# REFERENCE
# ONLY

# MAMMAL
# SPECIES
## OF THE
# WORLD

# MAMMAL SPECIES OF THE WORLD

## A Taxonomic and Geographic Reference

**THIRD EDITION**
**Volume 1**

*Edited by* Don E. Wilson
and DeeAnn M. Reeder

The Johns Hopkins University Press
*Baltimore*

The editors assume full responsibility for the contents and form
of Volumes 1 and 2.

The Johns Hopkins University Press
2715 North Charles Street
Baltimore, Maryland 21218-4363
www.press.jhu.edu

Library of Congress Cataloging-in-Publication Data

Mammal species of the world : a taxonomic and geographic
reference / edited by Don E. Wilson and DeeAnn M. Reeder. —
3rd ed.
    p. cm.
  Includes bibliographical references and index.
  ISBN 0-8018-8221-4 (hardcover set (vols. 1 and 2) : alk. paper)
  1. Mammals—Classification. 2. Mammals—Geographical
distribution. I. Wilson, Don E. II. Reeder, DeeAnn M.
QL708.M35 2005
599'.012—dc22        2005001870

A catalog record for this book is available from the British Library.

*In memory of*
*Peter Cannell, Charles O. Handley, Jr., and Karl F. Koopman*

# Contents
## Volume 1

## CHECKLIST OF MAMMAL SPECIES OF THE WORLD (CLASS MAMMALIA)

# Volume 2

# Contributors

GARY N. BRONNER
Small Mammal Research Unit
Department of Zoology
University of Cape Town
South Africa

ROBERT L. BROWNELL, JR.
Southwest Fisheries Center
National Marine Fisheries Service
Pacific Grove, CA 93950
USA

MICHAEL D. CARLETON
Division of Mammals
National Museum of Natural History
Washington, DC 20013-7012
USA

FRITZ DIETERLEN
Staatliches Museum für Naturkunde
Rosenstein 1
70191 Stuttgart
Germany

ALFRED L. GARDNER
U.S. Geological Survey
National Museum of Natural History
Washington, DC 20013-7012
USA

COLIN P. GROVES
Department of Archaeology and
    Anthropology
The Australian National University
GPO Box 4
Canberra ACT 2601
Australia

PETER GRUBB
35 Downhills Park Road
London N17 6PE
UK

KRISTOFER M. HELGEN
School of Earth and Environmental Sciences
University of Adelaide
Adelaide, SA 5005
Australia

ROBERT S. HOFFMANN
Division of Mammals
National Museum of Natural History
Washington, DC 20013-7012
USA

MARY ELLEN HOLDEN
Division of Vertebrate Zoology
American Museum of Natural History
Correspondence Address:
1714 Henley Lane
Charleston, SC 29412
USA

RAINER HUTTERER
Zoologisches Forschunginsmuseum
    Alexander Koenig
Adenauerallee 160
53113 Bonn
Germany

PAULINA D. JENKINS
Mammal Group, Department of Zoology
The Natural History Museum
Cromwell Road
London SW7 5BD
UK

C. WILLIAM KILPATRICK
Department of Biology
University of Vermont
Burlington, VT 05405-0086
USA

JAMES G. MEAD
Division of Mammals
National Museum of Natural History
Washington, DC 20013-7012
USA

GUY G. MUSSER
Division of Vertebrate Zoology
American Museum of Natural History
Correspondence Address:
1714 Henley Lane
Charleston, SC 29412
USA

JAMES L. PATTON
Museum of Vertebrate Zoology
University of California
Berkeley, CA 94720
USA

DEEANN M. REEDER
Department of Biology
Bucknell University
Lewisburg, PA 17837
USA

DUANE A. SCHLITTER
Texas Parks and Wildlife Department
3000 South IH-35, Suite 100
Austin, TX 78704
USA

JEHESKEL SHOSHANI
Department of Biology
University of Asmara
PO Box 1220 Asmara
Eritrea (Horn of Africa)
Correspondence Address:
Elephant Research Foundation
106 East Hickory Grove Road
Bloomfield Hills, Michigan 48304-1729
USA

NANCY B. SIMMONS
Division of Vertebrate Zoology
American Museum of Natural History
New York, NY 10024
USA

ANDREW T. SMITH
School of Life Sciences
Box 874501
Arizona State University
Tempe, AZ85287-4501
USA

BRIAN J. STAFFORD
Division of Mammals
National Museum of Natural History
Washington, DC 20013-7012
USA

RICHARD W. THORINGTON, JR.
Division of Mammals
National Museum of Natural History
Washington, DC 20013-7012
USA

DON E. WILSON
Division of Mammals
National Museum of Natural History
Washington, DC 20013-7012
USA

CHARLES A. WOODS
Department of Biology
University of Vermont
Burlington, VT 05405-0086
USA

W. CHRISTOPHER WOZENCRAFT
Division of Natural Science
Bethel College
1001 W. McKinley Avenue
Mishawaka, IN 46545
USA

# Preface

"A checklist of species is an invaluable tool for both researchers and the interested public." Thus began the first edition of this work, published by the Association of Systematics Collections (ASC) and Allen Press in 1982. That first edition was prepared by 189 professional mammalogists from twenty-three countries and was coordinated by a special Checklist Committee of the American Society of Mammalogists. During the ensuing decade, it became the industry standard for mammalian taxonomy, providing an authoritative reference for nonspecialists and establishing an overall taxonomic hypothesis for testing by systematic mammalogists.

The American Society of Mammalogists anticipated the need for revision and established a Standing Checklist Committee under the chairmanship of Karl F. Koopman in June 1982, concurrent with publication of the first edition. Duane A. Schlitter joined Koopman as co-chair in 1985, and they coordinated the committee's efforts until 1990. At that time, Don E. Wilson assumed chairmanship of the committee, with a mandate to expand the committee and produce a second edition of the checklist. With support from the Office of Biodiversity Programs at the Smithsonian Institution's National Museum of Natural History and with additional funding from the Seidell Fund of the Smithsonian Institution, DeeAnn M. Reeder joined the project in August 1991. At that time, we shifted from the egalitarian approach of multiple authors per section and assigned authorship of the various taxonomic groups to specialists in the field. In 1993, the second edition of *Mammal Species of the World* was published, and the database from which it was derived became available to the public at www.nmnh.si.edu/msw.

In 2002, various authors, some of whom were new to the project, began in earnest to update the text for the third edition, which is significantly enhanced by the inclusion of common names, recognition of subspecies, and inclusion of authorities for all synonyms. This additional information, coupled with the virtual explosion in taxonomic literature over the past decade, has resulted in the near doubling in size of the text between the second and third editions. Students of mammalian taxonomy have made significant advances in recent years, especially with the advent and refinement of additional molecular techniques. Beyond the additions due to revisions of known mammals that have occurred over the decades, a significant number of new mammalian species have been described, totaling 171 new species between the first and second editions of *Mammal Species of the World* (1982-1992) and 260 new species between the second and third editions (1993-2003).

Because of the inherent fluidity of mammalian taxonomy, with dramatic changes occurring in relatively short periods of time because of new data and interpretations and new species discoveries, we anticipate continued changes to the arrangement presented here.

The checklist will be maintained and updated in a Web-based database that will be freely available to the public in a variety of ways and usable formats. We anticipate updating the database regularly. We welcome your suggestions, comments, and additions and would particularly appreciate receiving copies of pertinent literature for preparation of future editions.

*Don E. Wilson*
*National Museum of Natural History*
*Smithsonian Institution*


*DeeAnn M. Reeder*
*Bucknell University*

# Acknowledgments

When a work as complex as *Mammal Species of the World* is finished, it is almost impossible to identify all of the many contributions deserving mention that have made the feat possible. We are ultimately most grateful to the hundreds of professional mammalogists who devoted their careers to the systematic study of Mammalia. Each has made a particular contribution to this synthesis.

The project owes much to Stephen R. Edwards and Robert S. Hoffmann, who initiated the effort that led to the first edition. James H. Honacki, Kenneth E. Kinman, and James W. Koeppl capably edited that first edition. The original Checklist Committee, chaired by Alfred L. Gardner and including Robert L. Brownell, Jr., Robert S. Hoffmann, Karl F. Koopman, Guy G. Musser, and Duane A. Schlitter, worked intensively to complete the first edition.

Elaine Hoagland, who succeeded Steve Edwards as executive director of the Association of Systematics Collections, facilitated transfer of the copyright to the American Society of Mammalogists. Duane Schlitter and Karl Koopman served as co-chairmen of the Checklist Committee for many years, keeping the project functioning during the decade between the first and second editions.

The production of the second edition was greatly facilitated by the creation of a database system for housing and searching the data from the text, which has been available on the Internet at www.nmnh.si.edu/msw. A number of Smithsonian Institution Information Resource Management staff were involved in this initiative, including Joe Russo and Barbara Weitbrecht. W. Christopher Wozencraft managed the thousands of references for the second edition. Ralph Walker of Information Resource Management and Peter Cannell of Smithsonian Institution Press diligently worked with the editors to facilitate final production of the second edition. The Web version of the second edition was created and maintained by Don Gourley and subsequently by Dennis Hasch, both Webmasters for the National Museum of Natural History. This has proved to be one of the museum's most popular Web sites and has been a valuable resource to the research community and the public at large.

The Division of Mammals at the National Museum of Natural History, Smithsonian Institution, provided an intellectual home for the editors for the second and third editions and the assistance of the curatorial, collections, and library staff is much appreciated. Kristofer M. Helgen reviewed most of the text and contributed significantly to the integrity of the data. A myriad of professional mammalogists and others from around the world made significant contributions by reviewing the text, offering advice, examining and/or providing specimens, providing unpublished data, and otherwise contributing intellectually, including H. Abe, Alexei Abramov, John Aguilar, Ticul Alvarez-Castaneda, Robert P. Anderson, Ken Aplin, Robert Asher, J.-C. P. Auffray, Stephane Aulagnier, Patrick Barriere, Kurt Bauer, Simon Bearder, Susan K. Bell, Wim Bergmans, Mary Boise, Robert Bradley, Doug Brandon-Jones, W. G. Breed, Norman Bridges, Patricia Brunauer, Leslie Carraway, Francois Catzeflis, Chris Chimimba, Gordon B. Corbet, Woody Cotterill, João Crawford Cabral, Gabor Csorba, Nick Czaplewski,

Daryl P. Domning, Judith Eger, Guillermo d'Elia, Louise Emmons, James W. Demastes, Christiane Denys, Jerry Dragoo, Jean-Francois Ducroz, Jakob Fahr, Katherine E. Ferrell, Robert D. Fisher, Tim Flannery, Larry J. Flynn, Fred Ford, Nikolai A. Formozov, Rosa García-Perea, Alfred L. Gardner, Philippe Gaubert, Valerius Geist, F. N. Golenishchev, Enrique Gonzalez, Steve Goodman, Antonia Gorog, Laurent Granjon, David J. Hafner, Mark S. Hafner, Meredith Happold, Rudolf Haslauer, Larry Heaney, Robert Hoffmann, M. Hugueney, Rainer Hutterer, Peter Jackson, Sharon Jansa, Paula Jenkins, Kate Jones, Y. Kaneko, Julian Kerbis Peterhans, Andrew Kitchener, Dieter Kock, Karl Koopman, Boris Kryštufek, Marcia Lara, Leonid A. Lavrenchenko, Emilie Lecompte, Yuri Leite, Georges Lenglet, Enrique P. Lessa, Burton Lim, Alicia Linzey, Darrin Lunde, Milos Macholan, V. G. Malikov, Joe T. Marshall, Jr., Erik Meijaard, Pierre Mein, Jin Meng, J. Michaux, A. Miljutin, Alvaro Mones, Brian Mould, Masaharu Motokawa, Guy Musser, J. Obuch, Jose Ochoa, Link Olson, Ulyses Pardinas, Bruce Patterson, James L. Patton, I. Ya. Pavlinov, Caroline Pollock, E. G. Potapova, Fiona Reid, Dale Rice, Ken Richardson, Eric Rickart, Luis A. Ruedas, Sue Ruff, Anthony Rylands, Eric Sargis, Erik Seiffert, Chad Schennum, Duane Schlitter, G. Shenbrot, Chris Smeenk, Angela Smith, William Stanley, Gerhard Storch, Peter Taylor, Adrian Tejedor, M. Tranier, Victor Van Cakenberghe, Harry Van Rompaey, Erik Van der Straeten, Walter N. Verheyen, Geraldine Veron, Vitaly Volobouev, John H. Wahlert, Robert Wayne, Lars Werdelin, Daniel F. Williams, John Wible, Neal Woodman, Jiong Wozencraft, Xie Yan, and Mikhail Zaitsev.

Finally, we would like to express our gratitude to the National Museum of Natural History for making funds available for the completion of this project.

# Introduction

The dynamic, rapidly changing state of mammalian taxonomy, documented by an enormous literature, long hampered the compilation of a detailed, complete world checklist. It was only about a century ago that Trouessart (1898-99, 1904-5) produced the first complete appraisal of all mammals of the world. A more modern compilation was provided when E. P. Walker and colleagues brought out, in 1964, the first edition of *Mammals of the World.* This compendium, now in its sixth edition (Nowak, 1999), is arranged systematically (with considerable supplementary natural history data) to the generic level, and later editions list the species in each genus in addition to furnishing an illustration of at least one member of each genus. V. E. Sokolov based his *Systematics of Mammals,* published in Russian (1973-79), on Walker's *Mammals of the World.* He provided a list of species with a brief summary of geographic distribution in each generic account. Corbett and Hill (1980, 1991) listed the species of the world, abbreviated distributions, common names, literature citations to major regional distributional works, and some additional revisionary works where appropriate. McKenna and Bell (1997) provided a complete phylogeny of mammals above the species level, including fossil as well as recent forms. That work provided a starting point for this edition of *Mammal Species of the World,* and deviations from their arrangement are noted in the comment sections of the accounts that follow. In the short time since the publication of McKenna and Bell (1997), an explosion of literature based on new techniques of molecular systematics has resulted in wholesale changes in our thinking about mammalian phylogeny. Those changes are reflected in the following pages, but this work is primarily a checklist at the species level, and higher-level relationships are used primarily to provide structure rather than to reflect phylogeny.

This volume, like previous editions (1982, 1993), will undoubtedly be used by many readers who are not systematic mammalogists. Do not be alarmed or disheartened by the debate over definition of species limits within many groups of mammals. Differences of opinion are aired in the comments sections to emphasize areas needing additional taxonomic study. Mammals are no worse off in this regard than other groups of animals and in fact are probably better known than most, with the possible exception of birds. One recurring suggestion from users of previous editions spurred us to include common names in this edition. The publication of the first complete list of common names of mammal species of the world (Wilson and Cole, 2000) made this possible. Contributors to this edition used those names as a starting point but were urged to adopt alternatives if there were compelling reasons to do so. As a result, this volume can be viewed as a second edition of Wilson and Cole (2000).

## THE PROCESS OF COMPILATION

Knowledge of the systematics of mammals is distributed over an extensive assortment of works. The process of compiling and editing the information contained in this edition drew heavily on lessons learned from prior editions.

### First Edition

The first edition (Honacki et al., 1982) evolved from three sources: a manuscript written by Kenneth E. Kinman, a preliminary list developed by the U.S. Fish and Wildlife Service, and the International Species Inventory (ISIS) List (Seal and Makey, 1974). A draft checklist developed from those sources was made available to the professional mammalogy community. Members of the American Society of Mammalogists were invited to contribute to the list at whatever level they wished. As a result, appropriate portions of the draft checklist were sent to 255 individuals. These reviewers were asked to provide the following information: author of the scientific name of the species and citation, type locality, distribution, citations of revisions or reviews, important synonyms, and explanatory comments if necessary. One hundred fifty mammalogists provided reviews, covering 85 percent of the species in the list. Additional reviewers were eventually found for the remaining species. Compiling the various drafts meant comparing the various contributions and including information on which there was agreement. In cases where there was not agreement, errors may have been retained. To determine valid names of taxa in cases of disagreement, the most recent reviewer was followed. Some of these names have stood the test of time, and others have not.

The penultimate draft was forwarded to the Checklist Committee of the American Society of Mammalogists. This group (Robert L. Brownell, Jr., Alfred L. Gardner [chair], Robert S. Hoffmann, Karl F. Koopman, Guy G. Musser, and Duane A. Schlitter) furnished the final review of the text and provided much useful additional information. The committee detected many problems resulting from the egalitarian effort, not all of which were satisfactorily resolved in the first edition.

Two categories of information were added after the review process. The protected status of affected taxa was listed based on information from the Federal Register. The ISIS numbers for each species were listed based on Seal and Makey (1974). In the second edition, the protected status was updated and retained, but the ISIS numbers were not because of their limited usefulness to most users.

### Second Edition

The second edition was compiled and edited in a distinctly different fashion from the first. Although the original decision to consult many professional mammalogists was theoretically sensible, it introduced a practical information management dilemma for the first edition. The wealth of material and the challenge of uniting numerous different, frequently conflicting opinions caused weaknesses in the original volume. Nevertheless, the taxonomic hypotheses outlined in that volume provided the basis for the refinements apparent in the second edition. The second edition was envisioned as the second step in organizing a continuing taxonomic database for mammals.

Because comprehensive survey of the literature is now prohibitive for a single mammalogist (or perhaps because they don't make us like they used to), the labor of composing the second edition was partitioned among twenty authors, specialists on their respective taxa. Although some authors had been maintaining records for their groups of interest for some time, preparation of the second edition began in earnest in the fall of 1990 when electronic and paper copies of the appropriate portions of the original text were distributed to the members of the committee along with a mandate to produce an up-to-date revision. Most of 1991 was spent in the production of that updated text. The authors examined the literature on mammals, from the initial works to the most recent, and provided references published in many languages. As each portion was received from the authors, it was converted to an electronic database, edited, regenerated in word-processing format, and returned to the authors for revisions. After a second round of editing, reviews were sought from other members of the community of systematic mammalogists. Although both editors and reviewers made suggestions to the authors, each section was the product of an individual author's scholarship and represented that author's best hypothesis of the relationships of a particular group at that time.

The information leading to that edition was compiled with computer technology that permitted rapid manuscript revision and organization. This allowed automatic indexing of the many names used and standardization of literature citations, as well as considerably more consistency in citations, punctuation, format, and spelling of place names. In spite of this, infelicities crept in, and the third printing of the second edition contained corrections of some typographical mistakes, misspellings, incorrect dates, and overlooked synonyms. Some literature citations were further standardized at that time.

Details on the 4,629 species of mammals covered by the second edition (up from 4,170 in the first edition) were the product of myriad former scholars united by a common curiosity about the diversity of mammals (Table 1). A major feature of the second edition, like the original, was that it identified gaps in our knowledge in need of further study and served as a starting point for the third edition.

## Third Edition

The third edition was prepared in much the same manner as the second, although the number of contributors was increased slightly. During the past decade, information on the systematic relationships of mammals has continued to increase. The advent of modern molecular techniques has allowed increasingly detailed comparisons of species limits and evolutionary relationships. The number of species represented in the current edition is 5,416, up from 4,629 in the second edition. Although most of this increase is due to taxonomic revision, a significant proportion is due to newly described species. In addition to currently recognized species, this edition contains 37,378 synonyms and full references to 9,373 scientific publications.

*Table 1*

*Comparisons of Genera and Species since the Second Edition*

| | Number of Genera | Number of Species | Number of New Species[a] |
|---|---|---|---|
| **Class Mammalia** | 1,229 | 5,416 | 260 |
| **Order Monotremata** | 3 | 5 | 1 |
| Family Tachyglossidae | 2 | 4 | 1 |
| Family Ornithorhynchidae | 1 | 1 | — |
| **Order Didelphimorphia** | 17 | 87 | 2 |
| Family Didelphidae | 17 | 87 | 2 |
| **Order Paucituberculata** | 3 | 6 | 1 |
| Family Caenolestidae | 3 | 6 | 1 |
| **Order Microbiotheria** | 1 | 1 | — |
| Family Microbiotheriidae | 1 | 1 | — |
| **Order Notoryctemorphia** | 1 | 2 | — |
| Family Notoryctidae | 1 | 2 | — |
| **Order Dasyuromorphia** | 22 | 71 | 5 |
| Family Thylacinidae | 1 | 1 | — |
| Family Myrmecobiidae | 1 | 1 | — |
| Family Dasyuridae | 20 | 69 | 5 |
| **Order Paramelemorphia** | 8 | 21 | — |
| Family Thylacomyidae | 1 | 2 | — |
| Family Chaeropodidae | 1 | 1 | — |
| Family Peramelidae | 6 | 18 | — |
| **Order Diprotodontia** | 39 | 143 | 8 |
| Suborder Vombatiformes | 3 | 4 | — |
| Family Phascolarctidae | 1 | 1 | — |
| Family Vombatidae | 2 | 3 | — |
| Suborder Phalangeriformes | 20 | 63 | 3 |
| Superfamily Phalangeroidea | 8 | 32 | 3 |
| Family Burramyidae | 2 | 5 | — |
| Family Phalangeridae | 6 | 27 | 3 |
| Superfamily Petauroidea | 12 | 31 | — |
| Family Pseudocheiridae | 6 | 17 | — |
| Family Petauridae | 3 | 11 | — |
| Family Tarsipedidae | 1 | 1 | — |
| Family Acrobatidae | 2 | 2 | — |
| Suborder Macropodiformes | 16 | 76 | 5 |
| Family Hypsiprymnodontidae | 1 | 1 | — |
| Family Potoroidae | 4 | 10 | — |
| Family Macropodidae | 11 | 65 | 5 |
| **Order Afrosoricida** | 19 | 51 | 6 |
| Suborder Tenrecomorpha | 10 | 30 | 5 |
| Family Tenrecidae | 10 | 30 | 5 |
| Suborder Chrysochloridea | 9 | 21 | 1 |
| Family Chrysochloridae | 9 | 21 | 1 |
| **Order Macroscelidea** | 4 | 15 | — |
| Family Macroscelididae | 4 | 15 | — |
| **Order Tubulidentata** | 1 | 1 | — |
| Family Orycteropodidae | 1 | 1 | — |
| **Order Hyracoidea** | 3 | 4 | — |
| Family Procaviidae | 3 | 4 | — |

*Table 1 (Continued)*

|  | Number of Genera | Number of Species | Number of New Species[a] |
|---|---|---|---|
| **Order Proboscidea** | 2 | 3 | — |
| Family Elephantidae | 2 | 3 | — |
| **Order Sirenia** | 3 | 5 | — |
| Family Dugongidae | 2 | 2 | — |
| Family Trichechidae | 1 | 3 | — |
| **Order Cingulata** | 9 | 21 | 1 |
| Family Dasypodidae | 9 | 21 | 1 |
| **Order Pilosa** | 5 | 10 | 1 |
| Suborder Folivora | 2 | 6 | 1 |
| Family Bradypodidae | 1 | 4 | 1 |
| Family Megalonychidae | 1 | 2 | — |
| Suborder Vermilingua | 3 | 4 | — |
| Family Cyclopedidae | 1 | 1 | — |
| Family Myrmecophagidae | 2 | 3 | — |
| **Order Scandentia** | 5 | 20 | — |
| Family Tupaiidae | 4 | 19 | — |
| Family Ptilocercidae | 1 | 1 | — |
| **Order Dermoptera** | 2 | 2 | — |
| Family Cynocephalidae | 2 | 2 | — |
| **Order Primates** | 69 | 376 | 24 |
| Suborder Strepsirrhini | 23 | 88 | 9 |
| Infraorder Lemuriformes | 14 | 59 | 8 |
| Superfamily Cheirogaleoidea | 5 | 21 | 6 |
| Family Cheirogaleidae | 5 | 21 | 6 |
| Superfamily Lemuroidea | 9 | 38 | 1 |
| Family Lemuridae | 5 | 19 | — |
| Family Lepilemuridae | 1 | 8 | — |
| Family Indridae | 3 | 11 | 1 |
| Infraorder Chiromyiformes | 1 | 1 | — |
| Family Daubentoniidae | 1 | 1 | — |
| Infraorder Lorisiformes | 8 | 28 | 2 |
| Family Lorisidae | 5 | 9 | 1 |
| Family Galagidae | 3 | 19 | 1 |
| Suborder Haplorrhini | 46 | 288 | 14 |
| Infraorder Tarsiiformes | 1 | 7 | — |
| Family Tarsiidae | 1 | 7 | — |
| Infraorder Simiiformes | 45 | 281 | 14 |
| Parvorder Platyrrhini | 16 | 128 | 11 |
| Family Cebidae | 6 | 56 | 8 |
| Family Aotidae | 1 | 8 | — |
| Family Pitheciidae | 4 | 40 | 3 |
| Family Atelidae | 5 | 24 | — |
| Parvorder Catarrhini | 29 | 153 | 3 |
| Superfamily Cercopithecoidea | 21 | 132 | 4 |
| Family Cercopithecidae | 21 | 132 | 4 |
| Superfamily Hominoidea | 8 | 21 | — |
| Family Hylobatidae | 4 | 14 | — |
| Family Hominidae | 4 | 7 | — |

*Table 1 (Continued)*

| | Number of Genera | Number of Species | Number of New Species[a] |
|---|---|---|---|
| **Order Rodentia** | **481** | **2,277** | **128** |
| Suborder Sciuromorpha | 61 | 307 | 1 |
| Family Aplondontiidae | 1 | 1 | — |
| Family Sciuridae | 51 | 278 | — |
| Family Gliridae | 9 | 28 | 1 |
| Suborder Castorimorpha | 13 | 102 | 2 |
| Family Castoridae | 1 | 2 | — |
| Family Heteromyidae | 6 | 60 | 2 |
| Family Geomyidae | 6 | 40 | — |
| Suborder Myomorpha | 326 | 1,569 | 96 |
| Superfamily Dipodoidea | 16 | 51 | — |
| Family Dipodidae | 16 | 51 | — |
| Superfamily Muroidea | 310 | 1,518 | 96 |
| Family Platacanthomyidae | 2 | 2 | — |
| Family Spalacidae | 6 | 36 | 4 |
| Family Calomyscidae | 1 | 8 | — |
| Family Nesomyidae | 21 | 61 | 6 |
| Family Cricetidae | 130 | 681 | 53 |
| Family Muridae | 150 | 730 | 33 |
| Suborder Anomaluromorpha | 4 | 9 | — |
| Family Anomaluridae | 3 | 7 | — |
| Family Pedetidae | 1 | 2 | — |
| Suborder Hystricomorpha | 77 | 290 | 29 |
| Infraorder Ctenodactylomorphi | 4 | 5 | — |
| Family Ctenodactylidae | 4 | 5 | — |
| Infraorder Hystricognathi | 73 | 285 | 29 |
| Family Bathyergidae | 5 | 16 | 2 |
| Family Hystricidae | 3 | 11 | — |
| Family Petromuridae | 1 | 1 | — |
| Family Thryonomyidae | 1 | 2 | — |
| Family Erethizontidae | 5 | 16 | 2 |
| Family Chinchillidae | 3 | 7 | — |
| Family Dinomyidae | 1 | 1 | — |
| Family Caviidae | 6 | 18 | 2 |
| Family Dasyproctidae | 2 | 13 | — |
| Family Cuniculidae | 1 | 2 | — |
| Family Ctenomyidae | 1 | 60 | 5 |
| Family Octodontidae | 8 | 13 | 3 |
| Family Abrocomidae | 2 | 10 | 2 |
| Family Echimyidae | 21 | 90 | 13 |
| Family Myocastoridae | 1 | 1 | — |
| Family Capromyidae | 8 | 20 | — |
| Family Heptaxodontidae | 4 | 4 | — |
| **Order Lagomorpha** | **13** | **92** | **5** |
| Family Ochotonidae | 1 | 30 | 2 |
| Family Prolagidae | 1 | 1 | — |
| Family Leporidae | 11 | 61 | 3 |
| **Order Erinaceomorpha** | **10** | **24** | **1** |
| Family Erinaceidae | 10 | 24 | 1 |

*Table 1 (Continued)*

| | Number of Genera | Number of Species | Number of New Species[a] |
|---|---|---|---|
| **Order Soricomorpha** | 45 | 428 | 18 |
| Family Nesophontidae | 1 | 9 | — |
| Family Solenodontidae | 1 | 4 | 1 |
| Family Soricidae | 26 | 376 | 17 |
| Family Talpidae | 17 | 39 | — |
| **Order Chiroptera** | 202 | 1,116 | 49 |
| Family Pteropodidae | 42 | 186 | 6 |
| Family Rhinolophidae | 1 | 77 | 4 |
| Family Hipposideridae | 9 | 81 | 7 |
| Family Megadermatidae | 4 | 5 | — |
| Family Rhinopomatidae | 1 | 4 | — |
| Family Craseonycteridae | 1 | 1 | — |
| Family Emballonuridae | 13 | 51 | 2 |
| Family Nycteridae | 1 | 16 | — |
| Family Myzopodidae | 1 | 1 | — |
| Family Mystacinidae | 1 | 2 | — |
| Family Phyllostomidae | 55 | 160 | 9 |
| Family Mormoopidae | 2 | 10 | — |
| Family Noctilionidae | 1 | 2 | — |
| Family Furipteridae | 2 | 2 | — |
| Family Thyropteridae | 1 | 3 | 1 |
| Family Natalidae | 3 | 8 | — |
| Family Molossidae | 16 | 100 | 2 |
| Family Vespertilionidae | 48 | 407 | 18 |
| **Order Pholidota** | 1 | 8 | — |
| Family Manidae | 1 | 8 | — |
| **Order Carnivora** | 126 | 286 | 1 |
| Suborder Feliformia | 54 | 121 | 1 |
| Family Felidae | 14 | 40 | — |
| Family Viverridae | 15 | 35 | 1 |
| Family Eupleridae | 7 | 8 | — |
| Family Nandiniidae | 1 | 1 | — |
| Family Herpestidae | 14 | 33 | — |
| Family Hyaenidae | 3 | 4 | — |
| Suborder Caniformia | 72 | 165 | — |
| Family Canidae | 13 | 35 | — |
| Family Ursidae | 5 | 8 | — |
| Family Otariidae | 7 | 16 | — |
| Family Odobenidae | 1 | 1 | — |
| Family Phocidae | 13 | 19 | — |
| Family Mustelidae | 22 | 59 | — |
| Family Mephitidae | 4 | 12 | — |
| Family Procyonidae | 6 | 14 | — |
| Family Ailuridae | 1 | 1 | — |
| **Order Perissodactyla** | 6 | 17 | — |
| Family Equidae | 1 | 8 | — |
| Family Tapiridae | 1 | 4 | — |
| Family Rhinocerotidae | 4 | 5 | — |

*Table 1 (Continued)*

| | Number of Genera | Number of Species | Number of New Species[a] |
|---|---|---|---|
| **Order Artiodactyla** | 89 | 240 | 8 |
| Family Suidae | 5 | 19 | 1 |
| Family Tayassuidae | 3 | 3 | — |
| Family Hippopotamidae | 2 | 2 | — |
| Family Camelidae | 3 | 4 | — |
| Family Tragulidae | 3 | 8 | — |
| Family Moschidae | 1 | 7 | — |
| Family Cervidae | 19 | 51 | 5 |
| Family Antilocapridae | 1 | 1 | — |
| Family Giraffidae | 2 | 2 | — |
| Family Bovidae | 50 | 143 | 2 |
| **Order Cetacea** | 40 | 84 | 1 |
| Suborder Mysticeti | 6 | 13 | — |
| Family Balaenidae | 2 | 4 | — |
| Family Balaenopteridae | 2 | 7 | — |
| Family Eschrichtiidae | 1 | 1 | — |
| Family Neobalaenidae | 1 | 1 | — |
| Suborder Odontoceti | 34 | 71 | 1 |
| Family Delphinidae | 17 | 34 | — |
| Family Monodontidae | 2 | 2 | — |
| Family Phocoenidae | 3 | 6 | — |
| Family Physeteridae | 2 | 3 | — |
| Family Platanistidae | 1 | 2 | — |
| Family Iniidae | 3 | 3 | — |
| Family Ziphiidae | 6 | 21 | 1 |

[a]Described since *Mammal Species of the World,* second edition.

## ORGANIZATION OF THE BOOK

Species recognized herein are limited to existing or recently extinct species (possibly alive during the preceding five hundred years); in instances where the persistence of a species is doubtful, the comment section so indicates. Five hundred years is an arbitrary span of time, selected to better allow us to judge the "recent" distribution of mammals on Earth. It has nothing to do with Western humans' arrival into the New World or with any other historical event. Obviously, determining the exact time of extinction of any species is difficult at best, and undoubtedly we have included some species that push our arbitrary limit.

### List of Museum Abbreviations

Abbreviations for specific museums are used in the text to indicate specimen repositories in some accounts. See page xxxv for full museum names.

## Taxonomic Arrangement

Various workers have reviewed the higher categories of mammals since Simpson (1945) produced his definitive classification, including Anderson and Jones (1967, 1984), McKenna (1975), Corbet and Hill (1980, 1986, 1991), and McKenna and Bell (1997). We used McKenna and Bell (1997) as the starting point for this edition. Modifications to their arrangement have been duly noted and documented in the text. In addition to a wealth of changes at the species level since the second edition, our understanding of the phylogenetic relationships of mammals at the higher category level is changing rapidly, and currently presents a moving target. The only mandatory categories are order, family, genus, and species. We are recognizing twenty-nine orders in this edition, up from twenty-six in the second edition. Xenarthra is split into the orders Cingulata and Pilosa, and Insectivora is split into Afrosoricida, Erinaceomorpha, and Soricomorpha. Within each order, individual authors have chosen the higher categories that best represent our knowledge of the group. Comment fields provide additional justification for the use or nonuse of additional categories (e.g., suborder, superfamily, subfamily, tribe, etc.) within each order. Generic names are ordered alphabetically within subfamilies (or families, if no subfamilies are recognized), and species names are ordered alphabetically within genera, without exception. If subgenera are thought to be useful, they are listed in the comments section of each species. Most authors have indicated subspecies as well by boldface type within the synonyms field. Readers are urged to review the introductory comments for each section.

## Scientific Name and Authority

Each currently used scientific name is followed by the name of the author(s) and the year in which it was described, e.g., *Vulpes vulpes* (Linnaeus, 1758). A species originally named by its author in a genus other than the one in which it is now placed has parentheses surrounding the author and date. In this example, Linnaeus, when he first named the red fox *vulpes*, placed it in the genus *Canis* instead of *Vulpes*. Should it be returned to *Canis*, the parentheses would be removed, i.e., *Canis vulpes* Linnaeus, 1758. Following the species name, authority, and date is the citation for the work in which the original description appeared. In some instances, especially in older literature, the date printed on a publication was not the actual date of publication. For example, many of the "parts" of the *Proceedings of the Zoological Society of London* were published in the year following that on the text. The actual date of publication is used as the date of authority and is found in brackets after the citation.

Usually, only the first page on which the species name appears follows the title citation. In some cases, for a variety of reasons, more than one page may be cited, as well as references to figures or plates. We have attempted to cite the first page on which the name appears, but if it is not listed unequivocally, the reader is advised to consult the original source.

Family names follow the principle of coordination outlined in Article 36 of the International Code of Zoological Nomenclature (1999). Generic accounts include the type species for which the generic name was proposed. In general, the name appears as it was originally described, including authority. When the type species is no longer a recognized species, the species with which it is currently synonymized is listed in parentheses in its original form with its authority.

## Common Name

Unlike previous editions, we have provided a common name for each recognized species. The starting point for these names is Wilson and Cole (2000), but each author was encouraged to examine those names and to provide a different one if there was good reason to do so. Thus, this list can be viewed as a second edition of Wilson and Cole (2000). There are no rules governing vernacular names, but Wilson and Cole (2000) outlined several reasons for adopting a single such name for each species of mammal.

## Type Locality

The type locality is the geographical site where the type material of a species was obtained. Type localities quoted exactly from the original description are enclosed in quotation marks. Information not surrounded by quotation marks has been arranged where possible with the current country name followed by state, province, or district, and specific locality. Elevation above sea level has been included when available, as have global coordinates in some cases. When appropriate, restrictions of the type locality made by revisers have been included as well.

## Distribution

The geographical range of each species is summarized using contemporary political units or, in some cases, geographical names. However, geographical names are usually used only when the entire area is included in the range of the species. We have attempted to standardize usage and spelling, but some inconsistencies may remain, particularly in transliterations from other alphabets. Country names have changed drastically during the time the text was being prepared; we have attempted to give the most current name whenever possible. Current country names have been standardized according to the list of Independent States of the World, produced by the U.S. Department of State as of February 7, 2003. With the exception of the Republic of Congo and the Democratic Republic of Congo (herein abbreviated as Dem. Rep. Congo), the official "short forms" of country names were used. Geographic names are subject to rapid changes in many parts of the world. Readers are urged to consult geographic thesauruses and local governmental lists when questions arise. Distribution records resulting from human introduction are sometimes noted. Maps of the distributions of many species are provided in the cited literature. If a species is known only from the type locality, that is noted.

## Status

Mammal species covered by the regulations for the U.S. Endangered Species Act (U.S. ESA) as of February 12, 2004; those listed in the 2003 International Union for Conservation of Nature and Natural Resources (IUCN) Red List of Threatened Animals, and those listed in the appendices of the Convention on International Trade in Endangered Species of Wild Fauna and Flora (CITES) as of March 2, 2004, are noted in the text in this category. In addition, some authors have included local governmental regulations in this field as well as general comments regarding the conservation status of particular species. For U.S. ESA listings, the categories of Endangered and Threatened (and supplemental categories such as proposed listings, emergency listings, and delisted taxa) are included. For the IUCN listings, the categories of Extinct, Extinct in the Wild, Critically Endangered, Endangered, Vulnerable, Lower Risk—near threatened (nt), Lower Risk—least concern (lc), Lower Risk—conservation dependent (cd), and Data Deficient are included. IUCN listings for species either officially "not evaluated" or not yet on the IUCN list are not included. For the CITES listings, Appendix I, II, and III are listed, where Appendix I includes species threatened with extinction that are or may be affected by trade, Appendix II includes species that although not necessarily threatened may become so unless trade in them is strictly controlled, as well as nonthreatened species that must be subject to regulation in order to control threatened species, and Appendix III includes species that any party identifies as being subject to regulation within its jurisdiction for purposes of preventing or restricting exploitation, and for which it needs the cooperation of other parties in controlling trade. Readers are cautioned against interpreting the lack of information in a status field as an indication that a particular species is not threatened. In many cases, recent taxonomic changes that are not yet reflected in the evaluations of the major conservation agencies will result in a taxon not being listed that is in fact endangered. Readers should consult the following Web sites for updates to the status fields: U.S. ESA (http://endangered.fws.gov/), IUCN (www.redlist.org), and CITES (www.cites.org).

## Synonyms

Authors have attempted to provide taxonomic lists consistent with recent literature, tempered by their own individual judgments. Considerable effort has been expended to compile a complete list of synonyms that have been used in the scientific literature for each taxon. These are usually either names of later origin than that used (junior synonyms) or names that are invalid systematically, for various reasons. Also included here are subspecies names, including any that might be currently recognized. Currently recognized subspecies are listed in boldface type, followed by their junior synonyms. Names listed before a boldfaced subspecies are synonyms of the nominate form. Subsequent emendations, misspellings, incorrect allocations, and partial synonyms are not included, for the most part. Theoretically, any scientific name used for a mammal should be found in this volume, either as a currently recognized species or as a synonym. One addition to this edition is the inclusion of the authority and date for each synonym, which should make it easier to find the original description of the name.

## Comments

Taxonomic and nomenclatorial alternatives are accompanied by appropriate documentation, including opinions of the International Commission on Zoological Nomenclature

(ICZN); revisions and additional literature sources are also cited. In the interest of brevity, secondary reference sources are sometimes cited to document taxonomic evidence, and reference to the primary sources can be found therein. Personal opinions of the author(s) and unpublished information are sometimes included here as well. When appropriate, other data such as references discussing type locality, occurrence of hybridization, and species known only from a single or few specimens are also included in the comments section. Absence of a comments section may indicate either that the species is taxonomically uncontroversial or that it is too poorly known to require comment.

**Bibliographic Treatment**

The literature cited section contains the works consulted in the compilation of this text. Works appearing after 2003 are generally not cited, although several works in press at the end of 2003 are included. References cited as authorities for original descriptions are not included in the literature cited but are given after each currently recognized name in the text in sufficient detail to allow the reader to locate them. Author and date citations in the synonym section are not included in the literature cited because they are provided only as authorities for the names.

**Index**

The index contains all taxonomic names contained in this volume, including synonyms. Page references to all currently recognized generic and species names employed in this volume are in italic boldface type. Species names are individually listed in alphabetical order.

# Museum Abbreviations

| | |
|---|---|
| AM | Australian Museum, Sydney |
| AMNH | American Museum of Natural History, New York |
| ANSP | Philadelphia Academy of Natural Sciences, Philadelphia |
| BBM | Bernice P. Bishop Museum, Hawaii |
| BMNH | The Natural History Museum, London |
| CIB | Centro de Investigaciones Biológicas del Noroeste, S. C., Mexico |
| CM | Carnegie Museum of Natural History, Pittsburgh |
| CMNH | Cincinnati Museum of Natural History |
| FMNH | Field Museum of Natural History, Chicago |
| HUNHM | Hokkaido University Natural History Museum, Hokkaido |
| IEBR | Institute of Ecology and Biological Resources, Hanoi |
| IRSNB | Institute de Royale de Sciences Naturelles Belgique, Brussels |
| IZAC | Institute of Zoology, Academia Sinica (Chinese Academy of Sciences), Beijing |
| KNMB | Koninklijk Belgisch Instituut voor Natuurwetenschappen, Brussels |
| MCZ | Museum of Comparative Zoology, Harvard University, Cambridge |
| MNHN | Museum National d'Histoire Naturelle, Paris |
| MRAC | Musée royal de l'Afrique centrale, Tervuren |
| MVZ | Museum of Vertebrate Zoology, University of California, Berkeley |
| MZB | Museum Zoologicum Bogoriense, Bogor (now Balai penelitian Zoologi-Pusat Penelitian Biologi-LIPI) |
| NMW | Naturhistorisches Museum Wien, Wien |
| OMNH | Oklahoma Museum of Natural History, University of Oklahoma |
| RMBR | Raffles Museum of Biodiversity Research, Department of Biological Sciences, National University of Singapore, Singapore |
| RMNH | Rijksmuseum van Natuurlijke Historie, Leiden (now National Natuurhistorisch Museum, Leiden) |
| ROM | Royal Ontario Museum, Toronto |
| TCWC | Texas Cooperative Wildlife Collection, Texas A&M University |
| UCA | Universitair Centrum, Antwerpen |
| UMMZ | University of Michigan Museum of Zoology, Ann Arbor |
| USNM | United States National Museum of Natural History, Smithsonian Institution |
| ZFMK | Zoologisches Forschungsinstitut und Museum Alexander Koenig, Bonn |
| ZMA | Zoological Museum of the University of Amsterdam, Amsterdam |
| ZMB | Zoologisches Museum und Institut für Spezielle Zoologie, Museum für Naturkunde der Humboldt-Universität zu Berlin, Berlin |
| ZMO | Zoological Museum, Oslo |
| ZMVNU | Zoological Museum, Vietnam National University, Hanoi |
| ZSI | Zoological Survey of India Collections, Calcutta |

# CHECKLIST OF MAMMAL SPECIES OF THE WORLD
## (CLASS MAMMALIA)

# ORDER MONOTREMATA
by Colin P. Groves

**ORDER MONOTREMATA** Bonaparte, 1837.
> COMMENTS: Reviewed by Griffiths (1978). The order is the sole extant representative of the Subclass Prototheria (all other living mammals belong to subclass Theria). McKenna and Bell (1997) divided the order into two (Platypoda and Tachyglossa); the date of divergence of the two living families is unknown, and conservatively they are retained here in a single order.

**Family Tachyglossidae** Gill, 1872. Smithson. Misc. Coll., 11:27.

*Tachyglossus* Illiger, 1811. Prodr. Syst. Mammal. Avium., p. 114.
> TYPE SPECIES: *Echidna novaehollandiae* Lacépède, 1799 (= *Myrmecophaga aculeatus* Shaw, 1792).
> SYNONYMS: *Acanthonotus* Goldfuss, 1809; *Echidna* G. Cuvier, 1797; *Echinopus* G. Fischer, 1814; *Syphonia* Rafinesque, 1815.

*Tachyglossus aculeatus* (Shaw, 1792). Nat. Misc., 3, pl. 109.
> COMMON NAME: Short-beaked Echidna.
> TYPE LOCALITY: Australia, New South Wales, New Holland (= Sydney).
> DISTRIBUTION: S and E New Guinea; Australia, including Kangaroo Isl (off South Australia) and Tasmania.
> STATUS: IUCN – Lower Risk (nt) as *T. a. multiaculeatus*; otherwise Lower Risk (lc). Abundant throughout its range.
> SYNONYMS: *australiensis* (Lesson, 1827); *australis* (Lesson, 1836); *corealis* (Krefft, 1872); *eracinius* (Mudie, 1829); *hystrix* (Home, 1802); *longiaculeata* (Tiedemann, 1808); *myrmecophagus* (Goldfuss, 1809); *novaehollandiae* (Lacépède, 1799); *orientalis* (Krefft, 1872); *sydneiensis* (Kowarzik, 1909); *typica* (Thomas, 1885); **acanthion** (Collett, 1884); *ineptus* Thomas, 1906; **lawesii** Ramsay, 1877; **multiaculeatus** (W. Rothschild, 1905); **setosus** (E. Geoffroy St. Hilaire, 1803); *breviaculeata* (Tiedemann, 1808); *hobartensis* (Kowarzik, 1909); *longirostrus* (Perry, 1810).
> COMMENTS: Includes *lawesii* and *setosus,* see Ride (1970:231). Subspecies are unclear, and revision is needed. Species name commonly attributed to Shaw and Nodder, but Nodder was the publisher, not an author.

*Zaglossus* Gill, 1877. Ann. Rec. Sci. Indus., May:171.
> TYPE SPECIES: *Tachyglossus bruijni* Peters and Doria, 1876.
> SYNONYMS: *Acanthoglossus* Gervais, 1877; *Bruynia* Dubois, 1882; *Proechidna* Dubois, 1884; *Prozaglossus* Kerbert, 1913.
> COMMENTS: Revised by Flannery and Groves (1998).

*Zaglossus attenboroughi* Flannery and Groves, 1998. Mammalia, 62:387.
> COMMON NAME: Sir David's Long-beaked Echidna.
> TYPE LOCALITY: Indonesia, Prov. of Papua (= Irian Jaya), Cyclops Mtns, "Oost-top, Berg Rara, 1600 m".
> DISTRIBUTION: Known only from type locality.
> STATUS: CITES - Appendix II.

*Zaglossus bartoni* (Thomas, 1907). Ann. Mag. Nat. Hist., 20:294.
> COMMON NAME: Eastern Long-beaked Echidna.
> TYPE LOCALITY: Papua New Guinea, Albert Edward Range, Mount Victoria, 8,000 ft (2438 m).
> DISTRIBUTION: Interior New Guinea, east of Paniai Lakes, 600 to 3200 m.
> STATUS: CITES - Appendix II; IUCN – Endangered as *Z. bruijni* (part).
> SYNONYMS: *bubuensis* Laurie, 1952; **clunius** Thomas and W. Rothschild, 1922; **diamondi** Flannery and Groves, 1998; **smeenki** Flannery and Groves, 1998.
> COMMENTS: The subspecies are all highly distinctive, and may represent distinct species.

*Zaglossus bruijni* (Peters and Doria, 1876). Ann. Mus. Civ. Stor. Nat. Genova, 9:183.
> COMMON NAME: Western Long-beaked Echidna.
> TYPE LOCALITY: Indonesia, Prov. of Papua (= Irian Jaya), Vogelkop, Manokwari Div., Arfak Mtns.
> DISTRIBUTION: Western New Guinea, west of Paniai Lakes; Salawati Isl (Indonesia).
> STATUS: CITES - Appendix II; IUCN – Endangered.
> SYNONYMS: *goodfellowi* (Thomas, 1907); *gularis* W. Rothschild, 1922; *nigro-aculeatus* (W. Rothschild, 1892); *pallidus* W. Rothschild, 1922; *tridactyla* (Dubois, 1882); *villosissima* (Dubois, 1884).
> COMMENTS: Does not include *bartoni* and *bubuensis*; see Flannery and Groves (1998).

**Family Ornithorhynchidae** Gray, 1825. Ann. Philos., n.s., 10:343.

*Ornithorhynchus* Blumenbach, 1800. Gotting. Gelehrt. Anz., 1:609.
> TYPE SPECIES: *Ornithorhynchus paradoxus* Blumenbach, 1800 (= *Platypus anatinus* Shaw, 1799).
> SYNONYMS: *Dermipus* Wiedermann, 1800; *Platypus* Shaw, 1799 [not of Herbst, 1793].
> COMMENTS: *Platypus* Shaw, 1799 was preoccupied by *Platypus* Herbst, 1793, a genus of Coleoptera. *Ornithorynchus* is the next available name.

*Ornithorhynchus anatinus* (Shaw, 1799). Nat. Misc., 10, pl. 385-386.
> COMMON NAME: Platypus.
> TYPE LOCALITY: Australia, New South Wales, New Holland (= Sydney).
> DISTRIBUTION: Queensland, New South Wales, SE South Australia, Victoria, and Tasmania (Australia).
> STATUS: IUCN – Lower Risk (lc); common but vulnerable to local extinction.
> SYNONYMS: *brevirostris* Ogilby, 1832; *crispus,* Macgillivray, 1827; *fuscus* Péron, 1807; *laevis* Macgillivray, 1827; *novaehollandiae* Lacépède, 1800; *paradoxus* Blumenbach, 1800; *phoxinus* Thomas, 1923; *rufus* Péron, 1807; *triton* Thomas, 1923.
> COMMENTS: Whether there are any valid subspecies is unclear; revision is needed. Species name is commonly attributed to Shaw and Nodder, but Nodder was the publisher, not an author.

# ORDER DIDELPHIMORPHIA
by Alfred L. Gardner

**ORDER DIDELPHIMORPHIA** Gill, 1872.
SYNONYMS: Didelphidia (*sensu* Hershkovitz, 1992*a*).
COMMENTS: Traditionally included in Marsupialia; included in Ameridelphia (see Aplin and Archer, 1987; Marshall et al., 1990; and Szalay, 1982); but not Microbiotheriidae (Marshall et al., 1990; *contra* Reig et al., 1987).

**Family Didelphidae** Gray, 1821. London Med. Repos., 15:308.
COMMENTS: Placed in the order Polyprotodontia by Kirsch (1977); also see Aplin and Archer (1987). Does not include *Dromiciops*; see Kirsch and Calaby (1977). Includes Caluromyidae, Glironiidae, and Marmosidae *sensu* Hershkovitz (1992*a*).

**Subfamily Caluromyinae** Kirsch, 1977. Aust. J. Zool., Suppl. ser., 52:111.
COMMENTS: Included in Caluromyidae along with Caluromysiopsinae by Hershkovitz (1992*a*); also includes Glironiidae.

*Caluromys* J. A. Allen, 1900. Bull. Am. Mus. Nat. Hist., 13:189.
TYPE SPECIES: *Didelphis philander* Linnaeus, 1758, by original designation.
SYNONYMS: *Didelphis* Linnaeus, 1758 [part]; *Gamba* Liais, 1872 [part]; *Mallodelphys* Thomas, 1920 [type species *Didelphis laniger* Desmarest, 1820, by original designation; valid as a subgenus]; *Philander* Burmeister, 1856 [preoccupied by *Philander* Brisson, 1762, and *Philander* Tiedemann, 1808]; *Sarigua* Muirhead, 1819 [part].
COMMENTS: *Calurosymys* Avila-Pires, 1964, is an incorrect subsequent spelling of *Caluromys* J. A. Allen. Comparatively uncommon to rare in collections, perhaps due to nocturnal and arboreal habits; but probably common in suitable habitat. Vulnerable to loss of tropical forest habitat.

*Caluromys derbianus* (Waterhouse, 1841). Jardine's Natur. Libr., 11:97.
COMMON NAME: Derby's Woolly Opossum.
TYPE LOCALITY: None given; restricted to Colombia, Cauca, Cauca Valley (Cabrera, 1958:2).
DISTRIBUTION: Mexico, Central America, W Colombia, and W Ecuador.
STATUS: IUCN – Vulnerable.
SYNONYMS: *antioquiae* (Matschie, 1917); *guayanus* (Thomas, 1899); *pictus* (Thomas, 1913); *pyrrhus* Thomas, 1901; *senex* (Thomas, 1913); **aztecus** (Thomas, 1913); **centralis** (Hollister, 1914); *canus* (Matschie, 1917); **fervidus** (Thomas, 1913); **nauticus** (Thomas, 1913); **pallidus** (Thomas, 1899); **unassigned:** *pulcher* (Matschie, 1917).
COMMENTS: Subgenus *Mallodelphys*. Reviewed by Bucher and Hoffmann (1980, Mammalian Species, 140). The name *pulcher* Matschie is based on a zoo specimen of unknown origin.

*Caluromys lanatus* (Olfers, 1818). *In* W. L. Eschwege, J. Brasilien, Neue Bibliothek. Reisen., 15:206.
COMMON NAME: Brown-eared Woolly Opossum.
TYPE LOCALITY: "Paraguay;" restricted to Caazapá, Caazapá (Cabrera, 1916).
DISTRIBUTION: N and C Colombia, NW and S Venezuela, E Ecuador, E Perú, E Bolivia, E and S Paraguay, N Argentina (Provincia Misiones), and W and S Brazil.
STATUS: IUCN – Lower Risk (nt).
SYNONYMS: *cahyensis* (Matschie, 1917); *lanigera* (Desmarest, 1820); **cicur** (Bangs, 1898); *meridensis* (Matschie, 1917); **nattereri** (Matschie, 1917); *modesta* (Miranda-Ribeiro, 1936); **ochropus** (Wagner, 1842); **ornatus** (Tschudi, 1845); *bartletti* (Matschie, 1917); *jivaro* (Thomas, 1913); *juninensis* (Matschie, 1917); **vitalinus** (Miranda-Ribeiro, 1936); **unassigned:** *hemiurus* (Miranda-Ribeiro, 1936).
COMMENTS: Subgenus *Mallodelphys*. The type locality of *hemiurus* (Miranda-Ribeiro) is Brazil, specific locality unknown. The name *calmensis* Vieira, 1955, is an incorrect subsequent spelling of *cahyensis* Matschie.

*Caluromys philander* (Linnaeus, 1758). Syst. Nat., 10th ed., 1:54.
>  COMMON NAME: Bare-tailed Woolly Opossum.
>  TYPE LOCALITY: "America;" restricted to Surinam (Thomas, 1911*a*).
>  DISTRIBUTION: Venezuela (including Margarita Isl), Trinidad, Guyana, Surinam, French Guiana, Brazil, and E Bolivia.
>  STATUS: IUCN – Lower Risk (nt).
>  SYNONYMS: *cajopolin* (Müller, 1776); *cayopollin* (Schreber, 1777); *leucurus* Thomas, 1904; **affinis** (Wagner, 1842); **dichurus** (Wagner, 1842); **trinitatis** (Thomas, 1894); *venezuelae* Thomas, 1903.
>  COMMENTS: Subgenus *Caluromys*. *Didelphys longicaudata* Pelzeln, 1883, and *D. macrura* Pelzeln, 1883 (not *D. macrura* Olfers, 1818 [= *Thylamy macrurus*]) are *nomina nuda*; *dichrura* Schinz, 1844, *dichrura* Cabrera, 1919, and *dichrurus* Vieira, 1953, are incorrect subsequent spellings of *dichura* Wagner.

*Caluromysiops* Sanborn, 1951. Fieldiana Zool., 31:474.
>  TYPE SPECIES: *Caluromysiops irrupta* Sanborn, 1951.
>  COMMENTS: Monotypic.

*Caluromysiops irrupta* Sanborn, 1951. Fieldiana Zool., 31:474.
>  COMMON NAME: Black-shouldered Opossum.
>  TYPE LOCALITY: Perú, Cuzco, "Quincemil, Province of Quispicanchis."
>  DISTRIBUTION: SE Colombia, SE Perú, and W Brazil.
>  STATUS: IUCN – Vulnerable.
>  COMMENTS: Uncommon to rare; see review by Izor and Pine (1987).

*Glironia* Thomas, 1912. Ann. Mag. Nat. Hist., ser. 8, 9:239.
>  TYPE SPECIES: *Glironia venusta* Thomas, 1912, by original designation.

*Glironia venusta* Thomas, 1912. Ann. Mag. Nat. Hist., ser. 8, 9:240.
>  COMMON NAME: Bushy-tailed Opossum.
>  TYPE LOCALITY: Perú, Pasco, "Pozuzo."
>  DISTRIBUTION: Amazonian Brazil, Ecuador, Perú, and Bolivia.
>  STATUS: IUCN – Vulnerable; rare.
>  SYNONYMS: *aequatorialis* Anthony, 1926; *criniger* Anthony, 1926.
>  COMMENTS: Reviewed by Marshall (1978*c*, Mammalian Species, 107).

**Subfamily Didelphinae** Gray, 1821. London Med. Repos., 15:308.
>  SYNONYMS: Chironectinae, Lestodelphyinae, Lutreolininae, Marmosidae, Marmosinae, Metachirinae, Monodelphinae, Thylamyinae.
>  COMMENTS: Hershkovitz (1992*a*) restricted the Didelphinae to the genera *Chironectes*, *Didelphis*, *Philander*, and *Lutreolina*. Hershkovitz (1997) further restricted the Didelphinae by establishing Chironectinae and Lutreolininae for *Chironectes* and *Lutreolina*, respectively.

*Chironectes* Illiger, 1811. Prodr. Syst. Mammal. Avium., p. 76.
>  TYPE SPECIES: *Lutra minima* Zimmermann, 1780, by monotypy.
>  SYNONYMS: *Gamba* Liais, 1872 [part]; *Memina* G. Fischer, 1814; *Sarigua* Muirhead, 1819 [part].
>  COMMENTS: *Memina* G. Fischer, 1813, is a *nomen nudum*. *Cheronectis* Fleming, 1822, *Cheironectes* Gray, 1827, *Cheironectes* Jentink, 1888, and *Chironeytes* Goeldi and Hagmann, 1904, are incorrect subsequent spellings of *Chironectes* Illiger.

*Chironectes minimus* (Zimmermann, 1780). Geogr. Gesch. Mensch. Vierf. Thiere, 2:317.
>  COMMON NAME: Water Opossum.
>  TYPE LOCALITY: "Gujana;" restricted to Cayenne, French Guiana (Cabrera, 1958:44).
>  DISTRIBUTION: Oaxaca and Tabasco, México, south through Central America to Colombia, Ecuador, Brazil, Perú, Venezuela, the Guianas, Paraguay, and NE Argentina.
>  STATUS: IUCN – Lower Risk (nt)
>  SYNONYMS: *cayennensis* Turton, 1800; *guianensis* Kerr, 1792; *gujanensis* Link, 1795; *palmata* Daudin *in* Lacépède, 1802; *paraguensis* Kerr, 1792; *sarcovienna* Shaw, 1800; *variegatus*

Olfers, 1818; *yapock* Desmarest, 1820; **argyrodytes** Dickey, 1928; **langsdorffi** Boitard, 1845; *bresslaui* Pohle, 1927; **panamensis** Goldman, 1914.

COMMENTS: Reviewed by Marshall (1978*d*, Mammalian Species, 109). The spelling *variegatus* Illiger, 1815, is a *nomen nudum*. The spellings *memia* Desmarest, 1803, *memina* Boddaert, 1784, *memina* F. Cuvier, 1825, *memmina* Desmarest, 1804, and *memmina* Muirhead, 1819, are incorrect subsequent spellings of *minimus* Zimmermann.

*Didelphis* Linnaeus, 1758. Syst. Nat., 10th ed., 1:54.

TYPE SPECIES: *Didelphis marsupialis* Linnaeus, 1758, by subsequent selection (Thomas, 1888).

SYNONYMS: *Dasyurotherium* Liais, 1872 [replacement name for *Thylacotherium* Lund]; *Dimerodon* Ameghino, 1889; *Gamba* Liais, 1872 [part]; *Gambatherium* Liais, 1872 [replacement name for *Thylacotherium* Lund]; *Leucodidelphis* Ihering, 1914; *Opossum* Schmid, 1818 [part]; *Sarigua* Muirhead, 1819 [part]; *Thylacotherium* Lund, 1839 [preoccupied].

COMMENTS: *Didelphys* Schreber, 1778, is an invalid emendation of *Didelphis* Linnaeus, and *Leucodidelphys* Krumbiegel, 1941, is an invalid emendation of *Leucodidelphis* Ihering, 1914. *Didelphus* Lapham, 1853, is an incorrect subsequent spelling of *Didelphis* Linnaeus.

*Didelphis albiventris* Lund, 1840. K. Dansk. Vid. Selsk. Afhandl., p. 20 [preprint of Lund, 1841].

COMMON NAME: White-eared Opossum.

TYPE LOCALITY: Brazil, Minas Gerais, "Rio das Velhas," Lagoa Santa.

DISTRIBUTION: Colombia, Ecuador, Perú, Brazil, Bolivia, Paraguay, Uruguay, and the northern half of Argentina.

STATUS: IUCN – Lower Risk (lc).

SYNONYMS: *antigua* Ameghino, 1889; *bonariensis* (Marelli, 1930); *brasiliensis* (Liais, 1872) (part); *dennleri* (Marelli, 1930); *lechei* (Ihering, 1892); *leucotis* (Wagner, 1847); *paraguayensis* (J. A. Allen, 1902) [not available from Oken, 1816]; *poecilonota* (Schinz, 1844); *poecilotis* Wagner, 1842.

COMMENTS: Formerly known as *D. azarae*; see Hershkovitz (1969). *Didelphis paraguayensis* J. A. Allen, 1902, ex Oken, 1816 (= *Didelphis albiventris* Lund) is the type species of *Leucodidelphis* Ihering, 1914.

*Didelphis aurita* (Wied-Neuwied, 1826). Beitr. Naturgesch. Brasil., 2:395.

COMMON NAME: Big-eared Opossum.

TYPE LOCALITY: Brazil, Bahia, "Villa Viçosa am Flusse Paruhype."

DISTRIBUTION: E Brazil, SE Paraguay, and NE Argentina.

STATUS: IUCN – Lower Risk (lc).

SYNONYMS: *azarae* Temminck, 1824 [see comments]; *brasiliensis* (Liais, 1872) [part]; *koseritzi* Ihering, 1892; *longipilis* Miranda-Ribeiro, 1935; *melanoidis* Miranda-Ribeiro, 1935; *typica* (Thomas, 1888) [part].

COMMENTS: Previously considered a disjunct population of *D. marsupialis* (see Cerqueira, 1985). The senior synonym is *D. azarae* Temminck, 1824 (see Hershkovitz, 1969); however, the name had been misapplied to *D. albiventris* for over 160 years. The name *leucoprymnus* Matschie, 1916, is a *nomen nudum*; *longigilis* Ávila-Pires, 1968, is an incorrect subsequent spelling of *longipilis* Miranda-Ribeiro.

*Didelphis imperfecta* Mondolfi and Pérez-Hernández, 1984. Acta Cient. Venezolana 35:407.

COMMON NAME: Guianan White-eared Opossum.

TYPE LOCALITY: Venezuela, Bolívar, "km 125, Carretera El Dorado-Santa Elena."

DISTRIBUTION: Venezuela (south of the Orinoco), SW Suriname, French Guiana, and N Brazil.

COMMENTS: Previously included in *D. albiventris* (see Gardner, 1993); revised by Lemos and Cerqueira (2002).

*Didelphis marsupialis* Linnaeus, 1758. Syst. Nat., 10th ed., 1:54.

COMMON NAME: Common Opossum.

TYPE LOCALITY: "America;" restricted to Surinam (Thomas, 1911*a*).

DISTRIBUTION: Tamaulipas, México, south throughout Central and South America to Perú, Bolivia, and Brazil.

STATUS: IUCN – Lower Risk (lc).

SYNONYMS: *austroamericana* J. A. Allen, 1902 [not available from Oken, 1816]; *brasiliensis* (Liais, 1872) [part]; *cancrivora* Gmelin, 1788; *karkinophaga* Zimmermann, 1780; *typica* (Thomas, 1888); *caucae* J. A. Allen, 1900; *battyi* J. A. Allen, 1902; *colombica* J. A. Allen, 1900; *etensis* J. A. Allen, 1902; *insularis* J. A. Allen, 1902; *mesamericana* J. A. Allen, 1902 [part; not available from Oken, 1816]; *particeps* Goldman, 1917; *richmondi* J. A. Allen, 1901; *tabascensis* J. A. Allen, 1901.

COMMENTS: Middle American populations were revised by Gardner (1973); Cerqueira (1985) treated the species as monotypic; *carcinophaga* Boddaert, 1784, is an invalid emendation of *karkinophaga* Zimmermann.

*Didelphis pernigra* J. A. Allen, 1900. Bull. Amer. Mus. Nat. Hist. 13:191.

COMMON NAME: Andean White-eared Opossum.

TYPE LOCALITY: Perú, Puno, "Juliaca;" corrected to "Inca Mines" (= Santo Domingo) by J. A. Allen (1901).

DISTRIBUTION: Andes of Colombia, Venezuela, Ecuador, Perú, and Bolivia.

SYNONYMS: *andina* J. A. Allen, 1902; *meridensis* J. A. Allen, 1902.

COMMENTS: Previously in *albiventris* (see Gardner, 1993); revised by Lemos and Cerqueira (2002).

*Didelphis virginiana* Kerr, 1792. *In* Linnaeus, Anim. Kingdom, p. 193.

COMMON NAME: Virginia Opossum.

TYPE LOCALITY: "Virginia, Louisiana, Mexico, Brasil, and Peru;" restricted to Virginia (J. A. Allen, 1901:160c).

DISTRIBUTION: S Canada; E and C United States (with introduced populations in Pacific states); México; and in Central America south into N Costa Rica.

STATUS: IUCN – Lower Risk (lc).

SYNONYMS: *boreoamericana* J. A. Allen, 1902 [not available from Oken, 1816]; *illinensium* Link, 1795; *pilosissima* Link, 1795; *pruinosa* Wagner, 1843; *typica* (Thomas, 1888) [part]; *woapink* Barton, 1806; *californica* Bennett, 1833; *breviceps* Bennett, 1833; *mesamericana* J. A. Allen, 1902 [part; not available from Oken, 1816]; *pigra* Bangs, 1898; *texensis* J. A. Allen, 1901; *yucatanensis* J. A. Allen, 1901; *cozumelae* Merriam, 1901.

COMMENTS: Revised by Gardner (1973, 1982); reviewed by McManus (1974, Mammalian Species, 40).

*Gracilinanus* Gardner and Creighton, 1989. Proc. Biol. Soc. Wash., 102:4.

TYPE SPECIES: *Didelphys microtarsus* Wagner, 1842, by original designation.

SYNONYMS: *Didelphis* Linnaeus, 1758 [part]; *Grymaeomys* Burmeister, 1854 [part].

COMMENTS: Previously in *Marmosa* (*sensu lato*; see Gardner and Creighton, 1989).

*Gracilinanus aceramarcae* (Tate, 1931). Am. Mus. Novit., 493:12.

COMMON NAME: Aceramarca Gracile Opossum.

TYPE LOCALITY: Bolivia, La Paz, "Rio Aceramarca, tributary of Rio Unduavi, Yungas."

DISTRIBUTION: Bolivia (type locality) and SE Perú.

STATUS: IUCN – Critically Endangered.

COMMENTS: Included under *Thylamys* by Reig et al. (1987).

*Gracilinanus agilis* (Burmeister, 1854). Syst. Uebers. Thiere Bras., 1:139.

COMMON NAME: Agile Gracile Opossum.

TYPE LOCALITY: Brazil, Minas Gerais, "Lagoa Santa."

DISTRIBUTION: Brazil, E Perú, E Bolivia, Paraguay, Uruguay, and adjacent Argentina.

STATUS: IUCN – Lower Risk (nt).

SYNONYMS: *beatrix* (Thomas, 1910); *blaseri* (Miranda-Ribeiro, 1936); *buenavistae* (Tate, 1931); *chacoensis* (Tate, 1931); *peruana* (Tate, 1931); *rondoni* (Miranda-Ribeiro, 1936); *unduaviensis* (Tate, 1931).

COMMENTS: Anderson (1997) recognized three subspecies in Bolivia, and Flores et al. (2000) recognized *G. a. chacoensis* in Argentina; however, limits between subspecies are unclear and the species is treated here as monotypic pending revision. Costa et al.

(2003) detailed differences between *G. agilis* and *G. microtarsus*, two taxa that Gardner (1993) suggested could prove conspecific.

*Gracilinanus agricolai* (Moojen, 1943). Bol. Mus. Nac., Rio de Janeiro, Nova Sér., Zool. 5:2.
COMMON NAME: Agricola's Gracile Opossum.
TYPE LOCALITY: Brazil, Ceará, "Crato."
DISTRIBUTION: Known only from the type locality.
COMMENTS: Included in *emiliae* (Thomas) by Gardner (1993; see review of *emiliae* by Voss et al., 2001); placement in *Gracilinanus* tentative and problematic.

*Gracilinanus dryas* (Thomas, 1898). Ann. Mag. Nat. Hist., ser. 7, 1:456.
COMMON NAME: Wood Sprite Gracile Opossum.
TYPE LOCALITY: Venezuela, Mérida, "Culata."
DISTRIBUTION: Andes of W Venezuela.
STATUS: IUCN –Vulnerable.

*Gracilinanus emiliae* (Thomas, 1909). Ann. Mag. Nat. Hist., ser. 8, 3:379.
COMMON NAME: Emilia's Gracile Opossum.
TYPE LOCALITY: Brazil, "Para."
DISTRIBUTION: Colombia, Venezuela, Guyana, Suriname, French Guiana, NE Brazil.
STATUS: IUCN – Vulnerable; Listed as Data Deficient as *G. longicaudis*, but *G. longicaudis* represents a misidentification by Hershkovitz and has no standing as a species or as a population.
SYNONYMS: *longicaudis* Hershkovitz, 1992.
COMMENTS: See review by Voss et al., (2001).

*Gracilinanus formosus* (Shamel, 1930). J. Mammal., 11:311.
COMMON NAME: Pygmy Opossum.
TYPE LOCALITY: Argentina, Formosa, "Riacho Pilago, 10 miles [16 km] northwest of Kilometro 182.
DISTRIBUTION: Known only from the type locality in Formosa Province, Argentina.
SYNONYMS: *Marmosa muscula* Shamel, 1930 [preoccupied].
COMMENTS: Included in *agilis* (Burmeister) by Gardner (1993); placement in *Gracilinanus* tentative and problematic.

*Gracilinanus ignitus* Díaz, Flores, and Barquez, 2002. J. Mammal., 83:825.
COMMON NAME: Red-bellied Gracile Opossum.
TYPE LOCALITY: Argentina, Jujuy, "Yuto."
DISTRIBUTION: Known only from type locality.

*Gracilinanus marica* (Thomas, 1898) Ann. Mag. Nat. Hist., ser. 7, 1:455.
COMMON NAME: Northern Gracile Opossum.
TYPE LOCALITY: Venezuela, Mérida, "R. Albarregas."
DISTRIBUTION: N Colombia and Venezuela.
STATUS: IUCN – Lower Risk (nt); Data Deficient as *G. perijae*.
SYNONYMS: *perijae* Hershkovitz, 1992.

*Gracilinanus microtarsus* (Wagner, 1842). Arch. Naturgesch., 8(1):359.
COMMON NAME: Brazilian Gracile Opossum.
TYPE LOCALITY: Brazil, São Paulo, "Ypanema."
DISTRIBUTION: SE Brazil.
STATUS: IUCN – Lower Risk (nt).
SYNONYMS: **guahybae** (Tate, 1931); *herhardti* Miranda-Ribeiro, 1936.
COMMENTS: Costa et al. (2003) detailed differences between *G. agilis* and *G. microtarsus*, two taxa that Gardner (1993) suggested could prove conspecific.

*Hyladelphys* Voss, Lunde, and Simmons, 2001. Bull. Am. Mus. Nat. Hist., 263:30.
TYPE SPECIES: *Gracilinanus kalinowskii* Hershkovitz, 1992, by original designation.
COMMENTS: Monotypic.

*Hyladelphys kalinowskii* (Hershkovitz, 1992). Fieldiana Zool., n.s., 70:37.
　　COMMON NAME: Kalinowski's Mouse Opossum.
　　TYPE LOCALITY: Perú, Cuzco, "Hacienda Cadena, Marcapata."
　　DISTRIBUTION: S Guyana, N French Guiana, west to E Perú.
　　STATUS: IUCN – Data Deficient as *Gracilinanus kalinowskii*.
　　COMMENTS: Described as *Gracilinanus kalinowskii*; see Voss et al. (2001).

*Lestodelphys* Tate, 1934. J. Mammal., 15:154.
　　TYPE SPECIES: *Notodelphys halli* Thomas, 1921, by original designation.
　　SYNONYMS: *Notodelphys* Thomas, 1921 [preoccupied].
　　COMMENTS: Monotypic.

*Lestodelphys halli* (Thomas, 1921). Ann. Mag. Nat. Hist., ser. 9, 8:137.
　　COMMON NAME: Patagonian Opossum.
　　TYPE LOCALITY: Argentina, Santa Cruz, "Cabo Tres Puntas;" subsequently emended to
　　　　"Estancia Madujada [= Estancia La Madrugada], not far from Puerto Deseado"
　　　　(Thomas, 1929:45).
　　DISTRIBUTION: Provincia Mendoza south to Provincia de Santa Cruz, Argentina.
　　STATUS: IUCN – Vulnerable.
　　COMMENTS: Reviewed by L. G. Marshall (1977, Mammalian Species, 81).

*Lutreolina* Thomas, 1910. Ann. Mag. Nat. Hist., ser. 8, 5:247.
　　TYPE SPECIES: *Didelphis crassicaudata* Desmarest, 1804, by monotypy.
　　SYNONYMS: *Didelphis* Linnaeus, 1758 [part]; *Sarigua* Muirhead, 1819 [part].

*Lutreolina crassicaudata* (Desmarest, 1804). Tabl. Méth. Hist. Nat., *in* Nouv. Dict. Hist. Nat., 24:19.
　　COMMON NAME: Lutrine Opossum.
　　TYPE LOCALITY: Paraguay, Asunción, by subsequent restriction (Cabrera, 1958:39).
　　DISTRIBUTION: South America in two populations: E Colombia, Venezuela, and W Guyana;
　　　　E Bolivia, SE Brazil, Paraguay, Uruguay, and Argentina south to Provincia de Buenos
　　　　Aires.
　　STATUS: IUCN – Lower Risk (lc).
　　SYNONYMS: *bonaria* Thomas, 1923; *crassicaudis* (Olfers, 1818); *ferruginea* Larrañaga, 1923;
　　　　*lutrilla* Thomas, 1923; *paranalis* Thomas, 1923; *travassosi* Miranda-Ribeiro, 1936;
　　　　**turneri** (Günther, 1879).
　　COMMENTS: Reviewed by Marshall (1978*a*, Mammalian Species, 91). The names *crassicaudis*
　　　　Olfers, 1815, and *mustelina* Waterhouse, 1846, are *nomina nuda*.

*Marmosa* Gray, 1821. London Med. Repos., 15:308.
　　TYPE SPECIES: *Didelphis marina* Gray, 1821, by monotypy (incorrect subsequent spelling of
　　　　*Didelphis murina* Linnaeus, 1758).
　　SYNONYMS: *Asagis* Gloger, 1841; *Cuica* Liais, 1872; *Didelphis* Linnaeus, 1758 [part]; *Grayium*
　　　　Kretzoi and Kretzoi, 2000; *Grymaeomys* Burmeister, 1854; *Notagogus* Gloger, 1841;
　　　　*Opossum* Schmid, 1818 [part]; *Sarigua* Muirhead, 1819 [part]; *Stegomarmosa* Pine, 1972.
　　COMMENTS: *Ouica* Cabrera, 1958, is an incorrect subsequent spelling of *Cuica* Liais.

*Marmosa andersoni* Pine, 1972. J. Mammal., 53:279.
　　COMMON NAME: Heavy-browed Mouse Opossum.
　　TYPE LOCALITY: Perú, Cuzco, "Hda. Villa Carmen, Cosñipata."
　　DISTRIBUTION: Known only from the type locality.
　　STATUS: IUCN – Critically Endangered.
　　COMMENTS: Type species of *Stegomarmosa* Pine.

*Marmosa lepida* (Thomas, 1888). Ann. Mag. Nat. Hist., ser. 6, 1:158.
　　COMMON NAME: Rufous Mouse Opossum.
　　TYPE LOCALITY: "Peruvian Amazons;" identified as Perú, Loreto, Santa Cruz, Huallaga River,
　　　　by Thomas (1888*a*:348).
　　DISTRIBUTION: Surinam and E Colombia, Ecuador, Perú, Brazil, and Bolivia.
　　STATUS: IUCN – Lower Risk (nt).

SYNONYMS: *grandis* Tate, 1931.

*Marmosa mexicana* Merriam, 1897. Proc. Biol. Soc. Wash., 11:44.
COMMON NAME: Mexican Mouse Opossum.
TYPE LOCALITY: México, Oaxaca, "Juquila."
DISTRIBUTION: Tamaulipas, México to W Panamá.
STATUS: IUCN – Lower Risk (lc).
SYNONYMS: *mayensis* Osgood, 1913; *savannarum* Goldman, 1917; *zeledoni* Goldman, 1917.

*Marmosa murina* (Linnaeus, 1758). Syst. Nat., 10th ed., 1:55.
COMMON NAME: Linnaeus's Mouse Opossum.
TYPE LOCALITY: "Asia, America;" restricted to Surinam by Thomas (1911*a*).
DISTRIBUTION: Colombia, Venezuela, Trinidad and Tobago, Guyana, Surinam, French
   Guiana, Brazil, E Ecuador, E Perú, and Bolivia.
STATUS: IUCN – Lower Risk (lc).
SYNONYMS: *bombascarae* Anthony, 1922; *chloe* Thomas, 1907; *dorsigera* (Linnaeus, 1758);
   *duidae* Tate, 1931; *guianensis* (Kerr, 1792); *klagesi* J. A. Allen, 1900; *macrotarsus* (Wagner,
   1842); *madeirensis* Cabrera, 1913; *maranii* Thomas, 1924; *meridionalis* Miranda-Ribeiro,
   1936; *moreirae* Miranda-Ribeiro, 1936; *muscula* (Cabanis, 1848); *parata* Thomas, 1911;
   *roraimae* Tate, 1931; *tobagi* Thomas, 1911; *waterhousii* (Tomes, 1860).
COMMENTS: Tate (1933) recognized 8 subspecies and Cabrera (1959) recognized 7; however,
   limits between subspecies are unclear and the species is treated here as monotypic
   pending revision. Work by Patton and Costa (2003) suggests that, as currently
   understood, *murina* is composite. The names *marina* E. Geoffroy, 1803, and *marina*
   Gray, 1821, are incorrect subsequent spellings of *murina* Linnaeus; *waterhausi*
   Miranda-Ribeiro, 1936, is an incorrect subsequent spelling of *waterhousii* (Tomes).
   *Didelphis murina* Linnaeus, 1758, is the type species of *Asagis* Gloger, 1841; *Notagogus*
   Gloger, 1841; *Grymaeomys* Burmeister, 1854; and *Cuica* Liais, 1872.

*Marmosa quichua* Thomas, 1899. Ann. Mag. Nat. Hist., ser. 7, 3:43.
COMMON NAME: Quechuan Mouse Opossum.
TYPE LOCALITY: "Ocabamba" identified as Valle de Occobamba by Ceballos Bendezu (1981),
   Cuzco, Perú.
DISTRIBUTION: E Perú.
SYNONYMS: *musicola* Osgood, 1913.
COMMENTS: Included in *murina* by Gardner (1993); see Voss et al. (2001:41).

*Marmosa robinsoni* Bangs, 1898. Proc. Biol. Soc. Wash., 12:95.
COMMON NAME: Robinson's Mouse Opossum.
TYPE LOCALITY: Venezuela, Nueva Esparta, "Margarita Island."
DISTRIBUTION: Belize, Honduras (Isla Ruatán), Panamá, Colombia, W Ecuador, NW Perú,
   N Venezuela, and Grenada (Lesser Antilles).
STATUS: IUCN – Lower Risk (lc).
SYNONYMS: *casta* Thomas, 1911; *mitis* Bangs, 1898; *pallidiventris* Osgood, 1912; *chapmani*
   J. A. Allen, 1900; *nesaea* Thomas, 1911; *fulviventer* Bangs, 1901; *grenadae* Thomas,
   1911; *isthmica* Goldman, 1912; *luridavolta* Goodwin, 1961; *ruatanica* Goldman,
   1911; *simonsi* Thomas, 1899; *mimetra* Thomas, 1921.
COMMENTS: Reviewed by O'Connell (1983, Mammalian Species, 203).

*Marmosa rubra* Tate, 1931. Am. Mus. Novit., 493:6.
COMMON NAME: Red Mouse Opossum.
TYPE LOCALITY: Perú, Loreto, "mouth of Rio Curaray."
DISTRIBUTION: E Ecuador and Perú.
STATUS: IUCN –Lower Risk (lc).

*Marmosa tyleriana* Tate, 1931. Am. Mus. Novit., 493:6.
COMMON NAME: Tyleria Mouse Opossum.
TYPE LOCALITY: Venezuela, Amazonas, "Central Camp, Mt. Duida Plateau, Upper Rio
   Orinoco."
DISTRIBUTION: Guayanan Highland tepuis of Venezuela.

STATUS: IUCN – Data Deficient.
SYNONYMS: *phelpsi* Tate, 1939.

*Marmosa xerophila* Handley and Gordon, 1979. *In* J. F. Eisenberg (ed.), Vertebrate ecology in the northern Neotropics, Smithson. Inst. Press, p. 68.
COMMON NAME: Guajira Mouse Opossum.
TYPE LOCALITY: Colombia, Guajira, "La Isla, 15 m, near Cajoro, 37 km NNE Paraguaipoa."
DISTRIBUTION: NE Colombia and NW Venezuela.
STATUS: IUCN – Endangered.

*Marmosops* Matschie, 1916. Sitzb. Ges. Naturf. Fr., Berlin, 1916(1):267.
TYPE SPECIES: *Didelphis incana* Lund, 1840, by original designation.
SYNONYMS: *Didelphis* Linnaeus, 1758 [part]; *Grymaeomys* Burmeister, 1854 [part]; *Thylamys* Gray, 1821 [part].
COMMENTS: Previously in *Marmosa* (*sensu lato*; see Gardner and Creighton, 1989).

*Marmosops bishopi* (Pine, 1981). Mammalia, 45:63.
COMMON NAME: Bishop's Slender Opossum.
TYPE LOCALITY: "264 km N (by road) Xavantina (locality is at 12°51'S, 51°46'W), Serra do Roncador, Mato Grosso, Brazil."
DISTRIBUTION: Brasil, Perú, and Bolivia.
COMMENTS: Described as a subspecies of *M. parvidens*; treated as a species by Voss et al. (2001:48).

*Marmosops cracens* (Handley and Gordon, 1979). *In* J. F. Eisenberg (ed.), Vertebrate ecology in the northern Neotropics, Smithson. Inst. Press., p. 66.
COMMON NAME: Narrow-headed Slender Opossum.
TYPE LOCALITY: Venezuela, Falcón, "near La Pastora (11°12'N, 68°37'W), 150 m, 14 km ENE Mirimire."
DISTRIBUTION: Known only from the type locality.
STATUS: IUCN – Endangered.

*Marmosops dorothea* (Thomas, 1911). Ann. Mag. Nat. Hist., ser. 8, 7:516.
COMMON NAME: Dorothys' Slender Opossum.
TYPE LOCALITY: Bolivia, La Paz, "Rio Solocame, 67°W., 16°S."
DISTRIBUTION: NE Bolivia.
STATUS: IUCN – Vulnerable (B1+2c).
SYNONYMS: *yungasensis* (Tate, 1931).

*Marmosops fuscatus* (Thomas, 1896). Ann. Mag. Nat. Hist., ser. 6, 18:313.
COMMON NAME: Dusky Slender Opossum.
TYPE LOCALITY: Venezuela, Mérida, "Rio Abbarregas [= Río Alvarregas]."
DISTRIBUTION: E Andes of Colombia, N Venezuela, and Isl of Trinidad.
STATUS: IUCN – Lower Risk (nt).
SYNONYMS: ***carri*** (J. A. Allen and Chapman, 1897); ***perfuscus*** (Thomas, 1924).

*Marmosops handleyi* (Pine, 1981). Mammalia, 45:67.
COMMON NAME: Handley's Slender Opossum.
TYPE LOCALITY: Colombia, Antioquia, "9 km S Valdivia."
DISTRIBUTION: Known from the vicinity of the type locality.
STATUS: IUCN – Critically Endangered.
COMMENTS: Known from only two specimens.

*Marmosops impavidus* (Tschudi, 1845). Fauna Peruana, 1:149.
COMMON NAME: Tschudi's Slender Opossum.
TYPE LOCALITY: "Der mittleren und tiefen Waldregion;" interpreted by Cabrera (1958:16) as Perú, Junín, "Montaña de Vitoc, cerca de Chanchamayo."
DISTRIBUTION: Darién of Panamá to Venezuela, Colombia, Ecuador, Perú, W Brazil, and E Bolivia.
STATUS: IUCN – Lower Risk (nt).
SYNONYMS: *albiventris* (Tate, 1931); *caucae* (Thomas, 1900); *celicae* (Anthony, 1922);

*madescens* (Osgood, 1913); *ocellatus* (Tate, 1931); *oroensis* (Anthony, 1922); *purui* (Miller, 1913); *sobrina* (Thomas, 1913); *ucayaliensis* (Tate, 1931).
COMMENTS: This taxon is too poorly known to assign names to subspecies with confidence.

*Marmosops incanus* (Lund, 1840). K. Dansk. Vid. Selsk. Afhandl., p. 21 [preprint of Lund, 1841].
COMMON NAME: Gray Slender Opossum.
TYPE LOCALITY: Brazil, Minas Gerais, "Rio das Velhas," Lagoa Santa.
DISTRIBUTION: E Brazil from the states of Bahia south to São Paulo.
STATUS: IUCN – Lower Risk (nt).
SYNONYMS: *bahiensis* (Tate, 1931); *scapulatus* (Burmeister, 1856).
COMMENTS: Reviewed by Mustrangi and Patton (1997).

*Marmosops invictus* (Goldman, 1912). Smithson. Misc. Coll., 60(2):3.
COMMON NAME: Panama Slender Opossum.
TYPE LOCALITY: Panamá, Darien, "Cana."
DISTRIBUTION: Panamá.
STATUS: IUCN – Lower Risk (nt).
COMMENTS: Reviewed by Pine (1981).

*Marmosops juninensis* (Tate, 1931). Am. Mus. Novit., 493:13.
COMMON NAME: Junin Slender Opossum.
TYPE LOCALITY: "Utcuyacu. Between Tarma and Chanchamayo," Junín, Perú.
DISTRIBUTION: Chanchamayo Valley of Perú.
COMMENTS: Included under *parvidens* by Pine (1981) and Gardner (1993); considered a species by Voss et al. (2001:47).

*Marmosops neblina* Gardner, 1990. *In* K. H. Redford and J. F. Eisenberg, eds., Advances in Neotropical Mammalogy, The Sandhill Crane Press, Inc., p. 414.
COMMON NAME: Neblina Slender Opossum.
TYPE LOCALITY: Venezuela, Amazonas, "Camp VII (00°50'40"N, 65°58'10"W), 1800 m, Cerro de la Neblina."
DISTRIBUTION: S Venezuela, E Ecuador, E Perú, and W Brazil (Patton et al., 2000).
COMMENTS: Described as a subspecies of *M. impavidus*; reviewed by Mustrangi and Patton (1997) and Patton et al. (2000). The spelling *neblinae* Linares, 1998, is an incorrect subsequent spelling of *neblina* (Gardner). One of the several names grouped under *M. impavidus* may prove to be a senior synonym for *neblina*.

*Marmosops noctivagus* (Tschudi, 1844). Fauna Peruana, 1:148.
COMMON NAME: White-bellied Slender Opossum.
TYPE LOCALITY: "Der mittleren und tiefen Waldregion"; restricted by Tate (1933) to Perú, Junín, Montaña de Vitoc, near Chanchamayo, Río Perené drainage.
DISTRIBUTION: Amazonian Ecuador, Perú, Bolivia, and adjacent Brazil.
STATUS: IUCN – Lower Risk (lc).
SYNONYMS: *collega* (Thomas, 1920); *keaysi* (J. A. Allen, 1900); *leucastrus* (Thomas, 1927); *lugenda* (Thomas, 1927); *neglectus* (Osgood, 1915); *politus* (Cabrera, 1913); *stollei* (Miranda-Ribeiro, 1936).
COMMENTS: Anderson (1997) recognized *M. noctivagus keaysi* in Bolivia. Nevertheless, *noctivagus* is too poorly known to assign names to subspecies with confidence.

*Marmosops parvidens* (Tate, 1931). Am. Mus. Novit., 493:13.
COMMON NAME: Delicate Slender Opossum.
TYPE LOCALITY: Guyana, East Demerara-West Coast Berbice, "Hyde Park, 30 miles [48 km] up the Demerara River."
DISTRIBUTION: E Colombia, N Venezuela, Guyana, French Guiana and N Brazil.
STATUS: IUCN – Lower Risk (nt).
COMMENTS: Voss et al. (2001) considered Pine's (1981) five subspecies to represent four species. The name *parvidentata* Tate, 1933, is an incorrect subsequent spelling of *parvidens* (Tate).

*Marmosops paulensis* (Tate, 1931). Am. Mus. Novit., 493:8.
COMMON NAME: Brazilian Slender Opossum.

TYPE LOCALITY: Brazil, "Therezopolis, Rio de Janeiro, São Paulo [sic]."
DISTRIBUTION: SE Brazil in states of Minas Gerais, Rio de Janeiro, São Paulo, and Paraná.
COMMENTS: Reviewed by Mustrangi and Patton (1997).

*Marmosops pinheiroi* (Pine, 1981). Mammalia, 45:61.
COMMON NAME: Pinheiro's Slender Opossum.
TYPE LOCALITY: Brazil, Amapá, "Rio Amapari, Serra do Navio (0°59'N, 52°03'W)."
DISTRIBUTION: Venezuela, Guyana, Suriname, French Guiana, and Brazil.
SYNONYMS: *woodalli* (Pine, 1981).
COMMENTS: Not a subspecies of *M. parvidens*; see Voss et al. (2001).

*Metachirus* Burmeister, 1854. Syst. Uebers. Thiere Bras., 1:135.
TYPE SPECIES: *Didelphis myosuros* Temminck, 1824, by subsequent designation (Thomas, 1888).
SYNONYMS: *Cuica* Liais, 1872 [part].
COMMENTS: Hall (1981) followed Pine (1973) in using the name *Philander*; see Hershkovitz (1976, 1981) who reaffirmed the use of *Philander* for the gray four-eyed opossums. *Metacherius* Sanderson, 1949, is an incorrect subsequent spelling of *Metachirus* Burmeister.

*Metachirus nudicaudatus* (É. Geoffroy, 1803). Cat. Mam. Mus. Natl. Hist. Nat., Paris, 142.
COMMON NAME: Brown Four-eyed Opossum.
TYPE LOCALITY: French Guiana, "Cayenne."
DISTRIBUTION: Mexico to Paraguay and N Argentina.
STATUS: IUCN – Lower Risk (lc).
SYNONYMS: **colombianus** J. A. Allen, 1900; *antioquiae* J. A. Allen, 1916; *dentaneus* Goldman, 1912; *imbutus* Thomas, 1923; *phaeurus* Thomas, 1901; **modestus** Thomas, 1923; **myosuros** (Temminck, 1824); *personatus* Miranda-Ribeiro, 1936; **tschudii** J. A. Allen, 1900; *bolivianus* J. A. Allen, 1901; *infuscus* Thomas, 1923.
COMMENTS: É. Geoffroy St.-Hilaire's (1803) catalog of the mammals in the Paris Museum was previously considered an unpublished work (Wilson and Reeder, 1993); therefore, Gardner (1993) dated *nudicaudatus* from Desmarest, 1817. Geoffroy's catalog was subsequently placed on the Official List of Works Approved as Available for Zoological Nomenclature (Opinion 2005 of the International Commission on Zoological Nomenclature, 2002*b*). The names *nudicauda* Lesson, 1827, and *nudicaudus* Gray, 1843, are incorrect subsequent spellings of *nudicaudatus* (É. Geoffroy). The name *colombica* J. A. Allen, 1901, is an incorrect subsequent spelling of *colombianus* J. A. Allen.

*Micoureus* Lesson, 1842. Nouv. Tabl. Regn. Anim. Mammifères, p. 186.
TYPE SPECIES: *Didelphis cinerea* Temminck, 1824, by subsequent designation (Thomas, 1888*a*).
SYNONYMS: *Grymaeomys* Burmeister, 1854 [part].
COMMENTS: Previously in *Marmosa* (*sensu lato*; see Gardner and Creighton, 1989). *Micoures* Reig et al. (1985) and *Micoures* Massoia, 1988, are incorrect subsequent spellings of *Micoureus* Lesson.

*Micoureus alstoni* (J. A. Allen, 1900). Bull. Am. Mus. Nat. Hist., 13:189.
COMMON NAME: Alston's Mouse Opossum.
TYPE LOCALITY: Costa Rica, Cartago, "Tres Rios."
DISTRIBUTION: E Central America from Belize to Panamá and adjacent Caribbean islands.
STATUS: IUCN – Lower Risk (nt).
SYNONYMS: *nicaraguae* Thomas, 1905.
COMMENTS: The species also may occur in Colombia.

*Micoureus constantiae* (Thomas, 1904). Proc. Zool. Soc. Lond., 1903(2):243 [1904].
COMMON NAME: White-bellied Woolly Mouse Opossum.
TYPE LOCALITY: Brazil, Mato Grosso, "Chapada."
DISTRIBUTION: E Bolivia and adjacent Brazil south into N Argentina.
STATUS: IUCN – Lower Risk (nt).
SYNONYMS: *budini* (Thomas, 1920).
COMMENTS: Anderson recognized *M. c. constantiae* and *M. c. budini* in Bolivia.

*Micoureus demerarae* (Thomas, 1905). Ann. Mag. Nat. Hist., ser. 7, 16:313.
COMMON NAME: Woolly Mouse Opossum.
TYPE LOCALITY: Guyana, East Demerara-West Coast Berbice, "Comaccka, 80 miles [129 km] up Demerara River."
DISTRIBUTION: Colombia, Venezuela, French Guiana, Guyana, Surinam, Brazil, E Perú, and E Bolivia.
STATUS: IUCN – Lower Risk (lc).
SYNONYMS: *areniticola* (Tate, 1931); *dominus* (Thomas, 1920); *limae* (Thomas, 1920); *esmeraldae* (Tate, 1931); *meridae* (Tate, 1931).
COMMENTS: Previously included under *Marmosa cinerea* (Temminck, 1824).

*Micoureus paraguayanus* (Tate, 1931). Amer. Mus. Novit. 493:1.
COMMON NAME: Tate's Woolly Mouse Opossum.
TYPE LOCALITY: Paraguay, Guairá, "Villa Rica."
DISTRIBUTION: E Brazil from Minas Gerais and S Bahia south to Rio Grande do Sul and E Paraguay.
SYNONYMS: *travassosi* (Miranda-Ribeiro, 1936); *cinereus* (Temminck, 1824) [preoccupied]; *pfrimeri* (Miranda-Ribeiro, 1936).
COMMENTS: Previously included under *Marmosa cinerea*; original name *Didelphis cinerea* Temminck, 1824, preoccupied by *D. cinerea* Goldfuss, 1812. Included in *M. demerarae* by Gardner (1993); see review by Patton and Costa (2003).

*Micoureus phaeus* (Thomas, 1899). Ann. Mag. Nat. Hist., ser. 7, 3:44.
COMMON NAME: Little Woolly Mouse Opossum.
TYPE LOCALITY: Colombia, Nariño, "San Pablo."
DISTRIBUTION: W slopes of the Andes from SW Colombia to SW Ecuador.
SYNONYMS: *perplexus* (Anthony, 1922).
COMMENTS: *pahea* Ceballos Bendezu, 1981, is an incorrect subsequent spelling of *phaea* (Thomas).

*Micoureus regina* (Thomas, 1898). Ann. Mag. Nat. Hist., ser. 7, 2:274.
COMMON NAME: Bare-tailed Woolly Mouse Opossum.
TYPE LOCALITY: Colombia, Cundinamarca, "West Cundinamarca (Bogotá Region)."
DISTRIBUTION: Colombia, Ecuador, Perú, Brazil, and Bolivia.
STATUS: IUCN – Lower Risk (lc).
SYNONYMS: *germanus* (Thomas, 1904); *parda* (Tate, 1931); *rutteri* (Thomas, 1924); *rapposa* (Thomas, 1899); *mapiriensis* (Tate, 1931).

*Monodelphis* Burnett, 1830. Quart. J. Sci. Lit. Art., 1829:351 [1830].
TYPE SPECIES: *Monodelphis brachyura* Burnett, 1830 (= *Didelphys brachyuros* Schreber, 1777, = *Didelphis brevicaudata* Erxleben, 1777) by subsequent designation (Matschie, 1916).
SYNONYMS: *Hemiurus* Gervais, 1855; *Microdelphys* Burmeister, 1856; *Minuania* Cabrera, 1919; *Monodelphiops* Matschie, 1916; *Peramys* Lesson, 1842.
COMMENTS: *Microdidelphys* Trouessart, 1898, is an incorrect subsequent spelling of *Microdelphys* Burmeister.

*Monodelphis adusta* (Thomas, 1897). Ann. Mag. Nat. Hist., ser. 6, 20:219.
COMMON NAME: Sepia Short-tailed Opossum.
TYPE LOCALITY: Colombia, "W. Cundinamarca, in the low-lying hot regions."
DISTRIBUTION: E Panamá, Colombia, Ecuador, Perú, and Bolivia.
STATUS: IUCN – Lower Risk (lc).
SYNONYMS: *melanops* (Goldman, 1912); *peruvianus* (Osgood, 1913).
COMMENTS: Emmons (1990) recognized *peruvianus* as a separate species.

*Monodelphis americana* (Müller, 1776). Linné's Vollständ. Natursyst., Suppl., p. 36.
COMMON NAME: Northern Three-striped Opossum.
TYPE LOCALITY: "Brasilien," restricted to Brazil, Pernambuco, Pernambuco, by Cabrera (1958:7).
DISTRIBUTION: E Brazil from the states of Pará south to Santa Catarina.
STATUS: IUCN – Lower Risk (nt).

SYNONYMS: *brasiliensis* (Erxleben, 1777); *brasiliensis* (Daudin *in* Lacépède, 1802); *trilineata* (Lund, 1840); *tristriata* (Olfers, 1818).

COMMENTS: The name *tristriata* Illiger, 1815, is a *nomen nudum*. *Microdelphys tristriatus* as used by Burmeister, 1856 (= *Didelphys tristriata* Olfers, 1818 = *Sorex americanus* Muller, 1776), is the type species of *Microdelphys* Burmeister, 1856.

*Monodelphis brevicaudata* (Erxleben, 1777). Syst. Regni Anim., 1:80.
COMMON NAME: Northern Red-sided Opossum.
TYPE LOCALITY: "In Americae australis silvis;" restricted to Surinam by Matschie (1916).
DISTRIBUTION: Venezuela, the Guianas, and adjacent Brasil.
STATUS: IUCN – Lower Risk (lc).
SYNONYMS: *brachyuros* (Schreber, 1777); *dorsalis* (J. A. Allen, 1904); *hunteri* (Waterhouse, 1841); *orinoci* (Thomas, 1899); *sebae* (Gray, 1827); *surinamensis* (Zimmermann, 1780); *touan* (Bechstein, 1800); *touan* (Shaw, 1800); *touan* (Daudin *in* Lacépède, 1802); *tricolor* (E. Geoffroy, 1803).
COMMENTS: Reviewed by Voss et al. (2001), who restricted the name to populations in the Guiana Region of northern South America. The name *orinoci* (Thomas) has been applied to populations in the Venezuelan Llanos north and west of the Orinoco by several authors (e.g., Reig et al., 1977; Linares, 1998). Apparently, that population lacks a name. The name *brachyura* Burnett, 1830, is an incorrect subsequent spelling of *brachyuros* (Schreber).

*Monodelphis dimidiata* (Wagner, 1847). Abh. Akad. Wiss., München, 5(1):151, footnote.
COMMON NAME: Yellow-sided Opossum.
TYPE LOCALITY: Uruguay, Maldonado, "Maldonado am la Plata."
DISTRIBUTION: Uruguay, SE Brazil, and NE Argentina.
STATUS: IUCN – Lower Risk (nt).
SYNONYMS: *fosteri* Thomas, 1924.
COMMENTS: See Pine et al. (1985). *Peramys brachyurus* as used by Lesson, 1842 (= *Didelphys dimidiata* Wagner, 1847), is the type species of *Peramys* Lesson, 1842. *Didelphys dimidiata* Wagner is the type species of *Minuania* Cabrera, 1919; therefore *Minuania* Cabrera is a junior objective synonym of *Peramys* Lesson.

*Monodelphis domestica* (Wagner, 1842). Arch. Naturgesh., 8:359.
COMMON NAME: Gray Short-tailed Opossum.
TYPE LOCALITY: Brazil, Mato Grosso, "Cuyaba."
DISTRIBUTION: Brazil, Bolivia, Paraguay, and N Argentina.
STATUS: IUCN – Lower Risk (lc).
SYNONYMS: *concolor* Gervais, 1856.

*Monodelphis emiliae* (Thomas, 1912). Ann. Mag. Nat. Hist., ser. 8, 9:89.
COMMON NAME: Emilia's Short-tailed Opossum.
TYPE LOCALITY: Brazil, Pará, "Boim, R. Tapajoz."
DISTRIBUTION: Amazon Basin of Perú, Brazil, and N Bolivia.
STATUS: IUCN – Vulnerable.
COMMENTS: Previously treated as a subspecies of *M. touan*. Reviewed by Pine and Handley (1984).

*Monodelphis glirina* (Wagner, 1842). Arch. Naturgesch. 8(1):359.
COMMON NAME: Amazonian Red-sided Opossum.
TYPE LOCALITY: Brazil, Rondônia, "Mamoré."
DISTRIBUTION: Central Brazil to SE Peru and N Bolivia.
COMMENTS: Considered a species by Voss et al. (2001).

*Monodelphis iheringi* (Thomas, 1888). Ann. Mag. Nat. Hist., ser. 6, 1:159.
COMMON NAME: Ihering's Three-striped Opossum.
TYPE LOCALITY: "Rio Grande do Sul;" identified as Brazil, Rio Grande do Sul, Taquara, by Thomas (1888a).
DISTRIBUTION: SE Brazil (Espírito Santo, São Paulo, Santa Catarina, and Rio Grande do Sul).
STATUS: IUCN – Lower Risk (nt).

COMMENTS: Previously considered a subspecies of *M. americana*. Reviewed by Pine (1977).

*Monodelphis kunsi* Pine, 1975. Mammalia, 39:321.
COMMON NAME: Pygmy Short-tailed Opossum.
TYPE LOCALITY: Bolivia, Beni, "La Granja, W bank of Río Itonamas, 4 k N Magdalena."
DISTRIBUTION: Known from only four localities, two in Bolivia and two in Brazil.
STATUS: IUCN – Endangered.
COMMENTS: See Anderson (1982, Mammalian Species, 190). To be expected in N Argentina.

*Monodelphis maraxina* Thomas, 1923. Ann. Mag. Nat. Hist., ser. 9, 12:157.
COMMON NAME: Marajó Short-tailed Opossum.
TYPE LOCALITY: Brazil, Pará, "Caldeirão."
DISTRIBUTION: Brazil, Pará, Marajó Isl.
STATUS: IUCN – Vulnerable.
COMMENTS: Eight specimens known. See Pine (1980*a*).

*Monodelphis osgoodi* Doutt, 1938. J. Mammal., 19:100.
COMMON NAME: Osgood's Short-tailed Opossum.
TYPE LOCALITY: Bolivia, Cochabamba, "Incachaca."
DISTRIBUTION: E Perú and C Bolivia.
STATUS: IUCN – Vulnerable.
COMMENTS: Previously included in *M. adusta* by Gardner (1993).

*Monodelphis palliolata* (Osgood, 1914). Field Mus. Nat. Hist. Publ., Zool Ser., 10:135.
COMMON NAME: Hooded Red-sided Opossum.
TYPE LOCALITY: "San Juan de Colon, State of Tachira, Venezuela."
DISTRIBUTION: NE Colombia and W Venezuela.
COMMENTS: Previously included in *M. brevicaudata* by Gardner (1993).

*Monodelphis rubida* (Thomas, 1899). Ann. Mag. Nat. Hist., ser. 7, 4:155.
COMMON NAME: Chestnut-striped Opossum.
TYPE LOCALITY: Brazil, "Bahia."
DISTRIBUTION: E Brazil from Goiás south to São Paulo.
STATUS: IUCN – Vulnerable.

*Monodelphis scalops* (Thomas, 1888). Ann. Mag. Nat. Hist., ser. 6, 1:158.
COMMON NAME: Long-nosed Short-tailed Opossum.
TYPE LOCALITY: "Brazil;" restricted to Rio de Janeiro, Therezôpolis, by Vieira (1949).
DISTRIBUTION: From Espírito Santo, SE Brazil, south Alto Paraguay, Paraguay (Contreras and Silvera Avalos, 1995).
STATUS: IUCN – Vulnerable
COMMENTS: Reviewed by Pine and Abravaya (1978).

*Monodelphis sorex* (Hensel, 1872). Abh. König. Akad. Wiss. Berlin, 1872:122.
COMMON NAME: Southern Red-sided Opossum.
TYPE LOCALITY: "Provinz Rio Grande do Sul;" restricted to Brazil, Rio Grande do Sul, Taquara, by Cabrera (1958).
DISTRIBUTION: SE Brazil, S Paraguay, and NE Argentina.
STATUS: IUCN – Vulnerable.
SYNONYMS: *henseli* (Thomas, 1888); *itatiayae* (Miranda-Ribeiro, 1936); *lundi* (Matschie, 1916); *paulensis* Vieira, 1950.
COMMENTS: See Pine et al. (1985). *Monodelphis sorex* may be the animal Azara (1801) described as "micouré cinquième, ou micouré à queue corte," in which case *brevicaudis* Olfers, 1818, and its objective synonym *wagneri* (Matschie, 1916) would be synonyms and the name *sorex* (Hensel, 1872) would be replaced by *brevicaudis* Olfers, 1818, which has 52 years priority over *sorex*. However, the true identity of Azara's animal cannot be determined with reasonable certainty. Lacking any compelling evidence to the contrary, the names *brevicaudis* Olfers, and *wagneri* Matschie are best treated as *nomina oblita* and do not affect the validity of *sorex* Hensel or any other *Monodelphis* sp. that may be found in the region. *Microdelphys sorex* Hensel, 1872, is the type species of *Monodelphiops* Matschie, 1916.

*Monodelphis theresa* Thomas, 1921. Ann. Mag. Nat. Hist., ser. 9, 8:441.
  COMMON NAME: Southern Three-striped Opossum.
  TYPE LOCALITY: Brazil, Rio de Janeiro, "Theresopolis, Organ Mtns."
  DISTRIBUTION: E Brazil.
  STATUS: IUCN – Vulnerable.
  COMMENTS: The Peruvian distribution given by Gardner (1993) was based on a similar, but
    undescribed species from Departamento Huánuco.

*Monodelphis umbristriata* (Miranda-Ribeiro, 1936). Rev. Mus. Paulista, São Paulo 20:422.
  COMMON NAME: Red Three-striped Opossum.
  TYPE LOCALITY: Brazil, Goiás, Veadeiros (Ávila Pires, 1968).
  DISTRIBUTION: Known from the Brazilian states of Goiás and Minas Gerais.
  SYNONYMS: *goyana* (Miranda-Ribeiro, 1936).
  COMMENTS: Previously included under *M. rubida*; see Pine (1976a); reviewed by Lemos et al.
    (2000).

*Monodelphis unistriata* (Wagner, 1842). Arch. Naturgesch., 8(1):360.
  COMMON NAME: One-striped Opossum.
  TYPE LOCALITY: Brazil, São Paulo, "Ytarare" (= Itararé).
  DISTRIBUTION: State of São Paulo, Brazil, and Provincía Misiones, Argentina.
  STATUS: IUCN – Vulnerable.
  COMMENTS: Apparently known from only two specimens.

*Philander* Brisson, 1762. Regnum animale, p. 13.
  TYPE SPECIES: *Didelphis opossum* Linnaeus, 1758, by plenary action (Opinion 1894 of the
    International Commission on Zoological Nomenclature, 1998).
  SYNONYMS: *Gamba* Liais, 1872 [part]; *Hylothylax* Cabrera, 1919; *Metachirops* Matschie, 1916;
    *Philander* Tiedemann, 1808; *Sarigua* Muirhead, 1919 [part].
  COMMENTS: Pine (1973) used *Metachirops* for this genus, as did Hall (1981), Husson (1978), and
    Corbet and Hill (1980; but not 1991 when they used *Philander*).

*Philander andersoni* (Osgood, 1913). Field Mus. Nat. Hist. Publ., Zool Ser., 10:95.
  COMMON NAME: Anderson's Four-eyed Opossum.
  TYPE LOCALITY: Perú, Loreto, "Yurimaguas."
  DISTRIBUTION: S Venezuela, E Colombia, Ecuador, and Andean foothills of Perú.
  STATUS: IUCN – Lower Risk (lc).
  SYNONYMS: *nigratus* Thomas, 1923.
  COMMENTS: Reviewed by Patton and Silva (1997).

*Philander frenatus* (Olfers, 1818). *In* W. L. Eschwege, J. Brasilien, Neue Bibliothek. Reisen., 15:204.
  COMMON NAME: Southeastern Four-eyed Opossum.
  TYPE LOCALITY: "Südamerica;" restricted to Bahia, Brasil, by J. A. Wagner (1843).
  DISTRIBUTION: E Brasil southward to Paraguay and adjacent Argentina.
  SYNONYMS: *azaricus* (Thomas, 1923); *quica* Temminck, 1824; *superciliaris* (Olfers, 1818).
  COMMENTS: Previously included in *P. opossum*; reviewed by Patton and Silva (1997); *frenata*
    Illiger, 1815, and *superciliaris* Illiger, 1815, are *nomina nuda*. The spelling *quichua*
    Krumbiegel, 1941, is an incorrect subsequent spelling of *quica* Temminck. *Didelphis*
    *quica* Temminck, 1924 (= *P. frenatus*) is the type species of *Metachirops* Matschie, 1916.

*Philander mcilhennyi* Gardner and Patton, 1972. Occas. Papers Mus. Zool., Louisiana State Univ.,
    43:2.
  COMMON NAME: McIlhenny's Four-eyed Opossum.
  TYPE LOCALITY: Perú, Ucayali, "Balta (10°08'S, 17°13'W), Río Curanja, *ca.* 300 meters."
  DISTRIBUTION: Amazon Basin of C Perú and W Brazil.
  COMMENTS: Previously included in *P. andersoni*; reviewed by Patton and Silva (1997). The
    spelling *mcilhenyi* Pérez-Hernández, 1990, is an incorrect subsequent spelling of
    *mcilhennyi* Gardner and Patton.

*Philander opossum* (Linnaeus, 1758). Syst. Nat., 10th ed., 1:55.
  COMMON NAME: Gray Four-eyed Opossum.

TYPE LOCALITY: "America;" restricted to Surinam by J. A. Allen (1900) and further restricted
   to Paramaribo, Surinam, by Matschie (1916).
DISTRIBUTION: Tamaulipas, México, through Central and South America to Bolivia and
   SC Brazil.
STATUS: IUCN – Lower Risk (lc).
SYNONYMS: *austroamericana* O. Thomas, 1923; *virginianus* Tiedemann, 1808; **canus** (Osgood,
   1913); *crucialis* (Thomas, 1923); **fuscogriseus** (J. A. Allen, 1900); *grisescens* (J. A. Allen,
   1901); *melantho* (Thomas, 1923); *pallidus* (J. A. Allen, 1901); **melanurus** (Thomas, 1899).
COMMENTS: Corbet and Hill (1980), Hall (1981), Husson (1978), and Pine (1973) used
   *Metachirops opossum* for this species. Reviewed by Castro-Arellano et al. (2000,
   Mammalian Species, 638). The name *D. larvata* Jentink, 1888, is a *nomen nudum*.
   *Didelphis opossum* Linnaeus, 1758, is the type species for *Holothylax* Cabrera, 1919.

*Thylamys* Gray, 1843. List Specimens Mamm. Coll. Brit. Mus., p. 101.
   TYPE SPECIES: *Didelphis elegans* Waterhouse, 1839, by monotypy.
   SYNONYMS: *Grymaeomys* Burmeister, 1854 [part]; *Microdelphys* Burmeister, 1856 [part]; *Sarigua*
      Muirhead, 1819 [part].
   COMMENTS: Previously a subgenus under *Marmosa*. *Thulamys* Reig et al. (1985) is an incorrect
      subsequent spelling of *Thylamys* Gray.

*Thylamys cinderella* (Thomas, 1902). Ann. Mag. Nat. Hist., ser. 7, 10:159.
   COMMON NAME: Cinderella Fat-tailed Mouse Opossum.
   TYPE LOCALITY: Argentina, "Tucuman."
   DISTRIBUTION: N Argentina and probably S Bolivia.
   COMMENTS: Reviewed by Flores et al. (2000). Previously included under *T. elegans* by
      Gardner (1993).

*Thylamys elegans* (Waterhouse, 1839). Zool. H.M.S. "Beagle," Mammalia, p. 95.
   COMMON NAME: Elegant Fat-tailed Mouse Opossum.
   TYPE LOCALITY: Chile, Coquimbo, "Valparaiso."
   DISTRIBUTION: Chile, on Pacific side of Andes between 32° and 38° S.
   STATUS: IUCN – Lower Risk (lc).
   SYNONYMS: *coquimbensis* (Tate, 1931); *soricinus* (Philippi, 1894).
   COMMENTS: Reviewed by Palma (1997, Mammalian Species 572; Palma's concept of the
      species further restricted by Solari (2003). The name *hortenis* Reid, 1837, is a *nomen
      nudum*.

*Thylamys karimii* (Petter, 1968). Mammalia, 32:313.
   COMMON NAME: Karimi's Fat-tailed Mouse Opossum.
   TYPE LOCALITY: Brazil, Pernambuco, "région d'Exu."
   DISTRIBUTION: Known only from type locality and Serra do Roncador, Mato Grosso.
   COMMENTS: Previously included under *T. pusillus* by Gardner (1993), and under *T. velutinus*
      by Palma (1995).

*Thylamys macrurus* (Olfers, 1818). *In* W. L. Eschwege, J. Brasilien, Neue Bibliothek. Reisen. 15:205.
   COMMON NAME: Paraguayan Fat-tailed Mouse Opossum.
   TYPE LOCALITY: "Sudamerica;" restricted to Paraguay, Presidente Hayes, "Tapoua" (= Tapua);
      see comments.
   DISTRIBUTION: Paraguay and S Brazil.
   STATUS: IUCN – Lower Risk (nt).
   SYNONYMS: *grisea* (Desmarest, 1827).
   COMMENTS: Based on Azara's (1801:290) "Micouré à queue longue;" therefore, the type
      locality is Tapuá. Previously called *T. griseus*. The name *macroura* Illiger, 1815, is a
      *nomen nudum*.

*Thylamys pallidior* (Thomas, 1902). Ann. Mag. Nat. Hist., ser. 7, 10:161.
   COMMON NAME: White-bellied Fat-tailed Mouse Opossum.
   TYPE LOCALITY: Bolivia, Oruro, "Challapata."
   DISTRIBUTION: W and S Peru, N Chile, and S Bolivia as far south as Peninsula Valdéz,
      Argentina.

STATUS: IUCN – Lower Risk (lc).
SYNONYMS: *bruchi* (Thomas, 1921); *fenestrae* (Marelli, 1932); *pulchellus* (Cabrera, 1934).
COMMENTS: Identification of Peruvian and Chilean populations based on Solari (2003). Subspecies are not recognized here; needs revision.

*Thylamys pusillus* (Desmarest, 1804). Tabl. Méth. Hist. Nat., *in* Nouv. Dict. Hist. Nat., 24:19.
COMMON NAME: Common Fat-tailed Mouse Opossum.
TYPE LOCALITY: Not stated; restricted to Paraguay, Misiones, "San Ignacio," by Tate (1933).
DISTRIBUTION: Paraguay, SE Bolivia, and N Argentina.
STATUS: IUCN – Lower Risk (lc).
SYNONYMS: *citellus* (Thomas, 1912); *marmotus* (Thomas, 1902); *nanus* (Olfers, 1818); *verax* (Thomas, 1921).
COMMENTS: The name *nana* Illiger, 1815, is a *nomen nudum* and was not validated by Oken, 1816. The name *marmota* Oken, 1816, is not available; however its usage by Thomas (1902) is valid.

*Thylamys sponsorius* (Thomas, 1921). Ann. Mag. Nat. Hist., ser. 9, 7:186.
COMMON NAME: Argentine Fat-tailed Mouse Opossum.
TYPE LOCALITY: Argentina, Jujuy, "Sunchal, 1200 m."
DISTRIBUTION: N Argentina and S Bolivia.
COMMENTS: Reviewed by Flores et al. (2000). Previously included under *T. elegans* by Gardner (1993).

*Thylamys tatei* (Handley, 1957). J. Washington Acad. Sci., 1956, 46:402 [1957].
COMMON NAME: Tate's Fat-tailed Mouse Opossum.
TYPE LOCALITY: "Chasquitambo 710 m, lat. 10°18'48" S., long. 77°37'20" W.), Ancash, Peru."
DISTRIBUTION: Departments of Ancash and Lima, Peru.
COMMENTS: Reviewed by Solari (2003). Previously included under *T. elegans* by Gardner (1993).

*Thylamys velutinus* (Wagner, 1842). Archiv Naturgesch., 8(1):360.
COMMON NAME: Dwarf Fat-tailed Mouse Opossum.
TYPE LOCALITY: Brazil, São Paulo, "Ypanema."
DISTRIBUTION: SE Brazil.
STATUS: IUCN – Lower Risk (lc).
SYNONYMS: *pimelurus* (Reinhardt, 1851).

*Thylamys venustus* (Thomas, 1902). Ann. Mag. Nat. Hist., ser. 7, 10:159.
COMMON NAME: Buff-bellied Fat-tailed Mouse Opossum.
TYPE LOCALITY: Bolivia, Cochabamba, "Paratani, W. of Cochabamba."
DISTRIBUTION: Bolivia south to Provincia Neuquén, Argentina.
SYNONYMS: *janetta* (Thomas, 1926).
COMMENTS: Previously included under *T. elegans* by Gardner (1993).

*Tlacuatzin* Voss and Jansa, 2003. Bull. Amer. Mus. Nat. Hist., 276:61.
TYPE SPECIES: *Didelphis (Micoureus) canescens* J. A. Allen, 1893, by original designation.
COMMENTS: Monotypic; previously included under *Marmosa*.

*Tlacuatzin canescens* (J. A. Allen, 1893). Bull. Amer. Mus. Nat. Hist., 5:235.
COMMON NAME: Gray Mouse Opossum.
TYPE LOCALITY: México, Oaxaca, "Santo Domingo de Guzman, Isthmus of Tehuantepec."
DISTRIBUTION: México from S Sonora to Oaxaca, Yucatán, and Tres Marías Isls.
STATUS: IUCN – Data Deficient as *Marmosa canescens*.
SYNONYMS: *insularis* (Merriam, 1908); *oaxacae* (Merriam, 1897); *sinaloae* (J. A. Allen, 1898); **gaumeri** (Osgood, 1913).
COMMENTS: Previously known as *Marmosa canescens*. A mandible recovered from an owl pellet suggests *T. canescens* also occurs in Baja California Sur (López-Forment and Urbano, 1977).

# ORDER PAUCITUBERCULATA
by Alfred L. Gardner

**ORDER PAUCITUBERCULATA** Ameghino, 1894.
> COMMENTS: Formerly included in Marsupialia.

**Family Caenolestidae** Trouessart, 1898. Cat. Mamm. Viv. Foss., 2(5):1205.
> COMMENTS: Reviewed by Marshall (1980) and Bublitz (1987).

*Caenolestes* Thomas, 1895. Ann. Mag. Nat. Hist., ser. 6, 16:367.
> TYPE SPECIES: *Hyracodon fuliginosus* Tomes, 1863, by monotypy.
> SYNONYMS: *Hyracodon* Tomes, 1863 [preoccupied].
> COMMENTS: *Coenolestes* Thomas, 1917, is an incorrect subsequent spelling of *Caenolestes* Thomas.

*Caenolestes caniventer* Anthony, 1921. Am. Mus. Novit., 20:6.
> COMMON NAME: Gray-bellied Caenolestid.
> TYPE LOCALITY: Ecuador, El Oro, "El Chiral."
> DISTRIBUTION: SW Ecuador and NW Perú.
> STATUS: IUCN – Lower Risk (lc).

*Caenolestes condorensis* Albuja and Patterson, 1996. J. Mammal., 77:42
> COMMON NAME: Andean Caenolestid.
> TYPE LOCALITY: Ecuador, Morona-Santiago, "Achupallas."
> DISTRIBUTION: Cordillera del Cóndor of SE Ecuador.

*Caenolestes convelatus* Anthony, 1924. Am. Mus. Novit., 120:1.
> COMMON NAME: Northern Caenolestid.
> TYPE LOCALITY: Ecuador, Pichincha, "Las Maquinas, Western Andes 7000 feet [2134 m] altitude, on trail from Aloag to Santo Domingo de los Colorados."
> DISTRIBUTION: W Colombia and NW Ecuador.
> STATUS: IUCN – Lower Risk (lc).
> SYNONYMS: ***barbarensis*** Bublitz, 1987.

*Caenolestes fuliginosus* (Tomes, 1863). Proc. Zool. Soc. Lond., 1863:51.
> COMMON NAME: Dusky Caenolestid.
> TYPE LOCALITY: "Ecuador."
> DISTRIBUTION: Colombia, Ecuador, and NW Venezuela.
> STATUS: IUCN – Lower Risk (lc).
> SYNONYMS: *tatei* Anthony, 1923; ***centralis*** Bublitz, 1987; ***obscurus*** Thomas, 1895.

*Lestoros* Oehser, 1934. J. Mammal., 15:240.
> TYPE SPECIES: *Orolestes inca* Thomas, 1917, by original designation.
> SYNONYMS: *Cryptolestes* Tate, 1934 [preoccupied]; *Orolestes* Thomas, 1917 [preoccupied].
> COMMENTS: *Lestoros* Oehser is a replacement name for *Cryptolestes* Tate, which in turn was a replacement name for *Orolestes* Thomas, and automatically takes the same type species as that designated by Thomas for *Orolestes*.

*Lestoros inca* (Thomas, 1917). Smithson. Misc. Coll., 68(4):3.
> COMMON NAME: Incan Caenolestid.
> TYPE LOCALITY: Perú, Cuzco, "Torontoy."
> DISTRIBUTION: S Andean Perú and adjacent Bolivia.
> STATUS: IUCN – Lower Risk (lc).
> SYNONYMS: *gracilis* (Bublitz, 1987).

*Rhyncholestes* Osgood, 1924. Field Mus. Nat. Hist. Publ., Zool. Ser., 14:170.
> TYPE SPECIES: *Rhyncholestes raphanurus* Osgood, 1924, by original designation.

*Rhyncholestes raphanurus* Osgood, 1924. Field Mus. Nat. Hist. Publ., Zool. Ser., 14:170.

    COMMON NAME: Long-nosed Caenolestid.

    TYPE LOCALITY: Chile, Biobio, "mouth of Rio Inio, south end of Chiloé Island."

    DISTRIBUTION: SC Chile including Chiloé Isl and adjacent Argentina.

    STATUS: IUCN – Vulnerable.

    SYNONYMS: *continentalis* Bublitz, 1987.

    COMMENTS: See Patterson and Gallardo (1987, Mammalian Species, 286).

# ORDER MICROBIOTHERIA
by Alfred L. Gardner

**ORDER MICROBIOTHERIA** Ameghino, 1889.
SYNONYMS: Gondawanadelphia [part]; Dromiciopsia.
COMMENTS: Traditionally included in Marsupialia. Included in order Polyprotodonta by Reig et al. (1987); considered a separate order by Aplin and Archer (1987) and Marshall et al. (1990). Phylogenetically allied with australidelphian marsupials (e.g., Spotorno et al., 1997; Burk et al., 1999).

**Family Microbiotheriidae** Ameghino, 1887. Bol. Mus. la Plata, 1:6.
COMMENTS: Previously considered a subfamily of the Didelphidae.

*Dromiciops* Thomas, 1894. Ann. Mag. Nat. Hist., ser. 6, 14:186.
TYPE SPECIES: *Dromiciops gliroides* Thomas, 1894, by monotypy.

*Dromiciops gliroides* Thomas, 1894. Ann. Mag. Nat. Hist., ser. 6, 14:187.
COMMON NAME: Monito del Monte.
TYPE LOCALITY: Chile, Biobío, "Huite, N.E. Chiloe Island."
DISTRIBUTION: Chile and adjacent Argentina from about 36°S to near 43°S.
STATUS: IUCN – Vulnerable.
SYNONYMS: *australis* F. Philippi, 1893 [preoccupied by *Didelphys australis* Goldfuss, 1812].
COMMENTS: Reviewed by Marshall (1978b, Mammalian Species, 99, as *D. australis*); also see review by Hershkovitz (1999).

# ORDER NOTORYCTEMORPHIA
by Colin P. Groves

**ORDER NOTORYCTEMORPHIA** Kirsch, *in* Hunsaker, 1977.
SYNONYMS: Syndactyliformes.
COMMENTS: Recognized as an order by Aplin and Archer (1987) who proposed a new syncretic classification of the marsupials.

**Family Notoryctidae** Ogilby, 1892. Cat. Aust. Mammalia, p. 5.
COMMENTS: Relationships unknown. Support for a relationship with dasyuromorphs suggested by Springer et al. (1998).

*Notoryctes* Stirling, 1891. Trans. R. Soc. S. Aust., 14:154.
TYPE SPECIES: *Psammoryctes typhlops* Stirling, 1889.
SYNONYMS: *Neoryctes* Stirling, 1891; *Psammoryctes* Stirling, 1889.
COMMENTS: *Psammoryctes* is preoccupied (see Iredale and Troughton, 1934).

*Notoryctes caurinus* Thomas, 1920. Ann. Mag. Nat. Hist., ser. 9, 6:111.
COMMON NAME: Northern Marsupial Mole.
TYPE LOCALITY: Australia, Western Australia, Wollal (= Wallal), Ninety Mile Beach.
DISTRIBUTION: NW Western Australia.
STATUS: IUCN – Endangered. Unknown.
COMMENTS: Separated from *N. typhlops* by Walton (1988:47).

*Notoryctes typhlops* (Stirling, 1889). Trans. R. Soc. S. Aust., 12:158.
COMMON NAME: Southern Marsupial Mole.
TYPE LOCALITY: Australia, Northern Territory, Indracowrie, 100 mi. (161 km) from Charlotte Waters.
DISTRIBUTION: Western deserts from Ooldea (South Australia) to Charlotte Waters and NW Western Australia, Northern Territory.
STATUS: IUCN – Endangered. Thought to be rare, but no real data exist.

# ORDER DASYUROMORPHIA
## by Colin P. Groves

**ORDER DASYUROMORPHIA** Gill, 1872.
> SYNONYMS: Dasyuroidea, Dasyuriformes, Creatophaga.
> COMMENTS: Recognized as an order by Aplin and Archer (1987) who proposed a new syncretic classification of the marsupials. Includes the Australian component of Marsupicarnivora (see Ride, 1964*b*). Some authors include Thylacinidae and Myrmecobiidae in the Dasyuridae (Vaughan, 1978:39); but Ride (1964*b*), Archer and Kirsch (1977), Kirsch and Calaby (1977:15), and Archer (1982) retained three families; their monophyly is supported by DNA sequence studies (Krajewski et al., 2000).

**Family Thylacinidae** Bonaparte, 1838. Nuovi Ann. Sci. Nat., 2(1):112.
> COMMENTS: Some authors include this family in the Dasyuridae, see Vaughan (1978:39), but also see Ride (1964*b*), Archer and Kirsch (1977), and Kirsch and Calaby (1977:15) who retained this family.

*Thylacinus* Temminck, 1824. Monogr. Mamm., 1:23.
> TYPE SPECIES: *Didelphis cynocephala* Harris, 1808.
> SYNONYMS: *Lycaon* Wagler, 1830 [preoccupied by *Lycaon* Brooks, 1827, a canid]; *Paracyon* Gray, 1843; *Peralopex* Gloger, 1841.

> *Thylacinus cynocephalus* (Harris, 1808). Trans. Linn. Soc. London, 9:174.
>> COMMON NAME: Thylacine.
>> TYPE LOCALITY: Australia, Tasmania.
>> DISTRIBUTION: Tasmania.
>> STATUS: CITES – Appendix I [Possibly Extinct]; U.S. ESA – Endangered; IUCN – Extinct.
>> SYNONYMS: *breviceps* Krefft, 1868; *communis* Anon., 1859; *harrisii* Temminck, 1824; *lucocephalus* (Grant, 1831); *striatus* Warlow, 1833.
>> COMMENTS: Probably extinct; but tracks and sightings continue to be reported; see Ride (1970:201) and Rounsvell and Smith (1982). Species reviewed by Guiler (1986) and Paddle (2000).

**Family Myrmecobiidae** Waterhouse, 1841. Nat. Hist. Marsup. or Pouched Animals (Naturalist's Libr., 10):60.
> COMMENTS: This citation is usually listed as "Cat. Mamm. Mus. Zool. Soc., 1838" but Myrmecobiidae is not used in this catalogue; see Palmer (1904). Some authors (including McKenna and Bell, 1997) include this family in the Dasyuridae; see Vaughan (1978:39); but also see Ride (1964*b*), Archer and Kirsch (1977), and Kirsch and Calaby (1977:15), who retained this family.

*Myrmecobius* Waterhouse, 1836. Proc. Zool. Soc. Lond., 1836:69.
> TYPE SPECIES: *Myrmecobius fasciatus* Waterhouse, 1836.

> *Myrmecobius fasciatus* Waterhouse, 1836. Proc. Zool. Soc. Lond., 1836:69.
>> COMMON NAME: Numbat.
>> TYPE LOCALITY: Australia, Western Australia, Mt. Kokeby, south of Beverley.
>> DISTRIBUTION: SW Western Australia; formerly in NW South Australia and SW New South Wales.
>> STATUS: U.S. ESA – Endangered; IUCN – Vulnerable.
>> SYNONYMS: *rufus* Wood Jones, 1923.

**Family Dasyuridae** Goldfuss, 1820. Handb. Zool., II:447.
> COMMENTS: Kirsch et al. (1997) and Krajewski et al. (2000) confirmed the division of the Dasyuridae into two subfamilies, Dasyurinae (with two tribes, Dasyurini and Phascogalini) and Sminthopsinae (with two tribes, Sminthopsini and Planigalini).

**Subfamily Dasyurinae** Goldfuss, 1820. Handb. Zool., II:447.

**Tribe Dasyurini** Goldfuss, 1820. Handb. Zool., II:447.

*Dasycercus* Peters, 1875. Sitzb. Ges. Naturf. Fr. Berlin, 1875:73.
    TYPE SPECIES: *Chaetocercus cristicauda* Krefft, 1867.
    SYNONYMS: *Amperta* Cabrera, 1919; *Chaetocercus* Krefft, 1867 [preoccupied by Gray, 1855].
    COMMENTS: Combined with *Dasyuroides* by Mack (1961) and Mahoney and Ride (*in* Walton,
        1988:18).

*Dasycercus cristicauda* (Krefft, 1867). Proc. Zool. Soc. Lond., 1866:435 [1867].
    COMMON NAME: Mulgara.
    TYPE LOCALITY: Australia, South Australia, probably Lake Alexandrina.
    DISTRIBUTION: Arid Australia from NW Western Australia to SW Queensland, N South
        Australia.
    STATUS: IUCN – Vulnerable as *D. cristicauda*, Endangered as *D. hillieri*. Rare or indeterminate.
    SYNONYMS: *blighi* (Woodward, 1902); *blythi* (Waite, 1904); *hillieri* (Thomas, 1905).

*Dasykaluta* Archer, 1982. *In* M. Archer, Carnivorous Marsupials, 2:434.
    TYPE SPECIES: *Antechinus rosamondae* Ride, 1964.
    COMMENTS: Formerly included in *Antechinus*, and this allocation was continued by McKenna
        and Bell (1997), but Archer (1982) and Krajewski et al. (2000) showed that they are not
        sister-groups.

*Dasykaluta rosamondae* (Ride, 1964). W. Aust. Nat., 9:58.
    COMMON NAME: Little Red Kaluta.
    TYPE LOCALITY: Australia, Western Australia, Woodstock Station (via Marble Bar), 21°35'S,
        119°E.
    DISTRIBUTION: NW Western Australia.
    STATUS: IUCN – Lower Risk (lc). Rare.
    COMMENTS: Formerly included in *Antechinus*; separated by Archer (1982). Its closest
        affinities are to *Parantechinus apicalis* (Krajewski et al., 1997a).

*Dasyuroides* Spencer, 1896. Proc. Roy. Soc. Victoria, N. S., VIII:5.
    TYPE SPECIES: *Dasyuroides byrnei*. Spencer, 1896.
    COMMENTS: Included in *Dasycercus* by Mack (1961), Mahoney and Ride (*in* Walton, 1988) and
        Groves (1993), but Kirsch et al. (1997) considered this "premature"; see also N. K. Cooper
        et al. (2000), who also favored making them congeneric.

*Dasyuroides byrnei* Spencer, 1896. Proc. Roy. Soc. Victoria, N. S., VIII:6.
    COMMON NAME: Kowari.
    TYPE LOCALITY: Australia, Northern Territory, Charlotte Waters.
    DISTRIBUTION: Junction of Northern Territory, South Australia, and Queensland (C Australia).
    STATUS: IUCN – Vulnerable as *Dasycercus byrnei*. Rare.
    SYNONYMS: *pallidior* Thomas, 1906.

*Dasyurus* É. Geoffroy, 1796. Mag. Encyclop., ser. 2, 3:469.
    TYPE SPECIES: *Didelphis maculata* Anon., 1791 (= *Didelphis viverrina* Shaw, 1800).
    SYNONYMS: *Dasyurinus* Matschie, 1916; *Dasyurops* Matschie, 1916; *Nasira* Harvey, 1841;
        *Notoctonus* Pocock, 1926; *Satanellus* Pocock, 1926; *Stictophonus* Pocock, 1926.
    COMMENTS: See Haltenorth (1958:20). Archer (1982) revived *Satanellus* (for *albopunctatus* and
        *hallucatus*), as he considered that the remaining species might belong in a clade with
        *Sarcophilus*; but Krajewski et al. (1997a), using molecular data, confirmed that *Dasyurus* is
        monophyletic.

*Dasyurus albopunctatus* Schlegel, 1880. Notes Leyden Mus., 2:51.
    COMMON NAME: New Guinean Quoll.
    TYPE LOCALITY: Indonesia, Prov. of Papua (= Irian Jaya), Vogelkop, Manokwari Div., Arfak
        Mtns, Sapoea.

DISTRIBUTION: New Guinea, sea level to 3500 m, possibly Yapen Isl: see Flannery (1995*a*).
STATUS: IUCN – Vulnerable.
SYNONYMS: *daemonellus* Thomas, 1904; *fuscus* Milne-Edwards, 1880.

*Dasyurus geoffroii* Gould, 1841. Proc. Zool. Soc. Lond., 1840:151 [1841].
COMMON NAME: Western Quoll.
TYPE LOCALITY: Australia, New South Wales, Liverpool Plains.
DISTRIBUTION: Western Australia; formerly in South Australia, Northern Territory, S Queens-
land, W New South Wales, and NW Victoria (Archer, *in* Tyler, 1979; Waithman, 1979).
STATUS: IUCN – Vulnerable. Rare.
SYNONYMS: *fortis* Thomas, 1906.
COMMENTS: Formerly included in *Dasyurinus*. The New Guinea records of this species
actually refer to *D. spartacus*.

*Dasyurus hallucatus* Gould, 1842. Proc. Zool. Soc. Lond., 1842:41.
COMMON NAME: Northern Quoll.
TYPE LOCALITY: Australia, Northern Territory, Port Essington.
DISTRIBUTION: Australia: N Northern Territory, N and NE Queensland, and N Western
Australia.
STATUS: IUCN – Lower Risk (nt). Common.
SYNONYMS: *exilis* Thomas, 1909; *nesaeus* Thomas, 1926; *predator* Thomas, 1926; *quoll*
(Zimmermann, 1783).
COMMENTS: Sometimes assigned to *Satanellus*. The original name *Mustela quoll*
Zimmermann, 1783, was suppressed under Article 80 of the International Code of
Zoological Nomenclature (International Commission on Zoological Nomenclature,
1999), now correctly *Dasyurus hallucatus*.

*Dasyurus maculatus* (Kerr, 1792). *In* Linnaeus, Anim. Kingdom, p. 170.
COMMON NAME: Tiger Quoll.
TYPE LOCALITY: Australia, New South Wales, Port Jackson.
DISTRIBUTION: Australia: E Queensland, E New South Wales, E and S Victoria, SE South
Australia, Tasmania. Formerly occurred in South Australia.
STATUS: IUCN – Endangered as *D. m.gracilis*, Vulnerable as *D. maculatus* and *D. m. maculatus*.
Widespread but rare; locally common in Tasmania.
SYNONYMS: *gracilis* Ramsay, 1888; *macrourus* E. Geoffroy St. Hilaire, 1803; *novaehollandiae*
(Meyer, 1793); *ursinus* Giebel, 1874.
COMMENTS: Formerly included in *Dasyurops*; see Haltenorth (1958). See Firestone et al.
(1999) for possible subspecies boundaries; the Tasmanian form is rather distinctive.

*Dasyurus spartacus* Van Dyck, 1987. Aust. Mamm., 11:145.
COMMON NAME: Bronze Quoll.
TYPE LOCALITY: Papua New Guinea, Trans-fly Plains, Marehead, 8°41'S, 141°39'E.
DISTRIBUTION: Fly Plains, Papua New Guinea.
STATUS: IUCN – Vulnerable.
COMMENTS: Formerly included in *D. geoffroii*.

*Dasyurus viverrinus* (Shaw, 1800). Gen. Zool. Syst. Nat. Hist., 1(2), Mammalia, p. 491.
COMMON NAME: Eastern Quoll.
TYPE LOCALITY: Australia, New South Wales, Sydney.
DISTRIBUTION: Probably survives only in Tasmania; formerly South Australia, New South
Wales, and Victoria (Archer, *in* Tyler, 1979).
STATUS: U.S. ESA – Endangered; IUCN – Lower Risk (nt).
SYNONYMS: *alboguttata* (Burmeister, 1854); *guttatus* Desmarest, 1804; *maculata* (Anon, 1791);
*maugei* E. Geoffroy St. Hilaire, 1803; *quoll* Zimmermann, 1777.
COMMENTS: *Dasyurus quoll* Zimmermann, 1777 (not *Mustela quoll* Zimmermann, 1783), is
invalid: this work was rejected by Opinion 257 of the International Commission on
Zoological Nomenclature (1954*a*). Also the name *Dasyurus maculata* Anon., 1791, was
suppressed under Article 80 of the International Code of Zoological Nomenclature
(International Commission on Zoological Nomenclature, 1999). "Original" name is
now *Didelphis viverrina* Shaw, 1800.

*Myoictis* Gray, 1858. Proc. Zool. Soc. Lond., 1858:112.
   TYPE SPECIES: *Myoictis wallacii* Gray, 1858.

   *Myoictis melas* (Müller, 1840). *In* Temminck, Verh. Nat. Ges. Ned. Overz. Bezitt., Land-en
       Volkenkunde, p. 20[1840], see comments.
       COMMON NAME: Three-striped Dasyure.
       TYPE LOCALITY: "Nieuw-Guinea, in de triton's baai (op 3°39'Z. breedte)" = Indonesia, Prov. of
           Papua (= Irian Jaya), Fakfak Div., Lobo Dist., near Triton Bay, Mt. Lamantsjieri.
       DISTRIBUTION: Prov. of Papua (= Irian Jaya) and NE Papua New Guinea; Salawati Isl. Also
           Waigeo, Yapen, possibly Batanta.
       STATUS: IUCN – Lower Risk (lc). Secure.
       SYNONYMS: *bruijni* (Peters, 1875); *buergersi* (Stein, 1932); *senex* (Stein, 1932); *thorbeckiana*
           (Schlegel, 1866); *wavicus* Tate, 1947.
       COMMENTS: This species was further described by Müller and Schlegel, *in* Temminck, Verh.
           Nat. Gesch. Nederland. Overz. Bezitt., Zool., Mammalia, p. 149[1845], pl. 25[1843].

   *Myoictis wallacii* Gray, 1858. Proc. Zool. Soc. Lond., 1858:112.
       COMMON NAME: Wallace's Dasyure.
       TYPE LOCALITY: Aru Isl.
       DISTRIBUTION: Southern New Guinea and Aru Isls (Indonesia).
       STATUS: Unknown, but rare.
       SYNONYMS: *pilicauda* (Peters and Doria, 1881).
       COMMENTS: Usually considered a subspecies of *M. melas*, but separated as a species by Archer
           (1982:417), and molecular data suggest that they may have separated several million
           years ago (Krajewski et al., 1997*a*).

*Neophascogale* Stein, 1933. Z. Säugetierk., 8:87.
   TYPE SPECIES: *Phascogale venusta* Thomas, 1921 (= *Phascogale lorentzi* Jentink, 1911).

   *Neophascogale lorentzi* (Jentink, 1911). Notes Leyden Mus., 33:234.
       COMMON NAME: Speckled Dasyure.
       TYPE LOCALITY: Indonesia, Prov. of Papua (= Irian Jaya), Djajawidjaja (= Jayawijaya)
           Division, Helwig Mtns, south of Mt. Wilhelmina, 2600 m.
       DISTRIBUTION: C New Guinea (highlands).
       STATUS: IUCN – Lower Risk (lc). Uncommon.
       SYNONYMS: *rubrata* (Thomas, 1922); *venusta* (Thomas, 1921).
       COMMENTS: Previously included *Phascogale nouhuysi* Jentink, 1911, but Husson (1955)
           showed that it is based on a specimen of *Phascolosorex doriae*.

*Parantechinus* Tate, 1947. Bull. Am. Mus. Nat. Hist., 88:137.
   TYPE SPECIES: *Phascogale apicalis* Gray, 1842.
   COMMENTS: Formerly included in *Antechinus*, and this allocation was continued by McKenna
       and Bell (1997), but Archer (1982) and Krajewski et al. (2000) showed that they are not
       sister-groups. Kitchener and Caputi (1988) restricted this genus to *P. apicalis*, which
       according to Krajewski et al. (1997*a*) is closely related to *Dasykaluta*.

   *Parantechinus apicalis* (Gray, 1842). Ann. Mag. Nat. Hist., [ser. 1], 9:518.
       COMMON NAME: Southern Dibbler.
       TYPE LOCALITY: Australia, SW Western Australia.
       DISTRIBUTION: Inland periphery of SW Western Australia.
       STATUS: U.S. ESA and IUCN – Endangered.

*Phascolosorex* Matschie, 1916. Mitt. Zool. Mus. Berlin, 8:263.
   TYPE SPECIES: *Phascogale dorsalis* Peters and Doria, 1876.

   *Phascolosorex doriae* (Thomas, 1886). Ann. Mus. Civ. Stor. Nat. Genova, 4:208.
       COMMON NAME: Red-bellied Marsupial Shrew.
       TYPE LOCALITY: Indonesia, Prov. of Papua (= Irian Jaya), Vogelkop, Manokwari Div., Arfak
           Mtns, Mori.
       DISTRIBUTION: W interior New Guinea, 100-2000 m.

STATUS: IUCN – Data Deficient. Uncommon.
SYNONYMS: *nouhuysii* (Jentink, 1911); *pan* (Stein, 1932); *umbrosa* (Dollman, 1930).
COMMENTS: *Phascogale nouhuysi* Jentink, 1911 is usually placed in synonymy of *Neophascogale lorentzii*, but Husson (1955) showed that it is based on a specimen of *P. doriae*.

**Phascolosorex dorsalis** (Peters and Doria, 1876). Ann. Mus. Civ. Stor. Nat. Genova, 8:335.
COMMON NAME: Narrow-striped Marsupial Shrew.
TYPE LOCALITY: Indonesia, Prov. of Papua (= Irian Jaya), Vogelkop, Manokwari Div., Arfak Mtns, Hatam.
DISTRIBUTION: W and E interior New Guinea (not known from central region).
STATUS: IUCN – Lower Risk (lc). Common.
SYNONYMS: **brevicaudata** (Rothschild and Dollman, 1932); **whartoni** (Tate and Archbold, 1936).

*Pseudantechinus* Tate, 1947. Bull. Am. Mus. Nat. Hist., 88:139.
TYPE SPECIES: *Phascogale macdonnellensis* Spencer, 1896.
COMMENTS: Separated from *Antechinus* by Archer (1982:434), but retained in it by McKenna and Bell (1997). According to molecular evidence, it is not even closely related to *Antechinus*, but forms a distinctive genus of the Dasyurini (Krajewski et al., 1997a).

**Pseudantechinus bilarni** (Johnson, 1954). Proc. Biol. Soc. Wash., 67:77.
COMMON NAME: Sandstone Dibbler.
TYPE LOCALITY: Australia, Northern Territory, Oenpelli (12°20'S, 133°3'E).
DISTRIBUTION: Northern Territory (Australia), known from region of type locality.
STATUS: IUCN – Lower Risk (lc). Rare.
COMMENTS: Included in *Antechinus* (= *Pseudantechinus*) *macdonnellensis* by Ride (1970:116), but see Kirsch and Calaby (1977:15). Archer (1982) placed it in *Parantechinus*. Kitchener and Caputi (1988) suggested that this species should be transferred to *Pseudantechinus*; Krajewski et al. (1997a) agreed that it is not close to *P. apicalis* and placed it on a clade by itself. N. K. Cooper et al. (2000) definitively removed it from *Parantechinus* but admitted that its inclusion (and that of *P. woolleyae*) in *Pseudantechinus* might make the genus paraphyletic.

**Pseudantechinus macdonnellensis** (Spencer, 1896). Rept. Horn Sci. Exped. Cent. Aust., Zool., 2:27.
COMMON NAME: Fat-tailed False Antechinus.
TYPE LOCALITY: Australia, Northern Territory, south of Alice Springs.
DISTRIBUTION: Uplands of Western Australia, about 24-28°S, 124-130°E, and southern desert region of Northern Territory, from type locality north to about 19°S (see N. K. Cooper et al., 2000:Fig. 12).
STATUS: IUCN – Lower Risk (lc).
COMMENTS: Does not include *mimulus* (D. J. Kitchener, 1991).

**Pseudantechinus mimulus** (Thomas, 1906). Proc. Zool. Soc. Lond., 1906, 2:540.
COMMON NAME: Alexandria False Antechinus.
TYPE LOCALITY: Australia, Northern Territory, near Alexandria (19°03'S, 136°42'E).
DISTRIBUTION: Northern Territory: known only from type locality and North Isl, Sir Edward Pellew Group.
STATUS: IUCN – Vulnerable.
COMMENTS: Not a synonym of *P. macdonnellensis* (D. J. Kitchener, 1991).

**Pseudantechinus ningbing** Kitchener, 1988. Rec. W. Aust. Mus., 14:62.
COMMON NAME: Ningbing False Antechinus.
TYPE LOCALITY: Australia, Western Australia, Kimberley region, Mitchell Plateau, *ca.* 220 m, 14°53'40"S, 125°45'20"E.
DISTRIBUTION: Kimberley region, Western Australia.
STATUS: IUCN – Lower Risk (lc).

**Pseudantechinus roryi** Cooper, Aplin and Adams, 2000. Rec. W. Aust. Mus., 20:125.
COMMON NAME: Rory Cooper's False Antechinus.

TYPE LOCALITY: Australia, Western Australia, Woodstock Station, 500 m north of the homestead, 21°36'42"S, 117°57'20"E.

DISTRIBUTION: Western Australia: Northern Pilbara, north of the Hamersley Range, into Great Sandy Desert as far east as Clutterbuck Hills; Cape Range Peninsula; probably Barrow Isl (N. K. Cooper et al., 2000).

*Pseudantechinus woolleyae* Kitchener and Caputi, 1988. Rec. W. Aust. Mus., 14:39.
COMMON NAME: Woolley's False Antechinus.
TYPE LOCALITY: Australia, Western Australia, near Newligunn bore, 10 km from Errabiddy Homestead, 25°33'00"S, 117°08'00"E.
DISTRIBUTION: Western Australia: Pilbara region and further south - between *ca.* 20° and 30°S, 115° and 123°E (see N. K. Cooper et al., 2000:Fig. 12).
STATUS: IUCN – Lower Risk (lc).
COMMENTS: N. K. Cooper et al. (2000) admitted that the inclusion of this species (and that of *P. bilarni*) in *Pseudantechinus* might make the genus paraphyletic.

*Sarcophilus* F. G. Cuvier, 1837. *In* E. Geoffroy and F. G. Cuvier, Hist. Nat. Mammifères, pt. 4, 7(70):1-6, "Sarcophile oursin."
TYPE SPECIES: *Didelphis ursina* Harris, 1808 (= *Ursinus harrisii* Boitard, 1841).
SYNONYMS: *Diabolus* Gray, 1841; *Ursinus* Boitard, 1841.

*Sarcophilus harrisii* (Boitard, 1841). *Le Jardin des Plantes*, p. 290.
COMMON NAME: Tasmanian Devil.
TYPE LOCALITY: Tasmania.
DISTRIBUTION: Australia: Tasmania; known as a subfossil in S Victoria (Mt. Hamilton and Lake Corangamite).
STATUS: IUCN – Lower Risk (lc). Common.
SYNONYMS: *satanicus* Thomas, 1903; *ursina* (Harris, 1808) [not of Shaw, 1800]; **dixonae** Werdelin, 1987.
COMMENTS: Groves (1993) called this species *S. laniarius* Owen, 1838, which was based on Pleistocene specimens from Wellington Caves, New South Wales. Werdelin (1987:9) argued that the Pleistocene and recent *Sarcophilus* were only subspecifically distinct, and as Owen's name antedates Boitard's (a replacement name for *Didelphis ursina* Harris, 1808 [preoccupied by *Didelphis ursina* Shaw, 1800]) by three years, *laniarius* must take precedence. Inspection of Werdelin (1987, Tables 1 and 2) shows, however, that recent and fossil ranges do not overlap in many variables, so they may be retained as different species. The Victorian subfossil *dixonae* remains as a subspecies of *S. harrisii*; though it is distinctive its measurements overlap with those of the living Tasmanian form.

**Tribe Phascogalini** Gill, 1872. Smithson. Misc. Collect., 11:26.
COMMENTS: Treated as a distinct subfamily by Krajewski et al. (1996).

*Antechinus* Macleay, 1841. Ann. Mag. Nat. Hist., [ser. 1], 8:242.
TYPE SPECIES: *Antechinus stuartii* Macleay, 1841.
COMMENTS: For the exclusion of *Parantechinus* and *Pseudantechinus* see Haltenorth (1958:18) and Ride (1964a); and of *Dasykaluta* see Archer (1982:434). Formerly included *habbema*, *melanura*, *naso*, and *wilhelmina*, which were transferred to *Murexia* by Armstrong et al. (1998), and have been transferred to a series of new genera (*Micromurexia, Murexechinus,* and *Phascomurexia*) by Van Dyck (2002).

*Antechinus adustus* (Thomas, 1923). Ann. Mag. Nat. Hist., ser.9, 11:175.
COMMON NAME: Tropical Antechinus.
TYPE LOCALITY: Australia, Queensland, Dinner Creek, now Charmillan Creek (17°42'S, 145°31'E), 885 m, near Ravenshoe.
DISTRIBUTION: Australia: Dense tropical vine forests from Paluma (19°00'S, 146°12'E) to Mt. Spurgeon (16°25' S, 145°12'E).
STATUS: Unknown.
COMMENTS: Van Dyck (1982) stated that what was then known as *Antechinus stuartii adustus*

is probably a valid species. Van Dyck and Crowther (2000) definitively raised it to species rank.

*Antechinus agilis* Dickman, Parnaby, Crowther and King, 1998. Aust. J. Zool., 46:5.
   COMMON NAME: Agile Antechinus.
   TYPE LOCALITY: Australia, Australian Capital Territory, Brindabella Range, on the south side of Warks Road and Blundells Creek Road, near Lees Creek (35°21′45″S, 148°50′17″E), 740 m.
   DISTRIBUTION: Australia: Victoria (SW, and C, E and NE districts), and SE New South Wales north on the coast to Kioloa (35°32′S, 150°23′E), and inland north to Mt. Canobolas (33°10′S, 149°00′E).
   STATUS: Common.
   COMMENTS: Dickman et al. (1988) first showed that *A. stuartii* in E New South Wales is actually divided into two quite distinct species: *A. stuartii* north of about 35°S, and an undescribed species mainly south of this latitude. The new species was not described for another ten years (Dickman et al., 1998). The two species are sympatric at Kioloa, New South Wales, and other places.

*Antechinus bellus* (Thomas, 1904). Nov. Zool., 11:229.
   COMMON NAME: Fawn Antechinus.
   TYPE LOCALITY: Australia, Northern Territory, South Alligator River.
   DISTRIBUTION: N Northern Territory (Australia).
   STATUS: IUCN – Lower Risk (lc).

*Antechinus flavipes* (Waterhouse, 1838). Proc. Zool. Soc. Lond., 1837:75 [1838].
   COMMON NAME: Yellow-footed Antechinus.
   TYPE LOCALITY: Australia, New South Wales, north of Hunter River.
   DISTRIBUTION: Cape York Peninsula (Queensland); Victoria and SE South Australia; SW Western Australia. In SE Australia, bioclimatic modeling predicts its range as (somewhat discontinuously) from about 23°N to S Victoria, and west to about 136°E on the South Australian coast. Inland, it may reach as far as about 145°E in S New South Wales (Sumner and Dickman, 1998).
   STATUS: Lower Risk (lc).
   SYNONYMS: *leucogaster* (Gray, 1841); *rufogaster* (Gray, 1841); **rubeculus** Van Dyck, 1982.
   COMMENTS: No direct comparisons between Western Australian and eastern forms of the species have been made. The Cape York Peninsula form appears to be only subspecifically distinct.

*Antechinus godmani* (Thomas, 1923). Ann. Mag. Nat. Hist., ser. 9, 11:174.
   COMMON NAME: Atherton Antechinus.
   TYPE LOCALITY: Australia, Queensland, Ravenshoe, Dinner Creek, 2900 ft. (884 m), 17°40′S, 145°30′E.
   DISTRIBUTION: NE Queensland (Australia).
   STATUS: IUCN – Lower Risk (nt).
   COMMENTS: Included in *flavipes* by Haltenorth (1958:18); but see Kirsch and Calaby (1977:15).

*Antechinus leo* Van Dyck, 1980. Aust. Mamm., 3:1.
   COMMON NAME: Cinnamon Antechinus.
   TYPE LOCALITY: Australia, Queensland, Cape York Penninsula, Nesbit River, Buthen Buthen (13°21′S, 143°28′E).
   DISTRIBUTION: Cape York Peninsula from the Iron Range to the southern limit of the McIlwraith Range.
   STATUS: IUCN – Lower Risk (nt).

*Antechinus minimus* (É. Geoffroy, 1803). Bull. Sci. Soc. Philom. Paris, 81:159.
   COMMON NAME: Swamp Antechinus.
   TYPE LOCALITY: Australia, Tasmania; probably Waterhouse Isl, Bass Strait (see Wakefield and Warneke, 1963:209-210).
   DISTRIBUTION: Coastal SE South Australia to Tasmania.

STATUS: IUCN – Lower Risk (lc) as *A. minimus*, Lower Risk (nt) as *A. m. maritimus*.
SYNONYMS: *affinis* (Gray, 1841); *concinnus* Higgins and Petterd, 1884; *maritima* (Finlayson, 1958); *rolandensis* Higgins and Petterd, 1883.

*Antechinus stuartii* Macleay, 1841. Ann. Mag. Nat. Hist., [ser. 1], 8:242.
COMMON NAME: Brown Antechinus.
TYPE LOCALITY: Australia, New South Wales, Manly (Spring Cove, Sydney Harbour); neotype from Waterfall, Royal National Park.
DISTRIBUTION: SE Queensland, E New South Wales south to Kioloa (35°32'S, 150°23'E). Bioclimatic modeling predicts an almost entirely coastal distribution, north to about 26°S (Sumner and Dickman, 1998).
STATUS: IUCN – Lower Risk (lc).
SYNONYMS: *adusta* (Thomas, 1923); *burrelli* (Le Souef and Burrel, 1926); *unicolor* Gould, 1854.
COMMENTS: Dickman et al. (1988) showed that what had been thought to be the single species *A. stuartii* in E New South Wales is actually divided into a northern and a southern species: *A. stuartii* and what is now (Dickman et al., 1998) named *A. agilis*; the two have been found together at Kioloa, in S New South Wales.

*Antechinus subtropicus* Van Dyck and Crowther, 2000. Mem. Qld. Mus., 45:613.
COMMON NAME: Subtropical Antechinus.
TYPE LOCALITY: Australia, Queensland, Emu Creek, 38 km E of Warwick (23°13'03"S, 152°24'54"E).
DISTRIBUTION: SE Queensland, south from Gympie (26°11'S, 152°40'E) into NE NSW, in subtropical vine forests.
STATUS: Unknown.
COMMENTS: Related to *A. stuartii*, with which it is sympatric in SE Queensland at Wallangarra (28°55'S, 151°55'E) and Pyramid Creek, Wyberba (28°50S, 151°57'E).

*Antechinus swainsonii* (Waterhouse, 1840). Mag. Nat. Hist. [Charlesworth's], 4:299.
COMMON NAME: Dusky Antechinus.
TYPE LOCALITY: Australia, Tasmania.
DISTRIBUTION: SE Queensland, E New South Wales, E and SE Victoria, coastal SE Australia, and Tasmania.
STATUS: IUCN – Lower Risk (lc).
SYNONYMS: *assimilis* Higgins and Petterd, 1884; *mimetes* (Thomas, 1924); *moorei* Higgins and Petterd, 1884; *niger* Higgins and Petterd, 1883.

*Micromurexia* Van Dyck, 2002. Mem. Queensl. Mus., 48:246.
TYPE SPECIES: *Antechinus habbema* Tate and Archbold, 1941.
COMMENTS: Separated from *Murexia* by Van Dyck, 2002.

*Micromurexia habbema* (Tate and Archbold, 1941). Amer. Mus. Novit., 1101:8.
COMMON NAME: Habbema Dasyure.
TYPE LOCALITY: Indonesia, Prov. of Papua (= Irian Jaya), 9 km N of Lake Habbema, north slope of Mt. Wilhelmina, 2800 m.
DISTRIBUTION: Central Cordillera of New Guinea, 4°05'-8°03'S, 138°50'-146°53'E, at 1600-3660 m (Van Dyck, 2002).
STATUS: IUCN – Data Deficient as *Antechinus habbema*.
SYNONYMS: *hageni* (Laurie, 1952).
COMMENTS: Formerly considered a synonym of *Antechinus naso* (here reallocated as *Phascomurexia naso*), but shown to be distinct by Woolley (1989) and Flannery (1995a); it is actually the most distinct species of the *Murexia* clade according to Krajewski et al. (1996) and was referred to a new genus by Van Dyck (2002).

*Murexechinus* Van Dyck, 2002. Mem. Queensl. Mus., 48:300.
TYPE SPECIES: *Phascogale melanura* Thomas, 1899.
COMMENTS: Separated from *Murexia* by Van Dyck, 2002.

*Murexechinus melanurus* (Thomas, 1899). Ann. Mus. Civ. Stor. Nat. Genova, 20:191.
COMMON NAME: Black-tailed Dasyure.

TYPE LOCALITY: Papua New Guinea, Central Prov., Astrolabe Range, Moroka, 9°24'S, 147°32'E, 1300 m.
DISTRIBUTION: New Guinea, from 134°00'E to 151°01'E (Normanby Isl), sea level to 2800 m.
STATUS: IUCN – Lower Risk (lc) as *Antechinus melanurus*, Data Deficient as *A. wilhelmina*.
SYNONYMS: *mayeri* (Dollman, 1930); *modesta* (Thomas, 1912); *wilhelmina* Tate, 1947.
COMMENTS: Usually included in *Antechinus*, but reassigned to *Murexia* by Armstrong et al. (1998) and to a new genus by Van Dyck (2002). Synonymy after Van Dyck (2000).

*Murexia* Tate and Archbold, 1937. Bull. Am. Mus. Nat. Hist., 73:335 (footnote), 339.
TYPE SPECIES: *Phascogale murex* Thomas, 1913 (= *Phascogale longicaudata* Schlegel, 1866).
COMMENTS: Some of these species have traditionally been assigned to *Antechinus*, but so-called "New Guinea *Antechinus*" were transferred to *Murexia* by Armstrong et al. (1998). The undescribed species from Normanby Isl in the D'Entrecasteaux group, Papua New Guinea (Flannery, 1995a) is actually *Murexechinus melanurus* according to Van Dyck (2002).

*Murexia longicaudata* (Schlegel, 1866). Ned. Tijdschr. Dierk., 3:356.
COMMON NAME: Short-furred Dasyure.
TYPE LOCALITY: Indonesia, Aru Islands, Wonumbai.
DISTRIBUTION: New Guinea, sea level to 1800 m; Aru Isls; Yapen.
STATUS: IUCN – Lower Risk (lc).
SYNONYMS: *aspera* (Thomas, 1913); *maxima* (Stein, 1932); *murex* (Thomas, 1913).

*Paramurexia* Van Dyck, 2002. Mem. Queensl. Mus., 48:293.
TYPE SPECIES: *Phascogale (Murexia) rothschildi* Tate, 1938.
COMMENTS: Separated from *Murexia* by Van Dyck (2002).

*Paramurexia rothschildi* (Tate, 1938). Nov. Zool., 41:58.
COMMON NAME: Broad-striped Dasyure.
TYPE LOCALITY: Papua New Guinea, Central Prov., head of Aroa River, 8°50'S, 147°06'E, about 1220 m.
DISTRIBUTION: SE New Guinea, between 9°56'-10°02'S and 147°00'-149°43'E, 600 to 1400 m (Van Dyck, 2002).
STATUS: IUCN – Data Deficient as *Murexia rothschildi*. Vulnerable.

*Phascomurexia* Van Dyck, 2002. Mem. Queensl. Mus., 48:257.
TYPE SPECIES: *Phascogale naso* Jentink, 1911.
COMMENTS: Separated from *Murexia* by Van Dyck, 2002.

*Phascomurexia naso* (Jentink, 1911). Notes Leyden Mus., 33:236.
COMMON NAME: Long-nosed Dasyure.
TYPE LOCALITY: Indonesia, Prov. of Papua (= Irian Jaya), Jayawijaya Div., Helwig Mtns, south of Mt. Wilhelmina, about 2000 m, 4°32'S, 138°41'E.
DISTRIBUTION: Interior New Guinea, 3°32'S, 139°10'E to 8°35'S, 147°09'E, 1400-2800 m (Van Dyck, 2002).
STATUS: IUCN – Data Deficient as *Antechinus naso*.
SYNONYMS: *centralis* (Tate and Archbold, 1941); *misim* (Tate, 1947); *parva* Laurie, 1952; *tafa* (Tate and Archbold, 1936).
COMMENTS: Usually included in *Antechinus*, but reassigned to *Murexia* by Armstrong et al. (1998) and awarded a separate genus by Van Dyck (2002), who recognised no subspecies.

*Phascogale* Temminck, 1824. Monogr. Mamm., 1:23, 56.
TYPE SPECIES: *Didelphis penicillata* Shaw, 1800 (= *Vivera tapoatafa* Meyer, 1793).
SYNONYMS: *Ascogale* Gloger, 1841; *Phascologale* Lenz, 1831; *Phascoloictis* Matschie, 1916; *Tapoa* Lesson, 1842.

*Phascogale calura* Gould, 1844. Proc. Zool. Soc. Lond., 1844:104.
COMMON NAME: Red-tailed Phascogale.
TYPE LOCALITY: Australia, Western Australia, Williams River, Military Station.

DISTRIBUTION: Inland SW Western Australia, formerly in Northern Territory, South
Australia, NW Victoria, SW New South Wales, but probably extinct in all places except
the Western Australian wheat belt.
STATUS: IUCN – Endangered.

*Phascogale tapoatafa* (Meyer, 1793). Zool. Entdeck., p. 28.
COMMON NAME: Brush-tailed Phascogale.
TYPE LOCALITY: Australia, New South Wales, Sydney.
DISTRIBUTION: SW Western Australia, SE South Australia, S Victoria, E New South Wales,
SE and N Queensland, Northern Territory.
STATUS: IUCN – Lower Risk (nt).
SYNONYMS: *penicillata* (Shaw, 1800); *pirata* (Thomas, 1904); *tafa* (White, 1803).

**Subfamily Sminthopsinae** Archer, 1982. *Carnivorous Marsupials*, 2:439.
COMMENTS: Kirsch et al. (1997) divided the subfamily into two tribes, and Krajewski et al.
(1997b) concurred in finding a deep division between *Planigale* and others.

**Tribe Sminthopsini** Archer, 1982. *Carnivorous Marsupials*, 2:439.

*Antechinomys* Krefft, 1867. Proc. Zool. Soc. Lond., 1866:434.
TYPE SPECIES: *Phascogale lanigera* Gould, 1856.
COMMENTS: Included in *Sminthopsis* by Archer (1979:329, 1981:187) and by McKenna and Bell
(1997); but Lidicker (1983:1317) considered *Antechinomys* a distinct genus, and Krajewski
et al. (1997b) and Blacket et al. (1999) separated it from *Sminthopsis* on the basis of their
molecular findings.

*Antechinomys laniger* (Gould, 1856). Mamm. Aust., 1, pl. 33.
COMMON NAME: Kultarr.
TYPE LOCALITY: Australia, interior New South Wales.
DISTRIBUTION: Western Australia, S Northern Territory, N Victoria, W New South Wales,
SW Queensland, N South Australia.
STATUS: U.S. ESA – Endangered; IUCN – Data Deficient.
SYNONYMS: *spenceri* Thomas, 1906.
COMMENTS: Includes *spenceri*; see Archer (1977:19).

*Ningaui* Archer, 1975. Mem. Queensl. Mus., 17(2):239.
TYPE SPECIES: *Ningaui timealeyi* Archer, 1975.
COMMENTS: An undescribed species of *Ningaui* occurs in Northern Territory (Australia); see
Johnson and Roff (1980). According to Blacket et al. (1999), *Ningaui* clusters within
*Sminthopsis*.

*Ningaui ridei* Archer, 1975. Mem. Queensl. Mus., 17(2):246.
COMMON NAME: Wongai Ningaui.
TYPE LOCALITY: Australia, Western Australia, 38.6 km ENE Laverton (28°30′S, 122°47′E).
DISTRIBUTION: Northern Territory, South Australia, and Western Australia (deserts).
STATUS: IUCN – Lower Risk (lc). Common.

*Ningaui timealeyi* Archer, 1975. Mem. Queensl. Mus., 17(2):244.
COMMON NAME: Pilbara Ningaui.
TYPE LOCALITY: Australia, Western Australia, 32.2 km SE Mt. Robinson.
DISTRIBUTION: NW Western Australia.
STATUS: IUCN – Lower Risk (lc). Common.

*Ningaui yvonnae* Kitchener, Stoddart, and Henry, 1983. Aus. J. Zool., 31:366.
COMMON NAME: Southern Ningaui.
TYPE LOCALITY: "Mt. Manning Area, Western Australia Goldfields, 29°58′S, 119°32′E".
DISTRIBUTION: Australia: Western Australia to New South Wales, Victoria.
STATUS: IUCN – Lower Risk (lc). Common.

*Sminthopsis* Thomas, 1887. Ann. Mus. Civ. Stor. Nat. Genova, ser. 2, 4:503.

TYPE SPECIES: *Phascogale crassicaudata* Gould, 1844.

SYNONYMS: *Podabrus* Gould, 1845 [not of Westwood, 1840].

COMMENTS: Original name *Podabrus*, Gould, 1845, is preoccupied. *Antechinomys* was considered a subgenus by Archer (1979:329); see also Kirsch and Calaby (1977:15), but on the basis of other anatomic and isozymic data, Lidicker (1983:1317) considered *Antechinomys* a distinct genus, despite similarities in dental morphology with *Sminthopsis*. Blacket et al. (1999) confirmed that *Antechinomys* forms a clade separate from *Sminthopsis* plus *Ningaui*. Revised by Archer (1981); according to Krajewski et al. (1997*b*) and Blacket al. (1999), *Sminthopsis* may be paraphyletic relative to *Ningaui*. The species-groups recognized in these three revisions are: *S. crassicaudata* group (monotypic); *S. macroura* group (containing *S. bindi*, *S. butleri*, *S. douglasi*, *S. macroura* and *S. virginiae*); *S. granulipes* group (monotypic); *S. griseoventer* group (*S. aitkeni*, *S. boullangerensis*, *S. griseoventer*); *S. longicaudata* group (monotypic); *S. murina* group (*S. archeri*, *S. dolichura*, *S. fuliginosus*, *S. gilberti*, *S. leucopus*, *S. murina*); and *S. psammophila* group (*S. hirtipes*, *S. ooldea*, *S. psammophila*, *S. youngsoni*). The species groups are strongly distinct, and some of them may ultimately be given generic rank.

*Sminthopsis aitkeni* Kitchener, Stoddart and Henry, 1984. Rec. W. Aust. Mus., 11:204.

COMMON NAME: Kangaroo Island Dunnart.

TYPE LOCALITY: Australia, South Australia, Kangaroo Isl, Section 46, Cassini.

DISTRIBUTION: Kangaroo Isl (South Australia).

STATUS: IUCN - Endangered.

COMMENTS: *S. griseoventer* species-group. Separated from *S. murina* by Kitchener et al. (1984*b*:204).

*Sminthopsis archeri* Van Dyck, 1986. Aust. Mamm., 9:112.

COMMON NAME: Chestnut Dunnart.

TYPE LOCALITY: Papua New Guinea, Trans-Fly Plains, Morehead (8°04'S, 141°39'E).

DISTRIBUTION: Lowland S Papua New Guinea; Northern Gulf, Queensland (Australia).

STATUS: IUCN – Data Deficient. Common?

COMMENTS: *S. murina* species-group.

*Sminthopsis bindi* Van Dyck, Woinarski and Press, 1994. Mem. Qld. Mus., 37:312.

COMMON NAME: Kakadu Dunnart.

TYPE LOCALITY: Australia, Northern Territory, Eva Valley Station, Stage 3, Kakadu National Park (14°30'S, 132°45'E).

DISTRIBUTION: Known from Stuart Highway (12°51'S, 131°08'E) southeast to Roper Valley (14°55'S, 133°54'E), but predicted from bioclimatic records to occur in much of C and S Arnhem Land (Australia).

STATUS: IUCN – Lower Risk (lc).

COMMENTS: *S. macroura* species-group. Formerly misidentified as *S. macroura*.

*Sminthopsis boullangerensis* Crowther, Dickman and Lynam, 1999. Aust. J. Zool., 47:220.

COMMON NAME: Boullanger Island Dunnart.

TYPE LOCALITY: Australia, Western Australia, Boullanger Isl (30°18'S, 115°02'E), Jurien Bay.

DISTRIBUTION: Boullanger Isl, and on the mainland at Lesueur, near Jurien; subfossil material near Hastings Cave may belong to this species according to the describers.

STATUS: Threatened.

COMMENTS: Described as a subspecies of *S. griseoventer*, but Crowther et al. (1999) described fixed differences in both morphology and allozymes from what they provisionally referred to as *S. g. griseoventer*, and considered their placement of *boullangerensis* as "conservative".

*Sminthopsis butleri* Archer, 1979. Aust. Zool., 20(2):329.

COMMON NAME: Carpentarian Dunnart.

TYPE LOCALITY: Australia, Western Australia, Kalumburu (14°15'S, 126°40'E).

DISTRIBUTION: In Australia known only from the type locality; also in Papua New Guinea.

STATUS: IUCN – Vulnerable.

COMMENTS: Name first mentioned by Kirsch (1977:47), but first made available by Archer

(1979). Not studied by Krajewski et al. (1999, 2000), but Archer (1981) noted that its affinities are probably with *S. macroura*.

*Sminthopsis crassicaudata* (Gould, 1844). Proc. Zool. Soc. Lond., 1844:105.
    COMMON NAME: Fat-tailed Dunnart.
    TYPE LOCALITY: Australia, Western Australia, Williams River.
    DISTRIBUTION: South Australia, SW Queensland, SE Northern Territory, S Western Australia, W New South Wales, W Victoria.
    STATUS: IUCN – Lower Risk (lc).
    SYNONYMS: *centralis* Thomas, 1902; *ferruginea* Finlayson, 1933.
    COMMENTS: Forms a species-group by itself. See Archer (1979:329, 1981:176). See S. J. B. Cooper et al. (2000) for possible subspecies boundaries.

*Sminthopsis dolichura* Kitchener, Stoddart and Henry, 1984. Rec. West. Aust. Mus., 11:204.
    COMMON NAME: Little Long-tailed Dunnart.
    TYPE LOCALITY: Western Australia, 6 km SSE of Buningonia Spring, 32°28′S, 123°36′E.
    DISTRIBUTION: Western Australia, South Australia.
    STATUS: IUCN – Lower Risk (lc).
    COMMENTS: *S. murina* species-group.

*Sminthopsis douglasi* Archer, 1979. Aust. Zool., 20(2):337.
    COMMON NAME: Julia Creek Dunnart.
    TYPE LOCALITY: Australia, Queensland, Cloncurry River Watershed, Julia Creek (20°40′S, 141°40′E).
    DISTRIBUTION: Known from 19-25°30′S, 141-143°E, in the "downs country" of NW Queensland; and possibly Mitchell Plateau, Western Australia.
    STATUS: IUCN – Endangered.
    COMMENTS: *S. macroura* species-group. For additions to range see Woolley (1992).

*Sminthopsis fuliginosus* (Gould, 1852). Mamm. Aust., 1, pl. 41.
    COMMON NAME: Sooty Dunnart
    TYPE LOCALITY: Western Australia, King George Sound.
    DISTRIBUTION: SW Western Australia.
    STATUS: IUCN – Data Deficient.
    COMMENTS: *S. murina* species-group. Separated from *S. murina* by Kitchener et al. (1984b). A little-known species; comparative notes given by Crowther et al. (1999).

*Sminthopsis gilberti* Kitchener, Stoddart, and Henry, 1984. Rec. W. Aust. Mus., 11:204.
    COMMON NAME: Gilbert's Dunnart.
    TYPE LOCALITY: Western Australia, Mt. Saddleback, 32°58′S, 116°20′E.
    DISTRIBUTION: SW Western Australia.
    STATUS: IUCN – Lower Risk (lc).
    COMMENTS: *S. murina* species-group.

*Sminthopsis granulipes* Troughton, 1932. Rec. Aust. Mus., 18:350.
    COMMON NAME: White-tailed Dunnart.
    TYPE LOCALITY: Australia, Western Australia, King George Sound (Albany).
    DISTRIBUTION: SW Western Australia.
    STATUS: IUCN – Lower Risk (lc).
    COMMENTS: Forms a species-group by itself.

*Sminthopsis griseoventer* Kitchener, Stoddart, and Henry, 1984. Rec. W. Aust. Mus., 11:204.
    COMMON NAME: Gray-bellied Dunnart.
    TYPE LOCALITY: Western Australia, Bindoon, 31°18′S, 116°01′E.
    DISTRIBUTION: SW Western Australia.
    STATUS: IUCN – Lower Risk (lc).
    SYNONYMS: *caniventer* Baverstock, Adams, and Archer, 1984.
    COMMENTS: *S. griseoventer* species-group.

*Sminthopsis hirtipes* Thomas, 1898. Nov. Zool., 5:3.
    COMMON NAME: Hairy-footed Dunnart.

TYPE LOCALITY: Australia, Northern Territory, Charlotte Waters.

DISTRIBUTION: Central deserts in Northern Territory and Western Australia; also coastal scrub 500 km N of Perth.

STATUS: IUCN – Lower Risk (lc).

COMMENTS: *S. psammophila* species-group.

*Sminthopsis leucopus* (Gray, 1842). Ann. Mag. Nat. Hist., [ser. 1], 10:261.

COMMON NAME: White-footed Dunnart.

TYPE LOCALITY: Australia, Tasmania.

DISTRIBUTION: S and SE Victoria, Tasmania, New South Wales, and Queensland (Australia).

STATUS: IUCN – Data Deficient. Common?

SYNONYMS: *ferruginifrons* (Gould, 1854); *leucogenys* Higgins and Petterd, 1883; *mitchelli* (Krefft, 1867).

COMMENTS: *S. murina* species-group. See Archer (1979:329, 1981:102).

*Sminthopsis longicaudata* Spencer, 1909. Proc. R. Soc. Victoria, 21 (n.s.):449.

COMMON NAME: Long-tailed Dunnart.

TYPE LOCALITY: Australia, Western Australia.

DISTRIBUTION: Western Australia.

STATUS: CITES – Appendix I; U.S. ESA – Endangered; IUCN – Lower Risk (lc).

COMMENTS: Forms a species-group by itself. Known from only four specimens; see Ride (1970:201).

*Sminthopsis macroura* (Gould, 1845). Proc. Zool. Soc. Lond., 1845:79.

COMMON NAME: Stripe-faced Dunnart.

TYPE LOCALITY: Australia, Queensland, Darling Downs.

DISTRIBUTION: Australia: NW New South Wales, W Queensland, S Northern Territory, N South Australia, N Western Australia.

STATUS: IUCN – Lower Risk (lc).

SYNONYMS: *froggatti* Ramsay, 1887; *larapinta* Spencer, 1896; *monticola* Troughton, 1965; *stalkeri* Thomas, 1906.

COMMENTS: *S. macroura* species-group. See Archer (1979:329, 1981:148). According to Blacket et al. (2001), this is probably a species complex rather than a single species, including *froggatti* and *stalkeri*.

*Sminthopsis murina* (Waterhouse, 1838). Proc. Zool. Soc. Lond., 1837:76 [1838].

COMMON NAME: Slender-tailed Dunnart.

TYPE LOCALITY: Australia, New South Wales, N of Hunter River.

DISTRIBUTION: SW Western Australia, SE South Australia, Victoria, New South Wales, E Queensland.

STATUS: IUCN – Lower Risk (nt) as *S. m. tatei*; otherwise Lower Risk (lc). Common.

SYNONYMS: *albipes* (Waterhouse, 1842); *tatei* Troughton, 1965.

COMMENTS: *S. murina* species-group. See Archer (1979:329, 1981:94-99).

*Sminthopsis ooldea* Troughton, 1965. Proc. Linn. Soc. N.S.W., 1964, 89:316 [1965].

COMMON NAME: Ooldea Dunnart.

TYPE LOCALITY: Australia, South Australia, Ooldea.

DISTRIBUTION: Edge of Nullarbor Plain (South Australia), Western Australia, S Northern Territory.

STATUS: IUCN – Lower Risk (lc).

COMMENTS: *S. psammophila* species-group. Originally described as a subspecies of *murina*, but considered a distinct species by Archer (1975:243) and Kirsch and Calaby (1977:15).

*Sminthopsis psammophila* Spencer, 1895. Proc. R. Soc. Victoria, 7 (n.s.):223.

COMMON NAME: Sandhill Dunnart.

TYPE LOCALITY: Australia, Northern Territory, Lake Amadeus.

DISTRIBUTION: Australia: SW Northern Territory (vicinity of Ayers Rock) and Eyre Peninsula (South Australia).

STATUS: CITES – Appendix I; U.S. ESA – Endangered; IUCN – Endangered.

COMMENTS: *S. psammophila* species-group. Known only from five specimens; see Archer
   (1981:215).

*Sminthopsis virginiae* (de Tarragon, 1847). Rev. Zool. Paris, p. 177.
   COMMON NAME: Red-cheeked Dunnart.
   TYPE LOCALITY: None given; Archer (1981:132) designated Australia, Queensland, Herbert
      Vale.
   DISTRIBUTION: N Queensland, N Northern Territory (Australia); Aru Isls (Indonesia);
      lowlands of S New Guinea.
   STATUS: IUCN – Lower Risk (lc).
   SYNONYMS: **nitela** Collett, 1897; *lumholtzi* Iredale and Troughton, 1934; **rufigenis** Thomas,
      1922; *rona* (Tate and Archbold, 1936).
   COMMENTS: *S. macroura* species-group. See Archer (1979:329, 1981:132) and Kirsch and
      Calaby (1977:15). De Tarragon's 1847 description of *S. virginiae* did not specify a type
      locality and his type specimen, now lost, had no known locality. Collett (1886[1887]:
      548) named *S. nitela* from Herbert Vale and it was subsequently renamed *S. lumholtzi*,
      both being referable to *virginiae* according to Archer (1981:136); but *nitela* and
      probably *rufigenis* may be distinct species according to Blacket et al. (2001).

*Sminthopsis youngsoni* McKenzie and Archer, 1982. Austr. Mamm., 5:267.
   COMMON NAME: Lesser Hairy-footed Dunnart.
   TYPE LOCALITY: Western Australia, Edgar Ranges, 18°50′S, 123°05′E.
   DISTRIBUTION: Western Australia, Northern Territory.
   STATUS: IUCN – Lower Risk (lc).
   COMMENTS: *S. psammophila* species-group.

**Tribe Planigalini** Archer, 1982. *Carnivorous Marsupials*, 2:439.

*Planigale* Troughton, 1928. Rec. Aust. Mus., 16:282.
   TYPE SPECIES: *Planigale brunneus* Troughton, 1928 (= *Phascogale ingrami* Thomas, 1906).
   COMMENTS: Revised by Archer (1976) on the basis of morphology and by Painter et al. (1995)
      on the basis of mitochondrial cytochrome *b* sequencing.

*Planigale gilesi* Aitken, 1972. Rec. S. Aust. Mus., 16(10):1.
   COMMON NAME: Paucident Planigale.
   TYPE LOCALITY: Australia, South Australia, Ann Creek Station (No. 3 bore) (28°18′S,
      136°29′40″E).
   DISTRIBUTION: NE South Australia, NW New South Wales, and SW Queensland (Australia).
   STATUS: IUCN – Lower Risk (lc).

*Planigale ingrami* (Thomas, 1906). Abstr. Proc. Zool. Soc. Lond., 1906(32):6.
   COMMON NAME: Long-tailed Planigale.
   TYPE LOCALITY: Australia, Northern Territory, Alexandria.
   DISTRIBUTION: Australia: N and E Queensland, NE Northern Territory, NE Western Australia.
   STATUS: U.S. ESA – Endangered as *P. i. subtilissima*; IUCN – Lower Risk (lc)..
   SYNONYMS: **brunnea** Troughton, 1928; **subtilissima** Lönnberg, 1913.
   COMMENTS: See Archer (1976:351). Woolley (1974) suggested that Western Australian
      *P. i. subtilissima* may be specifically distinct. A related but distinct species, referred to
      simply as *Planigale* 2 by Painter at al. (1995), is known from the Pilbara, Western
      Australia.

*Planigale maculata* (Gould, 1851). Mamm. Aust., 1, pl. 44.
   COMMON NAME: Pygmy Planigale.
   TYPE LOCALITY: Australia, New South Wales, Clarence River.
   DISTRIBUTION: E Queensland, NE New South Wales, and N Northern Territory (Australia).
   STATUS: IUCN – Lower Risk (lc).
   SYNONYMS: *minutissimus* (Gould, 1852); **sinualis** (Thomas, 1926).
   COMMENTS: Transferred to *Planigale* from *Antechinus* by Archer (1976:346). According to
      Painter et al. (1995), it is the most distinctive species in the genus, and may be closer to
      *Sminthopsis*. The subspecies *P. m. sinualis* (Arnhem Land and Groote Eylandt) is widely

divergent in mitochondrial cytochrome *b* sequences, as well as morphologically, and may be specifically distinct (Painter et al., 1995). A related but distinct species, referred to simply as *Planigale* 1 by Painter et al. (1995), is known from the Pilbara, Western Australia.

*Planigale novaeguineae* Tate and Archbold, 1941. Am. Mus. Novit., 1101:7.
 COMMON NAME: New Guinean Planigale.
 TYPE LOCALITY: Papua New Guinea, Central Prov., Rona Falls, Laloki River (vicinity of Port Moresby), 250 m.
 DISTRIBUTION: Lowlands of S New Guinea.
 STATUS: IUCN – Vulnerable.

*Planigale tenuirostris* Troughton, 1928. Rec. Aust. Mus., 16:285.
 COMMON NAME: Narrow-nosed Planigale.
 TYPE LOCALITY: Australia, "collected at Bourke or Wilcannia, New South Wales".
 DISTRIBUTION: NW New South Wales, and SC Queensland (Australia).
 STATUS: U.S. ESA – Endangered; IUCN – Lower Risk (lc).

# ORDER PERAMELEMORPHIA
by Colin P. Groves

**ORDER PERAMELEMORPHIA** Ameghino, 1889.
SYNONYMS: Peramelia, Perameliformes, Perameloidea.
COMMENTS: Recognized as an order by Aplin and Archer (1987) who proposed a new syncretic classification of the marsupials. McKenna and Bell (1997) used the name Peramelia.

**Family Thylacomyidae** Bensley, 1903. Trans. Linn. Soc. Lond. (Zool.), ser.2, 9:110.
COMMENTS: Included in the family Peramelidae by Vaughan (1978:39) and Groves and Flannery (1990); but more distinct than Peroryctidae according to Kirsch et al. (1997). McKenna and Bell (1997) included it in family Peramelidae, subfamily Chaeropodinae, with *Chaeropus*.

*Macrotis* Reid, 1837. Proc. Zool. Soc. Lond., 1836:131 [1837].
TYPE SPECIES: *Perameles lagotis* Reid, 1837.
SYNONYMS: *Paragalia* Gray, 1841; *Peragale* Lydekker, 1887; *Phalacomys* Anon., 1854; *Thalaconus* Richardson, Dallas, Cobbold, Baird and White, 1862; *Thylacomys* Blyth, 1840.
COMMENTS: Not preoccupied by *Macrotis* Dejean, 1833, a *nomen nudum* (Troughton, 1932*b*). Archer and Kirsch (1977) placed *Macrotis* (including its junior synonym *Thylacomys*) in a separate family (the name available being Thylacomyidae), rather than in Peramelidae. Groves and Flannery (1990) placed *Macrotis* back in Peramelidae.

*Macrotis lagotis* (Reid, 1837). Proc. Zool. Soc. Lond., 1836:129 [1837].
COMMON NAME: Greater Bilby.
TYPE LOCALITY: Australia, Western Australia, Swan River.
DISTRIBUTION: Formerly in Western Australia, South Australia, Northern Territory, W New South Wales, SW Queensland. Survives only in SW Queensland, Northern Territory/Western Australia border region and Kimberleys.
STATUS: CITES – Appendix I; U.S. ESA – Endangered; IUCN – Vulnerable.
SYNONYMS: *cambrica* Troughton, 1932; *grandis* Troughton, 1932; *interjecta* Troughton, 1932; *nigripes* (Wood Jones, 1923); *sagitta* (Thomas, 1905).

*Macrotis leucura* (Thomas, 1887). Ann. Mag. Nat. Hist., ser. 5, 19:397.
COMMON NAME: Lesser Bilby.
TYPE LOCALITY: Undesignated; as the specimen was sent by the South Australian Museum's taxidermist to London, Thomas (1887*a*) thought it might have originated near Adelaide or in the northern part of South Australia, whence others in the same collection had come.
DISTRIBUTION: C Australia.
STATUS: CITES – Appendix I; U.S. ESA – Endangered; IUCN – Extinct.
SYNONYMS: *minor* (Spencer, 1897); *miseliae* Tate, 1948; *miselius* (Finlayson, 1932).
COMMENTS: Probably extinct; see Ride (1970:200).

**Family Chaeropodidae** Gill, 1872. Smithson. Misc. Collect., 11:26.
SYNONYMS: Chaeropini Szalay, 1994.
COMMENTS: A subfamily of Peramelidae according to McKenna and Bell (1997), who included Thylacomyidae as a synonym. Molecular data (Westerman et al., 1999, 2001) do not support this arrangement, but indicate that *Chaeropus* is the sister-group of all other Peramelemorphia, so it is given family rank here.

*Chaeropus* Ogilby, 1838. Proc. Zool. Soc. Lond., 1838:26.
TYPE SPECIES: *Perameles ecaudatus* Ogilby, 1838.
SYNONYMS: *Choeropus* Waterhouse, 1841.
COMMENTS: Molecular data indicate that *Chaeropus* is the sister group to all other bandicoots (Westerman et al., 1999, 2001).

*Chaeropus ecaudatus* (Ogilby, 1838). Proc. Zool. Soc. Lond., 1838:25.
    COMMON NAME: Pig-footed Bandicoot.
    TYPE LOCALITY: Australia, New South Wales, banks of Murray River, south of the junction
        with Murrumbridge River.
    DISTRIBUTION: Australia: SW New South Wales, Victoria, S Northern Territory, N South
        Australia, Western Australia.
    STATUS: CITES – Appendix I [Possibly Extinct]; U.S. ESA – Endangered; IUCN – Extinct.
    SYNONYMS: *castanotis* Gray, 1842; *occidentalis* Gould, 1845.
    COMMENTS: Probably extinct, last taken in 1907; see Ride (1970:200).

**Family Peramelidae** Gray, 1825. Ann. Philos., n.s., 10:336.
    SYNONYMS: Peroryctidae, Echymiperinae.
    COMMENTS: Formerly considered to include the family Thylacomyidae; see Vaughan (1978:39)
        and Groves and Flannery (1990); but also see Archer and Kirsch (1977). Revised by Tate
        (1948*b*). Divided by McKenna and Bell (1997) into two families: Peramelidae, with two
        subfamilies, Chaeropodinae (for *Chaeropus* and *Macrotis*) and Peramelinae; and
        Peroryctidae, with two subfamilies, Peroryctinae and Echymiperinae. Molecular data
        (Westerman et al., 1999, 2001) do not support the monophyly of Chaeropodinae in the
        sense of McKenna and Bell (1997), show that *Chaeropus* is very distinct from other
        bandicoots, and do not support the monophyly of Peroryctidae.

**Subfamily Peramelinae** Gray, 1825. Ann. Philos., n.s., 10:336.

*Isoodon* Desmarest, 1817. Nouv. Dict. Hist. Nat., Nouv. ed., 16:409.
    TYPE SPECIES: *Didelphis obesula* Shaw, 1797.
    SYNONYMS: *Thylacis* Haltenorth, 1958 [not of Illiger, 1811 (=*Perameles*)].
    COMMENTS: Includes *Thylacis* of Haltenorth, 1958, which was an incorrect usage; see Van
        Deusen and Jones (1967:74) and Lidicker and Follett (1968). Revised by Lyne and Mort
        (1981).

*Isoodon auratus* (Ramsay, 1887). Proc. Linn. Soc. N.S.W., ser. 2, 2:551.
    COMMON NAME: Golden Bandicoot.
    TYPE LOCALITY: Australia, Western Australia, Derby.
    DISTRIBUTION: Australia: Formerly Northern Territory and N Western Australia, survives in
        NW of Western Australia and on Barrow Isl.
    STATUS: IUCN – Vulnerable as *I. auratus, I. a. auratus,* and *I. a. barrowensis.* Rare.
    SYNONYMS: **arnhemensis** Lyne and Mort, 1981; **barrowensis** (Thomas, 1901).
    COMMENTS: See Ride (1970:96) and Lyne and Mort (1981), who recognized *arnhemensis* and
        *barrowensis* as distinct species. A revision based on wider material is needed.

*Isoodon macrourus* (Gould, 1842). Proc. Zool. Soc. Lond., 1842:41.
    COMMON NAME: Northern Brown Bandicoot.
    TYPE LOCALITY: Australia, Northern Territory, Port Essington.
    DISTRIBUTION: NE Western Australia, N Northern Territory, E Queensland, and NE New
        South Wales (Australia); S and E New Guinea.
    STATUS: IUCN – Lower Risk (lc). Common.
    SYNONYMS: *macrura* (Wagner, 1853); **moresbyensis** (Ramsay, 1877); *torosa* (Ramsay, 1877).

*Isoodon obesulus* (Shaw, 1797). Nat. Misc., 8:298.
    COMMON NAME: Southern Brown Bandicoot.
    TYPE LOCALITY: Australia, New South Wales, Sydney, Ku-ring-gai Chase Natl. Park, 33°36'S,
        151°16'E, see Dixon (1981).
    DISTRIBUTION: SE New South Wales, S Victoria, SE South Australia, N Queensland,
        SW Western Australia, Nuyts Arch. (Great Australian Bight, S Australian coast), and
        Tasmania.
    STATUS: IUCN – Vulnerable as *I. o. nauticus,* Lower Risk (nt) as *I. o. fusciventer, I. o. obesulus,*
        and *I. o. peninsulae;* otherwise Lower Risk (lc). Locally common.
    SYNONYMS: *affinis* (Waterhouse, 1846); *fusciventer* (Gray, 1841); *peninsulae* Thomas, 1922;
        **nauticus** Thomas, 1922.

*Perameles* É. Geoffroy, 1804. Ann. Mus. Hist. Nat. Paris, 4:56.
  TYPE SPECIES: *Perameles nasuta* É. Geoffroy, 1804.
  SYNONYMS: *Thylacis* Illiger, 1811.
  COMMENTS: *Perameles* was also used by Geoffroy, 1804, Bull. Sci. Soc. Philom. Paris, 3(80):249,
      which may take priority (dates uncertain). Molecular data (Westerman et al., 2001)
      suggest that this genus may be paraphyletic with respect to *Isoodon*. Species boundaries in
      the genus are confused and a revision is needed.

*Perameles bougainville* Quoy and Gaimard, 1824. *In* de Freycinet, Voy. autour du monde . . . l'Uranie
      et al Physicienne, Zool., p. 56.
  COMMON NAME: Western Barred Bandicoot.
  TYPE LOCALITY: Australia, Western Australia, Shark Bay, Peron Peninsula.
  DISTRIBUTION: Formerly in S South Australia, NW Victoria, W New South Wales, S Western
      Australia, Bernier and Dorre Isls, survives only on Bernier and Dorre Isls (off Western
      Australia).
  STATUS: CITES – Appendix I; U.S. ESA – Endangered; IUCN – Extinct as *P. b. fasciata*,
      Endangered as *P. b. bougainville*. Rare.
  SYNONYMS: *arenaria* Gould, 1844; *fasciata* Gray, 1841; *myosuros* Wagner, 1841; *notina*
      Thomas, 1922.
  COMMENTS: See Ride (1970:100).

*Perameles eremiana* Spencer, 1897. Proc. R. Soc. Victoria, 9 (n.s.):9.
  COMMON NAME: Desert Bandicoot.
  TYPE LOCALITY: Australia, Northern Territory, Burt Plain (N of Alice Springs).
  DISTRIBUTION: N South Australia, S Northern Territory, Great Victoria Desert (Western
      Australia).
  STATUS: U.S. ESA – Endangered; IUCN – Extinct.
  COMMENTS: Possibly extinct; see Ride (1970:200).

*Perameles gunnii* Gray, 1838. Ann. Nat. Hist., 1:107.
  COMMON NAME: Eastern Barred Bandicoot.
  TYPE LOCALITY: Australia, Tasmania.
  DISTRIBUTION: Australia: S Victoria, where restricted to Hamilton, and Tasmania.
  STATUS: IUCN – Vulnerable as *P. g. gunnii*. Common in Tasmania, Endangered in Victoria.

*Perameles nasuta* É. Geoffroy, 1804. Ann. Mus. Hist. Nat. Paris, 4:62.
  COMMON NAME: Long-nosed Bandicoot.
  TYPE LOCALITY: Australia, New South Wales, Sydney.
  DISTRIBUTION: Australia: E Queensland, E New South Wales, E Victoria.
  STATUS: IUCN – Lower Risk (lc). Common.
  SYNONYMS: *lawson* Quoy and Gaimard, 1824; *major* Schinz, 1825; *musei* (Boitard, 1841);
      *pallescens* Thomas, 1923.

**Subfamily Peroryctinae** Groves and Flannery, 1990. *In* Seebeck et al. (eds.), Bandicoots and Bilbies,
      p. 2.
  COMMENTS: McKenna and Bell (1997) attributed this name to a publication of Archer et al.
      (1989), but this is a *nomen nudum,* referring to a manuscript name of Groves and Flannery.
      Kirsch et al. (1997) reduced this to a subfamily under Peramelidae. Westerman et al.
      (1999) considered the family, *sensu* Groves and Flannery, probably polyphyletic, and that
      *Echymipera* and *Microperoryctes* form a sister clade to genera of the Peramelidae; further
      material confirms this, and adds *Peroryctes* as a third, probably sister clade to the rest
      (Westerman et al, 2001). A peroryctid, probably *Echymipera* sp., occurred on Halmahera
      until 1870 B. P. (Flannery et al., 1995a).

*Peroryctes* Thomas, 1906. Proc. Zool. Soc. Lond., 1906:476.
  TYPE SPECIES: *Perameles raffrayana* Milne-Edwards, 1878.
  COMMENTS: May not be close to the other genera of this family (Westerman et al., 1999).

*Peroryctes broadbenti* (Ramsay, 1879). Proc. Linn. Soc. N.S.W., 3:402, pl. 27.
  COMMON NAME: Giant Bandicoot.

TYPE LOCALITY: Papua New Guinea, Central Prov., banks of Goldie River (a tributary of the Laloki River) inland from Port Moresby.

DISTRIBUTION: SE New Guinea, probably lowlands.

STATUS: IUCN – Data Deficient. Rare.

COMMENTS: Included in *raffrayana* by Laurie and Hill (1954:10), but considered a distinct species by Van Deusen and Jones (1967:74).

*Peroryctes raffrayana* (Milne-Edwards, 1878). Ann. Sci. Nat. (Paris), 7(11):1.

COMMON NAME: Raffray's Bandicoot.

TYPE LOCALITY: Indonesia, Prov. of Papua (= Irian Jaya), Vogelkop, Manokwari Div., Amberbaki.

DISTRIBUTION: New Guinea, sea level to 4000 m; Yapen Isl.

STATUS: IUCN – Lower Risk (lc). Secure.

SYNONYMS: *rothschildi* (Förster, 1913); *mainois* (Förster, 1913).

COMMENTS: Laurie and Hill (1954:10) included *broadbenti* in this species; but see Van Deusen and Jones (1967:74).

**Subfamily Echymiperinae** McKenna and Bell, 1997. Class. Mamm. Above Species Level, p. 57.

*Echymipera* Lesson, 1842. Nouv. Tabl. Regn. Anim. Mammifères, p. 192.

TYPE SPECIES: *Perameles kalubu* Fischer, 1829.

SYNONYMS: *Peramelopsis* Heude, 1896; *Suillomeles* Allen and Barbour, 1909.

COMMENTS: See Groves and Flannery (1990).

*Echymipera clara* Stein, 1932. Z. Säugetierk., 7:256.

COMMON NAME: Clara's Echymipera.

TYPE LOCALITY: Indonesia, Prov. of Papua (= Irian Jaya), Tjenderawasih Div., Japen (=Yapen) Isl.

DISTRIBUTION: NC New Guinea, 300-1700 m; Yapen Isl (Indonesia).

STATUS: IUCN – Data Deficient. Rare.

COMMENTS: See Flannery (1990*a*) for an assessment of its affinities.

*Echymipera davidi* Flannery, 1990. *In* Seebeck et al. (eds.), Bandicoots and Bilbies, p. 29.

COMMON NAME: David's Echymipera.

TYPE LOCALITY: Papua New Guinea, Trobriand Isls, Kiriwina Isl (08°30'S, 151°00'E).

DISTRIBUTION: Kiriwina Isl (Papua New Guinea).

STATUS: IUCN – Data Deficient. Unknown.

COMMENTS: Not closely related to any other species of *Echymipera* (Flannery, 1990*a*).

*Echymipera echinista* Menzies, 1990. Science in New Guinea, 16:92.

COMMON NAME: Menzies' Echymipera.

TYPE LOCALITY: Papua New Guinea, Western (Fly River) Province, Wipim, near Iamega (08°51'S, 142°58'E).

DISTRIBUTION: Known only from Western (Fly River) Province, Papua New Guinea.

STATUS: IUCN – Data Deficient. Rare.

*Echymipera kalubu* (Fischer, 1829). Synopsis Mammal., p. 274.

COMMON NAME: Common Echymipera.

TYPE LOCALITY: Indonesia, Prov. of Papua (= Irian Jaya), Sorong Div., Waigeo Isl.

DISTRIBUTION: New Guinea and adjacent small islands including Bismarck Arch., Yapen, Biak-Supiori, Waigeo, Misool and Salawati Isls.

STATUS: IUCN – Lower Risk (lc). Common.

SYNONYMS: *alticeps* (Cohn, 1910); *breviceps* (Cohn, 1910); *doreyanus* (Quoy and Gaimard, 1830); *garagassi* (Miklouho-Maclay, 1884); *hispida* (Allen and Barbour, 1909); *rufiventris* (Heller, 1897); *cockerelli* (Ramsay, 1877); *myoides* (Günther, 1883); *oriomo* Tate and Archbold, 1936; *philipi* Troughton, 1954.

COMMENTS: The name *kalubu* has been attributed to Lesson, 1828, Dict. Class. Hist. Nat., 13:200; but see Husson (1955:290). According to Westerman et al. (2001), the molecular relationships of *E. k. cockerelli* (from New Britain and Duke of York Isl) are with *E. rufescens* rather than with other *E. kalubu*, but it is morphologically similar to *kalubu*.

*Echymipera rufescens* (Peters and Doria, 1875). Ann. Mus. Civ. Stor. Nat. Genova, 7:541.
COMMON NAME: Long-nosed Echymipera.
TYPE LOCALITY: Indonesia, Kai Isls.
DISTRIBUTION: Cape York Peninsula (Queensland, Australia); New Guinea and
  D'Entrecasteaux Isls; Kai and Aru Isls, Yapen Isl, and Misool Isl (Indonesia).
STATUS: IUCN – Lower Risk (lc). Uncommon.
SYNONYMS: *aruensis* (Peters and Doria, 1875); *gargantua* Thomas, 1914; *keiensis* (Cohn,
  1910); *welsianus* (Heude, 1896); ***australis*** Tate, 1948.

*Microperoryctes* Stein, 1932. Z. Säugetierk., 7:256.
TYPE SPECIES: *Microperoryctes murina* Stein, 1932.
SYNONYMS: *Ornoryctes* Tate and Archbold, 1937.
COMMENTS: For synonymy of *Ornoryctes* with *Microperoryctes* instead of with *Peroryctes*, see
  Groves and Flannery (1990).

*Microperoryctes longicauda* (Peters and Doria, 1876). Ann. Mus. Civ. Stor. Nat. Genova, 8:335.
COMMON NAME: Striped Bandicoot.
TYPE LOCALITY: Indonesia, Prov. of Papua (= Irian Jaya), Vogelkop, Manokwari Div., Arfak
  Mtns, Hatam, 1520 m.
DISTRIBUTION: Interior New Guinea, 1000-4000 m.
STATUS: IUCN – Lower Risk (lc). Common.
SYNONYMS: ***dorsalis*** (Thomas, 1922); ***ornatus*** (Thomas, 1904); *magnus* Laurie, 1952.
COMMENTS: Formerly included in *Peroryctes*, but see Groves and Flannery (1990).

*Microperoryctes murina* Stein, 1932. Z. Säugetierk., 7:257.
COMMON NAME: Mouse Bandicoot.
TYPE LOCALITY: Indonesia, Prov. of Papua (= Irian Jaya), Paniai Div., Weyland Mtns, Sumuri
  Mtn, 2500 m.
DISTRIBUTION: W interior New Guinea., Vogelkop and Weyland Range, around 2000 m.
STATUS: IUCN – Data Deficient. Rare.
COMMENTS: The two isolates of this species are actually distinct species (Helgen and
  Flannery, work in progress).

*Microperoryctes papuensis* (Laurie, 1952). Bull. Brit. Mus. (Nat. Hist.), Zool., 1:291.
COMMON NAME: Papuan Bandicoot.
TYPE LOCALITY: Papua New Guinea, Milne Bay Prov., Mt. Mura, (30 mi [48 km] NW Mt.
  Simpson), Boneno, 1220-1525 m.
DISTRIBUTION: SE interior New Guinea, 1200-2650 m.
STATUS: IUCN – Data Deficient. Rare.
COMMENTS: Formerly included in *Peroryctes*, but see Groves and Flannery (1990).

*Rhynchomeles* Thomas, 1920. Ann. Mag. Nat. Hist., ser. 9, 6:429-430.
TYPE SPECIES: *Rhynchomeles prattorum* Thomas, 1920.

*Rhynchomeles prattorum* Thomas, 1920. Ann. Mag. Nat. Hist., ser. 9, 6:429-430.
COMMON NAME: Seram Bandicoot.
TYPE LOCALITY: Indonesia, Ceram (= Seram) Isl, Mt. Manusela, 1800 m.
DISTRIBUTION: Seram Isl (Indonesia), restricted to high elevations.
STATUS: IUCN – Data Deficient. Rare.
COMMENTS: See Groves and Flannery (1990), A. C. Kitchener et al. (1993).

# ORDER DIPROTODONTIA
by Colin P. Groves

**ORDER DIPROTODONTIA** Owen, 1866.
SYNONYMS: Phalangeriformes.
COMMENTS: Recognized as an order by Aplin and Archer (1987) who proposed a new syncretic classification of the marsupials. Divided into three suborders by Kirsch et al. (1997); their arrangement is followed here.

**SUBORDER VOMBATIFORMES** Burnett, 1830.

**Family Phascolarctidae** Owen, 1839. Proc. Zool. Soc. Lond., 1839:19.
COMMENTS: Formerly included in the Phalangeridae; see Ride (1970:225).

*Phascolarctos* de Blainville, 1816. Nouv. Bull. Sci. Soc. Philom. (Paris), p. 108.
TYPE SPECIES: *Lipurus cinereus* Goldfuss, 1817 (seen by de Blainville in ms., published 1817).
SYNONYMS: *Draximenus* Lay, 1825; *Koala* Schinz, 1821; *Lipurus* Goldfuss, 1817; *Liscurus* McMurtie, 1834; *Morodactylus* Goldfuss, 1820.

*Phascolarctos cinereus* (Goldfuss, 1817). Die Säugethiere, pt. 65, pl. 155, Aa, Ac.
COMMON NAME: Koala.
TYPE LOCALITY: Australia, New South Wales.
DISTRIBUTION: Australia: SE Queensland, E New South Wales, SE South Australia, and Victoria. Introduced on Kangaroo Isl, South Australia and at Yanchep, Western Australia.
STATUS: U.S. ESA – Threatened; IUCN – Lower Risk (nt). Vulnerable.
SYNONYMS: *adustus* Thomas, 1923; *flindersii* Lesson, 1827; *fuscus* Desmarest, 1820; *koala* Gray, 1827; *subiens* (Burnett, 1830); *victor* Troughton, 1935.
COMMENTS: Subspecies are uncertain.

**Family Vombatidae** Burnett, 1830. Quart. J. Lit. Sci. Art., 1829:351 [1830].
SYNONYMS: Phascolomyidae Goldfuss, 1820.
COMMENTS: Phascolomyidae Goldfuss, 1820, is based on *Phascolomis*, a junior synonym (Haltenorth, 1958:32). Because Phascolomyidae was replaced with Vombatidae before 1961, and because Vombatidae has won general acceptance, it is to be maintained (Art. 40.2 of the International Code of Zoological Nomenclature, International Commission on Zoological Nomenclature, 1999).

*Lasiorhinus* Gray, 1863. Ann. Mag. Nat. Hist., ser. 3, 11:458.
TYPE SPECIES: *Lasiorhinus mcoyi* Gray, 1863 (= *Phascolomys latifrons* Owen, 1845).
SYNONYMS: *Wombatula* Iredale and Troughton, 1934.
COMMENTS: This genus needs revision; *krefftii* may be better restricted to a Pleistocene species, and neither *barnardi* nor *gillespiei* may belong to it.

*Lasiorhinus krefftii* (Owen, 1873). Philos. Trans. R. Soc. London, 162:178, pl. 17, 20.
COMMON NAME: Northern Hairy-nosed Wombat.
TYPE LOCALITY: Australia, New South Wales, Wellington Caves, Breccia Cavern.
DISTRIBUTION: Australia: SE and E Queensland, Deniliquin (New South Wales).
STATUS: CITES – Appendix I; U.S. ESA Endangered; IUCN – Critically Endangered.
SYNONYMS: *gillespiei* (DeVis, 1900); *barnardi* Longman, 1939.
COMMENTS: Includes *gillespiei* and *barnardi* according to Kirsch and Calaby (1977:23), who stated that only a single remnant population of *krefftii* remained at the type locality of *barnardi*. However, populations historically known as *barnardi* may not be referable to *krefftii*.

*Lasiorhinus latifrons* (Owen, 1845). Proc. Zool. Soc. Lond., 1845:82.
COMMON NAME: Southern Hairy-nosed Wombat.

TYPE LOCALITY: Australia, South Australia.
DISTRIBUTION: S South Australia, SE Western Australia.
STATUS: IUCN – Lower Risk (lc). Locally common.
SYNONYMS: *lasiorhinus* (Gould, 1863); *mcoyi* Gray, 1863.

*Vombatus* É. Geoffroy, 1803. Bull. Sci. Soc. Philom. Paris, 72:185.
TYPE SPECIES: *Didelphis ursina* Shaw, 1800.
SYNONYMS: *Amblotis* Illiger, 1811; *Opossum* Perry, 1810; *Phascolomis* É. Geoffroy, 1803;
*Phascolomus* Rafinesque, 1815; *Phascolomys* Duméril, 1806; *Wombatus* Desmarest, 1804.

*Vombatus ursinus* (Shaw, 1800). Gen. Zool. Syst. Nat. Hist., 1(2), Mammalia, p. 504.
COMMON NAME: Common Wombat.
TYPE LOCALITY: Australia, Tasmania, Bass Strait, Cape Barren Isl.
DISTRIBUTION: E New South Wales, S Victoria, SE South Australia, Tasmania, islands in the
Bass Strait, and extreme SE Queensland (Australia).
STATUS: IUCN – Vulnerable as *V. u. ursinus*, Lower Risk (lc) as *V. usrinus*. Common.
SYNONYMS: *angasii* (Gray, 1863); *assimilis* (Krefft, 1872); *bassii* (Lesson, 1827); *fossor*
(Desmarest, 1804); *fuscus* (Tiedemann, 1808); *hirsutum* (Perry, 1810); *mitchelli* (Owen,
1838); *niger* (Gould, 1863); *platyrhinus* (Owen, 1853); *setosus* (Gray, 1863); *tasmaniensis*
(Spencer and Kershaw, 1910); *vombatus* (Leach, 1815); *wombat* (Voigt, 1802).
COMMENTS: Subspecies uncertain; marked differences in size exist between insular and
mainland wombats, but there are wide overlaps.

# SUBORDER PHALANGERIFORMES Szalay, *in* Archer (ed.), 1982.

**Superfamily Phalangeroidea** Thomas, 1888. Cat. Marsup. Monotr. Brit. Mus., p. 126.

**Family Burramyidae** Broom, 1898. Proc. Linn. Soc. N.S.W., 10:564.

*Burramys* Broom, 1896. Proc. Linn. Soc. N.S.W., 10:564.
TYPE SPECIES: *Burramys parvus* Broom, 1896.

*Burramys parvus* Broom, 1896. Proc. Linn. Soc. N.S.W., 10:564 [fig. in pl. 25, p. 273].
COMMON NAME: Mountain Pygmy Possum.
TYPE LOCALITY: Australia, New South Wales, Taralga (fossil).
DISTRIBUTION: Mountains of NE Victoria and S New South Wales (Australia).
STATUS: U.S. ESA and IUCN – Endangered.

*Cercartetus* Gloger, 1841. Gemein Hand.-Hilfsbuch. Nat., 1:85.
TYPE SPECIES: *Phalangista nana* Desmarest, 1818.
SYNONYMS: *Dromicia* Gray, 1841; *Dromiciella* Matschie, 1916; *Dromiciola* Matschie, 1916;
*Eudromicia* Mjöberg, 1916.
COMMENTS: Includes *Eudromicia* (see Kirsch and Calaby, 1977:16).

*Cercartetus caudatus* (Milne-Edwards, 1877). C. R. Acad. Sci. Paris, 85:1079.
COMMON NAME: Long-tailed Pygmy Possum.
TYPE LOCALITY: Indonesia, Prov. of Papua (= Irian Jaya), Vogelkop, Manokwari Div., Arfak
Mtns.
DISTRIBUTION: Interior New Guinea, above 1500 m; Fergusson Isl (Papua New Guinea);
NE Queensland (Australia).
STATUS: IUCN – Lower Risk (nt) as *C. c. macrurus*, Lower Risk (lc) as *C. c. caudatus*. Common.
SYNONYMS: **macrura** (Mjöberg, 1916).
COMMENTS: Includes *macrura* (see Ride, 1970:224). Formerly included in *Eudromicia*; see
comment under genus.

*Cercartetus concinnus* (Gould, 1845). Proc. Zool. Soc. Lond., 1845:2.
COMMON NAME: Southwestern Pygmy Possum.
TYPE LOCALITY: Australia, Western Australia, Swan River.

DISTRIBUTION: SW Western Australia, S and SE South Australia including Kangaroo Isl,
  W Victoria.
STATUS: IUCN – Lower Risk (lc). Common.
SYNONYMS: *neillii* (Waterhouse, 1846); ***minor*** Wakefield, 1963.

*Cercartetus lepidus* (Thomas, 1888). Cat. Marsup. Monotr. Brit. Mus., p. 142.
  COMMON NAME: Tasmanian Pygmy Possum.
  TYPE LOCALITY: Australia, Tasmania.
  DISTRIBUTION: Australia: Tasmania, NW Victoria/South Australia border, and Kangaroo Isl
    (South Australia).
  STATUS: IUCN – Lower Risk (lc). Common.
  COMMENTS: Formerly included in *Eudromicia*; see comment under genus.

*Cercartetus nanus* (Desmarest, 1818). Nouv. Dict. Hist. Nat., Nouv. ed., 25:477.
  COMMON NAME: Eastern Pygmy Possum.
  TYPE LOCALITY: Australia, Tasmania, Ile Maria.
  DISTRIBUTION: Australia: SE South Australia, E New South Wales to SE Queensland, Victoria,
    and Tasmania.
  STATUS: IUCN – Lower Risk (lc). Common.
  SYNONYMS: *gliriformis* (Bell, 1828); ***unicolor*** (Krefft, 1863); *britta* (Wood Jones, 1925).

**Family Phalangeridae** Thomas, 1888. Cat. Marsup. Monotr. Brit. Mus., p. 126.
  COMMENTS: Distinct from Phascolarctidae (see Ride, 1970:22) and does not include
    Pseudocheiridae, Petauridae, Burramyidae, or Acrobatidae (see Aplin and Archer, 1987). A
    provisional classification was given by Flannery et al. (1987), and this was modified by
    Norris (1994).

**Subfamily Ailuropinae** Flannery, Archer and Maynes, 1987. In M. Archer (ed.), *Possums and
  Opossums: Studies in Evolution*, 2: 482.

*Ailurops* Wagler, 1830. Naturliches Syst. Amphibien, p. 26.
  TYPE SPECIES: *Phalangista ursina* Temminck, 1824.
  SYNONYMS: *Ceonix* Temminck, 1827; *Eucuscus* Gray, 1862.
  COMMENTS: McKenna and Bell (1997) did not separate this genus from *Phalanger*, but Flannery
    et al. (1987) had placed it in its own subfamily, Ailuropinae, citing evidence that it is the
    sister-group of the rest of the Phalangeridae.

*Ailurops melanotis* (Thomas, 1898). Novit. Zool., 5:2.
  COMMON NAME: Talaud Bear Cuscus.
  TYPE LOCALITY: Indonesia, Talaud Isls, Salebabu Isl, Lirung.
  DISTRIBUTION: Only found on Salebabu Isl in the Talaud Isls (Indonesia).
  STATUS: Unknown.
  COMMENTS: Formerly included in *A. ursinus*; but measurements of the type (and only known
    preserved) specimen given by Feiler (1977) fall strongly outside those from Sulawesi
    and offshore islands, and the distinctive colouration and patterning of the type is seen
    in a second (living) specimen, whose photograph I examined, courtesy of R. Wirth.

*Ailurops ursinus* (Temminck, 1824). Monogr. Mamm., 1:10.
  COMMON NAME: Sulawesi Bear Cuscus.
  TYPE LOCALITY: Indonesia, Sulawesi, Sulawesi Utara, Minahasa, Manado.
  DISTRIBUTION: Sulawesi, Peleng Isl, Muna and Butung Isls, Togian Isls.
  STATUS: IUCN – Date Deficient. Common.
  SYNONYMS: ***flavissimus*** (Feiler, 1977); ***furvus*** (Miller and Hollister, 1922); *intermedius*
    (Hooijer, 1952); ***togianus*** (Tate, 1945).
  COMMENTS: Formerly included in *Phalanger*. May be a species complex.

**Subfamily Phalangerinae** Thomas, 1888. *Cat. Marsup. Monotr. Brit. Mus.*, p. 126.
  COMMENTS: Norris (1994), studying the anatomy of the periotic, and Kirsch et al. (1997), on
    molecular grounds, divided this subfamily into two tribes, Phalangerini and Trichosurini.

**Tribe Phalangerini** Thomas, 1888. *Cat. Marsup. Monotr. Brit. Mus.*, p. 126.

*Phalanger* Storr, 1780. Prodr. Meth. Mammal., p. 38.
    TYPE SPECIES: *Didelphis orientalis* Pallas, 1766.
    SYNONYMS: *Balantia* Illiger, 1811; *Coescoes* Lacépède, 1799; *Cuscus* Lesson, 1826; *Phalangista*
        E. Geoffroy St. Hilaire and G. Cuvier, 1795; *Sipalus* G. Fischer, 1813.
    COMMENTS: Does not include *Spilocuscus* (see Ride, 1970:248). Revised by Tate (1945), Feiler
        (1978*a-c*), and G. G. George (1979).

*Phalanger alexandrae* Flannery and Boeadi, 1995. Aust. Mammal., 18:42.
    COMMON NAME: Gebe Cuscus.
    TYPE LOCALITY: Indonesia, North Moluccas, Gebe Isl, near the airport (0°05'S, 129°25'E).
    DISTRIBUTION: Pulau (Isl) Gebe.
    STATUS: IUCN – Data Deficient. Very restricted distribution.
    COMMENTS: Related to *P. ornatus* and *P. rothschildi*.

*Phalanger carmelitae* Thomas, 1898. Ann. Mus. Civ. Stor. Nat. Genova, 19:5.
    COMMON NAME: Mountain Cuscus.
    TYPE LOCALITY: Papua New Guinea, Central Prov., upper Vanapa River.
    DISTRIBUTION: Interior New Guinea.
    STATUS: IUCN – Lower Risk (lc). Common.
    SYNONYMS: **coccygis** Thomas, 1922.
    COMMENTS: Formerly included in *vestitus* (see G. G. George, 1979:94).

*Phalanger gymnotis* Peters and Doria, 1875. Ann. Mus. Civ. Stor. Nat. Genova, 7:543.
    COMMON NAME: Ground Cuscus.
    TYPE LOCALITY: Indonesia, Aru Isls, Gialnhegen Isl (restricted by Van der Feen, 1962:40).
    DISTRIBUTION: New Guinea; Aru Isls and other small Indonesian islands.
    STATUS: IUCN – Data Deficient. Common.
    SYNONYMS: **leucippus** Thomas, 1898.
    COMMENTS: Distribution poorly known. Included in *Strigocuscus* by Flannery et al. (1987),
        but Norris (1994) and Kirsch et al. (1997) found that it is part of the *Phalanger* clade.

*Phalanger intercastellanus* Thomas, 1895. Novit. Zool., 2:165.
    COMMON NAME: Eastern Common Cuscus.
    TYPE LOCALITY: Papua New Guinea, D'Entrecasteaux group, Fergusson Isl.
    DISTRIBUTION: SE Papua New Guinea, east of Markham valley; Sariba, Itoh, Goodenough,
        Fergusson, Normanby, Kiriwina, Misima, Sudest and Rossel Isl.
    STATUS: IUCN – Lower Risk (nt). CITES – Appendix II as *P. orientalis*.
    SYNONYMS: *brevinasus* Thomas, 1895; *kiriwinae* Thomas, 1896; *matsika* Tate and Archbold,
        1935; *meeki* Thomas, 1898.
    COMMENTS: Flannery et al. (1987) suggested that what were hitherto regarded as the
        southern races of *P. orientalis* may prove to be specifically, even generically distinct, as
        *Strigocuscus mimicus*, and Flannery (1994*a*) definitively separated them but used the
        earlier name *intercastellanus*. Norris and Musser (2001) returned this species to
        *Phalanger*, and separated *P. mimicus* from it (see below). The species needs revision; at
        present no subspecies are recognized.

*Phalanger lullulae* Thomas, 1896. Novit. Zool., 3:528.
    COMMON NAME: Woodlark Cuscus.
    TYPE LOCALITY: Papua New Guinea, Milne Bay Prov., Woodlark Isl.
    DISTRIBUTION: Woodlark Isl (Papua New Guinea).
    STATUS: IUCN – Lower Risk (lc). Rare.
    COMMENTS: Formerly included in *orientalis* (see G. G. George, 1979:97). Reviewed by Norris
        (1999).

*Phalanger matabiru* Flannery and Boeadi, 1995. Aust. Mammal., 18:40.
    COMMON NAME: Blue-eyed Cuscus.
    TYPE LOCALITY: Indonesia, North Moluccas, Ternate, Tege Tege (0°50'N, 127°20'E), 400 m.

DISTRIBUTION: Ternate and Tidore Isls (Indonesia).

STATUS: Restricted distribution.

COMMENTS: Described as a subspecies of *P. ornatus* by Flannery and Boeadi (1995), but differs absolutely (diagnostically), and so raised to specific rank here.

*Phalanger matanim* Flannery, 1987. Rec. Aust. Mus., 39:183.

COMMON NAME: Telefomin Cuscus.

TYPE LOCALITY: Papua New Guinea, West Sepik Prov., Telefomin area, Upper Sol River, 5°06'S, 141°42'E, 2600 m.

DISTRIBUTION: Telefomin area, W Papua New Guinea.

STATUS: IUCN – Endangered. Rare.

*Phalanger mimicus* Thomas, 1922. Ann. Mag. Nat. Hist., ser. 9, 9:679.

COMMON NAME: Southern Common Cuscus.

TYPE LOCALITY: Indonesia, Prov. of Papua (= Irian Jaya), Mimika River, Parimau, 4°31'S, 136°36'E, 250 ft (76 m).

DISTRIBUTION: S New Guinea, from Mimika River (Prov. of Papua) east to Mt. Bosavi and Oriomo River; perhaps Aru Isls; Cape York Peninsula (Australia).

STATUS: CITES – Appendix II as *P. orientalis*; otherwise Lower Risk (lc).

SYNONYMS: *microdon* Tate and Archbold, 1935; **peninsulae** Tate, 1945.

COMMENTS: Separated from *P. intercastellanus* by Norris and Musser (2001). It is unclear whether this is the species on the Aru Isls.

*Phalanger orientalis* (Pallas, 1766). Misc. Zool., p. 61.

COMMON NAME: Northern Common Cuscus.

TYPE LOCALITY: Indonesia, Amboina (= Ambon) Isl, Maluku.

DISTRIBUTION: Timor, Sanana (Sulu Isls), Buru, Halmahera and Seram Isls (Indonesia) to northern New Guinea, Karkar Isl, Schouten group, Bismarck Arch., Solomon Isls; Sanana (Sula Isls), Buru, and Halmahera.

STATUS: IUCN – Lower Risk (lc); CITES – Appendix II.

SYNONYMS: *alba* (E. Geoffroy, 1803); *amboinensis* (Lacépède, 1799); *cavifrons* Temminck, 1824; *fusca* (Oken, 1816) [unavailable]; *indica* (Müller, 1776); *kori* Menzies and Pernetta, 1986; *minor* (Oken, 1816) [unavailable]; *molucca* (Gmelin, 1789); *moluccensis* (Oken, 1816) [unavailable]; *rufa* (E. Geoffroy, 1803); *timoriensis* Menzies and Pernetta, 1986; *vulpecula* (Förster, 1913); **breviceps** Thomas, 1888; *albidus* Feiler, 1978; *ducatoris* Thomas, 1922.

COMMENTS: Formerly included *intercastellanus*, *interpositus* (= *vestitus*) and *lullulae* (see G. G. George, 1979). Flannery et al. (1987) first suggested that the southern races may prove to be specifically distinct; Flannery (1990) did not adopt this course, but Flannery (1994a) separated them. Two undescribed species, related to *P. orientalis*, occur on Mt. Karimui. The species needs revision; Flannery (1995b) provisionally recognized only Solomons/Bismarcks *breviceps* as a subspecies.

*Phalanger ornatus* (Gray, 1860). Proc. Zool. Soc. Lond., 1860:374.

COMMON NAME: Ornate Cuscus.

TYPE LOCALITY: Indonesia: Bachian (= Bacan or Batjan) Isl.

DISTRIBUTION: Halmahera, Bacan, possibly Morotai Isls (Indonesia).

STATUS: IUCN – Lower Risk (lc). Apparently common.

COMMENTS: For status see Groves (1987). Provisionally allotted to *Strigocuscus* by Flannery et al. (1987) although they did not examine specimens.

*Phalanger rothschildi* Thomas, 1898. Novit. Zool., 5:433.

COMMON NAME: Rothschild's Cuscus.

TYPE LOCALITY: Moluccas, Pulau Obi, Loiwuj.

DISTRIBUTION: Pulau (Isl) Obi, and Bisa.

STATUS: IUCN – Vulnerable. Common.

COMMENTS: Commonly included in *Strigocuscus celebensis*, but see Groves (1987).

*Phalanger sericeus* Thomas, 1907. Ann. Mag. Nat. Hist., ser. 7, 20:74.

COMMON NAME: Silky Cuscus.

TYPE LOCALITY: Papua New Guinea, Angabunga Range, Owgarra, 6,000 ft (= 1829 m).
DISTRIBUTION: C and E New Guinea, higher elevations, above 1500 m.
STATUS: IUCN – Lower Risk (lc). Common.
SYNONYMS: *occidentalis* Menzies and Pernetta, 1986.
COMMENTS: Called *P. vestitus* by most authors, but see Menzies and Pernetta (1986:594).

*Phalanger vestitus* (Milne-Edwards, 1877). C. R. Acad. Sci. Paris, 85:1080.
COMMON NAME: Stein's Cuscus.
TYPE LOCALITY: Indonesia, Prov. of Papua (= Irian Jaya), Vogelkop, Sorong Div., Tamrau Range, Karons Mtns.
DISTRIBUTION: Interior New Guinea, 1200-1500 m. (in western range, up to 2200 m.).
STATUS: IUCN – Vulnerable.
SYNONYMS: *interpositus* Stein, 1933; *permixtio* Menzies and Pernetta, 1986.
COMMENTS: Formerly known as *P. interpositus* (see Flannery, 1990).

*Spilocuscus* Gray, 1862. Proc. Zool. Soc. Lond., 1861:316 [1862].
TYPE SPECIES: *Phalangista maculata* Desmarest, 1818.
COMMENTS: Separated from *Phalanger* by G. G. George (1979), and recognized by subsequent workers, though the separation was not mentioned by McKenna and Bell (1997).

*Spilocuscus kraemeri* (Schwarz, 1910). Sitzb. Ges. Naturf. Fr. Berlin, p. 406.
COMMON NAME: Admiralty Island Cuscus.
TYPE LOCALITY: Papua New Guinea, Admiralty Isls, Manus Isl.
DISTRIBUTION: Manus and Lou Isls.
STATUS: IUCN – Lower Risk (lc).
SYNONYMS: *minor* Cohn, 1914.
COMMENTS: Separated as a species by Flannery (1994a), who noted that it is very distinctive, but that it does not occur in archaeological deposits, so may possibly be derived from introductions.

*Spilocuscus maculatus* (E. Geoffroy, 1803). Cat. Mamm. Mus. Hist. Nat., Paris., p. 149.
COMMON NAME: Common Spotted Cuscus.
TYPE LOCALITY: Indonesia, Prov. of Papua (= Irian Jaya), Vogelkop, Manokwari Div., Manokwari.
DISTRIBUTION: New Guinea and adjacent small islands; Aru and Kei Isls, Seram, Ambon and Selayar Isls (Indonesia); Cape York Peninsula (Queensland, Australia). On New Guinea, found mainly above 1200 m.
STATUS: IUCN – Lower Risk (lc); CITES – Appendix II; but "common" in New Guinea (Flannery, 1990).
SYNONYMS: *variegata* (Oken, 1816) [unavailable]; *chrysorrhous* (Temminck, 1824); *goldiei* (Ramsay, 1876); *nudicaudatus* (Gould, 1850); *brevicaudatus* (Gray, 1858); *ochropus* (Gray, 1866).
COMMENTS: Feiler (1978a) included *atrimaculatus* in this species, but G. G. George (1979:98) placed it in *rufoniger*. Usually includes *papuensis* and *kraemeri*, but these were separated by Flannery (1994a).

*Spilocuscus papuensis* (Desmarest, 1822). Encycl. Méth. Mamm. Suppl., p. 541.
COMMON NAME: Waigeou Cuscus.
TYPE LOCALITY: Indonesia, Prov. of Papua (= Irian Jaya), Waigeou Isl.
DISTRIBUTION: Waigeou.
STATUS: IUCN – Data Deficient.
SYNONYMS: *macrourus* (Lesson and Garnot, 1826), *quoy* (Gaimard, 1824).
COMMENTS: Separated as a species by Flannery (1994a).

*Spilocuscus rufoniger* (Zimara, 1937). Anz. Akad. Wiss. Wien, 74:35.
COMMON NAME: Black-spotted Cuscus.
TYPE LOCALITY: Papua New Guinea, Morobe Prov., Sattelberg.
DISTRIBUTION: N New Guinea, above 1200 m.
STATUS: IUCN – Endangered.
SYNONYMS: *atrimaculatus* (Tate, 1945).

COMMENTS: Includes *atrimaculatus* (see G. G. George, 1979:98), but also see Feiler (1978*a*), who placed it in *maculatus*.

**Tribe Trichosurini** Flynn, 1911. Papers and Proceedings Roy. Soc. Tasmania, 1911:120.

*Strigocuscus* Gray, 1862. Proc. Zool. Soc. Lond., 1861:319 [1862].
    TYPE SPECIES: *Cuscus celebensis* Gray, 1858.
    COMMENTS: Not distinguished from *Phalanger* by McKenna and Bell (1997), but when Flannery et al. (1987) resurrected this genus for *S. celebensis*, and provisionally for *S. gymnotis*, they gave evidence that it is sister-group to *Trichosurus*. Norris (1994) retained *gymnotis* in *Phalanger* and thus showed that it belongs in Phalangerini.

*Strigocuscus celebensis* (Gray, 1858). Proc. Zool. Soc. Lond., 1858:105.
    COMMON NAME: Sulawesi Dwarf Cuscus.
    TYPE LOCALITY: Indonesia, Sulawesi, Sulawesi Selatan, Ujung Pandang (= Macassar).
    DISTRIBUTION: Sulawesi, Peleng Isl, Sanghir Isls. Records from Taliabu and Obi Isl (Indonesia) may be erroneous (T. F. Flannery, pers. comm.).
    STATUS: IUCN – Data Deficient. Common.
    SYNONYMS: *callenfelsi* (Hooijer, 1950); *feileri* (Groves, 1987); **sangirensis** Meyer, 1896.

*Strigocuscus pelengensis* (Tate, 1945). Am. Mus. Novit., 1283:3.
    COMMON NAME: Banggai Cuscus.
    TYPE LOCALITY: Indonesia: Peleng Isl.
    DISTRIBUTION: Peleng and Sula Isls (Indonesia).
    STATUS: IUCN – Lower Risk (lc). Unknown.
    SYNONYMS: **mendeni** (Feiler, 1978).
    COMMENTS: For status see Groves (1987). Flannery et al. (1987) doubted that *pelengensis* really belongs to *Phalanger*, and Flannery (1994*a*) placed it in *Strigocuscus*.

*Trichosurus* Lesson, 1828. *In* Bory de Saint-Vincent (ed.), Dict. Class. Hist. Nat. Paris, 13:333.
    TYPE SPECIES: *Didelphis vulpecula* Kerr, 1792.
    SYNONYMS: *Cercaertus* Burmeister, 1837; *Psilogrammurus* Gloger, 1841; *Tapoa* Owen, 1839; *Trichurus* Wagner, 1843.

*Trichosurus arnhemensis* Collett, 1897. Proc. Zool. Soc. Lond., 1897:328.
    COMMON NAME: Northern Brushtail.
    TYPE LOCALITY: Australia, Northern Territory, Daly River.
    DISTRIBUTION: N Northern Territory, NE Western Australia, Barrow Isl (Australia).
    STATUS: Common.

*Trichosurus caninus* (Ogilby, 1836). Proc. Zool. Soc. Lond., 1835:191 [1836].
    COMMON NAME: Short-eared Possum.
    TYPE LOCALITY: Australia, New South Wales, Hunter River.
    DISTRIBUTION: Australia: C Queensland (near Gladstone) south to C New South Wales, and possibly to the Victorian border (Lindenmayer et al., 2002).
    STATUS: IUCN – Lower Risk (lc). Common.
    SYNONYMS: *nigrans* Le Souef, 1916.
    COMMENTS: What were previously thought to be populations of this species in Victoria have been shown to be a separate species, *T. cunninghami* (Lindemayer et al., 2002). The species and *T. cunninghami* are together sometimes known as Bobuck; Lindemayer et al. (2002) suggested new vernacular names to distinguish them.

*Trichosurus cunninghami* Lindenmayer, Dubach and Viggers, 2002. Aust. J. Zool., 50:17.
    COMMON NAME: Mountain Brushtail Possum.
    TYPE LOCALITY: Australia, C Victoria, Cambarville region.
    DISTRIBUTION: Australia: C to NE Victoria, and possibly into S New South Wales (Lindenmayer et al., 2002).
    STATUS: Common.
    COMMENTS: Formerly thought to be the southern populations of *T. caninus*, but distinguished as a new species by Lindenmayer et al. (2002).

*Trichosurus johnstonii* (Ramsay, 1888). Proc. Linn. Soc. N.S.W., ser. 2, 8:1297.
   COMMON NAME: Coppery Brushtail.
   TYPE LOCALITY: Australia, Queensland, Atherton Tableland.
   DISTRIBUTION: Australia: rainforests of NE Queensland.
   STATUS: Common.
   COMMENTS: Separated as a species by Flannery (1994*a*).

*Trichosurus vulpecula* (Kerr, 1792). *In* Linnaeus, Anim. Kingdom, 1:198.
   COMMON NAME: Common Brushtail.
   TYPE LOCALITY: Australia, New South Wales, Sydney.
   DISTRIBUTION: Australia: E Queensland, E New South Wales, Victoria, Tasmania, SE and
      N South Australia, SW Western Australia; introduced to New Zealand (Wodzicki, 1950).
   STATUS: IUCN – Lower Risk (nt) as *T. v. hypoleuca*, otherwise Lower Risk (lc). Common, in
      many cities, lives commensally.
   SYNONYMS: *bougainvillei* (J. B. Fischer, 1829); *cookii* (G. Cuvier, 1824); *cuvieri* (Waterhouse,
      1841); *eburacensis* Lönnberg, 1916 [status *fide* Flannery, 1994a]; *felina* (Wagner, 1842);
      *fuliginosa* (Ogilby, 1831); *grisea* (Gray, 1841); *hypoleucus* (Wagner, 1855); *lemurina*
      (Shaw, 1800); *melanura* (Wagner, 1842); *mesurus* Thomas, 1926; *novaehollandiae*
      (Bechstein, 1800); *raui* Finlayson, 1963; *ruficollis* Schwarz, 1909; *selma* (Gervais, 1847);
      *tapouaru* (F. Meyer, 1793); *vulpina* (F. Meyer, 1793); *xanthopus* (Ogilby, 1831).

*Wyulda* Alexander, 1918. J. R. Soc. West. Aust. (1917-1918), 4:31.
   TYPE SPECIES: *Wyulda squamicaudata* Alexander, 1918.
   COMMENTS: Provisionally combined with *Trichosurus* by Flannery et al. (1987). Norris (1994)
      confirmed, on examination of the type of *W. squamicaudata*, that the two genera are
      closely related.

*Wyulda squamicaudata* Alexander, 1918. J. R. Soc. West. Aust. (1917-1918), 4:31.
   COMMON NAME: Scaly-tailed Possum.
   TYPE LOCALITY: Australia, Western Australia, Wyndham.
   DISTRIBUTION: NE Western Australia, Kimberleys.
   STATUS: U.S. ESA – Endangered; IUCN – Lower Risk (nt).
   COMMENTS: Provisionally placed in *Trichosurus* by Flannery et al. (1987).

**Superfamily Petauroidea** Bonaparte, 1838. Nuovi Ann. Sci. Nat., 2(1):112.

**Family Pseudocheiridae** Winge, 1893. Med. Udsigt over Pungdyrenes Slaegtskab. E. Mus. Lundii,
   11. pt. 2:89.
   COMMENTS: Separated from Petauridae by Archer (1984); not recognized as a family by
      McKenna and Bell (1997), though retained by molecular workers (Kirsch et al., 1997), but
      Osborne and Christidis (2001) again questioned the separation. McKenna and Bell (1997)
      retained an old scheme whereby *Hemibelideus, Pseudochirops, Pseudochirulus* and
      *Petropseudes* are included in *Pseudocheirus*, and recognized only one other genus
      (*Schoinobates* [sic], which correctly should be *Petauroides*); but Kirsch et al. (1997) showed
      that this does not reflect the true pattern of relationships, and is greatly over-lumped.

**Subfamily Hemibelideinae** Kirsch, Lapointe and Springer, 1997. Aust. J. Zool., 45:245.

*Hemibelideus* Collett, 1884. Proc. Zool. Soc. Lond., 1884:385.
   TYPE SPECIES: *Phalangista (Hemibelideus) lemuroides* Collett, 1884.
   COMMENTS: Formerly included in *Pseudocheirus*, and retained in that genus by McKenna and
      Bell (1997), but it was shown by Kirsch et al. (1997) that the two are very distinct, and
      *Hemibelideus* is sister-group to *Petauroides*.

*Hemibelideus lemuroides* (Collett, 1884). Proc. Zool. Soc. Lond., 1884:385.
   COMMON NAME: Lemur-like Ringtail.
   TYPE LOCALITY: Australia, North Queensland.
   DISTRIBUTION: NE Queensland (Australia).
   STATUS: IUCN – Lower Risk (nt). Localized but not endangered where it occurs.

SYNONYMS: *cervinus* (Longman, 1915).

*Petauroides* Thomas, 1888. Cat. Marsup. Monotr. Brit. Mus., p. 163.
    TYPE SPECIES: *Didelphis volans* Kerr, 1792.
    SYNONYMS: *Petaurista* Desmarest, 1820; *Schoinobates* Iredale and Troughton, 1934; *Volucella*
        Bechstein, 1800.
    COMMENTS: Formerly known as *Schoinobates* Lesson, 1842. This name was used by Lesson only
        for *Petaurista leucogenys* Temminck, 1823, a giant flying squirrel (McKay, 1982); its use for
        this genus dates only from Iredale and Troughton (1934), though McKenna and Bell
        (1997) continue to use *Schoinobates* for it without citing McKay's (1982) paper. The names
        *Volucella* and *Petaurista* were both preoccupied.

    *Petauroides volans* (Kerr, 1792) *In* Linnaeus, Anim. Kingdom, 1:199.
        COMMON NAME: Greater Glider.
        TYPE LOCALITY: Australia, New South Wales, Sydney.
        DISTRIBUTION: E Australia, from Dandenong Ranges (Victoria) to Rockhampton
            (Queensland).
        STATUS: Lower Risk (lc). Common.
        SYNONYMS: *didelphoides* (G. Cuvier, 1825); *incanus* Thomas, 1923; *macroura* (Shaw, 1794);
            *maximus* (Partington, 1837); *peronii* (Desmarest, 1818); *taguanoides* (Desmarest, 1818);
            *voluccella* (F. Meyer, 1793); **minor** (Collett, 1887); *armillatus* Thomas, 1923; *cinereus*
            Ramsay, 1890.

**Subfamily Pseudocheirinae** Winge, 1893. Med. Udsigt over Pungdyrenes Slaegtskab. E. Mus. Lundii,
    11. pt. 2:89.

*Petropseudes* Thomas, 1923. Ann. Mag. Nat. Hist., ser. 9, 11:250.
    TYPE SPECIES: *Pseudochirus dahli* Collett, 1895.
    COMMENTS: Separated from *Pseudocheirus* by McKay (1988:94).

    *Petropseudes dahli* (Collett, 1895). Zool. Anz., 18(490):464.
        COMMON NAME: Rock-haunting Ringtail.
        TYPE LOCALITY: Australia, Northern Territory, Mary River.
        DISTRIBUTION: N Northern Territory, NW Western Australia.
        STATUS: Lower Risk (lc). Localized.

*Pseudocheirus* Ogilby, 1837. Mag. Nat. Hist. (Charlesworth), 1:457.
    TYPE SPECIES: *Phalangista cookii* Desmarest, 1818 (= *Didelphis peregrinus* Boddaert, 1785), see
        Thomas (1888).
    SYNONYMS: *Hepoona* Gray, 1841; *Pseudochirus* Ogilby, 1836; *Ptenos* Gray, 1843.
    COMMENTS: *Pseudochirus* Ogilby, 1836. Proc. Zool. Soc. Lond., 1836:26 is a *nomen nudum*, see
        Palmer (1904). Does not include *Pseudochirulus* Matschie, 1915, see Flannery (1994*a*).

    *Pseudocheirus peregrinus* (Boddaert, 1785). Elench. Anim., p. 78.
        COMMON NAME: Common Ringtail.
        TYPE LOCALITY: Australia, Queensland, Endeavour River.
        DISTRIBUTION: Australia: Cape York Peninsula (Queensland) to SE South Australia and
            SW Western Australia, Tasmania, islands of the Bass Straits.
        STATUS: IUCN – Vulnerable as *P. occidentalis,* Lower Risk (lc) as *P. peregrinus.* Common.
        SYNONYMS: *banksii* (Gray, 1838); *caudivolvula* (Kerr, 1792); *cookii* (Desmarest, 1818); *incanens*
            (Thomas, 1923); *laniginosa* (Gould, 1858); *modestus* (Thomas, 1926); *notialis* (Thomas,
            1923); *novaehollanidiae* (Bechstein, 1800); *oralis* (Thomas, 1926); *victoriae* (Matschie,
            1915); **occidentalis** (Thomas, 1888); **pulcher** (Matschie, 1915); *rubidus* (Troughton and
            Le Souef, 1929); **convolutor** Schinz, 1821; *bassianus* (Le Souef, 1929); *incana* (Schinz,
            1844); *viverrina* Ogilby, 1838.
        COMMENTS: Probably a species complex rather than a single species (see Flannery, 1994*a*).

*Pseudochirulus* Matschie, 1915. Sitzb. Ges. Naturf. Fr. Berlin, p. 91.
    TYPE SPECIES: *Phalangista (Pseudocheirus) canescens* Waterhouse, 1845.

COMMENTS: Separated from *Pseudocheirus* by Flannery (1994*a*).

*Pseudochirulus canescens* (Waterhouse, 1846). Nat. Hist. Mamm., 1:306.
COMMON NAME: Lowland Ringtail.
TYPE LOCALITY: Indonesia, Prov. of Papua (= Irian Jaya), Fakfak Div., Triton Bay.
DISTRIBUTION: New Guinea and Salawati Isl, below 1300 m; Yapen Isl.
STATUS: IUCN – Data Deficient. Uncommon.
SYNONYMS: *grisescenti* (Peters, 1874); *avarus* Thomas, 1906); **bernsteini** (Schlegel, 1866); **dammermani** (Thomas, 1922); *gyrator* (Thomas, 1904).

*Pseudochirulus caroli* (Thomas, 1921). Ann. Mag. Nat. Hist., ser. 9, 8:357.
COMMON NAME: Weyland Ringtail.
TYPE LOCALITY: Indonesia, Prov. of Papua (= Irian Jaya), Paniai Div., Weyland Range, Menoo Valley, Mt. Kunupi, 1830 m.
DISTRIBUTION: WC New Guinea.
STATUS: IUCN – Data Deficient. Rare.
SYNONYMS: **versteegi** (Thomas, 1922).

*Pseudochirulus cinereus* Tate, 1945. Amer. Mus. Novit., 1287:17.
COMMON NAME: Cinereous Ringtail.
TYPE LOCALITY: Australia, Queensland, Mt. Spurgeon.
DISTRIBUTION: North Queensland, tablelands from Mt. Carbine north to Thornton Peak.
STATUS: IUCN – Lower Risk (nt). Localized but not uncommon.
COMMENTS: Separated from *P. herbertensis*, and perhaps not even a sister species, by Kirsch et al. (1997).

*Pseudochirulus forbesi* (Thomas, 1887). Ann. Mag. Nat. Hist., ser. 5, 19:146.
COMMON NAME: Painted Ringtail.
TYPE LOCALITY: Papua New Guinea, Central Prov., Astrolabe Range, near Port Moresby, Sogeri, 458 m.
DISTRIBUTION: SE New Guinea.
STATUS: IUCN – Lower Risk (lc). Common.
SYNONYMS: *longipilis* (Tate and Archbold, 1935).
COMMENTS: Does not include *larvatus* (see below).

*Pseudochirulus herbertensis* (Collett, 1884). Proc. Zool. Soc. Lond., 1884:383.
COMMON NAME: Herbert River Ringtail.
TYPE LOCALITY: Australia, Queensland, Herbert Vale.
DISTRIBUTION: NE Queensland (Australia), Ingham north to Atherton Tablelands.
STATUS: IUCN – Lower Risk (nt). Localized but not uncommon.
SYNONYMS: *colletti* (Waite, 1899); *mongan* (De Vis, 1887).

*Pseudochirulus larvatus* (Förster and Rothschild, 1911). Ann. Mag. Nat. Hist., ser. 6, 7:337.
COMMON NAME: Masked Ringtail.
TYPE LOCALITY: Papua New Guinea, Huon Peninsula, Rawlinson Mtns.
DISTRIBUTION: NE New Guinea (Huon Peninsula and eastern part of Central Cordillera).
STATUS: Common.
SYNONYMS: *barbatus* Matschie, 1915; *capistratus* Matschie, 1915.
COMMENTS: Diagnostically distinct from SE *forbesi*, and separated by a sharp boundary between Wau and the Kratke Mtns (Flannery, 1994*a*; Musser and Sommer, 1992), which strongly suggests that *larvatus* is specifically distinct.

*Pseudochirulus mayeri* (Rothschild and Dollman, 1932). Abstr. Proc. Zool. Soc. Lond., 1932(353):15.
COMMON NAME: Pygmy Ringtail.
TYPE LOCALITY: Indonesia, Prov. of Papua (= Irian Jaya), Paniai Div., Weyland Range, Gebroeders Mtns, 1830 m.
DISTRIBUTION: C interior New Guinea.
STATUS: IUCN – Lower Risk (lc). Common.
SYNONYMS: *pygmaeus* Stein, 1932
COMMENTS: See Laurie and Hill (1954: 21).

*Pseudochirulus schlegeli* (Jentink, 1884). Notes Leyden Mus., 6:110.
  COMMON NAME: Vogelkop Ringtail.
  TYPE LOCALITY: Indonesia, Prov. of Papua (= Irian Jaya), Vogelkop, Manokwari Div., Arfak
    Mtns.
  DISTRIBUTION: Extreme NW New Guinea.
  STATUS: IUCN – Data Deficient. Rare.
  SYNONYMS: *lewisi* (Dollman, 1930).
  COMMENTS: The putative taxon *P. forbesi lewisi* is actually a colour variant of this species
    (Flannery, 1995*a*).

**Subfamily Pseudochiropsinae** Kirsch, Lapointe and Springer, 1997. Aust. J. Zool., 45:245.

*Pseudochirops* Matschie, 1915. Sitzb. Ges. Naturf. Fr. Berlin, 4:86.
  TYPE SPECIES: *Phalangista* (*Pseudochirus*) *albertisii* Peters, 1874.
  COMMENTS: Separated from *Pseudocheirus* by McKay (1988). According to Kirsch et al. (1997), it
    is at least as distinct cladistically as are *Petauroides* and *Hemibelideus*.

*Pseudochirops albertisii* (Peters, 1874). Ann. Mus. Civ. Stor. Nat. Genova, 6:303.
  COMMON NAME: D'Albertis' Ringtail.
  TYPE LOCALITY: Indonesia, Prov. of Papua (= Irian Jaya), Vogelkop, Manokwari Div., Arfak
    Mtns, Hatam, 1520 m.
  DISTRIBUTION: N and W New Guinea, including Yapen Isl (Indonesia).
  STATUS: IUCN – Vulnerable. Uncommon.
  SYNONYMS: ***insularis*** Stein, 1933; ***schultzei*** Matschie, 1915

*Pseudochirops archeri* (Collett, 1884). Proc. Zool. Soc. Lond., 1884:381.
  COMMON NAME: Green Ringtail.
  TYPE LOCALITY: Australia, N Queensland, Herbert River District.
  DISTRIBUTION: NE Queensland (Australia).
  STATUS: IUCN – Lower Risk (nt). Locally common, but restricted.

*Pseudochirops corinnae* (Thomas, 1897). Ann. Mus. Civ. Stor. Nat. Genova, 18:142.
  COMMON NAME: Plush-coated Ringtail.
  TYPE LOCALITY: Papua New Guinea, Central Prov., upper Vanapa River.
  DISTRIBUTION: Interior New Guinea.
  STATUS: IUCN – Vulnerable. Common.
  SYNONYMS: *buergersi* Matchie, 1915; *caecias* (Thomas, 1922); ***argenteus*** (Förster, 1913);
    ***fuscus*** (Laurie, 1952).

*Pseudochirops coronatus* (Thomas, 1897). Ann. Mus. Civ. Stor. Nat. Genova, 18:144.
  COMMON NAME: Reclusive Ringtail.
  TYPE LOCALITY: Indonesia, Prov. of Papua (= Irian Jaya), Vogelkop, Arfak Mtns, 2000 m.
  DISTRIBUTION: Arfak Mtns (Prov. of Papua = Irian Jaya, Indonesia), above 1000 m.
  STATUS: IUCN – Lower Risk (lc). Unknown.
  SYNONYMS: *paradoxus* (Dollman, 1930)
  COMMENTS: Separated from *P. albertisii*, with which it is sympatric, by Flannery (1994*a*).

*Pseudochirops cupreus* (Thomas, 1897). Ann. Mus. Civ. Stor. Nat. Genova, 18:145.
  COMMON NAME: Coppery Ringtail.
  TYPE LOCALITY: Papua New Guinea, Owen Stanley Range.
  DISTRIBUTION: Interior New Guinea.
  STATUS: IUCN – Lower Risk (lc). Common.
  SYNONYMS: *beauforti* (Thomas, 1922); *obscurior* Tate and Archbold, 1935.

**Family Petauridae** Bonaparte, 1838. Nuovi Ann. Sci. Nat., 2(1):112.
  COMMENTS: May not be monophyletic (Osborne and Christidis, 2001).

*Dactylopsila* Gray, 1858. Proc. Zool. Soc. Lond., 1858:109.
  TYPE SPECIES: *Dactylopsila trivirgata* Gray, 1858.
  SYNONYMS: *Dactylonax* Thomas, 1910.

COMMENTS: See Haltenorth (1958:28).

*Dactylopsila megalura* Rothschild and Dollman, 1932. Abstr. Proc. Zool. Soc. Lond., 1932(353):14.
   COMMON NAME: Great-tailed Triok.
   TYPE LOCALITY: Indonesia, Prov. of Papua (= Irian Jaya), Paniai Div., Weyland Range,
      Gebroeders Mtns.
   DISTRIBUTION: Interior New Guinea.
   STATUS: IUCN – Vulnerable. Rare.
   COMMENTS: Considered a subspecies of *trivirgata* by Ziegler (*in* Stonehouse and Gilmore,
      1977:131).

*Dactylopsila palpator* Milne-Edwards, 1888. Mem. Cent. Soc. Philom. Paris, p. 174.
   COMMON NAME: Long-fingered Triok.
   TYPE LOCALITY: "South coast of New Guinea".
   DISTRIBUTION: Interior New Guinea.
   STATUS: IUCN – Lower Risk (lc). Common.
   SYNONYMS: *ernstmayri* (Stein, 1932).
   COMMENTS: Formerly included in *Dactylonax* (see Haltenorth, 1958).

*Dactylopsila tatei* Laurie, 1952. Bull. Br. Mus. (Nat. Hist.), Zool., 1:278.
   COMMON NAME: Tate's Triok.
   TYPE LOCALITY: Papua New Guinea, Milne Bay Prov., Fergusson Isl, Faralulu Dist.,
      mountains above Taibutu Village, 610-915 m.
   DISTRIBUTION: Fergusson Isl; Papua New Guinea.
   STATUS: IUCN – Endangered. Rare.
   COMMENTS: Considered a subspecies of *trivirgata* by Ziegler (*in* Stonehouse and Gilmore,
      1977:131); considered a distinct species by G. G. George (1979:94).

*Dactylopsila trivirgata* Gray, 1858. Proc. Zool. Soc. Lond., 1858:111.
   COMMON NAME: Striped Possum.
   TYPE LOCALITY: Indonesia, Aru Isls.
   DISTRIBUTION: New Guinea, Yapen and Waigeo; Aru Isls; NE Queensland (Australia).
   STATUS: IUCN – Lower Risk (lc). Common.
   SYNONYMS: *albertisii* Peters and Doria, 1875; *angustivittis* (Peters and Doria, 1880); *arfakensis*
      Matschie, 1916; *hindenburgi* Ramme, 1914; *occidentalis* Matschie, 1916; ***kataui***
      Matschie, 1916; ***melampus*** Thomas, 1908; *biedermanni* Matschie, 1916; ***picata***
      Thomas, 1908; *infumata* Tate, 1945.

*Gymnobelideus* McCoy, 1867. Ann. Mag. Nat. Hist., ser. 3, 20:287.
   TYPE SPECIES: *Gymnobelideus leadbeateri* McCoy, 1867.

*Gymnobelideus leadbeateri* McCoy, 1867. Ann. Mag. Nat. Hist., ser. 3, 20:287.
   COMMON NAME: Leadebeater's Possum.
   TYPE LOCALITY: Australia, Victoria, Bass River.
   DISTRIBUTION: NE Victoria (Australia).
   STATUS: U.S. ESA – Endangered; IUCN – Endangered.

*Petaurus* Shaw, 1791. Nat. Misc., 2, pl. 60.
   TYPE SPECIES: *Petaurus australis* Shaw, 1791.
   SYNONYMS: *Belideus* Waterhouse, 1839; *Petaurella* Matschie, 1916; *Petaurula* Matschie, 1916;
      *Ptilotus* G. Fischer, 1814; *Xenochirus* Gloger, 1841.

*Petaurus abidi* Ziegler, 1981. Austr. Mamm., 4:81.
   COMMON NAME: Northern Glider.
   TYPE LOCALITY: Papua New Guinea, West Sepik Prov., Mt. Somoro, 3°25'S, 142°05'E.
   DISTRIBUTION: NC New Guinea.
   STATUS: IUCN – Vulnerable. Rare.

*Petaurus australis* Shaw, 1791. Nat. Misc., 2, pl. 6.
   COMMON NAME: Yellow-bellied Glider.
   TYPE LOCALITY: Australia, New South Wales, Sydney.

DISTRIBUTION: Coastal Queensland, New South Wales, and Victoria (Australia).
STATUS: IUCN – Lower Risk (nt) as *P. a. australis* and as *P. australis*. Common locally.
SYNONYMS: *cunninghami* Gray, 1843; *flaviventer* Desmarest, 1818; *petaurus* (Shaw, 1800); *reginae* Thomas, 1923.

*Petaurus biacensis* Ulmer, 1940. Notul. Nat. Philad., 52:1.
COMMON NAME: Biak Glider.
TYPE LOCALITY: Indonesia, Prov. of Papua (= Irian Jaya), Biak Isl, Korrido.
DISTRIBUTION: Biak, Supiori and Owi Isls.
STATUS: IUCN – Lower Risk (lc). Unknown.
SYNONYMS: *kohlsi* Troughton, 1945.
COMMENTS: Regarded as a species by Flannery (1994*a*).

*Petaurus breviceps* Waterhouse, 1839. Proc. Zool. Soc. Lond., 1838:152 [1839].
COMMON NAME: Sugar Glider.
TYPE LOCALITY: Australia, New South Wales.
DISTRIBUTION: SE South Australia to Cape York Peninsula (Queensland), Tasmania (introduction), N Northern Territory, NE Western Australia; New Guinea and adjacent small islands, including Bismarck Arch.; Aru Isls and N Moluccas (Indonesia).
STATUS: IUCN – Lower Risk (lc). Common.
SYNONYMS: *notatus* Peters, 1859; **ariel** (Gould, 1842); *arul* (Gervais, 1869); **longicaudatus** Longman, 1924; **papuanus** Thomas, 1888; *flavidus* Tate and Archbold, 1935; *tafa* Tate and Archbold, 1935.
COMMENTS: See Smith (1973, Mammalian Species, 30). McAllan and Bruce (1989) argued that the original publication of this name was in *The Athenaeum*, 580:880 [8 Dec 1838]. An undescribed form from Tifalmin, west of the Sepik, is very distinct (Colgan and Flannery, 1992). Gliders from Goodenough, Fergusson and Normanby Isls (D'Entrecasteaux group), Papua New Guinea, usually identified as belonging to this species, are very distinct morphologically, if not electrophoretically (Flannery, 1994*a*).

*Petaurus gracilis* (de Vis, 1883). Abstr. Proc. Linn. Soc. N.S.W., 20 Dec. 1882, ii.
COMMON NAME: Ebony Glider.
TYPE LOCALITY: Australia, Queensland, Cardwell region.
DISTRIBUTION: From Wharps Holding (18°41'18"S, 146°04'25"E) to Hall River (17°58'29"S, 146°02'02"E), NE Queensland.
STATUS: IUCN – Endangered.
COMMENTS: History of description given by Van Dyck (1990). Species resurrected from synonymy with *P. norfolcensis* by Van Dyck (1991). Full description given by Van Dyck (1993), who also gave results of field surveys to determine its distribution.

*Petaurus norfolcensis* (Kerr, 1792). *In* Linnaeus, Anim. Kingdom, 1:270.
COMMON NAME: Squirrel Glider.
TYPE LOCALITY: Australia, New South Wales, Sydney.
DISTRIBUTION: Australia: E Queensland, E New South Wales, E Victoria.
STATUS: IUCN – Lower Risk (nt). Common.
SYNONYMS: *sciurea* (Shaw, 1794).

**Family Tarsipedidae** Gervais and Verreaux, 1842. Proc. Zool. Soc. Lond., 1842:1.

*Tarsipes* Gervais and Verreaux, 1842. L'Institut, l'ere Section, Sci., Math, Phys., Nat., 427:75.
TYPE SPECIES: *Tarsipes rostratus* Gervais and Verreaux, 1842.
COMMENTS: For correct authorship see Mahoney (1981).

*Tarsipes rostratus* Gervais and Verreaux, 1842. L'Institut, l'ere Section, Sci., Math, Phys., Nat., 427:75.
COMMON NAME: Honey Possum.
TYPE LOCALITY: Australia, Western Australia, King George Sound (Albany), see Gray (1842).
DISTRIBUTION: SW Western Australia.
STATUS: IUCN – Lower Risk (lc). Rare.
SYNONYMS: *spencerae* Ride, 1970; *spenserae* Gray, 1842.

COMMENTS: The name *T. spenserae* is considered a misspelling because it was presented as a patronym for Spencer (Gray, 1842:40). Ride (1970) emended the name to *spencerae*; see Mahoney (1981) for details. Mahoney (1981) presented evidence that *Tarsipes rostratus* Gervais and Verreaux, 1842 predates *T. spenserae* Gray, 1842.

**Family Acrobatidae** Aplin, 1987. *In* M. Archer (ed.), Possums and Opossums, xxii.
COMMENTS: Separated from Burramyidae by Aplin (*in* Aplin and Archer, 1987).

*Acrobates* Desmarest, 1818. Nouv. Dict. Hist. Nat., Nouv. ed., 25:405.
TYPE SPECIES: *Didelphis pygmaea* Shaw, 1793.
SYNONYMS: *Ascobates* Anon., 1839; *Cercoptenus* Gloger, 1841.

*Acrobates pygmaeus* (Shaw, 1793). Zool. New Holland, 1:5.
COMMON NAME: Feather-tailed Glider.
TYPE LOCALITY: Australia, New South Wales, Sydney.
DISTRIBUTION: Australia: E Queensland to SE South Australia, inland to Deniliquin (New South Wales).
STATUS: IUCN – Lower Risk (lc). Common.
SYNONYMS: *frontalis* De Vis, 1887; *pulchellus* W. Rothschild, 1892.
COMMENTS: Tate (1938:60) believed the single specimen (of *A. pulchellus*; which is considered a synonym of *pygmaeus*) obtained in NW New Guinea was probably an introduction as a pet.

*Distoechurus* Peters, 1874. Ann. Mus. Civ. Stor. Nat. Genova, 6:303.
TYPE SPECIES: *Phalangista* (*Distoechurus*) *pennata* Peters, 1874.

*Distoechurus pennatus* (Peters, 1874). Ann. Mus. Civ. Stor. Nat. Genova, 6:303.
COMMON NAME: Feather-tailed Possum.
TYPE LOCALITY: Indonesia, Prov. of Papua (= Irian Jaya), Vogelkop, Manokwari Div., "Andai" (Probably = Arfak Mtns, Hatam, 1520 m). See Van der Feen (1962:52).
DISTRIBUTION: New Guinea.
STATUS: IUCN – Lower Risk (lc). Abundant.
SYNONYMS: *amoenus* Thomas, 1920; *dryas* Thomas, 1920; *neuhassi* Matschie, 1916.

## SUBORDER MACROPODIFORMES Ameghino, 1889.

**Family Hypsiprymnodontidae** Collett, 1877. Zool. Jb., 2:906.
COMMENTS: Retained in Potoroinae [sic] by McKenna and Bell (1997); considered a family distinct from Potoroidae by Burk et al. (1998) and Burk and Springer (2000), on the ground that it diverged from Potoroidae and Macropodidae long before they diverged from each other.

*Hypsiprymnodon* Ramsay, 1876. Proc. Linn. Soc. N.S.W., 1:33.
TYPE SPECIES: *Hypsiprymnodon moschatus* Ramsay, 1876.
SYNONYMS: *Pleopus* Owen, 1877.
COMMENTS: Formerly included in Macropodidae.

*Hypsiprymnodon moschatus* Ramsay, 1876. Proc. Linn. Soc. N.S.W., 1:34.
COMMON NAME: Musky Rat-kangaroo.
TYPE LOCALITY: Australia, Queensland, Rockingham Bay.
DISTRIBUTION: NE Queensland (Australia).
STATUS: IUCN – Lower Risk (lc). Rare, localized.
SYNONYMS: *nudicaudatus* (Owen, 1877).

**Family Potoroidae** Gray, 1821. London Med. Repos., 15:308.
COMMENTS: Separated from Macropodidae by Archer and Bartholamai (1978), but Burk et al. (1998) proposed to reunite them on the basis that they form a clade relative to *Hypsiprymnodon*. Burk and Springer (2000) found that *Potorous* is strongly distinct from

the other genera and separated from them little, if at all, later than the divergence of the Macropodidae; consequently, they considered the monophyly of the Potoroidae in doubt.

*Aepyprymnus* Garrod, 1875. Proc. Zool. Soc. Lond., 1875:59.
   TYPE SPECIES: *Bettongia rufescens* Gray, 1837.

   *Aepyprymnus rufescens* (Gray, 1837). Mag. Nat. Hist. [Charlesworth's], 1:584.
       COMMON NAME: Rufous Rat-kangaroo.
       TYPE LOCALITY: Australia, New South Wales.
       DISTRIBUTION: Australia: NE Victoria, E New South Wales, E Queensland.
       STATUS: IUCN – Lower Risk (lc). Rare, localized.
       SYNONYMS: *melanotis* (Ogilby, 1838).

*Bettongia* Gray, 1837. Mag. Nat. Hist. [Charlesworth's], 1:584.
   TYPE SPECIES: *Bettongia setosa* Gray, 1837 (= *Kangurus gaimardi* Desmarest, 1822).
   SYNONYMS: *Bettongiops* Matschie, 1916.
   COMMENTS: Formerly included in Macropodidae.

   *Bettongia gaimardi* (Desmarest, 1822). Mammalogie. *In* Encycl. Méth., 2(Suppl.):542.
       COMMON NAME: Eastern Bettong.
       TYPE LOCALITY: Australia, New South Wales, Port Jackson.
       DISTRIBUTION: Formerly coastal SE Queensland and N New South Wales, south to
           SW Victoria; now extinct on mainland Australia; survives in Tasmania.
       STATUS: CITES – Appendix I; U.S. ESA – Endangered; IUCN – Extinct as *B. g. gaimardi*,
           otherwise Lower Risk (nt).
       SYNONYMS: *cuniculus* (Ogilby, 1838); *formosus* (Ogilby, 1838); *hunteri* (Owen, 1841); *lepturus*
           (Quoy and Gaimard, 1824); *minimus* (Boitard, 1841); *phillippi* (Ogilby, 1838); *setosa*
           Gray, 1837; *white* (Quoy and Gaimard, 1824); *whitei* Gray, 1841.
       COMMENTS: See Corbet and Hill (1980:16).

   *Bettongia lesueur* (Quoy and Gaimard, 1824). *In* de Freycinet, Voy. autour du monde . . . Uranie et al
           Physicienne, Zool., p. 64.
       COMMON NAME: Boodie.
       TYPE LOCALITY: Australia, Western Australia, Dirk Hartog Isl (Shark Bay).
       DISTRIBUTION: Formerly in Dampier Land (Western Australia), South Australia, Dirk Hartog
           Isl, Barrow Isl, Bernier and Dorre Isls, Northern Territory, and SW New South Wales
           (Australia); now extinct except on W Australian Isls.
       STATUS: CITES – Appendix I; U.S. ESA – Endangered; IUCN – Extinct as *B. l. graii*, otherwise
           Vulnerable. Rare.
       SYNONYMS: *anhydra* Finlayson, 1957; *graii* (Gould, 1841); *harveyi* (Waterhouse, 1842).
       COMMENTS: Commonly misspelt "*lesueuri*", but the original spelling is *lesueur*, with no
           indication that it is an error.

   *Bettongia penicillata* Gray, 1837. Mag. Nat. Hist. [Charlesworth's], 1:584.
       COMMON NAME: Woylie.
       TYPE LOCALITY: Australia, New South Wales.
       DISTRIBUTION: SW Western Australia, S South Australia including St. Francis Isl,
           NW Victoria, C New South Wales.
       STATUS: CITES – Appendix I; U.S. ESA – Endangered; IUCN – Extinct as *B. p. penicillata*,
           otherwise Lower Risk (conservation dependent).
       SYNONYMS: *francisca* Finlayson, 1957; *gouldii* Waterhouse, 1845; *ogilbyi* (Waterhouse, 1841).
       COMMENTS: Does not include *tropica*, contra Sharman et al. (1980).

   *Bettongia tropica* Wakefield, 1967. Victorian Nat., 84:15.
       COMMON NAME: Northern Bettong.
       TYPE LOCALITY: Australia, Queensland, Mt. Spurgeon.
       DISTRIBUTION: E Queensland: Windsor and Carbine Tablelands, Lamb Range, and Paluma.
       STATUS: CITES– Appendix I; U.S. ESAand IUCN – Endangered.
       COMMENTS: Sharman et al. (1980) believed that *tropica* is not a distinct species but rather

that it barely differs from *B. penicillata*.However, it differs in several reproductive parameters (M. J. Smith, 1998), and is now always treated as a distinct species. Its distribution, habitat parameters and phylogeography are given by Pope et al. (2000).

*Caloprymnus* Thomas, 1888. Cat. Marsup. Monotr. Brit. Mus., p. 114.
> TYPE SPECIES: *Bettongia campestris* Gould, 1843.

*Caloprymnus campestris* (Gould, 1843). Proc. Zool. Soc. Lond., 1843:81.
> COMMON NAME: Desert Rat-kangaroo.
> TYPE LOCALITY: Australia, South Australia.
> DISTRIBUTION: South Australia/Queensland border country.
> STATUS: CITES – Appendix I pe [Possibly Extinct]; U.S. ESA – Endangered; IUCN – Extinct, see Ride (1970:198); not recorded or sighted since 1935.

*Potorous* Desmarest, 1804. Tabl. Méth Hist. Nat., *in* Nouv. Dict. Hist. Nat., 24:20.
> TYPE SPECIES: *Didelphis murina* Cuvier, 1798 (= *Didelphis tridactyla* Kerr, 1792).
> SYNONYMS: *Hypsiprymnus* Illiger, 1811; *Patoroo* Partington, 1839; *Potoroiis* Rafinseque, 1815; *Potoroo* Berthold, 1827; *Potoroops* Matschie, 1916.
> COMMENTS: Formerly included in Macropodidae.

*Potorous gilbertii* (Gould, 1841). Monograph of Macropodidae, 1:[unnumbered plate].
> COMMON NAME: Gilbert's Potoroo.
> TYPE LOCALITY: Australia, Western Australia, King George Sound.
> DISTRIBUTION: Southern tip of Western Australia.
> STATUS: IUCN – Critically Endangered.
> COMMENTS: Considered a species separate from *P. tridactylus* by Sinclair and Westerman (1997).

*Potorous longipes* Seebeck and Johnston, 1980. Aust. J. Zool., 28:121.
> COMMON NAME: Long-footed Potoroo.
> TYPE LOCALITY: Australia, Victoria, Bellbird Creek, 32 km E Orbost.
> DISTRIBUTION: NE Victoria (Australia).
> STATUS: IUCN – Endangered.
> COMMENTS: Known from very few specimens; first collected in 1968.

*Potorous platyops* (Gould, 1844). Proc. Zool. Soc. Lond., 1844:103.
> COMMON NAME: Broad-faced Potoroo.
> TYPE LOCALITY: Australia, Western Australia, Swan River, L. Walyormouring.
> DISTRIBUTION: Formerly SW Western Australia and Kangaroo Isl, South Australia.
> STATUS: IUCN – Extinct.
> SYNONYMS: *morgani* Finlayson, 1938.
> COMMENTS: Probably extinct (Ride, 1970:199); there are no records after 1875.

*Potorous tridactylus* (Kerr, 1792). *In* Linnaeus, Anim. Kingdom, 1:198.
> COMMON NAME: Long-nosed Potoroo.
> TYPE LOCALITY: Australia, New South Wales, Sydney.
> DISTRIBUTION: SE Queensland, coastal New South Wales, NE Victoria, SE South Australia, Tasmania, and King Isl (Australia).
> STATUS: IUCN – Vulnerable as *P. t. tridactylus*, otherwise Lower Risk (lc). Endangered on mainland Australia, common in Tasmania.
> SYNONYMS: *micropus* (Waterhouse, 1841); *minor* (Shaw, 1800); *murina* (G. Cuvier, 1798); *muscola* (Perry, 1810); *myosurus* (Ogilby, 1838); *peron* (Quoy and Gaimard, 1824); *potoru* (F. Meyer, 1793); *setosus* (Ogilby, 1832); *trisulcatus* (McCoy, 1865); *tuckeri* (Gray, 1840); **apicalis** (Gould, 1851); *benormi* Courtney, 1963; *rufus* Higgins and Petterd, 1884.
> COMMENTS: See Kirsch and Calaby (1977:21).

**Family Macropodidae** Gray, 1821. London Med. Repos., 15:308.
> COMMENTS: Revised by Tate (1948*a*).

**Subfamily Sthenurinae** Glauert, 1926. Geol. Survey W. A. Bull., 88:36-71.

*Lagostrophus* Thomas, 1887. Proc. Zool. Soc. Lond., 1886:544 [1887].
   TYPE SPECIES: *Kangurus fasciatus* Peron and Lesueur, 1807.
   COMMENTS: McKenna and Bell (1997) placed this in Macropodinae, but it was convincingly argued to be a member of subfamily Sthenurinae by Flannery (1983). The other Sthenurinae are giant fossil kangaroos.

   *Lagostrophus fasciatus* (Peron and Lesueur, 1807). *In* Péron, Voy. Decouv. Terres. Austral., Atlas, pl. 27, 1:114.
      COMMON NAME: Banded Hare-wallaby.
      TYPE LOCALITY: Australia, Western Australia, Bernier Isl (Shark Bay).
      DISTRIBUTION: Survives only on Bernier and Dorre Isls (Western Australia); formerly in SW Western Australia and South Australia.
      STATUS: CITES – Appendix I; U.S. ESA – Endangered; IUCN – Extinct as *L. f. albipilis,* otherwise Vulnerable.
      SYNONYMS: *albipilis* (Gould, 1842); *elegans* (G. Cuvier, 1816); *striatus* (Lesson, 1842); **baudinettei** Helgen and Flannery, 2003.

**Subfamily Macropodinae** Gray, 1821. London Med. Repos., 15:308.

*Dendrolagus* Müller, 1840. *In* Temminck, Verh. Nat. Gesch. Nederland Overz. Bezitt., Land-en Volkenkunde, p. 20, footnote [1840].
   TYPE SPECIES: *Dendrolagus ursinus* Müller, 1840 (*recte Hypsiprymnus ursinus* Temminck, 1836; designated by Thomas, 1888).
   COMMENTS: Groves (1982) and Flannery et al. (1995*b*) divided the genus into a plesiomorphic group (*D. lumholtzi, D. inustus*) and a derived group, among which *D. dorianus* and *D. scottae*, and perhaps *D. mbaiso*, are the most strongly derived. *D. bennettianus* occupies an isolated position in the genus.

   *Dendrolagus bennettianus* De Vis, 1887. Proc. R. Soc. Queensl., 3(1886):11 [1887].
      COMMON NAME: Bennett's Tree-kangaroo.
      TYPE LOCALITY: Australia, Queensland, Daintree River.
      DISTRIBUTION: NE Queensland (Australia): north of Daintree River as far as Mt. Amos.
      STATUS: IUCN – Lower Risk (nt).
      COMMENTS: Considered a subspecies of *dorianus* by Haltenorth (1958); but see Ride (1970:223) and Kirsch and Calaby (1977:17); occupies an isolated position in the genus (Groves, 1982; Flannery et al., 1995*b*).

   *Dendrolagus dorianus* Ramsay, 1883. Proc. Linn. Soc. N.S.W., 8:17.
      COMMON NAME: Doria's Tree-kangaroo.
      TYPE LOCALITY: Papua New Guinea, "ranges behind Mt. Astrolabe."
      DISTRIBUTION: Interior New Guinea: Wondiwoi Peninsula, Prov. of Papua (= Irian Jaya); Papua New Guinea/Indonesian border to extreme SE of mainland Papua New Guinea, 600 to 3650 m.
      STATUS: IUCN – Vulnerable; uncommon.
      SYNONYMS: *aureus* Rothschild and Dollman, 1936; *palliceps* Troughton and Le Souef, 1936; *profugus* Troughton and Le Souef, 1936; **mayri** Rothschild and Dollman, 1933; **notatus** Matschie, 1916.
      COMMENTS: Does not include *bennettianus* (see Ride, 1970:223, Kirsch and Calaby, 1977:17), or *stellarum* (see below); *mayri* may also be a distinct species, but is known by only one specimen.

   *Dendrolagus goodfellowi* Thomas, 1908. Ann. Mag. Nat. Hist., ser. 8, 2:452.
      COMMON NAME: Goodfellow's Tree-kangaroo.
      TYPE LOCALITY: Papua New Guinea, Owen Stanley Range, vic. Mt. Obree, 8000 ft. (2438 m).
      DISTRIBUTION: Mainland of Papua New Guinea, sea level to 2860 m, excluding Torricelli Mtns.
      STATUS: IUCN – Endangered; uncommon.
      SYNONYMS: **buergersi** Matschie, 1912; *shawmayeri* Rothschild and Dollman, 1936.

COMMENTS: Does not include *spadix*. Groves (1982*d*) regarded this species as a subspecies of the earlier named *matschiei*; but see Flannery (1990:100-104), also Ganslosser (1980). Does not include *pulcherrimus* (see below).

*Dendrolagus inustus* Müller, 1840. *In* Temminck, Verh. Nat. Gesch. Nederland Overz. Bezitt., Land-en Volkenkunde, p. 20, footnote [1840], see comments.
COMMON NAME: Grizzled Tree-kangaroo.
TYPE LOCALITY: Indonesia, Prov. of Papua (= Irian Jaya), Fakfak Div., Lobo Dist., near Triton Bay, Mt. Lamantsjieri.
DISTRIBUTION: New Guinea: Vogelkop and Fakfak peninsulas, along north coast as far east as Wewak; Yapen Isl; possibly present on Salawati and Waigeou (Flannery, 1995*a*).
STATUS: CITES – Appendix II; IUCN – Data Deficient.
SYNONYMS: *maximus* Rothschild and Rothschild, 1898; *sorongensis* Matschie, 1916; **finschi** Matschie, 1916; *keiensis* Matschie, 1916; *schoedei* Matschie, 1916.
COMMENTS: This species was further described by Schlegel and Müller, *in* Temminck, Verh. Nat. Gesch. Nederland. Overz. Bezitt., Zool., Mammalia, p. 131, 143[1845], pl. 20, 22, 23[1841]. Considered a subspecies of *ursinus* by Haltenorth (1958); but see Kirsch and Calaby (1977:17) and Groves (1982*d*).

*Dendrolagus lumholtzi* Collett, 1884. Proc. Zool. Soc. Lond., 1884:387.
COMMON NAME: Lumholtz's Tree-kangaroo.
TYPE LOCALITY: Australia, Queensland, Herbert Vale.
DISTRIBUTION: NE Queensland (Australia), Kirrima north to Mt. Spurgeon, 300 to 1622 m.
STATUS: IUCN – Lower Risk (nt).
SYNONYMS: *fulvus* De Vis, 1888.

*Dendrolagus matschiei* Forster and Rothschild, 1907. Nov. Zool., 14:506.
COMMON NAME: Huon Tree-kangaroo.
TYPE LOCALITY: Papua New Guinea, Morobe Prov., Rawlinson Mtns.
DISTRIBUTION: Extreme NE interior New Guinea (Huon Peninsula), 1000 to 3300 m; Umboi Isl and Mt. Agulupella in W New Britain (introduced).
STATUS: IUCN – Endangered. Uncommon.
SYNONYMS: *deltae* Troughton and Le Souef, 1936; *flavidior* Matschie, 1912; *xanthotis* Rothschild and Dollman, 1936.
COMMENTS: See Kirsch and Calaby (1977:21) and Lidicker and Ziegler (1968). See also comments under *goodfellowi*.

*Dendrolagus mbaiso* Flannery, Boeadi and Szalay, 1995. Mammalia, 59:66.
COMMON NAME: Dingiso.
TYPE LOCALITY: Indonesia, Prov. of Papua (= Irian Jaya), Tembagapura area, 3250-3500 m on the south slopes of Gunung (=Mt.) Ki (04°S05'S, 137°06'E).
DISTRIBUTION: Southern and western slopes of Sudirman Range, from Paniai Lakes region in west to Baliem Gorge in east, 136°30' to 139°10'E, at 3250 to 4200 m, but perhaps down to 2700 m in suitable low mossy forest or scrub; Prov. of Papua, Indonesia.
STATUS: IUCN – Vulnerable. Locally common but restricted in distribution.
COMMENTS: Some communities hold this species sacred, and its hunting is forbidden; so, in some parts of its range, said to be remarkably tame.

*Dendrolagus pulcherrimus* Flannery, 1993. Rec. Aust. Mus., 45:38.
COMMON NAME: Golden-mantled Tree-kangaroo.
TYPE LOCALITY: Papua New Guinea, West Sepik Prov., near Sibilanga, Kukumbau area on Mt. Sapau, 1120 m (3°32'S, 142°31'E).
DISTRIBUTION: Sibilanga district, Toricelli Mtns only, 680-1120 m (Papua New Guinea).
STATUS: Endangered.
COMMENTS: Described as a subspecies of *D. goodfellowi*, but differs absolutely (diagnostically) from that species.

*Dendrolagus scottae* Flannery and Seri, 1990. Rec. Aust. Mus., 42:237.
COMMON NAME: Tenkile.

TYPE LOCALITY: Papua New Guinea, West Sepik Prov., Torricelli Mtns, Sweipini, 1400 m
(3°23'S, 142°06'E).
DISTRIBUTION: Torricelli Mtns and Mt. Menawa, 900-2000 m (Papua New Guinea).
STATUS: IUCN – Endangered.
COMMENTS: Closely related to *D. dorianus*.

*Dendrolagus spadix* Troughton and Le Souef, 1936. Aust. Zool., 8:194.
COMMON NAME: Lowlands Tree-kangaroo.
TYPE LOCALITY: Papua New Guinea, Western Division, between Bamu, upper Awarra and
Strickland Rivers.
DISTRIBUTION: S New Guinea, Gulf of Papua east of Fly River, sea level to 800 m.
STATUS: IUCN – Data Deficient. Rare.
COMMENTS: A subspecies of *D. matschiei* according to Groves (1982*d*), but raised to a full
species by Flannery (1990).

*Dendrolagus stellarum* Flannery and Seri, 1990. Rec. Aust. Mus., 42:180.
COMMON NAME: Seri's Tree-kangaroo.
TYPE LOCALITY: Papua New Guinea, Western Prov., western end of Dokfuma basin, Star
Mtns, 3000 m (5°01'S, 141°07'E).
DISTRIBUTION: New Guinea, from Paniai Lakes east to Hak-Om region, northeast of
Telefomin, 2600-3200 m (Papua New Guinea).
STATUS: Unknown.
COMMENTS: Described as a subspecies of *D. goodfellowi*, but differs absolutely (diagnostically)
from that species.

*Dendrolagus ursinus* (Temminck, 1836). Discours preliminaire destine a servir d'introduction al
faune du Japon, p. 6 (footnote 2).
COMMON NAME: Ursine Tree-kangaroo.
TYPE LOCALITY: Indonesia, Prov. of Papua (= Irian Jaya), Fakfak Div., Lobo Dist., near Triton
Bay, Mt. Lamantsjieri.
DISTRIBUTION: Volgelkop and Fakfak peninsulas, extreme NW New Guinea, sea level to
2500 m.
STATUS: CITES – Appendix II; IUCN – Data Deficient.
SYNONYMS: *leucogenys* Matschie, 1916.
COMMENTS: Does not include *inustus* (see Kirsch and Calaby, 1977:17). Correct original
citation presented by Husson (1955).

*Dorcopsis* Schlegel and Müller, 1845. *In* Temminck, Verh. Nat. Gesch. Nederland Overz. Bezitt., Zool.,
p. 130 [1845].
TYPE SPECIES: *Didelphis brunii* Quoy and Gaimard, 1830 (= *Macropus muelleri* Lesson, 1827).
COMMENTS: Does not include *Dorcopsulus* (see Flannery, 1990:89-92). The name *Conoyces*
Lesson, 1842 was used for this genus by Troughton (1937), but Tate (1948*a*) showed that
the type species of *Conoyces* is *Didelphis brunii* Gmelin, 1788 [= *Thylogale brunii* (Schreber,
1778)]; see under *Thylogale*.

*Dorcopsis atrata* Van Deusen, 1957. Am. Mus. Novit., 1826:5.
COMMON NAME: Black Dorcopsis.
TYPE LOCALITY: Papua New Guinea, Milne Bay Prov., Goodenough Isl, E slopes, near "Top
Camp", about 1600 m.
DISTRIBUTION: Goodenough Isl (Papua New Guinea).
STATUS: IUCN – Endangered. Very rare, localized.

*Dorcopsis hageni* Heller, 1897. Abh. Zool. Anthrop.-Ethnology Mus. Dresden, 6(8):7.
COMMON NAME: White-striped Dorcopsis.
TYPE LOCALITY: Papua New Guinea, Madang Prov., near Astrolabe Bay, Stefansort.
DISTRIBUTION: NC New Guinea.
STATUS: IUCN – Lower Risk (lc). Common.
SYNONYMS: *caurina* Thomas, 1922; *eitape* (Troughton, 1937)

*Dorcopsis luctuosa* (D'Albertis, 1874). Proc. Zool. Soc. Lond., 1874:110.
    COMMON NAME: Gray Dorcopsis.
    TYPE LOCALITY: "Southeast of New Guinea".
    DISTRIBUTION: S New Guinea.
    STATUS: IUCN – Lower Risk (lc). Common.
    SYNONYMS: *beccarii* Mikouho-Maclay, 1885; *chalmersi* Milkouho-Maclay, 1884; **phyllis**
        Groves and Flannery, 1989.
    COMMENTS: Usually included in *D. veterum* (= *D. muelleri*), but see Groves and Flannery
        (1989). The subspecies *phyllis* (Fly River district) is highly distinctive; it appears to fall
        outside the range of variation of nominotypical *luctuosa* (Port Moresby district), and is
        probably a distinct species.

*Dorcopsis muelleri* (Lesson, 1827). *In* Duperry (Lesson and Garnot, eds.), Voy. autour du Monde . . .
    la Coquille, Zool., 1:164.
    COMMON NAME: Brown Dorcopsis.
    TYPE LOCALITY: Indonesia, Prov. of Papua (= Irian Jaya), Vogelkop, Manokwari Div., Dorei
        (= Manokwari), Lobo Bay.
    DISTRIBUTION: W New Guinea; Misool and Salawati Isls, Aru Isls, and Yapen Isl (Indonesia);
        occurred on Gebe until about 2000 ybp and on Halmahera until about 1870 ybp
        (Flannery et al., 1998).
    STATUS: IUCN – Lower Risk (lc). Common.
    SYNONYMS: *brunii* (Quoy and Gaimard, 1830); *rufolateralis* Rothschild and Rothschild, 1908;
        *veterum* of sundry authors but not *Kangurus veterum* Lesson and Garnot (see below);
        **lorentzii** Jentink, 1908; **mysoliae** Thomas, 1913; **yapeni** Groves and Flannery, 1989.
    COMMENTS: *D. muelleri* was regarded as a junior synonym of *D. veterum* by Kirsch and Calaby
        (1977:21) and Husson (1955:299). George and Schuerer (1978) rejected *veterum* as
        based on a *Dendrolagus* (probably *inustus*), and employed *muelleri*; Groves and
        Flannery (1989) agreed. The original name *Didelphis brunii* Quoy and Gaimard was
        preoccupied and is now *Macropus muelleri*. The Yapen Isl form (*yapeni*) is strongly
        distinct, apparently outside the range of those from other parts of the range, though
        samples are small; it may be a distinct species.

*Dorcopsulus* Matschie, 1916. Sitzb. Ges. Naturf. Fr., Berlin, p. 57.
    TYPE SPECIES: *Dorcopsis macleayi* Miklouho-Maclay, 1885.
    COMMENTS: Formerly included in *Dorcopsis* but revived as a full genus by Flannery (1990).

*Dorcopsulus macleayi* (Miklouho-Maclay, 1885). Proc. Linn. Soc. N.S.W., 10:145, 149.
    COMMON NAME: Macleay's Dorcopsis.
    TYPE LOCALITY: Papua New Guinea, Central Prov., "inland from Port Moresby".
    DISTRIBUTION: Extreme SE New Guinea.
    STATUS: IUCN – Vulnerable.

*Dorcopsulus vanheurni* (Thomas, 1922). Ann. Mag. Nat. Hist., ser. 9, 9:264.
    COMMON NAME: Small Dorcopsis.
    TYPE LOCALITY: Indonesia, Prov. of Papua (= Irian Jaya), Djajawidjaja Div., Doormanpad-
        bivak (3°30'S, 138°30'E), 1410 m.
    DISTRIBUTION: Interior New Guinea.
    STATUS: IUCN – Lower Risk (lc). Common.
    SYNONYMS: *rothschildi* Thomas, 1922.
    COMMENTS: Regarded as conspecific with *macleayi* by Kirsch and Calaby (1977:21).

*Lagorchestes* Gould, 1841. Monogr. Macropodidae, pt. 1, pl. 12 (text).
    TYPE SPECIES: *Macropus leporides* Gould, 1841.
    SYNONYMS: *Lagocheles* Owen, 1842 [*nomen nudum*].
    COMMENTS: This genus is probably polyphyletic.

*Lagorchestes asomatus* Finlayson, 1943. Trans. R. Soc. South Aust., 67:319.
    COMMON NAME: Lake Mackay Hare-wallaby.
    TYPE LOCALITY: Australia, Northern Territory, between Mt. Farewell and Lake Mackay.

DISTRIBUTION: Known only from the type locality.

STATUS: IUCN – Extinct. Probably extinct.

COMMENTS: Known from a single unsexed skull (Kirsch and Calaby, 1977:22).

*Lagorchestes conspicillatus* Gould, 1842. Proc. Zool. Soc. Lond., 1841:82 [1842].

COMMON NAME: Spectacled Hare-wallaby.

TYPE LOCALITY: Australia, Western Australia, Barrow Isl.

DISTRIBUTION: N Western Australia and adjacent islands, N Northern Territory, N and
        W Queensland (Australia).

STATUS: IUCN – Vulnerable as *L. c. conspicillatus*, otherwise Lower Risk (nt). Common locally.

SYNONYMS: *leichardti* Gould, 1853; *pallidior* Thomas and Dollman, 1909.

COMMENTS: May consist of two or three distinct species.

*Lagorchestes hirsutus* Gould, 1844. Proc. Zool. Soc. Lond., 1844:32.

COMMON NAME: Rufous Hare-wallaby.

TYPE LOCALITY: Australia, Western Australia, York district.

DISTRIBUTION: C Western Australia, C Australia, Dorre Isl and Bernier Isl (Western Australia).
        Survives only on Bernier and Dorre Isls, and a tiny area NW of Alice Springs (Northern
        Territory, Australia).

STATUS: CITES – Appendix I; U.S. ESA – Endangered; IUCN – Extinct as *L. h. hirsutus*,
        otherwise Vulnerable. Rare.

SYNONYMS: *bernieri* Thomas, 1907; *dorreae* Thomas, 1907.

*Lagorchestes leporides* (Gould, 1841). Proc. Zool. Soc. Lond., 1840:93 [1841].

COMMON NAME: Eastern Hare-wallaby.

TYPE LOCALITY: Australia, interior New South Wales.

DISTRIBUTION: Formerly W New South Wales, E South Australia, NW Victoria.

STATUS: IUCN – Extinct.

COMMENTS: Almost certainly extinct; not recorded for more than a century (Kirsch and
        Calaby, 1977:22).

*Macropus* Shaw, 1790. Nat. Misc., 1, pl. 23 (text).

TYPE SPECIES: *Macropus giganteus* Shaw, 1790.

SYNONYMS: *Boriogale* Owen, 1874; *Dendrodorcopsis* W. Rothschild, 1903; *Gerboides* Gervais,
        1855; *Gigantomys* Link, 1794; *Halmatopus* Wagner, 1841 [error]; *Halmaturus* Illiger, 1811;
        *Kalmaturus* Gervais, 1835 [error]; *Kanguroo* Lacépède, 1799; *Kangurus* E. Geoffroy St.
        Hilaire and G. Cuvier, 1795; *Megaleia* Gistel, 1848; *Notamacropus* Dawson and Flannery,
        1985; *Osphranter* Gould, 1842; *Phascolagus* Owen, 1874; *Prionotemmus* Stirton, 1955.

COMMENTS: Includes *Megaleia* and *Protemnodon* (*sensu* Haltenorth, 1958); see Kirsch and Calaby
        (1977:17). Rationale for present usage of *Macropus* given by Calaby (1966). Ride (1962)
        discussed generic nomenclature for all Macropodinae. Van Gelder (1977b) included
        *Thylogale* and *Wallabia* in this genus, but see Kirsch and Calaby (1977:17) and Corbet and
        Hill (1980:17-18). Dawson and Flannery (1985) divided this genus into three subgenera:
        *Macropus, Notamacropus* and *Osphranter*; almost certainly *Notamacropus* is paraphyletic.

*Macropus agilis* (Gould, 1842). Proc. Zool. Soc. Lond., 1841:81 [1842].

COMMON NAME: Agile Wallaby.

TYPE LOCALITY: Australia, Northern Territory, Port Essington.

DISTRIBUTION: NE Western Australia, Northern Territory, Queensland; S New Guinea;
        Kiriwina and D'Entrecasteaux Isls.

STATUS: IUCN – Lower Risk (lc). Common.

SYNONYMS: *aurescens* Schwarz, 1910; *binoe* (Gould, 1842); **jardinii** (De Vis, 1884); **nigrescens**
        Lönnber, 1913; **papuanus** Peters and Doria, 1875; *aurantiacus* Rothschild and
        Rothschild, 1898; *crassipes* (Ramsay, 1876); *papuensis* Sclater, 1875.

COMMENTS: Subgenus *Notamacropus*.

*Macropus antilopinus* (Gould, 1842). Proc. Zool. Soc. Lond., 1841:80 [1842].

COMMON NAME: Antilopine Kangaroo.

TYPE LOCALITY: Australia, Northern Territory, Port Essington.

DISTRIBUTION: N Queensland, Northern Territory, NE Western Australia.

STATUS: IUCN – Lower Risk (lc). Common.
COMMENTS: Subgenus *Osphranter*.

*Macropus bernardus* W. Rothschild, 1904. Nov. Zool., 10:543.
COMMON NAME: Woodward's Wallaroo.
TYPE LOCALITY: Australia, Northern Territory, head of South Alligator River.
DISTRIBUTION: Interior of N Northern Territory.
STATUS: IUCN – Lower Risk (nt). Rare.
SYNONYMS: *woodwardi* (W. Rothschild, 1903).
COMMENTS: Subgenus *Osphranter*. The original name *Dendrodorcopsis woodwardi* was preoccupied.

*Macropus dorsalis* (Gray, 1837). Mag. Nat. Hist. [Charlesworth's], 1:583.
COMMON NAME: Black-striped Wallaby.
TYPE LOCALITY: Australia, New South Wales, probably interior (Namoi Hills), according to Iredale and Troughton (1934).
DISTRIBUTION: Australia: E Queensland, E New South Wales.
STATUS: IUCN – Lower Risk (lc). Common.
COMMENTS: Subgenus *Notamacropus*.

*Macropus eugenii* (Desmarest, 1817). Nouv. Dict. Hist. Nat., Nouv. ed., 17:38.
COMMON NAME: Tammar Wallaby.
TYPE LOCALITY: Australia, South Australia, Nuyt's Arch., St. Peter's Isl.
DISTRIBUTION: SW Western Australia, South Australia, Kangaroo Isl, Wallaby Isl and other islands.
STATUS: IUCN – Extinct in the Wild as *M. e. eugenii*, otherwise Lower Risk (nt).
SYNONYMS: *bedfordi* (Thomas, 1900); *dama* (Gould, 1844); *decres* (Troughton, 1841); *derbianus* (Gray, 1837); *emiliae* (Gray, 1843); *flindersi* (Wood Jones, 1924); *gracilis* Gould, 1844; *houtmanni* (Gould, 1844); *obscurior* (Gray, 1841).
COMMENTS: Subgenus *Notamacropus*.

*Macropus fuliginosus* (Desmarest, 1817). Nouv. Dict. Hist. Nat., Nouv. ed., 17:35.
COMMON NAME: Western Grey Kangaroo.
TYPE LOCALITY: Australia, South Australia, Kangaroo Isl.
DISTRIBUTION: SW New South Wales, NW Victoria, South Australia, SW Western Australia, Tasmania, King Isl, and Kangaroo Isl (Australia).
STATUS: U.S. ESA – Delisted Taxon (recovered); IUCN – Lower Risk (nt) as *M. f. fuliginosus*, otherwise Lower Risk (lc). Abundant.
SYNONYMS: **melanops** Gould, 1842; **ocydromus** Gould, 1842.
COMMENTS: Subgenus *Macropus*; see Kirsch and Poole (1972) for discussion of specific limits and subspecies included in this taxon and in *giganteus*.

*Macropus giganteus* Shaw, 1790. Nat. Misc., 1, pl. 33 (text).
COMMON NAME: Eastern Grey Kangaroo.
TYPE LOCALITY: Australia, Queensland, Cooktown (= "New Holland"), King's Plains.
DISTRIBUTION: E and C Queensland, Victoria, New South Wales, SE South Australia, and Tasmania (Australia).
STATUS: U.S. ESA – Endangered as *M. g. tasmaniensis*; otherwise Delisted Taxon (recovered); IUCN – Lower Risk (nt) as *M. g. tasmaniensis*, otherwise Lower Risk (lc). Abundant throughout E Australia.
SYNONYMS: *griseofuscus* (Goldfuss, 1819); *labiatus* (Desmarest, 1817); *major* Shaw, 1800; *tridactylus* (Perry, 1810); **tasmaniensis** Le Souef, 1923.
COMMENTS: Subgenus *Macropus*. Opinion 760 of the International Commission on Zoological Nomenclature (1966) placed this name on the Official List of Specific Names in Zoology, see Calaby et al. (1963) for discussion. Revised by Kirsch and Poole (1972) who discussed specific limits and the subspecies included in this taxon. See Poole (1982, Mammalian Species, 187).

*Macropus greyi* Waterhouse, 1846. Nat. Hist. Mamm., 1:122.
COMMON NAME: Toolache Wallaby.

TYPE LOCALITY: Australia, South Australia, Coorong.

DISTRIBUTION: Formerly SE South Australia and adjacent Victoria.

STATUS: IUCN – Extinct.

COMMENTS: Subgenus *Notamacropus*. Almost certainly extinct (Kirsch and Calaby, 1977:22; Ride, 1970:47).

*Macropus irma* (Jourdan, 1837). C. R. Acad. Sci. Paris, 5:523.

COMMON NAME: Western Brush Wallaby.

TYPE LOCALITY: Australia, Western Australia, Swan River.

DISTRIBUTION: SW Western Australia.

STATUS: IUCN – Lower Risk (nt). Rare.

SYNONYMS: *manicatus* Gould, 1841.

COMMENTS: Subgenus *Notamacropus*.

*Macropus parma* Waterhouse, 1846. Nat. Hist. Mamm., 1:149.

COMMON NAME: Parma Wallaby.

TYPE LOCALITY: Australia, New South Wales.

DISTRIBUTION: E New South Wales; introduced to Kawau Isl (New Zealand), see Wodzicki and Flux (1967).

STATUS: U.S. ESA – Endangered; IUCN – Lower Risk (nt).

COMMENTS: Subgenus *Notamacropus* (Dawson and Flannery, 1985).

*Macropus parryi* Bennett, 1835. Proc. Zool. Soc. Lond., 1834:151 [1835].

COMMON NAME: Pretty-faced Wallaby.

TYPE LOCALITY: Australia, New South Wales, Stroud (near Port Stephens).

DISTRIBUTION: Australia: E Queensland, NE New South Wales.

STATUS: IUCN – Lower Risk (lc). Common.

SYNONYMS: *pallida* (Gray, 1837).

COMMENTS: Subgenus *Notamacropus* (Dawson and Flannery, 1985). Formerly included in *Protemnodon*, see Haltenorth (1958:39); but also see Kirsch and Calaby (1977).

*Macropus robustus* Gould, 1841. Proc. Zool. Soc. Lond., 1840:92 [1841].

COMMON NAME: Wallaroo.

TYPE LOCALITY: Australia, New South Wales, interior (summit of mountains).

DISTRIBUTION: Australia: Western Australia, South Australia, S Northern Territory, Queensland, New South Wales, Barrow Isl.

STATUS: IUCN – Vulnerable as *M. r. isabellinus,* otherwise Lower Risk (lc). Abundant.

SYNONYMS: *reginae* Schwarz, 1910; **erubescens** Sclater, 1870; *alexandriae* Schwarz, 1910; *argentatus* Rothschild, 1905; *cervinus* Thomas, 1900; *hagenbecki* W. Rothschild, 1907; *magnus* W. Rothschild, 1905 [preoccupied by Owen, 1874]; *rubens* Schwarz, 1910; **isabellinus** (Gould, 1842); **woodwardi** Thomas, 1901; *alligatoris* Thomas, 1904; *bracteator* Thomas, 1904.

COMMENTS: Subgenus *Osphranter*. See Richardson and Sharman (1976). McAllan and Bruce (1989) would date *robustus* from: The Athenaeum, 670:685 [29 August 1840].

*Macropus rufogriseus* (Desmarest, 1817). Nouv. Dict. Hist. Nat., Nouv. ed., 17:36.

COMMON NAME: Red-necked Wallaby.

TYPE LOCALITY: Australia, Tasmania, King Isl.

DISTRIBUTION: SE South Australia, Victoria, SE Queensland, E New South Wales, Tasmania, King Isl and adjacent islands (Australia); introduced in England (Corbet and Hill, 1980:18).

STATUS: IUCN – Lower Risk (lc). Common.

SYNONYMS: *griseorufus* (Goldfuss, 1819); *griseus* (Gray, 1827); *kingii* (Illiger, 1815) [*nomen nudum*]; *rutilans* (Illiger, 1815) [*nomen nudum*]; *rutilus* (Lichtenstein, 1818) [*nomen nudum*]; *vinosus* (Boitard, 1841); **banksianus** (Quoy and Gaimard, 1825); **fruticus** Ogilby, 1838; *bennetti* Waterhouse, 1838; **unassigned:** *leptonyx* (Wagner, 1842).

COMMENTS: Subgenus *Notamacropus*.

*Macropus rufus* (Desmarest, 1822). Mammalogis. *In* Encycl. Méth., 2(Suppl.):541.
COMMON NAME: Red Kangaroo.
TYPE LOCALITY: Australia, New South Wales, Blue Mtns.
DISTRIBUTION: Mainland, mid-latitude Australia.
STATUS: U.S. ESA – Delisted Taxa (recovered); IUCN – Lower Risk (lc). Very abundant.
SYNONYMS: *dissimulatus* W. Rothschild, 1905; *griseolanosus* (Quoy and Gaimard, 1825); *laniger* (Gaimard, 1823); *lanigerus* Gray, 1825; *lanosus* (Gray, 1827) [error]; *occidentalis* Cahn, 1906; *pallidus* Schwarz, 1910; *pictus* Gould, 1861 [*nomen nudum*]; *ruber* Crisp, 1862 [error].
COMMENTS: Maintained in a monotypic genus *Megaleia* by McKenna and Bell (1997), but Dawson and Flannery (1985) showed that it belongs in genus *Macropus*, subgenus *Osphranter*.

*Onychogalea* Gray, 1841. Appendix C. *In* J. Two Exped. Aust., 2:402.
TYPE SPECIES: *Macropus unguifer* Gould, 1841.

*Onychogalea fraenata* (Gould, 1841). Proc. Zool. Soc. Lond., 1840:92 [1841].
COMMON NAME: Bridled Nail-tail Wallaby.
TYPE LOCALITY: Australia, New South Wales, interior.
DISTRIBUTION: Formerly in S Queensland, interior New South Wales; survives only near Taunton, Queensland (Australia).
STATUS: CITES – Appendix I; U.S. ESA and IUCN – Endangered.
COMMENTS: McAllan and Bruce (1989) argued that the original publication of this name was in *The Athenaeum*, 670:685 [29 August 1840], as [*Macropus*] *frenatus*.

*Onychogalea lunata* (Gould, 1841). Proc. Zool. Soc. Lond., 1840:93 [1841].
COMMON NAME: Crescent Nail-tail Wallaby.
TYPE LOCALITY: Australia, Western Australia, coast.
DISTRIBUTION: SC and SW Western Australia, S Northern Territory.
STATUS: CITES – Appendix I; U.S. ESA – Endangered; IUCN – Extinct; probably extinct.
COMMENTS: Extinct throughout most or all of its former range. McAllan and Bruce (1989) argued that the original publication of this name was in *The Athenaeum*, 670:685 [29 August 1840].

*Onychogalea unguifera* (Gould, 1841). Proc. Zool. Soc. Lond., 1840:93 [1841].
COMMON NAME: Northern Nail-tail Wallaby.
TYPE LOCALITY: Australia, Western Australia, Derby (King Sound).
DISTRIBUTION: N Australia: Western Australia, Northern Territory, Queensland.
STATUS: IUCN – Lower Risk (lc). Secure.
SYNONYMS: *annulicauda* De Vis, 1884.
COMMENTS: McAllan and Bruce (1989) argued that the original description of this name was in *The Athenaeum*, 670:685 [29 August 1840].

*Petrogale* Gray, 1837. Mag. Nat. Hist. [Charlesworth's], 1:583.
TYPE SPECIES: *Kangurus penicillatus* Gray, 1827.
SYNONYMS: *Heteropus* Jourdan, 1837 [not of Fitzinger, 1826]; *Peradorcas* Tomas, 1904.
COMMENTS: Revision of this genus is needed; see Poole (1979) and Briscoe et al. (1982). Kitchener and Sanson (1978) considered this genus as probably congeneric with *Peradorcas*, though McKenna and Bell (1997) continued to recognize *Peradorcas* as a genus. Species-groups (Briscoe et al., 1982; Eldredge and Close, 1997) are: *P. brachyotis* group (*brachyotis*, *burbidgei* and *concinna*), *P. xanthopus* group (*xanthopus*, *rothschildi* and *persephone*), and *P. lateralis*/*penicillata* group, containing the remainder.

*Petrogale assimilis* Ramsay, 1877. Proc. Linn. Soc. N.S.W., 1:360.
COMMON NAME: Allied Rock-wallaby.
TYPE LOCALITY: Australia, Queensland, Palm Isl.
DISTRIBUTION: C Queensland, from Townsville southward to lower Burdekin-Bowen Rivers, northwest to Croydon, southwest to Hughenden and Mt. Hope; Palm Isl, Magnetic Isl.
STATUS: IUCN – Lower Risk (lc). Locally common.

SYNONYMS: *puella* Thomas, 1926.
COMMENTS: *P. lateralis/penicillata* species-group.

*Petrogale brachyotis* (Gould, 1841). Proc. Zool. Soc. Lond., 1840:128 [1841].
COMMON NAME: Short-eared Rock-wallaby.
TYPE LOCALITY: Australia, Western Australia, Hanover Bay.
DISTRIBUTION: Coast of NW Australia, N Northern Territory.
STATUS: IUCN – Lower Risk (lc). Common.
SYNONYMS: *longmani* Thomas, 1926; *signata* Thomas, 1926; *venustula* Thomas, 1926; *wilkinsi* Thomas, 1926.
COMMENTS: *P. brachyotis* species-group.

*Petrogale burbidgei* Kitchener and Sanson, 1978. Rec. W. Aust. Mus., 6:269.
COMMON NAME: Monjon.
TYPE LOCALITY: Australia, Western Australia, Mitchell Plateau, Crystal Creek (14°30'S, 125°47'20"E).
DISTRIBUTION: Kimberleys (Western Australia), Bonaparte Arch., and adjacent islands.
STATUS: IUCN – Lower Risk (nt). Rare.
COMMENTS: *P. brachyotis* species-group.

*Petrogale coenensis* Eldredge and Close, 1992. Aust. J. Zool., 40:621.
COMMON NAME: Cape York Rock-wallaby.
TYPE LOCALITY: Australia, Queensland, "Twin Humps", 13°47'S, 143°04'E, north of Coen.
DISTRIBUTION: Musgrave north to Pascoe River, N Queensland (Australia).
STATUS: IUCN – Lower Risk (nt). Localized.
COMMENTS: *P. lateralis/ penicillata* species-group.

*Petrogale concinna* Gould, 1842. Proc. Zool. Soc. Lond., 1842:57.
COMMON NAME: Nabarlek.
TYPE LOCALITY: Australia, Western Australia, Wyndham.
DISTRIBUTION: Australia: NE and NW Northern Territory, NE Western Australia.
STATUS: IUCN – Lower Risk (nt).
SYNONYMS: *canescens* (Thomas, 1909); *monastria* (Thomas, 1926).
COMMENTS: *P. brachyotis* species-group. Formerly included in a separate genus *Peradorcas*, and this is retained by McKenna and Bell (1997), but see Kitchener and Samson (1978).

*Petrogale godmani* Thomas, 1923. Abstr. Proc. Zool. Soc. Lond., 1923(235):13.
COMMON NAME: Godman's Rock-wallaby.
TYPE LOCALITY: Australia, Queensland, Cooktown (Black Mtn).
DISTRIBUTION: Near Mt. Carbine and Mitchell River, north to Bathurst Head, west to "Pinnacles" (Australia).
STATUS: IUCN – Lower Risk (lc). Threatened by genetic introgression from *P. assimilis*.
COMMENTS: *P. lateralis/penicillata* species-group. Included in *penicillata* in a preliminary account by Poole (1979:21). Separated as a full species by Eldredge and Close (1992).

*Petrogale herberti* Thomas, 1926. Ann. Mag. Nat. Hist., ser. 9, 17:626.
COMMON NAME: Herbert's Rock-wallaby.
TYPE LOCALITY: Australia, Queensland, Eidsvold, Burnett River.
DISTRIBUTION: S Queensland, from Nanango, 100 km NW of Brisbane, north to Fitzroy River and northwest to Mt. Ball, near Rubyvale, and Mt. Donneybrook, near Clermont (Australia).
STATUS: IUCN – Lower Risk (lc). Common.
COMMENTS: *P. lateralis/penicillata* species-group. Considered a species separate from *P. penicillata* by Eldridge and Close (1992).

*Petrogale inornata* Gould, 1842. Monogr. Macropodidae, pt. 2, pl. 25.
COMMON NAME: Unadorned Rock-wallaby.
TYPE LOCALITY: Australia, Queensland, Cape Upstart.
DISTRIBUTION: Coastal C Queensland, from Fitzroy River north to lower Burdekin-Bowen Rivers; Whitsunday Isl (Australia).
STATUS: IUCN – Lower Risk (lc). Localized.

COMMENTS: *P. lateralis/penicillata* species-group.

*Petrogale lateralis* Gould, 1842. Monogr. Macropodidae, pt. 2, pl. 24.
   COMMON NAME: Black-flanked Rock-wallaby.
   TYPE LOCALITY: Australia, Western Australia, Swan River.
   DISTRIBUTION: Australia: Western Australia (Barrow Isl, Northwest Cape, West Kimberley, Recherche Arch.), South Australia (Pearson Isl, Mann, Musgrave, Everard and Davenport Ranges), Northern Territory (Uluru and Macdonnell Ranges), W Queensland (Dajarra and Selwyn Range).
   STATUS: IUCN – Vulnerable as *P. l. hacketti, P. l. lateralis,* and *P. l. pearsoni.* Common.
   SYNONYMS: **hacketti** Thomas, 1905; **pearsoni** Thomas, 1922.
   COMMENTS: *P. lateralis/penicillata* species-group. McAllan and Bruce (1989) argued that the original publication of this name is in *The Athenaeum*, 670:685 [29 August 1840]. Eldridge and Close (1995) noted the existence of two undescribed races from Central Australia and West Kimberley.

*Petrogale mareeba* Eldredge and Close, 1992. Aust. J. Zool., 40:619.
   COMMON NAME: Mareeba Rock-wallaby.
   TYPE LOCALITY: Australia, Queensland, Mungana Trucking Yards, 16 km west of Chillagoe, 17°06'S, 144°23'E.
   DISTRIBUTION: Mareeba north to Mitchell River and near Mt. Carbine, west to Mungana and south to Burdekin River, N Queensland (Australia).
   STATUS: Localized.
   COMMENTS: *P. lateralis/penicillata* species-group.

*Petrogale penicillata* (Gray, 1827). *In* Griffith et al., Anim. Kingdom, Mamm., 3, plate only.
   COMMON NAME: Brush-tailed Rock-wallaby.
   TYPE LOCALITY: Australia, New South Wales, Sydney.
   DISTRIBUTION: SE Australia, from East Gippsland (Victoria) to 100 km NW of Bisbane (Queensland).
   STATUS: IUCN – Vulnerable. Locally common but declining.
   SYNONYMS: *albogularis* (Jourdan, 1837); *longicauda* Krefft, 1865.
   COMMENTS: *P. lateralis/penicillata* species-group. Does not include *herberti* (Eldridge and Close, 1992).

*Petrogale persephone* Maynes, 1982. Aust. Mamm., 5:47.
   COMMON NAME: Proserpine Rock-wallaby.
   TYPE LOCALITY: Australia, Queensland, 9.6 km north of Proserpine, base of Mt. Dryander, 20°19'S, 148°33'E.
   DISTRIBUTION: Restricted to district around Proserpine.
   STATUS: IUCN – Endangered.
   COMMENTS: *P. xanthopus* species-group.

*Petrogale purpureicollis* Le Souef, 1924. Aust. Zool., 3:274.
   COMMON NAME: Purple-necked Rock-wallaby.
   TYPE LOCALITY: Australia, NW Queensland, Dajarra.
   DISTRIBUTION: Australia: Dajarra district, NW Queensland.
   STATUS: Uncertain.
   COMMENTS: *P. lateralis/penicillata* species-group. Considered probably a species distinct from *P. lateralis* by Eldridge et al. (1991).

*Petrogale rothschildi* Thomas, 1904. Nov. Zool., 11:166.
   COMMON NAME: Rothschild's Rock-wallaby.
   TYPE LOCALITY: Australia, Western Australia, Cossack.
   DISTRIBUTION: NW Western Australia.
   STATUS: IUCN – Lower Risk (lc). Vulnerable.
   COMMENTS: *P. xanthopus* species-group.

*Petrogale sharmani* Eldredge and Close, 1992. Aust. J. Zool., 40:618.
   COMMON NAME: Mt. Claro Rock-wallaby.
   TYPE LOCALITY: Australia, Queensland, Mt. Claro, 18°52'S, 145°44'S.

DISTRIBUTION: Seaview and Coane Ranges, west of Ingham, N Queensland (Australia).
STATUS: IUCN – Lower Risk (nt).
COMMENTS: *P. lateralis/penicillata* species-group.

*Petrogale xanthopus* Gray, 1855. Proc. Zool. Soc. Lond., 1854:259 [1855].
COMMON NAME: Yellow-footed Rock-wallaby.
TYPE LOCALITY: Australia, South Australia, Flinders Range.
DISTRIBUTION: Australia: SW Queensland, South Australia, NW New South Wales.
STATUS: U.S. ESA – Endangered; IUCN – Vulnerabler as *P. x. xanthopus,* Lower Risk (nt) as
    *P. x. celeris.*
SYNONYMS: *xanthopygus* (Giebel, 1874) [error]; *celeris* Le Souef, 1924.
COMMENTS: *P. xanthopus* species-group.

*Setonix* Lesson, 1842. Nouv. Tabl. Regn. Anim. Mammifères, p. 194.
TYPE SPECIES: *Kangurus brachyurus* Quoy and Gaimard, 1830.

*Setonix brachyurus* (Quoy and Gaimard, 1830). *In* Dumont d'Urville, Voy . . . de Astrolabe, Zool.,
    1(L'Homme, Mamm. Oiseaux):114.
COMMON NAME: Quokka.
TYPE LOCALITY: Australia, Western Australia, King George Sound (Albany).
DISTRIBUTION: SW Western Australia, Rottnest Isl, and Bald Isl (Australia).
STATUS: U.S. ESA – Endangered; IUCN – Vulnerable.
SYNONYMS: *brevicaudatus* (Gray, 1838).

*Thylogale* Gray, 1837. Mag. Nat. Hist. [Charlesworth's], 1:583.
TYPE SPECIES: *Halmaturus* (*Thylogale*) *eugenii* Gray, 1837 (= *Halmaturus thetis* Lesson, 1828).
COMMENTS: Included in *Macropus* by Van Gelder (1977*b*), but see Kirsch and Calaby (1977:17).
    See comments under *Dorcopsis.*

*Thylogale billardierii* (Desmarest, 1822). Mammalogie. *In* Encycl. Méth., 2(Suppl.):542.
COMMON NAME: Tasmanian Pademelon.
TYPE LOCALITY: Australia, Tasmania.
DISTRIBUTION: Australia: SE South Australia, Victoria, Tasmania, islands in Bass Strait;
    probably survives only in Tasmania.
STATUS: IUCN – Lower Risk (lc).
SYNONYMS: *brachytarsus* (Wagner, 1842); *rufiventer* Ogilby, 1838; *tasmanei* (Gray, 1838).

*Thylogale browni* (Ramsay, 1887). Die Säugethiere, 3:551.
COMMON NAME: Brown's Pademelon.
TYPE LOCALITY: Papua New Guinea, New Ireland.
DISTRIBUTION: E New Guinea and adjacent small islands, sea level to 2000 m; Bismarck Arch.
    (Papua New Guinea).
STATUS: IUCN – Vulnerable.
SYNONYMS: *keysseri* (Förster and Rothschild, 1914); *lauterbachi* Matschie, 1916; *lugens*
    (Alston, 1877); *tibol* (Miklouho-Maclay, 1885).
COMMENTS: Formerly considered a subspecies of *T. brunii*, but see Flannery (1992).

*Thylogale brunii* (Schreber, 1778). Die Säugethiere, 3:551.
COMMON NAME: Dusky Pademelon.
TYPE LOCALITY: Indonesia, Aru Isls.
DISTRIBUTION: Southern New Guinea (Trans-Fly Plains and Port Moresby district); Aru Isls;
    Palau (= Isl) Kei Besar.
STATUS: IUCN – Vulnerable.
SYNONYMS: *gracilis* Miklouho-Maclay, 1884; *jukesii* (Miklouho-Maclay, 1884).
COMMENTS: *T. bruijni* is a later spelling (Haltenorth, 1958:38).

*Thylogale calabyi* Flannery, 1992. Aust. Mammal., 15:18.
COMMON NAME: Calaby's Pademelon.
TYPE LOCALITY: Papua New Guinea, Central Province, Mt. Albert Edward, 3000 m on south
    side of Neon Basin.

DISTRIBUTION: Known from Mt. Albert Edward and Mt. Giluwe, Eastern highlands of Papua New Guinea, about 3000 m.

STATUS: IUCN – Endangered.

*Thylogale lanatus* (Thomas, 1922). Ann. Mag. Nat. Hist., ser. 9, 9:670.

COMMON NAME: Mountain Pademelon.

TYPE LOCALITY: Papua New Guinea, Saruwaged Mtns, 3600 m.

DISTRIBUTION: Subalpine grasslands of Huon Peninsula.

STATUS: Vulnerable.

COMMENTS: Close to *T. browni*, and considered a subspecies of it, with some misgivings, by Flannery (1992), who however noted some consistent differences, so considered a full species here.

*Thylogale stigmatica* (Gould, 1860). Mamm. Aust., 2, pt. 12, pl. 33-34.

COMMON NAME: Red-legged Pademelon.

TYPE LOCALITY: Australia, Queensland, Point Cooper (N of Rockingham Bay).

DISTRIBUTION: E Queensland, E New South Wales (Australia); SC lowland New Guinea.

STATUS: IUCN – Lower Risk (lc). Uncommon.

SYNONYMS: **coxenii** (Gray, 1866); *gazella* (De Vis, 1884); **oriomo** (Tate and Archbold, 1935); *temporalis* (De Vis, 1884); **wilcoxi** (McCoy, 1866)

COMMENTS: Citation for original description given as Proc. Zool. Soc. Lond., 1860:375, by some authors, but this is dated Nov. 13, while Mammal. Aust., Part 12 was published Nov. 1.

*Thylogale thetis* (Lesson, 1828). Monogr. Mamm., p. 229.

COMMON NAME: Red-necked Pademelon.

TYPE LOCALITY: Australia, New South Wales, Sydney.

DISTRIBUTION: E Queensland, E New South Wales (Australia).

STATUS: IUCN – Lower Risk (lc). Uncommon.

SYNONYMS: *eugenii* (Gray, 1837); *nuchalis* (Wagner, 1842).

*Wallabia* Trouessart, 1905. Cat. Mamm. Viv. Foss., Suppl. fasc., 4:834.

TYPE SPECIES: *Kangurus ualabatus* Lesson and Garnot, 1826 (= *Kangurus bicolor* Desmarest, 1804).

COMMENTS: Included in *Macropus* by Van Gelder (1977*b*), but see Kirsch and Calaby (1977:17).

*Wallabia bicolor* (Desmarest, 1804). Tabl. Méth. Hist. Nat., *in* Nouv. Dict. Hist. Nat., 24:357.

COMMON NAME: Swamp Wallaby.

TYPE LOCALITY: Unknown.

DISTRIBUTION: Australia: E Queensland, E New South Wales, Victoria, SE South Australia, Stradbroke Isl, Fraser Isl.

STATUS: IUCN – Lower Risk (lc). Common.

SYNONYMS: *apicalis* (Gunther, 1874); *ingrami* (Thomas and Dollman, 1909); *lessonii* (Gray, 1837); *mastersii* (Krefft, 1867) [*nomen nudum*, made available in Krefft, 1871]; *nemoralis* (Wagner, 1842); *ualabatus* (Lesson and Garnot, 1826); *welsbyi* Longman, 1922.

# ORDER AFROSORICIDA
by Gary N. Bronner and Paulina D. Jenkins

## ORDER AFROSORICIDA Stanhope, 1998.

COMMENTS: Traditionally included in the Lipotyphla (= Insectivora *sensu stricto*). Various molecular studies (Madsen et al., 2001; Murphy et al., 2001*a, b*; Springer et al., 1999) and syntheses of morphological and molecular data (Asher et al., 2003; Liu et al., 2001) support a clade containing tenrecs and golden moles, which Stanhope et al. (1998) named Afrosoricida. This name is inappropriate since this clade does not include soricids, and could lead to confusion with the soricid subgenus *Afrosorex* Hutterer, 1986. Noting that Tenrecomorpha Butler, 1972 may be a prior, and more explicit name for this clade following Simpson's (1945) guidelines for naming superfamial taxa, Bronner et al. (2003) nevertheless accepted Afrosoricida because this name is entrenched in the recent literature.

While Afrosoricida is widely used as a name for a tenrec-golden mole clade (e.g., Cao et al., 2000; de Jong et al., 2003; Douady et al., 2002*a,b*; Hedges, 2001; Helgen, 2003*a*; Scally et al., 2001; van Dijk et al., 2001; Waddell et al., 2001), some authors have argued that Tenrecoidea McDowell, 1958 is the prior valid name for this taxon (e.g., Archibald, 2003; Asher, 2000; Malia et al., 2002; Mouchaty, 1999). Tenrecoidea, however, was first used by Simpson (1931) as a corrected superfamily name for Cententoidea, and comprised various "zalambdodont" taxa (Solenodontidae, Potamogalidae, Tenrecidae and extinct Palaeoryctidae and Apternodontidae) but not chrysochlorids (assigned to a separate superfamily Chrysochloridea). McDowell's (1958) restriction of Tenrecoidea to include only golden moles and tenrecs (thus identical in composition to Afrosoricida) implied a fundamentally different grouping concept to that of Simpson (1931); usage of this name arguably violates Simpson's (1945:33) guidelines (29-30) for reasonable emendation, and also his recommendation that superfamily names (ending in –oidea) should be avoided. Priority, which Simpson (1945:33) advocated as a deciding criterion only "..*when other things are about equal.*" is thus insufficient to justify acceptance of McDowell's (1958) revised "Tenrecoidea". Even if accepted, this name should be ascribed to Simpson, 1931 (Simpson 1945:32 – guideline 23).

Following Simpson (1931), the term "tenrecoid" has also been widely misused as a general name for a vaguely defined grouping of (extinct and extant) taxa characterized by zalambdodonty, even though it has become clear that some of these were not technically zalambdodont, and that zalambdodonty may have arisen independently several times (e.g. Broom, 1916). This further militates against its stricter nomenclatorial use, even at taxonomic ranks below order.

Tenrecomorpha is also a problematic name for the tenrec-golden mole clade (*c.f.* Bronner et al., 2003). Butler (1972) used this name for a suborder of the Lipotyphla (Insectivora *sensu stricto*) to accommodate only the Tenrecidae, and assigned chrysochlorids to the separate suborder Chrysochloridea Broom, 1915. Butler (1988) tentatively accepted a common origin of these families, but argued for their early separation. Eisenberg (1981:63) showed Tenrecomorpha as a separate clade from Chrysochloridea, but elsewhere (p. 113) included chrysochlorids in Tenrecomorpha without giving any character support. While Mouchaty et al. (2000*b*) and Waddell and Shelley (2003) recently used Tenrecomorpha in this context, this name has most consistently been used to include only tenrecs (e.g. MacPhee and Novecek, 1993) and consequently is better applied at the subordinal level to separate tenrecids from chrysochlorids, in accordance with both morphological and molecular data suggesting early phylogenetic divergence of these taxa. Following Bronner et al. (2003), we therefore reluctantly accept Afrosoricida as the most specific name for a tenrecid-chrysochlorid clade.

The supraordinal affiliation of the Afrosoricida remains controversial. Molecular data strongly support an affinity within the Afrotheria, whereas morphological data suggest a closer relationship to lipotyphlans. Lipotyphlan monophyly, however, is only weakly supported by cladistic analyses of morphological data (Asher, 1999) and phylogenetic analyses of combined anatomical and molecular data strongly support the inclusion of afrosoricids within Afrotheria (Asher et al., 2003).

**SUBORDER TENRECOMORPHA** Butler, 1972.
>   COMMENTS: Included in suborder Soricomorpha of the Lipotyphla by Butler (1988) and by
>   MacPhee and Novacek (1993). McKenna and Bell (1997:293) use the name Tenrecoidea
>   (attributing it to Gray, 1821:301) as a superfamily with the note "Proposed as suborder of
>   Lipothyphla, coordinate with Erinaceomorpha, Soricomorpha, and Chrysochlorida";
>   however, usage of this name is undesirable (see above).

**Family Tenrecidae** Gray, 1821. London Med. Repos. Rec., 15:301.
>   COMMENTS: Includes Potamogalinae (see Corbet, 1974).

**Subfamily Geogalinae** Trouessart, 1881. Revue. Mag. Zool., ser. 3, 7:275.

*Geogale* Milne-Edwards and A. Grandidier, 1872. Ann. Sci. Nat. Zool., 15 (art 19):2.
>   TYPE SPECIES: *Geogale aurita* Milne-Edwards and A. Grandidier, 1872.
>   SYNONYMS: *Cryptogale* G. Grandidier, 1928 (see Genest and Petter, 1975).

>   *Geogale aurita* Milne-Edwards and A. Grandidier, 1872. Ann. Sci. Nat. Zool., 15 (art. 19):2.
>   >   COMMON NAME: Large-eared Tenrec.
>   >   TYPE LOCALITY: Madagascar, Mouroundava [Morondava].
>   >   DISTRIBUTION: Western deciduous forest and spiny bush of S and W Madagascar and
>   >   Fénérive, NE Madagascar.
>   >   STATUS: IUCN – Lower Risk (lc).
>   >   SYNONYMS: *australis* (G. Grandidier, 1928); ***orientalis*** G. Grandidier and Petit, 1930.
>   >   COMMENTS: The subspecific status of *G. a. orientalis* is uncertain. This species was not found
>   >   during a recent survey of the environs of Fénérive (Rakotondravony et al., 1998). The
>   >   easternmost occurrence of this species was recorded recently in SE Madagascar
>   >   (Goodman et al., 1999*a*).

**Subfamily Oryzorictinae** Dobson, 1882. Monograph of the Insectivora, 1:71.

*Limnogale* Major, 1896. Ann. Mag. Nat. Hist., ser. 6, 18:318.
>   TYPE SPECIES: *Limnogale mergulus* Major, 1896.

>   *Limnogale mergulus* Major, 1896. Ann. Mag. Nat. Hist., ser. 6, 18:319.
>   >   COMMON NAME: Web-footed Tenrec.
>   >   TYPE LOCALITY: Madagascar, Imasindrary, NE Betsileo.
>   >   DISTRIBUTION: Freshwater streams of eastern humid forest and central highlands of
>   >   E Madagascar.
>   >   STATUS: IUCN – Endangered.
>   >   COMMENTS: Morphological evidence suggests that *Limnogale* and the African Potamogalinae
>   >   are sister taxa (Asher, 1999).

*Microgale* Thomas, 1882. J. Linn. Soc., Zool., 16:319.
>   TYPE SPECIES: *Microgale longicaudata* Thomas, 1882.
>   SYNONYMS: *Leptogale* Thomas, 1918; *Nesogale* Thomas, 1918; *Paramicrogale* G. Grandidier and
>   Petit, 1931.
>   COMMENTS: Revised by MacPhee (1987*a*); reviewed in part by Jenkins et al. (1996, 1997).

>   *Microgale brevicaudata* G. Grandidier, 1899. Bull. Mus. Hist. Nat. Paris, 5:349.
>   >   COMMON NAME: Short-tailed Shrew Tenrec.
>   >   TYPE LOCALITY: "Environs de Mahanara, à 75 km environ au sud de Vohémar, sur la côte nord-
>   >   est de Madagascar" [Maharana River, 78 km S of Iharana [Vohimarina], Antsiranana,
>   >   Antalaha, Madagascar (MacPhee, 1987*a*)].
>   >   DISTRIBUTION: Northern highlands, sambirano, western deciduous forest and eastern humid
>   >   forest of N and NW Madagascar.
>   >   STATUS: IUCN – Lower Risk (lc).
>   >   SYNONYMS: *breviceps* Kaudern, 1918; *occidentalis* (G. Grandidier and Petit, 1931).
>   >   COMMENTS: There are only two records from the eastern humid forest, the type locality and

Parc National de Marojejy (Goodman and Jenkins, 2000); similarly there is a single report of this species from the sambirano of Réserve Speciale de Manongarivo (Raxworthy and Rakotondraparany, 1988).

*Microgale cowani* Thomas, 1882. J. Linn. Soc., Zool., 16:320.

COMMON NAME: Cowan's Shrew Tenrec.
TYPE LOCALITY: Ankáfana forest, E Betsileo [Ankafina, Fianarantsoa, Fianarantsoa Province, 21°12'S 47°12'E (Carleton and Schmidt, 1990; MacPhee, 1987*a*)].
DISTRIBUTION: Eastern humid forest and central highlands of N, E and SE Madagascar.
STATUS: IUCN – Lower Risk (lc).
SYNONYMS: *crassipes* Milne-Edwards, 1893; *longirostris* Major, 1896; *nigrescens* Elliot, 1905.

*Microgale dobsoni* Thomas, 1884. Ann. Mag. Nat. Hist., ser. 5, 14:337.

COMMON NAME: Dobson's Shrew Tenrec.
TYPE LOCALITY: Nandésen forest, C Betsileo. [Nandihizana, c. 20 miles (30 km) SSW of Ambositra, see MacPhee, 1987*a*; c. 20°50'S 47°10'E, see Jenkins et al., 1996.]
DISTRIBUTION: Eastern humid forest of N, E and SE, and central highlands, Madagascar.
STATUS: IUCN – Lower Risk (lc).
COMMENTS: Formerly included in *Nesogale,* see Thomas (1918*a*), Genest and Petter (1975).

*Microgale drouhardi* G. Grandidier, 1934. Bull. Mus. Nat. Hist. Nat. Paris, ser. 2, 6:474.

COMMON NAME: Drouhard's Shrew Tenrec.
TYPE LOCALITY: "Environs de Diego-Suarez, extrême-nord de Madagascar". [Antsiranana, c. 12°16'S 49°18'E (MacPhee, 1987*a*)].
DISTRIBUTION: Eastern humid forest of N and E Madagascar.
SYNONYMS: *melanorrhachis* Morrison-Scott, 1948.
COMMENTS: Considered a synonym of *M. cowani* by MacPhee (1987*a*), specific status recognised by Jenkins et al. (1997).

*Microgale dryas* Jenkins, 1992. Bull. Br. Mus. Nat. Hist. (Zool.), 58:53.

COMMON NAME: Dryad Shrew Tenrec.
TYPE LOCALITY: Site 1, Ambatovaky Special Reserve, NE Madagascar, in primary rainforest, between 600–750 m, 16°51'S, 49°08'E.
DISTRIBUTION: Known only from the type locality and from an owl pellet collected at Réserve Spéciale d'Anjanharibe-Sud, NE Madagascar.
STATUS: IUCN – Critically Endangered.

*Microgale fotsifotsy* Jenkins, Raxworthy and Nussbaum, 1997. Bull. Br. Mus. Nat. Hist. (Zool.), 63:2.

COMMON NAME: Pale Shrew Tenrec.
TYPE LOCALITY: Camp 2, Antomboka River Fitsahana, Parc National de la Montagne d'Ambre, Antsiranana Fivondronana, Antsiranana Province 12°29'S 49°10'E, altitude 650 m, Madagascar.
DISTRIBUTION: Eastern humid rainforests of N, E and SE Madagascar.

*Microgale gracilis* (Major, 1896). Ann. Mag. Nat. Hist., ser. 6, 18:321.

COMMON NAME: Gracile Shrew Tenrec.
TYPE LOCALITY: Ambohimitombo forest, Madagascar [Ambohimitombo town, 43 km by road SE of Ambositra, 10 km into eastern forest, Fianarantsoa, 20°43'S 47°26'E (MacPhee, 1987*a*)].
DISTRIBUTION: Eastern humid forest and central highlands of N, E and SE Madagascar.
STATUS: IUCN – Vulnerable.
COMMENTS: Formerly included in *Oryzorictes* (see Major, 1896) and *Leptogale* (see Thomas, 1918*a*).

*Microgale gymnorhyncha* Jenkins, Goodman and Raxworthy, 1996. Fieldiana Zool. n.s., 85:211.

COMMON NAME: Naked-nosed Shrew Tenrec.
TYPE LOCALITY: E Madagascar, 38 km S [of] Ambalavao, Réserve Naturelle Intégrale d'Andringitra, on ridge E of Volotsangana River, Fianarantsoa Province, 22°11'39"S 46°58'16"E, altitude 1625 m.
DISTRIBUTION: Eastern humid forest of N, E and SE and central highlands, Madagascar.

*Microgale longicaudata* Thomas, 1882. J. Linn. Soc., Zool., 16:320.
  COMMON NAME: Lesser Long-tailed Shrew Tenrec.
  TYPE LOCALITY: Ankáfana forest, E Betsileo, Madagascar [Ankafina, Fianarantsoa,
      Fianarantsoa Province, 21°12′S 47°12′E (see MacPhee, 1987a; Carleton and Schmidt,
      1990)].
  DISTRIBUTION: Eastern humid forest and central highlands of N, E and SE, and western
      deciduous forest of W Madagascar.
  STATUS: IUCN – Lower Risk (lc).
  SYNONYMS: *majori* Thomas, 1918; *prolixacaudata* G. Grandidier, 1937.

*Microgale monticola* Goodman and Jenkins, 1998. Fieldiana Zool. n.s., 90:149.
  COMMON NAME: Montane Shrew Tenrec.
  TYPE LOCALITY: Madagascar, 11 km WSW of Befingitra, Réserve Spéciale d'Anjanaharibe-
      Sud, 14°44′S, 49°26′E, 1550 m.
  DISTRIBUTION: Northern highlands of N Madagascar.
  COMMENTS: Recorded to date only at high altitudes in the mountains surrounding the
      Andapa Basin (see Goodman and Jenkins, 2000).

*Microgale nasoloi* Jenkins and Goodman, 1999. Bull. Nat. Hist. Mus. Lond. (Zool.), 65:156.
  COMMON NAME: Nasolo's Shrew Tenrec.
  TYPE LOCALITY: Vohibasia Forest [Forêt de Vohibasia], 59 km northeast of Sakaraha, Province
      de Toliara, SW Madagascar, 22°27.5′S 44°50.5′E, 780 m.
  DISTRIBUTION: Known only from transitional dry deciduous forest at the type locality and
      Analavelona Forest, SW Madagascar.

*Microgale parvula* G. Grandidier, 1934. Bull. Mus. Nat. Hist. Nat. Paris, ser. 2, 6:476.
  COMMON NAME: Pygmy Shrew Tenrec.
  TYPE LOCALITY: "Environs de Diego Suarez, extrême-nord de Madagascar." [Antsiranana,
      c. 12°16′S 49°18′E (MacPhee, 1987a)].
  DISTRIBUTION: Eastern humid forest and central highlands of N, E and SE Madagascar.
  STATUS: Revised by Jenkins et al. (1997). Although recorded as Endangered as *M. parvula* and
      Vulnerable as *M. pulla* in the IUCN Red List 2003, this species is currently common
      over a wide distributional range (Goodman and Jenkins, 1998, 2000; Goodman et al.,
      1999a; Jenkins et al., 1996, 1997).
  SYNONYMS: *pulla* Jenkins, 1988.

*Microgale principula* Thomas, 1926. Ann. Mag. Nat. Hist., ser. 9, 17:251.
  COMMON NAME: Greater Long-tailed Shrew Tenrec.
  TYPE LOCALITY: "Midongy-du-Sud, South-east Madagascar" [Midongy Atsimo, Fianarantsoa,
      Farafangana, 23°35′S 47°01′E (MacPhee, 1987a)].
  DISTRIBUTION: Eastern humid forest of N, E and SE Madagascar.
  STATUS: IUCN – Endangered.
  SYNONYMS: *decaryi* G. Grandidier, 1928; *sorella* Thomas, 1926.
  COMMENTS: Distribution may be disjunct.

*Microgale pusilla* Major, 1896. Ann. Mag. Nat. Hist., ser. 6, 18:462.
  COMMON NAME: Least Shrew Tenrec.
  TYPE LOCALITY: Forest of the Independent Tanala of Ikongo, in the neighbourhood of
      Vinanitelo [50 km SE of Fianarantsoa town and 10 km SSE of Vohitrafeno town,
      W margin of E forest, Fianarantsoa, Fianarantsoa Province, Madagascar, 21°45′S
      47°17′E (MacPhee, 1987a)].
  DISTRIBUTION: Eastern humid forest of E and SE, and central highlands, Madagascar.
  STATUS: IUCN – Lower Risk (lc).

*Microgale soricoides* Jenkins, 1993. Am. Mus. Novit., 3067:2.
  COMMON NAME: Shrew-toothed Shrew Tenrec.
  TYPE LOCALITY: Mantady National Park, c. 15 km north of Perinet, Madagascar, 18°51′S
      48°27′E, in primary rainforest, between 1100 and 1150 m elevation.
  DISTRIBUTION: Northern highlands, eastern humid forest and central highlands of N, E and
      SE Madagascar.

*Microgale taiva* Major, 1896. Ann. Mag. Nat. Hist., ser. 6, 18:461.
>COMMON NAME: Taiva Shrew Tenrec.
>TYPE LOCALITY: Ambohimitombo forest, Tanala Country, Madagascar [Ambohimitombo town, 43 km by road SE of Ambositra, 10 km into eastern forest, Fianarantsoa, 20°43'S 47°26'E (MacPhee, 1987*a*)].
>DISTRIBUTION: Southern regions of eastern humid forest of Madagascar.
>COMMENTS: Synonymised with *M. cowani* by MacPhee (1987*a*), specific status recognised by Jenkins et al. (1996).

*Microgale talazaci* Major, 1896. Ann. Mag. Nat. Hist., ser. 6, 18:320.
>COMMON NAME: Talazac's Shrew Tenrec.
>TYPE LOCALITY: "Forest of the Independent Tanala of Ikongo, in the neighbourhood of Vinanitelo, one day's journey south of Fianarantsoa" [50 km SE of Fianarantsoa town and 10 km SSE of Vohitrafeno town, W margin of E forest, Fianarantsoa, Fianarantsoa Province, Madagascar, 21°45'S 47°17'E (MacPhee, 1987*a*)].
>DISTRIBUTION: Northern highlands, eastern humid forest and central highlands of N, E and SE Madagascar.
>STATUS: IUCN – Lower Risk (lc).
>COMMENTS: Formerly included in *Nesogale*, see Thomas (1918*a*), Genest and Petter (1975).

*Microgale thomasi* Major, 1896. Ann. Mag. Nat. Hist., ser. 6, 18:320.
>COMMON NAME: Thomas's Shrew Tenrec.
>TYPE LOCALITY: Ampitambè forest (NE Betsileo, Madagascar) [Locality uncertain, discussed by MacPhee (1987*a*) and Carleton and Schmidt (1990); c. 20°22'S 47°46'E, see Carleton and Schmidt (1990)].
>DISTRIBUTION: Eastern humid forest, and central highlands, E and SE Madagascar.
>STATUS: IUCN – Vulnerable.

*Oryzorictes* A. Grandidier, 1870. Revue Mag. Zool., 22:50.
>TYPE SPECIES: *Oryzorictes hova* A. Grandidier, 1870.
>SYNONYMS: *Oryzoryctes* Trouessart, 1879 [invalid emendation]; *Nesoryctes* Thomas, 1918.

*Oryzorictes hova* A. Grandidier, 1870. Revue Mag. Zool., 22:50.
>COMMON NAME: Mole-like Rice Tenrec.
>TYPE LOCALITY: "Ankaye et Antsianak" [Ankay, along the Mangoro River, near Lac Alaotra and Antsianaka, region E of Lac Alaotra, Madagascar (Viette, 1991)].
>DISTRIBUTION: Northern highlands, eastern humid forest and central highlands of N, NW, E and S Madagascar.
>STATUS: IUCN – Lower Risk (lc).
>SYNONYMS: *talpoides* G. Grandidier and Petit, 1930.
>COMMENTS: Subgenus *Oryzorictes*. Includes *talpoides* (see Goodman et al., 1999*a*).

*Oryzorictes tetradactylus* Milne-Edwards and A. Grandidier, 1882. Le Naturaliste, 4:55.
>COMMON NAME: Four-toed Rice Tenrec.
>TYPE LOCALITY: "Du plateau d'Emirne" [Imerina, Madagascar (Viette, 1991)].
>DISTRIBUTION: Eastern humid forest and central highlands of C and E Madagascar.
>STATUS: IUCN – Lower Risk (lc).
>SYNONYMS: *niger* Major, 1896.
>COMMENTS: Subgenus *Nesoryctes* (see Heim de Balsac [1972] and Genest and Petter [1975]). *O. niger* is considered a melanistic form of *tetradactylus*, see Thomas (1918*a*).

**Subfamily Potamogalinae** Allman, 1865. Proc. Zool. Soc. Lond., 1865:467.

*Micropotamogale* Heim de Balsac, 1954. C.R. Acad. Sci. Paris, 239:102.
>TYPE SPECIES: *Micropotamogale lamottei* Heim de Balsac, 1954.
>SYNONYMS: *Kivugale* Kretzoi, 1961; *Mesopotamogale* Heim de Balsac, 1956 [see Corbet, 1974].

*Micropotamogale lamottei* Heim de Balsac, 1954. C.R. Acad. Sci. Paris, 239:103.
>COMMON NAME: Nimba Otter Shrew.

TYPE LOCALITY: "Ziéla, dans une savane au pied du Nimba, altitude 550 m." [Ziéla, Mount Nimba, Guinea].

DISTRIBUTION: Environs of Mt. Nimba in Guinea, Liberia and Côte d'Ivoire.

STATUS: IUCN – Endangered.

COMMENTS: For a survey of the distribution and ecology, see Vogel (1983).

*Micropotamogale ruwenzorii* (de Witte and Frechkop, 1955). Bull. Inst. r. Sci. nat. Belg., 31(84):9.

COMMON NAME: Ruwenzori Otter Shrew.

TYPE LOCALITY: "Rivière Talya, à Mutsora (station du Parc National Albert), contreforts occidentaux du massif Ruwenzori, altitudes 1100 – 1200 m." [Mutsora, Talya River, W slopes of Mt. Ruwenzori, Dem. Rep. Congo].

DISTRIBUTION: Ruwenzori region (Uganda, Dem. Rep. Congo), and W of Lake Edward and Lake Kivu (Dem. Rep. Congo).

STATUS: IUCN – Endangered.

*Potamogale* Du Chaillu, 1860. Proc. Boston Soc. Nat. Hist., 7:363.

TYPE SPECIES: *Cynogale velox* Du Chaillu, 1860.

SYNONYMS: *Bayonia* Bocage, 1865; *Mystomys* Gray, 1861; *Mythomys* Gray, 1862.

*Potamogale velox* (Du Chaillu, 1860). Proc. Boston Soc. Nat. Hist., 7:361.

COMMON NAME: Giant Otter Shrew.

TYPE LOCALITY: "Mountains of the interior, or in the hilly country north and south of the equator, Equatorial Africa" [Gabon].

DISTRIBUTION: Tropical Africa: Angola, Cameroon, Central African Republic, Chad, Republic of Congo, Dem. Rep. Congo, Equatorial Guinea, Gabon, Kenya, Nigeria, Sudan, Tanzania, Uganda.

STATUS: IUCN – Endangered.

SYNONYMS: *allmani* Jentink, 1895; *argens* Thomas, 1915.

**Subfamily Tenrecinae** Gray, 1821. London Med. Repos. Rec., 15:301.

*Echinops* Martin, 1838. Proc. Zool. Soc. Lond., 1838:17.

TYPE SPECIES: *Echinops telfairi* Martin, 1838.

SYNONYMS: *Echinogale* Wagner, 1841.

*Echinops telfairi* Martin, 1838. Proc. Zool. Soc. Lond., 1838:17.

COMMON NAME: Lesser Hedgehog Tenrec.

TYPE LOCALITY: Madagascar.

DISTRIBUTION: Western deciduous forest and spiny bush of S Madagascar.

STATUS: IUCN – Lower Risk (lc).

SYNONYMS: *miwarti* A. Grandidier, 1869; *nigrescens* Petit, 1931; *pallescens* Thomas, 1892.

COMMENTS: The easternmost occurrence of this species was recorded recently in SE Madagascar (Goodman et al., 1999*a*).

*Hemicentetes* Mivart, 1871. Proc. Zool. Soc. Lond., 1871:58, 72.

TYPE SPECIES: *Ericulus semispinosus* G. Cuvier, 1798.

SYNONYMS: *Centetes* Schinz and Brodtmann, 1827 [not Illiger, 1811]; *Echinodes* Pomel, 1848 [*nomen nudum*]; *Ericius* Giebel, 1871 [not Tilesius, 1813; not Sundevall, 1814]; *Ericus* Bergroth, 1902; *Eteocles* Gray, 1821.

*Hemicentetes nigriceps* Günther, 1875. Ann. Mag. Nat. Hist., ser. 4, 16:125.

COMMON NAME: Highland Streaked Tenrec.

TYPE LOCALITY: Fienerentova [? = Fianarantsoa, Madagascar].

DISTRIBUTION: Eastern edge of central highlands, E Madagascar.

SYNONYMS: *buffoni* Jentink, 1879.

COMMENTS: Distinguished from *H. semispinosus* by craniodental features (Butler, 1941; Dobson, 1882*a*) and treated as a distinct species by Eisenberg and Gould (1970). Considered to be a subspecies of *H. semispinosus* (Genest and Petter, 1975; Hutterer, 1993*a*). Recorded sympatrically with *H. semispinosus* (Goodman et al., 2000).

*Hemicentetes semispinosus* (G. Cuvier, 1798). Tableau élémentaire de l'histoire naturelle des animaux, p. 108.
    COMMON NAME: Lowland Streaked Tenrec.
    TYPE LOCALITY: Madagascar.
    DISTRIBUTION: Eastern humid forest, central highlands of E Madagascar.
    STATUS: IUCN – Lower Risk (lc).
    SYNONYMS: *madagascariensis* (Shaw, 1800) [not Zimmermann]; *variegatus* E. Geoffroy, 1803.

*Setifer* Froriep, 1806. In Dumeril, Analytische Zoologie .. mit Zusatzen, p. 15.
    TYPE SPECIES: *Erinaceus setosus* Schreber, 1778.
    SYNONYMS: *Dasogale* G. Grandidier, 1930; *Ericulus* I. Geoffroy, 1837; *Hericulus* Gloger, 1841.
    COMMENTS: Includes *Ericulus*, see Eisenberg and Gould (1970:49); and *Dasogale*, see Poduschka and Poduschka (1982:253).

*Setifer setosus* (Schreber, 1778). Die Säugethiere .. mit Beschreibungen, 3:590, pl. 164.
    COMMON NAME: Greater Hedgehog Tenrec.
    TYPE LOCALITY: Madagascar.
    DISTRIBUTION: All latitudes and phytographic zones: eastern humid forest, central highlands, northern highlands, sambirano, western deciduous forest, spiny bush, Madagascar.
    STATUS: IUCN – Lower Risk (lc).
    SYNONYMS: *acanthurus* (Boddaert, 1785); *fontoynonti* (G. Grandidier, 1930); *melantho* (Thomas, 1926); *nigrescens* (I. Geoffroy, 1839); *nigricans* (Bartlett, 1875) [*nomen nudum*]; *spinosus* (Desmarest, 1820).
    COMMENTS: *Dasogale fontoynonti* was based on a juvenile *Setifer setosus*, see Poduschka and Poduschka (1982:253) and MacPhee (1987b:133).

*Tenrec* Lacépède, 1799. Tableau des divisions, .. Mammifères, p.7.
    TYPE SPECIES: *Erinaceus ecaudatus* Schreber, 1778.
    SYNONYMS: *Tenrecus* Desmarest, 1820.

*Tenrec ecaudatus* (Schreber, 1778). Die Säugethiere.. mit Beschreibungen, 3:590, pl. 165.
    COMMON NAME: Tail-less Tenrec.
    TYPE LOCALITY: Madagascar.
    DISTRIBUTION: All latitudes and phytographic zones: eastern humid forest, central highlands, northern highlands, sambirano, western deciduous forest, spiny bush, Madagascar; Comoro Isls. Introduced on Reunion, Mauritius, and the Seychelle Isls.
    STATUS: IUCN – Lower Risk (lc).
    SYNONYMS: *armatus* (I. Geoffroy, 1837); *tanrec* (Boddaert, 1785).

# SUBORDER CHRYSOCHLORIDEA Broom, 1915.
    COMMENTS: MacPhee and Novacek (1993) erected the suborder Chrysochloromorpha for golden moles, but following Simpson's (1945:32-33) nomenclatural principles for categories above superfamilies, Chrysochloridea is the senior synonym.

**Family Chrysochloridae** Gray, 1825. Ann. Philos., n.s., 10:335.
    COMMENTS: For widely divergent treatments see Simonetta (1968), Meester (1974), Meester et al. (1986) and Petter (1981a). The treatment below follows Bronner (1995a, 1996).

**Subfamily Chrysochlorinae** Gray, 1825. Ann. Philos., n.s., 10:335.

*Carpitalpa* Lundholm, 1955. Ann. Transvaal Mus., 22:285.
    TYPE SPECIES: *Chlorotalpa* (*Carpitalpa*) *arendsi* Lundholm, 1955.
    COMMENTS: Lundholm (1955a:285) described *Carpitalpa* and *Kilimatalpa* (here included in *Chrysochloris*) as subgenera within *Chlorotalpa*. Simonetta (1968) afforded *Carpitalpa* generic rank. Included in *Amblysomus* by Petter (1981a:50) and in *Chlorotalpa* by Meester (1974) and Meester et al. (1986:20).

*Carpitalpa arendsi* (Lundholm, 1955). Ann. Transvaal Mus., 22:285.
COMMON NAME: Arend's Golden Mole.
TYPE LOCALITY: E escarpment of Zimbabwe, Inyanga, Pungwe Falls.
DISTRIBUTION: E Zimbabwe and adjacent Mozambique.
STATUS: IUCN – Lower Risk (lc).

*Chlorotalpa* Roberts, 1924. Ann. Transvaal Mus., 10:64.
TYPE SPECIES: *Chrysochloris duthieae* Broom, 1907.
COMMENTS: Included in *Amblysomus* by Ellerman et al. (1953) and by Petter (1981a). Meester (1974) and Meester et al. (1986) included *leucorhina* (here referred to *Calcochloris*) and *arendsi* (here treated as *Carpitalpa*) in this genus. Revised by Bronner (1995a).

*Chlorotalpa duthieae* (Broom, 1907). Trans. S. Afr. Philos. Soc., 18:292.
COMMON NAME: Duthie's Golden Mole.
TYPE LOCALITY: South Africa, Western Cape Prov., Knysna.
DISTRIBUTION: Coastal belt of Western and Eastern Cape Prov., South Africa.
STATUS: IUCN – Vulnerable.

*Chlorotalpa sclateri* (Broom, 1907). Ann. Mag. Nat. Hist., ser. 7, 19:263.
COMMON NAME: Sclater's Golden Mole.
TYPE LOCALITY: South Africa, Western Cape Prov., Beaufort West.
DISTRIBUTION: Western Cape Prov., E Free State, and S Mpumalanga (South Africa); Lesotho.
STATUS: IUCN – Vulnerable.
SYNONYMS: **guillarmodi** Roberts, 1936; **montana** Roberts, 1924; **shortridgei** Broom, 1950.
COMMENTS: Included in *Amblysomus* by Petter (1981a).

*Chrysochloris* Lacépède, 1799. Tabl. Mamm., p. 7.
TYPE SPECIES: *Chrysochloris capensis* Lacépède, 1799 (= *Talpa asiatica* Linnaeus, 1758).
SYNONYMS: *Kilimatalpa* Lundholm, 1955.
COMMENTS: Includes *Kilimatalpa* as a subgenus, see comments under *C. stuhlmanni* below.

*Chrysochloris asiatica* (Linnaeus, 1758). Syst. Nat., 10th ed., 1:53.
COMMON NAME: Cape Golden Mole.
TYPE LOCALITY: "In Sibiria"; usually taken as Cape of Good Hope, South Africa. See Ellerman et al. (1953).
DISTRIBUTION: Western Cape Prov. and Robben Isl northwards along coastal plain to Orange River (South Africa).
STATUS: IUCN – Lower Risk (lc).
SYNONYMS: *auratus* (Vosmaer, 1787); *aurea* (Pallas, 1778); *bayoni* De Beaux, 1921; *calviniae* Shortridge, 1942; *capensis* Lacépède, 1799; *concolor* Shortridge and Carter, 1938; *damarensis* Ogilby, 1838; *dixoni* Broom, 1946; *elegans* Broom, 1946; *inaurata* (Pallas, 1777); *minor* Roberts, 1919; *namaquensis* Broom, 1907; *rubra* Lacépède, 1799; *shortridgei* Broom, 1946; *taylori* Broom, 1950; *tenuis* Broom, 1907; *visserae* Broom, 1950.
COMMENTS: Subgenus *Chrysochloris*. Geographic variation in size and colour appears to be clinal, hence no subspecies are recognized (see Meester et al., 1986), but further study may reveal some valid taxa. The only known specimen of *damarensis* may have been incorrectly labelled as no *Chrysochloris* have subsequently been collected in Damaraland, Namibia (Meester 1974).

*Chrysochloris stuhlmanni* Matschie, 1894. Sitzb. Ges. Naturf. Fr. Berlin, p. 123.
COMMON NAME: Stuhlmann's Golden Mole.
TYPE LOCALITY: Uganda, Ruwenzori region, "Ukondjo und Kinyawanga".
DISTRIBUTION: Burundi, Cameroon, Dem. Rep. Congo, Uganda, Kenya, Rwanda, Tanzania.
STATUS: IUCN – Lower Risk (lc).
SYNONYMS: **balsaci** (Lamotte and Petter, 1981); **stuhlmanni** (Matschie, 1894); *fosteri* (St. Leger, 1931); *tropicalis* (Allen and Loveridge, 1927); *vermiculus* (Thomas, 1910).
COMMENTS: Subgenus *Kilimatalpa,* which Lundholm (1955a) included in *Chlorotalpa.* Simonetta (1968:31) treated it as a subgenus of *Carpitalpa* (including *stuhlmanni* and

*fosteri*) but referred *tropicalis* to *Chlorotalpa*. Meester (1974) placed *stuhlmanni* in *Chrysochloris* based on malleus morphology, a treatment followed by Bronner (1995*a*) who argued that subgeneric distinction from *Chrysochloris* is warranted by anagenetic divergence in cranial shape. Validity and limits of subspecies are uncertain owing to the few specimens available. Lamotte and Petter (1981) described *balsaci* from Mt. Oku, Cameroon, an allopatric form that may deserve full specific status. In addition, the isolated *tropicalis* should be re-studied.

*Chrysochloris visagiei* Broom, 1950. Ann. Transvaal Mus., 21:238.
    COMMON NAME: Visagie's Golden Mole.
    TYPE LOCALITY: South Africa, Western Cape Prov., Gouna (54 mi. [87 km] E Calvinia).
    DISTRIBUTION: Known only from the holotype.
    STATUS: IUCN – Critically Endangered.
    COMMENTS: Subgenus *Chrysochloris*. Possibly an aberrant *asiatica*; see Meester (1974). Simonetta (1968:31) listed it as a subspecies of *asiatica*.

*Chrysospalax* Gill, 1883. Standard Nat. Hist., 5 (Mamm.):137.
    TYPE SPECIES: *Chrysochloris trevelyani* Günther, 1875.
    SYNONYMS: *Bematiscus* Cope, 1892 (see Ellerman et al., 1953).

*Chrysospalax trevelyani* (Günther, 1875). Proc. Zool. Soc. Lond., 1875:311.
    COMMON NAME: Giant Golden Mole.
    TYPE LOCALITY: South Africa, Eastern Cape Prov., Pirie Forest, near King William's Town.
    DISTRIBUTION: Eastern Cape Prov. (South Africa).
    STATUS: IUCN – Endangered.

*Chrysospalax villosus* (A. Smith, 1833). S. Afr. Quart. J., 2:81.
    COMMON NAME: Rough-haired Golden Mole.
    TYPE LOCALITY: "Towards Natal", near Durban, South Africa; see Roberts (1951:121).
    DISTRIBUTION: Eastern Cape Prov., KwaZulu-Natal, Gauteng and S Mpumalanga (South Africa).
    STATUS: IUCN – Vulnerable.
    SYNONYMS: ***dobsoni*** (Broom, 1918); ***leschae*** (Broom, 1918); ***rufopallidus*** (Roberts, 1924); ***rufus*** (Meester, 1953); ***transvaalensis*** (Broom, 1913); *pratensis* Roberts, 1913.
    COMMENTS: Validity of subspecies unclear, treatment here follows Meester et al. (1986:16)

*Cryptochloris* Shortridge and Carter, 1938. Ann. S. Afr. Mus., 32:284.
    TYPE SPECIES: *Cryptochloris zyli* Shortridge and Carter, 1938.
    COMMENTS: Simonetta (1968:31) considered *Cryptochloris* a synonym of *Chrysochloris*.

*Cryptochloris wintoni* (Broom, 1907). Ann. Mag. Nat. Hist., ser. 7, 19:264.
    COMMON NAME: De Winton's Golden Mole.
    TYPE LOCALITY: South Africa, Northern Cape Prov., Little Namaqualand, Port Nolloth.
    DISTRIBUTION: Little Namaqualand, Northern Cape Prov., South Africa.
    STATUS: IUCN – Vulnerable.

*Cryptochloris zyli* Shortridge and Carter, 1938. Ann. S. Afr. Mus., 32:284.
    COMMON NAME: Van Zyl's Golden Mole.
    TYPE LOCALITY: South Africa, Western Cape Prov., Compagnies Drift, 16 km inland from Lamberts Bay.
    DISTRIBUTION: Known only from the type locality.
    STATUS: IUCN – Critically Endangered.
    COMMENTS: Considered a subspecies of *wintoni* by Ellerman et al. (1953); however, Meester et al. (1986:18) and Helgen and Wilson (2001) argued for specific status.

*Eremitalpa* Roberts, 1924. Ann. Transvaal Mus., 10:63.
    TYPE SPECIES: *Chrysochloris granti* Broom, 1907.

*Eremitalpa granti* (Broom, 1907). Ann. Mag. Nat. Hist., ser. 7, 19:265.
    COMMON NAME: Grant's Golden Mole.

TYPE LOCALITY: South Africa: Garies, south of Kamiesberg, Little Namaqualand, Northern Cape Prov.
DISTRIBUTION: Coastal dunes from Western and Northern Cape Prov., South Africa, to Namib Desert, Namibia.
STATUS: IUCN – Vulnerable.
SYNONYMS: *cana* Broom, 1950; **namibensis** Bauer and Niethammer, 1959.
COMMENTS: Revised by Meester (1964).

**Subfamily Amblysominae** Simonetta, 1957. Arch. Ital. Anat. Embriol., 62:77.

*Amblysomus* Pomel, 1848. Arch. Sci. Phys. Nat. Geneve, 9:247.
TYPE SPECIES: *Chrysochloris hottentotus* A. Smith, 1829.
COMMENTS: Revised by Bronner (1995a, 1996).

*Amblysomus corriae* Thomas, 1905. Abstr. Proc. Zool. Soc. Lond., 1905(20):5; Proc. Zool. Soc. Lond., 2:57.
COMMON NAME: Fynbos Golden Mole.
TYPE LOCALITY: South Africa, Western Cape Prov., Knysna.
DISTRIBUTION: Western Cape Prov. from Stellenbosch/Paarl eastwards to Knysna and George (South Africa).
SYNONYMS: **devilliersi** Roberts, 1946; *swellendamensis* Roberts, 1946.
COMMENTS: *A. corriae* was previously treated as a subspecies of *A. iris*, and *devilliersi* as a subspecies of *A. hottentotus*, see Meester et al. (1986:23). Bronner (1996) showed that *iris* represents only a subspecies of *A. hottentotus*, and elevated *corriae* to species rank to include *devilliersi*.

*Amblysomus hottentotus* (A. Smith, 1829). Zool. J., 4:436.
COMMON NAME: Hottentot Golden Mole.
TYPE LOCALITY: "Interior parts of South Africa", Grahamstown, Eastern Cape Prov., South Africa.
DISTRIBUTION: Eastern Cape Prov., KwaZulu-Natal, NE Free State and Mpumalanga (South Africa); Lesotho; possibly Swaziland.
STATUS: IUCN – Lower Risk (lc) for *A. hottentotus* and *A. iris*.
SYNONYMS: *affinis* (Wagner, 1841); *albirostris* (Wagner, 1841) [*nomen dubium*]; *holosericea* (Lichtenstein, 1981); *rutilans* (Wagner, 1841); **iris** Thomas and Schwann, 1905; *littoralis* Roberts, 1946; **longiceps** (Broom, 1907); *albifrons* (Broom, 1907); **pondoliae** Thomas and Schwann, 1905; *albirostris* (Broom, 1908) [*nomen dubium*]; *natalensis* Roberts, 1946; **meesteri** Bronner, 2000.
COMMENTS: Bronner (1995b, 1996, 2000) demonstrated the existence of three cryptic species (*marleyi*, *septentrionalis* and *robustus*) in this species, as traditionally constituted (Meester 1974).

*Amblysomus marleyi* Roberts, 1931. Ann. Transvaal Mus., 14:225.
COMMON NAME: Marley's Golden Mole.
TYPE LOCALITY: South Africa, KwaZulu-Natal, Zululand, Ubombo.
DISTRIBUTION: Ubombo to Ingwavuma, KwaZulu-Natal (South Africa).
COMMENTS: Separated from *hottentotus* by Bronner (1995b, 1996, 2000).

*Amblysomus robustus* Bronner, 2000. Mammalia 64(1):42.
COMMON NAME: Robust Golden Mole.
TYPE LOCALITY: South Africa, Mpumalanga, Dullstroom, Verloren-Vallei Nat. Res.
DISTRIBUTION: Belfast to Dullstroom (Mpumalanga, South Africa).
COMMENTS: Separated from *hottentotus* by Bronner (1995b, 1996, 2000).

*Amblysous septentrionalis* Roberts, 1913. Ann. Transvaal Mus., 4:73.
COMMON NAME: Highveld Golden Mole.
TYPE LOCALITY: South Africa, Mpumalanga, Wakkerstroom.
DISTRIBUTION: Helibron and Parys (NE Free State) to Wakkerstroom and Ermelo (Mpumalanga, South Africa); possibly also Swaziland.
SYNONYMS: *drakensbergensis* Roberts, 1946; *garneri* Roberts, 1917; *orangensis* Roberts, 1946.

COMMENTS: Separated from *hottentotus* by Bronner (1995*b*, 1996, 2000).

*Calcochloris* Mivart, 1867. J. Anat. Physiol., London, 2:133.
TYPE SPECIES: *Chrysochloris obtusirostris* Peters, 1851.
SYNONYMS: *Chrysotricha* Broom, 1907; *Huetia* Forcart, 1942.
COMMENTS: Includes *Chrysotricha*, see Meester et al. (1986:23). Ellerman et al. (1953) included *Calcochloris* in *Amblysomus*. Bronner (1995*a*) included *leucorhinus* and *tytonis* in this genus.

*Calcochloris leucorhinus* (Huet, 1885). Nouv. Arch. Mus. Hist. Nat. Paris, Bull., 8:8.
COMMON NAME: Congo Golden Mole.
TYPE LOCALITY: "Gulf of Guinea Coast, Congo."
DISTRIBUTION: N Angola, Dem. Rep. Congo, Cameroon, Central African Republic.
STATUS: IUCN – Lower Risk (lc) as *Chlorotalpa leucorhina*.
SYNONYMS: *congicus* (Thomas, 1910); *luluanus* (Forcart, 1942); **cahni** (Schwarz and Mertens, 1922).
COMMENTS: Subgenus *Huetia* (see Bronner 1995*a*). Included in *Chrysochloris* by Allen (1939); included in *Amblysomus* by Simonetta (1968) and Petter (1981*a*).

*Calcochloris obtusirostris* (Peters, 1851). Bericht. Verhandl. K. Preuss. Akad. Wiss. Berlin, 16:467.
COMMON NAME: Yellow Golden Mole.
TYPE LOCALITY: Coastal Mozambique, Inhambane, 24°S.
DISTRIBUTION: Maputaland (KwaZulu-Natal) and Kruger Nat. Park (Northern Prov., South Africa); S Zimbabwe and S Mozambique.
STATUS: IUCN – Lower Risk (lc).
SYNONYMS: **chrysillus** (Thomas and Schwann, 1905); **limpopoensis** (Roberts, 1946).
COMMENTS: Subgenus *Calcochloris*.

*Calcochloris tytonis* (Simonetta, 1968). Monitore Zool. Ital., n.s., 2(suppl.):31.
COMMON NAME: Somali Golden Mole.
TYPE LOCALITY: Somalia, Giohar (= Villaggio Duca degli Abruzzi).
DISTRIBUTION: Known only from the type specimen.
STATUS: IUCN – Critically Endangered as *Chlorotalpa tytonis*.
COMMENTS: Subgenus *incertae sedis* (Bronner 1995*a*). Assigned to *Amblysomus* by Simonetta (1968:31) and Petter (1981*a*); Meester (1974) placed this species in *Chlorotalpa*.

*Neamblysomus* Roberts, 1924. Ann. Transvaal Mus., 10:64.
TYPE SPECIES: *Chrysochloris gunningi* Broom, 1908.
COMMENTS: Included in *Amblysomus* by Simonetta (1968), Meester (1974), Petter (1981*a*) and Meester et al. (1986). Bronner (1995*a*, *b*) elevated it to generic rank on the basis of cytogenetic and cranial divergence from *Amblysomus* species.

*Neamblysomus gunningi* (Broom, 1908). Ann. Transvaal Mus., 1:14.
COMMON NAME: Gunning's Golden Mole.
TYPE LOCALITY: South Africa, Limpopo Province, Woodbush Hill.
DISTRIBUTION: Woodbush Forest and New Agatha Forest Reserve, Limpopo Province, South Africa.
STATUS: IUCN – Vulnerable as *Amblysomus gunningi*.

*Neamblysomus julianae* Meester, 1972. Ann. Transvaal Mus., 28(4):35.
COMMON NAME: Juliana's Golden Mole.
TYPE LOCALITY: South Africa, Gauteng, Pretoria, The Willows.
DISTRIBUTION: Pretoria (Gauteng), Nylstroom (Limpopo Prov.) and Kruger Nat. Park (Mpumalanga, South Africa).
STATUS: IUCN – Critically Endangered as *Amblysomus julianae*.
COMMENTS: Consistent dental and fur differences between the western (Pretoria and Nylstroom) and eastern (Kruger Nat. Park) populations allude to distinct subspecies (Meester, 1972; Bronner 1990), but study of more specimens is needed to confirm this.

# ORDER MACROSCELIDEA
by Duane A. Schlitter

**ORDER MACROSCELIDEA** Butler, 1956.

**Family Macroscelididae** Bonaparte, 1838. Nuovi Ann. Sci. Nat., 2:111.
   COMMENTS: Revised by Corbet and Hanks (1968).

*Elephantulus* Thomas and Schwann, 1906. Abst. Proc. Zool. Soc. Lond., 1906(33):10.
   TYPE SPECIES: *Macroscelides rupestris* A. Smith, 1831.
   SYNONYMS: *Elephantomys* Broom, 1937; *Nasilio* Thomas and Schwann, 1906.
   COMMENTS: Includes *Nasilio*; see Corbet and Hanks (1968), and *Elephantomys*; see Meester et al.
      (1986). A key to the species was presented in Koontz and Roeper (1983) and another to
      southern African species in Meester et al. (1986).

*Elephantulus brachyrhynchus* (A. Smith, 1836). Rept. Exped. Exploring Central Africa, 1834:42
      [1836].
   COMMON NAME: Short-snouted Elephant Shrew.
   TYPE LOCALITY: "The country between Lake Latakoo and the Tropic" (= South Africa,
      Northern Cape Province, Kuruman, to S Botswana).
   DISTRIBUTION: N South Africa; NE Namibia; E and N Botswana; Angola; Zimbabwe; Malawi;
      Zambia; S Dem. Rep. Congo; Mozambique; Tanzania; Kenya and Uganda.
   STATUS: IUCN – Lower Risk (lc).
   SYNONYMS: *albiventer* (Osgood, 1910); *brachyura* (Bocage, 1882); *brevirostris* (Schinz, 1844);
      *delamerei* (Thomas, 1901); *langi* (Roberts, 1929); *luluae* (Matschie, 1926); *mababiensis*
      (Roberts, 1932); *selindensis* (Roberts, 1937); *shortridgei* (Roberts, 1929); *tzaneenensis*
      (Roberts, 1929).
   COMMENTS: Corbet (1974b:5) and Meester et al. (1986:311) regarded variation as clinal and
      did not recognize any subspecies.

*Elephantulus edwardii* (A. Smith, 1839). Illustr. Zool. S. Afr. Mamm., pl. 14.
   COMMON NAME: Cape Elephant Shrew.
   TYPE LOCALITY: South Africa, Western Cape Province, Oliphants River.
   DISTRIBUTION: W and SC South Africa.
   STATUS: IUCN – Least Concern.
   SYNONYMS: *capensis* Roberts, 1924; *edwardsii* (Sclater, 1901); *karoensis* Roberts, 1938.
   COMMENTS: Corbet and Hanks (1968:97) suggested there might be a western and eastern
      subspecies.

*Elephantulus fuscipes* (Thomas, 1894). Ann. Mag. Nat. Hist., ser. 6, 13:68.
   COMMON NAME: Dusky-footed Elephant Shrew.
   TYPE LOCALITY: Dem. Rep. Congo, Niam-Niam country, N'doruma.
   DISTRIBUTION: Uganda; NE Dem. Rep. Congo; S Sudan.
   STATUS: IUCN – Lower Risk (lc).

*Elephantulus fuscus* (Peters, 1852). Reise nach Mossambique, Säugethiere, p. 87.
   COMMON NAME: Dusky Elephant Shrew.
   TYPE LOCALITY: Mozambique, near Quelimane, Boror.
   DISTRIBUTION: Mozambique; S Malawi; SE Zambia.
   STATUS: IUCN – Lower Risk (lc).
   SYNONYMS: *malosae* (Thomas, 1898).
   COMMENTS: Regarded as distinct by Corbet (1974b:5).

*Elephantulus intufi* (A. Smith, 1836). Rept. Exped. Exploring Central Africa, 1834:42 [1836].
   COMMON NAME: Bushveld Elephant Shrew.
   TYPE LOCALITY: South Africa, North West Province, Marico District, flats beyond
      Kurrichaine.
   DISTRIBUTION: SW Angola; Namibia; Botswana; N South Africa.

STATUS: IUCN – Lower Risk (lc).

SYNONYMS: *alexandri* (Ogilby, 1838); *campbelli* Roberts, 1938; *canescens* Lundholm, 1955; *kalaharicus* Roberts, 1932; *mchughi* Roberts, 1946; *mossamedensis* Hill and Carter, 1937; *namibensis* Roberts, 1938; *omahekensis* Lehmann, 1955; *schinzi* (Noack, 1889).

COMMENTS: Corbet (1974*b*:5) and Meester et al. (1986:313) declined to recognize subspecies although Corbet and Hanks (1968:90) considered that two might be recognized with *alexandri* applicable to all but the eastern nominate population.

*Elephantulus myurus* Thomas and Schwann, 1906. Proc. Zool. Soc. Lond., 1906:586.

COMMON NAME: Eastern Rock Elephant Shrew.

TYPE LOCALITY: South Africa, Limpopo Province, Woodbush.

DISTRIBUTION: Zimbabwe; E Botswana; N, C and E South Africa; Lesotho; W Mozambique.

STATUS: IUCN – Lower Risk (lc).

SYNONYMS: *centralis* Roberts, 1946; *fitzsimonsi* Lundholm, 1955; *jamesoni* Chubb, 1909; *mapogonensis* Roberts, 1917.

COMMENTS: Corbet (1974*b*:6) and Meester et al. (1986:314) did not recognize any subspecies.

*Elephantulus revoili* (Huet, 1881). Bull. Sci. Soc. Philom. Paris, ser. 7, 5:96.

COMMON NAME: Somali Elephant Shrew.

TYPE LOCALITY: Somalia, Medjourtine.

DISTRIBUTION: N Somalia.

STATUS: IUCN – Endangered.

*Elephantulus rozeti* (Duvernoy, 1833). Mem. Soc. Hist. Nat. Strasbourg, 1(2), art. M:18.

COMMON NAME: North African Elephant Shrew.

TYPE LOCALITY: Algeria, near Oran.

DISTRIBUTION: Morocco; Algeria; Tunisia; W Libya.

STATUS: IUCN – Lower Risk (lc).

SYNONYMS: *atlantis* Thomas, 1913; *moratus* Thomas, 1913; **deserti** (Thomas, 1901); *clivorum* Thomas, 1913.

COMMENTS: Corbet and Hanks (1968:81) and Corbet (1974*b*:6) recognized two subspecies but Kawalski and Rzebik-Kawalska (1991:299) did not recognize any subspecies based on Algerian material.

*Elephantulus rufescens* (Peters, 1878). Monatsb. K. Preuss. Akad. Wiss., Berlin, 1878:198.

COMMON NAME: Rufous Elephant Shrew.

TYPE LOCALITY: Kenya, Taita, Ndi.

DISTRIBUTION: S and E Ethiopia; Kenya; E Uganda; S Sudan; N, C and W Tanzania; N and S Somalia.

STATUS: IUCN – Lower Risk (lc).

SYNONYMS: *mariakanae* Heller, 1912; **boranus** (Thomas, 1901); **dundasi** Dollman, 1910; *delicatus* Dollman, 1911; *hoogstraali* Setzer, 1956; *phaeus* Heller, 1910; *rendilis* Lönnberg, 1912; **peasei** (Thomas, 1901); **pulcher** (Thomas, 1894); *ocularis* Kershaw, 1921; *renatus* Kershaw, 1923; **somalicus** (Thomas, 1901).

COMMENTS: Corbet (1974*b*:5) provisionally recognized six subspecies based on considerable but discontinuous variation. See Koontz and Roeper (1983, Mammalian Species, 204).

*Elephantulus rupestris* (A. Smith, 1831). Proc. Zool. Soc. Lond., 1831:11.

COMMON NAME: Western Rock Elephant Shrew.

TYPE LOCALITY: South Africa or Namibia, mountains near mouth of Orange River.

DISTRIBUTION: W Namibia; SW and SC South Africa.

STATUS: IUCN – Least Concern.

SYNONYMS: *barlowi* Roberts, 1938; *gordoniensis* Roberts, 1946; *kobosensis* Roberts, 1938; *montanus* Lundholm, 1955; *okombahensis* Roberts, 1946; *tarri* Roberts, 1938; *typus* (Lesson, 1830); *vandami* Roberts, 1924.

COMMENTS: No subspecies are recognized.

*Macroscelides* A. Smith, 1829. Zool. J. Lond., 4:435.

TYPE SPECIES: *Macroscelides typus* A. Smith, 1829 (= *Sorex proboscideus* Shaw, 1800).

SYNONYMS: *Eumerus* I. Geoffroy, 1829; *Macroscelis* Fisher, 1830; *Rhinomys* Lichtenstein, 1831.

*Macroscelides proboscideus* (Shaw, 1800). Gen. Zool. Syst. Nat. Hist., 1(2), Mammalia, p. 536.
COMMON NAME: Short-eared Elephant Shrew.
TYPE LOCALITY: South Africa, Western Cape Province, Oudtshoorn Div., Roodeval.
DISTRIBUTION: W, NW and SC South Africa; S Namibia.
STATUS: IUCN – Least Concern.
SYNONYMS: *ausensis* Roberts, 1938; *brandvleiensis* Roberts, 1938; *calviniensis* Roberts, 1938; *chiversi* Roberts, 1933; *harei* Roberts, 1938; *hewitti* Roberts, 1929; *isabellinus* Shortridge and Carter, 1938; *jaculus* (Lichtenstein, 1831); *langi* Roberts, 1933; *melanotis* Ogilby, 1838; *typicus* A. Smith, 1839; *typus* A. Smith, 1829; **flavicaudatus** Lundholm, 1955.
COMMENTS: Meester et al. (1986:310) listed two subspecies.

*Petrodromus* Peters, 1846. Bericht Verhandl. K. Preuss. Akad. Wiss. Berlin, 11:258.
TYPE SPECIES: *Petrodromus tetradactylus* Peters, 1846.
SYNONYMS: *Cercoctenus* Hollister, 1916; *Mesoctenus* Thomas, 1918.

*Petrodromus tetradactylus* Peters, 1846. Bericht Verhandl. K. Preuss. Akad. Wiss. Berlin, 11:258.
COMMON NAME: Four-toed Elephant Shrew.
TYPE LOCALITY: Mozambique, Tette.
DISTRIBUTION: Mozambique; Tanzania (including Mafia and Zanzibar); SE Kenya; S Uganda; Zambia; Malawi; SE Zimbabwe; Dem. Rep. Congo; E Republic of Congo; NE Angola; E South Africa.
STATUS: IUCN – Not Evaluated as *P. t. sangi*; otherwise IUCN – Not listed.
SYNONYMS: *matschiei* Neumann 1900; *occidentalis* Roberts, 1913; *robustus* Thomas, 1918; *venustus* Thomas, 1903; **beirae** Roberts, 1913; *rovumae* Thomas, 1897; *mossambicus* Thomas, 1918; *nigriseta* Neumann, 1900; **schwanni** Thomas and Wroughton, 1907; **sultani** Thomas, 1897; *sangi* Heller, 1912; **swynnertoni** Thomas, 1918; **tordayi** Thomas, 1910; *tumbanus* Kershaw, 1923; **warreni** Thomas, 1918; **zanzibaricus** Corbet and Neal, 1965.
COMMENTS: Corbet (1974*b*:2) and Meester et al. (1986:309) recognized nine subspecies. Corbet (1974*b*:3) suggested that *matschiei* and *robustus* might be distinct subspecies. It is possible that *schwanni* and *tordayi* are separate species.

*Rhynchocyon* Peters, 1847. Bericht Verhandl. K. Preuss. Akad. Wiss. Berlin, 12:36.
TYPE SPECIES: *Rhynchocyon cirnei* Peters, 1847.
SYNONYMS: *Rhinonax* Thomas, 1918.
COMMENTS: A key to the species was presented in Rathbun (1979).

*Rhynchocyon chrysopygus* Günther, 1881. Proc. Zool. Soc. Lond., 1881:164.
COMMON NAME: Golden-rumped Elephant Shrew.
TYPE LOCALITY: Kenya, Mombasa.
DISTRIBUTION: E Kenya.
STATUS: IUCN – Endangered.
COMMENTS: Kingdon (1974*a*:41) considered *chrysopygus* as a subspecies of *cirnei*. See Rathbun (1979, Mammalian Species, 117).

*Rhynchocyon cirnei* Peters, 1847. Bericht Verhandl. K. Preuss. Akad. Wiss. Berlin, 12:37.
COMMON NAME: Checkered Elephant Shrew.
TYPE LOCALITY: Mozambique, Bororo Dist., Quelimane.
DISTRIBUTION: N Mozambique; Malawi, S and SW Tanzania, NE Zambia, N and E Dem. Rep. Congo, Uganda.
STATUS: IUCN – Not Evaluated as *R. c. cirnei* and *R. c. hendersoni*; otherwise Vulnerable.
SYNONYMS: **hendersoni** Thomas, 1902; **macrurus** Günther, 1881; *melanurus* Neumann, 1900; **reichardi** Reichenow, 1886; *swynnertoni* Kershaw, 1923; **shirensis** Corbet and Hanks, 1968; **stuhlmanni** Matschie, 1893; *claudi* Thomas and Wroughton, 1907; *nudicaudata* Lydekker, 1906.
COMMENTS: Six isolated forest-inhabiting subspecies are recognized by Corbet and Hanks (1968:57) and Corbet (1974*b*:2). Includes *stuhlmanni*, which could be a distinct species according to Corbet and Hanks (1968:63).

*Rhynchocyon petersi* Bocage, 1880. J. Sci. Math. Phys. Nat. Lisboa, ser. 1, 7:159.

COMMON NAME: Black and Rufous Elephant Shrew.

TYPE LOCALITY: Tanzania, mainland opposite Zanzibar.

DISTRIBUTION: E Tanzania (including Mafia and Zanzibar); SE Kenya.

STATUS: IUCN – Endangered.

SYNONYMS: *fischeri* Neumann, 1900; *usambarae* Neumann, 1900; ***adersi*** Dollman, 1912.

COMMENTS: Two subspecies are listed by Corbet and Hanks (1968:64) and Corbet (1974*b*:2). Kingdon (1974*a*:41) considered *petersi* a subspecies of *cirnei*.

# ORDER TUBULIDENTATA
by Duane A. Schlitter

**ORDER TUBULIDENTATA** Huxley, 1872.

**Family Orycteropodidae** Gray, 1821. London Med. Repos., 15:305.

*Orycteropus* G. Cuvier, 1798. Tabl. Elem. Hist. Nat. Anim., 1798:144.
   TYPE SPECIES: *Myrmecophaga capensis* Gmelin, 1788 (= *Myrmecophaga afra* Pallas, 1766).
   COMMENTS: Sherborn (1902:701) and Allen (1939:270) gave *Orycteropus* "Geoffroy, Decad. Phil. et Litt. XXVIII. 1795", but it is untraceable. Meester et al. (1986:182) reviewed and assigned the correct generic name.

*Orycteropus afer* (Pallas, 1766). Misc. Zool., p. 64.
   COMMON NAME: Aardvark.
   TYPE LOCALITY: South Africa, Western Cape Prov., Cape of Good Hope.
   DISTRIBUTION: Savannah zones of West Africa to E Sudan, Ethiopia and Eritrea; Kenya; Somalia; N and W Uganda to Tanzania; Rwanda; N, E, and C Dem. Rep. Congo; W Angola; Namibia; Botswana; Zimbabwe; Zambia; Mozambique; South Africa.
   STATUS: IUCN – Least Concern.
   SYNONYMS: *albicaudus* Rothschild, 1907; *capensis* (Gmelin, 1788); *adametzi* Grote, 1921; *aethiopicus* Sundevall, 1843; *angolensis* Zukowsky and Haltenorth, 1957; *erikssoni* Lönnberg, 1906; *faradjius* Hatt, 1932; *haussanus* Matschie, 1900; *kordofanicus* Rothschild, 1927; *lademanni* Grote, 1911; *leptodon* Hirst, 1906; *matschiei* Grote, 1921; *observandus* Grote, 1921; *ruvanensis* Grote, 1921; *senegalensis* Lesson, 1840; *senegalensis,* Schinz, 1845; *somalicus* Lydekker, 1908; *wardi* Lydekker, 1908; *wertheri* Matschie, 1898.
   COMMENTS: Reviewed by Melton (1976), Pocock (1924), and Shoshani et al. (1988, Mammalian Species, 300). Seventeen poorly defined subspecies are recognized (Meester, 1972b; Meester et al., 1986).

# ORDER HYRACOIDEA
by Jeheskel Shoshani

**ORDER HYRACOIDEA** Huxley, 1869.

SYNONYMS: Hyracea Haeckel, 1895; Hyraciformes Kinman, 1994; Laminungula Gray, 1869; Lamnungia Van der Hoeven, 1858; Lamnunguia Illiger, 1811 [*nomen oblitum*]; Procaviata Imamura, 1961.

COMMENTS: Traditionally (since Huxley, 1869) the category or rank of "Order" has been used for Hyracoidea. In 1997, McKenna and Bell, following cladistic classification, proposed a new category – "Suborder" for Hyracoidea, in the "Order" Uranotheria McKenna and Bell (1997). For stability I retain the ordinal category, recognizing the cladistic message implied by the McKenna and Bell arrangement.

**Family Procaviidae** Thomas, 1892. Proc. Zool. Soc. Lond., 1892:51.

SYNONYMS: Hyracida Haeckel, 1866; Hyracidae Gray, 1821; Procaviinae Whitworth, 1954; Procavioidea Kalandadze and Rautian, 1992.

COMMENTS: Hyracidae Gray, 1821, is a group name based on *Hyrax* Hermann, 1783. Revised by Hahn (1934:207). Roche (1972) retained only *Procavia* and *Dendrohyrax* but Hoeck (1978) and Meester et al. (1986:178) retained *Procavia*, *Heterohyrax*, and *Dendrohyrax* as separate genera. A modern key to the genera was developed by Meester et al. (1986).

*Dendrohyrax* Gray, 1868. Ann. Mag. Nat. Hist., ser. 4, 1:48.

TYPE SPECIES: *Hyrax arboreus* A. Smith, 1827.

COMMENTS: A key to species was provided by C. Jones (1978). Schlitter (1993) followed previous workers in recognizing three species of Tree Hyraxes. The validity of *D. validus* as a species distinct from *D. arboreus* was questioned by Bothma (1971, see comments in Schlitter, 1993:373). I briefly examined skeletal and skin specimens of *D. arboreus, D. dorsalis,* and *D. validus* at the National Museum of Natural History (Washington D.C.) and at the American Museum of Natural History (New York). Until a detailed study is conducted to evaluate the validity of *Dendrohyrax* species, I believe that *D. arboreus* and *D. dorsalis* may be valid separate species, but *D. validus* is not.

*Dendrohyrax arboreus* (A. Smith, 1827). Trans. Linn. Soc. Lond., 15:468.

COMMON NAME: Southern Tree Hyrax.

TYPE LOCALITY: South Africa, Western Cape Prov., forests of Cape of Good Hope.

DISTRIBUTION: Western Cape Prov., Eastern Cape Prov., and KwaZulu-Natal (South Africa); Mozambique; Zambia; Malawi; Dem. Rep. Congo; Tanzania to Kenya and Sudan.

STATUS: IUCN – Vulnerable as *D. validus* and as *D. arboreus* South African subpopulation only, otherwise Lower Risk (lc).

SYNONYMS: *adersi* Kershaw, 1924; *adolfi-friederici* (Brauer, 1913); *bettoni* (Thomas and Schwann, 1904); *braueri* Hahn, 1933; *crawshayi* (Thomas, 1900); *helgei* (Lönnberg and Gyldenstolpe, 1925); *laikipia* Dollman, 1911; *mimus* (Thomas, 1900); *neumanni* (Matschie, 1893); *ruwenzorii* (Neumann, 1902); *scheelei* (Matschie, 1895); *scheffleri* (Brauer, 1913); *schubotzi* (Brauer, 1913); *schusteri* Brauer, 1917; *stuhlmanni* (Matschie, 1892); *terricola* Mollison, 1905; *validus* True, 1890; *vilhelmi* (Lönnberg, 1916); *vosseleri* Brauer, 1917.

COMMENTS: Includes *validus*; no subspecies are recognized until detailed study is conducted.

*Dendrohyrax dorsalis* (Fraser, 1855). Proc. Zool. Soc. Lond., 1854:99 [1855].

COMMON NAME: Western Tree Hyrax.

TYPE LOCALITY: Equatorial Guinea, Bioko.

DISTRIBUTION: West and Central Africa from Senegal, Gambia to N Angola; Bioko (Equatorial Guinea); C and NE Dem. Rep. Congo; N Uganda.

STATUS: IUCN – Lower Risk (lc).

SYNONYMS: **emini** Thomas, 1887; *beniensis* Brauer, 1917; *brevimaculatus* Brauer, 1917; *congoensis* Brauer, 1917; *rubriventer* Brauer, 1917; **latrator** (Thomas, 1910); **marmota** (Thomas, 1901); **nigricans** (Peters, 1879); *adametzi* (Brauer, 1912); *tessmanni* (Brauer,

1912); *zenkeri* Brauer, 1914; **sylvestris** (Temminek, 1853); *aschantiensis* (Brauer, 1914); *stampflii* (Jentink, 1886).
COMMENTS: See C. Jones (1978, Mammalian Species, 113).

*Heterohyrax* Gray, 1868. Ann. Mag. Nat. Hist., ser. 4, 1:50.
TYPE SPECIES: *Dendrohyrax blainvillii* Gray, 1868 (= *Hyrax brucei* Gray, 1868).
COMMENTS: Included as a subgenus of *Dendrohyrax* by Roche (1972).

*Heterohyrax brucei* (Gray, 1868). Ann. Mag. Nat. Hist., ser. 4, 1:44.
COMMON NAME: Yellow-spotted Rock Hyrax.
TYPE LOCALITY: Ethiopia (= Abyssinia).
DISTRIBUTION: Egypt to Somalia to southern Africa to WC Angola, with 'pockets' in C Sahara.
STATUS: IUCN – Vulnerable as *H. antineae* and *H. chapini,* Lower Risk (lc) as *H. brucei.*
SYNONYMS: *blainvillii* (Gray, 1868); *irroratus* (Gray, 1869); **albipes** Hollister, 1922; **antineae** Heim de Balsac and Bégouen, 1932; **bakeri** (Gray, 1874); **bocagei** (Gray, 1869); *grayi* (Bocage, 1889); **chapini** (Hatt, 1933); **dieseneri** Brauer, 1917; **frommi** (Brauer, 1913); **granti** (Wroughton, 1910); **hindei** (Wroughton, 1910); *maculata* (Osgood, 1910); **hoogstraali** Setzer, 1956; **kempi** (Thomas, 1910); **lademanni** Brauer, 1917; **manningi** (Wroughton, 1910); **mossambicus** (Peters, 1870); **muenzneri** (Brauer, 1913); *ruckwaensis* Brauer, 1917; **princeps** (Thomas, 1910); *arboricola* Brauer, 1917; **prittwitzi** Brauer, 1917; **pumilus** (Thomas, 1910); **ruddi** (Wroughton, 1910); *rhodesiae* Roberts, 1946; **rudolfi** (Thomas, 1910); *borana* (Lönnberg, 1912); **somalicus** (Thomas, 1892); *hararensis* Brauer, 1917; *webensis* Brauer, 1917; **ssongeae** Brauer, 1917; **thomasi** (Neumann, 1901); **victorianjansae** Brauer, 1917.
COMMENTS: See Barry and Shoshani (2000, Mammalian Species, 645).

*Procavia* Storr, 1780. Prodr. Meth. Mamm., p. 40.
TYPE SPECIES: *Cavia capensis* Pallas, 1766.
SYNONYMS: *Euhyrax* Gray, 1868; *Hyrax* Hermann, 1783; *Procauia* Storr, 1780.
COMMENTS: See Bothma (1971), McKenna and Bell (1997).

*Procavia capensis* (Pallas, 1766). Misc. Zool., p. 30.
COMMON NAME: Rock Hyrax.
TYPE LOCALITY: South Africa, Western Cape Prov., Cape of Good Hope.
DISTRIBUTION: Sub-Saharan and NE Africa (a line from Senegal through S Algeria and Libya, Egypt to southern most tip of Africa), portion of the Levant (Syria, Lebanon, Turkey, Israel), and the Arabian peninsula (Saudi Arabia, Yemen); isolated mountains in Algeria and Libya.
STATUS: IUCN – Lower Risk (lc).
SYNONYMS: *albaniensis* Roberts, 1946; *chiversi* Roberts, 1937; *coombsi* Roberts, 1924; *griquae* Roberts, 1946; *klaverensis* Roberts, 1946; *letabae* Roberts, 1937; *marlothi* Brauer 1914; *natalensis* Roberts, 1924; *orangiae* Roberts, 1937; *reuningi* Brauer 1914; *schultzei* Brauer 1914; *semicircularis* (Gray, 1869); *vanderhorsti* Roberts, 1946; *waterbergensis* Brauer 1914; *windhuki* Brauer 1914; **bamendae** Brauer, 1913; **capillosa** Brauer, 1917; **erlangeri** Neumann, 1901; *comata* Brauer, 1917; **habessinicus** (Hemprich and Ehrenberg, 1832); *abyssinicus* (Gray, 1868); *alpini* (Gray, 1868); *ferrugineus* (Gray, 1869); *luteogaster* (Gray, 1869); *meneliki* Neumann, 1902; **jacksoni** Thomas, 1900; *daemon* Thomas, 1910; *varians* Granvik, 1925; **jayakari** Thomas, 1892; **johnstoni** Thomas, 1894; **kerstingi** Matschie, 1899; *elberti* Brauer, 1917; *goslingi* Thomas, 1905; *ituriensis* Brauer, 1917; *kamerunensis* Brauer, 1913; *lopesi* Thomas and Wroughton, 1907; *naumanni* Brauer, 1917; *oweni* Thomas, 1911; **mackinderi** Thomas, 1900; *zelotes* Osgood, 1910; **matschiei** Neumann, 1900; **pallida** Thomas, 1891; *minor* Thomas, 1892; **ruficeps** (Hemprich and Ehrenberg, 1832); *bounhioli* Kollman, 1912; *buchanani* Thomas and Hinton, 1921; *burtonii* (Gray, 1868); *dongolanus* (Blanford, 1870); *ebneri* Wettstein, 1916; *latastei* Thomas, 1892; *marrensis* Thomas and Hinton, 1923; *slatini* Sassi, 1906; **scioanus** (Giglioli, 1888); *butleri* Wroughton, 1911; **shoana** Thomas 1892; **sharica** Thomas and Wroughton, 1907; *melfica* Mertens, 1929; **syriacus** (Schreber, 1784);

*ehrenbergi* Brauer, 1917; *schmitzi* Brauer 1917; *sinaiticus* (Gray, 1868); **welwitschii** (Gray, 1868); *flavimaculata* Brauer, 1917; *otjiwarongensis* (Roberts, 1946); *tsumebensis* (Roberts, 1946); *volkmanni* Brauer, 1914.

COMMENTS: See Allen (1939), Meester et al. (1986), Olds and Shoshani (1982, Mammalian Species, 171).

# ORDER PROBOSCIDEA
by Jeheskel Shoshani

## ORDER PROBOSCIDEA Illiger, 1811.

SYNONYMS: Proboscidiae Gray, 1821; Probosciformes Kinman, 1994.

COMMENTS: Traditionally (since Illiger, 1811) the category or rank of "Order" has been used for Proboscidea. In 1997 McKenna, Bell et al., following cladistic classification, proposed a new category – "Parvorder" for PROBOSCIDEA. For 'stability' I retained the ordinal category, even though the category parvorder conveys a cladistic message.

**Family Elephantidae** Gray, 1821. London Med. Repos., 15:305.

SYNONYMS: Elephantida Haeckel, 1866.

COMMENTS: Revised by Maglio (1973).

*Elephas* Linnaeus, 1758. Syst. Nat., 10th ed., 1:33.

TYPE SPECIES: *Elephas maximus* Linnaeus, 1758.

SYNONYMS: *Elephantus* É. Geoffroy Saint-Hilaire and Cuvier, 1795 [not Cuvier and Geoffroy Saint-Hilaire].

*Elephas maximus* Linnaeus, 1758. Syst. Nat., 10th ed., 1:33.

COMMON NAME: Asian Elephant.

TYPE LOCALITY: "Zeylonae" [Sri Lanka].

DISTRIBUTION: Thirteen countries in SE Asia from India in the west to Borneo in the east.

STATUS: CITES – Appendix I; U.S. ESA and IUCN – Endangered.

SYNONYMS: *asiaticus* Blumenbach, 1797; *ceylanicus* de Blainville, 1845; *vilaliya* Deraniyagala, 1939; *zeylanicus* Lydekker 1907; ***indicus*** Cuvier, 1798; *asurus* Deraniyagala, 1950; *bengalensis* de Blainville, 1845; *birmanicus* Deraniyagala, 1951; *borneensis* Deraniyagala, 1950; *dakhunensis* Deraniyagala, 1950; *dauntela* Falconer and Cautley 1847; *gigas* Perry, 1811; *heterodactylus* Hodgson, 1841 [*nomen nudum*]; *hirsutus* Lydekker 1914; *isodactylus* Hodgson, 1841 [*nomen nudum*]; *mukna* Falconer and Cautley, 1847; *persicus* Deraniyagala, 1950; *ruber* Deraniyagala, 1951; *rubridens* Deraniyagala 1950; *sichiaoshanensis* Wang J-k, 1978; *sondaicus* Deraniyagala, 1953; ***sumatranus*** Temminck, 1847.

COMMENTS: See Shoshani and Eisenberg (1982, Mammalian Species, 182), who identified three subspecies of the Asian elephant: *E. m. sumatranus* from the island of Sumatra, *E. m. indicus* from mainland Asia, and *E. m. maximus* from the island of Sri Lanka. See also Deraniyagala (1955). Colin Groves (pers. comm., 2002) suggested that based on small measurements and restricted ear depigmentation, the Malay elephant (*hirsutus* Lydekker 1914) and the Borneo elephant (*borneensis* Deraniyagala, 1950) should be synonyms of *sumatranus* Temminck, 1847. Similarly, based on geographic grounds, the Javan elephant (*sondaicus* Deraniyagala, 1953) should be a synonym of *sumatranus* Temminck, 1847. This is not followed because the Sumatran elephant is distinguished from other Asian subspecies by its 20 instead of 19 pairs of ribs. In addition, the elephants of Borneo are believed to be feral descendants introduced in the 1750's (details in Shoshani and Eisenberg, 1982). *E. m. sondaicus* was designated by subfossil tooth from Java (Deraniyagala, 1955:41), no other data such as number of ribs is given. Based on DNA isolated from dung, Fernando et al. (2003) concluded that elephants from Sabah and Sarawak (Borneo) are "genetically distinct, with molecular divergence indicative of a Pleistocene colonization of Borneo and subsequent isolation." These authors suggested, however "that a formal reinstatement of the subspecies *E. m. borneensis* await a detailed morphological analysis of Borneo elephants and their comparison with other populations." *E. m. borneensis* was first described by Deraniyagala in 1950. I concur with Fernando et al.'s (2003) opinion that there should also be morphological differences among the recognized Asian elephant subspecies. Characters suggested by Deraniyagala (1955) and by Shoshani (2000) include: overall body size, ear size, tusk size and shape (e.g., straight vs. curved), number of ribs (20 vs. 19 pairs), amount of bodily depigmentation, and habitat (forest vs. savanna).

*Loxodonta* Anonymous, 1827 (not Cuvier, 1825; see comments). Zoology J., 3:140.

TYPE SPECIES: *Elephas africanus* Blumenbach, 1797.

SYNONYMS: *Loxodon* Falconer, 1857.

COMMENTS: The spelling in the original publication was "Loxodonte" [F. Cuvier, 1825, *in* E. Geoffroy St.-Hilaire and F. Cuvier, *Hist. Nat. Mammifères*, 3(52):2]. "Loxodonte" was latinized in 1827 (author unknown) to read *Loxodonta,* and has been accepted in this form. Following Article 11 of the International Code of Zoological Nomenclature (International Commission on Zoological Nomenclature, 1999), the format of "*Loxodonta* Anonymous, 1827" is accepted. See Laursen and Bekoff (1978, Mammalian Species, 92) and Deraniyagala (1955).

*Loxodonta africana* (Blumenbach, 1797). Handb. Naturgesch., 5th ed., p. 125.

COMMON NAME: African Bush Elephant.

TYPE LOCALITY: Restricted to the Orange River, South Africa by Pohle (1926; see Allen, 1939).

DISTRIBUTION: Sub-Saharan, except C and W coast of Africa, including 30 countries from Senegal in the west to Somalia in the east.

STATUS: CITES – Appendix II for Botswana, Namibia, South Africa, and Zimbabwe, Appendix I for other African countries; U.S. ESA – Threatened; IUCN – Endangered.

SYNONYMS: *angolensis* Frade, 1928; *berbericus* Seurat, 1930 [*nomen nudum*]; *capensis* (G. Cuvier, 1798); *cavendishi* (Lydekker, 1907); *cornaliae* (Aradas, 1870); *hannibali* Deraniyagala, 1953 [*nomen nudum*]; *knochenhaueri* (Matschie, 1900); *mocambicus* (Frade, 1924); *orleansi* (Lydekker, 1907); *oxyotis* Matschie, 1900; *peeli* (Lydekker, 1907); *pharaohensis* Deraniyagala, 1948; *rothschildi* (Lydekker, 1907); *selousi* (Lydekker, 1907); *toxotis* (Lydekker, 1907); *typicus* Blumenbach, 1797; *zukowskyi* Strand, 1924.

COMMENTS: See Laursen and Bekoff (1978, Mammalian Species, 92) and Deraniyagala (1955). The name *cornaliae* (Aradas, 1870) is based on a *Loxodonta* molar from Catania, Sicily, and inferentially was a Carthaginian import (C. Groves, pers comm., 2002). The North African names (*berbericus, hannibali, pharaohensis*) were placed in this synonymy instead of under *L. cyclotis* per suggestion of Colin Groves (pers. comm., 2002).

*Loxodonta cyclotis* (Matschie, 1900). Sitzb. Ges. Naturf. Fr. Berlin, p. 194.

COMMON NAME: African Forest Elephant.

TYPE LOCALITY: Yaunde, S Cameroon.

DISTRIBUTION: C and W coast of Africa, including 21 countries from Senegal in the west to Uganda in the east.

STATUS: CITES – Appendix I (as included in *L. africana*); U.S. ESA – Threatened (as included in *L. africana*); IUCN – Endangered (as included in *L. africana*).

SYNONYMS: *albertensis* (Lydekker, 1907); *cottoni* (Lydekker, 1908); *fransseni* (Schouteden, 1914); *pumilio* (Noack, 1906).

COMMENTS: See Laursen and Bekoff (1978, Mammalian Species, 92), where *cyclotis* was treated as a subspecies of *L. africana*. Grubb et al. (2000) presented morphological and some molecular data in support of upgrading *cyclotis* to a species, separate species from *africana*, corroborating earlier hypothesis that *Loxodonta cyclotis* and *L. africana* are distinct species. Further, Grubb et al. (2000) supported the hypothesis that *L. cyclotis* is morphologically more primitive than *L. africana*. Roca et al. (2001) provided genetic evidence for two species in Africa. Evidence provided by Debruyne (2003), however, suggests that the African Forest Elephant and the African Bush Elephant are only subspecifically distinct – this taxonomic question has not yet been resolved.

# ORDER SIRENIA
by Jeheskel Shoshani

**ORDER SIRENIA** Illiger, 1811.
SYNONYMS: Halobioidea Ameghino, 1889; Herbivorae Gray, 1821; Phycoceta Haeckel, 1866; Sirenoidea van Beneden, 1855; Sireniformes Kinman, 1994; Trichechiformes Hay, 1923.
COMMENTS: Traditionally (since Illiger, 1811) the category or rank of "Order" has been used for Sirenia. In 1997 McKenna and Bell, following cladistic classification, proposed a new category – "Infraorder" for Sirenia, in the "Suborder" Tethytheria McKenna, 1975, "Order" Uranotheria McKenna and Bell, 1997. For 'stability' I retained the ordinal category, even though infraorder category conveys a cladistic message. For thorough index, complete synonymy, and bibliography of Sirenia, see Domning (1996).

**Family Dugongidae** Gray, 1821. London Med. Repos., 15:309.
SYNONYMS: Halicoridae Gray, 1825. See McKenna and Bell (1997).

**Subfamily Dugonginae** Gray, 1821. London Med. Repos., 15:309.

*Dugong* Lacépède, 1799. Tab. Div. Subd. Orders Genres Mammiféres, 14:17.
TYPE SPECIES: *Dugong indicus* Lacépède, 1799 (= *Trichechus dugon* Müller, 1776).
SYNONYMS: *Amblychilus* Fischer von Waldheim, 1814; *Dugongidus* Gray, 1821; *Dugungus* Tiedemann, 1808; *Halicore* Illiger, 1811; *Platystomus* G. Fischer, 1803.
COMMENTS: See Husar (1978, Mammalian Species, 88), McKenna and Bell (1997). *Halicora* Fleming, 1822 is an incorrect subsequent spelling of *Halicore* Illiger.

*Dugong dugon* (Müller, 1776). Linne's Vollstand. Natursyst. Suppl., p. 21.
COMMON NAME: Dugong.
TYPE LOCALITY: Cape of Good Hope to the Philippines.
DISTRIBUTION: Tropical coastal waters of Indian and W Pacific Oceans.
STATUS: CITES – Appendix I; U.S. ESA – Endangered except in Palau, where it is Proposed Endangered; IUCN – Vulnerable.
SYNONYMS: *australis* (Retzius, 1794); *cetacea* (Illiger, 1815); *dugong* (Gmelin, 1788); *dugung* (Erxleben, 1777); *hemprichii* (Ehrenberg, 1832); *indicus* (Boddaert, 1785); *lottum* (Ehrenberg, 1832); *malayana* (Owen, 1875) [*nomen nudum*; *lapsus*?, see Domming, 1996]; *syren* (Brookes, 1828); *tabernaculi* (Rüppell, 1834).
COMMENTS: Reviewed by Husar (1978, Mammalian Species, 88).

**Subfamily Hydrodamalinae** Palmer, 1895. Science, n.s.2(40):450.

*Hydrodamalis* Retzius, 1794. K. Svenska Vet.-Akad. Handl. Stockholm, 15, p. 292.
TYPE SPECIES: *Hydrodamalis stelleri* Retzius, 1794 (= *Manati gigas* Zimmermann, 1780).
SYNONYMS: *Haligyna* Billberg, 1827 [see Doming, 1996]; *Manati* Zimmermann, 1780; *Nepus* Fischer von Waldheim, 1814; *Rytina* Illiger, 1811; *Sirene* Link, 1794; *Stellera* Bowdich, 1821; *Stellerus* Desmarest, 1822.
COMMENTS: See Domning (1978, 1996), McKenna and Bell (1997). *Rhytina* Berthold, 1827 is an unjustified emendation of *Rytina* Illiger, and *Rhytine* Burmeister, 1837 is an emendation of *Rhytina*.

*Hydrodamalis gigas* (Zimmermann, 1780). Geogr. Gesch. Mensch. Vierf. Thiere, 2:426.
COMMON NAME: Steller's Sea Cow.
TYPE LOCALITY: Bering Sea, Commander Isls, Bering Isl.
DISTRIBUTION: Known only from the Commander Isls, Bering Sea.
STATUS: IUCN – Extinct.
SYNONYMS: *balaenurus* (Boddaert, 1785); *borealis* (Gmelin, 1788); *cetacea* (Illiger, 1815); *stelleri* Retzius, 1794.
COMMENTS: See Forsten and Youngman (1982, Mammalian Species, 165), Domning (1978, 1996).

**Family Trichechidae** Gill, 1872. Smithson. Misc. Coll., 11(1):14.
    SYNONYMS: Manatida Haeckel, 1866; Manatidae Gray, 1821; Manatina C. L. Bonaparte, 1837; Manatoidea Gill, 1872; Trichechoidea Giebel, 1847; Trichecida Haeckel, 1866.
    COMMENTS: See Domning (1996). McKenna and Bell (1997:496) listed "Giebel, 1847" as the author and year for Trichechidae; Susan Bell (pers. comm., 2002) confirmed that Gill, 1872 is the correct entry.

*Trichechus* Linnaeus, 1758. Syst. Nat., 10th ed., 1:34.
    TYPE SPECIES: *Trichechus manatus* Linnaeus, 1758.
    SYNONYMS: *Halipaedisca* Gistel, 1848; *Manatus* Brünnich, 1771; *Oxystomus* G. Fischer von Waldheim, 1803; *Trichecus* Oken, 1816.
    COMMENTS: Revised by Hatt (1934*a*); evolutionary history summarized by Domning (1982). See McKenna and Bell (1997); Susan Bell (pers. comm., 2002) confirmed that Brünnich, 1771 is the correct authority for *Manatus*.

*Trichechus inunguis* (Natterer, 1883). *In* Pelzeln, Verh. Zool.-Bot. Ges. Wien, 33:89.
    COMMON NAME: Amazonian Manatee.
    TYPE LOCALITY: Brazil, Amazonas, Rio Madeira, Borba.
    DISTRIBUTION: Amazon basin of Brazil, Colombia, Ecuador, Guyana, and Peru.
    STATUS: CITES – Appendix I; U.S. ESA – Endangered; IUCN – Vulnerable.
    SYNONYMS: None.
    COMMENTS: Reviewed by Husar (1977, Mammalian Species, 72).

*Trichechus manatus* Linnaeus, 1758. Syst. Nat., 10th ed., 1:34.
    COMMON NAME: West Indian Manatee.
    TYPE LOCALITY: "Mari Americano"; restricted by Thomas (1911*a*) to "West Indies."
    DISTRIBUTION: Caribbean coastal areas and river systems from Virginia, USA to Espírito Santo, Brazil.
    STATUS: CITES – Appendix I; U.S. ESA – Endangered; IUCN – Vulnerable.
    SYNONYMS: *amazonius* Shaw, 1800; *americanus* Link, 1795; *antillarum* Link, 1795; *clusii* (Pennant, 1793); *fluviatilis* (Olfers, 1818); *guyannensis* (Bechstein, 1800); *koellikeri* (Kükenthal, 1887); *latirostris* (Harlan, 1824); *minor* (Daudin, 1802); *oronocensis* (Bechstein, 1800).
    COMMENTS: See Domning (1981, 1996); reviewed by Husar (1978, Mammalian Species, 93).

*Trichechus senegalensis* Link, 1795. Beitr. Naturgesch., 1(2):209.
    COMMON NAME: African Manatee.
    TYPE LOCALITY: Senegal.
    DISTRIBUTION: Coastal W Africa including river systems from Angola to Senegal.
    STATUS: CITES – Appendix II; U.S. ESA – Threatened; IUCN – Vulnerable.
    SYNONYMS: *australis* Gmelin, 1788; *nasutus* (Wyman, 1848); *oweni* (Du Chaillu, 1861); *sphaerurus* (Illiger, 1815); *stroggylonurus* (Bechstein, 1800); *vogelii* (Owen, 1856).
    COMMENTS: Reviewed by Husar (1978, Mammalian Species, 89). *Trichechus manatus australis* Gmelin, 1788 is a partial synonym, that was restricted to the African manatee by Shaw (1800) and Hatt (1934*a*); see Domning (1996).

# ORDER CINGULATA
by Alfred L. Gardner

**ORDER CINGULATA** Illiger, 1811.
> COMMENTS: Included in Xenarthra by Gardner (1993); reviewed as part of Xenarthra by Kraft (1995).

**Family Dasypodidae** Gray, 1821. London Med. Repos., 15:305.
> COMMENTS: Wetzel (1985*b*) divided the Dasypodidae into two subfamilies: Chlamyphorinae (monotypic) and Dasypodinae (with four tribes: Dasypodini, Euphractini, Priodontini, and Tolypeutini). McKenna and Bell (1997) divided Wetzel's Dasypodidae into three subfamilies: Dasypodinae (monotypic), Euphractinae (with tribes Chlamyphorini and Euphractini), and Tolypeutinae (with tribes Tolypeutini and Priodontini).

**Subfamily Dasypodinae** Gray, 1821. London Med. Repos., 15:305.

*Dasypus* Linnaeus, 1758. Syst. Nat., 10th ed., 1:50.
> TYPE SPECIES: *Dasypus novemcinctus* Linnaeus, 1758, by Linnaean tautonomy.
> SYNONYMS: *Cachicamus* McMurtrie, 1831; *Cataphractus* Storr, 1780; *Cryptophractus* Fitzinger, 1856; *Hyperoambon* Peters, 1864; *Loricatus* Desmarest, 1804; *Muletia* Gray, 1874; *Praopus* Burmeister, 1854; *Tatu* Blumenbach, 1779; *Tatusia* Lesson, 1827; *Zonoplites* Gloger, 1841.
> COMMENTS: Reviewed by Wetzel (1985*b*) and Wetzel and Mondolfi (1979). *Cachicama* Gervais *in* I. Geoffroy, 1835, is an invalid emendation of *Cachicamus* McMurtrie; *Tatus* Olfers, 1818, and *Tatua* Robinson and Lyon, 1901, are incorrect subsequent spellings of *Tatu* Blumenbach.

*Dasypus hybridus* (Desmarest, 1804). Tabl. Meth. Hist. Nat., *in* Nouv. Dict. Hist. Nat., 24:28.
> COMMON NAME: Southern Long-nosed Armadillo.
> TYPE LOCALITY: Paraguay, Misiones, San Ignacio (as restricted by Cabrera, 1958).
> DISTRIBUTION: Argentina, Paraguay, and S Brazil south to Río Negro, Argentina.
> STATUS: IUCN – Lower Risk (lc).
> SYNONYMS: *auritus* (Olfers, 1818); *brevicaudus* Larrañaga, 1923.
> COMMENTS: Wetzel and Mondolfi (1979) included the Paraguayan paratype of *mazzai* Yepes, 1933; however, Vizcaíno (1995) included this specimen as a paratype of *D. yepesi.* Tamayo (1968) said *Dasypus undecimcinctus* Molina, 1782, was based on a composite of an animal known as "mulita' and *Cabassous unicinctus*. The name *undecimcintus* Molina is best considered a *nomen oblitum.*

*Dasypus kappleri* Krauss, 1862. Archiv Naturgesch., 28(1):20.
> COMMON NAME: Greater Long-nosed Armadillo.
> TYPE LOCALITY: "Den Urwäldern des Marowiniflusse in Surinam;" restricted to the neighborhood of Albina near the mouth of the Marowijne River by Husson (1978).
> DISTRIBUTION: Colombia (east of the Andes), Venezuela (south of the Orinoco), Guyana, Surinam, and south through the Amazon Basin of Brazil, Ecuador and Perú to NE Bolivia.
> STATUS: IUCN – Lower Risk (lc).
> SYNONYMS: *pentadactylus* Peters, 1864; **pastasae** (Thomas, 1901); *beniensis* Lönnberg, 1942; *peruvianus* Lönnberg, 1928.

*Dasypus novemcinctus* Linnaeus, 1758. Syst. Nat., 10th ed., 1:51.
> COMMON NAME: Nine-banded Armadillo.
> TYPE LOCALITY: "America Meridionali;" restricted to Pernambuco, Brazil, by Cabrera (1958).
> DISTRIBUTION: S USA, México, Central and South America to N Argentina, the Lesser Antilles (Grenada), and Trinidad and Tobago.
> STATUS: IUCN – Lower Risk (lc).
> SYNONYMS: *boliviensis* Gray, 1873; *brevirostris* (Gray, 1873); *leptocephalus* (Gray, 1873); *longicaudatus* Kerr, 1792; *longicaudatus* Daudin, 1799; *longicaudus* Wied, 1826; *longicaudus* Larrañaga, 1923; *lundii* Fitzinger, 1871; *mazzai* Yepes, 1933; *niger*

(Desmarest, 1804); *niger* (Olfers, 1818) [preoccupied]; *niger* Lichtenstein, 1818 [preoccupied]; *octocintus* Schreber, 1774; *peba* Desmarest, 1822; *platycercus* (Hensel, 1872); *serratus* G. Fischer, 1814; *uroceras* Lund, 1839; **aequatorialis** Lönnberg, 1913; **fenestratus** Peters, 1864; *granadiana* (Gray, 1873); **hoplites** G. M. Allen, 1911; **mexianae** (Hagmann, 1908); **mexicanus** Peters, 1864; *davisi* Russell, 1953; *leptorhynchus* Gray, 1873; *texanum* (Bailey, 1905).

COMMENTS: Reviewed by McBee and Baker (1982, Mammalian Species, 162). The following are *nomina nuda*: *decumanus* Illiger, 1815; *decumanus* Olfers, 1818; *longicaudus* Schinz, 1824; *cucurbitinus* Gaumer, 1917. The name *novenxinctus* Peal and Palisot de Beauvois, 1796, is an incorrect subsequent spelling of *novemcinctus* Linnaeus; *pepa* Krauss, 1862, is an incorrect subsequent spelling of *peba* Desmarest; *longicaudatus* Peters, 1864, is an incorrect subsequent spelling of *longicaudus* Wied; *leptorhinus* Gray, 1874, is an incorrect subsequent spelling of *leptorhynchus* Gray. The names *Tatus cucurbitalis* Fermin, 1765, and *Tatus minor* Fermin, 1769 (2:110), are not available because Fermin did not apply the principles of binominal nomenclature.

*Dasypus pilosus* (Fitzinger, 1856). Versamml. Deutsch. Nat. Arzte, Wien, Tageblatt, 32:123.
COMMON NAME: Hairy Long-nosed Armadillo.
TYPE LOCALITY: "Peru;" restricted to montane Perú by Wetzel and Mondolfi (1979).
DISTRIBUTION: Known only from the Peruvian Andes in the departments of San Martín, La Libertad, Huánuco, and Junín.
STATUS: IUCN – Vulnerable.
SYNONYMS: *hirsutus* (Burmeister, 1862).

*Dasypus sabanicola* Mondolfi, 1968. Mem. Soc. Cienc. Nat. La Salle, 27:151.
COMMON NAME: Llanos Long-nosed Armadillo.
TYPE LOCALITY: Venezuela, Apure, "Hato Macanillal."
DISTRIBUTION: Llanos of Venezuela and Colombia.
STATUS: IUCN – Data Deficient.

*Dasypus septemcinctus* Linnaeus, 1758. Syst. Nat., 10th ed., 1:51.
COMMON NAME: Seven-banded Armadillo.
TYPE LOCALITY: "Indiis;" restricted to Pernambuco, Brazil, by Hamlett (1939).
DISTRIBUTION: Lower Amazon Basin of Brazil to the Gran Chaco of Bolivia, Paraguay, and N Argentina.
STATUS: IUCN – Lower Risk (lc).
SYNONYMS: *megalolepis* (Cope, 1889); *propalatum* (Rhoads, 1894).
COMMENTS: The name *megalolepe* Yepes, 1928, is an incorrect subsequent spelling of *megalolepis* Cope.

*Dasypus yepesi* Vizcaíno, 1995. Mastozool. Trop., 2:7.
COMMON NAME: Yepes's Mulita.
TYPE LOCALITY: Argentina, Salta, San Andrés.
DISTRIBUTION: Gran Chaco of Paraguay, and N Argentina.
COMMENTS: Known from few specimens (Vizcaíno and Giallombardo, 2001), including the paratype of *Dasypus mazzai* Yepes that Wetzel and Mondolfi (1979) mistakenly said was from Puerto Guaraní, Alto Paraguay, Paraguay.

**Subfamily Euphractinae** Winge, 1923. Pattedyr-Slćgter, 1:304.

*Calyptophractus* Fitzinger, 1871. Sitzungsber. Kaiserl. Akad. Wiss., Wien, 64:388.
TYPE SPECIES: *Chlamyphorus retusus* Burmeister, 1863, by monotypy.
SYNONYMS: *Burmeisteria* Gray, 1865 [preoccupied].
COMMENTS: Recognized by Cabrera (1958) as the genus *Burmeisteria* Gray, which is preoccupied. The next available name is *Calyptophractus*, as pointed out by Wetzel (1985*b*); included under *Chlamyphorus* by Gardner (1993).

*Calyptophractus retusus* (Burmeister, 1863). Abhandl. Gesd. Naturf. Halle, 7:167.
COMMON NAME: Greater Fairy Armadillo.
TYPE LOCALITY: Bolivia, Santa Cruz, "Sta. Cruz de la Sierra."

DISTRIBUTION: Gran Chaco of N Argentina, W Paraguay, and SE Bolivia.
STATUS: IUCN – Vulnerable as *Chlamyphorus retusus*.
SYNONYMS: *clorindae* (Yepes, 1939).
COMMENTS: Commonly listed under *Burmeisteria retusa*; listed under *Chlamyphorus retusus* in
Wetzel's (1985*b*) review and by Gardner (1993).

*Chaetophractus* Fitzinger, 1871. Sitzb. Kaiserl. Akad. Wiss., Wein, 64(1):268.
TYPE SPECIES: *Dasypus villosus* (Desmarest, 1804) by subsequent designation (Yepes, 1928).
SYNONYMS: *Dasyphractus* Fitzinger, 1871.
COMMENTS: Formerly included in *Euphractus* (see Wetzel, 1985*b*).

*Chaetophractus nationi* (Thomas, 1894). Ann. Mag. Nat. Hist., ser. 6, 13:70.
COMMON NAME: Andean Hairy Armadillo.
TYPE LOCALITY: Bolivia, Oruro, "Orujo."
DISTRIBUTION: Bolivian departments of La Paz, Oruro, and Potosí.
STATUS: CITES – Appendix II; IUCN – Vulnerable.
COMMENTS: Distribution and status uncertain, may be a subspecies of *vellerosus* (see Wetzel,
1985*b*). The name *sajama* Suárez and Morales, 1986 (fide Anderson, 1997), is a *nomen
nudum*.

*Chaetophractus vellerosus* (Gray, 1865). Proc. Zool. Soc. Lond., 1865:376.
COMMON NAME: Screaming Hairy Armadillo.
TYPE LOCALITY: Bolivia, Santa Cruz, "Santa Cruz de la Sierra."
DISTRIBUTION: Chaco Boreal of Bolivia and Paraguay south to C Argentina and west to the
Puna de Tarapacá of Chile.
STATUS: IUCN – Lower Risk (lc).
SYNONYMS: *boliviensis* (Grandidier and Neveu-Lemaire, 1908); *brevirostris* (Fitzinger, 1871);
**pannosus** (Thomas, 1902); *desertorum* (Krumbiegel, 1940).
COMMENTS: The names *vallerosus* Thomas, 1923, and *villerosus* Grandidier and Neveu-
Lemaire, 1908, are incorrect subsequent spellings of *vellerosus* Gray; *brevirostris*
Fitzinger, 1860, is a *nomen nudum*.

*Chaetophractus villosus* (Desmarest, 1804). Tabl. Meth. Hist. Nat., *in* Nouv. Dict. Hist. Nat., 24:28.
COMMON NAME: Big Hairy Armadillo.
TYPE LOCALITY: Argentina, Buenos Aires, "Les Pampas" south of Río de la Plata between 35°
and 36° south (Azara, 1801:164).
DISTRIBUTION: Gran Chaco of Bolivia, Paraguay, and Argentina south to Santa Cruz,
Argentina, and Magallanes, Chile.
STATUS: IUCN – Lower Risk (lc).
SYNONYMS: *octocinctus* (Molina, 1782) [preoccupied]; *pilosus* (Larrañaga, 1923)
[preoccupied].

*Chlamyphorus* Harlan, 1825. Ann. Lyc. Nat. Hist., 1:235.
TYPE SPECIES: *Chlamyphorus truncatus* Harlan, 1825, by monotypy.
COMMENTS: *Chlamydophorus* Wagler, 1830, is an unjustified emendation of *Chlamyphorus*
Harlan; the names *Chlamiphorus* Contreras, 1973, *Chlamydephorus* Lenz, 1831, and
*Chlamydiphorus* Bonaparte, 1831, are incorrect subsequent spellings of *Chlamyphorus*
Harlan.

*Chlamyphorus truncatus* Harlan, 1825. Ann. Lyc. Nat. Hist., 1:235.
COMMON NAME: Pink Fairy Armadillo.
TYPE LOCALITY: "Mendoza . . . interior of Chili, on the east of the Cordilleras, in lat. 33°25′
and long. 69°47′, in the province of Cuyo;" restricted to Río Tunuyán, 33°25′S,
69°45′W, Mendoza, Argentina, by Cabrera (1958).
DISTRIBUTION: Argentina.
STATUS: U.S. ESA – Endangered; IUCN – Endangered.
SYNONYMS: *minor* (Lahille, 1895); *ornatus* (Lahille, 1895); *patquiensis* Yepes, 1931; *typicus*
(Lahille, 1895).

*Euphractus* Wagler, 1830. Naturliches Syst. Amphibien, p. 36.
> TYPE SPECIES: *Dasypus sexcinctus* Linnaeus, 1758, by subsequent designation (Palmer, 1904).
> SYNONYMS: *Encoubertus* McMurtrie, 1831; *Pseudotroctes* Gloger, 1841; *Scleropleura* Milne-Edwards, 1871.
> COMMENTS: Moeller (1968) included *Chaetophractus* and *Zaedyus* in this genus, *contra* Wetzel (1985*b*) whose usage is followed here. *Scelopleura* Trouessart, 1898, is an incorrect subsequent spelling of *Scleropleura* Milne-Edwards.

*Euphractus sexcinctus* (Linnaeus, 1758). Syst. Nat., 10th ed., 1:51.
> COMMON NAME: Six-banded Armadillo.
> TYPE LOCALITY: "America meridionale;" restricted to Pará, Brazil, by Thomas (1907*b*).
> DISTRIBUTION: S Surinam and adjacent Brazil as a northern isolated segment; E Brazil to Bolivia, Paraguay, Uruguay, and N Argentina as the main population.
> STATUS: IUCN – Lower Risk (lc).
> SYNONYMS: *mustelinus* Fitzinger, 1871; **boliviae** (Thomas, 1907); **flavimanus** (Desmarest, 1804); *encoubert* (Desmarest, 1822); *flavipes* (G. Fischer, 1814); *gilvipes* (Lichtenstein, 1818); *poyu* (Larrañaga, 1923); **setosus** (Wied, 1826); *bruneti* (Milne-Edwards, 1871); **tucumanus** (Thomas, 1907).
> COMMENTS: Reviewed by Wetzel (1985*b*) and Redford and Wetzel (1985, Mammalian Species, 252). The names *gilvipes* Illiger, 1815, and *pilosus* Olfers, 1818, are *nomina nuda*. The name *gylvipes* Minoprio, 1945, is an incorrect subsequent spelling of *gilvipes* Lichtenstein.

*Zaedyus* Ameghino, 1889. Acta Acad. Nac. Cienc. Cordoba, 6:867.
> TYPE SPECIES: *Dasypus minutus* Desmarest, 1822 (= *Loricatus pichiy* Desmarest, 1804), by original designation.
> COMMENTS: Sometimes considered a subgenus of *Euphractus*; reviewed by Wetzel (1985*b*). The names *Zaedypus* Lydekker, 1890, *Zaedius* Lydekker, 1894, and *Zaedius* Krumbiegel, 1940, are incorrect subsequent spellings of *Zaedyus* Ameghino.

*Zaedyus pichiy* (Desmarest, 1804). Tabl. Meth. Hist. Nat., *in* Nouv. Dict. Hist. Nat., 24:28.
> COMMON NAME: Pichi.
> TYPE LOCALITY: Argentina, Buenos Aires, Bahia Blanca, as restricted by Cabrera (1958).
> DISTRIBUTION: Mendoza, San Luis, and Buenos Aires, Argentina, south through Argentina and E Chile to the Straits of Magellan.
> STATUS: IUCN – Data Deficient.
> SYNONYMS: *australis* (Larrañaga, 1923); *ciliatus* (G. Fischer, 1814); *fimbriatus* (Olfers, 1818); *marginatus* (Wagler, 1830); *patagonicus* (Desmarest, 1819); *quadricinctus* (Molina, 1782) [preoccupied]; **caurinus** Thomas, 1928; *minutus* (Desmarest, 1822).
> COMMENTS: The name *cilliatus* J. A. Allen, 1901, is an incorrect subsequent spelling of *ciliatus* (G. Fischer).

**Subfamily Tolypeutinae** Gray, 1865. Proc. Zool. Soc. Lond., 1865:365.

*Cabassous* McMurtrie, 1831. Anim. Kingdom, 1:164.
> TYPE SPECIES: *Dasypus unicinctus* Linnaeus, 1758, by monotypy.
> SYNONYMS: *Arizostus* Gloger, 1841; *Lysiurus* Ameghino, 1891; *Tatoua* Gray, 1865; *Xenurus* Wagler, 1830 [preoccupied]; *Ziphila* Gray, 1873.
> COMMENTS: Revised by Wetzel (1980).

*Cabassous centralis* (Miller, 1899). Proc. Biol. Soc. Wash., 13:4.
> COMMON NAME: Northern Naked-tailed Armadillo.
> TYPE LOCALITY: Honduras, Cortés, "Chamelecon."
> DISTRIBUTION: México (Chiapas) to N Colombia and NW Venezuela.
> STATUS: CITES – Appendix III (Costa Rica); IUCN – Data Deficient.

*Cabassous chacoensis* Wetzel, 1980. Ann. Carnegie Mus., 49(2):335.
> COMMON NAME: Chacoan Naked-tailed Armadillo.
> TYPE LOCALITY: Paraguay, Presidente Hayes, "5-7 km W Estancia Juan de Zalazar."

DISTRIBUTION: Gran Chaco of W Paraguay and NW Argentina. Known from Mato Grosso, Brazil, based on one zoological park specimen (Wetzel, 1980).
STATUS: IUCN – Data Deficient.

*Cabassous tatouay* (Desmarest, 1804). Tabl. Meth. Hist. Nat., *in* Nouv. Dict. Hist. Nat., 24:28.
COMMON NAME: Greater Naked-tailed Armadillo.
TYPE LOCALITY: Paraguay, restricted to "a 27° de lat. Sur" by Cabrera (1958).
DISTRIBUTION: Uruguay, S Brazil, SE Paraguay, and NE Argentina.
STATUS: CITES – Appendix III (Uruguay); IUCN – Lower Risk (nt).
SYNONYMS: *dasycercus* (G. Fischer, 1814); *gymnurus* (Olfers, 1818); *lugubris* (Gray, 1873) [part, see Wetzel, 1985*b*]; *nudicaudus* (Lund, 1839).

*Cabassous unicinctus* (Linnaeus, 1758). Syst. Nat., 10th ed., 1:50.
COMMON NAME: Southern Naked-tailed Armadillo.
TYPE LOCALITY: "Africa;" restricted to "l'Amérique" by Buffon (1763), and to Surinam by Thomas (1911*a*).
DISTRIBUTION: South America east of the Andes from Colombia to Bolivia and Mato Grosso do Sul, Brazil.
STATUS: IUCN – Lower Risk (lc).
SYNONYMS: *duodecimcinctus* (Schreber, 1774); *multicinctus* (Thunberg, 1818); *octodecimcinctus* (Erxleben, 1777); *verrucosus* (Wagner, 1844); **squamicaudis** (Lund, 1845); *hispidus* (Burmeister, 1854); *latirostris* (Gray, 1873); *loricatus* (Wagner, 1855); *lugubris* (Gray, 1873) [part, see Wetzel, 1985*b*].
COMMENTS: Tamayo (1968) said *Dasypus undecimcinctus* Molina, 1782, was based on a composite of an animal known as "mulita" and *Cabassous unicinctus*. It is doubtful that *unicinctus* occurred in NW Argentina (formerly part of Chile) and the name *undecimcinctus* Molina is best considered a *nomen oblitum*. The name *undecimcinctus* Illiger, 1815, is a *nomen nudum*.

*Priodontes* F. Cuvier, 1825. Dentes des Mamm., p. 257.
TYPE SPECIES: *Dasypus gigas* G. Cuvier, 1817 (= *Dasypus maximus* Kerr, 1792), by monotypy.
SYNONYMS: *Cheloniscus* Wagler, 1830; *Polygomphius* Gloger, 1841; *Priodon* McMurtrie, 1831; *Prionodos* Gray, 1865.
COMMENTS: Reviewed by Wetzel (1985*b*). The name *Prionodon* attributed to Gray (1843) by Palmer (1904), Yepes (1928), and Cabrera (1958) is a *nomen nudum*.

*Priodontes maximus* (Kerr, 1792). *In* Linnaeus, Anim. Kingdom, p. 112.
COMMON NAME: Giant Armadillo.
TYPE LOCALITY: French Guiana, "Cayenne."
DISTRIBUTION: South America east of the Andes from N Venezuela and the Guianas south to Bolivia, Paraguay, and N Argentina.
STATUS: CITES – Appendix I; U.S. ESA – Endangered; IUCN – Endangered.
SYNONYMS: *giganteus* (G. Fischer, 1814); *gigas* (G. Cuvier, 1817); *grandis* (Olfers, 1818).

*Tolypeutes* Illiger, 1811. Prodr. Syst. Mamm. Avium., p. 111.
TYPE SPECIES: *Dasypus tricinctus* Linnaeus, 1758, by subsequent designation (Yepes, 1928).
SYNONYMS: *Apara* McMurtrie, 1831; *Cheloniscus* Gray, 1873 [preoccupied]; *Sphaerocormus* Fitzinger, 1871; *Tolypoides* Grandidier and Neveu-Lemaire, 1905.
COMMENTS: Reviewed by Wetzel (1985*b*). *Matacus* Rafinesque, 1815, is a *nomen nudum*; *Tolypeutis* Olfers, 1818 and *Tolypentes* Matschie, 1894, are incorrect subsequent spellings of *Tolypeutes* Illiger, 1811.

*Tolypeutes matacus* (Desmarest, 1804). Tabl. Meth. Hist. Nat., *in* Nouv. Dict. Hist. Nat., 24:28.
COMMON NAME: Southern Three-banded Armadillo.
TYPE LOCALITY: No locality mentioned; restricted to Argentina, Tucumán, Tucumán, by Sanborn (1930).
DISTRIBUTION: E Bolivia and SW Brazil south through the Gran Chaco of Paraguay to Argentina (Buenos Aires).
STATUS: IUCN – Lower Risk (nt).

SYNONYMS: *apar* (Desmarest, 1822); *bicinctus* (Grandidier and Neveu-Lemaire, 1905); *brachyurus* (G. Fischer, 1814); *conurus* I. Geoffroy, 1847; *muriei* Garrod, 1878; *octodecimcinctus* (Molina, 1782) [preoccupied].

COMMENTS: The name *aparoides* Gervais, 1869, is a *nomen nudum*.

*Tolypeutes tricinctus* (Linnaeus, 1758). Syst. Nat., 10th ed., 1:56.

COMMON NAME: Brazilian Three-banded Armadillo.

TYPE LOCALITY: "In India orientali;" redefined as Pernambuco, Brazil, by Sanborn (1930).

DISTRIBUTION: Brazilian states of Bahia, Ceará, Maranhão, Piauí, and Pernambuco; and expected in Goiás (Santos et al., 1994; Olmos, 1995; Oliveira, 1995; Marinho-Filho et al., 1997).

STATUS: IUCN – Vulnerable; extremely rare (see Silva and Oren, 1993; Santos et al., 1994).

SYNONYMS: *globulus* (Olfers, 1818); *quadricinctus* (Linnaeus, 1758); *quadricinctus* Olfers, 1818 [preoccupied].

COMMENTS: *globulus* Illiger, 1815, is a *nomen nudum*.

# ORDER PILOSA
## by Alfred L. Gardner

**ORDER PILOSA** Flower, 1883.
    SYNONYMS: Edentata Vicq-d'Azyr, 1792 [part].
    COMMENTS: Reviewed by Wetzel (1985a) and by Kraft (1995) as part of Xenarthra; formerly included in Xenarthra, which was elevated to magnorder rank by McKenna and Bell (1997).

**SUBORDER FOLIVORA** Delsuc, Catzeflis, Stanhope, and Douzery, 2001.
    COMMENTS: Use of Folivora follows recommendation by Delsuc et al. (2001:1606). McKenna and Bell (1997) used Phyllophaga Owen, 1842, for this suborder.

**Family Bradypodidae** Gray, 1821. London Med. Repos., 15:304.

*Bradypus* Linnaeus, 1758. Syst. Nat., 10th ed., 1:34.
    TYPE SPECIES: *Bradypus tridactylus* Linnaeus, 1758, by subsequent designation (Miller and Rehn, 1901).
    SYNONYMS: *Acheus* F. Cuvier, 1825; *Arctopithecus* Gray, 1850 [preoccupied]; *Eubradypus* Lönnberg, 1942; *Hemibradypus* Anthony, 1906; *Ignavus* Blumenbach, 1779; *Scaeopus* Peters, 1864.
    COMMENTS: Avila-Pires (*in* Wetzel and Avila-Pires, 1980) considered *Scaeopus* a separate genus. *Achaeus* Erman, 1835, and *Achaeus* Gray, 1843, are incorrect subsequent spellings of *Acheus* F. Cuvier; *Arctopithecus* Gray, 1843, and *Neobradypus* Lönnberg, 1942, are *nomina nuda*; *Tardigradus* Brisson, 1762, is an unavailable name and later uses for sloths are preoccupied.

*Bradypus pygmaeus* Anderson and Handley, 2001. Proc. Biol. Soc. Wash., 114:17.
    COMMON NAME: Pygmy Three-toed Sloth.
    TYPE LOCALITY: Panamá, Bocas del Toro, "Isla Escudo de Veraguas, West Point."
    DISTRIBUTION: Known only from Isla Escudo de Veraguas.
    STATUS: Vulnerable.

*Bradypus torquatus* Illiger, 1811. Prodr. Syst. Mamm. Avium, p. 109.
    COMMON NAME: Maned Sloth.
    TYPE LOCALITY: "Brasilia;" restricted to the Atlantic drainage of the Brazilian states of Bahia, Espírito Santo, and Rio de Janeiro, by Wetzel and Avila-Pires (1980).
    DISTRIBUTION: Coastal forests of SE Brazil.
    STATUS: U.S. ESA – Endangered; IUCN – Endangered.
    SYNONYMS: *affinis* Gray, 1850; *crinitus* Gray, 1850; *cristatus* Hamilton-Smith, 1827; *mareyi* (Anthony, 1907); *melanotis* Swainson, 1835.
    COMMENTS: Some authors believe erroneously that *B. torquatus* Illiger, 1811, is a *nomen nudum* and attribute the name to Desmarest (1816a).

*Bradypus tridactylus* Linnaeus, 1758. Syst. Nat., 10th ed., 1:34.
    COMMON NAME: Pale-throated Sloth.
    TYPE LOCALITY: "Americae meridionalis arboribus;" restricted to Surinam by Thomas (1911a).
    DISTRIBUTION: Guyana, Suriname, French Guiana, Venezuela south of the Orinoco, and N Brazil (south to the Amazonas/Solimões).
    STATUS: IUCN – Lower Risk (lc).
    SYNONYMS: *ai* (Lesson, 1827); *blainvillii* (Gray, 1850); *cuculliger* Wagler, 1831; *cummunis* Lesson, 1841; *dysonii* (Gray, 1869); *flaccidus* (Gray, 1850); *guianensis* Blainville, 1840; *gularis* Rüppell, 1842; *smithii* (Gray, 1869).
    COMMENTS: The name *flaccions* Sanderson, 1949, is an incorrect subsequent spelling of *flaccidus* Gray.

*Bradypus variegatus* Schinz, 1825. Das Thierreich, 4:510.
    COMMON NAME: Brown-throated Sloth.

TYPE LOCALITY: "Sudamerika;" restricted to Brazil by Mertens (1925) who suggested that the type may have come from Bahia.

DISTRIBUTION: Honduras to Colombia, Ecuador, Brazil, W Venezuela, E Perú and Bolivia, Paraguay, and N Argentina.

STATUS: CITES – Appendix II; IUCN – Lower Risk (lc).

SYNONYMS: *dorsalis* Fitzinger, 1871; **boliviensis** (Gray, 1871); *beniensis* Lönnberg, 1942; **brasiliensis** Blainville, 1840; *ai* Wagler, 1831 [preoccupied]; *ustus* Lesson, 1840; *pallidus* Wagner, 1844; **ephippiger** Philippi, 1870; *castaneiceps* (Gray, 1871); *ecuadorianus* Spillmann, 1927; *griseus* (Gray, 1871); *ignavus* Goldman, 1913; *nefandus* Spillmann, 1927; *violeta* Thomas, 1917; **gorgon** Thomas, 1926; **infuscatus** Wagler, 1831; *brachydactylus* Wagner, 1855; *codajazensis* Lönnberg, 1942; *macrodon* Thomas, 1917; *subjuruanus* Lönnberg, 1942; **trivittatus** Cornalia, 1849; *marmoratus* (Gray, 1850); *miritibae* Lönnberg, 1942; *problematicus* (Gray, 1850); *tocantinus* Lönnberg, 1942; *unicolor* Fitzinger, 1871.

COMMENTS: The literature contains the following unavailable names: *braziliensis* Sanderson, 1949 [incorrect subsequent spelling of *brasiliensis* Blainville]; *infumatus* Tschudi, 1845 [incorrect subsequent spelling of *infuscatus* Wagler]; *rifuscatus* Cornelia, 1849 [incorrect subsequent spelling of *infuscatus* Wagler]; *speculiger* Fitzinger, 1871 [*nomen nudum*].

**Family Megalonychidae** Ameghino, 1889. Acta Acad. Nac. Cienc. Cordoba, Buenos Aires, 6:690.
COMMENTS: Includes *Choloepus* and approximately 12 genera of extinct sloths, some of which survived to the Holocene on Caribbean islands. *Choloepus* was formerly included in Bradypodidae (see Hoffstetter, 1969; Patterson and Pascual, 1968a) or Choloepidae (see Honacki et al., 1982). Placed in Megalonychidae by Webb (1985) and Wetzel (1985a).

*Choloepus* Illiger, 1811. Prodr. Syst. Mamm. Avium., p. 108.
TYPE SPECIES: *Bradypus didactylus* Linnaeus, 1758, by subsequent designation (Gray, 1827).
SYNONYMS: *Unaues* Rafinesque, 1815; *Unaus* Gray, 1821.
COMMENTS: Reviewed by Wetzel (1985a). The original spelling of *Unaues* Rafinesque was *Unaüs*, which required a substitution of *ue* for *ü*. The spellings *Cholaepus*, *Chaelopus*, and *Choelopus* of various authors are all incorrect subsequent spellings of *Choloepus* Illiger. *Cholopus* Agassiz, 1847, is an invalid emendation of *Choloepus* Illiger.

*Choloepus didactylus* (Linnaeus, 1758). Syst. Nat., 10th ed., 1:35.
COMMON NAME: Linnaeus's Two-toed Sloth.
TYPE LOCALITY: "Zeylona;" corrected to Surinam by Thomas (1911a). Not British Guiana as stated by Tate (1939).
DISTRIBUTION: Guianas and Venezuela (delta and south of Río Orinoco) south into Brazil (Maranhão west along Rio Amazonas/Solimões) and west into upper Amazon Basin of Ecuador and Perú.
STATUS: IUCN – Data Deficient.
SYNONYMS: *brasiliensis* Fitzinger, 1871; *columbianus* Gray, 1871; *curi* (Link, 1795); *florenciae* J. A. Allen, 1913; *guianensis* Fitzinger, 1871; *kouri* (Daudin, 1802); *napensis* Lönnberg, 1922; *unau* (Link, 1795).
COMMENTS: Reviewed by Wetzel and Avila-Pires (1980). Treated here as monotypic; needs revision. The name *didaetylus* Thomas, 1893, is an incorrect subsequent spelling of *didactylus* Linnaeus.

*Choloepus hoffmanni* Peters, 1858. Monatsb. K. Preuss. Akad. Wiss. Berlin, 1858:128.
COMMON NAME: Hoffmann's Two-toed Sloth.
TYPE LOCALITY: "Costa Rica;" restricted to Escazú, San José, by Goodwin (1946); corrected to Heredia, Volcán Barba, by Wetzel and Avila-Pires (1980).
DISTRIBUTION: Central America (Nicaragua) into South America east to W Venezuela and south to Brazil (Mato Grosso) and E Bolivia.
STATUS: CITES – Appendix III (Costa Rica); IUCN – Data Deficient.
SYNONYMS: **capitalis** J. A. Allen, 1913; **florenciae** J. A. Allen, 1913; *andinus* J. A. Allen, 1913; *augustinus* J. A. Allen, 1913; **juruanus** Lönnberg, 1942; **pallescens** Lönnberg, 1928; *peruvianus* Menegaux, 1906.

## SUBORDER VERMILINGUA Illiger, 1811.

**Family Cyclopedidae** Pocock, 1924. Proc. Zool. Soc. Lond., 1924:1030.

*Cyclopes* Gray, 1821. London Med. Repos., 15:305.
  TYPE SPECIES: *Myrmecophaga didactyla* Linnaeus, 1758, by monotypy.
  SYNONYMS: *Cyclothurus* Lesson, 1842; *Didactyla* Liais, 1872; *Didactyles* F. Cuvier, 1829;
     *Eurypterna* Gloger, 1841; *Myrmydon* Wagler, 1830.
  COMMENTS: *Cyclothurus* Gray, 1825, is a *nomen nudum*. *Cycloturus* Goeldi and Hagmann, 1904,
     is an incorrect subsequent spelling of *Cyclothurus* Lesson.

*Cyclopes didactylus* (Linnaeus, 1758). Syst. Nat., 10th ed., 1:35.
  COMMON NAME: Silky Anteater.
  TYPE LOCALITY: "America australi;" restricted to Surinam by Thomas (1911*a*).
  DISTRIBUTION: México (Veracruz and Oaxaca) to Colombia and west of Andes to S Ecuador,
     east of Andes to Venezuela, Trinidad, Guyana, Surinam, French Guiana, and
     S Colombia and Venezuela, south to Bolivia (La Paz and Santa Cruz) and Brazil (Acre
     east to Alagoas).
  STATUS: IUCN – Lower Risk (lc).
  SYNONYMS: *monodactyla* (Kerr, 1792); *unicolor* (Desmarest, 1822); **catellus** Thomas, 1928;
     *codajazensis* Lönnberg, 1942; **dorsalis** (Gray, 1865); *eva* Thomas, 1902; **ida** Thomas,
     1900; *juruanus* Lönnberg, 1942; **melini** Lönnberg, 1928; **mexicanus** Hollister, 1914.
  COMMENTS: The name *dydactyla* Brongniart, 1792, is an incorrect subsequent spelling of
     *didactylus* Linnaeus. The name *jurnanus* Cabrera, 1958, is an incorrect subsequent
     spelling of *juruanus* Lönnberg.

**Family Myrmecophagidae** Gray, 1825. Ann. Philos., n.s., 10:343.

*Myrmecophaga* Linnaeus, 1758. Syst. Nat., 10th ed., 1:35.
  TYPE SPECIES: *Myrmecophaga tridactyla* Linnaeus, 1758, by subsequent selection (Thomas,
     1901*a*).
  SYNONYMS: *Falcifer* Rehn, 1900.
  COMMENTS: *Myrmecopha* Fisher von Waldheim, 1803, and *Myrmecophagus* Gray, 1825, are
     invalid emendations of *Myrmecophaga* Linnaeus.

*Myrmecophaga tridactyla* Linnaeus, 1758. Syst. Nat., 10th ed., 1:35.
  COMMON NAME: Giant Anteater.
  TYPE LOCALITY: "America *meridionali*;" restricted to Brazil, Pernambuco, Pernambuco, by
     Thomas (1911*a*).
  DISTRIBUTION: Belize and Guatemala through South America to Uruguay and the Gran
     Chaco of Bolivia, Paraguay, Argentina.
  STATUS: CITES – Appendix II; IUCN – Vulnerable.
  SYNONYMS: *jubata* Linnaeus, 1758; **artata** Osgood, 1912; **centralis** Lyon, 1906.
  COMMENTS: The name *iubata* Wied-Neuwied, 1826, is an incorrect subsequent spelling of
     *jubata* Linnaeus.

*Tamandua* Gray, 1825. Ann. Philos., n.s., 10:343.
  TYPE SPECIES: *Myrmecophaga tamandua* G. Cuvier, 1798 (= *Tamandua tetradactyla* Linnaeus,
     1758), by monotypy.
  SYNONYMS: *Dryoryx* Gloger, 1841; *Tamanduas* F. Cuvier, 1829; *Uroleptes* Wagler, 1830.
  COMMENTS: *Uropeltes* Alston, 1880, is an incorrect subsequent spelling of *Uroleptes* Wagler.
     Revised by Wetzel (1975).

*Tamandua mexicana* (Saussure, 1860). Rev. Mag. Zool. Paris, ser. 2, 12:9.
  COMMON NAME: Northern Tamandua.
  TYPE LOCALITY: México, "Tabasco."
  DISTRIBUTION: E México (Tamaulipas), Central America, South America to NW Perú and
     NW Venezuela.
  STATUS: CITES – Appendix III (Guatemala); IUCN – Lower Risk (lc).

SYNONYMS: *hesperia* Davis, 1955; *tenuirostris* J. A. Allen, 1904; ***instabilis*** J. A. Allen, 1904; ***opistholeuca*** Gray, 1873; *chiriquensis* J. A. Allen, 1904; *sellata* (Cope, 1889); *tambensis* Lönnberg, 1937; ***punensis*** J. A. Allen, 1916.

COMMENTS: The name *leucopygia* Gray, 1873, is a *nomen nudum*. The name *quadridactyla* True, 1884, is a *lapsus* for *tetradactyla* of authors (not *tetradactyla* Linnaeus).

*Tamandua tetradactyla* (Linnaeus, 1758). Syst. Nat., 10th ed., 1:35.

COMMON NAME: Southern Tamandua.

TYPE LOCALITY: "America meridionali;" restricted to Brazil, Pernambuco, Pernambuco (= Recife), by Thomas (1911*a*).

DISTRIBUTION: South America east of the Andes from Colombia, Venezuela, Trinidad, and the Guianas, south to Uruguay and N Argentina.

STATUS: IUCN – Lower Risk (lc).

SYNONYMS: *bivittata* (Desmarest, 1817); *brasiliensis* Liais, 1872; *myosura* (Pallas, 1766); ***nigra*** (Geoffroy, 1803); *crispa* (Rüppell, 1842); *longicaudata* (Wagner, 1844); ***quichua*** Thomas, 1927; ***straminea*** (Cope, 1889); *chapadensis* J. A. Allen, 1904; *kriegi* Krumbiegel, 1940; **unassigned:** *longicaudata* (Turner, 1853) [preoccupied]; *nigra* Beaux, 1908 [preoccupied]; *opisthomelas* Gray, 1873; *tamandua* (G. Cuvier, 1798).

COMMENTS: The name *mexianae* Cabrera, 1958, is a *nomen nudum*. The names *longicauda* Vesey-FitzGerald, 1936, and *longicauda* Rode, 1937, are incorrect subsequent spellings of *longicaudata* (Wagner). The names *longicaudata, nigra, opisthomelas,* and *tamandua,* cannot be assigned to subspecies with certainty because their type localities are too general.

# ORDER SCANDENTIA
by Kristofer M. Helgen

**ORDER SCANDENTIA** Wagner, 1855.

SYNONYMS: Scandentiformes, Tupaii, Tupaioidea, Tupayae.

COMMENTS: In the past, treeshrews have commonly been considered basal members of the order Primates, or united with macroscelidids in the "insectivoran" clade Menotyphla. However, as a group they have no immediate living relatives and are best classified at ordinal rank (Butler, 1972, 1980; Dene et al., 1978; Luckett, 1980; McKenna and Bell, 1997). At a deeper phylogenetic level, scandentians apparently form a natural group with dermopterans and primates (Murphy et al., 2001*b*). Representatives of the order are confined to southern, eastern, and SE Asia both currently and in the fossil record, which extends back to the Middle Eocene in east Asia (McKenna and Bell, 1997). Most previous workers have arranged Scandentia as a monofamilial order, but recognition of two families (Tupaiidae and Ptilocercidae) more aptly conveys the anatomical disparity evident among the living treeshrews (see below).

Despite the attention paid to the higher-level phylogenetic relationships of treeshrews, a modern revision of species-level taxonomy in the group is still unavailable; the most recent comprehensive review remains that of Lyon (1913), a thorough but now long-outdated work. Chasen (1940), Ellerman and Morrison-Scott (1966), and Corbet (in Corbet and Hill, 1992) produced regional lists of named forms, but not critical systematic treatments, and the latter two listings are beset by overlumping. This account is likewise no substitute for a comprehensive systematic review of the order, but in its preparation I have examined all treeshrew specimens (including types) in the collections of the American Museum of Natural History, Field Museum of Natural History, Museum of Comparative Zoology, and National Museum of Natural History, as well as a number of type specimens stored in European collections. Starting points for further research are noted below.

**Family Tupaiidae** Gray, 1825. Ann. Philos., n.s., 10:339.

SYNONYMS: Cladobatae, Cladobatidina, Cladobatida, Cladobatina, Glisoricina, Glisoricinae, Tupaina, Tupaiadae, Tupajidae, Tupayae, Tupayidae.

COMMENTS: See Elliott (1971), Luckett (1980), and Emmons and Greene (2000) for references pertaining to tree shrew biology. For more detailed data on the distribution of tupaiid species on small islands on the Sunda Shelf, see Corbet (in Corbet and Hill, 1992).

*Anathana* Lyon, 1913. Proc. U. S. Natl. Mus., 45:120.

TYPE SPECIES: *Tupaia ellioti* Waterhouse, 1850.

COMMENTS: *Anathana* and *Tupaia* are closely allied.

*Anathana ellioti* (Waterhouse, 1850). Proc. Zool. Soc. Lond., 1849:107 [1850].

COMMON NAME: Madras Treeshrew.

TYPE LOCALITY: India, Andhra Pradesh, "hills between Cuddapah and Nellox" (= Velikanda Range).

DISTRIBUTION: India south of the Ganges River.

STATUS: CITES – Appendix II; IUCN – Lower Risk (nt).

SYNONYMS: *pallida* Lyon, 1913; *wroughtoni* Lyon, 1913.

COMMENTS: Museum material of *Anathana* is very limited. Lyon (1913) named *pallida* and *wroughtoni* as full species based on minor pelage differences; Corbet (in Corbet and Hill, 1992) is probably correct in his assumption that these three "intergrade without definable boundaries."

*Dendrogale* Gray, 1848. Proc. Zool. Soc. Lond., 1848:23.

TYPE SPECIES: *Hylogalea murina* Schlegel and Müller, 1843.

*Dendrogale melanura* (Thomas, 1892). Ann. Mag. Nat. Hist., ser. 6, 9:251.

COMMON NAME: Bornean Smooth-Tailed Treeshrew.

TYPE LOCALITY: Borneo, Sarawak, Mt. Dulit, 5,000 ft. (1,524 m).

DISTRIBUTION: Restricted to higher altitudes (above 900 m) in Malaysian N Borneo, including the mountains of NE Sarawak and Mts. Kinabalu and Trus Madi in Sabah.

STATUS: CITES –Appendix II; IUCN – Vulnerable.

SYNONYMS: **baluensis** Lyon, 1913.

COMMENTS: Endemic to montane Borneo.

*Dendrogale murina* (Schlegel and Müller, 1843). *In* Temminck, Verh. Nat. Gesch. Nederland. Overz. Bezitt., Zool., p. 167[1845], pls. 26, 27[1843].

COMMON NAME: Northern Smooth-Tailed Treeshrew.

TYPE LOCALITY: Given as "Pontianak" (Kalimantan, Borneo); with little doubt actually collected by Diard at Cochin Chine, Vietnam (C. Smeenk, in litt.).

DISTRIBUTION: From E Thailand (Chatraburi and Trat Provinces) through Cambodia to S Vietnam.

STATUS: CITES – Appendix II; IUCN – Lower Risk (lc).

SYNONYMS: *frenata* (Gray, 1860).

COMMENTS: Lyon (1913) and Ellerman and Morrison-Scott (1951) discussed the doubtful validity of the original type locality. Archival research at the Naturalis Museum, Leiden corroborates a Vietnamese origin for the holotype of *murina* (see type locality above).

*Tupaia* Raffles, 1821. Trans. Linn. Soc. Lond., 13:256.

TYPE SPECIES: *Tupaia ferruginea* Raffles, 1821 (= *Sorex glis* Diard, 1820).

SYNONYMS: *Chladobates* Schinz, 1824; *Cladobates* F. Cuvier, 1825; *Gladobates* Schinz, 1824; *Glipora* Jentink, 1888; *Glirisorex* Scudder, 1882; *Glisorex* Desmarest, 1822; *Glisosorex* Giebel, 1855; *Hylogale* Temminck, 1827; *Hylogalea* Schlegel and Mueller, 1843; *Lyonogale* Conisbee, 1953; *Palaeotupaia* Chopra and Vasishat, 1979; *Sorex-glis* É. Geoffroy and F. Cuvier, 1822; *Tana* Lyon, 1913; *Tapaia* Gray, 1860; *Tupaja* Haeckel, 1866; *Tupaya* É. Geoffroy and F. Cuvier, 1822.

*Tupaia belangeri* (Wagner, 1841). Schreber's Die Säugethiere, Suppl., 2:42.

COMMON NAME: Northern Treeshrew.

TYPE LOCALITY: Burma, Pegu, Siriam (near Yangon).

DISTRIBUTION: S and SE Asia north of and including the Isthmus of Kra: Thailand, Burma, Bangladesh, far E India and Nepal, S China, Cambodia, Laos, Vietnam, and associated coastal islands, including Hainan. Probably also Preparis Isl north of the Andaman Isls (Lyon, 1913:61).

STATUS: CITES – Appendix II; IUCN – Lower Risk (lc).

SYNONYMS: *brunetta* Thomas, 1923; *clarissa* Thomas, 1917; *peguanus* Lesson, 1842; *tenaster* Thomas, 1917; **chinensis** Anderson, 1879; *annamensis* Robinson and Kloss, 1922; *assamensis* Wroughton, 1921; *cambodiana* Kloss, 1919; *cochinchinensis* Robinson and Kloss, 1922; *concolor* Bonhote, 1907; *dissimilis* (Ellis, 1860); *gaoligongensis* Wang, 1987; *gongshanensis* Wang, 1987; *kohtauensis* Shamel, 1930; *laotum* Thomas, 1914; *lepcha* Thomas, 1922; *modesta* J. A. Allen, 1906; *olivacea* Kloss, 1919; *operosa* Robinson and Kloss, 1914; *pingi* Ho, 1936; *siccata* Thomas, 1914; *sinus* Kloss, 1916; *tonquinia* Thomas, 1925; *ultima* Robinson and Kloss, 1914; *versurae* Thomas, 1922; *yaoshanensis* Wang, 1987; *yunalis* Thomas, 1914.

COMMENTS: Often included in *T. glis*, but *belangeri* differs from *glis* in pelage coloration, mammae formula, and craniodental aspects. Toder et al (1992) and Hirai et al. (2002) discussed chromosomal differences between *belangeri* and *glis*, and Endo et al. (2000) reported their syntopic occurrence at Hat-Yai in S Thailand (south of the Isthmus of Kra). All forms north of this contact zone are referred here to *T. belangeri*. A careful revision of geographic variation within *T. belangeri* is needed; I have divided the named forms into *belangeri* and *chinensis* groups (which might be better recognized as closely-related parapatric species), but this simplistic arrangement no doubt masks a good deal of taxonomic complexity within these two groups. For additional discussion see Lyon (1913), Agrawal (1975), Lekagul and McNeely (1977), and Wang (1987).

*Tupaia chrysogaster* Miller, 1903. Smithson. Misc. Coll., 45:58.

COMMON NAME: Golden-bellied Treeshrew.

TYPE LOCALITY: N Pagai Isl, Mentawai Isls.

DISTRIBUTION: N and S Pagai Isls, and Sipora (Mentawai Isls, Indonesia). Recorded erroneously from Nias by Lyon (1913:36, 39).

STATUS: CITES – Appendix II; IUCN – Vulnerable.

COMMENTS: A distinctive species readily distinguished from populations of *T. glis* by pelage coloration and mammae formula. Part of the distinctive endemic mammal fauna of the Mentawai Isls (see account of *Leopoldamys siporanus*). Does not include *siberu* and *tephrura* (from Siberut and the Batu Isls, respectively), synonyms of *T. glis*.

*Tupaia dorsalis* Schlegel, 1857. Handl. Beoef. Dierk., 1:59, 447, pl. 3.

COMMON NAME: Striped Treeshrew.

TYPE LOCALITY: W Borneo, lower Kapuas River.

DISTRIBUTION: Borneo: Sabah, Sarawak (Malaysia), Brunei, and Kalimantan (Indonesia) except SE, at low to moderate elevations (below 1,000 m).

STATUS: CITES – Appendix II; IUCN – Lower Risk (lc).

COMMENTS: Endemic to Borneo. United with *T. tana* in the genus *Tana* by Lyon (1913), but this arrangement is probably not natural (see Dene et al., 1978).

*Tupaia glis* (Diard, 1820). Asiat. J. Mon. Reg., 10:478.

COMMON NAME: Common Treeshrew.

TYPE LOCALITY: Malaysia, Penang (= Pinang) Isl (fixed by Lyon, 1913:45).

DISTRIBUTION: SE Asia south of about 10° N latitude, from the vicinity of Hat-Yai, S Thailand through mainland Malaysia (and adjacent coastal isls) to Singapore; also Indonesia, including Siberut, Batu Isls, Sumatra, Java, Bangka, and the Riau, Lingga, and Anambas Isls.

STATUS: CITES – Appendix II; IUCN – Lower Risk (lc).

SYNONYMS: *anambae* Lyon, 1913; *batamana* Lyon, 1907; *castanea* Miller, 1903; *chrysomalla* Miller, 1900; *cognata* Chasen, 1940; *demissa* Thomas, 1904; *discolor* Lyon, 1906; *ferruginea* Raffles, 1821; *hypochrysa* Thomas, 1895; *jacki* Robinson and Kloss, 1918; *lacernata* Thomas and Wroughton, 1909; *longicanda* Lyon, 1913 [*nomen nudum*]; *longicauda* Kloss, 1911; *obscura* Kloss, 1911; *pemangilis* Lyon, 1911; *penangensis* Robinson and Kloss, 1911; *phaeura* Miller, 1902; *phoeniura* Thomas, 1923; *press* (É. Geoffroy and F. Cuvier, 1822); *pulonis* Miller, 1903; *raviana* Lyon, 1911; *redacta* Robinson, 1916; *siaca* Lyon, 1908; *siberu* Chasen and Kloss, 1928; *sordida* Miller, 1900; *tephrura* Miller, 1903; *umbratilis* Chasen, 1940; *wilkinsoni* Robinson and Kloss, 1911.

COMMENTS: See comments under *T. belangeri*. In the past many additional taxa have been included in the synonymy of *T. glis* (e.g. see Chasen, 1940, and Corbet, in Corbet and Hill, 1992). Even with the separation of *belangeri, chrysogaster, longipes, palawanensis, moellendorffi*, and their synonyms, *T. glis* still retains many forms of uncertain rank and validity. Pending a detailed study, no subspecies are listed, but many if not most of the named insular forms are distinctive (see Lyon, 1913).

*Tupaia gracilis* Thomas, 1893. Ann. Mag. Nat. Hist., ser. 6, 12:53.

COMMON NAME: Slender Treeshrew.

TYPE LOCALITY: Borneo, Sarawak, Baram Dist., Apoh River at base of Mt. Batu Song.

DISTRIBUTION: Borneo below 1,200 m, including Sabah and Sarawak (Malaysia) and Kalimantan (Indonesia) except SE; west to islands of Karimata, Belitung, and Bangka, and north to Banggi Isl.

STATUS: CITES – Appendix II; IUCN – Lower Risk (lc).

SYNONYMS: **edarata** Lyon, 1913; **inflata** Lyon, 1906.

*Tupaia javanica* Horsfield, 1822. Zool. Res. Java, pt. 3 (pages unno.).

COMMON NAME: Horsfield's Treeshrew.

TYPE LOCALITY: Java, probably near Banjuwangi (far eastern Java; see Lyon, 1913:106).

DISTRIBUTION: Indonesia: Bali, Java, W Sumatra, and Nias.

STATUS: CITES – Appendix II; IUCN – Lower Risk (lc).

SYNONYMS: *balina* Thomas, 1913; *bogoriensis* Sody, 1937; *occidentalis* Robinson and Kloss, 1918; *tjibruniensis* Sody, 1937.

COMMENTS: Chasen (1940) and Hill (1960) offered opinions on subspecific taxonomy, but more study is needed. Known from Nias by a single specimen (Lyon, 1913:106).

*Tupaia longipes* (Thomas, 1893). Ann. Mag. Nat. Hist., ser. 6, 11:343.
    COMMON NAME: Long-footed Treeshrew.
    TYPE LOCALITY: Borneo, Sarawak.
    DISTRIBUTION: Borneo, including Sabah and Sarawak (Malaysia), Kalimantan (Indonesia),
        and Brunei.
    STATUS: CITES – Appendix II; IUCN – Endangered.
    SYNONYMS: *salatana* Lyon, 1913.
    COMMENTS: A distinctive but variable species endemic to Borneo.

*Tupaia minor* Günther, 1876. Proc. Zool. Soc. Lond., 1876:426.
    COMMON NAME: Pygmy Treeshrew.
    TYPE LOCALITY: Borneo, Sabah, mainland "opposite the island of Labuan."
    DISTRIBUTION: S peninsular Thailand, Malaysia, (Malay Peninsula, Sabah, Sarawak, and
        Laut), and Indonesia (Kalimantan, Sumatra, Lingga Isls, Banggi and Balambangan).
    STATUS: CITES – Appendix II; IUCN – Lower Risk (lc).
    SYNONYMS: *caedis* Chasen and Kloss, 1932; *humeralis* Robinson and Kloss, 1919; *malaccana*
        Anderson, 1879; *sincipis* Lyon, 1911.
    COMMENTS: Arrangement of subspecies follows Corbet (in Corbet and Hill, 1992).

*Tupaia moellendorffi* Matschie, 1898. Sitzb. Ges. Naturf. Fr. Berlin, p. 39.
    COMMON NAME: Calamian Treeshrew.
    TYPE LOCALITY: Philippines, Culion Isl.
    DISTRIBUTION: Calamian Isls (Busuanga, Culion) and Cuyo in the Philippines.
    STATUS: CITES – Appendix II; IUCN – Vulnerable (as included in *T. palawanensis*).
    SYNONYMS: *busuangae* Sanborn, 1952; *cuyonis* Miller, 1910.
    COMMENTS: The three named forms included here appear more closely related to each
        another than any is to *T. palawanensis* (where they are usually arranged; e.g. Heaney
        et al., 1998) and I provisionally separate them here as a distinctive complex; see also
        comments by Lyon (1913). This arrangement is somewhat more consistent with
        species-boundaries traditionally recognized in squirrels from the Palawan region (see
        accounts of *Sundasciurus*).

*Tupaia montana* Thomas, 1892. Ann. Mag. Nat. Hist., ser. 6, 9:252.
    COMMON NAME: Mountain Treeshrew.
    TYPE LOCALITY: Borneo, Sarawak, Mt. Dulit, 5,000 ft. (1,524 m).
    DISTRIBUTION: Mountains of Sarawak and W Sabah (Malaysia); probably N Kalimantan
        (Indonesia).
    STATUS: CITES – Appendix II; IUCN – Lower Risk (lc).
    SYNONYMS: *baluensis* Lyon, 1913.
    COMMENTS: A Bornean montane endemic with two well-marked subspecies.

*Tupaia nicobarica* (Zelebor, 1869). Reise Oesterr. Fregatte Nov. Zool., 1(Wirbelth.), 1(Säugeth.):17,
    pl. 1.
    COMMON NAME: Nicobar Treeshrew.
    TYPE LOCALITY: Nicobar Isls, Great Nicobar Isl.
    DISTRIBUTION: Great and Little Nicobar Isls (India).
    STATUS: CITES – Appendix II; IUCN – Endangered.
    SYNONYMS: *surda* Miller, 1902.
    COMMENTS: A distinctive species with two slightly-differentiated subspecies. Part of the
        small assemblage of mammal species endemic to the Nicobar Isls that also includes
        *Pteropus faunulus, Rattus palmarum* and *Crocidura nicobarica*.

*Tupaia palawanensis* Thomas, 1894. Ann. Mag. Nat. Hist., ser. 6, 13:367.
    COMMON NAME: Palawan Treeshrew.
    TYPE LOCALITY: Philippines, Palawan Isl.
    DISTRIBUTION: Palawan and Balabac in the Philippines.
    STATUS: CITES – Appendix II; IUCN – Vulnerable.
    COMMENTS: Included in *T. glis* by Corbet (in Corbet and Hill, 1992) and many earlier
        authors, but specific separation is supported by pelage coloration differences,
        craniodental features (K. H. Han et al., 2000), karyotypic data (Arrighi et al., 1969), and

immunological distances (Dene et al., 1978). *T. moellendorffi* is provisionally separated from *palawanensis* here (see account above).

*Tupaia picta* Thomas, 1892. Ann. Mag. Nat. Hist., ser. 6, 9:251.
    COMMON NAME: Painted Treeshrew.
    TYPE LOCALITY: Borneo, Sarawak, Baram Dist., Apoh.
    DISTRIBUTION: Borneo: N Sarawak (Malaysia), E Kalimantan (Indonesia), and Brunei.
    STATUS: CITES – Appendix II; IUCN – Lower Risk (lc).
    SYNONYMS: *fuscior* Medway, 1965.
    COMMENTS: Endemic to Borneo. Medway (1977:25) recognized two subspecies.

*Tupaia splendidula* Gray, 1865. Proc. Zool. Soc. Lond., 1865:322, pl. 12.
    COMMON NAME: Ruddy Treeshrew.
    TYPE LOCALITY: Borneo (no specific locality).
    DISTRIBUTION: Malaysia and Indonesia, including S Borneo (*splendidula*) and Karimata Isl (*carimatae*), Bunguran (*natunae*) and Laut (*lucida*) in the N Natuna Isls, and Riabu (*riabus*) in the Anambas Isls.
    STATUS: CITES – Appendix II; IUCN – Lower Risk (lc).
    SYNONYMS: *muelleri* Kohlbrugge, 1896; *ruficaudata* Mivart, 1867; ***carimatae*** Miller, 1906; ***lucida*** Thomas and Hartert, 1895; ***natunae*** Lyon, 1911; *typica* Thomas and Hartert, 1895 [*nomen nudum*]; ***riabus*** Lyon, 1913.
    COMMENTS: *T. s. lucida* and *T. s. natunae* of the Natuna Isls are very distinctive forms; although traditionally included in *T. splendidula* they may be more closely allied to *T. glis*. Because of its probable affinity to Natuna treeshrews, *T. riabus* from Riabu (Anambas Isls) is provisionally placed here rather than with *T. glis*.

*Tupaia tana* Raffles, 1821. Trans. Linn. Soc. Lond., 13:257.
    COMMON NAME: Large Treeshrew.
    TYPE LOCALITY: Indonesia, Sumatra, Bencoolen (= Bengkulu).
    DISTRIBUTION: Malaysia (Sabah, Sarawak, Banggi) and Indonesia (Kalimantan, Sumatra, Batu Isls, Lingga Isls, Bangka, Belitung, Tambelan, Serasan).
    STATUS: CITES – Appendix II; IUCN – Lower Risk (lc).
    SYNONYMS: *nainggolani* (Sody, 1936); ***banguei*** Chasen and Kloss, 1932; ***besara*** (Lyon, 1913); ***bunoae*** Miller, 1900; ***cervicalis*** Miller, 1903; ***chrysura*** Günther, 1876; ***kelabit*** Davis, 1958; ***kretami*** Davis, 1962; ***lingae*** (Lyon, 1913); ***masae*** (Lyon, 1913); ***nitida*** Chasen, 1933; ***paitana*** (Lyon, 1913); *griswoldi* (Coolidge, 1938); ***sirhassenensis*** Miller, 1901; ***speciosa*** (Wagner, 1841); *tuancus* (Lyon, 1913); ***utara*** (Lyon, 1913).
    COMMENTS: Arrangement of subspecies is based on accounts by Chasen (1940), Lyon (1913), and Medway (1977), and on examination of specimens at USNM.

*Urogale* Mearns, 1905. Proc. U. S. Natl. Mus., 28:435.
    TYPE SPECIES: *Urogale cylindrura* Mearns, 1905 (= *Tupaia everetti* Thomas, 1892).
    COMMENTS: *Urogale* and *Tupaia* are closely allied.

*Urogale everetti* (Thomas, 1892). Ann. Mag. Nat. Hist., ser. 6, 9:250.
    COMMON NAME: Mindanao Treeshrew.
    TYPE LOCALITY: Philippines, Mindanao Isl, Zamboanga.
    DISTRIBUTION: Philippines: Mindanao, Siargao, and Dinagat.
    STATUS: CITES – Appendix II; IUCN – Vulnerable.
    SYNONYMS: *cylindrura* Mearns, 1905.
    COMMENTS: Variation in *U. everetti* was reviewed by Angst and Mann (1971). No subspecies are recognized here, but an undescribed subspecies occurs on Dinagat.

**Family Ptilocercidae** Lyon, 1913. Proc. U. S. Natl. Mus., 45:4.
    COMMENTS: Given full familial rank here and by Shoshani and McKenna (1998), but not by McKenna and Bell (1997). *Ptilocercus* exhibits many plesiomorphic and autapomorphic external, craniodental, and postcranial traits unique among treeshrews (including a terminally distichous tail); see Lyon (1913:4), Le Gros Clark (1926) and Sargis (2001, 2002*a*, *b*).

*Ptilocercus* Gray, 1848. Proc. Zool. Soc. Lond., 1848:23.
> TYPE SPECIES: *Ptilocercus lowii* Gray, 1848.
> SYNONYMS: *Ptilocerus* Brehm, 1864.
> COMMENTS: For discussion regarding the date of publication, see McAllan and Bruce (1989).

*Ptilocercus lowii* Gray, 1848. Proc. Zool. Soc. Lond., 1848:24.
> COMMON NAME: Pen-tailed Treeshrew.
> TYPE LOCALITY: Borneo, Sarawak, Kuching.
> DISTRIBUTION: S Thailand, Malaysia (Malay Peninsula, Sabah, Sarawak, Labuan), Singapore, Brunei, and Indonesia (Sumatra, Kalimantan, and Riau, Batu, Siberut, Bangka, and Serasan Isls).
> STATUS: CITES – Appendix II; IUCN – Lower Risk (lc).
> SYNONYMS: *continentis* Thomas, 1910.

# ORDER DERMOPTERA
by Brian J. Stafford

**ORDER DERMOPTERA** Illiger, 1811.

**Family Cynocephalidae** Simpson, 1945. Bull. Am. Nat. Hist., 85:54.
SYNONYMS: Colugidae Miller, 1906; Galeopithecidae Gray, 1821; Galeopteridae Thomas, 1908.

*Cynocephalus* Boddaert, 1768. Dierk. Meng., 2:8.
TYPE SPECIES: *Lemur volans* Linnaeus, 1758.
SYNONYMS: *Colugo* Gray, 1870; *Dermopterus* Burnett, 1829; *Galeolemur* Lesson, 1840; *Galeopithecus* Pallas, 1783; *Galeopus* Rafinesque, 1815; *Pleuropterus* Burnett, 1829.
COMMENTS: See Corbet and Hill (1992), Melville (1977), Stafford and Szalay (2000), and Thomas (1908*a*).

*Cynocephalus volans* (Linnaeus, 1758). Syst. Nat., 10th ed., 1:30.
COMMON NAME: Philippine Flying Lemur.
TYPE LOCALITY: Philippine Isls.
DISTRIBUTION: Philippine Isls: Dinagat, Mindanao, Basilan, Samar, Siargao, Leyte, and Bohol.
STATUS: IUCN – Vulnerable.
SYNONYMS: *philippinensis* Waterhouse, 1838.
COMMENTS: Corbet and Hill (1992) correctly noted that although Pampanga has been given as the type locality, this species has never been known to occur on Luzon Isl. Cabrera (1925) listed *C. philippensis* (Lesson, 1840) as a synonym. Unable to locate this description.

*Galeopterus* Thomas, 1908. Ann. Mag. Nat. Hist., ser. 8, 1:252.
TYPE SPECIES: *Galeopithecus temminckii* Waterhouse, 1838.
COMMENTS: See Thomas (1908*a*), Stafford and Szalay (2000).

*Galeopterus variegatus* (Audebert, 1799). Hist Nat. Singes Makis, sig. Rr. Java.
COMMON NAME: Sunda Flying Lemur.
TYPE LOCALITY: Indonesia, Java.
DISTRIBUTION: Indochina to Java (Indonesia), Borneo, and most associated islands.
STATUS: IUCN – Lower Risk (lc) as *Cynocephalus variegatus*.
SYNONYMS: *abbotti* Lyon, 1911; *aoris* Miller, 1903; *borneanus* Lyon, 1911; *chombolis* Lyon, 1909; *gracilis* Miller, 1903; *hantu* Cabrera, 1924; *lautensis* Lyon, 1911; *lecheyi* Gyldenstolpe, 1919; *natunae* Miller, 1903; *peninsulae* Thomas, 1908; *perhentianus* Chasen and Kloss, 1929; *pumilis* Miller, 1903; *rufus* Desmarest, 1820; *saturatus* Miller, 1903; *taylori* Thomas, 1908; *tellonis* Lyon, 1908; *temminckii* Waterhouse, 1838; *ternatensis* Desmarest, 1817; *terutaus* Chasen and Kloss, 1929; *tuancus* Miller, 1903; *undatus* Wagner, 1839; *varius* Desmarest, 1817.
COMMENTS: Cabrera (1925) gave Geoffroy (1796) as the author of this *nomen*. I was unable to locate this reference. However, reading Geoffroy (1803) suggests that he only used French common names for this species in 1796. Desmarest (1820) supported this interpretation. Also, Cabrera (1925) listed *G. marmoratus* (Temminck, 1829) as a synonym of *G. variegatus*. I was unable to locate this description. Not listed in Sherborn (1927, 1931), not contained in Temminck (1824-1841). Desmarest (1820) incorrectly gave the localities of *G. rufus* and *G. ternatensis* as Pelew Isl and Ternate respectively. Here, these *nomina* are synonymized with *G. variegatus* because the descriptions in Desmarest (1820) seem closer to this species, and because his plate 22, fig. 2, illustrates an animal with the characteristically narrow rostrum of *G. variegatus*. Chasen and Kloss (1929*a*) described *G. v. terutaus* and *G. v. perhentianus* as subspecies. See Chasen and Kloss (1929*b*), Chasen (1940), Corbet and Hill (1992), and Stafford and Szalay (2000).

# ORDER PRIMATES
## by Colin P. Groves

**ORDER PRIMATES** Linnaeus, 1758.
> COMMENTS: Fully reviewed by Groves (2001*c*), whose arrangement is followed here, with the addition of some subsequently described species. McKenna and Bell (1997) placed all living Primates in a suborder Euprimates, and reduced Strepsirrhini and Haplorrhini to infraorders; they regarded the Dermoptera as a second suborder of Primates. If Dermoptera are retained as a separate order, as in this volume, the need for Euprimates (in a classification of living taxa) disappears and Strepsirrhini and Haplorrhini revert to suborders.

**SUBORDER STREPSIRRHINI** É. Geoffroy Saint-Hilaire, 1812.
> COMMENTS: McKenna and Bell (1997) divide the Strepsirrhini into superfamilies Daubentonioidea, Lemuroidea, Loroidea (including Cheirogaleidae) and Indroidea. Evidence that Cheirogaleidae is not related to lorises, and that Indriidae is sister-group to Lemuridae, is given by Groves (2001*c*).

**INFRAORDER LEMURIFORMES** Gray, 1821.
> COMMENTS: All Malagasy lemur families, with the possible exception of Daubentoniidae, form a monophyletic clade (reviewed in Groves, 2001*c*). As the families within this clade themselves fall into two groups, it is convenient to recognize infraorders as well as superfamilies.

**Superfamily Cheirogaleoidea** Gray, 1873. Proc. Zool. Soc. Lond. 1872:849 [1873].

**Family Cheirogaleidae** Gray, 1873. Proc. Zool. Soc. Lond. 1872:849 [1873].
> COMMENTS: Formerly included in Lemuridae. For status of this taxon, see Rumpler (1975) and Groves (2001*c*), who reviewed and rejected the hypothesis that they may be more closely related to (non-Malagasy) Lorisiformes than to (Malagasy) Lemuriformes.

*Allocebus* Petter-Rousseaux and Petter, 1967. Mammalia, 31:574.
> TYPE SPECIES: *Cheirogaleus trichotis* Günther, 1875.
> COMMENTS: Previously included in *Cheirogaleus*, but very distinct.

*Allocebus trichotis* (Günther, 1875). Proc. Zool. Soc. Lond., 1875:78.
> COMMON NAME: Hairy-eared Dwarf Lemur.
> TYPE LOCALITY: Madagascar, between Tamatave and Morondava.
> DISTRIBUTION: E Madagascar, vicinity of Morondava Bay.
> STATUS: CITES – Appendix I; U.S. ESA and IUCN – Endangered.

*Cheirogaleus* É. Geoffroy, 1812. Ann. Mus. Hist. Nat. Paris, 19:172.
> TYPE SPECIES: *Cheirogaleus major* É. Geoffroy, 1812; fixed by Elliot (1907*b*:548).
> SYNONYMS: *Altililemur* Elliot, 1913; *Cebugale* Lesson, 1840; *Mioxocebus* Lesson, 1840; *Myspithecus* F. Cuvier, 1842; *Opolemur* Gray, 1873.
> COMMENTS: Revised by Groves (2000*a*).

*Cheirogaleus adipicaudatus* Grandidier, 1868. Ann. Sci. Nat., 5th ser., Zool. Paléont., 10:378.
> COMMON NAME: Spiny Desert Dwarf Lemur.
> TYPE LOCALITY: Madagascar, Fort Dauphin.
> DISTRIBUTION: W and S Madagascar.
> STATUS: CITES – Appendix I; U.S. ESA – Endangered.
> SYNONYMS: *thomasi* Forsyth Major, 1894.

*Cheirogaleus crossleyi* A. Grandidier, 1870. Rev. Zool. pur et appliquée, 22:49.
> COMMON NAME: Crossley's Dwarf Lemur.

TYPE LOCALITY: Madagascar, forests east of Antsianak.
DISTRIBUTION: E Madagascar: inland regions from Vohima south to Imerima.
STATUS: CITES – Appendix I; U.S. ESA – Endangered.
SYNONYMS: *melanotis* Forsyth Major, 1894.

*Cheirogaleus major* É. Geoffroy, 1812. Ann. Mus. Hist. Nat. Paris, 19:172.
COMMON NAME: Greater Dwarf Lemur.
TYPE LOCALITY: Madagascar, Fort Dauphin.
DISTRIBUTION: E Madagascar: Antongil Bay south to nearly 23°S.
STATUS: CITES – Appendix I; U.S. ESA – Endangered; IUCN – Lower Risk (lc).
SYNONYMS: *commersonii* Wolf, 1822; *griseus* Lesson, 1840; *milii* É. Geoffroy St. Hilaire, 1828;
*typicus* A. Smith, 1833; *typus* F. Cuvier, 1842.

*Cheirogaleus medius* É. Geoffroy, 1812. Ann. Mus. Hist. Nat. Paris, 19:172.
COMMON NAME: Lesser Dwarf Lemur.
TYPE LOCALITY: Madagascar, supposedly from Fort Dauphin; fixed as Tsidsibon River by
Groves (2000*a*).
DISTRIBUTION: W Madagascar.
STATUS: CITES – Appendix I; U.S. ESA – Endangered; IUCN – Lower Risk (lc).
SYNONYMS: *minor* É. Geoffroy, 1812; *samati* Grandidier, 1867.

*Cheirogaleus minusculus* Groves, 2000. Int. J. Primatol., 21:960.
COMMON NAME: Small Iron-gray Dwarf Lemur.
TYPE LOCALITY: Madagascar, central plateau, Ambositra, ca. 20°S, 47°E.
DISTRIBUTION: Known only from type locality.
STATUS: CITES – Appendix I; U.S. ESA – Endangered.

*Cheirogaleus ravus* Groves, 2000. Int. J. Primatol., 21:960.
COMMON NAME: Large Iron-gray Dwarf Lemur.
TYPE LOCALITY: Madagascar, Tamatave, ca. 18°S, 14°E.
DISTRIBUTION: E Madagascar, about 17°-18°S.
STATUS: CITES – Appendix I; U.S. ESA – Endangered.

*Cheirogaleus sibreei* Forsyth Major, 1896. Ann. Mag. Nat. Hist., ser. 6, 18: 325.
COMMON NAME: Sibree's Dwarf Lemur.
TYPE LOCALITY: Madagascar, Ankeramadinika ("one day's journey east of Antananarivo").
DISTRIBUTION: E Madagascar: Ankeramadinika, Imerima and Pasandava.
STATUS: CITES – Appendix I; U.S. ESA – Endangered.

*Microcebus* É. Geoffroy, 1834. Cours Hist. Nat. Mamm., lecon 11, 1828:24.
TYPE SPECIES: *Lemur pusillus* É. Geoffroy, 1795 (= *Lemur murinus* J. F. Miller, 1777).
SYNONYMS: *Azema* Gray, 1870; *Gliscebus* Lesson, 1840; *Murilemur* Gray, 1870; *Myocebus* Wagner,
1841; *Myscebus* Lesson, 1840; *Scartes* Swainson, 1835
COMMENTS: Revised by Rasoloarison et al. (2000).

*Microcebus berthae* Rasoloarison, Goodman and Ganzhorn, 2000. Int. J. Primatol., 21:1001.
COMMON NAME: Berthe's Mouse Lemur.
TYPE LOCALITY: Madagascar, Province of Toliara, Kirindy/CFPF Forest, 60 km NE of
Morondava, 20°04′S, 44°39′E, about 40 m.
DISTRIBUTION: Madagascar: region surrounding Kirindy/CFPF Forest, perhaps south to
RS d'Andranomena and north to region of Analabe.
STATUS: CITES – Appendix I; U.S. ESA – Endangered.

*Microcebus griseorufus* Kollman, 1910. Bull. Mus. Hist. Nat. Paris, 16:304.
COMMON NAME: Red-and-gray Mouse Lemur.
TYPE LOCALITY: Madagascar, Province of Toliara, north of RS Beza Mahafaly, Forest of
Ihazoara, 23°41′S, 44°38′E, about 130 m.
DISTRIBUTION: Madagascar: region around RS Beza Mahafaly, north at least to
Lamboharana.
STATUS: CITES – Appendix I; U.S. ESA – Endangered.

*Microcebus murinus* (J. F. Miller, 1777). Cimelia Physica, p. 25.
>    COMMON NAME: Gray Mouse Lemur.
>    TYPE LOCALITY: Madagascar: Province of Toliara, S Andranomena, 20 km NNE of
>         Morondava, 20°09'S, 44°33'E, 40 m (fixed by Rasoloarison et al., 2000).
>    DISTRIBUTION: W Madagascar, apparently from the far southwest to the Sambirano region.
>    STATUS: CITES – Appendix I; U.S. ESA – Endangered; IUCN – Lower Risk (lc).
>    SYNONYMS: *gliroides* A. Grandidier, 1868; *madagascarensis* É. Geoffroy, 1812; *minima*
>         Boddaert, 1785; *minor* Gray, 1842; *palmarum* Lesson, 1840; *prehensilis* Kerr, 1792;
>         *pusillus* É. Geoffroy, 1795.

*Microcebus myoxinus* Peters, 1852. Reise nach Mozambique, Zool., 1:14.
>    COMMON NAME: Peters's Mouse Lemur.
>    TYPE LOCALITY: Madagascar, between Tsiribihina River and Soalala Peninsula (restricted by
>         Rasoloarison et al., 2000).
>    DISTRIBUTION: Madagascar: between Tsiribihna River and Soalala Peninsula.
>    STATUS: CITES – Appendix I; U.S. ESA and IUCN – Endangered.

*Microcebus ravelobensis* Zimmermann, Ehresmann, Zietemann, Radespiel, Randrianambinina, and
>    Rakotoarison, 1997. Primate Eye, 63:26.
>    COMMON NAME: Ravelobe Mouse Lemur.
>    TYPE LOCALITY: Madagascar, Province of Mahajanga, RF Ankarafantsika, Station Forestière
>         d'Ampijoroa, 16°35'S, 46°52'E, ca. 200 m.
>    DISTRIBUTION: Madagascar: Ankarafantsika region.
>    STATUS: CITES – Appendix I; U.S. ESA and IUCN – Endangered.
>    COMMENTS: A full description appears in Zimmermann et al. (1998), but the briefer
>         description in Zimmermann et al. (1997) satisfies the requirements of the Code.

*Microcebus rufus* É. Geoffroy, 1834. Cours Hist. Nat. Mamm., lecon 11, 1828:24.
>    COMMON NAME: Eastern Rufous Mouse Lemur.
>    TYPE LOCALITY: Madagascar.
>    DISTRIBUTION: E and N Madagascar.
>    STATUS: CITES – Appendix I; U.S. ESA – Endangered; IUCN – Lower Risk (lc).
>    SYNONYMS: *smithii* Gray, 1842.
>    COMMENTS: Separated from *murinus* by Petter et al. (1977:30).

*Microcebus sambiranensis* Rasoloarison, Goodman and Ganzhorn, 2000. Int. J. Primatol., 21:982.
>    COMMON NAME: Sambirano Mouse Lemur.
>    TYPE LOCALITY: Madagascar, Province of Mahajanga, RS Manongarivo, Bekolosy Forest,
>         14°02'S, 48°16'E, ca. 360 m.
>    DISTRIBUTION: Madagascar: Manongarivo Reserve.
>    STATUS: CITES – Appendix I; U.S. ESA – Endangered.

*Microcebus tavaratra* Rasoloarison, Goodman and Ganzhorn, 2000. Int. J. Primatol., 21:977.
>    COMMON NAME: Northern Rufous Mouse Lemur.
>    TYPE LOCALITY: Madagascar, Province of Antsiranana, RS Ankarana, Campement des
>         Anglais, 13°05'S, 49°06'E, 180 m.
>    DISTRIBUTION: Madagascar: Ankarana Reserve.
>    STATUS: CITES – Appendix I; U.S. ESA – Endangered.

*Mirza* Gray, 1870. Cat. Monkeys, Lemurs, Fruit-eating Bats Brit. Mus., p. 131.
>    TYPE SPECIES: *Cheirogaleus coquereli* A. Grandidier, 1867.
>    COMMENTS: Recognised as a genus separate from *Microcebus* by Schwartz and Tattersall (1985);
>         not recognized by McKenna and Bell (1997).

*Mirza coquereli* (A. Grandidier, 1867). Rev. Mag. Zool. Paris, ser. 2, 19:85.
>    COMMON NAME: Giant Mouse Lemur.
>    TYPE LOCALITY: Madagascar, Morondava.
>    DISTRIBUTION: W Madagascar.
>    STATUS: CITES – Appendix I; U.S. ESA – Endangered; IUCN – Vulnerable.

*Phaner* Gray, 1870. Cat. Monkeys, Lemurs, Fruit-eating Bats Brit. Mus., p. 135.
    TYPE SPECIES: *Lemur furcifer* Blainville, 1839.
    COMMENTS: Placed in a separate subfamily, Phanerinae, by Rumpler (1974) and by McKenna
        and Bell (1997); but the evidence that they are the sister clade to other Cheirogaleidae is
        incomplete. Groves and Tattersall (1991) described three new subspecies, which were
        raised to species rank by Groves (2001*c*).

*Phaner electromontis* Groves and Tattersall, 1991. Folia Primatol., 56:47.
    COMMON NAME: Mt. d'Ambre Fork-crowned Lemur.
    TYPE LOCALITY: Madagascar, Mt. d'Ambre, 12°40'S, 49°10'E.
    DISTRIBUTION: N Madagascar, Amber Mountain (Ambohitra).
    STATUS: CITES – Appendix I; U.S. ESA – Endangered; IUCN – Vulnerable as *P. furcifer*
        *electromontis.*

*Phaner furcifer* (Blainville, 1839). Osteogr. Mamm., Primates, p. 35.
    COMMON NAME: Masoala Fork-crowned Lemur.
    TYPE LOCALITY: Madagascar, Morondava.
    DISTRIBUTION: Madagascar, Masoala Peninsula.
    STATUS: CITES – Appendix I; U.S. ESA – Endangered; IUCN – Lower Risk (nt).

*Phaner pallescens* Groves and Tattersall, 1991. Folia Primatol., 56:47.
    COMMON NAME: Pale Fork-crowned Lemur.
    TYPE LOCALITY: Madagascar, Tabika, 22°10'S, 44°15'E, ca. 20 km NW of Ankazoabo.
    DISTRIBUTION: W Madagascar, discontinuously from Soalala to somewhat south of
        Fiherenana River.
    STATUS: CITES – Appendix I; U.S. ESA – Endangered; IUCN – Vulnerable as *P. furcifer*
        *pallescens.*

*Phaner parienti* Groves and Tattersall, 1991. Folia Primatol., 56:47.
    COMMON NAME: Pariente's Fork-crowned Lemur.
    TYPE LOCALITY: NW Madagascar, Djangoa, 13°50'S, 48°20'E, ca. 20 km southwest of Ambanja.
    DISTRIBUTION: NW Madagascar, Sambirano region.
    STATUS: CITES – Appendix I; U.S. ESA – Endangered; IUCN – Vulnerable as *P. furcifer parienti.*

**Superfamily Lemuroidea** Gray, 1821. London Med. Repos., 15:296.
    COMMENTS: Includes Indridae (Groves, 2001*c*).

**Family Lemuridae** Gray, 1821. London Med. Repos., 15:296.
    COMMENTS: Reviewed by Petter et al. (1977) and Groves (2001*c*).

*Eulemur* Simons and Rumpler, 1988. C. R. Acad. Sci. Paris, III, 307:547 (15 September).
    TYPE SPECIES: *Lemur mongoz* Linnaeus, 1766.
    SYNONYMS: *Petterus* Groves and Eaglen, 1988 (6 October).

*Eulemur albifrons* (É. Geoffroy, 1796). Mag. Encyclop., 1:20.
    COMMON NAME: White-headed Lemur.
    TYPE LOCALITY: Madagascar.
    DISTRIBUTION: N Madagascar, except extreme south; Mayotte (Comoro Isls).
    STATUS: CITES – Appendix I; U.S. ESA – Endangered; IUCN – Lower Risk (nt) as *E. fulvus*
        *albifrons.*
    SYNONYMS: *frederici* (Lesson, 1840).
    COMMENTS: Regarded as a distinct species by Groves (2001*c*).

*Eulemur albocollaris* (Rumpler, 1975). *In* Tattersall and Sussman (eds.), Lemur Biology, p. 29.
    COMMON NAME: White-collared Lemur.
    TYPE LOCALITY: Madagascar, Tsimbazaza.
    DISTRIBUTION: Madagascar: restricted to forest between Manampatrana and Mananara
        Rivers.
    STATUS: CITES – Appendix I; U.S. ESA – Endangered; IUCN – Critically Endangered as
        *E. fulvus albocollaris.*

COMMENTS: Regarded as a distinct species by Groves (2001c).

*Eulemur cinereiceps* (A. Grandidier and Milne-Edwards, 1890). Hist. Nat. Madagascar, Mamm., 5:147, pl. 140.
COMMON NAME: Gray-headed Lemur.
TYPE LOCALITY: Madagascar, Frafafangana.
DISTRIBUTION: Known only from type locality.
STATUS: CITES – Appendix I; U.S. ESA – Endangered.
COMMENTS: Discussed by Groves (2001c), who recognized it with a query. The two known specimens (both females) are somewhat different, and both are unlike any known taxon, although females of some taxa, such as *E. albocollaris*, are poorly known and proper comparison is difficult.

*Eulemur collaris* (É. Geoffroy, 1812). Ann. Mus. Hist. Nat. Paris, 19:161.
COMMON NAME: Red-collared Lemur.
TYPE LOCALITY: Madagascar.
DISTRIBUTION: Madagascar: from Mananara River south to near Tolagnaro.
STATUS: CITES – Appendix I; U.S. ESA – Endangered; IUCN – Vulnerable as *E. fulvus collaris*.
SYNONYMS: *melanocephala* (Gray, 1863); *xanthomystax* (Gray, 1863).
COMMENTS: Regarded as a distinct species by Groves (2001c).

*Eulemur coronatus* (Gray, 1842). Ann. Mag. Nat. Hist., [ser. 1], 10:257.
COMMON NAME: Crowned Lemur.
TYPE LOCALITY: Madagascar.
DISTRIBUTION: Mt. Ambre (N Madagascar).
STATUS: CITES – Appendix I; U.S. ESA – Endangered; IUCN – Vulnerable.
SYNONYMS: *chrysampyx* (Scheuermans, 1846).
COMMENTS: Separated from *mongoz* by Petter et al. (1977:151).

*Eulemur fulvus* (É. Geoffroy, 1796). Mag. Encyclop., 1:47.
COMMON NAME: Brown Lemur.
TYPE LOCALITY: Madagascar.
DISTRIBUTION: W Madagascar from Akarafantsika north to the Sambirano, perhaps to the Tsaratanana; E Madagascar, some way northeast of Antananarivo; Mayotte (Comoro Isls).
STATUS: CITES – Appendix I; U.S. ESA – Endangered; IUCN – Lower Risk (nt).
SYNONYMS: *bruneus* (van der Hoeven, 1844); *mayottensis* (Schlegel, 1866).
COMMENTS: Does not include *albifrons, albocollaris, collaris, rufus* or *sanfordi*; Groves (2001c) treated these forms, generally classed as subspecies, as distinct species.

*Eulemur macaco* (Linnaeus, 1766). Syst. Nat., 12th ed., 1:34.
COMMON NAME: Black Lemur.
TYPE LOCALITY: Madagascar.
DISTRIBUTION: Nosi Be and NW Madagascar.
STATUS: CITES – Appendix I; U.S. ESA – Endangered; IUCN – Critically Endangered as *E. m. flavifrons*, otherwise Vulnerable.
SYNONYMS: *leucomystax* (Bartlett, 1863); *niger* (Schreber, 1775); **flavifrons** (Gray, 1867); *nigerrimus* (Sclater, 1880).

*Eulemur mongoz* (Linnaeus, 1766). Syst. Nat., 12th ed., 1:44.
COMMON NAME: Mongoose Lemur.
TYPE LOCALITY: Comoros, Anjouan Isl.
DISTRIBUTION: NW Madagascar, between Majunga and Betsiboka; Anjouan, Moheli (Comoro Isls).
STATUS: CITES – Appendix I; U.S. ESA – Endangered; IUCN – Vulnerable.
SYNONYMS: *albimanus* (É. Geoffroy, 1812); *anjuanensis* (É. Geoffroy, 1812); *brissonii* (Lesson, 1840); *bugi* (Lesson, 1840); *cuvieri* (Fitzinger, 1870); *dubius* (F. Cuvier, 1834); *johannae* (Trouessart, 1904); *macromongoz* (Lesson, 1840); *micromongoz* (Lesson, 1840); *nigrifrons* (É. Geoffroy, 1812); *noussardii* (Boitard, 1842); *ocularis* (Lesson, 1840).

*Eulemur rubriventer* (I. Geoffroy, 1850). C. R. Acad. Sci. Paris, 31:876.
  COMMON NAME: Red-bellied Lemur.
  TYPE LOCALITY: Madagascar, Tamatave.
  DISTRIBUTION: E Madagascar, from Tsaratanana Mtns to Ivohibé.
  STATUS: CITES – Appendix I; U.S. ESA – Endangered; IUCN – Vulnerable.
  SYNONYMS: *flaviventer* (I. Geoffroy, 1850); *rufipes* (Gray, 1871); *rufiventer* (Gray, 1870).

*Eulemur rufus* (Audebert, 1799). Hist. Nat. Singes et Makis, p. 12.
  COMMON NAME: Red-fronted Lemur.
  TYPE LOCALITY: Madagascar.
  DISTRIBUTION: E Madagascar, probably from Mangoro River to Manampatrana River;
      W Madagascar, from Betsiboka River to south of Fiherenana River.
  STATUS: CITES – Appendix I; U.S. ESA – Endangered; IUCN – Lower Risk (nt) as *E. fulvus*
      *rufus*.
  SYNONYMS: *rufifrons* (Bennett, 1833).
  COMMENTS: Regarded as a distinct species, not a subspecies of *E. fulvus*, by Groves (2001*c*).

*Eulemur sanfordi* (Archbold, 1932). Am. Mus. Novit., 518:1.
  COMMON NAME: Sanford's Lemur.
  TYPE LOCALITY: Madagascar, Mt. d'Ambre.
  DISTRIBUTION: N Madagascar, from Ampasindava peninsula south to Mahavavy and
      Manambato Rivers.
  STATUS: CITES – Appendix I; U.S. ESA – Endangered; IUCN – Vulnerable as *E. fulvus sanfordi*.
  COMMENTS: Regarded as a distinct species, not a subspecies of *E. fulvus*, by Groves (2001*c*).

*Hapalemur* I. Geoffroy, 1851. L'Inst. Paris, 19(929):341.
  TYPE SPECIES: *Lemur griseus* É. Geoffroy, 1812 (= *Lemur griseus* Link, 1795).
  SYNONYMS: *Hapalolemur* Giebel, 1855; *Myoxicebus* Elliot, 1913.
  COMMENTS: Placed in Lepilemuridae (= Megaladapidae) by Tattersall (1982).

*Hapalemur alaotrensis* Rumpler, 1975. *In* Tattersall and Sussman (eds.), Lemur Biology, p. 28.
  COMMON NAME: Bandro.
  TYPE LOCALITY: Madagascar, Lake Alaotra.
  DISTRIBUTION: Reed-beds around Lake Alaotra, E Madagascar.
  STATUS: CITES – Appendix I; U.S. ESA – Endangered; IUCN – Critically Endangered as
      *H. griseus alaotrensis*.
  COMMENTS: Groves (2001*c*) treated this as a full species.

*Hapalemur aureus* Meier, Albignac, Peyriéras, Rumpler, and Wright, 1987. Folia Primatol., 48:211.
  COMMON NAME: Golden Bamboo Lemur.
  TYPE LOCALITY: Madagascar, 6.25 km from the village of Ronomafana, 21°16′38″S,
      47°23′50″E.
  DISTRIBUTION: Between Namorona River and Bevoahazo Village, SE Madagascar.
  STATUS: CITES – Appendix I; U.S. ESA – Endangered; IUCN – Critically Endangered.

*Hapalemur griseus* (Link, 1795). Beytr. Naturg., 1:65.
  COMMON NAME: Gray Bamboo Lemur.
  TYPE LOCALITY: Madagascar.
  DISTRIBUTION: E Madagascar from Tsaratanana Massif south to Tolagnaro.
  STATUS: CITES – Appendix I; U.S. ESA – Endangered; IUCN – Lower Risk (nt) as *H. g. griseus*,
      otherwise Lower Risk (lc).
  SYNONYMS: *cinereus* (Desmarest, 1820); *olivaceus* I. Geoffroy, 1851; *schlegeli* Pocock, 1917;
      **meridionalis** Warter, Randrianosolo, Dutrillaux and Rumpler, 1987.
  COMMENTS: Groves (1989:87) suggested that *H. g. alaotrensis* may be a full species, and both
      *alaotrensis* and *occidentalis* were definitively separated as species by Groves (2001*c*).

*Hapalemur occidentalis* Rumpler, 1975. *In* Tattersall and Sussman (eds.), Lemur Biology, p. 28.
  COMMON NAME: Sambirano Bamboo Lemur.
  TYPE LOCALITY: North and West of Madagascar.
  DISTRIBUTION: Sambirano region and perhaps elsewhere in W Madagscar.

STATUS: CITES – Appendix I; U.S. ESA – Endangered; IUCN – Vulnerable as *H. griseus occidentalis*.

COMMENTS: Groves (2001*c*) treated this as a full species.

*Lemur* Linnaeus, 1758. Syst. Nat., 10th ed., 1:24.

TYPE SPECIES: *Lemur catta* Linnaeus, 1758.

SYNONYMS: *Catta* Link, 1806; *Maki* Muirhead, 1819; *Mococo* Lesson, 1878; *Odorlemur* Bolwig, 1961; *Procebus* Storr, 1780; *Prosimia* Boddaert, 1785.

COMMENTS: Revised by Petter et al. (1977:128-213). Restricted to *L. catta* by Groves and Eaglen (1988:533) and Simons and Rumpler (1988:547).

*Lemur catta* Linnaeus, 1758. Syst. Nat., 10th ed., 1:30.

COMMON NAME: Ring-tailed Lemur.

TYPE LOCALITY: Madagascar.

DISTRIBUTION: S Madagascar.

STATUS: CITES – Appendix I; U.S. ESA – Endangered; IUCN – Vulnerable.

SYNONYMS: *mococo* (Muirhead, 1819).

*Prolemur* Gray, 1871. Proc. Zool. Soc. Lond., 1870:828 [1871].

TYPE SPECIES: *Hapalemur* (*Prolemur*) *simus* Gray, 1871.

SYNONYMS: *Prohapalemur* Lamberton, 1936.

COMMENTS: A synonym of *Hapalemur* according to McKenna and Bell (1997); regarded as a full genus by Groves (2001*c*).

*Prolemur simus* (Gray, 1871). Cat. Monkeys, Lemurs, Fruit-eating Bats Brit. Mus., p. 133.

COMMON NAME: Greater Bamboo Lemur.

TYPE LOCALITY: Madagascar.

DISTRIBUTION: Madagascar: confined to Ranomafana district; in historic times occurred around Bay of Antongil.

STATUS: CITES – Appendix I; U.S. ESA – Endangered; IUCN – Critically Endangered as *Hapalemur simus*.

SYNONYMS: *gallieni* (Standing, 1905).

COMMENTS: Includes *H. gallieni* Standing, 1905 (see Vuillaume-Randriamanantena et al., 1985).

*Varecia* Gray, 1863. Proc. Zool. Soc. Lond., 1863:135.

TYPE SPECIES: *Lemur varius* É. Geoffroy (= *Lemur macaco variegatus* Kerr, 1792).

COMMENTS: Separated from *Lemur* by J.-J. Petter (1962).

*Varecia rubra* (É. Geoffroy, 1812). Ann. Mus. Hist. Nat. Paris, 19:159.

COMMON NAME: Red Ruffed Lemur.

TYPE LOCALITY: Madagascar.

DISTRIBUTION: Madagascar: Masoala Peninsula.

STATUS: CITES – Appendix I; U.S. ESA – Endangered; IUCN – Critically Endangered.

SYNONYMS: *erythromela* (Lesson, 1840).

COMMENTS: Recognized as a full species by Groves (2001*c*).

*Varecia variegata* (Kerr, 1792). *In* Linnaeus, Anim. Kingdom, p. 85.

COMMON NAME: Black-and-white Ruffed Lemur.

TYPE LOCALITY: Madagascar.

DISTRIBUTION: East coast of Madagascar, to 19°S latitude.

STATUS: CITES – Appendix I; U.S. ESA and IUCN – Endangered.

SYNONYMS: *vari* (Muirhead, 1819); *varius* (I. Geoffroy, 1851); **editorum** (Osman Hill, 1953); *subcincta* (A. Smith, 1833).

**Family Lepilemuridae** Gray, 1870. Cat. Monkeys, Lemurs, Fruit-eating Bats Brit. Mus., p. 132.

SYNONYMS: Megaladapidae Major, 1893.

COMMENTS: Includes many extinct (subfossil) genera, as well as *Lepilemur*, according to Schwartz and Tattersall (1985:20) and Groves (1989:92). Called Megaladapidae by Groves

(1993, 2001c), but Lepilemuridae takes precedence. Considered a subfamily (Lepilemurinae) of Lemuridae by McKenna and Bell (1997), but it is not clearly more related to Lemuridae than to Indriidae.

*Lepilemur* I. Geoffroy, 1851. Cat. Meth. Coll. Mamm. Ois. (Mus. Hist. Nat. Paris), Primates, p. 75.
    TYPE SPECIES: *Lepilemur mustelinus* I. Geoffroy, 1851.
    SYNONYMS: *Galeocebus* Wagner, 1855; *Lepidilemur* Giebel, 1859; *Mixocebus*, Peters, 1874.
    COMMENTS: Revised by Petter et al. (1977:274-318). Seven of the eight species are known to be karyotypically distinct. Rumpler (1975) and Corbet and Hill (1980:83) placed this genus in a subfamily of Lemuridae. Petter and Petter (1977:6) placed it in its own family Lepilemuridae. Yoder et al. (1999) considered that *Lepilemur* is probably not, in fact, a member of the Megaladapidae, which is based on a subfossil genus.

*Lepilemur ankaranensis* Rumpler and Albignac, 1975. Am. J. Phys. Anthropol., 42:425.
    COMMON NAME: Ankarana Sportive Lemur.
    TYPE LOCALITY: Madagascar, Ankarana Forest.
    DISTRIBUTION: Montagne d'Ambre, Ankarana, Andrafiamena and Analamera regions, extreme N Madagascar.
    STATUS: CITES – Appendix I; U.S. ESA – Endangered; IUCN – Vulnerable as included in *L. septentrionalis*.
    SYNONYMS: *andrafiamenensis* Rumpler and Albignac, 1975.
    COMMENTS: Shown by Rumpler et al. (2001) to be a distinct species, not a subspecies of *L. septentrionalis*; they used the name *L. andrafiamenensis* for it, but Groves (1989:95), who synonymised *andrafiamenensis* and *ankaranensis* as a subspecies of *L. septentrionalis*, selected *ankaranensis* to have priority.

*Lepilemur dorsalis* Gray, 1870. Cat. Monkeys, Lemurs, Fruit-eating Bats Brit. Mus., p. 135.
    COMMON NAME: Black-striped Sportive Lemur.
    TYPE LOCALITY: NW Madagascar.
    DISTRIBUTION: Nosi Bé, Nosy Komba, and Sambirano region (NW Madagascar).
    STATUS: CITES – Appendix I; U.S. ESA – Endangered; IUCN – Vulnerable.
    SYNONYMS: *grandidieri* (Forsyth Major, 1894).

*Lepilemur edwardsi* (Forbes, 1894). Handbook of Primates, 1:87.
    COMMON NAME: Milne-Edwards's Sportive Lemur.
    TYPE LOCALITY: Madagascar, Betsaka Bay, Bombetoka.
    DISTRIBUTION: E Madagascar, from 15°15′S perhaps to the Tsiribihina River.
    STATUS: CITES – Appendix I; U.S. ESA – Endangered; IUCN – Lower Risk (nt).
    SYNONYMS: *rufescens* (Lorenz, 1898).
    COMMENTS: Includes *rufescens*, which Petter and Petter (1977:7) considered a distinct species.

*Lepilemur leucopus* (Major, 1894). Ann. Mag. Nat. Hist., ser. 6; 13:211.
    COMMON NAME: White-footed Sportive Lemur.
    TYPE LOCALITY: Madagascar, Fort Dauphin (Bevilany).
    DISTRIBUTION: Arid zone of S Madagascar.
    STATUS: CITES – Appendix I; U.S. ESA – Endangered; IUCN – Lower Risk (nt).
    SYNONYMS: *globiceps* (Forsyth Major, 1894).

*Lepilemur microdon* (Forsyth Major, 1894). *In* Forbes, Handbook of Primates, 1:88.
    COMMON NAME: Small-toothed Sportive Lemur.
    TYPE LOCALITY: Madagascar, E Betsileo, Ankafana Forest.
    DISTRIBUTION: E Madagascar, from 18°S to 24°50′S.
    STATUS: CITES – Appendix I; U.S. ESA – Endangered; IUCN – Lower Risk (nt).

*Lepilemur mustelinus* I. Geoffroy, 1851. Cat. Meth. Coll. Mamm. Ois. (Mus. Hist. Nat. Paris), Primates, p. 76.
    COMMON NAME: Weasel Lemur.
    TYPE LOCALITY: Madagascar, Tamatave.
    DISTRIBUTION: E Madagascar, from about 13°45′S to 20°S.

STATUS: CITES – Appendix I; U.S. ESA – Endangered; IUCN – Lower Risk (nt).
SYNONYMS: *caniceps* (Peters, 1875).

*Lepilemur ruficaudatus* A. Grandidier, 1867. Rev. Mag. Zool. Paris, ser. 2, 19:256.
COMMON NAME: Red-tailed Sportive Lemur.
TYPE LOCALITY: Madagascar, Morondava.
DISTRIBUTION: SW Madagascar, from about 19°45'S to 23°S, and along the Onilahy River.
STATUS: CITES – Appendix I; U.S. ESA – Endangered; IUCN – Lower Risk (nt).
SYNONYMS: *pallidicauda* Gray, 1873.

*Lepilemur septentrionalis* Rumpler and Albignac, 1975. Am. J. Phys. Anthropol., 42:425.
COMMON NAME: Northern Sportive Lemur.
TYPE LOCALITY: Madagascar, Sahafary Forest.
DISTRIBUTION: Sahafary Forest, extreme NE Madagascar.
STATUS: CITES – Appendix I; U.S. ESA – Endangered; IUCN – Vulnerable.
SYNONYMS: *sahafarensis* Rumpler and Albignac, 1975.
COMMENTS: Does not include *L. andrafiamenensis* (= *L. ankaranensis*; Rumpler et al., 2001).

**Family Indriidae** Burnett, 1828. Quart. J. Lit. Sci. Arts Lond., 2:306-307.
COMMENTS: Reviewed by Groves (2001c). Properly Indridae (see Jenkins,1987:43), but Indriidae was conserved as the correct spelling by Opinion 1995 of the International Commission on Zoological Nomenclature (2002a). The Indriid clade includes two recently-extinct families or subfamilies, Archaeolemuridae and Palaeopropithecidae (McKenna and Bell, 1997).

*Avahi* Jourdan, 1834. L'Institute, Paris, 2:231.
TYPE SPECIES: *Lemur laniger* Gmelin, 1788.
SYNONYMS: *Habrocebus* Wagner, 1839; *Iropocus* Gloger, 1841; *Microrhynchus* Jourdan, 1834 [not of Megerle, 1823 (Coleoptera)]; *Semnocebus* Lesson, 1840.
COMMENTS: *Lichanotus* has commonly been used for this genus, but see Jenkins (1987:55). Revised in part by Thalmann and Geissmann (2000).

*Avahi laniger* (Gmelin, 1788). Syst. Nat., 13th ed., 1:44.
COMMON NAME: Eastern Woolly Lemur.
TYPE LOCALITY: Madagascar.
DISTRIBUTION: E coast and Ankarafantsika Dist. in NW Madagascar.
STATUS: CITES – Appendix I; U.S. ESA – Endangered; IUCN – Lower Risk (nt).
SYNONYMS: *avahi* (van der Hoeven, 1844); *lanatus* (Wagner, 1840); *longicaudatus* (E. Geoffroy, 1796); *orientalis* (von Lorenz-Liburnau, 1898) .

*Avahi occidentalis* von Lorenz-Liburnau, 1898. Abh. Senckenb. Naturf. Ges., 21:452.
COMMON NAME: Western Woolly Lemur.
TYPE LOCALITY: Madagascar, Ambondrobe, NE of Bombetoka Bay, ca. 15°38'S, 46°24'E (according to lectotype selection by Thalmann and Geissmann, 2000).
DISTRIBUTION: Madagascar: Betsiboka River north as far as Mahajamba or Sofia Rivers, perhaps to Maevarano River.
STATUS: CITES – Appendix I; U.S. ESA – Endangered; IUCN – Vulnerable.
COMMENTS: Rumpler et al. (1990) suggested that *occidentalis* may be a distinct species, and they were followed by Thalmann and Geissmann (2000), who drew attention to a population in the Bemaraha district at 18°59'S, 44°45'E, which they suggested is likely to represent a further, undescribed species.

*Avahi unicolor* Thalmann and Geissmann, 2000. Int. J. Primatol., 21:934.
COMMON NAME: Sambirano Woolly Lemur.
TYPE LOCALITY: Madagascar, Cacamba (=Kakamba), Ampasindava peninsula, ca. 13°35'S, 47°57'E.
DISTRIBUTION: Madagascar: probably restricted to Sambirano region.
STATUS: CITES – Appendix I; U.S. ESA – Endangered.

*Indri* É. Geoffroy and G. Cuvier, 1796. Mag. Encyclop., 1:46.
    TYPE SPECIES: *Lemur indri* Gmelin, 1788.
    SYNONYMS: *Indris* G. Cuvier, 1805; *Lichanotus* Illiger, 1811; *Pithelemur* Lesson, 1840.

  *Indri indri* (Gmelin, 1788). Syst. Nat., 13th ed., 1:42.
    COMMON NAME: Indri.
    TYPE LOCALITY: Madagascar.
    DISTRIBUTION: NE to EC Madagascar.
    STATUS: CITES – Appendix I; U.S. ESA and IUCN – Endangered.
    SYNONYMS: *ater* (I. Geoffroy, 1825); *brevicaudatus* É. Geoffroy and G. Cuvier, 1796; *mitratus* (Peters, 1871); *niger* Lacépède, 1799; **variegatus** (Gray, 1872).

*Propithecus* Bennett, 1832. Proc. Zool. Soc. Lond., 1832:20.
    TYPE SPECIES: *Propithecus diadema* Bennett, 1832.
    SYNONYMS: *Macromerus* A. Smith, 1833.
    COMMENTS: Revised by Groves (2001*c*).

  *Propithecus coquereli* (A. Grandidier, 1867). Rev. Mag. Zool. Paris, ser. 2, 19:314.
    COMMON NAME: Coquerel's Sifaka.
    TYPE LOCALITY: Madagascar, Morondava.
    DISTRIBUTION: NW Madagascar, from Ambato-Boéni region to Antsohihy.
    STATUS: CITES – Appendix I; U.S. ESA – Endangered; IUCN – Endangered as *P. verreauxi coquereli*.
    SYNONYMS: *damonis* Gray, 1870.
    COMMENTS: Recognized as a full species by Groves (2001*c*).

  *Propithecus deckenii* A. Grandidier, 1867. Rev. Mag. Zool. Paris, ser. 2, 19:84.
    COMMON NAME: Van der Decken's Sifaka.
    TYPE LOCALITY: Madagascar, Kanatsy.
    DISTRIBUTION: NC to SW Madagascar.
    STATUS: CITES – Appendix I; U.S. ESA – Endangered; IUCN – Vulnerable as *P. verreauxi deckenii*, Critically Endangered as *P. v. coronatus*.
    SYNONYMS: **coronatus** Milne-Edwards, 1871; *damanus* Pollen *in* Schlegel, 1876.
    COMMENTS: Recognized as a full species by Groves (2001*c*).

  *Propithecus diadema* Bennett, 1832. Proc. Zool. Soc. Lond., 1832:20.
    COMMON NAME: Diademed Sifaka.
    TYPE LOCALITY: Madagascar.
    DISTRIBUTION: N and E Madagascar, south to Mangoro River.
    STATUS: CITES – Appendix I; U.S. ESA – Endangered; IUCN – Critically Endangered as *P. diadema diadema* and *P. d. candidus*.
    SYNONYMS: *albus* Vinson, 1862; *typicus* A. Smith, 1833; **candidus** Grandidier, 1871; *sericeus* Milne-Edwards and A. Grandidier, 1872.

  *Propithecus edwardsi* Grandidier, 1871. C. R. Acad. Sci. Paris, 72:232.
    COMMON NAME: Milne-Edwards's Sifaka.
    TYPE LOCALITY: Madagascar, west of Mananjary.
    DISTRIBUTION: Between Mangoro and Mananara Rivers.
    STATUS: CITES – Appendix I; U.S. ESA and IUCN – Endangered as *P. diadema edwardsi*.
    SYNONYMS: *bicolor* Gray, 1872; *holomelas* Günther, 1875.
    COMMENTS: Recognized as a full species by Groves (2001*c*).

  *Propithecus perrieri* Lavauden, 1931. C. R. Acad. Sci. Paris, 193:77.
    COMMON NAME: Perrier's Sifaka.
    TYPE LOCALITY: Madagascar, Forest of Analamera, southeast of Diego Suarez.
    DISTRIBUTION: E Madagascar, between Ankarana and the coast.
    STATUS: CITES – Appendix I; U.S. ESA and IUCN – Critically Endangered as *P. diadema perrieri*.
    COMMENTS: Recognized as a full species by Groves (2001*c*).

*Propithecus tattersalli* Simons, 1988. Folia Primatol., 50:146.
   COMMON NAME: Golden-crowned Sifaka.
   TYPE LOCALITY: Madagascar, 6-7 km NE of Daraina, Antseranana Prov., 13°9'S, 49°41'E.
   DISTRIBUTION: Ampandraha, Madirabe and Daraina districts, Madagascar.
   STATUS: CITES – Appendix I; U.S. ESA – Endangered; IUCN – Critically Endangered.

*Propithecus verreauxi* A. Grandidier, 1867. Rev. Mag. Zool. Paris, ser. 2, 19:84.
   COMMON NAME: Verreaux's Sifaka.
   TYPE LOCALITY: Madagascar, Tsifanihy (N of Cape Ste.-Marie).
   DISTRIBUTION: Madagascar: xerophytic bush zone south of Tsiribihina River.
   STATUS: CITES – Appendix I; U.S. ESA – Endangered; IUCN – Vulnerable.
   SYNONYMS: *majori* Rothschild, 1894; *verreauxoides* Lamberton, 1936.

**INFRAORDER CHIROMYIFORMES** Anthony and Coupin, 1931.
   COMMENTS: Retained as an infraorder (i.e. of equal status to Lemuriformes and Lorisiformes) by
      Groves (1989:65, 74-78; 2001*c*), because it does not certainly form a clade with other
      Malagasy taxa.

**Family Daubentoniidae** Gray, 1863. Proc. Zool. Soc. Lond., 1863:151.
   SYNONYMS: Cheiromyidae I. Geoffroy St. Hilaire, 1851; Chiromyidae Bonaparte, 1850.
   COMMENTS: Groves (1989:65, 74-78) proposed separating this family to its own infraorder,
      Chiromyiformes.

*Daubentonia* É. Geoffroy, 1795. Decad. Philos. Litt., 28:195.
   TYPE SPECIES: *Sciurus madagascariensis* Gmelin, 1788.
   SYNONYMS: *Aye-aye* Lacépède, 1799; *Cheiromys* G. Cuvier, 1817; *Cheyromys* É. Geoffroy, 1803;
      *Chiromys* Illiger, 1811; *Myslemur* Anon. [?de Blainville], 1846; *Myspithecus* de Blainville,
      1839; *Psilodactylus* Oken, 1816 [unavailable]; *Scolecophagus* É. Geoffroy, 1795.

*Daubentonia madagascariensis* (Gmelin, 1788). Syst. Nat., 13th ed., 1:152.
   COMMON NAME: Aye-aye.
   TYPE LOCALITY: NW Madagascar.
   DISTRIBUTION: NE and NW Madagascar (discontinuous).
   STATUS: CITES – Appendix I; U.S. ESA and IUCN – Endangered.
   SYNONYMS: *daubentonii* (Shaw, 1800); *laniger* (G. Grandidier, 1930); *psilodactylus* (Schreber,
      1800).
   COMMENTS: Type locality of *laniger* is "forest of the east"; this seems to be a potentially
      available name for an eastern subspecies, if recognized. The name *D. robusta*
      Lamberton, 1934 was given to subfossil remains; it is not known whether these
      represent the living aye-aye or a separate, extinct species.

**INFRAORDER LORISIFORMES** Gregory, 1915.

**Family Lorisidae** Gray, 1821. London Med. Repos., 15:298.
   COMMENTS: The original form of the name was Loridae (see Jenkins [1987:1] and McKenna and
      Bell [1997]), but the more commonly used form Lorisidae was conserved by Opinion
      1995 of the International Commission on Zoological Nomenclature (2002*a*). Goodman
      et al. (1998) recognized two subfamilies, Lorinae and Perodicticinae, for the Asian and
      African genera respectively, and included Galagoninae as a third subfamily. McKenna and
      Bell (1997) included Galagoninae and Lorinae as subfamilies, with Lorini (*Arctocebus* and
      *Loris*) and Nycticebini (*Nycticebus* and *Perodicticus*) as tribes; but their tribes are not
      monophyletic (Groves, 2001*c*).

*Arctocebus* Gray, 1863. Proc. Zool. Soc. Lond., 1863:150.
   TYPE SPECIES: *Perodicticus calabarensis* J. A. Smith, 1860.

*Arctocebus aureus* de Winton, 1902. Ann. Mag. Nat. Hist., ser. 7, 9:48.
   COMMON NAME: Calabar Angwantibo.
   TYPE LOCALITY: Equatorial Guinea, 50 mi. (80 km) up Benito River.

DISTRIBUTION: C Africa, south of Sanaga River, W and N of Congo/Oubangui River system.
STATUS: CITES – Appendix II; IUCN – Lower Risk (nt).
SYNONYMS: *ruficeps* Thomas, 1913.
COMMENTS: Formerly classified as a subspecies of *Arctocebus calabarensis*, but considered a full species by Maier (1980:567) and Groves (1989:100-101; 2001*c*).

*Arctocebus calabarensis* (J. A. Smith, 1860). Proc. Roy. Phys. Soc. Edinburgh, 2:177.
COMMON NAME: Golden Angwantibo.
TYPE LOCALITY: Nigeria, Old Calabar.
DISTRIBUTION: C Africa, between Niger and Sanaga Rivers.
STATUS: CITES – Appendix II; IUCN – Lower Risk (nt).

*Loris* É. Geoffroy, 1796. Mag. Encyclop., 1:48.
TYPE SPECIES: *Lemur tardigradus* Linnaeus, 1758.
SYNONYMS: *Stenops* Illiger, 1811; *Tardigradus* Boddaert, 1785.

*Loris lydekkerianus* Cabrera, 1908. Bol. Soc. Esp. Hist. Nat., Madrid, 1908:139.
COMMON NAME: Gray Slender Loris.
TYPE LOCALITY: India: Madras.
DISTRIBUTION: Dry and hill zones of Sri Lanka; S India.
STATUS: CITES – Appendix II; IUCN – Endangered as *L. tardigradus grandis*, *L. t. nyctoceboides* (sic), and *L. t. nordicus*, Data Deficient as *L. t. lydekkerianus* and *L. t. malabaricus*.
SYNONYMS: **grandis** Hill and Phillips, 1932; *nordicus* Hill, 1933; **malabaricus** Wroughton, 1917; **nycticeboides** Hill, 1942.
COMMENTS: Separated as a species by Groves (2001*c*); but still more than one species may be concealed under this name.

*Loris tardigradus* (Linnaeus, 1758). Syst. Nat., 10th ed., 1:29.
COMMON NAME: Red Slender Loris.
TYPE LOCALITY: Sri Lanka.
DISTRIBUTION: Rainforest zone of SW Sri Lanka.
STATUS: CITES – Appendix II; IUCN – Endangered as *L. t. tardigradus*.
SYNONYMS: *ceylonicus* Fischer, 1804; *gracilis* É. Geoffroy, 1796; *zeylanicus* Lydekker, 1905.

*Nycticebus* É. Geoffroy, 1812. Ann. Mus. Hist. Nat. Paris, 19:163.
TYPE SPECIES: *Tardigradus coucang* Boddaert, 1785.

*Nycticebus bengalensis* (Lacépède, 1800). Tabl. Mamm. Oiseaux, p. 68.
COMMON NAME: Bengal Slow Loris.
TYPE LOCALITY: India, Bengal.
DISTRIBUTION: Assam (India) to Vietnam and S Thailand (Isthmus of Kra); Yunnan, perhaps Kwangsi (China).
STATUS: CITES – Appendix II; IUCN – Data Deficient.
SYNONYMS: *cinereus* Milne-Edwards, 1867; *incanus* Thomas, 1921; *tenasserimensis* Elliot, 1913.
COMMENTS: Recognized as a full species by Groves (2001*c*).

*Nycticebus coucang* (Boddaert, 1785). Elench. Anim., p. 67.
COMMON NAME: Slow Loris.
TYPE LOCALITY: Malaysia, Malacca.
DISTRIBUTION: Sulu Arch. (S Philippines); Malay Peninsula, Tioman and offshore islands, Sumatra, Bangka, Java, Borneo, Natuna Isl.
STATUS: CITES – Appendix II; IUCN - Data Deficient as *N. c. menagensis* and *N. javanicus*, Lower Risk (lc) as *N. c. coucang*.
SYNONYMS: *brachycephalus* Sody, 1949; *buku* Robinson, 1917; *hilleri* Stone and Rehn, 1902; *insularis* Robinson, 1917; *malaiana* Anderson, 1881; *natunae* Stone and Rehn, 1902; *sumatrensis* Ludeking, 1867; *tardigradus* (Raffles, 1821); **menagensis** Trouessart, 1898; *bancanus* Lyon, 1906; *borneanus* Lyon, 1906; *philippinus* Cabrera, 1908; **javanicus** É. Geoffroy, 1812; *ornatus* Thomas, 1921.

*Nycticebus pygmaeus* Bonhote, 1907. Abstr. Proc. Zool. Soc. Lond., 1907(38):2.
COMMON NAME: Pygmy Slow Loris.
TYPE LOCALITY: Vietnam, Nhatrang.
DISTRIBUTION: Laos; Cambodia; Vietnam, east of Mekong River; S Yunnan (China).
STATUS: CITES – Appendix II; U.S. ESA – Threatened; IUCN – Vulnerable.
SYNONYMS: *intermedius* Dao Van Tien, 1960.
COMMENTS: Includes *intermedius*; see Groves (1971c) and Lekagul and McNeely (1977:270).

*Perodicticus* Bennett, 1831. Proc. Zool. Soc. Lond., 1830:109 [1831].
TYPE SPECIES: *Lemur potto* Müller, 1766.
SYNONYMS: *Potto* Lesson, 1840.

*Perodicticus potto* (Müller, 1766). *In* Linnaeus, Vollstand Natursyst. Suppl., p. 12.
COMMON NAME: Potto.
TYPE LOCALITY: Ghana, Elmina.
DISTRIBUTION: Cameroon to Guinea; Republic of Congo; Gabon; Dem. Rep. Congo to
W Kenya.
STATUS: CITES – Appendix II; IUCN – Lower Risk (lc).
SYNONYMS: *bosmannii* (Lesson, 1840); *geoffroyi* Bennett, 1831; *guineensis* (Desmarest, 1820);
*ju-ju* Thomas, 1910; **edwardsi** Bouvier, 1879; *batesi* de Winton, 1902; *faustus* Thomas,
1910; **ibeanus** Thomas, 1910; *arrhenii* (Lönnberg, 1917); *nebulosus* (Lorenz, 1917).

*Pseudopotto* Schwartz, 1996. Anthropol. Pap. Amer. Mus. Nat. Hist., 78:8.
TYPE SPECIES: *Pseudopotto martini* Schwartz, 1996.

*Pseudopotto martini* Schwartz, 1996. Anthropol. Pap. Amer. Mus. Nat. Hist., 78:8.
COMMON NAME: False Potto.
TYPE LOCALITY: "Equatorial Africa".
DISTRIBUTION: unknown; one specimen is said to be from Cameroon.
STATUS: CITES – Appendix II.
COMMENTS: A controversial taxon, based on only two specimens, both of uncertain
provenance. Groves (1998, 2001c) argued that the coalescence of a number of unique
features sets it apart from *Perodicticus*; but Sarmiento (1998) considered that it falls
within the range of *Perodicticus*. McKenna and Bell (1997) recognized the genus, but
did not assign it to either of their two subfamilies (Lorinae, Galagoninae).

**Family Galagidae** Gray, 1825. Ann. Philos., n.s., 10:338.
COMMENTS: Formerly considered a subfamily of Lorisidae; see Hill and Meester (1977:2) and
Jenkins (1987:85). The correct form of the name is Galagonidae (see Jenkins, 1987:1), but
Galagidae was conserved by Opinion 1995 of the International Commission on
Zoological Nomenclature (2002a).

*Euoticus* Gray, 1863. Proc. Zool. Soc. Lond., 1863:140.
TYPE SPECIES: *Otogale pallida* Gray, 1863.
COMMENTS: Recognized as a genus by Groves (1989:103; 2001c), and by McKenna and Bell
(1997).

*Euoticus elegantulus* (Le Conte, 1857). Proc. Acad. Nat. Sci. Phila., 9:10.
COMMON NAME: Southern Needle-clawed Bushbaby.
TYPE LOCALITY: Gabon, Ogooué River, Njola.
DISTRIBUTION: Gabon, Republic of Congo, Rio Muni, Cameroon south of Sanaga River.
STATUS: CITES – Appendix II; IUCN – Lower Risk (nt).
SYNONYMS: *apicalis* (du Chaillu 1860); *tonsor* (Dollman, 1910).

*Euoticus pallidus* (Gray, 1863). Proc. Zool. Soc. Lond., 1863:140.
COMMON NAME: Northern Needle-clawed Bushbaby.
TYPE LOCALITY: Equatorial Guinea, Bioko.
DISTRIBUTION: Bioko (Equatorial Guinea); Korup and Cross River region, on both sides of
Nigeria-Cameroon border.

STATUS: CITES – Appendix II; IUCN – Endangered as *E. p. pallidus*, otherwise Lower Risk (nt).
SYNONYMS: ***talboti*** Dollman, 1910.
COMMENTS: Accepted as a species by Groves (1989:104; 2001c).

*Galago* É. Geoffroy, 1796. Mag. Encyclop., 1:49.
TYPE SPECIES: *Galago senegalensis* É. Geoffroy, 1796.
SYNONYMS: *Chirosciurus* Grevais, 1836; *Galagoides* A. Smith, 1833; *Hemigalago* Dahlbom, 1857; *Macropus* Fischer, 1811 [not of Shaw, 1790]; *Otolicnus* Illiger, 1811; *Sciurocheirus* Gray, 1872.
COMMENTS: McKenna and Bell (1997) included *Otolemur* as a synonym. Groves (2001c) accepted the following species groups: (1) *C. senegalensis* group (*senagalensis, moholi, gallarum*), (2) *G. matschiei* group (monotypic), (3) *G. alleni* group (*alleni, cameronensis, gabonensis*), (4) *G. zanzibaricus* group (*zanzibaricus, granti, nyasae*), (5) *G. orinus* group (*orinus, rondoensis*), and (6) *G. demidoff* group (*demidoff, thomasi*). DelPero et al. (2000) suggested that the *G. alleni* group should be reallocated to *Otolemur*.

*Galago alleni* Waterhouse, 1838. Proc. Zool. Soc. Lond., 1837:87 [1838].
COMMON NAME: Bioko Allen's Bushbaby.
TYPE LOCALITY: Equatorial Guinea, Bioko.
DISTRIBUTION: Bioko (Equatorial Guinea).
STATUS: CITES – Appendix II; IUCN – Endangered as *G. a. alleni*.
COMMENTS: *G. alleni* species group. Divided into three species by Groves (2001c), so does not include *cameronensis, gabonensis*.

*Galago cameronensis* (Peters, 1876). Monatsb. K. Preuss. Akad. Wiss. Berlin, 1876:472.
COMMON NAME: Cross River Bushbaby.
TYPE LOCALITY: Cameroon: Duala.
DISTRIBUTION: Cameroon northwest of the lower Sanaga River; SE Nigeria.
STATUS: CITES – Appendix II.
COMMENTS: *G. alleni* species group. Separated from *G. alleni* by Groves (2001c).

*Galago demidoff* G. Fischer, 1806. Mem. Soc. Imp. Nat. Moscow, 1:24.
COMMON NAME: Prince Demidoff's Bushbaby.
TYPE LOCALITY: Senegal.
DISTRIBUTION: Senegal to E Dem. Rep. Congo; Bioko (Equatorial Guinea).
STATUS: CITES – Appendix II; IUCN – Lower Risk (lc).
SYNONYMS: *anomurus* de Pousargues, 1893; *demidovii* Fischer, 1808; *medius* (Thomas, 1915); *murinus* Murray, 1859; *peli* (Temminck, 1853); *phasma* (Cabrera and Ruxton, 1926); *poensis* Thomas, 1904; *pusillus* (Peters, 1876).
COMMENTS: *G. demidoff* species group. Bearder et al. (1995) placed this group in genus *Galagoides*, but Groves (2001c) and Delpero et al. (2000) considered this premature, as it is unclear how species should be assorted between *Galagoides* and *Galago*. Nash et al. (1989) recognized *G. thomasi* Elliot, 1907, as a full species, partially sympatric with *demidoff*; followed by Groves (2001c), who considered that a number of other species would eventually be recognized in addition. For use of *demidoff* in place of *demidovii*, see Jenkins (1987:98).

*Galago gabonensis* Gray, 1863. Proc. Zool. Soc. Lond., 1863:146.
COMMON NAME: Gabon Bushbaby.
TYPE LOCALITY: Gabon.
DISTRIBUTION: Gabon, Cameroon south of the Sanaga River, Rio Muni, Republic of Congo.
STATUS: CITES – Appendix II.
SYNONYMS: *batesi* Elliot, 1907.
COMMENTS: *G. alleni* species group. Separated from *G. alleni* by Groves (2001c).

*Galago gallarum* Thomas, 1901. Ann. Mag. Nat. Hist., ser. 7, 8:27.
COMMON NAME: Somali Bushbaby.
TYPE LOCALITY: Ethiopia, Webi Dau, Boran County.
DISTRIBUTION: Between Tana River (Kenya) and Webi Shebele River (Somalia), to Lake Turkana and Ethiopian rift lakes.

STATUS: CITES – Appendix II; IUCN – Lower Risk (nt).
COMMENTS: *G. senegalensis* species group. Recognized as a species by Nash et al. (1989) and by Groves (1989, 2001*c*).

*Galago granti* Thomas and Wroughton, 1907. Proc. Zool. Soc. Lond., 1907:286.
COMMON NAME: Grant's Bushbaby.
TYPE LOCALITY: Mozambique, Inhambane, Coguno.
DISTRIBUTION: Mozambique north to Ulugurus in S Tanzania.
STATUS: CITES – Appendix II; IUCN – Data Deficient.
SYNONYMS: *mertensi* Frade, 1924.
COMMENTS: *G. zanzibaricus* species group. Recognized as a species by Groves (2001*c*).

*Galago matschiei* Lorenz, 1917. Ann. K. K. Naturhist. Hofmus, Wien, 31:237.
COMMON NAME: Dusky Bushbaby.
TYPE LOCALITY: Dem. Rep. Congo: Moera, Ituri River.
DISTRIBUTION: E Dem. Rep. Congo; perhaps Uganda.
STATUS: CITES – Appendix II; IUCN – Lower Risk (nt).
SYNONYMS: *inustus* Schwarz, 1931.
COMMENTS: *G. matschiei* species group. Placed in *Euoticus* by Petter and Petter-Rousseaux (1979). This name predates the more commonly used *inustus* (Groves, 1989:102; Nash et al., 1989:69-70).

*Galago moholi* A. Smith, 1836. Rept. Exped. Exploring Central Africa, 1834:42 [1836].
COMMON NAME: Moholi Bushbaby.
TYPE LOCALITY: South Africa, Limpopo, Marico-Limpopo confluence.
DISTRIBUTION: KwaZulu-Natal and N Namibia north to Lake Victoria.
STATUS: CITES – Appendix II; IUCN – Lower Risk (lc).
SYNONYMS: *australis* (Wagner, 1855); *bradfieldi* Roberts, 1931; *conspicillatus* I. Geoffroy, 1851; *intontoi* Monard, 1931; *mossambicus* (Peters, 1876), *tumbolensis* Monard, 1931.
COMMENTS: *G. senegalensis* species group. Separated from *senegalensis* by Jenkins (1987), Nash et al. (1989), and Groves (1989).

*Galago nyasae* Elliot, 1907. Ann. Mag. Nat. Hist., ser. 7, 20:188.
COMMON NAME: Malawi Bushbaby.
TYPE LOCALITY: Mozambique, mountains south of Lake Malawi.
DISTRIBUTION: S Malawi and neighbouring region of Mozambique.
STATUS: CITES – Appendix II.
COMMENTS: *G. zanzibaricus* species group. An undescribed taxon occurs on Mt. Cholo, Malawi; vocalizations of the "Kalwe small" galago (Bearder et al., 1995) may pertain to either of these (Groves, 2001*c*:113).

*Galago orinus* Lawrence and Washburn, 1936. Occas. Pap. Boston Soc. Nat. Hist., 8:259.
COMMON NAME: Uluguru Bushbaby.
TYPE LOCALITY: Tanzania, Uluguru Mtns, Bagilo.
DISTRIBUTION: High elevations in Uluguru and probably Usambara Mtns.
STATUS: CITES – Appendix II; IUCN – Data Deficient.
COMMENTS: *G. orinus* species group. Bearder et al. (1995) place this group in genus *Galagoides*, but Groves (2001*c*) and Delpero et al. (2000) considered this premature, as it is unclear how species should be assorted between *Galagoides* and *Galago*. Separated from *G. demidoff* by Bearder et al. (1995), and accepted as a distinct species by Groves (2001*c*).

*Galago rondoensis* (Honess, 1997). *In* Kingdon, Kingdon Field Guide to African Mammals, p. 106.
COMMON NAME: Rondo Bushbaby.
TYPE LOCALITY: Tanzania, Lindi district, Rondo plateau, Rondo Forest Reserve, 10°07'S, 39°23'E.
DISTRIBUTION: Rondo, Litipo, Ziwani and Pugu Forests, SE Tanzania.
STATUS: CITES – Appendix II; IUCN – Endangered.
COMMENTS: *G. zanzibaricus* species group. Recognised as a species by Groves (2001*c*). More fully described by Honess and Bearder (1996 [1997]).

*Galago senegalensis* É. Geoffroy, 1796. Mag. Encyclop., 1:38.
>    COMMON NAME: Senegal Bushbaby.
>    TYPE LOCALITY: Senegal.
>    DISTRIBUTION: Senegal to Somalia, south to the Mwanza and Ankole districts on Lake Victoria.
>    STATUS: CITES – Appendix II; IUCN – Lower Risk (lc).
>    SYNONYMS: *acaciarum* Lesson, 1840; *albipes* Dollman, 1909; *calago* (Shaw, 1800); *camerounensis* Monard, 1951; *galago* (G. Cuvier, 1798); *geoffroyi* Fischer, 1806; *pupulus* Elliot, 1910; *sennariensis* Gray, 1863; *teng* (Sundevall, 1843); **braccatus** Elliot, 1907; **dunni** Dollman, 1910; **sotikae** Hollister, 1920.
>    COMMENTS: *G. senegalensis* species group. Hill and Meester (1977:2) included *granti* in this species; but see Smithers and Wilson (1979).

*Galago thomasi* Elliot, 1907. Ann. Mag. Nat. Hist., ser. 7, 20:189.
>    COMMON NAME: Thomas's Bushbaby.
>    TYPE LOCALITY: Dem. Rep. Congo, Beni.
>    DISTRIBUTION: W Uganda and Kivu district of Dem. Rep. Congo.
>    STATUS: CITES – Appendix II; IUCN – Lower Risk (lc).
>    COMMENTS: *G. demidoff* species group. Nash et al. (1989) recognized *G. thomasi* as a full species, partially sympatric with *demidoff*; followed by Groves (2001c). Known only from specimens from W Uganda and Kivu district of Dem. Rep. Congo (including Idjwi Isl), but vocalizations ascribed to this species are recorded from all over the same range as *G. demidoff*.

*Galago zanzibaricus* Matschie, 1893. Sitzb. Ges. Naturf. Fr. Berlin, p. 111.
>    COMMON NAME: Zanzibar Bushbaby.
>    TYPE LOCALITY: Tanzania, Zanzibar, Yambiani.
>    DISTRIBUTION: E African coast from Tana River, south to S Mozambique; Zanzibar.
>    STATUS: CITES – Appendix II; IUCN – Lower Risk (nt).
>    SYNONYMS: *cocos* Heller, 1912; *udzungwensis* Honess, 1997.
>    COMMENTS: *G. zanzibaricus* species group. Separated from *senegalensis* by Kingdon (1971:309), Groves (1974b:463, 1989:103, 2001c), Jenkins (1987:118), and Nash et al. (1989). Bearder et al. (1995) placed this group in genus *Galagoides*, but Groves (2001c) considered this premature, as it is unclear how species should be assorted between *Galagoides* and *Galago*. For allocation of *udzungwensis* to this species, see Groves (2001c:116).

*Otolemur* Coquerel, 1859. Rev. Zool. Paris, 11:458.
>    TYPE SPECIES: *Otolemur agyisymbanus* Coquerel, 1859 (= *Otolicnus garnetti* Ogilby, 1838).
>    SYNONYMS: *Callotus* Gray, 1863; *Otogale* Gray, 1863.
>    COMMENTS: Included in *Galago* by McKenna and Bell (1997), but recognized as a full genus by Groves (1974b:461-463, 1989) and Jenkins (1987:122). May include the *Galago alleni* species group (DelPero et al., 2000).

*Otolemur crassicaudatus* (É. Geoffroy, 1812). Ann. Mus. Hist. Nat. Paris, 19:166.
>    COMMON NAME: Brown Greater Galago.
>    TYPE LOCALITY: Mozambique, Quelimane (see Thomas, 1917b).
>    DISTRIBUTION: Kenya, Tanzania and Rwanda to KwaZulu-Natal (South Africa) and Angola.
>    STATUS: CITES – Appendix II; IUCN – Lower Risk (lc).
>    SYNONYMS: *zuluensis* (Elliot, 1907); **kirkii** (Gray, 1865); *badius* Matschie, 1905; *lonnbergi* (Schwarz, 1930); *umbrosus* (Thomas, 1917).

*Otolemur garnettii* (Ogilby, 1838). Proc. Zool. Soc. Lond., 1838:6.
>    COMMON NAME: Northern Greater Galago.
>    TYPE LOCALITY: Zanzibar (designated by Thomas, 1917b:48).
>    DISTRIBUTION: S Somalia to SE Tanzania (including Zanzibar, Pemba and Mafia Isls) and perhaps N Mozambique.
>    STATUS: CITES – Appendix II; IUCN – Lower Risk (lc).
>    SYNONYMS: *agyisymbanus* Coquerel, 1859; **lasiotis** (Peters, 1876); *hindei* (Elliot, 1907); *hindsi* Elliot, 1913 [*lapsus*]; **kikuyuensis** (Lönnberg, 1912); **panganiensis** Matschie, 1905.

COMMENTS: Olson (1979, in litt.[Unpubl. Ph.D. Dissertation, Univ. London]) was the first to treat *garnettii* as a distinct species. It is partly sympatric with both *O. crassicaudatus* and *O. monteiri*.

*Otolemur monteiri* (Bartlett, 1863). Proc. Zool. Soc. Lond., 1863:145.
COMMON NAME: Silvery Greater Galago.
TYPE LOCALITY: Angola: Cuvo Bay.
DISTRIBUTION: *Brachystegia* woodland, from Angola to Tanzania to W Kenya and Rwanda.
STATUS: CITES – Appendix II.
SYNONYMS: **argentatus** (Lönnberg, 1913); *lestradei* (Schouteden, 1953).
COMMENTS: Separated from *crassicaudatus* by Groves (2001c), with some misgivings. The zone of intermediacy with *O. crassicaudatus* is very wide, extending through S Malawi, N Mozambique, and part of Zimbabwe. Kingdon (1997) recognized *Otolemur argentatus* from the Lake Victoria region, while retaining southerly examples (here *Otolemur m. monteiri*) in *O. crassicaudatus*.

# SUBORDER HAPLORRHINI Pocock, 1918.
COMMENTS: Recognised as an infraorder by McKenna and Bell (1997; see comments above, under Strepsirrhini); Tarsiiformes and Anthropoidea were regarded as Parvorders.

# INFRAORDER TARSIIFORMES Gregory, 1915.
COMMENTS: Almost invariably placed in a group with Simiiformes (=Anthropoidea), but according to Murphy et al. (2001c), may be closer to Strepsirrhini.

**Family Tarsiidae** Gray, 1825. Ann. Philos., n.s., 10:338.

*Tarsius* Storr, 1780. Prodr. Meth. Mamm., p. 33.
TYPE SPECIES: *Lemur tarsier* Erxleben, 1777.
SYNONYMS: *Cephalopachus* Swainson, 1835; *Hypsicebus* Lesson, 1840; *Macrotarsus* Link, 1795; *Rabienus* Gray, 1821.
COMMENTS: This genus, according to M. Goodman et al. (2001) and to work in progress by C. P. Groves and M. Shekelle, should probably be split into two or three separate genera (for Sulawesi, Philippine, and Western groups).

*Tarsius bancanus* Horsfield, 1821. Zool. Res. Java, 2, pl.
COMMON NAME: Horsfield's Tarsier.
TYPE LOCALITY: Indonesia, SE Sumatra, Bangka Isl.
DISTRIBUTION: Indonesia: Bangka Isl, Sumatra, Karimata Isl, Billiton Isl, and Sirhassen Isl (South Natuna Isls); Borneo.
STATUS: CITES – Appendix II; IUCN – Data Deficient as *T. b. borneanus*, *T. b. natunensis*, and *T. b. saltator*, otherwise Lower Risk (lc).
SYNONYMS: **borneanus** Elliot, 1910; *natunensis* Chasen, 1940; **saltator** Elliot, 1910.

*Tarsius dentatus* Miller and Hollister, 1921. Proc. Biol. Soc. Wash., 34:103.
COMMON NAME: Dian's Tarsier.
TYPE LOCALITY: C Sulawesi, Labua Sore.
DISTRIBUTION: C Sulawesi (Indonesia).
STATUS: CITES – Appendix II; IUCN – Lower Risk (conservation dependent) as *T. dianae*. Status unknown.
SYNONYMS: *dianae* Niemitz, Nietsch, Water, and Rumpler, 1991.
COMMENTS: Shekelle and Groves (in prep.) could find no difference between specimens from the type localities of *dentatus* (Labua Sore) and *dianae* (Kamarora) in either morphology or vocalizations; the two localities are very close.

*Tarsius pelengensis* Sody, 1949. Treubia, 20:143.
COMMON NAME: Peleng Tarsier.
TYPE LOCALITY: Indonesia, Peleng Isl.

DISTRIBUTION: Peleng Isl (Indonesia).
STATUS: CITES – Appendix II; IUCN – Data Deficient.
COMMENTS: Recognized as a species by Groves (2001*c*).

*Tarsius pumilus* Miller and Hollister, 1921. Proc. Biol. Soc. Wash., 34:103.
COMMON NAME: Pygmy Tarsier.
TYPE LOCALITY: C Sulawesi, Rano Rano.
DISTRIBUTION: Known only from type locality and Latimojong Mtns, C Sulawesi (Indonesia).
STATUS: CITES – Appendix II; IUCN – Data Deficient.
COMMENTS: Separated from *T. tarsier* (*T. spectrum*) by Musser and Dagosto (1987); recognized as a full species by Groves (1989, 2001*c*) and others.

*Tarsius sangirensis* Meyer, 1897. Abh. Zool. Anthrop.-Ethnology. Mus. Dresden, 6, 6:9.
COMMON NAME: Sangihe Tarsier.
TYPE LOCALITY: Indonesia, Sanghir (=Sangihe) Isls.
DISTRIBUTION: Restricted to Pulau (=Isl) Sangihe Besar (Indonesia).
STATUS: CITES – Appendix II; IUCN – Data Deficient.
COMMENTS: Recognised as a species by Feiler (1990:85) and Groves (2001*c*).

*Tarsius syrichta* (Linnaeus, 1758). Syst. Nat., 10th ed., 1:29.
COMMON NAME: Philippine Tarsier.
TYPE LOCALITY: Philippine Isls, Samar Isl.
DISTRIBUTION: Mindanao, Bohol Isl, Samar Isl, Leyte Isl (Philippines).
STATUS: CITES – Appendix II; U.S. ESA – Threatened; IUCN – Data Deficient.
SYNONYMS: *carbonarius* Heude, 1898; *fraterculus* Miller, 1911; *philippinensis* Meyer, 1894.

*Tarsius tarsier* (Erxleben, 1777). Syst. Regni Anim., I. Mammalia, p. 71.
COMMON NAME: Spectral Tarsier.
TYPE LOCALITY: Unknown. Based on the "Tarsier or Woolly Jerboa" of Buffon, 1749, *Histoire Naturelle*, 13, N°MCCXXXV. Shekelle and Groves (in prep.) restricted the type locality to Makassar (Sulawesi, Indonesia), following Hill's (1953) previous restriction of the type locality of *spectrum* to that region.
DISTRIBUTION: Sulawesi lowlands and Selayar Isl (Indonesia).
STATUS: CITES – Appendix II; IUCN – Lower Risk (nt) as *T. spectrum*.
SYNONYMS: *buffonii* (Link, 1795); *daubentonii* Fischer, 1804; *fischerii* Desmarest, 1804; *fuscomanus* Fischer, 1804; *fuscus* Fischer, 1804; *macrotarsos* Schreber, 1778; *pallassii* É. Geoffroy, 1796; *podje* (Kerr, 1792); *spectrum* (Pallas, 1778).
COMMENTS: The description and plate of Buffon's (1849) "Tarsier or Woolly Jerboa" clearly indicates a Sulawesi tarsier; the name therefore takes precedence over *spectrum* Pallas; it is not a *nomen oblitum* because it was used by Chasen (1940). Niemitz et al. (1991) implied that this species, which they called *T. spectrum*, may be restricted to N Sulawesi; but the N Sulawesi tarsier may not be conspecific with that from Makassar. There are probably a number of undescribed species, sharply distinguished by their vocalizations (Shekelle et al., 1998) and by morphological features (Groves, in prep.), in different regions of Sulawesi, and the Selayar tarsier is also distinct.

## INFRAORDER SIMIIFORMES Haeckel, 1866.
COMMENTS: Usually called Anthropoidea; for the potential confusion surrounding this name, and why it is best avoided, see Hoffstetter (1982). McKenna and Bell (1997) divided living members into two superfamilies, Callitrichoidea and Cercopithecoidea, corresponding to parvorders Platyrrhini and Catarrhini used here.

## PARVORDER PLATYRRHINI É. Geoffroy St. Hilaire, 1812.
COMMENTS: Reduced to superfamily rank (as Callitrichoidea) by McKenna and Bell (1997). See comments under Catarrhini (below).

**Family Cebidae** Bonaparte, 1831. Saggio Dist. Metod. Anim. Vert., p. 6.
COMMENTS: Groves (1993) divided Platyrrhini into two families, Callitrichidae and Cebidae, the traditional approach (see for example Martin, 1990; see also McKenna and Bell (1997), who used the earlier name Atelidae instead of Cebidae). But the evidence (reviewed by Groves, 2001c) is that the family Cebidae, in this traditional sense, is paraphyletic, and Nyctipithecidae, Pitheciidae and Atelidae must be extracted from it, while the marmoset group (here called Callitrichinae) is closer to the core Cebidae and should be united with them.

**Subfamily Callitrichinae** Gray, 1821. London Med. Repos., 15:298.
SYNONYMS: Callithricidae Thomas, 1903; Callitrichidae Napier and Napier, 1967; Hapalidae Wagner, 1840.
COMMENTS: Retained as a full family by McKenna and Bell (1997) and Rylands et al. (2000). Goodman et al. (1998) divided the subfamily into subtribes Saguinina, Leontopithecina, Callimiconina and Callitrichina. Groves (2001c:126-7) argued that the correct name for the family/subfamily is Hapalidae/Hapalinae, but this was discussed by Brandon-Jones and Groves (2002), who found that the argument was insecure and re-established Callitrichidae.

*Callimico* Miranda-Ribeiro, 1912. Brasil. Rundsch., p. 21.
TYPE SPECIES: *Callimico snethlageri* Miranda-Ribeiro, 1912 (= *Hapale goeldii* Thomas, 1904).
COMMENTS: Placed in a separate family, Callimiconidae, by Hershkovitz (1977), but recognized as a member of the Callitrichidae (=Hapalidae) by Pocock (1920a), Napier (1976), and Groves (1989); McKenna and Bell (1997) placed it in a separate subfamily, Callimiconinae, of their family Callitrichidae.

*Callimico goeldii* (Thomas, 1904). Ann. Mag. Nat. Hist., ser. 7, 14:189.
COMMON NAME: Goeldi's Marmoset.
TYPE LOCALITY: Brazil, Acre, Rio Yaco.
DISTRIBUTION: W Brazil, N Bolivia, E Peru, Colombia: Upper Amazon Rainforests.
STATUS: CITES – Appendix I; U.S. ESA – Endangered; IUCN – Near Threatened.
SYNONYMS: *snethlageri* Miranda-Ribeiro, 1912.

*Callithrix* Erxleben, 1777. Syst. Regni Anim., p. 55.
TYPE SPECIES: *Simia jacchus* Linnaeus, 1758.
SYNONYMS: Listed for the four subgenera separately, because they are ranked as full genera by some: (1) Subgenus *Callithrix* Erxleben, 1777: *Anthopithecus* F. Cuvier, 1829; *Arctopithecus* G. Cuvier, 1819; *Hapale* Illiger, 1811; *Hapales* F. Cuvier, 1829; *Harpale* Gray, 1821; *Iacchus* Spix, 1823; *Jacchus* É. Geoffroy, 1812; *Midas* É. Geoffroy, 1828 [not of Latreille, 1796]; *Ouistitis* Burnett, 1826; *Sagoin* Desmarest, 1804; *Sagoinus* Kerr, 1792; *Sagouin* Lacépède, 1799; *Saguin* Fischer, 1803. (2) Subgenus *Mico* Lesson, 1840: *Liocephalus* Wagner, 1840; *Micoella* Gray, 1870. (3) Subgenus *Cebuella* Gray, 1866: no synonyms. (4) Subgenus *Calibella* van Roosmalen and van Roosmalen, 2003: no synonyms.
COMMENTS: Includes *Mico* (see Cabrera, 1958:185), *Cebuella* (see Groves, 1989:110-115), and *Callithrix* as subgenera, which Rylands et al. (2000) regarded as full genera. McKenna and Bell (1997) recognized *Cebuella* as a separate genus, but it forms a clade with *Mico* (Groves, 2001c). The fourth subgenus, *Calibella*, was described by van Roosmalen and van Roosmalen (2003) as a full genus.

*Callithrix acariensis* M. van Roosmalen, T. van Roosmalen, Mittermeier and Rylands, 2000. Neotropical Primates, 8(1):7.
COMMON NAME: Rio Acari Marmoset.
TYPE LOCALITY: Brazil, Amazonas, "a small settlement on the right bank of the lower Rio Acari close to the confluence with the Rios Sucunduri and Canuma", 05°07'08"S, 60°01'14"W.
DISTRIBUTION: Presumed to be the entire interfluvium of the Rios Acari and Sucunduri (Brazil).
STATUS: CITES – Appendix II; IUCN – Least Concern as *Mico acariensis*.
COMMENTS: Subgenus *Mico*.

*Callithrix argentata* (Linnaeus, 1771). Mantissa Plantarum, 2, Appendix:521.
    COMMON NAME: Silvery Marmoset.
    TYPE LOCALITY: Brazil, Pará, Cametá, on banks of Rio Tocantins.
    DISTRIBUTION: N and C Brazil, E Bolivia.
    STATUS: CITES – Appendix II; IUCN – Least Concern as *Mico argentata*.
    COMMENTS: Subgenus *Mico*. Includes *emiliae*, *melanura* and *leucippe* according to Hershkovitz
        (1977:436), and *intermedius* (Ávila-Pires, 1985). De Vivo (1985) recognized *emiliae*
        (Thomas, 1920) as a full species; Coimbra-Filho (1990) recognized *emiliae* and
        *intermedius* as distinct species; Groves (2001*c*) recognized all these and others,
        subsequently described, as distinct species.

*Callithrix aurita* (É. Geoffroy, 1812). Ann. Mus. Hist. Nat. Paris, 19:119.
    COMMON NAME: Buffy-tufted Marmoset.
    TYPE LOCALITY: Brazil, Rio de Janeiro.
    DISTRIBUTION: SE Brazilian coast.
    STATUS: CITES – Appendix I; U.S. ESA and IUCN – Endangered.
    SYNONYMS: *chrysopyga* (Burmeister, 1854); *coelestis* (Miranda Ribeiro, 1924); *itatiayae* Avila-
        Pires, 1959; *petronius* (Miranda Ribeiro, 1924).
    COMMENTS: Subgenus *Callithrix*. Accepted as a species by Mittermeier et al. (1988) and
        Groves (1989).

*Callithrix chrysoleuca* (Wagner, 1842). Arch. Naturg., 8, 1ˢᵗ part, p. 357.
    COMMON NAME: Gold-and-white Marmoset.
    TYPE LOCALITY: Brazil, Borba, lower Rio Madeira.
    DISTRIBUTION: Between the Aripuanã-Madeira and Canuma-Uraria, south to about 8°S,
        north to the Amazon (Brazil).
    STATUS: CITES – Appendix II; IUCN – Data Deficient as *Mico chrysoleucus*.
    SYNONYMS: *melanoleucus* (Miranda Ribeiro, 1955); *sericeus* (Gray, 1868).
    COMMENTS: Subgenus *Mico*. Recognized as a full species by Coimbra-Filho (1990).

*Callithrix emiliae* (Thomas, 1920). Ann. Mag. Nat. Hist., ser. 9, 6:209.
    COMMON NAME: Emilia's Marmoset.
    TYPE LOCALITY: Brazil, Rio Curua, Maloca.
    DISTRIBUTION: The interfluvium of the Rios Tapajos and Iriri, north to Maica, on the lower
        Tapajos, where perhaps sympatric with *C. argentata* (Brazil).
    STATUS: CITES – Appendix II; IUCN – Least Concern as *Mico emiliae*.
    COMMENTS: Subgenus *Mico*. A different species, so far undescribed, from between the Rios
        Ji-Parana and Beni, in Rondonia, has been referred to as *C. emiliae* but is likely to be
        distinct (Rylands et al., 1993). Recognized as a full species by De Vivo (1985) and
        Coimbra-Filho (1990).

*Callithrix flaviceps* (Thomas, 1903). Ann. Mag. Nat. Hist., ser. 7, 12:240.
    COMMON NAME: Buffy-headed Marmoset.
    TYPE LOCALITY: Brazil, Rive, Espírito Santo.
    DISTRIBUTION: S Espírito Santo (Brazil).
    STATUS: CITES – Appendix I; U.S. ESA and IUCN – Endangered.
    SYNONYMS: *flavescente* Miranda Ribeiro, 1924.
    COMMENTS: Subgenus *Callithrix*. Accepted as a species by Mittermeier et al. (1988) and
        Groves (1989). Regarded as a subspecies of *aurita* by Coimbra-Filho (1990).

*Callithrix geoffroyi* (Humboldt, 1812). Rec. Obs. Zool. Anat. Comp., p. 360.
    COMMON NAME: White-headed Marmoset.
    TYPE LOCALITY: Brazil, Victoria, between Rios Espirito Santo and Jucu.
    DISTRIBUTION: EC Brazil (coast of Bahia).
    STATUS: CITES – Appendix II; IUCN – Vulnerable.
    SYNONYMS: *albifrons* (Thunberg, 1819); *leucocephalus* (Humboldt, 1812); *leucogenys* (Gray,
        1870) [*nomen nudum*]; *maximiliani* (Reichenbach, 1862); *melanotis* (Lesson, 1840)
        [part].
    COMMENTS: Subgenus *Callithrix*. Accepted as a species by Mittermeier et al. (1988) and
        Groves (1989).

*Callithrix humeralifera* (É. Geoffroy, 1812). Ann. Mus. Hist. Nat. Paris, 19:120.
COMMON NAME: Santarem Marmoset.
TYPE LOCALITY: Brazil, left bank of Rio Tapajós, Paricatuba.
DISTRIBUTION: Brazil, south of the Amazon between the Maues-Açu and Tapajós Rivers.
STATUS: CITES – Appendix II; IUCN – Least Concern as *Mico humeralifer*.
SYNONYMS: *santaremensis* Matschie, 1893.
COMMENTS: Subgenus *Mico*. Hershkovitz (1977:595-598), who called this species
    *C. humeralifer*, included *intermedia* and *chrysoleuca* in it; *intermedia* was transfered to
    *C. argentata* by Ávila-Pires (1985). Coimbra-Filho (1990), who pointed out that the
    correct name for this species is *humeralifera*, listed *chrysoleuca* and *intermedia* as distinct
    species.

*Callithrix humilis* M. van Roosmalen, T. van Roosmalen, Mittermeier and de Fonseca, 1998.
    Goeldiana Zoologia, 22:8.
COMMON NAME: Roosmalens' Dwarf Marmoset.
TYPE LOCALITY: Brazil, Amazonas, west bank of lower Rio Aripuanã, 1 km south of Nova
    Olinda, 41 km southwest of Novo Aripuanã, 5°30'63"S, 60°24'61"W.
DISTRIBUTION: Between the Rios Aripuanã and Madeira, to about 6°S (Brazil).
STATUS: CITES – Appendix II; IUCN – Least Concern as *Mico humilis*.
COMMENTS: Subgenus *Calibella*.

*Callithrix intermedia* Hershkovitz, 1977. Living New World Monkeys, p. 1020.
COMMON NAME: Hershkovitz's Marmoset.
TYPE LOCALITY: Brazil, Amazonas, near mouth of Rio Guariba, left bank of Rio Aripuanã.
DISTRIBUTION: Interfluvium of the Rios Aripuanã and Roosevelt (Brazil).
STATUS: CITES – Appendix II; IUCN – Least Concern as *Mico intermedius*.
COMMENTS: Subgenus *Mico*. Recognized as a full species by de Vivo (1991).

*Callithrix jacchus* (Linnaeus, 1758). Syst. Nat., 10th ed., 1:27.
COMMON NAME: Common Marmoset.
TYPE LOCALITY: Brazil, Pernambuco.
DISTRIBUTION: Brazilian coast: Piauí, Ceará, and Pernambuco Provinces.
STATUS: CITES – Appendix II; IUCN – Least Concern.
SYNONYMS: *albicollis* Spix, 1823; *communis* (South, 1845); *hapale* (Gray, 1870); *leucotis*
    (Lesson, 1840); *moschatus* (Kerr, 1792); *rufus* Fischer, 1829; *vulgaris* Humboldt, 1812.
COMMENTS: Subgenus *Callithrix*. Includes *aurita*, *flaviceps*, *geoffroyi*, and *penicillata* according
    to Hershkovitz (1977:489-527), but refuted by Mittermeier et al. (1988) who
    recognized all of these species.

*Callithrix kuhlii* Coimbra-Filho, 1985. Fundaçao Brasileira Conservaçao Natu. Bol. Inform., 9:5.
COMMON NAME: Wied's Marmoset.
TYPE LOCALITY: Brazil, Bahia, north of Rio Belmonte.
DISTRIBUTION: Between Rio de Contas and Rio Jequitinhonha, SW Brazil.
STATUS: CITES – Appendix II; IUCN – Least Concern.
COMMENTS: Subgenus *Callithrix*. Accepted as a species by Mittermeier et al. (1988) and
    Groves (1989), but ascribed by these authors to Wied-Neuwied, 1826, who in fact was
    not describing a new taxon; the earliest available reference to the species is as above
    (see Groves, 2001*c*).

*Callithrix leucippe* (Thomas, 1922). Ann. Mag. Nat. Hist., ser. 9, 9:199.
COMMON NAME: White Marmoset.
TYPE LOCALITY: Brazil, Pimental, near mouth of Rio Jamanxim, right bank of Rio Tapajos.
DISTRIBUTION: The interfluvium of the Rios Tapajos and Cupari (Brazil).
STATUS: CITES – Appendix II; IUCN – Data Deficient as *Mico leucippe*.
COMMENTS: Subgenus *Mico*. Separated from *C. argentata* de Vivo (1991).

*Callithrix manicorensis* M. van Roosmalen, T. van Roosmalen, Mittermeier and Rylands, 2000.
    Neotropical Primates, 8(1):3.
COMMON NAME: Manicore Marmoset.
TYPE LOCALITY: Brazil, Amazonas, near Manicoré, Seringal São Luis, 05°50'28"S, 61°18'19"W.

DISTRIBUTION: The interfluvium of the Rios Aripuanã and Manicoré, from the Rio Madeira south to the Rio Roosevelt (Brazil).
STATUS: CITES – Appendix II; IUCN – Least Concern as *Mico manicorensis*.
COMMENTS: Subgenus *Mico*.

*Callithrix marcai* Alperin, 1993. Bol. Mus. Para. Emilio Goeldi, ser. Zool., 9(2):325.
COMMON NAME: Marca's Marmoset.
TYPE LOCALITY: Brazil, Amazonas, Foz do Rio Castanho.
DISTRIBUTION: Known only from region of type locality.
STATUS: CITES – Appendix II; IUCN – Data Deficient as *Mico marcai*.
COMMENTS: Subgenus *Mico*.

*Callithrix mauesi* Mittermeier, M. Schwarz and Ayres, 1992. Goeldiana Zoologia, 14:6.
COMMON NAME: Maués Marmoset.
TYPE LOCALITY: Brazil, Amazonas, west bank of Rio Maués-Açu, opposite Maués town, 3°23'S, 57°46'W.
DISTRIBUTION: Interfluvium of the Rios Uraria-Abacaxis and Maues-Açu (Brazil).
STATUS: CITES – Appendix I; IUCN – Least Concern as *Mico mauesi*.
COMMENTS: Subgenus *Mico*.

*Callithrix melanura* (É. Geoffroy, 1812). Rec. Observ. Zool., 1:361.
COMMON NAME: Black-tailed Marmoset.
TYPE LOCALITY: Brazil, Mato Grosso, Cuiaba.
DISTRIBUTION: S Brazil, between the Rios Aripuanã and Juruena, southwest to the Rio Beni in Bolivia.
STATUS: CITES – Appendix II.
SYNONYMS: *leucomerus* (Gray, 1846); *leukeurin* (Natterer in Pelzeln, 1883).
COMMENTS: Subgenus *Mico*. Separated from *C. argentata* by de Vivo (1991).

*Callithrix nigriceps* Ferrari and Lopes, 1992. Goeldiana Zoologia, 12:4.
COMMON NAME: Black-headed Marmoset.
TYPE LOCALITY: Brazil, Amazonas, Lago dos Reis, 7°31'S, 62°52'W.
DISTRIBUTION: Interfluvium of the Rios Marmelos and Madeira, north of the Ji-Paranã River (Brazil).
STATUS: CITES – Appendix II; IUCN – Data Deficient as *Mico nigriceps*.
COMMENTS: Subgenus *Mico*.

*Callithrix penicillata* (É. Geoffroy, 1812). Ann. Mus. Hist. Nat. Paris, 19:119.
COMMON NAME: Black-tufted Marmoset.
TYPE LOCALITY: Brazil, Lamarão, Bahia.
DISTRIBUTION: Brazilian coast: Bahia to São Paulo, inland to Goiás.
STATUS: CITES – Appendix II; IUCN – Least Concern.
SYNONYMS: *jordani* Thomas, 1904; *melanotis* (Lesson, 1840) [part]; *trigonifer* Reichenbach, 1862.
COMMENTS: Subgenus *Callithrix*. Accepted as a species by Mittermeier et al. (1988) and Groves (1989, 2001c).

*Callithrix pygmaea* (Spix, 1823). Sim. Vespert. Brasil., p. 32.
COMMON NAME: Pygmy Marmoset.
TYPE LOCALITY: Brazil, Amazonas, Solimões River, Tabatinga.
DISTRIBUTION: N and W Brazil, N Peru, Ecuador.
STATUS: CITES – Appendix II; IUCN – Least Concern as *Cebuella pygmaea*.
SYNONYMS: *nigra* (Schinz, 1844); *leoninus* (Bates, 1864); **niveiventris** (Lönnberg, 1940).
COMMENTS: Subgenus *Cebuella* (recognized by McKenna and Bell (1997) as a full genus; but it is not the most divergent clade of marmosets). Revised by Hershkovitz (1977:462-464), placed in *Callithrix* by Groves (1989, 2001c). Reviewed by Townsend (2001).

*Callithrix saterei* Silva and Noronha, 1998. Goeldiana Zoologia, 21:6.
COMMON NAME: Satéré Marmoset.
TYPE LOCALITY: Brazil, Amazonas, Foz do Canuma, right bank of lower Rio Canumã, in front of its confluence with the Parana Uraria, 03°59'50.8"S, 59°05'36.7"W.

DISTRIBUTION: Presumed to be the interfluvium of the Rios Abacaxis and Canumã-Sucunduri (Brazil).

STATUS: CITES – Appendix II; IUCN – Data Deficient as *Mico saterei*.

COMMENTS: Subgenus *Mico*.

*Leontopithecus* Lesson, 1840. Spec. Mamm. Bim. Quadrum., 1844:184, 200.

TYPE SPECIES: *Leontopithecus marikina* Lesson, 1840 (= *Simia rosalia* Linnaeus, 1766).

SYNONYMS: *Leontideus* Cabrera, 1956; *Leontocebus* Elliot, 1913.

COMMENTS: For synonyms see Hershkovitz (1977:807-808).

*Leontopithecus caissara* Lorini and Persson, 1990. Bol. Mus. Nac., Rio de Janeiro, n.s., 338:2.

COMMON NAME: Superagui Lion Tamarin.

TYPE LOCALITY: Brazil, Superagui Isl (south of São Paulo), Guaraquecaba, 25°18'S, 48°11'W.

DISTRIBUTION: Superagui Isl and a small region on the opposite mainland (Brazil).

STATUS: CITES – Appendix I; U.S. ESA – Endangered; IUCN – Critically Endangered.

COMMENTS: Considered a subspecies of *chrysopygus* by Coimbra-Filho (1990).

*Leontopithecus chrysomelas* (Kuhl, 1820). Beitr. Zool. Vergl. Aust., p. 51.

COMMON NAME: Golden-headed Lion Tamarin.

TYPE LOCALITY: Brazil, Ribeirao das Minhocas, S Bahia.

DISTRIBUTION: Brazil, coastal Bahia.

STATUS: CITES – Appendix I; U.S. ESA and IUCN – Endangered.

SYNONYMS: *chrysurus* (I. Geoffroy St. Hilaire, 1827).

COMMENTS: Recognized as a full species by Rosenberger and Coimbra-Filho (1984).

*Leontopithecus chrysopygus* (Mikan, 1823). Delectus florae et faunae Brasiliensis, Vienna, 3, plate.

COMMON NAME: Black Lion Tamarin.

TYPE LOCALITY: Brazil, Ipanema, São Paulo.

DISTRIBUTION: Brazil, São Paulo region.

STATUS: CITES – Appendix I; U.S. ESA – Endangered; IUCN – Critically Endangered.

SYNONYMS: *ater* (Lesson, 1840).

COMMENTS: Recognized as a full species by Rosenberger and Coimbra-Filho (1984).

*Leontopithecus rosalia* (Linnaeus, 1766). Syst. Nat., 12th ed., 1:41.

COMMON NAME: Golden Lion Tamarin.

TYPE LOCALITY: Brazil, coast, Rio São João (right bank), between 22° and 23°S, see Wied-Neuwied (1826) and de Carcalho (1965).

DISTRIBUTION: SE Brazil: Rio Doce (Espírito Santo) south into Rio de Janeiro and Guanabara.

STATUS: CITES – Appendix I; U.S. ESA and IUCN – Endangered.

SYNONYMS: *aurora* Elliot, 1913; *brasiliensis* (Fischer, 1829); *guyannensis* (Fischer, 1829); *leoninus* (Pocock, 1914); *marikina* Lesson, 1840.

COMMENTS: See Kleiman (1981, Mammalian Species, 148).

*Saguinus* Hoffmannsegg, 1807. Mag. Ges. Naturf. Fr., 1:101.

TYPE SPECIES: *Saguinus ursula* Hoffmannsegg, 1807 (= *Simia midas* Linnaeus, 1758).

SYNONYMS: *Hapanella* Gray, 1870; *Leontocebus* Wagner, 1840; *Marikina* Lesson, 1840; *Midas* E. Geoffroy St. Hilaire, 1812; *Mystax* Gray, 1870; *Oedipomidus* Reichenbach, 1862; *Oedipus* Lesson, 1840; *Seniocebus* Gray, 1870; *Tamarin* Gray, 1870; *Tamarinus* Trouessart, 1904.

COMMENTS: See Hershkovitz (1977:601-603). For a summary of interrelationships between species, using external, cranial, vocalization and DNA evidence, see Groves (2001c:137-8), who divided the genus into the following species groups: (1) *S. midas* group (*midas, niger*), (2) *S. nigricollis* group (*nigricollis, graellsi, fuscicollis, melanoleucus, tropartitus*), (3) *S. mystax* group (*mystax, pileatus, labiatus, imperator*), (4) *S. bicolor* group (*bicolor, martinsi*), (5) *S. oedipus* group (*oedipus, geoffroyi, leucopus*), (6) *S. inustus* group (*inustus* only).

*Saguinus bicolor* (Spix, 1823). Sim. Vespert. Brasil., p. 30.

COMMON NAME: Pied Tamarin.

TYPE LOCALITY: Brazil, Manaus, Barra de Rio Negro.

DISTRIBUTION: N Brazil; perhaps NE Peru.

STATUS: CITES – Appendix I; U.S. ESA – Endangered; IUCN – Critically Endangered.

COMMENTS: *S. bicolor* species group. See Hershkovitz (1977:744).

*Saguinus fuscicollis* (Spix, 1823). Sim. Vespert. Brasil., p. 27.
    COMMON NAME: Brown-mantled Tamarin.
    TYPE LOCALITY: Brazil, between Solimões River and Içá River, São Paulo de Olivença.
    DISTRIBUTION: N and W Brazil, N Bolivia, E Peru, E Ecuador, SW Colombia.
    STATUS: CITES – Appendix II; IUCN – Data Deficient as *S. f. cruzlimai*, otherwise Least
        Concern (including listing as *S. fuscus*).
    SYNONYMS: *flavifrons* (I. Geoffroy and Deville, 1848); ***avilapiresi*** Hershkovitz, 1966;
        ***cruzlimai*** Hershkovitz, 1966; *fuscus* (Lesson, 1840); *leonina* (Humboldt, 1805) [not of
        Shaw, 1800]; ***illigeri*** (Pucheran, 1845); *bluntschlii* (Matschie, 1915); *devillei* (I. Geoffroy,
        1850); *mounseyi* (Thomas, 1920); ***leucogenys*** (Gray, 1866); *micans* (Thomas, 1928);
        *pacator* (Thomas, 1914); ***lagonotus*** (Jiménez de la Espada, 1870); *apiculatus* (Thomas,
        1904); ***nigrifrons*** (I. Geoffroy, 1851); *pebilis* (Thomas, 1928); ***primitivus*** Hershkovitz,
        1977; ***weddelli*** (Deville, 1849); *imberbis* (Lönnberg, 1940); *purillus* (Thomas, 1914).
    COMMENTS: *S. nigricollis* species group. See Hershkovitz (1977:640-642). Coimbra-Filho
        (1990) regarded *melanoleucus* as a distinct species, with subspecies of *acrensis* and
        *crandalli*, and this was followed by Groves (2001*c*) and is followed here.

*Saguinus geoffroyi* (Pucheran, 1845). Rev. Mag. Zool. Paris, 8:336.
    COMMON NAME: Geoffroy's Tamarin.
    TYPE LOCALITY: Panama, Canal Zone.
    DISTRIBUTION: SE Costa Rica to NW Colombia.
    STATUS: CITES – Appendix I; IUCN – Least Concern.
    SYNONYMS: *salaguiensis* (Elliot, 1912); *spixii* (Reichenbach, 1862).
    COMMENTS: *S. oedipus* species group. Distinct from *S. oedipus*; see Natori (1988).

*Saguinus graellsi* (Jimenez de la Espada, 1870). Bol. Rev. Univers. Madrid, p. 19.
    COMMON NAME: Graells's Tamarin.
    TYPE LOCALITY: Peru, Loreto, Tarapoto, opposite mouth of Rio Curaray, Rio Napo.
    DISTRIBUTION: Peru, Ecuador, Colombia, west of Rio Napo, from Rio Putumayo south to
        Rio Marañon, west to Rio Santiago.
    STATUS: CITES – Appendix II.
    COMMENTS: *S. nigricollis* species group. Raised to rank of full species by Groves (2001*c*),
        contra Hershkovitz (1977:628).

*Saguinus imperator* (Goeldi, 1907). Proc. Zool. Soc. Lond., 1907:93.
    COMMON NAME: Emperor Tamarin.
    TYPE LOCALITY: Brazil, Rio Acre.
    DISTRIBUTION: W Brazil, E Peru, Bolivia (see Anderson, 1997).
    STATUS: CITES – Appendix II; IUCN – Data Deficient as *S. i. imperator*, otherwise Least
        Concern.
    SYNONYMS: ***subgrisescens*** (Lönnberg, 1940).
    COMMENTS: *S. mystax* species group.

*Saguinus inustus* (Schwartz, 1951). Am. Mus. Novit., 1508:1.
    COMMON NAME: Mottle-faced Tamarin.
    TYPE LOCALITY: Brazil, Amazonas, Tabocal.
    DISTRIBUTION: NW Brazil, SW Colombia.
    STATUS: CITES – Appendix II; IUCN – Least Concern.
    COMMENTS: *S. inustus* species group.

*Saguinus labiatus* (É. Geoffroy, 1812). Rec. Observ. Zool., 1:361.
    COMMON NAME: White-lipped Tamarin.
    TYPE LOCALITY: Brazil, Amazonas, Lake Joanacan.
    DISTRIBUTION: W Brazil, E Peru, Bolivia (see Anderson, 1997).
    STATUS: CITES – Appendix II; IUCN – Least Concern.
    SYNONYMS: ?*elegantulus* (Slack, 1862); ?*erythrogaster* (Reichenbach, 1862); *griseovertex*
        (Goeldi, 1907); ***rufiventer*** (Gray, 1843); ***thomasi*** (Goeldi, 1907).
    COMMENTS: *S. mystax* species group.

*Saguinus leucopus* (Günther, 1877). Proc. Zool. Soc. Lond., 1876:746 [1877].
COMMON NAME: White-footed Tamarin.
TYPE LOCALITY: Colombia, Antioquia, Medellin.
DISTRIBUTION: N Colombia.
STATUS: CITES – Appendix I; U.S. ESA – Threatened; IUCN – Vulnerable.
SYNONYMS: *pegasis* (Elliot, 1913).
COMMENTS: *S. oedipus* species group.

*Saguinus martinsi* (Thomas, 1912). Ann. Mag. Nat. Hist., ser. 8, 11:85.
COMMON NAME: Martins's Tamarin.
TYPE LOCALITY: Brazil, Para, Faro.
DISTRIBUTION: N Brazil: a very small area north of the Amazon, on either side of the
    Rio Nhamunda.
STATUS: CITES – Appendix I as included in *S. bicolor*; U.S. ESA – Endangered (as included in
    *S. bicolor*) ; IUCN – Least Concern.
SYNONYMS: **ochraceus** Hershkovitz, 1966.
COMMENTS: *S. bicolor* species group. See Hershkovitz (1977:744). Separated from *bicolor* by
    Groves (2001*c*).

*Saguinus melanoleucus* (Miranda Ribeiro, 1912). Brasil. Rundsch., p. 22.
COMMON NAME: White-mantled Tamarin.
TYPE LOCALITY: Brazil, Amazonas, Santo Antonio, Rio Eiru, upper Rio Jurua.
DISTRIBUTION: Brazil, between Rios Jurua and Tarauca.
STATUS: CITES – Appendix II; IUCN – Data Deficient as *S. fuscicolis crandalli*, otherwise not
    listed.
SYNONYMS: *acrensis* (Carvalho, 1957); *crandalli* Hershkovitz, 1966; *hololeucus* (Pinto, 1937).
COMMENTS: *S. nigricollis* species group. Coimbra-Filho (1990) regarded *melanoleucus* as a
    species distinct from *S. fuscicollis*, and this was followed by Groves (2001*c*).

*Saguinus midas* (Linnaeus, 1758). Syst. Nat., 10th ed., 1:28.
COMMON NAME: Red-handed Tamarin.
TYPE LOCALITY: Surinam.
DISTRIBUTION: Brazil, Guyana, Cayenne, Surinam, north of the Amazon, east of the Rio
    Negro.
STATUS: CITES – Appendix II; IUCN – Least Concern.
SYNONYMS: *egens* (Thomas, 1912); *lacepedii* (Fischer, 1806); *rufimanus* (É. Geoffroy, 1812);
    *tamarin* (Link, 1795).
COMMENTS: *S. midas* species group.

*Saguinus mystax* (Spix, 1823). Sim. Vespert. Brasil., p. 29.
COMMON NAME: Moustached Tamarin.
TYPE LOCALITY: Brazil, Amazonas, between Solimões River and Iça River.
DISTRIBUTION: W Brazil, Peru, south of Amazon-Solimoes-Marañon, between lower Rio
    Huallaga and Rio Madeira.
STATUS: CITES – Appendix II; IUCN – Least Concern.
SYNONYMS: **pluto** (Lönnberg, 1926).
COMMENTS: *S. mystax* species group. See Hershkovitz (1977:700); but does not include
    *pileatus* (see Groves, 2001*c*:143-4).

*Saguinus niger* (É. Geoffroy, 1803). Cat. Mamm. Mus. Nat. Hist. Paris, p. 13.
COMMON NAME: Black Tamarin.
TYPE LOCALITY: Brazil: Belem.
DISTRIBUTION: Brazil, south of the Amazon, east of the Rio Xingu, including Marajo Isl.
STATUS: CITES – Appendix II; IUCN – Least Concern.
SYNONYMS: *gracilis* (G. Fischer, 1813); *umbratus* (Thomas, 1922); *ursula* Hoffmannsegg, 1807.
COMMENTS: *S. midas* species group. Formerly called *tamarin*, but this is a synonym of *S. midas*:
    see Hershkovitz (1977:711). Separated from *S. midas* by Natori and Hanihara (1988).

*Saguinus nigricollis* (Spix, 1823). Sim. Vespert. Brasil., p. 28.
COMMON NAME: Black-mantled Tamarin.

TYPE LOCALITY: Brazil, Amazonas, São Paulo de Olivença.

DISTRIBUTION: W Brazil, E Peru, E Ecuador.

STATUS: CITES – Appendix II; IUCN – Least Concern.

SYNONYMS: *rufoniger* (I. Geoffroy and Deville, 1848); **hernandezi** Hershkovitz, 1982.

COMMENTS: *S. nigricollis* species group. Does not include *graellsi*, contra Hershkovitz (1977:628).

*Saguinus oedipus* (Linnaeus, 1758). Syst. Nat., 10th ed., 1:28.

COMMON NAME: Cottontop Tamarin.

TYPE LOCALITY: Colombia, Bolivar, lower Rio Sinu.

DISTRIBUTION: N Colombia, Panama.

STATUS: CITES – Appendix I; U.S. ESA – Endangered; IUCN – Endangered.

SYNONYMS: *doguin* (Griffith, 1821); *meticulous* (Elliot, 1912); *titi* (Lesson, 1840).

COMMENTS: *S. oedipus* species group. Does not include *geoffroyi*; see Natori (1988).

*Saguinus pileatus* (I. Geoffroy and Deville, 1848). C. R. Acad. Sci. Paris, 27:490.

COMMON NAME: Red-capped Tamarin.

TYPE LOCALITY: Brazil, eastern margin of Lago de Téfé (Hershkovitz, 1977; further refined by Groves, 2001*c*:143).

DISTRIBUTION: W Brazil, east of Rio Téfé, west of Rio Purus.

STATUS: CITES – Appendix II.

SYNONYMS: *juruanus* Ihering, 1904.

COMMENTS: *S. mystax* species group. See Hershkovitz (1977:700). Separated from *S. mystax* by Groves (2001*c*:143-4).

*Saguinus tripartitus* (Milne-Edwards, 1878). Nour. Arch. Mus. Hist. Nat. Paris, Ser. 2, 1:161.

COMMON NAME: Golden-mantled Tamarin.

TYPE LOCALITY: Equador, Rio Napo.

DISTRIBUTION: East of Rio Curaray, Brazil-Colombia border; sympatric with *S. fuscicollis* around Curaray-Napo Confluence.

STATUS: CITES – Appendix II; IUCN – Least Concern.

COMMENTS: *S. nigricollis* species group. Given specific rank by Thorington (1988) on the evidence of marginal sympatry with *S. fuscicollis*.

**Subfamily Cebinae** Bonaparte, 1831. Saggio. Dist. Metod. Anim. Vert., p. 6.

*Cebus* Erxleben, 1777. Syst. Regni Anim., p. 44.

TYPE SPECIES: *Simia capucina* Linnaeus, 1758.

SYNONYMS: *Agipan* Rafinesque, 1815; *Calyptrocebus* Reichenbach, 1862; *Eucebus* Reichenbach, 1862; *Otocebus* Reichenbach, 1862; *Pseudocebus* Reichenbach, 1862; *Sapajus* Kerr, 1792.

COMMENTS: Species groups are (1) *C. capucinus* group (*capucinus, albifrons, olivaceus, kaapori*), (2) *C. apella* group (*apella, libidinosus, nigritus, xanthosternos*). The following specific names probably refer to this genus, but are regarded by Groves (2001*c*) as not certainly identifiable to species: *albulus* (Kerr, 1792); *albus* É. Geoffroy, 1812; *barbatus* É. Geoffroy, 1812; *flavia* (Schreber, 1774); *flavus* (Goldfuss, 1809); *lugubris* Erxleben, 1777; *pucheranii* Dahlbom, 1856; *paraguayanus* Reichenbach, 1862.

*Cebus albifrons* (Humboldt, 1812). Rec. Observ. Zool., 1:324.

COMMON NAME: White-fronted Capuchin.

TYPE LOCALITY: Venezuela, Orinoco River.

DISTRIBUTION: Venezuela, Colombia, Ecuador, N Peru, NW Brazil, Trinidad, Bolivia (see Anderson, 1997).

STATUS: CITES – Appendix II; IUCN – Critically Endangered as *C. a. trinitatis*, Data Deficient as *C. a. adustus, C. a. aequatorialis, C. a. cesarae, C. a. cuscinus, C. a. leucocephalus, C. a. malitiosus, C. a. versicolor*, and *C. a. yuracus*, otherwise IUCN – Least Concern.

SYNONYMS: **aequatorialis** J. A. Allen, 1914; **cuscinus** Thomas, 1901; *yuracus* Hershkovitz, 1949; **trinitatis** Pusch, 1942; **unicolor** Spix, 1823; *chrysopus* Lesson, 1827; *flavescens* Gray, 1865; *gracilis* Spix, 1823; **versicolor** Pucheran, 1845; *adustus* Hershkovitz, 1849;

*cesarae* Hershkovitz, 1949; *leucocephalus* Gray, 1865; *malitiosus* Elliot, 1909; *pleei* Hershkovitz, 1949.

COMMENTS: *C. capucinus* species group. Intermediates between this species and *C. capucinus* occur in Colombia (middle San Jorge Valley; Lower Cauca River), see Hernández-Camacho and Cooper (1976:58).

*Cebus apella* (Linnaeus, 1758). Syst. Nat., 10th ed., 1:28.

COMMON NAME: Tufted Capuchin.

TYPE LOCALITY: French Guiana.

DISTRIBUTION: N and W South America, from Guyana, Venezuela (south from the Río Orinoco delta) and Colombia south across the Amazon in Brazil to about 5°S in the east and to the headwaters of the upper tributaries in the west, nearly to 10°S.

STATUS: CITES – Appendix II; IUCN – Critically Endangered as *C. a. margaritae*, otherwise Least Concern (including listing as *C. macrocephalus*).

SYNONYMS: *avus* Pusch, 1840; *barbatus* Humboldt, 1812; *buffonii* Lesson, 1840; *fulvus* (Kerr, 1792); *griseus* Desmarest, 1820; *trepida* (Linnaeus, 1766); **fatuellus** (Linnaeus, 1766); **macrocephalus** Spix, 1823; **margaritae** Hollister, 1914; **peruanus** Thomas, 1901; *magnus* Pusch, 1941; *maranonis* Pusch, 1941; **tocantinus** Lönnberg, 1939. The following names belong to this species or another species of the same species group, but seem impossible to allocate to a subspecies, according to Groves (2001c): *capillatus* Gray, 1865; *cirrifer* É. Geoffroy, 1812; *crassiceps* Pucheran, 1857; *cristatus* G. Cuvier, 1829; *fallax* Schlegel, 1876; *fistulator* Reichenbach, 1862; *frontatus* Kuhl, 1820; *hypomelas* (Pucheran, 1854); *leucogenys* Gray, 1866; *lunatus* Kuhl, 1820; *monachus* F. Cuvier, 1820; *niger* É. Geoffroy, 1812; *subcristatus* Gray, 1865; *variegata* (Humboldt, 1812).

COMMENTS: *C. apella* species group. Mittermeier et al. (1988:13-75) suggested that *C. xanthosternos* is a distinct species; followed by Groves (2001c), who also separated *C. libidinosus* and *C. nigritus* as distinct species.

*Cebus capucinus* (Linnaeus, 1758). Syst. Nat., 10th ed., 1:29.

COMMON NAME: White-headed Capuchin.

TYPE LOCALITY: N Colombia.

DISTRIBUTION: W Ecuador to Honduras.

STATUS: CITES – Appendix II; IUCN – Vulnerable as *C. c. curtus*, otherwise Lower Risk (lc).

SYNONYMS: *albulus* Pusch, 1942; *curtus* Bangs, 1905; *hypoleucus* É. Geoffroy, 1812; *imitator* Thomas, 1903; *limitaneus* Hollister, 1914; *nigripectus* Elliot, 1909.

COMMENTS: *C. capucinus* species group.

*Cebus kaapori* Queiroz, 1992. Goeldiana Zoologia, 15:4.

COMMON NAME: Kaapori Capuchin.

TYPE LOCALITY: Brazil, Maranhao, near right bank of Rio Gurupi, Quadrant 7, 10 km southwest of the Chaga-Tudo Prospection (0°30'S, 47°30'W).

DISTRIBUTION: Brazil, between Rios Gurupi and Pindaré, or may extend a few kms west of the Gurupi.

STATUS: CITES – Appendix II; IUCN – Vulnerable *C. olivaceus kaapori*.

COMMENTS: *C. capucinus* species group.

*Cebus libidinosus* Spix, 1823. Sim. Vespert. Brasil, 1823, p. 5.

COMMON NAME: Black-striped Capuchin.

TYPE LOCALITY: Brazil, Rio Carinhanha, north of Minas Gerais.

DISTRIBUTION: Highland region of S Brazil, to Bolivia and Paraguay.

STATUS: CITES – Appendix II.

SYNONYMS: *elegans* I. Geoffroy, 1850; **juruanus** Lönnberg, 1939; **pallidus** Gray, 1866; *sagitta* Pusch, 1941; **paraguayanus** Fischer, 1829; *azarae* Rengger, 1830; *chacoensis* Pusch, 1941; *morrulus* Pusch, 1941; *versuta* Elliot, 1910.

COMMENTS: *C. apella* species group. Separated from *C. apella* by Groves (2001c).

*Cebus nigritus* (Goldfuss, 1809). Vergl. Naturbeschr., Säug., 1:74.

COMMON NAME: Black Capuchin.

TYPE LOCALITY: Brazil: Rio de Janeiro, Sierra dos Orgaos.

DISTRIBUTION: Brazilian coast, Atlantic forests, 16°-30°S.

STATUS: CITES – Appendix II; IUCN – Vulnerable as *C. robustus*.

SYNONYMS: *xanthocephalus* Spix, 1823; **cucullatus** Spix, 1823; *caliginosus* Elliot, 1910; *vellerosus* I. Geoffroy, 1851; **robustus** Kuhl, 1820.

COMMENTS: *C. apella* species group. Separated from *C. apella* by Groves (2001*c*).

*Cebus olivaceus* Schomburgk, 1848. Reise Brit. Guiana, 2:247.

COMMON NAME: Weeper Capuchin.

TYPE LOCALITY: Venezuela, Bolivar, southern base of Mt. Roraima, 930 m.

DISTRIBUTION: Guyana, French Guiana, Surinam, N Brazil, Venezuela, perhaps N Colombia.

STATUS: CITES – Appendix II; IUCN – Lower Risk (lc).

SYNONYMS: *annellatus* Gray, 1865; *apiculatus* Elliot, 1907; *brunneus* J. A. Allen, 1914; *castaneus* I. Geoffroy, 1851; *leporinus* Pusch, 1941; *nigrivittatus* Wagner, 1848; *pucheranii* Dahlbom, 1856.

COMMENTS: *C. capucinus* species group. Replaces *nigrivittatus*; see Husson (1978:223). Mittermeier and Coimbra-Filho (1981) queried the distinction of this species from *C. capucinus*.

*Cebus xanthosternos* Wied-Neuwied, 1826. Reis. Brasil., 1:371.

COMMON NAME: Golden-bellied Capuchin.

TYPE LOCALITY: Brazil, Rio Belmonte.

DISTRIBUTION: Brazil, formerly between Rio São Francisco and Rio Jequitinhonha or even further south; now much reduced.

STATUS: CITES – Appendix II; IUCN – Critically Endangered.

COMMENTS: *C. apella* species group. Mittermeier et al. (1988:13-75) suggested that *C. xanthosternos* is a species distinct from *C. apella*; this was followed by Groves (2001*c*).

**Subfamily Saimiriinae** Miller, 1812. Bull. U.S. Nat. Mus., 79:380.

SYNONYMS: Chrysotrichinae Cabrera, 1900.

COMMENTS: Goodman et al. (1998) reduced the subfamily to a tribe of Cebinae. Groves (2001*c*) pointed out that the name is antedated by Chrysotrichinae Cabrera, 1900, and Brandon-Jones and Groves (2002) agreed, but considered that it would be preferable not to upset "prevailing usage" (Art. 40.2.1; International Code of Zoological Nomenclature, 1999).

*Saimiri* Voigt, 1831. *In* Cuvier, Das Thierreich, 1:95.

TYPE SPECIES: *Simia sciurea* Linnaeus, 1758.

SYNONYMS: *Chrysothrix* Kaup, 1835; *Pithesciurus* Lesson, 1840.

COMMENTS: Placed in a separate subfamily Saimirinae by Hershkovitz (1970*a*) and Napier (1976); according to Groves (2001*c*), Chrysotrichinae Cabrera, 1900, has priority. The species fall into two groups, *S. boliviensis* species group (*boliviensis*, *vanzolinii*) and *S. sciureus* species group (*sciureus*, *ustus*, *oerstedti*).

*Saimiri boliviensis* (I. Geoffroy and Blainville, 1834). Nouv. Ann. Mus. Hist. Nat. Paris, 3:89.

COMMON NAME: Black-capped Squirrel Monkey.

TYPE LOCALITY: Bolivia, Santa Cruz, Rio San Miguel, Guarayos Mission.

DISTRIBUTION: Upper Amazon in Peru; SW Brazil; Bolivia.

STATUS: CITES – Appendix II; IUCN – Least Concern.

SYNONYMS: *entomophagus* (d'Orbigny, 1835); *jaburuensis* Lönnberg, 1940; *nigriceps* Thomas, 1902; *pluvialis* Lönnberg, 1940; **peruviensis** Hershkovitz, 1984.

COMMENTS: *S. boliviensis* species group. Originally separated from *S. sciureus* by Hershkovitz (1984); not recognized as a species by Thorington (1985) or Costello et al. (1993), but belongs to a different species group according to Groves (2001*c*).

*Saimiri oerstedii* (Reinhardt, 1872). Vidensk. Medd. Nat. Hist. Kjobenhaven, p. 157.

COMMON NAME: Central American Squirrel Monkey.

TYPE LOCALITY: Panama, Chiriquí, vicinity of David.

DISTRIBUTION: Panama, Costa Rica.

STATUS: CITES – Appendix I; U.S. ESA – Endangered; IUCN – Critically Endangered as

*S. o. citrinellus*, Endangered as *Saimiri o. oerstedii*.
SYNONYMS: **citrinellus** Thomas, 1904.
COMMENTS: *S. sciureus* species group. Hershkovitz (1972*a*) considered *oerstedii* a subspecies of *sciureus*; but see Hershkovitz (1984).

*Saimiri sciureus* (Linnaeus, 1758). Syst. Nat., 10th ed., 1:29.
COMMON NAME: Common Squirrel Monkey.
TYPE LOCALITY: Guyana, Kartabo.
DISTRIBUTION: N Brazil north of the Amazon-Jurua system, and south of the Amazon east of the Rio Xingu or the Rio Iriri; Marajo Isl (Brazil), Guyana, French Guiana, Surinam, Venezuela, Colombia, E Ecuador, NE Peru.
STATUS: CITES – Appendix II; IUCN – Least Concern.
SYNONYMS: *apedia* (Linnaeus, 1758); *collinsi* Osgood, 1916; *leucopsis* (Hermann, 1804); *morta* (Linnaeus, 1758); *nigrivittata* (Wagner, 1848); *saimiri* (Lacépède, 1803); **albigena** Pusch, 1942; **cassiquiarensis** (Lesson, 1840); *codajazensis* Lönnberg, 1940; *lunulatus* (I. Geoffroy, 1843); **macrodon** Elliot, 1907; *caquetensis* J. A. Allen, 1916; *?juruana* Lönnberg, 1940; *?petrina* Thomas, 1927.
COMMENTS: *S. sciureus* species group.

*Saimiri ustus* (I. Geoffroy, 1843). C. R. Acad. Sci. Paris, 16(21):1157.
COMMON NAME: Bare-eared Squirrel Monkey.
TYPE LOCALITY: Brazil, Amazonas, Rio Madeira, Humaitu.
DISTRIBUTION: S Brazil: south of Rio Amazon, probably from Rio Xingu to Lage Tefé, approx. 61°30′W.
STATUS: CITES – Appendix II; IUCN – Least Concern.
SYNONYMS: *madeirae* Thomas, 1908.
COMMENTS: *S. sciureus* species group. Thorington (1985) and Costello et al. (1993) argued that *ustus* is a synonym of *S. sciureus* (so that the correct name should be *madeirae*), but this is not the case (Groves, 2001*c*:159). Costello et al. (1993) regarded this species as not specifically distinct from *S. sciureus*.

*Saimiri vanzolinii* Ayres, 1985. Papeis Avulsos de Zoologia, São Paulo, 36:148.
COMMON NAME: Black Squirrel Monkey.
TYPE LOCALITY: Brazil, Amazonas, mouth of Rio Japura, left bank of Lago Mamirauá, 2°59′S, 64°55′W.
DISTRIBUTION: Between Rios Japura, Solimões and (probably) Paranado Jaraua (Brazil); Tarara and Capucho Isls (Brazil).
STATUS: CITES – Appendix II; IUCN – Vulnerable.
COMMENTS: *S. boliviensis* species group. A subspecies of *S. boliviensis* according to Hershkovitz (1987*c*:22, fn.).

**Family Aotidae** Elliot, 1913. Review of the Primates, 1:xxiv.
COMMENTS: Groves (1989) first suggested that this is a full family and used the name Aotidae Poche, 1908; Groves (2001*c*) pointed out that Nyctipithecidae Gray, 1870 has precedence, but overlooked the fact that Simpson (1945) had already replaced Nyctipithecinae by Aotinae (Brandon-Jones and Groves, 2002). Goodman et al. (1998) placed it as a third subfamily of Cebidae (with Cebinae and Callitrichinae), and McKenna and Bell (1997) placed *Aotus*, *Cebus* and *Saimiri* as three equal genera in their tribe Cebini (of subfamily Cebinae which also included Callicebini).

*Aotus* Illiger, 1811. Prodr. Syst. Mamm. Avium., p. 71.
TYPE SPECIES: *Simia trivirgata* Humboldt, 1811.
SYNONYMS: *Nocthora* F. Cuvier, 1824; *Nyctipithecus* Spix, 1823.
COMMENTS: Hershkovitz (1983) divided the species into two groups: Gray-necked and Red-necked.

*Aotus azarae* (Humboldt, 1811). Rec. Observ. Zool., 1:359.
COMMON NAME: Azara's Night Monkey.
TYPE LOCALITY: Argentina, right bank of Rio Paraguay.

DISTRIBUTION: Bolivia south of Amazon, between Rios Tocantins and Tapajos-Juruena, south to Paraguay and N Argentina.

STATUS: CITES – Appendix II; IUCN – Lower Risk (lc).

SYNONYMS: *miriquouina* (É. Geoffroy, 1812); **boliviensis** Elliot, 1907; *bidentatus* Lönnberg, 1941; **infulatus** (Kuhl, 1820); *roberti* Dollman, 1909.

COMMENTS: Red-necked species group. Pieczarka and Nagamuchi (1988) proposed that *infulatus* may be conspecific with *A. azarae*; followed by Ford (1994) and Groves (2001*c*). Groves (1993) changed the form of the name to *azarai* in accordance with Art. 31 of the International Code of Zoological Nomenclature (International Commission on Zoological Nomenclature, 1985*d*), but D. Brandon-Jones (pers. comm.) pointed out that Art. 31.1.3 recommends that the original spelling be preserved.

*Aotus hershkovitzi* Ramirez-Cerquera, 1983. IX Cong. Latinoamer. Zool. [abstracts], Arequipa, Peru, p. 148.

COMMON NAME: Hershkovitz's Night Monkey.

TYPE LOCALITY: Colombia, Dept. of Meta, east side of Cordillera Oriental.

DISTRIBUTION: Known from the type locality only.

STATUS: CITES – Appendix II; IUCN – Data Deficient.

COMMENTS: Gray-necked species group. According to Defler et al. (2001), probably a synonym of *Aotus lemurinus lemurinus*, but more work (especially on karyotypes) needs to be done.

*Aotus lemurinus* (I. Geoffroy, 1843). C. R. Acad. Sci. Paris, 16:1151.

COMMON NAME: Gray-bellied Night Monkey.

TYPE LOCALITY: Colombia, Santa Fe de Bogotá (see Defler et al., 2001).

DISTRIBUTION: Panama, Equador and Colombia west of Cordillera Oriental.

STATUS: CITES – Appendix II; IUCN – Endangered as *A. l. griseimembra*, Data Deficient as *A. l. zonalis*, otherwise Vulnerable as *A. l. brumbacki* and *A. l. lemurinus*.

SYNONYMS: *aversus* Elliot, 1913; *hirsutus* (Gray, 1847); *lanius* Dollman, 1909; *pervigilis* Elliot, 1913; *villosus* (Gray, 1847); **brumbacki** Hershkovitz, 1983; **griseimembra** Elliot, 1912; *bipunstatus* Bole, 1937; **zonalis** Goldman, 1914.

COMMENTS: Gray-necked species group. Ford (1994) suggested that *brumbacki* is a subspecies of *A. lemurinus*; followed by Groves (2001*c*). Defler et al. (2001) argued that the described subspecies of *A. lemurinus* are probably full species, but more work (especially on karyotypes) needs to be done.

*Aotus miconax* Thomas, 1927. Ann. Mag. Nat. Hist., ser. 9, 19:365.

COMMON NAME: Peruvian Night Monkey.

TYPE LOCALITY: Peru, Amazonas, San Nicolas, 4500 ft. (1372 m).

DISTRIBUTION: A small area in Peru between Rio Ucayali and the Andes, south of Rio Marañon.

STATUS: CITES – Appendix II; IUCN – Vulnerable.

COMMENTS: Red-necked species group.

*Aotus nancymaae* Hershkovitz, 1983. Am. J. Primatol., 4:223.

COMMON NAME: Nancy Ma's Night Monkey.

TYPE LOCALITY: Peru, Loreto, right bank of Rio Samiria, above Estacion Pithecia, 130 m.

DISTRIBUTION: Loreto Dept. (Peru) to Rio Jandiatuba, south of Rio Solimões (Brazil); and enclave between Rios Tigre and Pastaza (Peru).

STATUS: CITES – Appendix II; IUCN – Lower Risk (lc).

COMMENTS: Red-necked species group. The form of the name is changed in accordance with the International Code of Zoological Nomenclature, Art. 31.1.2 (International Commission on Zoological Nomenclature, 1999).

*Aotus nigriceps* Dollman, 1909. Ann. Mag. Nat. Hist., ser. 8, 4:200.

COMMON NAME: Black-headed Night Monkey

TYPE LOCALITY: Peru, Chanchamayo, 1000 m.

DISTRIBUTION: Brazil, south of Rio Solimões, west of Rio Tapajós Juruena, west into Peru; Bolivia (see Anderson, 1997).

STATUS: CITES – Appendix II; IUCN – Lower Risk (lc).

SYNONYMS: *senex* Dollman, 1909.
COMMENTS: Red-necked species group.

*Aotus trivirgatus* (Humboldt, 1811). Rec. Observ. Zool., 1:306.
COMMON NAME: Three-striped Night Monkey.
TYPE LOCALITY: Venezuela, Duida Range, Rio Casiquiare.
DISTRIBUTION: Venezuela, south of Rio Orinoco, south to Brazil north of Rios Negro and Amazon.
STATUS: CITES – Appendix II; IUCN – Lower Risk (lc).
SYNONYMS: *commersonii* (Vigors and Horsfield, 1829); *duruculi* (Lesson, 1840); *felinus* (Spix, 1823); *humboldti* Illiger *in* Humboldt, 1812; *rufus* (Lesson, 1840).
COMMENTS: Gray-necked species group. Long the only species recognized in the genus; divided into 9 species by Hershkovitz (1983), revised further by Ford (1994) and Groves (2001*c*).

*Aotus vociferans* (Spix, 1823). Sim. Vespert. Brasil., p. 25.
COMMON NAME: Spix's Night Monkey.
TYPE LOCALITY: Brazil, upper Marañon, Tabatinga.
DISTRIBUTION: Colombia, east of Cordillera Oriental, west of Rio Negro, south to Brazil (north of Amazon-Solimões Rivers).
STATUS: CITES – Appendix II; IUCN – Lower Risk (lc).
SYNONYMS: *gularis* Dollman, 1909; *microdon* Dollman, 1909; *oseryi* (I. Geoffroy and Deville, 1848); *rufipes* (Sclater, 1872); *spixi* (Pucheran, 1857).
COMMENTS: Gray-necked species group.

**Family Pitheciidae** Mivart, 1865. Proc. Zool. Soc. Lond., 1865:547.

**Subfamily Callicebinae** Pocock, 1925. Proc. Zool. Soc. Lond., 1925:45.
COMMENTS: Regarded as a full family by Groves (1989), but its affinities with Pitheciinae were emphasised by Horovitz et al. (1998). McKenna and Bell (1997) placed it as a tribe Callicebini in subfamily Cebinae, separating it from Pitheciinae.

*Callicebus* Thomas, 1903. Ann. Mag. Nat. Hist., ser. 7, 12:456.
TYPE SPECIES: *Simia personatus* É. Geoffroy, 1812.
SYNONYMS: *Torquatus* Goodman et al., 1998.
COMMENTS: Revised by Hershkovitz (1990*b*) and van Roosmalen et al. (2002). Affinities among the species are disputed; the *C. torquatus* group is very distinct, and provisionally two subgenera may be recognized: *Torquatus* for the *C. torquatus* group, and *Callicebus* for the rest, following Goodman et al. (1998). Apart from the *C. torquatus* group (here, subgenus *Torquatus*), van Roosmalen et al. (2002) divide the species into four species groups: the *C. donacophilus*, *C cupreus*, *C. moloch* and *C. personatus* groups.

*Callicebus baptista* Lönnberg, 1939. Ark. f. Zool., 31A, 13:7.
COMMON NAME: Baptista Lake Titi.
TYPE LOCALITY: Brazil, Amazona, Lago de Baptista.
DISTRIBUTION: C Brazil, north of the Parana do Uraria and Parana do Ramos and south of the Amazon and lowermost Rio Madeira; and a small wedge between the Rio Uira-Curupa and Rio Andira (van Roosmalen et al., 2002:23-24).
STATUS: CITES – Appendix II; IUCN – Lower Risk (lc).
COMMENTS: Subgenus *Callicebus*. *C. moloch* species group. Hershkovitz (1990b) placed this as a subspecies of *C. hoffmannsi*; Groves (1992) suggested that both could be subspecies of *moloch*, but see Groves (2001*c*:170, 173) and van Roosmalen et al. (2002:23-24), who also separated it from *hoffmannsi*.

*Callicebus barbarabrownae* Hershkovitz, 1990. Fieldiana Zool., n.s., 55:77.
COMMON NAME: Barbara Brown's Titi.
TYPE LOCALITY: Brazil, Bahia, Lamarao.
DISTRIBUTION: E Brazil, between Rio Paraguaçu and Rio Itapicuru, except where *C. coimbrai* is found.

STATUS: CITES – Appendix II; IUCN – Critically Endangered.

COMMENTS: Subgenus *Callicebus*. *C. personatus* species group. Regarded as full species, separate from *C. personatus*, by Kobayashi and Langguth (1999).

*Callicebus bernhardi* van Roosmalen, van Roosmalen and Mittermeier, 2002. Neotropical Primates, 10 (Suppl.):24.

COMMON NAME: Prince Bernhard's Titi.

TYPE LOCALITY: Brazil, Amazonas State, west bank of the lower Rio Aripuana, at the edge of Nova Olinda, 41km SW of Novo Aripuana.

DISTRIBUTION: Brazil, Amazonas and Rodonia states, between Rios Madeira-Ji-Parana and Rios Aripuana-Roosevelt.

STATUS: CITES - Appendix II.

COMMENTS: Subgenus *Callicebus*. *C. moloch* species group.

*Callicebus brunneus* (Wagner, 1842). Arch. Naturgesch., 8:357.

COMMON NAME: Brown Titi.

TYPE LOCALITY: Brazil, Rondônia, upper Rio Madeira, Cachoeira da Bananeira.

DISTRIBUTION: Middle to upper Madeira basin in Peru and Brazil, to upper Rio Purús (Brazil) and Ucayali (Peru); Bolivia (see van Roosmalen et al., 2002).

STATUS: CITES – Appendix II; IUCN – Lower Risk (lc).

COMMENTS: Subgenus *Callicebus*. *C. moloch* species group. Not a subspecies of *moloch*, contra Groves (1992).

*Callicebus caligatus* (Wagner, 1842). Arch. Naturg, 8:257.

COMMON NAME: Chestnut-bellied Titi.

TYPE LOCALITY: Brazil, Rio Madeira, Borba.

DISTRIBUTION: South of the Rio Solimões from Rio Purús to Rio Madeira, Brazil.

STATUS: CITES – Appendix II; IUCN – Lower Risk (lc).

SYNONYMS: *castaneoventris* (Gray, 1866); *usto-fuscus* Elliot, 1907.

COMMENTS: Subgenus *Callicebus*. *C. cupreus* species group. A synonym of *cupreus* according to Groves (2001*c*), who thought it was a colour morph, but shown to be a distinct, allopatric species by van Roosmalen et al. (2002).

*Callicebus cinerascens* (Spix, 1823). Sim. Vespert. Brasil., p. 20.

COMMON NAME: Ashy Black Titi.

TYPE LOCALITY: "Río Putumayo or Içá, Peruvian border of Amazonas, Brazil"; Hershkovitz (1990*b*) doubted its accuracy., and van Roosmalen et al. (2002) postulated it was from the Rio Madeira basin south of the Amazon.

DISTRIBUTION: Rio Madeira basin (SW Brazil).

STATUS: CITES – Appendix II; IUCN – Lower Risk (lc).

COMMENTS: Subgenus *Callicebus*. *C. moloch* species group.

*Callicebus coimbrai* Kobayashi and Langguth, 1999. Rev. Bras. Zool., 16:534.

COMMON NAME: Coimbra Filho's Titi.

TYPE LOCALITY: Brazil, Sergipe, Santansa dos Frades, about 11 km SW of Pacatuba, Aragao, 10°32'S, 36°41W, alt. 90 m.

DISTRIBUTION: NE Brazil, between Rio São Francisco and Rio Real.

STATUS: CITES – Appendix II; IUCN – Critically Endangered.

COMMENTS: Subgenus *Callicebus*. *C. personatus* species group.

*Callicebus cupreus* (Spix, 1823). Sim. Vespert. Brasil., p. 23.

COMMON NAME: Coppery Titi.

TYPE LOCALITY: Brazil, Rio Solimões, Tabatinga.

DISTRIBUTION: South of the Amazon from Rio Purús to Rio Ucayali, Brazil and Peru; probably Bolivia.

STATUS: CITES – Appendix II; IUCN – Lower Risk (lc).

SYNONYMS: *acreanus* Vieira, 1952; *egeria* Thomas, 1908; *toppinii* Thomas, 1914.

COMMENTS: Subgenus *Callicebus*. *C. cupreus* species group. Includes *caligata*, *discolor* and *dubius* according to Groves (1992, 2001*c*), who argued that, if the distributions are sympatric, as mapped by Hershkovitz (1990), then all these are likely to be color morphs; but according to van Roosmalen et al. (2002), the distributions given by Hershkovitz are incorrect, and they are vicariant and hence represent differentiated taxa.

*Callicebus discolor* (I. Geoffroy and Deville, 1848). C. R. Acad. Sci. Paris, 27:498.
COMMON NAME: White-tailed titi.
TYPE LOCALITY: Brazil, Rio Solimões, Tabatinga.
DISTRIBUTION: Upper Amazonian region in Peru, Ecuador and Colombia, and possibly into Brazil, between the Rios Ucayali and Huallaga and north of Rio Marañon across the Rio Napo to the Rio Putumayo and Rio Guamés.
STATUS: CITES – Appendix II; IUCN – Lower Risk (lc).
SYNONYMS: *leucometopa* (Cabrera, 1900); *napoleon* Lönnberg, 1922; *paenulatus* Elliot, 1909; *rutteri* Thomas, 1923; *subrufus* Elliot, 1907.
COMMENTS: Subgenus *Callicebus*. *C. cupreus* species group. Separated as a species from *cupreus* by van Roosmalen et al. (2002).

*Callicebus donacophilus* (d'Orbigny, 1836). Voy. Am. Merid., Atlas Zool., pl. 5.
COMMON NAME: White-eared Titi.
TYPE LOCALITY: Bolivia, Moxos Prov., Río Marmoré.
DISTRIBUTION: WC Bolivia, El Beni and Santa Cruz Provs., Upper Rios Marmoré-Grande and San Miguel basins.
STATUS: CITES – Appendix II; IUCN – Lower Risk (lc).
COMMENTS: Subgenus *Callicebus*. *C. donacophilus* species group. Does not include *pallescens* (see Groves, 2001*c*:172, van Roosmalen et al., 2002:8).

*Callicebus dubius* Hershkovitz, 1988. Fieldiana Zool., ns, 55:66
COMMON NAME: Hershkovitz's Titi.
TYPE LOCALITY: Brazil, east bank of R. Purus, opposite Lago Ayapua, ca.4°20′ S, 62°00′ W, according to Hershkovitz (1990).
DISTRIBUTION: Brazil, Ituxi River or the Mucuim River, east to the Madeira River south of Humaita, and west to the Purus River (van Roosmalen et al., 2002).
STATUS: CITES – Appendix II; IUCN – Vulnerable.
COMMENTS: Subgenus *Callicebus*. *C. cupreus* species group. A synonym of *cupreus* according to Groves (2001*c*), who thought it was a color morph, but shown to be consistently different, with a distinct geographic distribution by van Roosmalen et al. (2002).

*Callicebus hoffmannsi* Thomas, 1908. Ann. Mag. Nat. Hist., ser. 8, 2:89.
COMMON NAME: Hoffmanns's Titi.
TYPE LOCALITY: Brazil, Pará, Urucurituba, Rio Tapajós.
DISTRIBUTION: C Brazil, south of Amazon, between Rios Canuma and Tapajós-Jurena, south to the Rio Sucunduri.
STATUS: CITES – Appendix II; IUCN – Lower Risk (lc).
COMMENTS: Subgenus *Callicebus*. *C. moloch* species group. Groves (1992) suggested that this may be a subspecies of *moloch*, but see Groves (2001*c*:170). Does not include *baptista* (see Groves, 2001*c*:170, 173).

*Callicebus lucifer* Thomas, 1914. Ann. Mag. Nat. Hist., ser. 8, 13:345.
COMMON NAME: Lucifer Titi.
TYPE LOCALITY: Peru, Department of Loreto, Yahauas Territory, vicinity of Pebas.
DISTRIBUTION: Peru, Ecuador and Brazil, between Rios Caqueta-Japua and Rios Napo-Solimoes.
STATUS: CITES – Appendix II; IUCN – Lower Risk (lc).
SYNONYMS: *ignitus* Thomas, 1927.
COMMENTS: Subgenus *Torquatus*. Separated as a species from *C. torquatus* by van Roosmalen et al. (2002).

*Callicebus lugens* (Humboldt, 1811). Rec. Observ. Zool., p. 319.
COMMON NAME: Black Titi.
TYPE LOCALITY: Venezuela, Amazonas state, near San Fernando de Atabapo, at confluence of Rio Orinoco and Rio Guaviare.

DISTRIBUTION: Brazil, Colombia and Venezuela, west of Rio Branco and north of Rios Negro/Uaupes/Vaupes; then west of Rio Apaporis and north of Rio Caqueta, east of Andes north to Rio Tomo, possibly to Rio Orinoco but known to reach the Orinoco only between Rio Caura and Rio Caroni (van Roosmalen et al., 2002).

STATUS: CITES – Appendix II; IUCN – Lower Risk (lc).

SYNONYMS: *duida* Allen, 1914; *vidua* (Lesson, 1840).

COMMENTS: Subgenus *Torquatus*. Separated as a species from *C. torquatus* by van Roosmalen et al. (2002).

*Callicebus medemi* Hershkovitz, 1963. Mammalia, 27:52.

COMMON NAME: Colombian Black-handed Titi.

TYPE LOCALITY: Colombia, Putumayo, right bank of Rio Caqueta, Rio Mecaya.

DISTRIBUTION: Amazonian region of Colombia.

STATUS: CITES – Appendix II; IUCN – Vulnerable.

COMMENTS: Subgenus *Torquatus*. Separated as a species from *C. torquatus* by Groves (2001*c*).

*Callicebus melanochir* (Wied-Neuwied, 1820). Reise nach Brasil., 1:258, fn.

COMMON NAME: Coastal Black-handed Titi.

TYPE LOCALITY: Brazil, Bahia, Morro d'Arara or Fazenda Arara.

DISTRIBUTION: E Brazil, between Rio Mucuri and Rio Itapicuru.

STATUS: CITES – Appendix II; IUCN – Vulnerable.

SYNONYMS: *canescens* (Kuhl, 1820); *gigot* Spix, 1823.

COMMENTS: Subgenus *Callicebus*. *C. personatus* species group. Regarded as full species, separate from *C. personatus*, by Kobayashi and Langguth (1999).

*Callicebus modestus* Lönnberg, 1939. Ark. f. Zool., 31A:17.

COMMON NAME: Rio Beni Titi.

TYPE LOCALITY: Bolivia, Beni, El Consuelo.

DISTRIBUTION: Upper Río Beni basin (Bolivia).

STATUS: CITES – Appendix II; IUCN – Lower Risk (lc).

COMMENTS: Subgenus *Callicebus*. *C. donacophilus* species group.

*Callicebus moloch* (Hoffmannsegg, 1807). Magas. Ges. Nativf. Fr., 9:97.

COMMON NAME: Red-bellied Titi.

TYPE LOCALITY: Brazil, Pará; right bank of lower Rio Tapajós according to Hershkovitz (1963).

DISTRIBUTION: C Brazil, south of Amazon, between Rios Tapajós and Tocantins-Araguaia.

STATUS: CITES – Appendix II; IUCN – Lower Risk (lc).

SYNONYMS: *emiliae* Thomas, 1911; *geoffroyi* Miranda Ribeiro, 1914; *hypokantha* (Olfers, 1819); *remulus* Thomas, 1908; *sakir* (Giebel, 1855).

COMMENTS: Subgenus *Callicebus*. *C. moloch* species group. See Jones and Anderson (1978, Mammalian Species, 112).

*Callicebus nigrifrons* (Spix, 1823). Sim. Vespert. Brasil., p. 21.

COMMON NAME: Black-fronted Titi.

TYPE LOCALITY: Brazil, Rio de Janeiro, Rio Onças, Municipio Campos (see Hershkovitz, 1990:72-73).

DISTRIBUTION: SE Brazil, states of Rio de Janeiro, São Paulo (north of Rio Tietê), and S Minas Gerais.

STATUS: CITES – Appendix II; IUCN – Vulnerable.

SYNONYMS: *brunello* Thomas, 1913; *chlorocnemius* (Lund, 1840); *crinicaudus* (Lund, 1841); *grandis* (Lund, 1841); *melanops* (Vigors, 1829).

COMMENTS: Subgenus *Callicebus*. *C. personatus* species group. Regarded as full species, separate from *C. personatus*, by Kobayashi and Langguth (1999).

*Callicebus oenanthe* Thomas, 1924. Ann. Mag. Nat. Hist., ser. 9, 14:286.

COMMON NAME: Rio Mayo Titi.

TYPE LOCALITY: Peru, San Martín Dept., Moyobamba, 840 m.

DISTRIBUTION: Rio Mayo valley, 750-950 m (N Peru).

STATUS: CITES – Appendix II; IUCN – Vulnerable.

COMMENTS: Subgenus *Callicebus*. *C. donacophilus* species group.

*Callicebus olallae* Lönnberg, 1939. Ark. f. Zool., 31A:16.
    COMMON NAME: Ollala Brothers' Titi.
    TYPE LOCALITY: Bolivia, El Beni Prov., La Laguna (5 km from Santa Rosa).
    DISTRIBUTION: Known only from the type locality.
    STATUS: CITES – Appendix II; IUCN – Data Deficient.
    COMMENTS: Subgenus *Callicebus*. *C. donacophilus* species group.

*Callicebus ornatus* (Gray, 1866). Ann. Mag. Nat. Hist., ser. 4, 17:57.
    COMMON NAME: Ornate Titi.
    TYPE LOCALITY: Colombia, Villavicencia, Rio Meta.
    DISTRIBUTION: Colombia, headwaters of Rio Meta and Rio Guiviare.
    STATUS: CITES – Appendix II; IUCN – Vulnerable.
    COMMENTS: Subgenus *Callicebus*. *C. cupreus* species group. Separated from *C. cupreus* by Groves (2001c) and van Roosmalen et al. (2002).

*Callicebus pallescens* Thomas, 1907. Ann. Mag. Nat. Hist., ser. 7; 20:161.
    COMMON NAME: White-coated Titi.
    TYPE LOCALITY: Paraguay, Chaco, 48 km north of Concepcion.
    DISTRIBUTION: Paraguay, W of Rio Paraguay to about 23°S and 61°30'W in Gran Chaco; Mato Grosso do Sul, in the Pantanal (Brazil); probably Bolivia (see Anderson, 1997).
    STATUS: CITES – Appendix II; IUCN – Lower Risk (lc).
    COMMENTS: Subgenus *Callicebus*. *C. donacophilus* species group. Separated from *C. donacophilus* by Groves (2001c:172) and van Roosmalen et al. (2002).

*Callicebus personatus* (É. Geoffroy, 1812). *In* Humboldt, Rec. Observ. Zool., 1:357.
    COMMON NAME: Atlantic Titi.
    TYPE LOCALITY: Brazil, Espírito Santo, Rio Doce (see Hershkovitz, 1990b).
    DISTRIBUTION: SE Brazil, Espirito Santo, possibly into NW Minas Gerais.
    STATUS: CITES – Appendix II; IUCN – Vulnerable.
    SYNONYMS: *incanescens* (Kuhl, 1820).
    COMMENTS: Subgenus *Callicebus*. *C. personatus* species group. Does not include *melanochir*, *nigrifrons* or *barbarabrownae* (see Kobayashi and Langguth, 1999).

*Callicebus purinus* Thomas, 1927. Ann. Mag. Nat. Hist., ser. 9, 19:510.
    COMMON NAME: Rio Purus Titi.
    TYPE LOCALITY: Brazil, state of Amazonas, Lago Ayapua, left bank of lower Rio Purus.
    DISTRIBUTION: Brazil south of the Rio Solimoes between the Rio Tapaua and Rio Jurua.
    STATUS: CITES – Appendix II; IUCN – Lower Risk (lc).
    COMMENTS: Subgenus *Torquatus*. Separated as a species from *C. torquatus* by van Roosmalen et al. (2002).

*Callicebus regulus* Thomas, 1927. Ann. Mag. Nat. Hist., ser. 9, 19:510.
    COMMON NAME: Red-headed Titi.
    TYPE LOCALITY: Brazil, state of Amazonas, Fonte Boa, right bank of upper Rio Solimoes.
    DISTRIBUTION: Brazil, between Rios Javari/Solimoes and Rio Jurua.
    STATUS: CITES – Appendix II; IUCN – Lower Risk (lc).
    COMMENTS: Subgenus *Torquatus*. Separated as a species from *C. torquatus* by van Roosmalen et al. (2002).

*Callicebus stephennashi* van Roosmalen, van Roosmalen and Mittermeier, 2002. Neotropical Primates, 10, suppl.:15.
    COMMON NAME: Stephen Nash's Titi.
    TYPE LOCALITY: Brazil: somewhere along the middle to upper Rio Purus.
    DISTRIBUTION: Brazil: probably along the right bank of the Rio Purus, in between the distributions of *C. caligatus* and *C. dubius* (van Roosmalen et al., 2002).
    STATUS: CITES – Appendix II.
    COMMENTS: Subgenus *Callicebus*. *C. cupreus* species group.

*Callicebus torquatus* (Hoffmannsegg, 1807). Magas. Ges. Nativf. Fr., 10:86.
  COMMON NAME: Collared Titi.
  TYPE LOCALITY: Brazil, Codajás, north bank of Rio Solimões.
  DISTRIBUTION: Brazil, between Rios Negro/Uaupes and Rios Solimoes/Japura/Apaporis.
  STATUS: CITES – Appendix II; IUCN – Lower Risk (lc).
  SYNONYMS: *amicta* (É. Geoffroy, 1812).
  COMMENTS: Subgenus *Torquatus*. Does not include *lucifer*, *lugens*, *purinus* or *regulus* according
    to van Roosmalen et al. (2002).

**Subfamily Pitheciinae** Mivart, 1865. Proc. Zool. Soc. Lond., 1865:547.

*Cacajao* Lesson, 1840. Spec. Mamm. Bim. et Quadrum., p. 181.
  TYPE SPECIES: *Simia melanocephalus* Humboldt, 1812.
  SYNONYMS: *Brachyurus* Spix, 1823; *Cercoptochus* Gloger, 1842; *Cothurus* Palmer, 1899 [not of
    Champion, 1891 (Coleoptera)]; *Neocothurus* Palmer, 1903; *Ouakaria* Gray, 1849.
  COMMENTS: Revised by Hershkovitz (1987c). Reduced to a subgenus of *Chiropotes* by Goodman
    et al. (1998).

*Cacajao calvus* (I. Geoffroy, 1847). C. R. Acad. Sci. Paris, 24:576.
  COMMON NAME: Bald Uacari.
  TYPE LOCALITY: Brazil, Amazonas, Fonte Boa.
  DISTRIBUTION: NW Brazil, E Peru.
  STATUS: CITES – Appendix I; U.S. ESA – Endangered; IUCN – Endangered as *C. c. calvus*,
    *C. c. novaesi*, and *C. c. rubicundus*, Vulnerable as *C. c. ucayalii*.
  SYNONYMS: *alba* (Schlegel, 1876); **novaesi** Hershkovitz, 1987; **rubicundus** (I. Geoffroy St.
    Hilaire and Deville, 1848); **ucayalii** Thomas, 1928.
  COMMENTS: Includes *rubicundus*; see Hershkovitz (1972a), but also see Szalay and Delson
    (1979:290) who listed it as a distinct species.

*Cacajao melanocephalus* (Humboldt, 1812). Rec. Observ. Zool., 1:317.
  COMMON NAME: Black-headed Uacari.
  TYPE LOCALITY: Venezuela, Casiquiare Forests, Mision de San Francisco Solano.
  DISTRIBUTION: SW Venezuela, NW Brazil.
  STATUS: CITES – Appendix I; U.S. ESA – Endangered; IUCN – Lower Risk (lc).
  SYNONYMS: *ouakary* (Spix, 1823); *spixii* (Gray, 1849).

*Chiropotes* Lesson, 1840. Spec. Mamm. Bim. et Quadrum., p. 178.
  TYPE SPECIES: *Pithecia* (*Chiropotes*) *couxio* Lesson, 1840 (= *Cebus satanas* Hoffmannsegg, 1807).
  SYNONYMS: *Cheiropotes* Reichenbach, 1862; *Saki* Schlegel, 1876.
  COMMENTS: Reviewed by Hershkovitz (1985) and Bonvicino et al. (2003b).

*Chiropotes albinasus* (I. Geoffroy and Deville, 1848). C. R. Acad. Sci. Paris, 27:498.
  COMMON NAME: White-nosed Saki.
  TYPE LOCALITY: Brazil, Pará, Santarém.
  DISTRIBUTION: NC Brazil.
  STATUS: CITES – Appendix I; U.S. ESA – Endangered; IUCN – Lower Risk (lc).
  SYNONYMS: *roosevelti* J. A. Allen, 1914.

*Chiropotes chiropotes* (Humboldt, 1811). Rec. Obs. zool. Anat. Comp., 1:311.
  COMMON NAME: Red-backed Bearded Saki.
  TYPE LOCALITY: Reported to be Venezuela, Amazonas, upper Rio Orinoco south of the
    cataracts; but according to Bonvicino et al. (2003b) does not occur in Venezuela.
  DISTRIBUTION: Guyana, French Guiana, Surinam, Brazil east of the R.Branco.
  STATUS: CITES – Appendix II; status not known.
  SYNONYMS: *couxio* (Lesson, 1840); *fulvo-fusca* (Trouessart, 1897); *sagulata* (Traill, 1821).
  COMMENTS: A distinct species, not a subspecies of *C. satanas*, and does not include *israelita*
    as a synonym, according to Bonvicino et al. (2003b).

*Chiropotes israelita* (Spix, 1823). Sim. et Vesp. Brasil., 11.
  COMMON NAME: Brown-backed Bearded Saki.

TYPE LOCALITY: Brazil, Rio Negro.

DISTRIBUTION: Brazil north of the Amazon and east of the Rio Branco, S Venezuela east of the Rio Orinoco.

STATUS: CITES – Appendix II; status not known.

COMMENTS: Regarded as a distinct species, not a synonym of *chiropotes*, by Bonvicino et al. (2003*b*).

*Chiropotes satanas* (Hoffmannsegg, 1807). Mag. Ges. Naturf. Fr., 10:93.

COMMON NAME: Black Bearded Saki.

TYPE LOCALITY: Brazil, Pará, lower Rio Tocantins, Cametá.

DISTRIBUTION: Brazil south of Amazon estuary, between Rios Tocantins and Gurupi.

STATUS: CITES – Appendix II; U.S. ESA – Endangered as *C. s. satanas*; IUCN – Endangered as *C. s. satanas*.

SYNONYMS: *ater* Gray, 1870; *nigra* (Trouessart, 1897).

COMMENTS: The subspecies assigned to this species by Hershkovitz (1985) were regarded as distinct species by Bonvicino et al. (2003*b*).

*Chiropotes utahickae* Hershkovitz, 1985. Fieldiana (Zool.), N.S. 27:17.

COMMON NAME: Uta Hick's Bearded Saki.

TYPE LOCALITY: Brazil, Tapará, right (east) bank of Rio Xingu, near mouth.

DISTRIBUTION: N Brazil, south of Amazon, between Rios Xingu and Tocantins, south to Serra dos Carajás and Rio Itacaiuna.

STATUS: CITES – Appendix II; IUCN – Vulnerable as *C. satanus utahickae*.

COMMENTS: Not a subspecies of *C. satanas* (Bonvicino et al., 2003*b*).

*Pithecia* Desmarest, 1804. Tabl. Meth. Hist. Nat., *in* Nouv. Dict. Hist. Nat., 24:8.

TYPE SPECIES: *Simia pithecia* Linnaeus, 1766.

SYNONYMS: *Calletrix* Fleming, 1822 [*lapsus*]; *Yarkea* Lesson, 1840.

COMMENTS: Revised by Hershkovitz (1987*d*).

*Pithecia aequatorialis* Hershkovitz, 1987. Am. J. Primatol., 12:429.

COMMON NAME: Equatorial Saki.

TYPE LOCALITY: Peru, Loreto, lower Rio Nanay, Santa Luisa.

DISTRIBUTION: Napo (Ecuador) to Loreto (Peru).

STATUS: CITES – Appendix II; IUCN – Lower Risk (lc).

COMMENTS: Based upon specimens included by Hershkovitz (1979) in *P. monachus*.

*Pithecia albicans* Gray, 1860. Proc. Zool. Soc. Lond., 1860:231.

COMMON NAME: White-footed Saki.

TYPE LOCALITY: Brazil, Amazonas, Tefé (south bank of Solimões River).

DISTRIBUTION: South bank of Amazon, between lower Jurua and lower Purús Rivers.

STATUS: CITES – Appendix II; IUCN – Lower Risk (lc).

COMMENTS: Separated from *monachus* by Hershkovitz (1979, 1987*d*).

*Pithecia irrorata* Gray, 1842. Ann. Mag. Nat. Hist., [ser. 1], 10:256.

COMMON NAME: Rio Tapajos Saki.

TYPE LOCALITY: Brazil, Pará, West bank of Rio Tapajós, Parque Nacional de Amazônia.

DISTRIBUTION: South of the Amazon in SW Brazil; SW Peru; E Bolivia.

STATUS: CITES – Appendix II; IUCN – Lower Risk (lc).

SYNONYMS: *vanzolinii* Hershkovitz, 1987.

COMMENTS: Includes some specimens identified by Hershkovitz (1979) as *P. hirsuta*. May prove to be a subspecies of *P. monachus* (Hershkovitz, 1987*d*).

*Pithecia monachus* (É. Geoffroy, 1812). Rec. Observ. Zool. 1:359.

COMMON NAME: Monk Saki.

TYPE LOCALITY: Brazil, left bank of Rio Solimões between Tabatinga and Rio Tocantins.

DISTRIBUTION: West of Rio Jurua and Rio Japura-Caqueta (in Brazil), Colombia, Ecuador and Peru.

STATUS: CITES – Appendix II; IUCN – Vulnerable as *P. m. milleri*, Data Deficient as *P. m. napensis*, otherwise Lower Risk (lc).

SYNONYMS: *guapo* Schinz, 1844; *hirsuta* Spix, 1823; *inusta* Spix, 1823; *napensis* Lönnberg, 1938; **milleri** J. A. Allen, 1914.

COMMENTS: Does not include *albicans*, see Hershkovitz (1979). Includes *hirsuta*, see Hershkovitz (1987*d*).

*Pithecia pithecia* (Linnaeus, 1766). Syst. Nat., 12th ed., 1:40.

COMMON NAME: White-faced Saki.

TYPE LOCALITY: French Guiana, Cayenne.

DISTRIBUTION: Guyana; French Guiana; Surinam; N Amazon, east of Río Negro and Rio Orinoco (N Brazil, S Venezuela).

STATUS: CITES – Appendix II; IUCN – Lower Risk (lc).

SYNONYMS: *adusta* Olfers, 1818; *capillamentosa* Spix, 1823; *leucocephala* (Audebert, 1797); *ochrocephala* Kuhl, 1820; *pogonias* Gray, 1842; *rufibarbata* Kuhl, 1820; *rufiventer* (É. Geoffroy, 1812); *saki* Muirhead, 1819; **chrysocephala** I. Geoffroy, 1850; *lotichiusi* Mertens, 1925.

**Family Atelidae** Gray, 1825. Ann. Philos., n.s., 10:338.

**Subfamily Alouattinae** Trouessart, 1897. Cat. Mamm. Viv. Foss., p. 32.

SYNONYMS: Mycetinae Gray, 1825.

COMMENTS: Groves (2001*c*) used Mycetinae, but had overlooked the fact that Simpson (1945) had replaced Mycetinae with Alouattinae: see Brandon-Jones and Groves (2003).

*Alouatta* Lacépède, 1799. Tabl. Div. Subd. Orders Genres Mammifères, p. 4.

TYPE SPECIES: *Simia belzebul* Linnaeus, 1766.

SYNONYMS: *Mycetes* Illiger, 1811; *Stentor* É. Geoffroy, 1812.

COMMENTS: Includes the following species groups according to Groves (2001*c*): (1) *A. palliata* group (*palliata, pigra, coibensis*), (2) *A. seniculus* group (*seniculus, macconnelli, sara, belzebul, nigerrima, guariba*), and (3) *A. caraya* group (*caraya* only).

*Alouatta belzebul* (Linnaeus, 1766). Syst. Nat., 12th ed., 1:37.

COMMON NAME: Red-handed Howler.

TYPE LOCALITY: Brazil, Pará, Rio Capim.

DISTRIBUTION: N Brazil (mainly south of Lower Amazon, east of Rio Madeira); Mexiana Isl (Brazil); in Pará Prov. (Brazil), north of Amazon.

STATUS: CITES – Appendix II; IUCN – Critically Endangered as *A. b. ululata*, otherwise Lower Risk (lc) as *A. belezebul* (sic).

SYNONYMS: *beelzebub* (Bechstein, 1800); *discolor* (Spix, 1823); *flavimanus* (Bates, 1863); *mexianae* (Hagmann, 1908); *rufimanus* (Kuhl, 1820); *tapojozensis* Lönnberg, 1941; *ululata* Elliot, 1912.

COMMENTS: *A. seniculus* species group. Does not include *nigerrima* (Groves, 2001*c*).

*Alouatta caraya* (Humboldt, 1812). Rec. Observ. Zool., 1:355.

COMMON NAME: Black Howler.

TYPE LOCALITY: Paraguay.

DISTRIBUTION: N Argentina to Mato Grosso (Brazil), Bolivia (see Anderson, 1997).

STATUS: CITES – Appendix II; IUCN – Lower Risk (lc).

SYNONYMS: *barbatus* (Spix, 1823); *niger* (É. Geoffroy, 1812); *straminea* (Humboldt, 1812).

COMMENTS: *A. caraya* species group. For the inclusion of *straminea* in this species, not in *A. seniculus*, see Rylands and Brandon-Jones (1988).

*Alouatta coibensis* Thomas, 1902. Novit. Zool., 9:135.

COMMON NAME: Coiba Island Howler.

TYPE LOCALITY: Panama, Coiba Isl.

DISTRIBUTION: Coiba Isl and Azuero Penninsula, Panama.

STATUS: CITES – Appendix I; IUCN – Critically Endangered as *A. c. trabeata*, otherwise Endangered.

SYNONYMS: **trabeata** Lawrence, 1933.

COMMENTS: *A. palliata* species group. Recognized as a full species by Froehlich and Froehlich (1987).

*Alouatta guariba* (Humboldt, 1812). Rec. Observ. Zool., 1:pl. 30.
COMMON NAME: Brown Howler.
TYPE LOCALITY: Brazil.
DISTRIBUTION: N Bolivia (?- *beniensis*); SE and EC Brazil, north to the Rio São Francisco.
STATUS: CITES – Appendix II; IUCN – Critically Endangered as *A. g. guariba*, otherwise Vulnerable.
SYNONYMS: ?*beniensis* Lönnberg, 1941; *bicolor* (Gray, 1845); *fusca* (É. Geoffroy, 1812); *ursinus* (É. Geoffroy, 1812: plate only, not text); **clamitans** Cabrera, 1940; *iheringi* Lönnberg, 1941.
COMMENTS: *A. seniculus* species group. *A. guariba* (Humboldt, 1812) is the correct name for this species, not *fusca* (Rylands and Brandon-Jones, 1988). *A. "fusca" beniensis* may actually be *A. seniculus* (Mittermeier et al., 1988:13-75).

*Alouatta macconnelli* (Linnaeus, 1766). Syst. Nat., 12th ed., 1:37.
COMMON NAME: Guyanan Red Howler.
TYPE LOCALITY: Guyana: coast region.
DISTRIBUTION: Trinidad; Guyana, French Guiana, and Brazil north of the lower and Middle Amazon.
STATUS: CITES – Appendix II; IUCN – Vulnerable as *A. seniculus insulanus*.
SYNONYMS: *insulanus* Elliot, 1910.
COMMENTS: *A. seniculus* species group. Separated from *A. seniculus* by Groves (2001*c*).

*Alouatta nigerrima* Lönnberg, 1941. Ark. f. Zool., 33A, 10:33.
COMMON NAME: Amazon Black Howler.
TYPE LOCALITY: Brazil, Amazonas, Patinga.
DISTRIBUTION: N Brazil, east of the Rio Trombetas to the Rio Tapajos, perhaps to the Rio Tocantins.
STATUS: CITES – Appendix II.
COMMENTS: *A. seniculus* species group. Separated from *A. belzebul* by Groves (2001*c*).

*Alouatta palliata* (Gray, 1849). Proc. Zool. Soc. Lond., 1848:138 [1849].
COMMON NAME: Mantled Howler.
TYPE LOCALITY: Nicaragua, Lake Nicaragua.
DISTRIBUTION: W Ecuador to Veracruz and Oaxaca (Mexico).
STATUS: CITES – Appendix I; U.S. ESA – Endangered; IUCN – Vulnerable as *A. p. mexicana*, otherwise Lower Risk (lc).
SYNONYMS: *aequatorialis* Festa, 1903; *inclamax* Thomas, 1913; *inconsonans* Goldman, 1913; *matagalpae* J. A. Allen, 1908; *mexicana* Merriam, 1902; *niger* (Thomas, 1880); *quichua* Thomas, 1913.
COMMENTS: *A. palliata* species group. Subspecies are probably recognizable in this species, but further study is needed (Groves, 2001*c*).

*Alouatta pigra* Lawrence, 1933. Bull. Mus. Comp. Zool., 75:333.
COMMON NAME: Guatemalan Black Howler.
TYPE LOCALITY: Guatemala, Petén, Uaxactun.
DISTRIBUTION: Yucatan and Chiapas (Mexico) to Belize and Guatemala.
STATUS: CITES – Appendix I; U.S. ESA – Threatened; IUCN – Lower Risk (lc).
SYNONYMS: *luctuosa* Lawrence, 1833.
COMMENTS: *A. palliata* species group. The name *villosa* (Gray, 1845), has been applied to this species (see Napier, 1976:76) but Lawrence (1933) regarded it as a *nomen dubium*; see Smith (1970:366) and Hall (1981:260, 263).

*Alouatta sara* Elliot, 1910. Ann. Mag. Nat. Hist., ser. 8, 5:283.
COMMON NAME: Bolivian Red Howler.
TYPE LOCALITY: Bolivia, Sara Prov., Santa Cruz.
DISTRIBUTION: Bolivia (Sara Province), Peru, and Brazil to the Rio Negro and Rondônia.
STATUS: CITES – Appendix II; IUCN – Lower Risk (lc).

COMMENTS: *A. seniculus* species group. Separated from *A. seniculus* by Minezawa et al. (1985).

*Alouatta seniculus* (Linnaeus, 1766). Syst. Nat., 12th ed., 1:37.
COMMON NAME: Venezuelan Red Howler.
TYPE LOCALITY: Colombia, Bolivar, Rio Magdalena, Cartagena.
DISTRIBUTION: Colombia to Venezuela and NW Brazil.
STATUS: CITES – Appendix II; IUCN – Data Deficient at *A. s. amazonica*, *A. s. juara*, and *A. s. puruensis*, otherwise Lower Risk (lc).
SYNONYMS: ?*auratus* (Gray, 1845); *bogotensis* J. A. Allen, 1914; *caquetensis* J. A. Allen, 1914; *caucensis* J. A. Allen, 1904; *chrysurus* (I. Geoffroy, 1829); ?*laniger* (Gray, 1845); *rubicunda* J. A. Allen, 1904; **arctoidea** Cabrera, 1940; *ursina* (Humboldt, 1815) [not of Kerr 1792]; **juara** Elliot, 1910; *amazonica* Lönnberg, 1941; *juruana* Lönnberg, 1941; *puruensis* Lönnberg, 1941.
COMMENTS: *A. seniculus* species group. Does not include *sara* (Minezawa et al., 1985), *straminea* (Rylands and Brandon-Jones, 1998), or *macconnelli* (Groves, 2001*c*).

**Subfamily Atelinae** Gray, 1825. Ann. Philos., n.s., 10:338.
COMMENTS: Combined with Alouattinae as family Atelidae by Rosenberger (1977). Divided into subtribes Atelina (*Ateles* only) and Brachytelina (*Lagothrix*, *Brachyteles*) by Goodman et al. (1998), who did not study *Oreonax*.

*Ateles* É. Geoffroy, 1806. Ann. Mus. Hist. Nat. Paris, 1:262.
TYPE SPECIES: *Simia paniscus* Linnaeus, 1758.
SYNONYMS: *Ameranthropoides* Montandon, 1929; *Montaneia* Ameghino, 1911; *Paniscus* Rafinesque, 1815; *Sapajou* Lacépède, 1799.
COMMENTS: All *Ateles* were considered conspecific by Hernández-Camacho and Cooper (1976:66).

*Ateles belzebuth* É. Geoffroy, 1806. Ann. Mus. Hist. Nat. Paris, 7:27.
COMMON NAME: White-fronted Spider Monkey.
TYPE LOCALITY: Venezuela, Esmeralda.
DISTRIBUTION: Cordillera Oriental, Colombia to Venezuela and N Peru.
STATUS: CITES – Appendix II; IUCN – Vulnerable.
SYNONYMS: *bartlettii* Gray, 1867; *braccatus* Pelzeln, 1883; *brissonii* (Fischer, 1829); *chuva* Schlegel, 1876; *fuliginosus* Kuhl, 1820; *variegatus* Wagner, 1840.
COMMENTS: Does not include *hybridus* or *marginatus* (Groves, 2001*c*).

*Ateles chamek* (Humboldt, 1812). Rec. Observ. Zool., 1:353.
COMMON NAME: Peruvian Spider Monkey.
TYPE LOCALITY: Peru, Cuzco, Rio Comberciato.
DISTRIBUTION: NE Peru, E Bolivia to Brazil west of Rio Juruá and south of Rio Solimões.
STATUS: CITES – Appendix II; IUCN – Lower Risk (lc).
SYNONYMS: *longimembris* J. A. Allen, 1914; *peruvianus* Lönnberg, 1940.
COMMENTS: Separated from *paniscus* by Groves (1989).

*Ateles fusciceps* Gray, 1866. Proc. Zool. Soc. Lond., 1865:733 [1866].
COMMON NAME: Black-headed Spider Monkey.
TYPE LOCALITY: NW Ecuador, Imbabura Prov., Hacienda Chinipamba, 1500 m.
DISTRIBUTION: SE Panama to Ecuador, Colombia to W Cordillera (Paraguay).
STATUS: CITES – Appendix II; IUCN – Critically Endangered as *A. geoffroyi fusciceps*, Vulnerable as *A. g. rufiventris*.
SYNONYMS: **rufiventris** Sclater, 1872; *dariensis* Goldman, 1915; *robustus* J. A. Allen, 1914.

*Ateles geoffroyi* Kuhl, 1820. Beitr. Zool. Vergl. Anat., 1:26.
COMMON NAME: Geoffroy's Spider Monkey.
TYPE LOCALITY: Nicaragua, San Juan del Norte.
DISTRIBUTION: S Mexico to Panama.
STATUS: CITES – Appendix I and U.S. ESA – Endangered as *A. g. frontatus* and *A. g. panamensis* only, otherwise CITES – Appendix II; IUCN – Critically Endangered as *A. g. azuerensis*,

Endangered as *A. g. grisescens* and *A. g. panamensis*, Vulnerable as *A. g. ornatus*, *A. g. yucatanensis*, and *A. g. frontatus*, Lower Risk (lc) as *A. g. vellerosus*.

SYNONYMS: *frontatus* (Gray, 1842); *melanochir* Desmarest, 1820; *trianguligera* Weinland, 1862; **grisescens** Gray, 1866; *cucullatus* Gray, 1866; **ornatus** Gray, 1870; *azuerensis* Bole, 1937; *panamensis* Kellogg and Golman, 1944; **vellerosus** Gray, 1866; *neglectus* Reinhardt, 1873; *pan* Schlegel, 1876; *tricolor* Hollister, 1914; **yucatanensis** Kellogg and Goldman, 1944.

*Ateles hybridus* I. Geoffroy, 1829. Mem. Mus. Paris, 17:168.
COMMON NAME: Brown Spider Monkey.
TYPE LOCALITY: Colombia, valley of Rio Magdalena, La Gloria.
DISTRIBUTION: N Colombia and NW Venezuela.
STATUS: CITES – Appendix II; IUCN – Endangered as *A. h. hybridus* and *A. h. brunneus*.
SYNONYMS: *albifrons* Gray, 1870; *brunneus* Gray, 1870; *loysi* (Montandon, 1929).
COMMENTS: Separated from *belzebuth* by Froehlich et al. (1991). Rylands et al. (2000) recognized *brunneus* as a subspecies.

*Ateles marginatus* É. Geoffroy, 1809. Ann. Mus. Hist. Nat. Paris, 13:97.
COMMON NAME: White-cheeked Spider Monkey.
TYPE LOCALITY: Brazil, Rio Tocantins, Cametá.
DISTRIBUTION: South of Lower Amazon, Rio Tapajós to Rio Tocantins (Brazil).
STATUS: CITES – Appendix II; IUCN – Endangered.
SYNONYMS: *frontalis* Bennett, 1831.
COMMENTS: Separated from *belzebuth* by Groves (1989).

*Ateles paniscus* (Linnaeus, 1758). Syst. Nat., 10th ed., 1:26.
COMMON NAME: Red-faced Spider Monkey.
TYPE LOCALITY: French Guiana.
DISTRIBUTION: Guianas and Brazil, north of the Amazon (east of Rio Negro).
STATUS: CITES – Appendix II; IUCN – Lower Risk (lc).
SYNONYMS: *ater* F. Cuvier, 1823; *cayennensis* (Fischer, 1829); *pentadactylus* É. Geoffroy, 1806; *subpentadactylus* Desmarest, 1820; *surinamensis* (Fischer, 1829).

*Brachyteles* Spix, 1823. Sim. Vespert. Brasil., p. 36.
TYPE SPECIES: *Brachyteles macrotarsus* Spix, 1823 (= *Ateles arachnoides* É. Geoffroy, 1806).
SYNONYMS: *Eriodes* I. Geoffroy, 1829.

*Brachyteles arachnoides* (É. Geoffroy, 1806). Ann. Mus. Hist. Nat. Paris, 7:271.
COMMON NAME: Southern Muriqui.
TYPE LOCALITY: Brazil, Rio de Janeiro.
DISTRIBUTION: SE Brazil: states of Rio de Janeiro and São Paulo.
STATUS: CITES – Appendix I; U.S. ESA – Endangered; IUCN – Critically Endangered.
SYNONYMS: *eriodes* Brehm, 1876; *macrotarsus* Spix, 1823; *tuberifer* (I. Geoffroy, 1829).

*Brachyteles hypoxanthus* (Kuhl, 1820). Beitr. Zool., 1820:25.
COMMON NAME: Northern Muriqui.
TYPE LOCALITY: Brazil, Bahia.
DISTRIBUTION: E Brazil: Bahia, Minas Gerais, Espiritu Santo.
STATUS: CITES – Appendix I as included in *B. arachnoides*; U.S. ESA – Endangered as included in *B. arachnoides*; IUCN – Critically Endangered.
SYNONYMS: *hemidactylus* (I. Geoffroy, 1829).
COMMENTS: Separated from *B. arachnoides* by Rylands et al. (1995).

*Lagothrix* É. Geoffroy, 1812. *In* Humboldt, Rec. Observ. Zool., 1:356.
TYPE SPECIES: *Lagothrix humboldtii* É. Geoffroy, 1812 (= *Simia lagothricha* Humboldt, 1812).
SYNONYMS: *Gastrimargus* Spix, 1823.
COMMENTS: Does not include *flavicauda*; divided into four species by Groves (2001*c*).

*Lagothrix cana* (É. Geoffroy, 1812). Rec. Observ. Zool., 1:354.
COMMON NAME: Gray Woolly Monkey.

TYPE LOCALITY: Brazil, S bank of Rio Solimoes near mouth of Rio Tefé.

DISTRIBUTION: Brazil, south of Amazon; southern highlands of Peru; an isolated population in northern Bolivia.

STATUS: CITES – Appendix II; IUCN – Vulnerable as *L. c. cana* and *L. c. tschudii*.

SYNONYMS: *olivaceus* (Spix, 1823); *puruensis* Lönnberg, 1940; *ubericola* Elliot, 1909; **tschudii** Pucheran, 1857; *nigra* J. A. Allen, 1900; *thomasi* Elliot, 1909.

COMMENTS: Wallace and Painter (1999) recorded a southern range extension for this species in Madidi National Park, Bolivia; the description corresponds to *L. c. tschudii* according to Rylands et al. (2000).

*Lagothrix lagotricha* (Humboldt, 1812). Rec. Observ. Zool., 1:322.

COMMON NAME: Brown Woolly Monkey.

TYPE LOCALITY: Colombia, Úapes, Rio Guaviare.

DISTRIBUTION: Brazil N of Rio Napo-Amazon system, SE Colombia, extreme N Peru and NE Ecuador.

STATUS: CITES – Appendix II; IUCN – Lower Risk (lc).

SYNONYMS: *barrigo* Natterer, 1883; *caparro* Lesson, 1840; *caroarensis* Lönnberg, 1931; *geoffroyi* Pucheran, 1857; *humboldtii* É. Geoffroy, 1812; *infumatus* (Spix, 1823).

*Lagothrix lugens* Elliot, 1907. Ann. Mag. Nat. Hist., ser. 7, 20:193.

COMMON NAME: Colombian Woolly Monkey.

TYPE LOCALITY: Colombia, Upper Rio Magdalena valley, Tolima, 2°20′N, 1500-2100 m.

DISTRIBUTION: Colombia, headwaters of Orinoco tributaries; Venezuela, Sarare River drainage.

STATUS: CITES – Appendix II; IUCN – Vulnerable.

*Lagothrix poeppigii* Schinz, 1844. Syst. Verz. Säug., 1:72.

COMMON NAME: Silvery Woolly Monkey.

TYPE LOCALITY: Peru, Loreto, lower Rio Huallaga, north of Yurimaguas.

DISTRIBUTION: Highlands of E Ecuador and N Peru, to about 70°W, 5°S in Brazil.

STATUS: CITES – Appendix II; IUCN – Vulnerable.

SYNONYMS: *castelnaui* I. Geoffroy and Deville, 1848.

*Oreonax* Thomas, 1927. Ann. Mag. Nat. Hist., ser. 9, 19:156.

TYPE SPECIES: *Lagothrix (Oreonax) hendeei* Thomas, 1927 (= *Simia flavicauda* Humboldt, 1812).

COMMENTS: Separated from *Lagothrix* by Groves (2001c).

*Oreonax flavicauda* (Humboldt, 1812). Rec. Observ. Zool., 1:363.

COMMON NAME: Yellow-tailed Woolly Monkey.

TYPE LOCALITY: Peru, San Martin, Puca Tambo, 5100 ft. (1555 m).

DISTRIBUTION: E Andes in San Martin (Peru) and Amazonas (Brazil).

STATUS: CITES – Appendix I as *Lagothrix flavicauda*; U.S. ESA – Endangered as *L. flavicauda*; IUCN – Critically Endangered.

SYNONYMS: *hendeei* (Thomas, 1927).

**PARVORDER CATARRHINI** É. Geoffroy Saint-Hilaire, 1812.

COMMENTS: Not recognized by McKenna and Bell (1997), who regarded all included taxa as a single superfamily, Cercopithecoidea (including what are usually separated as Hominoidea). Goodman et al. (1998) also regarded the catarrhines as over-split.

**Superfamily Cercopithecoidea** Gray, 1821. London Med. Repos., 15:297.

**Family Cercopithecidae** Gray, 1821. London Med. Repos., 15:297.

COMMENTS: Hill (*in* Honacki et al., 1982:230) and Groves (1989) divided this family into the Colobidae and Cercopithecidae.

**Subfamily Cercopithecinae** Gray, 1821. London Med. Repos., 15:297.

SYNONYMS: Cercocebini Jolly, 1966; Cynopithecinae Osman Hill, 1966; Macacidae Owen, 1843; Papinae Chiarelli, 1966; Papioninae Burnett, 1828; Theropithecini Jolly, 1966.

COMMENTS: Divided by Groves (2001c) into two tribes: Cercopithecini (*Allenopithecus*, *Cercopithecus*, *Chlorocebus*, *Erythrocebus*, *Miopithecus*) and Papionini (*Cercocebus*, *Lophocebus*, *Macaca*, *Mandrillus*, *Papio*, *Theropithecus*).

*Allenopithecus* Lang, 1923. Am. Mus. Novit., 87:1.
TYPE SPECIES: *Cercopithecus nigroviridis* Pocock, 1907.
COMMENTS: Separated from *Cercopithecus* by Thorington and Groves (1970:638), Szalay and Delson (1979), and Groves (1989, 2001c).

*Allenopithecus nigroviridis* (Pocock, 1907). Proc. Zool. Soc. Lond., 1907:739.
COMMON NAME: Allen's Swamp Monkey.
TYPE LOCALITY: Dem. Rep. Congo, upper Congo River.
DISTRIBUTION: NW Dem. Rep. Congo, NE Angola.
STATUS: CITES – Appendix II; IUCN – Lower Risk (nt).

*Cercocebus* É. Geoffroy, 1812. Ann. Mus. Hist. Nat. Paris, 19:97.
TYPE SPECIES: *Cercocebus fuliginosus* É. Geoffroy, 1812 (= *Simia* (*Cercopithecus*) *aethiops torquatus* Kerr, 1792).
SYNONYMS: *Aethiops* Martin, 1841; *Leptocebus* Trouessart, 1904.
COMMENTS: Van Gelder (1977b:8) included *Cercocebus* in *Cercopithecus*. Goodman et al. (1998) included *Mandrillus* as a subgenus of *Cercocebus*. McKenna and Bell (1997) continued to include *Lophocebus* in *Cercocebus*, although the combined genus is clearly not monophyletc (Groves, 2001c).

*Cercocebus agilis* Milne-Edwards, 1886. Rev. Scient., 12:15.
COMMON NAME: Agile Mangabey.
TYPE LOCALITY: Dem. Rep. Congo, Republic Poste des Ouaddas (junction of Oubangui and Congo Rivers).
DISTRIBUTION: Equatorial Guinea (Rio Muni), Cameroon, NE Gabon, Central African Republic, N Republic of Congo, Dem. Rep. Congo N of Congo River to Garamba and Semliki River.
STATUS: CITES – Appendix II.
SYNONYMS: *fumosus* Matschie, 1914; *hagenbecki* Lydekker, 1900; *oberlaenderi* Lorenz, 1915.
COMMENTS: Does not include *chrysogaster*. Separated from *galeritus* by Groves (1978b).

*Cercocebus atys* (Audebert, 1797). Hist. Nat. Singes Makis, fam. 4, sect. 2, p. 13.
COMMON NAME: Sooty Mangabey.
TYPE LOCALITY: West Africa.
DISTRIBUTION: Senegal to Ghana.
STATUS: CITES – Appendix II; U.S. ESA – Endangered as included in *C. torquatus*; IUCN – Critically Endangered as *C. a. lunulatus*, otherwise Lower Risk (nt).
SYNONYMS: *aethiopicus* F. Cuvier, 1821; *aethiops* (Schreber, 1775) [not of Linnaeus, 1758]; *fuliginosus* É. Geoffroy, 1812; **lunulatus** (Temminick, 1853).
COMMENTS: The subspecies *lunulatus* appears strongly distinct, and may be a separate species (Groves, 2001c:242-3). Separated from *torquatus* by Groves (2001c:242).

*Cercocebus chrysogaster* Lydekker, 1900. Novit. Zool., 7:279.
COMMON NAME: Golden-bellied Mangabey.
TYPE LOCALITY: Dem. Rep. Congo: Upper Congo.
DISTRIBUTION: Dem. Rep. Congo, south of Congo River.
STATUS: CITES – Appendix II; IUCN – Data Deficient as *C. galeritus chrysogaster*.
COMMENTS: Separated from *agilis* by Groves (2001c:243).

*Cercocebus galeritus* Peters, 1879. Monatsb. K. Preuss. Akad. Wiss. Berlin, 1879:830.
COMMON NAME: Tana River Mangabey.
TYPE LOCALITY: Kenya, Tana River, Mitole (2°10'S, 40°10'E).
DISTRIBUTION: Lower Tana River (Kenya).
STATUS: CITES – Appendix I as *C. g. galeritus*; U.S. ESA – Endangered as *C. g. galeritus*; IUCN – Critically Endangered as *C. g. galeritus*.
COMMENTS: Formerly included *agilis*; but see Groves (1978b).

*Cercocebus sanjei* Mittermeier, 1986. *In* Else and Lee (eds.), Primate Ecology and Conservation, p. 338.
> COMMON NAME: Sanje Mangabey.
> TYPE LOCALITY: Tanzania, Mwanihana Forest, Sanje Waterfall.
> DISTRIBUTION: Tanzania, Mwanihana Forest and eastern slopes of Uzungwa Mtns.
> STATUS: CITES – Appendix II; IUCN – Endangered as *C. galeritus sanjei*.

*Cercocebus torquatus* (Kerr, 1792). *In* Linnaeus, Anim. Kingdom, p. 67.
> COMMON NAME: Collared Mangabey.
> TYPE LOCALITY: West Africa.
> DISTRIBUTION: W Nigeria to Gabon.
> STATUS: CITES – Appendix II; U.S. ESA – Endangered; IUCN – Lower Risk (nt).
> SYNONYMS: *collaris* Gray, 1843; *crossi* Gray, 1843.
> COMMENTS: Does not include *atys*, pace Dandelot (1974:12) and Groves (1978*b*): see Groves (2001*c*:242).

*Cercopithecus* Linnaeus, 1758. Syst. Nat., 10th ed., 1:26.
> TYPE SPECIES: *Simia diana* Linnaeus, 1758.
> SYNONYMS: *Allochrocebus* Elliot, 1913; *Cercocephalus* Temminck, 1853; *Diademia* Reichenbach, 1862; *Diana* Trouessart, 1878 [not of Risso, 1826]; *Insignicebus* Elliot, 1913; *Lasiopyga* Illiger, 1811; *Melanocebus* Elliot, 1913; *Mona* Reichenbach, 1862; *Monichus* Oken, 1816 [unavailable]; *Neocebus* Elliot, 1913; *Otopithecus* Trouessart, 1897; *Petaurista* Reichenbach, 1862 [not of Link, 1795]; *Pogonocebus* Trouessart, 1904; *Rhinosticteus* Trouessart, 1897; *Rhinostigma* Elliot, 1913.
> COMMENTS: Dandelot (1974:14) included *Allenopithecus*, *Erythrocebus*, and *Miopithecus* in this genus; but see Groves (1978*b*) and Corbet and Hill (1980:89) who considered them to be distinct genera; Szalay and Delson (1979) and McKenna and Bell (1997) considered *Miopithecus* a subgenus of *Cercopithecus*. Van Gelder (1977*b*:8) included *Cercocebus*, *Papio*, and *Theropithecus* in this genus, but see Groves (1978*b*), Ansell (1978:33), and Cronin and Meikle (1979:259). *Chlorocebus* was separated by Groves (1989). Designated as a subgroup of *Simia* by Linnaeus; type species *S. diana* designated by Stiles and Orleman (1926:52). *Simia* was suppressed by Opinion 114 of the International Commision on Zoological Nomenclature (1929*b*). The species groups are: (1) *C. dryas* group (monotypic), (2) *C. diana* group (*diana*, *roloway*), (3) *C. mitis* group (*nictitans*, *mitis*, *doggetti*, *kandti*, *albogularis*), (4) *C. mona* group (*mona*, *campbelli*, *lowei*, *pogonias*, *wolfi*, *denti*), (5) *C. cephus* group (*petaurista*, *erythrogaster*, *sclateri*, *erythrotis*, *cephus*, *ascanius*), (6) *C. lhoesti* group (*lhoesti*, *preussi*, *solatus*), (7) *C. hamlyni* group (monotypic), and (8) *C. neglectus* group (monotypic).

*Cercopithecus albogularis* (Sykes, 1831). Proc. Committee Sci. Correspondance Zool. Soc. Lond., 1:106.
> COMMON NAME: Sykes' Monkey.
> TYPE LOCALITY: Angola.
> DISTRIBUTION: Ethiopia to South Africa, S and E Dem. Rep. Congo, NW Angola.
> STATUS: CITES – Appendix II; IUCN – Endangered as *C. mitis labiatus*, Data Deficient as *C. m. albotorquatus*, Lower Risk (lc) as *C. m. albogularis*, otherwise not listed.
> SYNONYMS: **albotorquatus** Pousargues, 1896; *rufotinctus* Pocock, 1907; **erythrarchus** Peters, 1852; *beirensis* Pocock, 1907; *mossambicus* Pocock, 1907; *nyasae* Schwarz, 1928; *stairsi* Sclater, 1892; *stevensoni* Roberts, 1948; **francescae** Thomas, 1902; **kibonotensis** Lönnberg, 1908; *kima* (Heller, 1913); *maritima* (Heller, 1913); **kolbi** Neumann, 1902; *hindei* Pocock, 1907; *nubilus* Dollman, 1910; **labiatus** I. Geoffroy, 1842; *chimango* Temminck, 1853; *samango* Wahlberg, 1845; **moloneyi** Sclater, 1893; **monoides** I. Geoffroy, 1841; *rufilatus* Pocock, 1907; **phylax** Schwarz, 1927; **schwarzi** Roberts, 1931; **zammaranoi** de Beaux, 1924.
> COMMENTS: *C. mitis* species group. Separated from *C. mitis* by Dandelot (1974:19).

*Cercopithecus ascanius* (Audebert, 1799). Hist. Nat. Singes Makis, 4(2):13.
> COMMON NAME: Red-tailed Monkey.
> TYPE LOCALITY: Angola (NW, by lower Congo River). See Machado (1969).

DISTRIBUTION: Uganda, Dem. Rep. Congo, Zambia, Angola, marginally in Central African Republic; W Kenya.

STATUS: CITES – Appendix II; IUCN – Lower Risk (lc).

SYNONYMS: *histrio* Reichenbach, 1863; *melanogenys* Gray, 1845; *picturatus* Santos, 1886; **atrinasus** Machado, 1965; **katangae** Lönnberg, 1919; **schmidti** Matschie, 1892; *enkamer* Matschie, 1913; *ituriensis* Lorenz, 1914; *kaimosae* (Heller, 1913); *montanus* Lorenz, 1914; *mpangae* Matschie, 1913; *orientalis* Lorenz, 1919; *rutschuricus* Lorenz, 1917; *sassae* Matschie, 1913; **whitesidei** Thomas, 1909; *cirrhorhinus* Matschie, 1913; *kassaicus* Matschie, 1913; *omissus* Matschie, 1913; *pelorhinus* Matschie, 1913.

COMMENTS: *C. cephus* species group.

*Cercopithecus campbelli* Waterhouse, 1838. Proc. Zool. Soc. Lond., 1838:61.
COMMON NAME: Campbell's Mona Monkey.
TYPE LOCALITY: Sierra Leone.
DISTRIBUTION: Senegal to Cavally River (Liberia – Côte d'Ivoire border).
STATUS: CITES – Appendix II; IUCN – Lower Risk (lc).
SYNONYMS: *burnettii* Gray, 1842; *monella* Gray, 1870; ?*temminickii* Ogilby, 1838.
COMMENTS: *C. mona* species group. McAllan and Bruce (1989) argued that the original publication of this species should be: *The Analyst*, 24:298-299 [publ. 2 July 1838]. Does not include *lowei*: see Kingdon (1997).

*Cercopithecus cephus* (Linnaeus, 1758). Syst. Nat., 10th ed., 1:27.
COMMON NAME: Moustached Guenon.
TYPE LOCALITY: Africa.
DISTRIBUTION: Gabon, Republic of Congo, S Cameroon, Equatorial Guinea, SW Central African Republic, NW Angola.
STATUS: CITES – Appendix II; IUCN – Data Deficient as *C. c. ngottoensis*, otherwise Lower Risk (lc).
SYNONYMS: *buccalis* Leconte, 1857; *inobservatus* Elliot, 1927; *pulcher* Lorenz, 1915; **cephodes** Pocock, 1907; *gabonensis* Maclatchy and Malbrant, 1947; **ngottoensis** Colyn, 1999.
COMMENTS: *C. cephus* species group. Includes *erythrotis* according to Struhsaker (1970:374-376); but see Dandelot (1974:23).

*Cercopithecus denti* Thomas, 1907. Abstr. Proc. Zool. Soc. Lond., 1907(38):1.
COMMON NAME: Dent's Mona Monkey.
TYPE LOCALITY: Dem. Rep. Congo, between Mawambi and Avakubi, Ituri Forest.
DISTRIBUTION: Dem. Rep. Congo north and east of Congo-Lualaba system, Rwanda, W Uganda, Central African Republic.
STATUS: CITES – Appendix II.
SYNONYMS: *liebrechtsi* Dubois and Matschie, 1912.
COMMENTS: *C. mona* species group. Included in *C. wolfi* by Dandelot (1974:25); provisionally separated by Groves (2001c).

*Cercopithecus diana* (Linnaeus, 1758). Syst. Nat., 10th ed., 1:26.
COMMON NAME: Diana Monkey.
TYPE LOCALITY: Liberia.
DISTRIBUTION: Sierra Leone to Sassandra River, Côte d'Ivoire.
STATUS: CITES – Appendix I; U.S. ESA – Endangered; IUCN – Endangered.
SYNONYMS: *faunus* (Linnaeus, 1766); *ignita* Gray, 1870.
COMMENTS: *C. diana* species group. Type species; see comment under *Cercopithecus*. Does not include *roloway*, pace Dandelot (1974:25); see Groves (2001c:205).

*Cercopithecus doggetti* Pocock, 1907. Proc. Zool. Soc. Lond., 1907:691.
COMMON NAME: Silver Monkey.
TYPE LOCALITY: Uganda: SW Ankole.
DISTRIBUTION: Highlands of Dem. Rep. Congo west of Lake Albert and Lake Tanganyika, to S Burundi, NW Tanzania (Bukoba), Rwanda, S Uganda (Ankole, Busenya, Kaiso).
STATUS: CITES – Appendix II.
SYNONYMS: *sibatoi* (Lorenz, 1913).
COMMENTS: *C. mitis* species group. Separated from *C. mitis* by Groves (2001c:208).

*Cercopithecus dryas* Schwartz, 1932. Rev. Zool. Bot. Afr., 21:251.
  COMMON NAME: Dryas Monkey.
  TYPE LOCALITY: Dem. Rep. Congo, Ikela Zone, Yapatsi. (See Thys van den Audenaerde, 1977:1006).
  DISTRIBUTION: Known only from a few localities in C Dem. Rep. Congo (Wamba Dist., 22°31'-33°E, and 0°01'N-0°01'S); see Kuroda et al. (1985, as *Cercopithecus salongo*).
  STATUS: CITES – Appendix II; IUCN – Data Deficient.
  SYNONYMS: *salongo* Thys van den Audenaerde, 1977.
  COMMENTS: *C. dryas* species group. Not a subspecies of *diana*; possibly related to *Chlorocebus aethiops*; see Thys van den Audenaerde (1977:1007), Groves (2001c:204). *C. salongo* is an age-variant of this species, see Colyn et al. (1991).

*Cercopithecus erythrogaster* Gray, 1866. Proc. Zool. Soc. Lond., 1866:169.
  COMMON NAME: White-throated Guenon.
  TYPE LOCALITY: "West Africa"; restricted by Groves (2001c:214) to Benin, Lama Forest.
  DISTRIBUTION: S Nigeria, both west and east of the Niger in the delta region; Benin.
  STATUS: CITES – Appendix II; U.S. ESA – Endangered; IUCN – Endangered as *C. e. erythrogaster* and *C. e. pococki*.
  SYNONYMS: **pococki** Grubb, Lernould and Oates, 1999.
  COMMENTS: *C. cephus* species group.

*Cercopithecus erythrotis* Waterhouse, 1838. Proc. Zool. Soc. Lond., 1838:59.
  COMMON NAME: Red-eared Guenon.
  TYPE LOCALITY: Equatorial Guinea, Bioko.
  DISTRIBUTION: S and E Nigeria, Cameroon coast, Bioko.
  STATUS: CITES – Appendix II; U.S. ESA – Endangered; IUCN – Endangered as *C. c. erythrotis*, otherwise Vulnerable.
  SYNONYMS: **camerunensis** Hayman, 1940.
  COMMENTS: *C. cephus* species group. Considered a subspecies of *cephus* by Struhsaker (1970:374-376); but also see Dandelot (1974:23). McAllan and Bruce (1989) argued that the original publication of this species should be: *The Analyst*, 24:298-299 [publ. 2 July 1838].

*Cercopithecus hamlyni* Pocock, 1907. Ann. Mag. Nat. Hist., ser. 7, 20:521.
  COMMON NAME: Hamlyn's Monkey.
  TYPE LOCALITY: Dem. Rep. Congo, Ituri Forest.
  DISTRIBUTION: E Dem. Rep. Congo, Rwanda.
  STATUS: CITES – Appendix II; IUCN – Endangered as *C. h. kahuziensis*, otherwise Lower Risk (nt).
  SYNONYMS: *aurora* Thomas and Wroughton, 1910; **kahuziensis** Colyn and Verheyen, 1988.
  COMMENTS: *C. hamlyni* species group.

*Cercopithecus kandti* Matschie, 1905. Sitzb. Ges. Naturf. Fr. Berlin, p. 264.
  COMMON NAME: Golden Monkey.
  TYPE LOCALITY: Dem. Rep. Congo, Virunga Volcanoes.
  DISTRIBUTION: Virunga Volcanoes on Dem. Rep. Congo – Uganda – Rwanda borders; Nyungwe Forest, Rwanda.
  STATUS: CITES – Appendix II; IUCN – Endangered as *C. mitis kandti*.
  SYNONYMS: *insignis* Elliot, 1909.
  COMMENTS: *C. mitis* species group. Separated from *C. mitis* by Groves (2001c).

*Cercopithecus lhoesti* P. Sclater, 1899. Proc. Zool. Soc. Lond., 1898:586 [1899].
  COMMON NAME: L'Hoest's Monkey.
  TYPE LOCALITY: Dem. Rep. Congo, Tschepo River, near Stanleyville.
  DISTRIBUTION: E Dem. Rep. Congo, W Uganda, Rwanda, Burundi.
  STATUS: CITES – Appendix II; U.S. ESA – Endangered; IUCN – Lower Risk (nt).
  SYNONYMS: *rutschuricus* Lorenz, 1915; *thomasi* Matschie, 1905.
  COMMENTS: *C. lhoesti* species group. Does not include *preussi*; see Harrison (1988:562).

*Cercopithecus lowei* Thomas, 1923. Ann. Mag. Nat. Hist., ser. 9, 11:608.
　　COMMON NAME: Lowe's Mona Monkey.
　　TYPE LOCALITY: Côte d'Ivoire: Bandama.
　　DISTRIBUTION: Côte d'Ivoire (Cavally River) to Ghana (Volta River).
　　STATUS: CITES – Appendix II.
　　COMMENTS: *C. mona* species group. Separated from *campbelli* by Kingdon (1997).

*Cercopithecus mitis* Wolf, 1822. Abbild. Beschreib. Merkw. Naturgesch. Gegenstandes, 2:145.
　　COMMON NAME: Blue Monkey.
　　TYPE LOCALITY: Angola.
　　DISTRIBUTION: Congo-Oubangui River system (probably Itimbiri River) to East African Rift
　　　　Valley, N Angola and NW Zambia.
　　STATUS: CITES – Appendix II; IUCN – Lower Risk (lc) as *C. m. stuhlmanni* and as *C. mitis*.
　　SYNONYMS: *diadematus* I. Geoffroy, 1834; *dilophos* Ogilby, 1838; *leucampyx* (Fischer, 1829);
　　　　*nigrigenis* Pocock, 1907; *pluto* Gray, 1848; **boutourlinii** Giglioli, 1887; *omensis* Thomas,
　　　　1901; **elgonis** Lönnberg, 1919; **heymansi** Colyn and Verheyen, 1987; **opisthostictus**
　　　　Sclater, 1894; **stuhlmanni** Matschie, 1893; *carruthersi* Pocock, 1907; *maesi* Lönnberg,
　　　　1919; *mauae* (Heller, 1913); *neumanni* Matschie, 1906; *otoleucus* Sclater, 1902; *princeps*
　　　　Elliot, 1909; *schubotzi* Matschie, 1913.
　　COMMENTS: *C. mitis* species group. Does not include *albogularis*; pace Booth (1968); see
　　　　Dandelot (1974:19). Does not include *doggetti* or *kandti* (see Groves, 2001c).

*Cercopithecus mona* (Schreber, 1774). Die Säugethiere, 1:103.
　　COMMON NAME: Mona Monkey.
　　TYPE LOCALITY: "Guinea".
　　DISTRIBUTION: Ghana to Cameroon; introduced into Lesser Antilles (Caribbean).
　　STATUS: CITES – Appendix II; IUCN – Lower Risk (lc).
　　SYNONYMS: *monacha* (Schreber, 1804); *monella* (Schreber, 1804).
　　COMMENTS: *C. mona* species group.

*Cercopithecus neglectus* Schlegel, 1876. Mus. Hist. Nat. Pays-Bas. Simiae, p. 70.
　　COMMON NAME: De Brazza's Monkey.
　　TYPE LOCALITY: Sudan, "White Nile".
　　DISTRIBUTION: SE Cameroon to Uganda and N Angola, W Kenya, SW Ethiopia, and S Sudan.
　　STATUS: CITES – Appendix II; IUCN – Lower Risk (lc).
　　SYNONYMS: *brazzae* Milne-Edwards, 1886; *brazziformis* Pocock, 1907; *ezrae* Pocock, 1908;
　　　　*uellensis* Lönnberg, 1919.
　　COMMENTS: *C. neglectus* species group.

*Cercopithecus nictitans* (Linnaeus, 1766). Syst. Nat., 12th ed., 1:40.
　　COMMON NAME: Greater spot-nosed Monkey.
　　TYPE LOCALITY: Equatorial Guinea, Benito River.
　　DISTRIBUTION: Liberia; Côte d'Ivoire; Nigeria apparently to Itimbiri River in NW Dem. Rep.
　　　　Congo, Central African Republic; Rio Muni and Bioko (Equatorial Guinea).
　　STATUS: CITES – Appendix II; IUCN – Critically Endangered as *C. n. stampflii*, Endangered as
　　　　*C. n. martini*, otherwise Lower Risk (lc).
　　SYNONYMS: *laglaizei* Pocock, 1907; *sticticeps* Elliot, 1909; **martini** Waterhouse, 1838; *insolitus*
　　　　Elliot, 1909; *ludio* Gray, 1849; *stampflii* Jentink, 1888.
　　COMMENTS: *C. mitis* species group. *C. signatus* Jentink, 1886, probably represents a hybrid
　　　　between this species and one of the *cephus* group (Oates, 1985).

*Cercopithecus petaurista* (Schreber, 1774). Die Säugethiere, 1:97, 185.
　　COMMON NAME: Lesser Spot-nosed Monkey.
　　TYPE LOCALITY: "Guinea".
　　DISTRIBUTION: Gambia to Togo.
　　STATUS: CITES – Appendix II; IUCN – Lower Risk (lc).
　　SYNONYMS: *albinasus* (Reichenbach, 1863); *fantiensis* Matschie, 1893; *pygrius* Thomas, 1923;
　　　　**buettikoferi** Jentink, 1886.
　　COMMENTS: *C. cephus* species group.

*Cercopithecus pogonias* Bennett, 1833. Proc. Zool. Soc. Lond., 1833:67.
COMMON NAME: Crested Mona Monkey.
TYPE LOCALITY: Equatorial Guinea, Bioko.
DISTRIBUTION: SE Nigeria, Cameroon, Bioko and Rio Muni (Equatorial Guinea), N and
W Gabon, W Dem. Rep. Congo, Republic of Congo.
STATUS: CITES – Appendix II; IUCN – Endangered as *C. p. pogonias*, otherwise Lower Risk (lc).
SYNONYMS: **grayi** Fraser, 1850; *erxlebeni* Dallbet and Pucheran, 1856; *pallidus* Elliot, 1909;
*petronellae* Büttikofer, 1911; **nigripes** du Chaillu, 1860; **schwarzianus** Schouteden,
1946; *schwarzi* Schouteden, 1944 [not of Roberts, 1931].
COMMENTS: *C. mona* species group.

*Cercopithecus preussi* Matschie, 1898. Sitzb. Ges. Naturf. Fr. Berlin, p. 76.
COMMON NAME: Preuss's Monkey.
TYPE LOCALITY: Cameroon, Victoria.
DISTRIBUTION: Region of Mt. Cameroon; Bioko (Equatorial Guinea).
STATUS: CITES – Appendix II; IUCN – Endangered as *C. p. preusii* and *C. p. insularis*.
SYNONYMS: *crossi* Forbes, 1905; **insularis** Thomas, 1910.
COMMENTS: *C. lhoesti* species group. Not a subspecies of *C. lhoesti*; see Harrison (1988:562).

*Cercopithecus roloway* (Schreber, 1774). Die Säugethiere, 1:186.
COMMON NAME: Roloway Monkey.
TYPE LOCALITY: "Guinea" (= West Africa in general).
DISTRIBUTION: Sassandra River (Côte d'Ivoire) to Pra River, Ghana.
STATUS: CITES – Appendix I as included in *C. diana*; U.S. ESA – Endangered as included in
*C. diana*; IUCN – Critically Endangered as *C. diana roloway*.
SYNONYMS: *palatinus* (Wagner, 1855).
COMMENTS: *C. diana* species group. Separated from *C. diana* by Groves (2001c).

*Cercopithecus sclateri* Pocock, 1904. Abstr. Proc. Zool. Soc. Lond., 1904(5):18.
COMMON NAME: Sclater's Guenon.
TYPE LOCALITY: Nigeria, Benin City.
DISTRIBUTION: SE Nigeria, between Niger and Cross Rivers.
STATUS: CITES – Appendix II; IUCN – Endangered.
COMMENTS: *C. cephus* species group. Recognized as a full species by Kingdon (1980:661).

*Cercopithecus solatus* M. J. S. Harrison, 1988. J. Zool. Lond., 215:562.
COMMON NAME: Sun-tailed Monkey.
TYPE LOCALITY: C Gabon, SE of Booue, Forêt des Abeilles, River Bali, 0°14'S, 12°15'E.
DISTRIBUTION: C Gabon.
STATUS: CITES – Appendix II; IUCN – Vulnerable. Endangered.
COMMENTS: *C. lhoesti* species group.

*Cercopithecus wolfi* A. Meyer, 1891. Notes Leyden Mus., 13:63.
COMMON NAME: Wolf's Mona Monkey.
TYPE LOCALITY: "Central West Africa."
DISTRIBUTION: Dem. Rep. Congo, NE Angola, south of the Congo River.
STATUS: CITES – Appendix II; IUCN – Lower Risk (lc) as *C. pogonius wolfi*.
SYNONYMS: **elegans** Dubois and Matschie, 1912; *pyrogaster* Lönnberg, 1919.
COMMENTS: *C. mona* species group. *C. denti* provisionally separated as a species by Groves
(2001c).

*Chlorocebus* Gray, 1870. Cat. Monkeys, Lemurs, Fruit-eating Bats Brit. Mus., p. 5.
TYPE SPECIES: *Simia sabaea* Linnaeus, 1766 (= *Simia aethiops* Linnaeus, 1758).
SYNONYMS: *Callithrix* Reichenbach, 1862 [not of Erxleben, 1777]; *Cynocebus* Gray, 1870.
COMMENTS: Not recognized by McKenna and Bell (1997). Recognized as a full genus distinct
from *Cercopithecus* by Groves (1989, 2001c).

*Chlorocebus aethiops* (Linnaeus, 1758). Syst. Nat., 10th ed., 1:28.
COMMON NAME: Grivet.
TYPE LOCALITY: Sudan, Sennaar.

DISTRIBUTION: Sudan east of the White Nile, Eritrea, Ethiopia east to the Rift Valley.

STATUS: CITES – Appendix II; IUCN – Lower Risk (lc) as *Cercopithecus aethiops*.

SYNONYMS: *calliaudi* (Wettstein, 1918); *cano-viridis* (Gray, 1843); *?cinereo-viridis* (Gray, 1843); *engytithia* (Hermann, 1804); *griseo-viridis* (Desmarest, 1820); *griseus* (F. Cuvier, 1819); *matschiei* (Neumann, 1902); *subviridis* (F. Cuvier, 1821); *toldti* (Wettstein, 1916); *weidholzi* (Lorenz, 1922); *zavattarii* (de Beaux, 1943).

COMMENTS: Does not include *pygerythrus*, *sabaeus*, *djamdjamensis* or *tantalus* (Kingdon, 1997; Groves, 2001c) or *cynosuros* (Groves, 2001c). According to Napier (1981), *zavattarii* is based on a hybrid with *pygerythrus*.

*Chlorocebus cynosuros* (Scopoli, 1786). Del. Faun. Flor. Insub., 1:44.

COMMON NAME: Malbrouck.

TYPE LOCALITY: Dem. Rep. Congo, Lower Congo, Banana.

DISTRIBUTION: S Dem. Rep. Congo to N Namibia, Zambia west of Luangwa River.

STATUS: CITES – Appendix II.

SYNONYMS: *helvescens* (Thomas, 1926); *katangensis* (Lönnberg, 1919); *lukonzolwae* (Matschie, 1912); *tephrops* (Bennett, 1833); *tholloni* (Matschie, 1912); *weynsi* (Dubois and Matschie, 1912).

COMMENTS: Separated from *aethiops* by Groves (2001c).

*Chlorocebus djamdjamensis* (Neumann, 1902). Sitzb. Ges. Naturf. Fr. Berlin, p. 51.

COMMON NAME: Bale Mountains Vervet.

TYPE LOCALITY: Ethiopia, bamboo forest near Abera, east of Lake Abaya, 3300 m.

DISTRIBUTION: Ethiopia, highlands east of Lakes Abiata, Shalla and Zway.

STATUS: CITES – Appendix II; IUCN – Data Deficient as *Cercopithecus aethiops djamdjamensis*.

COMMENTS: A distinctive species, first recognized (as a subspecies) as distinct from *aethiops* by Dandelot and Prévost (1972); raised to species rank by Kingdon (1977).

*Chlorocebus pygerythrus* (F. Cuvier, 1821). Hist. Nat. Mamm., 24:2.

COMMON NAME: Vervet Monkey.

TYPE LOCALITY: "Africa".

DISTRIBUTION: Ethiopia (east of Rift Valley), Somalia, to Zambia east of the Luangwa, and South Africa.

STATUS: CITES – Appendix II.

SYNONYMS: *cloetei* (Roberts, 1931); *erythropyga* (G. Cuvier, 1829); *glaucus* (Lichtenstein, 1811) [*nomen nudum*]; *lalandii* (I. Geoffroy, 1841); *marjoriae* (Bradfield, 1936); *ngamiensis* (Roberts, 1932); *pusillus* (Desmoulins, 1825); **excubutor** (Schwarz, 1926); *voeltzkowi* Matschie, 1923 [*nomen nudum*]; **hilgerti** (Neumann, 1902); *arenaria* (Heller, 1913); *callida* (Heller, 1912); *contigua* (Heller, 1920); *ellenbecki* (Neumann, 1902); *johnstoni* (Pocock, 1907); *luteus* (Elliot, 1910); *rubellus* (Elliot, 1909); *tumbili* (Heller, 1913); **nesiotes** (Schwarz, 1926); *pembae* Matschie, 1923 [*nomen nudum*]; **nifoviridis** (I. Geoffroy, 1843); *centralis* (Neumann, 1900); *?circumcinctus* (Reichenbach, 1862); *flavidus* (Peters, 1852); *rufoniger* Gray, 1870; *whytei* (Pocock, 1907); *silaceus* (Elliot, 1909).

COMMENTS: Separated from *aethiops* as a species by Dandelot (1959), Kingdon (1997) and Groves (2001c).

*Chlorocebus sabaeus* (Linnaeus, 1766). Syst. Nat., 12th ed., 1:38.

COMMON NAME: Green Monkey.

TYPE LOCALITY: Cape Verde Isls.

DISTRIBUTION: Senegal to the Volta River; introduced to Cape Verde Isls and to St. Kitts, Nevis, and Barbados (West Indies).

STATUS: CITES – Appendix II.

SYNONYMS: *callitrichus* (I. Geoffroy, 1851); *chrysurus* (Blyth, 1845); *werneri* (I. Geoffroy, 1850).

COMMENTS: Separated from *aethiops* as a full species by Kingdon (1997) and Groves (2001c). Napier (1981) regarded *chrysurus* (usually placed in synonymy of *C. tantalus*) as a synonym of *sabaeus*.

*Chlorocebus tantalus* (Ogilby, 1841). Proc. Zool. Soc. Lond., 1841:33.

COMMON NAME: Tantalus Monkey.

TYPE LOCALITY: No locality.

DISTRIBUTION: Volta River (Ghana) east to White Nile (Sudan) and Lake Turkana (Kenya).

STATUS: CITES – Appendix II.

SYNONYMS: *alexandri* (Pocock, 1909); *graueri* (Lorenz, 1914); *passargei* (Matschie, 1897); *pousarguei* (Mitchell, 1905); *viridis* (Schultze, 1910); **budgetti** (Pocock, 1907); *beniana* (Lorenz, 1914); *griseistictus* (Elliot, 1909); *itimbiriensis* (Matschie and Dubois, 1912); **marrensis** (Thomas and Hinton, 1923).

COMMENTS: Separated from *aethiops* by Dandelot (1959), Kingdon (1997) and Groves (2001c).

*Erythrocebus* Trouessart, 1897. Cat. Mamm. Viv. Foss., 1:19.

TYPE SPECIES: *Simia patas* Schreber, 1775.

COMMENTS: Recognized as a distinct genus by Thorington and Groves (1970:638-639) and Szalay and Delson (1979).

*Erythrocebus patas* (Schreber, 1775). Die Säugethiere, 1:98.

COMMON NAME: Patas Monkey.

TYPE LOCALITY: Senegal.

DISTRIBUTION: Savannahs, from W Africa to Ethiopia, Kenya, and Tanzania.

STATUS: CITES – Appendix II; IUCN – Lower Risk (lc).

SYNONYMS: *albigenus* Elliot, 1909; *albo-fasciatus* (Kerr, 1792); *albosignatus* (Matschie, 1912); *baumstarki* Matschie, 1905; *circumcinctus* (Reichenbach, 1863); *formosus* Elliot, 1909; *kerstingi* (Matschie, 1906); *langheldi* Matschie, 1905; *nigro-fasciatus* (Kerr, 1792); *poliomystax* (Matschie, 1912); *poliophaeus* (Heuglin, 1877); *pyrrhonotus* (Hemprich and Ehrenberg, 1829); *rubra* (Gmelin, 1788); *rufa* (Wagner, 1839); *sannio* (Thomas, 1906); *villiersi* Dekeyser, 1950; *whitei* Hollister, 1910; *zechi* Matschie, 1905.

COMMENTS: Subspecies may exist, but at least some of the features supposed to characterize them were based on changes to the female's facial pattern during pregnancy (Groves, 2001c).

*Lophocebus* Palmer, 1903. Science, n.s., 17:873.

TYPE SPECIES: *Presbytis albigena* Gray, 1850.

SYNONYMS: *Cercolophocebus* Matschie, 1914; *Semnocebus* Gray, 1870.

COMMENTS: Formerly included in *Cercocebus* (for example, by McKenna and Bell [1997]); a subgenus of *Cercocebus*, according to Szalay and Delson (1979), but see Groves (1979, 1989, 2001c). Included in *Papio* as a subgenus by M. Goodman et al. (2001).

*Lophocebus albigena* (Gray, 1850). Proc. Zool. Soc. Lond., 1850:77.

COMMON NAME: Gray-cheeked Mangabey.

TYPE LOCALITY: Dem. Rep. Congo, Mayombe.

DISTRIBUTION: SE Nigeria (Cross River), Cameroon, Republic of Congo, Gabon, Equatorial Guinea, NE Angola, Central African Republic, Dem. Rep. Congo north and east of Congo-Lualaba system, W Uganda (to Busoga), Burundi.

STATUS: CITES – Appendix II; IUCN – Lower Risk (lc).

SYNONYMS: *weynsi* (Matschie, 1913); *zenkeri* (Schwarz, 1910); **johnstoni** (Lydekker, 1900); *ituricus* (Matschie, 1913); *jamrachi* (Pocock, 1906); *mawambicus* (Lorenz, 1917); *ugandae* Matschie, 1913; **osmani** Groves, 1978.

COMMENTS: Does not include *aterrimus* and *opdenboschi*; see Groves (2001c).

*Lophocebus aterrimus* (Oudemans, 1890). Zool. Garten, 31:267.

COMMON NAME: Black Crested Mangabey.

TYPE LOCALITY: Dem. Rep. Congo, Stanley Falls.

DISTRIBUTION: Dem. Rep. Congo, south of the Congo River, in rainforest.

STATUS: CITES – Appendix II; IUCN – Lower Risk (nt).

SYNONYMS: *coelognathus* Matschie, 1914; *congicus* Sclater, 1900; *hamlyni* Pocock, 1906; *rothschildi* Lydekker, 1900.

COMMENTS: Does not include *opdenboschi*; see Groves (2001c).

*Lophocebus opdenboschi* (Scouteden, 1944). Rev. Zool. Bot. Afr., 38:192.
COMMON NAME: Opdenbosch's Mangabey.
TYPE LOCALITY: Dem. Rep. Congo, Mwiliambongo.
DISTRIBUTION: Dem. Rep. Congo, gallery forests along the Kwilu, Wamba and Kwango Rivers, into Angola.
STATUS: CITES – Appendix II; IUCN – Data Deficient as *L. atterimus opdenboschi*.
COMMENTS: Separated from *aterrimus* by Groves (2001*c*).

*Macaca* Lacépède, 1799. Tabl. Div. Subd. Orders Genres Mammifères, p. 4.
TYPE SPECIES: *Simia inuus* Linnaeus, 1766 (= *Simia sylvanus* Linnaeus, 1758).
SYNONYMS: *Aulaxinus* Cocchi, 1872; *Cynamolgus* Reichenbach, 1862; *Cynomacaca* Khajuria, 1953; *Cynopithecus* É. Geoffroy, 1835; *Gymnopyga* Gray, 1866; *Inuus* É. Geoffroy, 1812; *Lyssodes* Gistel, 1848; *Magotus* Ritgen, 1824; *Magus* Lesson, 1827; *Maimon* Wagner, 1839; *Nemestrinus* Reichenbach, 1862 [not of Latreille, 1802]; *Ouanderou* Lesson, 1840; *Pithes* Burnett, 1828; *Rhesus* Lesson, 1840; *Salmacis* Gloger, 1841; *Silenus* Goldfuss, 1820; *Sylvanus* Oken, 1816 [unavailable]; *Vetulus* Reichenbach, 1862; *Zati* Reichenbach, 1862.
COMMENTS: Placed in a separate subtribe, Macacina, from other members of the Papionini, by McKenna and Bell (1997). Species groups are: (1) *M. sylvanus* group (monotypic), (2) *M. nemestrina* group (*silenus, leonina, nemestrina, pagensis, siberu, maura, ochreata, tonkeana, hecki, nigrescens, nigra*), (3) *M. fascicularis* group (*fascicularis, arctoides*), (4) *M. mulatta* group (*mulatta, cyclopis, fuscata*), (5) *M. sinica* group (*sinica, radiata, assamensis, thibetana*).

*Macaca arctoides* (I. Geoffroy, 1831). *In* Belanger (ed.), Voy. Indes Orient., Mamm., 3(Zool.):61.
COMMON NAME: Stump-tailed Macaque.
TYPE LOCALITY: "Cochin-China" (Indochina).
DISTRIBUTION: Assam (India) to S China and N Malay Peninsula.
STATUS: CITES – Appendix II; U.S. ESA – Threatened; IUCN – Vulnerable.
SYNONYMS: *brunneus* (Anderson, 1871); *harmandi* (Trouessart, 1897); *melanotus* (Ogilby, 1839); *melli* (Matschie, 1912); *rufescens* (Anderson, 1872); *speciosus* (Murie, 1875) [not of I. Geoffroy, 1826 = *M. fuscata*]; *ursinus* (Gervais, 1854).
COMMENTS: *M. fascicularis* species group according to Groves (2001*c*), but probably derived from an early Pleistocene hybridization between *M. fascicularis*, which it resembles in its mtDNA, and *M. assamensis/thibetana*, which it resembles in its Y chromosome. Reviewed by Fooden et al. (1985).

*Macaca assamensis* (M'Clelland, 1840). Proc. Zool. Soc. Lond., 1839:148 [1840].
COMMON NAME: Assam Macaque.
TYPE LOCALITY: India, Assam.
DISTRIBUTION: Nepal to N Vietnam, S China.
STATUS: CITES – Appendix II; IUCN – Vulnerable as *M. a. assamensia* and *M. a. pelops*.
SYNONYMS: *coolidgei* Osgood, 1932; *rhesosimilis* (Sclater, 1872); **pelops** (Hodgson, 1840); *macclellandii* (Gray, 1846); *problematicus* (Gray, 1870); *sikimensis* (Hodgson, 1867).
COMMENTS: *M. sinica* species group. Reviewed by Fooden (1982).

*Macaca cyclopis* (Swinhoe, 1863). Proc. Zool. Soc. Lond., 1862:350 [1863].
COMMON NAME: Formosan Rock Macaque.
TYPE LOCALITY: Taiwan, Jusan, Takao Pref.
DISTRIBUTION: Taiwan.
STATUS: CITES – Appendix II; U.S. ESA – Threatened; IUCN – Vulnerable.
SYNONYMS: *affinis* (Blyth, 1863).
COMMENTS: *M. mulatta* species group.

*Macaca fascicularis* (Raffles, 1821). Trans. Linn. Soc. Lond., 13:246.
COMMON NAME: Crab-eating Macaque.
TYPE LOCALITY: Indonesia, Sumatra, Bengkulen.
DISTRIBUTION: S Indochina and Burma to Borneo and Timor (Indonesia); Philippine Isls; Nicobar Isls (India).
STATUS: CITES – Appendix II; IUCN – Data Deficient as *M. f. atriceps*, *M. f. condorensis*,

*M. f. fusca*, *M. f. karimondjawae*, *M. f. lasiae*, *M. f. tua*, and *M. f. umbrosa*, Lower Risk (nt)
as *M. f. fascicularis*, *M. f. aurea*, and *M. f. philippensis*.

SYNONYMS: *agnatus* (Elliot, 1910); *alacer* (Elliot, 1909); *argentimembris* Kloss, 1911; *aygula*
(Linnaeus, 1758) [suppressed by International Commission on Zoological
Nomenclature (1986), Opinion 1400]; *baweanus* (Elliot, 1910); *bintangensis* (Elliot,
1909); *buku* (Martin, 1838); *cagayanus* (Mearns, 1905); *capitalis* (Elliot, 1910);
*carbonarius* (F. Cuvier, 1825); *carimatae* (Elliot, 1910); *cupidus* (Elliot, 1910);
*cynocephalus* (Reichenbach, 1862) [not of Linnaeus, 1766]; *cynomolgus* (Schreber, 1775)
[not of Linnaeus, 1758]; *dollmani* (Elliot, 1909); *impudens* (Elliot, 1910); *irus* I. Geoffroy,
1826; *karimoni* (Elliot, 1909); *kra* (Lesson, 1830); *laetus* (Elliot, 1909); *lapsus* (Elliot,
1910); *lautensis* (Elliot, 1910); *limitis* (Schwarz, 1913); *lingae* (Elliot, 1910); *lingungensis*
(Elliot, 1910); *mandibularis* (Elliot, 1910); *mansalaris* (Lyon, 1916); *mordax* Thomas and
Wroughton, 1909; *phaeura* (Miller, 1903); *pumilus* (Miller, 1900); *resima* Thomas and
Wroughton, 1909; *sihassensis* (Elliot, 1910); *sublimitus* Sody, 1932; *submordax* Sody,
1949; *suluensis* (Mearns, 1905); *sumbae* Sody, 1933; *validus* (Elliot, 1909); **atriceps**
Kloss, 1919; **aureus** (É. Geoffroy, 1831); *vitiis* (Elliot, 1910); **condorensis** Kloss, 1926;
**fuscus** (Miller, 1903); **karimondjawae** Sody, 1949; **lasiae** (Lyon, 1916); **philippensis**
(I. Geoffroy, 1843); *apoensis* (Mearns, 1905); *cumingii* (Gray, 1870); *fur* (Slack, 1867);
*mindanensis* (Mearns, 1905); *mindorus* (Hollister, 1913); *palpebrosus* (I. Geoffroy, 1851);
**tua** Kellogg, 1944; **umbrosus** (Miller, 1902).

COMMENTS: *M. fascicularis* species group. Includes *irus*; see Medway (1977:70-71). Includes
*cynomolgos*; see W. C. O. Hill (1974:476-477). Revised by Fooden (1995).

*Macaca fuscata* (Blyth, 1875). J. Asiat. Soc. Bengal, 44:6.
COMMON NAME: Japanese Macaque.
TYPE LOCALITY: Japan.
DISTRIBUTION: Honshu, Shikoku, Kyushu, and adjacent small islands (Japan); Yaku Isl
(Ryukyu Isls, Japan).
STATUS: CITES – Appendix II; U.S. ESA – Threatened; IUCN – Endangered as *M. f. yakui*,
otherwise Data Deficient.
SYNONYMS: *japanensis* (Schweyer, 1909); *speciosus* (F. Cuvier, 1825); **yakui** Kuroda, 1941.
COMMENTS: *M. mulatta* species group. Includes *speciosus* F. Cuvier, 1825 (not *speciosa* Blyth,
1875) which was suppressed by Opinion 920 of the International Commision on
Zoological Nomenclature (1970); see Fooden (1976).

*Macaca hecki* (Matschie, 1901). Abh. Senckenb. Naturf. Ges., 25:257.
COMMON NAME: Heck's Macaque.
TYPE LOCALITY: Indonesia, Sulawesi Tengah, Buol.
DISTRIBUTION: N Sulawesi, from the base of the northern peninsula northeast to Gorontalo
(Indonesia).
STATUS: CITES – Appendix II; IUCN – Lower Risk (nt).
COMMENTS: *M. nemestrina* species group. Separated from *tonkeana* by Groves (2001c).

*Macaca leonina* (Blyth, 1863). Cat. Mamm. Mus. As. Soc., p. 7.
COMMON NAME: Northern Pig-tailed Macaque.
TYPE LOCALITY: Burma, N Arakan.
DISTRIBUTION: Burma, coast (including Mergui Arch.), Thailand north of about 8°N,
S Yunnan (China), Laos, Bangladesh, India north to Brahmaputra River.
STATUS: CITES – Appendix II; IUCN – Vulnerable.
SYNONYMS: *adusta* Miller, 1906; *andamanensis* (Bartlett, 1869); *blythii* Pocock, 1931; *coininus*
(Kloss, 1903) [*lapsus* for *leoninus*]; *insulana* Miller, 1906; *indochinensis* Kloss, 1919.
COMMENTS: *M. nemestrina* species group. Considered a species separate from *nemestrina* by
Groves (2001c:223).

*Macaca maura* (H. R. Schinz, 1825). *In* Cuvier, Das Thierreich, p. 257.
COMMON NAME: Moor Macaque.
TYPE LOCALITY: Indonesia, Sulawesi Selatan.
DISTRIBUTION: S Sulawesi, south of Tempe Depression (Indonesia).
STATUS: CITES – Appendix II; IUCN – Endangered.

SYNONYMS: *cuvieri* (Fischer, 1829); *fusco-ater* (Schinz, 1844); *hypomelas* (Matschie, 1901); *inornatus* (Gray, 1866); *majuscula* Hooijer, 1950.

COMMENTS: *M. nemestrina* species group. Type species of *Gymnopyga*; see Fooden (1969:79). Included in *nigra* by Corbet and Hill (1980:87).

*Macaca mulatta* (Zimmermann, 1780). Geogr. Gesch. Mensch. Vierf. Thiere, 2:195.

COMMON NAME: Rhesus Monkey.

TYPE LOCALITY: India, Nepal Terai.

DISTRIBUTION: Afghanistan and India to N Thailand, China, and Hainan Isl (China).

STATUS: CITES – Appendix II; IUCN – Lower Risk (nt).

SYNONYMS: *brachyurus* (Elliot, 1909) [not of Hamilton Smith, 1842]; *brevicaudatus* (Elliot, 1913); *erythraea* (Shaw, 1800); *fulvus* (Kerr, 1792); *lasiotus* (Gray, 1868); *littoralis* (Elliot, 1909); *mcmahoni* Pocock, 1932; *nipalensis* Hodgson, 1840; *oinops* Hodgson, 1840; *rhesus* (Audebert, 1798); *sancti-johannis* (Swinhoe, 1866); *siamica* Kloss, 1917; *tcheliensis* (Milne-Edwards, 1872); *vestita* (Milne-Edwards, 1892); *villosa* (True, 1894).

COMMENTS: *M. mulatta* species group. Revised by Fooden (2000).

*Macaca nemestrina* (Linnaeus, 1766). Syst. Nat., 12th ed., 1:35.

COMMON NAME: Southern Pig-tailed Macaque.

TYPE LOCALITY: Indonesia, Sumatra.

DISTRIBUTION: Malay Peninsula, Borneo, Sumatra and Bangka Isl (Indonesia), Thailand north to about 7°30'N.

STATUS: CITES – Appendix II; IUCN – Vulnerable.

SYNONYMS: *brachyurus* (Hamilton Smith, 1842); *broca* Miller, 1906; *carpolegus* (Raffles, 1821); *fusca* (Shaw, 1800); *libidinosus* I. Geoffroy, 1826; *longicruris* (Link, 1795); *maimon* (de Blainville, 1839); *nucifera* Sody, 1936; *platypygos* (Schreber, 1774).

COMMENTS: *M. nemestrina* species group. Includes *pagensis* according to Fooden (1975:67, 1980:7) and Szalay and Delson (1979); but Wilson and Wilson (1977:216) considered *pagensis* a distinct species, and this was followed by Groves (2001*c*).

*Macaca nigra* (Desmarest, 1822). Mammalogie, *in* Encyclop. Meth., 2(Suppl.):534.

COMMON NAME: Celebes Crested Macaque.

TYPE LOCALITY: Indonesia, Sulawesi, Maluku, Bacan Isl.

DISTRIBUTION: Sulawesi, east of Onggak Dumoga River, Lembeh Isl, Bacan Isl (Indonesia).

STATUS: CITES – Appendix II; IUCN – Endangered.

SYNONYMS: *lembicus* (Miller, 1931); *malayanus* (Desmoulins, 1824).

COMMENTS: *M. nemestrina* species group. Type species of *Cynopithecus*; see Fooden (1969). Includes *nigrescens* according to Groves (1980*c*, 1993), but separated again by Groves (2001*c*).

*Macaca nigrescens* (Temminck, 1849). Coup d'Oeil Possess. Neerd., 3:111.

COMMON NAME: Gorontalo Macaque.

TYPE LOCALITY: Indonesia, Sulawesi ("Celebes").

DISTRIBUTION: Sulawesi, east of Gorontalo, to Onggak Dumoga River (Indonesia).

STATUS: CITES – Appendix II; IUCN – Lower Risk (conservation dependent).

COMMENTS: *M. nemestrina* species group. Separated from *M. nigra* by Groves (2001*c*).

*Macaca ochreata* (Ogilby, 1841). Proc. Zool. Soc. Lond., 1841:56.

COMMON NAME: Booted Macaque.

TYPE LOCALITY: Unknown.

DISTRIBUTION: SE Sulawesi, Kabaena, Muna, and Butung (Indonesia).

STATUS: CITES – Appendix II; IUCN – Vulnerable as *M. brunnescens*, otherwise Data Deficient.

SYNONYMS: **brunnescens** (Matschie, 1901).

COMMENTS: *M. nemestrina* species group. Fooden (1969) recognized *brunnescens* as a species; but Groves (1980*c*:1-9) included it in *ochreata*.

*Macaca pagensis* (Miller, 1903). Smithson. Misc. Collect., 45:61.

COMMON NAME: Pagai Island Macaque.

TYPE LOCALITY: Indonesia, South Pagai Isl.

DISTRIBUTION: Mentawai group: Islands of Sipura, North Pagai and South Pagai (Indonesia).
STATUS: CITES – Appendix II; IUCN – Critically Endangered as *M. pagensis pagensis*.
SYNONYMS: *mentaveensis* de Beaux, 1923.
COMMENTS: *M. nemestrina* species group. Included in *nemestrina* by Fooden (1975:67, 1980:7) and Szalay and Delson (1979); but Wilson and Wilson (1977:216) and Groves (2001c:224) considered it a distinct species.

*Macaca radiata* (É. Geoffroy, 1812). Ann. Mus. Hist. Nat. Paris, 19:98.
COMMON NAME: Bonnet Macaque.
TYPE LOCALITY: India; see W. C. O. Hill (1974).
DISTRIBUTION: S India.
STATUS: CITES – Appendix II; IUCN – Lower Risk (lc).
SYNONYMS: *diluta* Pocock, 1931.
COMMENTS: *M. sinica* species group. Revised by Fooden (1981).

*Macaca siberu* Fuentes and Olson, 1995. Asian Primates, 4(4):1.
COMMON NAME: Siberut Macaque.
TYPE LOCALITY: Indonesia, Siberut Isl.
DISTRIBUTION: Siberut (Mentawai group).
STATUS: CITES – Appendix II; IUCN – Critically Endangered as *M. pagensis siberu*.
COMMENTS: *M. nemestrina* species group. Considered a distinct species by Kitchener and Groves (2002).

*Macaca silenus* (Linnaeus, 1758). Syst. Nat., 10th ed., 1:26.
COMMON NAME: Lion-tailed Macaque.
TYPE LOCALITY: "Ceylon" India, Western Ghats; see W. C. O. Hill (1974:652).
DISTRIBUTION: SW India, Western Ghats.
STATUS: CITES – Appendix I; U.S. ESA and IUCN – Endangered.
SYNONYMS: *albibarbatus* (Kerr, 1792); *ferox* (Shaw, 1792); *veter* (Audebert, 1798); *vetulus* (Erxleben, 1777).
COMMENTS: *M. nemestrina* species group.

*Macaca sinica* (Linnaeus, 1771). Mantissa Plantarum, 2, Appendix:521.
COMMON NAME: Toque Macaque.
TYPE LOCALITY: Probably Sri Lanka; see Fooden (1979).
DISTRIBUTION: Sri Lanka.
STATUS: CITES – Appendix II; U.S. ESA – Threatened; IUCN – Endangered as *M. s. opisthomelas*, otherwise Vulnerable as *M. s. aurifrons* and *M. s. sinica*.
SYNONYMS: *audeberti* (Reichenbach, 1862); *inaurea* Pocock, 1931; *longicaudata* Deraniyagala, 1965; *opisthomelas* Osman Hill, 1942; *pileatus* (Ogilby, 1838); **aurifrons** Pocock, 1931.
COMMENTS: *M. sinica* species group. See Fooden (1979).

*Macaca sylvanus* (Linnaeus, 1758). Syst. Nat., 10th ed., 1:25.
COMMON NAME: Barbary Macaque.
TYPE LOCALITY: North Africa, "Barbary coast".
DISTRIBUTION: Morocco, Algeria, Gibraltar (introduced).
STATUS: CITES – Appendix II; IUCN – Vulnerable.
SYNONYMS: *ecaudatus* (É. Geoffroy, 1812); *inuus* (Linnaeus, 1766); *pithecus* (Schreber, 1799); *pygmaeus* (Reichenbach, 1863).
COMMENTS: *M. sylvanus* species group. See Fooden (1976:226) for the use of this name.

*Macaca thibetana* (Milne-Edwards, 1870). C. R. Acad. Sci. Paris, 70:341.
COMMON NAME: Milne-Edwards's Macaque.
TYPE LOCALITY: China, Szechwan, Moupin.
DISTRIBUTION: E Tibet, Szechwan to Kwangtung (China).
STATUS: CITES – Appendix II; IUCN – Lower Risk (conservation dependent).
SYNONYMS: **esau** (Matschie, 1912); *pullus* (Howell, 1928); **guiahouensis** Wang and Jiang, 1996; **huangshanensis** Jiang and Wang, 1996.
COMMENTS: *M. sinica* species group. Reviewed by Fooden (1983) and Jiang et al. (1996).

*Macaca tonkeana* (Meyer, 1899). Abh. Zool. Anthrop.-Ethnology. Mus. Dresden, 7(7):3.
> COMMON NAME: Tonkean Macaque.
> TYPE LOCALITY: Indonesia, Sulawesi Tengah, Tonkean.
> DISTRIBUTION: C Sulawesi, south to Latimojong, north to the base of the northern peninsula, between Palu and Parigi (Indonesia); Togian Isls (Indonesia).
> STATUS: CITES – Appendix II; IUCN – Lower Risk (nt).
> SYNONYMS: *hypomelanus* (Matschie, 1901); *tonsus* (Matschie, 1901); *togeanus* (Sody, 1949).
> COMMENTS: *M. nemestrina* species group. Formerly included in *Cynopithecus*; see Fooden (1969:106-115). According to Froehlich et al. (1998), the macaques of the Balantak Mtns on the E peninsula of Sulawesi constitute a separate species from those of the main part of the range; in this case, the name *tonkeana* would apply to the Balantak form. Whether one of the other synonyms applies to the better-known form from the western part of Central Sulawesi is unclear. Froehlich et al. (1998) said that *togeanus* (from the Togian Isl) is a hybrid swarm between the two. Does not include *hecki*; see Groves (2001c).

*Mandrillus* Ritgen, 1824. Natureichen Eintheilung der Säugethiere, p. 33.
> TYPE SPECIES: *Simia maimon* Linnaeus, 1766; *Simia mormon* Alstromer, 1766 (= *Simia sphinx* Linnaeus, 1758).
> SYNONYMS: *Chaeropithecus* Gray, 1870 [not of de Blainville, 1839]; *Drill* Reichenbach, 1862; *Maimon* Trouessart, 1904; *Mandril* Voigt, 1831; *Mormon* Wagner, 1839 [not of Illiger, 1811]; *Papio* P.L.S. Müller, 1773 [Suppressed under Opinion 1199 of Int. Commission on Zool. Nomenclature].
> COMMENTS: Not a synonym of *Papio* (see Groves, 1989), pace McKenna and Bell (1997). Delson and Napier (1976:46) considered these two species in genus *Papio*, subgenus *Papio*; placed in subgenus *Mandrillus* by Dandelot (1974:9). *Mandrillus* considered a full genus by Groves (1989, 2001c); placed as a subgenus of *Cercocebus* by Goodman et al. (1998).

*Mandrillus leucophaeus* (F. Cuvier, 1807). Ann. Mus. Hist. Nat. Paris, 9:477.
> COMMON NAME: Drill.
> TYPE LOCALITY: Africa.
> DISTRIBUTION: SE Nigeria; Cameroon, north of the Sanaga River and just south of it; Bioko (Equatorial Guinea). See Grubb (1973) for details.
> STATUS: CITES – Appendix I; U.S. ESA – Endangered; IUCN – Endangered as *M. l. leucophaeus* and *M. l. poensis*.
> SYNONYMS: *cinerea* (Kerr, 1792); *drill* (Lesson, 1838); *livea* (Kerr, 1792); *mundamensis* (Hilzheimer, 1906); *sylvestris* (Link, 1795); *sylvicola* (Kerr, 1792); ?*variegata* (Kerr, 1792); *poensis* Zukowsky, 1922.
> COMMENTS: The names *sylvicola*, *variegata*, *cinerea*, *livea* and *sylvestris* were suppressed by Opinion 935 of the International Commission on Zoological Nomenclature (1970).

*Mandrillus sphinx* (Linnaeus, 1758). Syst. Nat., 10th ed., 1:25.
> COMMON NAME: Mandrill.
> TYPE LOCALITY: Cameroon, Ja River, Bitye.
> DISTRIBUTION: Cameroon, south of the Sanaga River; Rio Muni (Equatorial Guinea); Gabon; Republic of Congo. See Grubb (1973) for details.
> STATUS: CITES – Appendix I; U.S. ESA – Endangered; IUCN – Vulnerable.
> SYNONYMS: *burlacei* Rothschild, 1922; *ebolowae* Matschie and Zukowsky, 1917; *escherichi* Matschie and Zukowsky, 1917; *hagenbecki* Matschie and Zukowsky, 1917; *insularis* Zukowsky, 1922; *latidens* (Bechstein, 1799); *madarogaster* (Zimmermann, 1780); *maimon* (Linnaeus, 1766); *mormon* (Alströmer, 1766); *pennanti* (Griffith, 1827); *planirostris* (Elliot, 1909); *schreberi* Matschie, 1917; *suilla* (Kerr, 1792); *tessmanni* Matschie and Zukowsky, 1917; *zenkeri* Matschie and Zukowsky, 1917.

*Miopithecus* I. Geoffroy, 1862. C. R. Acad. Sci. Paris, 15:720.
> TYPE SPECIES: *Simia talapoin* Schreber, 1774.
> COMMENTS: A subgenus of *Cercopithecus* according to Szalay and Delson (1979), but see Groves (1978b, 1989). See also van der Kuhl et al. (2001).

*Miopithecus ogouensis* Kingdon, 1997. The Kingdon Field Guide to African Mammals, p. 55.
   COMMON NAME: Gabon Talapoin.
   TYPE LOCALITY: "Endemic to the equatorial coastal watersheds between Cabinda and the
      River Nyong".
   DISTRIBUTION: S Cameroon, Rio Muni, Gabon, Angola (Cabinda).
   STATUS: CITES – Appendix II.
   COMMENTS: Named on the evidence of descriptions by Machado (1969). See also van der
      Kuhl et al. (2001).

*Miopithecus talapoin* (Schreber, 1774). Die Säugethiere, 1:101, 186, pl. 17.
   COMMON NAME: Angolan Talapoin.
   TYPE LOCALITY: Angola.
   DISTRIBUTION: Angola, SW Dem. Rep. Congo.
   STATUS: CITES – Appendix II; IUCN – Lower Risk (lc).
   SYNONYMS: *ansorgei* (Pocock, 1907); *capillatus* I. Geoffroy, 1842; *melarhinus* (Schinz, 1844);
      *niger* (Kerr, 1792); *pileatus* (É. Geoffroy, 1812); *pilettei* Lorenz, 1919; *vlesschouwersi*
      (Poll, 1940).

*Papio* Erxleben, 1777. Systema Regni Animalis, 1, Mammalia:xxx, 15.
   TYPE SPECIES: *Cynocephalus papio* Desmarest, 1820 (= *Simia hamadryas* Linnaeus, 1758).
   SYNONYMS: *Chaeropitheus* Gervais, 1839; *Comopithecus* J. A. Allen, 1925; *Cynocephalus* G. Cuvier
      and É. Geoffroy, 1795; *Hamadryas* Lesson, 1840.
   COMMENTS: Opinion 1199 of the International Commision on Zoological Nomenclature
      (1982) fixed this as the first available name, and fixed the type species. Includes
      *Theropithecus* according to Goodman et al. (1998), and also *Lophocebus* according to M.
      Goodman et al. (2001). Includes *Mandrillus* according to McKenna and Bell (1997).

*Papio anubis* (Lesson, 1827). Man. Mamm., p. 27.
   COMMON NAME: Olive Baboon.
   TYPE LOCALITY: Upper Nile.
   DISTRIBUTION: Mali to Ethiopia, Kenya, NW Tanzania.
   STATUS: CITES – Appendix II; IUCN – Lower Risk (lc).
   SYNONYMS: *choras* (Ogilby, 1843); *doguera* (Pucheran and Schimper, 1836); *furax* Elliot, 1907;
      *graueri* Lorenz, 1915; *heuglini* Matschie, 1898; *lestes* Heller, 1913; *lydekkeri* Rothschild,
      1902; *neumanni* Matschie, 1897; *nigeriae* Elliot, 1909; *niloticus* Roth, 1965 [*nomen
      nudum*]; *olivaceus* de Winton, 1902; *silvestris* Lorenz, 1915; *tesselatum* Elliot, 1909;
      *tibestianus* Dekeyser and Derivot, 1960; *vigilis* Heller, 1913; *werneri* Wettstein, 1916;
      *yokoensis* Matschie, 1900.

*Papio cynocephalus* (Linnaeus, 1766). Syst. Nat., 12th ed., 1:38.
   COMMON NAME: Yellow Baboon.
   TYPE LOCALITY: Kenya, inland from Mombasa.
   DISTRIBUTION: Somalia, coastal Kenya, Tanzania, to Zambezi River.
   STATUS: CITES – Appendix II; IUCN – Lower Risk (lc).
   SYNONYMS: *antiquorum* (Schinz, 1821); *babouin* (Desmarest, 1820); *basiliscus* (Schreber,
      1800); *flavidus* (Peters, 1852); *jubilaeus* Schwarz, 1928; *langheldi* (Matschie, 1892);
      *ochraceus* (Peters, 1852); *pruinosus* Thomas, 1897; ?*rhodesiae* (Hagner, 1918); *strepitus*
      Elliot, 1907; *sublutea* (Shaw, 1800); *thoth* (Ogilby, 1843); ?*variegata* (Kerr, 1792);
      ***ibeanus*** Thomas, 1893; *ruhei* Zukowsky, 1942; ***kindae*** Lönnberg, 1919.

*Papio hamadryas* (Linnaeus, 1758). Syst. Nat., 10th ed., 1:27.
   COMMON NAME: Hamadryas Baboon.
   TYPE LOCALITY: Egypt.
   DISTRIBUTION: Arid zone of Red Sea coast of Sudan, Eritrea, Ethiopia, N Somalia, Yemen,
      Saudi Arabia.
   STATUS: CITES – Appendix II; IUCN – Lower Risk (nt).
   SYNONYMS: *aegyptiaca* (Gray, 1870); *arabicus* Thomas, 1900; *brockmani* Elliot, 1909;
      *chaeropitheus* (Lesson, 1840); *cynamolgus* (Linnaeus, 1758); *nedjo* (Reichenbach, 1863);
      *wagleri* (Agassiz, 1828).

COMMENTS: Includes *anubis*, *cynocephalus*, *papio*, and *ursinus* according to Szalay and Delson (1979:336), but see Jolly and Brett (1973:85), Dandelot (1974:9), Corbet and Hill (1980:88), Groves (2001*c*) and others who recognized these as distinct species.

*Papio papio* (Desmarest, 1820). Encyclop. Méthodique, Mammalogie, 1:69.
COMMON NAME: Guinea Baboon.
TYPE LOCALITY: "Coast of Guinea".
DISTRIBUTION: Senegal, Guinea and Guinea-Bissau to Mauretania, Mali.
STATUS: CITES – Appendix II; IUCN – Lower Risk (nt).
SYNONYMS: *olivaceus* (I. Geoffroy, 1851); *rubescens* Temminck, 1853; *sphinx* Erxleben, 1777 [not of Linnaeus, 1758].

*Papio ursinus* (Kerr, 1792). Anim. Kingd., p. 63.
COMMON NAME: Chacma Baboon.
TYPE LOCALITY: South Africa, Western Cape Prov., Cape of Good Hope.
DISTRIBUTION: South of Zambezi River, to S Angola, SW Zambia.
STATUS: CITES – Appendix II; IUCN – Lower Risk (lc).
SYNONYMS: *capensis* (A. Smith, 1826); *comatus* É. Geoffroy, 1812; *nigripes* Roberts, 1932; *occidentalis* Goldblatt, 1926; *orientalis* Goldblatt, 1926; *porcaria* (Boddaert, 1787) [not of Brünnich, 1782]; *sphingiola* (Hermann, 1804); **griseipes** Pocock, 1911; *chobiensis* Roberts, 1932; *ngamiensis* Roberts, 1932; *transvaalensis* (Zukowsky, 1927); **ruacana** Shortridge, 1942; *chacamensis* Roth, 1965 [*nomen nudum*].

*Theropithecus* I. Geoffroy, 1843. Arch. Mus. Hist. Nat. Paris, 1841, 2:576.
TYPE SPECIES: *Macacus gelada* Rüppell, 1835.
SYNONYMS: *Gelada* Gray, 1843; *Simopithecus* Andrews, 1916.
COMMENTS: Considered a distinct genus by Cronin and Meikle (1979:259), but Van Gelder (1977*b*:8) included this genus in *Cercopithecus*; Goodman et al. (1998) included it in *Papio* as a subgenus. McKenna and Bell (1997) placed it in a subtribe, Theropithecina, separate from Macacina (*Macaca*) and Papionina (all other genera of Papionini).

*Theropithecus gelada* (Rüppell, 1835). Neue Wirbelt. Fauna Abyssin. Gehörig. Säugeth., p. 5.
COMMON NAME: Gelada.
TYPE LOCALITY: Ethiopia, Semyen (Simien).
DISTRIBUTION: N Ethiopia, highlands.
STATUS: CITES – Appendix II; U.S. ESA – Threatened; IUCN – Data Deficient as *T. g. obscurus*, otherwise Lower Risk (lc).
SYNONYMS: *ruppelli* (Gray, 1843); *senex* Pucheran, 1857; **obscurus** Heuglin, 1863.

**Subfamily Colobinae** Jerdon, 1867. Mammals of India, p. 3.
SYNONYMS: Presbytinae Gray, 1825; Semnopithecinae Owen, 1843.
COMMENTS: Separated provisionally as a full family (Colobidae) by Groves (1989); a subfamily of Cercopithecidae according to Groves (2001*c*). On the name of this subfamily, see Delson (1976), and Brandon-Jones (1978). Divided into two tribes, Colobini (African taxa) and Presbytini (Asian taxa) by McKenna and Bell (1997).

*Colobus* Illiger, 1811. Prodr. Syst. Mamm. Avium., p. 69.
TYPE SPECIES: *Simia polycomos* Schreber, 1800 (= *Cebus polykomos* Zimmerman, 1780).
SYNONYMS: *Colobolus* Gray, 1821; *Guereza* Gray, 1870; *Pterycolobus* Rochebrune, 1887; *Stachycolobus* Rochebrune, 1887.
COMMENTS: Does not include *Procolobus* or *Piliocolobus*, pace McKenna and Bell (1997); see Groves (1989).

*Colobus angolensis* P. Sclater, 1860. Proc. Zool. Soc. Lond., 1860:245.
COMMON NAME: Angola Colobus.
TYPE LOCALITY: Angola, 300 mi. (483 km) inland from Bembe.
DISTRIBUTION: NE Angola, S and E Dem. Rep. Congo, Rwanda, Burundi, NE Zambia, SE Kenya, E Tanzania.

STATUS: CITES – Appendix II; IUCN – Data Deficient as *C. a. palliatus* and *C. a. prigonginei*, Vulnerable as *C. a. ruwenzorii*, otherwise Lower Risk (lc).

SYNONYMS: *benamakimae* Matschie, 1914; *maniemae* Matschie, 1914; *sandbergi* Lönnberg, 1908; *weynsi* Matschie, 1913; **cordieri** Rahm, 1959; **cottoni** Lydekker, 1905; *mawambicus* Matschie, 1913; *nahani* Matschie, 1914; **palliates** Peters, 1868; *langheldi* Matschie, 1914; *sharpie* Thomas, 1902; **prigoginei** Verheyen, 1959; **ruwenzorii** Thomas, 1901; *adolfi-friederici* Matschie, 1914.

COMMENTS: Thorington and Groves (1970:629-647), Dandelot (1974:37), and Corbet and Hill (1980:89) listed *angolensis* as a distinct species.

*Colobus guereza* Rüppell, 1835. Neue Wirbelt. Fauna Abyssin. Gehörig. Säugeth., p. 1.
COMMON NAME: Mantled Guereza.
TYPE LOCALITY: Ethiopia, Gojjam and Kulla.
DISTRIBUTION: Nigeria to Ethiopia; Kenya; Uganda; Tanzania.
STATUS: CITES – Appendix II; IUCN – Endangered as *C. g. percivali*, Data Deficient as *C. g. gallarum*, otherwise Lower Risk (lc).
SYNONYMS: *abyssinicus* (Oken, 1816) [unavailable]; *managaschae* Matschie, 1913; *poliurus* Thomas, 1901; *ruppelli* (Gray, 1870); **caudatus** Thomas, 1885; *albocaudatus* Lydekker, 1906; **dodingae** Matschie, 1913; *gallarum* Neumann, 1902; **kikuyuensis** Lönnberg, 1912; *laticeps* Matschie, 1913; *thikae* Matschie, 1913; **matschiei** Neumann, 1899; *elgonis* Granvik, 1925; *roosevelti* Heller, 1913; **occidentalis** (Rochebrune, 1887); *brachychaites* Matschie, 1913; *dianae* Matschie, 1913; *escherichi* Matschie, 1914; *ituricus* Matschie, 1913; *rutschuricus* Lorenz, 1914; *terrestris* Heller, 1913; *uellensis* Matschie, 1913; **percivali** Heller, 1913.

*Colobus polykomos* (Zimmermann, 1780). Geogr. Gesch. Mensch. Vierf. Thiere, 2:202.
COMMON NAME: King Colobus.
TYPE LOCALITY: Sierra Leone.
DISTRIBUTION: Gambia to the Nzo-Sassandra system in Côte d'Ivoire.
STATUS: CITES – Appendix II; IUCN – Lower Risk (nt).
SYNONYMS: *comosa* (Shaw, 1800); *polycomos* (Schreber, 1800); *regalis* (Kerr, 1792); *tetradactyla* (Link, 1795).
COMMENTS: Does not include *vellerosus* (Groves, 2001*c*; Oates and Trocco, 1983).

*Colobus satanas* Waterhouse, 1838. Proc. Zool. Soc. Lond., 1837:87 [1838].
COMMON NAME: Black Colobus.
TYPE LOCALITY: Equatorial Guinea, Bioko.
DISTRIBUTION: SW Gabon, Rio Muni and Bioko (Equatorial Guinea), SW Cameroon; possibly Republic of Congo (Carpaneto, 1995).
STATUS: CITES – Appendix II; U.S. ESA – Endangered; IUCN – Endangered as *C. s. satanus*, Data Deficient as *C. s. antracinus*, otherwise Vulnerable.
SYNONYMS: *metternichi* Krumbiegel, 1943; **anthracinus** (Leconte, 1857); *limbarenicus* (Matschie, 1917); *municus* (Matschie, 1917); *zenkeri* (Matschie, 1917).
COMMENTS: McAllan and Bruce (1989) argued that the original publication of this species is: *The Analyst*, 24:298-299 [publ. 2 July 1838].

*Colobus vellerosus* (I. Geoffroy, 1834). *In* Bélanger, Voy. Indes-Orientales, Zool., p. 37.
COMMON NAME: Ursine Colobus.
TYPE LOCALITY: "Africa".
DISTRIBUTION: Nzi-Bandama system (Côte d'Ivoire) to W Nigeria.
STATUS: CITES – Appendix II; IUCN – Vulnerable.
SYNONYMS: *bicolor* (Wesmael, 1835); *dollmani* Schwarz, 1927; *leucomeros* Ogilby, 1838; *ursinus* Ogilby, 1835.
COMMENTS: Strictly, *dollmani* is hybrid swarm between *vellerosus* and *polykomos*, but phenetically much closer to *vellerosus*. Separated from *polykomos* by Oates and Trocco (1983) and Groves (2001*c*).

*Nasalis* É. Geoffroy, 1812. Ann. Mus. Hist. Nat. Paris, 19:89.
TYPE SPECIES: *Cercopithecus larvatus* Wurmb, 1787.

SYNONYMS: *Hanno* Gray, 1821; *Rhinolazon* Gloger, 1841; *Rhynchopithecus* Dahlbohm, 1856.

COMMENTS: *Simias* was included in this genus by Groves (1970:639) and McKenna and Bell (1997), and by Szalay and Delson (1979) and Delson (1975:217) who considered *Simias* a subgenus; but also see Krumbiegel (1978) and Napier (1985), who considered it as a distinct genus.

*Nasalis larvatus* (Wurmb, 1787). Verh. Batav. Genootsch., 3:353.

COMMON NAME: Proboscis Monkey.

TYPE LOCALITY: Indonesia, W Kalimantan, Pontianak.

DISTRIBUTION: Borneo.

STATUS: CITES – Appendix I; U.S. ESA and IUCN – Endangered.

SYNONYMS: *capistratus* (Kerr, 1792); *nasica* (Lacépède, 1799); *orientalis* Chasen 1940; *recurvus* Vigors and Horsfield, 1828.

*Piliocolobus* (Rochebrune, 1877). Faune de Sénégambie, Suppl. Vert., Mamm., 1:96.

TYPE SPECIES: *Simia* (*Cercopithecus*) *badius* Kerr, 1792.

SYNONYMS: *Tropicolobus* Rochebrune, 1887.

COMMENTS: Separate from *Colobus*, see Corbet and Hill (1980:90) and Groves (1989); separate from *Procolobus* (Groves, 2001*c*).

*Piliocolobus badius* (Kerr, 1792). *In* Linnaeus, Anim. Kingdom, p. 74.

COMMON NAME: Western Red Colobus.

TYPE LOCALITY: Sierra Leone.

DISTRIBUTION: Senegal to Ghana.

STATUS: CITES – Appendix II; IUCN – Critically Endangered as *Procolobus b. waldroni* (probably extinct: Oates et al., 2000), Endangered as *P. b. badius* and *P. b. temminckii*.

SYNONYMS: *ferriginea* (Shaw, 1800); *ferruginosus* (É. Geoffroy, 1812); *rufoniger* (Ogilby, 1838); **temminckii** (Kuhl, 1820); *fuliginosus* (Ogilby, 1835); *rufo-fuliginus* (Ogilby, 1838); **waldronae** (Hayman, 1936).

COMMENTS: Includes *waldronae* and *temmincki*; see Dandelot (1974:33); but also see Rahm (1970).

*Piliocolobus foai* (de Pousargues, 1899). Bull. Mus. Hist. Nat. Paris, 5:278.

COMMON NAME: Central African Red Colobus.

TYPE LOCALITY: Dem. Rep. Congo, between SW of Lake Tanganyika and upper Congo, Ouroua.

DISTRIBUTION: Republic of Congo (Sangha, Oubangui), Dem. Rep. Congo (north of Congo, east of Lualaba), Central African Republic (Ngotto), Sudan (southernmost forests).

STATUS: CITES – Appendix II; IUCN – Data Deficient as *Procolobus badius foai*, *P. b. ellioti*, *P. b. langi*, *P. b. lulidicus*, and *P. b. parmentierorum*.

SYNONYMS: *graueri* (Dollman, 1909); *kabambarei* Matschie, 1914; *lulidicus* Matschie, 1914; **ellioti** (Dollman, 1909); *anzeliusi* Matschie, 1914; *langi* (J. A. Allen, 1925); *melanochir* Matschie, 1914; *multicolor* (Lorenz, 1914); *variabilis* (Lorenz, 1914); **oustaleti** (Trouessart, 1906); *brunneus* (Lönnberg, 1919); *nigrimanus* (Trouessart, 1906); *powelli* (Matschie, 1913); *schubotzi* (Matschie, 1914); *umbrinus* (Matschie, 1914); **parmentierorum** (Colyn and Verheyen, 1987); **semlikiensis** (Colyn, 1991).

COMMENTS: A rather heterogeneous species, but subspecies are variable and hard to separate. Considered a separate species by Groves (2001*c*).

*Piliocolobus gordonorum* Matschie, 1900. Sitzb. Ges. Naturf. Fr. Berlin, p. 186.

COMMON NAME: Uzungwa Red Colobus.

TYPE LOCALITY: Tanzania, Uzungwa Mtns.

DISTRIBUTION: Tanzania, Uzungwa Mtns and forests between Little Ruaha and Ulanga Rivers.

STATUS: CITES – Appendix II; IUCN – Vulnerable.

COMMENTS: Considered a separate species by Groves (2001*c*).

*Piliocolobus kirkii* (Gray, 1868). Proc. Zool. Soc. Lond., 1868:180.

COMMON NAME: Zanzibar Red Colobus.

TYPE LOCALITY: Tanzania, Zanzibar.

DISTRIBUTION: Zanzibar.

STATUS: CITES – Appendix I and U.S. ESA – Endangered as *Procolobus pennantii kirki*; IUCN – Endangered.

COMMENTS: Considered a separate species by Dandelot (1974) and Groves (2001c).

*Piliocolobus pennantii* (Waterhouse, 1838). Proc. Zool. Soc. Lond., 1838:57.

COMMON NAME: Pennant's Red Colobus.

TYPE LOCALITY: Equatorial Guinea, Bioko.

DISTRIBUTION: Bioko (Equatorial Guinea), Niger Delta (Nigeria); Sangha-Likouala confluence (Republic of Congo).

STATUS: CITES – Appendix II; IUCN – Critically Endangered as *Procolobus. pennantii bouvieri*, Endangered as *P. p. pennantii* and *P. p. epieni*.

SYNONYMS: **bouvieri** Rochebrune, 1887; *likualae* Matschie, 1914; **epieni** (Grubb and Powell, 1999).

COMMENTS: Does not include *foai, gondonorum, kirki, tephrosceles*, or *tholloni*; see Groves (2001c). For discussion of original publication see McAllan and Bruce (1989).

*Piliocolobus preussi* (Matschie, 1900). Sitzb. Ges. Naturf. Fr. Berlin, p. 183.

COMMON NAME: Preuss's Red Colobus.

TYPE LOCALITY: Cameroon, Barombi (on Elephant Lake).

DISTRIBUTION: Yabassi Dist. (Cameroon).

STATUS: CITES – Appendix II; U.S. ESA – Endangered as *Procolobus preussi*; and IUCN – Endangered as *Procolobus pennantii preussi*.

COMMENTS: Considered by Rahm (1970) to be a subspecies of *badius*, but see Dandelot (1974:37).

*Piliocolobus rufomitratus* (Peters, 1879). Monatsb. K. Preuss. Akad. Wiss. Berlin, 1879:829.

COMMON NAME: Tana River Red Colobus.

TYPE LOCALITY: Kenya, Tana River, Muniuni.

DISTRIBUTION: Lower Tana River (Kenya).

STATUS: CITES – Appendix I and U.S. ESA – Endangered as *Procolobus rufomitratus*; IUCN – Critically Endangered as *Procolobus rufomitratus*.

*Piliocolobus tephrosceles* (Elliot, 1907). Ann. Mag. Nat. Hist., ser. 7, 20:195.

COMMON NAME: Ugandan Red Colobus.

TYPE LOCALITY: Uganda, Toro, Ruahara River, east side of Rwenzoris.

DISTRIBUTION: Uganda, Rwanda, Burundi, W Tanzania to Lake Rukwa.

STATUS: CITES – Appendix II.

SYNONYMS: *gudoviusi* (Matschie, 1914).

COMMENTS: Considered a full species by Groves (2001c).

*Piliocolobus tholloni* (Milne-Edwards, 1886). *In* Rivière, Rev. Scient (3) 12:15.

COMMON NAME: Thollon's Red Colobus.

TYPE LOCALITY: Dem. Rep. Congo, Lower Congo.

DISTRIBUTION: South of Congo River, west of Lomami River (Dem. Rep. Congo).

STATUS: CITES – Appendix II.

SYNONYMS: *lovizettii* (Matschie, 1913).

COMMENTS: Dandelot (1974:35) considered *tholloni* a distinct species; followed by Groves (2001c).

*Presbytis* Eschscholtz, 1821. Reise (Kotzebue), 3:196.

TYPE SPECIES: *Presbytis mitrata* Eschscholtz, 1821 (= *Simia melalophos* Raffles, 1821).

SYNONYMS: *Corypithecus* Trouessart, 1879; *Lophopitheus* Trouessart, 1879; *Presbypitheus* Trouessart, 1879.

COMMENTS: Does not include *Semnopithecus* and *Trachypithecus*, pace McKenna and Bell (1997; see Brandon-Jones [1984]; Groves [1989, 2001c]; Hooijer [1962:20-24]). Szalay and Delson (1979:402) included these and *Kasi* as subgenera.

*Presbytis chrysomelas* (Müller, 1838). Tijdschr. Nat. Gesch. Physiol., 5:138.

COMMON NAME: Sarawak Surili.

TYPE LOCALITY: Indonesia, Pontianak.

DISTRIBUTION: Kalimantan north of Kapuas River (Indonesia), Sarawak, Sabah (Borneo, Malaysia).

STATUS: CITES – Appendix II; IUCN – Data Deficient as *Presbytis femoralis chrysomelas* and *P. f. cruciger*.

SYNONYMS: ***cruciger*** (Thomas, 1892); *arwasca* Miller, 1934.

COMMENTS: Separated from *P. femoralis* by Groves (2001*c*).

*Presbytis comata* (Desmarest, 1822). Mammalogie, *in* Encycl. Meth., 2(Suppl.):533.

COMMON NAME: Javan Surili.

TYPE LOCALITY: Indonesia, W Java.

DISTRIBUTION: W and C Java (Indonesia).

STATUS: CITES – Appendix II; IUCN – Endangered as *P. comata*, Data Deficient as *P. fredericae*.

SYNONYMS: *aygula*, various authors; ***fredericae*** (Sody, 1930).

COMMENTS: Formerly called *P. aygula*, but see Napier and Groves (1983) who showed that *aygula* is a *nomen oblitum* for *Macaca fascicularis*. Brandon-Jones (1984) considered *fredericae* to be a separate species.

*Presbytis femoralis* (Martin, 1838). Mag. Nat. Hist. [Charlesworth's], 2:436.

COMMON NAME: Banded Surili.

TYPE LOCALITY: Singapore.

DISTRIBUTION: Far south and northwest of Malay Peninsula; peninsular part of Thailand and Burma; Singapore; NE Sumatra, between Rokan and Siak Rivers.

STATUS: CITES – Appendix II; IUCN – Data Deficient as *P. f. percura* and *P. f. robinsoni*, otherwise Lower Risk (nt).

SYNONYMS: *australis* Miller, 1913; *neglectus* (Schlegel, 1876); ***percura*** Lyon, 1908; ***robinsoni*** Thomas, 1910; *keatii* Robinson and Kloss, 1911.

COMMENTS: Separated from *P. melalophos* by Wilson and Wilson (1977:217-222); recognized as a species by Aimi et al. (1986). Does not include *chrysomelas* or *natunae* (see Groves, 2001*c*), or *siamensis*; see Brandon-Jones (1974).

*Presbytis frontata* (Müller, 1838). Tijdschr. Nat. Gesch. Physiol., 5:136.

COMMON NAME: White-fronted Langur.

TYPE LOCALITY: Indonesia, SE Kalimantan: Murung and "Pulu Lampy", near Banjarmasin, Pematang, Kuala (Medway, 1965:82).

DISTRIBUTION: C and E Borneo, from C Sarawak to S coast.

STATUS: CITES – Appendix II; IUCN – Data Deficient.

SYNONYMS: *nudifrons* Elliot, 1909.

*Presbytis hosei* (Thomas, 1889). Proc. Zool. Soc. Lond., 1889:159.

COMMON NAME: Hose's Langur.

TYPE LOCALITY: Malaysia, Sarawak, Niah.

DISTRIBUTION: N and E Borneo: Brunei, E Sarawak, Sabah (Malaysia), south to Karangan River in Kalimantan (Indonesia).

STATUS: CITES – Appendix II; IUCN – Vulnerable as *P. h. sabana*, Data Deficient as *P. h. hosei*, *P. h. canicrus* and *P. h. everetti*. Nominotypical *hosei* perhaps extinct.

SYNONYMS: ***canicrus*** Miller, 1934; ***everetti*** (Thomas, 1892); ***sabana*** (Thomas, 1893).

COMMENTS: Separated from "*aygula*" (= *comata*) by Medway (1970:544). Possibly *canicrus* and *sabana* are distinct species (Groves, 2001*c*).

*Presbytis melalophos* (Raffles, 1821). Trans. Linn. Soc. Lond., 13:245.

COMMON NAME: Sumatran Surili.

TYPE LOCALITY: Indonesia, Sumatra, Bengkulen.

DISTRIBUTION: Sumatra (Indonesia).

STATUS: CITES – Appendix II; IUCN – Data Deficient as *P. femoralis batuana*, Lower Risk (nt), not evaluated as *bicolor*, *mitrata*, and *nobilis*.

SYNONYMS: *aurata* (Müller and Schlegel, 1861); *ferrugineus* (Schlegel, 1876); *flavimanus* (I. Geoffroy, 1831); *nobilis* (Gray, 1842); ***bicolor*** Aimi and Bakar, 1992; ***mitrata*** Eschscholtz, 1821; *fluviatilis* (Chasen, 1940); *fusco-murina* Elliot, 1906; ***sumatranus*** (Müller and Schlegel, 1841); *batuanus* Miller, 1903; *margae* Hooijer, 1948.

COMMENTS: Does not include *P. femoralis*, which was regarded as a separate species by Wilson and Wilson (1977:217-222), or *siamensis*, which was separated by Brandon-Jones (1984), or *natunae* or *chrysomelas*, which were separated by Groves (2001c).

*Presbytis natunae* (Thomas and Hartert, 1894). Novit. Zool., 1:652.
    COMMON NAME: Natuna Island Surili.
    TYPE LOCALITY: Indonesia, North Natuna Isl, Bunguran.
    DISTRIBUTION: Bunguran Isl (Indonesia).
    STATUS: CITES – Appendix II.
    COMMENTS: Separated from *P. siamensis* by Groves (2001c).

*Presbytis potenziani* (Bonaparte, 1856). C. R. Acad. Sci. Paris, 43:412.
    COMMON NAME: Mentawai Langur.
    TYPE LOCALITY: Indonesia, W Sumatra, Sipora Isl.
    DISTRIBUTION: Mentawai Isls (Indonesia).
    STATUS: CITES – Appendix I; U.S. ESA – Threatened; IUCN – Vulnerable as *P. p. potenziani* and *P. p. siberu*.
    SYNONYMS: *chrysogaster* (Peters, 1867); **siberu** (Chasen and Kloss, 1927).

*Presbytis rubicunda* (Müller, 1838). Tijdschr. Nat. Gesch. Physiol., 5:137.
    COMMON NAME: Maroon Leaf-Monkey.
    TYPE LOCALITY: Indonesia, S Kalimantan, Mt. Sekumbang (SE of Banjermasin).
    DISTRIBUTION: Borneo; Karimata Isl (Indonesia).
    STATUS: CITES – Appendix II; IUCN – Data Deficient as *P. r. chrysea*, otherwise Lower Risk (lc), not evaluated as *carimatae* and *ignita*.
    SYNONYMS: **carimatae** Miller, 1906; **chrysea** Davis, 1962; **ignita** Dollman, 1909; **rubida** (Lyon, 1911).

*Presbytis siamensis* (Müller and Schlegel, 1841). *In* Temminck, Verh. Nat. Ges. Overz. Bezitt. Zool. (Mamm.), p. 60.
    COMMON NAME: White-thighed Surili.
    TYPE LOCALITY: Malaya, Melaka.
    DISTRIBUTION: Malay Peninsula, except far south and northwest; E Sumatra between Siak and Inderagiri Rivers, between Rokan and Barimun Rivers, Lake Toba region, and perhaps Jambi district; Kundur, Bintang, and probably Batam and Galang Isls, Riau Arch. (Indonesia).
    STATUS: CITES – Appendix II; IUCN – Lower Risk (lc).
    SYNONYMS: *dilecta* Elliot, 1909; *nigrimanus* (I. Geoffroy, 1843); *nubigena* Elliot, 1909; **cana** Miller, 1906; *amsiri* Kawamura, 1984 [*nomen nudum*]; *catemana* Lyon, 1908; **paenulata** (Chasen, 1940); **rhionis** Miller, 1903.
    COMMENTS: Separated from *P. femoralis* by Brandon-Jones (1984).

*Presbytis thomasi* (Collett, 1893). Proc. Zool. Soc. Lond., 1892:613 [1893].
    COMMON NAME: Thomas's Langur.
    TYPE LOCALITY: Indonesia, Sumatra, Aceh, Langkat.
    DISTRIBUTION: Sumatra: Aceh, south to about 3°50′N.
    STATUS: CITES – Appendix II; IUCN – Lower Risk (nt).
    SYNONYMS: *nubilus* Miller, 1942.
    COMMENTS: Separated from "*aygula*" (= *comata*) by Medway (1970:544). Considered a subspecies of *comata* by Brandon-Jones (1984).

*Procolobus* Rochebrune, 1877. Faune de Sénégambie, Suppl. Vert., Mamm., 1:95.
    TYPE SPECIES: *Colobus verus* Van Beneden, 1838.
    SYNONYMS: *Lophocolobus* de Pousargue, 1895.
    COMMENTS: Separate from *Colobus*, see Corbet and Hill (1980:90) and Groves (1989).

*Procolobus verus* (Van Beneden, 1838). Bull. Acad. Sci. Belles-Letters Bruxelles, 5:347.
    COMMON NAME: Olive Colobus.
    TYPE LOCALITY: Africa.
    DISTRIBUTION: Sierra Leone to Togo; Idah Dist. (E Nigeria, see Menzies, 1970, for comments).

STATUS: CITES – Appendix II; IUCN – Lower Risk (nt).

SYNONYMS: *chrysurus* (Gray, 1866); *cristatus* (Gray, 1866); *olivaceus* (Wagner, 1840).

COMMENTS: Separated as a genus from *Colobus* by Dandelot (1974:37) and Corbet and Hill (1980:90).

*Pygathrix* É. Geoffroy, 1812. Ann. Mus. Hist. Nat. Paris, 19:90.

TYPE SPECIES: *Simia nemaeus* Linnaeus, 1771.

SYNONYMS: *Daunus* Gray, 1821.

COMMENTS: *Rhinopithecus* included in *Pygathrix* by Groves (1970) and Szalay and Delson (1979:404) and McKenna and Bell (1997); but see Jablonski and Peng (1993) and Groves (2001c), who recognized that the two are sister-groups but retained them as separate genera.

*Pygathrix cinerea* Nadler, 1997. Zool. Garten, NF, 67:165.

COMMON NAME: Gray-shanked Douc Langur.

TYPE LOCALITY: Vietnam, Gia Lai Province, Play Ku, 13°59′N, 108°00′E.

DISTRIBUTION: C Vietnam, 13°59′-14°46′N.

STATUS: CITES – Appendix I; IUCN – Endangered as *P. nemaeus cinerea*.

*Pygathrix nemaeus* (Linnaeus, 1771). Mantissa Plantarum, p. 521.

COMMON NAME: Red-shanked Douc Langur.

TYPE LOCALITY: Cochin-China (Indo-China).

DISTRIBUTION: C Vietnam, E Laos, from 20°N to about 14°N, perhaps as far as 13°N.

STATUS: CITES – Appendix I; U.S. ESA – Endangered; IUCN – Endangered as *P. n. nemaeus*.

COMMENTS: Does not include *nigripes* (see Groves, 2001c).

*Pygathrix nigripes* (Milne-Edwards, 1871). Bull. Nouv. Arch. Mus. Hist. Nat. Paris, 6: 7.

COMMON NAME: Black-shanked Douc Langur.

TYPE LOCALITY: Vietnam, Saigon.

DISTRIBUTION: S Vietnam, from about 10°30′N to 14°30′N; Cambodia east of the Mekong River.

STATUS: CITES – Appendix I; IUCN – Endangered.

SYNONYMS: *moi* (Kloss, 1926).

COMMENTS: Separated from *nemaeus* by Nadler (1997) and Groves (2001c).

*Rhinopithecus* Milne-Edwards, 1872. Rech. Nat. Hist. Mamm., p. 233.

TYPE SPECIES: *Semnopithecus roxellana* Milne-Edwards, 1872.

SYNONYMS: *Presbytiscus* Pocock, 1924.

COMMENTS: Included in *Pygathrix* by Groves (1970) and Szalay and Delson (1979:404); but see Jablonski and Peng (1993) and Groves (2001c), who recognized that the two are sister-groups but retained them as separate genera. Jablonski and Peng (1993) recognized *Presbytiscus* as a valid subgenus.

*Rhinopithecus avunculus* (Dollman, 1912). Abstr. Proc. Zool. Soc. Lond., 1912(106):18.

COMMON NAME: Tonkin Snub-nosed Monkey.

TYPE LOCALITY: Vietnam, Songkoi River, Yen Bay.

DISTRIBUTION: NW Vietnam.

STATUS: CITES – Appendix I; U.S. ESA – Endangered; IUCN – Critically Endangered.

COMMENTS: Placed in subgenus *Presbytiscus* by Jablonski and Peng (1993).

*Rhinopithecus bieti* Milne-Edwards, 1897. Bull. Mus. Hist. Nat. Paris, 3:157.

COMMON NAME: Black Snub-nosed Monkey.

TYPE LOCALITY: China, Yunnan, left bank of upper Mekong, Kiape, 28°25′N, 98°55′E, "a day's journey south of Atentse."

DISTRIBUTION: Ridge of Mekong-Salween divide, Yunnan (China).

STATUS: CITES – Appendix I; U.S. ESA and IUCN – Endangered.

COMMENTS: Considered a subspecies of *roxellana* by Groves (1970:569), but regarded as a full species by Peng et al. (1988).

*Rhinopithecus brelichi* (Thomas, 1903). Proc. Zool. Soc. Lond., 1903(1):224.
    COMMON NAME: Gray Snub-nosed Monkey.
    TYPE LOCALITY: China, N Kweichow, Van Gin Shan Range.
    DISTRIBUTION: Van Gin Shan (Fanjinshan) Range (Guizhou, China).
    STATUS: CITES – Appendix I; U.S. ESA and IUCN – Endangered.
    COMMENTS: Considered a valid species by Groves (1970:569).

*Rhinopithecus roxellana* (Milne-Edwards, 1870). C. R. Acad. Sci. Paris, 70:341.
    COMMON NAME: Golden Snub-nosed Monkey.
    TYPE LOCALITY: China, Sichuan, Moupin (= Baoxing, 30°26'N, 102°50'E).
    DISTRIBUTION: Mountains of Sichuan, S Ganssu, Hubei, Shaanxi (China).
    STATUS: CITES – Appendix I; U.S. ESA – Endangered; IUCN – Vulnerable as *R. r. hubeiensis*,
        *R. r. qinlingensis*, and *R. r. roxellana*.
    SYNONYMS: *roxellanae* (Milne-Edwards, 1872); **hubeiensis** Y. Wang, Jiang and Li, 1998;
        **qinlingensis** Y. Wang, Jiang and Li, 1998.

*Semnopithecus* Desmarest, 1822. Mammalogie, *in* Encycl. Meth, 2(Suppl.):532.
    TYPE SPECIES: *Simia entellus* Dufresne, 1797.
    COMMENTS: Considered a subgenus of *Presbytis* by Szalay and Delson (1979); separated from
        *Presbytis* by Groves (1989). All taxa have generally been placed in a single species
        (*entellus*), but Groves (2001c) divided them into seven species.

*Semnopithecus ajax* (Pocock, 1928). J. Bombay Nat. Hist. Soc., 32:480.
    COMMON NAME: Kashmir Gray Langur.
    TYPE LOCALITY: India, Chamba, Deolah, 1800 m.
    DISTRIBUTION: India, Dehra Dun west into Pakistani Kashmir, 2000-3000 m.
    STATUS: CITES – Appendix I and U.S. ESA – Endangered as included in *S. entellus*; IUCN –
        Lower Risk (nt) as *S. entellus ajax*.

*Semnopithecus dussumieri* I. Geoffroy, 1843. C. R. Acad. Sci. Paris, 15:719.
    COMMON NAME: Southern Plains Gray Langur.
    TYPE LOCALITY: India, Malabar coast, Mahé.
    DISTRIBUTION: SW and WC India.
    STATUS: CITES – Appendix I and U.S. ESA – Endangered as included in *S. entellus*; IUCN –
        Data Deficient as *S. entellus dussumieri*, *S. e. achises*, and *S. e. elissa*.
    SYNONYMS: *achates* (Pocock, 1928); *anchises* (Blyth, 1844); *elissa* (Pocock, 1928); *iulus*
        (Pocock, 1928); *priamellus* (Pocock, 1928).

*Semnopithecus entellus* (Dufresne, 1797). Bull. Sci. Soc. Philom. Paris, ser. 1, 7:49.
    COMMON NAME: Northern Plains Gray Langur.
    TYPE LOCALITY: India, Bengal.
    DISTRIBUTION: Pakistan and India, lowlands north of Godavari and Krishna Rivers, south of
        Ganges.
    STATUS: CITES – Appendix I; U.S. ESA – Endangered; IUCN – Lower Risk (nt).

*Semnopithecus hector* (Pocock, 1928). J. Bombay Nat. Hist. Soc., 32:481.
    COMMON NAME: Tarai Gray Langur.
    TYPE LOCALITY: India, Kumaun, Ramnagar, Sitabani, 600 m.
    DISTRIBUTION: India (Kumaun) to Nepal (Hazaria district), 600-1800 m.
    STATUS: CITES – Appendix I and U.S. ESA – Endangered as included in *S. entellus*; IUCN –
        Lower Risk (nt) as *S. entellus hector*.

*Semnopithecus hypoleucos* Blyth, 1841. J. Asiat. Soc. Bengal, 10:839.
    COMMON NAME: Black-footed Gray Langur.
    TYPE LOCALITY: India, Travancore.
    DISTRIBUTION: India, Kerala, South Coorg region.
    STATUS: CITES – I and U.S. ESA – Endangered as included in *S. entellus*; IUCN – Data
        Deficient as *S. entellus hypoleucos*.
    SYNONYMS: *aeneas* (Pocock, 1928).

*Semnopithecus priam* Blyth, 1844. Ann. Mag. Nat. Hist., ser. 2, 13:312.
   COMMON NAME: Tufted Gray Langur.
   TYPE LOCALITY: India, Coromandel Coast.
   DISTRIBUTION: SE India; Sri Lanka.
   STATUS: CITES – Appendix I and U.S. ESA – Endangered as included in *S. entellus*; IUCN –
      Vulnerable as *S. entellus thersites*, Data Deficient as *S. entellus priam*.
   SYNONYMS: *pallipes* Blyth, 1844 [*nomen nudum*]; *priamus* Blyth, 1847; *thersites* (Blyth, 1847).

*Semnopithecus schistaceus* Hodgson, 1840. J. Asiat. Soc. Bengal, 9:1212.
   COMMON NAME: Nepal Gray Langur.
   TYPE LOCALITY: Nepal.
   DISTRIBUTION: Nepal, east of Gorkha, to Sikkim and parts of southernmost Tibet (China),
      1500-3500 m.
   STATUS: CITES – Appendix I and U.S. ESA – Endangered as included in *S. entellus*; IUCN –
      Lower Risk (nt) as *S. entellus schistaceus*.
   SYNONYMS: *achilles* (Pocock, 1928); *lania* (Elliot, 1909); *nipalensis* Hodgson, 1840.

*Simias* Miller, 1903. Smithson. Misc. Coll., 45:67.
   TYPE SPECIES: *Simias concolor* Miller, 1903.
   COMMENTS: Included in *Nasalis* by Groves (1970:639); Szalay and Delson (1979) and Delson
      (1975:217) considered *Simias* a subgenus; but also see Krumbiegel (1978) and Napier
      (1985), who restored it to generic rank.

*Simias concolor* Miller, 1903. Smithson. Misc. Coll., 45:67.
   COMMON NAME: Simakobou.
   TYPE LOCALITY: Indonesia, W Sumatra, S Pagai Isl.
   DISTRIBUTION: Mentawai Isls (Indonesia).
   STATUS: CITES – Appendix I and U.S. ESA – Endangered as *Nasalis concolor*; IUCN –
      Endangered as *S. c. concolor* and *S. c. siberu*.
   SYNONYMS: *siberu* Chasen and Kloss, 1927.

*Trachypithecus* Reichenbach, 1862. Vollständ. Nat. Affen, p. 89.
   TYPE SPECIES: *Semnopithecus pyrrhus* Horsfield, 1823 (= *Cercopithecus auratus* É. Geoffroy, 1812).
   SYNONYMS: *Kasi* Reichenbach, 1862.
   COMMENTS: Separated from *Presbytis* by Hooijer (1962) and Groves (1989, 2001c). Includes
      subgenus *Kasi* (for *T. vetulus* group). Included in *Semnopithecus* by Brandon-Jones (1984,
      1995); provisionally retained as a genus by Groves (2001c), but its monophyletic status is
      not confirmed. The species groups are: (1) *T. vetulus* group (*vetulus, johnii*), (2) *T. cristatus*
      group (*auratus, cristatus, germaini, barbei*), (3) *T. obscurus* group (*obscurus, phayrei*),
      (4) *T. pileatus* group (*pileatus, shortridgei, geei*), (5) *T. francoisi* group (*francoisi, hatinhensis,
      poliocephalus, laotum, delacouri, ebenus*).

*Trachypithecus auratus* (É. Geoffroy, 1812). Ann. Mus. Hist. Nat. Paris, 19:93.
   COMMON NAME: Javan Lutung.
   TYPE LOCALITY: Java, Semarang (Müller, 1840:16).
   DISTRIBUTION: Java, Bali, and Lombok (Indonesia).
   STATUS: CITES – Appendix II; IUCN – Endangered as *T. a. auratus* and *T. a. mauritius*.
   SYNONYMS: *kohlbruggei* (Sody, 1931); *maurus* (Horsfield, 1823); *pyrrhus* (Horsfield, 1823);
      *sondaicus* (Robinson and Kloss, 1919); *stresemanni* Pocock, 1934; **mauritius** (Griffith,
      1821).
   COMMENTS: *T. cristatus* species group. Separated from *T. cristatus* by Weitzel and Groves
      (1985). Brandon-Jones (1995) placed this species, together with *T. johnii* and all the
      *T. francoisi* species group, in a separate (*auratus*) species group.

*Trachypithecus barbei* (Blyth, 1847). J. Asiat. Soc. Bengal, 16:734.
   COMMON NAME: Tenasserim Lutung.
   TYPE LOCALITY: Burma, Ye south of Moulmein.
   DISTRIBUTION: N penisular Burma and Thailand, 14°20'-15°10'N, 98°30'-98°55'E.
   STATUS: CITES – Appendix II.

SYNONYMS: *atrior* (Pocock, 1928).
COMMENTS: *T. cristatus* species group. On the question of the identity of *barbei*, see Groves (2001c:266).

*Trachypithecus cristatus* (Raffles, 1821). Trans. Linn. Soc. Lond., 13:244.
COMMON NAME: Silvery Lutung.
TYPE LOCALITY: Indonesia, Sumatra, Bengkulen (Bengkulu).
DISTRIBUTION: Borneo, Natuna Isl, Bangka, Belitung, Sumatra, Riau Archipelago, and W coast of Malay Peninsula.
STATUS: CITES – Appendix II.
SYNONYMS: *pruinosus* (Desmarest, 1822); *pullata* (Thomas and Wroughton, 1909); *rutledgii* (Anderson, 1878); *ultima* (Elliot, 1910); **vigilans** (Miller, 1913).
COMMENTS: *T. cristatus* species group. Does not include *barbei* and *germaini*, see Groves (2001c).

*Trachypithecus delacouri* (Osgood, 1932). Field Mus. Nat. Hist. Zool., 18:205.
COMMON NAME: Delacour's Langur.
TYPE LOCALITY: Vietnam, Hoi Xuan.
DISTRIBUTION: Vietnam south of Red River, 18°-21°36′N.
STATUS: CITES – Appendix II; U.S. ESA – Endangered as included in *T. francoisi*; IUCN – Critically Endangered.
COMMENTS: *T. francoisi* species group. Considered a species separate from *francoisi* by Brandon-Jones (1995).

*Trachypithecus ebenus* (Brandon-Jones, 1995). Raffles Bulletin of Zoology, 43:15.
COMMON NAME: Indochinese Black Langur.
TYPE LOCALITY: "Indo China": probably either Lai Chau or Fan Si Pan chain (ca. 22°30′N, 103°50′E) according to Brandon-Jones (1995).
DISTRIBUTION: Unknown. May be restricted to region of type locality (Brandon-Jones, 1995); but apparently occurs in Hin Namno National Biodiversity Conservation area, Laos, on Vietnam border at about 17°30′N according to Nadler (1998).
STATUS: CITES – Appendix II; IUCN – Data Deficient as *T. francoisi ebenus*.
COMMENTS: *T. francoisi* species group. Described as a subspecies of *T. auratus* by Brandon-Jones (1995), but raised to species rank and transferred to the *T. francoisi* group by Groves (2001c).

*Trachypithecus francoisi* (Pousargues, 1898). Bull. Mus. Hist. Nat. Paris, 4:319.
COMMON NAME: François's Langur.
TYPE LOCALITY: China, Kwangsi, Lungchow.
DISTRIBUTION: N Vietnam, C Laos, Kwangsi (China).
STATUS: CITES – Appendix II; U.S. ESA – Endangered; IUCN – Vulnerable.
COMMENTS: *T. francoisi* species group. The taxa *poliocephalus*, *hatinhensis*, *laotum* and *delacouri*, generally placed as subspecies of *francoisi*, were raised to specific rank by Brandon-Jones (1995).

*Trachypithecus geei* Khajuria, 1956. Ann. Mag. Nat. Hist., ser. 12, 9:86.
COMMON NAME: Gee's Golden Langur.
TYPE LOCALITY: India, Assam, Goalpara Dist., Jamduar Forest Rest House, east bank of Sankosh River.
DISTRIBUTION: Between Sankosh and Manas Rivers, Indo-Bhutan border (on both sides).
STATUS: CITES – Appendix I; U.S. ESA and IUCN – Endangered.
SYNONYMS: **bhutanensis** Wangchuk, Inouye and Hare, 2003.
COMMENTS: *T. pileatus* species group. For authorship of *geei*, see Biswas (1967).

*Trachypithecus germaini* (Milne-Edwards, 1876). Bull. Soc. Philom., (6)11:8.
COMMON NAME: Indochinese Lutung.
TYPE LOCALITY: Cochin-china and Cambodia.
DISTRIBUTION: Thailand and Burma (north of the peninsula), Cambodia, Vietnam, to 15°N.
STATUS: CITES – Appendix II; IUCN – Data Deficient as *T. villosus* ? *germaini* and *T. v. caudalis*.
SYNONYMS: *koratensis* (Kloss, 1919); *mandibularis* (Kloss, 1916); *margarita* (Elliot, 1909); **caudalis** (Dao, 1977).

COMMENTS: *T. cristatus* species group. Separated from *cristatus* by Groves (2001*c*).

*Trachypithecus hatinhensis* (Dao, 1970). Mitt. Zool. Mus. Berlin, 46:61.
    COMMON NAME: Hatinh Langur.
    TYPE LOCALITY: Vietnam, Ha-tinh, Xom-cuc.
    DISTRIBUTION: Vietnam, Quang Binh and neighbouring regions.
    STATUS: CITES – Appendix II; U.S. ESA – Endangered as included in *T. francoisi*; IUCN –
        Endangered as *T. francoisi hatinhensis*.
    COMMENTS: *T. francoisi* species group. Considered a species separate from *francoisi* by
        Brandon-Jones (1995).

*Trachypithecus johnii* (J. Fischer, 1829). Synopsis Mamm., p. 25.
    COMMON NAME: Nilgiri Langur.
    TYPE LOCALITY: India, Tellicherry.
    DISTRIBUTION: S India.
    STATUS: CITES – Appendix II; IUCN – Vulnerable.
    SYNONYMS: *cucullatus* (Fischer, 1829); *jubatus* (Wagner, 1839); ?*leonina* (Shaw, 1800).
    COMMENTS: *T. vetulus* species group. Often referred to subgenus *Kasi*, see Szalay and Delson
        (1979:402).

*Trachypithecus laotum* (Thomas, 1911). Ann. Mag. Nat. Hist., ser. 9, 7:181.
    COMMON NAME: Laotian Langur.
    TYPE LOCALITY: Laos, Ban Na São, on the Mekong at 17°30′N.
    DISTRIBUTION: C Laos.
    STATUS: CITES – Appendix II; U.S. ESA– Endangered as included in *T. francoisi*; IUCN – Data
        Deficient.
    COMMENTS: *T. francoisi* species group. Considered a species separate from *francoisi* by
        Brandon-Jones (1995).

*Trachypithecus obscurus* (Reid, 1837). Proc. Zool. Soc. Lond., 1837:14.
    COMMON NAME: Dusky Leaf-monkey.
    TYPE LOCALITY: Malaysia, Malacca.
    DISTRIBUTION: S Thailand and Malay Peninsula, and small adjacent islands.
    STATUS: CITES – Appendix II; IUCN – Lower Risk (lc), subspecies not evaluated.
    SYNONYMS: *leucomystax* (Müller and Schlegel, 1841); ***carbo*** (Thomas and Wroughton, 1909);
        *corvus* (Miller, 1913); ***flavicauda*** (Elliot, 1910); *corax* Pocock, 1934; *ruhei* (Knottnerus-
        Meyer, 1933); *smithii* (Kloss, 1916); ***halonifer*** (Cantor, 1845); ***sanctorum*** (Elliot, 1910);
        ***seimundi*** (Chasen, 1940); ***styx*** (Kloss, 1911).
    COMMENTS: *T. obscurus* species group.

*Trachypithecus phayrei* (Blyth, 1847). J. Asiat. Soc. Bengal, 16:733.
    COMMON NAME: Phayre's Leaf-monkey.
    TYPE LOCALITY: Burma, Arakan.
    DISTRIBUTION: Laos, Burma, C Vietnam, C and N Thailand, Yunnan (China).
    STATUS: CITES – Appendix II; IUCN – Lower Risk (lc).
    SYNONYMS: *barbei* (Blyth, 1863) [not of Blyth, 1847]; *holotephreus* (Anderson, 1878);
        *melamera* (Elliot, 1909); ***crepuscula*** (Elliot, 1909); *argenteus* (Kloss, 1919); *wroughtoni*
        (Elliot, 1909); ***shanicus*** (Wroughton, 1917).
    COMMENTS: *T. obscurus* species group.

*Trachypithecus pileatus* (Blyth, 1843). J. Asiat. Soc. Bengal, 12:174.
    COMMON NAME: Capped Langur.
    TYPE LOCALITY: India, Assam.
    DISTRIBUTION: Assam, NW Burma (west of Chindwin River); E Bangladesh.
    STATUS: CITES – Appendix I; U.S. ESA – Endangered; IUCN – Endangered as *T. p. pileatus*,
        *T. p. brahma*, *T. p. durga*, and *T. p. tenebricus*.
    SYNONYMS: *argentatus* (Horsfield, 1851); ***brahma*** (Wroughton, 1916); ***durga*** (Wroughton,
        1916); *saturatus* (Hinton, 1923); ***tenebricus*** (Wroughton, 1915).
    COMMENTS: *T. pileatus* species group. Does not include *shortridgei* (see Groves, 2001*c*).

*Trachypithecus poliocephalus* (Pousargues, 1898). Bull. Mus. Hist. Nat. Paris, 4:319.
>    COMMON NAME: White-headed Langur.
>    TYPE LOCALITY: Vietnam, Cat Ba Isl (restricted by Brandon-Jones, 1995).
>    DISTRIBUTION: Cat Ba Isl (Vietnam); Guangxi (China).
>    STATUS: CITES – Appendix II; U.S. ESA– Endangered as included in *T. francoisi*; IUCN –
>        Critically Endangered as *T. p. poliocephalus* and *T. p. leucocephalus*.
>    SYNONYMS: **leucocephalus** Tan, 1955.
>    COMMENTS: *T. francoisi* species group. Considered a species separate from *francoisi* by
>        Brandon-Jones (1995). Subspecies *leucocephalus* was considered to be a partially
>        albinistic population of *francoisi* by Brandon-Jones (1995), but (provisionally) a
>        subspecies of *poliocephalus* by Groves (2001c).

*Trachypithecus shortridgei* (Wroughton, 1915). J. Bombay Nat. Hist. Soc., 24:56.
>    COMMON NAME: Shortridge's Langur.
>    TYPE LOCALITY: Burma, Homalin (upper Chindwin).
>    DISTRIBUTION: Burma, east of Chindwin River; Gongshan (Yunnan, China).
>    STATUS: CITES – Appendix I; U.S. ESA – Endangered as *T. pileatus shortridgei*.
>    SYNONYMS: *belliger* (Wroughton, 1915).
>    COMMENTS: *T. pileatus* species group. Separated from *pileatus* by Groves (2001c).

*Trachypithecus vetulus* (Erxleben, 1777). Syst. Regni Anim., p. 24.
>    COMMON NAME: Purple-faced Langur.
>    TYPE LOCALITY: Sri Lanka, Hill country of South.
>    DISTRIBUTION: Sri Lanka.
>    STATUS: CITES – Appendix II; U.S. ESA – Threatened as *Presbytis senex*; IUCN – Endangered as
>        *T. v. vetulus*, *T. v. monticola*, *T. v. nestor*, and *T. v. philbricki*.
>    SYNONYMS: *cephalopterus* (Boddaert, 1785); *fulvogriseus* (Desmoulins, 1825); *kelaarti*
>        (Schlegel, 1876); *kephalopterus* (Zimmermann, 1780); *latibarba* (Temminck, 1807);
>        *latibarbatus* (É. Geoffroy, 1812); *leucoprymnus* (Otto, 1825); *porphyrops* (Link, 1795);
>        *purpuratus* (Kerr, 1792); *veter* (Shaw, 1800); **monticola** (Kelaart, 1850); *ursinus* (Blyth,
>        1851); **nestor** (Bennett, 1833); *phillipsi* (Hinton, 1923); **philbricki** (Phillips, 1927); *harti*
>        (Deraniyagala, 1955); **unassigned:** *albinus* (Kelaart, 1851); *senex* (Erxleben, 1777).
>    COMMENTS: *T. vetulus* species group. Type of subgenus *Kasi*; see Szalay and Delson
>        (1979:402). On the use of *vetulus*, instead of the previously more commonly used
>        *senex*, for the species, see Napier (1985:72).

**Superfamily Hominoidea** Gray, 1825. Ann. Philos., n.s., 10:344.

**Family Hylobatidae** Gray, 1871. Cat. Monkeys, Lemurs, Fruit-eating Bats Brit. Mus., p. 4.
>    COMMENTS: Vaughan (1978:39-40) included this family in Pongidae (which is here considered a
>        part of Hominidae); but see Delson and Andrews (1975:441) and Thenius (1981). Szalay
>        and Delson (1979:461), McKenna and Bell (1997), and Goodman et al. (1998) included
>        Hylobatidae in Hominidae. The family is usually awarded a single genus (*Hylobates*),
>        which is divided into four subgenera, but Groves (2001c:289) and Roos and Geissmann
>        (2001) considered that they should probably be elevated to full genera, but Groves
>        (2001c) did not take this final step because of the nomenclature problem of *Bunopithecus*.
>        It seems, however, undesirable that a problem of nomenclature should be allowed to
>        obstruct a desirable taxonomic change. Goodman et al. (1998) separated *Symphalangus*
>        and *Hylobates* as full genera (they had no material for the other two genera/subgenera).

*Bunopithecus* Matthew and Granger, 1923. Bull. Amer. Mus. Nat. Hist., 48:588.
>    TYPE SPECIES: *Bunopithecus sericus* Matthew and Granger, 1923 (a fossil species).
>    SYNONYMS: *Hoolock* Haimoff et al., 1984 [*nomen nudum*].
>    COMMENTS: This genus is being renamed and redefined (Mootnick and Groves, in prep.), as
>        *Bunopithecus sericus* is outside the modern gibbon clade and the Hoolock gibbon does not
>        belong to the same genus (Groves, 2001c); this was the reason why Groves (2001c) did
>        not take the step of elevating the subgenera of *Hylobates* to generic rank.

*Bunopithecus hoolock* (Harlan, 1834). Trans. Am. Philos. Soc., 4:52.
    COMMON NAME: Hoolock Gibbon.
    TYPE LOCALITY: India, Assam, Garo Hills.
    DISTRIBUTION: Between the Brahmaputra and Salween Rivers in Assam (India), Burma, and
        Yunnan (China).
    STATUS: CITES – Appendix I; U.S. ESA – Endangered as included in *Hylobates*; IUCN –
        Endangered as *B. h. hoolock* and *B. h. leuconedys*.
    SYNONYMS: *choromandus* Ogilby, 1827; *fuscus* Winslow Lewis, 1834; *golock* (Bechstein, 1795)
        [suppressed by Int. Comm. Zool. Nomencl. (1982), Opinion 1219]; *hulock* Lesson,
        1840; *scyritus* Ogilby, 1840; **leuconedys** Groves, 1967.

*Hylobates* Illiger, 1811. Prodr. Syst. Mamm. Avium., p. 67.
    TYPE SPECIES: *Homo lar* Linnaeus, 1771.
    SYNONYMS: *Brachiopithecus* Sénéchal, 1839 [in part]; *Brachitanytes* Schultz, 1932; *Cheiron*
        Burnett, 1829; *Gibbon* Zimmermann, 1777 [Rejected by Int. Comm. Zool. Nomencl.
        (1954), Opinion 257]; *Laratus* Gray, 1821; *Methylobates* Ameghino, 1882 [in part].
    COMMENTS: Includes *Symphalangus* according to Anderson (1967:175); and *Nomascus*
        according to Corbet and Hill (1980:91); but these should be given generic rank (Groves,
        2001*c*). Revised by Groves (1972*b*). Reviewed by Marshall and Marshall (1976).

*Hylobates agilis* F. Cuvier, 1821. *In* É. Geoffroy and F. Cuvier, Hist. Nat. Mammifères, pt. 2, 3(32):1-3,
    "Wouwou".
    COMMON NAME: Agile Gibbon.
    TYPE LOCALITY: Indonesia, W Sumatra.
    DISTRIBUTION: Malay Peninsula from the Mudah and Thepha Rivers on the north to the
        Perak and Kelanton Rivers on the south; Sumatra (Indonesia), SE of Lake Toba and the
        Singkil River.
    STATUS: CITES – Appendix I; U.S. ESA – Endangered; IUCN – Lower Risk (nt) as *H. a. agilis*
        and *H. a. unko*.
    SYNONYMS: *albo griseus* Ludeking, 1862; *albo nigrescens* Ludeking, 1862; *rafflei* É. Geoffroy,
        1828; *unko* Lesson, 1829.
    COMMENTS: Not a subspecies of *lar*. Includes *albibarbis* according to Marshall and Marshall
        (1976), but this was considered a separate species by Groves (2001*c*).

*Hylobates albibarbis* Lyon, 1911. Proc. U. S. Nat. Mus., 40:142.
    COMMON NAME: Bornean White-bearded Gibbon.
    TYPE LOCALITY: Indonesia, W Borneo, Sukadana.
    DISTRIBUTION: SW Borneo, south of Kapuas River and W of Barito River.
    STATUS: CITES – Appendix I; U.S. ESA – Endangered; IUCN – Lower Risk (nt) as *H. agilis*
        *albibarbis*.
    COMMENTS: A subspecies of *agilis* according to Marshall and Marshall (1976), a subspecies of
        *muelleri* according to Groves (1974), a distinct species according to Groves (2001*c*).

*Hylobates klossii* (Miller, 1903). Smithson. Misc. Coll., 45:70.
    COMMON NAME: Kloss's Gibbon.
    TYPE LOCALITY: Indonesia, West Sumatra, S Pagai Isl.
    DISTRIBUTION: Mentawai Isls (Indonesia).
    STATUS: CITES – Appendix I; U.S. ESA – Endangered; IUCN – Vulnerable.

*Hylobates lar* (Linnaeus, 1771). Mantissa Plantarum, p. 521.
    COMMON NAME: Lar Gibbon.
    TYPE LOCALITY: Malaysia, Malacca (restricted by Kloss, 1929).
    DISTRIBUTION: Between the Salween and Mekong Rivers from S Yunnan (China) south to
        the Mun River (Thailand) and the Mudah and Thepha Rivers on the Malay Peninsula;
        S Malay Peninsula south of the Perak and Kelantan Rivers; Sumatra (Indonesia) NW of
        Lake Toba and the Singkil River; E and S Burma.
    STATUS: CITES – Appendix I; U.S. ESA – Endangered; IUCN – Critically Endangered as *H. l.*
        *yunnanensis*, Lower Risk (nt) as *H. l. lar*, *H. l. carpenteri*, *H. l. entelloides*, and *H. l. vestitus*.
    SYNONYMS: *albimana* (Vigors and Horsfield, 1828); *longimana* (Schreber, 1774); *variegatus*

(É. Geoffroy, 1812); *varius* (Latreille, 1801); ***carpenteri*** Groves, 1968; ***entelloides***
I. Geoffroy, 1842; ***vestitus*** Miller, 1942; *yunnanensis* Ma and Y. Wang, 1986.

*Hylobates moloch* (Audebert, 1798). Hist. Nat. Singes Makis, 1st fasc., sect. 2, pl. 2.
COMMON NAME: Silvery Javan Gibbon.
TYPE LOCALITY: Indonesia, W Java, Mt. Salak (restricted by Sody, 1949*b*).
DISTRIBUTION: Java (Indonesia).
STATUS: CITES – Appendix I; U.S. ESA – Endangered; IUCN – Critically Endangered as
*H. m. moloch* and *H. m. pongoalsoni*.
SYNONYMS: *cinereus* Latreille, 1804; *javanicus* Matschie, 1893; *leucisca* (Schreber, 1799);
*pongoalsoni* Sody, 1949.
COMMENTS: See Andayani et al. (2001), who reviewed the phylogeography of the species.

*Hylobates muelleri* Martin, 1841. Nat. Hist. Mamm. Anim., p. 444.
COMMON NAME: Müller's Bornean Gibbon.
TYPE LOCALITY: Indonesia, Kalimantan, "Southeast Borneo"; restricted by Lyon (1911:142).
DISTRIBUTION: Borneo from the N bank of the Kapuas River clockwise around the island to
the east bank of the Barito River.
STATUS: CITES – Appendix I; U.S. ESA – Endangered; IUCN – Lower Risk (nt) as
*H. m. muelleri*, *H. m. abbotti*, and *H. m. funereus*.
SYNONYMS: ***abbotti*** Kloss, 1929; ***funereus*** I. Geoffroy, 1850.
COMMENTS: There is a wide hybrid zone with *H. albibarbis* in C Borneo (Marshall and
Sugardjito, 1986).

*Hylobates pileatus* (Gray, 1861). Proc. Zool. Soc. Lond., 1861:136.
COMMON NAME: Pileated Gibbon.
TYPE LOCALITY: Cambodia.
DISTRIBUTION: SE Thailand and Cambodia south of the Mun and Takhrong Rivers and west
of the Mekong River.
STATUS: CITES – Appendix I; U.S. ESA – Endangered; IUCN – Vulnerable.

*Nomascus* Miller, 1933. J. Mamm., 14:159.
TYPE SPECIES: *Hylobates leucogenys* Ogilby, 1840.
COMMENTS: Separated as a genus distinct from *Hylobates* (Groves, 2001*c*).

*Nomascus concolor* (Harlan, 1826). J. Acad. Nat. Sci. Philadelphia, ser. 5, 4:231.
COMMON NAME: Black Crested Gibbon.
TYPE LOCALITY: Vietnam, Tonkin.
DISTRIBUTION: E of the Mekong River, and a small enclave W of the Mekong, in S Yunnan
(China), N and WC Laos, and in Vietnam to Red River; and an isolated region round
Ban Nam Khueung, Laos.
STATUS: CITES – Appendix I; U.S. ESA – Endangered; IUCN – Critically Endangered as *N. c.*
*furvogaster*, *N. c. jingdongensis*, and *N. nasutus*, Endangered as *N. c. concolor* and *N. c. lu*.
SYNONYMS: *harlani* Lesson, 1827; *henrici* de Pousargues, 1897; *niger* Ogilby, 1840; ***furvogaster***
Ma and Y. Wang, 1986; ***jingdongensis*** Ma and Y. Wang, 1986; ***lu*** Delacour, 1951; ?
***nasutus*** Kunkel d'Herculais, 1884.
COMMENTS: Does not include *hainanus*, see Groves (2001*c*). The status of *nasutus* is unclear;
it may be a senior synonym for *hainanus*.

*Nomascus gabriellae* Thomas, 1909. Ann. Mag. Nat. Hist., ser. 8, 4:112.
COMMON NAME: Red-cheeked Gibbon.
TYPE LOCALITY: Vietnam, Langbian.
DISTRIBUTION: S Laos, S Vietnam from 15°30′N, E Cambodia.
STATUS: CITES – Appendix I; U.S. ESA – Endangered; IUCN – Vulnerable.
COMMENTS: Separated from *leucogenys* by Groves and Y. Wang (1989). Does not include *siki*,
see Groves (2001*c*).

*Nomascus hainanus* (Thomas, 1892). Ann. Mag. Nat. Hist., ser. 6, 9:145.
COMMON NAME: Hainan Gibbon.
TYPE LOCALITY: China: Hainan Isl.

DISTRIBUTION: Hainan Isl (China); Hoa Binh and Cao Bang Provs., Vietnam.
STATUS: CITES – Appendix I; U.S. ESA – Endangered; IUCN – Critically Endangered as
       *N. nasutus hainanus*.
COMMENTS: Separated from *concolor* by Groves (2001c). It is possible that *nasutus* may be an
       earlier name for the mainland population of this species.

*Nomascus leucogenys* Ogilby, 1840. Proc. Zool. Soc. Lond., 1840:20.
   COMMON NAME: Northern White-cheeked Gibbon.
   TYPE LOCALITY: Laos, Muang Khi (Fooden, 1987).
   DISTRIBUTION: SW Yunnan (China) to 19°N in Vietnam and Laos.
   STATUS: CITES – Appendix I; U.S. ESA and IUCN – Endangered.
   COMMENTS: Separated from *concolor* by Dao (1983) and Ma and Wang (1986).

*Nomascus siki* Delacour, 1951. Mammalia, 15:122.
   COMMON NAME: Southern White-cheeked Gibbon.
   TYPE LOCALITY: Vietnam, Thua Luu.
   DISTRIBUTION: C Vietnam and Laos, from 15°45' to 20°N.
   STATUS: CITES – Appendix I; U.S. ESA – Endangered; IUCN – Data Deficient as *N. leucogenys*
       *siki*.
   COMMENTS: Separated from *leucogenys* and *gabriellae* by Groves (2001c).

*Symphalangus* Gloger, 1841. Gemeinn. Naturg., 1:34.
   TYPE SPECIES: *Simia syndactylus* Raffles, 1821.
   SYNONYMS: *Siamanga* Gray, 1843.
   COMMENTS: Probably a genus distinct from *Hylobates* (Goodman et al., 1998; Groves, 2001c).

*Symphalangus syndactylus* (Raffles, 1821). Trans. Linn. Soc. Lond., 13:241.
   COMMON NAME: Siamang.
   TYPE LOCALITY: Indonesia, W Sumatra, Bengkeulen.
   DISTRIBUTION: Barisan Mountains of Sumatra (Indonesia); mountains of Malay Peninsula
       south of Perak River.
   STATUS: CITES – Appendix I; U.S. ESA – Endangered; IUCN – Lower Risk (nt) as *S. s.*
       *syndactylus* and *S. s. continentis*.
   SYNONYMS: *continentis* Thomas, 1908; *gibbon* (C. Miller, 1779); *subfossilis* Hooijer, 1960; *volzi*
       (Pohl, 1911).

**Family Hominidae** Gray, 1825. Ann. Philos., n.s., 10:344.
   SYNONYMS: Pongidae Elliot, 1913.
   COMMENTS: For combining all genera in one family, see Groves (1989). The genera are placed in
       two subfamilies by Groves (2001c): Ponginae (*Pongo* alone), and Homininae (*Gorilla*,
       *Homo*, *Pan*). McKenna and Bell (1997) included Hylobatidae in addition, as a subfamily,
       and within the Homininae recognized two living tribes, Pongini and Hominini;
       Goodman et al. (1998) recognized gibbons only as a tribe (Hylobatini), with the other
       three genera as a separate tribe (Hominini), divided into subtribes Pongina and
       Hominina; they included *Pan* in *Homo*, and Watson et al. (2001) included both *Pan* and
       *Gorilla* in *Homo*.

*Gorilla* I. Geoffroy, 1852. C. R. Acad. Sci. Paris, 36:933.
   TYPE SPECIES: *Troglodytes gorilla* Savage, 1847.
   SYNONYMS: *Pseudogorilla* Elliot, 1913.
   COMMENTS: A subgenus of *Pan* according to Tuttle (1967); but see Groves (1989) who stated
       that they are not closely related. Included in *Homo* by Watson et al. (2001). Included in
       Hominini by McKenna and Bell (1997), but as a subtribe Gorillina, separate from
       Hominina (=*Pan* and *Homo*).

*Gorilla beringei* Matschie, 1903. Sber. Ges. naturf. Fr. Berlin, 257.
   COMMON NAME: Eastern Gorilla.
   TYPE LOCALITY: Rwanda, Mt. Sabinyo.
   DISTRIBUTION: N and E Dem. Rep. Congo, SW Uganda, N Rwanda.

STATUS: CITES – Appendix I and U.S. ESA – Endangered as included in *G. gorilla*; IUCN – Critically Endangered as *G. g. beringei*, Endangered as *G. g. graueri*.

SYNONYMS: *beringeri* Matschie, 1903 [*lapsus*]; *mikenensis* Lönnberg, 1917; **graueri** Matschie, 1914; *manyema* Rothschild, 1908 [*lapsus*]; *rex-pygmaeorum* Schwarz, 1927.

COMMENTS: Separated as a full species by Groves (2001*c*).

*Gorilla gorilla* (Savage, 1847). Boston J. Nat. Hist., 5:417.

COMMON NAME: Western Gorilla.

TYPE LOCALITY: Gabon, Gabon Estuary, Mpongwe country.

DISTRIBUTION: SE Nigeria, Cameroon, Rio Muni (Equatorial Guinea), Republic of Congo, SW Central African Republic, Gabon.

STATUS: CITES – Appendix I; U.S. ESA – Endangered; IUCN – Critically Endangered as *G. g. diehli*, Endangered as *G. g. gorilla*.

SYNONYMS: *adrotes* (Mayer, 1856); *africanus* (Mayer, 1856); *castaneiceps* Slack, 1862; *ellioti* (Frechkop, 1943); *gigas* Haeckel, 1903; *gina* I. Geoffroy, 1855; *halli* Rothschild, 1927; *hansmeyeri* Matschie, 1914; *jacobi* Matschie, 1905; *matschiei* Rothschild, 1905; *mayêma* Alix and Bouvier, 1877; *savagei* (Owen, 1848); *schwartzi* Fritze, 1912; *uellensis* Schouteden, 1927; *zenkeri* Matschie, 1914; **diehli** Matschie, 1904.

COMMENTS: The author of the name is Savage, not Savage and Wyman (see Groves, 2001*c*:301). The status of *diehli* was reviewed in detail by Sarmiento and Oates (2000).

*Homo* Linnaeus, 1758. Syst. Nat., 10th ed., 1:20.

TYPE SPECIES: *Homo sapiens* Linnaeus, 1758.

SYNONYMS: *Africanthropus* Weinert, 1938; *Anthropus* Boyd Dawkins, 1926; *Archanthropus* Arldt, 1915; *Atlanthropus* Arambourg, 1954; *Cyphanthropus* Pycraft, 1928; *Europanthropus* Wüst, 1950; *Javanthropus* Oppenoorth, 1932; *Maueranthropus* Montandon, 1943; *Meganthropus* Weidenreich, 1944; *Nipponanthropus* Hasebe, 1948; *Notanthropus* Sergi, 1911; *Palaeanthropus* Bonarelli, 1907; *Pithecanthropus* Dubois, 1894; *Praehomo* von Eickstedt, 1932; *Proanthropus* Wilser, 1900; *Pseudhomo* Ameghino, 1909; *Sinanthropus* Black and Zdansky, 1927; *Tchadanthropus* Coppens, 1965; *Telanthropus* Broom and Robinson, 1949.

COMMENTS: Included with *Pan* and the fossil taxon *Australopithecus* in subtribe Hominina by McKenna and Bell (1997). Includes *Pan* as a subgenus according to Goodman et al. (1998); according to M. Goodman et al. (2001), all fossil representatives of the human lineage (including *Australopithecus*) would be synonyms of *Homo*.

*Homo sapiens* Linnaeus, 1758. Syst. Nat., 10th ed., 1:20.

COMMON NAME: Human.

TYPE LOCALITY: Sweden, Uppsala.

DISTRIBUTION: Cosmopolitan.

STATUS: CITES – Appendix II as Order Primates; absolutely not endangered.

SYNONYMS: *aethiopicus* Bory de St. Vincent, 1825; *americanus* Bory de St. Vincent, 1825; *arabicus* Bory de St. Vincent, 1825; *aurignacensis* Klaatsch and Hauser, 1910; *australasicus* Bory de St. Vincent, 1825; *cafer* Bory de St. Vincent, 1825; *capensis* Broom, 1917; *columbicus* Bory de St. Vincent, 1825; *cro-magnonensis* Gregory, 1921; *drennani* Kleinschmidt, 1931; *eurafricanus* (Sergi, 1911); *grimaldiensis* Gregory, 1921; *grimaldii* Lapouge, 1906; *hottentotus* Bory de St. Vincent, 1825; *hyperboreus* Bory de St. Vincent, 1825; *indicus* Bory de St. Vincent, 1825; *japeticus* Bory de St. Vincent, 1825; *melaninus* Bory de St. Vincent, 1825; *monstrosus* Linnaeus, 1758 [unavailable]; *neptunianus* Bory de St. Vincent, 1825; ?*palestinus* McCown and Keith, 1932; *patagonus* Bory de St. Vincent, 1825; *priscus* Lapouge, 1899; *proto-aethiopicus* Giuffrida-Ruggeri, 1915; *scythicus* Bory de St. Vincent, 1825; *sinicus* Bory de St. Vincent, 1825; *spelaeus* Lapouge, 1899; *troglodytes* Linnaeus, 1758 [*nomen oblitum*]; *wadjakensis* Dubois, 1921.

COMMENTS: Most of the synonyms have fossil specimens as their type specimens; Bory de St. Vincent's names refer to living geographic varieties of modern humans.

*Pan* Oken, 1816. Lehrb. Naturgesch., ser. 3, 2:xi.

TYPE SPECIES: *Simia troglodytes* Blumenbach, 1775.

SYNONYMS: *Anthropithecus* Lesson, 1840; *Anthropopithecus* de Blainville, 1838; *Bonobo* Tratz and

Heck, 1954; *Chimpansee* Voigt, 1831; *Engeco* Haeckel, 1866; *Fsihego* de Pauw, 1905; *Hylanthropus* Gloger, 1841; *Mimetes* Anon, 1820; *Pongo* Haeckel, 1866 [not of Lacépède, 1799]; *Pseudanthropos* Reichenbach, 1860; *Satyrus* Mayer, 1856; *Theranthropus* Brookes, 1828; *Troglodytes* É. Geoffroy, 1812 [not of Vieillot, 1806].

COMMENTS: Included with *Homo* in subtribe Hominina by McKenna and Bell (1997). A subgenus of *Homo* according to M. Goodman et al. (1998, 2001); Watson et al. (2001) also included it in *Homo*. In accordance with Opinion 1368 (International Commission on Zoological Nomenclature, 1985c), *Pan* is used instead of *Chimpansee*. Reviewed by Hill (1969).

*Pan paniscus* Schwartz, 1929. Rev. Zool. Bot. Afr., 16:4.

COMMON NAME: Bonobo.
TYPE LOCALITY: Dem. Rep. Congo, south of the upper Maringa River, 30 km south of Befale.
DISTRIBUTION: Congo Basin of Dem. Rep. Congo, on south side of Congo River.
STATUS: CITES – Appendix I; U.S. ESA and IUCN – Endangered.

*Pan troglodytes* (Blumenbach, 1775). De generis humani varietate nativa, p. 37.

COMMON NAME: Common Chimpanzee.
TYPE LOCALITY: Gabon, Mayoumba.
DISTRIBUTION: S Cameroon; Gabon; S Republic of Congo; Uganda; W Tanzania; E and N Dem. Rep. Congo; W Central African Republic; Guinea to W Nigeria, south to Congo River in W Africa.
STATUS: CITES – Appendix I; U.S. ESA – Endangered in the wild, Threatened in captivity; IUCN – Endangered as *P. t. verus*, *P. t. troglodytes*, *P. t. schweinfurthii*, and *P. t. vellerosus*.
SYNONYMS: *africanus* Oken, 1816 [unavailable]; *angustimanus* (Brehm, 1876) [*nomen nudum*]; *aubryi* (Gratiolet and Alix, 1866); *calvus* (du Chaillu, 1860); *chimpanse* (Mayer, 1856); *fuliginosus* (Schaufuss, 1870); *fuscus* (Meyer, 1895); *heckii* (Koch, 1932); *?jocko* (Kerr, 1792); *koolookamba* (du Chaillu, 1860); *lagaros* (Mayer, 1856); *leucoprymnus* (Lesson, 1831); *mafuca* (Haeckel, 1903) [*nomen nudum*]; *niger* (É. Geoffroy, 1812); *ochroleucus* (Matschie, 1914); *pan* (Lesson, 1840); *?pongo* (Kerr, 1792); *pusillus* (Matschie, 1919); *raripilosus* (Rothschild, 1905); *reuteri* (Matschie, 1914); *satyrus* (Linnaeus, 1758) [in part; suppressed by Opinion 114 of the Int. Comm. Zool. Nomencl., 1929c]; *schneideri* (Matschie, 1919); *tschego* (Duvernoy, 1855); **schweinfurthii** (Giglioli, 1872); *adolfi-friederici* (Matschie, 1912); *calvescens* (Matschie, 1914); *castanomale* (Matschie, 1914); *cottoni* (Matschie, 1912); *graueri* (Matschie, 1914); *ituricus* (Matschie, 1912); *ituriensis* (de Pauw, 1905); *livingstonii* (Selenka, 1899) [*nomen nudum*]; *marungensis* (Noack, 1887); *nahani* (Matschie, 1912); *pfeifferi* (Matschie, 1914); *purschei* (Matschie, 1914); *schubotzi* (Matschie, 1914); *steindachneri* (Lorenz, 1914); *yambuyae* (Matschie, 1912); **vellerosus** (Gray, 1862); *ellioti* (Matschie, 1914); *oertzeni* (Matschie, 1914); *papio* (Matschie, 1919); **verus** Schwarz, 1934.
COMMENTS: For authorship of this name, see the International Commission on Zoological Nomenclature (1988). Watson et al. (2001:315) used the name *Homo niger*, on the grounds that *Homo troglodytes* would be a secondary homonym of *Homo troglodytes* Linnaeus, 1758 (=*Homo sapiens*); but the new edition of the *Code* (in force since January 1st, 2000) permits a senior homonym to be rejected if unused since 1899 (Art. 23.9.), so *Homo troglodytes* would be an acceptable combination for the chimpanzee. The species has been most recently reviewed by Jones et al. (1996).

*Pongo* Lacépède, 1799. Tabl. Div. Subd. Orders Genres Mammifères, p. 4.

TYPE SPECIES: *Pongo borneo* Lacépède, 1799 (= *Simia pygmaeus* Linnaeus, 1760).
SYNONYMS: *Faunus* Oken, 1816 [unavailable]; *Lophotus* Fischer, 1813; *Macrobates* Bilberg, 1828; *Satyrus* Lesson, 1840.

*Pongo abelii* Lesson, 1827. Man. Mamm. 32.

COMMON NAME: Sumatran Orangutan.
TYPE LOCALITY: Indonesia, Sumatra.
DISTRIBUTION: Sumatra, NW of Lake Toba (Indonesia).
STATUS: CITES – Appendix I and U.S. ESA as included in *P. pygmaeus*; IUCN – Critically Endangered.

SYNONYMS: *abongensis* (Selenka, 1896); *bicolor* (I.Geoffroy, 1843); *deliensis* (Selenka, 1896); *gigantica* (Pearson, 1841); *langkatensis* (Selenka, 1896).

COMMENTS: On the nomenclature, see Groves and Holthuis (1985). Reviewed by Groves (1971*a*, Mammalian Species, 4). Considered a species separate from *P. pygmaeus* by Groves (2001*c*), who summarized several previous sources suggesting this.

*Pongo pygmaeus* (Linnaeus, 1760). Amoenit. Acad., 6:68.

COMMON NAME: Bornean Orangutan.

TYPE LOCALITY: Indonesia, Kalimantan, Landak River.

DISTRIBUTION: Borneo, except the southeast.

STATUS: CITES – Appendix I; U.S. ESA and IUCN – Endangered as *P. pygmaeus*, *P. p. pygmaeus*, and *P. p. wurmbii*.

SYNONYMS: *agris* (Schreber, 1799); *batangtuensis* (Selenka, 1896); *borneensis* Röhrer-Ertl, 1983; *borneo* (Lacépède, 1799); *dadappensis* (Selenka, 1896); *genepaiensis* (Selenka, 1896); *landakkensis* (Selenka, 1896); *rantaiensis* (Selenka, 1896); *rufus* (Lesson, 1840); *satyrus* (Linnaeus, 1766) [in part; suppressed by Opinion 114 of the Int. Comm. Zool. Nomenclature, 1929*c*]; *skalauensis* (Selenka, 1896); *sumatranus* (Mayer, 1856); *tuakensis* (Selenka, 1896); *wallichii* (Gray, 1871); **morio** (Owen, 1837); *brookei* (Blyth, 1853); *curtus* (Blyth, 1855); *owenii* (Blyth, 1853); **wurmbii** (Tiedemann, 1808).

COMMENTS: On the nomenclature, see Groves and Holthuis (1985). Reviewed by Groves (1971*a*, Mammalian Species, 4).

# ORDER LAGOMORPHA
by Robert S. Hoffmann and Andrew T. Smith

**ORDER LAGOMORPHA** Brandt, 1855.
SYNONYMS: Duplicidentata Illiger, 1811; Leporida Averianov, 1999; Neolagomorpha Averianov, 1999; Ochotonida Averianov, 1999; Palarodentia Haeckel, 1895.
COMMENTS: Relationships between this order and Rodentia have been disputed for over a century. Its early history was discussed by Simpson (1945), while Landry (1999) provided an overview of more recent literature on the subject, discussing many synapomorphies, mostly morphological, that support the concept of a Cohort Glires including both orders. Molecular sequence data also support the concept (Huchon et al., 1999).

**Family Ochotonidae** Thomas, 1897. Proc. Zool. Soc. Lond., 1896:1026 [1897].
SYNONYMS: Lagomina Gray, 1825; Lagomyidae Lilljeborg, 1866; Prolaginae Gureev, 1960.
COMMENTS: Revisions of the family include Gureev (1964), Corbet (1978c), and Erbajeva (1988, 1994). Other useful treatments include Allen (1938), Ellerman and Morrison-Scott (1951), Ognev (1940), Hall (1981), A. T. Smith et al. (1990), and Yu et al. (2000).

*Ochotona* Link, 1795. Beitr. Naturgesch., 2:74.
TYPE SPECIES: *Ochotona minor* Link, 1795 (= *Lepus dauuricus* Pallas, 1776).
SYNONYMS: *Abra* Gray, 1863 [not Lamarck, 1818]; *Abrama* Strand, 1928 [for *Abra* Gray]; *Argyrotona* Rekovetz, 1988; *Buchneria* Erbajeva, 1988; *Conothoa* Lyon, 1904; *Lagomys* G. Cuvier, 1800 [not Storr, 1780]; *Lagotona* Kretzoi, 1941; *Ogotoma* Gray, 1867; *Pika* Lacèpde, 1799 [= *Pica* Fischer, 1803]; *Tibetholagus* Argyropulo and Pidoplichko, 1939 [*nomen nudum*]; *Tibetolagus* Argyropulo, 1948 [for *Tibetholagus*].
COMMENTS: There have been no grounds for recognizing subgenera due to lack of a phylogenetic analysis of specific relationships within the genus. The subgeneric classifications published (e.g., Allen, 1938; Ellerman and Morrison-Scott, 1951; Erbajeva, 1988; Ognev, 1940) differ dramatically, even when based on the same distinguishing characteristics (morphology of dentition and of incisive and palatal foramina). However, a recent phylogenetic analysis (Yu et al., 2000) based on molecular sequencing, divides the genus into three groups. These were termed "shrub-steppe", "mountain" and "northern". The shrub steppe group includes 7 species that had previously been placed in subgenus *Ochotona*; those in the northern group (N=5) in subgenus *Pika*; and 7 species in the mountain group had been placed in either subgenus, though predominately in *Ochotona*. The oldest available name for this third subgenus might appear to be *Lagomys* Cuvier, 1800, but the name is unavailable since Palmer (1904:361) fixed the type as *Ochotona (Pika) alpina* Pallas. Moreover, both Cuvier's and Olgilby's (1839) *Lagomys* are junior homonyms of *Lagomys* Storr, 1780, which was a renaming of *Arctomys* Schreber, 1780, which in turn became a junior synonym of *Marmota* Blumenbach, 1779 (I. Ya. Pavlinov, pers. comm.). The earliest available name for the third subgenus thus becomes *Conothoa* Lyon, 1904, type species *O. roylei*. The name is a rearrangement of the letters in "Ochotona". Reviewed by Sokolov et al. (1994). Considerable confusion remains as to the placement of subspecies within species and the independence of some species.

*Ochotona alpina* (Pallas, 1773). Reise Prov. Russ. Reichs., 2:701.
COMMON NAME: Alpine Pika.
TYPE LOCALITY: "in Alpinus, rupestribus Sibiriae". Restricted by Ognev (1940:23) to Kazakhstan, Altai Mtns, Vostocho-Kazakhstansk Obl., Tigiretskoe Range, vic. of Tigiretskoe [110 km NNW Ust-Kamenogorsk]. Not Tigiretskoe, ESE Minusinsk, Krasnoyarsk Krai, Russia.
DISTRIBUTION: Sayan and Altai Mtns; Khangai, Kentei and associated ranges; upper Amur drainage (NW Kazakhstan, S Russia, NW Mongolia); N Xinjiang (China).
STATUS: IUCN – Lower Risk (lc). Isolated montane populations in Mongolia may be threatened (A. T. Smith et al., 1990).
SYNONYMS: *ater* Eversmann, 1842; *nitida* Hollister, 1912; ***changaica*** Ognev, 1940; ***cinereofusca*** (Schrenk, 1858); *scorodumovi* Skalon, 1935; ***sushkini*** Thomas, 1924.

COMMENTS: Subgenus *Pika*. Formerly included *hyperborea*; but see Ivanitskaya (1985) and Pavlinov and Rossolimo (1987). Sokolov and Orlov (1980:79) considered *hyperborea* a distinct species with a distribution overlapping that of *alpina* in the Khangai and Kentei Mtns, Mongolia. Separate specific status was supported by differences in chromosome numbers (Vorontsov and Ivanitskaya, 1973). The race *sushkini*, formerly assigned to *O. pallasi*, is a subspecies of *alpina* (see A. T. Smith et al., 1990, and references therein; see Niu et al., 2001, for contrary view). Does not include *collaris* or *princeps*, see Weston (1981). Formerly included *argentata*, but see Erbajeva (1997), Formozov (1997), and Formozov et al. (2004) who gave it full species status on the basis of morphology, chromosome number and acoustic behavior.

*Ochotona argentata* Howell, 1928. Proc. Biol. Soc. Wash., 41:116.
> COMMON NAME: Silver Pika.
> TYPE LOCALITY: "15 miles [24 km] north-northwest of Ninghsia [Yinchuan], northern Kansu [Gansu, now Ningxia Auton. Reg.], China"
> DISTRIBUTION: Restricted to the Helan Shan range, Ningxia, China (Formozov, 1997; A. T. Smith et al., 1990).
> STATUS: IUCN – Critically Endangered (see Formozov, 1997; A. T. Smith et al., 1990).
> SYNONYMS: *helanshanensis* Zheng, 1987 [in Wang, 1990].
> COMMENTS: Subgenus *Pika*. Formerly considered a subspecies of *O. alpina* (see comments therein). There is confusion as to whether *argentata* is the same as or different from *O. helanshanensis*. The two forms come from the same area, share similar pelage descriptions, and have similar body and skull measurements, thus appear to be synonyms (Formozov et al., 2004). See also *O. pallasi*.

*Ochotona cansus* Lyon, 1907. Smithson. Misc. Coll., 50:136.
> COMMON NAME: Gansu Pika.
> TYPE LOCALITY: "Taocheo, Kan-su, China" [Lintan, Gannan A.D., Gansu, China].
> DISTRIBUTION: C China (Gansu, Qinghai, Sichuan); isolated populations in Shaanxi and Shanxi.
> STATUS: IUCN – Lower Risk (lc); but the Shanxi subspecies *sorella*, isolated in the extreme NW of the species range (Yunshung Shan) is IUCN – Endangered and the subspecies *morosa* is IUCN – Data Deficient.
> SYNONYMS: **morosa** Thomas, 1912; **sorella** Thomas, 1908; **stevensi** Osgood, 1932.
> COMMENTS: Subgenus *Ochotona*. Büchner (1890) originally included this species in the quite different *O. roylei*, but in recent years it has usually been assigned to *O. thibetana* (Allen, 1938; Argyropulo, 1948; Corbet, 1978c; Ellerman and Morrison-Scott, 1951; Gureev, 1964; Honacki et al., 1982; Weston, 1982). Additional studies showed that *cansus* and *thibetana* are broadly sympatric, with distinct ecological niches, and morphological characters that do not intergrade (Feng and Kao, 1974; Feng and Zheng, 1985). The latter authors, without access to holotypes, assigned the race *morosa* to *thibetana*, but it is an isolated subspecies of *cansus* that is sympatric with *O. thibetana* in the Tsing Ling Shan, Shaanxi Province (A. T. Smith et al., 1990, and references therein). Recent phylogenetic analyses based on molecular sequencing also show that *morosa* is a synonym of *O. cansus* rather than *O. thibetana*, and that the two species are independent (Yu et al., 1997; Yu et al., 2000). May also include *annectens* Miller, 1911, which is usually considered a subspecies of *O. dauurica*; but Yu et al. (2000) treat *annectens* as an independent sister species of *O. cansus*; see comments under that species. Until more data become available, we prefer to leave *annectens* as a subspecies of *dauurica* (Pavlinov et al., 1995b).

*Ochotona collaris* (Nelson, 1893). Proc. Biol. Soc. Wash., 8:117.
> COMMON NAME: Collared Pika.
> TYPE LOCALITY: "about 200 miles [322 km] south of Fort Yukon, Alaska near the head of the Tanana River." [USA].
> DISTRIBUTION: WC Mackenzie, S Yukon, NW British Columbia (Canada); SE Alaska (USA).
> STATUS: IUCN – Lower Risk (lc) (MacDonald and Jones, 1987).
> COMMENTS: Subgenus *Pika*. Broadbooks (1965) and Youngman (1975) considered *collaris* and *princeps* conspecific. Corbet (1978c), following Argyropulo (1948) and Gureev

(1964), included *collaris* in *alpina*. A statistical reevaluation of craniometric data by Weston (1981) indicated that *collaris*, *princeps* and *alpina* are separate species; Hall (1981:286) also recognized *collaris* as a distinct species. *O. collaris* and *O. princeps* share similar chromosome numbers that differ sharply from those of *alpina* and *hyperborea* (Vorontsov and Ivanitskaya, 1973), but are similar to *pusilla* (Erbajeva, 1994).

*Ochotona curzoniae* (Hodgson, 1858). J. Asiat. Soc. Bengal, 1857, 26:207 [1858].

COMMON NAME: Plateau Pika

TYPE LOCALITY: "district of Chumbi", Chumbi Valley, Tibet, China.

DISTRIBUTION: Tibetan Plateau; adjacent Gansu, Qinghai, Sichuan (China), Sikkim (India) and E Nepal.

STATUS: IUCN – Lower Risk (lc); this species is the focus of widespread control efforts throughout its range, and has been eliminated locally (A. T. Smith et al., 1990; Smith and Foggin, 1999).

SYNONYMS: *melanostoma* (Büchner, 1890).

COMMENTS: Subgenus *Ochotona*. Includes *melanostoma*, but not *seiana* from Iran (*contra* Corbet, 1978c : 69); see A. T. Smith et al. (1990). Treated as a subspecies of *dauurica* by Ellerman and Morrison-Scott (1955) and Mitchell (1978), but it is considered a distinct species by the Chinese; see however Feng and Zheng (1985) and Feng et al. (1986). *O. curzoniae* and *O. dauurica* occur in geographic sympatry in Hainan County, Qinghai Province, China, and differ both morphologically (Feng and Zheng, 1985; Feng et al., 1986), chromosomally (Vorontsov and Ivanitskaya, 1973), electrophoretically (Zhou and Xia, 1981), and in their mitochondrial DNA (Yu et al., 1997; Yu et al., 2000). Its sister species is likely *O. nubrica* (Yu et al., 2000).

*Ochotona dauurica* (Pallas, 1776). Reise Prov. Russ. Reichs., 3:692.

COMMON NAME: Daurian Pika.

TYPE LOCALITY: "Vivit in campis, montiumque declivibus arenosis apricis, per totam Dauuriam . . ." Restricted by Ellerman and Morrison-Scott (1951:452) to "Kulusutai, Onon River, Eastern Siberia" [Chitinsk. Obl. Russia].

DISTRIBUTION: Steppes from Altai, Tuva, and Transbaikalia (Russia) through N China and Mongolia, south to Qinghai Province, China. Zhang et al. (1997), listed it from Henan and Hebei provinces, but this may be a *lapsus*.

STATUS: IUCN – Lower Risk (lc). Considered a pest, is intensively controlled in China; control in Russia has been much less intensive. Isolated populations around the margins of the Gobi Desert in China and Mongolia are very vulnerable (A. T. Smith et al., 1990).

SYNONYMS: *altaina* Thomas, 1911; *minor* Link, 1795; *ogotona* (Pallas, 1862); **annectens** Miller, 1911; **bedfordi** Thomas, 1908; *shaanxiensis* Xu and Wang, 1992; **mursavi** Bannikov, 1951.

COMMENTS: Subgenus *Ochotona*. The spelling of *dauurica* conforms to that of the original description. Formerly included *curzoniae* and *melanostoma*; see *curzoniae*, above. The retention of *annectens* as a subspecies is conservative; see comments under *cansus*. Inclusion of *shaanxiensis* as a synonym of *O. d. bedfordi* is provisional (see Wang and Xu, 1992). Ellerman and Morrison-Scott's type restriction is dubious because modern Kulusutai is south of the Onon River, at the NE end of Lake Baron-Torei. See Ognev (1940:62) and Allen (1938:551) for alternate type localities in the same general area.

*Ochotona erythrotis* (Büchner, 1890). Wiss. Result. Przewalski Cent. Asien Reisen. Zool. Th., B. I: Säugeth., p. 165.

COMMON NAME: Chinese Red Pika.

TYPE LOCALITY: Not specified; restricted by Allen (1938:535) to "Burchan-Budda", East Tibet, China.

DISTRIBUTION: E Qinghai, W Gansu, possibly N Sichuan, and Tibet (China).

STATUS: IUCN – Lower Risk (lc).

SYNONYMS: *vulpina* Howell, 1928.

COMMENTS: Subgenus *Conothoa*, but see Niu et al. (2001), who considered the *erythrotis* species group (*brookei*, *gloveri*, *muliensis*) to be subgenus *Pika*. Formerly included *gloveri*; see Corbet (1978c:68). Feng and Zheng (1985) provided evidence that *gloveri* (including *brookei*) is a distinct species. Formerly included in *rutila* (Ellerman and Morrison-Scott, 1951), but now regarded as distinct (Weston, 1982). The distribution of this species is poorly known, as is its relationship with the apparently allopatric *gloveri*. See also under *forresti*.

*Ochotona forresti* Thomas, 1923. Ann. Mag. Nat. Hist., ser. 9, 11:662.
    COMMON NAME: Forrest's Pika.
    TYPE LOCALITY: "N.W. Flank.Li-kiang Range, 13,000'" [Yunnan, China, approx. 27°N, 100°30'°, 3962 m].
    DISTRIBUTION: NW Yunnan, SE Tibet (China); N Burma; Assam, Sikkim (India); Bhutan.
    STATUS: IUCN – Lower Risk (nt).
    COMMENTS: Subgenus *Conothoa*. Formerly included in *pusilla* (Ellerman and Morrison-Scott, 1951), *roylei* (Corbet, 1978*c*), and *thibetana* (Feng and Kao, 1974; Gureev, 1964; Weston, 1982), but now considered distinct (A. T. Smith et al., 1990, and references therein). *O. forresti* is poorly known, but appears to be morphologically similar to *O. gaoligongensis* (Yu et al., 1992) and a sister species of *O. erythrotis* (Yu et al., 2000). It is thought to be geographically sympatric with *O. gloveri* and/or *O. thibetana* in Yunnan (China), Burma, and Sikkim (India). See also *nigritia*.

*Ochotona gaoligongensis* Wang, Gong, and Duan, 1988. Zool. Res., 9:201, 206.
    COMMON NAME: Gaoligong Pika.
    TYPE LOCALITY: "Dongsao-fang [Mount Gaoligong], (27°45'N, 98°27'E), Gongshan Co., Northwest Yunnan, alt. 2950 m." [China].
    DISTRIBUTION: Known only from the type locality.
    STATUS: IUCN – Data Deficient.
    COMMENTS: Subgenus *Conothoa*. From the original description, this taxon is likely to prove to be a synonym or sister species of *O. forresti*, which is known to occur in the same area (Yu et al., 1992).

*Ochotona gloveri* Thomas, 1922. Ann. Mag. Nat. Hist., ser. 9, 9:190.
    COMMON NAME: Glover's Pika.
    TYPE LOCALITY: "Nagchuka [= Nyagquka (Yajiang), W Sichuan, China], 10,000' [3048 m]."
    DISTRIBUTION: W Sichuan, NW Yunnan, NE Tibet, SW Qinghai (China).
    STATUS: IUCN – Lower Risk (lc).
    SYNONYMS: *kamensis* Argyropulo, 1948 [not 1941; see Honacki et al., 1982]; **brookei** Allen, 1937; **calloceps** Pen et al., 1962.
    COMMENTS: Subgenus *Conothoa*; but see Niu et al. (2001). Formerly included in *erythrotis*; see comments therein. Whether *gloveri* and *erythrotis* are sym-, para-, or allopatric in distribution in Sichuan and/or Qinghai is unknown. The taxon *brookei* may be a separate sister species to *gloveri+muliensis* (Niu, 2002).

*Ochotona himalayana* Feng, 1973. Acta Zool. Sinica, 19:69, 73.
    COMMON NAME: Himalayan Pika.
    TYPE LOCALITY: "Qu-xiang, Bo-qu Valley, Nei-la-mu [= Nyalam] District, alt. 3500 m." [Xigaze (Shigatse) County, Xizang (Tibet), China].
    DISTRIBUTION: Mt. Jolmolungma (Everest) area, S Xizang, China; probably adjacent Nepal.
    STATUS: IUCN – Lower Risk (lc).
    COMMENTS: Subgenus *Conothoa*. This taxon was considered a synonym of *O. roylei* by Corbet (1978*c*), Weston (1982) and Formozov (1997). Additional data (Feng and Zheng, 1985; Feng et al., 1986) suggested that it might be an independent species, but its range was within that of the similar *O. roylei nepalensis*. Additional studies (Yu et al., 2000) have now confirmed its specific distinctness.

*Ochotona hoffmanni* Formosov et al., 1996. Byul. Mosk. Ob-va Ispytatelei Prirody. Otd. Biol., 101(1):29.
    COMMON NAME: Hoffmann's Pika.
    TYPE LOCALITY: "Mongoliya, Khenteiskii aimak, Delger-Khan somon, 47°20's.sh. [N lat.],108°40'v.d. [E long]. [Mongolia, Khenteisk district, Delger village]."
    DISTRIBUTION: Restricted to the subalpine zone of the Hentiyn Nuruu ridge, Bayan-Ulan

mountains, Mongolian People's Republic, and Erman range, Russia (Formozov and
Baklushinskaya, 1999).

STATUS: IUCN – Vulnerable.

COMMENTS: Subgenus *Pika*. Originally described as a subspecies of *alpina*, but elevated to full
species status by Formozov and Baklushinskaya (1999) on the basis of morphological,
bioacoustical and chromosomal evidence.

*Ochotona huangensis* (Matschie, 1908). Wiss. Ergebn. der Exped. Filchner nach China und Tibet
1903-05,10(1):214.

COMMON NAME: Tsing-Ling Pika.

TYPE LOCALITY: Not specified; fixed by Allen (1938:544) as " . . . from the Tsingling
[mountains] in the vicinity of Sianfu [Xian]." [China].

DISTRIBUTION: In the mountains of C China, including Shaanxi, Gansu, Qinghai and
Sichuan provinces.

STATUS: IUCN – Endangered as *O. thibetana huangensis*.

SYNONYMS: *syrinx* Thomas, 1911; *xunhuaensis* Shou and Feng, 1984.

COMMENTS: Subgenus *Ochotona*. Formerly considered a subspecies of *thibetana* (Allen, 1938;
Ellerman and Morrison-Scott, 1955; Feng and Zheng, 1985; A. T. Smith et al., 1990). Yu
and Zheng (1992*b*) first suggested full species status for *huangensis*, based on differing
morphology, and later (Yu et al., 1997, 2000) supported this with molecular studies.
*O. huangensis* appears to be broadly sympatric with *O. thibetana*, but largely allopatric
in distribution with *O. cansus* and *O. thomasi*.

*Ochotona hyperborea* (Pallas, 1811). Zoogr. Rosso-Asiat., 1:152.

COMMON NAME: Northern Pika.

TYPE LOCALITY: " . . . e terris Tschuktschicis," [Chukotsk peninsula (Ognev, 1940:41),
Chukotsk AO, Russia].

DISTRIBUTION: Ural, Putorana, Sayan Mtns, east of Lena River to Chukotka, Koryatsk and
Kamchatka; upper Yenesei, Transbaikalia, and Amur regions, Sakhalin Isl (Russia);
NC Mongolia; NE China; N Korea; Hokkaido (Japan).

STATUS: IUCN – Lower Risk (lc); the subspecies *yesoensis* is considered to be rare on
Hokkaido, Japan.

SYNONYMS: *kolymensis* Allen, 1903; *litoralis* Peters, 1882; **cinereoflava** (Schrenk, 1858);
**coreana** Allen and Andrews, 1913; **ferruginea** (Schrenk, 1858); *kamtschaticus*
Dybowski, 1922; **mantchurica** Thomas, 1909; **normalis** (Schrenk, 1858); *davanica*
(Sokolov et al., 1994); *svatoshi* Turov, 1924; **uralensis** Flerov, 1927; *yesoensis* Kishida,
1930; *ornata* Kishida, 1930; *sadaki* Kishida, 1933; **yoshikurai** Kishida, 1932.

COMMENTS: Subgenus *Pika*. Formerly included in *alpina*; see A. T. Smith et al. (1990), and
references therein. Differences in morphology and vocalizations are noticeable where
*hyperborea* and *alpina* are sympatric in the W Sayan Mtns, Khangai Mtns, and
Transbaikalia, and character displacement in size is also evident in some populations
(Lissovsky and Lissovskaya, 2000; A. T. Smith et al., 1990). The original Pallas citation
was printed and privately circulated in 1811, but not published for general
distribution until 1826. Subspecies follow Lissovsky; formerly included
*O. turuchanensis* Naumov 1934, which is considered a separate species (Lissovsky,
2002).

*Ochotona iliensis* Li and Ma, 1986. Acta Zool. Sinica, 32:375, 379.

COMMON NAME: Ili Pika.

TYPE LOCALITY: "Tienshan Mountain [Borokhoro Shan], Nilka [County], Xinjiang, China,
alt. 3200 m."

DISTRIBUTION: Tien shan Mts., Xinjiang, China (Li et al., 1988).

STATUS: IUCN – Vulnerable; but may be endangered (Formozov, 1997; Li Weidong, pers.
comm).

COMMENTS: Subgenus *Conothoa*. Perhaps related to the *roylei-macrotis* group (Yu and Zheng,
1992*a*), or to the *erythrotis* group; poorly known.

*Ochotona koslowi* (Büchner, 1894). Wiss. Reisen. Przewalski Cent. Asien Zool. Th. I: Säugeth., p. 187.

COMMON NAME: Kozlov's Pika.

TYPE LOCALITY: "Dolina Vetrov" [Valley of the Winds; pass between Guldsha Valley and valley of Dimnalyk River, tributary of Chechen, Tarim Basin, Xinjiang, China, 14,000' (37°55'N, 87°50'E; 4267 m)].

DISTRIBUTION: Arkatag Range, Kunlun Mtns (China), and S shore of Aru-Tso Lake, E of Lungdo, Ngari, Xizang.

STATUS: IUCN – Endangered.

COMMENTS: Subgenus *Conothoa*. According to the molecular phylogeny of Yu et al. (2000), the sister species of *koslowi* is *ladacensis*.

*Ochotona ladacensis* (Günther, 1875). Ann. Mag. Nat. Hist., ser. 4, 16:231.

COMMON NAME: Ladak Pika.

TYPE LOCALITY: "Chagra, 14000 feet above the sea" [Changra, Ladak, Kashmir, India; 4267 m].

DISTRIBUTION: SW Xinjiang, Qinghai, E Tibet (China); Kashmir (India); Pakistan.

STATUS: IUCN – Lower Risk (lc).

COMMENTS: Subgenus *Conothoa*; but see Niu et al. (2001), who placed it in subgenus *Pika* as a sister to the *erythrotis* group. See also *koslowi* above. Broadly sympatric with *curzoniae* on the Tibetan Plateau, though not so widely distributed.

*Ochotona macrotis* (Günther, 1875). Ann. Mag. Nat. Hist., ser. 4, 16:231.

COMMON NAME: Large-eared Pika.

TYPE LOCALITY: "Doba" [C Tibet, (31°N, 87°E), China] Ognev (1940:86). Not "Duba... N side Kuenlun . . . on road . . . via Kugiar" [Kakyar, . 37°45'N, 77°05'E, W Xinjiang, China] *contra* Blanford (1879:76). Not Dobo, Qinghai [(36°41'N, 101°30'E) (Vaurie, 1972:352)].

DISTRIBUTION: Mountainous regions including the Himalayas (Nepal, India) from Bhutan through Tibet, Kunlun (Qinghai, Xinjiang, Sichuan and Yunnan [China]), Karakorum (Pakistan), Hindu Kush (Afghanistan), Pamir, and W Tien Shan Mtns (Kyrgyzstan, Tajikistan, SE Kazakhstan).

STATUS: IUCN – Lower Risk (lc).

SYNONYMS: *griseus* Blanford, 1875; **auritus** Blanford, 1875; *baltina* Thomas, 1922; **chinensis** Thomas, 1911; *sinensis* Lydekker, 1912 [*lapsus calami* according to Ellerman and Morrison-Scott, 1951]; **sacana** Thomas, 1914; **wollastoni** Thomas and Hinton, 1922.

COMMENTS: Subgenus *Conothoa*. Included in *roylei* by Gureev (1964), Roberts (1977), Corbet (1978c:68), and Gromov and Baranova (1981:72). Morphological and ecological differences in the area of sympatry first documented by Kawamichi (1971) and Abe (1971), and confirmed by Mitchell (1978, 1981). Weston (1982), Feng and Zheng (1985), and Feng et al. (1986) indicated that *macrotis* is a distinct species. Yu et al. (2000) consider it a sister species to *roylei*. Whether the co-type from "Doba" in the Natural History Museum (London) is from C Tibet or W Xinjiang is uncertain.

*Ochotona muliensis* Pen and Feng, 1962. *In* Pen et al., Acta Zool. Sinica, 14 (supplement):120, 132.

COMMON NAME: Muli Pika.

TYPE LOCALITY: "Ting-Tung-Niu-Chang, southeastern Muli (alt. 3600m) Szechuan" [Muli A.D., Xichang County, Sichuan, China].

DISTRIBUTION: Known only from the vicinity of the type locality.

STATUS: IUCN – Data Deficient.

COMMENTS: Subgenus *Conothoa*; but see Niu et al. (2001), who placed it in subgenus *Pika* as sister to *brookei* in the *erythrotis* group. This taxon was originally described as a subspecies of *O. gloveri*, but is now thought to be specifically distinct (Feng and Zheng, 1985; Niu et al., 2001). It differs from *gloveri* in certain cranial characters, and in habitat (A. T. Smith et al., 1990), but its extreme rarity in collections makes its independent status difficult to demonstrate. Formozov (1997) claimed it to be a junior synonym of *gloveri*.

*Ochotona nigritia* Gong et al., 2000. Zool. Research, 21:204.

COMMON NAME: Black Pika.

TYPE LOCALITY: Piyanma, Yunnan.

DISTRIBUTION: Known only from the vicinity of the type locality.

STATUS: Not Evaluated.

COMMENTS: Subgenus *Conothoa*. Apparently close to *O. forresti*; may be only a melanistic individual.

*Ochotona nubrica* Thomas, 1922. Ann. Mag. Nat. Hist., ser. 9, 9:187.

COMMON NAME: Nubra Pika.

TYPE LOCALITY: "Tuggur, Nubra Valley, alt. 10,000" [Ladak, Kashmir, India].

DISTRIBUTION: Southern edge of Tibetan Plateau from Ladak (India, China) through Nepal to E Tibet (China).

STATUS: IUCN – Lower Risk (lc).

SYNONYMS: *aliensis* Zheng, 1979; *hodgsoni* (Blyth, 1841); **lhasaensis** Feng and Kao, 1974; *lama* Mitchell and Punzo, 1975.

COMMENTS: Subgenus *Ochotona*. Assigned to *pusilla* by Ellerman and Morrison-Scott (1951), to *roylei* (as *O. lama*) by Corbet (1978c) and to *thibetana* (as *O. t. lama*) by Feng et al. (1986). Recognized as distinct by A. T. Smith et al. (1990), followed by Yu and Zheng (1992a). Its closest relations were thought to be with *O. thibetana*, but Yu et al. (2000) demonstrated that they are with *O. curzoniae*. Also see comment under *O. roylei*.

*Ochotona pallasi* (Gray, 1867). Ann. Mag. Nat. Hist., ser. 3, 20:220.

COMMON NAME: Pallas's Pika.

TYPE LOCALITY: " . . . said to come from 'Asiatic Russia-Kirgisen' " (Thomas, 1908b:109). Restricted by Heptner (1941:328) to "southern parts . . . Karkaralinsk Mountains . . . north of Lake Balkhash" [Karagandinsk Obl., (Kazakhstan 49°N, 75°E)].

DISTRIBUTION: Discontinuous in arid areas (mtns and high steppes) in Kazakhstan; Altai Mtns, Tuva (Russia), and Mongolia, to Xinjiang, Inner Mongolia and Ningxia (China).

STATUS: IUCN – Lower Risk (lc); but isolated populations of *O. p. hamica* (IUCN – Critically Endangered), and *O. p. sunidica* (IUCN – Endangered) are threatened.

SYNONYMS: *ogotona* (Waterhouse, 1848) [not Pallas, 1778]; *ogotona* Bohnhote, 1905 [not Pallas, 1778]; *opaca* Argyropulo, 1939 [not Vinogradov and Argyropulo, 1948; see Ellerman and Morrison-Scott, 1951] [not Argyropulo, 1941, see comment under *gloveri kamensis*]; **hamica** Thomas, 1912; **pricei** Thomas, 1911; **sunidica** Ma et al., 1980.

COMMENTS: Subgenus *Pika*. Includes *pricei* (Corbet, 1978c) as it is commonly referred to in Russian literature. However, marked difference in reproduction, habitat, behavior, and vocalization suggest that *pallasi* and *pricei* may prove to be specifically distinct (A. T. Smith et al., 1990). The name *helanshanensis* (Zheng, 1987; in X. T. Wang, 1990) is restricted to the Helan Shan (Ningxia, China), and its range is congruent with the only known distribution of *O. argentata* (all known specimens of *argentata* and *helanshanensis* originate from the same 2 X 1.5 km forest patch), and here we consider it to be a synonym of *argentata* (Formozov, 1997; Formozov et al., 2004). On the basis of molecular evidence, Yu et al. (2000) and Niu et al. (2001) felt that *helanshanensis* (= *argentata*) should be treated as a subspecies of *O. pallasi*, but its molecular distance is too great and at the level of species, and its karyotype differs from *pallasi*, which is a sister species to the *alpina-hyperborea* group (Yu et al., 2000), which includes *argentata*.

*Ochotona princeps* (Richardson, 1828). Zool. J., 3:520.

COMMON NAME: American Pika.

TYPE LOCALITY: "Rocky Mountains"; restricted by Preble (1908) to "near the sources of Elk (Athabasca) River," [Athabasca Pass, head of Athabasca River, Alberta, Canada].

DISTRIBUTION: Mountains of W North America from C British Columbia (Canada) to N New Mexico, Utah, C Nevada, and EC California (USA).

STATUS: IUCN – Lower Risk (lc); isolated subspecies in the Great Basin: IUCN – Vulnerable (*goldmani, lasalensis, nevadensis, nigrescens, obscura, sheltoni, tutelata*; A. T. Smith et al., 1990).

SYNONYMS: Allozyme studies revealed 4-5 main groups of populations in this species (Hafner and Sullivan, 1995). Synonyms listed here follow these groupings: (1) Northern Rockies: *cuppes* Bangs, 1899; *goldmani* Howell, 1924; *levis* Hollister, 1912; *lutescens* Howell, 1919; *obscura* Long, 1965; *saturata* Cowan, 1955; (2) Central Rockies: **figginsi** Allen, 1912; *clamosa* Hall and Bowlus, 1938; *fuscipes* Howell, 1919; *howelli* Borell, 1931; *lemhi* Howell, 1919; *uinta* Hollister, 1912; *ventorum* Howell, 1919; *wasatchensis* Durrant and Lee, 1955; (3) Southern Rockies: *saxatilis* Bangs, 1899; *barnesi* Durrant and Lee, 1955; *incana* Howell, 1919; *lasalensis* Durrant and Lee, 1955; *moorei* Gardner, 1950; *nevadensis* Howell, 1919; *nigrescens* Bailey, 1913; *utahensis* Hall and Hayward, 1941; (4) Sierra Nevada-Great Basin: *schisticeps* (Merriam, 1889); *albata* Grinnell, 1912; *cinnamomea* Allen, 1905; *muiri* Grinnell and Storer, 1916; *sheltoni* Grinnell, 1918; *tutelata* Hall, 1934; (5) Cascades: *taylori* Grinnell, 1912; *brooksi* Howell, 1924; *brunnescens* Howell, 1919; *fenisex* Osgood, 1913; *fumosa* Howell, 1919; *jewetti* Howell, 1919; *littoralis* Cowan, 1955; *minimus* (Lord, 1863) [not Schinz, 1821]; *septentrionalis* Cowan and Racey, 1947.

COMMENTS: Subgenus *Pika*. Broadbooks (1965) and Youngman (1975) considered *princeps* and *collaris* conspecific. Corbet (1978c), following Gureev (1964), included *princeps* in *alpina*. A statistical reevaluation of craniometric data by Weston (1981) indicated that *princeps*, *collaris*, and *alpina* are separate species. Hafner and Sullivan (1995) analyzed allozymic variation from 56 populations of *princeps*, with *collaris* as outgroup (see above). All but eight of the populations were placed in one of five regional populations, those eight being geographically intermediate. Reviewed by Smith and Weston (1990, Mammalian Species, 352).

*Ochotona pusilla* (Pallas, 1769). Nova Comm. Imp. Acad. Sci. Petropoli, 13:531.

COMMON NAME: Steppe Pika.

TYPE LOCALITY: "in campis circa Volgam . . ."; restricted by Ognev (1940) to Samarsk Steppe, near Buzuluk, left bank of Samara River [Orenburgsk Obl. Russia].

DISTRIBUTION: Steppes from middle Volga (Russia), east and south through N Kazakhstan to upper Irtysh River and Chinese border. Not yet recorded in China.

STATUS: IUCN – Vulnerable (classification refers to subspecies *O. p. pusilla* only; Formozov, pers. comm).

SYNONYMS: *minutus* (Pallas, 1771); **angustifrons** Argyropulo, 1932

COMMENTS: Subgenus *Ochotona* (Ellerman and Morrison-Scott, 1951). Formerly included *O. nubrica* (= *lama*), *O. forresti*, and *O. osgoodi* (a subspecies of *thibetana*); see comments therein.

*Ochotona roylei* (Ogilby, 1839). Royle's Illus. Botany . . . Himalaya, vol. 2, 69, pl. 4.

COMMON NAME: Royle's Pika.

TYPE LOCALITY: "Choor Mountain, Lat. 30. Elev. 11,500[ft]", [60 mi. (96 km) N of Saharanpur; 3505 m], Punjab, India.

DISTRIBUTION: Himalayan Mtns in NW Pakistan and India to Nepal; adjacent Tibet (China).

STATUS: IUCN – Lower Risk (lc).

SYNONYMS: *angdawai* Biswas and Khajuria, 1955; *mitchelli* Agrawal and Chakraborty, 1971; *wardi* Bonhote, 1904; **nepalensis** Hodgson, 1841.

COMMENTS: Subgenus *Conothoa*. Includes *angdawai* and *mitchelli*, but not *forresti* and *himalayana*, which are here provisionally considered distinct; see comments therein. *O. hodgsoni* Blyth, 1841 is traditionally placed here, but based on the original description it is a synonym of *nubrica*.

*Ochotona rufescens* (Gray, 1842). Ann. Mag. Nat. Hist., [ser. 1], 10:266.

COMMON NAME: Afghan Pika.

TYPE LOCALITY: "India, Cabul, Rocky Hills near Baker Tomb at about 6000 or 8000 feet [ca. 1829 or 2438 m] elevation" [Baber's (?) Tomb, Kabul, Afghanistan].

DISTRIBUTION: Afghanistan, Baluchistan (Pakistan), Iran, Armenia, and SW Turkmenistan.

STATUS: IUCN – Lower Risk (lc); considered a crop pest and controlled in parts of its range (A. T. Smith et al., 1990).

SYNONYMS: *seiana* Thomas, 1922; *vizier* Thomas, 1911; *vulturina* Thomas, 1920; **regina** Thomas, 1911; **shukurovoi** Heptner, 1961.

COMMENTS: Subgenus *Ochotona*. Includes *seiana*; see A. T. Smith et al. (1990) and comments under *O. curzoniae*.

*Ochotona rutila* (Severtzov, 1873). Izv. Obshch. Lyubit. Estestvozn., 8(2):83

COMMON NAME: Turkestan Red Pika.

TYPE LOCALITY: " . . . in mountains near Vernyi[Alma-ata] . . . 7000-8000 ft." [2134-2438 m]. Restricted by Shnitnikov (1936) to valley of Maly Alma-atinsk River, Zailisk Alatau Mtns, Kazakhstan (43°05'N, 77°10'E).

DISTRIBUTION: Isolated ranges from the Pamirs (Tajikistan) to Tien Shan (SE Uzbekistan, Kyrgyzstan, SE Kazakhstan); perhaps N Afghanistan and E Xinjiang (China). Zhang et al. (1997) recorded it from Batang, Sichuan province, but this seems unlikely, and the record may be based on a misidentified *gloveri*.

STATUS: IUCN – Lower Risk (lc); however, sporadically distributed throughout its range, and common in only a few localities (A. T. Smith et al., 1990).

COMMENTS: Subgenus *Conothoa*. Apparently an allospecies of *O. erythrotis*, which has sometimes been included in *rutila*; see comments therein.

*Ochotona thibetana* (Milne-Edwards, 1871). Nouv. Arch. Mus. Hist. Nat. Paris, Bull., 7:93.

COMMON NAME: Moupin Pika.

TYPE LOCALITY: "mountain near Moupin" [Baoxing, Ya'an County, Sichuan, China].

DISTRIBUTION: Shanxi, Shaanxi, W Hubei, Yunnan, Sichuan, S Tibet (China); N Burma; Sikkim (India); perhaps adjacent Bhutan and India.

STATUS: IUCN – Lower Risk (lc); however, *O. t. sikimaria* of Sikkim is IUCN – Critically Endangered (A. T. Smith et al., 1990).

SYNONYMS: *hodgsoni* Bonhote, 1905; *zappeyi* Thomas, 1922; **nangqenica** Zheng et al., 1980; **osgoodi** Anthony, 1941; **sacraria** Thomas, 1923; **sikimaria** Thomas, 1922.

COMMENTS: Subgenus *Ochotona*. Formerly included *cansus, forresti, huangensis,* and *nubrica*; see comments therein. The taxon *aliensis*, originally described as a subspecies of *thibetana*, is now considered a synonym of *nubrica* (Feng et al., 1986; A. T. Smith et al., 1990). *O. osgoodi*, described as a distinct species by Anthony (1941), was listed as a subspecies of *O. pusilla* by Ellerman and Morrison-Scott (1951), and subsequently allocated to *thibetana* by Corbet (1978c) and Weston (1982). The isolated subspecies *sikimaria* was assigned to *cansus* by Feng and Kao (1974), Feng and Zheng (1985), and Zhang et al. (1997), but transferred to *thibetana* by A. T. Smith et al. (1990); it may deserve full species status. Erbajeva (1988:190-191) considered *cansus* and *sikimaria* subspecies of *thibetana*, as well as *lhasaensis*, here placed in *nubrica*; moreover, she thought *hodgsoni* Bonhote was probably a distinct species (based on examination of a skull photograph). She later retracted this (Erbajeva, 1994).

*Ochotona thomasi* Argyropulo, 1948. Trudy Zool. Inst. Leningrad, 7:127.

COMMON NAME: Thomas's Pika.

TYPE LOCALITY: "Valley of Alyk-nor." Restricted by Formosov (in A. T. Smith et al., 1990) to Alang-nor Lake, NE Qinghai, China (35°35'N, 97°25'E); see also Corbet (1978c).

DISTRIBUTION: NE Qinghai, Gansu, and Sichuan (China).

STATUS: IUCN – Lower Risk (nt).

SYNONYMS: *ciliana* Bannikov, 1940.

COMMENTS: Subgenus *Ochotona*. Widely sympatric with the similar *O. cansus*.

*Ochotona turuchanensis* Naumov, 1934. Trudy Polyarnoe Komissii, Akad. Nauk, 17:38.

COMMON NAME: Turuchan Pika.

TYPE LOCALITY: Uchami, Nizhnyaya Tunguska River, Krasnoyarskii Krai, Russia.

DISTRIBUTION: From middle to lower Yenesei River eastward to middle Lena River and Lake Baikal; the Middle Siberian Plateau and adjacent Lena River basin (Lissovsky, 2002).

STATUS: Not Evaluated.

COMMENTS: Subgenus *Pika*. This newly recognized species is widely sympatric with *O. hyperborea* over much of its range.

**Family Prolagidae** Gureev, 1964. Fauna SSSR. Mlekopitayushchie, Zaitseobraznye, III (10):49.

*Prolagus* Pomel, 1853. Cat. Meth. Desc. Vert. Foss. dans le Bassin Hydro. Super. de la Loire et Surt. la Val. Aff. Prin., l'Allier. Paris. p. 43.

TYPE SPECIES: *Anoema aeningensis* König, 1825 (fossil).

SYNONYMS: *Anoema* König, 1825; *Archaeomys* Fraas, 1856; *Lagomys* G. Cuvier, 1800; *Myolagus* Hensel, 1856.

COMMENTS: Previously considered a subfamily (Gureev, 1964), but elevated by Erbajeva (1988, 1994) to family Prolagidae. Reviewed by Tobien (1975).

*Prolagus sardus* (Wagner, 1832). Abh. Bayer. Akad. Wiss., 1:763-767.
COMMON NAME: Sardinian Pika.
TYPE LOCALITY: Italy, Sardinia.
DISTRIBUTION: Mediterranean Isles of Corsica (France) and Sardinia (Italy); adjacent small islands.
STATUS: IUCN – Extinct.
COMMENTS: Described from fossils, but apparently survived until historic times (Vigne, 1983), perhaps as late as 1774 (Kurtén, 1968). Reviewed by Dawson (1969).

**Family Leporidae** Fischer, 1817. Mém. Soc. Imp. Nat. Moscow, 5:372.
SYNONYMS: Bunolagini Averianov, 1999; Lagidae Schultze, 1897; Leporidae Gray, 1821; Leporinorum Fischer, 1817; Oryctolaginae Gureev, 1948; Pentalaginae Gureev, 1948.
COMMENTS: Often divided into subfamilies Paleolaginae (*Pentalagus, Pronolagus, Romerolagus*) and Leporinae (remaining genera) (Dice, 1929; Simpson, 1945), but no subfamilies were recognized by Ellerman and Morrison-Scott (1951). For basis of genera recognized here, see Corbet (1983); see also Averianov (1999).

*Brachylagus* Miller, 1900. Proc. Biol. Soc. Wash., 13:157.
TYPE SPECIES: *Lepus idahoensis* Merriam, 1891.

*Brachylagus idahoensis* (Merriam, 1891). N. Am. Fauna, 5:76.
COMMON NAME: Pygmy Rabbit.
TYPE LOCALITY: "Pahsimeroi Valley [near Goldburg, Custer County], Idaho." [USA].
DISTRIBUTION: SW Oregon to EC California, SW Utah, N to SW Montana (USA). Isolated population in WC Washington (USA).
STATUS: U. S. ESA – Endangered for isolated population in Washington (Columbia River Basin); IUCN – Lower Risk (nt), species-wide; yet recently species is in decline.
COMMENTS: Formerly included in *Sylvilagus*; but see Corbet (1983). Placed in the monotypic genus *Brachylagus* by Dawson (1967) and, together with *bachmani*, in the genus *Microlagus* by Gureev (1964:170-173); but also see Hall (1981:294), who recognized *Brachylagus* as a subgenus. This species is widely sympatric with *Sylvilagus nuttallii*, and perhaps overlaps narrowly with *S. audubonii*. It has been interpreted as either a primitive rabbit (Hibbard, 1963), or as derived from *Sylvilagus* (Corbet, 1983). Reviewed by Green and Flinders (1980, Mammalian Species, 125).

*Bunolagus* Thomas, 1929. Proc. Zool. Soc. Lond., 1929:109.
TYPE SPECIES: *Lepus monticularis* Thomas, 1903.

*Bunolagus monticularis* (Thomas, 1903). Ann. Mag. Nat. Hist., ser. 7, 11:78.
COMMON NAME: Riverine Rabbit.
TYPE LOCALITY: "Deelfontein, Cape Colony," South Africa.
DISTRIBUTION: C Karoo (31E22'S, 22EE), Western Cape Prov. (South Africa).
STATUS: IUCN – Critically Endangered.
COMMENTS: Reviewed by Petter (1972b). Formerly in *Lepus* (Ellerman and Morrison-Scott, 1951), but returned to *Bunolagus* by Angermann (1966). Karyological evidence supports the separation of *Bunolagus* (2n=44) from *Lepus* (2n=48) (Robinson and Dippenaar, 1987; Robinson and Skinner, 1983); its closest relatives are probably *Pronolagus* (Corbet, 1983).

*Caprolagus* Blyth, 1845. J. Asiat. Soc. Bengal, 14:247.
TYPE SPECIES: *Lepus hispidus* Pearson, 1839.

*Caprolagus hispidus* (Pearson, 1839). *In* M'Clelland, Proc. Zool. Soc. Lond., 1838:152 [1839].
COMMON NAME: Hispid Hare.

TYPE LOCALITY: " . . . Assam, . . . base of the Boutan [Bhutan] mountains" [India].

DISTRIBUTION: S Himalaya foothills from Uttar Pradesh (India) through Nepal and West Bengal to Assam (India), and south through NW Bangladesh. Since 1951 there have been very few reports from Uttar Pradesh and Assam; see Santapau and Humayun (1960), Mallinson (1971), and Ghose (1978). Presently known distribution summarized by Bell et al. (1990).

STATUS: CITES – Appendix I; U.S. ESA – Endangered, and IUCN – Endangered.

*Lepus* Linnaeus, 1758. Syst. Nat., 10th ed., 1:57.

TYPE SPECIES: *Lepus timidus* Linnaeus, 1758.

SYNONYMS: *Allolagus* Ognev, 1929; *Boreolagus* Barrett-Hamilton, 1911; *Chionobates* Kaup, 1829; *Eulagos* Gray, 1867; *Eulepus* Acloque, 1899; *Indolagus* Gureev, 1953; *Lagos* Palmer, 1904; *Macrotolagus* Mearns, 1895; *Poecilolagus* Lyon, 1904; *Proeulagus* Gureev, 1964; *Sabanolagus* Averianov, 1998; *Sinolagus* Averianov, 1998; *Tarimolagus* Gureev, 1947 [Pavlinov et al., 1995*b*].

COMMENTS: Formerly included *Bunolagus*; see Petter (1972*b*); and originally all other genera (*Brachylagus, Caprolagus, Macrotolagus, Poelagus, Pronolagus, Romerolagus, Sylvilagus*) in Leporidae except *Pentalagus*. The taxonomy of this genus remains controversial. *L. crawshayi* (including *whytei* ), *peguensis, ruficaudatus*, and *siamensis* have been variously treated as separate species or have been included in *nigricollis*. *L. europaeus, corsicanus, granatensis, mediterraneus, tolai*, and *tibetanus* have been placed in *capensis* or treated as distinct species; see comments therein.

*Lepus alleni* Mearns, 1890. Bull. Am. Mus. Nat. Hist., 2:294.

COMMON NAME: Antelope Jackrabbit.

TYPE LOCALITY: "Rillito Station [Pima Co.] Arizona" [USA].

DISTRIBUTION: SC Arizona (USA) to N Nayarit and Tiburon Isl (Mexico).

STATUS: IUCN – Lower Risk (lc).

SYNONYMS: *palitans* Bangs, 1900; **tiburonensis** Townsend, 1912.

COMMENTS: Subgenus *Macrotolagus* (Gureev (1964:155). Probably related to *callotis*, but recognized as a distinct species by Hall (1981:331). Reviewed by Best and Henry (1993*a*, Mammalian Species, 424).

*Lepus americanus* Erxleben, 1777. Syst. Regni Anim., 1:330.

COMMON NAME: Snowshoe Hare.

TYPE LOCALITY: "in America boreeli, ad fretum Hudsonis copiosissimus." Restricted by Nelson (1909:87) to Fort Severn, Ontario, Canada.

DISTRIBUTION: S and C Alaska (USA) to S and C coasts of Hudson Bay to Newfoundland and Anacosti Isl (introduced) (Canada), south to S Appalachians, S Michigan, North Dakota, NC New Mexico, SC Utah, and EC California (USA).

STATUS: IUCN – Lower Risk (lc).

SYNONYMS: *bishopi* J. A. Allen, 1899; *columbianus* Rhoads, 1895; *hudsonius* Pallas, 1778; *nanus* Schreber, 1790 [composite]; *pallidus* Cowan, 1938; *phaeonotus* J. A. Allen, 1899; **bairdii** Hayden, 1869; *pineus* Dalquest, 1942; *seclusus* Baker and Hankins, 1950; *setzeri* Baker, 1959; **cascadensis** Nelson, 1907; *klamathensis* Merriam, 1899; *oregonus* Orr, 1934; *tahoensis* Orr, 1933; **dalli** Merriam, 1900; *macfarlani* Merriam, 1900; *niediecki* Matschie, 1907; *saliens* Osgood, 1900; **struthopus** Bangs, 1898; **virginianus** Harlan, 1825; *borealis* Schinz, 1845; *wardi* Schinz, 1825; *washingtoni* Baird, 1855.

COMMENTS: Subgenus *Poecilolagus* (Gureev, 1964; Averianov, 1998). Distinctive small species, but subgeneric separation not supported Hall (1981:314).

*Lepus arcticus* Ross, 1819. Voy. Discovery, II; ed. 2, App. IV, p. 170.

COMMON NAME: Arctic Hare.

TYPE LOCALITY: "Southeast of Cape Bowen" (Nelson, 1909:61) [Possession Bay, Bylot Island, lat. 73E37'N, Canada].

DISTRIBUTION: Greenland and Canadian arctic islands southward in open tundra to WC shore of Hudson Bay, thence northwest to the west of Fort Anderson on coast of Arctic Ocean. Isolated populations in tundra of N Quebec and Labrador, and on Newfoundland (Canada).

STATUS: IUCN – Lower Risk (lc).

SYNONYMS: *andersoni* Nelson, 1934; *banksicola* Manning and Macpherson, 1958; *canus* Preble, 1902; *glacialis* Leach, 1819; **bangsii** Rhoads, 1896; *labradorius* Miller, 1899; **groenlandicus** Rhoads, 1896; *hyperboreus* Pedersen, 1930 [not Pallas, 1811]; *persimilis* Nelson, 1934; *porsildi* Nelson, 1934; **monstrabilis** Nelson, 1934; *hubbardi* Handley, 1952.

COMMENTS: Subgenus *Lepus* (Averianov, 1998). Formerly included in *timidus* by Gureev (1964), Angermann (1967), Honacki et al. (1982), and Dixon et al. (1983), but considered distinct by Corbet (1978c), Hall (1981), A. J. Baker et al. (1983), and Flux and Angermann (1990). Reviewed by Best and Henry (1994a, Mammalian Species, 457).

*Lepus brachyurus* Temminck, 1845. *In* Siebold, Fauna Japonica, 1(Mamm.), p. 44, pl. 11.
COMMON NAME: Japanese Hare.
TYPE LOCALITY: " . . . tout l'Empire mais surtout dans l'île de Jezo", Nagasaki, Kyushu, Japan.
DISTRIBUTION: Honshu, Shikoku, Kyushu, Oki Isls and Sado Isl (Japan).
STATUS: IUCN – Lower Risk (lc).
SYNONYMS: **angustidens** Hollister, 1912; *etigo* Abe, 1918; **lyoni** Kishida, 1937; **okiensis** Thomas, 1906.
COMMENTS: Subgenus uncertain. Gureev (1964:150) and Gromov and Baranova (1981:63) placed this species in genus *Caprolagus*, subgenus *Allolagus*, but Averianov (1998) in subgenus *Eulagos*; see also comment under *mandshuricus*. Hirikawa et al. (1992) analyzed variation in the insular populations of *L. b. okiensis*. Reviewed by Imaizumi (1970b).

*Lepus californicus* Gray, 1837. Mag. Nat. Hist. [Charlesworth's], 1:586.
COMMON NAME: Black-tailed Jackrabbit.
TYPE LOCALITY: "St. Antoine" [probably near Mission of San Antonio, California, USA]. Discussed by Hall (1981:326).
DISTRIBUTION: Hidalgo and S Queretaro to N Sonora and Baja California (Mexico), north to SW Oregon and C Washington, S Idaho, E Colorado, S South Dakota, W Missouri, and NW Arkansas (USA). Apparently isolated population in SW Montana.
STATUS: IUCN – Lower Risk (lc).
SYNONYMS: *bennettii* Gray, 1843; *martirensis* Stowell, 1895; *richardsonii* Bachman, 1839; *tularensis* Merriam, 1904; *vigilax* Dice, 1926; **deserticola** Mearns, 1896; *depressus* Hall and Witlow, 1932; *wallawalla* Merriam, 1904; **insularis** Bryant, 1891; *edwardsi* St. Loup, 1895; **magdalenae** Nelson, 1907; *sheldoni* Burt, 1933; *xanti* Thomas, 1898; **melanotis** Mearns, 1890; *altamirae* Nelson, 1907; *curti* Hall, 1951; *merriami* Mearns, 1896; **texianus** Waterhouse, 1848; *asellus* Miller, 1899; *eremicus* J. Allen, 1894; *festinus* Nelson, 1904; *griseus* Mearns, 1896; *micropus* J. Allen, 1903.
COMMENTS: Subgenus *Proeulagus* (Gureev, 1964:193), or *Eulagos* (Averianov, 1998). Chromosomes described by Cervantes et al. (1999-2000). Reviewed by Best (1996, Mammalian Species, 530).

*Lepus callotis* Wagler, 1830. Naturliches Syst. Amphibien, p. 23.
COMMON NAME: White-sided Jackrabbit.
TYPE LOCALITY: "Mexico"; restricted by Nelson (1909:122) to southern end of Mexican Tableland.
DISTRIBUTION: C Oaxaca (Mexico) north discontinuously to SW New Mexico (USA).
STATUS: IUCN – Lower Risk (nt).
SYNONYMS: *mexicanus* Lichtenstein, 1830; *nigricaudatus* Bennett, 1833; **gaillardi** Mearns, 1896; *battyi* J. Allen, 1903.
COMMENTS: Subgenus *Proeulagus* (Gureev, 1964), or *Macrotolagus* (Averianov, 1998). Includes *gaillardi* and *mexicanus*; see Anderson and Gaunt (1962) and Hall (1981:328-330); but see also Gureev (1964:192, 195). Range allopatric with *L. alleni*, to which it is probably related. Karyotype reported by Gonzalez and Cervantes (1996). Reviewed by Best and Henry (1993b, Mammalian Species, 442).

*Lepus capensis* Linnaeus, 1758. Sys. Nat., 10th ed., 1:58.
COMMON NAME: Cape Hare.

TYPE LOCALITY: "ad Cap. b. Spei" [South Africa, Cape of Good Hope].

DISTRIBUTION: As construed in the past, a single species (*capensis sensu lato*) inhabits Africa and the Near East in two separate, non-forested areas: South Africa, Namibia, Botswana, Zimbabwe, S Angola, S Zambia (?), Mozambique; and to the north, Tanzania, Kenya, Somalia, Ethiopia, countries of the Sahel and Sahara, and N Africa; thence eastward through the Sinai to the Arabian Peninsula, Jordan, S Syria, S Israel and W and S Iraq, west of the Euphrates River (Harrison and Bates, 1991; Hufnagl, 1972; Kingdon, 1997; Kowalski and Rezebik-Kowalska, 1991; Smith, 1985). However, there is no evidence of gene flow between the South African populations, and those "capensis" in East, West and North Africa, and the intervening areas are inhabited by other species of *Lepus,* particularly *L. microtis*). Herein, the name *capensis* will be restricted to the South African hare, and other names applied to East and North African, and Arabian-Near Eastern hares. This informal subdivision of *capensis sensu lato* creates four groups that might be considered as distinct species. However, pending sufficient data at this point to support a formal revision, this arrangement best reflects the poorly known relationships of the taxa.

STATUS: IUCN – Lower Risk (lc).

SYNONYMS: South Africa: *arenarius* Geoffroy, 1826; *centralis* Thomas, 1903; *ochropoides* Roberts, 1929; *ochropus* Wagner, 1844; **aquilo** Thomas and Wroughton, 1907; *bedfordi* Roberts, 1932; *hartensis* Roberts, 1932; *ermeloensis* Roberts, 1932; *vernayi* Roberts, 1932; **carpi** Lundholm, 1955; *salai* Jentink, 1880; **granti** Thomas and Schwann, 1904; *kalaharicus* Dollman, 1910; *langi* Roberts, 1932; *major* Grill, 1860; *mandatus* Thomas, 1926; *narranus* Thomas, 1926. East Africa: **aegyptius** Desmarest, 1822; *abbotti* Hollister, 1918; *chadensis* Thomas and Wroughton, 1907; *dinderus* Setzer, 1956; **hawkeri** Thomas, 1901; **isabellinus** Cretzschmar, 1826; *aethiopicus* Hemprich and Ehrenberg, 1833; **sinaiticus** Ehrenberg, 1833; *innesi* de Winton, 1902; *rothschildi* de Winton, 1902. Arabia and Near East: **arabicus** Ehrenberg, 1833; *atallahi* Harrison, 1972; *cheesmani* Thomas, 1921; *jefferyi* Harrison, 1980; *omanensis* Thomas, 1894. Northwest Africa (Mahgreb): **schlumbergeri** Remy-St. Loup, 1894; *harterti* Thomas, 1903; *kabylicus* de Winton, 1898; *pallidior* Barrett-Hamilton, 1898; *pediaeus* Cabrera, 1923; *sefranus* Thomas, 1913; *tunetae* de Winton, 1898; **atlanticus** de Winton, 1898; *maroccanus* Cabrera, 1906; *sherif* Cabrera, 1906; **whitakeri** Thomas, 1902. **Unassigned**: *barcaeus* Ghigi, 1920.

COMMENTS: Subgenus *Proeulagus* (Gureev, 1964:202) or *Eulagos* (Averianov, 1998). Includes *arabicus*; formerly included *europaeus*, *corsicanus*, *granatensis*, and *tolai*; see Corbet (1978c: 71), Angermann (1983:20), and Harrison and Bates (1991). Formerly included *habessinicus*, but Azzaroli-Puccetti (1987a, b) considered *habessinicus* distinct. The enigmatic form *connori*, often placed in *capensis* (Corbet, 1978c; Harrison and Bates, 1991) is provisionally placed in *europaeus* on the basis of pelage characteristics; see Angermann (1983:19). Most Russian authors consider *tolai* (including *tibetanus*) a distinct species; see Gromov and Baranova (1981:65); but also see Pavlinov and Rossolimo (1987:229). Sludskii et al. (1980:58, 85) indicated an area of sympatry between *europaeus* and *tolai* in Kazakhstan. Sokolov and Orlov (1980:85) considered *tibetanus* a distinct species. Arabian forms may be specifically distinct (Flux and Angermann, 1990); Angermann (1983:19) noted pronounced "size" groups within *arabicus*. These are *arabicus* (largest, gray), *cheesmani* (with insular *atallahi*) (smaller, buffy), and *omanensis* (with insular *jeffreyi*)(smallest, gray). See also *Lepus tibetanus.*

*Lepus castroviejoi* Palacios, 1977. Doñana, Acta Vertebr., 1976, 3(2):205 [1977].

COMMON NAME: Broom Hare.

TYPE LOCALITY: "Puerto de la Ventana, San Emiliano (León [Province])" [= Puerto Ventanas, Spain, 1500 m].

DISTRIBUTION: Cantabrian Mtns between Sierra de Ancares and Sierra de Peña Labra (N Spain).

STATUS: IUCN – Vulnerable.

COMMENTS: Subgenus *Eulagos* (Averianov, 1998). Reviewed by Palacios (1983, 1989) and Bonhomme et al. (1986).

*Lepus comus* Allen, 1927. Am. Mus. Novit., No 284:9.
COMMON NAME: Yunnan Hare.
TYPE LOCALITY: "Teng-yueh [Tengueh], Yunnan Province, China, 5,500 feet [1676 m] altitude."
DISTRIBUTION: Yunnan, W Guizhou (China).
STATUS: IUCN – Lower Risk (lc).
SYNONYMS: *peni* Wang and Luo, 1985; *pygmaeus* Wang and Feng, 1985.
COMMENTS: Subgenus *Eulagos* (Averianov, 1998). Formerly included in *oiostolus*; see Corbet (1978c). Elevated to specific status by Cai and Feng (1982) and Wang et al. (1985), on the basis of morphological and ecological differences. May be allo- or parapatric with *oiostolus*. Possibly related to *nigricollis* (Flux and Angermann, 1990).

*Lepus coreanus* Thomas, 1892. Ann. Mag. Nat. Hist., ser. 6, 9:146.
COMMON NAME: Korean Hare.
TYPE LOCALITY: "Söul" [Seoul], Korea.
DISTRIBUTION: Korea; S Kirin, S Liaoning, E Heilungjiang (China).
STATUS: IUCN – Lower Risk (lc).
COMMENTS: Subgenus *Eulagos* (Averianov, 1998). Formerly included in *sinensis* (Corbet, 1978c) or in *brachyurus* (Kim and Kim, 1974); here considered distinct, following Flux and Angermann (1990) and Jones and Johnson (1965).

*Lepus corsicanus* de Winton, 1898. Ann. Mag. Nat. Hist., ser. 7, 1:155.
COMMON NAME: Corsican Hare.
TYPE LOCALITY: "Bastia," [Corsica, Italy].
DISTRIBUTION: Italy from the Abruzzo Mtns southward; Sicily; introduced into Corsica no later than 16th Century (Vigne, 1988).
STATUS: Not Evaluated; likely to be listed as one of the IUCN threatened categories upon evaluation, due to probable reduction in numbers and range due to over-hunting and introduction of *L. europaeus* (Palacios et al., 1989; Pierpaoli et al., 1999).
COMMENTS: Subgenus probably *Eulagos*. Formerly included in *capensis* or *europaeus*; see Ellerman and Morrison-Scott (1951) and Petter (1961a); but see also Palacios et al. (1989) and Pierpaoli et al. (1999) who provided evidence of their specific distinctness.

*Lepus europaeus* Pallas, 1778. Nova Spec. Quad. Glir. Ord., p. 30.
COMMON NAME: European Hare.
TYPE LOCALITY: Not stated; restricted by Trouessart (1910), to Poland (see discussion in Ognev, 1940:140, who further restricted it to SW Poland).
DISTRIBUTION: Open woodland, steppe and sub-desert: from S Sweden and Finland to Britain, throughout Europe (not Iberian Penin. south of Cantabria and the Ebro River, or south of Siena in Italy), to W Siberian lowlands; south to N Israel, N Syria, N Iraq, the Tigris-Euphrates valley and W Iran. SE border of range (Iran) from S Caspian Sea south to Persian Gulf (54°E); see Angermann (1983:19). Introduced to Ireland, SE Canada-NE USA, S South America, Australia, New Zealand and several islands, including Barbados, Réunion, and the Falklands.
STATUS: IUCN – Lower Risk (lc); however, this once common species is now declining across Europe and a reevaluation of its status is likely to place it in one of the IUCN threatened categories (Schneider, 1997).
SYNONYMS: *alba* Bechstein, 1801; *argenteogrisea* König-Warthausen, 1875; *cyanotus* Blanchard, 1957; *flavus* Bechstein, 1801; *niger* Bechstein, 1801; *pyrenaicus* Hilzheimer, 1906; **caspicus** Hemprich and Erhenberg, 1832; *kalmykorum* Ognev, 1929; **connori** Robinson, 1918; *astaricus* Baloutch, 1978; *iranensis* Goodwin, 1939; **creticus** Barrett-Hamilton, 1903; **cyprius** Barrett-Hamilton 1903; *cyrensis* Satunin, 1905; *caucasicus* Ognev, 1929; *ghigi* de Beaux, 1927; **hybridus** Desmarest, 1822; *aquilonius* Blasius, 1842; *biarmicus* Heptner, 1948; *borealis* Kuznetsov, 1944 (not Pallas, 1778; not Nilsson, 1820); *campestris* Bogdanov, 1871; *hyemalis* Tumac, 1850; *tesquorum* Ognev and Worobiev, 1923; *tumac* Tichomirov and Kortchagin, 1889; **judeae** Gray, 1867; **karpathorum** Hilzheimer, 1906; **medius** Nilsson, 1820; **occidentalis** de Winton, 1898; **parnassius** Miller, 1903; *niethammeri* Wettstein, 1943; **ponticus** Ognev, 1929; **rhodius** Festa, 1914; **syriacus** Hemprich and Ehrenberg, 1832; **transsylvanicus** Matschie, 1901;

*campicola* Gervais, 1859 [*nomen nudum*]; *cinereus* Fitzinger, 1867 [*nomen nudum*]; *coronatus* Fitzinger, 1867 [*nomen nudum*]; *laskerewi* Khomenko, 1916; *maculatus* Fitzinger, 1867 [*nomen nudum*]; *meridiei* Hilzheimer, 1906 [*nomen nudum*]; *meridionalis* Gervais, 1859 [*nomen nudum*]; *nigricans* Fitzinger, 1867 [*nomen nudum*]; *rufus*, Fitzinger, 1867 [*nomen nudum*]; *transsylvaticus* Hilzheimer, 1906.

COMMENTS: Subgenus *Eulagus* (Averianov, 1998; Gromov and Baranova, 1981; Gureev, 1964). This species was earlier placed in *capensis* by Petter (1961*a*) based on what was interpreted as a cline in morphological characters (mainly size) from NE Africa eastward across the N Arabian peninsula and the Middle East, and northward through Israel to Turkey. Sympatry between large "*europaeus*" and small "*capensis*" (= *tolai*) in Kazakhstan, without evidence of hybridization (Sludskii et al., 1980) was interpreted as overlapping ends of a Rassenkreis. Angermann's (1983) re-analysis indicated a marked discontinuity between smaller "capensis" (incl. *arabicus*) and larger *europaeus* running from the E Mediterranean coast (C Israel) through Iran, and on this basis we separate *europaeus* from *capensis* and *tolai*. East of the border of the range of *europaeus* in Iran, *tolai* occurs, apparently in allo- or parapatry with *europaeus*. Insular populations in the E Mediterranean were assigned to this species (Ellerman and Morrison-Scott, 1951), but need to be reviewed. See also comments under *tolai* and *tibetanus*.

*Lepus fagani* Thomas, 1903. Proc. Zool. Soc. Lond., 1902(2):315 [1903].
COMMON NAME: Ethiopian Hare.
TYPE LOCALITY: "Zegi, Lake Tsana [Tana, Ethiopia] 4000 feet. [1219 m]"
DISTRIBUTION: N and W Ethiopia, and adjacent SE Sudan, south to extreme NW Kenya.
STATUS: IUCN – Data Deficient.
COMMENTS: Subgenus *Sabanalagus* (Avarianov, 1998). Formerly included in *crawshayi* (= *victoriae* = *microtis*) by Gureev (1964:204), but Azzaroli-Puccetti (1987*a*, *b*) maintained its specific identity. Its known distribution is largely allo- or parapatric to that of *microtis*; may be a highland allospecies (Flux and Angermann, 1990).

*Lepus flavigularis* Wagner, 1844. *In* Schreber, Die Säugethiere . . . , Suppl. 4:106.
COMMON NAME: Tehuantepec Jackrabbit.
TYPE LOCALITY: "Mexico" Restricted by Elliot (1905:543) to San Mateo del Mar, Tehuantepec [City, Oaxaca, Mexico].
DISTRIBUTION: Coastal plains and bordering foothills on south end of Isthmus of Tehuantepec (Oaxaca, Mexico), along Pacific coast to Chiapas (Mexico); now restricted to small area between Salina Cruz, Oaxaca, and extreme W Chiapas.
STATUS: IUCN – Endangered; likely to be reclassified as IUCN – Critically Endangered.
COMMENTS: Subgenus *Proeulagus* (Gureev, 1964:193) or *Macrotolagus* (Averianov, 1998). Closely related to *callotis*, with which it has an isolated allopatric distribution; see Anderson and Gaunt (1962); also see Hall (1981:330). Karyotype reported by Uribe-Alcocer et al. (1989). Reviewed by Cervantes (1993, Mammalian Species, 423).

*Lepus granatensis* Rosenhauer, 1856. Die Thiere Andalusiens, 3.
COMMON NAME: Granada Hare.
TYPE LOCALITY: "bei Graneda" [Granada, Analusia Prov., Spain].
DISTRIBUTION: Iberian Peninsula, except NE and NC parts (Spain, Portugal); Mallorca (Balearic Isl, Spain).
STATUS: Not Evaluated.
SYNONYMS: *hispanicus* Fitzinger, 1867; *iturissius*, Miller, 1907; *lilfordi* de Winton, 1898; *meridionalis* Graells, 1897; **gallaecius** Miller, 1907; **solisi** Palacios and Fernández, 1992. **Unassigned:** *mediterraneus* Wagner, 1841 [see comments]; *mediterraneus* Machado, 1869; *typicus* Hilzheimer, 1906.
COMMENTS: Subgenus probably *Eulagus*. Formerly included in *europaeus* or *capensis*; but see Palacios (1983, 1989), and Bonhomme et al. (1986). The Majorcan population (*solisi*) is thought to have been introduced by humans (Palacios and Fernández, 1992). The population in Sardinia, to which the names *mediterraneus* Wagner, 1841) and *typicus* Hilzheimer, 1906, are applied, is assigned to this species based on Miller (1912*a*), who regared it as closest to *granatensis,* though he retained it as a " . . . very distinct species"

(pg. 514) because of its small size, but its status needs investigation, as do populations from the NW African coast that have been assigned to "*capensis*". To date there appears not to be any definitive study of these populations. However, if in future *mediterraneus* is confirmed as a synonym of *granatensis*, it has priority over *granatensis*.

*Lepus habessinicus* Hemprich and Ehrenberg, 1832. Symbolae Physicae, Mammalia, dec. 2, folio p, page 2, plate 15, f. 2.
COMMON NAME: Abyssinian Hare.
TYPE LOCALITY: "East coast of Abyssinia, Near Arkiko." (G. M. Allen, 1939:275.)
DISTRIBUTION: Djibouti, E Ethiopia, Somalia, perhaps NE Kenya.
STATUS: IUCN – Lower Risk (lc).
SYNONYMS: *abyssinicus* Lefebvre, 1850; *berberanus* Heuglin, 1861; *cordeauxi* Drake-Brockman, 1911; *crispii* Drake-Brockman, 1911; *somalensis* Heuglin, 1861; *tigrensis* Blanford, 1869.
COMMENTS: Subgenus undetermined. "This hare apparently replaces the Cape hare in the open grassland, steppe, savanna and desert habitats . . ." (Flux and Angermann, 1990) in the Horn of Africa. Azzaroli-Puccetti (1987 *a*, *b*) provides evidence of full species status. " . . . appears to be sympatric with '*capensis*' [*hawkeri*] and '*crawshayi*' [*microtus*] in . . . Somalia and Ethiopia." (Petter, 1972:4).

*Lepus hainanus* Swinhoe, 1870. Proc. Zool. Soc. Lond., 1870:233.
COMMON NAME: Hainan Hare.
TYPE LOCALITY: Hainan Isl, "in the neighbourhood of the capital city" [Hainan Province, China].
DISTRIBUTION: Lowlands of Hainan Isl (China).
STATUS: IUCN – Vulnerable.
COMMENTS: Subgenus *Indolagus*. Placed in *Caprolagus* (*Indolagus*) by Gureev (1964:146). Considered a subspecies of *peguensis* by Ellerman and Morrison-Scott (1951); Flux and Angermann (1990) recommended provisional specific status, as did Gureev (1964).

*Lepus insularis* W. Bryant, 1891. Proc. California Acad. Sci., ser. 2, 3:92.
COMMON NAME: Black Jackrabbit.
TYPE LOCALITY: "Espiritu Santo Island, [near La Paz], Gulf of California [Baja California del Sur], Mexico."
DISTRIBUTION: Restricted to the type locality.
STATUS: IUCN – Lower Risk (nt).
SYNONYMS: *edwardsi* Saint-Loup, 1895.
COMMENTS: Subgenus *Proeulagus* (Gureev, 1964:195). Insular melanic allospecies, related to *californicus*; see Hall (1981:328) and Dixon et al. (1983). Chromosomes, which differ from those of *L. californicus*, are described by Cervantes et al. (1999-2000). Reviewed by Cervantes et al. (1996*a*), and Thomas and Best (1994*b*, Mammalian Species, 465).

*Lepus mandshuricus* Radde, 1861. Melanges Biol. Acad. St. Petersbourg, 3:684.
COMMON NAME: Manchurian Hare.
TYPE LOCALITY: "Im Chy (Gebirge)" Bureya Mtns [Khabarovskii Krai, Russia].
DISTRIBUTION: Ussuri region (Russia); NE China; extreme NE Korea.
STATUS: IUCN – Lower Risk (lc).
SYNONYMS: *melainus* Li and Luo, 1979; *melanonotus* Ognev, 1922.
COMMENTS: Subgenus *Eulagos* (Averianov, 1998). Distinct from *brachyurus*; see Angermann (1966, 1983); placed in *Caprolagus* (*Allolagus*) *brachyurus* by Gureev (1964:150); followed by Gromov and Baranova (1981:63). Melanic individuals known since at least the time of Sowerby (1923) have been given the specific designation *melainus* (Li and Luo, 1979). The range of this taxon is entirely within that of *mandshuricus*, and we provisionally retain them in that species, although Flux and Angermann (1990) recognized *melainus*. *L. mandshuricus* and *L. coreanus* are parapatric in distribution in NE Korea/SE Heilungjiang, but are described as occupying different habitats; the former, mixed forest in hilly country, the latter, both forest and cultivated land, primarily in the plains (Flux and Angermann, 1990). Moreover, *mandshuricus* is sympatric with another forest species, *timidus*, and with the plains species, *tolai*; as forest is cleared, *tolai* tends to replace *mandshuricus* (Flux and Angermann, 1990).

L. *mandshuricus*, *L. timidus* and *L. tolai* all occur in the area occupied by the taxon *melainus*; four species of sympatric hares, three of them forest-dwellers, is unprecedented in hare ecology, and supports the view that *melainus* is not a distinct species.

*Lepus microtis* Heuglin, 1865. Leopoldiana, 5:32, in Nova Acta Acad. Caes. Leop.-Carol., Halle, 24.
 COMMON NAME: African Savanna Hare.
 TYPE LOCALITY: "Lande der Ridj," [Bahr-el-Ghazal, Sudan].
 DISTRIBUTION: From Atlantic coast of NW Africa (Senegal, south to Guinea and Sierra Leone) eastward across Sahel to Sudan and extreme W Ethiopia; southward through E Africa (E Republic of Congo, W Kenya) to NE Namibia, Botswana, and KwaZulu-Natal (South Africa). Small isolated population in W Algeria.
 STATUS: IUCN – Lower Risk (lc); isolated population around Beni Abbés, Algeria, "deserve[s] attention" (Flux and Angermann, 1990).
 SYNONYMS: *crawshayi* de Winton, 1899; *kakumegae* Heller, 1912; *raineyi* Heller, 1912; *victoriae* Thomas, 1893; **angolensis** Thomas, 1904; *ansorgei* Thomas and Wroughton, 1905; *canopus* Thomas and Hinton, 1921; *meridionalis* Monard, 1933; *zairensis* Hatt, 1935; **senegalensis** Rochebrune, 1883; *zechi* Matschie, 1899; **whytei** Thomas, 1894; *herero* Thomas, 1926; *micklemi* Chubb, 1908; *zuluensis* Thomas and Schwann, 1905.
 COMMENTS: Placed (as *crawshayi*) in subgenus *Proeulagus* by Gureev (1964), and in subgenus *Sabanalagus* by Averianov (1998). Gureev recognized both *crawshayi* and *whytei* as distinct species, as did Azzaroli-Puccetti (1987a). Formerly included in *saxatilis*; see comments under that species. This species has been known under several different names (*saxatilis, crawshayi, whytei, victoriae,* and now *microtis*). Angermann and Feiler (1988) thought that the oldest available name for this species was *victoriae* Thomas, 1893, but apparently did not consider *microtus* Heuglin 1865. The species is widely sympatric with *capensis*, but allo- to parapatric with *saxatilis* (which is also sympatric with *capensis* sensu stricto), and with the small *L. habessinicus*.

*Lepus nigricollis* F. Cuvier, 1823. Dict. Sci. Nat., 26:307.
 COMMON NAME: Indian Hare.
 TYPE LOCALITY: "Malabar" [Madras, India].
 DISTRIBUTION: Pakistan; India; Bangladesh, except Sunderbands; Sri Lanka; introduced into Java (?) and Mauritius, Gunnera Quoin, Anskya, Réunion and Cousin Isls in the Indian Ocean. Considered native to Java by McNeely (1981:931).
 STATUS: IUCN – Lower Risk (lc); if *nigricollis* is native to Java (rather than an introduced population), its numbers are now very low there.
 SYNONYMS: **aryabertensis** Hodgson, 1844; **dayanus** Blanford, 1874; *cutchensis* Kloss, 1918; *joongshaiensis* Murray, 1884; *rajput* Wroughton, 1918; **ruficaudatus** Geoffroy, 1826; *macrotus* Hodgson, 1840; *tytleri* Tytler, 1854; **sadiya** Kloss, 1918; **simcoxi** Wroughton, 1912; *mahadeva* Wroughton and Ryley, 1913; **singhala** Wroughton, 1915.
 COMMENTS: Subgenus *Indolagus* (Gureev, 1964). Placed in genus *Caprolagus* (*Indolagus*) by Gureev (1964:139). Includes *ruficaudatus*; see Prater (1980) and Angermann (1983), but see Gureev (1964:142); *ruficaudatus* is closer to *capensis* according to Petter (1961a), and *nigricollis* may include *whytei, crawshayi, peguensis* and *siamensis*; but also see comments under *microtis, peguensis* and *saxatilis*. Includes *dayanus*, given specific status by Gureev (1964:139).

*Lepus oiostolus* Hodgson, 1840. J. Asiat. Soc. Bengal, 9:1186.
 COMMON NAME: Woolly Hare.
 TYPE LOCALITY: " . . . the snowy region of the Hemalaya, and perhaps also Tibet." Restricted by Kao and Feng (1964), to "Southern Tibet" [Xizang, China].
 DISTRIBUTION: Tibetan Plateau, from Ladak to Sikkim (India) Nepal, and eastward through Xizang (Tibet) and Qinghai, Gansu and Sichuan (China).
 STATUS: IUCN – Lower Risk (lc).
 SYNONYMS: *illuteus* Thomas, 1914; *oemodias* Gray, 1847; *qusongensis* Cai and Feng, 1982; **hypsibius** Blanford, 1875; **pallipes** Hodgson, 1842; *grahami* Howell, 1928; *kozlovi* Satunin, 1907; *sechuenensis* de Winton, 1899; **przewalskii** Satunin, 1907; *qinghaiensis* Cai and Feng, 1982; *tsaidamensis* Hilzheimer, 1910.

COMMENTS: Placed in subgenus *Proeulagus* by Gureev (1964) and *Eulagos* by Averianov (1998); *przewalskii* was assigned to *capensis* (= *tolai*) by Corbet (1978*c*), but is placed in *oiostolus* following Cai and Feng (1982).

*Lepus othus* Merriam, 1900. Proc. Wash. Acad. Sci., 2:28.

COMMON NAME: Alaskan Hare.

TYPE LOCALITY: "St. Michaels, [Norton Sound], Alaska." [USA].

DISTRIBUTION: W and SW Alaska (USA); formerly perhaps northwestward to Pt. Barrow; as here interpreted, also E Chukotsk (Russia); if this is verified, *tschuktschorum* Nordquist, 1883 has priority.

STATUS: IUCN – Lower Risk (lc).

SYNONYMS: *poadromus* Merriam, 1900; **tschuktschorum** Nordquist, 1883 [see comments].

COMMENTS: Subgenus *Lepus* (Averianov, 1998). Formerly included in *arcticus* or *timidus* (see comments therein). Regarded as distinct by Hall (1981) and by Flux and Angermann (1990), who, however, followed A. J. Baker et al. (1983) in allying populations from Eastern Siberia (Chukotka) (*tschuktschorum* Nordquist, 1883) with Alaskan populations; but see also Pavlinov and Rossolimo (1987). Reviewed by Anderson (1974) who found that *othus* was closer morphologically to *townsendii* than to other northern hares. More work is required to determine whether Eastern Siberian populations are linked to *L. othus* or to *L. timidus*. If the former relationship is supported, then *tschuktschorum* Nordquist, 1883 has priority over *othus* Merriam, 1900. Reviewed by Best and Henry (1994*b*, Mammalian Species, 458).

*Lepus peguensis* Blyth, 1855. J. Asiat. Soc. Bengal, 24:471.

COMMON NAME: Burmese Hare.

TYPE LOCALITY: "Pegu" [Upper Pegu, Burma].

DISTRIBUTION: C, S Burma from Chindwin River valley east through Thailand; Cambodia; S Laos, S Vietnam; south in upper Malay Peninsula (Burma, Thailand) to 120EN.

STATUS: IUCN – Lower Risk (lc).

SYNONYMS: *siamensis* Bonhote, 1902; **vassali** Thomas, 1906.

COMMENTS: Subgenus *Indolagus* by Gureev (1964:144) in *Caprolagus* (*Indolagus*); he ranked *siamensis* as a distinct species; but see Lekagul and McNeely (1977:333) and Flux and Angermann (1990). Petter (1961*a*) suggested that *peguensis* might be conspecific with *nigricollis* because of its close resemblance to *L. n. ruficaudatus*. However, *L. n. ruficaudatus* appears to be allopatric with respect to *peguensis* in E India-W Burma. Suchentrunk (2004:28) considered the separate species status of *L. peguensis* still open. Formerly included *hainanus* (Ellerman and Morrison-Scott, 1955), which is here considered a full species.

*Lepus saxatilis* F. Cuvier, 1823. Dict. Sci. Nat., 26:309.

COMMON NAME: Scrub Hare.

TYPE LOCALITY: "il habite les contrées qui se trouvent à trois journées au nord du cap de Bonne-Espérance," [Cape of Good Hope, South Africa].

DISTRIBUTION: South Africa (former Cape Province [and Zululand north to C KwaZulu-Natal?]) and S Namibia.

STATUS: IUCN – Lower Risk (lc).

SYNONYMS: *albaniensis* Roberts, 1932; *aurantii* Thomas and Hinton, 1923; *chiversi* Roberts, 1929; *fumigatus* Wagner, 1844; *longicaudatus* Gray, 1837; *megalotis* Thomas and Schwann, 1905; **subrufus** Roberts, 1913; *bechuanae* Roberts, 1932; *chobiensis* Roberts, 1932; *damarensis* Roberts, 1926; *gungunyanae* Roberts, 1914; *khanensis* Roberts, 1946; *ngamiensis* Roberts, 1932; *nigrescens* Roberts, 1932; *orangensis* Kolbe, 1948; *rufinucha* A. Smith, 1829; *timidus* A. Smith, 1826 [not Linnaeus, 1758].

COMMENTS: Placed by Gureev (1964:203) in subgenus *Proeulagus*, and in *Sabanalagus* by Averianov, 1998). Formerly included *crawshayi* and *whytei*, see Ansell (1978:67), Swanepoel et al. (1980:159), and Robinson and Dippenaar (1983*b*, 1987); but see also Petter (1961*a*, 1972*b*). Angermann (1983) considered *whytei* a distinct species that includes *crawshayi*; Flux and Angermann (1990) placed both as subspecies of *victoriae* (= *microtis*); see comments therein. The range of *saxatilis* completely overlaps the range of *capensis* sensu stricto, except in northern Southwest Africa, Botswana and

Mozambique, where the smaller northern subspecies (*subrufus*) *is* allopatric with respect to both large *capensis* and the southern race of equally large *L. s. saxatilis* (Flux and Angermann, 1990; Smithers, 1983).

*Lepus sinensis* Gray, 1832. Illustr. Indian Zool., 2, pl. 20.
COMMON NAME: Chinese Hare.
TYPE LOCALITY: "China". Restricted by G. Allen (1938:559) to "more or less in the region of Canton." [Guangzhou, Guandong Province, China].
DISTRIBUTION: SE China from Yangtse River southward; Taiwan; disjunct in NE Vietnam.
STATUS: IUCN – Lower Risk (lc).
SYNONYMS: *flaviventris* G. Allen, 1927; *formosus* Thomas, 1908; *yuenshanensis* Shih, 1930.
COMMENTS: Subgenus *Sinolagus* (Averianov, 1998). Placed in *Caprolagus* (*Indolagus*) by Gureev (1964:143). Formerly included *coreanus*; see Corbet (1978c: 73); here considered distinct, following Flux and Angermann (1990).

*Lepus starcki* Petter, 1963. Mammalia, 27:239.
COMMON NAME: Ethiopian Highland Hare.
TYPE LOCALITY: "Jeldu-Liban-Shoa, 2,740 mtres, 40 km W. Addis Abeba," [Ethiopia].
DISTRIBUTION: Central highlands of Ethiopia.
STATUS: IUCN – Lower Risk (lc).
COMMENTS: Subgenus *Eulagos* (Averianov, 1998). Formerly included in *capensis* (Petter, 1963a), or *europaeus* (Azzaroli-Puccetti, 1987a, b); but see Angermann (1983) and Flux and Angermann (1990). Considered by Azzaroli-Puccetti (1987a, b) to be closely related to *europaeus*, with a relict distribution dating back to the Pleistocene.

*Lepus tibetanus* Waterhouse, 1841. Proc. Zool. Soc. London, Part IX: 7.
COMMON NAME: Desert Hare.
TYPE LOCALITY: "Little Thibet." Fixed as Baltistan, Kashmir, by Ellerman and Morrison-Scott (1955).
DISTRIBUTION: Afghanistan and Baluchistan eastward through N Pakistan and Kashmir to the E Pamir, NW Xinjiang and the Altai Mountains, thence eastward across S Mongolia to Gansu and Ningxia (China). The distribution of *tibetanus* relative to *tolai* is allo-to parapatric, but in the Tien Shan mountains they may be sympatric.
STATUS: Not Evaluated; widespread, but population levels not studied.
SYNONYMS: *biddulphi* Blanford, 1877; *centrasiaticus* Satunin, 1907; *gansuicus* Satunin, 1907; *craspedotis* Blanford, 1875; *pamirensis* Günther, 1875; *stoliczkanus* Blanford, 1875; *kashgaricus* Satunin, 1907; *quercerus* Hollister, 1912; *zaisanicus* Satunin, 1907.
COMMENTS: Subgenus *Proeulagus* (Gromov, 1964). Until the 1930's *tibetanus* was considered a distinct species. The first major revision (Heptner, 1934) united *europaeus, tolai* and *tibetanus* in a single species, but Ognev (1966:154) rejected this concept, stating that " . . . there is much evidence against considering the common hare, the Tolai and desert hares as one species . . .". Next, Ellerman (in Ellerman and Morrison-Scott, 1955) placed *tibetanus* as a subspecies of *capensis*, along with *tolai*; he was supported by Petter (1959, 1961a). Then Harrison (1972) added *arabicus* to *capensis*; see also Corbet (1978), and comments under *tolai*. Some, however, continued to follow Ognev. Bannikov (1954), Sokolov and Orlov (1980), and Shou (1962) provided details of distribution in Mongolia and China respectively. Luo (1981) performed a cluster analysis that he interpreted as supporting Ellerman, et al., but was strongly criticized by Zhao et al. (1983) for his methodology. Qui (1989) then re-analyzed the data, and found that three races of *tibetanus* were clearly separated from four races of *tolai* (although Qui continued to employ *capensis* as the species name). *L. tibetanus* shares certain characteristics with *L. oiostolus* (but not *capensis* or *tolai* ) of the adjacent Tibetan Plateau, most notably the relatively long premaxillary and short nasal bones, combined with greater procumbency of the incisors, as well as other cranial and pelage characters described by Ognev (1966). Evaluation of these characters across the zones of potential contact between the ten taxon pairs comprising *L. capensis* (*sensu lato*) is necessary before the taxonomy of these hares can be resolved (Hoffmann, 1998).

*Lepus timidus* Linnaeus, 1758. Syst. Nat., 10th ed., 1:57.

> COMMON NAME: Mountain Hare.
>
> TYPE LOCALITY: "in Europa" [Uppsala, Sweden].
>
> DISTRIBUTION: Palearctic from Scandinavia to E Siberia, except E Chukotsk (Russia), south to Sakhalin and Sikhote-Alin Mtns (Russia); Hokkaido (Japan); Heilungjiang, N Xinjiang (China); N Mongolia; Altai, N Tien Shan Mtns; N Ukraine, E Poland, and Baltics; isolated populations in the Alps, Scotland, Wales and Ireland. Introduced into England, Faeros and Scottish Isles.
>
> STATUS: IUCN – Lower Risk (lc).
>
> SYNONYMS: *abei* Kuroda, 1938; *alpinus* Erxleben, 1777 [not Pallas, 1773]; *algidus* Pallas, 1778; *borealis* Pallas, 1778; *canescens* Nilsson, 1844; *collinus* Nilsson, 1831; *septentrionalis* Link, 1795; *sylvaticus* Nilsson, 1831; *typicus* Barrett-Hamilton, 1900; *variabilis* Pallas, 1778; **ainu** Barrett-Hamilton, 1900; *albus* Leach, 1816 [*nomen nudum*]; **begitschevi** Koljuschev, 1936; **gichiganus** J. Allen, 1903; **hibernicus** Bell, 1837; *lutescens* Barrett-Hamilton, 1900; **kamtschaticus** Dybowski, 1922; **kolymensis** Ognev, 1923; **kozhevnikovi** Ognev, 1929; **lugubris** Kastschenko, 1899; *altaicus* Barrett-Hamilton, 1900; **mordeni** Goodwin, 1933; **orii**, Kuroda, 1928; *saghaliensis* Abe, 1931; *rubustus* Urita, 1935 [*nomen nudum*]; **scoticus** Hilzheimer, 1906; **sibiricorum** Johanssen, 1923; **transbaicalicus** Ognev, 1929; **varronis** Miller, 1901; **breviauritus** Hilzheimer, 1906.
>
> COMMENTS: Subgenus *Lepus* (Gureev, 1964; Averianov, 1998). Formerly included *arcticus* and *othus*; see Corbet (1978c: 73); but also see comments under those species. A. J. Baker et al. (1983) found Scottish and Alpine populations morphologically distinct, as well as geographically isolated, from other populations, and Flux (1983) remarked that *L. t. scoticus* and *L. t. hibernicus* (from Scotland and Ireland, respectively), both introduced on the island of Mull (Hewson, 1991) still do not interbreed after 50 years. Reviewed by Angerbjorn and Flux (1995, Mammalian Species, 495).

*Lepus tolai* Pallas, 1778. Nova Spec. Quad. Glir. Ord., p. 17.

> COMMON NAME: Tolai Hare.
>
> TYPE LOCALITY: "Caeterum in montibus aprecis campisque rupestribus vel arenosis circa Selengam . . ." Restricted by Ognev (1940:162) to " . . . valley of the Selenga River. . . ." [Russia]. According to Ellerman and Morrison-Scott (1951:430) the type locality is "Adinscholo Mountain, near Tchinden [Chinden = Chindant], on Borsja [Boriya] River, a tributary of the Onon River, Eastern Siberia." This locality is more than 700 km east of the Selenga River.
>
> DISTRIBUTION: Steppes north of Caspian Sea southward along eastern shore of Caspian to E Iran; eastward through Afghanistan; Kazakhstan and S Siberia, Middle Asian republics to Mongolia; and W, C, and NE China.
>
> STATUS: Not Evaluated.
>
> SYNONYMS: *butlerowi* Bogdanov, 1882 [*nomen nudum*]; *gobicus* Satunin, 1907; *huangshuiensis* Luo, 1982; *kessleri* Bogdanov, 1882 [*nomen nudum*]; **aurigineus** Hollister, 1912; **buchariensis** Ognev, 1922; *desertorum* Ognev and Heptner, 1928; *habibi* Baloutch, 1978; **cheybani** Baloutch, 1978; *petteri* Baloutch, 1978; **cinnamomeus** Shamel, 1940; **filchneri** Matschie, 1907; *brevinasus* J. Allen, 1909; *sowerbyae* Hollister, 1912; *subluteus* Thomas, 1908; **lehmanni** Severtsov, 1873; *aralensis* Severtsov, 1861 [*nomen nudum*]; *turcomanus* Heptner, 1934; **swinhoei** Thomas, 1894; *stegmanni* Matschie, 1907.
>
> COMMENTS: Subgenus *Proeulagus* (Gureev, 1964:198). Formerly included in *capensis* or *europaeus*; see comments therein. Formerly included *tibetanus*; but also see Bannikov, 1965, Sokolov and Orlov (1980:85), and Shou et al. (1962); Qui (1989) also provided evidence of differentiation of *tibetanus* but did not address specific status. Formerly included *przewalskii*, now assigned to *L. oiostolus*; see Cai and Feng (1982). "The situation in [southern] Iraq [and SW Iran] deserves a more detailed analysis" (Angermann, 1983:19). *L. tolai cheybani* occurs westward to about 55E-56EE, while *L. c. arabicus* occurs eastward to SE Iraq. Whether the two forms come into contact is not known, but their ranges may be separated by that of *L. europaeus connori* in SW Iran; see Baloutch (1978) and Angermann (1983).

*Lepus townsendii* Bachman, 1839. J. Acad. Nat. Sci. Philadelphia, 8(1):90.
COMMON NAME: White-tailed Jackrabbit.
TYPE LOCALITY: " . . . on the Walla-walla . . . river"; restricted by Nelson (1909:78) to Fort
Walla Walla, [near present town of Wallula, Walla Walla Co., Washington].
DISTRIBUTION: C Alberta and Saskatchewan east to extreme SW Ontario (Canada), south to
SW Wisconsin, Iowa, NW Missouri, west through C Kansas to NC New Mexico, west to
C Nevada, EC California (USA) and north to SC British Columbia (Canada).
STATUS: IUCN – Lower Risk (lc).
SYNONYMS: *sierrae* Merriam, 1904; **campanius** Hollister, 1915; *campestris* Bachman, 1837
[preoccupied by *campestris* Meyer, 1790].
COMMENTS: Placed (as *campestris*) in subgenus *Proeulagus* by Gureev (1964), or in *Eulagos*
(Averianov, 1998). Reviewed by Lim (1987, Mammalian Species, 288).

*Lepus yarkandensis* Günther, 1875. Ann. Mag. Nat. Hist., ser. 4, 16:229.
COMMON NAME: Yarkand Hare.
TYPE LOCALITY: "neighbourhood of Yarkand" [15 mi (24 km) E, Xinjiang, China].
DISTRIBUTION: Steppes of Tarim Basin, S Xinjiang (China), around edge of Takla Makan
desert.
STATUS: IUCN – Lower Risk (nt).
COMMENTS: Placed in subgenus *Tarimolagus* by Gureev (1964); and also by Averianov (1998),
but see Xu (1986). Reviewed by Angermann (1967) and Gao (1983).

*Nesolagus* Forsyth-Major, 1899. Trans. Linn. Soc. London, 7:493.
TYPE SPECIES: *Lepus netscheri* Schlegel, 1880.

*Nesolagus netscheri* (Schlegel, 1880). Notes Leyden Mus., II:59.
COMMON NAME: Sumatran Striped Rabbit.
TYPE LOCALITY: "Sumatra: Padang-Padjang . . . about 2000 feet" [Padangpanjang, Sumatera
Barat, Indonesia; ca. 610 m].
DISTRIBUTION: Sumatra [Indonesia].
STATUS: IUCN – Critically Endangered; only a dozen museum specimens exist, collected
between 1880 and 1916. Since these early collections there has been only one
confirmed sighting (in 1972; Flux, 1990) and two photographic records (different
individuals, captured in an automatic camera trap in 1998; Surridge et al., 1999).

*Nesolagus timminsi* Averianov, Abramov and Tikhonov, 2000. Cont. Zool. Inst., St. Petersburg,
No. 3:3.
COMMON NAME: Annamite Striped Rabbit.
TYPE LOCALITY: "Vietnam, Ha Tinh Province, Huong Son District, Son Kim Community,
about 10 km south from village Nuoc Sot, 18°22′N,105°13′E, altitude 200m."
DISTRIBUTION: Known only from the vicinity of the type locality.
STATUS: IUCN – Data Deficient; presumed rare and potentially endangered.
COMMENTS: Little is known about this recently described species, except that morphologi-
cally it is very similar to *N. netscheri* (Averianov et al., 2000; Surridge et al., 1999).

*Oryctolagus* Lilljeborg, 1873. Sverig. Og Norges Ryggradsdjur, 1:417.
TYPE SPECIES: *Lepus cuniculus* Linnaeus, 1758.
SYNONYMS: *Cuniculus* Meyer, 1790.

*Oryctolagus cuniculus* (Linnaeus, 1758). Syst. Nat., 10th ed., 1:58.
COMMON NAME: European Rabbit.
TYPE LOCALITY: "in Europa australis" [= Germany; Ellerman and Morrison-Scott, 1951].
DISTRIBUTION: W and S Europe through the Mediterranean region to Morocco and
N Algeria; original post-Pleistocene range probably limited to S France, Iberia and
NW Africa, but Late Pleistocene records occur from Ireland to Italy, Hungary, and even
W Siberia (Kurtén, 1968); introduced on all continents except Antarctica and Asia; see
Gibb (1990). Worldwide as domesticated forms.
STATUS: IUCN – Lower Risk (lc); considered a pest species in most areas where it has been
introduced.

SYNONYMS: *fodiens* Gray, 1867; *kreyenbergi* Honigmann 1913; *vermicula* Gray, 1843 [*nomen nudum*]; *vernicularis* Thompson, 1837 [*nomen nudum*]; *algirus* (Loche, 1858); **brachyotus** Trouessart, 1917; *cnossius* Bate, 1906; **habetensis** Cabrera, 1923; *oreas* Cabrera, 1922; **huxleyi** Haeckel, 1874. **Unassigned**: *borkumensis* Harrison, 1952; *campestris* (Meyer, 1790); *nigripes* Bartlett, 1857.

COMMENTS: The specific name may be based on a feral specimen (Gibb, 1990).

*Pentalagus* Lyon, 1904. Smithson. Misc. Coll., 45:428.
TYPE SPECIES: *Caprolagus furnessi* Stone, 1900.

*Pentalagus furnessi* (Stone, 1900). Proc. Acad. Nat. Sci. Philadelphia, 52:460.
COMMON NAME: Amami Rabbit.
TYPE LOCALITY: "Liu Kiu Islands" [Amami-Oshima, Ryukyu Isls, Japan].
DISTRIBUTION: Amami Isls (Amami-Oshima and Tokuno-shima) (S Japan).
STATUS: U.S. ESA – Endangered; IUCN – Endangered.

*Poelagus* St. Leger, 1932. Proc. Zool. Soc. Lond., 1932(1):119.
TYPE SPECIES: *Lepus marjorita* St. Leger, 1929.
COMMENTS: Originally spelled *Poëlagus*, but this is a diaeresis and not an umlaut and thus the correct spelling is *Poelagus* (see Art. 32.5.2 of the International Code of Zoological Nomenclature, International Commission on Zoological Nomenclature, 1999). Formerly placed as subgenus of *Pronolagus*; see Ellerman and Morrison-Scott (1951:425); but see also Petter (1972b:5). Formerly placed as subgenus of *Caprolagus*; see Gureev (1964:152).

*Poelagus marjorita* (St. Leger, 1929). Ann. Mag. Nat. Hist., ser. 10, 4:292.
COMMON NAME: Bunyoro Rabbit.
TYPE LOCALITY: "Near Masindi, Bunyoro [Bunyuru] Uganda, 4000 ft. [1219 m]", Africa.
DISTRIBUTION: S Sudan, Uganda, Ruanda, Burundi, NE Dem. Rep. Congo, Central African Republic, S Chad, disjunct population in Angola.
STATUS: IUCN – Lower Risk (lc).
SYNONYMS: *larkeni* St. Leger, 1935; *oweni* Setzer, 1956.
COMMENTS: This savanna-woodland species, like *Pronolagus*, is associated with rocky outcrops.

*Pronolagus* Lyon, 1904. Smithson. Misc. Coll., 45:416.
TYPE SPECIES: *Lepus crassicaudatus* I. Geoffroy, 1832.
COMMENTS: From one (Peddie, 1975) to six (Roberts, 1951) species have been recognized in this genus (Robinson, 1982). Three species are now generally recognized (Duthie and Robinson, 1990).

*Pronolagus crassicaudatus* (I. Geoffroy, 1832). Mag. Zool. Paris, 2:cl. 1, pl. 9 and text.
COMMON NAME: Natal Red Rock Hare.
TYPE LOCALITY: "Port Natal" [Durban, KwaZulu-Natal, South Africa].
DISTRIBUTION: SE South Africa; extreme S Mozambique.
STATUS: IUCN – Lower Risk (lc).
SYNONYMS: *kariegae* Hewitt, 1927; **ruddi** Thomas and Schwann, 1905; *lebombo* Roberts, 1936; *lebomboensis* Roberts, 1936 [*lapsus*].
COMMENTS: Formerly included *randensis*, see Lundholm (1955a). The relationship of *crassicaudatus* and *randensis* is unclear, see Petter (1972b:6). Distribution allopatric to that of *randensis*, but sympatric in western half of range with *rupestris*.

*Pronolagus randensis* Jameson, 1907. Ann. Mag. Nat. Hist., ser. 7, 20:404.
COMMON NAME: Jameson's Red Rock Hare.
TYPE LOCALITY: "Observatory Kopje . . . Johannesburg . . . Witwatersrand Range, Transvaal . . . 5,900 ft." [1,798 m] [South Africa].
DISTRIBUTION: Two disjunct areas: NE South Africa, E Botswana to extreme W Mozambique, Zimbabwe; and W Namibia, perhaps SW Angola.
STATUS: IUCN – Lower Risk (lc).
SYNONYMS: *capricornis* Roberts, 1926; *makapani* Roberts, 1924; *powelli* Roberts, 1924;

*caucinus* Thomas, 1929; *ekmani* Lundholm, 1955; *fitzsimonsi* Roberts, 1938; *kaokoensis* Roberts, 1946; *kobosensis* Roberts, 1938; *waterbergensis* Hoesch and Von Lehmann, 1956; **whitei** Roberts, 1938.

COMMENTS: Formerly included in *crassicaudatus* by Lundholm (1955*a*); but see Petter (1972*b*: 6). The systematic position of the two widely disjunct populations needs clarification (Duthie and Robinson, 1990).

*Pronolagus rupestris* (A. Smith, 1834). S. Afr. Quart. J., 2:174.

COMMON NAME: Smith's Red Rock Hare.

TYPE LOCALITY: "South Africa, rocky situations" [probably Van Rhynsdorp District, Western Cape Prov., South Africa].

DISTRIBUTION: Two disjunct areas: S and C South Africa, S Namibia; and E Africa, from N Malawi and E Zambia north through C Tanzania to SW Kenya.

STATUS: IUCN – Lower Risk (lc).

SYNONYMS: *australis* Roberts, 1933; *melanurus* (Rüppell, 1842); *mülleri* Roberts, 1938; **curryi** (Thomas, 1902); **nyikae** (Thomas, 1902); **saundersiae** (Hewitt, 1927); *barretti* Roberts, 1949; *bowkeri* Hewitt, 1927; **vallicola** Kershaw, 1924.

COMMENTS: Formerly included in *crassicaudatus*, see Gureev (1964:174) and Peddie (1975); see also Robinson and Dippenaar (1983*a*). The systematic relationships of the two widely disjunct populations should be examined (Duthie and Robinson, 1990).

*Romerolagus* Merriam, 1896. Proc. Biol. Soc. Wash., 10:173.

TYPE SPECIES: *Lepus diazi* Ferrari-Peréz, 1893.

SYNONYMS: *Lagomys* Herrera, 1897 [not *Lagomys* Cuvier, 1800].

COMMENTS: Whether this monotypic genus represents "the most primitive of the living rabbits and hares" (Fa and Bell, 1990), or is closer to the more specialized leporids (*Sylvilagus*, *Oryctolagus*, *Lepus*) (Corbet, 1983), or is intermediate (Hibbard, 1963), remains controversial.

*Romerolagus diazi* (Ferrari-Pérez, 1893). *In* Diaz, Cat. Comision Geogr.-Expl. República Mexicana, Exposicion Intern, Columbia de Chicago, pl. 42.

COMMON NAME: Volcano Rabbit.

TYPE LOCALITY: "near San Martín Texmelusán, northeastern slope of Volcán Iztaccíhuatl [Ixtaccíhuatl, Puebla], Mexico."

DISTRIBUTION: Distrito Federal, Mexico, and W Puebla (Mexico), in three discontinuous areas on the slopes of Volcán Pelado, Tlaloc, Popocatépetl, and Ixtaccíhuatl.

STATUS: CITES – Appendix I; U.S. ESA – Endangered, IUCN – Endangered.

SYNONYMS: *nelsoni* Merriam, 1896.

COMMENTS: Reviewed by Cervantes et al. (1990, Mammalian Species, 360) and Velazquez et al. (1993).

*Sylvilagus* Gray, 1867. Ann. Mag. Nat. Hist., ser. 3, 20:221.

TYPE SPECIES: *Lepus sylvaticus* Bachman, 1837 (= *Lepus sylvaticus floridanus* J. Allen, 1890). Bachman's original name was preoccupied by *Lepus borealis sylvaticus* Nillson, 1832, which Allen did not realize when he used the name combination *L. sylvaticus floridanus*. The first use of the name combination *Sylvilagus floridanus* was Lyon, 1904.

SYNONYMS: *Hydrolagus* Gray, 1867; *Limnolagus* Mearns, 1897; *Microlagus* Trouessart, 1897; *Paludilagus* Hershkovitz, 1950; *Tapeti* Gray, 1867.

COMMENTS: Formerly included *Brachylagus* as a subgenus; see Hall (1981:294); but see also Corbet (1983:14).

*Sylvilagus aquaticus* (Bachman, 1837). J. Acad. Nat. Sci. Philadelphia, 7:319.

COMMON NAME: Swamp Rabbit.

TYPE LOCALITY: " . . . western parts of that state" [Alabama]. Restricted by Nelson (1909:272) to "Western Alabama".

DISTRIBUTION: S Illinois and SW Indiana, SW Missouri to SE Kansas southward through extreme W Kentucky and W Tennessee to E Oklahoma, E Texas, Louisiana, Alabama, Mississippi and NW South Carolina (USA).

STATUS: IUCN – Lower Risk (lc).

SYNONYMS: *attwateri* (J. Allen, 1895); *telmalemonus* (Elliott, 1899); ***littoralis*** Nelson, 1909.
COMMENTS: Subgenus *Tapeti* (Gureev, 1964:162). Reviewed by Chapman and Feldhamer (1981, Mammalian Species, 151).

*Sylvilagus audubonii* (Baird, 1858). Mammalia, *in* Repts. U.S. Expl. Surv., 8(8):608.
COMMON NAME: Desert Cottontail.
TYPE LOCALITY: "San Francisco" [San Francisco Co., California, USA].
DISTRIBUTION: NE Puebla and W Veracruz (Mexico) to NC Montana and SW North Dakota, NC Utah, C Nevada, and NC California (USA), south to Baja California and C Sinaloa (Mexico).
STATUS: IUCN – Lower Risk (lc).
SYNONYMS: *vallicola* Nelson, 1907; ***arizonae*** (Mearns, 1896); *laticinctus* (Elliot, 1904); *major* (Mearns, 1896); *rufipes* (Elliot, 1904); *sanctidiegi* (Miller, 1899); ***baileyi*** (Merriam, 1897); *neomexicanus* Nelson, 1907; ***confinis*** J. Allen, 1898; ***goldmani*** (Nelson, 1904); ***minor*** (Mearns, 1896); *parvulus* (J. Allen, 1904); ***warreni*** Nelson, 1907; *cedrophilus* Nelson, 1907.
COMMENTS: Subgenus *Sylvilagus* (Gureev, 1964:169). Revised by Hoffmeister and Lee (1963*b*). Reviewed by Chapman and Willner (1978, Mammalian Species, 106).

*Sylvilagus bachmani* (Waterhouse, 1839). Proc. Zool. Soc. Lond., 1839:103.
COMMON NAME: Brush Rabbit.
TYPE LOCALITY: "Between Monterey and Santa Barbara". Type locality restricted by Nelson (1909:247) to San Luis Obispo, California, USA.
DISTRIBUTION: W Oregon (USA) S of Columbia River to Baja California (Mexico), E to Cascade-Sierra Nevada Range (USA).
STATUS: U.S. ESA – Endangered as *S. b. riparius*; IUCN – Lower Risk (lc).
SYNONYMS: *macrorhinus* Orr, 1935; *riparius* Orr, 1935; *trowbridgii* (Baird, 1855); *virgulti* Dice, 1926; ***cerrosensis*** (J. Allen, 1898); ***cinerascens*** (J. Allen, 1890); *mariposae* Grinnell and Storer, 1916; ***exiguus*** Nelson, 1907; ***howelli*** Huey, 1927; *peninsularis* (J. Allen, 1898); *rosaphagus* Huey, 1940; ***ubericolor*** (Miller, 1899); *tehamae* Orr, 1935. See also *S. mansuetus*.
COMMENTS: Subgenus *Microlagus* (Lyon, 1904). Placed in genus *Microlagus* together with *idahoensis* by Gureev (1964:171). This is the only species of *Sylvilagus* known to have retained the putative ancestral karyotype (2n=48) shared by all known *Lepus*, and by *Romerolagus* (Robinson et al., 1981, 1984). Reviewed by Chapman (1974, Mammalian Species, 34).

*Sylvilagus brasiliensis* (Linnaeus, 1758). Syst. Nat., 10th ed., 1:58.
COMMON NAME: Tapeti.
TYPE LOCALITY: "in America meridionali"; type locality restricted by Thomas (1911*a*), to Pernambuco, Brazil.
DISTRIBUTION: S Tamaulipas (Mexico) southward through Central and South America as far as Peru, Bolivia, N Argentina and S Brazil.
STATUS: IUCN – Lower Risk (lc).
SYNONYMS: South of Isthmus of Panama: *braziliensis* (Waterhouse, 1848); *nigricaudatus* (Lesson, 1842); *tapeti* (Pallas, 1778); ***andinus*** (Thomas, 1897); ***canarius*** Thomas, 1913; *carchensis* Hershkovitz, 1938; *chimbanus* Thomas, 1913; *ecaudatus* Trouessart, 1910; *nivicola* Cabrera, 1912; ***apollinaris*** Thomas, 1920; ***capsalis*** Thomas, 1913; ***caracasensis*** Mondolfi and Méndez Aroche, 1957; ***chillae*** Anthony, 1957; ***chotanus*** Hershkovitz, 1938; ***defilippi*** (Cornalia, 1850); *defilippii* (Thomas, 1897); *dephilippii* Cabrera, 1912 [*lapsus*]; ***fulvescens*** J. Allen, 1912; *fuscescens* J. Allen, 1916 [*lapsus*]; *nicefori* Thomas, 1921; *salentus* J. Allen, 1913; ***gibsoni*** Thomas, 1918; ***inca*** Thomas, 1913; ***kelloggi*** Anthony, 1923; ***meridensis*** Thomas, 1904; ***minensis*** Thomas, 1901; ***paraguensis*** Thomas, 1901; *chapadae* Thomas, 1904; *chapadensis* Thomas, 1913; *paraguensis* Yepes, 1938; ***peruanus*** Hershkovitz, 1950; ***sanctaemartae*** Hershkovitz 1950; ***surdaster*** Thomas, 1901; *daulensis* J. Allen, 1914; *messorius* Goldman, 1912; ***tapetillus*** Thomas, 1913. North of Isthmus of Panama: *gabbi* (J. Allen, 1877); *consobrinus* Anthony, 1917; *incitatus* (Bangs, 1901); *tumacus* (J. Allen, 1908); ***truei*** (J. Allen, 1890).

COMMENTS: Subgenus *Tapeti* (Gureev, 1964:160); he also considered *gabbi*, which is included here, a distinct species. Formerly included *dicei*; revised by Diersing (1981). Two different karyotypes reported, by Guerena-Gandara et al. (1983) (2n=36; FN=68) and by Lorenzo and Cervantes (1995) (2n=40; FN=76). Allozymes described by Cervantes et al. (1999*a*).

*Sylvilagus cognatus* Nelson, 1907. Proc. Biol. Soc. Wash., 20:82.
   COMMON NAME: Manzano Mountain Cottontail.
   TYPE LOCALITY: "10,000 feet [3048 m] altitude, near summit of Manzano Mountains, New Mexico." [USA]. Restricted by Frey, et al. (1997) to " . . . vicinity of Rea Ranch . . . 1.9 km N and 13.4 km W of Tajique (T6N, NE 1/4 of NW 1/4 Sec. 9, N34°45'05.39", W106°25'18.04") on the northeast side of Bosque Peak, at 2880m (=9450 ft.) elevation."
   DISTRIBUTION: Restricted to the Manzano Mountains, New Mexico, USA.
   STATUS: Not Evaluated; likely endangered.
   COMMENTS: Subgenus *Sylvilagus*. Formerly included in *S. floridanus*, but see Ruedas (1998).

*Sylvilagus cunicularius* (Waterhouse, 1848). Nat. Hist. Mamm., 2:132.
   COMMON NAME: Mexican Cottontail.
   TYPE LOCALITY: "Mexico." Restricted by Goodwin (1969:125) to "Sacualpan" = Zacualpan.
   DISTRIBUTION: S Sinaloa to E Oaxaca and Veracruz (Mexico).
   STATUS: IUCN – Lower Risk (nt).
   SYNONYMS: *pacificus* (Nelson, 1904); *verae-crucis* (Thomas, 1890); *insolitus* (J. Allen, 1890). See also *S. graysoni*.
   COMMENTS: Subgenus *Sylvilagus* (Gureev, 1964:167). Reviewed by Cervantes et al. (1992, Mammalian Species, 412).

*Sylvilagus dicei* Harris, 1932. Occas. Pap. Mus. Zool. Univ. Mich., 248:1.
   COMMON NAME: Dice's Cottontail.
   TYPE LOCALITY: "El Copey de Dota, in the Cordillera de Talamanca, Costa Rica . . . 6000 feet. [1829 m]"
   DISTRIBUTION: Cordillera de Talamanca (SE Costa Rica, NW Panama).
   STATUS: IUCN – Endangered.
   COMMENTS: Subgenus *Tapeti*? Formerly included in *brasiliensis* (Hall, 1981:295); revised by Diersing (1981).

*Sylvilagus floridanus* (J. A. Allen, 1890). Bull. Am. Mus. Nat. Hist., 3:160.
   COMMON NAME: Eastern Cottontail.
   TYPE LOCALITY: "Sebastian River, Brevard Co.," [Florida, USA].
   DISTRIBUTION: N, C, and W Venezuela (including adjacent islands) and adjacent Colombia through Central America (disjunct in part); to NW Mexico, Arizona, north and east to North Dakota, Minnesota, N Michigan, New York and Massachusetts, Atlantic Coast south and Florida Gulf Coast (USA) west to Mexico; also S Saskatchewan, S Ontario and SC Quebec (C Canada).
   STATUS: IUCN – Lower Risk (lc).
   SYNONYMS: North of Mexico: *ammophilus* Howell, 1939; *paulsoni* Schwartz, 1956; **alacer** (Bangs, 1896); *mearnsi* (J. Allen, 1894); *similis* Nelson, 1907; **chapmani** (J. Allen, 1899); *caniclunis* (Miller, 1899); *llanensis* Blair, 1938; *simplicicanus* (Miller, 1902); **holzneri** (Mearns, 1896); *durangae* J. Allen, 1903; *hesperius* Hoffmeister and Lee, 1963; *rigidus* (Mearns, 1896); **mallurus** (Thomas, 1898); *hitchensi* Mearns, 1911; *sylvaticus* (Bachman, 1837) [preoccupied by Nillson, 1832]. Mexico and Central America: **aztecus** (J. Allen, 1890); *chiapensis* (Nelson, 1904); **connectens** (Nelson, 1904); *russatus* (J. Allen, 1904); **hondurensis** Goldman 1932; *costaricensis* Harris, 1933; **macrocorpus** Diersing and Wilson, 1980; *orizabae* (Merriam, 1893); *persultator* Elliot, 1903; *restrictus* Nelson, 1907; *subcinctus* (Miller, 1899); **yucatanicus** (Miller, 1899). South of Isthmus of Panama: **avius** Osgood, 1910; **cumanicus** (Thomas, 1897); *continentis* Osgood, 1912; *valenciae* Thomas, 1914; **margaritae** Miller, 1898; **nigronuchalis** (Hartert, 1894); **orinoci** Thomas, 1900; **purgatus** Thomas, 1920; **superciliaris** (J. Allen, 1899); *boylei* Allen, 1916.

COMMENTS: Subgenus *Sylvilagus* (Gureev, 1964:164). Widely introduced in North America (Hall, 1981:301) and Europe (Flux et al., 1990). Formerly included *robustus* (Bailey, 1905) and *cognatus* Nelson, 1907. Ruedas (1998) stated " . . . at least three of the subspecies [of *floridanus*] (*cognatus, holzneri,* and *robustus*) traditionally ascribed to *S. floridanus* are of species rank." However, he did not formally raise the first two to full species, as he did for *robustus*. Reviewed by Chapman et al. (1980, Mammalian Species, 136). See also *cognatus* and *robustus; holzneri* is provisionally retained as a subspecies of *floridanus,* since its relationship to Mexican populations of *floridanus* has not been reported yet. Allozymes described by Cervantes et al. (1999*a*).

*Sylvilagus graysoni* (J. A. Allen, 1877). *In* Coues and Allen, Monog. N. Amer. Rodentia (U.S. Geol. Geograph. Survey Terr., Rep., 11:347).
COMMON NAME: Tres Marias Cottontail.
TYPE LOCALITY: According to Nelson (1899*a*:16), "Tres Marias Islands," "undoubtedly from Maria Madre" Isl, Nayarit, Mexico.
DISTRIBUTION: Tres Marías Isls, Nayarit (Mexico).
STATUS: IUCN – Endangered.
SYNONYMS: *badistes* Diersing and Wilson, 1980.
COMMENTS: Subgenus *Sylvilagus* (Gureev, 1964:168). An insular species probably derived from *cunicularius* of the adjacent mainland; see Diersing and Wilson (1980) and Hall (1981:314). Reviewed by Cervantes (1997, Mammalian Species, 559).

*Sylvilagus insonus* Nelson, 1904. Proc. Biol. Soc. Wash., 17:103.
COMMON NAME: Omilteme Cottontail.
TYPE LOCALITY: "Omilteme, Guerrero," [Mexico].
DISTRIBUTION: Appears restricted to Sierra Madre del Sur, C Guerrero (Mexico) between 2300-5280 ft. (701-1609 m) elevation.
STATUS: IUCN – Critically Endangered; known from fewer than 10 records.
COMMENTS: Subgenus *Tapeti* (Gureev, 1964:164), or *Sylvilagus* (Hershkovitz, 1950:335). Reviewed by Cervantes and Lorenzo (1997, Mammalian Species, 568).

*Sylvilagus mansuetus* Nelson, 1907. Proc. Biol. Soc. Wash., 20:83.
COMMON NAME: San Jose Brush Rabbit.
TYPE LOCALITY: "San José Island, Gulf of California, Mexico" [Baja California del Sur, Mexico].
DISTRIBUTION: Known only from the type locality.
STATUS: IUCN – Lower Risk (nt).
COMMENTS: An insular allospecies closely related to *bachmani* (Chapman and Ceballos, 1990); a subspecies of *bachmani* according to Gureev (1964:171). Karyotype reported by Cervantes et al. (1996*b*). Reviewed by Thomas and Best (1994*a*, Mammalian Species, 464).

*Sylvilagus nuttallii* (Bachman, 1837). J. Acad. Nat. Sci. Philadelphia, 7:345.
COMMON NAME: Mountain Cottontail.
TYPE LOCALITY: " . . . west of the Rocky Mountains, . . . streams which flow into the Shoshonee and Columbia rivers"; restricted by Nelson (1909:201) to "eastern Oregon, near mouth of Malheur River." Listed by Bailey (1936) as "near Vale."
DISTRIBUTION: Intermountain area of North America from S British Columbia to S Saskatchewan (Canada), south to E California, Nevada, C Arizona, and NW New Mexico (USA).
STATUS: IUCN – Lower Risk (lc).
SYNONYMS: *artemesia* (Bachman, 1839); **grangeri** (J. Allen, 1895); *perplicatus* Elliott, 1904; **pinetis** (J. Allen, 1894).
COMMENTS: Subgenus *Sylvilagus* (Gureev, 1964:168). This species is closely allopatric with *S. floridanus* where the two species ranges meet across the N and C Great Plains (see map 223, Hall, 1981), and in the Southwestern USA (Hoffmeister and Lee, 1963*b*). Reviewed by Chapman (1975*a*, Mammalian Species, 56).

*Sylvilagus obscurus* Chapman et al., 1992. Proc. Biol. Soc. Wash., 105(4):858.
COMMON NAME: Appalachian Cottontail.

TYPE LOCALITY: "Dolly Sods Scenic Area, Grant Co. West Virginia."

DISTRIBUTION: N Pennsylvania south and west along the Appalachian Mtns to N Alabama.

STATUS: IUCN – Lower Risk (lc).

COMMENTS: Subgenus *Sylvilagus*. Ruedas (1986), and Ruedas et al. (1989) were the first to compare chromosomes of *S. transitionalis* (see below) and to discover the existence of two cytotypes, 2n=52 in the N Appalachians (*transitionalis* proper), and 2n=46 in the S Appalachians; the latter was eventually named *obscurus* (Chapman et al., 1992).

*Sylvilagus palustris* (Bachman, 1837). J. Acad. Nat. Sci. Philadelphia, 7:194.

COMMON NAME: Marsh Rabbit.

TYPE LOCALITY: "South Carolinan ever . . . more than forty miles [64 km] from the sea coast"; restricted by Miller and Rehn (1901:183) to E South Carolina [USA].

DISTRIBUTION: Florida to SE Virginia (Dismal Swamp) (USA) in coastal lowlands.

STATUS: U.S. ESA – Endangered as *S. p. hefneri*; IUCN – Lower Risk (lc); IUCN – Endangered as *S. p. hefneri*, otherwise Lower Risk (lc).

SYNONYMS: *douglasii* (Gray, 1837); **hefneri** Lazell, 1984; **paludicola** (Miller and Bangs, 1894).

COMMENTS: Subgenus *Tapeti* (Gureev, 1964:162). *S. aquaticus* and *S. palustris* share a derived karyotype, 2n=38 (Robinson et al., 1983, 1984). Reviewed by Chapman and Willner (1981, Mammalian Species, 153).

*Sylvilagus robustus* (Bailey, 1905). North American Fauna, 25:159.

COMMON NAME: Robust Cottontail.

TYPE LOCALITY: " . . . from Davis Mountains, Texas, 6,000 feet altitude." [USA; 1829 m]

DISTRIBUTION: Chisos, Davis and Guadalupe Mountains of Texas and New Mexico, and Sierra de la Madera of adjacent Coahuila (Mexico). Perhaps also in the Sierra del Carmen (Ruedas, 1998).

STATUS: Not Evaluated; likely endangered.

SYNONYMS: *nelsoni* Baker, 1955; *pinetis robustus* (Bailey, 1905).

COMMENTS: Subgenus *Sylvilagus*. Formerly considered a subspecies of *S. floridanus*, but see Ruedas (1998).

*Sylvilagus transitionalis* (Bangs, 1895). Proc. Boston Soc. Nat. Hist., 26:405.

COMMON NAME: New England Cottontail.

TYPE LOCALITY: "Liberty Hill, Conn." [New London Co., Connecticut, USA].

DISTRIBUTION: Boreal habitats from S Maine to S New York, mostly east of the Hudson River.

STATUS: IUCN – Vulnerable.

COMMENTS: Subgenus *Sylvilagus* (Gureev, 1964:166). Reviewed by Chapman (1975, Mammalian Species, 55). Also see comments under *S. obscurus*.

*Sylvilagus varynaensis* Durant and Guevara, 2001. Revista de Biologia Tropical, 49(1):370.

COMMON NAME: Venezuelan Lowland Rabbit.

TYPE LOCALITY: "Fundo Millano (8E46′LN and 69E56′LW), 146 m elevation, 18 km NE of the town of Sabaneta, Distrito Obispos, state of Barinas [Venezuela]."

DISTRIBUTION: Presently known only from the states of Barinas, Guarico, and Portuguesa, Venezuela.

STATUS: Not Evaluated; likely endangered.

COMMENTS: Subgenus *Tapeti*. Probably sympatric with *S. brasiliensis* and *S. floridanus,* but larger than either (Durant and Guevara, 2001).

# ORDER ERINACEOMORPHA
by Rainer Hutterer

**ORDER ERINACEOMORPHA** Gregory, 1910.
>COMMENTS: Formerly included in the Insectivora (as in the last edition; Hutterer, 1993*a*) or Lipotyphla, but treated here as a separate order in consequence of the obvious paraphyletic nature of the Insectivora clade (Asher et al., 2002; Stanhope et al., 1998). Various genetic studies (Emerson et al., 1999; Liu et al., 2001; Mouchaty et al., 2000*a*, *b*; Nikaido et al., 2001*a*) demonstrated that hedgehogs and soricomorphs keep distant positions in phylogenetic trees. Such results reflect ideas earlier expressed by paleontologists (Butler, 1988; McKenna, 1975) and are corroborated by a careful study of the morphology and relationships of fossil and extant zalambdodont mammals by Asher et al. (2002). The name Erinaceomorpha was proposed by Gregory (1910) and has since been widely used in the paleontological literature. It is adopted here in the sense of McKenna (1975) and Butler (1988). MacPhee and Novacek (1993) used it as a name for a suborder of Lipotyphla of unresolved relationships to other clades such as soricomorphs and chrysochloromorphs.

**Family Erinaceidae** G. Fischer, 1814. Zoognosia tabulis synopticis illustrata, 3:ix.
>COMMENTS: Name often accredited to Bonaparte, 1838. Reviewed by Corbet (1988), Frost et al. (1991), and Gould (1995). Bannikova et al. (2002) reviewed the phylogenetic relations of most extant genera using the fingerprinting method. Includes Tupaiodontinae Butler, 1988 and Brachyericinae Butler, 1948 as extinct subfamilies (Lopatin and Zazhigin, 2003).

**Subfamily Erinaceinae** G. Fischer, 1814. Zoognosia tabulis synopticis illustrata, 3:ix.
>COMMENTS: Reviewed by Robbins and Setzer (1985) and Corbet (1988). For a more general review, see Reeve (1994).

*Atelerix* Pomel, 1848. Arch. Sci. Phys. Nat. Geneve, 9:251.
>TYPE SPECIES: *Erinaceus albiventris* Wagner, 1841.
>
>SYNONYMS: *Aethechinus* Thomas, 1918; *Peroechinus* Fitzinger, 1866.
>
>COMMENTS: Formerly in *Erinaceus*, but see Robbins and Setzer (1985) and Corbet (1988:149). Some authors (Poduschka, 1990) retain *Aethechinus* as a genus.

*Atelerix albiventris* (Wagner, 1841). *In* Schreber, Die Säugethiere, Suppl. 2:22.
>COMMON NAME: Four-toed Hedgehog.
>
>TYPE LOCALITY: Probably Senegal or Gambia; see Allen (1939:20).
>
>DISTRIBUTION: Savanna and steppe zones from Senegal to Eritrea and Somalia and south to the Zambezi River.
>
>STATUS: IUCN – Lower Risk (lc).
>
>SYNONYMS: *adansoni* (Rochebrune, 1883); *atratus* (Rhoads, 1896); *diadematus* (Fitzinger, 1867); *faradjius* J.A. Allen, 1922; *heterodactylus* (Sundevall, 1842); *hindei* (Thomas, 1910); *kilimanus* Thomas, 1918; *langi* J.A. Allen, 1922; *lowei* Setzer, 1956; *oweni* (Setzer, 1953); *pruneri* (Wagner, 1841); *sotikae* (Heller, 1910); *spiculus* (Thomas and Wroughton, 1907); *spinifex* Thomas, 1918 [see Corbet, 1988:149 and Ansell, 1974*b*].
>
>COMMENTS: No convincing subspecific arrangement has been proposed for this species.

*Atelerix algirus* (Lereboullet, 1842). Mem. Soc. Hist. Nat. Strasbourg, 3(2), art. QQ:4.
>COMMON NAME: North African Hedgehog.
>
>TYPE LOCALITY: Algeria, "provient de Oran".
>
>DISTRIBUTION: Coastal Western Sahara to Algeria, Tunisia, and N Libya; introduced into Canary Isls, Balearic Isls, Malta, and Mediterranean France and Spain; one historical record from Puerto Rico.
>
>STATUS: IUCN – Lower Risk (lc).
>
>SYNONYMS: *caniculus* (Thomas, 1915); *diadematus* (Dobson, 1882) [not of Fitzinger, 1867]; *fallax* (Dobson, 1882); *lavaudeni* (Cabrera, 1928); **girbanensis** Vesmanis, 1980; **vagans** Thomas, 1901; ? *krugi* (Peters, 1877).

COMMENTS: Authorship is often credited to Duvernoy and Lereboullet, 1842, but Saint-Girons (1972) showed that Lereboullet was the only author. Hutterer (1983c) recognized *vagans* and *girbanensis* as subspecies.

*Atelerix frontalis* A. Smith, 1831. S. Afr. Quart. J., 2:10, 29.
COMMON NAME: Southern African Hedgehog.
TYPE LOCALITY: "Cape Colony"; restricted to South Africa, Eastern Cape Prov., northern parts of the Graaff Reinet district, by Ellerman et al. (1953).
DISTRIBUTION: S South Africa to E Botswana and W Zimbabwe; and Namibia to SW Angola.
STATUS: IUCN – Lower Risk (lc).
SYNONYMS: *capensis* (A. Smith, 1831) [*nomen nudum*]; *fractilis* (Peters, 1877); **angolae** (Thomas, 1918); *angolensis* (Roberts, 1951); *diadematus* (Dobson, 1882) [not of Fitzinger, 1867].
COMMENTS: Genus allocation uncertain. Results of Bannikova et al. (2002) obtained with the fingerprinting method indicate that *frontalis* does not form a monophyletic group with other species of *Atelerix* and thus may deserve a genus of its own. Meester et al. (1986:15) listed *angolae* as a valid subspecies.

*Atelerix sclateri* Anderson, 1895. Proc. Zool. Soc. Lond., 1895:415.
COMMON NAME: Somali Hedgehog.
TYPE LOCALITY: [Somalia], "Taf in Central Somaliland."
DISTRIBUTION: N Somalia.
STATUS: IUCN – Lower Risk (lc).
COMMENTS: Closely related to *albiventris* and might be only a subspecies, see Corbet (1988:152). Reviewed by Poduschka (1990).

*Erinaceus* Linnaeus, 1758. Syst. Nat., 10th ed., 1:52.
TYPE SPECIES: *Erinaceus europaeus* Linnaeus, 1758.
SYNONYMS: *Herinaceus* Mina-Palumbo, 1868.
COMMENTS: Formerly included *Atelerix* and *Aethechinus*; see Corbet (1988) and comments under *Atelerix*. Does not include *Mesechinus*, see comments therein. The genetic relationships among European populations were studied by Filippucci and Simson (1996), Santucci et al. (1998), and Seddon et al. (2002).

*Erinaceus amurensis* Schrenk, 1859. Reisen im Amur-Lande, 1, pl. 4, fig. 2:100.
COMMON NAME: Amur Hedgehog.
TYPE LOCALITY: Russia, E Siberia, "In der Nähe der Stadt Aigun, im mandschurischen Dorfe Gulssoja am Amur".
DISTRIBUTION: Russia; Amur River and tributaries, from Zeya eastward, then south through E China to Hunan Prov.; Korea.
STATUS: IUCN – Lower Risk (lc).
SYNONYMS: *chinensis* Satunin, 1907; *dealbatus* Swinhoe, 1870; *hanensis* Matschie, 1907; *koreanus* Lönnberg, 1922; *koreensis* Mori, 1922; *kreyenbergi* Matschie, 1907; *orientalis* J. Allen, 1903; *tschifuensis* Matschie, 1907; *ussuriensis* Satunin, 1907.
COMMENTS: Formerly included in *europaeus* (see Corbet, 1978c, Gromov and Baranova, 1981, also Zhang et al., 1997); but considered distinct by Corbet (1984), Zaitsev (1984), and Bannikova et al. (1996). Range and subspecific boundaries uncertain, partly due to confusion with *Hemiechinus*, see Corbet (1988:144). Indomalayan range mapped by Corbet and Hill (1992), Chinese range by Zhang et al. (1997).

*Erinaceus concolor* Martin, 1838. Proc. Zool. Soc. Lond., 1837:103 [1838].
COMMON NAME: Southern White-Breasted Hedgehog.
TYPE LOCALITY: Turkey, near Trabzon.
DISTRIBUTION: Asia Minor to Israel, Syria, Lebanon, Iraq and Iran; S Caucasus.
STATUS: IUCN – Lower Risk (lc).
SYNONYMS: *carmelitus* Bate, 1932; *ponticus* Satunin, 1907; *sacer* Thomas, 1918; *sharonis* Bate, 1937; **rhodius** Festa, 1914; **transcaucasicus** Satunin, 1905.
COMMENTS: Formerly included in *europaeus*; but see Kratochvíl (1975), Kral (1967), Orlov (1969), Suchentrunk et al. (1998), among others. Recently two genotypes have been

discovered (Filippucci and Simson, 1996, Santucci et al., 1998, Seddon et al., 2002) that correspond to two morphotypes (Kryštufek, 2002b; Kryštufek and Vohralik, 2001). As suggested by Filippucci and Simson (1996), we distinguish here a southern (*E. concolor*) and a northern (*E. roumanicus*) species. This step has also been taken by Bannikova et al. (2002). The map in Mitchell-Jones et al. (1999) refers mainly to *E. roumanicus*. The inclusion of *rhodius* follows a suggestion of B. Kryštufek (in litt., 2003) but still needs closer study.

*Erinaceus europaeus* Linnaeus, 1758. Syst. Nat., 10th ed., 1:52.
    COMMON NAME: West European Hedgehog.
    TYPE LOCALITY: Sweden, S Gothland Isl.
    DISTRIBUTION: W Europe; Spain to Italy and Istrian Peninsula; north to Poland, Scandinavia and NW European Russia. Islands of Ireland, Britain, Corsica, Sardinia, Sicily, Azores (Mathias et al., 1998), and many smaller islands. European range mapped by Holz and Niethammer (1990:37) and Mitchell-Jones et al. (1999). Introduced to New Zealand, see King (1990).
    STATUS: IUCN – Lower Risk (lc).
    SYNONYMS: *caniceps* H. Smith, 1845; *caninus* Geoffroy, 1803; *consolei* Barrett-Hamilton, 1900; *centralrossicus* Ognev, 1926; *echinus* Schulze, 1897; *erinaceus* (Blumenbach, 1779); *hispanicus* Barrett-Hamilton, 1900; *italicus* Barrett-Hamilton, 1900; *meridionalis* Altobello, 1920; *occidentalis* Barrett-Hamilton, 1900; *suillus* Geoffroy, 1803; *typicus* Barrett-Hamilton, 1900.
    COMMENTS: Formerly included *amurensis, concolor*, and *roumanicus*, see comments therein. Reviewed by Holz and Niethammer (1990). Subspecific boundaries are unresolved (Corbet, 1988:137) but studies of allozyme variation (Filippucci and Simson, 1996), and of mitochondrial DNA variation (Kretteck et al., 1995, Santucci et al., 1998) indicate a strong east-west geographical partitioning of the European populations, with Spain, France and Great Britain on one side, and Italy, Corsica, Germany and Sweden on the other side. A single sample from Sicily clustered with the western group and obscured the otherwise clear pattern (Santucci et al., 1998). Filippucci and Simson (1996) suggested that *E. hispanicus* could represent a distinct species, an assumption principally supported by the genetic study of Santucci et al. (1998). However, other than in the case of *E. concolor* and *E. roumanicus*, the geographic sampling of the *E. europaeus* group is still insufficient, the morphological variation has not been assessed yet, and taxonomic problems remain to be solved. Five available names with type localities in Spain, France and the UK must be evaluated before the correct species name can be fixed.

*Erinaceus roumanicus* Barrett-Hamilton, 1900. Ann. Mag. Nat. Hist., ser. 7, 5:365.
    COMMON NAME: Northern White-Breasted Hedgehog.
    TYPE LOCALITY: Rumania, Prahova, Gageni.
    DISTRIBUTION: E Europe from Poland to Austria and Slovenia; the Balkan states, Greek and Adriatic isls including Crete, Corfu, and Rhodes; Turkish Thrace; eastwards through Russsia and Ukraine to N Caucasus, W Siberia and River Ob.
    SYNONYMS: *abasgicus* Satunin, 1907; *cabardinicus* Tembotov, Dzuev and Khemykhov, 1984; *danubicus* Matschie, 1901; *dissimilis* Stein, 1930; *kievensis* Charlemagne, 1915; **bolkayi** V. Martino, 1930; **drozdovskii** V. and E. Martino, 1933; **nesiotes** Bate, 1906; **pallidus** Stroganov, 1957.
    COMMENTS: Formerly included in *europaeus*, but see Kratochvíl (1975), Kral (1967), Orlov (1969), Suchentrunk et al. (1998), among others. Subsequently included in *concolor*, but new genetic (Bannikova et al., 2002; Filippucci and Simson, 1996; Santucci et al., 1998) and morphological data (Kryštufek, 2002b; Kryštufek and Vohralik, 2001) suggest that *concolor* and *roumanicus* are two distinct species with parapatric distributions. Geographic variation of *roumanicus* was studied by Corbet (1988), Giagia and Ondrias (1980), Doğramaci and Gündüz (1993), Giagia-Athanasopoulou and Markakis (1996), and Kryštufek and Vohralik (2001). Some forms may deserve subspecies status, but the above arrangement is still tentative. The European range of the species is shown under *E. concolor* in Mitchell-Jones et al. (1999).

*Hemiechinus* Fitzinger, 1866. Sitzb. Akad. Wiss. Wien, 54, 1:565.
    TYPE SPECIES: *Erinaceus platyotis* Sundevall, 1842 (= *Erinaceus auritus* Gmelin, 1770).
    SYNONYMS: *Ericius* Sundevall, 1842 [not of Tilesius, 1813]; *Erinaceolus* Ognev, 1928;
        *Macroechinus* Satunin, 1907.
    COMMENTS: Regarded as a subgenus of *Erinaceus* by Gureev (1979:168) and Gromov and
        Baranova (1981:9). Corbet (1978c:15) considered *Hemiechinus* a distinct genus, later
        reviewed by Corbet (1988), who included *Mesechinus*, see comments therein. Pavlinov
        and Rossolimo (1987:12-13) included *Paraechinus* in *Hemiechinus* as a valid subgenus, as
        did Frost et al. (1991:27), while Corbet (1988) argued for a generic separation of
        *Paraechinus*. Morshed and Patton (2002) analyzed a part of the mitochondrial
        cytochrome *b* gene of *Hemiechinus auritus*, *Paraechinus aethiopicus* and *P. hypomelas* and
        found that *Hemiechinus* is paraphyletic with respect to *Paraechinus*. Bannikova et al.
        (2002) used the fingerprinting method to analyze the relationships between *Erinaceus*,
        *Hemiechinus*, *Paraechinus*, *Atelerix* and *Neotetracus*. They also found that *Hemiechinus* and
        *Paraechinus* do not form a monophyletic group. The present list therefore returns to the
        former practice of keeping *Hemiechinus* and *Paraechinus* as distinct genera, contra Frost
        et al. (1991).

*Hemiechinus auritus* (Gmelin, 1770). Nova Comm. Acad. Sci. Petropoli, 14:519.
    COMMON NAME: Long-eared Hedgehog.
    TYPE LOCALITY: S Russia, "in regione Astrachanensi", (= Astrakhan, 46°21′N, 48°03′E).
    DISTRIBUTION: Steppe zone from E Ukraine to Mongolia in the north and from Libya to
        W Pakistan in the south.
    STATUS: IUCN – Lower Risk (lc).
    SYNONYMS: **albulus** (Stoliczka, 1872); **aegyptius** (E. Geoffroy, 1803); *alaschanicus* Satunin,
        1907; *brachyotis* Satunin, 1908; *calligoni* (Satunin, 1901); *caspicus* (Sundevall, 1842);
        *chorassanicus* Laptev, 1926; *dorotheae* Spitzenberger, 1978; *frontalis* (Dobson, 1882)
        [not of Smith, 1831]; *holdereri* Matschie, 1922; *homalacanthus* Stroganov, 1944;
        *insularis* Timofeyev, 1934; **libycus** (Ehrenberg, 1833); *major* Ognev and Heptner, 1928;
        **megalotis** (Blyth, 1845); *metwallyi* Setzer, 1957; *microtis* Laptev, 1925; *minor* Satunin,
        1907; *persicus* Satunin, 1907; *platyotis* (Sundevall, 1842); *russowi* Satunin, 1907; *syriacus*
        (Wood, 1876); *turanicus* (Satunin, 1905); *turfanicus* Matschie, 1911; *turkestanicus*
        Ognev, 1928 [see Corbet, 1988, and Frost et al., 1991].
    COMMENTS: Corbet (1988:159) accepted *albulus*, *auritus*, and *megalotis* as valid subspecies;
        *megalotis* was formerly regarded as a distinct species but intergrades with *auritus* in
        Afghanistan; see Niethammer (1973) and Morshed and Patton (2002), who found a
        small genetic distance between the forms *auritus* and *megalotis* within Iran. Osborn
        and Helmy (1980:57-64) recognized two subspecies within Egypt, *aegyptius* and *libycus*.
        The form of Cyprus (*dorotheae*) may be also distinct but was probably introduced by
        man (Boye, 1991:115). Arabian records reviewed by Harrison and Bates (1991) and
        supplemented by Benda and Obuch (2001). Turkish long-eared hedgehogs have
        2N=48 chromosomes, with a NFa ranging from 90 to 92 (Colak et al., 1998; Kefelioglu,
        1998).

*Hemiechinus collaris* (Gray, 1830). *In* Hardwicke, Illust. Indian Zool., 1, pl.8.
    COMMON NAME: Indian Long-eared Hedgehog.
    TYPE LOCALITY: India, "Doab"; restricted to "between Jumna and Ganges Rivers"; see
        discussion in Wroughton (1910:81).
    DISTRIBUTION: Pakistan and NW India.
    STATUS: IUCN – Lower Risk (lc).
    SYNONYMS: *grayi* (Bennett, 1832); *indicus* (Royle, 1833); *spatangus* (Bennett, 1832).
    COMMENTS: Formerly included in *auritus*, but Roberts (1977) indicated that there is
        discontinuity in distribution and morphology between *collaris* (which he called *auritus
        collaris*) and *auritus* (which he called *megalotis*).

*Mesechinus* Ognev, 1951. Byull. Moskow. Ova. Ispyt. Prir. Otd. Biol., 56:8.
    TYPE SPECIES: *Erinaceus dauuricus* Sundevall, 1842.
    COMMENTS: Pavlinov and Rossolimo (1987:11) proposed to place *Mesechinus* as subgenus in

*Erinaceus* while Corbet (1988:163) included it in *Hemiechinus*, as did Bannikova et al. (1996) based on molecular data. In a new dendrogram based on fingerprinting data, Bannikova et al. (2002) included *dauuricus* also in *Hemiechinus*; the species plotted next though rather distant to four *H. auritus* samples. Frost et al. (1991:30) concluded that *Mesechinus* deserves full generic status, a conclusion supported by Gould (1995) in his comprehensive re-analysis of morphological hedgehog characters, and by Korablev et al. (1996) on the basis of chromosomal data.

*Mesechinus dauuricus* (Sundevall, 1842). K. Svenska Vetensk.-Akad. Handl. Stockholm, 1841:237 [1842].
    COMMON NAME: Daurian Hedgehog.
    TYPE LOCALITY: Russia, Transbaikalia, "Dauuria" = Dauryia (49°57'N, 116°55'E).
    DISTRIBUTION: NE Mongolia east to upper Amur Basin in Russia and adjacent parts of Inner Mongolia and W Manchuria, China.
    STATUS: IUCN – Lower Risk (lc).
    SYNONYMS: *manchuricus* (Mori, 1926); *przewalskii* (Satunin, 1907); ? *sibiricus* (Erxleben, 1777).
    COMMENTS: Species sometimes included in *Hemiechinus* (Corbet and Hill, 1992), but see Frost et al. (1991) and Gould (1995). Includes and has precedence over *sibiricus*; see Corbet (1978c:15; 1988). A considerable confusion of names has occurred in the literature; see Corbet (1988:163). Possibly includes *miodon*, see comments under *M. hughi*. Chinese range mapped by Zhang et al. (1997).

*Mesechinus hughi* (Thomas, 1908). Abstr. Proc. Zool. Soc. Lond., 1908(63):44.
    COMMON NAME: Hugh's Hedgehog.
    TYPE LOCALITY: China, Shaanxi Prov., "Paochi, Shen-si" = Baoji.
    DISTRIBUTION: Known from around two localities in Shaanxi and Shanxi Prov., C China.
    STATUS: IUCN – Vulnerable.
    SYNONYMS: *miodon* (Thomas, 1908); *sylvaticus* Ma, 1964.
    COMMENTS: Formerly included in *Erinaceus europaeus* by Ellerman and Morrison-Scott (1951:21), and in *Hemiechinus dauuricus* (here called *Mesechinus dauuricus*) by Corbet (1978c:15) and Corbet and Hill (1992:22). Includes *H. sylvaticus* described by Ma (1964:35). The form *miodon*, known from an isolated population in the Ordos desert, Shaanxi, has been alternatively assigned to *M. dauuricus* or to *M. hughi*, see discussion in Frost et al. (1991) for tentative placement here.

*Paraechinus* Trouessart, 1879. Rev. Zool. Paris, 7:242.
    TYPE SPECIES: *Erinaceus micropus* Blyth, 1846.
    SYNONYMS: *Macroechinus* Satunin, 1907.
    COMMENTS: Reviewed by Corbet (1988). Pavlinov and Rossolimo (1987:12-13) included *Paraechinus* in *Hemiechinus* as a valid subgenus, as did Frost et al. (1991:27), while Corbet (1988) argued for a generic separation of *Paraechinus*. Morshed and Patton (2002) provided DNA sequence data for animals from Iran and the Arabian Peninsula, representing *Paraechinus aethiopicus* and *P. hypomelas*. Their results show that *Hemiechinus* and *Paraechinus* are not sister taxa and are therefore best kept as distinct genera, but also show that *P. aethiopicus* and *P. hypomelas* are highly divergent (15.7%) from one another. Further support for a generic distinction of *Paraechinus* and *Hemiechinus* based on fingerprinting data was recently provided by Bannikova et al. (2002). Further molecular studies of the species united here under *Paraechinus* will add to our understanding of the phylogeny of this group, and may also lead to a further revision of the taxonomic arrangement of the species.

*Paraechinus aethiopicus* (Ehrenberg, 1832). Symb. Phys. Mamm., 2, sig. k, footnote.
    COMMON NAME: Desert Hedgehog.
    TYPE LOCALITY: Sudan, "In desertis dongolanis habitat".
    DISTRIBUTION: Sahara from Mauritania to Egypt and Awash, Ethiopia; Arabian deserts from Syria to Yemen; insular populations on Djerba (Tunisia), Bahrain and Tanb (Persian Gulf).
    STATUS: IUCN – Lower Risk (lc) as *Hemiechinus aethiopicus*.

SYNONYMS: *albatus* Thomas, 1922; *albior* Pocock, 1934; *blancalis* Thomas, 1921; *brachydactylus* (Wagner, 1841); *deserti* (Loche, 1858); *ludlowi* Thomas, 1919; *oniscus* Thomas, 1922; *pallidus* (Fitzinger, 1867); *pectoralis* (Heuglin, 1861); *dorsalis* (Anderson and de Winton, 1901); *sennaariensis* (Hedenborg, 1839) [*nomen nudum*]; *wassifi* Setzer, 1957.

COMMENTS: Species and subspecies arrangement unclear; Corbet (1988:153-154) retained Arabian *dorsalis* (= *pectoralis*) as a subspecies, while Osborn and Helmy (1980) regarded *aethiopicus*, *deserti* and *dorsalis* as distinct species. Reviewed by Harrison and Bates (1991), who considered *pectoralis*, *ludlowi* and *albatus* as valid subspecies. Nader and Al-Safadi (1993), Kock and Ebenau (1996), and Benda and Obuch (2001) provided new distribution records. Morshed and Patton (2002) provided DNA sequence data from Saudi Arabia.

*Paraechinus hypomelas* (Brandt, 1836). Bull. Sci. Acad. Imp. Sci. St. Petersbourg, 1:32.

COMMON NAME: Brandt's Hedgehog.

TYPE LOCALITY: "Pays de Turcomans", somewhere in S Kazakhstan. See Ognev (1927) for discussion.

DISTRIBUTION: Arid steppe and desert zones from Iran and Turkmenistan east almost to Tashkent (Uzbekistan), to the Indus River and N Pakistan; isolates in Saudi Arabia, Oman, Yemen and on the islands of Tanb and Kharg in the Persian Gulf.

STATUS: IUCN – Lower Risk (lc) as *Hemiechinus hypomelas*.

SYNONYMS: *amir* Thomas, 1918; *blanfordi* (Anderson, 1878); *eversmanni* Ognev, 1927; *jerdoni* (Anderson, 1878); *macracanthus* (Blanford, 1875); *niger* (Blanford, 1878); *sabaeus* Thomas, 1922; *seniculus* Thomas, 1922.

COMMENTS: Type species of *Macroechinus* Satunin. Includes *eversmanni*, *sabaeus* and *seniculus* as possible and *blanfordi* as a distinct subspecies; see Corbet (1988:155). Species reviewed by Nader (1991) and Harrison and Bates (1991). DNA sequence data of animals from Iran diverge strongly from *P. aethiopicus* (Morshed and Patton, 2002).

*Paraechinus micropus* (Blyth, 1846). J. Asiat. Soc. Bengal, 15:170.

COMMON NAME: Indian Hedgehog.

TYPE LOCALITY: Pakistan, Punjab, "Bhawulpore" = Bahawalpur.

DISTRIBUTION: The arid zones of Pakistan and NW India.

STATUS: IUCN – Lower Risk (lc) as *Hemiechinus micropus*.

SYNONYMS: *intermedius* Biswas and Ghose, 1970; *kutchicus* Biswas and Ghose, 1970; *mentalis* (Fitzinger, 1867); *pictus* (Stoliczka, 1872).

COMMENTS: Biswas and Ghose (1970) regarded *intermedius* as a species but Corbet (1988:156-157) included it in *Paraechinus micropus* as a synonym. Range mapped by Corbet and Hill (1992).

*Paraechinus nudiventris* (Horsfield, 1851). Cat. Mamm. Mus. E. India Co., p. 136.

COMMON NAME: Bare-bellied Hedgehog.

TYPE LOCALITY: India, "Madras" = Madras city or Tamil Nadu province.

DISTRIBUTION: Few records from the S Indian provinces Madras (= Tamil Nadu) and Travancore (= Kerala).

STATUS: IUCN – Vulnerable as *Hemiechinus nudiventris*.

COMMENTS: Biswas and Ghose (1970) gave *nudiventris* specific rank while Corbet (1988:156-157) regarded it as a distinct subspecies of *micropus*. Provisionally listed as a species, following Frost et al. (1991:29).

**Subfamily Galericinae** Pomel, 1848. Arch. Sci. Phys. Nat., 9:249.

SYNONYMS: Echinosoricinae Cabrera, 1925; Gymnurinae Gill, 1872; Hylomyinae Anderson, 1879.

COMMENTS: Also known as Echinosoricinae or Hylomyinae. Frost et al. (1991:23) rejected the use of Galericinae, a view not shared by Corbet and Hill (1992) and McKenna and Bell (1997), among others. Reviewed (in part) by Frost et al. (1991), Ruedi et al. (1994), and Jenkins and Robinson (2002). Their views are not always congruent. Hoek Ostende (2001) reviewed the fossil record of the Galericini which dates back into the Oligocene of France, and Mein and Ginsburg (1997) documented Miocene records from Thailand.

*Echinosorex* Blainville, 1838. C.R. Acad. Sci. Paris, 6:742.
> TYPE SPECIES: *Viverra gymnura* Raffles, 1822.
> SYNONYMS: *Gymnura* Lesson, 1827 [preoccupied by *Gymnura* Kuhl, 1824 (a fish); see Ellerman and Morrison-Scott (1951:17) and Medway (1977:15)].

*Echinosorex gymnura* (Raffles, 1822). Trans. Linn. Soc. London, 13:272.
> COMMON NAME: Moonrat.
> TYPE LOCALITY: Not given; "Sumatra" [Indonesia] implied.
> DISTRIBUTION: Malayan Peninsula, Borneo and Sumatra, Labuan Isl.
> STATUS: IUCN – Lower Risk (lc).
> SYNONYMS: *birmanica* (Trouessart, 1879); *minor* (Lyon, 1909); *rafflesii* (Lesson, 1827); *alba* (Giebel, 1863); *borneotica* (Fitzinger, 1868); *candida* (Günther, 1876).
> COMMENTS: Two subspecies, *gymnura* (Sumatra and Malay Peninsula) and *alba* (Borneo) are recognized; see Corbet (1988:128). The common spelling of the specific epithet as *gymnurus* is incorrect; see Frost et al. (1991:24).

*Hylomys* Müller, 1840. *In* Temminck, Verh. Nat. Gesch. Nederland Overz. Bezitt., Zool., Zoogd. Indisch. Archipel, p. 50 [1840].
> TYPE SPECIES: *Hylomys suillus* Müller, 1840.
> COMMENTS: Fossil *Hylomys* have been documented in Miocene sediments of Thailand (Mein and Ginsburg, 1997). The taxonomy of the extant species was partly revised by Ruedi et al. (1994). Frost et al. (1991) and Jenkins and Robinson (2002) included *Neohylomys* and *Neotetracus* in *Hylomys*, but both genera were retained as separate taxa by Corbet (1988) and Mein and Ginsburg (1997).

*Hylomys megalotis* Jenkins and M. F. Robinson, 2002. Bull. Nat. Hist. Mus. Lond. (Zool.), 68:2.
> COMMON NAME: Long-eared Gymnure.
> TYPE LOCALITY: Laos, Khammouan Province, Khammouan Limestone National Biodiversity Conservation Area, Thakhek district, c. 18 km N Thakhek, environs of Ban Muang and Ban Doy, 17°33′15″N, 104°49′30″E.
> DISTRIBUTION: Known only from the type locality.
> COMMENTS: A very distinct species with long ears and a long skull. Jenkins and Robinson (2002) performed a phylogenetic analysis of skeletal characters of the species of Galericinae and included five species in *Hylomys*. They did not, however, consider paleontological data presented by Mein and Ginsburg (1997) who retained *Neotetracus* and *Neohylomys* as separate genera. A genetic study is warranted to test the phylogenetic hypothesis presented by Jenkins and Robinson (2002).

*Hylomys parvus* Robinson and Kloss, 1916. J. Straits Branch Roy. Asiat. Soc., 73:269.
> COMMON NAME: Dwarf Gymnure.
> TYPE LOCALITY: Indonesia, W Sumatra, Mt. Kerinchi.
> DISTRIBUTION: Restricted to the highlands of Mt. Kerinchi, Sumatra.
> STATUS: IUCN – Critically Endangered.
> COMMENTS: Revised by Ruedi et al. (1994).

*Hylomys suillus* Müller, 1840. *In* Temminck, Verh. Nat. Gesch. Nederland Overz. Bezitt., Zool., Zoogd. Indisch. Archipel., p. 50 [1840].
> COMMON NAME: Short-tailed Gymnure.
> TYPE LOCALITY: Indonesia, "Java en het andere van Sumatra".
> DISTRIBUTION: Peninsular Malaysia to Indochina and the Yunnan/Burma border; islands of Borneo, Java, Sumatra and Tioman.
> STATUS: IUCN – Lower Risk (lc).
> SYNONYMS: *dorsalis* Thomas, 1888; *maxi* Sody, 1933; *microtinus* Thomas, 1925; *pegunensis* Blyth, 1859; *siamensis* Kloss, 1916; *tionis* Chasen, 1940.
> COMMENTS: Revised by Ruedi et al. (1994) who considered *suillus*, *maxi*, *dorsalis*, *siamensis* and probably *peguensis*, *microtinus*, and *tionis* as valid subspecies. Further genetic studies by Ruedi and Fumagalli (1996) suggest that *siamensis*, *maxi* and *dorsalis* may even be distinct species.

*Neohylomys* Shaw and Wong, 1959. Acta Zool. Sinica, 11:422.
    TYPE SPECIES: *Neohylomys hainanensis* Shaw and Wong, 1959.
    COMMENTS: Reviewed by Corbet (1988) and Frost et al. (1991). The latter authors and Jenkins
        and Robinson (2002) included it in *Hylomys* as a synonym or subgenus, but see Main and
        Ginsburg (1997).

*Neohylomys hainanensis* Shaw and Wong, 1959. Acta Zool. Sinica, 11:422.
    COMMON NAME: Hainan Gymnure.
    TYPE LOCALITY: China, "Pai-sa Hsian, Hainan Island" [= Baisha Xian, an administrative unit
        at 19°13′N, 109°26′E].
    DISTRIBUTION: Hainan Isl (China).
    STATUS: IUCN – Endangered as *Hylomys hainanensis*.
    COMMENTS: Corbet (1988:127) and Mein and Ginsburg (1997) retained the genus
        *Neohylomys* for this species. Jenkins and Robinson (2002) included it in *Hylomys*.

*Neotetracus* Trouessart, 1909. Ann. Mag. Nat. Hist., ser. 8, 4:389.
    TYPE SPECIES: *Neotetracus sinensis* Trouessart, 1909.
    COMMENTS: Reviewed by Corbet (1988) and Frost et al. (1991). The latter authors and Jenkins
        and Robinson (2002) included it in *Hylomys* as a synonym or subgenus, but Corbet (1988)
        retained it, and Mein and Ginsburg (1997) described a Miocene species, *Neotetracus butleri*
        Mein and Ginsburg, 1997.

*Neotetracus sinensis* Trouessart, 1909. Ann. Mag. Nat. Hist., ser. 8, 4:390.
    COMMON NAME: Shrew Gymnure.
    TYPE LOCALITY: "Ta-tsien-lou, province of Se-tchouen (China Occidental) at an altitude of
        2454 meters" [= Kangding, Sichuan Sheng, 30°07′N, 102°02′E].
    DISTRIBUTION: S China in Sichuan and Yunnan, and adjacent parts of Burma and N Vietnam.
    STATUS: IUCN – Lower Risk (nt) as *Hylomys sinensis*.
    SYNONYMS: *cuttingi* Anthony, 1941; *fulvescens* Osgood, 1932.
    COMMENTS: Corbet (1988:127) retained the genus *Neotetracus* for this species, as did Mein
        and Ginsburg (1997). In a comparative genetic study of six hedgehog genera by
        Bannikova et al. (2002), *Neotetracus* (and *Hylomys*) formed the most basal taxa in the
        dendrograms.

*Podogymnura* Mearns, 1905. Proc. U.S. Natl. Mus., 28:436.
    TYPE SPECIES: *Podogymnura truei* Mearns, 1905.
    COMMENTS: Reviewed by Heaney and Morgan (1982), Corbet (1988), and Frost et al. (1991).

*Podogymnura aureospinula* Heaney and Morgan, 1982. Proc. Biol. Soc. Wash., 95:14.
    COMMON NAME: Dinagat Gymnure.
    TYPE LOCALITY: Philippines, "Plaridel, Albor Municipality, Dinagat Island, Surigao del Norte
        Province".
    DISTRIBUTION: Dinagat Isl (Philippines).
    STATUS: IUCN – Endangered.
    COMMENTS: Heaney and Morgan (1982) suggested "golden-spined gymnure" as an English
        name, but Poduschka and Poduschka (1985) argued that the stiff dorsal hairs are not
        always spiny and golden, and Corbet and Hill (1991:27) suggested "spiny moonrat" as
        a common name. Heaney and Morgan (1982) considered that generic rank might be
        justified for this species but decided to include it in *Podogymnura* in order to emphasize
        the close relationship between the two species of Philippine gymnures; see also Corbet
        (1988:130-131).

*Podogymnura truei* Mearns, 1905. Proc. U.S. Natl. Mus., 28:437.
    COMMON NAME: Mindanao Gymnure.
    TYPE LOCALITY: Philippines, Mindanao, Mount Apo, Davao.
    DISTRIBUTION: Mindanao Isl (Philippines).
    STATUS: IUCN – Endangered.
    SYNONYMS: *minima* Sanborn, 1954.
    COMMENTS: Includes *minima* Sanborn, 1953; see data of Heaney and Morgan (1982).

# ORDER SORICOMORPHA
by Rainer Hutterer

**ORDER SORICOMORPHA** Gregory, 1910.
COMMENTS: Commonly included in the Insectivora (as in the last edition; Hutterer, 1993*a*) or Lipotyphla, but provisionally treated here as a separate order because of accumulating evidence for the paraphyletic nature of the former Insectivora clade (Asher, 1999, 2001; Stanhope et al., 1998). Various genetic studies (Emerson et al., 1999; Liu et al., 2001; Malia et al., 2002; Mouchaty et al., 2000*a*, *b*; Nikaido et al., 2001*a*) demonstrated that soricomophs and hedgehogs, sometimes also moles, keep distant positions in phylogenetic trees. Such results are reflected by ideas earlier expressed by Butler (1988) and McKenna (1975), and are corroborated by the careful study of fossil and extant zalambdodont mammals by Asher et al. (2002). The name Soricomorpha was proposed by Gregory (1910) and has since been widely used in the paleontological literature. It is adopted here in the sense of McKenna (1975) and Butler (1988). MacPhee and Novacek (1993) used it as a name for a suborder of Lipotyphla of unresolved relationships to other clades such as erinaceomorphs (now Erinaceomorpha) and chrysochloromorphs (now Tenrecomorpha or Afrosoricida). The results of Stanhope et al. (1998) offer weak support for a relationship between "mole, shrew and solenodon", e.g., Soricomorpha in the sense applied here. Other authors, however, suggest that even this clade may be polyphyletic. Malia et al. (2003) constructed a consensus tree for a set of 47 mammalian taxa including *Sorex*, *Talpa*, *Scalopus*, and *Erinaceus*. The shrews clustered with the bats, while the two moles and the hedgehog formed a separate trichotomy. Corneli (2002), who compared complete mitochondrial genomes, found that moles sister shrews and that hedgehogs are distantly related to both, which is in accordance with the Soricomorpha/Erinaceomorpha concept adopted here. L.-K. Lin et al. (2002*a*) studied four mitochondrial genomes and found some support for a mole-shrew-hedgehog clade. Waddell et al. (1999) called this group Eulipotyphla. A further problem is the allocation of *Solenodon* and *Nesophontes*. Emerson et al. (1999) compared a 12 S rRNA sequence of *Solenodon* with various mammals and in a strict consensus tree placed the genus next to myomorph rodents. On the other hand, Asher (1999), Whidden and Asher (2001), and Asher et al. (2002) discussed possible relationships of *Nesophontes* and *Solenodon* to tenrecomorphs. At this stage there exist many conflicting views and no consistent phylogeny of the members of the former Insectivora.

Soricomorpha in the present sense were reviewed by Cabrera (1925). Van Valen (1967) treated the phylogeny of many living and fossil insectivores. Grenyer and Purvis (2003) used the available phylogenetic literature to construct a supertree. Basic data on brain structure and evolution were presented by Stephan et al. (1991). For a synopsis of karyotype data see Reumer and Meylan (1986).

**Family Nesophontidae** Anthony, 1916. Bull. Am. Mus. Nat. Hist., 35:725.
COMMENTS: Known mainly from sub-Recent fossils from the Greater Antilles. Remains of a *Nesophontes*-sized mammal were found in a piece of late Oligocene/early Miocene Dominican amber (MacPhee and Grimaldi, 1996). One genus and nine taxa have been named; Hall (1981) recognized six and Morgan and Woods (1986) eight species. Efforts to locate surviving populations have been unsuccessful (Woods et al., 1985). MacPhee et al. (1999) concluded that the genus has probably been extinct for five hundred years or more, although Fischer (1977) speculated about a more recent survival. The relationships of the Nesophontidae to the Solenodontidae are not well established. Van Valen (1967) placed the Nesophontidae near the Soricidae in the order Insectivora, and the Solenodontidae near the Tenrecidae in the order Deltatheria. Whidden and Asher (2001) reviewed the relevant literature and discussed biogeographical hypotheses to explain the presence of *Nesophontes* and *Solenodon* in the West Indies. Comprehensive cladistic analyses (Asher, 1999, Asher et al., 2002) of up to 118 morphological characters found support for affinities between *Solenodon* and *Nesophontes*, and also between the two Carribean taxa and tenrecs.

*Nesophontes* Anthony, 1916. Bull. Am. Mus. Nat. Hist., 35:725.

    TYPE SPECIES: *Nesophontes edithae* Anthony, 1916.

    COMMENTS: All species of *Nesophontes* appear to have survived the late Pleistocene extinction, at least five species are known to have existed into post-Columbian times. Morgan and Woods (1986) concluded that several species survived until the early part of the 18th century, and Fischer (1977) suggested an even more recent survival. However, recent attempts to locate living representatives in the Dominican Republic were unsuccessful (MacPhee et al., 1999). Includes undescribed species from the Cayman Isls which were found in post-Columbian deposits (Morgan, 1994, 1996; Morgan and Woods, 1986; Varona, 1974).

*Nesophontes edithae* Anthony, 1916. Bull. Am. Mus. Nat. Hist., 35:725.

    COMMON NAME: Puerto Rican Nesophontes.

    TYPE LOCALITY: Puerto Rico, Cueva Cathedral, near Morovis.

    DISTRIBUTION: Puerto Rico, Vieques, Vieques, St. Johns, St. Thomas.

    STATUS: Extinct.

    COMMENTS: Size variation documented by McFarlane (1999*a*). This species may have died out much earlier than 1500 AD. McFarlane (1999*b*) reported a date of 5410 +/- 80 yrs B.P.

*Nesophontes hypomicrus* Miller, 1929. Smithson. Misc. Coll., 81:4.

    COMMON NAME: Atalaya Nesophontes.

    TYPE LOCALITY: Haití, 4 mi. (6.4 km) east of St. Michel, cave near the Atalaya plantation.

    DISTRIBUTION: Haití and Gonave Isl.

    STATUS: IUCN – Extinct.

    COMMENTS: Bones sometimes found associated with *Rattus* bones. Remains of *N. hypomicrus* and *N. zamicrus* collected in a cave in Dominican Republic dated from the 13th century (MacPhee et al., 1999).

*Nesophontes longirostris* Anthony, 1919. Bull. Am. Mus. Nat. Hist., 41:633.

    COMMON NAME: Slender Cuban Nesophontes.

    TYPE LOCALITY: Cuba, Oriente, cave near the beach at Daiquirí.

    DISTRIBUTION: Cuba.

    STATUS: Extinct.

    COMMENTS: Time of extinction uncertain (MacPhee et al., 1999).

*Nesophontes major* Arredondo, 1970. Memoria, Soc. Cienc. Nat. La Salle, 30(86):126.

    COMMON NAME: Greater Cuban Nesophontes.

    TYPE LOCALITY: Cuba, Habana, Bacuranao, Cueva de la Santa.

    DISTRIBUTION: Cuba.

    STATUS: Extinct.

    COMMENTS: Time of extinction uncertain (MacPhee et al., 1999).

*Nesophontes micrus* G. M. Allen, 1917. Bull. Mus. Comp. Zool., 61:5.

    COMMON NAME: Western Cuban Nesophontes.

    TYPE LOCALITY: Cuba, Matanzas, Sierra de Hato Neuvo.

    DISTRIBUTION: Cuba, Haití, and Pines Isl.

    STATUS: IUCN – Extinct.

    COMMENTS: Bones from two sites in Cuba dated from the 13th and 14th century (MacPhee et al., 1999). Remains of this species were found together with bones of *Mus* and *Rattus*.

*Nesophontes paramicrus* Miller, 1929. Smithson. Misc. Coll., 81(9):3.

    COMMON NAME: St. Michael Nesophontes.

    TYPE LOCALITY: Haití, cave approximately 4 mi. (6.4 km) E St. Michel.

    DISTRIBUTION: Haití.

    STATUS: IUCN – Extinct.

    COMMENTS: Some analyzed bones from Haití dated from the 14th century (MacPhee et al., 1999). Found in association with *Rattus* bones.

*Nesophontes submicrus* Arredondo, 1970. Memoria, Soc. Cienc. Nat. La Salle, 30(86):137.

    COMMON NAME: Lesser Cuban Nesophontes.

TYPE LOCALITY: Cuba, Habana, Bacuranao, Cueva de la Santa.
DISTRIBUTION: Cuba.
STATUS: Extinct.
COMMENTS: Time of extinction uncertain (MacPhee et al., 1999).

*Nesophontes superstes* Fischer, 1977. Z. geol. Wiss., Berlin, 5:221.
COMMON NAME: Cuban Nesophontes.
TYPE LOCALITY: Cuba, Pinar del Rio Province, Sumerido, Cueva de la Ventana.
DISTRIBUTION: Cuba.
STATUS: Extinct.
COMMENTS: A large species known only from a mandible found on the surface of a cave in association with *Rattus*. Its relation to *N. major* needs to be studied. Fischer (1977) speculated that the species might still be extant and gave it the name *superstes* (surviving). Listed by MacPhee et al. (1999) as part of the endemic Antillean fauna.

*Nesophontes zamicrus* Miller, 1929. Smithson. Misc. Coll., 81:7.
COMMON NAME: Haitian Nesophontes.
TYPE LOCALITY: Haití, 4 mi. (6.4 km) east of St. Michel, cave near Atalaya plantation.
DISTRIBUTION: Haití.
STATUS: IUCN – Extinct.
COMMENTS: Formerly occurred together with *N. hypomicrus*, *N. micrus*, *Solenodon marcanoi* and *S. paradoxus* on Hispaniola. Found in association with *Rattus* bones (MacPhee et al., 1999).

**Family Solenodontidae** Gill, 1872. Smithson. Misc. Coll., 11(1):19.
COMMENTS: Dobson (1882a:82) was the first to raise Gill's subfamily to family level. The inclusion of this family in the Soricomorpha may not be justified; see Asher (1999), Whidden and Asher (2001), and Asher et al. (2002).

*Solenodon* Brandt, 1833. Mem. Acad. Imp. Sci., St. Petersbourg, ser. 6, 2:459.
TYPE SPECIES: *Solenodon paradoxus* Brandt, 1833.
SYNONYMS: *Antillogale* Patterson, 1962; *Atopogale* Cabrera, 1925.
COMMENTS: Reviewed by Ottenwalder (2001). Includes *Antillogale* and *Atopogale*; see Patterson (1962:2) and Varona (1974:6). Besides the two extant species, two extinct species have been described from Cuba (Giant Solenodon) and Hispaniola (*Antillogale marcanoi* Patterson, 1962); see Morgan and Woods (1986). Remains of *Solenodon marcanoi* have been found in a horizon of "Late Pleistocene or Recent" age (Patterson, 1962).

*Solenodon arredondoi* Morgan and Ottenwalder, 1993. Annls Carnegie Mus., 62:154.
COMMON NAME: Giant Solenodon.
TYPE LOCALITY: Cuba, La Habana Province, San Antonio de los Banos, 3 km SW Ceiba del Agua, Cueva Paredones.
DISTRIBUTION: Known from three Quaternary sites in W Cuba (Morgan and Ottenwalder, 1993).
STATUS: Extinct.
COMMENTS: The age of the fossils of the giant solenodon is unclear, but faunal associations indicate a Quaternary age. The species occurred contemporarily wih *Solenodon cubanus* and *Nesophontes micrus*. Morgan and Ottenwalder (1993:161) suggested that, "habitat destruction and predation by dogs, which were introduced into Cuba by pre-Columbian peoples, are more likely explanations for the extinction of the giant Cuban *Solenodon*."

*Solenodon cubanus* Peters, 1861. Monatsb. K. Preuss. Akad. Wiss. Berlin, 1861:169.
COMMON NAME: Cuban Solenodon.
TYPE LOCALITY: Cuba, Oriente Prov., Bayamo.
DISTRIBUTION: Extant only in SE Cuba; Late Quaternary and Amerindian sites all over the island (Ottenwalder, 2001:317).
STATUS: U.S. ESA – Endangered; IUCN – Endangered.
SYNONYMS: *poeyanus* Barbour, 1944.

COMMENTS: Sometimes placed in a distinct genus or subgenus, *Atopogale*, see Hall and Kelson (1959:22) and Hall (1981:22), but see Poduschka and Poduschka (1983:225-238) who regarded *Atopogale* as a synonym of *Solenodon*. For biological information see Varona (1983*b*) and Ottenwalder (2001).

*Solenodon marcanoi* (Patterson, 1962). Breviora, 165:2.
COMMON NAME: Marcano's Solenodon.
TYPE LOCALITY: Dominican Republic, San Rafael Prov., Hondo Valle Mun.; unnamed cave 2 km SW of Rancho La Guardia.
DISTRIBUTION: Known from three Quaternary localities in S Haítí and one Pleistocene site in W Dominican Republic (Ottenwalder, 2001:318).
STATUS: IUCN – Extinct.
COMMENTS: Originally described in genus *Antillogale* by Patterson (1962). Reviewed by Ottenwalder (2001).

*Solenodon paradoxus* Brandt, 1833. Mem. Acad. Imp. Sci., St. Petersbourg, ser. 6, 2:459.
COMMON NAME: Hispaniolan Solenodon.
TYPE LOCALITY: Dominican Republic (Hispaniola), Port-au-Prince (Baranova et al., 1981:4).
DISTRIBUTION: S Haiti and Dominican Republic (Hispaniola).
STATUS: U.S. ESA – Endangered; IUCN – Endangered.
SYNONYMS: **woodi** Ottenwalder, 2001.
COMMENTS: Systematics, biogeography and ecology reviewed by Ottenwalder (1999, 2001) who also described a distinctly smaller form from S Hispaniola as subspecies *woodi*. Allard et al. (2001) found a considerable genetic variability in the mitochondrial control region in samples from Hispaniola which may correspond to the morphological variation recognized by Ottenwalder (2001).

**Family Soricidae** G. Fischer, 1814. Zoognosia tabulis synopticis illustrata, 3:x.
SYNONYMS: Sorexineae Lesson, 1842; Soricinorum G. Fischer, 1814.
COMMENTS: Shrews form a coherent group which leaves little doubt about its monophyly. One problem still under discussion is the inclusion or exclusion of the extinct Heterosoricinae Viret and Zapfe, 1951. Many authors include them in Soricidae as a subfamily (Engesser, 1975; Jammot, 1983; McKenna and Bell, 1997; Repenning, 1967; Storch and Qiu, 1991; Storch et al., 1998), others prefer family level for the Heterosoricinae (Harris, 1998; Reumer, 1987, 1998; Rzebik-Kowalska, 1998). If one compares the concepts expressed by Repenning (1967), Gureev (1971, 1979), Jammot (1983), George (1983), Reumer (1987, 1998), or Hutterer et al. (2002*b*), the subfamiliar and tribal subdivision of the Soricidae is not well resolved. Here I adopt a slightly modified system which is principally based on Reumer (1998). A strict phylogenetic classification however could include all extant shrews within a subfamily Soricinae, with Crocidurini, Myosoricini, and Soricini as tribes (Hutterer et al., 2002*b*; Stanley and Hutterer, 2000). Recognition of these three clades is supported by allozyme data (Maddalena and Bronner, 1992) and by RNA sequence data (Querouil et al., 2001). Yet, for practical reasons, I retain these clades at the subfamily level and use tribal subdivisions where appropriate. Lopatin (2002) recently named another new subfamily Soricolestinae for "the earliest and most primitive shrew" from the Middle Eocene of Mongolia.

Family-wide reviews are available on natural history (Churchfield, 1990), hair structure (Ducommun et al., 1994), external morphology (Hutterer, 1985), dental adaptations (Dannelid, 1998), and chromosomal evolution (Zima et al., 1998), protein evolution (Ruedi, 1998), life histories and energetic strategies (Innes, 1994; J. R. E. Taylor, 1998), and social systems (Rychlik, 1998). Congress volumes of interest covering a wide array of aspects include Findley and Yates (1991), Merritt et al. (1994), Zima et al. (1994), Hanski and Pankakoski (1998), and Wojcik and Wolsan (1998). Conservation aspects are dealt with in the IUCN action plans for Africa (Nicoll and Rathbun, 1990) and Eurasia (Stone, 1995). All African shrew species were re-assessed in January 2004 at a Global Mammal Assessment workshop in London, resulting in new IUCN ratings for many species. For the most current conservation status, the reader should consult the IUCN website (www.redlist.org).

The fossil history of shrews was reviewed by Repenning (1967) and Reumer (1994,

1995); excellent regional reviews cover Africa (Butler, 1998), Asia (Storch et al., 1998), Europe (Rzebik-Kowalska, 1998), and North America (Harris, 1998; see also Hutchison and Harington, 2002). Asher et al. (2002) convincingly identified the Eocene Parapternodontidae as the sister taxon of the Soricidae. Some genetic studies suggest a closer relationship of shrews to bats (e.g., Narita et al., 2001), but these studies did not include relevant fossil taxa identified by Asher et al. (2002). The earliest remains of shrews are known from the Eocene of North America (Harris, 1998), Oligocene of Eurasia (Rzebik-Kowalska, 1998; Storch et al., 1998), and Miocene of Africa (Butler, 1998). Presumedly older taxa such as *Cretasorex* (Nesov and Gureev, 1981) from the Upper Cretaceous of Uzbekistan and *Ernosorex* (Wang and Li, 1990) from the Eocene of China were assigned to this family, but the first was probably based on a contaminant (Nessov et al., 1994), and the second subsequently transferred to the family Changlelestidae (Tong and Wang, 1993). Fossil shrews are known from the High Arctic in the north (Hutchison and Harington, 2002) to South Africa in the south (Butler, 1998). The fossil and recent diversity is highest in Africa, Eurasia, and North America, but northern South America and the Indomalayan Region also house a rich, though younger radiation.

Illustrated textbooks or reviews are available for a number of larger geographical units, such as North America (Hall, 1981; Wilson and Ruff, 1999), Europe (Mitchell-Jones et al., 1999; Niethammer and Krapp, 1990), Russia (Nesterenko, 1999; Yudin, 1989), Japan (Imaizumi, 1970), Arabia and Asia Minor (Harrison and Bates, 1991; Kryštufek and Vohralík, 2001), Pakistan (Roberts, 1997), the Indomalayan Region (Corbet and Hill, 1992), or southern Africa (Mills and Hes, 1997; Smithers, 1983). Sources for common names are Wolsan and Hutterer (1998) and Wilson and Cole (2000).

**Subfamily Crocidurinae** Milne-Edwards, 1872. Rech. Hist. Nat. Mammifères, p. 256.
    SYNONYMS: Scutisoricinae Allen, 1917.
    COMMENTS: Previously included *Congosorex*, *Myosorex* and *Surdisorex*, which have been shifted into subfamily Myosoricinae based on anatomical and genetical evidence (Hutterer et al., 2002*b*; Maddalena and Bronner, 1992; Querouil et al., 2001).

*Crocidura* Wagler, 1832. Isis, p. 275.
    TYPE SPECIES: *Sorex leucodon* Herman, 1780.
    SYNONYMS: *Afrosorex* Hutterer, 1986; *Heliosorex* Heller, 1910; *Leucodon* Fatio, 1869; *Paurodus* Schulze, 1897; *Praesorex* Thomas, 1913; *Rhinomus* Murray, 1860 [see Allen, 1939; Heim de Balsac and Meester, 1977; and Hutterer, 1986*a*].
    COMMENTS: Eurasian species revised by Jenkins (1976), Indomalayan and Philippine species by Heaney and Ruedi (1994) and Ruedi (1995), Chinese species by Jiang and Hoffmann (2001). Phenetic and Phylogenetic relationships of African and Palearctic species studied by Butler et al. (1989), Maddalena (1990) and McLellan (1994), and of Asian and Indomalayan species by Heaney and Ruedi (1994), Ruedi (1996), and Ruedi et al. (1998). Karyotypes of SE Asian species described by Ruedi and Vogel (1995), those of Mediterranean isls shrews by Vogel et al. (1990), and of African shrews by Schlitter et al. (1999). No formal subgenera are recognized here, although the type species of proposed subgenera are indicated.

*Crocidura aleksandrisi* Vesmanis, 1977. Bonn. Zool. Beitr., 28:3.
    COMMON NAME: Cyrenaica Shrew.
    TYPE LOCALITY: Libya, Cyrenaica, 5 km W Tocra.
    DISTRIBUTION: Restricted to Cyrenaica, Libya.
    STATUS: IUCN – Lower Risk (lc).
    COMMENTS: Sometimes included in *C. suaveolens* but currently regarded as a valid species (Hutterer, 1991).

*Crocidura allex* Osgood, 1910. Field Mus. Nat. Hist. Publ., Zool. Ser., 10(3):20.
    COMMON NAME: East African Highland Shrew.
    TYPE LOCALITY: Kenya, "Naivasha, British East Africa".
    DISTRIBUTION: Higlands of SW Kenya; Mt. Kilimanjaro, Meru and Ngorogoro, N Tanzania.
    STATUS: IUCN – Vulnerable.

SYNONYMS: *alpina* Heller, 1910; *zinki* Heim de Balsac, 1957 [see Heim de Balsac and Meester, 1977].

COMMENTS: Gureev (1979) listed *alpina* as a distinct species without comment. Common in the afro-alpine zone at 3500 m on Mt. Kilimanjaro (Shore and Garbett, 1991).

*Crocidura andamanensis* Miller, 1902. Proc. U.S. Natl. Mus., 24:777.

COMMON NAME: Andaman Shrew.

TYPE LOCALITY: India, Andaman Isls, South Andaman Isl.

DISTRIBUTION: Andaman Isls, Bay of Bengal.

STATUS: IUCN – Data Deficient.

COMMENTS: Erroneously attributed to genus *Suncus* by Krumbiegel (1978:71). The species was recently collected on Mt. Harriet, Andaman Isls (Das, 1999).

*Crocidura ansellorum* Hutterer and Dippenaar, 1987. Bonn. Zool. Beitr., 38:1, 269.

COMMON NAME: Ansell's Shrew.

TYPE LOCALITY: Zambia, Mwinilunga Distr., Kasombu stream (= Isombu River), 4100 ft. (1250 m).

DISTRIBUTION: N Zambia.

STATUS: IUCN – Critically Endangered.

SYNONYMS: *anselli* Hutterer and Dippenaar, 1987.

COMMENTS: Known from only two specimens. Species regarded as endemic to the Ikelenge Pedicle, NW Zambia (Cotterill, 2002).

*Crocidura arabica* Hutterer and Harrison, 1988. Bonn. Zool. Beitr., 39:64.

COMMON NAME: Arabian Shrew.

TYPE LOCALITY: Oman, Dhofar, Khadrafi [16°42′N, 53°09′E].

DISTRIBUTION: Coastal plains of S Arabian Peninsula (Yemen, Oman).

STATUS: IUCN – Lower Risk (lc).

COMMENTS: Previous to the recognition of *arabica*, specimens have been assigned to *russula* or *suaveolens*; see Harrison and Bates (1991).

*Crocidura arispa* Spitzenberger, 1971. Ann. Naturhistor. Museum Wien, 75:547.

COMMON NAME: Jackass Shrew.

TYPE LOCALITY: S Turkey, Vil. Nigde, ca. 20 km ESE Ulukisla, mountains S Madenköy.

DISTRIBUTION: Taurus Mtns of S Turkey; known only from two localities (Kryštufek and Vohralík, 2001:91).

COMMENTS: A rare shrew living in rocky areas. First described as a subspecies of *pergrisea* (Spitzenberger, 1971a), then assigned to *serezkyensis* (Hutterer, 1993a), and finally given species rank by Kryštufek and Vohralík (2001).

*Crocidura armenica* Gureev, 1963. *In* Mammal Fauna of the U.S.S.R., 1:118.

COMMON NAME: Armenian Shrew.

TYPE LOCALITY: Armenia, 14 km down river from Garni.

DISTRIBUTION: Armenia, Caucasus.

STATUS: IUCN – Data Deficient.

COMMENTS: Revised by Gureev (1979), who considered *armenica* as distinct from *pergrisea*; but see Dolgov and Yudin (1975), who considered it a subspecies; Gromov and Baranova (1981) and Zaitsev (1991) listed it as a distinct species.

*Crocidura attenuata* Milne-Edwards, 1872. Rech. Hist. Nat. Mamm., p. 263.

COMMON NAME: Asian Gray Shrew.

TYPE LOCALITY: China, Szechuan, Moupin (= Sichuan, Baoxing).

DISTRIBUTION: Assam, Sikkim (India), Nepal, Bhutan, Burma, Thailand, Vietnam, Peninsular Malaysia, S China; a doubtful record from Batan Isl (Philippines).

STATUS: IUCN – Lower Risk (lc).

SYNONYMS: *grisea* Howell, 1926; *grisescens* Howell, 1928; *kingiana* Anderson, 1877; *rubricosa* Anderson, 1877.

COMMENTS: Reviewed by Heaney and Timm (1983b), Jenkins (1976), and Jiang and Hoffmann (2001:1069, Fig. 7), who included *tanakae,* which is now elevated to species level. Jenkins (1982) and Corbet and Hill (1992) included the long-tailed *aequicaudata*, but see under *C. paradoxura*. Motokawa et al. (2001*b*) contrasted the karyotype of Chinese mainland *attenuata* (2n = 35-38, FN = 54) against the karyotype of Taiwanese *tanakae* (2n = 40, FN = 56) and suggested species status for the latter. A karyotype from Thailand identified as *attenuata* (Tsuchiya et al., 1979) with 2n = 50 was probably based on a misidentification; it probably represented *C. hilliana*.

*Crocidura attila* Dollman, 1915. Ann. Mag. Nat. Hist., ser. 8, 15:141.
    COMMON NAME: Hun Shrew.
    TYPE LOCALITY: Cameroon, Bitye.
    DISTRIBUTION: Gotel Mtns (Nigeria) and Cameroon Mtns to E Dem. Rep. Congo.
    STATUS: IUCN – Vulnerable (not justified).
    COMMENTS: Formerly included in *buettikoferi*, but separated by Hutterer and Joger (1982). Recorded from SE Nigeria by Hutterer et al. (1992*a*). Karyotype (2n = 50, FN = 66) identical to *greenwoodi*, *hirta*, and *olivieri* group (Schlitter et al., 1999).

*Crocidura baileyi* Osgood, 1936. Field Mus. Nat. Hist. Publ., Zool. Ser., 20:225.
    COMMON NAME: Bailey's Shrew.
    TYPE LOCALITY: Ethiopia, Simien Mtns, Ras Dashan (= Mt. Geech).
    DISTRIBUTION: Ethiopian highlands west of the Rift Valley.
    STATUS: IUCN – Vulnerable.
    COMMENTS: Revised by Dippenaar (1980).

*Crocidura baluensis* Thomas, 1898. Ann. Mag. Nat. Hist., ser. 7, 2:247.
    COMMON NAME: Kinabalu Shrew.
    TYPE LOCALITY: Sabah, Mt. Kinabalu.
    DISTRIBUTION: High altitudes (ca. 1600-3700 m) of Gunung Kinabalu, Sabah; perhaps also Sarawak.
    COMMENTS: Regarded a distinct species by Corbet and Hill (1992) and Ruedi (1995). Replaced at lower altitudes by *C. foetida*.

*Crocidura batesi* Dollman, 1915. Ann. Mag. Nat. Hist., ser. 8, 15:143.
    COMMON NAME: Bate's Shrew.
    TYPE LOCALITY: "Como River, Gabon."
    DISTRIBUTION: Lowland forest in S Cameroon and Gabon.
    STATUS: IUCN – Lower Risk (lc).
    COMMENTS: Often included in *poensis*; specimens from Cameroon and Gabon have been reported as *wimmeri*; but see Brosset (1988). Karyotype (2n = 50, FN = 76) identical to that of *nigeriae* (Schlitter et al., 1999). This complex requires a pan-African revision.

*Crocidura beatus* Miller, 1910. Proc. U.S. Natl. Mus., 38:392.
    COMMON NAME: Mindanao Shrew.
    TYPE LOCALITY: Philippines, Mindanao, Summit of Mt. Bliss, 1,461 m.
    DISTRIBUTION: Philippines: forested areas in Mindanao Faunal Region (Biliran, Bohol, Leyte, Maripipi, Mindanao) and on Camiguin.
    STATUS: IUCN 2000 – Vulnerable.
    SYNONYMS: *parvacauda* Taylor, 1934.
    COMMENTS: Includes *parvacauda* as a synonym; see Heaney et al. (1987:36). Distribution reviewed by Heaney (1986) and Heaney and Ruedi (1994).

*Crocidura beccarii* Dobson, 1887. Ann. Mus. Civ. Stor. Nat. Genova, ser. 2, 4:556.
    COMMON NAME: Beccari's Shrew.
    TYPE LOCALITY: Indonesia, Sumatra, Mt. Singalang.
    DISTRIBUTION: Mountain ranges in N and W Sumatra.
    STATUS: IUCN – Endangered.
    SYNONYMS: *weberi* Jentink in Weber, 1890.
    COMMENTS: Revised by Ruedi (1995). Karyotype has 2n = 38, FN = 56 (Ruedi and Vogel, 1995).

*Crocidura bottegi* Thomas, 1898. Ann. Mus. Civ. Stor. Nat. Genova, ser. 2, 18:677.
    COMMON NAME: Bottego's Shrew.
    TYPE LOCALITY: Ethiopia, north-east of Lake Turkana, "between Badditu and Dime".

DISTRIBUTION: Ethiopia and N Kenya; West African records doubtful.

STATUS: IUCN – Lower Risk (lc).

COMMENTS: Heim de Balsac and Meester (1977) included *obscurior* and *eburnea* as subspecies; but see under *C. obscurior*.

*Crocidura bottegoides* Hutterer and Yalden, 1990. *In* Peters and Hutterer (eds.), Vertebrates in the Tropics, Bonn, p. 67.

COMMON NAME: Bale Shrew.

TYPE LOCALITY: Ethiopia, Bale Mtns, Harenna Forest, Katcha Camp, 2400 m.

DISTRIBUTION: Bale Mtns and Mt. Albasso, Ethiopia (Yalden and Largen, 1992).

STATUS: IUCN – Vulnerable.

COMMENTS: Lavrenchenko et al. (1997) described the karyotype (2n = 36, FN = 48) which is very different from that of the similar *"bottegi"* (now *obscurior*) (2n = 40, FN = 60) from Côte d'Ivoire.

*Crocidura brunnea* Jentink, 1888. Notes Leyden Mus., 10:164.

COMMON NAME: Thick-tailed Shrew.

TYPE LOCALITY: Indonesia, W Java.

DISTRIBUTION: Java and Bali.

SYNONYMS: *brevicauda* Jentink *in* Weber, 1890; *melanorhyncha* Jentink, 1910; ***pudjonica*** Sody, 1936.

COMMENTS: Formerly included in *C. fuliginosa* (see Jenkins, 1982). Revised by Ruedi (1995), who considered *brunnea* and *pudjonica* as distinct subspecies. Bali records by Kitchener et al. (1994*a*). Karyotype has 2n = 38, FN = 56 (Ruedi and Vogel, 1995).

*Crocidura buettikoferi* Jentink, 1888. Notes Leyden Mus., 10:47.

COMMON NAME: Buettikofer's Shrew.

TYPE LOCALITY: Robertsport, Liberia.

DISTRIBUTION: West African high forest; S Guinea (Heim de Balsac, 1958) to Liberia, Ghana (Decher et al., 1997), and Nigeria (Hutterer and Happold, 1983).

STATUS: IUCN – Lower Risk (lc).

COMMENTS: Formerly included *attila*; see Hutterer and Joger (1982). Grubb et al. (1998) discussed a number of possible records from Ghana and Sierra Leone that require confirmation.

*Crocidura caliginea* Hollister, 1916. Bull. Am. Mus. Nat. Hist., 35:664.

COMMON NAME: African Dusky Shrew.

TYPE LOCALITY: Dem. Rep. Congo, Medje.

DISTRIBUTION: NE Dem. Rep. Congo.

STATUS: IUCN – Critically Endangered.

COMMENTS: The species was recently rediscovered in the Masako Forest in NE Dem. Rep. Congo (Hutterer and Dudu, 1990).

*Crocidura canariensis* Hutterer, Lopez-Jurado and Vogel, 1987. J. Nat. Hist., 21:1354.

COMMON NAME: Canarian Shrew.

TYPE LOCALITY: Spain, Canary Isls, Fuerteventura, Tiscamanita.

DISTRIBUTION: E Canary Islands (Lanzarote, Lobos, Fuerteventura, Graciosa, Mtna Clara).

STATUS: IUCN – Vulnerable.

COMMENTS: Related to *sicula*; see Maddalena and Vogel (1990) and Hutterer et al. (1992*b*). Sarà (1995, 1996) studied the variation of mandibular measurements and found no difference between *C. canariensis* and *C. sicula*; as a consequence, he included *canariensis* as a subspecies in *C. sicula*. However, this author disregarded biological, ecological, and paleontological evidence provided by Hutterer et al. (1992*b*) and Michaux et al. (1991). Genetic distances suggest a separation of *C. canariensis* and *C. sicula* for 5 million years (Vogel et al., 2003).

*Crocidura caspica* Thomas, 1907. Ann. Mag. Nat. Hist., ser. 7, 20:197.

COMMON NAME: Caspian Shrew.

TYPE LOCALITY: Iran, S coast of Caspian Sea.

DISTRIBUTION: Iran, S coast of Caspian Sea.

COMMENTS: A distinct, chocolate-brown shrew formerly included in *C. russula* (Ellerman and Morrison-Scott, 1966) or *C. leucodon*. Zaitsev (1993) redefined the species and mapped its range.

*Crocidura cinderella* Thomas, 1911. Ann. Mag. Nat. Hist., ser. 8, 8:119.
COMMON NAME: Cinderella Shrew.
TYPE LOCALITY: "Gemenjulla, French Gambia."
DISTRIBUTION: Senegal and Gambia, Mali and Niger.
STATUS: IUCN – Lower Risk (lc).
COMMENTS: May be related to *tarfayaensis* of Morocco and Mauritania; see Hutterer (1987).

*Crocidura congobelgica* Hollister, 1916. Bull. Am. Mus. Nat. Hist., 35:670.
COMMON NAME: Congo White-toothed Shrew.
TYPE LOCALITY: Dem. Rep. Congo, "Lubila, near Bafwasende".
DISTRIBUTION: NE Dem. Rep. Congo.
STATUS: IUCN – Vulnerable.
COMMENTS: For a discussion of relationships, see Heim de Balsac (1968*a*).

*Crocidura crenata* Brosset, Dubost, and Heim de Balsac, 1965. Mammalia, 29:268.
COMMON NAME: Long-footed Shrew.
TYPE LOCALITY: Gabon, Belinga.
DISTRIBUTION: High forest in S Cameroon, N Gabon, and E Dem. Rep. Congo.
STATUS: IUCN – Lower Risk (lc).
COMMENTS: The specific epithet obviously was choosen because the species has extremily long feet and tail; Brosset (1988) observed that they aid in jumping rather than climbing.

*Crocidura crossei* Thomas, 1895. Ann. Mag. Nat. Hist., ser. 6, 16:53.
COMMON NAME: Crosse's Shrew.
TYPE LOCALITY: Nigeria, "Asaba, 150 mi. [241 km] up the Niger River".
DISTRIBUTION: Lowland forest from Guinea to W Cameroon.
STATUS: IUCN – Lower Risk (lc).
SYNONYMS: *ingoldbyi* Heim de Balsac, 1956.
COMMENTS: Formerly included *ebriensis* and *jouvenetae*; see Heim de Balsac and Meester (1977). *C. crossei* occurs almost sympatrically with *C. jouvenetae* from Guinea to Côte d'Ivoire.

*Crocidura cyanea* (Duvernoy, 1838). Mem. Soc. Hist. Nat. Strasbourg, 2:2.
COMMON NAME: Reddish-gray Musk Shrew.
TYPE LOCALITY: "La riviere des Elephants, au sud de l'Afrique" = Citrusdal, South Africa *fide* Shortridge (1942:27).
DISTRIBUTION: South Africa, Namibia, Angola, Botswana, Mozambique, Zimbabwe; records further north uncertain.
STATUS: IUCN – Lower Risk (lc).
SYNONYMS: *argentatus* Sundevall in Grill, 1860; *capensis* (Smuts, 1832) [not (E. Geoffroy, 1811)]; ? *capensoides* (A. Smith, 1833); ? *concolor* (A. Smith, 1836); *electa* Dollman, 1910; *infumata* (Wagner, 1841); *martensii* Dobson, 1890; *pondoensis* Roberts, 1913; *vryburgensis* Roberts, 1946.
COMMENTS: The species concept applied by Heim de Balsac and Meester (1977) included a number of names which evidently do not belong to *cyanea* but to species such as *parvipes* and *smithii*; see Hutterer (1986*a*) and Hutterer and Joger (1982). The limits of distribution of *cyanea* have not yet been established; Meester et al. (1986) distinguished *cyanea* and *infumata* as subspecies in South Africa. The taxon *erica* which has been included in *cyanea* may be related to *hirta*.

*Crocidura denti* Dollman, 1915. Ann. Mag. Nat. Hist., ser. 8, 16:377.
COMMON NAME: Dent's Shrew.
TYPE LOCALITY: Dem. Rep. Congo, Ituri Forest, between Mawambi and Avakubi.
DISTRIBUTION: African Congo Basin (Dem. Rep. Congo, Gabon, Cameroon) and isolated records in West Africa (Guinea, Sierra Leone).

STATUS: IUCN – Lower Risk (lc).

COMMENTS: Considered a distinct species by Heim de Balsac (1959:216). It remains to be determined whether the West African records (Ziegler et al., 2002) are conspecific.

*Crocidura desperata* Hutterer, Jenkins and Verheyen, 1991. Oryx, 25:165.

COMMON NAME: Desperate Shrew.

TYPE LOCALITY: S Tanzania, Rungwa Mtns, mountain bamboo zone above 2000 m.

DISTRIBUTION: Relict forest patches at Rungwa and Udzungwa Mtns, S Tanzania.

STATUS: IUCN – Critically Endangered.

COMMENTS: The relationships of this shrew are still unresolved.

*Crocidura dhofarensis* Hutterer and Harrison, 1988. Bonn. Zool. Beitr., 39:68.

COMMON NAME: Dhofar Shrew.

TYPE LOCALITY: Oman, Dhofar, Khadrafi, 620 m.

DISTRIBUTION: Known only from the type locality.

STATUS: IUCN – Critically Endangered.

COMMENTS: Originally described as a subspecies of *C. somalica*, but Hutterer et al. (1992c) provided arguments for full specific status.

*Crocidura dolichura* Peters, 1876. Monatsb. K. Preuss. Akad. Wiss. Berlin, 1876:475.

COMMON NAME: Long-tailed Musk Shrew.

TYPE LOCALITY: Cameroon, Bonjongo.

DISTRIBUTION: High forest in Nigeria, S Cameroon, Bioko, Gabon, Central African Republic, Republic of Congo, Dem. Rep. Congo, and adjacent Uganda and Burundi.

STATUS: IUCN – Lower Risk (lc).

COMMENTS: Does not include *latona*, *ludia*, *muricauda*, and *polia*; see under these species.

*Crocidura douceti* Heim de Balsac, 1958. Mem. Inst. Fr. Afr. Noire, 53:329.

COMMON NAME: Doucet's Musk Shrew.

TYPE LOCALITY: Côte d'Ivoire, Adiopodoume.

DISTRIBUTION: Forest-savanna border of Guinea, Côte d'Ivoire, and Nigeria.

STATUS: IUCN – Lower Risk (lc).

COMMENTS: Reviewed by Hutterer and Happold (1983).

*Crocidura dsinezumi* (Temminck, 1842). *In* Siebold, Fauna Japonica, 2(Mamm.):pl. IV figs c, c, pl. V, figs 3.

COMMON NAME: Dsinezumi Shrew.

TYPE LOCALITY: "Japan"; restricted to Kyushu by Abe (1967).

DISTRIBUTION: Japan including N Ryukyus.

STATUS: IUCN – Lower Risk (lc).

SYNONYMS: *chisai* Thomas, 1906; *intermedia* Kuroda, 1924, *kinczumi* Temminck, 1842; *kinezumi* Temminck, 1842; *okinoshimae* Kuroda and Uchida, 1959; *umbrinus* Temminck, 1844.

COMMENTS: The spelling of the name was clarified by Corbet (1978b) and Motokawa (1999); *dsinezumi* was placed on the Official List of Specific Names; see the International Commission on Zoological Nomenclature (1983). Includes *chisai*, but not *quelpartis* and *orii*; see Corbet (1978c) and Iwasa et al. (2001). The taxon *hosletti* described by Jameson and Jones (1977) from Taiwan is now included in *C. shantungensis* (see Jiang and Hoffmann, 2001). Geographic variation of Japanese populations studied by Motokawa (2003b). Allozyme data studied by Ruedi et al. (1993) place *C. dsinezumi* close to *C. fuliginosa*, *C. malayana*, and *C. grayi*.

*Crocidura eisentrauti* Heim de Balsac, 1957. Zool. Jahrb. Abt. Syst. Oekol. Geogr. Tiere, 85:616.

COMMON NAME: Eisentraut's Shrew.

TYPE LOCALITY: Cameroon, Mt. Cameroon, "Johann-Albrecht-Hütte, 2900 m".

DISTRIBUTION: Higher elevations of Mt. Cameroon (Cameroon).

STATUS: IUCN – Critically Endangered.

COMMENTS: Only known from Mt. Cameroon. Not conspecific with *C. vulcani*; see under *C. virgata*.

*Crocidura elgonius* Osgood, 1910. Ann. Mag. Nat. Hist., ser. 8, 5:369.
COMMON NAME: Elgon Shrew.
TYPE LOCALITY: Kenya, Mt. Elgon, Kirui.
DISTRIBUTION: Mountains in W Kenya and NE Tanzania.
STATUS: IUCN – Vulnerable.
COMMENTS: Regarded as a distinct species by Heim de Balsac and Meester (1977) and
        Hutterer (1983*b*).

*Crocidura elongata* Miller and Hollister, 1921. Proc. Biol. Soc. Wash., 34:101.
COMMON NAME: Elongated Shrew.
TYPE LOCALITY: Indonesia, Sulawesi, Temboan (SW from Tondano Lake).
DISTRIBUTION: N and C Sulawesi.
STATUS: IUCN – Lower Risk (lc).
COMMENTS: See Musser (1987*a*) for ecological notes and a photograph. Ruedi and Vogel
        (1995) described two different karyotypes: 2n = 30, FN = 56 for a male from lowland
        forest, and 2n = 34, FN = 60 for a smaller female from montane forest. Ruedi (1995)
        suggested that one of the specimens may represent a cryptic new species.

*Crocidura erica* Dollman, 1915. Ann. Mag. Nat. Hist., ser. 8, 15:145.
COMMON NAME: Heather Shrew.
TYPE LOCALITY: Angola, Pungo Andongo.
DISTRIBUTION: W Angola.
STATUS: IUCN – Vulnerable.
COMMENTS: Resembles *hirta* in cranial dimensions; see Heim de Balsac and Meester (1977).
        Related to *nigricans*, according to Crawford-Cabral (1987).

*Crocidura fischeri* Pagenstecher, 1885. Jb. Hamburger Wiss. Anst., 2:34.
COMMON NAME: Fischer's Shrew.
TYPE LOCALITY: "Nguruman"; northwest of Lake Natron, close to Mt. Sambo, Kenya (near
        border to Tanzania); see discussion by Moreau et al. (1946) and Aggundey and Schlitter
        (1986).
DISTRIBUTION: Nguruman (Kenya), and Himo (Tanzania).
STATUS: IUCN – Vulnerable.
COMMENTS: Type species of subgenus *Afrosorex*. Revised by Hutterer (1986*a*).

*Crocidura flavescens* (I. Geoffroy, 1827). Dict. Class. Hist. Nat., 11:324.
COMMON NAME: Greater Red Musk Shrew.
TYPE LOCALITY: "La Cafrerie et le pays des Hottentots" = King William's Town, South Africa.
DISTRIBUTION: South Africa.
STATUS: IUCN – Vulnerable.
SYNONYMS: *capensis* (A. Smith, 1833) [not (Smuts, 1832), not (E. Geoffroy, 1811)];
        *cinnamomeus* (Lichtenstein, 1829); *knysnae* Roberts, 1946; *rutilus* (Sundevall, 1846).
COMMENTS: For correct original citation see Ellerman et al. (1953). Does not include *olivieri*;
        see Maddalena et al. (1987) and comments under that species. Reviewed by Meester
        (1963). Karyotype (2n = 50, FN = 74) described by Maddalena et al. (1987).

*Crocidura floweri* Dollman, 1915. Ann. Mag. Nat. Hist., ser. 8, 15:515.
COMMON NAME: Flower's Shrew.
TYPE LOCALITY: "Giza, Egypt."
DISTRIBUTION: Environs of Upper Nile valley and Wadi el Natrun, Egypt (Goodman, 1989).
STATUS: IUCN – Endangered.
COMMENTS: Mummified shrews from Ancient Egypt have been identified as *C. floweri*; see
        Heim de Balsac and Mein (1971). Possibly related to *crossei* and *arabica*; see Hutterer
        and Harrison (1988).

*Crocidura foetida* Peters, 1870. Mber. K. Preuss. Acad. Wiss., 1870:586.
COMMON NAME: Bornean Shrew.
TYPE LOCALITY: Borneo.
DISTRIBUTION: Borneo (Kalimantan, Sarawak, and Sabah).
SYNONYMS: **doriae** Peters, 1870; **kelabit** Medway, 1965.

COMMENTS: Revised by Ruedi (1995). Karyotype has 2n = 38, FN = 56-58 (Ruedi and Vogel, 1995).

*Crocidura foxi* Dollman, 1915. Ann. Mag. Nat. Hist., ser. 8, 15:514.
COMMON NAME: Fox's Shrew.
TYPE LOCALITY: Nigeria, Panyam.
DISTRIBUTION: Jos Plateau, Nigeria, and Sudan savanna zone of West Africa from Senegal to S Sudan.
STATUS: IUCN – Vulnerable (not justified).
SYNONYMS: *tephra* Setzer, 1956.
COMMENTS: A member of the *poensis* group; may be conspecific with *theresae*, which it antedates; see Hutterer and Happold (1983). A series from Owerri, S Nigeria, referred to *foxi* by these authors, was later, upon re-examination, identified as a dark form of *lamottei*. The holotype of *tephra* Setzer, 1956 has been recently examined and is regarded as representing *foxi* in S Sudan; a previous allocation to *viaria* (Hutterer, 1984) was based upon examination of a paratype skin; however, the holotype represents a different species.

*Crocidura fuliginosa* (Blyth, 1855). J. Asiat. Soc. Bengal, ser. 2, 24:362.
COMMON NAME: Southeast Asian Shrew.
TYPE LOCALITY: Burma, Schwegyin, near Pegu.
DISTRIBUTION: N India, Burma, adjacent China, Malaysian Peninsula and adjacent isls; exact distribution unknown.
STATUS: IUCN – Lower Risk (lc).
SYNONYMS: *fuliginosus* (Blyth, 1855); **dracula** Thomas, 1912; *mansumensis* Carter, 1942; *praedax* Thomas, 1923.
COMMENTS: A large S Asian shrew with a complex taxonomic history. Ruedi et al. (1990) demonstrated unrecognized sympatry of two cryptic but chromosomally distinct forms, one of which was provisionally labeled *C.* cf. *malayana*. For taxa formerly and actually included in *fuliginosa* see Jenkins (1976, 1982) and Ruedi (1995). Medway (1977) and Heaney and Timm (1983b) included *dracula*, which Lekagul and McNeely (1977) considered a distinct species. The list of synonyms is provisional; see also under *malayana*. Specimens from Zhejiang, E China assiged to *C. fuliginosa* (Zhuge, 1993) may represent an undescribed taxon (Jiang and Hoffmann, 2001). Karyotypes from Indochina and the Malay Peninsula count 2n = 40, FN = 54-58 (Ruedi and Vogel, 1995).

*Crocidura fulvastra* (Sundevall, 1843). K. Svenska Vetensk-Akad. Handl. Stockholm, 1842:172 [1843].
COMMON NAME: Savanna Shrew.
TYPE LOCALITY: Sudan, Bahr el Abiad.
DISTRIBUTION: Sudan savanna from Kenya to Mali.
STATUS: IUCN – Lower Risk (lc).
SYNONYMS: *arethusa* Dollman, 1915; *beta* Dollman, 1915; *diana* Dollman, 1915; *fulvaster* Sundevall, 1843; *macrodon* Dobson, 1980; *marrensis* Thomas and Hinton, 1923; *sericeus* Sundevall, 1843; *strauchii* Dobson, 1890 [see Hutterer, 1984, Hutterer and Kock, 1983, and Hutterer and Happold, 1983].
COMMENTS: Gureev (1979) listed *beta* as a distinct species without comment.

*Crocidura fumosa* Thomas, 1904. Ann. Mag. Nat. Hist., ser. 7, 14:238.
COMMON NAME: Smoky White-toothed Shrew.
TYPE LOCALITY: Kenya, "Camp 18, western slope of Mt. Kenya, 2,600 m".
DISTRIBUTION: Mt. Kenya and Aberdare Range (Kenya).
STATUS: IUCN – Vulnerable.
SYNONYMS: *alchemillae* Heller, 1910.
COMMENTS: Dippenaar and Meester (1989) revised the species.

*Crocidura fuscomurina* (Heuglin, 1865). Leopoldina, 5, *in* Nouv. Acta Acad. Caes. Leop.-Carol., 32:36.
COMMON NAME: Bicolored Musk Shrew.

TYPE LOCALITY: Sudan, Bahr-el-Ghazal, Meshra-el-Req.

DISTRIBUTION: Sudan and Guinea savanna from Senegal to Ethiopia, and south to South Africa.

STATUS: IUCN – Lower Risk (lc).

SYNONYMS: *bicolor* Bocage, 1889; *bovei* Dobson, 1887; *cuninghamei* Thomas, 1904; *fuscomurinus* (Heuglin, 1865); *? glebula* Dollmann, 1916; *hendersoni* Dollman, 1915; *marita* Thomas and Hinton, 1923; *sansibarica* Neumann, 1900; *tephragaster* Setzer, 1956; *tephronotus* Heim de Balsac, 1968 [*lapsus*]; *woosnami* Dollman, 1915.

COMMENTS: Revised by Hutterer (1983*b*). Karyotype from Burundi has 2n = 56, FN = 86 (Maddalena and Ruedi, 1994). *C. planiceps* may belong here but relationships are yet unsolved.

*Crocidura glassi* Heim de Balsac, 1966. Mammalia, 30:448.

COMMON NAME: Glass's Shrew.

TYPE LOCALITY: Ethiopia, "Camp in Gara Mulata Mts, Harar".

DISTRIBUTION: Ethiopian highlands east of Rift Valley (Yalden et al., 1997).

STATUS: IUCN – Vulnerable.

COMMENTS: Often confused with *fumosa* or *thalia*; see Dippenaar (1980). Karyotype (2n = 36, FN = 52) described by Lavrenchenko et al. (1997), genetic relations by Bannikova et al. (2001*b*).

*Crocidura gmelini* (Pallas, 1811). Zoogr. Rosso-Asiat., 1:134, pl. 10, fig. 3.

COMMON NAME: Gmelin's White-toothed Shrew.

TYPE LOCALITY: "Hyrcania"; restricted to "Iran, Khorassan prov., Bujnurd distr., 85 km W Bujnurd, Dasht, 3200 ft. [975 m]" by Goodwin (1940) and Hoffmann (1996*a*).

DISTRIBUTION: Israel through Iran, Turkmenistan, Uzbekistan, Kazakhstan, Afghanistan and Pakistan to W China and Mongolia.

STATUS: IUCN – Lower Risk (lc).

SYNONYMS: *hyrcania* Goodwin, 1940; *ilensis* Miller, 1901; *lar* G. Allen, 1928; *lignicolor* Miller, 1900; *mordeni* Goodwin, 1934; *portali* Thomas, 1920.

COMMENTS: Hoffmann (1996*a*, *b*) designated a neotype, re-defined the species, and provided tentative distribution maps. The species is still unsufficiently known, particularly the relations to *C. suaveolens* and *C. katinka* must be studied (Hutterer and Kock, 2002). The taxon *gmelini* was previously regarded as a synomyn of *Sorex minutus*. *C. portali* was given species rank by Kryštufek and Vohralík (2001), but without regard to Hoffmann's papers.

*Crocidura goliath* Thomas, 1906. Ann. Mag. Nat. Hist., ser. 7, 17:177.

COMMON NAME: Goliath Shrew.

TYPE LOCALITY: Cameroon, Efulen.

DISTRIBUTION: High forest of S Cameroon, Gabon, and Dem. Rep. Congo; an isolated population in West Africa around Mt. Nimba (Guinea, Côte d'Ivoire).

STATUS: IUCN – Lower Risk (lc).

SYNONYMS: **nimbasilvanus** Hutterer, 2003; *guineensis* Heim de Balsac, 1968 [not Cabrera, 1903].

COMMENTS: Type species of subgenus *Praesorex* Thomas, 1913. Often included in *flavescens* or *olivieri*, but apparently represents a distinct species which lives in sympatry with *C. olivieri* in the Central and West African rainforest; see Hutterer (*in* Colyn, 1986:22) and Goodman et al. (2000). *C. goliath nimbasilvanus* replaces Heim de Balsac's *C. odorata guineensis* (Ziela, Mt. Nimba, Guinea), a name preoccupied by *C. occidentalis guineensis* Cabrera, currently a subspecies of *C. olivieri* (Hutterer, 2003). The West African population (*nimbasilvanus*) of *C. goliath* is characterized by smaller size and a shorter pelage, and is genetically distinct from the Central African population (Querouil et al., In Press).

*Crocidura gracilipes* Peters, 1870. Monatsb. K. Preuss. Akad. Wiss. Berlin, 1870:584.

COMMON NAME: Peters's Musk Shrew.

TYPE LOCALITY: "Auf der Reise nach dem Kilimandscharo"; unidentifiable but usually taken as "Kilimanjaro, Tanzania"; see Moreau et al. (1946:395).

DISTRIBUTION: Known only from the type specimen with unknown origin.

STATUS: IUCN – Critically Endangered.

COMMENTS: Does not include *hildegardeae*; see Demeter and Hutterer (1986:201). A recent examination of the type specimen indicates it might be conspecific with *C. cyanea*, however, that species is not known from as far north as Tanzania.

*Crocidura grandiceps* Hutterer, 1983. Rev. Suisse Zool., 90:699.

COMMON NAME: Large-headed Shrew.

TYPE LOCALITY: Ghana, Sefwi-Wiawso, Krokosua Hills, N of Asempanaya (Asampaniye).

DISTRIBUTION: High forest regions of Guinea, Côte d'Ivoire, Ghana, Nigeria, and possibly Cameroon (Hutterer and Schlitter, 1996).

STATUS: IUCN – Lower Risk (lc).

COMMENTS: Genetically allied to the *turba-poensis* complex (Querouil et al., In Press). Karyotype has 2n = 46, FN = 68 (Schlitter et al., 1999).

*Crocidura grandis* Miller, 1911. Proc. U.S. Natl. Mus., 38:393.

COMMON NAME: Greater Mindanao Shrew.

TYPE LOCALITY: Philippines, Mindanao, Grand Malindang Mt., 6100 ft. (1859 m).

DISTRIBUTION: Known only from Mt. Malindang, Mindanao, Philippines.

STATUS: IUCN – Endangered.

COMMENTS: Status unknown; probably confined to primary forest (Heaney et al., 1987:38).

*Crocidura grassei* Brosset, Dubost, and Heim de Balsac, 1965. Biologia Gabonica, 1:165.

COMMON NAME: Grasse's Shrew.

TYPE LOCALITY: Gabon, Belinga.

DISTRIBUTION: Recorded from high forest regions in Gabon, Central African Republic, Cameroon, and Equatorial Guinea (S. M. Goodman et al., 2001*a*; Heim de Balsac, 1968*c*; Lasso et al., 1996).

STATUS: IUCN – Lower Risk (lc).

COMMENTS: A large and distinctly gray shrew, related to the *maurisca-littoralis* group (Querouil et al., In Press).

*Crocidura grayi* Dobson, 1890. Ann. Mag. Nat. Hist., ser. 6, 6:494.

COMMON NAME: Luzon Shrew.

TYPE LOCALITY: Philippines, Luzon.

DISTRIBUTION: Luzon and Mindoro faunal regions, Philippines, in primary forest from 250 to 2400 m.

STATUS: IUCN – Vulnerable.

SYNONYMS: *halconus* Miller, 1910.

COMMENTS: Heaney et al. (1987) included *halconus* as a synonym. Revised by Heaney and Ruedi (1994).

*Crocidura greenwoodi* Heim de Balsac, 1966. Monitore Zool. Ital., 74(suppl.):215.

COMMON NAME: Greenwood's Shrew.

TYPE LOCALITY: Somalia, "Gelib".

DISTRIBUTION: S Somalia.

STATUS: IUCN – Vulnerable.

COMMENTS: Species confined to the Horn of Africa; apparently related to *fulvastra* and *hirta*, with which *greenwoodi* shares the same karyotype (2n = 50, FN = 66) (Schlitter et al., 1999).

*Crocidura harenna* Hutterer and Yalden, 1990. *In* Peters and Hutterer (eds.), Vertebrates in the Tropics, Bonn, p. 64.

COMMON NAME: Harenna Shrew.

TYPE LOCALITY: Ethiopia, Bale Mtns, Harenna Forest.

DISTRIBUTION: Known only from the type locality.

STATUS: IUCN – Critically Endangered.

COMMENTS: Related to *C. phaeura*. Karyotype has 2n = 36, FN = 50 (Lavrenchenko et al., 1997).

*Crocidura hildegardeae* Thomas, 1904. Ann. Mag. Nat. Hist., ser. 7, 14:240.
COMMON NAME: Hildegarde's Shrew.
TYPE LOCALITY: Kenya, Fort Hall.
DISTRIBUTION: Forests in C and E Africa south to Tanzania.
STATUS: IUCN – Lower Risk (lc).
SYNONYMS: *altae* Heller, 1912; *ibeana* Dollman, 1915; *lutreola* Heller, 1912; *maanjae* Heller, 1910; *phaios* Setzer, 1956; *procera* Heller, 1912; *rubecula* Dollman, 1915 [see Heim de Balsac and Meester, 1977].
COMMENTS: Does not include *gracilipes*; see Dieterlen and Heim de Balsac (1979) and Demeter and Hutterer (1986). Gureev (1979) listed *ibeana*, *lutreola*, and *maanjae* as distinct species without comment. Does not include *virgata*; see under that name. The remaining synonyms are in need of revision. A karyotype from Burundi has 2n = 52, FN = 76 (Maddalena and Ruedi, 1994).

*Crocidura hilliana* Jenkins and Smith, 1995. Bull. Nat. Hist. Mus. Lond. (Zool.), 61:103.
COMMON NAME: Hill's Shrew.
TYPE LOCALITY: NE Thailand, Loei Province, 48 km S Loei, Ban Nong Hin, Wat Tham Maho Lan, 17°06'N, 101°53'E, 575 m.
DISTRIBUTION: NE and C Thailand, Laos.
COMMENTS: A shrew somewhat smaller than *C. fuliginosa* described on the basis of skulls removed from owl pellets (Jenkins and Smith, 1995; A. L. Smith et al., 2000). Also recorded from Laos (Smith et al., 1998). Motokawa and Harada (1998) described the animal and its karyotype (2n = 50, NF = 60). This karyotype is similar to that of "*C. attenuata*" from Thailand (2n = 50, FN = 66), as published by Tsuchiya et al. (1979).

*Crocidura hirta* Peters, 1852. Reise nach Mossambique, Säugethiere, p. 78.
COMMON NAME: Lesser Red Musk Shrew.
TYPE LOCALITY: Mozambique, Tette, 17°S.
DISTRIBUTION: Angola, Dem. Rep. Congo, Uganda, Kenya, Somalia, Tanzania, Malawi, Zimbabwe, Zambia, Mozambique, Botswana, Namibia, South Africa.
STATUS: IUCN – Lower Risk (lc).
SYNONYMS: *annellata* Peters, 1852; *beirae* Dollman, 1915; *bloyeti* Dekeyser, 1943; *canescens* Peters, 1852; *deserti* Schwann, 1906; *flavidula* Thomas and Schwann, 1905; *langi* Cabrera, 1925; *luimbalensis* Hill and Carter, 1937; *velutina* Thomas, 1904 [see Heim de Balsac and Meester, 1977].
COMMENTS: Gureev (1979) listed *beirae* and *deserti* as distinct species; the latter may well be separable. *C. bloyeti*, formerly listed as a species, is included here because it was based on a juvenile *hirta*. The Angolan *erica* may also belong here. The karyotype of *hirta* (2n = 50, FN = 66) is shared by *olivieri*, *viaria*, *greenwoodi*, and *attila* (Schlitter et al., 1999).

*Crocidura hispida* Thomas, 1913. Ann. Mag. Nat. Hist., ser. 8, 11:468.
COMMON NAME: Andaman Spiny Shrew.
TYPE LOCALITY: India, Andaman Isls, Middle Andaman Isl (northern end).
DISTRIBUTION: Middle Andaman Isl (Andaman Isls, India).
STATUS: IUCN – Endangered.
COMMENTS: A rare and little-known shrew with a spiny dorsal fur.

*Crocidura horsfieldii* (Tomes, 1856). Ann. Mag. Nat. Hist., ser. 2, 17:23.
COMMON NAME: Horsfield's Shrew.
TYPE LOCALITY: Sri Lanka.
DISTRIBUTION: Sri Lanka and Indian Peninsula (Mysore and Ladak), perhaps Nepal (see below).
STATUS: IUCN – Lower Risk (lc).
SYNONYMS: ? *myoides* Blanford, 1875; *retusa* Peters, 1870.
COMMENTS: Subspecies or synonyms discussed by Jenkins (1976) and Jameson and Jones (1977). Usually spelled *horsfieldi* but Corbet and Hill (1991) correctly used *horsfieldii*. Formerly also included *indochinensis* and *wuchihensis*, but see under those species. Formerly included also *kurodai* and *tadae* from Taiwan, but they show a karyotype

(2n = 40, FN = 54; Fang et al., 1997) different from [*horsfieldii*] *indochinensis* (2n = 38, FN = 48; Rao and Aswathanarayana, 1978). Here they are both included in *rapax*. The distribution of *horsfieldii* sensu strictu is still a matter of disagreement. Lunde et al. (2003*b*) restricted its distribution to Sri Lanka and adjacent peninsular India.

*Crocidura hutanis* Ruedi and Vogel, 1995. Experientia, 51:175, Fig. 1.
COMMON NAME: Hutan Shrew.
TYPE LOCALITY: Indonesia, N Sumatra, Aceh, Gunung Leuser N.P., Alas Valley (300 m), Ketambe (03°31'N, 97°46'E).
DISTRIBUTION: Lowland forest in N and W Sumatra.
COMMENTS: A report by Ruedi and Vogel (1995:175) presenting the new name, the diagnostic karyotype (2n = 36-38, FN = 54), the type locality (Ketambe, N Sumatra), specimen catalog numbers, and a comparison with other SE Asian species appeared earlier in print (February, 1995) than the full taxonomic description by Ruedi (November, 1995:227), and thus constitutes the first valid description.

*Crocidura ichnusae* Festa, 1912. Boll. Mus. Zool. Anat. Comp. Torino, 27, 684:1.
COMMON NAME: North African White-toothed Shrew.
TYPE LOCALITY: Italy, Sardinia, Lanusei, Piscina.
DISTRIBUTION: Pantelleria Isl, Sardinia (Italy), Ibiza (Spain), and E North Africa (Tunisia, E Algeria).
STATUS: IUCN – Lower Risk (lc) as *C. cossyrensis*.
SYNONYMS:? *agilis* (Loche, 1867) [*nomen dubium*]; ? *anthonyi* Heim de Balsac, 1940; *cossyrensis* Contoli, 1989; *ibicensis* Vericad and Balcells, 1965.
COMMENTS: The species was first reported from Pantelleria as *russula*; see Contoli and Amori (1986); then named *cossyrensis* in a footnote (Contoli et al., 1989) and later redescribed by Contoli (1990). Closely related to *russula* if not conspecific (Sarà et al., 1990). However, morphological (Sarà and Vogel, 1996; Sarà and Zanca, 1992) as well as chromosomal differences, restricted fertility in the F2 generation (Vogel et al., 1992), and genetic data (Vogel, pers. comm.) suggest that *C. russula* includes two species; *cossyrensis* has been used for the E North African taxon (Contoli and Aloise, 2001; Hutterer, 1993*a*), but *ichnusae* has priority. Sarà and Vogel (1996) considered *agilis* as unidentifiable taxon. The names *mauritanicus*, *pigmaea* and *heljanensis* are listed here under *russula* on geographical grounds, while *ichnusae* and *ibicensis* belong to the E North African taxon, based on a genetical study of these island forms (Cosson et al., in prep.). The type locality of *anthonyi* is Gafsa, Tunisia, and thus this name has to be considered; however, Heim de Balsac (1968) cast some doubt on the reliability of the type locality, as did Contoli and Aloise (2001). By comparing the coefficient of variation of cranial measurements in various North African populations of *Crocidura*, these authors identified a region W of Algers as a possible zone of sympatry between *russula* and *ichnusae*.

*Crocidura indochinensis* Robinson and Kloss, 1922. Ann. Mag. Nat. Hist., Ser. 9:88.
COMMON NAME: Indochinese Shrew.
TYPE LOCALITY: Vietnam, Langbian Plateau, Dalat.
DISTRIBUTION: Burma, N Thailand to Vietnam; Yunnan and Fujian (China); exact limits unknown.
COMMENTS: This species is usually included in *horsfieldii* (Corbet and Hill, 1992; Jameson and Jones, 1977; Jenkins, 1976; Jiang and Hoffmann, 2001), but the restriction of *horsfieldii* to Sri Lanka and India and the removal of *wuchinensis* leaves *indochinensis* as the available name for the populations from Burma to China. Heaney and Timm (1983) reported this species from Vietnam under *C. horsfieldii*.

*Crocidura jacksoni* Thomas, 1904. Ann. Mag. Nat. Hist., ser. 7, 14:238.
COMMON NAME: Jackson's Shrew.
TYPE LOCALITY: Kenya, "Ravine Station".
DISTRIBUTION: E Dem. Rep. Congo, Uganda, Kenya, N Tanzania.
STATUS: IUCN – Lower Risk (lc).
SYNONYMS: *amalae* Dollman, 1915.

COMMENTS: Includes *amalae*; see Heim de Balsac and Meester (1977:17).

*Crocidura jenkinsi* Chakraborty, 1978. Bull. Zool. Surv. India, 1:303.
   COMMON NAME: Jenkin's Shrew.
   TYPE LOCALITY: India, South Andaman Isl, Wright Myo.
   DISTRIBUTION: Known only from the type locality and from Mt. Harriet.
   STATUS: IUCN – Data Deficient.
   COMMENTS: Included in *nicobarica* by Corbet and Hill (1991), without comment. Das (1999)
      found *C. jenkinsi* in sympatry with *C. andamanensis* on Mt. Harriet, Andaman Isl.

*Crocidura jouvenetae* Heim de Balsac, 1958. Mém. Inst. fr. d'Afr. noire, 53:331.
   COMMON NAME: Jouvenet's Shrew.
   TYPE LOCALITY: Guinea, Mt. Nimba, Ziéla.
   DISTRIBUTION: S Guinea, Liberia, and Côte d'Ivoire.
   SYNONYMS: *ebriensis* Heim de Balsac and Aellen, 1958.
   COMMENTS: Heim de Balsac and Meester (1977) included *ebriensis*, *ingoldbyi* and *jouvenetae* in
      *crossei*, which was followed by Hutterer (1993a). However, *jouvenetae* and *ebriensis*
      represent much larger animals than *crossei* and *ingoldbyi*. The exact ranges of both
      species remain to be determined, but both forms occur almost sympatrically from
      Guinea to Côte d'Ivoire. *C. ebriensis* was listed as a separate species by Maddalena and
      Ruedi (1994), Wolsan and Hutterer (1998), and Schlitter et al. (1999), but *jouvenetae*
      has priority.

*Crocidura katinka* Bate, 1937. Ann. Mag. Nat. Hist., ser. 10, 20:398.
   COMMON NAME: Katinka's Shrew.
   TYPE LOCALITY: Israel, Tabun Cave, Levels E to D (Pleistocene).
   DISTRIBUTION: Israel and Palestine, Syria, SW Iran (Hutterer and Kock, 2002, and
      unpublished).
   COMMENTS: The brief diagnosis of Bate (1937a) was detailed by Bate (1937b). The species was
      previously known only from Pleistocene fossils, but Hutterer and Kock (2002)
      allocated remains from fresh owl pellets collected in Syria to this species. A yet
      unreported specimen from SW Iran appears to represent the same species.

*Crocidura kivuana* Heim de Balsac, 1968. Biologia Gabonica, 4:319.
   COMMON NAME: Kivu Shrew.
   TYPE LOCALITY: Dem. Rep. Congo, Kivu, Tschibati.
   DISTRIBUTION: Kahuzi-Biega National Park (Dem. Rep. Congo).
   STATUS: IUCN – Vulnerable.
   COMMENTS: Very localized species occurring in montane swamps; see Dieterlen and Heim de
      Balsac (1979).

*Crocidura lamottei* Heim de Balsac, 1968. Mammalia, 32:386.
   COMMON NAME: Lamotte's Shrew.
   TYPE LOCALITY: Côte d'Ivoire, Lamto (savane).
   DISTRIBUTION: Sudan and Guinea savanna from Senegal to W Cameroon.
   STATUS: IUCN – Lower Risk (lc).
   SYNONYMS: **elegans** Hutterer, 1986.
   COMMENTS: Includes *elegans* as a subspecies; see Hutterer (1986a). Karyotype has 2n = 52,
      FN = 68 (Meylan, 1971).

*Crocidura lanosa* Heim de Balsac, 1968. Biologia Gabonica, 4:309.
   COMMON NAME: Kivu Long-haired Shrew.
   TYPE LOCALITY: Dem. Rep. Congo, Kivu, Lemera.
   DISTRIBUTION: Uinka (Rwanda); Kivu, Lemera and Irangi (Dem. Rep. Congo).
   STATUS: IUCN – Lower Risk (lc).
   COMMENTS: Present knowledge summarized by Dieterlen and Heim de Balsac (1979).

*Crocidura lasiura* Dobson, 1890. Ann. Mag. Nat. Hist., ser. 6, 5:31.
   COMMON NAME: Ussuri White-toothed Shrew.
   TYPE LOCALITY: NE China, (Manchuria), Ussuri River.

DISTRIBUTION: Ussuri Region (Russia) and NE China to Korea; Kiangsu (China). Range mapped by Zaitsev (1993) and Jiang and Hoffmann (2001).

STATUS: IUCN – Lower Risk (lc).

SYNONYMS: *campuslincolnensis* Sowerby, 1945; *lasiura* Giglioli and Salvadori, 1887 [*nomen nudum*]; *lizenkani* Kishida, 1931 [*nomen nudum*]; *neglecta* Kuroda, 1934 [not Jentink, 1888]; *sodyi* Kuroda, 1935; *thomasi* Sowerby, 1917; *yamashinai* Kuroda, 1934 [see Corbet, 1978*c*:29].

COMMENTS: Karyotype has 2n = 40, FN = 56 (Zima et al., 1998).

*Crocidura latona* Hollister, 1916. Bull. Am. Mus. Nat. Hist., 35:667.

COMMON NAME: Latona's Shrew.

TYPE LOCALITY: Dem. Rep. Congo, Medje.

DISTRIBUTION: Lowland rainforest of NE Dem. Rep. Congo.

STATUS: IUCN – Vulnerable.

COMMENTS: Known by a few specimens only. Recently found in the Masako Forest, near Kisangani (Dudu et al., In Press).

*Crocidura lea* Miller and Hollister, 1921. Proc. Biol. Soc. Wash., 34:102.

COMMON NAME: Sulawesi Shrew.

TYPE LOCALITY: Indonesia, Sulawesi, Temboan.

DISTRIBUTION: N and C Sulawesi, tropical rain forest (Musser, 1987*a*).

STATUS: IUCN – Lower Risk (lc).

COMMENTS: Smallest in an assemblage of endemic shrews (*C. elongata*, *C. lea*, *C. musseri*, *C. nigripes*, *C. rhoditis*) that occur together in C Sulawesi (Ruedi, 1995). Only *C. levicula* is smaller.

*Crocidura lepidura* Lyon, 1908. Proc. U. S. Nat. Mus., 34:662.

COMMON NAME: Sumatran Giant Shrew.

TYPE LOCALITY: Indonesia, E Sumatra, Kateman River.

DISTRIBUTION: E Sumatra.

SYNONYMS: *villosa* Robinson and Kloss, 1918.

COMMENTS: Formerly included in *C. fuliginosa*, but resurrected by Ruedi (1995). Karyotype has 2n = 37-38, FN = 54 (Ruedi and Vogel, 1995).

*Crocidura leucodon* (Hermann, 1780). *In* Zimmermann, Geogr. Gesch. Mensch. Vierf. Thiere, 2:382.

COMMON NAME: Bicolored Shrew.

TYPE LOCALITY: France, Bas Rhin, vicinity of Strasbourg.

DISTRIBUTION: France to the Volga and Caucasus; Elburz Mtns; Asia Minor; Israel; Lebanon; Lesbos Isl (Aegean Sea).

STATUS: IUCN – Lower Risk (lc).

SYNONYMS: *albipes* (Kerr, 1792); *avicennai* Stroganov, 1960; *hydruntina* Costa, 1844; *judaica* Thomas, 1919; *lasia* Thomas, 1906; *leucodus* Schulze, 1897; *microurus* (Fatio, 1869); *narentae* Bolkay, 1925; *persica* Thomas, 1907; *volgensis* Stroganov, 1960.

COMMENTS: Reviewed by Richter (1970) and Gureev (1979). Includes *persica*; see Dolgov (1979). Gureev (1979) and Gromov and Baranova (1981) listed *persica* as a distinct species without comment. Includes *lasia*; see Catzeflis et al. (1985), Gureev (1979), and Jenkins (1976); but see also Corbet (1978*c*). Does not include *caspica* from Iran, but *judaica* from Palestine. European range reviewed by Krapp (1990), Arabian range by Harrison and Bates (1991). The assignment of *hydruntina* follows Nappi and Maio (2000). Karyotype has 2n = 28, FN = 56 (Zima et al., 1998).

*Crocidura levicula* Miller and Hollister, 1921. Proc. Biol. Soc. Wash., 34:103.

COMMON NAME: Sulawesi Tiny Shrew.

TYPE LOCALITY: Indonesia, Sulawesi, Pinedapa.

DISTRIBUTION: Tropical rain forest of C and SE Sulawesi (Musser, 1987*a*).

STATUS: IUCN – Lower Risk (lc).

COMMENTS: Smallest mammal of Sulawesi (Ruedi, 1995). Karyotype has 2n = 34, FN = 52 (Ruedi and Vogel, 1995).

*Crocidura littoralis* Heller, 1910. Smithson. Misc. Coll., 56(15):5.
COMMON NAME: Naked-tail Shrew.
TYPE LOCALITY: Uganda, Butiaba, east shore of Lake Albert.
DISTRIBUTION: Rain forest of Cameroon, Central African Republic, Dem. Rep. Congo, Uganda and Kenya.
STATUS: IUCN – Lower Risk (lc).
SYNONYMS: *oritis* Hollister, 1916.
COMMENTS: This species was included in *monax*, but is a distinct species (Dieterlen and Heim de Balsac, 1979; Hutterer 1993*a*).

*Crocidura longipes* Hutterer and Happold, 1983. Bonn. Zool. Monogr., 18:53.
COMMON NAME: Savanna Swamp Shrew.
TYPE LOCALITY: Nigeria, "Dada, 11°34′N, 04°29′E".
DISTRIBUTION: Known from two swamps in Guinea savanna in W Nigeria.
STATUS: IUCN – Endangered.
COMMENTS: May be related to *foxi*.

*Crocidura lucina* Dippenaar, 1980. Ann. Transvaal Mus., 32:134-138.
COMMON NAME: Lucina's Shrew.
TYPE LOCALITY: Ethiopia, "Web River, near Dinshu".
DISTRIBUTION: Montane moorlands of E Ethiopia.
STATUS: IUCN – Vulnerable.
COMMENTS: Part of the Ethiopian radiation of *Crocidura* with 2n = 36 and FN = 52 (Lavrenchenko et al., 1997: Fig. 5, as *Crocidura* sp. A., Lavrenchenko, pers. comm.). Species confined to the Afro-Alpine moorland (Hutterer and Yalden, 1990).

*Crocidura ludia* Hollister, 1916. Bull. Am. Mus. Nat. Hist., 35:668.
COMMON NAME: Ludia's Shrew.
TYPE LOCALITY: Dem. Rep. Congo, Medje.
DISTRIBUTION: Medje and Tandala (N Dem. Rep. Congo); Dzanga-Sangha FR (Central African Republic).
STATUS: IUCN – Vulnerable.
COMMENTS: Included in *dolichura* by Heim de Balsac and Meester (1977), but regarded as a full species by Hutterer and Dippenaar (1987). Ray and Hutterer (1996) identifed this species in carnivore scats from Central African Republic.

*Crocidura luna* Dollman, 1910. Ann. Mag. Nat. Hist., ser. 8, 5:175.
COMMON NAME: Moonshine Shrew.
TYPE LOCALITY: Dem. Rep. Congo, "Bunkeya River, Shaba Province".
DISTRIBUTION: Mozambique, Zambia, Zimbabwe, E Angola, Dem. Rep. Congo, Malawi, Tanzania, Kenya, Uganda, Rwanda.
STATUS: IUCN – Lower Risk (lc).
SYNONYMS: *electa* Dollmann, 1910; *garambae* Heim de Balsac and Verschuren, 1968; *inyangai* Lundholm, 1955; *johnstoni* Dollmann, 1915; *schistacea* Osgood, 1910; *umbrosa* Dollman, 1915.
COMMENTS: A uniform group of bluish-gray shrews that exhibit a considerable geographic size variation; probably a composite group. Revised by Dippenaar and Meester (1989). Does not include *macmillani*, *raineyi*, and *selina*. In a biochemical comparison, specimens from Rwanda grouped outside all other African *Crocidura* studied (Maddalena, 1990).

*Crocidura lusitania* Dollman, 1915. Ann. Mag. Nat. Hist., ser. 8, 15:516.
COMMON NAME: Mauritanian Shrew.
TYPE LOCALITY: Mauritania, "Trarza country".
DISTRIBUTION: Sahelian zone from S Morocco to Senegal, Mauritania, Mali, Nigeria, Sudan and Ethiopia; a Saharan record from Mali.
STATUS: IUCN – Lower Risk (lc).
COMMENTS: Included in subgenus *Afrosorex* by Hutterer (1986*a*), but a recent genetic study (Querouil et al., In Press) does not support this action. For a summary of distributional

records, see Hutterer (1986*a*) and Sidiyene (1989). Karyotype has 2n = 38, FN = 64 (Maddalena and Ruedi, 1994).

*Crocidura macarthuri* St. Leger, 1934. Ann. Mag. Nat. Hist., ser. 10, 13:559.
COMMON NAME: MacArthur's Shrew.
TYPE LOCALITY: Kenya, Tana River, Merifano (32 km from mouth of Tana River).
DISTRIBUTION: Savanna plains of Kenya and Somalia.
STATUS: IUCN – Lower Risk (lc).
COMMENTS: The species has been recorded from Somalia as *smithi* (e.g., Heim de Balsac, 1966*a*); see Hutterer (1986*a*).

*Crocidura macmillani* Dollman, 1915. Ann. Mag. Nat. Hist., ser. 8, 16:361.
COMMON NAME: MacMillan's Shrew.
TYPE LOCALITY: Ethiopia, "Kotelee, Walamo".
DISTRIBUTION: Ethiopia, Western Plateau.
STATUS: IUCN – Critically Endangered.
COMMENTS: A rare species that was known from the holotype only, but recently rediscovered in the Middle Godjeb Valley, Western Plateau (Bannikova et al., 2001*b*). Formerly included in *fumosa* (Yalden et al., 1976) or *luna* (Heim de Balsac and Meester, 1977; Hutterer, 1981*b*), but Dippenaar (1980) showed that two endemic Ethiopian species, *macmillani* and *thalia*, were covered under these names. Karyotype (2n = 28) and restriction DNA analysis data indicate that *macmillani* is not part of the Ethiopian *glassi-thalia* group (Bannikova et al., 2001*b*).

*Crocidura macowi* Dollman, 1915. Ann. Mag. Nat. Hist., ser. 8, 16:378.
COMMON NAME: Nyiro Shrew.
TYPE LOCALITY: Kenya, Mt. Nyiro, south of Lake Rudolf [Lake Turkana].
DISTRIBUTION: Known only from the type locality.
STATUS: IUCN – Critically Endangered.
COMMENTS: Regarded as a synonym of *hildegardeae* by Osgood (1936), but retained as a species by Heim de Balsac and Meester (1977), who noticed similarities to *niobe*.

*Crocidura malayana* Robinson and Kloss, 1911. J. Fed. Malay St. Mus., 4:241-247.
COMMON NAME: Malayan Shrew.
TYPE LOCALITY: Malaysia, Perak, Maxwell's Hill.
DISTRIBUTION: Malay Peninsula and offshore islands south of the Isthmus of Kra (Pulau Aor and Pulau Redang).
STATUS: IUCN – Endangered.
SYNONYMS: *aagaardi* Kloss, 1917; *aoris* Robinson, 1912; *gravida* Kloss, 1917; *klossi* Robinson, 1912; *major* Kloss, 1911 [not Wagler, 1832].
COMMENTS: This species was included in *fuliginosa* by Jenkins (1976, 1982), but Ruedi et al. (1990) reported two different karyotypes from sympatric populations in Peninsular Malaysia. They provisionally used *malayana* for the sibling species with 2n = 38-40, and FN = 62-68. The listing of synonms follows Ruedi (1995) who subsequently reviewed the species.

*Crocidura manengubae* Hutterer, 1982. Bonn. Zool. Beitr., 32:242.
COMMON NAME: Manenguba Shrew.
TYPE LOCALITY: Cameroon, "Lager III, 1800m, Manenguba-See, Bamenda-Hochland".
DISTRIBUTION: Bamenda, Adamaoua, and Yaounde highlands, Cameroon.
STATUS: IUCN – Lower Risk (lc).
COMMENTS: Resembles *C. littoralis* and *C. maurisca* morphologically; see Hutterer (1994) and S. M. Goodman et al. (2001*a*).

*Crocidura maquassiensis* Roberts, 1946. Ann. Transvaal Mus., 20:312.
COMMON NAME: Makwassie Musk Shrew.
TYPE LOCALITY: South Africa, Northwest Prov., Maquassi, Klipkuil.
DISTRIBUTION: Mpumalanga Prov. and Northwest Prov. (South Africa); Nyamaziwa Falls, and Matopo Hills (Zimbabwe).
STATUS: IUCN – Lower Risk (lc).

SYNONYMS: *malani* Lundholm, 1955.

COMMENTS: Includes *malani*; and may be related to *pitmani*; see Meester (1963) and Meester et al. (1986). Geometrics of the mandible distinguish this species and the similar *cyanea* and *silacea* (Taylor and Contrafatto, 1997).

*Crocidura mariquensis* (A. Smith, 1844). Illustr. Zool. S. Afr. Mamm., pl. 44, fig. 1.
COMMON NAME: Swamp Musk Shrew.
TYPE LOCALITY: South Africa, "A wooded ravine near the tropic of Capricorn" = Marico River, near its junction with Limpopo.
DISTRIBUTION: Swamps and forest from South Africa to Mozambique, W Zimbabwe, and Zambia; NW Botswana and NE Namibia to SC Angola; perhaps SE Dem. Rep. Congo.
STATUS: IUCN – Lower Risk (lc).
SYNONYMS: *pilosa* Dobson, 1890; *sylvia* Thomas and Schwann, 1906; **neavei** Wroughton, 1907; **shortridgei** St. Leger, 1932.
COMMENTS: Includes *pilosa* and *sylvia* as synonyms and *shortridgei* and *neavei* as subspecies; see Dippenaar (1977, 1979), who reviewed the species and selected a lectotype. May also include *nigricans*, which Crawford-Cabral (1987) considered distinct.

*Crocidura maurisca* Thomas, 1904. Ann. Mag. Nat. Hist., ser. 7, 14:239.
COMMON NAME: Gracile Naked-tailed Shrew.
TYPE LOCALITY: Uganda, Entebbe.
DISTRIBUTION: Swamps and primary forest in Uganda, Rwanda, and Burundi; a single record from Gabon.
STATUS: IUCN – Lower Risk (lc).
COMMENTS: Part of the *C. maurisca-littoralis* species group; the extinct *balsamifera* from Ancient Egypt is probably related (Hutterer, 1994). A record from Gabon (S. M. Goodman et al., 2001a) is far outside of the known range of the species.

*Crocidura maxi* Sody, 1936. Natuurk. Tijdschr. Ned.-Ind., 96:53.
COMMON NAME: Javanese Shrew.
TYPE LOCALITY: Indonesia, Java, East Java.
DISTRIBUTION: Java, Lesser Sunda Isls (Bali, Sumbawa, Komodo, Sumba, Flores, Alor, Roti, Timor), and Ambon (Moluccas, Indonesia).
STATUS: IUCN – Lower Risk (lc).
COMMENTS: Occurs sympatrically with *monticola* in Java; see Jenkins (1982) and Kitchener et al. (1994a). Ruedi (1995) questioned the validity of *maxi* but stated (p. 243) that "Eastern representatives, which are usually larger, with a flatter skull profile than nominal and central forms, could be treated as a distinct subspecies, *C. monticola maxi.*"

*Crocidura mindorus* Miller, 1910. Proc. U.S. Natl. Mus., 38:392.
COMMON NAME: Mindoro Shrew.
TYPE LOCALITY: Philippines, Mindoro, Mt. Halcon, 1,938 m.
DISTRIBUTION: Mindoro and Sibuyan, uncommon in forest from 325 m to 1,325 m.
STATUS: IUCN – Endangered.
COMMENTS: Goodman and Ingle (1993) reported six specimens from Sibuyan, the first examples since the description of the species; see also Heaney and Ruedi (1994) and Heaney et al. (1998).

*Crocidura miya* Phillips, 1929. Spolia Zeylan., 15:113.
COMMON NAME: Sri Lankan Long-tailed Shrew.
TYPE LOCALITY: Sri Lanka, Kandyan Hills, Nilambe Dist., Moolgama, 3,000 ft. (914 m).
DISTRIBUTION: Highlands of C Sri Lanka.
STATUS: IUCN – Endangered.
COMMENTS: A very distinctive species, resembling *C. elongata* of Sulawesi, or *C. dolichura* of Africa. Known by a handful of specimens; see Phillips (1980) for further information.

*Crocidura monax* Thomas, 1910. Ann. Mag. Nat. Hist., ser. 8, 6:310.
COMMON NAME: Kilimanjaro Shrew.
TYPE LOCALITY: Tanzania, Mt. Kilimanjaro, Rombo, 6,000 ft. (1829 m).

DISTRIBUTION: Montane forests in N Tanzania and possibly W Kenya.

STATUS: IUCN – Vulnerable.

COMMENTS: Part of the *littoralis* group; see Dieterlen and Heim de Balsac (1979). Does not includes *oritis* (part of *littoralis*) and *ultima* (treated as full species here) as suggested by Heim de Balsac and Meester (1977). The real distribution of *C. monax* is unknown; Jenkins (in Burgess et al., 2000) restricted *monax* to the type locality, while Stanley et al. (2000b) listed it from a number of mountains in N Tanzania. Dollman (1914) recorded the species from W Kenya, but these records need to be confirmed.

*Crocidura monticola* Peters, 1870. Monatsb. K. Preuss. Akad. Wiss. Berlin, 1870:584.

COMMON NAME: Sunda Shrew.

TYPE LOCALITY: Indonesia, Java, Mount Lawu, near Surakarta.

DISTRIBUTION: Borneo, Sumatra, Java, doubtfully Peninsular Malaysia.

STATUS: IUCN – Lower Risk (lc) as *C. monticola*, Data Deficient as *C. minuta*.

SYNONYMS: *bartelsii* Jentink, 1910; *minuta* Otten, 1917 [not Lyddeker, 1902]; *neglecta* Jentink, 1888.

COMMENTS: Revised by Jenkins (1982) and Ruedi (1995); latter author included also *maxi* as a synonym. However, Kitchener et al. (1994a) recognized *maxi* and *monticola* and defined the range as given above. *C. minuta* was listed as a species in the previous edition (Hutterer, 1993a); the name is most probably a synonym of *monticola*; see Jenkins (1982) and Ruedi (1995).

*Crocidura montis* Thomas, 1906. Ann. Mag. Nat. Hist. ser. 7, 18:138.

COMMON NAME: Montane White-toothed Shrew.

TYPE LOCALITY: Uganda, "Ruwenzori East, 12,500' [3809 m]" = Bujongolo, Mubuku Valley, eastern slope of Mt. Ruwenzori.

DISTRIBUTION: Montane forest in C and E Africa; Mt. Ruwenzori and Mt. Elgon (Uganda), Kilimanjaro and Mt. Meru (Tanzania), Imatong Mtns (Sudan), Mt. Kenya (Kenya).

STATUS: IUCN – Lower Risk (lc).

COMMENTS: Formerly a subspecies of *fumosa* but see Demeter and Hutterer (1986) and Dippenaar and Meester (1989), who revised the species. Habitat and ecology on Mt. Elgon described by Clausnitzer et al. (2003).

*Crocidura muricauda* (Miller, 1900). Proc. Wash. Acad. Sci., 2:645.

COMMON NAME: West African Long-tailed Shrew.

TYPE LOCALITY: Liberia, Mount Coffee.

DISTRIBUTION: West African high forest from Guinea to Ghana.

STATUS: IUCN – Lower Risk (lc).

COMMENTS: Often included in *dolichura* as a subspecies but constantly differs in its hairy tail while *dolichura* never shows any pilosity of the tail. Genetically not even closely related to *dolichura* (Querouil et al., In Press).

*Crocidura musseri* Ruedi and Vogel, 1995. Experientia, 51:175, Fig. 1.

COMMON NAME: Mossy Forest Shrew.

TYPE LOCALITY: Indonesia, C Sulawesi, Gunung Rorekatimbo [01°16'S, 120°15'E], 2230 m.

DISTRIBUTION: Known only from type locality.

COMMENTS: Unique karyotype (2n = 32, FN = 54) described by Ruedi and Vogel (1995) who formally named the species (February, 1995). Full description subsequently published (in November) by Ruedi (1995:254). This species is a member of the moss forest guild of small mammals described by Musser (1982, 1987a).

*Crocidura mutesae* Heller, 1910. Smithson. Misc. Coll., 56(15):3.

COMMON NAME: Ugandan Musk Shrew.

TYPE LOCALITY: Uganda, Kampala.

DISTRIBUTION: Uganda; perhaps more widely distributed.

STATUS: IUCN – Lower Risk (lc).

COMMENTS: A large species, alternatively assigned to *hirta* (Allen, 1939) or *suahelae* (Heim de Balsac and Meester, 1977). Recorded as *mutesae* from Central African Republic (Ray and Hutterer, 1996), but its taxonomic status is far from being settled. Possibly a synonym of *C. olivieri*.

*Crocidura nana* Dobson, 1890. Ann. Mag. Nat. Hist., ser. 6, 5:225.
> COMMON NAME: Somali Dwarf Shrew.
> TYPE LOCALITY: Somalia, Dollo.
> DISTRIBUTION: Somalia, Ethiopia.
> STATUS: IUCN – Lower Risk (lc).
> COMMENTS: The name *nana* has been applied to various small shrews of Somalia, Ethiopia, and Egypt, leading to the proposal (Setzer, 1957) that *nana* is conspecific with *religiosa* (which it does not antedate); a conclusion followed by Heim de Balsac and Mein (1971) and Osborn and Helmy (1980). Personal examination of the holotype of *nana* revealed that it represents a juvenile (skull inside the skin) of a species larger that *religiosa*; this conclusion was supported by better preserved topotypical specimens from Somalia in the BMNH, which were also compared with the neotype of *religiosa* (Corbet, 1978c:27). The proposed conspecificy can therefore not be accepted, and *religiosa* remains an endemic of the Nile valley in Egypt (Hutterer, 1994). The relation of *nana* with other small species has yet to be studied.

*Crocidura nanilla* Thomas, 1909. Ann. Mag. Nat. Hist., ser. 8, 4:99.
> COMMON NAME: Savanna Dwarf Shrew.
> TYPE LOCALITY: Uganda, probably Entebbe.
> DISTRIBUTION: Dry and moist savanna from West Africa (Mauritania) to Kenya, Uganda, and Tanzania.
> STATUS: IUCN – Lower Risk (lc).
> SYNONYMS: *denti* St. Leger, 1932 [not Dollman, 1915]; *nancilla* St. Leger, 1932 [*lapsus*]; *rudolfi* St. Leger, 1932.
> COMMENTS: Includes *rudolfi*; see Heim de Balsac and Meester (1977). Often confused with other small species such as *fuscomurina* and *pasha*. For a discussion of "small *Crocidura*", see Heim de Balsac (1968d). Karyotype from Côte d'Ivoire has 2n = 42, FN = 74 (Maddalena and Ruedi, 1994).

*Crocidura negligens* Robinson and Kloss, 1914. Ann. Mag. Nat. Hist., 13:232.
> COMMON NAME: Peninsular Shrew.
> TYPE LOCALITY: Thailand, Koh Samui.
> DISTRIBUTION: Malay Peninsula and some adjacent islands (Koh Samui, Pulau Tioman, Pulau Mapor).
> STATUS: IUCN – Lower Risk (lc).
> SYNONYMS: *maporensis* Robinson and Kloss in Robinson, 1916; *tionis* Kloss, 1917.
> COMMENTS: Formerly included in *C. fuliginosa*, but given species rank by Ruedi (1995). Karyotype (2n = 38, FN = 62) described by Ruedi and Vogel (1995). May be sympatric with *C. malayana* in some areas.

*Crocidura negrina* Rabor, 1952. Chicago Acad. Sci. Nat. Hist. Misc., 96:6.
> COMMON NAME: Negros Shrew.
> TYPE LOCALITY: Philippines, Negros Isl, Cuernos de Negros Mtn, Dayongan, 1,300 m.
> DISTRIBUTION: Primary forest at 500 to 1450 m on S Negros Isl (Philippines).
> STATUS: IUCN – Critically Endangered.
> COMMENTS: Reviewed by Heaney and Ruedi (1994).

*Crocidura nicobarica* Miller, 1902. Proc. U.S. Natl. Mus., 24:776.
> COMMON NAME: Nicobar Shrew.
> TYPE LOCALITY: India, Nicobar Isls, Great Nicobar Isl.
> DISTRIBUTION: Great Nicobar Isl (Nicobar Isls, India).
> STATUS: IUCN – Data Deficient.
> COMMENTS: Not a species of *Suncus*, as suggested by Krumbiegel (1978:71). Corbet and Hill (1991) included *jenkinsi* which is retained as distinct until more evidence is presented.

*Crocidura nigeriae* Dollman, 1915. Ann. Mag. Nat. Hist., ser. 8, 15:524.
> COMMON NAME: Nigerian Shrew.
> TYPE LOCALITY: Nigeria, Asaba, 150 mi. (241 km) up the Niger.
> DISTRIBUTION: Rainforest in Nigeria, Cameroon, and Bioko; exact distribution unknown.
> STATUS: IUCN – Lower Risk (lc).

COMMENTS: Formerly included in *poensis*; but see Heim de Balsac (1957), Meylan and Vogel (1982), and Hutterer and Happold (1983). The karyotype of *nigeria* (2n = 50, FN = 76) is identical to that of *batesi* from Cameroon (Schlitter et al., 1999), and it is possible that both are conspecifics.

*Crocidura nigricans* Bocage, 1889. J. Sci. Math. Phys. Nat. Lisboa, ser. 2, 1:28.
COMMON NAME: Blackish White-toothed Shrew.
TYPE LOCALITY: Angola, Benguela Dist., Quindumbo.
DISTRIBUTION: Angola.
STATUS: IUCN – Lower Risk (lc).
COMMENTS: Regarded unidentifiable by Heim de Balsac and Meester (1977), but specific status upheld by Crawford-Cabral (1987).

*Crocidura nigripes* Miller and Hollister, 1921. Proc. Biol. Soc. Wash., 34:101.
COMMON NAME: Black-footed Shrew.
TYPE LOCALITY: Indonesia, Sulawesi, Temboan, SW from Tondano Lake.
DISTRIBUTION: N and C Sulawesi, in tropical rain forest (Musser, 1987a).
STATUS: IUCN – Lower Risk (lc).
SYNONYMS: **lipara** Miller and Hollister, 1921.
COMMENTS: Ruedi (1995) recognized *lipara* as a distinct subspecies. Karyotype has 2n = 38, FN = 56 (Ruedi and Vogel, 1995).

*Crocidura nigrofusca* Matschie, 1895. Säugethiere Deutsch-Ost-Afrikas, p. 33.
COMMON NAME: African Black Shrew.
TYPE LOCALITY: Dem. Rep. Congo, Semliki Valley, "Wukalala, Kinyawanga im Westen des Semliki".
DISTRIBUTION: S Ethiopia and Sudan through E Africa to Zambia and Angola, Dem. Rep. Congo, perhaps Cameroon.
STATUS: IUCN – Lower Risk (lc).
SYNONYMS: *ansorgei* Dollman, 1915; *cabrerai* Morales Agacino, 1935; *kempi* Dollman, 1915; *lakiundae* Heller, 1912; *luluae* Matschie, 1926; *nilotica* Heller, 1910; *nyikae* Dollman, 1915; *provocax* Thomas, 1910; *zaodon* Osgood, 1910; *zena* Dollman, 1915 [see Heim de Balsac and Meester, 1977].
COMMENTS: Includes *luluae* Matschie, 1926 (Luluabourg, Dem. Rep. Congo) and *zaodon* Osgood, 1910 (Nairobi, Kenya) which were listed as separate species by Heim de Balsac and Meester (1977) and Dippenaar and Meester (1989); see Hutterer et al. (1987b). The holotypes of *nigrofusca*, *luluae*, and *zaodon* have been studied. Gureev (1979) listed *ansorgei*, *nilotica*, and *zena* as distinct species without comment. The species requires a careful systematic revision.

*Crocidura nimbae* Heim de Balsac, 1956. Mammalia, 20:131.
COMMON NAME: Nimba Shrew.
TYPE LOCALITY: Guinea, Mt. Nimba, baraque de Zouguépo.
DISTRIBUTION: Submontane and primary lowland forest in Sierra Leone, Guinea, Liberia, and Côte d'Ivoire; in the Tai National Park (Côte d'Ivoire) it occurs only in undisturbed primary forest (Churchfield et al., 2004).
STATUS: IUCN – Lower Risk (lc).
COMMENTS: A very distinct species; not conspecific with *wimmeri* as previously suggested (see Hutterer, 1983a).

*Crocidura niobe* Thomas, 1906. Ann. Mag. Nat. Hist., ser. 7, 18:138.
COMMON NAME: Niobe's Shrew.
TYPE LOCALITY: Uganda, Ruwenzori East, 6,000 ft. (= Mubukee Valley, 1829 m).
DISTRIBUTION: Montane forests of EC Africa (Uganda, Dem. Rep. Congo); perhaps Ethiopia.
STATUS: IUCN – Lower Risk (lc).
COMMENTS: Ethiopian records (Corbet and Yalden, 1972; Yalden et al., 1976) uncertain; see Hutterer and Yalden (1990).

*Crocidura obscurior* Heim de Balsac, 1958. Mém. Inst. fr. Afr. noire, 53:328.
COMMON NAME: West African Pygmy Shrew.

TYPE LOCALITY: Guinea, Mt. Nimba, montane prairie.

DISTRIBUTION: Sierra Leone to Côte d'Ivoire; possibly Nigeria.

STATUS: IUCN – Lower Risk (lc).

SYNONYMS: *eburnea* Heim de Balsac, 1958.

COMMENTS: Described as a subspecies of *bottegi* (see Heim de Balsac and Meester, 1977), but its longer skull (Hutterer and Happold, 1983) and West African distribution distinguish it. Two different karyotypes have been described from Côte d'Ivoire, 2n = 36, FN = 56 (Maddalena and Ruedi, 1994), and 2n = 40, FN = 60 (Meylan, 1971; under *bottegi*). The problem needs to be analyzed.

*Crocidura olivieri* (Lesson, 1827). Manuel de Mammalogie, p. 121.

COMMON NAME: African Giant Shrew.

TYPE LOCALITY: Egypt, Sakkara; the neotype designated by Corbet (1978c:30) was collected "near Giza".

DISTRIBUTION: Egypt; Mauretania to Ethiopia, and southwards to N South Africa.

STATUS: IUCN – Lower Risk (lc).

SYNONYMS: *deltae* Heim de Balsac and Barloy, 1966; ***anchietae*** Bocage, 1889; ***bueae*** Heim de Balsac and Barloy, 1966; ***cara*** Dollmann, 1915; ***cinereoaenea*** (Rüppell, 1842); *doriana* Dobson, 1887; ***darfurea*** Thomas and Hinton, 1923; ***giffardi*** de Winton, 1898; ***guineensis*** Cabrera, 1903 [not Heim de Balsac, 1968]; *atlantis* Heim de Balsac and Barloy, 1966; ***hansruppi*** Hutterer, 1981; ***hedenborgiana*** (Sundevall, 1843); *ferruginea* Heuglin, 1865; *fuscosa* Thomas, 1913; *hedenborgi* (Sundevall, 1843); *hera* Dollmann, 1915; *kijabae* J. A. Allen, 1909; ***kivu*** Osgood, 1910; *luluana* Cabrera, 1925; ***manni*** Peters, 1878; ***martiensseni*** Neumann, 1900; ***nyansae*** Neumann, 1900; *daphnia* Hollister, 1916; ***occidentalis*** (Pucheran, 1855); *aequatorialis* (Pucheran, 1855); *petersii* Dobson, 1890; ***odorata*** (Leconte, 1857); ***spurelli*** Thomas, 1910; ***sururae*** Heller, 1910; *tatiana* Dollman, 1915; ***toritensis*** Setzer, 1956; ***zuleika*** Dollman, 1915; *herero* St. Leger, 1932.

COMMENTS: *Crocidura olivieri* is the valid and widely used name for large African shrews previously known as *flavescens* (which is now the valid name for a species restricted to South Africa; see Maddalena et al., 1987). Chitaukali et al. (2001) recently proposed to use *occidentalis* instead of *olivieri*, which they regard as a *nomen dubium*, despite the neotype designation by Corbet (1978). The description of *C. olivieri* was based on a large mummified shrew from Ancient Egypt. The same species occurs in Egypt today (Hutterer, 1994), and there is no reason to believe that the name *olivieri* does not refer to the extant species.

This group of giant shrews was reviewed by Heim de Balsac and Barloy (1966). Well known subspecies names are *anchietae*, *doriana*, *ferruginea*, *fuscosa*, *giffardi*, *guineeensis*, *hansruppi*, *hedenborgiana*, *kivu*, *manni*, *martiensseni*, *nyansae*, *occidentalis*, *odorata*, *spurelli*, and *sururae*. Some of these were considered allospecies of a *flavescens* superspecies by Hutterer and Happold (1983). Many authors also distinguished pale (*occidentalis*, *manni*, *spurelli*) and black (*giffardi*, *hedenborgiana*, *martiensseni*, *odorata*) color morphs as different species but biochemical evidence showed that they are merely color morphs of a single and highly variable species (Maddalena, 1990). *Crocidura olivieri* may also include *zaphiri*; see Yalden et al. (1976).

*Crocidura orientalis* Jentink in Weber, 1890. Zool. Ergebn. Reis. Ned. Ost. Ind., I:124.

COMMON NAME: Oriental Shrew.

TYPE LOCALITY: Indonesia, W Java, Gedeh.

DISTRIBUTION: Mountains of Java.

STATUS: IUCN – Vulnerable.

SYNONYMS: ***lawuana*** Sody, 1936.

COMMENTS: Sometimes referred to *fuliginosa* (Heaney et al., 1987; Jenkins, 1982), but represents a distinct species (Corbet and Hill, 1992; Ruedi, 1995). In Java *C. orientalis* is confined to the mountains and replaced by *C. brunnea* in the lowlands. Ruedi (1995) distinguished *lawuana* as a subspecies. Karyotype has 2n = 38, FN = 56 (Ruedi and Vogel, 1995).

*Crocidura orii* Kuroda, 1924. [New Mammals from the Ryukyu Islands], p. 3.
COMMON NAME: Ryukyu Shrew.
TYPE LOCALITY: Japan, Ryukyu Islands, Amamioshima, Komi.
DISTRIBUTION: Amami Group of Ryukyu Isls, Japan.
STATUS: IUCN – Endangered.
COMMENTS: Provisionally included in *dsinezumi* (Corbet, 1978c); but regarded as a separate
     species by Imaizumi (1961, 1970b), Abe (1967), Jenkins (1976), Corbet and Hill (1992),
     and Hutterer (1993a), among others. The species was first described by Kuroda (1924)
     in a publication which, although privately published, has been regarded as available
     by all subsequent authors. See Motokawa (1998) for a review of the species.

*Crocidura palawanensis* Taylor, 1934. Monogr. Bur. Sci. Manila, 30:88.
COMMON NAME: Palawan Shrew.
TYPE LOCALITY: Philippines, Palawan, Sir J. Brooke Point.
DISTRIBUTION: Palawan and Balabac Isl, Philippines.
STATUS: IUCN – Vulnerable.
COMMENTS: May belong to *C. fuliginosa* (Heaney et al., 1987) or to *C. foetida* (Ruedi, 1995).
     Reviewed by Heaney and Ruedi (1994).

*Crocidura paradoxura* Dobson, 1886. Ann. Mus. Civ. Stor. Nat. Genova, 4:566.
COMMON NAME: Sumatran Long-tailed Shrew.
TYPE LOCALITY: Indonesia, Sumatra, Mt. Singalang, 2,000 m.
DISTRIBUTION: Mountains of N and W Sumatra.
STATUS: IUCN – Endangered.
SYNONYMS: *aequicauda* Robinson and Kloss, 1918.
COMMENTS: A large species with a long tail. Revised by Ruedi (1995). Corbet and Hill (1992)
     also referred a specimen from Java which may represent *paradoxura*, or a different
     species (Ruedi, 1995).

*Crocidura parvipes* Osgood, 1910. Field Mus. Nat. Hist. Publ., Zool. Ser., 10:19.
COMMON NAME: Small-footed Shrew.
TYPE LOCALITY: Kenya, "Voi, British East Africa".
DISTRIBUTION: Africa; Guinea and Sudan savanna from Cameroon to S Sudan, Ethiopia
     (Hutterer and Yalden, 1990), Kenya, Tanzania, S Dem. Rep. Congo, Zambia to Angola
     (Hutterer, 1986a:31).
STATUS: IUCN – Lower Risk (lc).
SYNONYMS: *boydi* Dollman, 1915; *chitauensis* Hill and Carter, 1937; *cuanzensis* Hill and
     Carter, 1937; *katharina* Kershaw, 1922; *lutrella* Heller, 1910; *nisa* Hollister, 1916.
COMMENTS: A cranially and externally distinct group of medium-sized savanna shrews.
     Included in subgenus *Afrosorex* by Hutterer (1986a). Considerable size variation,
     however, may indicate that more than one species is included. Karyotypes from
     Ethiopia (2n = 50; Bannikova et al., 2001b) and from Cameroon (2n = 52, FN = 66;
     Schlitter et al., 1999) are different. Specimens recorded as *Crocidura butleri percivali*
     from Machakos, Kenya (Harrison and Bates, 1986) represent *C. parvipes*.

*Crocidura pasha* Dollman, 1915. Ann. Mag. Nat. Hist., ser. 8, 15:517.
COMMON NAME: Sahelian Tiny Shrew.
TYPE LOCALITY: Sudan, Atbara River.
DISTRIBUTION: Sudan and Sahelian savanna of Sudan and Mali; a single record from
     Ethiopia (Demeter, 1982).
STATUS: IUCN – Lower Risk (lc).
COMMENTS: Often confused with *nanilla* and *lusitania*; does not include *glebula* which is a
     synonym of *fuscomurina* or *planiceps*; see Hutterer and Kock (1983) and Hutterer and
     Happold (1983). Dobigny et al. (2001b) found this tiny shrew to be common in the
     Adrar des Iforas Massif in N Mali.

*Crocidura pergrisea* Miller, 1913. Proc. Biol. Soc. Wash., 26:113.
COMMON NAME: Pale Gray Shrew.
TYPE LOCALITY: Kashmir, Baltistan, Shigar, Skoro Loomba, 9,500 ft. (2,896 m).
DISTRIBUTION: Mountains of W Himalaya (Kashmir).

STATUS: IUCN – Vulnerable.

COMMENTS: Some authors have included *armenica*, *serezkyensis*, and *zarudnyi* (see Spitzenberger, 1971*a*, and Corbet, 1978*c*, for a review of literature); but all are now considered separate species. A considerable diversity of opinions exists in the literature on the allocation of the different forms. Following Jenkins (1976), the name *pergrisea* is applied only to the largest species, as represented by the type series from Baltistan.

*Crocidura phaeura* Osgood, 1936. Field Mus. Nat. Hist. Publ., Zool. Ser., 20:228.

COMMON NAME: Guramba Shrew.

TYPE LOCALITY: Ethiopia, Sidamo, west base of Mt. Guramba, NE of Allata.

DISTRIBUTION: Montane and riverine forest along the Ethiopian Rift Valley (Duckworth et al., 1993).

STATUS: IUCN – Critically Endangered.

COMMENTS: Considered a full species by Dippenaar and Meester (1989). Related to *harenna*; see discussion in Hutterer and Yalden (1990).

*Crocidura picea* Sanderson, 1940. Trans. Zool. Soc. Lond., 24:682.

COMMON NAME: Cameroonian Shrew.

TYPE LOCALITY: Cameroon, Mamfe Div., Assumbo, Tinta.

DISTRIBUTION: W Cameroon, Bamenda Highlands.

STATUS: IUCN – Critically Endangered.

COMMENTS: Holotype figured by Heim de Balsac and Hutterer (1982:142, fig. 3). New locality reported and karyotype (2n = 58, FN = 66) described by Schlitter et al. (1999).

*Crocidura pitmani* Barclay, 1932. Ann. Mag. Nat. Hist., ser. 10, 10:440.

COMMON NAME: Pitman's Shrew.

TYPE LOCALITY: Zambia, Maluwe-Serenje Distr., 3800 ft. (1158 m).

DISTRIBUTION: C and N Zambia.

STATUS: IUCN – Vulnerable.

COMMENTS: Mapped under the name *gracilipes* by Ansell (1978).

*Crocidura planiceps* Heller, 1910. Smithson. Misc. Coll., 56(15):5.

COMMON NAME: Flat-headed Shrew.

TYPE LOCALITY: Uganda, Lado Enclave, Rhino Camp.

DISTRIBUTION: Ethiopia, Uganda, Sudan, Dem. Rep. Congo, Nigeria.

STATUS: IUCN – Lower Risk (lc).

COMMENTS: Closely related to *fuscomurina*, if not conspecific; see Heim de Balsac (1968*d*) and Hutterer (1983*b*).

*Crocidura poensis* (Fraser, 1843). Proc. Zool. Soc. Lond., 1842:200 [1843].

COMMON NAME: Fraser's Musk Shrew.

TYPE LOCALITY: Equatorial Guinea, Bioko (Fernando Po), Clarence.

DISTRIBUTION: West Africa (Guinea to Cameroon), Bioko, and Principe Isl.

STATUS: IUCN – Lower Risk (lc).

SYNONYMS: *calabarensis* Sanderson, 1940; *pamela* Dollman, 1915; *schweitzeri* Peters, 1877; *soricoides* (Murray, 1860); *stampflii* Jentink, 1888 [see Heim de Balsac and Meester, 1977, and Hutterer and Happold, 1983].

COMMENTS: The definition of this large black shrew is still problematic. The karyotype (2n = 52, 53, FN = 70, 72) is only known from the Côte d'Ivoire population (*pamela*) (Meylan and Vogel, 1982), but not from the type locality.

*Crocidura polia* Hollister, 1916. Bull. Am. Mus. Nat. Hist., 35:669.

COMMON NAME: Polia's Shrew.

TYPE LOCALITY: Dem. Rep. Congo, Medje.

DISTRIBUTION: Known only from the type locality.

STATUS: IUCN – Critically Endangered.

COMMENTS: Included in *dolichura* by Heim de Balsac and Meester (1977) but represents a distinct species known only from the holotype.

*Crocidura pullata* Miller, 1911. Proc. Biol. Soc. Wash., 24:241.

COMMON NAME: Kashmir White-toothed Shrew.

TYPE LOCALITY: Kashmir, Kotihar, 7,000 ft. (2134 m).

DISTRIBUTION: N India/Pakistan; Kashmir and Ladak (Jiang and Hoffmann 2001), otherwise unknown.

STATUS: IUCN – Lower Risk (lc).

COMMENTS: The name *pullata* has been provisonally (and partly erroneously) used by Hutterer (1993*a*) as a label to include Asian populations formerly called *russula* by Jenkins (1976) and many other authors. It can be seen from the measurements provided by Jameson and Jones (1977) that the forms *pullata*, *rapax* and *vorax* differ from the European *russula* by a longer tail; all have been assigned to the West European species; see Lekagul and McNeely (1977), among others. All three are now regarded as separate species (Jiang and Hoffmann, 2001).

*Crocidura raineyi* Heller, 1912. Smithson. Misc. Coll., 60(12):7-8.

COMMON NAME: Rainey's Shrew.

TYPE LOCALITY: Kenya, Mt. Garguez, North Creek.

DISTRIBUTION: Known only from the type locality.

STATUS: IUCN – Critically Endangered.

COMMENTS: Since its description *C. raineyi* has been considered a valid species, but was synonymized by Heim de Balsac and Meester (1977) with *C. luna*, an error corrected by Dippenaar and Meester (1989).

*Crocidura ramona* Ivanitskaya, Shenbrot, and Nevo, 1996. Z. Säugetierk., 61:97.

COMMON NAME: Negev Shrew.

TYPE LOCALITY: Israel, Negev, NE region of Makhtesh Ramon (30°40′N, 34°56′E).

DISTRIBUTION: Israel, Negev Highlands and edge of Judean Desert.

COMMENTS: A light silver-gray shrew with a unique karyotype of 2n = 28, FN = 46 (Ivanitskaya et al., 1996*a*). Possibly related to the *arispa*, *armenica*, *pergrisea*, *serezkiensis*, *zarudnyi* group of rock shrews. Kryštufek and Vohralík (2001) raised the question whether *portali* might represent an earlier name for *ramona*, however, *portali* was discussed as part of *gmelini* by Hutterer and Kock (2002).

*Crocidura rapax* G. Allen, 1923. Amer. Mus. Novit., 100:9.

COMMON NAME: Chinese White-toothed Shrew.

TYPE LOCALITY: China, Yunnan, Mekong River, Yinpankai.

DISTRIBUTION: S China and adjacent countries (Jiang and Hoffmann, 2001), NE India (specimens in FMNH).

SYNONYMS: **kurodai** Jameson and Jones, 1977; **lutaoensis** Fang and Lee, 2002; **tadae** Tokuda and Kano, 1936.

COMMENTS: Formerly included in *russula* or *pullata*, but see Jiang and Hoffmann (2001). Fang et al. (1997) retained *kurodai* as a separate species, but did not study *rapax*. Fang and Lee (2002) demonstrated that allopatric populations from Taiwan (*kurodai*), Orchid Isl (*tadae*) and Green Isl (*lutaoensis*) share the same karyotype (2n = 40, FN = 54 or 64), but differ in size and morphology and should therefore be considered as valid subspecies.

*Crocidura religiosa* (I. Geoffroy, 1827). Mem. Mus. Hist. Nat. Paris, 15:128.

COMMON NAME: Egyptian Pygmy Shrew.

TYPE LOCALITY: Egypt, Giza.

DISTRIBUTION: Nile Valley (Egypt).

STATUS: IUCN – Data Deficient.

COMMENTS: Described from embalmed specimens from Ancient Egyptian tombs at Thebes; holotype not preserved. Corbet (1978*c*:27) selected a neotype from Giza.

*Crocidura rhoditis* Miller and Hollister, 1921. Proc. Biol. Soc. Wash., 34:102.

COMMON NAME: Sulawesi White-handed Shrew.

TYPE LOCALITY: Indonesia, Sulawesi, Temboan.

DISTRIBUTION: Tropical rainforest of N, C, and SW Sulawesi (Musser, 1987*a*).

STATUS: IUCN – Lower Risk (lc).

COMMENTS: Karyotype has 2n = 30, FN = 50 (Ruedi and Vogel, 1995).

*Crocidura roosevelti* (Heller, 1910). Smithson. Misc. Coll., 56(15):6.
    COMMON NAME: Roosevelt's Shrew.
    TYPE LOCALITY: Uganda, Lado Enclave, Rhino Camp.
    DISTRIBUTION: Forest-savanna margin of the Central African forest block; records from
        Angola, Cameroon, Central African Republic, Dem. Rep. Congo, Uganda, Rwanda,
        and Tanzania (Hutterer, 1981*a*).
    STATUS: IUCN – Lower Risk (lc).
    COMMENTS: Type species of subgenus *Heliosorex* Heller, 1910. Genetic data suggest that this
        species is not closely related to *C. dolichura* and allies (Querouil et al., In Press).

*Crocidura russula* (Hermann, 1780). *In* Zimmermann, Geogr. Gesch. Mensch. Vierf. Thiere, 2:382.
    COMMON NAME: Greater White-toothed Shrew.
    TYPE LOCALITY: France, Bas Rhin, near Strasbourg.
    DISTRIBUTION: S and W Europe including some Atlantic isls off France and Great Britain;
        N Africa (Morocco; Algeria; Canary Isls).
    STATUS: IUCN – Lower Risk (lc) as *C. russula*, Vulnerable as *C. osorio*.
    SYNONYMS: *albiventris* de Selys Longchamps, 1839; *araneus* Schreber, 1778 [not of Linnaeus,
        1758]; *candidus* Bechstein, 1801; *chrysothorax* Dehne, 1855; *cinereus* Bechstein, 1801;
        *constrictus* Hermann, 1780; *fimbriatus* Wagler, 1832; *inodorus* de Selys Longchamps,
        1839; *leucurus* Shaw, 1800; *major* Wagler, 1832; *moschata* Wagler, 1832; *musaraneus*
        Cuvier, 1798; *pigmaea* Loche, 1867; *poliogastra* Wagler, 1832; *rufa* Wagler, 1832;
        *thoracicus* Savi, 1832; *unicolor* Kerr, 1792; **cintrae** Miller, 1907; **osorio** Molina and
        Hutterer, 1989; **peta** Montagu and Pickford, 1923; **pulchra** Cabrera, 1907; **yebalensis**
        Cabrera, 1913; *chaouianensis* Vesmanis and Vesmanis, 1980; *foucauldi* Morales
        Agacino, 1943; *heljanensis* Vesmanis, 1975; *mauritanicus* Pomel, 1856; *safii* Vesmanis
        and Vesmanis, 1980.
    COMMENTS: Reviewed by Genoud and Hutterer (1990). The species is confined to W Europe
        and N Africa. Many populations from Asia and Africa have been erroneously assigned
        to *russula* (see Ellermann and Morrison-Scott, 1951). Allozyme and karyotype analyses
        by Catzeflis et al. (1985) have shown that animals from E Europe, Asia Minor, and
        Israel formerly identified as *russula* instead belong to *suaveolens*. This may also be true
        for other populations further east. Does not include *hosletti, rapax*, or *vorax* (as in
        Ellerman and Morrison-Scott, 1966:81; Jameson and Jones, 1977:465); see under
        species *shantungensis, rapax*, and *vorax*. Does not include *ichnusae* (= *cossyrensis*); see
        under that species. Recent morphological (Hutterer, unpubl.) and genetic studies
        (Vogel et al., 2003) suggest that *osorio* is a peripheral population of *C. russula*, possibly
        introduced by man, although differences in size, ecology, and behavior (Hutterer et al.,
        1992*b*) characterize it as a distinct island form.

*Crocidura selina* Dollman, 1915. Ann. Mag. Nat. Hist., ser. 8, 16:371-372.
    COMMON NAME: Ugandan Lowland Shrew.
    TYPE LOCALITY: Uganda, Mabira Forest, Chagwe.
    DISTRIBUTION: Known only from three lowland forests in Uganda.
    STATUS: IUCN – Endangered.
    COMMENTS: Previously included in *fumosa* or *luna*, but considered a distinct species by
        Dippenaar and Meester (1989).

*Crocidura serezkyensis* Laptev, 1929. Opred. Mlekopitay. Sredney Asyy, Tashkent, 1:16.
    COMMON NAME: Lesser Rock Shrew.
    TYPE LOCALITY: Tajikistan, Pamir Mtns, Lake Sarezskoye.
    DISTRIBUTION: Azerbaijan, Turkmenistan, Tajikistan and Kazakhstan.
    STATUS: IUCN – Lower Risk (lc).
    COMMENTS: Previously included in *pergrisea* (Jenkins, 1976; Spitzenberger, 1971*a*), but
        considered a distinct species by Stogov and Bondar (1966) and Stogov (1985). Records
        from Kazakhstan, Tajikistan, Azerbaijan (Graphodatsky et al., 1988) and Turkmenistan
        (Stogov and Bondar, 1966) presumably belong to the same species. Graphodatsky et al.
        (1988) reported on the karyotype of a specimen from Dzhulfa, SW Azerbaijan (under
        the name *pergrisea*); with 2n = 22, FN = 34, *serezkyensis* has the lowest chromosome
        number ever recorded for a shrew. Populations in Asia Minor (*arispa* Spitzenberger,

1971) were included in *serezkyensis* in the former edition (Hutterer, 1993a), but Kryštufek and Vohralík (2001) have demonstrated that both represent distinct species.

*Crocidura shantungensis* Miller, 1901. Proc. Biol. Soc. Wash., 14:158.
COMMON NAME: Asian Lesser White-toothed Shrew.
TYPE LOCALITY: China, Shantung (=Shandong), Chimeh.
DISTRIBUTION: SE Siberia, E China and Korea incl. Taiwan, Cheju and Tsuchima Isl.
STATUS: IUCN – Data Deficient.
SYNONYMS: *coreae* Thomas, 1907; *longicaudata* Mori, 1927 [not *longicaudata* Tichomirov and Kortchagin, 1889]; ? *orientis* Ognev, 1922; *phaeopus* G. Allen, 1923; *utsuryoensis* Mori, 1937; **quelpartis** Kuroda, 1934; *hosletti* Jameson and Jones, 1977.
COMMENTS: Revised by Jiang and Hoffmann (2001). Previously included in *C. suaveolens*. Iwasa et al. (2001) reported a karyotype of 2n = 40, FN = 46 for specimens from Cheju Isl (= Quelpart Isl) which is indistinguishable from that of *C. suaveolens*. Kuroda (1934) included *quelpartis* in *C. dsinezumi*, while Jameson and Jones (1977) included it in *C. russula*. Motokawa et al. (2003) analyzed the morphometric geographic variation of *shantungensis* and found that populations of Cheju Isl (*quelpartis*) and Taiwan (*hosletti*) average larger. Possibly *quelpartis* can be recognized as a subspecies.

*Crocidura sibirica* Dukelsky, 1930. Zool. Anz., 88:75.
COMMON NAME: Siberian Shrew.
TYPE LOCALITY: Russia, Siberia, S Krasnoyarsky Krai, upper Yenisei River, 96 km S of Minusinsk, Oznatchenoie.
DISTRIBUTION: C Asia from Lake Issyk Kul to Upper Ob River; Lake Baikal; perhaps also Sinkiang (China) and Mongolia (see Sokolov and Orlov, 1980:50).
STATUS: IUCN – Lower Risk (lc).
SYNONYMS: *ognevi* Stroganov, 1956.
COMMENTS: Includes *ognevi*; see Yudin (1989). Species reviewed by Zaitsev (1993). Genetic data show that *sibirica* is related to *suaveolens* and *shantungensis* (Han et al., 2002).

*Crocidura sicula* Miller, 1900. Proc. Biol. Soc. Wash., 14:41.
COMMON NAME: Sicilian Shrew.
TYPE LOCALITY: Italy, Sicily, Palermo.
DISTRIBUTION: Sicily, Egadi Isls and Ustica (Italy), and Gozo (Malta); extinct in Malta.
STATUS: IUCN – Lower Risk (lc).
SYNONYMS: *caudata* Miller, 1900; **aegatensis** Hutterer, 1991; **calypso** Hutterer, 1991; **esuae** Kotsakis, 1984.
COMMENTS: Revised by Hutterer (1991), who recognized one extinct and three extant subspecies. Formerly included in *leucodon*, *russula*, or *suaveolens*; but the species has a distinct karyotype (Sarà and Vitturi, 1996; Vogel, 1988) and morphology (Vogel et al., 1989). Temporal and geographic variation studied by Hutterer (1991), Sarà (1995, 1996), and Sarà and Vitturi (1996).

*Crocidura silacea* Thomas, 1895. Ann. Mag. Nat. Hist., ser. 6, 16:53.
COMMON NAME: Lesser Gray-brown Musk Shrew.
TYPE LOCALITY: South Africa, Mpumalanga Prov., Barberton dist., De Kaap, Figtree Creek.
DISTRIBUTION: Occurs in most of South Africa, and parts of Botswana, Angola; Mozambique, Zambia and S Malawi.
STATUS: IUCN – Lower Risk (lc).
SYNONYMS: *holobrunneus* Roberts, 1931.
COMMENTS: This species was formerly assigned to *gracilipes* or *hildegardeae*, but is not conspecific with either of these; see Meester et al. (1986) for a discussion.

*Crocidura smithii* Thomas, 1895. Ann. Mag. Nat. Hist., ser. 6, 15:51.
COMMON NAME: Desert Musk Shrew.
TYPE LOCALITY: Ethiopia, Webi Shebeli, near Finik.
DISTRIBUTION: Arid regions of Senegal, Ethiopia, and probably Somalia.
STATUS: IUCN – Lower Risk (lc).
SYNONYMS: **debalsaci** Hutterer, 1981.
COMMENTS: Revised by Hutterer (1986a). Specimens reported from Somalia by Heim de

Balsac (1966a) represent *macarthuri*; see under that species. Includes *debalsaci* as a distinct subspecies; see Hutterer (1981b).

*Crocidura somalica* Thomas, 1895. Ann. Mag. Nat. Hist., ser. 6, 16:52.
COMMON NAME: Somali Shrew.
TYPE LOCALITY: Ethiopia, Middle Webi Shebeli (about 5°30′N, 44°E) near Geledi (Galadi).
DISTRIBUTION: Dry savannas and semi-desert areas of Ethiopia, Sudan, and probably Somalia; Mali.
STATUS: IUCN – Lower Risk (lc).
COMMENTS: Revised by Hutterer and Jenkins (1983). Recently recorded from the Sahara (Mali) by Hutterer et al. (1992c), who regarded the subspecies *dhofarensis* from Oman as specifically distinct; see under *dhofarensis*.

*Crocidura stenocephala* Heim de Balsac, 1979. Säugetierkdl. Mitt., 27:258.
COMMON NAME: Kahuzi Swamp Shrew.
TYPE LOCALITY: E Dem. Rep. Congo, Kahuzi-Biega N.P.
DISTRIBUTION: Montane *Cyperus* swamps at Mt. Kahuzi, E Dem. Rep. Congo (Dieterlen and Heim de Balsac, 1979) and Bwindi Impenetrable N.P., SW Uganda (Kasangaki et al., 2003).
STATUS: IUCN – Vulnerable.
COMMENTS: Described as a subspecies of *littoralis* but regarded as a full species by Hutterer (1982a) and Dippenaar (pers. comm.).

*Crocidura suaveolens* (Pallas, 1811). Zoogr. Rosso-Asiat., 1:133.
COMMON NAME: Lesser White-toothed Shrew.
TYPE LOCALITY: Russia, Crimea, Khersones, near Sevastopol.
DISTRIBUTION: Palearctic from Spain to Siberia; Atlantic isls (Scilly, Jersey, Sark, Ushant, Yeu, and others; see Cosson et al., 1996); many Mediterranean isls including Corsica, Crete, Cyprus, and Menorca.
STATUS: IUCN – Lower Risk (lc) as *C. suaveolens* and *C. gueldenstaedtii*.
SYNONYMS: *antipae* Matschie, 1901; *aralychensis* Satunin, 1914; *ariadne* Pieper, 1979; *astrabadensis* Goodwin, 1940; *balcanica* Ondrias, 1970; *balearica* Miller, 1907; *bogdanowii* Tichomirov and Kortchagin, 1889; *bruecheri* Lehmann, 1977; *caneae* Miller, 1909; *cantabra* Cabrera, 1908; *cassiteridum* Hinton, 1924; *corsicana* Heim de Balsac and Reynaud, 1940; *cypria* Bate, 1904; *cyrnensis* Miller, 1907; *debeauxi* Dal Piaz, 1925; *dinnicki* Ognev, 1922; *enezsizunensis* Heim de Balsac and Beaufort, 1966; *fumigatus* di Filippi, 1863; *gueldenstaedtii* (Pallas, 1811); *iculisma* Mottaz, 1908; *italica* Cavazza, 1912; *longicaudata* Tichomirov and Kortchagin, 1889 [not Mori, 1927]; *mimula* Miller, 1901; *mimuloides* Cavazza, 1912; *minor* de Sélys Longchamps, 1839; *minuta* Lydekker, 1902; *monacha* Thomas, 1906; *oyaensis* Heim de Balsac, 1940; ? *pamirensis* Ognev, 1928; *praecypria* Reumer and Oberli, 1988; ? *sarda* Cavazza, 1912; *tristami* (Bodenheimer, 1935); *uxantisi* Heim de Balsac, 1951.
COMMENTS: A widespread and variable species which has often been confused with *russula*; the taxonomic status of many E Asian forms has recently been discussed; see also under *rapax* and *shantungensis*. The European, Arabian, and Asian ranges were reviewed by Vlasák and Niethammer (1990), Harrison and Bates (1991), and Jiang and Hoffmann (2001), respectively. The name *gueldenstaedtii* has produced much confusion. Ellerman and Morrison-Scott (1966) listed it as a subspecies of *russula* and were followed by Corbet (1978c), among others. Richter (1970) applied *gueldenstaedtii* even to Mediterranean populations of *suaveolens* and was followed in that action by Kahmann and Vesmanis (1976). Hutterer (1981d) suggested that all these populations represent *suaveolens*; this was supported by karyological (2n = 40, FN = 50) and biochemical data (Catzeflis et al., 1985). Despite convincing evidence, some Russian authors (e.g., Graphodatsky et al., 1988) still claim the existence of *gueldenstaedtii* as a species in the Caucasus. Bannikova et al. (2001a) however found no genetic differences between populations from N and S Caucasus. Vogel et al. (2003) compared the mtDNA of specimens from Georgia (type locality of *gueldenstaedtii*) with specimens from Greece and Italy (*suaveolens*) and found no big difference. A specimen from Spain was more distant than samples from Europe and Georgia.

*Crocidura susiana* Redding and Lay, 1978. Z. Säugetierk., 43:307.
    COMMON NAME: Iranian Shrew.
    TYPE LOCALITY: Iran, Khuzistan Province, 8 km SSW of Dezful (32°19′N, 48°21′E).
    DISTRIBUTION: Known only from the vicinity of Dezful (SW Iran), but may have a wider
        distribution.
    STATUS: IUCN – Endangered.

*Crocidura tanakae* Kuroda, 1938. A list of the Japanese mammals, p. 81.
    COMMON NAME: Taiwanese Gray Shrew.
    TYPE LOCALITY: C Taiwan, Taichusiu, Horigai, Shohosha.
    DISTRIBUTION: Taiwan, from sea level up to 2,200 m (Fang et al., 1997).
    COMMENTS: Originally described as a species, *C. tanakae* was subsequently sunk into
        synonymy of *C. attenuata* (Fang et al., 1997). Differences in the karyotype (Motokawa
        et al., 1997, 2001*b*) led to a renewed recognition of the species (Fang and Lee, 2002).

*Crocidura tansaniana* Hutterer, 1986. Bonn. Zool. Beitr., 37:27.
    COMMON NAME: Tanzanian Shrew.
    TYPE LOCALITY: Tanzania, Tanga Region, E Usambara Mtns, Amani.
    DISTRIBUTION: Usambara Mtns (Tanzania).
    STATUS: IUCN – Vulnerable.
    COMMENTS: Previously known only by the holotype, but recently more specimens from the
        E and W Usambaras have been identified (Stanley et al., 2000*b*).

*Crocidura tarella* Dollman, 1915. Ann. Mag. Nat. Hist., ser. 8, 17:135.
    COMMON NAME: Tarella Shrew.
    TYPE LOCALITY: "Chaya, near Ruchuru, Congo Belge."
    DISTRIBUTION: Uganda and adjacent Dem. Rep. Congo.
    STATUS: IUCN – Lower Risk (lc).
    COMMENTS: Formerly a subspecies of *turba* but Dippenaar (1980) regarded it a distinct
        species. Occurs in the Bwindi Impenetrable N.P., SW Uganda (Kasangaki et al., 2003).

*Crocidura tarfayensis* Vesmanis and Vesmanis, 1980. Zool. Abh. Mus. Tierk. Dresden, 36:47.
    COMMON NAME: Saharan Shrew.
    TYPE LOCALITY: Morocco, Agadir Prov., 8 km south Tarfaya, 27°50′N, 12°30′W.
    DISTRIBUTION: Atlantic coast of Sahara; south of Agadir (Morocco) through Western Sahara
        into Mauritania.
    STATUS: IUCN – Lower Risk (lc).
    SYNONYMS: *agadiri* Vesmanis and Vesmanis, 1980; *gouliminensis* Vesmanis and Vesmanis,
        1980; *tiznitensis* Vesmanis and Vesmanis, 1980.
    COMMENTS: Recorded as *whitakeri* from Western Sahara by Heim de Balsac (1968*e*). Reviewed
        by Hutterer (1987).

*Crocidura telfordi* Hutterer, 1986. Bonn. Zool. Beitr., 37:28.
    COMMON NAME: Telford's Shrew.
    TYPE LOCALITY: Tanzania, Uluguru Mtns, Morningside, 1150 m.
    DISTRIBUTION: Uluguru and Udzungwa Mtns, in montane forest.
    STATUS: IUCN 2000 – Critically Endangered CR B1+2c.
    COMMENTS: Part of the endemic fauna of the Eastern Arc Mountains in Tanzania. Stanley
        et al. (2000*b*) reported this species also from the Udzungwa Mtns.

*Crocidura tenuis* (Müller, 1840). *In* Temminck, Verh. Nat. Gesch. Nederland Overz. Bezitt., Zool.,
    Zoogd. Indisch. Archipel, p. 26, 50[1840].
    COMMON NAME: Timor Shrew.
    TYPE LOCALITY: Timor.
    DISTRIBUTION: Timor (Indonesia).
    STATUS: IUCN – Vulnerable.
    SYNONYMS: *macklotii* Jentink, 1888.
    COMMENTS: Jenkins (1982:273) considered conspecificy of *tenuis* with *fuliginosa* but stated
        that present evidence is not sufficient. Hutterer (1993*a*) and Ruedi (1995) retained it as

a separate species. In case of conspecificy with *fuliginosa*, *tenuis* would be the earliest name for the group.

*Crocidura thalia* Dippenaar, 1980. Ann. Transvaal Mus., 32:138-147.
  COMMON NAME: Thalia's Shrew.
  TYPE LOCALITY: Ethiopia, NW Bale Province, Gedeb Mtns, SE Dodola, 2,600 m (06°55′N, 39°10′E).
  DISTRIBUTION: Forest and moorland of the Ethiopian highlands on both sides of the Rift Valley (Yalden et al., 1997).
  STATUS: IUCN – Lower Risk (nt).
  COMMENTS: Previous to its description, *thalia* was known as *C. luna macmillani* (e.g., Hutterer, 1981*c*) or *C. fumosa*; see Yalden (1988), who studied the altitudinal distribution. Chromosomally (2n = 36) and genetically close to *C. glassi* (Bannikova et al., 2001*b*; Lavrenchenko et al., 1997).

*Crocidura theresae* Heim de Balsac, 1968. Mammalia, 32:398.
  COMMON NAME: Therese's Shrew.
  TYPE LOCALITY: Guinea, Nzerekore.
  DISTRIBUTION: Guinea savanna from Ghana to Guinea.
  STATUS: IUCN – Lower Risk (lc).
  COMMENTS: May be a subspecies of *foxi*, but *theresae* from Côte d'Ivoire are distinctly smaller and grayer. Karyotype has 2n = 50, FN = 82-84 (Meylan, 1971).

*Crocidura thomensis* (Bocage, 1887). J. Sci. Math. Phys. Nat., Lisboa, 11:212.
  COMMON NAME: São Tomé Shrew.
  TYPE LOCALITY: São Tomé Isl.
  DISTRIBUTION: Endemic to São Tomé.
  STATUS: IUCN – Vulnerable.
  COMMENTS: For description of the species and designation of a neotype, see Heim de Balsac and Hutterer (1982). Distribution, ecology, and status reviewed by Dutton and Haft (1996).

*Crocidura trichura* Dobson in Thomas, 1889. Proc. Zool. Soc. Lond., 1888:532 [1889].
  COMMON NAME: Christmas Island Shrew.
  TYPE LOCALITY: Australia, Christmas Isl.
  DISTRIBUTION: Christmas Isl.
  STATUS: IUCN – not listed, but should be Endangered [EN B1+2c].
  COMMENTS: Included in *C. attenuata* until recently. Ruedi (1995) studied 11 specimens and concluded that they were different from *C. attenuata*, a conclusion with which I concur. Surveys conducted on Christmas Isl to determine the status of the shrew (Meek, 2000) were not successful; the most recent specimens were found in 1985. *C. trichura* was common in 1900 (Andrews, 1900) but already rare in 1909 (Andrews, 1909). It is most likely that *C. trichura* formed part of the endemic mammal fauna of the Christmas Isl, along with *Rattus nativitatis* and *R. macleari*, both of which are now extinct (Meek, 2000).

*Crocidura turba* Dollman, 1910. Ann. Mag. Nat. Hist., ser. 8, 5:176.
  COMMON NAME: Turbo Shrew.
  TYPE LOCALITY: "Chilui Island, Lake Bangweolo", = Chilubi Isl, Zambia.
  DISTRIBUTION: Angola, Zambia, Dem. Rep. Congo, Malawi, Tanzania, Kenya, Uganda, Cameroon.
  STATUS: IUCN – Lower Risk (lc).
  SYNONYMS: *angolae* Dollman, 1915.
  COMMENTS: Includes *angolae*; see Heim de Balsac and Meester (1977:24-25). Range not exactly known, due to confusion with *zaodon* (= *nigrofusca*).

*Crocidura ultima* Dollman, 1915. Ann. Mag. Nat. Hist., ser. 8, 15:517.
  COMMON NAME: Ultimate Shrew.
  TYPE LOCALITY: Kenya, Nyeri District, Jombeni Range.
  DISTRIBUTION: Known only from the type locality.

STATUS: IUCN – Critically Endangered.
COMMENTS: Dippenaar (1980:126), following Allen (1939:46), recognized *ultima* as a full
    species within the *littoralis-monax* group.

*Crocidura usambarae* Dippenaar, 1980. Ann. Transvaal Mus., 32:128.
COMMON NAME: Usambara Shrew.
TYPE LOCALITY: Tanzania, Western Usambara Mtns, Shume, 16 mi. (26 km) N. Lushoto.
DISTRIBUTION: Western and Eastern Usambara Mtns and South Pare Mountains,
    NE Tanzania. perhaps also Ngozi Crater, SW Tanzania (Stanley et al., 1996, 2000*b*).
STATUS: IUCN – Vulnerable.
COMMENTS: Dippenaar (1980:126) included the species into the *monax* group. A skin
    recorded by this author from the Poroto Mtns (S Tanzania) may belong to a different
    species.

*Crocidura viaria* (I. Geoffroy, 1834). *In* Zool. Voy. de Belanger Indes-Orient., p. 127.
COMMON NAME: Savanna Path Shrew.
TYPE LOCALITY: "Senegal", restricted to region between Dakar and St. Luis by Hutterer
    (1984).
DISTRIBUTION: Sahelien and Sudan savanna from S Morocco to Senegal and east to Sudan,
    Ethiopia and Kenya; perhaps further south.
STATUS: IUCN – Lower Risk (lc).
SYNONYMS: *bolivari* Morales Agacino, 1934; *hindei* Thomas, 1904; *suahelae* Heller, 1912;
    *tamrinensis* Vesmanis and Vesmanis, 1980.
COMMENTS: Revised by Hutterer (1984); Possibly includes *suahelae*, which may alternatively
    belong to *zaphiri*. A member of the *flavescens* species group (Maddalena, 1990), as
    shown by the karyotype (2n = 50, FN = 66; Maddalena and Ruedi, 1994). Isolated
    Maroccan population reviewed by Vogel et al. (2000).

*Crocidura virgata* Sanderson, 1940. Trans. zool. Soc. Lond., 24:682.
COMMON NAME: Mamfe Shrew.
TYPE LOCALITY: Cameroon, Mamfe Division, Assumbo, Tinta [06°15′N, 09°31′E].
DISTRIBUTION: Nigeria (Meylan and Vogel, 1982) and highlands of W Cameroon.
SYNONYMS: *vulcani* Heim de Balsac, 1956.
COMMENTS: Previously included in *hildegardeae* (see Heim de Balsac and Meester, 1977), but
    has a different karyotype (2n = 52, FN = 86; Schlitter et al., 1999).

*Crocidura voi* Osgood, 1910. Field Mus. Nat. Hist. Publ., Zool. Ser., 10:18.
COMMON NAME: Voi Shrew.
TYPE LOCALITY: Kenya, "Voi, British East Africa".
DISTRIBUTION: Sudan savanna from Kenya and Somalia to Ethiopia and Sudan; single
    records from Nigeria and Mali.
STATUS: IUCN – Lower Risk (lc).
SYNONYMS: *aridula* Thomas and Hinton, 1923; *butleri* Thomas, 1911; *percivali* Dollman, 1915
    [see Hutterer, 1986*a*].
COMMENTS: Included in subgenus *Afrosorex* by Hutterer (1986*a*). New material from Kenya
    suggest a strong relation, if not conspecificy, with *C. fischeri* (Oguge and Hutterer, in
    prep.).

*Crocidura vorax* G. Allen, 1923. Amer. Mus. Novit., 100:8.
COMMON NAME: Voracious Shrew.
TYPE LOCALITY: China, Yunnan, Li-kiang (=Lijiang) Valley.
DISTRIBUTION: India, Thailand, Laos, Vietnam to S and C China.
COMMENTS: Revised by Jiang and Hoffmann (2001). Recorded (as *pullata vorax*) from Laos
    and NE Thailand by A. L. Smith et al. (1998, 2000).

*Crocidura vosmaeri* Jentink, 1888. Notes Leyden Mus., 10:165.
COMMON NAME: Banka Shrew.
TYPE LOCALITY: Indonesia, SE Sumatra, Banka Isl.
DISTRIBUTION: Banka Isl and perhaps also Sumatra.
COMMENTS: Revised by Ruedi (1995). May be related to *C. beccarii*.

*Crocidura watasei* Kuroda, 1924. [New Mammals from the Ryukyu Islands], p. 1.
   COMMON NAME: Lesser Ryukyu Shrew.
   TYPE LOCALITY: Japan, Ryukyu Archipelago, Amamioshima, Komi.
   DISTRIBUTION: C Ryukyu Isls.
   COMMENTS: Endemic to the C Ryukyus (Motokawa et al., 1996). Formerly a subspecies of
      *horsfieldii* (e.g., Jameson and Jones, 1977), but differs in size and karyotype (2n = 26,
      FN = 52; Hattori et al., 1990).

*Crocidura whitakeri* de Winton, 1898. Proc. Zool. Soc. Lond., 1897:954 [1898].
   COMMON NAME: Whitaker's Shrew.
   TYPE LOCALITY: Morocco, between Morocco City and Mogador, Sierzet.
   DISTRIBUTION: Atlantic and Mediterranean parts of Morocco, Algeria and Tunisia; one
      record from coastal Egypt (Hutterer, 1994). Range in Morocco mapped by Aulagnier
      and Thévenot (1987); in Algeria by Rzebik-Kowalska (1988); and in Tunisia by Sara and
      Zanca (1992).
   STATUS: IUCN – Lower Risk (lc).
   SYNONYMS: *essaouiranensis* Vesmanis and Vesmanis, 1980; *matruhensis* Setzer, 1960;
      *mesatanensis* Vesmanis and Vesmanis, 1980; *zaianensis* Vesmanis and Vesmanis, 1980
      [see Hutterer, 1987, 1991].

*Crocidura wimmeri* Heim de Balsac and Aellen, 1958. Rev. Suisse Zool., 65:952.
   COMMON NAME: Wimmer's Shrew.
   TYPE LOCALITY: Côte d'Ivoire, Adiopodoume.
   DISTRIBUTION: S Côte d'Ivoire.
   STATUS: IUCN – Endangered.
   COMMENTS: Has been assigned to *nimbae*; but see Hutterer (1983a). Records outside Côte
      d'Ivoire are based on misidentifcations; specimen recorded from Cameroon and
      Gabon refer to *batesi*; see Brosset (1988). Karyotype has 2n = 50, FN = 84 (Meylan and
      Vogel, 1982).

*Crocidura wuchihensis* Wang, 1966. Acta Zootax. Sin., 261.
   COMMON NAME: Hainan Island Shrew.
   TYPE LOCALITY: China, Hainan Isl, Mt. Wuchih.
   DISTRIBUTION: China (Hainan) and Vietnam; limits unknown.
   COMMENTS: Until recently included in *C. horsfieldii* (see Jiang and Hoffmann, 2001).
      However, very small specimens recently collected in Vietnam (Feiler and Ziegler, 1999;
      Kuznetsov, unpubl.) suggest that further species of *Crocidura* occur in Indochina.
      Lunde et al. (2003b) identified one taxon as *C. wuchihensis*, while a smaller specimen
      reported by Feiler and Ziegler (1999) may represent another (new) species.

*Crocidura xantippe* Osgood, 1910. Field Mus. Nat. Hist. Publ., Zool. Ser., 10:19.
   COMMON NAME: Xanthippe's Shrew.
   TYPE LOCALITY: Kenya, Voi.
   DISTRIBUTION: Nyiru, Voi, Tsavo (SE Kenya); Usambara Mtns (Tanzania).
   STATUS: IUCN – Vulnerable.
   SYNONYMS: *xanthippe* Dollman, 1915.
   COMMENTS: Status uncertain; probably related to *hirta*. Not to be confused with *Crocidura*
      *xanthippe* Bate, 1937 (replaced by *C. samaritana* Bate, 1937), a Pleistocene shrew from
      Palestine.

*Crocidura yankariensis* Hutterer and Jenkins, 1980. Bull. Brit. Mus. (Nat. Hist.) Zool., 39:305.
   COMMON NAME: Yankari Shrew.
   TYPE LOCALITY: Nigeria, Bauchi State, 16 km E of Yankari Game Reserve boundary, Futuk
      [9°50′N, 10°55′E].
   DISTRIBUTION: Sudan savanna zone in Cameroon, Nigeria, Sudan, Ethiopia, Kenya, and
      Somalia.
   STATUS: IUCN – Lower Risk (lc).
   COMMENTS: Previously confused with *somalica*; see Hutterer and Jenkins (1983). Karyotype
      (2n = 68, FN = 122) described by Schlitter et al. (1999).

*Crocidura zaphiri* Dollman, 1915. Ann. Mag. Nat. Hist., ser. 8, 15:509.
COMMON NAME: Zaphir's Shrew.
TYPE LOCALITY: Ethiopia, Kaffa, Charada Forest.
DISTRIBUTION: Kaffa Prov. (S Ethiopia); Kaimosi, Kisumu (Kenya).
STATUS: IUCN – Lower Risk (lc).
SYNONYMS: *simiolus* Hollister, 1916.
COMMENTS: Includes *simiolus*; see Osgood (1936:224). May also include *mutesae* and
    *suahelae* (here questionably listed in *viaria*), in which case it would be a widely
    distributed species; see Hutterer and Yalden (1990:70).

*Crocidura zarudnyi* Ognev, 1928. [Mammals of Eastern Europe and Northern Asia], 1:341.
COMMON NAME: Zarudny's Rock Shrew.
TYPE LOCALITY: Iran, Baluchistan (border).
DISTRIBUTION: SE Iran, SE Afghanistan, SW Pakistan (Spitzenberger, 1971a).
STATUS: IUCN – Lower Risk (lc).
SYNONYMS: *streetorum* Hassinger, 1970; *tatianae* Ognev, 1922 [not *tatiana* Dollman, 1915].
COMMENTS: The species was first named *tatianae* by Ognev (1922), but later (1928) replaced
    by *zarudnyi*; Ognev argued that *tatianae* was preoccupied by *tatiana* Dollman, 1915
    (now a synonym of the African *olivieri*), an action covered by the 4th edition of the
    International Code of Zoological Nomenclature (Art. 58, International Commission
    on Zoological Nomenclature, 1999). The definition of *zarudnyi* follows Spitzenberger
    (1971a) and Hassinger (1970), but not Jenkins (1976) who included *arispa* which is
    now regarded as a separate species; see that account. As Spitzenberger (1971a) pointed
    out, *zarudnyi* has a shorter rostrum and a heavier mandible than both *pergrisea* and
    *serezkyensis*. The status of *streetorum* is not clear although it is included here as
    suggested by Hassinger (1970). The distribution and morphology of *pergrisea*,
    *serezkyensis*, and *zarudnyi* should be carefully studied in the Hindukush, Karakoram
    and Pamir where their ranges may overlap.

*Crocidura zimmeri* Osgood, 1936. Field Mus. Nat. Hist. Publ., Zool. Ser., 20:223.
COMMON NAME: Upemba Shrew.
TYPE LOCALITY: Dem. Rep. Congo, Katanga Prov., near Bukama, "Lualaba River, Katobwe".
DISTRIBUTION: Environs of Upemba National Park, Dem. Rep. Congo.
STATUS: IUCN – Vulnerable.
COMMENTS: A large and striking species that is known only by the type series.

*Crocidura zimmermanni* Wettstein, 1953. Z. Säugetierk., 17:12.
COMMON NAME: Cretan Shrew.
TYPE LOCALITY: Greece, Crete, Ida Mtns, Nida plateau.
DISTRIBUTION: Highlands of the island of Crete.
STATUS: IUCN – Vulnerable.
COMMENTS: A Pleistocene relict (Reumer, 1986). Formerly regarded as a subspecies of *russula*
    but differs in morphology and karyotype (2n = 34, FN = 44); see Vesmanis and
    Kahmann (1978), Vogel (1986), Vogel et al. (1986), and Pieper (1990).

*Diplomesodon* Brandt, 1852. Beitr. Kenntn. Russ. Reiches, 17:299.
TYPE SPECIES: *Sorex pulchellus* Lichtenstein, 1823.
COMMENTS: For placement in Crocidurinae see Repenning (1967:15). A study of protein
    evolution (Maddalena, in Ruedi, 1998) showed that *Diplomesodon* is closely related to
    *Crocidura*. Repenning (1965) named a species from the Pleistocene of South Africa in this
    genus which, however, does not belong here (Butler, 1998).

*Diplomesodon pulchellum* (Lichtenstein, 1823). *In* Eversmann, Reise von Orenburg nach Bokhara,
    Berlin, p. 124.
COMMON NAME: Piebald Shrew.
TYPE LOCALITY: Kazakhstan, E bank of Ural River sands "Bolshie Barsuki".
DISTRIBUTION: W and S Kazakhstan, Uzbekistan, Turkmenistan.
STATUS: IUCN – Lower Risk (lc).
SYNONYMS: *pallidus* Heptner, 1938.

COMMENTS: Biology and distribution reviewed by Heptner (1939), who also specified the type locality. Karyotype of specimen from Turkmenistan (2n = 44, FN = 54) described by Ivanitskaya (1975).

*Feroculus* Kelaart, 1852. Prodr. Faun. Zeylanica, p. 31.
TYPE SPECIES: *Sorex macropus* Blyth, 1851 (= *Sorex feroculus* Kelaart, 1850).
COMMENTS: For placement in Crocidurinae see Repenning (1967:15).

*Feroculus feroculus* (Kelaart, 1850). J. Ceylon Branch Asiat. Soc., 2(5):211.
COMMON NAME: Kelaart's Long-clawed Shrew.
TYPE LOCALITY: Sri Lanka, C mountains at 6,000 ft. (1829 m), Nuwara Eliya.
DISTRIBUTION: Primary swamps and forests in the C highlands of Sri Lanka, and montane swamps and marshes between 2200 and 2400 m at the border between Kerala and Tamil Nadu, S India.
STATUS: IUCN – Endangered.
SYNONYMS: *macropus* (Blyth, 1851); *newera* (Wagner, 1855); *newera-ellia* (Kelaart, 1851).
COMMENTS: A rare and little-known species; available information summarized by Phillips (1980). Indian records confirmed by Pradhan et al. (1997). Their specimens showed smaller bodies and longer tails compared to specimens from Sri Lanka. Blanford (1888) also noted the presence of this shrew in the Palni Hills, S India.

*Paracrocidura* Heim de Balsac, 1956. Rev. Zool. Bot. Afr., 54:137.
TYPE SPECIES: *Paracrocidura schoutedeni* Heim de Balsac, 1956.
COMMENTS: Revised by Hutterer (1986c). A study of 16s rRNA sequences placed *Paracrocidura* in one clade with *Crocidura* (Querouil et al., 2001). However, unique external and cranial features distinguish the species of *Paracrocidura* from all other genera.

*Paracrocidura graueri* Hutterer, 1986. Bonn. Zool. Beitr., 37:81.
COMMON NAME: Grauer's Large-headed Shrew.
TYPE LOCALITY: "Urwald hinter den Randbergen des Nord-Westufers des Tanganjika" = Sibatwa, 2,000 m, Itombwe Mtns, Dem. Rep. Congo.
DISTRIBUTION: Known only from the type locality.
STATUS: IUCN – Critically Endangered.
COMMENTS: Known only from the holotype which was collected in 1908. Like *Myosorex schalleri*, the species forms part of the endemic fauna of the Itombwe Mtns.

*Paracrocidura maxima* Heim de Balsac, 1959. Rev. Zool. Bot. Afr., 59:26.
COMMON NAME: Greater Large-headed Shrew.
TYPE LOCALITY: Dem. Rep. Congo, Tshibati.
DISTRIBUTION: Dem. Rep. Congo, Rwanda, Uganda.
STATUS: IUCN – Lower Risk (lc).
COMMENTS: Regarded as a full species by Hutterer (1986c:79). Kasangaki et al. (2003) recorded the species from the Bwindi Impenetrable N.P. in SW Uganda.

*Paracrocidura schoutedeni* Heim de Balsac, 1956. Rev. Zool. Bot. Afr., 54:137.
COMMON NAME: Lesser Large-headed Shrew.
TYPE LOCALITY: Dem. Rep. Congo, Kasai, Lubondaie (75 km south of Luluabourg), Tshimbulu (Dibaya).
DISTRIBUTION: Lowland primary forest in S Cameroon, Gabon, Republic of Congo, Dem. Rep. Congo, and Central African Republic.
STATUS: IUCN – Lower Risk (lc).
SYNONYMS: *camerunensis* Heim de Balsac, 1968.
COMMENTS: A subspecies *camerunensis* was named by Heim de Balsac (1968b), based on a specimen from Mt. Cameroon.

*Ruwenzorisorex* Hutterer, 1986. Z. Säugetierk., 51:260.
TYPE SPECIES: *Sylvisorex suncoides* Osgood, 1936.
COMMENTS: Data on the brain structure support generic separation; see Stephan et al. (1991). Genetic data show that *Ruwenzorisorex* and *Suncus* s. str. are sister taxa (Querouil et al., 2001).

*Ruwenzorisorex suncoides* (Osgood, 1936). Field Mus. Nat. Hist. Publ., Zool. Ser., 20:217.
COMMON NAME: Ruwenzori Shrew.
TYPE LOCALITY: Dem. Rep. Congo, western slope of Ruwenzori Mtns, Kalongi.
DISTRIBUTION: Montane forest in W Dem. Rep. Congo, Uganda, Rwanda, and Burundi.
STATUS: IUCN – Vulnerable.
COMMENTS: A semi-aquatic shrew that occurs also in Burundi (Kerbis, pers. comm.) and
Uganda (Kasangaki et al., 2003).

*Scutisorex* Thomas, 1913. Ann. Mag. Nat. Hist., ser. 8, 11:321.
TYPE SPECIES: *Sylvisorex somereni* Thomas, 1910.
COMMENTS: For placement in Crocidurinae see Repenning (1967:15). Genetic data place
*Scutisorex* next to *Sylvisorex* (Querouil et al., 2001). The unique vertebral column (Ahmed
and Klima, 1978) and a characteristic skull roof distinguish the genus, however.

*Scutisorex somereni* (Thomas, 1910). Ann. Mag. Nat. Hist., ser. 8, 6:113.
COMMON NAME: Armored Shrew.
TYPE LOCALITY: Uganda, near Kampala, Kyetume.
DISTRIBUTION: Tropical rainforest of the Dem. Rep. Congo Basin and adjacent mountains in
Uganda, Rwanda, and Burundi.
STATUS: IUCN – Lower Risk (lc).
SYNONYMS: *congicus* Thomas, 1915.
COMMENTS: Includes *congicus*; see Heim de Balsac and Meester (1977:7).

*Solisorex* Thomas, 1924. Spolia Zeylan., 13:94.
TYPE SPECIES: *Solisorex pearsoni* Thomas, 1924.
COMMENTS: For placement in Crocidurinae see Repenning (1967:15).

*Solisorex pearsoni* Thomas, 1924. Spolia Zeylan., 13:94.
COMMON NAME: Pearson's Long-clawed Shrew.
TYPE LOCALITY: Sri Lanka, Central Province, near Nuwara Eliya, Hakgala.
DISTRIBUTION: C highlands of Sri Lanka.
STATUS: IUCN – Endangered.
COMMENTS: A very little known species that inhabits "virgin forest" in the mountains of
C Sri Lanka (Phillips, 1980).

*Suncus* Ehrenberg, 1832. *In* Hemprich and Ehrenberg, Symb. Phys. Mamm., 2:k.
TYPE SPECIES: *Suncus sacer* Ehrenberg, 1832 (= *Sorex murinus* Linnaeus, 1766).
SYNONYMS: *Pachyura* de Selys-Longchamps, 1839 [not *Pachyurus* Agassiz, 1829, a genus of
fishes]; *Paradoxodon* Wagner, 1855; *Plerodus* Schulze, 1897; *Podihik* Deraniyagala, 1958;
*Sunkus* Sundevall, 1843.
COMMENTS: For placement in Crocidurinae see Repenning (1967:15). Occasionally regarded as
part of *Crocidura* (e.g., Lekagul and McNeely, 1977:35), based on morphology (McLellan,
1994), molecular data (Motokawa et al., 2000), or chromosome homology (Biltueva et al.,
2001), but accepted as a full genus by most authors. Fons et al. (1994) stated that the
fauna of parasitic helminths is completely different between *Suncus* and *Crocidura* (but the
fauna of *S. murinus* and *S. etruscus* were also different). Querouil et al. (2001) compared
16s rRNA sequences of six African and Asian species of *Suncus*. Their results suggest
paraphyly; *Suncus dayi* clustered next to *Sylvisorex megalura* (transfered here to *Suncus*) and
the smaller species (*S. etruscus*, *S. infinitesimus*, *S. remyi*), while *S. murinus* and *S. montanus*
formed a separate cluster, along with *Ruwenzorisorex suncoides*. With further knowledge it
may be warranted to break up *Suncus* into two subgenera or genera. *Suncus* is available for
*S. murinus* and allies, while *Paradoxodon* is available for *S. etruscus* and other small species.
However, for several unstudied species a correct allocation is not possible yet. The name
*Pachyura*, which had been in use for a long time, is preoccupied. Meester and Lambrechts
(1971) revised the southern African species. Jenkins et al. (1998) discussed relationships
among Asian species.

*Suncus aequatorius* (Heller, 1912). Smithsonian Misc. Coll., 60, 12:4.
COMMON NAME: Taita Shrew.

TYPE LOCALITY: Kenya, Taita Hills, summit of Mt. Sagalla.
DISTRIBUTION: SE Kenya and N Tanzania.
COMMENTS: Included in *S. lixus* by Heim de Balsac and Meester (1977), but specimens of *aequatorius* are considerably larger (Oguge et al., 2004).

*Suncus ater* Medway, 1965. J. Malay. Branch R. Asiat. Soc., 36:38.
COMMON NAME: Black Shrew.
TYPE LOCALITY: Malaysia, Sabah, Gunong (= Mt.) Kinabalu, Lumu-Lumu, 5,500 ft. (1,676 m).
DISTRIBUTION: Known only from the type locality.
STATUS: IUCN – Critically Endangered.
COMMENTS: Reviewed by Medway (1977:16-17).

*Suncus dayi* (Dobson, 1888). Ann. Mag. Nat. Hist., ser. 6, 1:428.
COMMON NAME: Day's Shrew.
TYPE LOCALITY: India, Cochin, Trichur.
DISTRIBUTION: Montane evergreen forest of S India (Trichur and Nilgiri Hills).
STATUS: IUCN – Vulnerable.
COMMENTS: New records provided and relationships discussed by Jenkins et al. (1998). Based on the analysis of RNA sequences, *S. dayi* is sister of the African *S. megalura* (Querouil et al., 2001).

*Suncus etruscus* (Savi, 1822). Nuovo Giorn. de Letterati, Pisa, 1:60.
COMMON NAME: Etruscan Shrew.
TYPE LOCALITY: Italy, Pisa.
DISTRIBUTION: S Europe and N Africa (Morocco, Algeria, Tunisia, Egypt); Arabian Peninsula and Asia Minor to Iraq, Turkmenistan, Afghanistan, Pakistan, Nepal, Bhutan, Burma, Thailand, Laos, Vietnam and Yunnan (China); also India and Sri Lanka. West and East African records (Guinea, Nigeria, Ethiopia) are doubtful and need confirmation.
STATUS: IUCN – Lower Risk (lc).
SYNONYMS: *assamensis* (Anderson, 1873); *atratus* (Blyth, 1855); *bactrianus* Stroganov, 1958; *hodgsoni* (Blyth, 1855); *kura* (Deraniyagala, 1958); *macrotis* (Anderson, 1877); *melanodon* (Blyth, 1855) [not Wagler, 1832]; *micronyx* (Blyth, 1855); *nanula* (Stroganov, 1941); *nilgirica* (Anderson, 1877); *nitidofulva* (Anderson, 1877); *nudipes* (Blyth, 1855); *pachyurus* (Küster, 1835); *perrotteti* (Duvernoy, 1842); *pygmaeoides* (Anderson, 1877); *pygmaeus* (Hodgson, 1845); *suaveolens* (Blasius, 1857) [not Pallas, 1811]; *travancorensis* (Anderson, 1877).
COMMENTS: European and Asian range reviewed by Spitzenberger (1970, 1990c); N African distribution mapped by Vesmanis (1987). Heim de Balsac and Meester (1977) discussed the African records south of the Sahara. Probably includes *Podihik kura*; see Nowak and Paradiso (1983:141). The records east of Afghanistan, particularly from S India (*macrotis*, *nilgirica*) are only tentatively included; Corbet (1978c:31) expressed doubt on the conspecificy of the Indian forms. The same applies to records from further east (Feiler and Nadler, 1997). Many authors included *fellowesgordoni*, *hosei*, *madagascariensis*, and *malayanus* in *etruscus*, however, in the present list they are all treated as valid species.

*Suncus fellowesgordoni* Phillips, 1932. Spolia Zeylan., 17:124.
COMMON NAME: Sri Lankan Shrew.
TYPE LOCALITY: Sri Lanka, Central Province, Ohiya, West Haputale Estate (6,000 ft. = 1829 m).
DISTRIBUTION: C highlands of Sri Lanka.
STATUS: IUCN – Endangered.
COMMENTS: Although usually included in *S. etruscus*, this taxon represents a species endemic to Sri Lanka (see also Jenkins et al., 1998). *Podihik kura* Deraniyagala, 1958, which was included in this species by Phillips (1980), does not represent *fellowesgordoni*, but is more similar to *etruscus* and included therein.

*Suncus hosei* (Thomas, 1893). Ann. Mag. Nat. Hist., ser. 6, 11:343.
COMMON NAME: Bornean Pgymy Shrew.
TYPE LOCALITY: Sarawak, Bakong River.
DISTRIBUTION: Lowland forest of Borneo and Sarawak.

STATUS: IUCN – Vulnerable.

COMMENTS: Often included in *etruscus* (e.g., Medway, 1977) but represents a distinct forest species.

*Suncus infinitesimus* (Heller, 1912). Smithson. Misc. Coll., 60(12):5.
COMMON NAME: Least Dwarf Shrew.
TYPE LOCALITY: Kenya, Laikipia Plateau, Rumruti, 7,000 ft. (2,134 m).
DISTRIBUTION: South Africa, Kenya, Central African Republic, Cameroon.
STATUS: IUCN – Lower Risk (lc).
SYNONYMS: *chriseos* (Kershaw, 1921); *ubanguiensis* Petter and Chippaux, 1962.
COMMENTS: Includes *chriseos* and *ubanguiensis*; see Heim de Balsac and Meester (1977). Gureev (1979:383) listed *chriseos* as a distinct species without comment.

*Suncus lixus* (Thomas, 1898). Proc. Zool. Soc. Lond., 1897:930 [1898].
COMMON NAME: Greater Dwarf Shrew.
TYPE LOCALITY: Malawi, Nyika Plateau (between 10 and 11°S and 33°40' to 34°10'E).
DISTRIBUTION: Savanna zones of Tanzania, Malawi, Dem. Rep. Congo, Zambia, Angola, Botswana, and South Africa (KwaZulu-Natal, Northwest Prov., Mpumalanga, and Limpopo).
STATUS: IUCN – Lower Risk (lc).
SYNONYMS: *gratulus* (Thomas and Schwann, 1907).
COMMENTS: Heim de Balsac and Meester (1977) included *aequatorius* and *gratulus* in *lixus*; the former is treated as a separate species here. Gureev (1979:383) listed *gratulus* as a distinct species without comment.

*Suncus madagascariensis* (Coquerel, 1848). Ann. Sci. Nat., Zool. (Paris), ser. 3, 9:194, pl. 11, fig. 1.
COMMON NAME: Madagascan Pgymy Shrew.
TYPE LOCALITY: Madagascar, Nossi-Bé.
DISTRIBUTION: Madagascar and Comores Isls, Socotra (Yemen).
STATUS: IUCN – Lower Risk (lc).
SYNONYMS: *coquerelii* (Trouessart, 1880).
COMMENTS: This species is often included in *etruscus* but treated as a full species in most reports on the fauna of Madagascar (e.g., Eisenberg and Gould, 1984). The population of Socotra has been included in *S. etruscus* by Hutterer and Harrison (1988), but unpublished genetic data show that it is closer to *S. madagascariensis*.

*Suncus malayanus* (Kloss, 1917). J. Nat. Hist. Soc. Siam, 2:282.
COMMON NAME: Malayan Pgymy Shrew.
TYPE LOCALITY: Thailand, "Bang Nara, Patani, Peninsular Siam".
DISTRIBUTION: Malaysian peninsula.
STATUS: IUCN – Data Deficient.
COMMENTS: Commonly included in *etruscus* but inhabits tropical forest and does not fit morphologically with the diagnosis of that species; *malayanus* was therefore regarded as a species by Corbet and Hill (1991:36), and Hutterer (1993*a*).

*Suncus megalura* (Jentink, 1888). Notes Leyden Mus., 10:48.
COMMON NAME: Climbing Shrew.
TYPE LOCALITY: Liberia, Junk River, Schieffelinsville.
DISTRIBUTION: Tropical forest and Guinea savanna zone of Africa from Upper Guinea to Ethiopia and south to Mozambique and Zimbabwe.
STATUS: IUCN – Lower Risk (lc) as *Sylvisorex megalura*.
SYNONYMS: *angolensis* (Roberts, 1929); *gemmeus* (Heller, 1910); *infuscus* (Thomas, 1915); *irene* (Thomas, 1915); *phaeopus* (Osgood, 1936); *sheppardi* (Kershaw, 1921); *sorella* (Thomas, 1898); *sorelloides* (Lönnberg, 1912) [see Heim de Balsac and Meester, 1977:7-8].
COMMENTS: Until recently placed in genus *Sylvisorex*, but genetic data (Querouil et al., 2001) show that *S. megalura* is the sister species of the Indian *S. dayi*. Karyotype (2n = 48, FN = 96) described by Meylan (1975). A common African species of forest edges and forested savannas; range mapped by Hutterer et al. (1987*b*). Gureev (1979:381) listed *sorella* as a distinct species without comment. Some geographic variation exists, the Central African forest populations being smallest and darkest.

*Suncus mertensi* Kock, 1974. Senckenbergiana Biol., 55:198.
> COMMON NAME: Flores Shrew.
> TYPE LOCALITY: Indonesia, Flores, Rana Mese.
> DISTRIBUTION: Flores Isl, Indonesia.
> STATUS: IUCN – Critically Endangered.
> COMMENTS: A distinct, long-tailed forest shrew, similar to *S. dayi* and *S. megalura*.

*Suncus montanus* (Kelaart, 1850). J. Ceylon Br. Asiat. Soc., 2:211.
> COMMON NAME: Asian Highland Shrew.
> TYPE LOCALITY: Sri Lanka, Nuwara Eliya, Pidurutalagala.
> DISTRIBUTION: Forested highlands in Sri Lanka and S India.
> STATUS: IUCN – Vulnerable.
> SYNONYMS: *ferrugineus* Kelaart, 1850; *kelaarti* Blyth, 1855; **niger** Horsfield, 1851.
> COMMENTS: Commonly included in *murinus* (Ellerman and Morrison-Scott, 1966:66), but represents a much smaller and always blackish species of primary forest habitats. Also genetically distinct from *murinus* (Ruedi et al., 1996). Listed as a species by Corbet and Hill (1991:36). The Indian population is recognized as a valid subspecies (*niger*).

*Suncus murinus* (Linnaeus, 1766). Syst. Nat., 12th ed., 1:74.
> COMMON NAME: Asian House Shrew.
> TYPE LOCALITY: Indonesia, Java.
> DISTRIBUTION: Afghanistan, Pakistan, India, Sri Lanka, Nepal, Bhutan, Burma, China, Taiwan, Japan, continental and peninsular Indomalayan Region; introduced into Guam, the Maldive Isls, Philippines, and probably many other islands; introduced in historical times into coastal Africa (Egypt to Tanzania), Madagascar, the Comores, Mauritius, and Réunion, and into coastal Arabia (Iraq, Bahrain, Oman, Yemen, Saudi Arabia).
> STATUS: IUCN – Lower Risk (lc).
> SYNONYMS: *albicauda* (Peters, 1866); *albinus* (Blyth, 1860); *andersoni* (Trouessart, 1879); *auriculata* (Fitzinger, 1868); *beddomei* (Anderson, 1881); *blanfordii* (Anderson, 1877); *blythii* (Anderson, 1877); *caerulaeus* (Kerr, 1792); *caerulescens* (Shaw, 1800); *caeruleus* (Kerr, 1792); *celebensis* (Revilliod, 1911); *ceylanica* (Peters, 1870); *crassicaudus* (Lichtenstein, 1834); *duvernoyi* (Fitzinger, 1868); *edwardsiana* (Trouessart, 1880); *fulvocinerea* (Anderson, 1877); *fuscipes* (Peters, 1870); *geoffroyi* (J. B. Fischer, 1830); *giganteus* (Geoffroy, 1831); *grayii* Motley and Dillwyn, 1855; *griffithi* (Horsfield, 1851); *heterodon* (Blyth, 1855); *indicus* (Geoffroy, 1811); *kandianus* (Kelaart, 1852); *kelaarti* (Blyth, 1855); *kroonii* (Kohlbrugge, 1896); *kuekenthali* (Matschie, 1901); *leucura* (Matschie, 1894); *luzoniensis* (Peters, 1870); *malabaricus* (Lindsay, 1929); *mauritiana* (Reichenbach, 1834); *media* (Peters, 1870); *microtis* (Peters, 1970); *muelleri* (Jentink, 1888); *muschata* (Hatori, 1915); *myosurus* (Pallas, 1785); *nemorivagus* (Hodgson, 1845); *occultidens* (Hollister, 1913); *palawanensis* (Taylor, 1934); *pealana* (Anderson, 1877); *pilorides* (Shaw, 1796); *riukiuana* (Kuroda, 1924); *rubicunda* (Anderson, 1877); *sacer* (Ehrenberg, 1832); *saturatior* (Hodgson, 1855); *semmelicki* Tate, 1944 [*lapsus*]; *semmeliki* Koller, 1930 [*lapsus*]; *semmelinki* (Jentink, 1888); *seramensis* Kitchener, 1994; *serpentarius* (I. Geoffroy in Bélanger, 1831); *sindensis* (Anderson, 1877); *soccatus* (Hodgson, 1845); *sonneratii* (I. Geoffroy, 1827); *sumatranus* (Peters, 1870); *swinhoei* (Blyth, 1859); *tytleri* (Blyth, 1859); *unicolor* (Jentink, 1888); *viridescens* (Blyth, 1859); *waldemarii* (Peters, 1870).
> COMMENTS: A highly variable species with a number of genetically distinct populations that almost behave like semispecies (Hasler et al., 1977; Rogatcheva et al., 2000; Yamagata et al., 1987; Yoshida, 1985). Chromosomes show Robertsonian polymorphism and vary geographically from $2n = 30$ to $2n = 40$ (Yosida, 1985). Forms with lower numbers are found in S India, Sri Lanka and peninsular Malaya. A number of laboratory strains have been established (Oda et al., 1985). Much of the present distribution is the result of human agency (Hutterer and Tranier, 1990). A clear allocation of all listed taxa to subspecies is not possible at this moment. Kitchener et al. (1994*b*) discussed subspecies in the Sunda Isls and recognized *murinus*, *muelleri* and *seramensis* as distinct. However, they did not consider names such as *edwardsiana* from S Philippines that may have

priority. African synonyms include *albicauda*, *auriculata*, *crassicaudus*, *duvernoyi*, *leucura*, *mauritiana*, *sacer*, and *geoffroyi*; see Heim de Balsac and Meester (1977).

*Suncus remyi* Brosset, Dubost and Heim de Balsac, 1965. Biologia Gabonica, 1:170.
    COMMON NAME: Remy's Pygmy Shrew.
    TYPE LOCALITY: Gabon, Makokou.
    DISTRIBUTION: Rain forest in Gabon, Central African Republic, Republic of Congo.
    STATUS: IUCN – Critically Endangered (not justified).
    COMMENTS: One of the smallest shrews. Genetically related to *S. infinitesimus* (Querouil et al., 2001).

*Suncus stoliczkanus* (Anderson, 1877). J. Asiat. Soc. Bengal, 46:270.
    COMMON NAME: Anderson's Shrew.
    TYPE LOCALITY: India, Bombay.
    DISTRIBUTION: Deserts and arid country in Pakistan, Nepal, India, and Bangladesh.
    STATUS: IUCN – Lower Risk (lc).
    SYNONYMS: *bidianus* (Anderson, 1877); *leucogenys* (Dobson, 1888); *subfulvus* (Anderson, 1877).

*Suncus varilla* (Thomas, 1895). Ann. Mag. Nat. Hist., ser. 6, 16:54.
    COMMON NAME: Lesser Dwarf Shrew.
    TYPE LOCALITY: South Africa, Eastern Cape Prov., East London.
    DISTRIBUTION: Savannahs from the Cape (South Africa) to Zimbabwe, Zambia, Tanzania, E Dem. Rep. Congo, Malawi; an isolated record from Nigeria.
    STATUS: IUCN – Lower Risk (lc).
    SYNONYMS: *meesteri* Butler and Greenwood, 1979; *minor* G. M. Allen and Loveridge, 1933; *natalensis* Roberts, 1946; *orangiae* (Roberts, 1924); *tulbaghensis* Roberts, 1946; *warreni* Roberts, 1929 [see Heim de Balsac and Meester, 1977:6].
    COMMENTS: Closely associated with termite mounds (Lynch, 1986). Gureev (1979:383) listed *orangiae* and *warreni* as distinct species without comment. Common in the Pleistocene of Kenya and South Africa (Butler and Greenwood, 1979).

*Suncus zeylanicus* Phillips, 1928. Spolia Zeylan., 14:313.
    COMMON NAME: Jungle Shrew.
    TYPE LOCALITY: Sri Lanka, Gonagama Estate, Kitulgala, 900 ft. (274 m).
    DISTRIBUTION: Higlands of Sri Lanka.
    STATUS: IUCN – Endangered.
    COMMENTS: Phillips (1980) stressed that *zeylanicus* differs distinctly from *murinus* in the flesh, particularly by its long and almost naked tail, and that it lives in primary forest. However, its relation to *montanus* has still to be studied.

*Sylvisorex* Thomas, 1904. Abstr. Proc. Zool. Soc. Lond., 1904(10):12.
    TYPE SPECIES: *Crocidura morio* Gray, 1862.
    COMMENTS: For placement in Crocidurinae see Repenning (1967:15). The genus was regarded as part of *Suncus* by Smithers and Tello (1976), but was retained by Ansell (1978). Querouil et al. (2001) studied 16s rRNA sequences of four species and found the genus, as currently understood, polyphyletic. As a consequence, *S. megalura* is removed from *Sylvisorex* and included in *Suncus*. Further taxonomic action will require a careful analysis of all taxa. Heim de Balsac (1968) and Jenkins (1984) figured and discussed most of the species listed below. Maddalena and Ruedi (1994) and Schlitter et al. (1999) described karyotypes of five species.

*Sylvisorex camerunensis* Heim de Balsac, 1968. Bonn. Zool. Beitr., 19:35.
    COMMON NAME: Cameroonian Forest Shrew.
    TYPE LOCALITY: Cameroon, Lake Manengouba, 1800 m.
    DISTRIBUTION: Montane forests of W Cameroon (Mt. Oku, Lake Manengouba) and SE Nigeria (Gotel Mtns).
    COMMENTS: Formerly included in *granti*, but given species rank by Hutterer et al. (1992*a*).

*Sylvisorex granti* Thomas, 1907. Ann. Mag. Nat. Hist., ser. 7, 19:118.
  COMMON NAME: Grant's Forest Shrew.
  TYPE LOCALITY: Uganda, Ruwenzori East, Mubuku Valley, 10,000 ft. (3,048 m).
  DISTRIBUTION: Mountain forests of C (Dem. Rep. Congo, Uganda, Rwanda) and E Africa
      (Kenya, Tanzania).
  STATUS: IUCN – Lower Risk (lc).
  SYNONYMS: *mundus* Osgood, 1910.
  COMMENTS: The isolated East African populations require a careful comparison with typical
      *granti* from the Rift Valley.

*Sylvisorex howelli* Jenkins, 1984. Bull. Brit. Mus. (Nat. Hist). Zool., 47:65.
  COMMON NAME: Howell's Forest Shrew.
  TYPE LOCALITY: Tanzania, Uluguru Mtns, Morningside.
  DISTRIBUTION: Eastern Arc Mtns (Tanzania): W and E Usambara, Nguru and Uluguru Mtns
      (Stanley et al., 2000*b*).
  STATUS: IUCN – Vulnerable.
  SYNONYMS: *usambarensis* Hutterer, 1986.
  COMMENTS: Includes *usambarensis*, which may represent a distinct species; see Hutterer
      (1986*b*). However, the variation among the five populations has to be studied before
      taxonomic conclusions can be drawn.

*Sylvisorex isabellae* Heim de Balsac, 1968. Bonn. Zool. Beitr., 19:31.
  COMMON NAME: Bioko Forest Shrew.
  TYPE LOCALITY: Equatorial Guinea, Bioko (Fernando Po), Pic Santa Isabel, Refugium, 2000 m.
  DISTRIBUTION: Bioko; a similar form of unsolved taxonomic status occurs in the Bamenda
      Highlands, Cameroon.
  STATUS: IUCN – Vulnerable.
  COMMENTS: Included in *morio* by Heim de Balsac and Meester (1977), but represents a
      distinctly smaller species. Karyotype (2n = 36, FN = 50) of the Bamenda population
      described by Schlitter et al. (1999).

*Sylvisorex johnstoni* (Dobson, 1888). Proc. Zool. Soc. Lond., 1887:577 [1888].
  COMMON NAME: Johnston's Forest Shrew.
  TYPE LOCALITY: Cameroon, Rio del Rey.
  DISTRIBUTION: Lowland forest of the Dem. Rep. Congo Basin, SW Cameroon, Gabon, Bioko,
      Republic of Congo, Dem. Rep. Congo, Uganda, NW Tanzania, Burundi.
  STATUS: IUCN – Lower Risk (lc).
  SYNONYMS: *dieterleni* Hutterer, 1986.
  COMMENTS: Species reviewed by Hutterer (1986*b*); recently found in the Republic of Congo
      (Dowsett and Granjon, 1991) and Burundi (Kerbis, pers. comm.). A peculiar karyotype
      (2n = 30, FN = 38; Schlitter et al., 1999) and 16s rRNA sequence data (Querouil et al.,
      2001) set this species apart from other species of the genus. Querouil et al. (2003)
      studied the phylogeography of the species across the Congo Basin. They found great
      genetic distances between populations in S Gabon and W Congo, and suggested that
      cryptic species may be involved.

*Sylvisorex konganensis* Ray and Hutterer, 1996. Ecotropica, 1:93 [1995].
  COMMON NAME: Kongana Shrew.
  TYPE LOCALITY: SW Central African Republic, Dzanga-Sangha Forest Reserve, unlogged
      mixed species forest near Kongana Camp [02°47'N, 16°25'E).
  DISTRIBUTION: High forest in Central African Republic and Republic of Congo (unpubl.).
  COMMENTS: A comparison of 16s rRNA sequences showed a closer relationship to *S. ollula*
      (Querouil et al., 2001).

*Sylvisorex lunaris* Thomas, 1906. Ann. Mag. Nat. Hist., ser. 7, 18:139.
  COMMON NAME: Moon Forest Shrew.
  TYPE LOCALITY: Uganda, Ruwenzori East, Mubuku Valley, 12,000 ft. (3,810 m).
  DISTRIBUTION: The high mountain zone of C Africa up to 4,500 m; Ruwenzori (Uganda,
      Dem. Rep. Congo), Virunga Volcanoes (Rwanda), and on both sides of Lake Kivu
      (Dem. Rep. Congo, Burundi).

STATUS: IUCN – Lower Risk (lc).

SYNONYMS: *ruandae* Lönnberg and Gyldenstolpe, 1925.

COMMENTS: Includes *ruandae* but not *oriundus*; both were listed as distinct species by Gureev (1979:380-381). Karyotype (2n = 58, FN = 80) described by Maddalena and Ruedi (1994).

*Sylvisorex morio* (Gray, 1862). Proc. Zool. Soc. Lond., 1862:180.

COMMON NAME: Mt. Cameroon Forest Shrew.

TYPE LOCALITY: Cameroon Mountains.

DISTRIBUTION: Confined to Mt. Cameroon (Cameroon).

STATUS: IUCN – Endangered.

COMMENTS: Does not include *isabellae*; see under that species. Karyotype (2n = 38) different from that of *isabellae* (2n = 36), see Schlitter et al. (1999).

*Sylvisorex ollula* Thomas, 1913. Ann. Mag. Nat. Hist., ser. 8, 11:321.

COMMON NAME: Greater Forest Shrew.

TYPE LOCALITY: Cameroons, Bitye, Ja River, 2,000 feet (610 m).

DISTRIBUTION: W Cameroon and adjacent Nigeria to S Cameroon, Gabon, Central African Republic, Equatorial Guinea, Republic of Congo, and Dem. Rep. Congo.

STATUS: IUCN – Lower Risk (lc).

COMMENTS: The largest species of the genus; discussed in some detail by Dieterlen and Heim de Balsac (1979). Querouil et al. (2003) studied the phylogeography of the species in the Congo Basin; they found little genetic differentiation between populations from SW Cameroon to W Congo. Karyotype (2n = 38, FN = 64) described by Schlitter et al. (1999). For biolgical and distributional data, see Brosset (1988), Lasso et al. (1996), and Ray and Hutterer (1996).

*Sylvisorex oriundus* Hollister, 1916. Bull. Am. Mus. Nat. Hist., 35:672.

COMMON NAME: Lesser Forest Shrew.

TYPE LOCALITY: Dem. Rep. Congo, Medje.

DISTRIBUTION: NE Dem. Rep. Congo.

STATUS: IUCN – Vulnerable.

COMMENTS: Often included in *ollula* but regarded as distinct by Dieterlen and Heim de Balsac (1979), a view supported by personal examination of the holotype.

*Sylvisorex pluvialis* Hutterer and Schlitter, 1996. Contributions in Mammalogy: A Memorial Volume Honoring Dr. J. Knox Jones, Jr., p. 61.

COMMON NAME: Rain Forest Shrew.

TYPE LOCALITY: Cameroon, SW Province, Korup N. P., Ikenge Research Station, 160 m [05°16'N, 09°08'E].

DISTRIBUTION: Known only from the type locality and from Kongana, Central African Republic.

COMMENTS: Ray and Hutterer (1996) identified this species in carnivore scats collected in forest around Kongana (Central African Republic), the type locality of *S. konganensis*.

*Sylvisorex vulcanorum* Hutterer and Verheyen, 1985. Z. Säugetierk., 50:266.

COMMON NAME: Volcano Shrew.

TYPE LOCALITY: Rwanda, Parc National des Volcans, Karisoke (0°28'S, 29°29'E, 3,100 m).

DISTRIBUTION: High altitude rainforest of E Dem. Rep. Congo, Uganda, Rwanda, and Burundi.

STATUS: IUCN – Lower Risk (lc); of conservation concern (Nicoll and Rathbun, 1990:21).

COMMENTS: One of the smallest species in the genus; rather similar to *S. granti*.

**Subfamily Myosoricinae** Kretzoi, 1965. Vertebrata Hungarica, 7:124.

SYNONYMS: Crocidosoricinae Reumer, 1987; Oligosoricini Gureev, 1971.

COMMENTS: Kretzoi (1965) based his tribe Myosoricini on *Myosorex*, an extant genus known to exhibit a number of ancestral characters (Heim de Balsac, 1966b). Maddalena and Bronner (1992) confirmed its isolated position in a study of allozymes. Jammot (1983) included also fossil taxa and applied the name Myosoricina to an evolutionary lineage. The dental characters that he and Reumer (1987) used to characterise this lineage (which

Reumer called Crocidosoricinae) are all present in extant *Myosorex*, *Congosorex*, and *Surdisorex* (Hutterer, 1993*a*; Hutterer et al., 2002*b*). It remains to be analyzed whether all the fossil and extant genera are related, or whether they only share a few ancestral characters.

*Congosorex* Heim de Balsac and Lamotte, 1956. Mammalia, 20:167.
   TYPE SPECIES: *Myosorex polli* Heim de Balsac and Lamotte, 1956.
   COMMENTS: Described as a subgenus of *Myosorex* by Heim de Balsac and Lamotte (1956), but differs in its tooth formula and was therefore treated as a full genus by Heim de Balsac (1967), Hutterer (1993*a*), and Hutterer et al. (2002*b*). Genetically, *Congosorex* is the sister taxon of *Myosorex* (Querouil et al., 2001). Reviewed by Hutterer et al. (2002*b*).

*Congosorex polli* (Heim de Balsac and Lamotte, 1956). Mammalia, 20:155.
   COMMON NAME: Greater Congo Shrew.
   TYPE LOCALITY: Dem. Rep. Congo, Kasai Prov., Lubondai.
   DISTRIBUTION: Known only from type locality in S Dem. Rep. Congo.
   STATUS: IUCN – Critically Endangered.
   COMMENTS: Since the discovery in 1955, this distinct species has not been collected again.

*Congosorex verheyeni* Hutterer, Barriere and Colyn, 2002. Bull. Inst. Roy. Sci. Nat. Belg., Biol., 72, Suppl.:10.
   COMMON NAME: Lesser Congo Shrew.
   TYPE LOCALITY: Republic of Congo, Parc National d'Odzala, Mbomo, 00°24′N, 14°44′E.
   DISTRIBUTION: Known from three sites in the lowland forest of the Western Congo Basin (Central African Republic, Republic of Congo, and Dem. Rep. Congo).
   COMMENTS: A 16s rRNA sequence analyzed by Querouil et al. (2001) as *Congosorex* sp. refers to this species.

*Myosorex* Gray, 1838. Proc. Zool. Soc. Lond., 1837:124 [1838].
   TYPE SPECIES: *Sorex varius* Smuts, 1832.
   COMMENTS: Repenning (1967) grouped *Myosorex* in the Crocidurinae; Reumer's (1987) Crocidosoricinae would fit as well. Kretzoi (1965) based the tribe Myosoricini on this genus; this name is available and used here for the subfamily. Generic status sometimes questioned; but see Meester (1954). *Surdisorex* and *Congosorex* are often included as subgenera but are treated here as full genera, following Thomas (1906*b*), Hollister (1918), Meester (1953), Heim de Balsac (1966*b*), Hutterer (1993*a*) and Hutterer et al. (2002*b*). Partial reviews of *Myosorex* were provided by Heim de Balsac (1967, 1968*b*), Heim de Balsac and Lamotte (1956), Meester and Dippenaar (1978), and Stanley and Hutterer (2000). The formerly recognized *Myosorex preussi* (Matschie, 1893), described from "Mount Cameroun", is based on mismatched parts of three different genera (*Crocidura*, *Sorex*, *Sylvisorex*), and does not represent a biological species (Hutterer, 1993*a*).

*Myosorex babaulti* Heim de Balsac and Lamotte, 1956. Mammalia, 20:150.
   COMMON NAME: Babault's Mouse Shrew.
   TYPE LOCALITY: Dem. Rep. Congo, Kivu.
   DISTRIBUTION: Mountains west and east of Lake Kivu, including Idjwi Isl (Dem. Rep. Congo, Uganda, Rwanda, Burundi).
   STATUS: IUCN – Lower Risk (lc).
   COMMENTS: Formerly included in *blarina*; but see Dieterlen and Heim de Balsac (1979). The species does not co-occur with *blarina* in the Bwindi Impenetrable NP, Uganda, as suggested by Kasangaki et al. (2003) (Kerbis Peterhans, pers. comm., 2004).

*Myosorex blarina* Thomas, 1906. Ann. Mag. Nat. Hist., ser. 7, 18:139.
   COMMON NAME: Montane Mouse Shrew.
   TYPE LOCALITY: Uganda, Ruwenzori East, Mubuku Valley, 10,000 ft. (3048 m).
   DISTRIBUTION: Montane forest at Mt. Ruwenzori (Uganda, Dem. Rep. Congo).
   STATUS: IUCN – Vulnerable.
   COMMENTS: Does not occur together with *babaulti* in the Bwindi Impenetrable NP, Uganda, as suggested by Kasangaki et al. (2003).

*Myosorex cafer* (Sundevall, 1846). Ofv. Kongl. Svenska Vet.-Akad. Forhandl. Stockholm, 3:119.
COMMON NAME: Dark-footed Mouse Shrew.
TYPE LOCALITY: South Africa, "E Caffraria interiore et Port-Natal".
DISTRIBUTION: South Africa, eastern escarpment from Eastern Cape Prov. north to Limpopo and Mpumalanga Provinces; extreme W Mozambique and E Zimbabwe, in higher elevations above 1,000 m.
STATUS: IUCN – Lower Risk (lc).
SYNONYMS: *swinnyi* Chubb, 1908.
COMMENTS: Meester (1958) described the geographic variation of the species. Heim de Balsac and Meester (1977) included *affinis, sclateri, swinnyi, talpinus,* and *tenuis* in *cafer,* while Wolhuter (in Smithers, 1983:3) and Dippenaar et al. (1983) regarded *sclateri* and *tenuis* as distinct, partly based on new karyotype information (*M. cafer*: 2n = 38). Although no additonal data have yet been published, this view is provisionally accepted here as it better reflects existing variation within the southern African representatives of the genus.

*Myosorex eisentrauti* Heim de Balsac, 1968. Bonn. Zool. Beitr., 19:20.
COMMON NAME: Eisentraut's Mouse Shrew.
TYPE LOCALITY: Equatorial Guinea, Bioko, Pic Santa Isabel, 2400 m.
DISTRIBUTION: Montane forest of Bioko (Fernando Po).
STATUS: IUCN – Endangered.
COMMENTS: The forms *okuensis* and *rumpii* were included in *eisentrauti* by Heim de Balsac and Meester (1977); both were regarded as distinct species by Hutterer (1993*a*).

*Myosorex geata* (Allen and Loveridge, 1927). Proc. Boston Soc. Nat. Hist., 38:417.
COMMON NAME: Geata Mouse Shrew.
TYPE LOCALITY: Tanzania, Uluguru Mtns, Nyingwa.
DISTRIBUTION: Uluguru Mtns in Tanzania; other localities questionable.
STATUS: IUCN – Endangered.
COMMENTS: Formerly in *Crocidura*; see Heim de Balsac (1967:610).

*Myosorex kihaulei* Stanley and Hutterer, 2000. Bonn. Zool. Beitr., 49:20.
COMMON NAME: Kihaule's Mouse Shrew.
TYPE LOCALITY: Tanzania, Udzungwa Scarp Forest Reserve, 19.5 km W Chita, 8°20'50"S, 35°56'20"E, 2000 m.
DISTRIBUTION: Udzungwa Mtns, Tanzania.
COMMENTS: Populations of the Rungwe Mtns may belong to this species or represent a different taxon (Stanley and Hutterer, 2000).

*Myosorex longicaudatus* Meester and Dippenaar, 1978. Ann. Transvaal Mus., 31:30.
COMMON NAME: Long-tailed Forest Shrew.
TYPE LOCALITY: South Africa, Western Cape Prov., 14 km NNE Knysna, Diepwalle State Forest Station, 33°57'S, 23°10'E.
DISTRIBUTION: Endemic to South Africa. Occurs in escarpment forests of SE Western Cape Prov., South Africa, between 2000-3600 m.
STATUS: IUCN – Vulnerable.
SYNONYMS: **boosmani** Dippenaar, 1995.
COMMENTS: Dippenaar (1995) described a distinct population from the Langeberg Mtns whose conservation status he considered as "vulnerable".

*Myosorex okuensis* Heim de Balsac, 1968. Bonn. Zool. Beitr., 19:20.
COMMON NAME: Oku Mouse Shrew.
TYPE LOCALITY: Cameroon, Bamenda Highlands, Mt. Oku, Lake Oku, 2100 m.
DISTRIBUTION: Forested mountains of the Bamenda plateau, Cameroon (Lake Manenguba, Mt. Oku, Mt. Lefo).
STATUS: IUCN – Vulnerable.
COMMENTS: Formerly included in *eisentrauti* (see Heim de Balsac and Meester, 1977), but cranially very distinct.

*Myosorex rumpii* Heim de Balsac, 1968. Bonn. Zool. Beitr., 19:20.
COMMON NAME: Rumpi Mouse Shrew.
TYPE LOCALITY: Cameroon, "Rumpi-Hills, 1100 mètres".
DISTRIBUTION: Known only from the type locality.
STATUS: IUCN – Critically Endangered.
COMMENTS: The holotype and only known specimen is so unique (Heim de Balsac, 1968*b*, fig. 4) that it is considered to represent a valid species. Heim de Balsac (1968*b*) himself was uncertain about the status of this taxon; while he formally named it *M. eisentrauti rumpii*, he labeled all figures and the map with "*Myosorex rumpii*".

*Myosorex schalleri* Heim de Balsac, 1966. C.R. Acad. Sci. Paris, 263:889.
COMMON NAME: Schaller's Mouse Shrew.
TYPE LOCALITY: E Dem. Rep. Congo, Itombwe Mtns, Nzombe (Mwenga).
DISTRIBUTION: Known only from the type locality.
STATUS: IUCN – Critically Endangered.
COMMENTS: Provisionally named by Heim de Balsac (1966*b*); full description by Heim de Balsac (1967). The type locality was later erroneously shifted to the "Albert N. P." (Heim de Balsac and Meester, 1977); Nzombe is located in the Itombwe Mtns (Hutterer, 1986*c*).

*Myosorex sclateri* Thomas and Schwann, 1905. Abstr. Proc. Zool. Soc. Lond., 1905(15):10.
COMMON NAME: Sclater's Mouse Shrew.
TYPE LOCALITY: South Africa, KwaZulu-Natal Prov., Zululand, Ngoye hills, 250 m.
DISTRIBUTION: Wet habitats in KwaZulu-Natal (South Africa).
STATUS: IUCN – Vulnerable.
SYNONYMS: *affinis* Thomas and Schwann, 1905; *talpinus* Thomas and Schwann, 1905.
COMMENTS: Provisionally regarded as a distinct species by Wolhuter (in Smithers, 1983:3); occurs in sympatry with *cafer* and has a different karyotype (2n = 38). Meester et al. (1986) included *sclateri* in *cafer*.

*Myosorex tenuis* Thomas and Schwann, 1905. Proc. Zool. Soc. Lond., 1905:131-132.
COMMON NAME: Thin Mouse Shrew.
TYPE LOCALITY: South Africa, Mpumalanga Prov., near Wakkerstroom, Zuurbron.
DISTRIBUTION: Mpumalanga Prov. (South Africa) and possibly W Mozambique.
STATUS: IUCN – Vulnerable.
COMMENTS: Provisionally regarded as a distinct species by Wolhuter (in Smithers, 1983:3) because of sympatry with *cafer* and a different karyotype (2n = 40). Meester et al. (1986) included *tenuis* in *cafer*.

*Myosorex varius* (Smuts, 1832). Enumer. Mamm. Capensium, p. 108.
COMMON NAME: Forest Shrew.
TYPE LOCALITY: South Africa, Cape of Good Hope, Algoa Bay (Port Elizabeth).
DISTRIBUTION: South Africa, from West and Eastern Cape across Northern Cape, KwaZulu-Natal and Free State to Mpumalanga and Gauteng Provinces; also Lesotho and Swaziland (Baxter, in litt., 2004).
STATUS: IUCN – Lower Risk (lc).
SYNONYMS: *herpestes* (Duvernoy, 1838); *pondoensis* Roberts, 1946; *transvaalensis* Roberts, 1924 [see Heim de Balsac and Meester, 1977].
COMMENTS: Revised by Meester (1958). Karyotype (2n = 42) differs from all other southern African *Myosorex* (Wolhuter, in Smithers, 1983).

*Myosorex zinki* Heim de Balsac and Lamotte, 1956. Mammalia, 20:148.
COMMON NAME: Kilimanjaro Mouse Shrew.
TYPE LOCALITY: Tanzania, Mt. Kilimanjaro, SE slope, 3,700 m.
DISTRIBUTION: High altitudes of Mt. Kilimanjaro, Tanzania, 2470-4000 m (Stanley et al., In Press).
COMMENTS: In the last edition of this checklist (Hutterer, 1993*a*) this species was ommitted by mistake. This is a large and distinct species that resembles *M. eisentrauti* from Bioko (Stanley and Hutterer, 2000). Grimshaw et al. (1997) listed it as endemic to Mt.

Kilimanjaro. Shore and Garbett (1991) collected one specimen at 3500 m. Recently it was found to be common on that mountain (W. T. Stanley, pers. comm., 2002).

*Surdisorex* Thomas, 1906. Ann. Mag. Nat. Hist., ser. 7, 18:223.
 TYPE SPECIES: *Surdisorex norae* Thomas, 1906.
 COMMENTS: This genus was commonly included in *Myosorex* but retained as a full genus by
  Hollister (1918), Meester (1953), Heim de Balsac (1966*b*), Hutterer (1993*a*), and Hutterer
  et al. (2002*b*).

*Surdisorex norae* Thomas, 1906. Ann. Mag. Nat. Hist., ser. 7, 18:223.
 COMMON NAME: Aberdare Mole Shrew.
 TYPE LOCALITY: Kenya, east side of Aberdare Range, near Nyeri.
 DISTRIBUTION: Aberdare Range (Kenya).
 STATUS: IUCN – Vulnerable.
 COMMENTS: Formerly in *Myosorex*; see Heim de Balsac and Meester (1977). Ecology and
  distribution described by Duncan and Wrangham (1971).

*Surdisorex polulus* Hollister, 1916. Smithson. Misc. Coll., 66(1):1.
 COMMON NAME: Mt. Kenya Mole Shrew.
 TYPE LOCALITY: Kenya, west side of Mt. Kenya, 10,700 ft. (3,261 m).
 DISTRIBUTION: Mt. Kenya (Kenya).
 STATUS: IUCN – Vulnerable.
 COMMENTS: Included in genus *Myosorex* and regarded as a subspecies of *norae* by Heim de
  Balsac and Meester (1977); however, both species form a quite distinct clade (Hutterer
  et al., 2002*b*). For ecology and distribution see Duncan and Wrangham (1971).

**Subfamily Soricinae** G. Fischer, 1814. Zoognosia tabulis synopticis illustrata, 3:x.
 SYNONYMS: Sorexineae Lesson, 1842; Soricinorum G. Fischer, 1814.
 COMMENTS: The contents of Soricinae have been principally defined by Repenning (1967) and
  modified by Reumer (1984, 1998). All extant Nearctic shrews belong in this subfamily.

**Tribe Anourosoricini** Anderson, 1879. Anatom. and zool. Res., 1:159.
 SYNONYMS: Amblycoptini Kormos, 1926.
 COMMENTS: McKenna and Bell (1997) included this tribe in Nectogalini, but see Reumer (1998).

*Anourosorex* Milne-Edwards, 1872. Rech. Hist. Nat. Mamm., p. 264.
 TYPE SPECIES: *Anourosorex squamipes* Milne-Edwards, 1872.
 SYNONYMS: *Anaurosorex* Günther, 1871; *Anurosorex* Anderson, 1875; *Pygmura* Anderson, 1873.
 COMMENTS: Sole living representative of Tribe Anourosoricini (see Reumer, 1998). Reumer
  (1984:17) placed the genus in the tribe Amblycoptini Kormos, 1926, but this was
  antedated by Anourosoricini Anderson, 1879. Geographic variation of extant species
  studied by Motokawa and Lin (2002) and Motokawa et al. (2004). The fossil history of the
  genus was reviewed by Zheng (1985) and Storch and Qiu (1991).

*Anourosorex assamensis* Anderson, 1875. Ann. Mag. nat. Hist., ser. 4, 16:282.
 COMMON NAME: Assam Mole Shrew.
 TYPE LOCALITY: Assam, Subsasugu.
 DISTRIBUTION: NE India (Assam, Meghalaya, Manipur, Nagaland).
 COMMENTS: Blanford (1888) and Allen (1938) considered *A. assamensis* as a species distinct
  from *A. squamipes*. Mandal and Das (1969) reported on a series of mole shrews from
  Assam and noted their larger size in comparison to the Chinese population, an
  observation corroborated by my study of specimens in the FMNH. These differences
  were neatly shown in a principal component analysis performed by Motokawa and
  Lin (2002). Records from Meghalaya and Nagaland are based on specimens in the
  FMNH. It is not known whether the ranges of *A. assamensis* and *A. squamipes* overlap
  in S Assam or Manipur, but the occurrence of the latter in Mizoram (Mandal et al.,
  1995) may indicate the presence of a contact zone.

*Anourosorex schmidi* Petter, 1963. Mammalia, 27:444.

    COMMON NAME: Giant Mole Shrew.

    TYPE LOCALITY: India, Arunachal Pradesh, Bombdila, 2700 m.

    DISTRIBUTION: NE India (Sikkim, Arunachal Pradesh) and Bhutan.

    COMMENTS: Petter (1963) described this taxon as a subspecies of *A. squamipes*, a decision followed by most authors (Hutterer, 1993*a*; Mandal and Das, 1969; Motokawa and Lin, 2002). A study of the holotype (MNHN Paris) and the discovery (in FMNH Chicago) of another specimen from Sikkim however reveals that *A. schmidi* is a giant form that deserves species status. Saha (1978) reported a similarly sized specimen from Gomchu, Bhutan. This mole shrew has a very large skull (condylo-incisive length 29.1-30.5 mm) and a bulbous dentition not found in any other population. The species appears to be confined to the SE slopes of the Himalayas.

*Anourosorex squamipes* Milne-Edwards, 1872. Rech. Hist. Nat. Mamm., p. 264.

    COMMON NAME: Chinese Mole Shrew.

    TYPE LOCALITY: China, Sichuan Prov., probably Moupin (= Baoxing).

    DISTRIBUTION: Shaanxi, Hubei, Sichuan and Yunnan (China); N and W Burma; E India (Mizoram); North Vietnam; Thailand.

    STATUS: IUCN – Lower Risk (lc).

    SYNONYMS: *capito* G. Allen, 1923; *capnias* G. Allen, 1923.

    COMMENTS: Formerly also included *assamensis, schmidi,* and *yamashinai* as subspecies (see Petter, 1963*b*; Jameson and Jones, 1977; Motokawa and Lin, 2002), but these three are given species rank here. The Indian records are based on measurements of specimens from Mizoram provided by Mandal et al. (1995).

*Anourosorex yamashinai* Kuroda, 1935. J. Mammal., 16:288.

    COMMON NAME: Taiwanese Mole Shrew.

    TYPE LOCALITY: N Taiwan, Taiheizan, Taihoku-siu, 5500 ft. (1676 m).

    DISTRIBUTION: C highlands of Taiwan from 1500-2500 m.

    COMMENTS: Described as a subspecies of *A. squamipes* and kept as such until recently; see Petter (1963*b*), Jameson and Jones (1977), Hutterer (1993*a*), and Motokawa and Lin (2002). However, this island population differs from the Chinese mainland forms by a smaller body, shorter tail, and a distinct karyotype (Harada and Takada, 1985), and represents a species endemic to Taiwan (Motokawa et al., 2004). The ecology, distribution, and sub-population structure of *A. yamashinai* was studied by Alexander et al. (1987), Yu (1993, 1994) and Yu et al. (2001), making it the best-studied species of the genus.

**Tribe Blarinellini** Reumer, 1998. In Wójcik, J. M., and M. Wolsan, Evolution of shrews, p. 19.

*Blarinella* Thomas, 1911. Proc. Zool. Soc. Lond., 1911:166.

    TYPE SPECIES: *Sorex quadraticauda* Milne-Edwards, 1872.

    COMMENTS: Formerly in Soricini; see Repenning (1967:61). The genus is known from the Late Miocene of China (Storch and Qiu, 1991); supposed records from the Pleistocene of Europe (Reumer, 1984, Rzebik-Kowalska, 1989) refer to *Alloblarinella* Storch, 1995; see Storch (1995*a*) and Reumer (1998). Extant taxa revised by Jiang et al. (2003).

*Blarinella griselda* Thomas, 1912. Ann. Mag. Nat. Hist., ser. , 10:400.

    COMMON NAME: Indochinese Short-tailed Shrew.

    TYPE LOCALITY: China, Gansu, 42 mi. (68 km) SE Taochou, 10,000 ft. (3048 m).

    DISTRIBUTION: China (Gansu, Yunnan and Hubei), and N Vietnam.

    COMMENTS: Since Allen (1938), included in *quadraticauda*, but given species status by Jiang et al. (2003). Occurs also in N Vietnam (Lunde et al., 2003*b*).

*Blarinella quadraticauda* (Milne-Edwards, 1872). Rech. Hist. Nat. Mamm., p. 261.

    COMMON NAME: Asiatic Short-tailed Shrew.

    TYPE LOCALITY: China, Sichuan, "Moupin, Thibet oriental".

    DISTRIBUTION: Montane taiga forest of W Sichuan (China).

    STATUS: IUCN – Lower Risk (lc).

COMMENTS: Species range and status revised by Jiang et al. (2003).

*Blarinella wardi* Thomas, 1915. Ann. Mag. Nat. Hist., ser. 8, 15:336.
COMMON NAME: Burmese Short-tailed Shrew.
TYPE LOCALITY: Burma, "Hpimaw, Upper Burma, about 26°N., 98°35'E. Alt. 8000' [2400 m]."
DISTRIBUTION: N Burma and NW Yunnan (China).
STATUS: IUCN – Lower Risk (nt).
COMMENTS: Included in *quadraticauda* by Allen (1938), Ellerman and Morrison-Scott (1951) and subsequent authors, but the species has a much smaller and narrower skull (see measurements in Hoffmann, 1987:134) and was therefore regarded as distinct (Hutterer, 1993*a*). These differences were also recognized by Corbet (1978*c*:26) and Lunde et al. (2003*b*). Revised by Jiang et al. (2003).

**Tribe Blarinini** Kretzoi, 1965. Vertebrata Hungarica, 7:126.
COMMENTS: Repenning (1967) and Reumer (1998) recognized and re-defined this tribe.

*Blarina* Gray, 1838. Proc. Zool. Soc. Lond., 1837:124 [1838].
TYPE SPECIES: *Corsira* (*Blarina*) *talpoides* Gray, 1838 (= *Sorex talpoides* Gapper, 1830 = *Sorex brevicaudus* Say, 1823).
SYNONYMS: *Anotus* Wagner, 1855; *Blaria* Gray, 1843; *Brachysorex* Duvernoy, 1842; *Mamblarinaus* Herrera, 1899; *Talposorex* Pomel, 1848 [not Lesson].
COMMENTS: Type genus of Blarinini (Repenning, 1967:37). Reviewed by George et al. (1982, 1986). Phylogeny analyzed by Brandt and Ortí (2002).

*Blarina brevicauda* (Say, 1823). *In* James, Account Exped. Pittsburgh to Rocky Mtns, 1:164.
COMMON NAME: Northern Short-tailed Shrew.
TYPE LOCALITY: USA, Engineer cantonment, west bank of the Missouri River; restricted to Nebraska, Washington Co., approximately 2 mi. (3.2 km) east Ft. Calhoun by Jones (1964:68).
DISTRIBUTION: S Canada west to C Saskatchewan and east to SE Canada, south to Nebraska and N Virginia (USA).
STATUS: IUCN – Lower Risk (lc).
SYNONYMS: *angusticeps* Baird, 1858; *costaricensis* J. A. Allen, 1891; *dekayi* (Bachmann, 1837); *fossilis* Hibbard, 1943; *micrurus* (Pomel, 1848); *ozarkensis* Brown, 1908; *simplicidens* Cope, 1899; *aloga* Bangs, 1902; *angusta* Anderson, 1943; *churchi* Bole and Moulthrop, 1942; *compacta* Bangs, 1902; *hooperi* Bole and Moulthrop, 1942; *kirtlandi* Bole and Moulthrop, 1942; *manitobensis* Anderson, 1947; *pallida* R. W. Smith, 1940; *talpoides* (Gapper, 1830); *telmalestes* Merriam, 1895.
COMMENTS: Includes *telmalestes* (see review by George et al., 1986, Mammalian Species No. 261), which Hall (1981:57) listed as a distinct species. Karyotype has 2n = 48-50, FN = 52 (Zima et al., 1998).

*Blarina carolinensis* (Bachman, 1837). J. Acad. Nat. Sci. Philadelphia, 7:366.
COMMON NAME: Southern Short-tailed Shrew.
TYPE LOCALITY: USA, "in the upper and maritime districts of South Carolina".
DISTRIBUTION: S Illinois east to N Virginia, and south through E Texas and N Florida (USA).
STATUS: IUCN – Lower Risk (lc).
SYNONYMS: *minima* Lowery, 1943; *shermani* Hamilton, 1955.
COMMENTS: For specific status see Genoways and Choate (1972) and Tate et al. (1980), for a general review Genoways and Choate (1998) and McCay (2001, Mammalian Species No. 673). Hall (1981:54) listed *carolinensis* as a subspecies of *brevicauda*. Karyotype has 2n = 36-46, FN = 48-49 (Genoways and Benedict, in Wilson and Ruff, 1999).

*Blarina hylophaga* Elliot, 1899. Field Columb. Mus. Publ., Zool. Ser., 1:287.
COMMON NAME: Elliot's Short-tailed Shrew.
TYPE LOCALITY: USA, Oklahoma, Murray Co., Dougherty.
DISTRIBUTION: USA: S Nebraska and SW Iowa south to S Texas; east to Missouri and NW Arkansas; Oklahoma; extending into Louisiana.
STATUS: IUCN – Lower Risk (lc).

SYNONYMS: *hulophaga* Elliott, 1899; ***plumbea*** Davis, 1941.

COMMENTS: Original spelling *hulophaga* Elliot, 1899, corrected to *hylophaga* by Elliot (1905). Formerly included in *carolinensis*, but separated as a distinct species by George et al. (1982). Karyotype 2n = 52, FN = 64-66 (Zima et al., 1998). Molecular data suggest that this is the basal taxon of the genus (Brandt and Ortí, 2002).

*Blarina peninsulae* Merriam, 1895. N. Amer. Fauna, 10:14.

COMMON NAME: Everglades Short-tailed Shrew.

TYPE LOCALITY: USA, Florida, Dade Co., Miami River.

DISTRIBUTION: Peninsular Florida (USA).

COMMENTS: Formerly listed under *brevicauda* or *carolinensis*. The population of short-tailed shrews in S Florida has a distinct karyotype (2n = 50-52) and a distinct morphology (Genoways and Choate, 1998; George et al., 1982). In addition, the presence of a contact zone with *carolinensis* (2n = 36-46) suggests that *peninsulae* represents a valid species (Genoways and Benedict, in Wilson and Ruff, 1999).

*Cryptotis* Pomel, 1848. Arch. Sci. Phys. Nat. Geneve, 9:249.

TYPE SPECIES: *Sorex cinereus* Bachman, 1837 (= *Sorex parvus* Say, 1823).

SYNONYMS: *Soriciscus* Coues, 1877; *Xenosorex* Schaldach, 1966.

COMMENTS: See Repenning (1967:37) and Reumer (1998:19) for placement in Blarinini. North and Central American species revised in part by Choate (1970) and Choate and Fleharty (1974); Central and South American species revised by Woodman (1996, 2002, 2003), Woodman and Timm (1993, 1999, 2000), and Vivar et al. (1997). Gureev (1979:433-437) listed many species that Choate (1970) considered synonyms. Formerly included *C. surinamensis* which was transferred to *Sorex araneus* by Husson (1963). Woodman (1993) argued that *Cryptotis* is feminine in gender, a conclusion followed here.

*Cryptotis alticola* (Merriam, 1895). N. Amer. Fauna, 10:27.

COMMON NAME: Central Mexican Broad-clawed Shrew.

TYPE LOCALITY: Mexico, Volcán Popocatépetl, 11,500 ft. (3505 m).

DISTRIBUTION: Highlands above 2000 m in the Mexican states of Colima, Hidalgo, Jalisco, Michoacán, México, and Puebla, Morelos, and in the Distrito Federal (Carraway, ms; Woodman and Timm, 1999).

SYNONYMS: *euryrhynchis* Genoways and Choate, 1967.

COMMENTS: *C. mexicanus* group, *goldmani* subset. Formerly a member of *goldmani*; revised by Woodman and Timm (1999).

*Cryptotis brachyonyx* Woodman, 2003. Proc. Biol. Soc. Wash., 116:855.

COMMON NAME: Eastern Cordillera Small-footed Shrew.

TYPE LOCALITY: Colombia, Cundinamarca, La Selva, near Bogotá, 8900 ft. (2740 m).

DISTRIBUTION: C Eastern Cordillera of Colombia.

COMMENTS: *C. nigrescens* group; most similar to *C. colombiana* (Woodman, 2003).

*Cryptotis colombiana* Woodman and Timm, 1993. Fieldiana: Zoology, n.s., 74:24.

COMMON NAME: Colombian Small-eared Shrew.

TYPE LOCALITY: Colombia, Central Cordillera, Antioquia Dept., Río Negrito; 15 km E of Sonsón, 1750 m.

DISTRIBUTION: Colombia, Central Cordillera.

COMMENTS: *C. nigrescens* group (Woodman and Timm, 1993). Woodman (1996) reported a second specimen from the Cordillera Oriental that he later described as a new species related to *C. colombiana* (Woodman, 2003); see under *C. brachyonyx*.

*Cryptotis endersi* Setzer, 1950. J. Wash. Acad. Sci., 40:300.

COMMON NAME: Enders's Small-eared Shrew.

TYPE LOCALITY: Panama, Cylindro.

DISTRIBUTION: Highlands of W Panama.

STATUS: IUCN – Endangered.

COMMENTS: Choate (1970) recognized this taxon as a relict species known only from the type locality. In 1980, a second specimen was collected 70 km further east (Pine et al., 2002).

*Cryptotis equatoris* (Thomas, 1912). Ann. Mag. Nat. Hist., ser. 8, 9:409.
COMMON NAME: Ecuadorean Small-eared Shrew.
TYPE LOCALITY: Ecuador, Sinchig, Guaranda, 4000 m.
DISTRIBUTION: Western Andes of Ecuador.
SYNONYMS: *osgoodi* Stone, 1914.
COMMENTS: *C. thomasi* group. Revised by Vivar et al. (1997) who listed *osgoodi* as a
subspecies.

*Cryptotis goldmani* (Merriam, 1895). N. Am. Fauna, 10:25.
COMMON NAME: Goldman's Broad-clawed Shrew.
TYPE LOCALITY: Mexico, Guerrero, mountains near Chilpancingo, 10,000 ft. (3505 m).
DISTRIBUTION: Mexico; highlands above 1500 m in Oaxaca and Guerrero.
STATUS: IUCN – Lower Risk (lc).
SYNONYMS: *guerrerensis* Jackson, 1933; **machetes** Merriam, 1895; *fossor* Merriam, 1895;
*frontalis* Miller, 1911.
COMMENTS: *C. mexicana* group, *goldmani* subset. Choate (1970) recognized two distinct
subspecies, *alticola* and *goldmani*; the former is now considered a species. Revised by
Woodman and Timm (1999), and Carraway (ms), who recognized the Oaxaca and
Guerrero populations as distinct subspecies.

*Cryptotis goodwini* Jackson, 1933. Proc. Biol. Soc. Wash., 46:81.
COMMON NAME: Goodwin's Broad-clawed Shrew.
TYPE LOCALITY: Guatemala, Calel, 10,200 ft. (3108 m).
DISTRIBUTION: Highlands above 1100 m in S Mexico (Chiapas) and S Guatemala;
N El Salvador and W Honduras.
STATUS: IUCN – Lower Risk (lc).
SYNONYMS: *magnimana* Woodman and Timm, 1999.
COMMENTS: *C. mexicana* group, *goldmani* subset. Reviewed by Choate and Fleharty (1974,
Mammalian Species No. 44) and by Woodman and Timm (1999), who described
*magnimana* as a subspecies from Honduras. Woodman (pers. comm., 2003) considered
this name as a synonym of *goodwini*.

*Cryptotis gracilis* Miller, 1911. Proc. Biol. Soc. Wash., 24:221.
COMMON NAME: Talamancan Small-eared Shrew.
TYPE LOCALITY: Costa Rica, Talamanca [= Limón], near base of Pico Blanco, head of Lari
River, altitude about 6000 ft. (1800 m).
DISTRIBUTION: SE Costa Rica and W Panama.
STATUS: IUCN – Vulnerable.
SYNONYMS: *jacksoni* Goodwin, 1944.
COMMENTS: Considered a relict species by Choate (1970). Specimens from Honduras
previously included in *gracilis* were described as a new species, *C. hondurensis*, by
Woodman and Timm (1992).

*Cryptotis griseoventris* Jackson, 1933. Proc. Biol. Soc. Wash., 46:80.
COMMON NAME: Guatemalan Broad-clawed Shrew.
TYPE LOCALITY: Mexico, Chiapas, San Cristóbal de las Casas, 9500 ft. (2900 m).
DISTRIBUTION: Highlands above 2000 m in S Mexico (Chiapas) and Guatemala.
COMMENTS: *C. mexicana* group, *goldmani* subset (Woodman and Timm, 1999). Formerly
included in *goldmani* by Choate (1970), but given species rank by Woodman and
Timm (1999).

*Cryptotis hondurensis* Woodman and Timm, 1992. Proc. Biol. Soc. Wash., 105:2.
COMMON NAME: Honduran Small-eared Shrew.
TYPE LOCALITY: Honduras, Francisco Morazán Department, 12 km WNW of El Zamorano,
W slope of Cerro Oyuca [ca. 14°05′N, 87°06′W], 1680 m.
DISTRIBUTION: Pine, mixed pine, and oak forests on highlands east of Tegucicalpa,
Honduras; possibly also in adjacent regions of Guatemala, El Salvador, and Nicaragua.
STATUS: IUCN – Vulnerable.
COMMENTS: *C. nigrescens* group. Formerly included in *gracilis*, see comments therein.

*Cryptotis magna* (Merriam, 1895). N. Am. Fauna, 10:28.
COMMON NAME: Big Mexican Small-eared Shrew.
TYPE LOCALITY: Mexico, Oaxaca, Totontepec, 6800 ft. (2000 m).
DISTRIBUTION: NC Oaxaca (Mexico) from ca. 1500 to 2500 m elevation (Carraway, ms).
STATUS: IUCN – Lower Risk (lc).
COMMENTS: Reviewed by Robertson and Rickart (1975, Mammalian Species No. 61). A relict
    species, according to Choate (1970).

*Cryptotis mayensis* (Merriam, 1901). Proc. Wash. Acad. Sci., 3:559.
COMMON NAME: Yucatan Small-eared Shrew.
TYPE LOCALITY: Mexico, Yucatán, Chichén Itzá (from a Maya ruin).
DISTRIBUTION: Yucatan Peninsula of Mexico and adjacent Belize and Guatemala. Also
    known from owl pellets collected in Guerrero.
COMMENTS: *C. nigrescens* group. Species redefined by Woodman and Timm (1993). Also
    recorded from the Pleistocene of the Yucatan Peninsula, Mexico (Woodman, 1995).

*Cryptotis medellinia* Thomas, 1921. Ann. Mag. Nat. Hist., ser. 9, 8 :354.
COMMON NAME: Medellín Small-eared Shrew.
TYPE LOCALITY: Colombia, Antioquia Dept., 30 km N Medellín, San Pedro.
DISTRIBUTION: Northern portions of the Central and Western Cordilleras, Colombia.
COMMENTS: *C. thomasi* group. Revised by Vivar et al. (1997) and Woodman (2002).

*Cryptotis mera* Goldman, 1912. Smiths. Misc. Coll., 60(2):17.
COMMON NAME: Darién Small-eared Shrew.
TYPE LOCALITY: Panama, Darién Province, Cerro Pirre, near head of Río Limón, 4500 ft.
    (1400 m).
DISTRIBUTION: Highlands along the Panama-Colombia border.
SYNONYMS: *merus* Goldman, 1912.
COMMENTS: *C. nigrescens* group. Revised by Woodman and Timm (1993).

*Cryptotis meridensis* Thomas, 1898. Ann. Mag. Nat. Hist., ser. 7, 1:457.
COMMON NAME: Merida Small-eared Shrew.
TYPE LOCALITY: Venezuela, Merida, alt. 2165 m.
DISTRIBUTION: Venezuela; cloud forest and páramo in the Cordillera de los Andes in Trujillo,
    Mérida, and E Táchira; probably also mountains near Caracas, Venezuela, see Tello
    (1979).
STATUS: IUCN – Lower Risk (lc).
COMMENTS: *C. thomasi* group. This species was commonly included in *thomasi* (Handley,
    1976; Eisenberg, 1989) but is much làrger and has a more robust dentition. Choate
    (pers. comm., 1983) and Hutterer (1986d) therefore considered *meridensis* a valid
    species, an action followed by Vivar et al. (1997) and Woodman (1996). Specimens
    from Páramo de Tamá previously referred to *meridensis* have been described as a new
    species, *C. tamensis*, by Woodman (2002). A specimen from coastal highlands near
    Caracas reported by Ojasti and Mondolfi (1968) will require a detailed study.

*Cryptotis merriami* Choate, 1970. Univ. Kansas Publ. Mus. Nat. Hist., 19:277.
COMMON NAME: Merriam's Small-eared Shrew.
TYPE LOCALITY: Guatemala, Huehuetenango, Jacaltenango, 5400 ft. (1646 m).
DISTRIBUTION: Highlands of S Mexico (Chiapas), Guatemala, Honduras, El Salvador,
    N Nicaragua, and N Costa Rica.
COMMENTS: *C. nigrescens* group. Formerly included in *nigrescens*, but redefined by Woodman
    and Timm (1993). Occurs syntopically with *nigrescens* in Costa Rica (Woodman, 2000).

*Cryptotis mexicana* (Coues, 1877). Bull. U.S. Geol. Geogr. Surv. Terr., 3:652.
COMMON NAME: Mexican Small-eared Shrew.
TYPE LOCALITY: Mexico, Veracruz, Jalapa, ca. 1520 m.
DISTRIBUTION: Humid upper tropical zone in Chiapas, Oaxaca, Puebla, and Veracruz
    (Mexico); altitudinal range 520 to 2600 m (Carraway, ms; Fa, 1989).
STATUS: IUCN – Lower Risk (lc).
COMMENTS: *C. mexicana* group. Reviewed by Choate (1970; 1973, Mammalian Species No.

28), who recognized four subspecies, *mexicana*, *nelsoni*, *obscura*, and *peregrina*; the latter three were regarded as separate species by Woodman and Timm (1999).

*Cryptotis montivaga* (Anthony, 1921). Am. Mus. Novit., 20:5.
  COMMON NAME: Wandering Small-eared Shrew.
  TYPE LOCALITY: Ecuador, Prov. del Azuay, Bestion, 10,000 ft. (3049 m).
  DISTRIBUTION: Andean zone of S Ecuador.
  STATUS: IUCN – Lower Risk (lc).
  COMMENTS: *C. thomasi* group. Revised by Vivar et al. (1997), who called it *montivagus*.

*Cryptotis nelsoni* (Merriam, 1895). N. Amer. Fauna, 10:26.
  COMMON NAME: Nelson's Small-eared Shrew.
  TYPE LOCALITY: Mexico, Veracruz, Volcán Tuxtla, 4800 ft. (1460 m).
  DISTRIBUTION: Known only from type locality.
  COMMENTS: *C. mexicana* group. Included in *mexicana* by Choate (1970) and Hall (1981), but given species rank by Woodman and Timm (1999).

*Cryptotis nigrescens* (J. A. Allen, 1895). Bull. Am. Mus. Nat. Hist., 7:339.
  COMMON NAME: Blackish Small-eared Shrew.
  TYPE LOCALITY: Costa Rica, San José Province, San Isidro.
  DISTRIBUTION: Highlands above 800 m in Costa Rica and W Panama.
  STATUS: IUCN – Lower Risk (lc).
  SYNONYMS: *tersus* Goodwin, 1954; *zeteki* Setzer, 1950.
  COMMENTS: *C. nigrescens* group. Choate (1970) included *mayensis* and *merriami* as subspecies, but see Woodman and Timm (1993) who revised the *nigrescens* group.

*Cryptotis obscura* (Merriam, 1895). N. Amer. Fauna, 10:23.
  COMMON NAME: Grizzled Mexican Small-eared Shrew.
  TYPE LOCALITY: Mexico, Hidalgo, Tulancingo, 8500 ft. (2600 m).
  DISTRIBUTION: Mexican highlands from 1040 to 2500 m in Hidalgo, México, Querétaro, San Luis Potosí, Tamaulipas, and Veracruz (Carraway, ms).
  SYNONYMS: *madrea* Goodwin, 1954.
  COMMENTS: *C. mexicana* group. Included in *mexicana* by Choate (1970) and Hall (1981), but given species rank by Woodman and Timm (1999).

*Cryptotis orophila* (J. A. Allen, 1895). Bull. Amer. Mus. Nat. Hist., 7:340.
  COMMON NAME: Central American Least Shrew.
  TYPE LOCALITY: Costa Rica, Cartago Prov., Volcán Irazú.
  DISTRIBUTION: Highlands and mid-elevations from Honduras and El Salvador south to C Costa Rica (Woodman, pers. comm., 2003).
  SYNONYMS: *olivaceus* J. A. Allen, 1908.
  COMMENTS: *C. parva* group. Until recently a subspecies of *parva*, but given species rank by Woodman (2002, and pers. comm.).

*Cryptotis parva* (Say, 1823). *In* James, Account Exped. Pittsburgh to Rocky Mtns, 1:163.
  COMMON NAME: North American Least Shrew.
  TYPE LOCALITY: "Engineer Cantonment," west bank of Missouri River; restricted by Jones (1964:68) to USA, Nebraska, Washington Co., approximately 2 mi. (3.2 km) east Ft. Calhoun.
  DISTRIBUTION: Extreme SE Canada through EC and SW USA and Mexico S to Chiapas and W to Nayarit.
  STATUS: IUCN – Lower Risk (lc).
  SYNONYMS: *cinereus* Bachmann, 1837 [not Kerr, 1792]; *elasson* Bole and Moulthrop, 1942; *exilipes* Baird, 1858; *eximius* Baird, 1858; *harlani* Duvernoy, 1842; *parvus* Say, 1823; **berlandieri** Baird, 1858; *macer* Miller, 1911; *nayaritensis* Jackson, 1933; *pergracilis* Elliot, 1903; **floridana** Merriam, 1895; **pueblensis** Jackson, 1933; *celatus* Goodwin, 1956; **soricina** Merriam, 1895.
  COMMENTS: Reviewed by Whitaker (1974, Mammalian Species No. 43), who recognized 9 subspecies, 5 of which occur in Middle America (Choate, 1970). Woodman (2002) concluded that *orophila* is a diagnosable species; see under that name. The remaining

taxa are tentatively assigned to the five subspecies noted here (Carraway, pers. comm., 2003; Woodman, pers. comm., 2002). Some of these taxa may in fact represent good species. Karyotypes from Missouri and Texas count 2n = 52, FN = 54 (Genoways et al., 1977).

*Cryptotis peregrina* (Merriam, 1895). N. Amer. Fauna, 10:24.
COMMON NAME: Oaxacan Broad-clawed Shrew.
TYPE LOCALITY: Mexico, Oaxaca, Mtns near 15 mi. (24 km) southwest of Oaxaca and just pass Santa Inéz, 9500 ft. (2900 m). Further specified by Woodman and Timm (2000).
DISTRIBUTION: Mexico, Oaxaca, Sierra de Cuatro Venados and Sierra Yucuyacua.
COMMENTS: *C. mexicana* group, *goldmani* subset. Formerly included in *mexicana* (see Choate, 1970), but raised to species level by Woodman and Timm (1999). Further restricted and defined by Woodman and Timm (2000).

*Cryptotis peruviensis* Vivar, Pacheco and Valqui, 1997. Am. Mus. Novit., 3202:7.
COMMON NAME: Peruvian Small-eared Shrew.
TYPE LOCALITY: Peru, Department Cajamarca, Las Ashitas, 3150 m, about 42 km W Jaén (05°42′S, 79°08′W).
DISTRIBUTION: Only known from elfin forest in N Peru, E and W Andes, at 3150 and 2050 m.
COMMENTS: *C. thomasi* group. The populations of *peruviensis* document the southernmost occurrence of *Cryptotis* in South America (Vivar et al., 1997).

*Cryptotis phillipsii* (Schaldach, 1966). Säugetierkdl. Mitt., 4:289.
COMMON NAME: Phillips' Small-eared Shrew.
TYPE LOCALITY: Mexico, S Oaxaca, 3 km SW San Miguel Suchixtepec, 2250 m, Río Molino.
DISTRIBUTION: Mexico, S Oaxaca, Sierra de Miahuatlán.
COMMENTS: *C. mexicana* group. Described as *Notiosorex (Xenosorex) phillipsii* but later included in *Cryptotis mexicana*; see Choate (1969). Recently resurrected by Woodman and Timm (2000).

*Cryptotis squamipes* (J. A. Allen, 1912). Bull. Am. Mus. Nat. Hist., 31:93.
COMMON NAME: Western Colombian Small-eared Shrew.
TYPE LOCALITY: Colombia, Cauca, 40 mi. (64 km) west of Popayan, crest of Western Andes, 10,340 ft. (3150 m).
DISTRIBUTION: S Cordillera Occidental of Colombia and Ecuador.
STATUS: IUCN – Lower Risk (lc).
COMMENTS: *C. thomasi* group.

*Cryptotis tamensis* Woodman, 2002. Proc. Biol. Soc. Wash., 115:254.
COMMON NAME: Tamá Small-eared Shrew.
TYPE LOCALITY: Venezuela, Táchira, Buena Vista, 07°27′N, 72°26′W, 2415 m.
DISTRIBUTION: Montane forest and pasture margins in the Tamá highlands, W Venezuela, and adjacent highlands in Colombia between 2385 and 3329 m.
COMMENTS: *C. thomasi* group; most similar to *meridensis* and *thomasi* (Woodman, 2002).

*Cryptotis thomasi* (Merriam, 1897). Proc. Biol. Soc. Wash., 11:227.
COMMON NAME: Thomas' Small-eared Shrew.
TYPE LOCALITY: Colombia, Plains of Bogota, near city of Bogota, 9000 ft. (2740 m), "on G. O. Child's estate".
DISTRIBUTION: Colombia, highlands above 2700 m around Bogotá in the Eastern Cordillera.
STATUS: IUCN – Lower Risk (lc).
SYNONYMS: *avia* G. M. Allen, 1923 as *C. thomasi* and *C. avia*.
COMMENTS: Woodman (1996) redefined the species and included *avia* as a synonym.

*Cryptotis tropicalis* (Merriam, 1895). N. Amer. Fauna, 10:21.
COMMON NAME: Tropical Small-eared Shrew.
TYPE LOCALITY: Guatemala, Cobán.
DISTRIBUTION: Eastern highlands of Chiapas (Mexico) east and south into highlands of Belize and Guatemala (Carraway, ms; Choate, 1970).
SYNONYMS: *micrurus* (Tomes, 1861) [not Pomel, 1848]; *tropicalis* (Gray, 1843) [*nomen nudum*].
COMMENTS: *C. parva* group. Until recently a subspecies of *parva*, but species rank suggested

by Carraway (ms) and Woodman (pers. comm., 2003). Relations to *C. orophila* remain to be studied.

**Tribe Nectogalini** Anderson, 1879. Anat. and zool. Res., 1:149.
SYNONYMS: Crossopinae A. Milne-Edwards, 1872; Hydrosoridae Anonymous, 1838; Neomyini Matschie, 1909; Soriculi Winge, 1917; Soriculini Kretzoi, 1965.
COMMENTS: Better known as Neomyini (Repenning, 1967), but Nectogalini has priority, as pointed out by McKenna and Bell (1997).

*Chimarrogale* Anderson, 1877. J. Asiat. Soc. Bengal, 46:262.
TYPE SPECIES: *Crossopus himalayicus* Gray, 1842.
SYNONYMS: *Chimmarogale* Ellermann and Morrison-Scott, 1951; *Crossogale* Thomas, 1921.
COMMENTS: Formerly in tribe Neomyini; see Repenning (1967:61). Because of the presence of white teeth the genus was occasionally included in subfamily Crocidurinae, but since Repenning (1967), overwhelming evidence has accumulated showing that *Chimarrogale* is a soricine shrew (Mori et al., 1991; Vogel and Besancon, 1979). Gureev (1971:226) included *Chimarrogale* in his subtribe Nectogalina within the Blarinini, while Reumer (1984:14) included it in the tribe Soriculini; see comments under genus *Neomys*. Includes *Crossogale*; see Harrison (1958), who also revised the genus. His arrangement was found to be more realistic than the common practice of lumping all forms together in one or two species. Hutterer (1993a) recognized the six species listed below.

*Chimarrogale hantu* Harrison, 1958. Ann. Mag. Nat. Hist., ser. 13, 1:282.
COMMON NAME: Malayan Water Shrew.
TYPE LOCALITY: "banks of a stream at low altitude (under 1,000 ft. [305 m]) in the Ulu Langat Forest Reserve, Selangor, Malaya, about 20 km. east of Kuala Lumpur."
DISTRIBUTION: Tropical forest of the Malaysian peninsula.
STATUS: IUCN – Critically Endangered.
COMMENTS: Included in *himalayica* by Medway (1977) and other authors but retained by Jones and Mumford (1971). The species differs considerably in its morphology and ecology from the species that inhabit the Himalayan region. The photograph of a live animal in Nowak (1991:156) depicts this species.

*Chimarrogale himalayica* (Gray, 1842). Ann. Mag. Nat. Hist., [ser. 1], 10:261.
COMMON NAME: Himalayan Water Shrew.
TYPE LOCALITY: "India", Punjab, Chamba.
DISTRIBUTION: Kashmir through SE Asia to Indochina; C and S China; Taiwan.
SYNONYMS: *himalayicus* (Gray, 1842); *leander* Thomas, 1902; *varennei* Thomas, 1927.
COMMENTS: Corbet (1978c) included *leander*, *platycephala*, *varennei*, and probably *hantu* in *himalayica*. Gureev (1979) listed *leander*, *hantu*, *platycephala*, and *varennei* as distinct species without comment; both views are only partially accepted here. Species reviewed by Jones and Mumford (1971) and Hoffmann (1987).

*Chimarrogale phaeura* Thomas, 1898. Ann. Mag. Nat. Hist., ser. 7, 2:246.
COMMON NAME: Bornean Water Shrew.
TYPE LOCALITY: Malaysia, Sabah, "Saiap, Mount Kina Balu".
DISTRIBUTION: Streams in tropical forest of island of Borneo (Malaysia and Indonesia).
STATUS: IUCN – Endangered.
COMMENTS: Medway (1977) considered *phaeura* as a subspecies of *himalayica* but Corbet (1978c) and Jones and Mumford (1971) maintained *styani* and *phaeura* as separate species. Ellerman and Morrison-Scott (1966:87) included *sumatrana* in this species, but Gureev (1979:458) listed it as a distinct species, a view followed by Hutterer (1993a).

*Chimarrogale platycephalus* (Temminck, 1842). Fauna Japon., 1(Mamm.), p. 23, pl. V, fig. 1.
COMMON NAME: Japanese Water Shrew.
TYPE LOCALITY: Japan, Kyushu, near Nagasaki and Bungo.
DISTRIBUTION: Most of the Japanese Isls.
STATUS: IUCN – Lower Risk (lc).
COMMENTS: Included in *himalayica* since Ellerman and Morrison-Scott (1951), but retained

as a separate species by Harrison (1958), Hutterer and Hürter (1981), Hoffmann (1987), and Corbet and Hill (1991). Arai et al. (1985) reported on clinal size variation in Japan. For date of publication see Holthuis and Sakai (1970). The species name is often spelled *platycephala* (as such in the second edition, 1993*a*); however, *platycephalus* is to be regarded as a noun, in which case the specific name does not change with the gender of the genus (C. Smeenk, in litt., 2002). Detailed observations on distribution and habitat provided by Abe (2003). Karyotype has 2n = 52, FN = 104 (Obara and Tada, 1985).

*Chimarrogale styani* de Winton, 1899. Proc. Zool. Soc. Lond., 1899:574.
COMMON NAME: Chinese Water Shrew.
TYPE LOCALITY: China, "Yangl-iu-pa, N.W. Sechuen [= Sichuan]."
DISTRIBUTION: Shensi and Sichuan (China), and N Burma.
STATUS: IUCN – Lower Risk (lc).
COMMENTS: Certainly a distinct species, and regarded as such by Jones and Mumford (1971), Corbet (1978*c*), and Hoffmann (1987). Occurs nearly sympatrically with *himalayica* in N Burma.

*Chimarrogale sumatrana* (Thomas, 1921). Ann. Mag. Nat. Hist., ser. 9, 7:244.
COMMON NAME: Sumatran Water Shrew.
TYPE LOCALITY: Indonesia, Sumatra, "Pager Alam, Padang Highlands".
DISTRIBUTION: Streams in tropical forest of Sumatra.
STATUS: IUCN – Critically Endangered.
COMMENTS: Regarded as a race of *phaeura* by Ellerman and Morrison-Scott (1966:87), but considered distinct by Harrison (1958) and Gureev (1979), a view supported by its cranial anatomy.

*Chodsigoa* Kastchenko, 1907. Ann. Mus. Zool. Acad. St. Pétersb., 10:251.
TYPE SPECIES: *Soriculus salenskii* Kastchenko, 1907 (by subsequent designation of Allen, 1938:104).
COMMENTS: Often included in *Soriculus* as a subgenus (Hoffmann, 1985*b*). Hutterer (1994*b*) and Motokawa (1997*b*, 1998) proposed generic status for *Chodsigoa* and *Episoriculus*.

*Chodsigoa caovansunga* Lunde, Musser and Son, 2003. Mammal Study, 28:37.
COMMON NAME: Van Sung's Shrew.
TYPE LOCALITY: Vietnam, Ha Giang province, Vi Xuyen distr., Cao Bo Commune, Mt. Tay Con Linh II, 1500 m (22°45'27"N, 104°49'49"E).
DISTRIBUTION: Known only from the type locality.
COMMENTS: Species contrasted against *parca* and *parva* by Lunde et al. (2003*b*).

*Chodsigoa hypsibia* (de Winton, 1899). Proc. Zool. Soc. Lond., 1899:574.
COMMON NAME: De Winton's Shrew.
TYPE LOCALITY: China, Sichuan, "Yang-liu-pa".
DISTRIBUTION: SW and C China, Yunnan, Sichuan and Shaanxi; apparently disjunct population (*larvarum*) in Hebei.
STATUS: IUCN – Lower Risk (lc) as *Soriculus hypsibius*.
SYNONYMS: *beresowskii* (Kastchenko, 1907); *larvarum* Thomas, 1911.
COMMENTS: Does not include *parva* and *lamula*; see Hoffmann (1985*b*).

*Chodsigoa lamula* Thomas, 1912. Ann. Mag. Nat. Hist., ser. 8, 10:399.
COMMON NAME: Lamulate Shrew.
TYPE LOCALITY: China, Gansu, "40 miles [64 km] S.E. of Tao-chou [Lintan]. Alt. 9500' [2896 m]".
DISTRIBUTION: C China, from Yunnan, Sichuan, and Gansu to Fujian.
STATUS: IUCN – Lower Risk (lc) as *Soriculus lamula*.
COMMENTS: Hoffmann (1985) included *parva* in *C. lamula*, but Lunde et al. (2003*b*) retained *parva* as a distinct species. Formerly in *hypsibia* but occurs sympatrically with that species; see Hoffmann (1985*b*).

*Chodsigoa parca* G. M. Allen, 1923. Am. Mus. Novit., 100:6.
COMMON NAME: Lowe's Shrew.
TYPE LOCALITY: "Ho-mu-shu Pass, Western Yunnan, China, 8000 feet [2438 m]".
DISTRIBUTION: SW China, N Burma, Thailand and N Vietnam.
STATUS: IUCN – Lower Risk (lc) as *Soriculus parca*.
SYNONYMS: **furva** Anthony, 1941; **lowei** Osgood, 1932.
COMMENTS: Formerly included in *smithii*, but retained as a separate species by Hoffmann (1985*b*), with *lowei* and *furva* as tentative subspecies.

*Chodsigoa parva* G. M. Allen, 1923. Am. Mus. Novit., 100:5.
COMMON NAME: Pygmy Brown-toothed Shrew.
TYPE LOCALITY: China, W Yunnan, Likiang Range, Ssushancheng.
DISTRIBUTION: Known only from the type locality.
COMMENTS: Hoffmann (1985*b*) synonymized *C. parva* with *C. lamula* and was followed by Corbet and Hill (1992), but Lunde et al. (2003*b*) demonstrated that *C. parva* is much smaller than *C. lamula* and represents a good species known only by the type series, a conclusion already drawn by Allen (1938).

*Chodsigoa salenskii* (Kastschenko, 1907). Ann. Mus. Zool. Acad. Sci. St. Petersbourg, 10:253.
COMMON NAME: Salenski's Shrew.
TYPE LOCALITY: China, Sichuan, "Lun-ngan'-fu" (= Liangfu).
DISTRIBUTION: Known only from the type locality in N Sichuan.
STATUS: IUCN – Critically Endangered as *Soriculus salenskii*.
COMMENTS: Related to *C. smithii*.

*Chodsigoa smithii* Thomas, 1911. Abstr. Proc. Zool. Soc. Lond., 1911(90):4.
COMMON NAME: Smith's Shrew.
TYPE LOCALITY: China, Sichuan, "Ta-tsien-lu".
DISTRIBUTION: C Sichuan to W Shaanxi (China).
STATUS: IUCN – Lower Risk (lc) as *Soriculus smithii*.
COMMENTS: Regarded as a subspecies of *salenskii* by Ellerman and Morrison-Scott (1951:60), but as a separate species by Corbet (1978*c*:24). Formerly included *parca* and *furva*; but see Hoffmann (1985*b*).

*Chodsigoa sodalis* Thomas, 1913. Ann. Mag. nat. Hist., 11:217.
COMMON NAME: Lesser Taiwanese Shrew.
TYPE LOCALITY: Central Taiwan, Mt. Arizan, 8,000 ft. (2438 m).
DISTRIBUTION: Mountains of C Taiwan.
COMMENTS: Formerly included as a synonym in *Episoriculus fumidus* (Hoffmann, 1984), but Yu (1994) and Motokawa et al. (1997*b*, 1998) re-defined *C. sodalis* and described its characteristic karyotype (2n = 44, FN = 88). Kuroda (1935) already correctly distinguished between *Chodsigoa sodalis* and *Episoriculus fumidus*.

*Episoriculus* Ellermann and Morrison-Scott, 1966. Checklist of Palaearctic and Indian Mammals 1758 to 1946, second ed., p. 56.
TYPE SPECIES: *Sorex caudatus* Horsfield, 1851.
COMMENTS: Described as a subgenus of *Soriculus* and often included in that genus (Hoffmann, 1985*b*). Hutterer (1994*b*) presented evidence that *Soriculus*, *Episoriculus* and *Chodsigoa* are not closely related to each other, and that Pleistocene and Pliocene shrews from Europe referred to *Episoriculus* by many authors (e.g., Rzebik-Kowalska, 1981) belong to the extinct genus *Asoriculus*, a conclusion widely accepted by paleontologists (Reumer, 1998; Rzebik-Kowalska, 2002).

*Episoriculus caudatus* (Horsfield, 1851). Cat. Mamm. Mus. E. India Co., p. 135.
COMMON NAME: Hodgsons's Brown-toothed Shrew.
TYPE LOCALITY: "Sikkim", no exact locality.
DISTRIBUTION: Kashmir to N Burma and SW China.
STATUS: IUCN – Lower Risk (lc) as *Soriculus caudatus*.
SYNONYMS: *gracilicauda* (Anderson, 1877); *homourus* (Gray, 1863) [*nomen nudum*]; **sacratus** (Thomas, 1911); *soluensis* (Gruber, 1969); **umbrinus** (G. Allen, 1923).

COMMENTS: Includes *sacratus* and *umbrinus* as subspecies; see Gruber (1969) and Hoffmann (1985*b*).

*Episoriculus fumidus* (Thomas, 1913). Ann. Mag. Nat. Hist., ser. 8, 11:216.
COMMON NAME: Taiwanese Brown-toothed Shrew.
TYPE LOCALITY: Taiwan, Chiai Hsien, "Mt. Arisan (= Alishan); Central Formosa. Alt. 8,000' [2438 m]".
DISTRIBUTION: Montane forests of Taiwan.
STATUS: IUCN – Lower Risk (lc) as *Soriculus fumidus*.
COMMENTS: Formerly included in *caudatus* by Ellerman and Morrison-Scott (1951:59), but see Jameson and Jones (1977:474) and Hoffmann (1985*b*), who included *sodalis* (now in *Chodsigoa*) in *fumidus*. Motokawa et al. (1998) described the karyotype (2n = 64, FN = 116) and distribution of *Episoriculus fumidus* that occurs symatrically with *Chodsigoa sodalis* in Taiwan. A RNA sequence was published by Querouil et al. (2001) and Corneli (2002) under the name *Soriculus fumidus*.

*Episoriculus leucops* (Horsfield, 1855). Ann. Mag. Nat. Hist., ser. 2, 16:111.
COMMON NAME: Long-tailed Brown-toothed Shrew.
TYPE LOCALITY: Nepal.
DISTRIBUTION: C Nepal, Sikkim and Assam to S China, N Burma and N Vietnam.
STATUS: IUCN – Lower Risk (lc) as *Soriculus leucops*.
SYNONYMS: **baileyi** (Thomas, 1914); *gruberi* (Weigel, 1969).
COMMENTS: Includes *baileyi* as a subspecies; see Hoffmann (1985*b*).

*Episoriculus macrurus* (Blanford, 1888). Fauna Brit. India, 1:231.
COMMON NAME: Long-tailed Mountain Shrew.
TYPE LOCALITY: India, Darjeeling.
DISTRIBUTION: C Nepal to W and S China and to N Burma and Vietnam.
STATUS: IUCN – Lower Risk (lc) as *Soriculus macrurus*.
SYNONYMS: *irene* (Thomas, 1911).
COMMENTS: Formerly confused with *leucops*, but shown to be a distinct species by Hoffmann (1985*b*).

*Nectogale* Milne-Edwards, 1870. C.R. Acad. Sci. Paris, 70:341.
TYPE SPECIES: *Nectogale elegans* Milne-Edwards, 1870.
COMMENTS: For placement in Soricinae see Vogel and Besancon (1979); formerly in tribe Neomyini; see Repenning (1967:45). Gureev (1971:226) placed *Nectogale* in a subtribe Nectogalina, but all authors prior to McKenna and Bell (1997) overlooked the fact that Nectogalini Anderson, 1879 was available.

*Nectogale elegans* Milne-Edwards, 1870. C.R. Acad. Sci. Paris, 70:341.
COMMON NAME: Elegant Water Shrew.
TYPE LOCALITY: China, Sichuan, Moupin (= Baoxing).
DISTRIBUTION: Cold mountain streams across the Himalayas and in W and C China; Tibet (Xizang Aut. Region), Nepal, Sikkim (India), Bhutan, N Burma, and Yunnan, Sichuan and Shaanxi (China).
STATUS: IUCN – Lower Risk (lc).
SYNONYMS: *sikhimensis* de Winton, 1899.
COMMENTS: A species highly adapted for a semi-aquatic life (Hutterer, 1985). Includes *sikhimensis*, see Ellerman and Morrison-Scott (1951) and Hoffmann (1987). Reviewed by Hutterer (1993*b*).

*Neomys* Kaup, 1829. Skizz. Entwickel.-Gesch. Nat. Syst. Europ. Thierwelt, 1:117.
TYPE SPECIES: *Sorex daubentonii* Erxleben, 1777 (= *Sorex fodiens* Pennant, 1771).
SYNONYMS: *Amphisorex* Duvernoy, 1835; *Crossopus* Wagler, 1832; *Hydrogale* Kaup, 1829 [not Pomel, 1848]; *Hydrosorex* Duvernoy, 1835; *Leucorrhynchus* Kaup, 1829; *Myosictis* Pomel, 1854; *Pinalia* Gray, 1838.
COMMENTS: Type genus of tribe Neomyini Repenning, 1967, for which Reumer (1984:14) used the name Soriculini Kretzoi, 1965. However, both are antedated by Neomyini Matschie,

1909 and Nectogalini Anderson, 1879. Biogeography and phylogeny of the genus reviewed by Kryštufek et al. (2000*a*).

*Neomys anomalus* Cabrera, 1907. Ann. Mag. Nat. Hist., ser. 7, 20:214.

COMMON NAME: Mediterranean Water Shrew.

TYPE LOCALITY: Spain, Madrid Prov., Jarama River, San Martin de la Vega.

DISTRIBUTION: Temperate woodlands of Europe, from Portugal to Poland and east to Voronesh, Russia; N Asia Minor and N Iran.

STATUS: IUCN – Lower Risk (lc).

SYNONYMS: *amphibius* Brehm, 1826; *josti* Martino, 1940; *milleri* Mottaz, 1907; *mokrzeckii* Martino, 1917; *rhenanus* Lehmann, 1976; *soricoides* Ognev, 1922.

COMMENTS: Reviewed by Spitzenberger (1990*b*); European range mapped by Mitchell-Jones et al. (1999). *Sorex amphibius* Brehm, 1826 is probably an earlier name for the species (von Knorre, pers. comm.), although it has to be treated as a *nomen oblitum*. Karyotype from Spain to Turkey uniformly 2n = 52, FN = 98 (Zima et al., 1998).

*Neomys fodiens* (Pennant, 1771). Synopsis Quadrupeds, p. 308.

COMMON NAME: Eurasian Water Shrew.

TYPE LOCALITY: Germany, Berlin.

DISTRIBUTION: Most of Europe including the British Isls and eastwards to Lake Baikal, Yenisei River (Russia), Tien Shan (China), and NW Mongolia; disjunct in Sakhalin Isl and adjacent Siberia, Jilin (China), and N Korea.

STATUS: IUCN – Lower Risk (lc).

SYNONYMS: *albiventris* (de Sélys Longchamps, 1839) [*nomen nudum*]; *albus* (Bechstein, 1800); *alpestris* Burg, 1924 [*nomen nudum*]; *aquaticus* (Müller, 1776) [not of Linnaeus, 1758]; *argenteus* Ognev, 1922; *bicolor* (Shaw, 1791); *brachyotus* Ognev, 1922; *canicularius* (Bechstein, 1800); *carinatus* (Hermann, 1780); *ciliatus* (Sowerby, 1805); *collaris* (Desmarest, 1818); *dagestanicus* Heptner and Formozov, 1928; *daubentonii* (Erxleben, 1777); *eremita* (Meyer, 1793); *fimbriatus* (Fitzinger, 1868); *fluviatilis* (Bechstein, 1793); *griseogularis* (Fitzinger, 1868); *hermanni* (Duvernoy, 1835); *hydrophilus* (Pallas, 1811); *ignotus* Fatio, 1905; *intermedius* (Cornalia, 1870); *intermedius* Brunner, 1952 [not Cornalia, 1870]; *leucotis* (de Sélys Longchamps, 1839) [*nomen nudum*]; *limchunhunii* Won, 1954; *lineatus* (E. Geoffroy, 1811); *linneana* (Gray, 1838); *liricaudatus* (Kerr, 1792); *longobarda* (Sordelli, 1899); *macrourus* (Lehmann, 1822); *minor* Miller, 1901; *musculus* (Wagler, 1832); *naias* Barrett-Hamilton, 1905; *nanus* Lydekker, 1906; *natans* (Brehm, 1826); *niethammeri* Bühler, 1963; *nigricans* (Nilsson, 1845) [*nomen nudum*]; *nigripes* (Melchior, 1834); *orientalis* Hinton, 1915; *orientis* Thomas, 1914; *pennantii* (Gray, 1838); *psilurus* (Wagler, 1832); *remifer* (E. Geoffroy, 1811); *rivalis* (Brehm, 1830); *sowerbyi* (Bonaparte, 1840); *stagnatilis* (Brehm, 1826); *stresemanni* Stein, 1931; *watasei* Kishida, 1930 [*nomen nudum*]; *watasei* Kuroda, 1941.

COMMENTS: Includes *orientis* and *watasei* as possible subspecies (Hoffmann, 1987; Ognev, 1928; Yudin, 1989). Many of the listed synonyms have never been properly studied and identified and a meaningful subspecific division of the species is not possible yet. Recently, Lehmann (1983) referred *constrictus* to *Crocidura russula*. The form *niethammeri* from NE Spain may represent a valid species (Bühler, 1996; López-Fuster et al., 1990). *Neomys newtoni* Hinton, 1911 is regarded as an extinct Pleistocene species; see Rzebik-Kowalska (1998). *Neomys intermedius* Brunner, 1952 was identified as a chronosubspecies of *N. fodiens* by Schaefer (1973); the name, however, is not available. Karyotypes of *N. fodiens* from Sweden to Mongolia were consistantly 2n = 52, FN = 98 (Zima et al., 1998).

*Neomys teres* Miller, 1908. Ann. Mag. Nat. Hist., ser. 8, 1:68.

COMMON NAME: Transcaucasian Water Shrew.

TYPE LOCALITY: NE Turkey, 25 mi. (40 km) N Erzerum, 7,000 ft. (2134 m).

DISTRIBUTION: Caucasus (Armenia, Azerbaijan, Georgia); and adjacent Turkey (Kryštufek and Vohralík, 2001) and Iran (specimen in ZFMK); eastern Palearctic limits unknown.

STATUS: IUCN – Lower Risk (lc) as *N. schelkovnikovi*.

SYNONYMS: *balkaricus* Ognev, 1926; *leptodactylus* Satunin, 1914; *schelkovnikovi* Satunin, 1913.

COMMENTS: Better known as *N. schelkovnikovi* as reviewed by Sokolov and Tembotov (1989).

Kryštufek et al. (1998) reviewed the species in Turkey and showed that *teres* is the corect name for the species. Karyotype (2n = 52, FN = 98) identical to that of the other two species (Graphodatsky et al., 1993).

*Nesiotites* Bate, 1945. Ann. Mag. Nat. Hist., 11, 11:741.
TYPE SPECIES: *Nesiotites hidalgo* Bate, 1945.
COMMENTS: A genus endemic to the Balearic Isls that went extinct during historical times. Vigne (1987) provided a summary of the extinction process. Further information in Vigne and Alcover (1985), and Vigne and Marinval-Vigne (1990, 1991). The morphological separation of *Nesiotites* from the extinct genus *Asoriculus* is weak and requires further study (Masini and Sarà, 1998).

*Nesiotites hidalgo* Bate, 1945. Ann. Mag. Nat. Hist., 11, 11:742.
COMMON NAME: Balearic Shrew.
TYPE LOCALITY: Spain, Balearic Isls, E Mallorca, Cap Ferrutx.
DISTRIBUTION: Mallorca, Menorca, and Menorca (Reumer, 1982).
STATUS: Extinct.
COMMENTS: Vanished in historical times (Reumer, 1980*b*). The morphology of the species was described in detail by Reumer (1980*a*). The epithet *hidalgoi*, as used by Alcover et al. (2000) in the combination *Asoriculus hidalgoi*, is an unjustified emendation.

*Nesiotites similis* (Hensel, 1855). Z. Dtsch. Geol. Ges., 7:459.
COMMON NAME: Sardinian Shrew.
TYPE LOCALITY: Italy, Sardinia, Montereale, near Cagliari.
DISTRIBUTION: Sardinia.
STATUS: Extinct.
COMMENTS: This species probably vanished from Sardinia during the Middle Ages (Vigne, 1992).

*Soriculus* Blyth, 1854. J. Asiatic Soc. Bengal, 23:733.
TYPE SPECIES: *Corsira nigrescens* Gray, 1842.
COMMENTS: Formerly most authors included *Chodsigoa* and *Episoriculus* as subgenera, but both are treated as distinct genera here, following Hutterer (1994*b*). Genus reviewed by Hoffmann (1985*b*).

*Soriculus nigrescens* (Gray, 1842). Ann. Mag. Nat. Hist., [ser. 1], 10:261.
COMMON NAME: Himalayan Shrew.
TYPE LOCALITY: "India", West Bengal, Darjeeling.
DISTRIBUTION: Middle altitudes of the Himalaya from Tibet and Nepal to Bhutan, Assam (India) and SW China.
STATUS: IUCN – Lower Risk (lc).
SYNONYMS: *aterrimus* (Blyth, 1842) [*nomen nudum*]; *caurinus* Hinton, 1922; *centralis* Hinton, 1922; *holosericeus* (Gray, 1863) [*nomen nudum*]; *oligurus* (Gray, 1863) [*nomen nudum*]; *pahari* Hinton, 1922; *sikimensis* (Hodgson, 1849) [*nomen nudum*]; **minor** Dobson, 1890; *radulus* Thomas, 1922 (see Ellerman and Morrison-Scott, 1951).
COMMENTS: The form *radulus* is distinctly smaller (Hoffmann, 1985*b*), however, Motokawa (2003*a*) has shown that *minor* antedates *radulus* and should be used as the valid name for the small form (or species) occurring in Bhutan and Assam.

**Tribe Notiosoricini** Reumer, 1984. Scripta Geol., 73:18.
COMMENTS: Includes also the fossil genera *Hesperosorex* Hibbard, 1957, and *Beckiasorex* Dalquest, 1972. Diagnosis specified by Reumer (1989).

*Megasorex* Hibbard, 1950. Contrib. Mus. Paleontol. Univ. Michigan, 8:129.
TYPE SPECIES: *Notiosorex gigas* Merriam, 1897.
COMMENTS: Repenning (1967) and George (1986) placed the genus in Neomyini, but see Reumer (1998).

*Megasorex gigas* (Merriam, 1897). Proc. Biol. Soc. Wash., 11:227.
    COMMON NAME: Mexican Shrew.
    TYPE LOCALITY: Mexico, Jalisco, near San Sebastián, mountains at Milpillas.
    DISTRIBUTION: Nayarit to Oaxaca (Mexico).
    STATUS: IUCN – Lower Risk (lc).
    COMMENTS: Formerly included in *Notiosorex* (Hall, 1981:65); but Repenning (1967:56) and
        Armstrong and Jones (1972*a*, Mammalian Species No. 16) considered *Megasorex* a
        distinct genus; a view supported by George (1986) on the basis of allozyme data.

*Notiosorex* Coues, 1877. Bull. U.S. Geol. Geogr. Surv. Terr., 3:646.
    TYPE SPECIES: *Sorex* (*Notiosorex*) *crawfordi* Coues, 1877.
    COMMENTS: Repenning (1967:45) placed the genus in Neomyini. Reumer (1984:14) created a
        new tribe Notiosoricini to include *Notiosorex*, but George (1986:160) could find no
        evidence to support this separation. Hall (1981:65) included also *Megasorex gigas*, but
        Repenning (1967:56), Armstrong and Jones (1972), and George (1986) considered
        *Megasorex* a distinct genus. *Notiosorex* (*Xenosorex*) *phillipsii* belongs to *Cryptotis*; see Choate
        (1969) and Woodman and Timm (2000). Lindsay and Jacobs (1985) described an extinct
        species from Pliocene sediments of Chihuahua, Mexico. Extant species revised by
        Carraway and Timm (2000) and Baker et al. (2003*a*).

*Notiosorex cockrumi* Baker, O'Neill, and McAliley, 2003. Occas. Pap., Mus. Texas Tech Univ.,
    no. 222:2.
    COMMON NAME: Cockrum's Gray Shrew.
    TYPE LOCALITY: USA, Arizona, Cochise Co., Leslie Canyon National Wildlife Refuge, T21S,
        R28E Section NW 1/4 20, 4460 ft. (1359 m).
    DISTRIBUTION: Arizona (USA) to C Sonora (Mexico).
    COMMENTS: A cryptic species indentified on the basis of cytochrome *b* gene fragments.
        Carraway and Timm (2000) included this new taxon in *crawfordi* and stated that they
        "found no identifyable morphological differences." Sympatric with *crawfordi* in
        SE Arizona. A karyotype of 2n = 62, FN = 94 reported earlier from Pima County,
        Arizona (Baker and Hsu, 1970), may refer to this species, while 2n= 68 and FN = 102
        may refer to *crawfordi*. Molecular data also indicate that a third species occurs in Baja
        California (Baker et al., 2003*a*).

*Notiosorex crawfordi* (Coues, 1877). Bull. U. S. Geol. Geogr. Surv. Terr., 3:631.
    COMMON NAME: Crawford's Gray Shrew.
    TYPE LOCALITY: USA, Texas, El Paso Co., 2 mi. (3.2 km) above El Paso, "near Fort Bliss, New
        Mexico (Practically El Paso Texas)." (Merriam, 1895*b*:32).
    DISTRIBUTION: SW and SC USA to Baja California and N and C Mexico.
    STATUS: IUCN – Lower Risk (lc).
    COMMENTS: Formerly included *evotis*; but see under that species. Reviewed by Armstrong
        and Jones (1972*b*, Mammalian Species No. 17) and Carraway and Timm (2000).
        Further Mexican records by Vela (1999) and Alvarez and González-Ruíz (2001). In
        Arizona the species occurs in sympatry with *N. cockrumi* (Baker et al., 2003*a*).

*Notiosorex evotis* (Coues, 1877). Bull. U.S. Geol. Geogr. Surv. Terr., 3:652.
    COMMON NAME: Large-eared Gray Shrew.
    TYPE LOCALITY: Mexico, Sinaloa, area of Mazatlan.
    DISTRIBUTION: WC Mexico (Colima, Jalisco, Michoacan, Nayarit, and Sinaloa).
    COMMENTS: Formerly included in *crawfordi* (Armstrong and Jones, 1971*a*), but given species
        rank by Carraway and Timm (2000).

*Notiosorex villai* Carraway and Timm, 2000. Proc. Biol. Soc. Wash., 113:307.
    COMMON NAME: Villa's Gray Shrew.
    TYPE LOCALITY: Mexico, Tamaulipas, Jaumave, 23°34'N, 99°23'W, 2400 ft. (732 m).
    DISTRIBUTION: C mountains of Tamaulipas, Mexico.
    COMMENTS: Differs from *crawfordi* and *evotis* by subtle cranial features (Carraway and Timm,
        2000).

**Tribe Soricini** G. Fischer, 1814. Zoognosia tabulis synopticis illustrata, 3:x.

*Sorex* Linnaeus, 1758. Syst. Nat., 10th ed., 1:53.
>    TYPE SPECIES: *Sorex araneus* Linnaeus, 1758.
>    SYNONYMS: *Amphiosorex* Hall, 1959 [*lapsus*]; *Amphisorex* Duvernoy, 1835; *Asorex* Mezhzherin, 1965; *Atophyrax* Merriam, 1884; *Corsira* Gray, 1838; *Dolgovia* Vorontsov and Kral, 1986 [? *nomen nudum*]; *Eurosorex* Stroganov, 1952; *Fredgia* Vorontsov and Kral, 1986 [? *nomen nudum*]; *Homalurus* Schulze, 1890; *Hydrogale* Pomel, 1848 [not Kaup, 1829]; *Kratochvilia* Vorontsov and Kral, 1986 [? *nomen nudum*]; *Microsorex* Coues, 1877; *Musaraneus* Brisson, 1762; *Neosorex* Baird, 1858; *Ognevia* Heptner and Dolgov, 1967; *Otisorex* De Kay, 1842; *Oxyrhin* Kaup, 1829; *Soricidus* Altobello, 1927; *Stroganovia* Yudin, 1989; *Yudinia* Vorontsov and Kral, 1986 [? *nomen nudum*].
>    COMMENTS: Type genus of Soricidae. The systematic relationships of a large number of Holarctic species were studied by George (1988); her proposals for subgeneric allocation are mainly followed here. Keys and/or reviews are available for the species of various geographical areas: Canada (van Zyll de Jong, 1983*a*); North and Middle America (Carraway, 1990, 1995; Junge and Hoffmann, 1981); China (Hoffmann, 1987); Siberia (Yudin, 1989); and Europe (Niethammer and Krapp, 1990). *Microsorex* was formerly regarded as a full genus, then reduced to a subgenus of *Sorex* by Diersing (1980*b*), and is now regarded as a synonym of subgenus *Otisorex* (see George, 1988). Besides subgenera a number of species groups have been distinguished such as the *araneus-arcticus* group (Hausser et al., 1985; Meylan and Hausser, 1973), the *cinereus* group (van Zyll de Jong, 1991*b*), and the *vagrans* group (Carraway, 1990), the boundaries and contents of which are still highly controversial (see Demboski and Cook, 2003). Old World species of *Sorex* were reviewed by Dannelid (1991*b*) who provided a phylogenetic hypothesis of relationships. Fumagalli et al. (1996, 1999), Ohdachi et al. (1997*a*, 2001), and Demboski and Cook (2001, 2003), among others, provided important genetic studies of the phylogeny and geographical variation of certain species complexes. Zima et al. (1998) reviewed data on chromosomal variation in *Sorex* and suggested species groups that are not always congruent with clusters based on genetical or morphological data.
>
>    Not all species can be allocated to subgenera or species groups. There exist various proposals for species groups based on chromosomes, genetics, or morphology, and not all are congruent. Many species have never been studied properly, and the genus offers a wide and promising field for further research. Not all taxon names can be allocated to species. The enigmatic *Sorex fimbripes* Bachman, 1837 cannot be identified in the absence of the holotype specimen or new material from the type locality (Handley and Varn, 1994). Also the identities of the taxa *Sorex pusillus* Gmelin, 1774; *Sorex dorichurus* Kishida, 1937; *Sorex longiusculus* Kishida, 1928 (? *nomen nudum*); *Sorex longicaudatus* Kishida, 1930 (*nomen nudum*, not Okhotina, 1993); and *Sorex longicaudatus* Yoshikura, 1956 (not Okhotina, 1993) remain unsolved.

*Sorex alaskanus* Merriam, 1900. Proc. Wash. Acad. Sci., 2:18.
>    COMMON NAME: Glacier Bay Water Shrew.
>    TYPE LOCALITY: USA, Alaska, Glacier bay, Point Gustavus.
>    DISTRIBUTION: Known only from the type locality.
>    STATUS: IUCN – Lower Risk (lc).
>    COMMENTS: Subgenus *Otisorex*; *S. vagrans* complex. The species was tentatively included in *palustris* by Junge and Hoffmann (1981) and Harris (in Wilson and Ruff, 1999), but retained as a species by Hall (1981), Jones et al. (1982), and George (1988); a view supported by the skull figures and measurements given by Jackson (1928) and Carraway (1995). Apparently the species has not been collected again since 1899.

*Sorex alpinus* Schinz, 1837. Neue Denkschr. Allgem. Schweiz. Gesell. Naturwiss. Neuchatel, 1:13.
>    COMMON NAME: Alpine Shrew.
>    TYPE LOCALITY: Switzerland, Canton Uri, St. Gotthard Pass.
>    DISTRIBUTION: Montane forests of C Europe; including Pyrenees, Carpathians, Tatra, Sudeten, Harz, and Jura Mtns.
>    STATUS: IUCN – Lower Risk (lc).

SYNONYMS: *hercynicus* Miller, 1909; ? *intermedius* Cornalia, 1870; *longobarda* Sordelli, 1899; *tatricus* Kratochvil and Rosicky, 1952.

COMMENTS: Subgenus *Sorex*, *S. alpinus* group. Type species of subgenus *Homalurus*; see Hutterer (1982*b*). Reviewed by Spitzenberger (1990*a*). No clear subspecific variation; three karyotypic forms (2n = 54-58, FN = 68) have been described (Dannelid 1994; Lukácová et al. 1996). European range shown in Mitchell-Jones et al. (1999). Peripheral populations in the Pyrenees and the Harz Mtns (Gahsche, 1994) now probably extinct.

*Sorex antinorii* Bonaparte, 1840. Iconogr. Faun. Ital., 1:29.

COMMON NAME: Valais Shrew.

TYPE LOCALITY: No exact locality given; restricted to N Italy, Lake Lugano, Porlezza, by Lehmann (1963).

DISTRIBUTION: SE France, S Switzerland, and Italy.

SYNONYMS: *crassicaudatus* Fatio, 1905 [not Hemprich and Ehrenberg, 1834]; *silanus* Lehmann, 1961; *valaicus* Zagorodnyuk and Khazan, 1996 [*nomen nudum*].

COMMENTS: Subgenus *Sorex*, *S. araneus* group. Formerly known as the Valais chromosome race of the common shrew (2n = 24/25, FN = 40). Given species rank, diagnosed and reviewed by Brünner et al. (2002*a*). A contact zone in the Alps between *araneus* and *antinorii* was investigated by Brünner et al. (2002*b*). *Sorex arunchi* may be related, if not conspecific.

*Sorex araneus* Linnaeus, 1758. Syst. Nat., 10th ed., 1:53.

COMMON NAME: Common Shrew.

TYPE LOCALITY: "*in Europe cryptis*"; restricted to Uppsala, Sweden by Thomas (1911*a*:143).

DISTRIBUTION: C, E, and N Europe including the British Isls (with some isolated populations in France, Italy and Spain), east to Siberia.

STATUS: IUCN – Lower Risk (lc).

SYNONYMS: *alticola* Miller, 1901; *bergensis* Miller, 1909; *bohemicus* Stepanek, 1944; *bolkayi* Martino, 1930; *carpathicus* Barrett-Hamilton, 1905; *castaneus* Jenys, 1838; *concinnus* Wagler, 1832; *csikii* Ehik, 1928; *daubentonii* Cuvier, 1829 [not Erxleben, 1777]; *eleonorae* Wettstein, 1927; *grantii* Barrett-Hamilton and Hinton, 1913 [not Okhotina, 1993]; *hermanni* Duvernoy, 1834; *huelleri* Lehmann, 1966; *ignotus* Fatio, 1905; *iochanseni* Ognev, 1933; *labiosus* Jenys, 1839; *macrotrichus* de Sélys Longchamps, 1839; *marchicus* Passarge, 1984; *melanodon* Wagler, 1832; *mollis* Fatio, 1900; *nigra* Fatio, 1869; *novyensis* Schaefer, 1975; *nuda* Fatio, 1869; ? *pallidus* Fitzinger, 1868 [*nomen dubium*]; *personatus* Millet, 1828 [not Geoffroy, 1827]; *petrovi* Martino, 1939; *peucinius* Thomas, 1913; *preussi* (Matschie, 1893); *pulcher* Zalesky, 1937; *pyrenaicus* Miller, 1909; *pyrrhonota* (Jentink, 1910); *quadricaudatus* Kerr, 1792; *rhinolophus* Wagler, 1832; *ryphaeus* Yudin, 1989; *sultanae* Simsek, 1986; ? *surinamensis* Gmelin, 1788; *tetragonurus* Hermann, 1780; *uralensis* Ognev, 1933; *vulgaris* Nilsson, 1847; *wettsteini* Bauer, 1960.

COMMENTS: Subgenus *Sorex*, *S. araneus* group. *S. araneus* is the preferred Palearctic species for studies in ecology and evolution; see Hausser et al. (1990), Hausser (1991), and Wójcik et al. (2002) for reviews. Karyotype variable (2n = 20-33, FN = 40). The species is well known for its Robertsonian chromosome polymorphism (Meylan, 1964) and for the tendency to establish local karyotype races (Hausser et al., 1985; Searle, 1984; Searle and Wójcik, 1998, 2000; Volobouev, 2003; Zima and Král, 1984*b*; Zima et al., 1994). In Switzerland two karyotype races occur that behave like parapatric species (Hausser et al., 1986); see under *S. antinorii*. Includes *Blarina pyrrhonota* Jentink, 1910, a name assigned to *Cryptotis surinamensis* by Cabrera (1958); however, Husson (1963) showed that the locality information was incorrect and that it was based on a *Sorex araneus*. The holotype skin (skull lost) of *Myosorex preussi* Matschie, 1893, formerly thought to represent an endemic species of Mt. Cameroon (Heim de Balsac, 1968*b*), is a *Sorex araneus* and therefore included as a synonym. *Sorex isodon marchicus*, described from E Germany (Passarge, 1984), is included in *araneus* as no clear morphological characters distinguish it from the latter (Turni et al., 2001), and because its karyotype represents the Laska race of the common shrews (Brünner et al., 2002*c*).

*Sorex arcticus* Kerr, 1792. Animal Kingdom, p. 206.
>COMMON NAME: Arctic Shrew.
>TYPE LOCALITY: Canada, Ontario, settlement on Severn River (now Fort Severn), Hudson Bay.
>DISTRIBUTION: Yukon and Northwest Territory to Quebec (Canada); Dakota, Wisconsin, Michigan, and Minnesota (USA).
>STATUS: IUCN – Lower Risk (lc).
>SYNONYMS: *belli* Merriam, 1892 [*nomen nudum*]; *pachyurus* Baird, 1858 [not Küster, 1835]; *richardsonii* Bachman, 1837; *spagnicola* Coues, 1877; **laricorum** Jackson, 1925.
>COMMENTS: Subgenus *Sorex*. *S. arcticus* group. Reviewed by Kirkland and Schmidt (1996, Mammalian Species No. 524). Karyotype has 2n = 28/29, FN = 38. Palearctic species currently referred to *arcticus* (Gromov and Baranova, 1981:18) represent *tundrensis* (Ivanitskaya et al., 1986; Junge et al., 1983); see also Sokolov and Orlov (1980), Okhotina (1983), and Hoffmann (1985*a*). Formerly included *maritimensis*, but see Stewart et al. (2002).

*Sorex arizonae* Diersing and Hoffmeister, 1977. J. Mammal., 58:329.
>COMMON NAME: Arizona Shrew.
>TYPE LOCALITY: USA, "upper end of Miller Canyon, 15 mi S [= 10 mi. (16.1 km) S, 4¾ mi. (7.6 km) E] Fort Huachuca [near spring at lower edge of Douglas fir zone, Huachuca Mts.] Cochise County, Arizona".
>DISTRIBUTION: Disjunct mountains in SE Arizona and SW New Mexico (USA; see Conway and Schmitt, 1978 and Hoffmeister, 1986); one specimen from Sierra Madre Occidental of Chihuahua (Mexico; see Caire et al., 1978).
>STATUS: IUCN – Vulnerable.
>COMMENTS: Refered to unnamed subgenus by George (1988). Close to *emarginatus* (see Diersing and Hoffmeister, 1977). Species reviewed by Simons and Hoffmeister (2003, Mammalian Species No. 732).

*Sorex arunchi* Lapini and Testone, 1998. Gortania, 20:246.
>COMMON NAME: Udine Shrew.
>TYPE LOCALITY: NE Italy, Udine Prov., Muzzana del Turgnano, Bosco Baredi-Selva di Arvonchi.
>DISTRIBUTION: NE Italy, Udine Province, and probably adjacent Slovenia (Lapini and Testone, 1998).
>COMMENTS: Subgenus *Sorex*, *S. araneus* group. Possibly related or conspecific with *Sorex antinorii* (Brünner et al., 2002*a*), although its karyotype has not been studied yet. Genetically well distinguished from *araneus* (Lapini et al., 2001), but morphological differences between both species weak (Breda, 2002).

*Sorex asper* Thomas, 1914. Ann. Mag. Nat. Hist., ser. 8, 13:565.
>COMMON NAME: Tien Shan Shrew.
>TYPE LOCALITY: "Thian-shan [Tien-shan], Tekes Valley". Note on type specimen tag says "Jigalong" (= Dzhergalan?, see Hoffmann, 1987:119); Narynko'skii r-n., Alma-Ata Obl., Kazakhstan.
>DISTRIBUTION: Tien Shan Mountains (Kazakhstan and Sinkiang, China).
>STATUS: IUCN – Lower Risk (lc).
>COMMENTS: Subgenus *Sorex*. *S. tundrensis* group. Karyotype has 2n = 32/33, FN = 58. Type locality discussed by Hoffmann (1987) and Pavlinov and Rossolimo (1987). Does not include *excelsus* as suggested by Corbet (1978*c*); see under that species. Reviewed by Hoffmann (1987), who discussed the relationship between *asper* and *tundrensis*. Genetically, they are not closely related (Fumagalli et al., 1999).

*Sorex averini* Zubko, 1937. Kharkov A. Gorky-State Univ., Proc. Zool.-Biol. Inst., 4:300.
>COMMON NAME: Dneper Common Shrew.
>TYPE LOCALITY: Ukraine, Kherson Region, Golaja Pristan.
>DISTRIBUTION: Recorded from the Lower Dneper Region, Ukraine.
>COMMENTS: Subgenus *Sorex*. Zagorodnyuk (1996*c*) defined this larger form as an allospecies of *S. araneus*. The species differs in larger external and cranial measurements from the allopatric common shrew (Zubko, 1937).

*Sorex bairdi* Merriam, 1895. N. Am. Fauna, 10:77.
COMMON NAME: Baird's Shrew.
TYPE LOCALITY: USA, Oregon, [Clatsop Co.], Astoria.
DISTRIBUTION: NW Oregon (USA).
STATUS: IUCN – Lower Risk (lc).
SYNONYMS: *bairdii* Hennings and Hoffmann, 1977; **permiliensis** Jackson, 1918.
COMMENTS: Subgenus *Otisorex*; *S. vagrans* complex (Carraway, 1990; Demboski and Cook, 2001). This taxon has been alternatively referred to *obscurus*, *vagrans*, and *monticulus*, but was given specific rank by Carraway (1990). Includes *permiliensis* as a valid subspecies (Alexander, 1996; Carraway, 1990).

*Sorex bedfordiae* Thomas, 1911. Abstr. Proc. Zool. Soc. Lond., 1911(90):3.
COMMON NAME: Lesser Striped Shrew.
TYPE LOCALITY: "Omi-san, Sze-chwan" [= China, Sichuan, Emei Shan].
DISTRIBUTION: Montane forests of S Gansu and W Shensi to Yunnan (China); adjacent Burma and Nepal.
STATUS: IUCN – Lower Risk (lc).
SYNONYMS: *fumeolus* Thomas, 1911; *gomphus* G. Allen, 1923; *nepalensis* Weigel, 1969; *wardi* Thomas, 1911.
COMMENTS: Subgenus *Sorex*. Formerly a subspecies of *cylindricauda* but recognized as a full species by Corbet (1978c) and Hoffmann (1987).

*Sorex bendirii* (Merriam, 1884). Trans. Linnean Soc. New York, 2:217.
COMMON NAME: Marsh Shrew.
TYPE LOCALITY: USA, "Klamath Basin, Oregon" = Oregon, Klamath Co., l mi. (1.6 km) from Williamson River, l8 mi. (29 km) SE of Fort Klamath.
DISTRIBUTION: A narrow coastal area from NW California to Oregon and Washington (USA); a few records from SE British Columbia (Canada).
STATUS: IUCN – Lower Risk (lc).
SYNONYMS: **albiventer** Merriam, 1895; **palmeri** Merriam, 1895.
COMMENTS: Originally described in the monotypic genus *Atophyrax* Merriam; now in subgenus *Otisorex*. *S. vagrans* complex (Carraway, 1990; Demboski and Cook, 2001). Karyotype has 2n = 54, FN = 70. Reviewed by Pattie (1973, Mammalian Species No. 27).

*Sorex bucharensis* Ognev, 1922. Ann. Mus. Zool. Acad. Sci. St. Petersbourg, 22:320.
COMMON NAME: Buchara Shrew.
TYPE LOCALITY: Tajikistan, Pamir Mtns, Davan-su River Valley, "Gornaya Bukhara, drevyaya morena lednika Oshanina, dol. p. Davan-Su (khrebet' Petra Velikavo)" [Montane Bukhara, ancient moraine of Oshanin glacier, Peter the Great range].
DISTRIBUTION: Pamir Mtns (Tajikistan).
STATUS: IUCN – Lower Risk (lc).
COMMENTS: Subgenus *Sorex*, *S. minutus* group. Referred to subgenus *Eurosorex* by Yudin (1989). Considered a subspecies of *thibetanus* by Dolgov and Hoffmann (1977) and Hoffmann (1987, 1996a), but retained as a distinct species by Ivanitskaya et al. (1977), Hutterer (1979), Zaitsev (1988), and Yudin (1989). Karyotypes of two specimens from Tajikistan (2n = 40, FN = 60) were similar to that of *volnuchini* (Ivanitskaya et al., 1977).

*Sorex caecutiens* Laxmann, 1788. Nova Acta Acad. Sci. Petropoli, 1785, 3:285 [1788].
COMMON NAME: Laxmann's Shrew.
TYPE LOCALITY: Russia, Buryatskaya ASSR, SW shore of Lake Baikal (Pavlinov and Rossolimo, 1987:17).
DISTRIBUTION: Taiga and tundra zones from E Europe to E Siberia, south to C Ukraine, N Kazakhstan, Altai Mtns, Mongolia, Gansu and NE China, Korea, Sakhalin, and Japan (Hokkaido).
STATUS: IUCN – Lower Risk (lc).
SYNONYMS: *altaicus* Ognev, 1922; *annexus* Thomas, 1907; *araneoides* Ognev, 1922; *buxtoni* J. Allen, 1903; *caecutienoides* Stroganov, 1967; *centralis* Thomas, 1911; *insularis* Okhotina, 1984 [*nomen nudum*]; *insularis* Okhotina, 1993 [not Cowan, 1944]; *karpinskii* Dehnel, 1949; *koreni* G. Allen, 1914; *kurilensis* Okhotina, 1984 [*nomen*

*nudum*]; *lapponicus* Melander, 1942; *longicaudatus* Okhotina, 1984 [*nomen nudum*]; *longicaudatus* Okhotina, 1993 [not *longicaudatus* Kishida, 1930, *nomen nudum*, not *longicaudatus* Yoshikuru, 1956]; *macropygmaeus* Miller, 1901; *orii* Kuroda, 1933; *pleskei* Ognev, 1922; *rozanovi* Ognev, 1922; *saevus* Thomas, 1907; *tasicus* Ognev, 1933; *tungussensis* Naumoff, 1933.

COMMENTS: Subgenus *Sorex*, *S. caecutiens* group. Karyotype has 2n = 42, FN = 68-70. This species still offers many unsolved problems, along with the species of the *tundrensis* and *arcticus* groups. Names like *cansulus*, *granarius*, and *shinto* have been included in *caecutiens* in the past but are presently included in other species or treated as separate species; see Hoffmann (1987) for a discussion of problems. The European range was reviewed by Sulkava (1990). Okhotina (1993) revised the subspecies taxonomy in the Far East; two of the names that she had proposed were replaced by Hutterer and Zaitsev (2004). Ohdachi et al. (2001) studied the mtDNA variation of the species and found distinct clusters in Hokkaido (= *saevus*) and Eurasian Far East.

*Sorex camtschatica* Yudin, 1972. Teriologiya, 1:48.
COMMON NAME: Kamchatka Shrew.
TYPE LOCALITY: Russia, "Kamchatka, Kambal'naya Bay".
DISTRIBUTION: Russia, S Kamchatka Peninsula.
STATUS: IUCN – Lower Risk (lc).
COMMENTS: Subgenus *Otisorex*; *S. cinereus* group (Demboski and Cook, 2003). Formerly included in *cinereus* (van Zyll de Jong, 1982) but now recognized as a full species (Ivanitskaya and Kozlovsky, 1983; van Zyll de Jong, 1991*b*).

*Sorex cansulus* Thomas, 1912. Ann. Mag. Nat. Hist., ser. 8, 10:398.
COMMON NAME: Gansu Shrew.
TYPE LOCALITY: China, Gansu, "46 miles [74 km] south-east of SE Taochou" (= Lintan).
DISTRIBUTION: Known only from the type locality.
STATUS: IUCN – Critically Endangered.
COMMENTS: Subgenus *Sorex*, related to *tundrensis*. The species was recognized by Hoffmann (1987); no specimens other than the type series are known.

*Sorex cinereus* Kerr, 1792. Animal Kingdom, p. 206.
COMMON NAME: Cinereus Shrew.
TYPE LOCALITY: Canada, Ontario, Fort Severn.
DISTRIBUTION: North America throughout Alaska and Canada and southward along the Rocky and Appalachian Mtns to 45°.
STATUS: IUCN – Lower Risk (lc).
SYNONYMS: **acadicus** Gilpin, 1867; **fontinalis** Hollister, 1911; **hollisteri** Jackson, 1900; **lesueurii** (Duvernoy, 1842); **miscix** Bangs, 1899; **ohionensis** Bole and Moulthrop, 1942; **streatori** Merriam, 1895. **Unassigned**: *cooperi* Bachman, 1837; *forsteri* Richardson, 1828; *frankstounensis* Peterson, 1926; *idahoensis* Merriam, 1891; *nigriculus* Green, 1932; *personatus* I. Geoffroy, 1827; *platyrhinus* (De Kay, 1842).
COMMENTS: Type species of subgenus *Otisorex*, and of *S. cinereus* group. Does not occur in Siberia as previously suggested; the taxa *haydeni, jacksoni, ugyunak, portenkoi, leucogaster, beringianus,* and *camtschatica* have been included previously but are now considered as separate species; see comments under these taxa and Junge and Hoffmann (1981, and references therein), van Zyll de Jong (1982, 1991*b*), van Zyll de Jong and Kirkland (1989), and Pavlinov and Rossolimo (1987). *S. fontinalis* was separated from *cinereus* by Kirkland (1977), Junge and Hoffmann (1981), and Jones et al. (1992), but is considered, together with *lesueurii*, as a subspecies (van Zyll de Jong and Kirkland, 1989). However, George's (1988) data indicate it is a sister taxon to both *cinereus* and *haydeni*. Paraphyly and/or mtDNA introgression of *cinereus* and *haydeni* documented by Stewart and Baker (1994, 1997) and Brunet et al. (2002). Genetic relations of the entire group studied by Demboski and Cook (2003). Karyotype has 2n = 60, FN = 70. The name *fimbripes* Bachman, 1837 is usually included in the synonymy of *cinereus*, but Handley and Varn (1994) pointed out that Bachman's description resembles *Nectogale elegans* rather than *Sorex cinereus*.

*Sorex coronatus* Millet, 1828. Faune de Maine-et-Loire, I, p. 18.
  COMMON NAME: Millet's Shrew.
  TYPE LOCALITY: France, Main-et-Loire, Blou.
  DISTRIBUTION: W Europe from the Netherlands and NW Germany to France and
      Switzerland, south to N Spain; also in Jersey (Channel Isls), Liechtenstein and
      westernmost tip of Austria.
  STATUS: IUCN – Lower Risk (lc).
  SYNONYMS: *euronotus* Miller, 1901; *fretalis* Miller, 1909; *gemellus* Ott, 1968; *personatus* Millet,
      1828 [not I. Geoffroy, 1827]; *santonus* Mottaz, 1906.
  COMMENTS: Subgenus *Sorex*, *S. araneus* group. A sibling species of *araneus* (Meylan and
      Hausser, 1978), characterized mainly by the karyotype (2n = 22/24, FN = 44). Its
      distribution broadly overlaps with that of *araneus* in Germany. Revised by Hausser
      (1990), range map in Mitchell-Jones et al. (1999).

*Sorex cylindricauda* Milne-Edwards, 1872. Nouv. Arch. Mus. Hist. Nat. Paris, Bull. for 1871, 7:92
      [1872].
  COMMON NAME: Stripe-backed Shrew.
  TYPE LOCALITY: China, Sichuan, Moupin (= Baoxing).
  DISTRIBUTION: Montane forests of N Sichuan.
  STATUS: IUCN – Endangered.
  COMMENTS: Subgenus *Sorex*. Revised by Hoffmann (1987). The species is sympatric with
      *S. bedfordiae* in C Sichuan.

*Sorex daphaenodon* Thomas, 1907. Proc. Zool. Soc. Lond., 1907:407.
  COMMON NAME: Siberian Large-toothed Shrew.
  TYPE LOCALITY: Russia, Sakhalin Isl, "Dariné, 25 miles [40 km] N.W. of Korsakoff, Saghalien".
  DISTRIBUTION: Ural Mountains to the Kolyma River (Siberia); Sakhalin Isl; Kamchatka
      Peninsula; Paramushir Isl (N Kuriles); Jilin and Nei Mongol Aut. Region (China).
  STATUS: IUCN – Lower Risk (lc).
  SYNONYMS: *? megalotis* Kuroda, 1933; *sanguinidens* G. Allen, 1914; *scaloni* Ognev, 1933.
  COMMENTS: Subgenus *Sorex*, *S. araneus* group. Karyotype has 2n = 26-29, FN = 46. Type
      species of subgenus *Stroganovia*, see Yudin (1989), who recognized three subspecies,
      *daphaenodon*, *sanguinidens*, and *scaloni*. Okhotina (1993) transferred *orii* to *caecutiens*.

*Sorex dispar* Batchelder, 1911. Proc. Biol. Soc. Wash., 24:97.
  COMMON NAME: Long-tailed Shrew.
  TYPE LOCALITY: USA, "Beede's (sometimes called Keene Heights), in the township of Keene,
      Essex county, New York". Redescribed by Martin (1966:131) as 0.6 mi. (1 km) S, 0.5 mi.
      (0.8 km) E St. Huberts, Essex Co., New York, lat. 44°09′, long. 73°46′.
  DISTRIBUTION: Appalachian Mtns from W Virginia to N Carolina and Tenessee; New
      England, S New Brunswick and adjacent Nova Scotia (Canada).
  STATUS: IUCN – Lower Risk (lc).
  SYNONYMS: *macrurus* Batchelder, 1896 [not Hodgson, 1863, not *macrourus* Lehmann, 1822];
      *blitchi* Schwartz, 1956.
  COMMENTS: Subgenus *Otisorex*. For comparison with *gaspensis* see Kirkland and Van Deusen
      (1979). Reviewed by Kirkland (1981, Mammalian Species No. 155).

*Sorex emarginatus* Jackson, 1925. Proc. Biol. Soc. Wash., 38:129.
  COMMON NAME: Zacatecas Shrew.
  TYPE LOCALITY: "Sierra Madre, near Bolanos, altitude 7,600 feet (2316 m), State of Jalisco,
      Mexico".
  DISTRIBUTION: Durango, Zacatecas, and Jalisco (Mexico).
  STATUS: IUCN – Lower Risk (lc).
  COMMENTS: Referred to unnamed subgenus by George (1988:456). Findley (1955a)
      considered this a subspecies of *oreopolus*; however, *oreopolus* belongs to subgenus
      *Otisorex* (Diersing and Hoffmeister, 1977). For biological and distributional
      information, see Alvarez and Polaco (1984) and Matson and Baker (1986).

*Sorex excelsus* G. M. Allen, 1923. Am. Mus. Novit., 100:4.
  COMMON NAME: Chinese Highland Shrew.

TYPE LOCALITY: "summit of Ho-shan (=Xue Shan), Pae-tai, 30 miles (48 km) south of Chung-tien (=Zhongdian), Yunnan, China, altitude 13000 feet [3962 m]."

DISTRIBUTION: Yunnan and Sichuan (China), and possibly Nepal.

STATUS: IUCN – Data Deficient.

COMMENTS: Subgenus *Sorex*. Considered as a possible subspecies of *asper* (Corbet, 1978*c*) but retained as a full species related to *tundrensis* by Hoffmann (1987) who also suggested that a specimen from Nepal recorded by Agrawal and Chakraborty (1971) may represent *excelsus*. However, mtDNA data indicate that *excelsus* is neither related to *asper* nor to *tundrensis* (Fumagalli et al., 1999).

*Sorex fumeus* G. M. Miller, 1895. N. Am. Fauna, 10:50.

COMMON NAME: Smoky Shrew.

TYPE LOCALITY: USA, "Peterboro [Madison Co.], New York."

DISTRIBUTION: S Ontario, S Quebec, New Brunswick, and Nova Scotia (Canada); all of New England and Appalachian Mtns and adjacent areas to NE Georgia (USA).

STATUS: IUCN – Lower Risk (lc).

SYNONYMS: **umbrosus** Jackson, 1917.

COMMENTS: Subgenus *Otisorex*. Reviewed by Owen (1984, Mammalian Species No. 215). Karyotype has 2n = 66, FN = 98. Overlaps in distribution and may be easily confused with *arcticus* in part of its range (Junge and Hoffmann, 1981).

*Sorex gaspensis* Anthony and Goodwin, 1924. Am. Mus. Novit., 109:1.

COMMON NAME: Gaspé Shrew.

TYPE LOCALITY: Canada, "Mt. Albert, Gaspé Peninsula, Quebec, 2000 feet [= 610 m] elevation".

DISTRIBUTION: Gaspe Peninsula, N New Brunswick, Nova Scotia, and Cape Breton Isl (Canada).

STATUS: IUCN – Lower Risk (lc).

COMMENTS: Subgenus *Otisorex*. For comparison with *dispar*, see Kirkland and Van Deusen (1979). Reviewed by Kirkland (1981, Mammalian Species No. 155).

*Sorex gracillimus* Thomas, 1907. Proc. Zool. Soc. Lond., 1907:408.

COMMON NAME: Slender Shrew.

TYPE LOCALITY: Russia, Sakhalin Isl, "Dariné, 25 miles [40 km] N.W. of Korsakoff, Saghalien".

DISTRIBUTION: SE Siberia from S shore of the Sea of Okhotsk to N Korea and probably Manchuria; Sakhalin Isl; Hokkaido (Japan).

STATUS: IUCN – Lower Risk (lc).

SYNONYMS: **granti** Okhotina, 1993 [not Barrett-Hamilton and Hinton, 1913]; **hyojironis** Kuroda, 1939; **minor** Okhotina, 1993; *minor* Okhotina, 1984 [*nomen nudum*]; **natalae** Okhotina, 1993.

COMMENTS: Subgenus *Sorex*, *S. gracillimus* group. Karyotype has 2n = 36, FN = 62 (Sakhalin). This species has long been included in *minutus* but its specific status is now widely accepted on the basis of penial (Dolgov and Lukanova, 1966) and cranial (Hutterer, 1979) morphology, karyotype (Orlov and Bulatova, 1983), and allozyme data (George, 1988). The inclusion of *hyojironis* follows Corbet (1978*c*) and is tentative. Okhotina (1993) studied the geographic variation of the species. One of the subspecies that she distinguished requires renaming (Hutterer and Zaitsev, 2004). A study of mtDNA variation over most of its range was performed by Ohdachi et al. (2001).

*Sorex granarius* Miller, 1910. Ann. Mag. Nat. Hist., ser. 8, 6:458.

COMMON NAME: Iberian Shrew.

TYPE LOCALITY: Spain, Segovia, La Granja.

DISTRIBUTION: W to C Iberian Peninsula (Portugal and Spain).

STATUS: IUCN – Lower Risk (lc).

COMMENTS: Subgenus *Sorex*, *S. araneus* group. Karyotype has 2n = 36/37, FN 38-40. Afforded specific rank by Hausser et al. (1975); reviewed by Hausser (1990) and García-Perea et al. (1997, Mammalian Species No. 554).

*Sorex haydeni* Baird, 1857. Mammalia, *in* Repts. U.S. Expl. Surv., 8(1):29.

COMMON NAME: Prairie Shrew.

TYPE LOCALITY: USA, "Fort Union, Nebraska" (later Fort Buford, now Mondak, Montana, near Buford, Williams Co., North Dakota).

DISTRIBUTION: SE Alberta, S Saskatchewan, SW Manitoba (Canada); NW Montana southeast to Kansas, east to W and S Minnesota (USA).

STATUS: IUCN – Lower Risk (lc).

COMMENTS: Subgenus *Otisorex*; *S. cinereus* group (Demboski and Cook, 2003). Karyotype has 2n = 64, FN = 66. Formerly included in but now separated from *cinereus* by van Zyll de Jong (1980) and Junge and Hoffmann (1981); both species are closely related (George, 1988). *S. haydeni* occurs in grassy habitats while *S. cinereus* prefers forest and woodland (van Zyll de Jong, 1980). In Minnesota, Stewart and Baker (1994, 1997) and Brunet et al. (2002) found evidence of introgression between *haydeni* and *cinereus*.

*Sorex hosonoi* Imaizumi, 1954. Bull. Natl. Sci. Mus. Tokyo, 35:94.

COMMON NAME: Azumi Shrew.

TYPE LOCALITY: "Tokiwa Mura (Maneki, about 900 m altitude, foot of Mt. Gaki, Japan Alps), Kita-Azumi Gun, Nagano Pref., Central Honsyû [= Honshu], Japan".

DISTRIBUTION: Montane forests of C Honshu (Japan).

STATUS: IUCN – Vulnerable.

SYNONYMS: *shiroumanus* Imaizumi, 1954.

COMMENTS: Subgenus *Sorex*. Imaizumi (1970*b*) reported that *hosonoi* occurs sympatrically with *shinto* and therefore should be considered as separate species (Corbet, 1978*c*). Ohdachi et al. (1997*a*, 2001) demonstrated that the species is genetically related to *minutissimus*.

*Sorex hoyi* Baird, 1857. Mammalia, *in* Repts. U.S. Expl. Surv., 8(1):32.

COMMON NAME: American Pygmy Shrew.

TYPE LOCALITY: USA, Wisconsin, Racine.

DISTRIBUTION: N taiga zone of Alaska, Canada and the USA, with S outliers in the montane forests of the Appalachian and Rocky Mtns.

STATUS: IUCN – Lower Risk (lc).

SYNONYMS: *intervectus* (Jackson, 1925); *washingtoni* (Jackson, 1925); **alnorum** Preble, 1902; **eximius** Osgood, 1901; **montanus** (Brown, 1966) [not (Kelaart, 1850), not Skalon and Rajevsky, 1940, infrasubspecific name]; **thompsoni** Baird, 1858; **winnemana** (Preble, 1910).

COMMENTS: Formerly in *Microsorex*, which is a synonym of subgenus *Otisorex*, according to George (1988). Includes *thompsoni* since Diersing (1980*b*), but D. T. Stewart (in litt.) suggested that it may be a distinct species, as previously postulated by Long (1972*b*). Reviewed by Long (1974, Mammalian Species No. 33) and Junge and Hoffmann (1981). *Sorex browni* George, 1988 [replacement name for *minutus* (Brown, 1908), not Linnaeus, 1766] is the name for a Wisconsinan shrew related, if not conspecific, with *S. hoyi*. The name *montanus* used by Long (in Wilson and Ruff, 1999) for the Colorado and Wyoming population is a secondary homonym of *S. montanus* Kelaart (now *Suncus montanus*), and may therefore persist (Hutterer and Zaitsev, 2004). Karyotype insufficently known; Meylan (1968) provisionally identified 2n = 62, FN = 72.

*Sorex isodon* Turov, 1924. C.R. Acad. Sci. Paris, p. 111.

COMMON NAME: Taiga Shrew.

TYPE LOCALITY: Russia, Siberia, NE of Lake Baikal, Barguzinsk taiga, River Sosovka.

DISTRIBUTION: SE Norway and Finland through Siberia to the Pacific coast; Kamchatka; Sakhalin Isl; Kurile Isls; also NE China and Korea (S.-H. Han et al., 2000).

STATUS: IUCN – Lower Risk (lc).

SYNONYMS: *gravesi* Goodwin, 1933; *isodon* Stroganov, 1936; *montanus* Skalon and Rajevsky, 1940 [infrasubspecific name]; *montanus* Pavlinov, Borissenko, Kruskop and Yahontov, 1995; *princeps* Skalon and Rajevsky, 1940; *ruthenus* Stroganov, 1936; *sachalinensis* Okhotina, 1984 [*nomen nudum*]; *sachalinensis* Okhotina, 1993.

COMMENTS: Subgenus *Sorex*, *S. caecutiens* group. Karyotype has 2n = 42, FN = 70. Probably not conspecific with *sinalis* as suggested by Corbet (1978*c*) and Dolgov (1985); see Siivonen (1965) and Hoffmann (1987). Because the well established name *isodon* Turov is formally not available, and because *isodon* Stroganov is antedated by *gravesi*

Goodwin, Hoffmann (1987) suggested that *isodon* be declared the valid name; the case still needs to be submitted to the International Commission on Zoological Nomenclature. The species was reviewed by Sulkava (1990). The recently described *isodon marchicus* (Passarge, 1984) belongs to *araneus*; see Brünner et al. (2002c).

*Sorex jacksoni* Hall and Gilmore, 1932. Univ. California Publ. Zool., 38:392.
COMMON NAME: St. Lawrence Island Shrew.
TYPE LOCALITY: USA, "Sevoonga, 2 miles [3.2 km] east of North Cape, St. Lawrence Island, Bering Sea, Alaska."
DISTRIBUTION: Known only from St. Lawrence Isl (Bering Sea).
STATUS: IUCN – Endangered.
COMMENTS: Subgenus *Otisorex*; *S. cinereus* group (Demboski and Cook, 2003). Placed in the *arcticus* species group by Hall and Gilmore (1932) and in the *cinereus* species group by Hoffmann and Peterson (1967). Separated from *cinereus* by Junge and Hoffmann (1981). Van Zyll de Jong (1982) included *leucogaster* (= *beringianus*), *portenkoi*, and *ugyunak* in this species, but van Zyll de Jong (1991b) retained all three as distinct. Rausch and Rausch (1995) included *jacksoni* as a subspecies in *S. cinereus* based on identical karyotypes (2n = 66, FN = 70), but Demboski and Cook (2003) found a nonsister relationship between *S. cinereus* and *S. jacksoni* in their genetic analysis of the group.

*Sorex kozlovi* Stroganov, 1952. Byull. Moscow Ova. Ispyt. Prir. Otd. Biol., 57:21.
COMMON NAME: Kozlov's Shrew.
TYPE LOCALITY: China, E Tibet (= Qinghai), Dzechu (Za Qu) River, tributary of Mekong River (= Lancang Jiang).
DISTRIBUTION: SE Tibet (China).
STATUS: IUCN – Critically Endangered.
COMMENTS: Subgenus *Sorex*, *S. minutus* group. Type species of subgenus *Eurosorex* Stroganov, 1952. Known for long from a single specimen, until a near-topotype (USNM 449080) was obtained in 1987 (Hoffmann, 1996a). Regarded as a subspecies of *thibetanus* by some authors (Dolgov and Hoffmann, 1977; Hoffmann, 1987, 1996a, b) or included in *buchariensis* by others (Corbet, 1978c; Gureev, 1979). Hutterer (1979) recognized inconsistencies in the various published figures and descriptions of the same holotype specimen and regarded *kozlovi* as a doubtful taxon; see also under *buchariensis* and *thibetanus*.

*Sorex leucogaster* Kuroda, 1933. Bull. Biogeogr. Soc. Japan, 3,3:155.
COMMON NAME: Paramushir Shrew.
TYPE LOCALITY: Russia, Paramushir Isl; given by Ellerman and Morrison-Scott (1951:48) as "Nasauki, Amamu-shiru, 200 ft. [61 m], North Kurile Islands".
DISTRIBUTION: Probably confined to Paramushir Isl, south of Kamchatka Peninsula.
STATUS: IUCN – Vulnerable.
SYNONYMS: *beringianus* Yudin, 1967.
COMMENTS: Subgenus *Otisorex*, *S. cinereus* group (Demboski and Cook, 2003). Karyotype has 2n = 66, FN = 70. Formerly included in *cinereus* or *gracillimus* (Corbet, 1978c); includes *beringianus* Yudin, 1967. On the status, authorship and valid date of publication see Pavlinov and Rossolimo (1987). Related to *jacksoni* and *ugyunak* (van Zyll de Jong, 1982, 1991b).

*Sorex longirostris* Bachman, 1837. J. Acad. Nat. Sci. Philadelphia, 7:370.
COMMON NAME: Southeastern Shrew.
TYPE LOCALITY: USA, "in the swamps of Santee [River], South Carolina"; restricted to Hume Plantation (Cat Isl in the mouth of Santee River) by Jackson (1928:85).
DISTRIBUTION: SE USA, Florida west to Louisiana, Arkansas, Missouri, Illinois, and Indiana; Virginia and N Carolina.
STATUS: U.S. ESA – Delisted Taxon as *S. l. fisheri*; IUCN – Lower Risk (lc).
SYNONYMS: *bachmani* (Pomel, 1848); *wagneri* Fitzinger, 1868; **eonis** Davis, 1957; **fisheri** Merriam, 1895.
COMMENTS: Subgenus *Otisorex*; *S. cinereus* species group (Demboski and Cook, 2003). As

pointed out by Junge and Hoffmann (1981), this species is inappropriately named because it has one of the shortest rostra of North American *Sorex*. Junge and Hoffmann (1981) also suggested that shrews of the Great Dismal Swamp described as *fisheri* and traditionally included in *longirostris* as a subspecies are much larger and may represent a valid species. Reviewed by French (1980, Mammalian Species No. 143). Part of range mapped in detail by Pagels and Handley (1989) and Pagels et al. (1982).

*Sorex lyelli* Merriam, 1902. Proc. Biol. Soc. Wash., 15:75.
   COMMON NAME: Mt. Lyell Shrew.
   TYPE LOCALITY: USA, "Mt. Lyell, Tuolumne Co., California".
   DISTRIBUTION: Altitudes above 2000 m in the Sierra Nevada, California (USA).
   STATUS: IUCN – Lower Risk (lc).
   COMMENTS: Subgenus *Otisorex*; member of the *cinereus* species group. Related to *milleri*, according to van Zyll de Jong (1991*b*).

*Sorex macrodon* Merriam, 1895. N. Am. Fauna, 10:82.
   COMMON NAME: Large-toothed Shrew.
   TYPE LOCALITY: "Orizaba, Veracruz, Mexico (altitude 4,200 feet [1280 m])."
   DISTRIBUTION: Oaxaca, Puebla, and Veracruz, in mountains from 4000-9500 ft (1219-2896 m). See Heaney and Birney (1977).
   STATUS: IUCN – Lower Risk (nt).
   COMMENTS: Subgenus *Otisorex*. Similar to, and possibly conspecific with, *veraepacis* (see Junge and Hoffmann, 1981).

*Sorex maritimensis* Smith, 1939. J. Mammal., 20:244.
   COMMON NAME: Maritime Shrew.
   TYPE LOCALITY: Canada, "Nova Scotia, Kings County, Wolfville".
   DISTRIBUTION: Canada, Nova Scotia and New Brunswick.
   STATUS: Conservation status requires evaluation (Stewart et al., 2002).
   COMMENTS: Subgenus *Sorex*, *S. arcticus* group. Karyotype has 2n = 28/29, FN = 34. Van Zyll de Jong (1983*b*) and Volobouev and van Zyll de Jong (1988) suggested that *maritimensis* may be an independent species. Genetic data provided by Stewart et al. (2002) support that view. They suggest that *S. arcticus* and *S. maritimensis* shared a common ancestor approximately 2.4 million years ago.

*Sorex merriami* Dobson, 1890. Monogr. Insectivora, pt. 3 (Soricidae), fasc. l, pl. 23.
   COMMON NAME: Merriam's Shrew.
   TYPE LOCALITY: USA, "Fort Custer, Montana" = Bighorn Co., Little Bighorn River, ca. l mi. (1.6 km) above Fort Custer (= Hardin).
   DISTRIBUTION: Xeric habitats in EC Washington to N and E California, Arizona, northeastward to Nebraska, Wyoming and Montana (USA).
   STATUS: IUCN – Lower Risk (lc).
   SYNONYMS: *leucogenys* Osgood, 1909.
   COMMENTS: Referred to unnamed subgenus by George (1988:456). Reviewed by Armstrong and Jones (1971*b*, Mammalian Species No. 2).

*Sorex milleri* Jackson, 1947. Proc. Biol. Soc. Wash., 60:131.
   COMMON NAME: Carmen Mountain Shrew.
   TYPE LOCALITY: "Madera Camp, altitude 8,000 feet [2438 m], Carmen Mountains, Coahuila, Mexico".
   DISTRIBUTION: Restricted to the Sierra Madre Oriental of Coahuila and Nuevo Leon, Mexico.
   STATUS: IUCN – Vulnerable.
   COMMENTS: Subgenus *Otisorex*. Controversial opinions on the systematic status of *milleri* exist; Findley (1955*a*) regarded it as a morphologically distinct relict population allied to *cinereus* and accepted its specific status, as did Hall (1981) and Junge and Hoffmann (1981); while van Zyll de Jong and Kirkland (1989) suggested that *milleri* may not merit full specific status.

*Sorex minutissimus* Zimmermann, 1780. Geogr. Gesch. Mensch. Vierf. Thiere, 2:385.
   COMMON NAME: Eurasian Least Shrew.

TYPE LOCALITY: "Yenisei"; given by Stroganov (1957:176) as "iz raiona sela Kiiskow chto na r. Kie (nyne g. Mariinsk Kemerovskoi oblasti)" [= Russia, Kemerovsk. Obl., Mariinsk (= Kiiskoe), bank of Kiia River (near Yenesei River)]; Restricted by Pavlinov and Rossolimo (1987:23) to "Krasnoyarskii kr., Krasnoyarsk."

DISTRIBUTION: Taiga zone from Norway, Sweden and Estonia to E Siberia; Sakhalin; Hokkaido, and perhaps Honshu (Japan); Mongolia; China; South Korea.

STATUS: IUCN – Lower Risk (lc).

SYNONYMS: *abnormis* Stroganov, 1949; *barabensis* Stroganov, 1956; *burneyi* Thomas, 1915; *caudata* Yudin, 1964; *czekanovskii* Naumoff, 1933; *exilis* Gmelin, 1788; *hawkeri* Thomas, 1906; *ishikawai* Yoshiyuki, 1988; *karelicus* Stroganov, 1949; *minimus* Gmelin, 1793; *neglectus* Ognev, 1922; *praeminutus* Heller, 1963; *stroganovi* Yudin, 1964; *tscherskii* Ognev, 1913; *tschuktschorum* Stroganov, 1949; *ussuriensis* Ognev, 1922.

COMMENTS: Subgenus *Sorex* or *Eurosorex*, *S. caecutiens* group. Karyotype has 2n= 38 (Finland) or 42 (Siberia), FN = 74. Yoshiyuki (1988*a*) recognized nine subspecies, but the evidence seems to be weak, and none is recognized here. Ohdachi et al. (1997*b*) identified a sister relationship between *S. minutissimus* and *S. hosonoi*.

*Sorex minutus* Linnaeus, 1766. Syst. Nat., 12th ed., l:73.

COMMON NAME: Eurasian Pygmy Shrew.

TYPE LOCALITY: "Yenisei"; restricted by Pavlinov and Rossolimo (1987:15) to "Krasnoyarskii kr., Krasnoyarsk." According to Ellerman and Morrison-Scott (1951:47), Linnaeus' name is based on Laxmann's ms. of *Sibir. Briefe*, and the type locality is Barnaul, Russia.

DISTRIBUTION: Europe to Yenesei River and Lake Baikal, south to Altai and Tien Shan Mtns; populations of Nepal and China have been alternatively identified as *minutus* or *thibetanus*; populations of Turkey and the Caucasus as *minutus* or *volnuchini*; populations of Kashmir and N Pakistan as *minutus*, *planiceps*, or *thibetanus*.

STATUS: IUCN – Lower Risk (lc).

SYNONYMS: *becki* Lehmann, 1963; *canaliculatus* Ljungh, 1806; *carpetanus* Rey, 1971; *exiguus* Brink, 1952; *exilis* Gmelin, 1788; *gymnurus* Chaworth-Musters, 1932; *heptapotamicus* Stroganov, 1956; *hibernicus* Jenys, 1838; *insulaebellae* Heim de Balsac, 1940; *kastchenkoi* Johansen, 1923; *lucanius* Miller, 1909; *melanderi* Ognev, 1928; *minimus* Geoffroy, 1811; *pumilio* Wagler, 1832; *pumilus* Nilsson, 1844; *pygmaeus* Laxmann, 1769; *rusticus* Jenys, 1838.

COMMENTS: Subgenus *Sorex*, *S. caecutiens* group. Formerly included *gracillimus* and *volnuchini*, which are each now accepted as specifically distinct; see comments therein. Corbet (1978*c*) included also *planiceps* and *thibetanus*; but see Dolgov and Hoffmann (1977), Hoffmann (1996*a*, *b*) and Hutterer (1979). The European populations of *minutus* were revised by Hutterer (1990, 1999). Karyotype has 2n = 42, FN = 56 (Zima et al., 1998), but 2n = 40 and 36 in the Baltic islands Öland and Gotland (Fredga et al., 1995).

*Sorex mirabilis* Ognev, 1937. Byull. Moscow Ova. Ispyt. Prir. Otd. Biol., 46(5):268.

COMMON NAME: Ussuri Shrew.

TYPE LOCALITY: Russia, Primorskii Krai, Ussuriiskii r-n., Kamenka River (specified by Pavlinov and Rossolimo (1987).

DISTRIBUTION: N and S Korea, NE China, and Ussuri region (Russia).

STATUS: IUCN – Lower Risk (lc).

SYNONYMS: *kutscheruki* Stroganov, 1956.

COMMENTS: Placed in monotypic subgenus *Ognevia* by Heptner and Dolgov (1967), who demonstrated that *mirabilis* is not conspecific with *pacificus*, as had been suggested earlier (Bobrinskii et al., 1965). Hutterer (1982*b*) suggested a closer relationship with *Sorex* (*Homalurus*) *alpinus* because of shared derived features of genital morphology. Karyotype has 2n = 38, FN = 66 (Zima et al., 1998).

*Sorex monticolus* Merriam, 1890. N. Am. Fauna, 3:43.

COMMON NAME: Dusky Shrew.

TYPE LOCALITY: USA, Arizona, Coconino Co., San Francisco Mountain, altitude 3500 meters.

DISTRIBUTION: Montane boreal and coastal coniferous forest and alpine areas from Alaska to California and New Mexico, east to Montana, Wyoming, and Colorado (USA) and to W Manitoba (Canada); Chihuahua, Durango (Mexico).

STATUS: IUCN – Lower Risk (lc).

SYNONYMS: *durangae* Jackson, 1928; *melanogenys* Hall, 1932; **alascensis** Merriam, 1895; *glacialis* Merriam, 1900; **calvertensis** Cowan, 1941; **elassodon** Osgood, 1901; **insularis** Cowan, 1941 [not Okhotina, 1993]; **isolatus** Jackson, 1922; **longicaudus** Merriam, 1895; **malitiosus** Jackson, 1919; **obscurus** Merriam, 1891; *longiquus* Findley, 1955; *obscuroides* Findley, 1955; *similis* Merriam, 1891 [not of Hensel, 1855]; **parvidens** Jackson, 1921; **prevostensis** Osgood, 1901; **setosus** Elliot, 1899; *mixtus* Hall, 1938; **shumaginensis** Merriam, 1900; **soperi** Anderson and Rand, 1945.

COMMENTS: Subgenus *Otisorex*. *S. vagrans* complex (Carraway, 1990) or "dusky shrew complex" (Demboski and Cook, 2001). Includes *obscurus* and *durangae*, which were previously included in *vagrans* and *saussurei* respectively; see Hennings and Hoffmann (1977) and map in Junge and Hoffmann (1981); other synonyms follow van Zyll de Jong (1983a) and George and Smith (1991). Related to *pacificus* (see George, 1988). Reviewed by Smith and Belk (1996, Mammalian Species No. 528), and by Alexander (1996). Phylogeography of the entire complex analyzed by Demboski and Cook (2001).

*Sorex nanus* Merriam, 1895. N. Am. Fauna, 10:81.

COMMON NAME: Dwarf Shrew.

TYPE LOCALITY: USA, "Estes Park [Larimer Co.], Colorado".

DISTRIBUTION: Rocky Mountains from Montana to New Mexico; South Dakota; Arizona (USA).

STATUS: IUCN – Lower Risk (lc).

COMMENTS: Subgenus *Otisorex*. Very similar to, and perhaps conspecific with *tenellus*, see review by Hoffmann and Owen (1980, Mammalian Species No. 131); but George (1988) retained both as distinct species.

*Sorex neomexicanus* Bailey, 1913. Proc. Biol. Soc. Wash., 26:133.

COMMON NAME: New Mexico Shrew.

TYPE LOCALITY: USA, New Mexico, Sacramento Mtns, Cloudcroft.

DISTRIBUTION: USA, New Mexico, Capitan and Sacramento Mtns.

COMMENTS: Subgenus *Otisorex*. *S. vagrans* complex (Demboski and Cook, 2001). Formerly included in *S. monticolus* but given species rank by Alexander (1996) based on morphometric differences.

*Sorex oreopolus* Merriam, 1892. Proc. Biol. Soc. Wash., 7:173.

COMMON NAME: Mexican Long-tailed Shrew.

TYPE LOCALITY: "Sierra de Colima, Jalisco, Mexico (altitude 10,000 feet) [= 3,048 m]".

DISTRIBUTION: Endemic to Mexico; in Distrito Federal, Jalisco, México, Morelos, Puebla, and Tlaxcala (Carraway, ms).

STATUS: IUCN – Lower Risk (nt).

COMMENTS: Subgenus *Otisorex*. Contrary to Findley (1955a), this species does not include *emarginatus* or *ventralis*, which Diersing and Hoffmeister (1977) placed in the subgenus *Sorex*. *S. orizabae* was included in *vagrans* by Hennings and Hoffmann (1977:8), and in *oreopolus* by Junge and Hoffmann (1981:43), but considered a separate species by Carraway (ms).

*Sorex orizabae* Merriam, 1895. North American Fauna, 10:71.

COMMON NAME: Orizaba Long-tailed Shrew.

TYPE LOCALITY: Mexico, "Falda oeste del Pico de Orizaba, 9500 ft. [2895 m], Puebla".

DISTRIBUTION: Endemic to Mexico; in Distrito Federal, México, Michoacán, Morelos, Puebla, Tlaxcala, and Veracruz (Carraway, ms).

STATUS: IUCN – Lower Risk (nt) as included in *S. oreopolus*.

COMMENTS: Subgenus *Otisorex*. Contrary to Findley (1955b), this species is not conspecific with *S. Vagrans*. *S. orizabae* was included in *vagrans* by Hennings and Hoffmann (1977:8), later included in *oreopolus* by Junge and Hoffmann (1981:43), and most recently considered a separate species by Carraway (ms).

*Sorex ornatus* Merriam, 1895. N. Am. Fauna, 10:79.
COMMON NAME: Ornate Shrew.
TYPE LOCALITY: USA, "San Emigdio Canyon, Mt. Piños [Kern Co.], California".
DISTRIBUTION: California coastal ranges from N of San Francisco Bay to N part and S tip of
Baja California; Santa Catalina Isl.
STATUS: U.S. ESA – Endangered as *S. o. relictus*; IUCN – Lower Risk (lc).
SYNONYMS: *californicus* Merriam, 1895; *oreinus* Elliot, 1903; **juncensis** Nelson and Goldman,
1909; **lagunae** Nelson and Goldman, 1909; **relictus** Grinnell, 1932; **salarius**
von Bloeker, 1939; **salicornicus** von Bloeker, 1932; **sinuosus** Grinnell, 1913; **willetti**
von Bloeker, 1942.
COMMENTS: Subgenus *Otisorex*. Genetically related to *S. vagrans* (Demboski and Cook, 2001).
Reviewed by Owen and Hoffmann (1983, Mammalian Species No. 212), who
recognized 9 subspecies. For further distributional information see Williams (1979)
and Junge and Hoffmann (1981); for karyotype (2n = 54, FN = 76) and allozyme data
see Brown and Rudd (1981) and George (1988).

*Sorex pacificus* Coues, 1877. Bull. U.S. Geol. Geogr. Surv. Terr., 3(3):650.
COMMON NAME: Pacific Shrew.
TYPE LOCALITY: USA, "Fort Umpqua [mouth Umpqua River, Douglas County], Oregon."
DISTRIBUTION: Coastal forests and Cascade Mountains, Oregon (USA).
STATUS: IUCN – Lower Risk (lc).
SYNONYMS: *yaquinae* Jackson, 1918; **cascadensis** Carraway, 1990.
COMMENTS: Subgenus *Otisorex*. *S. vagrans* complex (Carraway, 1990). Karyotype has 2n = 54,
FN = 62. Not conspecific with *mirabilis*; see Yudin (1969) and Hoffmann (1971).
Related to *monticolus*; see Findley (1955b), Junge and Hoffmann (1981), and George
(1988). Reviewed by Carraway (1985, Mammalian Species No. 231), who later (1990)
removed *sonomae* from synonymy; see comments under that species.

*Sorex palustris* Richardson, 1828. Zool. J., 3:517.
COMMON NAME: American Water Shrew.
TYPE LOCALITY: Canada, "marshy places, from Hudson's Bay to the Rocky Mountains."; not
specified.
DISTRIBUTION: Montane and boreal areas of North America below the tree line from Alaska
to the Sierra Nevada, Rocky and Appalachian Mtns.
STATUS: IUCN – Lower Risk (lc).
SYNONYMS: *acadicus* (Allen, 1915) [not Gilpin, 1867]; **albibarbis** (Cope, 1862); **brooksi**
Anderson, 1934; **gloveralleni** Jackson, 1926; **hydrobadistes** Jackson, 1926;
**labradorensis** Burt, 1938; **navigator** (Baird, 1858); **punctulatus** Hooper, 1942; **turneri**
Johnson, 1951.
COMMENTS: Formerly placed in genus *Neosorex* Baird; now in *Sorex*, subgenus *Otisorex*,
*S. vagrans* complex (Demboski and Cook, 2001; Fumagalli et al., 1999). Reviewed by
Beneski and Stinson (1987, Mammalian Species No. 296), who recognized
9 subspecies. They did not include *alaskanus* as suggested by Junge and Hoffmann
(1981:28) and Hall (1981:43); George (1988) and Carraway (1995) also treated
*alaskanus* as distinct.

*Sorex planiceps* Miller, 1911. Proc. Biol. Soc. Wash., 24:242.
COMMON NAME: Kashmir Pygmy Shrew.
TYPE LOCALITY: India, "Dachin, Khistwar, Kashmir (altitude, 9000 feet [2743 m])".
DISTRIBUTION: Kashmir (India) and N Pakistan.
STATUS: IUCN – Lower Risk (lc).
COMMENTS: Considered a subspecies of *thibetanus* by Dolgov and Hoffmann (1977) and
Hoffmann (1987, 1996a, b), but retained by Hutterer (1979) because of larger skull
measurements. The problem remains unresolved.

*Sorex portenkoi* Stroganov, 1956. Proc. Inst. Biol. W. Siberian Branch Acad. Sci. USSR, Zool., 1:11-14.
COMMON NAME: Portenko's Shrew.
TYPE LOCALITY: Russia, Koryaksk. Auv. Okr. "bliz pos Anadyr', poberejhe Anadyrsk limana
[near Anadyr' settlement, shore of Anadyr' estuary]."

DISTRIBUTION: NE Siberia.

STATUS: IUCN – Lower Risk (lc).

COMMENTS: Subgenus *Otisorex*; *S. cinereus* group (Demboski and Cook, 2003). Originally described as a subspecies of *cinereus* and treated as such by Yudin (1972) and Okhotina (1977), then included in *ugyunak* (Ivanitskaya and Kozlovsky, 1985), but recently recognized as a distinct species by Zaitsev (1988), and van Zyll de Jong (1991*b*), who, however, pointed out its close relationship to *jacksoni* and *ugyunak*.

*Sorex preblei* Jackson, 1922. J. Wash. Acad. Sci., 12:263.

COMMON NAME: Preble's Shrew.

TYPE LOCALITY: USA, "Jordan Valley, altitude 4,200 feet [1280 m], Malheur County, Oregon."

DISTRIBUTION: Columbia Plateau of Washington, Oregon, California, Idaho, Nevada to W Great Plains of Montana, Utah, Wyoming, Colorado, and New Mexico (USA). For reviews of distributional records, see Tomasi and Hoffmann (1984) and Long and Hoffmann (1992).

STATUS: IUCN – Lower Risk (lc).

COMMENTS: Subgenus *Otisorex*; *S. cinereus* group (Demboski and Cook, 2003). Reviewed by Cornely et al. (1992, Mammalian Species No. 416).

*Sorex pribilofensis* Merriam, 1895. N. Amer. Fauna, 10:87.

COMMON NAME: Pribilof Island Shrew.

TYPE LOCALITY: USA, Alaska, Pribilof Isls, St. Paul Isl.

DISTRIBUTION: Recently known only from St. Paul in the Bering Sea Isls (Pribilof Isls).

STATUS: IUCN – Endangered as *S. hydrodromus*.

SYNONYMS: ? *hydrodromus* Dobson, 1889.

COMMENTS: Subgenus *Otisorex*; *S. cinereus* group (Demboski and Cook, 2003). There exists some discrepancy in the literature on the correct name for this species. Dobson's *hydrodromus* would have priority, but the type locality (Unalaska Isl) may be incorrect, and the holotype specimen shows dental characters of the *Sorex araneus* group (Rausch and Rausch, 1997). Hoffmann and Peterson (1967) proposed to suppress *hydrodromus* in favour of *pribilofensis*, a suggestion followed by van Zyll de Jong (1991*b*). However, Yudin (1969), Baranova et al. (1981), Hall (1981), Junge and Hoffmann (1981), Honacki et al. (1982), and Hutterer (1993*a*) retained *hydrodromus*, while Gureev (1979) listed both *hydrodromus* and *pribilofensis* as species. Rausch and Rausch (1997) discussed in detail the complex taxonomic history of this species and described its unique karyotype (2n = 55, FN = 67). Demboski and Cook (2003) included it in the *S. cinereus* group on the basis of mtDNA data.

*Sorex raddei* Satunin, 1895. Arch. Naturgesch., l:109.

COMMON NAME: Radde's Shrew.

TYPE LOCALITY: Georgia, near Kutais.

DISTRIBUTION: Transcaucasia and N Turkey.

STATUS: IUCN – Lower Risk (lc).

SYNONYMS: *batis* Thomas, 1913; *caucasicus* Satunin, 1913.

COMMENTS: Subgenus *Sorex*; *S. raddei* group (Fumagalli et al., 1999). Includes *batis* (Corbet, 1978*c*) and *caucasicus* which in turn now must be called *satunini* (see comments therein and Pavlinov and Rossolimo, 1987). Karyotype has 2n = 36, FN = 68 (Zima et al., 1998).

*Sorex roboratus* Hollister, 1913. Smithson. Misc. Coll., 60(24):2.

COMMON NAME: Flat-skulled Shrew.

TYPE LOCALITY: Russia, Gorno-Altaisk A.O., "5 mi [8 km] S Dapuchu [Altai Mtns, Tapucha]".

DISTRIBUTION: Russia east of River Ob to Ussuri River, south to Altai Mtns, N Mongolia, and Primorsk Krai.

STATUS: IUCN – Lower Risk (lc).

SYNONYMS: *dukelskiae* Ognev, 1933; *jacutensis* Dudelsky, 1928; *platycranius* Ognev, 1922; *thomasi* Ognev, 1922; *tomensis* Ognev, 1922; *turuchanensis* Naumoff, 1931; *vir* G. Allen, 1914.

COMMENTS: Subgenus *Sorex*, *S. caecutiens* group. Formerly known as *vir* but *roboratus* has priority (Hoffmann, 1985*a*; Zaitsev, 1988). Taxonomy and distribution revised by Hoffmann (1985*a*). Karyotype has 2n = 42, FN = 70 (Orlov and Kozlovsky, 1971).

*Sorex samniticus* Altobello, 1926. Bol. Inst. Zool. Univ. Roma, 3:102.
COMMON NAME: Apennine Shrew.
TYPE LOCALITY: Italy, Campobasso Prov., Molise, 700 m.
DISTRIBUTION: Italy.
STATUS: IUCN – Lower Risk (lc).
SYNONYMS: *garganicus* Pasa, 1953; *monsvairani* (Altobello, 1927).
COMMENTS: Subgenus *Sorex*. *S. samniticus* group (Zima et al., 1998). Karyotype has 2n = 52, FN = 52. Formerly included in *araneus*; considered a distinct species by Graf et al. (1979). Reviewed by Hausser (1990).

*Sorex satunini* Ognev, 1922. Ann. Zool. Mus. Russ. Acad. Sci., 22:311.
COMMON NAME: Caucasian Shrew.
TYPE LOCALITY: Turkey, Kars, Goele, "Gel'skaya kotlovina [depression], Mvuzaret".
DISTRIBUTION: N Turkey and Caucasus.
STATUS: IUCN – Lower Risk (lc).
COMMENTS: Subgenus *Sorex*; *S. araneus* group (Fumagalli et al., 1999). Karyotype has 2n = 24/25, FN = 46. Formerly referred to as *caucasicus* Satunin, which is now synonymized with *raddei* Satunin; see Pavlinov and Rossolimo (1987) and Zaitsev (1988). Sokolov and Tembotov (1989), who reviewed the distribution in Caucasus, used *caucasicus* for this species. Considered a distinct species by Graf et al. (1979).

*Sorex saussurei* Merriam, 1892. Proc. Biol. Soc. Wash., 7:173.
COMMON NAME: Saussure's Shrew.
TYPE LOCALITY: "Sierra de Colima, Jalisco, Mexico, (altitude 8000 feet [2438 m])".
DISTRIBUTION: Colima, Distrito Federal, Guerrero, Jalisco, México, Michoacán, Morelos, Puebla, from 2100 to 3650 m or more (Mexico); Guatemala.
STATUS: IUCN – Lower Risk (lc).
SYNONYMS: **godmani** Merriam, 1897; *salvini* Merriam, 1897.
COMMENTS: Referred to unnamed subgenus by George (1988:456). Populations from Guatemala provisionally included by Junge and Hoffmann (1981) may be distinct and should be carefully studied.

*Sorex sclateri* Merriam, 1897. Proc. Biol. Soc. Wash., 11:228.
COMMON NAME: Sclater's Shrew.
TYPE LOCALITY: "Tumbala, Chiapas, Mexico (alt. 5000 ft. [1524 m])"
DISTRIBUTION: Endemic to Mexico; known from two localities in Chiapas (Carraway, ms).
STATUS: IUCN – Endangered.
COMMENTS: Referred to unnamed subgenus by George (1988:456).

*Sorex shinto* Thomas, 1905. Abstr. Proc. Zool. Soc. Lond., 1905(23):19.
COMMON NAME: Shinto Shrew.
TYPE LOCALITY: Japan, N Honshu, "Makado, near Nohechi".
DISTRIBUTION: Honshu, Shikoku, and Sado (Japan).
STATUS: IUCN – Endangered as *S. sadonis*; otherwise Lower Risk (lc).
SYNONYMS: *chouei* Imaizumi, 1954; **sadonis** Yoshiyuki and Imaizumi, 1986; **shikokensis** Abe, 1967.
COMMENTS: Subgenus *Sorex*. *S. caecutiens* group. Karyotype has 2n = 42, FN = 70. Included in *caecutiens* by Abe (1967) and Corbet (1978*c*), but Imaizumi (1970*b*) treated *shinto* as a separate species, a view supported by Pavlinov and Rossolimo (1987), and by the allozyme data of George (1988). Ohdachi et al. (1997*a*) used mitochondrial gene sequences to show that both species occur in Japan: *caecutiens* in Hokkaido, and *shinto* in Honshu, Shikoku, and Sado. These authors also provided evidence that *sadonis* and *shikokensis* should be included in *shinto*; *S. sadonis* had been treated as a separate species before. Yoshiyuki and Imaizumi (1986) assigned the species to "the *caecutiens-arcticus* section of the *minutus* group." Dokuchaev et al. (1999) studied the

geographical variation of *shinto* and found that all three island populations are morphometrically distinct.

*Sorex sinalis* Thomas, 1912. Ann. Mag. Nat. Hist., ser. 8, 10:398.
COMMON NAME: Chinese Shrew.
TYPE LOCALITY: China, Shaanxi, "45 miles [72 km] S.E. of Feng-siang-fu [Feng Xian], Shen-si, 10,500′ [3200 m]".
DISTRIBUTION: C and W China.
STATUS: IUCN – Vulnerable.
COMMENTS: Subgenus *Sorex*. Formerly regarded as conspecific with *isodon*; see Corbet (1978c) and Hoffmann (1987) for discussion and specific boundaries.

*Sorex sonomae* Jackson, 1921. J. Mammal., 2:162.
COMMON NAME: Fog Shrew.
TYPE LOCALITY: USA, "Sonoma Country side of Gualala River, Gualala, California".
DISTRIBUTION: Pacific coast from Oregon to N California (USA).
STATUS: IUCN – Lower Risk (lc).
SYNONYMS: *tenelliodus* Carraway, 1990.
COMMENTS: Subgenus *Otisorex*. *S. vagrans* complex (Carraway, 1990). The small and morphologically distinct subspecies *tenelliodus* may well be a good species.

*Sorex stizodon* Merriam, 1895. N. Am. Fauna, 10:98.
COMMON NAME: San Cristobal Shrew.
TYPE LOCALITY: "San Cristobal, Chiapas, Mexico, [9,000 ft.= 2,743 m]".
DISTRIBUTION: Endemic to Mexico; known only from the type locality and the Reserva Ecológica Huitepec in Chiapas (Carraway, ms).
STATUS: IUCN – Endangered.
COMMENTS: Referred to unnamed subgenus by George (1988:456). Similar to *ventralis* (Junge and Hoffmann, 1981).

*Sorex tenellus* Merriam, 1895. N. Am. Fauna, 10:81.
COMMON NAME: Inyo Shrew.
TYPE LOCALITY: USA, "summit of Alabama Hills near Lone Pine, Owens Valley [Inyo Co.], Calif[ornia, about 4500 ft. = 1372 m]."
DISTRIBUTION: Mountains of WC Nevada and EC California (USA).
STATUS: IUCN – Lower Risk (lc).
SYNONYMS: *myops* Merriam, 1902.
COMMENTS: Subgenus *Otisorex*. Similar to *nanus* with which it may form an allospecies (Hoffmann and Owen, 1980, Mammalian Species No. 131), but George (1988) retained both as separate species on the basis of allozyme frequencies.

*Sorex thibetanus* Kastschenko, 1905. Izv. Tomsk. Univ., 27:93.
COMMON NAME: Tibetan Shrew.
TYPE LOCALITY: "Tsaidam" [NE Tibet, China].
DISTRIBUTION: Himalyas and NE Tibet.
STATUS: IUCN – Lower Risk (lc).
COMMENTS: The pygmy shrews of the Himalayas are still a subject of controversy. The original description of *thibetanus* (as a subspecies of *minutus*) is not very informative; the holotype in the Tomsk Academy was considered to be lost (Yudin, pers. comm. 1977), which is why Hutterer (1979) regarded *thibetanus* as a *nomen dubium*. Dolgov and Hoffmann (1977) and later Hoffmann (1987) used *thibetanus* to define a Himalayan species in which they included *buchariensis*, *kozlovi*, *planiceps*, and specimens from Nepal and China reported as *minutus* by various authors. Hutterer (1979) instead recognized three species, *buchariensis*, *planiceps*, and *minutus* as occurring in the Himalayas and regarded *kozlovi* and *thibetanus* as indeterminable. Zaitsev (1988) pointed out differences between *buchariensis* and *thibetanus*. Surprisingly, the holotype of *thibetanus* turned up in the Zoological Museum of Moscow (Baranova et al., 1981) and Hoffmann (1987) reported on its measurements. Hoffmann (1996a, b) provided a distribution map for all known specimens of *thibetanus*, *kozlovi*, *buchariensis*, and *planiceps*, which he understood as subspecies of

*S. thibetanus.* However, in the light of drastic size and tooth shape differences among these taxa, and in the light of subtle differences between better-known small *Sorex* species (*minutus* and *volnuchini*, *caecutiens* and *shinto*), I prefer to list *buchariensis*, *kozlovi*, *planiceps*, and *thibetanus* as separate species until a more complete analysis is available.

*Sorex trowbridgii* Baird, 1857. Mammalia, *in* Repts. U.S. Expl. Surv., 8(1):13.
    COMMON NAME: Trowbridge's Shrew.
    TYPE LOCALITY: USA, "Astoria [mouth of the Columbia River, Clatsop Co.], Oregon".
    DISTRIBUTION: Coastal ranges from Washington (including Destruction Isl) to California (USA); SW British Columbia (Canada).
    STATUS: IUCN – Lower Risk (lc).
    SYNONYMS: **destructioni** Scheffer and Dalquest, 1942; **humboldtensis** Jackson, 1922; **mariposae** Grinnell, 1913; **montereyensis** Merriam, 1895.
    COMMENTS: Referred to unnamed subgenus by George (1988:456). Karyotype variable, 2n = 31-42, FN = 56-60 (Zima et al., 1998). Karyotype data analyzed by Ivanitskaya (1994) also place *S. trowbridgii* on a distinct branch in her cladogram. Reviewed by George (1989, Mammalian Species No. 337).

*Sorex tundrensis* Merriam, 1900. Proc. Wash. Acad. Sci., 2:16.
    COMMON NAME: Tundra Shrew.
    TYPE LOCALITY: USA, Alaska, St. Michaels.
    DISTRIBUTION: Sakhalin Isl; Siberia, from the Pechora River to Chukotka, south to the Altai Mtns; Mongolia and NE China; Alaska (USA); Yukon, Northwest Territories (Canada).
    STATUS: IUCN – Lower Risk (lc).
    SYNONYMS: *amasari* Ognev, 1922; *amazari* Ognev, 1928; *baikalensis* Ognev, 1913; *borealis* Kastchenko, 1905; *centralis* Thomas, 1911; *irkutensis* Ognev, 1933; *jenissejensis* Dudelski, 1930; *khankae* Baranova and Zaitsev, 2003 [replacement name for *stroganovi* Okhotina, 1984 (not Yudin, 1964, not Yudin, 1989)]; *margarita* Fetisov, 1950; *middendorffii* Ognev, 1933; *parvicaudatus* Okhotina, 1976; *petschorae* Ognev, 1922; *schnitnikovi* Ognev, 1922; *sibiriensis* Ognev, 1922; *transrypheus* Stroganov, 1956; *ultimus* G. Allen, 1914; *ussuriensis* Okhotina, 1983 [not Ognev, 1922].
    COMMENTS: Subgenus *Sorex*, *S. tundrensis* group (Fumagalli et al., 1999). Youngman (1975) provided evidence that *tundrensis* is specifically distinct from *arcticus*. Palearctic populations formerly referred to *arcticus* were included in *tundrensis* by Junge et al. (1983) and Okhotina (1983). Hoffmann (1987) and van Zyll de Jong (1991*b*) discussed additional aspects of its taxonomy and distribution. Karyotype variable: 2n = 31-41, FN 56-60 in Siberia, 2n = 32/33, FN = 58 in Yukon, and 2n = 32/33, FN = 62 in C Alaska. Kozlovsky (1976) found *irkutensis* and *sibiriensis* to be karyotypically distinct; possibly two sibling species occur throughout the Palearctic range. Meylan and Hausser (1991) described a karyotype from Canada that was identical to some in Siberia.

*Sorex ugyunak* Anderson and Rand, 1945. Canadian Field Nat., 59:62.
    COMMON NAME: Barren Ground Shrew.
    TYPE LOCALITY: "Tuktuk (Tuktuyaktok), northeast side of Mackenzie River delta, south of Toker Point, Mackenzie District, Northwest Territories, Canada."
    DISTRIBUTION: Mainland tundra west of Hudson Bay (Canada), and N Alaska (USA).
    STATUS: IUCN – Lower Risk (lc).
    COMMENTS: Subgenus *Otisorex*; *S. cinereus* group (Demboski and Cook, 2003). Karyotype has 2n = 60, FN = 62. Formerly included in *cinereus*, but van Zyll de Jong (1976, 1991*b*) provided arguments for a specific destinction of *ugyunak*; genetically related to *jacksoni*, *portenkoi*, *camtschatica*, and *pribilofensis* (Demboski and Cook, 2003). See Junge and Hoffmann (1981) and van Zyll de Jong (1983*a*, and in Wilson and Ruff, 1999) for further information.

*Sorex unguiculatus* Dobson, 1890. Ann. Mag. Nat. Hist., ser. 6, 5:155.
    COMMON NAME: Long-clawed Shrew.
    TYPE LOCALITY: Russia, "Saghalien [Sakhalin] Island; Nikolajewsk, at the mouth of the Amur

River." Ognev (1928:204) and Ellerman and Morrison-Scott (1951:52) both restricted the type locality to Sakhalin Isl.

DISTRIBUTION: Pacific coast of Siberia from Vladivostok to the Amur, and the islands of Sakhalin (Russia) and Hokkaido (Japan); from Corbet (1978c).

STATUS: IUCN – Lower Risk (lc).

SYNONYMS: *yesoensis* Kishida, 1924.

COMMENTS: Subgenus *Sorex*, *S. caecutiens* group. Karyotype has 2n = 42, FN = 68-70. The inclusion of *yesoensis* follows Abe (1967). Skaren (1964) suggested a relationship with *obscurus* (= *monticolus*) but this was rejected by Siivonen (1965) and Hoffmann (1971).

**Sorex vagrans** Baird, 1857. Mammalia, *in* Repts. U.S. Expl. Surv., 8(1):15.

COMMON NAME: Vagrant Shrew.

TYPE LOCALITY: USA, "Shoalwater Bay, W.T. [= Willapa Bay, Pacific Co., Washington]."

DISTRIBUTION: Riparian and montane areas of the N Great Basin and Columbia Plateau, north to S British Columbia and Vancouver Isl (Canada); east to W Montana, W Wyoming, and Wasatch Mtns (Utah); C Nevada to Sierra Nevada (California).

STATUS: IUCN – Lower Risk (lc).

SYNONYMS: *amoenus* Merriam, 1895; *dobsoni* Merriam, 1891; *nevadensis* Merriam, 1895; *shastensis* Merriam, 1899; *sukleyi* Baird, 1858; *trigonirostris* Jackson, 1922; *vancouverensis* Merriam, 1895; **halicoetes** Grinnell, 1913; **paludivagus** von Bloeker, 1939.

COMMENTS: Subgenus *Otisorex*. *S. vagrans* complex. Karyotype has 2n = 53-54, FN = 62-67 (Brown, 1974). Findley's (1955b) wide concept of the *vagrans* group was substantially modified by Hennings and Hoffmann (1977) and Junge and Hoffmann (1981). The group was partly revised by Carraway (1990). Demboski and Cook (2001) analyzed the phylogeography of the "dusky shrew complex" which overlaps with the *S. vagrans* complex of Carraway (1990).

**Sorex ventralis** Merriam, 1895. N. Am. Fauna, 10:75.

COMMON NAME: Chestnut-bellied Shrew.

TYPE LOCALITY: "Cerro San Felipe, Oaxaca, Mexico (altitude 1000 feet [305 m])."

DISTRIBUTION: Endemic to Mexico; in Distrito Federal, México, Oaxaca, Puebla, and Tlaxcala (Carraway, ms).

STATUS: IUCN – Lower Risk (lc).

COMMENTS: Referred to unnamed subgenus by George (1988:456). Similar to *saussurei* but smaller; see Junge and Hoffmann (1981), who allocated the species to subgenus *Sorex*. Hall (1981) included *ventralis* in *oreopolus*.

**Sorex veraecrucis** Jackson, 1925. Proc. Biol. Soc. Wash., 38:128.

COMMON NAME: Veracruz Shrew.

TYPE LOCALITY: [Mexico], "Xico, 6000 ft. [1829 m], Veracruz".

DISTRIBUTION: Mexico, from Coahuila to Chiapas at elevations ranging from 1600 to 3650 m or more (Carraway, ms).

SYNONYMS: **cristobalensis** Jackson, 1925; **oaxacae** Jackson, 1925.

COMMENTS: Referred to unnamed subgenus by George (1988:456, under *S. saussurei*). Previously included in *S. saussurei* (e.g., Junge and Hoffmann, 1981) but resurrected by Carraway (ms, 2003).

**Sorex veraepacis** Alston, 1877. Proc. Zool. Soc. Lond., 1877:445.

COMMON NAME: Verapaz Shrew.

TYPE LOCALITY: "Coban (Vera Paz) [Alta Verapaz], Guatemala."

DISTRIBUTION: Montane forests of C Guerrero, Puebla, and Veracruz, south through the highlands of Oaxaca and Chiapas (Mexico), to SW Guatemala.

STATUS: IUCN – Lower Risk (lc).

SYNONYMS: *teculyas* (Alston, 1877); *verae-pacis* Alston, 1877; **chiapensis** Jackson, 1925; **mutabilis** Merriam, 1898; *caudatus* Merriam, 1895 [not Hodgson, 1849, not Horsfield, 1851].

COMMENTS: Subgenus *Otisorex*. Mexican subspecies according to Carraway (ms).

*Sorex volnuchini* Ognev, 1922. Ann. Mus. Zool. Akad. St. Petersbourg, 22:322.
    COMMON NAME: Caucasian Pygmy Shrew.
    TYPE LOCALITY: Russia, Krasnodarskii kr., Adygeiskaya A.O. [middle course], r. Kisha (see
        Pavlinov and Rossolimo, 1987).
    DISTRIBUTION: S Russia and Caucasus States; Turkey and N Iran. Perhaps also Crimea,
        Ukraine.
    STATUS: IUCN – Lower Risk (lc).
    SYNONYMS: *colchica* Sokolov and Tembotov, 1989; ***dahli*** Zagorodnyuk, 1996.
    COMMENTS: Subgenus *Sorex*, *S. minutus* group. Formerly included in *minutus* but specimens
        from Caucasus have a slightly different karyotype (2n = 40, FN = 60) which led
        Kozlovsky (1973) and Sokolov and Tembotov (1989) to regard *volnuchini* as a full
        species. The karyotype of *S. buchariensis* is very similar (Ivanitskaya et al., 1977).
        Zaitsev and Osipova (2003) were able to distinguish *volnuchini* morphologically from
        *minutus*; they also documented Pleistocene records for *volnuchini*. Zagorodnyuk
        (1996c) described the population of Crimea as subspecies *dahli*, in contrast to the
        smaller *minutus* inhabiting mainland Ukraine. Kryštufek and Vohralík (2001)
        demonstrated that Turkish populations (except those from Thrace) represent
        *S. volnuchini*.

*Sorex yukonicus* Dokuchaev, 1997. J. Mammal., 78:814.
    COMMON NAME: Alaskan Tiny Shrew.
    TYPE LOCALITY: United States, Alaska, near Galena, Crown Creek, 1.75 mi. (2.83 km) N,
        2.25 mi. (3.62 km) W Beaver Creek (64°44′N, 156°50′W).
    DISTRIBUTION: C to SW Alaska.
    COMMENTS: Subgenus *Sorex*. Formerly included in *minutissimus* (Dokuchaev, 1994) but later
        described as a new species (Dokuchaev, 1997). Subsequently reported from SW Alaska
        by Peirce and Peirce (2000).

**Family Talpidae** G. Fischer, 1814. Zoognosia tabulis synopticis illustrata, 3:x.
    SYNONYMS: Desmaninae Thomas, 1912; Myaladae Gray, 1821; Myogalidae A. Milne-Edwards,
        1868; Myogalina Bonaparte, 1845; Scalopidae Cope, 1889; Talpinorum G. Fischer, 1814.
    COMMENTS: Subfamily systematics controversal. Cabrera (1925) proposed a division into five
        subfamilies, Yates (1984), Hutterer (1993a), and others used three subfamilies. Includes
        also the extinct Gaillardiinae Hutchison, 1968. Scalopinae is used here as a subfamily,
        contra Hutchison (1968) and others. *Desmana* and *Galemys* were often placed in a
        separate family, Desmanidae, or subfamily, Desmaninae; see Bobrinskii et al. (1965), and
        McKenna and Bell (1997). The present arrangement of subfamilies and tribes is mainly
        based on the genetic study of Shinohara et al. (2003). It deviates from the previous edition
        (Hutterer, 1993a) and from most other sources. Family reviewed by Gureev (1979); see
        also Gorman and Stone (1990). Relationships of recent moles discussed by Ziegler (1971),
        Yates and Moore (1990), Whidden (1990), and Shinohara et al. (2003). Systematics of
        North American forms reviewed by Yates and Greenbaum (1982); of Palearctic forms by
        Corbet (1978c); of Siberian forms by Yudin (1989); of Japanese forms by Abe (1988),
        Motokawa and Abe (1996), Motokawa et al. (2001b), Okamoto (1999), and Tsuchiya et al.
        (2000). Phylogeny of Urotrichini discussed by Storch and Qiu (1983). Skull morphology
        described by Koppers (1990), myology by Whidden (1990, and references therein).
        Karyotype data reviewed by Yates and Schmidly (1975) and Kawada et al. (2002b).

**Subfamily Scalopinae** Gill, 1875. Bull. Geol. Geogr. Surv., 1, 2:106.
    SYNONYMS: Scalopeae Trouessart, 1879; Scalopes Gill, 1875; Scalopes Dobson, 1883.
    COMMENTS: Formerly included in the Talpinae, but Shinohara et al. (2003) convincingly
        showed that *Condylura, Parascalops, Scalopus* and *Scapanus* form a monophyletic clade
        distinct from the Talpinae, a conclusion reached earlier by Thomas (1912c).

**Tribe Condylurini** Gill, 1875. Bull. Geol. Geogr. Surv., 1, 2:106.

*Condylura* Illiger, 1811. Prodr. Syst. Mamm. Avium, p. 125.
    TYPE SPECIES: *Sorex cristatus* Linnaeus, 1758.

SYNONYMS: *Astromycter* Harris, 1825; *Astromyctes* Gray, 1843; *Astromydes* Blyth, 1863; *Rhinaster* Wagler, 1830; *Talpasorex* Schinz, 1821 [not Lesson, 1827].

COMMENTS: Reviewed by Peterson and Yates (1980). Recorded from the Pliocene of Europe; see Skoczen (1976).

*Condylura cristata* (Linnaeus, 1758). Syst. Nat., 10th ed., 1:53.
   COMMON NAME: Star-nosed Mole.
   TYPE LOCALITY: USA, Pennsylvania.
   DISTRIBUTION: Georgia and NW South Carolina (USA) to Nova Scotia and Labrador (Canada); Great Lakes region to SE Manitoba.
   STATUS: IUCN – Lower Risk (lc).
   SYNONYMS: *caudata* (Zimmermann, 1777); *longicaudata* (Erxleben, 1777); *macroura* Harlan, 1825; *prasinata* Harris, 1825; *prasinatus* (Harris, 1825); *radiata* (Shaw, 1800); *radiatus* (Shaw, 1800); *nigra* Smith, 1940; *parva* Paradiso, 1959.
   COMMENTS: Reviewed by Peterson and Yates (1980, Mammalian Species No. 129) who included *parva* as a subspecies. However, Hartman (in Wilson and Ruff, 1999) stated that *nigra* is a synonym of *parva*, in which case *nigra* has priority. Karyotype has 2n=34, FN = 64 (Yates and Schmidly, 1975).

**Tribe Scalopini** Gill, 1875. Bull. Geol. Geogr. Surv., 1, 2:106.
   SYNONYMS: Parascalopina Hutchison, 1968.

*Parascalops* True, 1894. Diagnoses New N. Am. Mamm., p. 2. (preprint of Proc. U.S. Natl. Mus., 17:242).
   TYPE SPECIES: *Scalops breweri* Bachman, 1842.
   COMMENTS: Reviewed by Hallett (1978).

*Parascalops breweri* (Bachman, 1842). Boston J. Nat. Hist., 4:32.
   COMMON NAME: Hairy-tailed Mole.
   TYPE LOCALITY: "Martha's Vineyard."; restricted to "E. North America" by Hall and Kelson (1959).
   DISTRIBUTION: NE United States and SE Canada.
   STATUS: IUCN – Lower Risk (lc).
   COMMENTS: Reviewed by Hallett (1978, Mammalian Species No. 98). Karyotype has 2n = 34, FN = 62 (Yates and Schmidly, 1975).

*Scalopus* E. Geoffroy, 1803. Cat. mamm. Mus. Nat. Hist. Nat.:77.
   TYPE SPECIES: *S[calopus]. virginanus* E. Geoffroy, 1803 = *Sorex aquaticus* Linnaeus, 1758.
   SYNONYMS: *Hesperoscalops* Hibbard, 1941; *Scalops* Illiger, 1811; *Scalpos* Brooks, 1910; *Talpasorex* Lesson, 1827 [not Schinz, 1821].
   COMMENTS: Reviewed by Yates and Schmidly (1978). For authorship of the genus, see Grubb (2001) and Opinion 2005 of the International Commission on Zoological Nomenclature (2002b).

*Scalopus aquaticus* (Linnaeus, 1758). Syst. Nat., 10th ed., 1:53.
   COMMON NAME: Eastern Mole.
   TYPE LOCALITY: E USA; fixed by Jackson (1915:33) to Pennsylvania, Philadelphia.
   DISTRIBUTION: N Tamaulipas and N Coahuila (Mexico) through E USA to Massachusetts and Minnesota.
   STATUS: IUCN – Lower Risk (lc).
   SYNONYMS: *argentatus* (Audubon and Bachmann, 1842); *cryptus* Davis, 1942; *cupreata* (Rafinesque, 1814); *intermedius* Elliot, 1899; *pennsylvanica* (Harlan, 1825); *aereus* (Bangs, 1896); *pulcher* Jackson, 1914; *sericea* (Rafinesque, 1832); *virginianus* E. Geoffroy, 1803; *alleni* Baker, 1951; *anastasae* (Bangs, 1898); *australis* (Chapman, 1893); *bassi* Howell, 1939; *caryi* Jackson, 1914; *howelli* Jackson, 1914; *inflatus* Jackson, 1914; *machrinoides* Jackson, 1914; *machrinus* (Rafinesque, 1832); *montanus* Baker, 1951; *nanus* Davis, 1942; *parvus* (Rhoads, 1894); *porteri* Schwartz, 1952; *texanus* (J. A. Allen, 1891).
   COMMENTS: Subspecies taxonomy revised by Yates and Schmidly (1977). Yates and Schmidly (1978, Mammalian Species No. 105) and Yates (in Wilson and Ruff, 1999) listed sixteen

subspecies. Gureev (1979:254) listed *aereus* and *inflatus* as distinct species without comment. Hall (1981:72) included *aereus* and *inflatus* in *aquaticus*. Karyotype has 2n = 34, FN = 64 (Yates and Schmidly, 1975).

**Scapanulus** Thomas, 1912. Ann. Mag. Nat. Hist., ser. 8, 10:396.
    TYPE SPECIES: *Scapanulus oweni* Thomas, 1912.
    COMMENTS: For placement in Scalopini see Storch and Qiu (1983:118).

*Scapanulus oweni* Thomas, 1912. Ann. Mag. Nat. Hist., ser. 8, 10:397.
    COMMON NAME: Gansu Mole.
    TYPE LOCALITY: China, Kansu, "23 miles (37 km) S.E. of Tao-chou, 9000' " (2,743 m).
    DISTRIBUTION: Montane forest in C China: Kansu, Shensi and Sichuan.
    STATUS: IUCN – Lower Risk (lc).

**Scapanus** Pomel, 1848. Arch. Sci. Phys. Nat. Geneve, 9:247.
    TYPE SPECIES: *Scalops townsendii* Bachman, 1839.
    SYNONYMS: *Xeroscapheus* Hutchison, 1968.
    COMMENTS: Revised by Jackson (1915:54-76) and Hutchison (1987).

*Scapanus latimanus* (Bachman, 1842). Boston J. Nat. Hist., 4:34.
    COMMON NAME: Broad-footed Mole.
    TYPE LOCALITY: Santa Clara, Santa Clara Co., California, USA; *fide* Osgood (1907:52).
    DISTRIBUTION: SC Oregon (USA) to N Baja California (Mexico).
    STATUS: IUCN – Lower Risk (lc).
    SYNONYMS: *californicus* (Ayres, 1856); *townsendii* Peters, 1863; **anthonyi** J. A. Allen, 1893;
        **campi** Grinnell and Storer, 1916; **caurinus** F. G. Palmer, 1937; **dilatus** True, 1894;
        *alpinus* Merriam, 1897; *truei* Merriam, 1894; **grinnelli** Jackson, 1914; **insularis**
        F. G. Palmer, 1937; **minusculus** Bangs, 1899; **monoensis** Grinnell, 1918; **occultus**
        Grinnell and Swarth, 1912; **parvus** F. G. Palmer, 1937; **sericatus** Jackson, 1914.
    COMMENTS: Hall (1981:69-70) listed 12 subspecies. Reviewed by Verts and Carraway (2001,
        Mammalian Species No. 666). Ceballos and Navarro (1991) listed *anthonyi* as a distinct
        species without further reference. Karyotype has 2n = 34, FN = 64 (Yates and Schmidly,
        1975).

*Scapanus orarius* True, 1896. Proc. U.S. Natl. Mus., 19:52.
    COMMON NAME: Coast Mole.
    TYPE LOCALITY: USA, Washington, Pacific Co., Shoalwater Bay (= Willapa Bay).
    DISTRIBUTION: SW British Columbia (Canada) to NW California, WC Idaho, N Oregon,
        C and SE Washington (USA).
    STATUS: IUCN – Lower Risk (lc).
    SYNONYMS: **schefferi** Jackson, 1915; *yakimensis* Dalquest and Scheffer, 1944.
    COMMENTS: Includes *schefferi* as a subspecies; see Hartman and Yates (1985, Mammalian
        Species No. 253).

*Scapanus townsendii* (Bachman, 1839). J. Acad. Nat. Sci. Philadelphia, 8:58.
    COMMON NAME: Townsend's Mole.
    TYPE LOCALITY: USA, Washington, Clark Co., vicinity of Vancouver.
    DISTRIBUTION: SW British Columbia (Canada) to NW California (USA).
    STATUS: IUCN – Lower Risk (lc).
    SYNONYMS: *aeneus* (Cassin, 1853); *laeniata* (Le Conte, 1853); *towsendii* Pomel, 1848;
        **olympicus** Johnson and Yates, 1980.
    COMMENTS: Reviewed by Carraway et al. (1993, Mammalian Species No. 434).

**Subfamily Talpinae** G. Fischer, 1814. Zoognosia tabulis synopticis illustrata, 3:x.
    SYNONYMS: Desmaninae Thomas, 1912; Urotrichi Dobson, 1883.
    COMMENTS: Hutchison (1968), Storch and Qiu (1983), Hutterer (1993a), and other authors
        included also Condylurini and Scalopini, which are shifted here into Scalopinae, based on
        the the mtDNA phylogeny of Shinohara et al. (2001). The former Desmaninae are
        downgraded to tribal level, and a new tribe is proposed for the *Neurotrichus* lineage.

**Tribe Desmanini** Thomas, 1912. Ann. Mag. Nat. Hist., ser. 8, 10:397.

SYNONYMS: Myaladae Gray, 1821; Myogalina Bonaparte, 1837; Mygalina Pomel, 1848; Myogalidae A. Milne-Edwards, 1868; Myogalinae Gill, 1875; Myogalini Winge, 1917.

COMMENTS: Commonly regarded as a subfamily, sometimes even as a separate family; see Barabasch-Nikiforow (1975). Hutchinson (1974) concluded that Desmaninae and Talpidae were separated since the Eocene. Taxonomy of fossil and extant taxa reviewed by Rümke (1985). Distribution and status reviewed by Queiroz et al. (1996).

*Desmana* Güldenstaedt, 1777. Beschaft. Berliner Ges. Naturforsch. Fr., 3:108.

TYPE SPECIES: *Castor moschatus* Linnaeus, 1758.

SYNONYMS: *Caprios* Wagler, 1830; *Desman* Lacepède, 1799; *Desmanus* Rafinesque, 1815; *Galemodesmana* Topachevskii and Pashkov, 1983; *Myale* Gray, 1821; *Mygale* Cuvier, 1800 [not Latreille, 1802]; *Myogale* Brandt, 1836; *Myogalea* J. B. Fischer, 1829; *Palaeospalax* Owen, 1846; *Pliodesmana* Topachevskii and Pashkov, 1983; *Praedesmana* Topachevskii and Pashkov, 1983.

*Desmana moschata* (Linnaeus, 1758). Syst. Nat., 10th ed., 1:59.

COMMON NAME: Desman.

TYPE LOCALITY: "Habitat in Russiae aquosis."

DISTRIBUTION: Russia, Ukraine, and Kazakhstan; Don, Volga, and Ural Rivers and their tributaries; introduced into Tachan and Tartas Rivers (Ob basin) and Dnepr River. Almost extinct in Belarus.

STATUS: IUCN – Vulnerable.

SYNONYMS: *moschatus* Linnaeus, 1958; *moscovitica* (Geoffroy, 1811).

COMMENTS: Borodin (1963), Barabash-Nikiforov (1968, 1975), Khakhin and Ivanov (1990), and Queiroz et al. (1996) reviewed the morphology, distribution and ecology of the species. Dental formula discussed by Kawada et al. (2002c).

*Galemys* Kaup, 1829. Skizz. Entwickel.-Gesch. Nat. Syst. Europ. Thierwelt, 1:119.

TYPE SPECIES: *Mygale pyrenaica* E. Geoffroy, 1811.

SYNONYMS: *Galomys* Agassiz, 1846; *Mygalina* I. Geoffroy in Gervais, 1835.

*Galemys pyrenaicus* (E. Geoffroy St. Hilaire, 1811). Ann. Mus. Hist. Nat. Paris, 17:193.

COMMON NAME: Pyrenean Desman.

TYPE LOCALITY: France, "Les montagnes pres de Tarbes (Hautes-Pyrenees)".

DISTRIBUTION: Streams of the Pyrenees and the northern and central mountains of the Iberian Peninsula (France, Andorra, Spain and Portugal).

STATUS: IUCN – Vulnerable.

SYNONYMS: *pyrenaica* (Geoffroy, 1811); **rufulus** (Graells, 1897).

COMMENTS: Includes *rufulus* as a possible subspecies; reviewed by Palmeirim and Hoffmann (1983, Mammalian Species No. 207) and Juckwer (1990). Karyotype has 2n = 42, FN = 68 (Peyre, 1957).

**Tribe Neurotrichini**, new tribe.

COMMENTS: Type genus – *Neurotrichus* Günther, 1880. Definition – Extant species represents smallest New World talpid; semi-fossorial moles that spend much time and nest above ground; tail about half the length of head-and-body, thick, constricted at base, scaled, annulated, covered sparsely with long hairs; pelage black to blue-black; eyes rudimentary, pinnae absent; digitigrade, pentadactyle feet scaly; forefoot longer than broad, equipped with long curved claws; 6 weak tubercles on sole of hindfoot (Hall, 1981; Carraway and Verts, 1991, Mammalian Species 387); humerus shrew-like (Storch and Qiu, 1983); bullae incomplete; skull with broad braincase and wide interorbital constriction; zygoma short; 36 teeth, six upper and seven lower molariform teeth on each side; for details of the dentition and differences to *Urotrichus*, see Storch and Qiu (1983). Contents – *Neurotrichus* Günther, 1880; *Quyania* Storch and Qiu, 1983.

A molecular study by Shinohara et al. (2003) clearly revealed that *Neurotrichus* has no sister relationship to *Urotrichus* and *Dymecodon* and therefore cannot be included in the Urotrichini. Storch and Qiu (1983) described similarities in the dentition and humerus of

*Neurotrichus* and *Quyania* Storch and Qiu, 1983 from the Neogene of China, and postulated a long separation of Urotrichini and *Neurotrichus* (plus *Quyania*). *Neurotrichus* is the only known vertebrate to possess a pigmented layer covering the anterior surface of the eye lens (Lewis, 1983), and the only talpid known to possess a pair of ampullary glands and have external lobulation of the two bodies of the prostate gland (Eadie, 1951).

*Neurotrichus* Günther, 1880. Proc. Zool. Soc. Lond., 1880:441.
> TYPE SPECIES: *Urotrichus gibbsii* Baird, 1858.
> COMMENTS: Formerly in Urotrichini; see comments above, Hutchison (1968), Storch and Qiu (1983:100), among many others.

*Neurotrichus gibbsii* (Baird, 1858). Mammalia, *in* Repts. U.S. Expl. Surv., 8(1):76.
> COMMON NAME: Shrew-mole.
> TYPE LOCALITY: USA, Washington, Pierce Co., "Naches Pass, 4,500 ft." (1,372 m).
> DISTRIBUTION: SW British Columbia (Canada) to WC California (USA).
> STATUS: IUCN 2000 – Not listed.
> SYNONYMS: *major* Merriam, 1899; **hyacinthinus** Bangs, 1897; **minor** Dalquest and Burgner, 1941.
> COMMENTS: Hall (1981:67) and Yates (in Wilson and Ruff, 1999) listed *hyacinthinus* and *minor* as subspecies. Reviewed by Carraway and Verts (1991*b*, Mammalian Species No. 387). Karyotype has 2n = 38, FN = 72 (Yates and Schmidly, 1975).

**Tribe Scaptonychini** Van Valen, 1967. Bull. Amer. Mus. Nat. Hist., 135:263.

*Scaptonyx* Milne-Edwards, 1872. *In* David, Nouv. Arch. Mus. Hist. Nat. Paris, Bull. 7, p. 92.
> TYPE SPECIES: *Scaptonyx fusicauda* Milne-Edwards, 1872.
> COMMENTS: For placement in Scaptonychini see Van Valen (1967).

*Scaptonyx fusicaudus* Milne-Edwards, 1872. *In* David, Nouv. Arch. Mus. Hist. Nat. Paris, Bull. 7, p. 92.
> COMMON NAME: Long-tailed Mole.
> TYPE LOCALITY: "Frontière du Kokonoor", vicinity of Kukunor (Lake), China.
> DISTRIBUTION: N Burma; S China, Tsinghai, Shensi, Sichuan and Yunnan; N Vietnam.
> STATUS: IUCN – Lower Risk (lc).
> SYNONYMS: *affinis* Thomas, 1912; *fusicaudatus* Milne-Edwards, 1872.
> COMMENTS: Includes *affinis*; see Ellerman and Morrison-Scott (1966:35). Recently recorded from Mt. Tay Con Linh II in N Vietnam (Lunde et al., 2003*b*).

**Tribe Talpini** G. Fischer, 1814. Zoognosia tabulis synopticis illustrata, 3:x.
> SYNONYMS: Talpae Gill, 1875; Talpae Dobson, 1883.

*Euroscaptor* Miller, 1940. J. Mammal., 21:443.
> TYPE SPECIES: *Talpa klossi* Thomas, 1929.
> SYNONYMS: *Eoscalops* Stroganov, 1941.
> COMMENTS: Corbet (1978*c*:32), and subsequent work, included *Euroscaptor* in *Talpa* while Russian and Japanese authors retained it as a genus; most recently Abe et al. (1991). Species allocations and limits are tentative.

*Euroscaptor grandis* Miller, 1940. J. Mammal., 21:444.
> COMMON NAME: Greater Chinese Mole.
> TYPE LOCALITY: China, Sichuan, "Mount Omei, alt. 5000 feet [1524 m]", = Omei-Shan.
> DISTRIBUTION: N and S Bakbo and Cha-pa (Vietnam); S China.
> STATUS: IUCN – Lower Risk (lc).
> COMMENTS: Often included in *Talpa*; but see Gureev (1979:272). Regarded as a synonym of [*E*]. *micrura longirostris* by Ellerman and Morrison-Scott (1966:40).

*Euroscaptor klossi* (Thomas, 1929). Ann. Mag. Nat. Hist., ser. 10, 3:206.
> COMMON NAME: Kloss's Mole.
> TYPE LOCALITY: Thailand, Tonkin.

DISTRIBUTION: Highlands of Thailand, Laos and Peninsular Malaysia.
STATUS: IUCN – Lower Risk (lc).
SYNONYMS: *malayana* Chasen, 1940.
COMMENTS: Corbet (1978*c*:33) and Corbet and Hill (1991:38) included *klossi* in *micrura*; but see Yoshiyuki (1988*b*). May include *malayana*, which Harrison (1974:57) included in *micrura*.

*Euroscaptor longirostris* (Milne-Edwards, 1870). C.R. Acad. Sci. Paris, 70:341.
COMMON NAME: Long-nosed Mole.
TYPE LOCALITY: China, Sichuan, Moupin.
DISTRIBUTION: S China.
STATUS: IUCN – Lower Risk (lc).
COMMENTS: Type species of *Eoscalops* Stroganov, 1941. Formerly included in *micrura* by Ellerman and Morrison-Scott (1966:40) and Corbet (1978*c*:35). In the *Euroscaptor* group of *Talpa*; see Gureev (1979:272).

*Euroscaptor micrura* (Hodgson, 1841). Calcutta J. Nat. Hist., 2:221.
COMMON NAME: Himalayan Mole.
TYPE LOCALITY: Nepal, C and N hills.
DISTRIBUTION: E Himalaya and Peninsular Malaysia.
STATUS: IUCN – Lower Risk (lc).
SYNONYMS: *cryptura* Blyth, 1843.
COMMENTS: Does not include *klossi*; see Yoshiyuki (1988*b*). Does not include *malayana*, which Harrison (1974:57) included in *micrura*. Kawada et al. (2003) identified moles from the Cameron Highlands (Peninsular Malaysia) as belonging to *E. micrura*.

*Euroscaptor mizura* (Gunther, 1880). Proc. Zool. Soc. Lond., 1880:441.
COMMON NAME: Japanese Mountain Mole.
TYPE LOCALITY: Japan, Honshu, "In the neighbourhood of Yokohama".
DISTRIBUTION: Mountains of Honshu (Japan).
STATUS: IUCN – Vulnerable.
SYNONYMS: *hiwaensis* (Imaizumi, 1955); **othai** (Imaizumi, 1955).
COMMENTS: Imaizumi (1970*b*) and Abe et al. (1991) included this species in the genus *Euroscaptor*, while Corbet (1978*c*) placed it in *Talpa*. Three populations have been named, of which *othai* represents "probably a distinct species", according to Imaizumi (1970*b*); a view supported by Yoshiyuki (1988*b*). Karyotype (2n = 36, FN = 52) described by Kawada et al. (2001).

*Euroscaptor parvidens* (Miller, 1940). J. Mammal., 21:203.
COMMON NAME: Small-toothed Mole.
TYPE LOCALITY: Vietnam, Di Linh, Blao Forest Station.
DISTRIBUTION: Known from type locality and Rakho on the Chinese border.
STATUS: IUCN – Critically Endangered.
COMMENTS: Ellerman and Morrison-Scott (1966:40) included this species in [*E*]. *micrura leucura*. Corbet (1978*c*:33) mentioned *leucura* as a species, but Gureev (1979:274) also listed *parvidens* in the *Euroscaptor* group of *Talpa*, where Miller (1940*b*:444) put his species soon after description. Corbet and Hill (1991:38) did not list *parvidens* and one may assume that they included it in *micrura*.

*Mogera* Pomel, 1848. Arch. Sci. Phys. Nat. Geneve, 9:246.
TYPE SPECIES: *Talpa wogura* Temminck, 1842.
SYNONYMS: *Nesoscaptor* Abe, Shiraishi and Arai, 1991.
COMMENTS: Formerly included in *Talpa* by Corbet (1978*c*); but see Imaizumi (1970*b*), Gureev (1979), Yudin (1989), and Abe et al. (1991). Revised by Abe (1995). *Nesoscaptor uchidai* (Abe et al., 1991) was included in *Mogera* by Motokawa et al. (2001*b*).

*Mogera imaizumii* (Kuroda, 1957). J. Mammal. Soc. Japan, 1:74.
COMMON NAME: Small Japanese Mole.
TYPE LOCALITY: Japan, Honshu, Tochigi Pref., Shiobara.
DISTRIBUTION: Shikoku and N and SC Honshu, Japan.

STATUS: IUCN – Lower Risk (lc) as *M. minor*.

SYNONYMS: *minor* (Kuroda, 1936) [not Freudenberg, 1914].

COMMENTS: Included in *Talpa* [*Euroscaptor*] *micrura* by Ellerman and Morrison-Scott (1966); but retained as a separate species by Yoshiyuki (1986). Renamed *Talpa wogura imaizumii* by Kuroda (1957) for homonymy with *Talpa europaea* var. *minor*. In a revision of Japanese moles Abe (1995) reintroduced the name *M. minor*, but Motokawa and Abe (1996) subsequently corrected this view. Karyotype from Honshu has 2n = 36, FN = 54 (Kawada et al., 2001). Populations from E Honshu and W Honshu plus Shikoku exhibit considerable genetic differences (Tsuchiya et al., 2000).

*Mogera insularis* (Swinhoe, 1863). Proc. Zool. Soc. Lond., 1862:356 [1863].

COMMON NAME: Insular Mole.

TYPE LOCALITY: "Formosa (China)" = Taiwan.

DISTRIBUTION: Taiwan, Hainan, SE China.

STATUS: IUCN – Lower Risk (lc).

SYNONYMS: *montana* Kano, 1940 [*nomen nudum*]; **hainana** Thomas, 1910; **latouchei** Thomas, 1907.

COMMENTS: Includes *latouchei*; see Corbet and Hill (1991*c*:38). Included in *Talpa* [*Euroscaptor*] *micrura* by Ellerman and Morrison-Scott (1951:40); but see Corbet (1978*c*:33), Abe (1995) and Motokawa and Abe (1996). Karyotype of a specimen from Taiwan determined as 2n = 32, FN = 58 (L.-K. Lin et al., 2002*a*). The status of the mainland (*latouchei*) and Hainan (*hainana*) populations still has to be determined (Abe, 1995). Skulls of *latouchei* are smaller than in *insularis*.

*Mogera tokudae* Kuroda, 1940. [A monograph of Japanese mammals . . . ], Tokyo and Osaka, p. 196.

COMMON NAME: Sado Mole.

TYPE LOCALITY: Japan, Sado Isl.

DISTRIBUTION: Sado Isl and Echigo Plain, Honshu, C Japan.

STATUS: IUCN – Endangered as *M. tokudae* and *M. etigo*.

SYNONYMS: **etigo** Yoshiyuki and Imaizumi, 1991.

COMMENTS: Overlooked by Ellerman and Morrison-Scott (1966); included in *Talpa robusta* by Corbet (1978*c*); but retained as a separate species by Yoshiyuki (1986) and Abe et al. (1991). Yoshiyuki and Imaizumi (1991) named the Honshu population as a separate species, but Abe (1995, 1996) ascribed the differences between the Sado and Honshu population to geographical size variation. However, considerable genetic (Tsuchiya et al., 2000) and karyological differences exist: the karyotype from Sado is 2n = 36, FN = 60, and from Honshu 2n = 36, FN = 54 (Kawada et al., 2001; Tsuchiya, 1988).

*Mogera uchidai* (Abe, Shiraishi and Arai, 1991). J. Mammal. Soc. Japan, 15:53.

COMMON NAME: Senkaku Mole.

TYPE LOCALITY: Japan, Ryukyu Isls, Senkaku Isls, west coast of Uotsuri-jima.

DISTRIBUTION: Known only from the type locality.

STATUS: IUCN – Endangered.

comments: Originally described as *Nesoscaptor uchidai*, but Motokawa et al. (2001*b*) recognized a close relationship to *Mogera insularis*. The species is endemic to the Senkaku Isls (Motokawa, 2000) and seriously threatened by habitat degradation (Yokohata, 1999).

*Mogera wogura* (Temminck, 1842). *In* Siebold, Fauna Japonica, 1(Mamm.), 1:19.

COMMON NAME: Japanese Mole.

TYPE LOCALITY: Japan; restricted to Yokohama, Honshu by Thomas (1905*b*), but believed to have come from W or S Kyushu by Abe (1995).

DISTRIBUTION: Japan (Honshu, Kyushu, Shikoku, Senkaku, Tane, Amakusa, Tsushima and other Isls), Korea to NE China and adjacent Siberia (Abe, 1995, 1996).

STATUS: IUCN – Lower Risk (lc) as *M. wogura*, *M. kobeau*, and *M. robusta*.

SYNONYMS: *aquilonaris* Kishida, 1950 [*nomen nudum*]; *gracilis* Kishida, 1936; *kanai* Thomas, 1905; *kiusiuana* Kuroda, 1940; *kobeae* Thomas, 1905; *moogura* Temminck, 1842; **robusta** Nehring, 1891; *coreana* Thomas, 1907.

COMMENTS: For a taxonomic discussion see Corbet (1978*c*), who treated *robusta* as a different species. European authors often included *kobeae* and *tokudae*; however, Japanese authors (Imaizumi 1970*b*; Yoshiyuki 1988*b*) treated them as separate species. Formerly included in *Talpa*; but see Imaizumi (1970*b*), Gureev (1979), and Gromov and Baranova (1981). The present arrangement follows Abe (1995). However, moles from Japan have a different karyotype (2n = 36, FN = 52) than moles from the Korean mainland (2n = 36, FN = 58) (Kawada et al., 2001). Mitochondrial cytochrome *b* gene sequences studied by Tsuchiya et al. (2000) revealed three clades in Japan (Honshu, Shikoku, and Kyushu) and two distinct clades on the mainland of Korea and E Russia.

*Parascaptor* Gill, 1875. Bull. U.S. Geol. Geogr. Surv. Terr., I, 2:110.
    TYPE SPECIES: *Talpa leucura* Blyth, 1850.
    COMMENTS: Included in *Talpa* by Corbet and Hill (1991:38); but retained as a genus by Abe et al. (1991).

*Parascaptor leucura* (Blyth, 1850). J. Asiat. Soc. Bengal, 19:215, pl. 4.
    COMMON NAME:White-tailed Mole.
    TYPE LOCALITY: India, Assam, Khasi Hills, Cherrapunji.
    DISTRIBUTION: Burma, Assam (India), and Yunnan (China).
    STATUS: IUCN – Lower Risk (lc).
    COMMENTS: Formerly included in *T. micrura*; see Ellerman and Morrison-Scott (1951:40); but also see Corbet (1978*c*:33).

*Scaptochirus* Milne-Edwards, 1867. Ann. Sci. Nat. Zool. (Paris), 7:375.
    TYPE SPECIES: *Scaptochirus moschatus* Milne-Edwards, 1867.
    SYNONYMS: *Chiroscaptor* Heude, 1898.
    COMMENTS: Included in *Talpa* by Corbet (1978*c*:36) and Corbet and Hill (1991:38). Retained as a genus by Abe et al. (1991) and Gureev (1979:282).

*Scaptochirus moschatus* Milne-Edwards, 1867. Ann. Sci. Nat. Zool. (Paris), 7:375.
    COMMON NAME: Short-faced Mole.
    TYPE LOCALITY: "En Mongolie"; Swanhwafu, 100 mi. (= 161 km) NW of Peking, China.
    DISTRIBUTION: NE China: Hopei, Shantung, Shansi, Shensi.
    STATUS: IUCN – Lower Risk (lc).
    SYNONYMS: *davidianus* Swinhoe, 1879 [accidental renaming, *nomen oblitum*, not Milne-Edwards, 1884]; *gilliesi* Thomas, 1910; *grandidens* (Stroganov, 1941); *leptura* (Thomas, 1881); *moschiferus* Heude, 1898; *sinensis* (Heude, 1898).
    COMMENTS: Included in *Talpa micrura* by Ellerman and Morrison-Scott (1966:40); but see Corbet (1978*c*:36). Grulich (1982) pointed out that *Scaptochirus davidianus* Milne-Edwards, 1884, often regarded as a synonym of *moschatus*, is a species of *Talpa;* see under *Talpa davidiana*. Karyotype (2n = 48, FN = 54-56) described by Kawada et al. (2002*b*).

*Talpa* Linnaeus, 1758. Syst. Nat., 10th ed., 1:52.
    TYPE SPECIES: *Talpa europaea* Linnaeus, 1758.
    SYNONYMS: *Asioscalops* Stroganov, 1941; *Asioscaptor* Schwarz, 1948; *Heterotalpa* Peters, 1863; *Talpops* Gervais, 1868.
    COMMENTS: Includes *Asioscalops* which was retained as a full genus by Yudin (1989:60). Schwarz (1948), Corbet (1978*c*:36), and Corbet and Hill (1991:38) included *Euroscaptor*, *Parascaptor*, *Mogera*, and *Scaptochirus*; but these are retained here as full genera, see Abe et al. (1991) and Gureev (1979:256-285). For phylogenetic considerations based on morphology, see Stein (1960) and Grulich (1971); based on allozyme variation, see Filippucci et al. (1987). For a review of European species, see Niethammer and Krapp (1990).

*Talpa altaica* Nikolsky, 1883. Trans. Soc. Nat. St. Petersburg, 14:165.
    COMMON NAME: Altai Mole.
    TYPE LOCALITY: Russia, Siberia, Altai Mtns, Valley of Tourak.
    DISTRIBUTION: Taiga zone of Siberia between Ob and Lena Rivers; south to N Mongolia.
    STATUS: IUCN – Lower Risk (lc).

SYNONYMS: *gusevi* (Fetisov, 1956); *irkutensis* Dybowski, 1922; *saianensis* Bielovusev, 1921; *salairica* Egorin, 1936; *salairici* Corbet, 1978 [*lapsus*]; *sibirica* Egorin, 1937; *suschkini* Kastschenko, 1905; *tymensis* Egorin, 1937.

COMMENTS: Placed by Yudin (1989:52) in genus *Asioscalops*; but see Corbet (1978*c*:33). Kratochvíl and Kral (1972) provided karyological evidence for a separation of *altaica* from the remaining *Talpa* species.

*Talpa caeca* Savi, 1822. Nuovo Giorn. de Letterati, Pisa, 1:265.

COMMON NAME: Blind Mole.

TYPE LOCALITY: Italy, Pisa.

DISTRIBUTION: S Europe and (doubtfully) Asia Minor; Alps, Apennines, Balkan, Thrazia.

STATUS: IUCN – Lower Risk (lc).

SYNONYMS: *dobyi* Grulich, 1971; *minor* Freudenberg, 1914; **augustana** Capolongo and Panasci, 1978; **hercegovinensis** Bolkay, 1925; *beaucournui* Grulich, 1971; *olympica* Chaworth-Musters, 1932; **steini** Grulich, 1971 [see Niethammer, *in* Niethammer and Krapp, 1990].

COMMENTS: Kryštufek (1994) revised the taxonomy of European populations of *T. caeca* and demonstrated morphometric differences to *stankovici* and to the Caucasus moles (*caucasica*, *levantis*). Species reviewed by Niethammer (*in* Niethammer and Krapp, 1990). In the paleontological literature, *Talpa minor* Freudenberg, 1914 is either regarded as a fossil ancestor of, or as conspecific with the extant *caeca* (Rabeder, 1972). However, Cleef-Roders and Hoek Ostende (2001) pointed out clear differences in the dentition of *minor* and *caeca* and regarded both as different species. Karyotype has 2n = 36, FN = 64 (Ticino) or 68 (Balkans) (see Niethammer, *in* Niethammer and Krapp, 1990).

*Talpa caucasica* Satunin, 1908. Mitt. Kaukas. Mus., 4:5.

COMMON NAME: Caucasian Mole.

TYPE LOCALITY: Russia, Stavropol Krai, Stavropol.

DISTRIBUTION: NW Caucasus (Russia and Georgia, NE Turkey), Talysh Mtns (Iran).

STATUS: IUCN – Lower Risk (lc).

SYNONYMS: **ognevi** Stroganov, 1944; **orientalis** Ognev, 1926.

COMMENTS: Included in *europaea* by Ellerman and Morrison-Scott (1951), but considered a distinct species by Gromov et al. (1963). Reviewed by Sokolov and Tembotov (1989) and Zaitsev (1999). New records from Turkey (Kefelioglu and Gencoglu, 1996) and Iran (Kryštufek and Benda, 2002) considerably extend the range of the species. Karyotype has 2n = 38, FN = 64 (Sokolov and Tembotov, 1989).

*Talpa europaea* Linnaeus, 1758. Syst. Nat., 10th ed., 1:52.

COMMON NAME: European Mole.

TYPE LOCALITY: Sweden, Kristianstad, Engelholm.

DISTRIBUTION: Temperate Europe including Britain to the Ob and Irtysh Rivers (Russia) in the east.

STATUS: IUCN – Lower Risk (lc).

SYNONYMS: *alba* Gmelin, 1788; *albida* Reichenbach, 1852; *albo-maculata* Erxleben, 1777; *brauneri* Satunin, 1909; *caudata* Boddaert, 1772; *ehiki* Czajlik, 1987; *flavescens* Reichenbach, 1836; *frisius* Müller, 1776; *friseus* Corbet, 1978 [*lapsus*]; *kratochvili* Grulich, 1969; *kratochvilli* Corbet, 1978 [*lapsus*]; *lutea* Reichenbach, 1852; *maculata* Fitzinger, 1869; *major* Bechstein, 1800 [not of Altobello, 1920]; *nigra* Kerr, 1792; *obensis* Skalon and Rajevsky, 1940; *pancici* Martino, 1930; *scalops* Schulze, 1897; *transuralensis* Stroganov, 1956; *uralensis* Ognev, 1925; *variegata* Gmelin, 1788; *vulgaris* Boddaert, 1785; **cinerea** Gmelin, 1788; *grisea* Fitzinger, 1869; *rufa* Borkhausen, 1797; **velessiensis** Petrov, 1941.

COMMENTS: Husson and Heurn (1959) recognized and named twelve color morphs from the Netherlands that are not listed here. Does not include *altaica*, *caucasica*, *romana*, and *stankovici*; see comments under these species. Only *europaea* and *cinerea* were regarded as subspecies by Niethammer (*in* Niethammer and Krapp, 1990). Doğramaci (1989) considered *velessiensis* as the valid subspecies for Turkish Thrace. A biological review of the species was provided by Witte (1997). Karyotype has 2n = 34, FN = 68 (Kratochvil and Kral, 1972).

*Talpa davidiana* (Milne-Edwards, 1884). Compt. Rend. Acad. Sci., Paris, 99:1141.
    COMMON NAME: Père David's Mole.
    TYPE LOCALITY: SE Turkey, "environs d'Akbès, sur les confins de la Syrie et de l'Asie Mineure"
        = Meydanekbez, SW Gaziantep.
    DISTRIBUTION: SE Turkey, NW Iran.
    STATUS: IUCN – Critically Endangered as *Talpa streeti*.
    SYNONYMS: *streeti* Lay, 1965; *streetorum* Lay, 1967.
    COMMENTS: Includes *Talpa streeti*, as already suggested by Grulich (1982). Kryštufek et al.
        (2001) revised the species, discussed the morphological variation, and mapped all
        known localities.

*Talpa levantis* Thomas, 1906. Ann. Mag. Nat. Hist., ser. 7, 17:416.
    COMMON NAME: Levant Mole.
    TYPE LOCALITY: Turkey, S Trabzon, Scalita (= Altindere).
    DISTRIBUTION: Bulgaria, Thracia and N Anatolia (Turkey), and adjacent Caucasus.
    STATUS: IUCN – Lower Risk (lc).
    SYNONYMS: **minima** Deparma, 1960; **talyschensis** Vereschchagin, 1945; **transcaucasica**
        Dahl, 1944.
    COMMENTS: On specific status, see Grulich (1972) and Felten et al. (1973). Reviewed by
        Sokolov and Tembotov (1989), Zaitsev (1999), and Kryštufek (2001a, b). European
        records by Vohralík (1991). Karyotype has 2n = 34, FN = 68 (Kefelioglu and Gencoglu,
        1996).

*Talpa occidentalis* Cabrera, 1907. Ann. Mag. Nat. Hist., ser. 7, 20:212.
    COMMON NAME: Spanish Mole.
    TYPE LOCALITY: C Spain, Guadarrama Mtns, 1200-1300 m, "La Granja, Segovia".
    DISTRIBUTION: W and C Iberian Peninsula (Portugal, Spain).
    STATUS: IUCN – Lower Risk (lc).
    COMMENTS: Formerly regarded as a subspecies of *caeca*, but see Ramalhinho (1985) and
        Filippucci et al. (1987). Reviewed by Niethammer (*in* Niethammer and Krapp, 1990).
        Karyotype has 2n = 34, FN = 68. Dental morphology of *occidentalis* and *europaea*
        described by Cleef-Roders and Hoek Ostende (2001).

*Talpa romana* Thomas, 1902. Ann. Mag. Nat. Hist., ser. 7, 10:516.
    COMMON NAME: Roman Mole.
    TYPE LOCALITY: Italy, Ostia near Rome.
    DISTRIBUTION: Apennines, Italy, and extreme SE France; a historical record from Sicily.
    STATUS: IUCN – Lower Risk (lc).
    SYNONYMS: *major* Altobello, 1920 [not Bechstein, 1800]; **adamoi** Capolongo et Panasci,
        1976; **aenigmatica** Capolongo and Panasci, 1976; **brachycrania** Capolongo and
        Panasci, 1976; **montana** Cabrera, 1925; **wittei** Capolongo, 1986 [see Niethammer, *in*
        Niethammer and Krapp, 1990].
    COMMENTS: Does not include *stankovici*, see comments therein. Karyotype as in *europaea*
        (Capanna, 1981).

*Talpa stankovici* V. Martino and E. Martino, 1931. J. Mammal., 12:53.
    COMMON NAME: Balkan Mole.
    TYPE LOCALITY: Serbia and Montenegro, Pelister Mtns, "Magarevo Mts., Perister, S. Serbia
        (Macedonia). Alt. 1000 m."
    DISTRIBUTION: European Balkans, Greece including Corfu Isl, S Serbia and Montenegro,
        Macedonia; probably Albania.
    STATUS: IUCN – Lower Risk (lc).
    SYNONYMS: **montenegrina** Kryštufek, 1994.
    COMMENTS: Formerly included in *romana*; specific status supported by Filippucci et al.
        (1987). Reviewed by Niethammer (*in* Niethammer and Krapp, 1990) and Kryštufek
        (1994).

**Tribe Urotrichini** Dobson, 1883. Monogr. Insectivora, 2:128.
    SYNONYMS: Urotrichi Dobson, 1883.

*Dymecodon* True, 1886. Proc. U.S. Natl. Mus., 9:97.
>        TYPE SPECIES: *Dymecodon pilirostris* True, 1886.
>        SYNONYMS: *Dimecodon* Coues, 1889.
>        COMMENTS: Often included in *Urotrichus,* as in the last edition (Hutterer, 1993*a*), but
>                morphological (Imaizumi, 1970*b*) and genetical data (Shinohara et al., 2003) support
>                generic status.

*Dymecodon pilirostris* True, 1886. Proc. U.S. Natl. Mus., 9:97.
>        COMMON NAME: True's Shrew Mole.
>        TYPE LOCALITY: Japan, Honshu, Enoshima (Yenosima), at mouth of Bay of Yeddo.
>        DISTRIBUTION: Montane forests of Honshu, Shikoku, Kyushu (Japan).
>        STATUS: IUCN – Lower Risk (lc) as *Urotricus pilirostris.*
>        SYNONYMS: *dewanus* Kishida, 1950.
>        COMMENTS: Often included in *Urotrichus*; see under genus.

*Urotrichus* Temminck, 1841. Het. Instit. K. Ned. Inst., p. 212.
>        TYPE SPECIES: *Urotrichus talpoides* Temminck, 1841.
>        COMMENTS: For placement in Urotrichini see Storch and Qiu (1983:100). Includes sometimes
>                *Dymecodon*; see Ellerman and Morrison-Scott (1951:33-34). See also Imaizumi (1970*b*:123)
>                who considered *Dymecodon* a distinct genus.

*Urotrichus talpoides* Temminck, 1841. Het. Instit. K. Ned. Inst., p. 215.
>        COMMON NAME: Japanese Shrew Mole.
>        TYPE LOCALITY: Japan, Kyushu, Nagasaki.
>        DISTRIBUTION: Grassland and forest of Honshu, Shikoku, Kyushu (Japan); Dogo Isl,
>                N Tsushima Isl (Japan).
>        STATUS: IUCN – Lower Risk (lc).
>        SYNONYMS: **adversus** Thomas, 1908; **centralis** Thomas, 1908; **hondoensis** Thomas, 1918;
>                *yokohamanis* Kanda, 1929; **minutus** Tokuda, 1932; *shinanensis* Yagi, 1927.
>        COMMENTS: Imaizumi (1970*b*:128) recognized five taxa as valid subspecies. Harada et al.
>                (2001) studied the karyotype (2n = 34, FN = 64) of the species and discovered two
>                parapatric karyotype races in C Honshu.

**Subfamily Uropsilinae** Dobson, 1883. Monogr. Insectivora, 2:128.
>        COMMENTS: Species belonging to this subfamily form the basal branch in a mtDNA study of all
>                major mole taxa (Shinohara et al., 2003).

*Uropsilus* Milne-Edwards, 1871. *In* David, Nouv. Arch. Mus. Hist. Nat. Paris, Bull. 7, pp. 92-93.
>        TYPE SPECIES: *Uropsilus soricipes* Milne-Edwards, 1871.
>        SYNONYMS: *Nasillus* Thomas, 1911; *Rhynchonax* Thomas, 1912.
>        COMMENTS: *Nasillus* and *Rhynchonax* were included by Ellerman and Morrison-Scott (1966:31)
>                and Corbet and Hill (1980:33); but Gureev (1979:201-204), listed both as distinct genera.
>                Reviewed by Hoffmann (1984).

*Uropsilus andersoni* (Thomas, 1911). Abstr. Proc. Zool. Soc. Lond., 1911(100):49.
>        COMMON NAME: Anderson's Shrew Mole.
>        TYPE LOCALITY: China, Sichuan, "Omi-san" = Emei-Shan.
>        DISTRIBUTION: C Sichuan (China).
>        STATUS: IUCN – Lower Risk (lc).
>        COMMENTS: Type species of *Rhynchonax*. Formerly included in *soricipes*, but see Hoffmann
>                (1984).

*Uropsilus gracilis* (Thomas, 1911). Abstr. Proc. Zool. Soc. Lond., 1911(100):49.
>        COMMON NAME: Gracile Shrew Mole.
>        TYPE LOCALITY: China, Sichuan, near Nan-chwan (Nanchuan), Mt. Chin-fu-san
>                (Jingfu Shan).
>        DISTRIBUTION: Sichuan and Yunnan (China) and N Burma.
>        STATUS: IUCN – Lower Risk (lc).
>        SYNONYMS: *atronates* (Allen, 1923); *nivatus* (Allen, 1923).

COMMENTS: Type species of *Nasillus*. Formerly included in *soricipes*; but see Hoffmann (1984). The shrew-like humerus was figured by Storch and Dahlmann (2000).

*Uropsilus investigator* (Thomas, 1922). Ann. Mag. Nat. Hist., ser. 9, 10:393.
    COMMON NAME: Inquisitive Shrew Mole.
    TYPE LOCALITY: China, Yunnan, Kui-chiang-Salween divide at 28°N, 11,000 ft. (3353 m).
    DISTRIBUTION: Yunnan (China).
    STATUS: IUCN – Endangered.
    COMMENTS: Hoffmann (1984) included *investigator* in *gracilis* but on morphological and distributional grounds Wang and Yang (1989) and Storch (pers. comm.) concluded that both are sympatric in Yunnan and must therefore be regarded as distinct species.

*Uropsilus soricipes* Milne-Edwards, 1871. *In* David, Nouv. Arch. Mus. Hist. Nat. Paris, Bull. 7, p. 92.
    COMMON NAME: Chinese Shrew Mole.
    TYPE LOCALITY: China, Sichuan, Moupin.
    DISTRIBUTION: C Sichuan (China).
    STATUS: IUCN – Endangered.
    COMMENTS: Formerly included *andersoni, gracilis*, and *investigator*, according to Ellerman and Morrision-Scott (1966), but see Hoffmann (1984). Gureev (1979) listed these as distinct species without comment.

# ORDER CHIROPTERA
## By Nancy B. Simmons

**ORDER CHIROPTERA** Blumenbach, 1779.

STATUS: As is the case for all species in this book, the conservation status for each bat species is reported below based upon listings of the Convention on International Trade in Endangered Species of Wild Fauna and Flora (CITES), the United States Endangered Species Act (U.S. ESA), and the 2003 International Union for Conservation of Nature and Natural Resources Redlist (here cited "IUCN 2003"). In addition, two IUCN/SSC Action Plans provide more detailed information. The Action Plan for Old World Fruit Bats (Pteropodidae; cited below as "IUCN/SSC Action Plan, 1992", compiled by Mickleburgh et al., 1992) is over a decade old, but provides detailed information on the conservation status of subspecies as well as species of pteropodids, including the status of taxa not thought to be at risk. This publication also summarizes considerable information on ecology and population biology of pteropodids. The conservation status of other families of bats was assessed more recently in the Global Status Survey and Conservation Action Plan for Microchiropteran Bats (cited below as "IUCN/SSC Action Plan, 2001", compiled by Hutson et al., 2001). This work summarized the status of each microchiropteran species recognized by Koopman (1993) as well as some additional species described or revised subsequent to that publication. Threat categories listed in this work are identical to those found in the IUCN 2003 Redlist (which for bats is identical to the 2000 Redlist), but additional regional and taxon-specific conservation status information is included. Although both of the Action Plans were very comprehensive, many of the conservation assessments need to be revised in light of the new classification presented here, which includes 191 species not listed in Koopman (1993). Some of these are new species discovered in the last decade, but many were previously considered to be subspecies of other taxa. The majority of these newly recognized species have restricted geographic ranges, and may therefore be at risk. In addition, many long-recognized species are now believed to have much smaller geographic ranges than previously thought (a result of "taxonomic pruning" of populations now considered to be distinct species). The conservation status of these taxa should also be reevaluated.

COMMENTS: The higher-level classification of Chiroptera is in a state of flux due to incongruence among results of phylogenetic studies based on different data sets, and a rapidly emerging body of molecular data that strongly suggests that many traditionally recognized groups are not monophyletic. Prior to the late 1990s, most higher-level classifications were based on morphological data (see Simmons and Geisler [1998] for a review). Simmons (1998) and Simmons and Geisler (1998) conducted phylogenetic analyses of family-level relationships based on morphological data, and proposed a new classification that retained many traditional groups. They recognized two suborders (Megachiroptera and Microchiroptera), with Microchiroptera comprising two infraorders (Yinochiroptera and Yangochiroptera) and seven superfamilies (Emballonuroidea, Rhinopomatoidea, Rhinolophoidea, Noctilionoidea, Nataloidea, Molossoidea, and Vespertilionoidea). Jones et al. (2002) conducted a supertree analysis based on phylogenies and classifications published between 1970 and 2000, and their results were largely congruent with the Simmons and Geisler (1998) classification. However, recent molecular studies have strongly contradicted many of these groupings (Hutcheon et al., 1988; Hoofer and Van Den Bussche, 2001; Kirsch et al., 1998; Murphy et al., 2001; Springer et al., 2001; Teeling et al., 2000, 2002, 2003; Van Den Bussche and Hoofer, 2000). Several studies have seriously challenged the monophyly of Microchiroptera, suggesting instead that Yinochiroptera is the sister-group of Megachiroptera (Hutcheon et al., 1988; Teeling et al., 2000, 2002, 2003). Springer et al. (2001) created a new suborder Yinpterochiroptera for this clade, which now appears to include Pteropodidae, Rhinolophidae, Hipposideridae, Megadermatidae, Rhinopomatidae, and Craseonycteridae (Hulva and Horacek, 2002; Teeling et al., 2002, 2003). Nycteridae, usually included in Yinochiroptera within the superfamily Rhinolophoidea, now appears more closely related to Yangochiropteran bats, as does Emballonuridae (Teeling et al., 2002, 2003). Monophyly of other superfamilies recognized by Simmons (1998) and Simmons and Geisler (1998) has also been questioned based on molecular data (e.g., Nataloidea and Molossoidea; Van Den Bussche and Hoofer, 2000, 2001; Hoofer et al., 2001; Van Den Bussche et al., 2002b, 2003; Hoofer et al., 2003; Teeling et al., 2003). Most data sets, including morphology, now agree that Mystacinidae belongs in Noctilionoidea (Simmons and Conway, 2001; Teeling et al., 2003; Van Den Bussche and Hoofer, 2000, 2001; Van den Bussche et al., 2002a). However, relationships of Mystacinidae to the families traditionally placed in this superfamily (Noctilionidae, Mormoopidae, Phyllostomidae) and to others recently allied with Noctilionoidea based on molecular data (Thyropteridae and Furipteridae; Van Den Bussche and Hoofer [2000, 2001], Van Den Bussche et al. [2002b, 2003]) remain uncertain. Similarly, relationships among Myzopodidae, Emballonuridae, Nycteridae, and the two large Yangochiropteran superfamilies (Noctilionoidea and Vespertilionoidea) recognized in the molecular studies remain somewhat unclear because no single study has included representatives of all of these clades. The strength of the molecular sequence data supporting many of the novel clades noted above (e.g., Yinpterochiroptera) is increasingly compelling, and it seems likely that a new consensus view of higher-level classification of bats that contradicts most traditional arrangements will soon emerge. Teeling et al. (2002, 2003) provided a conservative classification based on an simultaneous analysis of five nuclear genes, but this classification did not include all families; Craseonycteridae, Myzopodidae, Thyropteridae, and Furipteridae were omitted as they have not been sampled for the genes in question. No complete classification of bat families based on molecular data yet exists, and those complete classifications that are available (e.g., McKenna and Bell, 1997; Simmons, 1998; Simmons and Geisler, 1998) were based on morphology and are not at all congruent with the new molecular data. More comprehensive analyses that include molecular and morphological data from all families are needed. In this context it seems premature to propose a new complete higher-level classification, while at the same time it would be counterproductive to use an older classification that is clearly out of date. As a compromise I have chosen to include no groups above the level of families in the present classification. However, the families are listed in an order which is consistent with the classifications proposed by Teeling et al. (2002, 2003) and Hoofer et al. (2003) as supplemented with information from Simmons and Geisler (1998) and Hulva and Horacek (2002).

**Family Pteropodidae** Gray, 1821. London Med. Repos., 15:299.

SYNONYMS: Cephalotidae Gray, 1821; Harpyidae H. Smith, 1842.

COMMENTS: Various workers have recognized between two and six subfamilies of Pteropodidae including: Cynopterinae Andersen, 1912, Epomophorinae K. Andersen, 1912, Harpionycterinae Miller, 1907, Nyctimeninae Miller, 1907, Macroglossinae Gray, 1866, Rousettinae Andersen, 1912, and Pteropodinae Gray, 1821 (Bergmans, 1997; Corbet and Hill, 1980, 1992; Hill and Smith, 1984; Koopman, 1993, 1994; McKenna and Bell, 1997). Recent phylogenetic studies agree that Macroglossinae and Pteropodinae sensu Koopman (1993, 1994) and McKenna and Bell (1997) are not monophyletic (Alvarez et al., 1999; Giannini and Simmons, 2003; Hollar and Springer, 1997; Hood, 1989; Juste et al., 1997; Kirsch et al., 1995; Romagnoli and Springer, 2000; Springer et al., 1995). Monophyly of cynopterines and empomophorines has also been questioned (Alvarez et al., 1999; Hollar and Springer, 1997; Kirsch et al., 1995; Romagnoli and Springer, 2000). Instead of supporting traditional taxonomic groupings, phylogenetic studies based on DNA hybridization and DNA sequences have found support for a large clade of endemic African taxa including genera previously placed in several different subfamilies/tribes (Alvarez et al., 1999; Giannini and Simmons, 2003; Hollar and Springer, 1997; Kirsch et al., 1995; Romagnoli and Springer, 2000). Relationships among pteropodid genera are not yet fully resolved, however, and questions remain concerning the position of *Nyctimene*, *Paranyctimene*, *Eidolon*, and several SE Asian endemic genera. Existing subfamilial and tribal classifications are not adequately congruent with recent phylogenies. Accordingly, no subfamilial or tribal groups are recognized here *pending* a thorough reevaluation of pteropodid classification.

*Acerodon* Jourdan, 1837. L'Echo du Monde Savant, 4, No. 275, p. 156.
> TYPE SPECIES: *Pteropus jubatus* Eschscholtz, 1831.
> COMMENTS: Very closely related to and possibly congeneric with *Pteropus*; see Musser et al.
> (1982*a*) and Corbet and Hill (1992).

*Acerodon celebensis* (Peters, 1867). Monatsb. K. Preuss. Akad. Wiss. Berlin, 1867:333.
> COMMON NAME: Sulawesi Fruit Bat.
> TYPE LOCALITY: Indonesia, Sulawesi.
> DISTRIBUTION: Sulawesi, Saleyer Isl, Sangihe Isls, Sula Isls (Indonesia).
> STATUS: CITES – Appendix II. IUCN/SSC Action Plan (1992) – No data. IUCN 2003 – Lower
> Risk (nt).
> SYNONYMS: *arquatus* Miller and Hollister, 1921.
> COMMENTS: Includes *arquatus* and Sulawesi specimens formerly in *Pteropus argentatus*
> (see Musser et al., 1982*a*). Also see Flannery (1995*b*).

*Acerodon humilis* K. Andersen, 1909. Ann. Mag. Nat. Hist., ser. 7, 3:24-25.
> COMMON NAME: Talaud Fruit Bat.
> TYPE LOCALITY: Indonesia, Talaud Isls, Lirong.
> DISTRIBUTION: Talaud Isls (Indonesia).
> STATUS: CITES – Appendix II. IUCN/SSC Action Plan (1992) – No Data: Limited Distribution.
> IUCN 2003 – Vulnerable.
> COMMENTS: Previously known only from the holotype, which Flannery (1995*b*) suggested
> might be a chimera consisting of a mismatched skull from an individual of *celebensis*
> and a skin of a *Pteropus hypomelanus* However, Feiler (1990) described two additional
> museum specimens, and a living population of this taxon has recently been
> rediscovered (Riley, 2001). It appears to represent a distinct species.

*Acerodon jubatus* (Eschscholtz, 1831). Zool. Atlas, Part 4:1.
> COMMON NAME: Golden-capped Fruit Bat.
> TYPE LOCALITY: Philippines, Luzon, Manila.
> DISTRIBUTION: Philippines except Palawan region.
> STATUS: CITES – Appendix I (and possibly extinct) as *A. lucifer*, Appendix I as *A. jubatus*;
> otherwise Appendix II; IUCN/SSC Action Plan (1992) and IUCN 2003 – Endangered as
> *A. jubatus*, *A. lucifer* listed as Extinct.
> SYNONYMS: *aurinuchalis* Elliot, 1896; *pyrrhocephalus* Meyen, 1833; **lucifer** Elliot, 1896;
> **mindanensis** K. Andersen, 1909.
> COMMENTS: Includes *lucifer*; see Ingle and Heaney (1992) and Heaney et al. (1998).

*Acerodon leucotis* (Sanborn, 1950). Proc. Biol. Soc. Wash., 63:189.
> COMMON NAME: Palawan Fruit Bat.
> TYPE LOCALITY: Philippines, Calamianes Isls, Busuanga Isl, Singay.
> DISTRIBUTION: Balabac, Palawan, Busuanga Isl (Philippines).
> STATUS: CITES – Appendix II. IUCN/SSC Action Plan (1992) – No Data. IUCN 2003 –
> Vulnerable.
> SYNONYMS: **obscurus** Sanborn, 1950.
> COMMENTS: Formerly included in *Pteropus* (see Musser et al., 1982*a*).

*Acerodon mackloti* (Temminck, 1837). Monogr. Mamm., 2:69.
> COMMON NAME: Sunda Fruit Bat.
> TYPE LOCALITY: Indonesia, Timor.
> DISTRIBUTION: Lombok, Sumbawa, Flores, Alor Isl, Sumba, and Timor (Indonesia).
> STATUS: CITES – Appendix II. IUCN/SSC Action Plan (1992) – Not Threatened; IUCN 2003 –
> Lower Risk (lc).
> SYNONYMS: *ochraphaeus* Muller and Jentink, 1887; **alorensis** K. Andersen, 1909; **floresii** Gray,
> 1871; *floresianus* Heude, 1896; **gilvus** K. Andersen, 1909; **prajae** Sody, 1936.
> COMMENTS: The subspecies nomenclature of this taxon is in need of revision (Helgen and
> Wilson, 2002). This name is sometimes spelled *macklotii*.

*Aethalops* Thomas, 1923. Proc. Zool. Soc. Lond., 1923:178.
> TYPE SPECIES: *Aethalodes alecto* Thomas, 1923.

SYNONYMS: *Aethalodes* Thomas, 1923 [not *Atehalodes* Gahan, 1888, an insect].

*Aethalops aequalis* G. M. Allen, 1938. J. Mammal., 19:497.
COMMON NAME: Borneo Fruit Bat.
TYPE LOCALITY: Malaysia (N Borneo), Sabah, Mt. Kinabalu, 5,500 ft. (1,833 m).
DISTRIBUTION: Brunei, Sabah, and Sarawak (Borneo).
STATUS: IUCN/SSC Action Plan (1992) – Indeterminate as *A. alecto aequalis*. IUCN 2003 –
Not listed.
COMMENTS: Considered a subspecies of *alecto* by many authors, but see D. J. Kitchener
et al. (1990, 1993*a*).

*Aethalops alecto* (Thomas, 1923). Ann. Mag. Nat. Hist., ser. 9, 11:251.
COMMON NAME: Pygmy Fruit Bat.
TYPE LOCALITY: Indonesia, Sumatra, Indrapura Peak, 7,300 ft. (2,225 m).
DISTRIBUTION: W Malaysia, Sumatra, Java, Bali, and Lombok.
STATUS: IUCN/SSC Action Plan (1992) – Indeterminate. IUCN 2003 – Lower Risk (nt).
SYNONYMS: **boeadii** Kitchener, 1993; **ocypete** Boeadi and Hill, 1986.
COMMENTS: Reviewed by Boeadi and Hill (1986) and D. J. Kitchener et al. (1993*a*). Does not
include *aequalis*; see D. J. Kitchener et al. (1993*a*).

*Alionycteris* Kock, 1969. Senckenberg. Biol., 50:319.
TYPE SPECIES: *Alionycteris paucidentata* Kock, 1969.

*Alionycteris paucidentata* Kock, 1969. Senckenberg. Biol., 50:322.
COMMON NAME: Mindanao Pygmy Fruit Bat.
TYPE LOCALITY: Philippines, Mindanao, Bukidion Prov., Mt. Katanglad.
DISTRIBUTION: Known only from the type locality.
STATUS: IUCN/SSC Bat Action Plan (1992) – Rare: Limited Distribution. IUCN 2003 –
Vulnerable.

*Aproteles* Menzies, 1977. Aust. J. Zool., 25:330.
TYPE SPECIES: *Aproteles bulmerae* Menzies, 1977.

*Aproteles bulmerae* Menzies, 1977. Aust. J. Zool., 25:331.
COMMON NAME: Bulmer's Fruit Bat.
TYPE LOCALITY: Papua New Guinea, Chimbu Prov., 2 km SE Chuave Govt. Sta., 1,530 m.
DISTRIBUTION: Mainland Papua New Guinea.
STATUS: U.S. ESA – Endangered. IUCN/SSC Action Plan (1992) – Endangered: Limited
Distribution. IUCN 2003 – Critically Endangered.
COMMENTS: Originally described from fossil material, but since found living (Flannery and
Seri, 1993; Hyndman and Menzies, 1980). See also Flannery (1995*a*) and Bonaccorso
(1998).

*Balionycteris* Matschie, 1899. Flederm. Berliner Mus. Naturk., p. 72, 80.
TYPE SPECIES: *Cynopterus maculatus* Thomas, 1893.

*Balionycteris maculata* (Thomas, 1893). Ann. Mag. Nat. Hist., ser. 6, 11:341.
COMMON NAME: Spotted-winged Fruit Bat.
TYPE LOCALITY: Malaysia (N Borneo), Sarawak.
DISTRIBUTION: Thailand; W Malaysia; Borneo; Sumatra; Durian and Galang Isls (Riau Arch.,
Indonesia).
STATUS: IUCN/SSC Action Plan (1992) – Not Threatened. IUCN 2003 – Lower Risk (lc).
SYNONYMS: **seimundi** Kloss, 1921.

*Casinycteris* Thomas, 1910. Ann. Mag. Nat. Hist., ser. 8, 6:111.
TYPE SPECIES: *Casinycteris argynnis* Thomas, 1910.

*Casinycteris argynnis* Thomas, 1910. Ann. Mag. Nat. Hist., ser. 8, 6:111.
COMMON NAME: Golden Short-palated Fruit Bat.
TYPE LOCALITY: Cameroon, Ja River, Bitye.

DISTRIBUTION: Cameroon to E Dem. Rep. Congo.
STATUS: IUCN/SSC Action Plan (1992) – No Data: Limited Distribution. IUCN 2003 –
 Lower Risk (nt).
COMMENTS: Reviewed by Bergmans (1990).

*Chironax* K. Andersen, 1912. Cat. Chiroptera Brit. Mus., 2nd ed., p. 658.
 TYPE SPECIES: *Pteropus melanocephalus* Temminck, 1825.

 *Chironax melanocephalus* (Temminck, 1825). Monogr. Mamm., 1:190.
  COMMON NAME: Black-capped Fruit Bat.
  TYPE LOCALITY: Indonesia, W Java, Bantam, Gunung Karang (restricted by Bergmans and
   Rozendaal, 1988).
  DISTRIBUTION: Thailand, W Malaysia, Borneo, Sumatra, Java, Nias Isl, and Sulawesi.
  STATUS: IUCN/SSC Action Plan (1992) – Not Threatened. IUCN 2003 – Lower Risk (lc).
  SYNONYMS: **tumulus** Bergmans and Rozendaal, 1988.
  COMMENTS: Reviewed by Hill (1983) and Bergmans and Rozendaal (1988).

*Cynopterus* F. Cuvier, 1824. Dentes des Mammifères, p. 248.
 TYPE SPECIES: *Pteropus marginatus* E. Geoffroy, 1810 (= *Vespertilio sphinx* Vahl, 1797).
 SYNONYMS: *Niadius* Miller, 1906; *Pachysoma* Geoffroy, 1828 [not *Pachysoma* Macleay, 1821,
  an insect].
 COMMENTS: Genetic variation within the genus was discussed by Peterson and Heaney (1993)
  and Schmitt et al. (1995).

 *Cynopterus brachyotis* (Müller, 1838). Tijdschr. Nat. Gesch. Physiol., 5:146.
  COMMON NAME: Lesser Short-nosed Fruit Bat.
  TYPE LOCALITY: Borneo, Dewei (= Dewai) River.
  DISTRIBUTION: Sri Lanka, India, Nepal, Burma, Thailand, Cambodia, Vietnam, S China,
   Malaysia, Nicobar and Andaman Isls, Borneo, Sumatra, Sulawesi, Magnole, Sanana,
   Sangihe Isls, Talaud Isls and adjacent small islands. Perhaps present in the Palawan
   region of the Philippines (L. Heaney, pers. comm.)
  STATUS: IUCN/SSC Action Plan (1992) – Not Threatened. IUCN 2003 – Lower Risk (lc).
  SYNONYMS: *brevicaudatum* I. Geoffroy, 1828 [*nomen nudum*]; *duvaucelii* E. Geoffroy, 1828
   [*nomen nudum*]; *grandidieri* Peters, 1869; *minor* Revilliod, 1911 [not Trousseart or Lyon];
   *montanoi* Robin, 1881; *titthaecheilum* Waterhouse, 1843 [not Temminck; *nomen*
   *dubium*]; **altitudinis** Hill, 1961; **brachysoma** Dobson, 1871; *andamanensis* Dobson,
   1873; *ceylonensis* Gray, 1871; **concolor** Sody, 1940; **hoffeti** Bourret, 1944; **insularum**
   K. Andersen, 1910; **javanicus** K. Andersen, 1910.
  COMMENTS: This taxon is sometimes confused with *sphinx,* and the status of many
   populations is in doubt. Does not include *angulatus,* which was transferred to *sphinx*
   by Hill and Thonglongya (1972). Includes *minor*; see Hill (1983) and Corbet and Hill
   (1992). Does not include *luzoniensis* and *minutus*; see Kitchener and Maharadatunkamsi (1991). May include *scherzeri,* here included in *sphinx* following Kitchener and
   Maharadatunkamsi (1991) and Bates and Harrison (1997). Bates and Harrison (1997)
   also referred *brachysoma* and *andamanesis* to *sphinx* with some reservations. See
   Andersen (1912) for discussion of *duvaucelii* and *grandidieri.* Corbet and Hill (1992)
   included *babi* (here considered a subspecies of *sphinx*) in this species without
   comment. See discussion of diagnostic characters in Bates and Harrison (1997) and
   Mapatuna et al. (2002).

 *Cynopterus horsfieldii* Gray, 1843. List Specimens Mamm. Coll. Brit. Mus., p. 38.
  COMMON NAME: Horsfield's Fruit Bat.
  TYPE LOCALITY: Indonesia, Java.
  DISTRIBUTION: Thailand, Cambodia, W Malaysia, Borneo, Java, Sumatra, Lesser Sunda Isls,
   and adjacent small islands.
  STATUS: IUCN/SSC Action Plan (1992) – Not Threatened. IUCN 2003 – Lower Risk (lc).
  SYNONYMS: **harpax** Thomas and Wroughton, 1909; *lyoni* K. Andersen, 1912; *minor* Lyon,
   1908 [not Trouessart, 1878]; **persimilis** K. Andersen, 1912; **princeps** Miller, 1906.
  COMMENTS: Includes *harpax*; see Hill (1961*a*). This name is sometimes spelled *horsefieldi.*

*Cynopterus luzoniensis* (Peters, 1861). Monatsb. K. Preuss. Akad. Wiss. 1861:708
    COMMON NAME: Peters's Fruit Bat.
    TYPE LOCALITY: Philippines, Luzon, S Camarines, Iriga.
    DISTRIBUTION: Sulawesi, Philippines, and adjacent small islands.
    STATUS: Not evaluated in IUCN/SSC Action Plan (1992). IUCN 2003 – not evaluated.
    SYNONYMS: *archipelagus* Taylor, 1934; *cumingii* Gray, 1871; *philippensis* Gray, 1871.
    COMMENTS: Included in *brachyotis* by many authors, but see Kitchener and Maharadatun-
        kamsi (1991) and Schmitt et al. (1995). Heaney et al. (1987) placed *archipelagus* (known
        only from the juvenile holotype) in *brachyotis*, but see Kitchner and Maharadatunkamsi
        (1991). Specimens of *luzoniensis* from the Palawan region of the Philippines may
        actually represent *brachyotis* as used herein (L. Heaney, pers. comm.).

*Cynopterus minutus* Miller, 1906. Proc. Biol. Soc. Wash., 19:63.
    COMMON NAME: Minute Fruit Bat.
    TYPE LOCALITY: Sumatra, Nias Isl.
    DISTRIBUTION: Sumatra, Java, Borneo, Sulawesi.
    STATUS: IUCN/SSC Action Plan (1992) – No Data: Limited Distribution as *C. brachyotis*
        *minutus*. IUCN 2003 – Not listed.
    COMMENTS: Included in *brachyotis* by Hill (1983) and Koopman (1993, 1994), but see
        Kitchener and Maharadatunkamsi (1991).

*Cynopterus nusatenggara* Kitchener and Maharadatunkamsi, 1991. Rec. West. Aust. Mus., 15:312.
    COMMON NAME: Nusatenggara Short-nosed Fruit Bat.
    TYPE LOCALITY: Indonesia, Lesser Sunda Isls, W Sumbawa, Jerewah, Desa Belo (8°52'S,
        116°50'E), ca 40 m.
    DISTRIBUTION: Lombok, Moyo, Sumbawa, Sangeang, Komodo, Flores, Sumba, Adonara,
        Lembata, Pantar, Alor, and Wetar Isls (Indonesia).
    STATUS: Described after completion of IUCN/SSC Action Plan (1992). IUCN 2003 – Lower
        Risk (nt).
    SYNONYMS: **sinagai** Kitchener, 1996 [in Kitchener and Maharadatunkamsi, 1996];
        **wetarensis** Kitchener, 1996 [in Kitchener and Maharadatunkamsi, 1996].
    COMMENTS: Specimens of this species were tentatively included in *brachyotis* by Corbet and
        Hill (1992), but see Schmitt et al. (1995) and Kitchener and Maharadatunkamsi (1996).

*Cynopterus sphinx* (Vahl, 1797). Skr. Nat. Selsk. Copenhagen, 4(1):123.
    COMMON NAME: Greater Short-nosed Fruit Bat.
    TYPE LOCALITY: India, Madras, Tranquebar.
    DISTRIBUTION: Sri Lanka, Pakistan, Bangladesh, India, S China, SE Asia including Burma,
        Vietnam, and Cambodia, W Malaysia, Sumatra, adjacent small islands; perhaps
        Borneo.
    STATUS: IUCN/SSC Action Plan (1992) – Not Threatened. IUCN 2003 – Lower Risk (lc).
    SYNONYMS: *brevicaudatum* Temminck, 1837 [not I. Geoffroy]; *ellioti* Gray, 1870; *fibulatus*
        Vahl, 1797; *gangeticus* K. Andersen, 1910; *marginatus* E. Geoffroy, 1810; *pusillus*
        E. Geoffroy, 1803; *sphynx* Sody, 1933; **angulatus** Miller, 1898; **babi** Lyon, 1916;
        **pagensis** Miller, 1906; **scherzeri** Zelebor, 1869; **serasani** Paradiso, 1971.
    COMMENTS: This taxon is sometimes confused with *brachyotis,* and the status of many
        populations is in doubt. See discussion of diagnostic characters in Bates and Harrison
        (1997) and Mapatuna et al. (2002). Includes *angulatus*; see Hill and Thonglongya
        (1972). Does not include *titthaecheilus*; see Hill (1983). Apparently includes *babi*; see
        Kitchener and Maharadatunkamsi (1991), but also see Corbet and Hill (1992), who
        included *babi* in *brachyotis* without comment. May not include *scherzeri*; see Corbet
        and Hill (1992), but also see Bates and Harrison (1997), who retained *scherzeri* in *sphinx*
        but noted that it may represent a distinct species. May also include *brachysoma* and
        *andamanesis* (here listed as synonyms of *brachyotis*); see Bates and Harrison (1997).
        Some authors recognize *gangeticus* as a distinct subspecies; it is here grouped in the
        nominate subspecies following Koopman (1994). Clinal variation in size discussed by
        Storz et al. (2001). Also see Storz and Kunz (1999).

*Cynopterus titthaecheilus* (Temminck, 1825). Monogr. Mamm. 1:198.

COMMON NAME: Indonesian Short-nosed Fruit Bat.

TYPE LOCALITY: Indonesia, Java, Bogor (restricted by Andersen, 1912).

DISTRIBUTION: Sumatra, Java, Bali, Lombok, Timor, and adjacent small islands.

STATUS: IUCN/SSC Action Plan (1992) – Not Threatened. IUCN 2003 – Lower Risk (lc).

SYNONYMS: *diardii* E. Geoffroy, 1828; **major** Miller, 1906; **terminus** Sody, 1940.

COMMENTS: Formerly included in *sphinx*, but see Hill (1983); also see Corbet and Hill (1992). The position of *diardii* in this synonymy remains somewhat uncertain, see Corbet and Hill (1992) and Pavlinov et al. (1995*b*).

*Dobsonia* Palmer, 1898. Proc. Biol. Soc. Wash., 12:114.

TYPE SPECIES: *Cephalotes peroni* E. Geoffroy, 1810.

SYNONYMS: *Hypoderma* E. Geoffroy, 1828 [not *Hypoderma* Latreille, 1825, a Diptera]; *Pteronotus* Rafinesque, 1815 [*nomen nudum*]; ?*Tribonophorus* Burnett, 1829.

COMMENTS: Reviewed by Jong and Bergmans (1981). Species groups follow Koopman (1994).

*Dobsonia anderseni* Thomas, 1914. Ann. Mag. Nat. Hist., ser. 8, 13:435.

COMMON NAME: Andersen's Naked-backed Fruit Bat.

TYPE LOCALITY: Papua New Guinea, Admiralty Isls, Manus Isl.

DISTRIBUTION: Bismarck Archipelago including Admiralty Isls.

STATUS: IUCN/SSC Action Plan (1992) – No Data as *D. pannietensis anderseni*. IUCN 2003 – Not listed.

COMMENTS: *moluccensis* species group. Often included in *moluccensis* or *pannietensis*, but see Bergmans and Sarbini (1985), Flannery (1995*b*), and Bonaccorso (1998).

*Dobsonia beauforti* Bergmans, 1975. Beaufortia, 23(295):3.

COMMON NAME: Beaufort's Naked-backed Fruit Bat.

TYPE LOCALITY: Indonesia, Prov. of Papua, Sorong Div., Waigeo Isl, Njanjef.

DISTRIBUTION: Waigeo, Batanta, Salawati, Gebe, Gag, and Biak Isls (off Vogelkop Peninsula, New Guinea).

STATUS: IUCN/SSC Action Plan (1992) – No Data: Limited Distribution. IUCN 2003 – Endangered.

COMMENTS: *viridis* species group. Closely related to *viridis*; see Bergmans (1975). Also see Flannery (1995*b*).

*Dobsonia chapmani* Rabor, 1952. Nat. Hist. Misc., Chicago Acad. Sci., 96:2.

COMMON NAME: Negros Naked-backed Fruit Bat.

TYPE LOCALITY: Philippines, Negros, Bais, Pagabonin.

DISTRIBUTION: Cebu and Negros Isls (Philippines).

STATUS: IUCN/SSC Action Plan (1992) – Extinct? IUCN 2003 – Extinct. Previously thought to be extinct, but a living population was discovered in 2000 by S. Pedregosa (L. Heaney, pers. comm.).

COMMENTS: *moluccensis* species group. Listed by Corbet and Hill (1992) as a possible subspecies of *exoleta*; also see Bergmans (1978).

*Dobsonia crenulata* K. Andersen, 1909. Ann. Mag. Nat. Hist., ser. 8, 4:532.

COMMON NAME: Halmahera Naked-backed Fruit Bat.

TYPE LOCALITY: Indonesia, Maluku (Moluccas), Ternate.

DISTRIBUTION: N Moluccas, Togian Isls, Sangihe Isls, Talaud Isls, Pelang, Sulawesi (Indonesia).

STATUS: IUCN/SSC Action Plan (1992) – No Data as *D. viridis crenulata*. IUCN 2003 – Not listed.

COMMENTS: *viridis* species group. Included by Hill (1983) and Hill and Corbet (1992) as a subspecies of *viridis*, but see Bergmans and Rozendaal (1988). Also see Flannery (1995*b*). Non-Moluccan populations apparently represent an undescribed subspecies (K. Helgen, pers. comm.).

*Dobsonia emersa* Bergmans and Sarbini, 1985. Beaufortia, 34:185.

COMMON NAME: Biak Naked-backed Fruit Bat.

TYPE LOCALITY: Indonesia, Prov. of Papua, Biak, Sorido.

DISTRIBUTION: Biak and Owii Isls (in Geelvink Bay, New Guinea).

STATUS: IUCN/SSC Action Plan (1992) – No Data: Limited Distribution. IUCN 2003 –
    Vulnerable.
COMMENTS: *moluccensis* species group. See Flannery (1995*b*). There is a closely related,
    undescribed species on Numfoor Isl (K. Helgen, pers. comm.).

*Dobsonia exoleta* K. Andersen, 1909. Ann. Mag. Nat. Hist., ser. 8, 4:531, 533.
COMMON NAME: Sulawesi Naked-backed Fruit Bat.
TYPE LOCALITY: Indonesia, Sulawesi, Minahassa, Tomohon.
DISTRIBUTION: Sulawesi, Muna Togian Isls, Sula Isls (Indonesia).
STATUS: IUCN/SSC Action Plan (1992) – No Data. IUCN 2003 – Lower Risk (nt).
COMMENTS: *moluccensis* species group. May include *chapmani*; see Corbet and Hill (1992).
    Reviewed by Hill (1983); also see Flannery (1995*b*).

*Dobsonia inermis* K. Andersen, 1909. Ann. Mag. Nat. Hist., ser. 8, 4:532.
COMMON NAME: Solomons Naked-backed Fruit Bat.
TYPE LOCALITY: Solomon Isls, Makira (San Cristobal Isl).
DISTRIBUTION: Solomon Isls, including Bougainville Isl (Papua New Guinea).
STATUS: IUCN/SSC Action Plan (1992) – Not Threatened. IUCN 2003 – Lower Risk (lc).
SYNONYMS: *nesea* K. Andersen, 1909; **minimus** Phillips, 1968.
COMMENTS: *viridis* species group; see discussion in Bergmans (1978) and Hill (1983). See also
    Flannery (1995*b*) and Bonaccorso (1998).

*Dobsonia magna* Thomas, 1905. Ann. Mag. Nat. Hist., ser. 7, 16:423.
COMMON NAME: New Guinea Naked-backed Fruit Bat.
TYPE LOCALITY: Papua New Guinea, Mambare River, Tamata, 100 ft. (33 m.).
DISTRIBUTION: Waigeo, Yapen, Batanta, and Misool Isls through New Guinea to
    N Queensland (Australia); possibly the Aru Isls.
STATUS: IUCN/SSC Action Plan (1992) – Not Threatened as *D. moluccense magna*. IUCN 2003
    – Not listed.
COMMENTS: *moluccensis* species group. Often included in *moluccensis* (e.g., Koopman, 1979;
    Hill, 1983), but see Bergmans and Sarbini (1985). Also see Flannery (1995*a, b*).

*Dobsonia minor* (Dobson, 1879). Proc. Zool. Soc. Lond., 1878:875 [1879].
COMMON NAME: Lesser Naked-backed Fruit Bat.
TYPE LOCALITY: Indonesia, Prov. of Papua, Manokwari Div., Amberbaki.
DISTRIBUTION: C and W New Guinea and adjacent small islands; Sulawesi.
STATUS: IUCN/SSC Action Plan (1992) – Rare. IUCN 2003 – Lower Risk (nt).
COMMENTS: *minor* species group. Reviewed by Bergmans and Sarbini (1985) and Corbet and
    Hill (1992); also see Flannery (1995*a, b*) and Bonaccorso (1998).

*Dobsonia moluccensis* (Quoy and Gaimard, 1830). *In* d'Urville, Voy . . . de Astrolabe, Zool.,
    1(L'Homme, Mamm. Oiseaux):86.
COMMON NAME: Moluccan Naked-backed Fruit Bat.
TYPE LOCALITY: Indonesia, Maluku (Moluccas), Amboina Isl.
DISTRIBUTION: Molucca Isls including Bacan, Buru and Seram; Banda Isls, Aru Isls, Waigeo
    (Prov. of Papua, Indonesia).
STATUS: IUCN/SSC Action Plan (1992) – Not Threatened. IUCN 2003 – Lower Risk (lc).
COMMENTS: *moluccensis* species group. Does not include *pannietensis*; see Bergmans (1979).
    Koopman (1979, 1982) included *magna* and *anderseni* in *moluccensis*, but see Bergmans
    and Sarbini (1985) and Bonaccorso (1998).

*Dobsonia pannietensis* (De Vis, 1905). Ann. Queensl. Mus., 6:36.
COMMON NAME: Panniet Naked-backed Fruit Bat.
TYPE LOCALITY: Papua New Guinea, Louisiade Arch., Panniet Isl.
DISTRIBUTION: Louisiade Arch., D'Entrecasteaux Isls, and Trobriand Isls.
STATUS: IUCN/SSC Action Plan (1992) – No Data. IUCN 2003 – Lower Risk (lc).
SYNONYMS: **remota** Cabrera, 1920.
COMMENTS: *moluccensis* species group. Considered a subspecies of *moluccensis* by Laurie and
    Hill (1954), but apparently distinct; see Bergmans (1979) and Bonaccorso (1998).
    Includes *remota*; see Koopman (1982). A record of *remota* from Bougainville Isl is based

on a misidentified *inermis* (see Bergmans, 1979).

*Dobsonia peronii* (E. Geoffroy, 1810). Ann. Mus. Natn. Hist. Nat. Paris, 15:104.
    COMMON NAME: Western Naked-backed Fruit Bat.
    TYPE LOCALITY: Indonesia, Lesser Sunda Isls, Timor.
    DISTRIBUTION: Bali, Nusa Penida, Lombok, Moyo, Sangeang, Komodo, Sumbawa, Rinca,
        Flores, Lembata, Pantar, Alor, Wetar, Babar, Timor, Sematu, Roti, Savu, and Sumba Isls
        (Indonesia).
    STATUS: IUCN/SSC Action Plan (1992) – Indeterminate. IUCN 2003 – Vulnerable.
    SYNONYMS: *desmarestii* Burnett, 1829 [*nomen nudum*]; *paliatus* Geoffroy, 1810; *sumbanus*
        K. Andersen, 1909; ***grandis*** Bergmans, 1978.
    COMMENTS: *peronii* species group. Reviewed by Bergmans (1978) and Kitchener et al.
        (1997*a*). Sometimes spelled *peroni* (e.g., Andersen, 1912; Koopman, 1993).

*Dobsonia praedatrix* K. Andersen, 1909. Ann. Mag. Nat. Hist., ser. 8, 4:532.
    COMMON NAME: New Britain Naked-backed Fruit Bat.
    TYPE LOCALITY: Papua New Guinea, Bismarck Arch., "Duke of York group".
    DISTRIBUTION: Bismarck Arch. (Papua New Guinea).
    STATUS: IUCN/SSC Action Plan (1992) – No Data. IUCN 2003 – Lower Risk (nt).
    COMMENTS: *viridis* species group; see discussion in Bergmans (1978) and Hill (1983). See also
        Flannery (1995*b*) and Bonaccorso (1998).

*Dobsonia viridis* (Heude, 1896). Mem. Hist. Nat. Emp. Chin., 3:176.
    COMMON NAME: Greenish Naked-backed Fruit Bat.
    TYPE LOCALITY: Indonesia, Maluku (Moluccas), Kai Isls.
    DISTRIBUTION: C and S Moluccas including Seram, Ambon, and Buru; Banda, and Kai Isls
        (Indonesia). A closely related but undescribed species occurs in the Tanimbar Isls
        (K. Helgen, pers. comm.).
    STATUS: IUCN/SSC Action Plan (1992) – Not Threatened. IUCN 2003 – Lower Risk (lc).
    SYNONYMS: *umbrosa* Thomas, 1910.
    COMMENTS: *viridis* species group. Does not include *chapmani*; see Bergmans (1978). Does not
        include *crenulata*; see Bergmans and Rozendaal (1988). See also Hill (1983) and
        Flannery (1995*b*).

*Dyacopterus* K. Andersen, 1912. Cat. Chiroptera Brit. Mus., 1:651.
    TYPE SPECIES: *Cynopterus spadiceus* Thomas, 1890.

*Dyacopterus brooksi* Thomas, 1920. Ann. Mag. Nat. Hist., ser. 9, 5:284.
    COMMON NAME: Brooks's Dyak Fruit Bat.
    TYPE LOCALITY: Sumatra, ca. 100 mi. (150 km) N of Bencoolen, upper Ketuan River, Lebang
        Tandai.
    DISTRIBUTION: Sumatra; possibly Luzon and Mindanao (Philippines).
    STATUS: IUCN/SSC Action Plan (1992) – No Data as *D. spadiceus brooksi*. IUCN 2003 – Not
        listed.
    COMMENTS: Formerly included in *spadiceus* (Koopman, 1993, 1994), but see Peterson (1969)
        and Corbet and Hill (1992).

*Dyacopterus spadiceus* (Thomas, 1890). Ann. Mag. Nat. Hist., ser. 6, 5:235.
    COMMON NAME: Dyak Fruit Bat.
    TYPE LOCALITY: Malaysia, N Borneo, Sarawak, Baram.
    DISTRIBUTION: NW Borneo including Bunei, Luzon and Mindanao (Philippines), Malaya,
        possibly S Thailand.
    STATUS: IUCN/SSC Action Plan (1992) – Rare. IUCN 2003 – Lower Risk (nt).
    COMMENTS: Does not include *brooksi*; see Peterson (1969) and Corbet and Hill (1992).

*Eidolon* Rafinesque, 1815. Analyse de la Nature, p. 54.
    TYPE SPECIES: *Vespertilio vampirus helvus* Kerr, 1792.
    SYNONYMS: *Leiponyx* Jentink, 1881; *Liponyx* Forbes, 1882; *Pterocyon* Peters, 1861.
    COMMENTS: Revised by Bergmans (1990).

*Eidolon dupreanum* (Pollen, 1866 *In* Schlegel and Pollen, 1866). Proc. Zool. Soc. Lond., 1866:419.
COMMON NAME: Malagasy Straw-colored Fruit Bat.
TYPE LOCALITY: Madagascar, Nossi Bé.
DISTRIBUTION: Madagascar.
STATUS: IUCN/SSC Action Plan (1992) – Not Threatened. IUCN 2003 – Lower Risk (lc).
COMMENTS: See comments under *helvum*. Reviewed by Bergmans (1990) and Peterson et al. (1995).

*Eidolon helvum* (Kerr, 1792). *In* Linnaeus, Anim. Kingdom, 1(1):xvii, 91.
COMMON NAME: African Straw-colored Fruit Bat.
TYPE LOCALITY: Senegal (restricted by K. Andersen, 1907).
DISTRIBUTION: Mauritania, Senegal, and Gambia to Ethiopia to South Africa; SW Arabia and Oman; islands in the Gulf of Guinea and off E Africa.
STATUS: IUCN/SSC Action Plan (1992) – Not Threatened. IUCN 2003 – Lower Risk (lc).
SYNONYMS: *buettikoferi* Jentink, 1881; *leucomelas* Fitzinger, 1866; *mollipilosus* H. Allen, 1862; *paleaceus* Peters, 1862; *palmarum* Heuglin, 1877; *stramineus* E. Geoffroy, 1803; **annobonensis** Juste, Ibáñez, and Machordom, 2000; **sabaeum** K. Andersen, 1907.
COMMENTS: Includes *sabaeum,* see Hayman and Hill (1971), Bergmans (1990), and Harrison and Bates (1991). Does not include *dupreanum*; see Bergmans (1990) and Peterson et al. (1995), but also see Hayman and Hill (1971). See DeFrees and Wilson (1988), but note that they included *dupreanum* in *helvum*. African forms reviewed in part by Juste et al. (2000); Palearctic forms reviewed by Horácek et al. (2000). Distribution mapped by Taylor (2000*a*) and Cotterill (2001*e*). The taxonomic status of populations in the Aïr Mountains of Niger is unclear.

*Eonycteris* Dobson, 1873. Proc. Asiat. Soc. Bengal, p. 148.
TYPE SPECIES: *Macroglossus spelaeus* Dobson, 1871.
SYNONYMS: *Callinycteris* Jentink, 1889.

*Eonycteris major* K. Andersen, 1910. Ann. Mag. Nat. Hist., ser. 8, 6:625.
COMMON NAME: Greater Dawn Bat.
TYPE LOCALITY: Malaysia, N Borneo, Sarawak, Mt. Dulit.
DISTRIBUTION: Borneo, Mentawai Isls (Indonesia).
STATUS: IUCN/SSC Action Plan (1992) – Not Threatened. IUCN 2003 – Lower Risk (lc).
COMMENTS: Apparently does not include *robusta* and *longicauda*; see Heaney et al. (1987, 1998). Corbet and Hill (1992) suggested that the Mentawai Isls record may have been based on a large example of *spelaea*, but it appears that this material may actually represent an undescribed subspecies (K. Helgen, pers. comm.).

*Eonycteris robusta* Miller, 1913. Proc. Biol. Soc. Wash., 26:73-74.
COMMON NAME: Phillipine Dawn Bat.
TYPE LOCALITY: Phillipines, Luzon Isl, Rizal Prov., Montalban Caves.
DISTRIBUTION: Philippines except Palawan region.
STATUS: IUCN/SSC Action Plan (1992) – Rare as *E. major robusta*. IUCN 2003 – Not listed.
SYNONYMS: *longicauda* Taylor, 1934.
COMMENTS: Often included in *major* following Tate (1942*b*), but apparently distinct; see Heaney et al. (1987, 1998).

*Eonycteris spelaea* (Dobson, 1871). Proc. Asiat. Soc. Bengal, p. 105, 106.
COMMON NAME: Lesser Dawn Bat.
TYPE LOCALITY: Burma, Tenasserim, Moulmein, Farm Caves.
DISTRIBUTION: India, Burma, Nepal, S China, Thailand, Laos, Cambodia, Vietnam, W Malaysia, Borneo; Sula Isls, N Moluccas, Sumatra, Java, Sumba, Timor and Sulawesi (Indonesia); Philippines; Andaman Isls (India).
STATUS: IUCN/SSC Action Plan (1992) – Not Threatened. IUCN 2003 – Lower Risk (lc).
SYNONYMS: **glandifera** Lawrence, 1939; **rosenbergii** Jentink, 1889; *bernsteini* Tate, 1942; **winnyae** Maharadatunkamsi and Kitchener, 1997.
COMMENTS: Includes *rosenbergii*; see Bergmans and Rozendaal (1988). Reviewed in part by

Hill (1983), Flannery (1995*b*), Bates and Harrison (1997), Maharadatunkamsi and Kitchener (1997), and Maharadatunkamsi et al. (2003).

*Epomophorus* Bennett, 1836. Proc. Zool. Soc. Lond., 1835:149 [1836].
   TYPE SPECIES: *Pteropus gambianus* Ogilby, 1835.
   COMMENTS: Revised by Bergmans (1988), who transferred *Micropteropus grandis* to this genus. Key to this genus was presented in Boulay and Robbins (1989) and Claessen and De Vree (1991). Species groups follow Koopman (1994) with some modifications.

*Epomophorus angolensis* Gray, 1870. Cat. Monkeys, Lemurs, Fruit-eating Bats Brit. Mus., p. 125.
   COMMON NAME: Angolan Epauletted Fruit Bat.
   TYPE LOCALITY: Angola, Benguela.
   DISTRIBUTION: W Angola, NW Namibia.
   STATUS: IUCN/SSC Action Plan (1992) – Rare. IUCN 2003 – Lower Risk (nt).
   COMMENTS: *gambianus* species group. Distribution mapped by Taylor (2000*a*).

*Epomophorus crypturus* Peters, 1852. Naturwiss. Reise nach Mossambique, Säug., p. 26.
   COMMON NAME: Peters's Epauletted Fruit Bat.
   TYPE LOCALITY: Mozambique, Tete.
   DISTRIBUTION: Zambia, Tanzania, SE Dem. Rep. Congo, Mozambique, Malawi, Zimbabwe, Botswana, Namibia, South Africa.
   STATUS: IUCN/SSC Action Plan (1992) – Not Threatened as *Epomophorus gambianus crypturus*. IUCN 2003 – Not listed.
   COMMENTS: *gambianas* species group. Often included in *gambianus* (e.g., Bergmans, 1988, 1997), but see Claessen and De Vree (1990). Genetic studies and more collecting in the gap between the ranges of *crypturus* and *gambianus* may be necessary to more completely resolve the relationship of these taxa.

*Epomophorus gambianus* (Ogilby, 1835). Proc. Zool. Soc. Lond., 1835:100.
   COMMON NAME: Gambian Epauletted Fruit Bat.
   TYPE LOCALITY: Gambia, Banjul (restricted by Kock et al., 2002).
   DISTRIBUTION: Senegal and Gambia to Central African Republic, east to Sudan, Ethiopia, S to Malawi and Botswana.
   STATUS: IUCN/SSC Action Plan (1992) – Not Threatened. IUCN 2003 – Lower Risk (lc).
   SYNONYMS: *epomophorus* Bennett, 1836; *guineensis* Bocage, 1898; *macrocephalus* Ogilby, 1835; *megacephalus* Swainson, 1835; *reii* Aellen, 1950; *whitei* Bennett, 1836; *zechi* Matschie, 1899; ***pousarguesi*** Trouessart, 1904.
   COMMENTS: *gambianus* species group. Does not include *crypturus* and *angolensis*; see Claessen and De Vree (1990). See Boulay and Robbins (1989), but note that they included *crypturus* and *angolensis* in *gambianus*.

*Epomophorus grandis* (Sanborn, 1950). Publ. Cult. Comp. Diamantes Angola, 10:55.
   COMMON NAME: Sanborn's Epauletted Fruit Bat.
   TYPE LOCALITY: Angola, Lunda, Dundo.
   DISTRIBUTION: N Angola, S Dem. Rep. Congo.
   STATUS: IUCN/SSC Action Plan (1992) – Rare: Limited Distribution. IUCN 2003 – Data Deficient.
   COMMENTS: *grandis* species group. Transferred from *Micropteropus* to *Epomophorus* by Bergmans (1988).

*Epomophorus labiatus* (Temminck, 1837). Monogr. Mamm., 2:83.
   COMMON NAME: Little Epauletted Fruit Bat.
   TYPE LOCALITY: Sudan, Blue Nile Prov., Sennar.
   DISTRIBUTION: Saudi Arabia; Nigeria to Ethiopia and Djibouti, south to Republic of Congo and Malawi. Senegal records are probably erroneous (see Bergmans, 1988).
   STATUS: IUCN/SSC Action Plan (1992) – Not Threatened. IUCN 2003 – Lower Risk (lc).
   SYNONYMS: *anurus* Heuglin, 1864; *doriae* Matscheie, 1899; *schoensis* Rüppell, 1842; *schovanus* Heuglin, 1877.
   COMMENTS: *gambianus* species group. Includes *anurus*; see Kock (1969*a*), Bergmans (1988, 1997), and Claessen and De Vree (1991). Koopman (1994) recognized two subspecies

(*labiatus* and *anurus*), but this arrangement does not appear justified given the morphometric data presented by Claessen and De Vree (1991), who did not recognize subspecies. Apparently does not include *minor* contra Claessen and De Vree (1991), see discussion in Bergmans (1988, 1997). Middle Eastern forms reviewed by Horácek et al. (2000).

*Epomophorus minimus* Claessen and De Vree, 1991. Senckenberg. Biol., 71:216.
  COMMON NAME: Least Epauletted Fruit Bat.
  TYPE LOCALITY: Ethiopia, Shewa, Bahadu.
  DISTRIBUTION: Ethiopia, Somalia, Kenya, Uganda and Tanzania.
  STATUS: Described after completion of IUCN/SSC Action Plan (1992); IUCN 2003 – Lower Risk (lc).
  COMMENTS: *gambianus* species group. Included in *minor* by Bergmans (1988), but see Claessen and De Vree (1991).

*Epomophorus minor* Dobson, 1880. Proc. Zool. Soc. Lond., 1879:715 [1880].
  COMMON NAME: Minor Epauletted Fruit Bat.
  TYPE LOCALITY: Zanzibar.
  DISTRIBUTION: Ethiopia, Somalia, Sudan, Kenya, Rwanda, SE Dem. Rep. Congo, Zambia, Tanzania, Zanzibar, Uganda, Malawi.
  STATUS: IUCN/SSC Action Plan (1992) – Not Threatened. IUCN 2003 – Not listed.
  COMMENTS: *gambianus* species group. Included in *labiatus* by some authors (e.g., Claessen and De Vree, 1991), but see Bergmans (1988, 1997).

*Epomophorus wahlbergi* (Sundevall, 1846). Ofv. Kongl. Svenska Vet.-Akad. Forhandl. Stockholm, 3(4):118.
  COMMON NAME: Wahlberg's Epauletted Fruit Bat.
  TYPE LOCALITY: South Africa, KwaZulu-Natal Prov., near Durban.
  DISTRIBUTION: Cameroon to Sudan and Somalia, south to Malawi, Angola, and South Africa; Pemba and Zanzibar Isls. A Liberian record is probably erroneous (Koopman, 1993), and Cameroon and Equatorial Guinea records are of uncertain validity (Bergmans, 1988).
  STATUS: IUCN/SSC Action Plan (1992) – Not Threatened. IUCN 2003 – Lower Risk (lc).
  SYNONYMS: *haldemani* Hallowell, 1846; *neumanni* Matschie, 1899; *stuhlmanni* Matschie, 1899; *unicolor* Gray, 1870; *zenkeri* Matschie, 1899.
  COMMENTS: *wahlbergi* species group. Revised by Bergmans (1988), and reviewed in part by Volpers and Kumirai (1996); also see Acharya (1992). For an updated distribution map see Taylor (2000a). Some authors have recognized *haldemani* as a distinct subspecies, but this arrangement does not seem to be justified, see discussion in Bergmans (1988).

*Epomops* Gray, 1870. Cat. Monkeys, Lemurs, Fruit-eating Bats Brit. Mus., p. 126.
  TYPE SPECIES: *Epomophorus franqueti* Tomes, 1860.
  COMMENTS: Reviewed by Bergmans (1989).

*Epomops buettikoferi* (Matschie, 1899). Megachiroptera Berlin Mus., p. 45.
  COMMON NAME: Büttikofer's Epauletted Fruit Bat.
  TYPE LOCALITY: Liberia, Junk River, Schlieffelinsville.
  DISTRIBUTION: Guinea to Nigeria.
  STATUS: IUCN/SSC Action Plan (1992) – Vulnerable; IUCN 2003 – Vulnerable.

*Epomops dobsonii* (Bocage, 1889). J. Sci. Math. Phys. Nat. Lisboa, ser. 2, 1:1.
  COMMON NAME: Dobson's Epauletted Fruit Bat.
  TYPE LOCALITY: Angola, Benguela, Quindumbo.
  DISTRIBUTION: Angola to Rwanda, Tanzania, Malawi, and N Botswana.
  STATUS: IUCN/SSC Action Plan (1992) – Not Threatened. IUCN 2003 – Lower Risk (lc).
  COMMENTS: Neotype designated by Bergmans (1989). Distribution mapped by Taylor (2000a). This name has sometimes been spelled *dobsoni* (e.g., Koopman, 1993) but the original spelling is with a double "i".

*Epomops franqueti* (Tomes, 1860). Proc. Zool. Soc. Lond., 1860:54.
    COMMON NAME: Franquet's Epauletted Fruit Bat.
    TYPE LOCALITY: Gabon.
    DISTRIBUTION: Côte d'Ivoire to Sudan, Uganda, NW Tanzania, N Zambia, and Angola.
        Previous reports of this species from Guinea are in error (J. Fahr, pers. comm.).
    STATUS: IUCN/SSC Action Plan (1992) – Not Threatened. IUCN 2003 – Lower Risk (lc).
    SYNONYMS: *comptus* H. Allen, 1861; *strepitans* K. Andersen, 1910.
    COMMENTS: Reviewed by Bergmans (1989). No subspecies are presently recognized.

*Haplonycteris* Lawrence, 1939. Bull. Mus. Comp. Zool., 86:31.
    TYPE SPECIES: *Haplonycteris fischeri* Lawrence, 1939.

*Haplonycteris fischeri* Lawrence, 1939. Bull. Mus. Comp. Zool., 86:33.
    COMMON NAME: Philippine Pygmy Fruit Bat.
    TYPE LOCALITY: Philippines, Mindoro, Mt. Halcon.
    DISTRIBUTION: Philippines except Palawan region.
    STATUS: IUCN/SSC Action Plan (1992) – Vulnerable; IUCN 2003 – Vulnerable.
    COMMENTS: Genetic variation discussed by Peterson and Heaney (1993); a new species from
        Sibuyan Isl is currently being described (Heaney et al., 1998).

*Harpyionycteris* Thomas, 1896. Ann. Mag. Nat. Hist., ser. 6, 18:243.
    TYPE SPECIES: *Harpyionycteris whiteheadi* Thomas, 1896.

*Harpyionycteris celebensis* Miller and Hollister, 1921. Proc. Biol. Soc. Wash., 34:99.
    COMMON NAME: Sulawesi Harpy Fruit Bat.
    TYPE LOCALITY: Indonesia, Sulawesi, middle Sulawesi, Gimpoe.
    DISTRIBUTION: Sulawesi.
    STATUS: IUCN/SSC Action Plan (1992) – No Data as *H. whiteheadi celebensis*. IUCN 2003 –
        Not listed.
    COMMENTS: Considered a subspecies of *whiteheadi* by Laurie and Hill (1954) and Koopman
        (1994), but as a separate species by Peterson and Fenton (1970). Hill (1983), Bergmans
        and Rozendaal (1988), and Corbet and Hill (1992) retained *celebensis* as a separate
        species with some reservations.

*Harpyionycteris whiteheadi* Thomas, 1896. Ann. Mag. Nat. Hist., ser. 6, 18:244.
    COMMON NAME: Harpy Fruit Bat.
    TYPE LOCALITY: Philippines, Mindoro Isl, 5,000 ft. (1,524 m).
    DISTRIBUTION: Philippines except Palawan region.
    STATUS: IUCN/SSC Action Plan (1992) – Not Threatened. IUCN 2003 – Lower Risk (lc).
    SYNONYMS: **negrosensis** Peterson and Fenton, 1970.
    COMMENTS: Does not include *celebensis*; see comments under that species.

*Hypsignathus* H. Allen, 1861. Proc. Acad. Nat. Sci. Phil., p. 156.
    TYPE SPECIES: *Hypsignathus monstrosus* H. Allen, 1861.
    SYNONYMS: *Sphyrocephalus* A. Murray, 1862; *Zygaenocephalus* A. Murray, 1862.
    COMMENTS: Revised by Bergmans (1989).

*Hypsignathus monstrosus* H. Allen, 1861. Proc. Acad. Nat. Sci. Phil., p. 157.
    COMMON NAME: Hammer-headed Fruit Bat.
    TYPE LOCALITY: Gabon.
    DISTRIBUTION: Sierra Leone to W Kenya, south to Zambia and Angola; Bioko (Equatorial
        Guinea). Records from Gambia and Ethiopia are doubtful.
    STATUS: IUCN/SSC Action Plan (1992) – Not Threatened. IUCN 2003 – Lower Risk (lc).
    SYNONYMS: *labrosus* Murray, 1862; *macrocephalus* Peters, 1876.
    COMMENTS: See Langevin and Barclay (1990).

*Latidens* Thonglongya, 1972. J. Bombay Nat. Hist. Soc., 69:151.
    TYPE SPECIES: *Latidens salimalii* Thonglongya, 1972.

*Latidens salimalii* Thonglongya, 1972. J. Bombay Nat. Hist. Soc., 69:153.
    COMMON NAME: Salim Ali's Fruit Bat.
    TYPE LOCALITY: India, Madras, Madurai Dist., High Wavy Mtns, 2,500 ft. (762 m).
    DISTRIBUTION: S India.
    STATUS: IUCN/SSC Action Plan (1992) – Rare: Limited Distribution. IUCN 2003 – Critically
        Endangered.
    COMMENTS: Reviewed by Bates and Harrison (1997).

*Lissonycteris* K. Andersen, 1912. Catalogue Chir. Brit. Mus. I. Megachiroptera, 23:814.
    TYPE SPECIES: *Cynopterus angolensis* Bocage, 1898.
    COMMENTS: Originally named as a subgenus of *Rousettus*. Often considered a junior synonym
        of either *Rousettus* (see Koopman, 1975) or *Myonycteris* (see Peterson et al., 1995), but Juste
        et al. (1997) showed that *Lissonycteris* is distinct from the latter genera. Bergmans (1997)
        also treated *Lissonycteris* as distinct, although he noted that it appears very closely related
        to *Myonycteris* (a conclusion confirmed by Juste et al., 1997).

*Lissonycteris angolensis* (Bocage, 1898). J. Sci. Math. Phys. Nat. Lisboa, ser. 2, 5:133.
    COMMON NAME: Angolan Soft-furred Fruit Bat.
    TYPE LOCALITY: Angola, Quibula, Cahata, Pungo Andongo.
    DISTRIBUTION: Gambia, Senegal, Guinea Bissau, Guinea, Sierra Leone, Liberia, Côte d'Ivoire,
        Burkina Faso, Ghana, Togo, Nigeria, Cameroon, Central African Republic, Sudan,
        Ethiopia, Equatorial Guinea (Bioko only), Republic of Congo, Dem. Rep. Congo,
        Uganda, Rwanda, Kenya, Tanzania, Angola, Zambia, Zimbabwe, Mozambique.
    STATUS: IUCN/SSC Action Plan (1992) – Not Threatened as *Rousettus* (*Lissonycteris*)
        *angolensis*. IUCN 2003 – Lower Risk (lc) as *Rousettus angolensis*.
    SYNONYMS: *crypticola* Cabrera, 1920; **goliath** Bergmans, 1997; **petraea** Bergmans, 1997;
        **ruwenzorii** Eisentraut, 1965; **smithii** Thomas, 1908.
    COMMENTS: Some authors have split this complex into more than one species: *smithii* was
        recognized as distinct by Peterson et al. (1995) and Cotterill (2001*e*), and *goliath* and
        *petraea* were also treated as distinct species by Cotterill (2001*e*). However, the most
        recent comprehensive revision of this complex is that of Bergmans (1997), who treated
        these taxa and *ruwenzorii* as subspecies of *angolensis*. Ongoing work by Kock et al.
        (2002) and J. Fahr (pers. comm.) supports Bergmans (1997) treatment of *smithii* as a
        subspecies of *angolensis*; the status of *goliath*, *petraea*, and *ruwenzorii* remains unclear.
        Pending further study, which should include molecular comparisons, I have chosen to
        follow Bergmans (1997) although it seems likely that more than one species may be
        present in this complex.

*Macroglossus* F. Cuvier, 1824. Dentes des Mammifères, p. 248.
    TYPE SPECIES: *Pteropus minimus* E. Geoffroy, 1810.
    SYNONYMS: *Carponycteris* Lydekker, 1891; *Kiodotus* Blyth, 1840; *Odontonycteris* Jentink, 1902;
        *Rhynchocyon* Gistel, 1848 [not *Rhynchocyon* Peters, 1847, a macroscelidid].
    COMMENTS: Reviewed by Hill (1983).

*Macroglossus minimus* (E. Geoffroy, 1810). Ann. Mus. Natn. Hist. Nat. Paris, 15:97.
    COMMON NAME: Dagger-toothed Long-nosed Fruit Bat.
    TYPE LOCALITY: Indonesia, Java.
    DISTRIBUTION: Thailand to Philippines, Indonesia, Papua New Guinea, Solomon Isls, and
        N Australia. This species has also been reported from Cambodia but there are no
        vouchered records; see Hendrichsen et al. (2001*a*).
    STATUS: IUCN/SSC Action Plan (1992) – Not Threatened. IUCN 2003 – Lower Risk (lc).
    SYNONYMS: *horsfieldi* Lesson, 1827; *kiodotes* Lesson, 1827; *rostratus* Horsfield, 1822; **booensis**
        Kompanje and Moeliker, 2001; **lagochilus** Matschie, 1899; *fructivorus* Taylor, 1934;
        *meyeri* Jentink, 1902; **nanus** Matschie, 1899; *microtus* K. Andersen, 1911; *novaeguineae*
        Matschie, 1899 [*nomen nudum*]; *pygmaeus* K. Andersen, 1911.
    COMMENTS: Includes *lagochilus*; see Hill (1983). Includes *fructivorus*; see Heaney and Rabor
        (1982). See Bergmans (2001) and Kompanje and Moeliker (2001) for a review of

subspecies limits, some of which are unclear. Also see Flannery (1995*a*, *b*), and Bonaccorso (1998).

*Macroglossus sobrinus* K. Andersen, 1911. Ann. Mag. Nat. Hist., ser. 8, 3:641, 642.
    COMMON NAME: Greater Long-nosed Fruit Bat.
    TYPE LOCALITY: Malaysia, Perak, Gunong Igari (= Mt Igari), 2,000 ft. (610 m).
    DISTRIBUTION: NE India, Burma, C and S Thailand, S Laos, Vietnam, Sumatra, Java, Bali, and Sipora, Siberut, and Mentawai Isls (Indonesia). Reports of this species from Cambodia cannot be confirmed (Kock, 2000*a*).
    STATUS: IUCN/SSC Action Plan (1992) – Not Threatened. IUCN 2003 – Lower Risk (lc).
    SYNONYMS: *fraternus* Chasen and Kloss, 1928.
    COMMENTS: Reviewed by Bates and Harrison (1997); also see Bergmans (2001).

*Megaerops* Peters, 1865. Monatsb. K. Preuss. Akad. Wiss. Berlin, 1865:256.
    TYPE SPECIES: *Pachysoma ecaudatum* Temminck, 1837.
    SYNONYMS: *Megaera* Temminck, 1841 [not *Megaera* Robineau-Desvoidy, 1830, an insect, and *Megaera* Wagler, 1830, a reptile].

*Megaerops ecaudatus* (Temminck, 1837). Monogr. Mamm., 2:94.
    COMMON NAME: Temminck's Tailless Fruit Bat.
    TYPE LOCALITY: Indonesia, W Sumatra, Padang.
    DISTRIBUTION: Borneo, Sumatra, W Malaysia, Thailand, perhaps Vietnam.
    STATUS: IUCN/SSC Action Plan (1992) – Not Threatened. IUCN 2003 – Lower Risk (lc).
    COMMENTS: Some records of this species from India, Thailand, and Vietnam (Hill, 1983; Van Peenen et al., 1969) are referable to *niphanae*; see Corbet and Hill (1992). Reviewed by Maharandatunkamsi and Maryanto (2002).

*Megaerops kusnotoi* Hill and Boeadi, 1978. Mammalia, 42:427.
    COMMON NAME: Javan Tailless Fruit Bat.
    TYPE LOCALITY: Indonesia, W Java, S Sukabumi, Lengkong, Hanjuang Ciletuh, 700 m.
    DISTRIBUTION: Java, Bali, Lombok.
    STATUS: IUCN/SSC Action Plan (1992) – Rare. IUCN 2003 – Vulnerable.
    COMMENTS: Reviewed by Maharandatunkamsi and Maryanto (2002).

*Megaerops niphanae* Yenbutra and Felten, 1983. Senckenberg. Biol., 64:2.
    COMMON NAME: Ratanaworabhan's Fruit Bat.
    TYPE LOCALITY: Thailand, Nakhon Ratchasima Province, Amphoe Pak Thong Chai, Sakaerat Environmental Research Station.
    DISTRIBUTION: NE India, Thailand, Laos, Cambodia, Vietnam.
    STATUS: IUCN/SSC Action Plan (1992) – No Data. IUCN 2003 – Lower Risk (lc).
    COMMENTS: Reviewed by Bates and Harrison (1997).

*Megaerops wetmorei* Taylor, 1934. Monogr. Bur. Sci. Manila, p. 191.
    COMMON NAME: White-collared Fruit Bat.
    TYPE LOCALITY: Philippines, Mindanao Isl, Cotabato near Tatayan.
    DISTRIBUTION: Minanao Isl (Philippines), Borneo, W Malaysia, Sumatra.
    STATUS: IUCN/SSC Action Plan (1992) – Not Threatened. IUCN 2003 – Lower Risk (lc).
    SYNONYMS: *albicollis* Francis, 1989.
    COMMENTS: Reviewed by Maharandatunkamsi and Maryanto (2002).

*Megaloglossus* Pagenstecher, 1885. Zool. Anz., 8:245.
    TYPE SPECIES: *Megaloglossus woermanni* Pagenstecher, 1885.
    SYNONYMS: *Trygenycteris* Lydekker, 1891.

*Megaloglossus woermanni* Pagenstecher, 1885. Zool. Anz., 8:245.
    COMMON NAME: Woermann's Long-tongued Fruit Bat.
    TYPE LOCALITY: Gabon, Sibange farm.
    DISTRIBUTION: Guinea Bissau, Guinea, and Sierra Leone to Dem. Rep. Congo and Uganda, Equatorial Guinea (Bioko, Mbini), Gabon, Republic of Congo, and N Angola.
    STATUS: IUCN/SSC Action Plan (1992) – Not Threatened. IUCN 2003 – Lower Risk (lc).

SYNONYMS: *prigoginei* Hayman, 1966.
COMMENTS: Reviewed by Bergmans and van Bree (1972) and Bergmans (1997).

*Melonycteris* Dobson, 1877. Proc. Zool. Soc. Lond., 1877:119.
TYPE SPECIES: *Melonycteris melanops* Dobson, 1877.
SYNONYMS: *Nesonycteris* Thomas, 1887.
COMMENTS: Includes *Nesonycteris*; see Phillips (1968) and Flannery (1993*b*). Revised by
Flannery (1993*b*). Two subgenera are presently recognized, *Melonycteris* and *Nesonycteris*.

*Melonycteris fardoulisi* Flannery, 1993. Rec. Aust. Mus., 45:68.
COMMON NAME: Fardoulis's Blossom Bat.
TYPE LOCALITY: Solomon Isls., Makira (= San Cristobal Isl.), Sesena.
DISTRIBUTION: S and E Solomon Isls.
STATUS: IUCN 2003 – Vulnerable.
SYNONYMS: **maccoyi** Flannery, 1993; **mengermani** Flannery, 1993; **schouteni** Flannery, 1993.
COMMENTS: Subgenus *Nesonycteris*. See Flannery (1993*b*, 1995*b*).

*Melonycteris melanops* Dobson, 1877. Proc. Zool. Soc. Lond., 1877:119.
COMMON NAME: Black-bellied Fruit Bat.
TYPE LOCALITY: Given by Andersen (1912:790) as "New Ireland, coast adjacent to Duke of
York Isl." (Papua New Guinea, Bismarck Arch.).
DISTRIBUTION: Bismarck Arch.; a New Guinea record is highly questionable (Flannery,
1993*b*).
STATUS: IUCN/SSC Action Plan (1992) – Not Threatened. IUCN 2003 – Lower Risk (lc).
SYNONYMS: *alboscapulatus* Ramsay, 1877.
COMMENTS: Subgenus *Melonycteris*. Reviewed by Flannery (1993*b*); also see Flannery (1995*b*)
and Bonaccorso (1998).

*Melonycteris woodfordi* (Thomas, 1887). Ann. Mag. Nat. Hist., ser. 5, 19:147.
COMMON NAME: Woodford's Fruit Bat.
TYPE LOCALITY: Solomon Isls, Western Province, Alu Isl (near Shortland Isl)
DISTRIBUTION: Bougainville and Buka Isls (Papua New Guinea), N and W Solomon Isls.
STATUS: IUCN/SSC Action Plan (1992) – Not Threatened as *M. woodfordi*, No Data: Limited
Distribution as *M. aurantius*. IUCN 2003 – Lower Risk (lc) as *M. woodfordi*; Vulnerable
as *M. aurantius*.
SYNONYMS: **aurantius** Phillips, 1966.
COMMENTS: Subgenus *Nesonycteris*. Includes *aurantius*; see Flannery (1993*b*). Also see
Flannery (1995*b*) and Bonaccorso (1998).

*Micropteropus* Matschie, 1899. Megachiroptera Berlin Mus., p. 36, 57.
TYPE SPECIES: *Epomophorus pusillus* Peters, 1868.
COMMENTS: Revised by Bergmans (1989), who transferred *grandis* from this genus to
*Epomophorus*.

*Micropteropus intermedius* Hayman, 1963. Publ. Cult. Comp. Diamantes Angola, 66:100.
COMMON NAME: Hayman's Lesser Epauletted Fruit Bat.
TYPE LOCALITY: Angola, Lunda, Dundo.
DISTRIBUTION: N Angola, SE Dem. Rep. Congo.
STATUS: IUCN/SSC Action Plan (1992) – Rare. IUCN 2003 – Data Deficient.

*Micropteropus pusillus* (Peters, 1868). Monatsb. K. Preuss. Akad. Wiss. Berlin, 1867:870 [1868].
COMMON NAME: Peters's Lesser Epauletted Fruit Bat.
TYPE LOCALITY: Nigeria, Yoruba (see Bergmans [1989] and Kock et al. [2002]).
DISTRIBUTION: Senegal and Gambia east to Ethiopia and Sudan; south to Angola, Zambia,
Burundi, and Tanzania.
STATUS: IUCN/SSC Action Plan (1992) – Not Threatened. IUCN 2003 – Lower Risk (lc).
COMMENTS: See Owen-Ashley and Wilson (1998). For discussion of publication date, see
Kock et al. (2002).

*Myonycteris* Matschie, 1899. Megachiroptera Berlin Mus., p. 61, 63.
TYPE SPECIES: *Cynonycteris torquata* Dobson, 1878.
SYNONYMS: *Phygetis* K. Andersen, 1912; *Phylletis* Juste and Ibáñez, 1993.
COMMENTS: Revised by Bergmans (1976, 1997). Two subgenera are presently recognized, *Myonycteris* and *Phygetis*, following Koopman (1994).

*Myonycteris brachycephala* (Bocage, 1889). J. Sci. Math. Phys. Nat. Lisboa, ser. 2, 1:198.
COMMON NAME: São Tomé Collared Fruit Bat.
TYPE LOCALITY: São Tomé and Príncipe, São Tomé Isl.
DISTRIBUTION: São Tomé Isl (Gulf of Guinea).
STATUS: IUCN/SSC Action Plan (1992) – Vulnerable: Limited Distribution. IUCN 2003 – Endangered.
SYNONYMS: *brachycephalus* Seabra, 1898; *collaris* Andersen, 1907 [in part: the São Tomé specimen].
COMMENTS: Subgenus *Phygetis*. Reviewed by Bergmans (1997).

*Myonycteris relicta* Bergmans, 1980. Zool. Meded. Rijksmus. Nat. Hist. Leiden, 14:126.
COMMON NAME: Bergmans's Collared Fruit Bat.
TYPE LOCALITY: Kenya, Coast Prov., Shimba Hills, Lukore area, Mukanda River.
DISTRIBUTION: Kenya, Tanzania, Zimbabwe along border with Mozambique.
STATUS: IUCN/SSC Action Plan (1992) – Vulnerable: Limited Distribution. IUCN 2003 – Vulnerable.
COMMENTS: Subgenus *Myonycteris*. Reviewed by Bergmans (1997). Peterson et al. (1995) assigned this species to *Rousettus*.

*Myonycteris torquata* (Dobson, 1878). Cat. Chiroptera Brit. Mus., p. 71, 76.
COMMON NAME: Little Collared Fruit Bat.
TYPE LOCALITY: N Angola.
DISTRIBUTION: Guinea and Sierra Leone to Uganda, south to Angola and NW Zambia; Bioko (Equatorial Guinea).
STATUS: IUCN/SSC Action Plan (1992) – Not Threatened. IUCN 2003 – Lower Risk (lc).
SYNONYMS: *collaris* Andersen, 1907; *leptodon* Andersen, 1908; *wroughtoni* Andersen, 1908.
COMMENTS: Subgenus *Myonycteris*. Includes *leptodon* and *wroughtoni*; see Hayman and Hill (1971), Peterson et al. (1995), and Bergmans (1976, 1997). Koopman (1994) recognized *torquata*, *leptodon*, and *wroughtoni* as subspecies, but see Bergmans (1997).

*Nanonycteris* Matschie, 1899. Megachiroptera Berlin Mus., p. 36, 58.
TYPE SPECIES: *Epomophorus veldkampii* Jentink, 1888.
COMMENTS: Revised by Bergmans (1989).

*Nanonycteris veldkampii* (Jentink, 1888). Notes Leyden Mus., 10:51.
COMMON NAME: Veldkamp's Dwarf Epauletted Fruit Bat.
TYPE LOCALITY: Liberia, Fisherman Lake, Buluma.
DISTRIBUTION: Guinea to Central African Republic.
STATUS: IUCN/SSC Action Plan (1992) – Not Threatened. IUCN 2003 – Lower Risk (lc).
COMMENTS: Sometimes misspelled *veldkampi*, but the original spelling is *veldkampii*.

*Neopteryx* Hayman, 1946. Ann. Mag. Nat. Hist., ser. 11, 12:569.
TYPE SPECIES: *Neopteryx frosti* Hayman, 1946.
SYNONYMS: *Neoptryx* Van der Zon, 1979.

*Neopteryx frosti* Hayman, 1946. Ann. Mag. Nat. Hist., ser. 11, 12:571.
COMMON NAME: Small-toothed Fruit Bat.
TYPE LOCALITY: Indonesia, W Sulawesi, Tamalanti, 3,300 ft. (1,006 m).
DISTRIBUTION: W and N Sulawesi.
STATUS: IUCN/SSC Action Plan (1992) – Rare: Limited Distribution. IUCN 2003 – Vulnerable.
COMMENTS: Known from only 7 specimens; see Bergmans and Rozendaal (1988) and Bergmans (2001).

*Notopteris* Gray, 1859. Proc. Zool. Soc. Lond., 1859:36.
TYPE SPECIES: *Notopteris macdonaldi* Gray, 1859.

*Notopteris macdonaldi* Gray, 1859. Proc. Zool. Soc. Lond., 1859:38.
COMMON NAME: Fijian Long-tailed Fruit Bat.
TYPE LOCALITY: Fiji Isls, Viti Levu.
DISTRIBUTION: Vanuatu (= New Hebrides), Fiji Isls. A record from the Caroline Isls is
probably incorrect (K. Helgen, pers. comm.).
STATUS: IUCN/SSC Action Plan (1992) and IUCN 2003 – Vulnerable.
COMMENTS: Apparently does not include *neocaledonica*, here considered a distinct species
following Flannery (1995*b*).

*Notopteris neocaledonica* Trouessart, 1908. Bull. Mus. Hist. Nat., Paris, 14:257.
COMMON NAME: New Caledonia Long-tailed Fruit Bat.
TYPE LOCALITY: New Caledonia, Nekliai Valley near Poya, Adio Caves.
DISTRIBUTION: New Caledonia.
STATUS: IUCN/SSC Action Plan (1992) – No Data: Limited Distribution as *N. macdonaldi
neocaledonica*. IUCN 2003 – Not listed.
COMMENTS: Included in *macdonaldi* by Sanborn and Nicholson (1950) and Hill (1983), but
distinguished by non-overlapping measurements in most dimensions; see Sanborn
and Nicholson (1950) and Flannery (1995*b*).

*Nyctimene* Borkhausen, 1797. Deutsche Fauna, 1:86.
TYPE SPECIES: *Vespertilio cephalotes* Pallas, 1767.
SYNONYMS: *Bdelygma* Matschie, 1899; *Cephalotes* E. Geoffroy, 1810; *Gelasinus* Temminck, 1837
[not *Gelasinus* Van der Hoeven, 1827, a crustacean]; *Harpyia* Illiger, 1811 [not *Harpyia*
Ochsenheimer, 1810, a lepidopteran]; *Uronycteris* Gray, 1863.
COMMENTS: Reviewed by Smith and Hood (1983) and Bergmans (2001). Bergmans (2001)
proposed inclusion of *Paranyctimene* as a subgenus of *Nyctimene*, but I retain
*Paranyctimene* as a distinct genus pending phylogenetic studies of relationships of these
taxa (see comments under *Paranyctimene*). Species groups follow Bergmans (2001).

*Nyctimene aello* (Thomas, 1900). Ann. Mag. Nat. Hist., ser. 7, 5:216.
COMMON NAME: Broad-striped Tube-nosed Fruit Bat.
TYPE LOCALITY: Papua New Guinea, Milne Bay Prov., Milne Bay.
DISTRIBUTION: Mainland New Guinea; Kairiru and Admosin Isls (Papua New Guinea);
Misool and Salawati Isl (Prov. of Papua, Indonesia).
STATUS: IUCN/SSC Action Plan (1992) – Rare as *N. aello*, No Data: Limited Distribution as
*N. celaeno*. IUCN 2003 – Lower Risk (nt) as *N. aello*; Vulnerable as *N. celaeno*.
SYNONYMS: *celaeno* Thomas, 1922.
COMMENTS: *aello* species group. Includes *celaeno*, see Flannery (1995*a, b*) and Bergmans
(2001). Also see Bonaccorso (1998).

*Nyctimene albiventer* (Gray, 1863). Proc. Zool. Soc. Lond., 1862:262 [1863].
COMMON NAME: Common Tube-nosed Fruit Bat.
TYPE LOCALITY: Indonesia, Maluku, Morotai Isl.
DISTRIBUTION: New Guinea, Molucca Isls.
STATUS: IUCN/SSC Action Plan (1992) – Not Threatened. IUCN 2003 – Lower Risk (lc).
SYNONYMS: **papuanus** K. Andersen, 1910.
COMMENTS: *albiventer* species group. Does not include *draconilla*; see Hill (1983). Does not
include *bougainville*, which Smith and Hood (1983) placed in *vizcaccia*. Includes
*papuanus*, see Kitchener et al. (1995*c*). Peterson (1991) treated *papuanus* as a distinct
species but provided no comparisons with *albiventer*. Does not include *keasti*; see
Kitchener et al. (1995*c*). Also see Flannery (1995*a, b*). Aru Isl population has not been
allocated to subspecies; see Kitchener et al. (1995*c*). Reviewed by Bergmans (2001).

*Nyctimene cephalotes* (Pallas, 1767). Spicil. Zool., 3:10.
COMMON NAME: Pallas's Tube-nosed Fruit Bat.
TYPE LOCALITY: Indonesia, Maluku, Ambon Isl; see Andersen (1912) for discussion.
DISTRIBUTION: Indonesia: Sulawesi, Sula Isls; Seram, Boano, Ambon, and Buru Isls (Molucca

Isls); extreme S New Guinea and Moa Isl (Australia). Records reported from Timor probably represent *keasti*; see Kitchener et al. (1995*c*). A record from Numfor Isl (off N coast New Guinea) represents an undescribed species, and another undescribed species occurs in the Sangihe Isls (K. Helgen, pers. comm.).

STATUS: IUCN/SSC Action Plan (1992) – Not Threatened. IUCN 2003 – Lower Risk (lc).

SYNONYMS: *melinus* Kerr, 1792; *pallasi* E. Geoffroy, 1810; **aplini** Kitchener, 1995 [*in* Kitchener et al., 1995*c*].

COMMENTS: *cephalotes* species group. Does not include *vizcaccia*; see Smith and Hood (1983). Revised by Kitchener et al. (1995*c*). Also see Heaney and Peterson (1984), Flannery (1995*b*), Bonaccorso (1998), Bergmans (2001), and Kompanje and Moeliker (2001).

*Nyctimene certans* K. Andersen, 1912. Ann. Mag. Nat. Hist., ser. 9, 8:95.

COMMON NAME: Mountain Tube-nosed Fruit Bat.

TYPE LOCALITY: New Guinea, Prov. of Papua, Mount Goliath.

DISTRIBUTION: New Guinea.

STATUS: IUCN/SSC Action Plan (1992) – Rare as *N. cyclotis certans*. IUCN 2003 – Lower Risk (nt).

COMMENTS: *cyclotis* species group. Formerly included in *cyclotis* but see Peterson (1991) and Flannery (1995*a*). Also see Bonaccorso (1998), who included *certans* in *cyclotis*. The relationship between these forms remains unclear and they may be conspecific (K. Helgen, pers. comm.).

*Nyctimene cyclotis* K. Andersen, 1910. Ann. Mag. Nat. Hist., ser. 7, 6:623.

COMMON NAME: Round-eared Tube-nosed Fruit Bat.

TYPE LOCALITY: Indonesia, Prov. of Papua, Manokwari Div., Arfak Mtns.

DISTRIBUTION: Arfak Mtns. (New Guinea). Specimens from Mansuar Isl. (Prov. of Papua, Indonesia) may also represent *cyclotis* (Meinig, 2002). Specimens from New Britain formerly assigned to this species apparently represent *vizcaccia* (Bonaccorso, 1998).

STATUS: IUCN/SSC Action Plan (1992) – No Data. IUCN 2003 – Lower Risk (nt).

COMMENTS: *cyclotis* species group. Apparently does not include *certans*; see Peterson (1991) and Flannery (1995*a, b*), though also see Bonaccorso (1998) and comments under *certans*.

*Nyctimene draconilla* Thomas, 1922. Nova Guinea, 13:725.

COMMON NAME: Dragon Tube-nosed Fruit Bat.

TYPE LOCALITY: Indonesia, Prov. of Papua, Southern Div., Lorentz River, Bivak Isl.

DISTRIBUTION: New Guinea.

STATUS: IUCN/SSC Action Plan (1992) – Rare. IUCN 2003 – Vulnerable.

COMMENTS: *albiventer* species group. Considered a subspecies of *albiventer* by Laurie and Hill (1954), but see Hill (1983), Koopman (1982), Flannery (1995*a*), and Bonaccorso (1998), all of whom treated it as distinct, though with some reservations. Bergmans (2001) questioned the validity of this species but did not revise it.

*Nyctimene keasti* Kitchener, 1993 (*in* Kitchener, Packer, and Maryanto). Rec. West. Aust. Mus., 16:408.

COMMON NAME: Keast's Tube-nosed Fruit Bat.

TYPE LOCALITY: Indonesia, Maluku, Pulau Dullah (closely associated with Pulau Kai Kecil), 12 km N Tual, near Taman Anggrek, 5°38′S, 132°44′E, sea level.

DISTRIBUTION: Babar, Tanimbar, and Kai Isls (Molucca Isls, Indonesia); probably Timor and Flores (K. Helgen, pers. comm.).

STATUS: Described after completion of IUCN/SSC Action Plan (1992); IUCN 2003 – not evaluated.

SYNONYMS: **babari** Bergmans, 2001; **tozeri** Kitchener, 1995 [*in* Kitchener et al., 1995*c*].

COMMENTS: *cephalotes* species group. Originally described as a subspecies of *albiventer*, but recognized as a distinct species by Kitchener et al. (1995*c*) and Bergmans (2001). Revised by Bergmans (2001).

*Nyctimene major* (Dobson, 1877). Proc. Zool. Soc. Lond., 1877:117.

COMMON NAME: Island Tube-nosed Fruit Bat.

TYPE LOCALITY: Papua New Guinea, Bismarck Arch., Duke of York Isl.

DISTRIBUTION: D'Entrecasteaux Isls, Trobriand Isls, Bismarck and Louisiade Archs.
(Papua New Guinea), Solomon Isls, and small islands off the north coast of New Guinea.
A New Guinea mainland record is almost certainly erroneous; see Koopman (1979).
STATUS: IUCN/SSC Action Plan (1992) – Not Threatened. IUCN 2003 – Lower Risk (lc).
SYNONYMS: *geminus* K. Andersen, 1910; *lullulae* Thomas, 1904; *scitulus* K. Andersen, 1910.
COMMENTS: *cephalotes* species group. Reviewed in part by Koopman (1979) and Hill (1983).
See also Flannery (1995*b*) and Bonaccorso (1998).

*Nyctimene malaitensis* Phillips, 1968. Univ. Kansas Publ. Mus. Nat. Hist., 16:822.
COMMON NAME: Malaita Tube-nosed Fruit Bat.
TYPE LOCALITY: Solomon Isls, Malaita Isl.
DISTRIBUTION: Malaita and Makira Isls (Solomon Isls).
STATUS: IUCN/SSC Action Plan (1992) – No Data: Limited Distribution. IUCN 2003 –
Vulnerable.
COMMENTS: *albiventer* species group. Possibly a synonym of *vizcaccia*; see Flannery (1995*b*)
but note that he used the name *bougainville* for the latter taxon.

*Nyctimene masalai* Smith and Hood, 1983. Occas. Pap. Mus. Texas Tech Univ., 81:1.
COMMON NAME: Demonic Tube-nosed Fruit Bat.
TYPE LOCALITY: Papua New Guinea, New Ireland, Ralum.
DISTRIBUTION: New Ireland (Bismarck Arch.).
STATUS: IUCN/SSC Action Plan (1992) – No Data: Limited Distribution. IUCN 2003 –
Vulnerable.
COMMENTS: *albiventer* species group. Apparently included in *vizcaccia* by Bonaccorso (1998).
Bergmans (2001) questioned the validity of this species but continued to list it as a
separate taxon pending a revision. Treated as distinct by Emmons and Kinbag (2002).

*Nyctimene minutus* K. Andersen, 1910. Ann. Mag. Nat. Hist., ser. 7, 6:622.
COMMON NAME: Lesser Tube-nosed Fruit Bat.
TYPE LOCALITY: Indonesia, N Sulawesi, Minahassa, Tondano.
DISTRIBUTION: Sulawesi, C Moluccas.
STATUS: IUCN/SSC Action Plan (1992) – Not Threatened. IUCN 2003 – Vulnerable.
SYNONYMS: *varius* K. Andersen, 1910.
COMMENTS: *albiventer* species group. The status of *minutus* and *varius* is unclear, and is
currently under review by K. Helgen (pers. comm.).

*Nyctimene rabori* Heaney and Peterson, 1984. Occas. Pap. Mus. Zool. Univ. Michigan, 708:3.
COMMON NAME: Philippine Tube-nosed Fruit Bat.
TYPE LOCALITY: Philippines, Negros Isl, Negros Oriental Prov., Sibulan Municipality, 6 km W
of Dumaguete City, Balinsasayo, (9°21'N, 123°10'E), 835 m.
DISTRIBUTION: Negros, Cebu, and Sibuyan Isls (Philippines); Karakelang Isl (Talaud Isls,
Indonesia).
STATUS: IUCN/SSC Action Plan (1992) – Endangered: Limited Distribution. IUCN 2003 –
Critically Endangered.
COMMENTS: *cephalotes* species group. Corbet and Hill (1992) noted that *rabori* might be
conspecific with *cephalotes*, but see Bergmans (2001).

*Nyctimene robinsoni* Thomas, 1904. Ann. Mag. Nat. Hist., ser. 7, 14:196.
COMMON NAME: Queensland Tube-nosed Fruit Bat.
TYPE LOCALITY: Australia, Queensland, Cooktown.
DISTRIBUTION: E Queensland (Australia).
STATUS: IUCN/SSC Action Plan (1992) – Not Threatened. IUCN 2003 – Lower Risk (lc).
SYNONYMS: *tryoni* Longman, 1921.
COMMENTS: *cephalotes* species group. See Churchill (1998).

*Nyctimene sanctacrucis* Troughton, 1931. Proc. Linn. Soc. N.S.W., 56:206.
COMMON NAME: Nendo Tube-nosed Fruit Bat.
TYPE LOCALITY: Solomon Isls, Temotu Province, Santa Cruz Isls.
DISTRIBUTION: Santa Cruz Isls.

STATUS: IUCN/SSC Action Plan (1992) and IUCN 2003 – Extinct.

COMMENTS: *cephalotes* species group. Known only from the holotype; see Flannery (1995*b*).

*Nyctimene vizcaccia* Thomas, 1914. Ann. Mag. Nat. Hist., ser. 8, 13:436.

COMMON NAME: Umboi Tube-nosed Fruit Bat.

TYPE LOCALITY: Papua New Guinea, Morobe Province, Umboi Isl.

DISTRIBUTION: Bismarck Arch., Bougainville Isl, Solomon Isls (N of Malaita only).

STATUS: IUCN/SSC Action Plan (1992) – Not Threatened. IUCN 2003 – Lower Risk (lc).

SYNONYMS: **bougainville** Troughton, 1936; *minor* Phillips, 1968.

COMMENTS: *albiventer* species group. Includes *bougainville*; see Smith and Hood (1983), but also see Flannery (1995*b*), who considered the latter to be a distinct species while treating *vizcacaia* as a synonym of *albiventer*. Includes *minor*, but see Peterson (1991). Formerly included in *cephalotes*, but see Smith and Hood (1983). Reviewed by Bergmans (2001); also see Bonaccorso (1998). May include *malaitensis*.

*Otopteropus* Kock, 1969. Senckenberg. Biol., 50:329.

TYPE SPECIES: *Otopteropus cartilagonodus* Kock, 1969.

SYNONYMS: *Otopterus* Sokolov, 1973 [*lapsus*, not *Otopterus* Lydekker, 1891].

*Otopteropus cartilagonodus* Kock, 1969. Senckenberg. Biol., 50:333.

COMMON NAME: Luzon Fruit Bat.

TYPE LOCALITY: Philippines, Luzon, Mountain Prov., Sitio Pactil.

DISTRIBUTION: Luzon (Philippines).

STATUS: IUCN/SSC Action Plan (1992) – Indeterminate. IUCN 2003 – Vulnerable.

*Paranyctimene* Tate, 1942. Am. Mus. Novit., 1204:1.

TYPE SPECIES: *Paranyctimene raptor* Tate, 1942.

COMMENTS: Revised by Bergmans (2001). Bergmans (2001) proposed that *Paranyctimene* be considered a subgenus of *Nyctimene*, but I retain *Paranyctimene* as a distinct genus pending studies of phylogenetic relationships of these taxa. Should *Paranyctimene* be shown to nest within a clade of *Nyctimene* species, I would support synonymizing these genera. However, I see little to be gained by this change if *Nyctimene* sensu stricto proves to be monophyletic with respect to *Paranyctimene* (and vice versa) given the long history of usage of these names. Electophoretic data published by Donnellan et al. (1995) suggests that these taxa are indeed reciprocally monophyletic.

*Paranyctimene raptor* Tate, 1942. Am. Mus. Novit., 1204:1.

COMMON NAME: Unstriped Tube-nosed Fruit Bat.

TYPE LOCALITY: Papua New Guinea, Western Province, Fly River, Oroville Camp. (ca. 4 mi. [6 km]) below Elavala River mouth); (6°13'S, 141°7'E).

DISTRIBUTION: Papua New Guinea; possibly Mainland Prov. of Papua (Indonesia) and Salawati Isl (Indonesia).

STATUS: IUCN/SSC Action Plan (1992) – Rare. IUCN 2003 – Lower Risk (nt).

COMMENTS: Reviewed and rediagnosed by Bergmans (2001); also see Flannery (1995*a*, *b*) and Bonaccorso (1998). Many of the published records of this species may represent *tenax*, see Bergmans (2001).

*Paranyctimene tenax* (Bergmans, 2001). Beaufortia, 51:146.

COMMON NAME: Steadfast Tube-nosed Fruit Bat.

TYPE LOCALITY: Papua New Guinea, Morobe Province, 32 km SSW of Wau, upstream of Anadea, about 07°36'S, 146°37'E, 850 m.

DISTRIBUTION: Mainland New Guinea; Waigeo Isl (Indonesia, Prov. of Papua).

STATUS: IUCN 2003 – not evaluated (new species).

SYNONYMS: **marculus** Bergmans, 2001.

COMMENTS: Originally placed in *Nyctimene* (subgenus *Paranyctimene*) by Bergmans (2001); see comments under those genera. Many published records referred to *raptor* may actually represent *tenax*, see Bergmans (2001).

*Penthetor* K. Andersen, 1912. Cat. Chiroptera Brit. Mus., p. 665.
 TYPE SPECIES: *Cynopterus* (*Ptenochirus*) *lucasi* Dobson, 1880.

 *Penthetor lucasi* (Dobson, 1880). Ann. Mag. Nat. Hist., ser. 5, 6:163.
  COMMON NAME: Lucas's Short-nosed Fruit Bat.
  TYPE LOCALITY: Malaysia, N Borneo, Sarawak.
  DISTRIBUTION: W Malaysia, Borneo, Sumatra, Riau Arch. (Indonesia).
  STATUS: IUCN/SSC Action Plan (1992) – Not Threatened. IUCN 2003 – Lower Risk (lc).

*Plerotes* K. Andersen, 1910. Ann. Mag. Nat. Hist., ser. 8, 5:97.
 TYPE SPECIES: *Epomophorus anchietae* Seabra, 1900.
 COMMENTS: Revised by Bergmans (1989).

 *Plerotes anchietae* (Seabra, 1900). J. Sci. Math. Phys. Nat. Lisboa, ser. 2, 6:116.
  COMMON NAME: Anchieta's Broad-faced Fruit Bat.
  TYPE LOCALITY: Angola, Benguela, Galanga.
  DISTRIBUTION: Angola, Zambia, S Dem. Rep. Congo, Malawi.
  STATUS: IUCN/SSC Action Plan (1992) – Rare. IUCN 2003 – Data Deficient.
  COMMENTS: Reviewed by Kock et al. (1998). Sometimes misspelled *anchietai*, but see Kock
   et al. (1998).

*Ptenochirus* Peters, 1861. Monatsb. K. Preuss. Akad. Wiss. Berlin, 1861:707.
 TYPE SPECIES: *Pachysoma* (*Ptenochirus*) *jagori* Peters, 1861.

 *Ptenochirus jagori* (Peters, 1861). Monatsb. K. Preuss. Akad. Wiss. Berlin, 1861:707.
  COMMON NAME: Greater Musky Fruit Bat.
  TYPE LOCALITY: Philippines, Luzon, Albay, Daraga.
  DISTRIBUTION: Philippines except Palawan region.
  STATUS: IUCN/SSC Action Plan (1992) – Not Threatened. IUCN 2003 – Lower Risk (lc).
  COMMENTS: Sometimes misspelled *jagorii* (e.g., Corbet and Hill, 1992).

 *Ptenochirus minor* Yoshiyuki, 1979. Bull. Natl. Sci. Mus. Tokyo, Ser. A (Zool.), 5:75.
  COMMON NAME: Lesser Musky Fruit Bat.
  TYPE LOCALITY: Philippines, Mindanao, Davao City Prov., Mt. Talomo, Baracatan.
  DISTRIBUTION: Philippines.
  STATUS: IUCN/SSC Action Plan (1992) – Not Threatened. IUCN 2003 – Lower Risk (lc).

*Pteralopex* Thomas, 1888. Ann. Mag. Nat. Hist., ser. 6, 1:155.
 TYPE SPECIES: *Pteralopex atrata* Thomas, 1888.
 COMMENTS: Revised by Parnaby (2002*b*); also see Hill and Beckon (1978). For a key to species
  see Parnaby (2002*b*).

 *Pteralopex acrodonta* Hill and Beckon, 1978. Bull. Brit. Mus. (Nat. Hist.) Zool., 34:68.
  COMMON NAME: Fijian Monkey-faced Fruit Bat.
  TYPE LOCALITY: Fiji Isls, Taveuni Isl, Des Voeux Peak, ca. 3,840 ft. (1,170 m).
  DISTRIBUTION: Taveuni Isl (Fiji Isls).
  STATUS: IUCN/SSC Action Plan (1992) – Endangered: Limited Distribution. IUCN 2003 –
   Critically Endangered.
  COMMENTS: Reviewed by Parnaby (2002*b*); also see Flannery (1995*b*).

 *Pteralopex anceps* K. Andersen, 1909. Ann. Mag. Nat. Hist., ser. 7, 3:266.
  COMMON NAME: Bougainville Monkey-faced Fruit Bat.
  TYPE LOCALITY: Papua New Guinea, Bougainville Isl.
  DISTRIBUTION: Buka, Bougainville Isls (Papua New Guinea); Choiseul Isl and Isabel Isl
   (Solomon Isls).
  STATUS: IUCN/SSC Action Plan (1992) – Endangered: Limited Distribution. IUCN 2003 –
   Critically Endangered.
  COMMENTS: Considered a subspecies of *atrata* by Phillips (1968), but clearly distinct; see Hill
   and Beckon (1978) Flannery (1991*b*, 1995*b*), and Parnaby (2002*b*). See also Bonaccorso
   (1998).

*Pteralopex atrata* Thomas, 1888. Ann. Mag. Nat. Hist., ser. 6, 1:155.
COMMON NAME: Guadalcanal Monkey-faced Fruit Bat.
TYPE LOCALITY: Solomon Isls, Guadalcanal, Aola.
DISTRIBUTION: Guadalcanal (Solomon Isls). A specimen from Isabel Isl formerly referred to this species has been reidentified as *anceps* (see Parnaby, 2002*b*).
STATUS: IUCN/SSC Action Plan (1992) – Endangered: Limited Distribution. IUCN 2003 – Critically Endangered.
COMMENTS: Does not include *anceps*; see Hill and Beckon (1978), Flannery (1991*b*, 1995*b*), and Parnaby (2002*b*).

*Pteralopex pulchra* Flannery, 1991. Rec. Aust. Mus., 43:125.
COMMON NAME: Montane Monkey-faced Fruit Bat.
TYPE LOCALITY: Solomon Isls, Guadacanal, Mount Makarakomburu, 1,230 m.
DISTRIBUTION: Montane Guadalcanal (Solomon Isls).
STATUS: Described after completion of IUCN/SSC Old World Fruit Bat Action Plan (1992). IUCN 2003 – Critically Endangered.
COMMENTS: Known only from the holotype. See Flannery (1991*b*, 1995*b*) and Parnaby (2002*b*).

*Pteralopex taki* Parnaby, 2002. Aust. Mammal., 23:146.
COMMON NAME: New Georgian Monkey-faced Bat.
TYPE LOCALITY: Solomon Isls, New Georgia Isl, Marovo Lagoon, 5 km N of Patutiva Village, Mt Javi, 8°31'S, 157°52'E, 50 m.
DISTRIBUTION: New Georgia Isl and Vangunu Isl (Solomon Isls). Apparently locally extinct on Kolombangara Isl.
STATUS: IUCN 2003 – Not Evaluated as *Pteralopex* sp. nov., but Parnaby (2000*b*) recommended that this species be classified in the IUCN threat category of "Critically Endangered."
COMMENTS: In addition to the original description by Parnaby (2002*b*), see Flannery (1995*b*), who discussed this species under its common name.

*Pteropus* Brisson, 1762. Regnum Animale, Ed. 2, pp. 13, 153.
TYPE SPECIES: *Vespertilio vampirus niger* Kerr, 1792, type species by designation under plenary powers of the International Commision on Zoological Nomenclature.
SYNONYMS: *Desmalopex* Miller, 1907; *Eunycteris* Gray, 1866; *Pselaphon* Gray, 1870 [not *Pselaphon* Herbst, 1792, a coleopteran]; *Sericonycteris* Matschie, 1899; *Spectrum* Lacépède, 1799 [not *Spectrum* Scopoli, 1777, a lepidopteran].
COMMENTS: The International Commission on Zoological Nomenclature ruled in favor of rejecting Brisson (1762), but conserved several of the generic names including *Pteropus* (Opinion 1894, ICZN 1998). Formerly included *arquatus* and *leucotis* which were transferred to *Acerodon* by Musser et al. (1982*a*). Species groups follow Koopman (1994).

*Pteropus admiralitatum* Thomas, 1894. Ann. Mag. Nat. Hist., ser. 6, 13:293.
COMMON NAME: Admiralty Flying Fox.
TYPE LOCALITY: Papua New Guinea, Bismarck Arch., Admiralty Isls.
DISTRIBUTION: Solomon Isls; Admiralty Isls, New Britain, and Tabar Isls (Bismarck Arch.).
STATUS: CITES – Appendix II. IUCN/SSC Action Plan (1992) – Not Threatened. IUCN 2003 – Lower Risk (lc).
SYNONYMS: *colonus* K. Andersen, 1908; *goweri* Tate, 1934; *solomonis* Thomas, 1904.
COMMENTS: *subniger* species group. Reviewed by Felten and Kock (1972); also see Flannery (1995*b*) and Bonaccorso (1998).

*Pteropus aldabrensis* True, 1893. Proc. U.S. Natl. Mus., 16:533.
COMMON NAME: Aldabra Flying Fox.
TYPE LOCALITY: Seychelles, Aldabra Isl.
DISTRIBUTION: Known only from the type locality.
STATUS: CITES – Appendix II. IUCN/SSC Action Plan (1992) – Vulnerable: Limited Distribution as *P. seychellensis aldabrensis*. IUCN 2003 – Vulnerable.
COMMENTS: *niger* species group. Included in *seychellensis* by Hill (1971*b*), but see Bergmans (1990).

*Pteropus alecto* Temminck, 1837. Monogr. Mamm., 2:75.

COMMON NAME: Black Flying Fox.

TYPE LOCALITY: Indonesia, N Sulawesi, Menado.

DISTRIBUTION: Sulawesi, Saleyer Isl, Lombok, Bawean Isl, Kangean Isls, Sumba Isl, and Savu Isl (Indonesia); N and E Australia; S New Guinea.

STATUS: CITES – Appendix II. IUCN/SSC Action Plan (1992) – Not Threatened. IUCN 2003 – Lower Risk (lc).

SYNONYMS: *nicobaricus* Heude, 1897 [not Zelebor, 1869]; ***aterrimus*** Matschie, 1899; *aterrimus* Temminck, 1846 [*nomen nudum*]; *baveanus* Miller, 1906; ***gouldi*** Peters, 1867; ***morio*** K. Andersen, 1908.

COMMENTS: *alecto* species group. Includes *gouldi*; see Tate (1942*b*); also see Bergmans and Rozendaal (1988). The synonymy of *nicobaricus* with *alecto* is uncertain (Corbet and Hill, 1992). See Webb and Tideman (1995) for discussion of cases of hybridization with *poliocephalus* and possible hybridization with *conspicillatus*. Also see Flannery (1995*a, b*).

*Pteropus anetianus* Gray, 1870. Cat. Monkeys, Lemurs, Fruit-eating Bats Brit. Mus., p. 101.

COMMON NAME: Vanuatu Flying Fox.

TYPE LOCALITY: Vanuatu, Aneiteum (= Aneityum).

DISTRIBUTION: Vanuatu including Banks Isls.

STATUS: CITES – Appendix II. IUCN/SSC Action Plan (1992) – Indeterminate. IUCN 2003 – Lower Risk (lc).

SYNONYMS: ***aorensis*** Lawrence, 1945; ***bakeri*** Thomas, 1925; ***banksiana*** Sanborn, 1930; ***eotinus*** K. Andersen, 1913; ***motalavae*** Felten and Kock, 1972; ***pastoris*** Felten and Kock, 1972.

COMMENTS: *samoensis* species group. Includes *eotinus*, *bakeri*, and *banksiana*; see Felten and Kock (1972). Also see Flannery (1995*b*).

*Pteropus aruensis* Peters, 1867. Monatsb. K. Preuss. Akad. Wiss. Berlin, 1867:330.

COMMON NAME: Aru Flying Fox.

TYPE LOCALITY: Indonesia, Aru Isls.

DISTRIBUTION: Aru Isls (Indonesia).

STATUS: CITES – Appendix II. IUCN/SSC Action Plan (1992) – No Data: Limited Distribution as *Pteropus melanopogon aruensis*. IUCN 2003 – Not listed. This species has not been collected since the nineteenth century, and probably should be listed as Critically Endangered (K. Helgen, pers. comm.).

SYNONYMS: *fumigatus* Rosenberg, 1867; *rubiginosus* Rosenberg, 1867.

COMMENTS: *melanopogon* species group. Often listed as a subspecies of *melanopogon* following Laurie and Hill (1954), but see Bergmans (2001), who argued that *aruensis* should be considered distinct pending additional review of this complex.

*Pteropus banakrisi* Richards and Hall, 2002. Australian Zool. 32:60.

COMMON NAME: Torresian Flying Fox.

TYPE LOCALITY: Australia, Torres Strait, Moa Isl, St. Pauls Mission.

DISTRIBUTION: Moa Isl (Australia).

STATUS: CITES – Appendix II. IUCN 2003 – not evaluated (new species).

COMMENTS: *alecto* species group. May be conspecific with *alecto* (K. Helgen, pers. comm.).

*Pteropus brunneus* Dobson, 1878. Cat. Chiroptera Brit. Mus., p. 37.

COMMON NAME: Dusky Flying Fox.

TYPE LOCALITY: Australia, Queensland, Percy Isl.

DISTRIBUTION: Known from the type locality only.

STATUS: CITES – Appendix II. IUCN/SSC Action Plan (1992) – Extinct? IUCN 2003 – Extinct.

COMMENTS: *subniger* species group. Known only from the holotype. It is not clear that this taxon represents a valid species, see Koopman (1984*c*).

*Pteropus caniceps* Gray, 1870. Cat. Monkeys, Lemurs, Fruit-eating Bats Brit. Mus., p. 107.

COMMON NAME: North Moluccan Flying Fox.

TYPE LOCALITY: Indonesia, Maluku, Halmahera, Batjan (= Batchian, Bacan).

DISTRIBUTION: Halmahera (Indonesia). Sula, Peleng, and Sangihe Isl records are erroneous,

and a single Sulawesi record (obtained from a dealer) is dubious (Bergmans and
Rozendaal, 1988; K. Helgen, pers. comm.; Koopman, 1993).
STATUS: CITES – Appendix II. IUCN/SSC Action Plan (1992) – Not Threatened. IUCN 2003 –
Lower Risk (lc).
SYNONYMS: *affinis* Gray, 1871; *batchiana* Gray, 1871; **dobsoni** Andersen, 1908; *fuscus*
Dobson, 1878 [not Geoffory, 1803, Desmarest, 1803, or Blainville, 1840].
COMMENTS: *caniceps* species group. Includes *dobsoni*; see Laurie and Hill (1954). Also see
Flannery (1995*b*).

*Pteropus capistratus* Peters, 1876. Monatsb. K. Preuss. Akad. Wiss. Berlin, 1876:316.
COMMON NAME: Bismark Masked Flying Fox.
TYPE LOCALITY: Papua New Guinea. The type locality was initially given as New Ireland Isl.,
but this is clearly incorrect (see Flannery and White, 1991). The type locality is
probably the Duke of York group or New Britain Isl.
DISTRIBUTION: Bismarck Arch. (Papua New Guinea).
STATUS: CITES – Appendix II. IUCN/SSC Action Plan (1992) – No Data as *P. temmincki
capistratus*. IUCN 2003 – Not listed.
SYNONYMS: **ennisae** Flannery and White, 1991.
COMMENTS: *Personatus* species group. Formerly regarded as a subspecies of *temmincki*, but
apparently distinct; see Flannery (1995*b*). See also Flannery and White (1991) and
Bonaccorso (1998).

*Pteropus chrysoproctus* Temminck, 1837. Monogr. Mamm., 2:67.
COMMON NAME: Moluccan Flying Fox.
TYPE LOCALITY: Indonesia, Maluku, Ambon.
DISTRIBUTION: Ambon, Buru, Seram, and small islands east of Seram (Indonesia). A Sangihe
Isl record is erroneous; see Bergmans and Rozendaal (1988).
STATUS: CITES – Appendix II. IUCN/SSC Action Plan (1992) – No Data. IUCN 2003 – Lower
Risk (nt).
SYNONYMS: *argentatus* Gray, 1843 [*nomen nudum*]; *argentatus* Gray, 1844.
COMMENTS: *chrysoproctus* species group. See Flannery (1995*b*). May not include *argentatus*, a
taxon based on a badly damaged immature specimen thought to be from Ambon (K.
Helgen, pers. comm.). Sulawesi specimens previously referred to *argentatus* were
allocated to *Acerodon celebensis* by Musser et al. (1982*a*). This complex includes several
undescribed species (K. Helgen, pers. comm.).

*Pteropus cognatus* K. Andersen, 1908. Ann. Mag. Nat. Hist., ser. 8, 2:365.
COMMON NAME: Makira Flying Fox.
TYPE LOCALITY: Solomon Isl, Makira ("San Cristoval" = San Cristobal Isl).
DISTRIBUTION: Makira and Uki Ni Masi Isls (Solomon Isls).
STATUS: CITES – Appendix II. IUCN/SSC Action Plan (1992) – No Data: Limited Distribution
as *P. rayneri cognatus*. IUCN 2003 – Not listed.
COMMENTS: *Chrysoproctus* species group. Often considered a subspecies of *rayneri* (e.g., Hill,
1962*a*), but apparently distinct; see Flannery (1995*b*). May be conspecific with *rennelli*
(K. Helgen, pers. comm.).

*Pteropus conspicillatus* Gould, 1850. Proc. Zool. Soc. Lond., 1849:109 [1850].
COMMON NAME: Spectacled Flying Fox.
TYPE LOCALITY: Australia, Queensland, Fitzroy Isl.
DISTRIBUTION: N Moluccas (Indonesia); New Guinea and West Papuan Isls (Raja Ampat Isl,
off NW coast of New Guinea); NE Queensland (Australia).
STATUS: CITES – Appendix II. IUCN/SSC Action Plan (1992) – Not Threatened. IUCN 2003 –
Lower Risk (lc).
SYNONYMS: **chrysauchen** Peters, 1862; *mysolensis* Gray, 1871.
COMMENTS: *conspicillatus* species group. See Webb and Tideman (1995) for discussion of
possible hybridization with *alecto*. Also see Flannery (1995*a*, *b*), Bonaccorso (1998),
and Bergmans (2001).

*Pteropus dasymallus* Temminck, 1825. Monogr. Mamm., 1:180.
COMMON NAME: Ryukyu Flying Fox.
TYPE LOCALITY: Japan, Ryukyu Isls, Kuchinoerabu Isl (restricted by Kuroda, 1933).

DISTRIBUTION: Taiwan; Ryukyu Isls, Daito Isls and extreme S Kyushu (Japan); Batan, Dalupiri, and Fuga Isls (Philippines).

STATUS: CITES – Appendix II. IUCN/SSC Action Plan (1992) – Endangered: Limited Distribution. IUCN 2003 – Endangered.

SYNONYMS: *rubricollis* Siebold, 1824 [not Geoffroy, 1810]; *yamagatai* Kishida, 1929; **daitonensis** Kuroda, 1921; **formosus** Sclater, 1873; **inopinatus** Kuroda, 1933; **yayeyamae** Kuroda, 1933.

COMMENTS: *subniger* species group. Includes *daitoensis*; see Kuroda (1933) and Yoshiyuki (1989). Reviewed in part by Yoshiyuki (1989) and Horácek et al. (2000); see also Ingle and Heaney (1992).

*Pteropus faunulus* Miller, 1902. Proc. U.S. Natl. Mus., 24:785.

COMMON NAME: Nicobar Flying Fox.

TYPE LOCALITY: India, Nicobar Isls, Car Nicobar Isl.

DISTRIBUTION: Nicobar Isls (India).

STATUS: CITES – Appendix II. IUCN/SSC Action Plan (1992) – No Data: Limited Distribution. IUCN 2003 – Vulnerable.

COMMENTS: *subniger* species group. Reviewed by Bates and Harrison (1997).

*Pteropus fundatus* Felten and Kock, 1972. Senckenberg. Biol., 53:186.

COMMON NAME: Banks Flying Fox.

TYPE LOCALITY: Vanuatu, Banks Isls, Mota Isl.

DISTRIBUTION: Banks Isls (Vanuatu).

STATUS: CITES – Appendix II. IUCN/SSC Action Plan (1992) – No Data: Limited Distribution. IUCN 2003 – Vulnerable.

COMMENTS: *chrysoproctus* species group. See Flannery (1995b).

*Pteropus giganteus* (Brünnich, 1782). Dyrenes Historie, 1:43.

COMMON NAME: Indian Flying Fox.

TYPE LOCALITY: India, Bengal.

DISTRIBUTION: Maldive Isls, India (incl. Andaman Isls), Sri Lanka, Pakistan, Bangladesh, Nepal, Burma, Tsinghai (China). The Tsinghai record requires confirmation. Cambodian records are apparently erroneous; see Kock (2000).

STATUS: CITES – Appendix II. IUCN/SSC Action Plan (1992) – Not Threatened. IUCN 2003 – Lower Risk (lc).

SYNONYMS: *edwardsi* Geoffroy, 1828 [not Geoffroy, 1810]; *kelaarti* Gray, 1871 [skin, not skull]; *medius* Temminck, 1825; *ruvicollis* Ogilby, 1840 [not E. Geoffroy, 1810]; **ariel** G. M. Allen, 1908; **chinghaiensis** Wang and Wang, 1962; **leucocephalus** Hodgson, 1835; **assamensis** McClelland, 1839.

COMMENTS: *vampyrus* species group. Includes *ariel*; see Hill (1958). Possibly conspecific with *vampyrus*; see Corbet and Hill (1992). Reviewed in part by Bates and Harrison (1997) and Horácek et al. (2000).

*Pteropus gilliardorum* Van Deusen, 1969. Am. Mus. Novit., 2371:5.

COMMON NAME: Gilliard's Flying Fox.

TYPE LOCALITY: Papua New Guinea, Bismarck Arch., West New Britain Province, Whiteman Mtns, Wild Dog Ridge, ca. 1,600 m.

DISTRIBUTION: New Britain and New Ireland (Bismarck Arch., Papua New Guinea).

STATUS: CITES – Appendix II. IUCN/SSC Action Plan (1992) – No Data: Limited Distribution. IUCN 2003 – Vulnerable as *P. gilliardi*.

COMMENTS: *scapulatus* species group. See Flannery (1995b) and Bonaccorso (1998). Previously spelled *gilliardi*; ammended to *gilliardorum* by Flannery (1995b) following Article 31.1.2 of the International Code of Zoological Nomenclature (International Commission on Zoological Nomenclature, 1999). Misspelled *gailliardi* by Koopman (1994).

*Pteropus griseus* (E. Geoffroy, 1810). Ann. Mus. Natn. Hist. Nat. Paris, 15:94.

COMMON NAME: Gray Flying Fox.

TYPE LOCALITY: Indonesia, Lesser Sunda Isls, Timor.

DISTRIBUTION: Timor, Samao Isl, Dyampea Isl, Bonerato Isl, Saleyer Isl, Paternoster Isls, Pelang, Isl, Sulawesi, and Banda Isls (Indonesia).

STATUS: CITES – Appendix II. IUCN/SSC Action Plan (1992) – Not Threatened. IUCN 2003 – Lower Risk (lc).

SYNONYMS: *mimus* K. Andersen, 1908; *pallidus* Temminck, 1825.

COMMENTS: *subniger* species group. Includes *mimus*; see Laurie and Hill (1954) and Corbet and Hill (1992). May also include *speciosus* (here retained as a separate species); see Corbet and Hill (1992). Subspecies limits and allocation are uncertain; see Bergmans (2001).

*Pteropus howensis* Troughton, 1931. Proc. Linn. Soc. N.S.W., 56:204.

COMMON NAME: Ontong Java Flying Fox.

TYPE LOCALITY: Solomon Isls, Ontong Java Isl.

DISTRIBUTION: Ontong Java Isl (Solomon Isls).

STATUS: CITES – Appendix II. IUCN/SSC Action Plan (1992) – No Data: Limited Distribution. IUCN 2003 – Vulnerable.

COMMENTS: *subniger* species group. See Flannery (1995b).

*Pteropus hypomelanus* Temminck, 1853. Esquisses Zool. sur la Côte de Guine, p. 61.

COMMON NAME: Variable Flying Fox.

TYPE LOCALITY: Indonesia, Molucca Isls, Ternate Isl.

DISTRIBUTION: Andaman and Maldive Isls; New Guinea through Indonesia to Vietnam and Thailand, and adjacent islands; Philippines. Solomon Isls records are probably erroneous (K. Helgen, pers. comm.).

STATUS: CITES – Appendix II. IUCN/SSC Action Plan (1992) – Not Threatened. IUCN 2003 – Lower Risk (lc).

SYNONYMS: *tricolor* Gray, 1871; *annectens* K. Andersen, 1908; *cagayanus* Mearns, 1905; *canus* K. Andersen, 1908; *condorensis* Peters, 1869; *enganus* Miller, 1906; *fretensis* Kloss, 1916; *geminorum* Miller, 1903; *lepidus* Miller, 1900; *luteus* K. Andersen, 1908; *vulcanius* Thomas, 1915; *macassaricus* Heude, 1897; *maris* Allen, 1936; *robinsoni* K. Andersen, 1909; *satyrus* K. Andersen, 1908; *simalurus* Thomas, 1923; *tomesi* Peters, 1869.

COMMENTS: *subniger* species group. It is possible that *vociferus* Peale, 1848, is an older name for this taxon; see K. Andersen (1912). Formerly included *brunneus*; see Ride (1970); but see Koopman (1984c) and Corbet and Hill (1992). Includes *satyrus*; see Bates and Harrison (1997), but also see Hill (1971c), who included *satyrus* in *melanotus*. Also see Flannery (1995a, b) and Bonaccorso (1998). Validity of many subspecies is questionable. Does not include *mearnsi*; see Heaney et al. (1987) and Flannery (1995b), but also see Corbet and Hill (1992).

*Pteropus insularis* Hombron and Jacquinot, 1842. *In* d'Urville, Voy. Pole Sud. Mammifères, p. 24.

COMMON NAME: Ruck Flying Fox.

TYPE LOCALITY: Caroline Isls, Truk Isl, Hogoleu (Micronesia).

DISTRIBUTION: Truk Isls (Micronesia).

STATUS: CITES – Appendix I. IUCN/SSC Action Plan (1992) – Endangered: Limited Distribution. IUCN 2003 – Critically Endangered.

SYNONYMS: *laniger* H. Allen, 1890; *phaeocephalus* Thomas, 1882.

COMMENTS: *pselaphon* species group. Includes *phaeocephalus* (K. Helgen, pers. comm.). See Flannery (1995b).

*Pteropus intermedius* K. Andersen, 1908. Ann. Mag. Nat. Hist., ser. 8, 2:368.

COMMON NAME: Andersen's Flying Fox.

TYPE LOCALITY: S Burma, Amherst, near Moulmein.

DISTRIBUTION: S Burma and W Thailand.

STATUS: CITES – Appendix II. IUCN/SSC Action Plan (1992) – No Data as *P. vampyrus intermedius*. IUCN 2003 – Not listed.

COMMENTS: *vampyrus* species group. Included in *vampyrus* by Lekagul and McNeely (1977) and Koopman (1993, 1994), but see Corbet and Hill (1992).

*Pteropus keyensis* Peters, 1867. Monatsb. K. Preuss. Akad. Wiss. Berlin, 1867:330.

COMMON NAME: Kei Flying Fox.

TYPE LOCALITY: Indonesia, Key (= Kei) Isls.

DISTRIBUTION: Kai Isls (Indonesia).

STATUS: CITES – Appendix II. IUCN/SSC Action Plan (1992) – No Data: Limited Distribution as *Pteropus melanopogon keyensis*. IUCN 2003 – Not listed.

SYNONYMS: *chrysargyrus* Heude, 1897.

COMMENTS: *melanopogon* species group. Often listed as a subspecies of *melanopogon* following Laurie and Hill (1954), but see Bergmans (2001), who argued that *keyensis* should be considered distinct pending additional review of this complex.

*Pteropus leucopterus* Temminck, 1853. Esquisses Zool. sur la Côte de Guine, p. 60.

COMMON NAME: White-winged Flying Fox.

TYPE LOCALITY: Philippines.

DISTRIBUTION: Luzon, Catanduanes, and Dinagat Isls (Philippines).

STATUS: CITES – Appendix II. IUCN/SSC Action Plan (1992) – Vulnerable. IUCN 2003 – Endangered.

SYNONYMS: *chinensis* Gray, 1871.

COMMENTS: *pselaphon* species group.

*Pteropus livingstonii* Gray, 1866. Proc. Zool. Soc. Lond., 1866:66.

COMMON NAME: Comoro Flying Fox.

TYPE LOCALITY: Comoro Isls, Anjouan Isl.

DISTRIBUTION: Comoro Isls.

STATUS: CITES – Appendix II. IUCN/SSC Action Plan (1992) – Endangered: Limited Distribution. IUCN 2003 – Critically Endangered.

COMMENTS: *livingstonii* species group. Reviewed by Bergmans (1990). Misspelled *livingstonei* by Koopman (1993, 1994).

*Pteropus lombocensis* Dobson, 1878. Cat. Chiroptera Brit. Mus., p. 34.

COMMON NAME: Lombok Flying Fox.

TYPE LOCALITY: Indonesia, Lesser Sunda Isls, Lombok Isl.

DISTRIBUTION: Lombok, Sumbawa, Komodo, Flores, Lembata, Pantar, Alor and Timor Isls (Indonesia).

STATUS: CITES – Appendix II. IUCN/SSC Action Plan (1992) – Not Threatened. IUCN 2003 – Lower Risk (lc).

SYNONYMS: *temmincki* Hartert, 1898 [not Peters, 1867]; **heudei** Matschie, 1899; *tricolor* Heude, 1897 [not Gray, 1871]; *solitarius* K. Andersen, 1908; **salottii** Kitchener, 1995 [in Kitchener and Maryanto, 1995].

COMMENTS: *molossinus* species group. Revised by Kitchener et al. (1995*d*) and Kitchener and Maryanto (1995*b*).

*Pteropus loochoensis* Gray, 1870. Cat. Monkeys, Lemurs and Fruit-eating Bats, British Museum, p. 106.

COMMON NAME: Japanese Flying Fox.

TYPE LOCALITY: Japan, Okinawa, Liû-kiû Isls.

DISTRIBUTION: Okinawa Isl, Ryûkyû Isls (Japan).

STATUS: CITES – Appendix II. IUCN/SSC Action Plan (1992) – Endangered: Limited Distribution as *P. mariannus loochoensis*. IUCN 2003 – Extinct.

SYNONYMS: *keraudreni* Fritze, 1894; *loochooensis* Fritze, 1894; *luchuensis* Seitz, 1892.

COMMENTS: *mariannus* species group. Often included in *mariannus*, but see Corbet and Hill (1980) and Yoshiyuki (1989). Reviewed by Yoshiyuki (1989). Flannery (1995*b*) treated *loochoensis* as a subspecies of *mariannus* without comment.

*Pteropus lylei* K. Andersen, 1908. Ann. Mag. Nat. Hist., ser. 8, 2:367.

COMMON NAME: Lyle's Flying Fox.

TYPE LOCALITY: Thailand, Bangkok.

DISTRIBUTION: Thailand, Vietnam, Cambodia.

STATUS: CITES – Appendix II. IUCN/SSC Action Plan (1992) – Not Threatened. IUCN 2003 – Lower Risk (lc).

COMMENTS: *vampyrus* species group.

*Pteropus macrotis* Peters, 1867. Monatsb. K. Preuss. Akad. Wiss. Berlin, 1867:327.
>   COMMON NAME: Big-eared Flying Fox.
>   TYPE LOCALITY: Indonesia, Aru Isls, Wokam Isl.
>   DISTRIBUTION: New Guinea; Aru Isls (Indonesia); Boigu Isl (Australia).
>   STATUS: CITES – Appendix II. IUCN/SSC Old World Fruit Bat Action Plan (1992) – Not Threatened. IUCN 2003 – Lower Risk (lc).
>   SYNONYMS: *insignis* Rosenberg, 1867; ***epularius*** Ramsay, 1878.
>   COMMENTS: *poliocephalus* species group. See Flannery (1995*a*, *b*), and Bonaccorso (1998).

*Pteropus mahaganus* Sanborn, 1931. Field Mus. Nat. Hist. Publ., Zool. Ser., 2:19.
>   COMMON NAME: Sanborn's Flying Fox.
>   TYPE LOCALITY: Solomon Isls, Ysabel Isl, Tunnibul.
>   DISTRIBUTION: Bougainville Isl (Papua New Guinea); Ysabel Isl and Choiseul Isl (Solomon Isls).
>   STATUS: CITES – Appendix II. IUCN/SSC Action Plan (1992) – Vulnerable: Limited Distribution. IUCN 2003 – Vulnerable.
>   COMMENTS: *scapulatus* species group. See Flannery (1995*b*) and Bonaccorso (1998).

*Pteropus mariannus* Desmarest, 1822. Mammalogie, *in* Encycl. Méth., 2(Suppl.):547.
>   COMMON NAME: Marianas Flying Fox.
>   TYPE LOCALITY: West Pacific, Mariana Isls, Guam (USA).
>   DISTRIBUTION: S Mariana Isls through Guam to Ulithi Isl.
>   STATUS: CITES – Appendix I; U.S. ESA – Endangered (but proposed reclassification to Threatened) in Guam as *P. m. mariannus*; Proposed Threatened in the Aguijan, Tinian, Saipan populations. IUCN/SSC Action Plan (1992) – Endangered: Limited Distribution. IUCN 2003 – Endangered.
>   SYNONYMS: *keraudren* Quoy and Gaimard, 1824; ***paganensis*** Yamashima, 1932; ***ulthiensis*** Yamashima, 1932. **Unassigned**: *vanikorensis* Quoy and Gaimard, 1830.
>   COMMENTS: *mariannus* species group. Probably includes *vanikorensis*, see Troughton (1930). If *vanikorensis* is in fact from the Mariana Isls (rather than Vanikoro Isl in the Santa Cruz Isls), it would likely be a synonym of either the nominate subspecies or *paganensis*. Systematics of this complex is somewhat confused; some authors have included *pelewensis*, *ualanus*, and *yapensis* as subspecies of *mariannus* (e.g., Koopman, 1994), while others have treated them as distinct species without comment (e.g., Corbet and Hill, 1980). I follow Flannery (1995*b*) in provisionally recognizing *pelewensis*, *ualanus*, and *yapensis* as distinct species pending further study.

*Pteropus melanopogon* Peters, 1867. Monatsb. K. Preuss. Akad. Wiss. Berlin, 1867:330.
>   COMMON NAME: Black-bearded Flying Fox.
>   TYPE LOCALITY: Indonesia, Molucca Isls, Amboina.
>   DISTRIBUTION: Amboina, Buru, Seram, Banda Isls, Yamdena (= Timor Laut), and adjacent islands (Indonesia). A Sangihe Isl record is erroneous; see Bergmans and Rozendaal (1988).
>   STATUS: CITES – Appendix II. IUCN/SSC Action Plan (1992) – Not Threatened. IUCN 2003 – Lower Risk (lc).
>   SYNONYMS: *argentatus* Gray, 1858 [not Gray, 1844]; *phaiops* Temminck, 1837 [not Temminck, 1825].
>   COMMENTS: *melanopogon* species group. Does not include *sepikensis*, here considered a synonym of *neohibernicus;* see Koopman (1979). I follow Bergmans (2001) in recognizing *aruensis* and *keyensis* (often listed as subspecies of *melanopogon* following Laurie and Hill [1954]) as distinct species pending additional review of this complex. This complex includes several undescribed species (K. Helgen, pers. comm.).

*Pteropus melanotus* Blyth, 1863. Cat. Mamm. Mus. Asiat. Soc. Calcutta, p. 20.
>   COMMON NAME: Black-eared Flying Fox.
>   TYPE LOCALITY: India, Nicobar Isls.
>   DISTRIBUTION: Nicobar and Andaman Isls (India); Engano Isl and Nias Isl (Indonesia); Christmas Isl.

STATUS: CITES – Appendix II. IUCN/SSC Action Plan (1992) – Not Threatened. IUCN 2003 – Lower Risk (lc).

SYNONYMS: *edulis* Blyth, 1846 [not E. Geoffroy, 1810]; *nicobaricus* Fitzinger, 1861 [*nomen nudum*]; *nicobaricus* Zelebor, 1869; **modiglianii** Thomas, 1894; **natalis** Thomas, 1887; **niadicus** Miller, 1906; **tytleri** Dobson, 1874.

COMMENTS: *melanotus* species group. Does not include *satyrus*; see Bates and Harrison (1997).

*Pteropus molossinus* Temminck, 1853. Esquisses Zool. sur la Côte de Guine, p. 62.
COMMON NAME: Caroline Flying Fox.
TYPE LOCALITY: Caroline Isls, Ponape (Micronesia).
DISTRIBUTION: Pohnpei (= Ponape) and possibly Mortlock Isls (Caroline Isls, Micronesia).
STATUS: CITES – Appendix I. IUCN/SSC Action Plan (1992) – Endangered: Limited Distribution. IUCN 2003 – Critically Endangered.
SYNONYMS: *breviceps* Thomas, 1883.
COMMENTS: *molossinus* species group. See Flannery (1995*b*).

*Pteropus neohibernicus* Peters, 1876. Monatsb. K. Preuss. Akad. Wiss. Berlin, 1876:317.
COMMON NAME: Great Flying Fox.
TYPE LOCALITY: Papua New Guinea, Bismarck Arch., New Ireland Isl.
DISTRIBUTION: Bismarck Arch. and Admiralty Isls (Papua New Guinea), New Guinea, Misool and Gebi Isls, Gag Isl.
STATUS: CITES – Appendix II. IUCN/SSC Action Plan (1992) – Not Threatened. IUCN 2003 – Lower Risk (lc).
SYNONYMS: *coronatus* Thomas, 1888; *degener* Peters, 1876; *papuanus* Peters and Doria, 1881; *rufus* Ramsay, 1891 [not E. Geoffroy, 1803, or Tiedemann, 1808]; *sepikensis* Sanborn, 1931; **hilli** Felten, 1961.
COMMENTS: *neohibernicus* species group. Includes *sepikensis*; see Koopman (1979). Also see Flannery (1995*a*, *b*) and Bonaccorso (1998).

*Pteropus niger* (Kerr, 1792). *In* Linnaeus, Anim. Kingdom, 1:90.
COMMON NAME: Greater Mascarene Flying Fox.
TYPE LOCALITY: Mascarene Isls, Réunion Isl (France).
DISTRIBUTION: Mascarene Isls (Réunion Isl, Mauritius Isl, subfossil on Rodrigues Isl). Madagascar records are probably erroneous (Bergmans, 1990).
STATUS: CITES – Appendix II. IUCN/SSC Action Plan (1992) – Vulnerable: Limited Distribution. IUCN 2003 – Vulnerable. Extinct on Réunion Isl, see Cheke and Dahl (1981).
SYNONYMS: *fuscus* E. Geoffroy, 1803; *mauritianus* Hermann, 1804; *rufus* Tiedemann, 1808 [not E. Geoffroy, 1803]; *pteropus* Merriam ex Brisson, 1895; *vulgaris* E. Geoffroy, 1810.
COMMENTS: *niger* species group. Reviewed by Bergmans (1990).

*Pteropus nitendiensis* Sanborn, 1930. Am. Mus. Novit., 435:2.
COMMON NAME: Temotu Flying Fox.
TYPE LOCALITY: Solomon Isls, Santa Cruz Isls, Ndeni Isl (= Nendö Isl.).
DISTRIBUTION: Nendö and Tömotu Neo (in the Santa Cruz Isls, Solomon Isls).
STATUS: CITES – Appendix II. IUCN/SSC Action Plan (1992) – No Data: Limited Distribution as *P. nitendiensis* and as *P. sanctacrucis*. IUCN 2003 – Vulnerable as *P. nitendiensis* and as *P. sanctacrucis*.
SYNONYMS: *sanctacrucis* Troughton, 1930.
COMMENTS: *pselaphon* species group. Includes *sanctacrucis*; see Flannery (1995*b*).

*Pteropus ocularis* Peters, 1867. Monatsb. K. Preuss. Akad. Wiss. Berlin, 1867:326.
COMMON NAME: Seram Flying Fox.
TYPE LOCALITY: Indonesia, Molucca Isls, Seram Isl.
DISTRIBUTION: Seram and Buru (Indonesia).
STATUS: CITES – Appendix II. IUCN/SSC Action Plan (1992) – No Data. IUCN 2003 – Vulnerable.
SYNONYMS: *ceramensis* Gray, 1871.
COMMENTS: *conspicillatus* species group. See Flannery (1995*b*).

*Pteropus ornatus* Gray, 1870. Cat. Monkeys, Lemurs, Fruit-eating Bats Brit. Mus., p. 105.
COMMON NAME: Ornate Flying Fox.
TYPE LOCALITY: New Caledonia, Noumea (France).
DISTRIBUTION: New Caledonia and Loyalty Isls.
STATUS: CITES – Appendix II. IUCN/SSC Action Plan (1992) – Indeterminate. IUCN 2003 – Vulnerable.
SYNONYMS: **auratus** K. Andersen, 1909.
COMMENTS: *subniger* species group. Includes *auratus*; see Felten (1964*b*). Also see Sanborn and Nicholson (1950) and Flannery (1995*b*).

*Pteropus pelewensis* K. Andersen, 1908. Ann. Mag. Nat. Hist., ser. 8, 2:364.
COMMON NAME: Pelew Flying Fox.
TYPE LOCALITY: Micronesia, Caroline Isls, Palau Isl (= Pelew Isls).
DISTRIBUTION: Pelew Isls (Micronesia).
STATUS: CITES – Appendix II. IUCN/SSC Action Plan (1992) – Endangered: Limited Distribution as *P. mariannus pelewensis*. IUCN 2003 – Not listed.
COMMENTS: *mariannus* species group. Often treated as a subspecies of *mariannus*, but apparently distinct (Corbet and Hill, 1980; Flannery, 1995*b*). See comments under *mariannus*.

*Pteropus personatus* Temminck, 1825. Monogr. Mamm., 1:189.
COMMON NAME: Moluccan Masked Flying Fox.
TYPE LOCALITY: Indonesia, Molucca Isls, Ternate.
DISTRIBUTION: North Molucca Isls (Halmahera and Obi Isl Groups), and Gag. Sulawesi records are erroneous; see Bergmans and Rozendaal (1988).
STATUS: CITES – Appendix II. IUCN/SSC Action Plan (1992) – No Data. IUCN 2003 – Lower Risk (lc).
COMMENTS: *personatus* species group. See Flannery (1995*b*).

*Pteropus pilosus* K. Andersen, 1908. Ann. Mag. Nat. Hist., ser. 8, 2:369.
COMMON NAME: Large Pelew Flying Fox.
TYPE LOCALITY: Micronesia, Caroline Isls, Palau Isls (= Pelew Isls).
DISTRIBUTION: Pelew Isls (Micronesia).
STATUS: CITES – Appendix I. IUCN/SSC Action Plan (1992) and IUCN 2003 – Extinct.
COMMENTS: *pselaphon* species group. Known from only two specimens and presumed to be extinct; see Flannery (1995*b*).

*Pteropus pohlei* Stein, 1933. Z. Säugetierk., 8:93.
COMMON NAME: Geelvink Bay Flying Fox.
TYPE LOCALITY: Indonesia, Prov. of Papua, Tjenderawasih Div., Yapen Isl.
DISTRIBUTION: Yapen, Biak-Supiori, Numfoor, and Rani Isls (off NW New Guinea).
STATUS: CITES – Appendix II. IUCN/SSC Action Plan (1992) – No Data: Limited Distribution. IUCN 2003 – Vulnerable.
COMMENTS: *poliocephalus* species group. See Flannery (1995*b*). Includes at least one undescribed species (K. Helgen, pers. comm.).

*Pteropus poliocephalus* Temminck, 1825. Monogr. Mamm., 1:179.
COMMON NAME: Gray-headed Flying Fox.
TYPE LOCALITY: Australia.
DISTRIBUTION: E Australia, from S Queensland to Victoria.
STATUS: CITES – Appendix II. IUCN/SSC Action Plan (1992) – Not Threatened. IUCN 2003 – Lower Risk (lc).
COMMENTS: *poliocephalus* species group. See Webb and Tideman (1995) for discussion of cases of hybridization with *alecto*.

*Pteropus pselaphon* Lay, 1829. Zool. J., 4:457.
COMMON NAME: Bonin Flying Fox.
TYPE LOCALITY: Japan, Bonin Isls.
DISTRIBUTION: Bonin and Volcano Isls (Japan).

STATUS: CITES – Appendix II. IUCN/SSC Action Plan (1992) – Vulnerable: Limited Distribution. IUCN 2003 – Critically Endangered.

SYNONYMS: *ursinus* Temminck (ex Kittlitz), 1837.

COMMENTS: *pselaphon* species group. Reviewed by Yoshiyuki (1989).

*Pteropus pumilus* Miller, 1911. Proc. U.S. Natl. Mus., 38:394.

COMMON NAME: Little Golden-mantled Flying Fox.

TYPE LOCALITY: Indonesia, Miangas Isl (= Palmas Isl) between Talaud Isls and Mindanao.

DISTRIBUTION: Philippines (except Palawan region), Talaud Isls (Indonesia).

STATUS: CITES – Appendix II. IUCN/SSC Action Plan (1992) – Vulnerable. IUCN 2003 – Vulnerable.

SYNONYMS: *balutus* Hollister, 1913; *tablasi* Taylor, 1934.

COMMENTS: *subniger* species group. Includes *balutus* and *tablasi*; see Klingener and Creighton (1984).

*Pteropus rayneri* Gray, 1870. Cat. Monkeys, Lemurs, Fruit-eating Bats Brit. Mus., p. 108.

COMMON NAME: Solomons Flying Fox.

TYPE LOCALITY: Solomon Isls, Guadalcanal Isl.

DISTRIBUTION: Bougainville and Buka Isls (Papua New Guinea); Solomon Isls.

STATUS: CITES – Appendix II. IUCN/SSC Action Plan (1992) – Not Threatened. IUCN 2003 – Lower Risk (lc).

SYNONYMS: *grandis* Thomas, 1887; *lavellanus* K. Andersen, 1908; *monoensis* Lawrence, 1945; *rubianus* K. Andersen, 1908.

COMMENTS: *chrysoproctus* species group. Does not include *cognatus* and *rennelli*; see Flannery (1995*b*). Also see Bonaccorso (1998).

*Pteropus rennelli* Troughton, 1929. Rec. Aust. Mus., 17:193.

COMMON NAME: Rennell Flying Fox.

TYPE LOCALITY: Solomon Isls, Rennell Isl.

DISTRIBUTION: Rennell Isl (Solomon Isls).

STATUS: CITES – Appendix II. IUCN/SSC Action Plan (1992) – No Data: Limited Distribution. IUCN 2003 – Not listed.

COMMENTS: *Chryoproctus* species group. Known from only 5 specimens. Formerly included in *rayneri*, but apparently distinct; see Flannery (1995*b*). May be conspecific with *cognatus* (K. Helgen, pers. comm.).

*Pteropus rodricensis* Dobson, 1878. Cat. Chiroptera Brit. Mus., p. 36.

COMMON NAME: Rodrigues Flying Fox.

TYPE LOCALITY: Mascarene Isls, Rodrigues.

DISTRIBUTION: Rodrigues Isl, Round Isl near Mauritius Isl (Mascarene Isls).

STATUS: CITES – Appendix II. U.S. ESA – Endangered. IUCN/SSC Action Plan (1992) – Endangered: Limited Distribution. IUCN 2003 – Critically Endangered; extinct on Round Isl.

SYNONYMS: *mascarinus* Mason, 1907.

COMMENTS: *molossinus* species group. See Bergmans (1990).

*Pteropus rufus* E. Geoffroy, 1803. Cat. Mamm. Mus. Nat. Hist. Nat. Paris, p. 47.

COMMON NAME: Malagasy Flying Fox.

TYPE LOCALITY: Madagascar. Restricted to "N. and C. Madagascar" by K. Andersen (1908).

DISTRIBUTION: Madagascar.

STATUS: CITES – Appendix II. IUCN/SSC Action Plan (1992) – Not Threatened. IUCN 2003 – Lower Risk (lc).

SYNONYMS: *edwardsi* E. Geoffroy, 1810; *phaiops* Temmnick, 1825; *princeps* K. Andersen, 1908.

COMMENTS: *niger* species group. Reviewed by Peterson et al. (1995); also see Bergmans (1990). Some authors have recognized *princeps* as a subspecies, but this has not been supported in recent analyses; see Bergmans (1990) and Peterson et al. (1995). Because Wilson and Reeder (1993) did not treat names established in E. Geoffroy (1803) as available, Koopman (1993) attributed authorship of *rufus* to Tiedemann ("1808, Zool., v.1, Allgemeine Zool., Mensch Saugthiere, Landshut, p.535."), but *rufus* Tiedemann is

a junior synonym of *niger* Kerr; see Grubb (2001*a*) and Opinion 2005 of the
International Commission on Zoological Nomenclature (2002*b*).

*Pteropus samoensis* Peale, 1848. Mammalia *in* Repts. U.S. Expl. Surv., 8:20.
COMMON NAME: Samoan Flying Fox.
TYPE LOCALITY: Samoan Isls, Tutuila Isl (American Samoa).
DISTRIBUTION: Fiji Isls, Samoan Isls.
STATUS: CITES – Appendix I. IUCN/SSC Action Plan (1992) and IUCN 2003 – Vulnerable.
SYNONYMS: *vitiensis* Gray, 1870; *whitmeei* Alston, 1874; **nawaiensis** Gray, 1870; *fuscicollis*
    Nicoll, 1904 [*nomen nudum*]; *ruficollis* Nicoll, 1908 [*nomen nudum*].
COMMENTS: *samoensis* species group. Includes *nawaiensis*; see Hill and Beckon (1978) and
    Banack (2001).

*Pteropus scapulatus* Peters, 1862. Monatsb. K. Preuss. Akad. Wiss. Berlin, 1862:574.
COMMON NAME: Little Red Flying Fox.
TYPE LOCALITY: Australia, Queensland, Cape York.
DISTRIBUTION: Australia, S New Guinea, accidental on New Zealand.
STATUS: CITES – Appendix II. IUCN/SSC Action Plan (1992) – Not Threatened. IUCN 2003 –
    Lower Risk (lc).
SYNONYMS: *elseyi* Peters, 1862.
COMMENTS: *scapulatus* species group. See Flannery (1995*a*) and Bonaccorso (1998).

*Pteropus seychellensis* Milne-Edwards, 1877. Bull. Sci. Soc. Philom. Paris, ser. 7, 2:221.
COMMON NAME: Seychelles Flying Fox.
TYPE LOCALITY: Seychelle Isls, Mahe Isl.
DISTRIBUTION: Seychelle Isls, Comoros Isls, Mafia Isl (off Tanzania).
STATUS: CITES – Appendix II. IUCN/SSC Action Plan (1992) – Not Threatened. IUCN 2003 –
    Lower Risk (lc).
SYNONYMS: **comorensis** Nicoll, 1908; *comorensis* Wallace, 1880 [*nomen nudum*]; *comorensis*
    Keller, 1898 [*nomen nudum*].
COMMENTS: *niger* species group. Includes *comorensis*; see Hill (1971*b*) and Bergmans (1990).
    Does not include *aldabrensis*; see Bergmans (1990).

*Pteropus speciosus* K. Andersen, 1908. Ann. Mag. Nat. Hist., ser. 8, 2:364.
COMMON NAME: Philippine Gray Flying Fox.
TYPE LOCALITY: Philippines, Malanipa Isl (off west end of Zamboanga, Mindanao).
DISTRIBUTION: Philippines; Solombo Besar and Mata Siri (Java Sea); Talaud Isls.
STATUS: CITES – Appendix II. IUCN/SSC Action Plan (1992) – Rare as *P. speciosus*, No Data:
    Limited Distribution as *P. mearnsi*. IUCN 2003 – Vulnerable as *P. speciosus*, Data
    Deficient as *P. mearnsi*.
SYNONYMS: *mearnsi* Hollister, 1913.
COMMENTS: *subniger* species group. Included in *griseus* by Corbet and Hill (1992), but I
    follow Flannery (1995*b*) and Heaney et al. (1998) in treating it as distinct pending
    further study. Some Philippine records were erroneously based on subadult
    *hypomelanus*; see Heaney et al. (1998). Includes *mearnsi*; see Heaney et al. (1987) and
    Flannery (1995*b*), but also see Corbet and Hill (1992), who suggested that *mearnsi* may
    be a synonym of *hypomelanus*.

*Pteropus subniger* (Kerr, 1792). *In* Linnaeus, Anim. Kingdom, 1:91.
COMMON NAME: Dark Flying Fox.
TYPE LOCALITY: Mascarene Isls, Réunion Isl (France).
DISTRIBUTION: Réunion and Mauritius Isls (Mascarene Isls).
STATUS: CITES – Appendix II. IUCN/SSC Action Plan (1992) and IUCN 2003 – Extinct.
SYNONYMS: *collaris* Illiger, 1815; *fuscus* Desmarest, 1803 [not E. Geoffroy]; *ruber* E. Geoffroy,
    1803; *rubidum* Daudin, 1802; *rubricollis* E. Geoffroy, 1810; *torquatus* G. Fischer, 1814;
    *vulgaris* Temminck, 1837.
COMMENTS: *subniger* species group. Reviewed by Bergmans (1990). Probably extinct, see
    Cheke and Dahl (1981).

*Pteropus temminckii* Peters, 1867. Monatsb. K. Preuss. Akad. Wiss. Berlin, 1867:331.
    COMMON NAME: Temminck's Flying Fox.
    TYPE LOCALITY: Indonesia, Molucca Isls, Amboina Isl; see Andersen (1912:318) for
        clarification.
    DISTRIBUTION: Buru, Ambon, Seram (Indonesia); nearby small islands; perhaps Timor Isl
        (Indonesia).
    STATUS: CITES – Appendix II. IUCN/SSC Action Plan (1992) – No Data. IUCN 2003 – Lower
        Risk (nt).
    SYNONYMS: *griseus* Temminck, 1837 [not Geoffroy, 1810]; *petersi* Matschie, 1899; ***liops***
        Thomas, 1910.
    COMMENTS: *personatus* species group. Does not include *capistratus*; see Flannery (1995*b*) and
        Bonaccorso (1998). This name is variously spelled "*temmincki*" and "*temminckii*";
        I follow Bergmans (2001) in preferring the latter because it is the original spelling.

*Pteropus tokudae* Tate, 1934. Am. Mus. Novit., 713:1.
    COMMON NAME: Guam Flying Fox.
    TYPE LOCALITY: Mariana Isls, Guam (USA).
    DISTRIBUTION: Guam (Mariana Isls, USA).
    STATUS: CITES – Appendix II. U.S. ESA – Endangered. IUCN/SSC Action Plan (1992) –
        Extinct? IUCN 2003 – Extinct.
    COMMENTS: *pselaphon* species group. See Flannery (1995*b*).

*Pteropus tonganus* Quoy and Gaimard, 1830. *In* d'Urville, Voy . . . de Astrolabe, Zool., 1(L'Homme,
    Mamm., Oiseaux):74.
    COMMON NAME: Pacific Flying Fox.
    TYPE LOCALITY: Tonga Isls, Tongatapu Isl.
    DISTRIBUTION: Karkar Isl (off NE New Guinea) and Rennell Isl (Solomon Isls), south to
        New Caledonia, east to Cook Isls.
    STATUS: CITES – Appendix I. IUCN/SSC Action Plan (1992) – Not Threatened. IUCN 2003 –
        Lower Risk (lc).
    SYNONYMS: *flavicollis* Gray, 1870; ***basiliscus*** Thomas, 1915; ***geddiei*** MacGillivray, 1860;
        *heffernani* Troughton, 1930.
    COMMENTS: *mariannus* species group. Includes *geddiei*; see Sanborn (1931) and Felten and
        Kock (1972). Karkar Isl population (*basiliscus*) may actually be a subspecies of
        *conspicillatus* (K. Helgen, pers. comm.). It is possible that this species has been
        transported to some islands by humans; see Flannery (1995*b*). Also see Miller and
        Wilson (1997), Bonaccorso (1998), and Bergmans (2001).

*Pteropus tuberculatus* Peters, 1869. Monatsb. K. Preuss. Akad. Wiss. Berlin, 1869:393.
    COMMON NAME: Vanikoro Flying Fox.
    TYPE LOCALITY: Solomon Isls, Santa Cruz Isls, Vanikoro Isl.
    DISTRIBUTION: Vanikoro Isl (Santa Cruz Isls, Solomon Isls).
    STATUS: CITES – Appendix II. IUCN/SSC Action Plan (1992) – No Data: Limited Distribution.
        IUCN 2003 – Vulnerable.
    COMMENTS: *pselaphon* species group. Reviewed by Troughton (1927); also see Flannery
        (1995*b*).

*Pteropus ualanus* Peters, 1883. Ges. Nat. Fr., 1:1.
    COMMON NAME: Kosrae Flying Fox.
    TYPE LOCALITY: Ualan (= Kosrae; Micronesia)
    DISTRIBUTION: Kosrae (Micronesia).
    STATUS: CITES – Appendix II. IUCN/SSC Action Plan (1992) – Endangered: Limited
        Distribution as *P. mariannus ualanus*. IUCN 2003 – Not listed.
    SYNONYMS: *ualensis* Finsch, 1881 [*nomen nudum*].
    COMMENTS: *mariannus* species group. Often treated as a subspecies of *mariannus*, but clearly
        distinct (Corbet and Hill, 1980; K. Helgen, pers. comm.; Flannery, 1995*b*). See
        comments under *mariannus*.

*Pteropus vampyrus* (Linnaeus, 1758). Syst. Nat., 10th ed., 1:31.
    COMMON NAME: Large Flying Fox.

TYPE LOCALITY: Indonesia, Java (designated by K. Andersen, 1912).

DISTRIBUTION: Vietnam, Burma, Malay Peninsula, Borneo, Philippines, Sumatra, Java, and Lesser Sunda Isls, adjacent small islands including Anak Krakatau. Reports of this species from Cambodia cannot be verified (Kock, 2000).

STATUS: CITES – Appendix II. IUCN/SSC Action Plan (1992) – Not Threatened. IUCN 2003 – Lower Risk (lc).

SYNONYMS: *celaeno* Hermann, 1804; *caninus* Blumenbach, 1797; *javanicus* Desmarest, 1820; *kalou* E. Geoffroy, 1810; *kelaarti* Gray, 1870 [skull, not skin]; *nudus* Hermann, 1804; *phaiops* Gray, 1870 [not Temminck, 1825]; *pteronotus* Dobson, 1878; **edulis** E. Geoffroy, 1810; *funereus* Temminck, 1837; **lanensis** Mearns, 1905; **natunae** K. Andersen, 1908; **pluton** Temminck, 1853; *kopangi* Kuroda, 1933; **sumatrensis** Ludeking, 1862; *malaccensis* K. Andersen, 1908.

COMMENTS: *vampyrus* species group. Does not include *intermedius*; see Corbet and Hill (1992). Reviewed in part by Bates and Harrison (1997). See Kunz and Jones (2000). Subspecies are poorly defined.

*Pteropus vetulus* Jouan, 1863. Mem. Soc. Imp. Sci. Nat. Cherbourg, 9:90.

COMMON NAME: New Caledonian Flying Fox.

TYPE LOCALITY: New Caledonia (France).

DISTRIBUTION: New Caledonia (France).

STATUS: CITES – Appendix II. IUCN/SSC Action Plan (1992) – Rare: Limited Distribution. IUCN 2003 – Lower Risk (nt).

SYNONYMS: *germaini* Dobson, 1878; *macmillani* Tate, 1942.

COMMENTS: *pselaphon* species group. Includes *macmillani*; see Felten (1964b). Also see Flannery (1995b).

*Pteropus voeltzkowi* Matschie, 1909. Sitzb. Ges. Naturf. Fr. Berlin, p. 486.

COMMON NAME: Pemba Flying Fox.

TYPE LOCALITY: Tanzania, Pemba Isl, Fufuni.

DISTRIBUTION: Pemba Isl (off coast of Tanzania).

STATUS: CITES – Appendix II. IUCN/SSC Action Plan (1992) – Endangered: Limited Distribution. IUCN 2003 – Critically Endangered.

COMMENTS: *niger* species group. Reviewed by Bergmans (1990).

*Pteropus woodfordi* Thomas, 1888. Ann. Mag. Nat. Hist., ser. 6, 1:156.

COMMON NAME: Dwarf Flying Fox.

TYPE LOCALITY: Solomon Isls, Guadalcanal Isl, Aola.

DISTRIBUTION: New Georgia group, Russell and Florida Isls, Guadalcanal, Malaita (Solomon Isls).

STATUS: CITES – Appendix II. IUCN/SSC Action Plan (1992) – No Data. IUCN 2003 – Lower Risk (lc).

SYNONYMS: *austini* Lawrence, 1945.

COMMENTS: *scapulatus* species group. See Flannery (1995b).

*Pteropus yapensis* K. Andersen, 1908. Ann. Mag. Nat. Hist., ser. 8, 2:365.

COMMON NAME: Yap Flying Fox.

TYPE LOCALITY: W Carolines, Yap Isl.

DISTRIBUTION: Yap Isls.

STATUS: CITES – Appendix II. IUCN/SSC Action Plan (1992) – Endangered: Limited Distribution as *P. mariannus yapensis*. IUCN 2003 – Not listed.

COMMENTS: *mariannus* species group. Often treated as a subspecies of *mariannus*, but apparently distinct (Corbet and Hill, 1980; Flannery, 1995b). See comments under *mariannus*.

*Rousettus* Gray, 1821. London Med. Repos., 15:299.

TYPE SPECIES: *Pteropus aegyptiacus* E. Geoffroy, 1810.

SYNONYMS: *Boneia* Jentink, 1879; *Cercopterus* Burnett, 1829; *Cynonycteris* Peters, 1852; *Eleutherura* Gray, 1844; *Senonycteris* Gray, 1870; *Stenonycteris* Gray, 1871; *Xantharpyia* Gray, 1834.

COMMENTS: Does not include *Lissonycteris* (Bergmans, 1994, 1997; Juste et al., 1997; Peterson et al., 1995). Revised by Bergmans (1994); also see Peterson et al. (1995). A key to the genus was provided by Kwiecinski and Griffiths (1999), however, this genus includes at least one undescribed species. Three subgenera are often recognized (*Rousettus*, *Boneia*, and *Stenonycteris*), although see Bergmans (1994), who rejected use of subgenera for the African species.

*Rousettus aegyptiacus* (E. Geoffroy, 1810). Ann. Mus. Natn. Hist. Nat. Paris, 15:96.
  COMMON NAME: Egyptian Rousette.
  TYPE LOCALITY: Egypt, Giza (Great Pyramid).
  DISTRIBUTION: Senegal and Egypt south to South Africa; Cyprus, Turkey, Jordan, Lebanon, Israel, S Syria, Yemen, Saudi Arabia, S Iraq, S. Iran, Pakistan, NW India; islands in the Gulf of Guinea (São Tomé and Príncipe); adjacent small islands.
  STATUS: IUCN/SSC Action Plan (1992) – Not Threatened. IUCN 2003 – Lower Risk (lc).
  SYNONYMS: *aegyptiacus* E. Geoffroy, 1818 [emendation of *egyptiacus*]; *geoffroyi* Temminck, 1825; **arabicus** Anderson and de Winton, 1902; **leachii** Smith, 1892; *hottentotus* Temminck, 1832; *sjostedti* Lönnberg, 1908; **princeps** Juste and Ibañez, 1993; **tomensis** Juste and Ibañez, 1993; *thomensis* Feiler, Haft, and Widmann, 1993; **unicolor** Gray, 1870; *occidentalis* Eisentraut, 1960.
  COMMENTS: Subgenus *Rousettus*. Includes *leachii* and *arabicus*; see Hayman and Hill (1971), Corbet (1978c), Harrison and Bates (1991), and Bergmans (1994). Revised by Bergmans (1994); reviewed in part by Bates and Harrison (1997) and Horácek et al. (2000). Also see Kwiecinski and Griffiths (1999). Spelling changed from *aegyptiacus* to *egyptiacus* by Corbet and Hill (1992), but returned to *aegyptiacus* by Kock (2001a).

*Rousettus amplexicaudatus* (E. Geoffroy, 1810). Ann. Mus. Natn. Hist. Nat. Paris, 15:96.
  COMMON NAME: Geoffroy's Rousette.
  TYPE LOCALITY: Indonesia, Lesser Sunda Isls, Timor Isl.
  DISTRIBUTION: Cambodia, Thailand, Burma, and Laos; Peninsular Malaysia through Indonesia, Java, and Bali; Philippines; New Guinea; Bismarck Archipelago, Solomon Isls.
  STATUS: IUCN/SSC Action Plan (1992) – Not Threatened. IUCN 2003 – Lower Risk (lc).
  SYNONYMS: *philippinensis* Gray, 1871; *stresemanni* Stein, 1933; **brachyotis** Dobson, 1877; **hedigeri** Pohle, 1952; **infumatus** Gray, 1871; *bocagei* Seabra, 1898; **minor** Dobson, 1873.
  COMMENTS: Subgenus *Rousettus*. Revised by Rookmaaker and Bergmans (1981); also see Hill (1983), Bergmans and Rozendaal (1988), and Flannery (1995a, b). Peterson et al. (1995) suggested that *brachyotis* and *minor* may represent distinct species. Subspecies allocation of Sulawesi and Kasi Isl (Indonesia) populations is uncertain; see Koopman (1994) and Kompanje and Moeliker (2001).

*Rousettus bidens* (Jentink, 1879). Notes Leyden Mus., 1:117.
  COMMON NAME: Manado Rousette.
  TYPE LOCALITY: Indonesia, N Sulawesi, Boné (near Gerontalo).
  DISTRIBUTION: N Sulawesi (Indonesia).
  STATUS: IUCN/SSC Action Plan (1992) – No Data: Limited Distribution. IUCN 2003 – Lower Risk (nt).
  SYNONYMS: *menadensis* Thomas, 1896.
  COMMENTS: Subgenus *Boneia*. Placed in its own genus (*Boneia)* by some authors (e.g., Andersen, 1912; Koopman, 1993) but see Bergmans and Rozendaal (1988) and Bergmans (1994). Corbet and Hill (1992) referred this species to the subgenus *Boneia*, the arrangement followed here.

*Rousettus celebensis* K. Andersen, 1907. Ann. Mag. Nat. Hist., ser. 7, 19:503, 509.
  COMMON NAME: Sulawesi Rousette.
  TYPE LOCALITY: Indonesia, Sulawesi, Mt. Masarang, 3,500 ft. (1,067 m).
  DISTRIBUTION: Sulawesi; Mangole, Sanana, Sangihe Isls (Indonesia).
  STATUS: IUCN/SSC Action Plan (1992) – Not Threatened. IUCN 2003 – Lower Risk (lc).
  COMMENTS: Subgenus *Rousettus*. Reviewed by Rookmaaker and Bergmans (1981), Hill

(1983), Bergmans and Rozendaal (1988), and Maryanto and Yani (2003). Also see
Flannery (1995b).

*Rousettus lanosus* Thomas, 1906. Ann. Mag. Nat. Hist., ser. 7, 18:137.
   COMMON NAME: Long-haired Rousette.
   TYPE LOCALITY: Uganda, Ruwenzori East, Mubuku Valley, 13,000 ft. (3,962 m).
   DISTRIBUTION: E Dem. Rep. Congo, Uganda, Rwanda, Kenya, Tanzania, Malawi, S Ethiopia,
      S Sudan.
   STATUS: IUCN/SSC Action Plan (1992) – Not Threatened. IUCN 2003 – Lower Risk (lc).
   SYNONYMS: *kempi* Thomas, 1909.
   COMMENTS: Subgenus *Stenonycteris*. Revised by Bergmans (1994), who argued against
      recognition of subspecies.

*Rousettus leschenaultii* (Desmarest, 1820). Mammalogie, *in* Encyclop. Méthod., 1:110.
   COMMON NAME: Leschenault's Rousette.
   TYPE LOCALITY: India, Pondicherry.
   DISTRIBUTION: Sri Lanka; Pakistan to Vietnam and S China; Peninsular Malaysia; Sumatra,
      Java, Bali, and Mentawai Isls (Indonesia).
   STATUS: IUCN/SSC Action Plan (1992) – Not Threatened. IUCN 2003 – Lower Risk (lc).
   SYNONYMS: *affinis* Gray, 1843; *fuliginosa* Gray, 1871; *fusca* Gray, 1871; *infuscata* Peters, 1873;
      *marginatus* Gray, 1843 [not Geoffroy, 1810]; *pirivarus* Hodgson, 1841; *pyrivorus*
      Hodgson, 1835; **seminudus** Kelaart, 1850; **shortridgei** Thomas and Wroughton, 1909.
   COMMENTS: Subgenus *Rousettus*. Includes *seminudus*; see Sinha (1970). See Peterson et al.
      (1995) for a discussion of *shortridgei*. Kock et al. (2000b) treated *shortridgei* as a distinct
      species without comment. Reviewed in part by Bates and Harrison (1997) and Kock
      et al. (2000b). This name is sometimes spelled *leschenaulti* (e.g., Koopman, 1993, 1994),
      but I prefer the original spelling.

*Rousettus linduensis* Maryanto and Yani, 2003. Mammal Study, 28:113.
   COMMON NAME: Linduan Rousette.
   TYPE LOCALITY: Indonesia, Central Sulawesi, Lore Lindu National Park, Lundu lake enclave,
      Kenawu village, 1°19′8″S, 120°6′8″E, 930 m.
   DISTRIBUTION: C Sulawesi.
   STATUS: IUCN 2003 – not evaluated; not considered in IUCN/SSC Action Plan (2001).
   COMMENTS: Subgenus *Rousettus*.

*Rousettus madagascariensis* G. Grandidier, 1928. Bull. Acad. Malgache, N.S., ll:91.
   COMMON NAME: Malagasy Rousette.
   TYPE LOCALITY: Madagascar, Beforona (between Tananarive [= Antananarivo] and
      Andevoranto).
   DISTRIBUTION: Madagascar except SW region.
   STATUS: IUCN/SSC Action Plan (1992) – Not Threatened. IUCN 2003 – Lower Risk (nt).
   COMMENTS: Subgenus *Stenonycteris*. Considered a subspecies of *lanosus* by Hayman and Hill
      (1977), but see Bergmans (1977). Revised by Peterson et al. (1995). Does not include
      *obliviosus*; see Bergmans (1994), but also see Peterson et al. (1995).

*Rousettus obliviosus* Kock, 1978. Proc. 4th Int. Bat Res. Conf. Nairobi, p. 208.
   COMMON NAME: Comoro Rousette.
   TYPE LOCALITY: Comoro Isls, Grand Comoro, near Boboni, 640 m.
   DISTRIBUTION: Comoro Isls.
   STATUS: IUCN/SSC Action Plan (1992) – Not Threatened. IUCN 2003 – Lower Risk (nt).
   COMMENTS: Subgenus *Rousettus*. Considered a subspecies of *madagascarensis* by Peterson
      et al. (1995), but see Bergmans (1994).

*Rousettus spinalatus* Bergmans and Hill, 1980. Bull. Brit. Mus. (Nat. Hist.) Zool., 38:95.
   COMMON NAME: Bare-backed Rousette.
   TYPE LOCALITY: Indonesia, N Sumatra, near Medan or near Prapat.
   DISTRIBUTION: Sumatra, Borneo.
   STATUS: IUCN/SSC Action Plan (1992) – No Data. IUCN 2003 – Vulnerable.
   COMMENTS: Subgenus *Rousettus*.

*Scotonycteris* Matschie, 1894. Sitzb. Ges. Naturf. Fr. Berlin, p. 200.
    TYPE SPECIES: *Scotonycteris zenkeri* Matschie, 1894.
    COMMENTS: Reviewed by Bergmans (1990).

*Scotonycteris ophiodon* Pohle, 1943. Sitzb. Ges. Naturf. Fr. Berlin, p. 76.
    COMMON NAME: Pohle's Fruit Bat.
    TYPE LOCALITY: Cameroon, Bipindi.
    DISTRIBUTION: Liberia, Ghana, Cameroon, Republic of Congo.
    STATUS: IUCN/SSC Action Plan (1992) – Not Threatened. IUCN 2003 – Lower Risk (nt).
    SYNONYMS: *cansdalei* Hayman, 1946.

*Scotonycteris zenkeri* Matschie, 1894. Sitzb. Ges. Naturf. Fr. Berlin, p. 202.
    COMMON NAME: Zenker's Fruit Bat.
    TYPE LOCALITY: Cameroon, Yaunde (Yaoundé).
    DISTRIBUTION: Liberia to Republic of Congo and E Dem. Rep. Congo.
    STATUS: IUCN/SSC Action Plan (1992) – Not Threatened. IUCN 2003 – Lower Risk (lc).
    SYNONYMS: **bedfordi** Thomas, 1904; **occidentalis** Hayman, 1947.
    COMMENTS: Current subspecific nomenclature does not adequately describe the known range of variation in this species; see Bergmans (1990).

*Sphaerias* Miller, 1906. Proc. Biol. Soc. Wash., 19:83.
    TYPE SPECIES: *Cynopterus blanfordi* Thomas, 1891.

*Sphaerias blanfordi* (Thomas, 1891). Ann. Mus. Civ. Stor. Nat. Genova, ser. 2, 10:884, 921, 922.
    COMMON NAME: Blanford's Fruit Bat.
    TYPE LOCALITY: Burma, Karin Hills, Cheba, Leito.
    DISTRIBUTION: N India, Bhutan, Burma, N Thailand, Vietnam, SW China.
    STATUS: IUCN/SSC Action Plan (1992) – No Data. IUCN 2003 – Lower Risk (lc).
    SYNONYMS: *motuoensis* Cai and Zhang, 1980.
    COMMENTS: Reviewed in part by Bates and Harrison (1997).

*Styloctenium* Matschie, 1899. Megachiroptera Berlin Mus., p. 33.
    TYPE SPECIES: *Pteropus wallacei* Gray, 1866.

*Styloctenium wallacei* (Gray, 1866). Proc. Zool. Soc. Lond., 1866:65.
    COMMON NAME: Stripe-faced Fruit Bat.
    TYPE LOCALITY: Indonesia, Sulawesi, Macassar.
    DISTRIBUTION: Sulawesi, Tongian Isls.
    STATUS: IUCN/SSC Action Plan (1992) – No Data. IUCN 2003 – Lower Risk (nt).
    COMMENTS: See Bergmans and Rozendaal (1988).

*Syconycteris* Matschie, 1899. Megachiroptera Berlin Mus., p. 94, 95, 98.
    TYPE SPECIES: *Macroglossus minimus* var. *australis* Peters, 1867.
    COMMENTS: Reviewed by Ziegler (1982*a*).

*Syconycteris australis* (Peters, 1867). Monatsb. K. Preuss. Akad. Wiss. Berlin, 1867:13, footnote.
    COMMON NAME: Southern Blossom Bat.
    TYPE LOCALITY: Australia, Queensland, Rockhampton.
    DISTRIBUTION: E Queensland and New South Wales (Australia); New Guinea, Aru Isl, Trobriand Isls, D'Entrecasteaux Isls, Kai Isls, Ambon, Seram, Haruku, and Boano Isls. (Indonesia), Bismarck Arch., including Manus (Papua New Guinea).
    STATUS: IUCN/SSC Action Plan (1992) – Not Threatened. IUCN 2003 – Lower Risk (lc).
    SYNONYMS: **crassa** Thomas, 1895; **finschi** Matschie, 1899; **keyensis** K. Andersen, 1911; **major** K. Andersen, 1911; **naias** K. Andersen, 1911; **papuana** Matschie, 1899.
    COMMENTS: Includes *naias* and *crassa*; see Lidicker and Ziegler (1968) and Koopman (1982). Reviewed by Hill (1983); also see Kitchener et al. (1994*d*), Flannery (1995*a*, *b*), Bonaccorso (1998), and Kompanje and Moeliker (2001). Subspecies limits are somewhat unclear, particularly the status of the Kai Isl form (*keyensis*); see Kitchener et al. (1994*d*). It is possible that *major* from Ambon and Seram Isls represents a distinct species; see Kitchener et al. (1994*d*). Material from Haruku and Boano Isls differs from

typical *major* and may require recognition as a distinct subspecies; see Kompanje and Moeliker (2001). There is also an undescribed subspecies from Biak-Supiori (K. Helgen, pers. comm.). High-altitude specimens of *australis* from mainland New Guinea are also in need of systematic revision (Kompanje and Moeliker, 2001).

*Syconycteris carolinae* Rozendaal, 1984. Zoologische Mededelingen, 58(13):200.
COMMON NAME: Halmaheran Blossom Bat.
TYPE LOCALITY: Indonesia, Moluccas, Halmahera Isl, S base of Gamkunora (01°20'N, 127°31'E), ca 180 m.
DISTRIBUTION: Bacan and Halmahera Isls (Moluccas).
STATUS: IUCN/SSC Action Plan (1992) – No Data: Limited Distribution. IUCN 2003 – Vulnerable.
COMMENTS: See Flannery (1995*b*).

*Syconycteris hobbit* Ziegler, 1982. Occas. Pap. Bernice P. Bishop Mus., 25(5):5.
COMMON NAME: Moss-forest Blossom Bat.
TYPE LOCALITY: Papua New Guinea, Morobe Prov., Mt. Kaindi.
DISTRIBUTION: Mountains of C New Guinea.
STATUS: IUCN/SSC Action Plan (1992) – Rare. IUCN 2003 – Vulnerable.
COMMENTS: See Flannery (1995*a*) and Bonaccorso (1998).

*Thoopterus* Matschie, 1899. Megachiroptera Berlin Mus., p. 72, 73, 77.
TYPE SPECIES: *Cynopterus marginatus* var. *nigrescens* Gray, 1870.
COMMENTS: This genus is presently monotypic, but contains at least one undescribed species.

*Thoopterus nigrescens* (Gray, 1870). Cat. Monkeys, Lemurs, Fruit-eating Bats Brit. Mus., p. 123.
COMMON NAME: Swift Fruit Bat.
TYPE LOCALITY: Indonesia, Molucca Isls, Morotai.
DISTRIBUTION: Sulawesi, Sula Isls, Sangihe Isls, Karakelang (Talaud Isls), and Morotai (Indonesia).
STATUS: IUCN/SSC Action Plan (1992) – No Data. IUCN 2003 – Lower Risk (nt).
SYNONYMS: *latidens* Dobson, 1878.
COMMENTS: See Flannery (1995*b*).

**Family Rhinolophidae** Gray, 1825. Zool. Journ., 2(6):242.
SYNONYMS: Histiorhina Van der Hoeven, 1855.
COMMENTS: Monogeneric. Does not include Hipposideridae; see discussion under that taxon.

*Rhinolophus* Lacépède, 1799. Tabl. Div. Subd. Orders Genres Mammifères, p. 15.
TYPE SPECIES: *Vespertilio ferrum-equinum* Schreber, 1774. Conserved in ICZN Opinion 91 (1926) and Direction 24 (1955).
SYNONYMS: *Aquias* Gray, 1847; *Coelophyllus* Peters, 1867; *Euryalus* Matschie, 1901; *Phyllorhina* Leach, 1816; *Phyllotis* Gray, 1866 [not Waterhouse, 1837]; *Rhinocrepis* Gervais, 1836; *Rhinomegalophus* Bourret, 1951; *Rhinophyllotis* Troughton, 1941.
COMMENTS: For a comprehensive review of the genus (including detailed accounts for each species) see Csorba et al. (2003). Includes *Rhinomegalophus*; see Thonglongyai (1973). For partial phylogenies see Qumsiyeh et al. (1988*b*), Bogdanowicz and Owen (1992), and Maree and Grant (1997); also see Guillén-Servent (2001). Species groups follow Csorba et al. (2003)

*Rhinolophus acuminatus* Peters, 1871. Monatsb. K. Preuss. Akad. Wiss. Berlin, 1871:308.
COMMON NAME: Accuminate Horseshoe Bat.
TYPE LOCALITY: Indonesia, Java, Gadok.
DISTRIBUTION: Thailand; Laos; Cambodia; Peninsular Malaysia and Sabah; Borneo; Sumatra (including Nias and Engano Isls); Java, Krakatau, Lombok, and Bali (Indonesia); Palawan, Balabac, Busuanga (Philippines).
STATUS: IUCN 2003 and IUCN/SSC Action Plan (2001) – Lower Risk (lc).
SYNONYMS: ***audax*** K. Andersen, 1905; ***calypso*** K. Andersen, 1905; ***circe*** K. Andersen, 1906; ***sumatranus*** K. Andersen, 1905.

COMMENTS: *pusillus* species group. Subspecific allocations of mainland and Philippine populations are uncertain.

*Rhinolophus adami* Aellen and Brosset, 1968. Rev. Suisse Zool., 75:443.
   COMMON NAME: Adam's Horseshoe Bat.
   TYPE LOCALITY: Republic of Congo, Kouilou.
   DISTRIBUTION: Republic of Congo.
   STATUS: IUCN 2003 and IUCN/SSC Action Plan (2001) – Data Deficient.
   COMMENTS: *adami* species group. Reviewed by Kock et al. (2000*a*).

*Rhinolophus affinis* Horsfield, 1823. Zool. Res. Java, 6, pl. figs. a, b.
   COMMON NAME: Intermediate Horseshoe Bat.
   TYPE LOCALITY: Indonesia, Java.
   DISTRIBUTION: India and Nepal to S China and Vietnam, through Malaysia to Borneo and Lesser Sunda Isls; Andaman Isls (India); perhaps Sri Lanka. Reports of this species from Cambodia cannot be confirmed (Kock, 2000*a*).
   STATUS: IUCN 2003 and IUCN/SSC Action Plan (2001) – Lower Risk (lc).
   SYNONYMS: *andamanensis* Dobson, 1872; *hainanus* Allen, 1906; *himalayanus* K. Andersen, 1905; *macrurus* K. Andersen, 1905; *nesites* K. Andersen, 1905; *princeps* K. Andersen, 1905; *superans* K. Andersen, 1905; *tener* K. Andersen, 1905.
   COMMENTS: *megaphyllus* species group. Includes *andamanensis*; see Sinha (1973). Reviewed in part by Bergmans and van Bree (1986) and Bates and Harrison (1997). Csorba (2002) designated a lectotype for this species; also see Csorba et al. (2003).

*Rhinolophus alcyone* Temminck, 1853. Esquisses Zool. sur la Côte de Guine, p. 80.
   COMMON NAME: Halcyon Horseshoe Bat.
   TYPE LOCALITY: Ghana, Boutry River.
   DISTRIBUTION: Senegal to Uganda, SW Sudan, N Dem. Rep. Congo, and Gabon; Bioko (Equatorial Guinea).
   STATUS: IUCN 2003 and IUCN/SSC Action Plan (2001) – Lower Risk (nt).
   COMMENTS: *landeri* species group.

*Rhinolophus arcuatus* Peters, 1871. Monatsb. K. Preuss. Akad. Wiss. Berlin, 1871:305.
   COMMON NAME: Arcuate Horseshoe Bat.
   TYPE LOCALITY: Philippines, Luzon.
   DISTRIBUTION: Sumatra to Philippines, New Guinea, and South Molucca Isls.
   STATUS: IUCN 2003 and IUCN/SSC Action Plan (2001) – Lower Risk (lc) as *Rhinolophus arcuatus*; Data Deficient as *R. anderseni*.
   SYNONYMS: *aequalis* Allen, 1922; *anderseni* Cabrera, 1909; *angustifolius* Sanborn, 1939; *beccarii* K. Andersen, 1907; *exiguus* K. Andersen, 1905; *mcintyrei* Hill and Schlitter, 1982; *proconsulis* Hill, 1959; *toxopeusi* Hinton, 1925.
   COMMENTS: *euryotis* species group. Includes *toxopeusi*; see Hill and Schlitter (1982). Includes *anderseni*; see Csorba et al. (2003). Also see Ingle and Heaney (1992), Flannery (1995*a*, *b*), and Bonaccorso (1998). The relationships of *aequalis* (originally described as a subspecies of *anderseni*) are uncertain; see Heaney et al. (1998) and Csorba et al. (2003). Koopman (1993) incorrectly spelled *proconsulis* as "*proconsularis*."

*Rhinolophus beddomei* K. Andersen, 1905. Ann. Mag. Nat. Hist., ser. 7, 16:253.
   COMMON NAME: Bedomme's Horseshoe Bat.
   TYPE LOCALITY: India, Madras [= Kerala], Wynaad.
   DISTRIBUTION: S India, Sri Lanka.
   STATUS: IUCN 2003 and IUCN/SSC Action Plan (2001) – Lower Risk (nt).
   SYNONYMS: *sobrinus* K. Andersen, 1918.
   COMMENTS: *trifoliatus* species group. Distinct from *luctus*; see Topál and Csorba (1992), Bates and Harrison (1997), and Hendrichsen et al. (2001*a*).

*Rhinolophus blasii* Peters, 1867. Monatsb. K. Preuss. Akad. Wiss. Berlin, 1866:17 [1867].
   COMMON NAME: Blasius's Horseshoe Bat.
   TYPE LOCALITY: SE Europe; restricted to Italy by Ellerman et al. (1953:59).
   DISTRIBUTION: NE South Africa to S Dem. Rep. Congo; Ethiopia; Somalia; Morocco; Algeria;

Tunisia; Turkey; Yemen; Israel; Jordan; Syria; Iran; Serbia and Montenegro; Albania; Bulgaria; Romania; Transcaucasia and Turkmenistan; Afghanistan; Pakistan; Italy; Greece; Cyprus.

STATUS: IUCN 2003 and IUCN/SSC Action Plan (2001) – Lower Risk (nt).

SYNONYMS: *blasiusi* Trouessart, 1910; *clivosus* Blasius, 1857 [not Cretzschmar, 1828]; ***andreinii*** Senna, 1905; *brockmani* Thomas, 1910; ***empusa*** K. Andersen, 1904; ***meyeroehmi*** Felten, 1977 [in Felten, Spitzenberger, and Storch, 1977].

COMMENTS: *landeri* species group. Includes *brockmani*; see Koopman (1975). Reviewed in part by Paz (1995), Harrison and Bates (1991), Bates and Harrison (1997), Zagorodnyuk (1999), and Horácek et al. (2000).

*Rhinolophus bocharicus* Kastchenko and Akimov, 1917. Annu. Mus. Zool. Acad. St. Petersb., 22:221.

COMMON NAME: Central Asian Horseshoe Bat.

TYPE LOCALITY: Turkmenistan, Murgab River.

DISTRIBUTION: Kyrgyzstan, W Tajikistan, NE Iran, Uzbekistan, Turkmenistan, Afghanistan, possibly N Pakistan.

STATUS: IUCN 2003 and IUCN/SSC Action Plan (2001) – Lower Risk (lc).

COMMENTS: *ferrumequinum* species group. Included in *clivosus* by Aellen (1959), but see Hanák (1969), Felten (1977), DeBlase (1980), Gromov and Baranova (1981), Pavlinov and Rossolimo (1987), and Horácek et al. (2000). Apparently does not include *rubiginosus*, which is here placed in *ferrumequinum* following Csorba et al. (2003).

*Rhinolophus borneensis* Peters, 1861. Monatsb. K. Preuss. Akad. Wiss. Berlin, 1861:709.

COMMON NAME: Bornean Horseshoe Bat.

TYPE LOCALITY: Malaysia, N Borneo, Sabah, Labuan Isl.

DISTRIBUTION: Borneo; Labuan and Banguey Isls (Malaysia); Java, Karimata Isls, and South Natuna Isls (Indonesia); Cambodia, Laos, and Vietnam.

STATUS: IUCN 2003 and IUCN/SSC Action Plan (2001) – Lower Risk (lc).

SYNONYMS: ***chaseni*** Sanborn, 1939; ***importunus*** Chasen, 1939; ***spadix*** Miller, 1901.

COMMENTS: *megaphyllus* species group. Includes *chaseni* and *importunus*; see Hill (1983). Formerly included *javanicus*, *celebensis*, *madurensis*, and *parvus* (e.g., by Goodwin [1979] and Hill and Thonglongya [1972]) but see Hill (1983) and Kitchener et al. (1995a). Type material discussed by Csorba (2002). Subspecies limits are somewhat unclear, and the relationships of various forms to *celebensis* and *malayanus* remains problematic; see Csorba et al. (2003). This complex may include more than one species, see Csorba et al. (2003).

*Rhinolophus canuti* Thomas and Wroughton, 1909. Abstr. Proc. Zool. Soc. Lond., 1909(68):18.

COMMON NAME: Canut's Horseshoe Bat.

TYPE LOCALITY: Indonesia, South Java, Tji-Tangoi river, Kallipoet-jang.

DISTRIBUTION: Java, Bali, Timor (Indonesia).

STATUS: IUCN 2003 and IUCN/SSC Action Plan (2001) – Lower Risk (nt).

SYNONYMS: *conuti* Schwartz, 1914; ***timoriensis*** Goodwin, 1979.

COMMENTS: *euryotis* species group. Formerly included in *creaghi*; see Hill and Schlitter (1982).

*Rhinolophus capensis* Lichtenstein, 1823. Verz. Doblet. Mus. Univ. Berlin, p. 4.

COMMON NAME: Cape Horseshoe Bat.

TYPE LOCALITY: South Africa, Western Cape Prov., Cape of Good Hope.

DISTRIBUTION: South Africa, Zimbabwe, Mozambique. Occurence outside South Africa is doubtful; records from Zambia and Malawi are definitely erroneous (Koopman, 1993).

STATUS: IUCN 2003 and IUCN/SSC Action Plan (2001) – Vulnerable.

SYNONYMS: *auritus* Sundevall, 1860.

COMMENTS: *capensis* species group. See Taylor (2000a) for distribution map.

*Rhinolophus celebensis* K. Andersen, 1905. Proc. Zool. Soc. Lond., 1905(2):83.

COMMON NAME: Sulawesi Horseshoe Bat.

TYPE LOCALITY: Sulawesi, Macassar (= Ujung Pandang).

DISTRIBUTION: Java, Bali, Timor, Sulawesi, Sangihe, Kangean, and Talaud Isls (Indonesia).

STATUS: IUCN 2003 and IUCN/SSC Action Plan (2001) – Lower Risk (nt).

SYNONYMS: *javanicus* K. Andersen, 1918.

COMMENTS: *megaphyllus* species group. Closely related to *virgo*; see Corbet and Hill (1992). Does not include *parvus*; see Bergmans and van Bree (1986) and Kitchener et al. (1995a), but also see Csorba et al. (2003). Does not include *madurensis*; see Bergmans and van Bree (1986), but also see Csorba et al. (2003).

*Rhinolophus clivosus* Cretzschmar, 1828. *In* Rüppell, Atlas Reise Nordl. Afr., Zool. Säugeth., p. 47.

COMMON NAME: Geoffroy's Horseshoe Bat.

TYPE LOCALITY: Saudi Arabia, Red Sea Coast, Muwaylih (= Mohila), (approx. 27°49′N, 35°30′E).

DISTRIBUTION: Israel, Jordan, Saudi Arabia, Oman, Yemen, Egypt, Libya, Algeria, Sudan, Ethiopia, Eritrea, Djibouti, Somalia, Kenya, Uganda, Dem. Rep. Congo, Rwanda, Burundi, Tanzania, Malawi, Angola, Zambia, Mozambique, Zimbabwe, South Africa, Swaziland, Namibia.

STATUS: IUCN 2003 and IUCN/SSC Action Plan (2001) – Lower Risk (lc).

SYNONYMS: *andersoni* Thomas, 1904; *acrotis* Heuglin, 1861; *augur* K. Andersen, 1904; *brachygnathus* K. Andersen, 1905; *keniensis* Hollister, 1916; *schwarzi* Heim de Balsac, 1934; *zuluensis* K. Andersen, 1904; *zambesiensis* K. Andersen, 1904.

COMMENTS: *ferrumequinum* species group. Does not include *bocharicus*; see Hanák (1969), DeBlase (1980), Gromov and Baranova (1981), and Pavlinov and Rossolimo (1987). Does not include *deckenii* or *silvestris*; see Koopman (1975), Cotterill (2002), and Csorba et al. (2003). Does not include *hillorum*, see Cotterill (2002). Also see Harrison and Bates (1991). Reviewed in part by Horácek et al. (2000).

*Rhinolophus coelophyllus* Peters, 1867. Proc. Zool. Soc. Lond., 1866:426 [1867].

COMMON NAME: Croslet Horseshoe Bat.

TYPE LOCALITY: Burma, Salaween (= Salween) River.

DISTRIBUTION: W Malaysia, Thailand, Burma, Laos.

STATUS: IUCN 2003 and IUCN/SSC Action Plan (2001) – Lower Risk (lc).

COMMENTS: *euryotis* species group. Does not include *shameli*; see Hill and Thonglongya (1972). Reviewed in part by Yoshiyuki (1990).

*Rhinolophus cognatus* K. Andersen, 1906. Ann. Mus. Civ. Stor. Nat. Genova, ser. 3, 2:181.

COMMON NAME: Andaman Horseshoe Bat.

TYPE LOCALITY: India, Andaman Isls, S Andaman Isl, Port Blair.

DISTRIBUTION: Andaman Isls (India).

STATUS: IUCN 2003 and IUCN/SSC Action Plan (2001) – Vulnerable.

SYNONYMS: *famulus* K. Andersen, 1918.

COMMENTS: *pusillus* species group. Reviewed by Bates and Harrison (1997); see also Csorba (1997).

*Rhinolophus convexus* Csorba, 1997. J. Mammal. 78:343.

COMMON NAME: Convex Horseshoe Bat.

TYPE LOCALITY: Malaysia, Pahang State, Cameron Highlands, Tanah Rata (= Tana Rata), Gunung Jasar, 4°28′N, 101°22′E, 1,600 m.

DISTRIBUTION: Peninsular Malaysia, Laos.

STATUS: IUCN 2003 and IUCN/SSC Action Plan (2001) – Critically Endangered.

COMMENTS: *pusillus* species group. Reviewed by Csorba et al. (2003).

*Rhinolophus cornutus* Temminck, 1834. Tijdschrift Natuurl. Gesch. Physiol., 1:30.

COMMON NAME: Little Japanese Horseshoe Bat.

TYPE LOCALITY: Japan.

DISTRIBUTION: Japan.

STATUS: IUCN 2003 and IUCN/SSC Action Plan (2001) – Lower Risk (nt).

SYNONYMS: *miyakonis* Kuroda, 1924; *orii* Kuroda, 1924; *perditus* K. Andersen, 1918; *pumilus* K. Andersen, 1905.

COMMENTS: *pusillus* species group. Does not include *blythi*; see Hill and Yoshiyuki (1980). Includes *pumilus*, *perditus*, and *miyakonis*; see Corbet and Hill (1992) and Csorba et al. (2003). May be conspecific with *pusillus*; see Corbet and Hill (1992), but also see Yoshiyuki (1989, 1990) and Csorba et al. (2003). Does not include *monoceros*, see

Csorba et al. (2003), but also see Koopman (1994) and Csorba (1997). Reviewed by Horácek et al. (2000). Csorba et al. (2003) designated a lectotype for *cornutus*.

*Rhinolophus creaghi* Thomas, 1896. Ann. Mag. Nat. Hist., ser. 6, 18:244.
 COMMON NAME: Creagh's Horseshoe Bat.
 TYPE LOCALITY: Malaysia, N Borneo, Sabah, Sandakan.
 DISTRIBUTION: Borneo; Madura Isl, Kalimantan, (Indonesia); Sabah, Sarawak (Malaysia).
 STATUS: IUCN 2003 and IUCN/SSC Action Plan (2001) – Lower Risk (nt).
 SYNONYMS: *pilosus* K. Andersen, 1918.
 COMMENTS: *euryotis* species group. Includes *pilosus* but not *canuti*; see Hill and Schlitter (1982).

*Rhinolophus darlingi* K. Andersen, 1905. Ann. Mag. Nat. Hist., ser. 7, 15:70.
 COMMON NAME: Darling's Horseshoe Bat.
 TYPE LOCALITY: Zimbabwe, Mazoe.
 DISTRIBUTION: NE South Africa, Namibia, S Angola, N and W Botswana, Zimbabwe, Malawi, Mozambique, Tanzania, Nigeria.
 STATUS: IUCN 2003 and IUCN/SSC Action Plan (2001) – Lower Risk (lc).
 SYNONYMS: *barbertonensis* Roberts, 1924; *damarensis* Roberts, 1946.
 COMMENTS: *ferrumequinum* species group. Includes *barbertonensis*; see Hayman and Hill (1971). See Taylor (2000*a*) for distribution map.

*Rhinolophus deckenii* Peters, 1868. Monatsb. K. Preuss. Akad. Wiss. Berlin, 1867:705 [1868].
 COMMON NAME: Decken's Horseshoe Bat.
 TYPE LOCALITY: Tanzania, "Zanzibar coast" (mainland opposite Zanzibar).
 DISTRIBUTION: Uganda, Kenya, Tanzania, Zanzibar and Pemba.
 STATUS: IUCN 2003 and IUCN/SSC Action Plan (2001) – Data Deficient.
 COMMENTS: *ferrumequinum* species group. Treated as a subspecies of *clivosus* by Hayman and Hill (1971), but see Koopman (1975), Cotterill (2002), and Csorba et al. (2003). May include *silvestris*, see Csorba et al. (2003).

*Rhinolophus denti* Thomas, 1904. Ann. Mag. Nat. Hist., ser. 7, 13:386.
 COMMON NAME: Dent's Horseshoe Bat.
 TYPE LOCALITY: South Africa, Northern Cape Prov., Kuruman.
 DISTRIBUTION: Northern Cape Prov. (South Africa), Namibia, Angola, Botswana, Zimbabwe, Mozambique, Guinea-Bissau, Guinea, Ghana. A Côte d'Ivoire record is incorrect (it actually represents *landeri*; J. Fahr, pers. comm.), and reports from Gambia similarly seem to represent misidentified *landeri* (Kock et al., 2002).
 STATUS: IUCN 2003 and IUCN/SSC Action Plan (2001) – Lower Risk (lc).
 SYNONYMS: *knorri* Eisentraut, 1960.
 COMMENTS: *capensis* species group. May include *swinnyi*, see discussion in Csorba et al. (2003).

*Rhinolophus eloquens* K. Andersen, 1905. Ann. Mag. Nat. Hist., ser. 7, 15:74.
 COMMON NAME: Eloquent Horseshoe Bat.
 TYPE LOCALITY: Uganda, Entebbe.
 DISTRIBUTION: Uganda, S Somalia, S Sudan, NE Dem. Rep. Congo, Kenya, Rwanda, N Tanzania, Zanzibar and Pemba.
 STATUS: IUCN 2003 and IUCN/SSC Action Plan (2001) – Lower Risk (lc).
 SYNONYMS: *perauritus* De Beaux, 1922.
 COMMENTS: *fumigatus* species group. Includes *perauritus*; see Koopman (1975).

*Rhinolophus euryale* Blasius, 1853. Arch. Naturgesch., 19(1):49.
 COMMON NAME: Mediterranean Horseshoe Bat.
 TYPE LOCALITY: Italy, Milan.
 DISTRIBUTION: Transcaucasia to Turkey, Israel, and Jordan; S Europe from Portugal, C France to S Slovakia, Hungary, Slovenia, and Romania; Turkmenistan; Iran; Algeria; Morocco; Tunisia; various Mediterranean islands; perhaps Egypt.
 STATUS: IUCN 2003 and IUCN/SSC Action Plan (2001) – Vulnerable.
 SYNONYMS: *atlanticus* K. Andersen and Matschie, 1904; *barbarus* K. Andersen and Matschie,

1904; *cabrerae* K. Andersen and Matschie, 1904; *meridionalis* K. Andersen and Matschie, 1904; *nordmanni* Satunin, 1911; *toscanus* K. Andersen and Matschie, 1904; **judaicus** K. Andersen and Matschie, 1904. **Unassigned:** *algirus* Loche, 1867.

COMMENTS: *euryale* species group. Revised by DeBlase (1972); also see Harrison and Bates (1991), Paz (1995), Zagorodnyuk (1999), Horácek et al. (2000), and Gaisler (2001*b*). Does not include *tuneti*; see Cockrum (1976*b*).

*Rhinolophus euryotis* Temminck, 1835. Monogr. Mamm., 2:26.
COMMON NAME: Broad-eared Horseshoe Bat.
TYPE LOCALITY: Indonesia, Molucca Isls, Amboina Isl.
DISTRIBUTION: Aru Isls, Buru, Bacan, Amboina, Seram, and Tanimbar Isls, Kai Isls, Halmahera, and Sulawesi (Indonesia); New Guinea; Bismarck Arch.; adjacent small islands.
STATUS: IUCN 2003 and IUCN/SSC Action Plan (2001) – Lower Risk (lc).
SYNONYMS: **aruensis** K. Andersen, 1907; **burius** Hinton, 1925; **praestens** K. Andersen, 1905; **tatar** Bergmans and Rozendaal, 1982; **timidus** K. Andersen, 1905.
COMMENTS: *euryotis* species group. Subspecies, some of which are of dubious validity, were discussed by Hill (1983); also see Flannery (1995*a*, *b*) and Bonaccorso (1998).

*Rhinolophus ferrumequinum* (Schreber, 1774). Die Säugethiere, 1:174, pl. 62.
COMMON NAME: Greater Horseshoe Bat.
TYPE LOCALITY: France.
DISTRIBUTION: Algeria, Morocco, and Tunisia; S Europe from Portugal to Greece and north to S England, the Netherlands, S Germany, Austria, Czech Republic, Slovakia, and Bulgaria; Turkey, Cyprus, Georgia, and Azerbaijan; Urkrain, Crimea, and Caucacus regions; the Mediterranean coast from Turkey to Israel and Jordan; NE Iraq, Iran, Turkmenistan, Uzbekistan, S Kazakhstan, Afganistan, Pakistan, N India, Nepal, Sikkim, China, Korea, and Japan; adjacent small islands. Records at some localities in northern Europe (e.g., the Netherlands) apparently reflect temporary northern range extensions (Glas and Voûte, 1992*a*).
STATUS: IUCN 2003 and IUCN/SSC Action Plan (2001) – Lower Risk (nt).
SYNONYMS: *colchicus* Satunin, 1912; *equinus* Müller, 1776; *germanicus* Koch, 1865; *hippocrepis* Schrank, 1798; *homodorensis* Daday, 1887; *homorodalmasiensis* Daday, 1885 [*nomen nudum*]; *insulanus* Barrett-Hamilton, 1910; *italicus* Koch, 1865; *major* E Geoffroy, 1803 [not Kerr, 1792]; *major* Kerr, 1792:99 [not Kerr, 1792:97]; *martinoi* Petrov, 1940; *obscurus* Cabrera, 1904; *perspicillatus* Blumenbach, 1779; *solea* Zimmermann, 1777 [unavailable; see Bull. Zool. Nomen. (1950)4:547]; *typicus* K. Andersen, 1905; *ungula* Boddaert, 1785; *unihastatus* E. Geoffroy, 1803; **creticum** Iliopoulou-Georgudaki and Ondrias, 1985; **irani** Cheesman, 1921; *rubiginosus* Gubareff, 1941; **korai** Kuroda, 1938; *pachyodontus* Kishida, 1931 [*nomen nudum*]; *quelpartis* Mori, 1933; **nippon** Temminck, 1835; *fudisanus* Kishida, 1940; *kosidianus* Kishida, 1940; *mikadoi* Ognev, 1927; *norikuranus* Kishida, 1940; *ogasimanus* Kishida, 1940; **proximus** K. Andersen, 1905; **tragatus** Hodgson, 1835; *brevitarsus* Blyth, 1863 [*nomen nudum*]; *regulus* K. Andersen, 1905.
COMMENTS: *ferrumequinum* species group. Revised by Strelkov et al. (1978). Reviewed in part by Yoshiyuki (1989), Harrison and Bates (1991), Paz (1995), Kock (1996), Bates and Harrison (1997), Sinha (1999), Zagorodnyuk (1999), Horácek et al. (2000) and Gaisler (2001*a*). Subspecies limits are somewhat unclear and there may be more than one species present in this complex; see discussion in Csorba et al. (2003).

*Rhinolophus formosae* Sanborn, 1939. Publ. Field. Mus. Nat. Hist., Zool., 24:41.
COMMON NAME: Formosan Woolly Horseshoe Bat.
TYPE LOCALITY: Taiwan.
DISTRIBUTION: Taiwan.
STATUS: IUCN 2003 – not evaluated; not considered in IUCN/SSC Action Plan (2001).
COMMENTS: *trifoliatus* species group. Formerly considered a subspecies of *luctus*, but apparently distinct; see Yoshiyuki and Harada (1995) and Csorba et al. (2003).

*Rhinolophus fumigatus* Rüppell, 1842. Mus. Senckenbergianum, 3:132, 155.
>    COMMON NAME: Rüppell's Horseshoe Bat.
>    TYPE LOCALITY: Ethiopia, Shoa.
>    DISTRIBUTION: Somalia, Ethiopia, Eritrea, Sudan, Kenya, Uganda, Tanzania, Rwanda, Burundi, Dem. Rep. Congo, Nigeria, Niger, Sierra Leone, Côte d'Ivoire, Togo, Benin, Senegal, Gambia, Guinea, Mali, Burkina Faso, Ghana, Cameroon, Gabon, Republic of Congo, Central African Republic, Zambia, Malawi, Zimbabwe, Mozambique, Angola, Namibia, South Africa.
>    STATUS: IUCN 2003 and IUCN/SSC Action Plan (2001) – Lower Risk (lc).
>    SYNONYMS: *antinorii* Dobson, 1885; *macrocephalus* Heuglin, 1877; ***abae*** J. A. Allen, 1917; ***aethiops*** Peters, 1869; ***diversus*** Sanborn, 1939; ***exsul*** K. Andersen, 1905; *acrotis* G. M. Allen, 1914 [not Heuglin, 1861]; *foxi* Thomas, 1913.
>    COMMENTS: *fumigatus* species group. Does not include *eloquens* or *perauritus*, but does include *aethiops*; see Koopman (1975). Subspecies boundaries are not well delimited. See Taylor (2000*a*) for distribution map.

*Rhinolophus guineensis* Eisentraut, 1960. Stuttg. Beitr. Naturk., 39:1.
>    COMMON NAME: Guinean Horseshoe Bat.
>    TYPE LOCALITY: Guinea, Tahiré (foot of Kelesi Plateau).
>    DISTRIBUTION: Senegal, Guinea, Sierra Leone, Liberia.
>    STATUS: IUCN 2003 and IUCN/SSC Action Plan (2001) – Lower Risk (nt).
>    COMMENTS: *landeri* species group. Originally described as a subspecies of *landeri*, but see Böhme and Hutterer (1979), who demonstrated that it is a separate species.

*Rhinolophus hildebrandtii* Peters, 1878. Monatsb. K. Preuss. Akad. Wiss. Berlin, 1878:195.
>    COMMON NAME: Hildebrandt's Horseshoe Bat.
>    TYPE LOCALITY: Kenya, Taita, Ndi.
>    DISTRIBUTION: NE South Africa and Mozambique to Ethiopia, S Sudan, and NE Dem. Rep. Congo; Nigeria.
>    STATUS: IUCN 2003 and IUCN/SSC Action Plan (2001) – Lower Risk (lc).
>    COMMENTS: *fumigatus* species group. See Taylor (2000*a*) for distribution map. See Fahr et al. (2002) for information on Nigerian record. Sometimes spelled "*hildebranti*," but the original spelling is "*hildebrantii*."

*Rhinolophus hilli* Aellen, 1973. Period. Biol. Zagreb, 75:101.
>    COMMON NAME: Hill's Horseshoe Bat.
>    TYPE LOCALITY: Rwanda, Cyangugu, Uwinka, 2,512 m.
>    DISTRIBUTION: Rwanda.
>    STATUS: IUCN 2003 – not evaluated; not considered in IUCN/SSC Action Plan (2001). Fahr et al. (2002) suggested that this species be listed as Data Deficient.
>    COMMENTS: *maclaudi* species group. Previously included in *maclaudi* (e.g., Smith and Hood, 1980) or *ruwenzorii* (e.g., Csorba et al., 2003), but distinct from both of these species (Fahr et al., 2002).

*Rhinolophus hillorum* Koopman, 1989. Amer. Mus. Novit., 2946:45.
>    COMMON NAME: Upland Horseshoe Bat.
>    TYPE LOCALITY: Liberia, Lofa County, ca. 2 mi. (3 km) SW Voinjama, near Zozoma, John Hegbe Farm, 8°25'N, 9°35'W, ca. 500 m.
>    DISTRIBUTION: Guinea, Liberia, Nigeria, Cameroon.
>    STATUS: IUCN 2003 – not evaluated; not considered in IUCN/SSC Action Plan (2001).
>    COMMENTS: *ferrumequinum* species group. Originally named as a subspecies of *clivosus*, but apparently distinct; see Cotterill (2002).

*Rhinolophus hipposideros* (Bechstein, 1800). *In* Pennant, Allgemeine Ueber. Vierfüss. Thiere, 2:629.
>    COMMON NAME: Lesser Horseshoe Bat.
>    TYPE LOCALITY: France.
>    DISTRIBUTION: Ireland, N Europe to Iberia and Morocco, through S Europe and N Africa to Kyrgystan and Kashmir; Bulgaria; Israel and Jordan; Arabia; Sudan; Ethiopia; Djibouti. Records at some localities in N Europe (e.g., the Netherlands) apparently reflect temporary northern range extensions (Glas and Voûte, 1992*b*).

STATUS: IUCN 2003 and IUCN/SSC Action Plan (2001) – Vulnerable.

SYNONYMS: *alpinus* Koch, 1865; *anomalus* Soderland, 1920; *bifer* Kaup, 1829 [*nomen nudum*]; *bifer* Blainville, 1840 [replacement for *bifer* Kaup, 1829]; *bihastatus* E. Geoffroy, 1813; *eggenhoeffner* Fitzinger, 1870; *helvetica* Bretschner, 1904; *intermedius* Soderland, 1920; *kisnyiresiensis* Daday, 1885; *minor* Kerr 1792:99 [not Kerr 1792:97]; *minuta* Leach, 1816 [*nomen nudum*]; *moravicus* Kostron, 1943; *trogophilus* Daday, 1887; *typicus* K. Andersen, 1905; *typus* Koch, 1865; **escalerae** K. Andersen, 1918; *vespa* Laurent, 1937; **majori** K. Andersen, 1918; *billanyani* DeBlase, 1972; **midas** K. Andersen, 1905; **minimus** Heuglin, 1861; *pallidus* Koch, 1865; *phasma* Cabrera, 1904; **minutus** Montagu, 1808.

COMMENTS: *hipposideros* species group. Revised by Felten et al. (1977). Reviewed by Paz (1995) and Bates and Harrison (1997); also see Harrison and Bates (1991) and Horácek et al. (2000). It is possible that *minimus* represents a distinct species; see Zagorodnyuk (1999).

*Rhinolophus imaizumii* Hill and Yoshiyuki, 1980. Bull. Natl. Sci. Mus. Tokyo, ser. A (Zool.), 6:180.
COMMON NAME: Imaizumi's Horseshoe Bat.
TYPE LOCALITY: Japan, Ryukyu Isls, Yayeyama Isls, Iriomote Isl, Otomi-do cave.
DISTRIBUTION: Iriomote Isl and Yaeyama Isl (Japan: Ryukyu Isls).
STATUS: IUCN 2003 and IUCN/SSC Action Plan (2001) – Endangered.
COMMENTS: *pusillus* species group. Reviewed by Yoshiyuki (1989).

*Rhinolophus inops* K. Andersen, 1905. Ann. Mag. Nat. Hist., ser. 7, 16:284, 651.
COMMON NAME: Philippine Forest Horseshoe Bat.
TYPE LOCALITY: Philippines, Mindanao, Davao, Mt. Apo, Todaya (= Jodaya), 1,325 m.
DISTRIBUTION: Philippines except Palawan region.
STATUS: IUCN 2003 and IUCN/SSC Action Plan (2001) – Data Deficient.
COMMENTS: *euryotis* species group. This taxon may include more than one species; see Ingle and Heaney (1992) and Heaney et al. (1998).

*Rhinolophus keyensis* Peters, 1871. Monatsb. K. Preuss. Akad. Wiss. Berlin, 1871:307.
COMMON NAME: Kai Horseshoe Bat.
TYPE LOCALITY: Key-Inseln (= Kai Isls).
DISTRIBUTION: Many islands in Indonesia; see Kitchener et al. (1995a).
STATUS: IUCN 2003 and IUCN/SSC Action Plan (2001) – Endangered.
SYNONYMS: *nanus* K. Andersen, 1905; *truncatus* Peters, 1871; **amiri** Kitchener, 1995 [in Kitchener et al., 1995a]; **parvus** Goodwin, 1979; *annectens* Sanborn, 1939; **simplex** K. Andersen, 1905.
COMMENTS: *megaphyllus* species group. Revised by Kitchener (1995a), who apparently overlooked the fact that *keyensis* is the oldest name for this complex (not *simplex*). Not included in *megaphyllus*, although see Corbet and Hill (1992). Includes *parvus*; see Kitchener et al. (1995a), but see also Bergmans and van Bree (1986). The holotype of *annectans* is a damaged skull that is difficult to assign with any certainty, but may represent *parvus*; see Kitchener et al. (1995a).

*Rhinolophus landeri* Martin, 1838. Proc. Zool. Soc. Lond., 1837:101 [1838].
COMMON NAME: Lander's Horseshoe Bat.
TYPE LOCALITY: Equatorial Guinea, Bioko.
DISTRIBUTION: Senegal and Gambia to Ethiopia and Somalia, south to South Africa and Namibia; Bioko (Equatorial Guinea); Zanzibar.
STATUS: IUCN 2003 and IUCN/SSC Action Plan (2001) – Lower Risk (lc).
SYNONYMS: **angolensis** Seabra, 1898; **lobatus** Peters, 1852; *axillaris* Allen, Lang, and Chapin, 1917; *dobsoni* Thomas, 1904.
COMMENTS: *landeri* species group. According to Hayman and Hill (1971) and Koopman (1975), this species includes *angolensis*, *dobsoni*, and *guineensis*, but not *brockmani*; but see Böhme and Hutterer (1979) who correctly treated *guineensis* as a separate species. See Brown and Dunlop (1997); also see Kock et al. (2002), who discussed differences between *landeri* and *denti*.

*Rhinolophus lepidus* Blyth, 1844. J. Asiat. Soc. Bengal, 13:486.
COMMON NAME: Blyth's Horseshoe Bat.

TYPE LOCALITY: India, Bengal, Calcutta (uncertain); see Das (1986).

DISTRIBUTION: Afghanistan, Pakistan, N India, Nepal, Burma, Thailand, Szechwan and
Yunnan (China), Peninsular Malaysia, Sumatra (Indonesia).

STATUS: IUCN 2003 and IUCN/SSC Action Plan (2001) – Lower Risk (lc).

SYNONYMS: *cuneatus* K. Andersen, 1918; *feae* K. Andersen, 1907; *monticola* K. Andersen,
1905; *refulgens* K. Andersen, 1905.

COMMENTS: *pusillus* species group. Includes *feae*, *monticola*, and *refulgens*; see Hill and
Yoshiyuki (1980) and Corbet and Hill (1992). Does not include *osgoodi* or *shortridgei*;
see Csorba et al. (2003), although also see Corbet and Hill (1992). Reviewed in part by
Bates and Harrison (1997).

*Rhinolophus luctus* Temminck, 1834. Tijdschrift Natuurl. Gesch. Physiol., 1:23.

COMMON NAME: Woolly Horseshoe Bat.

TYPE LOCALITY: Indonesia, Java, Tapos.

DISTRIBUTION: India, Nepal, Burma, Sri Lanka, S China, Vietnam, Cambodia, Laos,
Thailand, Peninsular Malaysia; Borneo, Sumatra, Java, and Bali (Indonesia).

STATUS: IUCN 2003 and IUCN/SSC Action Plan (2001) – Lower Risk (lc).

SYNONYMS: *geminus* K. Andersen, 1905; *foetidus* K. Andersen, 1918; *lanosus* K. Andersen,
1905; *morio* Gray, 1842; *perniger* Hodgson, 1843; *spurcus* Allen, 1928.

COMMENTS: *trifoliatus* species group. Includes *lanosus*; see Ellerman and Morrison-Scott
(1951). Does not include *beddomei*; see Topál and Csorba (1992) and Bates and
Harrison (1997). Does not include *formosae*; see Yoshiyuki and Harada (1995) and
Csorba et al. (2003).

*Rhinolophus maclaudi* Pousargues, 1897. Bull. Mus. Natn. Hist. Nat. Paris, 3:358.

COMMON NAME: Maclaud's Horseshoe Bat.

TYPE LOCALITY: Guinea, Conakry.

DISTRIBUTION: Guinea. A record from Nigeria was based on misidentified specimens
(Fahr et al., 2002).

STATUS: IUCN 2003 and IUCN/SSC Action Plan (2001) – Lower Risk (nt). Fahr et al. (2002)
proposed that this be changed to Endangered.

COMMENTS: *maclaudi* species group. Smith and Hood (1980) included *ruwenzorii* in this
taxon, but morphological differences and a major range disjunction indicate that
*maclaudi* and *ruwenzorii* are distinct species; see Csorba et al. (2003) and Fahr et al.
(2002). Revised by Fahr et al. (2002), who provided a key to species in the *maclaudi*
species group.

*Rhinolophus macrotis* Blyth, 1844. J. Asiat. Soc. Bengal, 13:485.

COMMON NAME: Big-eared Horseshoe Bat.

TYPE LOCALITY: Nepal.

DISTRIBUTION: Pakistan, N India, Nepal to S China, Burma, Thailand, Laos, Vietnam, and
Peninsualr Malaysia; Sumatra (Indonesia); Philippines.

STATUS: IUCN 2003 and IUCN/SSC Action Plan (2001) – Lower Risk (lc).

SYNONYMS: *caldwelli* Allen, 1923; *dohrni* K. Andersen, 1907; *episcopus* Allen, 1923;
*hirsutus* K. Andersen, 1905; *topali* Csorba and Bates, 1995.

COMMENTS: *philippinensis* species group. Includes *episcopus* and *hirsutus*; see Ellerman and
Morrison-Scott (1951), Tate (1943), Corbet and Hill (1992), and Bates and Harrison
(1997), but also see Ingle and Heaney (1992), who suggested that *hirsutus* may deserve
recognition as a distinct species. Does not include *siamensis*, see Francis et al. (1999*b*)
and Hendrichsen et al. (2001*b*).

*Rhinolophus madurensis* K. Andersen, 1918. Ann. Mag. Nat. Hist., ser. 9, 2:375.

COMMON NAME: Madura Horseshoe Bat.

TYPE LOCALITY: Indonesia, off NE Java, Madura Isl, E. Madura, Soemenep.

DISTRIBUTION: Madura and Kangean Isls (Indonesia).

STATUS: IUCN 2003 – not evaluated; not considered in IUCN/SSC Action Plan (2001).

COMMENTS: *megaphyllus* species group. Previously included in *celebensis*, but see Bergmans
and van Bree (1986). Does not include *parvus*; see Kitchener et al. (1995*a*), but also see

Bergmans and van Bree (1986). Csorba et al. (2003) retained both *madurensis* and *parvus* in *celebensis*.

*Rhinolophus maendeleo* Kock, Csorba, and Howell, 2000. Senkenbergiana Biol., 80:234.
COMMON NAME: Maendeleo Horseshoe Bat.
TYPE LOCALITY: Tanzania, Tanga Dist., 2.5 km W of Tanga, Mkulumuzi River Gorge, Amboni Cave Forest, 05°05'S, 39°02'E, 0-80 m.
DISTRIBUTION: NE Tanzania.
STATUS: IUCN 2003 – not evaluated; not considered in IUCN/SSC Action Plan (2001).
COMMENTS: *adami* species group. Known from only two specimens.

*Rhinolophus malayanus* Bonhote, 1903. *In* N. Annandale, Fasciculi Malayenses, Zool., 1:15.
COMMON NAME: Malayan Horseshoe Bat.
TYPE LOCALITY: Thailand, Jalor, Biserat.
DISTRIBUTION: Thailand, Burma, Cambodia, Laos, Vietnam, Peninsular Malaysia.
STATUS: IUCN 2003 and IUCN/SSC Action Plan (2001) – Lower Risk (lc).
COMMENTS: *megaphyllus* species group. McFarlane and Blood (1986) suggested that characters used by Lekagul and McNeely (1977) to separate *stheno* and *malayanus* may not be reliable, but see Corbet and Hill (1992), Csorba and Jenkins (1998), and Hendrichsen (2001*a*, *b*).

*Rhinolophus marshalli* Thonglongya, 1973. Mammalia, 37:590.
COMMON NAME: Marshall's Horseshoe Bat.
TYPE LOCALITY: Thailand, Chantaburi, Amphoe Pong Nam Ron, foothills of Khao Soi Dao Thai.
DISTRIBUTION: Thailand, Burma, Vietnam, Laos, Peninsular Malaysia.
STATUS: IUCN 2003 – and IUCN/SSC Action Plan (2001) – Lower Risk (nt).
COMMENTS: *philippinensis* species group. Reviewed by Yoshiyuki (1990) and Hill and Topál (1990); also see Bates et al. (2001) and Hendrichsen et al. (2001*b*).

*Rhinolophus megaphyllus* Gray, 1834. Proc. Zool. Soc. Lond., 1834:52.
COMMON NAME: Smaller Horseshoe Bat.
TYPE LOCALITY: Australia, New South Wales, Murrumbidgee River.
DISTRIBUTION: E New Guinea; Misima Isl (Louisiade Arch.), Goodenough Isl (D'Entrecasteaux Isls), and Bismarck Arch. (Papua New Guinea); Moluccas, Lesser Sundas; E Queensland, E New South Wales, and E Victoria (Australia).
STATUS: IUCN 2003 and IUCN/SSC Action Plan (2001) – Lower Risk (lc).
SYNONYMS: **fallax** K. Andersen, 1906; **ignifer** Allen, 1933; **monachus** K. Andersen, 1905; **vandeuseni** Koopman, 1982.
COMMENTS: *megaphyllus* species group. Does not include *keyensis* and *amiri*; see Kitchener et al. (1995*a*), although also see discussion in Csorba et al. (2003). May be closely related to *philippinensis*, and both taxa as presently recognized may be polyphyletic; see Cooper et al. (1998). Does not include *robinsoni* and *simplex*, although see discussion in Corbet and Hill (1992) and Csorba et al. (2003). Also see Flannery (1995*a*, *b*) and Bonaccorso (1998).

*Rhinolophus mehelyi* Matschie, 1901. Sitzb. Ges. Naturf. Fr. Berlin, p. 225.
COMMON NAME: Mehely's Horseshoe Bat.
TYPE LOCALITY: Romania, Bucharest.
DISTRIBUTION: Portugal, Spain, France, Romania, Bulgaria, Greece, Serbia and Montenegro, Transcaucasia; Morocco, Tunisia, Egypt, Algeria, and Libya; Mediterranean islands, Turkey, Cyprus, Iran, Iraq, Israel, and Jordan; Afghanistan.
STATUS: IUCN 2003 and IUCN/SSC Action Plan (2001) – Vulnerable.
SYNONYMS: *carpetanus* Cabrera, 1904; **tuneti** Deleuil and Labbe, 1955.
COMMENTS: *euryale* species group. Revised by DeBlase (1972); also see Harrison and Bates (1991), Paz (1995), Zagorodnyuk (1999), Horácek et al. (2000), and Gaisler (2001*c*). Includes *tuneti*; see Cockrum (1976*b*).

*Rhinolophus mitratus* Blyth, 1844. J. Asiat. Soc. Bengal, 13:483.
COMMON NAME: Mitred Horseshoe Bat.

TYPE LOCALITY: India, Orissa, Chaibassa.
DISTRIBUTION: Known only from the type locality.
STATUS: IUCN 2003 and IUCN/SSC Action Plan (2001) – Data Deficient.
COMMENTS: *trifoliatus* species group. Known only from the holotype; see Sinha (1973), Corbet and Hill (1992), and Bates and Harrison (1997) for discussion of morphology and possible affinities.

*Rhinolophus monoceros* K. Andersen, 1905. Proc. Zool. Soc. Lond., 1905, 2:131.
COMMON NAME: Formosan Lesser Horseshoe Bat.
TYPE LOCALITY: Taiwan, Baksa.
DISTRIBUTION: Taiwan, possibly S China.
STATUS: IUCN 2003 and IUCN/SSC Microchiropteran Bats Action Plan (2001) – Lower Risk (nt).
COMMENTS: *pusillus* species group. May be conspecific with *cornutus* and/or *pusillus*; see Corbet and Hill (1992), Koopman (1994), and Csorba (1997). Reviewed by Yoshiyuki (1989).

*Rhinolophus montanus* Goodwin, 1979. Bull. Amer. Mus. Nat. Hist., 163:112.
COMMON NAME: Timorese Horseshoe Bat.
TYPE LOCALITY: Timor, 5 mi. (8 km) S of Ermera, near Village of Lequi Mia, Quoto Lou Caves.
DISTRIBUTION: Known only from the type locality.
STATUS: IUCN 2003 – not evaluated; not considered in IUCN/SSC Action Plan (2001).
COMMENTS: *philippinensis* species group. Formerly considered a subspecies of *philippinensis*, but apparently distinct; see Csorba (2002) and Csorba et al. (2003).

*Rhinolophus nereis* K. Andersen, 1905. Proc. Zool. Soc. Lond., 1905:90.
COMMON NAME: Anamban Horseshoe Bat.
TYPE LOCALITY: Indonesia, Anamba Isls, Siantan Isl.
DISTRIBUTION: Anamba and North Natuna Isls (Indonesia).
STATUS: IUCN 2003 and IUCN/SSC Action Plan (2001) – Lower Risk (nt).
COMMENTS: *megaphyllus* species group.

*Rhinolophus osgoodi* Sanborn, 1939. Field Mus. Nat. Hist. Publ., Zool. Ser., 24:40.
COMMON NAME: Osgood's Horseshoe Bat.
TYPE LOCALITY: China, Yunnan, N of Likiang, Nguluko, (27°05′N, 100°15′E).
DISTRIBUTION: Yunnan (China).
STATUS: IUCN 2003 and IUCN/SSC Action Plan (2001) – Data Deficient.
COMMENTS: *pusillus* species group. Possibly conspecific with *lepidus* (Corbet and Hill, 1992), but treated as distinct following Csorba et al. (2003).

*Rhinolophus paradoxolophus* (Bourret, 1951). Bull. Mus. Natn. Hist. Nat. Paris, ser. 2, 33:607.
COMMON NAME: Bourret's Horseshoe Bat.
TYPE LOCALITY: Vietnam, Tonkin, Lao Key Prov., near Chapa, Rochepercée cave, 1,700 m.
DISTRIBUTION: Vietnam, Thailand, Laos, and China.
STATUS: IUCN 2003 and IUCN/SSC Action Plan (2001) – Vulnerable.
COMMENTS: *philippinensis* species group. Redescribed by Hill (1972*b*); also see Hendrichsen et al. (2001*b*). Corbet and Hill (1992) suggested that paradoxolophus may be conspecific with *rex*, but see Eger and Fenton (2003).

*Rhinolophus pearsonii* Horsfield, 1851. Cat. Mamm. Mus. E. India Co., p. 33.
COMMON NAME: Pearson's Horseshoe Bat.
TYPE LOCALITY: India, W Bengal, Darjeeling.
DISTRIBUTION: N India; Nepal; Bhutan; Burma; Tibet, Szechwan, Anhwei, and Fukien (China) to Vietnam; Laos; Thailand; Peninsular Malaysia.
STATUS: IUCN 2003 and IUCN/SSC Action Plan (2001) – Lower Risk (lc).
SYNONYMS: *larvatus* Milne-Edwards, 1872 [not Horsfield, 1823]; ***chinensis*** K. Andersen, 1905.
COMMENTS: *pearsonii* species group. Reviewed in part by Bates and Harrison (1997). Subspecies are of questionable validity (Corbet and Hill, 1992). Sometimes spelled *pearsoni* (e.g., Koopman, 1993).

*Rhinolophus philippinensis* Waterhouse, 1843. Proc. Zool. Soc. Lond., 1843:68.
> COMMON NAME: Large-eared Horseshoe Bat.
> TYPE LOCALITY: Philippines, Luzon.
> DISTRIBUTION: Phillipines; Kai Isls, Sabah, Sarawak, and Sulawesi (Indonesia); Borneo; New Guinea; NE Queensland (Australia).
> STATUS: IUCN 2003 and IUCN/SSC Action Plan (2001) – Lower Risk (nt).
> SYNONYMS: **achilles** Thomas, 1900; **alleni** Lawrence, 1939; **maros** Tate and Archbold, 1939; **robertsi** Tate, 1952; **sanborni** Chasen, 1940.
> COMMENTS: *philippinensis* species group. Variation discussed by Goodwin (1979). May be closely related to *megaphyllus*, and both taxa as presently recognized may be polyphyletic; see Cooper et al. (1998). Does not include *montanus*, see Csorba et al. (2003). Two morphologically distinct populations occur on the Cape York peninisula of Australia; see Flannery (1995*a*, *b*), Churchill (1998), and Csorba et al. (2003). Flannery (1995*a*, *b*) referred the smaller of these forms to the subspecies *maros* (which he considered to be a senior synonym of *alleni* and *sanborni*) and the larger-bodied form to *achilles*. The only name based on an Australian holotype, *robertsi*, was treated as a junior synonym of *achilles* by Flannery (1995*b*). Flannery (1995*a*, *b*) referred all New Guinea populations to *maros*, but Bonaccorso (1998) referred the New Guinea and Cape York populations to *robertsi* while recognizing the Kai Isl form (*achilles*) as a distinct subspecies. Based on sympatry of two forms of "*philippinensis*" on the Cape York peninsula, it seems clear that at least two species are present in this complex, but taxonomic limits and the appropriate names for each population remain unclear. I follow Koopman (1994) and Csorba et al. (2003) in recognizing each of the named forms as a distinct subspecies pending a thorough revision of this complex.

*Rhinolophus pusillus* Temminck, 1834. Tijdschr. Nat. Gesch. Physiol., 1:29.
> COMMON NAME: Least Horseshoe Bat.
> TYPE LOCALITY: Indonesia, Java.
> DISTRIBUTION: India; Nepal; Thailand; Burma; Laos; S China; Peninsular Malaysia; Mentawai Isls, Java and Lesser Sunda Isls (Indonesia), small adjacent islands. Reports of this species from Cambodia cannot be confirmed (Kock, 2000*a*).
> STATUS: IUCN 2003 and IUCN/SSC Action Plan (2001) – Lower Risk (lc).
> SYNONYMS: *minor* Horsfeld, 1823 [not Kerr, 1792]; **blythi** K. Andersen, 1918; **calidus** G. M. Allen, 1923; **gracilis** K. Andersen, 1905; **lakkhanae** Yoshiyuki, 1990; **minutillus** Miller, 1906; *minutus* Miller, 1900 [not Montague, 1808]; **pagi** Tate and Archbold, 1939; **parcus** Allen, 1928; **szechwanus** K. Andersen, 1918.
> COMMENTS: *pusillus* species group. Includes *blythi*, *minutillus*, and *pagi*; see Hill and Yoshiyuki (1980) and Corbet and Hill (1992). Contains *gracilis*; see Corbet and Hill (1992), but also see Sinha (1973). May include *cornutus*, *pumilus*, and *perditus*; see Corbet and Hill (1992). Reviewed in part by Yoshiyuki (1990), Kock (1996), and Bates and Harrison (1997). Lectotype designated by Csorba (2002). See Corbet and Hill (1992) and Hendrichsen et al. (2001*b*) for a discussion of usage of the name *minor*, which was preoccupied by *minor* Kerr, 1792.

*Rhinolophus rex* G. M. Allen, 1923. Am. Mus. Novit., 85:3.
> COMMON NAME: King Horseshoe Bat.
> TYPE LOCALITY: China, Szechwan, Wanhsien.
> DISTRIBUTION: SW China.
> STATUS: IUCN 2003 and IUCN/SSC Action Plan (2001) – Vulnerable.
> COMMENTS: *philippinensis* species group. Redescribed by Hill (1972*b*). May be conspecific with *paradoxolophus*; see Corbet and Hill (1992).

*Rhinolophus robinsoni* K. Andersen, 1918. Ann. Mag. Nat. Hist., ser. 9, 2:375.
> COMMON NAME: Peninsular Horseshoe Bat.
> TYPE LOCALITY: Thailand, Surat Thani, Bandon, Kaho Nawng.
> DISTRIBUTION: W Malaysia, Thailand, adjacent small islands.
> STATUS: IUCN 2003 – not evaluated; not considered in IUCN/SSC Action Plan (2001).
> SYNONYMS: **klossi** K. Andersen, 1918; **thaianus** Hill, 1992 [in Corbet and Hill (1992),

replacement name for *siamensis* McFarlane and Blood, 1986]; *siamensis* McFarlane and Blood, 1986 [not *siamensis* Gyldenstolpe, 1917].
COMMENTS: *megaphyllus* species group. Includes *klossi*; see Medway (1969). Not included in *megaphyllus*, although see Corbet and Hill (1992) and Csorba et al. (2003).

*Rhinolophus rouxii* Temminck, 1835. Monogr. Mamm., 2:306.
COMMON NAME: Rufous Horseshoe Bat.
TYPE LOCALITY: India, Pondicherry and Calcutta.
DISTRIBUTION: Sri Lanka, peninsular India to S Burma and Vietnam. Reports of this species from Cambodia are likely erroneous; see Kock (2000*a*).
STATUS: IUCN 2003 and IUCN/SSC Action Plan (2001) – Lower Risk (lc).
SYNONYMS: *cinerascens* Kelaart, 1852; *fulvidus* Blyth, 1851; *petersii* Dobson, 1872; *rammanika* Kelaart, 1852; *rubidus* Kelaart, 1850.
COMMENTS: *rouxii* species group. Includes *petersii*; see Sinha (1973). Does not include *sinicus*; see Thomas (2000). Reviewed in part by Bates and Harrison (1997) and Horáček et al. (2000); revised by Thomas (2000). Sometimes spelled *rouxi* (e.g., Horáček et al., 2000; Koopman, 1993).

*Rhinolophus rufus* Eydoux and Gervais, 1836. *In* Laplace, Voy. autour du monde par les mers de l'Inde . . . la Favorite, 5(Zoologie), pt. 2:9.
COMMON NAME: Large Rufous Horseshoe Bat.
TYPE LOCALITY: Philippines, Luzon, Manila.
DISTRIBUTION: Philippines.
STATUS: IUCN 2003 and IUCN/SSC Action Plan (2001) – Lower Risk (nt).
SYNONYMS: *eudoxii* Fitzinger, 1870.
COMMENTS: *euryotis* species group. Name revived by Lawrence (1939); also see Corbet and Hill (1992).

*Rhinolophus ruwenzorii* J. Eric Hill, 1942. Amer. Mus. Novit., 1180:1-2.
COMMON NAME: Ruwenzori Horseshoe Bat.
TYPE LOCALITY: Dem. Rep. Congo, Kivu, W slope of Mount Ruwenzori, Buhatu Valley, 7,500 ft. (2,500 m).
DISTRIBUTION: E Dem. Rep. Congo, Rwanda, W Uganda.
STATUS: IUCN 2003 – Not evaluated; not considered in IUCN/SSC Action Plan (2001). Fahr et al. (2002) suggested that this species be listed as Vulnerable.
COMMENTS: *maclaudi* species group. Smith and Hood (1980) considered this taxon to be a junior synonym of *maclaudi*, but morphological differences and a major range disjunction indicate that *maclaudi* and *ruwenzorii* are distinct species; see Csorba et al. (2003) and Fahr et al. (2002). Does not include *hilli*, a taxon sometimes considered a junior synonym of *ruwenzorii* (Fahr et al., 2002).

*Rhinolophus sakejiensis* Cotterill, 2002. J. Zool., 256:166.
COMMON NAME: Sakeji Horseshoe Bat.
TYPE LOCALITY: Zambia, Mwinilunga District, Ikelenge Pedicle between the Sakeji and Zambezi Rivers, approx. 11 km NNE of source of Zambezi River, Kavunda, 11°17′S, 24°21′E, 1,388 m.
DISTRIBUTION: Known only from the type locality.
STATUS: IUCN 2003 – Not evaluated (new species).
COMMENTS: *ferrumequinum* species group.

*Rhinolophus sedulus* K. Andersen, 1905. Ann. Mag. Nat. Hist., ser. 7, 16:244, 247.
COMMON NAME: Lesser Wooly Horseshoe Bat.
TYPE LOCALITY: Malaysia, Sarawak.
DISTRIBUTION: Peninsular Malaysia, Sarawak and Sabah (Malaysia), Borneo (Indonesia).
STATUS: IUCN 2003 and IUCN/SSC Action Plan (2001) – Lower Risk (lc).
COMMENTS: *trifoliatus* species group. Does not include *edax*; see Tate (1943) and Corbet and Hill (1992), but also see Chasen (1940).

*Rhinolophus shameli* Tate, 1943. Am. Mus. Novit., 1219:3.
COMMON NAME: Shamel's Horseshoe Bat.

TYPE LOCALITY: Thailand, off SE Thailand, Koh Chang Isl.

DISTRIBUTION: Burma, Thailand, Laos, Cambodia, Peninsular Malaysia.

STATUS: IUCN 2003 and IUCN/SSC Action Plan (2001) – Lower Risk (nt).

COMMENTS: *euryotis* species group. Described as a subspecies of *coelophyllus*, but see Hill and Thonglongya (1972).

*Rhinolophus shortridgei* K. Andersen, 1918. Ann. Mag. Nat. Hist., ser. 9, 2:376.

COMMON NAME: Shortridge's Horseshoe Bat.

TYPE LOCALITY: Burma, Irrawaddy River, Pagan (= Bagan).

DISTRIBUTION: N India, Burma

STATUS: IUCN 2003 – Not evaluated; not considered in IUCN/SSC Action Plan (2001).

COMMENTS: *pusillus* species group. Formerly considered a subspecies of *lepidus*, but recently captured in sympatry with that species and thus clearly distinct; see Csorba (2002) and Csorba et al. (2003).

*Rhinolophus siamensis* Gyldenstolpe, 1917. Kungliga Svenska VetenskAkad. Handl., 57:12.

COMMON NAME: Thai Horseshoe Bat.

TYPE LOCALITY: Thailand, NW Thailand, Doi Par Sakang.

DISTRIBUTION: Thailand, Laos, Vietnam.

STATUS: IUCN 2003 – Not evaluated; not considered in IUCN/SSC Action Plan (2001).

COMMENTS: Formerly included in *macrotis* but apparently distinct, see Francis et al. (1999*b*) and Hendrichsen et al. (2001*b*).

*Rhinolophus silvestris* Aellen, 1959. Arch. Sci. Phys. Nat. Geneve, 12:228.

COMMON NAME: Forest Horseshoe Bat.

TYPE LOCALITY: Gabon, Latoursville, N'Dumbu Cave.

DISTRIBUTION: Gabon, Republic of Congo.

STATUS: IUCN 2003 and IUCN/SSC Action Plan (2001) – Lower Risk (nt).

COMMENTS: *ferrumequinum* species group. Considered a subspecies of *clivosus* by Hayman and Hill (1971), but see Koopman (1975), Cotterill (2002), and Csorba et al. (2003). The relationships of *silvestris* and *deckeni* are unclear, these forms may be conspecific; see Csorba et al. (2003).

*Rhinolophus simulator* K. Andersen, 1904. Ann. Mag. Nat. Hist., ser. 7, 14:384.

COMMON NAME: Bushveld Horseshoe Bat.

TYPE LOCALITY: Zimbabwe, Mazoe.

DISTRIBUTION: South Africa to S Sudan and Ethiopia; Cameroon; Liberia; Nigeria; Guinea.

STATUS: IUCN 2003 and IUCN/SSC Action Plan (2001) – Lower Risk (lc).

SYNONYMS: *bembanicus* Senna, 1914; **alticolus** Sanborn, 1936.

COMMENTS: *capensis* species group. Includes *alticolus* and *bembanicus*; see Koopman (1975) and Hayman and Hill (1971).

*Rhinolophus sinicus* K. Andersen, 1905. Proc. Royal Soc. Lond. B., 2:98.

COMMON NAME: Chinese Rufous Horseshoe Bat.

TYPE LOCALITY: China, Anhwei (= Anhui), Chinteh.

DISTRIBUTION: S China, Nepal, N India, Vietnam.

STATUS: IUCN 2003 and IUCN/SSC Action Plan (2001) – Lower Risk (lc).

SYNONYMS: **septentrionalis** Sanborn, 1939.

COMMENTS: *rouxii* species group. Previously included in *rouxii*, but see Thomas (2000). Includes *septentrionalis*, see Csorba (2002) and Csorba et al. (2003).

*Rhinolophus stheno* K. Andersen, 1905. Proc. Zool. Soc. Lond., 1905:91.

COMMON NAME: Lesser Brown Horseshoe Bat.

TYPE LOCALITY: Malaysia, Selangor.

DISTRIBUTION: Vietnam, Thailand, Laos, Peninsular Malaysia, Sumatra and Java (Indonesia).

STATUS: IUCN 2003 and IUCN/SSC Action Plan (2001) – Lower Risk (lc).

SYNONYMS: **microglobosus** Csorba and Jenkins, 1998.

COMMENTS: *megaphyllus* species group. McFarlane and Blood (1986) suggested that characters used by Lekagul and McNeely (1977) to separate *stheno* and *malayanus* may not be reliable, but see Corbet and Hill (1992), Csorba and Jenkins (1998), and Hendrichsen et al. (2001b).

*Rhinolophus subbadius* Blyth, 1844. J. Asiat. Soc. Bengal, 13:486.
COMMON NAME: Little Nepalese Horseshoe Bat.
TYPE LOCALITY: Nepal.
DISTRIBUTION: NE India, Nepal, Vietnam, Burma.
STATUS: IUCN 2003 and IUCN/SSC Action Plan (2001) – Data Deficient.
SYNONYMS: *garoensis* Dobson, 1872; *subbadius* Hodgson, 1841 [*nomen nudum*].
COMMENTS: *pusillus* species group. Reviewed by Bates and Harrison (1997) and Csorba (1997).

*Rhinolophus subrufus* K. Andersen, 1905. Ann. Mag. Nat. Hist., ser. 7, 16:283.
COMMON NAME: Small Rufous Horseshoe Bat.
TYPE LOCALITY: Philippines, Luzon, Manila.
DISTRIBUTION: Philippines except Palawan region.
STATUS: IUCN 2003 and IUCN/SSC Action Plan (2001) – Vulnerable.
SYNONYMS: *rufus* Peters, 1861 [not Eydoux and Gervais, 1836]; **bunkeri** Taylor, 1934.
COMMENTS: *euryotis* species group. Includes *bunkeri*; see Lawrence (1939) and Corbet and Hill (1992). Also see Ingle and Heaney (1992).

*Rhinolophus swinnyi* Gough, 1908. Ann. Transvaal Mus., 1:72.
COMMON NAME: Swinny's Horseshoe Bat.
TYPE LOCALITY: South Africa, Eastern Cape Prov., Pondoland, Ngqeleni Dist.
DISTRIBUTION: South Africa, Zimbabwe, Mozambique, Malawi, Zambia, S Dem. Rep. Congo, Tanzania, Zanzibar.
STATUS: IUCN 2003 and IUCN/SSC Action Plan (2001) – Lower Risk (lc).
SYNONYMS: *piriensis* Hewitt, 1913; *rhodesiae* Roberts, 1946.
COMMENTS: *capensis* species group. Possibly a subspecies of *denti* (Koopman, 1993). See Taylor (2000a) for distribution map.

*Rhinolophus thomasi* K. Andersen, 1905. Proc. Zool. Soc. Lond., 1905:100.
COMMON NAME: Thomas's Horseshoe Bat.
TYPE LOCALITY: Burma, Karin Hills.
DISTRIBUTION: Burma, Vietnam, Thailand, Laos.
STATUS: IUCN 2003 and IUCN/SSC Action Plan (2001) – Lower Risk (nt).
SYNONYMS: **latifolius** Sanborn, 1939.
COMMENTS: *rouxii* species group. Does not include *septentrionalis*, see Csorba et al. (2003). Also see Corbet and Hill (1992) and Hendrichsen et al. (2001b).

*Rhinolophus trifoliatus* Temminck, 1834. Tijdschr. Nat. Gesch. Physiol., 1:24.
COMMON NAME: Trefoil Horseshoe Bat.
TYPE LOCALITY: Indonesia, W Java, Bantam.
DISTRIBUTION: NE India, SW Thailand, and Burma; Peninsular Malaysia, Sarawak, and Sabah (Malaysia); Singapore; Borneo, Sumatra, Riau Archipelago, Banguey Isl, Java, Banka Isl and Nias Isl (Indonesia).
STATUS: IUCN 2003 and IUCN/SSC Action Plan (2001) – Lower Risk (lc).
SYNONYMS: **edax** K. Andersen, 1918; **niasensis** K. Andersen, 1906; **solitarius** K. Andersen, 1905.
COMMENTS: *trifoliatus* species group. Includes *edax*; see Tate (1943) and Corbet and Hill (1992); but also see Chasen (1940). Reviewed by Bates and Harrison (1997).

*Rhinolophus virgo* K. Andersen, 1905. Proc. Zool. Soc. Lond., 1905:88.
COMMON NAME: Yellow-faced Horseshoe Bat.
TYPE LOCALITY: Philippines, Luzon, Camarines Sur, Pasacao.
DISTRIBUTION: Philippines.
STATUS: IUCN 2003 and IUCN/SSC Action Plan (2001) – Lower Risk (nt).
COMMENTS: *megaphyllus* species group. Closely related to *celebensis*; see Corbet and Hill (1992).

*Rhinolophus yunanensis* Dobson, 1872. J. Asiat. Soc. Bengal, 41:336.
    COMMON NAME: Dobson's Horseshoe Bat.
    TYPE LOCALITY: China, Yunnan, Hotha.
    DISTRIBUTION: Yunnan (China), Burma, Thailand, NE India.
    STATUS: IUCN 2003 and IUCN/SSC Action Plan (2001) – Lower Risk (nt).
    COMMENTS: *pearsonii* species group. Formerly included in *pearsonii*, but see Lekagul and
        McNeely (1977) and Yoshiyuki (1990). Reviewed by Bates and Harrison (1997).

*Rhinolophus ziama* Fahr, Vierhaus, Hutterer, and Kock, 2002. Myotis, 40:109.
    COMMON NAME: Ziama Horseshoe Bat.
    TYPE LOCALITY: Guinea, Guinée Forestière, Réserve de la Biosphère du Massif du Ziama,
        western edge of Sérédou near park station.
    DISTRIBUTION: SE Guinea, NW Liberia.
    STATUS: IUCN 2003– Not evaluated (new species). Fahr et al. (2002) suggested that this
        species be listed as Data Deficient.
    COMMENTS: *maclaudi* species group. See Fahr et al. for a key to species of the *maclaudi* species
        group.

**Family Hipposideridae** Lydekker, 1891. *In* Flower and Lydekker, Mamm., Living and Extinct,
    p. 657.
    SYNONYMS: Coelopsinae Tate, 1941.
    COMMENTS: Treated as a subfamily of Rhinolophidae by Koopman (1993, 1994), McKenna and
        Bell (1997), Simmons (1998), Simmons and Geisler (1998), and Teeling et al. (2002), but
        returned to family rank here following Corbet and Hill (1992), Bates and Harrison (1997),
        Bogdanowicz and Owen (1998), Hand and Kirsch (1998), and numerous other authors.
        McKenna and Bell (1997) used the name Rhinonycterinae Gray, 1866 for this group, but
        this has not been accepted by other authors. Although Rhinonycteridae (= Rhinony-
        cterina Gray, 1866) has priory over Hipposideridae as a family-group name, nobody
        other than Gray (1866) used the former name until it was resurrected by McKenna and
        Bell (1997). Miller (1907) used the name Hipposideridae for this group because *Hipposi-*
        *deros* Gray, 1831 has priority over *Rhinonycteris* Gray, 1866 (= *Rhinonicteris* Gray, 1847). All
        subsequent authors have followed Miller's (1907) usage of Hipposideridae/inae, and I
        believe that there is little to be gained by replacing it with an unknown name. I therefore
        retain Hipposideridae for this group pending action by the International Commission on
        Zoological Nomenclature. McKenna and Bell (1997) proposed a tribal classification for
        hipposiderids (which they treated as a subfamily), but many of the groups they defined
        have subsequently been shown to be paraphyletic (Bogdanowicz and Owen, 1998; Hand
        and Kirsch, 1998); accordingly, I do not recognize subfamilies or tribes within
        Hipposideridae at this time.

*Anthops* Thomas, 1888. Ann. Mag. Nat. Hist., ser. 6, 1:156.
    TYPE SPECIES: *Anthops ornatus* Thomas, 1888.

*Anthops ornatus* Thomas, 1888. Ann. Mag. Nat. Hist., ser. 6, 1:156.
    COMMON NAME: Flower-faced Bat.
    TYPE LOCALITY: Solomon Isls, Guadalcanal Isl, Aola.
    DISTRIBUTION: Solomon Isls, Bougainville Isl (Papua New Guinea).
    STATUS: IUCN 2003 and IUCN/SSC Action Plan (2001) – Vulnerable.
    COMMENTS: See Flannery (1995*b*) and Bonaccorso (1998).

*Asellia* Gray, 1838. Mag. Zool. Bot., 2:493.
    TYPE SPECIES: *Rhinolophus tridens* E. Geoffroy, 1813.

*Asellia patrizii* DeBeaux, 1931. Ann. Mus. Civ. Stor. Nat. Genova, 55:186.
    COMMON NAME: Patrizi's Trident Leaf-nosed Bat.
    TYPE LOCALITY: Ethiopia, Dancalia, Gaare.
    DISTRIBUTION: N Ethiopia, Saudi Arabia, and islands in the Red Sea.
    STATUS: IUCN 2003 and IUCN/SSC Action Plan (2001) – Vulnerable.
    COMMENTS: See Moeschler et al. (1990) and Horácek et al. (2000).

*Asellia tridens* (E. Geoffroy, 1813). Ann. Mus. Natn. Hist. Nat. Paris, 20:265.
    COMMON NAME: Geoffroy's Trident Leaf-nosed Bat.
    TYPE LOCALITY: Egypt, Qena, near Luxor.
    DISTRIBUTION: Pakistan and Afganistan to Israel and Jordan, Iran, Iraq, Syria, Saudi Arabia,
        Sinai peninsula (NE Egypt), Socotra (Yemen) and Oman; Egypt to Morocco including
        S Lybia, Tunisia, and Algeria; Senegal, Mauritania, Gambia, Burkina Faso, Mali, Niger,
        Chad, Sudan, S Somalia, and Eritrea; perhaps Zanzibar.
    STATUS: IUCN 2003 and IUCN/SSC Action Plan (2001) – Lower Risk (lc).
    SYNONYMS: *diluta* Anderson, 1881; *pallida* Laurent, 1937; *italosomalica* De Beaux, 1931;
        *murraiana* Anderson, 1881.
    COMMENTS: Reviewed in part by Owen and Qumiseyeh (1987), Harrison and Bates (1991),
        Bates and Harrison (1997), and Horácek et al. (2000). Subspecies are poorly delimited,
        see Owen and Qumiseyeh (1987) and Kock et al. (2002).

*Aselliscus* Tate, 1941. Am. Mus. Novit., 1140:2.
    TYPE SPECIES: *Rhinolophus tricuspidatus* Temminck, 1835.
    COMMENTS: Reviewed by Corbet and Hill (1992).

*Aselliscus stoliczkanus* (Dobson, 1871). Proc. Asiat. Soc. Bengal, p. 106.
    COMMON NAME: Stoliczka's Asian Trident Bat.
    TYPE LOCALITY: Malaysia, West, Penang Isl.
    DISTRIBUTION: Burma, S China, Thailand, Laos, Vietnam, W Malaysia.
    STATUS: IUCN 2003 and IUCN/SSC Action Plan (2001) – Lower Risk (lc).
    SYNONYMS: *trifidus* Peters, 1871; *wheeleri* Osgood, 1932.
    COMMENTS: See Sanborn (1952*b*).

*Aselliscus tricuspidatus* (Temminck, 1835). Monogr. Mamm., 2:20.
    COMMON NAME: Temminck's Asian Trident Bat.
    TYPE LOCALITY: Indonesia, Molucca Isls, Amboina.
    DISTRIBUTION: Molucca Isls, New Guinea, Bismarck Arch., Solomon Isls (including Santa
        Cruz Isls), Vanuatu (New Hebrides), adjacent small islands.
    STATUS: IUCN 2003 and IUCN/SSC Action Plan (2001) – Lower Risk (lc).
    SYNONYMS: *koopmani* Schlitter, Williams, and Hill, 1983; *novaeguinae* Schlitter, Williams,
        and Hill, 1983; *novehebridensis* Sanborn and Nicholson, 1950.
    COMMENTS: Revised by Schlitter et al. (1983); also see Hill (1983), Flannery (1995*a*, *b*), and
        Bonaccorso (1998).

*Cloeotis* Thomas, 1901. Ann. Mag. Nat. Hist., ser. 7, 8:28.
    TYPE SPECIES: *Cloetis percivali* Thomas, 1901.
    COMMENTS: Reviewed by Hill (1982).

*Cloeotis percivali* Thomas, 1901. Ann. Mag. Nat. Hist., ser. 7, 8:28.
    COMMON NAME: Percival's Short-eared Trident Bat.
    TYPE LOCALITY: Kenya, Coast Prov., Takaungu.
    DISTRIBUTION: Kenya, Tanzania, S Dem. Rep. Congo, Mozambique, Zambia, Zimbabwe,
        SE Botswana, Swaziland, NE South Africa.
    STATUS: IUCN 2003 and IUCN/SSC Action Plan (2001) – Lower Risk (nt).
    SYNONYMS: *australis* Roberts, 1917.

*Coelops* Blyth, 1848. J. Asiat. Soc. Bengal, 17:251.
    TYPE SPECIES: *Coelops frithii* Blyth, 1848.
    SYNONYMS: *Chilophylla* Miller, 1910.
    COMMENTS: Includes *Chilophylla*; see Ellerman and Morrison-Scott (1951), also Corbet and Hill
        (1992).

*Coelops frithii* Blyth, 1848. J. Asiat. Soc. Bengal, 17:251.
    COMMON NAME: East Asian Tailless Leaf-nosed Bat.
    TYPE LOCALITY: Bangladesh, Sunderbans.

DISTRIBUTION: Bangladesh and NE India to S China, Thailand, Burma, Laos, Vietnam, south to W Malaysia, Sumatra, and Java and Bali; Taiwan.

STATUS: IUCN 2003 and IUCN/SSC Action Plan (2001) – Lower Risk (lc).

SYNONYMS: *bernsteini* Peters, 1862; *formosanus* Horikawa, 1928; *inflatus* Miller, 1928; *sinicus* Allen, 1928.

COMMENTS: Reviewed in part by Bates and Harrison (1997). Malay material has not been allocated to subspecies. Sometimes spelled *frithi* (e.g., Koopman, 1993).

*Coelops robinsoni* Bonhote, 1908. J. Fed. Malay St. Mus., 3:4.

COMMON NAME: Malayan Tailless Leaf-nosed Bat.

TYPE LOCALITY: Malaya, Pahang, foot of Mt. Tahan.

DISTRIBUTION: W Malaysia, Borneo, Philippines. The record from Thailand is in error; see Hill (1983).

STATUS: IUCN 2003 and IUCN/SSC Action Plan (2001) – Lower Risk (nt) as *C. robinsoni*; Data Deficient as *C. hirsutus*.

SYNONYMS: *hirsutus* Miller, 1910.

COMMENTS: Includes *hirsutus*; see Hill (1972*a*, 1983) and Corbet and Hill (1992).

*Hipposideros* Gray, 1831. Zool. Misc., 1:37.

TYPE SPECIES: *Vespertilio speoris* Schneider, 1800.

SYNONYMS: *Chrysonycteris* Gray, 1866; *Cyclorhina* Peters, 1871; *Gloionycteris* Gray, 1866; *Phyllorhina* Bonaparte, 1837 [not Leach, 1816]; *Ptychorhina* Peters, 1871; *Rhinophylla* Gray, 1866 [not Peters, 1865]; *Speorifera* Gray, 1866; *Syndesmotis* Peters, 1871; *Syndesmotus* Waterhouse, 1902 [objective synonym of *Syndesmotis* Peters]; *Thyreorhina* Peters, 1871.

COMMENTS: Revised by Hill (1963*b*). The genus is apparently paraphyletic, but alternative phylogenies (e.g., those of Bogdanowicz and Owen [1998] and Hand and Kirsch [1998]) disagree about genus and species relationships. Accordingly, I have retained the traditional contents of *Hipposideros* pending a thorough revision. Species groups follow Koopman (1994) with some modifications.

*Hipposideros abae* J. A. Allen, 1917. Bull. Am. Mus. Nat. Hist., 37:432.

COMMON NAME: Aba Leaf-nosed Bat.

TYPE LOCALITY: Dem. Rep. Congo, Oriental, Aba.

DISTRIBUTION: Guinea-Bissau to SW Sudan and Uganda.

STATUS: IUCN 2003 and IUCN/SSC Action Plan (2001) – Lower Risk (lc).

COMMENTS: *speoris* species group.

*Hipposideros armiger* (Hodgson, 1835). J. Asiat. Soc. Bengal, 4:699.

COMMON NAME: Great Leaf-nosed Bat.

TYPE LOCALITY: Nepal.

DISTRIBUTION: N India, Nepal, Burma, S and SE China, Vietnam, Laos, Cambodia, Thailand, Malay Peninsula, Taiwan.

STATUS: IUCN 2003 and IUCN/SSC Action Plan (2001) – Lower Risk (lc).

SYNONYMS: *debilis* K. Andersen, 1906; *swinhoei* Peters, 1871; *fujianensis* Zhen, 1987; *terasensis* Kishida, 1924; *tranninhensis* Bourret, 1942.

COMMENTS: *armiger* species group. Includes *terasensis*, but see Yoshiyuki (1991*a*) and Pavlinov et al. (1995*b*). Reviewed in part by Kock (1996) Bates and Harrison (1997), Sinha (1999), and Hendrichsen et al. (2001*b*).

*Hipposideros ater* Templeton, 1848. J. Asiat. Soc. Bengal, 17:252.

COMMON NAME: Dusky Leaf-nosed Bat.

TYPE LOCALITY: Sri Lanka, Western Prov., Colombo.

DISTRIBUTION: Sri Lanka; India to W Malaysia, through Philippines, Indonesia, and New Guinea to N Queensland, N Northern Territory, and N Western Australia (Australia).

STATUS: IUCN 2003 and IUCN/SSC Action Plan (2001) – Lower Risk (lc).

SYNONYMS: *atratus* Kelaart, 1850; *amboinensis* Peters, 1871; *antricola* Peters, 1861; *aruensis* Gray, 1858; *albanensis* Gray, 1866; *gilberti* Johnson, 1959; *nicobarulae* Miller, 1902; *saevus* K. Andersen, 1918; *toala* Shamel, 1940.

COMMENTS: *bicolor* species group. Formerly included in *bicolor*, but see Hill (1963*b*). Does not include *wrighti* (here considered a subspecies of *cineraceus*); see Hill and Francis (1984). Reviewed in part by Bates and Harrison (1997); also see Flannery (1995*a*, *b*) and Bonaccorso (1998).

*Hipposideros beatus* K. Andersen, 1906. Ann. Mag. Nat. Hist., ser. 7, 17:279.
    COMMON NAME: Benito Leaf-nosed Bat.
    TYPE LOCALITY: Equatorial Guinea, Rio Muni, 15 mi. (24 km) from Benito River.
    DISTRIBUTION: Sierra Leone, Liberia, Ghana, Côte d'Ivoire, Nigeria, Cameroon, Rio Muni (Equatorial Guinea), Gabon, N Dem. Rep. Congo. A previous report of this species from Guinea-Bissau is in error (J. Fahr, pers. comm.).
    STATUS: IUCN 2003 and IUCN/SSC Action Plan (2001) – Lower Risk (lc).
    SYNONYMS: *maximus* Verschuren, 1957.
    COMMENTS: *bicolor* species group.

*Hipposideros bicolor* (Temminck, 1834). Tijdschr. Nat. Gesch. Physiol., 1:19.
    COMMON NAME: Bicolored Leaf-nosed Bat.
    TYPE LOCALITY: Indonesia, Java, Anjer coast. Lectotype designated and type locality restricted by Tate (1941).
    DISTRIBUTION: Laos, Vietnam, S Thailand, and Malaysia to Borneo and the Philippines; Java, Sumbawa, Seralu, Sumba, Savu, Roti, and Timor Isls (Indonesia), and adjacent small islands. A Cambodian record was rejected by Kock (2000*a*) and a Bali record was rejected by Kock and Dobat (2000); a Taiwan record is doubtful, see Corbet and Hill (1992).
    STATUS: IUCN 2003 and IUCN/SSC Action Plan (2001) – Lower Risk (lc).
    SYNONYMS: *javanicus* Sody, 1937; *atrox* K. Andersen, 1918; *erigens* Lawrence, 1939; *hilli* Kitchener, 1996 [in Kitchener et al., 1996]; *major* K. Andersen, 1918; *selatan* Kitchener, 1996 [in Kitchener et al., 1996]; *tanimbarensis* Kitchener, 1996 [in Kitchener et al., 1996].
    COMMENTS: *bicolor* species group. Includes *erigens*; see Hill (1963*b*). Does not include *pomona*, *gentilis*, or *macrobullatus*; see Hill et al. (1986). Reviewed in part by Hill (1983), Bergmans and van Bree (1986), Corbet and Hill (1992), Kitchener and Maharadatun-kamsi (1995), and Kitchener et al. (1996). Sumbawa specimens have not been allocated to subspecies; see Kitchener et al. (1996). Probably includes more than one species, including cryptic species distinguishable primarily by echolocation call frequencies (see Kingston et al., 2001).

*Hipposideros breviceps* Tate, 1941. Bull. Am. Mus. Nat. Hist., 78:358.
    COMMON NAME: Short-headed Leaf-nosed Bat.
    TYPE LOCALITY: Indonesia, Sumatra, Mentawai Isls, N Pagi Isl.
    DISTRIBUTION: Mentawai Isls (Indonesia).
    STATUS: IUCN 2003 and IUCN/SSC Action Plan (2001) – Vulnerable.
    COMMENTS: *bicolor* species group.

*Hipposideros caffer* (Sundevall, 1846). Öfv. Kongl. Svenska Vet.-Akad. Forhandl. Stockholm, 3(4):118.
    COMMON NAME: Sundevall's Leaf-nosed Bat.
    TYPE LOCALITY: South Africa, KwaZulu-Natal Prov., near Durban.
    DISTRIBUTION: SW Arabian Peninsula including Yemen; most of subsaharan Africa except the central forested region; Morocco; Zanzibar and Pemba.
    STATUS: IUCN 2003 and IUCN/SSC Action Plan (2001) – Lower Risk (lc).
    SYNONYMS: *aurantiaca* De Beaux, 1924; *bicornis* Heuglin, 1861; *gracilis* Peters, 1825; *angolensis* Seabra, 1898; *nanus* J. A. Allen, 1917; *tephrus* Cabrera, 1906; *braima* Monard, 1939.
    COMMENTS: *bicolor* species group. Includes *tephrus*; see Hayman and Hill (1971). Reviewed in part by Harrison and Bates (1991) and Horácek et al. (2000). See Taylor (2000*a*) for distribution map. Subspecies limits are somewhat unclear, and it is possible that this complex includes more than one species.

*Hipposideros calcaratus* (Dobson, 1877). Proc. Zool. Soc. Lond., 1877:122.
    COMMON NAME: Spurred Leaf-nosed Bat.

TYPE LOCALITY: Papua New Guinea, Bismarck Archipelago, Duke of York Isl.

DISTRIBUTION: New Guinea, Bismarck Arch., Solomon Isls, adjacent small islands.

STATUS: IUCN 2003 and IUCN/SSC Action Plan (2001) – Lower Risk (lc).

SYNONYMS: *cupidus* K. Andersen, 1918.

COMMENTS: *bicolor* species group. Includes *cupidus*; see Smith and Hill (1981). Does not include *maggietaylorae*; see Smith and Hill (1981). Also see Flannery (1995*a*, *b*) and Bonaccorso (1998).

*Hipposideros camerunensis* Eisentraut, 1956. Zool. Jahrb. Abt. Syst. Oekol. Geogr. Tiere, 84:526.

COMMON NAME: Cameroon Leaf-nosed Bat.

TYPE LOCALITY: Cameroon, near Buea.

DISTRIBUTION: Cameroon, E Dem. Rep. Congo, W Kenya.

STATUS: IUCN 2003 and IUCN/SSC Action Plan (2001) – Lower Risk (nt).

COMMENTS: *cyclops* species group.

*Hipposideros cervinus* (Gould, 1854). Mamm. Austr., 3: pl. 34.

COMMON NAME: Fawn-colored Leaf-nosed Bat.

TYPE LOCALITY: Australia, Queensland, Cape York and Albany Isl.

DISTRIBUTION: W Malaysia, Sumatra, and Mindanao (Philippines) to the Mollucca Isls, Vanuatu, and NE Australia. Specimens from Mansuar Isl (Prov. of Papua, Indonesia) may represent *cyclotis* (Meinig, 2002).

STATUS: IUCN 2003 and IUCN/SSC Action Plan (2001) – Lower Risk (lc).

SYNONYMS: *celebensis* Sody, 1936; *batchianus* Matschie, 1901; *labuanensis* Tomes, 1859; *schneidersi* Thomas, 1904; *misoriensis* Peters, 1906.

COMMENTS: *bicolor* species group. Distinct from *galeritus*; see Flannery (1995*a*, *b*). Also see Jenkins and Hill (1981) and Hill (1983).

*Hipposideros cineraceus* Blyth, 1853. J. Asiat. Soc. Bengal, 22:410.

COMMON NAME: Ashy Leaf-nosed Bat.

TYPE LOCALITY: Pakistan, Punjab, Salt Range, near Pind Dadan Khan.

DISTRIBUTION: Pakistan and India to Burma, Thailand, Laos, Vietnam, Sumatra and Borneo; adjacent small islands including Kangean Isls (Indonesia); probably the Philippines.

STATUS: IUCN 2003 and IUCN/SSC Action Plan (2001) – Lower Risk (lc).

SYNONYMS: *micropus* Peters, 1872; *wrighti* Taylor, 1934.

COMMENTS: *bicolor* species group. Does not includes *durgadasi*; see Topál (1975), Khajuria (1982), Corbet and Hill (1992), and Pavlinov et al. (1995*b*). Includes *wrighti*, see Hill and Francis (1984). Reviewed in part by Bates and Harrison (1997); also see Bonaccorso (1998).

*Hipposideros commersoni* (E. Geoffroy, 1813). Ann. Mus. Natn. Hist. Nat. Paris, 20:263.

COMMON NAME: Commerson's Leaf-nosed Bat.

TYPE LOCALITY: Madagascar, Fort Dauphin (= Tolagnaro).

DISTRIBUTION: Madagascar.

STATUS: IUCN 2003 and IUCN/SSC Action Plan (2001) – Lower Risk (lc) (including *gigas*, *vittata*, and *thomensis*, which are now regarded as distinct species).

COMMENTS: *commersoni* species group. Does not include *gigas*, *vittatus* (including *marungensis)* or *thomensis*, which are now recognized as distinct species based on differences in morphology and echolocation calls (J. Fahr and D. Kock, pers. comm.; D. Lunde, pers. comm.; McWilliam, 1982; Pye, 1972). Reviewed by Peterson et al. (1995).

*Hipposideros coronatus* (Peters, 1871). Monatsb. K. Preuss. Akad. Wiss. Berlin, 1871:327.

COMMON NAME: Large Mindanao Leaf-nosed Bat.

TYPE LOCALITY: Philippines, Mindanao, Surigao, Mainit.

DISTRIBUTION: NE Mindanao (Philippines).

STATUS: IUCN 2003 and IUCN/SSC Action Plan (2001) – Lower Risk (nt).

COMMENTS: *bicolor* species group. See Ingle and Heaney (1992).

*Hipposideros corynophyllus* Hill, 1985. Mammalia, 49:527.

COMMON NAME: Telefomin Leaf-nosed Bat.

TYPE LOCALITY: Papua New Guinea, W Sepik, 3 km ENE Telefomin, 1,800 m.
DISTRIBUTION: C New Guinea.
STATUS: IUCN 2003 and IUCN/SSC Action Plan (2001) – Vulnerable.
COMMENTS: *cyclops* species group. Reviewed by Flannery and Colgan (1993); also see
Flannery (1995*a*) and Bonaccorso (1998).

*Hipposideros coxi* Shelford, 1901. Ann. Mag. Nat. Hist., ser. 7, 8:113.
COMMON NAME: Cox's Leaf-nosed Bat.
TYPE LOCALITY: Malaysia, Borneo, Sarawak, Mt. Penrisen, 4,200 ft. (1,280 m).
DISTRIBUTION: Sarawak (Borneo, Malaysian part).
STATUS: IUCN 2003 and IUCN/SSC Action Plan (2001) – Vulnerable.
COMMENTS: *bicolor* species group.

*Hipposideros crumeniferus* (Lesueur and Petit, 1807). *In* Peron, Voyage Decouv. Terres Australes,
Atlas, pl. 35.
COMMON NAME: Timor Leaf-nosed Bat.
TYPE LOCALITY: Indonesia, Timor.
DISTRIBUTION: Timor (Indonesia).
STATUS: IUCN 2003 and IUCN/SSC Action Plan (2001) – Data Deficient.
COMMENTS: *bicolor* species group. Based on palate only; not certainly determinable; see
Laurie and Hill (1954) and Hill (1963*b*). It may represent *cervinus*, see Corbet and Hill
(1992) and Pavlinov et al. (1995*b*).

*Hipposideros curtus* G. M. Allen, 1921. Rev. Zool. Afr., 9:194.
COMMON NAME: Short-tailed Leaf-nosed Bat.
TYPE LOCALITY: Cameroon, Sakbayeme.
DISTRIBUTION: Cameroon, Bioko (Equatorial Guinea).
STATUS: IUCN 2003 and IUCN/SSC Action Plan (2001) – Lower Risk (nt).
SYNONYMS: *sandersoni* Sanderson, 1937.
COMMENTS: *bicolor* species group. Includes *sandersoni*; see Hill (1963*b*).

*Hipposideros cyclops* (Temminck, 1853). Esquisses Zool. sur la Côte de Guine, p. 75.
COMMON NAME: Cyclops Leaf-nosed Bat.
TYPE LOCALITY: Ghana, Boutry River.
DISTRIBUTION: Kenya and S Sudan to Senegal and Guinea-Bissau; Bioko (Equatorial Guinea).
STATUS: IUCN 2003 and IUCN/SSC Action Plan (2001) – Lower Risk (lc).
SYNONYMS: *langi* J. A. Allen, 1917; *micaceus* de Winton, 1897.
COMMENTS: *cyclops* species group.

*Hipposideros demissus* K. Andersen, 1909. Ann. Mag. Nat. Hist., ser. 8, 3:268.
COMMON NAME: Makira Leaf-nosed Bat.
TYPE LOCALITY: East Solomon Isls, San Cristoval (= San Cristobal) Isl, Yanuta.
DISTRIBUTION: San Cristobal Isl (Solomon Isls).
STATUS: IUCN 2003 and IUCN/SSC Action Plan (2001) – Vulnerable.
COMMENTS: *diadema* species group. Formerly considered a subspecies of *diadema*, but
apparently distinct; see Kitchener et al. (1992*b*) and Flannery (1995*b*).

*Hipposideros diadema* (E. Geoffroy, 1813). Ann. Mus. Natn. Hist. Nat. Paris, 20:263.
COMMON NAME: Diadem Leaf-nosed Bat.
TYPE LOCALITY: Indonesia, Lesser Sunda Isls, Timor Isl.
DISTRIBUTION: Burma and Vietnam through Thailand, Laos, W Malaysia and Indonesia
(including Sumatra, Borneo, and Bali) to New Guinea, Bismarck Arch., Solomon Isls
and NE Australia; Philippines; Nicobar Isls. Reports of this species from Cambodia
cannot be confirmed (Kock, 2000*a*).
STATUS: IUCN 2003 and IUCN/SSC Action Plan (2001) – Lower Risk (lc).
SYNONYMS: **ceramensis** Laurie and Hill, 1954; **custos** K. Andersen, 1918; **enganus**
K. Andersen, 1907; **euotis** K. Andersen, 1905; **griseus** Meyen, 1883; *anderseni* Taylor,
1934; *pullatus* K. Andersen, 1905; **masoni** Dobson, 1872; **mirandus** Thomas, 1914;
**natunensis** Chasen, 1940; **nicobarensis** Dobson, 1871; **nobilis** Horsfield, 1823; *vicarius*

K. Andersen, 1905; *oceanitis* K. Andersen, 1905; *malaitensis* Phillips, 1967; **reginae** Troughton, 1937; **speculator** K. Andersen, 1918; **trobrius** Troughton, 1937.
COMMENTS: *diadema* species group. Many subspecies are of dubious validity, and several island populations have not been assigned to subspecies. Reviewed in part by Laurie and Hill (1954) and Kitchener et al. (1992*b*). Does not include *demissus* and *inornatus*; see Kitchener et al. (1992*b*). Also see Flannery (1995*a*, *b*) and Bonaccorso (1998). May include *ornatus*, a name listed as a synonym of *diadema* by Koopman (1993) but which I have been unable to trace (although it may be a *lapsus* for *inornatus*).

*Hipposideros dinops* K. Andersen, 1905. Ann. Mag. Nat. Hist., ser. 7, 16:502.
COMMON NAME: Fierce Leaf-nosed Bat.
TYPE LOCALITY: Solomon Isls, New Georgia Group, Rubiana Isl.
DISTRIBUTION: Solomon Isls; Bougainville Isl (Papua New Guinea).
STATUS: IUCN 2003 and IUCN/SSC Action Plan (2001) – Lower Risk (nt).
COMMENTS: *diadema* species group. Does not include *pelingensis*; see discussion under that species. See Flannery (1995*b*) and Bonaccorso (1998).

*Hipposideros doriae* (Peters, 1871). Monatsb. K. Preuss. Akad. Wiss. Berlin, 1871:326.
COMMON NAME: Bornean Leaf-nosed Bat.
TYPE LOCALITY: Malaysia, Sarawak.
DISTRIBUTION: W Malaysia, Sarawak and Sabah (Malaysia), Borneo and Sumatra (Indonesia).
STATUS: IUCN 2003 and IUCN/SSC Action Plan (2001) – Lower Risk (lc) as *H. sabanus*; Data Deficient as *H. doriae*.
SYNONYMS: *sabanus* Thomas, 1898.
COMMENTS: *bicolor* species group. Includes *sabanus*, see Hill (1963*b*) and Benda (2000). Lectotype designated by Benda (2000).

*Hipposideros durgadasi* Khajuria, 1970. Mammalia, 64:623.
COMMON NAME: Durga Das's Leaf-nosed Bat.
TYPE LOCALITY: India, Madhya Pradesh, Jabalpur Dist., near Katungi village.
DISTRIBUTION: C India.
STATUS: IUCN 2003 and IUCN/SSC Action Plan (2001) – Vulnerable.
COMMENTS: *bicolor* species group. Formerly included in *cineraceus*, but see see Topál (1975), Khajuria (1982), and Corbet and Hill (1992). Reviewed by Bates and Harrison (1997).

*Hipposideros dyacorum* Thomas, 1902. Ann. Mag. Nat. Hist., ser. 7, 9:271.
COMMON NAME: Dayak Leaf-nosed Bat.
TYPE LOCALITY: Malaysia, Sarawak, Baram, Mt. Mulu.
DISTRIBUTION: Borneo (including Sarawak, Malaysia), Peninsular Thailand.
STATUS: IUCN 2003 and IUCN/SSC Action Plan (2001) – Lower Risk (lc).
COMMENTS: *bicolor* species group.

*Hipposideros edwardshilli* Flannery and Colgan, 1993. Rec. Aust. Mus., 45:45.
COMMON NAME: Hill's Leaf-nosed Bat.
TYPE LOCALITY: Papua New Guinea, W Sepik, Bewani Mtns., Imonda Sta.
DISTRIBUTION: NW Papua New Guinea.
STATUS: IUCN 2003 and IUCN/SSC Action Plan (2001) – Lower Risk (nt).
COMMENTS: *cyclops* species group. See Flannery (1995*a*) and Bonaccorso (1998).

*Hipposideros fuliginosus* (Temminck, 1853). Esquisses Zool. sur la Côte de Guine, p. 77.
COMMON NAME: Sooty Leaf-nosed Bat.
TYPE LOCALITY: Ghana.
DISTRIBUTION: Sierra Leone and Liberia to Dem. Rep. Congo. Ethiopian records represent another, possibly undescribed, species (J. Fahr, pers. comm.).
STATUS: IUCN 2003 and IUCN/SSC Action Plan (2001) – Lower Risk (nt).
COMMENTS: *bicolor* species group.

*Hipposideros fulvus* Gray, 1838. Mag. Zool. Bot., 2:492.
COMMON NAME: Fulvus Leaf-nosed Bat.
TYPE LOCALITY: India, Karnatika, Dharwar.
DISTRIBUTION: Afganistan, India, Sri Lanka, Pakistan to Vietnam.

STATUS: IUCN 2003 and IUCN/SSC Action Plan (2001) – Lower Risk (lc).
SYNONYMS: *atra* Fitzinger, 1870 [not Templeton, 1848]; *aurita* Tomes, 1859; *fulgens* Elliot, 1839; *murinus* Gray, 1838; **pallidus** K. Andersen, 1918.
COMMENTS: *bicolor* species group. Reviewed by Bates and Harrison (1997).

*Hipposideros galeritus* Cantor, 1846. J. Asiat. Soc. Bengal, 15:183.
COMMON NAME: Cantor's Leaf-nosed Bat.
TYPE LOCALITY: Malaysia, Penang Isl.
DISTRIBUTION: Sri Lanka and India through SE Asia (including Burma, Thailand, and Peninsular Malaysia) to Java and Borneo; Sanana Isl (Sula Group, Moluccas Isls). A record from Bali is possibly erroneous; see Kock and Dobat (2000).
STATUS: IUCN 2003 and IUCN/SSC Action Plan (2001) – Lower Risk (lc).
SYNONYMS: **brachyotis** Dobson, 1874; **insolens** Lyon, 1911; **longicauda** Peters, 1861.
COMMENTS: *bicolor* species group. Includes *longicauda*; see Hill (1963*b*). Formerly included *cervinus*; but see Jenkins and Hill (1981). Reviewed in part by Bates and Harrison (1997); also see Flannery (1995*b*).

*Hipposideros gigas* (Wagner, 1845). Arch. Naturgesch., 11(1); 148.
COMMON NAME: Giant Leaf-nosed Bat.
TYPE LOCALITY: Angola, Benguela.
DISTRIBUTION: Kenya, Tanzania, Angola, Central African Republic, Uganda, Dem. Rep. Congo, Gabon, Equatorial Guinea (incl. Bioko), Cameroon, Nigeria and west to Senegal. The range of this taxon may be more extensive and is currently under review (J. Fahr, pers. comm.)
STATUS: IUCN 2003 – Not evaluated; not considered in IUCN/SSC Action Plan (2001).
SYNONYMS: *gambiensis* K. Andersen, 1906; *niangarae* J. A. Allen, 1917.
COMMENTS: *commersoni* species group. Formerly included in *commersoni*, but clearly distinct based on differences in morphology and echolocation calls (J. Fahr and D. Kock, pers. comm.; D. Lunde, pers. comm.; McWilliam, 1982; Pye, 1972). Reviewed in part by Peterson et al. (1995). Some West African specimens identified as *gigas* may represent *vittatus* (J. Fahr, pers. comm.).

*Hipposideros grandis* G. M. Allen, 1936. Rec. Indian Mus., 38:345.
COMMON NAME: Grand Leaf-nosed Bat.
TYPE LOCALITY: Burma, Upper Chindwin, Akanti, 300 ft. (91 m).
DISTRIBUTION: Burma, Thailand, and Vietnam.
STATUS: IUCN 2003 – Not evaluated; not considered in IUCN/SSC Action Plan (2001).
COMMENTS: *larvatus* species group. Distinct from *larvatus*; see Kitchener and Maryanto (1993*a*).

*Hipposideros halophyllus* Hill and Yenbutra, 1984. Bull. Brit. Mus. (Nat. Hist.) Zool., 47:77.
COMMON NAME: Thailand Leaf-nosed Bat.
TYPE LOCALITY: Thailand, Lop Buri, Tha Woong, Khao Sa Moa Khan.
DISTRIBUTION: Thailand.
STATUS: IUCN 2003 and IUCN/SSC Action Plan (2001) – Lower Risk (nt).
COMMENTS: *bicolor* species group.

*Hipposideros hypophyllus* Kock and Bhat, 1994. Senk. Biol., 73:26.
COMMON NAME: Leafletted Leaf-nosed Bat.
TYPE LOCALITY: India, Karnataka, Bangalore Region, 15 km E Kolar Town, Hanumanhalli Village (13°09'N, 78°07'E).
DISTRIBUTION: S India.
STATUS: IUCN 2003 and IUCN/SSC Action Plan (2001) – Vulnerable.
COMMENTS: *bicolor* species group. Reviewed by Bates and Harrison (1997).

*Hipposideros inexpectatus* Laurie and Hill, 1954. List of land mammals of New Guinea, Celebes, and adjacent islands, p. 60.
COMMON NAME: Crested Leaf-nosed Bat.
TYPE LOCALITY: Indonesia, N Sulawesi, Poso (= Posso).
DISTRIBUTION: N Sulawesi (Indonesia).

STATUS: IUCN 2003 and IUCN/SSC Action Plan (2001) – Vulnerable.
COMMENTS: *diadema* species group.

*Hipposideros inornatus* McKean, 1970. West. Aust. Nat., 11(6):138.
COMMON NAME: McKean's Leaf-nosed Bat.
TYPE LOCALITY: Australia, Northern Territory, 55 mi. (85 km) S of Oenpelli, Deaf Adder
    Creek, where it emerges from the Arnhem Land Plateau, 13°06'S, 132°56'E.
DISTRIBUTION: Northern Territory (Australia).
STATUS: IUCN 2003 – Not evaluated; not considered in IUCN/SSC Action Plan (2001).
COMMENTS: *diadema* species group. Formerly considered a subspecies of *diadema*, but
    apparently distinct; see Kitchener et al. (1992*b*). Also see Churchill (1998).

*Hipposideros jonesi* Hayman, 1947. Ann. Mag. Nat. Hist., ser. 11, 14:71.
COMMON NAME: Jones's Leaf-nosed Bat.
TYPE LOCALITY: Sierra Leone, Makeni.
DISTRIBUTION: Sierra Leone and Guinea to Mali, Burkina Faso and Nigeria.
STATUS: IUCN 2003 and IUCN/SSC Action Plan (2001) – Lower Risk (nt).
COMMENTS: *bicolor* species group.

*Hipposideros lamottei* Brosset, 1985. Mammalia, 48:548.
COMMON NAME: Lamotte's Leaf-nosed Bat.
TYPE LOCALITY: Guinea, Mt. Nimba, Pierre Richaud.
DISTRIBUTION: Mt. Nimba on Guinea-Liberia border.
STATUS: IUCN 2003 and IUCN/SSC Action Plan (2001) – Data Deficient.
COMMENTS: *bicolor* species group. Distinction from *ruber* is not entirely clear.

*Hipposideros lankadiva* Kelaart, 1850. J. Sri Lanka Branch Asiat. Soc., 2(2):216.
COMMON NAME: Indian Leaf-nosed Bat.
TYPE LOCALITY: Sri Lanka, Kandy.
DISTRIBUTION: Sri Lanka, S and C India.
STATUS: IUCN 2003 and IUCN/SSC Action Plan (2001) – Data Deficient as *H. schistaceus*;
    Lower Risk (lc) as *H. lankadiva*.
SYNONYMS: *indus* K. Andersen, 1918; *mixtus* K. Andersen, 1918; *schistaceus* K. Andersen,
    1918; *unitus* K. Andersen, 1918.
COMMENTS: *diadema* species group. Includes *schistaceus*; see Bates and Harrison (1997) and
    Srivinasulu and Srivinasulu (2001). Multiple subspecies have been recognized by some
    authors, but these do not appear justified; see Sinha (1999), although also see
    Srinivasulu and Srinivasulu (2001).

*Hipposideros larvatus* (Horsfield, 1823). Zool. Res. Java, 6: *Rhinolophus larvatus*, pl. and 10 unno. pp.
COMMON NAME: Intermediate Leaf-nosed Bat.
TYPE LOCALITY: Indonesia, Java.
DISTRIBUTION: N and E India and Bangladesh; Yunnan, Kwangsi and Hainan (China);
    Burma, Thailand, Cambodia, Laos, and Vietnam; W Malaysia to Sumatra, Java,
    Borneo, and adjacent small islands including Kangean Isls (Indonesia).
STATUS: IUCN 2003 and IUCN/SSC Action Plan (2001) – Lower Risk (lc).
SYNONYMS: *deformis* Horsfield, 1823; *insignis* Horsfield, 1823; *vulgaris* Horsfield, 1823;
    **barbensis** Miller, 1900; **leptophyllus** Dobson, 1874; **neglectus** Sody, 1936; **poutensis**
    Allen, 1906.
COMMENTS: *larvatus* species group. Does not include *grandis* and *sumbae*; see Kitchner and
    Maryantu (1993), who revised this complex. See also Hill (1963) and Sinha (1999).
    Subspecies limits and validity are uncertain. Does not include *alongensis*, see Topál
    (1993).

*Hipposideros lekaguli* Thonglongya and Hill, 1974. Mammalia, 38:286.
COMMON NAME: Large Asian Leaf-nosed Bat.
TYPE LOCALITY: Thailand, Saraburi, Kaeng Khoi, Phu Nam Tok Tak Kwang, (c 14°34'N,
    101°09'E).
DISTRIBUTION: Thailand; peninsular Malaysia; Luzon (Philippines).
STATUS: IUCN 2003 and IUCN/SSC Action Plan (2001) – Lower Risk (nt).

COMMENTS: *diadema* species group.

*Hipposideros lylei* Thomas, 1913. Ann. Mag. Nat. Hist., ser. 8, 12:88.
    COMMON NAME: Shield-faced Leaf-nosed Bat.
    TYPE LOCALITY: Thailand, 50 mi. (80 km) N Chiengmai (= Chiang Mai), Chiengdao Cave,
        350 m.
    DISTRIBUTION: Burma, Vietnam, Thailand, W Malaysia.
    STATUS: IUCN 2003 and IUCN/SSC Action Plan (2001) – Lower Risk (nt).
    COMMENTS: *pratti* species group. Reviewed by Hendrichsen et al. (2001*b*) and Robinson et al.
        (2003).

*Hipposideros macrobullatus* Tate, 1941. Bull Am. Mus. Nat. Hist., 78:357.
    COMMON NAME: Big-eared Leaf-nosed Bat.
    TYPE LOCALITY: Indonesia, Sulawesi, Talassa (near Maros).
    DISTRIBUTION: Sulawesi, Seram (Molucca Isls) and Kangean Isls (Indonesia).
    STATUS: IUCN 2003 and IUCN/SSC Action Plan (2001) – Lower Risk (nt).
    COMMENTS: *bicolor* species group. Formerly included in *bicolor*, but see Hill et al. (1986) and
        Bergmans and van Bree (1986). May be conspecific with *pomona*; see Corbet and Hill
        (1992).

*Hipposideros madurae* Kitchener and Maryanto, 1993. Rec. West. Aust. Mus., 16:132.
    COMMON NAME: Maduran Leaf-nosed Bat.
    TYPE LOCALITY: Indonesia, Madura Isl, Pulau, Sampang.
    DISTRIBUTION: Madura Isl, C Java (Indonesia).
    STATUS: IUCN 2003 and IUCN/SSC Action Plan (2001) – Lower Risk (nt).
    SYNONYMS: *jenningsi* Kitchener and Maryanto, 1993.
    COMMENTS: *larvatus* species group.

*Hipposideros maggietaylorae* Smith and Hill, 1981. Los Angeles Cty. Mus. Contrib. Sci., 331:9.
    COMMON NAME: Maggie Taylor's Leaf-nosed Bat.
    TYPE LOCALITY: Papua New Guinea, Bismarck Arch., New Ireland, 1.3 km S, 3 km E,
        Lakuramau Plantation.
    DISTRIBUTION: New Guinea (possibly extending as far west as Waigeo Isl.), Bismarck Arch.
    STATUS: IUCN 2003 and IUCN/SSC Action Plan (2001) – Lower Risk (lc).
    SYNONYMS: *erroris* Smith and Hill, 1981.
    COMMENTS: *bicolor* species group. Formerly confused with *calcaratus*; see Smith and Hill
        (1981). Also see Flannery (1995*a*, *b*), Bonaccorso (1998), and Meinig (2002).

*Hipposideros marisae* Aellen, 1954. Rev. Suisse Zool., 61:474.
    COMMON NAME: Aellen's Leaf-nosed Bat.
    TYPE LOCALITY: Côte d'Ivoire, Duékoué, White Leopard Rock.
    DISTRIBUTION: Côte d'Ivoire, Liberia, Guinea.
    STATUS: IUCN 2003 and IUCN/SSC Action Plan (2001) – Vulnerable.
    COMMENTS: *bicolor* species group.

*Hipposideros megalotis* (Heuglin, 1862). Nova Acta Acad. Caes. Leop.-Carol., Halle, 29(8):4, 8.
    COMMON NAME: Large-eared Leaf-nosed Bat.
    TYPE LOCALITY: Eritrea, Bogos Land, Keren.
    DISTRIBUTION: Saudi Arabia, Ethiopia, Eritrea, Djibouti, and Kenya. A record from Somalia is
        erroneous (M. Happold, pers. comm.)
    STATUS: IUCN 2003 and IUCN/SSC Action Plan (2001) – Lower Risk (nt).
    COMMENTS: *megalotis* species group. Sometimes placed in the subgenus *Syndesmotis*; see
        Legendre (1982) and Gaucher and Brosset (1990). Also see Hill (1963*b*).

*Hipposideros muscinus* (Thomas and Doria, 1886). Ann. Mus. Civ. Stor. Nat. Genova, 4:201.
    COMMON NAME: Fly River Leaf-nosed Bat.
    TYPE LOCALITY: Papua New Guinea, Western Prov., Fly River.
    DISTRIBUTION: New Guinea.
    STATUS: IUCN 2003 and IUCN/SSC Action Plan (2001) – Vulnerable.
    COMMENTS: *cyclops* species group. See Flannery (1995*a*) and Bonaccorso (1998).

*Hipposideros nequam* K. Andersen, 1918. Ann. Mag. Nat. Hist., ser. 9, 2:380.
    COMMON NAME: Malayan Leaf-nosed Bat.
    TYPE LOCALITY: Malaysia, Selangor, Klang.
    DISTRIBUTION: Known only from the type locality.
    STATUS: IUCN 2003 and IUCN/SSC Action Plan (2001) – Critically Endangered.
    COMMENTS: *bicolor* species group. Known only from the holotype; see Hill (1963*b*).

*Hipposideros obscurus* (Peters, 1861). Monatsb. K. Preuss. Akad. Wiss. Berlin, 1861:707.
    COMMON NAME: Philippine Forest Leaf-nosed Bat.
    TYPE LOCALITY: Philippines, Luzon, Camarines, Paracale.
    DISTRIBUTION: Philippines except Palawan region.
    STATUS: IUCN 2003 and IUCN/SSC Action Plan (2001) – Lower Risk (nt).
    COMMENTS: *bicolor* species group.

*Hipposideros orbiculus* Francis, Kock, and Habersetzer, 1999. Senkenbergiana Biol., 79:259.
    COMMON NAME: Orbiculus Leaf-nosed Bat.
    TYPE LOCALITY: Indonesia, Sumatra, Sumatera Barat, SE Kota Baru, Abai Siat, 01°02′S, 101°43′E.
    DISTRIBUTION: Sumatra (Indonesia); Peninsular Malaysia.
    STATUS: IUCN 2003 – Not evaluated; not considered in IUCN/SSC Action Plan (2001).
    COMMENTS: *bicolor* species group.

*Hipposideros papua* (Thomas and Doria, 1886). Ann. Mus. Civ. Stor. Nat. Genova, 4:204.
    COMMON NAME: Biak Leaf-nosed Bat.
    TYPE LOCALITY: Indonesia, Prov. of Papua, Tjenderawasih Div. (= Geelvinck Bay), Misori Isl (= Biak Isl = Schouten Isl), Korido.
    DISTRIBUTION: Biak and Numfoor Isls, W New Guinea, and N Molucca Isls.
    STATUS: IUCN 2003 and IUCN/SSC Action Plan (2001) – Vulnerable.
    COMMENTS: *bicolor* species group. See Hill and Rozendaal (1989) and Flannery (1995*a*, *b*).

*Hipposideros pelingensis* Shamel, 1940. J. Mammal., 21:353.
    COMMON NAME: Peleng Leaf-nosed Bat.
    TYPE LOCALITY: Indonesia, Peling (= Peleng) Isl east of Sulawesi.
    DISTRIBUTION: Peleng Isl and Sulawesi (Indonesia).
    STATUS: IUCN 2003 – Not evaluated; not considered in IUCN/SSC Action Plan (2001).
    COMMENTS: *diadema* species group. Hill (1963*b*, 1983) followed Tate (1941) in treating *pelingensis* as a subspecies of *dinops*, but these taxa (which are separated by 1,800 km with no known populations on the many islands in between) are diagnosably distinct; see Flannery (1995*b*). *H. pelingensis* is therefore provisionally treated here as a separate species pending further study.

*Hipposideros pomona* K. Andersen, 1918. Ann. Mag. Nat. Hist., ser. 9, 2:380, 381.
    COMMON NAME: Pomona Leaf-nosed Bat.
    TYPE LOCALITY: India, Mysore, N Coorg, Haleri (a few miles N of Mercara, Coorg Dist., Karnataka).
    DISTRIBUTION: Bangladesh and India to Burma, Thailand, Laos, Cambodia, Vietnam, S China and W Malaysia.
    STATUS: IUCN 2003 and IUCN/SSC Action Plan (2001) – Lower Risk (lc).
    SYNONYMS: *gentilis* K. Andersen, 1918; *sinensis* K. Andersen, 1918.
    COMMENTS: *bicolor* species group. Formerly included in *bicolor* but see Hill et al (1986). May be conspecific with *macrobullatus*; see Corbet and Hill (1992). Some specimens from peninsular India previously referred to this species were subsequently removed to form the type series of *hypophyllus*. Reviewed in part by Bates and Harrison (1997) and Hendrichsen et al. (2001*b*).

*Hipposideros pratti* Thomas, 1891. Ann. Mag. Nat. Hist., ser. 6, 7:527.
    COMMON NAME: Pratt's Leaf-nosed Bat.
    TYPE LOCALITY: China, Szechwan, Kiatingfu.
    DISTRIBUTION: S China, Burma, Thailand, Vietnam, W Malaysia.
    STATUS: IUCN 2003 and IUCN/SSC Action Plan (2001) – Lower Risk (nt).

COMMENTS: *pratti* species group. Reviewed by Hendrichsen et al. (2001*b*) and Robinson et al. (2003).

*Hipposideros pygmaeus* (Waterhouse, 1843). Proc. Zool. Soc. Lond., 1843:67.
COMMON NAME: Philippine Pygmy Leaf-nosed Bat.
TYPE LOCALITY: Philippines.
DISTRIBUTION: Philippines except Palawan region.
STATUS: IUCN 2003 and IUCN/SSC Action Plan (2001) – Lower Risk (nt).
COMMENTS: *bicolor* species group.

*Hipposideros ridleyi* Robinson and Kloss, 1911. J. Fed. Malay St. Mus., 4:241.
COMMON NAME: Ridley's Leaf-nosed Bat.
TYPE LOCALITY: Singapore, Botanic Gardens.
DISTRIBUTION: Penninsular Malaysia, Singapore, N Borneo.
STATUS: U.S. ESA – Endangered. IUCN 2003 and IUCN/SSC Action Plan (2001) – Vulnerable.
COMMENTS: *bicolor* species group. Reviewed by Francis et al. (1999*a*).

*Hipposideros rotalis* Francis, Kock, and Habersetzer, 1999. Senkenbergiana Biol., 79:266.
COMMON NAME: Laotian Leaf-nosed Bat.
TYPE LOCALITY: Laos, Bolikhamxai Prov., Nam (River) Kading, Ban Keng Bit, 18°15′N, 104°34′E.
DISTRIBUTION: Laos.
STATUS: IUCN 2003 – Not evaluated; not considered in IUCN/SSC Action Plan (2001).
COMMENTS: *bicolor* species group.

*Hipposideros ruber* (Noack, 1893). Zool. Jahrb. Abt. Syst. Oekol. Geogr. Tiere, 7:586.
COMMON NAME: Noack's Leaf-nosed Bat.
TYPE LOCALITY: Tanzania, Eastern Province, Ngerengere River.
DISTRIBUTION: Senegal and Gambia to Ethiopia, south to Angola, Zambia, Malawi, and Mozambique; Bioko (Equatorial Guinea); São Tomé and Principe.
STATUS: IUCN 2003 and IUCN/SSC Action Plan (2001) – Lower Risk (lc).
SYNONYMS: *centralis* K. Andersen, 1906; *niapu* J. A. Allen, 1917; **guineensis** K. Andersen, 1906.
COMMENTS: *bicolor* species group. Included in *caffer* by Hill (1963*b*), but clearly distinct; see Lawrence (1964), Kock (1969*a*), Heller (1992), Jones et al. (1993), and Cotterill (2001*f*). Subspecies limits are problematic, and it is possible that this complex includes more than one species.

*Hipposideros scutinares* Robinson, Jenkins, Francis, and Fulford, 2003. Acta. Chiropt., 5:33.
COMMON NAME: Shield-nosed Leaf-nosed Bat.
TYPE LOCALITY: Laos, Khammouan Limestone NBCA, Bolikhamsai Province, along the upper Nam Hinboun, Ban Khankeo, 17°58′N, 104°49′E.
DISTRIBUTION: Laos, Vietnam.
STATUS: IUCN 2003 – Not evaluated; not considered in IUCN/SSC Action Plan (2001).
COMMENTS: *pratti* species group.

*Hipposideros semoni* Matschie, 1903. Denks. Med. Nat. Ges. Jena (Semon Zool. Forsch. Austr.), 8:774 (Heft 6:132).
COMMON NAME: Semon's Leaf-nosed Bat.
TYPE LOCALITY: Australia, Queensland, Cooktown.
DISTRIBUTION: N Queensland (Australia), E New Guinea.
STATUS: IUCN 2003 and IUCN/SSC Action Plan (2001) – Lower Risk (nt).
COMMENTS: *cyclops* species group. See Flannery (1995*a*) and Bonaccorso (1998).

*Hipposideros sorenseni* Kitchener and Maryanto, 1993. Rec. West. Aust. Mus., 16:142.
COMMON NAME: Sorensen's Leaf-nosed Bat.
TYPE LOCALITY: Indonesia, W Java, Pangandaran, Gua Karmat (= holy cave)(c. 7°41′S, 108°40′E).
DISTRIBUTION: C and W Java (Indonesia).
STATUS: IUCN 2003 and IUCN/SSC Action Plan (2001) – Lower Risk (nt).
COMMENTS: *larvatus* species group.

*Hipposideros speoris* (Schneider, 1800). *In* Schreber, Die Säugethiere, pl. 59b.
COMMON NAME: Schneider's Leaf-nosed Bat.
TYPE LOCALITY: India, Madras, Tranquebar.
DISTRIBUTION: India, Sri Lanka.
STATUS: IUCN 2003 and IUCN/SSC Action Plan (2001) – Lower Risk (lc).
SYNONYMS: *apiculatus* Gray, 1838; *aureus* Kelaart, 1853; *blythi* Kelaart, 1953; *dukhunensis*
Sykes, 1831; *marsupialis* Desmarest, 1820; *penicillatus* Gray, 1838; *pulchellus*
K. Andersen, 1918; *templetonii* Kelaart, 1850.
COMMENTS: *speoris* species group. Reviewed by Bates and Harrison (1997).

*Hipposideros stenotis* Thomas, 1913. Ann. Mag. Nat. Hist., ser. 8, 12:206.
COMMON NAME: Narrow-eared Leaf-nosed Bat.
TYPE LOCALITY: Australia, Northern Territory, Mary River.
DISTRIBUTION: Northern Territory, N Western Australia and N Queensland (Australia).
A New Guinea record is probably erroneous, see Hill (1963*b*:87).
STATUS: IUCN 2003 and IUCN/SSC Action Plan (2001) – Lower Risk (nt).
COMMENTS: *cyclops* species group.

*Hipposideros sumbae* Oei, 1960. Hemera Zoa, 67:28.
COMMON NAME: Sumban Leaf-nosed Bat.
TYPE LOCALITY: Indonesia, E Sumba, Nusa Tenggara, from cave (c. 9°55′S, 120°41′E).
DISTRIBUTION: Sumba, Roti, Sumbawa, Flores, Semau, and Savu Isls (Indonesia).
STATUS: IUCN 2003 and IUCN/SSC Action Plan (2001) – Lower Risk (nt).
SYNONYMS: **rotiensis** Kitchener and Maryanto, 1993; **sumbawae** Kitchener and Maryanto,
1993.
COMMENTS: *larvatus* species group. Distinct from *larvatus*; see Kitchener and Maryanto
(1993*a*). Lectotype designated by van Bree (1961).

*Hipposideros thomensis* (Bocage, 1891). J. Sci. Math. Phys. Nat. Lisboa, 2(2):88.
COMMON NAME: Saõ Tomé leaf-nosed Bat.
TYPE LOCALITY: Sao Tome and Princepe, Saõ Tomé Isl.
DISTRIBUTION: Saõ Tomé Isl.
STATUS: IUCN 2003 – Not evaluated; not considered in IUCN/SSC Action Plan (2001).
COMMENTS: *commersoni* species group. Formerly included in *commersoni*, but apparently
distinct (J. Fahr and D. Kock, pers. comm.).

*Hipposideros turpis* Bangs, 1901. Am. Nat., 35:561.
COMMON NAME: Lesser Leaf-nosed Bat.
TYPE LOCALITY: Japan, Ryukyu Isls, Sakishima Isls, Ishigaki Isl.
DISTRIBUTION: Peninsular Thailand and Vietnam; Ryukyu Isls (Japan).
STATUS: IUCN 2003 and IUCN/SSC Action Plan (2001) – Endangered.
SYNONYMS: **alongensis** Bourret, 1942; **pendleburyi** Chasen, 1936.
COMMENTS: *armiger* species group. Distinct from *armiger*; see Hill (1963*b*), Yoshiyuki (1989),
and Hendrichsen et al. (2001*b*). Includes *alongensis*, see Topál (1993).

*Hipposideros vittatus* (Peters, 1852). Naturwiss. Reise Mossambique, Säugeth., p. 32.
COMMON NAME: Striped Leaf-nosed Bat.
TYPE LOCALITY: Mozambique, Cap Delgado group, Ibo Isl.
DISTRIBUTION: Ethiopia, Somalia, Kenya, Tanzania (incl. Pemba, Chumbwe and Zanzibar
Isl), Malawi, Mozambique (incl. Ibo Isl), Zambia, Zimbabwe, Botswana, Dem. Rep.
Congo, Angola, Namibia, South Africa, Guinea-Bissau. May occur throughout much of
West Africa in sympatry with *gigas*, but distribution is presently unclear; it is likely
considerably more extensive than given here (J. Fahr, pers. comm.)
STATUS: IUCN 2003 – Not evaluated; not considered in IUCN/SSC Action Plan (2001).
SYNONYMS: *marungensis* Noack, 1887; *mostellum* Thomas, 1904; *viegasi* Monard, 1939.
COMMENTS: *commersoni* species group. Includes *maurngensis* (J. Fahr, pers. comm.); also see
Hayman and Hill (1971). Formerly included in *commersoni*, but clearly distinct based
on differences in morphology and echolocation calls (J. Fahr and D. Kock, pers.
comm.; McWilliam, 1982; Pye, 1972). Reviewed in part by Peterson et al. (1995). The
status of *viegasi* is unclear, but it probably represents *vittatus* (J. Fahr, pers. comm.).

*Hipposideros wollastoni* Thomas, 1913. Ann. Mag. Nat. Hist., ser. 8, 12:205.
 COMMON NAME: Wollaston's Leaf-nosed Bat.
 TYPE LOCALITY: Indonesia, Prov. of Papua, Utakwa River, 2,500 ft. (762 m).
 DISTRIBUTION: W and C New Guinea.
 STATUS: IUCN 2003 and IUCN/SSC Action Plan (2001) – Lower Risk (nt).
 SYNONYMS: **fasensis** Flannery and Colgan, 1993; **parnabyi** Flannery and Colgan, 1993.
 COMMENTS: *cyclops* species group. Revised by Flannery and Colgan (1993); also see Flannery
   (1995*a*) and Bonaccorso (1998).

*Paracoelops* Dorst, 1947. Bull. Mus. Natn. Hist. Nat. Paris, ser. 2, 19:436.
 TYPE SPECIES: *Paracoelops megalotis* Dorst, 1947.

*Paracoelops megalotis* Dorst, 1947. Bull. Mus. Natn. Hist. Nat. Paris, ser. 2, 19:436.
 COMMON NAME: Vietnamese Leaf-nosed Bat.
 TYPE LOCALITY: Vietnam, Annam, Vinh.
 DISTRIBUTION: C Vietnam.
 STATUS: IUCN 2003 and IUCN/SSC Action Plan (2001) – Critically Endangered.
 COMMENTS: Known only from the badly damaged holotype.

*Rhinonicteris* Gray, 1847. Proc. Zool. Soc. Lond., 1847:16.
 TYPE SPECIES: *Rhinolophus aurantius* Gray, 1845.
 COMMENTS: *Rhinonicteris* is the original spelling, but *Rhinonycteris* Gray, 1866, Proc. Zool. Soc.
   Lond., 1866:81, is sometimes used. Reviewed by Hill (1982) who spelled it *Rhinonycteris*.

*Rhinonicteris aurantia* (Gray, 1845). *In* Eyre, Central Australia, 1:405.
 COMMON NAME: Orange Leaf-nosed Bat.
 TYPE LOCALITY: Australia, Northern Territory, Port Essington.
 DISTRIBUTION: N Western Australia, Northern Territory and NW Queensland (Australia).
 STATUS: IUCN 2003 and IUCN/SSC Action Plan (2001) – Vulnerable.
 COMMENTS: Sometimes spelled "*aurantius*", but "*aurantia*" is the correct spelling in
   combination with *Rhinonicteris*. Reviewed by Armstrong (2002).

*Triaenops* Dobson, 1871. J. Asiat. Soc. Bengal, 40:455.
 TYPE SPECIES: *Triaenops persicus* Dobson, 1871.
 COMMENTS: Reviewed by Hill (1982); see also Peterson et al. (1995).

*Triaenops auritus* Grandidier, 1912. Bull. Mus. Natl. Hist. Nat. Paris, 18:8.
 COMMON NAME: Grandidier's Trident Bat.
 TYPE LOCALITY: Madagascar, near Diégo-Suarez (=Antsiranana).
 DISTRIBUTION: Known only from the type locality.
 STATUS: IUCN 2003 and IUCN/SSC Action Plan (2001) – Data Deficient.
 COMMENTS: Known only from the holotype. Often included in *furculus* (e.g., Hayman and
   Hill, 1971; Koopman, 1993, 1994), but see Peterson et al. (1995). Originally spelled
   *aurita* but emended to *auritus* by Peterson et al. (1995), presumably to agree in gender
   with the generic epithet.

*Triaenops furculus* Trouessart, 1906. Bull. Mus. Natn. Hist. Nat. Paris, 1906, 7:446.
 COMMON NAME: Trouessart's Trident Bat.
 TYPE LOCALITY: Madagascar, near Tulear (= Toliara), St. Augustine Bay, Grotte de
   Sarondrano.
 DISTRIBUTION: N and W Madagascar, Aldabra and Cosmoledo Isls (Seychelles).
 STATUS: IUCN 2003 and IUCN/SSC Action Plan (2001) – Vulnerable.
 COMMENTS: Does not includes *auritus*; see Peterson et al. (1995), but also see Hayman and
   Hill (1971). Originally spelled *furcula* but emended to *furculus* by Hill (1982),
   presumably to agree in gender with the generic epithet. May include *furinea* Tate,
   1941, possibly a *lapsus* for *furcula* (see discussion in Hill, 1982).

*Triaenops persicus* Dobson, 1871. J. Asiat. Soc. Bengal, 40:455.
 COMMON NAME: Persian Trident Bat.
 TYPE LOCALITY: Iran, Shiraz, 4,750 ft. (1,448 m).

DISTRIBUTION: Somalia, Djibouti, Ethiopia, Kenya, Tanzania, Uganda, Angola, Zanzibar, Malawi, Mozambique, Zimbabwe, Yemen, Oman, Republic of Congo, Iran, Pakistan.

STATUS: IUCN 2003 and IUCN/SSC Action Plan (2001) – Lower Risk (lc).

SYNONYMS: *macdonaldi* Harrison, 1955; ***afer*** Peters, 1876; ***majusculus*** Aellen and Brosset, 1968.

COMMENTS: See Hayman and Hill (1971) and Hill (1982) for discussion of contents. Does not include *rufus*; see Peterson et al. (1995). It is possible that *majusculus* represents a distinct species; see Cotterill (2001a). Reviewed in part by DeBlase (1980), Harrison and Bates (1991), and Bates and Harrison (1997). See Taylor (2000a) for distribution map.

*Triaenops rufus* Milne-Edwards, 1881. C. R. Hebd. Séanc. Acad, Sci., Paris, 91:1035.

COMMON NAME: Rufous Trident Bat.

TYPE LOCALITY: E Madagascar.

DISTRIBUTION: E and C Madagascar.

STATUS: IUCN 2003 and IUCN/SSC Action Plan (2001) – Data Deficient.

SYNONYMS: *humbolti* Milne-Edwards, 1881.

COMMENTS: Often included in *persicus* (e.g., Koopman, 1993, 1994), but see Peterson et al. (1995).

**Family Megadermatidae** H. Allen, 1864. Monogr. Bats N. Am., pp. xxiii, 1.

COMMENTS: For discussion of the correct formation of the family name, see Handley (1980). Hand (1985, 1996) and Griffiths et al. (1992) have provided alternative phylogenies for the group. No subfamilies are presently recognized.

*Cardioderma* Peters, 1873. Monatsb. K. Preuss. Akad. Wiss. Berlin, 1873:488.

TYPE SPECIES: *Megaderma cor* Peters, 1872.

*Cardioderma cor* (Peters, 1872). Monatsb. K. Preuss. Akad. Wiss. Berlin, 1872:194.

COMMON NAME: Heart-nosed Bat.

TYPE LOCALITY: Ethiopia.

DISTRIBUTION: Ethiopia, Djibouti, Somalia, Kenya, Uganda, E Sudan, Tanzania, Zanzibar.

STATUS: IUCN 2003 and IUCN/SSC Action Plan (2001) – Lower Risk (nt).

*Lavia* Gray, 1838. Mag. Zool. Bot., 2:490.

TYPE SPECIES: *Megaderma frons* E. Geoffroy, 1810.

SYNONYMS: *Livia* Agassiz, 1846 [misspelling].

*Lavia frons* (E. Geoffroy, 1810). Ann. Mus. Natn. Hist. Nat. Paris, 15:192.

COMMON NAME: Yellow-winged Bat.

TYPE LOCALITY: Senegal.

DISTRIBUTION: Senegal and Gambia to Somalia, south to Namibia, Zambia, and Malawi; Zanzibar.

STATUS: IUCN 2003 and IUCN/SSC Action Plan (2001) – Lower Risk (lc).

SYNONYMS: ***affinis*** K. Andersen and Wroughton, 1907; ***rex*** Miller, 1905.

COMMENTS: See Vonhof and Kalcounis (1999).

*Macroderma* Miller, 1906. Proc. Biol. Soc. Wash., 19:84.

TYPE SPECIES: *Megaderma gigas* Dobson, 1880.

*Macroderma gigas* (Dobson, 1880). Proc. Zool. Soc. Lond., 1880:461.

COMMON NAME: Australian False Vampire Bat.

TYPE LOCALITY: Australia, Queensland, Wilson's River, Mt. Margaret.

DISTRIBUTION: N and C Australia.

STATUS: IUCN 2003 and IUCN/SSC Action Plan (2001) – Vulnerable.

SYNONYMS: *saturata* Douglas, 1962.

COMMENTS: See Hudson and Wilson (1986).

*Megaderma* E. Geoffroy, 1810. Ann. Mus. Natn. Hist. Nat. Paris, 15:197.
  TYPE SPECIES: *Vespertilio spasma* Linnaeus, 1758.
  SYNONYMS: *Eucheira* Hodgson, 1847 [not *Eucheira* Westwood, 1838, an insect]; *Lyroderma* Peters, 1872; *Spasma* Gray, 1866.
  COMMENTS: Includes *Lyroderma*, but see Hand (1985). Two subgenera (*Megaderma* and *Lyroderma*) are recognized here following Corbet and Hill (1992).

*Megaderma lyra* E. Geoffroy, 1810. Ann. Mus. Natn. Hist. Nat. Paris, 15:190.
  COMMON NAME: Greater False Vampire Bat.
  TYPE LOCALITY: India, Madras.
  DISTRIBUTION: Afghanistan to S China, Burma, Thailand, Cambodia, Laos, Vietnam; south to Sri Lanka and W Malaysia; Bangladesh.
  STATUS: IUCN 2003 and IUCN/SSC Action Plan (2001) – Lower Risk (lc).
  SYNONYMS: *carnatica* Elliot, 1839; *caurina* K. Andersen and Wroughton, 1907; *schistacea* Hodgson, 1847; *spectrum* Wagner, 1844; *sinensis* K. Andersen and Wroughton, 1907.
  COMMENTS: Subgenus *Lyroderma*. Reviewed in part by Bates et al. (1994) and Bates and Harrison (1997).

*Megaderma spasma* (Linnaeus, 1758). Syst. Nat., 10th ed., 1:32. (based on Seba, 1734, Locupletissimi rerum naturalium . . . p. 90).
  COMMON NAME: Lesser False Vampire Bat.
  TYPE LOCALITY: Indonesia, Molucca Isls, Ternate.
  DISTRIBUTION: Sri Lanka and India through SE Asia (including Vietnam) to Lesser Sundas, the Philippines and Molucca Isls, various adjacent islands.
  STATUS: IUCN 2003 and IUCN/SSC Action Plan (2001) – Lower Risk (lc).
  SYNONYMS: **abditum** Chasen, 1940; **carimatae** Miller, 1906; **celebensis** Shamel, 1940; **ceylonense** K. Andersen, 1918; **horsfieldii** Blyth, 1863; **kinabalu** Chasen, 1940; **lasiae** Lyon, 1916; **majus** K. Andersen, 1918; **medium** K. Andersen, 1918; **minus** K. Andersen, 1918; **naisense** Lyon, 1916; **natunae** K. Andersen and Wroughton, 1907; **pangandarana** Sody, 1936; **philippinensis** Waterhouse, 1843; **siumatis** Lyon, 1916; **trifolium** Geoffroy, 1810.
  COMMENTS: Subgenus *Megaderma*. See Bergmans and van Bree (1986) for discussion of subspecies limits in the Indonesian region. Boundaries of some subspecies are unclear. Reviewed in part by Hill (1983) and Bates and Harrison (1997); also see Flannery (1995*b*).

**Family Rhinopomatidae** Bonaparte, 1838. Syn. Vert. Syst., *in* Nuovi Ann. Sci. Nat., Bologna, 2:111.
  SYNONYMS: Rhinopomidae Miller, 1911.
  COMMENTS: Monogeneric.

*Rhinopoma* E. Geoffroy, 1818. Descrip. de L'Egypte, 2:113.
  TYPE SPECIES: *Vespertilio microphyllus* Brünnich, 1782.
  SYNONYMS: *Rhinopomus* Gervais, 1854.
  COMMENTS: Revised by Van Cakenberghe and De Vree (1994), who provided a key to species; also see Hill (1977*b*), Bates and Harrison (1997), and Kock et al. (2001).

*Rhinopoma hardwickii* Gray, 1831. Zool. Misc., 1:37.
  COMMON NAME: Lesser Mouse-tailed Bat.
  TYPE LOCALITY: India, resticted to Bengal by Qumsiyeh et al. (1992).
  DISTRIBUTION: Morocco to Burma, south to Mauritania, Senegal, Mali, Burkina Faso, Niger, and Kenya; Socotra Isl (Yemen).
  STATUS: IUCN 2003 and IUCN/SSC Action Plan (2001) – Lower Risk (lc).
  SYNONYMS: **arabium** Thomas, 1913; *ferox* Stresemann, 1954; *sennaariense* Fitzinger, 1866 [*nomen nudum*; validated by Kock, 1969]; *cystops* Thomas, 1903; *sondaicum* Van Cakenberghe and De Vree, 1994. **Unassigned:** *brevicaudatum* Gray, 1831 [not available; International Commission on Zoological Nomenclature, Opinion 417, 1956]; *longicaudatum* Fitzinger 1866 [*nomen nudum*].
  COMMENTS: See Qumsiyeh and Jones (1986), Harrison and Bates (1991), and Kock et al.

(2001). Does not include *macinnesi*; see Van Cakenberghe and De Vree (1994). Sometimes spelled *hardwickei* (because the species was named after Major General Hardwicke), but the original spelling is *hardwickii* (see Kock et al., 2001). I follow Corbet and Hill (1992) and Kock et al. (2001) in using the original spelling.

*Rhinopoma macinnesi* Hayman, 1937. Ann. Mag. Nat. Hist., ser. 10, 19:530.
COMMON NAME: MacInnes's Mouse-tailed Bat.
TYPE LOCALITY: Kenya, Lake Rudolf, near Central Isl, Bat Isl.
DISTRIBUTION: Kenya, Somalia, Eritrea, and Ethiopia.
STATUS: IUCN 2003 and IUCN/SSC Action Plan (2001) – Vulnerable.
COMMENTS: Considered a subspecies of *hardwickii* by Koopman (1975, 1993, 1994) but see Van Cakenberghe and De Vree (1994).

*Rhinopoma microphyllum* (Brünnich, 1782). Dyrenes Historie, 1:50.
COMMON NAME: Greater Mouse-tailed Bat.
TYPE LOCALITY: Egypt, restricted to Giza by Koopman (1975).
DISTRIBUTION: Morocco, Mauritania, Senegal, Burkina Faso, and Nigeria to Afghanistan, Pakista, and India; possibly Burma; Thailand; N Sumatra.
STATUS: IUCN 2003 and IUCN/SSC Action Plan (2001) – Lower Risk (lc).
SYNONYMS: *cordofanicum* Heuglin, 1877; *hadithaensis* Khajuria, 1988; *harrisoni* Schlitter and Deblase, 1974; *lepsianum* Peters, 1859; *tropicalis* Kock, 1969; *asirensis* Nader and Kock, 1982; *kinneari* Wroughton, 1912; *sumatrae* Thomas, 1903.
COMMENTS: Reviewed in part by Harrison and Bates (1991) and Kock et al. (2001). Includes *hadithaensis*, see Kock et al. (2001). Subspecies nomenclature revised by Van Cakenberghe and De Vree (1994); also see Pearch et al. (2001).

*Rhinopoma muscatellum* Thomas, 1903. Ann. Mag. Nat. Hist., ser. 7, 11:498.
COMMON NAME: Small Mouse-tailed Bat.
TYPE LOCALITY: Oman, Muscat, Wadi Bani Ruha.
DISTRIBUTION: United Arab Emirates, Oman, Yemen, SW Iran, S Afghanistan, W Pakistan, SW India.
STATUS: IUCN 2003 and IUCN/SSC Action Plan (2001) – Lower Risk (lc).
SYNONYMS: *pusillum* Thomas, 1920; *seianum* Thomas, 1913.
COMMENTS: Ethiopian specimens referred to this species actually represent *macinnesi*; see Van Cakenberghe and de Vree (1994). Reviewed by Kock et al. (2001). Also see Qumsiyeh and Jones (1986) and Harrison and Bates (1991).

**Family Craseonycteridae** Hill, 1974. Bull. Brit. Mus. (Nat. Hist.) Zool., 27:303.
COMMENTS: Monotypic.

*Craseonycteris* Hill, 1974. Bull. Brit. Mus. (Nat. Hist.) Zool., 27:304.
TYPE SPECIES: *Craseonycteris thonglongyai* Hill, 1974.

*Craseonycteris thonglongyai* Hill, 1974. Bull. Brit. Mus. (Nat. Hist.) Zool., 27:305.
COMMON NAME: Hog-nosed Bat.
TYPE LOCALITY: Thailand, Kanchanaburi, Ban Sai Yoke (= Yok), cave near Forestry Station (14°26′N, 98°51′E).
DISTRIBUTION: Thailand, Burma
STATUS: U.S. ESA – Endangered; IUCN 2003 and IUCN/SSC Action Plan (2001) – Endangered.
COMMENTS: See Hill and Smith (1981) and Bates et al. (2001).

**Family Emballonuridae** Gervais, 1855. *In* F. Comte de Castelnau, Exped. Partes Cen. Am. Sud., Zool.(Sec. 7), Vol. 1, pt. 2(Mammifères), p. 62 footnote.
COMMENTS: For alternative phylogenies see Barghoorn (1977), Robbins and Sarich (1988), Griffiths and Smith (1991), and Dunlop (1998).

**Subfamily Taphozoinae** Jerdon, 1867. Mammals of India, p. 30.
COMMENTS: Equivalent to Tribe Taphozoini of McKenna and Bell (1997).

*Saccolaimus* Temminck, 1838. Tijdschr. Nat. Gesch. Physiol., 5:14.
    TYPE SPECIES: *Taphozous saccolaimus* Temminck, 1838.
    SYNONYMS: *Taphonycteris* Dobson, 1876.
    COMMENTS: Considered a subgenus of *Taphozous* by Ellerman and Morrison-Scott (1951),
        Corbet and Hill (1980, 1992), and Bates and Harrison (1997), but see Barghoorn (1977),
        Robbins and Sarich (1988), and Chimimba and Kitchener (1991). Key to species provided
        by Chimimba and Kitchener (1991).

*Saccolaimus flaviventris* Peters, 1867. Proc. Zool. Soc. Lond., 1866:430 [1867].
    COMMON NAME: Yellow-bellied Pouched Bat.
    TYPE LOCALITY: Australia.
    DISTRIBUTION: Australia (except Tasmania), SE New Guinea.
    STATUS: IUCN 2003 and IUCN/SSC Action Plan (2001) – Lower Risk (nt).
    SYNONYMS: *hargravei* Ramsay, 1876; *insignis* Leche, 1884.
    COMMENTS: Revised by Chimimba and Kitchener (1991). Also see Flannery (1995*a*) and
        Bonaccorso (1998).

*Saccolaimus mixtus* Troughton, 1925. Rec. Aust. Mus., 14:322.
    COMMON NAME: Troughton's Pouched Bat.
    TYPE LOCALITY: Papua New Guinea, Central Prov., Port Moresby.
    DISTRIBUTION: SE New Guinea, NE Queensland (Australia).
    STATUS: IUCN 2003 and IUCN/SSC Action Plan (2001) – Vulnerable.
    COMMENTS: Reviewed by Chimimba and Kitchener (1991). Also see Flannery (1995*a*) and
        Bonaccorso (1998).

*Saccolaimus peli* (Temminck, 1853). Esquisses Zool. sur la Côte de Guiné, p. 82.
    COMMON NAME: Pel's Pouched Bat.
    TYPE LOCALITY: Ghana, Boutry River.
    DISTRIBUTION: Liberia to W Kenya south to Angola.
    STATUS: IUCN 2003 and IUCN/SSC Action Plan (2001) – Lower Risk (nt).

*Saccolaimus saccolaimus* (Temminck, 1838). Tijdschr. Nat. Gesch. Physiol., 5:14.
    COMMON NAME: Naked-rumped Pouched Bat.
    TYPE LOCALITY: Indonesia, Java.
    DISTRIBUTION: Bangladesh, India, and Sri Lanka through SE Asia (including Burma,
        Cambodia, Thailand, and the Nicobar Isls) to the Philippines, Sulawesi, and Borneo,
        Sumatra, Java, Bali, and Timor (Indonesia); New Guinea; New Britain and Bougainville
        Isls (Papua New Guinea); NE Queensland (Australia); Guadalcanal Isl (Solomon Isls).
    STATUS: IUCN 2003 and IUCN/SSC Action Plan (2001) – Lower Risk (lc).
    SYNONYMS: **affinis** Dobson, 1875; *flavimaculatus* Sody, 1931; **crassus** Blyth, 1844; *pulcher*
        Blyth, 1844; **nudicluniatus** De Vis, 1905; *granti* Thomas, 1911; **pluto** Miller, 1910;
        *capito* Hollister, 1913.
    COMMENTS: Corbet and Hill (1980) listed *nudicluniatus* as a distinct species without
        comment. Includes *pulcher*; see Medway (1977) and Goodwin (1979). Includes *pluto*;
        see Corbet and Hill (1992). Reviewed in part by Chimimba and Kitchener (1991) and
        Bates and Harrison (1997); also see Flannery (1995*b*) and Bonaccorso (1998).

*Taphozous* E. Geoffroy, 1818. Descrip. de L'Egypte, 2:113.
    TYPE SPECIES: *Taphozous perforatus* E. Geoffroy, 1818.
    SYNONYMS: *Liponycteris* Thomas, 1922.
    COMMENTS: Includes *Liponycteris* but not *Saccolaimus*; see Hayman and Hill (1971), Barghoorn
        (1977), Robbins and Sarich (1988), and Chimimba and Kitchener (1991), though also see
        Corbet and Hill (1992). Key to Australian species provided by Chimimba and Kitchener
        (1991). Two subgenera are recognized, *Taphozous* and *Liponycteris*.

*Taphozous achates* Thomas, 1915. J. Bombay Nat. Hist. Soc., 24:60.
    COMMON NAME: Indonesian Tomb Bat.
    TYPE LOCALITY: Indonesia, near Timor, Nusa Tenggara, Savu Isl.
    DISTRIBUTION: Kei, Savu, Roti, Semau, and Nusa Penida Isls (Indonesia); possibly Timor.
    STATUS: IUCN 2003 and IUCN/SSC Action Plan (2001) – Vulnerable.

SYNONYMS: *minor* Kitchener, 1995 [in Kitchener and Suyanto, 1995].
COMMENTS: Subgenus *Taphozous*. Formerly included in *melanopogon*, but see D. J. Kitchener et al. (1993*b*) and Kitchener and Suyanto (1995). Also see Flannery (1995*b*).

*Taphozous australis* Gould, 1854. Mamm. Aust., p. 3.
COMMON NAME: Coastal Tomb Bat.
TYPE LOCALITY: Australia, Queensland, Albany Isl (off Cape York).
DISTRIBUTION: N Queensland (Australia), Torres Strait Isls, SE New Guinea.
STATUS: IUCN 2003 and IUCN/SSC Action Plan (2001) – Lower Risk (nt).
SYNONYMS: *fumosus* De Vis, 1905.
COMMENTS: Subgenus *Taphozous*. Includes *fumosus*; see Troughton (1925) and Chimimba and Kitchener (1991). Tate (1952) included *georgianus* in this species, but see McKean and Price (1967) and Chimimba and Kitchener (1991). Also see Flannery (1995*a*) and Bonaccorso (1998).

*Taphozous georgianus* Thomas, 1915. J. Bombay Nat. Hist. Soc., 24:62.
COMMON NAME: Sharp-nosed Tomb Bat.
TYPE LOCALITY: Australia, Western Australia, King George Sound.
DISTRIBUTION: N and W Australia.
STATUS: IUCN 2003 and IUCN/SSC Action Plan (2001) – Lower Risk (lc).
COMMENTS: Subgenus *Taphozous*. McKean and Price (1967) and Koopman (1993, 1994) included *troughtoni* in this species, but see Chimimba and Kitchener (1991).

*Taphozous hamiltoni* Thomas, 1920. Ann. Mag. Nat. Hist., ser. 9, 5:142.
COMMON NAME: Hamilton's Tomb Bat.
TYPE LOCALITY: Sudan, Equatoria, Mongalla.
DISTRIBUTION: S Sudan, Chad, Kenya, possibly Somalia.
STATUS: IUCN 2003 and IUCN/SSC Action Plan (2001) – Vulnerable.
COMMENTS: Subgenus *Liponycteris*.

*Taphozous hildegardeae* Thomas, 1909. Ann. Mag. Nat. Hist., ser. 8, 4:98.
COMMON NAME: Hildegarde's Tomb Bat.
TYPE LOCALITY: Kenya, Coast Province, Rabai (near Mombassa).
DISTRIBUTION: Kenya, NE Tanzania, Zanzibar.
STATUS: IUCN 2003 and IUCN/SSC Action Plan (2001) – Vulnerable.
COMMENTS: Subgenus *Taphozous*. See Colket and Wilson (1998).

*Taphozous hilli* Kitchener, 1980. Rec. W. Aust. Mus., 8:162.
COMMON NAME: Hill's Tomb Bat.
TYPE LOCALITY: Australia, Western Australia, Hamersley range, near Mt. Bruce.
DISTRIBUTION: Western Australia, South Australia, and Northern Territory.
STATUS: IUCN 2003 and IUCN/SSC Action Plan (2001) – Lower Risk (lc).
COMMENTS: Subgenus *Taphozous*. Reviewed by Chimimba and Kitchener (1991).

*Taphozous kapalgensis* McKean and Friend, 1979. Vict. Nat., 96:239.
COMMON NAME: Arnhem Tomb Bat.
TYPE LOCALITY: Australia, Northern Territory, S Alligator River, near Rookery Point, Kapalga.
DISTRIBUTION: Northern Territory (Australia).
STATUS: IUCN 2003 and IUCN/SSC Action Plan (2001) – Vulnerable.
COMMENTS: Subgenus *Taphozous*. Reviewed by Chimimba and Kitchener (1991).

*Taphozous longimanus* Hardwicke, 1825. Trans. Linn. Soc. Lond., 14:525.
COMMON NAME: Long-winged Tomb Bat.
TYPE LOCALITY: India, Bengal, Calcutta.
DISTRIBUTION: Sri Lanka; India and Bangladesh to Burma, Cambodia, and Thailand; Peninsular Malaysia; Sumatra, Borneo, Java, Bali, Sumbawa, and Flores (Indonesia).
STATUS: IUCN 2003 and IUCN/SSC Action Plan (2001) – Lower Risk (lc).
SYNONYMS: *brevicaudus* Blyth, 1841; *cantorii* Blyth, 1842; *fulvidus* Blyth, 1841; *albipinnis* Thomas, 1898; *kampenii* Jentink, 1907; *leucopleurus* Dobson, 1875.
COMMENTS: Subgenus *Taphozous*. Reviewed in part by Bates et al. (1994) and Bates and Harrison (1997).

*Taphozous mauritianus* E. Geoffroy, 1818. Descrip. de L'Egypte, 2:127.
    COMMON NAME: Mauritian Tomb Bat.
    TYPE LOCALITY: Mauritius.
    DISTRIBUTION: South Africa to Sudan and Somalia to Senegal; Mauritius and Réunion Isls
        (Mascarene Isls); São Tomé and Princepe; Madagascar; Assumption Isl and Aldabra Isl.
    STATUS: IUCN 2003 and IUCN/SSC Action Plan (2001) – Lower Risk (lc).
    SYNONYMS: *cinerascens* Seabra, 1900; *dobsoni* Jentink, 1879; *leucopterus* Temminck, 1835.
    COMMENTS: Subgenus *Taphozous*. Reviewed in part by Peterson et al. (1995); also see Taylor
        (2000*a*).

*Taphozous melanopogon* Temminck, 1841. Monogr. Mamm., 2:287.
    COMMON NAME: Black-bearded Tomb Bat.
    TYPE LOCALITY: Indonesia, W Java, Bantam.
    DISTRIBUTION: Sri Lanka; India; Burma; Thailand; Laos; Cambodia; Vietnam; S China; Malay
        Peninsula and adjacent islands; Borneo; Sumatra, Java, Lombok, Sumbawa, Moyo,
        Alor, Timor, and Sulawesi (Indonesia), Philippines.
    STATUS: IUCN 2003 – Lower Risk (lc); IUCN/SSC Action Plan (2001) – Lower Risk (lc) as
        *T. melanopogon* and *T. solifer.*
    SYNONYMS: **bicolor** Temminck, 1841; **cavaticus** Hollister, 1913; **fretensis** Thomas, 1916;
        **phillipinensis** Waterhouse, 1845; *solifer* Hollister, 1913.
    COMMENTS: Subgenus *Taphozous*. Does not includes *achates*; see D. J. Kitchener et al.
        (1993*b*). Includes *phillipinensis*; see Heaney et al. (1987, 1998) and Corbet and Hill
        (1992). Reviewed in part by Bates et al. (1994) and Bates and Harrison (1997); also see
        Flannery (1995*b*). Rediagnosed by D. J. Kitchener et al. (1993*b*). Sulawesi and Kei
        populations have not been allocated to subspecies.

*Taphozous nudiventris* Cretzschmar, 1830. *In* Rüppell, Atlas Reise Nördl. Afr., Zool. Säugeth., p. 70.
    COMMON NAME: Naked-rumped Tomb Bat.
    TYPE LOCALITY: Egypt, Giza.
    DISTRIBUTION: Mauritania, Senegal, and Guinea-Bissau to Djibouti, Egypt, Jordan, and
        NE Turkey, south to Tanzania and east to Burma.
    STATUS: IUCN 2003 and IUCN/SSC Action Plan (2001) – Lower Risk (lc).
    SYNONYMS: *assabensis* Monticelli, 1885; *nudiventer* Temminck, 1841; **kachhensis** Dobson,
        1872; **magnus** Wettstein, 1913; *babylonicus* Thomas, 1915; **nudaster** Thomas, 1915;
        **zayidi** Harrison, 1955. **Unassigned**: *serratus* Heuglin, 1877 [see comments].
    COMMENTS: Subgenus *Liponycteris*. Includes *kachhensis*; see Felten (1962) and Bates and
        Harrison (1997). Formerly included in genus *Liponycteris*; see Hayman and Hill (1977).
        Reviewed in part by Bates et al. (1994) as *kachhensis*. Also see Harrison and Bates
        (1991). May include *serratus* Heuglin, 1877, an enigmatic taxon variously referred to
        either *Taphozous nudiventris* (e.g., G. M. Allen, 1939; Koopman, 1993) or *Scotophilus*
        *leucogaster* (e.g., G. M. Allen, 1939; Koopman, 1975) but which may not represent
        either of those species.

*Taphozous perforatus* E. Geoffroy, 1818. Descrip. de L'Egypte, 2:126.
    COMMON NAME: Egyptian Tomb Bat.
    TYPE LOCALITY: Egypt, Kom Ombo.
    DISTRIBUTION: Mauritania and Senegal to Botswana, Mozambique, Somalia, Djibouti and
        Egypt; S Arabia; Jordan; S Iran; Pakistan; NW India.
    STATUS: IUCN 2003 and IUCN/SSC Action Plan (2001) – Lower Risk (lc).
    SYNONYMS: *maritimus* Heuglin, 1877; **haedinus** Thomas, 1915; **senegalensis** Desmarest,
        1820; *swirae* Harrison, 1958 **sudani** Thomas, 1915; *australis* Harrison, 1962 [not
        Gould, 1854]; *rhodesiae* Harrison, 1964 [replacement name for *australis*].
    COMMENTS: Subgenus *Taphozous*. Includes *senegalensis* and *sudani*; see Hayman and Hill
        (1977). Reviewed by Bates et al. (1994) and Bates and Harrison (1997); see also Meester et
        al. (1986), Harrison and Bates (1991), and Taylor (2000*a*). Subspecies are poorly defined.

*Taphozous theobaldi* Dobson, 1872. Proc. Asiat. Soc. Bengal, p. 152.
    COMMON NAME: Theobald's Tomb Bat.

TYPE LOCALITY: Burma, Tenasserim.
DISTRIBUTION: C India to Vietnam; Java, Borneo and Sulawesi. A record from Malaysia
    appears to be in error; see Medway (1969).
STATUS: IUCN 2003 and IUCN/SSC Action Plan (2001) – Lower Risk (lc).
SYNONYMS: *secatus* Thomas, 1915.
COMMENTS: Subgenus *Taphozous*. Reviewed in part by Bates and Harrison (1997).

*Taphozous troughtoni* Tate, 1952. Bull. Am. Mus. Nat, Hist., 98:563.
COMMON NAME: Troughton's Tomb Bat.
TYPE LOCALITY: Australia, Queensland, 10 mi. (15 km) E of Mt. Isa, Rifle Creek.
DISTRIBUTION: NW Queensland (Australia).
STATUS: IUCN 2003 and IUCN/SSC Action Plan (2001) – Critically Endangered.
COMMENTS: Subgenus *Taphozous*. Included in *georgianus* by McKean and Price (1967) and
    Koopman (1993, 1994), but see Chimimba and Kitchener (1991).

**Subfamily Emballonurinae** Gervais, 1855. *In* F. Comte de Castelnau, Exped. Partes Cen. Am. Sud.,
    Zool. (Sec. 7), Vol. 1, pt. 2 (Mammifères), p. 62 footnote.
COMMENTS: McKenna and Bell (1997) divided this subfamily into two tribes, Emballonurini
    Gervais, 1855 (*Mosia, Emballonura, Coleura*) and Diclidurini Gray, 1866 (Neotropical
    emballonurids). However, both of these groups may be paraphyletic as so defined
    (Dunlop, 1998). Accordingly, I do not recognize tribes within Emballonurinae at this
    time. For a key to Neotropical species see Jones and Hood (1993).

*Balantiopteryx* Peters, 1867. Monatsb. K. Preuss. Akad. Wiss. Berlin, 1867:476.
TYPE SPECIES: *Balantiopteryx plicata* Peters, 1867.
COMMENTS: Revised by Hill (1987); also see Arroyo-Cabrales and Jones (1988*a*) and Jones and
    Hood (1993).

*Balantiopteryx infusca* (Thomas, 1897). Ann. Mag. Nat. Hist., ser. 6, 20:546.
COMMON NAME: Ecuadorian Sac-winged Bat.
TYPE LOCALITY: Ecuador, Esmeraldas, Cachabi.
DISTRIBUTION: W Ecuador, Colombia.
STATUS: IUCN 2003 and IUCN/SSC Action Plan (2001) – Endangered.
COMMENTS: Reviewed by Hill (1987), Arroyo-Cabrales and Jones (1988*b*), and McCarthy
    et al. (2000).

*Balantiopteryx io* Thomas, 1904. Ann. Mag. Nat. Hist., ser. 7, 13:252.
COMMON NAME: Thomas's Sac-winged Bat.
TYPE LOCALITY: Guatemala, Alta Verapaz, Río Dolores (near Coban).
DISTRIBUTION: S Veracruz and Oaxaca (Mexico) to EC Guatemala and Belize.
STATUS: IUCN 2003 and IUCN/SSC Action Plan (2001) – Lower Risk (nt).
COMMENTS: See Arroyo-Cabrales and Jones (1988*b*).

*Balantiopteryx plicata* Peters, 1867. Monatsb. K. Preuss. Akad. Wiss. Berlin, 1867:476.
COMMON NAME: Gray Sac-winged Bat.
TYPE LOCALITY: Costa Rica, Puntarenas.
DISTRIBUTION: Costa Rica to C Sonora and S Baja California (Mexico); N Colombia.
STATUS: IUCN 2003 and IUCN/SSC Action Plan (2001) – Lower Risk (lc).
SYNONYMS: *ochoterenai* Martínez and Villa, 1938; *pallida* Burt, 1948.
COMMENTS: See Arroyo-Cabrales and Jones (1988*a*).

*Centronycteris* Gray, 1838. Mag. Zool. Bot., 2:499.
TYPE SPECIES: *Vespertilio calcaratus* Schinz, 1821 (preoccupied by Rafinesque, 1818) (= *Vespertilio
    maximiliani*, J. Fischer, 1829).
COMMENTS: Revised by Simmons and Handley (1998). The two species have not yet been found
    in sympatry, but their ranges may overlap in NE Peru (Hice and Solari, 2002).

*Centronycteris centralis* Thomas, 1912. Ann. Mag. Nat. Hist., ser. 8, 10:638.
COMMON NAME: Thomas's Shaggy Bat.
TYPE LOCALITY: Panama, Chiriquí, Bogava.

DISTRIBUTION: S Mexico to SE Peru.

STATUS: IUCN 2003 – Not evaluated; not considered in IUCN/SSC Action Plan (2001).

COMMENTS: Formerly included in *maximiliani* but clearly distinct, see Simmons and Handley (1998).

*Centronycteris maximiliani* (J. Fischer, 1829). Synopsis Mamm., p. 122.
COMMON NAME: Shaggy Bat.
TYPE LOCALITY: Brazil, Espirito Santo, Rio Jucy, Fazenda do Coroaba.
DISTRIBUTION: NE Peru, S Venezuela, Brazil, Guyana, Surinam, French Guiana.
STATUS: IUCN 2003 and IUCN/SSC Action Plan (2001) – Lower Risk (lc).
SYNONYMS: *calcaratus* Schinz, 1821 [preoccupied by *calcaratus* Rafinesque, 1818]; *wiedi* Palmer, 1898.
COMMENTS: Does not include *centralis*, see Simmons and Handley (1998).

*Coleura* Peters, 1867. Monatsb. K. Preuss. Akad. Wiss. Berlin, 1867:479.
TYPE SPECIES: *Emballonura afra* Peters, 1852.

*Coleura afra* (Peters, 1852). Reise nach Mossambique, Säugethiere, p. 51.
COMMON NAME: African Sheath-tailed Bat.
TYPE LOCALITY: Mozambique, Tete.
DISTRIBUTION: Guinea-Bissau to Somalia and Djibouti, south to Angola, Dem. Rep. Congo, and Mozambique; Yemen.
STATUS: IUCN 2003 and IUCN/SSC Action Plan (2001) – Lower Risk (lc).
SYNONYMS: *gallarum* Thomas, 1915; *kummeri* Monard, 1939; *nilosa* Thomas, 1915.
COMMENTS: Includes *kummeri*, see Rosevear (1965). Also see Harrison and Bates (1991) and Dunlop (1997).

*Coleura seychellensis* Peters, 1868. Monatsb. K. Preuss. Akad. Wiss. Berlin, 1868:367.
COMMON NAME: Seychelles Sheath-tailed Bat.
TYPE LOCALITY: Seychelle Isls, Mahe Isl.
DISTRIBUTION: Seychelle Isls; possibly Zanzibar. The Zanzibar record is extremely dubious (Koopman, 1993).
STATUS: IUCN 2003 and IUCN/SSC Action Plan (2001) – Critically Endangered.
SYNONYMS: *silhouettae* Thomas, 1915.

*Cormura* Peters, 1867. Monatsb. K. Preuss. Akad. Wiss. Berlin, 1867:475.
TYPE SPECIES: *Emballonura brevirostris* Wagner, 1843.
SYNONYMS: *Myropteryx* Miller, 1906.
COMMENTS: Reviewed by Jones and Hood (1993).

*Cormura brevirostris* (Wagner, 1843). Arch. Naturgesch., ser. 9, 1:367.
COMMON NAME: Chestnut Sac-winged Bat.
TYPE LOCALITY: Brazil, Amazonas, Rio Negro, Marabitanas.
DISTRIBUTION: Nicaragua south to Peru and C Brazil.
STATUS: IUCN 2003 and IUCN/SSC Action Plan (2001) – Lower Risk (lc).
SYNONYMS: *pullus* Miller, 1906.
COMMENTS: See Bernard (2003).

*Cyttarops* Thomas, 1913. Ann. Mag. Nat. Hist., ser. 8, 11:134.
TYPE SPECIES: *Cyttarops alecto* Thomas, 1913.

*Cyttarops alecto* Thomas, 1913. Ann. Mag. Nat. Hist., ser. 8, 11:135.
COMMON NAME: Short-eared Bat.
TYPE LOCALITY: Brazil, Pará, Mocajatube.
DISTRIBUTION: Nicaragua, Costa Rica, Guyana, French Guiana, Amazonian Brazil.
STATUS: IUCN 2003 and IUCN/SSC Action Plan (2001) – Lower Risk (nt).
COMMENTS: Reviewed by Jones and Hood (1993); also see Starrett (1972). See Emmons (1997) for distribution map.

*Diclidurus* Wied-Neuwied, 1820. Isis von Oken, 1819:1629 [1820].
 TYPE SPECIES: *Diclidurus albus* Wied-Neuwied, 1820.
 SYNONYMS: *Depanycteris* Thomas, 1920.
 COMMENTS: Includes *Depanycteris*. Reviewed by Jones and Hood (1993); also see Ojasti and
  Linares (1971) and Ceballos and Medellín (1988). Two subgenera are recognized,
  *Diclidurus* and *Depanycteris*.

*Diclidurus albus* Wied-Neuwied, 1820. Isis von Oken, 1819:1630 [1820].
 COMMON NAME: Northern Ghost Bat.
 TYPE LOCALITY: Brazil, Bahia, Rio Pardo, Canavieiras.
 DISTRIBUTION: Nayarit (Mexico) to E Brazil and Trinidad.
 STATUS: IUCN 2003 and IUCN/SSC Action Plan (2001) – Lower Risk (lc).
 SYNONYMS: *freyreisii* Wied, 1838; *virgo* Thomas, 1903.
 COMMENTS: Subgenus *Diclidurus*. Includes *virgo*; see Goodwin (1969), but see also Ojasti and
  Linares (1971). Corbet and Hill (1980) listed *virgo* as a distinct species without
  comment. See Ceballos and Medellín (1988).

*Diclidurus ingens* Hernandez-Camacho, 1955. Caldasia, 7:87.
 COMMON NAME: Greater Ghost Bat.
 TYPE LOCALITY: Colombia, Caqueta, Río Putumayo, Puerto Leguizamo.
 DISTRIBUTION: Venezuela, SE Colombia, Guyana, NW Brazil.
 STATUS: IUCN 2003 and IUCN/SSC Action Plan (2001) – Vulnerable.
 COMMENTS: Subgenus *Diclidurus*.

*Diclidurus isabellus* (Thomas, 1920). Ann. Mag. Nat. Hist., ser. 9, 6:271.
 COMMON NAME: Isabelle's Ghost Bat.
 TYPE LOCALITY: Brazil, Amazonas, Manacapuru (lower Solimões River).
 DISTRIBUTION: NW Brazil, Venezuela, Guyana.
 STATUS: IUCN 2003 and IUCN/SSC Action Plan (2001) – Lower Risk (nt).
 COMMENTS: Subgenus *Depanycteris*. Formerly placed in its own genus (*Depanycteris*), see
  Ojasti and Linares (1971).

*Diclidurus scutatus* Peters, 1869. Monatsb. K. Preuss. Akad. Wiss. Berlin, 1869:400.
 COMMON NAME: Lesser Ghost Bat.
 TYPE LOCALITY: Brazil, Pará, Belem.
 DISTRIBUTION: Amazonian Brazil, Venezuela, Peru, Guyana, Surinam, French Guiana.
 STATUS: IUCN 2003 and IUCN/SSC Action Plan (2001) – Lower Risk (lc).
 COMMENTS: Subgenus *Diclidurus*.

*Emballonura* Temminck, 1838. Tijdschr. Nat. Gesch. Physiol., 5:22.
 TYPE SPECIES: *Emballonura monticola* Temminck, 1838.
 COMMENTS: Does not include *Mosia*; see Griffiths et al. (1991). Species groups follow Koopman
  (1994).

*Emballonura alecto* (Eydoux and Gervais, 1836). Mag. Zool. Paris, 6:7.
 COMMON NAME: Small Asian Sheath-tailed Bat.
 TYPE LOCALITY: Philippines, Luzon, Manila.
 DISTRIBUTION: Philippines, Borneo, Sulawesi, and Tanimbar (Indonesia), Moluccas, and
  adjacent small islands including Anambas Isl.
 STATUS: IUCN 2003 and IUCN/SSC Action Plan (2001) – Lower Risk (lc).
 SYNONYMS: *discolor* Peters, 1861; *anambensis* Miller, 1900; *palawanensis* Taylor, 1934;
  *rivalis* Thomas, 1915.
 COMMENTS: *alecto* species group. Includes *rivalis*; see Medway (1977). Includes *anambensis*;
  see Corbet and Hill (1992). Also see Flannery (1995b).

*Emballonura atrata* Peters, 1874. Monatsb. K. Preuss. Akad. Wiss. Berlin, 1874:693.
 COMMON NAME: Peters's Sheath-tailed Bat.
 TYPE LOCALITY: Madagascar, restricted to "interior of Madagascar" by Peterson et al. (1995).
 DISTRIBUTION: Madagascar except for S region.
 STATUS: IUCN 2003 and IUCN/SSC Action Plan (2001) – Vulnerable.

COMMENTS: *atrata* species group. Reviewed by Peterson et al. (1995).

*Emballonura beccarii* Peters and Doria, 1881. Ann. Mus. Civ. Stor. Nat. Genova, 16:693.
COMMON NAME: Beccari's Sheath-tailed Bat.
TYPE LOCALITY: Indonesia, Prov. of Papua, Tjenderawasih Div., Yapen Isl, Ansus.
DISTRIBUTION: New Guinea, Kai Isls, Biak, Ypen, Trobriand Isls, Bougainville, New Ireland
(Bismarck Arch.) and nearby smaller islands.
STATUS: IUCN 2003 and IUCN/SSC Action Plan (2001) – Lower Risk (lc).
SYNONYMS: *locusta* Thomas, 1920; **clavium** Thomas, 1915; **meeki** Thomas, 1896.
COMMENTS: *alecto* species group. See Flannery (1995*a*, *b*) and Bonaccorso (1998).

*Emballonura dianae* Hill, 1956. *In* Wolff, Nat. Hist. Rennell Isl, Brit. Solomon Isls, 1:74.
COMMON NAME: Large-eared Sheath-tailed Bat.
TYPE LOCALITY: Solomon Isls, Rennell Isl, near Tigoa, Te-Abagua Cave, about 35 m.
DISTRIBUTION: Rennell, Guadalcanal, Malaita, Choiseul and San Isabel Isls (Solomon Isls),
New Ireland (Bismarck Arch.), New Guinea.
STATUS: IUCN 2003 and IUCN/SSC Action Plan (2001) – Vulnerable.
SYNONYMS: **fruhstorferi** Flannery, 1994; **rickwoodi** Flannery, 1994.
COMMENTS: *raffrayana* species group. Revised by Flannery (1994*b*). See also Flannery (1995*a*,
*b*) and Bonaccorso (1998).

*Emballonura furax* Thomas, 1911. Ann. Mag. Nat. Hist., ser. 8, 7:384.
COMMON NAME: New Guinean Sheath-tailed Bat.
TYPE LOCALITY: Indonesia, Prov. of Papua, S of Charles Louis Range, Kapare River,
Whitewater Camp., 400 ft. (122 m).
DISTRIBUTION: Prov. of Papua (Indonesia); Papua New Guinea including Bismarck Arch.
STATUS: IUCN 2003 and IUCN/SSC Action Plan (2001) – Vulnerable.
COMMENTS: *raffrayana* species group. Revised by Flannery (1994*b*), who described *serii* based
on specimens from New Ireland Isl originally referred to *furax*. Also see Flannery
(1995*a*) and Bonaccorso (1998).

*Emballonura monticola* Temminck, 1838. Tijdschr. Nat. Gesch. Physiol., 5:25.
COMMON NAME: Lesser Sheath-tailed Bat.
TYPE LOCALITY: Indonesia, Java, Mt. Munara.
DISTRIBUTION: Burma and Thailand to W Malaysia; Borneo; Sumatra, Rhio Arch., Banka,
Billiton, Enggano, Babi Isls, Batu Isls, Nias Isl, Mentawai Isls, Java, Sulawesi.
STATUS: IUCN 2003 and IUCN/SSC Action Plan (2001) – Lower Risk (lc).
SYNONYMS: *peninsularis* Miller, 1898; *pusilla* Lyon, 1911.
COMMENTS: *alecto* species group.

*Emballonura raffrayana* Dobson, 1879. Proc. Zool. Soc. Lond., 1878:876 [1879].
COMMON NAME: Raffray's Sheath-tailed Bat.
TYPE LOCALITY: Indonesia, Prov. of Papua, Tjenderawasih Div., Geelvinck Bay, Numfor Isl
(= Mefor Isl = Noemfor Isl = Numfoor Isl); for clarification see Thomas (1914*b*).
DISTRIBUTION: Moluccas, New Guinea, Bismark Arch., Solomons, and Vanuatu.
STATUS: IUCN 2003 and IUCN/SSC Action Plan (2001) – Lower Risk (nt).
SYNONYMS: **cor** Thomas, 1915; **stresemanni** Thomas, 1914.
COMMENTS: *raffrayana* species group. See Flannery (1995*a*, *b*) and Bonaccorso (1998).

*Emballonura semicaudata* (Peale, 1848). Mammalia *in* Repts. U.S. Expl. Surv., 8:23.
COMMON NAME: Polynesian Sheath-tailed Bat.
TYPE LOCALITY: Samoa.
DISTRIBUTION: Mariana Isls and Caroline Isls (including Palau Isls), Vanuatu, Fiji Isls, Samoa.
STATUS: U. S. ESA – Candidate taxon (in Aguijan, American Samoa); IUCN 2003 and
IUCN/SSC Action Plan (2001) – Endangered.
SYNONYMS: *fuliginosa* Tomes, 1859; **palauensis** Yamashima, 1932; **rotensis** Yamashima,
1943; **sulcata** Miller, 1911.
COMMENTS: *semicaudata* species group. Includes *sulcata*, see Griffiths et al. (1991). Subspecies
reviewed by Koopman (1997).

*Emballonura serii* Flannery, 1994. Mammalia, 58:606.
    COMMON NAME: Seri's Sheath-tailed Bat.
    TYPE LOCALITY: Bismarck Archipelago, New Ireland (Papua New Guinea), Matapara Cave
        near Medina, 2°55'N, 151°23'E.
    DISTRIBUTION: Los Negros Isl, Manus Isl, New Ireland Isl (Bismarck Arch.).
    STATUS: IUCN 2003 and IUCN/SSC Action Plan (2001) – Data Defficient.
    COMMENTS: *raffrayana* species group. Described based on specimens orginally referred to
        *furax*. See Flannery (1995*b*) and Bonaccorso (1998).

*Mosia* Gray, 1843. Ann. Mag. Nat. Hist., [ser. 1], 11:117.
    TYPE SPECIES: *Mosia nigrescens* Gray, 1843.
    COMMENTS: Formerly included in *Emballonura*, but see Griffiths et al. (1991). Corbet and Hill
        (1992) recognized *Mosia* as a subgenus of *Emballonura*.

*Mosia nigrescens* Gray, 1843. Ann. Mag. Nat. Hist., [ser. 1], 11:117.
    COMMON NAME: Dark Sheath-tailed Bat.
    TYPE LOCALITY: Indonesia, Molucca Isls, Amboina Isl.
    DISTRIBUTION: New Guinea; New Ireland (Papua New Guinea); Kai Isls, Halmahera Isls,
        Schouten Isls, Sulawesi, Moluccas Isls; Waigeo Isl. (Prov. of Papua, Indonesia),
        Bismarck Arch. (Papua New Guinea); Solomon Isls; adjacent small islands.
    STATUS: IUCN 2003 and IUCN/SSC Action Plan (2001) – Lower Risk (lc).
    SYNONYMS: **papuana** Thomas, 1914; **solomonis** Thomas, 1904.
    COMMENTS: Includes *papuana*; see Laurie and Hill (1954) and Hill (1983). Includes *solomonis*,
        considered a distinct species by McKean (1972). See Flannery (1995*a*, *b*) and
        Bonaccorso (1998).

*Peropteryx* Peters, 1867. Monatsb. K. Preuss. Akad. Wiss. Berlin, 1867:472.
    TYPE SPECIES: *Vespertilio caninus* Wied-Neuwied, 1821 (preoccupied; = *Emballonura macrotis*
        Wagner, 1843).
    SYNONYMS: *Peronymus* Peters, 1868.
    COMMENTS: Includes *Peronymus*; see Griffiths and Smith (1991), Jones and Hood (1993),
        Dunlop (1998), and Simmons and Voss (1998). Reviewed by Jones and Hood (1993). Two
        subgenera recognized, *Peropteryx* and *Peronymus*.

*Peropteryx kappleri* Peters, 1867. Monatsb. K. Preuss. Akad. Wiss. Berlin, 1867:473.
    COMMON NAME: Greater Dog-like Bat.
    TYPE LOCALITY: Surinam.
    DISTRIBUTION: S Veracruz (Mexico) to the Guianas, E Brazil, Peru, and N Bolivia.
    STATUS: IUCN 2003 and IUCN/SSC Action Plan (2001) – Lower Risk (lc).
    SYNONYMS: **intermedia** Sanborn, 1951.
    COMMENTS: Subgenus *Peropteryx*.

*Peropteryx leucoptera* Peters, 1867. Monatsb. K. Preuss. Akad. Wiss. Berlin, 1867:474.
    COMMON NAME: White-winged Dog-like Bat.
    TYPE LOCALITY: Surinam.
    DISTRIBUTION: Peru, Colombia, N and E Brazil, Venezuela, Guianas.
    STATUS: IUCN 2003 and IUCN/SSC Action Plan (2001) – Lower Risk (lc).
    SYNONYMS: **cyclops** Thomas, 1924.
    COMMENTS: Subgenus *Peronymus*. Formerly placed in its own genus (*Peronymus*), but clearly
        a member of the *Peropteryx* clade, see Griffiths and Smith (1991), Jones and Hood
        (1993), Dunlop (1998), and Simmons and Voss (1998).

*Peropteryx macrotis* (Wagner, 1843). Arch. Naturgesch., ser. 9, 1:367.
    COMMON NAME: Lesser Dog-like Bat.
    TYPE LOCALITY: Brazil, Mato Grosso.
    DISTRIBUTION: Guerrero and Yucatán (Mexico) to Peru, Bolivia, Paraguay, and S and E Brazil.
    STATUS: IUCN 2003 and IUCN/SSC Action Plan (2001) – Lower Risk (lc).
    SYNONYMS: *brunnea* Gervais, 1855; *caninus* Schinz, 1821 [not Blumenbach, 1797].
    COMMENTS: Subgenus *Peropteryx*. Does not include *trinitatis*; see Brosset and Charles-

Dominique (1990) and Simmons and Voss (1998). Does not include *phaea*; see Genoways et al. (1998). This complex may include more than one species; see Reid et al. (2000). See Yee (2000), but note that they included *trinitatis* and *phaea* in this species.

*Peropteryx trinitatis* Miller, 1899. Bull. Amer. Mus. Nat. Hist., 12:178.
COMMON NAME: Trinidad Dog-like Bat.
TYPE LOCALITY: Trinidad, Port-of-Spain.
DISTRIBUTION: Trinidad and Tobago; Aruba Isl (Netherlands Antilles); Grenada; Venezuela; Margarita Isl (Venezuela); French Guiana.
STATUS: IUCN 2003 – Not evaluated; not considered in IUCN/SSC Microchiropteran Bats Action Plan (2001).
SYNONYMS: *phaea* G. M. Allen, 1911.
COMMENTS: Subgenus *Peropteryx*. Considered a subspecies of *macrotis* by many authors, but see Brosset and Charles-Dominique (1990) and Simmons and Voss (1998). Includes *phaea*, see discussion in Genoways et al. (1998).

*Rhynchonycteris* Peters, 1867. Monatsb. K. Preuss. Akad. Wiss. Berlin, 1867:477.
TYPE SPECIES: *Vespertilio naso* Wied-Neuwied, 1820.
SYNONYMS: *Proboscidea* Spix, 1823 [not Brugière, 1791]; *Rhynchoniscus* Miller, 1907.
COMMENTS: Reviewed by Jones and Hood (1993).

*Rhynchonycteris naso* (Wied-Neuwied, 1820). Reise nach Brasilien, 1:251.
COMMON NAME: Proboscis Bat.
TYPE LOCALITY: Brazil, Bahia, Rio Mucuri, near Morro d'Arara; for clarification see Avila-Pires (1965:9).
DISTRIBUTION: E Oaxaca and C Veracruz (Mexico) to C and E Brazil, Peru, Bolivia, French Guiana, Guyana, and Surinam; Trinidad.
STATUS: IUCN 2003 and IUCN/SSC Action Plan (2001) – Lower Risk (lc).
SYNONYMS: *lineata* Temminck, 1838; *priscus* G. M. Allen, 1914; *rivalis* Spix, 1823; *saxatilis* Spix, 1823; *villosa* Gervais, 1855.
COMMENTS: See Plumpton and Jones (1992); see Emmons (1997) for an updated distribution map.

*Saccopteryx* Illiger, 1811. Prodr. Syst. Mamm. Avium., p. 121.
TYPE SPECIES: *Vespertilio lepturus* Schreber, 1774.
SYNONYMS: *Urocryptus* Temminck, 1838.
COMMENTS: Reviewed by Jones and Hood (1993); also see Muñoz and Cuartas (2001).

*Saccopteryx antioquensis* Muñoz and Cuartas, 2001. Actual. Biol. 23:53.
COMMON NAME: Antioquian Sac-winged Bat.
TYPE LOCALITY: Colombia, Antioquia, Municipality of Sonsón, ca. 15 km along La Soledad road E of Sonsón; 5°40'N, 75°05'W; 1,200 m.
DISTRIBUTION: Known only from the Cordillera Central of N Colombia.
STATUS: IUCN 2003 and IUCN/SSC Action Plan (2001) – Not evaluated (new species).
COMMENTS: Most similar to *gymnura*; see Muñoz and Cuartas (2001).

*Saccopteryx bilineata* (Temminck, 1838). Tijdschr. Nat. Gesch. Physiol., 5:33.
COMMON NAME: Greater Sac-winged Bat.
TYPE LOCALITY: Surinam.
DISTRIBUTION: Jalisco and Veracruz (Mexico) to Bolivia, Guianas, and E Brazil south to Rio de Janiero; Trinidad and Tobago.
STATUS: IUCN 2003 and IUCN/SSC Action Plan (2001) – Lower Risk (lc).
SYNONYMS: *centralis* Thomas, 1904; *insignis* Wagner, 1855; *perspicillifer* Miller, 1899.
COMMENTS: Several subspecies have been recognized, but these do not appear justified; see Simmons and Voss (1998). See Yancey et al. (1998*a*).

*Saccopteryx canescens* Thomas, 1901. Ann. Mag. Nat. Hist., ser. 7, 7:366.
COMMON NAME: Frosted Sac-winged Bat.
TYPE LOCALITY: Brazil, Pará, Obidos.

DISTRIBUTION: Colombia, Venezuela, Guianas, N Brazil, Peru, Bolivia.

STATUS: IUCN 2003 and IUCN/SSC Action Plan (2001) – Lower Risk (lc).

SYNONYMS: *leptura* J. A. Allen, 1900 [preoccupied by *leptura* Schreber, 1774]; **pumila** Thomas, 1914.

COMMENTS: Includes *pumila*; see Husson (1962).

*Saccopteryx gymnura* Thomas, 1901. Ann. Mag. Nat. Hist., ser. 7, 7:367.

COMMON NAME: Amazonian Sac-winged Bat.

TYPE LOCALITY: Brazil, Pará, Santarem.

DISTRIBUTION: Amazonian Brazil, French Guiana, Guyana, perhaps Venezuela.

STATUS: IUCN 2003 and IUCN/SSC Action Plan (2001) – Vulnerable.

COMMENTS: Reviewed by Simmons and Voss (1998) and Lim and Engstrom (2001).

*Saccopteryx leptura* (Schreber, 1774). Die Säugethiere, 1(8):57.

COMMON NAME: Lesser Sac-winged Bat.

TYPE LOCALITY: Surinam.

DISTRIBUTION: Chiapas and Tabasco (Mexico) to SE Brazil, Peru, and N Bolivia; Guianas; Margarita Isl (Venezuela); Trinidad and Tobago.

STATUS: IUCN 2003 and IUCN/SSC Action Plan (2001) – Lower Risk (lc).

COMMENTS: See Yancey et al. (1998*b*). Reviewed in part by Nogueira et al. (2002).

**Family Nycteridae** Van der Hoeven, 1855. Handb. Dierkunde, 2nd ed., 2:1028.

COMMENTS: Monogeneric; see Griffiths (1994) for a phylogeny. Although some authors have indicated that family-group names based on the greek root -nycteris should be spelled -nycterididae (e.g., Russell and Sigé 1970; Habersetzer and Storch, 1987; Kock et al., 2002), I prefer to maintain the commonly accepted spelling (-nycteridae) for these names in the interests of stability (see discussion in Simmons and Geisler [1998: footnote 13]). Accordingly, I use "Nycteridae" for this family instead of "Nycterididae".

*Nycteris* E. Geoffroy and G. Cuvier, 1795. Mag. Encyclop., 2:186.

TYPE SPECIES: *Vespertilio hispidus* Schreber, 1774 [*nomen nudum*, validated by Opinion 111 of the International Commission, 1929].

SYNONYMS: *Petalia* Gray, 1838; *Pelatia* Gray, 1866.

COMMENTS: Hall (1981) disregarded ICZN Opinion 111 and used *Nycteris* Borkhausen, 1797 for the Nearctic genus commonly known as *Lasiurus* Gray, 1831, but few other authors followed this usage and *Nycteris* is now universally used for Slit-faced Bats of the Old World. Revised by Van Cakenberghe and De Vree (1985, 1993*a*, *b*, 1998). Thomas et al. (1994) summarized character variation and identified species groups, which we follow here with modifications based on Van Cakenberghe and De Vree (1993*a*). For a key to the genus see Gray et al. (1999), but note that they did not distinguish all species recognized here.

*Nycteris arge* Thomas, 1903. Ann. Mag. Nat. Hist., ser. 7, 12:633.

COMMON NAME: Bates's Slit-faced Bat.

TYPE LOCALITY: Cameroon, Efulen.

DISTRIBUTION: Sierra Leone to S and E Dem. Rep. Congo; W Kenya; SW Sudan; NE Angola; Bioko (Equatorial Guinea).

STATUS: IUCN 2003 and IUCN/SSC Action Plan (2001) – Lower Risk (lc).

COMMENTS: *arge* species group. Formerly included *intermedia*; see Hayman and Hill (1971), but see Van Cakenberghe and De Vree (1985).

*Nycteris aurita* (K. Andersen, 1912). Ann. Mag. Nat. Hist., ser. 8, 10:547.

COMMON NAME: Andersen's Slit-faced Bat.

TYPE LOCALITY: Kenya, Kitui.

DISTRIBUTION: Ethiopia, S Somalia, N + E Kenya, NE Tanzania.

STATUS: IUCN 2003 and IUCN/SSC Action Plan (2001) – Lower Risk (nt).

COMMENTS: *hispida* species group. Often considered a synonym or subspecies of *hispida*, but apparently distinct; see Van Cakenberghe and De Vree (1993*b*)

*Nycteris gambiensis* (K. Andersen, 1912). Ann. Mag. Nat. Hist., ser. 8, 10:548.
    COMMON NAME: Gambian Slit-faced Bat.
    TYPE LOCALITY: Senegal, Dialakoto (= Dialocote).
    DISTRIBUTION: Senegal, Gambia, Guinea Bissau, Guinea, Ghana, Côte d'Ivoire, Togo, Benin,
        Burkina Faso, Nigeria. A record from Sierra Leone is in error (J. Fahr, pers. comm.).
    STATUS: IUCN 2003 and IUCN/SSC Action Plan (2001) – Lower Risk (lc).
    COMMENTS: *thebaica* species group. Reviewed by Van Cakenberghe and De Vree (1998).

*Nycteris grandis* Peters, 1865. Monatsb. K. Preuss. Akad. Wiss. Berlin, 1865:358.
    COMMON NAME: Large Slit-faced Bat.
    TYPE LOCALITY: "Guinea".
    DISTRIBUTION: Senegal to Dem. Rep. Congo, Kenya, Zimbabwe, Malawi, and Mozambique;
        Zanzibar and Pemba.
    STATUS: IUCN 2003 and IUCN/SSC Action Plan (2001) – Lower Risk (lc).
    SYNONYMS: *baikii* Gray, 1866; *marica* Kershaw, 1923; *proxima* Lönnberg and Gyldenstolpe,
        1925.
    COMMENTS: *hispida* species group. Reviewed by Van Cakenberghe and De Vree (1993*b*).
        *N. marica* is sometimes recognized as a distinct savanna subspecies, but this does not
        seem justified based on morphology; see Van Cakenberghe and De Vree (1993*b*). See
        Hickey and Dunlop (2000).

*Nycteris hispida* (Schreber, 1775). Die Säugethiere, 1:169, 188.
    COMMON NAME: Hairy Slit-faced Bat.
    TYPE LOCALITY: Senegal.
    DISTRIBUTION: Senegal, Gambia, and extreme S Mauritania to Somalia and south to Angola,
        C Mozambique, Botswana, and Malawi; Zanzibar; Bioko (Equatorial Guinea). A South
        African record is dubious; see Cotterill (1996).
    STATUS: IUCN 2003 and IUCN/SSC Action Plan (2001) – Lower Risk (lc).
    SYNONYMS: *daubentoni* E. Geoffroy, 1813; *martini* Fraser, 1834; *pallida* J. A. Allen, 1917; *pilosa*
        Gray, 1866; *poensis* Gray, 1843; *villosa* Peters, 1852.
    COMMENTS: *hispida* species group. Revised by Van Cakenberghe and De Vree (1993*b*); also
        see Koopman (1975). Does not include *aurita*. Several subspecies are often recognized,
        but these do not seem justified; see Van Cakenberghe and De Vree (1993*b*).

*Nycteris intermedia* Aellen, 1959. Arch. Sci. Genève, 12:218.
    COMMON NAME: Intermediate Slit-faced Bat.
    TYPE LOCALITY: Côte d'Ivoire, Adiopodoume.
    DISTRIBUTION: Liberia to W Tanzania and south to Angola.
    STATUS: IUCN 2003 and IUCN/SSC Action Plan (2001) – Lower Risk (nt).
    COMMENTS: *arge* species group. Formerly included in *arge* but see Van Cakenberghe and
        De Vree (1985).

*Nycteris javanica* E. Geoffroy, 1813. Ann. Mus. Natn. Hist. Nat. Paris, 20:20.
    COMMON NAME: Javan Slit-faced Bat.
    TYPE LOCALITY: Indonesia, Java.
    DISTRIBUTION: Java, Nusa Penida (near Bali), and Kangean Isl (Indonesia). A record from Bali
        is in error (see Kock and Dobat, 2000), as is a record from Timor (see Corbet and Hill,
        1992).
    STATUS: IUCN 2003 and IUCN/SSC Action Plan (2001) – Vulnerable.
    SYNONYMS: **bastiani** Bergmans and van Bree, 1986.
    COMMENTS: *javanica* species group. Does not include *tragata*; see Ellerman and Morrison-
        Scott (1955) and Van Cakenberge and De Vree (1993*a*), but also see Corbet and Hill
        (1992). Reviewed by Bergmans and van Bree (1986) and Van Cakenberghe and De Vree
        (1993*a*).

*Nycteris macrotis* Dobson, 1876. Monogr. Asiat. Chiroptera, p. 80.
    COMMON NAME: Large-eared Slit-faced Bat.
    TYPE LOCALITY: Sierra Leone.
    DISTRIBUTION: Senegal and Gambia to Ethiopia, south to Zimbabwe, Malawi and
        Mozambique; Zanzibar.

STATUS: IUCN 2003 and IUCN/SSC Action Plan (2001) – Lower Risk (lc).

SYNONYMS: *aethiopica* Dobson, 1878; *aurantiaca* Monard, 1939; *guineensis* Monard, 1939; *luteola* Thomas, 1901; *oriana* Kershaw, 1922.

COMMENTS: *macrotis* species group. For discussion of synonyms see Koopman (1975, 1992), Van Cakenberghe and De Vree (1985), and Kock (1969a). Does not include *madagascariensis*; see Peterson et al. (1995). Does not include *vinsoni*; see Van Cakenberghe and De Vree (1998). See Taylor (2000a) for distribution map.

*Nycteris madagascariensis* Grandidier, 1937. Bull. Mus. Natn. Hist. Nat. Paris, ser. 2, 9:353.

COMMON NAME: Malagasy Slit-faced Bat.

TYPE LOCALITY: Madagascar, N of Ankarana, Vallé de la Rodo (= Irodo), 12°05'S, 49°05'E.

DISTRIBUTION: N Madagascar.

STATUS: IUCN 2003 and IUCN/SSC Action Plan (2001) – Data Deficient.

COMMENTS: *macrotis* species group. Known from only two specimens. Often included in *macrotis* (e.g., Koopman, 1993, 1994; Van Cakenberghe and De Vree, 1985) but see Peterson et al. (1995).

*Nycteris major* (K. Andersen, 1912). Ann. Mag. Nat. Hist., ser. 8, 10:547.

COMMON NAME: Dja Slit-faced Bat.

TYPE LOCALITY: Cameroon, Ja (= Dja) River.

DISTRIBUTION: Liberia, Côte d'Ivoire, Cameroon, Dem. Rep. Congo, and Zambia.

STATUS: IUCN 2003 and IUCN/SSC Action Plan (2001) – Vulnerable.

SYNONYMS: *avakubia* J. A. Allen, 1917.

COMMENTS: *arge* species group. Includes *avakubia*; see Koopman (1965) and Van Cakenberghe and De Vree (1985).

*Nycteris nana* (K. Andersen, 1912). Ann. Mag. Nat. Hist., ser. 8, 10:547.

COMMON NAME: Dwarf Slit-faced Bat.

TYPE LOCALITY: Equatorial Guinea, Rio Muni, Benito River.

DISTRIBUTION: Côte d'Ivoire to NE Angola, W Kenya, and SW Sudan. A record from Tanzania actually represents *intermedia* (J. Fahr, pers. comm.).

STATUS: IUCN 2003 and IUCN/SSC Action Plan (2001) – Lower Risk (lc).

SYNONYMS: *tristis* Allen and Lawrence, 1936.

COMMENTS: *arge* species group. Includes *tristis*; see Koopman (1975) and Van Cakenberghe and De Vree (1985).

*Nycteris parisii* De Beaux, 1924. Atti. Soc. Ital. Sci. Nat., 62:254.

COMMON NAME: Parisi's Slit-faced Bat.

TYPE LOCALITY: Somalia, Bali.

DISTRIBUTION: Cameroon; S Somalia; Ethiopia.

STATUS: IUCN 2003 – Not evaluated; not considered in IUCN/SSC Microchiropteran Bats Action Plan (2001).

SYNONYMS: *benuensis* Aellen, 1952.

COMMENTS: *macrotis* species group. Distinct from *woodi*; see Thomas et al. (1995), but see also Van Cakenberghe and De Vree (1985).

*Nycteris thebaica* E. Geoffroy, 1818. Descrip. de L'Egypte, 2:119.

COMMON NAME: Egyptian Slit-faced Bat.

TYPE LOCALITY: Egypt, Thebes (near Luxor).

DISTRIBUTION: Central Arabia, Israel, Sinai, Egypt, Morocco, Senegal, Guinea, Mali, Burkina Faso, Ghana, Benin, Niger, Nigeria, Somalia, Djibouti, and Kenya, south to South Africa in open country; Zanzibar and Pemba.

STATUS: IUCN 2003 and IUCN/SSC Action Plan (2001) – Lower Risk (lc).

SYNONYMS: *albiventer* Wagner, 1840; *geoffroyi* Desmarest, 1820; *senegalensis* Hartmann, 1868; *adana* K. Anderen, 1912; *angolensis* Peters, 1870; *brockmani* K. Andersen, 1912; *media* K. Andersen, 1912; *capensis* A. Smith, 1829; *affinis* A. Smith, 1829; *discolor* Wagner, 1840; *fuliginosa* Peters, 1852; *damarensis* Peters, 1870; *labiata* Heuglin, 1861; *aurantiaca* De Beaux, 1923; *revoilii* Robin, 1881; *najdiya* Nader and Kock, 1982. **Unassigned:** *aethiopicus* Heuglin and Fitzinger, 1866 [*nomen nudum*].

COMMENTS: *thebaica* species group. Reviewed by Van Cakenberghe and De Vree (1998). The status of *brockmani* and *damarensis* remains unclear; these forms may represent distinct species, see discussion in Van Cakenberghe and De Vree (1998). Also see Harrison and Bates (1991) and Gray et al. (1999).

*Nycteris tragata* (K. Andersen, 1912). Ann. Mag. Nat. Hist., ser. 8, 10:546.
    COMMON NAME: Malayan Slit-faced Bat.
    TYPE LOCALITY: Malaysia, Sarawak, Bidi caves.
    DISTRIBUTION: Burma, Thailand, W Malaysia, Sumatra, Borneo.
    STATUS: IUCN 2003 and IUCN/SSC Action Plan (2001) – Lower Risk (lc).
    COMMENTS: *javanica* species group. Distinct from *javanica*; see Ellerman and Morrison-Scott (1955) and Van Cakenberghe and De Vree (1993*a*), but also see Corbet and Hill (1992).

*Nycteris vinsoni* Dalquest, 1965. J. Mammal., 46:254.
    COMMON NAME: Vinson's Slit-faced Bat.
    TYPE LOCALITY: Mozambique, Zinave.
    DISTRIBUTION: Mozambique; known only from the type locality.
    STATUS: IUCN 2003 – not evaluated; not considered in IUCN/SSC Action Plan (2001).
    COMMENTS: *thebaica* species group. Distinct from *macrotis* and *thebaica*; see Van Cakenberghe and De Vree (1998).

*Nycteris woodi* K. Andersen, 1914. Ann. Mag. Nat. Hist., ser. 8, 13:563.
    COMMON NAME: Wood's Slit-faced Bat.
    TYPE LOCALITY: Zambia, Chilanga.
    DISTRIBUTION: Zambia and South Africa to NW Mozambique and SW Tanzania.
    STATUS: IUCN 2003 and IUCN/SSC Action Plan (2001) – Lower Risk (nt).
    SYNONYMS: **sabiensis** Roberts, 1946.
    COMMENTS: *macrotis* species group. For synonyms see Van Cakenberghe and De Vree (1985). Does not include *parisii* or *benuensis*; see Thomas et al. (1995).

**Family Myzopodidae** Thomas, 1904. Proc. Zool. Soc. Lond., 1904(2):5.
    COMMENTS: Monotypic.

*Myzopoda* Milne-Edwards and Grandidier, 1878. Bull. Sci. Soc. Philom. Paris, sér. 7, 2:220.
    TYPE SPECIES: *Myzopoda aurita* Milne-Edwards and Grandidier, 1878.

    *Myzopoda aurita* Milne-Edwards and Grandidier, 1878. Bull. Sci. Soc. Philom. Paris, sér. 7, 2:220.
    COMMON NAME: Sucker-footed Bat.
    TYPE LOCALITY: Madagascar.
    DISTRIBUTION: Madagascar.
    STATUS: IUCN 2003 and IUCN/SSC Action Plan (2001) – Vulnerable.
    COMMENTS: See Schliemann and Maas (1978) and Peterson et al. (1995).

**Family Mystacinidae** Dobson, 1875. Ann. Mag. Nat. Hist., ser. 4, 16:349.
    COMMENTS: Monogeneric.

*Mystacina* Gray, 1843. *In* Dieffenbach, Travels in New Zealand, 2:296.
    TYPE SPECIES: *Mystacina tuberculata* Gray, 1843 (by ICZN ruling, Opinion 1994 [2002]).
    SYNONYMS: *Mystacops* Lydeckker, 1891 [placed on the Offical List of Rejected and Invalid Generic Names in Zoology; ICZN Opinion 1994 (2002)].
    COMMENTS: Revised by Hill and Daniel (1985); see Lloyd (2001) for a review and key to species.

*Mystacina robusta* Dwyer, 1962. Zool. Publ. Victoria Univ., Wellington, 28:3.
    COMMON NAME: New Zealand Greater Short-tailed Bat.
    TYPE LOCALITY: New Zealand, Big South Cape Isl.
    DISTRIBUTION: New Zealand.
    STATUS: IUCN 2003 and IUCN/SSC Action Plan (2001) – Extinct.
    COMMENTS: See Flannery (1995*b*) and Lloyd (2001).

*Mystacina tuberculata* Gray, 1843. Mammalia, *in* Voy. "Sulphur," Zool., p. 23.
    COMMON NAME: New Zealand Lesser Short-tailed Bat.

TYPE LOCALITY: New Zealand.

DISTRIBUTION: New Zealand.

STATUS: IUCN 2003 and IUCN/SSC Action Plan (2001) – Vulnerable.

SYNONYMS: *aupourica* Hill and Daniel, 1985; *rhyacobia* Hill and Daniel, 1985; *velutina* Hutton, 1872 [see comments].

COMMENTS: Mayer et al. (1999) and Mayer and Kirsch (2000) have argued that the correct name for this species is *velutina* Hutton, 1872. However, Spencer and Lee (1999, 2000) disagreed, and filed a petition with the International Commission on Zoological Nomeclature to conserve *tuberculata* as the name for this species. This petition was upheld in Opinion 1994 of the International Commission on Zoological Nomenclature (2002), which conserved *tuberculata* Gray, 1843 as the name for this species and which placed *velutina* Hutton, 1872 on the Official List of Rejected and Invalid Specific Names in Zoology. Subspecies limits recognized previously (e.g., by Hill and Daniel, 1985) do not correspond to observed patterns of genetic variation (Lloyd, 2003); accordingly no subspecies are recognized here pending a thorough revision of this taxon.

**Family Phyllostomidae** Gray, 1825. Zool. Journ., 2(6):242.

COMMENTS: Includes Desmodontidae; see Jones and Carter (1976). For use of Phyllostomidae rather than Phyllostomatidae, see Handley (1980). The classification used here generally follows that of Wetterer et al. (2000), which was based on a phylogenetic analysis of morphological data, restriction sites, and sex chromosomes; all genera, tribes, and subfamilies appear to be monophyletic unless otherwise noted. See Baker et al. (2000) for an alternative phylogeny based on mtDNA sequence data; also see Baker et al. (1989). Carstens et al. (2002) provided a updated phylogeny of the nectar-feeding subfamilies (Brachyphyllinae, Phyllonycterinae, and Glossophaginae) based on combined analysis of morphological and molecular data.

**Subfamily Desmodontinae** Bonaparte, 1845. Cat. Met. Mamm. Europe, p. 5.

COMMENTS: Formerly treated as a separate family; see Jones and Carter (1976). See Emmons (1997) for distribution maps.

*Desmodus* Wied-Neuwied, 1826. Beitr. Naturgesch. Brasil, 2:231.

TYPE SPECIES: *Desmodus rufus* Wied-Neuwied, 1824 (= *Phyllostoma rotundus* E. Geoffroy, 1810).

SYNONYMS: *Desmodon* Elliot, 1905; *Edostoma* D'Orbigny, 1834-36.

*Desmodus rotundus* (E. Geoffroy, 1810). Ann. Mus. Natn. Hist. Nat. Paris, 15:181.

COMMON NAME: Common Vampire Bat.

TYPE LOCALITY: Paraguay, Asunción (restricted by Cabrera, 1958).

DISTRIBUTION: Uruguay, N Argentina, Paraguay, Bolivia, and N Chile north to Sonora, Nuevo León and Tamaulipas (Mexico); Margarita Isl (Venezuela); Trinidad.

STATUS: IUCN 2003 and IUCN/SSC Action Plan (2001) – Lower Risk (lc).

SYNONYMS: *cinerea* D'Orbigny, 1834; *dorbignyi* Waterhouse, 1838; *ecaudatus* Schinz, 1821; *fuscus* Burmeister, 1854; *mordax* Burmeister, 1879; *murinus* Wagner, 1840; *rufus* Wied-Neuwied, 1824.

COMMENTS: See Greenhall et al. (1983).

*Diaemus* Miller, 1906. Proc. Biol. Soc. Wash., 19:84.

TYPE SPECIES: *Desmodus youngi* Jentink, 1893.

COMMENTS: Included in *Desmodus* by Handley (1976) and Anderson (1997), but more often treated as a distinct genus; see Greenhall and Schutt (1996).

*Diaemus youngi* (Jentink, 1893). Notes Leyden Mus., 15:282.

COMMON NAME: White-winged Vampire Bat.

TYPE LOCALITY: Guyana, Berbice River, upper Canje Creek.

DISTRIBUTION: Tamaulipas (Mexico) south to N Argentina, Bolivia, Paraguay, and E Brazil; Trinidad; Margarita Isl (Venezuela).

STATUS: IUCN 2003 and IUCN/SSC Action Plan (2001) – Lower Risk (lc).

SYNONYMS: *cypselinus* Thomas, 1928.
COMMENTS: See Greenhall and Schutt (1996). Sometimes spelled *youngii*, but *youngi* is the
original spelling.

*Diphylla* Spix, 1823. Sim. Vespert. Brasil., p. 68.
TYPE SPECIES: *Diphylla ecaudata* Spix, 1823.
SYNONYMS: *Haematonycteris* H. Allen, 1896.

*Diphylla ecaudata* Spix, 1823. Sim. Vespert. Brasil., p. 68.
COMMON NAME: Hairy-legged Vampire Bat.
TYPE LOCALITY: Brazil, Bahia, San Francisco River.
DISTRIBUTION: S Tamaulipas (Mexico) to Venezuela, Peru, Bolivia, and E Brazil; a single
vagrant individual has also been reported from S Texas (USA).
STATUS: IUCN 2003 and IUCN/SSC Action Plan (2001) – Lower Risk (nt).
SYNONYMS: *centralis* Thomas, 1903; *diphylla* Fischer, 1829.
COMMENTS: See Greenhall et al. (1984).

**Subfamily Brachyphyllinae** Gray, 1866. Proc. Zool. Soc. Lond., 1866:115.
COMMENTS: Treated as tribe within Glossophaginae by McKenna and Bell (1997).

*Brachyphylla* Gray, 1834. Proc. Zool. Soc. Lond., 1833:122 [1834].
TYPE SPECIES: *Brachyphylla cavernarum* Gray, 1834.
COMMENTS: Revised by Swanepoel and Genoways (1978). A key to this genus was presented by
Swanepoel and Genoways (1983*a*).

*Brachyphylla cavernarum* Gray, 1834. Proc. Zool. Soc. Lond., 1833:123 [1834].
COMMON NAME: Antillean Fruit-eating Bat.
TYPE LOCALITY: St. Vincent (Lesser Antilles, UK).
DISTRIBUTION: Puerto Rico, Virgin Isls and throughout Lesser Antilles south to St. Vincent
and Barbados.
STATUS: IUCN 2003 and IUCN/SSC Action Plan (2001) – Lower Risk (lc).
SYNONYMS: ***intermedia*** Swanepoel and Genoways, 1978; ***minor*** Miller, 1913.
COMMENTS: Includes *minor*; see Swanepoel and Genoways (1978) and Varona (1974).
Reviewed by Swanepoel and Genoways (1983*a*) and Timm and Genoways (2003).

*Brachyphylla nana* Miller, 1902. Proc. Acad. Nat. Sci. Phil., 54:409.
COMMON NAME: Cuban Fruit-eating Bat.
TYPE LOCALITY: Cuba, Pinar del Río, El Guama.
DISTRIBUTION: Cuba, Hispaniola, Jamaica (extinct, known only from fossils), Grand
Cayman (Cayman Isls, UK), Middle Caicos (SE Bahamas).
STATUS: IUCN 2003 and IUCN/SSC Action Plan (2001) – Lower Risk (nt).
SYNONYMS: *pumila* Miller, 1918.
COMMENTS: Includes *pumila*; see Jones and Carter (1976) and Swanepoel and Genoways
(1978). Considered a subspecies of *cavernarum* by Buden (1977) and Hall (1981).
Reviewed by Swanepoel and Genoways (1983*b*) and Timm and Genoways (2003).

**Subfamily Phyllonycterinae** Miller, 1907. Bull. U.S. Natl. Mus., 57:171.
COMMENTS: Treated as tribe within Glossophaginae by McKenna and Bell (1997).

*Erophylla* Miller, 1906. Proc. Biol. Soc. Wash., 19:84.
TYPE SPECIES: *Phyllonycteris bombifrons* Miller, 1899.
COMMENTS: Revised by Buden (1976). Included as a subgenus of *Phyllonycteris* by Varona
(1974).

*Erophylla bombifrons* (Miller, 1899). Proc. Biol. Soc. Wash., 13:36.
COMMON NAME: Brown Flower Bat.
TYPE LOCALITY: Puerto Rico, cave near Bayamón.
DISTRIBUTION: Hispaniola and Puerto Rico.
STATUS: IUCN 2003 – Not evaluated; not considered in IUCN/SSC Action Plan (2001).

SYNONYMS: *santacristobalensis* Elliot, 1905.
COMMENTS: Included in *sezekorni* by Buden (1976), but see Varona (1974), Hall (1981), and
Koopman (1993). Reviewed by Timm and Genoways (2003).

*Erophylla sezekorni* (Gundlach, 1861). Monatsb. K. Preuss. Akad. Wiss. Berlin, 1860:818 [1861].
COMMON NAME: Buffy Flower Bat.
TYPE LOCALITY: Cuba, Pinar del Río, Santa Cruz de los Pinos, Rangel.
DISTRIBUTION: Cuba, Jamaica, Bahamas, and Cayman Isls.
STATUS: IUCN 2003 and IUCN/SSC Action Plan (2001) – Lower Risk (lc).
SYNONYMS: *mariguanensis* Shamel, 1931; *planifrons* Miller, 1899; *syops* G. M. Allen, 1917.
COMMENTS: Does not include *bombifrons*; see Varona (1974), Hall (1981), and Koopman
(1993). Reviewed by Baker et al. (1978), but note that they included *bombifrons* in this
taxon. Reviewed by Timm and Genoways (2003).

*Phyllonycteris* Gundlach, 1861. Monatsb. K. Preuss. Akad. Wiss. Berlin, 1860:817 [1861].
TYPE SPECIES: *Phyllonycteris poeyi* Gundlach, 1861.
SYNONYMS: *Reithronycteris* Miller, 1898; *Rhithronycteris* Elliot, 1904.
COMMENTS: Two subgenera are recognized, *Phyllonycteris* and *Reithronycteris*.

*Phyllonycteris aphylla* (Miller, 1898). Proc. Acad. Nat. Sci. Phil., 50:334.
COMMON NAME: Jamaican Flower Bat.
TYPE LOCALITY: Jamaica.
DISTRIBUTION: Jamaica.
STATUS: IUCN 2003 and IUCN/SSC Action Plan (2001) – Endangered.
COMMENTS: Subgenus *Reithronycteris*.

*Phyllonycteris major* Anthony, 1917. Bull. Amer. Mus. Nat. Hist., 37:567.
COMMON NAME: Puerto Rican Flower Bat.
TYPE LOCALITY: Puerto Rico, Cueva Catedral near Morovis.
DISTRIBUTION: Puerto Rico.
STATUS: IUCN 2003 and IUCN/SSC Action Plan (2001) – Extinct.
COMMENTS: Subgenus *Phyllonycteris*. Known only from subfossil skeletal material from the
type locality; see Hall (1981).

*Phyllonycteris poeyi* Gundlach, 1861. Monatsb. K. Preuss. Akad. Wiss. Berlin, 1860:817 [1861].
COMMON NAME: Cuban Flower Bat.
TYPE LOCALITY: Cuba, Matanzas, Canimar (cafetal "San Antonio el Fundador").
DISTRIBUTION: Cuba, Isle of Pines, Hispaniola.
STATUS: IUCN 2003 and IUCN/SSC Action Plan (2001) – Lower Risk (nt).
SYNONYMS: *obtusa* Miller, 1929.
COMMENTS: Subgenus *Phyllonycteris*. Includes *obtusa*; see Jones and Carter (1976) and
Klingener et al. (1978). Also see Hall (1981), but note that he was unaware that *obtusa*
was not extinct. Reviewed by Timm and Genoways (2003).

**Subfamily Glossophaginae** Bonaparte, 1845. Cat. Met. Mamm. Europe, p. 5.
SYNONYMS: Lonchophyllinae Griffiths, 1982.
COMMENTS: Includes two tribes, Glossophagini and Lonchophillini, which are recognized as
separate subfamilies by some authors. Baker et al. (2000) suggested that this subfamily
may not be monophyletic, but see Carstens et al. (2002), who recovered a monophyletic
Glossophaginae in a combined analysis of molecular and morphological data.
Morphology, biology, and evolution reviewed by Solmsen (1998).

**Tribe Glossophagini** Bonaparte, 1845. Cat. Met. Mamm. Europe, p. 5.
COMMENTS: Equivalent to Glossophaginae as used by Griffiths (1982). See discussion in
Wetterer et al. (2000); also see Carstens et al. (2002). Informally divided into to two
groups by Carstens et al. (2002); "choeronycterines" (*Anoura, Choeronycteris, Choeroniscus,
Hylonycteris, Lichonycteris, Musonycteris,* and *Scleronycteris*), and "glossophagines"
(*Glossophaga, Leptonycteris,* and *Monophyllus*).

*Anoura* Gray, 1838. Mag. Zool. Bot., 2:490.
    TYPE SPECIES: *Anoura geoffroyi* Gray, 1838.
    SYNONYMS: *Anura* Agassiz, 1846; *Glossonycteris* Peters, 1868; *Lonchoglossa* Peters, 1868.
    COMMENTS: Includes *Lonchoglossa*; see Cabrera (1958). Keys to species of *Anoura* were provided
        by Tamsitt and Nagorsen (1982) and Handley (1984), but usefulness of these keys has
        been reduced by subsequent descriptions of new species (i.e., by Handley [1984] and
        Molinari [1994]) and suggestions that other undescribed species exist (Emmons, 1997).

*Anoura caudifer* (E. Geoffroy, 1818). Mem. Mus. Natn. Hist. Nat. Paris, 4:418.
    COMMON NAME: Tailed Tailless Bat.
    TYPE LOCALITY: Brazil, Rio de Janeiro.
    DISTRIBUTION: Colombia, Venezuela, Guianas, Brazil, Ecuador, Peru, Bolivia, NW Argentina.
    STATUS: IUCN 2003 and IUCN/SSC Action Plan (2001) – Lower Risk (lc) as *Anoura caudifera*.
    SYNONYMS: *aequatoris* Lönnberg, 1921; *ecaudata* Geoffroy, 1818; *wiedii* Peters, 1869.
    COMMENTS: Some specimens previously referred to *caudifer* may represent *luismanueli*; see
        Molinari (1994); also see Cadena et al. (1998). Often spelled "*caudifera*" (see Handley,
        1984), but the correct spelling is *caudifer* according to Article 31.2.2 of the Code of the
        International Commission on Zoological Nomenclature (International Commission
        on Zoological Nomenclature, 1999).

*Anoura cultrata* Handley, 1960. Proc. U.S. Natl. Mus., 112:463.
    COMMON NAME: Handley's Tailless Bat.
    TYPE LOCALITY: Panama, Darién, Río Pucro, Tacarcuna Village, 3,200 ft. (1,033 m); 08°10′N,
        77°18′W.
    DISTRIBUTION: Costa Rica, Panama, Venezuela, Colombia, Ecuador, Peru, Bolivia.
    STATUS: IUCN 2003 and IUCN/SSC Action Plan (2001) – Lower Risk (lc).
    SYNONYMS: *brevirostrum* Carter, 1968; *werckleae* Starrett, 1969.
    COMMENTS: Includes *brevirostrum* and *werckleae*; see Nagorsen and Tamsitt (1981) and
        Molinari (1994). See Tamsitt and Nagorsen (1982).

*Anoura geoffroyi* Gray, 1838. Mag. Zool. Bot., 2:490.
    COMMON NAME: Geoffroy's Tailless Bat.
    TYPE LOCALITY: Brazil, Rio de Janeiro.
    DISTRIBUTION: Peru, Bolivia, SE Brazil, the Guianas and Ecuador to Tamaulipas and Sinaloa
        (Mexico); Trinidad; Grenada (Lesser Antilles).
    STATUS: IUCN 2003 and IUCN/SSC Action Plan (2001) – Lower Risk (lc).
    SYNONYMS: **lasiopyga** Peters, 1868; **peruana** Tschudi, 1844; *antricola* Anthony, 1921;
        *apolinari* J. A. Allen, 1916.
    COMMENTS: Subspecies reviewed by Sanborn (1933); also see Arroyo-Cabrales and Gardner
        (2003).

*Anoura latidens* Handley, 1984. Proc. Biol. Soc. Wash., 97:503.
    COMMON NAME: Broad-toothed Tailless Bat.
    TYPE LOCALITY: Venezuela, Distrito Federal, Pico Avila.
    DISTRIBUTION: Venezuela, Guyana, Colombia, Peru.
    STATUS: IUCN 2003 and IUCN/SSC Action Plan (2001) – Lower Risk (nt).
    COMMENTS: Clearly distinct from *geoffroyi*; see Solari et al. (1999).

*Anoura luismanueli* Molinari, 1994. Trop. Zool., 7:76.
    COMMON NAME: Luis Manuel's Tailless Bat.
    TYPE LOCALITY: Venezuela, Estado Mérida, 4 km E Bailadores, inside the Cueva del Salado,
        2,000 m.
    DISTRIBUTION: Andes of Venezuela.
    STATUS: IUCN 2003 and IUCN/SSC Action Plan (2001) – Data deficient.
    COMMENTS: Distinct from *caudifer*; see Molinari (1994), Cadena et al. (1998), and Lim and
        Engstrom (2001).

*Choeroniscus* Thomas, 1928. Ann. Mag. Nat. Hist., ser. 10, 1:122.
    TYPE SPECIES: *Choeronycteris minor* Peters, 1868.

*Choeroniscus godmani* (Thomas, 1903). Ann. Mag. Nat. Hist., ser. 7, 11:288.
  COMMON NAME: Godman's Long-tongued Bat.
  TYPE LOCALITY: Guatemala.
  DISTRIBUTION: Sinaloa (Mexico) to Colombia, Venezuela, Guyana, and Surinam.
  STATUS: IUCN 2003 and IUCN/SSC Action Plan (2001) – Lower Risk (nt).

*Choeroniscus minor* (Peters, 1868). Monatsb. K. Preuss. Akad. Wiss. Berlin, 1868:366.
  COMMON NAME: Lesser Long-tongued Bat.
  TYPE LOCALITY: Surinam.
  DISTRIBUTION: Guianas, Venezuela, Trinidad, Amazonian Brazil, C Colombia, Ecuador, Peru,
      Bolivia.
  STATUS: IUCN 2003 and IUCN/SSC Action Plan (2001) – Lower Risk (nt) as *C. intermedius*;
      Lower Risk (lc) as *C. minor*.
  SYNONYMS: *inca* Thomas, 1912; *intermedius* J. A. Allen and Chapman, 1893.
  COMMENTS: Includes *intermedius*; see Simmons and Voss (1998).

*Choeroniscus periosus* Handley, 1966. Proc. Biol. Soc. Wash., 79:84.
  COMMON NAME: Greater Long-tongued Bat.
  TYPE LOCALITY: Colombia, Valle, 27 km S Buenaventura, Río Raposo.
  DISTRIBUTION: NW Venezuela, W Colombia, W Ecuador.
  STATUS: IUCN 2003 and IUCN/SSC Action Plan (2001) – Vulnerable.
  SYNONYMS: **ponsi** Pirlot, 1967.
  COMMENTS: Includes *ponsi*; see Koopman (1994).

*Choeronycteris* Tschudi, 1844. Fauna Peruana, 1:70.
  TYPE SPECIES: *Choeronycteris mexicana* Tschudi, 1844.

*Choeronycteris mexicana* Tschudi, 1844. Fauna Peruana, 1:72.
  COMMON NAME: Mexican Long-tongued Bat.
  TYPE LOCALITY: Mexico.
  DISTRIBUTION: Honduras and El Salvador to S California, Nevada, Arizona, and New Mexico
      (USA); a single record from S Texas; perhaps Venezuela (Koopman, 1993).
  STATUS: IUCN 2003 and IUCN/SSC Action Plan (2001) – Lower Risk (nt).
  COMMENTS: Does not include *ponsi*, which is now recognized as a subspecies of *Choeroniscus
      periosus*; see Koopman (1994). See Arroyo-Cabrales et al. (1987).

*Glossophaga* E. Geoffroy, 1818. Mem. Mus. Natn. Hist. Nat. Paris, 4:418.
  TYPE SPECIES: *Vespertilio soricinus* Pallas, 1766.
  SYNONYMS: *Nicon* Gray, 1847; *Phyllophora* Gray, 1838.
  COMMENTS: Revised by Webster (1993); see Hoffmann and Baker (2001) for a phylogeny of the
      genus.

*Glossophaga commissarisi* Gardner, 1962. Los Angeles Cty. Mus. Contrib. Sci., 54:1.
  COMMON NAME: Commissaris's Long-tongued Bat.
  TYPE LOCALITY: Mexico, Chiapas, 10 km SE Tonala.
  DISTRIBUTION: Sinaloa (Mexico) to Panama; SE Colombia; E Ecuador; E Peru; NW Brazil.
  STATUS: IUCN 2003 and IUCN/SSC Action Plan (2001) – Lower Risk (lc).
  SYNONYMS: **bakeri** Webster and Jones, 1987; **hespera** Webster and Jones, 1982.
  COMMENTS: See Webster and Jones (1993).

*Glossophaga leachii* Gray, 1844. Mammalia, *in* Zool. Voy. "Sulfur," 1:18.
  COMMON NAME: Gray's Long-tongued Bat.
  TYPE LOCALITY: Nicaragua, Chinandega, Realejo.
  DISTRIBUTION: Costa Rica north to Jalisco, Michoacan, Morelos, Tlaxcala, and Colima
      (Mexico).
  STATUS: IUCN 2003 and IUCN/SSC Action Plan (2001) – Lower Risk (lc).
  SYNONYMS: *alticola* Davis, 1944; *caudifer* Gray, 1847.
  COMMENTS: Originally considered a subspecies of *soricina*; see Jones and Carter (1976).
      Includes *alticola*; see Webster and Jones (1980). See Webster and Jones (1984). Solmsen
      (1998) treated *morenoi* as a synonym of *leachii* with no comment.

*Glossophaga longirostris* Miller, 1898. Proc. Acad. Nat. Sci. Phil., 50:330.
    COMMON NAME: Miller's Long-tongued Bat.
    TYPE LOCALITY: Colombia, Magdalena, Sierra Nevada de Santa Marta.
    DISTRIBUTION: Colombia; Venezuela (including Margarita Isl); N Brazil; Guyana; Trinidad
        and Tobago; Grenada, St Vincent, Curaçao, Bonaire, and Aruba (Lesser Antilles). The
        records from Dominica and Ecuador are erroneous.
    STATUS: IUCN 2003 and IUCN/SSC Action Plan (2001) – Lower Risk (lc).
    SYNONYMS: **campestris** Webster and Handley, 1986; **elongata** Miller, 1900; **major** Goodwin,
        1958; **maricelae** Soriano, Fariñas, and Naranjo, 2000; **reclusa** Webster and Handley,
        1986; **rostrata** Miller, 1913.
    COMMENTS: Includes *elongata*; see Jones and Carter (1976) and Koopman (1958). Revised by
        Webster and Handley (1986); also see Handley and Webster (1987), Webster et al.
        (1998), Soriano et al. (2000), and Timm and Genoways (2003).

*Glossophaga morenoi* Martínez and Villa-R., 1938. Anal. Inst. Biol. Univ. Nac. Auto. Mexico, 9:347.
    COMMON NAME: Western Long-tongued Bat.
    TYPE LOCALITY: Mexico, Oaxaca, Río Guamol, 34 mi. (55 km) S (by Hwy. 190) La Ventosa Jct.
    DISTRIBUTION: Chiapas to Michoacan and Tlaxcala (Mexico).
    STATUS: IUCN 2003 and IUCN/SSC Action Plan (2001) – Lower Risk (nt).
    SYNONYMS: **brevirostris** Webster and Jones, 1984; **mexicana** Webster and Jones, 1980.
    COMMENTS: Includes *mexicana*; see Gardner (1986) and Webster (1993). See Webster and
        Jones (1985). Solmsen (1998) treated *morenoi* as a synonym of *leachii,* and *brevirostris* as
        a subspecies of *mexicana* (listed as a distinct species) with no comment.

*Glossophaga soricina* (Pallas, 1766). Misc. Zool., p. 48.
    COMMON NAME: Pallas's Long-tongued Bat.
    TYPE LOCALITY: Surinam.
    DISTRIBUTION: Tamaulipas, Sonora and Trés Marías Isls (Mexico) south to the Guianas,
        SE Brazil, N Argentina, Paraguay, Bolivia, and Peru; Margarita Isl (Venezuela); Trinidad;
        Grenada (Lesser Antilles); Jamaica; perhaps Bahama Isls.
    STATUS: IUCN 2003 and IUCN/SSC Action Plan (2001) – Lower Risk (lc).
    SYNONYMS: *amplexicaudata* Spix, 1823; *microtis* Miller, 1913; *nigra* Gray, 1844; *truei* H. Allen,
        1897; *villosa* H. Allen, 1896; **antillarum** Rehn, 1902; **handleyi** Webster and Jones,
        1980; **mutica** Merriam, 1898; **valens** Miller, 1913.
    COMMENTS: Reviewed by Alvarez et al. (1991); also see Timm and Genoways (2003).
        Phylogeography discussed by Ditchfield (2000) and Hoffmann and Baker (2001). May
        contain more than one species, see Hoffmann and Baker (2001).

*Hylonycteris* Thomas, 1903. Ann. Mag. Nat. Hist., ser. 7, 11:286.
    TYPE SPECIES: *Hylonycteris underwoodi* Thomas, 1903.

*Hylonycteris underwoodi* Thomas, 1903. Ann. Mag. Nat. Hist., ser. 7, 11:287.
    COMMON NAME: Underwood's Long-tongued Bat.
    TYPE LOCALITY: Costa Rica, San José, Rancho Redondo.
    DISTRIBUTION: W Panama to Nayarit and Veracruz (Mexico).
    STATUS: IUCN 2003 and IUCN/SSC Action Plan (2001) – Lower Risk (nt).
    SYNONYMS: **minor** Phillips and Jones, 1971.
    COMMENTS: Includes *minor*, but see Alvarez and Alvarez-Castañeda (1991). See Jones and
        Homan (1974). See Reid (1997) for distribution map.

*Leptonycteris* Lydekker, 1891. *In* Flower and Lydekker, Intro. Mamm. Living and Extinct, p. 674.
    TYPE SPECIES: *Ischnoglossa nivalis* Saussure, 1860.
    SYNONYMS: *Ischnoglossa* Saussure, 1860 [not Kraatz, 1856].
    COMMENTS: Revised by Arita and Humphrey (1988). A key for the genus was presented by
        Hensley and Wilkins (1988).

*Leptonycteris curasoae* Miller, 1900. Proc. Biol. Soc. Wash., 13:126.
    COMMON NAME: Curaçaoan Long-nosed Bat.
    TYPE LOCALITY: Curaçao, Willemstad (Netherlands).

DISTRIBUTION: NE Colombia, N Venezuela, Margarita Isl, Curaçao, Bonaire and Aruba (Netherlands Antilles).

STATUS: IUCN 2003 and IUCN/SSC Action Plan (2001) – Vulnerable.

SYNONYMS: *tarlosti* Pirlot, 1965.

COMMENTS: Does not include *yerbabuenae* (= *sanborni*); see Watkins et al. (1972), Hall (1981), Koopman (1994), and Simmons and Wetterer (2002).

*Leptonycteris nivalis* (Saussure, 1860). Rev. Mag. Zool. Paris, ser. 2, 12:492.

COMMON NAME: Mexican Long-nosed Bat.

TYPE LOCALITY: Mexico, Veracruz, Mt. Orizaba.

DISTRIBUTION: SE Arizona, S New Mexico, and W Texas (USA) to S Mexico and Guatemala.

STATUS: U.S. ESA – Endangered; IUCN 2003 and IUCN/SSC Action Plan (2001) – Endangered.

SYNONYMS: *longala* Stains, 1957.

COMMENTS: Does not include *yerbabuenae*; see Watkins et al. (1972) and Hall (1981). See Hensley and Wilkins (1988).

*Leptonycteris yerbabuenae* Martínez and Villa-R, 1940. Anal. Inst. Biol. Univ. Nac. Auto. Mexico, 11:313.

COMMON NAME: Lesser Long-nosed Bat.

TYPE LOCALITY: Mexico, Guerrero, Yerbabuena.

DISTRIBUTION: C California, S Arizona, and New Mexico (USA) to Honduras and El Salvador

STATUS: U.S. ESA – Endangered as *L. curasoae yerbabuenae*. IUCN 2003 – Not listed; not considered in IUCN/SSC Action Plan (2001).

SYNONYMS: *sanborni* Hoffmeister, 1957.

COMMENTS: Included in *curasoae* by Koopman (1993), but see Watkins et al. (1972), Hall (1981), Koopman (1994), and Simmons and Wetterer (2002). The name *sanborni* has been used widely in the literature for this species, but is a junior synonym. A neotype for *yerbabuenae* was designated by Arita and Humphrey (1988).

*Lichonycteris* Thomas, 1895. Ann. Mag. Nat. Hist., ser. 6, 16:55.

TYPE SPECIES: *Lichonycteris obscura* Thomas, 1895.

*Lichonycteris obscura* Thomas, 1895. Ann. Mag. Nat. Hist., ser. 6, 16:55.

COMMON NAME: Dark Long-tongued Bat.

TYPE LOCALITY: Nicaragua, Managua, Managua.

DISTRIBUTION: Guatemala and Belize south to Bolivia and SE Brazil.

STATUS: IUCN 2003 and IUCN/SSC Action Plan (2001) – Lower Risk (lc).

SYNONYMS: *degener* Miller, 1931.

COMMENTS: Includes *degener*; see Hill (1985).

*Monophyllus* Leach, 1821. Trans. Linn. Soc. Lond., 13:75.

TYPE SPECIES: *Monophyllus redmani* Leach, 1821.

COMMENTS: Reviewed by Schwartz and Jones (1967). A key to the genus was published by Homan and Jones (1975*a*).

*Monophyllus plethodon* Miller, 1900. Proc. Washington Acad. Sci., 2:35.

COMMON NAME: Insular Single-leaf Bat.

TYPE LOCALITY: Barbados (Lesser Antilles), St. Michael Parish.

DISTRIBUTION: Lesser Antilles from Anguilla to St. Vincent and Barbados. Fossils known from Puerto Rico.

STATUS: IUCN 2003 and IUCN/SSC Action Plan (2001) – Lower Risk (nt).

SYNONYMS: **frater** Anthony, 1917; **luciae** Miller, 1902.

COMMENTS: Includes *luciae* and *frater*; see Schwartz and Jones (1967), Hall (1981), and Timm and Genoways (2003). See Homan and Jones (1975*b*).

*Monophyllus redmani* Leach, 1821. Trans. Linn. Soc. Lond., 13:76.

COMMON NAME: Leach's Single-leaf Bat.

TYPE LOCALITY: Jamaica.

DISTRIBUTION: Cuba, Hispaniola, Puerto Rico, Jamaica, S Bahama Isls.

STATUS: IUCN 2003 and IUCN/SSC Action Plan (2001) – Lower Risk (lc).
SYNONYMS: *clinedaphus* Miller, 1900; *cubanus* Miller, 1902; *ferreus* Miller, 1918; *portoricensis* Miller, 1900.
COMMENTS: Reviewed by Schwartz and Jones (1967), Hall (1981), and Timm and Genoways (2003). See Homan and Jones (1975*a*).

*Musonycteris* Schaldach and McLaughlin, 1960. Los Angeles Co. Mus. Contrib. Sci., 37:2.
TYPE SPECIES: *Musonycteris harrisoni* Schaldach and McLaughlin, 1960.
COMMENTS: Included in *Choeronycteris* by Handley (1966*b*) and Hall (1981), but see Phillips (1971) and Webster et al. (1982). Wetterer et al. (2000) found that *Musonycteris* is more closely related to *Choeroniscus* than to *Choeronycteris*.

*Musonycteris harrisoni* Schaldach and McLaughlin, 1960. Los Angeles Cty. Mus. Contrib. Sci., 37:3.
COMMON NAME: Banana Bat.
TYPE LOCALITY: Mexico, Colima, 2 km SE Pueblo Juarez.
DISTRIBUTION: Jalisco, Colima, Michoacan and Guerrero (Mexico).
STATUS: IUCN 2003 and IUCN/SSC Action Plan (2001) – Vulnerable.
COMMENTS: See Tellez and Ortega (1999).

*Scleronycteris* Thomas, 1912. Ann. Mag. Nat. Hist., ser. 8, 10:404.
TYPE SPECIES: *Scleronycteris ega* Thomas, 1912.

*Scleronycteris ega* Thomas, 1912. Ann. Mag. Nat. Hist., ser. 8, 10:405.
COMMON NAME: Ega Long-tongued Bat.
TYPE LOCALITY: Brazil, Amazonas, Ega.
DISTRIBUTION: Amazonian Brazil, S Venezuela.
STATUS: IUCN 2003 and IUCN/SSC Action Plan (2001) – Vulnerable.
COMMENTS: See Emmons (1997) for distribution map.

**Tribe Lonchophyllini** Griffiths, 1982. Amer. Mus. Novit., 2742:43.
COMMENTS: Used at the tribal level for the first time by McKenna and Bell (1997); see also Wetterer et al. (2000) and Carstens et al. (2002).

*Lionycteris* Thomas, 1913. Ann. Mag. Nat. Hist., ser. 8, 12:270.
TYPE SPECIES: *Lionycteris spurrelli* Thomas, 1913.

*Lionycteris spurrelli* Thomas, 1913. Ann. Mag. Nat. Hist., ser. 8, 12:271.
COMMON NAME: Chestnut Long-tongued Bat.
TYPE LOCALITY: Colombia, Chocó, Condoto.
DISTRIBUTION: E Panama, Colombia, Venezuela, Guianas, Amazonian Peru and Brazil.
STATUS: IUCN 2003 and IUCN/SSC Action Plan (2001) – Lower Risk (lc).
COMMENTS: See Emmons (1997) for distribution map.

*Lonchophylla* Thomas, 1903. Ann. Mag. Nat. Hist., ser. 7, 12:458.
TYPE SPECIES: *Lonchophylla mordax* Thomas, 1903.
COMMENTS: Taddei et al. (1983) gave a key to the species. Includes at least one undescribed species from northern South America (L. Davalos, pers. comm.).

*Lonchophylla bokermanni* Sazima, Vizotto, and Taddei, 1978. Rev. Brasil. Biol., 38:82.
COMMON NAME: Bokermann's Nectar Bat.
TYPE LOCALITY: Brazil, Minas Gerais, Jaboticatubas, Serra do Cipo.
DISTRIBUTION: SE Brazil.
STATUS: IUCN 2003 and IUCN/SSC Action Plan (2001) – Vulnerable.

*Lonchophylla dekeyseri* Taddei, Vizotto, and Sazima, 1983. Ciencia e Cultura, 35:626.
COMMON NAME: Dekeyser's Nectar Bat.
TYPE LOCALITY: Brazil, D. F., 8 km N Brasilia.
DISTRIBUTION: E Brazil.
STATUS: IUCN 2003 and IUCN/SSC Action Plan (2001) – Vulnerable.

*Lonchophylla handleyi* Hill, 1980. Bull. Brit. Mus. (Nat. Hist.) Zool., 38:233.
COMMON NAME: Handley's Nectar Bat.
TYPE LOCALITY: Ecuador, Morona, Santiago, Los Tayos (03°07′S, 18°12′W).
DISTRIBUTION: Ecuador, Peru, possibly SW Colombia.
STATUS: IUCN 2003 and IUCN/SSC Action Plan (2001) – Vulnerable.
COMMENTS: Formerly confused with *robusta*; see Hill (1980a). Colombian record may
represent *robusta*; see Cadena et al. (1998).

*Lonchophylla hesperia* G. M. Allen, 1908. Bull. Mus. Comp. Zool., 52:35.
COMMON NAME: Western Nectar Bat.
TYPE LOCALITY: Peru, Tumbes, Zorritos.
DISTRIBUTION: N Peru, Ecuador.
STATUS: IUCN 2003 and IUCN/SSC Action Plan (2001) – Vulnerable.
COMMENTS: Known from only five specimens; see Gardner (1976).

*Lonchophylla mordax* Thomas, 1903. Ann. Mag. Nat. Hist., ser. 7, 12:459.
COMMON NAME: Goldman's Nectar Bat.
TYPE LOCALITY: Brazil, Bahia, Lamarao.
DISTRIBUTION: Costa Rica south to Ecuador, Peru, and perhaps Bolivia; E Brazil.
STATUS: IUCN 2003 and IUCN/SSC Action Plan (2001) – Lower Risk (lc).
SYNONYMS: *concava* Goldman, 1914.
COMMENTS: Includes *concava*; see Handley (1966a); but also see Jones and Carter (1976),
who provisionally recognized it as a distinct species.

*Lonchophylla robusta* Miller, 1912. Proc. U.S. Natl. Mus., 42:23.
COMMON NAME: Orange Nectar Bat.
TYPE LOCALITY: Panama, Canal Zone, Río Chilibrillo, near Alajuela.
DISTRIBUTION: Nicaragua to Venezuela, Ecuador, and Peru.
STATUS: IUCN 2003 and IUCN/SSC Action Plan (2001) – Lower Risk (lc).
COMMENTS: See Solari et al. (1999) for discussion of southern range.

*Lonchophylla thomasi* J. A. Allen, 1904. Bull. Am. Mus. Nat. Hist., 20:230.
COMMON NAME: Thomas's Nectar Bat.
TYPE LOCALITY: Venezuela, Bolivar, Ciudad Bolivar.
DISTRIBUTION: E Panama, Colombia, Venezuela, Guianas, Amazonian Brazil, Ecuador, Peru,
Bolivia.
STATUS: IUCN 2003 and IUCN/SSC Action Plan (2001) – Lower Risk (lc).
COMMENTS: Specimens of this species have frequently been confused with *concava*, *mordax*,
and *Lionycteris spurrelli*; see Taddei et al. (1978) and Koopman (1978b).

*Platalina* Thomas, 1928. Ann. Mag. Nat. Hist., ser. 10, 8:120.
TYPE SPECIES: *Platalina genovensium* Thomas, 1928.

*Platalina genovensium* Thomas, 1928. Ann. Mag. Nat. Hist., ser. 10, 8:121.
COMMON NAME: Long-snouted Bat.
TYPE LOCALITY: Peru, near Lima.
DISTRIBUTION: Peru, N Chile.
STATUS: IUCN 2003 and IUCN/SSC Action Plan (2001) – Vulnerable.
COMMENTS: Reviewed by Galaz et al. (1999).

**Subfamily Phyllostominae** Gray, 1825. Zool. Journ., 2(6):242.
COMMENTS: Possibly monophyletic as defined here; see Wetterer et al. (2000), though see also
Baker et al. (2000). Wetterer et al. (2000) recognized four tribes within Phyllostominae
(Lonchorhinini, Micronycterini, Phyllostomini, and Vampyrini), but monophyly of these
groups is uncertain (see Baker et al., 2000). Accordingly, I do not recognize tribes within
Phyllostominae at the present time.

*Chrotopterus* Peters, 1865. Monatsb. K. Preuss. Akad. Wiss. Berlin, 1865:505.
TYPE SPECIES: *Vampyrus auritus* Peters, 1856.
SYNONYMS: *Vampyrus* Peters, 1856 [not Leach, 1821].

*Chrotopterus auritus* (Peters, 1856). Abhandl. Akad. Wiss. Berlin, 1856:305.
>   COMMON NAME: Woolly False Vampire Bat.
>   TYPE LOCALITY: Mexico. The type locality was incorrectly changed to Brazil, Santa Catarina, by Carter and Dolan (1978), see remarks in Medellin (1989).
>   DISTRIBUTION: Veracruz (Mexico) south to the Guianas, S Brazil, Peru, Bolivia, and N Argentina.
>   STATUS: IUCN 2003 and IUCN/SSC Action Plan (2001) – Lower Risk (lc).
>   SYNONYMS: *australis* Thomas, 1905; *guianae* Thomas, 1905.
>   COMMENTS: Simmons and Voss (1998) discussed problems with previously recognized subspecies. See Medellín (1989). See Emmons (1997) for distribution map.

*Glyphonycteris* Thomas, 1896. Ann. Mag. Nat. Hist., ser. 6, 18:301.
>   TYPE SPECIES: *Glyphonycteris sylvestris* Thomas, 1896.
>   SYNONYMS: *Barticonycteris* Hill, 1964.
>   COMMENTS: Recognized as a subgenus of *Micronycteris* by Sanborn (1949) and Simmons (1996); raised to genus rank by Simmons and Voss (1998) following information later published in Wetterer et al. (2000).

*Glyphonycteris behnii* (Peters, 1865). Monatsb. K. Preuss. Akad. Wiss. Berlin, 1865:505.
>   COMMON NAME: Behn's Bat.
>   TYPE LOCALITY: Brazil, Mato Grosso, Cuiaba (= Cuyaba).
>   DISTRIBUTION: Known only from the holotype.
>   STATUS: IUCN 2003 and IUCN/SSC Action Plan (2001) – Vulnerable as *Micronycteris behnii*.
>   COMMENTS: Restricted to include only the holotype by Simmons (1996). This taxon may be a senior synonym of *sylvestis*, but the relationship of these taxa cannot be resolved without reexamination of the holotype of *behnii* in the context of what is now known about variation in *sylvestris*; see discussion in Simmons (1996).

*Glyphonycteris daviesi* (Hill, 1964). Mammalia, 28:557.
>   COMMON NAME: Graybeard Bat.
>   TYPE LOCALITY: Guyana, Essequibo Prov., Potaro road, 24 mi. (39 km) from Bartica.
>   DISTRIBUTION: Honduras south to Peru, the Guianas, Brazil, and Bolivia; Trinidad.
>   STATUS: IUCN 2003 and IUCN/SSC Action Plan (2001) – Lower Risk (nt) as *Micronycteris daviesi*.
>   COMMENTS: Formerly included in the monotypic genus *Barticonycteris*; see Koopman (1978*b*) and Simmons (1996). Reviewed by Pine et al. (1996).

*Glyphonycteris sylvestris* Thomas, 1896. Ann. Mag. Nat. Hist., ser. 6, 18:302.
>   COMMON NAME: Tricolored Bat.
>   TYPE LOCALITY: Costa Rica, Guanacaste, Hda. Miravalles, between 1,400 and 2,000 ft. (427-610 m).
>   DISTRIBUTION: Peru and SE Brazil north to Nayarit and Veracruz (Mexico); Trinidad.
>   STATUS: IUCN 2003 and IUCN/SSC Action Plan (2001) – Lower Risk (nt) as *Micronycteris sylvestris*.
>   COMMENTS: Probably a junior synonym of *behnii*; see Simmons (1996).

*Lampronycteris* Sanborn, 1949. Fieldiana Zool., 31:223.
>   TYPE SPECIES: *Micronycteris brachyotis* Dobson, 1879.
>   COMMENTS: Recognized as a subgenus of *Micronycteris* by Sanborn (1949) and Simmons (1996); raised to genus rank by Wetterer et al. (2000).

*Lampronycteris brachyotis* (Dobson, 1879). Proc. Zool. Soc. Lond., 1878:880 [1879].
>   COMMON NAME: Orange-throated Bat.
>   TYPE LOCALITY: French Guiana, Cayenne.
>   DISTRIBUTION: Oaxaca (Mexico) to Guyana, French Guiana and Brazil; Peru; Trinidad.
>   STATUS: IUCN 2003 and IUCN/SSC Action Plan (2001) – Lower Risk (lc) as *Micronycteris brachyotis*.
>   SYNONYMS: *platyceps* Sanborn, 1949.
>   COMMENTS: Includes *platyceps*; see Jones and Carter (1976). See Medellín et al. (1985).

*Lonchorhina* Tomes, 1863. Proc. Zool. Soc. Lond., 1863:81.
  TYPE SPECIES: *Lonchorhina aurita* Tomes, 1863.
  COMMENTS: Reviewed by Hernandez-Camacho and Cadena-G. (1978). A key to the genus was presented by Lassieur and Wilson (1989) and subsequently modified by Handley and Ochoa (1997). Species groups follow Handley and Ochoa (1997).

*Lonchorhina aurita* Tomes, 1863. Proc. Zool. Soc. Lond., 1863:83.
  COMMON NAME: Common Sword-nosed Bat.
  TYPE LOCALITY: Trinidad and Tobago, Trinidad.
  DISTRIBUTION: Oaxaca (Mexico) south to SE Brazil, Bolivia, Peru, and Ecuador; Trinidad; perhaps New Providence Isl (Bahama Isls), see Jones and Carter (1976).
  STATUS: IUCN 2003 and IUCN/SSC Action Plan (2001) – Lower Risk (lc).
  SYNONYMS: *occidentalis* Anthony, 1923.
  COMMENTS: *aurita* species group. Includes *occidentalis*; see Jones and Carter (1976). See Lassieur and Wilson (1989) and Handley and Ochoa (1997). Some specimens previously referred to this species actually represent *inusitata*; see Handley and Ochoa (1997).

*Lonchorhina fernandezi* Ochoa and Ibáñez, 1982. Mem. Soc. Cienc. Nat. La Salle, 42:147.
  COMMON NAME: Fernandez's Sword-nosed Bat.
  TYPE LOCALITY: Venezuela, Amazonas, 40-50 km (by road) NE Puerto Ayacucho.
  DISTRIBUTION: S Venezuela.
  STATUS: IUCN 2003 and IUCN/SSC Action Plan (2001) – Vulnerable.
  COMMENTS: *orinocensis* species group. Known only from the type locality. Some specimens previously referred to this species actually represent *inusitata*; see Handley and Ochoa (1997).

*Lonchorhina inusitata* Handley and Ochoa, 1997. Mem. Soc. Cien. Nat. La Salle, 57:73.
  COMMON NAME: Uncommon Sword-nosed Bat.
  TYPE LOCALITY: Venezuela, Amazonas, 84 km SSE Esmeralda, Boca Mavaca, 2°30'N, 56°13'W.
  DISTRIBUTION: S Venezuela, Guyana, Surinam, French Guiana, W Brazil.
  STATUS: IUCN 2003 and IUCN/SSC Action Plan (2001) – Data Deficient.
  COMMENTS: *aurita* species group. Specimens referred to this species were previously identified as *aurita*, *marinkellei*, or *fernandezi*; see Handley and Ochoa (1997) and Simmons et al. (2000).

*Lonchorhina marinkellei* Hernández-Camacho and Cadena-G., 1978. Caldasia, 12:229.
  COMMON NAME: Marinkelle's Sword-nosed Bat.
  TYPE LOCALITY: Colombia, Vaupes, near Mitu, Durania.
  DISTRIBUTION: SE Colombia.
  STATUS: IUCN 2003 and IUCN/SSC Action Plan (2001) – Vulnerable.
  COMMENTS: *aurita* species group. Some specimens previously referred to this species actually represent *inusitata*; see Handley and Ochoa (1997).

*Lonchorhina orinocensis* Linares and Ojasti, 1971. Novid. Cient. Contrib. Occas. Mus. Hist. Nat. La Salle, Ser. Zool., 36:2.
  COMMON NAME: Orinocoan Sword-nosed Bat.
  TYPE LOCALITY: Venezuela, Bolivar, 50 km NE Puerto Paez, Boca de Villacoa.
  DISTRIBUTION: Venezuela, SE Colombia.
  STATUS: IUCN 2003 and IUCN/SSC Action Plan (2001) – Lower Risk (nt).
  COMMENTS: *orinocensis* species group. See Handley and Ochoa (1997) for distribution map.

*Lophostoma* d'Orbigny, 1836. Voy. Amer. Merid. Atlas Zool., 4:11.
  TYPE SPECIES: *Lophostoma silvicolum* d'Orbigny, 1836.
  SYNONYMS: *Chrotopterus* J. A. Allen, 1910 [not Peters, 1865].
  COMMENTS: Formerly included *Tonatia*, but see Lee et al. (2002), who demonstrated that *Tonatia* as traditionally defined is not monophyletic. Those authors proposed restricting *Tonatia* to the type species and its close relative (*bidens* and *saurophila*), and using the next available generic name (*Lophostoma*) for the remaining species, which together form a clade that is not closely related to *Tonatia*. That recommendation is followed here. Keys to

species now included in *Lophostoma* were provided by Genoways and Williams (1984) and Medellín and Arita (1989).

*Lophostoma brasiliense* Peters, 1866. Monatsb. K. Preuss. Akad. Wiss. Berlin, 1866:674.
COMMON NAME: Pygmy Round-eared Bat.
TYPE LOCALITY: Brazil, Bahia.
DISTRIBUTION: Veracruz (Mexico) south to Peru, Bolivia, NE Brazil; Trinidad.
STATUS: IUCN 2003 and IUCN/SSC Action Plan (2001) – Lower Risk (lc) as *Tonata brasiliense*.
SYNONYMS: *minuta* Goodwin, 1942; *nicaraguae* Goodwin, 1942; *venezuelae* Robinson and Lyon, 1901.
COMMENTS: For synonyms see Jones and Carter (1979); but also see Gardner (1976). Hall (1981) listed *nicaraguae* (including *minuta*) as a distinct species.

*Lophostoma carrikeri* (J. A. Allen, 1910). Bull. Am. Mus. Nat. Hist., 28:147.
COMMON NAME: Carriker's Round-eared Bat.
TYPE LOCALITY: Venezuela, Bolivar, Río Mocho.
DISTRIBUTION: Colombia, Venezuela, Guianas, N Brazil, Bolivia, Peru.
STATUS: IUCN 2003 and IUCN/SSC Action Plan (2001) – Vulnerable as *Tonatia carrikeri*.
COMMENTS: See McCarthy et al. (1992).

*Lophostoma evotis* (Davis and Carter, 1978). Occas. Pap. Mus. Texas Tech Univ., 53:8.
COMMON NAME: Davis's Round-eared Bat.
TYPE LOCALITY: Guatemala, Izabál, 25 km S.S.W. Puerto Barrios.
DISTRIBUTION: S Mexico, Belize, Guatemala, Honduras.
STATUS: IUCN 2003 and IUCN/SSC Action Plan (2001) – Lower Risk (nt) as *Tonatia evotis*.
COMMENTS: Formerly included in *silvicola*. See Medellín and Arita (1989).

*Lophostoma schulzi* (Genoways and Williams, 1980). Ann. Carnegie Mus., 49:205.
COMMON NAME: Schulz's Round-eared Bat.
TYPE LOCALITY: Surinam, Brokopondo, 3 km SW Rudi Koppelvliegveld.
DISTRIBUTION: Guianas, N Brazil.
STATUS: IUCN 2003 and IUCN/SSC Action Plan (2001) – Vulnerable as *Tonatia schulzi*.

*Lophostoma silvicolum* d'Orbigny, 1836. Voy. Amer. Merid. Atlas Zool., 4:11, pl. 7.
COMMON NAME: White-throated Round-eared Bat.
TYPE LOCALITY: Bolivia, Yungas between Secure and Isiboro rivers.
DISTRIBUTION: Honduras to Bolivia, NE Argentina, Guianas, and E Brazil.
STATUS: IUCN 2003 and IUCN/SSC Action Plan (2001) – Lower Risk (lc) as *Tonatia silvicola*.
SYNONYMS: *amblyotis* Wagner, 1843; *auritus* Sanborn, 1923 [not Peters, 1865]; *colombianus* Anthony, 1920; *midas* Pelzeln, 1883; **centralis** Davis and Carter, 1978; **laephotis** Thomas, 1910; **occidentalis** Davis and Carter, 1978.
COMMENTS: Includes *laephotis* and *amblyotis*; see Davis and Carter (1978). See Medellín and Arita (1989). The species name for this taxon was formerly spelled *silvicola* when used in *Tonatia*, but must be spelled *silvicolum* when combined with *Lophostoma*, which is a Greek neuter noun. These names have sometimes been spelled *sylvicola* and *sylvicolum*, but I retain the original spelling here.

*Macrophyllum* Gray, 1838. Mag. Zool. Bot., 2:489.
TYPE SPECIES: *Macrophyllum nieuwiedii* Gray, 1838 (= *Phyllostoma macrophyllum* Schinz, 1821).
SYNONYMS: *Dolichophyllum* Lydekker, 1891; *Mesophyllum* Vieira, 1942.

*Macrophyllum macrophyllum* (Schinz, 1821). Das Thierreich, 1:163.
COMMON NAME: Long-legged Bat.
TYPE LOCALITY: Brazil, Bahia, Río Mucuri.
DISTRIBUTION: Tabasco (Mexico) south to Peru, Bolivia, SE Brazil, Paraguay, and NE Argentina.
STATUS: IUCN 2003 and IUCN/SSC Action Plan (2001) – Lower Risk (lc).
SYNONYMS: *neuwiedii* Gervais, 1855; *nieuwiedii* Gray, 1838.
COMMENTS: See Harrison (1975). See Emmons (1997) for distribution map.

*Macrotus* Gray, 1843. Proc. Zool. Soc. Lond., 1843:21.
  TYPE SPECIES: *Macrotus waterhousii* Gray, 1843.
  SYNONYMS: *Otopterus* Lydekker, 1891.
  COMMENTS: Revised by Anderson and Nelson (1965).

*Macrotus californicus* Baird, 1858. Proc. Acad. Nat. Sci. Phil., 10:116.
  COMMON NAME: Californian Leaf-nosed Bat.
  TYPE LOCALITY: USA, California, Imperial Co., Old Fort Yuma.
  DISTRIBUTION: N Sinaloa and SW Chihuahua (Mexico) north to S Nevada and S California
      (USA); Baja California and Tamaulipas (Mexico).
  STATUS: IUCN 2003 and IUCN/SSC Action Plan (2001) – Vulnerable.
  COMMENTS: For a comparison with *waterhousii*, see Davis and Baker (1974) and Greenbaum
      and Baker (1976). Considered to be a subspecies of *waterhousii* by Anderson (1969*a*)
      and Hall (1981).

*Macrotus waterhousii* Gray, 1843. Proc. Zool. Soc. Lond., 1843:21.
  COMMON NAME: Waterhouse's Leaf-nosed Bat.
  TYPE LOCALITY: Haiti.
  DISTRIBUTION: Sonora and Hidalgo (Mexico) south to Guatemala; Bahama Isls; Jamaica;
      Cuba; Cayman Isls (NW of Jamaica); Hispaniola and Beata Isls.
  STATUS: IUCN 2003 and IUCN/SSC Action Plan (2001) – Lower Risk (lc).
  SYNONYMS: *heberfolium* Shamel, 1931; ***bulleri*** H. Allen, 1890; ***compressus*** Rehn, 1904;
      ***jamaicensis*** Rehn, 1904; ***mexicanus*** Saussure, 1860; *bocourtianus* Dobson, 1876;
      ***minor*** Gundlach, 1864.
  COMMENTS: Includes *mexicanus*; see Anderson and Nelson (1965). See Anderson (1969*a*).
      Caribbean forms reviewed by Timm and Genoways (2003).

*Micronycteris* Gray, 1866. Proc. Zool. Soc. Lond., 1866:113.
  TYPE SPECIES: *Phyllophora megalotis* Gray, 1842.
  SYNONYMS: *Schizastoma* Gray, 1862 [*lapsus* for *Schizostoma*]; *Schizostoma* Gervais, 1856 [not
      Bronn, 1835]; *Vampirella* Reinhardt, 1872 [not Cienkowsky, 1865]; *Xenoctenes* Miller,
      1907.
  COMMENTS: Includes *Xenoctenes*; see Simmons (1996). Does not include *Barticonycteris*,
      *Glyphonycteris*, *Lampronycteris*, *Neonycteris*, or *Trinycteris*; see Simmons and Voss (1998)
      and Wetterer et al. (2000).

*Micronycteris brosseti* Simmons and Voss, 1998. Bull. Amer. Mus. Nat. Hist., 273:62.
  COMMON NAME: Brosset's Big-eared Bat.
  TYPE LOCALITY: French Guiana, Paracou near Sinnamary.
  DISTRIBUTION: E Peru, Guyana, French Guiana, SE Brazil.
  STATUS: IUCN 2003 and IUCN/SSC Action Plan (2001) – Data Deficient.

*Micronycteris hirsuta* (Peters, 1869). Monatsb. K. Preuss. Akad. Wiss. Berlin, 1869:397.
  COMMON NAME: Hairy Big-eared Bat.
  TYPE LOCALITY: Costa Rica, Guanacaste, Pozo Azul.
  DISTRIBUTION: Honduras to French Guiana, Trinidad, Amazonian Brazil, Peru, and Ecuador.
  STATUS: IUCN 2003 and IUCN/SSC Action Plan (2001) – Lower Risk (lc).
  COMMENTS: Placed in subgenus *Xenoctenes* by Sanborn (1949), but see Simmons (1996).

*Micronycteris homezi* Pirlot, 1967. Mammalia, 31:265.
  COMMON NAME: Pirlot's Big-eared Bat.
  TYPE LOCALITY: Venezuela, Zulia, Maracaibo Basis, Río Palmar, Hacienda El Cerro.
  DISTRIBUTION: NW Venezuela, Guyana, French Guiana, Brazil.
  STATUS: IUCN 2003 – Not evaluated; not considered in IUCN/SSC Action Plan (2001).
  COMMENTS: Described by Pirlot (1967) as a subspecies of *megalotis*, but clearly a distinct
      species; see Simmons and Voss (1998). Also see Bernard (2001) and Lim and Engstrom
      (2001).

*Micronycteris matses* Simmons, Voss, and Fleck, 2002. Amer. Mus. Novit., 3358:5.
  COMMON NAME: Matses Big-eared Bat.

TYPE LOCALITY: Peru, Departmento Loreto, SE bank of Río Gálvez, village of Nuevo San Juan, 5°17'30"S, 73°9'50"W, 150 m.

DISTRIBUTION: Known only from the type locality.

STATUS: IUCN 2003 and IUCN/SSC Action Plan (2001) – Not evaluated (new species).

*Micronycteris megalotis* (Gray, 1842). Ann. Mag. Nat. Hist., [ser. 1], 10:257.

COMMON NAME: Little Big-eared Bat.

TYPE LOCALITY: Brazil, São Paulo, Pereque.

DISTRIBUTION: Colombia to Peru, Bolivia, and Brazil; Venezuela and the Guianas; Trinidad and Tobago; Margarita Isl (Venezuela); Grenada; St. Vincent.

STATUS: IUCN 2003 and IUCN/SSC Action Plan (2001) – Lower Risk (lc).

SYNONYMS: *elongatum* Gray, 1842; *megalotes* Robinson, 1896; *scrobiculatum* Wagner, 1855; *typica* K. Andersen, 1906.

COMMENTS: Does not include *microtis*; see Brosset and Charles-Dominique (1990), Simmons (1996), and Simmons and Voss (1998). Does not include *mexicana*; see Simmons (1996). Does not include *homezi*; see Simmons and Voss (1998). See Alonso-Mejia and Medellín (1991), but note that these authors included *microtis*, *mexicana*, and *homezi* in *megalotis*.

*Micronycteris microtis* Miller, 1898. Proc. Acad. Nat. Sci. Phil., 50:328.

COMMON NAME: Common Big-eared Bat.

TYPE LOCALITY: Nicaragua, San Juan del Norte, Graytown.

DISTRIBUTION: Tamaulipas and Jalisco (Mexico) to northern Colombia, Venezuela, the Guianas, northern Brazil, and Bolivia.

STATUS: IUCN 2003 – Not evaluated; not considered in IUCN/SSC Action Plan (2001).

SYNONYMS: **mexicana** Miller, 1898; *pygmaeus* Rehn, 1904.

COMMENTS: Formerly included in *megalotis*, but see Brosset and Charles-Dominique (1990), Simmons (1996), and Simmons and Voss (1998). Simmons (1996) included *mexicana* as a subspecies of *microtis*, but noted that it may be a distinct species.

*Micronycteris minuta* (Gervais, 1856). *In* F. Comte de Castelnau, Exped. Partes Cen. Am. Sud., Zool. (Sec. 7), Vol. 1, pt. 2 (Mammifères):50.

COMMON NAME: Tiny Big-eared Bat.

TYPE LOCALITY: Brazil, Bahia, Capela Nova.

DISTRIBUTION: Honduras to S Brazil, Bolivia, and Peru; Guianas; Trinidad.

STATUS: IUCN 2003 and IUCN/SSC Action Plan (2001) – Lower Risk (lc).

SYNONYMS: *hypoleuca* J. A. Allen, 1900.

COMMENTS: See Simmons (1996), Simmons and Voss (1998), and López-González (1998).

*Micronycteris sanborni* Simmons, 1996. Amer. Mus. Novit., 3158:6.

COMMON NAME: Sanborn's Big-eared Bat.

TYPE LOCALITY: Brazil, Ceará, Itaitera, Sitio Luanda, 4 mi. (6 km) S of Crato.

DISTRIBUTION: NE Brazil, Bolivia.

STATUS: IUCN 2003 and IUCN/SSC Action Plan (2001) – Data Deficient.

*Micronycteris schmidtorum* Sanborn, 1935. Field Mus. Nat. Hist. Publ., Zool. Ser., 20:81.

COMMON NAME: Schmidt's Big-eared Bat.

TYPE LOCALITY: Guatemala, Izabal, Bobos.

DISTRIBUTION: S Mexico to Guianas; NE Peru; Brazil.

STATUS: IUCN 2003 and IUCN/SSC Action Plan (2001) – Lower Risk (lc).

COMMENTS: Ascorra et al. (1991) reviewed *schmidtorum* but referred some specimens to this species that were subsequently reidentified as either *sanborni* (Simmons, 1996) or *brosseti* (Simmons and Voss, 1998).

*Mimon* Gray, 1847. Proc. Zool. Soc. Lond., 1847:14.

TYPE SPECIES: *Phyllostoma bennettii* Gray, 1838.

SYNONYMS: *Chrotopterus* Elliot, 1904 [not Peters, 1865]; *Vampyrus* Saussure, 1860 [not Leach, 1821].

COMMENTS: Does not include *Anthorhina*, a name that is actually a junior synonym of *Tonatia* (see Gardner and Ferrell [1990] and Article 67.8 of the International Code of Zoological

Nomenclature [International Commission on Zoological Nomenclature, 1999]) despite its frequent use as a subgenus of *Mimon* following Handley (1960). The nominate subgenus of *Mimon* includes *bennettii* and *cozumelae*; the other subgenus (formerly called *Anthorhina*) includes *crenulatum* and *koepckeae*, but there is presently no valid name for this subgenus.

*Mimon bennettii* (Gray, 1838). Mag. Zool. Bot., 2:483.
  COMMON NAME: Southern Golden Bat.
  TYPE LOCALITY: Brazil, São Paulo, Ipanema (restricted by Hershkovitz, 1951).
  DISTRIBUTION: Guianas; SE Brazil.
  STATUS: IUCN 2003 and IUCN/SSC Action Plan (2001) – Lower Risk (lc).
  SYNONYMS: *auricularis* Saussure, 1860; *auritus* Elliot, 1904.
  COMMENTS: Does not include *cozumelae*; see McCarthy (1987), McCarthy et al. (1993), and Simmons and Voss (1998).

*Mimon cozumelae* Goldman, 1914. Proc. Biol. Soc. Wash., 27:75.
  COMMON NAME: Cozumelan Golden Bat.
  TYPE LOCALITY: Mexico, Quintana Roo, Cozumel Isl.
  DISTRIBUTION: S Mexico to Colombia
  STATUS: IUCN 2003 – Not evaluated; not considered in IUCN/SSC Action Plan (2001).
  COMMENTS: Formerly included in *bennettii* (see Hall, 1981; Schaldach, 1965; Villa-R., 1967), but recognized as a distinct species by McCarthy (1987), McCarthy et al. (1993), and Simmons and Voss (1998).

*Mimon crenulatum* (E. Geoffroy, 1803). Cat. Mamm. Mus. Nat. d'Hist. Nat., p. 61.
  COMMON NAME: Striped Hairy-nosed Bat.
  TYPE LOCALITY: Brazil, Bahia; see Handley (1960).
  DISTRIBUTION: Chiapas and Campeche (Mexico) to Guianas, E Brazil, Bolivia, Ecuador and E Peru; Trinidad.
  STATUS: IUCN 2003 and IUCN/SSC Action Plan (2001) – Lower Risk (lc).
  SYNONYMS: **keenani** Handley, 1960; **longifolium** Wagner, 1843; *peruanum* Thomas, 1923; **picatum** Thomas, 1903.
  COMMENTS: Does not include *koepckeae*; see Gardner and Patton (1972). Reviewed by Handley (1960), Jones and Carter (1979), and Koopman (1978*b*). Because Wilson and Reeder (1993) did not treat names established in E. Geoffroy (1803) as available, Koopman (1993) attributed authorship of *crenulatum* to a later work by E. Geoffroy, but this name was actually published in the 1803 volume, which is now accepted (Grubb, 2001*a*; Opinion 2005 of the International Commission on Zoological Nomenclature, 2002*b*).

*Mimon koepckeae* Gardner and Patton, 1972. Occas. Papers Mus. Zool. Louisiana State Univ., 43:7.
  COMMON NAME: Koepcke's Hairy-nosed Bat.
  TYPE LOCALITY: Peru, Departamento de Ayacucho, Huanhuachayo (12°44'S, 73°47'W), 1,660 m.
  DISTRIBUTION: Highlands of central Peru.
  STATUS: IUCN 2003 – Not evaluated; not considered in IUCN/SSC Action Plan (2001).
  COMMENTS: Koopman (1993, 1994) treated *koepckeae* as a subspecies of *crenulatum*, but this does not appear justified given the data presented by Gardner and Patton (1972).

*Neonycteris* Sanborn, 1949. Fieldiana Zool. 31:226.
  TYPE SPECIES: *Micronycteris pusilla* Sanborn, 1949.
  COMMENTS: Recognized as a subgenus of *Micronycteris* by Sanborn (1949) and Simmons (1996); raised to genus rank by Wetterer et al. (2000).

*Neonycteris pusilla* (Sanborn, 1949). Fieldiana Zool., 31:228.
  COMMON NAME: Least Big-eared Bat.
  TYPE LOCALITY: Brazil, Amazonas, Tahuapunta (Vaupes River).
  DISTRIBUTION: NW Brazil, E Colombia.
  STATUS: IUCN 2003 and IUCN/SSC Action Plan (2001) – Vulnerable as *Micronycteris pusilla*.
  COMMENTS: Known only from the type series.

*Phylloderma* Peters, 1865. Monatsb. K. Preuss. Akad. Wiss. Berlin, 1865:513.
    TYPE SPECIES: *Phylloderma stenops* Peters, 1865.
    SYNONYMS: *Guandira* Gray, 1866.
    COMMENTS: Included in *Phyllostomus* by Baker et al. (1988*b*) and Van Den Bussche and Baker
        (1993), but see Simmons and Voss (1998) and Wetterer et al. (2000).

*Phylloderma stenops* Peters, 1865. Monatsb. K. Preuss. Akad. Wiss. Berlin, 1865:513.
    COMMON NAME: Pale-faced Bat.
    TYPE LOCALITY: French Guiana, Cayenne.
    DISTRIBUTION: S Mexico to SE Brazil, Bolivia, and Peru.
    STATUS: IUCN 2003 and IUCN/SSC Action Plan (2001) – Lower Risk (lc).
    SYNONYMS: *cayenensis* Gray, 1866; **boliviensis** Barquez and Ojeda, 1979; **septentrionalis**
        Goodwin, 1940.
    COMMENTS: Includes *septentrionalis*; see Jones and Carter (1976). Bolivian form reviewed by
        Bárquez and Ojeda (1979). See Emmons (1997) for distribution map.

*Phyllostomus* Lacépède, 1799. Tabl. Div. Subd. Order Genres Mammifères, p. 16.
    TYPE SPECIES: *V(espertilio) hastatus* Pallas, 1767.
    SYNONYMS: *Alectops* Gray, 1866; *Phyllostoma* Cuvier, 1800.
    COMMENTS: Does not include *Phylloderma*, but see Baker et al. (1988*b*) and Van Den Bussche
        and Baker (1993). Phylogenetic relationships among species discussed by Baker et al.
        (1988*b*) and Van Den Bussche and Baker (1993).

*Phyllostomus discolor* Wagner, 1843. Arch. Naturgesch., 9(1):366.
    COMMON NAME: Pale Spear-nosed Bat.
    TYPE LOCALITY: Brazil, Mato Grosso, Cuiaba (=Cuyaba).
    DISTRIBUTION: Oaxaca and Veracruz (Mexico) to Guianas, SE Brazil, Bolivia, Paraguay,
        N Argentina and Peru; Trinidad; Margarita Isl (Venezeula).
    STATUS: IUCN 2003 and IUCN/SSC Action Plan (2001) – Lower Risk (lc).
    SYNONYMS: *angusticeps* Gervais, 1856; *innominatum* Tschudi, 1844; **verrucosus** Elliot, 1905.

*Phyllostomus elongatus* (E. Geoffroy, 1810). Ann. Mus. Natn. Hist. Nat. Paris, 15:182.
    COMMON NAME: Lesser Spear-nosed Bat.
    TYPE LOCALITY: Brazil, Mato Grosso, Rio Branco.
    DISTRIBUTION: Bolivia, E Peru, Ecuador, and Colombia to Guianas and E Brazil.
    STATUS: IUCN 2003 and IUCN/SSC Action Plan (2001) – Lower Risk (lc).
    SYNONYMS: *ater* Gray, 1866.

*Phyllostomus hastatus* (Pallas, 1767). Spicil. Zool., 3:7.
    COMMON NAME: Greater Spear-nosed Bat.
    TYPE LOCALITY: Surinam.
    DISTRIBUTION: Guatemala and Belize to the Guianas, Brazil, Paraguay, N Argentina, Bolivia,
        and Peru; Trinidad and Tobago; Margarita Isl (Venezuela).
    STATUS: IUCN 2003 and IUCN/SSC Action Plan (2001) – Lower Risk (lc).
    SYNONYMS: *aruma* Thomas, 1924; *curaca* Cabrera, 1912; *maximus* Wied, 1821; **panamensis**
        J. A. Allen, 1904; *caucae* J. A. Allen, 1916; *caurae* J. A. Allen, 1904; *paeze* Thomas, 1924.

*Phyllostomus latifolius* (Thomas, 1901). Ann. Mag. Nat. Hist., ser. 7, 8:142.
    COMMON NAME: Guianan Spear-nosed Bat.
    TYPE LOCALITY: Guyana, Essequibo Prov., Mt. Kanuku.
    DISTRIBUTION: Guianas, SE Colombia.
    STATUS: IUCN 2003 and IUCN/SSC Action Plan (2001) – Lower Risk (nt).
    COMMENTS: Clearly distinct from *elongatus*.

*Tonatia* Gray, 1827. *In* Griffith, Anim. Kingdom, 5:71.
    TYPE SPECIES: *Vampyrus bidens* Spix, 1823.
    SYNONYMS: *Anthorhina* Lydekker, 1891; *Phyllostoma* Gray, 1838 [not Cuvier, 1800]; *Tylostoma*
        Gervais, 1855; *Vampyrus* Spix, 1823 [not Leach, 1821].
    COMMENTS: *Tonatia* as traditionally defined was recently shown to be non-monophyletic (see
        Lee et al., 2002), and so has been restricted to include only the type species (*bidens*) and its

close relative (*saurophila*). The remaining species, which together form a clade that is not closely related to *Tonatia*, are here transferred to *Lophostoma* following the recommendation of Lee et al. (2002). Keys to *Tonatia* have been published by several authors, but only one publication – Williams et al. (1995) – describes distinctions between the species now included in the genus.

*Tonatia bidens* (Spix, 1823). Sim. Vespert. Brasil., p. 65.
    COMMON NAME: Greater Round-eared Bat.
    TYPE LOCALITY: Brazil, Bahia, Rio Sao Francisco.
    DISTRIBUTION: NE Brazil to N Argentina and Paraguay.
    STATUS: IUCN 2003 and IUCN/SSC Action Plan (2001) – Lower Risk (lc).
    SYNONYMS: *childreni* Gray, 1838.
    COMMENTS: Most material previously referred to *bidens* from Central America and northern South America is now recognized as *saurophila*; see Williams et al. (1995).

*Tonatia saurophila* Koopman and Williams, 1951. Amer. Mus. Novit., 1519:11.
    COMMON NAME: Stripe-headed Round-eared Bat.
    TYPE LOCALITY: Jamaica, St. Elizabeth Parish, Balaclava, Wallingford Roadside Cave.
    DISTRIBUTION: Chiapas (Mexico) and Belize to Peru, Bolivia, Venezuela, the Guianas, and NE Brazil; Trinidad.
    STATUS: IUCN 2003 – Lower Risk (lc); not considered in IUCN/SSC Action Plan (2001).
    SYNONYMS: **bakeri** Williams, Willig, and Reid, 1995; **maresi** Williams, Willig, and Reid, 1995.
    COMMENTS: Originally described from fossils from Jamaica. Reviewed by Williams et al. (1995).

*Trachops* Gray, 1847. Proc. Zool. Soc. Lond., 1847:14.
    TYPE SPECIES: *Trachops fuliginosus* Gray, 1865 (= *Vampyrus cirrhosus* Spix, 1823).
    SYNONYMS: *Istiophorus* Gray, 1825 [not Lacépède, 1802]; *Histiophorus* Agassiz, 1846; *Trachyops* Peters, 1865; *Tylostoma* Saussure, 1860 [not Gervais, 1855 or Gervais, 1856].

*Trachops cirrhosus* (Spix, 1823). Sim. Vespert. Brasil., p. 64.
    COMMON NAME: Fringe-lipped Bat.
    TYPE LOCALITY: Brazil, Pará (restricted by Husson, 1962).
    DISTRIBUTION: Oaxaca (Mexico) to Guianas, SE Brazil, Bolivia and Ecuador; Trinidad.
    STATUS: IUCN 2003 and IUCN/SSC Action Plan (2001) – Lower Risk (lc).
    SYNONYMS: *fuliginosus* Gray, 1865; **coffini** Goldman, 1925; **ehrhardti** Felten, 1956.
    COMMENTS: Reviewed by Cramer et al. (2001).

*Trinycteris* Sanborn, 1949. Fieldiana Zool. 31:228
    TYPE SPECIES: *Micronycteris nicefori* Sanborn, 1949.
    COMMENTS: Recognized as a subgenus of *Micronycteris* by Sanborn (1949) and Simmons (1996); raised to genus rank by Simmons and Voss (1998) following information later published in Wetterer et al. (2000).

*Trinycteris nicefori* (Sanborn, 1949). Fieldiana Zool., 31:230.
    COMMON NAME: Niceforo's Bat.
    TYPE LOCALITY: Colombia, Norte de Santander, Cucuta.
    DISTRIBUTION: Belize to N Colombia, Venezuela, Guianas, Amazonian Brazil, Ecuador, and Peru; Bolivia; Trinidad.
    STATUS: IUCN 2003 and IUCN/SSC Action Plan (2001) – Lower Risk (lc) as *Micronycteris nicefori*.
    COMMENTS: See Simmons and Voss (1998).

*Vampyrum* Rafinesque, 1815. Analyse de la Nature, p. 54.
    TYPE SPECIES: *Vespertilio spectrum* Linnaeus, 1758.
    SYNONYMS: *Vampirus* Lesson, 1827; *Vampyrus* Leach, 1821.

*Vampyrum spectrum* (Linnaeus, 1758). Syst. Nat., 10th ed., 1:31.
    COMMON NAME: Spectral Bat.

TYPE LOCALITY: Surinam.

DISTRIBUTION: Veracruz (Mexico) to Ecuador and Peru, Bolivia, N and SW Brazil, and Guianas; Trinidad; perhaps Jamaica.

STATUS: IUCN 2003 and IUCN/SSC Action Plan (2001) – Lower Risk (nt).

SYNONYMS: *guianensis* Lacépède, 1789; *maximus* E. Geoffroy, 1806; *nasutus* Shaw, 1800; *nelsoni* Goldman, 1917.

COMMENTS: See Navarro and Wilson (1982). See Emmons (1997) for distribution map.

**Subfamily Carolliinae** Miller, 1924. Bull. U.S. Natl. Mus., 128:53.

COMMENTS: May not be monophyletic; see Lim and Engstrom (1998), Wright et al. (1999), and Baker et al. (2000). Treated as a tribe within Stenodermatinae by McKenna and Bell (1997), but see Wetterer et al. (2000).

*Carollia* Gray, 1838. Mag. Zool. Bot., 2:488.

TYPE SPECIES: *Carollia braziliensis* Gray, 1838 (= *Vespertilio perspicillata* Linnaeus, 1758).

SYNONYMS: *Hemiderma* Gervais, 1856; *Rhinops* Gray, 1866.

COMMENTS: Revised by Pine (1972); phylogeny and geographic patterns discussed by Lim and Engstrom (1998), Wright et al. (1999), Baker et al. (2002), and Hoffman and Baker (2003). Also see Cloutier and Thomas (1992) and Cuartas et al. (2001). Keys were provided by many of these authors, but none included all of the species recognized here.

*Carollia brevicauda* (Schinz, 1821). Das Thierreich, 1:164.

COMMON NAME: Silky Short-tailed Bat.

TYPE LOCALITY: Brazil, Espirito Santo, Jucu River, Fazenda de Coroaba.

DISTRIBUTION: E Panama, Colombia, Venezuela, Guyana, Surinam, French Guiana, Ecuador, Peru, Bolivia, and N & E Brazil; Trinidad.

STATUS: IUCN 2003 and IUCN/SSC Action Plan (2001) – Lower Risk (lc).

SYNONYMS: *bicolor* Wagner, 1840; *grayi* Waterhouse, 1838; *lanceolatum* Natterer, 1843 [*nomen nudum*]; *minor* Gray, 1866.

COMMENTS: Long confused with *perspicillata* or *subrufa*; see Pine (1972). Range restricted to N South America and E Panama by Baker et al. (2002), who referred all Central American records (from W Panama north through Mexico) to *sowelli*. It is possible that these taxa occur in sympatry in Panama, but this has not yet been demonstrated.

*Carollia castanea* H. Allen, 1890. Proc. Am. Philos. Soc., 28:19.

COMMON NAME: Chestnut Short-tailed Bat.

TYPE LOCALITY: Costa Rica, Angostura.

DISTRIBUTION: Honduras to Peru, Bolivia, W Brazil and Venezuela.

STATUS: IUCN 2003 and IUCN/SSC Action Plan (2001) – Lower Risk (lc).

COMMENTS: This complex probably includes more than one species; see Hoffman and Baker (2003).

*Carollia colombiana* Cuartas, Muñoz, and González, 2001. Actual. Biol., 23(75):65.

COMMON NAME: Colombian Short-tailed Bat.

TYPE LOCALITY: Colombia, Antioquia, Municipality of Barbosa, La Cejita road, 6°25'N, 75°15'W.

DISTRIBUTION: Cordillera Central of N Colombia.

STATUS: IUCN 2003 and IUCN/SSC Action Plan (2001) – Not evaluated (new species).

COMMENTS: Known only from the type locality.

*Carollia perspicillata* (Linnaeus, 1758). Syst. Nat., 10th ed., 1:31.

COMMON NAME: Seba's Short-tailed Bat.

TYPE LOCALITY: Surinam.

DISTRIBUTION: Oaxaca, Veracruz and Yucatán Peninsula (Mexico) to Peru, Bolivia, Paraguay, SE Brazil and Guianas; Trinidad and Tobago; perhaps Jamaica, N Lesser Antilles. A record from Grenada (Lesser Antilles) is probably erroneous; see Genoways et al. (1998).

STATUS: IUCN 2003 and IUCN/SSC Action Plan (2001) – Lower Risk (lc).

SYNONYMS: *amplexicaudata* E. Geoffroy, 1818; *azteca* Saussure, 1860; *brachyotus* Schinz, 1821;

*braziliensis* Gray, 1838; *calcaratum* Wagner, 1843; *tricolor* Miller, 1902; *verrucata* Gray, 1844.

COMMENTS: Includes *tricolor*; see Pine (1972). Some authors have recognized subspecies, but see Pine (1972), McLellan (1984), and Koopman (1994). See also Cloutier and Thomas (1992). Phylogeography discussed by Ditchfield (2000).

*Carollia sowelli* Baker, Solari, and Hoffmann, 2002. Occ. Pap. Mus. Texas Tech. Univ., 217:4.
COMMON NAME: Sowell's Short-tailed Bat.
TYPE LOCALITY: Honduras, Comayagua, Cueva de Taulabe, 14°41'42"N, 87°57'07"W.
DISTRIBUTION: San Luis Potosi (Mexico) south to W Panama.
STATUS: IUCN 2003 and IUCN/SSC Action Plan (2001) – Not evaluated (new species).
COMMENTS: Specimens of this species were previously referred to *brevicauda*, which is now thought to be restricted to E Panama and South America; see Wright et al. (1999), Baker et al. (2002), and Hoffman and Baker (2003). It is possible that these taxa occur in sympatry in Panama, but this has not yet been demonstrated.

*Carollia subrufa* (Hahn, 1905). Proc. Biol. Soc. Wash., 18:247.
COMMON NAME: Gray Short-tailed Bat.
TYPE LOCALITY: Mexico, NW coast of Oaxaca, 8 mi (12 km) NW Tapanatepec, Sta. Efigenia.
DISTRIBUTION: Jalisco (Mexico) to NW Nicaragua. A report of this species from Guyana (Koopman, 1993) appears to be in error (B. Lim, pers. comm.).
STATUS: IUCN 2003 and IUCN/SSC Action Plan (2001) – Lower Risk (lc).
COMMENTS: Not a subspecies of *castanea*; see Pine (1972).

*Rhinophylla* Peters, 1865. Monatsb. K. Preuss. Akad. Wiss. Berlin, 1865:355.
TYPE SPECIES: *Rhinophylla pumilio* Peters, 1865.
COMMENTS: Phylogeny and geographic patterns discussed by Lim and Engstrom (1998) and Wright et al. (1999).

*Rhinophylla alethina* Handley, 1966. Proc. Biol. Soc. Wash., 79:86.
COMMON NAME: Hairy Little Fruit Bat.
TYPE LOCALITY: Colombia, Valle, 27 km S Buenaventura, Raposo River.
DISTRIBUTION: W Colombia, W Ecuador.
STATUS: IUCN 2003 and IUCN/SSC Action Plan (2001) – Lower Risk (nt).

*Rhinophylla fischerae* Carter, 1966. Proc. Biol. Soc. Wash., 79:235.
COMMON NAME: Fischer's Little Fruit Bat.
TYPE LOCALITY: Peru, Loreto, 61 mi. (98 km) SE Pucallpa.
DISTRIBUTION: Peru, Ecuador, SE Colombia, S Venezuela, Amazonian Brazil.
STATUS: IUCN 2003 and IUCN/SSC Action Plan (2001) – Lower Risk (nt).

*Rhinophylla pumilio* Peters, 1865. Monatsb. K. Preuss. Akad. Wiss. Berlin, 1865:355.
COMMON NAME: Dwarf Little Fruit Bat.
TYPE LOCALITY: Brazil, Bahia.
DISTRIBUTION: Colombia, Ecuador, Peru, and Bolivia to Guianas and E Brazil.
STATUS: IUCN 2003 and IUCN/SSC Action Plan (2001) – Lower Risk (lc).

**Subfamily Stenodermatinae** Gervais, 1856. *In* F. Comte de Castelnau, Exped. Partes Cen. Am. Sud., Zool.(Sec. 7), Vol. 1, pt. 2(Mammifères):32 footnote.
COMMENTS: McKenna and Bell (1997) included Carolliinae as a tribe (Carolliini) within Stenodermatinae, but the traditional usage of these group names is retained here; see discussion in Wetterer et al. (2000). Phylogenetic relationships have been discussed by Owen (1987, 1991), Lim (1993), Van Den Bussche et al. (1993a), Wetterer et al. (2000), and Baker et al. (2000).

**Tribe Sturnirini** Miller, 1907. Bull. U.S. Natl. Mus., 57:33.
COMMENTS: Monogeneric; equivalent to subtribe Sturnirina of McKenna and Bell (1997). See discussion in Wetterer et al. (2000). Relationships and biogeography reviewed by Pacheco and Patterson (1992).

*Sturnira* Gray, 1842. Ann. Mag. Nat. Hist., [ser. 1], 10:257.
    TYPE SPECIES: *Sturnira spectrum* Gray, 1842 (= *Phyllostoma lilium* E. Geoffroy, 1810).
    SYNONYMS: *Corvira* Thomas, 1915; *Nyctiplanus* Gray, 1849; *Stenoderma* Gray, 1847 [not Geoffroy, 1813]; *Sturnirops* Goodwin, 1938.
    COMMENTS: Includes *Corvira*; see Jones and Carter (1976). Davis (1980) gave a key to all but one of the species recognized here. Two subgenera are recognized, *Sturnira* and *Corvira*. Phylogenies of the genus have been proposed by Pacheco and Patterson (1991) and Villalobos and Valerio (2002).

*Sturnira aratathomasi* Peterson and Tamsitt, 1968. R. Ontario Mus. Life Sci. Occas. Pap., 12:1.
    COMMON NAME: Aratathomas's Yellow-shouldered Bat.
    TYPE LOCALITY: Colombia, Valle, 2 km S Pance (ca. 20 km SW Cali), 1,650 m.
    DISTRIBUTION: Colombia, Ecuador, NW Venezuela, Peru.
    STATUS: IUCN 2003 and IUCN/SSC Action Plan (2001) – Lower Risk (nt).
    COMMENTS: Subgenus *Sturnira*. See Soriano and Molinari (1987).

*Sturnira bidens* Thomas, 1915. Ann. Mag. Nat. Hist., ser. 8, 16:310.
    COMMON NAME: Bidentate Yellow-shouldered Bat.
    TYPE LOCALITY: Ecuador, Napo, Baeza, Upper Coca River, 6,500 ft. (1,981 m).
    DISTRIBUTION: Peru, Ecuador, Colombia, Venezuela, perhaps Amazonian Brazil.
    STATUS: IUCN 2003 and IUCN/SSC Action Plan (2001) – Lower Risk (nt).
    COMMENTS: Subgenus *Corvira*. Formerly placed in a distinct genus (*Corvira*); see Gardner and O'Neill (1969) and Jones and Carter (1976). See Molinari and Soriano (1987).

*Sturnira bogotensis* Shamel, 1927. Proc. Biol. Soc. Wash., 40:129.
    COMMON NAME: Bogotan Yellow-shouldered Bat.
    TYPE LOCALITY: Colombia, Cundinamarca, Bogota.
    DISTRIBUTION: Colombia, Ecuador, and Peru. Records from Venezuela, Bolivia, and Argentina are erroneous (see Pacheco and Patterson, 1992).
    STATUS: IUCN 2003 and IUCN/SSC Action Plan (2001) – Lower Risk (lc).
    COMMENTS: Subgenus *Sturnira*. Often confused with *erythromos*, *ludovici*, and *oporaphilum*, but see Handley (1976) and Pacheco and Patterson (1992).

*Sturnira erythromos* (Tschudi, 1844). Fauna Peruana, p. 64.
    COMMON NAME: Hairy Yellow-shouldered Bat.
    TYPE LOCALITY: Peru.
    DISTRIBUTION: Venezuela to Peru, Bolivia, and NW Argentina.
    STATUS: IUCN 2003 and IUCN/SSC Action Plan (2001) – Lower Risk (lc).
    COMMENTS: Subgenus *Sturnira*. Reviewed by Pacheco and Patterson (1992); also see Giannini and Barquez (2003).

*Sturnira lilium* (E. Geoffroy, 1810). Ann. Mus. Natn. Hist. Nat. Paris, 15:181.
    COMMON NAME: Little Yellow-shouldered Bat.
    TYPE LOCALITY: Paraguay, Asunción (restricted by Cabrera [1958]).
    DISTRIBUTION: Lesser Antilles; Sonora and Tamaulipas (Mexico) south to Bolivia, Paraguay, N Argentina, Uruguay, and E Brazil; Trinidad and Tobago; Grenada; perhaps Jamaica.
    STATUS: IUCN 2003 and IUCN/SSC Action Plan (2001) – Lower Risk (lc).
    SYNONYMS: *albescens* Wagner, 1847; *chilense* Gray, 1847; *chrysocomos* Wagner, 1855; *erythromas*, Tschudi, 1844; *excisum* Wagner, 1842; *fumarium* Wagner, 1847; *oporophilum* Tschudi, 1844; *rotundatus* Gray, 1849; *spectrum* Gray, 1842; *spiculatum* Illiger, 1825; *vampyrus* Schinz, 1845; **angeli** de la Torre, 1966; **luciae** Jones and Phillips, 1976; **parvidens** Goldman, 1917; **paulsoni** de la Torre, 1966; **serotinus** Genoways, 1998; **vulcanensis** Genoways, 1998; **zygomaticus** Jones and Phillips, 1976.
    COMMENTS: Subgenus *Sturnira*. Includes *angeli* and *paulsoni*; see Jones and Phillips (1976). Reviewed in part by Jones and Phillips (1976), Genoways (1998), and Timm and Genoways (2003); also see Jones (1989) and Gannon et al. (1989). Phylogeography discussed by Ditchfield (2000).

*Sturnira ludovici* Anthony, 1924. Am. Mus. Novit., 139:8.
    COMMON NAME: Highland Yellow-shouldered Bat.

TYPE LOCALITY: Ecuador, Pichincha, near Gualea, ca. 4,000 ft (1,333 m).

DISTRIBUTION: Ecuador and Guyana north to Sonora and Tamaulipas (Mexico).

STATUS: IUCN 2003 and IUCN/SSC Action Plan (2001) – Lower Risk (lc).

SYNONYMS: *hondurensis* Goodwin, 1940; *occidentalis* Jones and Phillips, 1964.

COMMENTS: Subgenus *Sturnira*. Includes *hondurensis*; see Jones and Carter (1976). Bolivian records probably pertain to *oporaphilum*; Peruvian ones definitely do; see Anderson et al. (1982) and Pacheco and Patterson (1992).

*Sturnira luisi* Davis, 1980. Occas. Pap. Mus. Texas Tech Univ., 70:1.

COMMON NAME: Luis's Yellow-shouldered Bat.

TYPE LOCALITY: Costa Rica, Alajuela, 11 mi. (18 km) NE Naranjo, Cariblanco, 3,000 ft. (914 m).

DISTRIBUTION: Costa Rica to Ecuador and NW Peru.

STATUS: IUCN 2003 and IUCN/SSC Action Plan (2001) – Lower Risk (lc).

COMMENTS: Subgenus *Sturnira*. The presence of this species in Colombia has not been verified; previously confused with *Sturnira ludovici* (Tamsitt, *in* Honacki et al., 1982). Brosset and Charles-Dominique (1990) suggested that *luisi* might be conspecific with *tildae*, but see Simmons and Voss (1998).

*Sturnira magna* de la Torre, 1966. Proc. Biol. Soc. Wash., 79:267.

COMMON NAME: Greater Yellow-shouldered Bat.

TYPE LOCALITY: Peru, Loreto, Iquitos, Río Maniti, Santa Cecilia.

DISTRIBUTION: Colombia, Ecuador, Peru, W Brazil, Bolivia.

STATUS: IUCN 2003 and IUCN/SSC Action Plan (2001) – Lower Risk (nt).

COMMENTS: Subgenus *Sturnira*. See Tamsitt and Häuser (1985).

*Sturnira mistratensis* Vega and Cadena, 2000. Rev. Acad. Colomb. Cienc., 24:286.

COMMON NAME: Mistratoan Yellow-shouldered Bat.

TYPE LOCALITY: Colombia, Risaralda, Mistrató, corregimiento de Puerto de Oro, 980 m.

DISTRIBUTION: W Andes of Colombia.

STATUS: IUCN 2003 – Not evaluated (new species); and not considered in IUCN/SSC Action Plan (2001).

COMMENTS: Subgenus *Sturnira*. Known only from the holotype.

*Sturnira mordax* (Goodwin, 1938). Am. Mus. Novit., 976:1.

COMMON NAME: Talamancan Yellow-shouldered Bat.

TYPE LOCALITY: Costa Rica, Cartago, El Sauce Peralta.

DISTRIBUTION: Costa Rica, Panama.

STATUS: IUCN 2003 and IUCN/SSC Action Plan (2001) – Lower Risk (nt).

COMMENTS: Subgenus *Sturnira*. Formerly included in *Sturnirops*; see Davis et al. (1964).

*Sturnira nana* Gardner and O'Neill, 1971. Occas. Pap. Mus. Zool. La. St. Univ., 42:1.

COMMON NAME: Lesser Yellow-shouldered Bat.

TYPE LOCALITY: Peru, Ayacucho, Huanhuachayo, 1,660 m.

DISTRIBUTION: S Peru.

STATUS: IUCN 2003 and IUCN/SSC Action Plan (2001) – Vulnerable.

COMMENTS: Subgenus *Corvira*.

*Sturnira oporaphilum* (Tschudi, 1844). Fauna Peruana, p. 64.

COMMON NAME: Tschudi's Yellow-shouldered Bat.

TYPE LOCALITY: Peru.

DISTRIBUTION: Ecuador, Peru, Bolivia, and NW Argentina.

STATUS: IUCN 2003 – Not evaluated; not considered in IUCN/SSC Action Plan (2001).

COMMENTS: Subgenus *Sturnira*. Often confused with *bogotensis*; see Pacheco and Patterson (1992).

*Sturnira thomasi* de la Torre and Schwartz, 1966. Proc. Biol. Soc. Wash., 79:299.

COMMON NAME: Thomas's Yellow-shouldered Bat.

TYPE LOCALITY: Guadeloupe (Lesser Antilles), Sofaia, 1,200 ft. (366 m) (France).

DISTRIBUTION: Guadeloupe and Montserrat (Lesser Antilles).

STATUS: IUCN 2003 and IUCN/SSC Action Plan (2001) – Endangered.

COMMENTS: Subgenus *Sturnira*. See Jones and Genoways (1975c) and Pedersen et al. (1996).

*Sturnira tildae* de la Torre, 1959. Chicago Acad. Sci. Nat. Hist. Misc., 166:1.
    COMMON NAME: Tilda's Yellow-shouldered Bat.
    TYPE LOCALITY: Trinidad and Tobago, Trinidad, Arima Valley.
    DISTRIBUTION: Brazil, Guianas, Venezuela, Trinidad, Colombia, Ecuador, Peru, Bolivia.
    STATUS: IUCN 2003 and IUCN/SSC Action Plan (2001) – Lower Risk (lc).
    COMMENTS: Subgenus *Sturnira*. Does not include *luisi*; see Simmons and Voss (1998).

**Tribe Stenodermatini** Gervais, 1856. *In* Comte de Castelnau, Exped. Partes Cen. Am. Sud., Zool.
    (Sec. 7), Vol. 1, pt. 2(Mammifères):32 footnote.
    COMMENTS: Equivalent to subtribe Stenodermatina of McKenna and Bell (1997). The subtribal
        classification used here follows Wetterer et al. (2000). The subtribe Ectophyllina may be
        paraphylletic (see Baker et al., 2000), but there is strong support for monophyly of the
        subtribe Stenodermatina from both morphology and DNA sequence data (Baker et al.,
        2000; Wetterer et al. 2000).

*Ametrida* Gray, 1847. Proc. Zool. Soc. Lond., 1847:15.
    TYPE SPECIES: *Ametrida centurio* Gray, 1847.
    COMMENTS: Subtribe Stenodermatina. Revised by Peterson (1965b).

*Ametrida centurio* Gray, 1847. Proc. Zool. Soc. Lond., 1847:15.
    COMMON NAME: Little White-shouldered Bat.
    TYPE LOCALITY: Brazil, Pará, Belem.
    DISTRIBUTION: Amazonian Brazil, Guianas, Panama, Venezuela, Trinidad, Bonaire Isl
        (Netherlands Antilles).
    STATUS: IUCN 2003 and IUCN/SSC Action Plan (2001) – Lower Risk (lc).
    SYNONYMS: *minor* H. Allen, 1894.
    COMMENTS: Includes *minor*; see Jones and Carter (1976). See Emmons (1997) for distribution
        map.

*Ardops* Miller, 1906. Proc. Biol. Soc. Wash., 19:84.
    TYPE SPECIES: *Stenoderma nichollsi* Thomas, 1891.
    COMMENTS: Subtribe Stenodermatina. Revised by Jones and Schwartz (1967). Included under
        *Stenoderma* by Varona (1974) and Simpson (1945), but see Jones and Carter (1976).

*Ardops nichollsi* (Thomas, 1891). Ann. Mag. Nat. Hist., ser. 6, 7:529.
    COMMON NAME: Tree Bat.
    TYPE LOCALITY: Dominica (Lesser Antilles).
    DISTRIBUTION: Lesser Antilles, from St. Eustatius to St. Vincent.
    STATUS: IUCN 2003 and IUCN/SSC Action Plan (2001) – Lower Risk (nt).
    SYNONYMS: **annectens** Miller, 1913; **koopmani** Jones and Schwartz, 1967; **luciae** Miller,
        1902; **montserratensis** Thomas, 1894.
    COMMENTS: For discussion of subspecies see Jones and Schwartz (1967) and Jones (1989). See
        also Jones and Genoways (1973).

*Ariteus* Gray, 1838. Mag. Zool. Bot., 2:491.
    TYPE SPECIES: *Istiophorus flavescens* Gray, 1831.
    SYNONYMS: *Peltorhinus* Peters, 1876.
    COMMENTS: Subtribe Stenodermatina. Included as a subgenus of *Stenoderma* by Varona (1974)
        and Simpson (1945); but see Jones and Carter (1976).

*Ariteus flavescens* (Gray, 1831). Zool. Misc., 1:37.
    COMMON NAME: Jamaican Fig-eating Bat.
    TYPE LOCALITY: Not designated in original publication (presumably Jamaica).
    DISTRIBUTION: Jamaica.
    STATUS: IUCN 2003 and IUCN/SSC Action Plan (2001) – Vulnerable.
    SYNONYMS: *achradophilus* Gosse, 1851.
    COMMENTS: See Timm and Genoways (2003).

*Artibeus* Leach, 1821. Trans. Linn. Soc. Lond., 13:75.
  TYPE SPECIES: *Artibeus jamaicensis* Leach, 1821.
  SYNONYMS: *Arctibeus* Gray, 1838; *Artibaeus* Gervais, 1856; *Artibius* Bonaparte, 1847; *Artobius*
      Winge, 1892; *Dermanura* Gervais, 1856; *Koopmania* Owen, 1991; *Medateus* Leach, 1821;
      *Pteroderma* Gervais, 1856.
  COMMENTS: Subtribe Ectophyllina. Includes three subgenera as recognized here: *Artibeus*,
      *Dermanura*, and *Koopmania*. Does not include *Enchisthenes*; see Lim (1993), Van Den
      Bussche et al. (1993*a*, 1998), Baker et al. (2000), and Wetterer et al. (2000). Some
      researchers regard *Dermanura* as a distinct genus, but I prefer to treat it as a subgenus of
      *Artibeus* in recognition of its close phylogenetic affinities with *Artibeus* sensu stricto
      (see phylogenies in Van Den Bussche et al. [1993*a*, 1998], Baker et al. [2000], and Wetterer
      et al. [2000]). Large-bodied species (subgenus *Artibeus*) were reviewed by Marques-Aguiar
      (1994); see also Lim and Wilson (1993). Handley (1987) reviewed the smaller species
      (subgenera *Dermanura* and *Koopmania*). Species relationships discussed by Marques-
      Aguiar (1994) and Van Den Bussche et al. (1998).

*Artibeus amplus* Handley, 1987. Fieldiana, Zool., n.s., 39:164.
  COMMON NAME: Large Fruit-eating Bat.
  TYPE LOCALITY: Venezuela, Zulia, Kasmera.
  DISTRIBUTION: Guyana, Venezuela, N Colombia.
  STATUS: IUCN 2003 and IUCN/SSC Action Plan (2001) – Lower Risk (nt).
  COMMENTS: Subgenus *Artibeus*. Reviewed by Lim and Wilson (1993).

*Artibeus anderseni* Osgood, 1916. Field Mus. Nat. Hist. Publ., Zool. Ser., 10:212.
  COMMON NAME: Andersen's Fruit-eating Bat.
  TYPE LOCALITY: Brazil, Rondonia, Porto Velho.
  DISTRIBUTION: W Brazil, Bolivia, Ecuador, Peru.
  STATUS: IUCN 2003 and IUCN/SSC Action Plan (2001) – Lower Risk (lc).
  COMMENTS: Subgenus *Dermanura*. Previously considered a subspecies of *cinereus*, but see
      Koopman (1978*b*) and Handley (1987).

*Artibeus aztecus* K. Andersen, 1906. Ann. Mag. Nat. Hist., ser. 7, 18:422.
  COMMON NAME: Aztec Fruit-eating Bat.
  TYPE LOCALITY: Mexico, Morelos, Tetela del Volcán.
  DISTRIBUTION: Michoacan and Oaxaca to Nuevo León and Sinaloa (Mexico), south to
      W Panama.
  STATUS: IUCN 2003 and IUCN/SSC Action Plan (2001) – Lower Risk (lc).
  SYNONYMS: ***major*** Davis, 1969; ***minor*** Davis, 1969.
  COMMENTS: Subgenus *Dermanura*. Not a subspecies of *cinereus*; see Jones and Carter (1976).
      Revised by Davis (1969). See Webster and Jones (1982*b*).

*Artibeus cinereus* (Gervais, 1856). *In* Comte de Castelnau, Exped. Partes Cen. Am. Sud., Zool.
      (Sec. 7), Vol. 1, pt. 2 (Mammifères):36.
  COMMON NAME: Gervais's Fruit-eating Bat.
  TYPE LOCALITY: Brazil, Pará, Belem.
  DISTRIBUTION: Guianas, Venezuela, N Brazil, Peru, Trinidad.
  STATUS: IUCN 2003 and IUCN/SSC Action Plan (2001) – Lower Risk (lc).
  SYNONYMS: *quadrivittatus* Peters, 1865.
  COMMENTS: Subgenus *Dermanura*. Does not include *gnomus* or *glaucus*; see Handley (1987).

*Artibeus concolor* Peters, 1865. Monatsb. K. Preuss. Akad. Wiss. Berlin, 1865:357.
  COMMON NAME: Brown Fruit-eating Bat.
  TYPE LOCALITY: Surinam, Paramaribo.
  DISTRIBUTION: Guianas, Venezuela, Colombia, N Brazil, Peru.
  STATUS: IUCN 2003 and IUCN/SSC Action Plan (2001) – Lower Risk (nt).
  COMMENTS: Subgenus *Koopmania*. A revised diagnosis was presented by Owen (1991), who
      placed this species in its own genus (*Koopmania*), which is here considered a subgenus
      of *Artibeus* based on phylogenetic results of Marques-Aguiar (1994), Van Den Bussche
      et al. (1998), Baker et al. (2000), and Wetterer et al. (2000). See Acosta and Owen
      (1993).

*Artibeus fimbriatus* Gray, 1838. Mag. Zool. Bot., 2:487.
COMMON NAME: Fringed Fruit-eating Bat.
TYPE LOCALITY: Brazil, Paraná, Serra do Mar, Morretes.
DISTRIBUTION: S Brazil, Paraguay.
STATUS: IUCN 2003 and IUCN/SSC Action Plan (2001) – Lower Risk (nt).
SYNONYMS: *grandis* Dobson, 1878 [*nomen nudum*].
COMMENTS: Subgenus *Artibeus*. Reviewed by Handley (1989) and Marques-Aguiar (1994).

*Artibeus fraterculus* Anthony, 1924. Am. Mus. Novit., 114:5.
COMMON NAME: Fraternal Fruit-eating Bat.
TYPE LOCALITY: Ecuador, El Oro, Portovelo, 2,000 ft. (610 m).
DISTRIBUTION: Ecuador, Peru.
STATUS: IUCN 2003 and IUCN/SSC Action Plan (2001) – Vulnerable.
COMMENTS: Subgenus *Artibeus*. Considered a subspecies of *jamaicensis* by Jones and Carter
    (1976), but see Koopman (1978*b*) and Marques-Aguiar (1994).

*Artibeus glaucus* Thomas, 1893. Proc. Zool. Soc. Lond., 1893:336.
COMMON NAME: Silvery Fruit-eating Bat.
TYPE LOCALITY: Peru, Junín, Chauchamayo.
DISTRIBUTION: S Mexico to Bolivia and S Brazil; Grenada (Lesser Antilles).
STATUS: IUCN 2003 and IUCN/SSC Action Plan (2001) – Lower Risk (lc).
SYNONYMS: *bogotensis* K. Andersen, 1906; *pumilio* Thomas, 1924; *rosenbergii* Thomas, 1897.
COMMENTS: Subgenus *Dermanura*. Does not include *gnomus* or *watsoni*; see Handley (1987).
    Koopman (1994) recognized several subspecies in this complex (which he referred to
    *cinereus*), but the boundaries among them are unclear. Caribbean records reviewed by
    Genoways et al. (1998).

*Artibeus gnomus* Handley, 1987. Fieldiana Zool. 39:167.
COMMON NAME: Dwarf Fruit-eating Bat.
TYPE LOCALITY: Venezuela, Bolívar, 59 km SE El Dorado, El Manaco.
DISTRIBUTION: Ecuador, Peru, Bolivia, Amazonian Brazil, Venezuela, Guianas.
STATUS: IUCN 2003 – Not listed (*lapsus*); IUCN/SSC Action Plan (2001) – Lower Risk (lc).
COMMENTS: Subgenus *Dermanura*. Distinct from *cinereus* and *glaucus*; see Handley (1987),
    Brosset and Charles-Dominique (1990), and Simmons and Voss (1998).

*Artibeus hirsutus* K. Andersen, 1906. Ann. Mag. Nat. Hist., ser. 7, 18:420.
COMMON NAME: Hairy Fruit-eating Bat.
TYPE LOCALITY: Mexico, Michoacan, La Salada.
DISTRIBUTION: Sonora to Guerrero (Mexico).
STATUS: IUCN 2003 and IUCN/SSC Action Plan (2001) – Vulnerable.
COMMENTS: Subgenus *Artibeus*. See Webster and Jones (1983) and Marques-Aguiar (1994).

*Artibeus incomitatus* Kalko and Handley, 1994. Z. Säugtierk., 59:260.
COMMON NAME: Solitary Fruit-eating Bat.
TYPE LOCALITY: Panama, Bocas del Toro, Isla Escudo de Veraguas, near West Point, 1 m.
DISTRIBUTION: Known only from type locality.
STATUS: IUCN 2003 and IUCN/SSC Action Plan (2001) – Data Deficient.
COMMENTS: Subgenus *Dermanura*. Closely related to *watsoni*; see Kalko and Handley (1994).

*Artibeus inopinatus* Davis and Carter, 1964. Proc. Biol. Soc. Wash., 77:119.
COMMON NAME: Honduran Fruit-eating Bat.
TYPE LOCALITY: Honduras, Choluteca, Choluteca, 10 ft. (3 m).
DISTRIBUTION: El Salvador, Honduras, Nicaragua.
STATUS: IUCN 2003 and IUCN/SSC Action Plan (2001) – Vulnerable.
COMMENTS: Subgenus *Artibeus*. Closely related to *hirsutus* but apparently distinct; see
    Marques-Aguiar (1994). See Webster and Jones (1983).

*Artibeus jamaicensis* Leach, 1821. Trans. Linn. Soc. Lond., 13:75.
COMMON NAME: Jamaican Fruit-eating Bat.
TYPE LOCALITY: Jamaica.
DISTRIBUTION: Michoacan, Sinaloa, and Tamaulipas (Mexico) to Ecuador, Peru, Bolivia,

N Argentina, and E Brazil; Trinidad and Tobago; Greater and Lesser Antilles, S Bahamas. Perhaps Florida Keys; see Lazell and Koopman (1985), but see also Humphrey and Brown (1986).

STATUS: IUCN 2003 and IUCN/SSC Action Plan (2001) – Lower Risk (lc) as *A. jamaicensis* and *A. planirostris*.

SYNONYMS: *carpolegus* Gosse, 1851; *coryi* J. A. Allen, 1890; *eva* Cope, 1889; *insularis* J. A. Allen, 1904; *lewisi* Leach, 1821; *praeceps* K. Andersen, 1906; **aequatorialis** K. Andersen, 1906; **fallax** Peters, 1865; *alidum* Elliot, 1907; **grenadensis** K. Andersen, 1906; **hercules** Rehn, 1902; **parvipes** Rehn, 1902; **paulus** Davis, 1970; **planirostris** Spix, 1823; **richardsoni** J. A. Allen, 1908; **schwartzi** Jones, 1978; **trinitatis** K. Andersen, 1906; **triomylus** Handley, 1966; **yucatanicus** J. A. Allen, 1904. **Unassigned**: *macleayii* Dobson, 1878 [*nomen nudum*].

COMMENTS: Subgenus *Artibeus*. Does not include *obscurus* (= *fuliginosus*); see Handley (1989), Brosset and Charles-Dominique (1990), Lim and Wilson (1993), and Simmons and Voss (1998). There is little agreement about whether *jamaicensis* includes *planirostris* (supported by Handley [1987, 1991] and Marques-Aguiar [1994]) or if *planirostris* (including *fallax* and *hercules*) represents a distinct species (supported by Koopman [1978b], Lim and Wilson [1993], and Lim [1997]). Pumo et al. (1996) treated *planirostris* and *jamaicensis* as separate species in their analysis of mtDNA sequences, but their data are more consistent with recognition of these as members of a single species. Accordingly, I have retained *planirostris* in *jamaicensis* pending further study. Includes *fallax*, *hercules*, and *praeceps*; see Koopman (1968), Handley (1987), and Marques-Aguiar (1994). Subspecies limits and relationships discussed by Jones and Phillips (1970), J. K. Jones (1978), Hall (1981), Handley (1987), and Genoways et al. (1998); also see Pumo et al. (1988, 1996), Phillips et al. (1991), Lim (1997), and Timm and Genoways (2003). Reviewed by Ortega and Castro-Arellano (2001), but note that they excluded *planirostris*, *fallax*, *hercules*, *aequatorialis*, and *schwartzi*. Ortega and Castro-Arellano (2001) mapped the range of *A. j. richardsoni* as including most of South America (extensively overlapping the ranges *planirostris*, *fallax*, *hercules*, and *aequatorialis*), implying that *jamaicensis* and of *planirostris* (if distinct) are broadly sympatric throughout much of South America. However, this was probably an error because such sympatry has never been proposed in the primary systematic literature (see Lim and Wilson [1993] and Lim [1997]).

*Artibeus lituratus* (Olfers, 1818). *In* Eschwege, J. Brasilien, Neue Bibliothek. Reisenb., 15:224.

COMMON NAME: Great Fruit-eating Bat.

TYPE LOCALITY: Paraguay, Asunción.

DISTRIBUTION: Michoacan, Sinaloa, and Tamaulipas (Mexico) south to S Brazil, N Argentina, and Bolivia; Trinidad and Tobago; S Lesser Antilles; Trés Marías Isls.

STATUS: IUCN 2003 and IUCN/SSC Action Plan (2001) – Lower Risk (lc) as *A. lituratus* and *A. intermedius*.

SYNONYMS: *frenatus* Illiger, 1815 [*nomen nudum*]; *frenatus* Olfers, 1818; *rusbyi* J. A. Allen, 1904; *superciliatum* Schinz, 1821; **koopmani** Wilson, 1991; **palmarum** J. A. Allen, 1897; *femurvillosum* Bangs, 1899; *intermedius* J. A. Allen, 1897. **Unassigned**: *dominicanus* Andersen, 1908 [*nomen nudum*].

COMMENTS: Subgenus *Artibeus*. Includes *palmarum* but not *fallax*, *hercules*, or *praeceps* (Koopman, 1968, 1978b), Handley (1987), and Marques-Aguiar (1994). Includes *intermedius*; see Jones and Carter (1976) and Marques-Aguiar (1994), but see also Davis (1984) and Wilson (1991). It is not appropriate to treat *intermedius* as a subspecies of *lituratus* because it supposedly co-occurs with other populations of *lituratus* (referred to *palmarum*, which has priority) at several Central American localities (Davis, 1984). Because there are no characters that unambiguously separate *palmarum* and *intermedius* (Davis, 1984; Marques-Aguiar, 1994; Rodrigo Medellin, pers. comm.), it seems most likely that *intermedius* simply represents individuals of *palmarum* that fall at the lower end of the normal range of size variation. Accordingly, I treat *intermedius* as a junior synonym of *A. lituratus palmarum*. Phylogeography discussed by Phillips et al. (1991) and Ditchfield (2000).

*Artibeus obscurus* (Schinz, 1821). *In* G. Cuvier, Das Tierreich, 1:164.
    COMMON NAME: Dark Fruit-eating Bat.
    TYPE LOCALITY: Brazil, Bahia, Rio Peruhype, Villa Vicosa.
    DISTRIBUTION: Colombia, Venezuela, Guianas, Ecuador, Peru, Bolivia, Brazil.
    STATUS: IUCN 2003 and IUCN/SSC Action Plan (2001) – Lower Risk (nt).
    SYNONYMS: *fuliginosus* Gray, 1838.
    COMMENTS: Subgenus *Artibeus*. Distinct from *jamaicensis*; see Handley (1989), Brosset and
        Charles-Dominique (1990), Lim and Wilson (1993), Marques-Aguiar (1994), and
        Simmons and Voss (1998). This species has often been referred to as *fuliginosus* in the
        literature, but *obscurus* is a senior synonym; see Handley (1989).

*Artibeus phaeotis* (Miller, 1902). Proc. Acad. Nat. Sci. Phil., 54:405.
    COMMON NAME: Pygmy Fruit-eating Bat.
    TYPE LOCALITY: Mexico, Yucatán, Chichén-Itzá.
    DISTRIBUTION: Veracruz, Sinaloa, and Michoacan (Mexico) south to Ecuador and Guyana.
    STATUS: IUCN 2003 and IUCN/SSC Action Plan (2001) – Lower Risk (lc).
    SYNONYMS: *turpis* K. Andersen, 1906; **nanus** K. Andersen, 1906; **palatinus** Davis, 1970;
        **ravus** Miller, 1902.
    COMMENTS: Subgenus *Dermanura*. Includes *nanus* and *turpis*; see Jones and Lawlor (1965)
        and Davis (1970). For including *ravus* and other synomyms see Timm (1985) and
        Handley (1987).

*Artibeus toltecus* (Saussure, 1860). Rev. Mag. Zool. Paris, ser. 2, 12:427.
    COMMON NAME: Toltec Fruit-eating Bat.
    TYPE LOCALITY: Mexico, Veracruz, Mirador.
    DISTRIBUTION: Panama to Nuevo León and Sinaloa (Mexico).
    STATUS: IUCN 2003 and IUCN/SSC Action Plan (2001) – Lower Risk (lc).
    SYNONYMS: **hesperus** Davis, 1969.
    COMMENTS: Subgenus *Dermanura*. Not a subspecies of *cinereus*; see Jones and Carter (1976).
        Revised by Davis (1969). Does not include *ravus*, see Handley (1987). See Webster and
        Jones (1982*c*).

*Artibeus watsoni* Thomas, 1901. Ann. Mag. Nat. Hist., ser. 7, 7:542.
    COMMON NAME: Thomas's Fruit-eating Bat.
    TYPE LOCALITY: Panama, Chiriquí, Bogava [Bugaba], 250 m.
    DISTRIBUTION: S Mexico to SW Colombia.
    STATUS: IUCN 2003 – Not listed (*lapsus*); IUCN/SSC Action Plan (2001) – Lower Risk (lc).
    SYNONYMS: *jucundum* Elliot, 1906.
    COMMENTS: Subgenus *Dermanura*. Distinct from *glaucus*; see Handley (1987). See also Kalko
        and Handley (1994).

*Centurio* Gray, 1842. Ann. Mag. Nat. Hist., [ser. 1], 10:259.
    TYPE SPECIES: *Centurio senex* Gray, 1842.
    SYNONYMS: *Trichocoryctes* Trouessart, 1897; *Trichocoryes* H. Allen, 1861; *Trichocorytes* Gray, 1866.
    COMMENTS: Subtribe Stenodermatina.

*Centurio senex* Gray, 1842. Ann. Mag. Nat. Hist., [ser. 1], 10:259.
    COMMON NAME: Wrinkle-faced Bat.
    TYPE LOCALITY: Nicaragua, Chinandega, Realejo.
    DISTRIBUTION: Venezuela to Tamaulipas and Sinaloa (Mexico); Trinidad and Tobago.
    STATUS: IUCN 2003 and IUCN/SSC Action Plan (2001) – Lower Risk (lc).
    SYNONYMS: *flavogularis* Lichtenstein and Peters, 1854; *mexicanus* Saussure, 1860; *mcmurtrii*
        H. Allen, 1861; *minor* Ward, 1891; **greenhalli** Paradiso, 1967.
    COMMENTS: Reviewed by Paradiso (1967) and Snow et al. (1980). Venezuelan populations
        have not been allocated to subspecies. See Emmons (1997) for distribution map.

*Chiroderma* Peters, 1860. Monatsb. K. Preuss. Akad. Wiss. Berlin, 1860:747.
    TYPE SPECIES: *Chiroderma villosum* Peters, 1860.
    SYNONYMS: *Mimetops* Gray, 1866.

COMMENTS: Subtribe Ectophyllina. Reviewed by Goodwin (1958). See Baker et al. (1994) for a phylogeny of the genus.

*Chiroderma doriae* Thomas, 1891. Ann. Mus. Civ. Stor. Nat. Genova, ser. 2, 10:881.
COMMON NAME: Brazilian Big-eyed Bat.
TYPE LOCALITY: Brazil, Minas Gerais.
DISTRIBUTION: Minas Gerais and São Paulo (SE Brazil), Paraguay.
STATUS: IUCN 2003 and IUCN/SSC Action Plan (2001) – Vulnerable.
SYNONYMS: *dorsale* Lund, 1842.

*Chiroderma improvisum* Baker and Genoways, 1976. Occas. Pap. Mus. Texas Tech Univ., 39:2.
COMMON NAME: Guadeloupean Big-eyed Bat.
TYPE LOCALITY: Guadeloupe (Lesser Antilles), Basse Terre, 2 km S and 2 km E Baie-Mahault (France).
DISTRIBUTION: Guadeloupe and Montserrat (Lesser Antilles).
STATUS: IUCN 2003 and IUCN/SSC Action Plan (2001) – Endangered.
COMMENTS: See Jones and Baker (1980) and Jones (1989).

*Chiroderma salvini* Dobson, 1878. Cat. Chiroptera Brit. Mus., p. 532.
COMMON NAME: Salvin's Big-eyed Bat.
TYPE LOCALITY: Costa Rica.
DISTRIBUTION: Peru, Bolivia, and Venezuela north to Michoacan, Hidalgo, and Chihuahua (Mexico).
STATUS: IUCN 2003 and IUCN/SSC Action Plan (2001) – Lower Risk (lc).
SYNONYMS: *scopaeum* Handley, 1966.

*Chiroderma trinitatum* Goodwin, 1958. Am. Mus. Novit., 1877:1.
COMMON NAME: Little Big-eyed Bat.
TYPE LOCALITY: Trinidad and Tobago, Trinidad, Cumaca, 1,000 ft. (305 m).
DISTRIBUTION: Panama south to Amazonian Brazil, Bolivia and Peru; Trinidad.
STATUS: IUCN 2003 and IUCN/SSC Action Plan (2001) – Lower Risk (lc).
SYNONYMS: *gorgasi* Handley, 1960.
COMMENTS: Includes *gorgasi*; see Jones and Carter (1976).

*Chiroderma villosum* Peters, 1860. Monatsb. K. Preuss. Akad. Wiss. Berlin, 1860:748.
COMMON NAME: Hairy Big-eyed Bat.
TYPE LOCALITY: Brazil; see Carter and Dolan (1978).
DISTRIBUTION: Hidalgo (Mexico) south to S Brazil, Bolivia and Peru; Trinidad and Tobago.
STATUS: IUCN 2003 and IUCN/SSC Action Plan (2001) – Lower Risk (lc).
SYNONYMS: *jesupi* J. A. Allen, 1900; *isthmicum* Miller, 1912.
COMMENTS: See Handley (1960).

*Ectophylla* H. Allen, 1892. Proc. U.S. Natl. Mus., 15:441.
TYPE SPECIES: *Ectophylla alba* H. Allen, 1892.
COMMENTS: Subtribe Ectophyllina. Some authors have treated *Mesophylla* as a junior synonym; see Goodwin and Greenhall (1962), Simmons and Voss (1998), and Wetterer et al. (2000). However, relationships of these taxa remain unclear (see Owen [1987] and Baker et al. [2000]), so *Mesophylla* is treated here as a distinct genus pending further study.

*Ectophylla alba* H. Allen, 1892. Proc. U.S. Natl. Mus., 15:442.
COMMON NAME: Honduran White Bat.
TYPE LOCALITY: Honduras (= Río Segovia) (McCarthy et al., 1993).
DISTRIBUTION: Honduras to W Panama.
STATUS: IUCN 2003 and IUCN/SSC Action Plan (2001) – Lower Risk (nt).
COMMENTS: See Timm (1982). See Emmons (1997) for distribution map. Koopman (1993) included W Colombia in the range of this species based on Cuervo-Diaz et al. (1986), but that specimen has been reidentified as *Vampyressa pusilla*.

*Enchisthenes* K. Andersen, 1906. Ann. Mag. Nat. Hist, ser. 7, 18:419.
TYPE SPECIES: *Artibeus hartii* Thomas, 1892.

COMMENTS: Subtribe Ectophyllina. Formerly included in *Artibeus* (e.g., Goodwin, 1969), but see Jones and Carter (1979), Van Den Bussche et al. (1993*a*, 1998), Baker et al. (2000), and Wetterer et al. (2000).

*Enchisthenes hartii* (Thomas, 1892). Ann. Mag. Nat. Hist., ser. 6, 10:409.
  COMMON NAME: Velvety Fruit-eating Bat.
  TYPE LOCALITY: Trinidad and Tobago, Trinidad, Port of Spain.
  DISTRIBUTION: Bolivia and Venezuela north to Michoacan, Jalisco, and Tamaulipas (Mexico); Trinidad. There is an extralimital record from Tucson, Arizona (Irwin and Baker, 1967).
  STATUS: IUCN 2003 and IUCN/SSC Action Plan (2001) – Lower Risk (lc) as *Artibeus hartii*.

*Mesophylla* Thomas, 1901. Ann. Mag. Nat. Hist., ser. 7, 8:143.
  TYPE SPECIES: *Mesophylla macconnelli* Thomas, 1901.
  COMMENTS: Subtribe Ectophyllina. Included in *Ectophylla* by Goodwin and Greenhall (1962), Simmons and Voss (1998), and Wetterer et al. (2000); included in *Vampyressa* by Owen (1987). Treated as distinct here pending further study; also see Baker et al. (2000).

*Mesophylla macconnelli* Thomas, 1901. Ann. Mag. Nat. Hist., ser. 7, 8:145.
  COMMON NAME: MacConnell's Bat.
  TYPE LOCALITY: Guyana, Essequibo Dist., Kanuku Mtns.
  DISTRIBUTION: Nicaragua south to Peru, Bolivia, and Amazonian Brazil; Trinidad.
  STATUS: IUCN 2003 and IUCN/SSC Action Plan (2001) – Lower Risk (lc).
  SYNONYMS: *flavescens* Goodwin and Greenhall, 1962.
  COMMENTS: See Kunz and Pena (1992).

*Phyllops* Peters, 1865. Monatsb. K. Preuss. Akad. Wiss. Berlin, 1865:356.
  TYPE SPECIES: *Phyllostoma albomaculatum* Gundlach, 1861 (= *Arctibeus falcatus* Gray, 1839).
  COMMENTS: Subtribe Stenodermatina. Included in *Stenoderma* by Varona (1974), Simpson (1945), and Silva-Taboada (1979); but see Jones and Carter (1976) and Corbet and Hill (1980). Two additional species of *Phyllops* have been described from cave fossils: *Phyllops silvai* Suárez and Díaz-Franco, 2003 and *Phyllops vetus* Anthony, 1917; see review by Suárez and Díaz-Franco (2003).

*Phyllops falcatus* (Gray, 1839). Ann. Nat. Hist., 4:1.
  COMMON NAME: Cuban Fig-eating Bat.
  TYPE LOCALITY: Cuba, Habana, Guanabacoa.
  DISTRIBUTION: Cuba; Hispaniola; as fossil, Isle of Pines (Cuba).
  STATUS: IUCN 2003 and IUCN/SSC Action Plan (2001) – Lower Risk (nt).
  SYNONYMS: *albomaculatum* Gundlach, 1861; *haitiensis* J. A. Allen, 1908.
  COMMENTS: Includes *haitiensis*; see Koopman (1989*c*) and Timm and Genoways (2003). Reviewed by Suárez and Díaz-Franco (2003).

*Platyrrhinus* Saussure, 1860. Rev. Mag. Zool., Paris, ser. 2, 12:429.
  TYPE SPECIES: *Phyllostoma lineatum* E. Geoffroy, 1810.
  SYNONYMS: *Vampyrops* Peters, 1865.
  COMMENTS: Subtribe Ectophyllina. For a history of the nomeclature of this genus and reasons for using *Platyrrhinus* in place of *Vampyrops*, see Gardner and Ferrell (1990) and Alberico and Velasco (1991). A key to the genus was provided by Ferrell and Wilson (1991).

*Platyrrhinus aurarius* (Handley and Ferris, 1972). Proc. Biol. Soc. Wash., 84:522.
  COMMON NAME: Eldorado Broad-nosed Bat.
  TYPE LOCALITY: Venezuela, Bolivar, 85 km SSE El Dorado, 1,000 m.
  DISTRIBUTION: S Venezuela, Guyana, Surinam. Specimens previously reported from Colombia appear to have been misidentified.
  STATUS: IUCN 2003 and IUCN/SSC Action Plan (2001) – Lower Risk (nt).
  COMMENTS: May be a synonym of *dorsalis*; see Jones and Carter (1976).

*Platyrrhinus brachycephalus* (Rouk and Carter, 1972). Occas. Pap. Mus. Texas Tech Univ., 1:1.
  COMMON NAME: Short-headed Broad-nosed Bat.

TYPE LOCALITY: Peru, Huanuco, 3 mi. (5 km) S Tingo Maria, 2,400 ft. (732 m).

DISTRIBUTION: N Brazil; Colombia to Guianas; Ecuador; Peru; Bolivia.

STATUS: IUCN 2003 and IUCN/SSC Action Plan (2001) – Lower Risk (lc).

SYNONYMS: *latus* Handley and Ferris, 1972; *saccharus* Handley and Ferris, 1972.

COMMENTS: Includes *latus*; see Jones and Carter (1976). May not be distinct from *helleri*; see Alberico (1990), but also see Anderson (1996).

*Platyrrhinus chocoensis* Alberico and Velasco, 1991. Bonn. zool. Beitr., 42:238.

COMMON NAME: Choco Broad-nosed Bat.

TYPE LOCALITY: Colombia, Departamento del Chocó, 12 km W Istmina (by road), Quebrada El Platinero, 5°00'N, 76°45'W, 100 m.

DISTRIBUTION: W Colombia, lowlands between the Western Cordillera of the Andes and the Pacific coast.

STATUS: IUCN 2003 and IUCN/SSC Action Plan (2001) – Vulnerable.

*Platyrrhinus dorsalis* (Thomas, 1900). Ann. Mag. Nat. Hist., ser. 7, 5:269.

COMMON NAME: Thomas's Broad-nosed Bat.

TYPE LOCALITY: Ecuador, Paramba, 1,100 m.

DISTRIBUTION: Panama to Peru and Bolivia.

STATUS: IUCN 2003 and IUCN/SSC Action Plan (2001) – Lower Risk (lc).

COMMENTS: Although the named forms *umbratus*, *oratus*, and *aquilus* were regarded as synonyms of *dorsalis* by Carter and Rouk (1973), these apparently represent a distinct species for which the oldest name is *umbratus*; see Handley (1976).

*Platyrrhinus helleri* (Peters, 1866). Monatsb. K. Preuss. Akad. Wiss. Berlin, 1866:392.

COMMON NAME: Heller's Broad-nosed Bat.

TYPE LOCALITY: Mexico.

DISTRIBUTION: Oaxaca and Veracruz (Mexico) to Peru, Bolivia, and Amazonian Brazil; Trinidad. A Paraguay record is erroneous.

STATUS: IUCN 2003 and IUCN/SSC Action Plan (2001) – Lower Risk (lc).

SYNONYMS: *zarhinus* H. Allen, 1891; *incarum* Thomas, 1912.

COMMENTS: Includes *zarhinus*; see Jones and Carter (1976) and Gardner and Carter (1972). May include *brachycephalus*, see Alberico (1990). Reviewed by Ferrell and Wilson (1991) and Anderson (1996).

*Platyrrhinus infuscus* (Peters, 1880). Monatsb. K. Preuss. Akad. Wiss. Berlin, 1880:259.

COMMON NAME: Buffy Broad-nosed Bat.

TYPE LOCALITY: Peru, Cajamarca, Hualgayoc, Hac. Ninabamba.

DISTRIBUTION: Colombia to Peru, Bolivia, and NW Brazil.

STATUS: IUCN 2003 and IUCN/SSC Action Plan (2001) – Lower Risk (nt).

SYNONYMS: *fumosus* Miller, 1902; *intermedius* Marinkelle, 1970.

COMMENTS: Includes *intermedius* and *fumosus*; see Gardner and Carter (1972), who also designated a neotype for *infuscus*.

*Platyrrhinus lineatus* (E. Geoffroy, 1810). Ann. Mus. Natn. Hist. Nat. Paris, 15:180.

COMMON NAME: White-lined Broad-nosed Bat.

TYPE LOCALITY: Paraguay, Asunción.

DISTRIBUTION: Colombia to Peru, Bolivia, Uruguay, N Argentina, and S and E Brazil; French Guyana; Surinam.

STATUS: CITES – Appendix III (Uruguay). IUCN 2003 and IUCN/SSC Action Plan (2001) – Lower Risk (lc).

SYNONYMS: *sacrillus* Thomas, 1924; *nigellus* Gardner and Carter, 1972.

COMMENTS: Includes *nigellus*; see Jones and Carter (1979). See Willig and Hollander (1987).

*Platyrrhinus recifinus* (Thomas, 1901). Ann. Mag. Nat. Hist., ser. 7, 8:192.

COMMON NAME: Recife Broad-nosed Bat.

TYPE LOCALITY: Brazil, Pernambuco, Recife.

DISTRIBUTION: E and SE Brazil. A Guyana record is erroneous, because the specimen was referred to *latus* (= *brachycephalus*) by Handley and Ferris (1972).

STATUS: IUCN 2003 and IUCN/SSC Action Plan (2001) – Vulnerable.

*Platyrrhinus umbratus* (Lyon, 1902). Proc. Biol. Soc. Wash., 15:151.
  COMMON NAME: Shadowy Broad-nosed Bat.
  TYPE LOCALITY: Colombia, Magdalena, San Miguel (Macotama River).
  DISTRIBUTION: Panama, N and W Colombia, N Venezuela.
  STATUS: IUCN 2003 and IUCN/SSC Action Plan (2001) – Lower Risk (nt).
  SYNONYMS: *aquilius* Handley and Ferris, 1972; *oratus* Thomas, 1914.
  COMMENTS: Formerly included in *dorsalis* by Carter and Rouk (1973), but see Handley (1976).

*Platyrrhinus vittatus* (Peters, 1860). Monatsb. K. Preuss. Akad. Wiss. Berlin, 1860:225.
  COMMON NAME: Greater Broad-nosed Bat.
  TYPE LOCALITY: Venezuela, Carabobo, Puerto Cabello.
  DISTRIBUTION: Costa Rica to Venezuela, Peru, and Bolivia.
  STATUS: IUCN 2003 and IUCN/SSC Action Plan (2001) – Lower Risk (lc).

*Pygoderma* Peters, 1863. Monatsb. K. Preuss. Akad. Wiss. Berlin, 1863:83.
  TYPE SPECIES: *Stenoderma microdon* Peters, 1863 (= *Phyllostoma bilabiatum* Wagner, 1843).
  COMMENTS: Subtribe Stenodermatina.

*Pygoderma bilabiatum* (Wagner, 1843). Arch. Naturgesch., 1:366.
  COMMON NAME: Ipanema Broad-nosed Bat.
  TYPE LOCALITY: Brazil, São Paulo, Ipanema.
  DISTRIBUTION: Bolivia, SE Brazil, Paraguay, N Argentina. Reported occurrences in North America and Surinam are erroneous (Jones and Carter, 1976; Voss and Emmons, 1996).
  STATUS: IUCN 2003 and IUCN/SSC Action Plan (2001) – Lower Risk (nt).
  SYNONYMS: *leucomus* Gray, 1848; *microdon* Peters, 1863; *magna* Owen and Webster, 1983.
  COMMENTS: See Webster and Owen (1984). See Emmons (1997) for distribution map.

*Sphaeronycteris* Peters, 1882. Sitzb. Preuss. Akad. Wiss., 45:988.
  TYPE SPECIES: *Sphaeronycteris toxophyllum* Peters, 1882.
  COMMENTS: Subtribe Stenodermatina.

*Sphaeronycteris toxophyllum* Peters, 1882. Sitzb. Preuss. Akad. Wiss., 45:989.
  COMMON NAME: Visored Bat.
  TYPE LOCALITY: Peru, Loreto, Pebas.
  DISTRIBUTION: Colombia to Venezuela, Peru, and Bolivia; Amazonian Brazil.
  STATUS: IUCN 2003 and IUCN/SSC Action Plan (2001) – Lower Risk (lc).
  COMMENTS: See Emmons (1997) for distribution map.

*Stenoderma* E. Geoffroy, 1818. Descrip. de L'Egypte, 2:114.
  TYPE SPECIES: "le sténoderme roux" (= *Stenoderma rufa* Desmarest, 1820).
  SYNONYMS: *Histiops* Peters, 1869.
  COMMENTS: Subtribe Stenodermatina. Some authors have included *Ardops*, *Phyllops*, and *Ariteus* in *Stenoderma*; see Varona (1974) and Simpson (1945), but most recent authors have followed the arrangement presented here; see also Jones and Carter (1976).

*Stenoderma rufum* Desmarest, 1820. Mammalogie, *in* Encycl. Méth., p. 117.
  COMMON NAME: Red Fruit Bat.
  TYPE LOCALITY: Not designated in original publication (probably Virgin Isls).
  DISTRIBUTION: Puerto Rico and Virgin Isls (St. John and St. Thomas).
  STATUS: IUCN 2003 and IUCN/SSC Action Plan (2001) – Vulnerable.
  SYNONYMS: *undatus* Gervais, 1855; *darioi* Hall and Tamsitt, 1968; *anthonyi* Choate and Birney, 1968.
  COMMENTS: See Genoways and Baker (1972) and Timm and Genoways (2003).

*Uroderma* Peters, 1866. Monatsb. K. Preuss. Akad. Wiss. Berlin, 1865:587 [1866].
  TYPE SPECIES: *Phyllostoma personatum* Peters, 1865 (preoccupied; = *Uroderma bilobatum* Peters, 1866).

COMMENTS: Subtribe Ectophyllina. Revised by Davis (1968).

*Uroderma bilobatum* Peters, 1866. Monatsb. K. Preuss. Akad. Wiss. Berlin, 1866:392.
    COMMON NAME: Common Tent-making Bat.
    TYPE LOCALITY: Brazil, São Paulo.
    DISTRIBUTION: Veracruz and Oaxaca (Mexico) south to Peru, Bolivia, the Guianas, and
        Brazil; Trinidad.
    STATUS: IUCN 2003 and IUCN/SSC Action Plan (2001) – Lower Risk (lc).
    SYNONYMS: *personatum* Peters, 1865; *thomasi* K. Andersen, 1906; *trinitatum* Davis, 1968;
        **convexum** Lyon, 1902; *molaris* Davis, 1968; **davisi** Baker and McDaniel, 1972.
    COMMENTS: See Baker and Clark (1987). There are three chromosomal races that are largely
        genetically distinct although some hybridization may occur; these may represent
        distinct species (Hoffman et al., 2003). I have chosen to treat these races as subspecies
        here pending further study of this complex.

*Uroderma magnirostrum* Davis, 1968. J. Mammal., 49:679.
    COMMON NAME: Brown Tent-making Bat.
    TYPE LOCALITY: Honduras, Valle, 10 km E San Lorenzo.
    DISTRIBUTION: Michoacan (Mexico) to south Venezuela, Peru, Bolivia, and Brazil.
    STATUS: IUCN 2003 and IUCN/SSC Action Plan (2001) – Lower Risk (lc).

*Vampyressa* Thomas, 1900. Ann. Mag. Nat. Hist., ser. 7, 5:270.
    TYPE SPECIES: *Phyllostoma pusillum* Wagner, 1843.
    SYNONYMS: *Metavampyressa* Peterson, 1968; *Vampyriscus* Thomas, 1900.
    COMMENTS: Subtribe Ectophyllina. Includes *Metavampyressa* and *Vampyriscus*, here recognized
        as subgenera along with *Vampyressa*; see Jones and Carter (1976). Probably not
        monophyletic; see Wetterer et al. (2000) and Baker et al. (2000). A key for this genus was
        presented in Lewis and Wilson (1987). See Lim et al. (2003) for a phylogeny of the genus.

*Vampyressa bidens* (Dobson, 1878). Cat. Chiroptera Brit. Mus., p. 535.
    COMMON NAME: Bidentate Yellow-eared Bat.
    TYPE LOCALITY: Peru, Loreto, Santa Cruz (Río Huallaga).
    DISTRIBUTION: Guianas to Colombia to Peru; N Bolivia; Amazonian Brazil.
    STATUS: IUCN 2003 and IUCN/SSC Action Plan (2001) – Lower Risk (nt).
    COMMENTS: Subgenus *Vampyriscus*. Formerly placed in its own genus (*Vampyriscus*); see
        Jones and Carter (1976). Reviewed by Lee et al. (2001).

*Vampyressa brocki* Peterson, 1968. R. Ontario Mus. Life Sci. Contrib., 73:1.
    COMMON NAME: Brock's Yellow-eared Bat.
    TYPE LOCALITY: Guyana, Rupununi, ca. 40 mi. (64 km) E Dadanawa, at Ow-wi-dy-wau
        (Oshi Wau head, near Marara Waunowa), Kuitaro River.
    DISTRIBUTION: Guianas, Amazonian Brazil, SE Colombia, Peru.
    STATUS: IUCN 2003 and IUCN/SSC Action Plan (2001) – Lower Risk (nt).
    COMMENTS: Subgenus *Metavampyressa*. Characters of this species have been reported
        inconsistently in the literature; see Simmons and Voss (1998).

*Vampyressa melissa* Thomas, 1926. Ann. Mag. Nat. Hist., ser. 9, 18:157.
    COMMON NAME: Melissa's Yellow-eared Bat.
    TYPE LOCALITY: Peru, Amazonas, Chachapoyas, Puca Tambo, 1,480 m.
    DISTRIBUTION: Peru, S Colombia. A record from French Guiana is apparently erroneous
        (Charles-Dominique et al., 2001).
    STATUS: IUCN 2003 and IUCN/SSC Action Plan (2001) – Lower Risk (nt).
    COMMENTS: Subgenus *Vampyressa*.

*Vampyressa nymphaea* Thomas, 1909. Ann. Mag. Nat. Hist., ser. 8, 4:230.
    COMMON NAME: Striped Yellow-eared Bat.
    TYPE LOCALITY: Colombia, Chocó, Novita (San Juan River).
    DISTRIBUTION: W Ecuador to Nicaragua. A record from SE Peru is suspect.
    STATUS: IUCN 2003 and IUCN/SSC Action Plan (2001) – Lower Risk (lc).
    COMMENTS: Subgenus *Metavmpyressa*.

*Vampyressa pusilla* (Wagner, 1843). Abh. Akad. Wiss., München, 5:173.
  COMMON NAME: Southern Little Yellow-eared Bat.
  TYPE LOCALITY: Brazil, Rio de Janeiro, Sapitiba.
  DISTRIBUTION: SE Brazil, Paraguay, and NE Argentina.
  STATUS: IUCN 2003 and IUCN/SSC Action Plan (2001) – Lower Risk (lc).
  SYNONYMS: *nattereri* Goodwin, 1963.
  COMMENTS: Subgenus *Vampyressa*. Does not include *thyone*, see Lim et al. (2003). Discussed
      by Jones and Carter (1976) and Lewis and Wilson (1987), but note that they included
      *thyone* in this taxon.

*Vampyressa thyone* Thomas, 1909. Ann. Mag. Nat. Hist., ser. 8, 4:231.
  COMMON NAME: Northern Little Yellow-eared Bat.
  TYPE LOCALITY: Ecuador, Bolívar, Chimbo; 1000 ft (305 m).
  DISTRIBUTION: Oaxaca and Veracruz (Mexico) to Bolivia, Peru, Venezuela, Guyana, and
      French Guiana.
  STATUS: IUCN 2003 – Not evaluated; not considered in IUCN/SSC Action Plan (2001).
  SYNONYMS: *minuta* Miller, 1912; *venilla* Thomas, 1924.
  COMMENTS: Subgenus *Vampyressa*. Previously included in *pusilla*, but clearly distinct; see
      Lim et al. (2003). Much of the account of *V. pusilla* provided in Lewis and Wilson
      (1987) actually applies to *thyone*.

*Vampyrodes* Thomas, 1900. Ann. Mag. Nat. Hist., ser. 7, 5:270.
  TYPE SPECIES: *Vampyrops caracciolae* Thomas, 1889.
  COMMENTS: Subtribe Ectophyllina.

*Vampyrodes caraccioli* (Thomas, 1889). Ann. Mag. Nat. Hist., ser. 6, 4:167.
  COMMON NAME: Great Stripe-faced Bat.
  TYPE LOCALITY: Trinidad and Tobago, Trinidad.
  DISTRIBUTION: Oaxaca (Mexico) to Peru, Bolivia, the Guianas, and N Brazil; Trinidad and
      Tobago.
  STATUS: IUCN 2003 and IUCN/SSC Action Plan (2001) – Lower Risk (lc).
  SYNONYMS: **major** G. M. Allen, 1908; *ornatus* Thomas, 1924.
  COMMENTS: Includes *major*; see Jones and Carter (1976), but also see Starrett and Casebeer
      (1968). See Willis et al. (1990). Originally spelled *caracciolae* but later emended to
      *caraccioli*; see discussion in Carter and Dolan (1978).

**Family Mormoopidae** Saussure, 1860. Revue et Mag. Zool., 2:286.
  COMMENTS: Revised by Smith (1972); see Lewis-Oritt et al. (2001*a*), Simmons and Conway
      (2001), Van Den Bussche et al. (2002), and Van den Bussche and Weyandt (2003) for
      phylogenies. See Smith (1972) for a discussion of authorship and priority of the name
      Mormoopidae.

*Mormoops* Leach, 1821. Trans. Linn. Soc. Lond., 13:76.
  TYPE SPECIES: *Mormoops blainvillii* Leach, 1821.
  SYNONYMS: *Aello* Leach, 1821.
  COMMENTS: This name is used instead of *Aello* following Opinion 462 of the International
      Commission on Zoological Nomenclature (1958*b*).

*Mormoops blainvillei* Leach, 1821. Trans. Linn. Soc. Lond., 13:77.
  COMMON NAME: Antillean Ghost-faced Bat.
  TYPE LOCALITY: Jamaica.
  DISTRIBUTION: Greater Antilles, adjacent small islands.
  STATUS: IUCN 2003 and IUCN/SSC Action Plan (2001) – Lower Risk (nt).
  SYNONYMS: *cinnamomeum* Gundlach, 1840; *cuvieri* Leach, 1821.
  COMMENTS: See Lancaster and Kalko (1996) and Timm and Genoways (2003). Often spelled
      *blainvillii*, but this was an incorrect original spelling; the correct spelling is *blainvillei*
      (see Opinion 462 of the International Commission on Zoological Nomenclature,
      1958*b*). The ICZN placed *blainvilli* on the Offical Index of Rejected and Invalid Specific

Names in Zoology, and placed *blainvillei* on the Offical List of Specific Names in Zoology in Opinion 462.

*Mormoops magna* Silva-Taboada, 1974. Acta Zool. Cracoviensia, 19:52.
  COMMON NAME: Giant Ghost-faced Bat.
  TYPE LOCALITY: Cuba, Las Villas Province, Trinidad, Cueva de los Masones.
  DISTRIBUTION: Known only from the type locality.
  STATUS: Extinct; IUCN 2003 – Not evaluated; not considered in IUCN/SSC Action Plan (2001).
  COMMENTS: Known only from subfossils, but found in the same sedimentary deposits as remains of many extant bat species (Silva-Taboada, 1974, 1979).

*Mormoops megalophylla* (Peters, 1864). Monatsb. K. Preuss. Akad. Wiss. Berlin, 1864:381.
  COMMON NAME: Peters's Ghost-faced Bat.
  TYPE LOCALITY: Mexico, Coahuila, Parras.
  DISTRIBUTION: S Texas, S Arizona (USA), and Baja California (Mexico) south to NW Peru and N Venezuela; Aruba, Curaçao, and Bonaire (Netherlands Antilles); Trinidad; Margarita Isl (Venezuela).
  STATUS: IUCN 2003 and IUCN/SSC Action Plan (2001) – Lower Risk (lc).
  SYNONYMS: *rufescens* Davis and Carter, 1962; *senicula* Rehn, 1902; *carteri* Smith, 1972; *intermedia* Miller, 1900; *tumidiceps* Miller, 1902.
  COMMENTS: See Rezsutek and Cameron (1993).

*Pteronotus* Gray, 1838. Mag. Zool. Bot., 2:500.
  TYPE SPECIES: *Pteronotus davyi* Gray, 1838.
  SYNONYMS: *Chilonycteris* Gray, 1839; *Dermonotus* Gill, 1901; *Lobostoma* Gundlach, 1840; *Phyllodia* Gray, 1843.
  COMMENTS: Includes *Chilonycteris* and *Phyllodia*, which are recognized as subgenera along with *Pteronotus*; see Smith (1972). Keys to this genus were presented by Herd (1983) and by Rodríguez-Durán and Kunz (1992).

*Pteronotus davyi* Gray, 1838. Mag. Zool. Bot., 2:500.
  COMMON NAME: Davy's Naked-backed Bat.
  TYPE LOCALITY: Trinidad and Tobago, Trinidad.
  DISTRIBUTION: NW Peru and N Venezuela to S Baja California, S Sonora, and Nuevo León (Mexico); Trinidad; S Lesser Antilles. A Brazilian record is erroneous, see Willig and Mares (1989).
  STATUS: IUCN 2003 and IUCN/SSC Action Plan (2001) – Lower Risk (lc).
  SYNONYMS: *fulvus* Thomas, 1892; *calvus* Goodwin, 1958; *incae* Smith, 1972.
  COMMENTS: Subgenus *Pteronotus*. See Adams (1989) and Timm and Genoways (2003).

*Pteronotus gymnonotus* Natterer, 1843. *In* Wagner, Arch. Naturgesch., 9:367.
  COMMON NAME: Big Naked-backed Bat.
  TYPE LOCALITY: Brazil, Mato Grosso, Cuiaba (= Cuyaba).
  DISTRIBUTION: S Veracruz (Mexico) south to Peru, NE and C Brazil, Bolivia, Guyana, and French Guiana.
  STATUS: IUCN 2003 and IUCN/SSC Action Plan (2001) – Lower Risk (lc).
  SYNONYMS: *centralis* Goodwin, 1942; *suapurensis* J. A. Allen, 1904.
  COMMENTS: Subgenus *Pteronotus*. Includes *suapurensis*; see Smith (1977).

*Pteronotus macleayii* (Gray, 1839). Ann. Nat. Hist., 4:5.
  COMMON NAME: MacLeay's Mustached Bat.
  TYPE LOCALITY: Cuba, Habana, Guanabacoa.
  DISTRIBUTION: Cuba, Jamaica.
  STATUS: IUCN 2003 and IUCN/SSC Action Plan (2001) – Vulnerable.
  SYNONYMS: *griseus* Gosse, 1851.
  COMMENTS: Subgenus *Chilonycteris*. Reviewed in part by Timm and Genoways (2003).

*Pteronotus parnellii* (Gray, 1843). Proc. Zool. Soc. Lond., 1843:50.
  COMMON NAME: Common Mustached Bat.

TYPE LOCALITY: Jamaica.

DISTRIBUTION: Peru, Bolivia, Brazil, Guianas, and Venezuela to S Sonora and S Tamaulipas (Mexico); Cuba; Jamaica; Puerto Rico; Hispaniola; St. Vincent; Trinidad and Tobago; Margarita Isl (Venezuela); La Gonave Isl (Haiti).

STATUS: IUCN 2003 and IUCN/SSC Action Plan (2001) – Lower Risk (lc).

SYNONYMS: *boothi* Gundlach, 1861; *osburni* Tomes, 1861; *fuscus* J. A. Allen, 1911; *gonavensis* Koopman, 1955; *mesoamericanus* Smith, 1972; *mexicanus* Miller, 1902; *paraguanensis* Linares and Ojasti, 1974; *portoricensis* Miller, 1902; *pusillus* G. M. Allen, 1917; *rubiginosus* Wagner, 1843.

COMMENTS: Subgenus *Phyllodia*. Hall (1981) reviewed the numerous Central American and Carribean subspecies; also see Timm and Genoways (2003). See Herd (1983). This complex probably includes more than one species (Lewis-Oritt et al., 2001*a*).

*Pteronotus personatus* (Wagner, 1843). Arch. Naturgesch., 9:367.

COMMON NAME: Wagner's Mustached Bat.

TYPE LOCALITY: Brazil, Mato Grosso, São Vicente.

DISTRIBUTION: Colombia, Peru, Brazil, Bolivia, and Surinam to S Sonora and S Tamaulipas (Mexico); Trinidad.

STATUS: IUCN 2003 and IUCN/SSC Action Plan (2001) – Lower Risk (lc).

SYNONYMS: *psilotis* Dobson, 1878; *continentis* Sanborn, 1938.

COMMENTS: Often placed in the subgenus *Chilonycteris* (e.g., Smith, 1972; Simmons and Conway, 2001), but recent molecular studies suggest that it represents an unnamed subgenus (Lewis-Oritt et al., 2001*a*; Van Den Bussche and Weyandt, 2003). Includes *psilotis*; see Smith (1972). This complex may include more than one species (Lewis-Oritt et al., 2001*a*).

*Pteronotus pristinus* Silva-Taboada, 1974. Acta Zool. Cracoviensia, 19:49.

COMMON NAME: Prinstine Mustached Bat.

TYPE LOCALITY: Cuba, Las Villas Province, Trinidad, Cueva de los Masones.

DISTRIBUTION: Cuba, possibly Florida (USA).

STATUS: Extinct; IUCN 2003 – Not evaluated; not considered in IUCN/SSC Action Plan (2001).

COMMENTS: Subgenus *Phyllodia*. Known only from subfossils, but found in the same sedimentary deposits as remains of many extant bat species (Silva-Taboada, 1974, 1979). Morphology and phylogenetic relationships discussed by Simmons and Conway (2001).

*Pteronotus quadridens* (Gundlach, 1840). Arch. Naturgesch., 6:357.

COMMON NAME: Sooty Mustached Bat.

TYPE LOCALITY: Cuba, Matanzas, Canimar.

DISTRIBUTION: Cuba, Jamaica, Hispaniola, Puerto Rico.

STATUS: IUCN 2003 and IUCN/SSC Action Plan (2001) – Lower Risk (nt).

SYNONYMS: *torrei* G. M. Allen, 1916; *fuliginosus* Gray, 1843; *inflata* Rehn, 1904.

COMMENTS: Subgenus *Chilonycteris*. Includes *torrei*; For use of *quadridens* in place of *fuliginosus*, see Silva-Taboada (1976). See Rodríguez-Durán and Kunz (1992) and Timm and Genoways (2003).

**Family Noctilionidae** Gray, 1821. London Med. Reposit., 15:299.

COMMENTS: Monogeneric.

*Noctilio* Linnaeus, 1766. Syst. Nat., 12th ed., 1:88.

TYPE SPECIES: *Noctilio americanus* Linnaeus, 1766 (= *Vespertilio leporinus* Linnaeus, 1758).

SYNONYMS: *Celaeno* Leach, 1821; *Dirias* Miller, 1906; *Noctileo* Tiedemann, 1808.

COMMENTS: Two subgenera are recognized, *Noctilio* and *Dirias*.

*Noctilio albiventris* Desmarest, 1818. Nouv. Dict. Hist. Nat., Nouv. ed., 23:15.

COMMON NAME: Lesser Bulldog Bat.

TYPE LOCALITY: Brazil, Bahia, Rio Sao Francisco.

DISTRIBUTION: S Mexico to Guianas, E Brazil, Peru, Bolivia, and N Argentina.

STATUS: IUCN 2003 and IUCN/SSC Action Plan (2001) – Lower Risk (lc).

SYNONYMS: *affinis* D'Orbigny, 1835; *albiventer* Spix, 1823; *irex* Thomas, 1920; *leporinus* Gervais, 1856 [not Linnaeus, 1758]; *ruber* Rengger, 1830; *zaparo* Cabrera, 1907; **cabrerai** Davis, 1976; **minor** Osgood, 1910.

COMMENTS: Subgenus *Dirias*. Formerly referred to as *labialis*; see Davis (1976). See Simmons and Voss (1998) for discussion of Amazonian subspecies. Also see Hood and Pitocchelli (1983). May include more than one species, see Lewis-Oritt et al. (2001*b*).

*Noctilio leporinus* (Linnaeus, 1758). Syst. Nat., 10th ed., 1:32.

COMMON NAME: Greater Bulldog Bat.

TYPE LOCALITY: Surinam (restricted by Thomas, 1911*a*).

DISTRIBUTION: Sinaloa (Mexico) to the Guianas, S Brazil, N Argentina, Paraguay, Bolivia, and Peru; Trinidad; Greater and Lesser Antilles; S Bahamas.

STATUS: IUCN 2003 and IUCN/SSC Action Plan (2001) – Lower Risk (lc).

SYNONYMS: *americanus* Linnaeus, 1766; *brooksiana* Leach, 1821; *dorsatus* Desmarest, 1818; *labialis* Kerr, 1792; *longipes* Pelzeln, 1883; *macropus* Pelzeln, 1883; *minor* Fermin, 1765; *rufus* Spix, 1823; *unicolor* Desmarest, 1818; *vittatus* Schinz, 1821; **mastivus** Vahl, 1797; *mexicanus* Goldman 1915; **rufescens** Pelzeln, 1883; *rufipes* D'Orbigny, 1835.

COMMENTS: Subgenus *Noctilio*. See Hood and Jones (1984). Antillean form reviewed by Timm and Genoways (2003)

**Family Furipteridae** Gray, 1866. Ann. Mag. Nat. Hist., ser. 3, 17:91.

*Amorphochilus* Peters, 1877. Monatsb. K. Preuss. Akad. Wiss. Berlin, 1877:185.

TYPE SPECIES: *Amorphochilus schnablii* Peters, 1877.

*Amorphochilus schnablii* Peters, 1877. Monatsb. K. Preuss. Akad. Wiss. Berlin, 1877:185.

COMMON NAME: Smoky Bat.

TYPE LOCALITY: Peru, Tumbes, Tumbes.

DISTRIBUTION: W Peru, W Ecuador, Puna Isl (Ecuador), N Chile.

STATUS: IUCN 2003 and IUCN/SSC Action Plan (2001) – Vulnerable.

SYNONYMS: *osgoodi* J. A. Allen, 1914.

*Furipterus* Bonaparte, 1837. Iconogr. Fauna Ital., 1, fasc. 21.

TYPE SPECIES: *Furia horrens* F. Cuvier, 1828.

SYNONYMS: *Furia* F. Cuvier, 1828 [not Linnaeus, 1758].

*Furipterus horrens* (F. Cuvier, 1828). Mem. Mus. Natn. Hist. Nat. Paris, 16:150.

COMMON NAME: Thumbless Bat.

TYPE LOCALITY: French Guiana, Mana River.

DISTRIBUTION: Costa Rica south to Peru, the Guianas, and E Brazil; Trinidad.

STATUS: IUCN 2003 and IUCN/SSC Action Plan (2001) – Lower Risk (lc).

SYNONYMS: *coerulescens* Tomes, 1856.

COMMENTS: See Emmons (1997) for distribution map.

**Family Thyropteridae** Miller, 1907. Bull. U.S. Natl. Mus., 57:84, 186.

*Thyroptera* Spix, 1823. Sim. Vespert. Brasil., p. 61.

TYPE SPECIES: *Thyroptera tricolor* Spix, 1823.

SYNONYMS: *Hyonycteris* Lichtenstein and Peters, 1854.

*Thyroptera discifera* (Lichtenstein and Peters, 1855). Monatsb. K. Preuss. Akad. Wiss. Berlin, 1855:335.

COMMON NAME: Peters's Disk-winged Bat.

TYPE LOCALITY: Venezuela, Carabobo, Puerto Cabello.

DISTRIBUTION: Nicaragua; Panama and Colombia to Guianas, Amazonian Brazil, Peru, and Bolivia.

STATUS: IUCN 2003 and IUCN/SSC Action Plan (2001) – Lower Risk (lc).

SYNONYMS: *major* Miller, 1931; **abdita** Wilson, 1976.

COMMENTS: See Pine (1993) and Wilson (1978).

*Thyroptera lavali* Pine, 1993. Mammalia, 57:213.
COMMON NAME: LaVal's Disk-winged Bat.
TYPE LOCALITY: Peru, Loreto, Río Javari-Mirim, Quebrada Esperanza.
DISTRIBUTION: Peru, Ecuador, Venezuela, Brazil.
STATUS: IUCN 2003 and IUCN/SSC Action Plan (2001) – Vulnerable.
SYNONYMS: *robusta* Czaplewski, 1996.
COMMENTS: Reviewed by Reid et al. (2000); see also Solari et al. (1999).

*Thyroptera tricolor* Spix, 1823. Sim. Vespert. Brasil., p. 61.
COMMON NAME: Spix's Disk-winged Bat.
TYPE LOCALITY: Brazil, Amazon River.
DISTRIBUTION: Veracruz (Mexico) to Guianas, E Brazil, Bolivia, and Peru; Trinidad.
STATUS: IUCN 2003 and IUCN/SSC Action Plan (2001) – Lower Risk (lc).
SYNONYMS: *bicolor* Cantraine, 1845; *thyropterus* Schinz 1844; **albiventer** Tomes, 1856;
*albigula* G. M. Allen, 1923; **juquiaensis** Vieira, 1942.
COMMENTS: See Wilson and Findley (1977) and Pine (1993).

**Family Natalidae** Gray, 1866. Ann. Mag. Nat. Hist., ser. 3, 17:90.
COMMENTS: Many recent authors have considered Natalidae to be monogeneric, but see
Morgan (1989*b*) and Morgan and Czaplewski (2003), who raised *Nyctiellus* and
*Chilonatalus* to genus rank.

*Chilonatalus* Miller, 1898. Proc. Acad. Nat. Sci. Phil., 1898:326.
TYPE SPECIES: *Natalus micropus* Dobson, 1880.
COMMENTS: Previously considered a subgenus of *Natalus*, but see Morgan and Czaplewski
(2003).

*Chilonatalus micropus* (Dobson, 1880). Proc. Zool. Soc. Lond., 1880:443.
COMMON NAME: Cuban Lesser Funnel-eared Bat.
TYPE LOCALITY: Jamaica, Kingston.
DISTRIBUTION: Cuba, Jamaica, Hispaniola, Providencia Isl (Colombia).
STATUS: IUCN 2003 and IUCN/SSC Action Plan (2001) – Lower Risk (lc) as *Natalus micropus*.
SYNONYMS: **brevimanus** Miller, 1898; **macer** Miller, 1914.
COMMENTS: Includes *brevimanus* and *macer*; see Varona (1974) and Timm and Genoways
(2003). Formerly included *tumidifrons*; but see Ottenwalder and Genoways (1982) who
revised both species; also see Hall (1981). Kerridge and Baker (1978) treated only the
nominate subspecies.

*Chilonatalus tumidifrons* Miller, 1903. Proc. Biol. Soc. Wash., 16:119.
COMMON NAME: Bahamian Lesser Funnel-eared Bat.
TYPE LOCALITY: Bahamas, Watling Isl (= San Salvador Isl).
DISTRIBUTION: Isls of the Bahamas.
STATUS: IUCN 2003 and IUCN/SSC Action Plan (2001) – Vulnerable as *Natalus tumidifrons*.
COMMENTS: Formerly included in *micropus* (e.g., Hall, 1981) but see Ottenwalder and
Genoways (1982) who revised both species.

*Natalus* Gray, 1838. Mag. Zool. Bot., 2:496.
TYPE SPECIES: *Natalus stramineus* Gray, 1838.
SYNONYMS: *Phodotes* Miller, 1906; *Spectrellum* Gervais, 1855.
COMMENTS: Revised by Goodwin (1959*b*). Does not include *Chilonatalus* and *Nyctiellus*; see
Morgan (1989*b*) and Morgan and Czaplewski (2003).

*Natalus jamaicensis* Goodwin, 1959. Amer. Mus. Novit., 1977: 910.
COMMON NAME: Jamaican Greater Funnel-eared Bat.
TYPE LOCALITY: Jamaica, St. Catherine Parish, St. Clair.
DISTRIBUTION: Jamaica.
STATUS: IUCN 2003 – Not evaluated; not considered in IUCN/SSC Action Plan (2001).
COMMENTS: Formerly included in *stramineus*, but clearly distinct from that species; see

Morgan (1989*b*) and Morgan and Czaplewski (2003). Also distinct from *major* and *primus* (A. Tejedor, pers. comm.). See Arroyo-Cabrales et al. (1997), who reviewed genetic variation and possible relationships of populations of *jamaicensis*, *major*, and *stramineus* (although note that all were treated as *stramineus*). Reviewed by Goodwin (1959*b*).

*Natalus major* Miller, 1902. Proc. Acad. Nat. Sci. Phil., 54:398.
COMMON NAME: Hispaniolan Greater Funnel-eared Bat.
TYPE LOCALITY: Dominican Republic, near Savaneta.
DISTRIBUTION: Dominican Republic, Haiti.
STATUS: IUCN 2003 – Not evaluated; not considered in IUCN/SSC Action Plan (2001).
COMMENTS: Formerly included in *stramineus*, but see Morgan (1989*b*) and Morgan and Czaplewski (2003), although also see Timm and Genoways (2003). Does not include *jamaicensis* or *primus* (A. Tejedor, pers. comm.). See Arroyo-Cabrales et al. (1997), who reviewed genetic variation and possible relationships of populations of *major*, *jamaicensis*, and *stramineus* (although note that all were treated as *stramineus*). Reviewed by Goodwin (1959*b*) and Hoyt and Baker (1980), but note that they included *jamaicensis* and *primus* in *major*.

*Natalus primus* Anthony, 1919. Bull. Amer. Mus. Nat. Hist., 61:612.
COMMON NAME: Cuban Greater Funnel-eared Bat.
TYPE LOCALITY: Cuba, Oriente, Daiquiri, Cuevos de los Indios.
DISTRIBUTION: Cuba, Isle of Pines.
STATUS: IUCN 2003 – Not evaluated; not considered in IUCN/SSC Action Plan (2001).
COMMENTS: Formerly included in *stramineus*, but clearly distinct from that species; see Morgan (1989*b*) and Morgan and Czaplewski (2003). Also distinct from *major* and *jamaicensis* (A. Tejedor, pers. comm.). Reviewed by Goodwin (1959*b*).

*Natalus stramineus* Gray, 1838. Mag. Zool. Bot., 2:496.
COMMON NAME: Mexican Greater Funnel-eared Bat.
TYPE LOCALITY: Specified as unknown in the original description. Cabrera (1958) restricted the type locality to Lagoa Sanata, Minas Gerais, Brazil, but Goodwin (1959*b*) disagreed. Based on measurements and cranial morphology, Goodwin (1959*b*) concluded that the holotype was probably from Antigua, Lesser Antilles. Handley and Gardner (1990) subsequently confirmed the identity of the holotype and confirmed restriction of the type locality to Antigua.
DISTRIBUTION: S Baja California, Nuevo León, and Sonora (Mexico) to N Colombia, Venezuela, the Guianas, C and E Brazil, Bolivia; Lesser Antilles.
STATUS: IUCN 2003 and IUCN/SSC Action Plan (2001) – Lower Risk (lc).
SYNONYMS: *dominicensis* Shamel, 1928; *splendidus* Wagner, 1845; **espiritosantensis** Ruschi, 1951; **mexicanus** Miller, 1902; **natalensis** Goodwin, 1959; **saturatus** Dalquest and Hall, 1949; **tronchonii** Linares, 1971.
COMMENTS: See Handley and Gardner (1990) for clarification of the holotype. For synonyms see Goodwin (1959*b*) and Varona (1974). Includes *espiritosantensis*; see Pine and Ruschi (1976). Does not include *major*, *jamaicensis*, or *primus*; see Morgan (1989*b*) and Morgan and Czaplewski (2003), but also see Linares (1971). Arroyo-Cabrales et al. (1997) reviewed genetic variation and possible relationships of populations of *major* and *stramineus*. Morphometrics and distribution within South America reviewed by Taddei and Uieda (2001); see Timm and Genoways (2003) for discussion of the Carbibbean form.

*Natalus tumidirostris* Miller, 1900. Proc. Biol. Soc. Wash., 13:160.
COMMON NAME: Trinidadian Greater Funnel-eared Bat.
TYPE LOCALITY: Curaçao, Hatto (Netherlands).
DISTRIBUTION: Venezuela, Colombia, Trinidad and Tobago, Curaçao and Bonaire (Netherlands Antilles), the Guianas.
STATUS: IUCN 2003 and IUCN/SSC Action Plan (2001) – Lower Risk (lc).
SYNONYMS: **continentis** Thomas, 1911; **haymani** Goodwin, 1959.
COMMENTS: Revised by Goodwin (1959*b*).

*Nyctiellus* Gervais, 1855. Expéd. du compte de Castelnau, Zool., Mamm., p. 84.
    TYPE SPECIES: *Vespertilio lepidus* Gervais, 1837.
    COMMENTS: Previously considered a subgenus of *Natalus*, but see Morgan (1989*b*) and Morgan
        and Czaplewski (2003),

*Nyctiellus lepidus* (Gervais, 1837). L'Inst. Paris, 5(218):253.
    COMMON NAME: Gervais's Funnel-eared Bat.
    TYPE LOCALITY: Cuba.
    DISTRIBUTION: Cuba, Bahama Isls.
    STATUS: IUCN 2003 and IUCN/SSC Action Plan (2001) – Lower Risk (nt) as *Natalus lepidus*.
    SYNONYMS: *barbatus* Gundlach, 1840; *macrurum* Gervais, 1855.
    COMMENTS: Reviewed by Morgan (1989*b*); also see Timm and Genoways (2003).

**Family Molossidae** Gervais, 1856. *In* Comte de Castelnau, Exped. Partes Cen. Am. Sud., Zool.
    (Sec. 7), Vol. 1, pt. 2 (Mammifères):53 footnote.
    COMMENTS: Reviewed by Freeman (1981) and Legendre (1984). Includes Tomopeatinae; see
        Barkley (1984), Sudman et al. (1994), Simmons (1998), and Simmons and Geisler (1998).
        South American species reviewed by Jones and Hood (1993).

**Subfamily Tomopeatinae** Miller, 1907. Bull. U.S. Natl. Mus., 57:237.
    COMMENTS: Not included in Vespertilionidae; see Barkley (1984), Sudman et al. (1994),
        McKenna and Bell (1997), Simmons (1998), and Simmons and Geisler (1998).

*Tomopeas* Miller, 1900. Ann. Mag. Nat. Hist., ser. 7, 6:570.
    TYPE SPECIES: *Tomopeas ravus* Miller, 1900.

*Tomopeas ravus* Miller, 1900. Ann. Mag. Nat. Hist., ser. 7, 6:571.
    COMMON NAME: Blunt-eared Bat.
    TYPE LOCALITY: Peru, Cajamarca, Yayan, 1,000 m.
    DISTRIBUTION: W Peru.
    STATUS: IUCN 2003 and IUCN/SSC Action Plan (2001) – Vulnerable.

**Subfamily Molossinae** Gervais, 1856. *In* Comte de Castelnau, Exped. Partes Cen. Am. Sud., Zool.
    (Sec. 7), Vol. 1, pt. 2 (Mammifères):53 footnote.
    SYNONYMS: Cheiromelinae Legendre, 1984; Tadaridinae Legendre, 1984.
    COMMENTS: Equivalent to Molossidae sensu Freeman (1981), Legendre (1984), Koopman (1993,
        1994), and Peterson et al. (1995). Some recent authors (e.g., Pavlinov et al., 1995*b*) have
        followed Legendre (1984) in subdividing this group into three subfamilies, but confusion
        concerning intergeneric relationships leads me to reject any such arrangement pending a
        thorough phylogenetic analysis. A key to Brazilian species was provided by Gregorin and
        Taddei (2002).

*Chaerephon* Dobson, 1874. J. Asiat. Soc. Bengal, 43:144.
    TYPE SPECIES: *Molossus* (*Nyctinomus*) *johorensis* Dobson, 1873.
    SYNONYMS: *Lophomops* J. A Allen, 1917; *Nyctinomus* E. Geoffroy, 1818.
    COMMENTS: Formerly included in *Tadarida* but apparently distinct, see Freeman (1981).
        Recognized as a subgenus of *Tadarida* by Hill (1983), Legendre (1984), Corbet and Hill
        (1992), and Peterson et al. (1995). Keys have been provided by a number of authors; see
        Taylor (1999) for a critical summary of those used for African species, and Corbet and Hill
        (1992) for SE Asian species. Also see Bouchard (1998), but note that her key apparently
        includes errors in the first two couplets (M. Happold, pers. comm.). Species groups follow
        Koopman (1994).

*Chaerephon aloysiisabaudiae* (Festa, 1907). Bol. Mus. Zool. Anat. Comp. Univ. Torino, 22(546):1.
    COMMON NAME: Duke of Abruzzi's Free-tailed Bat.
    TYPE LOCALITY: Uganda, Toro.
    DISTRIBUTION: Ghana, Gabon, Dem. Rep. Congo, Uganda. Koopman (1993) listed "perhaps
        Ethiopia" in the range for this species, but there are no substantiated records.
    STATUS: IUCN 2003 and IUCN/SSC Action Plan (2001) – Lower Risk (lc).

SYNONYMS: *cyclotis* Brosset, 1966.
COMMENTS: *plicatus* species group.

*Chaerephon ansorgei* (Thomas, 1913). Ann. Mag. Nat. Hist., ser. 8, 11:318.
COMMON NAME: Ansorge's Free-tailed Bat.
TYPE LOCALITY: Angola, Malange.
DISTRIBUTION: Nigeria and Cameroon to Ethiopia, south to Angola and KwaZulu-Natal
(South Africa).
STATUS: IUCN 2003 and IUCN/SSC Action Plan (2001) – Lower Risk (lc).
SYNONYMS: *rhodesiae* Roberts, 1946.
COMMENTS: *bivittatus* species group. Distinct from *bivittatus*; see Eger and Peterson (1979),
Taylor (1999), and Bouchard (2001).

*Chaerephon bemmeleni* (Jentink, 1879). Notes Leyden Mus., 1:125.
COMMON NAME: Gland-tailed Free-tailed Bat.
TYPE LOCALITY: Liberia.
DISTRIBUTION: Sierra Leone, Liberia, Cameroon, Sudan, Dem. Rep. Congo, Uganda, Kenya,
Tanzania.
STATUS: IUCN 2003 and IUCN/SSC Action Plan (2001) – Lower Risk (lc).
SYNONYMS: **cistura** Thomas, 1903.
COMMENTS: *bivittatus* species group. Includes *cistura*; see Koopman (1975) and Peterson
(1971). Revised by Peterson (1971).

*Chaerephon bivittatus* (Heuglin, 1861). Nova Acta Acad. Caes. Leop.-Carol., Halle, 29(8):413.
COMMON NAME: Spotted Free-tailed Bat.
TYPE LOCALITY: Eritrea, Keren.
DISTRIBUTION: Sudan, Ethiopia, Eritrea, Uganda, Kenya, Tanzania, Zambia, Zimbabwe,
Mozambique.
STATUS: IUCN 2003 and IUCN/SSC Action Plan (2001) – Lower Risk (lc).
SYNONYMS: *hepaticus* Heuglin, 1866.
COMMENTS: *bivittatus* species group. Revised by Eger and Peterson (1979); also see Taylor
(1999). Note that the correct spelling for the specific epithet in combination with
*Chaerephon* is *bivittatus* (not *bivittata*) because the genus name is masculine.

*Chaerephon bregullae* (Felten, 1964). Senkenberg. Biol., 45:9.
COMMON NAME: Fijian Mastiff Bat.
TYPE LOCALITY: New Hebrides (= Vanatu), Malo Isl.
DISTRIBUTION: Vanuatu, Fiji Isls.
STATUS: IUCN 2003 and IUCN/SSC Action Plan (2001) – Lower Risk (nt).
COMMENTS: *plicatus* species group. Often included in *jobensis* (e.g. Felten, 1964*a*; Hill, 1983),
but provisionally recognized as distinct following Flannery (1995*b*).

*Chaerephon chapini* J. A. Allen, 1917. Bull. Am. Mus. Nat. Hist., 37:461.
COMMON NAME: Pale Free-tailed Bat.
TYPE LOCALITY: Dem. Rep. Congo, Oriental, Faradje.
DISTRIBUTION: Ghana, N Dem. Rep. Congo, Sudan, Uganda, Kenya, Ethiopia.
STATUS: IUCN 2003 and IUCN/SSC Action Plan (2001) – Lower Risk (nt).
SYNONYMS: **lancasteri** Hayman, 1938.
COMMENTS: *plicatus* species group. Apparently does not include *shortridgei*, see Peterson et al.
(1995). See Fenton and Eger (2002), but note that they included *shortridgei* in this
species.

*Chaerephon gallagheri* (Harrison, 1975). Mammalia, 39:313.
COMMON NAME: Gallagher's Free-tailed Bat.
TYPE LOCALITY: Dem. Rep. Congo, Kivu, 30 km SW Kindu, Scierie Forest (3°10′S and
25°46′E).
DISTRIBUTION: Dem. Rep. Congo.
STATUS: IUCN 2003 and IUCN/SSC Action Plan (2001) – Critically Endangered.
COMMENTS: *plicatus* species group.

*Chaerephon jobensis* (Miller, 1902). Proc. Biol. Soc. Wash., 15:246.
 COMMON NAME: Northern Mastiff Bat.
 TYPE LOCALITY: Indonesia, Prov. of Papua, Tjenderawasih Div. [= Geelvinck Bay], Yapen Isl
  [= Jobi Isl], Ansus.
 DISTRIBUTION: Seram (Moluccas), Yapen Isl (Indonesia), New Guinea, N and C Australia.
 STATUS: IUCN 2003 and IUCN/SSC Action Plan (2001) – Lower Risk (lc).
 SYNONYMS: *colonicus* Thomas, 1906.
 COMMENTS: *plicatus* species group. Listed as a subspecies of *plicatus* by Laurie and Hill (1954),
  but subsequently recognized as distinct by most authors. Revised by Felten (1964*a*),
  who included *bregullae* and *solomonis*; also see Hill (1983). In contrast, Flannery
  (1995*a, b*) treated *bregullae* and *solomonis* as distinct species based on morphological
  differences. The latter arrangement is provisionally followed here.

*Chaerephon johorensis* (Dobson, 1873). Proc. Asiat. Soc. Bengal, p. 22.
 COMMON NAME: Northern Free-tailed Bat.
 TYPE LOCALITY: Malaysia, Johore.
 DISTRIBUTION: W Malaysia, Sumatra (Indonesia).
 STATUS: IUCN 2003 and IUCN/SSC Action Plan (2001) – Lower Risk (nt).
 COMMENTS: *plicatus* species group. Reviewed by J. E. Hill (1974*b*).

*Chaerephon leucogaster* A. Grandidier, 1870. Rev. Mag. Zool., ser. 2, 21:337.
 COMMON NAME: Grandidier's Free-tailed Bat.
 TYPE LOCALITY: Madagascar, Mahab (= Mahabo?) and Ménabé, E of Morondava.
 DISTRIBUTION: Ethiopia to Ghana, Nigeria, Dem. Rep. Congo, Mali, Madagascar.
 STATUS: IUCN 2003 and IUCN/SSC Action Plan (2001) – Data Deficient.
 SYNONYMS: *cristatus* J. A. Allen, 1917; *frater* J. A. Allen, 1917; *nigri* Hatt, 1928; *websteri*
  Dollman, 1908.
 COMMENTS: *plicatus* species group. Often included in *pumilus*, but apparently distinct;
  see Peterson et al. (1995).

*Chaerephon major* (Trouessart, 1897). Cat. Mamm. Viv. Foss., 1:146.
 COMMON NAME: Lappet-eared Free-tailed Bat.
 TYPE LOCALITY: N Sudan, 5th Cataract of the Nile.
 DISTRIBUTION: Senegal, Liberia, Mali, Burkina Faso, Ghana, Togo, Nigeria, Niger, Sudan,
  NE Dem. Rep. Congo, Uganda, Tanzania.
 STATUS: IUCN 2003 and IUCN/SSC Action Plan (2001) – Lower Risk (lc).
 SYNONYMS: *abae* J. A. Allen, 1917; *emini* de Winton, 1901.
 COMMENTS: *plicatus* species group. Specimens reported as *pumilus* by Happold (1967)
  actually represent *major*.

*Chaerephon nigeriae* Thomas, 1913. Ann. Mag. Nat. Hist., ser. 8, 11:319.
 COMMON NAME: Nigerian Free-tailed Bat.
 TYPE LOCALITY: Nigeria, Northern Region, Zaria Province.
 DISTRIBUTION: Guinea, Sierra Leone, Mali, Ghana, Togo, and Nigeria to Saudi Arabia and
  Yemen, Ethiopia south to Namibia, Botswana, Uganda, Malawi, and Zimbabwe.
 STATUS: IUCN 2003 and IUCN/SSC Action Plan (2001) – Lower Risk (lc).
 SYNONYMS: **spillmani** Monard, 1933.
 COMMENTS: *plicatus* species group. Reviewed in part by Nader and Kock (1979), Harrison and
  Bates (1991), and Taylor (1999); also see Willis et al. (2002).

*Chaerephon plicatus* (Buchannan, 1800). Trans. Linn. Soc. Lond., 5:261.
 COMMON NAME: Wrinkle-lipped Free-tailed Bat.
 TYPE LOCALITY: India, Bengal, Puttahaut (restricted to Puttahaut by G. M. Allen, 1939).
 DISTRIBUTION: India and Sri Lanka to S China, Hong Kong, Cambodia, and Vietnam,
  southeast through Malyasia to the Philippines, Borneo and Lesser Sunda Isls; Hainan
  (China); Cocos Keeling Isl (Indian Ocean).
 STATUS: IUCN 2003 and IUCN/SSC Action Plan (2001) – Lower Risk (lc).
 SYNONYMS: *bengalensis* Desmarest, 1820; *murinus* Gray, 1830; **dilatatus** Horsfield, 1822;
  **insularis** Phillips, 1932; **luzonus** Hollister, 1913; **tenuis** Horsfield, 1822; *adustus* Sody,
  1936.

COMMENTS: *plicatus* species group. Includes *luzonus*; see Hill (1961*b*) and Corbet and Hill (1992). Reviewed in part by Bates and Harrison (1997). Subspecies limits are problematic.

*Chaerephon pumilus* (Cretzschmar, 1830-1831). *In* Rüppell, Atlas Reise Nördl. Afr., Zool. Säugeth., 1:69.

COMMON NAME: Little Free-tailed Bat.

TYPE LOCALITY: Eritrea, Massawa.

DISTRIBUTION: Senegal to Yemen, south to South Africa; Bioko (Equatorial Guinea); São Tomé; Pemba and Zanzibar; Comoro Isls; Aldabra and Amirante Isls (Seychelles); Madagascar.

STATUS: IUCN 2003 and IUCN/SSC Action Plan (2001) – Lower Risk (lc) as *C. pumila*; Vulnerable as *C. pusilla*.

SYNONYMS: *dubius* Peters, 1852 [not A. Smith, 1833]; *elphicki* Roberts, 1926; *faini* Hayman, 1951; *gambianus* de Winton, 1901; *hindei* Thomas, 1904; *langi* Roberts, 1932; *limbata* Peters, 1852; *naivashae* Hollister, 1916; *pusillus* Miller, 1902.

COMMENTS: *plicatus* species group. Includes *pusillus*; see Hayman and Hill (1971). Does not include *leucogaster*; see Peterson et al. (1995). Koopman (1994) included *leucogaster* in *pumilus* and recognized 12 subspecies in the resulting complex. However, subspecies limits are poorly defined and many populations have not been allocated, rendering any subspecific classification useless. This complex probably includes more than one species; Peterson et al. (1995) recognized *hindei*, *limbata,* and *naivashae* as distinct, but did not diagnose or delimit them. Note that the correct spelling for the specific epithet in combination with *Chaerephon* is *pumilus* (not *pumila*) because the genus name is masculine. Northern records reviewed in part by Harrison and Bates (1991). See Bouchard (1998), but note that she included *leucogaster* in this species. Specimens reported as *pumilus* by Happold (1967) actually represent *major*.

*Chaerephon russatus* J. A. Allen, 1917. Bull. Am. Mus. Nat. Hist., 37:458.

COMMON NAME: Russet Free-tailed Bat.

TYPE LOCALITY: Dem. Rep. Congo, Oriental, Medje.

DISTRIBUTION: Ghana, Cameroon, Dem. Rep. Congo, Kenya.

STATUS: IUCN 2003 and IUCN/SSC Action Plan (2001) – Lower Risk (lc).

COMMENTS: *plicatus* species group. The correct spelling for the specific epithet in combination with *Chaerephon* is *russatus* (not *russata*) because the genus name is masculine.

*Chaerephon shortridgei* Thomas, 1926. Proc. Zool. Soc. Lond., 1926: 289.

COMMON NAME: Shortridge's Free-tailed Bat.

TYPE LOCALITY: Namibia, NW Ovamboland, Ukualukasi, 3400 ft. (1100 m).

DISTRIBUTION: S Dem. Rep. Congo, Angola, Namibia, Botswana, Zimbabwe, Zambia.

STATUS: IUCN 2003 – Not evaluated; not considered in IUCN/SSC Action Plan (2001).

COMMENTS: *plicatus* species group. Often considered a subspecies of *chapini*, but see Peterson et al. (1995), who treated these taxa as distinct species based on significant size differences and the large geographic gap apparently separating the southern populations (*shortridgei*) from northern populations (*chapini*). Fenton and Eger (2002) included *shortridgei* in *chapini* with no comment. Based on my own limited observations, I prefer to treat *chapini* and *shortridgei* as distinct species pending additional data.

*Chaerephon solomonis* (Troughton, 1931). Proc. Linn. Soc. N.S.W., 56:201.

COMMON NAME: Solomons Mastiff Bat.

TYPE LOCALITY: Solomon Isls, SW coast of Ysabel Isl, 6 mi. (9 km) W of Tuarugu Village, cave at Mufu Point.

DISTRIBUTION: Solomon Isls.

STATUS: IUCN 2003 and IUCN/SSC Action Plan (2001) – Lower Risk (nt).

COMMENTS: *plicatus* species group. Often included in *jobensis* (e.g. Felten, 1964*a*; Hill, 1983), but provisionally recognized as distinct following Flannery (1995*b*).

*Chaerephon tomensis* (Juste and Ibañez, 1993). J. Mammal., 74:901.
    COMMON NAME: São Tomé Free-tailed Bat.
    TYPE LOCALITY: Sao Tome and Principe, São Tomé Isl, 3 km NW Guadalupe, Praia das
        Conchas.
    DISTRIBUTION: São Tomé Isl.
    STATUS: IUCN 2003 and IUCN/SSC Action Plan (2001) – Vulnerable.
    COMMENTS: *plicatus* species group.

*Cheiromeles* Horsfield, 1824. Zool. Res. Java, Part 8: *Cheiromeles torquatus*, pl. and 10 unno. pp.
    TYPE SPECIES: *Cheiromeles torquata* Horsfield, 1824.
    SYNONYMS: *Chiropotes* Gloger, 1841.
    COMMENTS: Placed in its own subfamiliy, Cheiromelinae Legendre, 1984, by some authors.

*Cheiromeles parvidens* Miller and Hollister, 1921. Proc. Biol. Soc. Wash., 34:100.
    COMMON NAME: Lesser Naked Bat.
    TYPE LOCALITY: Indonesia, Sulawesi, Middle Sulawesi, Pinedapa.
    DISTRIBUTION: Sulawesi, Sanana Isl (Sula Isls; Indonesia); Mindanao, Minoro, and Negros
        (Philippines).
    STATUS: IUCN 2003 and IUCN/SSC Action Plan (2001) – Lower Risk (nt).
    COMMENTS: Formerly included in *torquatus*, but see Corbet and Hill (1992) and Ingle and
        Heaney (1992). Also see Flannery (1995*b*).

*Cheiromeles torquatus* Horsfield, 1824. Zool. Res. Java, Part 8: *Cheiromeles torquatus*, pl. and
    10 unno. pp.
    COMMON NAME: Greater Naked Bat.
    TYPE LOCALITY: Malaysia, Penang.
    DISTRIBUTION: Peninsular Malaysia, Terutau Isl (Thailand), Sumatra and Java, Borneo,
        Palawan Isl (Philippines).
    STATUS: IUCN 2003 and IUCN/SSC Action Plan (2001) – Lower Risk (nt).
    SYNONYMS: *cheiropus* Temminck, 1826; **caudatus** Temminck, 1841; **jacobsoni** Thomas, 1923.
    COMMENTS: Does not include *parvidens*; see Corbet and Hill (1992) and Ingle and Heaney
        (1992).

*Cynomops* Thomas, 1920. Ann. Mag. Nat. Hist., ser. 9, 5:189.
    TYPE SPECIES: *Molossus cerastes* Thomas, 1901 (= *Vespertilio abrasus* Temminck, 1827).
    COMMENTS: Often considered a subgenus of *Molossops*, but here treated as distinct at the genus
        level following Barquez et al. (1993), Peterson et al. (1995), Solari et al. (1999), Reid et al.
        (2000), Barquez and Diaz (2001), and Peters et al. (2002); also see Gardner (1977) and
        Freeman (1981). See Simmons and Voss (1998) and Peters et al. (2002) for diagnoses and
        reviews of species, but note that the former authors did not treat *mexicanus* as a species
        distinct from *greenhalli*.

*Cynomops abrasus* (Temminck, 1827). Monogr. Mamm., 1:232.
    COMMON NAME: Cinnamon Dog-faced Bat.
    TYPE LOCALITY: "Brazil."
    DISTRIBUTION: Venezuela, Guyana, Surinam, French Guiana, Peru, Brazil, Bolivia, Paraguay,
        N Argentina.
    STATUS: IUCN 2003 and IUCN/SSC Action Plan (2001) – Lower Risk (nt) as *Molossops*
        *abrasus*.
    SYNONYMS: **brachymeles** Peters, 1865; **cerastes** Thomas, 1901; **mastivus** Thomas, 1911.
    COMMENTS: Called *brachymeles* by Cabrera (1958) and Freeman (1981), but see Carter and
        Dolan (1978).

*Cynomops greenhalli* Goodwin, 1958. Am. Mus. Novit., 1877:3.
    COMMON NAME: Greenhall's Dog-faced Bat.
    TYPE LOCALITY: Trinidad and Tobago, Trinidad, Port of Spain, Botanic Gardens.
    DISTRIBUTION: Peru, Ecuador, Venezuela, Guianas, and NE Brazil; Trinidad.
    STATUS: IUCN 2003 and IUCN/SSC Action Plan (2001) – Lower Risk (lc) as *Molossops*
        *greenhalli*.

COMMENTS: Reviewed by Simmons and Voss (1998) and Peters et al. (2002); also see Freeman (1981). Does not include *mexicanus*, see Peters et al. (2002).

*Cynomops mexicanus* Jones and Genoways, 1967. Proc. Biol. Soc. Wash., 80:207.
COMMON NAME: Mexican Dog-faced Bat.
TYPE LOCALITY: Mexico, Jalisco, 7.5 mi (20 km) SE Tecomate, 1,500 ft. (500 m).
DISTRIBUTION: Nayarit to Chiapas (Mexico), Honduras, Costa Rica.
STATUS: IUCN 2003 – Not evaluated; not considered in IUCN/SSC Action Plan (2001).
COMMENTS: Formerly considered a subspecies of *greenhalli* (e.g., Koopman, 1994) but apparently distinct, see Peters et al. (2002), also see discussion in Simmons and Voss (1998).

*Cynomops paranus* (Thomas, 1901). Ann. Mag. Nat. Hist., ser. 7, 8:190.
COMMON NAME: Brown Dog-faced Bat.
TYPE LOCALITY: Brazil, Pará.
DISTRIBUTION: Panama, Colombia, Ecuador, Peru, Venezuela, Guyana, Surinam, French Guiana, Brazil, N Argentina. A record from C Mexico listed by Corbet and Hill (1980, 1991) is dubious.
STATUS: IUCN 2003 – Not evaluated; not considered in IUCN/SSC Action Plan (2001).
SYNONYMS: *milleri* Osgood, 1914.
COMMENTS: Distinct from *planirostris*; see Williams and Genoways (1980c), Barquez et al. (1993), and Simmons and Voss (1998). Includes *milleri*; see Simmons and Voss (1998).

*Cynomops planirostris* (Peters, 1866). Monatsb. K. Preuss. Akad. Wiss. Berlin, 1865:575 [1866].
COMMON NAME: Southern Dog-faced Bat.
TYPE LOCALITY: French Guiana, Cayenne.
DISTRIBUTION: Panama to Peru, Bolivia, N Argentina, Paraguay, Brazil, French Guiana, Surinam, Venezuela, probably Guyana.
STATUS: IUCN 2003 and IUCN/SSC Action Plan (2001) – Lower Risk (lc).
COMMENTS: Lectotype designated by Carter and Dolan (1978). Does not include *milleri* or *paranus*; see Williams and Genoways (1980c), Barquez et al. (1993), and Simmons and Voss (1998). Specimens previously reported from Ecuador apparently represent *paranus*; *planirostris* is not presently known from Ecuador (Reid et al., 2000).

*Eumops* Miller, 1906. Proc. Biol. Soc. Wash., 19:85.
TYPE SPECIES: *Molossus californicus* Merriam, 1890 (= *Molossus perotis* Schinz, 1821).
COMMENTS: Revised by Eger (1977). Probably includes *Molossus ater* E. Geoffroy, 1805, see Dolan (1989). Unfortunately, the type of *ater* has been lost and its affinities are unclear.

*Eumops auripendulus* (Shaw, 1800). Gen. Zool. Syst. Nat. Hist., 1(1):137.
COMMON NAME: Black Bonneted Bat.
TYPE LOCALITY: French Guiana.
DISTRIBUTION: Oaxaca and Yucatán (Mexico) to Peru, Bolivia, N Argentina, E Brazil, Venezuela, the Guianas, Trinidad, and Jamaica.
STATUS: IUCN 2003 and IUCN/SSC Action Plan (2001) – Lower Risk (lc).
SYNONYMS: *abrasus* Miller, 1906 [not Temminck, 1827]; *amplexicaudatus* Geoffroy, 1805; *barbatus* J. A. Allen, 1904; *leucopleura* Wagner, 1843; *longimanus* Wagner, 1843; *milleri* J. A. Allen, 1900; *oaxacensis* Goodwin, 1956; *major* Eger, 1974.
COMMENTS: Called *abrasus* in Hall and Kelson (1959), but see Husson (1962) and Hall (1981). Also see Best et al. (2002). Jamaican form reviewed by Timm and Genoways (2003).

*Eumops bonariensis* (Peters, 1874). Monatsb. K. Preuss. Akad. Wiss. Berlin, 1874:232.
COMMON NAME: Dwarf Bonneted Bat.
TYPE LOCALITY: Argentina, Buenos Aires.
DISTRIBUTION: Veracruz (Mexico) to NW Peru, NW Argentina, Parguay, Uruguay, and Brazil.
STATUS: IUCN 2003 and IUCN/SSC Action Plan (2001) – Lower Risk (lc).
SYNONYMS: *delticus* Thomas, 1923; *nanus* Miller, 1900.
COMMENTS: Does not include *patagonicus* or *beckeri*; see Barquez and Ojeda (1992), Barquez et al. (1993), and Saralegui (1996). See Hunt et al. (2003), but note that they included *patagonicus* in *bonariensis*.

*Eumops dabbenei* Thomas, 1914. Ann. Mag. Nat. Hist., ser. 8, 13:481.
> COMMON NAME: Big Bonneted Bat.
> TYPE LOCALITY: Argentina, Chaco.
> DISTRIBUTION: Colombia, Venezuela, Brazil, Paraguay, N Argentina.
> STATUS: IUCN 2003 and IUCN/SSC Action Plan (2001) – Lower Risk (lc).
> SYNONYMS: *mederai* Massoia, 1976.
> COMMENTS: Includes *mederai*, which was originally described as a subspecies of *underwoodi* (Koopman, 1993). See McWilliams et al. (2002).

*Eumops glaucinus* (Wagner, 1843). Arch. Naturgesch., 9(1):368.
> COMMON NAME: Wagner's Bonneted Bat.
> TYPE LOCALITY: Brazil, Mato Grosso, Cuiaba (= Cuyaba).
> DISTRIBUTION: Jalisco (Mexico) to Peru, Bolivia, Paraguay, N Argentina and Brazil; Jamaica; Cuba; Florida (USA).
> STATUS: IUCN 2003 and IUCN/SSC Action Plan (2001) – Lower Risk (lc). *E. g. floridanus* is classified as an Endangered Species by the Florida Game and Fresh Water Fish Commission.
> SYNONYMS: *ferox* Gundlach, 1861; *orthotis* H. Allen, 1889; *floridanus* G. M. Allen, 1932.
> COMMENTS: Includes *floridanus*; see Eger (1977). This complex may include more than one species, see Timm and Genoways (2003).

*Eumops hansae* Sanborn, 1932. J. Mammal., 13:356.
> COMMON NAME: Sanborn's Bonneted Bat.
> TYPE LOCALITY: Brazil, Santa Catarina, Joinville, Colonia Hansa.
> DISTRIBUTION: Chiapas (Mexico), NW Honduras, SW Costa Rica, Panama, Venezuela, Guianas, Ecuador, Peru, Bolivia, Brazil.
> STATUS: IUCN 2003 and IUCN/SSC Action Plan (2001) – Lower Risk (lc).
> SYNONYMS: *amazonicus* Handley, 1955.
> COMMENTS: Includes *amazonicus*; see Gardner et al. (1970) and Eger (1977). Also see Best et al. (2001*b*).

*Eumops maurus* (Thomas, 1901). Ann. Mag. Nat. Hist., ser. 7, 7:141.
> COMMON NAME: Guianan Bonneted Bat.
> TYPE LOCALITY: Guyana, Kanuku Mtns, about 59° W and 37°N, 240 ft. (80 m).
> DISTRIBUTION: Ecuador, Venezuela, Guyana, Surinam. Best et al. (2001*a*) included extreme N Brazil in the range of this species, but I am unaware of any records from that area.
> STATUS: IUCN 2003 and IUCN/SSC Action Plan – Vulnerable.
> SYNONYMS: *geijskesi* Husson, 1962.
> COMMENTS: Includes *geijskesi*; see Eger (1977). Reviewed by Best et al. (2001*a*); also see Reid et al. (2000).

*Eumops patagonicus* Thomas, 1924. Ann. Mag. Nat. Hist., ser. 9, 13:234.
> COMMON NAME: Patagonian Dwarf Bonneted Bat.
> TYPE LOCALITY: Argentina, Buenos Ayres (= Buenos Aires).
> DISTRIBUTION: Bolivia, Argentina, Uruguay.
> STATUS: IUCN 2003 – Not evaluated; not considered in IUCN/SSC Action Plan (2001).
> SYNONYMS: *beckeri* Sanborn, 1932.
> COMMENTS: Distinct from *bonariensis*; see Barquez and Ojeda (1992), Barquez et al. (1993), Mares et al. (1995), Saralegui (1996), Barquez and Diaz (2001), and Gregorin and Taddei (2002).

*Eumops perotis* (Schinz, 1821). *In* Cuvier, Das Thierreich, 1:870.
> COMMON NAME: Greater Bonneted Bat.
> TYPE LOCALITY: Brazil, Rio de Janeiro, Campos do Goita Cazes, Villa São Salvador.
> DISTRIBUTION: California to Texas (USA), south to Zacatecas and Hidalgo (Mexico); N Venezuela, W Ecuador and W Peru, Bolivia, N Argentina, Paraguay, and E Brazil; Cuba.
> STATUS: IUCN 2003 and IUCN/SSC Action Plan (2001) – Lower Risk (lc).
> SYNONYMS: *renatae* Pirlot, 1965; *californicus* Merriam, 1890; *gigas* Peters, 1864.
> COMMENTS: Does not include *trumbulli*; see Eger (1977). The large geographic gap between

the North American and South American ranges of this taxon suggests that this complex may include more than one species.

*Eumops trumbulli* (Thomas, 1901). Ann. Mag. Nat. Hist., ser. 8, 7:190.
COMMON NAME: Trumbull's Bonneted Bat.
TYPE LOCALITY: Brazil, Pará.
DISTRIBUTION: Colombia, W Peru, N Bolivia, S Venezuela, Guianas, Amazon basin of Brazil.
STATUS: IUCN 2003 – Not evaluated; not considered in IUCN/SSC Action Plan (2001).
COMMENTS: Included in *perotis* by Koopman (1971*b*, 1978*b*, 1993, 1994) but see Eger (1977).

*Eumops underwoodi* Goodwin, 1940. Am. Mus. Novit., 1075:2.
COMMON NAME: Underwood's Bonneted Bat.
TYPE LOCALITY: Honduras, La Paz, 6 km N Chinacia.
DISTRIBUTION: Arizona (USA) to Nicaragua.
STATUS: IUCN 2003 and IUCN/SSC Action Plan (2001) – Lower Risk (nt).
SYNONYMS: *sonoriensis* Benson, 1947.
COMMENTS: Does not include *mederai*, which has been transferred to *dabbenei* (Koopman, 1993).

*Molossops* Peters, 1866. Monatsb. K. Preuss. Akad. Wiss. Berlin, 1865:575 [1866].
TYPE SPECIES: *Dysopes temminckii* Burmeister, 1854.
SYNONYMS: *Cabreramops* Ibáñez, 1980; *Dysopes* Burmeister, 1854 [not Illiger, 1811]; *Myopterus* Peters, 1869 [not Geoffory, 1813]; *Neoplatymops* Peterson, 1965.
COMMENTS: Includes *Cabreramops* and *Neoplatymops* with *Molossops* as subgenera. *Cynomops* is here treated as distinct at the genus level following Barquez et al. (1993), Peterson et al. (1995), Solari et al. (1999), Reid et al. (2000), Barquez and Diaz (2001), and Peters et al. (2002); also see Gardner (1977) and Freeman (1981).

*Molossops aequatorianus* Cabrera, 1917. Trab. Mus. Nac. Cienc. Nat. Zool., 31:20.
COMMON NAME: Equatorial Dog-faced Bat.
TYPE LOCALITY: Ecuador, Los Rios, Babahoyo.
DISTRIBUTION: Ecuador.
STATUS: IUCN 2003 and IUCN/SSC Action Plan (2001) – Vulnerable.
COMMENTS: Subgenus *Cabreramops*. Placed in its own genus (*Cabreramops*) by Ibáñez (1980).

*Molossops mattogrossensis* Vieira, 1942. Argent. Zool. Sao Paulo, 3:430.
COMMON NAME: Mato Grosso Dog-faced Bat.
TYPE LOCALITY: Brazil, Mato Grosso, Juruena River, São Simao.
DISTRIBUTION: Venezuela, Guyana, C and NE Brazil.
STATUS: IUCN 2003 and IUCN/SSC Action Plan (2001) – Lower Risk (nt).
COMMENTS: Subgenus *Neoplatymops*; see Freeman (1981). Listed as a subspecies of *Molossops temminckii* by Cabrera (1958), but see Peterson (1965*a*), who considered *Neoplatymops* a distinct genus. See Willig and Jones (1985). See Emmons (1997) for distribution map.

*Molossops neglectus* Williams and Genoways, 1980. Ann. Carnegie Mus., 49(25):489.
COMMON NAME: Rufous Dog-faced Bat.
TYPE LOCALITY: Surinam, Surinam, 1 km S, 2 km E Powaka (5°25'N, 53°03'W).
DISTRIBUTION: Colombia, Venezuela, Guyana, Surinam, Amazonian Brazil, Peru, N Argentina. Also found in the Atlantic Forest of SE Brazil (B. Lim and R. Gregorin, pers. comm.).
STATUS: IUCN 2003 and IUCN/SSC Action Plan (2001) – Lower Risk (nt).
COMMENTS: Subgenus *Molossops*. See Lim and Engstrom (2001).

*Molossops temminckii* (Burmeister, 1854). Syst. Uebers. Thiere Bras., p. 72.
COMMON NAME: Dwarf Dog-faced Bat.
TYPE LOCALITY: Brazil, Minas Gerais, Lagoa Santa.
DISTRIBUTION: Guyana, Venezuela, Colombia, Ecuador, Peru, Bolivia, S Brazil, Paraguay, N Argentina, Uruguay.
STATUS: IUCN 2003 and IUCN/SSC Action Plan (2001) – Lower Risk (lc).
SYNONYMS: *hirtipes* Winge, 1892; *griseiventer* Sanborn, 1941; *sylvia* Thomas, 1924.

COMMENTS: Subgenus *Molossops*.

*Molossus* E. Geoffroy, 1805. Ann. Mus. Natn. Hist. Nat. Paris, 6:151.
TYPE SPECIES: *Vespertilio molossus* Pallas, 1766.
SYNONYMS: *Dysopes* Illiger, 1811.
COMMENTS: Central American species revised by Dolan (1989). Jennings et al. (2000) provided a
key to species modified from Hall (1981), but did not include many of the species
recognized here as distinct.

*Molossus aztecus* Saussure, 1860. Rev. Mag. Zool. Paris, Ser. 2, 12:285.
COMMON NAME: Aztec Mastiff Bat.
TYPE LOCALITY: Mexico, Tlaxcala, Amecameca, at the foot of Popocatepetl.
DISTRIBUTION: Jalisco (Mexico) to Nicaragua; Cozumel Isl (Mexico); S Venezuela.
STATUS: IUCN 2003 and IUCN/SSC Action Plan (2001) – Lower Risk (nt).
COMMENTS: Included in *molossus* by Koopman (1993, 1994), but see Dolan (1989). The
Venezuela record is from Lim and Engstrom (2001). Also see López-González and
Presley (2001).

*Molossus barnesi* Thomas, 1905. Ann. Mag. Nat. Hist., ser. 7, 15:584.
COMMON NAME: Barnes's Mastiff Bat.
TYPE LOCALITY: French Guiana, Cayenne.
DISTRIBUTION: French Guiana.
STATUS: IUCN 2003 – Not evaluated; not considered in IUCN/SSC Action Plan (2001).
COMMENTS: Placed in *coibensis* by Dolan (1989) and considered a subspecies of *molossus* by
Koopman (1993, 1994), but clearly distinct, see Simmons and Voss (1998). Sometimes
spelled *burnesi* (e.g., Freeman, 1981), but the correct spelling is *barnesi*; see Cabrera
(1958), Carter and Dolan (1978), and Simmons and Voss (1998).

*Molossus coibensis* J. A. Allen, 1904. Bull. Amer. Mus. Nat. Hist., 20:227.
COMMON NAME: Coiban Mastiff Bat.
TYPE LOCALITY: Panama, Coiba Isl.
DISTRIBUTION: Chiapas (Mexico) south to Venezuela, SW Guyana, Colombia, Ecuador, Peru,
Mato Grosso (Brazil).
STATUS: IUCN 2003 and IUCN/SSC Action Plan (2001) – Lower Risk (nt).
SYNONYMS: *cherriei* J. A. Allen, 1916; *lambi* Gardner, 1966.
COMMENTS: Included in *molossus* by Koopman (1993, 1994), but see Dolan (1989) and Reid
et al. (2000). Does not include *barnesi* but does include *cherriei* and *lambi*; see Dolan
(1989) and Simmons and Voss (1998). Also see Lim and Engstrom (2001).

*Molossus currentium* Thomas, 1901. Ann. Mag. Nat. Hist., ser. 7. 8:438.
COMMON NAME: Thomas's Mastiff Bat.
TYPE LOCALITY: Argentina, Corrientes, Goya.
DISTRIBUTION: Honduras to Costa Rica; E Panama, Colombia, Ecuador, and Venezuela;
Amazonian Brazil; Paraguay and N Argentina.
STATUS: IUCN 2003 and IUCN/SSC Action Plan (2001) – Lower Risk (lc) as *M. bondae*.
SYNONYMS: **bondae** J. A. Allen, 1904; **robustus** López-González and Presley, 2001
COMMENTS: This species was formerly known as *bondae*, but *currentium* (previously listed as a
junior synonym of *molossus*) is an earlier name; see López-González and Presley
(2001). Subspecies nomenclature revised by López-González and Presley (2001). Also
see Burnett et al. (2001).

*Molossus molossus* (Pallas, 1766). Misc. Zool., p. 49-50.
COMMON NAME: Pallas's Mastiff Bat.
TYPE LOCALITY: France, Martinique (Lesser Antilles).
DISTRIBUTION: Sinaloa and Coahuila (Mexico) to Peru, N Argentina, Paraguay, Uruguay,
Brazil and Guianas; Greater and Lesser Antilles; Florida Keys (USA); Margarita Isl
(Venezuela); Curaçao and Bonaire (Netherlands Antilles); Trinidad and Tobago.
STATUS: IUCN 2003 and IUCN/SSC Action Plan (2001) – Lower Risk (lc).
SYNONYMS: *acuticaudatus* Desmarest, 1820; *amplexicaudus* Wagner, 1850; *crassicaudatus*
Geoffroy, 1805; *currentium* Miller, 1913 [not Thomas, 1901]; *daulensis* J. A. Allen, 1916;

*fusciventer* Geoffroy, 1805; *longicaudatus* Geoffroy, 1805; *major* Kerr, 1792; *minor* Kerr, 1792; *moxensis* D'Orbigny, 1835; *obscurus* Geoffroy, 1805; *olivaceofuscus* Wagner, 1850; *velox* Temminck, 1827; **debilis** Miller, 1913; **fortis** Miller, 1913; **milleri** Johnson, 1952; *fuliginosus* Gray, 1838 [not Cooper, 1837]; **pygmaeus** Miller, 1900; **tropidorhynchus** Gray, 1839; **verrilli** J. A. Allen, 1908.

COMMENTS: Includes *fortis*, *milleri*, *debilis*, and *tropidorhynchus*; see Varona (1974). Called *major* by Hall and Kelson (1959) and Cabrera (1958) but see Husson (1962). Does not include *aztecus*, *barnesi*, *coibensis*, *cherriei*, and *lambi*; see Dolan (1989) and Simmons and Voss (1998). Includes *daulensis*, but see Albuja (1982). Antillean populations reviewed by Genoways et al. (1981) and Timm and Genoways (2003). Records from the Florida Keys may have resulted from transportation by humans; see Frank (1997). *M. pygmaeus* may represent a distinct species, possibly including populations from Guyana; see Lim and Engstrom (2001). This complex is desperately in need of revision.

*Molossus pretiosus* Miller, 1902. Proc. Acad. Nat. Sci. Phil., p. 396.
COMMON NAME: Miller's Mastiff Bat.
TYPE LOCALITY: Venezuela, Caracas, LaGuaira.
DISTRIBUTION: Guerrero, Oaxaca (Mexico); Nicaragua to Colombia, Venezuela, Guyana, and Brazil.
STATUS: IUCN 2003 and IUCN/SSC Action Plan (2001) – Lower Risk (lc).
COMMENTS: Listed as a synonym of *rufus* by Cabrera (1958), but see Jones et al. (1977) and Dolan (1989). Does not include *macdougalli*; see Dolan (1989). See Jennings et al. (2000).

*Molossus rufus* E. Geoffroy, 1805. Ann. Mus. Nat. Hist. Paris, 6:155.
COMMON NAME: Black Mastiff Bat.
TYPE LOCALITY: French Guiana, Cayenne by restriction (Miller, 1913*b*).
DISTRIBUTION: Tamaulipas, Michoacan, and Sinaloa (Mexico) to Peru, N Argentina, Brazil and Guianas; Trinidad.
STATUS: IUCN 2003 and IUCN/SSC Action Plan (2001) – Lower Risk (lc) as *M. ater*.
SYNONYMS: *albus* Wagner, 1843; *alecto* Temminck, 1827; *fluminensis* Lataste, 1891; *holosericeus* Wagner, 1843; *myosurus* Tschudi, 1844; *ursinus* Spix, 1823; **castaneus** Geoffroy, 1805; **nigricans** Miller, 1902; *macdougalli* Goodwin, 1956; *malagai* Villa-R., 1955.
COMMENTS: Called *ater* by many authors, but see Carter and Dolan (1978) and Dolan (1989), who argued, based on descriptions of head and ear shape of both taxa, and examination of the specimens labeled as types of *rufus* in the Muséum National d'Histoire Naturelle in Paris, that *Molossus ater* Geoffroy, 1805, is really an *Eumops*, and that *rufus* is really the correct name for the large *Molossus* often incorrectly called *ater*. Lectotype designated by Carter and Dolan (1978). Unfortunately, the type of *ater* has been lost and its relationships are unclear. Includes *malagai*; see Jones (1965). Includes *macdougalli*; see Jones et al. (1977) and Dolan (1989).

*Molossus sinaloae* J. A. Allen, 1906. Bull. Am. Mus. Nat. Hist., 22:236.
COMMON NAME: Sinaloan Mastiff Bat.
TYPE LOCALITY: Mexico, Sinaloa, Esquinapa.
DISTRIBUTION: Sinaloa and Michoacan (Mexico) to Colombia, Guyana, Surinam, and French Guiana; Trinidad.
STATUS: IUCN 2003 and IUCN/SSC Action Plan (2001) – Lower Risk (lc).
SYNONYMS: **trinitatus** Goodwin, 1959.
COMMENTS: Includes *trinitatus*, see Dolan (1989) and Simmons and Voss (1998). Reviewed by Jennings et al. (2002).

*Mops* Lesson, 1842. Nouv. Tabl. Regn. Anim. Mammifères, p. 18.
TYPE SPECIES: *Mops indicus* Lesson, 1842 (= *Molossus mops* de Blainville, 1840).
SYNONYMS: *Allomops* J. A. Allen, 1917; *Philippinopterus* Taylor, 1934; *Xiphonycteris* Dollman, 1911.
COMMENTS: Formerly included in *Tadarida*, often as a subgenus, but apparently distinct; see Freeman (1981), also see Legendre (1984). Dunlop (1999) provided a key to subgenera and species in this genus. Two subgenera are recognized, *Mops* and *Xiphonycteris*.

*Mops brachypterus* (Peters, 1852). Reise nach Mossambique, Säugethiere, p. 59.
COMMON NAME: Short-winged Free-tailed Bat.
TYPE LOCALITY: Mozambique, Mozambique Isl (15°S, 40°42′E).
DISTRIBUTION: Gambia to Kenya; Tanzania (including Zanzibar and Mafia Isl); Mozambique.
STATUS: IUCN 2003 and IUCN/SSC Action Plan (2001) – Lower Risk (lc).
SYNONYMS: *leonis* Thomas, 1908; *ochraceus* J. A. Allen, 1917.
COMMENTS: Subgenus *Xiphonycteris*. Includes *leonis*; see El-Rayah (1981).

*Mops condylurus* (A. Smith, 1833). S. Afr. Quart. J., 1:54.
COMMON NAME: Angolan Free-tailed Bat.
TYPE LOCALITY: South Africa, KwaZulu-Natal Prov., Durban.
DISTRIBUTION: Mauritania and Senegal to Somalia, south to Angola, Botswana, and KwaZulu-Natal (South Africa).
STATUS: IUCN 2003 and IUCN/SSC Action Plan (2001) – Lower Risk (lc).
SYNONYMS: *angolensis* Peters, 1870; *orientis* G. M. Allen and Loveridge, 1942; *osborni* J. A. Allen, 1917; *fulva* Monard, 1939; *occidentalis* Monard, 1939; *wonderi* Sanbron, 1936.
COMMENTS: Subgenus *Mops*. Does not include *leucostigma*; see Peterson et al., 1995. Distribution mapped by Taylor (2000a).

*Mops congicus* J. A. Allen, 1917. Bull. Am. Mus. Nat. Hist., 37:467.
COMMON NAME: Congo Free-tailed Bat.
TYPE LOCALITY: Dem. Rep. Congo, Oriental, Medje.
DISTRIBUTION: Cameroon, Dem. Rep. Congo, Uganda. Specimens reported from Ghana and Nigeria actually represent *trevori* (J. Fahr, pers. comm.). Koopman (1993) included "perhaps Gambia" in the distribution, but this was apparently a *lapsus* for a specimen of *demonstrator* taken at sea off the coast of Gambia (see Koopman, 1989b).
STATUS: IUCN 2003 and IUCN/SSC Action Plan (2001) – Lower Risk (nt).
COMMENTS: Subgenus *Mops*. Does not include *trevori*; see Peterson (1972).

*Mops demonstrator* (Thomas, 1903). Ann. Mag. Nat. Hist., ser. 7, 12:504.
COMMON NAME: Mongallan Free-tailed Bat.
TYPE LOCALITY: Sudan, Equatoria, Mongalla.
DISTRIBUTION: Sudan, Dem. Rep. Congo, Uganda, Burkina Faso, Ghana, perhaps Gambia (see Koopman, 1989).
STATUS: IUCN 2003 and IUCN/SSC Action Plan (2001) – Lower Risk (nt).
SYNONYMS: *faradjius* J. A. Allen, 1917.
COMMENTS: Subgenus *Mops*. Koopman (1993) suggested that *demonstrator* may include *niveiventer*.

*Mops leucostigma* G. M. Allen, 1918. Bull. Mus. Comp. Zool. Harvard, 61(4):513.
COMMON NAME: Malagasy White-bellied Free-tailed Bat.
TYPE LOCALITY: Madagascar, Tananarive (= Antananarivo).
DISTRIBUTION: E, N, and W Madagascar.
STATUS: IUCN 2003 and IUCN/SSC Action Plan (2001) – Data Deficient.
COMMENTS: Subgenus *Mops*. Formerly included in *condylurus*, but see Peterson et al. (1995).

*Mops midas* (Sundevall, 1843). Kongl. Svenska Vet.-Akad. Handl. Stockholm, 1842:207 [1843].
COMMON NAME: Midas' Free-tailed Bat.
TYPE LOCALITY: Sudan, Blue Nile (= Bahr-el-Abiad Prov.), White Nile River, West bank, Jebel el Funj.
DISTRIBUTION: Senegal to Saudi Arabia, south to Botswana, NE South Africa, and Zimbabwe; Madagascar.
STATUS: IUCN 2003 and IUCN/SSC Action Plan (2001) – Lower Risk (lc).
SYNONYMS: *unicolor* A. Grandidier, 1870; *miarensis* A. Grandidier, 1869.
COMMENTS: Subgenus *Mops*. Reviewed by Peterson et al. (1995) and Dunlop (1999); also see Harrison and Bates (1991).

*Mops mops* (de Blainville, 1840). Osteogr. Mamm., pt. 5 (Vespertilio), p. 101.
COMMON NAME: Malayan Free-tailed Bat.

TYPE LOCALITY: Indonesia, Sumatra.

DISTRIBUTION: W Malaysia, Sumatra, Borneo, perhaps Java.

STATUS: IUCN 2003 and IUCN/SSC Action Plan (2001) – Lower Risk (lc).

SYNONYMS: *indicus* Lesson, 1842 [*nomen nudum*]; *mops* F. Cuvier, 1824 [*nomen nudum*]; *tenuis* Temminck, 1827 [not Horsefield, 1822].

COMMENTS: Subgenus *Mops*. *Dysopes labiatus* Temminck, 1827, may be an older name for this taxon; see discussion in Hill (1961*b*) and Corbet and Hill (1992). Possibly includes *sarasinorum*; see Corbet and Hill (1992).

*Mops nanulus* J. A. Allen, 1917. Bull. Am. Mus. Nat. Hist., 37:477.

COMMON NAME: Dwarf Free-tailed Bat.

TYPE LOCALITY: Dem. Rep. Congo, Oriental, Niangara.

DISTRIBUTION: Sierra Leone to Ethiopia and Kenya. A previous report of this species from The Gambia is in error, probably based on a specimen of *brachypterus* (J. Fahr, pers. comm.)

STATUS: IUCN 2003 and IUCN/SSC Action Plan (2001) – Lower Risk (lc).

SYNONYMS: *calabarensis* Hayman, 1940.

COMMENTS: Subgenus *Xiphonycteris*.

*Mops niangarae* J. A. Allen, 1917. Bull. Am. Mus. Nat. Hist., 37:468.

COMMON NAME: Niangaran Free-tailed Bat.

TYPE LOCALITY: Dem. Rep. Congo, Oriental, Niangara.

DISTRIBUTION: Dem. Rep. Congo (known only from the holotype).

STATUS: IUCN 2003 and IUCN/SSC Action Plan (2001) – Critically Endangered.

COMMENTS: Subgenus *Mops*. Peterson (1972) included this species in *trevori*, while Hayman and Hill (1971) listed it as a subspecies of *Tadarida congica* (= *Mops congicus*). Freeman (1981) found that holotype skull differed significantly from skulls of both *trevori* and *congicus*, and therefore retained *niangarae* a distinct species. I follow this treatment pending a more formal revision of the *trevori/congicus* complex.

*Mops niveiventer* Cabrera and Ruxton, 1926. Ann. Mag. Nat. Hist., ser. 9, 17:594.

COMMON NAME: White-bellied Free-tailed Bat.

TYPE LOCALITY: Dem. Rep. Congo, Kasai Occidental, Luluabourg (= Kananga).

DISTRIBUTION: Dem. Rep. Congo, Rwanda, Burundi, Tanzania, Angola, Zambia, Mozambique. Records from Botswana and Madagascar are erroneous; Botswana records are now thought to represent *condylurus* while Madagascar records represent *leucostigma* (see Hayman and Hill [1971], Meester et al. [1986], and Peterson et al. [1995]).

STATUS: IUCN 2003 and IUCN/SSC Action Plan (2001) – Lower Risk (lc).

SYNONYMS: *chitauensis* Hill, 1937.

COMMENTS: Subgenus *Mops*. Clearly distinct from *condylurus*, see Ansell (1967), Hayman and Hill (1971), and Meester et al. (1986). Koopman (1993) suggested that *niveiventer* is possibly a subspecies of *demonstrator*. Reviewed in part by Van Cakenberghe et al. (1999).

*Mops petersoni* (El Rayah, 1981). R. Ontario Mus. Life Sci. Occas. Pap., 36:3.

COMMON NAME: Peterson's Free-tailed Bat.

TYPE LOCALITY: Cameroon, 15 km S Kumba (4°39'N, 9°26'E).

DISTRIBUTION: Cameroon and Ghana. Koopman (1993) included "perhaps Sierra Leone" in the distribution, but there are apparently no documented records from that country (J. Fahr, pers. comm.).

STATUS: IUCN 2003 and IUCN/SSC Action Plan (2001) – Lower Risk (nt).

COMMENTS: Subgenus *Xiphonycteris*.

*Mops sarasinorum* (A. Meyer, 1899). Abh. Zool. Anthrop.-Ethnolog. Mus. Dresden, 7(7):16.

COMMON NAME: Sulawesian Free-tailed Bat.

TYPE LOCALITY: Indonesia, Sulawesi, Batulappa (North of Lake Tempe).

DISTRIBUTION: Sulawesi (Indonesia) and adjacent small islands; Philippines.

STATUS: IUCN 2003 and IUCN/SSC Action Plan (2001) – Lower Risk (nt).

SYNONYMS: **lanei** Taylor, 1934.

COMMENTS: Subgenus *Mops*. Includes *lanei* (formerly included in *Philippinopterus*);
    see Freeman (1981) and Hill and Rozendaal (1989). Possibly conspecific with *mops*;
    see Corbet and Hill (1992).

*Mops spurrelli* (Dollman, 1911). Ann. Mag. Nat. Hist., ser. 8, 7:211.
    COMMON NAME: Spurrell's Free-tailed Bat.
    TYPE LOCALITY: Ghana, Bibianaha.
    DISTRIBUTION: Guinea to Rio Muni, Bioko (Equatorial Guinea), Central African Republic,
    and Dem. Rep. Congo.
    STATUS: IUCN 2003 and IUCN/SSC Action Plan (2001) – Lower Risk (lc).
    COMMENTS: Subgenus *Xiphonycteris*.

*Mops thersites* (Thomas, 1903). Ann. Mag. Nat. Hist., ser. 7, 12:634.
    COMMON NAME: Railer Free-tailed Bat.
    TYPE LOCALITY: Cameroon, Efulen.
    DISTRIBUTION: Sierra Leone to Rwanda; Bioko (Equatorial Guinea); perhaps Mozambique
    and Zanzibar.
    STATUS: IUCN 2003 and IUCN/SSC Action Plan (2001) – Lower Risk (lc).
    SYNONYMS: *occipitalis* J. A. Allen, 1917.
    COMMENTS: Subgenus *Xiphonycteris*.

*Mops trevori* J. A. Allen, 1917. Bull. Am. Mus. Nat. Hist., 37:468.
    COMMON NAME: Trevor's Free-tailed Bat.
    TYPE LOCALITY: Dem. Rep. Congo, Oriental, Faradje.
    DISTRIBUTION: NE Dem. Rep. Congo, Uganda, Sudan, Ghana, Nigeria.
    STATUS: IUCN 2003 and IUCN/SSC Action Plan (2001) – Lower Risk (nt).
    COMMENTS: Subgenus *Mops*. Formerly included *niangarae*; see Peterson (1972) and Freeman
    (1981). Specimens reported as *congicus* from Ghana and Nigeria actually represent
    *trevori* (J. Fahr, pers. comm.).

*Mormopterus* Peters, 1865. Monatsb. K. Preuss. Akad. Wiss. Berlin, 1865:258.
    TYPE SPECIES: *Nyctinomus* (*Mormopterus*) *jugularis* Peters, 1865.
    SYNONYMS: *Micronomus* Troughton, 1943.
    COMMENTS: Formerly included in *Tadarida* but apparently distinct; see Koopman (1975) and
    Legendre (1984), but also see Freeman (1981). Does not include *Platymops* and *Sauromys*;
    see Corbet and Hill (1992) and Peterson et al. (1995). Species groups follow Koopman
    (1994). This genus apparently includes at least seven undescribed species in Australia; see
    Adams et al. (1988), Churchill (1998), and Menkhorst and Knight (2001). These forms
    have already been given common names (Churchill, 1998): Eastern Freetail Bat, Inland
    Freetail Bat, Little Northern Freetail Bat, Little Western Freetail Bat, Southern Freetail Bat,
    Western Freetail Bat, and Hairy-nosed Freetail Bat.

*Mormopterus acetabulosus* (Hermann, 1804). Observ. Zool., p. 19.
    COMMON NAME: Mauritian Little Mastiff Bat.
    TYPE LOCALITY: Mauritius, Port Louis.
    DISTRIBUTION: Réunion and Mauritius (Mascarene Isls), and a single record from Ethiopia.
    A record from South Africa is questionable, and no specimens are known from
    Madagascar despite several reports to the contrary (Peterson et al., 1995).
    STATUS: IUCN 2003 and IUCN/SSC Action Plan (2001) – Vulnerable.
    SYNONYMS: *natalensis* A. Smith, 1847.
    COMMENTS: *acetabulosus* species group. Reviewed by Peterson et al. (1995).

*Mormopterus beccarii* Peters, 1881. Monatsb. K. Preuss. Akad. Wiss. Berlin, 1881:484.
    COMMON NAME: Beccari's Mastiff Bat.
    TYPE LOCALITY: Indonesia, Molucca Isls, Amboina Isl.
    DISTRIBUTION: Molucca Isls, New Guinea, adjacent small islands, N Australia.
    STATUS: IUCN 2003 and IUCN/SSC Action Plan (2001) – Lower Risk (lc).
    SYNONYMS: **astrolabiensis** Meyer, 1899.
    COMMENTS: *norfolkensis* species group. Includes *astrolabiensis*, see Freeman (1981) and Hill

(1983), also see Flannery (1995*a*, *b*) and Bonaccorso (1998). Peterson et al. (1995) listed *astrolabiensis* as a distinct species with no comment.

*Mormopterus doriae* K. Andersen, 1907. Ann. Mus. Civ. Stor. Nat. Genova, 3(38):42.
COMMON NAME: Sumatran Mastiff Bat.
TYPE LOCALITY: Indonesia, NW Sumatra, Deli, Soekaranda.
DISTRIBUTION: Sumatra.
STATUS: IUCN 2003 and IUCN/SSC Action Plan (2001) – Vulnerable.
COMMENTS: *acetabulosus* species group.

*Mormopterus jugularis* (Peters, 1865). *In* Sclater, Proc. Zool. Soc. Lond., 1865:468.
COMMON NAME: Peters's Wrinkle-lipped Bat.
TYPE LOCALITY: Madagascar, Tananarive (= Antananarivo).
DISTRIBUTION: Madagascar.
STATUS: IUCN 2003 and IUCN/SSC Action Plan (2001) – Vulnerable.
SYNONYMS: *albiventer* Dobson, 1877.
COMMENTS: *acetabulosus* species group. Reviewed by Peterson et al. (1995).

*Mormopterus kalinowskii* (Thomas, 1893). Proc. Zool. Soc. Lond., 1893:334.
COMMON NAME: Kalinowski's Mastiff Bat.
TYPE LOCALITY: "Central Peru."
DISTRIBUTION: Peru, N Chile.
STATUS: IUCN 2003 and IUCN/SSC Action Plan (2001) – Vulnerable.
COMMENTS: *kalinowskii* species group.

*Mormopterus loriae* Thomas, 1897. Ann. Mus. Civ. Stor. Nat. Genova, 18:609.
COMMON NAME: Loria's Mastiff Bat.
TYPE LOCALITY: Papua New Guinea, Kamali, mouth of Kemp Welch River, 10°10′S, 147°44′E.
DISTRIBUTION: N Australia; New Guinea.
STATUS: IUCN 2003 – Not evaluated; not considered in IUCN/SSC Action Plan (2001).
SYNONYMS: **cobourgiana** Johnson, 1959; **ridei** Felten, 1964.
COMMENTS: *norfolkensis* species group. Formerly included in *planiceps*, but see Flannery (1995*a*) and Bonaccorso (1998). Also see Hill (1961*b*) and Koopman (1984*c*). This complex may include at least two undescribed species; see Menkhorst and Knight (2001).

*Mormopterus minutus* (Miller, 1899). Bull. Am. Mus. Nat. Hist., 12:173.
COMMON NAME: Little Goblin Bat.
TYPE LOCALITY: Cuba, Las Villas, Trinidad, San Pablo.
DISTRIBUTION: Cuba.
STATUS: IUCN 2003 and IUCN/SSC Action Plan (2001) – Vulnerable.
COMMENTS: *kalinowskii* species group.

*Mormopterus norfolkensis* (Gray, 1840). Ann. Nat. Hist., 4:7.
COMMON NAME: Eastern Little Mastiff Bat.
TYPE LOCALITY: Australia, Norfolk Isl (S Pacific Ocean); uncertain.
DISTRIBUTION: Norfolk Isl?, SE Queensland, E New South Wales (Australia).
STATUS: IUCN 2003 and IUCN/SSC Action Plan (2001) – Lower Risk (lc).
SYNONYMS: *wilcoxii* Krefft, 1871.
COMMENTS: *norfolkensis* species group. There is considerable doubt as to the status of this species; see Hill (1961*b*) and Koopman (1984*c*). Freeman (1981) included *wilcoxii* in *planiceps*.

*Mormopterus phrudus* (Handley, 1956). Proc. Biol. Soc. Wash., 69:197.
COMMON NAME: Incan Little Mastiff Bat.
TYPE LOCALITY: Peru, Cuzco, Machu Picchu, Urubamba River, San Miguel Bridge.
DISTRIBUTION: Peru.
STATUS: IUCN 2003 and IUCN/SSC Action Plan (2001) – Endangered.
COMMENTS: *kalinowskii* species group.

*Mormopterus planiceps* (Peters, 1866). Monatsb. K. Preuss. Akad. Wiss. Berlin, 1866:23.
    COMMON NAME: Southern Free-tailed Bat.
    TYPE LOCALITY: Australia. Probably New South Wales, Sydney; see Iredale and Troughton
        (1934) for discussion.
    DISTRIBUTION: S and C Australia.
    STATUS: IUCN 2003 and IUCN/SSC Action Plan (2001) – Lower Risk (lc).
    SYNONYMS: *petersi* Leche, 1844.
    COMMENTS: *norfolkensis* species group. Formerly included *loriae*, but see Flannery (1995*a*)
        and Bonaccorso (1998). See Hill (1961*b*) and Koopman (1984*c*). This complex may
        includes at as many as three undescribed species; see Menkhorst and Knight (2001).

*Myopterus* E. Geoffroy, 1818. Descrip. de L'Egypte, 2:113.
    TYPE SPECIES: *Myopterus senegalensis* Oken, 1816 (not available) (= *Myopterus daubentonii*
        Desmarest, 1820).
    SYNONYMS: *Eomops* Thomas, 1905.
    COMMENTS: Includes *Eomops*; see Hayman and Hill (1971).

*Myopterus daubentonii* Desmarest, 1820. Mammalogie, *in* Encyclop. Méth., 1:132.
    COMMON NAME: Daubenton's Winged-mouse Bat.
    TYPE LOCALITY: Senegal.
    DISTRIBUTION: Senegal, Côte d'Ivoire, NE Dem. Rep. Congo, Central African Republic.
    STATUS: IUCN 2003 and IUCN/SSC Action Plan (2001) – Data Deficient.
    SYNONYMS: **albatus** Thomas, 1915.
    COMMENTS: Holotype lost; neotype designated by Adam et al. (1993). Includes *albatus*;
        see Koopman (1989*b*) and Adam et al. (1993).

*Myopterus whitleyi* (Scharff, 1900). Ann. Mag. Nat. Hist., ser. 7, 6:569.
    COMMON NAME: Bini Winged-mouse Bat.
    TYPE LOCALITY: Nigeria, Mid-Western Region, Benin City.
    DISTRIBUTION: Ghana, Nigeria, Cameroon, Dem. Rep. Congo, Uganda.
    STATUS: IUCN 2003 and IUCN/SSC Action Plan (2001) – Lower Risk (lc).
    COMMENTS: Reviewed by Adam et al. (1993).

*Nyctinomops* Miller, 1902. Proc. Acad. Nat. Sci. Phil., 54:393.
    TYPE SPECIES: *Nyctinomus femorosaccus* Merriam, 1889.
    COMMENTS: Formerly included in *Tadarida* but apparently distinct; see Hall (1981) and
        Legendre (1984), but also see Freeman (1981). A key to the species was presented by
        Kumirai and Jones (1990).

*Nyctinomops aurispinosus* (Peale, 1848). Mammalia, *in* Repts. U.S. Expl. Surv., 8:21.
    COMMON NAME: Peale's Free-tailed Bat.
    TYPE LOCALITY: Brazil, Rio Grande do Norte, 100 mi. (161 km) off Cape Sao Roque.
    DISTRIBUTION: Sonora and Tamaulipas (Mexico) to Peru, Bolivia, and Brazil.
    STATUS: IUCN 2003 and IUCN/SSC Action Plan (2001) – Lower Risk (lc).
    SYNONYMS: *similis* Sanborn, 1941.
    COMMENTS: Includes *similis*; see Jones and Arroyo-Cabrales (1990).

*Nyctinomops femorosaccus* (Merriam, 1889). N. Am. Fauna, 2:23.
    COMMON NAME: Pocketed Free-tailed Bat.
    TYPE LOCALITY: USA, California, Riverside Co., Palm Springs.
    DISTRIBUTION: Guerrero (Mexico) to New Mexico, Arizona, California (USA) and
        Baja California (Mexico).
    STATUS: IUCN 2003 and IUCN/SSC Action Plan (2001) – Lower Risk (lc).
    COMMENTS: See Kumirai and Jones (1990).

*Nyctinomops laticaudatus* (E. Geoffroy, 1805). Ann. Mus. Natn. Hist. Nat. Paris, 6:156.
    COMMON NAME: Broad-eared Free-tailed Bat.
    TYPE LOCALITY: Paraguay, Asunción.
    DISTRIBUTION: Tamaulipas and Jalisco (Mexico) to Venezuela and the Guianas, NW Peru,
        Bolivia, N Argentina, Paraguay, and Brazil; Trinidad; Cuba.

STATUS: IUCN 2003 and IUCN/SSC Action Plan (2001) – Lower Risk (lc).

SYNONYMS: *caecus* Rengger, 1830; *espiritosantensis* Ruschi, 1951 [see discussion of availability in Pine and Ruschi, 1976]; *gracilis* Wagner, 1843; *europs* H. Allen, 1889; *ferruginea* Goodwin, 1954; *macarenensis* Barriga-Bonilla, 1965; *yucatanicus* Miller, 1902.

COMMENTS: Includes *yucatanicus*, *europs*, and *gracilis*; see Silva-Taboada and Koopman (1964), Freeman (1981), and Avila-Flores et al. (2002). Includes *espiritosantensis*, see Zortéa and Taddei (1995) and Avila-Flores et al. (2002). Note that the correct spelling for the specific epithet in combination with *Nyctinomops* is *laticaudatus* (not *laticaudata*) because the generic name is masculine. Reviewed by Avila-Flores et al. (2002).

*Nyctinomops macrotis* (Gray, 1840). Ann. Nat. Hist., 4:5.
COMMON NAME: Big Free-tailed Bat.
TYPE LOCALITY: Cuba.
DISTRIBUTION: SW British Columbia and Iowa (USA) to SW Mexico; Colombia, Venezuela, Guyana, and Surinam to Peru, N Argentina and Uruguay; Cuba; Jamaica; Hispaniola.
STATUS: IUCN 2003 and IUCN/SSC Action Plan (2001) – Lower Risk (lc).
SYNONYMS: *aequatoralis* J. A. Allen, 1914; *affinis* J. A. Allen, 1900; *auritus* Wagner, 1843; *depressus* Ward, 1891; *megalotis* Dobson, 1876; *molossa* Hershkovitz, 1949 [not Pallas]; *nevadensis* H. Allen, 1894.
COMMENTS: Called *Tadarida molossa* by Hall and Kelson (1959), but see Husson (1962). See Milner et al. (1990) and Timm and Genoways (2003).

*Otomops* Thomas, 1913. J. Bombay Nat. Hist. Soc., 22:90.
TYPE SPECIES: *Nyctinomus wroughtoni* Thomas, 1913.
COMMENTS: Reviewed by Peterson et al. (1995).

*Otomops formosus* Chasen, 1939. Treubia, 17:186.
COMMON NAME: Java Giant Mastiff Bat.
TYPE LOCALITY: Indonesia, W Java, Tjibadak.
DISTRIBUTION: Java.
STATUS: IUCN 2003 and IUCN/SSC Action Plan (2001) – Vulnerable.
COMMENTS: Reviewed by Boeadi (1990), Kitchener et al. (1992*a*), and Walston and Bates (2001).

*Otomops johnstonei* Kitchener, How, and Maryanto, 1992. Rec. W. Aust. Mus., 15:730.
COMMON NAME: Johnstone's Giant Mastiff Bat.
TYPE LOCALITY: Indonesia, Nusa Tenggara, Alor Isl, Desa Apui, 08°15′S, 124°43′E.
DISTRIBUTION: Alor Isl (Indonesia).
STATUS: IUCN 2003 and IUCN/SSC Action Plan (2001) – Vulnerable.
COMMENTS: Known only from the holotype.

*Otomops madagascariensis* Dorst, 1953. Mém. Inst. Scient. Madagascar (A), 8:236.
COMMON NAME: Malagasy Giant Mastiff Bat.
TYPE LOCALITY: Madagascar, S of Soalala, Namoroka, Réserve naturelle intégrale (no. 8), 16°23′S, 45°28′E.
DISTRIBUTION: N, S, and W Madagascar.
STATUS: IUCN 2003 – Not evaluated; not considered in IUCN/SSC Action Plan (2001).
COMMENTS: Reviewed by Peterson et al. (1995).

*Otomops martiensseni* (Matschie, 1897). Arch. Naturgesch., 63(1):84.
COMMON NAME: Large-eared Giant Mastiff Bat.
TYPE LOCALITY: Tanzania, W of Tanga, SE Usambara Mtns, Magrotto Plantation.
DISTRIBUTION: Yemen; Djibouti and Central African Republic to Angola and KwaZulu-Natal (South Africa); Ghana.
STATUS: IUCN 2003 and IUCN/SSC Action Plan (2001) – Vulnerable.
SYNONYMS: *icarus* Chubb, 1917.
COMMENTS: Formerly included *madagascariensis* (e.g., Hayman and Hill, 1971; Koopman, 1993, 1994), but see Peterson et al. (1995). Reviewed by Al-Jumaily (1999).

*Otomops papuensis* Lawrence, 1948. J. Mammal., 29:413.
    COMMON NAME: Papuan Giant Mastiff Bat.
    TYPE LOCALITY: Papua New Guinea, Gulf Prov., Vailala River.
    DISTRIBUTION: SE New Guinea.
    STATUS: IUCN 2003 and IUCN/SSC Action Plan (2001) – Vulnerable.
    COMMENTS: Reviewed by Kitchener et al. (1992*a*); also see Hill (1983), Flannery (1995*a*), and
        Bonaccorso (1998).

*Otomops secundus* Hayman, 1952. *In* Laurie, Bull. Brit. Mus. (Nat. Hist.), Zool., 1:314.
    COMMON NAME: Mantled Giant Mastiff Bat.
    TYPE LOCALITY: Papua New Guinea, Madang Prov., Tapu.
    DISTRIBUTION: NE New Guinea.
    STATUS: IUCN 2003 and IUCN/SSC Action Plan (2001) – Vulnerable.
    COMMENTS: Distinct from *papuensis*; see Kitchener et al. (1992*a*). Also see Hill (1983),
        Flannery (1995*a*), and Bonaccorso (1998).

*Otomops wroughtoni* (Thomas, 1913). J. Bombay Nat. Hist. Soc., 22:87.
    COMMON NAME: Wroughton's Giant Mastiff Bat.
    TYPE LOCALITY: India, Mysore, Kanara, near Talewadi, Barapede Cave.
    DISTRIBUTION: S and NE India, Cambodia.
    STATUS: IUCN 2003 and IUCN/SSC Action Plan (2001) – Critically Endangered.
    COMMENTS: Reviewed by Bates and Harrison (1997), Walston and Bates (2001), and Thabah
        and Bates (2002).

*Platymops* Thomas, 1906. Ann. Mag. Nat. Hist., ser. 7, 7:499.
    TYPE SPECIES: *Platymops macmillani* Thomas, 1906 (= *Mormopterus setiger* Peters, 1878).
    COMMENTS: Included in *Mormopterus* by Freeman (1981) and Koopman (1993, 1994), but see
        Harrison and Fleetwood (1960), Corbet and Hill (1992), and Peterson et al. (1995).

*Platymops setiger* (Peters, 1878). Monatsb. K. Preuss. Akad. Wiss. Berlin, 1878:196.
    COMMON NAME: Peters's Flat-headed Bat.
    TYPE LOCALITY: Kenya, Taita.
    DISTRIBUTION: S Sudan, Ethiopia, Kenya.
    STATUS: IUCN 2003 and IUCN/SSC Action Plan (2001) – Lower Risk (lc) as *Mormopterus*
        *setiger*.
    SYNONYMS: *parkeri* Harrison and Fleetwood, 1960; **macmillani** Thomas, 1906; *barbatogularis*
        Harrison, 1956.

*Promops* Gervais, 1856. *In* Comte de Castelnau, Exped. Partes Cen. Am. Sud., Zool.(Sec. 7), Vol 1,
    pt. 2(Mammifères):58.
    TYPE SPECIES: *Promops ursinus* Gervais, 1856 (= *Molossus nasutus* Spix, 1823).

*Promops centralis* Thomas, 1915. Ann. Mag. Nat. Hist., ser. 8, 16:62.
    COMMON NAME: Big Crested Mastiff Bat.
    TYPE LOCALITY: Mexico, N Yucatán.
    DISTRIBUTION: Jalisco and Yucatán (Mexico) to Ecuador, Peru, W Brazil, Bolivia, Paraguay,
        N Argentina, Guianas; Trinidad.
    STATUS: IUCN 2003 and IUCN/SSC Action Plan (2001) – Lower Risk (lc).
    SYNONYMS: **davisoni** Thomas, 1921; **occultus** Thomas, 1915.
    COMMENTS: *davisoni* may actually be a subspecies of *nasutus*; see Genoways and Williams
        (1979*b*) and Freeman (1981). See also Ojasti and Linares (1971).

*Promops nasutus* (Spix, 1823). Sim. Vespert. Brasil., p. 58.
    COMMON NAME: Brown Mastiff Bat.
    TYPE LOCALITY: Brazil, Bahia, Sao Francisco River.
    DISTRIBUTION: Venezuela, Trinidad, Guyana, Surinam, Brazil, Ecuador, Peru, Bolivia,
        Paraguay, N Argentina.
    STATUS: IUCN 2003 and IUCN/SSC Action Plan (2001) – Lower Risk (lc).
    SYNONYMS: *fumarius* Spix, 1823; *rufocastaneus* Schinz, 1844; *ursinus* Gervais, 1855; **ancilla**
        Thomas, 1915; **downsi** Goodwin, 1962; **fosteri** Miller, 1907; **pamana** Miller, 1913.

COMMENTS: Includes *pamana*; see Goodwin and Greenhall (1962). May include *davisoni*; see Genoways and Williams (1979*b*) and Freeman (1981).

*Sauromys* Roberts, 1917. Ann. Transvaal Mus., 6:5.
TYPE SPECIES: *Platymops petrophilus* Roberts, 1917.
COMMENTS: Originally described as a subgenus of *Platymops*. Included in *Mormopterus* by Freeman (1981), Legendre (1984), and Koopman (1993, 1994), but see Peterson (1965*a*), Corbet and Hill (1992), and Peterson et al. (1995).

*Sauromys petrophilus* (Roberts, 1917). Ann. Transvaal Mus., 6:4.
COMMON NAME: Roberts's Flat-headed Bat.
TYPE LOCALITY: South Africa, Northwest Prov., near Rustenburg, Bleskap.
DISTRIBUTION: South Africa, Namibia, Botswana, Zimbabwe, Mozambique, perhaps Ghana.
STATUS: IUCN 2003 and IUCN/SSC Action Plan (2001) – Lower Risk (lc) as *Mormopterus petrophilus*.
SYNONYMS: ***erongensis*** Roberts, 1946; ***fitzsimonsi*** Roberts, 1946; ***haagneri*** Roberts, 1917; ***umbratus*** Shortridge and Carter, 1938.
COMMENTS: See Jacobs and Fenton (2002).

*Tadarida* Rafinesque, 1814. Precis Som., p. 55.
TYPE SPECIES: *Cephalotes teniotis* Rafinesque, 1814.
SYNONYMS: *Austronomus* Iredale and Troughton, 1934 [*nomen dubium*; later validated by Troughton, 1941]; *Dinops* Savi, 1825; *Dysops* Cretzschmar, 1830-1831 [proccupied by *Dysops* Illiger, 1911]; *Nictinomes* Gray, 1821; *Nyctinoma* Bowdich, 1821; *Nyctinomia* Fleming, 1822; *Rhizomops* Legendre, 1984.
COMMENTS: Formerly included *Chaerephon*, *Mops*, *Mormopterus*, *Nyctinomops*, *Platymops*, and *Sauromys*, which are here treated as distinct genera. Includes *Rhizomops*; see R. D. Owen et al. (1990), but also see Legendre (1984). Mahoney and Walton (1988) regarded *Nyctinomus* (here considered a junior synonym of *Chaerephon*) as an older name for this genus. Species groups follow Koopman (1994).

*Tadarida aegyptiaca* (E. Geoffroy, 1818). Descrip. de L'Egypte, 2:128.
COMMON NAME: Egyptian Free-tailed Bat.
TYPE LOCALITY: Egypt, Giza (resticted by Koopman, 1975).
DISTRIBUTION: South Africa to Nigeria, Algeria, and Egypt to Saudi Arabia, Yemen and Oman, east to India and Sri Lanka, N to Afganistan.
STATUS: IUCN 2003 and IUCN/SSC Action Plan (2001) – Lower Risk (lc).
SYNONYMS: *brunneus* Seabra, 1900; *geoffroyi* Temminck, 1826; *talpinus* Heuglin, 1877; *tongaensis* Wettstein, 1916; ***bocagei*** Seabra, 1900; *anchietae* Seabra, 1900; ***sindica*** Wroughton, 1919; ***thomasi*** Wroughton, 1919; *gossei* Wroughton, 1919; ***tragatus*** Dobson, 1874.
COMMENTS: *aegyptiaca* species group. Includes *tragata*; see Corbet (1978*c*) and Freeman (1981). Reviewed in part by Harrison and Bates (1991) and Bates and Harrison (1997). For African range see Taylor (2000*a*).

*Tadarida australis* (Gray, 1839). Mag. Zool. Bot., 2:501.
COMMON NAME: White-striped Free-tailed Bat.
TYPE LOCALITY: Australia, New South Wales.
DISTRIBUTION: S and C Australia.
STATUS: IUCN 2003 and IUCN/SSC Action Plan (2001) – Lower Risk (nt).
SYNONYMS: *albidus* Leche, 1884; *atratus* Thomas, 1924.
COMMENTS: *australis* species group. Does not include *kuboriensis*, although see Koopman (1982).

*Tadarida brasiliensis* (I. Geoffroy, 1824). Ann. Sci. Nat. Zool., 1:343.
COMMON NAME: Brazilian Free-tailed Bat (known as the Mexican Free-tailed Bat in North America).
TYPE LOCALITY: Brazil, Paraná, Curitiba (= Curityba).

DISTRIBUTION: S Brazil, Bolivia, Argentina, and Chile to Oregon, S Nebraska and Ohio (USA); Greater and Lesser Antilles.

STATUS: IUCN 2003 and IUCN/SSC Action Plan (2001) – Lower Risk (nt).

SYNONYMS: *multispinosus* Burmeister, 1861; *naso* Wagner, 1840; *nasutus* Temminck, 1827; *peruanus* J. A. Allen, 1914; *rugosus* D'Orbigny, 1837; **antillularum** Miller, 1902; **bahamensis** Rhen, 1902; **constanzae** Shamel, 1931; **cynocephala** Le Conte, 1831; *fuliginosus* Cooper, 1837; **intermedia** Shamel, 1931; **mexicana** Saussure, 1860; *californicus* H. Allen, 1894; *mohavensis* Merriam, 1889; *texana* Stager, 1942; **murina** Gray, 1827; **muscula** Gundlach, 1861.

COMMENTS: *aegyptiaca* species group. Placed in distinct genus (*Rhizomops*) by Legendre (1984), but see Freeman (1981) and R. D. Owen et al. (1990). See Hall (1981) and Wilkins (1989); also see Emmons (1997) for distribution map. Caribbean subspecies reviewed by Timm and Genoways (2003).

*Tadarida fulminans* (Thomas, 1903). Ann. Mag. Nat. Hist., ser. 7, 12:501.

COMMON NAME: Malagasy Free-tailed Bat.

TYPE LOCALITY: Madagascar, Betsilo, Fianarantsoa.

DISTRIBUTION: E Dem. Rep. Congo, Rwanda, Kenya, Tanzania, Zambia, Malawi, Zimbabwe, NE South Africa, Madagascar.

STATUS: IUCN 2003 and IUCN/SSC Action Plan (2001) – Lower Risk (nt).

SYNONYMS: **mastersoni** Roberts, 1946.

COMMENTS: *teniotis* species group. Reviewed by Peterson et al. (1995) and Cotterill (2001*b*).

*Tadarida insignis* Blyth, 1862. J. Asiat. Soc. Bengal, 30:90.

COMMON NAME: East Asian Free-tailed Bat.

TYPE LOCALITY: China, Fukien (= Fujian), Amoy.

DISTRIBUTION: Japan, Taiwan, Korea, S China.

STATUS: IUCN 2003 – Not evaluated; not considered in IUCN/SSC Action Plan (2001).

SYNONYMS: *chinensis* Westwood, 1874; *cinerea* Gubareff, 1939; *coecata* Thomas, 1922; *septentrionalis* Kishida, 1931 [*nomen nudum*].

COMMENTS: *teniotis* species group. Formerly included in *teniotis*, but see Yoshiyuki (1989), Yoshiyuki et al. (1989), and Funakoshi and Kunisaki (2000). Does not include *latouchei*; see Kock (1999*a*) and Funakoshi and Kunisaki (2000). Status of *coecata* from Yunnan (China) is somewhat unclear; see Kock (1999*a*), who suggested that it might represent either *teniotis* or *insignis*.

*Tadarida kuboriensis* McKean and Calaby, 1968. Mammalia, 32:375.

COMMON NAME: New Guinea Mastiff Bat.

TYPE LOCALITY: Papua New Guinea, Chimbu Prov., Kubor Range, Minj-nona Divide, 6°02'S, 144°45'E, 2,750 m.

DISTRIBUTION: New Guinea.

STATUS: IUCN 2003 – Not evaluated; not considered in IUCN/SSC Action Plan (2001).

COMMENTS: *australis* species group. Koopman (1982, 1994) treated *kuboriensis* as a subspecies of *australis*, but described character variation suggests that they represent distinct species. See Flannery (1995*a*) and Bonaccorso (1998).

*Tadarida latouchei* Thomas, 1920. Ann. Mag. Nat. Hist., ser. 9, 5:283.

COMMON NAME: La Touche's Free-tailed Bat.

TYPE LOCALITY: China, NE coast of Hopei [Hebei], Ching-wang Tao [= Qinhuangdao].

DISTRIBUTION: N. China, Thailand, Laos, Japan.

STATUS: IUCN 2003 and IUCN/SSC Action Plan (2001) – Data Deficient.

COMMENTS: *teniotis* species group. Clearly distinct from *teniotis* and *insignis*; see Kock (1999*a*), Funakoshi and Kunisaki (2000), and Helgen and Wilson (2002).

*Tadarida lobata* (Thomas, 1891). Ann. Mag. Nat. Hist., ser. 6, 7:303.

COMMON NAME: Big-eared Free-tailed Bat.

TYPE LOCALITY: Kenya, West Pokot, Turkwell Gorge.

DISTRIBUTION: Kenya, Zimbabwe.

STATUS: IUCN 2003 and IUCN/SSC Action Plan (2001) – Vulnerable.

COMMENTS: *teniotis* species group. Reviewed by Cotterill (2001*b*).

*Tadarida teniotis* (Rafinesque, 1814). Prícis Som., p. 12.
COMMON NAME: European Free-tailed Bat.
TYPE LOCALITY: Italy, Sicily.
DISTRIBUTION: France, Spain, and Portugal south to Morocco and Algeria, east through Tunisia, Libya, Israel, Jordan, W Saudi Arabia, Iran, Iraq, Azerbaijan, Turkmenistan, Tajikistan, Kyrgyzstan, and Afghanistan to W Bengal (India), Yunnan (China), and Flores (Indonesia); Madeira (Portugal) and Canary Isls (Spain).
STATUS: IUCN 2003 and IUCN/SSC Action Plan (2001) – Lower Risk (lc).
SYNONYMS: *cestoni* Savi, 1825; *nigrogriseus* Schneider, 1871; *savii* Schinz, 1840; *rueppelli* Temminck, 1826.
COMMENTS: *teniotis* species group. Revised by Aellen (1966) and Kock and Nader (1984), although both included *insignis* in this complex as a subspecies. Does not include *insignis*; see Yoshiyuki (1989), Yoshiyuki et al. (1989), and Funakoshi and Kunisaki (2000). Does not include *latouchei*; see Kock (1999a), Funakoshi and Kunisaki (2000), and Helgen and Wilson (2002). May include *coecata* from Yunnan (China), here considered a synonym of *insignis*; see Kock (1999). A specimen from India seems clearly referable to *teniotis*, see Funakoshi and Kunisaki (2000), though also see Kock. (1999a). Eastern-most records reviewed by Bates and Harrison (1997) and Helgen and Wilson (2002); Middle Eastern records reviewed by Harrison and Bates (1991); Palearctic forms reviewed by Horácek et al. (2000).

*Tadarida ventralis* (Heuglin, 1861). Nova. Acta. Acad. Caes. Leop.-Carol., 29(8):4, 11.
COMMON NAME: Giant Free-tailed Bat.
TYPE LOCALITY: Eritrea, Keren.
DISTRIBUTION: Eritrea to South Africa.
STATUS: IUCN 2003 and IUCN/SSC Action Plan (2001) – Lower Risk (nt).
SYNONYMS: *africana* Dobson, 1876.
COMMENTS: *teniotis* species group. Reviewed by Kock (1975) and Cotterill (2001b).

**Family Vespertilionidae** Gray, 1821. London Med. Repos., 15:299.
COMMENTS: Does not include Tomopeatinae; see Barkley (1984), Sudman et al. (1994), Simmons (1998), and Simmons and Geisler (1998). Includes Antrozoinae; see discussion under that subfamily. For a phylogeny including representatives of most genera, see Volleth and Heller (1994); also see Kawai et al. (2002) on possible relationships of subfamilies.

**Subfamily Vespertilioninae** Gray, 1821. London Med. Repos., 15:299.
SYNONYMS: Nyctophilinae Peters, 1865.
COMMENTS: May not be monophyletic; see Simmons (1998). Includes Nyctophilinae; see Koopman (1984a), Volleth and Tidemann (1991), and Volleth and Heller (1994). Does not include Myotinae; see Volleth and Heller (1994), Simmons (1998), and Simmons and Geisler (1998). Koopman (1994) proposed a tribal classification for the subfamily (subsequently reproduced by McKenna and Bell, 1997), but these groupings have not been supported in phylogenetic studies. Volleth and Tidemann (1991) and Volleth and Heller (1994) proposed a tribal classification based on a phylogenetic analysis of karyotypes but did not include all genera. The tribal classification adopted here follows Koopman (1994) with modifications suggested by Volleth and Tidemann (1991), Tumlinson and Douglas (1992), Frost and Timm (1992), Volleth and Heller (1994), Bogdanowicz et al. (1998), Hoofer and Van Den Bussche (2001), and Volleth et al. (2001).

**Tribe Eptesicini** Volleth and Heller, 1994. Z. Zool. Syst. Evolut.-forsch, 32:24.
COMMENTS: Includes *Arielulus*, *Eptesicus*, and *Hesperoptenus*; see Volleth and Heller (1994) and Volleth et al. (2001).

*Arielulus* Hill and Harrison, 1987. Bull. Br. Mus. Nat. Hist., 52:250.
TYPE SPECIES: *Vespertilio circumdatus* Temminck, 1840.
SYNONYMS: *Thainycteris* Kock and Storch, 1996.
COMMENTS: Named as a subgenus of *Pipistrellus* by Hill and Harrison (1987). Transferred to

*Eptesicus* by Heller and Volleth (1984) and Volleth and Heller (1994), but subsequently recognized as a distinct genus by Csorba and Lee (1999). Includes *Thainycteris*; see Csorba and Lee (1999).

*Arielulus aureocollaris* (Kock and Storch, 1996). Senkenberg. Biol. 76(1/2):2.
COMMON NAME: Collared Sprite.
TYPE LOCALITY: Thailand, Chiang Mai Prov., Amphoe (District) Mae Ai, Doi (Mount) Pha Hom Pok, 20°08'N, 99°10'E, 1,500 m.
DISTRIBUTION: Thailand, Cambodia, Vietnam.
STATUS: IUCN 2003 and IUCN/SSC Action Plan (2001) – Data Deficient.
COMMENTS: Originally placed in its own genus, *Thainycteris*, but see Csorba and Lee (1999). Also see Eger and Theberge (1999).

*Arielulus circumdatus* (Temminck, 1840). Monogr. Mamm., 2:214.
COMMON NAME: Bronze Sprite.
TYPE LOCALITY: Indonesia, Java, Tapos.
DISTRIBUTION: Java (Indonesia), W Malaysia, Cambodia, Thailand, Burma, NE India, Nepal, SW China.
STATUS: IUCN 2003 Lower Risk (lc) as *Arielulus circumdatus*; IUCN/SSC Action Plan (2001) – Lower Risk (lc) as *Pipistrellus circumdatus*.
SYNONYMS: *drungicus* Wang, 1982.
COMMENTS: Heller and Volleth (1984) included this taxon in *societatis*, but see Hill and Francis (1984) and Corbet and Hill (1992).

*Arielulus cuprosus* (Hill and Francis, 1984). Bull. Brit. Mus. (Nat. Hist.) Zool., 47:312.
COMMON NAME: Coppery Sprite.
TYPE LOCALITY: Malaysia, Borneo, Sabah, Sepilok (05°52'N, 117°56'E).
DISTRIBUTION: Borneo.
STATUS: IUCN 2003 and IUCN/SSC Action Plan (2001) – Vulnerable.

*Arielulus societatis* (Hill, 1972). Bull. Brit. Mus. (Nat. Hist.) Zool., 23:34.
COMMON NAME: Social Sprite.
TYPE LOCALITY: Malaysia, Pahang, Gunong Benom, Base Camp (03°51'N, 102°11'E), 800 ft. (266 m).
DISTRIBUTION: W Malaysia.
STATUS: IUCN 2003 and IUCN/SSC Action Plan (2001) – Data Deficient.
COMMENTS: Synonymized with *circumdatus* by Heller and Volleth (1984), but see Hill and Francis (1984) and Corbet and Hill (1992).

*Arielulus torquatus* Csorba and Lee, 1999. J. Zool., Lond., 248:364-366.
COMMON NAME: Necklace Sprite.
TYPE LOCALITY: Taiwan, Taichung County, Wu-ling Farm; 1,800 m; 24°24'N, 121°18'E.
DISTRIBUTION: Taiwan.
STATUS: IUCN 2003 and IUCN/SSC Action Plan (2001) – Data Deficient.

*Eptesicus* Rafinesque, 1820. Ann. Nature, p. 2.
TYPE SPECIES: *Eptesicus melanops* Rafinesque, 1820 (= *Vespertilio fuscus* Beauvois, 1796).
SYNONYMS: *Adelonycteris* H. Allen, 1891; *Amblyotis* Kolenati, 1858; *Cateorus* Kolenati, 1856; *Cnephaeus* Kaup, 1829; *Noctula* Bonaparte, 1837; *Nyctiptenus* Fitzinger, 1870; *Pachyomus* Gray, 1866; *Pareptesicus* Bianchi, 1917; *Rhyneptesicus* Bianchi, 1917; *Rhinopterus* Miller, 1906; *Scabrifer* G. M. Allen, 1908; *Tuitatus* Kishida and Mori, 1931.
COMMENTS: Middle and South American species reviewed by W. B. Davis (1965, 1966). Indomalayan species reviewed by Corbet and Hill (1992). Definition and content discussed by Horacek and Hanák (1985-1986), Hill and Harrison (1987), Menu (1987), Heller and Volleth (1994), and Kearney et al. (2002). Does not include *Vespadelus*; see Kitchener et al. (1987), Volleth and Tidemann (1991), and Volleth and Heller (1994). Does not include *Arielulus*; see Csorba and Lee (1991). Does not include *Neoromicia*; see Volleth et al. (2001) and Kearney et al. (2002). Two subgenera are recognized here, *Eptesicus* and *Rhinopterus*. Some authors have recognized several subgenera from among the taxa here included in the subgenus *Eptesicus* (e.g., Horácek et al. [2000] used

*Amblyotus* and *Rhyneptesicus* as subgenera), but I prefer to retain a more conservative usage pending a thorough revision of the genus.

*Eptesicus andinus* J. A. Allen, 1914. Bull. Amer. Mus. Nat. Hist. 33:382.
    COMMON NAME: Little Black Serotine.
    TYPE LOCALITY: Colombia, Valle de las Papas, 10,000 ft. (3,333 m).
    DISTRIBUTION: Colombia, Ecuador, Peru, Venezuela, Amazonian Brazil; possibly Bolivia. Also known from S Guyana (B. Lim and M. Engstrom, pers. comm.).
    STATUS: IUCN 2003 and IUCN/SSC Action Plan (2001) – Lower Risk (lc).
    SYNONYMS: *chiralensis* Anthony, 1926; *montosus* Thomas, 1920.
    COMMENTS: Subgenus *Eptesicus*. Included in *brasiliensis* by Koopman (1978b, 1993, 1994) but see W. B. Davis (1966) and Simmons and Voss (1998). Does not include *chiriquinus* and *inca* contra W. B. Davis (1966); see Simmons and Voss (1998). Anderson (1997) reported specimens of both *andinus* and *montosus* from Bolivia, but these records must be considered provisional until the specimens are reexamined in light of Simmons and Voss' (1998) revised diagnoses of *andinus* and *chiriquinus*.

*Eptesicus bobrinskoi* Kuzyakin, 1935. Bull. Soc. Nat. Moscow, 44:435.
    COMMON NAME: Bobrinski's Serotine.
    TYPE LOCALITY: Kazakhstan, 65 km E Aralsk, Tyulek Wells in Aral-Kara-Kum desert.
    DISTRIBUTION: Kazakhstan. Records from Caucasus, Uzbekistan, Turkmenistan, and Iran are apparently erroneous, based on juvenile *nilssonii* (Hanák and Horácek, 1986; Horácek et al., 2000).
    STATUS: IUCN 2003 and IUCN/SSC Action Plan (2001) – Lower Risk (lc).
    COMMENTS: Subgenus *Eptesicus*. Revised by Hanák and Gaisler (1971); see also Hanák and Horácek (1986) and Horácek et al. (2000). Placed in the subgenus *Amblyotus* by Horácek et al. (2000).

*Eptesicus bottae* (Peters, 1869). Monatsb. K. Preuss. Akad. Wiss. Berlin, 1869:406.
    COMMON NAME: Botta's Serotine.
    TYPE LOCALITY: Yemen.
    DISTRIBUTION: Rhodes (Greece), Turkey, Egypt, Yemen, Israel, Jordan, Iran, Iraq, Kazakhstan, Turkmenistan, Uzbekistan, Kyrgyzstan, Tajikistan, Afghanistan, east to Mongolia, NW China, and Pakistan.
    STATUS: IUCN 2003 and IUCN/SSC Action Plan (2001) – Lower Risk (lc).
    SYNONYMS: **anatolicus** Felten, 1971; **hingstoni** Thomas, 1919; **innesi** Lataste, 1887; **ognevi** Bobrinskii, 1918; **omanensis** Harrison, 1976; **taftanimontis** de Roguin, 1988.
    COMMENTS: Subgenus *Eptesicus*. Does not include *sodalis*; see Gaisler (1970). See also DeBlase (1971) for discussion of synonyms. Revised by Nader and Kock (1990). Reviewed in part by Bates and Harrison (1997).

*Eptesicus brasiliensis* (Desmarest, 1819). Nouv. Dict. Hist. Nat., Nouv. ed., 35:478.
    COMMON NAME: Brazilian Brown Bat.
    TYPE LOCALITY: Brazil, Goias (restricted by Cabrera, 1957).
    DISTRIBUTION: Veracruz (Mexico) south to N Argentina, Paraguay, and Uruguay; Trinidad and Tobago.
    STATUS: IUCN 2003 and IUCN/SSC Action Plan (2001) – Lower Risk (lc).
    SYNONYMS: *arctoideus* Wagner, 1855; *derasus* Burmeister, 1854; *ferrugineus* Temminck, 1839; *hilarii* I. Geoffroy 1824; *nitens* Wagner, 1855; **argentinus** Thomas, 1920; *arge* Cope, 1889; **melanopterus** Jentink, 1904; **thomasi** Davis, 1966.
    COMMENTS: Subgenus *Eptesicus*. Does not include *andinus*, *chiriquinus*, *inca*, or *montosus*; see W. B. Davis (1966) and Simmons and Voss (1998). W. B. Davis (1966) suggested that the holotype of *hilarii* may be referable to *fuscus*, but retained it in *brasiliensis* pending more comparisons. See Williams (1978c) for discussion of *hilarii* and *melanopterus*. Subspecies were delimited by W. B. Davis (1966), but additional specimens collected subsequently have made subspecies limits somewhat unclear.

*Eptesicus chiriquinus* Thomas, 1920. Ann. Mag. Nat. Hist., ser. 9, 5:362.
    COMMON NAME: Chiriquinan Serotine.
    TYPE LOCALITY: Panama, Chiriquí, Boquete, 4,000 ft. (1,333 m).

DISTRIBUTION: Costa Rica, Panama, Colombia, Ecuador, Peru, Venezuela, Guyana, French
      Guiana, Amazonian Brazil.
STATUS: IUCN 2003 – Not evaluated; not considered in IUCN/SSC Action Plan (2001).
SYNONYMS: *inca* Thomas, 1920.
COMMENTS: Subgenus *Eptesicus*. Distinct from *andinus* and *brasiliensis*; see Simmons and
      Voss (1998).

*Eptesicus diminutus* Osgood, 1915. Field Mus. Nat. Hist. Publ., Zool. Ser., 10:197.
COMMON NAME: Diminutive Serotine.
TYPE LOCALITY: Brazil, Bahia, Rio Preto, São Marcello.
DISTRIBUTION: Venezuela, E Brazil, Paraguay, Uruguay, N Argentina.
STATUS: IUCN 2003 and IUCN/SSC Action Plan (2001) – Lower Risk (lc).
SYNONYMS: *fidelis* Thomas, 1920.
COMMENTS: Subgenus *Eptesicus*. Includes *fidelis* but does not include *dorianus*; see Williams
      (1978c).

*Eptesicus dimissus* Thomas, 1916. J. Fed. Malay St. Mus., 7:1.
COMMON NAME: Surat Serotine.
TYPE LOCALITY: Thailand, Bandon, Kao Nawg, 3,500 ft. (1,166 m).
DISTRIBUTION: Peninsular Thailand, Nepal.
STATUS: IUCN 2003 and IUCN/SSC Action Plan (2001) – Vulnerable as *E. demissus* (sic).
COMMENTS: Subgenus *Eptesicus*. Reviewed by Myers et al. (2000). The correct spelling of this
      name is *dimissus*, not *demissus*; see Myers et al. (2000).

*Eptesicus floweri* (de Winton, 1901). Ann. Mag. Nat. Hist., ser. 7, 7:46.
COMMON NAME: Horn-skinned Serotine.
TYPE LOCALITY: Sudan, Khartoum, Wad Marium.
DISTRIBUTION: Sudan, Mali.
STATUS: IUCN 2003 and IUCN/SSC Action Plan (2001) – Lower Risk (nt).
SYNONYMS: *lowei* Thomas, 1915.
COMMENTS: Subgenus *Rhinopterus*. Includes *lowei*; see Braestrup (1935). Hayman and Hill
      (1971) listed *lowei* as a distinct species but expressed serious doubts about its validity,
      noting almost complete overlap with *floweri* in size and color.

*Eptesicus furinalis* (d'Orbigny, 1847). Voy. Am. Merid., Atlas Zool., 4:13.
COMMON NAME: Argentinian Brown Bat.
TYPE LOCALITY: Argentina, Corrientes.
DISTRIBUTION: N Argentina, Paraguay, Bolivia, Brazil, and the Guianas east to Peru and
      north to Jalisco and Tamaulipas (Mexico).
STATUS: IUCN 2003 and IUCN/SSC Action Plan (2001) – Lower Risk (lc).
SYNONYMS: *dorianus* Dobson, 1885; **carteri** Davis, 1965; **findleyi** Williams, 1978; **gaumeri**
      J. A. Allen, 1897; *chapmani* J. A. Allen, 1915.
COMMENTS: Subgenus *Eptesicus*. Reviewed by Williams (1978c). Apparently includes
      *dorianus*, but questions still remain about identity of the holotype; see Williams
      (1978c). Does not include *chiralensis* and *montosus*; see Simmons and Voss (1998).

*Eptesicus fuscus* (Beauvois, 1796). Cat. Raisonne Mus. Peale Phil., p. 18.
COMMON NAME: Big Brown Bat.
TYPE LOCALITY: USA, Pennsylvania, Philadelphia.
DISTRIBUTION: S Canada to Colombia and N Brazil; Greater Antilles; Bahamas; Dominica
      and Barbados (Lesser Antilles); Alaska.
STATUS: IUCN 2003 and IUCN/SSC Action Plan (2001) – Lower Risk (lc).
SYNONYMS: *arquatus* Say, 1823; *carolinensis* E. Geoffroy, 1806; *greenii* Gray, 1843 [*nomen
      nudum*]; *melanops* Rafinesque, 1820; *phaiops* Rafinesque, 1820; *ursinus* Temminck,
      1835-1841; **bahamensis** Miller, 1897; **bernardinus** Rhoads, 1902; *melanopterus* Rehn,
      1904 [not Jentink, 1904]; **dutertreus** P. Gervais, 1837; *cubensis* Gray, 1839; **hispaniolae**
      Miller, 1918; **lynni** Shamel, 1945; **miradorensis** H. Allen, 1866; *pelliceus* Thomas, 1920;
      **osceola** Rhoads, 1902; **pallidus** Young, 1908 [not Bobrinskii, 1929]; **peninsulae**
      Thomas, 1898; **petersoni** Silva Taboada, 1974; **wetmorei** Jackson, 1916.
COMMENTS: Subgenus *Eptesicus*. Very similar to *serotinus* with which it may be conspecific

according to Koopman (1993). Includes *lynni*; see Koopman (1989c). See Kurta and Baker (1990). Caribbean forms reviewed by Timm and Genoways (2003).

*Eptesicus gobiensis* Bobrinskii, 1926. Doklady Akad. Nauk SSSR A:96.
    COMMON NAME: Gobi Big Brown Bat.
    TYPE LOCALITY: Mongolia, Gobi Altai Mtns, Burchastei-tala.
    DISTRIBUTION: Iran, N Afghanistan, Kashmir, Pakistan, and Nepal, S Russia, Mongolia. Records from Tajikistan and W China including Tibet are uncertain (Horácek et al., 2000).
    STATUS: IUCN 2003 and IUCN/SSC Action Plan (2001) – Lower Risk (lc).
    SYNONYMS: *centrasiaticus* Bobrinskii, 1926; *kashgaricus* Bobrinskii, 1926.
    COMMENTS: Subgenus *Eptesicus*. Sometime considered conspecific with *nilssonii*, but see Strelkov (1986), Pavlinov and Rossolimo (1987), Yoshiyuki (1989), Corbet and Hill (1992), Bates and Harrison (1997), and Horácek et al. (2000). Placed in the subgenus *Amblyotus* by Horácek et al. (2000).

*Eptesicus guadeloupensis* Genoways and Baker, 1975. Occas. Pap. Mus. Texas Tech Univ., 34:1.
    COMMON NAME: Guadeloupean Big Brown Bat.
    TYPE LOCALITY: Guadeloupe (Lesser Antilles), Basse Terre, 2 km S and 2 km E Baiae-Mahault (France).
    DISTRIBUTION: Guadeloupe (Lesser Antilles).
    STATUS: IUCN 2003 and IUCN/SSC Action Plan (2001) – Endangered.
    COMMENTS: Subgenus *Eptesicus*. Probably closely related to *fuscus*.

*Eptesicus hottentotus* (A. Smith, 1833). S. Afr. J., 2:59.
    COMMON NAME: Long-tailed Serotine.
    TYPE LOCALITY: South Africa, Eastern Cape Prov., Uitenhage.
    DISTRIBUTION: South Africa to Angola and Kenya.
    STATUS: IUCN 2003 and IUCN/SSC Action Plan (2001) – Lower Risk (lc).
    SYNONYMS: *angusticeps* Shortridge and Carter, 1938; *megalurus* Temminck, 1840; *pallidior* Shortridge, 1942; *smithii* Wagner, 1855; *bensoni* Roberts, 1946; *portavernus* Schlitter and Aggundey, 1986.
    COMMENTS: Subgenus *Eptesicus*. Revised by Schlitter and Aggundey (1986).

*Eptesicus innoxius* (Gervais, 1841). In Vaillant, Voy. autour du monde . . . la Bonite, Zool. (Eydoux and Souleyet), 1:pl. 2.
    COMMON NAME: Harmless Serotine.
    TYPE LOCALITY: Peru, Piura, Amotape.
    DISTRIBUTION: NW Peru, W Ecuador, Puna Isl (Ecuador).
    STATUS: IUCN 2003 and IUCN/SSC Action Plan (2001) – Vulnerable.
    SYNONYMS: *espadae* Cabrera, 1901; *punicus* Thomas, 1920.
    COMMENTS: Subgenus *Eptesicus*. Reviewed by W. B. Davis (1966).

*Eptesicus japonensis* Imaizumi, 1953. Bull. Nat. Sci. Mus. Tokyo, 33:91.
    COMMON NAME: Japanese Short-tailed Bat.
    TYPE LOCALITY: Japan, C Honshû, Nagano Pref., Kita-Azumi-Gun, Hokujô-Mura (Shinden), 720 m.
    DISTRIBUTION: Honshû Isl (Japan).
    STATUS: IUCN 2003 – Not evaluated; not considered in IUCN/SSC Action Plan (2001).
    COMMENTS: Subgenus *Eptesicus*. Included in *nilssonii* by Corbet (1978c), but see Yoshiyuki (1989). See also Wallin (1969) and Rydell (1993).

*Eptesicus kobayashii* Mori, 1928. Zool. Mag. (Tokyo), 40:292.
    COMMON NAME: Kobayashi's Serotine.
    TYPE LOCALITY: Korea, Nando, Heian, Heijo.
    DISTRIBUTION: Korea.
    STATUS: IUCN 2003 and IUCN/SSC Action Plan (2001) – Data Deficient.
    COMMENTS: Subgenus *Eptesicus*. Status uncertain; see Corbet (1978c) and Horácek et al. (2000). Possibly a synonym of *bottae* (see Koopman, 1993, 1994) or *serotinus* (Horácek et al., 2000). Sometimes spelled *kobayashi*.

*Eptesicus matroka* (Thomas and Schwann, 1905). Proc. Zool. Soc. Lond., 1:258.
    COMMON NAME: Malagasy Serotine.
    TYPE LOCALITY: Madagascar, Ambositra, Betsileo, 1,100 m (20°31'S, 47°15'E).
    DISTRIBUTION: E Madagascar.
    STATUS: IUCN 2003 and IUCN/SSC Action Plan (2001) – Data Deficient.
    COMMENTS: Subgenus *Eptesicus*. Included in *Neoromicia capensis* by Hayman and Hill (1971)
        and Koopman (1993, 1994), but see Peterson et al. (1995).

*Eptesicus nasutus* (Dobson, 1877). J. Asiat. Soc. Bengal, 46 (2):311.
    COMMON NAME: Sind Bat.
    TYPE LOCALITY: Pakistan, Sind, Shikarpur, E of Rohri.
    DISTRIBUTION: Saudi Arabia, Oman, Yemen, Iraq, Iran, Afghanistan, Pakistan.
    STATUS: IUCN 2003 and IUCN/SSC Action Plan (2001) – Vulnerable.
    SYNONYMS: ***batinensis*** Harrison, 1968; ***matschiei*** Thomas, 1905; ***pellucens*** Thomas, 1906;
        *walli* Thomas, 1919.
    COMMENTS: Subgenus *Eptesicus*. Does not include *bobrinskoi*; see Harrison (1963) and Hanák
        and Gaisler (1971). Includes *walli*; see DeBlase (1980). Revised by Gaisler (1970) and
        DeBlase (1980); also see Harrison and Bates (1991) and Bates and Harrison (1997).
        Placed in the subgenus *Rhyneptesicus* by Horácek et al. (2000).

*Eptesicus nilssonii* (Keyserling and Blasius, 1839). Arch. Naturgesch., 5(1):315.
    COMMON NAME: Northern Bat.
    TYPE LOCALITY: Sweden.
    DISTRIBUTION: W and E Europe to E Siberia and NW China; north beyond Arctic Circle in
        Scandinavia, south to Bulgaria, Iraq, the Elburz Mtns (N Iran), The Pamirs and
        W China (not Tibet); Korea; Hokkaido (Japan); Sakhalin Isl (Russia).
    STATUS: IUCN 2003 and IUCN/SSC Action Plan (2001) – Lower Risk (lc).
    SYNONYMS: *atratus* Kolenati, 1858; *borealis* Nilsson 1838 [not Müller, 1776]; *kuhli* Nilsson
        1836 [not Kuhl, 1819]; *propinquus* Peters, 1872; ***parvus*** Kishida, 1932.
    COMMENTS: Subgenus *Eptesicus*. Includes *propinquus*; see W. B. Davis (1965). Revised by
        Wallin (1969). Does not include *japonensis*; see Yoshiyuki (1989). Does not include
        *gobiensis*; see Strelkov (1986), Pavlinov and Rossolimo (1987), and Corbet and Hill
        (1992). See Rydell (1993), but note that he included *japonensis* in this species. Closely
        related to *serotinus* and possibly paraphyletic with respect to that species; see Mayer
        and von Helversen (2001*a*). Specific epithet has often been spelled *nilssoni*, but the
        correct spelling is *nilssonii*. Placed in the subgenus *Amblyotus* by Horácek et al. (2000).

*Eptesicus pachyotis* (Dobson, 1871). Proc. Asiat. Soc. Bengal, p. 211.
    COMMON NAME: Thick-eared Bat.
    TYPE LOCALITY: India, Assam (= Meghalaya), Khasi Hills.
    DISTRIBUTION: Bangladesh, NE India, Tibet (China), N Burma, N Thailand.
    STATUS: IUCN 2003 and IUCN/SSC Action Plan (2001) – Lower Risk (nt).
    COMMENTS: Subgenus *Eptesicus*. Reviewed by Bates and Harrison (1997). Placed in the
        subgenus *Amblyotus* by Pavlinov et al. (1995*b*).

*Eptesicus platyops* (Thomas, 1901). Ann. Mag. Nat. Hist., ser. 7, 8:31.
    COMMON NAME: Lagos Serotine.
    TYPE LOCALITY: Nigeria, Western Region, Lagos.
    DISTRIBUTION: Nigeria, Senegal, Bioko (Equatorial Guinea).
    STATUS: IUCN 2003 IUCN/SSC Action Plan (2001) – Vulnerable.
    COMMENTS: Subgenus *Eptesicus*. Considered a subspecies of *serotinus* by Ibáñez and Valverde
        (1985), but no comparison with *bottae* was made. Also see Hayman and Hill (1971),
        who treated *platyops* as a distinct species based on morphological differences.

*Eptesicus serotinus* (Schreber, 1774). Die Säugethiere, 1:167.
    COMMON NAME: Common Serotine.
    TYPE LOCALITY: France.
    DISTRIBUTION: W Europe through Turkey and S Asiatic Russia to Himalayas, Thailand and
        China, north to Korea; Taiwan; S England; N Africa; most islands in Mediterranean.
        Koopman (1993) listed "perhaps Subsaharan Africa" under his account of the range of

this species, but there are no known records from that region (M. Happold, pers. comm.)

STATUS: IUCN 2003 and IUCN/SSC Action Plan (2001) – Lower Risk (lc).

SYNONYMS: *incisivus* Crespon, 1844; *insularis* Cabrera, 1904; *intermedius* Ognev, 1927; *mirza* de Filippi, 1865; *okenii* Brehm, 1827; *rufescens* Koch, 1865; *serotine* Müller, 1776; *sodalis* Barrett-Hamilton, 1910; *transsylvanus* Daday, 1885; *typus* Koch, 1865; *wiedii* Brehm, 1827; **andersoni** Dobson, 1871; **boscai** Cabrera, 1904; *meridionalis* Dal Piaz, 1926; **horikawai** Kishida, 1924; **isabellinus** Temminck, 1840; **pachyomus** Tomes, 1857; **pallens** Miller, 1911; *brachydigitatus* Mori, 1928; *pallidus* Bobrinskii, 1929 [not Young, 1908]; **pashtonus** Gaisler, 1970; **shirazensis** Dobson, 1871; **turcomanus** Eversmann, 1840; *albescens* Karelin, 1875 [*nomen nudum*]. **Unassigned**: *gabonensis* Trouessart, 1897 [see discussion in Hayman and Hill, 1971].

COMMENTS: Subgenus *Eptesicus*. Revised by Gaisler (1970), who noted that *shiraziensis* may be synonymous with *turcomanicus*. Includes *sodalis*; see Gaisler (1970) and Corbet (1978c). Includes *horikawai*; see Jones (1975). See additional comments under *fuscus* and *platyops*. Reviewed in part by Harrison and Bates (1991), Bates and Harrison (1997), Horácek et al. (2000), and Baagøe (2001c).

*Eptesicus tatei* Ellerman and Morrison-Scott, 1951. Checklist Palaearctic Indian Mammals, p. 158.

COMMON NAME: Sombre Bat.

TYPE LOCALITY: India, Darjeeling.

DISTRIBUTION: NE India.

STATUS: IUCN 2003 and IUCN/SSC Action Plan (2001) – Data Deficient.

SYNONYMS: *atratus* Blyth, 1863 [not Kolenati, 1858].

COMMENTS: Subgenus *Eptesicus*. Corbet and Hill (1992) noted that this species is known only from the holotype, but see Agrawal et al. (1992), who reported additional specimens.

*Hesperoptenus* Peters, 1868. Monatsb. K. Preuss. Akad. Wiss. Berlin, 1868:626.

TYPE SPECIES: *Vesperus* (*Hesperoptenus*) *doriae* Peters, 1868.

SYNONYMS: *Milithronycteris* Hill, 1976.

COMMENTS: Revised by Hill (1976); Indomalayan species reviewed by Corbet and Hill (1992) and Bates and Harrison (1997). Two subgenera are recognized, *Hesperoptenus* and *Milithronycteris*.

*Hesperoptenus blanfordi* (Dobson, 1877). J. Asiat. Soc. Bengal, 46:312.

COMMON NAME: Blanford's Bat.

TYPE LOCALITY: Burma, E of Moulmein, Tenasserim.

DISTRIBUTION: Burma, Thailand, Cambodia, Laos, Malay Peninsula, Borneo.

STATUS: IUCN 2003 and IUCN/SSC Action Plan (2001) – Lower Risk (lc).

COMMENTS: Subgenus *Milithronycteris*.

*Hesperoptenus doriae* (Peters, 1868). Monatsb. K. Preuss. Akad. Wiss. Berlin, 1868:626.

COMMON NAME: False Serotine Bat.

TYPE LOCALITY: Malaysia, Borneo, Sarawak.

DISTRIBUTION: Borneo, Malay Peninsula.

STATUS: IUCN 2003 and IUCN/SSC Action Plan (2001) – Endangered.

COMMENTS: Subgenus *Hesperoptenus*.

*Hesperoptenus gaskelli* Hill, 1983. Bull. Brit. Mus. (Nat. Hist.) Zool., 45:169.

COMMON NAME: Gaskell's False Serotine.

TYPE LOCALITY: Indonesia, Sulawesi, Central R. Ranu (01°51'S, 121°30'E).

DISTRIBUTION: Sulawesi (known only from the type locality).

STATUS: IUCN 2003 and IUCN/SSC Action Plan (2001) – Vulnerable.

COMMENTS: Subgenus *Milithronycteris*.

*Hesperoptenus tickelli* (Blyth, 1851). J. Asiat. Soc. Bengal, 20:157.

COMMON NAME: Tickell's Bat.

TYPE LOCALITY: India, Bihar, Chaibassa (restricted by J. Anderson, 1881).

DISTRIBUTION: India (including Andaman Isls), Sri Lanka, Nepal, Bhutan, Burma, Cambodia, Laos, Thailand, perhaps SW China.

STATUS: IUCN 2003 and IUCN/SSC Action Plan (2001) – Lower Risk (lc).
SYNONYMS: *isabellinus* Horsfield, 1851; *isabellinus* Kelaart, 1850 [*nomen nudum*].
COMMENTS: Subgenus *Milithronycteris*. Reviewed by Bates and Harrison (1997) and
    Hendrichsen et al. (2001*b*).

*Hesperoptenus tomesi* Thomas, 1905. Ann. Mag. Nat. Hist., ser. 7, 16:575.
COMMON NAME: Large False Serotine.
TYPE LOCALITY: Malaysia, Malacca.
DISTRIBUTION: Borneo, Malay Peninsula.
STATUS: IUCN 2003 and IUCN/SSC Action Plan (2001) – Lower Risk (lc).
COMMENTS: Subgenus *Milithronycteris*.

**Tribe Lasiurini** Tate, 1942. Bull. Amer. Mus. Nat. Hist., 80:290.
COMMENTS: Includes only *Lasiurus*.

*Lasiurus* Gray, 1831. Zool. Misc., 1:38.
TYPE SPECIES: *Vespertilio borealis* Müller, 1776.
SYNONYMS: *Atalapha* Peters, 1871 [not Rafinesque, 1814]; *Dasypterus* H. Allen, 1894; *Nycteris*
    Borkhausen, 1797 [not Cuvier and Geoffroy, 1795].
COMMENTS: Treated under the name *Nycteris* by Hall (1981). In Opinion 111 of the
    International Commission on Zoological Nomenclature (1929*b*), *Lasiurus* was adopted
    rather than *Nycteris*. *Atalapha* was used for this genus until the early 20th century, when
    application of the name to bats now included in *Lasiurus* was shown to date from Peters,
    not Rafinesque (Hall and Jones, 1961). Includes *Dasypterus*; see Hall and Jones (1961). Two
    subgenera are recognized, *Lasiurus* and *Dasypterus*. Keys to the genus were presented by
    Hall and Jones (1961) and Shump and Shump (1982*a*); see also Handley (1996). Species
    groups in the subgenus *Lasiurus* generally follow results of Morales and Bickham (1995).

*Lasiurus atratus* Handley 1996. Proc. Biol. Soc. Wash., 109:5.
COMMON NAME: Handley's Red Bat.
TYPE LOCALITY: Surinam, Zuid River, Kaiserberg Airport.
DISTRIBUTION: S and E Venezuela, Guyana, Surinam, French Guiana.
STATUS: IUCN 2003 and IUCN/SSC Action Plan (2001) – Data Deficient.
COMMENTS: Subgenus *Lasiurus*, *borealis* species group.

*Lasiurus blossevillii* (Lesson and Garnot, 1826). Ferussac's Bull. Sci. Nat. Geol., 8:95.
COMMON NAME: Red Bat (known as the Western Red Bat in North America).
TYPE LOCALITY: Uruguay, Montevideo.
DISTRIBUTION: Bolivia, N Argentina, Uruguay, and Brazil to W North America (but not
    E North America); Trinidad and Tobago; Galapagos (Ecuador).
STATUS: IUCN 2003 and IUCN/SSC Action Plan (2001) – Lower Risk (lc).
SYNONYMS: *bonariensis* Lesson, 1826; *enslenii* Lima, 1926; **brachyotis** J. A. Allen, 1882;
    **frantzii** Peters, 1871; **teliotis** H. Allen, 1891; *ornatus* Hall, 1951.
COMMENTS: Subgenus *Lasiurus*, *borealis* species group. Included in *borealis* by Koopman
    (1993, 1994) but see Schmidly and Hendricks (1984), Baker et al. (1988*a*), and Morales
    and Bickham (1995). Does not include *degelidus* (Baker et al., 1988*a*) but might include
    *minor*. Does not include *pfeifferi*; see Morales and Bickham (1995). Includes *brachyotis*;
    see Niethammer (1964) and McCracken et al. (1997). Does not include *varius*; see
    Barquez (1987), Barquez et al. (1993), and Mares et al. (1995). Does not include *salinae*,
    see Mares et al. (1995) and Tiranti and Torres (1998), but also see Barquez and Diaz
    (2001).

*Lasiurus borealis* (Müller, 1776). Linné's Vollstand. Natursystem, Suppl., p. 20.
COMMON NAME: Eastern Red Bat.
TYPE LOCALITY: USA, New York.
DISTRIBUTION: E North America, Bermuda.
STATUS: IUCN 2003 and IUCN/SSC Action Plan (2001) – Lower Risk (lc).
SYNONYMS: *funebris* Fitzinger, 1870; *lasiurus* Schreber, 1781; *monachus* Rafinesque, 1818;

*noveboracensis* Erxleben, 1777; *quebecensis* Yourans, 1930; *rubellus* Palisot de Beauvois, 1796; *rubra* Ord, 1815; *rufus* Wardern, 1820; *tesselatus* Rafinesque, 1818.

COMMENTS: Subgenus *Lasiurus*, *borealis* species group. Does not include *blossevillii*, *frantzii*, *teliotis*, and *varius*; see Schmidly and Hendricks (1984), Baker et al. (1988*a*), and Morales and Bickham (1995). Does not include *degelidus* (Baker et al., 1988*a*) but might include *minor*. Does not include *pfeifferi*; see Morales and Bickham (1995). See Shump and Shump (1982*a*) but note that they included *blossevillii* and its synonyms in *borealis*.

*Lasiurus castaneus* Handley, 1960. Proc. U.S. Natl. Mus., 112:468.
COMMON NAME: Tacarcunan Bat.
TYPE LOCALITY: Panama, Darien, Río Pucro, Tacarcuna Village, 3,200 ft. (1,066 m).
DISTRIBUTION: Panama, Costa Rica. A record from French Guiana was subsequently reidentified as *atratus* (Handley, 1996).
STATUS: IUCN 2003 and IUCN/SSC Action Plan (2001) – Vulnerable.
COMMENTS: Subgenus *Lasiurus*, *borealis* species group.

*Lasiurus cinereus* (Palisot de Beauvois, 1796). Cat. Raisonne Mus. Peale Phil., p. 18.
COMMON NAME: Hoary Bat.
TYPE LOCALITY: USA, Pennsylvania, Philadelphia.
DISTRIBUTION: Colombia and Venezuela to C Chile, Bolivia, Uruguay, and C Argentina; Hawaii (USA); Guatemala and Mexico throughout the USA to S British Columbia, SE Mackenzie, Hudson Bay and S Quebec (Canada); Galapagos Isls (Ecuador); Bermuda; accidental on Cuba, Hispaniola, Iceland, and the Orkney Isls (Scotland).
STATUS: U.S. ESA – Endangered as *L. c. semotus*. IUCN/SSC Action Plan (2001) – Not Evaluated as *L. c. semotus*; otherwise Lower Risk (lc).
SYNONYMS: *mexicana* Saussure, 1861; *pruinosus* Say, 1823; **semotus** H. Allen, 1890; **villosissimus** E. Geoffroy, 1806; *brasiliensis* Pira, 1905; *grayi* Tomes, 1857; *pallescens* Peters, 1871. **Unassigned**: *fossilis* Hibbard, 1950 [fossil].
COMMENTS: Subgenus *Lasiurus*, *cinereus* species group. Includes *villosissimus* and *semotus*; see Sanborn and Crespo (1957) and Morales and Bickham (1995). See Shump and Shump (1982*b*).

*Lasiurus degelidus* Miller, 1931. J. Mammal., 12:410.
COMMON NAME: Jamaican Red Bat.
TYPE LOCALITY: Jamaica, District of Vere, Sutton's.
DISTRIBUTION: Jamaica.
STATUS: IUCN 2003 – Not evaluated; not considered in IUCN/SSC Action Plan (2001).
COMMENTS: Subgenus *Lasiurus*, *borealis* species group. Closely related to *seminolis*, but apparently distinct; see Baker et al. (1988*a*).

*Lasiurus ebenus* Fazzolari-Corrêa, 1994. Mammalia, 58:119.
COMMON NAME: Blackish Red Bat.
TYPE LOCALITY: Brazil, São Paulo, Parque Estadual da Ilha do Cardoso, 25°05′S, 47°59′W.
DISTRIBUTION: SE Brazil.
STATUS: IUCN 2003 and IUCN/SSC Action Plan (2001) – Vulnerable.
COMMENTS: Subgenus *Lasiurus*, *borealis* species group. Known only from the holotype.

*Lasiurus ega* (Gervais, 1856). *In* F. Comte de Castelnau, Exped. Partes Cen. Am. Sud. Zool. (Sec. 7), Vol. 1, pt. 2 (Mammifères):73.
COMMON NAME: Southern Yellow Bat.
TYPE LOCALITY: Brazil, Amazonas, Ega.
DISTRIBUTION: S Texas, E and S Mexico south to Bolivia, Argentina, Paraguay, Uruguay, and Brazil; Trinidad.
STATUS: IUCN 2003 and IUCN/SSC Action Plan (2001) – Lower Risk (lc).
SYNONYMS: **argentinus** Thomas, 1901; **caudatus** Tomes, 1857; **fuscatus** Thomas, 1901; *punensis* J. A. Allen, 1914; **panamensis** Thomas, 1901.
COMMENTS: Subgenus *Dasypterus*. Does not include *xanthinus*, see Baker et al. (1988*a*) and Morales and Bickham (1995). For discussion of the ranges of *ega* and *xanthinus* see Baker and Patton (1967), Baker et al. (1971, 1988*a*), and Bickham (1987).

*Lasiurus egregius* (Peters, 1870). Monatsb. K. Preuss. Akad. Wiss. Berlin, 1870:275.
COMMON NAME: Big Red Bat.
TYPE LOCALITY: Brazil, Santa Catarina.
DISTRIBUTION: Brazil, French Guiana, Panama.
STATUS: IUCN 2003 and IUCN/SSC Action Plan (2001) – Lower Risk (nt).
COMMENTS: Subgenus *Lasiurus*, *borealis* species group.

*Lasiurus insularis* Hall and Jones, 1961. Univ. Kansas Mus. Nat. Hist. Publ., 14:85.
COMMON NAME: Cuban Yellow Bat.
TYPE LOCALITY: Cuba, Las Villas Province, Cienfuegos.
DISTRIBUTION: Cuba.
STATUS: IUCN 2003 – Not evaluated; not considered in IUCN/SSC Action Plan (2001).
COMMENTS: Subgenus *Dasypterus*. Named as a subspecies of *intermedius*, but clearly distinct; see Silva-Taboada (1976) and Morales and Bickham (1995).

*Lasiurus intermedius* H. Allen, 1862. Proc. Acad. Nat. Sci. Phil., 14:246.
COMMON NAME: Northern Yellow Bat.
TYPE LOCALITY: Mexico, Tamaulipas, Matamoros.
DISTRIBUTION: Honduras to Sinaloa (Mexico) and through Texas to Florida and New Jersey (USA); Cuba.
STATUS: IUCN 2003 and IUCN/SSC Action Plan (2001) – Lower Risk (lc).
SYNONYMS: *floridanus* Miller, 1902.
COMMENTS: Subgenus *Dasypterus*. Includes *floridanus*; see Hall and Jones (1961) and Morales and Bickham (1995). Does not include *insularis*; see Silva-Taboada (1976) and Morales and Bickham (1995). See Webster et al. (1980), but note that their account included *insularis*.

*Lasiurus minor* Miller, 1931. J. Mammal., 12:410.
COMMON NAME: Minor Red Bat.
TYPE LOCALITY: Haiti, Voûte l'Église, a cave near Jacmel road a few km N Trouin, 1,350 ft. (450 m).
DISTRIBUTION: Bahamas, Hispaniola, Puerto Rico.
STATUS: IUCN 2003 – Not evaluated; not considered in IUCN/SSC Action Plan (2001).
COMMENTS: Subgenus *Lasiurus*, *borealis* species group. Possibly conspecific with *seminolis*, *borealis*, or *blossevillii*.

*Lasiurus pfeifferi* (Gundlach, 1861). Monatsb. K. Preuss. Akad. Wiss., Berlin, 1861:152.
COMMON NAME: Pfeiffer's Red Bat.
TYPE LOCALITY: Cuba, Trinidad.
DISTRIBUTION: Cuba.
STATUS: IUCN 2003 – Not evaluated; not considered in IUCN/SSC Action Plan (2001).
COMMENTS: Subgenus *Lasiurus*, *borealis* species group. May represent a subspecies of *seminolus*; see Morales and Bickham (1995).

*Lasiurus salinae* Thomas, 1902. Ann. Mag. Nat. Hist., ser. 7, 9:238.
COMMON NAME: Saline Red Bat.
TYPE LOCALITY: Argentina, Córdoba Province, Cruz del Eje.
DISTRIBUTION: Argentina.
STATUS: IUCN 2003 – Not evaluated; not considered in IUCN/SSC Action Plan (2001).
COMMENTS: Subgenus *Lasiurus*, *borealis* species group. The status of this form is unclear. Formerly considered a subspecies or synonym of *borealis* or *blossevillii*, but apparently distinct; see Mares et al. (1995) and Tiranti and Torres (1998), but also see Barquez and Diaz (2001).

*Lasiurus seminolus* (Rhoads, 1895). Proc. Acad. Nat. Sci. Phil., 47:32.
COMMON NAME: Seminole Bat.
TYPE LOCALITY: USA, Florida, Pinellas Co., Tarpon Springs.
DISTRIBUTION: Florida and Texas to Oklahoma and Virginia; Pennsylvania and New York (USA); Bermuda. N Veracruz (Mexico) record unverified.
STATUS: IUCN 2003 and IUCN/SSC Action Plan (2001) – Lower Risk (lc).

SYNONYMS: *peninsularis* Coues, 1896.

COMMENTS: Subgenus *Lasiurus*, *borealis* species group. Formerly included in *borealis*, but see Hall (1981), Baker et al. (1988*a*), and Morales and Bickham (1995). May include *pfeifferi*, see Morales and Bickham (1995). See Wilkins (1987*a*).

*Lasiurus varius* Poeppig, 1835. Reis. Chilie, Peru, und Amaz., 1:451.

COMMON NAME: Cinnamon Red Bat.

TYPE LOCALITY: Chile, Antuco.

DISTRIBUTION: S Argentina, Chile.

STATUS: IUCN 2003 – Not evaluated; not considered in IUCN/SSC Action Plan (2001).

SYNONYMS: *poeppigii* Lesson, 1836.

COMMENTS: Subgenus *Lasiurus*, *borealis* species group. Often listed as synonym of *borealis* or *blossevillii*, but apparently distinct; see Barquez (1987), Barquez et al. (1993), and Mares et al. (1995).

*Lasiurus xanthinus* Thomas, 1897. Ann. Mag. Nat. Hist. ser. 6, 20:544.

COMMON NAME: Western Yellow Bat.

TYPE LOCALITY: Mexico, Baja California, Sierra Laguna.

DISTRIBUTION: S California, Arizona, and New Mexico south to Baja California, W and C Mexico.

STATUS: IUCN 2003 and IUCN/SSC Action Plan (2001) – Lower Risk (lc).

COMMENTS: Subgenus *Dasypterus*. Often considered a subspecies of *ega*, but see Baker et al. (1988*a*) and Morales and Bickham (1995). For discussion of the ranges of *ega* and *xanthinus* see Baker and Patton (1967), Baker et al. (1971, 1988*a*), and Bickham (1987).

**Tribe Nycticeiini** Gervais, 1855. *In* F. Comte de Castelnau, Exped. Partes Cen. Am. Sud., Zool. (Sec. 7), Vol. 1, pt. 2 (Mammifères), p. 71.

COMMENTS: May not be monophyletic; see Hoofer and Van Den Bussche (2001).

*Nycticeinops* Hill and Harrison, 1987. Bull. Br. Mus. Nat. Hist. 52:254.

TYPE SPECIES: *Nycticeius schlieffeni* Peters, 1859.

COMMENTS: Previously included in *Nycticeius*, but see Hill and Harrison (1987) and Hoofer and Van Den Bussche (2001).

*Nycticeinops schlieffeni* (Peters, 1859). Monatsb. K. Preuss. Akad. Wiss. Berlin, 1859:223.

COMMON NAME: Schlieffen's Twilight Bat.

TYPE LOCALITY: Egypt, Cairo.

DISTRIBUTION: Saudi Arabia, Yemen, and Egypt to Djibouti, Somalia, Mozambique, Mali, Botswana, South Africa, and Namibia; Mauritania and Ghana to Sudan and Tanzania.

STATUS: IUCN 2003 and IUCN/SSC Action Plan (2001) – Lower Risk (lc) as *Nycticeius schlieffeni*.

SYNONYMS: *adovanus* Heuglin, 1877; *africanus* Allen, 1911; *albiventer* Thomas and Wroughton, 1908; *australis* Thomas and Wroughton, 1908; *bedouin* Thomas and Wroughton, 1908; *cinnamomeus* Wettstein, 1916; *fitzsimonsi* Roberts, 1932; *minimus* Noack, 1887.

COMMENTS: Includes *cinnamomeus*; see Koopman (1975). Several poorly defined subspecies are often recognized, but there seems little justification for separation of these taxa. Reviewed in part by Harrison and Bates (1991); see Taylor (2000*a*) for distribution map.

*Nycticeius* Rafinesque, 1819. J. Phys. Chim. Hist. Nat. Arts Paris, 88:417.

TYPE SPECIES: *Vespertilio humeralis* Rafinesque, 1818.

SYNONYMS: *Nycticea* Le Conte, 1831; *Nycticejus* Temminck, 1827; *Nycticeus* Lesson, 1827; *Nycticeyx* Wagler, 1830.

COMMENTS: Does not include *Scotoecus*, see J. E. Hill (1974*c*). Does not include *Scotorepens* or *Scoteanax*; see Kitchener and Caputi (1985) and Volleth and Tidemann (1991). Does not includes *Nycticeinops*; see Hill and Harrison (1987) and Hoofer and Van Den Bussche (2001).

*Nycticeius aenobarbus* Temminck, 1840. Monographies de Mammalogie, 2:247.

COMMON NAME: Temminck's Mysterious Bat.

TYPE LOCALITY: "Amérique méridionale."

DISTRIBUTION: Unknown; Carter and Dolan (1978) have suggested that the type and only known specimen is probably not from South America.

STATUS: IUCN 2003 – Not evaluated; not considered in IUCN/SSC Action Plan (2001).

COMMENTS: Listed as a synonym of *Myotis albescens* by many authors following Miller and Allen (1928), but clearly distinct at both the genus and species level; see Husson (1962) and Carter and Dolan (1978). The latter authors suggested that this species probably belongs with *Nycticeius*, but its status remains unclear. If the holotype originated in the Old World, this taxon might be referable to *Scotoecus*, *Scotorepens*, or *Scoteanax*.

*Nycticeius cubanus* Gundlach, 1861. Monatsb. K. Preuss. Akad. Wiss. Berlin, 1861:150.

COMMON NAME: Cuban Evening Bat.

TYPE LOCALITY: Cuba, Matanzas, near Cárdenas.

DISTRIBUTION: Cuba.

STATUS: IUCN 2003 – Not evaluated; not considered in IUCN/SSC Action Plan (2001).

COMMENTS: Apparently distinct from *humeralis*; see Hall (1981), but also see Varona (1974).

*Nycticeius humeralis* (Rafinesque, 1818). Am. Mon. Mag., 3(6):445.

COMMON NAME: Evening Bat.

TYPE LOCALITY: USA, Kentucky.

DISTRIBUTION: N Veracruz (Mexico) to Nebraska, the Great Lakes, and Pennsylvania, south to Florida and the Gulf coast (USA).

STATUS: IUCN 2003 and IUCN/SSC Action Plan (2001) – Lower Risk (lc).

SYNONYMS: *creeks* F. Cuvier, 1832; *crepuscularis* Le Conte, 1831; **mexicanus** Davis, 1944; **subtropicalis** Schwartz, 1951.

COMMENTS: Does not include *cubanus*; see Hall (1981), but also see Varona (1974). See Watkins (1972).

*Rhogeessa* H. Allen, 1866. Proc. Acad. Nat. Sci. Phil., 18:285.

TYPE SPECIES: *Rhogeessa tumida* H. Allen, 1866.

SYNONYMS: *Baeodon* Miller, 1906.

COMMENTS: Includes *Baeodon*, here recognized with *Rhogeessa* as a subgenus; see Jones et al. (1977). Revised by LaVal (1973b) and Genoways and Baker (1996). For a partial phylogeny see Hoofer and Van Den Bussche (2001); also see Baker et al. (1985).

*Rhogeessa aeneus* Goodwin, 1958. Amer. Mus. Novit., 1923:6

COMMON NAME: Yucatan Yellow Bat.

TYPE LOCALITY: Mexico, Yucatán, Chichén-Itzâ, 10 m.

DISTRIBUTION: Yucatán (Mexico).

STATUS: IUCN 2003 – Not evaluated; not considered in IUCN/SSC Action Plan (2001).

COMMENTS: Subgenus *Rhogeessa*. Often included in *tumida* but see Audet et al. (1993) and Genoways and Baker (1996).

*Rhogeessa alleni* Thomas, 1892. Ann. Mag. Nat. Hist., ser. 6, 10:477.

COMMON NAME: Allen's Yellow Bat.

TYPE LOCALITY: Mexico, Jalisco, near Autlan, Santa Rosalia.

DISTRIBUTION: Oaxaca to Zacatecas (Mexico).

STATUS: IUCN 2003 – Endangered. IUCN/SSC Action Plan (2001) – Lower Risk (nt).

COMMENTS: Subgenus *Baeodon*.

*Rhogeessa genowaysi* Baker, 1984. Syst. Zool., 33:178.

COMMON NAME: Genoways's Yellow Bat.

TYPE LOCALITY: Mexico, Chiapas, 23.6 mi. (42 km) NW Huixtla.

DISTRIBUTION: Pacific lowlands of S Chiapas (Mexico).

STATUS: IUCN 2003 and IUCN/SSC Action Plan (2001) – Vulnerable.

COMMENTS: Subgenus *Rhogeessa*. Apparently morphologically inseparable from *tumida*, but with distinctive karyotype, see Baker (1984). Also see Roots and Baker (1998).

*Rhogeessa gracilis* Miller, 1897. N. Am. Fauna, 13:126.
COMMON NAME: Slender Yellow Bat.
TYPE LOCALITY: Mexico, Puebla, Piaxtla, 1,100 m.
DISTRIBUTION: Jalisco and Zacatecas to Oaxaca (Mexico).
STATUS: IUCN 2003 and IUCN/SSC Action Plan (2001) – Lower Risk (nt).
COMMENTS: Subgenus *Rhogeessa*. See J. K. Jones (1977).

*Rhogeessa hussoni* Genoways and Baker, 1996. Cont. Mammal.: a Memorial Vol. Honoring
    Dr. J. K. Jones, Mus. Texas Tech Univ., p. 85.
COMMON NAME: Husson's Yellow Bat.
TYPE LOCALITY: Surinam, Nickerie Dist., Sipaliwini Airstrip.
DISTRIBUTION: S Surinam, E Brazil.
STATUS: IUCN 2003 – not evaluated; not considered in IUCN/SSC Action Plan (2001).
COMMENTS: Subgenus *Rhogeessa*.

*Rhogeessa io* Thomas, 1903. Ann Mag. Nat. Hist., ser. 7, 11:382.
COMMON NAME: Thomas's Yellow Bat.
TYPE LOCALITY: Venezuela, Carabobo, Valencia.
DISTRIBUTION: C and S Nicaragua south to N Colombia and W Ecuador; Venezuela; Trinidad
    and Tobago; Guyana; N and C Brazil; N Bolivia.
STATUS: IUCN 2003 – Not evaluated; not considered in IUCN/SSC Action Plan (2001).
SYNONYMS: *bombyx* Thomas, 1913; *riparia* Goodwin, 1958; *velilla* Thomas, 1903.
COMMENTS: Subgenus *Rhogeessa*. Formerly included in *tumida* (e.g., Hall, 1981; Koopman,
    1993, 1994) but see Genoways and Baker (1996).

*Rhogeessa minutilla* Miller, 1897. Proc. Biol. Soc. Wash., 11:139.
COMMON NAME: Tiny Yellow Bat.
TYPE LOCALITY: Venezuela, Margarita Isl.
DISTRIBUTION: NE Colombia, coastal Venezuela (including Margarita Isl).
STATUS: IUCN 2003 and IUCN/SSC Action Plan (2001) – Lower Risk (nt).
COMMENTS: Subgenus *Rhogeessa*. Listed as a subspecies of *parvula* by Cabrera (1958), but see
    LaVal (1973b) and Genoways and Baker (1996).

*Rhogeessa mira* LaVal, 1973. Occas. Pap. Mus. Nat. Hist. Univ. Kansas, 19:26.
COMMON NAME: Least Yellow Bat.
TYPE LOCALITY: Mexico, Michoacan, 20 km N El Infernillo.
DISTRIBUTION: S Michoacan (Mexico).
STATUS: IUCN 2003 and IUCN/SSC Action Plan (2001) – Endangered.
COMMENTS: Subgenus *Rhogeessa*.

*Rhogeessa parvula* H. Allen, 1866. Proc. Acad. Nat. Sci. Phil., 18:285.
COMMON NAME: Little Yellow Bat.
TYPE LOCALITY: Mexico, Nayarit, Trés Marías Isls.
DISTRIBUTION: Oaxaca to Sonora (Mexico); Trés Marías Isls (Mexico).
STATUS: IUCN 2003 and IUCN/SSC Action Plan (2001) – Lower Risk (nt).
SYNONYMS: *major* Goodwin, 1958.
COMMENTS: Subgenus *Rhogeessa*. For scope of this species, see LaVal (1973b) and Genoways
    and Baker (1996).

*Rhogeessa tumida* H. Allen, 1866. Proc. Acad. Nat. Sci. Phil., 18:286.
COMMON NAME: Black-winged Little Yellow Bat.
TYPE LOCALITY: Mexico, Veracruz, Mirador.
DISTRIBUTION: Tamaulipas (Mexico) to N Nicaragua and NW Costa Rica.
STATUS: IUCN 2003 and IUCN/SSC Action Plan (2001) – Lower Risk (lc).
COMMENTS: Subgenus *Rhogeessa*. Listed as a subspecies of *parvula* by Hall and Kelson (1959),
    but see LaVal (1973b) and Hall (1981). Does not include *aeneus*; see Audet et al. (1993)
    and Genoways and Baker (1996). Does not include *io*; see Genoways and Baker (1996).
    See Vonhof (2000).

*Scoteanax* Troughton, 1943. Furred Animals of Australia, 1st ed., Sydney: Angus and Robertson, p. 353.
  TYPE SPECIES: *Oligotomus australis* Iredale (ex MacGillivray), 1937 (= *Nyticejus reuppellii* Peters, 1866).
  SYNONYMS: *Oligotomus* Iredale (ex MacGillivray), 1937 [preoccupied by *Oligotomus* Cope, 1843].
  COMMENTS: Often included in *Nycticeius*, but see Kitchener and Caputi (1985).

*Scoteanax rueppellii* (Peters, 1866). Monatsb. K. Preuss. Akad. Wiss. Berlin, 1866:21.
  COMMON NAME: Rüppell's Broad-nosed Bat.
  TYPE LOCALITY: Australia, New South Wales, Sydney.
  DISTRIBUTION: E Queensland and E New South Wales (Australia).
  STATUS: IUCN 2003 and IUCN/SSC Action Plan (2001) – Lower Risk (nt) as *Nycticeius ruppellii*.
  SYNONYMS: *australis* Iredale (ex MacGillivray), 1937.
  COMMENTS: Reviewed by Kitchener and Caputi (1985).

*Scotoecus* Thomas, 1901. Ann. Mag. Nat. Hist., ser. 7, 7:263.
  TYPE SPECIES: *Scotophilus albofuscus* Thomas, 1890.
  COMMENTS: Considered a subgenus of *Nycticeius* by Hayman and Hill (1971), but see J. E. Hill (1974c), who revised the genus.

*Scotoecus albigula* Thomas, 1909. Ann. Mag. Nat. Hist., ser. 8, 4:544.
  COMMON NAME: White-throated Lesser House Bat.
  TYPE LOCALITY: Kenya, Mount Elgon.
  DISTRIBUTION: Angola, Zambia, Mozambique, Uganda, Kenya, Malawi, Somalia.
  STATUS: IUCN 2003 – Not evaluated; not considered in IUCN/SSC Action Plan (2001).
  COMMENTS: Formerly included in *hirundo*, but apparently distinct; but see Happold et al. (1987), Happold and Happold (1989), Taylor and Van der Merwe (1998), and Cotterill (2001d).

*Scotoecus albofuscus* (Thomas, 1890). Ann. Mus. Civ. Stor. Nat. Genova, 29:84.
  COMMON NAME: Light-winged Lesser House Bat.
  TYPE LOCALITY: Gambia, Bathurst.
  DISTRIBUTION: Senegal and Gambia to Kenya, Tanzania, Mozambique, Malawi, KwaZulu-Natal, South Africa.
  STATUS: IUCN 2003 and IUCN/SSC Action Plan (2001) – Lower Risk (nt).
  SYNONYMS: **woodi** Thomas, 1917.
  COMMENTS: See discussion in Kearney and Taylor (1997).

*Scotoecus hindei* Thomas, 1901. Ann. Mag. Nat. Hist., ser. 7, 7:264.
  COMMON NAME: Hinde's Lesser House Bat.
  TYPE LOCALITY: Kenya, Kitui.
  DISTRIBUTION: Nigeria and Cameroon to S Sudan and Somalia; south to SE Dem. Rep. Congo, Kenya, Tanzania, Zambia, Mozambique, Malawi.
  STATUS: IUCN 2003 – Not evaluated; not considered in IUCN/SSC Action Plan (2001).
  SYNONYMS: **falabae** Thomas, 1915.
  COMMENTS: Formerly included in *hirundo*, but apparently distinct; see Happold et al. (1987), Happold and Happold (1989), Taylor and Van Der Merwe (1998), and Cotterill (2001d).

*Scotoecus hirundo* (de Winton, 1899). Ann. Mag. Nat. Hist., ser. 7, 4:355.
  COMMON NAME: Dark-winged Lesser House Bat.
  TYPE LOCALITY: Ghana, Gambaga.
  DISTRIBUTION: Senegal to Ethiopia.
  STATUS: IUCN 2003 and IUCN/SSC Action Plan (2001) – Lower Risk (lc).
  SYNONYMS: *artinii* de Winton, 1899.
  COMMENTS: Does not include *albigula* or *hindei*; see Happold et al. (1987), Happold and Happold (1989), and Taylor and Van der Merwe (1998); also see Happold and Happold (1997b).

*Scotoecus pallidus* (Dobson, 1876). Monogr. Asiat. Chiroptera, App. D:186.
COMMON NAME: Desert Yellow Lesser House Bat.
TYPE LOCALITY: Pakistan, Punjab, Lahore, Mian Mir.
DISTRIBUTION: Pakistan, N India.
STATUS: IUCN 2003 and IUCN/SSC Action Plan (2001) – Lower Risk (lc).
SYNONYMS: *noctulinus* I. Geoffroy, 1831 [see discussion in J. E. Hill (1974c)].
COMMENTS: Included in *Nycticeius* by Ellerman and Morrison-Scott (1951); but see J. E. Hill
(1974c). Reviewed by Bates and Harrison (1997). *S. noctulinus* may be an earlier name
for this species.

*Scotomanes* Dobson, 1875. Proc. Zool. Soc. Lond., 1875:371.
TYPE SPECIES: *Nycticejus ornatus* Blyth, 1851.
SYNONYMS: *Scoteinus* Dobson, 1875.
COMMENTS: Includes *Scoteinus*; see Sinha and Chakraborty (1971).

*Scotomanes ornatus* (Blyth, 1851). J. Asiat. Soc. Bengal, 20:511.
COMMON NAME: Harlequin Bat.
TYPE LOCALITY: India, Assam, Khasi Hills, Cherrapunji.
DISTRIBUTION: NE India (including Sikkim), Burma, S China, Thailand, Vietnam.
STATUS: IUCN 2003 and IUCN/SSC Action Plan (2001) – Lower Risk (nt) as *S. ornatus*; Data
Deficient as *S. emarginatus*.
SYNONYMS: *nivicolus* Hodgson, 1855; **imbrensis** Thomas, 1921; **sinensis** Thomas, 1921.
**Unassigned**: *emarginatus* Dobson, 1871 [locality unknown, although thought to be
from some part of India].
COMMENTS: Includes *emarginatus*; see Corbet and Hill (1992). Included in *Nycticeius* by
Ellerman and Morrison-Scott (1951); but see J. E. Hill (1974c). Reviewed by Bates and
Harrison (1997); also see Sinha (1999) and Hendrichsen et al. (2001b).

*Scotophilus* Leach, 1821. Trans. Linn. Soc. Lond., 13:69, 71.
TYPE SPECIES: *Scotophilus kuhlii* Leach, 1821.
SYNONYMS: *Pachyotus* Gray, 1831.
COMMENTS: Includes *Pachyotus*; see Walker et al. (1975). African species revised by Robbins et al.
(1985); also see Peterson et al. (1995).

*Scotophilus borbonicus* (E. Geoffroy, 1803). Cat. Mamm. Mus. Nat. d'Hist. Nat., p. 46.
COMMON NAME: Réunion House Bat.
TYPE LOCALITY: Réunion Isl (France).
DISTRIBUTION: Réunion Isl (Mascarene Isls). Records from Mauritius (Mascarene Isls) are
erroneous, see Cheke and Dahl (1981). Reports from Madagascar have not been
confirmed; see Peterson et al. (1995).
STATUS: IUCN 2003 and IUCN/SSC Action Plan (2001) – Critically Endangered. May be
extinct; see Cheke and Dahl (1981).
COMMENTS: Hill (1980b) considered African *viridis* and *damarensis*, and possibly *leucogaster*,
to be conspecific with *borbonicus*, and Koopman (1986) included *viridis, damarensis*,
and *nigritellus* (but not *leucogaster*) in this species. However, in a comprehensive
revision of the African forms C. B. Robbins et al. (1985) rejected any affinity of
*borbonicus* sensu stricto with African mainland species. I therefore restrict usage of the
name *borbonicus* to the Réunion Isl form, and follow C. B. Robbins et al. (1985) in
using *leucogaster* and *dinganii* for the smaller mainland species.

*Scotophilus celebensis* Sody, 1928. Natuurk. Tijdschr. Ned.-Ind., 88:90.
COMMON NAME: Sulawesi Yellow House Bat.
TYPE LOCALITY: Indonesia, N Sulawesi, Toli Toli.
DISTRIBUTION: Sulawesi (Indonesia).
STATUS: IUCN 2003 and IUCN/SSC Action Plan (2001) – Data Deficient.
COMMENTS: May represent a subspecies of *heathii*, see Tate (1942a) and Sinha (1980).

*Scotophilus collinus* Sody, 1936. Natuurk. Tijdschr. Ned.-Ind., 96:48.
COMMON NAME: Sody's Yellow House Bat.

TYPE LOCALITY: Indonesia, Bali, SW Bali, Djembrana, ca. 50 m.
DISTRIBUTION: Sabah, W Java, Bali, Lombok, Flores, Lembata, Timor, Semanu, and Roti Isls
    (Indonesia); probably also Sumba, Sawu, and Banda Isls (Indonesia).
STATUS: IUCN 2003 – Not evaluated; not considered in IUCN/SSC Action Plan (2001).
COMMENTS: Formerly included in *kuhlii*, but see Kitchener et al. (1997*b*). Kitchener et al.
    (1997*b*) recognized eastern and western forms of *collinus*, but did not name them as
    subspecies.

*Scotophilus dinganii* (A. Smith, 1833). S. Afr. Quart. J., 2:59.
COMMON NAME: Yellow-bellied House Bat.
TYPE LOCALITY: South Africa, KwaZulu-Natal Prov., Port Natal (= Durban).
DISTRIBUTION: Senegal, Guinea-Bissau, and Sierra Leone east to Somalia, Djibouti, and
    S Yemen, and south to South Africa and Namibia.
STATUS: IUCN 2003 and IUCN/SSC Action Plan (2001) – Lower Risk (lc).
SYNONYMS: *planirostris* Peters, 1852; *colias* Thomas, 1904; *herero* Thomas, 1906; *pondoensis*
    Roberts, 1946.
COMMENTS: Distinct from *nigrita* and *leucogaster*; see Schlitter et al. (1980) and C. B. Robbins
    et al. (1985). Includes *colias*; see C. B. Robbins et al. (1985). Also see Koopman (1975).
    Many literature records of this species are in error due to taxonomic confusion
    surrounding these names; see C. B. Robbins et al. (1985). Subspecies are poorly defined.

*Scotophilus heathii* (Horsfield, 1831). Proc. Zool. Soc. Lond., 1831:113.
COMMON NAME: Greater Asiatic Yellow House Bat.
TYPE LOCALITY: India, Madras.
DISTRIBUTION: Afghanistan to S China, including Hainan Isl, south to Sri Lanka, Vietnam,
    Cambodia, Thailand, and Burma.
STATUS: IUCN 2003 and IUCN/SSC Action Plan (2001) – Lower Risk (lc).
SYNONYMS: *belangeri* Geoffroy, 1834; *flaveolus* Horsfield, 1851; *luteus* Blyth, 1851; *insularis*
    Allen, 1906; *watkinsi* Sanborn, 1952.
COMMENTS: May include *celebensis*; see Tate (1942*a*) and Sinha (1980). Reviewed in part by
    Bates and Harrison (1997) and Hendrichsen et al. (2001*b*). Populations from Vietnam
    have not been allocated to subspecies. Sometimes spelled *heathi* (e.g., Koopman,
    1993).

*Scotophilus kuhlii* Leach, 1821. Trans. Linn. Soc. Lond., 13:71.
COMMON NAME: Lesser Asiatic Yellow House Bat.
TYPE LOCALITY: "India".
DISTRIBUTION: Bangladesh, Pakistan to Taiwan, south to Sri Lanka, Burma, Cambodia,
    W Malaysia, Java, Bali, Nusa Tenggara (Indonesia), southeast to Philippines and Aru
    Isls (Indonesia).
STATUS: IUCN 2003 and IUCN/SSC Action Plan (2001) – Lower Risk (lc).
SYNONYMS: *wroughtoni* Thomas, 1897; *castaneus* Horsfield, 1851; *castaneus* Gray, 1838
    [*nomen nudum*]; *sumatrana* Gray, 1838; *consobrinus* Allen, 1906; *swinhoei* Blyth, 1860;
    *gairdneri* Kloss, 1917; *panayensis* Sody, 1928; *solutatus* Sody, 1936; *temminckii*
    Horsfield, 1824; *fulvus* Gray, 1843.
COMMENTS: Often called *temminckii*, but see Hill and Thonglongya (1972). Does not include
    *collinus*; see Kitchener et al. (1997*b*). Reviewed in part by Bates and Harrison (1997)
    and Kitchener et al. (1997*b*); see also Tate (1942*a*). There is some confusion regarding
    the use of this name in S Asia, see Hendrichsen et al. (2001*b*).

*Scotophilus leucogaster* (Cretzschmar, 1830). *In* Rüppell, Atlas Reise Nördl. Afr., Zool. Säugeth., p. 71.
COMMON NAME: White-bellied House Bat.
TYPE LOCALITY: Sudan, Kordofan, Brunnen Nedger (Nedger Well = Bir Nedger).
DISTRIBUTION: Mauritania, Senegal, and Gambia to N Kenya and Ethiopia.
STATUS: IUCN 2003 and IUCN/SSC Action Plan (2001) – Lower Risk (lc).
SYNONYMS: *altilis* G. M. Allen, 1914; *flavigaster* Heuglin, 1861; *murinoflavus* Heuglin, 1861;
    *damarensis* Thomas, 1906.
COMMENTS: Does not include *nucella*, see C. B. Robbins et al. (1985). Also see Koopman
    (1994). Distinct from *dinganii*; see Schlitter et al. (1980) and C. B. Robbins et al. (1985),

but also see Koopman (1975) and Koopman et al. (1978). Includes *damarensis*, see Robbins et al. (1985). Many literature records of this species are in error due to taxonomic confusion surrounding the names *nigrita, dinganii, leucogaster,* and *borbonicus*; see Robbins (1978) and C. B. Robbins et al. (1985). May include *serratus* Heuglin, 1877, an enigmatic taxon variously referred to either *Taphozous nudiventris* (e.g., G. M. Allen, 1939; Koopman, 1993) or *Scotophilus leucogaster* (e.g., G. M. Allen, 1939; Koopman, 1975) but which may not represent either of those species.

*Scotophilus nigrita* (Schreber, 1774). Die Säugethiere, 1:171.
COMMON NAME: Giant House Bat.
TYPE LOCALITY: Senegal.
DISTRIBUTION: Senegal to Sudan, E Dem. Rep. Congo, Kenya, Zimbabwe, Malawi, and Mozambique.
STATUS: IUCN 2003 and IUCN/SSC Action Plan (2001) – Lower Risk (nt).
SYNONYMS: *gigas* Dobson, 1875; **alvenslebeni** Dalquest, 1965.
COMMENTS: Reviewed by Robbins (1978) and Cotterill (1996). The identity of this species is clear and *nigrita* is the senior synonym of *gigas*. Many literature records of this species are in error due to taxonomic confusion surrounding the names *nigrita, dinganii, leucogaster, and borbonicus*; see Robbins (1978) and C. B. Robbins et al. (1985).

*Scotophilus nucella* Robbins, 1973. Ann. Kon. Mus. Mid. Afr., Zool. Wetensch. and Ann. Mus Roy. Afr. Centr., Sc. Zool., 273:19. (Publication has Dutch and English titles)
COMMON NAME: Robbins's House Bat.
TYPE LOCALITY: Ghana, Eastern Region, 1 mi N Nkawkaw.
DISTRIBUTION: Côte d'Ivoire, Ghana, Uganda.
STATUS: IUCN 2003 – Not evaluated; not considered in IUCN/SSC Action Plan (2001).
COMMENTS: Sometimes considered a subspecies of *leucogaster*, but apparently distinct in both morphology and habitat preferences; see Koopman (1994).

*Scotophilus nux* Thomas, 1904. Ann. Mag. Nat. Hist., ser. 7, 13:208.
COMMON NAME: Nut-colored House Bat.
TYPE LOCALITY: Cameroon, Efulen.
DISTRIBUTION: High forest zones from Sierra Leone to Kenya.
STATUS: IUCN 2003 and IUCN/SSC Action Plan (2001) – Lower Risk (lc).
COMMENTS: Often treated as a subspecies of *dinganii* or *leucogaster* (or *nigrita*, when that name was misapplied to the former species), see G. M. Allen (1939), Rosevear (1965), and Hayman and Hill (1971), Koopman et al. (1978) and Koopman (1994). However, *nux* appears to be distinct from all of the above species; see C. B. Robbins et al. (1985).

*Scotophilus robustus* Milne-Edwards, 1881. C. R. Acad. Sci. Paris, 91:1035.
COMMON NAME: Robust House Bat.
TYPE LOCALITY: Madagascar.
DISTRIBUTION: N Madagascar.
STATUS: IUCN 2003 and IUCN/SSC Action Plan (2001) – Lower Risk (nt).
COMMENTS: Recognized as a subspecies of *nigrita* (when *nigrita* was used for the species now called *dinganii*) by Hayman and Hill (1971). However, C. B. Robbins et al. (1985) considered it specifically distinct from *dinganii* and *borbonicus*. Reviewed by Peterson et al. (1995), who also considered it to be distinct.

*Scotophilus viridis* (Peters, 1852). Reise nach Mossambique, Säugethiere, p. 67.
COMMON NAME: Green House Bat.
TYPE LOCALITY: Mozambique, Mozambique Isl, 15°S.
DISTRIBUTION: Senegal to Ethiopia south to Namibia and South Africa.
STATUS: IUCN 2003 and IUCN/SSC Action Plan (2001) – Lower Risk (lc).
SYNONYMS: **nigritellus** de Winton, 1899.
COMMENTS: Included in *leucogaster* by Hayman and Hill (1971), but see Koopman (1975, 1986) and Schlitter et al. (1980). Distinct from *dinganii*; see Schlitter et al. (1980). Includes *nigritellus* but does not include *damarensis*, see C. B. Robbins et al. (1985). Also see comments under *borbonicus*.

*Scotorepens* Troughton, 1943. Furred Animals of Australia, 1st ed., Syndey: Angus and Robertson, p. 354.
    TYPE SPECIES: *Scoteinus orion* Troughton, 1937.
    COMMENTS: Often included in *Nycticeius*, but see Kitchener and Caputi (1985) and Volleth and
        Tidemann (1991). The latter authors suggested that *Scotorepens* may be more closely
        related to Vespertilionini than Nycticeiini. Revised by Kichener and Caputi (1985), who
        provided a key to the species. An undescribed species of *Scotorepens* may be present in
        E Australia; see Menkhorst and Knight (2001).

*Scotorepens balstoni* (Thomas, 1906). Abstr. Proc. Zool. Soc. Lond., 1906(31):2.
    COMMON NAME: Western Broad-nosed Bat.
    TYPE LOCALITY: Australia, Western Australia, Laverton, North Pool, 503 m.
    DISTRIBUTION: Mainland Australia.
    STATUS: IUCN 2003 and IUCN/SSC Action Plan (2001) – Lower Risk (lc) as *Nycticeius balstoni*.
    SYNONYMS: ***influatus*** Thomas, 1924.
    COMMENTS: Includes *influatus*; see Kitchener and Caputi (1985). Does not include *orion* and
        *caprenus*; see Kitchener and Caputi (1985), but also see Koopman (1978a) and Hall and
        Richards (1979).

*Scotorepens greyii* (Gray, 1842). Zool. Voy. H.M.S. "Erebus" and "Terror," pl. 20.
    COMMON NAME: Little Broad-nosed Bat.
    TYPE LOCALITY: Australia, Northern Territory, Port Essington.
    DISTRIBUTION: Western Australia (excluding the south), Northern Territory, South Australia,
        New South Wales, and Queensland (Australia). Records from Victoria refer to *balstoni*.
    STATUS: IUCN 2003 and IUCN/SSC Action Plan (2001) – Lower Risk (lc) as *Nycticeius greyii*.
    SYNONYMS: *aqeilo* Troughton, 1937; *caprenus* Troughton, 1937.
    COMMENTS: Reviewed by Kitchener and Caputi (1985).

*Scotorepens orion* (Troughton, 1937). Aust. Zool. 8:211.
    COMMON NAME: Orion Broad-nosed Bat.
    TYPE LOCALITY: Australia, New South Wales, Sydney
    DISTRIBUTION: SE Australia.
    STATUS: IUCN 2003 and IUCN/SSC Action Plan (2001) – Lower Risk (lc).
    COMMENTS: Included in *balstoni* by Koopman (1978a, 1993, 1994) and Hall and Richards
        (1979), but see Kitchener and Caputi (1985).

*Scotorepens sanborni* (Troughton, 1937). Aust. Zool., 8:280.
    COMMON NAME: Northern Broad-nosed Bat.
    TYPE LOCALITY: Papua New Guinea, Milne Bay Prov., East Cape.
    DISTRIBUTION: W Timor; SE New Guinea; NE Queensland, Northern Territory, and
        N Western Australia (Australia).
    STATUS: IUCN 2003 and IUCN/SSC Action Plan (2001) – Lower Risk (lc) as *Nycticeius*
        *sanborni*.
    COMMENTS: Included in *balstoni* by Koopman (1978a), but see Kitchener and Caputi (1985).
        Reviewed by Kitchener et al. (1994c); also see Flannery (1995a, b) and Bonaccorso
        (1998).

**Tribe Nyctophilini** Peters, 1865. Monatsb. K. Preuss. Akad. Wiss. Berlin, 1865:524.
    COMMENTS: Volleth and Tidemann (1991) suggested on the basis of karyotype data that
        *Nyctophilus* may belong in Vespertilionini.

*Nyctophilus* Leach, 1821. Trans. Linn. Soc. Lond., 13:78.
    TYPE SPECIES: *Nyctophilus geoffroyi* Leach, 1821.
    SYNONYMS: *Lamingtona* McKean and Calaby, 1968.
    COMMENTS: Includes *Lamingtona*, see Hill and Koopman (1981). Australian species reviewed by
        Hall and Richards (1979).

*Nyctophilus arnhemensis* Johnson, 1959. Proc. Biol. Soc. Wash., 72:184.
    COMMON NAME: Northern Long-eared Bat.
    TYPE LOCALITY: Australia, Northern Territory, Cape Arnhem Peninsula, S of Yirkala, Rocky
        Bay. (12°13'S, 36°47'E).

DISTRIBUTION: N Australia.

STATUS: IUCN 2003 and IUCN/SSC Action Plan (2001) – Lower Risk (lc).

*Nyctophilus bifax* Thomas, 1915. Ann. Mag. Nat. Hist., ser. 8, 15:496.

COMMON NAME: Bifax Long-eared Bat.

TYPE LOCALITY: Australia, Queensland, Herberton.

DISTRIBUTION: N Western Australia, N Northern Territory, coastal Queensland, NE New South Wales (Australia); Papua New Guinea.

STATUS: IUCN 2003 and IUCN/SSC Action Plan (2001) – Lower Risk (lc) as *N. bifax*; IUCN/SSC Action Plan (2001) – Lower Risk (nt) as *N. daedalus*; but *N. daedalus* not listed in IUCN 2003 (*lapsus*).

SYNONYMS: *daedalus* Thomas, 1915.

COMMENTS: Included in *gouldi* by Koopman (1984*c*, 1993, 1994), but see Parnaby (1987, 2002*a*). Also see Flannery (1995*a*). The status of *daedalus* is uncertain; data presented by Parnaby (1987) suggested that it may represent a distinct species, but Bonaccoroso (1998) indicated that it may not be distinct from *bifax* even at the subspecies level.

*Nyctophilus geoffroyi* Leach, 1821. Trans. Linn. Soc. Lond., 13:78.

COMMON NAME: Lesser Long-eared Bat.

TYPE LOCALITY: Australia, Western Australia, King George Sound.

DISTRIBUTION: Australia (except NE) including Tasmania.

STATUS: IUCN 2003 and IUCN/SSC Action Plan (2001) – Lower Risk (lc).

SYNONYMS: *australis* Peters, 1861; *leachii* Dobson, 1878; *novaehollandiae* Gray, 1831; **pacificus** Gray, 1831; *geayi* Troussart, 1915; *unicolor* Tomes, 1858; **pallescens** Thomas, 1913.

COMMENTS: Reviewed in part by Kitchener et al. (1991*d*). The three subspecies are poorly defined.

*Nyctophilus gouldi* Tomes, 1858. Proc. Zool. Soc. Lond., 1858:31.

COMMON NAME: Gould's Long-eared Bat.

TYPE LOCALITY: Australia, Queensland, Moreton Bay.

DISTRIBUTION: E Queensland, E New South Wales, Victoria, SE South Australia, SW Western Australia; a Tasmanian record appears to be erroneous (Koopman, 1993).

STATUS: IUCN 2003 and IUCN/SSC Action Plan (2001) – Lower Risk (lc).

COMMENTS: Koopman (1984*c*, 1993, 1994) included *bifax* and *daedalus* in this species, but see Parnaby (1987, 2002*a*).

*Nyctophilus heran* Kitchener, How, and Maharadatunkamsi, 1991. Rec. West. Aust. Mus., 15:100.

COMMON NAME: Sundan Long-eared Bat.

TYPE LOCALITY: Indonesia, Lesser Sundas (Nusa Tenggara), Lembata Isl (= Lomblen Isl), Desa Hadakewa, Kampong Merdeka (08°22′S, 123°31′E; restricted by Corbet and Hill, 1992).

DISTRIBUTION: Known only from the type locality.

STATUS: IUCN 2003 and IUCN/SSC Action Plan (2001) – Endangered.

COMMENTS: Known only from the holotype; similar to *geoffroyi*, to which it may be related; see Kitchener et al. (1991*d*) and Corbet and Hill (1992).

*Nyctophilus howensis* McKean, 1975. Aust. Mammalogy, 1:330.

COMMON NAME: Lord Howe Island Long-eared Bat.

TYPE LOCALITY: Australia, Lord Howe Isl, North Bay.

DISTRIBUTION: Known only from the type locality.

STATUS: IUCN 2003 and IUCN/SSC Action Plan (2001) – Extinct.

COMMENTS: Known only from the holotype, a fossil found in a cave on Lord Howe Isl. McKean (1975) suggested that this species may have survived into historic times on the basis of Etheridge's (1889) statement that a bat larger than *Chalinolobus morio* was occasionally seen on the island.

*Nyctophilus microdon* Laurie and Hill, 1954. List of Land Mammals of New Guinea, Celebes, and adjacent Islands, p. 78.

COMMON NAME: Small-toothed Long-eared Bat.

TYPE LOCALITY: Papua New Guinea, Western Highlands (?) Prov., Welya (W of Hagen Range, 7,000 ft. (2,134 m)).

DISTRIBUTION: EC Papua New Guinea.
STATUS: IUCN 2003 and IUCN/SSC Action Plan (2001) – Vulnerable.
COMMENTS: See Flannery (1995a) and Bonaccorso (1998).

*Nyctophilus microtis* Thomas, 1888. Ann. Mag. Nat. Hist., ser. 6, 2:226.
COMMON NAME: New Guinea Long-eared Bat.
TYPE LOCALITY: Papua New Guinea, Central Prov., Astrolabe Range, Sogeri.
DISTRIBUTION: Papua New Guinea including New Ireland.
STATUS: IUCN 2003 and IUCN/SSC Action Plan (2001) – Lower Risk (lc).
SYNONYMS: *bicolor* Thomas, 1915; *lophorhina* McKean and Calaby, 1968.
COMMENTS: Includes *lophorhina*; see Hill and Koopman (1981). Hill and Koopman (1981)
    tentatively recognized the three named forms as subspecies, but Koopman (1994)
    rejected this arrangement and did not recognize subspecies. See Flannery (1995a, b)
    and Bonaccorso (1998).

*Nyctophilus nebulosus* Parnaby, 2002. Aust. Mammal., 23:116.
COMMON NAME: New Caledonian Long-eared Bat.
TYPE LOCALITY: New Caledonia, Nouméa, southwestern slopes of Mt. Koghis, 150 m N of
    Station d'Altitude car park, 22°10'37"S, 166°30'12"E, 430 m.
DISTRIBUTION: Known only from Nouméa area of New Caledonia.
STATUS: Not yet assessed by IUCN, but Parnaby (2000b) recommended that this species be
    classified in the IUCN threat category of Vulnerable (B1ab+2ab, D2).
COMMENTS: In addition to the original description by Parnaby (2002a), see Flannery
    (1995b), who discussed this species under its common name.

*Nyctophilus timoriensis* (E. Geoffroy, 1806). Ann. Mus. Natn. Hist. Nat. Paris, 8:200.
COMMON NAME: Greater Long-eared Bat.
TYPE LOCALITY: Indonesia, Timor (uncertain).
DISTRIBUTION: All of Australia including Tasmania; New Guinea; Timor (Indonesia).
STATUS: IUCN 2003 and IUCN/SSC Action Plan (2001) – Vulnerable as *N. timoriensis*;
    IUCN/SSC Action Plan (2001) – Lower Risk (nt) as *N. sherrini*; but *N. sherrini* not listed
    in IUCN 2003 (*lapsus*).
SYNONYMS: **major** Gray, 1844; **sherrini** Thomas, 1915.
COMMENTS: This bat has been confused with the smaller *gouldi* in coastal SE Queensland.
    Reviewed by Hall and Richards (1979) and Kitchener et al. (1991d). Corbet and Hill
    (1992) discussed the problems associated with the Timor record. See also Flannery
    (1995a).

*Nyctophilus walkeri* Thomas, 1892. Ann. Mag. Nat. Hist., ser. 6, 9:405.
COMMON NAME: Pygmy Long-eared Bat.
TYPE LOCALITY: Australia, Northern Territory, Adelaide River.
DISTRIBUTION: Northern Territory and N Western Australia (Australia).
STATUS: IUCN 2003 and IUCN/SSC Action Plan (2001) – Lower Risk (nt).

*Pharotis* Thomas, 1914. Ann. Mag. Nat. Hist., ser. 8, 14:381.
TYPE SPECIES: *Pharotis imogene* Thomas, 1914.

*Pharotis imogene* Thomas, 1914. Ann. Mag. Nat. Hist., ser. 8, 14:382.
COMMON NAME: Thomas's Big-eared Bat.
TYPE LOCALITY: Papua New Guinea, Central Prov., Lower Kemp Welch River, Kamali.
DISTRIBUTION: SE New Guinea.
STATUS: IUCN 2003 and IUCN/SSC Action Plan (2001) – Critically Endangered.
COMMENTS: See Flannery (1995a) and Bonaccorso (1998).

**Tribe Pipistrellini** Tate, 1942. Bull. Amer. Mus. Nat. Hist., 80:232.
COMMENTS: Includes *Pipistrellus*, *Glischropus*, *Nyctalus*, and *Scotozous*; see Volleth (1992), Volleth
    and Heller (1994), and Volleth et al. (2001); also see Mayer and von Helversen (2001a).

*Glischropus* Dobson, 1875. Proc. Zool. Soc. Lond., 1875:472.
TYPE SPECIES: *Vesperugo tylopus* Dobson, 1875.

COMMENTS: Menu (1987) considered this genus to be a synonym of *Pipistrellus*, but see Corbet and Hill (1992).

*Glischropus javanus* Chasen, 1939. Treubia, 17:189.
COMMON NAME: Javan Thick-thumbed Bat.
TYPE LOCALITY: Indonesia, Java, West Java, Mt. Pangeango.
DISTRIBUTION: W Java (Indonesia).
STATUS: IUCN 2003 and IUCN/SSC Action Plan (2001) – Endangered.
COMMENTS: May be conspecific with *tylopus*; see Corbet and Hill (1992), but also see Menu (1987).

*Glischropus tylopus* (Dobson, 1875). Proc. Zool. Soc. Lond., 1875:473.
COMMON NAME: Common Thick-thumbed Bat.
TYPE LOCALITY: Malaysia, N Boreneo, Sabah.
DISTRIBUTION: Burma, Thailand, W Malaysia, Borneo, Palawan (Philippines), Sumatra and N Molucca Isls.
STATUS: IUCN 2003 and IUCN/SSC Action Plan (2001) – Lower Risk (lc).
SYNONYMS: **batjanus** Marschie, 1901.
COMMENTS: See Corbet and Hill (1992) and Flannery (1995*b*).

*Nyctalus* Bowditch, 1825. Excursions in Madeira and Porto Santo, p. 36, footnote.
TYPE SPECIES: *Nyctalus verrucosus* Bowditch, 1825 (= *Vespertilio leisleri* Kuhl, 1817).
SYNONYMS: *Noctulina* Gray, 1842; *Panugo* Kolenati, 1856; *Pterygistes* Kaup, 1829.
COMMENTS: Members of Koopman's (1994) *stenopterus* species group are here included in *Pipistrellus* and *Hypsugo*.

*Nyctalus aviator* (Thomas, 1911). Ann. Mag. Nat. Hist., ser. 8, 8:380.
COMMON NAME: Birdlike Noctule.
TYPE LOCALITY: Japan, Honshu, Tokyo.
DISTRIBUTION: Hokkaido, Shikoku, Kyushu, Tsushima, Iki (Japan); Korea; E and C China. Possibly occurs in Russian Far East, see Tiunov (1997).
STATUS: IUCN 2003 and IUCN/SSC Action Plan (2001) – Lower Risk (nt).
SYNONYMS: *molossus* Temminck, 1840 [not Pallas, 1767].
COMMENTS: Listed as a subspecies of *lasiopterus* by Ellerman and Morrison-Scott (1951), but see Corbet (1978*c*) and Yoshiyuki (1989). Reviewed by Yoshiyuki (1989).

*Nyctalus azoreum* (Thomas, 1901). Ann. Mag. Nat. Hist., ser. 7, 8:34.
COMMON NAME: Azores Noctule.
TYPE LOCALITY: Portugal, Azores, St. Michael.
DISTRIBUTION: Azores Isls (Portugal).
STATUS: IUCN 2003 and IUCN/SSC Action Plan (2001) – Vulnerable.
COMMENTS: Listed as a subspecies of *leisleri* by Corbet (1978*c*), but see Palmeirim (1991) and Horácek et al. (2000).

*Nyctalus furvus* Imaizumi and Yoshiyuki, 1968. Bull. Nat. Sci. Mus. Tokyo, 11:127.
COMMON NAME: Japanese Noctule.
TYPE LOCALITY: Japan, Iwate Pref., Shimohei-gun, Iwaizumi-Machi, Kado, 300 m.
DISTRIBUTION: N Honshû Isl (Japan).
STATUS: IUCN 2003 – Not evaluated; not considered in IUCN/SSC Action Plan (2001).
COMMENTS: Included in *noctula* by Corbet (1978*c*) and Corbet and Hill (1992), but see Yoshiyuki (1989).

*Nyctalus lasiopterus* (Schreber, 1780). *In* Zimmermann, Geogr. Gesch. Mensch. Vierf. Thiere, 2:412.
COMMON NAME: Giant Noctule.
TYPE LOCALITY: Northern Italy, ?Pisa (uncertain).
DISTRIBUTION: W Europe to Urals, Caucasus, and Balkans, Asia Minor, Iran and Ust-Urt Plateau (Kazakhstan), Morocco, Libya, possbily Algeria.
STATUS: IUCN 2003 and IUCN/SSC Action Plan (2001) – Lower Risk (nt).
SYNONYMS: *ferrugineus* Brehm, 1827; *maxima* Fatio, 1869; *sicula* Mina-Palumbo, 1868.
COMMENTS: Reviewed by Corbet (1978*c*).

*Nyctalus leisleri* (Kuhl, 1817). Die Deutschen Fledermäuse. Hanau, p. 14, 46.
    COMMON NAME: Leisler's Noctule.
    TYPE LOCALITY: Germany, Hessen, Hanau.
    DISTRIBUTION: W Europe to Urals, Caucasus, and Turkey; Britain and Ireland; Sweden,
        S Finland, Baltic states; Madeira Isl; W Himalayas, Pakistan, E Afghanistan; NW Africa.
    STATUS: IUCN 2003 and IUCN/SSC Action Plan (2001) – Lower Risk (nt).
    SYNONYMS: *dasykarpos* Kuhl, 1819; *pachygnathus* Michahelles, 1839; **verrucosus** Bowditch,
        1825; *madeirae* Barrett-Hamilton, 1906.
    COMMENTS: Includes *verrucosus*, see Corbet (1978c), who also included *azoreum*; but see
        Palmeirim (1991). Reviewed in part by Bates and Harrison (1997). For discussion of
        correct spelling (*leisleri*) see Bogdanowicz and Kock (1998).

*Nyctalus montanus* (Barrett-Hamilton, 1906). Ann. Mag. Nat. Hist., ser. 7, 17:99.
    COMMON NAME: Mountain Noctule.
    TYPE LOCALITY: India, Uttar Pradesh, Dehra Dun, Mussooree.
    DISTRIBUTION: E Afghanistan, Pakistan, N India, Nepal.
    STATUS: IUCN 2003 and IUCN/SSC Action Plan (2001) – Lower Risk (nt).
    COMMENTS: Listed as a subspecies of *leisleri* by Ellerman and Morrison-Scott (1951), but see
        Gaisler (1970), Corbet (1978c), Corbet and Hill (1992), and Bates and Harrison (1997).

*Nyctalus noctula* (Schreber, 1774). Die Säugethiere, 1:166.
    COMMON NAME: Noctule.
    TYPE LOCALITY: France.
    DISTRIBUTION: Europe and S Scandinavia to Urals and Caucasus; Turkey to Israel and Oman;
        W Turkmenistan, W Kazakhstan, Uzbekistan, Kyrgyzstan, and Tajikistan to SW Siberia,
        Himalayas, south to Burma, Vietnam, and W Malaysia; possibly Algeria. A record from
        Mozambique is dubious (Koopman, 1993, 1994).
    STATUS: IUCN 2003 and IUCN/SSC Action Plan (2001) – Lower Risk (lc).
    SYNONYMS: *altivolans* White, 1789; *lardarius* Müller, 1776; *magnus* Berkenhout, 1789; *major*
        Leach, 1818; *minima* Fatio, 1869; *palustris* Crespon, 1844; *princeps* Ognev and
        Worobyev, 1923; *proterus* Kuhl, 1818; *rufescens* Brehm, 1829; **labiata** Hodgson, 1835;
        **lebanoticus** Harrison, 1962; **mecklenburzevi** Kuziakin, 1934; *montanus* Kishida, 1934
        [not Barrett-Hamilton, 1906]. **Unassigned:** *macuanus* Peters, 1852 [type locality =
        Mozambique, but this provenance is dubious; Koopman, 1994].
    COMMENTS: Formerly included *furvus* and *velutinus*, but these appear to be distinct; see
        Yoshiyuki (1989), but also see Corbet (1978c) and Corbet and Hill (1992). Does not
        include *sinensis*, which was recognized as a senior synonym of *Vespertilio superans* by
        Horácek (1997). Reviewed in part by Harrison and Bates (1991), Bates and Harrison
        (1997), and Horácek et al. (2000).

*Nyctalus plancyi* Gerbe, 1880. Bull. Soc. Zool. France, 5:71.
    COMMON NAME: Chinese Noctule.
    TYPE LOCALITY: China, Peking.
    DISTRIBUTION: E China, Taiwan.
    STATUS: IUCN 2003 – Not evaluated; not considered in IUCN/SSC Action Plan (2001).
    SYNONYMS: **velutinus** G. M. Allen, 1923.
    COMMENTS: Included in *noctula* by Corbet (1978c) and Corbet and Hill (1992), but see Tate
        (1942a), Yoshiyuki (1989), Zhang (1990), and L.-K. Lin et al. (2002b). This name is
        sometimes misspelled *plancei*, but the correct spelling is *plancyi* after M. V. Collin
        Plancy.

*Pipistrellus* Kaup, 1829. Skizz. Entwickel.-Gesch. Nat. Syst. Europ. Thierwelt, 1:98.
    TYPE SPECIES: *Vespertilio pipistrellus* Schreber, 1774.
    SYNONYMS: *Alobus* Peters, 1867 [not Le Conte, 1856]; *Attalepharca* Menu, 1987 [no type species
        designated, therefore not available]; *Eptesicops* Roberts, 1926; *Euvesperugo* Acloque, 1899;
        *Nannugo* Kolenati, 1856; *Perimyotis* Menu, 1984; *Romicia* Gray, 1838; *Romicius* Blyth,
        1840; *Vansonia* Roberts, 1946.
    COMMENTS: For discussion of synonyms see Ellerman and Morrison-Scott (1951), Hill (1976),
        Menu (1984), and Kitchener et al. (1986). Hill and Harrison (1987) reviewed the genus

and recognized seven subgenera (*Pipistrellus, Hypsugo, Falsistrellus, Perimyotis, Arielulus, Vespadelus, and Neoromicia*), but most of these groups are now recognized as distinct genera. Does not include *Hypsugo*; see Horácek and Hanák (1985-1986), Tiunov (1986), Menu (1987), Ruedi and Arlettaz (1991), Volleth and Heller (1994), Volleth et al. (2001), and Mayer and von Helversen (2001*a*). Does not include *Glischropus*; see Corbet and Hill (1992), but also see Menu (1987). Does not include *Scotozous*; see Corbet and Hill (1992). Does not include *Vespadelus*; see Volleth and Tidemann (1991) and Volleth and Heller (1994). Does not include *Falsistrellus*; see Kitchener et al. (1986) and Volleth and Heller (1994). Does not include *Arielulus*; see Heller and Volleth (1984) and Volleth and Heller (1994), who transferred this subgenus to *Eptesicus*, and Csorba and Lee (1999), who subsequently argued that it should be recognized as a distinct genus. Does not include *Neoromicia*; see Volleth et al. (2001) and Kearney et al. (2002). Only *Pipistrellus* and *Perimyotis* are retained here as subgenera. American species reviewed by Hall and Dalquest (1950); Indomalayan species reviewed by Corbet and Hill (1992). Also see Peterson et al. (1995) and Barratt et al. (1995).

*Pipistrellus abramus* (Temminck, 1838). Mongr. Mamm., Tome 2:232.
COMMON NAME: Japanese Pipistrelle.
TYPE LOCALITY: Japan, Kyushu, Nagasaki.
DISTRIBUTION: S Ussuri region (Russia and China), Taiwan, S and C Japan, Korea, Vietnam, Burma, India.
STATUS: IUCN 2003 – Not listed (*lapsus*); IUCN/SSC Action Plan (2001) – Lower Risk (lc).
SYNONYMS: *akokomuli* Temminck 1838; *irretitus* Cantor, 1842; *pomiloides*, Mell, 1922; *pumiloides* Tomes, 1857.
COMMENTS: Subgenus *Pipistrellus*. Often regarded as a subspecies of *javanicus*, but clearly separable; see Hill and Harrison (1987), Yoshiyuki (1989), Corbet and Hill (1992), and Tiunov (1997). Does not include *paterculus*; see Hill and Harrison (1987), Corbet and Hill (1992), Bates and Harrison (1997), Bates et al. (1997), and Hendrichsen et al. (2001*b*). Reviewed by Horácek et al. (2000) and Srinivasulu and Srinivasulu (2001).

*Pipistrellus adamsi* Kitchener, Caputi, and Jones, 1986. Rec. West. Aust. Mus., 12:463.
COMMON NAME: Adams's Pipistrelle.
TYPE LOCALITY: Australia, Queensland, Cape York, 40 km E Archer River Crossing, 13°27'S, 143°18'E.
DISTRIBUTION: Queensland and Northern Territory (Australia).
STATUS: IUCN 2003 – Not listed (*lapsus*); IUCN/SSC Action Plan (2001) – Lower Risk (lc).
COMMENTS: Subgenus *Pipistrellus*. Included in *tenuis* by Koopman (1993, 1994), but see Kitchener et al. (1986).

*Pipistrellus aero* Heller, 1912. Smithson. Misc. Coll., 60(12):3.
COMMON NAME: Mt. Gargues Pipistrelle.
TYPE LOCALITY: Kenya, Mathews Range, Mt. Gargues.
DISTRIBUTION: NW Kenya, perhaps Ethiopia.
STATUS: IUCN 2003 and IUCN/SSC Action Plan (2001) – Data Deficient.
COMMENTS: Subgenus *Pipistrellus*. The Ethiopian specimens in the British Museum are clearly *kuhlii*; see Hayman and Hill (1971).

*Pipistrellus angulatus* Peters, 1880. Sitz. Ges. Naturf. Freunde, p. 122.
COMMON NAME: Angulate Pipistrelle.
TYPE LOCALITY: Duke of York Isl, between New Britain and New Ireland (New Hebrides, = Vanuatu).
DISTRIBUTION: New Guinea; Bismarck Arch.; Bougainville Isl and Solomon Isls; adjacent small islands.
STATUS: IUCN 2003 – Not listed (*lapsus*); IUCN/SSC Action Plan (2001) – Lower Risk (lc).
SYNONYMS: **ponceleti** Troughton, 1936.
COMMENTS: Subgenus *Pipistrellus*. Included in *tenuis* by Koopman (1993, 1994), but see Kitchener et al. (1986). Also see Flannery (1995*a*, *b*) and Bonaccorso (1998).

*Pipistrellus ceylonicus* (Kelaart, 1852). Prodr. Faun. Zeylanica, p. 22.
COMMON NAME: Kelaart's Pipistrelle.

TYPE LOCALITY: Sri Lanka, Trincomalee.

DISTRIBUTION: Pakistan, India, Sri Lanka, Bangladesh, Burma, Kwangsi and Hainan (China), Vietnam, Borneo.

STATUS: IUCN 2003 and IUCN/SSC Action Plan (2001) – Lower Risk (lc).

SYNONYMS: *borneanus* Hill, 1963; *indicus* Dobson, 1878; *chrysothrix* Wroughton, 1899; *raptor* Thomas, 1904; *shanorum* Thomas, 1915; *subcanus* Thomas, 1915; *tongfangensis* Wang, 1966.

COMMENTS: Subgenus *Pipistrellus*. Reviewed in part by Bates and Harrison (1997).

*Pipistrellus collinus* Thomas, 1920. Ann. Mag. Nat. Hist., ser. 9, 6:533.

COMMON NAME: Greater Papuan Pipistrelle.

TYPE LOCALITY: British Papua (= Papua New Guinea), head of Mambare River, Bihagi, 8°04′S, 148°01′E.

DISTRIBUTION: Highlands of Papua New Guinea.

STATUS: IUCN 2003 – Not listed (*lapsus*); IUCN/SSC Action Plan (2001) – Lower Risk (lc).

COMMENTS: Subgenus *Pipistrellus*. Included in *angulatus* by Laurie and Hill (1954) and in *tenuis* by Koopman (1993, 1994), but see Kitchener et al. (1986). Also see Flannery (1995a) and Bonaccoroso (1998).

*Pipistrellus coromandra* (Gray, 1838). Mag. Zool. Bot., 2:498.

COMMON NAME: Indian Pipistrelle.

TYPE LOCALITY: India, Coromandel Coast, Pondicherry.

DISTRIBUTION: Afghanistan, Bangladesh, India (including Nicobar Isls), Sri Lanka, Pakistan, Nepal, Bhutan, Burma, Cambodia, Thailand, S China.

STATUS: IUCN 2003 and IUCN/SSC Action Plan (2001) – Lower Risk (lc).

SYNONYMS: *afghanus* Gaisler, 1970; *blythii* Wagner, 1855; *coromandelianus* Blyth, 1863; *coromandelicus* Blyth, 1851; *micropus* Peters, 1872; *nicobaricus* Fitzinger, 1861; *parvipes* Blyth, 1853.

COMMENTS: Subgenus *Pipistrellus*. Does not include *aladdin*; see Corbet (1978c). Does not include *portensis* and *tramatus*; see Corbet and Hill (1992). See comment under *pipistrellus*. Reviewed by Bates and Harrison (1997).

*Pipistrellus deserti* Thomas, 1902. Proc. Zool. Soc. Lond., 1902, II:4.

COMMON NAME: Desert Pipistrelle.

TYPE LOCALITY: Libya, Fezzan, Murzuk.

DISTRIBUTION: Egypt, N Sudan, Libya, Algeria, Burkina Faso, Ghana.

STATUS: IUCN 2003 and IUCN/SSC Action Plan (2001) – Lower Risk (lc) as *P. aegyptius*.

SYNONYMS: *aegyptius* J. Fischer, 1829 [*nomen dubium*].

COMMENTS: Subgenus *Pipistrellus*. Qumsiyeh (1985) proposed use of *aegyptius* for this species and many subsequent authors followed this usage, but see Kock (1999b), who showed *aegyptius* to be a *nomen dubium*. Reviewed by Horáček et al. (2000).

*Pipistrellus endoi* Imaizumi, 1959. Bull. Natl. Sci. Mus. Tokyo, 4:363.

COMMON NAME: Endo's Pipistrelle.

TYPE LOCALITY: Japan, Honshu, Iwate Pref., Ninohe-Gun, Ashiro-cho, Horobe.

DISTRIBUTION: Honshu (Japan).

STATUS: IUCN 2003 and IUCN/SSC Action Plan (2001) – Endangered.

COMMENTS: Subgenus *Pipistrellus*. Very similar to *javanicus* and *abramus* but apparently distinct, see Yoshiyuki (1989) and Horáček et al. (2000).

*Pipistrellus hesperidus* (Temminck, 1840). Monograph. Mammal . . . Musées de l'Europe, 2:211.

COMMON NAME: Dusky Pipistrelle.

TYPE LOCALITY: Not definitely identifiable, although known to be from the Red Sea coast of Africa; probably Ethiopia, probably Shewa Province [= Shoa] (see discussion in Kock, 2001b).

DISTRIBUTION: Cape Verde Isls, Canary Isls, Liberia, Chad, Bioko (Equatorial Guinea), Nigeria, Cameroon, Dem. Rep. Congo, Ethiopia, Eritrea, Kenya, Uganda, Rwanda, Burundi, Tanzania, Malawi, Zambia, Mozambique, Zimbabwe, Botswana, South Africa, Madagascar.

STATUS: IUCN 2003 – Not evaluated; not considered in IUCN/SSC Action Plan (2001).

SYNONYMS: *fuscatus* Thomas, 1901; *subtilis* Sundevall, 1846; *broomi* Roberts, 1948.
**Unassigned**: *platycephlus* Temminck, 1832 [*nomen dubium*].
COMMENTS: Subgenus *Pipistrellus*. Distinct from *kuhlii*, see Kock (2001*b*). Lectotype
designated by Kock (2001*b*). Chromosomal differences between populations in South
Africa/Madagascar and those in N Africa strongly suggest that the southern
populations (for which *subtilis* is apparently the oldest name) represent a distinct
species (Volleth et al., 2001). Similarly, differences in ectoparasites suggest that North
African and Afrotropical forms may represent different species (Kock, 2001*b*). It thus
seems clear that more than one species is present in this complex. However, allocation
of many populations is uncertain, taxonomic limits have not yet been adequately
described, and holotypes of several important forms (e.g., *subtilis*) have not been
reexamined (Kock et al., 2001*b*; Volleth et al., 2001). I therefore treat this complex as a
single taxon, recognizing the following subspecies (which may be shown to be distinct
species): *hesperidus* (Northeastern Africa), *fuscatus* (Afrotropical regions excluding
Southern Africa and Madagascar), and *subtilis* (Southern Africa and Madagascar).

*Pipistrellus hesperus* H. Allen, 1864. Smithson. Misc. Coll., 7:43.
COMMON NAME: Western Pipistrelle.
TYPE LOCALITY: USA, California, Imperial Co., Old Fort Yuma.
DISTRIBUTION: Washington to SW Oklahoma (USA), and Baja California, south to Hidalgo
and Guerrero (Mexico).
STATUS: IUCN 2003 and IUCN/SSC Action Plan (2001) – Lower Risk (lc).
SYNONYMS: *apus* Elliot, 1904; *australis* Miller, 1897; *merriami* Dobson, 1866; *maximus*
Hatfield, 1936; *oklahomae* Glass and Morse, 1959; *potosinus* Dalquest, 1951; *santarosae*
Hatfield, 1936.
COMMENTS: Subgenus *Pipistrellus*. See Hall (1981). Placed in *Hypsugo* by Koopman (1993),
but here retained in *Pipistrellus* pending further study.

*Pipistrellus inexspectatus* Aellen, 1959. Arch. Sci. Phys. Nat. Geneve, 12:226.
COMMON NAME: Aellen's Pipistrelle.
TYPE LOCALITY: Cameroon, Upper Benoue Valley, Ngaaouyanga.
DISTRIBUTION: Sierra Leone, Ghana, Benin, Cameroon, and Uganda. Specimens from Kenya
and Dem. Rep. Congo previously referred to this species are now thought to represent
*eisentrauti*; see Koopman et al. (1995). A possible record from Sudan cannot be
confirmed as the specimen is too immature to identify (M. Happold, pers. comm.)
STATUS: IUCN 2003 and IUCN/SSC Action Plan (2001) – Lower Risk (lc).
COMMENTS: Subgenus *Pipistrellus*. Often misspelled *inexpectatus*.

*Pipistrellus javanicus* (Gray, 1838). Mag. Zool. Bot., 2:498.
COMMON NAME: Javan Pipistrelle.
TYPE LOCALITY: Indonesia, Java.
DISTRIBUTION: E Afganistan, N Pakistan, N, C India, SE Tibet (China), Burma, Thailand,
Vietnam, through SE Asia to Lesser Sunda Isls and the Philippines; perhaps Australia.
Reports of this species from Cambodia cannot be confirmed (Kock, 2000*a*).
STATUS: IUCN 2003 and IUCN/SSC Action Plan (2001) – Lower Risk (lc) as *P. javanicus*; Data
Deficient as *P. peguensis*.
SYNONYMS: *bancanus* Sody, 1937; *tralatitius* Horsfield, 1824 [indeterminable; see comments];
*tralatitius* Thomas, 1928; *babu* Thomas, 1915; *camortae* Miller, 1902; *meyeni*
Waterhouse, 1845; *peguensis* Sinha, 1969.
COMMENTS: Subgenus *Pipistrellus*. Includes *meyeni*; see Laurie and Hill (1954), Ellerman and
Morrison-Scott (1951), Hill (1967), and Koopman (1973). Includes *camortae*; see Soota
Chaturverdi (1980) and Corbet and Hill (1992), but also see Das (1990). Includes *babu*
and *peguensis*; see Corbet and Hill (1992), Kock (1996), and Bates and Harrison (1997),
but also see Das (1990) and Sinha (1999). Does not include *paterculus* and *abramus*; see
Hill and Harrsion (1987), Corbet and Hill (1992), Bates et al. (1997), and Hendrichsen
et al. (2001*b*). For many years this species was known as *tralatitius* Horsfield, but Laurie
and Hill (1954) regarded this name as indeterminable.

*Pipistrellus kuhlii* (Kuhl, 1817). Die Deutschen Fledermäuse, Hanau, p. 14.
    COMMON NAME: Kuhl's Pipistrelle.
    TYPE LOCALITY: Italy, Friuli-Venezia Giulia, Trieste.
    DISTRIBUTION: C Europe, Near East through the Caucasus to Kazakhstan and Pakistan;
        SW Asia.
    STATUS: IUCN 2003 and IUCN/SSC Action Plan (2001) – Lower Risk (lc).
    SYNONYMS: *albicans* Monticelli, 1886; *albolimbatus* Küster, 1835; *alcythoe* Bonaparte, 1837;
        *marginatus* Cretzschmar, 1830; *marginatus* Bonaparte, 1841 [not Cretzschmar, 1830];
        *minuta* Loche, 1867; *pallidus* Heim de Balsac, 1936; *pullatus* Monticelli, 1886; *saharae*
        Heim de Balsac, 1936 [*nomen nudum*]; *ursula* Wagner, 1840; **ikhwanius** Cheesman and
        Hinton, 1924; *latastei* Laurent, 1937; **lepidus** Blyth, 1845; *canus* Blyth, 1863; *leucotis*
        Dobson, 1872; *lobatus* Jerdon, 1867; *vispistrellus* Bonaparte, 1837. **Unassigned**:
        *calcarata* Gray, 1838 [*nomen dubium*; locality unknown].
    COMMENTS: Subgenus *Pipistrellus*. Does not include African populations (here referred to
        *hesperidus*), see Kock (2001*b*). Canary Isls populations referred to *kuhlii* by Pestano
        et al. (2003) probably also represent *hesperidus* and also listed under that taxon.
        Reviewed in part by Harrsion and Bates (1991) and Bates and Harrison (1997). For
        discussion of correct spelling (*kuhlii*) and authorship (Kuhl not Natterer), see
        Bogdanowicz and Kock (1998).

*Pipistrellus maderensis* (Dobson, 1878). Cat. Chiroptera Brit. Mus., p. 231.
    COMMON NAME: Madeiran Pipistrelle.
    TYPE LOCALITY: Madeira Isls, Madeira Isl (Portugal).
    DISTRIBUTION: Madeira Isl (Portugal); Canary Isls (Spain).
    STATUS: IUCN 2003 and IUCN/SSC Action Plan (2001) – Vulnerable.
    COMMENTS: Subgenus *Pipistrellus*. Phylogeography investigated by Pestano et al. (2003).

*Pipistrellus minahassae* (Meyer, 1899). Abh. Zool. Anthrop.- Ethnology. Mus. Dresden, 7(7):14.
    COMMON NAME: Minahassa Pipistrelle.
    TYPE LOCALITY: Indonesia, Sulawesi, Minahassa, Tomohon.
    DISTRIBUTION: N Sulawesi.
    STATUS: IUCN 2003 and IUCN/SSC Action Plan (2001) – Data Deficient.
    COMMENTS: Subgenus *Pipistrellus*. Reviewed by Tate (1942*a*).

*Pipistrellus nanulus* Thomas, 1904. Ann. Mag. Nat. Hist., ser. 7, 14:198.
    COMMON NAME: Tiny Pipistrelle.
    TYPE LOCALITY: Cameroon, Efulen.
    DISTRIBUTION: Sierra Leone and Côte d'Ivoire to Kenya; Bioko (Equatorial Guinea).
    STATUS: IUCN 2003 and IUCN/SSC Action Plan (2001) – Lower Risk (lc).
    COMMENTS: Subgenus *Pipistrellus*.

*Pipistrellus nathusii* (Keyserling and Blasius, 1839). Arch. Naturgesch., 5(1):320.
    COMMON NAME: Nathusius's Pipistrelle.
    TYPE LOCALITY: Germany, Berlin.
    DISTRIBUTION: W Europe to Urals and Caucasus, and W Asia Minor; S England.
    STATUS: IUCN 2003 and IUCN/SSC Action Plan (2001) – Lower Risk (lc).
    SYNONYMS: *unicolor* Fatio, 1905.
    COMMENTS: Subgenus *Pipistrellus*.

*Pipistrellus papuanus* Peters and Doria, 1881. Ann. Mus. Stor. Nat. Genova, 16:696.
    COMMON NAME: Lesser Papuan Pipistrelle.
    TYPE LOCALITY: Indonesia, Prov. of Papua, Salawati Isl.
    DISTRIBUTION: Seram, Aru Isls, Baik-Supiori, New Guinea, New Ireland (Bismarck Arch.),
        adjacent small islands.
    STATUS: IUCN 2003 and IUCN/SSC Action Plan (2001) – Lower Risk (nt).
    SYNONYMS: *orientalis* Meyer, 1899.
    COMMENTS: Subgenus *Pipistrellus*. Included in *tenuis* by many authors, but see Kitchener
        et al. (1986); also see Flannery (1995*a, b*) and Bonaccorso (1998).

*Pipistrellus paterculus* Thomas, 1915. J. Bombay Nat. Hist. Soc., 24:32.
  COMMON NAME: Mount Popa Pipistrelle.
  TYPE LOCALITY: Burma, Mt. Popa.
  DISTRIBUTION: N India, Burma, Thailand, Vietnam, SW China.
  STATUS: IUCN 2003 and IUCN/SSC Action Plan (2001) – Lower Risk (nt).
  SYNONYMS: *yunnanensis* Wang, 1982.
  COMMENTS: Subgenus *Pipistrellus*. Included in *abramus* by Ellerman and Morrison-Scott
    (1951), but see Hill and Harrison (1987), Corbet and Hill (1992), Bates and Harrison
    (1997), and Bates et al. (1997), and Hendrichsen et al. (2001*b*). Also see Lunde et al.
    (2003*a*).

*Pipistrellus permixtus* Aellen, 1957. Rev. Suisse Zool., 64:200.
  COMMON NAME: Dar-es-Salaam Pipistrelle.
  TYPE LOCALITY: Tanzania, Dar-es-Salaam.
  DISTRIBUTION: NE Tanzania.
  STATUS: IUCN 2003 and IUCN/SSC Action Plan (2001) – Data Deficient.
  COMMENTS: Subgenus *Pipistrellus*.

*Pipistrellus pipistrellus* (Schreber, 1774). Die Säugethiere, 1:167.
  COMMON NAME: Common Pipistrelle.
  TYPE LOCALITY: France.
  DISTRIBUTION: British Isles, S Denmark, and W Europe to the Volga and Caucasus; Morocco;
    Greece, Turkey, Israel and Lebanon to Afghanistan, Kashmir, Kazakhstan, Pakistan,
    Burma, Sinkiang (China). Perhaps Korea, Japan and Taiwan.
  STATUS: IUCN 2003 and IUCN/SSC Action Plan (2001) – Lower Risk (lc).
  SYNONYMS: *brachyotos* Baillon, 1834; *flavescens* Koch, 1865; *genei* Bonaparte, 1845; *griseus*
    Gray, 1842; *limbatus* Koch, 1863; *macropterus* Jeitteles, 1862; *melanopterus* Schinz,
    1840; *minutissimus* Schinz, 1840; *murinus* Gray, 1838; *nigra* de Selys Longchamps, 1839
    [*nomen nudum*]; *nigricans* Bonaparte, 1845; *pipistrelle* Müller, 1776; *pusillus* Schinz,
    1840; *rufescens* de Selys Longchamps, 1839 [*nomen nudum*, not *rufescens* Brehm, 1829];
    *stenotus* Schinz, 1840 [not Noack, 1899, or LeConte, 1857]; *typus* Bonaparte, 1845;
    **aladdin** Thomas, 1905; *almatensis* Severtzov, 1873 [*nomen nudum*]; *bactrianus* Satunin,
    1905; *fulvus* Korelov, 1947; *kuzyakini* Korelov, 1947; *oxianus* Bogdanov, 1882 [*nomen
    nudum*]. **Unassigned**: *lacteus* Temminck, 1840 [locality unknown].
  COMMENTS: Subgenus *Pipistrellus*. A cryptic species previously confused with *pipistrellus* was
    recently identified based on echolocation call frequency and DNA sequence
    divergence; this taxon has been given the name *pygmaeus* Leach, 1825, see Jones and
    van Parijs (1993), Barratt et al. (1995, 1997), Jones and Barratt (1999), Häussler et al.
    (2000), Russo and Jones (2000), and Sendor et al. (2002). The International
    Commission on Zoological Nomenclature (2003*b*) placed both *pipistrellus* and
    *pygmaeus* on the Official List of Specific Names in Zoology, and designated neotypes
    for both species to prevent future confusion of these taxa. Includes *aladdin*; see Corbet
    (1978*c*) and Bates and Harrison (1997). Does not include *mediterraneus*, which is a
    synonym of *pygmaeus*; see Jones and Barratt (1999) and Häussler et al. (2000). Some of
    the synonyms listed above may actually represent *pygmaeus*; they are retained here
    pending reexamination. See Jones (1997) and Mayer and Helversen (2001*b*) for
    geographic range in Europe and Harrison and Bates (1991) for the Middle East.

*Pipistrellus pygmaeus* (Leach, 1825). Zool. J. 1:559.
  COMMON NAME: Soprano Pipistrelle.
  TYPE LOCALITY: England, Devonshire, Dartmoor.
  DISTRIBUTION: British Isles, S Scandinavia south to Spain, Portugal, Corsica, Sardinina, Italy,
    Slovenia, and Greece; east to Ukraine and W Russia (perhaps much further east);
    N Algeria, Tunisia, Libya (Cyrenaica only).
  STATUS: IUCN 2003 – Not evaluated; not considered in IUCN/SSC Action Plan (2001).
  SYNONYMS: *mediterraneus* Cabrera, 1904.
  COMMENTS: Subgenus *Pipistrellus*. Previously confused with *pipistrellus*, but clearly distinct;
    see Jones and van Parijs (1993), Barratt et al. (1995, 1997), Jones and Barratt (1999),
    Häussler et al. (2000), Russo and Jones (2000), Ziegler et al. (2001), and Sendor et al.

(2002). Conspecific with *mediterraneus*, see Jones and Barratt (1999) and Häussler et al. (2000). von Helversen et al. (2000) supported use of the name *mediterraneus* rather than *pygmaeus* for this species, but this was rejected by the International Commission on Zoological Nomenclature (2003*b*), which recently placed both *pipistrellus* and *pygmaeus* on the Official List of Specific Names in Zoology and designated neotypes for both species to prevent future confusion of these taxa. Some of the synonyms listed under *pipistrellus* may actually represent *pygmaeus*; they are retained under the former pending reexamination. See Jones (1997) and Mayer and Helversen (2001*b*) for geographic range in Europe. The eastern limits of the range of this species are presently unknown, as some Asian and Middle Eastern populations presently attributed to *Pipistrellus pipistrellus* may actually represent *pygmaeus*.

*Pipistrellus rueppellii* (J. Fischer, 1829). Synopsis Mamm., p. 109.
COMMON NAME: Rüppell's Pipistrelle.
TYPE LOCALITY: Sudan, Northern Province, Dongola.
DISTRIBUTION: Mauritania, Senegal, Algeria, Israel, Egypt, and Iraq, south to Botswana and NE South Africa; Zanzibar.
STATUS: IUCN 2003 and IUCN/SSC Action Plan (2001) – Lower Risk (lc).
SYNONYMS: *hypoleucus* Fitzinger, 1866; *temminckii* Cretzschmar, 1826 [not Horsfield, 1824]; **coxi** Thomas, 1919; **fuscipes** Thomas, 1913; **pulcher** Dobson, 1875; **senegalensis** Dorst, 1960; **vernayi** Roberts, 1932; *leucomelas* Monard, 1933.
COMMENTS: Subgenus *Pipistrellus*. Reviewed in part by Harrison and Bates (1991). See Taylor (2000*a*) for distribution map. Different authors have misspelled this name in a variety of ways, dropping the first "e", second "p", second "l", or second "i". The original, correct spelling is "rueppellii".

*Pipistrellus rusticus* (Tomes, 1861). Proc. Zool. Soc. Lond., 1861:35.
COMMON NAME: Rusty Pipistrelle.
TYPE LOCALITY: Namibia, Damaraland, Olifants Vlei.
DISTRIBUTION: Senegal, Gambia, Burkina Faso, Ghana, Nigeria, Central African Republic, and Ethiopia, south to Kenya, Tanzania, Malawi, Zambia, South Africa. A specimen from Liberia has been tentatively reidentified as *kuhlii* (see Koopman et al., 1995).
STATUS: IUCN 2003 and IUCN/SSC Action Plan (2001) – Lower Risk (lc).
SYNONYMS: **marrensis** Thomas and Hinton, 1923.
COMMENTS: Subgenus *Pipistrellus*. Includes *marrensis*; see Koopman (1975). Geographic range reviewed by Kock et al. (2002).

*Pipistrellus stenopterus* (Dobson, 1875). Proc. Zool. Soc. Lond., 1875:470.
COMMON NAME: Narrow-winged Pipistrelle.
TYPE LOCALITY: Malaysia, Borneo, Sarawak.
DISTRIBUTION: W Malaysia, Sumatra, Riau Arch., N Borneo, Mindanao (Philippines).
STATUS: IUCN 2003 and IUCN/SSC Action Plan (2001) – Lower Risk (lc).
COMMENTS: Subgenus *Pipistrellus*. Transferred from *Nyctalus* to *Pipistrellus* by Medway (1977) following Tate (1942*a*). Koopman (1989*a*, 1993) suggested that this species might best be returned to *Nyctalus*, but see Hill and Harrison (1987) and Corbet and Hill (1992), who instead placed it in *Hypsugo*. Volleth and Heller (1994) presented strong karyotypic evidence that *stenopterus* is a true *Pipistrellus* closely related to *javanicus* and *mimus* (the latter here considered a junior synonym of *tenuis*).

*Pipistrellus sturdeei* Thomas, 1915. Ann. Mag. Nat. Hist., ser. 8, 15:230.
COMMON NAME: Sturdee's Pipistrelle.
TYPE LOCALITY: Japan, Bonin Isls, Hillsboro (= Hahajima) Isl.
DISTRIBUTION: Bonin Isls (Japan).
STATUS: IUCN 2003 and IUCN/SSC Action Plan (2001) – Extinct.
COMMENTS: Subgenus *Pipistrellus*. Reviewed by Yoshiyuki (1989).

*Pipistrellus subflavus* (F. Cuvier, 1832). Nouv. Ann. Mus. Natn. Hist. Nat. Paris, 1:17.
COMMON NAME: Eastern Pipistrelle.
TYPE LOCALITY: USA, Georgia.

DISTRIBUTION: Nova Scotia, S Quebec (Canada), and Minnesota (USA), south to Florida (USA) and Honduras.

STATUS: IUCN 2003 and IUCN/SSC Action Plan (2001) – Lower Risk (lc).

SYNONYMS: *erythrodactylus* Temminck, 1835-1841; *monticola* Audubon and Bachman, 1841; *obscurus* Miller, 1897; *clarus* Baker, 1954; *floridanus* Davis, 1957; *veraecrucis* Ward, 1891.

COMMENTS: Subgenus *Perimyotis*. Transferred by Menu (1984) to its own genus *(Perimyotis)*, but see Hill and Harrison (1987). See Fujita and Kunz (1984).

*Pipistrellus tenuis* (Temminck, 1840). Monogr. Mamm., 2:229.

COMMON NAME: Least Pipistrelle.

TYPE LOCALITY: Indonesia, Sumatra.

DISTRIBUTION: Afghanistan to the Moluccas; S China, Laos, Vietnam; Cocos Keeling Isl and Christmas Isl (Indian Ocean).

STATUS: IUCN 2003 and IUCN/SSC Action Plan (2001) – Lower Risk (lc).

SYNONYMS: *mimus* Wroughton, 1899; *glaucillus* Wroughton, 1912; *principulus* Thomas, 1915; *murrayi* Andrews, 1900; *nitidus* Tomes, 1859; *ponceleti* Troughton, 1936; *portensis* Allen, 1906; *tramatus* Thomas, 1928; *sewelanus* Oei, 1960; *subulidens* Miller, 1901.

COMMENTS: Subgenus *Pipistrellus*. Does not include *adamsi*, *angulatus*, *collinus*, *orientalis*, *papuanus*, *wattsi*, or *westralis*; see Kitchener et al. (1986), but also see Koopman (1984c, 1994) and Corbet and Hill (1992). See also Koopman (1973) and McKean and Price (1978) for discussion of synonyms. Reviewed in part by Bates and Harrison (1997) and Hendrichsen et al. (2001b). This complex may include more than one species.

*Pipistrellus wattsi* Kitchener, Caputi, and Jones, 1986. Rec. West. Aust. Mus., 12:472.

COMMON NAME: Watts's Pipistrelle.

TYPE LOCALITY: Papua New Guinea, Tepala, 8°05'S, 146°12'E.

DISTRIBUTION: SE Papua New Guinea and Sanari Isl.

STATUS: IUCN 2003 and IUCN/SSC Action Plan (2001) – Lower Risk (nt).

COMMENTS: Subgenus *Pipistrellus*. Included in *tenuis* by Koopman (1993, 1994), but see Kitchener et al. (1986). Also see Flannery (1995a, b) and Bonaccorso (1998).

*Pipistrellus westralis* Koopman, 1984. Amer. Mus. Novit., 2778:13.

COMMON NAME: Koopman's Pipistrelle.

TYPE LOCALITY: Australia, Western Australia, Cape Bossut, 18°40'S, 121°30'E.

DISTRIBUTION: N Australia from Kimberly to E Gulf of Carpentaria.

STATUS: IUCN 2003 and IUCN/SSC Action Plan (2001) – Lower Risk (lc).

COMMENTS: Subgenus *Pipistrellus*. Included in *tenuis* by Koopman (1984c, 1993, 1994), but see Kitchener et al. (1986).

*Scotozous* Dobson, 1875. Proc. Zool. Soc. Lond., 1875:372.

TYPE SPECIES: *Scotozous dormeri* Dobson, 1875.

COMMENTS: Included in *Pipistrellus* by some authors (e.g., Bates and Harrison, 1997; Ellerman and Morrison-Scott, 1951; Koopman, 1993, 1994; Sinha, 1999), but see Corbet and Hill (1992). Considered congeneric with *Scotoecus* by Menu (1987), but see Hill and Harrison (1987) and Corbet and Hill (1992). Phylogenetic relationships were discussed by Volleth and Heller (1994).

*Scotozous dormeri* Dobson, 1875. Proc. Zool. Soc. Lond., 1875:373.

COMMON NAME: Dormer's Pipistrelle.

TYPE LOCALITY: India, Mysore, Bellary Hills.

DISTRIBUTION: India, Pakistan. A record from Taiwan is erroneous (Koopman, 1994).

STATUS: IUCN 2003 and IUCN/SSC Action Plan (2001) – Lower Risk (lc) as *Pipistrellus dormeri*.

SYNONYMS: *caurinus* Thomas, 1915.

COMMENTS: Reviewed by Bates and Harrison (1997).

**Tribe Plecotini** Gray, 1866. Ann. Mag. Nat. Hist., ser. 3, 17:90.
> COMMENTS: Apparently includes *Otonycteris*; see Qumsiyeh and Bickham (1993) and
> Bogdanowicz et al. (1998), although also see Pine et al. (1971) and Hoofer and Van Den
> Bussche (2001). For phylogenies also see Tumlison and Douglas (1992) and Frost and
> Timm (1992). See Frost and Timm (1992) for generic diagnoses, but note that they
> included *Idionycteris* in *Euderma*.

*Barbastella* Gray, 1821. London Med. Repos., 15:300.
> TYPE SPECIES: *Vespertilio barbastellus* Schreber, 1774.
> SYNONYMS: *Synotus* Keyserling and Blasius, 1839.
> COMMENTS: Corbet (1978c) provided a key separating the two species.

*Barbastella barbastellus* (Schreber, 1774). Die Säugethiere, 1:168.
> COMMON NAME: Western Barbastelle.
> TYPE LOCALITY: France, Burgundy.
> DISTRIBUTION: England and W Europe to Caucasus; Bulgaria; Turkey; Crimea (Ukraine);
> Morocco; larger Mediterranean islands; Canary Isls; perhaps Senegal.
> STATUS: IUCN 2003 and IUCN/SSC Action Plan (2001) – Vulnerable.
> SYNONYMS: *barbastelle* Müller, 1776; *communis* Gray, 1838; *daubentonii* Bell, 1836; ***guanchae***
> Trujillo, Ibáñez, and Juste, 2002.
> COMMENTS: Apparently does not include *leucomelas*, but see Qumsiyeh (1985) and Benda
> and Horácek (1998), who suggested that they might be conspecific (see discussion
> under *leucomelas*). Reviewed by Trujillo et al. (2002). Juste et al. (2003) discussed
> phylogeography of this species.

*Barbastella leucomelas* (Cretzschmar, 1826). *In* Ruppell, Atlas Reise Nordl. Afr., Zool. Säugeth., p. 73.
> COMMON NAME: Eastern Barbastelle.
> TYPE LOCALITY: Egypt, Sinai.
> DISTRIBUTION: Caucasus to The Pamirs, N Iran, Afghanistan, India, Nepal, and W China;
> Honshu, Hokkaido (Japan); Sinai (Egypt); Eritrea; perhaps Indo-China.
> STATUS: IUCN 2003 and IUCN/SSC Action Plan (2001) – Lower Risk (lc).
> SYNONYMS: ***darjelingensis*** Hodgson, 1855 [in Horsfield, 1855]; *blanfordi* Bianchi, 1917;
> *caspica* Satunin, 1908; *dargelinensis* Dobson, 1875; *walteri* Bianchi, 1916.
> COMMENTS: Reviewed in part by De Blase (1980), Qumsiyeh (1985), Yoshiyuki (1989),
> Harrison and Bates (1991), and Bates and Harrison (1997). Horácek et al. (2000)
> suggested that the western subspecies *leucomelas* may be conspecific with *barbastellus*,
> but retained these as separate species pending further study. If *leucomelas* is conspecific
> with *barbastellus*, the oldest name for the Eastern Barbastelle (widely regarded as a
> distinct species) would be *darjelingensis*. Japanese populations may also be distinct at
> the subspecies or species level (Horácek et al., 2000).

*Corynorhinus* H. Allen, 1865. Proc. Acad. Nat. Sci. Phildelphia, 17:173.
> TYPE SPECIES: *Plecotus macrotis* Le Conte, 1831 (= *Plecotus rafinesquii* Lesson, 1827).
> COMMENTS: Included in *Plecotus* by many authors, but see Tumlinson and Douglas (1992), Frost
> and Timm (1992), Bogdanowicz et al. (1998), and Hoofer and Van Den Bussche (2001).

*Corynorhinus mexicanus* G. M. Allen, 1916. Bull. Mus. Comp. Zool., 60:347.
> COMMON NAME: Mexican Big-eared Bat.
> TYPE LOCALITY: Mexico, Chihuahua, Pacheco.
> DISTRIBUTION: Sonora and Coahuila to Michoacan Yucatán (Mexico); Cozumel Isl (Mexico).
> STATUS: IUCN 2003 and IUCN/SSC Action Plan (2001) – Lower Risk (lc) as *Plecotus*
> *mexicanus*.
> COMMENTS: Listed as a subspecies of *townsendii* by Hall and Kelson (1959), but see Handley
> (1959b) and Hall (1981). See Tumlison (1992).

*Corynorhinus rafinesquii* (Lesson, 1827). Manuel de Mammalogie, p. 96.
> COMMON NAME: Rafinesque's Big-eared Bat.
> TYPE LOCALITY: USA, Illinois, Wabash Co., Mt. Carmel.
> DISTRIBUTION: SE USA from Virginia to Missouri, south to E Texas and Florida.

STATUS: IUCN 2003 and IUCN/SSC Action Plan (2001) – Vulnerable as *Plecotus rafinesquii*.
SYNONYMS: *megalotis* Rafinesque, 1818 [not Bechstein, 1800]; **macrotis** Le Conte, 1831; *leconteii* Cooper, 1837.
COMMENTS: See C. Jones (1977).

*Corynorhinus townsendii* (Cooper, 1837). Ann. Lyc. Nat. Hist., 4:73.
COMMON NAME: Townsend's Big-eared Bat.
TYPE LOCALITY: USA, Washington, Clark Co., Fort Vancouver.
DISTRIBUTION: S British Columbia (Canada) through W USA to Oaxaca (Mexico), east to Virginia.
STATUS: U.S. ESA – Endangered as *Plecotus ingens* and *Plecotus virginianus*. IUCN 2003 and IUCN/SSC Action Plan (2001) – Vulnerable as *Plecotus townsendii*.
SYNONYMS: **australis** Handley, 1955; **ingens** Handley, 1955; **pallescens** Miller, 1897; *intermedius* H. W. Grinnell, 1914; **virginianus** Handley, 1955.
COMMENTS: See Kunz and Martin (1982).

*Euderma* H. Allen, 1892. Proc. Acad. Nat. Sci. Phil., 43:467.
TYPE SPECIES: *Histiotus maculatus* J. A. Allen, 1891.
COMMENTS: Revised by Handley (1959b). Does not include *Idionycteris*; see comments under that genus.

*Euderma maculatum* (J. A. Allen, 1891). Bull. Am. Mus. Nat. Hist., 3:195.
COMMON NAME: Spotted Bat.
TYPE LOCALITY: USA, California, Los Angeles Co., Santa Clara Valley, Castac Creek mouth.
DISTRIBUTION: SW Canada and Montana (USA) to Queretaro (Mexico).
STATUS: IUCN 2003 and IUCN/SSC Action Plan (2001) – Lower Risk (lc).
COMMENTS: See Watkins (1977).

*Idionycteris* Anthony, 1923. Am. Mus. Novit., 54:1.
TYPE SPECIES: *Idionycteris mexicanus* Anthony, 1923 (= *Corynorhinus phyllotis* G. M. Allen, 1916).
COMMENTS: *Idionycteris* is considered a separate genus following Williams et al. (1970), Tumlinson and Douglas (1992), Bogdanowicz et al. (1998), and Hoofer and Van Den Bussche, but also see Handley (1959b), who retained it in *Plecotus*, and Frost and Timm (1992), who placed it in *Euderma*.

*Idionycteris phyllotis* (G. M. Allen, 1916). Bull. Mus. Comp. Zool., 60:352.
COMMON NAME: Allen's Big-eared Bat.
TYPE LOCALITY: Mexico, San Luis Potosi, probably near city of San Luis Potosi (see Hall, 1981).
DISTRIBUTION: Distrito Federal and Michoacan (Mexico) to S Utah and S Nevada (USA).
STATUS: IUCN 2003 and IUCN/SSC Action Plan (2001) – Lower Risk (lc).
SYNONYMS: *mexicanus* Anthony, 1923.
COMMENTS: See Czaplewski (1983).

*Otonycteris* Peters, 1859. Monatsb. K. Preuss. Akad. Wiss. Berlin, 1859:223.
TYPE SPECIES: *Otonycteris hemprichii* Peters, 1859.
COMMENTS: Often placed in Nycticeini (e.g., Koopman, 1994; McKenna and Bell, 1997), but recent phylogenetic analyses have grouped this taxon with plecotines (Bogdanowicz et al., 1998; Qumsiyeh and Bickham, 1993), although also see Pine et al. (1971) and Hoofer and Van Den Bussche (2001), who suggested a close relationship between *Otonycteris* and *Antrozous*.

*Otonycteris hemprichii* Peters, 1859. Monatsb. K. Preuss. Akad. Wiss. Berlin, 1859:223.
COMMON NAME: Hemprich's Desert Bat.
TYPE LOCALITY: Restricted by Kock (1969a) to the Nile Valley between north of Aswan, Egypt and Chondek, Sudan.
DISTRIBUTION: The desert zone from Morocco and Niger through Tunisia, Algeria, Libya, Egypt, Oman, Saudi Arabia, Jordan, Syria, and Iraq to Turkmenistan, Tajikistan, Uzbekistan, Kyrgyzstan, Afghanistan, and Kashmir.

STATUS: IUCN 2003 and IUCN/SSC Action Plan (2001) – Lower Risk (lc).

SYNONYMS: *brevimanus* Severtzov, 1873 [not Jenyns, 1829]; *cinerea* Satunin, 1909; *jin* Cheesman and Hinton, 1924; *leucophaeus* Severtzov, 1873; *petersi* Anderson and de Winton, 1902; *saharae* Laurent, 1936; *ustus* Fitzinger and Heuglin, 1866 [*nomen nudum*].

COMMENTS: Reviewed by Harrison and Bates (1991), Horácek (1991), and Bates and Harrison (1997). Several subspecies are sometimes recognized (e.g., Harrison and Bates, 1991; Koopman, 1994), but Horácek (1991) and Horácek et al. (2000) have argued that geographic variation in size and coloration is clinal and therefore does not support recognition of local populations as subspecies.

*Plecotus* E. Geoffroy Saint-Hilaire, 1818. Descrip. de L'Egypte, 2:112.

TYPE SPECIES: *Vespertilio auritus* Linnaeus, 1758.

SYNONYMS: *Macrotus* Leach, 1816 [*nomen nudum*; not Gray, 1842].

COMMENTS: Does not include *Idionycteris* or *Corynorhinus*; see Williams et al. (1970), Tumlinson and Douglas (1992), Frost and Timm (1993), Bogdanowicz et al. (1998), and Hoofer and Van Den Bussche (2001). See C. Jones (1977) for a key to species of *Plecotus* and *Corynorhinus*, but note that new species have been described since that publication.

*Plecotus alpinus* Kiefer and Veith, 2002. Myotis, 39:8. [dated 2001; issued April, 2002].

COMMON NAME: Alpine Long-eared Bat.

TYPE LOCALITY: France, Haute-Alpes, Ristolas, 44°46′N, 06°57′E, 1600 m.

DISTRIBUTION: France, Liechtenstein, Switzerland, Austria, Croatia, Greece.

STATUS: IUCN 2003 – Not evaluated (new species).

SYNONYMS: *microdontus* Spitzenberger, 2002 [in Spitsenberger et al., 2002].

COMMENTS: Morphologically similar to *auritus* and *austriacus* and probably confused with these taxa in some previous studies. See Kock (2002) for discussion of priority of the name *alpinus* over *microdontus* for this species. For comparisons with other European *Plecotus* species, see Mucedda et al. (2002) and Spitzenberger et al. (2002). Also see Kiefer et al. (2002). Garin et al. (2003) suggested that *macrobullaris* may be a senior synonym of *alpinus* (rather than a subspecies of *austriacus*), but the supporting data have not yet been published.

*Plecotus auritus* (Linnaeus, 1758). Syst. Nat., 10th ed., 1:32.

COMMON NAME: Brown Long-eared Bat.

TYPE LOCALITY: Sweden.

DISTRIBUTION: Norway, Ireland, and Spain to Sakhalin Isl (Russia), Korea, Japan, N China, Nepal, India.

STATUS: IUCN 2003 and IUCN/SSC Action Plan (2001) – Lower Risk (lc).

SYNONYMS: *bonapartii* Gray, 1838 [*nomen nudum*]; *brevimanus* Jenyns, 1829; *communis* Lesson, 1827; *cornutus* Faber, 1826; *europaeus* Leach, 1816 [*nomen nudum*]; *megalotos* Schinz, 1840; *montanus* Koch, 1865; *otus* Boie, 1825; *peronii* I. Geoffroy, 1832; *typus* Koch, 1865; *velatus* I. Geoffroy, 1832; *vulgaris* Desmarest, 1829; **begognae** de Paz, 1994; **homochrous** Hodgson, 1847; *puck* Barrett-Hamilton, 1907; **sacrimontis** G. M. Allen, 1908; *ognevi* Kishida, 1927; **uenoi** Imaizumi and Yoshiyuki, 1969.

COMMENTS: Reviewed in part by Yoshiyuki (1989, 1991*b*), de Paz (1994), Bates and Harrison (1997), Sinha (1999), and Spitzenberger et al. (2001); also see Kiefer and Veith (2001), Kiefer et al. (2002), and Mucedda et al. (2002). Subspecies allocation of populations from northern China, eastern Siberia, and Sakhalin is uncertain. This complex may include more than one species; *homochrous* may represent a distinct species (Horácek et al., 2000), and it is possible that other forms may also be distinct (see Mucedda et al., 2002).

*Plecotus austriacus* (J. Fischer, 1829). Synopsis Mamm., p. 117.

COMMON NAME: Gray Long-eared Bat.

TYPE LOCALITY: Austria, Vienna.

DISTRIBUTION: England and Spain to Mongolia and W China; N Africa from Morocco to Egypt and Sudan; Canary Isls (Spain) and Cape Verde Isls. A report of this species from Senegal is in error, see Grubb and Ansell (1996).

STATUS: IUCN 2003 and IUCN/SSC Action Plan (2001) – Lower Risk (lc).

SYNONYMS: *brevipes* Koch, 1865; *hispanicus* Bauer, 1957; *kirschbaumii* Koch, 1860; ***ariel*** Thomas, 1911; *kozlovi* Bobrinski, 1926; ***christii*** Gray, 1838; *aegyptius* Fischer, 1829:117 [not Fischer, 1829:105]; *meridionalis* Martino, 1940; ***macrobullaris*** Kuzyakin, 1965; ***turkmenicus*** Strelkov, 1988 [replacement for *turkmenicus* Strelkov, 1983, *nomen nudum*]; ***wardi*** Thomas, 1911; *mordax* Thomas, 1926.

COMMENTS: Included in *auritus* by Ellerman and Morrison-Scott (1951), but see Corbet (1978c). Does not include *teneriffae*; see Ibáñez and Fernández (1985), though also see Corbet (1978c). Does not include *kolombatovici*, see Mayer and von Helversen (2001a), Kiefer and Veith (2001), Spitzenberger et al. (2001), Kiefer et al. (2002), and Mucedda et al. (2002). Reviewed in part by Yoshiyuki (1991b), Kock (1996), Harrison and Bates (1991), Bates and Harrison (1997), Horácek et al. (2000), and Mucedda et al. (2002). See Horácek et al. (2000) for a summary of presumed subspecies limits, but note that subspecific allocation of many populations is uncertain. *P. wardi* is included here following Koopman (1993, 1994), Horácek et al. (2000) and other authors, but Sinha (1999) treated this taxon as a subspecies of *auritus* rather than *austriacus*. Garin et al. (2003) suggested that *macrobullaris* may be a senior synonym of *alpinus* (rather than a subspecies of *austriacus*), but the supporting data have not yet been published.

*Plecotus balensis* Kruskop and Lavrenchenko, 2000. Myotis 38:6.
COMMON NAME: Bale Long-eared Bat.
TYPE LOCALITY: Ethiopia, southern Ethiopia, Bale Mountains National Park, Harenna Forest, 6°45′N, 39°44′E, 2,760 m.
DISTRIBUTION: S Ethiopia.
STATUS: IUCN 2003 – Not evaluated (new species); not considered in IUCN/SSC Action Plan (2001).

*Plecotus kolombatovici* Dulic, 1980. Proc. 5th Internat. Bat Res. Conf., (D. E. Wilson and A. L. Gardner, eds.), Texas Tech Press., pg. 159.
COMMON NAME: Kolombatovic's Long-eared Bat.
TYPE LOCALITY: Croatia, Dalmatia, Korcula Isl., 2.5 km NW Zrnovo, 276 m.
DISTRIBUTION: Croatia and nearby islands in the Adriatic Sea.
STATUS: IUCN 2003 – Not evaluated; not considered in IUCN/SSC Action Plan (2001).
COMMENTS: Originally described as a subspecies of *austriacus*, but clearly distinct; see Mayer and von Helversen (2001a), Kiefer and Veith (2001), Spitzenberger et al. (2001), Kiefer et al. (2002), and Mucedda et al. (2002).

*Plecotus sardus* Mucedda, Kiefer, Pidinchedda, and Veith, 2002. Acta Chiropterol., 4:123.
COMMON NAME: Sardinian Long-eared Bat.
TYPE LOCALITY: Italy, Sardinia, Nuoro Province, Oliena District, Lanaitto's Valley, in a cave, 40°15′29″N, 09°29′13″E, 150 m.
DISTRIBUTION: Sardinia (Italy).
STATUS: IUCN 2003 – Not evaluated (new species).
COMMENTS: Closely related to *auritus* and *alpinus*, but clearly distinct.

*Plecotus taivanus* Yoshiyuki, 1991. Bull. Natl. Sci. Mus. Tokyo, ser. A(Zool.), 17:189.
COMMON NAME: Taiwan Long-eared Bat.
TYPE LOCALITY: Taiwan, Taichung Hsien, Hoping Hsiang, Mt. Anma Shan, 2,250 m.
DISTRIBUTION: Taiwan.
STATUS: IUCN 2003 and IUCN/SSC Action Plan (2001) – Vulnerable.
COMMENTS: Most similar to *homochrous* and *puck*, here included in *auritus*.

*Plecotus teneriffae* Barrett-Hamilton, 1907. Ann. Mag. Nat. Hist., ser. 7, 20:520.
COMMON NAME: Canary Long-eared Bat.
TYPE LOCALITY: Spain, Canary Isls, Teneriffe Isl.
DISTRIBUTION: Canary Isls (Spain).
STATUS: IUCN 2003 and IUCN/SSC Action Plan (2001) – Vulnerable.
COMMENTS: Synonymized with *austriacus* by Corbet (1978c), but see Ibáñez and Fernández (1985).

**Tribe Vespertilionini** Gray, 1821. London Med. Repos., 15:299.
> COMMENTS: Does not include *Arielulus, Eptesicus, Hesperoptenus, Glischropus, Nyctalus, Pipistrellus,* and *Scotozous*; see Volleth (1992), Volleth and Heller (1994), and Volleth et al. (2001). Includes *Hypsugo, Falsistrellus, Tylonycteris,* and *Vespadelus*; see Volleth and Tidemann (1991) and Volleth and Heller (1994). Includes *Neoromicia*; see Volleth et al. (2001). May include *Nyctophilus* (Volleth and Tideman, 1991), here placed in a separate tribe Nyctophilini along with *Pharotis*.

*Chalinolobus* Peters, 1867. Monatsb. K. Preuss. Akad. Wiss. Berlin, 1866:679 [1867].
> TYPE SPECIES: *Vespertilio tuberculatus* Forster, 1844 (by International Commision on Zoological Nomenclature ruling, Opinion 1994 [2002]).
> COMMENTS: Does not include *Glauconycteris*; see Hill and Harrison (1987) and Volleth and Heller (1994). Tate (1942*a*) reviewed all named forms, and Chruszez and Barclay (2002) provided a key to the genus.

*Chalinolobus dwyeri* Ryan, 1966. J. Mammal., 47:89.
> COMMON NAME: Large-eared Pied Bat.
> TYPE LOCALITY: Australia, New South Wales, 14 mi. (23 km) S Inverell, Copeton.
> DISTRIBUTION: New South Wales and adjacent part of Queensland (Australia).
> STATUS: IUCN 2003 and IUCN/SSC Action Plan (2001) – Vulnerable.

*Chalinolobus gouldii* (Gray, 1841). Appendix C *in* J. Two Exped. Aust., 2:401, 405.
> COMMON NAME: Gould's Wattled Bat.
> TYPE LOCALITY: Australia, Tasmania, Launceston.
> DISTRIBUTION: Australia but not Cape York Peninsula N of Cardwell; Tasmania, Norfolk Isl (Australia).
> STATUS: IUCN 2003 and IUCN/SSC Action Plan (2001) – Lower Risk (lc).
> SYNONYMS: *venatoris* Thomas, 1908.
> COMMENTS: Reviewed by Tidemann (1986) and Chruszez and Barclay (2002), although note that they included *neocaledonicus* in this species. Does not include *neocaledonicus*, see Flannery (1995*b*) and discussion under that species. The population from Norfolk Isl (as yet unnamed) may also represent a distinct species, see Flannery (1995*b*).

*Chalinolobus morio* (Gray, 1841). Appendix C *in* J. Two Exped. Aust., 2:400, 405.
> COMMON NAME: Chocolate Wattled Bat.
> TYPE LOCALITY: Australia, Tasmania.
> DISTRIBUTION: Southern Australia, Tasmania.
> STATUS: IUCN 2003 and IUCN/SSC Action Plan (2001) – Lower Risk (lc).
> SYNONYMS: *australis* Gray, 1841; *microdon* Tomes, 1860; *signifer* Dobson, 1876.

*Chalinolobus neocaledonicus* Revilliod, 1914. *In* Sarasin and Roux, Nova Caledonia, A. Zool., p. 355.
> COMMON NAME: New Caledonia Wattled Bat.
> TYPE LOCALITY: New Caledonia, Canala.
> DISTRIBUTION: New Caledonia.
> STATUS: IUCN 2003 and IUCN/SSC Action Plan (2001) – Endangered.
> COMMENTS: Often treated as a subspecies of *gouldii* (e.g., Koopman, 1971*a*, 1994; Tidemann, 1986), but evidence for synonymy is weak; I follow Flannery (1995*b*) in provisionally recognizing *neocaledonicus* as distinct pending further study.

*Chalinolobus nigrogriseus* (Gould, 1852). Mamm. Aust., pt. 4, vol. 3, pl. 43.
> COMMON NAME: Hoary Wattled Bat.
> TYPE LOCALITY: Australia, Queensland, vic. of Moreton Bay.
> DISTRIBUTION: N and E Australia; SE New Guinea and adjacent small islands.
> STATUS: IUCN 2003 and IUCN/SSC Action Plan (2001) – Lower Risk (lc).
> SYNONYMS: **rogersi** Thomas, 1909.
> COMMENTS: Includes *rogersi*; see Van Deusen and Koopman (1971), who revised the species. Also see Flannery (1995*a, b*) and Bonaccorso (1998).

*Chalinolobus picatus* (Gould, 1852). Mamm. Aust., pt. 4, vol. 3, pl. 43.
> COMMON NAME: Little Pied Bat.

TYPE LOCALITY: Australia, New South Wales, Capt. Sturt's Depot.
DISTRIBUTION: NW New South Wales, C and S Queensland, and South Australia (Australia).
STATUS: IUCN 2003 and IUCN/SSC Action Plan (2001) – Lower Risk (nt).
COMMENTS: Reviewed by Van Deusen and Koopman (1971).

*Chalinolobus tuberculatus* (Forster, 1844). Descrip. Animal. Itinere Maris Aust. Terras, 1772-74:62.
COMMON NAME: Long-tailed Wattled Bat.
TYPE LOCALITY: New Zealand.
DISTRIBUTION: New Zealand and adjacent small islands.
STATUS: IUCN 2003 and IUCN/SSC Action Plan (2001) – Vulnerable.
COMMENTS: See O'Donnell (2001). Placed on the Offical List Specific Names in Zoology;
    International Commission on Zoological Nomenclature (Opinion 1994 [2002]).

*Eudiscopus* Conisbee, 1953. Last names proposed genera subgenera Recent Mamm., p. 30.
TYPE SPECIES: *Discopus denticulus* Osgood, 1932.
SYNONYMS: *Discopus* Osgood, 1932 [not *Discopus* Thompson, 1864, a coleopteran].

*Eudiscopus denticulus* (Osgood, 1932). Field Mus. Nat. Hist. Publ., Zool. Ser., 18:236.
COMMON NAME: Disk-footed Bat.
TYPE LOCALITY: Laos, Phong Saly, 4,000 ft. (1,219 m).
DISTRIBUTION: Thailand, Laos, Vietnam, C Burma.
STATUS: IUCN 2003 and IUCN/SSC Action Plan (2001) – Lower Risk (nt).
COMMENTS: See Koopman (1972) and Kock and Kovac (2000).

*Falsistrellus* Troughton, 1943. Furred animals of Australia, 1st ed., Sydney: Angus and Robertson,
    p. 349.
TYPE SPECIES: *Vespertilio tasmaniensis* Gould, 1858.
COMMENTS: Included in *Pipistrellus* by many authors (e.g., Corbet and Hill, 1992; Hill and
    Harrison, 1987; Koopman, 1993, 1994), but see Kitchener et al. (1986) and Volleth and
    Heller (1994). See Kitchener et al. (1986) for diagnosis.

*Falsistrellus affinis* (Dobson, 1871). Proc. Asiat. Soc. Bengal, p. 213.
COMMON NAME: Chocolate Pipistrelle.
TYPE LOCALITY: Burma, Bhamo.
DISTRIBUTION: NE Burma, Yunnan (China), India, Nepal, Sri Lanka.
STATUS: IUCN 2003 and IUCN/SSC Action Plan (2001) – Lower Risk (lc) as *Pipistrellus affinus*.
COMMENTS: May include *petersi*; see Francis and Hill (1986) and Corbet and Hill (1992). May
    be conspecific with *mordax*; see Corbet and Hill (1992). Reviewed by Bates and
    Harrison (1997).

*Falsistrellus mackenziei* Kitchener, Caputi, and Jones, 1986. Rec. West. Aust. Mus., 12:451.
COMMON NAME: Mackenzie's False Pipistrelle.
TYPE LOCALITY: Australia, Donelly, 34°06′, 115°58′E.
DISTRIBUTION: SW Australia.
STATUS: IUCN 2003 and IUCN/SSC Action Plan (2001) – Vulnerable as *Pipistrellus mackenziei*
    (misspelled as *mckenziei* in 2001 Action Plan).
COMMENTS: Included in *tasmaniensis* by Koopman (1993, 1994), but see Kitchener et al.
    (1986).

*Falsistrellus mordax* (Peters, 1866). Monatsb. K. Preuss. Akad. Wiss. Berlin, 1866:402.
COMMON NAME: Pungent Pipistrelle.
TYPE LOCALITY: Indonesia, Java.
DISTRIBUTION: Java; records from India and Sri Lanka are erroneous, based on misidentified
    *affinis*, see Hill and Harrison (1987).
STATUS: IUCN 2003 and IUCN/SSC Action Plan (2001) – Lower Risk (nt) as *Pipistrellus
    mordax*.
SYNONYMS: *maderaspatanus* Gray, 1843 [*nomen nudum*].
COMMENTS: May include *petersi* and/or *affinis*; see Corbet and Hill (1992).

*Falsistrellus petersi* (A. Meyer, 1899). Abh. Zool. Anthrop.-Ethnology. Mus. Dresden, 7(7):13.
    COMMON NAME: Peters's Pipistrelle.
    TYPE LOCALITY: Indonesia, Sulawesi, N Sulawesi, Minahassa.
    DISTRIBUTION: Borneo; Sulawesi; Buru and Amboina (Molucca Isls); Philippines.
    STATUS: IUCN 2003 and IUCN/SSC Action Plan (2001) – Lower Risk (lc) as *Pipistrellus petersi*.
    COMMENTS: May be conspecific with *affinis*; see Francis and Hill (1986) and Corbet and Hill (1992).

*Falsistrellus tasmaniensis* (Gould, 1858). Mamm. Aust., 3, pl. 48.
    COMMON NAME: Eastern False Pipistrelle.
    TYPE LOCALITY: Australia, Tasmania.
    DISTRIBUTION: E and SE Australia, Tasmania.
    STATUS: IUCN 2003 and IUCN/SSC Action Plan (2001) – Lower Risk (lc) as *Pipistrellus tasmaniensis*.
    SYNONYMS: *krefftii* Peters, 1869.
    COMMENTS: Does not includes *mackenziei*; see Kitchener et al. (1986).

*Glauconycteris* Dobson, 1875. Proc. Zool. Soc. Lond., 1875:383.
    TYPE SPECIES: *Kerivoula poensis* Gray, 1842.
    COMMENTS: Formerly included in *Chalinolobus*, but see Hill and Harrison (1987) and Volleth and Heller (1994). Reviewed by Tate (1942*a*) and Ryan (1966).

*Glauconycteris alboguttata* J. A. Allen, 1917. Bull. Am. Mus. Nat. Hist., 37:449.
    COMMON NAME: Striped Butterfly Bat.
    TYPE LOCALITY: Dem. Rep. Congo, Oriental, Medje.
    DISTRIBUTION: Dem. Rep. Congo, Cameroon.
    STATUS: IUCN 2003 and IUCN/SSC Action Plan (2001) – Vulnerable as *Chalinolobus alboguttatus*.
    COMMENTS: See Eger and Schlitter (2001).

*Glauconycteris argentata* (Dobson, 1875). Proc. Zool. Soc. Lond., 1875:385.
    COMMON NAME: Common Butterfly Bat.
    TYPE LOCALITY: Cameroon, Western Province, Mt. Cameroon.
    DISTRIBUTION: Cameroon to Kenya, south to Angola, Tanzania, and N Malawi.
    STATUS: IUCN 2003 and IUCN/SSC Action Plan (2001) – Lower Risk (lc) as *Chalinolobus argentatus*.
    COMMENTS: See Peterson and Smith (1973) and Peterson (1982).

*Glauconycteris beatrix* Thomas, 1901. Ann. Mag. Nat. Hist., ser. 7, 8:256.
    COMMON NAME: Beatrix Butterfly Bat.
    TYPE LOCALITY: Equatorial Guinea, Rio Muni, Benito River, 15 mi. (24 km) from mouth.
    DISTRIBUTION: Equatorial Guinea, Côte d'Ivoire, Ghana, Nigeria, Cameroon, Gabon, Angola. A report of this species from Guinea-Bissau is in error (J. Fahr, pers. comm.)
    STATUS: IUCN 2003 and IUCN/SSC Action Plan (2001) – Lower Risk (nt) as *Chalinolobus beatrix*.
    COMMENTS: Does not include *humeralis*; see Hill and Harrison (1987) and Heller et al. (1994), but also see Eger and Schlitter (2001).

*Glauconycteris curryae* Eger and Schlitter, 2001. Acta Chiropterologica 3:2.
    COMMON NAME: Curry's Butterfly Bat.
    TYPE LOCALITY: Cameroon, 10 km W Bipindi, approximately 300 m above sea level; 03°05'N, 10°25'E.
    DISTRIBUTION: Cameroon; Dem. Rep. Congo.
    STATUS: IUCN 2003 – Not evaluated (new species); not considered in IUCN/SSC Action Plan (2001).
    COMMENTS: Originally spelled *curryi*, but emended to *curryae* by Eger (2001).

*Glauconycteris egeria* Thomas, 1913. Ann. Mag. Nat. Hist., ser. 8, 11:144.
    COMMON NAME: Bibundi Butterfly Bat.
    TYPE LOCALITY: Cameroon, Western Province, Bibundi.

DISTRIBUTION: Cameroon, Uganda, Central African Republic (Lunde et al., 2002).
STATUS: IUCN 2003 and IUCN/SSC Action Plan (2001) – Lower Risk (nt) as *Chalinolobus egeria*.

*Glauconycteris gleni* Peterson and Smith, 1973. R. Ont. Mus. Life Sci. Occas. Pap., 22:3.
COMMON NAME: Glen's Butterfly Bat.
TYPE LOCALITY: Cameroon, near Lomie.
DISTRIBUTION: Cameroon, Uganda.
STATUS: IUCN 2003 and IUCN/SSC Action Plan (2001) – Lower Risk (nt) as *Chalinolobus gleni*.
COMMENTS: See Peterson (1982).

*Glauconycteris humeralis* J. A. Allen, 1917. Bull. Am. Mus. Nat. Hist., 37:448.
COMMON NAME: Spotted Butterfly Bat.
TYPE LOCALITY: Dem. Rep. Congo, Oriental, Medje.
DISTRIBUTION: Dem. Rep. Congo, Uganda, Kenya.
STATUS: IUCN 2003 – Not evaluated; not considered in IUCN/SSC Action Plan (2001).
COMMENTS: Apparently distinct from *beatrix*; see Hill and Harrison (1987) and Heller et al. (1994), but also see Eger and Schlitter (2001).

*Glauconycteris kenyacola* Peterson, 1982. Canadian J. Zool., 60:2521.
COMMON NAME: Kenyacola Butterfly Bat.
TYPE LOCALITY: Kenya, Coast Prov., 8.5 km N Garsen.
DISTRIBUTION: Kenya.
STATUS: IUCN 2003 and IUCN/SSC Action Plan (2001) – Data Deficient as *Chalinolobus kenyacola*.
COMMENTS: Known only from the holotype.

*Glauconycteris machadoi* Hayman, 1963. Comp. Diamantes de Angola, Ser. Cult., 1963:107.
COMMON NAME: Machado's Butterfly Bat.
TYPE LOCALITY: Angola, Lac Calundo.
DISTRIBUTION: Angola; known only from the type locality.
STATUS: IUCN 2003 – Not evaluated; not considered in IUCN/SSC Action Plan (2001).
COMMENTS: Known only from the holotype. Koopman (1971a:6) treated *machadoi* as a subspecies of *variegata*, suggesting that it might be "simply a melanistic mutant individual" of the latter widespread species, which is typically pale creamy buff with a whitish head and pale wing membranes. Alternatively, Hayman and Hill (1971) treated *machadoi* as a separate species in recognition of its distinct coloration, which includes brown dorsal fur, a dark brown head, dark wing membranes, and a creamy-white underbelly. Peterson and Smith (1973) and Crawford-Cabral (1986b) also treated *machadoi* as distinct. I follow the latter authors pending additional evidence.

*Glauconycteris poensis* (Gray, 1842). Ann. Mag. Nat. Hist., [ser. 1], 10:258.
COMMON NAME: Abo Butterfly Bat.
TYPE LOCALITY: Nigeria, Abo (lower Niger River).
DISTRIBUTION: Senegal to Uganda; Bioko (Equatorial Guinea); Cameroon.
STATUS: IUCN 2003 and IUCN/SSC Action Plan (2001) – Lower Risk (nt) as *Chalinolobus poensis*.
SYNONYMS: *kraussii* Peters, 1868.
COMMENTS: Distinct from *beatrix* and *humeralis*; see Heller et al. (1994).

*Glauconycteris superba* Hayman, 1939. Ann. Mag. Nat. Hist., ser. 11, 3:219.
COMMON NAME: Pied Butterfly Bat.
TYPE LOCALITY: Dem. Rep. Congo, Oriental, Ituri Dist., Pawa.
DISTRIBUTION: Côte d'Ivoire, Ghana, NE Dem. Rep. Congo.
STATUS: IUCN 2003 and IUCN/SSC Action Plan (2001) – Vulnerable as *Chalinolobus superbus*.
SYNONYMS: *sheila* Hayman, 1947.
COMMENTS: Although *sheila* is sometimes recognized as a subspecies, it might represent only a color variant (Rosevear, 1965; J. Fahr, pers. comm.).

*Glauconycteris variegata* (Tomes, 1861). Proc. Zool. Soc. Lond., 1861:36.
    COMMON NAME: Variegated Butterfly Bat.
    TYPE LOCALITY: Namibia, Otjoro.
    DISTRIBUTION: Senegal to Somalia, south to South Africa.
    STATUS: IUCN 2003 and IUCN/SSC Action Plan (2001) – Lower Risk (lc) as *Chalinolobus*
        *variegatus.*
    SYNONYMS: *papilio* Thomas, 1915; ***phalaena*** Thomas, 1915.
    COMMENTS: Does not include *machadoi*, see Hayman and Hill (1971), but also see Koopman
        (1971*a*).

*Histiotus* Gervais, 1856. *In* F. Comte de Castelnau, Exped. Partes Cen. Am. Sud.(Sec. 7), Vol, 1, pt. 2
    (Mammifères):77.
    TYPE SPECIES: *Plecotus velatus* I. Geoffroy, 1824.
    COMMENTS: Species differences discussed by Handley (1996).

*Histiotus alienus* Thomas, 1916. Ann. Mag. Nat. Hist., ser. 8, 17:276.
    COMMON NAME: Strange Big-eared Brown Bat.
    TYPE LOCALITY: Brazil, Santa Catarina, Joinville.
    DISTRIBUTION: SE Brazil, Uruguay.
    STATUS: IUCN 2003 and IUCN/SSC Action Plan (2001) – Vulnerable.

*Histiotus humboldti* Handley, 1996. Proc. Biol. Soc. Wash., 109:2.
    COMMON NAME: Humboldt's Big-eared Brown Bat.
    TYPE LOCALITY: Venezuela, Distrito Federal, 4 km NNW Caracas, Los Venados, 1,498 m.
        10°32'N, 66°54'W.
    DISTRIBUTION: Colombia, W Venezuela.
    STATUS: IUCN 2003 and IUCN/SSC Action Plan (2001) – Data Deficient.

*Histiotus laephotis* Thomas, 1916. Ann. Mag. Nat. Hist., ser. 8, 17:275.
    COMMON NAME: Thomas's Big-eared Brown Bat.
    TYPE LOCALITY: Bolivia, Caiza.
    DISTRIBUTION: Argentina, S Bolivia, S Peru.
    STATUS: IUCN 2003 – Not evaluated; not considered in IUCN/SSC Action Plan (2001).
    COMMENTS: Treated as a subspecies of *montanus* by Anderson (1997) and as a subspecies of
        *macrotis* by Koopman (1994) and Barquez et al. (1993, 1999), but apparently distinct;
        see Autino et al. (1999) and Barquez and Diaz (2001).

*Histiotus macrotus* (Poeppig, 1835). Reise Chile Peru Amaz., 1:451.
    COMMON NAME: Big-eared Brown Bat.
    TYPE LOCALITY: Chile, Bio-BIo, Antuco.
    DISTRIBUTION: Chile, Argentina.
    STATUS: IUCN 2003 and IUCN/SSC Action Plan (2001) – Lower Risk (nt).
    SYNONYMS: *chilensis* Lesson, 1836; *poeppigii* Fitzinger, 1872.
    COMMENTS: Does not include *laephotis*; see Autino et al. (1999) and Barquez and Diaz (2001).

*Histiotus magellanicus* Philippi, 1866. Arch. Naturg., 1866:113.
    COMMON NAME: Southern Big-eared Brown Bat.
    TYPE LOCALITY: Chile.
    DISTRIBUTION: S Argentina, S Chile.
    STATUS: IUCN 2003 – Not evaluated; not considered in IUCN/SSC Action Plan (2001).
    SYNONYMS: *capucinus* Philippi, 1866.
    COMMENTS: Often treated as a subspecies of *montanus*, but apparently distinct; see Barquez
        et al. (1993) and Mares et al. (1995).

*Histiotus montanus* (Philippi and Landbeck, 1861). Arch. Naturgesch., p. 289.
    COMMON NAME: Small Big-eared Brown Bat.
    TYPE LOCALITY: Chile, Santiago Cordillera.
    DISTRIBUTION: N Chile, Argentina, Uruguay, W Bolivia, S Peru, Ecuador, Colombia,
        Venezuela, perhaps N Peru and S Brazil.
    STATUS: IUCN 2003 and IUCN/SSC Action Plan (2001) – Lower Risk (lc).

SYNONYMS: *segethii* Peters, 1864; ***colombiae*** Thomas, 1916; ***inambarus*** Anthony, 1920.

COMMENTS: Does not include *laephotis*; see Autino et al. (1999) and Barquez and Diaz (2001). Does not include *magellanicus*; see Barquez et al. (1993) and Mares et al. (1995).

*Histiotus velatus* (I. Geoffroy, 1824) Ann. Sci. Nat. Zool., 3:446.

COMMON NAME: Tropical Big-eared Brown Bat.

TYPE LOCALITY: Brazil, Parana, Curitiba.

DISTRIBUTION: E Brazil, Bolivia, Paraguay, NW Argentina

STATUS: IUCN 2003 and IUCN/SSC Action Plan (2001) – Lower Risk (lc).

SYNONYMS: *miotis* Thomas, 1916.

*Hypsugo* Kolenati, 1856. Allegemeine Deutsche Naturhist. Zeit. 2:131.

TYPE SPECIES: *Vespertilio savii* Bonaparte, 1837 (type species fixed by Wallin [1969]).

SYNONYMS: *Parastrellus* Horacek and Hanák, 1985.

COMMENTS: Often included in *Pipistrellus*, but see Horacek and Hanák (1985-1986), Tiunov (1986), Menu (1987), Ruedi and Arlettaz (1991), Volleth and Heller (1994), and Volleth et al. (1994).

*Hypsugo alaschanicus* Bobrinskii, 1926. C. R. Acad. sci. URSS, A, 1926:98.

COMMON NAME: Alashanian Pipistrelle.

TYPE LOCALITY: Mongolia, Alashan Range, Hotin Gol Pass.

DISTRIBUTION: Mongolia, China, Russian Far East to Korea and Tsushima Isl (Japan).

STATUS: IUCN 2003 – Not evaluated; not considered in IUCN/SSC Action Plan (2001).

SYNONYMS: *coreensis* Imazumi, 1955; *velox* Ognev, 1927.

COMMENTS: Formerly included in *savii*, but see Horácek et al. (2000). Horácek et al. (2000) suggested that *coreensis* might represent a separate subspecies, but also see Yoshiyuki (1989), who treated *coreensis* as a distinct species.

*Hypsugo anchietae* (Seabra, 1900). J. Sci. Math. Phys. Nat. Lisboa, ser. 2, 6:26, 120.

COMMON NAME: Anchieta's Pipistrelle.

TYPE LOCALITY: Angola, Cahata.

DISTRIBUTION: Angola, S Dem. Rep. Congo, Zambia, Zimbabwe, KwaZulu-Natal (South Africa).

STATUS: IUCN 2003 and IUCN/SSC Action Plan (2001) – Vulnerable as *Pipistrellus anchietai*.

COMMENTS: The oldest name for this species may be *bicolor*, here listed as a synonym of *Neoromicia tenuipinnis* following Hayman and Hill (1971); see discussion in Koopman (1975) and Hill and Harrison (1987). Reviewed by Cotterill (1996) and Kearney and Taylor (1997). Sometimes misspelled *anchieta* or *anchietai*, but the correct spelling is *anchietae*; see Kock (2001*a*).

*Hypsugo anthonyi* (Tate, 1942). Bull. Am. Mus. Nat. Hist., 80:252.

COMMON NAME: Anthony's Pipistrelle.

TYPE LOCALITY: Burma, Changyinku, 7,000 ft. (2,134 m).

DISTRIBUTION: Known only from the type locality.

STATUS: IUCN 2003 and IUCN/SSC Action Plan (2001) – Critically Endangered as *Pipistrellus anthonyi*.

SYNONYMS: *affinis* Anthony, 1941 [not Dobson, 1871].

COMMENTS: Known only by the holotype. May be referrable to *Nyctalus* or even *Philetor*; see Hill (1966), Koopman (1993), Hill and Harrison (1987), and Corbet and Hill (1992).

*Hypsugo arabicus* (Harrison, 1979). Mammalia, 43:575.

COMMON NAME: Arabian Pipistrelle.

TYPE LOCALITY: Oman, Wadi Sahtan (23°22'N, 57°18'E).

DISTRIBUTION: Oman, Iran.

STATUS: IUCN 2003 and IUCN/SSC Action Plan (2001) – Vulnerable as *Pipistrellus arabicus*.

COMMENTS: Reviewed by Harrison and Bates (1991). May be conspecific with *ariel*, see Benda et al. (2002).

*Hypsugo ariel* (Thomas, 1904). Ann. Mag. Nat. Hist., ser. 7, 14:157.

COMMON NAME: Fairy Pipistrelle.

TYPE LOCALITY: Sudan, Kassala Province, Wadi Alagi (22°N, 35°E), 2,000 ft. (610 m).
DISTRIBUTION: Israel, Jordan, N Sudan, possibly Egypt.
STATUS: IUCN 2003 and IUCN/SSC Action Plan (2001) – Vulnerable as *Pipistrellus ariel*.
COMMENTS: Reviewed by Harrison and Bates (1991). May include *arabicus* and *bodenheimeri*, see Benda et al. (2002).

*Hypsugo bodenheimeri* (Harrison, 1960). Durban Mus. Novit., 5:261.
COMMON NAME: Bodenheimer's Pipistrelle.
TYPE LOCALITY: Israel, 40 km N Eilat, Wadi Araba, Yotwata.
DISTRIBUTION: Israel, Saudi Arabia, S Yemen, Oman, perhaps Socotra Isl (Yemen).
STATUS: IUCN 2003 and IUCN/SSC Action Plan (2001) – Lower Risk (nt) as *Pipistrellus bodenheimeri*.
COMMENTS: Reviewed by Harrison and Bates (1991) and Gaucher and Harrison (1995); also see Riskin (2001). May be conspecific with *ariel*, see Benda et al. (2002).

*Hypsugo cadornae* (Thomas, 1916). J. Bombay Nat. Hist. Soc., 24:416.
COMMON NAME: Cadorna's Pipistrelle.
TYPE LOCALITY: India, Darjeeling, Pashok, 3,500 ft. (1,067 m).
DISTRIBUTION: NE India, Burma, Thailand, Vietnam, Laos.
STATUS: IUCN 2003 and IUCN/SSC Action Plan (2001) – Lower Risk (nt) as *Pipistrellus cadornae*.
COMMENTS: Listed as a subspecies of *savii* by Ellerman and Morrison-Scott (1951), but see Hill (1962b) and Bates and Harrison (1967). Reviewed in part by Bates et al. (1997), Hendrichsen et al. (2001b), and Lunde et al. (2003a).

*Hypsugo crassulus* (Thomas, 1904). Ann. Mag. Nat. Hist., ser. 7, 13:206.
COMMON NAME: Broad-headed Pipistrelle.
TYPE LOCALITY: Cameroon, Efulen.
DISTRIBUTION: Liberia, Côte d'Ivoire, Cameroon, Dem. Rep. Congo, N Angola, S Sudan.
STATUS: IUCN 2003 and IUCN/SSC Action Plan (2001) – Lower Risk (lc) as *Pipistrellus crassulus*.
SYNONYMS: *bellieri* De Vree, 1972.
COMMENTS: Includes *bellieri*; see Heller et al. (1994).

*Hypsugo eisentrauti* (Hill, 1968). Bonn. Zool. Beitr., 19:45.
COMMON NAME: Eisentraut's Pipistrelle.
TYPE LOCALITY: Cameroon, Western Province, Rumpi Highlands, Dikume-Balue.
DISTRIBUTION: Cameroon, Rwanda, Kenya, and Somalia.
STATUS: IUCN 2003 and IUCN/SSC Action Plan (2001) – Lower Risk (lc) as *Pipistrellus eisentrauti*.
COMMENTS: Formerly included *bellieri* (e.g., Koopman, 1989, 1993, 1994), which is here listed as a synonym of *crassulus* following Heller et al. (1994).

*Hypsugo imbricatus* (Horsfield, 1824). Zool. Res. Java, part 8, p. 5 (unno.) of *Vespertilio Temminckii* acct.
COMMON NAME: Brown Pipistrelle.
TYPE LOCALITY: Indonesia, Java.
DISTRIBUTION: Java, Kangean Isl, Bali, and Lesser Sunda Isls; Borneo.
STATUS: IUCN 2003 and IUCN/SSC Action Plan (2001) – Lower Risk (lc) as *Pipistrellus imbricatus*.
COMMENTS: Previous reports of *imbricatus* from the Philippines all appear to represent *javanicus*; see Heaney et al. (1998).

*Hypsugo joffrei* (Thomas, 1915). Ann. Mag. Nat. Hist., ser. 8, 15:225.
COMMON NAME: Joffre's Pipistrelle.
TYPE LOCALITY: Burma, Kachin Hills.
DISTRIBUTION: N Burma.
STATUS: IUCN 2003 and IUCN/SSC Action Plan (2001) – Critically Endangered as *Pipistrellus joffrei*.
COMMENTS: Transferred from *Nyctalus*; see Hill (1966). Koopman (1989a, 1993) suggested

that it might best be returned to *Nyctalus*, but see Hill (1966) and Hill and Harrison (1987).

*Hypsugo kitcheneri* (Thomas, 1915). Ann. Mag. Nat. Hist., ser. 8, 15:229.
COMMON NAME: Red-brown Pipistrelle.
TYPE LOCALITY: Borneo, Kalimantan Tengah, Barito River.
DISTRIBUTION: Borneo.
STATUS: IUCN 2003 and IUCN/SSC Action Plan (2001) – Lower Risk (nt) as *Pipistrellus kitcheneri*.
COMMENTS: Listed as a subspecies of *imbricatus* by Chasen (1940), but see Tate (1942*a*) and Medway (1977). May be conspecific with *lophurus*; see Francis and Hill (1986).

*Hypsugo lophurus* (Thomas, 1915). J. Bombay Nat. Hist. Soc., 23:413.
COMMON NAME: Burmese Pipistrelle.
TYPE LOCALITY: Burma, Tenasserim, Victoria Province, Maliwun.
DISTRIBUTION: Peninsular Burma.
STATUS: IUCN 2003 and IUCN/SSC Action Plan (2001) – Data Deficient as *Pipistrellus lophurus*.
COMMENTS: May be conspecific with *kitcheneri*; see Francis and Hill (1986). Corbet and Hill (1992) argued that *lophurus* would be considered the older name.

*Hypsugo macrotis* (Temminck, 1840). Monogr. Mamm., 2:218.
COMMON NAME: Big-eared Pipistrelle.
TYPE LOCALITY: Indonesia, Sumatra, Padang.
DISTRIBUTION: W Malaysia, Sumatra, Bali, adjacent small islands.
STATUS: IUCN 2003 and IUCN/SSC Action Plan (2001) – Lower Risk (nt) as *Pipistrellus macrotis*.
SYNONYMS: *curtatus* Miller, 1911.
COMMENTS: Listed as a subspecies of *imbricatus* by Medway (1969), but see Tate (1942*a*) and Corbet and Hill (1980). May include *vordermanni*; see Corbet and Hill (1992).

*Hypsugo musciculus* (Thomas, 1913). Ann. Mag. Nat. Hist., ser. 8, 11:316.
COMMON NAME: Mouse-like Pipistrelle.
TYPE LOCALITY: Cameroon, Ja River, Bitye, 2,000 ft. (610 m).
DISTRIBUTION: Cameroon, Dem. Rep. Congo, Gabon, possibly Ghana.
STATUS: IUCN 2003 and IUCN/SSC Action Plan (2001) – Lower Risk (nt) as *Pipistrellus musciculus*.

*Hypsugo pulveratus* (Peters, 1871). *In* Swinhoe, Proc. Zool. Soc. Lond., 1870:618 [1871].
COMMON NAME: Chinese Pipistrelle.
TYPE LOCALITY: China, Fukien, Amoy.
DISTRIBUTION: Szechwan, Yunnan, Hunan, Kiangsu, Fukien (China), Hong Kong; Thailand, Laos, Vietnam.
STATUS: IUCN 2003 and IUCN/SSC Action Plan (2001) – Lower Risk (nt) as *Pipistrellus pulveratus*.
COMMENTS: Reviewed in part by Bates et al. (1997) and Hendrichsen et al. (2001*b*).

*Hypsugo savii* (Bonaparte, 1837). Fauna Ital., 1, fasc. 20.
COMMON NAME: Savi's Pipistrelle.
TYPE LOCALITY: Italy, Pisa.
DISTRIBUTION: France, Portugal, Spain, Italy, S Switzerland, Austria, E. Hungary, Balkan Countries, Morocco, N Algeria, and the Canary Isls. (Spain) and Cape Verde Isls through the Crimea and Caucasus, Turkey, Lebanon, Syria, Israel, Iran, Kazakhstan, Turkmenistan, Uzbekistan, Kyrgyzstan, Tajikistan, Afghanistan to N India and Burma.
STATUS: IUCN 2003 and IUCN/SSC Action Plan (2001) – Lower Risk (lc) as *Pipistrellus savii*.
SYNONYMS: *agilis* Fatio, 1872; *aristippe* Bonaparte, 1837; *bonapartei* Savi, 1838; *darwini* Tomes, 1859; *leucippe* Bonaparte, 1837; *maurus* Blasius, 1853; *nigrans* Crespon, 1844; **austenianus** Dobson, 1871; **caucasicus** Satunin, 1901; *tauricus* Ognev, 1927; *pallescens* Bobrinskii, 1926; *tamerlani* Bobrinskii, 1918; **ochromixtus** Cabrera, 1904.
COMMENTS: Does not include *coreensis*, *alaschanicus*, or *velox*, see Yoshiyuki (1989) and

Horácek et al. (2000). Reviewed in part by Harrison and Bates (1991), Bates and
Harrison (1997), and Horácek et al. (2000). See Horácek et al. (2000) for discussion of
subspecies limits. Srinivasulu and Srinivasulu (2001) suggested that *austenianus* may be
a distinct species.

*Hypsugo vordermanni* (Jentink, 1890). Zool. Ergebnisse Reis. Niederlandische Ost-Indien, p. 152.
COMMON NAME: Vordermann's Pipistrelle.
TYPE LOCALITY: Indonesia, Billiton Isl (= Belitung).
DISTRIBUTION: Belitung Isl (Indonesia), Borneo (Sarawak, Malaysia).
STATUS: IUCN 2003 – Not evaluated; not considered in IUCN/SSC Action Plan (2001).
COMMENTS: Reviewed by Hill (1983) and Francis and Hill (1986). May be conspecific with
*macrotis*; see Corbet and Hill (1992).

*Ia* Thomas, 1902. Ann. Mag. Nat. Hist., ser. 7, 10:163.
TYPE SPECIES: *Ia io* Thomas, 1902.
SYNONYMS: *Parascotomanes* Bourret, 1942.
COMMENTS: Considered a subgenus of *Pipistrellus* by Ellerman and Morrison-Scott (1951); but
see Topál (1970*a*).

*Ia io* Thomas, 1902. Ann. Mag. Nat. Hist., ser. 7, 10:164.
COMMON NAME: Great Evening Bat.
TYPE LOCALITY: China, Hupeh, Chungyang.
DISTRIBUTION: Sichuan, Hubei, Yunnan, Guichow, Tibet (S China), Laos, N Vietnam,
N Thailand, NE India, Nepal.
STATUS: IUCN 2003 and IUCN/SSC Action Plan (2001) – Lower Risk (nt).
SYNONYMS: *beaulieui* Bourret, 1942; *longimana* Pen, 1962.
COMMENTS: Reviewed by Topál (1970*a*), Bates and Harrison (1997), Csorba (1998), and
Hendrichsen et al. (2001*b*).

*Laephotis* Thomas, 1901. Ann. Mag. Nat. Hist., ser. 7, 7:460.
TYPE SPECIES: *Laephotis wintoni* Thomas, 1901.
COMMENTS: Considered monotypic by Hayman and Hill (1971), but see J. E. Hill (1974*a*), who
revised the genus. Kearney et al. (2002) suggested that *Laephotis* may nest within
*Neoromicia* based on analysis of bacular structure, but this hypothesis needs to be tested
with additional data sets.

*Laephotis angolensis* Monard, 1935. Arch. Mus. Bocage, 6:45.
COMMON NAME: Angolan Long-eared Bat.
TYPE LOCALITY: Angola, Tyihumbwe, 15 km W. Dala.
DISTRIBUTION: Angola, Dem. Rep. Congo.
STATUS: IUCN 2003 and IUCN/SSC Action Plan (2001) – Lower Risk (nt).

*Laephotis botswanae* Setzer, 1971. Proc. Biol. Soc. Wash., 84:260, 263.
COMMON NAME: Botswanan Long-eared Bat.
TYPE LOCALITY: Botswana, 50 mi. (80 km) W and 12 mi. (19 km) S Shakawe.
DISTRIBUTION: Dem. Rep. Congo, Zambia, Malawi, Botswana, Zimbabwe, NE South Africa.
STATUS: IUCN 2003 and IUCN/SSC Action Plan (2001) – Lower Risk (nt).
COMMENTS: For a range map see Cotterill (1996).

*Laephotis namibensis* Setzer, 1971. Proc. Biol. Soc. Wash., 84:259.
COMMON NAME: Namibian Long-eared Bat.
TYPE LOCALITY: Namibia, Gobabeb, Kuiseb River.
DISTRIBUTION: Namibia, South Africa.
STATUS: IUCN 2003 and IUCN/SSC Action Plan (2001) – Endangered.

*Laephotis wintoni* Thomas, 1901. Ann. Mag. Nat. Hist., ser. 7, 7:460.
COMMON NAME: de Winton's Long-eared Bat.
TYPE LOCALITY: Kenya, Kitui, 1,150 m.
DISTRIBUTION: Ethiopia, Kenya, Tanzania, SW Cape Province (South Africa).
STATUS: IUCN 2003 and IUCN/SSC Action Plan (2001) – Lower Risk (nt).

*Mimetillus* Thomas, 1904. Abstr. Proc. Zool. Soc. Lond., 1904(10):12.
TYPE SPECIES: *Vesperugo* (*Vesperus*) *moloneyi* Thomas, 1891.

*Mimetillus moloneyi* (Thomas, 1891). Ann. Mag. Nat. Hist., ser. 6, 7:528.
COMMON NAME: Moloney's Mimic Bat.
TYPE LOCALITY: Nigeria, Western Region, Lagos.
DISTRIBUTION: Sierra Leone east to Ethiopia and Kenya, Tanzania south to Mozambique,
west to Zambia, S Dem. Rep. Congo, and Angola; no records have been documented in
the central Congo Basin.
STATUS: IUCN 2003 and IUCN/SSC Action Plan (2001) – Lower Risk (lc).
SYNONYMS: ***thomasi*** Hinton, 1920; *berneri* Monard, 1933.
COMMENTS: Cotterill (2001*c*) suggested that *thomasi* represents a distinct, large-bodied,
savanna-dwelling species (*moloneyi* being a strict forest-dwelling species characterized
by smaller size). However, additional locality and morphometric data indicate a more
complex pattern (J. Fahr, pers. comm.). Accordingly, I have chosen to treat these taxa
as conspecific pending a thorough revision.

*Neoromicia* Roberts, 1926. Annals Transvaal Mus., 11:245.
TYPE SPECIES: *Eptesicus zuluensis* Roberts, 1924.
COMMENTS: Often considered a subgenus of *Pipistrellus* (e.g, Hill and Harrison, 1987; Koopman,
1994) or *Eptesicus* (e.g., Koopman, 1993) but raised to generic rank and transferred to
Vespertilionini by Volleth et al. (2001) based on karyotype data. Also see Kearney et al.
(2002), who provided additional evidence in support of this arrangment. For a partial
phylogeny see Kearney et al. (2002), who also suggested that *Laephotis* might nest within
*Neoromicia*.

*Neoromicia brunneus* (Thomas, 1880). Ann. Mag. Nat. Hist., ser. 5, 6:165.
COMMON NAME: Dark-brown Pipistrelle.
TYPE LOCALITY: Nigeria, Eastern region, Calabar.
DISTRIBUTION: Liberia to Dem. Rep. Congo.
STATUS: IUCN 2003 and IUCN/SSC Action Plan (2001) – Lower Risk (nt) as *Eptesicus
brunneus*.

*Neoromicia capensis* (A. Smith, 1829). Zool. J., 4:435.
COMMON NAME: Cape Serotine.
TYPE LOCALITY: South Africa, Eastern Cape Prov., Grahamstown.
DISTRIBUTION: Guinea-Bissau to Ethiopia, south to South Africa.
STATUS: IUCN 2003 and IUCN/SSC Action Plan (2001) – Lower Risk (lc) as *Eptesicus capensis*.
SYNONYMS: ***damarensis*** Noack, 1889; ***garambae*** J. A. Allen, 1917; ***gracilior*** Thomas and
Schwann, 1905; ***grandidieri*** Dobson, 1876; ***nkatiensis*** Roberts, 1932; ***notius***
G. M. Allen, 1908.
COMMENTS: Includes *notius*; see Koopman (1975*b*). Does not include *matroka*; see Peterson
et al. (1995). May also include *minuta* Temminck, 1840 [not of Montgu, 1808; see
Koopman, 1975*b*]. Probably includes *melckorum* Roberts, 1919, but not all of the
material referred to that species (see account for *melckorum* below). Most West African
populations have not been allocated to subspecies. See Taylor (2000*a*) for distribution
map, but note that he apparently included *matroka* from Madagascar (here treated as a
species of *Eptesicus*) in *capensis*.

*Neoromicia flavescens* (Seabra, 1900). J. Sci. Math. Phys. Nat. Lisboa, ser. 2, 6:23.
COMMON NAME: Yellow Serotine.
TYPE LOCALITY: Angola, Galanga.
DISTRIBUTION: Angola, Burundi, Malawi; also Cameroon (Van Cakenberghe, pers. comm.).
STATUS: IUCN 2003 and IUCN/SSC Action Plan (2001) – Data Deficient as *Eptesicus
flavescens*.
SYNONYMS: *angolensis* Hill, 1937.

*Neoromicia guineensis* (Bocage, 1889). J. Sci. Math. Phys. Nat. Lisboa, ser. 2, 1:6.
COMMON NAME: Guinean Serotine.
TYPE LOCALITY: Guinea-Bissau, Bissau.

DISTRIBUTION: Senegal, Gambia, Guinea Bissau, and Guinea to Ethiopia and NE Dem. Rep. Congo; perhaps Tanzania.

STATUS: IUCN 2003 and IUCN/SSC Action Plan (2001) – Lower Risk (nt) as *Eptesicus guineensis*.

SYNONYMS: *rectitragus* Wettstein, 1916.

COMMENTS: This species was called *pusillus* by Hayman and Hill (1971), but see Koopman (1975). Tanzanian record may represent *somalicus*.

*Neoromicia helios* (Heller, 1912). Smiths. Misc. Coll., 60(12):3.

COMMON NAME: Samburu Pipistrelle.

TYPE LOCALITY: Kenya, 30 mi S Mt. Marsabit, Merelle Water.

DISTRIBUTION: Kenya, Somalia, Djibouti, NE Uganda, extreme S Sudan, N Tanzania. Maybe more widespread (Peterson, 1987).

STATUS: IUCN 2003 – Not evaluated; not considered in IUCN/SSC Action Plan (2001).

COMMENTS: Considered to be a subspecies of *nanus* by some authors, but differences in bacular morphology (Hill and Harrison, 1987), roosting and social behavior (O'Shea, 1980; Happold and Happold, 1996), habitat and pelage coloration, and the presence of a pair of glands on the interfemoral membrane in *helios* (O'Shea, 1980) that are rarely found in *nanus* indicate that *helios* is a distinct species (M. Happold, pers. comm.).

*Neoromicia melckorum* (Roberts, 1919). Ann. Transvaal Mus., 6:113.

COMMON NAME: Melcks' Serotine.

TYPE LOCALITY: South Africa, Western Cape Prov., Berg River, Kersfontein.

DISTRIBUTION: SW South Africa, Zimbabwe, Zambia, Mozambique, Kenya, Tanzania.

STATUS: IUCN 2003 and IUCN/SSC Action Plan (2001) – Lower Risk (lc) as *Eptesicus melckorum*.

COMMENTS: This species has not been clearly distinguished from *capensis*, see discussion in Rautenbach et al. (1993) and Kearney et al. (2002). Koopman (1994) noted that the type series from W Cape Province is probably conspecific with *capensis*; however, material from the northern part of the supposed range of *melckorum* (Kenya to Zambia and Transvaal) is clearly distinct from *capensis* and should be renamed.

*Neoromicia nanus* (Peters, 1852). Reise nach Mossambique, Säugethier, p. 63.

COMMON NAME: Banana Pipistrelle.

TYPE LOCALITY: Mozambique, Inhambane.

DISTRIBUTION: South Africa to Ethiopia, Eritrea, Sudan, Niger, Mali, and Senegal; Madagascar; Pemba and Zanzibar.

STATUS: IUCN 2003 and IUCN/SSC Action Plan (2001) – Lower Risk (lc) as *Pipistrellus nanus*.

SYNONYMS: *abaensis* J. A. Allen, 1917; *africanus* Rüppell, 1842; *culex* Thomas, 1911; *fouriei* Thomas, 1926; *meesteri* Kock, 2001 [replacement name for *australis* Roberts, 1913]; *australis* Roberts, 1913 [not Miller, 1897]; *pagenstecheri* Noack, 1889; *pusillulus* Peters, 1870; *pusillus* LeConte, 1857 [not Schinz, 1840, or Noack, 1889]; *pusillus* Noack, 1889 [not Schinz, 1840, or LeConte, 1857]; *minusculus* Miller, 1900; *stampflii* Jentink, 1888.

COMMENTS: Usually placed in *Pipistrellus* or *Hypsugo*, but recently transferred to *Neoromicia* based on analyses of karyotype data (Kearney et al., 2002; Volleth et al., 2001). The oldest name for this species is *africanus* (see Koopman [1975], Meester et al. [1986], and Kock [2001*b*]), but the name *nanus* has been applied to this taxon extensively in the literature for many decades. A petition has been filed with the International Commission on Zoological Nomenclature to conserve *nanus* in place of *africanus* (M. Happold, pers. comm.) Pending a ruling of the International Commission, I follow Koopman (1993) in retaining the name *nanus* for this species in the interest of stability. Does not include *helios*, see comments under that species. May not include *minusculus*, *culex*, or *fouriei*; see Peterson (1987), who suggested in an abstract that these taxa and *helios* represent two distinct species, *helios* (no junior synonyms) and *minusculus* (including *culex* and *fouriei*). May not include *pusillulus* (here treated as a syononym of *meesteri*), see Kock et al. (2002). It is probable that *africanus* and *nanus* represent different subspecies, but their limits have yet to be evaluated, see Kock (2001*b*).

*Neoromicia rendalli* (Thomas, 1889). Ann. Mag. Nat. Hist., ser. 6, 3:362.
  COMMON NAME: Rendall's Serotine.
  TYPE LOCALITY: Gambia, Bathurst.
  DISTRIBUTION: Senegal, Mali, and Gambia to Somalia, south to Botswana, Malawi,
    Mozambique, South Africa.
  STATUS: IUCN 2003 and IUCN/SSC Action Plan (2001) – Lower Risk (lc) as *Eptesicus rendalli*.
  SYNONYMS: *phasma* G. M. Allen, 1911; *faradjius* J. A. Allen, 1917.
  COMMENTS: Reviewed by Koopman (1975) and Kock et al. (2002). This complex is in need of
    review as it may include more than one species (see Kock et al., 2002).

*Neoromicia somalicus* (Thomas, 1901). Ann. Mag. Nat. Hist., ser. 7, 8:32.
  COMMON NAME: Somali Serotine.
  TYPE LOCALITY: Somalia, Northwest Province, Hargeisa.
  DISTRIBUTION: Senegal and Guinea-Bissau to Somalia, south to Uganda, Dem. Rep. Congo,
    Kenya, and Tanzania; Madagascar.
  STATUS: IUCN 2003 and IUCN/SSC Action Plan (2001) – Lower Risk (lc) as *Eptesicus
    somalicus*.
  SYNONYMS: *humbloti* Milne-Edwards, 1881; *malagasyensis* Peterson, Eger, and Mitchell,
    1995; *ugandae* Hollister, 1916.
  COMMENTS: Does not include *zuluensis*; see Peterson et al. (1995). Reviewed in part by
    Peterson et al. (1995). Subspecific allocations of West African and Tanzanian
    populations are uncertain.

*Neoromicia tenuipinnis* (Peters, 1872). Monatsb. K. Preuss. Akad. Wiss. Berlin, 1872:263.
  COMMON NAME: White-winged Serotine.
  TYPE LOCALITY: "Guinea".
  DISTRIBUTION: Senegal to Kenya and Ethiopia, south to Angola and Dem. Rep. Congo.
  STATUS: IUCN 2003 and IUCN/SSC Action Plan (2001) – Lower Risk (lc) as *Eptesicus
    tenuipinnis*.
  SYNONYMS: *ater* J. A. Allen, 1917; *bicolor* Bocage, 1889.
  COMMENTS: *bicolor* (known only from the type locality in Angola) was tentatively included
    here by Hayman and Hill (1971), but it may be an older name for *Hypsugo anchietae*;
    see Koopman (1975) and Hill and Harrison (1987).

*Neoromicia zuluensis* (Roberts, 1924). Ann. Transv. Mus., 15:15.
  COMMON NAME: Zulu Serotine.
  TYPE LOCALITY: South Africa, KwaZulu-Natal Prov., Zululand, White Umfolosi Game
    Reserve.
  DISTRIBUTION: Namibia, Botswana, Zambia, Natal, Malawi, N South Africa; also known from
    Kenya, Ethiopia, and Sudan (V. Van Cakenberghe, pers. comm.).
  STATUS: IUCN 2003 and IUCN/SSC Action Plan (2001) – Lower Risk (nt) as *Eptesicus
    zuluensis*.
  SYNONYMS: *vansoni* Roberts, 1932.
  COMMENTS: Often included in *somalicus* (e.g., Koopman, 1975, 1993, 1994), but apparently
    distinct; see Peterson et al. (1995).

*Philetor* Thomas, 1902. Ann. Mag. Nat. Hist., ser. 7, 9:220.
  TYPE SPECIES: *Philetor rohui* Thomas, 1902 (= *Vespertilio brachypterus* Temminck, 1840).

*Philetor brachypterus* (Temminck, 1840). Monogr. Mamm., 2:215.
  COMMON NAME: Rohu's Bat.
  TYPE LOCALITY: Indonesia, Sumatra, Padang Dist.
  DISTRIBUTION: Nepal, W Malaysia, Sumatra, Borneo, Philippines, Sulawesi, New Guinea,
    New Britain and New Ireland Isls (Bismarck Arch.). A record from Java is erroneous,
    and a record from Bangka Isl (Indonesia) does not appear to be authentic, see Corbet
    and Hill (1992).
  STATUS: IUCN 2003 and IUCN/SSC Action Plan (2001) – Lower Risk (lc).
  SYNONYMS: *rohui* Thomas, 1902; *verecundus* Chasen, 1940.
  COMMENTS: Reviewed by Hill (1966, 1971d, 1983); also see Corbet and Hill (1992), Bates and

Harrison (1997), Flannery (1995*a*, *b*), and Bonaccorso (1998). Three subspecies are often recognized, but the actual pattern of variation is too complex to fit this taxonomy.

*Tylonycteris* Peters, 1872. Monatsb. K. Preuss. Akad. Wiss. Berlin, 1872:703.
>TYPE SPECIES: *Vespertilio pachypus* Temminck, 1840.
>COMMENTS: Reviewed by Corbet and Hill (1992).

*Tylonycteris pachypus* (Temminck, 1840). Monogr. Mamm., 2:217.
>COMMON NAME: Lesser Bamboo Bat.
>TYPE LOCALITY: Indonesia, W Java, Bantam.
>DISTRIBUTION: Bangladesh, India, Burma, S China, Thailand, Burma, Laos, Cambodia, Vietnam to Peninsular Malaysia, Philippines, Sumatra, Java, Borneo, Bali (Indonesia); Andaman Isls (India).
>STATUS: IUCN 2003 and IUCN/SSC Action Plan (2001) – Lower Risk (lc).
>SYNONYMS: ***aurex*** Thomas, 1915; ***bhaktii*** Oei, 1960; ***fulvidus*** Blyth, 1859; *rubidus* Thomas, 1915; ***meyeri*** Peters, 1872.
>COMMENTS: Reviewed in part by Bates and Harrison (1997) and Hendrichsen et al. (2001*b*).

*Tylonycteris robustula* Thomas, 1915. Ann. Mag. Nat. Hist., ser. 8, 15:227.
>COMMON NAME: Greater Bamboo Bat.
>TYPE LOCALITY: Malaysia, Borneo, Sarawak, Upper Sarawak.
>DISTRIBUTION: NE India, Burma, Cambodia, Laos, Vietnam, S China to the Philippines, Sulawesi, Sumatra, Java, Bali, Borneo, and Ambon Isl (Moluccas).
>STATUS: IUCN 2003 and IUCN/SSC Action Plan (2001) – Lower Risk (lc).
>SYNONYMS: ***malayana*** Chasen, 1940.
>COMMENTS: Includes *malayana*; see Medway (1969). Reviewed in part by Bates and Harrison (1997) and Hendrichsen et al. (2001*b*).

*Vespadelus* Troughton, 1943. Furred animals of Australia, 1st ed., Sydney: Angus and Robertson, p. 349.
>TYPE SPECIES: *Scotophilus pumilus* Gray, 1841.
>SYNONYMS: *Registrellus* Troughton, 1943.
>COMMENTS: *Vespadelus* was first published as a *nomen nudum* (no accompanying diagnosis) by Iredale and Troughton (1934); the name was made available by Troughton (1943), who provided a diagnosis. Often included in *Pipistrellus* (e.g., Hill and Harrison, 1987) or *Eptesicus* (e.g., Adams et al., 1987; Kitchener et al., 1987; Koopman, 1993, 1994; McKean et al., 1978), but see Volleth and Tidemann (1991) and Volleth and Heller (1994). Revised by Kitchener et al. (1987); see also Adams et al. (1987) and Queale (1997). Species groups follow Kitchener et al. (1987).

*Vespadelus baverstocki* (Kitchener, Jones, and Caputi, 1987). Rec. West. Aust. Mus., 13:481.
>COMMON NAME: Baverstock's Forest Bat.
>TYPE LOCALITY: Australia, Western Australia, Yuinmery area.
>DISTRIBUTION: C and S Australia.
>STATUS: IUCN 2003 and IUCN/SSC Action Plan (2001) – Lower Risk (lc) as *Eptesicus baverstocki*.
>COMMENTS: *pumilus* species group. Included in *vulturnus* by Koopman (1994), but see Kitchener at al. (1987) and Queale (1997).

*Vespadelus caurinus* (Thomas, 1914). Ann. Mag. Nat. Hist., ser. 8, 13:439.
>COMMON NAME: Western Cave Bat.
>TYPE LOCALITY: Australia, Western Australia, Kimberley, Drysdale.
>DISTRIBUTION: N Australia.
>STATUS: IUCN 2003 and IUCN/SSC Action Plan (2001) – Lower Risk (nt) as *Eptesicus caurinus*.
>COMMENTS: *caurinus* species group. Included in *pumilus* by McKean et al. (1978) and Koopman (1993, 1994), but see Kitchener et al. (1987) and Adams et al. (1987).

*Vespadelus darlingtoni* (G. M. Allen, 1933). J. Mammal. 14:150.
>COMMON NAME: Large Forest Bat.

TYPE LOCALITY: Australia, New South Wales, 13 km NW Braidwood.

DISTRIBUTION: SE Australia, including Tasmania and Lord Howe Isl.

STATUS: IUCN 2003 and IUCN/SSC Action Plan (2001) – Lower Risk (lc) as *Eptesicus darlingtoni*.

SYNONYMS: *sagittula* McKean, Richards, and Price, 1978.

COMMENTS: *pumilus* species group. See Kitchener et al. (1987) for discussion of synonmy. Adams et al. (1987) and Koopman (1993) used the name *sagittula* for this taxon, which may include more than one species.

*Vespadelus douglasorum* (Kitchener, 1976). Rec. West. Aust. Mus., 4:295, 296.

COMMON NAME: Yellow-lipped Bat.

TYPE LOCALITY: Australia, Western Australia, Kimberley, Napier Range, Tunnel Creek.

DISTRIBUTION: Kimberley (N Western Australia).

STATUS: IUCN 2003 and IUCN/SSC Action Plan (2001) – Lower Risk (nt) as *Eptesicus douglasorum*.

COMMENTS: *caurinus* species group. Originally described as *douglasi* but emended by Kitchener et al. (1987).

*Vespadelus finlaysoni* (Kitchener, Jones, and Caputi, 1987). Rec. West, Aust. Mus., 13:456.

COMMON NAME: Finlayson's Forest Bat.

TYPE LOCALITY: Australia, Western Australia, Cossack.

DISTRIBUTION: Western and central Australia.

STATUS: IUCN 2003 and IUCN/SSC Action Plan (2001) – Lower Risk (lc) as *Eptesicus finlaysoni*.

COMMENTS: *caurinus* species group. Included in *pumilus* by Koopman (1993, 1994), but see Kitchener et al. (1987) and Queale (1997).

*Vespadelus pumilus* (Gray, 1841). Appendix C *in* J. Two Exped. Austr., 2:406.

COMMON NAME: Eastern Forest Bat.

TYPE LOCALITY: Australia, New South Wales, Yarrundi.

DISTRIBUTION: Eastern Australia, Lord Howe Isl.

STATUS: IUCN 2003 and IUCN/SSC Action Plan (2001) – Lower Risk (lc) as *Eptesicus pumilus*.

COMMENTS: *pumilus* species group. Does not include *caurinus, darlingtoni, finlaysoni* and *troughtoni;* see Kitchener et al. (1987), Adams et al. (1987), and Queale (1997). Many specimens attributed to this species by McKean et al. (1978) were later referred to *troughtoni* by Kitchener et al. (1987). See Flannery (1995b).

*Vespadelus regulus* (Thomas, 1906). Proc. Zool. Soc. Lond., 1906:470, 471.

COMMON NAME: Southern Forest Bat.

TYPE LOCALITY: Australia, Western Australia, King George Sound, King River (near Albany).

DISTRIBUTION: SW and SE Australia, including Tasmania.

STATUS: IUCN 2003 and IUCN/SSC Action Plan (2001) – Lower Risk (lc) as *Eptesicus regulus*.

COMMENTS: *pumilus* species group. See McKean et al. (1978), Kitchener et al. (1987), and Queale (1997).

*Vespadelus troughtoni* (Kitchener, Jones, and Caputi, 1987). Rec. West. Aust. Mus., 13:467.

COMMON NAME: Troughton's Forest Bat.

TYPE LOCALITY: Australia, Queensland, Mt. Surprise, Yarramulla Lava Tunnels.

DISTRIBUTION: E Australia.

STATUS: IUCN 2003 and IUCN/SSC Action Plan (2001) – Lower Risk (lc) as *Eptesicus troughtoni*.

COMMENTS: *caurinus* species group. Included in *pumilus* by Koopman (1993, 1994), but see Kitchener et al. (1987).

*Vespadelus vulturnus* (Thomas, 1914). Ann. Mag. Nat. Hist., ser. 8, 13:440.

COMMON NAME: Little Forest Bat.

TYPE LOCALITY: Australia, Tasmania.

DISTRIBUTION: SE Australia including Tasmania.

STATUS: IUCN 2003 and IUCN/SSC Action Plan (2001) – Lower Risk (lc) as *Eptesicus vulturnus*.

SYNONYMS: *pygmaeus* Becker, 1858 [not Leach, 1825].

COMMENTS: *pumilus* species group. Includes *pygmaeus*; see McKean et al. (1978). Does not include *baverstocki*; see Kitchener et al. (1987) and Queale (1997). This complex may include more than one species; see Adams et al. (1987).

*Vespertilio* Linnaeus, 1758. Syst. Nat., 10th ed., 1:31.

TYPE SPECIES: *Vespertilio murinus* Linnaeus, 1758.

SYNONYMS: *Aristippe* Kolenati, 1863; *Marsipolaemus* Peters, 1872; *Meteorus* Kolenati, 1856; *Vesperugo* Keyserling and Blasius, 1839; *Vesperus* Keyserling and Blasius, 1839 [not Latreille, 1829].

*Vespertilio murinus* Linnaeus, 1758. Syst. Nat., 10th ed., 1:32.

COMMON NAME: Particolored Bat.

TYPE LOCALITY: Sweden. Baagøe (2001*b*) indicated that the type locality is probably near Uppsala, Central Sweden.

DISTRIBUTION: E France, Britain, and Norway across C Russia, Caucasus, S Ural, S Siberia, Ussuri region (Russia), Mongolia, NE China, and Korea; Bulgaria, Turkey, Iran, Kazakhstan, Turkmenistan, Uzbekistan, Kyrgyzstan, Tajikistan, E Afghanistan and N Pakistan.

STATUS: IUCN 2003 and IUCN/SSC Action Plan (2001) – Lower Risk (lc).

SYNONYMS: *albigularis* Peters, 1872; *discolor* Kuhl, 1819; *krascheninnikovi* Eversmann, 1853; *luteus* Kastschenko, 1905; *michnoi* Kastschenko, 1913; *siculus* Daday, 1885; **ussuriensis** Wallin, 1969.

COMMENTS: Reviewed in part by Bates and Harrison (1997), Horácek et al. (2000), and Baagøe (2001*b*).

*Vespertilio sinensis* (Peters, 1880). Monatsber. K. Preuss. Acad. Wiss. Berlin, 1880:259.

COMMON NAME: Asian Particolored Bat.

TYPE LOCALITY: Peking (China).

DISTRIBUTION: China, Ussuri region (Russia), Korea, Japan, Taiwan.

STATUS: IUCN 2003 and IUCN/SSC Action Plan (2001) – Lower Risk (lc) as *Vespertilio superans*.

SYNONYMS: *aurijunctus* Mori, 1928; *montanus* Kishida, 1931 [not Barrett-Hamilton, 1906; substitute for *noctula* Namie, 1889]; *motoyoshii* Kuroda, 1934 [substitute for *montanus* Kishida, 1931]; *superans* Thomas, 1899; **andersoni** Wallin, 1963; **namiyei** Kuroda, 1920; **noctula** Namie, 1889 [not Schreber, 1774]; **orientalis** Wallin, 1969.

COMMENTS: Includes *namiyei* and *orientalis*; see Yoshiyuki (1989) and Horácek (1997). The name *superans* was commonly applied to this taxon until Horácek (1997) demonstrated that *sinensis* (erroneously grouped in *Nyctalus* in previous classifications) is the oldest name for the species.

**Subfamily Antrozoinae** Miller, 1897. North American Fauna, 13:41.

COMMENTS: Simmons (1998) raised this group to family level and moved it to Molossoidea, but recent studies based on DNA sequence data (e.g., Hoofer and Van Den Bussche, 2001) indicate that *Antrozous* belongs in Vespertilionidae, a placement in line with more traditional classifications (e.g., Hill and Smith, 1984; Koopman, 1993, 1994; McKenna and Bell, 1997; Miller, 1897). This group may nest within Vespertilioninae, but its placement remains unclear; accordingly, it is here retained as a distinct subfamily pending further study.

*Antrozous* H. Allen, 1862. Proc. Acad. Nat. Sci. Phil., 14:248.

TYPE SPECIES: *Vespertilio pallidus* Le Conte, 1856.

COMMENTS: Does not include *Bauerus*; see Engstrom and Wilson (1981) and Engstrom et al. (1987*b*), but also see Pine et al. (1971).

*Antrozous pallidus* (Le Conte, 1856). Proc. Acad. Nat. Sci. Phil., 7:437.

COMMON NAME: Pallid Bat.

TYPE LOCALITY: USA, Texas, El Paso Co., El Paso.

DISTRIBUTION: Queretaro and Baja California (Mexico) to Kansas (USA) and British Columbia (Canada); Cuba.

STATUS: IUCN 2003 and IUCN/SSC Action Plan (2001) – Lower Risk (lc).

SYNONYMS: *cantwelli* V. Bailey, 1936; *bunkeri* Hibbard, 1934; *koopmani* Orr and Silva-Taboada, 1960; *minor* Miller, 1902; *obscurus* Baker, 1967; *pacificus* Merriam, 1897; *packardi* Martin and Schmidly, 1982.

COMMENTS: Includes *bunkeri*; see Morse and Glass (1960). Includes *koopmani*; see Martin and Schmidly (1982). See Hermanson and O'Shea (1983).

*Bauerus* Van Gelder, 1959. Amer. Mus. Novit., 1973:1.

TYPE SPECIES: *Antrozous dubiaquercus* Van Gelder, 1959.

COMMENTS: For use of this name see Engstrom and Wilson (1981) and Engstrom et al. (1987), but also see Pine et al. (1971).

*Bauerus dubiaquercus* (Van Gelder, 1959). Am. Mus. Novit., 1973:2.

COMMON NAME: Van Gelder's Bat.

TYPE LOCALITY: Mexico, Nayarit, Trés Marias Isls, Maria Magdalena Isl.

DISTRIBUTION: Trés Marias Isls, Jalisco, Veracruz, Oaxaca, and Chiapas (Mexico); Belize; Honduras, Costa Rica.

STATUS: IUCN 2003 and IUCN/SSC Action Plan (2001) – Vulnerable as *Antrozous dubiaquercus*.

SYNONYMS: *meyeri* Pine, 1966.

COMMENTS: Includes *Baeodon meyeri*; see Pine (1967) and Engstrom and Wilson (1981). See Engstrom et al. (1987b). See Emmons (1997) for distribution map.

**Subfamily Myotinae** Tate, 1942. Bull. Amer. Mus. Nat. Hist., 80:229.

COMMENTS: Originally named as a tribe within Vespertilioninae; raised to subfamily level by Simmons (1998) following the suggestion of Volleth and Heller (1994). May not be monophyletic; see Hoofer and Van Den Bussche (2001).

*Cistugo* Thomas, 1912. Ann. Mag. Nat. Hist., ser. 8, 10:205.

TYPE SPECIES: *Myotis seabrai* Thomas, 1912.

COMMENTS: Formerly included in *Myotis* by most authors (e.g., Ellerman and Morrison-Scott, 1951; Hayman and Hill, 1971; Koopman, 1993, 1994), but see Rautenbach et al. (1993).

*Cistugo lesueuri* Roberts, 1919. Ann. Transv. Mus., 6:112.

COMMON NAME: Lesueur's Wing-gland Bat.

TYPE LOCALITY: South Africa, Western Cape Prov., Paarl Dist., Lormarins.

DISTRIBUTION: S South Africa; Lesotho.

STATUS: IUCN 2003 and IUCN/SSC Action Plan (2001) – Vulnerable as *Myotis lesueuri*.

COMMENTS: For distribution map see Taylor (2000a).

*Cistugo seabrae* Thomas, 1912. Ann. Mag. Nat. Hist., ser. 8, 10:205.

COMMON NAME: Angolan Wing-gland Bat.

TYPE LOCALITY: Angola, Mossamedes.

DISTRIBUTION: Northern Cape Prov. (South Africa), Namibia, SW Angola.

STATUS: IUCN 2003 and IUCN/SSC Action Plan (2001) – Vulnerable as *Myotis seabrai*.

COMMENTS: For distribution map see Taylor (2000a). Sometimes spelled "*seabrai*", but the original spelling is *seabrae*.

*Lasionycteris* Peters, 1866. Monatsb. K. Preuss. Akad. Wiss. Berlin, 1866:8.

TYPE SPECIES: *Vespertilio noctivagans* Le Conte, 1831.

SYNONYMS: *Vesperides* Coues, 1875.

*Lasionycteris noctivagans* (Le Conte, 1831). *In* McMurtie, Anim. Kingdom, 1(App.):431.

COMMON NAME: Silver-haired Bat.

TYPE LOCALITY: "Eastern United States".

DISTRIBUTION: S Canada, USA (including SE Alaska, and except extreme southern parts), NE Mexico, Bermuda.

STATUS: IUCN 2003 and IUCN/SSC Action Plan (2001) – Lower Risk (lc).

SYNONYMS: *pulverlentus* Temminck, 1840.

COMMENTS: See Kunz (1982).

*Myotis* Kaup, 1829. Skizz. Entwickel.-Gesch. Nat. Syst. Europ. Thierwelt, 1:106.
> TYPE SPECIES: *Vespertilio myotis* Borkhausen, 1797.
> SYNONYMS: *Aeorestes* Fitzinger, 1870; *Anamygdon* Troughton, 1929; *Brachyotis* Kolenati, 1856 [not Gould, 1837]; *Capaccinus* Bonaparte, 1841; *Chrysopteron* Jentink, 1910; *Comastes* Fitzinger, 1870; *Dichromyotis* Bianchi, 1916; *Euvespertilio* Acloque, 1899; *Exochurus* Fitzinger, 1870; *Hesperomyotis* Cabrera, 1958; *Isotus* Kolenate, 1856; *Leuconoe* Boie, 1830; *Megapipistrellus* Bianchi, 1917; *Nyctactes* Kaup, 1829; *Paramyotis* Bianchi, 1916; *Pizonyx* Miller, 1906; *Pternopterus* Peters, 1867; *Rickettia* Bianchi, 1916; *Selysius* Bonaparte, 1841; *Tralatitus* Gervais, 1849; *Trilatitus* Gray, 1842.
> COMMENTS: For discussion of synonyms see Findley (1972), Hayman and Hill (1971), and Phillips and Birney (1968). Neotropical species revised by LaVal (1973*a*). Apparently does not include *Cistugo*; see Rautenbach et al. (1993). Hall (1981) provided a key to North and Central American species; Corbet and Hill (1992) gave a key to Indomalayan species; Bates et al. (1999) provided a key to species found in Vietnam and adjoining countries. Also see Topál (1997) and Stormark (1998). For partial phylogenies see Mayer and von Helversen (2001*a*), Ruedi and Mayer (2001), and Kawai et al. (2003). These studies have convincingly demonstrated that the three subgenera of *Myotis* typically recognized (*Myotis*, *Leuconoe*, and *Selysius*) are not monophyletic, but instead represent ecomorphs characterized by convergent morphologies. Menu et al. (2002:320) argued that *Leuconoe* can be diagnosed as distinct based on dental morphology, and noted that it includes "most of the fossil and living species attributed to *Myotis* sensu lato within recent accounts and revisions." While this may be true, there is presently no agreement concerning relationships among species previously referred to *Myotis*, *Leuconoe*, and *Selysius*. Accordingly, no subgeneric classification is recognized here, and *Leuconoe* and *Selysius* are treated as junior synonyms of *Myotis*. Woodman (1993) argued that *Myotis* should be considered feminine in gender (thus requiring changes in the spelling of many specific epithets in *Myotis*), but Pritchard (1994) disagreed. Both of these authors appear to have overlooked a 1958 ruling by the International Commission on Zoological Nomenclature that fixed the gender of *Myotis* as masculine and placed the name as such on the Offical list of Generic Names in Zoology (International Commission on Zoological Nomenclature, 1958*a*). I follow this ruling, and thus retain the traditional spellings of specific epithets in combination with *Myotis*.

*Myotis abei* Yoshikura, 1944. Zool. Mag. (Tokyo), 56:6.
> COMMON NAME: Sakhalin Myotis.
> TYPE LOCALITY: Russia, S Sakhalin, Shirutoru.
> DISTRIBUTION: Known only from the type locality.
> STATUS: IUCN 2003 and IUCN/SSC Action Plan (2001) – Data Deficient.
> COMMENTS: Horácek et al. (2000) suggested that *abei* might be conspecific with *brandtii*, but retained these as separate taxa pending additional data.

*Myotis adversus* (Horsfield, 1824). Zool. Res. Java, part 8, p. 3(unno.) of *Vespertilio Temminckii* acct.
> COMMON NAME: Large-footed Myotis.
> TYPE LOCALITY: Indonesia, Java.
> DISTRIBUTION: Numerous islands in Indonesia (see Kitchener et al., 1995*b*); New South Wales; Taiwan; possibly Vietnam and peninsular Malaysia.
> STATUS: IUCN 2003 and IUCN/SSC Action Plan (2001) – Lower Risk (lc).
> SYNONYMS: **carimatae** Miller, 1906; **orientis** Hill, 1983; **taiwanensis** Ärnbäck-Christie Linde, 1908; **tanimbarensis** Kitchener, 1995 [in Kitchener et al., 1995*b*]; **wetarensis** Kitchener, 1995 [in Kitchener et al., 1995*b*].
> COMMENTS: Includes *taiwanensis*; see Ellerman and Morrison-Scott (1951); but see also Findley (1972). Includes *carimatae*; see Hill (1983). Does not include *macropus*, *moluccarum*, or *solomonis*; see Kitchener et al. (1995*b*), who revised this complex, but also see Churchill (1998). Vietnamese records are dubious; see Bates et al. (1999). Subspecies affinities of a specimen from New South Wales are unclear; see Kitchener et al. (1995*b*).

*Myotis aelleni* Baud, 1979. Rev. Suisse Zool., 86:268.
> COMMON NAME: Southern Myotis.

TYPE LOCALITY: Argentina, Chubut, El Hoyo de Epuyen.

DISTRIBUTION: SW Argentina.

STATUS: IUCN 2003 and IUCN/SSC Action Plan (2001) – Vulnerable.

COMMENTS: May not be distinct from *chiloensis*; see Pearson and Pearson (1989), but also see Barquez et al. (1993).

*Myotis albescens* (E. Geoffroy, 1806). Ann. Mus. Natn. Hist. Nat. Paris, 8:204.

COMMON NAME: Silver-tipped Myotis.

TYPE LOCALITY: Paraguay, Paraguari, Yaguaron (of neotype).

DISTRIBUTION: S Veracruz (Mexico), Guatemala, Honduras, Nicaragua, Panama, Colombia, Venezuela, Guyana, Surinam, Equador, Peru, Brazil, Uruguay, N Argentina, Paraguay, and Bolivia.

STATUS: IUCN 2003 and IUCN/SSC Action Plan (2001) – Lower Risk (lc).

SYNONYMS: *argentatus* Dalquest and Hall, 1947; *isidori* D'Orbigny and Gervais, 1847; *leucogaster* Schinz, 1821.

COMMENTS: Includes *argentatus*; see LaVal (1973*a*). Reviewed in part by López-González et al. (2001). Does not include *aenobarbus*, which is here placed in *Nycticeius* following Carter and Dolan (1978); also see Husson (1962). Although I follow Koopman (1993) in listing *isidori* as a synonym, there are serious problems with idenfication of the holotype; see Carter and Dolan (1978), who suggested that this name might actually belong in *Pipistrellus*. Does not include *mundus*; see LaVal (1973*a*). Apparently closely related to *nigricans*, *levis*, and *oxyotus*; see Ruedi and Mayer (2001).

*Myotis alcathoe* von Helversen and Heller, 2001. In von Helversen, Heller, Mayer, Nemeth, Volleth, and Gombkötö, Naturwissenschaften, 88:217

COMMON NAME: Alcathoe Myotis.

TYPE LOCALITY: Greece, Nomos Evritanias, near the village of Kleistos, over Fournikos Patomos stream, 39°05′N, 21°49′E.

DISTRIBUTION: Greece, Hungary, France. Specimens from Bulgaria, Romania, and Ukraine previously reported as *ikonnikovi* might represent *alcathoe* (von Helveren et al., 2001).

STATUS: IUCN 2003 – Not evaluated (new species); not considered in IUCN/SSC Action Plan (2001).

COMMENTS: See Ruedi et al. (2002).

*Myotis altarium* Thomas, 1911. Abstr. Proc. Zool. Soc. Lond., 1911(90):3.

COMMON NAME: Szechwan Myotis.

TYPE LOCALITY: China, Szechwan, Omi San (= Omei Shan).

DISTRIBUTION: Szechwan, Kweichow (China), Thailand.

STATUS: IUCN 2003 and IUCN/SSC Action Plan (2001) – Lower Risk (lc).

COMMENTS: Redescribed by Blood and McFarlane (1988).

*Myotis anjouanensis* Dorst, 1960. Bull. Mus. Nat. Hist. Nat., ser. 2, 31:476.

COMMON NAME: Anjouan Myotis.

TYPE LOCALITY: Comoro Isls, Anjouan Isl.

DISTRIBUTION: Anjouan Isl (Comoro Isls).

STATUS: IUCN 2003 – Not evaluated; not considered in IUCN/SSC Action Plan (2001).

COMMENTS: Usually included in *goudoti*, but appears to be distinct; see Peterson et al. (1995).

*Myotis annamiticus* Kruskop and Tsytsulina, 2001. Mammalia, 65:65.

COMMON NAME: Annamit Myotis.

TYPE LOCALITY: Vietnam, Qaun Binh prov., Minh Hoa district, ca. 35 km S Minh Hoa (Qui Dat), Yen Hop valley near Yen Hop.

DISTRIBUTION: Vietnam.

STATUS: IUCN 2003 – Not evaluated (new species); not considered in IUCN/SSC Action Plan (2001).

COMMENTS: Known only from the type locality. Most similar to *csorbai*.

*Myotis annectans* (Dobson, 1871). Proc. Asiat. Soc. Bengal, p. 213.

COMMON NAME: Hairy-faced Myotis.

TYPE LOCALITY: India, NE India, Assam, Naga Hills.

DISTRIBUTION: NE India to Burma, Thailand, Laos, Cambodia, and Vietnam.

STATUS: IUCN 2003 and IUCN/SSC Action Plan (2001) – Lower Risk (nt).

SYNONYMS: *primula* Thomas, 1920.

COMMENTS: Includes *primula*; see Topál (1970*b*), who transfered the species from *Pipistrellus*. Reviewed by Bates and Harrison (1997); also see Hendrichsen et al. (2001*a*) and Lunde et al. (2003*a*).

*Myotis atacamensis* (Lataste, 1892). Actes Soc. Sci. Chile, 1:80.

COMMON NAME: Atacaman Myotis.

TYPE LOCALITY: Chile, Antofogasta, San Pedro de Atacama.

DISTRIBUTION: S Peru, N Chile.

STATUS: IUCN 2003 and IUCN/SSC Action Plan (2001) – Vulnerable.

SYNONYMS: *nicholsoni* Sanborn, 1941.

COMMENTS: Listed as a subspecies of *chiloensis* by Cabrera (1958). Includes *nicholsoni*; see LaVal (1973*a*).

*Myotis ater* (Peters, 1866). Monatsb. K. Preuss. Akad. Wiss. Berlin, 1866:18.

COMMON NAME: Peters's Myotis.

TYPE LOCALITY: Moluccas, Ternate Isl.

DISTRIBUTION: Vietnam, W Sumatra, Peninsular Malaysia, Sulawesi, Togian Isl, N Borneo, Moluccas, Papua New Guinea, possibly Philippines and Australia.

STATUS: IUCN 2003 and IUCN/SSC Action Plan (2001) – Lower Risk (lc) as *Myotis atra* (misspelled).

SYNONYMS: *amboinensis* Peters, 1866; **nugax** Allen and Coolidge, 1940.

COMMENTS: Formerly included in *muricola*, but see Hill (1983), Corbet and Hill (1992), Bates et al. (1999), and Hendrichsen et al. (2001). Revised by Francis and Hill (1998). Specimens from Peninsular Malaysia were tentatively referred to *ater* by Francis and Hill (1998) but may represent another species. There is apparently only one species of "*muricola*-type" *Myotis* in the Philippines, but it is not yet clear if this taxon is *ater* or *muricola* (L. Heaney, pers. comm.); the same is probably also true of the Moluccas (K. Helgen, pers. comm.). May include *australis*; see Hill (1983). Also see Flannery (1995*b*).

*Myotis auriculus* Baker and Stains, 1955. Univ. Kansas Publ. Mus. Nat. Hist., 9:83.

COMMON NAME: Southwestern Myotis.

TYPE LOCALITY: Mexico, Tamaulipas, Sierra de Tamaulipas, 10 mi. (16 km) W, 2 mi. (3 km) S Piedra, 1,200 ft. (366 m).

DISTRIBUTION: Arizona and New Mexico (USA) to Jalisco and Veracruz (Mexico); Guatemala.

STATUS: IUCN 2003 and IUCN/SSC Action Plan (2001) – Lower Risk (lc).

SYNONYMS: **apache** Hoffmeister and Krutzsch, 1955.

COMMENTS: Listed as a subspecies of *evotis* by Hall and Kelson (1959), but see Genoways and Jones (1969*b*), Hall (1981), and Gannon (1998). See Warner (1982). Woodman (1993) argued that the correct spelling of the specific epithet is *auriculacea*, but see Pritchard (1994).

*Myotis australis* (Dobson, 1878). Cat. Chiroptera Brit. Mus., p. 317.

COMMON NAME: Australian Myotis.

TYPE LOCALITY: Australia, New South Wales.

DISTRIBUTION: New South Wales, possibly Western Australia (Australia).

STATUS: IUCN 2003 and IUCN/SSC Action Plan (2001) – Data Deficient.

COMMENTS: Poorly known, the holotype and only certain specimen possibly being incorrectly labelled, or a vagrant individual of *muricola* (Husson, 1970). A specimen from NW Australia may belong in this species (Koopman, 1984*c*). Hill (1983) considered *australis* a subspecies of *ater*.

*Myotis austroriparius* (Rhoads, 1897). Proc. Acad. Nat. Sci. Phil., 49:227.

COMMON NAME: Southeastern Myotis.

TYPE LOCALITY: USA, Florida, Pinellas Co., Tarpon Springs.

DISTRIBUTION: SE USA including Florida, north to Indiana and North Carolina, west to Texas and SE Oklahoma.

STATUS: IUCN 2003 and IUCN/SSC Action Plan (2001) – Lower Risk (lc).
SYNONYMS: *gatesi* Lowery, 1943; *mumfordi* Rice, 1955.
COMMENTS: Reviewed by LaVal (1970). See Jones and Manning (1989).

*Myotis bechsteinii* (Kuhl, 1817). Die Deutschen Fledermäuse. Hanau, p. 14, 30.
COMMON NAME: Bechstein's Myotis.
TYPE LOCALITY: Germany, Hessen, Hanau.
DISTRIBUTION: Europe to Caucasus and Iran; Bulgaria; England; S Sweden.
STATUS: IUCN 2003 and IUCN/SSC Action Plan (2001) – Vulnerable.
SYNONYMS: *favonicus* Thomas, 1906; *ghidinii* Fatio, 1902.
COMMENTS: For discussion of correct spelling (*bechsteinii*, not *bechsteini*) see Bogdanowicz
and Kock (1998). Apparently closely related to *daubentonii*; see Ruedi and Mayer
(2001). Reviewed by Horácek et al. (2000) and Baagøe (2001a).

*Myotis blythii* (Tomes, 1857). Proc. Zool. Soc. Lond., 1857:53.
COMMON NAME: Lesser Mouse-eared Myotis.
TYPE LOCALITY: India, Rajasthan, Nasirabad.
DISTRIBUTION: Turkey and Israel to Iraq and Iran; NW India and the Himalayas; NW Altai
Mtns; Inner Mongolia and Shensi (China).
STATUS: IUCN 2003 and IUCN/SSC Action Plan (2001) – Lower Risk (lc).
SYNONYMS: *africanus* Dobson, 1875 [actually from Kasmir, N India]; *dobsoni* Trouessart,
1878; *murinoides* Dobson, 1873 [not Lartet, 1851]; **ancilla** Thomas, 1910; **lesviacus**
Iliopoulou, 1984; **omari** Thomas, 1906; *risorius* Cheesman, 1921.
COMMENTS: For discussion of synonyms see Strelkov (1972), Felten et al. (1977), Corbet
(1978c), Bogan et al. (1978), and Horácek et al. (2000). Middle Eastern records
reviewed by Harrison and Bates (1991), Palearctic records by Horácek et al. (2000).
Does not include *oxygnathus* and *punicus*, which together with *blythii* form a
paraphyletic assemblage that includes *myotis*; see Ruedi and Mayer (2001). In order to
restrict all species to potentially monophyletic groups of populations, *oxygnathus* and
*punicus* are here treated as separate species. It is possible that *ancilla*, *lesviacus*, and/or
*omari* may also be distinct, but these are here retained in *blythii* pending further study.
Zhang Yongzu et al. (1997) included *ancilla* in *myotis*, but this is apparently incorrect;
see Horácek et al. (2000).

*Myotis bocagii* (Peters, 1870). J. Sci. Math. Phys. Nat. Lisboa, ser. 1, 3:125.
COMMON NAME: Rufous Myotis.
TYPE LOCALITY: Angola, Duque de Braganca.
DISTRIBUTION: Senegal and Liberia to S Yemen, south to Angola, Zambia, Malawi, and
NE South Africa.
STATUS: IUCN 2003 and IUCN/SSC Action Plan (2001) – Lower Risk (lc).
SYNONYMS: *hildegardeae* Thomas, 1904; **cupreolus** Thomas, 1904; **dogalensis** Monticelli,
1887.
COMMENTS: Includes *dogalensis*; see Corbet (1978c). Reviewed in part by Harrison and Bates
(1991). Misspelled *bocagei* by some authors, see Bogdanowicz and Kock (1998).

*Myotis bombinus* Thomas, 1906. Proc. Zool. Soc. Lond., 1905(2):337 [1906].
COMMON NAME: Far Eastern Myotis.
TYPE LOCALITY: Japan, Kiushiu, Miyasaki Ken, Tano.
DISTRIBUTION: Japan, Korea, SE Siberia, NE China.
STATUS: IUCN 2003 and IUCN/SSC Action Plan (2001) – Lower Risk (nt).
SYNONYMS: **amurensis** Ognev, 1927.
COMMENTS: Formerly included in *nattereri*, but see Horácek and Hanák (1984) and Kawai
et al. (2003). Includes *amurensis*; see Yoon (1990) and Horácek et al. (2000), but also see
Yoshiyuki (1989).

*Myotis brandtii* (Eversmann, 1845). Bull. Soc. Nat. Moscow, 18(1):505.
COMMON NAME: Brandt's Myotis.
TYPE LOCALITY: Russia, Orenburgsk. Obl., S. Ural, Bolshoi-Ik River, Spasskoie. Foothills of the
Ural Mountains.
DISTRIBUTION: Britain south to Italy, Greece, and Bulgaria; east to Kazakhstan and

Mongolia, E Siberia including Sakhalin Isls, Kamchatka Peninsula and Kurile Isls; Ussuri region (Russia); Korea.

STATUS: IUCN 2003 and IUCN/SSC Action Plan (2001) – Lower Risk (lc).

SYNONYMS: *aureus* Koch, 1865; *coluotus* Kostron, 1943; *sibiricus* Kastschenko, 1905; ***gracilis*** Ognev, 1927.

COMMENTS: Listed as a subspecies of *mystacinus* by Ellerman and Morrison-Scott (1951), but see Strelkov and Buntova (1982) and Benda and Tsytsulina (2000). Does not include *fujiensis*, see Yoshiyuki (1989), Benda and Tsytsulina (2000), and Horáček et al. (2000). Includes *gracilis*, see Benda and Tsytsulina (2000), but also see Yoshiyuki (1989). Horáček et al. (2000) provisionally treated *gracilis* as distinct from *brandtii* and *fujiensis* pending further study. See also Yoon and Son (1989). Sometimes misspelled *brandti*, but *brandtii* is the original spelling.

*Myotis bucharensis* Kuzyakin, 1950. Letuchieye myschi, Izd. Sovetskaya Nauk, Moscow, p. 286.

COMMON NAME: Bocharic Myotis.

TYPE LOCALITY: Tajikistan, Kurgan-Tjubinskaja obl., Ayvadj.

DISTRIBUTION: Uzbekistan, Tajikistan, and Afghanistan.

STATUS: IUCN 2003 – Not evaluated; not considered in IUCN/SSC Action Plan (2001).

COMMENTS: Formerly included in *frater*, but clearly distinct; see Horacek et al. (2000) and Tsytsulina and Strelkov (2001).

*Myotis californicus* (Audubon and Bachman, 1842). J. Acad. Nat. Sci. Phil., ser. 1, 8:285.

COMMON NAME: Californian Myotis.

TYPE LOCALITY: USA, California, Monterey.

DISTRIBUTION: S Alaska Panhandle (USA) to Baja California and higher elevations in the Sonoran and Chihuahuan deserts (Mexico); Guatemala.

STATUS: IUCN 2003 and IUCN/SSC Action Plan (2001) – Lower Risk (lc).

SYNONYMS: *exilis* H. Allen, 1866; *nitidus* H. Allen, 1862; *oregonensis* H. Allen, 1864; *quercinus* H. W. Grinnell, 1914; *tenuidorsalis* H. Allen, 1866; ***caurinus*** Miller, 1897; ***mexicanus*** Saussure, 1860; *agilis* H. Allen, 1866; ***stephensi*** Dalquest, 1946; *pallidus* Stephens, 1900 [not Blyth, 1863].

COMMENTS: See Miller and Allen (1928) for discussion of the holotype. Reviewed in part by Yancey (1997). Subspecies are poorly delimited.

*Myotis capaccinii* (Bonaparte, 1837). Fauna Ital., 1, fasc. 20.

COMMON NAME: Long-fingered Myotis.

TYPE LOCALITY: Italy, Sicily.

DISTRIBUTION: Mediterranean zone and islands of Europe and NW Africa; Bulgaria; Turkey; Israel; Iraq; Iran; Uzbekistan.

STATUS: IUCN 2003 and IUCN/SSC Action Plan (2001) – Vulnerable.

SYNONYMS: *blasii* Kolenati, 1860; *bureschi* Heinrich, 1936; *dasypus* de Selys Longchamps, 1841; *majori* Ninni, 1878; *megapodius* Temminck, 1840; *pellucens* Crespon, 1844.

COMMENTS: See comment under *macrodactylus*. Does not include *fimbriatus*; see Corbet (1978c) and Corbet and Hill (1992). Reviewed in part by Harrison and Bates (1991), Horáček et al. (2000), and Alayrak and Asan (2002).

*Myotis chiloensis* (Waterhouse, 1840). Zool. Voy. H.M.S. "Beagle", Mammalia, p. 5.

COMMON NAME: Chilean Myotis.

TYPE LOCALITY: Chile, Chiloe Isl, Islets on eastern side.

DISTRIBUTION: C and S Chile; Argentina.

STATUS: IUCN 2003 and IUCN/SSC Action Plan (2001) – Lower Risk (nt).

SYNONYMS: *arescens* Osgood, 1943; *atacamensis* Miller and Allen, 1928 [not Lataste, 1892]; *gayi* Lataste, 1892.

COMMENTS: See LaVal (1973a) for restriction of the scope of this species. May include *aelleni*; see Pearson and Pearson (1989), but also see Barquez et al. (1993).

*Myotis chinensis* (Tomes, 1857). Proc. Zool. Soc. Lond., 1857:52.

COMMON NAME: Large Myotis.

TYPE LOCALITY: "Southern China".

DISTRIBUTION: Szechwan and Yunnan to Kiangsu (China); Hong Kong; N Thailand; Burma; Vietnam.

STATUS: IUCN 2003 and IUCN/SSC Action Plan (2001) – Lower Risk (lc).

SYNONYMS: *luctuosus* Allen, 1923.

COMMENTS: Included in the species *myotis* by Ellerman and Morrison-Scott (1951) and Zhang Yongzu et al. (1997), but see Lekagul and McNeely (1977), Corbet (1978c), Horácek et al. (2000), and Kawai et al. (2003). Reviewed by Bates et al. (1999). Two subspecies are sometimes recognized, but these do not adequately correspond to known variation in the species; see Bates et al. (1999) and Hendrichsen et al. (2001b).

*Myotis ciliolabrum* Merriam, 1886. Proc. Biol. Soc. Wash., 4:2.

COMMON NAME: Western Small-footed Myotis.

TYPE LOCALITY: United States, Kansas, Trego Co., near Banner, about 1 mi. (1.5 km) from Castle Rock, bluff on Hackberry Creek.

DISTRIBUTION: S Alberta and Saskatchewan (Canada) south through E Colorado and W Kansas (USA).

STATUS: IUCN 2003 and IUCN/SSC Action Plan (2001) – Lower Risk (lc).

COMMENTS: Formerly included in *leibii* (for which Hall [1981] used the name *subulatus*), but see van Zyll de Jong (1984). Does not include *melanorhinus*; see van Zyll de Jong (1984). Reviewed by Holloway and Barclay (2001), but note that they included *melanorhinus* as a subspecies of *ciliolabrum*.

*Myotis cobanensis* Goodwin, 1955. Am. Mus. Novit., 1744:2.

COMMON NAME: Guatemalan Myotis.

TYPE LOCALITY: Guatemala, Alta Verapaz, Coban, 1,305 m.

DISTRIBUTION: C Guatemala.

STATUS: IUCN 2003 and IUCN/SSC Action Plan (2001) – Critically Endangered.

COMMENTS: Listed as a subspecies of *velifer* by Goodwin (1955a), but see de la Torre (1958) and Hall (1981).

*Myotis csorbai* Topál, 1997. Acta Zool. Acad. Scient. Hungaricae, 43(4):377.

COMMON NAME: Csorba's Mouse-eared Myotis.

TYPE LOCALITY: Nepal, Syangja District, 4 km E of Syangja, about 30 km S of Pokhara town, 1,300 m.

DISTRIBUTION: Nepal.

STATUS: IUCN 2003 and IUCN/SSC Action Plan (2001) – Data Deficient.

COMMENTS: Distinct from *longipes*; see Topál (1997).

*Myotis dasycneme* (Boie, 1825). Isis Jena, p. 1200.

COMMON NAME: Pond Myotis.

TYPE LOCALITY: Denmark, Jutland, Dagbieg (near Wiborg).

DISTRIBUTION: France and Sweden east to Yenisei River (Russia), south to Ukraine, NW Kazakhstan; a single record from Manchuria (China).

STATUS: IUCN 2003 and IUCN/SSC Action Plan (2001) – Vulnerable.

SYNONYMS: *ferrugineus* Temminck, 1840 [not Brehm, 1827]; *limnophilus* Temminck, 1839; *major* Ognev and Worobiev, 1923; *mystacinus* Boie, 1823 [not Kuhl, 1819]; *surinamensis* Husson, 1962 [replacement name for *ferrugineus* Temminck, 1840].

COMMENTS: Probably includes *surinamensis*; see Carter and Dolan (1978).

*Myotis daubentonii* (Kuhl, 1817). Die Deutschen Fledermäuse. Hanau, p. 14.

COMMON NAME: Daubenton's Myotis.

TYPE LOCALITY: Germany, Hessen, Hanau.

DISTRIBUTION: Europe (including Britain and Ireland; Scandinavia) east to Kamtschatka, Vladivostok, Sakhalin and Kurile Isls (Russia), Japan, Korea, Manchuria, N and E China (including Tibet), Vietnam.

STATUS: IUCN 2003 and IUCN/SSC Action Plan (2001) – Lower Risk (lc).

SYNONYMS: *aedilus* Jenyns, 1839; *albus* Fitzinger, 1871; *capucinellus* Fitzinger, 1871; *lanatus* Crespon, 1844; *minutellus* Fitzinger, 1871; *staufferi* Fatio, 1890; **chasanensis** Tiunov, 1997; **loukashkini** Shamel, 1942; **nathalinae** Tupinier, 1977; **petax** Hollister, 1912; **ussuriensis** Ognev, 1927; **volgensis** Eversmann, 1840.

COMMENTS: Includes *nathalinae*; see Horácek and Hanák (1984), Fairon (1985), Mayer and von Helversen (2001*a*), and Ruedi and Mayer (2001). Reviewed in part by Yoshiyuki (1989), Yoon (1990), Bates and Harrison (1997), Bates et al. (1999), and Horácek et al. (2000). Does not appear to include *laniger*; see Topál (1997) and Bates et al. (1999), though also see Corbet and Hill (1992). For discussion of correct spelling see Bogdanowicz and Kock (1998). See Bogdanowicz (1994), but note that *laniger* was included in *daubentonii* in that publication. Apparently closely related to *bechsteinii*; see Ruedi and Mayer (2001). Subspecies limits are problematic, see Bogdanowicz (1994), Horácek et al. (2000), and Kruskop (2002). Genetic studies suggest that this complex includes more than one species, with at least some Russian and Japanese specimens representing a taxon distinct from the European form (Kawai et al., 2003).

*Myotis davidii* Peters, 1869. Monatsb. K. Preuss. Akad. Wiss. Berlin, 1869:402.
COMMON NAME: David's Myotis.
TYPE LOCALITY: China, Hopei, Peiping.
DISTRIBUTION: N China.
STATUS: IUCN 2003 – Not evaluated; not considered in IUCN/SSC Action Plan (2001).
COMMENTS: Formerly included in *mystacinus* but apparently distinct; see Pavlinov et al. (1995*b*) and Kawai et al. (2003).

*Myotis dominicensis* Miller, 1902. Proc. Biol. Soc. Wash., 15:243.
COMMON NAME: Dominican Myotis.
TYPE LOCALITY: Dominica (Lesser Antilles).
DISTRIBUTION: Dominica, Guadeloupe.
STATUS: IUCN 2003 and IUCN/SSC Action Plan (2001) – Vulnerable.
COMMENTS: Listed as a subspecies of *nigricans* by Hall and Kelson (1959), but see LaVal (1973*a*) and Hall (1981). Reviewed by Masson and Breuil (1992). Apparently closely related to *velifer* and *yumanensis*; see Ruedi and Mayer (2001).

*Myotis elegans* Hall, 1962. Univ. Kansas Publ. Mus. Nat. Hist., 14:163-164.
COMMON NAME: Elegant Myotis.
TYPE LOCALITY: Mexico, Veracruz, 12.5 mi. (20 km) N Tihuatlan.
DISTRIBUTION: San Luis Potosi (Mexico) to Costa Rica.
STATUS: IUCN 2003 and IUCN/SSC Action Plan (2001) – Lower Risk (nt).

*Myotis emarginatus* (E. Geoffroy, 1806). Ann. Mus. Natn. Hist. Nat. Paris, 8:198.
COMMON NAME: Geoffroy's Myotis.
TYPE LOCALITY: France, Ardennes, Givet, Charlemont.
DISTRIBUTION: S Europe, north to Netherlands and S Poland, Crimea, Caucasus and Kopet Dag Mtns, east and south to Israel, Jordan, Syria, Lebanon, Saudi Arabia, Oman, E Iran, Kyrgyzstan, Tajikistan, Uzbekistan, and Afghanistan; Morocco, Algeria, and Tunisia.
STATUS: IUCN 2003 and IUCN/SSC Action Plan (2001) – Vulnerable.
SYNONYMS: *budapestiensis* Margo, 1880; *ciliatus* Blasius, 1853; *kuzyakini* Pavilnov, 1979 [replacement name for *saturatus* Kuzyakin, 1934]; *neglectus* Fatio, 1890; *rufescens* Crespon 1844 [not Brehm, 1829]; *saturatus* Kuzyakin, 1934 [not Miller, 1897]; *schrankii* Kolenati, 1856 [*nomen nudum*; not *schranki* Wagner, 1843]; **desertorum** Dobson, 1875; *lanaceus* Thomas, 1920; **turcomanicus** Bobrinskii, 1925.
COMMENTS: Reviewed in part by Harrison and Bates (1991); also see Gaucher (1995) and Horácek et al. (2000). Apparently closely related to *welwitschii*; see Ruedi and Mayer (2001).

*Myotis evotis* (H. Allen, 1864). Smithson. Misc. Coll., 7:48.
COMMON NAME: Long-eared Myotis.
TYPE LOCALITY: USA, California, Monterey.
DISTRIBUTION: S British Columbia, S Alberta, S Saskatchewan (Canada) to New Mexico (USA) and Baja California (Mexico).
STATUS: IUCN 2003 and IUCN/SSC Action Plan (2001) Lower risk (lc) as *M. evotis*; Endangered as *M. milleri*.
SYNONYMS: **chrysonotus** J. A. Allen, 1896; **jonesorum** Manning, 1993; **micronyx** Nelson and Goldman, 1909; **milleri** Elliot, 1903; **pacificus** Dalquest, 1943.

COMMENTS: See Genoways and Jones (1969*b*) and Manning and Jones (1989). Includes *milleri*; see Reducker et al. (1983) and Manning (1993). Does not include *auriculus*; see Genoways and Jones (1969*b*), Hall (1981), and Gannon (1998). Revised by Manning (1993).

*Myotis fimbriatus* (Peters, 1871). *In* R. Swinhoe, Catalogue of Mammals of China, Proc. Zool. Soc. Lond., 1870:617 [1871].
COMMON NAME: Fringed Long-footed Myotis.
TYPE LOCALITY: China, Fujian, Amoy.
DISTRIBUTION: SE China.
STATUS: IUCN 2003 and IUCN/SSC Action Plan (2001) – Lower Risk (nt).
SYNONYMS: *hirsutus* Howell, 1926.
COMMENTS: Not conspecific with *macrodactylus* or *capaccinii*; see Corbet (1978*c*) and Corbet and Hill (1992).

*Myotis findleyi* Bogan, 1978. J. Mammal., 59:524.
COMMON NAME: Findley's Myotis.
TYPE LOCALITY: Mexico, Nayarit, Trés Marías Isls, Maria Magdalena Isl.
DISTRIBUTION: Trés Marías Isls (Mexico).
STATUS: IUCN 2003 and IUCN/SSC Action Plan (2001) – Endangered.

*Myotis formosus* (Hodgson, 1835). J. Asiat. Soc. Bengal, 4:700.
COMMON NAME: Hodgson's Myotis.
TYPE LOCALITY: Nepal.
DISTRIBUTION: Afghanistan to N India, Nepal, Tibet, Kweichow, Kwangsi, Kiangsu and Fukien (China); Taiwan, Korea, Tsushima Isl (Japan), Malaysia, Philippines, Sumatra, Java, Sulawesi, and Bali.
STATUS: IUCN 2003 and IUCN/SSC Action Plan (2001) – Lower Risk (lc). IUCN 2003 – Not evaluated as *M. f. bartelsi*.
SYNONYMS: *andersoni* Trouessart, 1897; *auratus* Dobson, 1871; *dobsoni* Anderson, 1881 [not Trouessart, 1878]; *pallida* Blyth, 1863; **bartelsi** Jentink, 1910; **rufoniger** Tomes, 1858; **rufopictus** Waterhouse, 1845; *tsuensis* Kuroda, 1922; *chofukusei* Mori, 1928; **watasei** Kishida, 1924; *flavus* Shamel, 1944; **weberi** Jentink, 1890.
COMMENTS: For discussion of synonyms see Findley (1972). Does not include *hermani*; see Corbet and Hill (1992). Reviewed in part by Yoshiyuki (1989), Yoon (1990), and Bates and Harrison (1997). *M. f. rufopictus* may represent a distinct species; see Heaney et al. (1998).

*Myotis fortidens* Miller and Allen, 1928. Bull. U.S. Natl. Mus., 144:54.
COMMON NAME: Cinnamon Myotis.
TYPE LOCALITY: Mexico, Tabasco, Teapa.
DISTRIBUTION: Sonora and Veracruz (Mexico) to Guatemala.
STATUS: IUCN 2003 and IUCN/SSC Action Plan (2001) – Lower Risk (nt).
SYNONYMS: *cinnamomeus* Miller, 1902 [not Wagner, 1855]; **sonoriensis** Findley and Jones, 1967.

*Myotis frater* G. M. Allen, 1923. Am. Mus. Novit., 85:6.
COMMON NAME: Fraternal Myotis.
TYPE LOCALITY: China, SE China, Fukien (= Fujian), Yenping.
DISTRIBUTION: E Siberia, Ussuri Region, Krasnoyarsk Region (Russia) to Korea, Heilungkiang (China), SE China; Japan.
STATUS: IUCN 2003 and IUCN/SSC Action Plan (2001) – Lower Risk (nt).
SYNONYMS: **eniseensis** Tsytsulina and Strelkov, 2001; **kaguyae** Imaizumi, 1956; **longicaudatus** Ognev, 1927.
COMMENTS: Includes *longicaudatus*, see Corbet (1978*c*) and Tsytsulina and Strelkov (2001). Does not include *bucharensis*, see Horácek et al. (2000) and Tsytsulina and Strelkov (2001).

*Myotis gomantongensis* Francis and Hill, 1998. Mammalia, 62(2):248.
COMMON NAME: Gomantong Myotis.

TYPE LOCALITY: Malaysia, Borneo, Sabah, Gomantong Caves, 5°31′N, 118°04′E.
DISTRIBUTION: Sabah (Borneo, Malaysia).
STATUS: IUCN 2003 and IUCN/SSC Action Plan (2001) – Data Deficient.
COMMENTS: Based on specimens previously referred to *ater* by Hill and Francis (1984) and
  Payne et al. (1985).

*Myotis goudoti* (A. Smith, 1834). S. Afr. Quart. J., 2:244.
COMMON NAME: Malagasy Myotis.
TYPE LOCALITY: Madagascar.
DISTRIBUTION: Madagascar.
STATUS: IUCN 2003 and IUCN/SSC Action Plan (2001) – Lower Risk (nt).
SYNONYMS: *madagascariensis* Tomes, 1858; *sylvicola* A. Grandidier, 1870.
COMMENTS: Does not appear to include *anjouanensis*; see Peterson et al. (1995).

*Myotis grisescens* A. H. Howell, 1909. Proc. Biol. Soc. Wash., 22:46.
COMMON NAME: Gray Myotis.
TYPE LOCALITY: USA, Tennessee, Marion Co., Nickajack Cave, near Shellmound.
DISTRIBUTION: Florida Panhandle to Kentucky, Indiana, Illinois, E Kansas and NE Oklahoma
  (USA).
STATUS: U.S. ESA – Endangered; IUCN 2003 and IUCN/SSC Action Plan (2001) –
  Endangered.

*Myotis hajastanicus* Argyropulo, 1939. Zool. Sbornick 1 (Trudy Biol. Inst. 3):27.
COMMON NAME: Hajastan Myotis.
TYPE LOCALITY: Armenia, eastern bank of Sevan Lake, Sordza (= Nadezdino), 2,000 m.
DISTRIBUTION: Known only from the Sevan Lake basin in Armenia.
STATUS: IUCN 2003 – Not evaluated; not considered in IUCN/SSC Action Plan (2001).
COMMENTS: Formerly included in *mystacinus*, but see Benda and Tsytsulina (2000).

*Myotis hasseltii* (Temminck, 1840). Monogr. Mamm., 2:225.
COMMON NAME: Lesser Large-footed Myotis.
TYPE LOCALITY: Indonesia, Java, Bantam.
DISTRIBUTION: E India, Sri Lanka, Burma, Thailand, Cambodia, Vietnam, W Malaysia,
  Sumatra, Mentawai Isls, Riau Arch., Java, Borneo.
STATUS: IUCN 2003 and IUCN/SSC Action Plan (2001) – Lower Risk (lc).
SYNONYMS: **abboti** Lyon, 1916; **continentis** Shamel, 1942; *berdmorei* Blyth, 1863; **macellus**
  Temminck, 1840.
COMMENTS: Reviewed by Hill (1983), Bates and Harrison (1997), and Bates et al. (1999).
  Apparently closely related to *macrotarsus* and *horsfieldii*; see Ruedi and Mayer (2001).

*Myotis hermani* Thomas, 1923. Ann. Mag. Nat. Hist., ser. 9, 11:252.
COMMON NAME: Herman's Myotis.
TYPE LOCALITY: Indonesia, NW Sumatra, Sabang.
DISTRIBUTION: Sumatra (Indonesia).
STATUS: IUCN 2003 and IUCN/SSC Action Plan (2001) – Data Deficient.
COMMENTS: Included in *formosus* by Findley (1972), but see Corbet and Hill (1992).

*Myotis horsfieldii* (Temminck, 1840). Monogr. Mamm., 2:226.
COMMON NAME: Horsfield's Myotis.
TYPE LOCALITY: Indonesia, Java, Mount Gede.
DISTRIBUTION: India (including Andaman Isls, SE China, Thailand, Burma, Laos, Vietnam,
  W Malaysia, Java, Bali, Sulawesi, Borneo, Philippines.
STATUS: IUCN 2003 and IUCN/SSC Action Plan (2001) – Lower Risk (lc).
SYNONYMS: *lepidus* Thomas, 1915; **deignani** Shamel, 1942; **dryas** K. Andersen, 1907; **jeannei**
  Taylor, 1934; **peshwa** Thomas, 1915.
COMMENTS: Reviewed by Hill (1983), Bates and Harrison (1997), Bates et al. (1999), and
  Hendrichsen et al. (2001). Apparently closely related to *macrotarsus* and *hasseltii*; see
  Ruedi and Mayer (2001).

*Myotis hosonoi* Imaizumi, 1954. Bull. Natl. Sci. Mus. Tokyo, N. S., 1:44.
COMMON NAME: Hosono's Myotis.

TYPE LOCALITY: Japan, Honshu, Nagano Pref., Kita-azumi-Gun (about 30 km N Matsumotao City), Tokiwa-Mura, Koumito, 732 m.

DISTRIBUTION: Honshu (Japan).

STATUS: IUCN 2003 and IUCN/SSC Action Plan (2001) – Vulnerable.

COMMENTS: Reviewed by Yoshiyuki (1989), Horácek et al. (2000), and Tsytsulina (2000).

*Myotis ikonnikovi* Ognev, 1912. Ann. Mus. Zool. Acad. Imp. Sci. St. Petersbourg, 16:477.

COMMON NAME: Ikonnikov's Myotis.

TYPE LOCALITY: Russia, Primorsk. Krai (= Ussuri Region), Dalnerechen Dist., Euseevka.

DISTRIBUTION: Ussuri region and N Korea to Lake Baikal (Russia), the Altai Mtns, and Mongolia, NE China; Sakhalin Isl (Russia) and Honshû and Hokkaido Isls (Japan).

STATUS: IUCN 2003 and IUCN/SSC Action Plan (2001) – Lower Risk (lc).

SYNONYMS: *fujiensis* Imaizumi, 1954.

COMMENTS: Revised by Tsytsulina (2001); also see Corbet (1978c), Yoshiyuki (1989), and Benda and Tsytsulina (2000). Molecular sequence data support placement of *fujiensis* in *ikonnikovi* (Kawai et al., 2003).

*Myotis insularum* (Dobson, 1878). Cat. Chiroptera Brit. Mus., p. 313.

COMMON NAME: Insular Myotis.

TYPE LOCALITY: Samoa.

DISTRIBUTION: Samoa.

STATUS: IUCN 2003 and IUCN/SSC Action Plan (2001) – Data Deficient.

COMMENTS: Poorly known, the type and only specimen possibly being incorrectly labelled; see Koopman (1984c).

*Myotis keaysi* J. A. Allen, 1914. Bull. Am. Mus. Nat. Hist., 33:383.

COMMON NAME: Hairy-legged Myotis.

TYPE LOCALITY: Peru, Puno, Inca Mines.

DISTRIBUTION: Tamaulipas (Mexico) to Bolivia, N Argentina, Peru, Ecuador, Venezuela, and Trinidad.

STATUS: IUCN 2003 and IUCN/SSC Action Plan (2001) – Lower Risk (lc).

SYNONYMS: **pilosotibialis** LaVal, 1973.

COMMENTS: Revised by LaVal (1973a). Apparently closely related to *riparius* and *ruber*; see Ruedi and Mayer (2001).

*Myotis keenii* (Merriam, 1895). Am. Nat., 29:860.

COMMON NAME: Keen's Myotis.

TYPE LOCALITY: Canada, British Columbia, Queen Charlotte Isls, Graham Isl, Massett.

DISTRIBUTION: Alaska Panhandle to W Washington (USA).

STATUS: IUCN 2003 and IUCN/SSC Action Plan (2001) – Lower Risk (lc).

COMMENTS: Does not include *septentrionalis*; see van Zyll de Jong (1979) and Caceres and Barclay (2000). See Fitch and Shump (1979), but note that they included *septentrionalis*.

*Myotis laniger* Peters, 1871. Proc. Royal Soc. Lond., 3 (1870):617.

COMMON NAME: Chinese Water Myotis.

TYPE LOCALITY: China, Fujian, Amoy.

DISTRIBUTION: S China including Tibet, Vietnam, E India.

STATUS: IUCN 2003 – Not evaluated; not considered in IUCN/SSC Action Plan (2001).

COMMENTS: Included in *daubentonii* by many authors, but see Topál (1997) and Bates et al. (1999).

*Myotis leibii* (Audubon and Bachman, 1842). J. Acad. Nat. Sci. Phil., ser. 1, 8:284.

COMMON NAME: Eastern Small-footed Myotis.

TYPE LOCALITY: USA, Pennsylvania, Erie Co.

DISTRIBUTION: E North America from S Ontario, S Quebec (Canada), and S Maine (USA) south to Georgia and west to E Oklahoma (USA).

STATUS: IUCN 2003 and IUCN/SSC Action Plan (2001) – Lower Risk (lc).

SYNONYMS: *henshawii* H. Allen, 1894; *orinomus* Elliot, 1903; *winnemana* Nelson, 1913.

COMMENTS: Formerly included *ciliolabrum* and *melanorhinus*, but see van Zyll de Jong (1984).

An older name for this species may be *subulatus* Say, 1823; see Glass and Baker (1968). These authors recommended that *subulatus* should be supressed, but see Hall (1981), who used *subulatus* instead of *leibii* for this species. Koopman (1993) disagreed, and suggested that *subulatus* is probably an older name for *yumanensis*.

*Myotis levis* (I. Geoffroy, 1824). Ann. Sci. Nat. Zool., ser. 1, 3:444-445.
COMMON NAME: Yellowish Myotis.
TYPE LOCALITY: "Southern Brazil."
DISTRIBUTION: Bolivia, Argentina, SE Brazil, Uruguay.
STATUS: IUCN 2003 and IUCN/SSC Action Plan (2001) – Lower Risk (lc).
SYNONYMS: *alter* Miller and Allen, 1928; *nubilus* J. A. Wagner, 1855; *polythrix* I. Geoffroy, 1824; *dinellii* I Geoffroy, 1824.
COMMENTS: Included in *ruber* by Cabrera (1958), but see LaVal (1973a). Reviewed in part by López-González et al. (2001). Apparently closely related to *nigricans*; see Ruedi and Mayer (2001).

*Myotis longipes* (Dobson, 1873). Proc. Asiat. Soc. Bengal, p. 110.
COMMON NAME: Kashmir Cave Myotis.
TYPE LOCALITY: India, Kashmir, Bhima Devi Caves, 6,000 ft. (1,829 m).
DISTRIBUTION: Afghanistan, NE India, Nepal, possibly Vietnam.
STATUS: IUCN 2003 and IUCN/SSC Action Plan (2001) – Vulnerable.
SYNONYMS: *macropus* Dobson, 1872 [not Gould, 1854]; *megalopus* Dobson, 1875.
COMMENTS: Included in *capaccinii* by Ellerman and Morrison-Scott (1951), but considered a distinct species by Hanák and Gaisler (1969), Corbet (1978c), and Bates and Harrison (1997). Some specimens referred to *longipes* by Bates and Harrison (1997) subsequently formed the type series for *csorbai* (Topál, 1997). Vietnamese records are dubious; see Bates et al. (1999).

*Myotis lucifugus* (Le Conte, 1831). *In* McMurtie, Animal Kingdom, 1 (App.):431.
COMMON NAME: Little Brown Myotis.
TYPE LOCALITY: USA, Georgia, possibly Liberty Co., LeConte Plantation near Riceboro (but see Davis and Rippy, 1968).
DISTRIBUTION: Alaska (USA) to Labrador and Newfoundland (Canada), south to S California, N Arizona, N New Mexico (USA).
STATUS: IUCN 2003 and IUCN/SSC Action Plan (2001) – Lower Risk (lc).
SYNONYMS: *affinis* H. Allen, 1864; *brevirostris* Wied-Neuwied, 1862; *carolii* Temminck, 1840; *crassus* F. Cuvier, 1832; *domesticus* Green, 1832; *gryphus* F. Cuvier, 1832; *lanceolatus* Wied, 1839; *salarii* F. Cuvier, 1832; *virginianus* Audubon and Bachman, 1841; *alascensis* Miller, 1897; *carissima* Thomas, 1904; *albicinctus* G. M. Allen, 1919; *altipetens* H. W. Grinnell, 1916; *baileyi* Hollister, 1909; *pernox* Hollister, 1911; *relictus* Harris, 1974.
COMMENTS: Does not include *occultus*, see Piaggio et al. (2002). Hybridizes with *yumanensis* in some areas; see Parkinson (1979), but see Herd and Fenton (1983). See Fenton and Barclay (1980). Apparently closely related to *thysanodes*; see Ruedi and Mayer (2001). Status of the type and type locality was discussed by Davis and Rippy (1968).

*Myotis macrodactylus* (Temminck, 1840). Monogr. Mamm., 2:231.
COMMON NAME: Big-footed Myotis.
TYPE LOCALITY: Japan.
DISTRIBUTION: Japan, Kunashir Isl and Kurile Isls (Russia), SE Siberia, Korea.
STATUS: IUCN 2003 and IUCN/SSC Action Plan (2001) – Lower Risk (lc).
SYNONYMS: *continentalis* Tiunov, 1997; *insularis* Tiunov, 1997.
COMMENTS: Koopman (1993) included *fimbriatus* in this species and noted that it was probably conspecific with *capaccinii* (see Wallin, 1969), but see Corbet (1978c), Yoshiyuki (1989), and Corbet and Hill (1992), who argued that *fimbriatus* and *capaccinii* are distinct species. Reviewed in part by Yoshiyuki (1989) and Yoon (1990).

*Myotis macropus* (Gould, 1854). Mammals of Australia, unnumbered page of text.
COMMON NAME: Gould's Large-footed Myotis.
TYPE LOCALITY: South Australia.

DISTRIBUTION: S Australia, Victoria (Australia).

STATUS: IUCN 2003 – Not evaluated; not considered in IUCN/SSC Action Plan (2001).

COMMENTS: Distinct from *adversus* and *moluccarum*; see Kitchener et al. (1995*b*), who revised this complex.

*Myotis macrotarsus* (Waterhouse, 1845). Proc. Zool. Soc. Lond., 1845, 3:5.

COMMON NAME: Pallid Large-footed Myotis.

TYPE LOCALITY: Philippines.

DISTRIBUTION: Philippines, N Borneo.

STATUS: IUCN 2003 and IUCN/SSC Action Plan (2001) – Lower Risk (nt).

SYNONYMS: *saba* Davis, 1962.

COMMENTS: May include *stalkeri*; see Findley (1972) and Corbet and Hill (1992). Apparently closely related to *hasseltii* and *horsfieldii*; see Ruedi and Mayer (2001).

*Myotis martiniquensis* LaVal, 1973. Bull. Los Angeles Cty. Mus. Nat. Hist. Sci. Soc., 15:35.

COMMON NAME: Schwartz's Myotis.

TYPE LOCALITY: Martinique (Lesser Antilles), Tartane, 6 km E La Trinité (France).

DISTRIBUTION: Martinique, Barbados (Lesser Antilles).

STATUS: IUCN 2003 and IUCN/SSC Action Plan (2001) – Lower Risk (nt).

SYNONYMS: *nyctor* LaVal and Schwartz, 1975.

COMMENTS: See Masson and Breuil (1992) and Timm and Genoways (2003).

*Myotis melanorhinus* Merriam, 1890. N. Amer. Fauna, 3:46.

COMMON NAME: Dark-nosed Small-footed Myotis.

TYPE LOCALITY: United States, Arizona, Coconino Co., N base of San Francisco Mountain, Little Spring, 8,250 ft (2,750 m).

DISTRIBUTION: British Columbia (Canada) south to C Mexico and east to W Oklahoma (USA).

STATUS: IUCN 2003 – Not evaluated; not considered in IUCN/SSC Action Plan (2001).

COMMENTS: Included in *leibii* or *ciliolabrum* by various authors, but see van Zyll de Jong (1984). Reviewed by Holloway and Barclay (2001), who treated it as a subspecies of *ciliolabrum*.

*Myotis moluccarum* (Thomas, 1915). Ann. Mag. Nat. Hist., ser. 8, 15:170.

COMMON NAME: Maluku Myotis.

TYPE LOCALITY: Indonesia, Maluku Tenggara, Kei (=Kai) Isls, Ara.

DISTRIBUTION: Ambon and Kai Isls (Moluccas), N and W Australia, Seram, Waigeo Isl (West Palua, Indonesia), Papua New Guinea, Bismarck Arch., Solomon Isls.

STATUS: IUCN 2003 – Not evaluated; not considered in IUCN/SSC Action Plan (2001).

SYNONYMS: *richardsi* Kitchener, 1995 [in Kitchener et al., 1995*b*]; *solomonis* Troughton, 1929.

COMMENTS: Distinct from *adversus* and *macropus*; see Kitchener et al. (1995*b*), who revised this complex. Also see Flannery (1995*b*) and Bonaccorso (1998). Includes *Anamygdon solomonis* Troughton, 1929; see Phillips and Birney (1968) and Kitchener et al. (1995*b*). Kitchener et al. (1995*b*) tentatively retained *solomonis* as a synonym of *M. moluccarum moluccarum*, but morphological differences suggest that it is best considered as a distinct subspecies.

*Myotis montivagus* (Dobson, 1874). J. Asiat. Soc. Bengal, 43:237.

COMMON NAME: Burmese Whiskered Myotis.

TYPE LOCALITY: China, Yunnan, Hotha.

DISTRIBUTION: Yunnan to Fukien and Chihli (China), NE India, Burma, Vietnam, Laos, NE Thailand, W Malaysia, Borneo.

STATUS: IUCN 2003 and IUCN/SSC Action Plan (2001) – Lower Risk (nt).

SYNONYMS: *borneoensis* Hill and Francis, 1984; *federatus* Thomas, 1916; *peytoni* Wroughton and Ryley, 1913.

COMMENTS: Includes *peytoni*; see Hill (1962*b*), Corbet and Hill (1992), and Bates and Harrison (1997); but see also Findley (1972). Das (1987) reviewed the type series. Reviewed by Bates et al. (1999). Subspecies affinities of Vietnam specimen are unclear; see Bates et al. (1999) and Hendrichsen et al. (2001).

*Myotis morrisi* Hill, 1971. Bull. Brit. Mus. (Nat. Hist.) Zool., 21:43.
    COMMON NAME: Morris's Myotis.
    TYPE LOCALITY: Ethiopia, Walaga, Didessa River mouth.
    DISTRIBUTION: Ethiopia, Nigeria.
    STATUS: IUCN 2003 and IUCN/SSC Action Plan (2001) – Vulnerable.

*Myotis muricola* (Gray, 1846). Cat. Hodgson Coll. Brit. Mus., p. 4.
    COMMON NAME: Nepalese Whiskered Myotis.
    TYPE LOCALITY: Nepal.
    DISTRIBUTION: Afghanistan through N India and Nepal to Taiwan, Vietnam, Malaysia,
        Indonesia, and New Guinea; possibly the Philippines.
    STATUS: IUCN 2003 and IUCN/SSC Action Plan (2001) – Lower Risk (lc).
    SYNONYMS: *lobipes* Peters, 1867; *tralatitus* Temminck, 1840 [not Horsfield, 1824]; **browni**
        Taylor, 1934; **caliginosus** Tomes, 1859; *blanfordi* Dobson, 1871; **herrei** Taylor, 1934;
        **latirostris** Kishida, 1932; *orii* Kuroda, 1935; **moupinensis** Milne-Edwards, 1872;
        **niasensis** Lyon, 1916; **patriciae** Taylor, 1934. **Unassigned**: *muricola* Hodgson, 1841
        [*nomen nudum*]; *trilatitoides* Gray, 1843 [*nomen nudum*].
    COMMENTS: Includes *caliginosus*, *moupinensis*, and *latirostris*; see Findley (1972). Does not
        include *ater* or *nugax*; see Hill (1983) and Hill and Corbet (1992). Includes *browni*; see
        Hill and Rozendaal (1989). Includes *herrei* and *patriciae*; see Heaney et al. (1987), but
        also see Corbet and Hill (1992), who listed *patriciae* as a separate species with some
        reservations. Reviewed in part by Hill (1983), Kock (1996), Bates and Harrison (1997),
        Bates et al. (1999), Francis et al. (1999), and Hendrichsen et al. (2001). This complex
        may include more than one species, see Francis et al. (1999). There is apparently only
        one species of "*muricola*-type" *Myotis* in the Philippines, but it is not yet clear if this
        taxon is *ater* or *muricola* (L. Heaney, pers. comm.); the same is probably also true of the
        Moluccas (K. Helgen, pers. comm.).

*Myotis myotis* (Borkhausen, 1797). Deutsche Fauna, 1:80.
    COMMON NAME: Mouse-eared Myotis.
    TYPE LOCALITY: Germany, Thuringia.
    DISTRIBUTION: C and S Europe, east to Ukraine; S England; most Mediterranean islands;
        Azores (Portugal); Asia Minor; Lebanon, Syria, and Israel.
    STATUS: IUCN 2003 and IUCN/SSC Action Plan (2001) – Lower Risk (nt).
    SYNONYMS: *alpinus* Koch, 1865; *latipennis* Crespon, 1844; *myosotis* author unknown, date
        1797 or 1800 [see Ellerman and Morrison-Scott, 1951]; *spelaea* Bielz, 1886 [not Koch,
        1865]; *submurinus* Brehm, 1827; *typus* Koch, 1865; **macrocephalicus** Harrison and
        Lewis, 1961.
    COMMENTS: See Corbet (1978c) and Horácek et al. (2000) for discussion of synonyms. Zhang
        Yongzu et al. (1997) included *ancilla* and *chinensis* in *myotis*, but this is apparently
        incorrect; see Horácek et al. (2000). Closely related to *blythii*, *oxygnathus*, and *punicus*;
        see Castella et al. (2000), Mayer and von Helversen (2001a), and Ruedi and Mayer
        (2001). Middle Eastern records reviewed by Harrison and Bates (1991), Palearctic
        records by Horácek et al. (2000).

*Myotis mystacinus* (Kuhl, 1817). Die Deutschen Fledermäuse. Hanau, p. 15.
    COMMON NAME: Whiskered Myotis.
    TYPE LOCALITY: Germany.
    DISTRIBUTION: Ireland and Scandinavia to C Russia and the Ural Moutains, Kazakhstan,
        south to Syria, Israel, and Morocco.
    STATUS: IUCN 2003 and IUCN/SSC Action Plan (2001) – Lower Risk (lc).
    SYNONYMS: *aurascens* Kuzyakin, 1935; *bulgaricus* Heinrich, 1936; *collaris* Schinz, 1821;
        *humeralis* Baillon, 1834; *lugubris* Fatio, 1869; *nigricans* Koch, 1865 [not Schinz, 1821];
        *nigricans* Fatio, 1869 [not Schinz, 1821, or Koch, 1865]; *nigrofuscus* Fitzinger, 1871;
        *rufofuscus* Koch, 1865; *schinzii* Brehm, 1837; *schrankii* Wagner, 1843; **caucasicus**
        Tsytsulina, 2000 [in Benda and Tsytsulina, 2000]; **occidentalis** Benda, 2000 [in Benda
        and Tsytsulina, 2000].
    COMMENTS: Reviewed by Strelkov (1983), Bates and Harrison (1997), and Horácek et al.
        (2000), revised by Benda and Tsytsulina (2000) and Tsytsulina (2001). Does not

include *davidii*, *hajastanicus*, *nipalensis*, *przewalskii*, *sogdianus*, or *transcaspicus*; see Benda and Tsytsulina (2000) and Kawai et al. (2003). Includes *aurascens*; see Mayer and von Helversen (2001*a*), but also see Benda and Tsytsulina (2000). This complex may include at least one cryptic species in Europe; see Mayer and von Helversen (2001*a*). Japanese specimens previously referred to this species clearly represent an apparently unnamed taxon distinct from both *mystacinus* and *davidii*; see Kawai et al. (2003). Specimens from Vietnam originally identified as *mystacinus* may represent *muricola* (see Bates et al., 1999); alternatively, they might represent *davidii* or be conspecific with the unnamed Japanese form.

*Myotis nattereri* (Kuhl, 1817). Die Deutschen Fledermäuse. Hanau, p. 14, 33.
   COMMON NAME: Natterer's Myotis.
   TYPE LOCALITY: Germany, Hessen, Hanau.
   DISTRIBUTION: Ireland, Great Britain, Europe (except N Scandinavia), Morocco, N Algeria, Turkey, Israel, Jordan, Lebanon, Iraq, Iran, Bulgaria, Crimea and Caucasus to Turkmenistan.
   STATUS: IUCN 2003 and IUCN/SSC Action Plan (2001) – Lower Risk (lc).
   SYNONYMS: *escalerae* Cabrera, 1904; *hoveli* Harrison, 1964; *spelaeus* Koch, 1865; *typus* Koch, 1865; **tschuliensis** Kuzyakin, 1935.
   COMMENTS: Does not include *araxenus* or *bombinus*; see Horácek and Hanák (1984) and Kawai et al. (2003). Reviewed in part by Harrison and Bates (1991) and Horácek et al. (2000). For discussion of correct spelling see Bogdanowicz and Kock (1998).

*Myotis nesopolus* Miller, 1900. Proc. Biol. Soc. Wash., 13:123.
   COMMON NAME: Curaçao Myotis.
   TYPE LOCALITY: Curaçao, Willemstad (Netherlands).
   DISTRIBUTION: NE Venezuela; Curaçao and Bonaire (Netherlands Antilles).
   STATUS: IUCN 2003 and IUCN/SSC Action Plan (2001) – Lower Risk (nt).
   SYNONYMS: **larensis** LaVal, 1973.
   COMMENTS: Includes *larensis*; see Genoways and Williams (1979*a*). A single specimen reported from St. Martin probably represents *nigricans*; see Jones (1989).

*Myotis nigricans* (Schinz, 1821). Das Thierreich, 1:179.
   COMMON NAME: Black Myotis.
   TYPE LOCALITY: Brazil, Espírito Santo, between Itapemirin and Iconha Rivers, Fazenda de Aga.
   DISTRIBUTION: Nayarit and Tamaulipas (Mexico) to Peru, Bolivia, N Argentina, Paraguay, and S Brazil; Trinidad and Tobago; St. Martin, Montserrat, Grenada (Lesser Antilles).
   STATUS: IUCN 2003 and IUCN/SSC Action Plan (2001) – Lower Risk (lc).
   SYNONYMS: *arsinoe* Temminck, 1840; *bondae* J. A. Allen, 1914; *brasiliensis* Spix, 1823; *chiriquensis* J. A. Allen, 1904; *concinnus* H. Allen, 1866; *dalquesti* Hall and Alvarez, 1961; *esmeraldae* J. A. Allen, 1914; *exiguus* H. Allen, 1866; *hypothrix* D'Orbigny and Gervais, 1847; *maripensis* J. A. Allen, 1914; *mundus* H. Allen, 1866; *parvulus* Temminck, 1840; *punensis* J. A. Allen, 1914; *spixii* J. B. Fischer, 1829; *splendidus* J. A. Wagner, 1855; **carteri** LaVal, 1973; **extremus** Miller and Allen, 1928; **osculatii** Cornalia, 1849; *caucensis* Miller and G. M. Allen, 1928; *quixensis* Osculati, 1854.
   COMMENTS: Includes *carteri*; see Corbet and Hill (1980), but see Bogan (1978). Neotype designated by LaVal (1973*a*). See Wilson and LaVal (1974). Reviewed in part by López-González et al. (2001). Apparently closely related to *levis*; see Ruedi and Mayer (2001). More than one species may be represented in this complex.

*Myotis nipalensis* Dobson, 1871. Proc. Asiat. Soc. Bengal, 1871:214.
   COMMON NAME: Nepalese Myotis.
   TYPE LOCALITY: Nepal, Katmandu.
   DISTRIBUTION: Iran, Turkey, and Uzbekistan to Nepal, Mongolia, Tibet and NW China, Siberia.
   STATUS: IUCN 2003 – Not evaluated; not considered in IUCN/SSC Action Plan (2001).
   SYNONYMS: *kukunoriensis* Bobrinskii, 1929; *meinertzhageni* Thomas, 1926; *pallidiventris* Hodgson, 1844 [*nomen nudum*]; **przewalskii** Bobrinski, 1926; *mongolicus* Kruskop and

Borissenko, 1996; *transcaspicus* Ognev and Heptner, 1928; *pamirensis* Kuzyakin, 1935; *sogdianus* Kuzyakin, 1934.
COMMENTS: Often included in *mystacinus*, but see Benda and Tsytsulina (2000).

*Myotis occultus* Hollister, 1909. Proc. Biol. Soc. Wash., 22:43.
COMMON NAME: Arizona Myotis.
TYPE LOCALITY: California, San Bernadino Co., 10 mi above Needles, W side of Colorado River.
DISTRIBUTION: S California to Arizona, New Mexico, and Colorado (USA), south to Distrito Federal (Mexico); possibly W Texas (USA).
STATUS: IUCN 2003 – Not evaluated; not considered in IUCN/SSC Action Plan (2001).
COMMENTS: Included in *lucifugus* by Findley and Jones (1967) and most subsequent authors, but apparently distinct, see Piaggio et al. (2002).

*Myotis oreias* (Temminck, 1840). Monogr. Mamm., 2:270.
COMMON NAME: Singaporese Whiskered Myotis.
TYPE LOCALITY: Singapore.
DISTRIBUTION: Known only from the type locality.
STATUS: IUCN 2003 and IUCN/SSC Action Plan (2001) – Data Deficient.
COMMENTS: Known only from the holotype. Redescribed by Francis and Hill (1998), who noted that the type locality is questionable.

*Myotis oxygnathus* Monticelli, 1885. Ann. Accad. O. Costa de Aspir. Nat. Napoli, 1:82.
COMMON NAME: Monticelli's Myotis.
TYPE LOCALITY: Italy, Basilicata, Matera.
DISTRIBUTION: Mediterranean region from Spain to Italy and Greece; Bulgaria to Turkmenistan, Kyrgyzstan, and Afghanistan.
STATUS: IUCN 2003 – Not evaluated; not considered in IUCN/SSC Action Plan (2001).
COMMENTS: Formerly treated as a subspecies of *blythii* (e.g., Koopman, 1994), but shown to be more closely related to *myotis* based on molecular data; see Ruedi and Mayer (2001).

*Myotis oxyotus* (Peters, 1867). Monatsb. K. Preuss. Akad. Wiss. Berlin, 1867:19.
COMMON NAME: Montane Myotis.
TYPE LOCALITY: Ecuador, Mount Chimborazo, between 2,743 and 3,048 m.
DISTRIBUTION: Venezuela to Bolivia; Panama; Costa Rica.
STATUS: IUCN 2003 and IUCN/SSC Action Plan (2001) – Lower Risk (lc).
SYNONYMS: *thomasi* Cabrera, 1901; *gardneri* LaVal, 1973.
COMMENTS: Revised by LaVal (1973a). Subspecies allocation of populations from coastal Peru is uncertain. Apparently closely related to *nigricans* and *levis*; see Ruedi and Mayer (2001).

*Myotis ozensis* Imaizumi, 1954. Bull. Natl. Sci. Mus. Tokyo, N. S., 1:49.
COMMON NAME: Honshu Myotis.
TYPE LOCALITY: Japan, Honshu, Gunma Pref, Ozegahara, 1,400 m.
DISTRIBUTION: Honshu (Japan).
STATUS: IUCN 2003 and IUCN/SSC Action Plan (2001) – Endangered.
COMMENTS: Reviewed by Yoshiyuki (1989).

*Myotis peninsularis* Miller, 1898. Ann. Mag. Nat. Hist., ser. 7, 2:124.
COMMON NAME: Peninsular Myotis.
TYPE LOCALITY: Mexico, Baja California, San Jose del Cabo.
DISTRIBUTION: S Baja California (Mexico).
STATUS: IUCN 2003 and IUCN/SSC Action Plan (2001) – Vulnerable.
COMMENTS: Listed as a subspecies of *velifer* by Hall and Kelson (1959), but see Hayward (1970) and Hall (1981). See Alvarez-Castañeda and Bogan (1998).

*Myotis pequinius* Thomas, 1908. Proc. Zool. Soc. Lond., 1908:637.
COMMON NAME: Peking Myotis.
TYPE LOCALITY: China, Hopeh, 30 mi. (48 km) W Peking, 600 ft (183 m).
DISTRIBUTION: Hong Kong, Hopeh, Shantung, Honan and Kiangsu (China).
STATUS: IUCN 2003 and IUCN/SSC Action Plan (2001) – Lower Risk (nt).

COMMENTS: See Horácek et al. (2000).

*Myotis planiceps* Baker, 1955. Proc. Biol. Soc. Wash., 68:165.
COMMON NAME: Flat-headed Myotis.
TYPE LOCALITY: Mexico, Coahuila, 7 mi. (11 km) S and 4 mi. (6 km) E Bella Union, 7,200 ft. (2,195 m).
DISTRIBUTION: Coahuila, Nuevo León, and Zacatecas (Mexico).
STATUS: IUCN 2003 and IUCN/SSC Action Plan (2001) – Critically Endangered.
COMMENTS: See Matson (1975).

*Myotis pruinosus* Yoshiyuki, 1971. Bull. Natl. Sci. Mus. Tokyo, 14:305.
COMMON NAME: Frosted Myotis.
TYPE LOCALITY: Japan, NE Honshu, Iwate Pref., Waga-Gun, Waga-Machi, Geto Hot Spring.
DISTRIBUTION: Honshu and Shikoku (Japan).
STATUS: IUCN 2003 and IUCN/SSC Action Plan (2001) – Endangered.
COMMENTS: Reviewed by Yoshiyuki (1989), also see Horácek et al. (2000) and Kawai et al. (2003).

*Myotis punicus* Felten, Spitzenberger, and Storch, 1977. Senkenberg. Biol., 58:39.
COMMON NAME: Maghrebian Myotis.
TYPE LOCALITY: Tunisia, Cap Bon, El Haouaria Cave.
DISTRIBUTION: Tunisia, Algeria, Libya, Malta, Corsica (France), and Sardinia (Italy).
STATUS: IUCN 2003 – Not evaluated; not considered in IUCN/SSC Action Plan (2001).
COMMENTS: Originally described as a subspecies of *blythii*, but recently shown to lie outside a clade including *blythii*, *myotis*, and *oxygnathus*; see Ruedi and Mayer (2001). Also see Borg (1998) and Castella et al. (2000). Accordingly, *punicus* is treated as a separate species here.

*Myotis ricketti* (Thomas, 1894). Ann. Mag. Nat. Hist., ser. 6, 14:300.
COMMON NAME: Rickett's Big-footed Myotis.
TYPE LOCALITY: China, Fukien (= Fujian), Foochow.
DISTRIBUTION: Fukien, Anhwei, Kiangsu, Shantung, Yunnan (China); Hong Kong; Vietnam and Laos.
STATUS: IUCN 2003 and IUCN/SSC Action Plan (2001) – Lower Risk (nt).
COMMENTS: *Myotis pilosus* Peters, 1869 (type locality unknown) may be the oldest name for this species; see Ellerman and Morrison-Scott (1951) and Corbet and Hill (1992). Sometimes placed in its own subgenus *Rickettia*, see discussion in Findley (1972) and Corbet and Hill (1992). Reviewed in part by Bates et al. (1999) and Hendrichsen et al. (2001); also see Horácek et al. (2000), who discussed this taxon under the name *pilosus*.

*Myotis ridleyi* Thomas, 1898. Ann. Mag. Nat. Hist., ser. 7, 1:361.
COMMON NAME: Ridley's Myotis.
TYPE LOCALITY: Malaysia, Selangor (= Kepong).
DISTRIBUTION: W Malaysia, Sumatra, Borneo.
STATUS: IUCN 2003 and IUCN/SSC Action Plan (2001) – Lower Risk (nt).
COMMENTS: Transferred from *Pipistrellus*; see Medway (1978); also see Hill and Topál (1973).

*Myotis riparius* Handley, 1960. Proc. U.S. Natl. Mus., 112:466-468.
COMMON NAME: Riparian Myotis.
TYPE LOCALITY: Panama, Darien, Río Puero, Tacarcuna Village.
DISTRIBUTION: Honduras south to Uruguay, E Brazil, Argentina, Paraguay, and Bolivia; Trinidad.
STATUS: IUCN 2003 and IUCN/SSC Action Plan (2001) – Lower Risk (lc).
COMMENTS: Originally described as a subspecies of *simus*. Reviewed in part by López-González et al. (2001). LaVal (1973a) suggested that *guaycuru* may be the oldest name for this species, but López-González et al. (2001) have shown *guaycuru* to be a junior synonym of *simus*. Apparently closely related to *ruber*; see Ruedi and Mayer (2001).

*Myotis rosseti* (Oey, 1951). Beaufortia, 1(8):4.
COMMON NAME: Thick-thumbed Myotis.
TYPE LOCALITY: Cambodia.

DISTRIBUTION: Cambodia, Thailand, possibly Vietnam.

STATUS: IUCN 2003 and IUCN/SSC Action Plan (2001) – Lower Risk (nt).

COMMENTS: Originally described as a species of *Glischropus*; see Hill and Topál (1973). Vietnamese record is not well documented; see Bates et al. (1999).

*Myotis ruber* (E. Geoffroy, 1806). Ann. Mus. Natn. Hist. Nat. Paris, 8:204.

COMMON NAME: Red Myotis.

TYPE LOCALITY: Paraguay, Neembucu, Sapucay (neotype locality).

DISTRIBUTION: SE Brazil, SE Paraguay, NE Argentina.

STATUS: IUCN 2003 and IUCN/SSC Action Plan (2001) – Vulnerable.

SYNONYMS: *cinnamomeus* Wagner, 1855; *kinnamon* Gervais, 1856.

COMMENTS: Does not include *levis*; revised by LaVal (1973*a*), who with Miller and Allen (1928) discussed the type. Reviewed in part by López-González et al. (2001). Apparently closely related to *riparius*; see Ruedi and Mayer (2001).

*Myotis schaubi* Kormos, 1934. Földt Közl., Budapest, 64:310.

COMMON NAME: Schaub's Myotis.

TYPE LOCALITY: Hungary (Pliocene). See discussion in Horácek et al. (2000).

DISTRIBUTION: Extant populations limited to Armenia and W Iran.

STATUS: IUCN 2003 and IUCN/SSC Action Plan (2001) – Endangered.

SYNONYMS: **araxenus** Dahl, 1947; *kretzoii* Topál, 1981.

COMMENTS: The nominate subspecies is known only from the Pliocene and is presumably extinct. The living subspecies, *araxenus*, was formerly included in *nattereri*; see Horácek and Hanák (1984). Reviewed by Horácek et al. (2000).

*Myotis scotti* Thomas, 1927. Ann. Mag. Nat. Hist., ser. 9, 19:554.

COMMON NAME: Scott's Myotis.

TYPE LOCALITY: Ethiopia, Shoa, Djem-Djem Forest (ca. 40 mi. (64 km) W Addis Ababa), 8,000 ft. (2,438 m).

DISTRIBUTION: Ethiopia.

STATUS: IUCN 2003 and IUCN/SSC Action Plan (2001) – Vulnerable.

*Myotis septentrionalis* (Trouessart, 1897). Catalog. Mammal. Vivent., p. 131.

COMMON NAME: Northern Myotis.

TYPE LOCALITY: Canada, Nova Scotia, Halifax.

DISTRIBUTION: E United States and Canada west to British Columbia, E Montana, E Wyoming; south to Alabama, Georgia, and Florida Panhandle.

STATUS: IUCN 2003 and IUCN/SSC Action Plan (2001) – Lower Risk (lc).

COMMENTS: Formerly included in *keenii*, but see van Zyll de Jong (1979) and Caceres and Barclay (2000).

*Myotis sicarius* Thomas, 1915. J. Bombay Nat. Hist. Soc., 23:608.

COMMON NAME: Mandelli's Mouse-eared Myotis.

TYPE LOCALITY: India, N Sikkim.

DISTRIBUTION: Sikkim (NE India); Nepal.

STATUS: IUCN 2003 and IUCN/SSC Action Plan (2001) – Vulnerable.

COMMENTS: Reviewed by Bates and Harrison (1997).

*Myotis siligorensis* (Horsfield, 1855). Ann. Mag. Nat. Hist., ser. 2, 16:102.

COMMON NAME: Himalayan Whiskered Myotis.

TYPE LOCALITY: Nepal, Siligori.

DISTRIBUTION: N India to S China, Burma, Vietnam, and Laos; south to W Malaysia; Borneo.

STATUS: IUCN 2003 and IUCN/SSC Action Plan (2001) – Lower Risk (lc).

SYNONYMS: *darjilingensis* Horsfield, 1855; **alticraniatus** Osgood, 1932; **sowerbyi** Howell, 1926; **thaianus** Shamel, 1942.

COMMENTS: Reviewed by Bates and Harrison (1997), Bates et al. (1999), and Hendrichsen et al. (2001). Populations from Malaysia and Borneo have not been allocated to subspecies, and the subspecific status of Vietnamese populations is questionable (see Hendrichsen et al., 2001*b*).

*Myotis simus* Thomas, 1901. Ann. Mag. Nat. Hist., ser. 7, 7:541.
COMMON NAME: Velvety Myotis.
TYPE LOCALITY: Peru, Loreto, Sarayacu (Ucayali River).
DISTRIBUTION: Colombia, Ecuador, Peru, N Brazil, Bolivia, NE Argentina, and Paraguay.
STATUS: IUCN 2003 and IUCN/SSC Action Plan (2001) – Lower Risk (lc).
SYNONYMS: *guaycuru* Proença, 1943.
COMMENTS: Revised by LaVal (1973*a*), Baud and Menu (1993), and López-González et al., 2001. Includes *guaycuru*; see López-González et al. (2001).

*Myotis sodalis* Miller and Allen, 1928. Bull. U.S. Natl. Mus., 144:130.
COMMON NAME: Indiana Myotis.
TYPE LOCALITY: USA, Indiana, Crawford Co., Wyandotte Cave.
DISTRIBUTION: New Hampshire to Florida Panhandle, west to Wisconsin and Oklahoma (USA).
STATUS: U.S. ESA – Endangered; IUCN 2003 and IUCN/SSC Action Plan (2001) – Endangered.
COMMENTS: See Thomson (1982).

*Myotis stalkeri* Thomas, 1910. Ann. Mag. Nat. Hist., ser. 8, 5:384.
COMMON NAME: Kei Myotis.
TYPE LOCALITY: Indonesia, Molucca Isls, Kai Isl, Ara.
DISTRIBUTION: Kai and Gebe Isls (Molucca Isls), Waigeo Isl (Prov. of Papua, Indonesia)..
STATUS: IUCN 2003 and IUCN/SSC Action Plan (2001) – Endangered.
COMMENTS: May be conspecific with *macrotarsus*; see Findley (1972) and Corbet and Hill (1992). Also see Flannery (1995*b*) and Meinig (2002).

*Myotis thysanodes* Miller, 1897. N. Am. Fauna, 13:80.
COMMON NAME: Fringed Myotis.
TYPE LOCALITY: USA, California, Kern Co., Tehachapi Mountains, Old Fort Tejon.
DISTRIBUTION: Chiapas (Mexico) to SW South Dakota (USA) and SC British Columbia (Canada).
STATUS: IUCN 2003 and IUCN/SSC Action Plan (2001) – Lower Risk (lc).
SYNONYMS: **aztecus** Miller and G. M. Allen, 1928; **pahasapensis** Jones and Genoways, 1967; **vespertinus** Manning and Jones, 1988.
COMMENTS: Revised by Miller and Allen (1928). Also see O'Farrell and Studier (1980) and Manning and Jones (1988*b*). Apparently closely related to *lucifugus*; see Ruedi and Mayer (2001).

*Myotis tricolor* (Temminck, 1832). *In* Smuts, Enumer. Mamm. Capensium, p. 106.
COMMON NAME: Temminck's Myotis.
TYPE LOCALITY: South Africa, Western Cape Prov., Capetown.
DISTRIBUTION: Liberia, Ethiopia and Dem. Rep. Congo, south to South Africa.
STATUS: IUCN 2003 and IUCN/SSC Action Plan (2001) – Lower Risk (lc).
SYNONYMS: *loveni* Granvik, 1924.
COMMENTS: Includes *Eptesicus loveni*, see Schlitter and Aggundey (1986). See Taylor (2000*a*) for distribution map.

*Myotis velifer* (J. A. Allen, 1890). Bull. Am. Mus. Nat. Hist., 3:177.
COMMON NAME: Cave Myotis.
TYPE LOCALITY: Mexico, Jalisco, Guadalajara, Santa Cruz del Valle.
DISTRIBUTION: Honduras to Kansas and SE California (USA).
STATUS: IUCN 2003 and IUCN/SSC Action Plan (2001) – Lower Risk (lc).
SYNONYMS: *jaliscensis* Menegaux, 1901; **brevis** Vaughan, 1954; **grandis** Hayward, 1970; **incautus** J. A. Allen, 1896; **magnamolaris** Choate and Hall, 1967.
COMMENTS: See Hayward (1970), Hall (1981), Fitch et al. (1981). Includes *magnamolaris*; see Dalquest and Stangl (1984). Apparently closely related to *yumanensis*; see Ruedi and Mayer (2001).

*Myotis vivesi* Menegaux, 1901. Bull. Mus. Natn. Hist. Nat. Paris, 7:323.
COMMON NAME: Fish-eating Myotis.

TYPE LOCALITY: Mexico, Baja California, Partida Isl.
DISTRIBUTION: Coast of Sonora and Baja California (Mexico), chiefly on small islands.
STATUS: IUCN 2003 and IUCN/SSC Action Plan (2001) – Vulnerable.
COMMENTS: Often placed in its own genus, *Pizonyx*. See Blood and Clark (1998).

*Myotis volans* (H. Allen, 1866). Proc. Acad. Nat. Sci. Phil., 18:282.
COMMON NAME: Long-legged Myotis.
TYPE LOCALITY: Mexico, Baja California, Cabo San Lucas.
DISTRIBUTION: Jalisco to Veracruz (Mexico); Alaska Panhandle (USA) to Baja California
    (Mexico), east to N Nuevo León (Mexico), South Dakota (USA), and C Alberta
    (Canada).
STATUS: IUCN 2003 and IUCN/SSC Action Plan (2001) – Lower Risk (lc).
SYNONYMS: *capitaneus* Nelson and Goldman, 1909; *amotus* Miller, 1914; *interior* Miller,
    1914; *longicrus* True, 1886; *altifrons* Hollister, 1911; *ruddi* Silliman and von Bloeker,
    1938.
COMMENTS: Revised by Miller and Allen (1928). See Warner and Czaplewski (1984).
    Apparently closely related to *lucifugus* and *thysanodes*; see Ruedi and Mayer (2001).

*Myotis welwitschii* (Gray, 1866). Proc. Zool. Soc. Lond., 1866:211.
COMMON NAME: Welwitsch's Myotis.
TYPE LOCALITY: NE Angola.
DISTRIBUTION: South Africa, Mozambique, Zimbabwe, Angola, Zambia, Dem. Rep. Congo,
    Tanzania, Kenya, Uganda, Ethiopia.
STATUS: IUCN 2003 and IUCN/SSC Action Plan (2001) – Lower Risk (lc).
SYNONYMS: *venustus* Matschie, 1899.
COMMENTS: Reviewed by Kock (1967) and Ratcliffe (2002). Sometimes misspelled *welwitschi*
    but the original spelling is *welwitschii*. Apparently closely related to *emarginatus*; see
    Ruedi and Mayer (2001).

*Myotis yanbarensis* Maeda and Matsumura, 1998. Zool. Sci. 15:301.
COMMON NAME: Yanbaru Myotis.
TYPE LOCALITY: Japan, Okinawa Isl, Kunigami-mura, Aha, upper stream of Funga River.
DISTRIBUTION: Northern Okinawa Isl (Japan); known only from the type locality.
STATUS: IUCN 2003 and IUCN/SSC Action Plan (2001) – Data Deficient.
COMMENTS: Apparently related to *pruinosus* and *montivagus*; see Kawai et al. (2003).

*Myotis yesoensis* Yoshiyuki, 1984. Bull. Natl. Sci. Mus. Tokyo, ser. A(Zool.), 10:153.
COMMON NAME: Yoshiyuki's Myotis.
TYPE LOCALITY: Japan, Hokkaido, Hiddaka, Mt. Petegari, neighborhood of Petegari River,
    400 m.
DISTRIBUTION: Hokkaido (Japan).
STATUS: IUCN 2003 and IUCN/SSC Action Plan (2001) – Vulnerable.
COMMENTS: Closely related to *hosonoi*. Reviewed by Yoshiyuki (1989); also see Horácek et al.
    (2000).

*Myotis yumanensis* (H. Allen, 1864). Smithson. Misc. Coll., 7:58.
COMMON NAME: Yuma Myotis.
TYPE LOCALITY: USA, California, Imperial Co., Old Fort Yuma.
DISTRIBUTION: Hidalgo, Morelos and Baja California (Mexico) north to British Columbia
    (Canada), east to Montana and W Texas (USA).
STATUS: IUCN 2003 and IUCN/SSC Action Plan (2001) – Lower Risk (lc).
SYNONYMS: *durangae* J. A. Allen, 1903; *macropus* H. Allen, 1866 [not Gould, 1854]; *obscurus*
    H. Allen, 1866; *phasma* Miller and G. M. Allen, 1928; *lambi* Benson, 1947; *lutosus*
    Miller and G. M. Allen, 1928; *oxalis* Dalquest, 1947; *saturatus* Miller, 1897 [not
    Kuzyakin, 1934]; *sociabilis* H. W. Grinnell, 1914.
COMMENTS: An older name for this species may be *subulatus* Say, 1823; see Glass and Baker
    (1968). Those authors recommended that *subulatus* should be supressed, but see Hall
    (1981), who used *subulatus* for the species we recognize as *leibii*. See also comments
    under *leibii* and *lucifugus*. Apparently closely related to *velifer*; see Ruedi and Mayer
    (2001).

**Subfamily Miniopterinae** Dobson, 1875. Ann. Mag. Nat. Hist., ser. 4, 16:349.

*Miniopterus* Bonaparte, 1837. Fauna Ital., 1, fasc. 20.
TYPE SPECIES: *Vespertilio ursinii* Bonaparte, 1837 (= *Vespertilio schreibersii* Kuhl, 1817).
COMMENTS: Reviewed (in part) by Goodwin (1979), Peterson (1981), Maeda (1982), Hill (1983), Corbet and Hill (1992), and Peterson et al. (1995). These authors have often come to different conclusions regarding classification and synonymys; the arrangement given here generally follows Corbet and Hill (1992) and Peterson et al. (1995). See Maeda (1982) and Peterson et al. (1995) for a summary of authorship, type localities, and holotypes of most named forms.

*Miniopterus africanus* Sanborn, 1936. Field Mus. Nat. Hist., Publ., Zool. Ser., 20:111.
COMMON NAME: African Long-fingered Bat.
TYPE LOCALITY: Ethiopia, Shoa.
DISTRIBUTION: Kenya, Ethiopia, Eritrea, Tanzania, Botswana, Namibia.
STATUS: IUCN 2003 – Not evaluated; not considered in IUCN/SSC Action Plan (2001).
COMMENTS: Formerly included in *inflatus*, but see Peterson et al. (1995).

*Miniopterus australis* Tomes, 1858. Proc. Zool. Soc. Lond., 1858:125.
COMMON NAME: Little Long-fingered Bat.
TYPE LOCALITY: New Caledonia, Loyalty Isls, Lifu (21°S, 167°03'E) (France).
DISTRIBUTION: Philippines, Borneo, Java, Timor, Moluccas, southeast to Vanuatu and E Australia.
STATUS: IUCN 2003 and IUCN/SSC Action Plan (2001) – Lower Risk (lc).
SYNONYMS: **solomonensis** Maeda, 1982; **tibialis** Tomes, 1858.
COMMENTS: Reviewed by Peterson (1981), Maeda (1982), Hill (1983), Koopman (1989*a*), Flannery (1995*a*, *b*), and Bonaccorso (1998), but note that these authors included taxa now considered to be distinct species (i.e., *paululus*, *shortridgei*). Revised by Kitchener and Suyanto (2002), who recognized but did not name an additional subspecies from Kai Isl. Does not include *witkampi*, referred to *paululus* by Kitchener and Suyanto (2002).

*Miniopterus fraterculus* Thomas and Schwann, 1906. Proc. Zool. Soc. Lond., 1906:162.
COMMON NAME: Lesser Long-fingered Bat.
TYPE LOCALITY: South Africa, Western Cape Prov., Knysna.
DISTRIBUTION: South Africa, Malawi, Zambia, Angola, Mozambique, Madagascar.
STATUS: IUCN 2003 and IUCN/SSC Action Plan (2001) – Lower Risk (nt).
COMMENTS: Reviewed by Peterson et al. (1995).

*Miniopterus fuscus* Bonhote, 1902. Novit. Zool., 9:626.
COMMON NAME: Southeast Asian Long-fingered Bat.
TYPE LOCALITY: Japan, Ryukyu Isls, Okinawa.
DISTRIBUTION: Ryukyu Isls (Japan).
STATUS: IUCN 2003 and IUCN/SSC Action Plan (2001) – Vulnerable.
SYNONYMS: *yayeyamae* Kuroda, 1924.
COMMENTS: May include *medius*; see Hill (1983) and Corbet and Hill (1992). Does not include *fraterculus*; see Peterson et al. (1995). Reviewed by Yoshiyuki (1989). Corbet and Hill (1992) suggested that *yayeyamae* may merit recognition as a distinct subspecies.

*Miniopterus gleni* Peterson, Eger, and Mitchell, 1995. Faune de Madagascar, Chiroptères, 84:128.
COMMON NAME: Glen's Long-fingered Bat.
TYPE LOCALITY: Madagascar, 20 km S Tuléar (= Toliara), in a marine cave between Sarodrano and St. Augustin.
DISTRIBUTION: N, W, and S Madagascar.
STATUS: IUCN 2003 and IUCN/SSC Action Plan (2001) – Lower Risk (nt).

*Miniopterus inflatus* Thomas, 1903. Ann. Mag. Nat. Hist., ser. 7, 12:634.
COMMON NAME: Greater Long-fingered Bat.
TYPE LOCALITY: Cameroon, Efulen.

DISTRIBUTION: Kenya, Uganda, Burundi, E and S Dem. Rep. Congo, Cameroon, Gabon, Mozambique, Liberia, perhaps Nigeria. W African distribution uncertain because of confusion with *schreibersii*.

STATUS: IUCN 2003 and IUCN/SSC Action Plan (2001) – Lower Risk (lc).

SYNONYMS: ***rufus*** Sanborn, 1936.

COMMENTS: Koopman (1993, 1994) included *africanus* in this species, but see Peterson et al. (1995).

*Miniopterus macrocneme* Revilliod, 1914. *In* Sarasin and Roux, Nova Caledonia, A. Zool., 1:360.

COMMON NAME: Small Melanesian Long-fingered Bat.

TYPE LOCALITY: New Caledonia and Loyalty Isls.

DISTRIBUTION: New Guinea to Vanuatu and New Caledonia.

STATUS: IUCN 2003 – Not evaluated; not considered in IUCN/SSC Action Plan (2001).

COMMENTS: Listed a subspecies of *pusillus* by Koopman (1993, 1994), but see Sanborn and Nicholson (1950), Flannery (1995*a*, *b*), and Bonaccorso (1998). In a recent revision of the *pusillus/australis* complex, Kitchener and Suyanto (2002) treated *macrocneme* as a subspecies of *pusillus*, but did not examine specimens of *macrocneme* sensu stricto.

*Miniopterus magnater* Sanborn, 1931. Field Mus. Nat. Hist. Publ., Zool. Ser., 18:26.

COMMON NAME: Western Long-fingered Bat.

TYPE LOCALITY: Papua New Guinea, E Sepik, Marienberg.

DISTRIBUTION: NE India, SE China, Burma, Thailand, Laos, and Vietnam to Malaysia, Sumatra, Java, Timor (Indonesia), Borneo, Moluccas, and New Guinea including the Bismarck Arch.

STATUS: IUCN 2003 and IUCN/SSC Action Plan (2001) – Lower Risk (lc).

SYNONYMS: ***macrodens*** Maeda, 1982.

COMMENTS: Reviewed by Hill (1983) and Corbet and Hill (1992). May also include *bismarckensis*, here listed as a synonym of *tristis* following Koopman (1993); see discussion in Hill (1983). See also Flannery (1995*a*) and Bonaccorso (1998). Some specimens from SE Asia previously identified as *schrebersii* may represent *magnater*; see Hendrichsen et al. (2001*b*).

*Miniopterus majori* Thomas, 1906. Ann. Mag. Nat. Hist., ser. 7, 17:175.

COMMON NAME: Major's Long-fingered Bat.

TYPE LOCALITY: Madagascar, NE Betsileo, d'Imasindrary [Sahamananina].

DISTRIBUTION: Madagascar, Comoro Isls.

STATUS: IUCN 2003 and IUCN/SSC Action Plan (2001) – Data Deficient.

COMMENTS: Formerly included in *schreibersii*, but see Peterson et al. (1995).

*Miniopterus manavi* Thomas, 1906. Ann. Mag. Nat. Hist., ser. 7, 17:176.

COMMON NAME: Manavi Long-fingered Bat.

TYPE LOCALITY: Madagascar, E/NE of Betsileo, 20°17'S, 47°31'E [Fandriana region].

DISTRIBUTION: Madagascar, Comoro Isls.

STATUS: IUCN 2003 and IUCN/SSC Action Plan (2001) – Data Deficient as *M. menavi* (misspelled).

SYNONYMS: ***griveaudi*** Harrison, 1959.

COMMENTS: Formerly included in *minor*, but see Peterson et al. (1995); also see Juste and Ibáñez (1992). Includes *griveaudi*; see Peterson et al. (1995).

*Miniopterus medius* Thomas and Wroughton, 1909. Proc. Zool. Soc. Lond., 1909:382.

COMMON NAME: Intermediate Long-fingered Bat.

TYPE LOCALITY: Indonesia, W Java, Tji-Tandoei River, Kalipoetjang.

DISTRIBUTION: SE China, Thailand, W Malaysia, Borneo, Java, Sulawesi, Philippines, New Guinea, possibly the Solomon Isls.

STATUS: IUCN 2003 and IUCN/SSC Action Plan (2001) – Lower Risk (lc).

COMMENTS: May be conspecific with *fuscus*; see Hill (1983) and Corbet and Hill (1992). See also Flannery (1995*a*) and Bonaccorso (1998).

*Miniopterus minor* Peters, 1867. Monatsb. K. Preuss. Akad. Wiss. Berlin, 1866:885 [1867].

COMMON NAME: Least Long-fingered Bat.

TYPE LOCALITY: Tanzania, coast opposite Zanzibar Isl.

DISTRIBUTION: Kenya, Tanzania, Dem. Rep. Congo, Republic of Congo, São Tomé Isl.

STATUS: IUCN 2003 and IUCN/SSC Action Plan (2001) – Lower Risk (nt).

SYNONYMS: *newtoni* Bocage, 1889; *occidentalis* Juste and Ibáñez, 1992.

COMMENTS: Reviewed by Juste and Ibañez (1992), who designated a neotype for *newtoni*. Does not include *manavi* or *griveaudi* see Peterson et al. (1995).

*Miniopterus natalensis* (A. Smith, 1834). S. Afr. Quart. J., 2:59.

COMMON NAME: Natal Long-fingered Bat.

TYPE LOCALITY: Natal, Durban.

DISTRIBUTION: Sudan and SW Arabia to South Africa.

STATUS: IUCN 2003 – Not evaluated; not considered in IUCN/SSC Action Plan (2001).

SYNONYMS: *breyeri* Jameson, 1909; *scotinus* Sundevall, 1846; *vicinior* J. A. Allen, 1917; *arenarius* Heller, 1912.

COMMENTS: Formerly included in *schreibersii*, but apparently distinct; see O'Shea and Vaughan (1980), Koopman (1994), and Peterson et al. (1995).

*Miniopterus paululus* Hollister, 1913. Proc. U. S. Nat. Mus., 46:311.

COMMON NAME: Philippine Long-fingered Bat.

TYPE LOCALITY: Philippines, Guimarás Isls.

DISTRIBUTION: Majuyod, Negros, and Guimarás Isls (Philippines), Borneo, Selaru.

STATUS: IUCN 2003 – Not evaluated; not considered in IUCN/SSC Action Plan (2001).

SYNONYMS: *graysonae* Kitchener, 2002 [in Kitchener and Suyanto, 2002]; *witkampi* Sody, 1930.

COMMENTS: Revised by Kitchener and Suyanto (2002).

*Miniopterus pusillus* Dobson, 1876. Monogr. Asiatic Chiroptera, p. 162.

COMMON NAME: Small Long-fingered Bat.

TYPE LOCALITY: India, Nicobar Isls (NW of Sumatra).

DISTRIBUTION: India, Nepal, and Burma to Sumatra and Timor (Indonesia), Philippines, and Moluccas.

STATUS: IUCN 2003 and IUCN/SSC Action Plan (2001) – Lower Risk (lc).

COMMENTS: Reviewed by Hill (1983), Corbet and Hill (1992), Bates and Harrison (1997), and Kitchener and Suyanto (2002). Philippine records may actually represent *australis*; see Heaney et al. (1998). Does not seem to include *macrocneme*; see Sanborn and Nicholson (1950), Flannery (1995*a*, *b*), and Bonaccorso (1998), although also see Kitchener and Suyanto (2002), who treated *macrocneme* as a subspecies of *pusillus* but did not examine specimens of *macrocneme* sensu stricto. Some specimens from SE Asia previously identified as *schreibersii* may represent *pusillus*; see Hendrichsen et al. (2001*b*). Kitchener and Suyanto (2002) recognized but did not name a subspecies from Alor, Roti, Timor, Ambon, probably Seram, and possibly Sulawesi.

*Miniopterus robustior* Revilliod, 1914. *In* Sarasin and Roux, Nova Caledonia, A. Zool., 1:359.

COMMON NAME: Loyalty Long-fingered Bat.

TYPE LOCALITY: New Caledonia (France), Loyalty Isls, Lifu Isl, Quepenee (= Chépénéhé).

DISTRIBUTION: Loyalty Isls (E of New Caledonia).

STATUS: IUCN 2003 and IUCN/SSC Action Plan (2001) – Endangered.

COMMENTS: See Hill (1971*a*), Peterson (1981), and Flannery (1995*b*).

*Miniopterus schreibersii* (Kuhl, 1817). Die Deutschen Fledermäuse, Hanau, p. 14.

COMMON NAME: Schreibers's Long-fingered Bat.

TYPE LOCALITY: Romania, Mountains of Banat, Banat, near Coronini, Kolumbacs Cave (= Kulmbazer Cave = Columbäzar Cave).

DISTRIBUTION: S Europe and Morocco through the Caucasus, Iran, and Bulgaria to most of China and Japan; most of Indo-Malayan region; Philippines; New Guinea; Solomon Isls (including Bougainville Isl); Australia; subsaharan Africa; Bismarck Arch.

STATUS: IUCN 2003 and IUCN/SSC Action Plan (2001) – Lower Risk (nt).

SYNONYMS: *baussencis* Laurent, 1944; *inexpectatus* Heinrich, 1936; *italicus* Dal Piaz, 1926; *ursinii* Bonaparte, 1837; *bassanii* Cardinal and Christidis, 2000; *blepotis* Temminck, 1840; *ravus* Sody, 1930; *chinensis* Thomas, 1908; *dasythrix* Temminck, 1840;

*eschscholtzii* Waterhouse, 1845; *fuliginosus* Hodgson, 1835; *haradai* Maeda, 1982; *japoniae* Thomas, 1905; *oceanensis* Maeda 1982; *orianae* Thomas, 1922; *orsinii* Temminck, 1840; *pallidus* Thomas, 1907; *pulcher* Harrison, 1956; *parvipes* G. M. Allen, 1923; *smitianus* Thomas, 1927; *villiersi* Aellen, 1956.

COMMENTS: Formerly included *magnater*. Does not include *natalensis* or *arenarius*, see Koopman (1994). Does not include *majori*; see Peterson et al. (1995). Reviewed by Crucitti (1976); see also Maeda (1982), Hill (1983), Harrison and Bates (1991), Kock (1996), Bates and Harrison (1997), Cardinal and Christidis (2000), Conole (2000), Horácek et al. (2000), and Hendrichsen et al. (2001b). Subspecies boundaries are not always clear (e.g., see Hill [1983] and Yoshiyuki [1989]), and some populations have not been allocated to subspecies. Sometimes misspelled *schriebersi*, but see Bogdanowicz and Kick (1998) for correct spelling (*schreibersii*). This complex probably includes more than one species.

*Miniopterus shortridgei* Laurie and Hill, 1957. J. Mammal. 38: 128.
COMMON NAME: Shortridge's Long-fingered Bat.
TYPE LOCALITY: Indonesia, Java, south Java, Tji -Tandoei River, Kalipoetjang.
DISTRIBUTION: Java, Madura, Lombok, Sumbawa, Moyo, Alor, Wetar, Seralu, Timor, Semau, Roti, and Savu Isls (Indonesia).
STATUS: IUCN 2003 – Not evaluated; not considered in IUCN/SSC Action Plan (2001).
SYNONYMS: *minor* Hill, 1954 [in Laurie and Hill, 1954; not *minor* Peters, 1867].
COMMENTS: Revised by Kitchener and Suyanto (2002).

*Miniopterus tristis* (Waterhouse, 1845). Proc. Zool. Soc. Lond., 1845:3.
COMMON NAME: Great Long-fingered Bat.
TYPE LOCALITY: Philippine Isls.
DISTRIBUTION: Philippines; Sulawesi, Sanan Isl, New Guinea; Bismarck Arch., Solomon Isls, Vanuatu (= New Hebrides).
STATUS: IUCN 2003 and IUCN/SSC Action Plan (2001) – Lower Risk (lc).
SYNONYMS: *celebensis* Peterson, 1981; *grandis* Peterson, 1981; *insularis* Peterson, 1981; *bismarckensis* Maeda, 1982; *melanesiensis* Maeda, 1982; *propritristis* Peterson, 1981.
COMMENTS: Includes *propritristis*; see Koopman (1984c) and Hill (1983). Peterson (1981) and Maeda (1982) recognized more than one species in this complex, but did not agree on species limits; see Hill (1983), who argued convincingly that all of these forms should be regarded as subspecies of *tristis* pending further study. Koopman (1993) included *bismarckensis* in this complex, but also see Hill (1983), who suggested that this poorly-known taxon might be allied to *magnater*. Also see accounts in Flannery (1995a, b), Bonaccorso (1998), and Meinig (2002) under *propritristis*.

**Subfamily Murininae** Miller, 1907. Bull. U.S. Natl. Mus., 57:229.

*Harpiocephalus* Gray, 1842. Ann. Mag. Nat. Hist., [ser. 1], 10:259.
TYPE SPECIES: *Harpiocephalus rufus* Gray, 1842 (= *Vespertilio harpia* Temminck, 1840).

*Harpiocephalus harpia* (Temminck, 1840). Monogr. Mamm., 2:219.
COMMON NAME: Lesser Hairy-winged Bat.
TYPE LOCALITY: Indonesia, Java, NE side of Mt. Gede.
DISTRIBUTION: S and NE India, S China, Taiwan, Laos and Vietnam, Sumatra, Java, Borneo, S Moluccas, and the Philippines.
STATUS: IUCN 2003 and IUCN/SSC Action Plan (2001) – Lower Risk (lc).
SYNONYMS: *pearsonii* Horsfield, 1851; *rufus* Gray, 1842; *lasyurus* Hodgson, 1847; *madrassius* Thomas, 1923; *rufulus* G. M. Allen, 1913.
COMMENTS: Does not include *mordax*; see Hill and Francis (1984) Corbet and Hill (1992), and Hendrichsen et al. (2002b). Reviewed in part by Bates and Harrison (1997) and Hendrichsen et al. (2001b). The Taiwan record, if valid, has not been allocated to subspecies.

*Harpiocephalus mordax* Thomas, 1923. J. Bombay Nat. Hist. Soc., 29:88.
COMMON NAME: Greater Hairy-winged Bat.

TYPE LOCALITY: Burma, Mogok.

DISTRIBUTION: Burma, Thailand, Vietnam, Borneo. This species has also been reported from Cambodia but there are no vouchered records; see Hendrichsen et al. (2001a). Some specimens from India previously identified as *harpia* may represent this species, see Hendrichsen et al. (2001b)

STATUS: IUCN 2003 and IUCN/SSC Action Plan (2001) – Lower Risk (nt).

COMMENTS: Formerly included in *harpia* (e.g., Koopman, 1993, 1994), but apparently distinct, see Hill and Francis (1984), Corbet and Hill (1992), and Hendrichsen et al. (2001b).

*Murina* Gray, 1842. Ann. Mag. Nat. Hist., [ser. 1], 10:258.

TYPE SPECIES: *Vespertilio suillus* Temminck, 1840.

SYNONYMS: *Harpiola* Thomas, 1915; *Ocypetes* Lesson, 1841 [not Risso, 1826].

COMMENTS: Includes *Harpiola*, here recognized as a subgenus, see Corbet and Hill (1980, 1992) although also see Bhattacharyya (2002). The other recognized subgenus, *Murina*, is sometimes divided into two species groups but there is disagreement about membership; see Corbet and Hill (1992), Maeda and Matsumura (1998), and Kawai et al. (2002).

*Murina aenea* Hill, 1964. Fed. Mus. J., Kuala Lumpur, N.S., 8:57.

COMMON NAME: Bronze Tube-nosed Bat.

TYPE LOCALITY: Malaysia, Pahang, Bentong Dist., near Janda Baik, Ulu Chemperoh (c 03°18'N, 101°50'E).

DISTRIBUTION: W Malaysia, Borneo.

STATUS: IUCN 2003 and IUCN/SSC Action Plan (2001) – Lower Risk (nt).

COMMENTS: Subgenus *Murina*. Reviewed by Francis (1997).

*Murina aurata* Milne-Edwards, 1872. Rech. Hist. Nat. Mammifères, p. 250.

COMMON NAME: Little Tube-nosed Bat.

TYPE LOCALITY: China, Szechwan, Moupin.

DISTRIBUTION: NE India, Nepal to SW China (including E Tibet) and Burma, Thailand.

STATUS: IUCN 2003 and IUCN/SSC Action Plan (2001) – Lower Risk (nt).

SYNONYMS: *aurita* Miller, 1907; *feae* Thomas, 1891.

COMMENTS: Subgenus *Murina*. Formerly included *ussuriensis*; see Maeda (1980); see also comments under *silvatica*. Reviewed by Hill (1983) and Bates and Harrison (1997).

*Murina cyclotis* Dobson, 1872. Proc. Asiat. Soc. Bengal, p. 210.

COMMON NAME: Round-eared Tube-nosed Bat.

TYPE LOCALITY: India, Darjeeling.

DISTRIBUTION: Sri Lanka and India to Kwangtung and Hainan (China); Myanamar, Laos, and Vietnam, south to W Malaysia, Borneo, Sumatra, Philippines, and Lesser Sunda Isls. Records from Cambodia are erroneous (Kock, 2000a).

STATUS: IUCN 2003 and IUCN/SSC Action Plan (2001) – Lower Risk (lc).

SYNONYMS: *eileenae* Phillips, 1932; *peninsularis* Hill, 1964.

COMMENTS: Subgenus *Murina*. Reviewed in part by Hill (1983), Bates and Harrison (1997), Sinha (1999), and Hendrichsen et al. (2001b).

*Murina florium* Thomas, 1908. Ann. Mag. Nat. Hist., ser. 8, 2:371.

COMMON NAME: Flores Tube-nosed Bat.

TYPE LOCALITY: Indonesia, Lesser Sunda Isls, Flores.

DISTRIBUTION: Lesser Sunda Isls, Sulawesi, Moluccas, Seram, New Guinea including the Bismark Arch, and NE Australia.

STATUS: IUCN 2003 and IUCN/SSC Action Plan (2001) – Lower Risk (lc).

SYNONYMS: *lanosa* Thomas, 1910; *toxopei* Thomas, 1923.

COMMENTS: Subgenus *Murina*. See Flannery (1995a, b) and Bonaccorso (1998). The three subspecies are poorly delimited.

*Murina fusca* Sowerby, 1922. J. Mammal., 3:46.

COMMON NAME: Dusky Tube-nosed Bat.

TYPE LOCALITY: China, Manchuria, Kirin, Imienpo area.

DISTRIBUTION: Manchuria (China).

STATUS: IUCN 2003 and IUCN/SSC Action Plan (2001) – Data Deficient.
COMMENTS: Subgenus *Murina*. Listed as a subspecies of *leucogaster* by Ellerman and Morrison-Scott (1951) and Corbet (1978c), but see Wallin (1969). Wang (1959) suggested that *fusca* might be a synonym of *hilgendorfi*.

*Murina grisea* Peters, 1872. Monatsb. K. Preuss. Akad. Wiss. Berlin, 1872:258.
COMMON NAME: Peters's Tube-nosed Bat.
TYPE LOCALITY: India, Uttar Pradesh, Dehra Dun, Mussooree, Jeripanee, 5,500 ft. (1,676 m).
DISTRIBUTION: NW Himalayas, Mizoram (India).
STATUS: IUCN 2003 and IUCN/SSC Action Plan (2001) – Endangered.
COMMENTS: Subgenus *Harpiola*. Reviewed by Bates and Harrison (1997) and Bhattacharyya (2002).

*Murina hilgendorfi* Peters, 1880. Monatsb. K. Preuss. Akad. Wiss. Berlin, 1880:24.
COMMON NAME: Hilgendorf's Tube-nosed Bat.
TYPE LOCALITY: Japan, near Tokyo, Yedo.
DISTRIBUTION: N China; Upper Yenisei River (Russia); Altai Mtns (Russia, Kazakhstan and Mongolia); Korea; Ussur region (Russia); Sakhalin Isl (Russia); Honshu, Kyushu and Shikiku (Japan).
STATUS: IUCN 2003 – Not evaluated; not considered in IUCN/SSC Action Plan (2001).
SYNONYMS: *intermedia* Mori, 1933; *ognevi* Bianchi, 1916; *sibirica* Kastschenko, 1905.
COMMENTS: Subgenus *Murina*. Formerly included in *leucogaster*, but apparently distinct. May include more than one species; see Yoshiyuki (1989). Also see Wang (1959).

*Murina huttoni* (Peters, 1872). Monatsb. K. Preuss. Akad. Wiss. Berlin, 1872:257.
COMMON NAME: Hutton's Tube-nosed Bat.
TYPE LOCALITY: India, Uttar Pradesh, Kumaon, Dehra Dun.
DISTRIBUTION: Tibet, NE and S China, NW India to Vietnam, Thailand, W Malaysia.
STATUS: IUCN 2003 and IUCN/SSC Action Plan (2001) – Lower Risk (nt).
SYNONYMS: **rubella** Thomas, 1914.
COMMENTS: Subgenus *Murina*. Reviewed in part by Sinha (1999) and Hendrichsen et al. (2001b). Does not include *tubinaris*, see Hill (1963a, 1983), Bates and Harrison (1997), Sinha (1999), and Hendrichsen et al. (2001b). Some SE Asian specimens previously referred to *tubinaris* may represent *huttoni*, see Hendrichsen et al. (2001b). Sometimes spelled *huttonii*.

*Murina leucogaster* Milne-Edwards, 1872. Rech. Hist. Nat. Mammifères, p. 252.
COMMON NAME: Greater Tube-nosed Bat.
TYPE LOCALITY: China, Szechwan, Moupin Dist.
DISTRIBUTION: NE India, Nepal, S China, W Thailand.
STATUS: IUCN 2003 and IUCN/SSC Action Plan (2001) – Lower Risk (lc).
SYNONYMS: *leucogastra* Thomas, 1899; **rubex** Thomas, 1916.
COMMENTS: Subgenus *Murina*. Does not include *hilgendorfi*; see Yoshiyuki (1989). Reviewed in part by Bates and Harrison (1997).

*Murina puta* Kishida, 1924. Zool. Mag. (Tokyo), 36:127.
COMMON NAME: Taiwanese Tube-nosed Bat.
TYPE LOCALITY: Taiwan, Chang Hua, Erh-Shui.
DISTRIBUTION: Taiwan.
STATUS: IUCN 2003 and IUCN/SSC Action Plan (2001) – Vulnerable.
COMMENTS: Subgenus *Murina*. Closely related to and possibly conspecific with *huttoni*, see Yoshiyuki (1989).

*Murina rozendaali* Hill and Francis, 1984. Bull. Brit. Mus. Nat. Hist. (Zool.), 47:319.
COMMON NAME: Gilded Tube-nosed Bat.
TYPE LOCALITY: Borneo, Sabah, Gomantong (c 05°31'N, 118°04'E).
DISTRIBUTION: Peninsular Malaysia, Borneo.
STATUS: IUCN 2003 and IUCN/SSC Action Plan (2001) – Lower Risk (nt).
COMMENTS: Subgenus *Murina*. Reviewed by Francis (1997).

*Murina ryukyuana* Maeda and Matsumura, 1998. Zool. Sci. 15:303.

COMMON NAME: Ryukyu Tube-nosed Bat.
TYPE LOCALITY: Japan, Okinawa Isl, Kunigami-mura, Aha, upper stream of Funga River.
DISTRIBUTION: Northern Okinawa Isl (Japan); known only from the type locality.
STATUS: IUCN 2003 and IUCN/SSC Action Plan (2001) – Data Deficient.
COMMENTS: Subgenus *Murina*.

*Murina silvatica* Yoshiyuki, 1983. Bull. Natl. Sci. Mus. Tokyo, ser. A(Zool.), 9:141.

COMMON NAME: Forest Tube-nosed Bat.
TYPE LOCALITY: Japan, Honshu, Fukushima Prefecture, Minamiaiau-Gug, Hinoemata-Mura, Oze-Numa Lake.
DISTRIBUTION: Japan, including Tsushima Isls.
STATUS: IUCN 2003 and IUCN/SSC Action Plan (2001) – Lower Risk (nt).
COMMENTS: Subgenus *Murina*. Includes specimens formerly included in *aurata* or *ussuriensis*. Reviewed by Yoshiyuki (1989).

*Murina suilla* (Temminck, 1840). Monogr. Mamm., 2:224.

COMMON NAME: Brown Tube-nosed Bat.
TYPE LOCALITY: Indonesia, Java, Tapos.
DISTRIBUTION: Java, Sumatra, Borneo, W Malaysia, nearby small islands. Reports of this species from Sulawesi, Peleng Isl, and New Guinea are doubtfull, see discussion in Corbet and Hill (1992).
STATUS: IUCN 2003 and IUCN/SSC Action Plan (2001) – Lower Risk (lc).
SYNONYMS: *balstoni* Thomas, 1908; **canescens** Thomas, 1923.
COMMENTS: Subgenus *Murina*. Includes *balstoni* and *canescens*, see Koopman (1989*a*) and Corbet and Hill (1992). See also Francis (1997).

*Murina tenebrosa* Yoshiyuki, 1970. Bull. Natl. Sci. Mus. Tokyo, 13:195.

COMMON NAME: Gloomy Tube-nosed Bat.
TYPE LOCALITY: Japan, Tsushima Isls, Kamishima Isl, Sago.
DISTRIBUTION: Tsushima Isls (Japan), perhaps Yakushima (Ryukyu Isls, Japan).
STATUS: IUCN 2003 and IUCN/SSC Action Plan (2001) – Critically Endangered.
COMMENTS: Subgenus *Murina*. Reviewed by Yoshiyuki (1989).

*Murina tubinaris* (Scully, 1881). Proc. Zool. Soc. Lond., 1881:200.

COMMON NAME: Scully's Tube-nosed Bat.
TYPE LOCALITY: Pakistan, Kasmir, Gilgit.
DISTRIBUTION: Pakistan, N India, Burma, Thailand, Laos, Vietnam.
STATUS: IUCN 2003 and IUCN/SSC Action Plan (2001) – Lower Risk (lc).
COMMENTS: Subgenus *Murina*. Listed as a subspecies of *huttoni* by Ellerman and Morrison-Scott (1951), but apparently distinct, see Hill (1963*a*, 1983), Bates and Harrison (1997), Sinha (1999), and Hendrichsen et al. (2001*b*). Koopman and Danforth (1989) suggested that *tubinaris* may be conspecific with *suilla*, but this has not been supported by recent authors.

*Murina ussuriensis* Ognev, 1913. Ann. Mus. Zool. Acad. Imp. Sci. St. Petersbourg, 18:402.

COMMON NAME: Ussurian Tube-nosed Bat.
TYPE LOCALITY: Russia, SE Siberia, Ussuri, Imansky distr., Evseevka
DISTRIBUTION: Ussuri region, Kurile Isls, and Sakhalin (Russia); Korea.
STATUS: IUCN 2003 and IUCN/SSC Action Plan (2001) – Endangered.
COMMENTS: Subgenus *Murina*. Formerly included in *aurata*, see Maeda (1980) and Corbet (1978*c*). Japanese populations have been separated as *M. silvatica*.

**Subfamily Kerivoulinae** Miller, 1907. Bull. U.S. Natl. Mus., 57:232.

*Kerivoula* Gray, 1842. Ann. Mag. Nat. Hist., [ser. 1], 10:258.

TYPE SPECIES: *Vespertilio pictus* Pallas, 1767, by subsequent designation (Peters, 1866).
SYNONYMS: *Cerivoula* Blanford, 1891; *Nyctophylax* Fitzinger, 1861.
COMMENTS: Does not include *Phoniscus*. Koopman (1982, 1993, 1994) and Ryan (1965) considered *Phoniscus* to be congeneric with *Kerivoula*, but see Hill (1965) and Corbet and

Hill (1980, 1991, 1992). Characters separating these genera were summarized by Corbet and Hill (1992).

*Kerivoula africana* Dobson, 1878. Cat. Chiroptera Brit. Mus., p. 335.
COMMON NAME: Tanzanian Woolly Bat.
TYPE LOCALITY: Tanzania, coast opposite Zanzibar Isl.
DISTRIBUTION: Tanzania.
STATUS: IUCN 2003 and IUCN/SSC Action Plan (2001) – Data Deficient.
COMMENTS: See Burgess et al. (2000).

*Kerivoula agnella* Thomas, 1908. Ann. Mag. Nat. Hist., ser. 8, 2:372.
COMMON NAME: St. Aignan's Woolly Bat.
TYPE LOCALITY: Papua New Guinea, Louisiade Archipelago, Misima Isl.
DISTRIBUTION: Louisiade Arch., Woodlark and D'Entrecasteaux Isls (Papua New Guinea).
STATUS: IUCN 2003 and IUCN/SSC Action Plan (2001) – Vulnerable.
COMMENTS: See Flannery (1995*b*) and Bonaccorso (1998).

*Kerivoula argentata* Tomes, 1861. Proc. Zool. Soc. Lond., 1861:32.
COMMON NAME: Damara Woolly Bat.
TYPE LOCALITY: Namibia, Otjoro.
DISTRIBUTION: Uganda and S Kenya to Malawi, Angola, Namibia and KwaZulu-Natal (South Africa).
STATUS: IUCN 2003 and IUCN/SSC Action Plan (2001) – Lower Risk (lc).
SYNONYMS: *nidicola* Kirk, 1865; *zuluensis* Roberts, 1924.
COMMENTS: See Taylor (2000*a*) for distribution map.

*Kerivoula cuprosa* Thomas, 1912. Ann. Mag. Nat. Hist., ser. 8, 10:41.
COMMON NAME: Copper Woolly Bat.
TYPE LOCALITY: Cameroon, Ja River, Bitye.
DISTRIBUTION: N Dem. Rep. Congo, S Cameroon. A record of this species from Kenya was based on a specimen subsequently reidentified as *smithii* (J. Fahr, pers. comm.)
STATUS: IUCN 2003 and IUCN/SSC Action Plan (2001) – Lower Risk (nt).

*Kerivoula eriophora* (Heuglin, 1877). Reise Nordost-Afrika, 2:34.
COMMON NAME: Ethiopian Woolly Bat.
TYPE LOCALITY: Ethiopia, Belegaz Valley, between Semian and Wogara.
DISTRIBUTION: Ethiopia.
STATUS: IUCN 2003 – Not evaluated; not considered in IUCN/SSC Action Plan (2001).
COMMENTS: Very poorly known; may be conspecific with *africana* which it antedates; see Hayman and Hill (1971).

*Kerivoula flora* Thomas, 1914. Ann. Mag. Nat. Hist., ser. 8, 13:441.
COMMON NAME: Flores Woolly Bat.
TYPE LOCALITY: Indonesia, Lesser Sundas, S Flores.
DISTRIBUTION: Borneo, Lesser Sunda Isls, Bali, Sumbawa, and Sumba (Indonesia; see Corbet and Hill, 1992); possibly Vietnam and Thailand (see Hendrichsen et al., 2001*b*).
STATUS: IUCN 2003 and IUCN/SSC Action Plan (2001) – Lower Risk (lc).
COMMENTS: Reviewed by Hendrichsen et al. (2001*b*) and Vanitharani et al. (2003); also see Hill and Rozendaal (1989).

*Kerivoula hardwickii* (Horsfield, 1824). Zool. Res. Java, Part 8, p. 4(unno.) of *Vespertilio Temminckii* acct.
COMMON NAME: Hardwicke's Woolly Bat.
TYPE LOCALITY: Indonesia, Java.
DISTRIBUTION: India and Sri Lanka, Burma, Laos, Cambodia, Vietnam, Thailand, China, W Malaysia, Borneo, Java, Sumatra, Nusa Penida, Mentawai Isls, Sulawesi, Bali, Lesser Sundas, Kangean Isl and Talaud Isl (Indonesia), Philippines.
STATUS: IUCN 2003 and IUCN/SSC Action Plan (2001) – Lower Risk (lc).
SYNONYMS: *crypta* Wroughton and Ryley, 1913; *depressa* Miller, 1906; *engana* Miller, 1906; *fusca* Dobson, 1871; *malpasi* Phillips, 1932.
COMMENTS: Does not include *flora*; see Hill and Rozendaal (1989). Reviewed in part by Bates

and Harrison (1997) and Hendrichsen et al. (2001*b*). Multiple subspecies have been recognized in the past, but recent studies suggest that these are not justified; see Corbet and Hill (1992) and Sinha (1999). This taxon is sometimes spelled *hardwickei* or *hardwicki* but most recent authors (e.g., Corbet and Hill, 1992; Koopman, 1993; Sinha, 1999; Hendrichsen et al., 2001*b*) have used the spelling *hardwickii*.

*Kerivoula intermedia* Hill and Francis, 1984. Bull. Brit. Mus. Nat. Hist. (Zool.), 47:323.
COMMON NAME: Small Woolly Bat.
TYPE LOCALITY: Malaysia, Borneo, Sabah, Lumerao (05°12'N, 118°52'E).
DISTRIBUTION: Borneo, W Malaysia.
STATUS: IUCN 2003 and IUCN/SSC Action Plan (2001) – Lower Risk (nt).

*Kerivoula lanosa* (A. Smith, 1847). Illustr. Zool. S. Afr. Mamm., pl. 50.
COMMON NAME: Lesser Woolly Bat.
TYPE LOCALITY: South Africa, 200 mi. (322 km) E Capetown.
DISTRIBUTION: Guinea and Liberia to Ethiopia, south to South Africa.
STATUS: IUCN 2003 and IUCN/SSC Action Plan (2001) – Lower Risk (lc).
SYNONYMS: *brunnea* Dobson, 1878; **harrisoni** Thomas, 1901; **lucia** Hinton, 1920; *lueia* Kershaw, 1922; **muscilla** Thomas, 1906; *bellula* Aellen, 1959.
COMMENTS: Includes *harrisoni* and *muscilla*; see Hill (1977*a*). See Cotterill (1996) for range map.

*Kerivoula lenis* Thomas, 1916. J. Bombay Nat. Hist. Soc., 24:416.
COMMON NAME: Lenis Woolly Bat.
TYPE LOCALITY: India, Calcutta.
DISTRIBUTION: NE and S India, W Malaysia, Sabah.
STATUS: IUCN 2003 – Not evaluated; not considered in IUCN/SSC Action Plan (2001).
COMMENTS: Formerly included in *papillosa* but clearly distinct; see Vanitharani et al. (2003).

*Kerivoula minuta* Miller, 1898. Proc. Acad. Nat. Sci. Phil., 50:321.
COMMON NAME: Least Woolly Bat.
TYPE LOCALITY: Thailand, Trang Province, Lay Song Hong.
DISTRIBUTION: W Malaysia, S Thailand, Borneo.
STATUS: IUCN 2003 and IUCN/SSC Action Plan (2001) – Lower Risk (nt).

*Kerivoula muscina* Tate, 1941. Bull. Am. Mus. Nat. Hist., 78:586.
COMMON NAME: Fly River Woolly Bat.
TYPE LOCALITY: Papua New Guinea, Western Province, Lake Daviumbu, ca. 20 m.
DISTRIBUTION: C New Guinea.
STATUS: IUCN 2003 and IUCN/SSC Action Plan (2001) – Vulnerable.
COMMENTS: See Flannery (1995*a*) and Bonaccorso (1998).

*Kerivoula myrella* Thomas, 1914. Ann. Mag. Nat. Hist., ser. 8, 13:438.
COMMON NAME: Bismarck's Woolly Bat.
TYPE LOCALITY: Papua New Guinea, Bismarck Archipelago, Admiralty Isls, Manus Isl.
DISTRIBUTION: Bismarck Arch.; possibly Wetar Isl (Lesser Sunda Isls).
STATUS: IUCN 2003 and IUCN/SSC Action Plan (2001) – Vulnerable.
COMMENTS: Specimens from Wetar Isl reported by Hill and Rozendaal (1989) may represent *hardwickii*; see Bonaccorso (1998). See Flannery (1995*b*).

*Kerivoula papillosa* (Temminck, 1840). Monogr. Mamm., 2:220.
COMMON NAME: Papillose Woolly Bat.
TYPE LOCALITY: Indonesia, Java, Bantam (restricted by Tate, 1940).
DISTRIBUTION: Thailand, Cambodia, Vietnam, W Malaysia, Sumatra, Java, Sulawesi, Borneo.
STATUS: IUCN 2003 and IUCN/SSC Action Plan (2001) – Lower Risk (lc).
SYNONYMS: **malayana** Chasen, 1940.
COMMENTS: Does not include *lenis*; see Vanitharani et al. (2003). See Hill (1983) and Corbet and Hill (1992) for discussion of subspecies. Reviewed in part by Bates and Harrison (1997). Some specimens referred to this species may represent *lenis*, which appreas to be broadly sympatric with papillosa (Vanitharani et al., 2003).

*Kerivoula pellucida* (Waterhouse, 1845). Proc. Zool. Soc. Lond., 1845:6.
COMMON NAME: Clear-winged Woolly Bat.
TYPE LOCALITY: Philippines.
DISTRIBUTION: Borneo, Philippines, Java and Sumatra, W Malaysia.
STATUS: IUCN 2003 and IUCN/SSC Action Plan (2001) – Lower Risk (lc).
SYNONYMS: *bombifrons* Lyon, 1911.
COMMENTS: Includes *bombifrons*; see Hill (1965).

*Kerivoula phalaena* Thomas, 1912. Ann. Mag. Nat. Hist., ser. 8, 10:281.
COMMON NAME: Spurrell's Woolly Bat.
TYPE LOCALITY: Ghana, Bibianaha.
DISTRIBUTION: Liberia, Ghana, Cameroon, Republic of Congo, Dem. Rep. Congo.
STATUS: IUCN 2003 and IUCN/SSC Action Plan (2001) – Lower Risk (lc).

*Kerivoula picta* (Pallas, 1767). Spicil. Zool., 3:7.
COMMON NAME: Painted Woolly Bat.
TYPE LOCALITY: Indonesia, Molucca Isls, Ternate Isl. See discussion in Corbet and Hill (1992).
DISTRIBUTION: Sri Lanka; India and Nepal to Vietnam, W Malaysia, and S China; Borneo;
        Sumatra, Java, Bali, Lombok, and Molucca Isls.
STATUS: IUCN 2003 and IUCN/SSC Action Plan (2001) – Lower Risk (lc).
SYNONYMS: *kirivoula* F. Cuvier, 1832; *rubellus* Kerr, 1792; **bellissima** Thomas, 1906.
COMMENTS: Reviewed in part by Bates and Harrison (1997). Also see Flannery (1995*b*).

*Kerivoula smithii* Thomas, 1880. Ann. Mag. Nat. Hist., ser. 5, 6:166.
COMMON NAME: Smith's Woolly Bat.
TYPE LOCALITY: Nigeria, Calabar.
DISTRIBUTION: Nigeria, Cameroon, N and E Dem. Rep. Congo, Kenya. Previous records from
        Côte d'Ivoire and Liberia are apparently erroneous (J. Fahr, pers. comm.)
STATUS: IUCN 2003 and IUCN/SSC Action Plan (2001) – Lower Risk (nt).
COMMENTS: Sometimes mispelled *smithi* but the original spelling is *smithii*.

*Kerivoula whiteheadi* Thomas, 1894. Ann. Mag. Nat. Hist., ser. 6, 14:460.
COMMON NAME: Whitehead's Woolly Bat.
TYPE LOCALITY: Philippines, Luzon, Isabella, Molino.
DISTRIBUTION: Philippines, Borneo, S Thailand, W Malaysia.
STATUS: IUCN 2003 and IUCN/SSC Action Plan (2001) – Lower Risk (lc).
SYNONYMS: **bicolor** Thomas, 1904; **pusilla** Thomas, 1894.
COMMENTS: Reviewed by Hill (1965) and Corbet and Hill (1992).

*Phoniscus* Miller, 1905. Proc. Biol. Soc. Wash., 18:229.
TYPE SPECIES: *Phoniscus atrox* Miller, 1905.
COMMENTS: Distinct from *Kerivoula*. Koopman (1982, 1993, 1994) and Ryan (1965) considered
        *Phoniscus* to be congeneric with *Kerivoula*, but see Hill (1965) and Corbet and Hill (1980,
        1992). Characters separating these genera were summarized by Corbet and Hill (1992).

*Phoniscus aerosa* (Tomes, 1858). Proc. Zool. Soc. Lond., 1858:333.
COMMON NAME: Dubious Trumpet-eared Bat.
TYPE LOCALITY: "Eastern coast of South Africa."
DISTRIBUTION: Possibly South Africa, but more likely somewhere in SE Asia; known only
        from two syntypes that may have been incorrectly localized (Corbet and Hill, 1992).
STATUS: IUCN 2003 and IUCN/SSC Action Plan (2001) – Data Deficient as *Kerivoula aerosa*.
COMMENTS: See Hill (1965) and Corbet and Hill (1992) for discussion of the uncertain
        affinities of this species.

*Phoniscus atrox* Miller, 1905. Proc. Biol. Soc. Wash., 18:230.
COMMON NAME: Groove-toothed Trumpet-eared Bat.
TYPE LOCALITY: Indonesia, E Sumatra, near Kateman River.
DISTRIBUTION: S Thailand, W Malaysia, Sumatra, Borneo.
STATUS: IUCN 2003 and IUCN/SSC Action Plan (2001) – Lower Risk (lc) as *Kerivoula atrox*.
COMMENTS: Discussed by Hill and Francis (1984) and Corbet and Hill (1992).

*Phoniscus jagorii* (Peters, 1866). Monatsb. K. Preuss. Akad. Wiss. Berlin, 1866:399.
> COMMON NAME: Peters's Trumpet-eared Bat.
> TYPE LOCALITY: Philippines, Samar Isl.
> DISTRIBUTION: Laos; Peninsular Malaysia, Borneo, Java, Bali, Sulawesi, and Lesser Sunda Isls,
> Samar Isl (Philippines).
> STATUS: IUCN 2003 and IUCN/SSC Action Plan (2001) – Lower Risk (lc) as *Kerivoula jagori*.
> SYNONYMS: *javana* Thomas, 1880; *rapax* Miller, 1931.
> COMMENTS: See Hill (1965) and Kingston et al. (1997). Specimens from Laos are slightly
> smaller than those reported from elsewhere (Robinson and Webber, 2000), and may
> represent a distinct taxon. Sometimes spelled *jagori*.

*Phoniscus papuensis* (Dobson, 1878). Cat. Chiroptera Brit. Mus., p. 339.
> COMMON NAME: Golden-tipped Bat.
> TYPE LOCALITY: Papua New Guinea, Central Prov., Port Moresby.
> DISTRIBUTION: SE New Guinea, Biak-Supiori Isl, Queensland and New South Wales
> (Australia).
> STATUS: IUCN 2003 and IUCN/SSC Action Plan (2001) – Lower Risk (lc) as *Kerivoula
> papuensis*.
> COMMENTS: See Flannery (1995*a*, *b*) and Bonaccorso (1998).

# ORDER PHOLIDOTA
by Duane A. Schlitter

## ORDER PHOLIDOTA Weber, 1904.

**Family Manidae** Gray, 1821. London Med. Repos., 15:305.

*Manis* Linnaeus, 1758. Syst. Nat., 10th ed., 1:36.
   TYPE SPECIES: *Manis pentadactyla* Linnaeus, 1758.
   SYNONYMS: *Pangolin* Gray, 1873; *Pangolinus* Rafinesque, 1815; *Pangolinus* Rafinesque, 1821; *Paramanis* Pocock, 1824; *Phatages* Sundevall, 1843; *Phatagin* Gray, 1865; *Phataginus* Rafinesque, 1821; *Pholidotus* Brisson, 1762; *Smutsia* Gray, 1865; *Triglochinopolis* Fitzinger, 1872; *Uromanis* Pocock, 1924.
   COMMENTS: Family reviewed by Mohr (1961). Morphological evidence suggests a subdivision of the genus into two genera (*Manis* and *Phataginus*), see Corbet and Hill (1992) and Patterson (1978); or four subgenera (*Manis, Paramanis, Smutsia,* and *Uromanis*), see Meester (1972*a*), Meester et al. (1986), and Mohr (1961). Gaudin and Wible (1999) conducted a cladistical analysis of 67 cranial characters in extant pangolins plus one fossil genus and found that the Asian pangolins form a monophyletic clade while the African species form a paraphyletic assemblage.

*Manis crassicaudata* E. Geoffroy, 1803. Cat. Mamm. Mus. H. N. Paris, p. 213.
   COMMON NAME: Indian Pangolin.
   TYPE LOCALITY: India.
   DISTRIBUTION: E Pakistan; India; Bangladesh; Sri Lanka.
   STATUS: CITES – Appendix II; IUCN – Lower Risk (nt).
   SYNONYMS: *crassicaudata* Gray, 1827; *indicus* (Gray, 1865).
   COMMENTS: Subgenus *Manis*. Formerly erroneously called *pentadactyla*; see Ellerman and Morrison-Scott (1951), Emry (1970), and Corbet and Hill (1992). Probably includes *laticauda* Illiger, 1815, a *nomen nudum*. Chromosomes reported by Aswathanayana (2000).

*Manis culionensis* (de Elera, 1915). Cont. Fauna Filipina, Manila, Col. Santo Tomás, p. 274.
   COMMON NAME: Philippine Pangolin.
   TYPE LOCALITY: Philippines, Calamian Isl, Culion Isl.
   DISTRIBUTION: Palawan and adjacent islands, Philippines.
   SYNONYMS: *culionensis* (de Elera, 1895) [*nomen nudum*].
   COMMENTS: Subgenus *Paramanis*. A synonym of *javanica* according to Corbet and Hill (1992:19) but accorded specific rank by Lawrence (1939:70), Sanborn (1952*a*:114) and Feiler (1998:161).

*Manis gigantea* Illiger, 1815. Abh. Phys. Klasse K. Pruess Konigl. Akad. Wiss., p. 84.
   COMMON NAME: Giant Pangolin.
   TYPE LOCALITY: Not indicated.
   DISTRIBUTION: Senegal to W Kenya, south to Rwanda, C Dem. Rep. Congo and SW Angola.
   STATUS: CITES – Appendix II; IUCN – Lower Risk (lc).
   SYNONYMS: *africanus* (Gray, 1865); *wagneri* Fitzinger, 1872.
   COMMENTS: Subgenus *Smutsia*.

*Manis javanica* Desmarest, 1822. Mammalogie, *in* Encycl. Méth., 2:377.
   COMMON NAME: Sunda Pangolin.
   TYPE LOCALITY: Indonesia, Java.
   DISTRIBUTION: Burma; Thailand; S Laos; C and S Vietnam; Cambodia; Malaysia; Sumatra; Java; Borneo; adjacent islands.
   STATUS: CITES – Appendix II; IUCN – Lower Risk (nt).
   SYNONYMS: *aspera* Sundevall, 1843; *guy* Focillon, 1850; *labuanensis* (Fitzinger, 1872); *leptura* Blyth, 1842; *leucura* Blyth, 1847; *malaccensis* (Fitzinger, 1872); *sumatrensis* Ludeking, 1862.

COMMENTS: Subgenus *Paramanis*. Formerly included *culionensis* (Corbet and Hill, 1992:19).

***Manis pentadactyla*** Linnaeus, 1758. Syst. Nat., 10th ed., 1:36.
  COMMON NAME: Chinese Pangolin.
  TYPE LOCALITY: Taiwan.
  DISTRIBUTION: E Nepal; NE India; E Bangladesh; Burma; Thailand; N Cambodia; N Laos; N
      Vietnam; C and S China, including Hainan Isl; Taiwan.
  STATUS: CITES – Appendix II; IUCN – Lower Risk (nt).
  SYNONYMS: *brachyura* Erxleben, 1777; **auritus** Hodgson, 1836; *assamensis* (Fitzinger, 1872);
      *bengalensis* (Fitzinger, 1872); *dalmanni* Sundevall, 1843; *kreyenbergi* (Matschie, 1907);
      **pusilla** J. Allen, 1906.
  COMMENTS: Subgenus *Manis*. Ellerman and Morrison-Scott (1951:214) recognized three
      subspecies. Zhang and Shi (1991*b*) analyzed mtDNA in two scale color morphs and
      concluded they could represent two taxa but Su-Bing et al (1994) concluded from
      protein polymorphisms that the two morphs were indistinguishable.

***Manis temminckii*** Smuts, 1832. Enumer. Mamm. Capensium, p. 54.
  COMMON NAME: Ground Pangolin.
  TYPE LOCALITY: South Africa, Northern Cape Prov., Latakou (= Litakun), near Kuruman.
  DISTRIBUTION: N South Africa; N and E Namibia; Zimbabwe; Mozambique; Botswana; S
      Angola; S Zambia; SE Dem. Rep. Congo; S Rwanda; Malawi; Tanzania; E Uganda; W
      Kenya; S Sudan; S Chad.
  STATUS: CITES – Appendix II; U.S. ESA – Endangered; IUCN – Lower Risk (nt).
  SYNONYMS: *hedenborgii* (Fitzinger, 1872).
  COMMENTS: Subgenus *Smutsia*. Reviewed by Stuart (1980).

***Manis tetradactyla*** Linnaeus, 1766. Syst. Nat., 12th ed., 1:53.
  COMMON NAME: Long-tailed Pangolin.
  TYPE LOCALITY: West Africa.
  DISTRIBUTION: Equatorial Africa from Senegal and Gambia to W Uganda, south to SW
      Angola.
  STATUS: CITES – Appendix II; IUCN – Lower Risk (lc).
  SYNONYMS: *africana* Desmarest, 1822; *ceonyx* Rafinesque, 1820; *guineensis* Fitzinger, 1872;
      *hessi* Noack, 1889; *longicaudatus* (Brisson, 1756); *longicaudatus* (Brisson, 1762);
      *macroura* Erxleben, 1777; *senegalensis* Fitzinger, 1872.
  COMMENTS: Subgenus *Uromanis*. The name *longicaudatus* Brisson, 1756, is unavailable; see
      Mohr (1961) and Meester (1972*a*).

***Manis tricuspis*** Rafinesque, 1821. Ann. Sci. Phys. Brux., 7:215.
  COMMON NAME: Tree Pangolin.
  TYPE LOCALITY: West Africa, "Guinee."
  DISTRIBUTION: Equatorial Africa from Senegal to W Kenya, south to NW Zambia and SW
      Angola; NE Mozambique; Bioko (Equatorial Guinea).
  STATUS: CITES – Appendix II; IUCN – Lower Risk (lc).
  SYNONYMS: *multiscutata* Gray, 1843; *tridentata* Focillon, 1850; **mabirae** (Allen and Loveridge,
      1942).
  COMMENTS: Subgenus *Phataginus*. Meester (1972*a*:2) recognized two subspecies. Includes
      *tridentata* described from "coastal Mozambique"; see Ansell (1982:35).

# ORDER CARNIVORA
by W. Christopher Wozencraft

**ORDER CARNIVORA** Bowdich, 1821.
>    COMMENTS: Higher taxonomic arrangement follows that of McKenna and Bell (1997), except that Ailuridae, Eupleridae, Mephitidae, and Odobenidae are raised to Family rank.

**SUBORDER FELIFORMIA** Kretzoi, 1945.

**Family Felidae** Fischer de Waldheim, 1817. Mém. Soc. Imp. Nat. Moscow, 5:372.
>    SYNONYMS: Euailuroida Kretzoi, 1929; Felinoidea Brunet, 1979; Feloidae Hay, 1930; Feloidea Simpson, 1931; Lyncina Gray, 1867.
>    COMMENTS: Revised by Pocock (1917*a*, *b*, 1951), Weigel (1961), de Beaumont (1964), Hemmer (1978), Král and Zima (1980), Kratochvíl (1982*c*), Groves (1982*a*), Collier and O'Brien (1985), Salles (1992), Johnson and O'Brien (1997), McKenna and Bell (1997), Bininda-Emonds et al. (1999), and Mattern and McLennan (2000). Some (Honacki et al., 1982; McKenna and Bell, 1997; Van Gelder, 1977*b*) have followed Simpson (1945) and placed the majority of taxa in *Felis*, except for the large cats (i.e., *Panthera* and *Acinonyx*); however, this is not well supported by primary systematic studies and only poorly represents relationships below the family level. Most studies agree on the clear separation of the "big cats" (i.e., *Panthera*, *Neofelis*, *Uncia*) from the remainder. However, within the remaining group, there does not appear to be a clear consensus. Even the cheetah's (*Acinonyx*) traditional position has been called into question (Bininda-Emonds et al., 1999; Mattern and McLennan, 2000). For these reasons, only two subfamilies of cats are recognized, and taxa are listed alphabetically within each subfamily. Synonyms allocated according to McKenna and Bell (1997) and Kitchener (pers. comm.). Species distributions were suplemented by Kristin Nowell, IUCN/SSC Cat Specialist Group (pers. comm.). For an excellent review of the biology of the felids, see Sunquist and Sunquist (2002).

**Subfamily Felinae** Fischer de Waldheim, 1817. Mém. Soc. Imp. Nat. Moscow, 5:372.
>    SYNONYMS: Acinonychinae Pocock, 1917; Guepardina Gray, 1867; Lyncini Kalandadze and Rautian, 1992; Profelina Kalandadze and Rautian, 1992; Therailurini Kalandadze and Rautian, 1992.
>    COMMENTS: A comparison of four recent phylogenetic analyses of the non-pantherine cats shows little consensus at branch points other than those that might be recognized as genera. For this reason all non-pantherine cats are tentatively grouped together in the Felinae. Synonyms allocated according to McKenna and Bell (1997).

*Acinonyx* Brookes, 1828. Cat. Anat. Zool. Mus. Joshua Brookes, London, p. 16, 33.
>    TYPE SPECIES: *Acinonyx venator* Brookes, 1828 (= *Felis jubata* Schreber, 1775), by monotypy (International Commission on Zoological Nomenclature, 1956*a*; Melville and Smith, 1987).
>    SYNONYMS: *Acinomyx* de Beaumont, 1964; *Cynaelurus* Gloger, 1841; *Cynailurus* Wagner, 1830; *Cynofelis* Lesson, 1842; *Guepar* Boitard, 1842; *Gueparda* Gray, 1843; *Guepardus* Duvernoy, 1834; *Paracinonyx* Kretzoi, 1929.
>    COMMENTS: Wozencraft (1993) placed *Acinonyx* in the monophyletic subfamily Acinonychinae. Salles (1992), Johnson and O'Brien (1997), Bininda-Emonds et al. (1999), and Mattern and McLennan (2000) considered *Acinonyx*, *Puma concolor*, and *Puma* (= *Herpailurus*) *yagouaroundi* to represent close sister groups. Synonyms allocated according to McKenna and Bell (1997).

*Acinonyx jubatus* (Schreber, 1775). Die Säugethiere, 2(15):pl. 105[1775]; text 3(22):392 [1777].
>    COMMON NAME: Cheetah.
>    TYPE LOCALITY: "südliche Afrika; man bekömmt die Felle vom Vorgebirge der guten Hofnung" [South Africa, Western Cape Province, Cape of Good Hope].
>    DISTRIBUTION: Algeria, Angola, Benin, Botswana, Burkina Faso, Cameroon, Central African

Republic, Chad, Dem. Rep. Congo, Egypt, Eritrea, Ethiopia, Malawi, Mali, Mauritania, Mozambique, Namibia, Niger, Nigeria, Somalia, South Africa, Sudan, Swaziland, Tanzania, Togo, Zambia, Zimbabwe. Recently extinct: Afghanistan, Burundi, India, Iran, Iraq, Israel, Jordan, Kazakhstan, Kenya, Kuwait, Lebanon, Libya, Morocco, Pakistan, Saudi Arabia, Senegal, Syrian Arab Republic, Tunisia, Turkmenistan, Uganda, Uzbekistan, Western Sahara, Yemen.

STATUS: CITES – Appendix I; U.S. ESA – Endangered; IUCN – Critically Endangered as *A. j. venaticus*, Endangered as *A. j. hecki*, otherwise Vulnerable.

SYNONYMS: *guttata* (Hermann, 1804); *fearonii* (A. Smith, 1834); *fearonis* (Fitzinger, 1869); *lanea* (Sclater, 1877); *obergi* Hilzheimer, 1913; *rex* Pocock, 1927; *hecki* Hilzheimer, 1913; *senegalensis* (de Blainville, 1843) [preoccupied]; *raineyi* Heller, 1913; *ngorongorensis* Hilzheimer, 1913; *soemmeringii* (Fitzinger, 1855); *megabalica* (Heuglin, 1863); *wagneri* Hilzheimer, 1913; *velox* Heller, 1913; *venaticus* (Griffith, 1821); *raddei* Hilzheimer, 1913; *venator* Brookes, 1828.

COMMENTS: Placed in *Acinonyx* by Pocock (1917*b*), Weigel (1961), Hemmer (1978), Král and Zima (1980), Kratochvíl (1982*c*), and Groves (1982*a*). Subspecies and their synonyms allocated according to G. M. Allen (1939) and Ellerman et al. (1953).

*Caracal* Gray, 1843. List. Spec. Mamm. Coll. Brit. Mus., p. 46.

TYPE SPECIES: *Caracal melanotis* Gray, 1843 (= *Felis caracal* Schreber, 1776) by monotypy.

SYNONYMS: *Caracala* Gray, 1843 [*nomen nudum*]; *Urolynchus* Severtzov, 1858.

COMMENTS: Recent studies (Groves, 1982*a*; Král and Zima, 1980) emphasized the closeness of this taxon to *Felis* and the separation of it from *Lynx* (*sensu* Simpson, 1945). Weigel (1961), Hemmer (1978), and Werdelin (1981) placed *caracal* Schreber in the monotypic *Caracal* followed here (placed in *Urolynchus* by Kratochvíl, 1982*c*). Bininda-Emonds et al. (1999) and Mattern and McLennan (2000) considered *Caracal* the sister taxon to *Leptailurus serval*. However Mattern and McLennan then considered this pair related to *Profelis aurata* whereas Bininda-Emonds et al. (1999) placed it with more typical *Felis*. Johnson and O'Brien (1997) considered *C. caracal* and *P. aurata* to be sister groups. McKenna and Bell (1997) placed *Caracal* in *Felis* (*Lynx*).

*Caracal caracal* (Schreber, 1776). Die Säugethiere, 3(16):pl. 110[1776]; text 3(24):413, 587[1777].

COMMON NAME: Caracal.

TYPE LOCALITY: "Vorgebirge der guten Hofnung", restricted by J. A. Allen (1924:281) to "Table Mountain, near Cape Town, South Africa".

DISTRIBUTION: Afghanistan, Algeria, Angola, Benin, Botswana, Burkina Faso (?), Cameroon, Central African Republic, Chad, Côte d'Ivoire, Dem. Rep. Congo, Djibouti, Egypt, Eritrea, Ethiopia, Gambia, Ghana, Guinea, Guinea-Bissau, India, Iran, Iraq, Israel, Jordan, Kazakhstan, Kenya, Kuwait, Lebanon, Lesotho, Libya, Malawi, Mali (?), Mauritania, Morocco, Mozambique, Namibia, Niger, Nigeria, Oman, Pakistan, Saudi Arabia, Senegal, Somalia, South Africa, Sudan, Syrian Arab Republic, Tajikistan, Tanzania, Togo, Tunisia, Turkey, Turkmenistan, Uganda, United Arab Emirates, Uzbekistan, Yemen, Zambia, Zimbabwe.

STATUS: CITES – Appendix I (Asian population); otherwise, Appendix II; IUCN – Least Concern.

SYNONYMS: *coloniae* Thomas, 1926; *melanotis* Gray, 1843; *melanotix* Gray, 1843; *roothi* (Roberts, 1926); *algira* (Wagner, 1841); *berberorum* Matschie, 1892; *corylinus* (Matschie, 1912); *medjerdae* (Matschie, 1912); *spatzi* (Matschie, 1912); *damarensis* (Roberts, 1926); *limpopoensis* (Roberts, 1926); *lucani* (Rochebrune, 1885); *nubica* (J. B. Fischer, 1829); *poecilotis* Thomas and Hinton, 1921; *schmitzi* (Matschie, 1912); *aharonii* (Matschie, 1912); *bengalensis* (J. B. Fischer, 1829) [preoccupied]; *michaelis* Heptner, 1945.

COMMENTS: J. A. Allen (1924) discussed the authority. Subspecies allocated according to G. M. Allen (1939), Ellerman et al. (1953), and Smithers (1971).

*Catopuma* Severtzov, 1858. Rev. Mag. Zool. Paris, ser. 2, 10:387.

TYPE SPECIES: *Felis moormensis* Hodgson, 1831 (= *Felis temminckii* Vigors and Horsfield, 1827), by monotypy.

SYNONYMS: *Badiofelis* Pocock, 1932.

COMMENTS: *Catopuma* is the sister group to *Profelis aurata* according to Bininda-Emonds et al. (1999); however, Mattern and McLennan (2000) placed it next to more typical *Felis*. McKenna and Bell (1997) placed it in *Felis* (*Profelis*). Bininda-Emonds et al. (1999) and Mattern and McLennan (2000) demonstrated that *C. badia* and *C. temmincki* are sister taxa.

*Catopuma badia* (Gray, 1874). Proc. Zool. Soc. Lond., 1874:322.

COMMON NAME: Bay Cat.

TYPE LOCALITY: "Borneo, Sarawak" [Malaysia].

DISTRIBUTION: Brunei Darussalam, Indonesia (Kalimantan), Malaysia (Sabah, Sarawak).

STATUS: CITES – Appendix II; IUCN – Endangered.

COMMENTS: Placed in *Catopuma* by Hemmer (1978) and Groves (1982*a*). Placed in the monotypic *Badiofelis* by Pocock (1932*d*), and followed by Weigel (1961), and Johnson and O'Brien (1997).

*Catopuma temminckii* (Vigors and Horsfield, 1827). Zool. J., 3:451.

COMMON NAME: Asian Golden Cat.

TYPE LOCALITY: "Sumatra" [Indonesia].

DISTRIBUTION: Bangladesh, Bhutan, Cambodia, China, India, Indonesia (Sumatra), Laos, Malaysia, Burma, Nepal, Thailand, Vietnam.

STATUS: CITES – Appendix I; U.S. ESA – Endangered; IUCN – Vulnerable.

SYNONYMS: *aurata* (Blyth, 1863); *bainsei* (Sowerby, 1924); *moormensis* (Hodgson, 1831); *nigrescens* (Gray, 1863); **dominicanorum** (Sclater, 1898); *badiodorsalis* (Howell, 1926); *dominicorum* (Howell, 1929); *melli* (Matschie, 1922); *mitchelli* (Lydekker, 1908); **tristis** (Milne-Edwards, 1872); *semenovi* (Satunin, 1905).

COMMENTS: Placed in *Catopuma* by Hemmer (1978) and Groves (1982*a*). Placed in *Profelis* by Pocock (1932*d*), followed by Weigel (1961), Král and Zima (1980), Kratochvíl (1982*c*), and Johnson and O'Brien (1997). Includes *tristis* after Pocock (1932*d*). Subspecies allocated according to Ellerman et al. (1953).

*Felis* Linnaeus, 1758. Syst. Nat., 10th ed., 1:41.

TYPE SPECIES: *Felis catus* Linnaeus, 1758, by Linnean tautonymy (Melville and Smith, 1987).

SYNONYMS: *Avitofelis* Kretzoi, 1930; *Catolynx* Severtzov, 1858; *Catus* Frisch, 1775; *Chaus* Gray, 1843; *Eremaelurus* Ognev, 1927; *Mamfelisus* Herrera, 1899; *Microfelis* Roberts, 1926; *Otailurus* Severtzov, 1858; *Otocolobus* Brandt, 1842; *Poliailurus* Lönnberg, 1925; *Trichaelurus* Satunin, 1905.

COMMENTS: Revised by Schwangart (1943), Pocock (1951), and Haltenorth (1953). Subspecies allocated according to McKenna and Bell (1997) and Kitchener (pers. comm.). Opinion 91 (1926) and Direction 24 (1955*b*) of the International Commission on Zoological Nomenclature declared the type of *Felis* to be *Felis catus* Linnaeus, 1758.

*Felis bieti* Milne-Edwards, 1892. Rev. Gen. Sci. Pures Appl., 3:671.

COMMON NAME: Chinese Mountain Cat.

TYPE LOCALITY: "Batang Tatsien-Lou", restricted by Pousargues (1898:358) to "environ de Tongolo et de Ta-tsien-lou" [China, Sichuan].

DISTRIBUTION: China (E Qinghai and N Sichuan; see He et al., 2004).

STATUS: CITES – Appendix II; IUCN – Vulnerable.

SYNONYMS: *pallida* Büchner, 1892; *subpallida* Jacobi, 1923.

COMMENTS: Haltenorth (1953) suggested *chutuchta* and *vellerosa* belonged in *silvestris* (followed here). *F. bieti*, *margarita*, *nigripes*, and *chaus* are considered closely related by Hemmer (1978), Collier and O'Brien (1985), Salles (1992), and Bininda-Emonds et al. (1999).

*Felis catus* Linnaeus, 1758. Syst. Nat., 10th ed., 1:41.

COMMON NAME: Domestic Cat.

TYPE LOCALITY: Listed as "Sweden" in Pocock (1951:6).

DISTRIBUTION: Cosmopolitan. Specifically reported in: Albania, Belgium, Bosnia and Herzegovina, Bulgaria, Crete, Croatia, France, Germany, Greece, Hungary, Italy,

Poland, Portugal, Romania, Scotland, Serbia and Montenegro, Slovakia, Slovenia, Spain, Turkey.

STATUS: CITES – specifically excluded from protection.

SYNONYMS: *agria* Bate, 1906; *angorensis* Gmelin, 1788; *antiquorum* J. B. Fischer, 1829; *aureus* Kerr, 1792; *bouvieri* Rochebrune, 1883; *brevicaudata* Schinz, 1844; *caerulea* Erxleben, 1777; *cumana* Schinz, 1844; *daemon* Satunin, 1904; *domestica* Erxleben, 1777; *hispanica* Erxleben, 1777; *huttoni* Blyth, 1846; *inconspicua* Gray, 1837; *japonica* J. B. Fischer, 1829; *longiceps* Bechstein, 1800; *madagascariensis* Kerr, 1792; *megalotis* Müller, 1839; *pulchella* Gray, 1837; *rubra* Gmelin, 1788; *siamensis* Trouessart, 1904; *sinensis* Kerr, 1792; *striaas* Bechstein, 1800; *syriaca* J. B. Fischer, 1829; *tralatitia* J. B. Fischer, 1829; *vulgaris* J. B. Fischer, 1829.

COMMENTS: Also see comments under *Felis* and *Felis silvestris*. Synonyms allocated according to Pocock (1951) and should be considered provisional. There has been almost universal use of *F. catus* for the domestic cat and *silvestris* for wild cats. Several authors have treated the domestic cat as separate from the wildcats (Corbet and Hill, 1991; Daniels et al., 1998; A. C. Kitchener, 1991; Mattern and McLennan, 2000; Nowak, 1999; Pocock, 1951; Wiseman et al., 2000); however also see Randi and Ragni (1986), Essop et al. (1997), and Johnson and O'Brien (1997), who presented morphological and molecular evidence to support *catus*, *libyca*, and *silvestris* as conspecific. If conspecific, there would be a problem with the continued use of the name *Felis silvestris* (see comments therein).

*Felis chaus* Schreber, 1777. Die Säugethiere, 2(13):pl. 110.B[1777]; text, 3(24):414[1777].

COMMON NAME: Jungle Cat.

TYPE LOCALITY: "wohnt in den sumpfigen mit Schilf bewachsenen oder bewaldeten Gegenden der Steppen um das kaspische Meer, und die in selbiges fallenden Flüse. Auf der Nordseite des Terekflusses und der Festung Kislar . . . desto Hünfiger aber bey der Mündung der Kur . . .". Listed in Honacki et al. (1982) as "U.S.S.R., Dagestan, Terek River, N. of the Caucasus".

DISTRIBUTION: Afghanistan, Bangladesh, Bhutan, Cambodia, China, Egypt, India, Iran, Iraq, Israel, Jordan, Kazakhstan, Laos, Mongolia, Burma, Nepal, Pakistan, Russia, Sri Lanka, Syrian Arab Republic, Tajikistan, Thailand, Turkey, Turkmenistan, Uzbekistan, Vietnam.

STATUS: CITES – Appendix II; IUCN – Least Concern.

SYNONYMS: *catolynx* Pallas, 1811; *shawiana* Blanford, 1876; *typica* de Winton, 1898; **affinis** Gray, 1830; *erythrotus* (Hodgson, 1836); *jacquemontii* I. Geoffroy Saint-Hilaire, 1844; **fulvidina** Thomas, 1929; **furax** de Winton, 1898; *chrysomelanotis* (Nehring, 1902); **kelaarti** Pocock, 1939 [based on *Felis chaus* Kelaart, 1852]; **kutas** Pearson, 1832; **maimanah** Zukowsky, 1915; **nilotica** de Winton, 1898; *rüppelii* Brandt, 1832 [preoccupied]; **oxiana** Heptner, 1969; **prateri** Pocock, 1939.

COMMENTS: *F. chaus* Güldenstädt, 1776, is invalid (J. A. Allen, 1920). Subspecies allocated according to Pocock (1951) and Ellerman and Morrison-Scott (1951).

*Felis manul* Pallas, 1776. Reise Prov. Russ. Reichs, 3:692.

COMMON NAME: Pallas's Cat.

TYPE LOCALITY: "Frequens in rupestribus, apricis totius Tatariae Mongoliaeque desertae" [USSR, Chita Province, Borzya District, Kulusutai (Heptner and Sludskii, 1992)].

DISTRIBUTION: Afghanistan, Armenia, China, India, Iran, Kazakhstan, Kyrgyzstan, Mongolia, Pakistan, Russia, Tajikistan, Turkmenistan, Uzbekistan.

STATUS: CITES – Appendix II; IUCN – Near Threatened.

SYNONYMS: *mongolica* Satunin, 1905; *satuni* Lydekker, 1907; **ferruginea** Ognev, 1928; **nigripecta** Hodgson, 1842.

COMMENTS: Revised by Pocock (1907a), Birula (1913, 1916), Ognev (1935), and Schwangart (1936). Most consider *F. manul* to be included with *Felis* (*Felis*) and separation into *Otocolobus* following Wozencraft (1993) would make *Felis* paraphyletic (Bininda-Emonds et al., 1999; Collier and O'Brien, 1985; Hemmer, 1978; Mattern and McLennan, 2000; Salles, 1992; Weigel, 1961). McKenna and Bell (1997) placed in *Felis* (*Otocolobus*). Subspecies allocated according to Pocock (1951) and Ellerman and Morrison-Scott (1951).

*Felis margarita* Loche, 1858. Rev. Mag. Zool. Paris, ser. 2, 10:49.
> COMMON NAME: Sand Cat.
> TYPE LOCALITY: "environs de Négonca (Sahara)" [Algeria].
> DISTRIBUTION: Afghanistan (?), Algeria, Chad, Egypt, Iran, Iraq (?), Israel, Jordan, Kazakhstan, Kuwait, Libya (?), Mali (?), Mauritania (?), Morocco, Niger, Oman, Pakistan, Qatar, Saudi Arabia, Sudan (?), Syrian Arab Republic (?), Turkmenistan, United Arab Emirates, Uzbekistan, Western Sahara (?), Yemen.
> STATUS: CITES – Appendix II; U.S. ESA – Endangered as *F. margarita scheffeli*; IUCN – Near Threatened.
> SYNONYMS: *marginata* Gray, 1867; *margaritae* Trouessart, 1897; *marguerittei* Trouessart, 1905; ***airensis*** Pocock, 1951; ***harrisoni*** Hemmer, Grubb and Groves, 1976; ***meinertzhageni*** Pocock, 1938; ***scheffeli*** Hemmer, 1974; ***thinobia*** (Ognev, 1927).
> COMMENTS: Revised by Schauenberg (1974) and Hemmer et al. (1976). Pocock (1951) and Schauenberg (1974) included *Eremaelurus thinobia*, which was recognized as separate by Haltenorth (1953) and Weigel (1961), but see discussion by Hemmer et al. (1976). Král and Zima (1980) suggested this species was closely related to *F. manul*. Subspecies allocated according to G. M. Allen (1939), Pocock (1951), Ellerman and Morrison-Scott (1951), Schauenberg (1974), and Hemmer et al. (1976).

*Felis nigripes* Burchell, 1824. Travels Interior of Southern Africa, 2:592.
> COMMON NAME: Black-footed Cat.
> TYPE LOCALITY: Burchell (1824:509) implied the country of the "Bachapins", presumably in the capital, "the town of Litákun (Letárkoon) . . . 27°6'44". [S] . . . 24°39'27"[E]" [South Africa].
> DISTRIBUTION: Angola, Botswana, Namibia, South Africa, Zimbabwe.
> STATUS: CITES – Appendix I; U.S. ESA – Endangered; IUCN – Vulnerable.
> SYNONYMS: ***thomasi*** Shortridge, 1931.
> COMMENTS: Král and Zima (1980) noted a distinctly different karotype from other *Felis*. McKenna and Bell (1997) placed in *Felis* (*Microfelis*).

*Felis silvestris* Schreber, 1777. Die Säugethiere, 3(23):397.
> COMMON NAME: Wildcat.
> TYPE LOCALITY: Not given. Fixed by Haltenorth (1953) as "vielleicht Nordfrankreich". Listed by Pocock (1951) as "Germany".
> DISTRIBUTION: Afghanistan, Algeria, Angola, Armenia, Azerbaijan, Belarus, Benin, Botswana, Burkina Faso, Burundi, Cameroon, Central African Republic, Chad, China, Djibouti, Egypt, Eritrea, Ethiopia, Gambia, Georgia, Ghana, Guinea, Guinea-Bissau, India, Iran, Iraq, Israel, Jordan, Kazakhstan, Kenya, Kuwait, Kyrgyzstan, Lebanon, Lesotho, Libya, Luxembourg, Macedonia, Malawi, Mali, Mauritania, Morocco, Mozambique, Namibia, Niger, Nigeria, Oman, Pakistan, Republic of Congo, Russia, Saudi Arabia, Senegal, Serbia and Montenegro, Sierra Leone, Slovakia, Slovenia, Somalia, South Africa, Sudan, Swaziland, Switzerland, Syrian Arab Republic, Tajikistan, Tanzania, Togo, Tunisia, Turkey, Turkmenistan, Uganda, United Arab Emirates, USA, Uzbekistan, Western Sahara, Yemen, Zambia, Zimbabwe.
> STATUS: CITES – Appendix II; IUCN – Vulnerable as *F. s. grampia*, otherwise Least Concern.
> SYNONYMS: *euxina* Pocock, 1943; *ferox* Martorelli, 1896; *ferus* Erxleben, 1777; *foxi* Pocock, 1944; *hybrida* J. B. Fischer, 1829; *molisana* Altobello, 1921; *morea* Trouessart, 1904; *obscura* Desmarest, 1820; *tartessia* Miller, 1907; ***cafra*** Desmarest, 1822; *caffra* A. Smith, 1826; *caligata* Temminck, 1824; *obscura* Anderson and de Winton, 1902; *namaquana* Thomas, 1926; *rusticana* Thomas, 1928; ***caucasica*** Satunin, 1905; *trapezia* Blackler, 1916; ***caudata*** Gray, 1874; *griseoflava* Zukowsky, 1915; *issikulensis* Ognev, 1930; *kozlovi* Satunin, 1905; *longipilis* Zukowsky, 1915; *macrothrix* Zukowsky, 1915; *matschiei* Zukowsky, 1914; *murgabensis* Zukowsky, 1915; *schnitnikovi* Birula, 1915; ***chutuchta*** Birula, 1916; ***cretensis*** Haltenorth, 1953; ***foxi*** Pocock, 1944; ***gordoni*** Harrison, 1968; ***grampia*** Miller, 1907; ***griselda*** Thomas, 1926; *vernayi* Roberts, 1932; *xanthella* Thomas, 1926; ***haussa*** Thomas and Hinton, 1921; ***iraki*** Cheesman, 1921; ***jordansi*** Schwarz, 1930; ***lybica*** Forster, 1780; *bubastis* Hemprich and Ehrenberg, 1833; *cyrenarum* Ghigi, 1920; *dongolana* Hemprich and Ehrenberg, 1832 [not *V. dongolana* Hemprich and

Ehrenberg, 1832, a viverrid]; *cristata* Lataste, 1885; *libyca* Olivier, 1804; *lowei* Pocock, 1944; *lybiensis* Kerr, 1792; *lynesi* Pocock, 1944; *maniculata* Temminck, 1824; *mauritana* Cabrera, 1906; *mediterranea* Martorelli, 1896; *ruppelii* Schinz, 1824; *sarda* Lataste, 1885; **mellandi** Schwann, 1904; *pyrrhus* Pocock, 1944; **nesterovi** Birula, 1916; **ocreata** Gmelin, 1791; *brockmani* Pocock, 1944; *guttata* Hermann, 1804; *maniculata* Cretschmar, 1826; *nubiensis* Kerr, 1792; **ornata** Gray, 1832; *servalina* Jardine, 1834; *torquata* Blyth, 1863; **reyi** Lavauden, 1929; **rubida** Schwann, 1904; **tristrami** Pocock, 1944; *maniculata* Yerbury and Thomas, 1895; *syriaca* Tristram, 1867; **ugandae** Schwann, 1904; *nandae* Heller, 1913; *taitae* Heller, 1913; *vellerosa* Pocock, 1943.

COMMENTS: Also see comments under *Felis catus*. There is some confusion as to the correct species name. Schreber (1775) illustrated a plate as '*Felis Catus ferus*', and in 1777 the text listed '*Felis (Catus) silvestris*' and '*Felis Catus (domestica)*.' Opinion 465 of the International Commission on Zoological Nomenclature (1957*f*) declared *silvestris* as the specific name for the European wild cat (with the understanding that *F. catus* and *F. silvestris* are usually considered conspecific). Revised by Ragni and Randi (1986), who included *lybica*, and by Haltenorth (1953), who included *chutuchta*, *lybica*, and *vellerosa*. However, Pocock's (1951) revision considered *catus* as separate and placed *chutuchta* and *vellerosa* in *bieti*, and they probably should be considered *incertae sedis*. Does not include *F. catus* (worldwide), which was domesticated from this species (Corbet, 1978*c*). Ellerman and Morrison-Scott (1951) argued that *lybica* Forster (1780), was a *lapsus* for *libyca*; however, there is no clear internal evidence that the name was misspelled (Meester et al., 1986). Rosevear (1974), Ansell (1978), Smithers (1983), Meester et al. (1986), and Wiseman et al. (2000) retained *lybica* as separate from *silvestris*. Hemmer (1978), Collier and O'Brien (1985), Salles (1992), Johnson and O'Brien (1997) and Essop et al. (1997*b*) supported the inclusion of *silvestris*, *lybica* and the domestic cat (*catus*), however Mattern and McLennan (2000) considered *silvestris* closer to *margarita*, and consider *catus* as sister group to *lybica*. Placed in *Felis* (*Felis*) by McKenna and Bell (1997). Subspecies allocated according to Pocock (1951), Ellerman et al. (1953), Smithers (1971), and Kitchener (pers. comm.).

*Leopardus* Gray, 1842. Ann. Mag. Nat. Hist., [ser. 1], 10:260.

TYPE SPECIES: *Leopardus griseus* Gray, 1842 (= *Felis pardalis* Linnaeus, 1758), by subsequent designation by Pocock (1917*b*).

SYNONYMS: *Colocolo* Pocock, 1941; *Dendrailurus* Severtzov, 1858; *Lynchailurus* Severtzov, 1858; *Margay* Gray, 1867; *Montifelis* Schwangart, 1941; *Mungofelis* Antonius, 1933; *Noctifelis* Severtzov, 1858; *Oncifelis* Severtzov, 1858; *Oncilla* J. A. Allen, 1919; *Oncoides* Severtzov, 1858; *Oreailurus* Cabrera, 1940; *Pajeros* Gray, 1867; *Pardalina* Gray, 1867; *Pardalis* Gray, 1867; *Pseudolynx* Schwangart, 1941.

COMMENTS: There has been almost unanimous agreement that this group is monophyletic (Bininda-Emonds et al., 1999; Herrington, 1986; Johnson and O'Brien, 1997; Johnson et al., 1998; Mattern and McLennan, 2000; Pocock, 1917*a*; Salles, 1992; Weigel, 1961). However, the relationships within this genus are unclear. Most recognize *wiedii* and *pardalis* as a monophyletic group, however there is considerable controversy on the arrangement of the remaining species. Under phylogenies put forward by some recent genetic and molecular studies, the remaining taxa (*guigna*, *colocolo*, *jacobitus*, *tigrinus*, *geoffroyi*) would be paraphyletic if *wiedii* and *pardalis* were separated (Collier and O'Brien, 1985; Johnson and O'Brien, 1997; Johnson et al., 1998). All are provisionally included here in *Leopardus*. García-Perea (1994) revised the pampas cat group and found three clearly distinct allopatric populations. She argued that these populations had been separated for some time and should be considered full species (*braccatus*, *colocolo*, and *pajeros*). Although the distinguishing morphological features she found were in some cases variable, they would fall within the range of differences recognized elsewhere at the species level. I have provisionally followed García-Perea for these endangered populations.

*Leopardus braccatus* (Cope, 1889). Am. Nat., 23:144.

COMMON NAME: Pantanal Cat.

TYPE LOCALITY: "Chapada, matto Grosso," Brazil.

DISTRIBUTION: Brazil (Mato Grosso and mato Grosso do Sul), Paraguay, Uruguay (García-Perea, 1994).

STATUS: CITES – Appendix II.

SYNONYMS: *munoai* (Ximenez, 1961).

COMMENTS: See comments under genus.

*Leopardus colocolo* (Molina, 1782). Sagg. Stor. Nat. Chile, p. 295.

COMMON NAME: Colocolo.

TYPE LOCALITY: "che abitano i boschi del Chili," restricted by Osgood (1943) to "Province of Valparaiso" [Chile].

DISTRIBUTION: Chile.

STATUS: CITES – Appendix II, ; IUCN – Near Threatened as *Oncifelis colocolo*.

SYNONYMS: *albescens* (Fitzinger, 1869); *colocola* (Molina, 1782); *colorolla* (Bechstein, 1800); *huinus* (Pocock, 1941); *passerum* (Sclater, 1871); *wolffsohni* (García-Perea, 1994).

COMMENTS: The validity of *colocolo* was questioned by Osgood (1943) however, Wolffsohn (1908) and Cabrera (1940, 1958) defended the original description. Some have placed it in *Oncifelis* (but see comments under genus). Subspecies allocated according to Pocock (1941*b*) and García-Perea (1994).

*Leopardus geoffroyi* (d'Orbigny and Gervais, 1844). Bull. Sci. Soc. Philom. Paris, 1844:40.

COMMON NAME: Geoffroy's Cat.

TYPE LOCALITY: "des rives du Rio Negro, en Patagonie".

DISTRIBUTION: Argentina, Bolivia, Brazil, Chile, Paraguay, Uruguay.

STATUS: CITES – Appendix I as *Oncifelis geoffroyi*; IUCN – Near Threatened as *Oncifelis geoffroyi*.

SYNONYMS: *argenteus* (Schwangart, 1941); *flavus* (Schwangart, 1941); *geoffroyi* (Severtzow, 1858); *guigna* (Mivart, 1881); *himalayanus* Gray, 1843 [*nomen nudum*]; *macdonaldi* (Marelli, 1932); *pardoides* (Gray, 1867); *tigrina* (Larranaga, 1923); *warwickii* (Gray, 1867); *euxanthus* (Pocock, 1940); *leucobaptus* (Pocock, 1940); *paraguae* (Pocock, 1940); *melas* (Betoni, 1914) [preoccupied]; *salinarum* (Thomas, 1903).

COMMENTS: Subspecies allocated according to Cabrera (1958). Revised by Pocock (1940*c*) and reviewed by Ximenez (1975). Placed in *Oncifelis* by J. A. Allen (1919*a*), Weigel (1961), Hemmer (1978), Král and Zima (1980), and Kratochvíl (1982*c*). Ximeñez (1975) followed Cabrera (1957), and placed *geoffroyi* in *Felis* (*Leopardus*). Includes *F. pardoides* (Cabrera, 1957).

*Leopardus guigna* (Molina, 1782). Sagg. Stor. Nat. Chile, p. 295.

COMMON NAME: Kodkod.

TYPE LOCALITY: "Chili", restricted by Thomas (1903:240) to "Valdivia" [Chile].

DISTRIBUTION: Argentina and Chile.

STATUS: CITES – Appendix II; IUCN – Vulnerable as *Oncifelis guigna*.

SYNONYMS: *santacrucensis* (Artayeta, 1950); *tigrillo* (Schinz, 1844); *molinae* (Osgood, 1943).

COMMENTS: Placed in *Oncifelis* by Weigel (1961) and Hemmer (1978). Placed in subgenus *Leopardus* of *Felis* by Cabrera (1958). Synonyms allocated according to Cabrera (1957).

*Leopardus jacobitus* (Cornalia, 1865). Mem. Soc. Ital. Sci. Nat., 1:5.

COMMON NAME: Andean Mountain Cat.

TYPE LOCALITY: "Bolivia, circa Potosi et Humacuaca in montibus sat elevatis"; further clarified by Cabrera (1958:297): as "Sur del departamento boliviano de Potosi, cerca de la frontera argentina, entre Potosi y Humahuaca".

DISTRIBUTION: NW Argentina, SW Bolivia, NE Chile, S Peru.

STATUS: CITES – Appendix I as *Oreailurus jacobitus*; U. S. ESA – Endangered as *Felis jacobitus*; IUCN – Endangered as *Oreailurus jacobita* (sic).

SYNONYMS: *colocolo* (Philippi, 1869).

COMMENTS: Reviewed by García-Perea (2002), Yensen and Seymour (2000), and Gray (1867*b*). Placed in *Oreailurus* by Cabrera (1940), Weigel (1961), and Hemmer (1978). Later, Cabrera (1958) reconsidered *Oreailurus* as a subgenus of *Felis*.

*Leopardus pajeros* (Desmarest, 1816). Nouv. Dict. Hist. Nat., (2), 6:114.

COMMON NAME: Pampas cat.

TYPE LOCALITY: "Pampas de Buenos Ayres entre los 35 y 36 grados" [Argentina].

DISTRIBUTION: Argentina, Bolivia, Chile, Ecuador, Peru.

STATUS: CITES – Appendix II.

SYNONYMS: *pageros* (Lesson, 1827); *pajero* (Burmeister, 1879); *pampa* (Schinz, 1831); *pampanus* (Gray, 1867); *passerum* (Sclater, 1872); **budini** (Pocock, 1941); *crespoi* (Cabrera, 1957); *crucinus* (Thomas, 1901); **garleppi** (Matschie, 1912); *parleppi* (Lönnberg, 1913); **steinbachi** (Pocock, 1941); *thomasi* (Lönnberg, 1913); *garleppi* (Pocock, 1941) [preoccupied].

COMMENTS: See comments under genus, see also García-Perea (1994).

*Leopardus pardalis* (Linnaeus, 1758). Syst. Nat., 10th ed., 1:42.

COMMON NAME: Ocelot.

TYPE LOCALITY: "America", restricted to "Mexico", by Thomas (1911*a*:136), further restricted by J. A. Allen (1919*b*:345) to "State of Vera Cruz".

DISTRIBUTION: Argentina, Belize, Bolivia, Brazil, Colombia, Costa Rica, Ecuador, El Salvador, French Guiana, Guatemala, Guyana, Honduras, Mexico, Nicaragua, Panama, Paraguay, Peru, Suriname, Trinidad and Tobago, USA (Texas, Arizona), Uruguay, Venezuela.

STATUS: CITES – Appendix I; U. S. ESA – Endangered; IUCN – Endangered as *L. p. albescens*, otherwise Least Concern.

SYNONYMS: *buffoni* (Brass, 1911); *canescens* (Swainson, 1838); *griffithii* (J. B. Fischer, 1829); *griseus* Gray, 1842; *mexicanus* (Kerr, 1792); *ocelot* (Link, 1795); *pictus* Gray, 1842; **aequatorialis** (Mearns, 1903); *costaricensis* (Mearns, 1903); *mearnsi* (J. A. Allen, 1904); *pardalis* (Alston, 1882) [preoccupied]; **albescens** (Pucheran, 1855); *limitis* (Mearns, 1902); *ludovicianus* (Brass, 1911); **melanurus** (Ball, 1844); *chibigouazou* (Mearns, 1903); *maripensis* (J. A. Allen, 1904); *ocelot* (Osgood, 1916); *tumatumari* (J. A. Allen, 1915); **mitis** (F. G. Cuvier, 1820); *armillatus* (F. G. Cuvier, 1820); *brasiliensis* (Schinz, 1844); *chati* (Gray, 1827); *chibigouavou* (Ditmars, 1939); *chibigouazou* (Gray, 1827); *chibiguazu* (J. B. Fischer, 1829); *hamiltonii* (J. B. Fischer, 1829); *maracaya* (Wagner, 1841); *ocelot* (Smith, 1827); *smithii* (Swainson, 1838); *tigrinus* (Elliot, 1877); *pardalis* (Lahille, 1899); **nelsoni** (Goldman, 1925); **pseudopardalis** (Boitard, 1842); *sanctaemartae* (J. A. Allen, 1904); **pusaeus** (Thomas, 1914); *sonoriensis* (Goldman, 1925); **steinbachi** (Pocock, 1941).

COMMENTS: Placed in *Leopardus* by J. A. Allen (1919*b*), Weigel (1961), Hemmer (1978), and Kratochvíl (1982*c*). Reviewed by Murray and Gardner (1997). Synonyms allocated according to Cabrera (1958*b*), Hall (1981), and Murray and Gardner (1997).

*Leopardus tigrinus* (Schreber, 1775). Die Säugethiere, 2(15):pl. 106[1775]; text, 3(23):396[1777].

COMMON NAME: Oncilla.

TYPE LOCALITY: "südlichen Amerika", restricted by J. A. Allen (1919*b*:356), to "Cayenne" [French Guiana].

DISTRIBUTION: Argentina, Bolivia, Brazil, Colombia, Costa Rica, Ecuador, French Guiana, Guyana, Nicaragua (?), Panama, Paraguay (?), Peru, Suriname, Venezuela.

STATUS: CITES – Appendix I; U.S. ESA – Endangered; IUCN – Near Threatened.

SYNONYMS: *emiliae* (Thomas, 1914); *margay* (Müller, 1776); **guttulus** (Hensel, 1872); *guigna* (Hensel, 1872); *guttula* (Trouessart, 1897); *mitis* (Lahille, 1899); *pardinoides* (Thomas, 1903) [preoccupied]; **oncilla** (Thomas, 1903); *carrikeri* (J. A. Allen, 1904); **pardinoides** (Gray, 1867); *andinus* (Thomas, 1903); *caucensis* (J. A. Allen, 1915); *elenae* (J. A. Allen, 1915); *emeritus* (Thomas, 1912); *geoffroyi* (Elliot, 1872); *wiedi* (J. A. Allen, 1916).

COMMENTS: Placed in *Leopardus* by J. A. Allen (1919*b*), Weigel (1961), and Kratochvíl (1982*c*); placed in *Oncifelis* (with *guigna* and *geoffroyi*) by Hemmer (1978). *L. tigrinus* shares a derived chromosomal number with *pardalis* and *wiedii* (Wurster-Hill, 1973). Includes *Felis pardinoides* after Cabrera (1958); however, see J. A. Allen (1919*b*) and Weigel (1961) who considered *pardinoides* as distinct, but closely related to *tigrinus*. Synonyms allocated according to Pocock (1941*c*), Cabrera (1957), and Hall (1981).

*Leopardus wiedii* (Schinz, 1821). *In* Cuvier, Das Thierreich, 1:235.

COMMON NAME: Margay.

TYPE LOCALITY: "Brasilieri", restricted by J. A. Allen (1919*b*:357) to "northern Espirito Santo,

Brazil", and further restricted by Cabrera (1957:290), to "Brasil, restringida al Morro de Arará, sobre el rio Mucurí, estado de Baía".

DISTRIBUTION: Argentina, Belize, Bolivia, Brazil, Colombia, Costa Rica, Ecuador, El Salvador, French Guiana, Guatemala, Guyana, Honduras, Mexico, Nicaragua, Panama, Paraguay, Peru, Suriname, USA (Texas), Uruguay, Venezuela.

STATUS: CITES – Appendix I; U.S. ESA – Endangered (from Mexico southward); IUCN – Least Concern.

SYNONYMS: *elegans* (Lesson, 1830); *geoffroyi* (Rochebrune, 1895); *macroura* (Wied-Neuwied, 1823); *macrourus* (C. E. H. Smith, 1827); *macrura* (Hensel, 1872); *pardictis* Pocock, 1941; *tigrinoides* Gray, 1842; *venusta* (Reichenbach, 1836); *amazonicus* (Cabrera, 1917); *pirrensis* (Pocock, 1941); *boliviae* Pocock, 1941; *cooperi* (Goldman, 1943); *glauculus* (Thomas, 1903); *nicaraguae* (J. A. Allen, 1919); *oaxacensis* (Nelson and Goldman, 1931); *mexicana* (Saussure, 1860) [preoccupied]; *pirrensis* Goldman, 1914; *andina* (J. A. Allen, 1916); *ludovici* (Lönnberg, 1925); *salvinius* Pocock, 1941; *vigens* (Thomas, 1904); *catenata* (Cabrera, 1917); *macrura* (Goeldi and Hagmann, 1904); *yucatanicus* Nelson and Goldman, 1931.

COMMENTS: Placed in *Leopardus* by Weigel (1961), Hemmer (1978), and Kratochvíl (1982*c*). Allen (1919*b*) and Weigel (1961) suggested that *wiedii* (in part) may be conspecific with *tigrinus*; however, Hemmer (1978) considered differences between *wiedii* and *tigrinus* to warrant generic distinction. Synonyms allocated according to Pocock (1941*d*), Cabrera (1957), Hall (1981), and Oliveira (1998*a*).

*Leptailurus* Severtzov, 1858. Rev. Mag. Zool. Paris, ser. 2, 10:389.

TYPE SPECIES: *Felis serval* Schreber, 1776, by monotypy.

SYNONYMS: *Galeopardus* Heuglin and Fitzinger, 1866; *Serval* Brehm, 1864; *Servalina* Greve, 1894.

COMMENTS: Placed as a subgenus of *Felis* by McKenna and Bell (1997). There appears to be little agreement on the relationship of *Leptailurus* to other cats. Pocock (1917*a*) placed it with *Leopardus*; whereas Weigel (1961), Hemmer (1978), and Bininda-Emonds et al. (1999) placed with *Felis*, *Lynx*, and *Caracal*. Salles (1992) grouped it with *Prionailurus bengalensis*, and Johnson and O'Brien (1997) and Mattern and McLennan (2000) with *Caracal* and *Profelis*. Severtzov (1858), Groves (1982*a*), and McKenna and Bell (1997) considered *Leptailurus* a subgenus of *Felis*.

*Leptailurus serval* (Schreber, 1776). Die Säugethiere, 3(16):pl. 108[1776]; text 3(23):407[1777].

COMMON NAME: Serval.

TYPE LOCALITY: "Ostindien und Tibet in gebirgegen Gegenden, vielleicht auch am Vorgebirge der guten Hofnung und dem heissern Afrika"; restricted by J. A. Allen (1924) to the "Cape region of South Africa".

DISTRIBUTION: Angola, Benin, Botswana, Burkina Faso, Burundi, Cameroon, Central African Republic, Chad, Côte d'Ivoire, Dem. Rep. Congo, Ethiopia, Gabon, Gambia, Ghana, Guinea, Guinea-Bissau, Kenya, Lesotho (?), Liberia, Malawi, Mali, Morocco, Mozambique, Namibia, Niger, Nigeria, Rwanda, Senegal, Sierra Leone, Somalia, South Africa, Sudan, Swaziland, Tanzania, Togo, Uganda, Zambia, Zimbabwe. Believed to be extirpated in Algeria, and Tunisia.

STATUS: CITES – Appendix II; U.S. ESA – Endangered as *L. s. constantina* [sic]; IUCN – Endangered as *L. s. constantinus*, otherwise Least Concern.

SYNONYMS: *capensis* (Forster, 1781); *galeopardus* (Desmarest, 1820); *beirae* (Wroughton, 1910); *brachyurus* (Wagner, 1841); *ogilbyi* (Schinz, 1844); *servalinus* (Ogilby, 1839); *constantinus* (Forster, 1780); *algiricus* (J. B. Fischer, 1829); *faradjius* J. A. Allen, 1924; *ferrarii* (de Beaux, 1924); *hamiltoni* Roberts, 1931; *hindei* (Wroughton, 1910); *kempi* (Wroughton, 1910); *kivuensis* (Lönnberg, 1919); *lipostictus* (Pocock, 1907); *larseni* (Thomas, 1913); *lonnbergi* (Cabrera, 1910); *niger* (Lönnberg, 1897) [preoccupied]; *mababiensis* Roberts, 1932; *pantastictus* (Pocock, 1907); *poliotricha* (Pocock, 1907); *phillipsi* (G. M. Allen, 1914); *pococki* (Cabrera, 1910); *senegalensis* (Lesson, 1839) [preoccupied]; *robertsi* Ellerman, Morrison-Scott and Hayman, 1953; *togoensis* (Matschie, 1893).

COMMENTS: Synonyms allocated according to Smithers (1971).

*Lynx* Kerr, 1792. *In* Linnaeus, Anim. Kingdom, 1:155.
 TYPE SPECIES: *Felis lynx* Linnaeus, 1758, by absolute tautonymy (Melville and Smith, 1987).
 SYNONYMS: *Cervaria* Gray, 1867; *Eucervaria* Palmer, 1903; *Lynceus* Gray, 1821; *Lynchus* Jardine, 1834; *Lyncus* Gray, 1825; *Pardina* Kaup, 1829.
 COMMENTS: Revised by Matyushkin (1979), Werdelin (1981), and García-Perea (1992), who recognized the generic status of *Lynx*. Groves (1982a), Hemmer (1978), and McKenna and Bell (1997) considered *Lynx* a subgenus of *Felis*.

*Lynx canadensis* Kerr, 1792. *In* Linnaeus, Anim. Kingdom, 1:157.
 COMMON NAME: Canadian Lynx.
 TYPE LOCALITY: "Canada"; listed in Miller (1912a) as "Eastern Canada".
 DISTRIBUTION: Canada, USA (C Utah and SW Colorado, NE Nebraska, S Indiana, and West Virginia).
 STATUS: CITES – Appendix II; U.S. ESA – Threatened; IUCN – Least Concern.
 SYNONYMS: *mollipilosus* Stone, 1900; *subsolanus* Bangs, 1897.
 COMMENTS: Considered distinct from *L. lynx* by Kurtén and Anderson (1980), Matyushkin (1979), Werdelin (1981), and García-Perea (1992). Reviewed in part by Tumlison (1987) as *Felis lynx*.

*Lynx lynx* (Linnaeus, 1758). Syst. Nat., 10th ed., 1:43.
 COMMON NAME: Eurasian Lynx.
 TYPE LOCALITY: "Europe sylvis and desertis", subsequently restricted by Thomas (1911a:136) to "Wennersborg, S. Sweden".
 DISTRIBUTION: Afghanistan, Albania, Armenia, Austria, Azerbaijan, Belarus, Bhutan, Bosnia and Herzegovina (?), China, Croatia, Czech Republic, Estonia, Finland, France, Georgia, Germany, Greece, Hungary, India, Iran, Iraq (?), Italy, Kazakhstan, North Korea, Kyrgyzstan, Latvia, Lithuania, Macedonia, Mongolia, Nepal, Norway, Pakistan, Poland, Romania, Russia, Slovakia, Slovenia, Sweden, Switzerland, Tajikistan, Turkey, Turkmenistan, Uzbekistan, Serbia and Montenegro. Recently extinct in Bulgaria.
 STATUS: CITES – Appendix II; IUCN – Near Threatened.
 SYNONYMS: *albus* Kerr, 1792; *baicalensis* (Dybowsky, 1922); *borealis* (Thunberg, 1798); *carpathica* (Heptner, 1972); *cervarius* Temminck, 1824; *dinniki* Satunin, 1915; *guttatus* Smirnov, 1922; *kattlo* (Schrank, 1798); *lupulinus* (Thunberg, 1825); *lynculus* (Nilsson, 1820); *melinus* Kerr, 1792; *neglectus* Stroganov, 1962; *orientalis* Satunin, 1905; *virgata* (Nilsson, 1829); *vulgaris* Kerr, 1792; *vulpinus* (Thunberg, 1825); *wrangeli* Ognev, 1928; *isabellinus* (Blyth, 1847); *kamensis* (Satunin, 1905); *tibetanus* (Gray, 1863); *wardi* (Lydekker, 1904); *kozlovi* Fetisov, 1950; *sardiniae* Mola, 1908; *stroganovi* Heptner, 1969.
 COMMENTS: Does not include *L. canadensis* or *L. pardinus*, following Pocock (1917a), Hemmer (1978), Matyushkin (1979), García-Perea (1992), Werdelin (1981), Salles (1992), Johnson and O'Brien (1997), Bininda-Emonds et al. (1999), and Mattern and McLennan (2000). Includes *isabellinus* (Gao, 1987). Reviewed in part by Tumlison (1987) as *Felis lynx*. Synonyms allocated according to Ellerman and Morrison-Scott (1951) and Tumlison (1987).

*Lynx pardinus* (Temminck, 1827). Monogr. Mamm., 1:116.
 COMMON NAME: Iberian Lynx.
 TYPE LOCALITY: "Portugal, puisque le commerce reçoit des peaux préparées de Lisbonne, et que M. le baron de Vionénil tua, en 1818, sur les bords du Tage, à dix lieues de Lisbonne".
 DISTRIBUTION: Portugal, SW Spain.
 STATUS: CITES – Appendix I; U.S. ESA – Endangered as *Felis pardina*; IUCN – Critically Endangered.
 SYNONYMS: *pardella* Miller, 1907.
 COMMENTS: Given specific status by Matyushkin (1979), Werdelin (1981) and García-Perea (1992); however Weigel (1961) and Tumlison (1987) considered *pardinus* conspecific with *L. lynx*.

*Lynx rufus* (Schreber, 1777). Die Säugethiere, 3(25):pl. 109.B[1777]; text 3(24):412[1777].
> COMMON NAME: Bobcat.
> TYPE LOCALITY: "Provinz New York in Amerika".
> DISTRIBUTION: Canada (S British Columbia to Nova Scotia), Mexico (south to Oaxaca), USA.
> STATUS: CITES – Appendix II; U.S. ESA – Endangered as *L. rufus escuinapae*. IUCN – Least Concern.
> SYNONYMS: *montanus* Rafinesque, 1817; ***baileyi*** Merriam, 1890; *eremicus* Mearns, 1897; ***californicus*** Mearns, 1897; *oculeus* Bangs, 1899; ***escuinapae*** J. A. Allen, 1903; ***fasciatus*** Rafinesque, 1817; *fasciata* Elliot, 1901; ***floridanus*** Rafinesque, 1817; ***gigas*** Bangs, 1897; ***oaxacensis*** Goodwin, 1963; ***pallescens*** Merriam, 1899; *uinta* Merriam, 1902; ***peninsularis*** Thomas, 1898; ***superiorensis*** Peterson and Downing, 1952; ***texensis*** J. A. Allen, 1895; *maculata* Horsfield and Vigors, 1829 [preoccupied].
> COMMENTS: Reviewed by Larivière and Walton (1997). Mattern and McLennan (2000) demonstrated that a *rufus-lynx* clade would be paraphyletic. Synonyms allocated according to Hall (1981) and Larivière and Walton (1997).

*Pardofelis* Severtzov, 1858. Rev. Mag. Zool. Paris, ser. 2, 10:387.
> TYPE SPECIES: *Felis marmorata* Martin, 1837, by monotypy.
> COMMENTS: There is considerable controversy over the correct placement of this genus. Hemmer (1978), Král and Zima (1980), Groves (1982*a*), Kratochvíl (1982*c*), Collier and O'Brien (1985), and Bininda-Emonds et al. (1999) placed it as the first outgroup to the Pantherines. Pocock (1932*d*), Weigel (1961), Salles (1992), and Mattern and McLennan (2000) suggested a relationship with felines while recognizing similarities with *Panthera*. McKenna and Bell (1997) placed it as a subgenus of *Felis*. It perhaps should be considered *incertae sedis*.

*Pardofelis marmorata* (Martin, 1837). Proc. Zool. Soc. Lond., 1836:108 [1837].
> COMMON NAME: Marbled Cat.
> TYPE LOCALITY: "Java or Sumatra" [Indonesia], restricted by Robinson and Kloss (1919*a*:261), to "Sumatra".
> DISTRIBUTION: Bangladesh (?), Bhutan (?), Brunei Darussalam, Cambodia, China, India, Indonesia (Sumatra and Kalimantan), Laos, Malaysia, Burma, Nepal, Thailand, Vietnam.
> STATUS: CITES – Appendix I; U.S. ESA – Endangered; IUCN – Vulnerable.
> SYNONYMS: *diardii* (Jardine, 1834) [preoccupied]; *longicaudata* (de Blainville, 1843); ***charltonii*** (Gray, 1846); *dosul* Gray, 1863; *duvaucellii* (Hodgson, 1863); *ogilbii* (Hodgson, 1847).
> COMMENTS: Revised by Pocock (1932*d*). Placed in *Pardofelis* by Pocock (1932*d*), Weigel (1961), Král and Zima (1980), Kratochvíl (1982*c*), Hemmer (1978), and Groves (1982*a*). Synonyms allocated according to Ellerman and Morrison-Scott (1951).

*Prionailurus* Severtzov, 1858. Rev. Mag. Zool. Paris, ser. 2, 10:387.
> TYPE SPECIES: *Felis pardochrous* Hodgson, 1844 (= *Felis bengalensis* Kerr, 1792), by original designation.
> SYNONYMS: *Aelurina* Gill, 1871; *Ailurin* Gervais, 1855; *Ailurina* Trouessart, 1885; *Ailurogale* Fitzinger, 1869; *Ictailurus* Severtzov, 1858; *Mayailurus* Imaizumi, 1967; *Plethaelurus* Cope, 1882; *Priononfelis* Kretzoi, 1929; *Viverriceps* J. E. Gray, 1867; *Zibethailurus* Severtzov, 1858.
> COMMENTS: Weigel (1961), Hemmer (1978), Groves (1982*a*), Kratochvil (1982*c*), Collier and O'Brien (1985), and Bininda-Emonds et al. (1999), considered *bengalensis*, *planiceps*, *iriomotenisis*, *rubiginosus*, and *viverrinus* a monophyletic group. McKenna and Bell (1997) included *bengalensis*, *iriomotensis*, *rubiginosus*, and *viverrinus*. Salles (1992), Johnson and O'Brien (1997), and Mattern and McLennan (2000) separated *rubiginosus*, although there is no agreement nor consensus as to its correct placement. It is provisionally left in *Prionailurus* here, although perhaps it would be better to consider it as *incertae sedis*.

*Prionailurus bengalensis* (Kerr, 1792). *In* Linnaeus, Anim. Kingdom, 1:151.
> COMMON NAME: Leopard Cat.

TYPE LOCALITY: "Bengal" [India].

DISTRIBUTION: Afghanistan, Bangladesh, Burma, Cambodia, China, India, Indonesia, Laos, Malaysia, Nepal, North Korea, Pakistan, Philippines, Russia (Far East), Taiwan, Thailand, South Korea, and Vietnam.

STATUS: CITES – Appendix I as *P. b. bengalensis* (populations of India, Bangladesh, and Thailand); otherwise Appendix II; U.S. ESA – Endangered as *P. b. bengalensis*; IUCN – Least Concern.

SYNONYMS: *ellioti* (Gray, 1842); *herschelii* (Gray, 1869); *jerdoni* Blyth, 1863; *nipalensis* (Horsfield and Vigors, 1829); *servalinus* Gray, 1843; *tenasserimensis* (Gray, 1867); *undatus* Desmarest, 1816; *wagati* (Gray, 1867); **alleni** Sody, 1949; *hainanus* Xu and Liu, 1983; **borneoensis** Brongersma, 1936; **chinensis** (Gray, 1837); *anastasiae* (Satunin, 1905); *decoloratus* (Milne-Edwards, 1872); *ingrami* (Bonhote, 1903); *microtis* (Milne-Edwards, 1872); *minutus* (Temminck, 1824) [preoccupied]; *reevesii* (Gray, 1843); *ricketti* (Bonhote, 1903); *scriptus* (Milne-Edwards, 1870); *sinensis* (Shih, 1930); *undatus* (Radde, 1862); **euptilurus** (Elliot, 1871); *manchuricus* (Mori, 1922); *raddei* (Trouessart, 1904); **heaneyi** Groves, 1997; **horsfieldii** (Gray, 1842); *nipalensis* (Hodgson, 1832) [preoccupied]; *pardochrous* (Hodgson, 1844); **javanensis** Desmarest, 1816; *anguliferus* Fitzinger, 1868; *javensis* Elliot, 1882; **rabori** Groves, 1997; **sumatranus** Horsfield, 1821; *tingius* Lyon, 1908; **trevelyani** Pocock, 1939.

COMMENTS: Includes *euptilurus* following G. M. Allen (1939) and Gao (1987). Heptner (1971) and Gromov and Baranova (1981) considered *euptilurus* a distinct species; however, Gao (1987) pointed out that Heptner compared Russian specimens with those from SE Asia, whereas when intervening Chinese populations are included, his distinctions do not hold. Includes *minuta* following Chasen (1940). Excludes *iriomotensis* (see comments below). Synonyms allocated according to Ellerman and Morrison-Scott (1951) and Groves (1997*b*).

*Prionailurus iriomotensis* Imaizumi, 1967. J. Mamm. Soc. Japan, 3:75.

COMMON NAME: Iriomote Cat.

TYPE LOCALITY: "Haimida, Iriomote".

DISTRIBUTION: Japan (Iriomote Isl.).

STATUS: CITES – Appendix II; U.S. ESA and IUCN – Endangered as *P. bengalensis iriomotensis*.

COMMENTS: Petzsch (1970), Glass and Todd (1977), Hemmer (1978), Groves (1982*a*), Herrington (1986), and Johnson et al. (1999) argued that differences only warranted subspecific status and separation of *iriomotensis* from *bengalensis* may make some populations of *bengalensis* paraphyletic. Suzuki et al. (1994*a*), Masuda et al. (1994), and Leyhausen and Pfleiderer (1994, 1999) presented evidence that it should be considered distinct from other *bengalensis*. It should probably be best considered *incertae sedis*.

*Prionailurus planiceps* (Vigors and Horsfield, 1827). Zool. J., 3:449.

COMMON NAME: Flat-headed Cat.

TYPE LOCALITY: "Sumatra" [Indonesia].

DISTRIBUTION: Brunei Darussalam, Burma, Indonesia (Sumatra, Kalimantan), Malaysia, Singapore, Thailand.

STATUS: CITES – Appendix I; U.S. ESA – Endangered; IUCN – Vulnerable.

COMMENTS: Ellerman and Morrison-Scott (1951) listed this taxon as *incertae sedis*, but placed it with *Felis viverrina*. Grouped with *Prionailurus* by Weigel (1961), Hemmer (1978), Kratochvíl (1982*c*), Groves (1982*a*), Johnson and O'Brien (1997), Bininda-Emonds et al. (1999), and Mattern and McLennan (2000).

*Prionailurus rubiginosus* (I. Geoffroy Saint-Hilaire, 1831). *In* Bélanger (ed.), Voy. Indes Orient., Mamm., 3(Zoologie):140.

COMMON NAME: Rusty-Spotted Cat.

TYPE LOCALITY: "bois de lataniers qui couvrent une hauteur voisine de Pondichéry" [India, Pondicherry].

DISTRIBUTION: India, Sri Lanka (see Chakraborty, 1978).

STATUS: CITES – Appendix I (Indian population), otherwise Appendix II; IUCN – Vulnerable.

SYNONYMS: **phillipsi** Pocock, 1939; *koladivius* Deraniyagala, 1956.

COMMENTS: Placed in *Prionailurus* by Pocock (1917*a*), Weigel (1961), Hemmer (1978),

Kratochvíl (1982*c*), and Groves (1982*a*). Herrington (1986), Salles (1992), Johnson and O'Brien (1997), and Mattern and McLennan (2000) all considered *rubiginosus* distant from other *Prionailurus*, although there is little agreement where it should be placed. See comments under genus.

*Prionailurus viverrinus* (Bennett, 1833). Proc. Zool. Soc. Lond., 1833:68.
    COMMON NAME: Fishing Cat.
    TYPE LOCALITY: "from the continent of India".
    DISTRIBUTION: Bangladesh, Bhutan, Brunei, Cambodia, India, Indonesia (Java, Sumatra), Laos, Malaysia (Peninsular), Burma, Nepal, Pakistan, Sri Lanka, Thailand, Vietnam.
    STATUS: CITES – Appendix II; IUCN – Vulnerable.
    SYNONYMS: *bennettii* (Gray, 1867); *himalayanus* (Jardine, 1834); *rhizophoreus* Sody, 1936; *viverriceps* (Hodgson, 1836).
    COMMENTS: Grouped with *Prionailurus* by Weigel (1961), Hemmer (1978), Kratochvíl (1982*c*), Groves (1982*a*), Salles (1992), Johnson and O'Brien (1997), Bininda-Emonds et al. (1999), and Mattern and McLennan (2000). Synonyms allocated according to Ellerman and Morrison-Scott (1951).

*Profelis* Severtzov, 1858. Revue Mag. Zool. Paris, ser. 2, 10:386.
    TYPE SPECIES: *Felis celidogaster* Temminck, 1827 (= *Felis aurata* Temminck, 1827), by monotypy.
    SYNONYMS: *Chrysailurus* Severtzov, 1858.
    COMMENTS: Placed as a subgenus of *Felis* by McKenna and Bell (1997). There is considerable controversy with the relationship of *Profelis* to other cats. Some follow Pocock (1917*a*) and unite it with *Catopuma* (Bininda-Emonds et al., 1999; Collier and O'Brien, 1985; Herrington, 1986; Weigel, 1961). Others consider *Profelis* closely related to *Caracal* (Johnson and O'Brien, 1997; Mattern and McLennan, 2000).

*Profelis aurata* (Temminck, 1827). Monogr. Mamm., 1:120.
    COMMON NAME: African Golden Cat.
    TYPE LOCALITY: "Nous ne savons pas au juste dans quelle partie du globe a été trouvé"; fixed by Van Mensch and Van Bree (1969) to "probably the coastal region of Lower Guinea (Between Cross River and River Congo. . .)".
    DISTRIBUTION: Angola, Benin, Burkina Faso (?), Burundi, Cameroon, Central African Republic, Côte d'Ivoire, Dem. Rep. Congo, Equatorial Guinea, Ethiopia, Gabon, Gambia, Ghana, Guinea, Guinea-Bissau, Kenya, Liberia, Mali (?), Nigeria (?), Rwanda, Senegal (?), Sierra Leone, Togo, Uganda.
    STATUS: CITES – Appendix II; IUCN – Vulnerable.
    SYNONYMS: *celidogaster* (Temminck, 1827); *chrysothrix* (Temminck, 1827); *neglecta* (Gray, 1838); *rutila* (Waterhouse, 1843); **cottoni** (Lydekker, 1907).
    COMMENTS: Revised by Van Mensch and Van Bree (1969). Placed in *Profelis* by Pocock (1917*a*), Weigel (1961), Hemmer (1978), Kratochvíl (1982*c*), and Groves (1982*a*). Král and Zima (1980) placed it in *Felis*. Synonyms allocated according to G. M. Allen (1939).

*Puma* Jardine, 1834. Natur. Libr., pp. 266-267.
    TYPE SPECIES: *Felis concolor* Linnaeus, 1771, by original designation.
    SYNONYMS: *Herpailurus* Severtzov, 1858.
    COMMENTS: Placed as a subgenus in *Felis* by McKenna and Bell (1997) who separated *yagouaroundi* into *Felis* (*Herpailurus*). Salles (1992), Johnson and O'Brien (1997), Bininda-Emonds et al. (1999), and Mattern and McLennan (2000) considered *concolor* and *yagouaroundi* monophyletic, with *Acinonyx* as the sister group.

*Puma concolor* (Linnaeus, 1771). Mantissa Plantarum, 2:266.
COMMON NAME: Cougar.
TYPE LOCALITY: "Brassilia", restricted by Goldman (*In* Young and Goldman, 1946:200) to "Cayenne region, French Guiana".
DISTRIBUTION: Argentina, Belize, Bolivia, Brazil, Canada, Chile, Colombia, Costa Rica, Ecuador, El Salvador, French Guiana, Guatemala, Guyana, Honduras, Mexico, Nicaragua, Panama, Paraguay, Peru, Suriname, USA, Uruguay, Venezuela.

STATUS: CITES – Appendix I as *F. c. coryi, F. c. costaricensis*, and *F. c. couguar;* otherwise Appendix II; U.S. ESA – Endangered as *F. c. coryi, F. c. costaricensis*, and *F. c. couguar;* U.S. ESA – Similarity of Appearance to a Threatened Taxa (Florida); IUCN – Critically Endangered as *P. c. couguar* and *P. c. coryi*, otherwise Near Threatened.

SYNONYMS: *bangsi* (Merriam, 1901); *incarum* (Nelson and Goldman, 1929); *osgoodi* (Nelson and Goldman, 1929); *soasoaranna* (Lesson, 1842); *soderstromii* (Lönnberg, 1913); *sucuacuara* (Liais, 1872); *wavula* (Lesson, 1842); **anthonyi** (Nelson and Goldman, 1931); *acrocodia* (Goldman, 1943); *borbensis* (Nelson and Goldman, 1933); *capricornensis* (Goldman, 1946); *concolor* (Pelzeln, 1883) [preoccupied]; *greeni* (Nelson and Goldman, 1931); *nigra* Jardine, 1834 [preoccupied]; **cabrerae** Pocock, 1940; *hudsoni* (Cabrera, 1958); *puma* (Marcelli, 1922); **costaricensis** (Merriam, 1901); **couguar** (Kerr, 1792); *arundivaga* (Hollister, 1911); *aztecus* (Merriam, 1901); *browni* (Merriam, 1903); *californica* (May, 1896); *coryi* (Bangs, 1899); *floridana* (Cory, 1896); *hippolestes* (Merriam, 1897); *improcera* (Phillips, 1912); *kaibabensis* (Nelson and Goldman, 1931); *mayensis* (Nelson and Goldman, 1929); *missoulensis* (Goldman, 1943); *olympus* (Merriam, 1897); *oregonensis* (Rafinesque, 1832); *schorgeri* (Jackson, 1955); *stanleyana* (Goldman, 1938); *vancouverensis* (Nelson and Goldman, 1932); *youngi* (Goldman, 1936); **puma** (Molina, 1782); *araucanus* (Osgood, 1943); *concolor* (Gay, 1847); *patagonica* (Merriam, 1901); *pearsoni* (Thomas, 1901); *puma* (Trouessart, 1904).

COMMENTS: Reviewed by Currier (1983) as *Felis concolor*. Placed in *Puma* by Pocock (1917a), Weigel (1961), Hemmer (1978), and Kratochvíl (1982c). Synonyms allocated according to Culver et al. (2000).

*Puma yagouaroundi* (É. Geoffory Saint-Hilaire, 1803). Catal. Mam. Mus. Hist. Nat., p. 124.

    COMMON NAME: Jaguarundi.

    TYPE LOCALITY: "Paraguay", restricted by Hershkovitz, (1951) to "Cayenne, French Guiana".

    DISTRIBUTION: Argentina, Belize, Bolivia, Brazil, Colombia, Costa Rica, Ecuador, El Salvador, French Guiana, Guatemala, Guyana, Honduras, Mexico, Nicaragua, Panama, Paraguay, Peru, Suriname, USA (Arizona, Texas, Florida – introduced), Venezuela. Recently extinct in Uruguay.

    STATUS: CITES – Appendix I as *Herpailurus jagouaroundi* (North and Central American populations); otherwise Appendix II. U.S. ESA – Endangered as *H. y. cacomitli, H. y. fossata, H. y. panamensis*, and *H. y. tolteca;* IUCN – Endangered as *Herpailurus yagouaroundi cacomitli*, otherwise Least Concern as *H. yagouaroundi*.

    SYNONYMS: *jaguarondi* (G. Fischer, 1814); *jaguarondi* (Sanderson, 1949); *unicolor* (Traill, 1819); *yaguarondi* (Lacépède, 1809); *yaguarundi* (Goeldi and Hagmann, 1904); **ameghinoi** (Holmberg, 1898); *yaguarondi* (Thomas, 1920); **cacomitli** (Berlandier, 1859); *apache* (Mearns, 1901); **eyra** (G. Fischer, 1814); *eira* (Desmarest, 1816); *darwini* (Martin, 1837); *yaguarundi* (Lahille, 1899); **fossata** (Mearns 1901); **melantho** (Thomas, 1914); **panamensis** (J. A. Allen, 1904); *eyra* (Alfaro, 1897); **tolteca** (Thomas, 1898).

    COMMENTS: Others have used *yaguaroundi* Lacépède, 1809, or *yagouaroundi* Desmarest, 1816, however, the former is invalid and the latter is a junior synonym. Placed in *Herpailurus* by Weigel (1961), Hemmer (1978), and Kratochvíl (1982c). See comments under genus for its inclusion here. Reviewed by Oliveira (1998b). Synonyms allocated according to Cabrera (1957) and Oliveira (1998b).

**Subfamily Pantherinae** Pocock, 1917. Ann. Mag. Nat. Hist., ser. 8, 20:332.

    SYNONYMS: Neofelinae Kretzoi, 1929; Neofelina Kalandadze and Rautian, 1992; Pantherini Kalandadze, 1992.

    COMMENTS: Pocock's (1917a) original classification for this subfamily placed *Neofelis* in the Felinae. Most recent studies have considered these taxa as a monophyletic group (Bininda-Emonds et al., 1999; Hemmer, 1978; Herrington, 1986; Johnson and O'Brien, 1997; Mattern and McLennan, 2000; Salles, 1992; Weigel, 1961). Janczewski et al. (1995) argued that separation of *Neofelis* and *Uncia* made *Panthera* paraphyletic.

*Neofelis* Gray, 1867. Proc. Zool. Soc. Lond., 1867:265.

    TYPE SPECIES: *Felis macrocelis* Horsfield, 1825 (= *Felis nebulosa* Griffith, 1821), by subsequent designation by Pocock (1917a).

COMMENTS: Placed in *Pantherinae* by Hemmer (1978) and Weigel (1961). Placed in *Neofelinae* by Kratochvíl (1982c). Pocock (1917a), Weigel (1961), Collier and O'Brien (1985), Bininda-Emonds et al. (1999), and Mattern and McLennan (2000) considered *Neofelis* either as the most primitive member of the Pantherinae, or as the first outgroup. Placed in Felinae by McKenna and Bell (1997). Considered a synonym of *Pardofelis* by Corbet and Hill (1992).

*Neofelis nebulosa* (Griffith, 1821). Gen. Particular Descrip. Vert. Anim. (Carn.), p. 37, pl.
COMMON NAME: Clouded Leopard.
TYPE LOCALITY: "brought from Canton" [China, Guangdong: Guangzhou].
DISTRIBUTION: Bangladesh, Bhutan, Brunei, Cambodia, China, India, Indonesia (Sumatra, Kalimantan), Laos, Malaysia, Burma, Nepal, Taiwan, Thailand, Vietnam.
STATUS: CITES – Appendix I; U.S. ESA – Endangered; IUCN – Vulnerable.
SYNONYMS: *melli* (Matschie, *in* Mell, 1922); **brachyura** (Swinhoe, 1862); **diardi** (G. Cuvier, 1823); **macrosceloides** (Hodgson, 1853); *macrocelis* (Tickell, 1843) [preoccupied].
COMMENTS: Placed in *Neofelis* by Pocock (1917a), Weigel (1961), Hemmer (1978), and Kratochvíl (1982c). Groves (1982a) placed in *Panthera*. Synonyms allocated according to Ellerman and Morrison-Scott (1951).

*Panthera* Oken, 1816. Lehrb. Naturgesch, 3, 2:1052.
TYPE SPECIES: *Felis pardus* Linnaeus, 1758, by subsequent designation by J. A. Allen (1902).
SYNONYMS: *Jaguarius* Severtzov, 1858; *Leo* Frisch, 1775; *Leonina* Greve, 1894; *Leoninae* Wagner, 1841; *Pardotigris* Kretzoi, 1929, *Pardus* Fitzinger, 1868; *Tigrina* Greve, 1894; *Tigrinae* Wagner, 1841; *Tigris* Gray, 1843; *Tigris* Frisch, 1775.
COMMENTS: Synonyms allocated according to McKenna and Bell (1997). Revised by Hemmer (1966, 1968, 1974). *Panthera* Oken, 1816, has been ruled available (International Commission on Zoological Nomenclature, 1985c). Includes *Tigris* following Pocock (1916b, 1929). Van Gelder (1977b) included *Panthera* as a synonym of *Felis*.

*Panthera leo* (Linnaeus, 1758). Syst. Nat., 10th ed., 1:41.
COMMON NAME: Lion.
TYPE LOCALITY: "Africa", restricted by J. A. Allen (1924:222) to "the Barbary coast region of Africa, or, more explicitly, Constantine, Algeria".
DISTRIBUTION: Angola, Benin, Botswana, Burkina Faso, Burundi, Cameroon, Central African Republic, Chad, Côte d'Ivoire, Eritrea, Ethiopia, Gabon, Ghana, Guinea, Guinea-Bissau (?), India, Kenya, Lesotho, Malawi, Mali, Mozambique, Namibia, Niger, Nigeria, Senegal, Sierra Leone, Somalia, South Africa, Sudan, Swaziland, Tanzania, Togo, Uganda, Zambia, Zimbabwe. Recently extinct: Afghanistan, Algeria, Egypt, Gambia, Iraq, Israel, Jordan, Lebanon, Libya, Kuwait, Mauritania, Morocco, Pakistan, Republic of Congo, Rwanda, Saudi Arabia, Syrian Arab Republic, Turkey, Tunisia, Western Sahara.
STATUS: CITES – Appendix I as *P. l. persica*; otherwise Appendix II; U.S. ESA – Endangered as *P. l. persica*; IUCN – Critically Endangered as *P. l. persica*, otherwise Vulnerable.
SYNONYMS: *africana* (Brehm, 1829); *barbara* (J. B. Fisher, 1829); *barbarica* (Meyer, 1826); *nigra* (Loche, 1867); *nobilis* (Gray, 1867); *nubica* (de Blainville, 1843); *somaliensis* (Noack, 1891); **azandica** (J. A. Allen, 1924); **bleyenberghi** (Lönnberg, 1914); **hollisteri** (J. A. Allen, 1924); **kamptzi** (Matschie, 1900); **krugeri** (Roberts, 1929); **massaica** (Neumann, 1900); *roosevelti* (Heller, 1913); *sabakiensis* (Lönnberg, 1908); **melanochaita** (C. E. H. Smith, 1858); *capensis* (J. B. Fischer, 1830) [preoccupied]; **nyanzae** (Heller, 1913); **persica** (Meyer, 1826); *asiaticus* (Brehm, 1829); *bengalensis* (Bennett, 1829); *goojratensis* (Smee, 1833); *indica* (de Blainville, 1843); **senegalensis** (J. N. von Meyer, 1826); *gambiana* (Gray, 1843).
COMMENTS: Synonyms allocated according to Ellerman and Morrison-Scott (1951); G. M. Allen (1939). Revised by Pocock (1930c). Placed in *Panthera* by Pocock (1930c), Weigel (1961), Hemmer (1978), Kratochvíl (1982c), and Groves (1982a).

*Panthera onca* (Linnaeus, 1758). Syst. Nat., 10th ed., 1:42.
COMMON NAME: Jaguar.
TYPE LOCALITY: "America meridionali", fixed by Thomas (1911a:136) as "Pernambuco" [Brazil].

DISTRIBUTION: Argentina, Belize, Bolivia, Brazil, Colombia, Costa Rica, Ecuador, French Guiana, Guatemala, Guyana, Honduras, Mexico, Nicaragua, Panama, Paraguay, Peru, Suriname, Venezuela. Recently extinct in United States, El Salvador, and Uruguay.

STATUS: CITES – Appendix I; U. S. ESA – Endangered; IUCN – Near Threatened.

SYNONYMS: *boliviensis* (Nelson and Goldman, 1933); *coxi* (Nelson and Goldman, 1933); *jaguapara* (Liais, 1872); *jaguar* (Link, 1795); *jaguarete* (Liais, 1872); *jaguatyrica* (Liais, 1872); *madeirae* (Nelson and Goldman, 1933); *major* (J. B. Fischer, 1830); *mexianae* (Hagmann, 1908); *minor* (J. B. Fisher, 1830); *nigra* (Wagner, 1841); *onza* (Brehm, 1876); *ucayalae* (Nelson and Goldman, 1933); **arizonensis** (Goldman, 1932); **centralis** (Mearns, 1901); *onca* (Alfaro, 1897) [preoccupied]; **goldmani** (Mearns, 1901); *gikdnabu* (Goldman, 1932); **hernandesii** (J. E. Gray, 1857); **palustris** (Ameghino, 1888); *antiqua* (Ameghino, 1889); *fossilis* (Ameghino, 1889); *onssa* (Ihering, 1911); *proplatensis* (Ameghino, 1904); **paraguensis** (Hollister, 1914); *milleri* (Nelson and Goldman, 1933); *notialis* (Hollister, 1914); *paulensis* (Nelson and Goldman, 1933); *ramsayi* (Miller, 1930); **peruviana** (de Blainville, 1843); *onza* (Tschudi, 1844); *peruviana* (Hoffstetter, 1952); **veraecruscis** (Nelson and Goldman, 1933).

COMMENTS: Synonyms allocated according to Hall (1981), Seymour (1989) and Cabrera (1957). Revised by Nelson and Goldman (1933*a*), Pocock (1939*b*), and Larson (1997). Placed in *Panthera* by Pocock (1939*b*), Weigel (1961), Hemmer (1978), Kratochvíl (1982*c*), and Groves (1982*a*). Reviewed by Seymour (1989). A multivariate analysis of skull morphology could not discriminate among subspecies (Larson, 1997).

*Panthera pardus* (Linnaeus, 1758). Syst. Nat., 10th ed., 1:41.

COMMON NAME: Leopard.

TYPE LOCALITY: "Indiis", fixed by Thomas (1911*a*:135), as "Egypt"; see discussion by Pocock (1930).

DISTRIBUTION: Afghanistan, Algeria, Angola, Arabia, Armenia, Botswana, Burma, Cameroon, Central African Republic, Chad, China, Dem. Rep. Congo, Egypt, Ethiopia, Gabon, Guinea-Bissau, India, Indonesia (Java), Iran, Iraq, Kenya, Liberia, Laos, Malawi, Malaysia, Mauritania, Morocco, Mozambique, Namibia, Nepal, Niger, Nigeria, North and South Korea, Pakistan, Republic of Congo, Russia, Saudia Arabia, Senegal, Sierra Leone, Somalia, South Africa, Sri Lanka, Sudan, Tanzania, Thailand, Tunisia, Turkey, Turkmenistan, Uganda, Vietnam, Zambia, Zimbabwe.

STATUS: CITES – Appendix I; U.S. ESA – Endangered (except in Africa, in the wild, south of, and including Gabon, Republic of Congo, Dem. Rep. Congo, Uganda, and Kenya, where this species is Threatened). IUCN – Critically Endangered as *P. p. nimr*, *P. p. orientalis*, *P. p. panthera*, and *P. p. tulliana*, Endangered as *P. p. japonensis*, *P. p. kotiya*, *P. p. melas*, and *P. p. saxicolor*, otherwise Least Concern.

SYNONYMS: *adersi* Pocock, 1932; *adusta* Pocock, 1927; *antinorii* (de Beaux, 1924); *barbara* (de Blainville, 1843); *brockmani* Pocock, 1932; *centralis* (Lönnberg, 1917); *chui* (Heller, 1913); *fortis* (Heller, 1913); *leoparda* (Schreber, 1775); *melanosticta* (Lydekker, 1908); *melanotica* (Gunther, 1885); *minor* (Matschie, 1895); *nanoparda* (Thomas, 1904); *palearia* (F. G. Cuvier, 1832); *panthera* (Schreber, 1777); *poecilura* (Valenciennes, 1856); *puella* (Pocock, 1932); *reichenowi* Cabrera, 1918; *ruwenzorii* (Camerano, 1906); *shortridgei* Pocock, 1932; *suahelicus* (Neumann, 1900); *varia* (J. E. Gray, 1843); *vulgaris* (Oken, 1816); **delacouri** Pocock, 1930; *variegata* (Lydekker, 1914) [preoccupied]; **fusca** (Meyer, 1794); *antiquorum* Fitzinger, 1868; *centralis* (Lönnberg, 1917); *chinenesis* (Brass, 1904); *iturensis* J. A. Allen, 1924; *longicaudata* (Valenciennes, 1856); *melas* (Pousargues, 1896); *millardi* Pocock, 1930; *pernigra* (J. E. Gray, 1863); *variegata* (G. M. Allen, 1912); **japonensis** (J. E. Gray, 1862); *bedfordi* Pocock, 1930; *chinensis* (J. E. Gray, 1867); *fontanierii* (Milne-Edwards, 1867); *grayi* (Trouessart, 1904); *hanensis* Matschie, 1907; **kotiya** Deraniyagala, 1956; **melas** G. Cuvier, 1809; *variegata* (Wagner, 1841) [preoccupied]; **nimr** (Hemprich and Ehrenberg, 1833); *ciscaucasica* (Satunin, 1914); *dathei* Zukowsky, 1964; *jarvisi* Pocock, 1932; *leoparda* (Sclater, 1878); *saxicolor* Pocock, 1927; *sindica* Pocock, 1930; *tulliana* (Valenciennes, 1856); **orientalis** (Schlegel, 1857); *villosa* (Bonhote, 1903).

COMMENTS: Placed in *Panthera* by Pocock (1930*a*, *b*), Weigel (1961), Hemmer (1978),

Kratochvíl (1982c), and Groves (1982a). Revised by Pocock (1930a, b, 1932c) and Miththapala et al. (1996). Synonyms allocated according to Miththapala et al. (1996).

*Panthera tigris* (Linnaeus, 1758). Syst. Nat., 10th ed., 1:41.
COMMON NAME: Tiger.
TYPE LOCALITY: "Asia", fixed by Thomas (1911a:135) as "Bengal" [India].
DISTRIBUTION: Bangladesh, Bhutan, Cambodia, China, India, Indonesia (Sumatra only), Laos, Malaysia, Burma, Nepal, Russia, Thailand, Vietnam. Recently extinct in: Afghanistan, Georgia, Bali, Java, Iran, Iraq, Kazakhstan, North Korea, Kyrgyzstan, Pakistan, Singapore, Tajikistan, Turkmenistan, Uzbekistan.
STATUS: CITES – Appendix I; U.S. ESA – Endangered; IUCN – Extict as *P. t. balica*, *P. t. sondaica*, and *P. t. virgata*, Critically Endangered as *P. t. altaica*, *P. t. amoyensis*, and *P. t. sumatrae*, otherwise Endangered.
SYNONYMS: *fluviatilis* (Sterndale, 1884); *montana* (Sterndale, 1884); *regalis* (J. E. Gray, 1842); *striata* (Severtzov, 1858); **altaica** Temminck, 1844; *amurensis* (Dode, 1871); *coreensis* (Brass, 1904); *longipilis* (Fitzinger, 1868); *mandshurica* (Baykov, 1925); *mikadoi* (Satunin, 1915); **amoyensis** (Hilzheimer, 1905); *styani* (Pocock, 1929); **balica** Schwarz, 1912; **corbetti** Mazak, 1968; **sondaica** Temminck, 1844; **sumatrae** Pocock, 1929; **virgata** (Illiger, 1815); *septentrionalis* (Satunin, 1904); *trabata* (Schwarz, 1916).
COMMENTS: Synonyms allocated according to Mazák (1981). Revised by Pocock (1929) and Mazák (1979, 1981). Placed in *Panthera* by Pocock (1929), Weigel (1961), Hemmer (1978), Kratochvíl (1982c), and Groves (1982a).

*Uncia* Gray, 1854. Ann. Mag. Nat. Hist., ser. 2, 14:394.
TYPE SPECIES: *Felis irbis* Ehrenberg, 1830 (= *Felis uncia* Schreber, 1775), by subsequent designation (Palmer, 1904).
COMMENTS: Revised by Pocock (1916b). N. Yu et al. (1996) considered *Uncia* congeneric with *Panthera*.

*Uncia uncia* (Schreber, 1775). Die Säugethiere, 2(14):pl. 100[1775]; text, 3(22):386-7[1777].
COMMON NAME: Snow leopard.
TYPE LOCALITY: "Barbarey, Persien, Ostindien, und China", restricted by Pocock (1930c:332) to "Altai Mountains".
DISTRIBUTION: Afghanistan, Bhutan, China, India, Kazakhstan, Kyrgyzstan, Mongolia, Nepal, Pakistan, Russia, Tajikistan, Uzbekistan.
STATUS: CITES – Appendix I; U.S. ESA – Endangered; IUCN – Endangered.
SYNONYMS: *baikalensis-romanii* Medvedev, 2000; *irbis* (Ehrenberg, 1830); *schneideri* (Zukowsky, 1950); *uncioides* (Horsfield, 1855).
COMMENTS: Revised by Pocock (1930a, b). Placed in *Uncia* by Pocock (1930a, b), Weigel (1961), Kratochvíl (1982c), and Heptner et al. (1967a). Placed in *Uncia* and reviewed by Hemmer (1972). Status of proposed subspecies *baikalensisromanii* needs evaluation, see Medvedev (2000).

**Family Viverridae** Gray, 1821. London Med. Repos., 1821:301.
COMMENTS: Does not include (1) Herpestinae Bonaparte, 1845 or Galidiinae Gray, 1865 (Flynn et al., 1988; Gregory and Hellman, 1939; Hunt, 1987; Pocock, 1916c, 1919; Radinsky, 1975; Thenius, 1972; Wozencraft, 1989a, b; Wurster and Benirschke, 1968); (2) *Nandinia* (Flynn and Nedbal, 1998; Hunt, 2001; Veron and Heard, 2000; Yoder et al., 2003); and (3) *Cryptoprocta*, *Eupleres*, and *Fossa* (Veron, 1995; Veron and Catzeflis, 1993; Veron and Heard, 2000; Yoder et al., 2003). The viverrids are one of the most problematic families of carnivores. Hunt (2001) placed the members of this family into six subfamilies: Prionodontinae (incl. *Prionodon*, *Poiana*, and *Genetta*); Viverrinae (incl. *Viverra*, *Viverricula*, *Osbornictis*, and *Civettictis*); Euplerinae (incl. *Fossa* and *Eupleres*); Cryptoproctinae (*Cryptoprocta*); Hemigalinae (incl. *Hemigalus*, *Diplogale*, *Chrotogale*, and *Cynogale*); and Paradoxurinae (incl. *Paradoxurus*, *Paguma*, *Arctictis*, *Arctogalidia*, and *Macrogalidia*). Hunt's (2001) study gave a morphological basis for the separation of the Prionodontinae from the Viverrinae. However, Veron and Heard (2000) raised serious doubts about the monophyly of these subfamilies. They find that neither Hunt's Prionodontinae, nor the

more traditional system of Prionodontinae+Viverrinae would represent monophyletic groups. Gaubert et al. (2004) excluded *Prionodon* from other Viverrinae based on cytochrome *b* sequences (followed here). The position of the Malagasy carnivores was not addressed by Hunt's (2001) study. However, Veron and Catzeflis (1993), Veron (1995), Veron and Heard (2000), and Yoder et al. (2003) have consistently shown that these taxa do not belong within the Viverridae and probably represent a single monophyletic origin for the Malagasy Carnivora.

**Subfamily Paradoxurinae** Gray, 1865. Proc. Zool. Soc. Lond., 1864:508 [1865].
SYNONYMS: Arctictidina Gray, 1864; Arctictidae Cope, 1882; Arctogalidiinae Pocock, 1933; Arctogalidiini Simpson, 1945; Paradoxurida Gregory and Hellman, 1939; Paradoxurini Simpson, 1945.

*Arctictis* Temminck, 1824. Prospectus de Monographies de Mammifères, p. xxi. [issued March, 1824].
TYPE SPECIES: *Viverra binturong* Raffles, 1821, by monotypy (Melville and Smith, 1987).
SYNONYMS: *Ictides* Valenciennes, 1825.
COMMENTS: First placed in Paradoxurinae by Gray (1869).

*Arctictis binturong* (Raffles, 1821). Trans. Linn. Soc. Lond., 13:253.
COMMON NAME: Binturong.
TYPE LOCALITY: "Malacca".
DISTRIBUTION: Bangladesh, Bhutan, Burma, China (Yunnan), India (incl. Sikkim), Indonesia (Kalimantan, Java, Sumatra), Laos, Malaysia, Nepal, Philippine Isls (Palawan), Thailand, Vietnam.
STATUS: CITES – Appendix III (India); IUCN – Vulnerable as *A. b. whitei*, otherwise Lower Risk (lc).
SYNONYMS: *ater* (F. Cuvier and É. Geoffroy Saint-Hilaire, 1824); *gairdneri* Thomas, 1916; **albifrons** (F. G. Cuvier, 1822); *niasensis* Lyon, 1916; **kerkhoveni** Sody, 1936; **menglaensis** Wang and Li, 1987; **penicillatus** Temminck, 1835; *pageli* Schwarz, 1911; **whitei** J. A. Allen, 1910.
COMMENTS: Revised by Pocock (1933a). Synonyms allocated according to Pocock (1933a, 1941a), Ellerman and Morrison-Scott (1951), and D. D. Davis (1962).

*Arctogalidia* Merriam, 1897. Science, 5:302.
TYPE SPECIES: *Paradoxurus trivirgatus* Gray, 1832, by monotypy through the replaced name *Arctogale* Gray, 1865 (Melville and Smith, 1987).
SYNONYMS: *Arctogale* Gray, 1865.
COMMENTS: Gray's (1864[1865]) generic name stood until Merriam (1897) pointed out that the name *Arctogale* was preoccupied (= *Arctogale erminea* Kaup, 1829).

*Arctogalidia trivirgata* (Gray, 1832). Proc. Zool. Soc. Lond., 1832:68.
COMMON NAME: Small-toothed Palm Civet.
TYPE LOCALITY: "from a specimen in the Leyden Museum, sent from the Molúccas", restricted by Jentink (1887) to "Java, Buitenzorg" [= Indonesia, Java, Bogor] (see comments).
DISTRIBUTION: Bangladesh, Burma, China (Yunnan), India, Indonesia, Laos, Malaysia, Thailand, and Vietnam.
STATUS: IUCN – Endangered as *A. t. trilineata*, otherwise Lower Risk (lc).
SYNONYMS: **bancana** Schwarz, 1913; **fusca** Miller, 1906; *depressa* Miller, 1913; **inornata** Miller, 1901; **leucotis** Horsfield, 1851; *prehensilis* Sclater, 1877; **macra** Miller, 1913; **major** Miller, 1906; **millsi** Wroughton, 1921; **minor** Lyon, 1906; **simplex** Miller, 1902; *mima* Miller, 1913; **stigmaticus** (Temminck, 1853); *bicolor* Miller, 1913; **sumatrana** Lyon, 1908; *tingia* Lyon, 1908; **trilineata** Wagner, 1841.
COMMENTS: Revised by Pocock (1933a) and Van Bemmel (1952). Gray (1832) originally described the type from the "Moluccas"; later Temminck (1841) referred to the same specimen as being from "Java". Gray (1843), then corrected the presumed geographic error and listed the same type as from "Malacca". Jentink (1887) listed the same type from "Buitenzorg". However, Van Bemmel (1952) stated that the collector, Reinwardt,

was in the eastern part of the Indo-Australian Archipelago in 1821 and the type did not match other specimens from Java. Synonyms allocated according to Pocock (1933a) except that Pocock placed *leucotis*, *millsi* and *macra* in a separate species (=*leucotis*). Corbet and Hill (1992) proposed three subspecies: Mainland north of the Isthmus of Kra (*leucotis*); Malaya, Sumatra, and Borneo (*trivirgata*); and Java (*trilineata*).

*Macrogalidia* Schwarz, 1910. Ann. Mag. Nat. Hist., ser. 5, 8:423.
 TYPE SPECIES: *Paradoxurus musschenbroekii* Schlegel, 1877, by monotypy.
 COMMENTS: First placed in the Paradoxurinae by Pocock (1933a).

*Macrogalidia musschenbroekii* (Schlegel, 1877). Prosp. Mus. Publ., 1877: [unnumbered].
 COMMON NAME: Sulawesi Palm Civet.
 TYPE LOCALITY: "in the Northern parts of the isle of Celebes", restriced by Jentink (1887), to "Celebes, Menado-Kinilo" [Indonesia, Kinilou, 1°22'N, 124°51'E].
 DISTRIBUTION: Indonesia (Sulawesi).
 STATUS: IUCN – Vulnerable.
 COMMENTS: Schlegel (1877) circulated an unnumbered "Prospectus," for the "Annals of the Royal Zoological Museum of the Netherlands at Leyden", which contained the first mention of the new species. The prospectus was to preceed the first issue of the new "Annals" which was never printed. He republished the type description in the Notes of the Leyden Museum (1879).

*Paguma* Gray, 1831. Proc. Comm. Sci. Corres. Zool. Soc. London, 1831:94.
 TYPE SPECIES: *Gulo larvatus* C. E. H. Smith, 1827.
 SYNONYMS: *Ambliodon* Jourdan, 1837.
 COMMENTS: Reviewed by Pocock (1933c).

*Paguma larvata* (C. E. H. Smith, 1827). *In* Griffith et al., Anim. Kingdom, 2:281.
 COMMON NAME: Masked Palm Civet.
 TYPE LOCALITY: Not given. Fixed by Temminck (1841) as "Nepal". Gray (1864) discounted this because he knew of no specimens from Nepal, and reassigned the name to two specimens from Canton, China collected by J. R. Reeve (Pocock, 1934b).
 DISTRIBUTION: Bangladesh, Burma, Cambodia, China (Hainan north to Hopei, Shanxi and the vicinity of Beijing), India (and S Andaman Isls), Indonesia (Kalimantan, Sumatra), Japan (introduced), Laos, Malaysia (Sabah, Sarawak, West), Nepal, Pakistan, Singapore, Taiwan, Thailand, Vietnam.
 STATUS: CITES – Appendix III (India); IUCN – Lower Risk (lc).
 SYNONYMS: *reevesi* Matschie, 1907; *rivalis* Thomas, 1921; **chichingensis** Wang, 1981; **grayi** (Bennett, 1835); *nipalensis* (Hodgson, 1836); **hainana** Thomas, 1909; **intrudens** Wroughton, 1910; *vagans* Kloss, 1919; *yunalis* Thomas, 1921; **janetta** Thomas, 1928; **jourdanii** (J. E. Gray, 1837); *aurata* (de Blainville, 1842); *annectens* Robinson and Kloss, 1917; *dore* (Jourdan, 1837); **lanigera** (Hodgson, 1836); *grayi* Wroughton, 1918; *laniger* (Hodgson, 1841); **leucomystax** (J. E. Gray, 1837); **neglecta** Pocock, 1934; **nigriceps** Pocock, 1939; **ogilbyi** (Fraser, 1846); *leucocephala* J. E. Gray, 1864; *rubidus* (Blyth, 1858); **robusta** (Miller, 1906); **taivana** Swinhoe, 1862; **tytlerii** (Tytler, 1864); **wroughtoni** Schwarz, 1913.
 COMMENTS: Pocock (1934b) included *Paradoxurus tytlerii*. *P. lanigera*, the "imperfect, no doubt immature skin, without skull (B.M. no. 43.1.12.103)" provisionally recognized as separate by Pocock (1941a:416) does not contain diagnostic features that would definitively align the specimen with *Paguma* (Ellerman and Morrison-Scott, 1951). Synonyms allocated according to Pocock (1934b) and Ellerman and Morrison-Scott (1951).

*Paradoxurus* F. Cuvier, 1821. *In* É. Geoffroy Saint-Hilaire and F. Cuvier, Hist. Nat. Mammifères, pt. 2, 3(24): "Martre des Palmiers", 5 pp., 1 pl.
 TYPE SPECIES: *Paradoxurus typus* F. Cuvier, 1821 (= *Viverra hermaphrodita* Pallas, 1777), by indication (Melville and Smith, 1987).
 SYNONYMS: *Bondar* Gray, 1865; *Macrodus* Gray, 1865; *Platyschista* Otto, 1835.

COMMENTS: Revised by Pocock (1933c, 1934a).

*Paradoxurus hermaphroditus* (Pallas, 1777). *In* Schreber, Die Säugethiere, 3(25):426, [1777].
COMMON NAME: Asian Palm Civet.
TYPE LOCALITY: Uncertain. "Das Vaterland des beschreibenen Thieres ist die Barbarey".
Listed as "India?" by Corbet and Hill (1992).
DISTRIBUTION: Bhutan, Burma, Cambodia, China, India, Indonesia, Laos, Malaysia, Nepal,
New Guinea, Philippine Isls, Singapore, Sri Lanka, Thailand, Vietnam; scattered
records in Sulawesi, Moluccas, and Aru Isls, probably resulting from introductions.
STATUS: CITES – Appendix III (India); IUCN – Vulnerable as *P. h. lignicolor*, otherwise Lower
Risk (lc).
SYNONYMS: *felinus* Wagner, 1841; *fuliginosus* Gray, 1832; *laneus* Pocock, 1934; *niger* Blanford,
1885; *nigra* (Desmarest, 1820); *pallasii* (Otto, 1835); *typus* F. G. Cuvier and É. Geoffroy,
1821; ***balicus*** Sody, 1933; ***bondar*** (Desmarest, 1820); *crossi* Gray, 1832; *hirsutus*
Hodgson, 1836; *pennantii* Gray, 1832; *strictus* Horsfield (Hodgson, 1855 MS.);
***canescens*** Lyon, 1907; ***canus*** Miller, 1913; ***cochinensis*** Schwarz, 1911; *kutensis* Chasen
and Kloss, 1916; ***dongfangensis*** Corbet and Hill, 1992; *hainanus* Wang and Xu, 1981
[preoccupied]; ***enganus*** Lyon, 1916; ***exitus*** Schwarz, 1911; ***javanica*** Horsfield, 1824;
*dubius* Gray, 1832; *macrodus* Gray, 1864; ***kangeanus*** Thomas, 1910; ***laotum***
Gyldenstolpe, 1917; *birmanicus* Wroughton, 1917; ***lignicolor*** Miller, 1903; *siberu*
Chasen and Kloss, 1928; ***milleri*** Kloss, 1908; *fuscus* Miller, 1913; ***minor*** Bonhote, 1903;
*ravus* Miller, 1913; ***musanga*** (Raffles, 1821); *brunneipes* Miller, 1906; *cantori* Pocock,
1934; *fossa* (Marsden, 1811) [preoccupied]; *musangoides* Gray, 1837; *padangus* Lyon,
1908; *sumatrensis* Fischer, 1829; ***nictitans*** Taylor, 1891; ***pallasii*** Gray, 1832; *nigrifons*
Gray, 1864; *prehensilis* Desmarest, 1820 [preoccupied]; *quadriscriptus* Horsfield
(Hodgson, 1855 MS.); *strictus* Wroughton, 1917; *vicinus* Schwarz, 1910; ***pallens*** Miller,
1913; ***parvus*** Miller, 1913; *enganus* Lyon, 1916; ***philippinensis*** Jourdan, 1837;
*baritensis* Lönnberg, 1925; *minax* Thomas, 1909; *sabanus* Thomas, 1909; *torvus*
Thomas, 1909; ***pugnax*** Miller, 1913; ***pulcher*** Miller, 1913; ***sacer*** Miller, 1913; ***scindiae***
Pocock, 1934; ***senex*** Miller, 1913; ***setosus*** Jacquinot and Pucheran, 1853; *celebensis*
Schwarz, 1911; ***simplex*** Miller, 1913; ***sumbanus*** Schwarz, 1910; ***vellerosus*** Pocock,
1934.
COMMENTS: Two primary systematic studies – Chasen and Kloss (1927) and Pocock (1934a)
– considered *lignicolor* Miller, 1903 (an insular population) as conspecific. However,
Pocock did not directly study Miller's (1903b) type specimen. Corbet and Hill's (1992)
review separated *lignicolor* from other *hermaphroditus* and perhaps it should best be
listed as *incertae sedis*. They point out the pelage differences (a paler tail, and a more
uniform body color). Synonyms allocated according to Pocock (1933c, 1934a) and
Ellerman and Morrison-Scott (1951).

*Paradoxurus jerdoni* Blanford, 1885. Proc. Zool. Soc. Lond., 1885:613, 802.
COMMON NAME: Jerdon's Palm Civet.
TYPE LOCALITY: "Kodaikanal, on the Palni (or Pulney) hills in the Madura district, Madras
Presidency" [India, Tamil Nadu Province, Palni Hills, Kodaikanal; 10°15'N, 77°31'E].
DISTRIBUTION: S India.
STATUS: CITES – Appendix III (India); IUCN – Vulnerable.
SYNONYMS: ***caniscus*** Pocock, 1933.
COMMENTS: Synonyms allocated according to Pocock (1933c) and Ellerman and Morrison-
Scott (1951).

*Paradoxurus zeylonensis* Schreber, 1778. Die Säugethiere, 3(26):451.
COMMON NAME: Golden Palm Civet.
TYPE LOCALITY: "Ceylon" [= Sri Lanka].
DISTRIBUTION: Endemic to Sri Lanka.
STATUS: IUCN – Lower Risk (lc).
SYNONYMS: *aureus* F. G. Cuvier, 1822; *fuscus* Kelaart, 1852; *montanus* Kelaart, 1852; *zeylanica*
(Gmelin, 1788).

**Subfamily Hemigalinae** Gray, 1865. Proc. Zool. Soc. Lond., 1864:508 [1865].
> SYNONYMS: Cynogalina Gray, 1865; Cynogalidae Gray, 1869; Cynogalini Simpson, 1945;
> Hemigalida Gregory and Hellman, 1939; Hemigalini Simpson, 1945.
> COMMENTS: Simpson (1945) also included *Eupleres* and *Fossa*, following Pocock (1915*b*). Pocock
> (1933*d*) and Gregory and Hellman (1939) placed *Cynogale* in the monotypic
> Cynogalinae, although both recognized the close relationship of *Cynogale* to other
> hemigalines. Placed in Hemigalinae by Simpson (1945) and Ellerman and Morrison-Scott
> (1951).

*Chrotogale* Thomas, 1912. Abstr. Proc. Zool. Soc. Lond., 1912(106):17.
> TYPE SPECIES: *Chrotogale owstoni* Thomas, 1912, by monotypy.
> COMMENTS: Corbet and Hill (1992) suggested that *Chrotogale* and *Hemigalus* are congeneric.
> This has not been supported by morphological or molecular studies (Veron and Heard,
> 2000).

*Chrotogale owstoni* Thomas, 1912. Abstr. Proc. Zool. Soc. Lond., 1912(106):17.
> COMMON NAME: Owston's Palm Civet.
> TYPE LOCALITY: "Yen-bay, on the Song-koi River, Tonkin" [Vietnam: Yen Bay on the Songhoi
> River; 21°43′N 104°54′E].
> DISTRIBUTION: China (Yunnan, Guangxi), Laos, Vietnam.
> STATUS: IUCN – Vulnerable.
> COMMENTS: Revised by Pocock (1933*d*).

*Cynogale* Gray, 1837. Proc. Zool. Soc. Lond., 1836:88 [1837].
> TYPE SPECIES: *Cynogale bennettii* Gray, 1837, by monotypy (Melville and Smith, 1987).
> SYNONYMS: *Lamictis* de Blainville, 1837; *Potamophilus* Müller, 1838.

*Cynogale bennettii* Gray, 1837. Proc. Zool. Soc. Lond., 1836:88 [1837].
> COMMON NAME: Otter Civet.
> TYPE LOCALITY: "Sumatra" [Indonesia].
> DISTRIBUTION: Brunei, Indonesia (Kalimantan, Sumatra), Malaysia, Thailand, Vietnam.
> STATUS: CITES – Appendix II; IUCN – Endangered; *Cynogale* populations from Vietnam
> (*C. b. lowei*) considered by IUCN to have high conservation priority status for viverrids
> (Schreiber et al., 1989).
> SYNONYMS: *barbatus* Müller, 1838; *carcharias* de Blainville, 1837; *lowei* Pocock, 1933.
> COMMENTS: Revised by Pocock (1933*d*). Includes *C. lowei*, which is known only from the
> type, a poorly preserved juvenile skin from N Vietnam (Ellerman and Morrison-Scott,
> 1951). Schreiber et al. (1989) and Corbet and Hill (1992) recognized *lowei* as a separate
> species, however primary systematic studies are lacking. Synonyms allocated
> according to Pocock (1933*d*).

*Diplogale* Thomas, 1912. Abstr. Proc. Zool. Soc. Lond., 1912(106):18.
> TYPE SPECIES: *Hemigale hosei* Thomas, 1892, by original designation.

*Diplogale hosei* (Thomas, 1892). Ann. Mag. Nat. Hist., ser. 6, 9:250.
> COMMON NAME: Hose's Palm Civet.
> TYPE LOCALITY: "Mount Dulit, N. Borneo, 4000 ft" [Malaysia, Sarawak, Gunung Dulit,
> 3°15′N, 114°15′E, 1219 m].
> DISTRIBUTION: Malaysia (Sabah, Sarawak).
> STATUS: IUCN – Vulnerable.
> COMMENTS: Thomas published two accounts of the type description in 1892, one printed in
> August (Thomas, 1892*a*), and another printed in October (Thomas, 1892*b*). Although
> Pocock (1933*d*) and Corbet and Hill (1992) supported Thomas (1912*d*) in separating
> this species into *Diplogale*; Chasen (1940), Medway (1977), and Payne et al. (1985) did
> not.

*Hemigalus* Jourdan, 1837. C. R. Acad. Sci. Paris, 5:442.
> TYPE SPECIES: *Hemigalus zebra* Jourdan, 1837 (= *Paradoxurus derbyanus* Gray, 1837), by
> monotypy.

SYNONYMS: *Hemigale* Gray, 1865; *Hemigalea* I. Geoffroy Saint-Hilaire and de Blainville, 1837.
COMMENTS: Corbet and Hill (1992) included *Chrotogale* but excluded *Diplogale*. There is little
question that these three taxa represent a monophyletic group and few would question
that *Diplogale* is more closely related to *Hemigalus* than *Chrotogale*. Molecular studies by
Veron and Heard (2000) supported the separation of *Hemigalus* and *Chrotogale*.

*Hemigalus derbyanus* (Gray, 1837). Mag. Nat. Hist. [Charlesworth's], 1:579.
  COMMON NAME: Banded Palm Civet.
  TYPE LOCALITY: Not given. Fixed by Gray (1837) as "in Peninsulâ Malayanâ".
  DISTRIBUTION: Burma (peninsular), Indonesia (Sipora Isl, South Pagi Isl, Kalimantan,
    Sumatra), Malaysia, Thailand.
  STATUS: CITES – Appendix II; IUCN – Lower Risk (lc).
  SYNONYMS: *derbianus* Gray, 1838; *derbyi* Temminck, 1841; *incursor* Thomas, 1915; *invisus*
    Pocock, 1933; *zebra* Gray, 1837; **boiei** Müller, 1838; **minor** Miller, 1903; **sipora** Chasen
    and Kloss, 1927.
  COMMENTS: Gervais (1841) was the first to place in *Hemigalus*. Gray (1849) later considered
    *Paradoxurus derbyanus* a junior synonym of *Viverra hardwicki* Gray, 1830, and placed it
    also in the genus *Hemigalea* after Jourdan (1837). However, *V. hardwicki* is a junior
    synonym of *Prionodon linsang* Raffles, 1821. Synonyms allocated according to
    D. D. Davis (1962).

**Subfamily Prionodontinae** Pocock, 1933. Proc. Zool. Soc. Lond., 1933:970.
  COMMENTS: Pocock (1933*d*) and Gregory and Hellman (1939) placed *Poiana* and *Prionodon* in
    the Prionodontinae, considered a sister group to the remaining viverrines. This was not
    followed by Gill (1872), Simpson (1945), Ellerman and Morrison-Scott (1951), Rosevear
    (1974), and Wozencraft (1989*b*). Gaubert et al. (2004) demonstrated that *Prionodon*
    should be excluded from the Viverrinae but left *Poiana* in the Viverrinae (which is
    followed here); however, see Hunt (2001) who included *Poiana*.

*Prionodon* Horsfield, 1822. Zool. Res. Java, Part 5, p. 13(unno.) of *Mangusta javanica* acct.
  TYPE SPECIES: *Prionodon gracilis* (Horsfield, 1822) (= *Viverra* ? *linsang* Hardwicke, 1821).
  SYNONYMS: *Linsang* Müller, 1840; *Linsanga* Lydekker, 1896; *Pardictis* Thomas, 1925; *Priodontes*
    Lesson, 1842.
  COMMENTS: See comments under family and subfamily for the taxonomic position of this
    genus.

*Prionodon linsang* (Hardwicke, 1821). Trans. Linn. Soc. Lond., 13:236, pl. 24.
  COMMON NAME: Banded Linsang.
  TYPE LOCALITY: "Malaysia, Malacca", restricted by Robinson and Kloss (1920:264) to
    "Malacca".
  DISTRIBUTION: Burma (peninsular), Indonesia (Banka Isl; Java; Kalimantan, Billiton Isl);
    Malaysia (West) to Sumatra.
  STATUS: CITES – Appendix II; IUCN – Lower Risk (lc).
  SYNONYMS: *maculosus* Blanford, 1878; **fredericae** Sody, 1936; *interliniurus* Sody, 1949; **gracilis**
    (Horsfield, 1822); *hardwichii* (Lesson, 1827).
  COMMENTS: Synonyms allocated according to Pocock (1933*d*), Ellerman and Morrison-Scott
    (1951), and D. D. Davis (1962).

*Prionodon pardicolor* Hodgson, 1842. Calcutta J. Nat. Hist., 2:57.
  COMMON NAME: Spotted Linsang.
  TYPE LOCALITY: "Sikim. . . Sub-Hemalayan mountains". [India].
  DISTRIBUTION: Bhutan, Burma, China (Guizhou, Sichuan, Yunnan), India, Laos, Nepal,
    Thailand, Vietnam.
  STATUS: CITES – Appendix I; U.S. ESA – Endangered; IUCN – Lower Risk (lc).
  SYNONYMS: *pardochrous* Gray, 1863; *perdicator* (Schinz, 1844); **presina** (Thomas, 1925).

**Subfamily Viverrinae** Gray, 1821. London Med. Repos., 15:301.
  COMMENTS: Pocock (1933*c*) and Gregory and Hellman (1939) placed *Poiana* and *Prionodon* in
    the Prionodontinae, considered a sister group to the remaining viverrines. This was not

followed by Gill (1872), Simpson (1945), Ellerman and Morrison-Scott (1951), Rosevear (1974), and Wozencraft (1989*b*). Gaubert et al. (2004) excluded *Prionodon* from the Viverrinae (followed here).

*Civettictis* Pocock, 1915. Proc. Zool. Soc. Lond., 1915:134.
>    TYPE SPECIES: *Viverra civetta* Schreber, 1776, by monotypy (Melville and Smith, 1987).
>    COMMENTS: Included in *Viverra* by Coetzee (1977*b*); recognized as *Civettictis* by Rosevear (1974), Kingdon (1977), Ansell (1978), Smithers (1983), and Wozencraft (1989*b*).

*Civettictis civetta* (Schreber, 1776). Die Säugethiere, 3(16):pl. 111[1776]; text 3(24):418, 3:index, p. 587[1777].
>    COMMON NAME: African Civet.
>    TYPE LOCALITY: "Guinea, Kongo, das Vorgebirge der guten Hofnung und Aethiopien", restricted by Allen (1924:117) to "Guinea".
>    DISTRIBUTION: Angola, Benin, Botswana, Cameroon, Central African Republic, Côte d'Ivoire, Dem. Rep. Congo, Equatorial Guinea, Ethiopia, Gabon, Gambia, Guinea, Kenya, Liberia, Malawi, Mozambique, Namibia, Niger, Nigeria, Republic of Congo, Rwanda, Senegal, Sierra Leone, South Africa, Sudan, Tanzania, Uganda, Zambia, Zimbabwe.
>    STATUS: CITES – Appendix III (Botswana); IUCN – Lower Risk (lc).
>    SYNONYMS: *poortmanni* (Pucheran, 1855); ***australis*** Lundholm, 1955; ***congica*** Cabrera, 1929; ***pauli*** Kock, Künzel and Rayaleh, 2000; ***schwarzi*** Cabrera, 1929; *orientalis* (Matschie, 1891) [preoccupied]; *matschiei* (Pocock, 1933); *megaspila* (Noack, 1891) [preoccupied]; ***volkmanni*** Lundholm, 1955.
>    COMMENTS: Ray (1995) reviewed this species. Some authors have placed this species in *Viverra* (see Coetzee, 1977); most have followed Pocock (1915*b*), who placed this species in *Civettictis*. Rosevear (1974) noted differences in scent glands; G. Petter (1969) discussed dental differences. Synonyms allocated according to Kock et al. (2000*c*).

*Genetta* G.[Baron] Cuvier, 1816. Règne Anim., 1:156.
>    TYPE SPECIES: *Viverra genetta* Linnaeus, 1758, by designation.
>    SYNONYMS: *Odmaelurus* Gloger, 1841; *Paragenetta* Kuhn, 1960; *Pseudogenetta* Dekeyser, 1949.
>    COMMENTS: For reviews, see Crawford-Cabral (1966*a*, 1969, 1970, 1973, 1981*a, b*), Rosevear (1974), Coetzee (1977), Schlawe (1980*a*, 1981), Wozencraft (1984, 1989*b*), Gaubert et al. (2002*b*, 2003*a, b*, 2004). Synonyms for the species in this genus follow Gaubert (2003) and Gaubert et al. (2002*a*, 2003*a, b*). Gaubert et al. (2002*b*) suggested that subgenera within the genus should be abandoned.

*Genetta abyssinica* (Rüppell, 1836). Neue Wirbelt. Fauna Abyssin. Gehörig. Säugeth., 1:33.
>    COMMON NAME: Abyssinian Genet.
>    TYPE LOCALITY: "In Abyssinien, wo es sehr häufig vorkömmt, führt es beiden Landeseinge-bornen zu Gondar"; Ethiopia, Gondar (12°36′N, 37°28′E).
>    DISTRIBUTION: Djibouti, Eritrea, Ethiopia, Somalia, Sudan.
>    STATUS: IUCN – Data Deficient.

*Genetta angolensis* Bocage, 1882. J. Sci. Math. Phys. Nat. Lisboa, ser. 1, 9:29.
>    COMMON NAME: Angolan Genet.
>    TYPE LOCALITY: "Calcuimba" [placed in Angola, Caconda (13°47′S, 15°08′E)].
>    DISTRIBUTION: Angola, Dem. Rep. Congo, Malawi, Mozambique, Tanzania, Zambia, Zimbabwe.
>    STATUS: IUCN – Lower Risk (lc).
>    SYNONYMS: *hintoni* Schwarz, 1929; *mossambica* Matschie, 1902.
>    COMMENTS: Crawford-Cabral (1970) argued that *mossambica* should be considered a synonym of *angolensis*, but later (Crawford-Cabral and Pacheco, 1992) changed and placed *mossambica* in *zambesiana* (= *maculata*). Crawford-Cabral and Fernandes (1999) considered *"mossambica"* specimens—as identified by Roberts (1951)—distinct from *mossambica* Matschie, 1902, and therefore constituting a separate species closely related to *maculata*.

*Genetta bourloni* Gaubert, 2003. Mammalia, 67(1):95.
COMMON NAME: Bourlon's Genet.
TYPE LOCALITY: "Sérédou, Cercle de Macenta" [Guinea, 8°33′N, 9°28′W].
DISTRIBUTION: Guinea, Côte d'Ivoire, Liberia, Sierra Leone.

*Genetta cristata* Hayman, *In* Sanborn, 1940. Trans. Zool. Soc. Lond. 24:686.
COMMON NAME: Crested Servaline Genet.
TYPE LOCALITY: "Okoiyong, Mamfe Division, Cameroons"; [Nigeria, 5°45′N, 8°25′E].
DISTRIBUTION: Cameroon-Nigeria border region.
STATUS: IUCN – Endangered.
SYNONYMS: *bini* Rosevear, 1974.
COMMENTS: Considered conspecific with *servalina* by Hayman in the original description, and followed by Coetzee (1977b) and Wozencraft (1993). However, see Rosevear (1974), Crawford-Cabral (1981a), Powell and Van Rompaey (1998), and Van Rompaey and Colyn (1998) who considered it distinct. Synonyms allocated according to Gaubert et al. (2003a, b).

*Genetta genetta* (Linnaeus, 1758). Syst. Nat., 10th ed., 1:45.
COMMON NAME: Common Genet.
TYPE LOCALITY: "oriente juxta rivos", restricted by Linnaeus (1766) to "oriente juxta rivos, Hispania", later listed by Thomas (1911a) as "Spain". Cabrera (1914), synonymizing *G. peninsulae*, further restricted the type locality to "El Pardo, cerca de Madrid" [Spain, El Pardo, near Madrid (40°32′N, 3°46′W)].
DISTRIBUTION: Algeria, Angola, Arabia, Belgium, Benin, Botswana, Burkina Faso, Cameroon, Central African Republic, Chad, Egypt, Ethiopia, France, Ghana, Kenya, Liberia, Libya, Mali, Mauritania, Morocco, Mozambique, Namibia, Niger, Nigeria, Oman, Portugal, Senegal, Spain, Somalia, South Africa, Sudan, Tanzania, Togo, Tunisia, Uganda, Yemen, Zambia, Zimbabwe.
STATUS: IUCN – Vulnerable as *G. g. isabelae*, otherwise Lower Risk (lc).
SYNONYMS: *balearica* Thomas, 1902; *barbar* (Wagner, 1841); *communis* Burnett, 1830; *gallica* (Oken, 1816); *hispanica* (Oken, 1816); *isabelae* Delibes, 1977; *lusitanica* Seabra, 1924; *melas* Graells, 1897; *peninsulae* Cabrera, 1905; *pyrenaica* E. Bourdelle and De Zillière, 1951; *rhodanica* Matschie, 1902; *terraesanctae* Neumann, 1902; *vulgaris* (Lesson, 1827); *afra* F. G. Cuvier, 1825; *barbara* C. E. H. Smith, 1842; *bonapartei* Loche, 1857; *dongolana* Hemprich and Ehrenberg, 1832; *albipes* Trouessart, 1904; *grantii* Thomas, 1902; *guardafuensis* Neumann, 1902; *hararensis* Neumann, 1902; *neumanni* Matschie, 1902; *tedescoi* de Beaux; 1924; *felina* (Thunberg, 1811); *bella* Matschie, 1902; *ludia* Thomas and Schwann, 1906; *macrura* (Jentink, 1892); *pulchra* Matschie, 1902; *senegalensis* J. B. Fischer, 1829; *leptura* Reichenbach, 1836.
COMMENTS: Schlawe (1981) included *afra, bonapartei, barbar, barbara, balearica, lusitanica, melas, peninsulae, pyrenaica, terraesanctae, rhodanica,* and *isabelae*; and provisionally separated into *G. felina* the following: *guardafuensis, hararensis, leptura, senegalensis, dongolana, granti, neumanni, bella, pulchra,* and *ludia*, which are included here, following Crawford-Cabral (1966a, 1969; 1981a), Coetzee (1977b), Smithers (1983), and Wozencraft (1984, 1989b). Reviewed by Crawford-Cabral (1966a, 1969, 1981a), Schlawe (1981), and Larivière and Calzada (2001). Rosevear (1974) separated *senegalensis* from *genetta*. However, this was not followed by Coetzee (1977b), Kingdon (1977), Ansell (1978), Crawford-Cabral (1981a), or Wozencraft (1989b). Synonyms allocated according to Ellerman et al. (1953), Schlawe (1981), Wozencraft (1984), and Larivière and Calzada (2001).

*Genetta johnstoni* Pocock, 1908. Proc. Zool. Soc. Lond., 1907:1041 [1908].
COMMON NAME: Johnston's Genet.
TYPE LOCALITY: "in a district from fifteen to twenty miles [32 km] west of the Putu Mountains, which lie west of the Duobe and Cavally Rivers". [Liberia]
DISTRIBUTION: Côte d'Ivoire, Ghana, Guinea, Liberia.
STATUS: IUCN – Data Deficient.
SYNONYMS: *lehmanni* Kuhn, 1960.
COMMENTS: Reviewed by Gaubert et al. (2002a).

*Genetta maculata* (Gray, 1830). Spicil. Zool., 2:9.
    COMMON NAME: Rusty-spotted Genet.
    TYPE LOCALITY: "in Africa Boreali". Subsequently redefined by Gaubert et al. (2003*b*) as
        "6 km from Hirna (Harrar Road, 2180m), Ethiopia", following designation of a
        neotype.
    DISTRIBUTION: Angola, Botswana, Cameroon, Central African Republic, Chad, Dem. Rep.
        Congo, Eritrea, Equatorial Guinea (incl. Bioko), Ethiopia, Gabon, Ghana, Kenya,
        Malawi, Mozambique, Namibia, Nigeria, Republic of Congo, Rwanda, Somalia,
        South Africa, Sudan, Tanzania, Togo, Uganda, Zambia, Zimbabwe.
    STATUS: IUCN – Lower Risk (lc).
    SYNONYMS: *aequatorialis* Heuglin, 1866; *albiventris* Roberts, 1932; *deorum* Funaioli and
        Simonetta, 1960; *erlangeri* Matschie, 1902; *fieldiana* Du Chaillu, 1860; *gleimi* Matschie,
        1902; *insularis* Cabrera, 1921; *letabae* Thomas and Schwann, 1906; *matschiei*
        Neumann, 1902; *pumila* Hollister, 1916; *schoutedeni* Crawford-Cabral, 1970; *schraderi*
        Matschie, 1902; *soror* Schwarz, 1929; *stuhlmanni* Matschie, 1902; *suahelica* Matschie,
        1902; *zambesiana* Matschie, 1902; *zuluensis* Roberts, 1924.
    COMMENTS: Traditionally recognized as *G. rubiginosa* Pucheran, 1855; this form is attributed
        to *G. thierryi* (see Schlawe, 1980*a*, 1981; Crawford-Cabral and Pacheco, 1992). Gaubert
        et al. (2003*a*, *b*) proposed that *rubiginosa* Pucheran, 1855 is a *nomen oblitum*. Synonyms
        allocated according to Roberts (1951), Crawford-Cabral and Pacheco (1992) and
        Gaubert et al. (2003*a*, *b*). Rosevear (1974) believed that *V. maculata* Gray (1830) was
        invalid, but see Schlawe (1980*a*, 1981) who defended its use. See Gaubert et al. (2003*a*,
        *b*) for usage of the name *V. maculata* Gray, 1830. Crawford-Cabral (1981*a*) and Ansell
        (1978) placed genets west of the Dahomey Gap in *pardina* and southern and eastern
        populations in *rubiginosa* (=*maculata*) (except for the extreme southern *tigrina*), which
        is followed here. *G. schoutedeni* Crawford-Cabral, 1970 and *suahelica* Matschie, 1902
        should probably best be considered *incertae sedis*.

*Genetta pardina* I. Geoffroy Saint-Hilaire, 1832. Etudes zoologiques, fasc. 1:8.
    COMMON NAME: Pardine Genet.
    TYPE LOCALITY: "intérieur du Sénégal".
    DISTRIBUTION: Burkina Faso, Côte d'Ivoire, Gambia, Ghana, Guinea, Liberia, Mali, Niger,
        Senegal, Sierra Leone.
    SYNONYMS: *amer* Gray, 1843; *dubia* Matschie, 1902; *genettoides* Temminck, 1853; *pantherina*
        Hamilton-Smith, 1842.
    COMMENTS: The status of *genettoides* is perhaps best listed as *incertae sedis*. It may be a hybrid
        population between *pardina* and *maculata* (Gaubert et al., 2003*a*).

*Genetta piscivora* (J. A. Allen, 1919). J. Mammal., 1:25.
    COMMON NAME: Aquatic Genet.
    TYPE LOCALITY: "Niapu, Belgian Congo" [Dem. Rep. Congo, Niapu, 2°25′N, 26°28′E)].
    DISTRIBUTION: N and E Dem. Rep. Congo.
    STATUS: IUCN – Data Deficient as *Osbornictis piscivora*.
    COMMENTS: Some have placed in *Osbornictis* (Bininda-Edmonds et al., 1999; Van Rompaey,
        1988; Wozencraft, 1993). Reviewed by Van Rompaey (1988) and Gaubert (2003) and
        Gaubert et al. (2002*b*, 2004). Hunt (2001) placed *Osbornictis* in a subfamily separate
        from *Genetta*. Gaubert et al. (2004) demonstrated that *Osbornictis* and *Genetta* are
        congeneric, in agreement with Verheyen (1962) and Stains (1983).

*Genetta poensis* Waterhouse, 1838. Proc. Zool. Soc. Lond., 1838:59.
    COMMON NAME: King Genet.
    TYPE LOCALITY: "Fernando Po" [Bioko Isl, Equatorial Guinea].
    DISTRIBUTION: Ghana, Côte d'Ivoire, Equatorial Guinea (Bioko), Liberia, Republic of Congo.
    COMMENTS: Considered conspecific with *pardina-maculata* by Schlawe (1981), Wozencraft
        (1993) and Grubb et al. (1998). Rosevear (1974) and Crawford-Cabral (1981*a*)
        discussed its taxonomic status. Erroneously used for designating *cristata* by Jeannin
        (1936) and Happold (1987). Gaubert et al. (2003*a*) proposed specific status on the basis
        of coat pattern and hair ultrastructure, together with sympatric distribution with
        *pardina* and *maculata*.

*Genetta servalina* Pucheran, 1855. Rev. Mag. Zool. Paris, 7(2):154.

COMMON NAME: Servaline Genet.

TYPE LOCALITY: "Gabon".

DISTRIBUTION: Cameroon, Central African Republic, Dem. Rep. Congo, Equatorial Guinea, Gabon, Kenya, Republic of Congo, Tanzania, Uganda.

STATUS: IUCN – Lower Risk (lc).

SYNONYMS: *aubryana* Pucheran, 1855; **archeri** Van Rompaey and Colyn, 1998; **bettoni** Thomas, 1902; *intensa* Lönnberg, 1917; **lowei** Kingdon, 1977; **schwarzi** Crawford-Cabral, 1970.

COMMENTS: Includes *bettoni* and *aubryana*, but not *bini*, which is here considered a junior synonym of *cristata*, following Gaubert et al. (2003a). Does not include *cristata* after Rosevear (1974), Crawford-Cabral (1981a), and Van Rompaey and Colyn (1998). Subspecies allocated according to Van Rompaey and Colyn (1998).

*Genetta thierryi* Matschie, 1902. Verh. V. Internat. Zool. Congr., 1901:1142.

COMMON NAME: Haussa Genet.

TYPE LOCALITY: "Hinterlang von Togo von 9° n. Br. ab.", restricted to "Borogu = Borgou, (10.78 N., 0.65 E)" by Schlawe (1981:159).

DISTRIBUTION: Benin, Burkina Faso, Cameroon, Côte d'Ivoire, Gambia, Ghana, Mali, Nigeria, Niger, Sierra Leone, Senegal, Togo.

STATUS: IUCN – Lower Risk (lc).

SYNONYMS: *rubiginosa* Pucheran, 1855; *villiersi* (Dekeyser, 1949).

COMMENTS: Includes *Pseudogenetta villiersi* (see Crawford-Cabral, 1969, 1981a; Rosevear, 1974; and Schlawe, 1981). Schlawe (1981) pointed out that the type of *G. rubiginosa*, which traditionally has been considered a synonym of *G. maculata*, is actually a senior synonym of *G. thierryi*. This was later verified by Crawford-Cabral and Pacheco (1992), who nevertheless continued to use "*rubiginosa*" to represent the Rusty-spotted Genet (=*maculata*). See Gaubert et al. (2003a) for use of the name *G. thierryi* Matschie, 1902, under the status of *nomen protectum*.

*Genetta tigrina* (Schreber, 1776). Die Säugethiere, 3(17):pl. 115[1776]; text, 3(25):425 [1777].

COMMON NAME: Cape Genet.

TYPE LOCALITY: "von dem Vorgebirge der guten Hofnug" [South Africa, Western Cape Prov., Cape of Good Hope].

DISTRIBUTION: South Africa.

STATUS: IUCN – Lower Risk (lc).

SYNONYMS: **methi** Roberts, 1948.

COMMENTS: There is great confusion in the taxonomy of the *pardina-rubiginosa* (= *maculata*) -*tigrina* complex. Some authors (Coetzee, 1977b; Meester et al., 1986) considered *rubiginosa* as conspecific with *tigrina*. Others believed that *maculata* may be conspecific with *pardina* (Ansell, 1978; Crawford-Cabral, 1966a; Pringle, 1977). Crawford-Cabral (1981a) later reversed his earlier opinion and considered three types to exist as separate species; (1) *pardina*, (2) *rubiginosa* (= *maculata*), and (3) *tigrina*. This was followed by Gaubert (2003) and Gaubert et al. (2003a, b). Roberts (1951) considered *zambesiana*, *letabae*, *zuluensis*, and *albiventris* as subspecies of *maculata*; whereas he limited *tigrina* to only two subspecies: *tigrina* and *methi*. Subspecies allocated according to Roberts (1951) and Crawford-Cabral and Pacheco (1992).

*Genetta victoriae* Thomas, 1901. Proc. Zool. Soc. Lond., 1901(2):87.

COMMON NAME: Giant Forest Genet.

TYPE LOCALITY: "Entebbe, Uganda". Subsequently restricted by Moreau et al. (1946:410) to "Near Lupanzula's, ten miles [16 km] west of Beni, Ituri Forest, Congo Belge [Dem. Rep. Congo]". See Allen (1924) for discussion.

DISTRIBUTION: N and E Dem. Rep. Congo (N and E).

STATUS: IUCN – Lower Risk (lc).

*Poiana* Gray, 1865. Proc. Zool. Soc. Lond. 1864:507, 520 [1865].

TYPE SPECIES: *Genetta richardsonii* Thomson, 1842, by monotypy (Melville and Smith, 1987).

COMMENTS: There are two widely separated (approximately 1600 km) allopatric populations of

*Poiana*: the congo population (*P. richardsonii*), and the West African population, which Rosevear (1974) raised to the specific level (*P. leightoni*). *Poiana* records are few and scattered; however, recent reviews place *leightoni* as a separate species (de Beaufort, 1965; Gaubert et al., 2002*b*; Kingdon, 1977; Michaelis, 1972).

*Poiana leightoni* Pocock, 1908. Proc. Zool. Soc. Lond., 1907:1043 [1908].
   COMMON NAME: Leighton's Linsang.
   TYPE LOCALITY: "fifteen to twenty miles [24 to 32 km] west of the Putu Mountains, which lie west of the Duobe and Cavally Rivers. The Cavally River is the eastern boundary line between Liberia and the Côte d'Ivoire, and the Duobe is one of its tributaries joining the Cavally about seventy miles [113 km], as the crow flies, from its mouth, after running for over one hundred miles [161 km] nearly parallel to the main stream" [Liberia].
   DISTRIBUTION: Côte d'Ivoire, Liberia.
   STATUS: IUCN – Data Deficient as *P. richardsonii liberiensis*.
   SYNONYMS: *liberiensis* Pocock, 1908.
   COMMENTS: Pocock (1907*b*:1045) mentioned that there was one specimen in the British Museum labeled "Sierra Leone", but he believed that the "locality is probably errone-ous." Coetzee (1977*b*) considered *leightoni* a lapsus and replaced it with *liberiensis*.

*Poiana richardsonii* (Thomson, 1842). Ann. Mag. Nat. Hist., [ser. 1], 10:204.
   COMMON NAME: African Linsang.
   TYPE LOCALITY: "Fernando Po" [Equatorial Guinea: Bioko].
   DISTRIBUTION: Cameroon, Central African Republic, Dem. Rep. Congo, Equatorial Guinea, Gabon, Republic of Congo.
   STATUS: IUCN – Lower Risk (lc).
   SYNONYMS: *poensis* (Waterhouse, 1838) [preoccupied]; ***ochracea*** Thomas and Wroughton, 1907.

*Viverra* Linnaeus, 1758. Syst. Nat., 10th ed., 1:43.
   TYPE SPECIES: *Viverra zibetha* Linnaeus, 1758, by subsequent designation (Sclater, 1900; Melville and Smith, 1987).
   SYNONYMS: *Moschothera* Pocock, 1933; *Vivera* Gray, 1821.
   COMMENTS: Pocock (1933*a*) placed *civettina* and *megaspila* in *Moschothera*, which was not recognized by Ellerman and Morrison-Scott (1951), Wozencraft (1989*b*), and Corbet and Hill (1992). Does not include *Civettictis*; see Kingdon (1977) and Ansell (1978); but also see Coetzee (1977*b*).

*Viverra civettina* Blyth, 1862. J. Asiatic Soc. Bengal, 31:332.
   COMMON NAME: Malabar Large-spotted Civet.
   TYPE LOCALITY: "Southern Malabar" restricted by Pocock (1933*a*:446) to "Travancore" [India].
   DISTRIBUTION: Endemic to S India.
   STATUS: CITES – Appendix III (India); U.S. ESA – Endangered; IUCN – Critically Endangered.
   COMMENTS: Considered a subspecies of *V. megaspila* by Ellerman and Morrison-Scott (1951); however, considered at the specific level by Lindsay (1928), Pocock (1941*a*), and Wozencraft (1984, 1989*b*). Corbet and Hill (1992) raised doubts as to their separation.

*Viverra megaspila* Blyth, 1862. J. Asiatic Soc. Bengal, 31:331.
   COMMON NAME: Large-spotted Civet.
   TYPE LOCALITY: "vicinity of Prome" [Burma, Prome (= Pye) 18°49′N, 95°13′E].
   DISTRIBUTION: Burma, Cambodia, Laos, Malaysia (West), Thailand, Vietnam.
   STATUS: IUCN – Lower Risk (lc).
   COMMENTS: Does not include *V. civettina*; reviewed by Lindsay (1928), Pocock (1941*a*), and Wozencraft (1989*b*). Corbet and Hill (1992) raised doubts as to their separation.

*Viverra tangalunga* Gray, 1832. Proc. Zool. Soc. Lond., 1832:63.
   COMMON NAME: Malayan Civet.
   TYPE LOCALITY: Not given. Fixed by Gray (1843:48) as "Sumatra" [Indonesia].
   DISTRIBUTION: Cambodia, Indonesia (Sumatra, Rhio-Lingga Arch., Bangka Isl, Borneo,

Karimata Isl, Sulawesi, Amboina), Malaysia, Philippines, and Thailand. Introduced throughout the Moluccas.

STATUS: IUCN – Lower Risk (lc).

SYNONYMS: *lankavensis* Robinson and Kloss, 1920.

*Viverra zibetha* Linnaeus, 1758. Syst. Nat., 10th ed., 1:44.

COMMON NAME: Large Indian Civet.

TYPE LOCALITY: "Indiis", subsequently restricted by Thomas (1911*a*:137) to "Bengal".

DISTRIBUTION: Burma, Cambodia, China (Anhui, Shaanxi, Zhejiang and Jiangsu), India, Indonesia, Laos, Western Malaysia, Nepal, Thailand, and Vietnam.

STATUS: CITES – Appendix III (India); IUCN – Lower Risk (lc).

SYNONYMS: *civettoides* Hodgson, 1842; *melanurus* Hodgson, 1842; *orientalis* Hodgson, 1842; *undulata* Gray, 1830; *ashtoni* Swinhoe, 1864; *expectata* Corbert and Hooijer, 1953; *filchneri* Matschie, 1907; *hainana* Wang and Xu, 1983; *picta* Wroughton, 1915; *surdaster* Thomas, 1927; *tainguensis* Sokolov, Rozhnov and Pham Trong, 1997; *pruinosus* Wroughton, 1917; *sigillata* Robinson and Kloss, 1920.

COMMENTS: See Walston and Veron (2001) for inclusion of *V. tainguensis* here. Synonyms allocated according to Pocock (1933*a*) and Ellerman and Morrison-Scott (1951).

*Viverricula* Hodgson 1838. Ann. Mag. Nat. Hist., [ser. 1], 1:152.

TYPE SPECIES: *Civetta indica* É. Geoffroy Saint-Hilaire, 1803

SYNONYMS: *Viverrula* Hodgson, 1842.

*Viverricula indica* (É. Geoffroy Saint-Hilaire, 1803). Cat. Mamm. Mus. Nat. Hist. Nat., p. 113.

COMMON NAME: Small Indian Civet.

TYPE LOCALITY: "l'Inde" [India].

DISTRIBUTION: Afghanistan, Bangladesh, Burma, Cambodia, China, Hong Kong, India, Indonesia (Borneo, Sumatra, Java, Kangean Isl, Sumbawa, Bali), Laos, Malaysia, Nepal, Pakistan, Sri Lanka, Taiwan, Thailand, Vietnam. Introduced to Yemen, Zanzibar and Pemba Isl, Socotra Isl, Madagascar, the Comoro Isls, and the Philippines; scattered distribution on many SE Asian islands due to introductions.

STATUS: CITES – Appendix III (India); IUCN – Lower Risk (lc).

SYNONYMS: *rasse* (Horsfield, 1821); *atchinensis* Sody, 1931; *baliensis* (Sody, 1931); *baptistae* Pocock, 1933; *bengalensis* (Gray and Hardwicke, 1830); *deserti* Bonhote, 1898; *klossi* Pocock, 1933; *mayori* Pocock, 1933; *muriavensis* Sody, 1931; *pallida* (Gray, 1831); *hanensis* Matschie, 1907; *schlegelii* Pollen, 1866; *taivana* Schwarz, 1911; *thai* Kloss, 1919; *wellsi* Pocock, 1933.

COMMENTS: Subspecies arranged according to Pocock (1933*b*) and Ellerman and Morrison-Scott (1951).

**Family Eupleridae** Chenu, 1850. Ency. Hist. Nat., 21:165.

SYNONYMS: Euplerini Simpson, 1945. Including: Cryptoproctina Gray, 1864; Galidiina Gray, 1864; Cryptoproctidae Flower, 1869, Galidiinae Gill, 1872; Galidictinae Mivart, 1882; Cryptoproctinae Trouessart, 1885; Fossinae Pocock, 1915.

COMMENTS: The Malagasy carnivores have been problematic since their discovery. They have been placed in the Viverridae, the Herpestidae, and separated into monotypic families. Veron and Catzeflis (1993) and Yoder et al. (2003) provided the strongest evidence that all Malagasy carnivores represent a single radiation.

**Subfamily Euplerinae** Chenu, 1850. Ency. Hist. Nat., 21:165.

SYNONYMS: Cryptoproctina Gray, 1865; Cryptoproctidae Flower, 1869; Cryptoproctinae Trouessart, 1885.

COMMENTS: Bininda-Emonds et al. (1999) and Yoder et al. (2003) considered these taxa a monophyletic group (followed here). Few have questioned that *Fossa* and *Eupleres* are more closely related to each other than to *Cryptoprocta* and Hunt (2001) placed *Cryptoprocta* in a separate taxon.

*Cryptoprocta* Bennett, 1833. Proc. Zool. Soc. Lond., 1833:46.
>     TYPE SPECIES: *Cryptoprocta ferox* Bennett, 1833, by designation (Melville and Smith, 1987).
>     COMMENTS: Gregory and Hellman (1939), Beaumont (1964), and Veron (1995) argued that *Cryptoprocta* be placed in the Felidae because of morphological similarities, but Veron (1995) showed that the dental features shared by Felidae and Cryptoprocta are the result of a convergence. Albignac (1970), Thenius (1972), Radinsky (1975), Coetzee (1977*b*), Flynn et al. (1988), and Wozencraft (1989*a, b*) placed in the Viverridae; Hemmer (1978) suggested an intermediate position. Veron and Catzeflis (1993) and Yoder et al. (2003) suggested affinities with the Galidiinae (Herpestidae) rather than with the Viverridae.

*Cryptoprocta ferox* Bennett, 1833. Proc. Zool. Soc. Lond., 1833:46.
>     COMMON NAME: Fossa.
>     TYPE LOCALITY: "Madagascar".
>     DISTRIBUTION: Endemic to Madagascar.
>     STATUS: CITES – Appendix II; IUCN – Endangered.
>     SYNONYMS: *typicus* A. Smith, 1834.
>     COMMENTS: Reviewed by Köhncke and Leonhardt (1986).

*Eupleres* Doyère, 1835. Bull. Soc. Sci. Nat., 3:45.
>     TYPE SPECIES: *Eupleres goudotii* Doyère, 1835, by monotypy (Melville and Smith, 1987).
>     COMMENTS: Gregory and Hellman (1939) followed Chenu and Desmarest (1852), and suggested placing *Eupleres* in a separate family; however, Pocock (1915*a*), Albignac (1973, 1974), Petter (1974), Coetzee (1977*b*), and Wozencraft (1989*b*) included it in Viverridae. See comments under family for its inclusion here.

*Eupleres goudotii* Doyère, 1835. Bull. Soc. Sci. Nat., 3:45.
>     COMMON NAME: Falanouc.
>     TYPE LOCALITY: "Tamatave" [Madagascar, 18°10′S, 49°23′E].
>     DISTRIBUTION: Endemic to Madagascar.
>     STATUS: CITES – Appendix II; IUCN – Endangered as *E. g. goudotii* and *E. g. major*.
>     SYNONYMS: ***major*** Lavauden, 1929.

*Fossa* Gray, 1865. Proc. Zool. Soc. Lond., 1864:518 [1865].
>     TYPE SPECIES: *Fossa d'aubentonii* Gray, 1865 (= *Viverra Fossana* Müller, 1776).
>     COMMENTS: Veron (1995) and Gaubert et al. (2002*b*) suggested that *Fossa* could be separated from Viverridae based on morphological features. Molecular data (Yoder et al., 2003) showed close affinities with *Cryptoprocta*.

*Fossa fossana* (Müller, 1776). Linné's Vollstand, Natursyst. Suppl., p. 32.
>     COMMON NAME: Malagasy Civet.
>     TYPE LOCALITY: "Madagascar".
>     DISTRIBUTION: Endemic to Madagascar.
>     STATUS: CITES – Appendix II; IUCN – Vulnerable.
>     SYNONYMS: *daubentonii* Gray, 1865; *fossana* (Schreber, 1777); *majori* Dollman, 1909.
>     COMMENTS: The commonly used *Viverra fossa* Schreber, 1777, is a junior synonym (G. Petter, 1962, 1974).

**Subfamily Galidiinae** Gray, 1865. Proc. Zool. Soc. Lond., 1864:508 [1865].
>     SYNONYMS: Galidictinae Mivart, 1882.
>     COMMENTS: Veron and Catzeflis (1993), Yoder et al. (2003), and Gaubert et al. (in press) suggested that the Galidiinae and the Cryptoproctidae are a monophyletic group.

*Galidia* I. Geoffroy Saint-Hilaire, 1837. C. R. Acad. Sci. Paris, 5:580.
>     TYPE SPECIES: *Galidia elegans* I. Geoffroy Saint-Hilaire, 1837, by monotypy.

*Galidia elegans* I. Geoffroy Saint-Hilaire, 1837. C. R. Acad. Sci. Paris, 5:581.
>     COMMON NAME: Ring-tailed Mongoose.
>     TYPE LOCALITY: "Madagascar".
>     DISTRIBUTION: Endemic to Madagascar.

STATUS: IUCN – Vulnerable.

SYNONYMS: *afra* (Kerr, 1792) [preoccupied]; ***dambrensis*** Tate and Rand, 1941; ***occidentalis*** Albignac, 1971.

*Galidictis* I. Geoffroy Saint-Hilaire, 1839. Mag. Zool., Mamm Art., No. 5, p. 33, footnote, 37.
TYPE SPECIES: *Mustela striata* I. Geoffroy Saint-Hilaire, 1837 (= *Viverra fasciata* Gmelin, 1788) by original designation.
SYNONYMS: *Galictis* I. Geoffroy Saint-Hilaire, 1837 [preoccupied]; *Musanga* Coues, 1891.
COMMENTS: Gregory and Hellman (1939) separated *Galidictis* from other galidiines and placed it in the Viverridae.

*Galidictis fasciata* (Gmelin, 1788). *In* Linnaeus, Syst. Nat., 13th ed., 1:92.
COMMON NAME: Broad-striped Malagasy Mongoose.
TYPE LOCALITY: Unknown, erroneously listed by Gmelin as "in India".
DISTRIBUTION: Endemic to Madagascar.
STATUS: IUCN – Vulnerable.
SYNONYMS: *eximius* Pocock, 1915; *striata* (Desmarest, 1820); *vittata* Schinz, 1844; ***striatus*** G. Cuvier, 1829.

*Galidictis grandidieri* Wozencraft, 1986. J. Mammal., 67:561.
COMMON NAME: Grandidier's Mongoose.
TYPE LOCALITY: "Madagascar".
DISTRIBUTION: Known only from the spiny desert of SW Madagascar.
STATUS: IUCN – Endangered.
COMMENTS: *G. grandidiensis* Wozencraft, 1986 was emended to *G. grandidieri* by Wozencraft (1987).

*Mungotictis* Pocock, 1915. Ann. and Mag. Nat. Hist. ser. 8, 16:120.
TYPE SPECIES: *Galidictis vittatus* Gray, 1848 (= *Galidia decemlineata* Grandidier, 1867).

*Mungotictis decemlineata* (Grandidier, 1867). Rev. Mag. Zool. Paris, ser. 2, 19:85.
COMMON NAME: Narrow-striped Mongoose.
TYPE LOCALITY: "à la côte ouest de Madagascar" (pg. 84).
DISTRIBUTION: Endemic to Madagascar.
STATUS: IUCN – Endangered.
SYNONYMS: *rufa* Grandidier, 1869; *substriatus* Pocock, 1915; *vittata* Gray, 1848 [preoccupied]; ***lineatus*** Pocock, 1915.
COMMENTS: Synonyms after Hawkins et al. (2000).

*Salanoia* Gray, 1865. Proc. Zool. Soc. Lond., 1864:523 [1865].
TYPE SPECIES: *Galidia concolor* I. Geoffroy Saint-Hilaire, 1837.
SYNONYMS: *Hemigalidia* Mivart, 1882.

*Salanoia concolor* (I. Geoffroy Saint-Hilaire, 1837). C. R. Acad. Sci. Paris, 5:581.
COMMON NAME: Brown-tailed Mongoose.
TYPE LOCALITY: "Madagascar".
DISTRIBUTION: Endemic to Madagascar.
STATUS: IUCN – Vulnerable.
SYNONYMS: *olivacea* (I. Geoffroy Saint-Hilaire, 1839); *unicolor* (I. Geoffroy Saint-Hilaire, 1837).
COMMENTS: Geoffroy Saint-Hilaire (1839) noted that his first listed species name, *unicolor*, was a typographical error and should have been *concolor* (Coetzee, 1977*b*:35).

**Family Nandiniidae** Pocock, 1929. Ency. Brit., (ed. 14) 3:898.
COMMENTS: Listed in Nandiniinae by Gregory and Hellman (1939), and Coetzee (1977*b*). Hunt (1987, 1989, 1998), McKenna and Bell (1997), Hunt and Tedford (1993), Flynn and Nedbal (1998), and Veron and Heard (2000) argued that *Nandinia* should be placed in a monotypic family based on the plesiomorphic condition of its auditory bullae (Pohle, 1920*b*). This has been confirmed by molecular data (Flynn and Nedbal, 1998; Veron and Heard, 2000; Yoder et al., 2003). Bininda-Emonds et al. (1999) considered *Nandinia* most closely related to the Paradoxurine palm civets.

*Nandinia* Gray, 1843. List Spec. Mamm. Coll. Brit. Mus., p. 54.
    TYPE SPECIES: *Viverra binotata* Gray, 1830, by monotypy (Melville and Smith, 1987).
    COMMENTS: See comments under family.

*Nandinia binotata* (Gray, 1830). Spicil. Zool., 2:9.
    COMMON NAME: African Palm Civet.
    TYPE LOCALITY: "Africa, Ashantee" [Ghana; Ashanti Region; aproximately at 6°55′N 0°32′E].
    DISTRIBUTION: Angola, Benin, Burundi, Cameroon, Central African Republic, Côte d'Ivoire, Dem. Rep. Congo, Equatorial Guinea, Gabon, Ghana, Guinea, Guinea-Bissau, Kenya, Liberia, Malawi, Mozambique, Nigeria, Republic of Congo, Rwanda, Senegal, Sierra Leone, Sudan, Tanzania, Togo, Uganda, Zambia, Zimbabwe.
    STATUS: IUCN – Lower Risk (lc).
    SYNONYMS: *hamiltonii* (Gray, 1832); **arborea** Heller, 1913; **gerrardi** Thomas, 1893; **intensa** Cabrera and Ruxton, 1926.
    COMMENTS: Synonyms allocated according to G. M. Allen (1939) and Ellerman and Morrison-Scott (1953).

**Family Herpestidae** Bonaparte, 1845. Cat. Meth. Mamm. Europe, p. 3.
    SYNONYMS: Cynictidae Cope, 1882; Herpestoidei Winge, 1895; Mongotidae Pocock, 1919; Rhinogalidae Gray, 1869; Suricatidae Cope, 1882; Suricatinae Thomas, 1882.
    COMMENTS: Wozencraft (1989b) placed *Crossarchus, Cynictis, Dologale, Helogale, Liberiictis, Mungos, Paracynictis,* and *Suricata* in the Mungotinae but gave no supporting rationale. Fredga's (1972) analysis of chromosomes and the recent molecular work by Veron et al. (2004) would support Wozencraft's Mungotinae (with the inclusion of *Bdeogale* and *Ichneumia*). The phylogenetic analysis of allozyme data by Taylor et al. (1991) also supported *Cynictis, Suricata,* and *Helogale* as a monophyletic group.

*Atilax* F. G. Cuvier, 1826. *In* É. Geoffroy Saint-Hilaire and F. G. Cuvier, Hist. Nat. Mammifères, pt. 3, 5(54), "Vansire," 2 pp., 1 pl.
    TYPE SPECIES: *Herpestes paludinosus* G. [Baron] Cuvier (1829), by original designation (Melville and Smith, 1987).
    SYNONYMS: *Athylax* de Blainville, 1837.
    COMMENTS: Fredga's (1972) comparative chromosome study of mongooses suggested that recognition of *Atilax* as distinct from *Herpestes* would make *Herpestes* paraphyletic. However, allozyme data support *Atilax* as the first early offshoot of the main herpestine branch (Taylor et al., 1991).

*Atilax paludinosus* (G.[Baron] Cuvier, 1829). Regn. Anim., Nouv. ed., 1:158.
    COMMON NAME: Marsh Mongoose.
    TYPE LOCALITY: "une grand des marais du Cap" [South Africa, Western Cape Prov., Cape of Good Hope].
    DISTRIBUTION: Algeria, Angola, Botswana, Cameroon, Central African Republic, Côte d'Ivoire, Dem. Rep. Congo, Equatorial Guinea, Ethiopia, Gabon, Liberia, Malawi, Mozambique, Niger, Ruwanda, Senegal, Sierra Leone, Somalia, South Africa, Sudan, Tanzania, Uganda, and Zambia.
    STATUS: IUCN – Lower Risk (lc).
    SYNONYMS: *atilax* (Wagner, 1841); *paludosus* Gray, 1865; *urinatrix* (A. Smith, 1829); *vansire* (F. G. Cuvier, 1842); **macrodon** J. A. Allen, 1924; **mitis** (Thomas, 1903); **mordax** (Thomas, 1912); **nigerianus** (Thomas, 1912); **pluto** (Temminck, 1853); **robustus** (Gray, 1865); **rubellus** (Thomas and Wroughton, 1908); **rubescens** (Hollister, 1912); **spadiceus** Cabrera, 1921; **transvaalensis** Roberts, 1933.
    COMMENTS: Reviewed by Baker (1992).

*Bdeogale* Peters, 1850. Spenersche Z., 25 June, 1850 (unpaginated).
    TYPE SPECIES: *Bdeogale crassicauda* Peters, 1852; by subsequent designation by Thomas (1882) (Melville and Smith, 1987).
    SYNONYMS: *Beleogale* Marshall, 1873; *Galeriscus* Thomas, 1894.
    COMMENTS: Matschie (1895), Pocock (1916a), Coetzee (1977b), Kingdon (1977), and Meester

et al. (1986) included *Galeriscus* Thomas (1894). Rosevear (1974) believed that no one had advanced any "reasoned argument" for combining *Galeriscus* with *Bdeogale*, and followed Schoutenden (1945) and Hill and Carter (1941) who considered them distinct; all have agreed that *jacksoni* and *nigripes* are sister taxa.

*Bdeogale crassicauda* Peters, 1852. Monatsb. K. Preuss. Akad. Wiss. Berlin, 1852:81.
   COMMON NAME: Bushy-tailed Mongoose.
   TYPE LOCALITY: "Africa orient., Tette, Boror, 17-18° Lat. austr". (pg. 82). Restricted by Moreau et al. (1945:410) to "Tette" [Mozambique].
   DISTRIBUTION: Kenya, Malawi, C Mozambique, Tanzania (incl. Zanzibar), S and E Zambia, NE Zimbabwe.
   STATUS: IUCN – Endangered as *B. c. omnivora*, otherwise Lower Risk (lc).
   SYNONYMS: *nigrescens* Sale and Taylor, 1970; *omnivora* Heller, 1913; *puisa* Peters, 1852; *tenuis* Thomas and Wroughton, 1908.
   COMMENTS: Reviewed by Sale and Taylor (1970) and Taylor (1987).

*Bdeogale jacksoni* (Thomas, 1894). Ann. Mag. Nat. Hist., ser. 6, 13:522.
   COMMON NAME: Jackson's Mongoose.
   TYPE LOCALITY: "Mianzini, Masailand, 8000 feet [2438 m]" (pg. 523). Restricted by Moreau et al. (1945:410) to "Mianzini. . . a few miles E.S.E. of Naivasha and on the southern end of the Kinangop Plateau. . . 9000 ft [2743 m]" [Kenya].
   DISTRIBUTION: C Kenya, SE Uganda.
   STATUS: IUCN – Vulnerable.
   COMMENTS: Rosevear (1974) placed *jacksoni* in *Galeriscus*. Kingdon (1977) considered *jacksoni* conspecific with *nigripes*; however, Rosevear (1974) and Coetzee (1977*b*) noted skull and skin differences.

*Bdeogale nigripes* Pucheran, 1855. Rev. Mag. Zool Paris, 7(2):111.
   COMMON NAME: Black-footed Mongoose.
   TYPE LOCALITY: "Gubon" [Gabon].
   DISTRIBUTION: Nigeria to N Angola.
   STATUS: IUCN – Lower Risk (lc).
   COMMENTS: Rosevear (1974) placed *nigripes* in *Galeriscus*. Kingdon (1977) considered *jacksoni* conspecific with *nigripes*; however, Rosevear (1974) and Coetzee (1977*b*) noted skull and skin differences.

*Crossarchus* F. G. Cuvier, 1825. *In* É. Geoffroy Saint-Hilaire and F. G. Cuvier, Hist. Nat. Mammifères, pt. 3, 5(47), "Mangue," 3 pp., 1 pl.
   TYPE SPECIES: *Crossarchus obscurus* F. G. Cuvier, 1825, by original designation (Melville and Smith, 1987).
   COMMENTS: Revised by Goldman (1984) and Colyn and Van Rompaey (1994). Placed in *Mungos* by Hill and Carter (1941). Van Rompaey and Colyn (1992) presented a key to the species.

*Crossarchus alexandri* Thomas and Wroughton, 1907. Ann. Mag. Nat. Hist., ser. 7, 19:373.
   COMMON NAME: Alexander's Kusimanse.
   TYPE LOCALITY: "from Banzyville, Ubanghi" [= Mobayi, Zaire (=Dem. Rep. Congo), 4°N, 21°11'E (Goldman, 1984)].
   DISTRIBUTION: Central African Republic, Dem. Rep. Congo, Republic of Congo, Uganda.
   STATUS: IUCN – Lower Risk (lc).
   SYNONYMS: *minor* Goldman, 1984.
   COMMENTS: Colyn and Van Rompaey's (1994) study did not support the recognition of subspecies.

*Crossarchus ansorgei* Thomas, 1910. Ann. Mag. Nat. Hist., ser. 8, 5:195.
   COMMON NAME: Angolan Kusimanse.
   TYPE LOCALITY: "Dalla Tando" [= Angola, Ndala Tando, 9°18'S, 14°54'E (Goldman, 1984)].
   DISTRIBUTION: N Angola, SE Dem. Rep. Congo.
   STATUS: IUCN – Data Deficient as *C. a. ansorgei*, otherwise Lower Risk (lc).
   SYNONYMS: *nigricolor* Colyn and Van Rompaey, 1990.

*Crossarchus obscurus* F. G. Cuvier, 1825. *In* É. Geoffroy Saint-Hilaire and F. G. Cuvier, Hist. Nat. Mammifères, pt. 3, 5(47), "Mangue" 3 pp., 1 pl.
COMMON NAME: Common Kusimanse.
TYPE LOCALITY: "côtes occidentales de l'Afrique, et vraisemblablement des parties qui sont au midi de la Gambie" restricted by F. G. Cuvier (1829:158) to "Sierra Leone".
DISTRIBUTION: Côte d'Ivoire, Sierra Leone, Liberia, and Ghana (west of the Dahomey Gap).
STATUS: IUCN – Lower Risk (lc).
SYNONYMS: *punctatissimus* (Temminck, 1853).
COMMENTS: Goldman (1984) separated central (*C. platycephalus*) from western (*C. obscurus*) African populations based on phenetic differences in skull proportions. Wozencraft (1989b) argued that these populations are conspecific. Reviewed by Goldman (1987). The separation of *C. platycephalus* from *C. obscurus* has been generally supported by others (Colyn and Van Rompaey, 1994).

*Crossarchus platycephalus* Goldman, 1984. Can. J. Zool., 62(8):1624.
COMMON NAME: Flat-headed Kusimanse.
TYPE LOCALITY: "Eséka, Cameroon".
DISTRIBUTION: Benin, Nigeria, and Cameroon (east of the Dahomey Gap).
STATUS: IUCN – Lower Risk (lc).
COMMENTS: See comments under *C. obscurus*.

*Cynictis* Ogilby, 1833. Proc. Zool. Soc. Lond., 1833:48.
TYPE SPECIES: *Cynictis steedmanni* Ogilby, 1833 (= *Herpestes penicillatus* G. Cuvier, 1829).
COMMENTS: McKenna and Bell (1997) included *Paracynictis* Pocock, 1916, without discussion.

*Cynictis penicillata* (G.[Baron] Cuvier, 1829). Regn. Anim., Nouv. ed., 2 1:158.
COMMON NAME: Yellow Mongoose.
TYPE LOCALITY: "du Cap", restricted by Roberts (1951:151) to "Uitenhage, C.P." [South Africa].
DISTRIBUTION: S Angola, Botswana, Namibia, South Africa, SW Zimbabwe.
STATUS: IUCN – Lower Risk (lc).
SYNONYMS: *levaillantii* (A. Smith, 1829); *typicus* A. Smith, 1834; *steedmanni* Ogilby, 1833; **bechuanae** Roberts, 1932; **brachyura** Roberts, 1924; **bradfieldi** Roberts, 1924; **cinderella** Thomas, 1927; **coombsi** Roberts, 1929; **intensa** Schwann, 1906; **kalaharica** Roberts, 1932; **karasensis** Roberts, 1938; **lepturus** A. Smith, 1839; **ogilbyii** A. Smith, 1834; **pallidior** Thomas and Schwann, 1904.
COMMENTS: Synonyms allocated according to Ellerman and Morrison-Scott (1953) and Taylor and Meester (1993). Revised by Lundholm (1955b) and Taylor and Meester (1993).

*Dologale* Thomas, 1926. Ann. Mag. Nat. Hist., ser. 9, 17:183.
TYPE SPECIES: *Crossarchus dybowskii* Pousargues, 1893, by original designation.
COMMENTS: Revised by Hayman (1936). Although originally placed in *Crossarchus*, most since Hayman (1936) believed this genus to be the sister group to *Helogale*; Allen (1924) identified some specimens of this taxon as *Helogale hirtula robusta*. McKenna and Bell (1997) included it in *Helogale* without discussion.

*Dologale dybowskii* (Pousargues, 1893). Bull. Soc. Zool. Fr., 18:51.
COMMON NAME: Pousargues's Mongoose.
TYPE LOCALITY: "Ubangi, Congo Belge", restricted by Moreau et al. (1945:410) to "on the Upper Kemo, a tributary to the north of the Ubangui, about 6°17′N, 19°12′E" [Central African Republic].
DISTRIBUTION: Central African Republic, NE Dem. Rep. Congo, S Sudan, W Uganda.
STATUS: IUCN – Lower Risk (lc).
SYNONYMS: *nigripes* (Kershaw, 1924); *robusta* (J. A. Allen, 1924).
COMMENTS: Pousargues (1894) later redescribed the species in detail.

*Galerella* Gray, 1865. Proc. Zool. Soc. Lond., 1864:564 [1865].
TYPE SPECIES: *Herpestes ochraceus* Gray, 1849 by original designation.

COMMENTS: Revised by Lynch (1981), Watson and Dippenaar (1987), Watson (1990), and Taylor et al. (1991) who considered these taxa a monophyletic group. Crawford-Cabral (1989a:2) regarded this group as a "superspecies with several allospecies." These taxa are provisionally separated from *Herpestes* (*sensu latu*) following revisions and reviews by Rosevear (1974), Ansell (1978), Smithers (1983), and Meester et al. (1986) (see discussion under *Herpestes*). McKenna and Bell (1997) included in *Herpestes*.

*Galerella flavescens* (Bocage, 1889). J. Sci. Math. Phys. Nat. Lisboa, ser. 2, 3:179.
COMMON NAME: Angolan Slender Mongoose.
TYPE LOCALITY: "Benguella", [Angola].
DISTRIBUTION: S Angola, C and N Namibia.
STATUS: IUCN – Lower Risk (lc).
SYNONYMS: *annulatus* Lundholm, 1955; *nigratus* (Thomas, 1928); *shortridgei* (Roberts, 1932).
COMMENTS: Synonyms allocated after Crawford-Cabral (1989a, 1996). Included in *sanguinea* by Taylor (1975). The form *flavescens* was not mentioned in Meester et al. (1986) or Watson and Dippenaar's (1987) revision; Crawford-Cabral (1989a, 1996) considered *nigratus* conspecific with *flavescens*, the senior synonym. Watson (1990) considered these taxa in *nigratus*. Meester et al. (1986) listed *nigratus* as a synonym of *G. pulverulenta*. *G. shortridgei* and *annulatus* provisionally included here (see comments under *G. pulverulenta*).

*Galerella ochracea* (J. E. Gray, 1848). Proc. Zool. Soc. Lond., 1848:138.
common name: Somalian Slender Mongoose.
TYPE LOCALITY: "Abyssinia".
DISTRIBUTION: Somalia.
SYNONYMS: *bocagei* (Thomas and Wroughton, 1905); *fulvidior* (Thomas, 1904); *perfulvidus* (Thomas, 1904).
COMMENTS: Recognized as distinct by Azzaroli and Simonetta (1966) and Taylor and Goldman (1993).

*Galerella pulverulenta* (Wagner, 1839). Gelehrte. Anz. I. K. Bayer. Akad. Wiss. München, 9:426.
COMMON NAME: Cape Gray Mongoose.
TYPE LOCALITY: "Kap" [Cape of Good Hope, Western Cape Prov., South Africa].
DISTRIBUTION: South Africa, south of 27°S latitude (Bronner, 1990).
STATUS: IUCN – Lower Risk (lc).
SYNONYMS: *apiculatus* (Gray, 1865); *caffra* (A. Smith, 1826) [preoccupied]; *maritimus* (Roberts, 1919); *basuticus* (Roberts, 1936); *ruddi* (Thomas, 1903).
COMMENTS: Reviewed by Cavallini (1992). Revised by Lynch (1981) and Watson and Dippenaar (1987), who removed *annulata* and *shortridgei* (provisionally listed here under *G. flavescens*; perhaps best considered *incertae sedis*) and *nigrata* (placed also in *flavescens*) from *pulverulenta*; although Meester et al. (1986) did not. Crawford-Cabral (1989a) included these taxa with *flavescens*. Lynch (1981) argued that subspecies recognition was not warranted. Synonyms allocated accoding to Cavallini (1992) and Watson and Dippenaar (1987).

*Galerella sanguinea* (Rüppell, 1835). Neue Wirbelt. Fauna Abyssin. Gehörig. Säugeth., 1:27.
COMMON NAME: Slender Mongoose.
TYPE LOCALITY: "Kordofan" [Sudan].
DISTRIBUTION: Angola, Benin, Botswana, Burkina Faso, Cameroon, Cape Verde Isls, Central African Republic, Côte d'Ivoire, Dem. Rep. Congo, Equatorial Guinea, Ethiopia, Ghana, Kenya, Liberia, Malawi, Mauritana, Mozambique, Namibia, Niger, Republic of Congo, Rwanda, Senegal, Sierra Leone, Somalia, South Africa, Sudan, Tanzania, Togo, Uganda, Zambia, Zimbabwe.
STATUS: IUCN – Lower Risk (lc).
SYNONYMS: *canus* (Wroughton, 1907); *cauui* (A. Smith, 1836); *auratus* (Thomas and Wroughton, 1908); *badius* (A. Smith, 1838); *bradfieldi* (Roberts, 1932); *caldatus* (Thomas, 1927); *erongensis* (Roberts, 1946); *ignitoides* (Roberts, 1932); *kalaharicus* (Roberts, 1932); *kaokoensis* (Roberts, 1932); *khanensis* (Roberts, 1932); *ngamiensis* (Roberts, 1932); *okavangensis* (Roberts, 1932); *ornatus* (Peters, 1852); *punctulatus* (Gray,

1849); *ratlamuchi* (A. Smith, 1836); *upingtoni* (Shortridge, 1934); *venatica* (Gray, 1865); *zombae* (Wroughton, 1907); **dasilvai** (Roberts, 1938); **dentifer** (Heller, 1913); **fulvidior** Thomas, 1904; **galbus** (Wroughton, 1909); **gracilis** (Rüppell, 1835); *galinieri* (Gúerinr, 1847); *iodoprymnus* (Heuglin, 1861); *lefebvrii* (Prevost and Desmurs, 1850); *nigricaudatus* (I. Geoffroy Saint-Hilaire, 1839); *ochromelas* (Pucheran, 1855); *ruficauda* (Heuglin, 1877); **grantii** (Gray, 1865); **ibeae** (Wroughton, 1907); *elegans* (Matschie, 1914); *marae* (Matschie, 1914); **ignitus** (Roberts, 1913); **lancasteri** (Roberts, 1932); **melanura** (Martin, 1836); **mossambica** (Matschie, 1914); **mustela** Schwarz, 1935; **mutgigella** (Rüppell, 1835); *fuscus* (Rüppell, 1835); *mutscheltschela* (Heuglin, 1877); **orestes** (Heller, 1911); **parvipes** (Hollister, 1916); **perfulvidus** (Thomas, 1904); **phoenicurus** (Thomas, 1912); **proteus** (Thomas, 1907); **rendilis** (Lönnberg, 1912); **saharae** (Thomas, 1925); **swalius** (Thomas, 1926); **swinnyi** (Roberts, 1913); *ugandae* (Wroughton, 1909).

COMMENTS: This enigmatic group, reviewed by Taylor (1975) and Watson (1990), is represented by several allopatric populations (a situation similar to the *Genetta genetta* complex where they are recognized as conspecific). Watson and Dippenaar (1987), in their revision, argued for the separation of *nigratus* (placed here in *flavescens*), and *swalius* (considered here as conspecific), and considered *swinnyi* Roberts (1913), as *incertae sedis* (included here), although their study did not include representative samples from NE and W Africa. It is believed that *swinnyi* has been extirpated from the type locality (Watson, in litt.). Taylor (1989) suggested that the allopatric *ochraceus* from Somalia warrants full specific status (followed here). Certainly, these studies suggest that a thorough revision, inclusive of all of the African forms of *sanguinea* is badly needed. Taylor and Goldman (1993) present convincing evidence to suggest that *swalius* is conspecific with *sanguinea*, but also see Watson (1990).

*Helogale* Gray, 1862. Proc. Zool. Soc. Lond., 1861:308 [1862].
TYPE SPECIES: *Herpestes parvulus* Sundevall, 1847 by subsequent designation by Thomas (1882) (Melville and Smith, 1987).
COMMENTS: The number of taxa ascribed to this genus is provisional. Ellerman and Morrison-Scott (1953) recognized seven species, Allen (1939) listed eleven. The acceptance here of two follows Coetzee (1977*b*). The range of *H. hirtula* is included within that of *H. parvula*. McKenna and Bell (1997) included *Dologale* without discussion.

*Helogale hirtula* Thomas, 1904. Ann. and Mag. Nat. Hist., ser. 7, 14:97.
COMMON NAME: Ethiopian Dwarf Mongoose.
TYPE LOCALITY: "Gabridehari, 60 mi [96 km] West of Gerlogobi", restricted by Moreau et al. (1946:410) to "south-east Ethiopia (Ogaden) at about 7°0'N, 45°20'E". Further restricted by Yalden et al. (1980) to "Gabridehari (= Gabredarre, Kebridar) 6°45'N, 44°17'E".
DISTRIBUTION: S Ethiopia, N and C Kenya; S and C Somalia.
STATUS: IUCN – Lower Risk (lc).
SYNONYMS: **ahlselli** Lönnberg, 1912; **annulata** Drake-Brockman, 1912; **lutescens** Thomas, 1911; **powelli** Drake-Brockman, 1912.
COMMENTS: Synonyms allocated according to G. M. Allen (1939).

*Helogale parvula* (Sundevall, 1847). Ofv. K. Svenska Vet.-Akad. Forhandl, Stockholm, 1846. 3(4):121 [1847].
COMMON NAME: Common Dwarf Mongoose.
TYPE LOCALITY: "Caffraria superiore, juxta tropicum", restricted by Roberts (1951) to "Zoutpansberg" [South Africa].
DISTRIBUTION: Angola, Botswana, Dem. Rep. Congo, Ethiopia, Gambia, Kenya, Malawi, Mozambique, Namibia, Somalia, South Africa, Sudan, Tanzania, Uganda, Zambia.
STATUS: IUCN – Lower Risk (lc).
SYNONYMS: *brunnula* Thomas and Schwann, 1906; **ivori** Thomas, 1919; **mimetra** Thomas, 1926, *brunetta* Thomas, 1926; **nero** Thomas, 1928; *bradfieldi* Roberts, 1928; **ruficeps** Kershaw, 1922; **undulatus** (Peters, 1852); **varia** Thomas, 1902.
COMMENTS: Synonyms allocated according to Ellerman et al. (1953).

*Herpestes* Illiger, 1811. Prodr. Syst. Mamm. Avium., p. 135.

TYPE SPECIES: *Viverra ichneumon* Linnaeus, 1758, by absolute tautonomy, through the replaced name *Ichneumon* Lacépède, 1799 (Melville and Smith, 1987).

SYNONYMS: *Calogale* Gray, 1865; *Calictis* Gray, 1865; *Herpertes* Illiger, 1811; *Ichneumon* Frisch, 1775; *Mangusta* Horsfield, 1822; *Mesobema* Hodgson, 1841; *Onychogale* Gray, 1865; *Taeniogale* Gray, 1865; *Urva* Hodgson, 1837; *Xenogale* Allen, 1919.

COMMENTS: Revised by Pocock (1919, 1937, 1941*a*), Bechthold (1939), and Taylor and Matheson (1999). Coetzee (1977*b*) and Hayman (*in* Sanderson, 1940) included *Xenogale* (see discussion under *naso*). Allen (1924) included only *ichneumon* in this genus and separated *sanguinea* and *pulverulenta* into *Galerella*; for support, he contrasted the large *ichneumon* with the smaller *sanguine-pulverulenta* complex and reported proportion differences in measurements of skeleton and skull. His rationale has been repeated, in some cases verbatum, by Rosevear (1974), Ansell (1978), Smithers (1983), Meester et al. (1986), and Watson and Dippenaar (1987). Taylor et al. (1991) presented an allozyme analysis and argued for generic recognition, however, they did not include Asiatic *Herpestes*, and their consensus tree made the placement of the *sanguinea/pulverulenta* clade equivocal. Fredga's (1972) comparative chromosome analysis looked at variation including Asiatic and African *Herpestes*, and based on this, recognition of *Galerella* would make *Herpestes* paraphyletic. Comparison of measurements from Allen (1924), Rosevear (1974), and Smithers (1983) for African forms, and Bechthold (1939) and Pocock (1941*a*) for Asiatic forms reveals that the large morphological gaps originally identified by Allen (1924) dissolve when Asiatic species are included. Ellerman and Morrison-Scott (1953) and Wozencraft (1989*b*) suggested that differences between these taxa and other *Herpestes* are less than those found within *Herpestes*. Morphological criteria similar to that used by Allen (1924) have been used mostly at the specific level in other carnivores. Although the case is not strong, many recent authors have separated *Galerella* and this is provisionally followed here (see comments under *Galerella*) as nearly all agree that this is a monophyletic group. McKenna and Bell (1997) included *Galerella*. Taylor and Matheson's (1999) phenetic analysis illustrated interesting parallel developments between Asiatic and African mongooses. Veron et al. (2004) suggested that placement of *H. auropunctata* and *H. ichneumon* in the same genus group to the exclusion of *Ichneumia* would make *Herpestes* paraphyletic.

*Herpestes brachyurus* Gray, 1837. Proc. Zool. Soc. Lond., 1836:88 [1837].

COMMON NAME: Short-tailed Mongoose.

TYPE LOCALITY: "Indian Islands", restricted by Kloss (1917) to "Borneo", however, Thomas (1921*c*) believed it to be from "Malacca". Pocock (1941*a*) believed the type to represent a "Malayan Race".

DISTRIBUTION: Indonesia (Kalimantan, Sumatra), Malaysia, Philippine Isls, Singapore, Vietnam.

STATUS: IUCN – Lower Risk (lc).

SYNONYMS: **hosei** Jentink, 1903; *dyacorum* Thomas, 1921; *rajah* Thomas, 1921; **javanensis** Bechthold, 1936; **palawanus** Allen, 1910; **parvus** Jentink, 1895; **sumatrius** Thomas, 1921.

COMMENTS: Bechthold (1939) included *hosei* (followed here) and *fuscus* (here considered separate), and listed characteristics suggesting that in some respects, *semitorquatus* was intermediate between *brachyurus* and *urva*; this was followed by Medway (1977). Bechthold (1939) believed that *fuscus* (*sensu stricto*) is most closely related to far-eastern *brachyurus* forms and considered them conspecific (both forms are short tailed mongooses); however, he gave features of the skull and pelage (used elsewhere at the specific level, i.e., *edwardsii* vs. *javanicus*) that distinguished the S India/Sri Lankan populations from those of SE Asia. Here they are provisionally treated as separate. Schwarz (1947) believed *semitorquatus* to be a red color morph of the dark *brachyurus*, although he did not address the most distinguishing feature of the collared mongoose – the collar – present in *semitorquatus* and absent in *brachyurus*. Medway (1977), followed by Payne et al. (1985), recognized *hosei* based on differences in the shape of the coronoid process of the mandible. Fredga (1972) believed *fuscus* to be intermediate to *edwardsii* and *javanicus*, but believed that it should remain separate (followed by

Corbet and Hill, 1992). Synonyms allocated according to Pocock (1937) and Ellerman and Morrison-Scott (1951).

*Herpestes edwardsi* (É. Geoffroy Saint-Hilaire, 1818). Descrip. de L'Egypte, 2:139.
COMMON NAME: Indian Gray Mongoose.
TYPE LOCALITY: "Indes orientales".
DISTRIBUTION: Afghanistan, Bahrain, India, Indonesia, Iran, Japan, Kuwait, Nepal, Pakistan, Saudia Arabia, Sri Lanka. Populations believed to be introductions in Malaysia, Ryukyu Isls, Mauritius (Corbet and Hill, 1980; Wells, 1989).
STATUS: CITES – Appendix III (India); IUCN – Lower Risk (lc).
SYNONYMS: *carnaticus* Wroughton, 1921; *ellioti* (Wroughton, 1915); *fimbriatus* Temminck, 1853; *frederici* Desmarest, 1823; *griseus* (I. Geoffroy Saint-Hilaire, 1818); *malaccensis* (J. B. Fischer, 1829); *moerens* (Wroughton, 1915); *pallidus* Wagner, 1841; *pondiceriana* Gervais, 1841; **ferrugineus** Blanford, 1874; *andersoni* Murray, 1884; *pallens* (Ryley, 1914); **lanka** (Wroughton, 1915); *mungo* (Blanford, 1888) [preoccupied]; **montanus** Bechthold, 1936; **nyula** (Hodgson, 1836).

*Herpestes fuscus* Waterhouse, 1838. Proc. Zool. Soc. Lond., 1838:55.
COMMON NAME: Indian Brown Mongoose.
TYPE LOCALITY: "India".
DISTRIBUTION: SW India, Sri Landa.
STATUS: CITES – Appendix III (India) as *H. brachyurus fuscus*; IUCN – Vulnerable as *H. f. fuscus*.
SYNONYMS: **flavidens** Kelaart, 1850; *ceylanicus* Nevill, 1887; *ceylonicus* Thomas, 1924; *fulvescens* Kelaart, 1851; *phillipsi* Thomas, 1924; **maccarthiae** (Gray, 1851); **rubidior** Pocock, 1937; **siccatus** Thomas, 1924.
COMMENTS: Some have placed *fuscus* in *brachyurus* (Bechthold, 1936; Wenzel and Haltenorth, 1972). Fredga (1972) and Corbet and Hill (1992) argued that it should remain separate. Corbet and Hill (1992) recognized only four subspecies (*fuscus, phillipsi, siccatus, rubidior*).

*Herpestes ichneumon* (Linnaeus, 1758). Syst. Nat., 10th ed., 1:43.
COMMON NAME: Egyptian Mongoose.
TYPE LOCALITY: "in Ægypto ad ripas Nili, . . . in India primario; mansuescit", restricted by Thomas (1911a) to "Egypt".
DISTRIBUTION: Algeria, Angola, Botswana, Cameroon, Chad, Côte d'Ivoire, Dem. Rep. Congo, Egypt, Ethiopia, Gambia, Ghana, Gibraltar, Guinea, Israel, Jordan, Kenya, Lebanon, Liberia, Lybia, Malawi, Morocco, Mozambique, Niger, Portugal, Rwanda, Senegal, Sierra Leone, South Africa, Spain, Sudan, Syria, Tanzania, Togo, Tunisia, Turkey, Uganda, Zambia.
STATUS: IUCN – Lower Risk (lc).
SYNONYMS: *aegyptiae* (Tiedemann, 1808); *egypti* (Tiedemann, 1808); *major* (É. Geoffroy Saint-Hilaire, 1818); *pharaon* (Lacépède, 1799); **angolensis** Bocage, 1890; **cafra** (Gmelin, 1788); *bennettii* Gray, 1837; *dorsalis* Gray, 1865; *griseus* Smuts, 1832; *madagascarensis* A. Smith, 1834; *nems* (Kerr, 1792); **centralis** (Lönnberg, 1917); **funestus** (Osgood, 1910); **mababiensis** Roberts, 1932; **numidicus** (F. G. Cuvier, 1834); *numidianus* Gray, 1865; **parvidens** (Lönnberg, 1908); **sabiensis** Roberts, 1926; **sangronizi** Cabrera, 1924; **widdringtonii** Gray, 1842; *dorsalis* Seabra, 1909; *ferruginea* Seabra, 1909; *grisea* Seabra, 1909.

*Herpestes javanicus* (É. Geoffroy Saint-Hilaire, 1818). Descrip. de L'Egypte, 2:138.
COMMON NAME: Small Asian Mongoose.
TYPE LOCALITY: "Java".
DISTRIBUTION: Afghanistan, Bangladesh, Bhutan, Burma, Cambodia, China, India, Indonesia, Malaysia, Nepal, Pakistan, Thailand, and Vietnam. Introduced to: Bosnia and Herzegovina, Croatia, Cuba, Dominican Republic, Fiji Isls, Jamacia, Japan, Puerto Rico, Surinam, West Indies, USA (Hawaiian Isls), and many other tropical regions.
STATUS: CITES – Appendix III (India) as *H. javanicus auropunctatus*; IUCN – Endangered as *H. palustris*, Lower Risk (lc) as *H. javanicus*.

SYNONYMS: ***auropunctatus*** (Hodgson, 1836); *birmanicus* Thomas, 1886; *nepalensis* Gray, 1837; ***exilis*** Gervais, 1841; *rutilus* Gray, 1861; ***orientalis*** (Sody, 1936); ***pallipes*** (Blyth, 1845); *helvus* (Ryley, 1914); *persicus* Gray, 1865; ***palustris*** Ghose, 1965; ***peninsulae*** (Schwarz, 1910); *incertus* (Kloss, 1917); ***perakensis*** (Kloss, 1917); ***rafflesii*** Anderson, 1875; ***rubrifrons*** J. A. Allen, 1909; ***siamensis*** (Kloss, 1917); *tjerapai* Sody, 1949.

COMMENTS: Bechthold (1939), Pocock (1941*a*), and Lekagul and McNeeley (1977) included *auropunctatus*. Wells (1989) discussed the situation for the morphotypes in Indochina. Nellis (1989) considered *auropunctatus* as a separate species. Taylor and Matheson's (1999) skull morphometic study showed that the oriental subspecies can be distinguished from the northern and western subspecies based on a phenetic analysis of skull measurements. Ghose (1965) separated *palustris* from *javanicus*. Wenzel and Haltenorth (1972) and Corbet and Hill (1992) considered *palustris*, *auropunctatus*, and *javanicus* as conspecific, which is followed here.

*Herpestes naso* de Winton, 1901. Bull. Liverpool Mus., 3:35.
COMMON NAME: Long-nosed Mongoose.
TYPE LOCALITY: "Cameroon River, West Africa" [Cameroon].
DISTRIBUTION: Cameroon, Dem. Rep. Congo, Equatorial Guinea, Gabon, Kenya, Niger, Republic of Congo, Tanzania.
STATUS: IUCN – Lower Risk (lc).
SYNONYMS: *almodovari* Cabrera, 1902; *microdon* (J. A. Allen, 1919); *nigerianus* Thomas, 1912.
COMMENTS: Placed in *Xenogale* by Allen (1919*b*), and followed by Rosevear (1974), Ansell (1978) and Colyn and Van Rompaey (1994). This taxon, and *ichneumon*, which is generally recognized as its sister taxon (Allen, 1919*b*; Hayman, *in* Sanderson, 1940; Rosevear, 1974), can be distinguished principally by proportional differences of the interorbital region (Rosevear, 1974). Allen (1919*b*), Hayman (*in* Sanderson, 1940), Wenzel and Haltenorth (1972), Rosevear (1974), and Coetzee (1977*b*) did not feel these differences were sufficient to warrant generic distinction. Recognition of *Xenogale* would make *Herpestes* paraphyletic. Reviewed by Orts (1970) and Colyn and van Rompaey (1994) who provide an excellent summary of the taxonomic discussion and who could find no subspecies differentiation.

*Herpestes semitorquatus* Gray, 1846. Ann. Mag. Nat. Hist., [ser. 1], 18:211.
COMMON NAME: Collared Mongoose.
TYPE LOCALITY: "Borneo" [Mainland opposite Labuan = Brunei)].
DISTRIBUTION: Indonesia (Borneo and Sumatra).
STATUS: IUCN – Lower Risk (lc).
SYNONYMS: ***uniformis*** Robinson and Kloss, 1919.
COMMENTS: Schwarz (1947) concluded that *semitorquatus* was a red color morph of the dark *brachyurus*, however, see comments under *brachyurus*.

*Herpestes smithii* Gray, 1837. Mag. Nat. Hist. [Charlesworth's], 1:578.
COMMON NAME: Ruddy Mongoose.
TYPE LOCALITY: Not given. Thomas (1923) suggested that it was from the "Bombay Region" but this was questioned by Pocock (1937).
DISTRIBUTION: India, Sri Lanka.
STATUS: CITES – Appendix III (India); IUCN – Lower Risk (lc).
SYNONYMS: *ellioti* Blyth, 1851; *torquatus* Kelaart, 1852; ***thysanurus*** Wagner, 1839; *canens* Thomas, 1921; *jerdonii* Gray, 1865; *monticolus* Jerdon, 1867; *rusanus* Thomas, 1921; ***zeylanius*** Thomas, 1921.

*Herpestes urva* (Hodgson, 1836). J. Asiat. Soc. Bengal, 5:238.
COMMON NAME: Crab-eating Mongoose.
TYPE LOCALITY: "Central and Northern Regions" [Nepal].
DISTRIBUTION: Burma, China, India, Laos, Malaysia (Wells and Francis, 1988), Nepal, Taiwan, Thailand, Vietnam.
STATUS: CITES – Appendix III (India); IUCN – Lower Risk (lc).
SYNONYMS: *cancrivora* (Hodgson, 1837); *hanensis* (Matschie, 1907); ***annamensis*** Bechthold, 1936; ***formosanus*** Bechthold, 1936; ***sinensis*** Bechthold, 1936.

COMMENTS: Synonyms allocated after Bechthold (1939).

*Herpestes vitticollis* Bennett, 1835. Proc. Zool. Soc. Lond., 1835:67.
 COMMON NAME: Stripe-necked Mongoose.
 TYPE LOCALITY: "in forests about twenty miles [32 km] inland from Kolun or Quilon, in the Travancore country" [India].
 DISTRIBUTION: S India, Sri Lanka.
 STATUS: CITES – Appendix III (India); IUCN – Lower Risk (lc).
 SYNONYMS: *rubiginosus* (Wagner, 1841); ***inornatus*** Pocock, 1941.

*Ichneumia* I. Geoffroy Saint-Hilaire, 1837. Ann. Sci. Nat. Zool. (Paris), 8(2):251.
 TYPE SPECIES: *Herpestes albicaudus* G. [Baron] Cuvier, 1829, by designation of Geoffroy Saint-Hilaire (1839) (Melville and Smith, 1987).
 SYNONYMS: *Lasiopus* I. Geoffroy Saint-Hilaire, 1835.

*Ichneumia albicauda* (G.[Baron] Cuvier, 1829). Regn. Anim., Nouv., ed. 2, 1:158.
 COMMON NAME: White-tailed Mongoose.
 TYPE LOCALITY: "l'Afrique australe et le Sénégal".
 DISTRIBUTION: Angola, Botswana, Burkina Faso, Central African Republic, Côte d'Ivoire, Dem. Rep. Congo, Ghana, Kenya, Kenya, Mozambique, Namibia, Niger, Nigeria, Oman, Senegal, Sierra Leone, Somalia, South Africa, Sudan, Yemen, Zambia, Zimbabwe.
 STATUS: IUCN – Lower Risk (lc).
 SYNONYMS: *abuwudan* Fitzinger and Heuglin, 1866; *albescens* I. Geoffroy Saint-Hilaire, 1839; *leucurus* (Hemprich and Ehrenberg, 1832); ***dialeucos*** (Hollister, 1916); ***grandis*** (Thomas, 1890); ***haagneri*** Roberts, 1924; ***ibeanus*** (Thomas, 1904); *ferox* (Heller, 1913); ***loandae*** (Thomas, 1904); ***loempo*** (Temminck, 1853); *nigricauda* Pucheran, 1855.
 COMMENTS: Reviewed by Taylor (1972).

*Liberiictis* Hayman, 1958. Ann. Mag. Nat. Hist., ser. 13, 1:449.
 TYPE SPECIES: *Liberiictis kuhni* Hayman, 1958, by original designation.

*Liberiictis kuhni* Hayman, 1958. Ann. Mag. Nat. Hist., ser. 13, 1:449.
 COMMON NAME: Liberian Mongoose.
 TYPE LOCALITY: "Kpeaplay, north-east Liberia, about 6°36′N, 8°30′W".
 DISTRIBUTION: Liberia, Côte d'Ivoire.
 STATUS: IUCN – Endangered.
 COMMENTS: Reviewed by Goldman and Taylor (1990).

*Mungos* É. Geoffroy Saint-Hilaire and F. G. Cuvier, 1795. Mag. Encyclop., 2:184, 187.
 TYPE SPECIES: Not given; *Viverra mungo* Gmelin, 1788, designated by Muirhead (1819) (Melville and Smith, 1987). McKenna and Bell (1997) argued that *Herpestes fasciatus* (Desmarest, 1823) should be considered as the type.
 SYNONYMS: *Ariela* Gray, 1864.
 COMMENTS: Allen (1919*b*) discussed the nomenclatural history of this name.

*Mungos gambianus* (Ogilby, 1835). Proc. Zool. Soc. Lond., 1835:102.
 COMMON NAME: Gambian Mongoose.
 TYPE LOCALITY: "Gambia".
 DISTRIBUTION: Côte d'Ivoire, Gambia, Ghana, Niger, Nigeria, Senegal, Sierra Leone, Togo.
 STATUS: IUCN – Data Deficient.

*Mungos mungo* (Gmelin, 1788). *In* Linnaeus, Syst. Nat., 13th ed., 1:84.
 COMMON NAME: Banded Mongoose.
 TYPE LOCALITY: "Bengala, Persia, aliisque asiae", restricted by Ogilby (1835:101) to "Gambia". However, Thomas (1882) believed it to be in the eastern part of South Africa, [former] Cape Prov., as did Roberts (1929).
 DISTRIBUTION: Angola, Botswana, Burundi, Cameroon, Central African Republic, Chad, Dem. Rep. Congo, Ethiopia, Guinea-Bissau, Kenya, Malawi, Mozambique, Namibia,

Niger, Nigeria, Rwanda, Senegal, Somalia, South Africa, Sudan, Tanzania, Uganda, Zambia, Zimbabwe.

STATUS: IUCN – Lower Risk (lc).

SYNONYMS: *fasciatus* (Desmarest, 1823); *taenionotus* (A. Smith, 1834); **adailensis** (Heuglin, 1861); *gothneh* (Heuglin and Fitzinger, 1866); *leucostethicus* (Heuglin and Fitzinger, 1866); **bororensis** Roberts, 1929; **caurinus** Thomas, 1926; *colonus* (Heller, 1911); **grisonax** Thomas, 1926; **mandjarum** (Schwarz, 1915); **marcrurus** (Thomas, 1907); *macrosus* (Lydekker, 1908); *ngamiensis* Roberts, 1932; **pallidipes** Roberts, 1929; **rossi** Roberts, 1929; **senescens** (Thomas and Wroughton, 1907); **somalicus** (Thomas, 1895); **talboti** (Thomas and Wroughton, 1907); **zebra** (Rüppell, 1835); **zebroides** (Lönnberg, 1908).

*Paracynictis* Pocock, 1916. Ann. Mag. Nat. Hist., ser. 8, 17:177.

TYPE SPECIES: *Cynictis selousi* de Winton, 1896, by original designation (Melville and Smith, 1987).

*Paracynictis selousi* (de Winton, 1896). Ann. Mag. Nat. Hist., ser. 6, 18:469.

COMMON NAME: Selous' Mongoose.

TYPE LOCALITY: "found on a grassy heap under a tree, EssexVale, Matabeleland. . . near Bulawayo" [Zimbabwe].

DISTRIBUTION: Angola, Botswana, Malawi, Mozambique, Namibia, South Africa, Zambia, Zimbabwe.

STATUS: IUCN – Lower Risk (lc).

SYNONYMS: **bechuanae** Roberts, 1932; **ngamiensis** Roberts, 1932; **sengaani** Roberts, 1931.

COMMENTS: McKenna and Bell (1997) included in *Cynictis* without comment.

*Rhynchogale* Thomas, 1894. Proc. Zool. Soc. Lond., 1894:139.

TYPE SPECIES: *Rhinogale melleri* Gray, 1865, by monotypy through the replaced name *Rhinogale* Gray, 1865.

SYNONYMS: *Rhinogale* Gray, 1865.

*Rhynchogale melleri* (Gray, 1865). Proc. Zool. Soc. Lond., 1864:575 [1865].

COMMON NAME: Meller's Mongoose.

TYPE LOCALITY: "from a ravine on the outskirts of the Otto Estate, near Mbweni, about 2½ miles [4 km] west of Kilosa, Tanganyika Territory"

DISTRIBUTION: Dem. Rep. Congo, Malawi, Mozambique, South Africa, Tanzania, Zambia, Zimbabwe.

STATUS: IUCN – Lower Risk (lc).

SYNONYMS: *caniceps* (Kershaw, 1924); **langi** Roberts, 1938.

*Suricata* Desmarest, 1804. Tabl. Méth. Hist. Nat., *In*, Nouv. Dict. Hist. Nat., 24:15.

TYPE SPECIES: *Suricata capensis* Desmarest, 1804 (= *Viverra suricatta* Schreber, 1776), by monotypy (Melville and Smith, 1987).

SYNONYMS: *Rhyzaena* Wagner, 1841; *Rysaena* Lesson, 1827; *Ryzaena* Illiger, 1811; *Surricata* Gray, 1821.

*Suricata suricatta* (Schreber, 1776). Die Säugethiere pl. 117 [1776].

COMMON NAME: Meerkat.

TYPE LOCALITY: Listed as "Cape of Good Hope" by Meester et al. (1986), restricted by Thomas and Schwann (1905:133) to "Deelfontein" [South Africa].

DISTRIBUTION: Angola, S Botswana, Namibia, South Africa.

STATUS: IUCN – Lower Risk (lc).

SYNONYMS: *capensis* Desmarest, 1804; *hahni* Thomas, 1927; *hamiltoni* Thomas and Schwann, 1905; *lophurus* Thomas and Schwann, 1905; *namaquensis* Thomas and Schwann, 1905; *suraktta* (A. Smith, 1826); *tetradactyla* (Pallas, 1777); *typicus* (A. Smith, 1834); *viverrina* Desmarest, 1819; *zenik* (Scopoli, 1786); **iona** Cabral, 1971; **marjoriae** Bradfield, 1936.

COMMENTS: Reviewed by van Staaden (1994). Synonyms allocated according to Coetzee (1977b) and van Staaden (1994).

**Family Hyaenidae** Gray, 1821. London Med. Repos., 15:302.
SYNONYMS: Protelidae Flower, 1869.
COMMENTS: Reviewed by Ronnefeld (1969), Werdelin and Solounias (1991), and Jenks and Werdelin (1998), which is followed here.

*Crocuta* Kaup, 1828. Oken's Isis. Encyclop. Zeit 21(11), column 1145.
TYPE SPECIES: *Canis crocuta* Erxleben, 1777, by original designation.
SYNONYMS: *Crocotta* Kaup, 1829.
COMMENTS: Antedated by *Crocuta* Meigen, 1800 (an insect), but that name has been suppressed (International Commission on Zoological Nomenclature, 1962).

*Crocuta crocuta* (Erxleben, 1777). Syst. Regni Anim., 1:578.
COMMON NAME: Spotted Hyena.
TYPE LOCALITY: "Guinea, Aethiopia, ad caput bonae spei in terrae rupiumque caueis", restricted by Cabrera (1911:95) to "Senegambia".
DISTRIBUTION: Angola, Botswana, Cameroon, Dem. Rep. Congo, Ethiopia, Gabon, Gambia, Guinea, Kenya, Malawi, Mauritania, Mozambique, Namibia, Nigeria, Senegal, Sierra Leone, Somalia, South Africa, Sudan, Tanzania, Togo, Uganda, Zambia, Zimbabwe.
STATUS: IUCN – Lower Risk (cd).
SYNONYMS: *capensis* (Desmarest, 1817); *cuvieri* (Boitard, 1842); *fisi* Heller, 1914; *fortis* J. A. Allen, 1924; *gariepensis* (Matschie, 1900); *germinans* (Matschie, 1900); *habessynica* (de Blainville, 1844); *kibonotensis* (Lönnberg, 1908); *leontiewi* (Satunin, 1905); *maculata* (Thunberg, 1811); *noltei* (Matschie, 1900); *nyasae* Cabrera, 1911; *nzoyae* Cabrera, 1911; *panganensis* (Lönnberg, 1908); *rufa* (Desmarest, 1817); *rufopicta* Cabrera, 1911; *sivalensis* (Falconer and Cautley, *In* Falconer, 1868); *thierryi* (Matschie, 1900); *thomasi* Cabrera, 1911; *togoensis* (Matschie, 1900); *wissmanni* (Matschie, 1900).
COMMENTS: Revised by Matthews (*1939a, b*). Synonyms according to Matthews (1939a) and Jenks and Werdelin (1998) who demonstrated that subspecies are not justified.

*Hyaena* Brisson, 1762. Regnum Animale, ed. 2., pg. 168.
TYPE SPECIES: *Canis hyaena* Linnaeus, 1758, by original designation.
SYNONYMS: *Euhyaena* Falconer, 1868; *Hyena* Gray, 1821.
COMMENTS: Revised by Pocock (1934c). *Hyaena* Brisson, 1762, is available (International Commission on Zoological Nomenclature, 1955b, 1998), even though the work *Regnum Animale* has been rejected. The relationship of *H. hyaena* and *H. brunnea* has been problematic. McKenna and Bell (1997) followed Galiano and Frailey (1977) and placed *brunnea* in *Pachycrocuta*. Werdelin and Solounias (1991) argued that *brunnea* should be placed in *Parahyaena*. Jenks and Werdelin (1998) placed *brunnea* back in *Hyaena* and used subgenera to maintain Hendey's (1978) original intention (which is followed here). Synonyms according to Jenks and Werdelin (1998).

*Hyaena brunnea* Thunberg, 1820. K. Svenska Vet.-Acad. Handl. Stockholm, p. 59.
COMMON NAME: Brown Hyena.
TYPE LOCALITY: "Goda Hopps Udden; Södra Afrika" [South Africa, Western Cape Prov., Cape of Good Hope].
DISTRIBUTION: Botswana, Mozambique, Namibia, South Africa, Zimbabwe.
STATUS: U.S. ESA – Endangered as *Parahyaena* (=*Hyaena*) *brunnea*; IUCN – Lower Risk (nt).
SYNONYMS: *fusca* É. Geoffroy Saint-Hilaire, 1825; *melampus* Pocock, 1934; *striata* A. Smith, 1826; *villosa* A. Smith, 1827.
COMMENTS: Reviewed by Mills (1982). See comments under genus for inclusion of *brunnea* under *Hyaena*. Pocock (1934c) and Jenks and Werdelin (1998) argued that neither morphological nor molecular studies have supported the recognition of subspecies at present. Synonyms according to Jenks and Werdelin (1998).

*Hyaena hyaena* (Linnaeus, 1758). Syst. Nat., 10th ed. 1:40.
COMMON NAME: Striped Hyena.
TYPE LOCALITY: "India", restricted by Thomas (1911a:134) to "Benna Mts., Laristan, S. Persia".
DISTRIBUTION: Afghanistan, Algeria, Armenia, Azerbaijan, Egypt, Ethiopia, India, Iran, Iraq,

Israel, Kenya, Libya, Mali, Morocco, Nepal, Nigeria, Pakistan, Saudia Arabia, Sierra Leone, Somalia, South Africa, Sudan, Tanzania, Turkmenistan, Uzbekistan, Yemen.

STATUS: U.S. ESA and IUCN – Data Deficient as *H. hyaena barbara*, otherwise Lower Risk (nt).

SYNONYMS: *antiquorum* (Temminck, 1820); *barbara* de Blainville, 1844; *bergeri* Matschie, 1910; *bilkiewiczi* Satunin, 1905; *bokcharensis* Satunin, 1905; *dubbah* Meyer, 1793; *dubia* Schinz, 1821; *fasciata* Thunberg, 1820; *hienomelas* Matschie, 1900; *hyaenomelas* (Bruce, *In* Desmarest, 1820); *indica* de Blainville, 1844; *orientalis* Tiedemann, 1808; *rendilis* Lönnberg, 1912; *satunini* Matschie, 1910; *schillingsi* Matschie, 1900; *striata* Zimmermann, 1777; *suilla* Filippi, 1853; *sultana* Pocock, 1934; *syriaca* Matschie, 1900; *virgata* Ogilby, 1840; *vulgaris* Desmarest, 1820; *zarudnyi* Satunin, 1905.

COMMENTS: Reviewed by Rieger (1981) and Jenks and Werdelin (1998). Pocock (1934*d*) and Jenks and Werdelin (1998) argued that at present neither morphological nor molecular studies have supported the recognition of subspecies. Synonyms according to Rieger (1981) and Jenks and Werdelin (1998).

**Proteles** I. Geoffroy Saint-Hilaire, 1824. Bull. Sci. Soc. Philom. Paris, 1824:139.

TYPE SPECIES: *Proteles lalandii* I. Geoffroy Saint-Hilaire, 1824 (= *Viverra cristata* Sparrman, 1783), by original designation (Melville and Smith, 1987).

SYNONYMS: *Geocyon* Wagler, 1830.

**Proteles cristata** (Sparrman, 1783). Resa Goda-Hopps-Udden., I. 1:581.

COMMON NAME: Aardwolf.

TYPE LOCALITY: English translation (Sparrman, 1786) of original locality: "Agter-Bruntjes hoogte . . . which takes in the upper part of Kleine Visch-rivier, and is separated from Camdebo by Bruntjes hoogtens . . ."; listed in G. M. Allen (1939) as "Near Little Fish River, Somerset East, Cape Colony" [South Africa].

DISTRIBUTION: Angola, Botswana, Central African Republic, Egypt, Ethiopia, Kenya, Mozambique, Namibia, Somalia, South Africa, Sudan, Tanzania, Uganda, Zambia, Zimbabwe.

STATUS: CITES – Appendix III (Botswana); IUCN – Lower Risk (lc).

SYNONYMS: *canescens* Shortridge and Carter, 1938; *harrisoni* Rothschild, 1902; *hyenoides* (Desmarest, 1821); *lalandii* I. Geoffroy Saint-Hilaire, 1824; *pallidior* Cabrera, 1910; *septentrionalis* Rothschild, 1902; *termes* Heller, 1913; *transvaalensis* Roberts, 1932; *typicus* A. Smith, 1834.

COMMENTS: Reviewed by Koehler and Richardson (1990) and Jenks and Werdelin (1998) who demonstrated that subspecies are not well defined and probably should not be recognized. Synonyms according to Jenks and Werdelin (1998).

# SUBORDER CANIFORMIA Kretzoi, 1938.

**Family Canidae** Fischer, 1817. Mém. Soc. Imp. Nat. Moscow, 5:372.

COMMENTS: Conservation status and distribution reviewed by Ginsberg and Macdonald (1990). Reviewed by Langguth (1975), Stains (1975), Tedford et al. (1995), and Wayne et al. (1997). Revisions by Langguth (1969), Clutton-Brock et al. (1976), Van Gelder (1978), Berta (1985, 1988), Wayne and O'Brien (1987), Wayne (1993), and Wayne et al. (1987*a, b*, 1989, 1997) gave little support to the subfamilies recognized by Simpson (1945); therefore, no subfamilies are recognized here. There are considerable questions regarding the validity of the South American genera (Xiaoming Wang et al., 1999; Wayne et al., 1997). Van Gelder's (1978) hybridization criteria for generic classification resulted in the recognition of only a few genera, including some paraphyletic groups.

**Atelocynus** Cabrera, 1940. Notas Mus. La Plata, 5:14.

TYPE SPECIES: *Canis microtis* Sclater, 1883, by original designation.

SYNONYMS: *Canis* Sclater, 1883 (preoccupied by *Canis* Linneaus, 1758); *Carcinocyon* J. A. Allen, 1905.

COMMENTS: See comments under *Dusicyon*. Placed in *Atelocynus* by Cabrera (1931, 1957), Langguth (1975), Stains (1975), Berta (1985, 1986, 1988) and McKenna and Bell (1997).

Van Gelder (1978) considered *Atelocynus* a subgenus of *Canis*. Tedford et al. (1995) placed it as the sister taxon to *Speothos*.

*Atelocynus microtis* (Sclater, 1883). Proc. Zool. Soc. Lond., 1882:631 [1883].
     COMMON NAME: Short-eared Dog.
     TYPE LOCALITY: "Amazons," restricted by Hershkovitz (1957*a*) to "south bank of the
         Rio Amazonas, Pará, Brazil."
     DISTRIBUTION: Amazonian basin: Bolivia (see Anderson, 1997), Brazil, Colombia, Ecuador,
         Peru, Venezuela (?).
     STATUS: IUCN – Data Deficient.
     SYNONYMS: *sclateri* J. A. Allen, 1905.
     COMMENTS: Reviewed by Hershkovitz (1961*a*) and Berta (1986).

*Canis* Linnaeus, 1758. Syst. Nat., 10th ed., 1:38.
     TYPE SPECIES: *Canis familiaris* Linnaeus, 1758 (= *Canis lupus* Linnaeus, 1758), by Linnean
         tautonomy (Melville and Smith, 1987).
     SYNONYMS: *Alopedon* Hilzheimer, 1906; *Alopsis* Rafinesque, 1815; *Chaon* C. E. H. Smith, 1839;
         *Dasycyon* Krumbiegel, 1953; *Dieba* Bray, 1869; *Lupulella* Hilzheimer, 1906; *Lupulus*
         Gervais, 1855; *Lupus* Oken, 1816; *Lyciscus* C. E. H. Smith, 1839; *Mamcanisus* Herrera,
         1899; *Neocyon* Gray, 1868; *Oreocyon* Krumbiegel, 1949; *Oxygous* Hodgson, 1841; *Sacalius*
         C. E. H. Smith, 1839; *Schaeffia* Hilzheimer, 1906; *Simenia* Gray, 1868; *Thos* Oken, 1816;
         *Vulpicanis* de Blainville, 1837.
     COMMENTS: Van Gelder (1978) included *Alopex*, *Atelocynus*, *Cerdocyon*, *Pseudalopex*, *Lycalopex*,
         *Dusicyon*, and *Vulpes* as subgenera, however, this arrangement is not currently employed
         by most mammalogists (Berta, 1987, 1988; Corbet, 1978; Corbet and Hill, 1980; Gromov
         and Baranova, 1981; Hall, 1981; McKenna and Bell, 1997; Wozencraft, 1989). Synonyms
         allocated according to McKenna and Bell (1997).

*Canis adustus* Sundevall, 1847. Ofv. K. Svenska Vet.-Akad. Forhandl., Stockholm, 3:121.
     COMMON NAME: Side-striped Jackal.
     TYPE LOCALITY: "Caffraria Interiore"; listed as "Magaliesberg" [South Africa] by Sclater
         (1900).
     DISTRIBUTION: Angola, Botswana, Cameroon, Central African Republic, Dem. Rep. Congo,
         Ethiopia, Gabon, Kenya, Malawi, Mozambique, Namibia, Niger, Nigeria, Republic of
         Congo, Senegal, South Africa, Sudan, Tanzania, Uganda, Zambia, Zimbabwe,
     STATUS: IUCN – Lower Risk (lc).
     SYNONYMS: *holubi* Lorenz, 1895; *wunderlichi* Noack, 1897; **bweha** Heller, 1914; *centralis*
         Schwarz, 1915; **grayi** Hilzheimer, 1906; **kaffensis** Neumann, 1902; **lateralis**
         P. L. Sclater, 1870; **notatus** Heller, 1914.
     COMMENTS: Synonyms allocated according to G. M. Allen (1939) and Ellerman et al. (1953).

*Canis aureus* Linnaeus, 1758. Syst. Nat., 10th ed., 1:40.
     COMMON NAME: Golden Jackal.
     TYPE LOCALITY: "oriente", restricted by Thomas (1911*a*) to "Benná Mts., Laristan, S. Persia"
         [Iran].
     DISTRIBUTION: Afghanistan, Albania, Algeria, Bangladesh, Burma, Chad, Coatia, Egypt,
         Eritrea, Ethiopia, Greece, Iran, Iraq, Israel, Italy, Jordan, Kenya, Lebanon, Libya,
         Macedonia, Mali, Mauritania, Morocco, Niger, Nigeria, Oman, Pakistan, Saudia Arabia,
         Senegal, Slovenia, Somalia, Sri Lanka, Sudan, Syria, Tajikistan, Tanzania, Thailand,
         Tunisia, Turkey, Turkmenistan, United Arab Emirates, Uzbekistan, Western Sahara,
         Yemen.
     STATUS: CITES – Appendix III (India); IUCN – Lower Risk (lc).
     SYNONYMS: *balcanicus* Brusina, 1892; *caucasica* Kolenati, 1858; *dalmatinus* Wagner, 1841;
         *hadramauticus* Noack, 1896; *hungaricus* Ehik, 1938; *kola* Wroughton, 1916; *lanka*
         Wroughton, 1916; *maroccanus* (Cabrera, 1921); *typicus* Kolenati, 1858; *vulgaris* Wagner,
         1841; **algirensis** Wagner, 1841; *barbarus* (C. E. H. Smith, 1839) [preoccupied]; *grayi*
         Hilzheimer, 1906; *tripolitanus* Wagner, 1841; **anthus** F. Cuvier, 1820; *senegalensis*
         (C. E. H. Smith, 1839); **bea** Heller, 1914; **cruesemanni** Matschie, 1900; **ecsedensis**
         (Kretzoi, 1947); *minor* Mojsisovico, 1897 [preoccupied]; **indicus** Hodgson, 1833;

*lupaster* Hemprich and Ehrenberg, 1833; *sacer* Hemprich and Ehrenberg, 1833; ***moreotica*** I. Geoffroy Saint-Hilaire, 1835; *graecus* Wagner, 1841; ***naria*** Wroughton, 1916; ***riparius*** Hemprich and Ehrenberg, 1832; *hagenbecki* Noack, 1897; *mengesi* Noack, 1897; *somalicus* Lorenz, 1906; ***soudanicus*** Thomas, 1903; *doederleini* Hilzheimer, 1906; *nubianus* (Cabrera, 1921); *thooides* Hilzheimer, 1906; *variegatus* Cretzschmar, 1826 [preoccupied]; ***syriacus*** Hemprich and Ehrenberg, 1833.

COMMENTS: Synonyms allocated according to G. M. Allen (1939) and Ellerman and Morrison-Scott (1951).

*Canis latrans* Say, 1823. *In* James, Account Exped. Pittsburgh to Rocky Mtns, 1:168.

COMMON NAME: Coyote.

TYPE LOCALITY: "Engineer cantonment" reported at "latitude 41°25′N, and longitude . . . 95°47′30′W" (p. XVIII, vol. 2). Reported in Honacki et al. (1982) as "U.S.A., Nebraska, Washington Co., Engineer Cantonment, about 12 mi. (19.2 km) S. E. Blair".

DISTRIBUTION: Canada, Costa Rica, El Salvador, Guatemala, Honduras, Mexico, Nicaragua, USA. Introduced to Florida and Georgia and currently widespread throughout Northern and Central America (Beckoff, 1977, 1999).

STATUS: IUCN – Lower Risk (lc).

SYNONYMS: *nebracensis* Merriam, 1898; *pallidus* Merriam, 1897; ***cagottis*** C. E. H. Smith, 1839; ***clepticus*** Elliot, 1903; ***dickeyi*** Nelson, 1932; ***frustror*** Woodhouse, 1851; ***goldmani*** Merriam, 1904; ***hondurensis*** Goldman, 1936; ***impavidus*** J. A. Allen, 1903; ***incolatus*** Hall, 1934; ***jamesi*** Townsend, 1912; ***lestes*** Merriam, 1897; ***mearnsi*** Merriam, 1897; *estor* Merriam, 1897; ***microdon*** Merriam, 1897; ***ochropus*** Eschscholtz, 1829; ***peninsulae*** Merriam, 1897; ***texensis*** Bailey, 1905; ***thamnos*** Jackson, 1949; ***umpquensis*** Jackson, 1949; ***vigilis*** Merriam, 1897.

COMMENTS: Revised by Young (1951) and reviewed by Beckoff (1977). Synonyms allocated according to Beckoff (1977) and Hall (1981).

*Canis lupus* Linnaeus, 1758. Syst. Nat., 10th ed., 1:39.

COMMON NAME: Wolf.

TYPE LOCALITY: "Europae sylvis, etjam frigidioribus", restricted by Thomas (1911*a*) to "Sweden".

DISTRIBUTION: Throughout the N hemisphere: North America south to 20°N in Oaxaca (Mexico); Europe; Asia, including the Arabian Peninsula and Japan, excluding Indochina and S India. Extirpated from most of the continental USA, Europe, and SE China and Indochina (Ginsburg and Macdonald, 1990). Afghanistan, Albania, Armenia, Azerbaijan, Belarus, Bhutan, Bulgaria, Canada, China, Egypt (?), Estonia, Finland, France, Georgia, Greece, Greenland, Hungary, India, Iran, Iraq, Israel, Jordan, Kyrgyzstan, Latvia, Lebanon (?), Lithuania, Macedonia, Mexico, Mongolia, Nepal, Norway, Pakistan, Poland, Portugal, Romania, Russia, Saudia Arabia, Serbia and Montenegro, Slovakia, Spain, Sweden, Syria, Tajikistan, Turkey, Turkmenistan, Ukraine, USA (see status below), Uzbekistan.

STATUS: CITES – Appendix I (Indian, Pakistan, Bhutan, and Nepal populations); otherwise Appendix II. U.S. ESA – as *C. lupus* varies by population: 1) Endangered in Southwestern Distinct Population Segment – Mexico and USA (AZ, NM, CO south of Interstate Highway 70, UT south of U.S. Highway 50, OK and TX, except those parts of OK and TX east of Interstate Highway 35; except where listed as an experimental population); 2) Threatened in Western Distinct Population Segment – USA (CA, ID, MT, NV, OR, WA, WY, UT north of U.S. Highway 50, and CO north of Interstate Highway 70, except where listed as an experimental population); 3) Threatened in Eastern Distinct Population Segment – USA (CT, IA, IL, IN, KS, MA, ME, MI, MN, MO, ND, NE, NH, NJ, NY, OH, PA, RI, SD, VT, and WI); 4) Experimental populations in portions of USA (WY and portions of ID and MT; portions of AZ, NM, and TX); otherwise, U.S. ESA – Delisted Taxa in USA (Delaware, West Virginia, Virginia, Maryland, District of Columbia, Kentucky, Tennessee, North Carolina, South Carolina, Georgia, Florida, Alabama, Mississippi, Louisiana, Arkansas, parts of Oklahoma and Texas east of Interstate Highway 35; delisting of all other lower 48 states or portions of lower 48 states not otherwise included in the 3 distinct population segments). U.S. ESA – as *C. rufus* Endangered in entire range except in portions of NC and TN (USA), where listed as experimental populations. IUCN – Lower Risk (lc), except for Mexican subpopulation, which is Extinct in the Wild, Italian subpopulation, which is Vulnerable, Spanish-Portuguese subpopulation, which is Lower Risk (cd), and as *Canis rufus*, which is Critically Endangered.

SYNONYMS: *altaicus* (Noack, 1911); *argunensis* Dybowski, 1922; *canus* de Sélys Longchamps, 1839; *communis* Dwigubski, 1804; *deitanus* Cabrera, 1907; *desertorum* Bogdanov, 1882; *flavus* Kerr, 1792; *fulvus* de Sélys Longchamps, 1839; *italicus* Altobello, 1921; *kurjak* Bolkay, 1925; *lycaon* Trouessart, 1910; *major* Ogérien, 1863; *minor* Ogerien, 1863, *niger* Hermann, 1804; *orientalis* (Wagner, 1841); *orientalis* Dybowski, 1922; *signatus* Cabrera, 1907; **albus** Kerr, 1792; *dybowskii* Domaniewski, 1926; *kamtschaticus* Dybowski, 1922; *turuchanensis* Ognev, 1923; **alces** Goldman, 1941; **arabs** Pocock, 1934; **arctos** Pocock, 1935; **baileyi** Nelson and Goldman, 1929; **beothucus** G. M. Allen and Barbour, 1937; **bernardi** Anderson, 1943; *banksianus* Anderson, 1943; **campestris** Dwigubski, 1804; *bactrianus* Laptev, 1929; *cubanenesis* Ognev, 1923; *desertorum* Bogdanov, 1882; **chanco** Gray, 1863; *coreanus* Abe, 1923; *dorogostaiskii* Skalon, 1936; *karanorensis* (Matschie, 1907); [preoccupied]; *niger* Sclater, 1874; *tschiliensis* (Matschie, 1907); **columbianus** Goldman, 1941; **crassodon** Hall, 1932; **dingo** Meyer, 1793 [domestic dog]; *antarcticus* Kerr, 1792[suppressed, ICZN, O. 451]; *australasiae* Desmarest, 1820; *australiae* Gray, 1826; *dingoides*, Matschie, 1915; *macdonnellensis* Matschie, 1915; *novaehollandiae* Voigt, 1831; *papuensis* Ramsay, 1879; *tenggerana* Kohlbrugge, 1896; *hallstromi* Troughton, 1957; *harappensis* Prashad, 1936; **familiaris** Linnaeus, 1758 [domestic dog]; *aegyptius* Linnaeus, 1758; *alco* C. E. H. Smith, 1839; *americanus* Gmelin, 1792; *anglicus* Gmelin, 1792; *antarcticus* Gmelin, 1792; *aprinus* Gmelin, 1792; *aquaticus* Linnaeus, 1758; *aquatilis* Gmelin, 1792; *avicularis* Gmelin, 1792; *borealis* C. E. H. Smith, 1839; *brevipilis* Gmelin, 1792; *cursorius* Gmelin, 1792; *domesticus* Linnaeus, 1758; *extrarius* Gmelin, 1792; *ferus* C. E. H. Smith, 1839; *fricator* Gmelin, 1792; *fricatrix* Linnaeus, 1758; *fuillus* Gmelin, 1792; *gallicus* Gmelin, 1792; *glaucus* C. E. H. Smith, 1839; *graius* Linnaeus, 1758; *grajus* Gmelin, 1792; *hagenbecki* Krumbiegel, 1950; *haitensis* C. E. H. Smith, 1839; *hibernicus* Gmelin, 1792; *hirsutus* Gmelin, 1792; *hybridus* Gmelin, 1792; *islandicus* Gmelin, 1792; *italicus* Gmelin, 1792; *laniarius* Gmelin, 1792; *leoninus* Gmelin, 1792; *leporarius* C. E. H. Smith, 1839; *major* Gmelin, 1792; *major* Gmelin, 1792; *mastinus* Linnaeus, 1758; *melitacus* Gmelin, 1792; *melitaeus* Linnaeus, 1758; *minor* Gmelin, 1792; *molossus* Gmelin, 1792; *mustelinus* Linnaeus, 1758; *obesus* Gmelin, 1792; *orientalis* Gmelin, 1792; *pacificus* C. E. H. Smith, 1839; *plancus* Gmelin, 1792; *pomeranus* Gmelin, 1792; *sagaces* C. E. H. Smith, 1839; *sanguinarius* C. E. H. Smith, 1839; *sagax* Linnaeus, 1758; *scoticus* Gmelin, 1792; *sibiricus* Gmelin, 1792; *suillus* C. E. H. Smith, 1839; *terraenovae* C. E. H. Smith, 1839; *terrarius* C. E. H. Smith, 1839; *turcicus* Gmelin, 1792; *urcani* C. E. H. Smith, 1839; *variegatus* Gmelin, 1792; *venaticus* Gmelin, 1792; *vertegus* Gmelin, 1792; **filchnevi** (Matschie, 1907); *laniger* (Hodgson, 1847); **floridanus** Miller, 1912; **fuscus** Richardson, 1839; *gigas* (Townsend, 1850); **gregoryi** Goldman, 1937; **griseoalbus** Baird, 1858; *knightii* Anderson, 1945; **hattai** Kishida, 1931; *rex* Pocock, 1935; **hodophilax** Temminck, 1839; *hodopylax* Temminck, 1844; *japonicus* Nehring, 1885; **hudsonicus** Goldman, 1941; **irremotus** Goldman, 1937; **labradorius** Goldman, 1937; **ligoni** Goldman, 1937; **lycaon** Schreber, 1775; *canadensis* de Blainville, 1843; *ungavensis* Comeau, 1940; **mackenzii** Anderson, 1943; **manningi** Anderson, 1943; **mogollonensis** Goldman, 1937; **monstrabilis** Goldman, 1937; *niger* Bartram, 1791; **nubilus** Say, 1823; *variabilis* Wied-Neuwied, 1841; **occidentalis** Richardson, 1829; *ater* Richardson, 1829; *sticte* Richardson, 1829; **orion** Pocock, 1935; **pallipes** Sykes, 1831; **pambasileus** Elliot, 1905; **rufus** Audubon and Bachman, 1851; **tundrarum** Miller, 1912; **youngi** Goldman, 1937.

COMMENTS: Reviewed by Mech, 1974. Opinion 2027 of the International Commission on Zoological Nomenclature (March, 2003*a*) ruled that *lupus* is not invalid by virtue of being pre-dated by a name based on a domestic form. Includes the domestic dog as a subspecies, with the dingo provisionally separate—artificial variants created by domestication and selective breeding (Vilá et al., 1999; Wayne and Ostrander, 1999;

Savolainen et al., 2002). Although this may stretch the subspecies concept, it retains the correct allocation of synonyms. Corbet and Hill (1992) suggested treating the domestic dog as a separate species in SE Asia. Synonyms allocated according to Ellerman and Morrison-Scott (1951), Mech (1974), and Hall (1981). Provisionally includes *rufus* (recognized by Paradiso, 1968; Paradiso and Nowak, 1972; Atkins and Dillion, 1971; Paradiso and Nowak, 1972; Nowak, 1979, 1992, 2002) although this problematic group (*rufus*, *floridanus*, *gregoryi*) should probably be best listed as *incertae sedis*. The widely used name *C. niger* is invalid (International Commission on Zoological Nomenclature, 1957*a*). The validity of *rufus* as a full species was questioned by Clutton-Brock et al. (1976), and Lawrence and Bossert (1967, 1975), due to the existence of natural hybrids with *lupus* and *latrans*. Natural hybridization may be a consequence of habitat disruption by man (Paradiso and Nowak, 1972, 2002). All specimens examined by Wayne and Jenks (1991) had either a *lupus* or *latrans* mtDNA genotype and there appears to be a growing consensus that all historical specimens are a product of hybridization (Nowak, 2002; Reich et al., 1999; Roy et al., 1994, 1996; Wayne et al., 1992, 1998). Hybridization between wolf and coyote has long been recognized (Nowak, 2002). Two recent studies make the strongest case for separation. Wilson et al. (2000) argued for separation of the Eastern Canadian Wolf (as *Canis lycaon*) and the Red Wolf (as *Canis rufus*) as separate species based on mtDNA, but see Nowak (2002) who could not find support for this in a morphometric study. Nowak (2002) in an extensive analysis of tooth morphology concluded that there was a distinct population intermediate between traditionally recognized wolves and coyotes, which warranted full species recognition (*C. rufus*). The red wolf is here considered a hybrid after Wayne and Jenks (1991), Wayne (1992, 1995), and Wayne et al. (1992). Although hybrids are not normally recognized as subspecies, I have chosen as a compromise to retain *rufus* because of its uncertain status. Also see Roy et al. (1994, 1996), Vilá et al. (1999), and Nowak (2002) who provided an excellent review of the situation.

*Canis mesomelas* Schreber, 1775. Die Säugethiere, 2(14):pl. 95[1775]; text, 3(21):370[1776], 586[1777].
COMMON NAME: Black-backed Jackal.
TYPE LOCALITY: "Vorgebirge der guten Hofnung" [South Africa, Western Cape Prov., Cape of Good Hope].
DISTRIBUTION: Allopatric south and east African populations: Angola, Botswana, Ethiopia, Kenya, Mozambique, Namibia, Somalia, Sudan, Tanzania, Uganda, Zimbabwe.
STATUS: IUCN – Lower Risk (lc).
SYNONYMS: *achrotes* (Thomas, 1925); *arenarum* (Thomas, 1926); *variegatoides* A. Smith, 1833; **schmidti** Noack, 1897; *elgonae* Heller, 1914; *mcmillani*, Heller, 1914.
COMMENTS: Reviewed by Walton and Joly (2003).

*Canis simensis* Rüppell, 1840. Neue Wirbelt. Fauna Abyssin. Gehörig. Säugeth., 1:39, pl. 14.
COMMON NAME: Ethiopian Wolf.
TYPE LOCALITY: "Wir beobachteten diesen wolfsartigen Hund in den Bergen von Simen . . ." [Ethiopia, mountains of Simen].
DISTRIBUTION: C Ethiopia.
STATUS: U.S. ESA – Endangered; IUCN – Critically Endangered.
SYNONYMS: *crinensis* (Erlanger and Neumann, 1900); *semiensis* Heuglin, 1862; *simensis* (Gray, 1869); *walgi* Heuglin, 1862; **citernii** de Beaux, 1922.
COMMENTS: Sometimes placed in subgenus *Simenia* Gray, 1868. Reviewed by Sillero-Zubiri and Gottelli (1994).

*Cerdocyon* C. E. H. Smith, 1839. Jardine's Natur. Libr., 9:259-267.
TYPE SPECIES: *Canis azarae* Wied, 1824 (= *Canis Thous* Linnaeus, 1766) by subsequent designation (Thomas, 1914*a*).
SYNONYMS: *Carcinocyon* J. A. Allen, 1905; *Thous* J. E. Gray, 1868.
COMMENTS: Tedford et al. (1995) considered *Cerdocyon* and *Nyctereutes* to be sister taxa.

*Cerdocyon thous* (Linnaeus, 1766). Syst. Nat., 12th ed., 1:60.
 COMMON NAME: Crab-eating Fox.
 TYPE LOCALITY: "Surinamo" [Surinam].
 DISTRIBUTION: N Argentina, Bolivia, Brazil (except Amazonia), Colombia, Guyanas,
  Suriname, Peru, Paraguay, Uruguay, Venezuela.
 STATUS: CITES – Appendix II; IUCN – Lower Risk (lc).
 SYNONYMS: *brasiliensis* (Wied-Neuwied, 1824); *cancrivorus* (Brongniart, 1792); *lunaris*
  (Thomas, 1914); *melampus* (Wagner, 1841); *rudis* (Günther, 1879); *savannarum*
  (Thomas, 1901); *vetulus* (Studer, 1905); **aquilus** (Bangs, 1898); *apollinaris* Thomas,
  1914; **azarae** (Wied-Neuwied, 1824); *angulensis* (Thomas, 1903); *brachyteles*
  (de Blainville, 1843); *cancrivorus* (Winge, 1896); *guaraxa* C. E. H. Smith, 1839;
  *melanostomus* (Wagner, 1843); *robustior* (Lund, 1843); **entrerianus** (Burmeister, 1861);
  *affinis* Marcelli, 1931; *flavogriseus* (Zukowsky, 1950); *fronto* Lönnberg, 1919; *jucundus*
  Thomas, 1921; *mimax* Thomas, 1914; *riograndensis* (Ihering, 1911); *tucumanus*
  Thomas, 1921; **germanus** G. M. Allen, 1923; **soudanicus** (Thomas, 1903).
 COMMENTS: Reviewed by Berta (1982). Placed in *Cerdocyon* by Langguth (1975), Stains
  (1975), and Berta (1982); placed in subgenus *Canis (Cerdocyon)* by Van Gelder (1978).
  Synonyms allocated according to Berta (1982) and Cabrera (1957).

*Chrysocyon* C. E. H. Smith, 1839. Jardine's Natur. Libr., 9:241-247.
 TYPE SPECIES: *Canis jubatus* Desmarest, 1820 (= *Canis brachyurus* Illiger, 1815).
 COMMENTS: Recognized by Langguth (1975), Stains (1975), Van Gelder (1978), Berta (1988),
  and McKenna and Bell (1997).

*Chrysocyon brachyurus* (Illiger, 1815). Abh. Phys. Klasse K. Pruess. Akad. Wiss., 1804-1811:121.
 COMMON NAME: Maned Wolf.
 TYPE LOCALITY: Listed by Cabrera (1957) as "los esteros del Paraguay."
 DISTRIBUTION: NE Argentina, Paraguay; Bolivia (lowlands), Brazil (from Rio Grande do Sul
  to Minas Gerais, Goiás and Mato Grosso), Paraguay.
 STATUS: CITES – Appendix II; U.S. ESA – Endangered; IUCN – Lower Risk (nt).
 SYNONYMS: *campestris* (Wied-Neuwied, 1826); *cancrosa* (Oken, 1816); *isodactylus* (Ameghino,
  1906); *jubatus* (Desmarest, 1820); *vulpes* (Larrañaga, 1923).
 COMMENTS: Reviewed by Dietz (1985). Synonyms allocated according to Dietz (1985).

*Cuon* Hodgson, 1838. Ann. Mag. Nat. Hist., [ser. 1], 1:152.
 TYPE SPECIES: *Canis primaevus* Hodgson, 1838 (=*Canis alpinus* Pallas, 1811) by monotypy
  (Melville and Smith, 1987). Conserved by Opinion 384 (1956*a*).
 SYNONYMS: *Anurocyon* Heude, 1888; *Chrysaeus* C. E. H. Smith, 1839; *Cyon* Agassiz, 1842;
  *Primaevus* Gray, 1843; *Primoevus* Hodgson, 1842.
 COMMENTS: Placed in subfamily *Simocyoninae* Dawkins, 1868, by Simpson (1945) and Stains
  (1975). Tedford et al. (1995) considered *Cuon* and *Lycaon* to be sister taxa.

*Cuon alpinus* (Pallas, 1811). Zoogr. Rosso-Asiat., 1:34.
 COMMON NAME: Dhole.
 TYPE LOCALITY: "Udskoi Ostrog"; reported in Honacki et al. (1982) as "U.S.S.R., Amurskaya
  Obl., Udskii-Ostrog."
 DISTRIBUTION: China (Tibet and Xinjiang: Tian Shan and Altai-extinct); Indonesia
  (Java, Sumatra), Malaysia, India (montane forest), N Pakistan, Indochina, North and
  South ?Korea, N Mongolia, Russia (Ussuri region and S Siberia).
 STATUS: CITES – Appendix II; U. S. ESA – Endangered; IUCN – Vulnerable.
 SYNONYMS: **adustus** Pocock, 1941; *antiquus* (Matthew and Granger, 1923); *dukhunensis*
  (Sykes, 1831); **fumosus** Pocock, 1936; *infuscus* Pocock, 1936; *javanicus* (Desmarest,
  1820); **laniger** Pocock, 1936; *grayiformis* Hodgson, 1863; *primaevus* (Hodgson, 1833);
  **lepturus** Heude, 1892; *clamitans* (Hende, 1892); *rutilans* Müller, 1839; *sumatrensis*
  Hodgson, 1863; **hesperius** (Afanasjev and Zolotarev, 1935); *jason* Pocock, 1936;
  **sumatrensis** (Hardwicke, 1821).
 COMMENTS: Reviewed by Cohen (1978). Synonyms allocated according to Cohen (1978).

*Dusicyon* C. E. H. Smith, 1839. Jardine's Natur. Libr., 9:248.
>TYPE SPECIES: *Canis antarcticus* Bechstein, 1799 (= *C. australis* Kerr, 1792), by subsequent designation (Cabrera, 1931).
>COMMENTS: There has been general disagreement as to generic classification of the South American canids, with most of the disagreement centered on the species *australis*, *culpaeus*, *griseus*, *gymnocercus*, *microtis*, *sechurae*, *thous*, and *vetulus*. Van Gelder (1978) proposed placing these taxa into *Canis* and giving only subgeneric recognition. The other extreme arrangement is best represented by Cabrera (1931) who recognized 5 genera for this group. Langguth (1969) first followed Cabrera's classification, but later (1975) decided to group most taxa into *Canis*, because he felt differences were not sufficient to warrant generic distinctions. The phenetic approaches of Clutton-Brock et al. (1976) and Wayne and O'Brien (1987) confirmed the close similarities of these taxa. Berta's (1987, 1988) phylogenetic hypothesis is followed here. Placed in *Dusicyon* by McKenna and Bell (1997) who consider it to include only the extinct Falkland Island wolf. Tedford et al. (1995) considered a different arrangement of South American canids and recogned the following monophyletic groups: (1) *L. vetulus* + *Chrysocyon* + *Cerdocyon* + *Nyctereutes* + *Speothos* + *Atelocynus*; (2) *L. culpaeus* + *Dusicyon*; (3) *L. griseus* + *L. gymnocercus* + *L. sechurae*.

*Dusicyon australis* (Kerr, 1792). *In* Linnaeus, Anim. Kingdom, p. 144.
>COMMON NAME: Falkland Islands Wolf.
>TYPE LOCALITY: "America and Falkland islands."
>DISTRIBUTION: Falkland Isls.
>STATUS: IUCN – Extinct.
>SYNONYMS: *antarcticus* (Bechstein, 1799).
>COMMENTS: Placed in *Dusicyon* by Cabrera (1931) and Berta (1987, 1988), and considered as a subgenus separate from other "foxes" (i.e., *culpaeus*, *griseus*, *gymnocercus*, and *sechurae*) by Langguth (1975) and Van Gelder (1978).

*Lycalopex* Burmeister, 1854. Systematische Uebersicht der Thiere Brasiliens, p. 95-101.
>TYPE SPECIES: *Canis magellanicus* Gray, 1847 (= *Canis culpaeus* Molina, 1782), by subsequent designation (Cabrera, 1931).
>SYNONYMS: *Angusticeps* Hilzheimer, 1906; *Eunothocyon* J. A. Allen, 1905; *Lupulus* Trouessart, 1897; *Microcyon* Trouessart, 1906; *Nothocyon* Wortman and Matthew, 1899; *Procyon* Fischer, 1814 (not Storr, 1780); *Pseudalopex* Burmeister, 1856; *Pseudolopex* Philippi, 1903; *Pseudolycos* Philippi, 1903; *Thous* Gray, 1869; *Viverriceps* Hilzheimer, 1906; *Vulpes* Martin, 1837.
>COMMENTS: Revised by Zunino et al. (1995). Also see comments under *Dusicyon*. Although combining taxa included here with *Dusicyon* would not be in conflict with Berta (1987, 1988), her analyses suggested that other genera, now extinct, are more closely related to *Dusicyon*. Berta (1987, 1988) presented derived features that would support a single origin for those taxa recognized here in *Lycalopex* (=*Pseudalopex*), which would also agree with Cabrera (1957) and Stains (1975). A detailed comparative morphological study by Langguth (1969) caused him to conclude (1975) that *Pseudalopex* (=*Lycalopex*) merited generic rank. Synonyms allocated according to Zunino et al. (1995). Tedford et al. (1995) considered *Lycalopex* as recognized here as paraphyletic and proposed a different arrangement of South American canids according to the following monophyletic groups: (1) *L. vetulus* + *Chrysocyon* + *Cerdocyon* + *Nyctereutes* + *Speothos* + *Atelocynus*; (2) *L. culpaeus* + *Dusicyon*; (3) *L. griseus* + *L. gymnocercus* + *L. sechurae*.

*Lycalopex culpaeus* (Molina, 1782). Sagg. Stor. Nat. Chile, p. 293.
>COMMON NAME: Culpeo.
>TYPE LOCALITY: "Chili" restricted by Cabrera (1931) to "the Santiago Province."
>DISTRIBUTION: Argentina (Tierra del Fuego), Bolivia, Chile, Colombia, Ecuador, Peru.
>STATUS: CITES – Appendix II as *Pseudalopex culpaeus*; IUCN – Lower Risk (lc) as *P. culpaeus*.
>SYNONYMS: *albigula* (Philippi, 1903); *amblyodon* (Philippi, 1903); *chilensis* (Kerr, 1792); *ferrugineus* (Huber, 1925); *magellanicus* (Gray, 1847, not Gray, 1836); **andinus** (Thomas, 1914); *azarae* (Tschudi, 1844) [preoccupied]; *culpaeolus* (Thomas, 1914); *inca* (Thomas, 1914); *magellanicus* (Waterhouse, 1838) [preoccupied]; *reissii* (Osgood,

1914); *smithersi* (Kraglievich, 1930); **lycoides** (Philippi, 1896); **magellanicus** (Gray, 1837); *montanus* (Prichard, 1902); *prichardi* (Trouessart, 1904); *typicus* (Trouessart, 1910); **reissii** (Hilzheimer, 1906); *riveti* (Trouessart, 1906); **smithersi** (Thomas, 1914).
COMMENTS: Revised by Zunino et al. (1995). Placed in *Pseudalopex* by Berta (1987, 1988); and in *Dusicyon* by Cabrera (1957). Considered in *Canis* (*Pseudalopex*) by Langguth (1975), Clutton-Brock et al. (1976), and Van Gelder (1978). Includes *culpaeolus* (part) and *inca* (part) mismatched skin and skull (Langguth, 1967). Reviewed by Novaro (1997). Synonyms allocated according to Cabrera (1957), Zunino et al. (1995), and Novaro (1997).

*Lycalopex fulvipes* (Martin, 1837). Proc. Zool. Soc. Lond., 1837:11.
COMMON NAME: Darwin's Fox.
TYPE LOCALITY: "island of Chiloé".
DISTRIBUTION: Chile (Chiloé Isl, and Nahuelbuta National Park)(Medel et al., 1990).
STATUS: CITES Appendix II (as included in *Pseudalopex* (= *Lycalopex*) *griseus*).
SYNONYMS: *lagopus* (Molina, 1782).
COMMENTS: The distinctiveness of *fulvipes* is supported by mtDNA analyses (Yahnke et al., 1996).

*Lycalopex griseus* (Gray, 1837). Mag. Nat. Hist. [Charlesworth's], 1:578.
COMMON NAME: South American Gray Fox.
TYPE LOCALITY: "Magellan", listed in Cabrera (1957) as "Costa del Estrecho de Magallanes" [Chile].
DISTRIBUTION: Argentina (Santiago del Estero), Chile, Falkland Isls.
STATUS: CITES – Appendix II as *Pseudalopex griseus*; IUCN – Lower Risk (lc) as *P. griseus*.
SYNONYMS: *gracilis* (Burmeister, 1861).
COMMENTS: Placed in *Pseudalopex* by Berta (1988) and *Dusicyon* by Cabrera (1957). Considered in *Canis* (*Pseudalopex*) by Langguth (1975), and Van Gelder (1978). Placed in *Lycalopex gymnocercus* by Zunino et al. (1995).

*Lycalopex gymnocercus* (G. Fischer, 1814). Zoognosia, 3:xi, p. 178.
COMMON NAME: Pampas Fox.
TYPE LOCALITY: "Paraguay", restricted by Cabrera (1957) to "a los alrededores de Asunción."
DISTRIBUTION: Argentina (north of Rio Negro), E Bolivia, S Brazil, Paraguay, Uruguay.
STATUS: CITES – Appendix II as *Pseudalopex gymnocercus*; IUCN – Lower Risk (lc) as *P. gymnocercus*.
SYNONYMS: *argenteus* (Larrañaga, 1923); *attenuatus* (Kraglievich, 1930); *brasiliensis* (Schinz, 1821); *protalopex* (Lund, 1839); **antiquus** (Ameghino, 1889); *antiguus* (Ameghino, 1889); *azarai* (Lahille, 1898); *azarica* (Thomas, 1914); *fossilis* (Gervais and Ameghino, 1880) [preoccupied]; **domeykoanus** (Philippi, 1901); *azarae* (Gay, 1847) [preoccupied]; *domeycoanus* (Wolffsohn, 1918); *griseus* (Wolffsohn and Porter, 1908); **gracilis** (Burmeister, 1861); *patagonicus* (Philippi, 1866); *zorrula* (Thomas, 1921); **maullinicus** (Philippi, 1903); *torquatus* (Philippi, 1903); *trichodactylus* (Philippi, 1903).
COMMENTS: Placed in *Pseudalopex* by Berta (1988) and in *Dusicyon* by Cabrera (1957). Considered in *Canis* (*Pseudalopex*) by Langguth (1975) and Van Gelder (1978). Synonyms allocated according to Zunino et al. (1995) and Cabrera (1957).

*Lycalopex sechurae* Thomas, 1900. Ann. Mag. Nat. Hist., ser. 7, 5:148.
COMMON NAME: Sechuran Fox.
TYPE LOCALITY: "Desert of Sechura, N.W. Peru. . . Sullana". [=Piura, Perú by Sheffield and Thomas (1997)].
DISTRIBUTION: SW Ecuador, NW Peru.
STATUS: IUCN – Data Deficient as *Pseudalopex sechurae*.
COMMENTS: Placed in *Pseudalopex* by Berta (1988); in *Dusicyon* by Cabrera (1957). Considered in *Canis* (*Pseudalopex*) by Langguth (1975) and Van Gelder (1978).

*Lycalopex vetulus* (Lund, 1842). K. Dansk. Vid. Selsk. Naturv. Math. Afhandl., 9:4.
COMMON NAME: Hoary Fox.
TYPE LOCALITY: "Rio das Velhas's Floddal" Lagoa Santa, Minas Gerais, Brazil (Zunino et al., 1995).

DISTRIBUTION: Brazil (highlands in the States of Mato Grosso, Goiás, Minas Gerais, Bahia, and Sao Paulo).

STATUS: IUCN – Data Deficient as *Pseudalopex vetulus*.

SYNONYMS: *chilensis* Gray, 1868; *fulvicaudus* (Lund, 1843); *parvidens* (Mivart, 1890); *sladeni* (Thomas, 1904); *urostictus* (Mivart, 1890); *vitulus* (Huber, 1925).

COMMENTS: Placed in *Pseudalopex* by Berta (1987, 1988); in *Dusicyon* (*Lycalopex*) by Cabrera (1957) and implied by Stains (1975); in *Lycalopex* by Langguth (1975); and considered in *Canis* (*Lycalopex*) by Van Gelder (1978).

*Lycaon* Brookes, 1827. *In* Griffith et al., Anim. Kingdom, 5:151.

TYPE SPECIES: *Lycaon tricolor* Brookes, 1827 (= *Hyaena picta* Temminck, 1820) by monotypy (Melville and Smith, 1987).

SYNONYMS: *Cynhyaena* F. G. Cuvier, 1829; *Hyaenoides* Gervais, 1855; *Hyenoides* Boitard, 1842; *Kynos* Rüppell, 1842.

COMMENTS: Placed in *Simocyoninae* Dawkins, 1868, by Simpson (1945) and Stains (1975). Reviewed by Girman et al. (2001).

*Lycaon pictus* (Temminck, 1820). Ann. Gen. Sci. Phys., 3:54, pl. 35.

COMMON NAME: African wild dog.

TYPE LOCALITY: "á la côte de Mosambique" [Mozambique].

DISTRIBUTION: Angola, Botswana, Cameroun, Central African Republic, Chad, Côte d'Ivoire (?), Ethiopia, Gambia (?), Guinea, Kenya, Malawi, Mali, Moçambique, Namibia, Sénégal, Somalia, South Africa, Sudan, Tanzania, Uganda, Zambia, Zimbabwe. Recently extinct: Algeria (?), Benin, Burkina Faso (?), Burundi, Dem. Rep. Congo (?), Eritrea, Gabon, Ghana, Niger, Mauritania, Nigeria, Republic of Congo, Rwanda, Sierra Leone, Togo (Fanshawe et al, 1997).

STATUS: U. S. ESA – Endangered; IUCN – Endangered.

SYNONYMS: *cacondae* Matschie, 1915; *fuchsi* Matschie, 1915; *gobabis* Matschie, 1915; *krebsi* Matschie, 1915; *lalandei* Matschie, 1915; *tricolor* (Brookes, 1827); *typicus* A. Smith, 1833; *venatica* (Burchell, 1822); *windhorni* Matschie, 1915; *zuluensis* Thomas, 1904; **lupinus** Thomas, 1902; *dieseneri* Matschie, 1915; *gansseri* Matschie, 1915; *hennigi* Matschie, 1915; *huebneri* Matschie, 1915; *kondoae* Matschie, 1915; *lademanni* Matschie, 1915; *langheldi* Matschie, 1915; *prageri* Matschie, 1912; *richteri* Matschie, 1915; *ruwanae* Matschie, 1915; *ssongaeae* Matschie, 1915; *stierlingi* Matschie, 1915; *styxi* Matschie, 1915; *wintgensi* Matschie, 1915; **manguensis** Matschie, 1915; *mischlichi* Matschie, 1915; **sharicus** Thomas and Wroughton, 1907; *ebermaieri* Matschie, 1915; **somalicus** Thomas, 1904; *luchsingeri* Matschie, 1915; *ruppelli* Matschie, 1915; *takanus* Matschie, 1915; *zedlitzi* Matschie, 1915.

COMMENTS: Allocated according to G. M. Allen (1939) and Ellerman et al. (1953). Girman et al. (1993) presented molecular evidence concerning subspecies.

*Nyctereutes* Temminck, 1838. Tijdschr. Nat. Gesch. Physiol., 5:285.

TYPE SPECIES: *Canis viverrinus* Temminck, 1838 (= *Canis procyonoides* Gray, 1834).

*Nyctereutes procyonoides* (Gray, 1834). Illustr. Indian Zool, . 2:pl. 1.

COMMON NAME: Raccoon dog.

TYPE LOCALITY: Unknown; restricted to "vicinity of Canton, China" by G. M. Allen (1938).

DISTRIBUTION: China, Japan, Mongolia, North and South Korea, Russia. Introduced into Europe and now found in: Austria, Belarus, Bosnia and Herzegovina, Bulgaria, Denmark, Estonia, Finland, France, Germany, Hungary, Latvia, Lithuania, Moldova, Netherlands, Norway, Poland, Romania, Serbia, Slovenia, Sweden, Switzerland, Ukraine.

STATUS: IUCN – Lower Risk (lc).

SYNONYMS: *kalininensis* Sorokin, 1958; *sinensis* Brass, 1904; *stegmanni* Matschie, 1907; **koreensis** Mori, 1922; **orestes** Thomas, 1923; **ussuriensis** Matschie, 1907; *amurensis* Matschie, 1907; **viverrinus** Temminck, 1838; *albus* Hornaday, 1904.

COMMENTS: Reviewed by Ward and Wurster-Hill (1990). Synonyms allocated according to Ellerman and Morrison-Scott (1951). Introduced populations in Europe.

*Otocyon* Müller, 1836. Arch. Anat. Physiol., Jahresber. Fortschr. Wiss., 1835:1 [1836].
  TYPE SPECIES: *Otocyon caffer* Müller, 1836 (= *Canis megalotis* Desmarest, 1822), by monotypy
    (Melville and Smith, 1987).
  SYNONYMS: *Agrodius* C. E. H. Smith, 1840.

*Otocyon megalotis* (Desmarest, 1822). Mammalogie, *In* Encyclop. Meth., 2(Suppl.):538.
  COMMON NAME: Bat-eared Fox.
  TYPE LOCALITY: "le Cap de Bonne-Espérance" [South Africa, Western Cape Prov., Cape of
    Good Hope].
  DISTRIBUTION: Allopatric south and east African populations: Angola, Botswana, Ethiopia,
    Mozambique, Namibia, Somalia, South Africa, Sudan, Tanzania, Uganda, Zambia,
    Zimbabwe.
  STATUS: IUCN – Lower Risk (lc).
  SYNONYMS: *auritus* (C. E. H. Smith, 1840); *caffer* Müller, 1836; *lalandi* (Desmoulins, 1823);
    *steinhardti* Zukowsky, 1924; **canescens** Cabrera, 1910.
  COMMENTS: Synonyms allocated according to G. M. Allen (1939) and Ellerman et al. (1953).

*Speothos* Lund, 1839. Ann. Sci. Nat. Zool. (Paris) ser. 2, 11:224.
  TYPE SPECIES: *Speothos pacivorus* Lund, 1839 (extinct).
  SYNONYMS: *Abathmodon* Lund, 1843; *Cynalicus* Gray, 1846; *Cynalius* Gray, 1847; *Cynalycus* Gray,
    1869; *Cynogale* Lund, 1842; *Icticyon* Lund, 1843; *Melictis* Schinz, 1848.
  COMMENTS: Berta and Marshall (1978) included *Icticyon*. Placed in *Simocyoninae* Dawkins, 1868,
    by Simpson (1945) and Stains (1975). Synonyms allocated according to McKenna and Bell
    (1997).

*Speothos venaticus* (Lund, 1842). K. Dansk. Vid. Selsk. Naturv. Math. Afhandl, 9:67.
  COMMON NAME: Bush Dog.
  TYPE LOCALITY: "Lagoa Santa" [Minas Gerais, Brazil].
  DISTRIBUTION: Forested areas of Bolivia, Brazil (except the semiarid NE), Colombia, Ecuador,
    French Guiana, Guyana, Panama, Paraguay, E Peru, Surinam, Venezuela.
  STATUS: CITES – Appendix I; IUCN – Vulnerable.
  SYNONYMS: *baskii* (Schinz, 1849); *melanogaster* (Gray, 1846); **panamensis** Goldman, 1912;
    **wingei** Ihering, 1911.
  COMMENTS: Synonyms allocated according to Cabrera (1957) and Hall (1981).

*Urocyon* Baird, 1857. Mammals, *In* Repts. Expl. Surv., 8(1):121, 138.
  TYPE SPECIES: *Canis virginianus* Schreber, 1775 (= *Canis cinereo argenteus* Schreber, 1775) by
    subsequent designation (Elliot, 1901; Melville and Smith, 1987).
  COMMENTS: Considered a subgenus of *Vulpes* by Clutton-Brock et al. (1976).

*Urocyon cinereoargenteus* (Schreber, 1775). Die Säugethiere, 2(13):pl. 92[1775]; text: 21:361[1776].
  COMMON NAME: Gray Fox.
  TYPE LOCALITY: "Sein Vaterland ist Carolina und die Wärmeren Gegenden von
    Nordamerica, vielleicht auch Surinam."
  DISTRIBUTION: Belize, Canada (along USA border); Colombia, Costa Rica, El Salvador,
    Guatemala, Mexico, Nicaragua, Panama, USA (most states except Idaho, Washington,
    Montana, Wyoming), Venezuela.
  STATUS: IUCN – Lower Risk (lc).
  SYNONYMS: *pensylvanicus* (Boddaert, 1784); *virginianus* (Schreber, 1775); **borealis** Merriam,
    1903; **californicus** Mearns, 1897; **costaricensis** Goodwin, 1938; **floridanus** Rhoads,
    1895; **fraterculus** Elliot, 1896; **furvus** G. M. Allen and Barbour, 1923; **guatemalae**
    Miller, 1899; **madrensis** Burt and Hooper, 1941; **nigrirostris** (Lichtenstein, 1850);
    **ocythous** Bangs, 1899; **orinomus** Goldman, 1938; **peninsularis** Huey, 1928; **scottii**
    Mearns, 1891; *texensis* Mearns, 1897; *inyoensis* Elliot, 1904; **townsendi** Merriam, 1899;
    *sequoiensis* Dixon, 1910; **venezuelae** J. A. Allen, 1911.
  COMMENTS: Reviewed by Fritzell and Haroldson (1982). Placed in *Canis (Vulpes)* by
    Van Gelder (1978). Synonyms allocated according to Hall (1981) and Fritzell and
    Haroldson (1982).

*Urocyon littoralis* (Baird, 1858). Mammalia, *In* Repts. U.S. Expl. Surv., 8(1):143.
COMMON NAME: Island Fox.
TYPE LOCALITY: "island of San Miguel, on the coast of California."
DISTRIBUTION: USA (Islands off the Pacific coast of S California).
STATUS: U.S. ESA – Proposed Endangered as *U. littoralis catalinae*, *U. l. littoralis*,
*U. l. santacruzae*, and *U. l. santarosae*; IUCN – Lower Risk (cd).
SYNONYMS: **catalinae** Merriam, 1903; **clementae** Merriam, 1903; **dickeyi** Grinnell and
Linsdale, 1930; **santacruzae** Merriam, 1903; **santarosae** Grinnell and Linsdale, 1930.
COMMENTS: Reviewed by Moore and Collins (1995). Placed in *Canis (Vulpes)* by Van Gelder
(1978). Gilbert et al. (1990), George and Wayne (1991), Wayne et al. (1991*a, b*), and
Collins (1993) supported full species status. Synonyms allocated according to Moore
and Collins (1995).

*Vulpes* Frisch, 1775. Das Natur-System der Vierfüssigen Thiere, p. 15.
TYPE SPECIES: *Canis vulpes* Linnaeus, 1758, by designation under the plenary powers (Melville
and Smith, 1978).
SYNONYMS: *Alopex* Kaup, 1829; *Cynalopex* C. E. H. Smith, 1839; *Fennecus* Desmarest, 1804;
*Leucocyon* Gray, 1869; *Mamvulpesus* Herrera, 1899; *Megalotis* Illiger, 1811; *Vulpis* Gray,
1821.
COMMENTS: Although Frisch (1775) has been ruled a rejected work for nomenclatural purposes,
*Vulpes* has been retained (International Commission on Zoological Nomenclature, 1979).
Considered a subgenus of *Canis* by Van Gelder (1978); however, this arrangement is not
currently employed by most mammalogists (Corbet, 1978; Corbet and Hill, 1980;
Gromov and Baranova, 1981; Hall, 1981; Wozencraft, 1989). McKenna and Bell (1997)
included *Fennecus* and *Alopex* as congeneric taxa (followed here). Bobrinskii et al. (1965),
McKenna and Bell (1997) and Bininda-Emonds et al. (1999) placed *Alopex* as a subgenus
of *Vulpes*; Van Gelder (1978) considered it a subgenus of *Canis*. Wayne et al. (1987),
Wayne and O'Brien (1987), and Mercure et al. (1993) argued for the inclusion of *Alopex*
with other *Vulpes*. Synonyms allocated after McKenna and Bell (1997).

*Vulpes bengalensis* (Shaw, 1800). Gen. Zool. Syst. Nat. Hist., 1(2), p. 330.
COMMON NAME: Bengal Fox.
TYPE LOCALITY: "Bengal."
DISTRIBUTION: India, S Nepal, Pakistan.
STATUS: CITES – Appendix III (India); IUCN – Data Deficient.
SYNONYMS: *chrysurus* (Gray, 1837); *hodgsonii* Gray, 1837; *indicus* (Hodgson, 1833); *kokree*
(Sykes, 1831); *rufescens* (Gray, 1834); *xanthura* Gray, 1837.
COMMENTS: Synonyms allocated according to Pocock (1941*a*).

*Vulpes cana* Blanford, 1877. J. Asiat. Soc. Bengal, 2:321.
COMMON NAME: Blanford's Fox.
TYPE LOCALITY: "Gwadar, Baluchistan", [Pakistan].
DISTRIBUTION: Afghanistan, Egypt (Sinai), NE Iran, Israel, Oman, Pakistan, Saudi Arabia,
Tajikistan, Turkmenistan, Uzbekistan.
STATUS: CITES – Appendix II; IUCN – Data Deficient.
SYNONYMS: *nigricans* Shitkow, 1907.
COMMENTS: Reviewed by Geffen (1994).

*Vulpes chama* (A. Smith, 1833). S. Afr. J., 2:89.
COMMON NAME: Cape Fox.
TYPE LOCALITY: "Namaqualand and the country on both sides of the Orange river"
[Namibia]; fixed by Shortridge (1942) as "Port Nolloth, Little Namaqualand."
DISTRIBUTION: S Angola, Botswana, Namibia, South Africa.
STATUS: IUCN – Lower Risk (lc).
SYNONYMS: *caama* (C. E. H. Smith, 1839); *hodsoni* (Noack, 1910); *variegatoides* (Layard, 1861).
COMMENTS: Synonyms allocated according to Ellerman et al. (1953).

*Vulpes corsac* (Linnaeus, 1768). Syst. Nat., 12th ed., 3: appendix 223.
COMMON NAME: Corsac Fox.

TYPE LOCALITY: "in campis magi deserti ab Jaco fluvio verus Irtim"; listed by Honacki et al. (1982) as "U.S.S.R., N. Kazakhstan, steppes between Ural and Irtysh rivers, near Petropavlovsk."

DISTRIBUTION: N Afghanistan, NE China, Kazakhstan, Kyrgyzstan, Mongolia, Russia.

STATUS: IUCN – Data Deficient.

SYNONYMS: *corsak* Ognev, 1935; *nigra* Kastschenko, 1912; *skorodumovi* Dorogostaiski, 1935; **kalmykorum** Ognev, 1935; **turcmenicus** Ognev, 1935.

COMMENTS: Synonyms allocated according to Ellerman and Morrison-Scott (1951).

*Vulpes ferrilata* Hodgson, 1842. J. Asiat. Soc. Bengal, 11:278.

COMMON NAME: Tibetan Sand Fox.

TYPE LOCALITY: "brought from Lassa" [Tibet, China].

DISTRIBUTION: China (Tibet, Tsinghai, Kansu, and Yunnan), Nepal.

STATUS: IUCN – Lower Risk (lc).

SYNONYMS: *ekloni* (Przewalski, 1883).

COMMENTS: Baryshnikov and Abramov (1992) discussed the taxonomic position of "*ekloni*".

*Vulpes lagopus* (Linnaeus, 1758). Syst. Nat., 10th ed., 1:40.

COMMON NAME: Arctic Fox.

TYPE LOCALITY: "alpibus Lapponicis, Sibiria," restricted by Thomas (1911*a*) to "Sweden (Lapland)."

DISTRIBUTION: Circumpolar, entire tundra zone of the Holarctic, including most of the Arctic islands: Canada, Finland, Greenland, Iceland, Norway, Russia, Sweden, USA (Alaska).

STATUS: IUCN – Lower Risk (lc) as *Alopex lagopus*.

SYNONYMS: *arctica* Oken, 1816; *argenteus* (Billberg, 1827); *caerulea* (Nilsson, 1820); *hallensis* Merriam, 1900; *innuitus* Merriam, 1902; *kenaiensis* (Brass, 1911); *typicus* (Barrett-Hamilton and Bonhote, 1898); *ungava* Merriam, 1902; **beringensis** Merriam, 1902; *beringianus* (Cherski, 1920); *semenovi* (Ognev, 1931); **fuliginosus** (Bechstein, 1799); *groenlandicus* (Bechstein, 1799); *spitzbergenensis* (Barrett-Hamilton and Bonhote, 1898); **pribilofensis** Merriam, 1902.

COMMENTS: Viable hybrids have been recorded between *V. lagopus* and *V. vulpes* (Chiarelli, 1975). Synonyms allocated according to Audet et al. (2002).

*Vulpes macrotis* Merriam, 1888. Proc. Biol. Soc. Wash., 4:136.

COMMON NAME: Kit Fox.

TYPE LOCALITY: "Riverside, San Bernardino county, California"

DISTRIBUTION: USA (S and C California, Nevada, SE Oregon, SW Idaho, W Utah, Arizona, New Mexico and W Texas).

STATUS: U.S. ESA – Endangered as *V. macrotis mutica*.

SYNONYMS: *arizonensis* Goldman, 1931; *arsipus* Elliot, 1904; *devius* Nelson and Goldman, 1909; *muticus* Merriam, 1902; *neomexicanus* Merriam, 1902; *nevadensis* Goldman, 1931; *tenuirostris* Nelson and Goldman, 1931; *zinseri* Benson, 1938.

COMMENTS: Reviewed by Egoscue (1979) and McGrew (1979). Revised by Waithman and Roest (1977) and Dragoo et al. (1990). Blair et al. (1968), Lechleitner (1969), Bueler (1973), and Dragoo et al. (1990) considered *macrotis* and *velox* conspecific. Packard and Bowers (1970), Rohwer and Kilgore (1973), Thornton and Creel (1975) (who found hybrids between *velox* and *macrotis* but concluded they were of reduced viability) and Mercure et al. (1993) retained both as separate species. Mercure et al. (1993) argued that the genetic differences between *macrotis* and *velox* were similar to that of *Vulpes vulpes* and *V. lagopus* and therefore argued that they should be recognized at the species level (followed here). Synonyms allocated according to Mercure et al. (1993).

*Vulpes pallida* (Cretzschmar, 1826). *In* Rüppell, Atlas Reise Nordl. Afr., Zool. Säugeth., 1(2):33, pl. 11.

COMMON NAME: Pale Fox.

TYPE LOCALITY: "Kordofan" [Sudan].

DISTRIBUTION: Semiarid sahelian region of Africa: Burkina Faso, Cameroon, Chad, Eritrea, Ethiopia, Gambia, Mali, Mauritania, Niger, Nigeria, Senegal, Somalia, Sudan.

STATUS: IUCN – Data Deficient.

SYNONYMS: *sabbar* (Hemprich and Ehrenberg, 1832); ***cyrenaica*** Festa, 1921; ***edwardsi*** Rochebrune, 1883; ***harterti*** Thomas and Hinton, 1921; *oertzeni* (Matschie, 1910).

COMMENTS: Synonyms allocated according to G. M. Allen (1939).

*Vulpes rueppellii* (Schinz, 1825). *In* G. Cuvier, Das Thierreich, 4:508.
    COMMON NAME: Rüppell's Fox.
    TYPE LOCALITY: "Vatherland Dongola, Sudan".
    DISTRIBUTION: Afghanistan, Egypt (Sinai), Iran, Morocco, Pakistan, Saudia Arabia, Somalia.
    STATUS: IUCN – Data Deficient.
    SYNONYMS: *famelicus* (Cretzschmar, 1826); *somalize* Thomas, 1918; ***caesia*** Thomas and Hinton, 1921; ***cyrenaica*** Festa, 1921; ***sabaea*** Pocock, 1934; *zarudnyi* Birula, 1913.
    COMMENTS: Synonyms allocated according to G. M. Allen (1939), Ellerman and Morrison-Scott (1951), and Larivière and Seddon (2001).

*Vulpes velox* (Say, 1823). *In* James, Account of an Exped. from Pittsburgh to the Rocky Mtns, 1:487.
    COMMON NAME: Swift Fox.
    TYPE LOCALITY: "camp on the river Platte, at the fording place of the Pawnee Indians, twenty-seven miles [43 km] below the confluence of the North and South, or Paduca Forks." [Camp on 20 June 1820 reported to be at 40.59'15'N (vol. 2)].
    DISTRIBUTION: Canada (SE British Columbia, SC Alberta and SW Saskatchewan), USA (C North America to NW Texas panhandle and E New Mexico).
    STATUS: U.S. ESA – Endangered as *V. velox hebes* (Canada); IUCN – Lower Risk (cd).
    SYNONYMS: *hebes* Merriam, 1902.
    COMMENTS: See comments under *V. macrotis* for the separation of *macrotis* and *velox* as followed here.

*Vulpes vulpes* (Linnaeus, 1758). Syst. Nat., 10th ed., 1:40.
    COMMON NAME: Red Fox.
    TYPE LOCALITY: "Europa, Asia, Africa, antrafodiens," restricted by Thomas (1911*a*) to "Sweden (Upsala)."
    DISTRIBUTION: Afghanistan, Albania, Algeria, Armenia, Austria, Azerbaijan, Bangladesh, Belarus, Belgium, Bhutan, Bosnia and Herzegovina, Bulgaria, China, Croatia, Czech Republic, Denmark, Egypt, Estonia, Finland, France, Geogria, Germany, Great Britain, Greece, Hungary, Iceland, India, Iran, Iraq, Ireland, Israel, Italy, Japan, Jordan, Kazakhstan, Kyrgyzstan, Laos, Latvia, Lithuania, Macedonia, Moldova, Mongolia, Morocco, Nepal, Netherlands, North and South Korea, Norway, Pakistan, Portugal, Romania, Russia, Serbia and Montenegro, Slovakia, Slovenia, Spain, Sweden, Switzerland, Syria, Tunisia, Turkey, Turkmenistan, Ukraine, USA (Alaska, throughout most of the contiguous 48 states except central plains and SW deserts), Uzbekistan, Vietnam. Introduced to Australia (Corbet and Hill, 1980)
    STATUS: CITES – Appendix III (India) as *V. vulpes griffithi*, *V. v. montana* and *V. v. pusilla*; IUCN – Lower Risk (lc).
    SYNONYMS: *alopex* (Linnaeus, 1758); *communis* Burnett, 1829; *lineatus* (Billberg, 1827); *nigro-argenteus* (Nilsson, 1820); *nigrocaudatus* (Billberg, 1827); *septentrionalis* Brass, 1911; *variegates* (Billberg, 1827); *vulgaris* Oken, 1816; ***abietorum*** Merriam, 1900; *sitkaensis* Brass, 1911; ***alascensis*** Merriam, 1900; ***alpherakyi*** Satunin, 1906; ***anatolica*** Thomas, 1920; ***arabica*** Thomas, 1902; ***atlantica*** (Wagner, 1841); *algeriensis* Loche, 1858; ***bangsi*** Merriam, 1900; ***barbara*** (Shaw, 1800); *acaab* Cabrera, 1916; ***beringiana*** (Middendorff, 1875); *anadyrensis* J. A. Allen, 1903; *beringensis* Merriam, 1902; *kamtschadensis* Brass, 1911; *kamtschatica* Dybowski, 1922; *schantaricus* Yudin, 1986; ***cascadensis*** Merriam, 1900; ***caucasica*** Dinnik, 1914; ***crucigera*** (Bechstein, 1789); *alba* (Borkhausen, 1797); *cinera* (Bechstein, 1801); *diluta* Ognev, 1924; *europaeus* (Kerr, 1792); *hellenica* Douma-Petridou and Ondrias, 1980; *hypomelas* Wagner, 1841; *lutea* (Bechstein, 1801); *melanogaster* (Bonaparte, 1832); *meridionalis* Fitzinger, 1855; *nigra* (Borkhausen, 1797); *stepensis* Brauner, 1914; ***daurica*** Ognev, 1931; *ussuriensis* Dybowski, 1922; ***deletrix*** Bangs, 1898; ***dolichocrania*** Ognev, 1926; *ognevi* Yudin, 1986; ***dorsalis*** (J. E. Gray, 1838); ***flavescens*** J. E. Gray, 1843; *cinerascens* Birula, 1913; *splendens* Thomas, 1902; ***fulvus*** (Desmarest, 1820); *pennsylvanicus* [sic] Rhoads, 1894;

*griffithi* Blyth, 1854; *flavescens* Hutton, 1845 [preoccupied]; **harrimani** Merriam, 1900; **hoole** Swinhoe, 1870; *aurantioluteus* Matschie, 1907; *lineiventer* Swinhoe, 1871; **ichnusae** Miller, 1907; **indutus** Miller, 1907; **jakutensis** Ognev, 1923; *sibiricus* Dybowski, 1922 [*nomen nudum*]; **japonica** J. E. Gray, 1868; **karagan** (Erxleben, 1777); *ferganensis* Ognev, 1926; *melanotus* (Pallas, 1811); *pamirensis* Ognev, 1926; *tarimensis* Matschie, 1907; **kenaiensis** Merriam, 1900; **kurdistanica** Satunin, 1906; *alticola* Ognev, 1926; **macroura** Baird, 1852; **montana** (Pearson, 1836); *alopex* Blanford, 1888; *himalaicus* (Ogilby, 1837); *ladacensis* Matschie, 1907; *nepalensis* J. E. Gray, 1837; *waddelli* Bonhote, 1906; **necator** Merriam, 1900; **niloticus** (É. Geoffroy Saint-Hilaire, 1803); *aegyptiacus* (Sonnini, 1816); *anubis* (Hemprich and Ehrenberg, 1833); *vulpecula* (Hemprich and Ehrenberg, 1833); **ochroxantha** Ognev, 1926; **palaestina** Thomas, 1920; **peculiosa** Kishida, 1924; *kiyomassai* Kishida and Mori, 1929; **pusilla** Blyth, 1854; *leucopus* Blyth, 1854; *persicus* Blanford, 1875; **regalis** Merriam, 1900; **rubricosa** Bangs, 1898; *bangsi* Merriam, 1900; *deletrix* Bangs, 1898; *rubricos* Churcher, 1960; *vafra* Bangs, 1897 [preoccupied]; **schrenckii** Kishida, 1924; **silacea** Miller, 1907; **splendidissima** Kishida, 1924; **stepensis** Brauner, 1914; *crymensis* Brauner, 1914; *krymeamontana* Brauner, 1914; **tobolica** Ognev, 1926; **tschiliensis** Matschie, 1907; *huli* Sowerby, 1923.
COMMENTS: Reviewed by Larivière and Pasitschniak (1996). Synonyms allocated according to Larivière and Pasitschniak (1996).

*Vulpes zerda* (Zimmermann, 1780). Geogr. Gesch. Mensch. Vierf. Thiere, 2:247.
COMMON NAME: Fennec Fox.
TYPE LOCALITY: "Es bewohnt die Soara und andere Theile von Nordafrika hinter den Atlas, der Ritter Bruce behauptet, man fände es auch in tripolitanischen."
DISTRIBUTION: Chad, Egypt, Kuwait, Libya, Mali, Mauritania, Morocco, Niger, Saudia Arabia, Sudan, Tunisia.
STATUS: CITES – Appendix II; IUCN – Data Deficient.
SYNONYMS: *arabicus* (Desmarest, 1804); *aurita* (F. A. A. Meyer, 1793); *brucei* (Desmarest, 1820); *cerda* (Illiger, 1811); *cerdo* (Gmelin, 1788); *denhamii* Boitard, 1842; *fennecus* (Lesson, 1827); *saarensis* Skjoldebrand, 1777 [suppressed, ICZN, O. 1129]; *zaarensis* Gray, 1843.
COMMENTS: Reviewed by Larivière (2002a) who did not recognize subspecies. Placed in *Fennecus* by Ellerman and Morrison-Scott (1951) and Stains (1975).

**Family Ursidae** Fischer de Waldheim, 1817. Mém. Soc. Imp. Nat. Moscow, 5:372.
SYNONYMS: *Ailuropodidae* Pocock, 1916; *Ursinidae* Gray, 1821.
COMMENTS: *Ailuropoda* has been placed in a separate family by some; however, morphological and molecular evidence strongly supports the placement of *Ailuropoda* in this family (Chorn and Hoffmann, 1978; Davis, 1964; Sarich, 1973, 1976; Mayr, 1986; Goldman et al., 1989; Hendey, 1980a, b; O'Brien et al., 1985; Wozencraft, 1989a). Thenius (1979) placed *Ailuropoda* in the monotypic family Ailuropodidae. Morphological studies have supported the monophyly of three subfamilies (Hendey, 1980; Kurtén, 1966; Thenius, 1979), although this has not been corroborated by a recent molecular approach (Goldman et al., 1989). Subfamilies are not recognized here. Synonyms allocated according to McKenna and Bell (1997).

*Ailuropoda* Milne-Edwards, 1870. Ann. Sci. Nat. Zool. (Paris), ser. 5, 13(10):1.
TYPE SPECIES: *Ursus melanoleucus* David, 1869, by monotypy.
SYNONYMS: *Aeluropus* Lydekker, 1891; *Ailuropus* Milne-Edwards, 1871; *Pandarctos* Gervais, 1870.
COMMENTS: Revised by Davis (1964) and Hendey (1980b). Reviewed by Chorn and Hoffmann (1978).

*Ailuropoda melanoleuca* (David, 1869). Nouv. Arch. Mus. Hist. Nat. Paris, Bull., 5:12-13.
COMMON NAME: Giant Panda.
TYPE LOCALITY: "Mou-pin" [China, Sichuan Sheng, Baoxing (=Moupin) 30°23′N, 102°50′E].
DISTRIBUTION: China (Sichuan, Shensi, Gansu; perhaps Qinghai, on E edge of Tibetan plateau).
STATUS: CITES – Appendix I; U.S. ESA – Endangered; IUCN – Endangered.

COMMENTS: Regarded by Hendey (1980*a*, *b*) as the only surviving species in the subfamily Agriotheriinae. Placed in the monotypic family Ailuropodidae by Thenius (1979). Reviewed by Chorn and Hoffmann (1978).

*Helarctos* Horsfield, 1825. Zool. J., 2(6):221.

TYPE SPECIES: *Helarctos euryspilus* Horsfield, 1825 (= *Ursus malayanus* Raffles, 1821) by original designation (Melville and Smith, 1987).

SYNONYMS: *Helarctus* Gloger, 1841.

COMMENTS: Revised by Pocock (1932*b*). Van Gelder (1977*b*) placed *Helarctos* in *Melursus*. Pocock (1941*a*) and Bininda-Emonds et al. (1999) suggested a close relationship between *M. ursinus* and *H. malayanus*; however, this was not supported by Goldman et al. (1989).

*Helarctos malayanus* (Raffles, 1821). Trans. Linn. Soc. Lond., 13:254.

COMMON NAME: Sun bear.

TYPE LOCALITY: "Sumatra" [Indonesia].

DISTRIBUTION: Burma, China (Yunnan), India, Indonesia (Sumatra, Kalimantan), Kampuchea, Laos, Malaysia, Thailand, Vietnam.

STATUS: CITES – Appendix I; IUCN – Data Deficient.

SYNONYMS: *annamiticus* Heude, 1901; *wardi* (Lydekker, 1906); **euryspilus** Horsfield, 1825.

COMMENTS: Reviewed by Fitzgerald and Krausman (2002). Synonyms allocated according to Ellerman and Morrison-Scott (1951).

*Melursus* Meyer, 1793. Zool. Entdeck., pp. 155-160.

TYPE SPECIES: *Melursus lybius* Meyer, 1793 (= *Bradypus ursinus* Shaw, 1791), by monotypy (Melville and Smith, 1987).

SYNONYMS: *Arceus* Goldfuss, 1809; *Chondrorhynchus* Fischer de Waldheim, 1814; *Prochilus* Illiger, 1811; *Prochylus* Gray, 1821.

COMMENTS: Revised by Pocock (1932*b*). See comments under *Helarctos* concerning the relationship between these taxa. Synonyms allocated according to McKenna and Bell (1997).

*Melursus ursinus* (Shaw, 1791). Nat. Misc., 2 (unpaginated) pl. 58.

COMMON NAME: Sloth Bear.

TYPE LOCALITY: "Abinteriore Bengala"; restricted by Pocock (1941*a*) as "Patna, north of the Ganges, Bengal" [India].

DISTRIBUTION: India (north to the Indian desert and to the foothills of the Himalayas), Sri Lanka.

STATUS: CITES – Appendix I; IUCN – Vulnerable.

SYNONYMS: *labiatus* (de Blainville, 1817); *longirostris* (Tiedemann, 1820); *lybius* Meyer, 1793; *niger* (Goldfuss, 1809); **inornatus** Pucheran, 1855.

COMMENTS: Synonyms allocated according to Pocock (1941*a*).

*Tremarctos* Gervais, 1855. Hist. Nat. Mammifères, 2:20.

TYPE SPECIES: *Ursus ornatus* F. G. Cuvier, 1825.

SYNONYMS: *Nearctos* Gray, 1873.

*Tremarctos ornatus* (F. G. Cuvier, 1825). *In* É. Geoffroy Saint-Hilaire and F. G. Cuvier, Hist. Nat. Mammifères, pt. 3, 5(50), "Ours des cordiliéres du Chili," 2 pp.

COMMON NAME: Spectacled Bear.

TYPE LOCALITY: "cordiliéres du Chili," restricted by Cabrera (1957) to "los montañas al este de Trujillo, departamento de la Libertad, Perú."

DISTRIBUTION: Mountainous regions of W Bolivia, Colombia, Ecuador, Panama (?), Peru, W Venezuela.

STATUS: CITES – Appendix I; IUCN – Vulnerable.

SYNONYMS: *frugilegus* (Tschudi, 1844); *lasallei* Maria, 1924; *majori* Thomas, 1902; *nasutus* (Sclater, 1868); *thomasi* (Hornaday, 1911).

COMMENTS: Some authors have considered this genus as the only extant member of the subfamily Tremarctinae (Thenius, 1976).

*Ursus* Linnaeus, 1758. Syst. Nat., 10th ed., 1:47.

TYPE SPECIES: *Ursus arctos* Linnaeus, 1758, by tautonymy (Melville and Smith, 1987).

SYNONYMS: *Arcticonus* Pocock, 1917; *Danis* J. E. Gray, 1825; *Euarctos* Gray, 1864; *Mamursus* Herrera, 1899; *Melanarctos* Heude, 1898; *Mylarctos* Lönnberg, 1923; *Myrmarctos* J. E. Gray, 1864; *Selenarctos* Heude, 1901; *Thalassarctos* J. E. Gray, 1825; *Thalassarctus* Gloger, 1841; *Thalassiarchus* Kobelt, 1896; *Ursarctos* Heude, 1898; *Ursulus* Kretzoi, 1954; *Vetularctos* Merriam, 1918.

COMMENTS: The close relationship of the four species included herein has been generally recognized by morphological and molecular studies (Goldman et al., 1989; Hendey, 1980*a*; Kurtén and Anderson, 1980; Shields and Kocher, 1991). Allen (1938) proposed a close relationship between *thibetanus* and *americanus*. Thenius (1953), Goldman et al. (1989), and Shields and Kocher (1991) gave support to the monophyly of *arctos* with *maritimus*. Synonyms allocated according to McKenna and Bell (1997).

*Ursus americanus* Pallas, 1780. Spicil. Zool., 14:5.

COMMON NAME: American Black Bear.

TYPE LOCALITY: Not given. In Pallas' (1780) description, he refered to Brickell (1737) who implied North Carolina (USA) by stating they "are very common in this province." Palmer (1904) listed the locality as "eastern North America".

DISTRIBUTION: Canada, Mexico (N Nayarit and S Tamaulipas), USA.

STATUS: CITES – Appendix II; U.S. ESA – Threatened as *U. americanus luteolus*; all other subspecies – Similarity of Appearance to a Threatened Species; IUCN – Lower Risk (lc).

SYNONYMS: *hunteri* Anderson, 1945; *randi* Anderson, 1945; *schwenki* Shoemaker, 1913; *sornborgeri* Bangs, 1898; **altifrontalis** Elliot, 1903; **amblyceps** Baird, 1859; **californiensis** Miller, 1900; **carlottae** Osgood, 1901; **cinnamomum** Audubon and Bachman, 1854; **emmonsii** Dall, 1895; *glacilis* Kells, 1897; **eremicus** Merriam, 1904; **floridanus** Merriam, 1896; **hamiltoni** Cameron, 1957; **kermodei** Hornaday, 1905; **luteolus** Griffith, 1821; **machetes** Elliot, 1903; **perniger** J. A. Allen, 1910; *kenaiensis* J. A. Allen, 1910; **pugnax** Swarth, 1911; **vancouveri** Hall, 1928.

COMMENTS: Reviewed by Larivière (2001*b*). Synonyms allocated according to Hall (1981) and Larivière (2001*b*).

*Ursus arctos* Linnaeus, 1758. Syst. Nat., 10th ed., 1:47.

COMMON NAME: Brown Bear. In US, often known as Grizzly Bear (see Free-tailed bat).

TYPE LOCALITY: "sylvis Europae frigidae" restricted by Thomas (1911*a*) to "Northern Sweden."

DISTRIBUTION: Afghanistan, Albania, Armenia, Austria, Azerbaijan, Belarus, Bhutan, Bosnia and Herzegovina, Bulgaria, N and W China, Croatia, Estonia, Finland, France, Greece, India, Iran, Iraq, Israel, Italy, Japan (Hokkaido), Kazakhstan, Lebanon, Macdeonia, N Mexico, Mongolia, North Korea, Norway, N Pakistan, Poland, Romania, Russia, Serbia and Montenegro, Slovakia, Spain, Sweden, Syria, Turkey, Ukraine, W USA.

STATUS: CITES – Appendix I as *U. arctos* (Mexico, Bhutan, China, and Mongolia populations) and *U. a. isabellinus*; otherwise Appendix II. U. S. ESA – Endangered as *U. arctos pruinosus*, as *U. arctos* in Mexico, and as *U. a. arctos* in Italy. Threatened as *U. a. horribilis* in the USA (48 conterminous states) except where listed as Experimental Non Essential Populations in portions of Idaho and Montana; IUCN – Extinct as *U. a. nelsoni*, otherwise Lower Risk (lc).

SYNONYMS: *albus* Gmelin, 1788; *alpinus* G. Fischer, 1814; *annulatus* Billberg, 1827; *argenteus* Billberg, 1827; *aureus* Fitzinger, 1855; *badius* Schrank, 1798; *brunneus* Billberg, 1827; *cadaverinus* Eversmann, 1840; *euryrhinus* Nilsson, 1847; *eversmanni* (Gray, 1864); *falciger* Reichenbach, 1836; *formicarius* Billberg, 1828; *fuscus* Gmelin, 1788; *grandis* J. E. Gray, 1864; *griseus* Kerr, 1792; *gobiensis* Sokolov and Orlov, 1992; *longirostris* Eversmann, 1840; *major* Nilsson, 1820; *marsicanus* Altobello, 1921; *minor* Nilsson, 1820; *myrmephagus* Billberg, 1827; *niger* Gmelin, 1788; *normalis* Gray, 1864; *norvegicus* J. B. Fischer, 1829; *polonicus* J. E. Gray, 1864; *pyrenaicus* J. B. Fischer, 1829; *rossicus* J. E. Gray, 1864; *rufus* Borkhausen, 1797; *scandinavicus* Gray, 1864; *stenorostris* Gray, 1864; *ursus* Boddaert, 1772; **alascensis** Merriam, 1896; *alexandrae* Merriam, 1914; *cressonus* Merriam, 1916; *eximius* Merriam, 1916; *holzworthi* Merriam, 1929; *innuitus* Merriam, 1914; *internationalis* Merriam, 1914; *kenaiensis* Merriam, 1904; *kidderi*

Merriam, 1902; *nuchek* Merriam, 1916; *phaeonyx* Merriam, 1904; *sheldoni* Merriam, 1910; *toklat* Merriam, 1914; *tundrensis* Merriam, 1914; **beringianus** Middendorff, 1851; *kolymensis* Ognev, 1924; *mandchuricus* Heude, 1898; *piscator* Pucheran, 1855; **californicus** Merriam, 1896; *colusus* Merriam, 1914; *henshawi* Merriam, 1914; *klamathensis* Merriam, 1914; *magister* Merriam, 1914; *mendocinensis* Merriam, 1916; *tularensis* Merriam, 1914; **collaris** F. G. Cuvier, 1824; *jeniseensis* Ognev, 1924; *sibiricus* J. E. Gray, 1864; **crowtheri** Schinz, 1844; **dalli** Merriam, 1896; *nortoni* Merriam, 1914; *orgiloides* Merriam, 1918; *townsendi* Merriam, 1916; **gyas** Merriam, 1902; *merriami* J. A. Allen, 1902; **horribilis** Ord, 1815; *absarokus* Merriam, 1914; *andersoni* Merriam, 1918; *apache* Merriam, 1916; *arizonae* Merriam, 1916; *bairdi* Merriam, 1914; *bisonophagus* Merriam, 1918; *canadensis* Merriam, 1914; *candescens* C. E. H. Smith, 1827; *cinereus* Desmarest, 1820; *crassus* Merriam, 1918; *dusorgus* Merriam, 1918; *ereunetes* Merriam, 1918; *griseus* Choris, 1822; *horriaeus* Baird, 1858; *hylodromus* Elliot, 1904; *idahoensis* Merriam, 1918; *imperator* Merriam, 1914; *impiger* Merriam, 1918; *inopinatus* Merriam, 1918; *kennerleyi* Merriam, 1914; *kluane* Merriam, 1916; *latifrons* Merriam, 1914; *macfarlani* Merriam, 1918; *macrodon* Merriam, 1918; *mirus* Merriam, 1918; *navaho* Merriam, 1914; *nelsoni* Merriam, 1914; *ophrus* Merriam, 1916; *oribasus* Merriam, 1918; *pallasi* Merriam, 1916; *pellyensis* Merriam, 1918; *perturbans* Merriam, 1918; *planiceps* Merriam, 1918; *pulchellus* Merriam, 1918; *richardsoni* Swainson, 1838; *rogersi* Merriam, 1918; *rungiusi* Merriam, 1918; *russelli* Merriam, 1914; *sagittalis* Merriam, 1918; *selkirki* Merriam, 1916; *shoshone* Merriam, 1914; *texensis* Merriam, 1914; *utahensis* Merriam, 1914; *washake* Merriam, 1916; **isabellinus** Horsfield, 1826; *leuconyx* Severtzov, 1873; *pamirensis* Ognev, 1924; **lasiotus** Gray, 1867; *baikalensis* Ognev, 1924; *cavifrons* (Heude, 1901); *ferox* Temminck, 1844 [preoccupied]; *macneilli* Lydekker, 1909; *melanarctos* Heude, 1898; *yesoensis* Lydekker, 1897; **middendorffi** Merriam, 1896; *kadiaki* Kleinschmidt, 1911; **pruinosus** Blyth, 1854; *lagomyiarius* Przewalski, 1883; **sitkensis** Merriam, 1896; *caurinus* Merriam, 1914; *eltonclarki* Merriam, 1914; *eulophus* Merriam, 1904; *insularis* Merriam, 1916; *mirabilis* Merriam, 1916; *neglectus* Merriam, 1916; *orgilos* Merriam, 1914; *shirasi* Merriam, 1914; **stikeenensis** Merriam, 1914; *atnarko* Merriam, 1918; *chelan* Merriam, 1916; *chelidonias* Merriam, 1918; *crassodon* Merriam, 1918; *hoots* Merriam, 1916; *kwakiutl* Merriam, 1916; *pervagor* Merriam, 1914; *tahltanicus* Merriam, 1914; *warburtoni* Merriam, 1916; **syriacus** Hemprich and Ehrenberg, 1828; *caucasicus* Smirnov, 1919; *dinniki* Smirnov, 1919; *lasistanicus* Satunin, 1913; *meridionalis* Middendorff, 1851; *persicus* Lönnberg, 1925; *schmitzi* Matschie, 1917; *smirnovi* Lönnberg, 1925.

COMMENTS: Reviewed by Erdbrink (1953), Couturier (1954), Rausch (1963*a*), Kurtén (1973), Hall (1984) and Pasitschniak-Arts (1993). Ognev (1931) and Allen (1938) recognized *U. pruinosus* as distinct; not followed by Ellerman and Morrison-Scott (1951), Gao (1987), and Stroganov (1962). Lönnberg (1923*b*) believed that differences between *pruinosus* and *arctos* warranted subgeneric distinction as (*Mylarctos*) *pruinosus*; however, this was not supported by Pocock's (1932*b*) thorough revision. Synonyms allocated according to Ellerman and Morrison-Scott (1966) and Hall (1984).

*Ursus maritimus* Phipps, 1774. Voyage Towards North Pole, p. 185.
COMMON NAME: Polar Bear.
TYPE LOCALITY: "on the main land of Spitsbergen" [Norway].
DISTRIBUTION: Canada, Greenland, USA (Alaska), Russia. Circumpolar in the Arctic, S limits determined by ice pack.
STATUS: CITES – Appendix II; IUCN – Lower Risk (cd).
SYNONYMS: *eogroenlandicus* (Knottnerus-Meyer, 1908); *groenlandicus* (Birula, 1932); *jenaensis* (Knottnerus-Mayer, 1908); *labradorensis* (Knottnerus-Meyer, 1908); *marinus* Pallas, 1776; *polaris* Shaw, 1792.
COMMENTS: Revised by Wilson (1976). Reviewed by DeMaster and Stirling (1981). Placed in subgenus *Thalarctos* by Gromov and Baranova (1981). *U. maritimus* is considered the sister species to *arctos* (Goldman et al., 1989; Shields and Kocher, 1991). Synonyms allocated according to Ellerman and Morrison-Scott (1951) and Hall (1981).

*Ursus thibetanus* G.[Baron] Cuvier, 1823. Rech. Oss. Foss., Nouv. ed., 4:325.
    COMMON NAME: Asian Black Bear.
    TYPE LOCALITY: "Cet ours a été trouvé d'abord par M. Wallich dans les montagnes du
        Napaul, et je l'ai rencontré également dans celles du Sylhet" [India, Assam, Sylhet].
    DISTRIBUTION: Afghanistan, China, India, Indochina, Japan, North and ? South Korea, Laos,
        Nepal, Pakistan, Taiwan, Thailand, Russia (SE Primorski Krai), Vietnam.
    STATUS: CITES – Appendix I; U. S. ESA – Endangered as *U. t. gedrosianus*; IUCN – Critically
        Endangered as *U. t. gedrosinus*, otherwise Vulnerable.
    SYNONYMS: *labiatus* Blanford, 1876; *torquatus* Wagner, 1841; **formosanus** Swinhoe, 1864;
        *melli* (Matschie, 1922); **gedrosianus** Blanford, 1877; **japonicus** Schlegel, 1857; *rexi*
        Matschie, 1897; **laniger** (Pocock, 1932); **mupinensis** (Heude, 1901); *clarki* Sowerby,
        1920; *leuconyx* (Heude, 1901); *macneilli* Lydekker, 1909; **ussuricus** (Heude, 1901);
        *wulsini* (Howell, 1928).
    COMMENTS: Placed in subgenus *Selenarctos* by Gromov and Baranova (1981); and in
        subgenus *Euarctos* by Thenius (1979). Allen (1938) suggested a close relationship to
        *U. americanus*; Pocock (1932a) retained in a separate genus—there is molecular support
        for both positions (Goldman et al., 1989). Synonyms allocated according to Ellerman
        and Morrison-Scott (1951).

**Family Otariidae** Gray, 1825. Ann. Philos., n.s., 10:340.
    SYNONYMS: Arctocephalina Gray, 1837; Callorhinae Muizon, 1978; Callorhinina Gray, 1869;
        Eumetopiina Gray, 1869; Gypsophocina Gray, 1874; Otariadae Brookes, 1828; Otariarina
        J. E. Gray, 1843; Otarioidea Smirnov, 1908; Ouliphocacae J. A. Allen, 1880; Ouliphocinae
        J. A. Allen, 1870; Trichiphocinae J. A. Allen, 1870; Trichophocacae J. A. Allen, 1880;
        Zalophina Gray, 1869.
    COMMENTS: Reviewed by Allen (1880, 1892), Repenning et al. (1971), Mitchell and Tedford
        (1973), J. E. King (1983), Berta and Deméré (1986), Barnes (1989) and Wynen et al. (2001).
        Does not include *Odobenus*, which was included in a monotypic subfamily (Odobeninae
        within Otariidae) by Mitchell and Tedford (1973), Tedford (1976), Hall (1981), Barnes
        (1989), and Wozencraft (1989a, b); however, see Wyss (1987) and Berta (1991). Berta and
        Deméré (1986) separated *Arctocephalus* and *Callorhinus* into the Arctocephalinae.
        Repenning et al. (1971), Repenning and Tedford (1977), and Wynen et al. (2001) argued
        against the recognition of subfamilies. Distributional information for species based on
        Rice (1998).

*Arctocephalus* É. Geoffroy Saint-Hilaire and F. Cuvier, 1826. *In* F. Cuvier, Dict. Sci. Nat. 39:554 [1826].
    TYPE SPECIES: "*Phoca ursina*" (= *Phoca pusilla* Schreber, 1775; not *Phoca ursina* Linnaeus,
        1758)(International Commission on Zoological Nomenclature, 2000).
    SYNONYMS: *Arctophoca* Peters, 1866; *Euotaria* Gray, 1866; *Gypsophoca* Gray, 1866; *Halarctus* Gill,
        1866.
    COMMENTS: Reviewed by King (1954) and Repenning et al. (1971) who included *Arctophoca*
        Peters, 1866. Van Gelder (1977b) considered *Zalophus* and *Arctocephalus* congeneric.
        Nearly all species of *Arctocephalus* are distributed allopatrically (Rice, 1998). Synonyms
        allocated according to Gardner and Robbins (1998). Allen (1905) discussed confusion in
        designation of type species.

*Arctocephalus australis* (Zimmermann, 1783). Geogr. Gesch. Mensch. Vierf. Thiere, 3:276.
    COMMON NAME: South American Fur Seal.
    TYPE LOCALITY: Zimmermann (1783) based the name on the "Falkland Isle Seal" of Pennant
        (1781), and added that it "Wohnt um Juan Fernandez, und über haupt in dortigen
        Meeren." [Falkland Isls, UK].
    DISTRIBUTION: South America coasts of Argentina, Brazil (from Recife dos Tôrres south),
        Chile, Falkland Isls, Peru (from Isla Lobos de Tierra south), Uruguay.
    STATUS: CITES – Appendix II; IUCN – Lower Risk (lc).
    SYNONYMS: *argentata* (Philippi, 1871); *australis* (J. A. Allen, 1880); *brachydactyla* (Philippi,
        1892); *falclandica* (J. B. Fischer, 1829); *falklandica* (Shaw, 1800); *gracilis* Nehring, 1887;
        *grayii* Scott, 1873; *hauvillii* (Lesson, 1827); *laitirostros* J. E. Gray, 1874; *latirostris*

(J. E. Gray, 1872); *leucostoma* (Philippi, 1892); *lupina* (Molina, 1782); *nigrescens* (J. E. Gray, 1850); *shawii* (Lesson, 1828); *ursinus* J. E. Gray, 1843.
COMMENTS: Scheffer (1958) included *galapagoensis* Heller, 1904, but this was not followed by Repenning et al. (1971) or J. E. King (1983). Synonyms allocated according to Cabrera (1957) and Rice (1998).

*Arctocephalus forsteri* (Lesson, 1828). *In* Bory de Saint-Vincet (ed.), Dict. Class. Hist. Nat. Paris., 13:421.
COMMON NAME: Australasian Fur Seal.
TYPE LOCALITY: Scheffer (1958) restricted the type locality to "Dusky Sound, New Zealand."
DISTRIBUTION: Coastal regions of Australia (Eclipse Isl in the west to S end of Tasmania), New Zealand and nearby subantarctic isls.
STATUS: CITES – Appendix II; IUCN – Lower Risk (lc).

*Arctocephalus galapagoensis* Heller, 1904. Proc. Calif. Acad. Sci., ser. 3(7):245.
COMMON NAME: Galapagos Fur Seal.
TYPE LOCALITY: "Wenman Island" [Ecuador, Galapagos Isls].
DISTRIBUTION: Endemic to Ecuador (Galapagos Isls).
STATUS: CITES – Appendix II; IUCN – Vulnerable.
COMMENTS: Repenning et al. (1971) supported recognition at the specific level, followed by J. E. King (1983); however, Scheffer (1958) considered *galapagoensis* conspecific with *australis*, which would be the most closely related taxon. Reviewed by Clark (1975).

*Arctocephalus gazella* (Peters, 1875). Monatsb. K. Preuss. Akad. Wiss. Berlin, 1875:393, 396.
COMMON NAME: Antarctic Fur Seal.
TYPE LOCALITY: "von Seehunden aus Kerguelenland". Restriced by Scheffer (1958) to "Anse Betsy (49°09'S, 70°11'E)."
DISTRIBUTION: Islands south of Antarctic convergence (Kerguelen, S Sandwich, S Orkney, Heard, Bouver, S Georgia, S Shetland Isls).
STATUS: CITES – Appendix II; IUCN – Lower Risk (lc).
COMMENTS: Reviewed by King (1959*a*, *b*).

*Arctocephalus philippii* (Peters, 1866). Monatsb. K. Preuss. Akad. Wiss. Berlin, 1866:276, pl. 2a, b, c.
COMMON NAME: Juan Fernández Fur Seal.
TYPE LOCALITY: "Insel Juan Fernandez". Listed by Scheffer (1958) as "Isla Más a Tierra, Islas Juan Fernández, Chile".
DISTRIBUTION: Specimens recorded from Chile (Juan Fernandez and San Felix Isls), Peru (vagrant populations).
STATUS: CITES – Appendix II; IUCN – Vulnerable.

*Arctocephalus pusillus* (Schreber, 1775). Die Säugethiere, 2(13):pl. 85[1775]; text, 3(17):314 [1776].
COMMON NAME: Brown Fur Seal.
TYPE LOCALITY: Unknown. "Diese Gattung findet sich in den levantischen, und nach dem Herrn Grafen von Büffon, im indischen Meere"; see Allen (1880).
DISTRIBUTION: Two allopatric populations: (1) Southern African coastal regions of Angola (vagrant populations), Namibia (Cape Cross southward), South Africa (east to Algoa Bay) (2) Coastal regions of SE Australia, Tasmania.
STATUS: CITES – Appendix II; IUCN – Lower Risk (lc).
SYNONYMS: *antarctica* (Thunberg, 1811); *compressa* (Gray, 1874); *delalandii* (Lesson, 1827); *nivosus* (Gray, 1868); *parva* (Boddaert, 1785); *peronii* (Desmarest, 1817); *schist-hyperves* (Turner, 1868); ***doriferus*** Wood Jones, 1925; *tasmanicus* Scott and Lord, 1926.
COMMENTS: Repenning et al. (1971) and J. E. King (1983) included *doriferus* Wood-Jones, 1925; however, Scheffer (1958) considered it a distinct species.

*Arctocephalus townsendi* Merriam, 1897. Proc. Biol. Soc. Wash., 11:175.
COMMON NAME: Guadalupe Fur Seal.
TYPE LOCALITY: "Guadalupe Island, off Lower California. . . collected on the beach on west side of Guadalupe." [Mexico]
DISTRIBUTION: Mexico (Guadalupe Isl), USA (Channel Isls).
STATUS: CITES – Appendix I; U.S. ESA – Threatened; IUCN – Vulnerable.

COMMENTS: Reviewed by Belcher and Lee (2002). Formerly included in *Arctophoca*; see
Repenning et al. (1971). Considered conspecific with *philippii* by Scheffer (1958).

*Arctocephalus tropicalis* (J. E. Gray, 1872). Proc. Zool. Soc. Lond., 1872:653, 659.
COMMON NAME: Subantarctic Fur Seal.
TYPE LOCALITY: "North coast of Australia." This is in error, fixed by King (1959*a*) to
"'Australasian sea' . . . to include the islands of St. Paul and Amsterdam as these are the
islands nearest to Australia . . .".
DISTRIBUTION: Islands north of Antarctic Convergence (Tristan, Gough, Marion, Crozet,
Amsterdam, Macquarie Isls).
STATUS: CITES – Appendix II; IUCN – Lower Risk (lc).
SYNONYMS: *elegans* Peters, 1876.

*Callorhinus* J. E. Gray, 1859. Proc. Zool. Soc. Lond., 1859:359.
TYPE SPECIES: *Arctocephalus ursinus* Gray, 1859 (= *Phoca ursina* Linnaeus, 1758), by original
designation (International Commission on Zoological Nomenclature, 2000).
SYNONYMS: *Callirhinus* J. E. Gray, 1859; *Callorhynchus* Greve, 1896; *Callotaria* Palmer, 1892;
*Otaria* Péron, 1816; *Otoes* G. Fischer, 1817; *Phoca* Linnaeus, 1758.
COMMENTS: Synonyms allocated according to McKenna and Bell (1997), and Gardner and
Robbins (1998).

*Callorhinus ursinus* (Linnaeus, 1758). Syst. Nat., 10th ed., 1:37.
COMMON NAME: Northern Fur Seal.
TYPE LOCALITY: "in Camschatcć maritimus inter Asiam and Americam proximam, primario
in infula Beringri," restricted by Thomas (1911*a*) to "Bering Island."
DISTRIBUTION: North Pacific coastal regions in Canada, China (vagrant to Shandong), Japan,
Mexico (costs of Baja California), Russia (Okhotsk and Bering Seas, Commander and
Pribilof Isls), USA (Alaska, Washington, Oregon, S California).
STATUS: IUCN – Vulnerable.
SYNONYMS: *alascanus* (Jordan and Clark, 1898); *californianus* (Gray, 1866); *curilensis* (Jordan
and Clark, 1899); *cynocephala* (Walbaum, 1792); *krachenninikowii* (Lesson, 1828);
*mimica* (Tilesius, 1835); *nigra* (Pallas, 1811).
COMMENTS: Subspecies not recognized, following Taylor et al. (1955) and Rice (1998).

*Eumetopias* Gill, 1866. Proc. Essex Inst. Salem, 5:7.
TYPE SPECIES: *Arctocephalus monterienis* Gray, 1859 (= *Phoca jubata* Schreber, 1776) by monotypy
(International Commission on Zoological Nomenclature, 2000).
COMMENTS: For a discussion of the type, see Scheffer (1958).

*Eumetopias jubatus* (Schreber, 1776). Die Säugethiere text, 3(17):300[1776]; 3(17):pl. 83.B[1776].
COMMON NAME: Steller Sea Lion.
TYPE LOCALITY: " . . . Aufenthalt in dem nördlichen Theil des stillen Meeres . . . westlichen
Küste von Amerika . . . östlichen von Kamtschatka . . . Inseln . . . Küsten unter dem
56ten Grade der Breite liegen." [N part of the Pacific. Russia, Commander and Bering
Isls].
DISTRIBUTION: Northern Pacific coastal regions of Canada, China (vagrant populations to
Jiangsu), Japan (from Hokkaido N), Russia, USA (Alaska, Washington, Oregon,
California).
STATUS: U.S. ESA – Threatened, except population segment west of 144° W. Long, which is
Endangered; IUCN – Endangered.
SYNONYMS: *leonina* (Pallas, 1811); *monteriensis* (Gray, 1859); *stellerii* (Lesson, 1828).
COMMENTS: The type of *O. californiana* Lesson, 1828, was shown by Allen (1880) to actually
be *Zalophus*. *A. monteriensis* Gray, 1859, is based on *P. jubata* Schreber, 1776. Scheffer
(1958) pointed out that *jubata* Forster, 1775, is invalid. Reviewed by Loughlin et al.
(1987). Synonyms allocated according to Ellerman and Morrison-Scott (1951), Hall
(1981), and Loughlin et al. (1987).

*Neophoca* Gray, 1866. Ann. Mag. Nat. Hist., ser. 3, 18:231.
TYPE SPECIES: *Arctocephalus lobatus* Gray, 1828 (= *Otaria cinerea* Péron, 1816).

COMMENTS: Sivertsen (1954) and Scheffer (1958) considered *Neophoca* congeneric with *Phocarctos*. However, it was retained as separate by J. E. King (1960, 1983), Rice (1977), Barnes (1989), and Wynen et al. (2001).

*Neophoca cinerea* (Péron, 1816). Voy. Decouv. Terres. Austral., 2:54.
COMMON NAME: Australian Sealion.
TYPE LOCALITY: "L'ile Decrès" [Australia, South Australia, Kangaroo Isl].
DISTRIBUTION: Australia coastal regions (Houtmans Abrolhos in the west to Kangaroo Isl in the south).
STATUS: IUCN – Lower Risk (lc).
SYNONYMS: *albicollis* (Péron, 1816); *australis* (Quoy and Gaimard, 1830); *fosteri* (Wood Jones, 1922); *lobatus* (J. E. Gray, 1828); *stelleri* (Temminck, 1844); *williamsi* (McCoy, 1877).
COMMENTS: Allen (1880) questioned the validity of the type description. Reviewed by Ling (1992). Synonyms allocated according to Ling (1992).

*Otaria* Péron, 1816. Voy. Decouv. Terres. Austral., 2:37 (footnote), pp. 40-52.
TYPE SPECIES: *Phoca leonina Molina, 1782* (= *Phoca byronia* de Blainville, 1820) by designation (International Commission on Zoological Nomenclature, 2000).
SYNONYMS: *Otoes* Fischer de Waldheim, 1817; *Platyrhynchus* F. G. Cuvier, 1826; *Pontoleo* Gloger, 1841.

*Otaria flavescens* (Shaw, 1800). Gener. Zool., 1, 2a parte:260.
COMMON NAME: South American Sealion.
TYPE LOCALITY: "Strait of Magellan".
DISTRIBUTION: South American coasts of Argentina, Brazil (south from Recife dos Tôrres), Chile, Peru, Uruguay, Falkland Isls. Vagrant populations occasionally in Columbia, Ecuador (Galapagos Isls), Panama.
STATUS: IUCN – Lower Risk (lc).
SYNONYMS: *aurita* (Bechstein, 1800); *byronia* (de Blainville, 1820); *chilensis* Muller, 1841; *chonotica* Philippi, 1892; *fulva* Philippi, 1892; *godeffroyi* Peters, 1866; *hookeri* Schlater, 1866; *leoninus* (F. G. Cuvier, 1827); *minor* Gray, 1874; *molossina* Lesson and Garnot, 1826; *molossinus* (Lesson, 1827); *pernettyi* Lesson, 1828; *pygmaea* Gray, 1874; *rufa* Philippi, 1892; *ulloae* Tschudi, 1844; *uraniae* (Lesson, 1827); *velutina* Philippi, 1892.
COMMENTS: There is some controversy regarding the validity of *O. byronia* (de Blainville, 1920) or *O. flavescens* (Shaw, 1800) (King, 1978). Rodriguez and Bastida (1993) reviewed the information and concluded that *flavescens* was a valid name with priority. Also see Oliva (1988) who argued for *O. byronia*. Synonyms allocated according to Cabrera (1957).

*Phocarctos* Peters, 1866. Monatsb. K. Preuss. Akad. Wiss. Berlin, 1866:269.
TYPE SPECIES: *Otaria hookeri* (=*Arctocephalus hookeri* Gray, 1844).
COMMENTS: Sivertsen (1954) and Scheffer (1958) considered *Phocarctos* congeneric with *Neophoca*, however, it was retained as separate by Clark (1873a), J. E. King (1960, 1983), Rice (1977), Barnes (1989), and Wynen et al. (2001).

*Phocarctos hookeri* (Gray, 1844). Zool. Voy. H.M.S. "Erebus" and "Terror," 1:4.
COMMON NAME: New Zealand Sealion.
TYPE LOCALITY: "Falkland Islands and Cape Horn." Locality in error; fixed by Clark (1873b) as "Auckland Islands. . . between 800 and 900 miles [1287 and 1448 km] S. of Tasmania, in lat. 50°48'S., long. 166°42'E." [New Zealand].
DISTRIBUTION: New Zealand subantarctic islands.
STATUS: IUCN – Vulnerable.

*Zalophus* Gill, 1866. Proc. Essex Inst. Salem, 5:7.
TYPE SPECIES: "*Otaria Gilliespii* Macbain", 1858 (= *Otaria californiana* Lesson, 1828).
COMMENTS: Included in *Arctocephalus* by Van Gelder (1977b). Mohr (1952) reported successful matings between *Arctocephalus pusillus* and *Z. californianus*. Rice (1998), followed here, argued for the retention of *japonicus*, *californianus*, and *wollenbaeki* as distinct species. Itoo (1985) concluded that *japonicus* was distinct, and behavioral differences separate

*californianus* and *wollenbaeki* (Eibl-Eibesfeldt, 1984). But see Scheffer (1958) who recognized these populations at the subspecies level.

*Zalophus californianus* (Lesson, 1828). *In* Bory de Saint-Vincent (ed.), Dict. Class. Hist. Nat. Paris. 13:420.
COMMON NAME: California Sealion.
TYPE LOCALITY: "les rochers dans le voisinage de la baie San-Francisco sont ordinairement couverts de lion marins." [USA, California, San Francisco Bay].
DISTRIBUTION: Northern Pacific coastal regions of Canada (British Columbia), Mexico (Baja California, and throughout the Gulf of California), USA (Washington, Oregon, California).
STATUS: IUCN – Lower Risk (lc).
SYNONYMS: *gillespii* (MacBain, 1858).

*Zalophus japonicus* (Peters, 1866). Monatsb. K. Preuss. Akad. Wiss., 1866:668.
COMMON NAME: Japanese Sealion.
TYPE LOCALITY: 'Japan'.
DISTRIBUTION: Sea of Japan. Historical range included Japan, Russia (Kamchatka, Sakhalin), South Korea (E coast).
STATUS: IUCN – Extinct.
SYNONYMS: *lobatus* Jentink, 1892.
COMMENTS: Probably extinct (Rice, 1998).

*Zalophus wollebaeki* Sivertsen, 1953. K. Norske Vidensk. Selsk. Forh., 26:2.
COMMON NAME: Galapagos Sea Lion.
TYPE LOCALITY: "Floreana, (Sancta Maria), Galapagos Islands."
DISTRIBUTION: Ecuador (Galapagos Isls), vagrant populations to coastal regions of Columbia and Ecuador.
STATUS: IUCN – Vulnerable.

**Family Odobenidae** Allen, 1880. U.S. Geol. and Geog. Surv. Territ., 12:ix, 5.
SYNONYMS: Odobaeninae Orlov, 1931; Odontobenidae Elliot, 1905; Rosmaridae Gill, 1866; Thalattailurina Albrecht, 1879; Trichechoidea Giebel, 1855; Trichecidae J. E. Gray, 1821; Trichiphocinae J. A. Allen, 1870; Trichisina J. E. Gray, 1837; Trichophocacae J. A. Allen, 1880.
COMMENTS: Trichecidae Gray (1821) and Rosmaridae Gill (1866) are invalid (International Commission on Zoological Nomenclature, 1959). The enigmatic walruses have been placed as: (1) the sister group to the otariids (Árnason, 1977; Árnason et al., 1995; Couturier and Dutrillaux, 1986; Dragoo and Honeycutt, 1997; Repenning and Tedford, 1977; Sarich, 1969*a, b*; Vrana et al., 1994); (2) in the family Otariidae (Barnes, 1989; Mitchell, 1975*b*); (3) the sister group to the phocids (Berta, 1994; Berta and Wyss, 1994; Wyss and Flynn, 1993); and finally (4) McKenna and Bell (1997) considered them a subfamily of Phocidae. Lento et al. (1995) believed the best answer was to leave the walrus as an independent family. Because of the uncertainty of the placement of this taxon, I have followed Rice (1998) who provided an excellent discussion of the various arrangements.

*Odobenus* Brisson, 1762. Regne Anim., 2nd ed., p. 30.
TYPE SPECIES: *Odobenus odobenus* Brisson, 1762 (= *Phoca rosmarus* Linnaeus, 1758).
SYNONYMS: *Hodobaenus* Sundevall, 1860; *Odobaenus* Fee, 1830; *Odontobaenus* Steenstrup, 1860; *Rosmarus* Brünnich, 1772; *Trichechus* Linnaeus, 1766; *Trichecus* F. G. Cuvier, 1829.
COMMENTS: Although the names in Brisson (1762) are invalid, *Odobenus* has been retained (International Commission on Zoological Nomenclature, 1955*a*, 1957*e*, 1998). Placed in Subfamily Odobeninae, Family Phocidae by McKenna and Bell (1997). Placed in separate family by Rice (1998).

*Odobenus rosmarus* (Linnaeus, 1758). Syst. Nat., 10th ed., 1:38.
COMMON NAME: Walrus.
TYPE LOCALITY: "intra Zonam arcticam Europae, Asiae, Americae".

DISTRIBUTION: Arctic sea- coastal regions of Belgium (vagrant), Canada, Great Britain (vagrant), Greenland, Iceland (vagrant), Japan (Honshu), Netherlands (vagrant), Norway, Russia, USA (Alaska, New England-vagrant).

STATUS: CITES – Appendix III (Canada); IUCN – Data Deficient as *O. r. laptevi*, otherwise Lower Risk (lc).

SYNONYMS: *arcticus* (Pallas, 1811); ***divergens*** (Illiger, 1815); *cookii* (Fremery, 1831); *orientalis* (Dybowski, 1922); ***laptevi*** Chapskii, 1940.

COMMENTS: Reviewed by Fay (1985). Synonyms allocated according to Ellerman and Morrison-Scott (1951) and Fay (1985). Distributional information from Rice (1998).

**Family Phocidae** Gray, 1821. London Med. Repos., 15:297.

SYNONYMS: Amphibia Trouessart, 1879; Amphibiae Gray, 1821; Cystophorina Gray, 1837; Erignathini Chapskii, 1955; Eumetopiina Gray, 1869; Halichoerina Gray, 1869; Histriophocina Chapskii, 1955; Hydrurginae Trouessart, 1907; Lobodoninae Kellogg, 1922; Lobodontina Gray, 1869; Miroungini Muizon, 1982; Monachina Gray, 1869; Ogmorhininae Turner, 1888; Phocadae [sic] Gray, 1821; Phocae Trouessart, 1879; Phocomorpha Berta and Wyss, 1994; Pinnigrada Owen, 1857; Pinnigrades Owen, 1857; Pinnipedia Illiger, 1811; Sibiricopusidae Dybowski, 1929; Stemmotopina [sic] Gray, 1825; Stenorhynchina Gray, 1844; Stenorhyncina [sic] Gray, 1825; Stenorynchina [sic] Gray, 1843; Thalattailurina Albrecht, 1879.

COMMENTS: Reviewed by Chapskii (1955), Scheffer (1958), J. E. King (1966, 1983), Hendey (1972), Muizon (1982*b*), and Wyss (1989). Muizon (1982*b*), and Wyss's (1988) phylogenetic analyses agreed on three points: 1) The monophyletic nature of two groups they refer to as the Lobodontini (*Hydrurga*, *Leptonychotes*, *Lobodon*, and *Ommatophoca*), and the Phocinae (*Erignathus*, *Cystophora*, *Halichoerus*, and *Phoca*), 2) The lobodonts, along with *Monachus* and *Mirounga* traditionally have been referred to as the Monachinae (kept by Muizon), however, they both suggested that this group may be paraphyletic, 3) Because of the "unsettled" nature of these taxa, no subfamilies are recognized at this time. Distributions for species after Rice (1998).

*Cystophora* Nilsson, 1820. Skand. Faun. Dagg. Djur., 1:382.

TYPE SPECIES: *Cystophora borealis* Nilsson, 1820 (= *Phoca cristata* Erxleben, 1777).

SYNONYMS: *Semmatopis* Gloger, 1841; *Stemmatops* Van der Hoeven, 1855; *Stemmatopus* F. G. Cuvier, 1826.

COMMENTS: Revised by King (1966).

*Cystophora cristata* (Erxleben, 1777). Syst. Regni Anim., 1:590.

COMMON NAME: Hooded Seal.

TYPE LOCALITY: "Habitat in Groenlandia australiori et Newfoundland". [S Greenland and Newfoundland].

DISTRIBUTION: N Atlantic and Arctic ocean coastal regions of Canada (Newfoundland), Denmark (vagrant), France (vagrant), Great Britain (vagrant), Greenland, Iceland, Portugal (vagrant), Puerto Rico (vagrant), Russia (Svalbard and Novaya Zemlya), Spain (vagrant), USA (vagrant: California and Florida), Virgin Isls (vagrant).

STATUS: IUCN – Lower Risk (lc).

SYNONYMS: *borealis* Nilsson, 1820; *cristata* Nilsson, 1841; *cucullata* (Boddaert, 1785); *isidorei* (Lesson, 1843); *leucopla* (Thienemann, 1824); *mitrata* (G.[Baron] Cuvier, 1823).

COMMENTS: Reviewed by Kovacs and Lavigne (1986) who placed it in subfamily Phocinae. Synonyms allocated according to Ellerman and Morrison-Scott (1951) and Kovacs and Lavigne (1986). No subspecies are recognized.

*Erignathus* Gill, 1866. Proc. Essex Inst. Salem, 5:5.

TYPE SPECIES: *Phoca barbata* Fabricius, 1776 [*nomen nudum*] (= *Phoca barbata* Erxleben, 1777).

*Erignathus barbatus* (Erxleben, 1777). Syst. Regni Anim., 1:590.

COMMON NAME: Bearded Seal.

TYPE LOCALITY: "ad Scotiam atque Groelandiam australiorem, vulgaris circa Islandiam" [North Atlantic, S Greenland].

DISTRIBUTION: Circumpolar Arctic seas and coastal regions of Canada, China (Zhejiang - vagrant), France (vagrant), Great Britian (vagrant), Greenland, Iceland, Japan (south to Hokkaido), Norway, Portugal (vagrant), Russia, Spain (vagrant), USA (Alaska, Massachusetts - vagrant).

STATUS: IUCN – Lower Risk (lc).

SYNONYMS: *lepechenii* (Lesson, 1828); *leporina* (Lepechin, 1778); *parsonsii* (Lesson, 1828); **nautica** (Pallas, 1811); *albigena* (Pallas, 1811); *naurica* (J. E. Gray, 1871).

COMMENTS: Synonyms allocated according to Ellerman and Morrison-Scott (1951) and Hall (1981).

*Halichoerus* Nilsson, 1820. Skand. Faun. Dagg. Djur., 1:376.

TYPE SPECIES: *Halichoerus griseus* Nilsson, 1820 (= *Phoca grypus* Fabricius, 1791).

SYNONYMS: *Halychoerus* Boitard, 1842.

COMMENTS: Mohr (1952) described successful mating in captivity between *Pusa hispida* and *Halichoerus grypus*.

*Halichoerus grypus* (Fabricius, 1791). Skr. Nat. Selsk. Copenhagen, 1(2):167.

COMMON NAME: Gray Seal.

TYPE LOCALITY: Listed by Scheffer (1958) as "Greenland".

DISTRIBUTION: Temperate and subarctic waters around Canada (Newfoundland area), Denmark, Estonia, Finland, France, Germany, Great Britain, Greenland, Iceland, Ireland, Latvia, Lithuania, Netherlands, Norway, Portugal (vagrant), Russia (Kola Peninsula), Sweden, USA (Maine, New Hampshire, Massachusetts, New Jersey - vagrant).

STATUS: IUCN – Lower Risk (lc), except for Northeast Atlantic subpopulation, which is Endangered.

SYNONYMS: *atlantica* Nehring, 1886; *griseus* Nilsson, 1820; **macrorhynchus** Hornschuch and Schilling, 1851; *baltica* Nehring, 1886.

COMMENTS: Synonyms allocated according to Ellerman and Morrison-Scott (1951).

*Histriophoca* Gill, 1873. Am. Nat., 7:179.

TYPE SPECIES: *Phoca fasciata* Zimmermann, 1783.

SYNONYMS: *Callocephalus* Heuglin, 1874.

COMMENTS: Burns and Fay (1970), Rice (1977), McDermid and Bonner (1975), Gromov and Baranova (1981), J. E. King (1983), and Wyss (1988) considered *Phoca*, *Pusa*, *Histriophoca*, and *Pagophilus* a monophyletic group. Cladistic analysis based on morphology and mtDNA revealed two clades, *Pagophilus+Histriophoca* and *Phoca+Pusa+Halichoerus* (Carr and Perry, 1998; Mouchaty et al., 1995; Muizon, 1982b; Perry et al., 1995; Rice, 1998). Burns and Fay (1970) and McDermid and Bonner (1975) argued that these differences should be recognized only at the subgeneric level.

*Histriophoca fasciata* (Zimmermann, 1783). Geogr. Gesch. Mensch. Vierf. Thiere, 3:277.

COMMON NAME: Ribbon Seal.

TYPE LOCALITY: "Wohnt um die Kurilischen Inseln" [Russia, Kurile Isls].

DISTRIBUTION: Japan (N Hokkaido), Russia (Okhotsk, W Bering, Chukchi and Japan Seas), USA (Alaska, California-vagrant).

STATUS: IUCN – Lower Risk (lc).

SYNONYMS: *equestris* (Pallas, 1831).

COMMENTS: Reviewed by Burns and Fay (1970). Placed in *Histriophoca* Gill, 1873, by Scheffer (1958), Muizon (1982b), and Rice (1998).

*Hydrurga* Gistel, 1848. Naturgesch. des Thierreichs, p. xi.

TYPE SPECIES: *Phoca leptonyx* de Blainville, 1820.

SYNONYMS: *Ogmorhinus* Peters, 1875; *Stenorhynchotes* Turner, 1888.

COMMENTS: Synonyms allocated according to McKenna and Bell (1997).

*Hydrurga leptonyx* (de Blainville, 1820). J. Phys. Chim. Hist. Nat. Arts Paris, 91:298.

COMMON NAME: Leopard seal.

TYPE LOCALITY: "des environs des îles Falckland ou Malouines" [Falkland Isls (UK)].

DISTRIBUTION: Circumpolar pack-ice zone south to the shores of Antarctica. Also coastal regions of Australia, Chile, Falkland Isls, Kerguelen Isls, New Zealand), South Sandwich Isls, South Africa.
STATUS: IUCN – Lower Risk (lc).
SYNONYMS: *homei* (Lesson, 1828); *leptonyz* (de Blainville, 1820).
COMMENTS: Synonyms allocated according to Cabrera (1957).

*Leptonychotes* Gill, 1872. Smithson. Misc. Coll., 11:70.
TYPE SPECIES: *Otaria weddellii* Lesson, 1826, by monotypy.
SYNONYMS: *Poecilophoca* Lydekker, 1891.
COMMENTS: Replacement name for *Leptonyx* Gray (1837), which is preoccupied by *Leptonyx* Swainson (1821).

*Leptonychotes weddellii* (Lesson, 1826). Bull. Sci. Nat. Geol., 7:437.
COMMON NAME: Weddell Seal.
TYPE LOCALITY: "sur les côtes des Orcades australes, situées sour 60 degrés 37 minutes de lat" [South Orkney Isl (Br. Antarct. Trust Terr.)].
DISTRIBUTION: Coastal fast ice areas of Antarctic continent and adjacent islands. Vagrant populations: Argentina, Australia, Chile, Falkland Isls, Macquarie Isl, New Zealand, Uruguay.
STATUS: IUCN – Lower Risk (lc).
SYNONYMS: *leopardina* (C. E. H. Smith, 1839); *leopardinus* Wagner, 1946; *leptonyx* (Moseley, 1879).
COMMENTS: Reviewed by Stirling (1971) and Kooyman (1981).

*Lobodon* Gray, 1844. Zool. Voy. H. M. S. "Erebus" and "Terror," 1:2.
TYPE SPECIES: *Phoca carcinophaga* Hombron and Jacquinot, 1842.

*Lobodon carcinophaga* (Hombron and Jacquinot, 1842) *In* Dumont d'Uville, Voy. Pole Sud., Zool., Altas: Mammifères, pl. 10 [1842], vol. 3:Mammifères et Oiseaux, p. 27 [1853].
COMMON NAME: Crabeater Seal.
TYPE LOCALITY: "capturé sur les glaces du Pole Sud, entre les îles Sandwich et les îles Powels, à 150 lieues de distance de chacune de ces îles." [Scotia Sea (midway between South Orkney and South Sandwich Isls) (Br. Antarct. Trust Terr.)].
DISTRIBUTION: Antarctic seas, frequently on pack ice around Antarctic Continent. Vagrant populations: Argentina, Australia, Brazil, Falkland Isls, New Zealand, South Africa, Tasmania, Uruguay.
STATUS: IUCN – Lower Risk (lc).
SYNONYMS: *serridens* (Owen, 1843).
COMMENTS: This species has often been listed incorrectly as *L. carcinophagus* (Rice, 1998).

*Mirounga* Gray, 1827. *In* Griffith et al., Anim. Kingdom, 5:179.
TYPE SPECIES: *Phoca proboscidea* Péron, 1816 (= *Phoca leonina* Linnaeus, 1758).
SYNONYMS: *Macrorhinus* F. G. Cuvier, 1826 [preoccupied]; *Morunga* J. E. Gray, 1943; *Rhinophoca* Wagler, 1830.
COMMENTS: Revised by King (1966). Reviewed by Davidson (1929), Briggs and Morejohn (1976), and Ling and Bryden (1992).

*Mirounga angustirostris* (Gill, 1866). Proc. Essex Inst. Salem, 5:13.
COMMON NAME: Northern Elephant Seal.
TYPE LOCALITY: "California" restricted by Poole and Schantz (1942) to "St. Bartholomews Bay, lower California, Mexico." Clarified by Scheffer (1958) as "Bahía Tórtola (= Bahía San Bartolomé) 27°39'N, 114°51'W, Baja California, Mexico".
DISTRIBUTION: Mexico (Baja California), USA (SE Alaska to California).
STATUS: IUCN – Lower Risk (lc).
COMMENTS: Reviewed by Stewart and Huber (1993).

*Mirounga leonina* (Linnaeus, 1758). Syst. Nat., 10th ed., 1:37.
COMMON NAME: Southern Elephant Seal.

TYPE LOCALITY: "ad polum Antarcticum" restricted by Thomas (1911a) to "Juan Fernandez", further restricted by Hamilton (1940) as "Isla Mas a Tierra" [Chile].
DISTRIBUTION: Circumpolar mainly in the subantarctic zone, including Antarctica, Macquarie, Kerguelen, S Georgia Isls, and Argentina (Peninsula Valdez). Vagrant populations recorded at Australia, Brazil, Chile, Mauritius, Mozambique, Namibia, New Zealand, Oman, Uruguay.
STATUS: CITES – Appendix II; IUCN – Lower Risk (lc).
SYNONYMS: *crosetensis* (Lydekker, 1909); *falclandica* (Peters, 1875); *kerguilensis* (Peters, 1875); *macquariensis* (Lydekker, 1909); *proboscidea* (Péron, 1816); *typicus* (Lydekker, 1909).
COMMENTS: Reviewed by Ling and Bryden (1992). Synonyms allocated according to Bryden (1995). Subspecies not recognized, following Lönnberg (1910); however see Carrick et al. (1962).

*Monachus* Fleming, 1822. Philos. Zool., 2:187.
TYPE SPECIES: *Phoca monachus* Hermann, 1779.
SYNONYMS: *Heliophoca* Gray, 1854; *Mammonachus* Herrera, 1899; *Pelagias* J. E. Gray, 1837; *Pelagios* F. G. Cuvier, 1824; *Pelagius* F. G. Cuvier, 1826; *Pelagocyon* Gloger, 1841; *Pelagus* McMurtrie, 1834; *Rigoon* Gistel, 1854.
COMMENTS: Revised by King (1956). Wyss (1988) suggested that this might be a paraphyletic group, however see Bininda-Emonds et al. (1999) who supported monophyly (followed here).

*Monachus monachus* (Hermann, 1779). Beschaft. Berlin Ges. Naturforsch. Fr., 4:501, pls. 12, 13.
COMMON NAME: Mediterranean Monk Seal.
TYPE LOCALITY: "Dalmation Sea at Ossero." [Serbia and Montenegro] King (1956).
DISTRIBUTION: Coastal regions of Mediterranean and Black Seas and NW Africa to Cape Blanc: Algeria, Balearic Isls, Cape Verde, Crete, Cyprus, France, Gambia, Greece, Italy, Lebanon, Mauritania, Morocco, Portugal, Russia, Sardinia, Senegal, Sicily, Spain, Tunisia, Turkey.
STATUS: CITES – Appendix I; U.S. ESA – Endangered; IUCN – Critically Endangered.
SYNONYMS: *albiventer* (Boddaert, 1785); *atlantica* (Gray, 1854); *bicolor* (Shaw, 1800); *crinita* (Menis, 1848); *hermannii* (Lesson, 1828); *isidorei* (Lesson, 1843); *leucogaster* (Péron and Lesueur, 1816); *mediterraneus* Nilsson, 1838.
COMMENTS: Synonyms allocated according to G. M. Allen (1939) and Ellerman and Morrison-Scott (1951).

*Monachus schauinslandi* Matschie, 1905. Sitzb. Ges. Naturf. Fr. Berlin, 1905:258.
COMMON NAME: Hawaiian Monk Seal.
TYPE LOCALITY: "Laysan ist eine kleine Koralleinsel, nordwestlich der Sandwich-Inseln" [USA, Laysan Isl, 25°50'N, 171°50'W].
DISTRIBUTION: USA (NW Hawaiian Isls, from Nihoa to Kure).
STATUS: CITES – Appendix I; U.S. ESA – Endangered; IUCN – Endangered.

*Monachus tropicalis* (Gray, 1850). Cat. Spec. Mamm. Coll. Br. Mus., Part 2(Seals), p. 28.
COMMON NAME: Caribbean Monk Seal.
TYPE LOCALITY: "Jamaica" restricted by King (1956) to "Pedro Cays, 80 km. south of Jamaica".
DISTRIBUTION: Historical records include coastal regions of the Caribbean Sea and Yucatan: Mexico (Veracruz to Yucatan), Bahamas, Guadeloupe, Jamaica, Puerto Rico, USA (Florida).
STATUS: CITES – Appendix I; U.S. ESA – Endangered; IUCN – Extinct.
SYNONYMS: *antillarum* (J. E. Gray, 1849) [*nomen nudum*].
COMMENTS: Extinct since 1952 (Kenyon, 1977; Rice, 1998).

*Ommatophoca* Gray, 1844. Zool. Voy. H. M. S. "Erebus" and "Terror," 1:3.
TYPE SPECIES: *Ommatophoca rossi* Gray, 1844, by monotypy.

*Ommatophoca rossii* Gray, 1844. Zool. Voy. H. M. S. "Erebus" and "Terror," 1:3.
COMMON NAME: Ross Seal.

TYPE LOCALITY: "Antarctic ocean", restricted by Barrett-Hamilton (1902) to "pack ice, north of Ross Sea 68°S, 176°E".

DISTRIBUTION: Circumpolar, Antarctic pack ice, particularly King Haakon VII Sea. Vagrant populations on Heard Isl and S Australia.

STATUS: IUCN – Lower Risk (lc).

*Pagophilus* Gray, 1844. Zool. Voy. H. M. S. "Erebus" and "Terror," 1:3.

TYPE SPECIES: *Phoca groenlandica* Erxleben, 1777.

SYNONYMS: *Callocephalus* Heuglin, 1874; *Haliphilus* J. E. Gray, 1866; *Pagomys* Gray, 1864; *Pagophoca* Trouessart, 1904.

COMMENTS: Burns and Fay (1970), Rice (1977), McDermid and Bonner (1975), Gromov and Baranova (1981), J. E. King (1983), and Wyss (1988) considered *Phoca*, *Pusa*, *Histriophoca*, and *Pagophilus* a monophyletic group. Cladistic analysis based on morphology and mtDNA reveal two clades, *Pagophilus+Histriophoca* and *Phoca+Pusa+Halichoerus* (Carr and Perry, 1998; Mouchaty et al., 1995; Muizon, 1982*b*; Perry et al., 1995; Rice, 1998). Burns and Fay (1970) and McDermid and Bonner (1975) argued that these differences should be recognized only at the subgeneric level.

*Pagophilus groenlandicus* (Erxleben, 1777). Syst. Regni Anim., 1:588.

COMMON NAME: Harp Seal.

TYPE LOCALITY: "in Groenlandia et Newfoundland."

DISTRIBUTION: N Atlantic and Arctic oceans and coastal regions of Canada (Newfoundland), France (vagrant), Germany (vagrant), Greenland, Iceland, Norway, Russia from E Canada to the White Sea (Russia), Scotland (vagrant), USA (Virginia - vagrant).

STATUS: IUCN – Lower Risk (lc).

SYNONYMS: *albicauda* (Desmarest, 1822); *albini* (Alessandrini, 1851); *dorsata* (Pallas, 1811); *leucopla* (Thienemann, 1824); *oceanica* (Lepechin, 1778); *semilunaris* (Boddaert, 1785).

*Phoca* Linnaeus, 1758. Syst. Nat., 10th ed., 1:37.

TYPE SPECIES: *Phoca vitulina* Linnaeus, 1758, by tautonomy.

SYNONYMS: *Ambysus* Rafinesque, 1815; *Arctias* Rafinesque, 1815; *Calocephalus* F. G. Cuvier, 1826; *Caspiopusa* Dybowski, 1929.

COMMENTS: Burns and Fay (1970), Rice (1977), McDermid and Bonner (1975), Gromov and Baranova (1981), J. E. King (1983), and Wyss (1988) considered *Phoca*, *Pusa*, *Histriophoca*, and *Pagophilus* a monophyletic group. Cladistic analysis based on morphology and mtDNA reveal two clades, *Pagophilus+Histriophoca* and *Phoca+Pusa+Halichoerus* (Carr and Perry, 1998; Mouchaty et al., 1995; Muizon, 1982*b*; Perry et al., 1995; Rice, 1998). This was also supported by Bininda-Emonds et al.'s (1999) "complete data" phylogeny. Burns and Fay (1970) and McDermid and Bonner (1975) argued that these differences should be recognized only at the subgeneric level.

*Phoca largha* Pallas, 1811. Zoogr. Rosso-Asiat., 1:113.

COMMON NAME: Spotted Seal.

TYPE LOCALITY: "quam quod observetur tantum ad orientale littus Camtschatcae" [Eastern coast of Kamchatka, Russia (Shaughnessy and Fay, 1977)].

DISTRIBUTION: Associated with pack ice in coastal N Pacific of Canada, China (south to Fujian), Japan (south to Shikoku), Russia (Bering and Okhotsk Seas), USA (Alaska).

STATUS: IUCN – Lower Risk (lc).

SYNONYMS: *chorisii* Lesson, 1828; *macrodens* J. A. Allen, 1902; *nummularis* Temminck, 1844; *ochotensis* J. A. Allen, 1902; *pallasii* Naumov and Smirnov, 1936; *petersi* Mohr, 1941; *pribilofensis* Allen, 1902; *tigrina* Lesson, 1827.

COMMENTS: Scheffer (1958) considered *largha* as conspecific with *vitulina*; however, Shaughnessy and Fay (1977) and J. E. King (1983) separated the two.

*Phoca vitulina* Linnaeus, 1758. Syst. Nat., 10th ed., 1:38.

COMMON NAME: Harbor Seal.

TYPE LOCALITY: "in mari Europaeo" restricted by Thomas (1911*a*) to "Mari Bothnico et Baltico", however, presently it does not occur in the Gulf of Bothnia (Bobrinski et al., 1944).

DISTRIBUTION: Coastal regions of Canada, China (south to Kiangsu), Denmark, Germany, Great Britain, Greenland, Iceland, Ireland, Japan (Hokkaido), Mexico (Baja California, Isla Guadalupe-vagrant), Netherlands, Norway, Portugal, Russia (Kurile Isls and Kamchatka), Sweden, USA (Atlantic coast: Maine, Massachusetts, New Hampshire, vagrants: New York, Florida, Vermont. Pacific Coast: Alaska, Washington, Oregon, California).

STATUS: IUCN – Data Deficient as *P. v. mellonae*, otherwise Lower Risk (lc).

SYNONYMS: *canina* Pallas, 1811; *linnaei* Lesson, 1828; *littorea* Thienemann, 1824; *scopulicola* Thienemann, 1824; *thienemannii* Lesson, 1828; *variegata* Nilsson, 1820; *concolor* De Kay, 1842; *vitulina* Trouessart, 1904; *mellonae* Doutt, 1942; *richardii* (Gray, 1864); *geronimensis* J. A. Allen, 1902; *pribilofensis* J. A. Allen, 1902; *stejnegeri* J. A. Allen, 1902; *insularis* Belkin, 1964; *kurilensis* McLaren, 1966.

COMMENTS: The position of *stejnegeri* remains uncertain; Scheffer (1958) placed it in *largha*; J. E. King (1983) placed it in *vitulina*; and Shaughnessy and Fay (1977) suggested *incertae sedis*. Reviewed by Shaughnessy and Fay (1977), Burns et al. (1984), and Smith et al. (1994). Synonyms allocated according to Ellerman and Morrison-Scott (1951) and Rice (1998).

*Pusa* Scopoli, 1771. Introductio ad historiam naturalem, p. 490.

TYPE SPECIES: *Phoca foetica* Fabricius, 1776 (= *Phoca hispida* Schreber, 1775) by subsequent designation by Ellerman and Morrison-Scott (1951).

SYNONYMS: *Caspiopusa* Dybowski, 1929.

COMMENTS: Burns and Fay (1970), Rice (1977), McDermid and Bonner (1975), Gromov and Baranova (1981), J. E. King (1983), and Wyss (1988) considered *Phoca, Pusa, Histriophoca*, and *Pagophilus* a monophyletic group. Cladistic analysis based on morphology and mtDNA reveal two clades, *Pagophilus+Histriophoca* and *Phoca+Pusa+Halichoerus* (Carr and Perry, 1998; Mouchaty et al., 1995; Muizon, 1982*b*; Perry et al., 1995; Rice, 1998). Burns and Fay (1970) and McDermid and Bonner (1975) argued that these differences should be recognized only at the subgeneric level.

*Pusa caspica* (Gmelin, 1788). *In* Linnaeus, Syst. Nat., 13th ed., 1:64.

COMMON NAME: Caspian Seal.

TYPE LOCALITY: "in mari, praesertim septentrionali, etiam Pacifico et Caspico" [Caspian Sea].

DISTRIBUTION: Coastal regions of the Caspian Sea: Azerbaijan, Iran, Kazakhstan, Russia, Turkmenistan.

STATUS: IUCN – Vulnerable.

*Pusa hispida* (Schreber, 1775). Die Säugethiere, 2(13):pl. 86[1775]; text 3(17):312[1776].

COMMON NAME: Ringed Seal.

TYPE LOCALITY: "Man fängt ihn auf den Küsten von Grönland und Labrader".

DISTRIBUTION: Arctic Ocean and coastal regions of Açôres (vagrant), Canada (Nettilling Lake, Baffin Isl), China (vagrant: Jiangsu), Estonia, Finland (Saimaa Lake), Germany (vagrant), Greenland, Japan (Hokkaido, vagrant: Shikoku, Kyushu), Latvia, Lithuania, Norway, Portugal (vagrant), Russia (Okhotsk, Bering, and Baltic Seas), Sweden, USA (vagrant: New Jersey, California).

STATUS: U.S. ESA – Endangered as *Phoca hispida saimensis*; IUCN – Endangered as *P. h. saimensis*, Vulnerable as *P. h. botnica* and *P. h. ladogensis*, Lower Risk as *P. hispida* and as *P. h. ochotensis*.

SYNONYMS: *annellata* (Nilsson, 1820); *beaufortiana* (Anderson, 1943); *birulai* (Smirnov, 1929); *foetica* (Fabricius, 1776); *krascheninikovi* (Naumov and Smirnov, 1936); *pomororum* (Smirnov, 1929); *pygmaea* (Zukowsky, 1921); *rochmistrovi* (Smirnov, 1929); *soperi* (Anderson, 1943); *botnica* (Gmelin, 1788); *annellata* (Nilsson, 1820); *octonata* (Kutorga, 1839); *undulata* (Kutorga, 1839); *ladogensis* (Nordquiest, 1899); *ochotensis* (Pallas, 1811); *gichigensis* (J. A. Allen, 1902); *saimensis* (Nordquist, 1899).

COMMENTS: Placed in *Pusa* Scopoli, 1771, by Scheffer (1958) and Rice (1998). Mohr (1952) described successful mating in captivity between *Pusa hispida* and *Halichoerus grypus*.

*Pusa sibirica* (Gmelin, 1788). *In* Linnaeus, Syst. Nat., 13th ed., 1:64.

COMMON NAME: Baikal Seal.

TYPE LOCALITY: "Baikal et Orom" [Lake Baikal and Lake Oron (=Ozero Oron), Russia].
DISTRIBUTION: Endemic to Lake Baikal (Russia).
STATUS: IUCN – Lower Risk (nt).
SYNONYMS: *baicalensis* (Dybowski, 1873).

**Family Mustelidae** Fischer, 1817. Mém. Soc. Imp. Nat. Moscow, 5:372.
SYNONYMS: Mustelladae Gray, 1821.
COMMENTS: Reviewed by Pocock (1921*b*, *d*), Muizon (1982*a*), Van Zyll de Jong (1987), Bryant et al. (1993), Masuda and Yoshida (1994*a*, *b*), Abramov and Baryshnikov (1995), Dragoo and Honeycutt (1997), Baryshnikov and Abramov (1997, 1998), Koepfli and Wayne (1998), Bininda-Emonds et al. (1999), Ginsburg and Morales (2000), and Kurose et al. (2000). Traditionally consisted of four subfamilies (Mephitinae, Melinae, Mustelinae, and Lutrinae), but McKenna and Bell (1997) also recognized Mellivorinae and Guloninae. The Mephitinae are separated here (see Mephitidae) following Dragoo and Honeycutt (1997). Few have questioned the distinctiveness of the Lutrinae, but the traditionally recognized Melinae and Mustelinae appear to be paraphyletic (Bininda-Emonds et al., 1999; Bryant et al., 1993). For these reasons, only two subfamilies are recognized here, the Lutrinae and the Mustelinae (provisionally including taxa traditionally placed in Melinae, Guloninae, Taxidiinae, and Mellivorinae). Bryant et al. (1993) argued that *Melogale* may be ancestral to Lutrinae and Mustelinae.

**Subfamily Lutrinae** Bonaparte, 1838. Nuovi Ann Sci. Nat., 2:111.
COMMENTS: Revised by Pohle (1920*a*), Pocock (1940*b*), van Zyll de Jong (1972, 1987, 1991*a*), Koepfli and Wayne (1998), and Bininda-Emonds et al. (1999). Reviewed by Harris (1968) and Sokolov (1973). Foster-Turley et al. (1990) reviewed the conservation status and distribution of otters.

*Aonyx* Lesson, 1827. Manual de Mammalogie, p. 157.
TYPE SPECIES: *Aonyx delalandi* Lesson, 1827 (= *Lutra capensis* Schinz, 1821) by subsequent designation (Palmer, 1904).
SYNONYMS: *Amblonyx* Rafinesque, 1832; *Anahyster* Murray, 1861; *Leptonyx* Lesson, 1842; *Micraonyx* J. A. Allen, 1919; *Paraonyx* Hinton, 1921.
COMMENTS: *Aonyx* is considered congeneric with *Amblonyx* (Coetzee, 1977*b*; Davis, 1978; Ellerman and Morrison-Scott, 1951; Koepfli and Wayne, 1998; Osgood, 1932). However see Harris (1968), Medway (1977), and van Zyll de Jong (1972, 1987) who considered them separate. There is little question that *capensis* and *congica* are sister species. Koepfli and Wayne (1998) and Bininda-Emonds et al. (1999) supported monophyly of *Amblonyx* + *Aonyx*.

*Aonyx capensis* (Schinz, 1821). *In* G. Cuvier, Das Thierreich, 1:211.
COMMON NAME: African Clawless Otter.
TYPE LOCALITY: "Capischer otter. . Afrika" [South Africa, (former) Cape Province].
DISTRIBUTION: Angola, Benin, Botswana, Burkina Faso, Burundi (?), Cameroon, Central African Republic, Chad, Côte d'Ivoire, Dem. Rep. Congo, Ethiopia, Gabon, Ghana, Guinea-Bissau, Kenya, Liberia, Malawi, Mozambique, Namibia, Niger, Nigeria, Republic of Congo, Rwanda, Senegal, Sierra Leone, South Africa, Sudan, Tanzania, Togo, Uganda, Zambia, Zimbabwe.
STATUS: CITES – Appendix I (populations of Cameroon and Nigeria), otherwise Appendix II as *A. congicus*; U.S. ESA – Endangered as *A. congicus microdon*; IUCN – Lower Risk (lc) as *A. capensis*, Data Deficient as *A. congicus*.
SYNONYMS: *angolae* Thomas, 1908; *calaboricus* (Murray, 1861); *coombsi* Roberts, 1926; *delalandi* Lesson, 1827; *inunguis* (F. G. Cuvier, 1823); *lenoiri* (Rochebrune, 1888); **congica** Lönnberg, 1910; **hindei** (Thomas, 1905); *helios* (Heller, 1913); **meneleki** (Thomas, 1903); **microdon** Pohle, 1920; *philippsi* Hinton, 1921.
COMMENTS: Reviewed by Larivière (2001*c*, *d*). Synonyms allocated according to G. M. Allen (1939), and Ellerman et al. (1953). *A. congica* was considered a distinct species by Pohle (1920*a*) and van Zyll de Jong (1987). Perret and Aellen (1956) and Davis (1978) included *Paraonyx philippsi* and *Aonyx microdon*.

*Aonyx cinerea* (Illiger, 1815). Abh. Phys. Klasse K. Preuss. Akad. Wiss., 1804-1811:99 [1815].
COMMON NAME: Oriental Small-clawed Otter.
TYPE LOCALITY: "Batavia" [Indonesia, Java, Jakarta].
DISTRIBUTION: Bangladesh, Burma, S China (incl. Hainan Isl), India, Indonesia (Sumatra, Java, Kalimantan), Laos, Malaysia (West, Sarawak, Sabah), Philippines (Palawan Isl), Taiwan, Thailand, Vietnam.
STATUS: CITES – Appendix II; IUCN – Lower Risk (nt) as *Amblonyx cinereus*.
SYNONYMS: *barang* Lesson, 1842; *horsfieldii* Gray, 1843; *leptonyx* (Horsfield, 1823); *sernaria* Nelson, 1983; *swinhoei* (Gray, 1867); **concolor** Rafinesque, 1832; *indigitatus* (Hodgson, 1839); *sikimensis* (Horsfield, 1855); *fulvus* Pohle, 1920; *wurmbi* Sody, 1933; **nirnai** Pocock, 1940.
COMMENTS: Synonyms allocated according to Larivière (2003).

*Enhydra* Fleming, 1822. Philos. Zool., 2:187.
TYPE SPECIES: *Mustela lutris* Linnaeus, 1758, by monotypy (Melville and Smith, 1987).
SYNONYMS: *Enhydria* Zittel, 1893; *Enhydris* Temminck, 1838; *Enhydrus* Dahl, 1823; *Enhydrus* MacLeay, 1925; *Enydris* J. B. Fischer, 1829; *Enydris* Lichtenstein, 1827; *Euhydris* Jordan, 1888; *Latax* Gloger, 1827; *Pusa* Oken, 1816; *Sutra* Elliot, 1874.

*Enhydra lutris* (Linnaeus, 1758). Syst. Nat., 10th ed., 1:45.
COMMON NAME: Sea Otter.
TYPE LOCALITY: "Asia et America septentrionali," restricted by Thomas (1911*a*), to "Kamtchatka", then by Heptner et al. (1967:884) to "Commander Isls" [Russia]. Roest (1973) listed the type locality as "the east central coast of Kamchatka, opposite the Commander Islands" [Russia].
DISTRIBUTION: Canada, Russia (Sakhalin Isl, Kurile Isls, Commander Isls, Kamchatka), USA (Aleutian Isls, and S Alaska to California). Formerly in Japan (coastal Hokkaido) and Mexico (Baja California).
STATUS: CITES – Appendix I as *E. l. nereis*; otherwise Appedix II. U.S. ESA – Threatened as *E. lutris nereis* in all of its range except for areas subject to U.S. jurisdiction south of Pt. Conception, CA (34°26.9′ N. Lat.), where it is listed as an Experimental Non Essential Population; U.S. ESA – Candidate taxon as *E. l. kenyoni*; IUCN – Endangered.
SYNONYMS: *gracilis* (Bechstein, 1800); *kamtschatica* Dybowski, 1922; *marina* (Erxleben, 1777); *orientalis* (Oken, 1816); *stelleri* (Lesson, 1827); **kenyoni** Wilson, 1991; **nereis** (Merriam, 1904).
COMMENTS: Reviewed by Roest (1973), Davis and Lidicker (1975), Estes (1980), and Wilson et al. (1991). Synonyms allocated according to Wilson et al. (1991).

*Hydrictis* Pocock, 1921. Proc. Zool. Soc. Lond., 1921:543.
TYPE SPECIES: *Lutra maculicollis* Lichtenstein, 1835.
SYNONYMS: *Hydrogale* Gray, 1865.
COMMENTS: Commonly included in *Lutra*, separated here. See discussion under *Lutra* and under *H. maculicollis*, below.

*Hydrictis maculicollis* (Lichtenstein, 1835). Arch. Naturgesch., 1:89.
COMMON NAME: Spotted-necked Otter.
TYPE LOCALITY: "Kafferlandes am östlichen Abhange der Bambusberge."
DISTRIBUTION: Angola, Benin, Botswana, Burkina Faso, Cameroon, Central African Republic, Chad, Côte d'Ivoire, Dem. Rep. Congo, Ethiopia, Gabon, Kenya, Liberia, Malawi, Mozambique, Namibia, Nigeria, Rwanda, Sierra Leone, South Africa, Sudan, Tanzania, Togo, Uganda, Zambia.
STATUS: CITES – Appendix II; IUCN – Vulnerable as *Lutra maculicollis*.
SYNONYMS: *chobiensis* (Roberts, 1932); *concolor* (Neumann, 1902); *grayii* (Gerrard, 1862); *kivuana* (Pohle, 1920); *malculicollis* (Roberts, 1932); *matschiei* (Cabrera, 1903); *mutandae* (Hinton, 1921); *nilotica* Thomas, 1911; *poensis* (Waterhouse, 1838); *tenuis* (Pohle, 1919).
COMMENTS: Reviewed by Larivière (2002*c*). Pocock (1921*c*) and Cabrera (1929) placed *maculicollis* in the monotypic *Hydrictis*; however, Ansell (1978) and Harris (1968)

considered *Hydrictis* a subgenus of *Lutra* (see comments under *Lutra*). Two recent studies (Koepfli and Wayne, 1998, and Bininda-Emonds et al., 1999) concluded that inclusion in *Lutra* would make *Lutra* paraphyletic. Van Zyll de Jong (1987) and Bininda-Emonds et al. (1999) suggested that *Hydrictis* and *Lutrogale* may be sister taxa. Subspecies not recognized (Kingdon, 1997; Larivière, 2002c).

*Lontra* Gray, 1843. Ann. Mag. Nat. Hist. [ser. 1], 11:118.
TYPE SPECIES: *Lutra canadensis* Gray, 1843 ( = *Lutra canadensis* Shreber, 1777).
COMMENTS: Van Zyll de Jong (1972, 1987, 1991), Koepfli and Wayne (1998), and Bininda-Emonds et al. (1999) supported the separation of New World otters (except *Pteronura*) into *Lontra*.

*Lontra canadensis* (Schreber, 1777). Die Säugethiere, 3(18):pl. 126.B[1776], text:3(26):457, 588(index)[1777]). (First occurance of name on pg. 588).
COMMON NAME: North American River Otter.
TYPE LOCALITY: "Der Fischotter . . . Europa überall gemein, . . . Einwohner des nordlichen Theils von Asien, bis nach Kamatschatka hinaus, und . . . Persian hinunter, und von Nordamerika." Miller (1912b:113) listed the type locality as "Eastern Canada".
DISTRIBUTION: Canada, USA (except for arid SW desert regions), USA (historical distributions of Alaska and most of contiguous 48 states exclusive of the Central Plains).
STATUS: CITES – Appendix II; IUCN – Lower Risk (lc).
SYNONYMS: *americana* (Wyman, 1847); *chimo* (Anderson, 1945); *degener* (Bangs, 1898); *destructor* (Barnston, 1863); *hudsonica* (Desmarest, 1803); **kodiacensis** (Goldman, 1935); **lataxina** (Cuvier, 1823); *interior* (Swenk, 1920); *mollis* (Gray, 1843); *parviceps* (Gidley and Gazin, 1933); *rhoadsi* (Cope, 1897); *vaga* (Bangs, 1898); **mira** (Goldman, 1935); *vancouverensis* (Goldman, 1935); **pacifica** (J. A. Allen, 1898); *atterima* (Elliot, 1901); *brevipilosus* (Grinnell, 1914); *californica* (Baird, 1857); *evexa* (Goldman, 1935); *extera* (Goldman, 1935); *nexa* (Goldman, 1935); *optiva* (Goldman, 1935); *paranensis* (Elliot, 1901); *preblei* (Goldman, 1935), *yukonensis* (Goldman, 1935); **periclyzomae** (Elliot, 1905); **sonora** (Rhoads, 1898).
COMMENTS: Reviewed by Larivière and Walton (1998). Synonyms allocated according to Hall (1981), and Larivière and Walton (1998).

*Lontra felina* (Molina, 1782). Sagg. Stor. Nat. Chile, p. 284.
COMMON NAME: Marine Otter.
TYPE LOCALITY: "Chili" [Chile].
DISTRIBUTION: Argentina (extreme S), Chile, Peru.
STATUS: CITES – Appendix I; U.S. ESA – Endangered; IUCN – Endangered.
SYNONYMS: *brachydactyla* (Wagner, 1841); *californica* (Gray, 1837); *chilensis* (Kerr, 1792); *cinerea* (Thomas, 1908); *lutris* (Larrañaga, 1923); *montana* (Tschudi, 1844); *paranensis* (Burmeister, 1861); *peruensis* (Pohle, 1920); *peruviensis* (Gervais, 1841).
COMMENTS: Placed in *Lontra* by van Zyll de Jong (1972, 1987) and Larivière (1998). Reviewed by Larivière (1998). Synonyms allocated according to Cabrera (1957).

*Lontra longicaudis* (Olfers, 1818). *In* Eschwege, J. Brasilien, Neue Bibliothek Reisenb., 15(2):233.
COMMON NAME: Neotropical Otter.
TYPE LOCALITY: "Brasilien."
DISTRIBUTION: Argentina, Belize, Brazil, Columbia, Costa Rica, Ecuador, El Salvador, Guatemala, Honduras, Mexico, Nicaragua, Panama, Paraguay, Peru, Uruguay, Venezuela.
STATUS: CITES – Appendix I; U.S. ESA – Endangered as *L. longicaudis* (incl. *platensis*); IUCN – Data Deficient.
SYNONYMS: *latifrons* (Nehring, 1887); *platensis* (Waterhouse, 1838); *solitaria* (Wagner, 1842); **annectens** (Major, 1897); *colombiana* (J. A. Allen, 1904); *emerita* (Thomas, 1908); *latidens* (J. A. Allen, 1908); *mesopetes* (Cabrera, 1924); *parilina* (Thomas, 1914); *repanda* (Goldman, 1914); **enudris** (F. G. Cuvier, 1823); *incarum* (Thomas, 1908); *insularis* (F. G. Cuvier, 1823); *mitis* (Thomas, 1908).
COMMENTS: Van Zyll de Jong (1972) included *annectens, enudris, incarum, mesopetes,* and

*platensis*; however, Pohle (1920*a*), Cabrera (1957), and Harris (1968) recognized these as distinct species. These taxa (often referred to as the *annectens*-group) were distinguished primarily by variation in the shape of the rhinarium; van Zyll de Jong's (1972) analysis suggested that these should be considered conspecific. Reviewed by Larivière (1999*b*). Synonyms allocated according to Hall (1981) and Larivière (1999*b*).

*Lontra provocax* (Thomas, 1908). Ann. Mag. Nat. Hist., ser. 8, 1:391.
COMMON NAME: Southern River Otter.
TYPE LOCALITY: "south of Lake Nahuel Huapi, Patagonia" [Argentina].
DISTRIBUTION: Patagonia (C and S Chile, W Argentina), between 36°S and 52°S.
STATUS: CITES – Appendix I; U.S. ESA – Endangered; IUCN – Endangered.
SYNONYMS: *huidobria* (Gay, 1847) [*nomen nudum*]; *paranensis* (Thomas, 1908).
COMMENTS: Placed in *Lontra* by van Zyll de Jong (1987). Reviewed by Larivière (1999*a*).

*Lutra* Brisson, 1762. Regnum Animale, Ed. 2, 13:201.
TYPE SPECIES: *Mustela lutra* Linnaeus, 1758.
SYNONYMS: *Lutris* Duméril, 1806; *Lutrix* Rafinesque, 1815; *Lutronectes* Gray, 1867; *Mamlutraus* Herrera, 1899.
COMMENTS: Brisson (1762) was ruled unavailable (International Commission on Zoological Nomenclature, 1955*a*); however, *Lutra* was ruled as still available (ICZN, 1998). Pocock (1921*c*, 1941*a*) recognized *Lutrogale* Gray, 1865, and *Hydrictis* Pocock, 1921. Harris (1968) and van Zyll de Jong (1987, 1991*a*) considered *lutra*, *maculicollis*, and *sumatrana* to represent a single monophyletic group; furthermore, van Zyll de Jong's analysis supported separation of *perspicillata* from other *Lutra*, which is followed here. Van Zyll de Jong (1972) referred New World otters to *Lontra* (see comment therein). McKenna and Bell (1997) included *Lontra* in *Lutra*; otherwise allocated according to McKenna and Bell (1997).

*Lutra lutra* (Linnaeus, 1758). Syst. Nat., 10th ed., 1:45.
COMMON NAME: European Otter.
TYPE LOCALITY: "Europae aquis dulcibus, fluviis, flagnis, piscinis," subsequently restricted by Thomas (1911*a*) to "Upsala" [Sweden].
DISTRIBUTION: Afghanistan, Albania, Algeria, Austria, Azerbaijan, Bangladesh, Belarus, Belgium, Bosnia and Herzegovina, Bulgaria, China, Czech Republic, Croatia, Denmark, Estonia, Finland, France, Georgia, Germany, Great Britain, Greece, Hungary, India, Indonesia, Iran, Iraq, Ireland, Israel, Italy, Jordan, Kazakhstan, Kyrgyzstan, Laos, Latvia, Lithuania, Macedonia, Malaysia, Moldova, Mongolia, Montenegro, Morocco, North and South Korea, Norway, Pakistan, Poland, Portugal, Romania, Russia, Serbia and Montenegro, Slovakia, Slovenia, Spain, Sri Lanka, Sweden, Tajikistan, Taiwan, Tunisia, Turkey, Urkaine, Vietnam.
STATUS: CITES – Appendix I; IUCN – Vulnerable.
SYNONYMS: *amurensis* Dybowski, 1922; *baicalensis* Dybowski, 1922; *fluviatilis* Leach, 1816; *kamtschatica* Dybowski, 1922; *marinus* Billberg, 1827; *nudipes* Melchior, 1834; *piscatoria* (Kerr, 1792); *roensis* Ogilby, 1834; *stejnegeri* Goldman, 1936; *vulgaris* Erxleben, 1777; *whiteleyi* (Gray, 1867); **angustifrons** Lataste, 1885; *splendida* Cabrera, 1906; **aurobrunneus** Hodgson, 1839; *nepalensis* (Gray, 1865); **barang** F. G. Cuvier, 1823; **chinensis** Gray, 1837; *hanensis* Matschie, 1907; *sinensis* Trouessart, 1897; **hainana** Xu and Lu, 1983; **kutab** Schinz, 1844; **meridionalis** Ognev, 1931; **monticolus** Hodgson, 1839; **nair** F. G. Cuvier, 1823; *ceylonica* Phole, 1920; *indica* Gray, 1837; **seistanica** Birula, 1913; *oxiana* Birula, 1915.
COMMENTS: Imaizumi and Yoshiyuki (1989) considered Japanese otters a distinct species (*L. nippon*). Synonyms allocated according to Ellerman and Morrison-Scott (1951).

*Lutra nippon* Imaizumi and Yoshiyuki, 1989. Bull. Nat. Sci. Mus., Tokyo. Ser. A, 15(3):178.
COMMON NAME: Japanese Otter.
TYPE LOCALITY: "Nenokubi Seaside, Shimoda, Nakamura City, Kôchi Prefecture."
DISTRIBUTION: Formerly widely distributed in Japan, now probably extinct. May now only exist on Shikoku Isl (Abe et al., 1997).
STATUS: CITES – Appendix II.

COMMENTS: Distinctiveness of the Japanese otter was supported by Suzuki et al. (1996*b*).

*Lutra sumatrana* (Gray, 1865). Proc. Zool. Soc. Lond., 1865:123.
COMMON NAME: Hairy-nosed Otter.
TYPE LOCALITY: "Sumatra (Raffles); Malacca (B.M.)," restricted by Pocock (1941*a*) to "Sumatra."
DISTRIBUTION: Indonesia (Sumatra, Borneo), Cambodia, Malaysia, Thailand, Vietnam.
STATUS: CITES – Appendix II; IUCN – Data Deficient.
SYNONYMS: *brunnea* Pohle, 1920; *lovii* Gunther, 1877.

*Lutrogale* Gray, 1865. Proc. Zool. Soc. Lond., 1865:127.
TYPE SPECIES: *Lutra perspicillata* I. Geoffroy Saint-Hilaire, 1826.
COMMENTS: See comments under *Lutra*. Consideration here as a separate genus is consistent with Pohle (1920*a*), Pocock (1941*a*), and van Zyll de Jong (1987).

*Lutrogale perspicillata* (I. Geoffroy Saint-Hilaire, 1826). *In* Bory de Saint-Vincent (ed.), Dict. Class. Hist. Nat. Paris. 9:519.
COMMON NAME: Smooth-coated Otter.
TYPE LOCALITY: "Sumatra" [Indonesia].
DISTRIBUTION: Afghanistan, Bangladesh, China, India, Indonesia (Sumatra, Java, Kalimantan), Iraq, Malaysia, Nepal, Pakistan, Thailand, Vietnam.
STATUS: CITES – Appendix II; IUCN – Vulnerable.
SYNONYMS: *ellioti* (Anderson, 1879); *macrodus* (Gray, 1865); *simung* (Lesson, 1827); *tarayensis* (Hodgson, 1839); **sindica** Pocock, 1940; *maxwelli* Hayman, 1957.
COMMENTS: Pocock (1941*a*), van Zyll de Jong (1972), and Davis (1978) placed *perspicillata* in the monotypic *Lutrogale*, considered a subgenus by Pohle (1920*a*); see comments under *Lutra*. Van Zyll de Jong's (1987) analysis placed as sister groups *L. maculicollis* and *L. lutra*+*L. sumatrana*. Synonyms allocated according to Ellerman and Morrison-Scott (1951).

*Pteronura* Gray, 1837. Mag. Nat. Hist. [Charlesworth's], 1:580.
TYPE SPECIES: *Pteronura sambachii* Gray, 1837 (= *Mustela brasilinesis* Gmelin, 1788), by monotypy (Melville and Smith, 1987).
SYNONYMS: *Pterura* Wiegmann, 1839; *Saricovia* Lesson, 1842.

*Pteronura brasiliensis* (Gmelin, 1788). *In* Linnaeus, Syst. Nat., 13th ed., 1:93.
COMMON NAME: Giant Otter.
TYPE LOCALITY: "in fluviis americae meridionalis"; Cabrera (1957:274) restricted to "rió São Francisco, en la orilla correspondiente al estado de Alagoas", Brazil.
DISTRIBUTION: Argentina, Bolivia, Brazil, Colombia, Ecuador, Guyana, Peru, Suriname, Venezuela.
STATUS: CITES – Appendix I; U.S. ESA – Endangered; IUCN – Endangered.
SYNONYMS: *brasiliana* (Shaw, 1800); *lupina* (Schinz, 1821); *lupina* Thomas, 1889; *sambachii* Gray, 1837; *sanbachii* (Wiegmann, 1838); *sandbachii* Gray, 1865; **paraguensis** (Schinz, 1821); *brasiliensis* (Boitard, 1845); *paranensis* (Rengger, 1830).
COMMENTS: See lengthy comments by Harris (1968) concerning the correct identity of the type, the confusion in published synonymies, and the type locality. Synonyms allocated according to Cabrera (1957).

**Subfamily Mustelinae** Fischer, 1817. Mém. Soc. Imp. Nat. Moscow, 5:372.
COMMENTS: Includes taxa traditionally included elsewhere (Melinae, Guloninae, Taxidiinae, and Mellivorinae); see comments under family heading.

*Arctonyx* F. G. Cuvier, 1825. *In* E. Geoffroy Saint-Hilaire and F. G. Cuvier, Hist. Nat. Mammifères, pt. 3, 5(51), "Bali-saur", 2 pp., 1 pl.
TYPE SPECIES: *Arctonyx collaris* F. G. Cuvier, 1825.
SYNONYMS: *Syarchus* Gloger, 1841; *Synarchus* Gray, 1865; *Trichomanis* Hubrecht, 1891.
COMMENTS: Revised by Lönnberg (1923*a*) and Pocock (1940*a*).

*Arctonyx collaris* F. G. Cuvier, 1825. *In* É. Geoffroy Saint-Hilaire and F. G. Cuvier, Hist. Nat.
Mammifères, pt. 3, 5(51), "Bali-saur", 2 pp., 1 pl.
COMMON NAME: Hog Badger.
TYPE LOCALITY: "dans les montagnes qui séparent le Boutan de l'Indoustan."
DISTRIBUTION: Bangladesh, Bhutan, Burma, China, India, Indonesia (Sumatra), Laos,
W Malaysia, Thailand, Vietnam.
STATUS: IUCN – Lower Risk (lc).
SYNONYMS: *isonyx* Horsfield, 1856; *taraiyensis* (Gray, 1863); *taxoides* (Blyth, 1853);
***albogularis*** (Blyth, 1853); *incultus* Thomas, 1922; *obscurus* (Milne-Edwards, 1871);
*orestes* Thomas, 1911; ***consul*** Pocock 1940; ***dictator*** Thomas, 1910; *annaeus* Thomas,
1921; ***hoevenii*** (Hubrecht, 1891); ***leucolaemus*** (Milne-Edwards, 1867); *milne-edwardsii*
Lönnberg, 1923.
COMMENTS: Synonyms allocated according to Ellerman and Morrison-Scott (1951).

*Eira* C. E. H. Smith, 1842. Jardine's Natur. Libr., 35:201.
TYPE SPECIES: *Mustela barbara* Linnaeus, 1758.
SYNONYMS: *Eirara* Lund, 1839; *Eraria* Gray, 1843; *Galera* Gray, 1843; *Tayra* Palmer, 1904.

*Eira barbara* (Linnaeus, 1758). Syst. Nat., 10th ed., 1:46.
COMMON NAME: Tayra.
TYPE LOCALITY: "Brasilia," restricted by Lönnberg (1913) to "Pernambuco."
DISTRIBUTION: Argentina, Belize, Bolivia, Brazil, Colombia, Costa Rica, Ecuador, Guatemala,
Guyana, Honduras, Mexico (Sinaloa and Tamaulipas), Nicaragua, Panama, Peru,
Surinam, Trinidad, Venezuela.
STATUS: CITES – Appendix III (Honduras); IUCN – Vulnerable as *E. b. senex*, otherwise Lower
Risk (lc).
SYNONYMS: *barbatus* (Desmarest, 1820); *canescens* (Lichtenstein, 1825); *gulina* (Schinz, 1821);
*kriegi* (Krumbiegal, 1942); *tucumana* (Lönnberg, 1913); ***biologiae*** (Thomas, 1900);
***inserta*** (J. A. Allen, 1908); ***madeirensis*** (Lönnberg, 1913); *peruana* (Osgood, 1914);
***peruana*** (Tschudi, 1844); *brunnea* (Thomas, 1907); ***poliocephala*** (Traill, 1821); *ilya*
Smith, 1842; *leira* (Cuvier, 1849); ***senex*** (Thomas, 1900); ***senilis*** (J. A. Allen, 1913);
***sinuensis*** (Humboldt, 1812); *bimaculata* (Martinez, 1873); *irara* (J. A. Allen, 1901).
COMMENTS: Reviewed by Thomas (1900*a*), Lönnberg (1913), and Presley (2000). Synonyms
allocated according to Cabrera (1957), Hall (1981), and Presley (2000).

*Galictis* Bell, 1826. Zool. J., 2:552.
TYPE SPECIES: *Viverra vittata* Schreber, 1776, by original designation.
SYNONYMS: *Galictes* Bell, 1837; *Gallictis* Waterhouse, 1839; *Grison* Oken, 1816; *Grisonella*
Thomas, 1912; *Grisonia* Gray, 1865; *Gulo* Desmarest, 1820; *Huro* I. Geoffroy Saint-Hilaire,
1835; *Mustela* Bechstein, 1800.

*Galictis cuja* (Molina, 1782). Sagg. Stor. Nat. Chile, p. 291.
COMMON NAME: Lesser Grison.
TYPE LOCALITY: "Chili" restricted by Thomas (1912*e*) to "S. Chili (Temuco)"; Cabrera (1957)
restricted the locality to "alrededores de Santiago" [Chile]. Honacki et al. (1982) listed
the type locality for *G. c. furax*.
DISTRIBUTION: Argentina, Bolivia, Peru, Brazil, Chile, Paraguay.
STATUS: IUCN – Lower Risk (lc).
SYNONYMS: *chilensis* Ihering, 1886; *melina* (Thomas, 1912); *quiqui* (Molina, 1782); *ratellina*
(Thomas, 1921); *shiptoni* (Thomas, 1926); *vittata* Gay, 1847; ***furax*** (Thomas, 1907);
*albifrons* (Larrañaga, 1923); *brasiliensis* (d'Orbigny, 1838) [preoccupied]; *vittata*
(Schreber, 1776) [preoccupied]; ***huronax*** (Thomas, 1921); *barbara* Hudson, 1903; *furax*
(Thomas, 1907); *vittata* Burmeister, 1897 [preoccupied]; ***luteola*** (Thomas, 1907).
COMMENTS: Reviewed by Yensen and Tarifa (2003*b*).

*Galictis vittata* (Schreber, 1776). Säugethiere 3(18):pl. 124[1776], text, 3(26):418, 447[1777].
COMMON NAME: Greater Grison.
TYPE LOCALITY: "Surinam".

DISTRIBUTION: Argentina, Bolivia, Colombia, Costa Rica, Guatemala, Guyana, Mexico (San Luis Potosi and Veracruz), Panama, Peru, Venezuela.

STATUS: CITES – Appendix III (Costa Rica).

SYNONYMS: *allamandi* (Göldi and Hagmann, 1904); *gujanensis* (Bechstein, 1800); ***andina*** (Thomas, 1903); ***brasiliensis*** (Thunberg, 1820); *aliamandi* (Ihering, 1911); *allamandi* Bell, 1841; *crassidens* Nehring, 1885; *intermedia* Lund, 1845; ***canaster*** (Nelson, 1901).

COMMENTS: Krumbiegel (1942) included *allamandi*. Reviewed by Yensen and Tarifa (2003*a*).

*Gulo* Pallas, 1780. Spicil. Zool., 14:25.

TYPE SPECIES: *Gulo sibiricus* Pallas, 1780 (= [*Mustela*] *gulo* Linnaeus, 1758), by absolute tautonymy (Melville and Smith, 1987).

COMMENTS: Corbet (1978) attributed *Gulo* to Storr (1780). *Gulo* Frisch, 1775, is invalid (International Commission on Zoological Nomenclature, 1954*b*). Placed in subfamily Guloninae according to McKenna and Bell (1997).

*Gulo gulo* (Linnaeus, 1758). Syst. Nat., 10th ed., 1:45.

COMMON NAME: Wolverine.

TYPE LOCALITY: "alpibus Lapponiae, Ruffiae, Sibiriae, sylvis vastissimis", restricted by Thomas (1911*a*) to "Lapland".

DISTRIBUTION: Canada, China (Heilongiang, Xinjiang, Inner Mongolia), Finland, Mongolia, Norway, Russia, Sweden, USA (Alaska, Wyoming, Idaho, Montana)

STATUS: IUCN – Vulnerable.

SYNONYMS: *arcticus* Desmarest, 1820; *arctos* Kaup, 1829; *biedermanni* Matschie, 1918; *borealis* Nilsson, 1820; *kamtschaticus* Dybowsky, 1922; *luscus* Trouessart, 1910; *sibirica* Pallas, 1780; *vulgaris* Oken, 1816; *wachei* Matschie, 1918; ***albus*** (Kerr, 1702); ***katschemakensis*** Matschie, 1918; ***luscus*** (Linnaeus, 1758); *auduboni* Matschie, 1918; *bairdi* Matschie, 1918; *hylaeus* Elliot, 1905; *luscus* Sabine, 1823; *niediecki* Matschie, 1918; ***luteus*** Elliot, 1904; ***vancouverensis*** Goldman, 1935.

COMMENTS: Degerbøl (1935) and Kurtén and Rausch (1959) demonstrated that *gulo* and *luscus* are conspecific. Reviewed by Pasitschniak-Arts and Larivière (1995). Synonyms allocated according to Ellerman and Morrison-Scott (1951) and Pasitschniak-Arts and Larivière (1995).

*Ictonyx* Kaup, 1835. Das Thierreich in Seinen Hauptformen, 1:352.

TYPE SPECIES: *Ictonyx capensis* Kaup, 1835 (= *Bradypus striatus* Perry, 1810) (Melville and Smith, 1987).

SYNONYMS: *Ictidonyx* Agassiz, 1846; *Ictomys* Roberts, 1936; *Ozolictis* Gloger, 1841; *Poecilictis* Thomas and Hinton, 1920; *Rhabdogale* Wiegmann, 1838; *Zorilla* I. Geoffroy Saint-Hilaire, 1826.

COMMENTS: There is considerable controversy over the correct name for this genus (Hershkovitz, 1955*b*; Van Gelder, 1966). *Zorilla* Oken (1816) is invalid (International Commission on Zoological Nomenclature, 1956*b*). *Zorilla* I. Geoffroy Saint-Hilaire (1826), was suppressed under the plenary powers for the purposes of the Principle of Priority (International Commission on Zoological Nomenclature, 1967). Rosevear (1974) strongly suggested, and Dekeyser (1955) and Niethammer (1987*a*) argued that *Poecilictis* and *Ictonyx* are congeneric. The principal skull features used to erect the new genus by Thomas and Hinton (1920) could not be supported when a more extensive series of specimens was measured (Rosevear, 1974).

*Ictonyx libyca* (Hemprich and Ehrenberg, 1833). Symb. Phys. Mamm., vol. 1, pt. 2, sig. K, verso.

COMMON NAME: Saharan Striped Polecat.

TYPE LOCALITY: "Libyae" [Libya].

DISTRIBUTION: Burkina Faso, Chad, Egypt, Libya, Mali, Mauritania, Morocco, Niger, Nigeria, Sudan, Tunisia, Western Sahara.

STATUS: IUCN – Lower Risk (lc).

SYNONYMS: ***multivittata*** (Wagner, 1841); *frenata* Sundevall, 1842; *vaillantii* (Loche, 1856); ***oralis*** (Thomas and Hinton, 1920); ***rothschildi*** (Thomas and Hinton, 1920).

COMMENTS: Thomas and Hinton (1920), and Baryshnikov and Abramov (1997, 1998) placed it in *Poecilictis*. Synonyms allocated according to G. M. Allen (1939) and Ellerman et al. (1953).

*Ictonyx striatus* (Perry, 1810). Arcana, Mus. Nat. Hist. Signature Y, Fig. [41][1810].
>     COMMON NAME: Striped Polecat.
>     TYPE LOCALITY: "South America". This is clearly in error and Hollister (1918) fixed the type locality as "Cape of Good Hope". [South Africa].
>     DISTRIBUTION: Angola, Benin, Botswana, Burkina Faso, Cameroon, Central African Republic, Chad, Cote d'Ivoire, Ethiopia, Gambia, Ghana, Guinea, Guinea-Bissau, Kenya, Malawi, Mali, Mauritania, Mozambique, Namibia, Niger, Nigeria, Republic of Congo, Senegal, Somalia, South Africa, Sudan, Tanzania, Togo, Uganda, Zambia, Zimbabwe.
>     STATUS: IUCN – Lower Risk (lc).
>     SYNONYMS: *capensis* (A. Smith, 1826); *mustelina* (Wagner, 1841); *pondoensis* Roberts, 1924; *variegata* (Lesson, 1842); **albescens** Heller, 1913; **arenarius** Roberts, 1924; **elgonis** Granvik, 1924; **erythreae** de Winton, 1898; **ghansiensis** Roberts, 1932; *nigricaudus* Roberts, 1932; **giganteus** Roberts, 1932; **intermedius** Anderson and de Winton, 1902; **kalaharicus** Roberts, 1932; **lancasteri** Roberts, 1932; **limpopoensis** Roberts, 1917; **maximus** Roberts, 1924; **obscuratus** de Beaux, 1924; **orangiae** Roberts, 1924; **ovamboensis** Roberts, 1951; **pretoriae** Roberts, 1924; **senegalensis** (J. B. Fischer, 1829); **shoae** Thomas, 1906; **shortridgei** Roberts, 1932.
>     COMMENTS: Reviewed by Larivière (2002*b*). See Hollister (1918) on the use of this name. Synonyms allocated according to Ellerman et al. (1953).

*Lyncodon* Gervais, 1845. *In* d'Orbigny, Dict. Univ. Hist. Nat., 4:685.
>     TYPE SPECIES: *Mustela patagonica* de Blainville, 1842.

*Lyncodon patagonicus* (de Blainville, 1842). Osteogr. Mamm., pt. 10 (Viverra):1.
>     COMMON NAME: Patagonian Weasel.
>     TYPE LOCALITY: Listed in Cabrera (1957) as "cercanías del río Negro." [Argentina].
>     DISTRIBUTION: Argentina and S Chile.
>     STATUS: IUCN – Lower Risk (lc).
>     SYNONYMS: *anticola* (Burmeister, 1869); *quiqui* (Burmeister, 1861); *lujanensis* Ameghino, 1889; **thomasi** Cabrera, 1928.
>     COMMENTS: Synonyms allocated according to Cabrera (1957).

*Martes* Pinel, 1792. Actes Soc. Hist. Nat. Paris, 1:55.
>     TYPE SPECIES: *Martes domestica* Pinel, 1792 (= *Mustela foina* Erxleben, 1777).
>     SYNONYMS: *Charronia* Gray, 1865; *Foiana* Gray, 1865; *Lamprogale* Ognev, 1928; *Mustela* Blasius, 1857; *Pekania* Gray, 1865; *Zibellina* Kaup, 1829.
>     COMMENTS: Stone and Cook's (2002) cytochrome *b* data suggested that the recognition of *Martes* as here understood (to the exclusion of *Gulo*) would make the genus paraphyletic.

*Martes americana* (Turton, 1806). *In* Linnaeus, Gen. Syst. Nat., 1:60.
>     COMMON NAME: American Marten.
>     TYPE LOCALITY: "North America".
>     DISTRIBUTION: Canada, USA (Alaska to N California, south in the Sierra Nevada and Rocky Mtns to 35°N, Maine, Minnesota, New York).
>     STATUS: IUCN – Lower Risk (lc).
>     SYNONYMS: *huro* (F. Cuvier, 1823); *leucopus* (Kuhl, 1820); *martinus* (Ames, 1874); **abieticola** (Preble, 1902); **abietinoides** Gray, 1865; **actuosa** (Osgood, 1900); *boria* (Elliot, 1905); **atrata** (Bangs, 1897); *brumalis* (Bangs, 1898); **caurina** (Merriam, 1890); **humboldtensis** Grinnell and Dixon, 1926; **kenaiensis** (Elliot, 1903); **nesophila** (Osgood, 1901); **origensis** (Rhoads, 1902); **sierrae** Grinnell and Storer, 1916; **vancourverensis** Grinnell and Dixon, 1926; **vulpina** (Rafinesque, 1819).
>     COMMENTS: May be conspecific with *martes*, *melampus*, and *zibellina* (Anderson, 1970; Hagmeier, 1961). Reviewed by Clark et al. (1987) who considered New World forms as a distinct species agreeing with Youngman (1975). Carr and Hicks (1997) recognized two distinct species based on genetic data (*M. americana* – Eastern North America;

*M. caurina* – Pacific Northwest and the Great Plains). However, most have recognized that there are two distinct subspecies groups – *"caurina"* and *"americana"* (Anderson, 1970; Clark et al., 1987; Graham and Graham, 1994; Hagmeier, 1961; Stone and Cook, 2002). Synonyms allocated according to Hall (1981) and Clark et al. (1987). Stone and Cook (2002) placed in the subgenus *Martes*.

*Martes flavigula* (Boddaert, 1785). Elench. Anim., 1:88.
COMMON NAME: Yellow-throated Marten.
TYPE LOCALITY: Not given; fixed by Pocock (1941*a*) as "Nepal".
DISTRIBUTION: China, India, Indonesia (Sumatra, Java, and Borneo), North and South Korea, Pakistan, Russia, Taiwan, Vietnam.
STATUS: CITES – Appendix III (India); U.S. ESA – Endangered as *M. f. chrysospila*; IUCN – Endangered as *M. f. robinsoni*, otherwise Lower Risk (lc).
SYNONYMS: *chrysogaster* (C. E. H. Smith, 1842); *hardwickei* (Horsfield, 1828); *kuatunensis* (Bonhote, 1901); *leucotis* (Bechstein, 1800); *melina* (Kerr, 1792); *melli* (Matschie, 1922); *quadricolor* (Shaw, 1800); *szetchuensis* (Hilzheimer, 1910); *typica* (Bonhote, 1901); *yuenshanensis* (Shih, 1930); **aterrima** (Pallas, 1811); **borealis** (Radde, 1862); *koreana* (Mori, 1922); **chrysospila** Swinhoe, 1866; *xanthospila* Swinhoe, 1870; **hainana** Hsu and Wu, 1981; *henrici* (Schinz, 1845); *lasiotis* (Temminck, 1892); **indochinensis** Kloss, 1916; **peninsularis** (Bonhote, 1901); **robinsoni** (Pocock, 1936); **saba** Chasen and Kloss, 1931.
COMMENTS: Pocock (1936*a*) and Baryshnikov and Abramov (1997, 1998) separated *flavigula* from other *Martes* and placed in the genus *Lamprogale* based on bacular morphology. Rozhnov (1995) separated the subspecies *henrici*, *hainana*, and *peninsularis* into *M. lasiotis*. Allocated according to Pocock (1936*a*), Chasen (1940), Ellerman and Morrison-Scott (1951), and Rozhnov (1995). Stone and Cook (2002) supported placing in the subgenus *Charronia* Gray, 1865.

*Martes foina* (Erxleben, 1777). Syst. Regni Anim., 1:458.
COMMON NAME: Beech Marten.
TYPE LOCALITY: "Europa inque Persia", listed by Miller (1912*a*) as "Germany."
DISTRIBUTION: Afghanistan, Albania, Austria, Belgium, Bosnia and Herzegovina, Bulgaria, China, Crete, Croatia, Czech Republic, Denmark, Estonia, France, Germany, Greece, Hungary, Italy, Kazakhstan, Latvia, Lithuania, Luxembourg, Macedonia, Moldova, Mongolia, Netherlands, Poland, Portugal, Romania, Russia, Serbia and Montenegro, Slovakia, Slovenia, Spain (incl. Ibiza Isl), Switzerland, Ukraine.
STATUS: CITES – Appendix III (India) as *M. f. intermedia*; IUCN – Lower Risk (lc).
SYNONYMS: *alba* (Bechstein, 1801); *domestica* Pinel, 1792; *fagorum* (Fatio, 1869); **bosniaca** Brass, 1911; **bunites** (Bate, 1906); **kozlovi** Ognev, 1931; **intermedia** (Severtzov, 1873); *altaica* Satunin, 1914; *leucolachnaea* Blanford, 1879; **mediterranea** (Barrett-Hamilton, 1898); **milleri** Festa, 1914; **nehringi** (Satunin, 1906); **rosanowi** Martino and Martino, 1917; **syriaca** (Nehring, 1902); **toufoeus** (Hodgson, 1842).
COMMENTS: Synonyms allocated according to Ellerman and Morrison-Scott (1951). Stone and Cook (2002) placed it in the subgenus *Martes*.

*Martes gwatkinsii* Horsfield, 1851. Cat. Mamm. Mus. E. India Co., p. 90.
COMMON NAME: Nilgiri Marten.
TYPE LOCALITY: "Madras" [India].
DISTRIBUTION: S India.
STATUS: CITES – Appendix III (India); IUCN – Vulnerable.
COMMENTS: Included in *Martes flavigula* by Corbet (1978), Honacki et al. (1982), and Corbet and Hill (1992); however, separated by Bonhote (1901*b*), Pocock (1936*a*, 1941*a*), Ellerman and Morrison-Scott (1951), Anderson (1970) and Rozhnov (1995). Pocock (1936*a*) placed *gwatkinsii* and *flavigula* in the genus *Lamprogale*.

*Martes martes* (Linnaeus, 1758). Syst. Nat., 10th ed., 1:46.
COMMON NAME: European Pine Marten.
TYPE LOCALITY: "sylvis antiquis", restricted by Thomas (1911*a*) to "Upsala" [Sweden].
DISTRIBUTION: Albania, Austria, Belgium, Bosnia and Herzegovina, Bulgaria, Corsica, Croatia, Czech Republic, Denmark, Estonia, Finland, France, Germany, Great Britain,

Hungary, Iran, Iraq, Ireland, Italy, Latvia, Lithuania, Luxembourg, Macedonia, Moldova, Netherlands, Norway, Poland, Portugal, Romania, Russia, Sardinia, Sicily, Serbia and Montenegro, Slovakia, Slovenia, Spain (incl. Mallorca and Minorca Isls), Sweden, Switzerland, Turkey, Ukraine.

STATUS: IUCN – Lower Risk (lc).

SYNONYMS: *abietum* Gray, 1865; *sylvatica* (Nilsson, 1820); *sylvestris* (Oken, 1816); *vulgaris* Griffith, 1827; **borealis** Kuznetsov, 1944; *kuznetsovi* Pavlinov and Rossolimo, 1987; *sabaneevi* Jurgenson, 1947; **latinorum** (Barrett-Hamilton, 1904); **lorenzi** Ognev, 1926; **minoricensis** Alcover, Delibes, Gosálbez, and Nadal, 1987; **notialis** (Cavazza, 1912); **ruthena** Ognev, 1926; **uralensis** Kuznetsov, 1941.

COMMENTS: May be conspecific with *americana*, *melampus*, and *zibellina* (Anderson, 1970; Hagmeier, 1961). Synonyms allocated according to Ellerman and Morrison-Scott (1951). Stone and Cook (2002) suggested a close relationship with *zibellina*, and placed in the subgenus *Martes*.

*Martes melampus* (Wagner, 1840). *In* Schreber, Die Säugethiere. Suppl., 2:229.

COMMON NAME: Japanese Marten.

TYPE LOCALITY: "Japan."

DISTRIBUTION: Japan (Honshu, Kyushu, Shikoku, Tsushima, introduced on Sado and Hokkaido Isls); North and South Korea.

STATUS: IUCN – Vulnerable as *M. m. tsuensis*, otherwise Lower Risk (lc).

SYNONYMS: *bedfordi* (Thomas, 1905); *japonica* Gray, 1865; *melanopus* Gray, 1865; **coreensis** Kuroda and Mori, 1923; **tsuensis** (Thomas, 1897).

COMMENTS: May be conspecific with *americana*, *martes*, and *zibellina* (Anderson, 1970; Hagmeier, 1961). Heptner et al. (1967) included Japanese and Korean *melampus* in *zibellina*. Synonyms allocated according to Ellerman and Morrison-Scott (1951). Stone and Cook (2002) placed it in the subgenus *Martes*.

*Martes pennanti* (Erxleben, 1777). Syst. Regni Anim., 1:470.

COMMON NAME: Fisher.

TYPE LOCALITY: "in America boreali, vulgaris victitans quadrupedibus minoribus." Listed by Miller and Rehm (1901) as "Eastern Canada."

DISTRIBUTION: Canada (Yukon to E Quebec), USA (Sierra Nevadas, N Rocky Mtns; Minnesota, N Wisconsin, Michigan upper peninsula, New York, Maine).

STATUS: IUCN – Lower Risk (lc).

SYNONYMS: *alba* (Richardson, 1829); *canadensis* (Schreber, 1777); *godmani* (Fischer, 1829); *melanorhyncha* (Boddaert, 1784); *nigra* (Kerr, 1792); *piscatoria* (Lesson, 1827); **columbiana** Goldman, 1935; **pacifica** Rhoads, 1898.

COMMENTS: Reviewed by Powell (1981); Goldman (1935) recognized three subspecies; Hagmeier (1961) concluded that subspecies could not be recognized. Synonyms allocated according to Hall (1981) and Powell (1981). Stone and Cook (2002) placed it in the subgenus *Pekania* Gray, 1865, and suggested that *pennanti* and *Gulo gulo* may form a monophyletic group, which would make *Martes* paraphyletic.

*Martes zibellina* (Linnaeus, 1758). Syst. Nat., 10th ed., 1:46.

COMMON NAME: Sable.

TYPE LOCALITY: "asia septentrionali," restricted by Thomas (1911a) to "N. Asia." Restricted by Ognev (1931:562) to "Tobol'skaya gub. v ee severnoi chasti" ["northern part of Tobol'sk Province" (1962 translation)] [Russia].

DISTRIBUTION: China (Xinjiang to NE), Japan (Hokkaido); Mongolia, North Korea, Russia (Ural Mtns to Siberia, Kamchatka, Sakhalin).

STATUS: IUCN – Data Deficient as *M. z. brachyura*, otherwise Lower Risk (lc).

SYNONYMS: *alba* (Brandt, 1855); *asiatica* (Brandt, 1855); *fusco-flavescens* (Brandt, 1855); *maculata* (Brandt, 1855); *ochracea* (Brandt, 1855); *rupestris* (Brandt, 1855); *sylvestris* (Brandt, 1855); **angarensis** Timofeev and Nadeev, 1955; **arsenjevi** Kuznetsov, 1944; **averini** Bashanov, 1943; *altaica* Jurgenson, 1947; *jurgensoni* Rossolimo and Pavlinov, 1987; **brachyura** (Temminck, 1844); **ilimpiensis** Timofeev and Nadeev, 1955; **jakutensis** Novikov, 1956; **kamtschadalica** (Birula, 1919); *coreensis* Kishida, 1927; *hamgyenensis* Kishida, 1927; *kamtschatica* (Dybowski, 1922); **lin kouensis** Ma and Wu,

1981; ***obscura*** Timofeev and Nadeev, 1955; ***princeps*** (Birula, 1922); *baicalensis* (Dybowski, 1922); *vitimensis* Timofeev and Nadeev, 1955; ***sahalinensis*** Ognev, 1925; ***sajanensis*** Ognev, 1925; ***schantaricus*** Kuznetsov, 1941; ***tomensis*** Timofeev and Nadeev, 1955; ***tungussensis*** Kuznetsov, 1944; *yeniseensis* Ognev, 1925.

COMMENTS: Reviewed by Pavlinin (1966). May be conspecific with *americana*, *martes*, and *melampus* (Anderson, 1970; Hagmeier, 1961). Heptner et al. (1967) included Japanese and Korean *melampus* in *zibellina*. Synonyms allocated according to Heptner et al. (1967). Stone and Cook (2002) placed it in the subgenus *Martes*.

*Meles* Brisson, 1762. Regnum Animale, pg. 13.
TYPE SPECIES: *[Ursus] meles* Linnaeus, 1758.
SYNONYMS: *Eumeles* Gray, 1865; *Meledes* Kastschenko, 1925; *Melesium* Rafinesque, 1815; *Taxus* É. Geoffroy Saint-Hilaire and G. Cuvier, 1795.
COMMENTS: Many authors considered *Meles* monotypic (Ellerman and Morrison-Scott, 1951; Heptner et al., 1967; Long and Killingley, 1983; Novikov, 1956; Stroganov, 1962). However, others supported the position that European and Asian badgers are not conspecific (Aristov and Baryshnikov, 2001; Baryshnikov and Potapova, 1990; Kastschenko, 1902; Neal, 1948; Ognev, 1931; Satunin, 1914). Kurose et al. (2001) argued that *Meles* is heterogeneous. Recent morphological studies (Abramov, 2001, 2002) support the separation of *Meles* into several species. Species and subspecies allocated following Abramov (2001, 2002).

*Meles anakuma* Temminck, 1844. Fauna Japon., Mamm.: 30, pl. 6.
COMMON NAME: Japanese Badger.
TYPE LOCALITY: "Environs of Nagasaki et d'Awa," [Japan].
DISTRIBUTION: Japan (Honshu, Kyushu, Shikoku).
COMMENTS: Heptner et al. (1967) and Baryshnikov and Potapova (1990) considered *anakuma* a subspecies of Asian badger. However, Abramov (2001, 2002) supported the recognition of these forms as distinct.

*Meles leucurus* (Hodgson, 1847). J. Asiat. Soc. Bengal., 16:763.
COMMON NAME: Asian Badger.
TYPE LOCALITY: "Lhasa, Tibet," [China].
DISTRIBUTION: China, Kazakhstan, North and South Korea, Russia (From Volga River through Siberia).
SYNONYMS: *chinensis* Gray, 1868; *hanensis* Matschie, 1907; *leptorhynchus* Milne-Edwards, 1867; *siningensis* Matschie, 1907; *tsingtauensis* Matschie, 1907; ***amurensis*** Schrenck, 1859; *melanogenys* J. A. Allen, 1913; *schrenkii* Nehring, 1891; ***arenarius*** Satunin, 1895; *blandfordi* Matschie, 1907; ***sibiricus*** Kastschenko, 1900; *aberrans* Stroganov, 1962; *altaicus* Kastschenko, 1902; *enisseyensis* Petrov, 1953; *eversmanni* Petrov, 1953; *raddei* Kastschenko, 1902; ***tianschanensis*** Hoyningen-Huene, 1910; *talassicus* Ognev, 1931.
COMMENTS: Reviewed by Petrov (1953), Heptner et al. (1967), and Abramov (2001). Synonyms allocated according to Heptner et al. (1967) and Abramov (2001).

*Meles meles* (Linnaeus, 1758). Syst. Nat., 10th ed., 1:48.
COMMON NAME: European Badger.
TYPE LOCALITY: "Europa inter rimas rupium et lapidum," restricted by Thomas (1911*a*) to "Upsala" [Sweden].
DISTRIBUTION: Afghanistan, Albania, Austria, Belgium, Bosnia and Herzegovina, Bulgaria, China (Xinjiang), Crete, Croatia, Czech Republic, Denmark, Estonia, Finland, France, Germany, Great Britain, Greece, Hungary, Iran, Iraq, Ireland, Israel, Italy, Latvia, Lithuania, Luxembourg, Macedonia, Moldova, Netherlands, Norway, Poland, Portugal, Romania, Russia (eastward up to Volga River), Serbia and Montenegro, Slovakia, Slovenia, Spain, Sweden, Switzerland, Ukraine.
STATUS: IUCN – Lower Risk (lc).
SYNONYMS: *alba* (Gmelin, 1788); *britannicus* Satunin, 1905; *caninus* Billberg, 1827; *caucasicus* Ognev, 1926; *communis* Billberg, 1827; *danicus* Degerbøl, 1933; *europaeus* Desmarest, 1816; *maculata* (Gmelin, 1788); *tauricus* Ognev, 1926; *taxus* Boddaert, 1785; *typicus* Barrett-Hamilton, 1899; *vulgaris* (Tiedemann, 1808); ***arcalus*** Miller, 1907; ***canescens***

Blanford, 1875; *minor* Satunin, 1905; *ponticus* Blackler, 1916; **heptneri** Ognev, 1931; **marianensis** Graells, 1897; *mediterraneus* Barrett-Hamilton, 1899; **milleri** Baryshnikov, Puzachenko and Abramov, 2003; **rhodius** Festa, 1914; **severzovi** Heptner, 1940; *bokharensis* Petrov, 1953.

COMMENTS: Reviewed by Petrov (1953) and Heptner et al. (1967). Synonyms allocated according to Heptner et al. (1967) and Abramov (2001).

*Mellivora* Storr, 1780. Prodr. Meth. Mamm., p. 34, Tabl. A.

TYPE SPECIES: *Viverra ratel* Sparrman, 1777 (= *Viverra capensis* Schreber, 1776), by designation (Sclater, 1900; Melville and Smith, 1987).

SYNONYMS: *Lipotus* Sundevall, 1843; *Melitoryx* Gloger, 1841; *Melivora* Gray, 1847; *Ratellus* Gray, 1827; *Rattelus* Swainson, 1835; *Ratelus* Gray, 1825; *Ursitaxus* Hodgson, 1835; *Ursotaxus* Blyth, 1840.

COMMENTS: Placed in Mellivorinae by McKenna and Bell (1997).

*Mellivora capensis* (Schreber, 1776). Die Säugethiere, 3(18): pl. 125[1776]; text, 3(26):450[1777].

COMMON NAME: Honey Badger.

TYPE LOCALITY: "Vorgebirge der guten Hofnung" [South Africa, Western Cape Prov., Cape of Good Hope].

DISTRIBUTION: Nepal (Savanna and steppe), India, Turkmenistan, Lebanon, South Africa.

STATUS: CITES – Appendix III (Ghana and Botswana); IUCN – Lower Risk (lc).

SYNONYMS: *mellivorus* (G. [Baron] Cuvier, 1798); *ratel* (Sparrman, 1777); *typicus* (A. Smith, 1833); *vernayi* Roberts, 1932; **abyssinica** Hollister, 1910; **buechneri** Baryshnikov, 2000; **concisa** Thomas, and Wroughton, 1907; *brockmani* Wroughton and Cheesman, 1920; *buchanani* Thomas, 1925; **cottoni** Lydekker, 1906; *sagulata* Hollister, 1910; **inaurita** (Hodgson, 1836); **indica** (Kerr, 1792); *mellivorus* (Bennett, 1830); *ratel* Horsfield, 1851; *ratelus* Fraser, 1862; **leuconota** Sclater, 1867; **maxwelli** Thomas, 1923; **pumilio** Pocock, 1946; **signata** Pocock, 1909; **wilsoni** Cheesman, 1920.

COMMENTS: Synonyms allocated according to Baryshnikov (2000) and Vanderhaar and Hwang (2003).

*Melogale* I. Geoffroy Saint-Hilaire, 1831. *In* Bélanger (ed.), Voy. Indes Orient. 3(Zoologie): 129, pl. 5 [issued 13 March 1831].

TYPE SPECIES: *Melogale personata* I. Geoffroy Saint-Hilaire, 1831.

SYNONYMS: *Helictis* Gray, 1831; *Nesictis* Thomas, 1922; *Rhinogale* Gloger, 1841.

COMMENTS: There is uncertainty as to the total number of species in this genus. Most recent authors tend to consider *everetii* and/or *orientalis* as conspecific with *personata*; however, Pocock (1941a), Everts (1968), Long (1978, 1981), and Long and Killingley (1983) supported the recognition of these populations as distinct. Long and Killingley pointed out that there is no published information to refute Pocock's (1941a) revision.

*Melogale everetti* (Thomas, 1895). Ann. Mag. Nat. Hist. ser. 6, 15:331-332.

COMMON NAME: Bornean Ferret-badger.

TYPE LOCALITY: "Mount Kina Balu, N. Borneo, about 4000 ft. [1219 m]"

DISTRIBUTION: Indonesia (Kalimantan), Malaysia (Sabah, Sarawak).

STATUS: IUCN – Vulnerable.

COMMENTS: Medway (1977) included *everetti* in *orientalis*; however, see comments under genus.

*Melogale moschata* (Gray, 1831). Proc. Zool. Soc. Lond., 1831:94.

COMMON NAME: Chinese Ferret-badger.

TYPE LOCALITY: "China," restricted by Allen (1929) to "Canton, Kwangtung Province, South China."

DISTRIBUTION: China (C and SE, Hainan), India (Naga Hills near Manipur, Assam), N Laos, Taiwan, N Vietnam.

STATUS: IUCN – Lower Risk (lc).

SYNONYMS: **ferreogrisea** (Hilzheimer, 1905); **hainanensis** Zheng and Xu, 1983; **millsi** (Thomas, 1922); **sorella** (G. M. Allen, 1929); **subaurantiaca** (Swinhoe, 1862); *modesta* (Thomas, 1922); **taxilla** (Thomas, 1925).

COMMENTS: Synonyms allocated according to Ellerman and Morrison-Scott (1951) and Storz and Wozencraft (1999).

*Melogale orientalis* (Horsfield, 1821). Zool. Res. Java., plate and 4 pages of text.
COMMON NAME: Javan Ferret-badger.
TYPE LOCALITY: "limited . . . to . . . south of Mountain Prahu, between the two prinicpal cones of the central part of Java, the Mountain Sumbing, and . . . Teggal, . . . Baggulen and Banyumas . . . to Gowong in the east." [Indonesia, Java].
DISTRIBUTION: Indonesia (Java).
STATUS: IUCN – Lower Risk (nt).
SYNONYMS: *maccourus* (Temminck, 1824); **sundaicus** (Sody, 1937).

*Melogale personata* I. Geoffroy Saint-Hilaire, 1831. *In* Bélanger (ed.), Voy. Indes Orient. 3(Zoologie):137, pl. 5.[issued 13 March 1831].
COMMON NAME: Burmese Ferret-badger.
TYPE LOCALITY: "environs de Rangoun" [Burma].
DISTRIBUTION: Burma, China, Nepal, India (Assam), Malaysia (West), Thailand, Vietnam
STATUS: IUCN – Lower Risk (lc).
SYNONYMS: **laotum** Thomas, 1922; **nipalensis** (Hodgson, 1836); *orientalis* (Blanford, 1888); **pierrei** (Bonhote, 1903); **tonquinia** Thomas, 1922.
COMMENTS: Synonyms allocated according to Ellerman and Morrison-Scott (1951).

*Mustela* Linnaeus, 1758. Syst. Nat., 10th ed., 1:45.
TYPE SPECIES: [*Mustela*] *erminea* Linnaeus, 1758, by tautonomy (Miller, 1912).
SYNONYMS: *Arctogale* Kaup, 1829; *Cabreragale* Baryshnikov and Abramov, 1997; *Cryptomustela* Abramov, 2000; *Cynomyonax* Coues, 1877; *Eumustela* Acloque, 1899; *Foetorius* Keyserling and Blasius, 1840; *Gale* Wagner, 1841; *Grammogale* Cabrera, 1940; *Gymnopus* Gray, 1865; *Hydromustela* Bogdanov, 1871; *Ictis* Kaup, 1829; *Kolonocus* Satunin, 1914; *Kolonokus* Satunin, 1911; *Lutreola* Wagner, 1841; *Mustelina* Bogdanov, 1871; *Mustella* Scopoli, 1777; *Neogale* Gray, 1865; *Neovison* Baryshnikov and Abramov, 1997; *Plesiogale* Pocock, 1921; *Pocockictis* Kretzoi, 1947; *Putorius* Cuvier, 1817; *Vison* Gray, 1843.
COMMENTS: Revised by Hall (1951), Youngman (1982), Abramov (1999), and Kurose et al. (2000). Youngman (1982) recognized five and Abramov (1999) recognized nine subgenera. Some have chosen to elevate these groupings to generic status; because of the lack of any comprehensive phylogenetic approach to this problem, they are provisionally recognized here as valid subgenera. Synonyms for species and subspecies principally follow Abramov (1999, pers. comm.).

*Mustela africana* Desmarest, 1818. Nouv. Dict. Hist. Nat., Nouv., ed., 9:376.
COMMON NAME: Amazon Weasel.
TYPE LOCALITY: "Africa", type locality is in error, fixed by Cabrera (1957) as "arrabales de Belem, la antigua Pará." [Brazil]
DISTRIBUTION: Amazon Basin in Brazil, Ecuador, and Peru.
STATUS: IUCN – Data Deficient.
SYNONYMS: *paraensis* (Goeldi, 1897); **stolzmanni** Taczanowski, 1881.
COMMENTS: Youngman (1982) and Abramov (1999) placed *africana* in subgenus *Grammogale*. Izor and de la Torre (1978) suggested that *africana* and *felipei* form a monophyletic group. Cabrera (1957) considered *Grammogale* a valid genus.

*Mustela altaica* Pallas, 1811. Zoogr. Rosso-Asiat., I:98.
COMMON NAME: Mountain Weasel.
TYPE LOCALITY: "qui alpes altaicas adibunt" [Altai Mtns].
DISTRIBUTION: W and N China, Kashmir, E Kazakhstan, Kyrgyzstan, Mongolia, North Korea (?), Russia (S and SE Siberia, Primorski Krai), Sikkim, Tajikistan.
STATUS: CITES – Appendix III (India); IUCN – Lower Risk (lc).
SYNONYMS: *alpina* (Gebler, 1823); *sacana* Thomas, 1914; **birulai** (Ognev, 1928); **raddei** (Ognev, 1928); **temon** Hodgson, 1857; *astutus* (Milne-Edwards, 1870); *longstaffi* Wroughton, 1911; **tsaidamensis** (Hilzheimer, A10).
COMMENTS: Youngman (1982) placed *altaica* in the subgenus *Mustela*; however, Ognev

(1935) considered *sibirica* and *altaica* closely related. Abramov (1999) placed it in the subgenus *Gale*. Synonyms allocated according to Ellerman and Morrison-Scott (1951).

*Mustela erminea* Linnaeus, 1758. Syst. Nat., 10th ed., 1:46.

COMMON NAME: Ermine.

TYPE LOCALITY: "Europa and Asia frigidiore; hyeme praefertim in alpinis regionibus nivea".

DISTRIBUTION: Circumboreal, tundra and forested regions of Palearctic. Afghanistan, Algeria, Austria, Belarus, Belgium, Bosnia and Herzegovina, Canada, China, Croatia, Czech Republic, Denmark, Estonia, Finland, France, Germany, Great Britain, Hungary, Ireland, Italy, Japan (C Honshu), Kazakhstan, Kyrgyzstan, Latvia, Lithuania, Luxembourg, Mongolia, Netherlands, New Zealand (introduced), Norway, Poland, Romania, Russia, Serbia and Montenegro, Slovakia, Slovenia, Spain, Sweden, Switzerland, Tajikistan, USA (C California, N New Mexico, N Iowa and Maryland).

STATUS: CITES – Appendix III (India) as *M. erminea ferghanae*; IUCN – Lower Risk (lc).

SYNONYMS: *hyberna* Kerr, 1792; *maculata* Billberg, 1827; ***aestiva*** Kerr, 1792; *algiricus* Thomas, 1895; *alpestris* Burg, 1920; *giganteus* Burg, 1920; *major* Nilsson, 1820; ***alascensis*** Merriam, 1896; ***anguinae*** Hall, 1932; ***arctica*** Merriam, 1896; *audax* Barrett-Hamilton, 1904; *kadiacensis* Merriam, 1896; *kadiacensis* Osgood, 1901; *richardsonii* Bonaparte, 1838; ***augustidens*** Brown, 1908; ***bangsi*** Hall, 1945; *cicognani* Mearns, 1891; *pusillus* Aughey, 1880 [preoccupied]; ***celenda*** Hall, 1944; ***cigognanii*** Bonaparte, 1838; *pusilla* DeKay, 1842; *vulgaris* Griffith, 1827 [preoccupied]; ***fallenda*** Hall, 1945; ***ferghanae*** (Thomas, 1895); *shnitnikovi* Ognev, 1935; *whiteheadi* Wroughton, 1908; ***gulosa*** Hall, 1945; ***haidarum*** Preble, 1898; ***hibernica*** Thomas and Barrett-Hamilton, 1895; ***initis*** Hall, 1945; ***invicta*** Hall, 1945; ***kadiacensis*** Merriam, 1896; ***kaneii*** (Baird, 1857); *baturini* Ognev, 1929; *digna* Hall, 1944; *kamtschatica* (Dybowski, 1922); *kanei* G. Allen, 1914; *naumovi* Jurgenson, 1938; *orientalis* Ognev, 1928; *transbaikalica* Ognev, 1928; ***karaginensis*** Jurgenson, 1936; ***lymani*** Hollister, 1912; ***martinoi*** Ellerman and Morrison-Scott, 1951; *birulai* Martino and Martino, 1930 [preoccupied]; ***minima*** Cavazza, 1912; ***mongolica*** Ognev, 1928; ***muricus*** Bangs, 1899; *leptus* Merriam, 1903; ***nippon*** Cabrera, 1913; ***ognevi*** Jurgenson, 1932; ***olympica*** Hall, 1945; ***polaris*** Barrett-Hamilton, 1904; ***richardsonii*** Bonaparte, 1838; *imperii* Barrett-Hamilton, 1904; *microtis* J. A. Allen, 1903; *mortigena* Bangs, 1913; ***ricinae*** Miller, 1907; ***salva*** Hall, 1944; ***seclusa*** Hall, 1944; ***semplei*** Sutton and Hamilton, 1932; *labiata* Degerbøl, 1935; ***stabilis*** Barrett-Hamilton, 1904; ***streatori*** Merriam, 1896; ***teberdina*** Korneev, 1941; *balkarica* Basiev, 1962; ***tobolica*** Ognev, 1923.

COMMENTS: Revised by Eger (1990). Reviewed by C. M. King (1983). Youngman (1982) and Abramov (1999) placed *erminea* in the subgenus *Mustela*. Synonyms allocated according to Ellerman and Morrison-Scott (1951), Hall (1951, 1981), and C. M. King (1983).

*Mustela eversmanii* Lesson, 1827. Manuel de Mammalogie, p. 144.

COMMON NAME: Steppe Polecat.

TYPE LOCALITY: "trouvé . . . entre Orembourg et Bukkara," restricted by Stroganov (1962:338) to "bassein srednego techeniya r. Ileka, v raione vpadenya . . . r. Bol'shoi Khobdy" [Russia, Orenburg Obl., south of Orenburg, mouth of Bol'shaya Khobda River, a tributary of Ilek River].

DISTRIBUTION: Austria, Bulgaria, China, Czech Republic, Georgia, Hungary, Kazakhstan, Kyrgyzstan, Moldova, Mongolia, Poland, Romania, Russia, Serbia and Montenegro, Slovakia, Tajikistan, Turkmenistan, Ukraine, Uzbekistan.

STATUS: IUCN – Vulnerable as *M. e. amurensis*, otherwise Lower Risk (lc).

SYNONYMS: *aureus* (Pocock, 1936); *heptapotamicus* (Stroganov, 1960); *nobilis* (Stroganov, 1958); *pallidus* (Stroganov, 1958); ***admirata*** (Pocock, 1936); ***amurensis*** (Ognev, 1930); ***hungarica*** Éhik, 1928; *moravica* Kostroò, 1948; *occidentalis* (Brauner, 1929); *satunini* (Migulin, 1928); ***larvatus*** (Hodgson, 1849); *tibetanus* Horsfield, 1851; ***michnoi*** (Kastschenko, 1910); *dauricus* (Stroganov, 1958); *lineiventer* Hollister, 1913; *sibiricus* (Kastschenko, 1912); *triarata* Hollister, 1913; *tuvinicus* (Stroganov, 1958); ***talassicus*** Ognev, 1928.

COMMENTS: The correct spelling of this name is *eversmanii* (Mazák, 1971). Reviewed by

Kostroò (1948), Stroganov (1958, 1962), Heptner et al. (1967), and Anderson (1977). Youngman (1982) and Abramov (1999) placed *eversmanii* in the subgenus *Putorius*. Anderson (1977) and Kurtén and Anderson (1980) suggested that *nigripes* and *eversmanii* may be conspecific. Pocock (1936*b*) and Ellerman and Morrison-Scott (1966) considered *eversmanii* and *putorius* conspecific; however, Ognev (1931), Stroganov (1962), and Heptner et al. (1967) recognized them as distinct species.

*Mustela felipei* Izor and de la Torre, 1978. J. Mammal., 59:92.
COMMON NAME: Colombian Weasel.
TYPE LOCALITY: "Santa Marta, elevation 2,700 m, near San Agustin, Huila, Colombia".
DISTRIBUTION: Colombia, Ecuador.
STATUS: IUCN – Endangered.
COMMENTS: Izor and de la Torre (1978) suggested that *africana* and *felipei* form a monophyletic group. Youngman (1982) placed *felipei* in subgenus *Grammogale*; Abramov (1999) placed it in subgenus *Cabreragale*.

*Mustela frenata* Lichtenstein, 1831. Darst. Säugeth., text: "Das gezäumte Wiesel" [not paginated], and plate 42.
COMMON NAME: Long-tailed Weasel.
TYPE LOCALITY: "der Nähe von Mexico"[placed in Ciudad Mexico, Mexico].
DISTRIBUTION: Belize, Bolivia, S Canada, Columbia, Costa Rica, Ecuador, El Salvador, Guatemala, Honduras, Mexico, Nicaragua, Panama, Peru, USA (most states excluding SW deserts), Venezuela.
STATUS: IUCN – Lower Risk (lc).
SYNONYMS: *aequatorialis* (Coues, 1877); *brasiliensis* Sevastianoff, 1813 [preoccupied]; *mexicanus* (Coues, 1877); *affinis* Gray, 1874; *costaricensis* J. A. Allen, 1916; *macrurus* (J. A. Allen, 1912); *meridana* Hollister, 1914; *agilis* Tschudi, 1844; *macrura* J. A. Allen, 1916; *alleni* (Merriam, 1896); *altifrontalis* Hall, 1936; *saturata* Miller, 1912 [preoccupied]; *arizonensis* (Mearns, 1891); *arthuri* Hall, 1927; *aureoventris* Gray, 1864; *affinis* Lönnberg, 1913; *jelskii* Taczanowski, 1881; *macrura* Taczanowski, 1874; *boliviensis* Hall, 1938; *costaricensis* Goldman, 1912; *brasiliensis* Gray, 1874 [preoccupied]; *effera* Hall, 1936; *goldmani* (Merriam, 1896); *gracilis* (Brown, 1908); *helleri* Hall, 1935; *inyoensis* Hall, 1936; *latirostra* Hall, 1936; *arizonensis* Grinnell and Swarth, 1913 [preoccupied]; *leucoparia* Merriam, 1896; *longicauda* Bonaparte, 1838; *macrophonius* (Elliot, 1905); *munda* (Bangs, 1899); *neomexicanus* (Barber and Cockerell, 1898); *nevadensis* Hall, 1936; *longicauda* (Coues, 1891) [preoccupied]; *nicaraguae* J. A. Allen, 1916; *nigriauris* Hall, 1936; *xanthogenys* Gray, 1874 [preoccupied]; *notius* (Bangs, 1899); *noveboracensis* (Emmons, 1840); *fusca* DeKay, 1842; *richardsonii* (Baird, 1858); *occisor* Bangs, 1899; *olivacea* Howell, 1913; *oregonensis* (Merriam, 1896); *oribasus* (Bangs, 1899); *panamensis* Hall, 1932; *peninsulae* (Rhoads, 1894); *perda* (Merriam, 1902); *perotae* Hall, 1936; *primulina* Jackson, 1913; *pulchra* Hall, 1936; *saturata* (Merriam, 1896); *spadix* (Bangs, 1896); *texensis* Hall, 1936; *tropicalis* (Merriam, 1896); *frenatus* Coues, 1877; *noveboracensis* DeKay, 1840; *perdus* Merriam, 1902; *richardsoni* Bonaparte, 1838; *washingtoni* (Merriam, 1896); *xanthogenys* Gray, 1843.
COMMENTS: Youngman (1982) and Abramov (1999) placed *frenata* in the subgenus *Mustela*. Synonyms allocated according to Hall (1951) and Sheffield and Thomas (1997).

*Mustela itatsi* Temminck, 1844. Fauna Japonica, Mamm., 34, pl. vii, fig. 2.
COMMON NAME: Japanese Weasel.
TYPE LOCALITY: "Japan."
DISTRIBUTION: Japan. Introduced to Russia (Sakhalin).
SYNONYMS: *asaii* Kuroda, 1943; *katsurai* Kishida, 1931; *natsi* Temminck, 1844; *sho* Kuroda, 1924.
COMMENTS: Abramov (2000), Kurose et al. (2000), and Graphodatsky et al. (1976) supported separation of *itatsi* from *sibirica*. Abramov (1999) placed it in the subgenus *Kolonokus*.

*Mustela kathiah* Hodgson, 1835. J. Asiat. Soc. Bengal, 4:702.
COMMON NAME: Yellow-bellied Weasel.

TYPE LOCALITY: "Kachar region" [Nepal].

DISTRIBUTION: Burma, S and E China, Indochinese peninsula, Nepal, N Pakistan.

STATUS: CITES – Appendix III (India); IUCN – Lower Risk (lc).

SYNONYMS: *auriventer* Hodgson, 1837; *dorsalis* (Trouessart, 1895); *melli* (Matschie, 1922); *tsaidamensis* (Hilzheimer, 1910); **caporiaccoi** de Beaux, 1935.

COMMENTS: Abramov (1999) placed *kathiah* in the subgenus *Gale*. Synonyms allocated according to Ellerman and Morrison-Scott (1951).

*Mustela lutreola* (Linnaeus, 1761). Fauna Suecica, 2nd ed., p. 5.

COMMON NAME: European Mink.

TYPE LOCALITY: "Finlandiae aquolis", restricted by Matschie (1912) to "Südwest-Finnland."

DISTRIBUTION: Belarus, Estonia, France, Latvia, Romania, Russia (Europe to the Urals), NE Spain. Formerly Germany, Poland, Austria, Czech Republic, Serbia and Montenegro, Slovakia, Hungary, Bulgaria.

STATUS: IUCN – Endangered.

SYNONYMS: *alba* de Sélys Longchamps, 1839; *alpinus* (Ogérien, 1863); *europeae* (Homeyer, 1879); *fulva* Kerr, 1792; *minor* (Erxleben, 1777); *wyborgensis* Matschie, 1912; **biedermanni** Matschie, 1912; *armorica* Matschie, 1912; **binominata** Ellerman and Morrison-Scott, 1951; *caucasica* (Novikov, 1939) [preoccupied]; **cylipena** Matschie, 1912; *albica* Matschie, 1912; *budina* Matschie, 1912; *glogeri* Matschie, 1912; *varina* Matschie, 1912; **novikovi** Ellerman and Morrison-Scott, 1951; *borealis* (Novikov, 1939) [preoccupied]; **transsylvanica** Éhik, 1932; *ehiki* Kretzoi, 1942; *hungarica* Éhik, 1932; **turovi** Kuznetsov in Novikov, 1939.

COMMENTS: Youngman (1982, 1990) and Abramov (1999) placed *lutreola* in subgenus *Lutreola*. Revised by Matschie (1912), Novikov (1939), and Youngman (1982, 1990). Occasional hybrids occur between *lutreola* and *putorius* (Ognev, 1931; Tumanov and Abramov, 2002; Youngman, 1982). Synonyms allocated according to Ellerman and Morrison-Scott (1951).

*Mustela lutreolina* Robinson and Thomas, 1917. Ann. Mag. Nat. Hist., ser. 8, 20:261-262.

COMMON NAME: Indonesian Mountain Weasel.

TYPE LOCALITY: "Tjibodas, West Java, 5500′ [1676 m]"; identified by Van Bree and Boeadi (1978) as "6°44′S 107°00′E".

DISTRIBUTION: Indonesia (Java, Sumatra).

STATUS: IUCN – Endangered.

COMMENTS: Revised by Van Bree and Boeadi (1978). Ellerman and Morrison-Scott (1951) implied, and Heptner et al. (1967), Corbet (1978), and Lekagul and McNeely (1988) believed *lutreolina* and *sibirica* to be conspecific; however this has not been supported by primary studies (Brongersma, 1940; Van Bree and Boeadi, 1978). Youngman (1982) placed *lutreolina* in the subgenus *Lutreola*, Abramov (1999) placed it in the subgenus *Kolonokus*.

*Mustela nigripes* (Audubon and Bachman, 1851). Viviparous Quadrupeds of North America, 2: 297.

COMMON NAME: Black-footed Ferret.

TYPE LOCALITY: "lower waters of the Platte River", restricted by Hayden (1863:138) to "Fort Laramie" [Wyoming, USA].

DISTRIBUTION: Formerly, Canada (S Alberta and Saskatchewan), USA (south to Arizona, Oklahoma, and NW Texas). Viable populations now only in captivity (see status).

STATUS: CITES – Appendix I; U.S. ESA – Endangered, except where listed as an Experimental Non Essential Population in portions of Arizona, Colorado, Montana, South Dakota, Utah, and Wyoming (USA); IUCN – Extinct in the Wild.

COMMENTS: Reviewed by Hillman and Clark (1980) and Anderson (1977). Youngman (1982) and Abramov (1999) placed *nigripes* in the subgenus *Putorius*. Anderson (1977) and Kurtén and Anderson (1980) suggested that *nigripes* and *eversmanii* may be conspecific.

*Mustela nivalis* Linnaeus, 1766. Syst. Nat., 12th ed., 1:69.

COMMON NAME: Least Weasel.

TYPE LOCALITY: "Westrobothnia" [Sweden].

DISTRIBUTION: Albania, Austria, Belgium, Bosnia and Herzegovina, Bulgaria, Canada, China,

Corsica, Crete, Croatia, Cyprus, Czech Republic, Denmark, Estonia, Finland, France, Germany, Great Britain, Greece, Hungary, Italy, Japan (Hokkaido and Honshu), Latvia, Lithuania, Luxembourg, Macedonia, Moldova, Mongolia, Netherlands, New Zealand (introduced – Corbet and Hill, 1980), Norway, Poland, Portugal, Romania, Russia, Sardinia, Serbia and Montenegro, Sicily, Slovakia, Slovenia, Spain, Sweden, Switzerland, Taiwan, USA, (Alaska and most of the USA except SW), Ukraine.

STATUS: IUCN – Lower Risk (lc).

SYNONYMS: *caraftensis* Kishida, 1936; *kerulenica* Bannikov, 1952; *punctata* Domaniewski, 1926; *yesoidsuna* Kishida, 1936; **allegheniensis** (Rhoads, 1901); **aistoodonnivalis** Wu and Kao, 1991; **boccamela** Bechstein, 1800; *italicus* (Barrett-Hamilton, 1900); **campestris** Jackson, 1913; **caucasica** Barrett-Hamilton, 1900; *dinniki* (Satunin, 1907); **eskimo** (Stone, 1900); **heptneri** Morozova-Turova, 1953; **mosanensis** Mori, 1927; **namiyei** Kuroda, 1921; **numidica** Pucheran, 1855; *albipes* Mina Palumbo, 1868; *algiricus* Thomas, 1895; *atlas* (Barrett-Hamilton, 1904); *corsicanus* (Cavazza, 1908); *fulva* Mina Palumbo, 1868; *galinthias* (Bate, 1905); *ibericus* (Barrett-Hamilton, 1900); *meridionalis* (Costa, 1869); *siculus* (Barrett-Hamilton, 1900); **pallida** Barrett-Hamilton, 1900; **pygmaea** (J. A. Allen, 1903); *kamtschatica* (Dybowski, 1922); **rixosa** (Bangs, 1896); **rossica** Abramov and Baryshnikov, 2000; **russelliana** Thomas, 1911; **stoliczkana** Blanford, 1877; **tonkinensis** Björkegren, 1941; **vulgaris** Erxleben, 1777; *dombrowskii* Matschie, 1901; *hungarica* Vásárhelyi, 1942; *minutus* (Pomel, 1853); *monticola* (Cavazza, 1908); *nikolskii* Semenov, 1899; *occidentalis* Kratochvil, 1977; *trettaui* Kleinschmidt, 1937; *vasarhelyi* Kretzoi, 1942.

COMMENTS: Reviewed by Reichstein (1957), van Zyll de Jong (1992), Reig (1997), and Abramov and Baryshinikov (2000). Reig divided this problematic taxon into four species based on a skull morphometric analysis (*subpalmata*, *rixosa*, *eskimo*, and *vulgaris*). Abramov and Baryshinikov separated only *subpalmata*. Youngman (1982) placed *nivalis* in the subgenus *Mustela*; Abramov (1999) placed it in the subgenus *Gale*. Synonyms allocated according to Ellerman and Morrison-Scott (1951), Hall (1981), Sheffield and King (1994), and Abramov and Baryshnikov (1999).

*Mustela nudipes* Desmarest, 1822. Mammalogie, *In* Encycl. Meth., 2(Suppl.):537.

COMMON NAME: Malayan Weasel.

TYPE LOCALITY: "L'île de Java". Locality is in error, fixed by Robinson and Kloss (1919*b*) as "West Sumatra" [Indonesia].

DISTRIBUTION: Thailand, Malaysia, Indonesia (Sumatra, Java, Borneo).

STATUS: IUCN – Lower Risk (lc).

SYNONYMS: *hamakeri* Dammerman, 1940; **leucocephalus** (Gray, 1865).

COMMENTS: Pocock (1941*a*) believed *strigidorsa* and *nudipes* to be closely related. Youngman (1982) placed nudipes in the subgenus *Lutreola*, Abramov (1999) placed it in subgenus *Pocockictis*.

*Mustela putorius* Linnaeus, 1758. Syst. Nat., 10th ed., 1:46.

COMMON NAME: European Polecat.

TYPE LOCALITY: "inter Europae rupes et lapidum acervos", restricted by Thomas (1911*a*) to "Scania, S. Sweden."

DISTRIBUTION: Albania, Austria, Belgium, Bosnia and Herzegovina, Bulgaria, Croatia, Czech Republic, Denmark, Estonia, Finland, France, Germany, Great Britian, Hungary, Italy, Latvia, Lithuania, Luxembourg, Macedonia, Moldova, Morocco, Netherlands, Norway, Poland, Portugal, Romania, Serbia and Montenegro, Slovakia, Slovenia, Spain, Sweden, Switzerland, Ukraine.

STATUS: IUCN – Lower Risk (lc).

SYNONYMS: *flavicans* de Sélys Longchamps, 1839; *foetens* (Thunberg, 1798); *foetidus* (Gray, 1843); *iltis* Boddaert, 1785; *infectus* (Ogérien, 1863); *manium* (Barrett-Hamilton, 1904); *putorius* Blyth, 1842; *verus* (Brandt *In* Simashko, 1851); *vison* de Sélys Longchamps, 1839; *vulgaris* (Griffith, 1827); **anglia** (Pocock, 1936); **aureola** (Barrett-Hamilton, 1904); **caledoniae** (Tetley, 1939); **furo** Linnaeus, 1758 [domestic ferret]; *albus* (Bechstein, 1801); *furoputorius* Link, 1795; *subrufo* (Gray, 1865); **mosquensis** Heptner, 1966; *orientalis* Brauner, 1929 [preoccupied]; *orientalis* (Polushina, 1955) [preoccupied]; *ognevi* Kratochvil, 1952 [preoccupied]; **rothschildi** Pocock, 1932.

COMMENTS: Reviewed by Heptner et al. (1967). Probable ancestor of the domestic ferret, *M. p. furo* (Rempe, 1970; Volobuev et al., 1974). Youngman (1982) and Abramov (1999) placed it in the subgenus *Putorius*. Pocock (1936*b*) and Ellerman and Morrison-Scott (1966) considered *eversmanii* and *putorius* conspecific; however, Ognev (1931), Stroganov (1962), and Heptner et al. (1967), recognized them as distinct species. Synonyms allocated according to Heptner et al. (1967).

*Mustela sibirica* Pallas, 1773. Reise Prov. Russ. Reichs., 2:701.
    COMMON NAME: Siberian Weasel.
    TYPE LOCALITY: "Sibiriae montanis, sylvis densissimis", restricted by Pocock (1941*a*) to "Vorposten Tigerazkoi, near Usstkomengorsk, W. Altai," based on Pallas (1773:570) [U.S.S.R., E. Kazakhstan, vic. of Ust-Kamenogorsk, Tigeretskoie (Honacki et al., 1982)].
    DISTRIBUTION: N Burma, China, Japan (Hokkaido, introduced to Honshu), North Korea, South Korea, Pakistan, Russia (From Kirov Prov., Tataria and W Ural Mtns throughout Siberia to Far East), Taiwan, N Thailand.
    STATUS: CITES – Appendix III (India); IUCN – Lower Risk (lc).
    SYNONYMS: *australis* (Satunin, 1911); *miles* Barrett-Hamilton, 1904; ***canigula*** Hodgson, 1842; ***charbinensis*** Lowkashkin, 1934; ***coreanus*** (Domaniewski, 1926); *peninsulae* (Kishida, 1931); ***davidiana*** (Milne-Edwards, 1871); *melli* (Matschie, 1922); *noctis* (Barrett-Hamilton, 1904); ***fontanierii*** (Milne-Edwards, 1871); *stegmanni* (Matschie, 1907); ***hodgsoni*** Gray, 1843; ***manchurica*** Brass, 1911; ***moupinensis*** (Milne-Edwards, 1874); *hamptoni* Thomas, 1921; *major* (Hilzheimer, 1910); *tafeli* (Hilzheimer, 1910); ***quelpartis*** (Thomas, 1908); ***subhemachalana*** Hodgson, 1837; *horsfieldii* Gray, 1843; *humeralis* Blyth, 1842; ***taivana*** Thomas, 1913.
    COMMENTS: Youngman (1982) placed *sibirica* in the subgenus *Lutreola*, Abramov (1999) placed it in the subgenus *Kolonokus*. Ognev (1935) considered *altaica* and *sibirica* closely related. Ellerman and Morrison-Scott (1951) implied, and Heptner et al. (1967), Corbet (1978), and Lekagul and McNeely (1988) believed *lutreolina* and *sibirica* conspecific; however, this has not been supported by primary studies (Brongersma, 1940; Van Bree and Boeadi, 1978). Abramov (1999) and Kurose et al. (2000) considered *itatsi* and *sibirica* separate. Synonyms allocated according to Ellerman and Morrison-Scott (1951).

*Mustela strigidorsa* Gray, 1853. Proc. Zool. Soc. Lond., 1853:191.
    COMMON NAME: Back-striped Weasel.
    TYPE LOCALITY: Not given. Gray (1853) based the type description on a manuscript given to him by Hodgson. Horsfield (1855) later fixed the type locality as "the Sikim Hills of Tarai." [India, Sikkim].
    DISTRIBUTION: Burma, China (Yunnan, Guizhou), India, Laos, Nepal, Thailand, Vietnam.
    STATUS: IUCN – Vulnerable.
    COMMENTS: Youngman (1982) suggested that *strigidorsa* belonged in subgenus *Lutreola*. Abramov (1999) placed it in subgenus *Cryptomustela*.

*Mustela subpalmata* Hemprich and Ehrenberg, 1833. Symb. Phys. Icon., Mamm. 3(2). *In* "Herpestes leucurus"; k verso.
    COMMON NAME: Egyptian Weasel.
    TYPE LOCALITY: "In domibus aegyptiacis Cahirae et Alexandriae murium vulgaris socius"
    DISTRIBUTION: Egypt.
    COMMENTS: Recognized as a separate species by van Zyll de Jong (1992), Reig (1997), and Abramov and Baryshnikov (1999).

*Neovison* Baryshnikov and Abramov, 1997. Zool. Zhurnal, 76(12):1408.
    TYPE SPECIES: *Mustela vison* Schreber, 1777.
    COMMENTS: Commonly included in *Mustela*, separated accordingly to Abramov (1999). There are significant differences between the American mink and *Mustela* (and other Mustelidae) according to cytogenetic and biochemical data. The level of these differences is higher then differences among *Mustela* species (Belyaev et al., 1980; Brinck et al., 1983; Graphodatsky et al., 1976; Kurose et al., 2000; Lushnikova et al., 1989; Taranin et al., 1991). The analyses by Graphodatsky et al. (1976), Youngman (1982), and Kurose et al.

(2000) support significant divergence of *vison* from the *Mustela* lineage. Masuda and Yoshida (1994*a*) argued that inclusion of *vison* in *Mustela* would make *Mustela* paraphyletic.

*Neovison macrodon* (Prentis, 1903). Proc. U.S. Natl. Mus. 26:887.
common name: Sea mink.
TYPE LOCALITY: "Brooklin, Handcock County, Maine." [Brooklin archaeological site on Black Isl, Maine].
DISTRIBUTION: Formerly found along the coasts of Canada (New Brunswick) and USA (Maine).
STATUS: IUCN – Extinct as *Mustela macrodon*.
SYNONYMS: *antiquus* (Loomis, 1911)
COMMENTS: Last collected in 1894. Manville (1966) argued that recently extinct form *macrodon* is conspecific with *vison*, although Kurtén and Anderson (1980) and Mead et al. (2000) recognized it as a distinct species.

*Neovison vison* (Schreber, 1777). Die Säugethiere, 3(19):pl. 127.B [1777]; text, 3(26):463 [1777].
COMMON NAME: American Mink.
TYPE LOCALITY: "Man findet das Vison in Canada un Pensilvanien". Larivière (1999*c*) listed type locality as "Eastern Canada".
DISTRIBUTION: Canada, USA (Alaska and through all of USA except SW deserts). Introduced to Belarus, Belgium, China, Czech Republic, Denmark, Estonia, Finland, France, Germany, Great Britian, Iceland, Ireland, Italy, Japan (Hokkaido), Latvia, Lithuania, Netherlands, Norway, Poland, Portugal, Russia, Spain, Sweden.
STATUS: IUCN – Lower Risk (lc) as *Mustela vision* (sic).
SYNONYMS: *altaica* (Ternovskii, 1958); *borealis* (Brass, 1911); *nigrescens* (Audubon and Bachman, 1854); *tatarica* (Popov, 1949); *winingus* (Baird, 1858); **aestuarina** (Ginnell, 1916); **aniakensis** (Burns, 1964); **energumenos** (Bangs, 1896); **evagor** (Hall, 1932); **evergladensis** (Hamilton, 1948); **ingens** (Osgood, 1900); **lacustris** (Preble, 1902); **letifera** (Hollister, 1913); **lowii** (Anderson, 1945); **lutensis** (Bangs, 1898); **melampeplus** (Elliot, 1904); **mink** (Peale and Palisot de Beauvois, 1796); *lutreocephala* (Harlan, 1825); *rufa* (Hamilton-Smith, 1858); **nesolestes** (Heller, 1909); *vulgivaga* (Bangs, 1895).
COMMENTS: Synonyms allocated according to Hall (1981) and Larivière (1999).

*Poecilogale* Thomas, 1883. Ann. Mag. Nat. Hist., ser. 5, 11:370.
TYPE SPECIES: *Zorilla albinucha* Gray, 1864, by monotypy (Melville and Smith, 1987).
SYNONYMS: *Zorilla* Gray, 1864.

*Poecilogale albinucha* (Gray, 1864). Proc. Zool. Soc. Lond., 1864:69, plate X.
COMMON NAME: African Striped Weasel.
TYPE LOCALITY: "it was without any habitat". Fixed by Coetzee (1977*b*) as "Cape Colony".
DISTRIBUTION: Angola, Botswana, Burundi, Dem. Rep. Congo, Kenya, Malawi, Mozambique, Namibia, Republic of Congo, Rwanda, South Africa, Tanzania, Uganda, Zambia, Zimbabwe.
STATUS: IUCN – Lower Risk (lc).
SYNONYMS: *africana* (Peters, 1865); *flavistriata* (Bocage, 1865); **bechuanae** Roberts, 1931; **doggetti** Thomas and Schwann, 1904; **lebombo** Roberts, 1931; **transvaalensis** Roberts, 1926.
COMMENTS: Reviewed by Larivière (2001). Synonyms allocated according to Ellerman et al. (1953).

*Taxidea* Waterhouse, 1839. Proc. Zool. Soc. Lond., 1838:154 [1839].
TYPE SPECIES: *Ursus meles labradorius* Gmelin, 1788 (= *Ursus taxus* Schreber, 1777), by original designation (Melville and Smith, 1987).

*Taxidea taxus* (Schreber, 1777). Säugethiere, 3(26):pl. 142[1778], text, 3(26):520[1777].
COMMON NAME: American Badger.
TYPE LOCALITY: "Er wohnt in Labrador und um die Hudsonsbay," restricted by Long (1972*a*), to "Carman, Manitoba." [Canada].

DISTRIBUTION: Canada (British Columbia, Alberta, Saskatschewan, Manitoba, Ontario), Mexico (Baja California N and C Mexico), USA (Illinois, Indiana, Ohio, Michigan, Wisconsin, and most states west of the Mississippi River, except Louisiana, Arkansas).

STATUS: IUCN – Lower Risk (lc).

SYNONYMS: *americanus* (Boddaert, 1784); *dacotensis* Schantz, 1946; *iowae* Schantz, 1947; *kansensis* Schantz, 1950; *labradorius* (Gmelin, 1788); *merriami* Schantz, 1950; **berlandieri** Baird, 1858; *apache* Schantz, 1948; *californica* Gray, 1865; *halli* Schantz, 1951; *hallorani* Schantz, 1949; *infusca* Thomas, 1898; *littoralis* Schantz, 1949; *nevadensis* Schantz, 1949; *papagoensis* Skinner, 1943; *phippsi* Figgins, 1918; *robusta* Hay, 1921; *sonoriensis* Goldman, 1939; **jacksoni** Schantz, 1946; **jeffersonii** (Harlan, 1825); *montana* Schantz, 1950; *neglecta* Mearns, 1891; *sulcata* Cope, 1878; **marylandica** Gidley and Gaxin, 1933.

COMMENTS: Reviewed by Long (1972a, 1973). Synonyms allocated according to Hall (1981) and Long (1973).

*Vormela* Blasius, 1884. Ber. Naturforsch Ges. Bemberg, 13:9.

TYPE SPECIES: *Mustela sarmatica* Pallas, 1771 (= *Mustela peregusna* Güldenstädt, 1770), by original designation (Melville and Smith, 1987).

*Vormela peregusna* (Güldenstädt, 1770). Nova Comm. Imp. Acad. Sci. Petropoli, 14(1):441.

COMMON NAME: Marbled Polecat.

TYPE LOCALITY: "habitat in campis apricis desertis Tanaicensibus" [U.S.S.R., Rostov Obl., steppes at lower Don River (Honacki et al., 1982)].

DISTRIBUTION: Afghanistan, Bulgaria, NC and W China, Greece, Iran, Kazakhstan, Macedonia, S Mongolia, Pakistan, Romania, Russia, Serbia, Syria, Tajikistan, Turkey, Turkmenistan, Ukraine, Uzbekistan.

STATUS: IUCN – Vulnerable as *V. p. peregusna*, otherwise Lower Risk (lc).

SYNONYMS: *euxina* Pocock, 1936; *intermedia* Ognev, 1935; *sarmatica* (Pallas, 1771); **koshewnikowi** Satunin, 1910; *alpherakii* Birula, 1910; *obscura* Stroganov, 1948; *tedshenika* Satunin, 1910; **negans** Miller, 1910; *chinensis* Stroganov, 1962; **pallidior** Stroganov, 1948; *ornata* Pocock, 1936 [*nomen dubium*]; **syriaca** Pocock, 1936.

COMMENTS: Synonyms allocated according to Heptner et al. (1967).

**Family Mephitidae** Bonaparte, 1845. Cat. Meth. Mamm. Europe, p. 1.

SYNONYMS: Myadina Gray, 1825; Mydaina Gray, 1864.

COMMENTS: The traditional Mustelidae (including skunks) has always been a problematic group. Radinsky (1973) first proposed a relationship between the Mephitinae (*sensu* Simpson, 1945) and *Mydaus*. Morphological studies seem to provide some support for a monophyletic Mustelidae (Hunt, 1974; Wolsan, 1999; Wozencraft, 1989; Wyss and Flynn, 1993). However, Bryant et al. (1993), Ledje and Arnason (1996), and Bininda-Emonds et al. (1999) showed support for Radinsky's Mephitinae-*Mydaus* group. Dragoo and Honeycutt (1997) and Flynn et al. (2000) provided a thorough review of the relationships of these groups and followed others (Árnason and Widegren, 1986; Ledje and Árnason, 1996; Wayne et al., 1989) in the recognition of a paraphyletic traditional Mustelidae. I have chosen to follow Dragoo and Honeycutt (1997) and Flynn et al. (2000) in the separation of *Mephitis*, *Conepatus*, *Spilogale* and *Mydaus* from the remaining mustelids. But see Wolsan (1999) for support of a monophyletic Mustelidae.

*Conepatus* Gray, 1837. Mag. Nat. Hist. [Charlesworth's], 1:581.

TYPE SPECIES: *Conepatus humboldtii* Gray, 1837, by monotypy (Melville and Smith, 1987).

SYNONYMS: *Lycodon* d'Orbigny (*In* Gray 1865); *Mamconepatus* Herrera, 1899; *Marputius* Gray, 1837; *Oryctogale* Merriam, 1902; *Ozolictus* Gloger, 1842 (*In* Gray, 1865); *Thiosmus* Lichtenstein 1838.

COMMENTS: Revised by Kipp (1965), who studied an extensive series of southern South American specimens and could not recognize distinctive groups among them based on skull morphology, and found only two groups based on pelage coloration. Dragoo et al. (2003) reviewed the North American species and is followed here.

*Conepatus chinga* (Molina, 1782). Sagg. Stor. Nat. Chile, p. 288.
COMMON NAME: Molina's Hog-nosed Skunk.
TYPE LOCALITY: "Chili," restricted by Cabrera (1957) to "alrededores de Valparaíso." [Chile].
DISTRIBUTION: N Argentina, Bolivia, S Brazil, Chile, Peru, Uruguay.
STATUS: IUCN – Lower Risk (lc).
SYNONYMS: *americana* (Desmarest, 1818); *chilensis* (É. Geoffroy Saint-Hilaire, 1803); *chinensis* (Gerrard, 1862); *chingha* (Molina, 1786); *chinghe* (Bechstein, 1800); *dimidiata* (G. Fischer, 1814); *furcata* (Wagner, 1841); *molinae* (Lichtenstein, 1838); **budini** Thomas, 1919; *calurus* Thomas, 1919; *mendosus* Yepes, 1937; **gibsoni** Thomas, 1910; *pampanus* Thomas, 1921; *suffocans* (Burmeister, 1879) [preoccupied]; **inca** Thomas, 1900; *mapurito* (Tschundi, 1844) [preoccupied]; **mendosus** Thomas, 1921; *enuchus* Thomas, 1927; **rex** Thomas, 1898; *ajax* Thomas, 1913; *arequipae* Thomas, 1900; *chorensis* Thomas, 1902; *hunti* Thomas, 1903; *porcinus* Thomas, 1902; **suffocans** (Illiger, 1811); *americana* (Desmarest, 1820); *feuillei* (Gervais, 1841); *feuillei* Trouessart, 1897; *monzoni* Aplin, 1894; *vittata* (Larrañaga, 1923).
COMMENTS: Kipp (1965) considered *rex* as conspecific with *chinga*; however, it was listed as separate by Osgood (1943) and Cabrera (1957). Synonyms allocated according to Cabrera (1958).

*Conepatus humboldtii* Gray, 1837. Mag. Nat. Hist. [Charlesworth's], 1:581.
COMMON NAME: Humboldt's Hog-nosed Skunk.
TYPE LOCALITY: "Magellan Straits." [Chile].
DISTRIBUTION: Argentina, Paraguay.
STATUS: CITES – Appendix II; IUCN – Lower Risk (lc).
SYNONYMS: *chinga* Wolffsohn and Porter, 1908 [preoccupied]; *conepatl* Gmelin (*In* Gray, 1837) [preoccupied]; *mapurito* (Humbolt, *In* Coues, 1877) [preoccupied]; *patachonica* (Burmeister, 1869); *patagonica* (Lichtenstein, 1838); *suffocans* J. A. Allen, 1916 [preoccupied]; *westermannii* (Reinhardt, 1865) [preoccupied]; **castaneus** (d'Orbigny and Gervais, 1847); *gaucho* Thomas, 1927; *humboldtii* (d'Orbigny, 1838) [preoccupied]; **proteus** Thomas, 1902.
COMMENTS: Kipp (1965) considered *castaneus* as conspecific; Cabrera (1958) considered it separate. Synonyms allocated according to Cabrera (1958).

*Conepatus leuconotus* (Lichtenstein, 1832). Darst. Säugeth., text: "*Mephitis leuconota*" [not paginated], pl. 44. fig 1.
COMMON NAME: American Hog-nosed Skunk.
TYPE LOCALITY: "oberen Lauf des Rio Alvarado" [Mexico, Veracruz, Rio Alvarado].
DISTRIBUTION: Guatemala, Honduras, Mexico (from USA south along coast to Veracruz), Nicaragua, USA (S Gulf coast of Texas).
STATUS: IUCN – Extinct as *C. mesoleucus telmalestes*, Lower Risk (lc) as *C. mesoleucus* and *C. leuconotus*.
SYNONYMS: *chinga* (Molina, 1865); *filipensis* Merriam, 1902; *intermedia* (Saussure, 1861); *longicaudata* (Tomes, 1861); *marputio* (Gray, 1865); *mearnsi* Merriam, 1902; *mesoleucus* (Lesson, 1865); *molinae* (Lichenstein, 1865); *nasuta* (Bennett, 1833); *nelsoni* Goldman, 1922; *nicaraguae* J. A. Allen, 1910; *nicaraguus* Goodwin, 1946; *pediculus* Merriam, 1902; *putorius* (Mutis, 1865); *sonoriensis* Merriam, 1902; *texensis* Merriam, 1902; *venaticus* Goldman, 1922; **figginsi** F. W. Miller, 1925; *fremonti* F. W. Miller, 1933; **telmalestes** Bailey, 1905.
COMMENTS: Includes *mesoleucus* (Dragoo et al., 2003; Hall, 1981).

*Conepatus semistriatus* (Boddaert, 1785). Elench. Anim., 1:84.
COMMON NAME: Striped Hog-nosed Skunk.
TYPE LOCALITY: "Mexico"; Cabrera (1958) listed the type locality as "Minas de Montuosa, cerca de Pamplona, departamento del norte de Santander, Colombia".
DISTRIBUTION: Belize, Brazil, Guatemala, Honduras, Nicaragua, Mexico (Veracruz, Tabasco, and Yucatan), Peru.
STATUS: IUCN – Lower Risk (lc).
SYNONYMS: *gumillae* (Lichtenstein, 1838); *gumillaei* (Boitard, 1842); *mapurito* (Gmelin, 1788); *putorius* (Mutis, 1770) [preoccupied]; *semistriata* (Boddaert, 1785); *zorilla* (J. B. Fischer,

1829); *amazonicus* (Lichtenstein, 1838); *bahiensis* Ihering, 1911; *childensis* (Hensel, 1872); *chilensis* Gray, 1865; *conepatl* (Gmelin, 1788); *lichtensteinii* Gray, 1865; *quitensis* (Humboldt, 1812); *suffocans* (Winge, 1876); *tropicalis* Merriam, 1902; *westermanni* (Reinhardt, 1856); ***taxinus*** Thomas, 1924; *amazonica* (Tschundi, 1844) [preoccupied]; ***trichurus*** Thomas, 1905; *mapurito* Bangs, 1902 [preoccupied]; ***yucatanicus*** Goldman, 1943; ***zorrino*** Thomas, 1901; *zorilla* Thomas, 1900 [preoccupied].
COMMENTS: Synonyms allocated according to Cabrera (1957).

*Mephitis* É. Geoffroy Saint-Hilaire and F. G. Cuvier, 1795. Mag. Encyclop., 2:187.
TYPE SPECIES: *Viverra mephitis* Schreber, 1776.
SYNONYMS: *Chincha* Lesson, 1842; *Leucomitra* Howell, 1901; *Mammephitisus* Herrera, 1899; *Mephites* Gray, 1847; *Mephritis* Gray, 1821; *Spilogale* Gray, 1865; *Viverra* Schreber, 1776.

*Mephitis macroura* Lichtenstein, 1832. Darst. Säugeth. text: "Mephitis macroura," [not paginated], pl. 46.
COMMON NAME: Hooded Skunk.
TYPE LOCALITY: "Gebirgs-Gegenden nordwestlich von der Stadt Mexico." [Mexico, mountains NW of Mexico City].
DISTRIBUTION: Costa Rica, El Salvador, Guatemala, Honduras, Mexico, Nicaragua, USA (S Arizona, S New Mexico, and W Texas).
STATUS: IUCN – Lower Risk (lc).
SYNONYMS: *concolor* Gray, 1865; *edulis* Coues, 1877; *intermedia* Gray, 1869; *longicaudata* Tomes, 1862; *mexicana* Gray, 1837; *vittata* Lichtenstein, 1832; ***eximius*** Hall and Dalquest, 1950; ***milleri*** Mearns, 1897; *richardsoni* Goodwin, 1957.
COMMENTS: Reviewed by Hwang and Larivière (2001). Synonyms allocated according to Hall (1981) and Hwang and Larivière (2001).

*Mephitis mephitis* (Schreber. 1776). Die Säugethiere, 3(17):pl. 121[1776], text, 3(26):444, 588 (index)[1777].
COMMON NAME: Striped Skunk.
TYPE LOCALITY: "Amerika".
DISTRIBUTION: Canada (SW Northwest Territories to Hudson Bay and S Quebec), Mexico (N Tamaulipas, N Durango, and N Baja California), USA.
STATUS: IUCN – Lower Risk (lc).
SYNONYMS: *americana* Desmarest, 1818; *chinche* Fischer, 1829; *mephitica* Saw, 1792; *vulgaris* F. Cuvier, 1842; ***avia*** Bangs, 1898; *newtonensis* Brown, 1908; ***elongata*** Bangs, 1895; ***estor*** Merriam, 1890; ***holzneri*** Mearns, 1898; ***hudsonica*** Richardson, 1829; *americana* (Lesson, 1865); *chinga* Tiedemann, 1808 [preoccupied]; *minnesotoe* Brass, 1911; ***major*** Howell, 1901; ***mesomelas*** Lichtenstein, 1832; *mesomeles* Gerrard, 1862; *scrutator* Bangs, 1896; ***nigra*** (Peale and Palisot de Beauvois, 1796); *bivirgata* C. E. H. Smith, 1839; *dentata* Brass, 1911; *fetidissima* Boitard, 1842; *frontata* Coues, 1875; *olida* Boitard, 1842; *putida* Boitard, 1842; ***notata*** Hall, 1936; ***occidentalis*** Baird, 1858; *notata* Howell, 1901; *platyrhina* (Howell, 1901); ***spissigrada*** Bangs, 1898; *foetulenta* Elliot, 1899; ***varians*** Gray, 1837; *texana* Low, 1879.
COMMENTS: Synonyms allocated according to Hall (1981) and Wade-Smith and Verts (1982).

*Mydaus* F. G. Cuvier, 1821. *In* É. Geoffroy Saint-Hilaire and F. G. Cuvier, Hist. Nat. Mammifères, pt. 2, 3(27), "Telagon", 2 pp., 1 pl.
TYPE SPECIES: *Mydaus meliceps* F. G. Cuvier, 1821 (= *Mephitis javanensis* Desmarest, 1820).
SYNONYMS: *Mephitis* Desmarest, 1820; *Mydaon* Gloger, 1865; *Suillotaxus* Lawrence, 1939.
COMMENTS: Lawrence (1939) believed that differences in dentition and pelage warranted separation of these taxa into separate, monotypic genera (i.e., *Suillotaxus marchei*, *Mydaus javanensis*). These differences parallel those found within the genus *Melogale* (Long and Killingley, 1983). This genus is provisionally placed in this family (see comments therein), its position may be better considered as *incertae sedis*.

*Mydaus javanensis* (Desmarest, 1820). Mammalogie, In Encycl. Meth., I:187.
COMMON NAME: Sunda Stink Badger.
TYPE LOCALITY: "l'île de Java." [Indonesia, Java].

DISTRIBUTION: Indonesia (Java, Borneo, Sumatra and the Natuna Isls) and Malaysia (Sabah, Sarawak).

STATUS: IUCN – Lower Risk (lc).

SYNONYMS: *foetidus* (Gray, 1865); *meliceps* (Cuvier, 1821); **lucifer** Thomas, 1902; *luciferoides* Lönnberg and Mjöberg, 1925; *montanus* Moulton, 1921; **ollula** Thomas, 1902.

COMMENTS: Synonyms allocated according to Chasen (1940) and Hwang and Larivière (2003).

*Mydaus marchei* (Huet, 1887). Le Naturaliste, ser. 2, 9(13):149-151.

COMMON NAME: Palawan Stink Badger.

TYPE LOCALITY: "l'ile Palaouan" [Philippine Isls, Palawan].

DISTRIBUTION: Philippine Isls (Palawan and Calamian Isls).

STATUS: IUCN – Vulnerable.

SYNONYMS: *schadenbergii* (Jentink, 1895).

COMMENTS: Referred to the genus *Suillotaxus* by Lawrence (1939). *Suillotaxus* was considered a subgenus of *Mydaus* by Long (1978, 1981).

*Spilogale* Gray, 1865. Proc. Zool. Soc. Lond., 1865:150.

TYPE SPECIES: *Mephitis interrupta* Rafinesque, 1820 (= [*Viverra*] *putorius* Linnaeus, 1758).

COMMENTS: Mead (1968) argued that *S. p. gracilis* and "possibly" *leucoparia* are reproductively isolated from eastern populations and therefore should be considered distinct species. Preliminary genetic data (Dragoo et al., 1993) support Mead (1968). However, both taxa were included by Van Gelder (1959). Kinlaw (1995) restricted *putorius* to the eastern spotted skunk. Owen et al. (1996) provided karyotypic data to support recognition of the southern spotted skunk, *angustifrons*, as a distinct species. Verts et al. (2001) reported information on subspecies of *gracilis* which included western populations of spotted skunks. They did not include taxa regarded by Hall and Kelson (1959) as *pygmaea* or *angustifrons*. Synonyms allocated according to Hall and Kelson (1959), Kinlaw (1995), and Verts et al. (2001).

*Spilogale angustifrons* Howell, 1902. Proc. Biol. Soc. Wash., 15:242.

COMMON NAME: Southern Spotted Skunk.

TYPE LOCALITY: "Tlalpam, Valley of Mexico".

DISTRIBUTION: Belize, Costa Rica, El Salvador, Guatemala, Honduras, C Mexico, Nicaragua.

SYNONYMS: **celeris** Hall, 1938; **elata** Howell, 1906; **tropicalis** Howell, 1902; **yucatanensis** Burt, 1938.

*Spilogale gracilis* Merriam, 1890. N. Amer. Fauna, 3:83.

COMMON NAME: Western Spotted Skunk.

TYPE LOCALITY: "Grand Canon of the Colorado (altitude 3,500 feet), [Coconino County, 1067 m] Arizona, north of San Francisco Mountain."

DISTRIBUTION: Mexico (central plateau); USA (from the Puget Sound region in the west to an eastern boundary in Montana, Wyoming, Colorado, W Oklahoma and W Texas).

SYNONYMS: *saxatilis* Merriam, 1890; *tenuis* Howell, 1902; **amphialus** Dickey, 1929; **latifrons** Merriam, 1890; *olympica* Elliot, 1899; **leucoparia** Merriam, 1890; *ambigua* Mearns, 1897; *arizonae* Mearns, 1891; *texensis* Merriam, 1890; **lucasana** Merriam, 1890; **martirensis** Elliot, 1903; *microdon* Howell, 1906; **phenax** Merriam, 1890; *microrhina* Hall, 1926; *zorrilla* (Lichtenstein, 1838).

COMMENTS: Reviewed by Verts et al. (2001).

*Spilogale putorius* (Linnaeus, 1758). Syst. Nat., 10th ed., 1:44.

COMMON NAME: Eastern Spotted Skunk.

TYPE LOCALITY: "America septentrionali" restricted by Thomas (1911*a*), to "South Carolina." [USA].

DISTRIBUTION: USA (Florida N to Kentucky and W Virginia, W through the Great Plains, from Texas to North Dakota and Minnesota).

STATUS: IUCN – Lower Risk (lc).

SYNONYMS: *bicolor* (J. E. Gray, 1837); *mapurita* (P. L. S. Müller, 1776); *putida* (F. G. Cuvier, 1798); *ringens* Merriam, 1890; *striata* (Shaw, 1800); *zorilla* (Schreber, 1776); **ambarvalis**

Bangs, 1898; *interrupta* (Rafinesque, 1820); *indianola* Merriam, 1890; *quaterlinearis* (Winans, 1859).
COMMENTS: Reviewed by Kinlaw (1995).

*Spilogale pygmaea* Thomas, 1898. Proc. Zool. Soc. Lond., 1897:898 [1898].
COMMON NAME: Pygmy Spotted Skunk.
TYPE LOCALITY: "Rosario, Sinaloa, W. Mexico."
DISTRIBUTION: Mexico (West coastal regions from Sinaloa to Oaxaca).
STATUS: IUCN – Lower Risk (lc).
SYNONYMS: *australis* Hall, 1938; *albipes* Goodwin, 1956; *intermedia* López-Forment and Urbano, 1979.
COMMENTS: Reviewed by Medellín et al. (1998a), from which synonyms are allocated. Ewer (1973) argued that *pygmaea* is conspecific with *putorius*.

**Family Procyonidae** Gray, 1825. Ann. Philos., n.s., 10:339.
SYNONYMS: Bassaricyonidae Coues, 1887; Bassaridae Gray, 1869; Bassariscidae Gray, 1869; Cercoleptidae Bonaparte, 1838; Nasuidae Gray, 1869; Potidae Degland, 1854; Potosinae Trouessart, 1904.
COMMENTS: Revised by Hollister (1915a), Pocock (1921a), Baskin (1982, 1989), and Decker and Wozencraft (1991). Does not include *Ailurus* or *Ailuropoda*, following Davis (1964), Todd and Pressman (1968), Sarich (1976), Ginsburg (1982), Wozencraft (1989), Decker and Wozencraft (1991), Bininda-Emonds et al., (1999), and Flynn et al. (2000). However, Hollister (1915a), Gregory (1936), and Thenius (1979) considered *Ailurus* in the Procyonidae.

*Bassaricyon* J. A. Allen, 1876. Proc. Acad. Nat. Sci. Philadelphia, 28:20, pl. 1.
TYPE SPECIES: *Bassaricyon gabbi* J. A. Allen, 1876, by designation.
COMMENTS: Reviewed by Poglayen-Neuwall (1965). Several workers have suggested that the several named forms of *Bassaricyon* are conspecific (Decker and Wozencraft, 1991; Ewer, 1973; Hall and Kelson, 1959; Stains, 1967; Wozencraft, 1989), but supporting systematic work is lacking.

*Bassaricyon alleni* Thomas, 1880. Proc. Zool. Soc. Lond., 1880:397.
COMMON NAME: Allen's Olingo.
TYPE LOCALITY: "Sarayacu, on the Bobonasa river, Upper Pastasa river" [Ecuador].
DISTRIBUTION: Bolivia, Ecuador (east of the Andes), Peru (to Cuzco Prov.), Venezuela (?).
STATUS: IUCN – Lower Risk (lc).

*Bassaricyon beddardi* Pocock, 1921. Ann. Mag. Nat. Hist., ser. 9, 7:231.
COMMON NAME: Beddard's Olingo.
TYPE LOCALITY: "Bastrica woods, Essequibo River, British Guiana".
DISTRIBUTION: Guyana, and possibly adjacent Venezuela and Brasil.
STATUS: IUCN – Lower Risk (nt).
SYNONYMS: *alleni* Sclater, 1895 [preoccupied].
COMMENTS: Cabrera (1957) erroneously listed *Bassaricyon beddardi* as *Bassariscus beddardi*.

*Bassaricyon gabbii* J. A. Allen, 1876. Proc. Acad. Nat. Sci. Philad., 28:21.
COMMON NAME: Olingo.
TYPE LOCALITY: "Costa Rica" restricted by Allen (1908) to "Talamanca".
DISTRIBUTION: W Colombia, Costa Rica, W Ecuador, C Nicaragua, Panama.
STATUS: CITES – Appendix III (Costa Rica).
STATUS: IUCN – Lower Risk (nt).
SYNONYMS: *medius* Thomas, 1909; *orinomus* Goldman, 1912; *richardsoni* J. A. Allen, 1908; *siccatus* Thomas, 1927.
COMMENTS: Synonyms allocated according to Cabrera (1957) and Hall (1981).

*Bassaricyon lasius* Harris, 1932. Occas. Pap. Mus. Zool., Univ. Michigan, 248:3.
COMMON NAME: Harris's Olingo.
TYPE LOCALITY: "Estrella de Cartago, Costa Rica. This locality is six to eight miles [10 to 13 km] south of Cartago near the source of the Rio Estrella, at an altitude of

about 4,500 feet [1372 m]."
DISTRIBUTION: Costa Rica (known only from the type locality).
STATUS: IUCN – Endangered.

*Bassaricyon pauli* Enders, 1936. Proc. Acad. Nat. Sci. Philadelphia, 88:365.
COMMON NAME: Chiriqui Olingo.
TYPE LOCALITY: "Between Rio Chiriqui Viejo and Rio Colorado, on a hill known locally as Cerro Pando, elevation 4800 feet [1463 m], about ten miles [16 km] from El Volcan, Province de Chiriqui, R. de Panama."
DISTRIBUTION: Panama (known only from the type locality).
STATUS: IUCN – Endangered.

*Bassariscus* Coues, 1887. Science, 9:516.
TYPE SPECIES: *Bassariscus astutus*, by monotypy through the replaced name *Bassaris astuta* Lichtenstein, 1830 (Melville and Smith, 1987).
SYNONYMS: *Bassaris* Lichtenstein, 1830; *Jentinkia* Trouessart, 1904; *Mambassariscus* Herrera, 1899; *Wagneria* Jentink, 1886.
COMMENTS: Hollister (1915a) placed in monotypic Bassariscidae.

*Bassariscus astutus* (Lichtenstein, 1830). Abh. König. Akad. Wiss., Berlin, 1827:119 [1830].
COMMON NAME: Ringtail.
TYPE LOCALITY: "Mexico" [near city of Mexico].
DISTRIBUTION: Mexico (from USA to the Isthmus of Tehuantepec, Tiburon Isl and several other islands in the Gulf of California), USA (SW Oregon, N Nevada, Utah, SW Wyoming and W Colorado, south through California, Arizona, New Mexico and Texas).
STATUS: IUCN – Lower Risk (lc).
SYNONYMS: *albipes* Elliot, 1904; *arizonensis* Goldman, 1932; *bolei* Goldman, 1945; *consitus* Nelson and Goldman, 1932; *flavus* Rhoads, 1893; *insulicola* Nelson and Goldman, 1909; *macdougalli* Goodwin, 1956; *nevadensis* Miller, 1913; *octavus* Hall, 1926; *palmarius* Nelson and Goldman, 1909; *raptor* (Baird, 1859); *oregonus* Rhoads, 1894; *saxicola* Merriam, 1897; *willetti* Stager, 1950; *yumanensis* Huey, 1937.
COMMENTS: Revised by Rhoads (1893). Reviewed by Poglayen-Neuwall and Toweill (1988). Synonyms allocated according to Hall (1981) and Poglayen-Neuwall and Toweill (1988).

*Bassariscus sumichrasti* (Saussure, 1860). Rev. Mag. Zool. Paris, ser. 2, 12:7.
COMMON NAME: Cacomistle.
TYPE LOCALITY: "Cet animal habite les greniers dans la région chaude du Mexique." Hall and Kelson (1959) listed "Mexico, Veracruz, Mirador."
DISTRIBUTION: Belize, Costa Rica, El Salvador, Guatemala, Mexico (Guerrero and S), Nicaragua, W Panama.
STATUS: CITES – Appendix III (Costa Rica); IUCN – Lower Risk (nt).
SYNONYMS: *campechensis* (Nelson and Goldman, 1932); *monticola* (Cordero, 1875); *latrans* (Davis and Lukens, 1958); *notinus* Thomas, 1903; *oaxacensis* (Goodwin, 1956); *variabilis* (Peters, 1874).
COMMENTS: Synonyms allocated according to Hall (1981).

*Nasua* Storr, 1780. Prodr. Meth. Mamm., p. 35, tabl. A.
TYPE SPECIES: *Viverra nasua* Linnaeus, 1766, by absolute tautomy (Melville and Smith, 1987).
SYNONYMS: *Coati* Lacépède, 1799; *Mamnasuaus* Herrera, 1899; *Nasica* South, 1845.
COMMENTS: Revised by Allen (1879) and Decker (1991). Reviewed by Cabrera (1957).

*Nasua narica* (Linnaeus, 1766). Syst. Nat., 12th ed., 1:64.
COMMON NAME: White-nosed Coati.
TYPE LOCALITY: "America", restricted by Allen (1879) to "Veracruz, Mexico"; Hershkovitz (1951) further restricted it to "Achotal, Isthmus of Techuantpec, Vera Cruz."
DISTRIBUTION: Belize, Colombia (Gulf of Uraba), Costa Rica, El Salvador, Guatemala, Honduras, Mexico (except Baja California), Nicaragua, Panama, USA (S Arizona and

SW New Mexico).

STATUS: CITES – Appendix III (Honduras) as *Nasua narica*; IUCN – Endangered as *N. nelsoni*, Lower Risk (lc) as *N. narica*.

SYNONYMS: *bullata* J. A. Allen, 1904; *isthmica* Goldman, 1942; *mexicana* Weinland, 1860; *nasica* Winge, 1895; *panamensis* J. A. Allen, 1904; *richmondi* Goldman, 1932; *subfusca* Tiedemann, 1808; *vulpecula* (Erxleben, 1777); **molaris** Merriam, 1902; *pallida* J. A. Allen, 1904; *tamaulipensis* Goldman, 1942; **nelsoni** Merriam, 1901; *thersites* Thomas, 1901; **yucatanica** J. A. Allen, 1904.

COMMENTS: Includes *nelsoni* (Decker, 1991). Reviewed by Gompper (1995), from which synonyms are allocated.

*Nasua nasua* (Linnaeus, 1766). Syst. Nat., 12th ed., 1:64.

COMMON NAME: South American Coati.

TYPE LOCALITY: "America"; listed by Cabrera (1957) as "Pernambuco" [Brazil].

DISTRIBUTION: Argentina, Bolivia, Brazil, Colombia, Guyana, Paraguay, Peru, Surinam, Uruguay, Venezuela.

STATUS: CITES – Appendix III as *N. n. solitaria* (Uruguay); IUCN – Lower Risk (lc).

SYNONYMS: *annulata* (Desmarest, 1920); *fusca* Desmarest, 1820; *mexiana* Hagmann, 1908; *mexianae* Vieira, 1945; *nasua* (Cuvier, 1798); *quasje* (Gmelin, 1788); *rufa* Desmarest, 1820; *socialis* J. B. Fischer, 1829; *striata* (Shaw, 1800); *vulgaris* F. G. Cuvier, 1842; *vulpecula* (Erxleben, 1777); **aricana** Vieira, 1945; **boliviensis** Cabrera, 1956; **candace** Thomas, 1912; **judex** Thomas, 1914; **cinerascens** Lönnberg, 1921; **dorsalis** Gray, 1866; *jivaro* Thomas, 1914; *juruana* Ihering, 1911; *masua* Lönnberg, 1921; *mephisto* Thomas, 1927; *soederstroemmi* Lönnberg, 1921; **manium** Thomas, 1912; *gualeae* Lönnberg, 1921; **molaris** Merriam, 1902; *pallida* J. A. Allen, 1904; *tamaulipensis* Goldman, 1942; **montana** Tschundi, 1844; *monticola* Schinz, 1844; **quichua** Thomas, 1901; **solitaria** Schinz, 1823; *fulva* Wagner, 1841; *fusca* Desmarest, 1820 [*nomen nudum*]; *henseli* Lönnberg, 1921; *rufa* J. A. Allen, 1875; *sociabilis* Schinz, 1823; *socialis* Wied-Neuwied, 1826; **spadicea** Olfers, 1818; **vittata** Tschudi, 1844; *dichromatica* Tate, 1939; *phaeocephala* J. A. Allen, 1904.

COMMENTS: Reviewed by Gompper and Decker (1998). Synonyms allocated according to Cabrera (1957) and Hall (1981).

*Nasuella* Hollister, 1915. Proc. U.S. Natl. Mus., 49:148.

TYPE SPECIES: *Nasua olivacea meridensis* Thomas, 1901, by designation.

*Nasuella olivacea* (Gray, 1865). Proc. Zool. Soc. Lond., 1865:703.

COMMON NAME: Mountain Coati.

TYPE LOCALITY: "Santa Fé de Bogota" [Colombia], subsequently restricted by Cabrera (1957: 249) to "Bogotá, lo que debe interpretarse como las montañas próximas a esta capital"

DISTRIBUTION: Colombia, Ecuador, W Venezuela.

STATUS: IUCN – Data Deficient.

SYNONYMS: *lagunetae* (J. A. Allen, 1913); **meridensis** (Thomas, 1901); **quitensis** (Lönnberg, 1913).

COMMENTS: Synonyms allocated according to Cabrera (1957).

*Potos* É. Geoffroy Saint-Hilaire and F. G. Cuvier, 1795. Mag. Encyclop., 2:187.

TYPE SPECIES: *Viverra caudivolvula* Schreber, 1777 (= *Lemur flavus* Schreber, 1774), by original designation.

SYNONYMS: *Cercoleptes* Illiger, 1811; *Kinkajou* Lacépède, 1799; *Kinkaschu* G. Fischer de Waldheim, 1813; *Mamcercolepteus* Herrera, 1899.

COMMENTS: Hernández-Camacho (1977) placed *Potos* in Cercoleptidae Bonaparte, 1838.

*Potos flavus* (Schreber, 1774). Die Säugethiere 1(9):pl. 42[1774]; text, p. 187[189](index)[1774].

COMMON NAME: Kinkajou.

TYPE LOCALITY: "Er ist, der Sage nach, auf den Gebirgen in Jamaica einheimisch"; restricted by Thomas (1902*b*) to "Surinam". Ford and Hoffmann (1988) and Husson (1978) discussed the confusion over the name and type locality.

DISTRIBUTION: Belize, Bolivia, Brazil (Mato Grosso), Colombia, Costa Rica, Ecuador,

Guatemala, Guyana, Mexico (S Tamaulipas and Guerrero and possibly Michoacan), Nicaragua, Panama, Peru, Surinam, Venezuela.

STATUS: CITES – Appendix III (Hondurus); IUCN – Lower Risk (lc).

SYNONYMS: *brachyotos* (Schinz, 1844); *brachyotus* (Martin, 1836); *caudivolvula* (Schreber, 1777); *caudivolvulus* (Cuvier, 1798); *caudivolvulus* (Lacépède, 1799); *potto* (Müller, 1776); *simiasciurus* (Schreber, 1774); **chapadensis** J. A. Allen, 1904; *brasiliensis* Ihering, 1911; *caudivolvulus* (Pelzeln, 1883) [preoccupied]; *dugesii* Villa, 1944; **chiriquensis** J. A. Allen, 1904; *arborensis* Goodwin, 1938; *boothi* Goodwin, 1957; *campechensis* Nelson and Goldman, 1931; **megalotus** (Martin, 1836); *brachyotus* Trouessart, 1910; *caucensis* J. A. Allen, 1904; *isthmicus* Goldman, 1913; *mansuetus* Thomas, 1914; *modestus* Lönnberg, 1921; *tolimensis* J. A. Allen, 1913; **meridensis** Thomas, 1902; **modestus** Thomas, 1902; *caudivolvulus* (Thomas, 1880) [preoccupied]; **nocturnus** Wied-Neuwied, 1826; *aztecus* Thomas, 1902; *guerrerensis* Goldman, 1915; *prehensilis* (Kerr, 1792) [preoccupied].

COMMENTS: Reviewed by Cabrera (1957), Husson (1978), and Ford and Hoffmann (1988). Revised by Kortlucke (1973) and Hernández-Camacho (1977). Synonyms allocated according to Ford and Hoffmann (1988).

*Procyon* Storr, 1780. Prodr. Meth. Mamm., p. 35.

TYPE SPECIES: *Ursus lotor* Linnaeus, 1758, by designation by Elliot (1901).

COMMENTS: Reviewed by Goldman (1950) and Lotze and Anderson (1979). Hall (1981) listed *minor*, *gloveralleni*, *insularis*, *maynardi*, and *pygmaeus* as distinct, but gave no supporting rationale. Koopman et al. (1957) examined the type series of *maynardi* and showed them to be conspecific with *lotor*. Lotze and Anderson (1979) and Corbet and Hill (1986) have suggested that only *cancrivorus* and *lotor* are distinct, and other species are conspecific with *lotor*. Pons et al. (1999) showed that *minor* is conspecific with *lotor*, and Helgen and Wilson (2003) showed that *gloveralleni*, *maynardi*, and *minor* are introductions to the Caribbean from eastern United States.

*Procyon cancrivorus* (G.[Baron] Cuvier, 1798). Tabl. Elem. Hist. Nat. Anim., p. 113.

COMMON NAME: Crab-eating Raccoon.

TYPE LOCALITY: "se trouve à Cayenne" [French Guiana, Cayenne].

DISTRIBUTION: Argentina, Bolivia, Brazil, Colombia, Costa Rica, Guyana, Panama, Peru, Surinam, Trinidad and Tobago, Venezuela.

STATUS: IUCN – Lower Risk (lc).

SYNONYMS: **aequatorialis** J. A. Allen, 1915; **nigripes** Mivart, 1886; *brasiliensis* Ihering, 1911; **panamensis** (Goldman, 1913); *proteus* J. A. Allen, 1904 [preoccupied].

COMMENTS: Synonyms allocated according to Cabrera (1957).

*Procyon lotor* (Linnaeus, 1758). Syst. Nat., 10th ed., 1:48.

COMMON NAME: Raccoon.

TYPE LOCALITY: "Americae maritimis," restricted by Thomas (1911a) to "Pennsylvania" [USA].

DISTRIBUTION: S Canada, Mexico, Panama, USA (except parts of the Rocky Mtns).
Introductions into: Austria, Azerbaijan, Belarus, Czech Republic, Denmark, France, Germany, Russia, Switzerland, Uzbekistan.

STATUS: IUCN – Extinct as *P. gloveralleni*, Endangered as *P. insularis*, *P. maynardi*, and *P. minor*, Lower Risk (lc) as *P. lotor*.

SYNONYMS: *annulatus* G. Fischer, 1814; *brachyurus* Wiegmann, 1837; *fusca* Burmeister, 1850; *gularis* C. E. H. Smith, 1848; *melanus* J. E. Gray, 1864; *obscurus* Wiegmann, 1837; *rufescens* de Beaux, 1910; *vulgaris* (Tiedemann, 1808); **auspicatus** Nelson, 1930; **elucus** Bangs, 1898; **excelsus** Nelson and Goldman, 1930; **fuscipes** Mearns, 1914; **gloveralleni** Nelson and Goldman, 1930; *solutus* Nelson and Goldman, 1931; **grinnelli** Nelson and Goldman, 1930; **hernandezii** Wagler, 1831; *crassidens* Hollister, 1914; *dickeyi* Nelson and Goldman, 1931; *mexicana* Baird, 1858; *shufeldti* Nelson and Goldman, 1931; **hirtus** Nelson and Goldman, 1930; **incautus** Nelson, 1930; **inesperatus** Nelson, 1930; **insularis** Merriam, 1898; *vicinus* Nelson and Goldman, 1931; **litoreus** Nelson and Goldman, 1930; **marinus** Nelson, 1930; *maritimus* Dozier, 1948; **maynardi** Bangs,

1898; *flavidus* de Beaux, 1910; *minor* Miller, 1911; *varius* Nelson and Goldman, 1930; **megalodous** Lowery, 1943; **pacificus** Merriam, 1899; *proteus* Brass, 1911; **pallidus** Merriam, 1900; *ochraceus* Mearns, 1914; **psora** Gray, 1842; *californicus* Means, 1914; **pumilus** Miller, 1911; **simus** Gidley, 1906; **vancouverensis** Nelson and Goldman, 1930.

COMMENTS: Reviewed by Lotze and Anderson (1979). Includes the Caribbean introduced populations of *gloveralleni*, *minor*, and *maynardi* after Helgen and Wilson (2003); includes *insularis* after Helgen and Wilson (2005). Synonyms allocated according to Cabrera (1957), Lotze and Anderson (1979), and Helgen and Wilson (2003; 2005).

*Procyon pygmaeus* Merriam, 1901. Proc. Biol. Soc. Wash., 14:101.

COMMON NAME: Cozumel Raccoon.

TYPE LOCALITY: "Cozumel Island, Yucatan" [Mexico].

DISTRIBUTION: Known only from the type locality.

STATUS: IUCN – Endangered.

COMMENTS: Placed in *Procyon pygmaeus* according to Hall (1981), Lazell (1981), and Helgen and Wilson (In Press); although similar differences between Cozumel and mainland forms of *Nasua* were recognized at the subspecies level (Decker, 1991).

**Family Ailuridae** Gray, 1843. List Spec. Mamm. Coll. B.M. p. xxi.

SYNONYMS: Ailuridae Flower, 1869; Ailuridae Gray, 1869; Ailurina Gray, 1843; Ailurinae Trouessart, 1885.

COMMENTS: Biochemical and molecular evidence has suggested that the enigmatic *Ailurus* is either (1) intermediate between procyonids and ursids (O'Brien et al., 1985; Sarich, 1973; Tagle et al., 1986; Wayne et al., 1989; Wurster and Benirschke, 1968); (2) more closely related to ursids than to procyonids (Todd and Pressmann, 1968; Zhang and Shi, 1991a); (3) more closely related to procyonids than to ursids (Goldman et al., 1989; Pecon Slattery and O'Brien, 1995), or, finally; (4) more closely related to mephitids+procyonids (Bininda-Emonds et al., 1999; Flynn et al., 2000). Morphological studies have pointed out the lack of any shared derived features with the procyonids (Bugge, 1978; Decker and Wozencraft, 1991; Ginsburg, 1982; Hunt, 1974; Mayr, 1986; Schmidt-Kittler, 1981; Wozencraft, 1989a, b). Flynn et al. (1988) could not find any unambiguous features to place *Ailurus* with the procyonids, and only two characters to unite *Ailurus* with some procyonids, the loss of $M_3$ (shared also with mustelids) and the presence of a cusp on $P^4$ on some, but not all procyonids. Wozencraft (1989) found nine shared derived features with the Ursidae, and Decker and Wozencraft (1991) identified six unambiguous shared derived features of the procyonids, all of which are lacking in *Ailurus*. Flynn et al. (2000) showed strong support for a musteloid clade consisting of *Ailurus*, Mephitidae, Mustelidae, and Procyonidae. Synonyms allocated according to McKenna and Bell (1997).

*Ailurus* F. G. Cuvier, 1825. *In* É. Geoffroy Saint-Hilaire and F. G. Cuvier, Hist. Nat. Mammifères, pt. 3, 5(50), "Panda" 3 pp., 1 pl.

TYPE SPECIES: *Ailurus fulgens* F. G. Cuvier, 1825, by monotypy (Melville and Smith, 1987).

SYNONYMS: *Aelurus* Agassiz, 1846; *Aelurus* Flower, 1870; *Arctaelurus* Gloger, 1841.

COMMENTS: See comments under Family.

*Ailurus fulgens* F. G. Cuvier, 1825. *In* É. Geoffroy Saint-Hilaire and F. G. Cuvier, Hist. Nat. Mammifères, pt. 3, 5(50), "Panda" 3 pp., 1 pl.

COMMON NAME: Red Panda.

TYPE LOCALITY: "Indes orientales".

DISTRIBUTION: N Burma, China (Sichuan, Xizang, Yunnan. Recently extinct, or absent in Guizhou, Gansu, Shaanxi, and Qinghai; see Wei, et al., 1999), Nepal, Sikkim (India).

STATUS: CITES – Appendix I; IUCN – Endangered.

SYNONYMS: *ochraceus* Hodgson, 1847; **refulgens** Milne-Edwards, 1874; *styani* Thomas, 1902.

COMMENTS: Reviewed by Roberts and Gittleman (1984). Synonyms allocated according to Ellerman and Morrison-Scott (1951).

# ORDER PERISSODACTYLA
by Peter Grubb

**ORDER PERISSODACTYLA** Owen, 1848.

**Family Equidae** Gray, 1821. London Med. Repos., 15:307.

*Equus* Linnaeus, 1758. Syst. Nat., 10th ed., 1:73.
   TYPE SPECIES: *Equus caballus* Linnaeus, 1758.
   SYNONYMS: *Asinohippus* Trumler, 1961; *Asinus* Brisson, 1762 [unavailable]; *Asinus* Gray, 1824; *Caballus* Rafinesque, 1815; *Dolichohippus* Heller, 1912; *Grevya* Hilzheimer, 1912; *Hemionus* F. Cuvier, 1823 [unavailable]; *Hemionus* Stehlin and Graziosi, 1935; *Hemippus* Dietrich, 1959; *Hippotigris* C. H. Smith, 1841; *Ludolphozecora* Griffin, 1913; *Megacephalon* Hilzheimer, 1912; *Megacephalonella* Strand, 1943; *Microhippus* Matschie, 1924; *Onager* Brisson, 1762 [unavailable]; *Pseudoquagga* Hoffstetter, 1951; *Quagga* Shortridge, 1934; *Quaggoides* Willoughby, 1974; *Zebra* J. A. Allen, 1909.
   COMMENTS: Species-groups with potential subgeneric names, based on Groves and Willoughby (1981), are as follows: *E. asinus* or *Asinus* group; *E. caballus* or nominate *Equus* group (synonym: *Caballus*); *E. grevyi* or *Dolichohippus* group (synonyms: *Grevya, Ludolphozecora, Megacephalon,* and *Megacephalonella*); *E. hemionus* or *Hemionus* group (synonyms: *Asinohippus, Hemippus, Microhippus* and *Onager*), including also *E. kiang*; *E. quagga* or *Quagga* group (synonyms: *Pseudoquagga, Quaggoides,* and *Zebra*), including also *E. burchellii*; and *E. zebra* or *Hippotigris* group. Earliest generic name for the *E. quagga* group (*Zebra* J. A. Allen, 1909) is preoccupied and next available name (*Kraterohippus* van Hoepen, 1930) would have priority over *Quagga* Shortridge, 1934 if these nominal genera were treated as valid taxa.

*Equus asinus* Linnaeus, 1758. Syst. Nat., 10th ed., 1:73.
   COMMON NAME: Ass.
   TYPE LOCALITY: "Habitat in oriente" (= Middle East?).
   DISTRIBUTION: NE Sudan (now extinct), NE Ethiopia, and N Somalia; domesticated worldwide; feral or possibly wild in Oman, Hoggar (S Algeria), and Tibesti (N Chad); feral in Sudan, Saudi Arabia, Socotra Isl (Yemen), Sri Lanka, Australia, USA (including Hawaiian Isls), Galapagos Isls, Chagos Isls, and probably other oceanic islands.
   STATUS: CITES – Appendix I as *E. africanus*; U.S. ESA – Endangered as *E. asinus*; IUCN – Critically Endangered as *E. africanus africanus* and *E. a. somalicus*.
   SYNONYMS: *arabicus* (Fitzinger, 1860); *domesticus* Erxleben, 1777; *europaeus* Sanson, 1871; *germanicus* (Fitzinger, 1860); *grajus* (Fitzinger, 1860); *palaestinae* Ducos, 1968; *sardous* (Fitzinger, 1860); *vulgaris* (Gray, 1824); **africanus** Heuglin and Fitzinger, 1866; *africanus* (Fitzinger, 1858) [*nomen nudum*]; *dianae* (Dollman, 1935); *hippagrus* Schomber, 1963 [unavailable]; *nubianus* Peel, 1900; **somalicus** P. L. Sclater, 1885; *aethiopicus* Denman, 1957; *somaliensis* Noack, 1884 [*nomen nudum*]; *taeniopus* (Heuglin, 1861) [*nomen dubium*].
   COMMENTS: Revised by Groves et al. (1966) who with Ansell (1974a:6) recommended use of *africanus* as specific name, not wishing to use the name *asinus* because it was based upon domestic populations. Revised also by Schlawe (1980b) who indicated that *Asinus africanus* Fitzinger was named in 1858, not 1857, and by Groves (1986), who noted that *Asinus africanus* Fitzinger is a *nomen nudum*. The apparent senior name for wild asses is then *taeniopus* (Heuglin, 1861) but Groves (1986) regarded this as a *nomen dubium*. Gentry et al. (1996) proposed that majority usage be confirmed by adoption of the first available specific name based on a wild population for the wild taxon, in this case deemed to be *Equus africanus* Heuglin and Fitzinger, 1866, though it has not been demonstrated that most authors have termed the wild ass *E. africanus* rather than *E. asinus*. They asked the International Commission on Zoological Nomenclature to use its plenary power to rule that the name for the wild species is not invalid by virtue of being antedated by the name based on the domestic form. A ruling in favor of the proposal has now been made (International Commission on Zoological

Nomenclature, 2003). While it stipulates that *africanus* is not invalid, it does not
explicitly specify which species-group name is to be assigned to the whole species by
those who consider both domestic and wild populations to be conspecific (see Bock,
1997). Furthermore, the domestic ass may represent a subspecies *E. asinus asinus*, now
extinct in the wild and not synonymous with *E. a. africanus* (Groves et al., 1966;
Pocock, 1909a). Accordingly, *africanus* is here maintained as a subspecies of *E. asinus*.
The Linnean names *mulus* and *hinnus*, treated as varieties of the ass, refer to mule and
hinny respectively (horse-ass hybrids). There is no such name as *hippagrus* Schomber,
1963; this is merely the application in error to an ass of the name *hippagrus*
C. H. Smith, 1841, based on a horse.

*Equus burchellii* (Gray, 1824). Zool. J., 1:247.
    COMMON NAME: Burchell's Zebra.
    TYPE LOCALITY: "The flat parts near the Cape", now identified as South Africa, Northern
        Cape Prov., Kuruman Dist., Little Klibbolikhonni Fontein (Grubb, 1999:16).
    DISTRIBUTION: S and E Angola, N and E Botswana, SE Dem. Rep. Congo, Kenya, N Namibia,
        SE Sudan, SW Ethiopia, Malawi, Mozambique, S Somalia, South Africa (N KwaZulu-
        Natal, Limpopo, and Mpumalanga Provs.; formerly more widespread, S to Orange
        River), Swaziland, Tanzania, Uganda, Zambia, and Zimbabwe.
    STATUS: IUCN – Extinct as *E. b. burchellii*, Data Deficient as *E. b. chapmani*, *E. b. crawshayi*,
        and *E. b. zambeziensis*, Least Concern as *E. b. antiquorum* and *E. b. boehmi*.
    SYNONYMS: **antiquorum** C. H. Smith, 1841; *burschelii* Schinz, 1845; *campestris* (C. H. Smith,
        1841); *festivus* Wagner, 1835; *isabellinus* (C. H. Smith, 1841); *kaokensis* (Zukowsky,
        1924); *paucistriatus* Hilzheimer, 1912; *typicus* Selous, 1899; *zebroides* Lesson, 1827;
        **boehmi** Matschie, 1892; *cuninghamei* Heller, 1914; *goldfinchi* Ridgeway, 1911; *granti* de
        Winton, 1896; *isabella* Ziccardi, 1959; *jallae* (Camerano, 1902); *mariae* Trouessart,
        1898; *muansae* (Matschie, 1906); *zambeziensis* Trouessart, 1898; **borensis** Lönnberg,
        1921; **chapmanni** Layard, 1866; *chapmani* Trouessart, 1898; *kaufmanni* Matschie,
        1912; *markhami* Tichomirow, 1878; *pococki* Brasil and Pennetier, 1909; *selousi* Pocock,
        1897; *transvaalensis* Ewart, 1897; *wahlbergi* Pocock, 1871; **crawshaii** de Winston, 1896;
        *annectans* Lydecker, 1908; annectens W. Rothschild, 1906; *crawshayi* Pocock, 1897; *foai*
        Prazack and Trouessart, 1899; *tigrinus* Johnston, 1897.
    COMMENTS: Reviewed by Grubb (1981, Mammalian Species, 157). Species status
        controversial. A species separate from *E. quagga*; see Gentry (1975), Eisenmann and
        Turlot (1978), Bennett (1980), Klein and Cruz-Uribe (1999), and Eisenmann and Brink
        (2000). Many previous workers regarded *quagga* and *burchellii* as conspecific; see Rau
        (1978) – and they were recently regarded as conspecific by Groves (1985b). Subspecies
        partly based on Ansell (1974a) and L. Schlawe and W. Wozniak (in litt., 1991). *Equus
        wardi* Ridgeway, 1910 is a hybrid between *E. burchellii* and *E. zebra* (Barnaby, 2001;
        Pocock, 1909b; Rzasnicki, 1938).

*Equus caballus* Linnaeus, 1758. Syst. Nat., 10th ed., 1:73.
    COMMON NAME: Horse.
    TYPE LOCALITY: "Habitat in Europa" (= Sweden?); based on domestic horses.
    DISTRIBUTION: In the late 18th Century, from Poland and Russian Steppes east to Turkestan
        and Mongolia; wild population survived (at least until recently) in SW Mongolia and
        adjacent Gansu, Sinkiang, and Inner Mongolia (China); reintroduced into Mongolia.
        Domesticated worldwide; feral in Portugal, Spain, France, Greece, Iran, Sri Lanka,
        Lesser Sundas (Flores and Rintja), Australia, New Zealand, Colombia, Hispaniola,
        Canada, USA (incl. Hawaiian Isls), Galapagos and probably other oceanic islands.
    STATUS: CITES – Appendix I as *E. przewalskii*; U.S. ESA – Endangered as *E. przewalskii*; IUCN –
        Extinct in the Wild as *E. ferus* and *E. f. przewalskii*.
    SYNONYMS: *africanus* Sanson, 1878; *agilis* Ewart, 1910; *anglicus* Desmarest, 1822; *arabicus*
        Desmarest, 1822; *aryanus* Piétrement, 1875; *asiaticus* Sanson, 1878; *belgius* Sanson,
        1878; *bohemicus* Marchlewlski, 1924; *brittanicus* Sanson, 1878; *celticus* Ewart, 1903;
        *cracoviensis* Storkowski, 1946; *domesticus* Gmelin, 1788; *equuleus* C. H. Smith, 1841;
        *europaeus* Stegmann von Pritzwald, 1924; *ewarti* Storkowski, 1946; *frisius* Desmarest,
        1822; *gallicus* Fitzinger, 1858; *germanicus* Fitzinger, 1859; *gracilis* Ewart, 1909;

*gutsenensis* Skorkowski, 1946; *helveticus* Desmarest, 1822; *hibernicus* Fitzinger, 1859; *hippagrus* C. H. Smith, 1841; *italicus* Desmarest, 1822; *lalisio* C. H. Smith, 1841; *libycus* Ridgeway, 1905; *midlandensis* Quinn, 1957; *moldavicus* Desmarest, 1822; *mongolicus* Piétrement, 1875; *muninensis* Storkowski, 1946; *nehringi* Duerst, 1904; *nipponicus* Shikama and Onuki, 1962; *nordicus* Skorkowski, 1933; *pallas* Skorkowski, 1933; *parvus* Franck, 1875; *persicus* Desmarest, 1822; *pumpelli* Duerst, 1908; *robustus* Fitzinger, 1859; *sequanicus* Desmarest, 1822; *sequanius* Sanson, 1878; *silvaticus* Vetulani, 1927 [unavailable]; *sinensis* Fitzinger, 1858; *sylvestris* von den Brincken, 1828 [*nomen nudum*]; *tanghan* Gray, 1846 [*nomen nudum*]; *tataricus* Desmarest, 1822; *transylvanicus* Desmarest, 1822; *typicus* Ewart, 1904; *varius* S. D. W., 1836; **ferus** Boddaert, 1785; *equiferus* Pallas, 1811; *gmelini* Antonius, 1912; *tarpan* Pidoplichko, 1951 [*nomen nudum*]; **przewalskii** Poliakov, 1881; *hagenbecki* Matschie, 1903; *prjevalskii* Ewart, 1903.

COMMENTS: Reviewed by Bennett and Hoffman (1999, Mammalian Species, 628). Recent caballine horses have been assigned to two different species, *E. caballus* (or *ferus*) and *E. przewalskii*, but many authors now include *przewalskii* in *caballus*; see Corbet (1978c:194), Groves (1974a), Bennett (1980), and Bennett and Hoffman (1999). Gromov and Baranova (1981:333-334) continued to recognize two species, *gmelini* (= *ferus*) and *przewalskii*. Groves (1971b) and Corbet (1978c:194) proposed that *ferus* (the Tarpan) replace *caballus*, objecting to the use of specific names based on domestic animals. Gentry et al. (1996) proposed that majority usage be confirmed by adoption of the first available specific name based on a wild population for the wild taxon, in this case deemed to be *E. ferus*. It has not been demonstrated that most authors have termed wild horses *E. ferus* rather than *E. caballus* or *E. c. ferus* or other names. Azzaroli (1984), Bennett and Hoffman (1999), and Forsten (1988) are among those who have used the name *caballus* for the species. The case is complicated by the very much wider use of *przewalskii* as a name for wild horses, though *przewalskii* is commonly treated as a species separate from *E. caballus*. Gentry et al. (1996) asked the International Commission on Zoological Nomenclature to use its plenary power to rule that the name for the wild species is not invalid by virtue of being antedated by the name based on the domestic form. The Commission has ruled in favor of the proposal (International Commission on Zoological Nomenclature, 2003). It has stipulated that *ferus* is not invalid but has not specified explicitly what name is to be used for the species by those who consider *E. caballus* and *E. ferus* to be conspecific (see Bock, 1997). Material evidence that *ferus* is a distinct form of wild horse is limited to osteological material of two specimens and it has not been reliably identified with Pleistocene or Holocene local populations (Forsten, 1988). Its status as a wild rather than a feral form is disputed (e.g. Epstein, 1971) and it is not regarded as ancestral to domestic horses by Kuz'mina (1997). Accordingly *ferus* is here treated as a subspecies of *E. caballus*. The systematics of *ferus* needs to be more thoroughly reviewed. *Equus ferus* Boddaert, 1785 is preoccupied by *Equus asinus ferus* Erxleben, 1777 (= *Equus hemionus*), cited also by Kerr (1792) and G. Fischer [von Waldheim] (1814), which can however be regarded as a nomen oblitum. The next published name, *equiferus* Pallas, 1811 is available, contrary to Heptner et al. (1961). It is listed by Pallas (1811:510) as "EQUUS *Caballus* ß. Equiferus" in his synonymy, that is as an infraspecific category, now regarded as a nominal subspecies. For domesticated horses, Desmarest (1822) employed 25 subspecific names and Fitzinger (1858, 1859, 1860) used 160 names for species, subspecies and infrasubspecific categories. Most of these names have been ignored but the specific and subspecific names are available and take part in synonymy. Only those that have been used elsewhere as junior or senior synonyms are cited above, under "synonymy".

*Equus grevyi* Oustalet, 1882. La Nature (Paris), 10(2):12.

COMMON NAME: Grévy's Zebra.

TYPE LOCALITY: Described as "region de l'Afrique orientale qu'on appelle le pays des Gallas", i.e. Ethiopia, Gallaland, Shoa Prov. (Rzasnicki, 1951), restricted to Awash Valley (Yalden et al., 1986).

DISTRIBUTION: Dry desert regions of S and E Ethiopia, N Kenya, and S Somalia (extinct).

STATUS: CITES – Appendix I; U.S. ESA – Threatened; IUCN – Endangered.

SYNONYMS: *berberensis* Pocock, 1902; *faurei* Matschie, 1898.
COMMENTS: Reviewed by Churcher (1993, Mammalian Species, 453). Restriction of type
locality to Lake Zwai (Roosevelt and Heller, 1914) is erroneous, as there are no records
of the species from this lake region (Yalden et al, 1986).

*Equus hemionus* Pallas, 1775. Nova Comm. Imp. Acad. Sci. Petrop., 19:394.
COMMON NAME: Onager.
TYPE LOCALITY: "ad Lacum Tarei Dauuriae", i.e. Russia, Transbaikalia, S Chitinsk. Obl., Tarei-
Nor, 50°N, 115°E.
DISTRIBUTION: Formerly much of Mongolia, north to Transbaikalia (Russia); east to NE Inner
Mongolia (China) and possibly W Manchuria (China); and west to Dzhungarian Gate.
Survives in SW and SC Mongolia and adjacent China; see Sokolov and Orlov
(1980:248). Also formerly Kazakhstan north to upper Irtysh and Ural Rivers (Russia);
westward north of the Caucasus and Black Sea at least to Dniestr River (Ukraine);
Anatolia, Syria, and SE of Caspian Sea in N Iraq, Iran, Afghanistan, and Pakistan to
Thar Desert of NW India; survives as isolated populations in Rann of Kutch (India),
Badkhys Preserve, Turkmenistan, and C Iran; also reestablished on Barsa-Khelmes Isl
(Aral Sea, Uzbekistan); until 17th-18th centuries in Armenia and Azerbaidjan.
STATUS: CITES – Appendix I as *E. h. hemionus*, *E. onager* and *E. onager khur*; otherwise
Appendix II; U.S. ESA – Endangered; IUCN – Extinct *E. h. hemippus*, Critically
Endangered as *E. h. onager* and *E. h. kulan*, Endangered as *E. h. khur*, Vulnerable as
*E. hemionus*, *E. h. hemionus* and *E. h. luteus*.
SYNONYMS: *castaneus* Lydekker, 1904; *finschi* Matschie, 1911; *hemionos* Boddaert, 1785;
*typicus* Sclater, 1891; **blanfordi** (Pocock, 1947); **hemippus** I. Geoffroy Saint-Hilaire,
1855; *syriacus* Milne-Edwards, 1869; **khur** Lesson, 1827; *indicus* (Sclater, 1862) [*nomen
nudum*]; *indicus* George, 1869; **kulan** (Groves and Mazák, 1967); **luteus** Matschie,
1911; *bedfordi* Matschie, 1911; **onager** Boddaert, 1785; *bahram* (Pocock, 1947);
*dzigguetai* (Wood, 1879); *ferus* Erxleben, 1777 [*nomen oblitum*]; *hamar* C. H. Smith,
1841; *onager* Pallas, 1777 [unavailable]; *typicus* Sclater, 1891.
COMMENTS: Revised by Groves and Mazák (1967), Groves (1986), and Schlawe (1986), who
included *onager* in *hemionus*. Bennett (1980) considered *onager* (including *hemippus*,
*khur,* and *kulan*) to be a distinct species. Nominate subspecies is also known by
common name Kulan or Dzigetai; *luteus* doubtfully separable from it. Groves (2003)
considered *khur* as a species separate from *E. hemionus*.

*Equus kiang* Moorcroft, 1841. Travels in the Himalayan Provinces, 1:312.
COMMON NAME: Kiang.
TYPE LOCALITY: "Ladak" (India, Kashmir).
DISTRIBUTION: Ladak (India), Tibet, Tsinghai and Szechwan (China), adjacent Nepal and
Sikkim (India).
STATUS: CITES – Appendix I; IUCN – Data deficient as *E. k. kiang* and *E. k. polyodon*, Lower
Risk (lc) as *E. k. holdereri*.
SYNONYMS: *equioides* (Hodgson, 1842) [*nomen nudum*]; *kyang* (Kinloch, 1869); **holdereri**
Matschie, 1911; *tafeli* (Matschie, 1924); **polyodon** (Hodgson, 1847); *nepalensis*
(Trumler, 1959).
COMMENTS: Revised by Groves and Mazák (1967) and Groves (1986), who with Bennett
(1980) separated *kiang* from *hemionus*; Schlawe (1986) regarded *kiang* as a subspecies of
*hemionus*.

*Equus quagga* Boddaert, 1785. Elench. Anim., p. 160.
COMMON NAME: Quagga.
TYPE LOCALITY: "Caffrorum regione"; locality of paralectotype now identified as South
Africa, Northern Cape Prov., Colesburg Dist., Seekoei River (Grubb, 1999).
DISTRIBUTION: Formerly South Africa, south of the Orange-Vaal Rivers.
STATUS: IUCN – Extinct.
SYNONYMS: *couagga* Desmarest, 1822; *danielli* Pocock, 1904; *greyi* Lydekker, 1902; *lorenzi*
Lydekker, 1902; *qouagga* Lesson, 1827; *quaccha* Gray, 1827; *trouessarti* Camerano, 1908.
COMMENTS: See comments under *burchellii*. Last specimen, a captive, died in 1872.

*Equus zebra* Linnaeus, 1758. Syst. Nat., 10th ed., 1:74.
>COMMON NAME: Mountain Zebra.
>TYPE LOCALITY: "Habitat in India, Africa", since restricted to South Africa, Western Cape Prov., Ceres Dist., Perdekop.
>DISTRIBUTION: S Angola, Namibia, South Africa (Eastern and Western Cape Provs.; formerly in W Northern Cape Prov.). Now much reduced in numbers and, in South Africa, confined to a few nature reserves.
>STATUS: CITES – Appendix I as *E. z. zebra*, Appendix II as *E. z. hartmannae*; U.S. ESA – Endangered as *E. z. zebra*, Threatened as *E. z. hartmannae*; IUCN – Endangered as *E. z. zebra* and *E. z. hartmannae*.
>SYNONYMS: *campestris* (Gray, 1852); *frederici* Trouessart, 1826; *indica* Trouessart, 1898; *montanus* Burchell, 1822; **hartmannae** Matschie, 1898; *matschiei* (Zukowsky, 1924); *penricei* Thomas, 1900.
>COMMENTS: Reviewed by Penzhorn (1988, Mammalian Species, 314).

**Family Tapiridae** Gray, 1821. London Med. Repos., 15:306.
>SYNONYMS: Elasmognathinae Gray, 1867.

*Tapirus* Brisson, 1762. Regnum Animale, 2nd ed., pp. 12, 81.
>TYPE SPECIES: *Hippopotamus terrestris* Linnaeus, 1758.
>SYNONYMS: *Acrocodia* Goldman, 1913; *Chinchecus* Trouessart, 1898; *Cinchacus* Gray, 1873; *Elasmognathus* Gill, 1865; *Hydrochoerus* Gray, 1821 [*nomen nudum*]; *Pinchacus* Hershkovitz, 1954; *Rhinochoerus* Wagler, 1830; *Syspotamus* Billburg, 1827; *Tapir* Blumenbach, 1779; *Tapyra* Liais, 1872; *Tapirella* Palmer, 1903; *Tapirussa* Frisch, 1775 [unavailable].
>COMMENTS: The genus is available from Brisson, 1762 (Opinion 1894, International Commission on Zoological Nomenclature, 1998). Nominate *T. terrestris* group also includes *T. pinchaque*; analysis of mtDNA indicates that *T. terrestris* and *T. pinchaque* are sister species (Ashley et al., 1996); *T. bairdii* or *Tapirella* group is monotypic; *T. indicus* or *Acrocodia* group is monotypic among living forms; *Acrocodia* separated as a genus by Eisenberg et al. (1987).

*Tapirus bairdii* (Gill, 1865). Proc. Acad. Nat. Sci. Philadelphia, 17:183.
>COMMON NAME: Baird's Tapir.
>TYPE LOCALITY: "Isthmus of Panama", restricted to Panama, Canal Zone by Hershkovitz (1954).
>DISTRIBUTION: East from Isthmus of Tehuantepec, Mexico, through all other Central American states to Colombia west of the Rio Cauca and Ecuador west of the Andes to the Gulf of Guayaquil.
>STATUS: CITES – Appendix I; U.S. ESA and IUCN – Endangered.
>SYNONYMS: *bairdi* (Gray, 1868); *dowi* Alston, 1880; *dowii* Gill, 1870.
>COMMENTS: Revised by Hershkovitz (1954).

*Tapirus indicus* Desmarest, 1819. Nouv. Dict. Hist. Nat., Nouv. ed., 32:458,
>COMMON NAME: Malayan Tapir.
>TYPE LOCALITY: "la presqu'ile de Malacca", i.e. Malaysia, Malay Peninsula.
>DISTRIBUTION: Burma and Thailand south of 18°N, south through peninsular Malaya and Sumatra; listed as occurring in S Cambodia and possibly S Vietnam (Brooks et al., 1997); reliably recorded from Hongquan district, eastern Cochin China, Vietnam, in 1944 (Harper, 1945); authentic record from Laos in 1902 (Duckworth et al., 1999).
>STATUS: CITES – Appendix I; U.S. ESA – Endangered; IUCN – Vulnerable.
>SYNONYMS: *bicolor* Wagner, 1835; *malayanus* Raffles, 1821; *me* (Gray, 1869) [*nomen dubium*]; *sumatrensis* (Gray, 1821) [*nomen nudum*]; *sumatranus* Gray, 1843; **brevetianus** Kuiper, 1926.
>COMMENTS: The all-black form *brevetianus* known by two specimens from Palembang, S Sumatra is provisionally treated as a subspecies (Kuiper, 1926).

*Tapirus pinchaque* (Roulin, 1829). Ann. Sci. Nat. Zool., 18:46.
>COMMON NAME: Mountain Tapir.

TYPE LOCALITY: "une journée de cette ville [Bogota], dans le *Paramo* de *Suma-Paz*," i.e., Colombia, Cundinamarca, Páramo de Sumapaz.

DISTRIBUTION: Andes of Colombia and Ecuador; perhaps W Venezuela and N Peru.

STATUS: CITES – Appendix I; U.S. ESA and IUNC – Endangered.

SYNONYMS: *andicola* Gloger, 1842; *leucogenys* Gray, 1872; *pinchacus* de Blainville, 1846; *roulini* de Blainville, 1846; *roulinii* Fischer, 1830; *villosus* (Wagler, 1830).

COMMENTS: Revised by Hershkovitz (1954).

*Tapirus terrestris* (Linnaeus, 1758). Syst. Nat., 10th ed., 1:74.

COMMON NAME: South American Tapir.

TYPE LOCALITY: "Habitat in Brasilia", i.e., Brazil, Pernambuco.

DISTRIBUTION: East of the western cordillera of the Andes in N Argentina, Bolivia, Brazil, Colombia, E Ecuador, French Guiana, Guyana, Paraguay, Peru, Surinam, and Venezuela.

STATUS: CITES – Appendix II; U.S. ESA – Endangered; IUCN – Vulnerable.

SYNONYMS: *americanus* (Gmelin, 1788); *anta* (Zimmermann, 1780); *brasiliensis* Liais, 1872; *guianae* J. A. Allen, 1916; *laurillardi* Gray, 1868; *maypuri* (Roulin, 1829); *mexianae* Hagmann, 1908; *rufus* G. Fischer [von Waldheim], 1814; *sabatyra* (Liais, 1872); *suillus* (Blumenbach, 1779); *tapir* (Erxleben, 1777); *tapirus* Merriam, 1895; **aenigmaticus** Gray, 1872; *ecuadorensis* Gray, 1872; *peruvianus* Gray, 1872; **colombianus** Hershkovitz, 1954; **spegazzinii** Amhegino, 1916; *anulipes* Hermann, 1924; *obscura* Dennler, 1939.

COMMENTS: Revised by Hershkovitz (1954), who provisionally recognized only nominate *terrestris* and *colombianus* as subspecies but indicated that *tapir* was probably a valid Guiana subspecies, and by Cabrera (1961), whose classification is followed here. Reviewed by Padilla and Dowler (1994, Mammalian Species, 481).

**Family Rhinocerotidae** Gray, 1821. London Med. Repos., 15:306.

SYNONYMS: Ceratorhinae Osborn, 1896; Dicerorhinae Ringström, 1924; Dicerinae Ringström, 1924.

COMMENTS: Living species all allocated to nominate subfamily. Taxonomy and nomenclature revised by Rookmaaker (1983). The holotype of *Zygomaturus diahotensis* (Guerin et al., 1981), described as a zygomaturine diprotodontid from New Caledonia, seems to be a rhinoceros tooth, perhaps *Rhinoceros sondaicus* or *Dicerorhinus sumatrensis*; see Rich et al. (1987), Guerin and Faure (1987).

*Ceratotherium* Gray, 1868. Proc. Zool. Soc. Lond., 1867:1027 [1868].

TYPE SPECIES: *Rhinoceros simus* Burchell, 1817.

*Ceratotherium simum* (Burchell, 1817). Bull. Sci. Soc. Philom. Paris, p. 97.

COMMON NAME: White Rhinoceros.

TYPE LOCALITY: "L'interior de l'Afrique Méridionale vers le vingt-sixième degré de latitude"; since identified as South Africa, North West Prov., Chue Spring (= Heuningvlei), about 26°15'S, 23°10'E. See Grubb (1999).

DISTRIBUTION: Formerly north of Equator in S Chad, Central African Republic, S Sudan, NE Dem. Rep. Congo, and Uganda. Southern Africa in SE Angola, Botswana, NE Namibia, S Mozambique, South Africa (north of Orange-Vaal Rivers and in KwaZulu-Natal), Swaziland, Zimbabwe, and possibly also SW Zambia. Now much restricted in distribution; in south of range, extinct except in E KwaZulu-Natal (South Africa), but reintroduced into other parts of South Africa (KwaZulu-Natal, Limpopo Prov., Mpumalanga, Free State), Namibia, Swaziland, Mozambique, Zimbabwe, and Botswana; introduced into Zambia and Kenya. In north of range, now confined to NE Dem. Rep. Congo.

STATUS: CITES – Appendix II as *C. s. simum*; otherwise Appendix I; U.S. ESA – Endangered as *C. s. cottoni*; IUCN – Critically Endangered as *C. s. cottoni*, Near Threatened as *C. s. simum*.

SYNONYMS: *burchellii* (Lesson, 1827); *camperis* (Gray, 1827); *camptoceros* (Brandt, 1878); *camus* (Gray, 1827); *kiaboaba* (Murray, 1866); *kulumane* (Player, 1972); *oswelli* (Elliot, 1847); *prostheceros* (Brandt, 1878); **cottoni** Lydekker, 1908.

COMMENTS: Reviewed by Groves (1972*a*, Mammalian Species, 8). Revised by Groves (1975*b*).

*Dicerorhinus* Gloger, 1841. Gemein Hand.-Hilfsbuch. Nat., p. 125.
    TYPE SPECIES: *Rhinoceros sumatrensis* Fischer, 1814.
    SYNONYMS: *Ceratorhinus* Gray, 1867; *Didermocerus* Brookes, 1828 [suppressed].
    COMMENTS: *Didermocerus* Brookes, 1828, has been rejected, and *Dicerorhinus* validated
        (International Commission on Zoological Nomenclature, 1977*b*).

*Dicerorhinus sumatrensis* (G. Fischer [von Waldheim], 1814). Zoognosia, 3:301.
    COMMON NAME: Sumatran Rhinoceros.
    TYPE LOCALITY: "Sumatra", now known to be Indonesia, Sumatra, Bencoolen (= Bintuhan)
        Dist., Fort Marlborough (Groves, 1967c).
    DISTRIBUTION: Formerly Bangladesh (Chittagong Hills), Borneo, Burma, India (Assam),
        Laos, Malaysia (peninsular Malaya), Mergui Isl, Sumatra, Thailand, and Vietnam;
        probably also S China, and Cambodia. Survives in Tenasserim Range (Thailand-
        Burma), Petchabun Range (Thailand), and other scattered localities in Burma,
        peninsular Malaya, Sumatra, and Borneo.
    STATUS: CITES – Appendix I; U.S. ESA – Endangered; IUCN – Extinct as *D. s. lasiotis*,
        Critically Endangered as *D. s. harrisoni* and *D. s. sumatrensis*.
    SYNONYMS: *blythii* (Gray, 1873) [unavailable]; *crossii* (Gray, 1854); *malayanus* (Newman,
        1874) [*nomen nudum*]; *niger* (Gray, 1873); *sumatranus* (Raffles, 1822); **harrissoni** Groves,
        1965; *borniensis* Hose and McDougall, 1912 [*nomen nudum*]; **lasiotis** (Buckland, 1872).
    COMMENTS: Reviewed by Groves and Kurt (1972, Mammalian Species, 21). Revised by
        Groves (1967c).

*Diceros* Gray, 1821. London Med. Repos., 15:306.
    TYPE SPECIES: *Rhinoceros bicornis* Linnaeus, 1758.
    SYNONYMS: *Colobognathus* Brandt, 1878; *Keitloa* Gray, 1868; *Opsiceros* Gloger, 1841; *Rhinaster*
        Gray, 1862.

*Diceros bicornis* (Linnaeus, 1758). Syst. Nat., 10th ed., 1:56.
    COMMON NAME: Black Rhinoceros.
    TYPE LOCALITY: "Habitat in India", now identified as South Africa, Western Cape Prov.,
        Cape of Good Hope.
    DISTRIBUTION: Formerly in S Angola, Botswana, Burundi, N Cameroon, Central African
        Republic, S Dem. Rep. Congo, S Chad, N Eritrea, Ethiopia, Kenya, Malawi,
        Mozambique, Namibia, SE Niger, Nigeria, Rwanda, Somalia, South Africa, Sudan,
        Swaziland, Tanzania, Uganda, Zambia, and Zimbabwe; possibly more widespread in
        Niger, extending to Benin and Côte d'Ivoire, within historic times (Blancou, 1960;
        Sayer and Green, 1984). Very much reduced in numbers, particularly in recent decades
        of 20th century, and probably now extinct in many countries which it formerly
        occupied. Survives in reserves in Kenya, Tanzania, Namibia, Zambia, Zimbabwe and
        KwaZulu-Natal (South Africa), and possibly still in Cameroon, Chad, Central African
        Republic, Sudan, Rwanda, Malawi, Mozambique, Angola, and Botswana; widely
        reintroduced into parts of South Africa (Cumming et al., 1990).
    STATUS: CITES – Appendix I; U.S. ESA – Endangered; IUCN – Vulnerable as *D. b. bicornis*,
        Critically Endangered as *D. bicornis* and subspecifically as *D. b. longipes*, *D. b. michaeli*,
        and *D. b. minor*.
    SYNONYMS: *africanus* (Blumenbach, 1797); *camperi* (Schinz, 1845); *capensis* (Gray, 1868)
        [unavailable]; *capensis* (Trouessart, 1898); *gordoni* (Lesson, 1842); *keitloa* (A. Smith,
        1836); *ketloa* (A. Smith, 1837); *niger* (Schinz, 1845); *platyceros* (Brandt, 1878); *plesioceros*
        (Brandt, 1878); **brucii** (Lesson, 1842); *atbarensis* Zukowsky, 1965; *palustris* Benzon,
        1947; *porrhoceros* (Brandt, 1878); *somaliensis* (Potocki, 1897); **chobiensis** Zukowsky,
        1965; *somaliensis* J. Allen, 1914 [preoccupied]; **longipes** Zukowsky, 1949; **michaeli**
        Zukowsky, 1965; *rendilis* Zukowsky, 1965; **minor** (Drummond, 1876); *angolensis*
        Zukowsky, 1965; *holmwoodi* (Sclater, 1893); *ladoensis* Zukowsky, 1965 [unavailable];
        *ladoensis* Groves, 1967; *major* (Drummond, 1876); *nyasae* Zukowsky, 1965

[unavailable]; *occidentalis* (Zukowsky, 1922); *punyana* Potter, 1947; *rowumae* Zukowsky, 1965 [unavailable].

COMMENTS: Revised by Groves (1967*b*) and Prins (1990). Zukowsky's names are dated 1965 not 1964 (Rookmaaker, 1983). Reviewed by Hillman-Smith and Groves (1994, Mammalian Species, 455). *Rhinoceros kulumane* Player, 1972 is referable to *Ceratotherium*, not *Diceros* (Ansell, 1989). The type of *Rhinoceros cucullatus* Wagner, 1835 has been referred to this species but was regarded as an artefact by Zukowsky (1965).

*Rhinoceros* Linnaeus, 1758. Syst. Nat., 10th ed., 1:56.
    TYPE SPECIES: *Rhinoceros unicornis* Linnaeus, 1758.
    SYNONYMS: *Eurhinoceros* Gray, 1867; *Monocerorhinus* Wüst, 1922; *Monoceros* Rafinesque, 1815; *Naricornis* Frisch, 1775 [unavailable]; *Unicornus* Rafinesque, 1815.

*Rhinoceros sondaicus* Desmarest, 1822. Mammalogie, *in* Encycl. Meth., 2:399.
    COMMON NAME: Javan Rhinoceros.
    TYPE LOCALITY: "Sumatra" (Indonesia), later corrected to "Java" (Indonesia).
    DISTRIBUTION: Formerly Bangladesh, Burma, Thailand, Laos, Cambodia, Vietnam, and probably S China through peninsular Malaya to Sumatra and Java. Survives in Ujung Kulon (W Java) and in Vietnam; perhaps in small areas of Burma, Thailand, Laos, and Cambodia.
    STATUS: CITES – Appendix I; U.S. ESA – Endangered; IUCN – Critically Endangered.
    SYNONYMS: *camperii* Jardine, 1836; *camperis* Gray, 1827; *floweri* Gray, 1868; *frontalis* Von Martens, 1876; *javanicus* Geoffroy and Cuvier, 1824; *javanus* G. Cuvier, 1829; *nasalis* Gray, 1868; **annamiticus** Heude, 1892; **inermis** Lesson, 1838.
    COMMENTS: Revised by Groves (1967*c*). The type was said to have been obtained by Diard and Duvaucel who were thought to have collected together only on Sumatra, not Java (Sody, 1946) but Rookmaaker (1983) showed that Java is correctly the type locality.

*Rhinoceros unicornis* Linnaeus, 1758. Syst. Nat., 10th ed., 1:56.
    COMMON NAME: Indian Rhinoceros.
    TYPE LOCALITY: "Habitat in Africa, India", now identified as India, Assam, Terai.
    DISTRIBUTION: Within the present millennium, Indus Valley (Pakistan) east in N India to Assam and N Burma. Survives in India (Assam, West Bengal), Nepal, and possibly N Burma.
    STATUS: CITES – Appendix I; U.S. ESA – Endangered; IUCN – Endangered.
    SYNONYMS: *asiaticus* Blumenbach, 1797; *bengalensis* Kourist, 1970 [unavailable]; *indicus* G. Cuvier, 1816; *jamrachi* Jamrach, 1875; *rugosus* Blumenbah, 1779; *stenocephalus* Gray, 1868.
    COMMENTS: Reviewed by Laurie et al. (1983, Mammalian Species, 211).

# ORDER ARTIODACTYLA
by Peter Grubb

**ORDER ARTIODACTYLA** Owen, 1848.
  COMMENTS: Sequence of non-ruminant families follows Simpson (1945) and McKenna and Bell (1997); sequence of ruminant families based on Janis and Scott (1987).

**Family Suidae** Gray, 1821. London Med. Repos., 15:306.
  SYNONYMS: Babyrousini Thenius, 1970; Babirussina Gray, 1868 [unavailable]; Eurodontina Gray, 1873 [unavailable]; Hylochoerini Mekayev, 2002; Phacochoerini Gray, 1868; Potamochoerini Gray, 1873.
  COMMENTS: McKenna and Bell (1997) assigned all extant suids to the subfamily Suinae. Babirussina Gray, 1868 is based on *Babirussa* Frisch, 1775 which is unavailable. Includes as Tribes Babyrousini (including *Babyrousa*), Phacochoerini (including *Phacochoerus*), Potamochoerini (including *Hylochoerus* and *Potamochoerus*), and Suini (including *Sus*).

*Babyrousa* Perry, 1811. Arcana, Mus. Nat. Hist. (plate and 2 pages, unno.).
  TYPE SPECIES: *Babyrousa quadricornua* Perry, 1811 (= *Sus babyrussa* Linnaeus, 1758).
  SYNONYMS: *Babiroussa* F. Cuvier, 1825; *Babiroussous* Thomas, 1895; *Babiroussus* Gray, 1821; *Babirusa* Lesson, 1842; *Babirussa* Frisch, 1775 [unavailable]; *Babirussa* Rafinesque, 1815; *Babyrussa* Burnett, 1830; *Choerelaphus* Gloger, 1841; *Elaphochoerus* Gistel, 1848; *Porcus* Wagler, 1830; *Sukotyrus* Kerr, 1792 [*nomen oblitum*]; *Suckoteirus* Gray, 1843.
  COMMENTS: Revised by Groves (1980*b*).

*Babyrousa babyrussa* (Linnaeus, 1758). Syst. Nat., 10th ed., 1:50.
  COMMON NAME: Buru Babirusa.
  TYPE LOCALITY: "Habitat in Borneo Indiae orientalis"; identified as "Island of Boero" by Thomas (1911*a*) (Indonesia, Buru Isl).
  DISTRIBUTION: Indonesia, Buru (N Molucca Isls) and Sula Isls.
  STATUS: CITES – Appendix I; U.S. ESA – Endangered; IUCN – Vulnerable.
  SYNONYMS: *alfurus* (Lesson, 1827); *babirousa* (Jardine, 1836); *babirusa* Guillemard, 1889; *babirussa* (Quoy and Gaimard, 1830); *frosti* (Thomas, 1920); *indicus* (Kerr, 1792); *orientalis* (Brisson, 1762) [unavailable]; *quadricornua* Perry, 1811.
  COMMENTS: Former subspecies raised to species rank (Groves, 2001*a*, Meijaard and Groves, 2002). Probably introduced to Buru and the Sulu Isls; original distribution unknown (Groves, 1980*b*).

*Babyrousa bolabatuensis* Hoojer, 1950. Verh. Kon. Ned. Akad. Wetensch. Amsterdam (Afd. Natuurk.), 46 (2):121.
  COMMON NAME: Bola Batu Babirusa.
  TYPE LOCALITY: Indonesia, "Bola Batoe cave, near Badjo (Barebo district), ca. 20 km S.W. of Watampone in Central Bone, S. Celebes [Sulawesi]".
  DISTRIBUTION: Known by jaws and teeth of Holocene age from the type locality and one Recent skull from Gunung Malema, Moa, near Kulawi in C Sulawesi.

*Babyrousa celebensis* (Deninger, 1909). Ber. Naturf. Ges. Freiburg, 18:7.
  COMMON NAME: North Sulawesi Babirusa.
  TYPE LOCALITY: Indonesia, "Lembeh b. Celebes" (N Sulawesi, Lembeh Isl).
  DISTRIBUTION: Northern peninsula of Sulawesi, at least as far west as Bumbulan and including Lembeh Isl.
  SYNONYMS: *merkusi* De Beaufort, 1964 [*nomen nudum*].

*Babyrousa togeanensis* (Sody, 1949). Treubia, 20:187.
  COMMON NAME: Malenge Babirusa.
  TYPE LOCALITY: "Malengi island, Togean group, Res. Manado, N. Celebes" (Indonesia, N Sulawesi, Togean Isls, Malenge Isl).
  DISTRIBUTION: Known only from Malenge Isl.

*Hylochoerus* Thomas, 1904. Nature (London), 70:577.
    TYPE SPECIES: *Hylochoerus meinertzhageni* Thomas, 1904.

*Hylochoerus meinertzhageni* Thomas, 1904. Nature, 70:577.
    COMMON NAME: Giant Forest Hog.
    TYPE LOCALITY: Kenya, "Nandi Forest, near the Victoria Nyanza, at an altitude of 7000 feet";
        Nandi Forest, near Kaimosi [2134 m] (Allen and Lawrence, 1936).
    DISTRIBUTION: W Africa in Guinea, Côte d'Ivoire, and Ghana; not confirmed from Guinea
        Bissau, Sierra Leone, and Togo (Grubb et al., 1998). C Africa in W and SE Cameroon,
        Central African Republic, N and E Dem. Rep. Congo, SW Ethiopia, N Gabon, Kenya,
        E Nigeria, N Republic of Congo, Rwanda, S Sudan, and Uganda; not reliably recorded
        from Tanzania (Grimshaw, 1998; Kock and Howell, 2000).
    STATUS: IUCN – Vulnerable as *H. m. ivoriensis*, otherwise Lower Risk (lc).
    SYNONYMS: *gigliolii* Balducci, 1909; **rimator** Thomas, 1906; *ituriensis* Matschie, 1906;
        **ivoriensis** Bouet and Neuville, 1930.
    COMMENTS: See Thomas (1904*a*) for designation of the type specimen. *Hylochoerus schulzi*
        Zukowsky, 1921 is a synonym of *Potamochoerus larvatus hassama* (Grimshaw, 1998;
        Kock and Howell, 2000). Synonymy otherwise follows Grubb (1993).

*Phacochoerus* F. Cuvier, 1826. Dict. Sci. Nat., 39:383.
    TYPE SPECIES: *Aper aethiopicus* Pallas, 1766.
    SYNONYMS: *Aper* Pallas, 1766 [suppressed]; *Dinochoerus* Gloger, 1841 [suppressed]; *Eureodon*
        G. Fischer von Waldheim, 1817 [suppressed]; *Macrocephalus* Frisch, 1775 (unavailable);
        *Macrocephalus* Palmer, 1904 [suppressed]; *Phacellochaerus* Hemprich and Ehrenberg, 1832;
        *Phacellochoerus* Hemprich and Ehrenberg, 1832; *Phacochaeres* Gray, 1821 [suppressed];
        *Phacocherus* Fleming, 1822 [suppressed]; *Phacochoerus* G. Cuvier, 1816 (unavailable);
        *Phascochaeres* Cretzschmar, 1828 [suppressed]; *Phascochaerus* Desmarest, 1822
        [suppressed]; *Phascochoeres* Ranzani, 1821 [suppressed]; *Phascochoerus* Ranzani, 1821
        [suppressed].
    COMMENTS: Senior synonyms of *Phacochoerus* F. Cuvier, 1826 and most junior synonyms have
        been supressed by Opinion 466 (International Commission on Zoological Nomenclature,
        1957*g*), which, together with Morrison-Scott (1955) should be consulted for dates and
        authors cited here. Parapatric distribution of species of *Phacochoerus* in the Horn of Africa
        reviewed by d'Huart and Grubb (2001); genetic divergence in these species described by
        Randi et al. (2002).

*Phacochoerus aethiopicus* (Pallas, 1766). Misc. Zool., p. 16.
    COMMON NAME: Desert Warthog.
    TYPE LOCALITY: "Promontoria Bona Spei advectus"; between Kaffraria and Great
        Namaqualand (South Africa, Eastern Cape Prov.), two hundred leagues from the
        Cape of Good Hope according to Vosmaer (1766).
    DISTRIBUTION: Formerly in Cape Provinces, South Africa (extinct since ca. 1870-1890);
        NE Africa in E Ethiopia, N Kenya, and Somalia (Grubb, 1993; d'Huart and Grubb,
        2001).
    STATUS: IUCN – Extinct as *P. a. aethiopicus*, Vulnerable as *P. a. delamerei*, otherwise Lower Risk
        (lc).
    SYNONYMS: *angalla* (Boddaert, 1785); *edentatus* I. Geoffroy Saint-Hilaire, 1828; *pallasii*
        Van der Hoeven, 1839; *typicus* (A. Smith, 1834); **delamerei** Lönnberg, 1909.
    COMMENTS: For distinctions from *P. africanus*, see Ewer (1957) and Grubb (1993).

*Phacochoerus africanus* (Gmelin, 1788). *In* Linnaeus, Syst. Nat., 13th ed., 1:220.
    COMMON NAME: Common Wart-hog.
    TYPE LOCALITY: "Habitat in Africa a capite viridi ad caput bonae spei"; restricted to Senegal,
        "Cape Verd [Verde]" (Lydekker, 1915:373).
    DISTRIBUTION: Outside rainforest zone of Africa in Angola, Benin, Botswana, Burkina Faso,
        Burundi, Cameroon, Central African Republic, Chad, Côte d'Ivoire, Dem. Rep. Congo,
        Eritrea, Ethiopia, Gambia, Ghana, Guinea, Guinea Bissau, Kenya, Malawi, Mali,
        Mauritania, Mozambique, Nambia, Niger, Nigeria, Republic of Congo, Rwanda,

Senegal, Sierra Leone, N Somalia, South Africa, Sudan, Tanzania, Togo, Uganda, Zambia, and Zimbabwe.

STATUS: IUCN – Endangered as *P. a. aeliani*, otherwise Lower Risk (lc).

SYNONYMS: *barbatus* Gloger, 1841; *incisivus* I. Geoffroy Saint-Hilaire, 1828; **aeliani** (Cretzschmar, 1828); *barkeri* W. Rothschild, 1920; *haroia* (Hemprich and Ehrenberg, 1832); *sclateri* Gray, 1870; **massaicus** Lönnberg, 1908; *bufo* Heller, 1914; *centralis* Lönnberg, 1917; *fossor* Schwarz, 1913; **sundevallii** Lönnberg, 1908; *shortridgei* St Leger, 1932.

COMMENTS: Specifically distinct from *P. aethiopicus* (Cooke and Wilkinson, 1978; Ewer, 1957; Grubb, 1993). Synonymy tentative, based on Grubb (1993).

*Potamochoerus* Gray, 1854. Proc. Zool. Soc. Lond., 1852:129 [1854].

TYPE SPECIES: *Choiropotamus pictus* Gray, 1852 (= *Sus porcus* Linnaeus, 1758).

SYNONYMS: *Choiropotamus* Gray, 1843 (*nomen oblitum*); *Koiropotamus* Gray, 1843 (*nomen nudum*); *Nyctochoerus* Heuglin, 1863.

COMMENTS: Revised by de Beaux (1924).

*Potamochoerus larvatus* (F. Cuvier, 1822). Mem. Mus. Hist. Nat. Paris, 8:447.

COMMON NAME: Bush-pig.

TYPE LOCALITY: "Madagascar" (no precise locality) here selected.

DISTRIBUTION: Angola, N Botswana, Burundi, N Eritrea, Ethiopia, E and S Dem. Rep. Congo, Kenya, Malawi, Mozambique, Rwanda, S Somalia, NE and S South Africa, S Sudan, Swaziland, Tanzania, Uganda, Zambia, and Zimbabwe; Madagascar and Comoro Isls (introduced?).

STATUS: IUCN – Lower Risk (lc).

SYNONYMS: **edwardsi** A. Grandidier, 1867; *hova* Lönnberg, 1910; *madagascariensis* (A. Grandidier, 1867); **hassama** (Heuglin, 1863); *arrhenii* Lönnberg, 1917; *daemonis* Forsyth Major, 1897; *intermedius* Lönnberg, 1910; *keniae* Lönnberg, 1912; *schulzi* (Zukowsky, 1921); **koiropotamus** (Desmoulins, 1831); *africanus* (von Schreber, 1791) [preoccupied]; *capensis* (Gray, 1847) [*nomen nudum*]; *choeropotamus* Forsyth Major, 1897; **nyasae** Forsyth Major, 1897; *congicus* Lönnberg, 1910; *cottoni* Pinfold, 1928; *johnstoni* Forsyth Major, 1897; *maschona* Lönnberg, 1910; **somaliensis** De Beaux, 1924.

COMMENTS: Specifically distinct from *P. porcus* (de Beaux 1924; Grubb 1993). Syntypes from "Madagascar" and "Sitsikamma" (South Africa, Eastern Cape Prov., Humansdorp dist., Tsitsikamma); lectotype here designated as the skull from Madagascar illustrated in the original description. Synonymy modified from Grubb (1993). Evidence of domestication of species of *Potamochoerus* and transportation of *P. porcus* to Brazil (Simoons, 1953) suggest that *P. larvatus* could have been transported to Madagascar by humans.

*Potamochoerus porcus* (Linnaeus, 1758). Syst. Nat., 10th ed., 1:50.

COMMON NAME: Red River Hog.

TYPE LOCALITY: "Habitat in Africa" (West Africa); based on animals exported to Brazil (Simoons, 1953).

DISTRIBUTION: Rainforest zone of Africa from Benin, Cameroon, Central African Republic, Côte d'Ivoire, Dem. Rep. Congo, Equatorial Guinea (Mbini), Gabon, Ghana, Guinea, Guinea Bissau, Liberia, Nigeria, Republic of Congo, Senegal, Sierra Leone, and Togo; no reliable record from Gambia (Grubb et al., 1998) or Sudan (Grubb, 1993).

STATUS: IUCN – Lower Risk (lc).

SYNONYMS: *albifrons* Du Chaillu, 1860; *albinuchalis* Lönnberg, 1919; *guineensis* (Pallas, 1766); *mawambicus* Lorenz, 1923; *penicillatus* (Schinz, 1848); *pictus* (Gray, 1852); *ubangensis* Lönnberg, 1910.

COMMENTS: A monotypic species (Grubb, 1993).

*Sus* Linnaeus, 1758. Syst. Nat., 10th ed., 1:49.

TYPE SPECIES: *Sus scrofa* Linnaeus, 1758.

SYNONYMS: *Annamisus* Heude, 1892 [*nomen nudum*]; *Aulacochoerus* Gray, 1873; *Capriscus* Gloger, 1841; *Caprisculus* Strand, 1928; *Centuriosus* Gray, 1862; *Dasychoerus* Gray, 1873; *Euhys* Gray, 1869; *Eusus* Gray, 1868; *Gyrosus* Gray, 1862; *Indisus* Heude, 1899; *Microsus* Heude,

1899; *Nesosus* Heude, 1892; *Porcula* Hodgson, 1847; *Porculia* Jerdon, 1874; *Porcus* S.D.W., 1836; *Ptychochoerus* Fitzinger, 1864; *Rhinosus* Heude, 1894; *Scrofa* Gray, 1868; *Sinisus* Heude, 1892; *Taenisus* Heude, 1899; *Verrusus* Heude, 1894; *Vittatus* Heude, 1899 [*nomen nudum*].

COMMENTS: Revised by Groves (1981*a*). Can be partitioned into *S. barbatus* or *Euhys* group (possibly parapatric; including also *S. ahoenobarbus, S. bucculentus, S. cebifrons, S. celebensis, S. oliveri, S. philippensis,* and *S. verrucosus*), and *S. scrofa* or nominate *Sus* group (including also *S. salvanius*).

**Sus ahoenobarbus** Huet, 1888. Naturaliste, ser. 2, 2:5
COMMON NAME: Palawan Pig.
TYPE LOCALITY: Philippines, "Palauan" (Palawan Isl).
DISTRIBUTION: Philippines (Palawan Isl, Balabac Isl, and Calamian Isls).
STATUS: IUCN – Vulnerable as *S. barbatus ahoenobarbis.*
SYNONYMS: *balabacensis* Forsyth Major, 1897; *calamianensis* Heude, 1892; *palavensis* Nehring, 1889.
COMMENTS: A separate species from *S. barbatus* according to Groves (2001*b*).

**Sus barbatus** Müller, 1838. Tijdschr. Nat. Gesch. Physiol., 5:149.
COMMON NAME: Bearded Pig.
TYPE LOCALITY: Indonesia, "Borneo", Kalimantan; "Bornéo, [near] Banjermassing [Banjarmasin]" (Jentink, 1892:192); "the neighborhood of the village Poeloe-Lampej [Pululampei], not off [i.e. not far from?] the bank of the Moloekko-river [Molukko River], South-eastern Borneo" (Jentink, 1905:161).
DISTRIBUTION: Brunei, Indonesia (Banka Isl, Kalimantan, Rhio Arch., Sumatra), Malaysia (Malay Peninsula, Sarawak).
STATUS: IUCN – Lower Risk (nt) as *S. b. oi,* otherwise Lower Risk (lc).
SYNONYMS: *gargantua* Miller, 1906; *longirostris* Nehring, 1885; *oi* Miller, 1902; *branti* Kloss, 1921 [unavailable]; *edmondi,* Sody, 1937; *sumatranus* Kelm, 1939.

**Sus bucculentus** Heude, 1892. Mem. Hist. Nat. Emp. Chin., 2:pl. 20b, fig. 7.
COMMON NAME: Heude's Pig.
TYPE LOCALITY: Viet Nam, Cochin China, "sur les bords du Donnaï" (= Dong Nai River); the type is labelled "Bienhoa" (Viet Nam).
DISTRIBUTION: Vietnam, Laos.
STATUS: IUCN – Data Deficient.
COMMENTS: The species is more extensively described by Heude, Mem. Hist. Nat. Emp. Chin., 2:219, pl. 40 [1894]. Known from the lectotype and paralectotype skulls, now in the Institute of Zoology, Academia Sinica, Beijing, and a third skull recently obtained from the Annamite Range, Laos (Braun et al., 2001; Groves and Schaller, 2000; Groves et al., 1997).

**Sus cebifrons** Heude, 1888. Mem. Hist. Nat. Emp. Chin., 2, pl. 17, fig. 5.
COMMON NAME: Visayan Warty Pig.
TYPE LOCALITY: Philippines, "l'ile de Cebu".
DISTRIBUTION: Philippines (Cebu, Negros, Panay and probably Masbate Isls).
STATUS: IUCN – Extinct as *S. c. cebifrons,* otherwise Critically Endangered.
SYNONYMS: *negrinus* Sanborn, 1952.
COMMENTS: The species is more extensively described by Heude, Mem. Hist. Nat. Emp. Chin., 2,pl. 28 [1892], 2:218 [1894]. Specifically distinct from *S. barbatus* and *S. philippensis* (Groves and Grubb, 1993; Sanborn, 1952*a*). Revised by Groves (1997*a*).

**Sus celebensis** Müller and Schlegel, 1843. *In* Temminck, Verh. Nat. Gesch. Nederland. Overz. Bezitt., Zool., pp. 172, 177 [1845]; pl. 28 bis [1843].
COMMON NAME: Celebes Warty Pig.
TYPE LOCALITY: Indonesia, "Celebes"; the type is from "Célèbes, Ménado" (Sulawesi, Manado) according to Jentink (1892:193).
DISTRIBUTION: Indonesia (Sulawesi and neighboring small islands; feral on Halmahera and Simaleue Isls; possibly feral on Flores Isl as *floresianus* (= *heureni*) and on Timor Isl as *timoriensis*).

STATUS: IUCN – Lower Risk (lc).

SYNONYMS: *amboinensis* Forsyth Major, 1897; *macassaricus* (Heude, 1898); *maritanus* Raven, 1935 [*nomen nudum*]; *maritimus* (Heude, 1898); *mimus* Miller, 1906; *nehringii* Jentink, 1905; *niadensis* Miller, 1906 [*nomen dubium*]; *weberi* Jentink, 1905; *floresianus* (Heude, 1899); *heureni* Hardjasasmita, 1987; *heurni* Corbet and Hill, 1992; *timoriensis* Müller, 1840.

COMMENTS: Although this species is sometimes cited from "Müller, 1840. *In* Temminck, Verh. Nat. Gesch. Nederland. Overz. Bezitt., Zool., Zoogd. Indisch. Archipel., p. 42", it is not found on that page. A species distinct from *S. verrucosus* (Groves, 1981*a*). *Sus timoriensis* is a feral population of *S. celebensis* according to Groves (1981*a*), but a valid species according to Hardjasasmita (1987) and a synonym of *S. scrofa* according to Corbet and Hill (1992). Groves (1981*a*) regarded warty pigs from Flores as a feral population of *S. celebensis* but Hardjasasmita (1987) assigned them to a separate species, *Sus heureni* which was included with a query in the synonymy of *S. celebensis* by Corbet and Hill (1992). A prior name for *S. heureni* is *Microsus floresianus* Heude, 1899. Both *floresianus* and *timoriensis* are provisionally ranked here as subspecies.

*Sus oliveri* Groves, 1997. Zool. J. Linn. Soc., 170:186.

COMMON NAME: Oliver's Warty Pig

TYPE LOCALITY: "Mayapang, Rizal, Mindoro Occidental, Philippines".

DISTRIBUTION: Philippines, Mindoro.

COMMENTS: A separate species from *S. philippensis* according to Groves (2001*b*), known from four skulls and a head skin.

*Sus philippensis* Nehring, 1886. Sber. Ges. Naturf. Fr., Berlin, 1886:83.

COMMON NAME: Philippine Warty Pig.

TYPE LOCALITY: Philippines, Luzon Isl.

DISTRIBUTION: Philippines (Luzon, Mainit, Mindanao, Jolo, Catanduanis and Samar Isls and probably Balabac and Leyte Isls).

STATUS: IUCN – Vulnerable.

SYNONYMS: *arietinus* Heude, 1892; *conchyvorus* Heude, 1888; *crassidens* Heude, 1892; *effrenus* Heude, 1888; *frenatus* Heude, 1888; *jalaensis* Heude, 1888; *joloensis* Groves, 1981; *mainitensis* Heude, 1892; *marchei* Huet, 1888; *megalodontus* Heude, 1892; *microtis* Heude, 1888; *minutus* Heude, 1888; *mindanensis* Forsyth Major, 1897; *inconstans* Heude 1892.

COMMENTS: Regarded as a species distinct from *S. barbatus* (Groves and Grubb, 1993); revised by Groves (1997*a*); an undescribed subspecies or related species recorded from Tawitawi Isls, Sulu Archipelago (Karen Rose, pers. comm.).

*Sus salvanius* (Hodgson, 1847). J. Asiat. Soc. Bengal, 16:423.

COMMON NAME: Pygmy Hog.

TYPE LOCALITY: "Habitat, Saul forest" but no locality given; N India, "Sikhim Tarai [Sikkim Terai] The moist forest-tract at the base of the eastern Himalaya" (Lydekker, 1915:343).

DISTRIBUTION: Bhutan, S Nepal, N India (incl. Sikkim).

STATUS: CITES – Appendix I; U.S. ESA – Endangered; IUCN – Critically Endangered.

*Sus scrofa* Linnaeus, 1758. Syst. Nat., 10th ed., 1:49.

COMMON NAME: Wild Boar.

TYPE LOCALITY: "Habitat in Europa australiore"; shown to be Germany, from where wild boar had been introduced to Sweden, Oeland (Thomas, 1911*a*:140).

DISTRIBUTION: N Africa in Algeria, Morocco, Tunisia; anciently introduced into Egypt and N Sudan where now absent. All states of mainland Europe east to Armenia, Azerbaijan, Georgia, W Russia (European Russia and Caucasus Mtns), and Ukraine; extinct in Ireland, Scandinavia, and United Kingdom but reintroduced into England, S Finland, and S Sweden; anciently introduced into Corsica and Sardinia. In Asia present in Burma, Cambodia, China (but absent from Tibetan Plateau, Singkiang, Gansu, Inner Mongolia, and Ordos Plateau), India, Indonesia (Sumatra, Java east to Bali and Sumbawa Isls), Iran, Iraq, Israel, Japan (including Riukiu Isls), W Jordan, Kazakhstan, Kyrgyzstan, Lebanon, Laos, Malaysia (peninsular Malaya only), Mongolia, Pakistan,

Russia (S Siberia and Soviet Far East), Sri Lanka, Syria, Taiwan, Tajikistan, Thailand, Turkmenistan, Uzbekistan, and Vietnam. Widespread as feral populations in South Africa, Indonesia (Lesser Sunda Isls), Australia, USA, West Indies, Central and South America and numerous oceanic islands, including Andaman Isls and Mauritius (Indian Ocean) and Hawaiian, Galapagos and Fiji Isls (Pacific Ocean). Feral and domestic populations of Molucca Isls, New Guinea and Solomon Isls thought to originate from hybrids between *scrofa* and *celebensis*.

STATUS: IUCN – Vulnerable as *S. s. riukiuanus*, otherwise Lower Risk (lc).

SYNONYMS: *anglicus* Reichenbach, 1846; *aper* Erxleben, 1777; *asiaticus* Sanson, 1878; *bavaricus* Reichenbach, 1846; *campanogallicus* Reichenbach, 1846; *capensis* Reichenbach, 1846; *castilianus* Thomas, 1911; *celticus* Sanson, 1878; *chinensis* Linnaeus, 1758; *crispus* Fitzinger, 1858; *deliciosus* Reichenbach, 1846; *domesticus* Erxleben, 1777; *europaeus* Pallas, 1811; *fasciatus* von Schreber, 1790; *ferox* Moore, 1870; *ferus* Gmelin, 1788; *gambianus* Gray, 1847 [*nomen nudum*]; *hispidus* von Schreber, 1790; *hungaricus* Reichenbach, 1846; *ibericus* Sanson, 1878; *italicus* Reichenbach, 1846; *juticus* Fitzinger, 1858; *lusitanicus* Reichenbach, 1846; *macrotis* Fitzinger, 1858; *monungulus* G. Fischer [von Waldheim], 1814; *moravicus* Reichenbach, 1846; *nanus* Nehring, 1884; *palustris* Rütimeyer, 1862; *pliciceps* Gray, 1862; *polonicus* Reichenbach, 1846; *sardous* Reichenbach, 1846; *scropha* Gray, 1827; *sennaarensis* Fitzinger, 1858 [*nomen nudum*]; *sennaarensis* Gray, 1868; *sennaariensis* Fitzinger, 1860; *setosus* Boddaert, 1785; *siamensis* von Schreber, 1790; *sinensis* Erxleben, 1777; *suevicus* Reichenbach, 1846; *syrmiensis* Reichenbach, 1846; *turcicus* Reichenbach, 1846; *variegatus* Reichenbach, 1846; *vulgaris* (S. D. W., 1836); *wittei* Reichenbach, 1846; **algira** Loche, 1867; *barbarus* Sclater, 1860 [*nomen nudum*]; *sahariensis* Heim de Balzac, 1937; **attila** Thomas, 1912; *falzfeini* Matschie, 1918; **cristatus** Wagner, 1839; *affinis* Gray, 1847 [*nomen nudum*]; *aipomus* Gray, 1868; *aipomus* Hodgson, 1842 [*nomen nudum*]; *bengalensis* Blyth, 1860; *indicus* Gray, 1843 [*nomen nudum*]; *isonotus* Gray, 1868; *isonotus* Hodgson, 1842 [*nomen nudum*]; *jubatus* Miller, 1906; *typicus* Lydekker, 1900; *zeylonensis* Blyth, 1851; **davidi** Groves, 1981; **leucomystax** Temminck, 1842; *japonica* Nehring, 1885; *nipponicus* Heude, 1899; **libycus** Gray, 1868; *lybicus* Groves, 1981; *mediterraneus* Ulmansky, 1911; *reiseri* Bolkay, 1925; **majori** De Beaux and Festa, 1927; **meridionalis** Forsyth Major, 1882; *baeticus* Thomas, 1912; *sardous* Ströbel, 1882; **moupinensis** Milne-Edwards, 1871; *acrocranius* Heude, 1892; *chirodontus* Heude, 1888; *chirodonticus* Heude, 1899; *collinus* Heude, 1892; *curtidens* Heude, 1892; *dicrurus* Heude, 1888; *flavescens* Heude, 1899; *frontosus* Heude, 1892; *laticeps* Heude, 1892; *leucorhinus* Heude, 1888; *melas* Heude, 1892; *microdontus* Heude, 1892; *oxyodontus* Heude, 1888; *paludosus* Heude, 1892; *palustris* Heude, 1888; *planiceps* Heude, 1892; *scrofoides* Heude, 1892; *spatharius* Heude, 1892; *taininensis* Heude, 1888; **nigripes** Blanford, 1875; **riukiuanus** Kuroda, 1924; **sibiricus** Staffe, 1922; *raddeanus* Adlerberg, 1930; **taivanus** (Swinhoe, 1863); **ussuricus** Heude, 1888; *canescens* Heude, 1888; *continentalis* Nehring, 1889; *coreanus* Heude, 1897; *gigas* Heude, 1892; *mandchuricus* Heude, 1897; *songaricus* Heude, 1897; **vittatus** Boie, 1828; *andersoni* Thomas and Wroughton, 1909; *jubatulus* Miller, 1906; *milleri* Jentink, 1905; *pallidiloris* Mees, 1957; *peninsularis* Miller, 1906; *rhionis* Miller, 1906; *typicus* Heude, 1899; names based on domestic or feral populations possibly to be assigned to *S. s. vittatus*: *andamanensis* Blyth, 1858; *babi* Miller, 1906; *enganus* Lyon, 1916; *floresianus* Jentink, 1905; *natunensis* Miller, 1901; *nicobaricus* Miller, 1902; *tuancus* Lyon, 1916; names based on populations possibly originating from *scrofa/celebensis* hybrids (Groves, 1981a): *aruensis* Rosenberg, 1878; *ceramensis* Rosenberg, 1878; *goramensis* De Beaux, 1924; *niger* Finsch, 1886; *papuensis* Lesson and Garnot, 1826; *ternatensis* Rolleston, 1877.

COMMENTS: Revised by Genov (1999) and Groves (1981a, 2003). Treatment of *majori* as a subspecies follows Randi et al. (1996). The species can be partitioned into the following divisions (Genov, 1999; Groves and Grubb 1993): *cristatus* division (including also *davidi*), *leucomystax* division (including also *moupinensis, riukiuanus, sibiricus, taivanus,* and *ussuricus*), nominate *scrofa* division (including also *algira, attila, libycus, majori, meridionalis,* and *nigripes*), and *vittatus* division. For systematics, origin, and distribution of feral populations see Groves (1981a), Lever (1985), Uerpmann

(1987), and Vigne (1988). Hardjasasmita (1987) recognised *floresianus*, *milleri* and *papuensis* as subspecies, but *Sus scrofa floresianus* Jentink, 1905 is a junior secondary homonym of *Microsus floresianus* Heude, 1899, a subspecies of *Sus celebensis*. Corbet and Hill (1992) listed the domestic pig as a separate species, *Sus domesticus*, from *Sus scrofa* on grounds of utility.

*Sus verrucosus* Boie, 1832. Neues Statsb. Mag. Schleswig, 1:466.
    COMMON NAME: Java Warty Pig.
    TYPE LOCALITY: Indonesia, "Java"; "Java, Palang" (Jentink, 1892:191).
    DISTRIBUTION: Indonesia (Java, Madoera Isl, Bawean Isl).
    STATUS: IUCN – Endangered.
    SYNONYMS: *borneensis* Forsyth Major, 1897; *ceramica* Gray, 1868; *mystaceus* Gray, 1873; *olivieri* Sody, 1941; **blouchi** Groves, 1981.
    COMMENTS: This species is usually credited to Müller, 1840 *in* Temminck, Verh. Nat. Gesch. Nederland. Overz. Bezitt., Zool., Zoogd. Indisch. Archipel., p. 42, but an earlier citation is Temminck, 1836 *in* von Siebold, Temminck and Schlegel, Fauna Japonica, Coup d'Oeil Faune Iles Sonde Emp. Japan, pp. viii, and Corbet and Hill (1992) noted the still earlier designation cited above. It was further described by Müller and Schlegel, *in* Temminck, Verh. Nat. Gesch. Nederland. Overz. Bezitt., Zool., Mammalia, p. 172 (and also on p. 175, but not p. 107 as widely cited)[1845], pl. 28[1843]. Synonyms apparently from Borneo and Seram were based on wrongly located specimens (Groves, 1981*a*).

**Family Tayassuidae** Palmer, 1897. Proc. Biol. Soc. Wash., 11:174.
    SYNONYMS: Dicotylidae Turner, 1849.
    COMMENTS: Dicotylidae does not have priority over Tayassuidae (Article 40.2, International Code of Zoological Nomenclature; International Commission on Zoological Nomenclature, 1999). G. M. Roosmalen (in litt.) is preparing to name a fourth species of peccary from the Amazon basin.

*Catagonus* Ameghino, 1904. An. Mus. Soc. Cient. Argent., 58:188.
    TYPE SPECIES: *Catagonus metropolitanus* Ameghino, 1904 (extinct).

*Catagonus wagneri* (Rusconi, 1930). An. Mus. Nac. Hist. Nat. Bernardino Rivadavia, 36:231.
    COMMON NAME: Chacoan Peccary.
    TYPE LOCALITY: Argentina, "Llajta-Maiica, tres leguas al noreste de Melero provincia de Santiago del Estero. Época moderna (precolombiana)".
    DISTRIBUTION: Gran Chaco of Argentina, Bolivia, and Paraguay.
    STATUS: CITES – Appendix I; IUCN – Endangered.
    COMMENTS: Originally described from pre-Hispanic and subfossil remains; subsequently discovered alive (Wetzel et al., 1975; Wetzel, 1977, 1981). Reviewed by Mayer and Wetzel (1986, Mammalian Species, 259).

*Pecari* Reichenbach, 1835. Bildergalerie der Thierwelt, part 6, p. 1.
    TYPE SPECIES: *Dicotyles torquatus* Cuvier, 1816 (= *Sus tajacu* Linnaeus, 1758).
    SYNONYMS: *Adenonotus* Brookes, 1827 [*nomen oblitum*]; *Notophorus* G. Fischer [von Waldheim], 1817 [*nomen oblitum*]; *Notophorous* Wooodburne, 1968; *Tagassu* von Frisch, 1775 [unavailable].
    COMMENTS: The Collared Peccary should be assigned to a separate genus from the White-lipped species according to Woodburne (1968), Husson (1978:347-348), and Wright (1989). Use of appropriate generic names for these taxa is controversial. Genotypes of *Tayassu* and *Dicotyles* by subsequent designation are White-lipped Peccaries (Miller and Rehn, 1901:12; Miller, 1912*b*:384). *Notophorus* and *Adenonotus* are obscure names now categorized as *nomina oblita* (Article 23.9, International Code of Zoological Nomenclature; International Commission on Zoological Nomenclature, 1999), so the valid generic name for Collared Peccaries is *Pecari*, with type by monotypy *Dicotyles torquatus*.

*Pecari tajacu* (Linnaeus, 1758). Syst. Nat., 10th ed., 1:50.

> COMMON NAME: Collared Peccary.
>
> TYPE LOCALITY: "Habitat in Mexici, Panamae, Brasiliae montibus, sylvis". Mexico selected by Thomas (1911*a*:140); however, Linnaeus's name *Sus tajacu* is based on the tajacu of Marcgraf, from Brazil, Pernambuco (Cabrera, 1961:319; Hershkovitz, 1963, 1987*b*) and this restriction of the type locality is adopted here.
>
> DISTRIBUTION: USA (mainly in Arizona and Texas), Mexico (outside the Sierra Madre), and all other Central American states; South America in N Argentina, Bolivia, Brazil, Colombia, Ecuador, French Guiana, Guyana, Paraguay, Peru, Surinam, Trinidad, and Venezuela. Introduced to Cuba.
>
> STATUS: CITES – Appendix II (populations in the USA and Mexico not covered by CITES); IUCN – Lower Risk (lc).
>
> SYNONYMS: *caitetu* Liais, 1872; *tajassu* (Erxleben, 1777); ***angulatus*** (Cope, 1889); ***bangsi*** Goldman, 1917; *modestus* Cabrera, 1917; ***crassus*** (Merriam, 1901); ***crusnigrum*** (Bangs, 1902); ***humeralis*** (Merriam, 1901); ***nanus*** (Merriam, 1901); ***nelsoni*** Goldman, 1926; ***niger*** (J. A. Allen, 1913); ***nigrescens*** Goldman, 1926; ***patira*** (Kerr, 1792); *macrocephalus* Anthony, 1921; *minor* (Kerr, 1792); *torquatus* (G. Cuvier, 1816); ***sonoriensis*** (Mearns, 1897); ***torvus*** (Bangs, 1898); *yucatanensis* (Merriam, 1901).

*Tayassu* G. Fischer [von Waldheim], 1814. Zoognosia, 3:284.

> TYPE SPECIES: *Tayassu pecari* G. Fischer [von Waldheim], 1814 (= *Sus pecari* Link, 1795).
>
> SYNONYMS: *Dicotyles* G. Cuvier, 1817; *Olidosus* Merriam, 1901.
>
> COMMENTS: By subsequent designation of Miller and Rehn (1901:12), the type of *Tayassu* is *T. pecari* G. Fischer [von Waldheim], 1814 (= *Sus pecari* Link, 1795). By subsequent designation of Miller (1912*b*:384), the type of *Dicotyles* is *D. labiatus* G. Cuvier. *Sus pecari* and *Dicotyles labiatus* are synonyms of *Tayassu pecari* (Hershkovitz, 1963). Therefore, *Dicotyles* is a synonym of *Tayassu*. Husson (1978:347-348) and Woodburne (1968) held contrary views.

*Tayassu pecari* (Link, 1795). Beitr. Naturgesch., 2:104.

> COMMON NAME: White-lipped Peccary.
>
> TYPE LOCALITY: No locality cited; identified as French Guiana, Cayenne (Hershkovitz, 1963).
>
> DISTRIBUTION: Mexico (E from Oaxaca and Veracruz) and all other Central American states; South America in N Argentina, Bolivia, Brazil, Colombia, W Ecuador, French Guiana, Guyana, Paraguay, Peru, Surinam, and Venezuela. Introduced to Cuba.
>
> STATUS: CITES – Appendix II; IUCN – Lower Risk (lc).
>
> SYNONYMS: *beebei* Anthony, 1921; ***aequatoris*** (Lönnberg, 1921); *equatorius* Rusconi, 1929; ***albirostris*** (Illiger, 1815); *labiatus* (Cuvier, 1817); ***ringens*** Merriam, 1901; ***spiradens*** Goldman, 1912.
>
> COMMENTS: Includes *albirostris*; see Husson (1978:353). Reviewed by Mayer and Wetzel (1987, Mammalian Species, 293).

**Family Hippopotamidae** Gray, 1821. London Med. Repos., 15:306.

> SYNONYMS: Choeropsinae Gill, 1872.
>
> COMMENTS: *Hexaprotodon madagascariensis* (Guldberg, 1883), *Hippopotamus lemerlei* Grandidier, 1868, and *H. laloumena* Faure and Guerin, 1990 were present in the Holocene on Madagascar (Stuenes, 1989), but have not been shown to have survived into the last 500 years.

*Hexaprotodon* Falconer and Cautley, 1836. Asia. Res. Calcutta, 19:51.

> TYPE SPECIES: *Hippopotamus sivalensis* Falconer and Cautley, 1836 (extinct fossil Asiatic species).
>
> SYNONYMS: *Choerodes* Leidy, 1852; *Choeropsis* Leidy, 1853; *Diprotodon* Duvernoy, 1849.
>
> COMMENTS: Includes *Choeropsis* Leidy, 1853, following Coryndon (1977).

*Hexaprotodon liberiensis* (Morton, 1849). J. Acad. Nat. Sci. Phila., ser. 2, 1:232.

> COMMON NAME: Pygmy Hippopotamus.
>
> TYPE LOCALITY: Liberia, "the river St. Pauls, a stream that rises in the mountains of Guinea, and passing through the Dey country and Liberia, empties into the Atlantic to the north of Cape Messurado".

DISTRIBUTION: Sierra Leone to Côte d'Ivoire; SC Nigeria (extinct?).

STATUS: CITES – Appendix II; IUCN – Critically Endangered D1 as *H. l. heslopi*, otherwise Vulnerable.

SYNONYMS: *minor* (Morton, 1844) [preoccupied]; **heslopi** (Corbet, 1969).

COMMENTS: First described as *Hippopotamus minor* Morton, 1844 (Proc. Acad. Nat. Sci. Phila., 2:14).

*Hippopotamus* Linnaeus, 1758. Syst. Nat., 10th ed., 1:74.

TYPE SPECIES: *Hippopotamus amphibius* Linnaeus, 1758.

SYNONYMS: *Hippopothamus* Boddaert, 1785; *Tetraprotodon* Falconer and Cautley, 1836.

*Hippopotamus amphibius* Linnaeus, 1758. Syst. Nat., 10th ed., 1:74.

COMMON NAME: Common Hippopotamus.

TYPE LOCALITY: "Habitat in Nilo & Bambolo Africae et ad ostia fluviorum Asiae"; restricted to River Nile (Thomas, 1911*a*:155) in Egypt (G. M. Allen, 1939:457).

DISTRIBUTION: Rivers of savanna zone of Africa, and main rivers of forest zone in C Africa, in Angola, Benin, N Botswana, Burkina Faso, Burundi, Cameroon, Central African Republic, S Chad, Côte d'Ivoire, Dem. Rep. Congo, Egypt (extinct; formerly along Nile to its Delta), N Eritrea, Ethiopia, Equatorial Guinea (Mbini), Gabon, Gambia, Ghana, Guinea, Guinea Bissau, Kenya, Liberia (only 2 records), Rwanda, Senegal, Sierra Leone, Somalia, Sudan, Swaziland, Malawi, Mozambique, Namibia (Caprivi Strip, Okavango River), Niger, Nigeria, Republic of Congo, Sierra Leone, South Africa (now only in N and E Limpopo Prov. and E Mpumalanga Prov., and N KwaZulu-Natal), Tanzania, Togo, Uganda, Zambia, and Zimbabwe.

STATUS: CITES – Appendix II; IUCN – Vulnerable as *H. a. tschadensis*, otherwise Lower Risk (lc).

SYNONYMS: *abyssinicus* Lesson, 1842; *africanus* Lacépède, 1799; *senegalensis* Desmoulins, 1826; *tschadensis* Schwarz, 1914; *typus* Duvernoy, 1846; **capensis** Desmoulins, 1825; *australis* Duvernoy, 1846; *constrictor* Zukowsky, 1924; *constrictus* Miller, 1910; **kiboko** Heller, 1914.

**Family Camelidae** Gray, 1821. London Med. Repos., 15:307.

SYNONYMS: Aucheniini Bonaparte, 1845; Lamini Webb, 1965.

COMMENTS: Extant camelids all belong to the Camelinae (McKenna and Bell, 1997). Includes as Tribes Camelini (including *Camelus*) and Aucheniini (including *Lama*). Lamini does not have priority over Aucheniini (Aucheniinae Bonaparte) (Article 40.2, International Code of Zoological Nomenclature; International Commission on Zoological Nomenclature, 1999).

*Camelus* Linnaeus, 1758. Syst. Nat., 10th ed., 1:65.

TYPE SPECIES: *Camelus dromedarius* Linnaeus, 1758, designated by Hay (1902). Has been widely cited as *C. bactrianus* (see Gentry et al., 1996) and hence a change in desigation was proposed by Erridge (1988), but has not been supported.

SYNONYMS: *Camellus* Molina, 1782; *Dromedarius* Gloger, 1841.

COMMENTS: Essentially allopatric distribution of the domesticated populations of the two species may reflect adaptations to different habitats of ancestral wild populations.

*Camelus bactrianus* Linnaeus, 1758. Syst. Nat., 10th ed., 1:65.

COMMON NAME: Bactrian Camel.

TYPE LOCALITY: "Habitat in Africa"; identified as "Bactria" (Uzbekistan, Bokhara) by Thomas (1911*a*:150); based on domesticated stock.

DISTRIBUTION: Exists in the wild in SW Mongolia and China (Gansu, Tsinghai, and Sinkiang); domesticated in Iran, Afghanistan, and Pakistan, north to Kazakhstan, Mongolia, and China.

STATUS: U.S. ESA – Endangered; IUCN – Critically Endangered.

SYNONYMS: *bocharicus* Kolenati, 1847; *caucasicus* Kolenati, 1847; *orientalis* J. Fischer, 1829 [*nomen nudum*]; *tauricus* J. Fischer, 1829 [*nomen nudum*]; **ferus** Przewalski, 1878; *genuinus* Kolenati, 1847 [*nomen oblitum*].

COMMENTS: Includes *ferus* Przewalski, based on wild specimen; *bactrianus* Linnaeus, 1758, has priority. Abramov (1996) showed that *ferus* dates from Przewalski, 1878, not 1883, and is preoccupied by *Camelus dromedarius ferus* Falk, 1786, which is probably a *nomen oblitum*. A. Gentry et al. (1996) proposed that majority usage should be confirmed by adoption of *C. ferus* as the name for the wild taxon of Bactrian camels. Though it has not been demonstrated that most authors term the wild Bactrian camel *C. ferus* rather than *C. bactrianus* (or *C. b. ferus*), they asked the International Commission on Zoological Nomenclature to use its plenary powers to rule that the name for the wild species is not invalid by virtue of being antedated by the name based on the domestic form. A ruling has now been made in their favour (International Commission on Zoological Nomenclature, 2003*a*), but it might still be valid for those who consider *C. bactrianus* and *C. ferus* to be conspecific to employ the senior name for the name of the species (see Bock, 1997). Domestic and wild camels vary by three base substitutions in the gene fragments studied and it was thought they were genetically differentiated at a level large enough that they could qualify as subspecies (Schaller, 1998). By implication, the domestic form has originated from a taxon subspecifically distinct from the extant wild form, in which case the names *C. b. bactrianus* and *C. b. ferus* would be applicable to different taxa. Bactrian and one-humped camels produce viable hybrids (*C. dromedarius hybridus* J. Fischer, 1829, unavailable) but hybrid males are said to be sterile (A. P. Gray, 1972).

*Camelus dromedarius* Linnaeus, 1758. Syst. Nat., 10th ed., 1:65.
  COMMON NAME: One-humped Camel.
  TYPE LOCALITY: "Habitat in Africae desertis arenosis siticulosis", identified as "deserts of Libya and Arabia" by Thomas (1911*a*:150); based on domesticated stock.
  DISTRIBUTION: Extinct in the wild; domesticated from wild populations which presumably had become restricted to the S Arabian Peninsula; domesticated in Senegal and Mauritania to Somalia and Kenya, throughout N Africa, the Middle East, Arabia, and Iran to NW India; feral populations in Australia.
  SYNONYMS: *aegyptiacus* Kolenati, 1847; *africanus* (Gloger, 1841); *arabicus* Desmoulins, 1823; *dromas* Pallas, 1811; *dromos* Kerr, 1792; *ferus* Falk, 1786 [*nomen oblitum*]; *lukius* Kolenati, 1847; *polytrichus* Kolenati, 1847; *turcomanicus* J. Fischer, 1829 [*nomen nudum*]; *vulgaris* Kolenati, 1847.
  COMMENTS: Produces viable hybrids with *bactrianus* (see comments therein). Bohlken (1961) considered *dromedarius* a synonym of *bactrianus*. Reviewed by Köhler-Rollefson (1991, Mammalian Species, 375). Biology reviewed by Gauthier-Pilters and Innis Dagg (1981). For history of domestication, see R. T. Wilson (1984).

*Lama* G. Cuvier, 1800. Leçons Anat. Comp., I, tab. 1.
  TYPE SPECIES: *Camelus glama* Linnaeus, 1758.
  SYNONYMS: *Aucheria* F. Cuvier, 1830; *Auchenia* Illiger, 1811; *Auchenias* Wagner, 1843; *Dromedarius* Wagler, 1830; *Guanaco* Perry, 1811; *Lacma* Tiedemann, 1804; *Lama* Frisch, 1775 [unavailable]; *Llacma* Illiger, 1815; *Llama* Gray, 1852; *Neoauchenia* Ameghino, 1891; *Pacos* Gray, 1872; *Vicunia* Rafinesque, 1815.
  COMMENTS: Evolution of domesticated llama and alpaca from wild ancestors reviewed by Wheeler (1995) and Kadwell et al. (2001).

*Lama glama* (Linnaeus, 1758). Syst. Nat., 10th ed., 1:65.
  COMMON NAME: Guanaco.
  TYPE LOCALITY: "Habitat in America meridionali", identified as "Peru", Andes, by Thomas (1911*a*:150); based on domesticated stock.
  DISTRIBUTION: Cordilleras of the Andes, Patagonia, and Tierra del Fuego in Argentina, Bolivia, Chile (including Navarino Isl), NW Paraguay, and S Peru. Domesticated as the Llama in S Peru, W Bolivia, and NW Argentina.
  STATUS: CITES – Appendix II as *Lama guanicoe*; IUCN – Endangered as *L. guanicoe huanacus*, Vulnerable as *L. g. voglii* and *L. g. cacsilensis*, otherwise Lower Risk (lc).
  SYNONYMS: *ameghiniana* López Aranguren, 1930; *araucana* (Molina, 1782); *arucana* (Kerr, 1792); *arrucana* (Link, 1795); *castelnaudi* (Gervais, 1855); *chilihueque* Boitard, 1845;

*cordubensis* (Ameghino, 1889); *domestica* Fischer, 1829; *ensenadensis* (Ameghino, 1889); *intermedia* (Gervais, 1855); *lama* (Illiger, 1811); *llama* (Link, 1795); *llacma* (F. Cuvier, 1821); *lujanensis* (Ameghino, 1889); *moromoro* (Schinz, 1845); *paco* (Gmelin, 1788); *pacos* (Linnaeus, 1758); *peruana* (Tiedemann, 1804); *peruviana* Lesson, 1827; *vulgaris* Wagner, 1837; **cacsilensis** Lönnberg, 1913; **guanicoe** (Müller, 1776); *fera* Gray, 1843; *guanaco* (Perry, 1811); *guanacos* (Schinz, 1845); *guanacus* Gray, 1852; *huanaca* (C. H. Smith, 1827); *huanacha* Elliot, 1907; *huanachus* Thomas, 1891; *huanacos* Sclater, 1891; *huanacus* (Molina, 1782); *llama* (Waterhouse, 1839); *molinaei* Boitard, 1845; *voglii* Krumbiegel, 1944.

COMMENTS: Haltenorth (1963) recognized four subspecies (*cacsilensis, guanicoe, huanacus, voglii*) but Wheeler (1995) regarded them as poorly defined and did not diagnose them; subspecies are those recognised by Cabrera (1961). The Guanaco has previously been included with the Llama, *L. glama*, of which it is understood to be the wild ancestor (Hemmer, 1990; Kadwell et al., 2001; Lydekker, 1915). Gentry et al. (1996) proposed that majority usage be confirmed by adoption of *Lama guanicoe* as the name for the wild Guanaco and asked the International Commission on Zoological Nomenclature to use its plenary powers to rule that the name for this wild species is not invalid by virtue of being antedated by the name based on the domestic form. A ruling has now been made in their favour (International Commission on Zoological Nomenclature, 2003a). It might still be valid for those who consider *L. glama* and *L. guanicoe* to be conspecific to employ the senior name for the name of the species (see Bock, 1997). Provisionally, *guanicoe* is listed here as a subspecies of *glama*. The Alpaca (*Lama pacos* including *lujanensis* and *paco*) has been regarded as a synonym of *glama*; see Corbet and Hill (1991:126). It originated from hybrids between *Lama glama* and *Vicugna vicugna* according to Hemmer (1990). Wheeler (1995) favoured a primary origin from the Vicugna and post-Conquest genetic introgression from the Llama. Kadwell et al. (2001) found that there was a relatively low estimated admixture of Vicugna mtDNA in the Alpaca but a high proportion for microsatellites (from four loci studied), and inferred that the Vicugna is the ancestor of the Alpaca. Nevertheless, the Alpaca appears to be of biphyletic origin and its synonymy may be somewhat arbitrary, though it could yet be shown that *pacos* is best regarded as the same species as *vicugna*.

*Vicugna* Lesson, 1842. Nouv. Tabl. Regn. Anim. Mammifères, p. 167.
TYPE SPECIES: *Camelus vicugna* Molina, 1782.

*Vicugna vicugna* (Molina, 1782). Sagg. Stor. Nat. Chile, p. 313.
COMMON NAME: Vicugna.
TYPE LOCALITY: Chile, "abondano nella parte della Cordigliera spettante alle Provincie de Coquimbo, e di Copiapò" (cordilleras of Coquimbo and Copiapo).
DISTRIBUTION: NW Argentina, W Bolivia, N Chile, and S Peru.
STATUS: CITES – Appendix I [except for populations of Bolivia and Peru, and parts of the population in Argentina and Chile, which are included in Appendix II]; U.S. ESA – Endangered; IUCN – Lower Risk (cd).
SYNONYMS: *elfridae* (Krumbiegel, 1949); *frontosa* (H. Gervais and Ameghino, 1880); *gracilis* (H. Gervais and Ameghino, 1880); *mensalis* (Thomas, 1917); *minuta* (Burmeister, 1891); *pristina* (Amhegino, 1891); *provicugna* (Boule, 1920); *vicunia* (Tschudi, 1844); *vicunna* (Tiedemann, 1804).
COMMENTS: Systematics reviewed by Wheeler (1995). Regarded as polytypic (*mensalis* a distinct subspecies) by some authors (Haltenorth, 1963) but systematics here follows Cabrera (1961). Kadwell et al. (2001) suggested that the Alpaca should be assigned to *Vicugna*. Gentry et al. (1996) had already proposed that majority usage be confirmed by adoption of *Vicugna vicugna* as the name for the wild taxon of Vicugna on the assumption that the Alpaca is the domesticated descendent and asked the International Commission on Zoological Nomenclature to use its plenary powers to rule that the name for this wild species is not invalid by virtue of being antedated by *pacos,* the name based on the domestic form. A ruling has been made in their favour (International Commission on Zoological Nomenclature, 2003a).

**Family Tragulidae** Milne-Edwards, 1864. Ann. Sci. Nat. Zool. Paris, ser. 5, 2:157.
> COMMENTS: From the description, *Moschus leverianus* Kerr, 1792 may be a tragulid, but its synonymy has not been determined.

*Hyemoschus* Gray, 1845. Ann. Mag. Nat. Hist., [ser. 1], 16:350.
> TYPE SPECIES: *Moschus aquaticus* Ogilby, 1841.
> SYNONYMS: *Hyaemoschus* Zittel, 1893; *Hyeomoschus* Turner, 1850; *Hyomoschus* Blyth, 1865.

> *Hyemoschus aquaticus* (Ogilby, 1841). Proc. Zool. Soc. Lond., 1840:35 [1841].
>> COMMON NAME: Water Chevrotain.
>> TYPE LOCALITY: "Sierra Leone".
>> DISTRIBUTION: W Africa in Ghana, Côte d'Ivoire, Liberia, Sierra Leone; C Africa in Angola (Cabinda), Cameroon, Central African Republic, Dem. Rep. Congo, Equatorial Guinea (Mbini), Gabon, Nigeria, Republic of Congo, and Bwamba Forest of Semliki Valley, Uganda, where not known to survive according to East et al. (1999) but seems likely to be present according to Kingdon's (1979) account. Supposed occurrence in Benin, Gambia, Guinea Bissau, Senegal, and Togo unsupported by evidence.
>> STATUS: CITES – Appendix III (Ghana); IUCN – Data Deficient (DSG recommended).
>> SYNONYMS: *batesi* (Lydekker, 1906); *cottoni* (Lydekker, 1906); *typicus* (Lydekker, 1906).

*Moschiola* Gray, 1852. Cat. Mamm. Brit. Mus., part 3, Ungulata Furcipeda, p. 247.
> TYPE SPECIES: *Moschus meminna* Erxleben, 1777.
> SYNONYMS: *Meminna* Gray, 1836 [*nomen oblitum?*]; *Moschiola* Hodgson, 1843 [*nomen nudum*].
> COMMENTS: Treated as a full genus by Groves and Grubb (1987), following Flerov (1931). It has yet to be shown that *Meminna* Gray, 1836 fully qualifies as a *nomen oblitum* (Article 23.9.1.2, International Code of Zoological Nomenclature; International Commission on Zoological Nomenclature, 1999).

> *Moschiola meminna* (Erxleben, 1777). Syst. Regn. Anim., 1:322.
>> COMMON NAME: Indian Spotted Chevrotain.
>> TYPE LOCALITY: "in Ceylona" (Sri Lanka).
>> DISTRIBUTION: Sri Lanka and peninsular India. Supposed occurrence in Himalayan foothills of India and Nepal not confirmed (Champion, 1929).
>> STATUS: IUCN – Lower Risk (lc).
>> SYNONYMS: *ceylonensis* (Pallas, 1779) [*nomen nudum*]; *indica* (Gray, 1843); *malaccensis* (Gray, 1843); *memennoides* (Hodgson, 1841) [*nomen nudum*]; *mimenoides* (Hodgson, 1842) [*nomen nudum*].

*Tragulus* Brisson, 1762. Regn. Anim., 2nd ed., pp. 12, 65.
> TYPE SPECIES: *Cervus javanicus* Osbeck, 1765.
> SYNONYMS: *Lagonebrax* Gloger, 1841.
> COMMENTS: *Tragulus* was attributed to Brisson (1762) by many authors (A. Gentry, 1994) though some (e.g. Chasen, 1940; Lydekker, 1915) assigned it to Pallas (1779). Recent rejection of Brisson (1762) was on the assumption that it was unavailable. Brisson (1762:65-68) listed the species *T. indicus, T. guineensis* (= *Neotragus pygmaeus*), *T. surinamensis* (= *Mazama americana*), *T. africanus* (= *Sylvicapra grimmia*), and [*T.*] *moschus* (= *Moschus moschiferus*). Of these, only *T. indicus* is referred to *Tragulus* as currently used, according to A. Gentry (1994:141). Brisson's *Tragulus* was defined by lacking horns (or antlers). However, three species were included on the strength of females or immatures which lack horns or antlers, while adult males possess them, so the character can not help to confirm that *T. indicus* is a mouse-deer. This nominal species is also based on descriptions of specimens of *N. pygmaeus* by Linnaeus and Seba (and secondary citations by Klein) and on Kolbe's "Chevre de Congo". Brisson's own description is mostly undiagnostic but indicates a small animal with upperparts of head, neck and body reddish-yellow mixed with blackish; throat, belly, and inner parts of thighs whitish; and 26 teeth (so not with the whole tragulid complement of 34), including two upper canines, one on each side. The vernacular name is "Le Chevrotain des Indes". The streaked pelage suggests a species of *Tragulus* rather than the uniformly-coloured *N. pygmaeus*, and so

does the presence of upper canines, as noted by A. W. Gentry (1995). Although infants of
*N. pygmaeus* often have upper milk canines, canines are rarely present in subadults and
adults and are then very small. Reference to the Indies is suggestive. This is the evidence
for inferring that Brisson had studied a subadult specimen of mouse-deer, one of the
syntypes of *T. indicus*. *Tragulus indicus* Brisson could be identified as a mouse-deer and the
genus could apply to these mammals. Merriam (1895c:375) designated *Tragulus indicus* as
the type of *Tragulus* Brisson, but identified it as the same as "*Capra pygmea* Linn., which
becomes *Tragulus pygmeus* (Linn.) 1758" (i.e. *N. pygmaeus*), probably for reasons suggested
by A. W. Gentry (1995). He did not designate a lectotype for *T. indicus* that was
*N. pygmaeus* (for instance, Linnaeus' or one of Seba's specimens) and ignored the evidence
that a tragulid was among the syntypes: the species can still be regarded as a tragulid.
Gardner (1995:79, 81) preferred to treat *Tragulus* Brisson as unavailable and to date
*Tragulus* from Pallas (1767, fasc. 6, p. 6), but this would appear to be a *nomen nudum*, with
*Tragulus* not distinguished from *Moschus*. It is also the only place where Pallas cited
*Tragulus pygmaeus* (possibly a misidentification of *N. pygmaeus* as a tragulid), which
contrary to A. Gentry (1994:140) is not the single species included in the genus by which
*Tragulus* Pallas was made available. Elsewhere, Pallas (1779, fasc. 13, p. 28) provided a
diagnosis of *Tragulus* and in a footnote referred to *T. ceylonensis*, *nomen nudum*, and
*T. javensis*, a name validated by a reference to Pallas' earlier description of a tragulid from
Java (Pallas, 1777, fasc. 12, p. 18). Hopwood (1947:534) considered *Tragulus* Brisson to be
unavailable and by quoting selected text, denied that Pallas defined or formally adopted
the genus *Tragulus* (he did not refer to the diagnosis in fasc. 13, p. 28). He therefore
assigned *Tragulus* to Boddaert (1785:131), who included in the genus *T. moschus*
(= *Moschus moschiferus*), *T.* (= *Moschiola*) *meminna*, and *T. pygmaeus*. Hopwood (1947:534)
selected *T. pygmaeus* as the type and as this is *Neotragus pygmaeus*, *Tragulus* Boddaert
would be referred to the Bovidae (A. Gentry, 1994: 141). But because *Tragulus* Boddaert is
a replacement for *Moschus* Linnaeus, 1758 and thus an objective synonym, the type of the
genus must be *T. moschus* (= *Moschus moschiferus*) according to Ellerman and Morrison-
Scott (1951:350). This would have to be set aside if *T. meminna* = *Moschus meminna*
Erxleben, 1777 were to be treated as the type of *Tragulus*, should *Tragulus* Brisson or
*Tragulus* Pallas be regarded as unavailable, and should this genus be therefore referred to
Boddaert (Gardner, 1995:81). Brisson (1762) was rejected for nomenclatural purposes (but
only in 1998), so it might seem that *Tragulus* Pallas, 1779:29 is now the first available
designation of the genus; however, *Tragulus* Brisson, 1762 has been conserved (Opinion
1894, International Commission on Zoological Nomenclature, 1998), with *Cervus
javanicus* Osbeck, 1765 as type species. In the same Opinion, *Tragulus* Pallas, 1767
(though not *Tragulus* Pallas, 1779) and *Tragulus* Boddaert, 1785 were placed on the
Official Index of Rejected and Invalid Generic Names in Zoology. It has long been
assumed that *Tragulus* includes only two species, *T. javanicus* (*T. kanchil* in the older
literature) and *T. napu* (mistakenly given the name *T. javanicus* in the older literature); see
Van Bemmel (1949b). A revision by Meijaard and Groves (2004) recognised six species in
three species-groups: *napu* group (including *T. napu* and *T. nigricans*), *versicolor* group
(monotypic) and *javanicus* group (including *T. javanicus*, *T. kanchil*, and *T. williamsoni*).

*Tragulus javanicus* (Osbeck, 1765). Reise nach Ostindien und China, p. 357.

    COMMON NAME: Java Mouse-deer.

    TYPE LOCALITY: Indonesia, W Java, Udjung Kulon Peninsula, "Nieu Bay" (Meeuwenbaai or
        Muara Tjikuja), Jankolan (Djungkulan) kampong; identified by Van Bemmel (1949b)
        and Hoogerwerf (1970:353).

    DISTRIBUTION: Indonesia (Java).

    STATUS: IUCN – Lower Risk (lc).

    SYNONYMS: *focalinus* Miller, 1903; *indicus* Brisson, 1765 [not available]; *indicus* (Gmelin,
        1788); *jasanicus* (C. H. Smith, 1827); *javanicus* Gmelin, 1788; *javensis* Pallas, 1779
        [*nomen oblitum*].

    COMMENTS: Meijaard and Groves (2004) were not convinced that *Cervus javanicus* Osbeck,
        1765 is a mouse-deer and preferred to date the name from *Tragulus javanicus* (Gmelin,
        1788). Until their evidence is published, the older name is retained here. *Tragulus
        javensis* Pallas, 1779 is available and predates *Moschus javanicus* Gmelin, 1788. The two

names are objective synonyms because they are both based on the description of a specimen by Pallas (1777, fasc. 12, p. 18), but *javensis* has not been noticed and is a *nomen oblitum*.

*Tragulus kanchil* (Raffles, 1821). Trans. Linn. Soc. Lond., 13:239.
 COMMON NAME: Lesser Mouse-deer.
 TYPE LOCALITY: "Sumatra"; identified as Indonesia, Sumatra, Bengkulu (Meijaard and Groves, 2004).
 DISTRIBUTION: Indochina, Burma (isthmus of Kra), Brunei, Cambodia, China (S Yunnan), Indonesia (Kalimantan, Sumatra, and many small islands), Laos, Malaysia (peninsular Malaya, Sarawak, and many small islands), Singapore, Thailand, and Vietnam.
 SYNONYMS: *pelandoc* C. H. Smith, 1827; *abruptus* Chasen, 1935; *affinis* Gray, 1861; *pierrei* Bonhote, 1903; *anambensis* Chasen and Kloss, 1928; *angustiae* Kloss, 1918; *brevipes* Miller, 1903; *carimatae* Miller, 1906; *everetti* Bonhote, 1903; *natunae* Miller, 1903; *fulvicollis* Lyon, 1908; *fulviventer* Gray, 1836; *fuscatus* Blyth, 1858; *pumilus* Chasen, 1940; *hosei* Bonhote, 1903; *virgicollis* Miller, 1903; *insularis* Chasen, 1940; *klossi* Chasen, 1935; *lampensis* Miller, 1903; *lancavensis* Miller, 1903; *longipes* Lyon, 1908; *luteicollis* Lyon, 1906; *masae* Lyon, 1916; *mergatus* Thomas, 1923; *pallidus* Miller, 1901; *penangensis* Kloss, 1918; *pidonis* Chasen, 1940; *pinius* Lyon, 1916; *ravulus* Miller, 1903; *ravus* Miller, 1902; *rubeus* Miller, 1903; *russeus* Miller, 1903; *russulus* Miller, 1903; *siantanicus* Chasen and Kloss, 1928; *subrufus* Miller, 1903.

*Tragulus napu* (F. Cuvier, 1822). *In* É. Geoffroy Saint-Hilaire and F. Cuvier, Hist. Nat. Mammifères, 4, part 37, "Le chevrotain napu", p. 2, pl. 329.
 COMMON NAME: Greater Mouse-deer.
 TYPE LOCALITY: Indonesia, "Sumatra"; restricted to the southern part of Sumatra by Sody (1931:355).
 DISTRIBUTION: Indochina, Burma (isthmus of Kra), Brunei, Cambodia, Indonesia (Kalimantan, Sumatra, and many small islands), Laos, Malaysia (peninsular Malaya, Sarawak, and many small islands), Singapore, and Thailand.
 STATUS: IUCN – Lower Risk (lc).
 SYNONYMS: *abjectus* Chasen, 1935; *annae* Matschie, 1897; *borneanus* Miller, 1902; *canescens* Miller, 1900; *umbrinus* Miller, 1900; *amoenus* Miller, 1903; *jugularis* Miller, 1903; *bancanus* Lyon, 1906; *banguei* Chasen and Kloss, 1931; *batuanus* Miller, 1903; *billitonus* Lyon, 1906; *bunguranensis* Miller, 1901; *flavicollis* Miller, 1903; *hendersoni* Chasen, 1940; *lutescens* Miller, 1903; *neubronneri* Sody, 1931; *niasis* Lyon, 1916; *nigricollis* Miller, 1902; *nigrocinctus* Miller, 1906; *parallelus* Miller, 1911; *pretiellus* Miller, 1906; *rufulus* Miller, 1900; *formosus* Miller, 1903; *perflavus* Miller, 1906; *pretiosus* Miller, 1902; *sebucus* Lyon, 1911; *stanleyanus* Gray, 1836; *terutus* Thomas and Wroughton, 1909.

*Tragulus nigricans* Thomas, 1892. Ann. Mag. Nat. Hist. ser. 6, 9:254.
 COMMON NAME: Philippine Mouse-deer.
 TYPE LOCALITY: "Balabac, Philippine Islands".
 DISTRIBUTION: Philippines (Balabac, Bugsuc, and Ramos Isls).
 STATUS: IUCN – Endangered as *I. napu nigricans*.

*Tragulus versicolor* Thomas, 1910. Ann. Mag. Nat. Hist., ser. 8, 5:535.
 COMMON NAME: Vietnam Mouse-deer.
 TYPE LOCALITY: Vietnam, "Nhatrang, Annam".
 DISTRIBUTION: Vietnam.

*Tragulus williamsoni* Kloss, 1961. J. Nat. Hist. Soc. Siam, 2:88.
 COMMON NAME: Williamson's Mouse-deer.
 TYPE LOCALITY: "Me Song forest, Pre, North Siam" (N Thailand, Song forest, Muang Pre, Meh Lem, 18°25'N, 100°23'E, according to Meijaard and Groves, 2004).
 DISTRIBUTION: N Thailand.
 COMMENTS: Known only from the holotype.

**Family Moschidae** Gray, 1821. London Med. Repos., 15:307.
>    COMMENTS: A family separate from the Cervidae; see Flerov (1960), Webb and Taylor (1980), Groves and Grubb (1987), and Janis and Scott (1987).

*Moschus* Linnaeus, 1758. Syst. Nat., 10th ed., 1:66.
>    TYPE SPECIES: *Moschus moschiferus* Linnaeus, 1758.
>    SYNONYMS: *Odontodorcus* Gistel, 1848; *Tragulus* Boddaert, 1785 (preoccupied).
>    COMMENTS: Species limits in Himalayas are still uncertain; see Cai and Feng (1981), Groves (1976, 1980a), Groves et al. (1995), Groves and Grubb (1987), and Grubb (1982a). Su et al. (1999) recognised the following phylogeny from study of cytochrome *b* genes: (*moschiferus*) ((*berezovskii*) (*chrysogaster, fuscus, leucogaster*)), confirming affinity of at least some alpine taxa. Sokolov and Prikhod'ko (1996, 1997) recognised only one species in the genus.

*Moschus anhuiensis* Wang, Hu, and Yan, 1982. Acta Ther. Sin., 2:133.
>    COMMON NAME: Anhui Musk Deer.
>    TYPE LOCALITY: China, "Changling region (31°10′42″N, 115°53′48″E, altitude 500 m), Jinzhai county, Anhui province".
>    DISTRIBUTION: Known only from Anhui Prov., China.
>    STATUS: CITES – Appendix II.
>    COMMENTS: Originally described as a subspecies of *M. moschiferus*; included in *M. berezovskii* by Groves and Feng (1986); a valid species according to Su et al. (2001), inferred from mtDNA sequences to be the sister taxon of a group including *M. chrysogaster, M. fuscus, M. leucogaster,* and *M. berezovskii.*

*Moschus berezovskii* Flerov, 1929. C. R. Acad. Sci. U.S.S.R., 1928A:519 [1929].
>    COMMON NAME: Forest Musk Deer.
>    TYPE LOCALITY: "Mountain défilé [sic] Ho-tzi-how, environs of town Lun-ngan-fu, Sze-chuan, China" (China, Sichuan, near Lungan, Ho-tsi-how Pass).
>    DISTRIBUTION: S and C China (Shaanxi to Yunnan, and S Tibet) and N Vietnam.
>    STATUS: CITES – Appendix II; U.S. ESA – Endangered in Tibet and Yunnan (China); IUCN – Lower Risk (nt).
>    SYNONYMS: **bijiangensis** Wang and Li, 1993; *caobangis* Dao, 1969; **yanguiensis** Wang and Ma, 1993.
>    COMMENTS: Revised by Y. Wang et al (1993). A well-defined species sharply distinct from the parapatric or marginally sympatric *M. chrysogaster*; see Kao (1963), Groves (1976), and Grubb (1982a), yet treated as a synonym of *M. moschiferus sifanicus* (= *M. chrysogaster sifanicus*) by Sokolov and Prikhod'ko (1997).

*Moschus chrysogaster* (Hodgson, 1839). J. Asiat. Soc. Bengal, 8:203.
>    COMMON NAME: Alpine Musk Deer.
>    TYPE LOCALITY: "Cis and Trans Hemelayan regions"; "lofty mountains of the interior of Tibet, especially towards the Chinese frontier, where the first and loveliest, or Chrysogaster, is almost exclusively found I have specimens of all three species [*chrysogaster, leucogaster, saturatus*] from Lassa and Digurchee" (Hodgson, 1842, J. Asiatic Soc. Bengal, 6:285). (China, Tibetan Plateau).
>    DISTRIBUTION: Bhutan, S and C China (S Gansu, S Ningxia, Qinghai, W Sichuan, S Tibet, and N Yunnan), N India (Sikkim), and Nepal.
>    STATUS: CITES – Appendix I in Bhutan, India and Nepal; otherwise Appendix II; U.S. ESA – Endangered in Bhutan, China (Yunnan and Tibet), India and Nepal; IUCN – Lower Risk (nt).
>    SYNONYMS: **sifanicus** Büchner, 1891.
>    COMMENTS: A well defined species; see Groves (1976) and Gao (1963), under the name *sifanicus*, which should be included in *M. chrysogaster* (see Grubb, 1982a). However, "*chrysogaster*" of Cai and Feng (1981) is subspecifically or specifically distinct and available name for this taxon may be *leucogaster* Hodgson, 1839; see Grubb (1982a).

*Moschus cupreus* Grubb, 1982. Säugetier. Mitt., 30:133.
>    COMMON NAME: Kashmir Musk Deer.

TYPE LOCALITY: India or Pakistan, "Kashmir (no precise locality)".

DISTRIBUTION: Himalayas of India and Pakistan in Kashmir, and N Afghanistan.

STATUS: CITES – Appendix I; U.S. ESA – Endangered.

COMMENTS: Originally described as a subspecies of *chrysogaster*; very similar to *leucogaster*; Groves et al. (1995) suggested that *cupreus* might be a separate species.

*Moschus fuscus* Li, 1981. Zool. Res. Kunming, 2:159.
  COMMON NAME: Black Musk Deer.
  TYPE LOCALITY: China, "Bapo, Gongshan-Xian, Yunnan. Altitude 3,500 m".
  DISTRIBUTION: N Burma, China (NW Yunnan and SE Tibet), India (Assam), and Nepal.
  STATUS: CITES – Appendix I in Bhutan, Burma, India, and Nepal; otherwise Appendix II; U.S. ESA – Endangered; IUCN – Lower Risk (nt).
  COMMENTS: *Moschus saturatus* Hodgson, 1839 may be a prior name for this species. Gao (1985) treated *fuscus* as a subspecies of *chrysogaster*.

*Moschus leucogaster* Hodgson, 1839. J. Asiat. Soc. Bengal, 8:203.
  COMMON NAME: Himalayan Musk Deer.
  TYPE LOCALITY: "Cis and Trans Hemelayan regions"; "lofty mountains of the interior of Tibet . . . On the Tibetan slopes of the Himanchal, Saturatus chiefly resides . . . I have specimens of all three species [*chrysogaster, leucogaster, saturatus*] from Lassa and Digurchee, whilst my garden is seldom deprived of the ornament of several live families of the Saturatus of the Kachar [Alpine life-zone]" (Hodgson, 1842, J. Asiatic Soc. Bengal, 6:285) (Nepal, Himalayas).
  DISTRIBUTION: Himalayas of Bhutan, N India (incl. Sikkim), and Nepal.
  STATUS: CITES – Appendix I; ESA – Endangered; IUCN – Lower Risk (nt) as *M. chrysogaster leucogaster*.
  SYNONYMS: *cacharensis* Lydekker, 1915 [*nomen nudum*]; *saturatus* Hodgson, 1839; *zhangmu* Groves, Wang and Grubb, 1995 [*nomen nudum*].
  COMMENTS: Groves and Grubb (1987) and Groves et al. (1995) treated *leucogaster* as a separate species from *M. chrysogaster*, from which it differs in skull proportions; Grubb (1990) listed it as a Himalayan subspecies-group of *M. chrysogaster*.

*Moschus moschiferus* Linnaeus, 1758. Syst. Nat., 10th ed., 1:66.
  COMMON NAME: Siberian Musk Deer.
  TYPE LOCALITY: "Habitat in Tataria versus Chinam"; restricted to Russia, SW Siberia, Altai Mtns by Heptner et al. (1961).
  DISTRIBUTION: Forests of Russia (Sakhalin Isl and E Siberia), N China (N Sinkiang; Inner Mongolia to Shanxi), Korea, and N Mongolia.
  STATUS: CITES – Appendix II; IUCN – Vulnerable.
  SYNONYMS: *altaicus* Eschscholtz, 1830; *fasciatus* Gray, 1872; *maculatus* Gray, 1872; *moschus* (Boddaert, 1785); *sibiricus* Pallas, 1779; **arcticus** Flerov, 1929; **parvipes** Hollister, 1911; **sachalinensis** Flerov, 1929; **turowi** Zalkin, 1945.
  COMMENTS: Includes *sibiricus*; see Corbet (1978c:198). Revised by Sokolov and Prikhod'ko (1996, 1997).

**Family Cervidae** Goldfuss, 1820. Handb. Zool., 2:xx, 374.
  COMMENTS: Reviewed by Whitehead (1972) and Groves and Grubb (1987). For introduced populations, see Lever (1985). For revision of the whole family, see Geist (1998). The following names have not been identified: *Cervus anomalus* Kerr, 1792; *C. minutus* Kerr, 1792; *C. paludosus* Kerr, 1792; *C. squinaton* Kerr, 1792.

**Subfamily Capreolinae** Brookes, 1828. Cat. Anat. Zool. Mus. J. Brookes, p. 62.
  SYNONYMS: Alceini Brookes, 1828; Elaphalcedae Brookes, 1828 [*nomen oblitum*]; Mazamadae Brookes, 1828 [*nomen oblitum*]; Mazaminae Kraglievitch, 1932; Odocoileini Pocock, 1923; Pudinae Pocock, 1923; Rangiferini Brookes, 1828; Subulidae Brookes, 1828 [*nomen oblitum*].
  COMMENTS: For use of Capreolinae, see Lister et al. (1998). The widely used term Neocervinae Carette, 1922 includes Odocoileini and Rangiferini and as it is not based on any genus is an unavailable name. Tribe Alcini is now to be spelt Alceini, approved by Opinion 1081

(International Commission on Zoological Nomenclature, 1977c). Allocation of genera to tribes is as follows: Alceini (*Alces*), Capreolini (*Capreolus*), Odocoileini (*Blastocerus, Mazama, Odocoileus, Ozotoceros*), Odocoileini or Rangiferini (*Hippocamelus, Pudu*), Rangiferini (*Rangifer*). Webb (2000) transferred *Hippocamelus* and *Pudu* to Rangiferini.

*Alces* Gray, 1821. London Med. Repos., 15:307.

TYPE SPECIES: *Cervus alces* Linnaeus, 1758.

SYNONYMS: *Alce* Frisch, 1775 [unavailable]; *Alcelaphus* Gloger, 1841; *Paralces* J. Allen, 1902.

COMMENTS: Has been regarded as a monotypic genus by most recent workers but Boyeskorov (1999) treated the *alces* and *americanus* subspecies groups as species. They are said to be separated by the Yenisei River in Siberia. More information is required on the location and nature of the contact zone, particular in the upper Yenisei, Mongolia, and China. It has yet to be confirmed that both species occur or occurred in Mongolia and China. Genus revised by Peterson (1952); reviewed by Franzmann (1981, Mammalian Species, 154) and Geist (1998).

*Alces alces* (Linnaeus, 1758). Syst. Nat., 10th ed., 1:66.

COMMON NAME: Eurasian Elk.

TYPE LOCALITY: "Habitat in boroealibus Europae, Asiaeque Populetis"; identified as Sweden by Thomas (1911a:151).

DISTRIBUTION: N Eurasia from Scandinavia, Poland, N Austria, and S Czech Republic (vagrant in Croatia, Hungary, and Romania), east to the Yenisei River (Siberia) and south to Ukraine, N Kazakhstan, N China (N Sinkiang), and possibly adjacent parts of Mongolia; extinct in Caucasus region since 19th century.

STATUS: IUCN – Lower Risk (lc).

SYNONYMS: *aces* (Shaw, 1801); *albes* (Bowdich, 1821); *alce* (Boddaert, 1785); *angusticephalus* Zukowsky, 1915; *antiquorum* Rüppell, 1842; *europaeus* Burnett, 1830 [*nomen nudum*]; *jubatus* Fitzinger, 1860; *machlis* Ogilby, 1837; *malchis* Gray, 1850; *palmatus* Gray, 1843; *platycephalus* Pusch, 1840; *resupinatus* Rouillier, 1842; *tymensis* Zukowsky, 1915; *typicus* Ward, 1910; *uralensis* Matschie, 1913; *vulgaris* de Serres, 1835; **caucasicus** Vereshchagin, 1955.

COMMENTS: Differs from *A. americanus* in karyotype, body dimensions and proportions, form of premaxilla, colouration, and structure and dimensions of antlers (Boyeskorov, 1999; Geist, 1998).

*Alces americanus* (Clinton, 1822). Letters on the natural history . . . of New York, p. 193.

COMMON NAME: Moose.

TYPE LOCALITY: "Country north of Whitestown". USA, New York, probably in the western Adirondack region.

DISTRIBUTION: Russia (E Siberia), east of the Yenisei River east to Anadyr region (E Siberia) and south to N Mongolia and N China (N of Inner Mongolia and Manchuria). N America in Canada and N USA (including Alaska); introduced to New Zealand where now extinct.

STATUS: IUCN – Lower Risk (nt) as *A. alces cameloides*.

SYNONYMS: *andersoni* Peterson, 1950; *buturlini* Chernyavsky and Zhelesnov, 1982; *columbae* Lydekker, 1915; *gigas* Miller, 1899; *lobatus* (Agassiz, 1846); *meridionalis* Matschie, 1913; *muswa* Richardson, 1852; *pfizenmayeri* Zukowsky, 1910; *shirasi* Nelson, 1914; *yakutskensis* Millais, 1911; **cameloides** (Milne-Edwards, 1867); *bedfordiae* Lydekker, 1902.

COMMENTS: Subspecies limits follow Geist (1998). Characters diagnosing species not fully confirmed for *cameloides*. *Cervus americanus* Clinton, 1822 is preoccupied by *Cervus americanus* Erxleben, 1777 (= *Odocoileus virginianus*), a name used in the literature that has not been declared to be unavailable. It is probably a *nomen oblitum* and the familiar name *Alces americanus* (Clinton, 1822) continues to be used here. *Cervus coronatus* Lesson, 1827 (= *C. coronatus* É. Geoffroy Saint-Hilaire, 1803), based on a single rack, is usually cited as a synonym of *Alces alces* but is reputedly from America. It is much too small to be a Moose (C. H. Smith, 1827), and is possibly an aberrant *Rangifer tarandus*, according to Blyth (1860).

*Blastocerus* Wagner, 1844. *In* von Schreber, Die Säugetiere, 4:366.
> TYPE SPECIES: *Cervus paludosus* Desmarest, 1822 (= *Cervus dichotomus* Illiger, 1815).
> SYNONYMS: *Bezoarticus* Marelli, 1932; *Blastoceros* Fitzinger, 1873; *Blastoros* Knottnerus-Meyer, 1907; *Edoceros* Avila-Pires, 1957.
> COMMENTS: Included in *Odocoileus* by Haltenorth (1963:44-45), but generically distinct (Groves and Grubb, 1987). Hershkovitz (1958) argued that first valid use of generic name was Gray, 1850 but Grubb (2000*a*) provided evidence to support wide acceptance of Wagner, 1844 as author.

*Blastocerus dichotomus* (Illiger, 1815). Abh. Phys. Klasse K.-Preuss. Akad. Wiss., 1804-1811:117 [1815].
> COMMON NAME: Marsh Deer.
> TYPE LOCALITY: No locality given; based on the gouazoupoukou of Azara; restricted to Paraguay, Lake Ypoá, south of Asuncion (Cabrera, 1961:329).
> DISTRIBUTION: N Argentina, Bolivia, Brazil (S of Amazon River), Paraguay, E Peru, and Uruguay.
> STATUS: CITES – Appendix I; U.S. ESA – Endangered; IUCN – Critically Endangered (Paraná Brazilian Basin subpopulation), Endangered (Delta del Paraná subpopulation), otherwise Vulnerable.
> SYNONYMS: *ensenadensis* (Ameghino, 1888); *furcata* (Gray, 1843); *paludosus* (Desmarest, 1822); *palustris* (Desmoulins, 1823).
> COMMENTS: Reviewed by Pinder and Grosse (1991, Mammalian Species, 380).

*Capreolus* Gray, 1821. London Med. Repos., 15:307.
> TYPE SPECIES: *Cervus capreolus* Linnaeus, 1758.
> SYNONYMS: *Caprea* Ogilby, 1837; *Capreolus* Frisch, 1775 [unavailable].
> COMMENTS: Reviewed by Sokolov et al (1986*c*).

*Capreolus capreolus* (Linnaeus, 1758). Syst. Nat., 10th ed., 1:68.
> COMMON NAME: European Roe.
> TYPE LOCALITY: "Habitat in Europa, Asia"; identified as Sweden by Thomas (1911*a*:151).
> DISTRIBUTION: Europe (excluding Corsica, Ireland, Sardinia, and Sicily) to W Russia and Ukraine, Turkey, Caucasus region, NW Syria, N Iraq, N Iran; extinct in Lebanon and Israel; Protoneolithic record from Jordan (Jericho).
> STATUS: IUCN – Lower Risk (lc).
> SYNONYMS: *albicus* Matschie, 1910; *albus* (Kerr, 1792); *armenius* Blackler, 1916; *baleni* Martino, 1933; *balticus* Matschie, 1910; *capraea* Gray, 1843; *cistaunicus* (Matschie, 1913); *coxi* Cheesman and Hinton, 1923; *dorcas* Burnett, 1830 [*nomen nudum*]; *europaeus* Sundevall, 1846; *grandis* Bolkay, 1925; *illyricus* von Lehmann and Sägesser, 1986 [*nomen nudum*]; *joffrei* Blackler, 1916; *niger* Fitzinger, 1874; *plumbeus* (Reichenbach, 1845); *rhenanus* Matschie, 1910; *thotti* Lönnberg, 1910; *transsylvanicus* Matschie, 1907; *transvosagicus* (Matschie, 1913); *varius* Fitzinger, 1874; *vulgaris* Fitzinger, 1832; *warthae* Matschie, 1912; *whittalli* Barclay, 1936; *zedlitzi* Matschie, 1916; **canus** Miller, 1910; *decorus* Cabrera, 1916; *garganta* Meunier, 1983; **caucasicus** Dinnik, 1910; **italicus** Festa, 1925.
> COMMENTS: Reviewed by Sempéré et al. (1996, Mammalian Species, 538) and by Lister et al. (1998) whose identification of *caucasicus* as correct name for large-sized subspecies north of Caucasus Mtns is provisional. Treatment of *italicus* as a valid subspecies follows Lorenzini et al. (2002).

*Capreolus pygargus* (Pallas, 1771). Reise Prov. Russ. Reichs, 1:453.
> COMMON NAME: Siberian Roe.
> TYPE LOCALITY: "In campestribus et montanis fruticosis ultra Volgam"; identified as Russia, former Samar district or province (Orenburgskaia Obl.), source of River Sok (a left tributary of the Volga), Bugulma-Belebei uplands (Heptner et al. 1961; Rossolimo in litt.).
> DISTRIBUTION: S Ural Mtns (Russia), N and E Kazakhstan, Kyrgyzstan, and S Siberia (Russia) eastward to Pacific coast, south into N and C China (N Sinkiang and Inner Mongolia south to Sichuan), N Mongolia, and Korea; apparently formerly in E Ukraine and N Caucasus Mtns (Russia) but original natural distribution not well documented.

STATUS: IUCN – Lower Risk (lc).

SYNONYMS: *ahu* (Gmelin, 1780) [unavailable]; *ahu* (Lydekker, 1915); **bedfordi** Thomas, 1908; *melanotis* Miller, 1911; **mantschuricus** (Noack, 1889); *ferghanicus* Rasewig, 1909; *tianschanicus* Satunin, 1906; **ochraceus** Barclay, 1935.

COMMENTS: Year of publication stated to be 1773 by Heptner et al. (1961) but no evidence provided and bibliographies cite 1771. Now regarded by most Russian authors as a species distinct from *C. capreolus* (Hewison and Danilkin, 2001; Sokolov et al., 1985; Sokolov and Gromov, 1990). Reviewed by Danilkin (1995, Mammalian Species, 512); revised by Sokolov et al. (1986c). *Cervus pygargus mantschuricus* Noack, 1889 is not preoccupied by *Cervus mantchuricus* Swinhoe, 1864 (= *Cervus nippon*) as there is a one letter difference (Article 57.6, International Code of Zoological Nomenclature; International Commission on Zoological Nomenclature, 1999). Treatment of *ochraceus* as a valid subspecies follows Koh and Randi (2001).

*Hippocamelus* Leuckart, 1816. Diss. Inaug. de *Equo bisulco* Molinae, p. 24.

TYPE SPECIES: *Hippocamelus dubius* Leuckart, 1816 (= *Equus bisulcus* Molina, 1792).

SYNONYMS: *Anomalocera* Gray, 1869; *Cervequus* Lesson, 1842; *Creagroceros* Fitzinger, 1873; *Furcifer* Wagner, 1844; *Huamela* Gray, 1873; *Xenelaphus* Gray, 1869.

COMMENTS: Included in *Odocoileus* by Haltenorth (1963:44, 46). Sister genus of *Rangifer* according to Webb (1992).

*Hippocamelus antisensis* (d'Orbigny, 1834). Nouv. Ann. Mus. Hist. Nat. Paris, 3:91.

COMMON NAME: Taruca.

TYPE LOCALITY: "du versant oriental des Cordillieres"; Bolivian Andes, near La Paz, at about 3,500 m (Cabrera, 1961:333).

DISTRIBUTION: Andes of NW Argentina, Bolivia, Ecuador, and Peru.

STATUS: CITES – Appendix I; U.S. ESA – Endangered; IUCN – Data Deficient.

SYNONYMS: *anomalocera* (Gray, 1872); *antisiensis* (Wagner, 1844).

*Hippocamelus bisulcus* (Molina, 1782). Sagg. Stor. Nat. Chile, p. 320.

COMMON NAME: Guemal.

TYPE LOCALITY: Chile, "delle Andi [Andes]"; restricted to Colchagua Prov. (Cabrera, 1961:334).

DISTRIBUTION: Andes of S Chile and S Argentina.

STATUS: CITES – Appendix I; U.S. ESA – Endangered; IUCN – Endangered.

SYNONYMS: *andicus* (Lesson, 1842); *chilensis* (Gay and Gervais, 1846); *dubius* Leuckart, 1816; *equinus* (Treviranus, 1803); *huamel* (Gray, 1850); *huemel* (C. H. Smith, 1827); *leucotis* (Gray, 1849).

*Mazama* Rafinesque, 1817. Am. Mon. Mag., 1(5):363.

TYPE SPECIES: *Mazama pita* Rafinesque, 1817 (= *Moschus americanus* Erxleben, 1777).

SYNONYMS: *Azarina* Larrañaga, 1923; *Coassus* Gray, 1843; *Doryceros* Fitzinger, 1873; *Doratoceros* Lydekker, 1915; *Homelaphus* Gray, 1872; *Nanelaphus* Fitzinger, 1873; *Passalites* Gloger, 1841; *Subulo* C. H. Smith, 1827.

COMMENTS: Revised by Czernay (1987). The genus includes brown brockets (*gouazoubira*, *pandora*), red brockets (*americana*, *temama*), and small brockets (*bororo, bricenii, chunyi, nana, rufina*); it has not been established that brown or small brockets are monophyletic groups.

*Mazama americana* (Erxleben, 1777). Syst. Regni Anim., 1:324.

COMMON NAME: South American Red Brocket.

TYPE LOCALITY: "Habitat in Guiania et Brazilia" and also cited from Surinam; restricted to French Guiana, Cayenne (Cabrera, 1961:335).

DISTRIBUTION: N Argentina, Bolivia, Brazil, Colombia, Ecuador, French Guiana, Guyana, Paraguay, Peru, Surinam, Trinidad, Tobago, and Venezuela.

STATUS: IUCN – Data Deficient.

SYNONYMS: *baralou* (Kerr, 1792) [*nomen dubium*]; *delicatulus* (Shaw, 1800); *inornata* (Gray, 1872); *juruana* J. A. Allen, 1915; *nemorosus* (Kerr, 1792) [*nomen dubium*]; **carrikeri** Hershkovitz, 1959; **gualea** J. A. Allen, 1915; *fuscata* J. A. Allen, 1915; **jucunda** Thomas,

1913; *rosii* Lönnberg, 1919; *rufa* (Illiger, 1815); *dolichurus* (Wagner, 1844); *pita* Rafinesque, 1817; *toba* Lönnberg, 1919; *sarae* Thomas, 1925; *sheila* Thomas, 1913; *trinitatis* J.A. Allen, 1915; *whitelyi* (Gray, 1873); *zamora* J. A. Allen, 1915; *zetta* Thomas, 1913.

COMMENTS: *Mazama americana tumatumari* J. A. Allen, 1915 is a composite based on a skull of *Odocoileus virginianus* (the lectotype) and a skin of *Mazama americana*; see Tate (1939).

*Mazama bororo* Duarte, 1996. Guia de identificação de cervídeos Brasileiros, p. 7.
COMMON NAME: São Paulo Bororó.
TYPE LOCALITY: "Esta espécie, aparentemente, se distribui nos poucos fragmentos de Mata Atlântica existentes no sudeste do Estado de São Paulo e nordeste do Estado do Paraná"; type from Brazil, São Paulo, Capão Benito (J. M. Barbanti Duarte, in litt.).
DISTRIBUTION: Brazil (Atlantic Forest from SE São Paulo State to NE Paraná State).
STATUS: IUCN – Data Deficient (DSG recommended).
SYNONYMS: *bororo* Mirando Ribeiro, 1919 [*nomen nudum*].
COMMENTS: Stated to have not yet been formally described (Wemmer, 1998) but Duarte's (1996) description made the name available. Revised by Duarte and Jorge (2003). The name *bororo* Mirando Ribeiro, 1919, which is a *nomen nudum*, may refer to this species.

*Mazama bricenii* Thomas, 1908. Ann. Mag. Nat. Hist., ser. 8, 1:349.
COMMON NAME: Mérida Brocket.
TYPE LOCALITY: "Paramo de la Culata, Merida, Venezuela".
DISTRIBUTION: W Venezuela.
COMMENTS: A species distinct from *M. rufina* according to Czernay (1987).

*Mazama chunyi* Hershkovitz, 1959. Proc. Biol. Soc. Wash., 72:45.
COMMON NAME: Dwarf Brocket.
TYPE LOCALITY: "Cocopunco, a site on the eastern slope of the Cordillera Real on the road to Mapiri, La Paz, Bolivia; altitude, about 3200 meters".
DISTRIBUTION: Bolivian Andes, S Peru.
STATUS: IUCN – Data Deficient.
COMMENTS: Prior to 1959 this species was confused with *Pudu mephistophiles*; see Hershkovitz (1959c). Regarded as a subspecies of *M. bricenii* by Anderson (1997).

*Mazama gouazoubira* (G. Fischer [von Waldheim], 1814). Zoognosia, 3:465.
COMMON NAME: South American Brown Brocket.
TYPE LOCALITY: "Paraqu." (Paraguay); restricted to Asuncion region (Cabrera, 1961:339).
DISTRIBUTION: N Argentina, Bolivia, Brazil, Colombia, Ecuador, French Guiana, Guyana, Panama (San Jose Isl), Paraguay, Peru, Surinam, Uruguay, and Venezuela.
STATUS: IUCN – Data Deficient (DSG recommended).
SYNONYMS: *argentina* Lönnberg, 1919; *bira* Rafinesque, 1817; *fusca* (Larrañaga, 1923); *gouazoupira* (G. Fischer [von Waldheim], 1814) [incorrect original spelling]; *kozeritzi* Mirando Ribeiro, 1919; *namby* (Fitzinger, 1879); *simplicicornis* (Illiger, 1815); *cita* Osgood, 1912; *humboldtii* (Wiegmann, 1833) [*nomen nudum*]; *medemi* Barriga-Bonilla, 1966; *mexianae* (Hagmann, 1908); *murelia* J. A. Allen, 1915; *nemorivaga* (F. Cuvier, 1817); *permira* Kellogg, 1946; *sanctaemartae* J. A. Allen, 1915; *rondoni* Miranda Ribeiro, 1915; *superciliaris* (Gray, 1852); *tschudii* (Wagner, 1855).
COMMENTS: Although the specific name is based on the gouazoubira of Azara, the original spelling was "*gouazoupira*" not "*gouezoubira*". The latter has been conserved as the correct original spelling (Opinion 1985; International Commission on Zoological Nomenclature, 2001b). Cabrera (1961) and Czernay (1987) included *rondoni* in *superciliaris*; Duarte (1996) and Duarte and Merino (1997) listed it as a separate species, based on an anomalous karyotype; Pinder and Leeuwenberg (1997) listed it as a subspecies, and provisionally also listed *namby* as valid.

*Mazama nana* (Hensel, 1872). Abhandl. Preuss. Akad. Will., 1872:99.
COMMON NAME: Southern Bororó.
TYPE LOCALITY: Brazil, "Río Grande do Sul".
DISTRIBUTION: N Argentina, SE Brazil, and E Paraguay.

STATUS: IUCN – Data Deficient (DSG recommended).

SYNONYMS: *nana* (Lund, 1841) [*nomen nudum*].

COMMENTS: Hensel misidentified this species as *Cervus rufinus* Pucheran, but regarded *Cervus nanus* Lund as a synonym. The latter name becomes valid from Hensel's description, as *Mazama nana* (Hensel, 1872). The name is preoccupied by *Cervus nanus* Kaup, 1839, possibly a junior synonym of *Euprox dicranocerus* (Kaup, 1833), and probably a *nomen oblitum*. A species distinct from *M. rufina* according to Czernay (1987). "Bororo" has been used as a vernacular name for *M. nana* by Czernay (1987) but needs to be qualified in view of the recently described *M. bororo*.

*Mazama pandora* Merriam, 1901. Proc. Biol. Soc. Wash., 14:105.

COMMON NAME: Yucatan Brown Brocket.

TYPE LOCALITY: Mexico, "Tunkas, Yucatan".

DISTRIBUTION: Campeche and Yucatán, Mexico.

STATUS: IUCN – Data Deficient (DSG recommended).

COMMENTS: Restored to species status by Medellin et al. (1998*b*).

*Mazama rufina* (Pucheran, 1851). Rev. Mag. Zool. Paris, 3: 561.

COMMON NAME: Ecuador Red Brocket.

TYPE LOCALITY: Ecuador, "la vallée de Lloa, sur le versant occidental de la Cordillière du Pichincha" (Pichincha, Pichincha Mtns, Lloa valley).

DISTRIBUTION: Ecuador and S Columbia.

STATUS: IUCN – Lower Risk (nt).

COMMENTS: Bourcier and Pucheran were cited as the authors in the original publication, but Pucheran provided the description and is the sole author; Bourcier collected the syntypes.

*Mazama temama* (Kerr, 1792). The Animal Kingdom, p. 303.

COMMON NAME: Central American Red Brocket.

TYPE LOCALITY: No locality cited but reference made to "Hernand. hist. nat. mexic. p. 325". Restricted by Hershkovitz (1951) to Mexico, Veracruz, Mirador.

DISTRIBUTION: Belize, W Colombia, Costa Rica, El Salvador, Guatemala, Honduras, Mexico (SE from S Tamaulipas), Nicaragua, and Panama.

STATUS: CITES – Appendix III (Guatemala) as *M. americana cerasina*.

SYNONYMS: *sartorii* (Saussure, 1860); *tema* Rafinesque, 1817; **cerasina** Hollister, 1914; **reperticia** Goldman, 1913.

COMMENTS: Raised to species status by Geist (1998), following suggestions by Groves and Grubb (1987).

*Odocoileus* Rafinesque, 1832. Atlantic Journal and Friend of Knowledge, 1:109.

TYPE SPECIES: *Odocoileus speleus* Rafinesque, 1832 (= *Dama virginianus* Zimmermann, 1780).

SYNONYMS: *Aplacerus* Hall and Kelson, 1959; *Cariacus,* Lesson, 1842; *Dama* Zimmermann, 1780 [preoccupied]; *Dorcelaphus* Gloger, 1841; *Elaphalces* Brookes, 1828 [*nomen oblitum*]; *Eucervus* Gray, 1866; *Gymnotis* Fitzinger, 1879; *Macrotis* Wagner, 1855; *Mazama* C. H. Smith, 1827 [preoccupied]; *Odocoelus* G. M. Allen, 1901; *Odontocoelus* Sclater, 1902; *Oplacerus* Haldleman, 1842; *Otelaphus* Fitzinger, 1874; *Palaeodocoileus* Spillman, 1931; *Protomazama* Spillman, 1931; *Reduncina* Wagner, 1844; *Subulus* Brookes, 1828 [*nomen oblitum*].

COMMENTS: Hall (1981:1087) employed *Dama* Zimmermann, 1780, of which *Dama virginiana* (= *Odocoileus virginianus*) is the type, for this genus, but *Dama* Frisch, 1775, with *Cervus dama* (= *Dama dama*) as type has priority and thus preoccupies *Dama* Zimmermann, 1780 (International Commission on Zoological Nomenclature, 1960).

*Odocoileus hemionus* (Rafinesque, 1817). Am. Mon. Mag., 1:436.

COMMON NAME: Mule Deer.

TYPE LOCALITY: USA, South Dakota, mouth of Big Sioux River (Bailey, 1926:41).

DISTRIBUTION: W Canada, Mexico (Baja California and Sonora to N Tamaulipas), W USA east to Minnesota, and Alaskan Panhandle. Introduced to Kauai (Hawaiian Isls) and Argentina.

STATUS: U.S. ESA – Endangered as *O. h. cedrocensis* [sic; = *cerrosensis*]; IUCN – Endangered as
*O. h. cerrosensis*, otherwise Lower Risk (lc).

SYNONYMS: *auritus* (Warden, 1820); *macrotis* (Say, 1823); *montanus* (Caton, 1881); *virgultus*
(Hallock, 1899); **californicus** (Caton, 1876); **cerrosensis** Merriam, 1898; **columbianus**
(Richardson, 1829); *lewisii* (Peale, 1848); *punctulatus* (Gray, 1852); *pusillus* (Gray, 1873);
*richardsoni* (Audubon and Bachman, 1853); *scaphiotus* (Merriam, 1898); **eremicus**
(Mearns, 1897); *canus* Merriam, 1901; **fuliginatus** (Cowan, 1933); **inyoensis** (Cowan,
1933); **peninsulae** (Lydekker, 1898); **sheldoni** Goldman, 1939; **sitkensis** Merriam,
1898.

COMMENTS: Revised by Cowan (1936); reviewed by Anderson and Wallmo (1984,
Mammalian Species, 219) and Geist (1998). The species can be partitioned into the
*columbianus* division or Black-tailed Deer (including also *sitkensis*) and the nominate
*hemionus* division or Mule Deer *sensu stricto* (including also *californicus, cerrosensis,
eremicus, fuliginatus, inyoensis, peninsulae,* and *sheldoni*). *Dorcelaphus crooki* Mearns,
1897 is based on a hybrid between *O. virginianus* and *O. hemionus* (Heffelfinger, 2000).

*Odocoileus virginianus* (Zimmermann, 1780). Geogr. Gesch. Mensch. Vierf. Thiere, 2:129.
COMMON NAME: White-tailed Deer.
TYPE LOCALITY: "Bewohnt in grossen Heerden Carolina v), Virginien, Louisiana w), und geht
vielleicht bis Panama x) hinunter"; restricted by Hershkovitz (1948*c*:43) to USA,
Virginia.
DISTRIBUTION: S Canada extending N of 60°N in the North West Territory and in the Yukon,
USA (absent from California to W Colorado), and all nations of Central America;
South America in Bolivia, N Brazil, Colombia, French Guiana, Guyana, Peru, Surinam,
and Venezuela. Introduced to Czech Republic, Finland, New Zealand, and West Indies,
possibly surviving on Cuba, Curacao, St. Croix, and St. Thomas Isls.
STATUS: CITES – Appendix III (Guatemala) as *O. v. mayensis*; U.S. ESA – Endangered as
*O. v. clavium,* Endangered (but proposed delisting) as *O. v. leucurus*; IUCN –
Endangered as *O. v. clavium,* Lower Risk (nt) as *O. v. leucurus,* otherwise Lower Risk (lc).
SYNONYMS: *americanus* (Erxleben, 1777) [*nomen oblitum*]; *clavatus* (C. H. Smith, 1827) [*nomen
dubium*]; *ramosus* (de Blainville, 1822); *typicus* (Lydekker, 1898); *virginianus*
(Zimmermann, 1777) [unavailable]; *wisconsinensis* (Belitz, 1919); **acapulcensis** (Caton,
1877); **borealis** Miller, 1900; **cariacou** (Boddaert, 1784); *campestris* (F. Cuvier, 1817);
*mangivorus* (Schrank, 1819) [*nomen dubium*]; *mazame* (Kerr, 1792); *pratensis* (Kerr,
1792); *spinosus* (Gay and Gervais, 1846); *suacuapara* Mirando Ribeiro, 1919; *sylvaticus*
(Kerr, 1792); **carminis** Goldman and Kellog, 1940; *chiriquensis* J. A. Allen, 1910;
**clavium** Barbour and G. M. Allen, 1922; *couesi* (Coues and Yarrow, 1875); *baileyi*
Lydekker, 1915; *battyi* J. A. Allen, 1903; **curassavicus** (Hummelinck, 1940); **dacotensis**
Goldman and Kellog, 1940; **goudotii** (Gay and Gervais, 1846); *columbicus* (Fitzinger,
1879); *lasiotis* Osgood, 1914; **gymnotis** (Wiegmann, 1833); *savannarum* (Cabanis,
1848); *tumatumari* (J. A. Allen, 1915); *wiegmanni* (Fitzinger, 1879); **hiltonensis**
Goldman and Kellog, 1940; **leucurus** (Douglas, 1829); **macrourus** (Rafinesque, 1817);
*louisianae* G. M. Allen, 1901; **mcilhennyi** F. W. Miller, 1928; **margaritae** Osgood, 1910;
**mexicanus** (Gmelin, 1788); *indicus* (Kerr, 1792); *lichtensteini* (J. A. Allen, 1902);
**miquihuanensis** Goldman and Kellogg, 1940; **nelsoni** Merriam, 1898; *mayensis* CITES,
1992 [*nomen nudum*]; **nemoralis** (C. H. Smith, 1827); *clavatus* (True, 1889); *costaricensis*
Miller, 1901; *truei* Merriam, 1898; **nigribarbis** Goldman and Kellogg, 1940; **oaxacensis**
Goldman and Kellogg, 1940; **ochrourus** V. Bailey, 1932; *osceola* (Bangs, 1896),
*fraterculus* (Coues, 1896); **peruvianus** (Gray, 1874); *brachyceros* (Philippi, 1894);
*peruanus* Sanborn, 1953; *philippii* Trouessart, 1904; **rothschildi** (Thomas, 1902);
**seminolus** Goldman and Kellogg, 1940; *sinaloae* J. A. Allen, 1903; **taurinsulae**
Goldman and Kellogg, 1940; **texanus** (Mearns, 1898); *texensis* (Miller and Rehn, 1901);
**thomasi** Merriam, 1898; **toltecus** (Saussure, 1860); **tropicalis** Cabrera, 1918; *punensis*
Spillmann, 1948 [*nomen nudum*]; **ustus** Trouessart, 1910; *abeli* (Spillmann, 1931);
*aequatorialis* (Spillmann, 1931); *antonii* (Spillmann, 1931); *consul* Lönnberg, 1922;
*gracilis* (Spillmann, 1931); **venatorius** Goldman and Kellogg, 1940; **veraecrucis**
Goldman and Kellogg, 1940; *yucatanensis* (Hays, 1872).

COMMENTS: Reviewed by Smith (1991, Mammalian Species, 388) and Geist (1998). The species includes two divisions (Groves and Grubb, 1987; Grubb, 1990), the *cariacou* division or Cariacu (including also *acapulcensis, chiriquensis, curassavicus, goudotii, margaritae, mexicanus, miquihuanensis, nelsoni, nemoralis, oaxacensis, peruvianus, rothschildi, sinaloae, thomasi, toltecus, tropicalils, truei, ustus, veraecrucis,* and *yucatanensis*) and the nominate *virginianus* division, the White-tailed Deer *sensu stricto* (including also *borealis, carminus, clavium, couesi, dacotensis, hiltonensis, leucurus, macrourus, mcilhennyi, nigribarbis, ochrourus, osceola, seminolus, taurinsulae, texanus,* and *venatorius*). Three taxa in Venezuela (*goudotii, gymnotis, margaritae*) regarded as species by Molina and Molinari (1999) but not by Moscarella et al. (2003).

*Ozotoceros* Ameghino, 1891. Rev. Argent. Hist. Nat., 1:243.
   TYPE SPECIES: *Cervus bezoarticus* Linnaeus, 1758 misidentified as *Cervus campestris* F. Cuvier, 1817.
   SYNONYMS: *Ozelaphus* Knottnerus-Meyer, 1907; *Ozotoceras* Palmer, 1904.
   COMMENTS: *Ozotoceros* is the name to be used for *Blastoceros* Fitzinger, 1860, if *Blastoceros* is regarded as an invalid emendation of *Blastocerus*; see Hershkovitz (1958) and Grubb (2000). Included in *Odocoileus* by Haltenorth (1963:46), Bianchini and Delupi (1979), and Ximenez et al. (1972), but a distinct genus (Groves and Grubb, 1987).

*Ozotoceros bezoarticus* (Linnaeus, 1758). Syst. Nat., 10th ed., 1:67.
   COMMON NAME: Pampas Deer.
   TYPE LOCALITY: "Habitat in America australis"; identified as Brazil, Pernambuco (Thomas, 1911*a*:151).
   DISTRIBUTION: N Argentina, SE Bolivia, Brazil (S of Amazon), Paraguay, and Uruguay.
   STATUS: CITES – Appendix I; U.S. ESA – Endangered; IUCN – Critically Endangered as *O. b. uruguayensis* and *O. b. arerunguaenis*, Endangered as *O. b. celer*, Data Deficient as *O. b. bezoarticus*, Lower Risk (nt) as *O. b. leucogaster*.
   SYNONYMS: *caenosus* (Wagner, 1844); *comosus* (Wagner, 1844); *cuguapara* (Kerr, 1792); *cuguete* (Kerr, 1792); *sylvestris* (Gray, 1873); **arerunguaensis** González, Álvares-Valin and Maldonado, 2002; **celer** Cabrera, 1943; **leucogaster** (Goldfuss, 1817); *albus* (Fitzinger, 1879); *azarae* (Wiegmann, 1833) [*nomen nudum*]; *dickii* (Goeldi, 1912); **uruguayensis** González, Álvares-Valin and Maldonado, 2002.
   COMMENTS: Reviewed by Jackson (1987, Mammalian Species, 295). Widely known in the older literature as *Blastocerus* or *Cariacus campestris* (= *Cervus campestris* F. Cuvier, 1817), a name which is properly a junior synonym of *Odocoilelus virginianus cariacou*.

*Pudu* Gray, 1852. Proc. Zool. Soc. Lond., 1850:242 [1852].
   TYPE SPECIES: *Capra puda* Molina, 1782.
   SYNONYMS: *Pudella* Thomas, 1913; *Pudua* Garrod, 1877.
   COMMENTS: Included in *Mazama* by Haltenorth (1963:48); includes *Pudella*; revised by Hershkovitz (1982).

*Pudu mephistophiles* (de Winton, 1896). Proc. Zool. Soc. Lond., 1896:508.
   COMMON NAME: Northern Pudu.
   TYPE LOCALITY: Ecuador, Napo-Pastaza Prov., "Paramo of Papallacta east of Quito, only just south of the Equator about 11,000 feet [3353 m] above the sea".
   DISTRIBUTION: Andes of Colombia, Ecuador, and Peru.
   STATUS: CITES – Appendix II; IUCN – Lower Risk (nt).
   SYNONYMS: *fusca* (Spillmann, 1931); *mephistopheles* Thomas, 1908; *mephistophelis* Cabrera and Yepes, 1940; *wetmorei* Lehmann, 1945.

*Pudu puda* (Molina, 1782). Sagg. Stor. Nat. Chile, p. 308.
   COMMON NAME: Southern Pudu.
   TYPE LOCALITY: Chile, "Cordigliera delle Province Australi"; restricted to "los bosques del lago Todos los Santos, en la provincia de Chiloé" (Cabrera, 1961:343).
   DISTRIBUTION: S Chile and SW Argentina.
   STATUS: CITES – Appendix I; U.S. ESA – Endangered as *P. pudu*; IUCN – Vulnerable.

SYNONYMS: *chilensis* Gray, 1852; *dubia* (Afzelius, 1815); *humilis* (Bennett, 1831); *pudu* (Gmelin, 1788); *pudua* Carette, 1922.

COMMENTS: For original spelling of specific name, see Hershkovitz (1982).

**Rangifer** C. H. Smith, 1827. *In* Griffith et al., Anim. Kingdom, 5:304.

TYPE SPECIES: *Cervus tarandus* Linnaeus, 1758.

SYNONYMS: *Achlis* Reichenbach, 1845; *Rangifer* Frisch, 1775 [unavailable]; *Tarandus* Billberg, 1827.

COMMENTS: Revised by Banfield (1961), Geist (1998), and Markov et al. (1994).

**Rangifer tarandus** (Linnaeus, 1758). Syst. Nat., 10th ed., 1:67.

COMMON NAME: Reindeer.

TYPE LOCALITY: "Habitat in Alpibus Europae et Asiae maxime septentrionalibus"; identified as Sweden, Alpine Lapland by Thomas (1911*a*:151); based on domesticated stock.

DISTRIBUTION: Circumboreal in tundra and taiga from Svalbard, Norway, Finland, Russia, Alaska (USA) and Canada including most arctic islands, and Greenland, south to N Mongolia, China (Inner Mongolia; now only domesticated or feral?), Sakhalin Isl, and USA (N Idaho and Great Lakes region). Introduced to, and feral in, Iceland, Kerguelen Isls, South Georgia Isl, Pribilof Isls, St. Matthew Isl. Extinct in Sweden.

STATUS: U.S. ESA – Endangered as *R. t. caribou* in Canada (SE British Columbia at the Canadian-USA border, Columbia River, Kootenay Lake, and Kootenai River) and USA (Idaho, Washington); IUCN – Endangered as *R. t. pearyi*, otherwise Lower Risk (lc). The woodland *caribou* is highly endangered throughout its distribution right into Ontario (V. Geist, in litt.).

SYNONYMS: *borealis* (Rüppell, 1842); *cilindricornis* Camerano, 1902 [unavailable]; *furcifer* (Baird, 1852); *lapponum* (Billberg, 1827); *rangifer* (Gmelin, 1788); *typicus* Lydekker, 1898; **buskensis** (Millais, 1915); *angustirostris* Flerov, 1932; *dichotomus* Hilzheimer, 1936; *silvicola* Hilzheimer, 1936; *transuralensis* Hilzheimer, 1936; *valentinae* Flerov, 1933; **caboti** G. M. Allen, 1914; *labradorensis* (Millais, 1915); **caribou** (Gmelin, 1788); *coronatus* (É. Geoffroy Saint-Hilaire, 1803) [*nomen dubium*]; *fortidens* Hollister, 1912; *hastalis* (Agassiz, 1847); *keewatinensis* (Millais, 1915); *montanus* Thompson-Seton, 1899; *sylvestris* (Richardson, 1829); **dawsoni** Thompson-Seton, 1900; **fennicus** Lönnberg, 1909; **groenlandicus** (Borowski, 1780); *arcticus* (Richardson, 1829); *excelsifrons* Hollister, 1912; *granti* J. A. Allen, 1902; *grewensis* (de Blainville, 1822); *ramosus* (de Blainville, 1822); **osborni** J. A. Allen, 1902; *mcguirei* Figgins, 1919; *ogilvyensis* (Millais, 1915); *selousi* Barclay, 1935; *stonei* J. A. Allen, 1901; **pearsoni** Lydekker, 1903; **pearyi** J. A. Allen, 1902; *eogroenlandicus* Degerbøl, 1957; **phylarchus** Hollister, 1912; *setoni* Flerov, 1933; **platyrhynchus** (Vrolik, 1829); *spetsbergensis* (Andersen, 1862); *spitzbergensis* Murray, 1866; **sibiricus** Murray, 1866; *asiaticus* Jacobi, 1931; *chukchensis* (Millais, 1915); *lenensis* (Millais, 1915); *sibiricus* (von Schreber, 1784) [unavailable]; *taimyrensis* Michurin, 1965; *yakutskensis* (Millais, 1915); **terraenovae** Bangs, 1896.

COMMENTS: Subspecies have been placed in two divisions, compressicornis or Woodland Reindeer, and cylindricornis or Tundra Reindeer (Jacobi, 1931). These names of divisions are non-Linnean; *cilindricornis* Camerano, 1902 is a lapsus for cylindricornis and is not a Linnean name. An additional category has since been recognised for the Peary Caribou, due to marginal or seasonal sympatry between caribou in Arctic America, following Banfield (1963). Subspecies here considered valid are based on Banfield (1961), considerably modified by Geist (1998): *caribou* division or Woodland Caribou (includes also *buskensis, valentinae, dawsoni, fennicus,* and *phylarchus*); populations transitional between *caribou* and *tarandus* divisions (includes *osborni*); *tarandus* division, Barren-ground Caribou or Reindeer (includes also *caboti, groenlandicus, pearsoni, sibiricus,* and *terraenovae*); and *platyrhynchus* division (including *pearyi* or Peary Caribou and *platyrhynchus* or Svalbard Reindeer). The extinct insular *dawsoni* has been treated as a distinct species (Cowan and Guiguet, 1965) but does not differ from *caribou* or *granti* (= *groenlandicus*) in mtDNA sequences (Byun et al., 2002). Grouping the Svalbard Reindeer with the Peary Caribou is provisional (Groves and Grubb, 1987).

**Subfamily Cervinae** Goldfuss, 1820. Handb. Zool., 2:xx, 374.

SYNONYMS: Axidae Brookes, 1828; Cervulinae Sclater, 1870 [suppressed]; Elaphidae Brookes, 1828; Elaphodinae Knottnerus-Meyer, 1907; Eucervidae Bubenik, 1990 [unavailable]; Muntiacini Knottnerus-Meyer, 1907; Platycerinidae Brookes, 1828 [unavailable]; Rusidae Brookes, 1828; Stylocerinidae Brookes, 1828 [unavailable].

COMMENTS: Cervini includes *Axis, Cervus, Elaphurus, Przewalskium, Rucervus,* and *Rusa.* Muntiacini includes *Elaphodus* and *Muntiacus.* Muntiacini generally has been regarded as a subfamily (Haltenorth, 1963) and has usually been attributed to Pocock, 1923, Proc. Zool. Soc. Lond., 1923:207; treated as a full family by Groves and Grubb (1990); relegated to tribal status in Cervinae, by Groves and Grubb (1987) and Grubb (2000*b*) supported by evidence in Kraus and Miyamoto (1991).

*Axis* C. H. Smith, 1827. *In* Griffith et al., Anim. Kingdom, 5:312.

TYPE SPECIES: *Cervus axis* Erxleben, 1777.

SYNONYMS: *Hyelaphus* Sundevall, 1846.

COMMENTS: Treated as a full genus, not a subgenus of *Cervus,* by Groves and Grubb (1987). Subgenus *Axis* contains *axis* only; subgenus *Hyelaphus* contains *calamianensis, kuhlii* and *porcinus.*

*Axis axis* (Erxleben, 1777). Syst. Regn. Anim., 1:312.

COMMON NAME: Chital.

TYPE LOCALITY: "Habitat ad ripas Gangis; in Iana, Ceylona"; restricted to India, Bihar, banks of the Ganges River (Ellerman and Morrison-Scott, 1951:360).

DISTRIBUTION: India (incl. Sikkim), Nepal, and Sri Lanka; introduced to Andaman Isls, Argentina, Armenia, Australia, Brazil, Croatia, Moldavia, Pakistan, Papua New Guinea, Ukraine, Uruguay, and USA (Florida, Hawaiian Isls, and Texas).

STATUS: IUCN – Lower Risk (lc).

SYNONYMS: *ceylonensis* (J. B. Fischer, 1829); *indicus* (J. B. Fischer, 1829); *maculatus* (Kerr, 1792); *major* Hodgson, 1842; *minor* Hodgson, 1842; *nudipalpebra* (Ogilby, 1831); *zeylanicus* (Lydekker, 1905).

*Axis calamianensis* (Heude, 1888). Mem. Hist. Nat. Emp. Chin., 2:49.

COMMON NAME: Calamian Deer.

TYPE LOCALITY: Philippines, "l'ile Calamian et l'isle de la Paragua [= Palawan Isl.]"; restricted to Calamian Isls, Culion Isl (Lydekker, 1915:59).

DISTRIBUTION: Philippines, Calamian Isls (Busuanga, Calauit, Culion and some smaller Isls). Not recorded from Palawan and Heude did not have material from Palawan.

STATUS: CITES – Appendix I; U.S. ESA – Endangered as *A. porcinus calamianensis;* IUCN – Endangered.

SYNONYMS: *culionensis* (Elliot, 1897).

COMMENTS: Included in *A. porcinus* by Haltenorth (1963), but treated as a full species by Groves and Grubb (1987).

*Axis kuhlii* (Temminck, 1836). *In* von Siebold, Temminck and Schlegel, Fauna Japonica, Coup d'Oeil Faune Iles Sonde Emp. Japan, pp. viii, ix.

COMMON NAME: Bawean Deer.

TYPE LOCALITY: Indonesia, "les îles Bavian" (= Bawean Isl).

DISTRIBUTION: Indonesia, Bawean Isl. Specimen in Institute of Zoology, Beijing, labelled from Bangka Isl, off Sumatra, Indonesia.

STATUS: CITES – Appendix I; U.S. ESA – Endangered as *A. porcinus kuhli* [sic]; IUCN – Endangered.

COMMENTS: The name of this deer is widely attributed to Müller, 1840 *in* Temminck, Verh. Nat. Gesch. Nederland. Overz. Bezitt., Zool., Zoogd. Indisch. Archipel., p. 45, but Corbet and Hill (1992) indicated the earlier publication. This species was further described by Müller and Schlegel, *in* Temminck, Verh. Nat. Gesch. Nederland. Overz. Bezitt., Zool., Mammalia, p. 223[1845], pl. 44[1842]. Included in *A. porcinus* by Haltenorth (1963), but treated as a full species by Groves and Grubb (1987) and Geist (1998).

*Axis porcinus* (Zimmermann, 1780). Geogr. Gesch. Mensch. Vierf. Thiere, 2:131.
> COMMON NAME: Hog Deer.
> TYPE LOCALITY: No locality given, based on captive in Bengal; "Indo-Gangetic Plain of India" (Lydekker, 1915:56); here restricted to India, West Bengal.
> DISTRIBUTION: Bangladesh, Burma, Cambodia, China (Yunnan), N India, Laos, Nepal, Pakistan, Sri Lanka (introduced?), and S Vietnam; introduced to S Australia.
> STATUS: CITES – Appendix I as *Cervus porcinus annamiticus*; U.S. ESA – Endangered as *Axis porcinus annamiticus*; IUCN – Lower Risk (nt) as *A. p. porcinus*, Data Deficient as *A. p. annamiticus*.
> SYNONYMS: *dodur* (Royle, 1834) [*nomen dubium*]; *maculatus* (Kerr, 1792); *oryzus* Kelaart, 1852; *porcinus* (Zimmermann, 1777) [unavailable]; *pumilio* (C. H. Smith, 1827); **annamiticus** (Heude, 1888); *hecki* (Lydekker, 1908).
> COMMENTS: *Cervus porcinus* Zimmermann, 1777, is not an available name as it was published in an unavailable work (*Spec. Zool. Geogr.*, p. 532): see Hemming (1950:547).

*Cervus* Linnaeus, 1758. Syst. Nat., 10th ed., 1:66.
> TYPE SPECIES: *Cervus elaphus* Linnaeus, 1758.
> SYNONYMS: *Elaphoceros* Fitzinger, 1874; *Elaphus* C. H. Smith, 1827; *Eucervus* Acloque, 1899; *Harana* Hodgson, 1838; *Pseudaxis* Gray, 1872; *Pseudocervus* Hodgson, 1841; *Sica* Trouessart, 1898; *Sika* Sclater, 1870; *Sikelaphus* Heude, 1894; *Sikaillus* Heude, 1898.
> COMMENTS: Formerly included *Rusa*, *Rucervus*, and *Przewalskium* as subgenera, see Groves and Grubb (1987). Van Gelder (1977*b*) also included *Elaphurus, Axis, Dama* and *Hyelaphus*. Information from various sources suggests that *Cervus sensu lato* is polyphyletic or paraphyletic. Dendrograms derived from mitochondrial-DNA restriction-site maps suggest that *Axis axis* and *Rucervus duvauceli* form a sister-clade to *Elaphurus davidianus, Rusa unicolor* and *Cervus elaphus* (Cronin, 1991), whereas genetic distances obtained from protein analysis suggested that *Axis axis, Dama* species and *Rusa* species formed a sister clade to *Elaphurus davidianus, C. elaphus*, and *C. nippon* (Emerson and Tate, 1993). Phylogram of Randi et al. (2001) suggests *Rucervus eldi* and *Elaphurus davidianus* form a clade whose sister-group includes *Rusa* and *Cervus* species.

*Cervus elaphus* Linnaeus, 1758. Syst. Nat., 10th ed., 1:67.
> COMMON NAME: Red Deer (see comments).
> TYPE LOCALITY: "Habitat in Europa, Asia"; identified as S Sweden by Thomas (1911*a*:151).
> DISTRIBUTION: N Africa in NE Algeria and Tunisia. All states of continental Europe east to S Norway, S Sweden, Ukraine and Caucasus (Armenia, Azerbaijan, Georgia, and Russia); extinct in Albania, Moldavia, and Sicily; introduced but now extinct on Lampedusa Isl and islands off Sicily; in Corsica and Sardinia only since Neolithic; not in Finland; reintroduced into Belorussia, Estonia, Kaliningrad, Latvia, and Lithuania. Near and Middle East in Turkey, N Iran, and Iraq; extinct in Israel, Jordan, Lebanon, and Syria. C Asia in Kazakhstan, Kyrgyzstan, Tajikistan, Turkmenistan (extinct), Uzbekistan, N Afghanistan, N India (Kashmir Valley), N Pakistan (vagrant), east to Siberia, Mongolia, W and N China (Gansu, Inner Mongolia, Jilin, Liaoning, Manchuria, Ninxia, Shaanxi, Shanxi, Sichuan, and E Tibet including Qinghai), Korea, and Ussuri region (Russia). Canada and USA, where now restricted to western areas and reserves. Red Deer (*elaphus* division) introduced to Morocco, USA, Argentina, Chile, Australia, and New Zealand; Elk or Wapiti (*canadensis* division) introduced to Ural Mtns and Volga Steppe (Russia), and New Zealand.
> STATUS: CITES – Appendix I as *C. e. hanglu*; Appendix II as *C. e. bactrianus*; Appendix III (Tunisia) as *C. e. barbarus*. U.S. ESA – Endangered as *C. e. bactrianus, C. e. barbarus, C. e. corsicanus, C. e. hanglu, C. e. macneilli, C. e. wallichi*, and *C. e. yarkandensis*; IUCN – Endangered as *C. e. yarkandensis, C. e. corsicanus* and *C. e. hanglu*, Vulnerable as *C. e. bactrianus*, Lower Risk (nt) as *C. e. barbarus*, Data Deficient as *C. e. affinis, C. e. alashanicus, C. e. macneilli*, and *C. e. wallichi*, otherwise Lower Risk (lc).
> SYNONYMS: *albicus* Matschie, 1907; *albifrons* Reichenbach, 1845; *albus* Desmarest, 1822; *bajovaricus* Matschie, 1907; *balticus* Matschie, 1907; *debilis* Matschie, 1912; *germanicus* Desmarest, 1822; *hippelaphus* Erxleben, 1777; *montanus* Botezat, 1903 [preoccupied];

*neglectus* Matschie, 1912; *rhenanus* Matschie, 1907; *saxonicus* Matschie, 1912; *typicus*
Lydekker, 1898; *varius* Fitzinger, 1874; *visurgensis* Matschie, 1912; *vulgaris* Botezat,
1903; **alashanicus** Bobrinskii and Flerov, 1935; **atlanticus** Lönnberg, 1906; *scoticus*
Lönnberg, 1906; **barbarus** Bennett, 1833; **brauneri** Charlemagne, 1920;  *tauricus*
Fortunatov, 1925;  **canadensis** Erxleben, 1777; *asiaticus* Lydekker, 1898; *baicalensis*
Lydekker, 1915; *biedermanni* Matschie, 1907; *major* Ord, 1815; *manitobensis* Millais,
1915; *merriami* Nelson, 1902; *nelsoni* Bailey, 1935; *occidentalis* C. H. Smith, 1827;
*roosevelti* Merriam, 1897; *sibiricus* Severtzov, 1873; *strongyloceros* von Schreber, 1784;
*wachei* Noack, 1902; *wapiti* Barton, 1808; **corsicanus** Erxleben, 1777; *corsiniacus*
Gervais, 1848; *mediterraneus* de Blainville, 1822; *minor* Wagner, 1855; **hanglu** Wagner,
1844; *cashmeerianus* Falconer, 1868; *cashmeriensis* Adams, 1859; *cashmirianus* Fitzinger,
1874; *casperianus* Gray, 1847; **hispanicus** Hilzheimer, 1909; *bolivari* Cabrera, 1911;
**kansuensis** Pocock, 1912; *wardi* Lydekker, 1910; **macneilli** Lydekker, 1909; **maral**
Gray, 1850; *caspius* Radde, 1886; *caucasicus* Winans, 1914 [*nomen nudum*]; *maral*
Ogilby, 1840 [*nomen nudum*]; **nannodes** Merriam, 1905; **pannoniensis** Banwell, 1997;
*campestris* Botezat, 1903 [*nomen nudum*, preoccupied]; *carpathicus* Tatarinov, 1956
[*nomen nudum*]; **songaricus** Severtzov, 1873; *eustephanus* Blanford, 1876; **wallichii**
G. Cuvier, 1823; *affinis* Hodgson, 1841; *nariyanus* Hodgson, 1851; *tibetanus* Hodgson,
1850; **xanthopygus** Milne-Edwards, 1867; *bedfordianus* Lydekker, 1897; *isubra* Noack,
1889; *luehdorfi* Bolau, 1880; *typicus* de Pousargues, 1898; *ussuricus* (Heude, 1892);
**yarkandensis** Blanford, 1892; *bactrianus* Lydekker, 1900; *hagenbeckii* Shitkov, 1904.

COMMENTS: European and North American populations are known as Red Deer and Elk
(Wapiti) respectively; neither is suited as the name for the whole species; "maral", a
Mongolian name widely used for Asiatic members of the species could be selected but
is unlikely to be acceptable; for history and meaning of "maral", see Oswald (2002).
Reviewed by Dolan (1988) and Geist (1998). Following Geist (1998) in part, subspecies
modified from Groves and Grubb (1987) who recognised divisions of the species
including nominate *elaphus* division or Red Deer *sensu stricto* (including also *atlanticus,
barbarus, brauneri, corsicanus, hispanicus, maral, pannoniensis,* and *scoticus*), possibly
paraphyletic *wallichii* division (primitive Wapiti *alashanicus* and *kansuensis,* Hangul
*hanglu,* McNeill's Deer *macneilli,* Shou *wallichii,* and Bactrian or Yarkand Deer
*yarkandensis*), and *canadensis* division or Elk (including also *nannodes, songaricus,* and
*xanthopygus*). The name *pannoniensis* has priority over other names for SE European
Red Deer (Banwell, 1997, 1998, 2002) though Oswald (2002) and V. Geist (in litt.)
regarded *pannoniensis* as a synonym of *maral;* too many subspecies are recognised in
the *elaphus* division but a definitive synonymy is not yet available. Retention of
*nannodes* follows Schonewald (1994). Advanced North American Elk belong to a clade
including *C. nippon, Rusa timorensis* and *R. unicolor,* of which the sister group consists
of European Red Deer according to study of the mtDNA control region (Randi et al.,
2001), so *C. elaphus* appears to be polyphyletic (supported by Kuwayama and Ozawa,
2000 but not by Mahmut et al., 2002). The work of these authors and A. Lister and
I. Van Piljen (in litt.) distinguished western and eastern lineages in the species, the
western lineage including *atlanticus, corsicanus, elaphus, hanglu,* cf. *hippelaphus,
hispanicus,* and *yarkandensis* and the eastern lineage *alashanicus,* cf. *kansuensis,
macneilli,* cf. *manitobensis, nannodes,* cf. *nelsoni,* cf. *roosevelti,* cf. *sibiricus, songaricus,
wallichii,* and *xanthopygus:* the *wallichii* division appears to be paraphyletic, with
*alashanicus, macneilli,* and *wallichii* affined to *canadensis,* and *hanglu,* and *yarkandensis*
associated with nominate *elaphus.* Groves (2003) ranked *hanglu* and *wallichii* (and
*canadensis*) as species separate from *C. elaphus.*

*Cervus nippon* Temminck, 1838. *In* von Siebold, Temminck and Schlegel, Fauna Japonica, Coup
d'Oeil Faune Iles Sonde Emp. Japan, p. xxii.
COMMON NAME: Sika.
TYPE LOCALITY: "Les îles du domaine du Japon"; restricted to Japan, Kyushu, Nagasaki
(Groves and Smeenk, 1978).
DISTRIBUTION: China (Manchuria south to Guangxi, and Sichuan to Anhui), Korea (incl.
Cheju Isl), Japan (incl. Tsushima Isls), Russia (Soviet Far East), Taiwan (extinct but
reintroduced), and Vietnam. Apparently wild populations now very localized in

China. Presumably anciently introduced to Philippines (Solo Isl; still extant?). Introduced in 17th century to Kerama Isls (Ryukyu Isls). Introduced in 19th-20th centuries to British Isles, mainland Europe (Armenia, Austria, Azerbaijan, Czech Republic, Denmark, Finland, France, Germany, Kaliningrad, Lithuania, Poland, W Russia, and Ukraine), New Zealand, USA, and small islands off Japan.

STATUS: U.S. ESA – Endangered as *C. n. grassianus*, *C. n. keramae*, *C. n. kopschi*, *C. n. mandarinus*, and *C. n. taiouanus*; IUCN – Critically Endangered as *C. n. grassianus*, *C. n. keramae*, *C. n. mandarinus*, *C. n. taiouanus*, and *C. n. pseudaxis*, Endangered as *C. n. sichuanicus* and *C. n. kopschi*, Data Deficient as *C. n. aplodontus*, *C. n. mantchuricus*, *C. n. pulchellus*, and *C. n. yesoensis*, otherwise Lower Risk (lc).

SYNONYMS: *aceros* (Heude, 1888); *brachypus* (Heude, 1884); *consobrinus* (Heude, 1898); *daimius* (Heude, 1898); *dejardinus* (Heude, 1888); *euopis* Sclater, 1874; *fuscus* (Heude, 1884); *granulosus* (Heude, 1888); *hollandianus* (Heude, 1884); *infelix* (Heude, 1884); *japonicus* Sundevall, 1846; *kematoceros* (Heude, 1888); *latidens* (Heude, 1898); *mageshimae* Kuroda and Okada, 1950; *marmandianus* (Heude, 1888); *minor* Brooke, 1878; *minutus* (Heude, 1888); *modestus* (Heude, 1888); *orthopodicus* (Heude, 1897); *orthopus* (Heude, 1884); *paschalis* (Heude, 1888); *regulus* (Heude, 1898); *rex* (Heude, 1888); *schlegeli* (Heude, 1884); *sica* Lydekker, 1893; *sicarius* (Heude, 1898); *sika* Temminck, 1844; *sika* (Heude, 1898); *surdescens* (Heude, 1888); *typicus* Lydekker, 1897; **aplodontus** (Heude, 1884); *aplodonticus* (Heude, 1897); *centralis* Kishida, 1936; *elegans* (Heude, 1897); *ellipticus* (Heude, 1897); *minoensis* (Heude, 1897); *mitratus* (Heude, 1884); *schizodonticus* (Heude, 1897); *sendaiensis* (Heude, 1897); *xendaiensis* (Heude, 1884); **grassianus** (Heude, 1884); **hortulorum** Swinhoe, 1864; **keramae** (Kuroda, 1924); **kopschi** Swinhoe, 1873; *andreanus* Heude, 1882; *arietinus* (Heude, 1894); *brachyrhinus* (Heude, 1884); *cycloceros* (Heude, 1884); *cyclorhinus* Heude, 1882; *dugennianus* (Heude, 1894); *frinianus* Heude, 1882; *gracilis* Heude, 1882; *grilloanus* (Heude, 1884); *hyemalis* Heude, 1882; *ignotus* Heude, 1882; *joretianus* Heude, 1882; *lacrymosus* Heude, 1882; *microdontus* (Heude, 1884); *oxycephalus* (Heude, 1884); *pouvrelianus* (Heude, 1884); *riverianus* (Heude, 1894); *yuanus* (Heude, 1884); **mageshimae** Kuroda and Okada, 1950; **mandarinus** Milne-Edwards, 1871; **mantchuricus** Swinhoe, 1864; *dybovskii* (Heude, 1894); *dybowskii* Taczanowski, 1876; *imperialis* (Heude, 1894); *major* Noack, 1889; *microspilus* (Heude, 1884); *typicus* Ward, 1910; **pseudaxis** Gervais, 1841; **pulchellus** Imaizumi, 1970; **sichuanicus** Guo, Chen and Wang, 1978; *swinhoei* Glover, 1956 [preoccupied]; **soloensis** (Heude, 1888); **taiouanus** Blyth, 1860; *devilleanus* Heude, 1882; *dominicanus* (Heude, 1884); *morrisianus* (Heude, 1884); *novioninus* (Heude, 1884); *schulzianus* (Heude, 1884); *taevanus* Sclater, 1862; *taioranus* Heude, 1882; *taivanus* Gray, 1872; **yakushimae** Kuroda and Okada, 1950; **yesoensis** (Heude, 1884); *blakistoninus* (Heude, 1884); *dolichorhinus* (Heude, 1884); *legrandianus* (Heude, 1884); *matsumotei* Kishida, 1924; *rutilus* (Heude, 1897); *sylvanus* (Heude, 1884).

COMMENTS: Further described by Temminck *in* von Siebold, Temminck and Schlegel, Fauna Japonica, Aperçu Gén. Spéc. Mamm. Japon, p. 54, pl. 17 [1844] as *Cervus sika*. Includes *hortulorum*, *taiouanus* and *pulchellus*, which were considered species by Imaizumi (1970a). Includes *soloensis*, see Grubb and Groves (1983). Revised in part by Groves and Smeenk (1978) who included *mageshimae* and *yakushimae* in nominate *nippon*, and noted that *aplodontus* has priority over *centralis*. Reviewed by Feldhamer (1980, Mammalian Species, 128) and Banwell (1999). Native and introduced populations seriously threatened by genetic pollution; numerous populations are of uncertain provenance or have mixed ancestry; status of *hortulorum* is particularly uncertain.

*Dama* Frisch, 1775. Das Natur-System der Vierfüssigen Thiere, 3.

TYPE SPECIES: *Cervus dama* Linnaeus, 1758.

SYNONYMS: *Dactyloceros* Wagner, 1844; *Machlis* Zittel, 1894; *Palmatus* Lydekker, 1896; *Platyceros* Zimmermann, 1780.

COMMENTS: *Dama* as generic name for the Fallow Deer was conserved by Opinion 581, International Commission on Zoological Nomenclature (1960).

*Dama dama* (Linnaeus, 1758). Syst. Nat., 10th ed., 1:67.
>    COMMON NAME: Fallow Deer.
>    TYPE LOCALITY: "Habitat in Europa"; identified as "Habitat in vivariis Regis & Magnatum" by
>        Thomas (1911*a*:151), in Sweden to which it had been introduced.
>    DISTRIBUTION: Naturally wild populations of nominate form still present in S Turkey;
>        introduced into nearly all countries of Europe (incl. Lithuania and Ukraine), South
>        Africa, Australia, New Zealand, USA, Argentina, Chile, Peru, and Uruguay, as well as
>        islands in Fijian group, Lesser Antilles, and off W Canadian Coast. For present
>        distribution, see Chapman and Chapman (1980); for natural recent distribution see
>        Uerpmann (1987). Subspecies *mesopotamica* formerly in Iraq, Israel, Jordan, Lebanon,
>        E Turkey, and possibly Syria; survives in W Iran.
>    STATUS: CITES – Appendix I as *D. mesopotamica*; U.S. ESA – Endangered as *D. mesopotamica*
>        (= *D. d. mesopotamica*); IUCN – Endangered as *D. dama mesopotamica*, otherwise Lower
>        Risk (lc).
>    SYNONYMS: *albus* Fitzinger, 1874; *leucaethiops* (J. B. Fischer, 1829); *maura* (J. B. Fischer, 1829);
>        *mauricus* (Cuvier, 1816); *niger* Fitzinger, 1874; *platyceros* (Cuvier, 1798); *plinii*
>        (Zimmermann, 1780); *schaeferi* Hilzheimer, 1926; *varius* Fitzinger, 1874; *vulgaris*
>        (J. B. Fischer, 1829); **mesopotamica** (Brooke, 1875); *mesopotamiae* (Trouessart, 1905).
>    COMMENTS: Reviewed by Feldhamer et al. (1988, Mammalian Species, 317), who included
>        *mesopotamica* in this species. *Dama schaeferi* Hilzheimer, 1926 was supposedly from
>        Africa, but the name is now known to have been based on a specimen from Italy
>        (Kock, 2000*b*). The form *mesopotamica* has recently been regarded as a subspecies of
>        *D. clactoniana* (Falconer, 1868), treated as a separate species from *D. dama* by di Stefano
>        (1996), based on the resemblance of its antlers to a fossil antler of *clactoniana* from
>        Edelsheim, Germany. Since characters of fossil antlers are open to varying
>        interpretations, the evidence supporting this conclusion seems insufficient at present
>        (A. Lister, in litt.); *mesopotamica* has also been regarded as a separate species from
>        *D. dama* by Haltenorth (1959), Ferguson et al. (1985), Uerpmann (1987), and Harrison
>        and Bates (1991) but in Geist's (1998) revision has been restored to subspecies status.

*Elaphodus* Milne-Edwards, 1872. Nouv. Arch. Mus. Hist. Nat. Paris, Bull., 7:93.
>    TYPE SPECIES: *Elaphodus cephalophus* Milne-Edwards, 1872.
>    SYNONYMS: *Lophotragus* Swinhoe, 1874.
>    COMMENTS: For year of publication, see Ellerman and Morrison-Scott (1953).

*Elaphodus cephalophus* Milne-Edwards, 1872. Nouv. Arch. Mus. Hist. Nat. Paris, Bull., 7:93.
>    COMMON NAME: Tufted Deer.
>    TYPE LOCALITY: China, Sichuan, "la principauté de Moupin" (= Baoxing).
>    DISTRIBUTION: N Burma and S and C China (S Gansu to Yunnan).
>    STATUS: IUCN – Data Deficient.
>    SYNONYMS: **fociensis** Lydekker, 1904; **ichangensis** Lydekker, 1904; **michianus** (Swinhoe,
>        1874).
>    COMMENTS: Fully described by Milne-Edwards, Rech. Hist. Nat. Mamm., Faune Tibet-
>        Oriental, p. 356, pl.65-67 [1874].

*Elaphurus* Milne-Edwards, 1866. Ann. Sci. Nat. Zool. (Paris), ser. 5, 5:382.
>    TYPE SPECIES: *Elaphurus davidianus* Milne-Edwards, 1866.

*Elaphurus davidianus* Milne-Edwards, 1866. Ann. Sci. Nat. Zool. (Paris), ser. 5, 5:382.
>    COMMON NAME: Père David's Deer.
>    TYPE LOCALITY: China, "dans le parc impérial situé à quelque distance de Pékin [Beijing]".
>    DISTRIBUTION: Formerly NE China; extinct in wild since 3rd or 4th Century; now
>        reintroduced to its former range, near Beijing and near Shanghai.
>    STATUS: IUCN – Critically Endangered.
>    SYNONYMS: *menziesianus* (Sowerby, 1933); *tarandoides* (David, 1867).
>    COMMENTS: Included in *Elaphurus* by Corbet (1978*c*:201); but see Van Gelder (1977*b*).

*Muntiacus* Rafinesque, 1815. Analyse de la Nature, p. 56.
> TYPE SPECIES: *Cervus muntjak* Zimmermann, 1780.
> SYNONYMS: *Caninmuntiacus* Giao, Tuoc, Eric, Dung et al. [sic], 1997; *Cervulus* de Blainville, 1816; *Diopplon* Brookes, 1828; *Megamuntiacus* Tuoc, Dung, Dawson, Arctander and Mackinnon, 1994; *Muntjacus* Gray, 1825 [*nomen nudum*]; *Procops* Pocock, 1923; *Prox* Ogilby, 1836; *Stylocerus* C. H. Smith, 1827.
> COMMENTS: *Muntiacus* Rafinesque is a *nomen nudum*, but was conserved by Opinion 460 (International Commission on Zoological Nomenclature, 1957*b*). Revised by Groves and Grubb (1990). Six new species named from 1982 through 1999. *Megamuntiacus* treated as a synonym of *Muntiacus* by Amato et al. (2000), Giao et al. (1998), and Schaller and Vrba (1996). Studies of mtDNA sequences by Wang and Lan (2000) suggested the following phylogeny: (*reevesi, vuquangensis*) (((*feae*) (*gongshanensis, crinifrons*)) (*muntjac*)), indicating *Muntiacus* would be paraphyletic if *Megamuntiacus* were regarded as valid.

*Muntiacus atherodes* Groves and Grubb, 1982. Zool. Meded. Leiden, 56:210.
> COMMON NAME: Bornean Yellow Muntjac.
> TYPE LOCALITY: Malaysia, Borneo, "near forest camp 1, Cocoa Research Station, Tawau, Saba [= Sabah], 800 ft. [244 m]"
> DISTRIBUTION: Borneo.
> STATUS: IUCN – Lower Risk (lc).
> COMMENTS: Formerly included in *M. muntjak*, or in a separate species, *M. pleiharicus*; see Chasen (1940:203); however *pleiharicus* is a synonym of *muntjak*; see Groves and Grubb (1982).

*Muntiacus crinifrons* (Sclater, 1885). Proc. Zool. Soc. Lond., 1885:1, pl. 1.
> COMMON NAME: Black Muntjac.
> TYPE LOCALITY: "Vicinity of Ningpo, China" (= China, Zhejiang, near Ningpo).
> DISTRIBUTION: E China (S Anhui, N Fujian, Jiangxi, and Zhejiang; not reliably recorded from Yunnan).
> STATUS: CITES – Appendix I; IUCN – Vulnerable.
> COMMENTS: Included in *muntjak* by Haltenorth (1963:42). Former presumed occurrence from Yunnan and Guangdong to Jaingsu (China) (Shou, 1962:454) may involve confusion with *M. truongsongensis*. Records from N Burma (Rabinowitz and Saw Tun Khaing, 1998; Rabinowitz et al., 1998) are probably based on *M. gongshanensis*. Differs radically from that species in karyotype (Yang et al., 1995, 1997).

*Muntiacus feae* (Thomas and Doria, 1889). Ann. Mus. Civ. Stor. Nat. Genova, 27:92.
> COMMON NAME: Fea's Muntjac.
> TYPE LOCALITY: Burma, "Thagatà Juva, a S. E. del Monte Mooleyit [= Mt. Mulaiyit], Tenasserim".
> DISTRIBUTION: Peninsular Burma and Thailand; records from China are doubtful (SE Yunnan) or refer to *M. gonghanensis* (SE Tibet and W Yunnan).
> STATUS: U.S. ESA – Endangered; IUCN – Endangered; IUCN – Data Deficient.
> SYNONYMS: *feai* Grubb, 1977.
> COMMENTS: Included in *muntjak* by Haltenorth (1963:42). For spelling of the species name *feai*, see Grubb (1977); but see Article 31.1.1. (International Code of Zoological Nomenclature; International Commission on Zoological Nomenclature, 1999) for retention of original spelling *feae*.

*Muntiacus gongshanensis* Ma, 1990. *In* Ma et al., Zool. Res. Kunming, 11:47.
> COMMON NAME: Gongshan Muntjac.
> TYPE LOCALITY: China, "Mijiao (27°35′ N., 98°47′ E.), Puladi, Gongshan county, East slope of the northern sector of Gaoligong Mountain, north-western Yunnan".
> DISTRIBUTION: N Burma and China (SE Tibet and W Yunnan).
> STATUS: IUCN – Data Deficient.
> COMMENTS: Identified in the literature as *M. reevesi* (F. M. Bailey, 1914, 1915; Dollman, 1932), *M. feae* (Groves and Grubb, 1990; Zhang et al, 1984), or *M. crinifirons* (Amato et al., 1999; Rabinowitz and Saw Tun Khaing, 1998; Rabinowitz et al., 1998), from which it differs in the structure of the chromosomes (Yang et al., 1995).

*Muntiacus muntjak* (Zimmermann, 1780). Geogr. Gesch. Mensch. Vierf. Thiere, 2:131.
    COMMON NAME: Red Muntjac.
    TYPE LOCALITY: Indonesia, "Java".
    DISTRIBUTION: Bangladesh, Bhutan, Burma, Cambodia, S China (S Tibet and Yunnan to
        Guangdong), India, Laos, peninsular Malaya, Nepal, NE Pakistan, Sri Lanka, Vietnam,
        and Sunda Isls (Sumatra, Borneo, Java, Bali, Lombok, and many smaller Indonesian
        islands).
    STATUS: IUCN – Lower Risk (lc).
    SYNONYMS: *bancanus* Lyon, 1907; *hamatus* (de Blainville, 1816); *moschatus* (de Blainville,
        1816); *moschus* (Desmarest, 1822); *muntjac* (Gmelin, 1788); *muntjuc* (Link, 1795);
        *nainggolani* Sody, 1932; *peninsulae* Lydekker, 1915; *pleiharicus* (Kohlbrugge, 1896);
        *robinsoni* Lydekker, 1915; *rubidus* Lyon, 1911; *subcornutus* (de Blainville, 1816); *typicus*
        (Ward, 1910); **annamensis** Kloss, 1928; **aureus** (C. H. Smith, 1826); *albipes* (Wagner,
        1844); *tamulicus* (Gray, 1872); **curvostylis** (Gray, 1872); *grandicornis* (Lydekker, 1904);
        **guangdongensis** Li and Xu, 1996; **malabaricus** Lydekker, 1915; **menglalis** Wang and
        Groves, 1988; **montanus** Robinson and Kloss, 1918; **nigripes** G.M. Allen, 1930;
        **vaginalis** (Boddaert, 1785); *melas* (Ogilby, 1840); *muntjacus* (Kelaart, 1852); *ratva*
        (Sundevall, 1846); *ratwa* (Hodgson, 1833); *styloceros* (Schinz, 1845); **yunnanensis**
        Ma and Wang, 1988.
    COMMENTS: Includes *pleiharicus*, listed as a distinct species by Chasen (1940:203), and
        *vaginalis*. Haltenorth (1963:40) included *reevesi*, *feae*, *rooseveltorum* and *crinifrons*.
        Distinctive differences in karyotype between single peninsular Malayan specimen
        (2N = 8) and other mainland populations (2n = 6 or 7) suggest possible division
        between Malesian and Continental semispecies (Groves and Grubb, 1987). Groves
        (2003) treated the Continental *vaginalis* (including subspecies *aureus*, *malabaricus* and
        others) as a species separate from *muntjak*.

*Muntiacus puhoatensis* Trai, 1997. *In* Chau, Vietnam Economic News, 47:46.
    COMMON NAME: Puhoat Muntjac.
    TYPE LOCALITY: Vietnam, "Puhoat area in Que Phong District, Nghe An Province".
    DISTRIBUTION: Known only from the type locality.
    COMMENTS: Systematic status uncertain.

*Muntiacus putaoensis* Amato, Egan and Rabinowitz, 1999. Anim. Conserv., 2:4.
    COMMON NAME: Leaf Deer.
    TYPE LOCALITY: "purchased at Atanga village, 30 km east of Putao (27°21'N, 97°24'E),
        northern Myanmar [N Burma]".
    DISTRIBUTION: N Burma.
    COMMENTS: Names *putaoensis* and *puhoatensis* refer to localities in Burma and Vietnam
        respectively.

*Muntiacus reevesi* (Ogilby, 1839). Proc. Zool. Soc. Lond., 1838:105 [1839].
    COMMON NAME: Reeves' Muntjac.
    TYPE LOCALITY: "China"; "Near Canton, Kwantung [Guangdong], Southern China"
        (Ellerman and Morrison-Scott, 1951:357).
    DISTRIBUTION: SE China (S Gansu to Yunnan) and Taiwan; introduced to England
        (successfully) and France (no longer present).
    STATUS: IUCN – Lower Risk (lc).
    SYNONYMS: *bridgemani* (Lydekker, 1910); *lachrymans* (Milne-Edwards 1871); *pingshiangicus*
        (Hilzheimer, 1906); *sclateri* (Swinhoe, 1873); *sinensis* (Hilzheimer, 1903); *teesdalei*
        Lydekker, 1915; **jiangkouensis** Gu and Zu, 1998; **micrurus** (Sclater, 1875).
    COMMENTS: Included in *muntjak* by Haltenorth (1963:42); but see Corbet (1978c:199).

*Muntiacus rooseveltorum* Osgood, 1932. Field. Mus. Publ. Zool., 18:232.
    COMMON NAME: Roosevelt Muntjac.
    TYPE LOCALITY: "Muong Yo, Laos. Altitude 2,300 feet [701 m]."
    DISTRIBUTION: Known from the type locality, ca. 31°30'N, 102°00'E. Recently recorded from
        the Annamite Mtns in N Laos at 19°49'N, 103°45'E and observed in captivity at Lak
        Sao, N Laos, 18°20'N, 106°00'E (Amato et al., 1999).

STATUS: IUCN – Data Deficient as *M. feae rooseveltorum*.
COMMENTS: Included in *M. feae* by Groves and Grubb (1990) but now known to differ.

*Muntiacus truongsonensis* (Giao, Tuoc, Eric, Dung et al. [sic; apparently Giao, Tuoc, Dung, Wikramanayake, Amato, Arctander and Mackinnon], 1997). *In* Ha, Vietnam Economic News, 38:46.
COMMON NAME: Annamite Muntjac.
TYPE LOCALITY: Vietnam, "in the west of Quang Nam province"; "collected from four houses in three locations in Hien District, West Quang Nam Province, Vietnam The three locations are: Hien, the District capital, A Tin village, and A Plo village (15°56′59″N, 107°34′18″E)" (Giao et al., 1998). The type locality is one of these three places.
DISTRIBUTION: Upland forest in S Laos and C Vietnam; possibly S China (including SE Yunnan).
STATUS: IUCN – Not Evaluated, Data Deficient DSG recommended, as Truong Son muntjac.
SYNONYMS: *napensis* Tobias, 1997 [*nomen nudum*].
COMMENTS: First described as *Caninmuntiacus truongsonensis* Giao, Tuoc, Eric [Wikramanayake], Dung et al., in Ha (1997). Later named as *Muntiacus truongsonensis* Giao, Tuoc, Dung, Wikramanayake, Amato, Arctander and Mackinnon, 1998. These authors named a holotype, though technically it is probably a lectotype. The earlier publication satisfies the requirements for availability and the authorship appears to be correctly attributable to Giao et al. (Articles 9 and 10 and Article 50.1.1, respectively, International Code of Zoological Nomenclature; International Commission on Zoological Nomenclature, 1999). Known distribution discussed by Groves and Schaller (2000). A skin attributed to *M. feae* from SE Yunnan (Sokolov, 1957) may represent this species; presumed occurrence of *M. crinifrons* in Yunnan and Guangdong to Jaingsu (Shou, 1962:454) may also refer to *M. truongsonensis*.

*Muntiacus vuquangensis* (Tuoc, Dung, Dawson, Arctander and Mackinnon, 1994). Science and Technology news. Forest Inventory and Planning Institue (Hanoi), p. 5.
COMMON NAME: Large-antlered Muntjac.
TYPE LOCALITY: "Vu Quang Nature Reserve in Ha tinh province of Vietnam".
DISTRIBUTION: Upland forest in Laos and Vietnam.
STATUS: CITES – Appendix I as *Megamuntiacus vuquanghensis* [sic]. Locally relatively abundant but its restriction to upland forests suggests it should be classified as Potentially At Risk (Duckworth et al., 1993).
COMMENTS: Distribution and status in Laos reviewed by Timmins et al (1998). Placed in genus *Muntiacus* by Amato et al. (2000), Giao et al (1998), and Schaller and Vrba (1996).

*Przewalskium* Flerov, 1930. C. R. Acad. Sci. URSS, p. 115.
TYPE SPECIES: *Cervus albirostris* Przewalski, 1883.

*Przewalskium albirostris* (Przewalski, 1883). Third Journey in Central Asia, p. 124.
COMMON NAME: White-lipped Deer.
TYPE LOCALITY: China, Gansu, 3 km above mouth of Kokusu River, Humboldt Mtns, Nan Shan (Flerov, 1960).
DISTRIBUTION: China (Gansu, Sichuan, E Tibet including Qinghai, and N Yunnan).
STATUS: IUCN – Vulnerable as *C. albirostris*.
SYNONYMS: *dybowskii* (Sclater, 1889); *sellatus* (Przewalski, 1883); *thoroldi* (Blanford, 1893).

*Rucervus* Hodgson, 1838. Ann. Nat. Hist. 1:154.
TYPE SPECIES: *Cervus elaphoides* Hodgson, 1835 (= *Cervus duvaucelii* G. Cuvier, 1823).
SYNONYMS: *Panolia* Gray, 1843; *Procervus* Hodgson, 1847; *Thaocervus* Pocock, 1943.

*Rucervus duvaucelii* (G. Cuvier, 1823). Rech. Oss. Foss., Nouv. ed., 4:505.
COMMON NAME: Barasingha.
TYPE LOCALITY: "des Indes"; restricted to N India, Uttar Pradesh, Kumaun by Groves (1982*b*:624).
DISTRIBUTION: N and C India, SW Nepal; extinct in Pakistan.

STATUS: CITES – Appendix I as *Cervus duvaucelii*; U.S. ESA – Endangered as *C. duvauceli* [sic]; IUCN – Critically Endangered as *Cervus duvauceli ranjitsinhi*, Endangered as *C. d. branderi*, Vulnerable as *C. d. duvauceli*.

SYNONYMS: *bahrainja* (Hodgson, 1834) [*nomen nudum*]; *dimorphe* (Hodgson, 1843); *duvaucelli* (Sundevall, 1846); *elaphoides* (Hodgson, 1835); *euceros* (Gray, 1850); *eucladoceros* (Falconer, 1868); *euryceros* (Gray, 1850); *smithii* (Gray, 1837); **branderi** Pocock, 1943; **ranjitsinhi** (Groves, 1982).

COMMENTS: Revised by Groves (1982*b*).

*Rucervus eldii* (M'Clelland, 1842). Calcutta J. Nat. Hist., 2:417, pl. 12.

COMMON NAME: Eld's Deer.

TYPE LOCALITY: India, Assam, "the valley of Munipore" (Manipur).

DISTRIBUTION: Burma, Cambodia, China (Hainan Isl), N India (Manipur), Laos, Thailand, and Vietnam; now much reduced in numbers in several of these countries.

STATUS: CITES – Appendix I as *Cervus eldii*; U.S. ESA – Endangered as *C. eldi* [sic]; IUCN – Critically Endangered as *Cervus eldii eldii,* Lower Risk (nt) as *C. e. thamin*, Data Deficient as *C. e. siamensis*.

SYNONYMS: *acuticauda* (Blyth, 1864); *acuticornis* (Gray, 1843); *cornipes* (Lydekker, 1901); *frontalis* (M'Clelland, 1843); *lyratus* (Schinz, 1845); *typicus* (Lydekker, 1898); **siamensis** (Lydekker, 1915); *hainanus* Thomas, 1918; *platyceros* (Gray, 1843) [preoccupied]; **thamin** Thomas, 1918; *brucei* Thomas, 1918.

COMMENTS: Phylogeography studies support the recognition of three subspecies (Balakrishnan et al., 2003).

*Rucervus schomburgki* (Blyth, 1863). Proc. Zool. Soc. Lond., 1863:155.

COMMON NAME: Schomburgk's Deer.

TYPE LOCALITY: "probably inhabiting Siam [Thailand]"; occurrence in central plains of Thailand since confirmed.

DISTRIBUTION: Thailand (extinct), China (Yunnan), and possibly in N Laos.

STATUS: IUCN – Extinct (but see below).

COMMENTS: Included in *duvaucelii* by Haltenorth (1963:58) and Groves (1982*b*); but treated as a full species by Lekagul and McNeely (1977). Last Thailand specimen killed in 1932 (Harper, 1945); one record from Sanda Valley, Yunnan (Bentham, 1908; Sclater, 1891); present status in Yunnan unknown; recently observed antlers suggest another population may survive in N Laos (Schroering, 1995, and in litt.).

*Rusa* C. H. Smith, 1827. *In* Griffith et al., Anim. Kingdom, 4:105.

TYPE SPECIES: *Cervus unicolor* Kerr, 1792.

SYNONYMS: *Hippelaphus* Sundevall, 1846; *Melanaxis* Heude, 1888; *Sambur* Heude, 1888; *Ussa* Heude, 1888.

COMMENTS: Assumption that this is a monophyletic group has been challenged by Randi et al. (2001).

*Rusa alfredi* Sclater, 1870. Proc. Zool. Soc. Lond., 1870:381.

COMMON NAME: Visayan Spotted Deer.

TYPE LOCALITY: "transmitted from Singapore Hab. Malayan peninsula, or adjoining islands (?)"; "Philippines, the type specimen having been received from Manila" (Lydekker, 1915:63).

DISTRIBUTION: Philippines (Panay and Negros Isls; formerly also Guimaras and possibly Siquijor but almost certainly not Bohol, Cebu or any other Isls according to W. Oliver, in litt.).

STATUS: U.S. ESA and IUCN – Endangered as *Cervus alfredi*.

SYNONYMS: *breviceps* (Heude, 1888); *cinerea* (Heude, 1899); *masbatensis* (Heude, 1888).

COMMENTS: Included in *R. marianna* by Haltenorth (1963). Revised by Grubb and Groves (1983), where treated as a full species.

*Rusa marianna* (Desmarest, 1822). Mammalogie, *in* Encycl. Meth., 2:436.

COMMON NAME: Philippine Deer.

TYPE LOCALITY: "Les îles Mariannes" (Mariana Isls, Guam); introduced.

DISTRIBUTION: Philippines (Basilan, Catanduanes, Leyte, Luzon, Polillo, and Samar Isls, and possibly Bohol and other small Isls); introduced to Mariana, Caroline and Ogasawara (= Bonin) Isls (W Pacific Ocean).

STATUS: IUCN – Data Deficient as *Cervus mariannus*.

SYNONYMS: *ambrosiana* (Heude, 1888); *atheneensis* (Heude, 1899); *baryceros* (Heude, 1899); *boninensis* (Lydekker, 1905); *brachyceros* (Heude, 1888); *chrysotrichos* (Heude, 1888); *corteana* (Heude, 1888); *crassicornis* (Heude, 1888); *dailliardiana* (Heude, 1888); *elegans* (Heude, 1888); *elorzana* (Heude, 1888); *garciana* (Heude, 1888); *gonzalina* (Heude, 1888); *gorrichana* (Heude, 1888); *guevarana* (Heude, 1888); *guidoteana* (Heude, 1888); *hippolitiana* (Heude, 1888); *longicuspis* (Heude, 1888); *macariana* (Heude, 1888); *maraisiana* (Heude, 1888); *marzanina* (Heude, 1888); *michaelina* (Heude, 1899); *microdontus* (Heude, 1888); *nublana* (Heude, 1888); *philippina* (C. H. Smith, 1827); *ramosiana* (Heude, 1888); *rosariana* (Heude, 1888); *roxasiana* (Heude, 1888); *rubiginosa* (Heude, 1888); *spatharia* (Heude, 1888); *telesforiana* (Heude, 1888); *tuasonina* (Heude, 1888); *verzosana* (Heude, 1888); *vidalina* (Heude, 1899); *villemeriana* (Heude, 1899); **barandana** (Heude, 1888); **nigella** Hollister, 1913; *apoensis* Sanborn, 1952; **nigricans** (Brooke, 1876); *basilanensis* (Heude, 1888); *franciana* (Heude, 1888); *steerii* (Elliot, 1896).

COMMENTS: Treated as a separate species from *C. unicolor* by Haltenorth (1963), and by Grubb and Groves (1983), who revised this taxon. Brought to Ogasawara Isls in late 18th to early 19th centuries by Spanish ships; extinct there by about 1925; reintroduced from Guam after World War II but do not now survive (Miura and Yoshihara, 2002).

*Rusa timorensis* (de Blainville, 1822). J. Phys. Chim. Hist. Nat. Arts Paris, 94:267.

COMMON NAME: Javan Rusa.

TYPE LOCALITY: Indonesia, Lesser Sunda Isls, "Timor" Isl.

DISTRIBUTION: Indonesia, Sunda Isls; autochthonous on Bali, and Java; probably introduced in antiquity to Lesser Sunda Isls, Molucca Isls (including Buru and Seram), Sulawesi, and Timor; since 17th century, introduced to Borneo (Kalimantan; now extinct?), New Guinea, New Britain, Aru Isls, Mauritius, Comoro Isls, Madagascar (extinct?), Australia, New Zealand, New Caledonia and small islands in Indonesia and off the coast of Australia.

STATUS: IUCN – Lower Risk (lc).

SYNONYMS: *paradoxa* Brehm, 1865; *peronii* (G. Cuvier, 1825); *tavistocki* (Lydekker, 1900); *timoriensis* (Müller and Schlegel, 1845); **djonga** Van Bemmel, 1949; **floresiensis** (Heude, 1896); *sumbavana* (Heude, 1896); **macassaricus** (Heude, 1896); *celebensis* (Rörig, 1896); *menadensis* (Heude, 1896); **moluccensis** (Quoy and Gaimard, 1830); *buruensis* (Heude, 1896); *hoevelliana* (Heude, 1896); **renschi** (Sody, 1932); **russa** (Müller and Schlegel, 1845); *hippelaphus* (G. Cuvier, 1825) [preoccupied]; *hippolaphus* (Schinz, 1845); *javanica* (Müller and Schlegel, 1845); *laronesiotes* Van Bemmel, 1949; *lepida* (Sundevall, 1846); *tunjuc* (Horsfield, 1830) [*nomen nudum*].

COMMENTS: Revised by Van Bemmel (1949*a*). Includes *tavistocki*; see Grubb and Groves (1983).

*Rusa unicolor* (Kerr, 1792). *In* Linnaeus, Anim. Kingdom, p. 300.

COMMON NAME: Sambar.

TYPE LOCALITY: "Inhabits the dry hilly forests of Ceylon, Borneo, Celebes and Java"; restricted to Ceylon (Sri Lanka; Lydekker, 1915:73).

DISTRIBUTION: India and Sri Lanka east to S China (E Tibet and Sichuan to Yunnan; Hainan Isl) and Taiwan; south to Peninsular Malaysia, Sunda Isles (Sumatra, Borneo, Siberut, Sipora, and Pagi and Nias Isls); introduced to Australia and New Zealand.

STATUS: IUCN – Lower Risk (lc).

SYNONYMS: *albicornis* (Bechstein, 1799); *aristotelis* (G. Cuvier, 1823); *bengalensis* (Schinz, 1845); *heterocerus* (Hodgson, 1831); *hippelaphus* (C. H. Smith, 1827); *jarai* (Hodgson, 1831); *leschenauldii* (G. Cuvier, 1823); *leschenaulti* (Sundevall, 1846); *major* (Kerr, 1792); *maxima* (de Blainville, 1822); *nepalensis* (Hodgson, 1841); *nigra* (de Blainville, 1816); *pennantii* (Gray, 1843); *tarai* Hodgson, 1863 [*nomen nudum*]; *typica* (Lydekker,

1898); ***brookei*** (Hose, 1893); *hamiltoniana* (Heude, 1896); ***cambojensis*** (Gray, 1861); *brachyrhina* (Heude, 1888); *colombertina* (Heude, 1888); *combalbertina* (Heude, 1888); *curvicornis* (Heude, 1888); *errardiana* (Heude, 1888); *joubertiana* (Heude, 1888); *latidens* (Heude, 1888); *lemeana* (Heude, 1888); *lignaria* (Heude, 1888); *longicornis* (Heude, 1888); *officialis* (Heude, 1888); *outreyana* (Heude, 1888); *planiceps* (Heude, 1888); *planidens* (Heude, 1888); *simonina* (Heude, 1888); *veruta* (Heude, 1888); ***dejeani*** de Pousargues, 1896; ***equina*** (G. Cuvier, 1823); *malaccensis* (F. Cuvier, 1824); *oceana* (Chasen and Kloss, 1928); ***hainana*** (Xu, 1983); ***swinhoii*** (Sclater, 1862).

COMMENTS: Subspecies from Groves and Grubb (1987).

**Subfamily Hydropotinae** Trouessart, 1898. Cat. Mamm. Viv. Foss., new ed., fasc. 4:865.

COMMENTS: Affinities controversial; from cladistic analysis of skeletal characters, distanced from all other Cervidae by Gentry and Hooker (1988), making Cervidae paraphyletic if included; so perhaps to be placed in a different family; alternatively nested within Odocoileinae (= Capreolinae), close to *Capreolus*, by Randi et al. (1998), using mtDNA studies, and inferred to have lost antlers secondarily.

*Hydropotes* Swinhoe, 1870. Athenaeum, 2208:264.

TYPE SPECIES: *Hydropotes inermis* Swinhoe, 1870.

SYNONYMS: *Hydrelaphus* Lydekker, 1898.

COMMENTS: Original description usually given as Proc. Zool. Soc. Lond., 1870:90 [publ. June, 1870], but McAllan and Bruce (1989) showed that publication in The Athenaeum was earlier (19 Feb. 1870).

*Hydropotes inermis* Swinhoe, 1870. Athenaeum, 2208:264.

COMMON NAME: Chinese Water Deer.

TYPE LOCALITY: Syntypes purchased in Shanghai market, but based on the place where Swinhoe saw the species in the wild, type locality restricted to China, Kiangsu, Chingkiang, Yangtze River, Deer Isl (Ellerman and Morrison-Scott, 1951:354).

DISTRIBUTION: China (formerly from Liaoning to Guangxi including the lower Yangtze Basin) and Korea; introduced in England and France.

STATUS: IUCN – Lower Risk (nt) as *H. i. inermis*, Data Deficient as *H. i. argyropus*.

SYNONYMS: *affinis* Brooke, 1872; *kreyenbergi* Hilzheimer, 1905; ***argyropus*** Heude, 1884.

COMMENTS: Original description usually given as Proc. Zool. Soc. Lond., 1870:89 [publ. June, 1870], but McAllan and Bruce (1989) showed that publication in the Athenaeum was earlier (19 Feb. 1870).

**Family Antilocapridae** Gray, 1866. Ann. Mag. Nat. Hist., ser 3, 18:325-326, 468.

*Antilocapra* Ord, 1818. J. Phys. Chim. Hist. Nat. Arts Paris, 87:149.

TYPE SPECIES: *Antilope americana* Ord, 1815.

SYNONYMS: *Dicranocerus* C. H. Smith, 1827.

COMMENTS: Included in Bovidae by O'Gara and Matson (1975); but restored to separate family status by Janis and Scott (1987) and Soulounias (1988).

*Antilocapra americana* (Ord, 1815). *In* Guthrie, New Geogr., Hist. Coml. Grammar., Philadelphia, 2nd ed., 2:292, 308.

COMMON NAME: Pronghorn.

TYPE LOCALITY: USA, "On the plains and the highlands of the Missouri [River]".

DISTRIBUTION: S Alberta and S Saskatchewan (Canada) south through W USA to Hidalgo, Baja California, W Sonora (Mexico). Introduced to Lanai Isl (Hawaiian Isls).

STATUS: CITES – Appendix I (Mexican populations); U.S. ESA – Endangered as *A. a. peninsularis* and *A. a. sonoriensis*; IUCN – Critically Endangered as *A. a. peninsularis*, Endangered as *A. a. sonoriensis*, Lower Risk (cd) as *A. a. mexicana*, otherwise Lower Risk (lc).

SYNONYMS: *anteflexa* Gray, 1855; *furcifer* (C. H. Smith, 1821); *palmata* (C. H. Smith, 1821); ***mexicana*** Merriam, 1901; ***oregona*** V. Bailey, 1932; ***peninsularis*** Nelson, 1912; ***sonoriensis*** Goldman, 1945.

COMMENTS: Reviewed by O'Gara (1978, Mammalian Species, 90).

**Family Giraffidae** Gray, 1821. London Med. Repos., 15:307.
 SYNONYMS: Camelopardalina Bonaparte, 1837; Camelopardina Gray, 1825; Okapinae Bohlin,
  1926; Palaeotragini Pilgrim, 1911.
 COMMENTS: Placement of this family follows Janis and Scott (1987). According to McKenna and
  Bell (1997) there are two subfamilies, Giraffinae and the wholly extinct Sivatheriinae.
  Giraffini includes *Giraffa*; Palaeotragini includes *Okapia*.

*Giraffa* Brisson, 1762. Regn. Anim., 2nd ed., pp. 12, 37.
 TYPE SPECIES: *Cervus camelopardalis* Linnaeus, 1758.
 SYNONYMS: *Camelopardalis* von Schreber, 1784; *Orasius* Oken, 1816; *Trachelotherium* Gistel, 1848.
 COMMENTS: Brisson (1762) is rejected for nomenclatural purposes (and see Hopwood, 1947)
  but *Giraffa* Brisson, 1762 has been conserved (Opinion 1894, International Commission
  on Zoological Nomenclature, 1998).

*Giraffa camelopardalis* (Linnaeus, 1758). Syst. Nat., 10th ed., 1:66.
 COMMON NAME: Giraffe.
 TYPE LOCALITY: "Habitat in Æthiopia et Sennar"; identified as Egypt, in captivity at Cairo
  (Thomas, 1911*a*:150); restricted to Sudan, Sennar, by Harper (1940:322).
 DISTRIBUTION: Disjunct; W and C Africa in Burkina Faso (vagrant), N Cameroon, Central
  African Republic, S Chad, NE Dem. Rep. Congo, Eritrea (extinct), W and S Ethiopia,
  Gambia (extinct), Kenya, Mali (extinct), SE Mauritania (extinct), Niger, Nigeria
  (extinct, now a vagrant), Senegal (extinct), S Somalia, Sudan, Tanzania and Uganda; no
  reliable records from Ghana, Guinea, and Togo; may have occurred in Benin;
  introduced into Rwanda; S Africa in S Angola (extinct?), Botswana, Mozambique
  (extinct), Namibia, South Africa (originally mostly N of Orange River), Swaziland
  (extinct, reintroduced), Zambia (SW and Luangwa Valley), and Zimbabwe.
  Distribution now much restricted; in W Africa still present in Niger, and N Cameroon
  but extinct in Mali according to Ciofolo and Le Pendu (2002), apparently very
  recently; in southern Africa, now naturally distributed no farther south than
  N Namibia, Botswana and NE South Africa (E Limpopo and E Mpumalanga Provs.).
  Introduced beyond its former range in South Africa, including KwaZulu-Natal.
 STATUS: IUCN – Lower Risk (cd).
 SYNONYMS: *aethiopica* (Ogilby, 1837) [*nomen nudum*]; *aethiopica* Sundevall, 1846; *africana*
  (Lacépède, 1799), *antiquorum* (Jardine, 1835); *biturigum* (Duvernoy, 1844); *congoensis*
  Lydekker, 1903; *peralta* Thomas, 1898; *renatae* Krumbiegel, 1971; *reticulata* Weinland,
  1863 [suppressed]; *senaariensis* Trouessart, 1898; *typica* Bryden, 1899; **reticulata**
  de Winton, 1899; *hagenbecki* Knottnerus-Meyer, 1910; *nigrescens* Lydekker, 1911;
  **rothschildi** Lydekker, 1903; *cottoni* Lydekker, 1904; **thornicrofti** Lydekker, 1911;
  **tippelskirchi** Matschie, 1898; *schillingsi* Matschie, 1898; **giraffa** (von Schreber, 1784);
  *angolensis* Lydekker, 1903; *australis* (Swainson, 1835) [*nomen nudum*]; *australis* Rhoads,
  1896 [suppressed]; *capensis* (Lesson, 1842); *infumata* Noack, 1908; *maculata* (Weinland,
  1863); *wardi* Lydekker, 1904.
 COMMENTS: Subspecific synonymy modified from Ansell (1972:13). Reviewed by Dagg
  (1971, Mammalian Species, 5). Cotterill (2003*a*) listed *thornicrofti* as a species. The
  names *reticulata* Weinland, 1863 and *australis* Rhoads, 1896 have been suppressed,
  while *reticulata* de Winton, 1899 has been conserved (International Commission on
  Zoological Nomenclature 1971*a*, 1979*b*), even though *australis* is probably a junior
  synonym of *giraffa*, not a senior synonym of *reticulata* de Winton.

*Okapia* Lankester, 1901. Nature, 64:24.
 TYPE SPECIES: *Equus johnstoni* P. L. Sclater, 1901.

*Okapia johnstoni* (P. L. Sclater, 1901). Proc. Zool. Soc. Lond., 1901(1):50.
 COMMON NAME: Okapi.
 TYPE LOCALITY: Dem. Rep. Congo, "in sylvis fluvio Semliki adjacentibus" (= Semliki Forest,
  Mundala).

DISTRIBUTION: N and E Dem. Rep. Congo.

STATUS: IUCN – Lower Risk (nt).

SYNONYMS: *erikssoni* Lankester, 1902; *kibalensis* Gatti, 1936; *liebrechtsi* Forsyth Major, 1902; *tigrinum* (Johnston, 1901).

**Family Bovidae** Gray, 1821. London Med. Repos., 15:308.

COMMENTS: Distribution and status of introduced populations reviewed by Lever (1985). Distribution and status of African species reviewed by East (1988, 1989, 1990) and East et al. (1999). Systematics of African species reviewed by Ansell (1972) and Gentry (1972). Family-group names reviewed by Grubb (2001*b*). Evidence is accumulating that all Bovidae other than Bovinae constitute a monophyletic clade (Hassanin and Douzery, 1999*a*; Kingdon, 1982; Vrba and Schaller, 2000). Within this section of Bovidae some conventional subfamilies are probably paraphyletic if not polyphyletic and until monophyletic clades are defined and downgraded into tribes, Aepycerotinae, Alcelaphinae, Antilopinae, Caprinae, Cephalophinae, Hippotraginae, and Reduncinae are retained here as subfamilies. The synonymy of the following names remains undecided: *Ixalus* Ogilby, 1836, *Adenota mengesi* Neumann, 1900, *Antilope mazama* C. H. Smith, 1821, *Antilope koba* Erxleben, 1777, *Antilope temamazama* C. H. Smith, 1821, *Ixalus probaton* Ogilby, 1836. *Cervus guineensis* Linnaeus, 1758 is possibly a bovid if correctly reported from Guinea (= West Africa).

**Subfamily Aepycerotinae** Gray, 1872. Cat. Ruminant Mamm. Brit. Mus., p. 4, 42.

*Aepyceros* Sundevall, 1847. Kongl. Svenska Vet.-Akad. Handl. Stockholm, 1845:271 [1847].

TYPE SPECIES: *Antilope melampus* Lichtenstein, 1812.

*Aepyceros melampus* (Lichtenstein, 1812). Reisen Sudl. Africa, 2, pl. 4 opp. p. 544.

COMMON NAME: Impala.

TYPE LOCALITY: "Koossi-Thale", now identified as South Africa, Northern Cape Prov., Kuruman Dist., Khosis (Grubb, 1999).

DISTRIBUTION: S Angola, N and E Botswana, Burundi (extinct?), Dem. Rep. Congo (SE Shaba Prov.), Kenya, Malawi, Mozambique, N Namibia, Rwanda, South Africa (North-West, Limpopo, and Mpumalanga Provs. and KwaZulu-Natal; formerly in N Northern Cape Prov.), Swaziland, Tanzania, Uganda (marginally in NE and SW), Zambia, Zimbabwe.

STATUS: U.S. ESA – Endangered as *A. m. petersi*; IUCN – Vulnerable as *A. m. petersi*, otherwise Lower Risk (cd).

SYNONYMS: *pallah* (Gervais, 1841); *typicus* Thomas, 1893; **johnstoni** Thomas, 1893; *holubi* Lorenz, 1894; **katangae** Lönnberg, 1914; **petersi** Bocage, 1879; **rendilis** Lönnberg, 1912; **suara** Matschie, 1892.

COMMENTS: Synonymy and inclusion of *petersi* follows Ansell (1972:57). Nersting and Arctander (2001) found *petersi* haplotypes to be strongly isolated from haplotypyes of populations in Kenya, Uganda, Tanzania, Zambia, Zimbabwe and N Botswana. Cotterill (2003*a*) considered *petersi* to be an evolutionary species.

**Subfamily Alcelaphinae** Brooke, 1876. *In* Wallace. Geog. Distr. Anim. p. 224.

SYNONYMS: Bubalidinae Sclater and Thomas, 1894 [unavailable]; Bubalinae Trouessart, 1898 [unavailable]; Bubalinae Trouessart, 1905 [unavailable]; Damalidae Brookes, 1828 [*nomen oblitum*]; Damalidae Gray, 1872 [*nomen oblitum*]; Connochetidae Gray, 1872; Damaliscina Vrba, 1997.

COMMENTS: Living genera assigned to two subtribes by Vrba (1997), Alcelaphini (including *Alcelaphus, Beatragus,* and *Connochaetes*) and Damaliscini (includes only *Damaliscus*), but retention of flehmen behavior, lost in *Alcelaphus* and *Damaliscus* (Estes, 1999), and karyology suggests *Beatragus* is sister group of *Alcelaphus* plus *Damaliscus* (Kumamoto et al., 1996; Robinson et al., 1991).

*Alcelaphus* de Blainville, 1816. Bull. Sci. Soc. Philom. Paris, 1816:75.

TYPE SPECIES: *Antilope bubalis* Pallas, 1767 (= *Antilope buselaphus* Pallas, 1766).

SYNONYMS: *Acronotus* C. H. Smith, 1827; *Alcephalus* Brooke, 1876; *Bubalis* Goldfuss, 1820; *Bubalus* Ogilby, 1837; *Damalis* C. H. Smith, 1827; *Sigmoceros* Heller, 1912.

COMMENTS: Van Gelder (1977*b*) included *Damaliscus* in this genus, but has not been followed by recent authors; see Swanepoel et al. (1980:187). Phylogeographic studies (Arctander et al., 1999; Flagstad et al., 2001) suggest a tree of the following form: (*A. lichtensteinii, A. caama*)((*A. buselaphus buselaphus, A. b. major*)(*A. b. tora, A. b. swaynei, A. b. cokii, A. b. lelwel*)).

*Alcelaphus buselaphus* (Pallas, 1766). Misc. Zool., p. 7.
COMMON NAME: Hartebeest.
TYPE LOCALITY: No locality cited but the name is based on "Le bubale" of Buffon, "en Barbarie & dans toutes les parties septentrionales de l'Afrique", and on other sources. Restricted to Barbary by designation of the "Vache de Barbarie" of Perrault as the lectotype (Ruxton and Schwarz, 1929:575). Further restricted to Morocco (Lydekker, 1914*a*:5).
DISTRIBUTION: In N Africa, now extinct but within historic times occurred in N Algeria, Libya (marginally), N Morocco, and Tunisia. In West and Equatorial Africa in Benin, Burkina Faso, Cameroon, Central African Republic, S Chad, Côte d'Ivoire, N Dem. Rep. Congo, N Eritrea, Ethiopia (outside highlands), Gambia (extinct or vagrant), Ghana, E Guinea Bissau, Guinea, S Kenya, S Mali, Niger (marginal in SW), Nigeria, Senegal, NW Somalia (extinct), S Sudan, N Tanzania, Togo, and Uganda. No authentic records from Sierra Leone (Grubb et al., 1998).
STATUS: U.S. ESA – Endangered as *A. b. swaynei* and *A. b. tora*; IUCN – Extinct as *A. b. buselaphus*, Endangered as *A. b. swaynei* and *A. b. tora*, otherwise Lower Risk (cd).
SYNONYMS: *ambiguus* (Pomel, 1894); *boselaphus* (Trouessart, 1898); *bubalinus* Flower and Lydekker, 1891; *bubalis* (Pallas, 1766); *mauretanicus* (Ogilby, 1837); *montanus* (Perry, 1811); **cokii** Günther, 1884; *cokei* Johnston, 1886; *cookei* (Noack, 1905); *deckeni* (Matschie and Zukowsky, 1916); *oscari* (Matschie and Zukowsky, 1916); *sabakiensis* (Zukowsky, 1913); *schillingsi* (Zukowsky, 1913); *schulzi* (Zukowsky, 1914); *tanae* (Matschie and Zukowsky, 1913); *wembaerensis* (Zukowsky, 1913); **lelwel** (Heuglin, 1877); *heuglini* (Millais, 1924); *insignis* (Thomas, 1904); *jacksoni* (Thomas, 1892); *modestus* (Schwarz, 1914); *niediecki* (Neumann, 1905); *rooseveltti* (Heller, 1912); *tschadensis* (Schwarz, 1913); **major** (Blyth, 1869); *invadens* (Schwarz, 1914); *luzarchei* (G. Grandidier, 1914); *matschiei* (Schwarz, 1914); *tunisianus* (Gray, 1852); **swaynei** (P. L. Sclater, 1892); *noacki* (Neumann, 1905); **tora** Gray, 1873; names based on hybrids between *lelwel* and *cokii, swaynei* or *tora*: *digglei* (Rothschild, 1913); *keniae* (Heller, 1913); *kongoni* (Heller, 1912); *nakurae* (Heller, 1912); *neumanni* (Rothschild, 1897); *rahatensis* (Neumann, 1906); *ritchiei* (Ruxton, 1926); *rothschildi* (Neumann, 1905).
COMMENTS: More than one taxon is included among the syntypes of *Antilope buselaphus* Pallas, 1766, hence the designation of a lectotype was necessary. Six species recognised by earlier authors (*buselaphus, cokii, lelwel, major, swaynei, tora*) were all assigned to *A. buselaphus* once hybridization between some of them was recognised (Ruxton and Schwarz, 1929); *caama* later included (Ellerman et al., 1953:202); *lichtensteinii* has also been included (Haltenorth, 1963:102; Kingdon, 1997:429) but not by most workers. The species can be partitioned into nominate *buselaphus* division (including also *major*), *lelwel* division, and *tora* division (including also *cokii* and *swaynei*) on the basis of skull morphology, but cytochrome *b* and D-loop sequence data (Flagstad et al., 2000) suggest a close affinity between *lelwel* and *tora* divisions. Synonymy modified from Ansell (1972:53).

*Alcelaphus caama* (É. Geoffroy Saint-Hilaire, 1803). Cat. Mamm. Mus. Nation. Hist. Nat., p. 269.
COMMON NAME: Red Hartebeest.
TYPE LOCALITY: "Le cap de Bonne Esperance"; since restricted to syntype locality South Africa, Eastern Cape Prov., Steynsburg (Grubb, 1999).
DISTRIBUTION: S Angola, Botswana, Lesotho (extinct), Namibia, South Africa, Swaziland (introduced), and W Zimbabwe.
STATUS: IUCN – Lower Risk (cd) as *A. buselaphus caama*.
SYNONYMS: *cama* (Bryden, 1899); *dorcas* (Sparrman, 1783) [unavailable]; *evalensis* (Monard,

1933); *obscurus* (Frechkop, 1937); *selbornei* (Lydekker, 1913); *senegalensis* (G. Cuvier, 1816).

COMMENTS: Authorship and date of publication validated by Opinion 2005 of the International Commission on Zoological Nomenclature (2002*b*). *Alcelaphus buselaphus* would be paraphyletic if *A. caama* were included, as the latter is the sister-species of *A. lichtensteinii* (Arctander et al., 1999; Flagstad et al., 2000).

*Alcelaphus lichtensteinii* (Peters, 1849). Spenerschen Zeitung, 18 December, 1849, p. unknown; reprinted in 1912 in Gesellschaft Natuurforschender Freunde zu Berlin for 1839-59.

COMMON NAME: Lichtenstein's Hartebeest.

TYPE LOCALITY: No type locality indicated; since identified as Mozambique, Tette.

DISTRIBUTION: E Angola, SE Dem. Rep. Congo, Malawi, Mozambique, South Africa (KwaZulu-Natal and doubtfully Limpopo Prov.; extinct but reintroduced), Swaziland (extinct), Tanzania, Zambia, SE Zimbabwe.

STATUS: IUCN – Lower Risk (cd).

SYNONYMS: *bangae* (Matschie and Zukowsky, 1916); *basengae* (Matschie and Zukowsky, 1910); *basengae* (Matschie and Zukowsky, 1916) [preoccupied]; *dieseneri* (Matschie and Zukowsky, 1925); *frommi* (Matschie and Zukowsky, 1918); *gendagendae* (Matschie and Zukowsky, 1925); *godonga* (Matschie and Zukowsky, 1916); *godowiusi* (Matschie and Zukowsky, 1925); *gombensis* (Matschie and Zukowsky, 1910); *gorongozae* (Matschie and Zukowsky, 1916); *grotei* (Matschie and Zukowsky, 1925); *hennigi* (Matschie and Zukowsky, 1925); *heuferi* (Matschie and Zukowsky, 1916); *inkulanondo* (Matschie and Zukowsky, 1916); *janenschi* (Matschie and Zukowsky, 1925); *kangosa* (Matschie and Zukowsky, 1918); *konzi* (Matschie and Zukowsky, 1916); *lacrymalis* (Matschie and Zukowsky, 1925); *lademanni* (Matschie and Zukowsky, 1916); *leucoprymnus* (Matschie, 1892); *leupolti* (Matschie and Zukowsky, 1916); *lindicus* (Matschie and Zukowsky, 1925); *munzneri* (Matschie and Zukowsky, 1918); *niediecki* (Matschie and Zukowsky, 1916); *niedieckianus* (Matschie and Zukowsky, 1916); *petersi* (Matschie and Zukowsky, 1918); *prittwitzi* (Matschie and Zukowsky, 1925); *rendalli* (Matschie and Zukowsky, 1925); *rowumae* (Matschie and Zukowsky, 1925); *rukwae* (Matschie and Zukowsky, 1910); *saadanicus* (Matschie and Zukowsky, 1925); *schmitti* (Matschie and Zukowsky, 1925); *schusteri* (Matschie and Zukowsky, 1925); *senganus* (Matschie and Zukowsky, 1916); *shirensis* (Matschie and Zukowsky, 1910); *stierlingi* (Matschie and Zukowsky, 1916); *tendagurucus* (Matschie and Zukowsky, 1925); *ufipae* (Matschie and Zukowsky, 1910); *ugalae* (Matschie and Zukowsky, 1910); *ulangae* (Matschie and Zukowsky, 1925); *ungonicus* (Matschie and Zukowsky, 1925); *ungoniensis* (Matschie and Zukowsky, 1925); *uwendensis* (Matschie and Zukowsky, 1918); *wiesei* (Matschie and Zukowsky, 1916); *wintgensi* (Matschie and Zukowsky, 1925).

COMMENTS: Included in *Alcelaphus buselaphus* by Haltenorth (1963:102) and Kingdon (1997:429), but regarded as a distinct species by other authors and placed in a separate genus, *Sigmoceros* by Vrba (1979). Included in *Alcelaphus* by Gentry (1990).

*Beatragus* Heller, 1912. Smithsonian Misc. Coll., 60(8):8.

TYPE SPECIES: *Cobus hunteri* P. L. Sclater, 1889.

COMMENTS: Placed in *Alcelaphus* by Van Gelder (1977*b*:18); but see also Vrba (1979) and Gentry (1990).

*Beatragus hunteri* (P. L. Sclater, 1889). Proc. Zool. Soc. Lond., 1889:58.

COMMON NAME: Hunter's Hartebeest.

TYPE LOCALITY: Kenya, "Africam orientalem, in ripis fl. Tana" (E bank of Tana River).

DISTRIBUTION: N Kenya, S Somalia. Introduced into Tsavo National Park, Kenya.

STATUS: IUCN – Critically Endangered as *Damaliscus hunteri*.

COMMENTS: Included in *Damaliscus lunatus* by Haltenorth (1963:100). Formerly in *Beatragus*; see Ansell (1972:54); retained in *Beatragus* by Gentry and Gentry (1978) and Gentry (1990). Differs from *Damaliscus* (and *Alcelaphus*) in independent fusions of formerly acrocentric chromosomes (Kumamoto et al., 1996).

*Connochaetes* Lichtenstein, 1812. Mag. Ges. Naturf. Fr. Berlin, 6:152.
    TYPE SPECIES: *Antilope gnu* Gmelin, 1788 (= *Antilope gnou* Zimmermann, 1780).
    SYNONYMS: *Butragus* Gray, 1872; *Catablepas* Gray, 1821; *Cemas* Oken, 1816 [unavailable];
        *Gorgon* Gray, 1850.

*Connochaetes gnou* (Zimmermann, 1780). Geogr. Gesch. Mensch. Vierf. Thiere, 2:102.
    COMMON NAME: Black Wildebeest.
    TYPE LOCALITY: South Africa, "Die lander der Caffern, ziemlich tief ins land vom Cap
        gerechnet in grossen Waldern ohnmeit der Uchtermanns Brenjes hogde und
        Camdebo"; since selected as Eastern Cape Prov., Somerset East Dist.,
        Agterbruintjieshoogte (Grubb, 1999).
    DISTRIBUTION: Formally W Lesotho, South Africa (E of 22°E in Karoo and grassveld
        vegetation types), and W Swaziland; now only in captivity, or as reintroduced
        populations in Lesotho, South Africa (including introductions beyond its former
        range), and Swaziland.
    STATUS: IUCN – Least Concern.
    SYNONYMS: *capensis* (Gatterer, 1780); *connochaetes* (Forster, 1844); *gnou* (Zimmermann, 1777)
        [unavailable]; *gnu* (Gmelin, 1788); *operculatus* (Brookes, 1828).
    COMMENTS: Reviewed by Von Richter (1974, Mammalian Species, 50). *Catoblepas brookii*
        C. H. Smith, 1827 is not a synonym but is based on a horn probably of the domestic
        cattle of Bornu (Lydekker, 1912).

*Connochaetes taurinus* (Burchell, 1824). Travels in Interior of Southern Africa, 2:278(footnote)
    [1824].
    COMMON NAME: Blue Wildebeest.
    TYPE LOCALITY: Apparently "Kosi Fountain", but lectotype came from South Africa, North
        West Prov., Vryburg Dist., "Chue Spring, Maadji Mtn [Klein Heuningvlei]"; see Grubb
        (1999).
    DISTRIBUTION: Angola, Botswana, S Kenya, Malawi (extinct), Mozambique, Namibia,
        NE South Africa, Tanzania, Zambia, Zimbabwe.
    STATUS: IUCN – Lower Risk (cd).
    SYNONYMS: *borlei* Monard, 1933; *corniculatus* (Gray, 1872); *fasciatus* (Gray, 1872); *gorgon*
        (C. H. Smith, 1827); *mattosi* Blaine, 1925; *reichei* (Noack, 1893); **albojubatus** Thomas,
        1892; *hecki* Neumann, 1905; **cooksoni** Blaine, 1914; **johnstoni** P.L. Sclater, 1896;
        *rufijianus* De Beaux, 1911; **mearnsi** (Heller, 1913); *babaulti* Kollman, 1919; *henrici*
        Zukowsky, 1913; *lorenzi* Zukowsky, 1913; *schulzi* Zukowsky, 1913.
    COMMENTS: For year of publication see Ellerman et al. (1953:205). Synonymy follows Ansell
        (1972:51). Status of *babaulti* discussed by Scoazec (1996). Cotterill (2003*a*) listed
        *johnstoni* and *cooksoni* as species.

*Damaliscus* Sclater and Thomas, 1894. Book of Antelopes, 1(part 1):3, 51.
    TYPE SPECIES: *Antilope pygargus* Pallas, 1767.
    SYNONYMS: *Damalis* Gray, 1872 [preoccupied].
    COMMENTS: Placed in *Alcelaphus* by Van Gelder (1977*b*:18); but see also Vrba (1979) and Gentry
        (1990).

*Damaliscus korrigum* (Ogilby, 1837). Proc. Zool. Soc. Lond., 1836:103 [1837].
    COMMON NAME: Topi.
    TYPE LOCALITY: N Nigeria, "Bornou" (Borno Prov.).
    DISTRIBUTION: West and Equatorial Africa in N Benin, Burkina Faso, Burundi, N Cameroon,
        N Central African Republic, Chad, Dem. Rep. Congo (Rwindi-Rutshuru plain only),
        W Ethiopia, Gambia, N Ghana, Guinea-Bisau, Kenya, Mali, S Mauritania, S Niger,
        N Nigeria, Ruanda, Senegal, S Somalia, Sudan, Tanzania, Togo, Uganda. Former
        occurrence in Guinea unconfirmed. Now extinct in Burundi, Gambia, Guinea-Bissau,
        Mali, Mauritania, and Senegal.
    STATUS: CITES – Appendix III (Ghana); IUCN – Vulnerable as *D. lunatus korrigum*, Lower Risk
        (cd) as *D. l. jimela* and *D. l. topi*, Lower Risk (nt) as *D. l. tiang*,
    SYNONYMS: *corrigum* H.C.V. Hunter, 1899; *floweri* Matschie, 1913; *jonesi* Lydekker, 1907; *lyra*
        Schwarz, 1914; *purpurescens* Blaine, 1914; *senegalensis* (Children, 1826) [preoccupied];

*tiang* (Heuglin, 1863); *tiangriel* (Heuglin, 1863); ***jimela*** (Matschie, 1892); *eurus* Blaine, 1914; *phalius* Cabrera, 1911; *selousi* Lydekker, 1907; *ugandae* Blaine, 1914; ***topi*** Blaine, 1914.

COMMENTS: The vernacular name "Topi" applies to both *jimela* and *topi*, while the Korrigum or Tiang is *korrigum*. Synonymy modified from Ansell (1972:56).

*Damaliscus lunatus* (Burchell, 1824). Travels in Interior of Southern Africa, 2:334 [1824].

COMMON NAME: Common Tsessebe.

TYPE LOCALITY: South Africa, Northern Cape Prov., Kuruman Dist., "Makkwarin" (Matlhwareng) River.

DISTRIBUTION: Southern Africa in E Angola, N Botswana, Mozambique (extinct), NE Namibia, South Africa (extinct in Northern Cape, North-West, and Mpumalanga Provs., and N KwaZulu-Natal; survives in E Limpopo Prov.; reintroduced within former range), Swaziland (extinct; reintroduced), E and C Zambia, and Zimbabwe.

STATUS: IUCN – Lower Risk (cd).

SYNONYMS: *reclinis* Matschie, 1912.

COMMENTS: For date of publication see Ellerman et al. (1953:201), who included *korrigum* in this species. Cotterill (2003*c*) separated a Zambian population as a separate species and also regarded *korrigum* as a separate species.

*Damaliscus pygargus* (Pallas, 1767). Spicil. Zool., 1:10.

COMMON NAME: Bontebok.

TYPE LOCALITY: No locality. Since restricted to South Africa, Western Cape Prov., Caledon, Swart River (Bigalke, 1948).

DISTRIBUTION: Bontebok *sensu stricto* (*D. p. pygargus*) only in South Africa (Western Cape Prov.); Blesbok (*D. p. phillipsi*) formerly in SW Lesotho, South Africa (Northern Cape Prov. E of 23°E, Eastern Cape Prov., Free State, North-West Prov., Gautung, Mpumalanga, and NW and W KwaZulu-Natal), and Swaziland; now only in captivity, or as reintroduced populations in Lesotho, South Africa, and Swaziland. Introduced on private land in Botswana, Namibia, and Zimbabwe (East, 1999).

STATUS: CITES – Appendix II as *D. pygargus pygargus*; and U.S. ESA – Endangered as *D. pygarus* [sic] (= *dorcas*) *dorcas*; IUCN – Vulnerable as *D. p. pygargus*, Least Concern as *D. p. phillipsi*.

SYNONYMS: *albifrons* (Burchell, 1823); *dorcas* (Pallas, 1766) [preoccupied]; ***phillipsi*** Harper, 1939.

COMMENTS: The name Bontebok usually applies to *D. p. pygargus*, while *D. d. phillipsi* is the Blesbok. Includes *phillipsi* and *albifrons*; see Ansell (1972:55). Includes *dorcas*, a junior secondary homonym; *pygargus* is the valid name; see Rookmaaker (1991).

*Damaliscus superstes* Cotterill, 2003. Durban Mus. Novit., 28:20.

COMMON NAME: Bangweulu Tsessebe.

TYPE LOCALITY: "Muku Muku Flats, Luapala Province, north east Zambia 12°21'S; 30°00'E".

DISTRIBUTION: Southern Bangweulu Flats in NE Zambia and extinct in Katanga Pedicle of Dem. Rep. Congo.

## Subfamily Antilopinae Gray, 1821. London Med. Repos., 15:307.

SYNONYMS: Ammodorcini Haltenorth, 1962; Antidorcatinae Knottnerus-Meyer, 1907; Dorcatragini Haltenorth, 1963; Eudorcatinae Knottnerus-Meyer, 1907; Gacellidae Knottnerus-Meyer, 1907 [unavailable]; Gazellae Haeckel, 1866; Litocraniidae Knottnerus-Meyer, 1907; Madoquinae Pocock, 1910; Neotragini Sclater and Thomas, 1894; Nesotragidae Gray, 1872 [*nomen oblitum*]; Oreotraginae Pocock, 1910; Procaprinae Knottnerus-Meyer, 1907; Raphicerinae Knottnerus-Meyer, 1907; Rhynchotraginae Roosevelt and Heller, 1914.

COMMENTS: Antilopini includes *Ammodorcas, Antidorcas, Antilope, Eudorcas, Gazella, Litocranius, Nanger, Procapra,* and *Saiga*; Neotragini includes *Dorcatragus, Madoqua, Neotragus, Oreotragus, Ourebia,* and *Raphicerus*. Neotragini probably paraphyletic if not polyphyletic; *Oreotragus* and *Neotragus* distant from other genera, associated with *Cephalophus* or *Aepyceros*, respectively, in molecular phylogenies, according to Hassanin and Douzery (1999*b*) and Matthee and Davis (2001). The latter regarded *Antidorcas, Gazella, Litocranius,*

*Madoqua, Ourebia* and *Raphicerus* as members of a clade. Possibly *Oreotragus* and *Neotragus* to be excluded from Neotragini and remaining neotragine genera placed in Gazellini. *Antilope, Eudorcas, Gazella,* and *Nanger* may form a clade within Antilopini, sharing translocation of autosome to X chromosome, while within this clade, relationships suggest *Gazella* sensu stricto is still paraphyletic: (*Antidorcas*) ((*Eudorcas, Nanger*) ((*Gazella dorcas* group) (*Antilope, Gazella subgutturosa* group))) (Vassart et al, 1995). Using 74 morphological characters, Groves' (2000*b*) cladogram was (*Litocranius*)((*Saiga, Procapra*) (*Ammodorcas, Antidorcas*) (*Gazella, Eudorcas, Nanger, Antilope*)). In an earlier study without *Saiga* (Groves, 1997*c*), the cladogram was (*Procapra*) ((*Antidorcas*) ((*Ammodorcas, Litocranius*) (gazelles + *Antilope*))).

*Ammodorcas* Thomas, 1891. Proc. Zool. Soc. Lond., 1891:207, pl. 21, 22.
    TYPE SPECIES: *Cervicapra clarkei* Thomas, 1891.
    COMMENTS: Sister taxon of *Litocranius*; see Groves (1997*c*) but not Groves (2000*b*).

*Ammodorcas clarkei* (Thomas, 1891). Ann. Mag. Nat. Hist., ser. 6, 7:304.
    COMMON NAME: Dibatag.
    TYPE LOCALITY: "Northern Somali-land"; according to the collector Clarke (*in* Sclater and Thomas, 1898:220), "about three hours from 'Bairwell' or about one day from "Buroa Well, Habergerhagi's country' " (N Somalia, vicinity of Burao and Ber).
    DISTRIBUTION: E Ethiopia and N Somalia.
    STATUS: U.S. ESA – Endangered; IUCN – Vulnerable.
    COMMENTS: Reviewed by Schomber (1964). Sometimes placed in a separate tribe, Ammodorcadini (e.g. East et al., 1999), the correct form of "Ammodorcini".

*Antidorcas* Sundevall, 1847. Kongl. Svenska Vet.-Akad. Handl. Stockholm, 1845:271 [1847].
    TYPE SPECIES: *Antilope euchore* J. R. Forster, 1790 (= *Antilope marsupialis* Zimmermann, 1780).

*Antidorcas marsupialis* (Zimmermann, 1780). Geogr. Gesch. Mensch. Vierf. Thiere, 2:427.
    COMMON NAME: Springbok.
    TYPE LOCALITY: South Africa, "die Lander am Cap der guten Hoffnung", since restricted to "Cape Colony [Cape Prov.]" (Lydekker, 1914*b*:111).
    DISTRIBUTION: SW Angola, Botswana, Namibia, and South Africa (range here now much reduced).
    STATUS: IUCN – Lower Risk (cd).
    SYNONYMS: *centralis* Lydekker and Blaine, 1914; *dorsata* (Daudin in Buffon, 1802); *euchore* (J. R. Forster, 1790); *pygargus* (Thunberg, 1788); *saccata* (Boddaert, 1785); *saliens* (Daudin in Buffon, 1802); *saltans* (Kerr, 1792); **angolensis** Blaine, 1922; **hofmeyri** Thomas, 1926.
    COMMENTS: Revised by Groves (1981*b*). Assigned to a separate tribe, Antidorcini by Kingdon (1997), correctly Antidorcadini.

*Antilope* Pallas, 1766. Misc. Zool., p. 1.
    TYPE SPECIES: *Capra cervicapra* Linnaeus, 1758.
    SYNONYMS: *Antelopa* Perry, 1811; *Antelope* Forster, 1790; *Cervicapra* Sparrman, 1780.

*Antilope cervicapra* (Linnaeus, 1758). Syst. Nat., 10th ed., 1:69.
    COMMON NAME: Blackbuck.
    TYPE LOCALITY: "Habitat in India, Asia"; restricted by Zukowsky (1927:125) to "Trivandrum im südlichsten Vorderindien nahe Kap Comorin" (India, Travancore, inland of Trivandrum).
    DISTRIBUTION: Bangladesh (extinct), India (Punjab south to Madras and east to Bihar; formerly up to Assam; now localized), Nepal (Terai; now very localized), and E Pakistan (extinct but vagrants occur); introduced to Texas (USA), and Argentina.
    STATUS: CITES – Appendix III (Nepal); IUCN – Near Threatened.
    SYNONYMS: *bezoartica* (Gray, 1843); *bilineata* (Gray, 1830); *hagenbecki* Zukowsky, 1927; *rupicapra* Müller, 1776; *strepsiceros* (Oken, 1816); **rajputanae** Zukowsky, 1927; *centralis* Zukowsky, 1928.
    COMMENTS: Revised by Groves (1982*c*).

*Dorcatragus* Noack, 1894. Zool. Anz., 17:202.
    TYPE SPECIES: *Oreotragus megalotis* Menges, 1894.
    SYNONYMS: *Dorcotragus* P. L. Sclater and Thomas, 1898.

*Dorcatragus megalotis* (Menges, 1894). Zool. Anz., 17:130.
    COMMON NAME: Beira.
    TYPE LOCALITY: Somalia, "in den Schluchten des Hekebo" (ravine in the Hekebo region); 35 mi (56 km) SW of Berbera (Moreau et al., 1946:437).
    DISTRIBUTION: Djibouti, NE Ethiopia (Marmar Mtns only), and N Somalia.
    STATUS: IUCN – Vulnerable.
    COMMENTS: Present occurrence in Djibouti established by Künzel and Künzel (1998).

*Eudorcas* Fitzinger, 1869. Sitzb. K. K. Akad. Wiss., Wien, math.-nat. Cl., 59(sect. 1):159.
    TYPE SPECIES: *Gazella laevipes* Sundevall, 1847 (= *Gazella rufifrons* Gray, 1846).
    SYNONYMS: *Korin* Sclater and Thomas, 1898.
    COMMENTS: *Eudorcas* has been treated as a full genus as a result of cladistic analysis by Groves (2000*b*); may be sister taxon of *Nanger*, as shares translocation of an autosome to the Y chromosome (Vassart et al., 1995); genetically distinct according to Rebholz and Harley (1999).

*Eudorcas rufifrons* (Gray, 1846). Ann. Mag. Nat. Hist., ser. 1, 18:214.
    COMMON NAME: Red-fronted Gazelle.
    TYPE LOCALITY: "Senegal".
    DISTRIBUTION: Burkino Faso, N Cameroon, N Central African Republic, Chad, N Eritrea, Ethiopia (NW and Omo valley in SW), N Ghana (probably extinct), S Mali, S Mauritania, S Niger, N Nigeria, N Senegal, Sudan, and N Togo. Possibly occurred in Benin; possibly formerly a rare vagrant in The Gambia.
    STATUS: IUCN – Vulnerable as *G. rufifrons* and *G. rufifrons tilonura*, Lower Risk (nt) as *G. thomsonii albonotatus*.
    SYNONYMS: *senegalensis* Fitzinger, 1869; *typica* (Ward, 1910); **albonotata** (W. Rothschild, 1903); *albonota* Roosevelt and Heller, 1915; **kanuri** (Schwarz, 1914); *centralis* (Schwarz, 1914); **laevipes** (Sundevall, 1847); *hasleri* (Pocock, 1912); *salmi* (Lorenz, 1906); **tilonura** (Heuglin, 1869); *melanura* (Heuglin, 1863) [preoccupied].
    COMMENTS: Haltenorth (1963:112) excluded *tilonura* but Gentry (1972:90) provided evidence to include it. Revised as subspecies of *Gazella cuvieri* (Groves, 1969*a*), though later separated from *cuvieri* (Groves 1975*a*). Skull and horn proportions associate *albonotata* with *rufifrons* rather than *thomsonii*.

*Eudorcas rufina* (Thomas, 1894). Proc. Zool. Soc. Lond., 1894:467.
    COMMON NAME: Red Gazelle.
    TYPE LOCALITY: "Hab. Doubtful. Type bought at Algiers"; "probably the interior of Algeria" (Lydekker, 1914*b*:66).
    DISTRIBUTION: N Algeria.
    STATUS: Thought to have become extinct in 20th Century; see Corbet (1978*c*:210); IUCN – Extinct.
    SYNONYMS: *pallaryi* (Pomel, 1895).

*Eudorcas thomsonii* (Günther, 1884). Ann. Mag. Nat. Hist., ser. 5, 14:427.
    COMMON NAME: Thomson's Gazelle.
    TYPE LOCALITY: "the range of country from Kilimanjaro to Baringo and at various heights above 6000' [1829 m]"; restricted to "Kilimanjaro district" (Lydekker, 1914*b*:84) in Kenya (G. M. Allen, 1939:526).
    DISTRIBUTION: S and C Kenya, N Tanzania.
    STATUS: IUCN – Lower Risk (cd).
    SYNONYMS: *arushae* Zukowsky, 1914; *bergeri* Knottnerus-Meyer, 1910; *bergerinae* Zukowsky, 1914; *macrocephala* Zukowsky, 1914; *manyarae* Knottnerus-Meyer, 1910; *marwitzi* Zukowsky, 1914; *ndjiriensis* Knottnerus-Meyer, 1910; *sabakiensis* Knottnerus-Meyer, 1910; *schillingsi* Knottnerus-Meyer, 1910; *wembaerensis* Knottnerus-Meyer, 1910; **nasalis** (Lönnberg, 1908); *baringoensis* Knottnerus-Meyer, 1910; *behni* Zukowsky,

1914; *biedermanni* Knottnerus-Meyer, 1910; *dieseneri* Zukowsky, 1914; *dongilanensis* Zukowsky, 1914; *langheldi* Knottnerus-Meyer, 1910; *mundorosica* Knottnerus-Meyer, 1910; *nakuroensis* Knottnerus-Meyer, 1910; *ruwanae* Knottnerus-Meyer, 1910; *seringetica* Zukowsky, 1914.

COMMENTS: Revised by Brooks (1961). Groves (1969a) included *thomsonii* in *Gazella cuvieri*; but Gentry (1972:88, 90-91) gave reasons for rejecting this classification. Groves (1985a, 1988) and Rebholz and Harley (1999) included *thomsonii* in *E. rufifrons* but Gentry (1964) presented evidence to show they are distinct.

*Gazella* de Blainville, 1816. Bull. Sci. Soc. Philom. Paris, 1816:75.

TYPE SPECIES: *Capra dorcas* Linnaeus, 1758.

SYNONYMS: *Dorcas* Gray, 1821; *Leptoceros* Wagner, 1844; *Trachelocele* Ellerman and Morrison-Scott, 1951; *Tragops* Hodgson, 1847; *Tragopsis* Fitzinger, 1869.

COMMENTS: Revised by Groves (1969a), who recognised *Trachelocele* as a subgenus for *G. subgutturosa* only. Since then, *Eudorcas* has been revived (Groves, 2000b) and *G. cuvieri, G. leptoceros,* and *G. subgutturosa* have been recognised as a clade (*G. subgutturosa* group) distinct from *G. dorcas* and *G. erlangeri* in sharing 11 unique centric fusions of the autosomes (Vassart et al. 1995) and having similar mitochondrial genes (Rebholz and Harley, 1999). *Trachelocele* may seem redundant but *Gazella* may be paraphyletic, as the *subgutturosa* group could be a sister taxon of *Antilope*, sharing two unique centric fusions (Vassart et al. 1995), suggesting the possible need to revive *Trachelocele*. The other species of *Gazella* may be divided into two groups. In the *dorcas* group, *G. dorcas* and *G. erlangeri* share 9 unique homologous centric fusions, *G. spekei* is similar to them in the low number of chromosomes; *G. arabica* and *G. gazella* should probably be included: a group including *G. dorcas, G. gazella* and *G. spekei* is supported, if weakly, by studies of mitochondrial DNA sequences (Rebholz and Harley, 1999). In the *bennettii* group, *G. bennettii* and *G. saudiya* share up to 6 unique homologous centric fusions, with some fission/fusion polymorphism, and are said to be genetically closer to the *subgutturosa* group than the *dorcas* group (Kumamoto et al., 1995; Rebholz and Harley, 1999). Status of Palaearctic and Indian species reviewed by Mallon and Kingswood (2001).

*Gazella arabica* (Lichenstein, 1827). Darst. Säugeth., pl. 6 and associated unpaginated text.

COMMON NAME: Arabian Gazelle.

TYPE LOCALITY: Saudi Arabia, "Insel Farsan" (Farasan Isls).

DISTRIBUTION: Saudi Arabia (Farasan Isls; extinct) and Yemen (mountains near Ta'izz; possibly extinct).

STATUS: IUCN – Extinct as *G. arabica* and *G. bilkis.*

SYNONYMS: **bilkis** Groves and Lay, 1985.

COMMENTS: Treated as a separate species from *G. gazella* by Groves (1985a). Nominate subspecies known from only two specimens; see Groves (1983); even if formerly present on Farasan Isls, now replaced there by *G. gazella farasani*; see Thouless and Al Bassri (1991). Status of *bilkis* (known from five specimens collected in 1951) reviewed by Greth et al (1993); treated as a subspecies of *arabica* by Groves (1997c); type locality is Yemen, Wadi Maleh 5 mi (8 km) east of Ta'izz, El Hauban.

*Gazella bennettii* (Sykes, 1831). Proc. Zool. Soc. Lond., 1830-1831:104 [1831].

COMMON NAME: Indian Gazelle.

TYPE LOCALITY: India, "found on the rocky hills of Dukhun [the Deccan]".

DISTRIBUTION: S Afghanistan, Iran, India, and Pakistan.

STATUS: IUCN – Least Concern.

SYNONYMS: *hazenna* (I. Geoffroy Saint-Hilaire, 1843); **christii** Blyth, 1842; *christyi* Lydekker, 1914; **fuscifrons** Blanford, 1873; *hayi* Lydekker, 1911; *kennioni* Lydekker, 1908; **karamii** Groves, 1993; **salinarum** Groves, 2003; **shikarii** Groves, 1993.

COMMENTS: A species distinct from *G. gazella* according to Furley et al. (1988) and Groves (1985a, 1988). Revised by Groves (2003).

*Gazella cuvieri* (Ogilby, 1841). Proc. Zool. Soc. Lond., 1840:35 [1841].

COMMON NAME: Cuvier's Gazelle.

TYPE LOCALITY: Morocco, "Mogadore" (Mogador).

DISTRIBUTION: Morocco, N Algeria, Tunisia. No reliable record from Libya.

STATUS: CITES – Appendix III (Tunisia); U.S. ESA – Endangered; IUCN – Endangered.

SYNONYMS: *cineraceus* Temminck, 1853; *corinna* Lacépède and Cuvier, 1804 [preoccupied]; *kevella* Tristram, 1860; *vera* Gray, 1850.

COMMENTS: Assigned to *G. gazella* by Haltenorth (1963:111); but a distinct species according to Groves (1969*a*).

*Gazella dorcas* (Linnaeus, 1758). Syst. Nat., 10th ed., 1:69.

COMMON NAME: Dorcas Gazelle.

TYPE LOCALITY: "Habitat in Africa"; restricted to Lower Egypt by Blaine (1913:292), west of the Nile River (Osborn and Helmy, 1980:508).

DISTRIBUTION: Algeria, N Burkina Faso, Chad, Djibouti, Egypt, Eritrea, N Ethiopia, S Israel, W Jordan, Libya, Mali, Mauritania, Morocco, Niger, N Nigeria (vagrant), Senegal (seasonal; reintroduced), N Somalia, N Sudan, and Tunisia.

STATUS: CITES – Appendix III (Tunisia); U.S. ESA – Endangered as *G. d. massaesyla* and *G. d. pelzelni* (sic); IUCN – Vulnerable as *G. dorcas,* not evaluated as *G. d. pelzelnii.*

SYNONYMS: *corinna* (Pallas, 1766); *kevella* (Pallas, 1766); *sundevalli* Fitzinger, 1869; **beccarii** De Beaux, 1931; **isabella** Gray, 1846; *isidis* (Sundevall, 1847); *littoralis* Blaine, 1913; *rueppelli* Neumann, 1906; **massaesyla** Cabrera, 1928; *cabrerai* Joleaud, 1929; *maculata* (Oken, 1816) [unavailable]; **osiris** Blaine, 1913; *neglecta* Lavauden, 1926; **pelzelnii** Kohl, 1886.

COMMENTS: Revised by Groves (1981*c*), reviewed by Ferguson (1981) and Yom-Tov et al. (1995, Mammalian Species, 491). Includes *pelzelnii*; see Gentry (1972:89); Haltenorth (1963:112) regarded it as a separate species. Due to differences in spelling *Gazella dorcas massaesyla* Cabrera, 1928 is not preoccupied by *Antilope (Dorcas) massoessilia* Pomel, 1895 = *Gazella* (or *Nanger*) *atlantica* Bourguignat, 1870.

*Gazella erlangeri* Neumann, 1906. Sitzber. Ges. Nat. Freunde, 1906:244.

COMMON NAME: Neumann's Gazelle.

TYPE LOCALITY: Yemen, Lahej.

DISTRIBUTION: W Saudi Arabia and W Yemen.

STATUS: In need of evaluation.

COMMMENTS: Morphologically distinct from *G. gazella*; distribution records intervene between those of *G. g. cora* suggesting sympatry or parapatry, at least in the past (Groves, 1996*a*); treated as a separate species by Groves (1997*c*).

*Gazella gazella* (Pallas, 1766). Misc. Zool., p. 7.

COMMON NAME: Mountain Gazelle.

TYPE LOCALITY: No locality cited; based on "La Gazelle" of Buffon, from "Syrié" (Syria).

DISTRIBUTION: Israel, Jordan, Lebanon, Oman, Saudi Arabia, W Syria, United Arab Emirates, and Yemen; introduced to Farur Isl (Iran, Persian Gulf) and Farasan Isls (Saudi Arabia, Red Sea). Marginal occurrence in Sinai Peninsula (Egypt) based on old sightings only; not known to occur there now.

STATUS: U.S. ESA – Endangered; IUCN – Critically Endangered as *G. g. acaciae* and *G. g. muscatensis,* Endangered as *G. g. gazella,* otherwise Vulnerable as *G. gazella, G. g. cora* and *G. g. farasini.*

SYNONYMS: *merilli* Thomas, 1904; **acaciae** Mendelssohn, Groves and Shalmon, 1997; **cora** (C. H. Smith, 1827); *hanishi* Dollman, 1927; *typica* Ward, 1910; **darehshourii** Karami and Groves, 1993; **farasani** Thoulless and Al Basari, 1991; **muscatensis** Brooke, 1874.

COMMENTS: Reviewed by Mendelssohn et al. (1995, Mammalian Species, 490), revised by Groves (1996*a*). Subspecies *darehshourii* and *farasani* apparently based on introduced populations. Captive population in King Khalid Wildlife Research Center, Thumamah, Saudi Arabia may represent an undescribed subspecies (Groves, 1996*a*, 1997*c*).

*Gazella leptoceros* (F. Cuvier, 1842). *In* É. Geoffroy Saint-Hilaire and F. Cuvier, Hist. Nat. Mammifères, 7, part 72, "Antilope aux longues cornes", p. 2, pls. 373, 374.

COMMON NAME: Slender-horned Gazelle.

TYPE LOCALITY: "rapportés du Sennaar [Sudan, Sennar] par M. Burton"; corrected to "desert between Giza and Wadi Natron, lower Egypt, as the type-specimen was brought to Paris by James Burton, circa 1833" (Flower, 1932:438).

DISTRIBUTION: Algeria, S Tunisia, Libya, NW Egypt, Niger (Air Massif), and N Chad; apparently Mali and Sudan, though material evidence is lacking; not recorded from Mauritania.

STATUS: CITES – Appendix III (Tunisia); U.S. ESA – Endangered; IUCN – Endangered.

SYNONYMS: *abuharab* (Fitzinger, 1869); *cuvieri* (Fitzinger, 1869); *typica* P. L. Sclater and Thomas, 1898; ***loderi*** Thomas, 1894.

COMMENTS: Status reviewed by East (1988) and Mallon and Kingswood (2001).

*Gazella saudiya* Carruthers and Schwarz, 1935. Proc. Zool. Soc. Lond., 1935:155.

COMMON NAME: Saudi Gazelle.

TYPE LOCALITY: Saudi Arabia, "Dhalm, about 150 miles [241 km] north-east of Mecca, central Arabia, 3500 feet [1067 m]".

DISTRIBUTION: Formerly Saudi Arabia and Yemen; one record dubiously from Kuwait; single reported specimen from S Iraq is *G. subgutturosa marica* (Mallon and Kingswood, 2001); extinct in the wild.

STATUS: U.S. ESA – Endangered as *G. dorcas saudiya*; IUCN – Extinct in the Wild.

COMMENTS: A species distinct from *G. dorcas* according to Groves (1988). Status reviewed by Mallon and Kingswood (2001).

*Gazella spekei* Blyth, 1863. Cat. Mamm. Mus. Asiat. Soc. Calcutta, p. 172.

COMMON NAME: Speke's Gazelle.

TYPE LOCALITY: "Somâli-land" (= Somalia).

DISTRIBUTION: Somalia, E Ethiopia.

STATUS: IUCN – Vulnerable.

*Gazella subgutturosa* (Guldenstaedt, 1780). Acta Acad. Sci. Petropoli, for 1778, 1:251 [1780].

COMMON NAME: Goitered Gazelle.

TYPE LOCALITY: "Patria Antilopes subgutturosai Persia est. Inter mare caspicum & nigrum septentrionem versus usque ad pedem australem promontorii iugi alpini caucasici, vix ultra gradum latitudinis 42 procedit per Georgiam per Cardueliam et Cachetiam regis Teflisi" (Georgia, steppes of E Transcaucasia, near Tbilisi).

DISTRIBUTION: Afghanistan, Azerbaijan, Bahrain, China (Gansu, Inner Mongolia, Sinkiang, N Tibet), SE Georgia (extinct), Iran, Iraq, E Jordan, Kazakhstan, Kuwait (extinct), Kyrgyztan (extinct?), Mongolia, Oman, WC Pakistan, Saudi Arabia, Syria, Tajikistan, SE Turkey, Turkmenistan, United Arab Emirates, Uzbekistan, and Yemen (possibly extinct).

STATUS: U.S. ESA – Endangered as *G. s. marica*; IUCN – Vulnerable as *G. s. marica*, otherwise Near Threatened.

SYNONYMS: *gracilicornis* Stroganov, 1956; *persica* (Gray, 1843); *seistanica* Lydekker, 1910; *typica* Lydekker, 1900; ***hillieriana*** Heude, 1894; *mongolica* Heude, 1894; *reginae* Adlerberg, 1931; *sairensis* Lydekker, 1900; *marica* Thomas, 1897; ***yarkandensis*** Blanford, 1875.

COMMENTS: The type locality may once have been within the boundaries of Persia (Iran) (Heptner et al., 1961) but does not lie within its modern limits. Revised by Groves (1969*a*). Reviewed by Kingswood and Blank (1996, Mammalian Species, 518).

*Litocranius* Kohl, 1886. Ann. K. K. Naturhist. Hofmus. Wien, 1:79.

TYPE SPECIES: *Gazella walleri* Brooke, 1879.

SYNONYMS: *Lithocranius* Thomas, 1891.

COMMENTS: Revised by Schomber (1963) and Grubb (2002). Sister taxon of *Ammodorcas* according to Groves (1997*c*) but not Groves (2000*b*).

*Litocranius walleri* (Brooke, 1879). Proc. Zool. Soc. Lond., 1878:929, pl. 56 [1879].

COMMON NAME: Gerenuk.

TYPE LOCALITY: "Mainland of Africa, north of the island of Zanzibar, about lat. 30°S and long. 38°E" and therefore apparently in Kenya, but shown to be correctly "Somalia, coast near Juba River" by Sclater and Thomas (1898) and Moreau et al. (1946) and more specifically, the vicinity of Chisimayo (Grubb, 2002).

DISTRIBUTION: E Ethiopia, Somalia, Kenya, NE Tanzania.

STATUS: IUCN – Lower Risk (cd).

SYNONYMS: *sclateri* Neumann, 1988.

*Madoqua* Ogilby, 1837. Proc. Zool. Soc. Lond., 1836:137 [1837].

TYPE SPECIES: *Antilope saltiana* Desmarest, 1816.

SYNONYMS: *Rhynchotragus* Neumann, 1905.

COMMENTS: Includes *Rhynchotragus*; see Ansell (1972:61). Comprises two species-groups, *saltiana* or nominate *Madoqua* group, including also *piacentinii* (revised by Yalden 1978); and *kirkii* or *Rhynchotragus* group, including also *guentheri*.

*Madoqua guentheri* Thomas, 1894. Proc. Zool. Soc. Lond., 1894:324.

COMMON NAME: Günther's Dikdik.

TYPE LOCALITY: Ethiopia, "Central Ogaden, 3000 feet [914 m]"; identified as "District immediately north of Imi and Karanle on the Webi Shebeli, Ogaden, Ethiopia, about 6°30′N, 42°30′E" by Moreau et al. (1946:437).

DISTRIBUTION: S Ethiopia, N Kenya, S and C Somalia, SE Sudan, NE Uganda.

STATUS: IUCN – Lower Risk (lc).

SYNONYMS: *smithii* Thomas, 1901; *hodsoni* (Pocock, 1926); *nasoguttatus* Lönnberg, 1907; *wroughtoni* (Drake-Brockman, 1909).

COMMENTS: Synonymy modified from Ansell (1972:63-64) and review by Kingswood and Kumamato (1996, Mammalian Species, 539).

*Madoqua kirkii* (Günther, 1880). Proc. Zool. Soc. Lond., 1880:17.

COMMON NAME: Kirk's Dikdik.

TYPE LOCALITY: Somalia,"near Brava, in the South Somali country".

DISTRIBUTION: Kenya, N and C Tanzania, and S Somalia in East Africa; SW Angola and Namibia in Southern Africa.

STATUS: IUCN – Lower Risk (lc).

SYNONYMS: *minor* Lönnberg, 1912; *cavendishi* Thomas, 1898; *langi* J.A. Allen, 1909; *thomasi* (Neumann, 1905); *damarensis* (Günther, 1880); *hemprichianus* (Jentink, 1887); *variani* (Drake-Brockman, 1913); *hindei* Thomas, 1902; *nyikae* (Heller, 1913).

COMMENTS: Synonymy modified from Ansell (1972:64) who included *cavendishi*, *damarensis*, and *thomasi*; and from review by Kingswood and Kumamoto (1997, Mammalian Species, 569). Specimens whose karyology was studied by Ryder et al. (1989) and Kumamoto et al. (1994) may really represent three or more species, *M. kirkii* *sensu stricto* or *M. hindei*, *M. cavendishi* and *M. damarensis*. Regarded as four evolutionary species (*M. kirkii, M. cavendishi, M. thomasi,* and *M. damarensis*) by Cotterill (2003b), but *thomasi* grades into *cavendishi* and is not a separate taxon.

*Madoqua piacentinii* Drake-Brockman, 1911. Proc. Zool. Soc. Lond., 1911:981.

COMMON NAME: Piacentini's Dikdik.

TYPE LOCALITY: E Somalia, "Gharabwein, within a day's march of Obbia [5°25′N, 48°25′E], in the Mijertain country, Italian Somaliland".

DISTRIBUTION: E Somalia.

STATUS: IUCN – Vulnerable.

COMMENTS: Included in *swaynei* by Ansell (1972:62) but a distinct species according to Yalden (1978:262).

*Madoqua saltiana* (de Blainville, 1816). Bull. Sci. Soc. Philom. Paris, 1816:79.

COMMON NAME: Salt's Dikdik.

TYPE LOCALITY: "Abyssinie" (Ethiopia).

DISTRIBUTION: Djibouti, Eritrea, N Ethiopia, NE Sudan, Somalia.

STATUS: IUCN – Lower Risk (lc).

SYNONYMS: *cordeauxi* Drake-Brockman, 1909; *hemprichiana* (Ehrenberg, 1832); *hemprichii* (Rüppell, 1835); *madoka* (C. H. Smith, 1827); *madoqua* (Waterhouse, 1838); *hararensis* Neumann, 1905; *lawrancei* Drake-Brockman, 1926; *phillipsi* Thomas, 1894;

*gubanensis* Drake-Brockman, 1909; **swaynei** (Thomas, 1894); *citernii* (De Beaux, 1922); *erlangeri* Neumann, 1905.

COMMENTS: The author of the name is usually cited as Desmarest, 1816 (Nouv. Dict. Nat., Nouv. ed., 2:192), who however acknowledged de Blainville's paper and cited the page number where *saltiana* was named, confirming de Blainville is the author. Smaller species of dikdik were revised by Ansell (1972:62-63) who recognised *M. saltiana* (including *cordeauxi*), *M. swaynei* (including *hararensis* and *piacentinii*), and *M. phillipsi* (including *erlangeri, gubanensis,* and *lawrancei*). Revised by Yalden (1978) who synonymised several subspecies and included all these taxa in *M. saltiana* except for *piacentinii*, which was treated as a distinct species. Regarded as five evolutionary species (*M. saltiana, M. hararensis, M. lawrancei, M. phillipsi* and *M. swaynei*) by Cotterill (2003*b*).

*Nanger* Lataste, 1885. Actes Soc. Linn. Bordeaux, 39:183.

TYPE SPECIES: *Antilope mhorr* Bennett, 1833 (= *Antilope dama* Pallas, 1766).

SYNONYMS: *Matschiea* Knottnerus-Meyer, 1907.

COMMENTS: Status as a full genus restored by Groves (2000*b*). May be sister taxon of *Eudorcas*, as shares translocation of autosome to Y chromosome; *Nanger* species share 9 unique centric fusions (Vassart et al., 1995) and are genetically distinct (Rebholz and Harley, 1999).

*Nanger dama* (Pallas, 1766). Misc. Zool., p. 5.

COMMON NAME: Dama Gazelle.

TYPE LOCALITY: No locality cited; based on "Le Nanguer" of Buffon from "Sénégal"; see discussion in Harper (1940).

DISTRIBUTION: S and W Algeria, N Burkina Faso, Chad, Egypt (one Recent record from Western Desert), S Mali, S Mauritania (extinct), Morocco, S Niger, N Nigeria (extinct?), N Senegal (extinct but reintroduced), N Sudan (W of Nile), Tunisia.

STATUS: CITES – Appendix I as *Gazella dama*; U.S. ESA – Endangered as *G. d. lozanoi* and *G. d. mhorr*, otherwise Proposed Endangered as *G. dama*; IUCN – Endangered as *G. dama*.

SYNONYMS: *damergouensis* (W. Rothschild, 1921); *nanguer* (Bennett, 1833); *occidentalis* (Sundevall, 1847); *permista* (Neumann, 1906); *reducta* (K. Heller, 1907); *weidholzi* (Zimara, 1935); **mhorr** (Bennett, 1833); *lazoni* (Gentry, 1972); *lozanoi* (Morales Agacino, 1934); *mhoks* (Lesson, 1836); *mohr* Gray, 1846; **ruficollis** (C. H. Smith, 1827); *addra* (Bennett, 1833); *orientalis* (Sundevall, 1847).

COMMENTS: Revised by Andreae and Krumbiegel (1976), who retained *lozanoi* and *permista* (includes *reducta*) as separate subspecies, and Cano Perez (1984), who reduced them to synonymy. Wirth (1984) defended the status of these two subspecies. Status reviewed by Mallon and Kingswood (2001).

*Nanger granti* (Brooke, 1872). Proc. Zool. Soc. Lond., 1872:602.

COMMON NAME: Grant's Gazelle.

TYPE LOCALITY: Tanzania, "Western Kinyenye, in Ugogo".

DISTRIBUTION: S Ethiopia, Kenya, S Somalia, SE Sudan, NE Uganda, and N Tanzania.

STATUS: IUCN – Lower Risk (cd) as *Gazella granti*.

SYNONYMS: *roosevelti* (Heller, 1913); **brighti** (Thomas, 1901); *lacuum* (Neumann, 1906); *raineyi* (Heller, 1913); **notata** (Thomas, 1897); **petersii** (Günther, 1884); *gelidjiensis* (Noack, 1887); *serengetae* (Heller, 1913); **robertsi** (Thomas, 1903).

COMMENTS: Subspecies recognised follow Arctander et al (1995), and Grubb (1994, 2000*b*).

*Nanger soemmerringii* (Cretzschmar, 1828). *In* Rüppell, Atlas Reise Nordl. Afr., Zool. Säugeth., p. 49, pl. 19.

COMMON NAME: Soemmerring's Gazelle.

TYPE LOCALITY: "an dem östlichen Abhange Abyssiniens" (E Ethiopia); restricted by Lydekker (1914*b*:97) to "Tal E'Sabb, Abyssinia" (Ethiopia, El Shab Valley).

DISTRIBUTION: N Somalia, Eritrea, Ethiopia, EC Sudan.

STATUS: IUCN – Vulnerable as *Gazella soemmerringii*.

SYNONYMS: *casanovae* (Matschie, 1912); *erlangeri* (Matschie, 1912); *sibyllae* (Matschie, 1912);

*typica* (P. L. Sclater and Thomas, 1898); ***berberana*** (Matschie, 1893); ***butteri*** (Thomas, 1904).
  COMMENTS: Synonymy modified from G. M. Allen (1939) and Gentry (1972). Year of publication is usually cited as 1826, but 1828 according to J. E. Hill (ms notes based on Anon., 1829:1291-1292).

*Neotragus* C. H. Smith, 1827. *In* Griffith et al., Anim. Kingdom, 5:349.
  TYPE SPECIES: *Capra pygmea* Linnaeus, 1758.
  SYNONYMS: *Hylarnus* Thomas, 1916; *Memina* Gray 1821 [preoccupied]; *Meminna* Agassiz, 1842; *Minytragus* Gloger, 1841 [*nomen nudum*]; *Nanotragus* Sundevall, 1846; *Nesotragus* Von Dueben, 1846; *Spinigera* Lesson, 1842; *Tragulus* Boddaert, 1785 [preoccupied]; *Tragulus* Ogilby, 1837.
  COMMENTS: Includes *Nesotragus*; see Ansell (1972:68). *Neotragus* distant from other Neotragine genera, according to Hassanin and Douzery (1999*b*) and Matthee and Davis (2001), and should perhaps be only member of Neotragini.

*Neotragus batesi* de Winton, 1903. Proc. Zool. Soc. Lond., 1903(1):192.
  COMMON NAME: Bates's Dwarf Antelope.
  TYPE LOCALITY: "Efulen, Bulu Country, Kamarun [Cameroon], 1500 ft. [457 m] above sea".
  DISTRIBUTION: Forest zone of SE Cameroon, E Dem. Rep. Congo, NE Gabon, SE Nigeria, N Republic of Congo, and W Uganda. Occurrence south of the Ogôoué River in Gabon not confirmed by material evidence.
  STATUS: IUCN – Lower Risk (nt).
  SYNONYMS: *harrisoni* (Thomas, 1906).
  COMMENTS: Synonymy modified from Ansell (1972:68). Regarded as three evolutionary species (*N. batesi*, *N. harrisoni*, and *N.* "ogouensis" [*nomen nudum*]) by Cotterill (2003*b*).

*Neotragus moschatus* (Von Dueben, 1846). *In* Sundevall, Ofv. K. Svenska Vet.-Akad. Forhandl., Stockholm, 3(7):221.
  COMMON NAME: Suni.
  TYPE LOCALITY: "in Chapani (Anglis French island) occisa insula prope Zanzibar in Lat. Austr. 6°9', Long. Orient. 39°14' a Greenwich sita, fructibus dense tecta et fonte irrigata" (Tanzania, Chapani Isl, 3 km from Zanzibar).
  DISTRIBUTION: SE Kenya, Malawi, Mozambique, South Africa (KwaZulu-Natal, E Limpopo and E Mpumalanga Provs.), E Tanzania (including Zanzibar and Mafia Isls), NE Zimbabwe.
  STATUS: U.S. ESA – Endangered as *N. m. moschatus*; IUCN – Lower Risk (cd).
  SYNONYMS: *zanzibaricus* (Layard, 1861); *deserticola* (Heller, 1913); ***kirchenpaueri*** (Pagenstecher, 1885); *akeleyi* (Heller, 1913); ***livingstonianus*** (Kirk, 1865); *livingstonei* (Bryden, 1899); ***zuluensis*** (Thomas, 1898).
  COMMENTS: Includes *livingstonianus* (Ellerman et al., 1953). Revised by Grubb (1989) who recognized nominate *moschatus* division (including *kirchenpaueri*) and *livingstonianus* division (including *zuluensis*), differing in dimensions and separated by Zambezi River; representatives of these divisions (cf. *akeleyi* and *zuluensis*) differ in chromosome complement (2n = 52 and 56 respectively) (Kingswood et al., 1998*a*). Regarded as three evolutionary species (*N. moschatus, N. livingsonianus,* and *N. zanzibaricus*) by Cotterill (2003*b*), but *zanzibaricus* appears to be an objective synonym of *moschatus*.

*Neotragus pygmaeus* (Linnaeus, 1758). Syst. Nat., 10th ed., 1:69.
  COMMON NAME: Royal Antelope.
  TYPE LOCALITY: "Habitat in Guinea, India"; restricted to "Guinea" (West Africa) by Thomas (1911*a*:152).
  DISTRIBUTION: Forest zone of Côte d'Ivoire, Ghana, Guinea, Liberia, and Sierra Leone.
  STATUS: IUCN – Lower Risk (nt).
  SYNONYMS: *perpusillus* (Gray, 1851); *pygmaeus* (Pallas, 1777); *pygmeus* (Linnaeus, 1758) [incorrect original spelling]; *regius* (Erxleben, 1777); *spinigera* (Lesson, 1842).
  COMMENTS: Pallas (1767:6, 1777:18) recognized two different species, *Tragulus pygmaeus* = *Neotragus pygmaeus* (Linnaeus, 1758), misidentified as a tragulid, and *Antilope pygmaea* Pallas, 1777. Both have types that are Royal Antelopes and therefore are both

homonyms and synonyms. Gmelin in Linnaeus (1788:173, 191) recognised the same two species as *Moschus pygmaeus* and *Antilope pygmaea* and Erxleben (1777:278) called them *M. pygmaeus* and *A. regia*, speculating that they may be female and male respectively of the same species.

*Oreotragus* A. Smith, 1834. S. Afr. Quart. J., 2:212.
TYPE SPECIES: *Antilope oreotragus* Zimmermann, 1783.
SYNONYMS: *Oritragus* Gloger, 1841.
COMMENTS: *Oreotragus* distant from other Neotragine genera and associated with *Cephalophus* in molecular phylogenies, according to Hassanin and Douzery (1999*b*) and Matthee and Davis (2001). Not shown to share any synapomorphies with *Cephalophus* and perhaps should be restored to tribe Oreotragini.

*Oreotragus oreotragus* (Zimmermann, 1783). Geogr. Gesch. Mensch. Vierf. Thiere, 3:269.
COMMON NAME: Klipspringer.
TYPE LOCALITY: "Die Caffern"; now known to be South Africa, Western Cape Prov., Cape Dist., False Bay (Grubb, 1999).
DISTRIBUTION: SW Angola, E Botswana, Burundi (extinct?), Djibouti, Eritrea, Ethiopia, C Nigeria, Central African Republic (NE and NW only), Dem. Rep. Congo (SE Shaba Prov. and formerly in western Rift Valley), Kenya, Malawi, Mozambique, Namibia, Rwanda, N Somalia, South Africa, Swaziland, NE and SE Sudan, Tanzania, NE and SW Uganda, Zambia, Zimbabwe. Former or present occurrence in Lesotho unconfirmed (Lynch, 1994).
STATUS: IUCN –Endangered as *O. o. porteousi*, otherwise Lower Risk (cd),
SYNONYMS: *klippspringer* (Daudin, 1802); *saltator* (Boddaert, 1785); *typicus* A. Smith, 1834; **aceratos** Noack, 1899; *centralis* Hinton, 1921; *transvaalensis* Roberts, 1917; **saltatrixoides** (Wagner, 1855); *aureus* Heller, 1913; *hyatti* Hinton, 1921; *porteousi* Lydekker, 1911; *saltatricoides* Neumann, 1902; *saltatrixoides* (Temminck, 1853) [*nomen nudum*]; *schillingsi* Neuman, 1902; *somalicus* Neumann, 1902; **stevensoni** Roberts, 1946; **tyleri** Hinton, 1921; *cunenensis* Zukowsky, 1924; *steinhardti* Zukowsky, 1924.
COMMENTS: Systematics considerably modified from Ansell (1972:61). Kingdon (1982) synonymised *aureus* with *schillngsi* (here followed) and implied that *stevensoni*, *transvaalensis* and *tyleri* are synonymous with nominate *oreotragus*. Cotterill (2003*b*) recognised *porteousi* and *schillingsi* as evolutionary species.

*Ourebia* Laurillard, 1842. *In* d'Orbigny, Dict. Univ. D'Hist. Nat., 1:622.
TYPE SPECIES: *Antilope scoparia* von Schreber, 1799 (= *Antilope ourebi* Zimmermann, 1783).
SYNONYMS: *Oribia* Kirby, 1899; *Quadriscopa* Fitzinger, 1869; *Qurebia* Moore, 1947; *Scopophorus* Gray, 1846.

*Ourebia ourebi* (Zimmermann, 1783). Geogr. Gesch. Mensch. Vierf. Thiere, 3:268.
COMMON NAME: Oribi.
TYPE LOCALITY: "Bewohnt die Cafferen" (South Africa, Eastern Cape Prov., Kaffraria); since restricted to one of the syntypical localities: South Africa, Eastern Cape Prov., Somerset East Dist., Bruintjieshoogte (Grubb, 1999).
DISTRIBUTION: Angola, Benin, N Botswana, Burkina Faso, Burundi (extinct?), Cameroon, Central African Republic, S Chad, N Côte d'Ivoire, N and SE Dem. Rep. Congo, N Eritrea, W Ethiopia, Gambia, Ghana, Guinea, Guinea Bissau, Kenya, Lesotho, Malawi, S Mali, Mozambique, SW Niger, Nigeria, Rwanda, S Senegal, N Sierra Leone, S Somalia, E South Africa, Sudan, Swaziland, Tanzania, Togo, Uganda, Zambia, Zimbabwe.
STATUS: IUCN – Extinct as *O. o. keniae*, Vulnerable as *O. o. haggardi*, otherwise Lower Risk (cd).
SYNONYMS: *grayi* (Fitzinger, 1869); *melanura* (Bechstein, 1799); *scoparia* (von Schreber, 1836); **dorcas** Schwarz, 1914; *splendida* Schwarz, 1914; **gallarum** Blaine, 1913; **haggardi** (Thomas, 1895); **hastata** (Peters, 1852); *kenyae* Meinertzhagen 1905; *masakensis* Lönnberg and Gyldenstolpe, 1925; *pitmani* Ruxton, 1926; **montana** (Cretzschmar, 1826); *aequatoria* Heller, 1912; *brevicaudata* (Rüppell, 1835); *cottoni* Thomas and

Wroughton, 1908; *goslingi* Thomas and Wroughton, 1907; *microdon* Hollister, 1910; *ugandae* De Beaux, 1921; **quadriscopa** (C. H. Smith, 1827); *nigricaudata* (Brookes, 1873); *smithii* (Fitzinger, 1869); **rutila** Blaine, 1922; *leucopus* Monard, 1930.

COMMENTS: Synonymy modified from Ansell (1972:66). Cotterill (2003b) regarded *haggardi* and perhaps *hastata* as evolutionary species.

*Procapra* Hodgson, 1846. J. Asiat. Soc. Bengal, 15:334.

TYPE SPECIES: *Procapra picticaudata* Hodgson, 1846.

SYNONYMS: *Prodorcas* Pocock, 1918.

COMMENTS: Revised by Groves (1967a). Gromov and Baranova (1981:393) considered *Procapra* a subgenus of *Gazella*; but Groves (1985a) maintained its status as a genus. Genus comprises *P. picticaudata* or *Procapra* group (includes also *przewalskii*) and *P. gutturosa* or *Prodorcas* group.

*Procapra gutturosa* (Pallas, 1777). Spicil. Zool., 12:46.

COMMON NAME: Mongolian Gazelle.

TYPE LOCALITY: "Intra Siberiae limites maxime Dauuriam transmontanum, campos dico circa Ononem and Argunum, frequentat" (Russia, SE Transbaikalia, Chitinsk. Obl., upper Onon River).

DISTRIBUTION: Formerly China (Gansu, Heilongjiang, Hebei, Inner Mongolia, Jilin, Ningxia, Shanxi, Shaanxi), NE Kazakhstan, Mongolia (except mountains and SW desert), and Russia (Chuya Steppe, Transbaikalia and Tuva on Mongolian border). Now extinct in Kazakhstan and survives only in Inner Mongolia (China), Khomin Tal Steppe in W Mongolia, E Mongolia, and Transbaikalia (Russia).

STATUS: IUCN – Least Concern.

SYNONYMS: *altaica* Hollister, 1913; *orientalis* (Erxleben, 1777).

COMMENTS: Reviewed by Sokolov and Lushchekina (1997, Mammalian Species, 571). Distribution reviewed by Mallon and Kingswood (2001). Monotypic status follows Groves (1986).

*Procapra picticaudata* Hodgson, 1846. J. Asiat. Soc. Bengal, 15:334, pl. 2.

COMMON NAME: Tibetan Gazelle.

TYPE LOCALITY: "Habitat: the plains of Tibet, amid ravines and low bare hills", restricted to China, "Hundes district of Tibet" (Lydekker, 1914b:31) "but more likely the district north of Sikkim, where most of Hodgson's specimens were obtained after 1844" (Groves, 1967a:148).

DISTRIBUTION: China (Gansu, Sichuan, Tibetan Plateau including Qinghai) and India (Ladak and seasonally in Sikkim).

STATUS: IUCN – Least Concern.

SYNONYMS: *picticauda* Gray, 1867.

*Procapra przewalskii* (Büchner, 1891). Melanges Biol. Soc. St. Petersb., 13:161.

COMMON NAME: Przewalski's Gazelle.

TYPE LOCALITY: China, "im südlichen Ordos" (S Ordos desert); Groves (1967:149) stated that the type locality is the Chagrin Gol (or Steppe).

DISTRIBUTION: China (Gansu, Inner Mongolia, Ningxia, Sinkiang, Qinghai); may only survive in Qinghai.

STATUS: IUCN – Critically Endangered.

SYNONYMS: *cuvieri* (Przewalski, 1888) [preoccupied]; **diversicornis** (Stroganov, 1949).

COMMENTS: Considered a subspecies of *picticaudata* by G. M. Allen (1940).

*Raphicerus* C. H. Smith, 1827. *In* Griffith et al., Animal Kingdom, 5:342.

TYPE SPECIES: *Cerophorus acuticornis* de Blainville, 1816 (= *Antilope campestris* Thunberg, 1811).

SYNONYMS: *Calotragus* Sundevall, 1846; *Grysbock* Knottnerus-Meyer, 1907; *Nototragus* Thomas and Schwann, 1906; *Pediotragus* Fitzinger, 1861; *Raphiceros* Thomas, 1897; *Rhaphiceros* Lydekker, 1897; *Rhaphicerus* Lönnberg, 1908; *Rhaphocerus* Agassiz, 1846.

COMMENTS: Genus comprises *R. melanotis* or *Nototragus* group (a superspecies according to Ansell, 1972:68), including also *R. sharpei*; and *R. campestris* or nominate *Raphicerus* group.

*Raphicerus campestris* (Thunberg, 1811). Mem. Acad. Imp. Sci. St. Petersbourg, 3:313.
    COMMON NAME: Steenbok.
    TYPE LOCALITY: No locality cited; South Africa by implication; since restricted to Cape
        Colony (Lydekker, 1914a:148) or Cape of Good Hope (G. M. Allen, 1939:502); since
        selected as Western Cape Prov., Malmesbury Div., Swartland (Grubb, 1999:23).
    DISTRIBUTION: E Africa in S Kenya and N and C Tanzania; S Africa in S Angola, Botswana,
        S Mozambique, Namibia, South Africa, Swaziland, W Zambia, and Zimbabwe.
    STATUS: IUCN – Lower Risk (lc).
    SYNONYMS: *acuticornis* (de Blainville, 1816); *capensis* (Afzelius, 1815); *fulvorubescens*
        (Desmoulins, 1822); *grayi* (Fitzinger, 1869) [*nomen nudum*]; *horstockii* (Jentink, 1900);
        *ibex* (Afzelius, 1815); *natalensis* W. Rothschild, 1907; *pallida* (Lichtenstein, 1812);
        *pediotragus* (Afzelius, 1815); *rufescens* (C. H. Smith, 1827); *rupestris* (Lichtenstein, 1812);
        *stenbock* (de Blainville, 1816) [*nomen nudum*]; *subulata* (C. H. Smith, 1827); *tragulus*
        (Lichtenstein, 1812); ***capricornis*** Thomas and Schwann, 1906; *zuluensis* Roberts, 1946;
        ***kelleni*** (Jentink, 1900); *bourquii* Monard, 1930; *cunenensis* (Zukowsky, 1924);
        *hoamibensis* (Zukowsky, 1924); *steinhardti* (Zukowsky, 1924); *ugabensis* (Zukowsky,
        1924); *zukowskyi* (Zukowsky, 1924); ***neumanni*** (Matschie, 1894); *stigmatus* Lönnberg,
        1908.
    COMMENTS: Synonymy modified from Ansell (1972:67).

*Raphicerus melanotis* (Thunberg, 1811). Mem. Acad. Imp. Sci. St. Petersbourg, 3:312.
    COMMON NAME: Cape Grysbok.
    TYPE LOCALITY: No locality cited; South Africa by implication; Cape Colony (Lydekker,
        1914a:157) or Cape of Good Hope (G. M. Allen, 1939:504); since selected as Western
        Cape Prov., Cape Peninsula (Grubb,1999:23).
    DISTRIBUTION: South Africa (Western Cape, Eastern Cape).
    STATUS: IUCN – Lower Risk (cd).
    SYNONYMS: *grisea* (Cuvier, 1816); *rubroalbescens* (Desmoulins, 1822); *rufescens* (C. H. Smith,
        1827).

*Raphicerus sharpei* Thomas, 1897. Proc. Zool. Soc. Lond., 1896:796, pl. 34 [1867].
    COMMON NAME: Sharpe's Grysbok.
    TYPE LOCALITY: Malawi, "Southern Angoniland".
    DISTRIBUTION: N Botswana, SE Dem. Rep. Congo, Malawi, Mozambique, South Africa
        (Limpopo Prov.), Swaziland, Tanzania, Zambia, and Zimbabwe.
    STATUS: IUCN – Lower Risk (cd).
    SYNONYMS: *colonicus* Thomas and Schwann, 1906.
    COMMENTS: Included in *melanotis* by Haltenorth (1963:78) but was too distinct for this
        according to Ansell (1972:67). Examination of museum material indicates the species
        is monotypic.

*Saiga* Gray, 1843. List Specimens Mamm. Coll. Brit. Mus., p. xxvi.
    TYPE SPECIES: *Capra tatarica* Linnaeus, 1766.
    SYNONYMS: *Colus* Wagner, 1844.
    COMMENTS: This generic name is spelt *"Saiga"* on p. xxvi and *"Siaga"* on p. 160 of the original
        citation. Synonymy suggested in part by Baryshnikov and Tikhonov (1994) and Kahlke
        (1999).

*Saiga borealis* (Tschersky, 1876). Izvest. Sibir. Otdel. Russ. Geog. Obshchest., 7(4-5):14.
    COMMON NAME: Mongolian Saiga.
    TYPE LOCALITY: Russia, Siberia, Yakutsia, River Wiljui or Vilyuy.
    DISTRIBUTION: W Mongolia (*S. b. mongolica*).
    STATUS: U.S. ESA – Endangered and IUCN – Vulnerable as *S. tatarica mongolica*.
    SYNONYMS: ***mongolica*** Bannikov, 1946.
    COMMENTS: The Pleistocene mammoth-steppe Saiga is a distinct species including the living
        subspecies *mongolica* according to Baryshnikov and Tikhonov (1994). Reviewed in part
        as *S. tatarica mongolica* by Sokolov (1974, Mammalian Species, 38).

*Saiga tatarica* (Linnaeus, 1766). Syst. Nat., 12th ed., 1:97.
>    COMMON NAME: Steppe Saiga.
>    TYPE LOCALITY: "Habitat in summa Asia"; identified as W Kazakhstan, steppes along the
>       Ural River.
>    DISTRIBUTION: China (extinct; formerly in Dzungarian Basin of Sinkiang), Kazakhstan,
>       Moldavia (extinct), E Poland (extinct), S Russia (now restricted to Kalmykia,
>       occasionally entering Dagestan), Ukraine (Crimea, extinct), NW Uzbekistan
>       (seasonal).
>    STATUS: CITES – Appendix II; IUCN – Critically Endangered.
>    SYNONYMS: *colus* (Oken, 1816) [unavailable]; *imberbis* (Gmelin, 1760) [unavailable]; *saiga*
>       (Pallas, 1766); *sayga* (Forster, 1768); *scythica* (Pallas, 1767).
>    COMMENTS: Reviewed by Sokolov (1974, Mammalian Species, 38).

**Subfamily Bovinae** Gray, 1821. London Med. Repos., 15:308.
>    SYNONYMS: Bibovina Rütimeyer, 1865 [unavailable]; Bibovina Mekayev, 2002; Bisontina
>       Rütimeyer, 1865; Boselaphini Knottnerus-Meyer, 1907; Bubalina Rütimeyer, 1865;
>       Buffelinae Knottnerus-Meyer, 1907; Poephagina Mekayev, 2002; Pseudonovibovina
>       Kuznetsov, Kalikov, Petrov, Ivanova, Lomov, Kholodova and Poltaraus, 2002;
>       Pseudoryina Hassanin and Douzery, 1999; Strepsiceriae Gray, 1846; Syncerina Pilgrim,
>       1939; Taurina Rütimeyer, 1865 [unavailable]; Taurotragini Leakey, 1965; Tetracerotidae
>       Brookes, 1828; Tragelaphini Blyth, 1863; Taurotragidae Knottnerus-Meyer, 1907
>       [unavailable].
>    COMMENTS: Boselaphini includes *Boselaphus* and *Tetraceros*; Bovini includes *Bison, Bos, Bubalus,*
>       *Pseudoryx* and *Syncerus*; and Tragelaphini includes *Taurotragus* and *Tragelaphus*.
>       Tetracerotidae, as Tetracerotini, has priority over Boselaphini but has only been used since
>       1899 as junior to Boselaphini so should not replace it (Article 35.5, International Code of
>       Zoological Nomenclature; International Commission on Zoological Nomenclature,
>       1999). Strepsiceriae as Strepsicerotini has priority over Tragelaphini, but the junior
>       synonym is in general use and should continue to be used, until an appropriate
>       submission is made to the International Commission on Zoological Nomenclature.
>       Subtribe Pseudoryina is wrongly constructed and should be Pseudorygina; inadvertantly
>       redescribed as tribe Pseudorygini by Grubb (2001*b*). Eubovini Geraads, 1992, is
>       unavailable (not based on a recognised genus). Tribe Bovini revised by Groves (1981*d*)
>       and Geraads (1992).

*Bison* H. Smith, 1827. *In* Griffith et al., Animal Kingdom, 5:373.
>    TYPE SPECIES: *Bos bison* Linnaeus, 1758.
>    SYNONYMS: *Bonasus* Fitzinger, 1860; *Urus* Bojanus, 1827.
>    COMMENTS: Revised by Bohlken (1967), and McDonald (1981). A synonym of *Bos* according to
>       Groves (1981*d*).

*Bison bison* (Linnaeus, 1758). Syst. Nat., 10th ed., 1:72.
>    COMMON NAME: American Bison.
>    TYPE LOCALITY: "Habitat in Mexico, Florida"; identified as "Mexico" by Thomas
>       (1911*a*:154); restricted to USA, C Kansas, "Quivera" by Hershkovitz (1957*b*);
>       redesignated as USA, E New Mexico, Canadian River valley by McDonald (1981:62).
>    DISTRIBUTION: Formerly NW and C Canada, south through USA, to Chihuahua, Coahuila
>       (Mexico). Exterminated in the wild except in Yellowstone Park, Wyoming (USA) and
>       Wood Buffalo Park, Northwest Territory (Canada). Reintroduced widely within native
>       range and in C Alaska.
>    STATUS: CITES – Appendix II as *B. b. athabascae*; U.S. ESA – Endangered in Canada as *B. b.*
>       *athabascae*; IUCN – Lower Risk (cd).
>    SYNONYMS: *americanus* (Linnaeus, 1766); *athabascae* Rhoads, 1898; *haningtoni* Figgins, 1933;
>       *montanae* Krumbiegel, 1980; *oregonus* V. Bailey, 1932; *pennsylvanicus* Schoemaker, 1915
>       [unavailable]; *septemtrionalis* [sic] Figgins, 1933; *sylvestris* Hay, 1915.
>    COMMENTS: Reviewed by Meagher (1986, Mammalian Species, 266). *Bison bison athabascae*
>       treated as a distinct taxon by Geist and Karsten (1977) and Van Zyll de Jong (1986),
>       and assigned to *B. priscus* by Flerov (1979), but regarded as an ecotype by Geist (1991).

*Bison bonasus* (Linnaeus, 1758). Syst. Nat., 10th ed., 1:71.
>   COMMON NAME: European Bison.
>   TYPE LOCALITY: "Habitat in Africa, Asia"; restricted to "Bielowitza, Lithuania" (Poland, Bialowieza Forest) by Lydekker (1913:35).
>   DISTRIBUTION: Europe, surviving in Germany, Romania and W Russia into 18th Century, in Hungary until about 1790, and in W Caucasus Mtns (Armenia, Georgia, Russia) and Poland until early part of 20th Century (but extinct in East Prussia, now N Poland, in 1755); extinct in the wild but now reintroduced to E Poland, W Russia, and Caucasus Mtns.
>   STATUS: IUCN – Endangered.
>   SYNONYMS: *ferus* (Gmelin, 1785); *nostras* (Bojanus, 1827); *urus* (Erxleben, 1777); *armeniacus* Mejlumjan, 1972; *europaeus* Owen, 1849; **caucasicus** (Turkin and Satunin, 1904); *caucasia* Grevé, 1906; *caucasicus* (Satunin, 1903) [*nomen nudum*]; *kaukasikus* Hilzheimer, 1909; **hungarorum** Kretzoi, 1946.
>   COMMENTS: Considered conspecific with *bison* by Bohlken (1967) and Van Zyll de Jong (1986); but not included in *B. bison* by MacDonald (1981) or Meagher (1986). Reviewed by Flerov (1979).

*Bos* Linnaeus, 1758. Syst. Nat., 10th ed., 1:71.
>   TYPE SPECIES: *Bos taurus* Linnaeus, 1758.
>   SYNONYMS: *Bibos* Hodgson, 1837; *Bubalibos* Heude, 1901; *Gauribos* Heude, 1901; *Gaveus* Hodgson, 1847; *Microbos* Heude, 1901; *Novibos* Coolidge, 1940; *Poephagus* Gray, 1843; *Pseudonovibos* Peter and Feiler, 1994 [*nomen dubium*]; *Taurus* Rafinesque, 1814; *Uribos* Heude, 1901; *Urus* C. H. Smith, 1827.
>   COMMENTS: Includes *Bibos*, *Novibos*, and *Poephagus*; see Ellerman and Morrison-Scott (1951:380). Genus traditionally comprises *B. frontalis* or *Bibos* group (includes also *B. javanicus* and *B. sauveli*), *B. taurus* or nominate *Bos* group, and *B. grunniens* or *Poephagus* group. From cranial morphometrics Groves (1981*d*) suggested that *Bibos* group is paraphyletic, *B. sauveli* is related to *B. taurus*, and *Bison* should be included in *Bos*. From mtDNA sequences, the most parsimonious cladogram suggested that *Poephagus* plus *Bison* formed the sister-group of domestic cattle, but *Bibos* was not studied (Miyamoto et al., 1989). From cranial morphometrics, Geraads (1992) suggested *Bison* plus *Poephagus* is the sister-group of *Bibos* plus nominate *Bos*, with *Bibos* paraphyletic.Using restriction-site mapping of nuclear-ribosomal DNA regions, Wall et al. (1992) came to similar conclusions but with the position of *Poephagus* unresolved and "*Bibos*" species, *Bos javanicus* and *Bos gaurus* (= *frontalis*), sister species.

*Bos frontalis* Lambert, 1804. Trans. Linn. Soc. Lond., 7:57.
>   COMMON NAME: Gaur.
>   TYPE LOCALITY: Native "of the hills to the north-east and east of the Company's province of Chittagong in Bengal, inhabiting that range of hills which separate it from the country of Arracan" (Bangladesh, NE Chittagong).
>   DISTRIBUTION: Bangladesh, Burma, Cambodia, China (S Tibet and Yunnan), India, Laos, Malaysia (peninsular Malaya), Nepal, Sri Lanka (extinct), Thailand, and S Vietnam.
>   STATUS: CITES – Appendix I as *B. gaurus* (excluding domesticated form); U.S. ESA – Endangered as *B. gaurus*; IUCN – Vulnerable.
>   SYNONYMS: *domesticus* Fitzinger, 1860; *gavaeus* Colebrook, 1805; *sylhetanus* F. Cuvier, 1824; **laosiensis** (Heude, 1901); *annamiticus* (Heude, 1901); *brachyrhinus* (Heude, 1901); *diardii* Temminck, 1838 [*nomen nudum*]; *fuscicornis* (Heude, 1901) [*nomen dubium*]; *hubbacki* Lydekker, 1907; *leptoceros* Heude, 1901 [*nomen dubium*]; *mekongensis* (Heude, 1901); *platyceros* (Heude, 1901); *readei* Lydekker, 1903; *sylvanus* (Heude, 1901); **gaurus** C. H. Smith, 1827; *asseel* Horsfield, 1851; *cavifrons* Hodgson, 1837; *frontatus* Temminck, 1838 [*nomen nudum*]; *gaur* Sundevall, 1846; *gour* Traill, 1824 [unavailable]; *gour* Hardwicke, 1827; *guavera* Kerr, 1792 [suppressed]; *subhemachalus* (Hodgson, 1837); **sinhaleyus** (Deraniyagala, 1951); *sinhaleyus* Deraniyagala, 1939 [*nomen nudum*].
>   COMMENTS: The name *frontalis* was based on a Gyall or Gayal (also known as Mithan). These are wild animals recurrently taken into captivity and hence categorised as feral or domestic. Gayal derive from wild Gaur and differ in proportions but are uniform and

tend to breed true; interbreeding with domestic cattle appears to be relatively recent (Simoons, 1984). Includes *gaurus*; but see Corbet and Hill (1991:130). Formerly placed in *Bibos*. Gentry et al. (1996) proposed that majority usage be confirmed by adoption of *Bos gaurus* as the name for the wild taxon of Gaur and asked the International Commission on Zoological Nomenclature to use its plenary powers to rule that the name for this wild species is not invalid by virtue of being antedated by the name based on the domestic form. A ruling has now been made in their favor (International Commission on Zoological Nomenclature, 2003*a*). It may still be valid for those who consider *B. gaurus* and *B. frontalis* to be conspecific to employ the senior name for the name of the species (see Bock, 1997). Gayal possibly originated from the form *laosiensis*, hence *frontalis* could be a synonym of *laosiensis*. Nearly all authors have termed the Gaur *B. gaurus* rather than *B. frontalis* (or *B. f. gaurus*). Provisionally *gaurus* and *laosiensis* are here listed as subspecies of *frontalis*. Gaur that survived into historic times on Sri Lanka were named from fossil material as *sinhaleyus* (Deraniyagala, 1951).

*Bos grunniens* Linnaeus, 1766. Syst. Nat., 12th ed., 1:99.

COMMON NAME: Yak.

TYPE LOCALITY: "Habitat in Asia boreali"; "in regno Tibetano" according to Gmelin, in Linnaeus, 1788 (China, Tibetan Plateau); based on domesticated stock.

DISTRIBUTION: China (Gansu, Sichuan, Sinkiang, Tibet including Qinghai), N India (Ladak), and Nepal; apparently in Kazakhstan, Mongolia, and S Russia (Siberia) until 13th to 18th centuries; domesticated in C Asia; feral in China, Inner Mongolia, Helan Mtns (Wiener et al, 2003).

STATUS: CITES – Appendix I as *B. mutus* (excluding domesticated form); U.S. ESA – Endangered as *B. mutus* (= *grunniens m.*); IUCN – Vulnerable.

SYNONYMS: *corriculus* von Schreber, 1789; *domesticus* (Fitzinger, 1860); *ecornis* Kerr, 1792; *ghainouk* Kerr, 1792 [*nomen nudum*]; *gruniens* (Gray, 1833); *poephagus* Pallas, 1811; *sarlyk* Kerr, 1792 [*nomen nudum*]; **mutus** (Przewalski, 1883).

COMMENTS: Includes *mutus*; but see Corbet (1978*c*:206). Formerly placed in *Poephagus*. Reviewed by Olsen (1990). Gentry et al. (1996) proposed that majority usage be confirmed by adoption of *Bos mutus* as the name for the wild taxon of yak, though it has not been demonstrated that most authors have termed the wild yak *B. mutus* rather than *B. grunniens* (or *B. g. mutus*). Gentry et al. (1996) asked the International Commission on Zoological Nomenclature to use its plenary powers to rule that the name for this wild species is not invalid by virtue of being antedated by the name based on the domestic form. A ruling has now been made in their favor (International Commission on Zoological Nomenclature, 2003*a*). It may still be valid for those who consider *B. grunniens* and *B. mutus* to be conspecific to employ the senior name for the name of the species (see Bock, 1997); here *mutus* is provisionally treated as a subspecies of *grunniens*. Domestic and wild yaks have identical mitochondrial haplotypes in the gene fragments tested (Schaller, 1998). *Bos bunelli* Frick, 1937 is not a Pleistocene Alaskan yak but a domestic cow (Guthrie, 1990; Olsen, 1991).

*Bos javanicus* d'Alton, 1823. Die Skelete der Wiederkauer, abgebildt und verglichen, p. 7.

COMMON NAME: Banteng.

TYPE LOCALITY: Indonesia, Java.

DISTRIBUTION: Borneo, Burma, Cambodia, China (S Yunnan), Java, Laos, Malaysia (N peninsular Malaya), Thailand, and Vietnam; introduced to Australia, Bali Isl, Sangihe, and Enggano Isls; domesticated in SE Asia.

STATUS: U.S. ESA – Endangered; IUCN – Endangered.

SYNONYMS: *banteng* Wagner, 1844; *bantiger* Schlegel and Müller, 1845; *banting* Sundevall, 1846; *birmanicus* Lydekker, 1898; *butleri* Lydekker, 1905; *discolor* (Heude, 1901); *domesticus* Wilckens, 1905; *leucoprymnus* Quoy and Gaimard, 1830; *longicornis* (Heude, 1901); *porteri* Lydekker, 1909; *seleniceros* Heller, 1890; *seligniceros* Meyer, 1878 [*nomen nudum*]; *sondaicus* Blyth, 1842; **lowi** Lydekker, 1912.

COMMENTS: For use of *javanicus* instead of *banteng*, see Hooijer (1956). Synonymy from C. P. Groves (in litt.). Formerly placed in *Bibos*.

*Bos sauveli* Urbain, 1937. Bull. Soc. Zool. Fr., 62:307.
> COMMON NAME: Kouprey.
> TYPE LOCALITY: "Nord Cambodge" (Cambodia, near Tchep Village).
> DISTRIBUTION: Cambodia, S Laos, SE Thailand, and W Vietnam; possibly extinct.
> STATUS: CITES – Appendix I; U.S. ESA – Endangered; IUCN – Critically Endangered.
> COMMENTS: Included in *Novibos* by Coolidge (1940); but see Ellerman and Morrison-Scott (1951:380). Reviewed by MacKinnon and Stuart (1989).

*Bos taurus* Linnaeus, 1758. Syst. Nat., 10th ed., 1:71.
> COMMON NAME: Aurochs.
> TYPE LOCALITY: Linnaeus (1758) stated *"Habitat in* Poloniae *depressis graminosis ferus Urus"*. "Urus" applies to the aurochs because Linnaeus' only source was Caesar in his "Gallic Wars" where the aurochs is described (Lydekker, 1912). Other authors have used the name "urus" for the European Bison *Bison bonasus*, in the 18th Century thought to be the wild form of domestic cattle. Thomas (1911*a*:154) proposed to restrict the type locality to Sweden, Upsala, applying *taurus* to domestic cattle.
> DISTRIBUTION: Extinct in the wild, except in Jaktorowka Forest, Masovia, Poland, by commencement of 15th century; last wild individual reputed to have died in 1627. Distributed worldwide under domestication; feral populations in Spain, France, Australia, New Guinea, USA, Colombia, Argentina and many islands, including Hawaiian, Galapagos, Dominican Republic/Haiti, Tristan da Cunha, New Amsterdam and Juan Fernandez Isls.
> STATUS: IUCN – Endangered as *Pseudonovibos spiralis* (but see comments).
> SYNONYMS: *aceratos* Hilzheimer, 1926; *adelensis* Boddaert, 1785; *aegyptiacus* Lydekker, 1904; *akeratos* Arenander, 1898; *albus* Sundevall, 1846; *alpestris* Wagner, 1836; *alpinus* Sanson, 1878; *alpium* Fitzinger, 1860; *aquitanicus* Sanson, 1878; *arnei* Amschler, 1939; *arvernensis* Sanson, 1878; *asiaticus* Sanson, 1878; *balticus* Stegmann von Pritzwald, 1924; *batavicus* Sanson, 1878; *brachiceros* Brehm, 1864; *brachycephalus* Wilckens, 1878; *brachyceroides* Pohlig, 1912; *brachyceros* Owen, 1846; *britannicus* Sanson, 1878; *bunelli* Frick, 1937; *caledoniensis* Sanson, 1878; *collicerus* Rostafinski, 1933; *communis* S. D. W., 1836; *curvidens* Pomel, 1894; *desertorum* Fitzinger, 1860; *domesticus* Erxleben, 1777; *dunelmensis* Fitzinger, 1860; *ecornis* Wagner, 1836; *europaeus* Kerr, 1792; *friburgensis* Fitzinger, 1860; *frisius* Wagner, 1836; *frontosus* Nilsson, 1849; *hibernicus* Sanson, 1878; *hollandicus* Fitzinger, 1860; *hypselurus* Wagner, 1836; *ibericus* Sanson, 1878; *inermis* Boddaert, 1785; *jurassicus* Sanson, 1878; *ligeriensis* Sanson, 1878; *longifrons* Owen, 1844; *macroceros* Duerst, 1899; *minor* (Owen, 1846); *minutus* von der Malsburg, 1911; *mastodontis* Pohlig, 1912; *orthoceros* Stegmann von Pritzwald, 1912; *palustris* Cardas 1936 [*nomen nudum*]; *podolicus* Wagner, 1836; *polonicus* Cardas, 1936 [*nomen nudum*]; *scoticus* (C. H. Smith, 1827); *spiralis* (Peter and Feiler, 1994) [*nomen dubium*]; *tinianus* Boddaert, 1785; *tinianensis* J. B. Fischer, 1829; *vulgaris* Wagner, 1836; **indicus** Linnaeus, 1758; *abessinicus* Kerr, 1792; *abessynicus* J. B. Fischer, 1829; *aegyptiorum* Fitzinger, 1860; *aethiopicus* Fitzinger, 1860; *africanus* Kerr, 1792; *brookii* (C. H. Smith, 1827); *chinensis* Swinhoe, 1870; *dante* Link, 1794; *galla* Salt, 1814; *gibbosus* (Blyth, 1860); *harveyi* de Rochebrune, 1882; *hottentottus* Fitzinger, 1860; *hybridus* Fitzinger, 1860; *madagascariensis* Kerr, 1792; *major* Fitzinger, 1860; *medius* Fitzinger, 1860; *pusio* Swainson, 1835; *sanga* Fitzinger, 1860; *triceros* de Rochebrune, 1882; *zebu* Boddaert, 1785; **primigenius** Bojanus, 1827; *priscus* von Schlotheim, 1820; *sylvestris* Bonaparte, 1845; *urus* Linnaeus, 1758; *urus* C. H. Smith, 1827.
> COMMENTS: Includes *primigenius* (extinct wild ancestor surviving into 17th Century) and *indicus*; but see Corbet (1978*c*:206). Studies of mtDNA suggest two independent domestications of cattle (Loftus et al. 1994), *taurus* and *indicus*, originating presumably from Eurasiatic and Indian populations. Formal synonymy disputed. Gentry et al. (1996) proposed that majority usage be confirmed by adoption of *Bos primigenius* as the name for the wild taxon of Aurochsen. They asked the International Commission on Zoological Nomenclature to use its plenary powers to rule that the name for this wild species is not invalid by virtue of being antedated by the name based on the domestic form. A ruling has now been made in their favor (International Commission on Zoological Nomenclature, 2003*a*). It may still be valid for those who

consider *B. taurus* and *B. primigenius* to be conspecific to employ the senior name for the name of the species (see Bock, 1997). Provisionally, *indicus,* and *primigenius* are here listed as subspecies of *taurus.* Kretzoi (1942) noted that *urus* Linnaeus, 1758 and *priscus* von Schlotheim, 1820 antedate *primigenius* Bojanus, 1827. The citation in Linnaeus (1758:71) is as follows: "BOS Taurus . Urus. *Caesar bell. Gall. VI. C. 5. Habitat in* Poloniae *depressis graminosis ferus Urus* [Only distribution given by Linnaeus for *Bos taurus*]." The name *urus* Linnaeus is varietal and such names are regarded as available. Independently, Harper (1945) questioned whether *primigenius* was an available name but it is now conserved (International Commission on Zoological Nomenclature, 2003*a*). Until the International Commission on Zoological Nomenclature rules on Kretzoi's (1942) paper, *primigenius* is retained as the name for the Aurochs. Fitzinger (1860) listed 102 mostly new binomial or trinomial names for European domestic cattle. Only those cited by other authors are included above. *Pseudonovibos spiralis* was named from isolated horns (Peter and Feiler, 1994*a, b*); since known from frontlets with horns *in situ* (Dioli, 1995, 1997; Timm and Brandt, 2001), some of which had previously been mistaken for *Bos sauveli* (Hoffman, 1986). History and phylogenetic relationships discussed by Timm and Brandt (2001) who recommended the vernacular name Khting Vor (Khting = gaur, and Vor = spiral climbing plant). Further material examined has been found to consist of horns and associated frontlets of domestic cattle, with the horns modified by carving and twisting when softened by heat (Thomas et al., 2001), and this view is supported by evidence from DNA (Hassanin et al., 2001), but not all specimens, including the type, have been confirmed to be artefacts. A review of the evidence (Brandt et al., 2001) leaves the status of this name equivocal; Brandt, Dioli, Olson, and Timm insist that some specimens are not artefacts and accept assignment to *Bos;* Seveau suggests that the holotype consists of modified buffalo horns, in which case *Pseudonovibos spiralis* would be a synonym of *Bubalus bubalis* (a new name would be necessary for the genuine specimens, if that is what they are); Melville finds the circumstantial evidence for this bovine occurring in Indochina to be flawed. Genetic analysis of another specimen by Kuznetsov et al. (2002) suggested affinities with *Bubalis.* Further review by Galbreath and Melville (2003) suggests that *Pseudonovibos spiralis* should not be regarded as a valid species unless new incontrovertable evidence is obtained.

*Boselaphus* de Blainville, 1816. Bull. Sci. Soc. Philom. Paris, 1816:75.
    TYPE SPECIES: *Antilope tragocamelus* Pallas, 1766.
    SYNONYMS: *Bosephalus* Horsfield, 1851; *Oreades* Schinz, 1845; *Portax* C. H. Smith, 1827.

    *Boselaphus tragocamelus* (Pallas, 1766). Misc. Zool., p. 5.
        COMMON NAME: Nilgai.
        TYPE LOCALITY: No locality cited; restricted to "plains of Peninsular India" (Lydekker, 1914*b*:227).
        DISTRIBUTION: India, Nepal (Terai), and E Pakistan; introduced into Texas (USA).
        STATUS: IUCN – Least Concern.
        SYNONYMS: *albipes* (Erxleben, 1777); *hippelaphus* (Ogilby, 1837); *picta* (Pallas, 1777); *risia* (C. H. Smith, 1827); *tragelaphus* (Sundevall, 1846).
        COMMENTS: *Antilope tragocamelus* Pallas, 1766, was based on accounts of the "tragelaphus" by Caius, Gesner and Ray and on Parsons' (1745) description of a male nilgai in London (here designated the lectotype), which had first been "brought to Bengal, from a very remote part of the Mogul's Dominions".

*Bubalus* C. H. Smith, 1827. *In* Griffith et al., Animal Kingdom, 5:371.
    TYPE SPECIES: *Bos bubalis* Linnaeus, 1758.
    SYNONYMS: *Anoa* C. H. Smith, 1827; *Bubalus* Frisch, 1775 [unavailable]; *Buffelus* Rütimeyer, 1865; *Probubalus* Rütimeyer, 1865.
    COMMENTS: *Bubalus* includes the *B. depressicornis* or *Anoa* group (includes also *quarlesi*), revised by Groves (1969*b*); and the *B. bubalis* or nominate *Bubalus* group, (includes also *B. mephistopheles* and *B. mindorensis*).

*Bubalus bubalis* (Linnaeus, 1758). Syst. Nat., 10th ed., 1:72.

COMMON NAME: Water Buffalo.

TYPE LOCALITY: "Habitat in Asia, cultus in Italia". Restricted by Thomas (1911*a*:154) to Italy, Rome, but Linnaeus' (1758) comment indicates Asia (India?).

DISTRIBUTION: Bangladesh, Burma, Cambodia, India (survives in Assam and Orissa), Nepal, N Thailand, Vietnam, and possibly at least formerly in Laos; domesticated in N Africa, S Europe, and even England, east to Indonesia and in E South America; supposedly feral populations in Sri Lanka, Sumatra, Java, Borneo, Philippines and other parts of SE Asia; feral populations resulting from introductions in New Britain and New Ireland (Bismarck Arch., Papua New Guinea), and Australia.

STATUS: CITES – Appendix III (Nepal) as *B. arnee* (excludes domesticated forms - but see comments below; IUCN – Endangered.

SYNONYMS: *bubalus* (Gmelin, 1788); *buffelus* (Blumenbach, 1821); *domesticus* Fitzinger, 1860; *indicus* (von Schreber, 1789); *italicus* (Rütimeyer, 1865); *minor* J. B. Fischer, 1829; *moellendorffi* Nehring, 1894; *seminudus* Kerr, 1792; *vulgaris* Fitzinger, 1860; **arnee** (Kerr, 1792); *arna* Hodgson, 1841; *arni* (Blumenbach, 1807); *macroceros* Hodgson, 1842 [*nomen nudum*]; *macrocerus* Hodgson, 1847; *septentrionalis* Matschie, 1912; *spirocerus* Gray, 1852 [*nomen nudum*]; *speirocerus* Hodgson, 1842 [*nomen nudum*]; *typicus* Lydekker, 1898; **fulvus** (Blanford, 1891); **kerabau** Fitzinger, 1860; *carabanensis* Castillo, 1971; *ferus* Nehring, 1894 [*nomen nudum*]; *hosei* Lydekker, 1898; *kerabau* (Sundevall, 1846) [*nomen nudum*]; *mainitensis* Heude, 1894; *sondaicus* (Schlegel and Müller, 1845) [preoccupied]; *sunda* (Schlegel and Müller, 1843) [*nomen oblitum*]; **migona** Deraniyagala, 1952; **theerapati** Groves, 1996.

COMMENTS: Includes *arnee*, the name used for the species by those workers who do not employ specific names based on domestic mammals; *bubalis* is the senior synonym; see Ellerman and Morrison-Scott (1951:383); but see also Corbet and Hill (1991:130). Gentry et al. (1996) proposed that majority usage be confirmed by adoption of *Bubalus arnee* as the name for the wild taxon of water buffaloes, though it has not been demonstrated that most authors term the wild buffalo *B. arnee* rather than *B. bubalis* (or *B. b. arnee*). They asked the International Commission on Zoological Nomenclature to use its plenary powers to rule that the name for this wild species is not invalid by virtue of being antedated by the name based on the domestic form. A ruling has now been made in their favor (International Commission on Zoological Nomenclature, 2003*a*). It may still be valid for those who consider *B. bubalis* and *B. arnee* to be conspecific to employ the senior name for the name of the species (see Bock, 1997). Domestic buffaloes comprise Murrah or river buffaloes (the *Bos bubalis* of Linnaeus), with distinctive morphology (Cockrill, 1974), and swamp buffaloes, which resemble the wild populations. These two kinds differ not only in morphology but also in karyotype (Berardino and Iannuzzi, 1981; Fischer and Ulbrich, 1968) and DNA sequences that suggest two independent domestications of water buffalo (Tanaka et al., 1996), presumably from different infraspecific wild taxa. However, Kierstein et al. (2003) inferred that there was only a single domestication. A third taxon, *Bubalus mephistopheles* Hopwood, 1925, was also domesticated but is not known to have survived later than ca. 3000 yr BP (Olsen, 1993; Teilhard de Chardin and Young, 1936). Whatever name might apply to swamp buffaloes, it would appear that river buffaloes could be recognized as a separate taxon (*Bubalus bubalis bubalis*) from *B. b. arnee*. Status of insular populations unclear; some populations on Sumatra and Java have "wild" morphology (Dammerman, 1934) and are here provisionally assigned to *kerabau*; *kerabau* and *migona* are here treated as subspecies until more information becomes available. Mainland wild populations revised by Groves (1996*b*).

*Bubalus depressicornis* (C. H. Smith, 1827). *In* Griffith et al., Animal Kingdom, 4:293.

COMMON NAME: Anoa.

TYPE LOCALITY: Indonesia, "Island of Celebes" (Sulawesi).

DISTRIBUTION: Sulawesi.

STATUS: CITES – Appendix I; U.S. ESA – Endangered; IUCN – Endangered.

SYNONYMS: *anoa* (Kerr, 1792) [suppressed]; *celebensis* (Rütimeyer, 1865); *fergusoni* (Lydekker, 1905); *platyceros* (Temminck, 1853).

COMMENTS: Includes *anoa*; see Groves (1969*b*:3). Formerly included in *Anoa* but placed in genus *Bubalus*, subgenus *Anoa* by Groves (1969*b*:3).

*Bubalus mindorensis* (Heude, 1888). Mem. Hist. Nat. Emp. Chin., 2:50.
COMMON NAME: Tamarau.
TYPE LOCALITY: Philippines, "l'ile de Mindoro".
DISTRIBUTION: Philippines, Mindoro.
STATUS: CITES – Appendix I; U.S. ESA – Endangered; IUCN – Critically Endangered.
SYNONYMS: *mindorensis* (Steere, 1888).
COMMENTS: Described independently as *Bubalus mindorensis* Heude, 1888 and *Anoa mindorensis* Steere, 1888. A subspecies of *B. bubalis* according to Bohlken (1958), but restored to specific status, in subgenus *Bubalus* by Groves (1969*b*:10). Reviewed by Custodio et al. (1996, Mammalian Species, 520).

*Bubalus quarlesi* (Ouwens, 1910). Bull. Dept. Agric. Indes Neerl., 38:7.
COMMON NAME: Mountain Anoa.
TYPE LOCALITY: Indonesia, Sulawesi, "des bois des hautes montagnes de la région centrale de Toradja".
DISTRIBUTION: Mountains of Sulawesi.
STATUS: CITES – Appendix I; U.S. ESA – Endangered; IUCN – Endangered.
COMMENTS: Subgenus *Anoa*; see Groves (1969*b*). Formerly included in *A. depressicornis*; see Haltenorth (1963:131).

*Pseudoryx* Dung, Giao, Chinh, Tuoc, Arctander, and MacKinnon, 1993. Nature, 363:443.
TYPE SPECIES: *Pseudoryx nghetinhensis* Dung, Giao, Chinh, Tuoc, Arctander, and MacKinnon, 1993.

*Pseudoryx nghetinhensis* Dung, Giao, Chinh, Tuoc, Arctander, and MacKinnon, 1993. Nature, 363:443.
COMMON NAME: Siola.
TYPE LOCALITY: "Vu Quang Nature Reserve, Vietnam 105°25′E by 18°15′N".
DISTRIBUTION: Rainforest of Vietnam (Ha Tinh and Nghe An Prov.) and neighbouring parts of Laos.
STATUS: CITES – Appendix I; IUCN – Endangered.
COMMENTS: A recently discovered monotypic genus (Dung et al,. 1993, 1994; Shaller and Rabinowitz, 1995), possibly a member of the Caprinae (Thomas, 1994) but more probably a member of the Bovinae (Dung et al., 1993; Robichaud 1998). Recent studies place it in the Bovini as a subtribe Pseudorygina Hassanin and Douzery (1999*b*).

*Syncerus* Hodgson, 1847. J. Asiat. Soc. Bengal, ser. 2, 16:709.
TYPE SPECIES: *Bos brachyceros* Gray, 1837 (= *Bos caffer* Sparrman, 1779).
SYNONYMS: *Planiceros* Gray, 1872.
COMMENTS: A subgenus of *Bubalus* according to Haltenorth (1963:133). Reviewed by Grubb (1972).

*Syncerus caffer* (Sparrman, 1779). K. Svenska Vet.-Akad. Handl. Stockholm, 40:79.
COMMON NAME: African Buffalo.
TYPE LOCALITY: "Seecov Rivier" and "Akter Brunties hoogte", now restricted to South Africa, Eastern Cape Prov., Uitenhage district, Sunday River, Algoa Bay.
DISTRIBUTION: Rain forest and savanna of Angola, Benin, N and E Botswana, Burkina Faso, Burundi, Cameroon, Central African Republic, S Chad, Côte d'Ivoire, Dem. Rep. Congo, Equatorial Guinea (Mbini; extinct on Bioko), N Eritrea, Ethiopia, Gabon, Gambia (extinct), Ghana, Guinea, Guinea-Bissau, Kenya, Liberia, Malawi, S Mali, Mozambique, NE Namibia (Caprivi Strip), SW Niger, Nigeria, Republic of Congo, Rwanda, Senegal, Sierra Leone, S Somalia, South Africa, Sudan, Swaziland, Tanzania, Togo, Uganda, Zambia, and Zimbabwe.
STATUS: IUCN – Lower Risk (cd).
SYNONYMS: *athiensis* (Matschie, 1912); *bubuensis* (Matschie, 1912); *cafer* (Link, 1795); *cottoni* (Lydekker, 1907); *cubangensis* (Zukowsky, 1910); *cunenensis* (Zukowsky, 1910);

*gariepensis* (Matschie, 1906); *gazae* (Matschie, 1918); *limpopoensis* (Matschie, 1906); *lomamiensis* (Zukowsky, 1910); *massaicus* (Matschie, 1913); *neumanni* (Matschie, 1906); *niediecki* (Matschie, 1918); *pungwensis* (Matschie, 1918); *radcliffei* (Thomas, 1904); *ruahaensis* (Matschie, 1906); *rufuensis* (Zukowsky, 1910); *sankurrensis* (Zukowsky, 1910); *schillingsi* (Matschie, 1906); *tanae* (Matschie, 1912); *typicus* (Lydekker, 1898); *urundicus* (Matschie, 1913); *ussanguensis* (Matschie, 1910); *wembarensis* (Matschie, 1906); *wiesei* (Matschie, 1906); *wintgensi* (Matschie, 1913); **aequinoctialis** (Blyth, 1866); *azrakensis* (Matschie, 1906); *orientalis* (Brooke, 1873); *solvayi* (Matschie, 1911); **brachyceros** (Gray, 1837); *beddingtoni* (Lydekker, 1913); *bornouensis* (C. H. Smith, 1842); *centralis* (Gray, 1872); *geoffroyi* (de Rochebrune, 1885); *houyi* (Schwarz, 1914); *niger* ("In Tanoust" [= Carbou], 1935), *planiceros* (Blyth, 1863); *thierryi* (Matschie, 1906); **matthewsi** (Lydekker, 1904); **nanus** (Boddaert, 1785); *adamauae* (Schwarz, 1914); *adametzi* (Matschie, 1913); *adolfifriederici* (Matschie, 1918); *corniculatus* (H. Smith, 1842); *diehli* (Schwrz, 1913); *hunti* (Lydekker, 1913); *hylaeus* (Schwarz, 1914); *mayi* (Matschie, 1906); *nuni* (Matschie, 1913); *pumilus* (Kerr, 1792); *reclinis* (Blyth, 1863); *savanensis* Malbrant, 1935; *simpsoni* (Lydekker, 1911); *sylvestris* Malbrant, 1935.

COMMENTS: Ansell (1972:19) compared subspecific systematics of different authors, here modified from Schouteden (1945). The species can be partitioned into the nominate *caffer* division (including also *aequinoctialis*) and the *nanus* division (including also *brachyceros*); phylogeography indicates similar haplotypes for *nanus* and cf. *brachyceros*, which differ from those of nominate *caffer* (Van Hooft et al., 2002); *matthewsi* is probably of polyphyletic origin; *cottoni* is based on a specimen of nominate *caffer* showing characters reflecting gene flow between *caffer* and *nanus*. *Bos pegasus* C. H. Smith, 1827 has been identified as an African buffalo, but is probably a sheep (Blyth, 1871).

*Taurotragus* Wagner, 1855. *In* Schreber, Die Säugethiere, Suppl., 5:438.
   TYPE SPECIES: *Antilope oreas* Pallas, 1777 (= *Antilope oryx* Pallas, 1766).
   SYNONYMS: *Doratoceros* Lydekker, 1891; *Oreas* Desmarest, 1822 [preoccupied].
   COMMENTS: This genus has been included in *Tragelaphus*; see Van Gelder (1977*a, b*) and Ansell (1978:53). Generic rank was restored by Smithers (1983:679), Meester et al. (1986:216), and Ansell and Dowsett (1988:87).

*Taurotragus derbianus* (Gray, 1847). Ann. Mag. Nat. Hist., ser. 1, 20:286.
   COMMON NAME: Derby Eland.
   TYPE LOCALITY: "Western Africa, Gambia".
   DISTRIBUTION: Savanna of W Africa in Gambia (extinct), Guinea (extinct?), Guinea Bissau, SW Mali (extinct?), S Senegal, and Sierra Leone (formerly a vagrant); purported records from Ghana and Togo not accepted (Grubb et al., 1998). C Africa in N Cameroon, Central African Republic, S Chad (extinct), N Dem. Rep. Congo, E Nigeria (extinct), SW Sudan, and NW Uganda (extinct).
   STATUS: U.S. ESA – Endangered as *T. d. derbianus*; IUCN – Endangered as *Tragelaphus d. derbianus* (but also listed, from the same evaluation date, as Lower Risk (nt) for *Taurotragus derbianus*), Lower Risk (nt) as *Tragelaphus d. gigas*.
   SYNONYMS: *colini* (de Rochebrune, 1883); *typicus* Rowland Ward, 1910; **gigas** (Heuglin, 1863); *cameroonensis* Millais, 1924; *congolanus* W. Rothschild, 1913; *derbii* (Johnston, 1884).
   COMMENTS: Regarded as conspecific with *T. oryx* by Haltenorth (1963:86), but usually treated as a full species; see Ansell (1972:26), whose synonymy is followed here. "Giant eland" refers only to the subspecies *gigas*.

*Taurotragus oryx* (Pallas, 1766). Misc. Zool., p. 9.
   COMMON NAME: Common Eland.
   TYPE LOCALITY: Known to the Dutch "ad Promontorium B. Spei", restricted to South Africa, Western Cape Prov., near Cape Town by Shortridge (1934:607).
   DISTRIBUTION: Angola, Botswana, Burundi (extinct), S Dem. Rep. Congo, Ethiopia (seasonal in Omo Valley), Kenya, Lesotho (seasonal), Malawi, Mozambique, Namibia, Rwanda,

South Africa, SE Sudan, Swaziland (extinct, reintroduced), Tanzania, Uganda, S Zaire, Malawi, Zambia, and Zimbabwe.

STATUS: IUCN – Lower Risk (cd) as *Tragelaphus oryx*.

SYNONYMS: *alces* (Oken, 1816) [unavailable]; *canna* (C. H. Smith, 1827); *barbatus* (Kerr, 1792); *oreas* (Pallas, 1777); *typicus* Selous, 1899; **livingstonei** (P. L. Sclater, 1864); *billingae* Kershaw, 1923; *kaufmanni* (Matschie, 1912); *niediecki* (Matschie, 1913); *selousi* Lydekker, 1910; *triangularis* (Günther, 1889); **pattersonianus** Lydekker, 1906.

COMMENTS: Systematics modified from Ansell (1972:27).

*Tetracerus* Leach, 1825. Trans. Linn. Soc. Lond., 14:524.
TYPE SPECIES: *Antilope chickara* Hardwicke, 1825 (= *Cerophorus quadricornis* de Blainville, 1816).
SYNONYMS: *Tetraceros* Brookes, 1827.

*Tetracerus quadricornis* (de Blainville, 1816). Bull. Sci. Soc. Philom. Paris, 1816:75.
COMMON NAME: Four-horned Antelope.
TYPE LOCALITY: "native de l'Inde"; "plains of Peninsular India" (Lydekker, 1914b:222).
DISTRIBUTION: India, Nepal (Terai).
STATUS: CITES – Appendix III (Nepal); IUCN – Vulnerable.
SYNONYMS: *chicara* (Kaup, 1833); *chickara* (Hardwicke, 1825); *chikara* (J. B. Fischer, 1829); *labipes* (F. Cuvier, 1832); *striatocornis* (Brookes, 1828); *tetracornis* (Hodgson, 1836); *typicus* Sclater and Thomas, 1895; **iodes** Hodgson, 1847; *paccerois* (Hodgson, 1847); **subquadricornutus** (Elliot, 1839); *subquadricornis* Gray, 1843.
COMMENTS: The type of *Cervus labipes* F. Cuvier, 1832 is not a deer from the Philippines but a female Four-horned Antelope (Sundevall, 1846). The incorrect original spelling *Antilope sub-4-cornutus* Elliot was justifiably emended to *subquadricornutus* by Hodgson (1847; Calcutta Journal of Natural History, 8:89). Revised by Groves (2003).

*Tragelaphus* de Blainville, 1816. Bull. Sci. Soc. Philom. Paris, 1816:75.
TYPE SPECIES: *Antilope sylvatica* Sparrman, 1780 (= *Antilope scripta* Pallas, 1766).
SYNONYMS: *Ammelaphus* Heller, 1912; *Boocercus* Thomas, 1902; *Calliope* Ogilby, 1837; *Euryceros* Gray, 1850; *Hydrotragus* Gray, 1872; *Limnotragus* Pocock, 1900; *Nyala* Heller, 1912; *Strepsicerastes* Knottnerus-Meyer, 1903; *Strepsicerella* Zukowsky, 1910; *Strepsiceros* C. H. Smith, 1827; *Strepticeros* Blyth, 1869.
COMMENTS: Includes *Boocercus, Limnotragus, Nyala* and *Strepsiceros*; see Ansell (1972:20) and Van Gelder (1977a, b); except for *T. buxtoni*, all species have been made types of genera. Monophyletic lineages within the Tragelaphini have not been proposed on the basis of morphology, though Ansell (1972) allocated *angasii* and *spekii* to a superspecies. From gene analysis, there is a lack of evidence that *Tragelaphus* species form a clade excluding *Taurotragus*, and therefore *Tragelaphus* is regarded as paraphyletic if *Taurotragus* is excluded (Essop et al., 1997a; Gatesy et al., 1997; Geordiadis et al., 1990; Hassanin and Douzery, 1999a; Hassanin and Douzery, 1999a; Matthee and Robinson, 1999).

*Tragelaphus angasii* Angas, 1849. Proc. Zool. Soc. Lond., 1848:89 [1849].
COMMON NAME: Nyala.
TYPE LOCALITY: South Africa, KwaZulu-Natal, "Hills that border upon the northern shores of St. Lucia Bay, in the Zulu country, lat. 28° south".
DISTRIBUTION: S Malawi, Mozambique, South Africa (N and E Limpopo Prov., E Mpumalanga, and KwaZulu-Natal), Swaziland (extinct, reintroduced), and N and S Zimbabwe. Reintroduced or newly introduced to private land in South Africa and Namibia (East, 1999).
STATUS: IUCN – Lower Risk (cd).
COMMENTS: The name *angasii* is usually attributed to Gray, because Angas (1849) stated "Mr Gray has named this species after my father, George Fife Angas, Esq, of South Australia" but this is insufficient to make Gray the author (Article 50.1.1, International Code of Zoological Nomenclature; International Commission on Zoological Nomenclature, 1999).

*Tragelaphus buxtoni* (Lydekker, 1910). Nature (London), 84:397.
COMMON NAME: Mountain Nyala.

TYPE LOCALITY: Ethiopia, Bak Prov., "Arusi plateau of Gallaland, in the Sahatu Mountains, and south-east of Lake Zwei [Zwai], at an estimated height of 9000 feet [2,700 m] above sea level".

DISTRIBUTION: Ethiopia, east of Rift Valley.

STATUS: IUCN – Endangered.

*Tragelaphus eurycerus* (Ogilby, 1837). Proc. Zool. Soc. Lond., 1836:120 [1837].

COMMON NAME: Bongo.

TYPE LOCALITY: "Their origin is unknown, but I have reason to believe they [the syntypes] came from Western Africa".

DISTRIBUTION: Rain forest of W Africa in Benin, Côte d'Ivoire, Ghana, Guinea, Liberia, Sierra Leone, and Togo; C Africa in SE Cameroon, Central African Republic, Dem. Rep. Congo, NE Gabon, Republic of Congo, SW Sudan, Uganda (extinct); and in S Kenya. Occurrence in Equatorial Guinea (Mbini) questionable.

STATUS: CITES – Appendix III (Ghana); IUCN – Endangered as *T. e. isaaci*, otherwise Lower Risk (nt).

SYNONYMS: *albovirgatus* Du Chaillu, 1860; *cooperi* (W. Rothschild, 1928); *euryceros* Gray, 1850 [incorrect subsequent spelling]; *isaaci* (Thomas, 1902); *katanganus* (W. Rothschild, 1927).

COMMENTS: Formerly placed in *Boocercus*. Reviewed by Ralls (1978, Mammalian Species, 111).

*Tragelaphus imberbis* (Blyth, 1869). Proc. Zool. Soc. Lond., 1869:55.

COMMON NAME: Lesser Kudu.

TYPE LOCALITY: "Abyssinia"; now known to be Ethiopia, Shoa Prov.

DISTRIBUTION: SE Ethiopia, Kenya, Somalia, SE Sudan, E Tanzania, NE Uganda. Also apparently Yemen and SW Saudi Arabia.

STATUS: IUCN – Lower Risk (cd).

SYNONYMS: *australis* (Heller, 1913); *tendal* (Gray, 1873).

COMMENTS: The type specimen was procured in Shoa Prov. by W. C. Harris, possibly at Manyo or Taboo Forest (Yalden et al., 1984). Arabian records are based on only two specimens (Harrison and Bates, 1991:192).

*Tragelaphus scriptus* (Pallas, 1766). Misc. Zool., p. 8.

COMMON NAME: Bushbuck.

TYPE LOCALITY: No locality cited but based on "Le Guib" of Buffon, from "Sénégal".

DISTRIBUTION: Savanna and secondary forest in Angola, Benin, N and E Botswana, Burkina Faso, Burundi, Cameroon, Central African Republic, S Chad, Côte d'Ivoire, Dem. Rep. Congo, Equatorial Guinea (Mbini), N Eritrea, Ethiopia, Gabon, Gambia, Ghana, Guinea, Guinea-Bissau, Kenya, Liberia, Malawi, S Mali, S Mauritania, Mozambiqiue, NE Namibia (Caprivi Strip), SW Niger, Nigeria, Republic of Congo, Rwanda, Senegal, Sierra Leone, S Somalia, E and S South Africa, Sudan, Swaziland, Tanzania, Togo, Uganda, Zambia, and Zimbabwe. Not recorded from Lesotho (Lynch, 1994).

STATUS: IUCN – Lower Risk (lc).

SYNONYMS: *johannae* Schwarz, 1929; *obscurus* Trouessart, 1898; *phaleratus* (C. H. Smith, 1827); *pictus* Schwarz, 1914; *punctatus* Schwarz, 1914; *signatus* Schwarz, 1914; *typicus* Thomas, 1891; *uellensis* Schwarz, 1914; **bor** Heuglin, 1877; *cottoni* Matschie, 1912; *dodingae* Matschie, 1912; *makalae* Matschie, 1912; *meridionalis* Matschie, 1912; **decula** (Rüppell, 1835); *fulvoochraceus* Matschie, 1912; *nigrinotatus* Neumann, 1902; **fasciatus** Pocock, 1900; *olivaceus* Heller, 1913; *reidae* Babault, 1947; **knutsoni** Lönnberg, 1905; **meneliki** Neumann, 1902; *multicolor* Neumann, 1902; *powelli* Matschie, 1912; **ornatus** Pocock, 1900; **sylvaticus** (Sparrman, 1780); *barkeri* J. G. Millais, 1924; *brunneus* Matschie, 1912; *dama* Neumann, 1902; *delamerei* Pocock, 1900; *dianae* Matschie, 1912; *eldomae* Matschie, 1912; *haywoodi* Thomas, 1905; *heterochrous* Cabrera, 1918; *insularis* Zukowsky, 1961; *laticeps* Matschie, 1912; *locorinae* Matschie, 1912; *massaicus* Neumann, 1902; *meruensis* Lönnberg, 1908; *roualeynei* Gordon Cumming, 1850; *roualeyni* Thomas, 1891; *sassae* Matschie, 1912; *simplex* Matschie, 1912; *tjaderi* J. A. Allen, 1909; *typicus* Sclater and Thomas, 1900.

COMMENTS: Subspecific systematics modified from Ansell (1972:24) and Grubb (1985,

2000*b*); the species comprises *decula* division (includes also *meneliki*), nominate *scriptus* division (includes also *bor* and *knutsoni*), and *sylvaticus* division (includes also *fasciatus* and *ornatus*).

*Tragelaphus spekii* Speke, 1863. Journal of the Discovery of the Source of the Nile, p. 223 (footnote).
    COMMON NAME: Sitatunga.
    TYPE LOCALITY: Tanzania, Karagwe, E of Lake Victoria, at a lake named "Little Windermere" by Speke; identified as Bukoba district, Lake Lwelo, 2°S, 30°57'E by Moreau et al. (1946:441).
    DISTRIBUTION: Disjunct. Swamps in Gambia, W Guinea, Guinea Bissau, and S Senegal; not authentically recorded from Sierra Leone and doubtfully recorded from Côte d'Ivoire (Grubb et al., 1998). Rainforest and swamps in C and E Angola, S Benin, N Botswana, Burundi, Cameroon, Central African Republic, Chad (Lake Chad only), Dem. Rep. Congo, Equatorial Guinea (Mbini), Gabon, SE Ghana, W Kenya, Mozambique (W Tete Prov. only), NE Namibia (Caprivi Strip only), Niger (Lake Chad only; extinct), S Nigeria (and Lake Chad), Republic of Congo, Rwanda, S Sudan, W and NW Tanzania, Togo (extinct?), Uganda, Zambia, and Zimbabwe (extreme NW). Occurrence in Ghana only recently confirmed (East, 1998).
    STATUS: CITES – Appendix III (Ghana); IUCN – Lower Risk (nt).
    SYNONYMS: *speckei* Neumann, 1900; *spekei* Heuglin, 1869; *typicus* R. Ward, 1910; *ugallae* Matschie, 1913; *wilhelmi* (Lönnberg and Gyldenstolpe, 1924); **gratus** P.L. Sclater, 1880; *albonotatus* Neumann, 1905; **larkenii** (St Leger, 1931); **selousi** W. Rothschild, 1898; *anderessoni* (Leyland, 1866) [*nomen oblitum*]; *baumii* (Sokolowsky, 1903); *inornatus* (Cabrera, 1918); **sylvestris** (Meinertzhagen, 1916).
    COMMENTS: Speke (1863) described and illustrated the "nzoé' or "water-boc' and reported in a footnote that Sclater had named the species *Tragelaphus Spekii* [sic], but this is insufficient to make Sclater the author of the name (Article 50.1.1, International Code of Zoological Nomenclature; International Commission on Zoological Nomenclature, 1999). Subspecific systematics from Ansell (1972:22).

*Tragelaphus strepsiceros* (Pallas, 1766). Misc. Zool., p. 9.
    COMMON NAME: Greater Kudu.
    TYPE LOCALITY: "Prom. B. Spei" (Cape of Good Hope); restricted to South Africa, southeastern Cape Prov. [eastern part of Western Cape Prov.] by Grubb (1999:36).
    DISTRIBUTION: Angola, Botswana, N Central African Republic, S Chad, SE Dem. Rep. Congo, Djibouti, N Eritrea, Ethiopia, Kenya, Malawi, Mozambique, Namibia, Somalia (extinct?), South Africa, W and E Sudan, NE Uganda, Zambia, and Zimbabwe.
    STATUS: IUCN – Lower Risk (cd).
    SYNONYMS: *capensis* (A. Smith, 1834); *excelsus* (Sundevall, 1846); *koodoo* (C. H. Smith, 1836); *torticornis* (Hermann, 1804); *typicus* (Lydekker, 1910); **bea** (Heller, 1913); *frommi* (Matschie, 1914); **burlacei** Ansell, 1969; *cottoni* Dollman and Burlace, 1928 [preoccupied]; **chora** (Cretzschmar, 1826); *abyssinicus* (Fitzinger, 1869); **zambesiensis** (Lorenz, 1894); *hamiltoni* (Matschie, 1914).
    COMMENTS: Subspecific systematics from Ansell (1972:25), modified by Grubb (1999).

**Subfamily Caprinae** Gray, 1821. London Med. Repos., 15:307.
    SYNONYMS: Budorcatinae Brooke, 1876; Capricornini Duvernois and Guérin, 1989; Hircidae Brookes, 1828 [unavailable]; Naemorhedini Brooke, 1876; "Oegosceridae" (Aegoceroti-dae) Cobbold, 1869 [unavailable]; Ovibovini Gray, 1872; Ovidae Brookes, 1828; Pantholo-pini Gray, 1872; Pseudoinae Knottnerus-Meyer, 1907; Rupicapridae Brookes, 1828.
    COMMENTS: Caprini includes *Ammotragus, Capra, Hemitragus, Ovis, Oreamnos, Pseudois,* and *Rupicapra*; Naemorhedini includes *Capricornis* and *Naemorhedus*; Ovibovini includes *Budorcas* and *Ovibos*; Pantholopini includes *Pantholops*. Placing of *Pantholops* in the Caprinae is supported by morphological and molecular studies (Gatesy et al., 1997; Gentry, 1992; Hassanin et al., 1998; Vrba and Schaller, 2000). It may be the sister taxon of all other Caprinae. Relationships in the rest of the Caprinae are problematical. Nadler et al. (1973) noted identity of karyotypes in *Ammotragus* and *Ovis* (*O. aries arkar*). From electrophoresis of proteins, Hartl et al. (1990) obtained the following tree: (*Ovis*)

((*Rupicapra, Oreamnos*) ((*Hemitragus*) (*Ammotragus, Capra*))). Hassanin et al. (1998) recognised three clades on the basis of their studies of cytochrome *b* sequences, namely (1) *Capricornis, Naemorhedus, Ovibos;* (2) *Capra, Hemitragus, Pseudois;* and (3) *Budorcas, Ovis;* with the positions of *Ammotragus, Oreamnos* and *Rupicapra* less certain; Ovibovini hence appeared to be polyphyletic. The tree based on behavior, glands, skull and postcrania (Vrba and Schaller, 2000) is of the following form: (*Budorcas*) ((*Ovibos*) ((*Oreamnos*) (*Naemorhedus, Capricornis*)))) ((*Ovis*) ((*Pseudois*) (*Capra, Hemitragus*))). Distribution reviewed by Shackleton (1997).

*Ammotragus* Blyth, 1840. Proc. Zool. Soc. Lond., 1840:13.
    TYPE SPECIES: *Antilope lervia* Pallas, 1777.
    SYNONYMS: *Traguelaphus* Pomel, 1898.
    COMMENTS: Ansell (1972:70) included *Ammotragus* in *Capra*; but see comment under *Capra*.

*Ammotragus lervia* (Pallas, 1777). Spicil. Zool., 12:12.
    COMMON NAME: Barbary Sheep.
    TYPE LOCALITY: "Africae borealori propria"; restricted to Algeria, Department of Oran (Harper, 1940).
    DISTRIBUTION: Algeria, N Chad, Egypt, Libya, N Mali, Mauritania, Morocco (including Western Sahara), Niger, Sudan (west of Nile and east of Nile in Red Sea Hills), and Tunisia; introduced to USA, N Mexico and Spain.
    STATUS: CITES – Appendix II; IUCN – Extinct in the Wild as *A. l. ornatus*, otherwise Vulnerable.
    SYNONYMS: *barbatus* (Kerr, 1792); *tragelaphus* (Afzelius, 1815); **angusi** W. Rothschild, 1921; **blainei** (W. Rothschild, 1913); *jaela* (C. H. Smith, 1827) [*nomen oblitum*]; **fassini** Lepri, 1930; *ornatus* (I. Geoffroy Saint-Hilaire, 1827); **sahariensis** (W. Rothschild, 1913).
    COMMENTS: Subspecific systematics from Ansell (1972:71). Reviewed by Gray and Simpson (1980, Mammalian Species, 144).

*Budorcas* Hodgson, 1850. J. Asiat. Soc. Bengal, 19:65.
    TYPE SPECIES: *Budorcas taxicolor* Hodgson, 1850.
    COMMENTS: Doubtfully included in Ovibovini following analysis of skull characters (Gentry, 1992).

*Budorcas taxicolor* Hodgson, 1850. J. Asiat. Soc. Bengal, 19:65.
    COMMON NAME: Takin.
    TYPE LOCALITY: India, Assam, "Mishmi mountains [Mishmi Hills] in the Eastern Himalaya".
    DISTRIBUTION: Bhutan, N Burma, China (Gansu, Sichuan, Shaanxi, SE Tibet, and Yunnan), and NE India (Sikkim and Mishmi Hills).
    STATUS: CITES – Appendix II; IUCN – Endangered as *B. t. taxicolor* and *B. t. bedfordi*, Vulnerable as *B. t. tibetana* and *B. t. whitei*.
    SYNONYMS: *mitchelli* Lydekker, 1908; *sinensis* Lydekker, 1907; *taxicola* Gray, 1852; **bedfordi** Thomas, 1911; **tibetana** Milne-Edwards, 1874; **whitei** Lydekker, 1907.
    COMMENTS: Reviewed by Neas and Hoffmann (1987, Mammalian Species, 277).

*Capra* Linnaeus, 1758. Syst. Nat., 10th ed., 1:68.
    TYPE SPECIES: *Capra hircus* Linnaeus, 1758.
    SYNONYMS: *Aegoceros* Pallas, 1811 [suppressed]; *Aegocerus* Agassiz, 1846; *Aries* Link, 1795; *Eucapra* Camerano, 1916; *Euibex* Camerano, 1916; *Hilzheimeria* Kretzoi, 2000; *Hircus* Boddaert, 1785; *Ibex* Pallas, 1776; *Orthaegoceros* Trouessart, 1905; *Tragus* Schrank, 1798; *Turocapra* de Beaux, 1949; *Turus* Hilzheimer, 1916.
    COMMENTS: Reviewed by Coutourier (1962). Includes *Orthaegoceros*; see Heptner et al. (1961:593). Some authors have included *Ammotragus* and *Ovis*; see Ansell (1972:70) and Van Gelder (1977*b*). However, most authors have not followed this arrangement; see Gray and Simpson (1980), Gromov and Baranova (1981), Hall (1981), and Corbet and Hill (1991). There is no consensus concerning the number of species to be recognized in this genus; some would recognize only two (*hircus* and *falconeri*; see Haltenorth, 1963), while others would recognize up to nine. Heptner et al. (1961) are followed here except that

only one species of Tur is recognized. Suggested divisions within the genus according to Ellerman and Morrison-Scott (1951) are *C. caucasica* or *Hilzheimeria* group, *C. falconeri* or *Orthaegoceros* group, *C. hircus* or nominate *Capra* group, *C. ibex* or *Ibex* group (including also *nubiana, sibirica* and *walie* as subspecies, as well as "*C. ibex severtzovi*"), and *C. pyrenaica* or *Turocapra* group. Systematics of the genus has been inferred from mtDNA sequences. Hassanin et al. (1998) suggested the following tree: (*C. sibirica, Hemitragus jemlahicus*) ((*C. nubiana*) ((*C. hircus, C. falconeri*) ((*C. caucasica*) (*C. cylindricornis, C. aegagrus*)))). Manceau et al (1999*a*) concluded that *C. aegagrus* and *C. ibex sensu lato* were polyphyletic, that *C. cylindricornis* and *C. caucasica* were distinct, though they did not state whether their material included specimens of the intermediate *severtzovi*, and that *C. pyrenaica* was allied to *C. ibex*. From studies of fossil material, Crégut-Bonnoure (1992) concluded that there were two lineages in late Pleistocene Europe, the *ibex* lineage, and the *caucasica-cylindricornis-pyrenaica* lineage.

*Capra caucasica* Güldenstaedt and Pallas, 1783. Acta Acad. Sci. Petropoli, for 1779, 2:275 [1783].
  COMMON NAME: Tur.
  TYPE LOCALITY: Russia, "in summis Caucasi jugis, circa fluviorum Terek et Kubam summas origines, itemque in Ossetino tractu et Cachetia" (Caucasus Mtns, between Malka and Baksan Rivers, east of Mt. Elbrus).
  DISTRIBUTION: Caucasus Mtns (Azerbaijan, Georgia, Russia).
  STATUS: IUCN – Endangered as *C. caucasica,* Vulnerable as *C. cylindricornis.*
  SYNONYMS: *dinniki* Satunin, 1905; *raddei* Matschie, 1901; **cylindricornis** Blyth, 1841; *ammon* (Pallas, 1811) [preoccupied]; *pallasii* (Rouillier, 1841); **severtzovi** Menzbier, 1888.
  COMMENTS: Ellerman and Morrison-Scott (1951:407) recognised two taxa, *C. caucasica* with synonym *cylindricornis*, and *C. ibex severtzovi*; Heptner et al. (1961) recognised *C. caucasica* with synonym *severtzovi*, and *C. cylindricornis*; Sokolov and Tembotov (1993) recognised these three taxa as subspecies in a single species and their classification is followed here.

*Capra falconeri* (Wagner, 1839). Gelehrt. Anz. I. K. Bayer Akad. Wiss., München, 9:430.
  COMMON NAME: Markhor.
  TYPE LOCALITY: "Kaschmir"; restricted to Pakistan, Kashmir, Astor (Lydekker, 1913).
  DISTRIBUTION: NE Afghanistan, N India (SW Jammu and Kashmir), N and C Pakistan, S Tajikistan, and S Uzbekistan.a
  STATUS: CITES – Appendix I; U.S. ESA – Endangered as *C. f. jerdoni* and *C. f. megaceros*; IUCN – Critically Endangered as *C. f. heptneri*, Endangered as *C. f. falconeri* and *C. f. megaceros.*
  SYNONYMS: *cashmiriensis* Lydekker, 1898; *chitralensis* Cobb, 1958 [*nomen nudum*]; *gilgitensis* Cobb, 1958 [*nomen nudum*]; **heptneri** Zalkin, 1945; *ognevi* Zalkin, 1945; **megaceros** Hutton, 1842; *jerdoni* Hume, 1875.
  COMMENTS: Revised by Schaller (1977).

*Capra hircus* Linnaeus, 1758. Syst. Nat., 10th ed., 1:68.
  COMMON NAME: Goat.
  TYPE LOCALITY: "Habitat in montosis"; identified as Sweden (Thomas, 1911*a*:152), based on domesticated stock.
  DISTRIBUTION: Afghanistan, Caucasus region (Armenia, Azerbaijan, NE Georgia, and S Russia), Iraq, Iran, Israel (till Neolithic), Jordan (extinct), Lebanon (extinct), S Pakistan, Syria (extinct), Turkey, and S Turkmenistan; anciently introduced into Greek isls and probably Oman. Domesticated worldwide; feral populations in British Isles, islands in the Mediterranean, USA, Canada, Chile, Argentina, Venezuela, Australia, New Zealand and many oceanic islands including Bonin, Hawaiian, Galapagos, Seychelles, and Juan Fernandez Isls.
  STATUS: CITES – Appendix I; ESA – Endangered *C. falconeri* (=*aegagrus*) *chiltanensis* [sic]; IUCN – Critically Endangered as *C. aegagrus chialtanensis*; otherwise Vulnerable as *C. acera* and as *C. a. aegagrus, C. a. cretica,* and *C. a. blythi.*
  SYNONYMS: *acera* Desmarest, 1822; *adametzi* Kretzoi, 1942; *aegyptiaca* (Fitzinger, 1860); *aethiopica* Hartmann, 1864; *africana* (Sanson, 1878); *anatolica* (Fitzinger, 1859); *angolensis* Linnaeus, 1758; *angorensis* Linnaeus, 1766; *arietina* Desmarest, 1822; *asiatica*

(Sanson, 1878); *barbara* de Blainville, 1816; *barbarica* (Fitzinger, 1860); *brachyceros* (Fitzinger, 1859); *brachyotis* (Fitzinger, 1859); *buraitica* (Fitzinger, 1859); *calotus* Reichenbach, 1845; *capra* (Fitzinger, 1860); *capricornus* Erxleben, 1777; *chungra* Gray, 1846 [*nomen nudum*]; *cossus* de Blainville, 1816; *crispa* (Fitzinger, 1859); *depressa* Linnaeus, 1758; *domestica* von Schreber, 1787; *ecornis* von Schreber, 1788; *ensicornis* Angst, 1911; *europaea* (Sanson, 1878); *gazella* (Fitzinger, 1860); *girgentana* Magliano, 1930; *graeca* Fiedler, 1841; *guineensis* (Fitzinger, 1859); *hirsuta* (Fitzinger, 1859); *imberbis* de Blainville, 1816; *indorum* (Fitzinger, 1859); *iowensis* (Palmer, 1956); *jamaicensis* J. B. Fischer, 1829; *kelleri* Duerst, 1908; *laevipes* (Fitzinger, 1859); *lanigera* Wagner, 1836; *mambrica* Linnaeus, 1758; *montana* Reichenbach, 1845; *mutica* Kerr, 1892; *nana* Kerr, 1792; *nepalensis* Reichenbach, 1845; *prisca* Adametz, 1915; *promaza* (Pomel, 1898); *resima* Wagner, 1836; *reversa* Linnaeus, 1758; *rossica* (Fitzinger, 1859); *ruetimeyeri* Duerst, 1899; *sericea* (Fitzinger, 1859); *simus* Rautenbach, 1845; *stenotis* (Fitzinger, 1859); *tatarorum* (Fitzinger, 1859); *thebaica* Desmarest, 1811; *thibetana* Desmarest, 1822; *villosa* Wagner, 1836; *vulgaris* von Schreber, 1787; **aegagrus** Erxleben, 1777; *aegagros* Chollet, Dayot and Neuville, 1904; *aegergus* Zivanèević, 1960; *algagrus* Medvedeff, 1927; *bezoartica* Linnaeus, 1766 [*nomen oblitum*]; *blythi* Hume, 1875; *caucasica* Gray, 1843; *cilicica* Matschie, 1907; *fera* Desmarest, 1822; *florstedti* Matschie, 1907; *gazella* (Gmelin, 1788); *neglecta* Zarudny and Bilkevitsch, 1918; *oegagrus* Crespon, 1844; *oegagyrus* França, 1908; *persica* Matschie, 1905; *turcmenica* Tzalkin, 1950; **chialtanensis** Lydekker, 1913; **cretica** Schinz, 1838; *cretensis* Brisson, 1756 [unavailable]; *cretensis* von Lorenz-Liburnau, 1899; **jourensis** Ivrea, 1899; *aegaeica* Kretzoi, 1942; *dorcas* Reichenow, 1888 [preoccupied]; **picta** (Erhard, 1858).

COMMENTS: Includes *aegagrus*, but see Corbet (1978c:214). Gentry et al. (1996) proposed that majority usage be confirmed by adoption of *Capra aegagrus* as the name for the wild taxon of Goats and asked the International Commission on Zoological Nomenclature to use its plenary power to rule that the name for this wild species is not invalid by virtue of being antedated by the name based on the domestic form. A ruling has now been made in their favor (International Commission on Zoological Nomenclature, 2003a). It might still be valid for those who consider *C. hircus* and *C. aegagrus* to be conspecific to employ the senior name for the name of the species (see Bock, 1997). Both names have been used when referring to wild Goats. *Capra aegagrus* Erxleben, 1777 is a junior synonym of *C. bezoartica* Linnaeus, 1766 whose syntypes may have included other species (Blanford, 1875b); *bezoartica* may be regarded as a *nomen oblitum*. Populations anciently introduced to Greek islands include *cretica* on Crete and Theodorou, *jourensis* on Giura or Joura in the Northern Sporades, and *picta* on Antimilo or Erimomilos in the Cyclades. They have been regarded as synonyms of *hircus* or *aegagrus*, but their systematic status needs evaluation. Provisionally, *aegagrus*, *cretica*, *jourensis* and *picta* are listed here as subspecies of *C. hircus*. A DNA analysis of *cretica*, *aegagrus* and domestic Goats (Kahila Bar-Gal et al., 2002) found that *cretica* was closely allied to domestic Goats and an Iranian wild Goat while a wild Goat from Turkmenistan was distinct. *Capra hircus chialtanensis* is a population originating from hybrids between Goat and Markhor; Schaller (1977) identified it as most like *C. hircus* and did not consider it valid, while Manceau et al. (1999a) regarded it as a markhor or a hybrid, from mtDNA sequence data.

*Capra ibex* Linnaeus, 1758. Syst. Nat., 10th ed., 1:68.
    COMMON NAME: Alpine Ibex.
    TYPE LOCALITY: "Habitat in Wallesiae praeruptis inaccessis"; identified as Switzerland, Valais by Thomas (1911a:152).
    DISTRIBUTION: Formerly the Alps of Austria, France, Germany, N Italy, and Switzerland; extinct except in Italy but reintroduced into its former range.
    STATUS: IUCN – Lower Risk (lc).
    SYNONYMS: *alpina* Girtanner, 1786; *europea* (Hodgson, 1847); *graicus* Matschie, 1912.

*Capra nubiana* F. Cuvier, 1825. *In* É. Geoffroy Saint-Hilaire and F. Cuvier, Hist. Nat. Mammifères, 6, part 50, "Bouc sauvage de la Haute-Egypte", p. 2, pl. 397.
    COMMON NAME: Nubian Ibex.

TYPE LOCALITY: Egypt, "de la Haute-Égypte ou de Nubie"; Nubia (Lydekker, 1913:153; G. M. Allen, 1939:549) or Upper Egypt (Ellerman and Morrison-Scott, 1951:407), which are virtually synonymous; here restricted to Sudan, Northern Prov., Nubian Desert, east of Nile River.

DISTRIBUTION: Egypt east of the Nile, N Eritrea, Israel, W Jordan, Lebanon (extinct), SE Oman, Saudi Arabia, NE Sudan, Syria (extinct; no archaeological records), and SE Yemen.

STATUS: IUCN – Endangered.

SYNONYMS: *arabica* Rüppell, 1835; *beden* (Wagner, 1835); *mengesi* Noack, 1896; *sinaitica* Ehrenberg, 1833; *typica* Lydekker, 1908.

COMMENTS: Treated as a species distinct from *C. ibex* by Uerpmann (1987).

*Capra pyrenaica* Schinz, 1838. N. Denkschr. Schneiz. Ges. Natur. Wiss., 2:9.
COMMON NAME: Spanish Ibex.
TYPE LOCALITY: "In den spanischen Pyrenäen, auf den Gebirgen der Sierra de Randa und des Königreiches Granada"; restricted to Spain, Pyrenees Mtns, Huesca, near Maladetta Pass (Harper, 1940).
DISTRIBUTION: Iberian Peninsula; extinct in Portugal.
STATUS: U.S. ESA – Endangered as *C. p. pyrenaica*; IUCN – Extinct as *C. p. pyrenaica*, Vulnerable as *C. p. victoriae*, Lower Risk (cd) as *C. p. hispanica*, otherwise Lower Risk (nt).
SYNONYMS: *cabrerae* Camerano, 1916 [*nomen nudum*]; *cabrerae* Camerano, 1917; *hispanica* Schimper, 1848; *lusitanica* Schlegel, 1872; *nowaki* Wyrwoll, 1999; *pyrenica* Mottl, 1938; *typica* Lydekker, 1898; *victoriae* Cabrera, 1911.
COMMENTS: Validity of subspecies questioned by Coutourier (1962), Clouet (1979) and following mtDNA analysis by Manceau et al. (1999*b*).

*Capra sibirica* (Pallas, 1776). Spicil. Zool., 11:52.
COMMON NAME: Siberian Ibex.
TYPE LOCALITY: "sylvas inter Udae et Birjussae fluviorum fontes ad ipsam calcem Sajensis"; "northern slope of Sayansk Mountains, in the neighbourhood of Munku Sardyx, west of Lake Baikal" (Lydekker, 1913:143) (Russia, Siberia, Sayan Mtns, near Munku-Sardyk).
DISTRIBUTION: Mountain ranges of N Afghanistan, China (N Gansu, W Inner Mongolia, Sinkiang, N Tibet), N India (Himalayas of Jammu and Kashmir and Himachal Pradesh), E Kazakhstan, Kyrgyzstan, S and W Mongolia, N Pakistan, Russia (S Siberia), and Tajikistan.
STATUS: IUCN – Lower Risk (lc).
SYNONYMS: *alaiana* Noack 1902; *almasyi* Lorenz, 1906; *altaica* Noack, 1902; *dauvergnii* Sterndale, 1886; *dementievi* Tzalkin, 1949; *fasciata* Noack, 1902; *filippii* Camerano, 1911; *formosovi* Tzalkin, 1949; *hagenbecki* Noack, 1903; *hemalayana* Hodgson, 1841; *lorenzi* Satunin, 1905; *lydekkeri* W. Rothschild, 1900; *merzbacheri* (Leisewitz, 1906); *pallasii* Schinz, 1838; *pedri* Lorenz, 1906; *sacin* Lydekker, 1898; *sakeen* Blyth, 1842; *sakin* (Hodgson, 1847); *skyn* (Wagner, 1844); *transalaiana* Lorenz, 1906; *typica* Lorenz, 1906; *wardi* Lydekker, 1900.
COMMENTS: Treated as a species distinct from *C. ibex* by Heptner et al. (1961).

*Capra walie* Rüppell, 1835. Neue Wirbelt. Fauna Abyssin. Gehörig, Säugeth., 1:16.
COMMON NAME: Walia.
TYPE LOCALITY: "die höchsten felsigten Gebirge Abyssiniens in den Provinzen Simen und Godjam"; restricted to Ethiopia, mountains of Simien (Lydekker, 1913:156).
DISTRIBUTION: N Ethiopia.
STATUS: U.S. ESA – Endangered; IUCN – Critically Endangered.
SYNONYMS: *vali* Lydekker, 1898; *valie* Sundevall, 1846; *wali* Richters, 1894.
COMMENTS: Treated as a species distinct from *C. ibex* by Ansell (1972:70) and Yalden et al. (1984).

*Capricornis* Ogilby, 1837. Proc. Zool. Soc. Lond., 1836:139.
TYPE SPECIES: *Antilope thar* Hodgson, 1831.

SYNONYMS: *Austritragus* Heude, 1898; *Capricornus* Gray, 1862; *Capricornulus* Heude, 1898; *Lithotragus* Heude, 1898; *Nemotragus* Heude, 1898.

COMMENTS: Revised by Groves and Grubb (in prep.) who raise *milneedwardsii, rubidus,* and *thar* to species status.

*Capricornis crispus* (Temminck, 1836). *In* von Siebold, Temminck and Schlegel, Fauna Japonica, Coup d'Oeil Faune Iles Sonde Emp. Japan, p. xxii.
  COMMON NAME: Japanese Serow.
  TYPE LOCALITY: "Les îles du domaine du Japon"; "On ne la trouve que dans les parties de l'île de Nippon couvertes de hautes alpes, telle qu la partie, nommée Josino; puis sur les montagnes les plus élevées de l'île de Sikok [Shikoku]"; restricted to "Nippon (Hondo) [Honshu], Japan" (Lydekker, 1913:200).
  DISTRIBUTION: Honshu, Shikoku and Kyushu (Japan).
  STATUS: IUCN – Lower Risk (cd).
  SYNONYMS: *pryeri* (Lydekker, 1901); *pryerianus* Heude, 1894; *saxicola* Heude, 1898.
  COMMENTS: Widely cited from Temminck, 1844. *In* von Siebold, Temminck and Schlegel, Fauna Japonica, Aperçu Gén. Spéc. Mamm. Japon, p. 55, pls. 18,19 [1844]. Included in *sumatraensis* by Haltenorth (1963:119); but a valid species according to Dolan (1963).

*Capricornis milneedwardsii* David, 1869. Nouv. Arch. Mus. H. N. Paris, 5 bull.:10.
  COMMON NAME: Chinese Serow.
  TYPE LOCALITY: China, Sichuan, "Moupin" (Baoxing).
  DISTRIBUTION: S Burma, Cambodia, S and C China (Himalayas and E Tibet, S Gansu to Zhejiang and S to Yunnan), Laos, Thailand, and Vietnam.
  STATUS: IUCN – Vulnerable as *C. sumatraensis maritimus* and *C. s. milneedwardsi.*
  SYNONYMS: *argyrochaetes* Heude, 1888; *brachyrhinus* Heude, 1894; *chrysochaetes* Heude, 1894; *collasinus* Heude, 1899; *cornutus* Heude, 1894; *edwardsii* (David, 1871); *erythropygius* Heude, 1894; *fargesianus* Heude, 1894; *longicornis* Heude, 1894; *maxillaris* Heude, 1894; *microdonticus* Heude, 1894; *microdontus* Heude, 1894; *montinus* G. M. Allen, 1930; *nasutus* Heude, 1894; *osborni* Andrews, 1921; *platyrhinus* Heude, 1894; *pugnax* Heude, 1894; *ungulosus* Heude, 1894; *vidianus* Heude, 1894; **maritimus** Heude, 1888; *annectens* Kloss, 1919; *benetianus* Heude, 1894; *berthetianus* Heude, 1898; *gendrelianus* Heude, 1899; *marcolinus* Heude, 1897; *rocherianus* Heude, 1894; *venetianus* Lydekker, 1913.

*Capricornis rubidus* Blyth, 1863. Cat. Mamm. Mus. Asiat. Soc., p. 174.
  COMMON NAME: Red Serow.
  TYPE LOCALITY: Burma, "Arakan Hills".
  DISTRIBUTION: N Burma.
  STATUS: IUCN – Endangered as *C. sumatraensis rubidus.*

*Capricornis sumatraensis* (Bechstein, 1799). *In* Pennant, Allgemeine Ueber. Vierfüss. Thiere, 1:98.
  COMMON NAME: Sumatran Serow.
  TYPE LOCALITY: Indonesia, "Sumatra".
  DISTRIBUTION: Indonesia (Sumatra), Malaysia (peninsular Malaya), Thailand (isthmus of Kra).
  STATUS: CITES – Appendix I as *Naemorhedus sumatraensis*; U.S. ESA – Endangered as *Naemorhedus* (= *Capricornis*) *sumatraensis*; IUCN – Endangered as *C. s. sumatraensis,* otherwise Vulnerable.
  SYNONYMS: *interscapularis* (Lichtenstein, 1814); *robinsoni* Pocock, 1908; *swettenhami* (Butler, 1900).

*Capricornis swinhoei* Gray, 1862. Ann. Mag. Nat. Hist., ser. 3, 10:320.
  COMMON NAME: Formosan Serow.
  TYPE LOCALITY: "Formosa [Taiwan], on the central ridge of the Snowy Mountains".
  DISTRIBUTION: Taiwan.
  STATUS: IUCN – Vulnerable.
  COMMENTS: Regarded as a species distinct from *C. crispus* by Groves and Grubb (1985).

*Capricornis thar* (Hodgson, 1831).Gleanings Science, 3:324.

    COMMON NAME: Himalayan Serow.

    TYPE LOCALITY: "the central region, equidistant from the snows on one hand, and the plains of India on the other; between the Sutlege, west, and the Teesta, east, in Nepal proper" (Nepal, Himalayas).

    DISTRIBUTION: E and SE Bangladesh, Himalayas (Bhutan, N India including Sikkim, and Nepal), and NE India (provinces E of Bangladesh). Probably W Burma.

    STATUS: IUCN – Vulnerable as *C. sumatraensis thar*.

    SYNONYMS: *bubalina* (Hodgson, 1832); *humei* Pocock, 1908; *jamrachi* Pocock, 1908; *proclivus* (Hodgson, 1842) [*nomen nudum*]; *rodoni* Pocock, 1908.

    COMMENTS: Reddish specimens from Arunchal Pradesh, Assam, Bangladesh, and Meghalaya are not attributable to *C. rubidus* and may represent a distinct subspecies of *C. thar*.

*Hemitragus* Hodgson, 1841. Calcutta J. Nat. Hist., 2:218.

    TYPE SPECIES: *Capra jharal* Hodgson, 1833 (= *Capra jemlahica* C. H. Smith, 1826).

*Hemitragus hylocrius* (Ogilby, 1838). Proc. Zool. Soc. Lond., 1837:81 [1838].

    COMMON NAME: Nilgiri Tahr.

    TYPE LOCALITY: India, "Neilgherry Hills" (Nilgiri Hills).

    DISTRIBUTION: S India (Western Ghats along border between Kerala and Tamil Nadu).

    STATUS: IUCN – Endangered.

    SYNONYMS: *warryato* (Gray, 1842).

    COMMENTS: Included in *jemlahicus* by Haltenorth (1963:125) but generally regarded as a full species, for example by Corbet and Hill (1991).

*Hemitragus jayakari* Thomas, 1894. Ann. Mag. Nat. Hist., ser. 6, 13:365.

    COMMON NAME: Arabian Tahr.

    TYPE LOCALITY: Oman, "Jebel Taw, Jebel Akhdar Range".

    DISTRIBUTION: Oman; United Arab Emirates (extinct).

    STATUS: U.S. ESA – Endangered; IUCN – Endangered.

    COMMENTS: Included in *jemlahicus* by Haltenorth (1963:125) but see Harrison (1968:324).

*Hemitragus jemlahicus* (C. H. Smith, 1826). *In* Griffith et al., Animal Kingdom, 4, plate [1826] opp. p. 308 [1827].

    COMMON NAME: Himalayan Tahr.

    TYPE LOCALITY: Nepal, "the district of Jemlah, between the sources of the Sargew and Sampoo" (Jemla Hills).

    DISTRIBUTION: Himalayas including China (S Tibet), N India (Jammu and Kashmir; Sikkim), and Nepal. Introduced in New Zealand and Western Cape Prov. (South Africa).

    STATUS: IUCN – Vulnerable.

    SYNONYMS: *iharal* (Wagner, 1836); *jemlaicus* Gray, 1847; *jemlanica* (C. H. Smith, 1827); *jharal* (Hodgson, 1833); *quadrimammis* (Hodgson, 1836); *schaeferi* Pohle, 1944; *tubericornis* (Wagner, 1836).

    COMMENTS: Specific name is spelt "*jemlanica*" on p. 308 in the original description which was published in 1827 and "*jemlahica*" in legend to the plate on the opposite unnumbered page, dated 1826.

*Naemorhedus* C. H. Smith, 1827. *In* Griffith et al., Animal Kingdom, 5:352.

    TYPE SPECIES: *Antilope goral* Hardwicke, 1825.

    SYNONYMS: *Caprina* Wagner, 1844; *Kemas* Ogilby, 1837; *Naemorhaedus* Jardine, 1836; *Nemorhaedus* Hodgson, 1841; *Nemorhedus* Agassiz, 1842; *Nemorrhaedus* Trouessart, 1898; *Nemorrhedus* Gray, 1843; *Urotragus* Gray, 1871.

    COMMENTS: The original spelling is "*Naemorhedus*". *Naemorhaedus, Nemorhaedus, Nemorhedus, Nemorrhaedus,* and *Nemorrhedus* are later spellings. If one of these is in prevailing usage (Article 33.2.3.1, International Code of Zoological Nomenclature; International Commission on Zoological Nomenclature, 1999), it can be deemed to be a justified emendation, but this has yet to be demonstrated. Reviewed by Dolan (1963) and Groves and Grubb (1985) who included *Capricornis* in this genus; revised by Groves and Grubb (in prep.), who once again confined *Naemorhedus* to gorals and separated *griseus* from *caudatus*.

*Naemorhedus baileyi* Pocock, 1914. J. Bombay Nat. Hist. Soc., 23:32.
    COMMON NAME: Red Goral.
    TYPE LOCALITY: China, Tibet, "Dre on banks of Yigrong Tso (Lake) in Po Me [Bomi]. 9,000 ft
        [2743 m]".
    DISTRIBUTION: N Burma, China (SE Tibet and Yunnan), and NE India (Arunachal Pradesh).
    STATUS: CITES – Appendix I; IUCN – Vulnerable as *N. b. baileyi* and *N. b. cranbrooki*.
    SYNONYMS: *cranbrooki* Hayman, 1961.
    COMMENTS: Regarded as a valid species by Groves and Grubb (1985); and by Zhang (1987),
        under the name *cranbrooki*.

*Naemorhedus caudatus* (Milne-Edwards, 1867). Ann. Sci. Nat. Zool. Paris, ser. 5, 7:377.
    COMMON NAME: Long-tailed Goral.
    TYPE LOCALITY: Russia, "Sibérie" (Amurland, Bureja Mtns).
    DISTRIBUTION: E Russia (Primorsky and Khabarovsk Territories), NE China (Heilonjiang,
        Jilin), and Korea.
    STATUS: CITES – Appendix I; IUCN – Vulnerable as *N. c. caudatus* and *N. c. raddeanus*.
    SYNONYMS: *crispa* (Radde, 1862); *raddeanus* (Heude, 1894).
    COMMENTS: Fully described by Milne-Edwards, Rech. Hist. Nat. Mamm., Faune Chine, p. ,
        pl.23, 23a, 23b [1868]. Regarded as a species distinct from *N. goral* by Groves and
        Grubb (1985). Reviewed by J. I. Mead (1989, Mammalian Species, 335, as *Nemorhaedus
        goral*).

*Naemorhedus goral* (Hardwicke, 1825). Trans. Linn. Soc. Lond., 14:518.
    COMMON NAME: Himalayan Goral.
    TYPE LOCALITY: "a native of the Himalayah range and the mountains of the Nepaul frontier"
        (Nepal, Himalayas).
    DISTRIBUTION: Himalayas in Bhutan, China (S Tibet), N India (including Sikkim), Nepal, and
        N Pakistan.
    STATUS: CITES – Appendix I; U.S. ESA – Endangered; IUCN – Lower Risk (nt) as *N. g. bedfordi*
        and *N. g. goral*.
    SYNONYMS: *duvaucelii* (C. H. Smith, 1827); *hodgsoni* (Lydekker, 1905); **bedfordi** (Lydekker,
        1905).
    COMMENTS: Reviewed by J. I. Mead (1989, Mammalian Species, 335, as *Nemorhaedus goral*).

*Naemorhedus griseus* (Milne-Edwards, 1871). Nouv. Arch. Mus. Hist. Nat. Paris, 7, bull.: 93.
    COMMON NAME: Chinese Goral.
    TYPE LOCALITY: "du nord de la Chine"; China, Sichuan, Moupin (Baoxing).
    DISTRIBUTION: W Burma, E China (SW Inner Mongolia to Yunnan), NE India (Provinces E of
        Bangladesh), and NW Thailand. Also SE Burma and N Vietnam (Tonkin) (Groves and
        Grubb, in prep.). Occurrence in Laos as yet unconfirmed (Duckworth et al., 1999).
    STATUS:IUCN – Vulnerable as *N. caudatus evansi* and *N. c. griseus*.
    SYNONYMS: *aldridgeanus* (Heude, 1894); *arnouxianus* (Heude, 1888); *cinerea* (Milne-Edwards,
        1874); *curvicornis* (Heude, 1894); *fantozatianus* (Heude, 1894); *fargesianus* (Heude,
        1894); *galeanus* (Heude, 1894); *henryanus* (Heude, 1890); *initialis* (Heude, 1894); *iodinus*
        (Heude, 1894); *niger* (Heude, 1894); *pinchonianus* (Heude, 1894); *versicolor* (Heude,
        1894); *vidianus* Heude, 1894; *xanthodeiros* (Heude, 1894); **evansi** (Lydekker, 1902).
    COMMENTS: Fully described by Milne-Edwards, Rech. Hist. Nat. Mamm., Faune Tibet-
        Oriental, p. 361, pl.70, 71, 71a [1874]. Reviewed by J. I. Mead (1989, Mammalian
        Species, 335, as *Nemorhaedus goral*).

*Oreamnos* Rafinesque, 1817. Am. Mon. Mag., 2:44.
    TYPE SPECIES: *Mazama dorsata* Rafinesque, 1817 (= *R[upicapra]. americanus* de Blainville, 1816).
    SYNONYMS: *Aplocerus* C. H. Smith, 1827; *Haploceros* Flower and Garson, 1884; *Haplocerus*
        Wagner, 1844; *Oreamnus* Elliot, 1901.

*Oreamnos americanus* (de Blainville, 1816). Bull. Sci. Soc. Philom. Paris, 1816:80.
    COMMON NAME: Mountain Goat.
    TYPE LOCALITY: "d'Amerique"; restricted to the type locality of *Ovis montanus* Ord, 1815, the
        Columbia River area, where observed by Lewis and Clark, in "the Cascade Mountains,

and the nearby range" (Hollister, 1912:185) "near Mt Adams, Washington" (Dalquest, 1948:409) (USA, Washington, Cascade Mountains, Mt. Adams).

DISTRIBUTION: SE Alaska (USA), S Yukon and SW Mackenzie (Canada) to NC Oregon, C Idaho, and Montana (USA). Introduced to Kodiak, Chichagof, and Baranof Isls (Alaska), Olympic Peninsula (Washington), C Montana, Black Hills (South Dakota), and Colorado (USA).

STATUS: IUCN – Lower Risk (lc).

SYNONYMS: *columbiae* Hollister, 1912; *columbianus* (Desmoulins, 1823); *dorsatus* (Rafinesque, 1817); *kennedyi* (Elliot, 1900); *lanigera* (C. H. Smith, 1821); *missoulae* J. A. Allen, 1904; *montanus* (Ord, 1815) [preoccupied]; *sericeus* (Rafinesque, 1817).

COMMENTS: Revised by Cowan and McCrory (1970). Reviewed by Rideout and Hoffmann (1975, Mammalian Species, 63).

*Ovibos* de Blainville, 1816. Bull. Sci. Soc. Philom. Paris, 1816:76.

TYPE SPECIES: *Bos moschatus* Zimmermann, 1780.

SYNONYMS: *Bosovis* Kowarzik, 1911; *Criotaurus* Gloger, 1841.

*Ovibos moschatus* (Zimmermann, 1780). Geogr. Gesch. Mensch. Vierf. Thiere, 2:86.

COMMON NAME: Muskox.

TYPE LOCALITY: Canada, Manitoba, "Bewohnt anjezt hauptsachlich Neuwallis an der Hudsonsbay ban zwischen Seefalber- (Seals) und Churchill-Fluss zwischen dem 59 bis 61. Grad Breite" (between Seal and Churchill Rivers).

DISTRIBUTION: Formerly Point Barrow, Alaska (USA) east to NE Greenland, south to NE Manitoba (Canada). Range now much reduced. Introduced to Seward Peninsula and Nunivak Isl, Alaska (USA); Taimyr Peninsula and Wrangel Isl (Russia); and Svalbard (Norway), where it has since died out.

STATUS: IUCN – Lower Risk (lc).

SYNONYMS: *mackenzianus* Kowarzik, 1908; *melvillensis* Kowarzik, 1909; *niphoecus* Elliot, 1905; *pearyi* Allen, 1901 [unavailable]; *platycerus* (G. Fischer [von Waldheim], 1814); *wardi* Lydekker, 1900.

COMMENTS: Revised by Tener (1965). Reviewed by Lent (1988, Mammalian Species, 302).

*Ovis* Linnaeus, 1758. Syst. Nat., 10th ed., 1:70.

TYPE SPECIES: *Ovis aries* Linnaeus, 1758.

SYNONYMS: *Ammon* de Blainville, 1816; *Argali* Gray, 1852; *Aries* Link, 1795; *Caprovis* Hodgson, 1847; *Musimon* Pallas, 1776; *Musmon* Schrank, 1798; *Pachyceros* Gromova, 1936.

COMMENTS: Placed in *Capra* by Van Gelder (1977b); see comments under *Capra*. There is no consensus concerning the number of species to be recognized in this genus; some would recognize only one (*ammon*; see Haltenorth, 1963:126-128); others two (*ammon*, *canadensis*; see Corbet, 1978c:218); while others recognize up to seven, as do the most recent reviews (Korobitsyna et al., 1974; Nadler et al., 1973). Five species are listed here. Species-groups are nominate *Ovis* or *aries* group (includes also *ammon*) and *Pachyceros* or *canadensis* group (includes also *dalli* and *nivicola*).

*Ovis ammon* (Linnaeus, 1758). Syst. Nat., 10th ed., 1:70.

COMMON NAME: Argali.

TYPE LOCALITY: "Habitat in Siberia"; since identified as Kazakhstan, Vostochno-Kazakhstansk. Obl., Altai Mtns, Bukhtarma; near Ust-Kamenogorsk.

DISTRIBUTION: China (Gansu, Inner Mongolia, Ningxia, N and S Sinkiang, W Sichuan, Tibet including Qinghai), N India (Ladak, Sikkim, Spiti), E Kazakhstan, Kyrgyzstan, Mongolia, Nepal (Tibetan border), Pamir Range (NE Afghanistan, N Pakistan), SC Siberia (Altai Mntns), and Tajikistan.

STATUS: CITES – Appendix I as *O. a. hodgsoni* and *O. a. nigrimontana*, otherwise Appendix II; U.S. ESA – Endangered, except in Kyrgyzstan, Mongolia, and Tajikistan, where status is Threatened; IUCN – Critically Endangered as *O. a. nigrimontana* and *O. a. jubata* [= *comosa*], Endangered as *O. a. severtsovi* and *O. a. darwini*, Vulnerable as *O. a. hodgsonii*, *O. a. ammon*, *O. a. collium*, *O. a. karelini*, and *O. a. polii*.

SYNONYMS: *altaica* Severtzov, 1873; *argail* Walther, 1809; *argali* Boddaert, 1785; *asiatica*

(Pallas, 1776); *daurica* Severtzov, 1873 [preoccupied]; *fera* Pallas, 1776; *mongolica* Severtzov, 1873 [preoccupied]; *przevalskii* Nasonov, 1923; *typica* Lydekker, 1898; **collium** Severtzov, 1873; **comosa** Hollister, 1919; *jubata* Peters, 1876 [preoccupied]; **darwini** Przewalski, 1883; *intermedia* Gromova, 1936; *kozlovi* Nasonov, 1913; **hodgsonii** Blyth, 1841; *adametzi* Kowarzik, 1913; *ammonoides* Hodgson, 1841; *bambhera* (Gray, 1852); *blythi* Severtzov, 1873; *broockii* Glür, 1894; *brookei* Ward, 1874; *dalailamae* Przewalski, 1888; *henrii* Milne-Edwards, 1892; **karelini** Severtzov, 1873; *heinsii* Severtzov, 1873; *humei* Lydekker, 1913; *littledalei* Lydekker, 1902; *nassanovi* Laptev, 1929; *sairensis* Lydekker, 1898; **nigrimontana** Severtzov, 1873; **polii** Blyth, 1841; *typica* Lydekker, 1898; **severtzovi** Nasonov, 1914.

COMMENTS: Haltenorth (1963:121) and Corbet (1978c:218) included *orientalis* (= *aries*), *musimon* and *vignei*, but Nadler et al. (1973) and Corbet and Hill (1991:136) excluded them. The names *daurica, jubata* and *mongolica* are preoccupied by names of domestic sheep. Subspecies reviewed by Sopin (1982) and revised by Geist (1990 and in Shackleton, 1997). Diploid chromosome complement (2n = 56) and examination of mitochondrial region sequences indicate that *severtzovi* is not part of *vignei* division of *O. aries*, but a primitive argali, sister taxon to the rest of the species (Bunch et al, 1998; Wu et al., 2003). Wu et al. (2003) transferred *severtzovi* to *O. ammon*.

*Ovis aries* Linnaeus, 1758. Syst. Nat., 10th ed., 1:70.

COMMON NAME: Red Sheep.

TYPE LOCALITY: "Habitat in siccis apricis calidis"; identified as Sweden by Thomas (1911a:153); domesticated stock.

DISTRIBUTION: Urial or Arkar in Afghanistan, NW India (Kashmir), NE and SE Iran, SW Kazakhstan, Oman (introduced?), Pakistan, Tajikistan, Turkmenistan, and Uzbekistan. Intermediate Laristan sheep in S Iran. Red Sheep or Mouflon in Armenia, S Azerbaijan, N Iraq, W Iran, and S and E Turkey; transported populations (Mouflon) on Corsica and Sardinia, introduced from there to Europe, Ukraine (Crimea), USA (incl. Hawaiian Isls), Chile, Kerguelen Isls, and Tenerife (Canary Isls); and on Cyprus. Domesticated worldwide; feral populations on St. Kilda and other small islands off the British Isles; improved domestic stock feral in Norway, Sweden, USA, islands off coasts of United Kingdom and New Zealand, Kerguelen Isls, and probably other oceanic islands.

STATUS: CITES – Appendix I as *O. orientalis ophion* and *O. vignei vignei*, Appendix II as *O. vignei*; U.S. ESA – Endangered as *O. musimon ophion* and *O. vignei vignei*; IUCN – Endangered as *O. orientalis ophion*, *O. o. bochariensis*, *O. o. punjabiensis*, and *O. o. vignei*, Vulnerable as *O. o. isphahanica*, *O. o. laristanica*, *O. o. arkal*, *O. o. gmelini*, *O. o. musimon*, and *O. o. cycloceros*.

SYNONYMS: *adimain* Boddaert, 1785; *aegyptiaca* Fitzinger, 1860; *africana* Linnaeus, 1758; *anatolica* Fitzinger, 1860; *anglica* Linnaeus, 1758; *angolensis* Fitzinger, 1860; *antillarum* Fitzinger, 1860; *appendiculata* Gené, 1834; *arabica* Fitzinger, 1860; *arietina* Fitzinger, 1860; *arvernensis* Sanson, 1878; *asiatica* Sanson, 1878; *astracanica* Reichenbach, 1845; *astrachanica* Gené, 1834; *bakelensis* de Rochbrune, 1882; *barbarica* Reichenbach, 1845; *batavica* Sanson, 1878; *belgica* Walther, 1809; *bergamena* Fitzinger, 1860; *bohemica* Walther, 1809 [nomen nudum]; *borealis* Fitzinger, 1860; *brachyura* Pallas, 1776; *brittanica* Sanson, 1878; *bucharica* Gmelin, 1758; *buraetica* Fitzinger, 1860; *cabardinica* Fitzinger, 1860; *calmuccorum* Fitzinger, 1860; *calotis* Fitzinger, 1860; *campestris* Fitzinger, 1860; *capensis* Erxleben, 1777; *carnapi* Müller-Liebenweide, 1896; *colchica* Fitzinger, 1860; *congensis* Reichenbach, 1845; *corneri* Millais, 1906; *cornuta* Erxleben, 1777; *cretensis* Boddaert, 1785; *dacica* Rautenbach, 1845; *danica* Walther, 1809; *daurica* Fitzinger, 1860; *djalonensis* de Rochebrune, 1882; *dolichura* Pallas, 1776; *domestica* (G. Fischer [von Waldheim], 1814); *ecaudata* I. Geoffroy Saint Hilaire, 1827; *gallica* Walther, 1809; *germanica* Walther, 1809; *gothlandica* G. Fischer [von Waldheim], 1814; *guineensis* Linnaeus, 1758; *gutturosa* Wagner, 1836; *hebridica* Fitzinger, 1860; *hibernica* Fitzinger, 1860; *hispanica* Linnaeus, 1758; *hollandica* Walther, 1809 [nomen nudum]; *holsatica* Fitzinger, 1860; *hoonia* Hodgson, 1842; *indica* Reichenbach, 1845; *ingevonensis* Sanson, 1878; *iberica* Sanson, 1878; *islandica* Pallas, 1794; *italica* Walther, 1809; *jubata* Kerr, 1792; *kirgisica* Fitzinger, 1860; *laticauda* Boddaert, 1785; *laticaudata* Linnaeus, 1758;

*leptura* von Schreber, 1788; *libyca* Fitzinger, 1860; *ligeriensis* Sanson, 1878; *lombardica* Brüggemann, 1831; *longicauda* Gmelin, 1789; *longicaudata* Erxleben, 1777; *longipes* Desmarest, 1822; *macedonica* Fitzinger, 1860; *macrocerca* von Schreber, 1788; *macroura* Walther, 1809; *madagascariensis* Fitzinger, 1860; *melanocephalus* de Rochebrune, 1883; *monasteriensis* Fitzinger, 1860; *mongolica* Fitzinger, 1860; *nana* Kerr, 1792; *numida* Fitzinger, 1860; *orcadica* Fitzinger, 1860; *ovis* (Pallas, 1811); *pachycerca* Fitzinger, 1860; *paduana* Reichenbach, 1845; *palaeoaegyptica* Duerst and Gaillard, 1902; *palustris* Glür, 1895; *parnassica* Fitzinger, 1860; *pegasus* (C. H. Smith, 1827) [*nomen dubium*]; *parnassica* Reichenbach, 1845; *persica* Reichenbach, 1845; *platyura* Pallas, 1794; *polonica* Walther, 1809; *polycerata* Linnaeus, 1758; *polyceros* Boddaert, 1785; *quadricornis* Boddaert, 1785; *recurvicauda* Gené, 1834; *rustica* Linnaeus, 1758; *scotica* Fitzinger, 1860; *senegalensis* Fitzinger, 1860; *sodanica* Sanson, 1878; *steatinion* C. H. Smith, 1827; *steatopyga* Pallas, 1776; *strepsiceros* Linnaeus, 1758; *studeri* Duerst, 1904; *suecica* Walther, 1809; *syenitica* Fitzinger, 1855; *syriaca* Fitzinger, 1860; *tarentina* Fitzinger, 1860; *tatarica* Fitzinger, 1860; *taurica* Pallas, 1794; *torticornis* Reichenbach, 1845; *tscherkessica* Pallas, 1776; *tunetana* Fitzinger, 1860; *turcica* Walther, 1809; *ungarica* Walther, 1809; *zetlandica* Fitzinger, 1860; **arkal** Eversmann, 1850; *arkar* Brandt, 1852; *dolgopolovi* Nasonov, 1913; *varentsowi* Satunin, 1905; **cycloceros** Hutton, 1842; *arabica* Sopin and Harrison, 1986; *blanfordi* Hume, 1877; *bochariensis* Nasonov, 1914; *punjabiensis* Lydekker, 1913; **isphahanica** Nasonov, 1910; *isphaganica* Nasonov, 1910 [incorrect original spelling]; **laristanica** Nasonov, 1909; **musimon** (Pallas, 1811); *corsicosardinensis* Kowarzik, 1913; *corsica* Bourguignat, 1870; *europaea* Kerr, 1792 [*nomen oblitum*]; *faidherbi* Bourguignat, 1870; *lartetiana* Bourguignat, 1870; *matschiei* Duerst, 1905; *musmon* C. H. Smith, 1827; *musimon* von Schreber, 1782 [unavailable]; *occidentalis* Brandt and Ratzeburg, 1829; *occidentosardinensis* Kowarzik, 1913; *rouvieri* Bourguignat, 1870; *sinesella* Turcek, 1949; **ophion** Blyth, 1841; *cypria* Blasius, 1842; **orientalis** Gmelin, 1774; *anatolica* Valenciennes, 1856; *armeniana* Nasonov, 1919; *erskinei* Lydekker, 1904; *gmelinii* Blyth, 1841; *typica* Lydekker, 1898; *urmiana* Günther, 1899; **vignei** Blyth, 1841.

COMMENTS: Includes *orientalis*; see Nadler et al. (1973). Also includes *musimon* and *ophion*, introduced in Neolithic to Corsica, Sardinia and Cyprus; see Payne (1968), Vigne (1988), and Hemmer (1990). For correct authorship of *musimon*, see Uerpmann (1980); Pallas (1811) included the Mouflon of Sardinia and Corsica described by Cetti with the Arkar of Turkmenistan and Iran as syntypes of *musimon* (Heptner et al., 1961). To avoid *musimon* being a senior synonym of *arkal*, the lectotype of *musimon* must be designated as a Corsican or Sardinian Mouflon. Gentry et al. (1996) proposed that majority usage be confirmed by adoption of *Ovis orientalis* as the name for the wild taxon of Red Sheep and asked the International Commission on Zoological Nomenclature to use its plenary powers to rule that the name for this wild species is not invalid by virtue of being antedated by the name based on the domestic form. A ruling has now been made in their favor (International Commission on Zoological Nomenclature, 2003*a*). It might still be valid for those who consider *O. aries* and *O. orientalis* to be conspecific to employ the senior name for the name of the species (see Bock, 1997), though authors have not referred to this species as *O. aries* when discussing wild populations; they have used the names *O. ammon, O. gmelinii,* or *O. orientalis* instead. Status of *orientalis* disputed; reputed type skull said to combine characters of Red Sheep, *O. a.* cf. *armeniana* and Arkar, *O. a. arkal* (Nadler et al. 1973; Valdez et al. 1978); type locality lies between habitat of Armenian Sheep and hybrid zone (see below); here treated as a senior synonym of *armeniana*, but provisionally listed as a subspecies of *O. aries*. Populations of the species can be partitioned into Red Sheep or Mouflon *sensu stricto*, the *aries/orientalis* division (includes also *isphahanica, musimon,* and *ophion*); and Urial or Arkar, the *vignei* division (includes also *arkal, cycloceros,* and formerly *severtsovi*, now transferred to *O. ammon* by Wu et al, 2003); Laristan Sheep, *laristanica*, combine Urial-type morphology and Red Sheep karyotype. Hence *vignei* and *aries/orientalis* divisions treated as conspecific following Valdez (1982); multivariate morphometrics (Ludwig and Knoll, 1998) did not discriminate between the two. Hybrid zone between *O. a. orientalis* (2N = 54) and *O. a. arkal*

(2N = 58) in Elburz Mtns, N Iran and apparent hybrids between *O. a. laristanica* and *O. a. cycloceros* recorded in SE Iran (Nadler et al., 1971*b*; Valdez et al., 1978). Subspecies modified from Valdez (1982).

*Ovis canadensis* Shaw, 1804. Nat. Misc., 51, text to pl. 610.
 COMMON NAME: Bighorn Sheep.
 TYPE LOCALITY: "the interior parts of Canada"; identified as Canada, Alberta, Mountains on Bow River, near Exshaw (Anderson, 1947:184).
 DISTRIBUTION: S British Columbia and SW Alberta (Canada) to Coahuila, Chihuahua, Sonora and Baja California (Mexico).
 STATUS: CITES – Appendix II (Mexican population); U.S. ESA – Endangered as *O. c. californiana* in Sierra Nevada, California; Endangered as *O. canadensis* in peninsular ranges of Calfornia; IUCN – Critically Endangered as *O. c. weemsi*; Endangered as *O. c. cremnobates*; Vulnerable as *O. c. mexicana*; Lower Risk (cd) as *O. c. californiana* and *O. c. nelsoni*; otherwise Lower Risk (lc).
 SYNONYMS: *auduboni* Merriam, 1901; *cervina* Desmarest, 1804; *montana* von Schreber, 1804; *palmeri* (Cragin, 1900); *pygargus* C. H. Smith, 1827; ***californiana*** Douglas, 1829; *ellioti* Kowarzik, 1913 [*nomen nudum*]; *samilkameenensis* Millais, 1915; *sierrae* Grinnell, 1912; ***cremnobates*** Elliot, 1904; ***mexicana*** Merriam, 1901; *gaillardi* Mearns, 1907; *sheldoni* Merriam, 1916; *texianus* V. Bailey, 1912; ***nelsoni*** Merriam, 1897; ***weemsi*** Goldman, 1937.
 COMMENTS: For locality where Duncan McGillivray shot the type, see J. A. Allen (1912). Corbet (1978*c*:218) included *nivicola*; but see also Korobitsyna et al. (1974) and Corbet and Hill (1991:135). Revised by Cowan (1940) and Wehausen and Ramey (2000). Reviewed by Shackleton (1985, Mammalian Species, 230).

*Ovis dalli* Nelson, 1884. Proc. U.S. Natl. Mus., 7:13.
 COMMON NAME: Dall's Sheep.
 TYPE LOCALITY: USA, Alaska, "killed by the Indians on some mountains south of Fort Yukon, and on the west bank of the river"; interpreted as "Mountains west of Fort Reliance, Alaska, on divide between Tanana and Yukon Rivers" (Miller, 1924:497), or "Mountains south of Fort Yukon on west bank of Yukon River, Alaska; probably Tanana Hills" (Miller and Kellog, 1955:823).
 DISTRIBUTION: Alaska to N British Columbia and W Mackenzie (Canada).
 STATUS: IUCN – Lower Risk (lc).
 SYNONYMS: *kenaiensis* J. A. Allen, 1902; ***stonei*** J. A. Allen, 1897; *cowani* W. Rothschild, 1907; *fannini* Hornaday, 1901; *liardensis* Lydekker, 1898; *niger* Millais, 1915.
 COMMENTS: Hybrid zone between Dall's Sheep *sensu stricto* (*dalli*) and Stone's Sheep (*stonei*) (Sheldon, 1919). Revised by Cowan (1940). Reviewed by Bowyer and Leslie (1992, Mammalian Species, 393).

*Ovis nivicola* Eschscholtz, 1829. Zool. Atlas, Part 1, p. 1, pl. 1.
 COMMON NAME: Snow Sheep.
 TYPE LOCALITY: "den Bergen der Halbinsel Kamtschatka" (Russia, E Kamchatka).
 DISTRIBUTION: Russia (Putorana Mtns, NC Siberia; NE Siberia from Lena River east to Chukotka and Kamchatka).
 STATUS: IUCN – Vulnerable as *O. n. borealis*, Lower Risk (nt) as *O. n. nivicola*; Lower Risk (lc) as *O. n. alleni* and *O. n. lydekkeri*.
 SYNONYMS: *storcki* J. A. Allen, 1904; ***borealis*** Severtzov, 1872; *albula* Nasonov, 1923 [unavailable]; *alleni* Matschie, 1907; *lenaensis* Kowarzik, 1913 [*nomen nudum*]; *lydekkeri* Kowarzik, 1913; *middendorfi* Kowarzik, 1913; *obscura* Nasonov, 1923 [unavailable]; *potanini* Nasonov, 1915; ***kodarensis*** Medvedev, 1994; ***koriakorum*** Chernyavskii, 1962.
 COMMENTS: Corbet (1978*c*:218) and others included *nivicola* in *canadensis*; but see Korobitsyna et al. (1974) and Gromov and Baranova (1981:407). Subspecies reviewed by Heptner et al. (1961) and Valdez (1982) whose synonymy is followed here.

*Pantholops* Hodgson, 1834. Proc. Zool. Soc. Lond., 1834:81.
 TYPE SPECIES: *Antilope hodgsonii* Abel, 1826.

*Pantholops hodgsonii* (Abel, 1826). Calcutta Gov't Gazette., 68:234.
COMMON NAME: Chiru.
TYPE LOCALITY: China, Tibet, Kooti Pass in Arrun Valley, Tingri Maiden.
DISTRIBUTION: China (S Sinkiang, Sichuan, Tibet including Qinghai, Szechwan) and N India
(Ladak).
STATUS: CITES – Appendix I; IUCN – Endangered. Under severe pressure from hunting.
SYNONYMS: *chiru* (Lesson, 1827); *kemas* (C. H. Smith, 1827).

*Pseudois* Hodgson, 1846. J. Asiat. Soc. Bengal, 15:343.
TYPE SPECIES: *Ovis nayaur* Hodgson, 1833.
SYNONYMS: *Pseudovis* Gill, 1872.
COMMENTS: Revised by Groves (1978c). Reviewed by Wang and Hoffmann (1987).

*Pseudois nayaur* (Hodgson, 1833). Asiat. Res., 18(2):135.
COMMON NAME: Bharal.
TYPE LOCALITY: "The Himálaya"; restricted to "the Tibetan frontier of Nepal" (Lydekker,
1913:127).
DISTRIBUTION: Bhutan, N Burma (Rabinowitz, 1996), China (Gansu, S Inner Mongolia,
Ningxia, Shaanxi, Sichuan, S Sinkiang, Tibet including Qinghai, and N Yunnan),
Himalayas (N India, Nepal, N Pakistan), and SE Tajikistan (Pamir Range).
STATUS: IUCN – Lower Risk (nt) as *P. n. nayaur* and *P. n. szechuanensis*, otherwise Lower Risk (lc).
SYNONYMS: *barhal* (Hodgson, 1846); *burrhel* (Blyth, 1841); *caesia* Howell, 1928; *nahoor*
(Hodgson, 1835); *nahura* (Gray, 1843); *szechuanensis* W. Rothschild, 1922.
COMMENTS: See Wang and Hoffmann (1987, Mammalian Species, 278).

*Pseudois schaeferi* Haltenorth, 1963. Handb. Zool., 8(32):126.
COMMON NAME: Dwarf Bharal.
TYPE LOCALITY: China, "Obere Jangtse-Talschlucht bei Batang" (upper Yangtze Gorge,
Drupalong, south of Batang).
DISTRIBUTION: China (Upper Yangtze Gorge in W Sichuan and adjacent parts of Tibet and
N Yunnan).
STATUS: IUCN – Endangered.
COMMENTS: A separate species according to Groves (1978c:183). See Wang and Hoffmann
(1987, Mammalian Species, 278).

*Rupicapra* de Blainville, 1816. Bull. Sci. Soc. Philom. Paris, 1816:75.
TYPE SPECIES: *Capra rupicapra* Linnaeus, 1768.
SYNONYMS: *Rupicapra* Frisch, 1775 [unavailable]; *Capella* Keyserling and Blasius, 1840.
COMMENTS: Revised by Lovari and Scala (1980, 1984), Scala and Lovari (1984), and Nascetti
et al. (1985).

*Rupicapra pyrenaica* Bonaparte, 1845. Cat. Meth. Mamm. Europe, p. 17.
COMMON NAME: Pyrenean Chamois.
TYPE LOCALITY: Spain, "Mont. Pyren." (Pyrenees).
DISTRIBUTION: Appenine Mtns (Italy), Cantabrian Mtns (Spain), and Pyrenees (France and
N Spain).
STATUS: CITES – Appendix I as *R. pyrenaica ornata*; U.S. ESA – Endangered as *R. rupicapra
ornata*; IUCN – Endangered as *R. p. ornata*, Lower Risk (lc) as *R. pyrenaica*.
SYNONYMS: *parva* Cabrera, 1911; **ornata** Neumann, 1899.
COMMENTS: Revised by Lovari and Scala (1980) and Scala and Lovari (1984). Regarded as a
species distinct from *R. rupicapra* by Lovari (1985, 1987).

*Rupicapra rupicapra* (Linnaeus, 1758). Syst. Nat., 10th ed., 1:68.
COMMON NAME: Alpine Chamois.
TYPE LOCALITY: "Habitat in alpibus Helveticis summis inaccessis" (Switzerland).
DISTRIBUTION: Albania, Alps (of Austria, France, Germany, Italy, and Switzerland), Bosnia
and Herzegovina, Bulgaria, Caucasus Mtns (Azerbaijan, Georgia, Russia), Croatia,
Greece, Macedonia, Romania (Carpathians), Slovakia (Tatra Mtns), Slovenia, Turkey,
and Serbia and Montenegro; introduced to New Zealand.

STATUS: IUCN – Critically Endangered as *R. r. cartusiana* and *R. r. tatrica*, Vulnerable as *R. r. caucasica*, Data Deficient as *R. r. asiatica*, Lower Risk (lc) as *R. r. balcanica*, *R. r. carpatica*, *R. r. rupicapra*.

SYNONYMS: *alpina* (Sundevall, 1847); *capella* Bonaparte, 1845; *cartusiana* Coutourier, 1938; *dorcas* Schulze, 1807; *europea* Cornallia, 1871; *faesula* Miller, 1912; *hamulicornis* Burnett, 1830; *sylvatica* (Sundevall, 1847); *tragus* Gray, 1843; **asiatica** Lydekker, 1908; *caucasica* Lydekker, 1910; **balcanica** Bolkay, 1925; *olympica* Koller, 1929; **carpatica** Coutourier, 1938; **tatrica** Blahout, 1971.

COMMENTS: Revised by Hrabe and Koubek (1985), Lovari and Scala (1980, 1984), and Pemberton et al (1989).

**Subfamily Cephalophinae** Gray, 1871. Proc. Zool. Soc. Lond., 1871:588.

SYNONYMS: Sylvicaprina Sundevall, 1846.

COMMENTS: Reviewed extensively by Wilson (2002), with systematics revised by Grubb and Groves (2002). Jansen van Vuuren and Robinson (2001) studied phylogeny, using evidence from mtDNA. Sylvicaprina as Sylvicaprinae has only been used since 1899 as junior to Cephalophinae so should not replace it (Article 35.5, International Code of Zoological Nomenclature; International Commission on Zoological Nomenclature, 1999).

*Cephalophus* C. H. Smith, 1827. *In* Griffith et al., Animal Kingdom, 5:344.

TYPE SPECIES: *Antilope silvicultrix* Afzelius, 1815.

SYNONYMS: *Cephalolophus* Wagner, 1843; *Cephalophella* Knottnerus-Meyer, 1907; *Cephalophia* Knottnerus-Meyer, 1907; *Cephalophidium* Knottnerus-Meyer, 1907; *Cephalophops* Knottnerus-Meyer, 1907; *Cephalophorus* Gray, 1842; *Cephalophula* Knottnerus-Meyer, 1907; *Potamotragus* Gray, 1871; *Terpone* Gray, 1871.

COMMENTS: Van Gelder (1977*b*) included *Sylvicapra* but recent authors have not followed this arrangement (Meester et al. 1986; Swanepoel et al. 1980:188). Jansen van Vuuren and Robinson (2001) together with Grubb and Groves (2002) categorized species groups as follows: *C. ogilbyi* or *Cephalophorus* group (includes three subgroups: [i] *adersi*, [ii] *leucogaster, natalensis, nigrifrons, rufilatus*, [iii] *brookei, callipygus, niger, ogilbyi, weynsi*); *C. silvicultor* or nominate *Cephalophus* group (includes also *badius, jentinki*, and *spadix*); and *C. zebra* or *Cephalophula* group (monotypic).

*Cephalophus adersi* Thomas, 1918. Ann. Mag. Nat. Hist., ser. 9, 2:151.

COMMON NAME: Aders' Duiker.

TYPE LOCALITY: Tanzania, "Zanzibar".

DISTRIBUTION: Tanzania (Zanzibar) and Kenya (Sokoke Forest).

STATUS: IUCN – Endangered.

COMMENTS: Thought possibly to be conspecific with *natalensis* and/or *callipygus*; see Ansell (1972:33) but these presumed affinities were not supported by Jansen van Vuuren and Robinson (2001).

*Cephalophus brookei* Thomas, 1903. Ann. Mag. Nat. Hist., ser. 7, 11:290.

COMMON NAME: Brooke's Duiker.

TYPE LOCALITY: "Fanti[land]", now identified as Ghana, inland from Cape Coast, probably in Denkara (Grubb et al., 1998).

DISTRIBUTION: Sierra Leone, Liberia, W Côte d'Ivoire, W Ghana.

COMMENTS: Formerly included in *C. ogilbyi*; restored to species status by Grubb et al. (1998) and Grubb and Groves (2002).

*Cephalophus callipygus* Peters, 1876. Monatsb. K. Preuss. Akad. Wiss. Berlin, 1876:483.

COMMON NAME: Peters' Duiker.

TYPE LOCALITY: "Africa occidentalis (Gabun)" (Gabon, Gabon River).

DISTRIBUTION: West of Congo and Ubangi Rivers in S Cameroon, S Central African Republic, Republic of Congo, and Gabon.

STATUS: IUCN – Lower Risk (nt).

COMMENTS: Thought possibly to be conspecific with *natalensis* and/or *adersi* (Ansell 1972:33) but *callipygus* is not closely related to *natalensis* (Groves and Grubb, 1974).

*Cephalophus dorsalis* Gray, 1846. Ann. Mag. Nat. Hist., [ser. 1], 18:165.
  COMMON NAME: Bay Duiker.
  TYPE LOCALITY: "Sierra Leone".
  DISTRIBUTION: Disjunct; in rainforest zone; W Africa in Côte d'Ivoire, Ghana, Guinea,
      Guinea Bissau, Liberia, Sierra Leone, and Togo; C Africa in NE Angola, Cameroon,
      Dem. Rep. Congo, S Central African Republic, Equatorial Guinea (Mbini), Gabon,
      SE Nigeria, Republic of Congo, Uganda (one record; extinct).
  STATUS: CITES – Appendix II; IUCN – Lower Risk (nt).
  SYNONYMS: *badius* Gray, 1852; *breviceps* Gray, 1852; *typicus* Thomas, 1892; **castaneus**
      Thomas, 1892; *kuha* Lorenz, 1923; *leucochilus* Jentink, 1901; *orientalis* Schwarz, 1914.
  COMMENTS: Treated as two evolutionary species, *C. dorsalis* and *C. castaneus* by Cotterill
      (2003*b*).

*Cephalophus jentinki* Thomas, 1892. Proc. Zool. Soc. Lond., 1892:417.
  COMMON NAME: Jentink's Duiker.
  TYPE LOCALITY: "Liberia"; since identified as Junk River opposite Schieffelinsville, Sharp-Hill
      (Kühn, 1965).
  DISTRIBUTION: W Côte d'Ivoire, Liberia, and Sierra Leone.
  STATUS: CITES – Appendix I; U.S. ESA – Endangered; IUCN – Vulnerable.
  COMMENTS: Reviewed by Kuhn (1968).

*Cephalophus leucogaster* Gray, 1873. Ann. Mag. Nat. Hist., ser. 4, 12:43.
  COMMON NAME: White-bellied Duiker.
  TYPE LOCALITY: "West Africa, Gaboon" (Gabon).
  DISTRIBUTION: Rainforest zone in Cameroon, S Central African Republic, Dem. Rep. Congo,
      Equatorial Guinea (Mbini), Gabon, Republic of Congo.
  STATUS: IUCN – Lower Risk (nt).
  SYNONYMS: **arrhenii** Lönnberg, 1917; *seke* Lorenz, 1924.
  COMMENTS: *Cephalophus castaneus arrhenii* Lönnberg, 1917 is based on a skull of
      *C. leucogaster* (the lectotype) and a skin of *C. badius* (Grubb and Groves, 2002).

*Cephalophus natalensis* A. Smith, 1834. S. Afr. Quart. J., 2:217.
  COMMON NAME: Red Duiker.
  TYPE LOCALITY: South Africa, KwaZulu-Natal, "Port Natal" (Durban).
  DISTRIBUTION: E Kenya, Malawi, Mozambique, S Somalia, South Africa (KwaZulu-Natal),
      Swaziland, E and S Tanzania, and E Zambia; sight records from E Ethiopia.
  STATUS: IUCN – Lower Risk (cd) as *C. natalensis* and *C. harveyi*.
  SYNONYMS: *amoenus* Wroughton, 1911; *bradshawi* Wroughton, 1911; *lebombo* Roberts, 1936;
      *robertsi* W. Rothschild, 1906; *vassei* Trouessart, 1906; **harveyi** (Thomas, 1893); *bottegoi*
      De Beaux, 1924; *keniae* Lönnberg, 1912.
  COMMENTS: Ansell (1972:34) included *harveyi* and *weynsi*, and placed the extended
      *natalensis* with *adersi* and *callipygus* in a superspecies; Grubb and Groves (2002) and
      Van Vuuren and Robinson (2001) regarded *natalensis* + *harveyi* as distinct from
      *callipygus* + *weynsi* (and *adersi*), and treated *harveyi* as a subspecies of *natalensis*, though
      it was regarded as a species separate from *C. natalensis* by East et al. (1999), Kingdon
      (1982:297) and Cotterill (2003*b*).

*Cephalophus niger* Gray, 1846. Ann. Mag. Nat. Hist., [ser. 1], 18:165.
  COMMON NAME: Black Duiker.
  TYPE LOCALITY: "Guinea" but apparently Ghana, Shama.
  DISTRIBUTION: Rainforest zone, in Côte d'Ivoire, Ghana, Guinea, Liberia, Nigeria (west of
      lower Niger River), Sierra Leone, and Togo. No record from Benin.
  STATUS: IUCN – Lower Risk (nt).
  SYNONYMS: *pluto* (Temminck, 1853).
  COMMENTS: The type of *niger* came from the Leiden Museum, one of a series from Chama
      (= Shama) and Dabocrom, Ghana, including the syntypes of *pluto*. Only the specimens
      from Dabocrom were retained in Leiden (Jentink, 1892) so presumably the type is
      from Shama.

*Cephalophus nigrifrons* Gray, 1871. Proc. Zool. Soc. Lond., 1871:598.
    COMMON NAME: Black-fronted Duiker.
    TYPE LOCALITY: "Gaboon" (Gabon).
    DISTRIBUTION: Rainforest zone in N Angola, Cameroon, Central African Republic, Dem. Rep.
        Congo, Equatorial Guinea (Mbini), Gabon, S Nigeria, and Republic of Congo; upland
        forest in Albertine Rift including Ruwenzori Mtns (Burundi, Dem. Rep. Congo,
        Rwanda, Uganda), Mt. Elgon (Uganda-Kenya border), Aberdare Range (Kenya), and
        Mt. Kenya (Kenya).
    STATUS: IUCN – Endangered as *C. n. rubidus*, otherwise Lower Risk (nt) as *C. nigrifrons*.
    SYNONYMS: *apanbanga* Lorenz-Liburnau, 1923; *aureus* Gray, 1873; *claudi* Thomas and
        Wroughton, 1907; *emini* Noack, 1904; *lusumbi* Lönnberg, 1919; *mixtus* Lönnberg,
        1917; **fosteri** St Leger, 1934; **hooki** St Leger, 1934; **hypoxanthus** Grubb and Groves,
        2002; **kivuensis** Lönnberg, 1919; **rubidus** Thomas, 1901.
    COMMENTS: Treated as at least four evolutionary species (*C. nigrifrons, C. fosteri, C. hooki,* and
        *C. rubidus*) by Cotterill (2003*b*). *Cephalophus rubidus* also regarded as a species separate
        from *C. nigrifrons* by Kingdon (1982:292) and Jansen van Vuuren and Robinson
        (2001), but status revised by Grubb and Groves (2002 and in prep.).

*Cephalophus ogilbyi* (Waterhouse, 1838). Proc. Zool. Soc. Lond., 1838:60.
    COMMON NAME: Ogilby's Duiker.
    TYPE LOCALITY: Equatorial Guinea, "Fernando Po" (= Bioko).
    DISTRIBUTION: SE Nigeria, S Cameroon, Equatorial Guinea (Bioko), Gabon.
    STATUS: CITES – Appendix II; IUCN – Lower Risk (nt) as *C. ogilbyi* and *C. o. crusalbum*.
    SYNONYMS: **crusalbum** Grubb, 1978.
    COMMENTS: Treated as two evolutionary species (*C. ogilbyi, C. crusalbum*) by Cotterill
        (2003*b*).

*Cephalophus rufilatus* Gray, 1846. Ann. Mag. Nat. Hist., [ser. 1], 18:166.
    COMMON NAME: Red-flanked Duiker.
    TYPE LOCALITY: "Sierra Leone, Village of Waterloo".
    DISTRIBUTION: Savanna zone in Benin, Burkina Faso, Cameroon, Central African Republic,
        S Chad, Côte d'Ivoire, NE Dem. Rep. Congo, Gambia, Guinea, Guinea Bissau, S Mali,
        S Niger, Nigeria, Senegal, Sierra Leone, SW Sudan, Togo, and NW Uganda.
    STATUS: IUCN – Lower Risk (cd).
    SYNONYMS: *cuvieri* Fitzinger, 1869; *rubidior* Thomas and Wroughton, 1907.
    COMMENTS: Known from a specimen collected by Adanson in Senegal as "la grimme"
        (Buffon, 1764:307, 329), and described as "*Antilope grimmia*" by Desmarest (1816*c*:191)
        from Coast of Guinea, based on misidentifications of this species as *Sylvicapra grimmia*
        (Linnaeus, 1758).

*Cephalophus silvicultor* (Afzelius, 1815). Nova Acta Reg. Soc. Sci. Upsala, 7:265, pl. 8, fig. l.
    COMMON NAME: Yellow-backed Duiker.
    TYPE LOCALITY: "Habitat in montibus Sierrae Leone & regionibus susuensium fluvios Pongas
        & Quia adjacentibus frequens"; since restricted to Sierra Leone, vicinity of Freetown
        (Grubb et al., 1998).
    DISTRIBUTION: Dense vegetation in N Angola, Benin, S Burkina Faso, Burundi, Cameroon,
        Central African Republic, S Chad, Côte d'Ivoire, Dem. Rep. Congo, Equatorial Guinea
        (Mbini), Gabon, Gambia (former occurrence doubtful), Ghana, Guinea, Guinea-
        Bisssau, W Kenya, Liberia, Nigeria, Republic of Congo, Rwanda (extinct?), S Senegal,
        Sierra Leone, SW Sudan, Togo, W Uganda, and Zambia.
    STATUS: CITES – Appendix II; IUCN – Lower Risk (nt).
    SYNONYMS: *punctulatus* Gray, 1850; *sclateri* Jentink, 1901; *sylvicultor* Thomas, 1892; **curticeps**
        Grubb and Groves, 2002; **longiceps** Gray, 1865; *ituriensis* Rothschild and Neuville,
        1907; *melanoprymnus* Gray, 1871; *thomasi* Jentink, 1901; **ruficrista** Bocage, 1869; *coxi*
        Jentink, 1906.
    COMMENTS: Reviewed by Lumpkin and Kranz (1984, Mammalian Species, 225, as
        *Cephalophus sylvicultor*, an incorrect subsequent spelling).

*Cephalophus spadix* True, 1890. Proc. U.S. Natl. Mus., 13:227.
COMMON NAME: Abbott's Duiker.
TYPE LOCALITY: "High altitudes on Mt. Kilima-njaro, frequenting the highest points" (Tanzania, Mt. Kilimanjaro; at 2400 m according to Grimshaw et al., 1995).
DISTRIBUTION: Highlands of NE and C Tanzania.
STATUS: IUCN – Vulnerable.
COMMENTS: Possibly a subspecies of *silvicultor* (Haltenorth, 1963:71).

*Cephalophus weynsi* Thomas, 1901. Ann. Mus. Congo Zool., 2(1):15.
COMMON NAME: Weyns's Duiker.
TYPE LOCALITY: "district des Stanley-Falls" (Dem. Rep. Congo, near Stanley Falls).
DISTRIBUTION: S Central African Republic, Dem. Rep. Congo, W Kenya, Rwanda, S Sudan, and Uganda. Possibly in Omo basin, Ethiopia.
STATUS: IUCN – Lower Risk (nt).
SYNONYMS: *centralis* Rothschild and Neuville, 1907; *leopoldi* Rothschild and Neuville, 1907; **johnstoni** Thomas, 1901; *barbertoni* Kershaw, 1923; *ignifer* Thomas, 1903; *rutshuricus* Lönnberg, 1917; **lestradei** Groves and Grubb, 1974.
COMMENTS: Formerly included in *callipygus*, but separate according to Groves and Grubb (1974).

*Cephalophus zebra* Gray, 1838. Ann. Nat. Hist., 1:27.
COMMON NAME: Zebra Duiker.
TYPE LOCALITY: "Sierra Leone".
DISTRIBUTION: W Côte d'Ivoire, Liberia, and W Sierra Leone.
STATUS: CITES – Appendix II; IUCN – Vulnerable.
SYNONYMS: *doria* (Ogilby, 1836) [suppressed]; *doriae* Thomas, 1898; *zebrata* (Robert, 1836) [suppressed].
COMMENTS: For synonyms see Ansell (1980) and the International Commission on Zoological Nomenclature (1985*a*). Reviewed by Kuhn (1966).

*Philantomba* Blyth, 1840. *In* Cuvier's Animal Kingdom, p. 140.
TYPE SPECIES: *Antilope philantomba* C. H. Smith, 1827 (= *Antilope maxwellii* C. H. Smith, 1827).
SYNONYMS: *Guevei* Gray, 1852.
COMMENTS: Restored to generic rank (Grubb and Groves, 2002; Grubb et al., 1998; supported by Jansen van Vuuren and Robinson, 2001). For parapatric distribution of species in Nigeria, see Wilson (2002).

*Philantomba maxwellii* (C. H. Smith, 1827). *In* Griffith et al., Animal Kingdom, 4:267.
COMMON NAME: Maxwell's Duiker.
TYPE LOCALITY: "Sierra Leone".
DISTRIBUTION: Forested habitats in Benin, S Burkina Faso, Côte d'Ivoire, Gambia, Ghana, Guinea, Guinea-Bissau, Liberia, Nigeria west of Cross River, Senegal, Sierra Leone, and Togo.
STATUS: IUCN – Lower Risk (nt) as *Cephalophus maxwellii*.
SYNONYMS: *frederici* (Laurillard, 1842); *liberiensis* (Hinton, 1920); *lowei* (Hinton, 1920); *philantomba* (C. H. Smith, 1827); *whitfieldi* (Gray, 1850); **danei** (Hinton, 1920).
COMMENTS: Included in *P. monticola* by Haltenorth and Diller (1977:43). Reviewed by Ralls (1973, Mammalian Species, 31).

*Philantomba monticola* (Thunberg, 1789). Resa uti Europa Africa, Asia . . . , 2:66.
COMMON NAME: Blue Duiker.
TYPE LOCALITY: South Africa, "Lange Kloof"; since identified as borders of Western and Eastern Cape, Uniondale and Humansdorp Dist., Langkloof, 33°48'S, 23° to 24° 30'E; see Grubb (1999:21).
DISTRIBUTION: Forested habitats in N Angola, Cameroon, Central African Republic, Dem. Rep. Congo, Equatorial Guinea (Bioko, Mbini), Gabon, W and E Kenya, Malawi, Mozambique, Nigeria east of Cross River, Republic of Congo, South Africa (East Cape Prov., KwaZulu-Natal), S Sudan, Tanzania including Pemba and Zanzibar Isls, Uganda,

Zambia, E Zimbabwe. Former or present occurrence in Swaziland uncertain. No record from Lesotho (Lynch, 1994).

STATUS: CITES – Appendix II as *Cephalophus monticola*; IUCN – Lower Risk (lc) as *Cepalophus monticola*.

SYNONYMS: *caerula* (C. H. Smith, 1827); *caffer* (Fitzinger, 1869); *minuta* (Forster, 1844); *perpusilla* (C. H. Smith, 1827); *pygmea* (Schinz, 1821); ***aequatorialis*** (Matschie, 1892); *aequinoctialis* (Lydekker, 1893); *bakeri* (Rothschild and Neuville, 1907); ***anchietae*** (Bocage, 1879); ***bicolor*** (Gray, 1863); *fuscicolor* (Blaine, 1922); *ruddi* (Blaine, 1922); ***congicus*** (Lönnberg, 1908); *schultzei* (Schwarz, 1914); ***defriesi*** (W. Rothschild, 1904); *ludlami* (Blaine, 1922); ***hecki*** (Matschie, 1897); *nyasae* (Thomas, 1902); ***lugens*** (Thomas, 1898); *schusteri* (Matschie, 1914); ***melanorheus*** (Gray, 1846); ***musculoides*** (Heller, 1913); ***simpsoni*** (Thomas, 1910); ***sundevalli*** (Fitzinger, 1869); *pembae* (Kershaw, 1924).

COMMENTS: May include *maxwellii*; see Haltenorth and Diller (1977:43). Can be partitioned between grey-legged *melanorheus* division (includes also *aequatorialis, congicus, lugens, musculoides,* and *sundevalli*) and red-legged nominate *monticola* division (includes also *anchietae, bicolor, defriesi, hecki,* and *simpsoni*). Cotterill (2003b) recognized *pembae* and *melanorheus* as evolutionary species.

*Sylvicapra* Ogilby, 1837. Proc. Zool. Soc. Lond., 1836:138 [1837].
TYPE SPECIES: *Antilope mergens* Desmarest, 1816 (= *Capra grimmia* Linnaeus, 1758).
SYNONYMS: *Cephalophora* Gray, 1842; *Grimmia* Laurillard, 1842.
COMMENTS: Included in *Cephalophus* by Haltenorth (1963:71) and Van Gelder (1977b:18); but see Ansell (1978:57), Swanepool et al. (1980:188), and Meester et al. (1986).

*Sylvicapra grimmia* (Linnaeus, 1758). Syst. Nat., 10th ed., 1:70.
COMMON NAME: Bush Duiker.
TYPE LOCALITY: "Habitat in Africa"; based on a specimen seen by Grimm in the fort at Cape Town (Thomas, 1911a:153) so now known to be South Africa, Western Cape Prov., Cape Town.
DISTRIBUTION: Non-forested habitats in Angola, Benin, Botswana, Burkina Faso, Burundi, N Cameroon, Central African Republic, S Chad, Côte d'Ivoire, S, E, and N Dem. Rep. Congo, Eritrea, Ethiopia, S Gabon, Gambia, Ghana, Guinea, Guinea-Bissau, Kenya, Malawi, S Mali, Mozambique, Namibia, S Niger, Nigeria, Rwanda, Republic of Congo, Senegal, N Sierra Leone, S Somalia, South Africa, Swaziland, Sudan, Tanzania, Togo, Uganda, Zambia, and Zimbabwe.
STATUS: IUCN – Lower Risk (lc).
SYNONYMS: *burchellii* (C. H. Smith, 1827); *cana* (Oken, 1816) [unavailable]; *mergens* (Desmarest, 1816); *nictitans* (Thunberg, 1811); *platous* (C. H. Smith, 1827); *ptoox* (C. H. Smith, 1827); ***altivallis*** Heller, 1912; ***caffra*** Fitzinger, 1869; *irrorata* (Gray, 1871); *noomei* Roberts, 1926; *transvaalensis* Roberts, 1926; ***campbelliae*** (Gray, 1843); *roosevelti* Heller, 1912; ***coronata*** (Gray, 1842); ***hindei*** (Wroughton, 1910); ***lobeliarum*** Lönnberg, 1919; ***madoqua*** (Rüppell, 1836); *abyssinica* (Thomas, 1892); ***nyansae*** Neumann, 1905; *lutea* (Dollman, 1914); ***orbicularis*** (Peters, 1852); *altifrons* (Peters, 1852); *deserti* Heller, 1913; *ocularis* (Peters, 1852); *shirensis* (Wroughton, 1910); *walkeri* (Thomas, 1906); ***pallidior*** Schwarz, 1914; *platyotis* (Lesson, 1836); ***splendidula*** (Gray, 1891); *flavescens* (Lorenz, 1894); *leucoprosopus* (Neumann, 1899); *uvirensis* Lönnberg, 1919; ***steinhardti*** Zukowsky, 1924; *bradfieldi* Roberts, 1926; *cunenensis* Zukowsky, 1924; *omurambae* Zukowsky, 1924; *ugabensis* Zukowsky, 1924; *vernayi* Hill, 1926.

**Subfamily Hippotraginae** Sundevall, 1845. Öfversigt. Kongl.-Vetensk. Akad. Förhand. for 1845, parts 2 and 3, p. 31.
SYNONYMS: Adacina Pilgrim, 1939; Oryginae Brooke, 1876.
COMMENTS: Until the availability of *Hippotragus* Sundevall, 1845 was restored, the available name was Hippotraginae Sundevall, 1845 *in* Retzius and Lovén, Archiv. Skand. Bietr. Naturgesch., Greifswald, 1:445. Cladistic relations of genera: (*Hippotragus*)(*Oryx, Addax*) (Hassanin and Douzery, 1999a; Vrba and Gatesy, 1994).

*Addax* Laurillard, 1841. *In* d'Orbigny, Dict. Univ. Hist. Nat. I, p. 619.
  TYPE SPECIES: *Antelope suturosa* Otto, 1825 (= *Cerophorus nasomaculatus* de Blainville, 1816).
  SYNONYMS: *Addax* Rafinesque, 1815 [*nomen nudum*].

*Addax nasomaculatus* (de Blainville, 1816). Bull. Sci. Soc. Philom. Paris, 1816:75.
  COMMON NAME: Addax.
  TYPE LOCALITY: No locality given. Here selected as the Tunisian Sahara.
  DISTRIBUTION: Extinct in Algeria, Egypt, Libya, Morocco, Tunisia, and probably Sudan.
      Vagrants still enter Algeria and Sudan. Survives in Chad, N Mali, Mauritania, and
      Niger.
  STATUS: Nearly extinct in wild (East, 1990). CITES – Appendix I; U.S. ESA – Proposed
      Endangered; IUCN – Critically Endangered.
  SYNONYMS: *addax* (Cretzschmar, 1826); *gibbosa* (Savi, 1828); *mytilopes* (C. H. Smith, 1827);
      *suturosa* (Otto, 1825).
  COMMENTS: The syntypes were observed by de Blainville in Bullock's Pantherion or Museum
      and the Museum of the Royal College of Surgeons, both in London, UK. C. H. Smith
      (1827) suggested the specimens came from Guinea or Western Africa; Lydekker
      (1914*b*:148) stated that the type locality was probably Senegambia. These authors
      provided no evidence to support their conclusions and from the discussion in Sclater
      and Thomas (1898), it seems more probable that British hunters or collectors obtained
      Addax from the Tunisian Sahara, to which the type locality is here restricted.

*Hippotragus* Sundevall, 1845. Öfversigt. Kongl.-Vetensk. Akad. Förhand. for 1845, parts 2 and 3, p. 31.
  TYPE SPECIES: *Antilope equina* É. Geoffroy Saint-Hilaire, 1803.
  SYNONYMS: *Aegocera* Berthold, 1827 [suppressed]; *Aegocoerus* Gervais, 1859; *Aigererus* Harris,
      1838 [*nomen oblitum*]; *Aigocerus* C. H. Smith, 1827 [suppressed]; *Egocerus* Desmarest, 1822
      [suppressed]; *Oegocerus* Lesson, 1842 [suppressed]; *Ozanna* Reichenbach, 1845
      [suppressed].
  COMMENTS: *Hippotragus* is usually quoted from Sundevall, 1846, K. Svenska Vet.-Akad. Handl.
      Stockholm, for 1844, p. 196. The earlier name, cited above, was declared to be
      unavailable, and while it remained unavailable, *Hippotragus* Sundevall, 1845 (*in* Retzius
      and Lovén, Archiv. Skand. Beitr. Naturgesch., Griefswald, 1:445) was the available name.
      Grubb (2001*c*) applied to the International Commission on Zoological Nomenclature to
      conserve *Hippotragus* Sundevall, 1845, and this action has now been taken (International
      Commission on Zoological Nomenclature, 2003*c*). Sister species in the genus are said to
      be *leucophaeus* and *niger* (Groves and Westwood, 1995), *equinus* and *niger* (Robinson et al.,
      1996), or *equinus* and *leucophaeus* (Vrba and Gatesy, 1994).

*Hippotragus equinus* (É. Geoffroy Saint-Hilaire, 1803). Cat. Mamm. Mus. Nation. Hist. Nat. p. 259.
  COMMON NAME: Roan Antelope.
  TYPE LOCALITY: "Inconnue"; now thought to be South Africa, Western Cape Prov.,
      Plettenberg Bay (Grubb, 1999).
  DISTRIBUTION: Savanna woodland in Angola, Benin, Botswana, Burkina Faso, Burundi
      (extinct), N Cameroon, Central African Republic, S Chad, Côte d'Ivoire, N and S Dem.
      Rep. Congo, N Eritrea (extinct?), W Ethiopia, Gambia (extinct), Ghana, Guinea,
      Guinea-Bissau, Kenya, Malawi, S Mali, S Mauritania, Mozambique, Namibia, S Niger,
      Nigeria, Rwanda, Senegal, E South Africa, Swaziland (extinct, reintroduced), Sudan,
      Tanzania, Togo, Uganda, Zambia, and Zimbabwe. No evidence of occurrence in Sierra
      Leone (Grubb et al., 1998).
  STATUS: IUCN – Lower Risk (cd).
  SYNONYMS: *aethiopica* (Schinz, 1821); *aurita* (C. H. Smith, 1827); *barbata* (C. H. Smith, 1827);
      *jubata* (Goldfuss, 1824); *truteri* (J. B. Fischer, 1829); *typicus* Sclater and Thomas, 1899;
      **bakeri** Heuglin, 1863; *dogetti* de Beaux, 1921; **cottoni** Dollman and Burlace, 1928;
      **koba** (Gray, 1872); *docoi* (Gray, 1872); *gambianus* P. L. Sclater and Thomas, 1899;
      **langheldi** Matschie, 1898; *rufopallidus* Neumann, 1899; *scharicus* (Schwarz, 1913).
  COMMENTS: For dating the name to É. Geoffroy Saint-Hilaire, 1803 see Grubb (2001*a*) and
      Opinion 2005 of the International Commission on Zoological Nomenclature (2002*b*).
      The type locality was selected as South Africa, Northern Cape, Lataku (= Kuruman) by

Harper (1940), but there is evidence to show that the holotype was collected much farther south where the species no longer occurs, at Plettenberg Bay (Grubb, 1999). Subspecific systematics follows Ansell (1972:46).

*Hippotragus leucophaeus* (Pallas, 1766). Misc. Zool., p. 4.
COMMON NAME: Blaaubok.
TYPE LOCALITY: "Promontoriae bonae Spei missas"; since restricted to South Africa, Western Cape Prov., Swellendam Dist.
DISTRIBUTION: South Africa (Western Cape); extirpated about 1799.
STATUS: IUCN – Extinct.
SYNONYMS: *capensis* (P.L.S. Müller, 1776); *glauca* (Oken, 1816) [unavailable].
COMMENTS: Reviewed by Groves and Westwood (1995), Klein (1974), Mohr (1967), and Rookmaaker (1992).

*Hippotragus niger* (Harris, 1838). Athenaeum, 535:71.
COMMON NAME: Sable Antelope.
TYPE LOCALITY: "The great mountain range in the county of Mataveld", and "On the northern side of the Cashan range of mountiains, about a degree and a half south of the tropic of Capricorn", since specified as South Africa, North West Prov., Krugersdorp and Rustenburg, Magaliesberg (Grubb, 1999).
DISTRIBUTION: Savanna woodland in Africa; giant sable (*variani*) in C Angola (between Cuanza and Loando Rs.); other subspecies in E Angola, N Botswana, S Dem. Rep. Congo, SE Kenya, Malawi, Mozambique, NE Namibia (Caprivi Strip), NE South Africa, Tanzania, Zambia, and Zimbabwe.
STATUS: CITES – Appendix I as *H. niger variani*; U.S. ESA – Endangered as *H. n. variani*; IUCN – Critically Endangered as *H. n. variani*, otherwise Lower Risk (cd).
SYNONYMS: *harrisi* (Harris, 1839); *kaufmanni* Matschie, 1912; *kirkii* (Gray, 1872); **anselli** Groves, 1983; **roosevelti** (Heller, 1910); **variani** Thomas, 1916.
COMMENTS: Includes *variani*; see Ansell (1972:47). Original publication usually assumed to be Proc. Zool. Soc. Lond., 1838:2 (publ. July, 1838), but McAllan and Bruce (1989) showed that an earlier publication is The Athenaeum (publ. 27 Jan., 1838). Subspecific synonymy follows Ansell (1972:47). In phylogeographic studies, Matthee and Robinson (1999) distinguishsed *niger, kirkii* and *variani* from "*roosevelti*", and Pitra et al. (2002) recognised clade I ("*roosevelti*" of Matthee and Robinson, in W Tanzania and merged with clade II), "pure" clade II (*niger* including *kirkii*) and clade III (*roosevelti*). Cotterill (2003a) treated *anselli* (mtDNA not studied) as specifically distinct from *niger*.

*Oryx* de Blainville, 1816. Bull. Sci. Soc. Philom. Paris, 1816:75.
TYPE SPECIES: *Antilope oryx* Pallas, 1777 (= *Capra gazella* Linnaeus, 1758).
SYNONYMS: *Aegoryx* Pocock, 1918; *Onyx* Gray, 1821.

*Oryx beisa* (Rüppell, 1835). Neue Wirbelthiere z. d. Fauna Abyssinien gehörig, Säugeth., p. 14, pl. 5.
COMMON NAME: Beisa.
TYPE LOCALITY: Eritrea, "in den Niederungen der Küstenlandschaft bei Massaua" (Red Sea coast west of Massawa).
DISTRIBUTION: Djibouti, Eritrea, Ethiopia, Kenya, Somalia, NE and SE Sudan, NE Uganda (extinct?), and NE Tanzania.
STATUS: IUCN – Lower Risk (cd) as *O. gazella beisa*.
SYNONYMS: *annectens* Hollister, 1910; *gallarum* Neumann, 1902; *subcallotis* W. Rothschild, 1921; **callotis** Thomas, 1892.
COMMENTS: Included in *gazella* by Ansell (1972:49), whose subspecific synonymy is otherwise followed here. Restored to species status by East et al. (1999), Grubb (2000b), and Kingdon (1997).

*Oryx dammah* (Cretzschmar, 1827). *In* Rüppell, Atlas Reise Nördl. Afr., Zool. Säugeth., 1:22.
COMMON NAME: Scimitar-horned Oryx.
TYPE LOCALITY: Sudan, Northern Kordofan Prov., "bewohnen die grossen Steppen von Haraza [vicinity of Jebel Haraza]".
DISTRIBUTION: Extinct in Algeria, N Burkina Faso, Chad, Egypt, Libya, Mali, Mauritania,

Morocco, Niger, N Nigeria, N Senegal, Sudan, and Tunisia. Probably last occurred in the wild in the 1980s in Chad (East et al., 1999). Survives as captive populations.

STATUS: CITES – Appendix I; U.S. ESA – Proposed Endangered; IUCN – Extinct in the Wild.

SYNONYMS: *algazel* (Oken, 1816) [unavailable]; *bezoastica* C. H. Smith, 1827; *ensicornis* (Wagner, 1844); *nubica* (Wagner, 1855); *senegalensis* (Wagner, 1855); *tao* (C. H. Smith, 1827).

COMMENTS: Year of publication is 1827, not 1826, according to J. E. Hill (ms notes based on Anon., 1829:1291-1292). Includes *tao* (Ansell, 1972:48). The name *algazel* Oken, 1816, was declared invalid by Opinion 417 of the International Commission on Zoological Nomenclature (1956*b*).

*Oryx gazella* (Linnaeus, 1758). Syst. Nat., 10th ed., 1:69.
COMMON NAME: Gemsbok.
TYPE LOCALITY: "India"; understood to be South Africa (Thomas, 1911*a*:152).
DISTRIBUTION: SW Angola (extinct?), Botswana, Namibia, N South Africa, and W Zimbabwe.
STATUS: IUCN – Lower Risk (cd).
SYNONYMS: *aschenborni* Strand, 1924; *bezoartica* (Pallas, 1766); *blainei* W. Rothschild, 1921; *capensis* Ogilby, 1837; *onyx* Gray, 1821; *oryx* (Pallas, 1777); *pasan* (Daudin, 1802); *recticornis* (Erxleben, 1777).

*Oryx leucoryx* (Pallas, 1777). Spicil. Zool., 12:17.
COMMON NAME: Arabian Oryx.
TYPE LOCALITY: "Arabiae et forte Lybiae proprium animal"; restricted to Arabia (Lydekker, 1914*b*:130).
DISTRIBUTION: Extinct in Egypt (Sinai Peninsula), Iraq, Israel, Jordan, Kuwait, Oman, Saudi Arabia, Syria, United Arab Emirates, and Yemen. Maintained in captivity; Middle Eastern breeding stock in Bahrain, Israel, Leban, Oman, Qatar, Saudi Arabia, Syria, and United Arab Emirates. Recently reintroduced to the wild in Oman and Saudi Arabia.
STATUS: CITES – Appendix I; U.S. ESA and IUCN – Endangered.
SYNONYMS: *asiatica* (Wagner, 1855); *beatrix* Gray, 1857; *latipes* Pocock, 1934; *leucorix* Link, 1795; *oryx* (Oken, 1816) [unavailable]; *pallasii* Fitzinger, 1869.
COMMENTS: Included in *O. gazella* by Haltenorth (1963:88). Status reviewed in Mallon and Kingswood (2001).

**Subfamily Reduncinae** Knottnerus-Meyer, 1907. Arch. Naturgesch. 73:39.
SYNONYMS: Adenotinae Blyth, 1863 [*nomen oblitum*]; Cervicaprinae Brooke, 1876 [invalid]; Eleotragidae Gray, 1872 [*nomen oblitum*]; Kobinae Roosevelt and Heller, 1914; Peleini Gray, 1872.
COMMENTS: Recent work has suggested that *Pelea* should be included in the Reduncinae (Gatesy et al., 1997; Hassanin and Douzery, 1999*a*; Vrba and Schaller, 2000) although *Pelea* is strongly differentiated from *Kobus* and *Redunca* in morphology. Peleinae has priority over Reduncinae but its seniority is not recognised here: inclusion of *Pelea* in Reduncinae is provisional, Reduncinae is a much more familiar and widely used name, and a proposal will be made to the International Commission on Zoological Nomenclature for the conservation of Reduncinae. Reduncini includes *Kobus* and *Redunca*; Peleini includes *Pelea*.

*Kobus* A. Smith, 1840. Illustr. Zool. S. Afr. Mamm., Part 12, pl. 28 plus text.
TYPE SPECIES: *Antilope ellipsiprymnus* Ogilby, 1833.
SYNONYMS: *Adenota* Gray, 1847; *Cobus* Buckley, 1876; *Hydrotragus* Fitzinger, 1866; *Kolus* Gray, 1843; *Onotragus* Gray, 1872; *Onototragus* Heller, 1913; *Pseudokobus* Fitzinger, 1869; *Robus* Zittel, 1893.
COMMENTS: Includes *Adenota* and *Onotragus*; see Ansell (1972:40). Species groups are *K. kob* or *Adenota* group (includes also *K. vardonii*, forming a superspecies), *K. leche* or *Hydrotragus* group (includes also *K. megaceros*), and *K. ellipsiprymnus* or nominate *Kobus* group (Ansell, 1972). The most parsimonious tree based on morphological characters is (*K. vardonii*) ((*K. kob*) ((*K. leche, K. megaceros*) (*K. ellipsprymnus*))) with *kob* group paraphyletic (Vrba et al., 1994). Phylogeny inferred from mtDNA is (*K. leche, K. megaceros*) ((*K. kob*,

*K. vardonii*) (*K. ellipsprymnus*)) (Birungi and Arctander, 2001), with species groups monophyletic.

*Kobus ellipsiprymnus* (Ogilby, 1833). Proc. Zool. Soc. Lond., 1833:47.
    COMMON NAME: Waterbuck.
    TYPE LOCALITY: "about twenty-five days' journey north of the Orange River between Latakoo and the western coast of Africa". Since restricted to Botswana, Gaborone, the top reaches of the Notwani River; see Smithers (1971:233) and Grubb (1999:25-26).
    DISTRIBUTION: Mesic non-forested habitats in Angola, Benin, N and EC Botswana, Burkina Faso, Burundi, N Cameroon, Central African Republic, S Chad, Côte d'Ivoire, N and S Dem. Rep. Congo, N Eritrea, Ethiopia, S Gabon, Gambia (extinct), Ghana, Guinea, Guinea-Bissau, Kenya, Malawi, S Mali, S Mauritania, Mozambique, NE Namibia (Caprivi Strip), S Niger, Nigeria, S Republic of Congo, Rwanda, Senegal, E South Africa, Swaziland, Sudan, Tanzania, Togo, Uganda, Zambia, and Zimbabwe.
    STATUS: IUCN – Lower Risk (cd) as *K. ellipsiprymnus* and subspecifically as *K. e. defassa* and *K. e. ellipsiprymnus*.
    SYNONYMS: *ellipsiprymnos* (Sundevall, 1846); **adolfifriderici** Matschie, 1910; *fulvifrons* Matschie, 1910; *nzoiae* Matschie, 1910; *raineyi* Heller, 1913; **annectens** Schwarz, 1913; *schubotzi* Schwarz, 1913; **crawshayi** (P.L. Sclater, 1894); *frommi* Matschie, 1911; *muenzneri* Matschie, 1911; *uwendensis* Matschie, 1911; **defassa** (Rüppell, 1835); *abyssinica* (Wagner, 1855); *hawashensis* Matschie, 1910; *matschiei* Neumann, 1905; *singsing* Gray, 1843; **harnieri** (Murie, 1867); *albertensis* Matschie, 1910; *avellanifrons* Matschie, 1910; *breviceps* Matschie, 1910; *cottoni* Matschie, 1910; *dianae* Matschie, 1910; *griseotinctus* Matschie, 1910; *ladoensis* Matschie, 1910; *ugandae* Neumann, 1905; **kondensis** Matschie, 1911; *kulu* Matschie, 1911; *lipuwa* Matschie, 1911; **pallidus** Matschie, 1910; **penricei** (W. Rothschild, 1895); **thikae** Matschie, 1910, *canescens* Lönnberg, 1912; *kuru* Heller, 1913; **tjaederi** (Lönnberg, 1907); *angusticeps* Matschie, 1910; *powelli* Matschie, 1910; **tschadensis** Schwarz, 1913; **unctuosus** (Laurillard, 1842); *senegalensis* (Wagner, 1855); *togoensis* Schwarz, 1914.
    COMMENTS: Reviewed by Ansell (1972:42); *ellipsiprymnus* division (includes also *kondensis*, *pallidus* and *thikae*) and *defassa* division (includes all other nominal subspecies) formerly regarded as species; they hybridize along a zone of contact (G. Peters, 1986) and differ in their centric fusion polymorphisms and structure of the Y chromosome (Kingswood et al, 1998b). Cotterill (2003a) treated *crawshayi* as a species.

*Kobus kob* (Erxleben, 1777). Syst. Regni Anim., 1:293.
    COMMON NAME: Kob.
    TYPE LOCALITY: "Senegal".
    DISTRIBUTION: Benin, Burkina Faso, Cameroon, Central African Republic, S Chad, Côte d'Ivoire, N Dem. Rep. Congo, Ethiopia (Gambela Salient only), Gambia (extinct), Ghana, Guinea, Guinea Bissau, W Kenya (extinct), S Mali, S Mauritania, S Niger, Nigeria, Senegal, Sierra Leone (extinct), Sudan, NW Tanzania (extinct), Togo, and Uganda.
    STATUS: IUCN – Lower Risk (cd) as *K. kob* and subspecifically as *K. k. kob* and *K. k. thomasi*, Lower Risk (nt) as *K. k. leucotis*.
    SYNONYMS: *adansoni* A. Smith, 1840; *adenota* (C. H. Smith, 1827); *annulipes* (Gray, 1842); *buffonii* (Fitzinger, 1869); *forfex* (C. H. Smith, 1827); *fraseri* (Fitzinger, 1869); *loderi* (Lydekker, 1900); *nigricans* (Lydekker, 1899); *typicus* (Ward, 1910); **adolfi** Lydekker and Blaine, 1914; *adolfifriderici* (Schwarz, 1913) [preoccupied]; **bahrkeetae** (Schwarz, 1913); **leucotis** (Lichtenstein and Peters, 1853); *kul* (Heuglin, 1863); *nigroscapulatus* (Matschie, 1899); *notatus* (W.Rothschild, 1913); *vaughani* (Lydekker, 1906); *wuil* (Heuglin, 1863); **pousarguesi** (Neumann, 1905); **riparia** (Schwarz, 1914); **thomasi** (P.L. Sclater, 1896); *alurae* Heller, 1913; *neumanni* (W. Rothschild, 1913); **ubangiensis** (Schwarz, 1913).

*Kobus leche* Gray, 1850. Gleanings, Knowsley Menagerie, 2:23.
    COMMON NAME: Lechwe.
    TYPE LOCALITY: "Banks of the river Zoaga, lat. 21°"; since identified as Botswana, Botletle River, near Lake Ngami (Smithers, 1971:233).

DISTRIBUTION: SE Angola, N Botswana, SE Dem. Rep. Congo, NE Namibia (Caprivi Strip), and Zambia.

STATUS: CITES – Appendix II; U.S. ESA – Threatened; IUCN – Extinct as *K. l. robertsi*, Vulnerable as *K. l. kafuensis* and *K. l. smithemani*, Lower Risk (cd) as *K. l. leche*.

SYNONYMS: *amboellensis* (Sokolowsky, 1903); *lechee* (Gray, 1852); *leechi* (Buckley, 1876); *lechwe* (W. Rothschild, 1907); *notatus* (Matschie, 1912); **kafuensis** Haltenorth, 1963; *grandicornis* Ansell, 1964; **robertsi** (W. Rothschild, 1907); **smithemani** (Lydekker, 1900).

COMMENTS: Revised by Ansell and Banfield (1979). Consists of four evolutionary species according to Cotterill (2003a), namely *K. leche, K. kafuensis, K. robertsi,* and *K. smithemani.*

*Kobus megaceros* (Fitzinger, 1855). Sitzb. K. Akad. Wiss. Wien, 17:247.

COMMON NAME: Nile Lechwe.

TYPE LOCALITY: Sudan, Upper Nile Prov., "Bahr el abiad" or according to the collector, von Heuglin (Nov. Act. Acad. Caes. Leop. Carol. Germ. Nat. Curios., Abhand. 30, part 2, p. 14 [1863]), "am Sobat, Bhar ghasál und untern Kir" (Bahr el Abiad or White Nile, Bahr el Ghazal, Kir River or lower Bahr el Jebel, and Sobat River); restricted to "Mouth of the Bahr el Ghazal at its junction with the White Nile" (Roosevelt and Heller, 1914:519).

DISTRIBUTION: S Sudan, W Ethiopia.

STATUS: IUCN – Lower Risk (nt).

SYNONYMS: *maria* Gray, 1859.

*Kobus vardonii* (Livingstone, 1857). Missionary Travels and Researches in South Africa, p. 256.

COMMON NAME: Puku.

TYPE LOCALITY: "Thirty or forty miles above Libonta"; since identified as Zambia, Barotseland, Chobe Valley, near Libonda at 40°0'S, 23°15'E.

DISTRIBUTION: NE Angola, N Botswana, S Dem. Rep. Congo, Malawi, NE Namibia (Caprivi Strip), S Tanzania, Zambia, and N Zimbabwe (vagrant).

STATUS: IUCN – Lower Risk (cd).

SYNONYMS: *typicus* (Selous, 1899); **senganus** (P.L. Sclater and Thomas, 1897).

COMMENTS: Included in *kob* by Haltenorth (1963:92) but Ansell (1972:40), whose subspecies are followed here, treated *kob* and *vardoni* as separate species in a superspecies. Cotterill (2003a) treated *senganus* as an evolutionary species.

*Pelea* Gray, 1851. Proc. Zool. Soc. Lond., 1850:126 [1851].

TYPE SPECIES: *Antilopa* (sic) *capreolus* Forster, 1790.

SYNONYMS: *Antilopa* Forster, 1790 [*nomen oblitum*].

*Pelea capreolus* (Forster, 1790). *In* Levaillant, Erste Reise Afrika, p. 71.

COMMON NAME: Vaal Rhebok.

TYPE LOCALITY: "Ouwe-hoeck", now specified as South Africa, Western Cape, Caledon, Houhoek Pass; see Skead (1973:79).

DISTRIBUTION: Lesotho, South Africa, and Swaziland.

STATUS: IUCN – Least Concern.

SYNONYMS: *lanata* (Desmoulins, 1822); *villosa* (Burchell, 1823).

*Redunca* C. H. Smith, 1827. *In* Griffith et al., Animal Kingdom, 5:337.

TYPE SPECIES: *Antilope redunca* Pallas, 1767.

SYNONYMS: *Cervicapra* de Blainville, 1816; *Eleotragus* Gray, 1843; *Heleotragus* Kirk, 1865; *Nagor* Laurillard, 1842; *Oreodorcas* Heller, 1912.

COMMENTS: Phylogeny inferred from morphology (Vrba et al., 1994) or mtDNA (Birungi and Arctander, 2001) is (*R. fulvorufula*) (*R. redunca, R. arundinum*).

*Redunca arundinum* (Boddaert, 1785). Elench. Anim., 1:141.

COMMON NAME: Southern Reedbuck.

TYPE LOCALITY: "Habitat ad Cap. Bn. Sp." (Cape of Good Hope); since selected as South

Africa, Free State, Bethulie, based on known collecting localities of syntypes (Grubb, 1999:25).

DISTRIBUTION: Angola, N and E Botswana, S Gabon, S Dem. Rep. Congo, Lesotho (vagrant), Malawi, Mozambique, N Namibia, S Republic of Congo, E South Africa, Swaziland, Tanzania, Zambia, and Zimbabwe.

STATUS: IUCN – Lower Risk (cd).

SYNONYMS: *algoensis* Fitzinger, 1869; *arundinacea* (Bechstein, 1799); *caffra* Fitzinger, 1869; *cinerea* (Bechstein, 1800); *coerulescens* (Link, 1795) [*nomen nudum*]; *eleotragus* (von Schreber, 1787); *isabellina* (Afzelius, 1815); *multiannulata* Fitzinger, 1869; *occidentalis* (W. Rothschild, 1907); *oleotragus* (Desmoulins, 1822); *oreotragus* (Bechstein, 1799); *penricei* (Lydekker, 1910); *thomasinae* (P.L. Sclater, 1900).

COMMENTS: Roberts' (1951:292) restriction of the type locality to Bathurst Dist. was not based on a syntype locality; see Grubb (1999:25).

*Redunca fulvorufula* (Afzelius, 1815). Nova Acta Reg. Soc. Sci. Upsala, 7:250.

COMMON NAME: Mountain Reedbuck.

TYPE LOCALITY: No type locality originally identified; restricted to eastern Cape Colony (Lydekker, 1914a:221) (South Africa, Eastern Cape).

DISTRIBUTION: W Africa in E Nigeria and W Cameroon; E Africa in C Ethiopia, Kenya, SE Sudan, N Tanzania, and NE Uganda; S Africa in SE Botswana, Lesotho, S Mozambique, E South Africa, and Swaziland.

STATUS: IUCN – Endangered as *R. f. adamauae*, Lower Risk (cd) as *R. f. fulvorufula*, Lower Risk (nt) as *R. f. chanleri*, otherwise Least Concern.

SYNONYMS: *lalandia* (Desmoulins, 1822); *landiana* (Desmarest, 1822); *subalpina* (Kirby, 1898); **adamauae** Pfeffer, 1962; **chanleri** (W. Rothschild, 1895); *schoana* (Neumann, 1902).

COMMENTS: Subspecies from Ansell (1972:40).

*Redunca redunca* (Pallas, 1767). Spicil. Zool., 1:8.

COMMON NAME: Common Reedbuck.

TYPE LOCALITY: No locality cited; based on "Le Nagor" of Buffon from Senegal, "dans les terres voisines de l'île de Gorée" (mainland opposite Gori Isl).

DISTRIBUTION: Benin, Burkina Faso, Burundi, Cameroon, Central African Republic, S Chad, Côte d'Ivoire, N Dem. Rep. Congo, N Eritrea, Ethiopia, Gambia, Ghana, Guinea, Guinea Bissau, Kenya, S Mali, S Mauritania, S Niger, Nigeria, Rwanda, Senegal, Sierra Leone, Sudan, Tanzania, Togo, and Uganda.

STATUS: IUCN – Lower Risk (cd).

SYNONYMS: *nagor* Rüppell, 1842; *reversa* (Pallas, 1776); *ridunca* Shaw, 1801; *rufa* (Afzelius, 1815); *typica* (R. Ward, 1907); **bohor** Rüppell, 1842; *odrob* (Heuglin, 1877); **cottoni** (W. Rothschild, 1902); *donaldsoni* (W. Rothschild, 1902); **dianae** Schwarz, 1929; **nigeriensis** (Blaine, 1913); **tohi** Heller, 1913; **wardi** (Thomas, 1900); *bayoni* De Beaux, 1921; *ugandae* (Blaine, 1913).

COMMENTS: Subspecies from Ansell (1972:39).

# ORDER CETACEA
by James G. Mead and Robert L. Brownell, Jr.

**ORDER CETACEA** Brisson, 1762.
>    COMMENTS: The definition of oceanic water masses follows Briggs (1974). Includes as suborders
>    Mysticeti (Balaenidae, Balaenopteridae, Eschrichtiidae, and Neobalaenidae) and
>    Odontoceti (Delphinidae, Monodontidae, Phocoenidae, Physeteridae, Platanistidae,
>    Iniidae and Ziphiidae).

**SUBORDER MYSTICETI** Flower, 1864.

**Family Balaenidae** Gray, 1821. Lond. Med. Repos., 15:310.
>    COMMENTS: Commonly included *Caperea*, which is here put in a separate family,
>    Neobalaenidae, following Barnes and McLeod (1984).

*Balaena* Linnaeus, 1758. Syst. Nat., 10th ed., 1:75.
>    TYPE SPECIES: *Balaena mysticetus* Linnaeus, 1758.
>    SYNONYMS: *Leiobalaena* Eschricht, 1849.

>    *Balaena mysticetus* Linnaeus, 1758. Syst. Nat., 10th ed., 1:75.
>    >    COMMON NAME: Bowhead.
>    >    TYPE LOCALITY: "Habitat in Oceano Groenlandico" (= Greenland Sea).
>    >    DISTRIBUTION: Northern hemisphere: arctic waters. Strays have occured in Japan, Gulf of
>    >    St. Lawrence, and Massachusetts.
>    >    STATUS: CITES – Appendix I; U.S. ESA – Endangered; IUCN – Critically Endangered
>    >    (Spitzbergen population), Endangered (Okhotsk Sea subpopulation and Baffin Bay-
>    >    Davis Strait stock), Vulnerable (Hudson Bay-Foxe Basin stock), Lower Risk (cd) (Bering-
>    >    Chukchi-Beaufort Sea stock), otherwise listed as Lower Risk (cd) for *B. mysticetus*.
>    >    COMMENTS: Reviewed by Reeves and Leatherwood (1985).

*Eubalaena* Gray, 1864. Proc. Zool. Soc. Lond., 1864(2):201.
>    TYPE SPECIES: *Balaena australis* Desmoulins, 1822.
>    SYNONYMS: *Halibalaena* Gray, 1873; *Hunterius* Gray, 1866.
>    COMMENTS: Corbet and Hill (1980) used this genus.

>    *Eubalaena australis* (Desmoulins, 1822). *In* Bory de Saint-Vincent (ed.), Dict. Class. Hist. Nat. Paris,
>    2:161, pl.
>    >    COMMON NAME: Southern Right Whale.
>    >    TYPE LOCALITY: Algoa Bay, Cape of Good Hope, South Africa.
>    >    DISTRIBUTION: Southern hemisphere: Antarctic to temperate waters; occasionally along the
>    >    northern part of the Antarctic Peninsula.
>    >    STATUS: CITES – Appendix I; U.S. ESA – Endangered (included with *E. glacialis*); IUCN –
>    >    Lower Risk (cd).
>    >    SYNONYMS: *antarctica* (Lesson, 1828); *antipodarum* (Gray, 1843); *temminckii* (Gray, 1864).
>    >    COMMENTS: Reviewed by Cummings (1985*b*). Included in *glacialis* by some recent authors.

>    *Eubalaena glacialis* (Müller, 1776). Zool. Danicae Prodr., p. 7.
>    >    COMMON NAME: North Atlantic Right Whale.
>    >    TYPE LOCALITY: None given, listed as Norway, Finnmark, Nord Kapp (vicinity of
>    >    North Cape) by Eschricht and Reinhardt (1861).
>    >    DISTRIBUTION: North Atlantic: temperate to tropical waters.
>    >    STATUS: CITES – Appendix I; U.S. ESA and IUCN – Endangered.
>    >    SYNONYMS: *biscayensis* (Eschricht, 1860); *nordcaper* (Lacépède, 1804).
>    >    COMMENTS: Reviewed by Cummings (1985*b*); see Hershkovitz (1961*b*).

>    *Eubalaena japonica* (Lacépède, 1818). Mem. Mus. Hist. Nat., Paris, 4:469.
>    >    COMMON NAME: North Pacific Right Whale.

TYPE LOCALITY: Japan.

DISTRIBUTION: North Pacific: temperate to to tropical waters; one stray record from Hawaii (Scarff, 1986).

STATUS: CITES – Appendix I; U.S. ESA – Endangered (as included in *E. glacialis*); IUCN – Endangered.

SYNONYMS: *sieboldii* (Gray, 1864).

COMMENTS: Reinstated by Rosenbaum et al. (2000), Brownell et al. (2001); see Hershkovitz (1961*b*).

**Family Balaenopteridae** Gray, 1864. Proc. Zool. Soc. Lond., 1864:203.

*Balaenoptera* Lacépède, 1804. Hist. Nat. Cetacees, p. 114.
   TYPE SPECIES: *Balaenoptera gibbar* Lacépède, 1804 (= *Balaena physalus* Linnaeus, 1758).
   SYNONYMS: *Catoptera* Rafinesque, 1815; *Cuvierius* Gray, 1866; *Physalus* Gray, 1821; *Pterobalaena* Eschricht, 1849; *Rorqualus* Cuvier, 1836; *Sibbaldius* Gray, 1864.

*Balaenoptera acutorostrata* Lacépède, 1804. Hist. Nat. Cetacees, p. 134.
   COMMON NAME: Common Minke Whale.
   TYPE LOCALITY: France, "pris aux environs de la rade de Cherbourg", Mancha.
   DISTRIBUTION: Worldwide: arctic to tropical waters.
   STATUS: CITES – Appendix I (except the population of West Greenland, which is Appendix II); IUCN – Lower Risk (nt).
   SYNONYMS: *davidsoni* Cope, 1872; *minima* (Rapp, 1837); *rostrata* (Fabricius, 1780).
   COMMENTS: Reviewed by Stewart and Leatherwood (1985). Two forms have been described from SW Pacific waters (Arnold et al., 1987). May represent two or three species (Wada and Numachi, 1991). Deméré (1986) reassigned *Eschrichtius davidsonii* (Cope 1872) to *Balaenoptera,* making *B. davidson*i Scammon 1872 a junior synonym and erecting *Balaenoptera acutorostrata scammoni* Deméré, 1986 as a substitute. Rice (1998:70) recognized *B. a. scammoni* (Scammon's Minke Whale) and another un-named subspecies in the southern hemisphere (*B. a.* subsp.; Dwarf Minke Whale). We recognize that there are probably others, such as the population in the Sea of Japan.

*Balaenoptera bonaerensis* Burmeister, 1867. Actas Soc. Paleo., Buenos Aires, p. XXIV.
   COMMON NAME: Antarctic Minke Whale.
   TYPE LOCALITY: Near Belgrano, Prov. Buenos Aires, Argentina.
   DISTRIBUTION: Polar to tropical waters in the southern hemisphere.
   STATUS: CITES – Appendix I; IUCN – Lower Risk (cd).
   SYNONYMS: *huttoni* Gray, 1874.

*Balaenoptera borealis* Lesson, 1828. Hist. Nat. Gen. Part. Mamm. Oiseaux, 1:342.
   COMMON NAME: Sei Whale.
   TYPE LOCALITY: Germany, Schleswig-Holstein, Lubeck Bay, near Gromitz (see Rudolphi, 1822).
   DISTRIBUTION: Worldwide: cold-temperate to warm-temperate waters. Distributional records sometimes confused with *B. edeni.*
   STATUS: CITES – Appendix I; U.S. ESA and IUCN – Endangered.
   SYNONYMS: *rostrata* (Rudolphi, 1822); ***schlegelii*** (Flower, 1865).
   COMMENTS: Reviewed by Gambell (1985*a*). Rice (1998:75-76) separated this species into *B. borealis borealis* and *B. borealis schlegelii* Flower, 1865 (Northern Sei Whale) (Southern Sei Whale).

*Balaenoptera edeni* Anderson, 1879. Anat. Zool. Res., Yunnan, p. 551, pl. 44.
   COMMON NAME: Bryde's Whale.
   TYPE LOCALITY: Burma, "found its way into the Thaybyoo Choung, which runs into the Gulf of Martaban between the Sittang and Beeling Rivers, and about equidistant from each".
   DISTRIBUTION: Worldwide: warm-temperate to tropical waters. Distributional records sometimes confused with *B. borealis.*
   STATUS: CITES – Appendix I; IUCN – Data Deficient.

SYNONYMS: *brydei* Olsen, 1913; *omurai* Wada, Oishi, and Yamada 2003.

COMMENTS: Reviewed by Cummings (1985*a*). May represent more than one species (Wada and Numachi, 1991). Rice (1998:71) recognized *B. edeni* (Eden's or Sittang Whale; smaller in size) as being distinct from *B. brydei* (Bryde's whale; larger in size) but we used only one species because it is not clear if *B. edeni* is the larger or the smaller type whale and *B. edeni* has a type specimen. Wada, Oishi and Yamada (2003) described *Balaenoptera omurai*, related to *B. edeni* and recognized *B. brydei*. Due to the controversy regarding this step, we have recognized *B. omurai* as a synonym with the provision that it may be recognized as a full species when *Balaenoptera* is revised. See comment in Rice (1998:71-75).

*Balaenoptera musculus* (Linnaeus, 1758). Syst. Nat., 10th ed., 1:76.

COMMON NAME: Blue Whale.

TYPE LOCALITY: UK, Scotland, Firth of Forth ("Habitat in Mari Scotico").

DISTRIBUTION: Worldwide: arctic to tropical waters.

STATUS: CITES – Appendix I; U.S. ESA – Endangered; IUCN – Endangered as *B. m. intermedia*, Data Deficient as *B. m. brevicauda*; Endangered as *B. musculus* except – Vulnerable (North Atlantic stock), Lower Risk (cd) (North Pacific stock).

SYNONYMS: *gigas* (Van Beneden, 1861); *major* (Knox, 1870); *sibbaldii* (Gray, 1847); **brevicauda** Ichihara, 1966 [not Zemsky and Boronin, 1964, which is a *nomen nudum*, see Rice, 1977:6]; ***indica*** Blyth, 1859; ***intermedia*** Burmeister, 1871; **unassigned**: *sulfureus* Cope, 1869.

COMMENTS: Reviewed by Yochem and Leatherwood (1985). Rice (1998:78) recognized *B. m. musculus* (Northern Blue Whale), *B. m. indica* (Great Indian Rorqual), *B. m. brevicauda* (Pygmy Blue Whale) and *B. m. intermedia* (Southern Blue Whale).

*Balaenoptera physalus* (Linnaeus, 1758). Syst. Nat., 10th ed., 1:75.

COMMON NAME: Fin Whale.

TYPE LOCALITY: "Habitat in Oceano Europeao", restricted to Norway, near Svalbard, Spitsbergen Sea by Thomas (1911*a*).

DISTRIBUTION: Worldwide: arctic to tropical waters.

STATUS: CITES – Appendix I; U.S. ESA and IUCN – Endangered.

SYNONYMS: *antiquorum* (Fischer, 1829); *boops* (Linnaeus, 1758); *gibbar* Lacépède, 1804; *velifera* Cope, 1869; **quoyi** (Fischer, 1829); *patachonica* Burmeister, 1865.

COMMENTS: Reviewed by Gambell (1985*b*). Rice (1998:77) recognized *B. b. quoyi* (sic) (Southern Fin Whale).

*Megaptera* Gray, 1846. Ann. Mag. Nat. Hist., [ser. 1], 17:83.

TYPE SPECIES: *Megaptera longipinna* Gray, 1846 (= *Balaena novaeangliae* Borowski, 1781).

SYNONYMS: *Cyphobalaena* Marschall, 1873; *Kyphobalaena* Eschricht, 1849; *Perqualus* Gray, 1846; *Poescopia* Gray, 1864.

*Megaptera novaeangliae* (Borowski, 1781). Gemein. Naturgesch. Thier., 2(1):21.

COMMON NAME: Humpback Whale.

TYPE LOCALITY: USA, "de la nouvelle Angleterre" (= coast of New England).

DISTRIBUTION: Worldwide: cold-temperate to tropical waters.

STATUS: CITES – Appendix I; U.S. ESA – Endangered; IUCN – Vulnerable.

SYNONYMS: *braziliensis* Cope, 1867; *burmeisteri* Burmeister, 1866; *indica* Gervais, 1883; *lalandii* (Fischer, 1829); *longimana* (Rudolphi, 1832); *longipinna* Gray, 1846; *nodosa* (Bonnaterre, 1789); *osphya* Cope, 1865; *versabilis* Cope, 1869.

COMMENTS: Reviewed by Winn and Reichley (1985).

**Family Eschrichtiidae** Ellerman and Morrison-Scott, 1951. Checklist of Palearctic Indian Mammals, p. 713.

SYNONYMS: Rhachianectidae Weber, 1904.

*Eschrichtius* Gray, 1864. Ann. Mag. Nat. Hist., ser. 3, 14:350.

TYPE SPECIES: *Balaenoptera robusta* Lilljeborg, 1861.

SYNONYMS: *Cyphonotus* Rafinesque, 1815 [*nomen nudum*]; *Cyphonotus* Gray, 1850 [preoccupied by *Cyphonotus* Fischer, 1823 (genus of beetles)]; *Rhachianectes* Cope, 1869.

COMMENTS: The type species of *Agaphelus* Cope, 1868 is *Balaena gibbosa* Erxleben, 1777, which is a composite species consisting mainly of junior synonyms of *Megaptera novaeangliae*. *Agaphelus* has been linked to *Eschrichtius* by the supposed inclusion of the "scrag whale of Dudley" in its synonymy. The "scrag whale of Dudley"is actually listed as "*Species obscurae*" just following the section on *Agaphelus* in Erxleben (p. 611).

*Eschrichtius robustus* (Lilljeborg, 1861). Forh. Skand. Naturf. Ottende Mode, Kopenhagen, 1860, 8:602 [1861].

COMMON NAME: Gray Whale.

TYPE LOCALITY: Sweden, "på Gräsön i Roslagen"; "Benen lägo 840 fot från hafsstranden, ungefär 12 à 15 fot öfver hafvets yta" (= Uppland, Graso Isl).

DISTRIBUTION: North Pacific: warm temperate to arctic waters. Formerly present in the North Atlantic. Sometimes enters tropical water at the southern boundaries of its distribution; see Henderson (1990) for further details. The eastern and western North Pacific populations are separate. The eastern population is distributed from Baja California and the adjacent coast of Mexico to the Bering and Chukchi Seas. The western population is distributed from the East China Sea to the Sea of Okhotsk.

STATUS: CITES – Appendix I; U.S. ESA – Endangered, except for Eastern North Pacific Ocean—coastal and Bering, Beaufort, and Chukchi Seas populations, which are listed as Delisted Taxa (recovered); IUCN – Critically Endangered (Northwest Pacific (Asian) stock), otherwise Lower Risk (cd).

SYNONYMS: *gibbosus* (Erxleben, 1777); *glaucus* (Cope, 1868).

COMMENTS: See Rice and Wolman (1971), Jones et al. (1984), and Wolman (1985).

**Family Neobalaenidae** Gray, 1873. Ann. Mag. Nat. Hist., ser. 4, 11:108.

COMMENTS: See Barnes and McLeod (1984) for comments. Gray, 1874 (Trans. Proc. N. Z. Inst. 6(18):93-97) is cited by Barnes and McLeod for Neobalaenidae.

*Caperea* Gray, 1864. Proc. Zool. Soc. Lond. 1864(2):202.

TYPE SPECIES: *Balaena* (*Caperea*) *antipodarum* Gray, 1864 (= *Balaena marginata* Gray, 1846).

SYNONYMS: *Neobalaena* Gray, 1870.

*Caperea marginata* (Gray, 1846). Zool. Voy. H.M.S. "Erebus" and "Terror", 1:48.

COMMON NAME: Pygmy Right Whale.

TYPE LOCALITY: "Inhab. W. Australia" (= southern hemisphere, temperate waters; see Baker, 1985).

DISTRIBUTION: Southern hemisphere: cold-temperate waters.

STATUS: CITES – Appendix I; IUCN – Lower Risk (lc).

SYNONYMS: *antipodarum* Gray, 1864.

COMMENTS: Reviewed by Baker (1985).

## SUBORDER ODONTOCETI Flower, 1867.

**Family Delphinidae** Gray, 1821. London Med. Repos., 15(1):310.

SYNONYMS: Globicephalidae Gray, 1850; Grampidelphidae Nishiwaki, 1963; Stenidae Fraser and Purves, 1960; Orcinae Fraser and Purves, 1960; Lissodelphinae Fraser and Purves, 1960; Cephalorhynchinae Fraser and Purves, 1960.

COMMENTS: Includes Globicephalidae, Grampidelphidae, Stenidae, Orcinae, Lissodelphinae, Cephalorhynchinae and Delphininae (Fraser and Purves, 1960); see Kasuya (1973), Mead (1975), Barnes (1978). Also includes *Orcaella* (see Heyning, 1989*a* and Lint et al., 1990), sometimes put in the family Monodontidae.

*Cephalorhynchus* Gray, 1846. Zool. Voy. H.M.S. "Erebus" and "Terror", 1:36.

TYPE SPECIES: *Delphinus heavisidii* Gray, 1828.

SYNONYMS: *Eutropia* Gray, 1862.
COMMENTS: Revised by Harmer (1922).

*Cephalorhynchus commersonii* (Lacépède, 1804). Hist. Nat. Cétacées, p. 317.
COMMON NAME: Commerson's Dolphin.
TYPE LOCALITY: Chile, "de la terre de Feu et dans le détroit de Magellan" (= Tierra del Fuego, Straits of Magellan).
DISTRIBUTION: Argentina to Chile: Gulf of San Matias, Argentina, to the Chilean side of the Straits of Magellan; South Shetland, Falkland and Kerguelen Isls. See Brownell and Praderi (1985) for further discussion. Rice (1998:101) recognized that the Falkland and Kerguelen Island populations are disjunct and suggested that they differ in subspecies, the Falkland population being *C. c. commersonii* and the Kerguelen population being *C. c.* subsp.
STATUS: CITES – Appendix II; IUCN – Data Deficient.
SYNONYMS: *floweri* (Moreno, 1892).
COMMENTS: Reviewed by Goodall et al. (1988).

*Cephalorhynchus eutropia* Gray, 1846. Zool. Voy. H.M.S. "Erebus" and "Terror", 1:pl. 34.
COMMON NAME: Chilean Dolphin.
TYPE LOCALITY: None given, listed as Pacific Ocean, off the coast of Chile by Gray (1850:112).
DISTRIBUTION: Chile: coastal waters between Valparaiso and Navarino Isl, Tierra del Fuego.
STATUS: CITES – Appendix II; IUCN – Data Deficient.
SYNONYMS: *albiventris* (Perez Canto *in* Philippi, 1893); *obtusata* (Philippi, 1893).
COMMENTS: Reviewed by Goodall et al. (1988). *Tursio? panope* is not a synonym (see *Lagenorhynchus obscurus*).

*Cephalorhynchus heavisidii* (Gray, 1828). Spicil. Zool., 1:2.
COMMON NAME: Haviside's Dolphin.
TYPE LOCALITY: South Africa, Western Cape Prov., "Inhab. Cape of Good Hope".
DISTRIBUTION: South Africa to perhaps S Angola: coastal waters from Cape Town to 17°09'S (Namibia).
STATUS: CITES – Appendix II; IUCN – Data Deficient.
SYNONYMS: *hastatus* (F. Cuvier, 1836).
COMMENTS: Type specimen was brought to the Royal College of Surgeons by Captain Haviside, at about the same time that Captain Heaviside sold a collection of anatomical specimens to the Royal College. Gray confused the two (Rice 1998:101).

*Cephalorhynchus hectori* (van Beneden, 1881). Bull. R. Acad. Belg., ser. 3, 4:877, pl. 11.
COMMON NAME: Hector's Dolphin.
TYPE LOCALITY: "capturé sur la côte nord-est de la Nouvelle-Zélande." (= New Zealand, north coast).
DISTRIBUTION: New Zealand: coastal waters. Harrison's (1960) reference to the occurrence of this species around Sarawak is undocumented by specimens or photos.
STATUS: CITES – Appendix II; IUCN – Endangered, except for North Island subpopulation, which is Critically Endangered.
SYNONYMS: *albifrons* True, 1889; *maui* Baker, Smith and Pichler 2002.
COMMENTS: Baker, Smith and Pichler (2002) recognized that *C. h. maui* form a morphologically and genetically recognizable subspecies on the North Isl of New Zealand, while the nominal subspecies was restricted to the South Isl.

*Delphinus* Linnaeus, 1758. Syst. Nat., 10th ed., 1:77.
TYPE SPECIES: *Delphinus delphis* Linnaeus, 1758.
SYNONYMS: *Eudelphinus* Van Beneden and Gervais, 1880; *Rhinodelphis* Wagner, 1846.

*Delphinus capensis* Gray, 1828. Spicilegia Zoologica, 1:1-2, tab. 2.
COMMON NAME: Long-beaked Common Dolphin.
TYPE LOCALITY: Cape of Good Hope.
DISTRIBUTION: Near-shore tropical to temperate waters, world-wide.
STATUS: CITES – Appendix I; IUCN – Lower Risk (lc).

SYNONYMS: *bairdii* Dall, 1873; *major* Gray, 1866; *tropicalis* van Bree, 1971.
COMMENTS: See Heyning and Perrin (1994) for details. Rice (1998:112) recognized
*D. tropicalis* (Arabian Common Dolphin) as a separate species.

*Delphinus delphis* Linnaeus, 1758. Syst. Nat., 10th ed., 1:77.
COMMON NAME: Short-beaked Common Dolphin.
TYPE LOCALITY: E North Atlantic ("Oceano Europaeo").
DISTRIBUTION: Worldwide: temperate and tropical waters, including the Black Sea.
STATUS: CITES – Appendix II; IUCN – Lower Risk (lc), except for Mediterranian
subpopulation, which is Endangered.
SYNONYMS: *fulvofasciatus* Wagner, 1846; *janira* Gray, 1846; *pomeegra* Owen, 1866; *vulgaris*
Lacépède, 1804; **ponticus** Barabash, 1935.
COMMENTS: See Heyning and Perrin (1991, 1994). Rice (1998:112) recognized *D. d. ponticus*
(Black Sea Common Dolphin).

*Feresa* Gray, 1870. Proc. Zool. Soc. Lond., 1870(1):77.
TYPE SPECIES: *Delphinus intermedius* Gray, 1827 (= *Feresa attenuata* Gray, 1874).
COMMENTS: *Delphinus intermedius* Gray, 1827 was preoccupied by *Delphinus intermedius* Harlan,
1827 (= *Globicephala melas*). Gray subsequently changed generic designations of that
nominal taxon (*Grampus intermedius* Gray, 1843; *Orca intermedia* Gray, 1846).

*Feresa attenuata* Gray, 1874. Ann. Mag. Nat. Hist., ser. 4, 14:238-239.
COMMON NAME: Pygmy Killer Whale.
TYPE LOCALITY: "South Seas."
DISTRIBUTION: Worldwide: tropical to warm-temperate waters.
STATUS: CITES – Appendix II; IUCN – Data Deficient.
SYNONYMS: *intermedius* (Gray, 1843); *occulta* Jones and Packard, 1956.

*Globicephala* Lesson, 1828. Compl. Oeuvres Buffon Hist. Nat., 1:441.
TYPE SPECIES: *Delphinus globiceps* Cuvier, 1812 (= *Delphinus melas* Traill, 1809).
SYNONYMS: *Cetus* Wagler, 1830; *Globiceps* Flower, 1884; *Sphaerocephalus* Gray, 1864.
COMMENTS: Reviewed by Van Bree (1971).

*Globicephala macrorhynchus* Gray, 1846. Zool. Voy. H.M.S. "Erebus" and "Terror", 1:33.
COMMON NAME: Short-finned Pilot Whale.
TYPE LOCALITY: "South Seas".
DISTRIBUTION: Worldwide: tropical and warm-temperate waters; cold-temperate waters of
the N Pacific, where it appears to stray as far north as the Gulf of Alaska (Pike and
MacAskie, 1969).
STATUS: CITES – Appendix II; IUCN – Lower Risk (cd).
SYNONYMS: *brachypterus* Cope, 1876; *scammonii* Cope, 1869; *sieboldii* Gray, 1846.
COMMENTS: See Van Bree (1971).

*Globicephala melas* (Traill, 1809). Nicholson's J. Nat. Philos. Chem. Arts, 22:81.
COMMON NAME: Long-finned Pilot Whale.
TYPE LOCALITY: UK, Scotland, "in Scapay Bay, in Pomona, one of the Orkneys".
DISTRIBUTION: North Atlantic and southern Oceans: cold-temperate waters. Kasuya (1975)
described the historic distribution in the NW Pacific.
STATUS: CITES – Appendix II; IUCN – Lower Risk (lc).
SYNONYMS: *globiceps* (G. Cuvier, 1812); *intermedius* Harlan 1827; *svineval* Gray, 1846;
**edwardii** Smith, 1834; *leucosagmaphora* Rayner, 1939.
COMMENTS: See Van Bree (1971). Formerly called *G. melaena* but Article 31b of the third
edition of the International Code of Zoological Nomenclature (1985*d*) specifically
gave *melas* as an example of a Greek adjective that does not change its ending when
transferred to a genus of another gender (see Schevill, 1990*a*, *b*; Rice, 1990). Rice
(1998:119) recognized *G. m. melas* (North Atlantic Longfinned Pilot whale),
*G. m. edwardii* (Southern Longfinned Pilot Whale) and an un-named subfossil
subspecies of *G. melas* that occupied the North Pacific.

*Grampus* Gray, 1828. Spicil. Zool., 1:2.
    TYPE SPECIES: *Delphinus griseus* Cuvier, 1812.
    SYNONYMS: *Grampidelphis* Iredale and Troughton, 1933; *Grayius* Scott, 1873.

*Grampus griseus* (G. Cuvier, 1812). Ann. Mus. Hist. Nat. Paris, 19:13.
    COMMON NAME: Risso's Dolphin.
    TYPE LOCALITY: France, Finistere, "envoyé de Brest".
    DISTRIBUTION: Worldwide: temperate to tropical waters.
    STATUS: CITES – Appendix II; IUCN – Data Deficient.
    SYNONYMS: *rissoanus* (Desmarest, 1822); *stearnsii* Dall, 1873.
    COMMENTS: Corbet and Hill (1980:110) included *rectipinna* in this species but it belongs in
        *Orcinus orca*.

*Lagenodelphis* Fraser, 1956. Sarawak Mus. J., n.s., 8(7):496.
    TYPE SPECIES: *Lagenodelphis hosei* Fraser, 1956.

*Lagenodelphis hosei* Fraser, 1956. Sarawak Mus. J., n.s., 8(7):496.
    COMMON NAME: Fraser's Dolphin.
    TYPE LOCALITY: "Collected at the mouth of Lutong River, Baram, Borneo."
    DISTRIBUTION: Worldwide: warm-temperate to tropical waters.
    STATUS: CITES – Appendix II; IUCN – Data Deficient.

*Lagenorhynchus* Gray, 1846. Ann. Mag. Nat. Hist., [ser 1.], 17:84.
    TYPE SPECIES: *Delphinus albirostris* Gray, 1846.
    SYNONYMS: *Electra* Gray, 1866; *Leucopleurus* Gray 1866; *Sagmatius* Cope, 1866.
    COMMENTS: Reviewed by Fraser (1966).

*Lagenorhynchus acutus* (Gray, 1828). Spicil. Zool., 1:2.
    COMMON NAME: Atlantic White-sided Dolphin.
    TYPE LOCALITY: None given, listed as North Sea, Faeroe Isls (Denmark) (uncertain) by
        Gray (1846:36).
    DISTRIBUTION: North Atlantic: cold temperate waters; *L. acutus* tends to be distributed to the
        south of *L. albirostris*.
    STATUS: CITES – Appendix II; IUCN – Lower Risk (lc).
    SYNONYMS: *gubernator* Cope, 1876; *leucopleurus* (Rasch, 1843); *perspicillatus* Cope, 1876.

*Lagenorhynchus albirostris* (Gray, 1846). Ann. Mag. Nat. Hist., [ser. 1], 17:84.
    COMMON NAME: White-beaked Dolphin.
    TYPE LOCALITY: None given in original description, given by Gray (1846:35) as UK, England,
        "North Sea, coast of Norfolk.", and by Gray (1850) as Great Yarmouth.
    DISTRIBUTION: North Atlantic: cold-temperate waters; *L. albirostris* tends to be distributed to
        the north of *L. acutus*.
    STATUS: CITES – Appendix II; IUCN – Lower Risk (lc).
    SYNONYMS: *pseudotursio* (Reichenbach, 1846); *ibseni* (Eschricht, 1846).

*Lagenorhynchus australis* (Peale, 1848). Mammalia *in* Repts. U.S. Expl. Surv., 8:33, pl. 6.
    COMMON NAME: Peale's Dolphin.
    TYPE LOCALITY: "South Atlantic Ocean, off the coast of Patagonia", Argentina, 1 day's sail
        north of the Straits of LeMaire.
    DISTRIBUTION: Chile to Argentina: Valparaiso to Commodoro Rivadavia and Falkland Isls:
        Cold-temperate waters. One published (photograph) sighting in the tropical waters of
        the South Pacific, Cook Isls (Leatherwood et al., 1991).
    STATUS: CITES – Appendix II; IUCN – Data Deficient.
    SYNONYMS: *amblodon* (Cope, 1866); *chilensis* (Philippi, 1896).
    COMMENTS: Included in *cruciger* by Bierman and Slijper (1947) and Hershkovitz (1966a:67),
        but considered a distinct species by Fraser (1966), Rice (1977), Brownell (1974), and
        Mitchell (1975a).

*Lagenorhynchus cruciger* (Quoy and Gaimard, 1824). Voy. autour du Monde . . . l'Uranie et la Physicienne, Zool., p. 87, pl. 2.

COMMON NAME: Hourglass Dolphin.

TYPE LOCALITY: Pacific Ocean, "entre la Nouvelle-Hollande et le cap Horn [= between Australia and Cape Horn] . . . par 49 [S] de latitude".

DISTRIBUTION: Southern hemisphere: antarctic and cold-temperate waters.

STATUS: CITES – Appendix II; IUCN – Lower Risk (lc).

SYNONYMS: *albigena* (Quoy and Gaimard, 1824); *bivitattus* (Lesson and Garnot, 1826); *clanculus* Gray, 1846; *wilsoni* Lillie, 1915.

COMMENTS: Formerly included *australis* and *obscurus*, see Hershkovitz (1966*a*) and comments under *australis* and *obscurus*.

*Lagenorhynchus obliquidens* Gill, 1865. Proc. Acad. Nat. Sci. Philadelphia, 17:177.

COMMON NAME: Pacific White-sided Dolphin.

TYPE LOCALITY: USA, "obtained at San Francisco, California".

DISTRIBUTION: North Pacific: cold-temperate waters except warm-temperate waters of the ends of its range. Undocumented sighting from Hong Kong (Hammond and Leatherwood, 1984:495).

STATUS: CITES – Appendix II; IUCN – Lower Risk (lc).

SYNONYMS: *longidens* (Cope, 1866); *ognevi* Slensov, 1955.

COMMENTS: May be a northern hemisphere form of *L. obscurus*. *Lagenorhynchus thicolea* is not synonymous with *L. obliquidens* (see *Lissodelphis*).

*Lagenorhynchus obscurus* (Gray, 1828). Spicil. Zool., 1:2.

COMMON NAME: Dusky Dolphin.

TYPE LOCALITY: South Africa, Western Cape Prov., "Inhab. Cape of Good Hope".

DISTRIBUTION: Southern hemisphere: cold-temperate continental waters.

STATUS: CITES – Appendix II; IUCN – Data Deficient.

SYNONYMS: *similis* (Gray, 1868); *supercilliosus* (Lesson and Garnot, 1826); *fitzroyi* (Waterhouse, 1838); *breviceps* (Wagner, 1846); *panope* (Canto *in* Philippi, 1896).

COMMENTS: Included in *cruciger* by Hershkovitz (1966*a*:65), but considered a distinct species by Rice (1977), Brownell (1974), and Mitchell (1975*a*). Previously reported from Kerguelen Isls, reidentified as young specimen of *Cephalorhynchus commersonii* (Robineau, 1989). Rice (1998:114) recognized *L. o. obscurus* (South African Dusky Dolphin), *L. o. fitzroyi* (South American Dusky Dolphin) and an un-named subspecies (New Zealand Dusky Dolphin).

*Lissodelphis* Gloger, 1841. Gemein. Naturgesch. Thier., 1:169.

TYPE SPECIES: *Delphinus peronii* Lacépède, 1804.

SYNONYMS: *Delphinapterus* Gray, 1846 [part]; *Leucorhamphus* Lilljeborg, 1861; *Tursio* Wagler, 1830.

COMMENTS: This may be a monotypic genus. The holotype of *Lagenorhynchus thicolea*, previously associated with *Lagenorhynchus*, is a specimen of *Lissodelphis* sp.

*Lissodelphis borealis* (Peale, 1848). Mammalia *in* Repts. U.S. Expl. Surv., 8:35, pl. 8.

COMMON NAME: Northern Right Whale Dolphin.

TYPE LOCALITY: "North Pacific Ocean, latitude 46° 6' 50" N., 134° 5' W. from Greenwich." 10°W of Astoria, Oregon, USA.

DISTRIBUTION: North Pacific: cold-temperate waters.

STATUS: CITES – Appendix II; IUCN – Lower Risk (lc).

*Lissodelphis peronii* (Lacépède, 1804). Hist. Nat. Cetacees, p. 316.

COMMON NAME: Southern Right Whale Dolphin.

TYPE LOCALITY: Indian Ocean, "dans les environs du cap sud de la terre de Diémen, et par conséquent vers le quarante-quatrime degré de latitude australe." (= about 44°S, 141°E, south of Tasmania).

DISTRIBUTION: Southern hemisphere: cold-temperate waters, occasionally Antarctic waters south of Argentina.

STATUS: CITES – Appendix II; IUCN – Data Deficient.
SYNONYMS: *leucorhamphus* (Lacépède, 1804).

*Orcaella* Gray, 1866. Cat. Seals Whales Brit. Mus., p. 285.
   TYPE SPECIES: *Orca* (*Orcaella*) *brevirostris* Owen *in* Gray, 1866.
   COMMENTS: We follow Fordyce (1989), Heyning (1989a), and Lint et al. (1990) in including
      *Orcaella* in the Delphinidae, not in the Monodontidae as was recently proposed (Barnes
      et al., 1985; Kasuya, 1973).

*Orcaella brevirostris* (Owen *in* Gray, 1866). Cat. Seals Whales Brit. Mus., p. 285, fig. 57.
   COMMON NAME: Irrawady Dolphin.
   TYPE LOCALITY: "Inhab. East coast of India, the harbour of Vizagapatam" (= Vishakhapat-
      nam Harbor, in Bay of Bengal).
   DISTRIBUTION: SE Asia, N Australia and Papua New Guinea: tropical coastal waters and large
      rivers.
   STATUS: CITES – Appendix II; IUCN – Data Deficient, except for Mahakam subpopulation,
      which is Critically Endangered.
   SYNONYMS: *fluminalis* Gray, 1871.
   COMMENTS: Reviewed by Marsh et al. (1989). See Rice (1998:120) for details on species
      citation.

*Orcinus* Fitzinger, 1860. Wiss.-Pop. Naturgesch. Säugeth., 6:204.
   TYPE SPECIES: *Delphinus orca* Linnaeus, 1758.
   SYNONYMS: *Gladiator* Gray, 1870; *Grampus* Iredale and Troughton, 1933; *Ophysia* Gray, 1868;
      *Orca* Gray, 1846 [preoccupied].

*Orcinus orca* (Linnaeus, 1758). Syst. Nat., 10th ed., 1:77.
   COMMON NAME: Killer Whale.
   TYPE LOCALITY: E North Atlantic ("Oceano Europaeo").
   DISTRIBUTION: Worldwide: all seas and oceans.
   STATUS: CITES – Appendix II; IUCN – Lower Risk (cd).
   SYNONYMS: *ater* (Cope *in* Scammon, 1869); *capensis* (Gray, 1846); *glacialis* Berzin and
      Vladimirov 1983; *gladiator* (Bonnaterre, 1789); *nanus* Mikhalev and Ivashin, 1981;
      *rectipinna* (Cope *in* Scammon, 1869).
   COMMENTS: Reviewed by Heyning and Dahlheim (1988, Mammalian Species, 304). Rice
      (1998:118) recognized the possibility that *O. glacialis* represented a separate taxon, but
      did not assign specific or subspecific status to it. Rice (1998:118) considered *O. nanus* a
      *nomen nudum*, which we agree with.

*Peponocephala* Nishiwaki and Norris, 1966. Sci. Rep. Whales Res. Inst., 20:95.
   TYPE SPECIES: *Lagenorhynchus electra* Gray, 1846.
   SYNONYMS: *Electra* Gray, 1868.
   COMMENTS: Formerly included in *Lagenorhynchus*.

*Peponocephala electra* (Gray, 1846). Zool. Voy. H.M.S. "Erebus" and "Terror", 1:35.
   COMMON NAME: Melon-headed Whale.
   TYPE LOCALITY: None given, unknown.
   DISTRIBUTION: Worldwide: tropical to warm-temperate waters.
   STATUS: CITES – Appendix II; IUCN – Lower Risk (lc).
   SYNONYMS: *asia* (Gray, 1846); *fusiformis* (Owen, 1866); *pectoralis* (Peale, 1848).
   COMMENTS: Historically this species was included in the genus *Lagenorhynchus*.

*Pseudorca* Reinhardt, 1862. Overs. Danske Vidensk. Selsk. Forh., 1862:151.
   TYPE SPECIES: *Phocaena crassidens* Owen, 1846.
   SYNONYMS: *Neoorca* Gray, 1871.

*Pseudorca crassidens* (Owen, 1846). Hist. Brit. Foss. Mamm. Birds, p. 516, fig. 213.
   COMMON NAME: False Killer Whale.

TYPE LOCALITY: UK, England, "in the great fen of Lincolnshire beneath the turf, in the neighborhood of the ancient town of Stamford". (subfossil).
DISTRIBUTION: Worldwide: temperate to tropical waters.
STATUS: CITES – Appendix II; IUCN – Lower Risk (lc).
SYNONYMS: *destructor* (Cope, 1866); *meridionalis* (Flower, 1865).

*Sotalia* Gray, 1866. Cat. Seals Whales Brit. Mus., p. 401.
TYPE SPECIES: *Delphinus guianensis* Van Beneden, 1864 (= *Delphinus fluviatilis* Gervais and Deville, 1853).
SYNONYMS: *Tucuxa* Gray, 1866.

*Sotalia fluviatilis* (Gervais and Deville, 1853). *In* Gervais, Bull. Soc. Agric. Herault, p. 148.
COMMON NAME: Tucuxi.
TYPE LOCALITY: Peru, Loreto, Rio Maranon above Pebas.
DISTRIBUTION: Western Atlantic: coastal waters from Panama to Santos, São Paulo, Brazil: Amazon and Orinoco river systems. See Vidal (1990) and Borobia et al. (1991).
STATUS: CITES – Appendix I; IUCN – Data Deficient.
SYNONYMS: *guianensis* (Van Beneden, 1864); *pallida* (Gervais, 1855); *tucuxi* (Gray, 1856).
COMMENTS: Due to the difficulty in finding the original work, the full citation is included here: Gervais, F. L. P. [and Deville]. 1853. Sur les mammiféres marins qui fréquentent les côtes de la France et plus particulièrement sur une novelle espéce de dauphins propre a la Méditerranés. Bulletin Société Centrale d'Agriculture et des Comices Agricoles du Département de l'Herault, Montpellier, 40me année, pp. 140-155, 1 pl. Rice (1998:104) recognized *S. f. guianensis* (Guiane Dolphin) and Monteiro-Filho et al. (2002) suggested the use of *S. guianensis* for the marine dolphins and *S. fluviatilis* for the Amazonian dolphins. We are taking the conservative approach and only recognizing one specific name until the problem is resolved.

*Sousa* Gray, 1866. Proc. Zool. Soc. Lond., 1866(2):213.
TYPE SPECIES: *Steno lentiginosus* Gray, 1866 (= *Delphinus chinensis* Osbeck, 1765).
SYNONYMS: *Stenopontistes* Miranda Ribiero, 1936.
COMMENTS: Formerly included in *Sotalia* (Hershkovitz 1966a:18).

*Sousa chinensis* (Osbeck, 1765). Reise nach Ostind. China Rostock, 1:7.
COMMON NAME: Indo-Pacific Humpbacked Dolphin.
TYPE LOCALITY: China, Guangdong Prov., Zhujiang Kou (mouth of Canton River).
DISTRIBUTION: Indian Ocean: coastal waters and rivers from False Bay, South Africa, east to S China and Moreton Bay, Queensland (Australia, see Corkeron, 1990).
STATUS: CITES – Appendix I; IUCN – Data Deficient.
SYNONYMS: *borneensis* (Lydekker, 1901); *huangi* Wang Peilie, 1999; *lentiginosa* (Gray, 1866); *plumbea* (G. Cuvier, 1829); *zambezicus* (Miranda Ribiero, 1936).
COMMENTS: See Perrin (1975) who placed *Delphinus malayanus* in *Stenella attenuata*, as is done here; Pilleri and Gihr (1973-74) considered *borneensis*, *plumbea* and *lentiginosa* to be distinct species. Mitchell (1975a) combined those species into *S. chinensis*; Brownell (1975b) included *Stenopontistes zambezicus* as a synonym of *S. plumbea*. Wang Peilie (1999:299) described *Sousa huangi* on the basis of an immature specimen. Because we feel that immature specimens of *Sousa* do not show diagnostic characters, we tentatively reject *Sousa huangi*. Rice (1998:103) recognized *S. plumbea* (Indian Humpbacked Dolphin) as well as *S. teuszi* and *S. chinensis*, but gave conflicting accounts as to the number of species recognized in his account of the genus. Accordingly we maintain the more conservative approach of two species, while indicating that more detailed systematic work needs to be done.

*Sousa teuszii* (Kükenthal, 1892). Zool. Jahrb. Syst., 6:442, pl. 21.
COMMON NAME: Atlantic Humpbacked Dolphin.
TYPE LOCALITY: "aus Kamerun" (= Cameroon), Cameroun Oriental, Bay of Warships, near Douala.
DISTRIBUTION: E South Atlantic: coastal waters in river mouths from S Morocco (W Sahara; see Beaubrun, 1990) to Cameroon.

STATUS: CITES – Appendix I; IUCN – Data Deficient.
COMMENTS: Reviewed by Pilleri and Gihr (1972).

*Stenella* Gray, 1866. Proc. Zool. Soc. Lond., 1866:213.
TYPE SPECIES: *Steno attenuatus* Gray, 1846.
SYNONYMS: *Clymene* Gray, 1864; *Euphrosyne* Gray, 1866; *Fretidelphis* Iredale and Troughton, 1934; *Micropia* Gray, 1868; *Prodelphinus* Gervais in Van Beneden and Gervais, 1880.
COMMENTS: Reviewed, in part, by Perrin (l975) and Perrin et al. (1981, 1987). The International Commission on Zoological Nomenclature (1991) conserved *Stenella* Gray, 1846.

*Stenella attenuata* (Gray, 1846). Zool. Voy. H.M.S. "Erebus" and "Terror", 1:44.
COMMON NAME: Pantropical Spotted Dolphin.
TYPE LOCALITY: None given, unknown (possibly India, see Gray, 1843).
DISTRIBUTION: Worldwide: temperate to tropical waters.
STATUS: CITES – Appendix II; IUCN – Lower Risk (cd).
SYNONYMS: *albirostratus* (Peale, 1848); *brevimanus* (Wagner, 1846); *capensis* (Rapp, 1837); *consimilis* (Malm, 1871); *malayanus* (Lesson, 1826); *pseudodelphis* (Wiegmann, 1846); *punctata* (Gray, 1866); *velox* (Cuvier, 1829); *graffmani* (Lönnberg, 1934).
COMMENTS: Perrin et al. (1987) revised this species. *D. dubius* is a *nomen nudum* [sic *dubium*] (Perrin et al., 1987). Opinion 1660 of the International Commission on Zoological Nomenclature (1991) conserved *attenuatus* Gray, 1846 and suppressed *velox* Cuvier, 1829, *pseudodelphis* Schlegel, 1841, and *brevimanus* Wagner, 1846. Rice (1998:108) recognized *Stenella attenuata* subspecies A of Perrin (1975) (Eastern Tropical Pacific Offshore Spotted Porpoise), *S. a.* subspecies B of Perrin (1975) (Hawaiian Spotted Porpoise), and *S. a. graffmani* (Eastern Pacific Coastal Spotted Porpoise (Perrin, 1975)).

*Stenella clymene* (Gray, 1846). Zool. Voy. H.M.S. "Erebus" and "Terror", 1:39.
COMMON NAME: Clymene Dolphin.
TYPE LOCALITY: None given, unknown.
DISTRIBUTION: Atlantic Ocean including the Gulf of Mexico: warm-temperate to tropical waters.
STATUS: CITES – Appendix II; IUCN – Data Deficient.
SYNONYMS: *metis* (Gray, 1846); *normalis* (Gray, 1866).
COMMENTS: Recognized by Hershkovitz (1966a), but not by Mitchell (1975a) who included it in *longirostris*. See Perrin et al. (1981) for redescription.

*Stenella coeruleoalba* (Meyen, 1833). Nova Acta Acad. Caes. Nat. Curios., 16(2):609, pl. 43.
COMMON NAME: Striped Dolphin.
TYPE LOCALITY: "an der östlichen Küste von Südamerika; wir karpunirten ihn in der Gegend des Rio de la Plata." (= South Atlantic Ocean near Rio de la Plata, off coast of Argentina and Uruguay).
DISTRIBUTION: Worldwide: cold-temperate to tropical waters.
STATUS: CITES – Appendix II; IUCN – Lower Risk (cd).
SYNONYMS: *asthenops* (Cope, 1865); *crotaphiscus* (Cope, 1865); *euphrosyne* (Gray, 1846); *styx* (Gray, 1846); *tethyos* (Gervais, 1853).
COMMENTS: See Mitchell (1970:720). Perrin et al. (1981, 1987) gave a revised synonymy of this species.

*Stenella frontalis* (G. Cuvier, 1829). Règne Anim., Nouv. ed., 1:288.
COMMON NAME: Atlantic Spotted Dolphin.
TYPE LOCALITY: "découvert un aux îles du Cap-Vert". (= off Cape Verde Isls, West Africa).
DISTRIBUTION: Atlantic Ocean including the Gulf of Mexico: warm-temperate to tropical waters.
STATUS: CITES – Appendix II; IUCN – Data Deficient.
SYNONYMS: *doris* (Gray, 1846); *froenatus* (F. Cuvier, 1836); *plagiodon* (Cope, 1866).
COMMENTS: Perrin et al. (1987) revised this species. The International Commission on Zoological Nomenclature (1977a) suppressed *D. pernettensis* de Blainville, 1817 and *D. pernettyi* Desmarest, 1820, which Hershkovitz (1966a) used as a senior synonym for *S. plagiodon*.

*Stenella longirostris* (Gray, 1828). Spicil. Zool., 1:1.
    COMMON NAME: Spinner Dolphin.
    TYPE LOCALITY: None given, unknown.
    DISTRIBUTION: Worldwide: warm-temperate to tropical waters.
    STATUS: CITES – Appendix II; IUCN – Lower Risk (cd).
    SYNONYMS: *alope* (Gray, 1846); *microps* (Gray, 1846); *roseiventris* (Wagner, 1846);
        **centroamericana** Perrin, 1990; **orientalis** Perrin, 1990.
    COMMENTS: See Perrin (1975:206). Perrin (1990) established three subspecies
        (*centroamericana, longirostris,* and *orientalis*), which Rice (1998:109) recognized.

*Steno* Gray, 1846. Zool. Voy. H.M.S. "Erebus" and "Terror", 1:43.
    TYPE SPECIES: *Delphinus rostratus* Cuvier, 1833 (= *Delphinus bredanensis* Cuvier *in* Lesson, 1828).
    SYNONYMS: *Glyphidelphis* Gervais, 1859.

*Steno bredanensis* (G. Cuvier in Lesson, 1828). Hist. Nat. Gen. Part. Mamm. Oiseaux, 1:206.
    COMMON NAME: Rough-toothed Dolphin.
    TYPE LOCALITY: Coast of France.
    DISTRIBUTION: Worldwide: warm-temperate to tropical waters.
    STATUS: CITES – Appendix II; IUCN – Data Deficient.
    SYNONYMS: *compressus* (Gray, 1843); *frontatus* (Cuvier, 1823); *perspicillatus* Peters, 1876;
        *rostratus* (Desmarest, 1817).
    COMMENTS: *Stenopontistes zambezicus* is not a synonym, see comment under *Sousa chinensis*.
        See Schevill (1987*a*) for further taxonomic notes.

*Tursiops* Gervais, 1855. Hist. Nat. Mammifères, 2:323.
    TYPE SPECIES: *Delphinus truncatus* Montagu, 1821.
    SYNONYMS: *Gadamu* Gray, 1868; *Tursio* Gray, 1843.
    COMMENTS: Two species are provisionally recognized in this highly polymorphic genus.

*Tursiops aduncus* (Ehrenberg, 1833). Hemprich and Ehrenberg, Symbolae Physicae. Mammalia,
    decas II, folio k, ftn. 1.
    COMMON NAME: Indo-Pacific Bottlenose Dolphin.
    TYPE LOCALITY: Ethiopia, Dahlak Arch., Belhosse Isl.
    DISTRIBUTION: Indian Ocean; distribution in Pacific uncertain.
    STATUS: CITES – Appendix II; IUCN – Data Deficient.
    SYNONYMS: *nuuanu* Andrews, 1911.
    COMMENTS: Rice (1998:106) recognized this species based primarily on the work of LeDuc
        and Curry (1997). There is a growing consensus that this is one of the species of
        *Tursiops* that is valid. Its relationship to *Tursiops nuuanu* remains equivocal.

*Tursiops truncatus* (Montagu, 1821). Mem. Wernerian Nat. Hist. Soc., 3:75, pl. 3.
    COMMON NAME: Bottlenose Dolphin.
    TYPE LOCALITY: UK, England, Devonshire, "in Duncannon Pool, near Stoke Gabriel, about
        five miles up the River Dart".
    DISTRIBUTION: Worldwide: temperate to tropical waters, including the Black Sea.
    STATUS: CITES – Appendix II; IUCN – Data Deficient.
    SYNONYMS: *nesarnack* (Lacépède, 1804); **gillii** Dall, 1873; **ponticus** Barabash-Nikiforov, 1940;
        **unassigned**: *gephyreus* Lahille, 1908.
    COMMENTS: See Leatherwood and Reeves (1990). Ross and Cockroft (1990:124) considered
        *aduncus* to be synonymous with *truncatus*. Hall (1981:885-887) considered *nesarnack*
        and *gillii* distinct species, and synonymized *truncatus* with *nesarnack*. Opinion 1413 of
        the International Commision on Zoological Nomenclature (1986) conserved *truncatus*
        Montagu, 1821 and suppressed *nesarnack* Lacépède, 1804. Rice (1998:106)
        provisionally recognized *T. t. truncatus* (Bottlenose Dolphin), *T. t. ponticus* (Black Sea
        Bottlenose Dolphin) and *T. t. gillii* (Cowfish). The relationship of *Tursiops gephyreus*
        (South American Bottlenose Dolphin), to either *T. truncatus* or *T. aduncus* remains
        uncertain.

**Family Monodontidae** Gray, 1821. London Med. Repos., 15(1):310.
    COMMENTS: Does not include *Orcaella*, a delphinid.

*Delphinapterus* Lacépède, 1804. Hist. Nat. Cetacees, p. 241.
    TYPE SPECIES: *Delphinapterus beluga* Lacépède 1804 (= *Delphinus leucas* Pallas, 1776).
    SYNONYMS: *Argocetus* Gloger, 1842; *Beluga* Rafinesque, 1815.
    COMMENTS: Reviewed by Kleinenberg et al. (1969) and T. G. Smith et al. (1990).

    *Delphinapterus leucas* (Pallas, 1776). Reise Prov. Russ. Reichs, 3(1):85 [footnote].
        COMMON NAME: Beluga.
        TYPE LOCALITY: NE Siberia, "die im Obischen Meerbusen" (= mouth of Ob River).
        DISTRIBUTION: Circumpolar in Arctic seas; Okhotsk and Bering Seas; northern Gulf of Alaska
            (Cook Inlet); Gulf of St. Lawrence: arctic to cold-temperate waters; occasionally strays
            south to Honshu, Japan; France; and Massachusetts, USA.
        STATUS: CITES – Appendix II; IUCN – Vulnerable.
        SYNONYMS: *albicans* (Lacépède, 1804); *beluga* Lacépède, 1804; *catodon* (Gray, 1846); *dorofeevi*
            Barabash and Klumov, 1935; *marisalbi* Ostroumov, 1935.
        COMMENTS: Reviewed by Kleinenberg et al. (1969), T. G. Smith et al. (1990), Stewart and
            Stewart (1989, Mammalian Species, 336) and Brodie (1989).

*Monodon* Linnaeus, 1758 Syst. Nat., 10th ed., 1:75.
    TYPE SPECIES: *Monodon monoceros* Linnaeus, 1758.
    SYNONYMS: *Ceratodon* Brünnich, 1772; *Diodon* Storr, 1780; *Narwalus* Lacépède, 1804; *Tachynices*
        Brookes, 1828.
    COMMENTS: Reviewed by Reeves and Tracey (1980).

    *Monodon monoceros* Linnaeus, 1758. Syst. Nat., 10th ed., 1:75.
        COMMON NAME: Narwhal.
        TYPE LOCALITY: "Habitat in Oceano Septentrionali Americae, Europae." (= northern seas of
            Europe and America).
        DISTRIBUTION: Arctic Ocean; rarely in Beaufort, Bering, Chuckchi and East Siberian Seas;
            occasional strays as far south as Newfoundland, the Netherlands, British Isles and
            Japan.
        STATUS: CITES – Appendix II; IUCN – Data Deficient.
        SYNONYMS: *microcephalus* (Lacépède, 1804); *monodon* (Pallas, 1811); *narhval* Blumenbach,
            1788; *vulgaris* (Lacépède, 1804).
        COMMENTS: Reviewed by Reeves and Tracey (1980, Mammalian Species, 127) and Hay and
            Mansfield (1989).

**Family Phocoenidae** Gray, 1825. Ann. Philos., n.s., 10:340.
    COMMENTS: Formerly considered a subfamily of Delphinidae; see Gromov and Baranova
        (1981:222).

*Neophocaena* Palmer, 1899. Proc. Biol. Soc. Wash., 13:23.
    TYPE SPECIES: *Delphinus phocaenoides* Cuvier, 1829.
    SYNONYMS: *Meomeris* Gray, 1847; *Neomeris* Gray, 1846.
    COMMENTS: Includes *Neomeris*; see Rice (1977) and Pilleri and Chen (1980).

    *Neophocaena phocaenoides* (G. Cuvier, 1829). Règne Anim., Nouv. ed., 1:291.
        COMMON NAME: Finless Porpoise.
        TYPE LOCALITY: South Africa, Western Cape Prov., Cape of Good Hope ("à découvert au
            Cap,"). Almost certainly erroneous; unknown today from coast of Africa.
        DISTRIBUTION: Indo-Pacific: warm-temperate to tropical waters; Persian Gulf to Malaysia,
            north coast of Java (Tasàn and Leatherwood, 1984), China, and Japan: coastal waters
            and some rivers.
        STATUS: CITES – Appendix I; IUCN – Endangered as *N. p. asiaeorientalis*, otherwise Data
            Deficient.
        SYNONYMS: *melas* (Temminck, 1841) [not Traill]; ***asiaeorientalis*** Pilleri and Gihr, 1972;
            ***sunameri*** Pilleri and Gihr, 1975.

COMMENTS: Reviewed by Pilleri and Gihr (1972; 1975:657, 673; 1980b). Van Bree (1973) considered *asiaeorientalis* to be of subspecific rank and *sunameri* to be synonymous with *phocaenoides*. Rice (1998:123) recognized *N. p. phocaenoides*, *N. p. sunameri* (Sunameri) and *N. p. asiaeorientalis* (Yangtse River Porpoise).

*Phocoena* G. Cuvier, 1816. Règne Anim., Nouv. ed., 1:279.
　　TYPE SPECIES: *Delphinus phocoena* Linnaeus, 1758.
　　SYNONYMS: *Acanthodelphis* Gray, 1866; *Australophocoena* Barnes, 1985.
　　COMMENTS: *Phocaena* and *Phocena* are later spellings.

*Phocoena dioptrica* Lahille, 1912. Ann. Mus. Nat. Hist., Buenos Aires, 23:269.
　　COMMON NAME: Spectacled Porpoise.
　　TYPE LOCALITY: Argentina, Buenos Aires, "capturado en Punta Colares, cerca de Quilmes".
　　DISTRIBUTION: Southern hemisphere: cold-temperate waters; Uruguay, Argentina; Falkland, South Georgia, Heard, Macquarie and the Auckland Isls, perhaps Kerguelen Isls. Perhaps circumpolar, see Baker (1977).
　　STATUS: CITES – Appendix II; IUCN – Data Deficient.
　　SYNONYMS: *stornii* Marelli, 1922.
　　COMMENTS: Reviewed by Brownell (1975a, Mammalian Species, 66). Barnes (1985) proposed *Australophocoena* to house this species, but Rosel *et al.* (1995) did not find the generic separation warranted. *Phocaena obtusata* is synonymous with *Cephalorhynchus eutropia*. See Goodall et al. (1988).

*Phocoena phocoena* (Linnaeus, 1758). Syst. Nat., 10th ed., 1:77.
　　COMMON NAME: Harbor Porpoise.
　　TYPE LOCALITY: "Habitat in Oceano Europaeo, & Balthico." (= Baltic Sea, "Swedish Seas").
　　DISTRIBUTION: N Pacific and N Atlantic: arctic to cold-temperate waters, isolated population in Black Sea; extends south to Senegal in the E Atlantic.
　　STATUS: CITES – Appendix II; IUCN – Vulnerable.
　　SYNONYMS: *americana* Allen, 1869; *communis* Lesson, 1827; *lineata* Cope, 1876; **relicta** Abel, 1905; **vomerina** Gill, 1865.
　　COMMENTS: Reviewed by Gaskin et al. (1974, Mammalian Species, 42). Rice (1998:124) recognized *P. p. phocoena* (North Atlantic Harbor Porpoise), *P. p. vomerina* (Eastern North Pacific Harbor Porpoise) and an un-named subspecies in the western North Pacific, but did not recognize *P. p. relicta* (Black Sea Harbor Porpoise).

*Phocoena sinus* Norris and McFarland, 1958. J. Mammal., 39:22, pl. 1-4.
　　COMMON NAME: Vaquita.
　　TYPE LOCALITY: "from the northeast shore of Punta San Felipe, Baja California Norte, Gulf of California, Mexico".
　　DISTRIBUTION: North Pacific: warm-temperate waters; northern Gulf of California (Mexico); erroneously reported from the S Gulf of California, including Tres Marías Isls and N Jalisco (Brownell, 1986).
　　STATUS: CITES – Appendix I; U.S. ESA – Endangered; IUCN – Critically Endangered.
　　COMMENTS: Reviewed by Brownell (1983, Mammalian Species, 198).

*Phocoena spinipinnis* Burmeister, 1865. Proc. Zool. Soc. Lond., 1865:228, figs 1-5.
　　COMMON NAME: Burmeister's Porpoise.
　　TYPE LOCALITY: Argentina, Buenos Aires, "captured in the mouth of the River Plata".
　　DISTRIBUTION: Southern hemisphere: coastal temperate waters of South America, from Rio Urucanga, Santa Catarina, Brazil to Tierra del Fuego to Paita, Peru.
　　STATUS: CITES – Appendix II; IUCN – Data Deficient.
　　SYNONYMS: *philippii* Perez Canto in Philippi, 1896.
　　COMMENTS: Reviewed by Brownell and Praderi (1984, Mammalian Species, 217). A recent specimen referred to this species from Heard Isl has been reidentified as *Phocaena dioptrica* (Brownell et al., 1989).

*Phocoenoides* Andrews, 1911. Bull. Am. Mus. Nat. Hist., 30:31.
　　TYPE SPECIES: *Phocoenoides truei* Andrews, 1911 (= *Phocaena dalli* True, 1885).

*Phocoenoides dalli* (True, 1885). Proc. U.S. Natl. Mus., 8:95, pls. 2-5.
COMMON NAME: Dall's Porpoise.
TYPE LOCALITY: USA, Alaska, "in the strait west of Adakh [sic] Island, one of the Aleutian
group".
DISTRIBUTION: North Pacific: cold-temperate waters.
STATUS: CITES – Appendix II; IUCN – Lower Risk (cd).
SYNONYMS: *truei* Andrews, 1911.
COMMENTS: Reviewed by Jefferson (1988, Mammalian Species 319). Rice (1998:125)
recognized *P. d. dalli* and *P. d. truei* (True's Porpoise).

**Family Physeteridae** Gray, 1821. London Med. Repos., 15(1):310.
SYNONYMS: Kogiidae Gill, 1871
COMMENTS: *Kogia* is sometimes put in a separate family, Kogiidae.

*Kogia* Gray, 1846. Zool. Voy. H.M.S. "Erebus" and "Terror", 1:22.
TYPE SPECIES: *Physeter breviceps* Blainville, 1838.
SYNONYMS: *Callignathus* Gill, 1871; *Cogia* Wallace, 1876; *Euphysetes* Wall, 1851.
COMMENTS: Reviewed by Handley (1966c).

*Kogia breviceps* (Blainville, 1838). Ann. Franc. Etr. Anat. Phys., 2:337.
COMMON NAME: Pygmy Sperm Whale.
TYPE LOCALITY: South Africa, Western Cape Prov., "rapportée des mers du cap de Bonne-
Espérance" (= Cape of Good Hope).
DISTRIBUTION: Worldwide: temperate to tropical waters.
STATUS: CITES – Appendix II; IUCN – Lower Risk (lc).
SYNONYMS: *floweri* Gill, 1871; *goodei* True, 1884; *grayii* (Wall, 1851).
COMMENTS: Reviewed by Caldwell and Caldwell (1989).

*Kogia sima* (Owen, 1866). Trans. Zool. Soc. Lond., 6(1):30, pls. 10-14.
COMMON NAME: Dwarf Sperm Whale.
TYPE LOCALITY: India, Andhra Pradesh (= Madras Presidency), "taken at Waltair".
DISTRIBUTION: Worldwide: warm-temperate to tropical waters, occasionally strands in cold-
temperate areas.
STATUS: CITES – Appendix II; IUCN – Lower Risk (lc).
COMMENTS: Reviewed by Nagorsen (1985, Mammalian Species, 239) and Caldwell and
Caldwell (1989). The specific name *simus. -a, -um* is a Latin adjectival form and has to
agree in gender with the generic name *Kogia* (Rice, 1998:84).

*Physeter* Linnaeus, 1758. Syst. Nat., 10th ed., 1:76.
TYPE SPECIES: *Physeter macrocephalus* Linnaeus, 1758 (= *Physeter catodon* Linnaeus, 1758) by
subsequent selection (Palmer, 1904:5).
SYNONYMS: *Catodon* Linnaeus, 1761; *Cetus* Billberg, 1828; *Meganeuron* Gray, 1865; *Megistosaurus*
Anonymous *in* Harlan, 1828; *Physalus* Lacépède, 1804.

*Physeter catodon* Linnaeus, 1758. Syst. Nat., 10th ed., 1:76.
COMMON NAME: Sperm Whale.
TYPE LOCALITY: "Habitat in Oceano Septentrionali.", restricted to Netherlands, Middenpiat
by Husson and Holthuis (1974).
DISTRIBUTION: Worldwide: antarctic and cold-temperate waters (northern hemisphere) to
tropical waters.
STATUS: CITES – Appendix I; U.S. ESA – Endangered; IUCN – Vulnerable as *P. macrocephalus*.
SYNONYMS: *australasianus* Desmoulins, 1822; *australis* Gray, 1846; *macrocephalus* Linnnaeus,
1758.
COMMENTS: Neotype designated by Husson and Holthuis (1974:212). Linnaeus used both
*catodon* and *macrocephalus* in the 10th edition. *P. catodon* has line priority and,
according to Linnaeus' diagnoses, is the only name applicable. See Hershkovitz
(1966a:121), Schevill (1986, 1987b), Holthuis (1987), and Rice (1989, who also
reviewed the species).

**Family Platanistidae** Gray, 1846. Zool. Voy. H.M.S. "Erebus" and "Terror", 1:25.
  SYNONYMS: Susuidae Gray, 1868.
  COMMENTS: The family grouping of the river dolphins has always been a problem. Using molecular data, both Árnason and Gullberg (1996) and Yang *et al.* (2002) found no direct relationship between *Platanista* and the other river dolphins. Messenger and McGuire (1998) generated a phylogenetic tree supporting the idea that river dolphins are not monophyletic. In a recent retroposon analysis of the major cetacean lineages the authors found that platanistid dolphins, beaked whales and ocean dolphins diverged (in this order) after sperm whales (Nikaido et al., 2001*b*). Within the other river dolphins the other three genera were found to form a monophyletic group. Hamilton (*et al.* 2001), based on cytochrome *b*, also supports *Platanista* in a separate family. The only remaining disagreement is the placement of the other three genera. Yang *et al.* (2002) and Rice (1998) placed *Lipotes* in a separate family and Rice (1998:92-95) placed all three of the other genera in separate families as well. At the present time we favor a two family arrangement, with *Platanista* in the family Platanistidae and the other three genera in the family Iniidae.

*Platanista* Wagler, 1830. Naturliches Syst. Amphibien, p. 35.
  TYPE SPECIES: *Delphinus gangetica* Roxburgh, 1801.
  SYNONYMS: *Susu* Lesson, 1828.
  COMMENTS: Authorship reviewed by Pilleri (1978). The International Commission on Zoological Nomenclature (1989) conserved *Platanista* Wagler, 1830 and *gangeticus* Roxburgh, 1801 and suppressed *Susu* Lesson, 1828. The controversy over the original description still is active, with Kinze (2000) giving reasons why Lebeck, 1801 should be the authority.

*Platanista gangetica* (Roxburgh, 1801). Asiat. Res. Trans. Soc. (Calcutta ed.), 7:170, pl. 5.
  COMMON NAME: Ganges River Dolphin.
  TYPE LOCALITY: India, West Bengal, "in the Ganges. . . rivers, and creeks, which intersect in the delta of that river to the South, S. E. and east of Calcutta." (= Hooghly River, Ganges River delta).
  DISTRIBUTION: India, Nepal, and Bangladesh: Ganges, Bramaputra, Meghna, Karnaphuli, and Hooghly river systems.
  STATUS: CITES – Appendix I; IUCN – Endangered as *P. g. gangetica*.
  COMMENTS: Reviewed by Reeves and Brownell (1989). Formerly included *minor* (= *indi*), see Van Bree (1976), Pilleri and Gihr (1971), and Pilleri (1978).

*Platanista minor* Owen, 1853. Descrip. Cat. Osteol. R. Mus. Coll. Surgeons, 2:448.
  COMMON NAME: Indus River Dolphin.
  TYPE LOCALITY: Pakistan, "from the Indus" River.
  DISTRIBUTION: Pakistan, Indus River system.
  STATUS: CITES – Appendix I; U.S. ESA – Endangered; IUCN – Endangered as *P. gangetica minor*.
  SYNONYMS: *indi* Blyth, 1859.
  COMMENTS: Reviewed by Reeves and Brownell (1989). See Van Bree (1976). Formerly included in *gangetica*; see Pilleri and Gihr (1971) and Pilleri and Gihr (1976*b*, 1976*a*).

**Family Iniidae** Gray, 1846. Zool. Voy. H.M.S. "Erebus" and "Terror", 1:25.
  SYNONYMS: Lipotidae Zhou, Quian and Li, 1978; Pontoporiidae Gray, 1870; Stenodelphinidae Miller, 1923.
  COMMENTS: See comments in family Platanistidae.

*Inia* d'Orbigny, 1834. Nouv. Ann. Mus. Hist. Nat. Paris, 3:31.
  TYPE SPECIES: *Inia boliviensis* d'Orbigny 1834 (= *Delphinus geoffrensis* Blainville, 1817).
  COMMENTS: Reviewed by Pilleri and Gihr (1980*a*).

*Inia geoffrensis* (Blainville, 1817). Nouv. Dict. Hist. Nat., Nouv. ed., 9:151.
  COMMON NAME: Amazon River Dolphin.
  TYPE LOCALITY: "sur la côte du Brésil.", probably upper Amazon River.

DISTRIBUTION: Peru, Ecuador, Brazil, Bolivia, Venezuela, Columbia: Amazon, Negro, Mamore (Bolivia), and Orinoco River systems.

STATUS: CITES – Appendix II; IUCN – Vulnerable.

SYNONYMS: *amazonicus* (Spix and Martius, 1831); *pallida* (Sanborn, 1949) [not Gervais]; **boliviensis** d'Orbigny, 1834; **humboldtiana** Pilleri and Gihr, 1978; **unassigned**: *rostratus* (G. Cuvier, 1812) [part, not Shaw]; *frontatus* (G. Cuvier, 1823).

COMMENTS: Reviewed by Best and de Silva (1989). Includes *boliviensis*, see Casinos and Ocaña (1979); but also see Pilleri and Gihr (1977), who considered it a distinct species. Rice (1998:93) recognized *I. g. geoffrensis* (Amazon River Dolphin), *I. g. humboldtiana* (Orinoco River Dolphin) and *I. g. boliviensis* (Bolivian River Dolphin).

*Lipotes* Miller, 1918. Smithson. Misc. Coll., 68(9):1.
TYPE SPECIES: *Lipotes vexillifer* Miller, 1918.

*Lipotes vexillifer* Miller, 1918. Smithson. Misc. Coll., 68(9):1.
COMMON NAME: Baiji.
TYPE LOCALITY: "Tung Ting Lake, about 600 miles up the Yangtze River, [Hunan] China".
DISTRIBUTION: China: Chang Jiang (Yangtze) and Qiantang Jiang (mouth of Fuchun Jiang) river systems.
STATUS: CITES – Appendix I; U.S. ESA – Endangered; IUCN – Critically Endangered.
COMMENTS: Reviewed by Chen (1989), Zhou et al. (1978, 1979). Reviewed by Brownell and Herald (1972, Mammalian Species, 10).

*Pontoporia* Gray, 1846. Zool. Voy. H.M.S. "Erebus" and "Terror", 1:46.
TYPE SPECIES: *Delphinus blainvillei* Gervais and d'Orbigny, 1844.
SYNONYMS: *Stenodelphis* d'Orbigny and Gervais, 1847.

*Pontoporia blainvillei* (Gervais and d'Orbigny, 1844). Bull. Sci. Soc. Philom. Paris, 1844:39.
COMMON NAME: Franciscana.
TYPE LOCALITY: Uruguay, "qui a été pris à Montevideo" = mouth of the Rio de La Plata near Montevideo.
DISTRIBUTION: Brazil to Argentina: coastal waters from Doce River, Regencia, Espírito Santo to Peninsula Valdez, including the Rio de la Plata.
STATUS: CITES – Appendix II; IUCN – Data Deficient, except for the Rio Grande do Sul/Uruguay subpopulation, which is Vulnerable.
SYNONYMS: *tenuirostris* Malm, 1871.
COMMENTS: Reviewed by Brownell (1989).

**Family Ziphiidae** Gray, 1865. Proc. Zool. Soc. Lond., 1865:528.
SYNONYMS: Hyperoodontidae Gray, 1846.
COMMENTS: Although Hyperoodontidae Gray, 1846 has priority over Ziphiidae, we have chosen to use the latter name following Article 23.12 of the International Code of Zoological Nomenclature (1999) because Ziphiidae has been the name of choice for more than 100 years. Family reviewed by Moore (1968).

*Berardius* Duvernoy, 1851. Ann. Sci. Nat. Zool. (Paris), ser. 3, 15:41.
TYPE SPECIES: *Berardius arnuxii* Duvernoy, 1851.
COMMENTS: This may be a monotypic genus.

*Berardius arnuxii* Duvernoy, 1851. Ann. Sci. Nat. Zool. (Paris), ser. 3, 15:52, fig. 1.
COMMON NAME: Arnoux's Beaked Whale.
TYPE LOCALITY: "échoué sur la côte, dans le port d'Akaroa, presqu'île de Bancks, dans la Nouvelle-Zélande." (= New Zealand, Canterbury Prov., Akaroa).
DISTRIBUTION: Southern hemisphere: circumpolar, Antarctic to temperate waters.
STATUS: CITES – Appendix I; IUCN – Lower Risk (cd).
COMMENTS: Reviewed by Balcomb (1989).

*Berardius bairdii* Stejneger, 1883. Proc. U.S. Natl. Mus., 6:75.
COMMON NAME: Baird's Beaked Whale.

TYPE LOCALITY: Russia, Commander Isls, "found stranded in Stare Gavan, on the eastern shore of Bering Island".
DISTRIBUTION: North Pacific: temperate waters.
STATUS: CITES – Appendix I; IUCN – Lower Risk (cd).
SYNONYMS: *vegae* Malm, 1883.
COMMENTS: Reviewed by Balcomb (1989); possibly a subspecies of *arnuxii*, see Davies (1963) and McLachlan et al. (1966).

*Hyperoodon* Lacépède, 1804. Hist. Nat. Cetacees, xliv, 319.
TYPE SPECIES: *Hyperoodon butskopf* Lacépède, 1804 (= *Balaena ampullata* Forster, 1770).
SYNONYMS: *Anodon* Wagler, 1830; *Chaenodelphinus* Eschricht, 1843; *Frasercetus* Moore, 1968; *Heterodon* Blainville *in* Desmarest, 1817; *Lagenocetus* Gray, 1863; *Uranodon* Illiger, 1811.
COMMENTS: Includes *Frasercetus* Moore, 1968 as a subgenus.

*Hyperoodon ampullatus* (Forster, 1770). *In* Kalm, Travels into N. Am., 1:18.
COMMON NAME: Northern Bottlenose Whale.
TYPE LOCALITY: "See Mr. Pennant's [1769] British Zoology Vol. 3, p. 43, where it is called the beaked whale, and very well described;" Pennant (1769:43) gave Maldon (England) as the locality and 1717 as the date stranded.
DISTRIBUTION: North Atlantic: arctic to cold-temperate waters. The Mediterranean record represents a stray (J. G. Mead, 1989*b*).
STATUS: CITES – Appendix I; IUCN – Lower Risk (cd).
SYNONYMS: *butskopf* (Bonnaterre, 1789); *latifrons* Gray, 1846; *rostratus* (Müller, 1776).
COMMENTS: Subgenus *Hyperoodon*. Reviewed by J. G. Mead (1989*b*).

*Hyperoodon planifrons* Flower, 1882. Proc. Zool. Soc. Lond., 1882:392, figs. 1, 2.
COMMON NAME: Southern Bottlenose Whale.
TYPE LOCALITY: "found upon the sea-beach of Lewis Island in the Dampier Archipelago, north-western Australia."
DISTRIBUTION: Southern hemisphere: circumpolar, antarctic to temperate waters, rarely into tropical waters. May occur in the W North Pacific.
STATUS: CITES – Appendix I; IUCN – Lower Risk (cd).
SYNONYMS: *burmeisterei* Moreno, 1895.
COMMENTS: Reviewed by J. G. Mead (1989*b*). Moore (1968) erected the subgenus *Frasercetus* for this species.

*Indopacetus* Moore, 1968. Fieldiana Zool., 53(4):254.
TYPE SPECIES: *Mesoplodon pacificus* Longman, 1926.
COMMENTS: Considered by many authors to be included in *Mesoplodon*. Known only from two specimens.

*Indopacetus pacificus* (Longman, 1926). Mem. Queensl. Mus., 8(3):269, pl. 43.
COMMON NAME: Tropical Bottlenose Whale.
TYPE LOCALITY: Australia, Queensland, "found at Mackay".
DISTRIBUTION: Indian Ocean and W South Pacific: tropical waters.
STATUS: CITES – Appendix II; IUCN – Data Deficient.
COMMENTS: Reviewed by J. G. Mead (1989*c*). Commonly included in *Mesoplodon* (Heyning, 1989*a*; J. G. Mead, 1989*c*). We prefer 'Tropical Bottlenose Whale', as used by Pitman et al. (1999), to the older name 'Longman's beaked whale'.

*Mesoplodon* Gervais, 1850. Ann. Sci. Nat. Zool. (Paris), ser. 3, 14:16.
TYPE SPECIES: *Delphinus sowerbensis* de Blainville, 1817 (= *Physeter bidens* Sowerby, 1804).
SYNONYMS: *Aodon* Lesson, 1828; *Dioplodon* Gervais, 1850; *Dolichodon* Gray, 1866; *Micropterus* Wagner, 1846; *Oulodon* von Haast, 1876; *Nodus* Wagler, 1830; *Paikea* Oliver, 1922.
COMMENTS: *Mesoplodon* Gervais, 1850 and *Physeter bidens* Sowerby, 1804 were conserved; *Nodus, Micropteron,* and *Mikropteron* were suppressed by the International Commission on Zoological Nomenclature (1985*b*).

*Mesoplodon bidens* (Sowerby, 1804). Trans. Linn. Soc. Lond., 7:310.
COMMON NAME: Sowerby's Beaked Whale.
TYPE LOCALITY: UK, Scotland, "stranded on the estate of James Brodie, Esq. F. L. S., in the county of Elgin."
DISTRIBUTION: North Atlantic and Baltic Sea: temperate waters. Occurrence in the Mediterranean Sea was discussed by van Bree (1975), who considered the evidence unconvincing; however, Casinos and Filella (1981) supported a report from the Italian coast (Brunelli and Fasella, 1929). There is one report from the Gulf of Mexico (Bonde and O'Shea, 1989) that is also considered a stray.
STATUS: CITES – Appendix II; IUCN – Data Deficient.
SYNONYMS: *dalei* (Lesson, 1827); *micropterus* (G. Cuvier, 1829); *sowerbensis* (de Blainville, 1817); *sowerbiensis* (Gray, 1846); *sowerbyi* (Desmarest, 1822).
COMMENTS: Reviewed by J. G. Mead (1989*c*).

*Mesoplodon bowdoini* Andrews, 1908. Bull. Am. Mus. Nat. Hist., 24:203, figs. 1-5, pl. 13.
COMMON NAME: Andrew's Beaked Whale.
TYPE LOCALITY: "collected at New Brighton Beach, Canterbury Province, New Zealand".
DISTRIBUTION: Southern hemisphere, South Pacific and Indian oceans, cold-temperate waters of Australia and New Zealand. The record from Kerguelen Isls (Robineau, 1973) is erroneous and is the fossil rostrum commonly known as *Mesoplodon longirostris* (J. G. Mead, 1989*c*).
STATUS: CITES – Appendix II; IUCN – Data Deficient.
COMMENTS: Reviewed by J. G. Mead (1989*c*). McCann (see Mead et al., 1982) felt that *M. bowdoini* was synonymous with *M. stejnegeri*.

*Mesoplodon carlhubbsi* Moore, 1963. Am. Midl. Nat., 70:396, figs. 1-3, 7, 8, 13-15.
COMMON NAME: Hubbs' Beaked Whale.
TYPE LOCALITY: "La Jolla, California, 32° 51' 41" N. Lat., 117° 15' 19" W. Long."
DISTRIBUTION: North Pacific: temperate waters.
STATUS: CITES – Appendix II; IUCN – Data Deficient.
COMMENTS: Reviewed by J. G. Mead (1989*c*). Very closely related to *bowdoini*. Orr believed that this species was synonymous with *M. stejnegeri* (see Mead et al., 1982). Hubbs (1946) first identified the holotype of this species as *M. bowdoini*.

*Mesoplodon densirostris* (Blainville, 1817). Nouv. Dict. Hist. Nat., Nouv. ed., 9:178.
COMMON NAME: Blainville's Beaked Whale.
TYPE LOCALITY: None given, unknown.
DISTRIBUTION: World-wide: temperate to tropical waters.
STATUS: CITES – Appendix II; IUCN – Data Deficient.
SYNONYMS: *seychellensis* (Gray, 1846).
COMMENTS: Reviewed by J. G. Mead (1989*c*).

*Mesoplodon europaeus* (Gervais, 1855). Hist. Nat. Mammifères, 2:320.
COMMON NAME: Gervais' Beaked Whale.
TYPE LOCALITY: English Channel, "qui provient d'un individu harponné dans la Manche."
DISTRIBUTION: Aside from the type, one specimen from Ireland, one specimen from Guinea-Bissau, and three records from Ascension Isl, it is only known from the Western North Atlantic: temperate to tropical waters.
STATUS: CITES – Appendix II; IUCN – Data Deficient.
SYNONYMS: *gervaisi* (Deslongchamps, 1866).
COMMENTS: Reviewed by J. G. Mead (1989*c*) and Norman and Mead (2002, Mammalian Species, 688). The type was not harpooned, as stated by Gervais, but was found as a "cadavre" (Deslongschamps, 1866:177).

*Mesoplodon ginkgodens* Nishiwaki and Kamiya, 1958. Sci. Rep. Whales Res. Inst. (Tokyo), 13:53, 13 figs., 17 pls.
COMMON NAME: Ginkgo-toothed Beaked Whale.
TYPE LOCALITY: Japan, "Oiso Beach, Sagami Bay, near Tokyo."
DISTRIBUTION: North Pacific and Indian Oceans: warm-temperate to tropical waters; Japan, Taiwan, Baja California, Mexico, Sri Lanka, Indonesia and Australia.

STATUS: CITES – Appendix II; IUCN – Data Deficient.
SYNONYMS: *hotaula* Deraniyagala, 1963.
COMMENTS: Reviewed by J. G. Mead (1989*c*).

*Mesoplodon grayi* Von Haast, 1876. Proc. Zool. Soc. Lond., 1876:9.
COMMON NAME: Gray's Beaked Whale.
TYPE LOCALITY: New Zealand, "the Chatham Islands. . . from specimens stranded. . . on the Waitangi beach of the main island of that group."
DISTRIBUTION: Southern hemisphere: cold-temperate waters; one specimen found in the Netherlands (Boschma, 1950:779).
STATUS: CITES – Appendix II; IUCN – Data Deficient.
SYNONYMS: *australis* Flower, 1878; *haasti* Flower, 1878.
COMMENTS: Reviewed by J. G. Mead (1989*c*).

*Mesoplodon hectori* (Gray, 1871). Ann. Mag. Nat. Hist., ser. 4, 8:116.
COMMON NAME: Hector's Beaked Whale.
TYPE LOCALITY: New Zealand, Wellington, "killed in Tatai [sic] Bay, Cook's Straits" (= Titai Bay).
DISTRIBUTION: Southern hemisphere: temperate waters.
STATUS: CITES – Appendix II; IUCN – Data Deficient.
SYNONYMS: *knoxi* Hector, 1873.
COMMENTS: Reviewed by J. G. Mead (1989*c*). The North Pacific form was found to be a new species, *Mesoplodon perrini* Dalebout et al., 2002.

*Mesoplodon layardii* (Gray, 1865). Proc. Zool. Soc. Lond., 1865:357, fig.
COMMON NAME: Strap-toothed Whale.
TYPE LOCALITY: None given, probably South Africa.
DISTRIBUTION: Southern hemisphere: temperate waters.
STATUS: CITES – Appendix II; IUCN – Data Deficient.
SYNONYMS: *floweri* Haast, 1876; *guntheri* Krefft, 1871; *longirostris* Gray, 1873; *thomsoni* Ogilby, 1896.
COMMENTS: Reviewed by J. G. Mead (1989*c*). *Mesoplodon traversii*, a former synonym of *M. layardii*, was recently recognized as a valid species (Helden et al., 2002).

*Mesoplodon mirus* True, 1913. Smithson. Misc. Coll., 60(25):1.
COMMON NAME: True's Beaked Whale.
TYPE LOCALITY: USA, "stranded in the outer bank of Bird Island Shoal in the harbor of Beaufort, North Carolina".
DISTRIBUTION: North Atlantic, South Atlantic coast of South Africa, Australia: temperate waters.
STATUS: CITES – Appendix II; IUCN – Data Deficient.
COMMENTS: Reviewed by J. G. Mead (1989*c*).

*Mesoplodon perrini* Dalebout et al., 2002. Mar. Mam. Sci., 18(3):577.
COMMON NAME: Perrin's Beaked Whale.
TYPE LOCALITY: U.S.A., Carlsbad, California.
DISTRIBUTION: Cold temperate waters, eastern North Pacific.
STATUS: CITES – Appendix II.
COMMENTS: Was first described as *Mesoplodon hectori* by Mead (1981), subsequently recognized as a new species by Dalebout et al. (2002).

*Mesoplodon peruvianus* Reyes, Mead, and Van Waerebeek, 1991. Marine Mammal Sci., 7(1):1, 6 figs.
COMMON NAME: Pygmy Beaked Whale.
TYPE LOCALITY: "Playa Paraiso (11°12' S), Huacho, Lima, Peru."
DISTRIBUTION: E South Pacific, E North Pacific: cold-temperate to tropical waters. Known from the coast of Peru between Playa Paraiso (11°S) and San Juan de Marcona (15°S). Two specimens are known from near La Paz, Baja California, Mexico (Urban-Ramirez and Aurioles-Gamboa, 1992), and one from New Zealand.
STATUS: CITES – Appendix II; IUCN – Data Deficient.

*Mesoplodon stejnegeri* True, 1885. Proc. U.S. Natl. Mus., 8:584, pl. 25.
    COMMON NAME: Stejneger's Beaked Whale.
    TYPE LOCALITY: Russia, Commander Isls, "Bering Island".
    DISTRIBUTION: North Pacific: cold-temperate waters; isolated population in the Sea of Japan.
    STATUS: CITES – Appendix II; IUCN – Data Deficient.
    COMMENTS: Reviewed by Loughlin and Perez (1985, Mammalian Species, 250) and
        J. G. Mead (1989*c*).

*Mesoplodon traversii* (Gray, 1874) Trans. New Zealand Inst. (1873) 6:96.
    COMMON NAME: Spade-toothed Whale.
    TYPE LOCALITY: Chatham Isl, New Zealand.
    DISTRIBUTION: South Pacific: temperate waters; known from 3 specimens: Pitt Isl (Chatham
        Isls), White Isl (North Isl, N.Z.) and Robinson Crusoe Isl (Juan Fernadez Arch., Chile).
    STATUS: CITES – Appendix II.
    SYNONYMS: *bahamondi* Reyes, Van Warebeek, Cárdenas, and Yáñez, 1996.
    COMMENTS: Was considered synonymous with *Mesoplodon layardii* until Helden et al. (2002)
        compared the DNA sequence from the type and determined that it was the same
        species as *M. bahamondi*. This species is the rarest member of the Ziphiidae.

*Tasmacetus* Oliver, 1937. Proc. Zool. Soc. Lond., 107:371.
    TYPE SPECIES: *Tasmacetus shepherdi* Oliver, 1937.

*Tasmacetus shepherdi* Oliver, 1937. Proc. Zool. Soc. Lond., 107:371, pls. 1-5.
    COMMON NAME: Shepherd's Beaked Whale.
    TYPE LOCALITY: New Zealand, North Isl, "cast upon the beach at Ohawe, in the province of
        Taranaki."
    DISTRIBUTION: Southern hemisphere: cold-temperate waters, particularly off New Zealand,
        Chile, Argentina and Tristan de Cunha.
    STATUS: CITES – Appendix II; IUCN – Data Deficient.
    COMMENTS: Reviewed by J. G. Mead (1989*a*).

*Ziphius* G. Cuvier, 1823. Rech. Oss. Foss., Nouv. ed., 5:350.
    TYPE SPECIES: *Ziphius cavirostris* G. Cuvier, 1823.
    SYNONYMS: *Diodon* Lesson, 1828; *Hypodon* Haldeman, 1841; *Petrorhynchus* Gray, 1875;
        *Ziphiorhynchus* Burmeister, 1865.

*Ziphius cavirostris* G. Cuvier, 1823. Rech. Oss. Foss., Nouv. ed., 5(1):350.
    COMMON NAME: Cuvier's Beaked Whale.
    TYPE LOCALITY: France, "dans le département des Bouches-du-Rhône, entre de Fos
        et l'embouchure du Galégeon" (= between Fos and the mouth of the Galégeon River).
    DISTRIBUTION: Worldwide: cold-temperate to tropical waters.
    STATUS: CITES – Appendix II; IUCN – Data Deficient.
    SYNONYMS: *australis* (Burmeister, 1865); *capensis* (Gray, 1865); *chathamensis* (Hector, 1873);
        *indicus* Van Beneden, 1863.
    COMMENTS: Reviewed by Heyning (1989*b*).